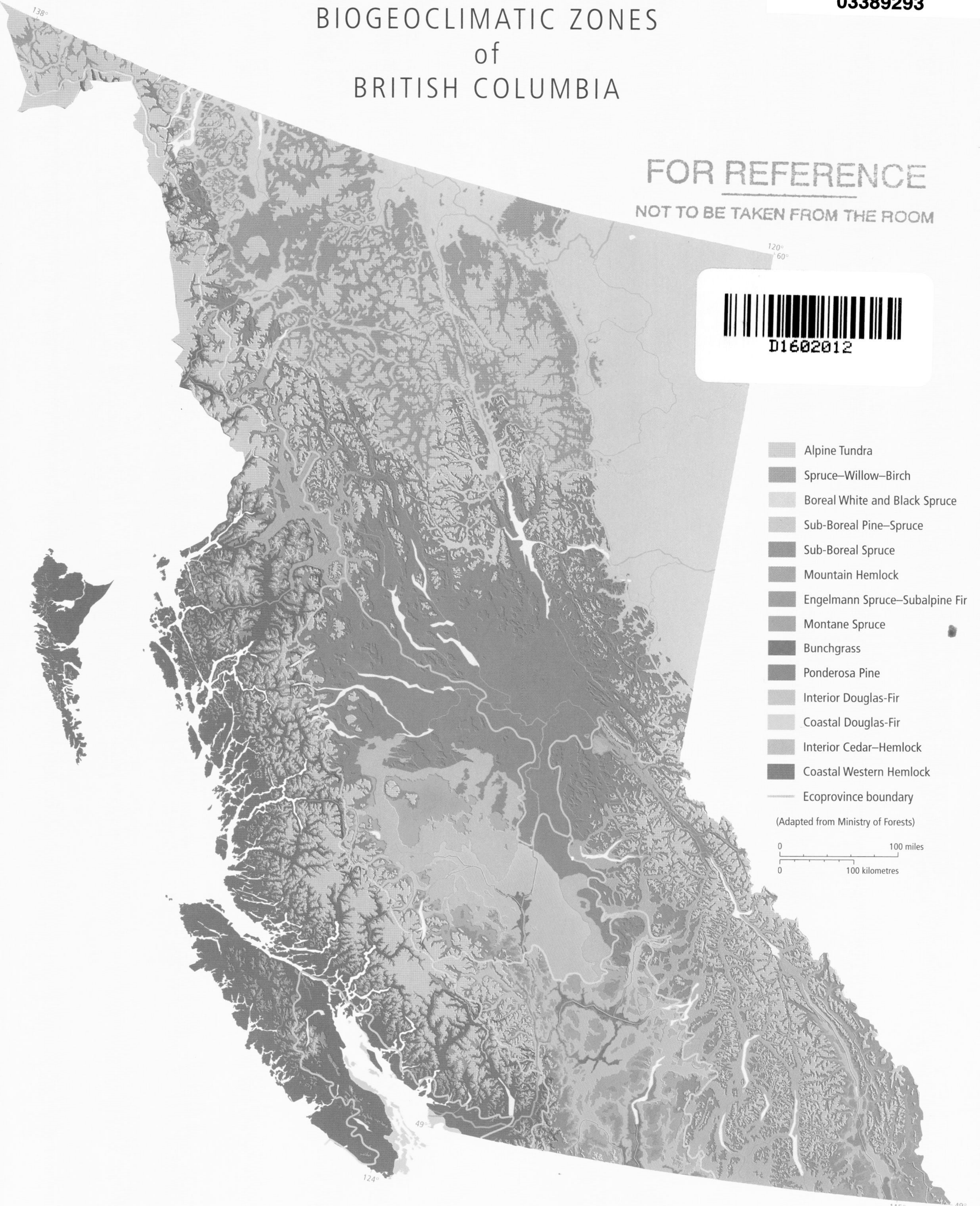
BIOGEOCLIMATIC ZONES
of
BRITISH COLUMBIA
60°
138°
120°
60°
49°
124°
115°
49°
Alpine Tundra
Spruce–Willow–Birch
Boreal White and Black Spruce
Sub-Boreal Pine–Spruce
Sub-Boreal Spruce
Mountain Hemlock
Engelmann Spruce–Subalpine Fir
Montane Spruce
Bunchgrass
Ponderosa Pine
Interior Douglas-Fir
Coastal Douglas-Fir
Interior Cedar–Hemlock
Coastal Western Hemlock
Ecoprovince boundary
(Adapted from Ministry of Forests)
0
100 miles
0
100 kilometres

THE Birds OF BRITISH COLUMBIA

VOLUME 4 PASSERINES

WOOD-WARBLERS
THROUGH
OLD WORLD SPARROWS

THE BIRDS OF BRITISH COLUMBIA

Volume 1	Nonpasserines: Introduction, Loons through Waterfowl
Volume 2	Nonpasserines: Diurnal Birds of Prey through Woodpeckers
Volume 3	Passerines: Flycatchers through Vireos
Volume 4	Passerines: Wood-Warblers through Old World Sparrows

Environment
Canada

Canadian Wildlife
Service

Ministry of Environment,
Lands and Parks
Wildlife Branch
Resources Inventory Branch

THE Birds OF BRITISH COLUMBIA

VOLUME 4 **PASSERINES**

WOOD-WARBLERS THROUGH OLD WORLD SPARROWS

Synopsis:
The Birds of British Columbia into the 21st Century

by
R. Wayne Campbell, Neil K. Dawe,
Ian McTaggart-Cowan, John M. Cooper,
Gary W. Kaiser, Andrew C. Stewart,
and Michael C.E. McNall

UBCPress · Vancouver · Toronto

Published in cooperation with Environment Canada (Canadian Wildlife Service), the British Columbia Ministry of Environment, Lands and Parks (Wildlife Branch and Resources Inventory Branch), and the Royal British Columbia Museum

Printed in Canada on acid-free paper ∞

ISBN 0-7748-0621-4

Canadian Cataloguing in Publication Data

Main entry under title:
The birds of British Columbia

"Published in cooperation with Environment Canada (Canadian Wildlife Service), the British Columbia Ministry of Environment, Lands and Parks (Wildlife Branch and Resources Inventory Branch), and the Royal British Columbia Museum."
Includes bibliographical references and index.
Contents: v. 1. Nonpasserines : introduction and loons through waterfowl – v. 2. Nonpasserines : diurnal birds of prey through woodpeckers – v. 3. Passerines : flycatchers through vireos – v. 4. Passerines : wood-warblers through old world sparrows.
ISBN 0-7748-0618-4 (v. 1) – ISBN 0-7748-0619-2 (v. 2) – ISBN 0-7748-0572-2 (v. 3) – ISBN 0-7748-0621-4 (v. 4)

1. Birds – British Columbia. I. Campbell, R. Wayne (Robert Wayne), 1942- II. British Columbia. Wildlife Branch. III. British Columbia. Ministry of Environment, Lands and Parks. Resources Inventory Branch. IV. Canadian Wildlife Service. V. Royal British Columbia Museum.

QL685.5.B7B574 1997 598'.09711 C96-910903-2

UBC Press acknowledges the financial support of the Government of Canada through the Book Publishing Industry Development Program (BPIDP) for our publishing activities.
Canadä

We also gratefully acknowledge the support of the Canada Council for the Arts for our publishing program, as well as the support of the British Columbia Arts Council.

Printed and bound in Canada by Friesens
Set in Palatino and Frutiger by Artegraphica Design Co. Ltd.
Cover design: Chris Tyrrell
Copy editor: Francis J. Chow
Proofreader: Deborah Kerr
Cartographer: Eric Leinberger
Bird illustrations on maps: Michael Hames
Front cover photograph: R. Wayne Campbell
Back cover photograph: Mark Nyhof

UBC Press
University of British Columbia
2029 West Mall
Vancouver, BC V6T 1Z2
(604) 822-5959
Fax: (604) 822-6083
E-mail: info@ubcpress.ca
www.ubcpress.ca

CONTENTS

We acknowledge the generous support
of the
following organizations
and companies:

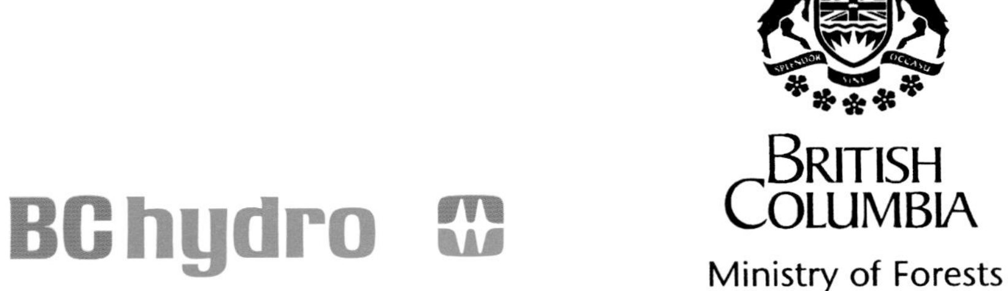

ACKNOWLEDGEMENTS

The completion of this project has been a truly cooperative effort between federal and provincial governments, professional biologists, natural history and conservation organizations, industry, and thousands of dedicated and informed birdwatchers. Support started at the top, with managers and administrators, and continued through the active field observers.

Over the past 3 years, we have been very fortunate to have had supervisors and colleagues who provided an unobstructed path to the completion of this volume. These people include Michael Dunn, Bob Elner, Trish Hayes, Arthur M. Martell, Richard W. McKelvey, and Peggy Ward (Canadian Wildlife Service); Nancy Bircher, Douglas Dryden, Donald S. Eastman, D. Ray Halladay, Greg Jones, Ted Lea, Jim S. Mattison, William T. Munro, Bruce A. Pendergast, and Rod S. Silver (British Columbia Ministry of Environment, Lands and Parks); William D. Barkley, Grant Hughes, Brent Cooke, Peter Newroth, and Jim Cosgrove (Royal British Columbia Museum); and Ron Erickson and Helen Torrance (The Nature Trust of British Columbia).

The completion of this complex project required the understanding and support of more than 11,000 contributors at many different levels. All have been acknowledged and those who contributed specifically to this volume are listed, beginning on page 723. In addition, many individuals provided support that merits special thanks.

CONTRIBUTORS TO DATA BASE

A total of 3,629 people provided information on the occurrence, distribution, numbers, and breeding biology of wood-warblers, sparrows, blackbirds, and finches by contributing to a variety of provincial data bases and participating in annual events such as Breeding Bird Surveys and Christmas Bird Counts.

Among the general contributors is a growing group of dedicated individuals whose passion for birds resulted in major contributions in the form of writing detailed, long-term field notes, coordinating annual birdwatching events, serving as regional contacts for information, compiling quarterly provincial reports for *American Birds/Field Notes*, and encouraging participation in our project. These contributors include: David Allinson, Errol Anderson, Jerry and Gladys Anderson, Gerry Ansell, Cathy Antoniazzi, Steve Baillie, E. Derek Beacham, Alice Beals, Alan Bear, Barbara Begg, Winnifred Bennie, Mike Bentley, Ed Beynon, Jack Bowling, Ken C. Boyce, Jan Bradshaw, Len Brown, Christopher A. Buis, Clyde Burton, Andreas Buttimer, Elmer Callin, Richard J. Cannings, Stephen R. and Jean Cannings, Sydney G. Cannings, Donald G. Cecile, Chris Charlesworth, Stewart Clow, George Clulow, John Comer, Evi Coulson, Linnea Cross-Tallman, Gary S. Davidson, Lyndis Davis, Brent Diakow, Adrian Dorst, Eva Durance, Linda Durrell, Michael C.R. Edgell, Kyle Elliott, Maurice Ellison, Anthony J. Erskine, Michael P. Force, Lee A. Foster, David F. Fraser, Jeff Gaskin, Bryan R. Gates, Les and Violet Gibbard, J.E. Victor and Margaret E. Goodwill, Hilary and Orville Gordon, B. Max Gotz, Al Grass, Tony Greenfield, Larry R. Halverson, Peter Hamel, Willie Haras, Margaret Harris, Robert B. Hay, Wendy Hayduk, W. Grant Hazelwood, Todd Heakes, Margo Hearne, Charles W. Helm, Joyce Henderson, R. Jerry Herzig, Werner and Hilde Hesse, Edward Hillary, Dennis Horwood, Richard R. Howie, Douglas and Marian Innes, John Ireland, Pat Janzen, Dale Jensen, Anne M.N. Jones, Elspeth Kerr, Joan Kerr, Frank Kime, Jeremy V. Kimm, Sandra Kinsey, Helen Knight, Burke Korol, W. Douglas Kragh, Nancy Krueger, Violet F. Lambie, Laird Law, Douglas Leighton, Betty and James Lunam, Jo Ann and Hue MacKenzie, Alan MacLeod, Nancy Mahony, Diana V. Maloff, Derrick Marven, Hylda Mayfield, Barb and Mike McGrenere, Karen McLaren, Martin K. McNicholl, William J. Merilees, Guy L. Monty, Elaine Moore, Art Morgan, Alexander Muir, Emily Müller, Gwen Nicol, Mark Nyhof, Elsie Nykyfork, Mary Pastrick, Aurora M. Paterson, Jim Perry, Connie Phillip, Mark Phinney, Douglas Powell, G. Allen Poynter, George and Bea Prehara, Al Preston, Michael I. Preston, D. Michael Price, Roy Prior, Nina Raginsky, Trish Reid, Diane Richardson, Dirk Rinehart-Pidcock, Anna Roberts, I. Laurie Rockwell, Michael S. Rodway, Manfred Roschitz, Glenn R. Ryder, Ron and Joy Satterfield, Barbara M. Sedgwick, Brian Self, Michael G. Shepard, Chris Siddle, Pam Sinclair, Marion Smith, Gail Spitler, David Stirling, Howard A. Telosky, Harvey Thommasen, John Toochin, Rick Toochin, Ruth Travers, Roger Tremblay, Jim Tuck, Danny Tyson, Linda M. Van Damme, Ronald P. Walker, Sydney and Emily Watts, Wayne C. Weber, Rita Wege, Bruce Whittington, Karen Willies, Douglas J. Wilson, Jim Wisnia, John G. Woods, and Kenneth G. Wright.

DATA ENTRY

The entry of the occurrence and breeding data into electronic data base files involved many people and organizations.

The Arrowsmith Naturalists (Parksville) obtained a series of grants from Employment and Immigration Canada that allowed most of the passerine data from the Wildlife Records Scheme to be entered into data base files. Pauline Tranfield and Betty Barnes of the Arrowsmith Naturalists helped administer the program.

The following people participated in this program over the 3½ years it took to complete the task; those with an asterisk after their names supervised the data entry: Lone Anderson, Anne Ashmore, Danielle Bras, Susan L. Bridge, Shirley Brittain, Kathryn J. Cole*, Heather Ann Colton, Deborah Cote, Avon Fitch*, Roseanne Marie Harrison, Michael E.J. Hayward, Karen Koehn, Renate L. Liddell*, Terri Martin, Jeanette W. Pryor, Dianna Reed, Marilyn Reed, Rosemarie Sather, Marlene Slawson, Carol J. Strynadka, and Jodi Waites*.

WBT Wild Bird Trust assisted with entry of current information by Maureen L. Funk, Laura Gretzinger, Scott Dickin, and Nicole Wallace.

The following people entered other portions of the data, particularly the British Columbia Nest Records Scheme, over the years: D. Sean Campbell, Eileen C. Campbell, Tessa N.

Campbell, Darren R. Copley, Jordan T. Dawe, Karen E. Dawe, Carolyn Hamilton, Renate L. Liddell, Andrew MacDonald, Edward L. Nygren, Charlene Pearce, Cynthia and Michael G. Shepard, and Bernice Smith.

Len Thomas, a graduate student at the University of British Columbia, organized and summarized computer files for trends in coastal and interior Breeding Bird Survey routes. Laura M. Kammermeier (Cornell University, Ithaca, New York) kindly provided us with Project FeederWatch data for several species accounts.

VOLUNTEERS

The following people assisted in searching journals and government reports for bird records, copying articles, and filing references: Mary Alpen, Gerry Anderson, Gladys Anderson, Tracy Anderson, Sheri Baker, Colin Barnfield, Marion Becevel, Tanya Bennett, Elizabeth Brooke, D. Sean Campbell, Eileen C. Campbell, Brian Chapel, Vivian Clark, Holly Clermont, Lyndis Davis, Jordan T. Dawe, Diana Demarchi, Scott Dickin, Peggy Dyson, John Elliott, Linda Funk, Maureen L. Funk, Ruth Gilson, Taiya Henderson, Joan Hooper, Tracey D. Hooper, Jennifer Hopkins, Jason Kimm, Jeremy V. Kimm, Joan Laharty, Wally Lee, John D. MacIntosh, Steve Madson, Alistar Marr, Jim W. McCammon, Faye L. McNall, Heather Melchior, Joan Mogenson, Cindy Moore, Angela Muellers, Phyllis Mundy, Geri Nishi, Andrea R. Norris, Meg Philpot, Karel Sars, Kelly Sendall, Bernice Smith, Terry Snye, Win Speechly, Irene I. Stewart, Laura E. Stewart, Doreen Sutherland, Calvin Tolkamp, Herbert Van Kampen, Margaret Wainwright, Nicole Wallace, Lillian Weston, Eldred Williams, and Mrs. J.M. Winterbottom.

Besides many regional reviewers, a small group of people hastened our writing by providing summaries of information on very short notice. These included Barbara Begg, Donald G. Cecile, Chris Charlesworth, Kimball L. Garrett, Hilary Gordon, Larry Halverson, Peter Hamel, Margo Hearne, Helen Knight, Douglas Leighton, Ron Mayo, Martin K. McNicholl, Michael S. Rodway, Manfred Roschitz, Barbara M. Sedgwick, Harvey Thommasen, Ron Walker, and John G. Woods.

OFFICE SUPPORT

Considerable assistance was provided the authors by Maureen L. Funk. Our day-to-day demands included photocopying, fielding telephone calls, coordinating volunteers, obtaining needed literature, preparing figures, and verifying electronic records with the original data. In the later stages of the project, she took over many of the tasks left by Jordan T. Dawe (see below) when he returned to university. The fact that Maureen stayed with us until the completion of the project is more a reflection of her abilities than of ours.

Jordan T. Dawe prepared most of the initial drafts of the figures, updated and readied the species dBASE files in preparation for writing the species accounts, maintained an orderly flow of sending draft accounts to reviewers and returning reviewed accounts to the respective authors, assisted the authors with their computer problems, and wrote the initial dBASE V program that summarized the data to prepare the species richness figures and many of the tables for the final chapters.

Eileen, Sean, and Tessa Campbell gave up most of the downstairs of their home and endured busy phones, strangers, piles of papers, and computers galore for much of the last 10 years of the project.

MEETING SUPPORT

A special thanks to Joyce McTaggart-Cowan for graciously opening her home to the monthly authors' meetings and providing tea and cookies, always at just the right moment.

FINANCIAL AND ADMINISTRATIVE SUPPORT

The following organizations and individuals provided important support to the project, either directly or indirectly: British Columbia Field Ornithologists; British Columbia Hydro and Power Authority; British Columbia Ministry of Attorney General (Jennifer Button), British Columbia Ministry of Environment, Lands and Parks, Wildlife Branch; British Columbia Ministry of Environment, Lands and Parks, Resources Inventory Branch; British Columbia Ministry of Forests, Research Branch (Allison Nicholson); British Columbia Waterfowl Society; Canada–British Columbia Partnership Agreement of Forest Resources: FRDA II; Centennial Wildlife Society of British Columbia; Ducks Unlimited Canada; Employment and Immigration Canada; Environment Canada, Canadian Wildlife Service, Pacific and Yukon Region; Environment Canada, Parks; Federation of British Columbia Naturalists; Frank M. Chapman Memorial Fund (American Museum of Natural History); Friends of Mount Revelstoke and Glacier National Parks; Habitat Conservation Trust Fund (Rod S. Silver); The Nature Conservancy of Canada (Jan Garnett); The Nature Trust of British Columbia; Northwest Wildlife Preservation Society; Royal British Columbia Museum; Saltspring Island Garden Club; Shearwater Scaling and Grading Limited (F. Wayne Diakow); James R. Slater; Bernice Smith; Win Speechly; William Taylor; Vancouver Natural History Society, Birding Section (George F. Clulow); Westcoast Energy; Weyerhaeuser (Ronald T. McLaughlin, Stan J. Coleman, and B.G. (Glen) Dunsworth); WBT Wild Bird Trust of British Columbia; and World Wildlife Fund (Canada).

PHOTOGRAPHS, MAPS, LITERATURE, AND SPECIMENS

We thank the following photographers and organizations who, in addition to the authors, contributed images to this volume: Michael D. Bentley, Jack Bowling, W. Sean Boyd, Stephen R. Cannings, Ernie Carlson, Donald G. Cecile, John K. Cooper, R. Curtis, Gary S. Davidson, Al Dawson, Adrian Dorst, Lance Goodwin, B. Henry, R. Jerry Herzig, Ted Lea, Douglas Leighton, Jo Ann MacKenzie, Brent Matsuda, Mas Matsushita, Ron Mayo, Bertha McHaffie-Gow, Anthony Mercieca, William J. Merilees, A. Morris, Nanny Mulder ten Kate, B. Murphy, National Air Photo Library, Mark Nyhof, Marie O'Shaughnessy, Aurora M. Paterson, Mark Phinney, Michael I. Preston, Chris Siddle, Brian E. Small, M. Stubblefield, D. Tipling, Triathlon Mapping Corporation, Linda M. Van Damme, T. Vezo, Douglas J. Wilson, S. Young, and Tim Zurowski.

Brian Low and Robin V. Quenet (Pacific Forestry Centre, Canadian Forest Service, Victoria) kindly helped in recreating the western spruce budworm map for the Evening Grosbeak account. G.E. John Smith (Canadian Wildlife Service, retired) wrote the computer programs we used to summarize some of the breeding data and produce the species distribution maps for this volume. Dwight McCulloch (Canadian Wildlife Service, Delta) produced draft maps for the diversity of birds in the province. Tony Hamilton and Kristin Karr (Wildlife Branch, B.C. Ministry of Environment, Lands and Parks, Victoria) provided provincial map area summaries for ecoprovinces and biogeoclimatic zones. John R. Sauer and Jane Fallon (Patuxent Wildlife Research Center, United States Geological Survey, Laurel, Maryland) graciously provided digital maps of the North American Christmas Bird Counts and Breeding Bird Surveys. Bryan Krueger (Resources Inventory Branch, B.C. Ministry of Environment, Lands and Parks, Victoria) translated these digital maps into a usable format for this volume. Ann Schau and Rod S. Silver helped obtain some of the early photographs of Ian McTaggart-Cowan for *The Last Word*.

We are grateful to librarians John Pinn and Kathy Neer (B.C. Ministry of Environment, Lands and Parks), and Paul Nystedt (B.C. Ministry of Forests), for help in tracking down many publications and theses that were not available in libraries in Victoria.

We also thank Michel Gosselin (Canadian Museum of Nature, Ottawa) and Christine Adkins (Cowan Vertebrate Museum, University of British Columbia, Vancouver) for their help with various specimen records.

CONTRIBUTING AUTHORS

In order to expedite this final volume, we enlisted the support of 3 people – Richard R. Howie, Chris Siddle, and Linda M. Van Damme – to help us prepare draft accounts for the following species: Black-throated Sparrow, Nelson's Sharp-tailed Sparrow, Harris's Sparrow, Bobolink, Western Meadowlark, Baltimore Oriole, Bullock's Oriole, Brambling, Common Redpoll, and Hoary Redpoll. Dennis A. Demarchi is acknowledged here for his significant contribution to the section, *The Environment*, in Volume 1 of *Birds of British Columbia*. Information about the contributing authors can be found on page 733.

REVIEWERS

All species accounts were updated, reviewed, and edited from a regional and provincial perspective by the following 32 individuals who reside in British Columbia: **Cathy Antoniazzi** (Prince George area), **Jack Bowling** (Prince George and Fort Nelson areas and B.C. general), **Jan Bradshaw** (Harrison and Shuswap areas), **Richard J. Cannings** (Okanagan valley and B.C. general), **Myke J. Chutter** (B.C. general), **Gary S. Davidson** (west Kootenay), **Dennis A. Demarchi** (B.C. general), **Adrian Dorst** (Tofino-Ucluelet area), **Kyle Elliott** (Greater Vancouver), **Bryan R. Gates** (southern Vancouver Island), **Tony Greenfield** (Sunshine Coast and B.C. general), **Margo Hearne** (Queen Charlotte Islands), **Charles Helm** (Tumbler Ridge), **R. Jerry Herzig** (Princeton area), **Richard R. Howie** (Thompson-Nicola valleys and east Kootenay), **Douglas W. Innes** (Comox-Courtenay area), **Sandra Kinsey** (Prince George area), **Nancy Krueger** (Prince George area), **Vi Lambie** (Mackenzie area), **Laird Law** (Prince George area), **Douglas Leighton** (Golden-Blaeberry valley area), **Mark Phinney** (Prince George and Dawson Creek areas), **Tom Plath** (Greater Vancouver), **Douglas Powell** (Revelstoke area), **Anna Roberts** (Cariboo and Chilcotin areas), **Chris Siddle** (Peace River region and B.C. general), **David Stirling** (B.C. general), **Rick Toochin** (Greater Vancouver), **Ruth E. Travers** (central Chilcotin and Cariboo areas), **Jim Tuck** (Mackenzie region), **Linda M. Van Damme** (west Kootenay and B.C. general), and **Ellen Zimmerman** (Golden area).

The following individuals reviewed at least 1 species account in their field of expertise: **Curtis S. Adkisson** (Virginia Tech, Blacksburg, Virginia – Pine Grosbeak), **Daniel A. Airola** (Sacramento, California – Gray-crowned Rosy-Finch), **Elisabeth M. Ammon** (University of Nevada, Reno – Lincoln's Sparrow), **Michael L. Avery** (United States Department of Agriculture, Gainesville, Florida – Rusty Blackbird), **Steve Baillie** (Nanaimo, British Columbia – Yellow-throated Warbler), **Myron C. Baker** (Colorado State University, Fort Collins – White-crowned Sparrow), **Christopher P. Bell** (Imperial College, London, United Kingdom – Fox Sparrow), **Laurence C. Binford** (Baton Rouge, Louisiana – Northern Waterthrush), **Anthony H. Bledsoe** (University of Pittsburgh, Pittsburgh, Pennsylvania – Common Redpoll), **Donald A. Blood** (Donald A. Blood and Associates, Nanaimo, British Columbia – Yellow-rumped Warbler), **Daniel W. Brauning** (Montgomery, Pennsylvania – Chestnut-sided and Canada warblers), **James V. Briske** (University of Canterbury, New Zealand – Yellow Warbler and Smith's Longspur), **Richard R. Buech** (St. Paul, Minnesota – Clay-colored Sparrow), **John S. Castrale** (Forest Wildlife Office, Mitchell, Indiana – Lapland Longspur), **Glen Chilton** (University of Calgary, Calgary, Alberta – White-crowned Sparrow), **George W. Cox** (Biosphere and Biosurvival, Santa Fe, New Mexico – Mourning and MacGillivray's warblers), **Barbara B. DeWolfe** (University of California, Santa Barbara – White-crowned Sparrow), **Therese M. Donovan** (University of Missouri-Columbia, Columbia – Ovenbird), **Stephen W. Eaton** (Allegany, New York – Magnolia Warbler and Northern Waterthrush), **J. Bruce Falls** (Don Mills, Ontario – White-throated Sparrow), **Thomas A. Gavin** (Cornell University, Ithaca, New York – Bobolink), **William M. Gilbert** (El Sobrante, California – Orange-crowned and Wilson's warblers), **Erick Greene** (University of Montana, Missoula – Lazuli Bunting), **Jon S. Greenlaw** (Cape Coral, Florida – Spotted Towhee and Nelson's Sharp-tailed Sparrow), **George A. Hall** (West Virginia University, Morgantown – Magnolia Warbler), **D. Paul Hendricks** (Missoula, Montana – Golden-crowned Sparrow), **Richard T. Holmes** (Dartmouth College, Hanover, New Hampshire – American Redstart), **John P. Hubbard** (Glenwood, New Mexico – Yellow-rumped Warbler), **Jocelyn Hudon** (Provincial Museum of Alberta, Edmonton – Western Tanager), **Peter Hunt** (Mascoma Lake Bird Observatory, Enfield, New Hampshire – Yellow-rumped Warbler), **Richard L. Hutto** (University of Montana, Missoula – MacGillivray's Warbler), **Ross D. James** (Sunderland, Ontario – Magnolia Warbler, White-throated Sparrow, and Rose-breasted Grosbeak), **Brina Kessel** (University of Alaska, Fairbanks – American Tree Sparrow), **Richard W. Knapton** (Fonthill, Ontario – Nashville Warbler, Chipping Sparrow, and Clay-colored Sparrow), **John C. Kricher** (Wheaton College, Norton, Massachusetts – Black-and-white Warbler), **Roger L. Kroodsma** (Oak Ridge, Tennessee – Black-headed Grosbeak), **Peter E. Lowther** (Field Museum of Natural History, Chicago, Illinois – Le Conte's Sparrow and Brown-headed Cowbird), **George A. Lozano** (McGill University, Montreal, Quebec – American Redstart), **Scott A. MacDougall-Shackleton** (University of Toronto, Toronto, Ontario – Gray-crowned Rosy-Finch), **Nancy Mahony** (University of British Columbia, Vancouver – Brewer's Sparrow), **Stephen G. Martin** (S.G. Martin and Associates, Inc., Wellington, Colorado – Bobolink), **Steven M. Matsuoka** (United States Geological Survey, Alaska Biological Science Center, Anchorage – Townsend's Warbler), **Douglas B. McNair** (Tall Timbers Research Station, Tallahassee, Florida – Grasshopper Sparrow), **Martin K. McNicholl** (Burnaby, British Columbia – Tennessee Warbler, Yellow-rumped Warbler, Townsend's Warbler, Wilson's Warbler, and Yellow-breasted Chat), **Alex L.A. Middleton** (University of Guelph, Guelph, Ontario – Chipping Sparrow and American Goldfinch), **Robert D. Montgomerie** (Queen's University, Kingston, Ontario – Snow Bunting), **Michael L. Morrison** (California State University, Sacramento – Orange-crowned and Townsend's warblers), **Douglass H. Morse** (Brown University, Providence, Rhode Island – Tennessee Warbler), **Thomas B. Mowbray** (Salem College, Winston-Salem, North Carolina – Swamp Sparrow), **Robert W. Nero** (Winnipeg, Manitoba – Red-winged Blackbird), **Erica Nol** (Trent University, Peterborough, Ontario – Ovenbird), **Christopher J. Norment** (Brockport State University of New York, Brockport – Harris's Sparrow), **Gordon H. Orians** (University of Washington, Seattle – Rusty Blackbird), **Cynthia A. Paszkowski** (University of Alberta, Edmonton – Black-and-white Warbler), **Robert B. Payne** (University of Michigan, Ann Arbor – Indigo Bunting), **Mark Phinney** (Louisiana-Pacific Corporation, Dawson Creek, British Columbia – Black-throated Green Warbler), **F. Jay Pitocchelli** (Saint Anselm College, Manchester, New Hampshire – Mourning Warbler and MacGillivray's Warbler), **Michael I. Preston** (Calgary, Alberta – Western Tanager), **J. Michael Reed** (Tufts University, Medford, Massachusetts – Vesper Sparrow), **Steven E. Reinert** (Barrington, Rhode Island – Swamp Sparrow), **Michael Richardson** (Brighton, Ontario – Chestnut-sided Warbler), **James D. Rising** (University of Toronto, Toronto, Ontario – Vesper Sparrow, Savannah Sparrow, Fox Sparrow, Song Sparrow, Lapland Longspur, Baltimore Oriole, and Bullock's Oriole), **Spencer G. Sealy** (University of Manitoba, Winnipeg – Yellow Warbler, Cape May Warbler, Bay-breasted Warbler, and Brown-headed Cowbird), **W. Dave Shuford** (Point Reyes Bird Observatory, Stinson Beach, California – Common Redpoll), **James N.M. Smith** (University of British Columbia, Vancouver – Song Sparrow and Brown-headed Cowbird), **Kimberly G. Smith** (University of Arkansas, Fayetteville – Dark-eyed Junco), **Navjot S. Sodhi** (National University of Singapore, Singapore – Black-and-white Warbler), **Charles F. Thompson** (Illinois State University, Normal – Yellow-breasted Chat), **Declan M. Troy** (Troy Ecological Research Associates, Anchorage, Alaska – Hoary Redpoll), **Peter D. Vickery** (Maine Audubon Society, Richmond – Grasshopper Sparrow), **George C. West** (Green Valley, Arizona – Lapland Longspur), **Nathaniel T. Wheelwright** (Bowdoin College, Brunswick, Maine – Savannah Sparrow), **John A. Wiens** (Colorado State University, Fort Collins – Grasshopper Sparrow), **Janet McI. Williams** (Swarthmore College, Swarthmore, Pennsylvania – Nashville and Bay-breasted warblers), **W. Herbert Wilson** (Colby College, Wareville, Maine – Palm Warbler), **John G. Woods** (Parks Canada, Revelstoke, British Columbia – Evening Grosbeak), **J. Timothy Wootton** (University of Chicago, Chicago, Illinois – Purple Finch), **Robert P. Yunick** (Schenectady, New York – Evening Grosbeak), and **Robert M. Zink** (University of Minnesota, Bell Museum, St. Paul – Fox Sparrow).

The final section – *Synopsis* – benefited greatly from information or editorial comments provided by the following: *Avian Biodiversity, Ecological Distribution, and Patterns of Change:* André Breault, Barry Booth, Les Gyug, Rosamond Pojar, James Quayle, G.G.E. Scudder, Anthony R.E. Sinclair (University of British Columbia), and Michaela Waterhouse. *What Lies in Store for the Birds of British Columbia?:* Ron Buechert, David W. Ehrenfeld (Rutgers, New Brunswick, New Jersey), Bob Elner, Susan Hannon (University of Alberta, Edmonton), Richard W. McKelvey, Kathleen Moore, Ken Ryan, and Peggy Ward.

APPENDICES

The regional migration dates listed in Appendix 1 were completed by the following people: **J.E. Victor Goodwill** and **Bryan R. Gates** (Victoria), **Adrian Dorst** (Tofino), **Wayne C. Weber** and **Kyle Elliott** (Vancouver), **Jan Bradshaw** (Harrison), **Tony Greenfield** (Sunshine Coast), **Harvey Thommasen** (Bella Coola), **Margo Hearne** and **Peter J. Hamel** (Queen Charlotte Islands), **Richard J. Cannings** (Okanagan), **Gary S. Davidson** (Nakusp and Fort Nelson), **Richard R. Howie** (Kamloops), **Anna Roberts** (Williams Lake), **Manfred Roschitz** (Quesnel), **Jack Bowling** (Prince George and Fort Nelson), **Vi** and **John Lambie** and **Jim Tuck** (Mackenzie), and **Chris Siddle** (Fort St. John).

More than 40 years of Christmas Bird Counts summarized in Appendix 2, were compiled onto species spreadsheets by Eileen C. Campbell, Tessa N. Campbell, Maureen L. Funk, Laura Gretzinger, Edward L. Nygren, and Pam Stacey.

PRODUCTION

Again, our affiliation with UBC Press and their associates made the onerous task of "putting the *albatross* to bed" a pleasant experience. Professional help and advice were provided by Peter Milroy (Director) and Holly Keller-Brohman (Managing Editor); Francis Chow (F & M Chow Consulting) copy-edited the book and taught the authors much about syntax; Deborah Kerr proofread the manuscript; Eric Leinberger designed and prepared the maps and technical figures; and Irma Rodriguez (Artegraphica Design Co. Ltd.) typeset and completed the layout of the book. Chris Tyrrell (Royal British Columbia Museum) designed the jacket.

INTRODUCTION

This is the fourth and final volume on the avifauna of British Columbia; it completes a discussion of the passerine birds of the province that began in Volume 3. Coming some 10 years following the publication of Volumes 1 and 2 (the non-passerine birds), it is the culmination of over 25 years of effort by the authors and significant contributions from colleagues around the world and thousands of volunteers (Fig. 1) throughout British Columbia.

The methodologies used in the preparation of this work have previously been described (Volume 1, page 146, and Volume 3, page 13) and will not be repeated here.

Because of the length of time between the publication of Volumes 1 and 2 (1990) and this final volume, we have included a chapter called "Additions to the Avifauna of British Columbia, 1987 through 1999." There we list 28 species new to the province that were reported over that period. Some species, such as Manx Shearwater, Crested Caracara, Xantus's Hummingbird, and Least Tern are new occurrence additions to the avifauna of the province. Others, such as Northern Fulmar, Baird's Sandpiper, Franklin's Gull, and Black-legged Kittiwake (Fig. 2), now breed in the province. One species, Pacific Golden-Plover, has been elevated from subspecies to full species status due to taxonomic changes.

Beyond the scope of the chapter are the many new changes in numbers, distribution, and results of research that contribute to our current knowledge and status of birds in the province. For example, the Flammulated Owl, Trumpeter Swan, and Black Tern (Fig. 3), have ranges in the province that are much greater than was previously known. And for others, such as the Barrow's Goldeneye and Marbled Murrelet, research has provided new information significant to the conservation and management of the species. In the case of the goldeneye, mature trembling aspen forest that provides nesting cavities is being harvested at an incredible rate to make hardwood products for the world market. In the case of the murrelet, an ornithological enigma was solved in 1991 with the discovery of the first murrelet nest for British Columbia in old-growth forest along Walbran Creek.

Figure 1. Over 11,000 volunteers contributed information to the data bases that were used to prepare *The Birds of British Columbia.* J.E. Victor Goodwill and Margaret E. Goodwill have been among our most consistent and supportive contributors (Victoria, 15 March 2000; R. Wayne Campbell).

Figure 2. Adult Black-legged Kittiwake with chick at nest ledge (Gjelpruvaer, Norway, 11 July 1994; R. Wayne Campbell).

Figure 3. Over the past 2 decades, the breeding distribution of the Black Tern has expanded northward and westward in British Columbia (south of Prince George, 13 June 1998; R. Wayne Campbell).

A final section, "Synopsis: The Birds of British Columbia into the 21st Century," summarizes significant information on the avifauna of the province from all 4 volumes. Here we discuss aspects of the biodiversity, ecological distribution, and patterns of change we have noted along with rare, threatened or endangered species and those species experiencing long-term changes in numbers. The national and international significance of select species and our responsibilities to ensure that viable populations survive for future generations are also covered, along with future management needs. That is followed by our thoughts on some new philosophies, concerns, and conservation challenges we believe will play a role in the future of the birds of the province.

Finally, we end much as we began, by respectfully reminding the reader

> that this [has truly been] a cooperative work; if he fails to find in these volumes anything that he knows about the birds, he can blame himself for not having sent the information to

THE AUTHORS

CHECKLIST OF BRITISH COLUMBIA BIRDS

Passerines: Wood-Warblers through Old World Sparrows

This phylogenetic list includes 101 species of birds, Wood-Warblers through Old World sparrows, that have been documented in British Columbia through 31 December 1999.

Family PARULIDAE: Wood-Warblers
- Tennessee Warbler
- Orange-crowned Warbler
- Nashville Warbler
- Northern Parula
- Yellow Warbler
- Chestnut-sided Warbler
- Magnolia Warbler
- Cape May Warbler
- Black-throated Blue Warbler
- Yellow-rumped Warbler
- Black-throated Gray Warbler
- Black-throated Green Warbler
- Townsend's Warbler
- Hermit Warbler
- Blackburnian Warbler
- Yellow-throated Warbler
- Prairie Warbler
- Palm Warbler
- Bay-breasted Warbler
- Blackpoll Warbler
- Black-and-white Warbler
- American Redstart
- Ovenbird
- Northern Waterthrush
- Connecticut Warbler
- Mourning Warbler
- MacGillivray's Warbler
- Common Yellowthroat
- Hooded Warbler
- Wilson's Warbler
- Canada Warbler
- Painted Redstart
- Yellow-breasted Chat

Family THRAUPIDAE: Tanagers
- Scarlet Tanager
- Western Tanager

Family EMBERIZIDAE: Towhees, Sparrows, Longspurs, and Allies
- Green-tailed Towhee
- Spotted Towhee
- American Tree Sparrow
- Chipping Sparrow
- Clay-colored Sparrow
- Brewer's Sparrow
- Vesper Sparrow
- Lark Sparrow
- Black-throated Sparrow
- Sage Sparrow
- Lark Bunting
- Savannah Sparrow
- Baird's Sparrow
- Grasshopper Sparrow
- Le Conte's Sparrow
- Nelson's Sharp-tailed Sparrow
- Fox Sparrow
- Song Sparrow
- Lincoln's Sparrow
- Swamp Sparrow
- White-throated Sparrow
- Harris's Sparrow
- White-crowned Sparrow
- Golden-crowned Sparrow
- Dark-eyed Junco
- McCown's Longspur
- Lapland Longspur
- Smith's Longspur
- Chestnut-collared Longspur
- Rustic Bunting
- Snow Bunting
- McKay's Bunting

Family CARDINALIDAE: Cardinals, Grosbeaks, and Allies
- Rose-breasted Grosbeak
- Black-headed Grosbeak
- Blue Grosbeak
- Lazuli Bunting
- Indigo Bunting
- Dickcissel

Family ICTERIDAE: Blackbirds, Orioles, and Allies
- Bobolink
- Red-winged Blackbird
- Western Meadowlark
- Yellow-headed Blackbird
- Rusty Blackbird
- Brewer's Blackbird
- Great-tailed Grackle
- Common Grackle
- Brown-headed Cowbird
- Orchard Oriole
- Hooded Oriole
- Baltimore Oriole
- Bullock's Oriole

Family FRINGILLIDAE: Cardueline Finches and Allies
- Brambling
- Gray-crowned Rosy-Finch
- Pine Grosbeak
- Purple Finch
- Cassin's Finch
- House Finch
- Red Crossbill
- White-winged Crossbill
- Common Redpoll
- Hoary Redpoll
- Pine Siskin
- Lesser Goldfinch
- American Goldfinch
- Evening Grosbeak

Family PASSERIDAE: Old World Sparrows
- House Sparrow

Regular Species

Tennessee Warbler TEWA

Vermivora peregrina (Wilson)

RANGE: Breeds from southeastern Alaska, southern Yukon, and northwestern and southern Mackenzie east across Canada from northern Alberta to northern Quebec, southern Labrador, and western Newfoundland; south to south-central British Columbia, southwestern Alberta, southern Saskatchewan, and southern Manitoba, northern Minnesota, northern Wisconsin, northern Michigan, southern Ontario, northeastern New York, northeastern Vermont, northern New Hampshire, central Maine, and Nova Scotia. Winters from Oaxaca and Veracruz south to Colombia and northern Venezuela.

STATUS: On the coast, *casual* in the Georgia Depression Ecoprovince; in the Coast and Mountains Ecoprovince, *casual* on the Southern Mainland Coast and Western Vancouver Island, *very rare* migrant and summer visitant on the Northern Mainland Coast, and absent from the Queen Charlotte Islands.

In the interior, *very rare* migrant and *casual* summer visitant in the Southern Interior and Central Interior ecoprovinces; *rare* migrant and irregular summer visitant in the Southern Interior Mountains Ecoprovince; *rare* to locally *uncommon* migrant and summer visitant in the Sub-Boreal Interior Ecoprovince; *fairly common* migrant and summer visitant in the Taiga Plains Ecoprovince; *very rare* migrant and summer visitant in the Boreal Plains and Northern Boreal Mountains ecoprovinces.

Breeds.

NONBREEDING: The Tennessee Warbler (Fig. 4) occurs very infrequently on the south coast, where it is known from various locations in the Lower Mainland and on the central west coast of Vancouver Island. Further north, the species occasionally reaches the coastal slope at Bella Coola, Kimsquit, Kitimat, and the lower Skeena valley, in each instance where a valley cuts through the mountains from the interior.

In the interior, the Tennessee Warbler occurs as an infrequent migrant in the Okanagan and Thompson valleys, and more frequently in the valleys of the west and east Kootenay regions. In the vicinity of Golden, most of the Tennessee Warblers appear to come from the east through various low-elevation routes such as the lower Blaeberry River valley, Athabasca Pass, and Yellowhead Pass, where there is an almost unbroken path of habitat from their range on the eastern slopes of the Rocky Mountains in Alberta (D. Leighton pers. comm.). This is the logical flow at the end of their spring route, curving north of the prairies then southwestward following suitable habitat down the eastern slope of the Rocky Mountains.

There are few records from the Cariboo and Chilcotin areas, but the Tennessee Warbler occurs more regularly in the valleys of the upper Fraser and Nechako rivers. It is relatively numerous from there northward into the Nechako Lowland and the valleys of the Skeena and Omineca Mountains ecoregions of the Sub-Boreal Interior, the Liard Basin of the

Figure 4. The Tennessee Warbler occurs most regularly in northeastern areas of British Columbia, especially east of the Rocky Mountains (Anthony Mercieca).

Northern Boreal Mountains, the Peace Lowland of the Boreal Plains, and the Fort Nelson Lowland of the Taiga Plains. In the latter area it has been recorded north to Kwokullie Lake. There are few records of this species in the Northern Boreal Mountains, but the species has been reported at Telegraph Creek, McDonald Creek, Muncho Lake, Liard River, and Atlin. Given the somewhat sporadic presence of bird observers in that vast region, the reports suggest a continuous distribution where habitat is suitable. The most northwesterly record is from Atlin.

During spring migration, the Tennessee Warbler occurs at elevations from near sea level to about 915 m, but in the autumn it has been recorded from elevations up to 1,200 m.

During spring and autumn migration, this species uses a variety of low-elevation habitats in which there is usually a mix of coniferous and deciduous trees. In the Okanagan valley, most records have been from gardens (Cannings et al. 1987). In Kootenay National Park, however, it prefers open and closed wetland forests dominated by spruce; the most important vegetation types include associations of spruce–buffalo berry–feather moss, or dwarf birch–shrubby cinquefoil–willow–brown moss, and spruce–Labrador tea–brown moss (Poll et al. 1984). Near Golden, it frequents mixed riparian spruce–trembling aspen–black cottonwood habitats (D. Leighton pers. comm.). As the spring migration moves closer to the nesting grounds, mixed forests of trembling aspen and white spruce with an open texture appear to be favoured. In the Peace Lowland, migrating flocks in late May moved rapidly through the greening tops of tall second-growth trembling aspens (Cowan 1939). At the western edge of the species' range near Smithers, it was present but scarce in sapling second-growth trembling aspen following clearcutting; it was not found in the young clearcuts or in the mature trembling aspen forest (Pojar 1995).

Brooks and Swarth (1925) note that the "line of migration is probably east of the Rocky Mountains," and the

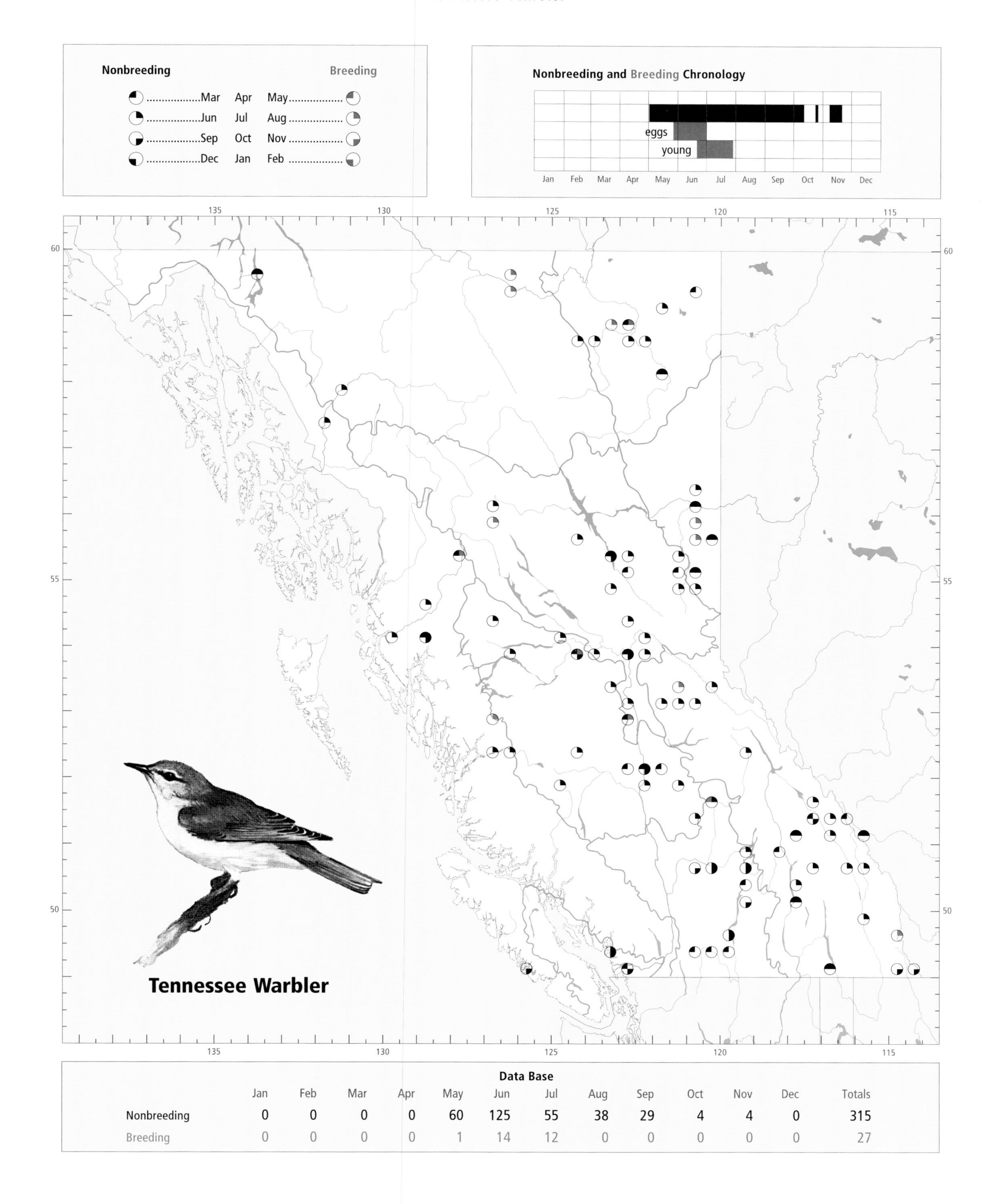

Data Base

	Jan	Feb	Mar	Apr	May	Jun	Jul	Aug	Sep	Oct	Nov	Dec	Totals
Nonbreeding	0	0	0	0	60	125	55	38	29	4	4	0	315
Breeding	0	0	0	0	1	14	12	0	0	0	0	0	27

evidence still supports this conclusion for the main movement. There are, however, data from the Southern Interior and Southern Interior Mountains pointing to a small but persistent population of Tennessee Warblers that migrates both north and south through the interior of the province. The species is known as a rare migrant in Washington and Oregon (Roberson 1980). In California it is a rare spring transient (about 25 reports per year), and rare to very common autumn transient (an average of 60 to 65 reports per year). Small numbers annually winter in California (Small 1994).

Spring migrants can arrive in southern portions of the interior of British Columbia as early as the first week of May (Figs. 5 and 6). The peak movement, however, occurs in the last 2 weeks of May. North of the latitude of Prince George (about 54°N), the dates of arrival of the first spring migrants and of the peak movement are similar to those of the southern portions of the interior, and suggest that this northern migration arises from a different migratory population. Birds arrive on the Northern Mainland Coast following a similar timetable and likely derive from the northern population. These arrival times are similar to those recorded in Alberta (Sadler and Myres 1976).

In some years, autumn migration from the northeastern nesting grounds begins as early as late July and is noticeable in all regions by mid-August (Figs. 5 and 6), soon after the young are independent. Most birds have left the area by mid-September. In the south, although there are too few records to provide much confidence, autumn records in the south Okanagan valley and in the Kootenays have been concentrated between the first week of August and the third week of September. The main movement is over by mid-September, after which sightings are unusual (Fig. 5). Most birds have left the Northern Mainland Coast by the end of August.

On the coast, the Tennessee Warbler has been recorded irregularly from 11 May to 20 November; in the interior, it has been recorded from 2 May to 10 October (Fig. 5).

BREEDING: In the interior, the Tennessee Warbler breeds in suitable habitats, from as far south as the vicinity of Sparwood in southeastern British Columbia and northwest through Quesnel and Nulki Lake (Munro 1949). However, it mainly breeds north of latitude 55°N, in the southern Sub-Boreal Interior, north through the Northern Boreal Mountains to the Liard River valley, and east throughout the Boreal Plains and Taiga Plains. East of the Rocky Mountains, localities of confirmed nesting include Dawson Creek, Jackfish Lake (near Chetwynd), Charlie Lake, Fort Nelson, and near Mile 335 on the Alaska Highway. There are isolated breeding records from the north coast at Kimsquit (Laing 1942) and near Hazelton.

Only 17 nests have been reported in the province, but Breeding Bird Surveys have identified large numbers of Tennessee Warblers on routes along the Alaska Highway from Dawson Creek to the Fort Nelson Lowland (see Appendix 3). Bennett and Enns (1996) considered the Tennessee Warbler to be an "abundant" species in their songbird study on the Liard River in the Taiga Plains. In fact, it was the most abundant bird they detected (452 birds); the next most abundant was the Swainson's Thrush (200 birds). Recent bird surveys in the early to mid-seral Engelmann spruce–subalpine fir forests of

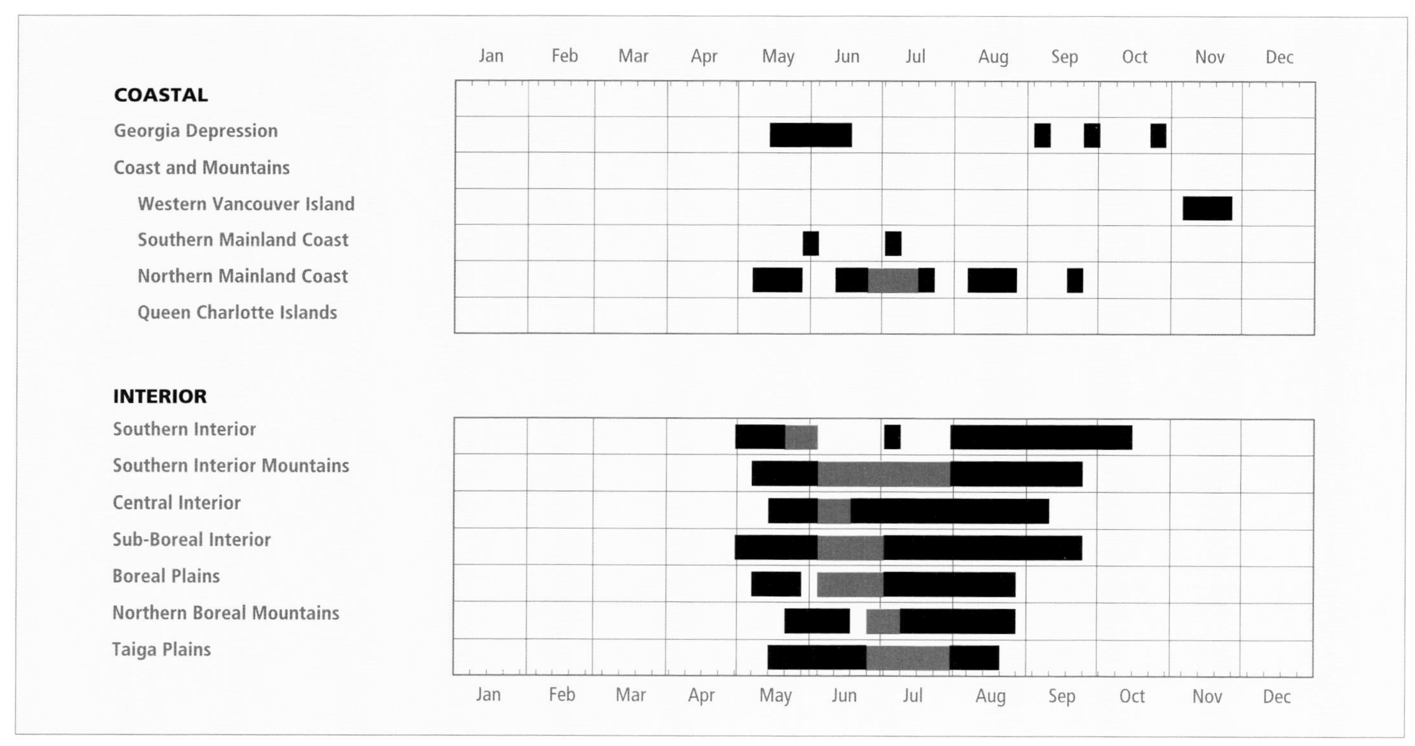

Figure 5. Annual occurrence (black) and breeding chronology (red) for the Tennessee Warbler in ecoprovinces of British Columbia. Records are shown for the week in which they occurred.

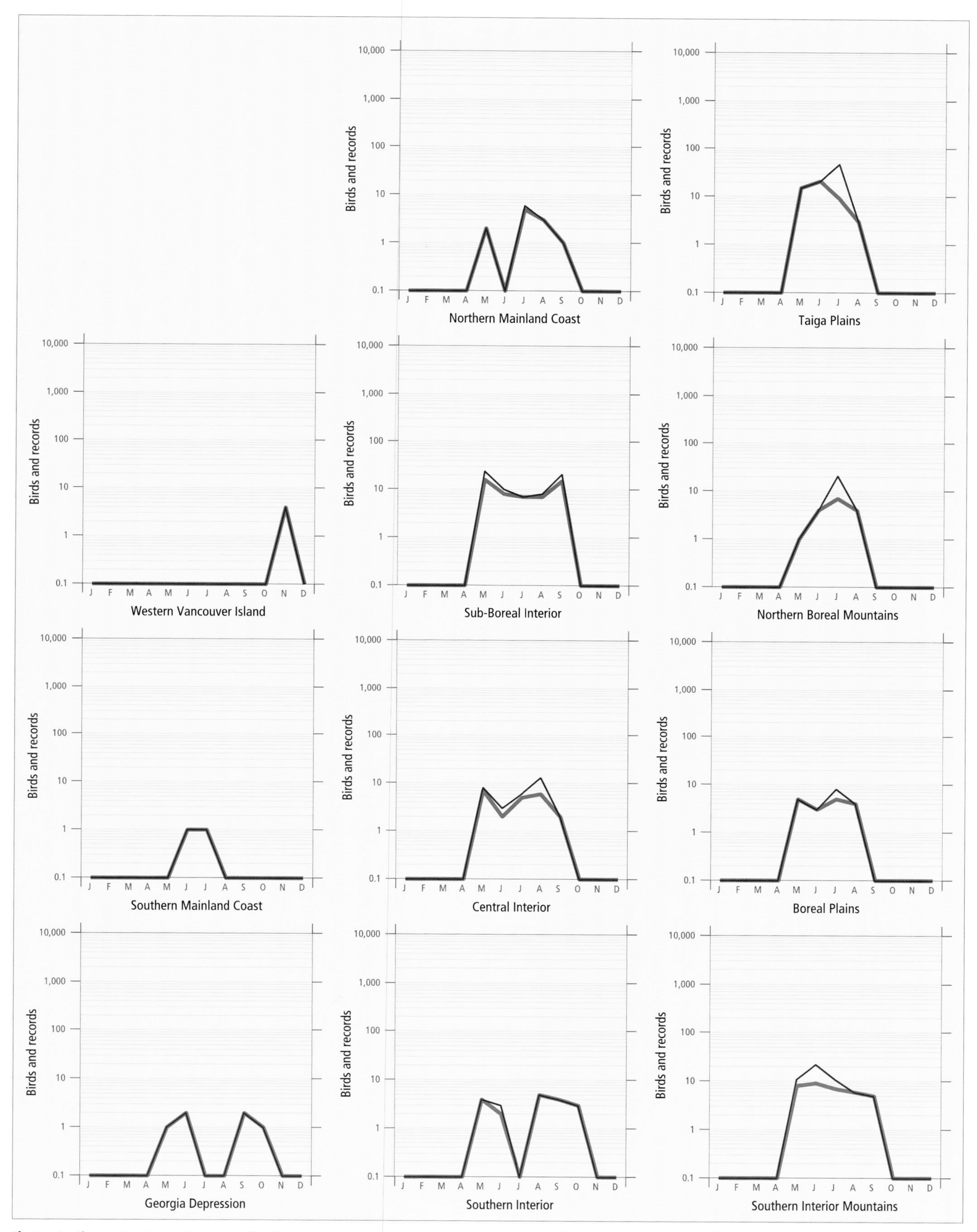

Figure 6. Fluctuations in total number of birds (purple line) and total number of records (green line) for the Tennessee Warbler in ecoprovinces of British Columbia. Breeding Bird Surveys and nest record data have been excluded.

the Quesnel Highlands near Barkerville suggest that this warbler is more common there than previously thought, and likely breeds (M.J. Waterhouse pers. comm.). There is little doubt that further work will extend the known breeding distribution of the species in the province (see REMARKS).

The Tennessee Warbler reaches its highest numbers in summer in the Fort Nelson Lowland of the Taiga Plains (Fig. 7). An analysis of Breeding Bird Surveys in British Columbia for the period 1968 through 1993 could not detect a net change in numbers on interior routes. Breeding Bird Surveys across Canada from 1980 through 1993 indicate that numbers declined at an average annual rate of 5.3% during this period ($P < 0.01$); continent-wide, numbers declined at an average annual rate of 5.1% during the same period ($P < 0.10$) (Sauer et al. 1997).

The Tennessee Warbler breeds near sea level on the coast and from 360 to 1,150 m elevation in the interior. The nesting habitat of this warbler in British Columbia includes a variety of forested habitats generally associated with moisture and a shrubby understorey or shrubby edge. In the Boreal Plains, habitats include associations of trembling aspen–balsam poplar, white spruce–trembling aspen, trembling aspen–tamarack–willow–birch–black spruce, pure trembling aspen, and alder-willow (Siddle, pers. comm.). Further north, in the Taiga Plains, the warbler frequents stands of old-growth and mature balsam poplars with a dense understorey of alders, trembling aspen, and rose (Fig. 8); mature white spruce–trembling aspen forest where occasional large openings are covered with alder, rose, and red-osier dogwood; and alder–scrub birch thickets bordering black spruce muskeg (C. Siddle pers. comm.). The Tennessee Warbler also utilizes thickets of trembling aspen, birch, and willow scrub on hillsides; willow and spruce with an understorey of shrubs; and mixed spruce–trembling aspen forests, where it frequents the understorey of alders and willows. In a study on the Liard River, Bennett and Enns (1996) found that the density of Tennessee Warblers was highest in mature conifer riparian habitat. This warbler has also been found nesting in relatively dry forests of trembling aspen in association with lodgepole pine (Brooks 1901). Nests were associated with edges around clearings, roadsides, ponds, sphagnum bogs (Fig. 8), and farmsteads.

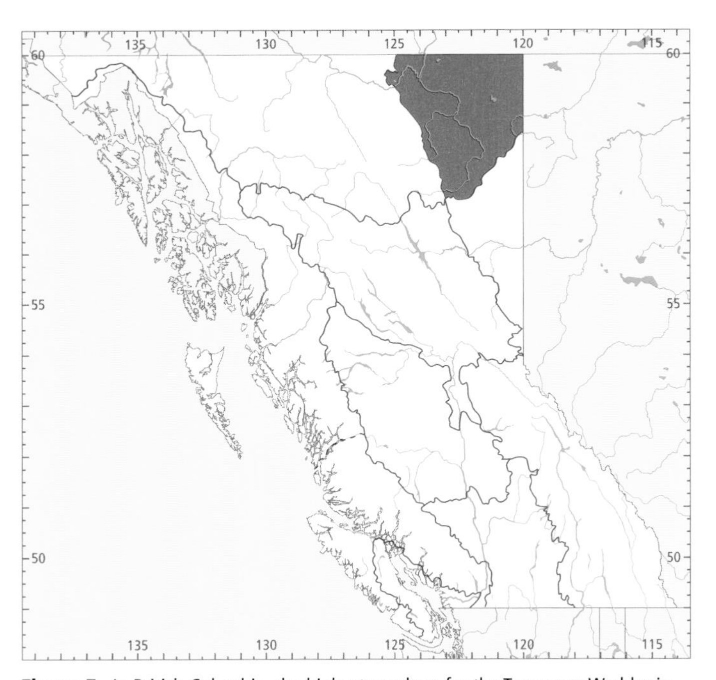

Figure 7. In British Columbia, the highest numbers for the Tennessee Warbler in summer occur in the Taiga Plains Ecoprovince.

Figure 8. Woodlands of mature balsam poplar with young stands of Sitka alder and a dense understorey of rose provide nesting habitat for the Tennessee Warbler in the Fort Nelson Lowland of the Taiga Plains Ecoprovince (3 km south of Fort Nelson, 29 June 1997; R. Wayne Campbell).

On the coast, the Tennessee Warbler has been recorded breeding from 29 June (calculated) to 14 July; in the interior, it has been recorded breeding from 27 May (calculated) to 28 July (Fig. 5).

Nests: The Tennessee Warbler almost always builds its nest on the ground. In the muskeg terrain that is characteristic of much of its breeding habitat in northern British Columbia (Fig. 8), it frequently uses sphagnum hummocks as a nest site. Other sites have been under a mossy overhanging bank and in a recess in a mossy stump. In dry trembling aspen–lodgepole pine stands in the Cariboo and Chilcotin areas, nests were placed on the ground, close to the base of a shrub and well sheltered by dry pine grass (Brooks 1901). Characteristically, ground nests are shielded from view by overhanging dry grasses. One nest was found about 3 m from the ground, on the trunk of a trembling aspen (M. Roschitz pers. comm.). Nests were constructed of dry grasses, mosses, and bark strips, and lined with fine grasses, rootlets, mosses, and sometimes mammal hair.

Eggs: Dates for 12 clutches ranged from 30 May to 30 June, with 50% recorded between 11 June and 21 June. Calculated dates suggest that nests can hold eggs as early as 27 May. Sizes of 8 clutches ranged from 2 to 6 eggs (2E-1, 3E-1, 4E-1, 5E-1, 6E-4). Elsewhere in North America, this species has been found to lay as many as 8 eggs, "more than any other warbler species" (Morse 1989). The incubation period is 11 to 12 days (Knight 1908; Baicich and Harrison 1997).

Young: Dates for 15 broods ranged from 21 June to 28 July, with 67% recorded between 7 July and 14 July. Sizes of 9 broods ranged from 1 to 6 young (1Y-3, 2Y-1, 3Y-1, 4Y-1, 5Y-1, 6Y-2). Rimmer and McFarland (1998) report the nestling period to be from 11 to 12 days. In British Columbia, the nestling period for 1 brood was 7 days (L. Law pers. comm.). There is no evidence that the Tennessee Warbler raises 2 broods a season.

Brown-headed Cowbird Parasitism: In British Columbia, only 1 instance of cowbird parasitism was reported from 11 nests found with eggs or young. In Maine, 8 nests located were without cowbird parasitism (Harrison 1984). Friedmann and Kiff (1985) note 8 occurrences of Brown-headed Cowbird parasitism of the Tennessee Warbler, and include the latter in their list of species that have been known to rear cowbird young.

Nest Success: One nest found with eggs and followed to a known fate produced 1 fledgling.

REMARKS: Territorial males singing throughout June and July in the lower Blaeberry River valley, 35 km north of Donald, strongly suggest that the Tennessee Warbler may breed locally in northern portions of the Southern Interior Mountains (D. Leighton pers. comm.).

In parts of eastern North America, the Tennessee Warbler shows cyclic fluctuations in numbers correlating with outbreaks of eastern spruce budworm (*Choristoneura fumiferana*) (Curson et al. 1994). Apparently it is one of the small group of forest warblers that prey heavily on both the larvae and adults of this moth. It is not known whether it responds similarly to outbreaks of other defoliating insects in its western range, although its numbers here vary locally and annually far more than those of any other warbler. For example, Holroyd and Van Tighem (1993) found the Tennessee Warbler to be "uncommon" in suitable habitat in the Athabasca valley in Jasper National Park, Alberta. Cowan (1955) stated that in 1944, 1945, and 1946 this species was one of the most abundant warblers in the Athabasca valley between Snaring and Pocahontas. On 16 June 1992, D. Leighton (pers. comm.) counted 32 Tennessee Warblers on a search of the same stretch, and again found it to be the most abundant warbler.

The status of this species on the coast requires clarification. Many records lack confirmatory details and have not been included in this account. They include the following published reports: Comox autumn 1970-1 (Taylor 1984); Cortes Island 3 June 1973-1 (Taylor 1984); Comox Lake 17 June 1973-1 on Breeding Bird Survey; Campbell River 1 September 1973-1, 26 January to February 1983-1, and 26 June 1983-1 (Taylor 1984); Victoria 24 August 1974-1 and 9 and 20 May 1982-1 (Taylor 1984); Chilliwack 26 June 1976-4 on Breeding Bird Survey; Saanich 9 May 1981-1 (Taylor 1984) and 17 September 1983-2 (Hunn and Mattocks 1984); Vancouver 7 September 1984-1 (Hunn and Mattocks 1985) and 18 May 1990-1 (Siddle 1990b); and Victoria 10 May 1993-1 (Siddle and Bowling 1993). In the interior of the province, Tennessee Warbler reports from Grand Forks, Kuskonook, and Chilkat Pass recorded on Breeding Bird Surveys have also been excluded from this account.

For more information on identification and a description of age classes of this warbler, see Bradshaw (1992) and Curson et al. (1994). For a general treatment of warbler biology, including this species, see Morse (1989), Dunn and Garrett (1997), and Rimmer and McFarland (1998).

NOTEWORTHY RECORDS

Spring: Coastal – Deer Lake (Burnaby) 18 May 1982-1 (Campbell 1982d; Weber 1985); Kitimat 11 May 1975-1 adult (Hay 1976). **Interior** – Creston 30 May 1984-1; 6 km e Princeton 20 May 1989-1 foraging in trembling aspen grove in open range at 914 m elevation; 11 km e Princeton 8 May 1994-1 gleaning leaves in mountain-ash on Old Hedley Rd; West Bench 6 May 1980-1 (Cannings et al. 1987); Vernon 15 May 1984-1 (Rogers 1984a); Nakusp 26 May 1990-1 (Siddle 1990b); Revelstoke 8 May 1979-1; Illecilewaet 20 May 1981-1; Vermilion Crossing 31 May 1983-1 singing male (Poll et al. 1984); Wapta Lake 27 May 1977-4; Williams Lake 16 May 1968-1; Cariboo 23 May 1901-1; Puntchesakut Lake 27 May 1944-1 (Munro and Cowan 1947); Nulki Lake 18 May 1945-1 (Munro 1949); Bullmoose Creek 2 and 3 May 1995-1; Tupper Creek 20 May 1938-1, first arrival (Cowan 1939); Helmut (Kwokullie Lake) 28 May 1982-1; Atlin 26 May 1934-1 (Swarth 1936).

Summer: Coastal – Vancouver 3 Jun 1987-1 in Jericho Park (Harrington-Tweit and Mattocks 1987), 17 Jun 1997-1 adult female at University of British Columbia; Bella Coola 3 Jun 1933-1 male collected (Dickinson 1953); Dean River 14 Jul 1939-female and nestling collected (Laing 1942); Kimsquit 18 Jul 1939-1 male collected (Laing 1942); Kitimat 10 Jul 1975-2, 23 Aug 1975-1 nr Eurocan dock (Hay 1976); Contact Creek 10 Jun 1972-4. **Interior** – North Galbraith Creek 23 Jul 1983-3 newly fledged young; Creston 31 Jun 1980-1; Michel 28 Jun 1980-3 nestlings; West Bench 1 Jun 1972-1 (Cannings et al. 1987), 5 Aug 1993-1 (Bowling 1994b), 27 Aug 1974-1 (Cannings et al. 1987); Nakusp 10 Jul 1985-1 (Rogers 1985); Brocklehurst 20 Aug 1989-1 (Weber and Cannings 1990); Nicholson 26 Jun 1976-1, 5 Aug 1993-1; 35 km n Donald 16 Jun 1995-4 males, 27 Jun 1995-1 male, all on territory, suggests nesting; Chezacut Lake 2 Aug 1931-1 (UBC 4684); Kleena Kleene 16 Jul 1932-1 (Dickinson 1953); Ten Mile Lake 1 Jul 1936-1; Bowron Lake Park 29 Jun 1990-9 (Siddle 1990c); Ootsa Lake 12 Jul 1936-2; Nulki Lake 11 Jun 1946-6 eggs (Munro 1949), 15 Aug 1946-1 (UBC 4682), 27 Aug 1946-1 (Munro 1949); Prince George 18 Jun 1984-3; nr Donna Creek 8 Jul 1993-adult feeding 3 fledged young; 23 km s Groundbirch 17 Jun 1993-6, 21 Jun 1993-4 newly hatched nestlings; nr Arras (Dawson Creek) 12 Jun 1992-3 eggs plus 2 Brown-headed Cowbird eggs; Beatton Park 10 Jul 1987-1 male feeding fledgling; Grindrod 13 Aug 1947-1 (UBC 7951); 8 km sw Telegraph Creek 12 Jun 1919-1 (Swarth 1922; MVZ 40155); 7 km wnw Clarke Lake 3 Jul 1986-2 females feeding 2 fledglings; Cecil Lake 3 Jul 1978-3 singing males (Campbell 1978b); Mile 335 Alaska Highway 28 Jul 1985-6; 1 km e Fort Nelson airport 8 Jul 1986-1 juvenile about 3 days fledged; Liard Hot Springs 21 Aug 1943-1, 2 Jul 1992-3 pairs feeding newly fledged young; Atlin 26 Jul 1929-1.

Breeding Bird Surveys: Coastal – Recorded from 2 of 27 routes and on 1% of all surveys. See REMARKS. Maxima: Kwinitsa 23 Jun 1968-3; Kitsumkalum 22 Jun 1975-1. **Interior** – Recorded from 25 of 73 routes and on 13% of all surveys. Maxima: Steamboat 14 Jun 1976-51; McLeod Lake 15 Jun 1968-37; Ferndale 9 Jun 1968-22.

Autumn: Interior – Mackenzie 2 Sep 1994-1; West Lake (Prince George) 21 Sep 1991-2; Nulki Lake 2 Sep 1945-1 (Munro 1949); Williams Lake 9 Sep 1965-1; lower Blaeberry River valley 10 Sep 1995-1, 19 Sep 1995-1; Grindrod 7 Oct 1949-1 (Munro 1953); Tranquille 21 Sep 1980-1; Brocklehurst 2 Oct 1993-1; Okanagan Landing 18 Sep 1927-1 (Munro and Cowan 1947), 10 Oct 1937-1 (Cannings et al. 1987). **Coastal** – Kitimat River 19 Sep 1974-1 adult (Hay 1976); Vancouver 30 Sep 1982-1 (Weber 1985), 25 Oct 1930-1 (UBC 4683); Langley 8 Sep 1990-1 (Siddle 1991a); Crescent Park (South Surrey) 2 Sep 1996-1; Tofino 9 Nov 1983-1 (Campbell 1984b); Radar Hill (Tofino) 20 Nov 1982-1 (Campbell 1983a).

Winter: No records.

Orange-crowned Warbler

Vermivora celata (Say) OCWA

RANGE: Breeds from western and northern Alaska, Yukon, northwestern and southern Mackenzie, northern Alberta, northern Saskatchewan, northern Manitoba, northern Ontario, central Quebec, and central Labrador; in the west, from southern Alaska, British Columbia, and the Pacific states to California and northern Baja California; south through the Rocky Mountains to Arizona, New Mexico, and extreme western Texas. Winters from California, Arizona, Texas, and the Gulf states south to Guatemala and Belize.

STATUS: On the coast, a *fairly common* to *common* migrant, *uncommon* summer visitant, and *rare* in winter in the Georgia Depression Ecoprovince; elsewhere on the coast, including Western Vancouver Island and the Queen Charlotte Islands, an *uncommon* to *fairly common* migrant and summer visitant; *casual* in winter.

In the interior, a *fairly common* to *common* migrant and summer visitant, and *casual* in winter in the Southern Interior and Southern Interior Mountains ecoprovinces; *fairly common* migrant and summer visitant in the Central Interior Ecoprovince; *uncommon* to *fairly common* migrant, and *fairly common* to locally *common* summer visitant, in the Sub-Boreal Interior and Northern Boreal Mountains ecoprovinces and in the Peace Lowland of the Boreal Plains Ecoprovince; *rare* in the Taiga Plains Ecoprovince.

Breeds.

NONBREEDING: The Orange-crowned Warbler (Fig. 9) is widely distributed throughout British Columbia, including the offshore islands, and is one of the few warblers recorded from all ecoprovinces. There are large areas, especially in the Northern Boreal Mountains and Taiga Plains, from which the species is not recorded, but where it may occur locally. These gaps may reflect the absence of suitable habitat, a local distribution, or inadequate exploration.

This warbler has been reported from sea level to about 1,700 m, but only rarely above 1,300 m outside the breeding season. Along the coast, it occurs during migration in mixed forests or deciduous woodlands, including red alder, bigleaf maple, birch, Garry oak, and arbutus, and, less frequently, in open coniferous stands including Douglas-fir and grand fir. In the interior, it is often seen in riparian black cottonwood or mixed coniferous stands, usually with shrub understoreys; in the edge habitat between young and old forests; in groves of trembling aspen along edges such as lakeshores and farm clearings; and in suburban gardens and parks. Inventories of forest birds in migration near Prince George and Fort Nelson suggest that this species prefers young seral trembling aspen stands on south-facing slopes of 15° to 30°.

On the south coast, spring migration begins with the arrival of a small number of warblers in late March. This species and the Yellow-rumped Warbler are consistently the earliest migrant warblers. The spring movement increases rapidly and reaches its height in mid to late April, by which time

Figure 9. The Orange-crowned Warbler is one of the most widely distributed warblers in British Columbia (Victoria, 31 August 1996; R. Wayne Campbell).

the first arrivals are reaching the Queen Charlotte Islands and the Northern Mainland Coast. The peak of the migration on Western Vancouver Island (Hatler et al. 1978) and the Southern Mainland Coast is in late April and early May, on the Northern Mainland Coast in early May, and on the Queen Charlotte Islands in early May to early June (Figs. 10 and 11).

In the Southern Interior, birds may arrive as early as 11 March. Cannings et al. (1987) found that first arrival dates over 25 years ranged from 17 March to 19 May, with a mean of 22 April. After late May, the numbers of Orange-crowned Warblers reported in the Okanagan valley decrease sharply. Further east, in the valleys of the Kootenay and Columbia rivers, the spring arrival is slightly later and shows a more rapid increase to peak numbers. The peak in numbers in southern portions of the interior occurs in the first half of May, while further north it occurs between late May and early June. In all regions, a drop in numbers is recorded after mid-June because singing migrant males have passed through.

The limited data from extreme northern regions of the province obscure the onset of autumn migration there, but the steady decline in numbers of birds reported in August and September in the northern half of the province certainly reflects the progress of the migration. Most of the Orange-crowned Warblers in the Northern Boreal Mountains have gone by the second week of September. In the Peace Lowland, there is a marked increase in the number of records from the last week of August to mid-September, but most birds have left by the end of September. Almost the same timetable is found in the Central Interior. In the southernmost portions of the interior, as represented by Vaseux Lake in 1994, the migration was at its height between 27 August and 14 September (J. Jones pers. comm.). This warbler is regularly present in the Okanagan valley into early October.

On the southern coast, the autumn movement reaches its peak from mid-August to mid-September; stragglers continue

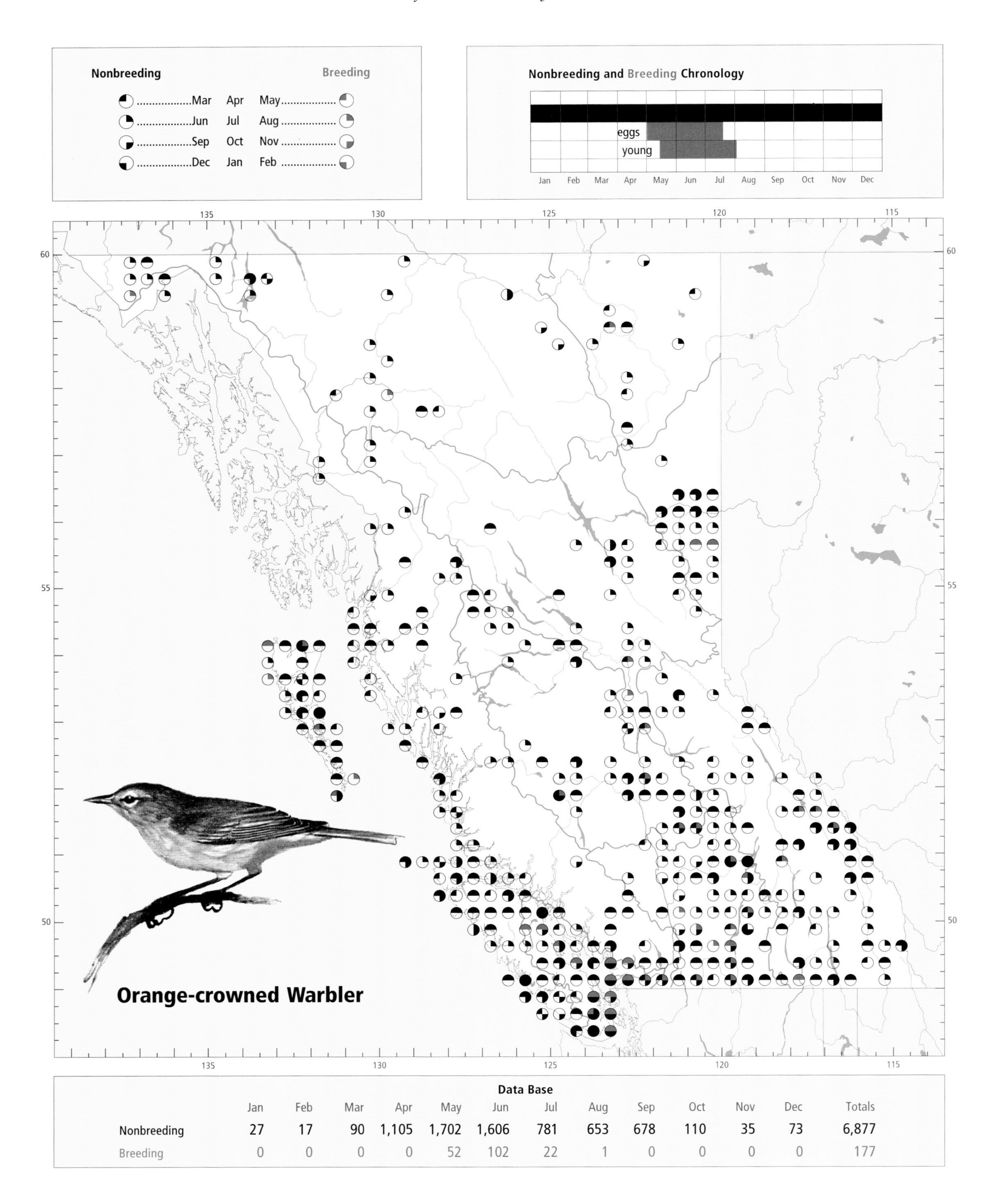

Data Base

	Jan	Feb	Mar	Apr	May	Jun	Jul	Aug	Sep	Oct	Nov	Dec	Totals
Nonbreeding	27	17	90	1,105	1,702	1,606	781	653	678	110	35	73	6,877
Breeding	0	0	0	0	52	102	22	1	0	0	0	0	177

into mid-October, with a few staying into December. For example, during banding of migrants at Rocky Point (near Victoria) over 3 years from 5 August to 23 October, 85% of the Orange-crowned Warblers (n = 66), were taken between 11 August and 18 September (R. Millikin pers. comm.). The latest was banded on 16 October. On Western Vancouver Island, this warbler is regularly present until the first week of October but there are scattered records as late as mid-December.

On the south coast, the Orange-crowned Warbler occurs throughout the year; in the interior, it has been reported regularly between mid-April and mid-September (Fig. 10).

BREEDING: The Orange-crowned Warbler probably breeds throughout most of British Columbia, including the offshore islands. It is a regular breeding species on both Vancouver Island and the Queen Charlotte Islands, and probably on other islands and the adjacent mainland as well, although there is only 1 nesting report (Goose Island – Guiguet 1953). In the southern portions of the interior, nesting records are widely scattered. Breeding has not been confirmed in the north-central region of the province, but we suspect that the species breeds locally in small patches of deciduous trees.

The Orange-crowned Warbler reaches its highest numbers in summer in the Georgia Depression and on the Queen Charlotte Islands in the Coast and Mountains (Fig. 12). It is also abundant in parts of the northern archipelago where Guiguet (1953) reported a singing male for every 182 m of the perimeter of one of the small islands in the Goose Island group.

An analysis of Breeding Bird Surveys for the period 1968 through 1993 could not detect a change in the mean number of birds on coastal routes; on interior routes, the mean number of birds increased at an average annual rate of 7% (Fig. 13). Canada-wide surveys and surveys across the bird's North American range covering the period 1966 through 1996 suggest a relatively stable population (Sauer et al. 1997). Continent-wide Breeding Bird Survey data indicate that British Columbia has some of the highest densities of the Orange-crowned Warbler in North America (Sauer et al. 1997).

The Orange-crowned Warbler breeds from sea level to 420 m elevation on the coast, and from 300 to 1,200 m in the interior. During the breeding season, this versatile warbler inhabits a wide variety of forest types, usually with associated dense thickets, bushes, or shrubs. It is numerous on the west coast of Vancouver Island and on the archipelago of the north coast, including the treeless islets that support colonies of nesting marine birds. There the shrubbery includes salmonberry, devil's club, salal, evergreen huckleberry, false azalea, copper-bush, Pacific crab apple, and wind-wracked western hemlock that backs the ocean shoreline and, in a different aspect, margins the river courses of the mainland and larger islands. The Orange-crowned Warbler is also frequently found on the margins of coniferous forests of yellow cedar, lodgepole pine, western hemlock, and Sitka spruce (Fig. 14), where they meet ocean shores, ponds, or muskegs. On the arid shorelines of the Gulf Islands in the Georgia Depression, the Orange-crowned Warbler is consistently present as a nesting species in open forest featuring arbutus, Garry oak, Douglas-fir, grand fir, Oregon-grape, and hairy honeysuckle. Elsewhere on southeastern Vancouver Island, it occupies Garry oak woodlands; riparian areas bordering lakes, ponds, and

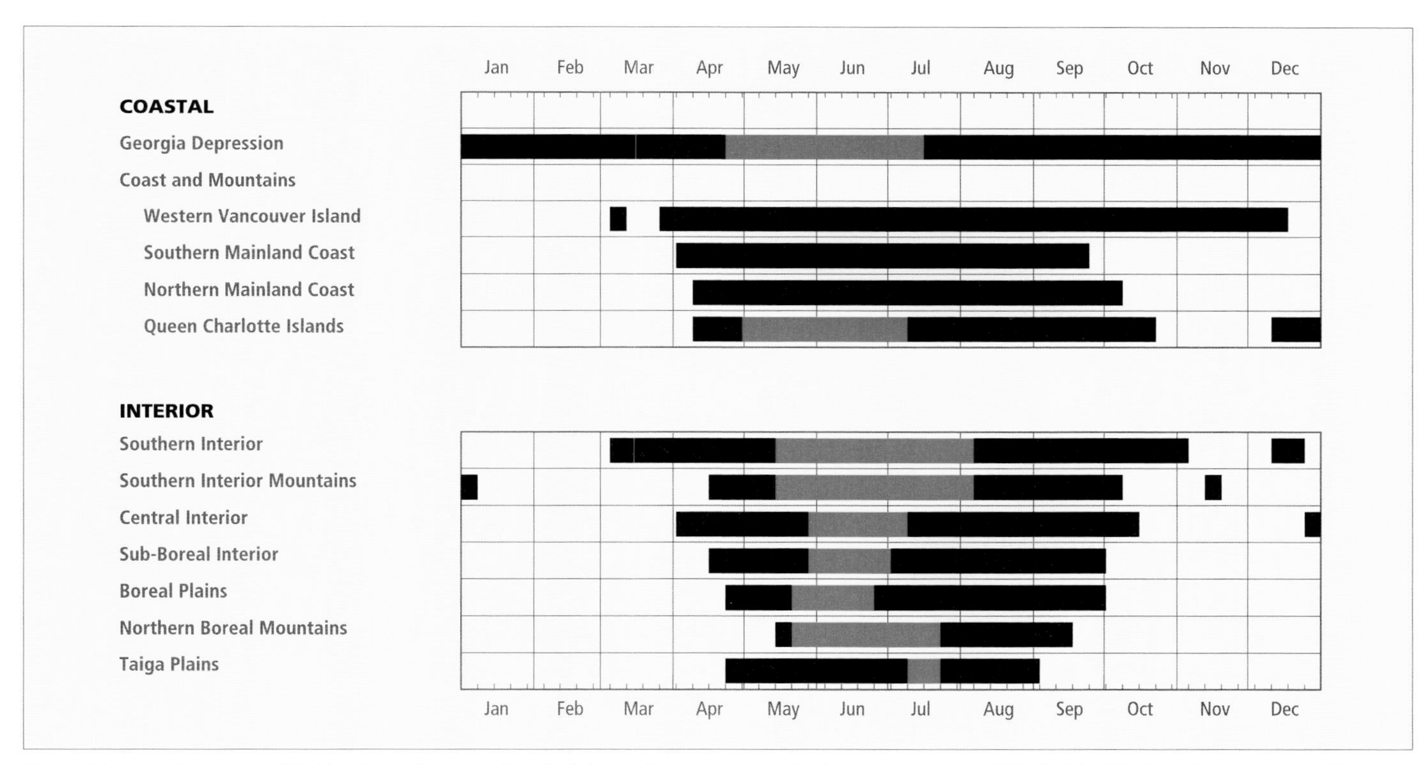

Figure 10. Annual occurrence (black) and breeding chronology (red) for the Orange-crowned Warbler in ecoprovinces of British Columbia. Records are shown for the week in which they occurred.

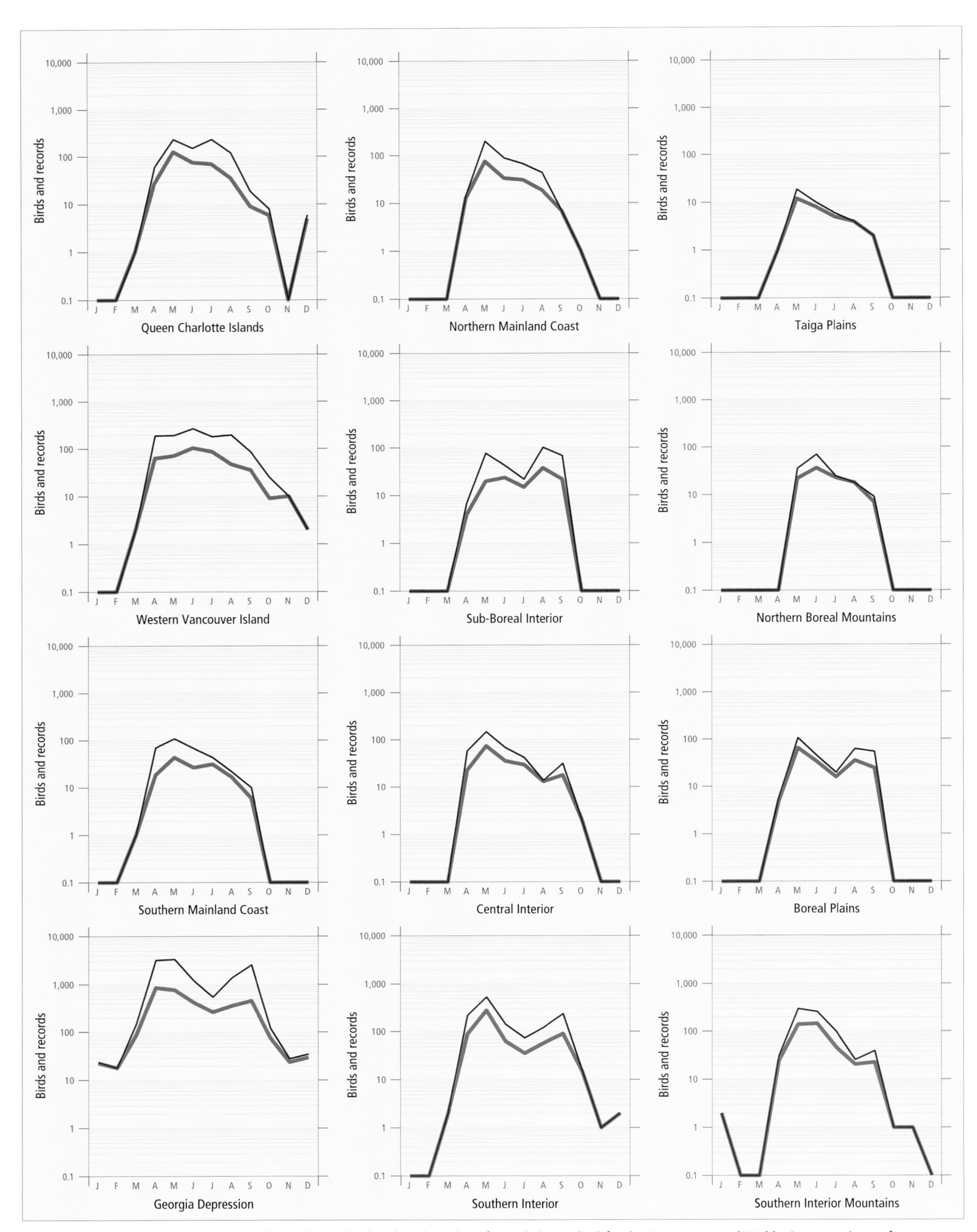

Figure 11. Fluctuations in total number of birds (purple line) and total number of records (green line) for the Orange-crowned Warbler in ecoprovinces of British Columbia. Christmas Bird Counts, Breeding Bird Surveys, and nest record data have been excluded.

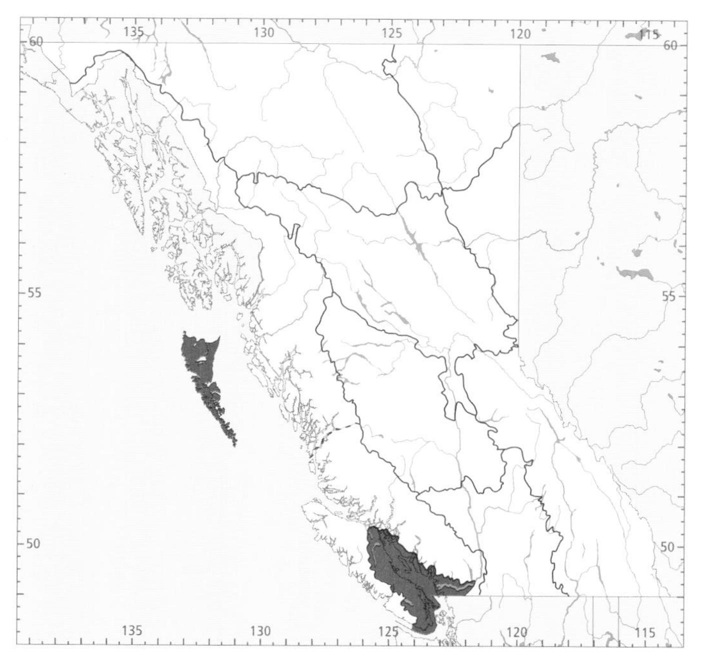

Figure 12. In British Columbia, the highest numbers for the Orange-crowned Warbler in summer occur in the Georgia Depression Ecoprovince and on the Queen Charlotte Islands in the Coast and Mountains Ecoprovince.

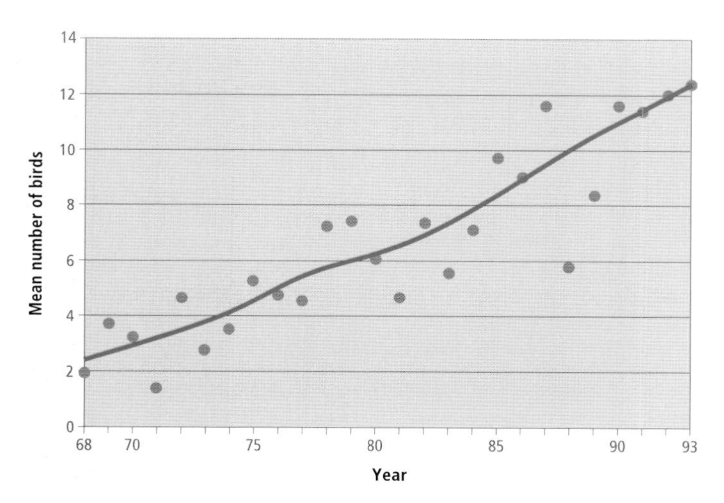

Figure 13. An analysis of Breeding Bird Surveys for the Orange-crowned Warbler in British Columbia shows that the mean number of birds on interior routes increased at an average annual rate of 7% over the period 1968 through 1993 ($P < 0.1$).

streams; second-growth forests, including red alder and willows along with young Douglas-fir; and edges of transmission line corridors and clearings.

In the southern portions of the interior, nesting habitat is equally varied and includes trembling aspen woodland and open montane forest with abundant shrubbery usually associated with the lower mountain slopes. The species seldom breeds in the ponderosa pine–bunchgrass communities of the valley bottoms. In the Bulkley River valley, a comparison of clearcuts with sapling trembling aspen and mature aspen woodland revealed that the Orange-crowned Warbler was most numerous in the sapling trembling aspen stands (16 to 23 singing males per 10 ha) and decreased with increasing density of the woodland (Pojar 1993). In northern regions of the province, especially in the Boreal Plains, nesting habitat includes second-growth trembling aspen stands on south-facing hillsides, and upper shrubby edges of trembling aspen groves with willow, saskatoon, and saplings that grow between grassy ridges on south- or northeast-facing slopes. It also breeds in tall red-osier dogwood and alder understorey in open mature balsam poplar riparian woodlands, in saplings and bushes growing in transmission corridors through mixed woodlands, in brushy edges and overgrown clearings in white spruce, balsam poplar, and trembling aspen woodlands, in swampy terrain surrounded by white spruce, green alder, and willow (Siddle 1997), as well as in subalpine terrain with subalpine fir and birch.

Figure 14. On the Queen Charlotte Islands, the Orange-crowned Warbler breeds among dense patches of red alder and salal bordering Sitka spruce forests (Rose Spit, 21 May 1996; R. Wayne Campbell).

The Orange-crowned Warbler has been recorded breeding in the province from 29 April (calculated) to 2 August. On the coast, it has been recorded from 29 April (calculated) to 14 July; in the interior, from 16 May to 2 August (Fig. 10).

Nests: Most nests of this warbler were neat cups, placed on the ground sheltered by nearby vegetation (81%, n = 104;

Figure 15. Clutch of 3 eggs of the Orange-crowned Warbler and 1 egg of the Brown-headed Cowbird in the warbler's typical ground nest of fine grasses (near McLean Lake, Hat Creek, 11 June 1998; R. Wayne Campbell).

Fig. 15) and in shrubs or trees (13%). The heights of 59 nests ranged from ground level to 6 m, with 81% between ground level and 0.3 m. Grasses were the most frequently used nesting material and occurred in 76% of nests (Fig. 15), followed by mosses (27%), mammal hair (20%), leaves (10%), and plant fibres (7%). An assortment of rootlets, lichens, twigs, sedges, and feathers was used less frequently.

Eggs: Dates for 113 clutches ranged from 2 May to 19 July, with 51% recorded between 22 May and 12 June. Calculated dates indicate nests may contain eggs as early as 29 April. Sizes of 74 clutches ranged from 1 to 6 eggs (1E-1, 2E-3, 3E-7, 4E-34, 5E-27, 6E-2), with 82% having 4 or 5 eggs. The incubation period is 12 to 14 days (Sogge et al. 1994) (Fig. 15).

On the coast, the first eggs may be found in early May, with a peak in late May. In the interior, eggs are normally not found until late May, with a peak about mid-June. Nests may contain eggs up to 2 weeks earlier on the coast than in the interior (Fig. 16).

Young: Dates for 53 broods ranged from 15 May to 2 August, with 51% recorded between 7 June and 27 June. Sizes of 35 broods ranged from 1 to 6 young (1Y-1, 2Y-4, 3Y-6, 4Y-17, 5Y-6, 6Y-1), with 50% having 4 young. The nestling period in a nest on Mayne Island was 10 days, although the young may have been prematurely disturbed. Sogge et al. (1994) give the nestling period as 11 to 13 days.

Brown-headed Cowbird Parasitism: In British Columbia, 4% of 118 nests found with eggs or young were parasitized by the cowbird (Fig. 15). In addition, 8 fledgling cowbirds, apparently from 7 nests, were seen being fed by the host parents. Friedmann et al. (1977) report just 12 instances of parasitism of the Orange-crowned Warbler by the Brown-headed Cowbird, 6 of them from southern Vancouver Island (Tatum 1973).

Nest Success: Of 16 nests found with eggs and followed to a known fate, 8 produced at least 1 fledgling.

REMARKS: Three subspecies of Orange-crowned Warbler are known to occur in British Columbia. *V. c. celata* breeds across the northeastern regions of the province. There is a nesting record for Horse Lake, in the Cariboo region, at or near the southwestern boundary of the range of this subspecies (Munro 1945a). *V. c. orestera* is described as breeding from southwestern Yukon south through mountains of the interior of British Columbia east of the coastal mountain ranges, and *V. c. lutescens* is the subspecies of the coastal region the length of the province (Sogge et al. 1994). The 3 subspecies occupy different winter ranges, with some overlap. *V. c. celata* winters mainly in the southeastern United States, *orestera* mainly in the southwestern states and Mexico, and *lutescens* mainly in southern California, Arizona, and northwestern New Mexico (Sogge et al. 1994).

The studies of Pojar (1993) in British Columbia and Morrison (1981) in Oregon relate the abundance of Orange-

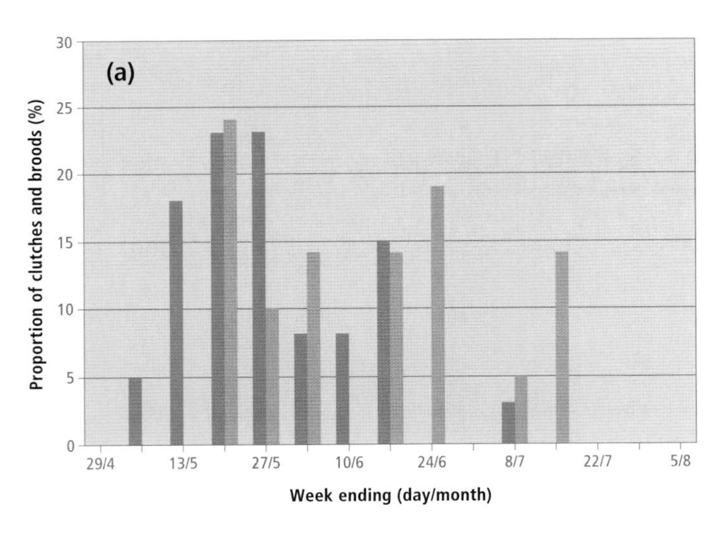

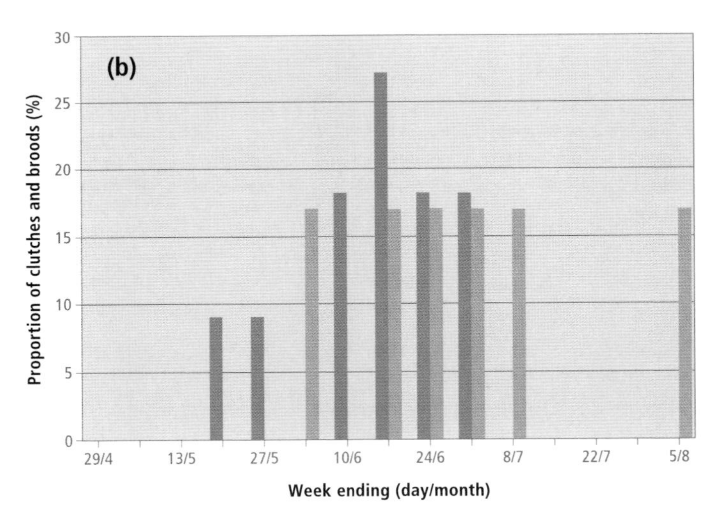

Figure 16. Weekly distribution of the proportion of clutches (dark bars) and broods (light bars) for the Orange-crowned Warbler in the (a) Georgia Depression (clutches: $n = 40$; 16: $n = 21$) and (b) Southern Interior (clutches: $n = 11$; broods: $n = 6$) ecoprovinces of British Columbia. The figures are based on the week eggs or young were first found in the nest. On the coast, nests may contain eggs 2 weeks earlier than in the interior; the figures also suggest that this warbler is double-brooded in the Georgia Depression.

crowned Warbler to regenerating forests in 2 different ecosystems. Pojar studied the relative densities of birds in clearcut forests regenerating to trembling aspen (see BREEDING for details). The Oregon studies were of habitat selection and foraging behaviour of Orange-crowned, Wilson's, and MacGillivray's warblers during the nesting season in clearcut forest areas in Oregon. Both studies revealed a strong preference of all 3 warblers for deciduous second-growth. In the absence of deciduous trees, however, the Orange-crowned Warbler, unlike the other 2, could maintain its population in young conifer stands if there was a good shrub layer. Densities of bird communities increased where patches of deciduous trees formed breaks in plant communities dominated by shrubs and conifers.

For additional information on this species, see Curson et al. (1994) and Morse (1989). For a summary of current information on the species, see Sogge et al. (1994).

NOTEWORTHY RECORDS

Spring: Coastal – Victoria 2 May 1946-4 eggs; Goldstream Park 25 Mar 1972-1; Beaver Lake (Saanich) 8 Apr 1973-30; Central Saanich 31 Mar 1987-20; Duncan 15 May 1972-4; e Cultus Lake 7 Apr 1968-5; Reifel Island 13 Apr 1985-60, high count of the year; Tofino 10 Mar 1992-2, 20 Apr 1982-9; Chilliwack Mountain 23 Apr 1982-9; Gabriola Island 19 Apr 1983-1; Vancouver 12 Mar 1970-2 (Campbell et al. 1972a); Vancouver 8 May 1984-200 in Stanley Park; Burquitlam 12 May 1935-5 eggs; Mount Seymour Park 16 Apr 1983-6; Hornby Island 14 Apr 1974-4; Powell River 8 Apr 1955-1; Miracle Beach Park 7 Apr 1963-18 (Westerborg and Stirling 1963); Port Alice 3 May 1973-20; Cape Scott Park 15 Apr 1980-1; Ninstints 17 May 1991-8; Skedans Islands, 10 May 1983-5 eggs; Queen Charlotte City 13 Apr 1987-1; Tlell 22 May 1951-11; Masset 27 May 1989-32; Langara Island 21 May 1947-5 eggs. **Interior** – Trail 24 May 1902-5 eggs; Vaseux Lake 5 Apr 1986-57; ne Naramata 24 May 1988-6 eggs; Nakusp 19 Apr 1980-1; Okanagan Landing 17 Mar 1962-1, first of year, 1 May 1911-15; Enderby 20 May 1945-5 eggs; Kootenay National Park 19 May 1983-20; Ottertail Valley 26 Apr 1977-1; lower Blaeberry River valley 25 Apr 1995-1; 100 Mile House 12 Apr 1986-4; Lunch Lake 28 Apr 1992-1; Kleena Kleene 19 May 1992-1; Williams Lake 19 May 1982-25; Spanish Lake 9 May 1992-1 singing male; Puntchesakut Lake 18 May 1944-35 (Munro 1947); Vanderhoof 7 May 1968-4; 24 km s Prince George 18 Apr 1983-1; nr Bullmoose Creek 27 Apr 1994-4; Chetwynd 3 May 1978-1; 5 km s Pouce Coupe 25 May 1992-5 eggs; Attachie 26 Apr 1986-1; Taylor 15 May 1983-8 territorial males evenly spaced along first 4 km Johnstone Rd; Fort St. John 22 May 1982-6; w side Boundary Lake (Goodlow) 30 Apr 1988-1 spring arrival; Hyland Post 29 May 1976-10; Fort Nelson 27 Apr 1987-1; Parker Lake 13 May 1975-3; Atlin 14 May 1930-1 (RBCM 5670), 18 May 1977-6.

Summer: Coastal – Rocky Point (Victoria) 20 Aug 1995-19; Sooke Hills 3 Jun 1984-35; Sea Island 31 Aug 1995-200 (Elliott and Gardner 1997); Campbell River 6 Jun 1980-20; Storm Islands 13 Jun 1976-32; Triangle Island 19 Jun 1974-6, 12 Jun 1976-4, adult pair feeding 2 young, 2 Aug 1994-21, 21 Aug 1974-40 migrating; Goose Group 29 Jun 1948-first fledgling (Guiguet 1953); Cape St. James 3 Jul 1982-8; Skedans Islands 12 Jul 1977-21; small islet off n end Hippa Island 21 Jun 1983-5 nestlings. **Interior** – Anarchist Mountain 11 Jul 1978-5; Apex Mountain (Penticton) 12 Aug 1978-20, elevation 1,700 m; Beaverdell 10 Jun 1987-5; ne Naramata 9 Jun 1988-6 feathered nestlings; Creighton Valley 21 Jun 1980-18; Revelstoke 26 Jun 1979-3 eggs; Spillimacheen 23 Jun 1985-22; Bridge Lake Jun 1959-4 eggs; Tatlayoko Lake 2 Jul 1994-5; Kleena Kleene 19 Jun 1963-5 nestlings, 3 days old; Topley 2 Jun 1956-5 eggs; 3.3 km ne Tabor Lake 18 Jun 1966-4 eggs; nr Gwillim Lake 26 Aug 1994-1; Tupper 20 Jun 1994-13; sw Fellers Heights 7 Jun 1992-2 eggs plus 2 Brown-headed Cowbird eggs; 55 km sw Dawson Creek 9 Jun 1993-5 nestlings; s Pink Mountain 18 Jun 1981-3; Todagin Creek to Burrage Creek 20 Jun 1975-12; Ealue Lake 16 Jul 1982-3 eggs; Steamboat 28 Jun 1980-3; Lower Tatshenshini River valley 5 Jun 1983-4 nestlings, probably 1 day old, 7 Jun 1983-6 eggs.

Breeding Bird Surveys: Coastal – Recorded from 27 of 27 routes and on 96% of all surveys. Maxima: Masset 25 Jun 1991-66; Queen Charlotte City 25 Jun 1994-56; Port Renfrew 27 Jun 1973-55. **Interior** – Recorded from 66 of 73 routes and on 72% of all surveys. Maxima: Bridge Lake 21 Jun 1992-53; Succour Creek 17 Jun 1993-48; Ferndale 7 Jun 1969-43.

Autumn: Interior – Alaska Highway (Mile 213 at Liard River) 2 Sep 1943-1 (Rand 1944); Atlin area 14 Sep 1972-1; St. John Creek 1 Sep 1985-9, 1 singing; Fort St. John 27 Sep 1984-2, last record; Mackenzie 2 Sep 1994-5; Chezacut 4 Oct 1932-1 (MCZ 284133); Riske Creek 10 Oct 1986-1; lower Blaeberry River valley 1 Oct 1997-1; Golden 29 Sep 1975-1; Bridge Lake 7 Sep 1960-13; Okanagan Landing 4 Nov 1935-1; Nakusp 16 Nov 1986-1 (Rogers 1987); Elk River, Michel Creek junction 11 Sep 1983-2 (Fraser 1984); Vaseux Lake 1 Sep 1994-13; Creston 14 Sep 1985-1; Osoyoos Lake 10 Oct 1975-1. **Coastal** – Cape St. James 18 Oct 1981-2; Vancouver 6 Sep 1972-55 (Campbell et al. 1974); Reifel Island 14 Sep 1983-93, on Alaksen National Wildlife Area; Port Renfrew 5 Oct 1974-10; Beacon Hill Park (Victoria) 12 Sep 1966-150 (Crowell and Nehls 1967c), 6 Oct 1981-6.

Winter: Interior – Indianpoint Lake 6 Jan 1934-1 (MCZ 284121); Sorrento 20 Dec 1987-1 (Rogers 1989); Kelowna 12 Dec 1994-1 (Bowling 1995b). **Coastal** – North Vancouver 3 Feb 1981-1; Vancouver 28 Feb 1938-1; Sea Island 22 Feb 1998-1; Ladner 23 Dec 1995-4 on survey; Tofino 14 Dec 1983-1; Swan Lake (Victoria) 13 Jan 1989-1, 10 Feb 1986-1; Victoria 1 Jan 1976-1.

Christmas Bird Counts: Interior – Recorded from 1 of 27 localities and on less than 1% of all counts. Maximum: Shuswap Lake Park 20 Dec 1988-1. **Coastal** – Recorded from 10 of 33 localities and on 8% of all counts. Maxima: Victoria 19 Dec 1992-3; Skidegate 14 Dec 1991-2; Ladner 14 Dec 1974-2; Vancouver 26 Dec 1957-2, 22 Dec 1996-2; Sooke 27 Dec 1986-2; Masset 18 Dec 1993-1; Port Clements 21 Dec 1993-1; Duncan 27 Dec 1970-1; White Rock 2 Jan 1994-1; Sunshine Coast 18 Dec 1993-1.

Nashville Warbler

NAWA

Vermivora ruficapilla (Wilson)

RANGE: Breeds in 2 geographically separated regions of North America. In the west, breeds from the southern interior of British Columbia and northern and northwestern Washington south through the Cascade Range into northern, western, and southwestern Idaho and northwestern Montana; also in central and southern Oregon, northern California, and west-central Nevada. In the east, breeds from central Saskatchewan and Manitoba across central Ontario and southern Quebec, and through the Maritime provinces, south to central Minnesota, east to Connecticut, New York, Maryland, and West Virginia. Winters from southern Texas south through Mexico to Honduras and El Salvador.

STATUS: On the coast, a *very rare* summer vagrant, *rare* transient, and *casual* winter visitant in the Georgia Depression Ecoprovince; in the Coast and Mountains Ecoprovince, *rare* to locally *uncommon* in southern portions of the Southern Mainland Coast, *casual* on Western Vancouver Island and on the Northern Mainland Coast, and absent from the Queen Charlotte Islands.

In the interior, a *fairly common* to occasionally *common* migrant and summer visitant in the Southern Interior Ecoprovince; *uncommon* to *fairly common* in the Southern Interior Mountains Ecoprovince; locally *uncommon* in the Central Interior Ecoprovince; *casual* in the Sub-Boreal Interior Ecoprovince.

Breeds.

NONBREEDING: The Nashville Warbler (Fig. 17) is widely distributed across the southern mainland of British Columbia. It is fairly widespread in the Okanagan, Thompson, and west Kootenay regions at moderate elevations, but is scarcer in the east Kootenay. It is more sparsely distributed and local in the Cariboo and Chilcotin areas, reaching its northern limit south of Prince George in the Sub-Boreal Interior. On the coast, transients occur infrequently along the west and east coasts of Vancouver Island, on the northern mainland coast, and on the southwestern mainland from the lower Fraser River valley north along the Sunshine Coast. The species appears to occur regularly in summer in the valley between Pemberton and Lillooet Lake. A single Nashville Warbler seen at Kitsumkalum Lake is the northernmost record for the province.

During its migration, the Nashville Warbler frequents low elevations on the coast, from near sea level to 200 m but sometimes as high as 600 m in autumn. In the interior, it migrates in fair numbers through the major valleys of the Southern Interior and Southern Interior Mountains. Most spring migrants are at valley bottom elevations but, as at the coast, the autumn migrants move over a wider band of elevation and have been recorded as high as 2,150 m.

In British Columbia, the Nashville Warbler prefers open forests, either mixed or deciduous, with abundant shrub growth. In the coastal regions, the small number of migrating Nashville Warblers have been seen in second-growth alder-willow stands as well as vine maple, black cottonwood, and related shrubbery, and in gardens and suburban parks where trees and shrubs are landscape features. In the interior, except for a greater use of suburban environments during migration, habitat use during the nonbreeding seasons is similar to that of the breeding season (see BREEDING).

Figure 17. The Nashville Warbler occurs mainly in southern portions of the interior of British Columbia (Anthony Mercieca).

On the coast, the first transients may arrive in the Georgia Depression in late March (Fig. 18). Most birds arrive in mid-April and numbers reach their peak a month later. There are few records from late June to late July (Fig. 18). By late August, a small southbound migration is in progress, and by early September observations at coastal localities are few (Fig. 18).

In the southern portions of the interior of the province, there are scattered records of migrants from mid to late April both in the south Okanagan valley and in the valleys of the Kootenay and Columbia rivers (Fig. 18). The dates of first arrival in the Okanagan valley over a period of 22 years ranged from 20 April to 19 May, with a median date of 27 April (Cannings et al. 1987). In both the Okanagan and Kootenay regions, the peak movement occurs about mid-May. The autumn migration is difficult to detect but appears to begin in early August, with the largest numbers moving between mid-August and mid-September. The last migrants usually leave the interior of the province by late September (Figs. 18 and 19).

On the coast, the Nashville Warbler has been recorded regularly from 11 April to 27 September; in the interior, it has been recorded regularly from 5 April to 1 October (Fig. 18).

BREEDING: The Nashville Warbler breeds mainly in the southern portions of the interior of the province from Osoyoos to Nelson and north to Squaam Bay on Adams Lake and the lower Blaeberry River valley near Golden, all in the Southern Interior or Southern Interior Mountains. There are no nesting records west of the Okanagan valley in the ecologically similar Similkameen valley. Small numbers may breed near

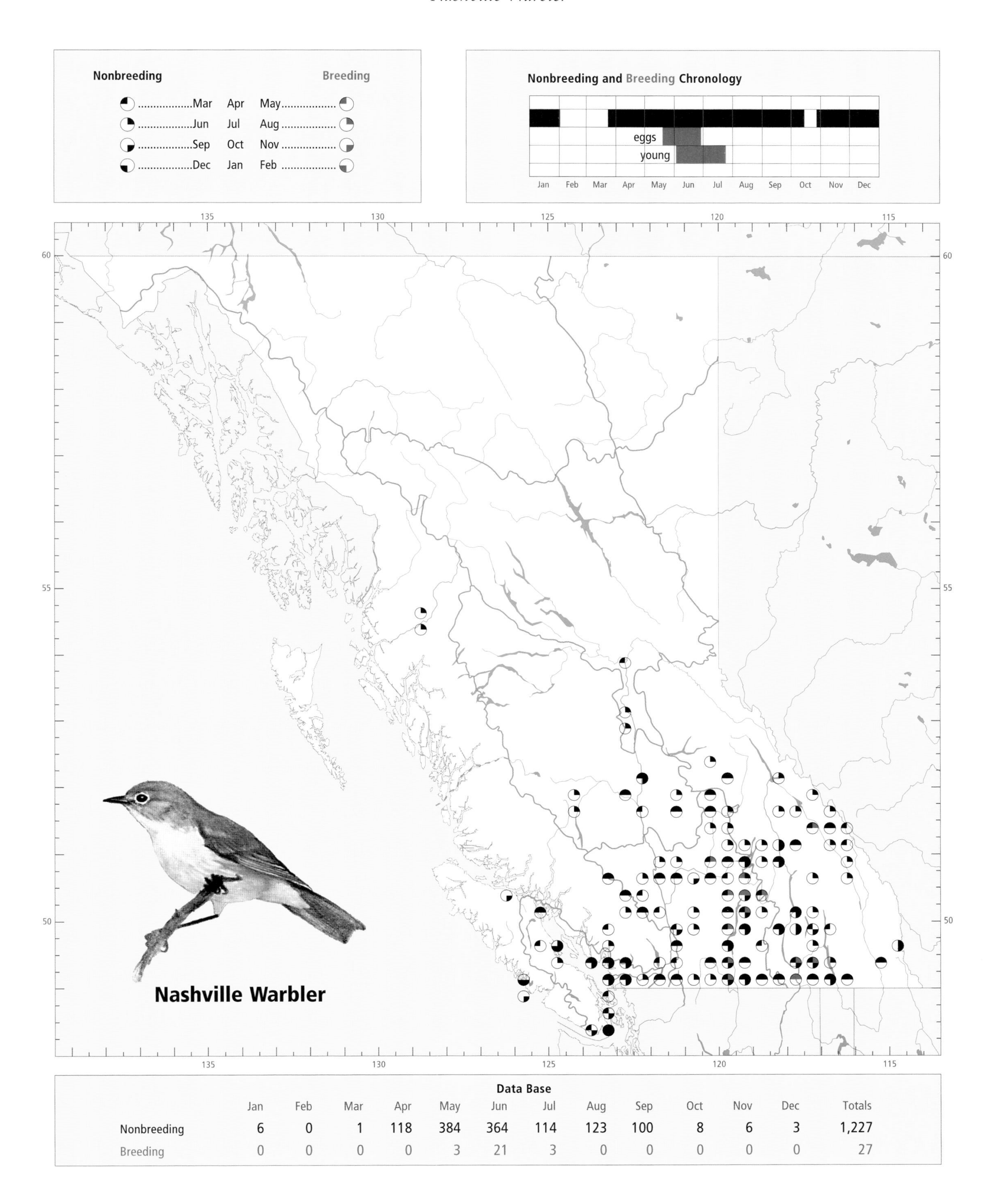

Data Base

	Jan	Feb	Mar	Apr	May	Jun	Jul	Aug	Sep	Oct	Nov	Dec	Totals
Nonbreeding	6	0	1	118	384	364	114	123	100	8	6	3	1,227
Breeding	0	0	0	0	3	21	3	0	0	0	0	0	27

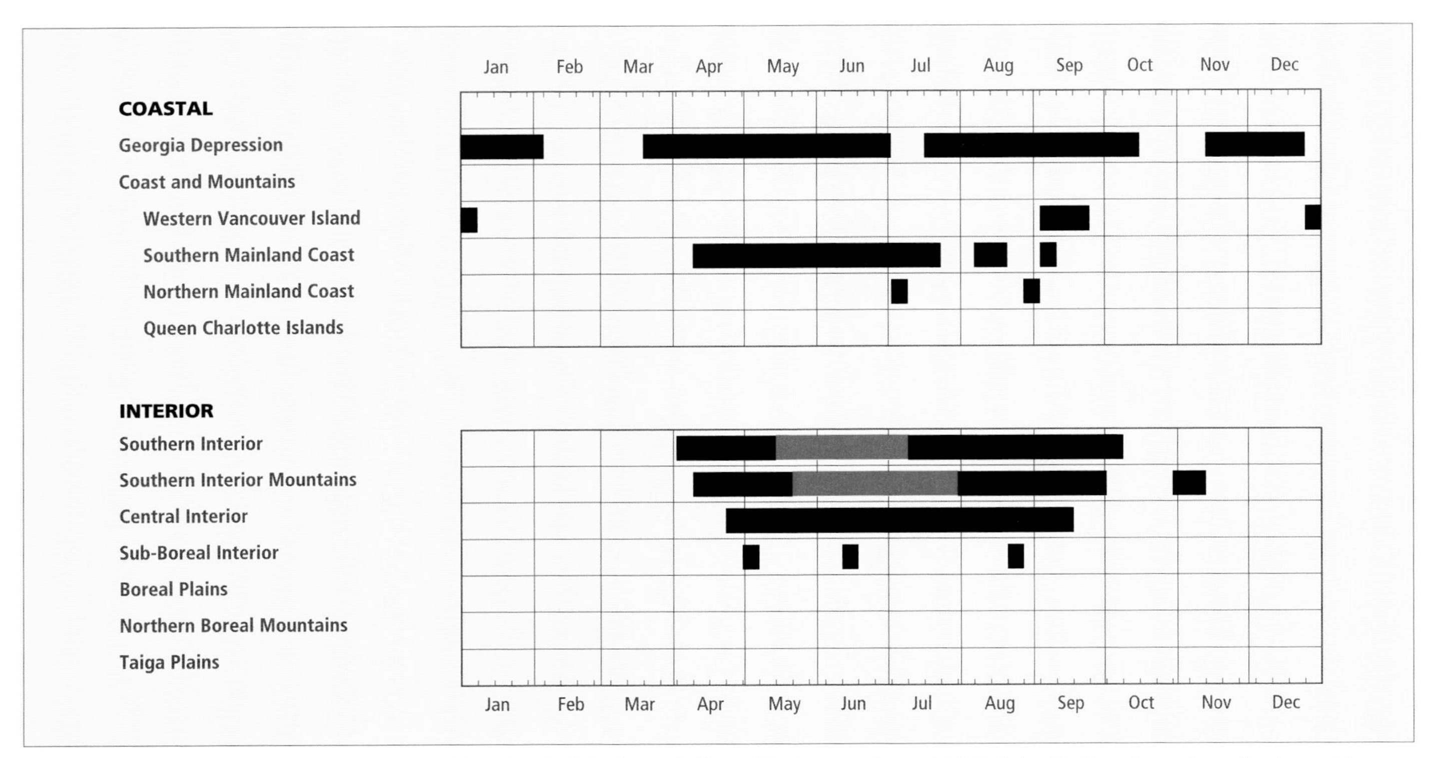

Figure 18. Annual occurrence (black) and breeding chronology (red) for the Nashville Warbler in ecoprovinces of British Columbia. Records are shown for the week in which they occurred.

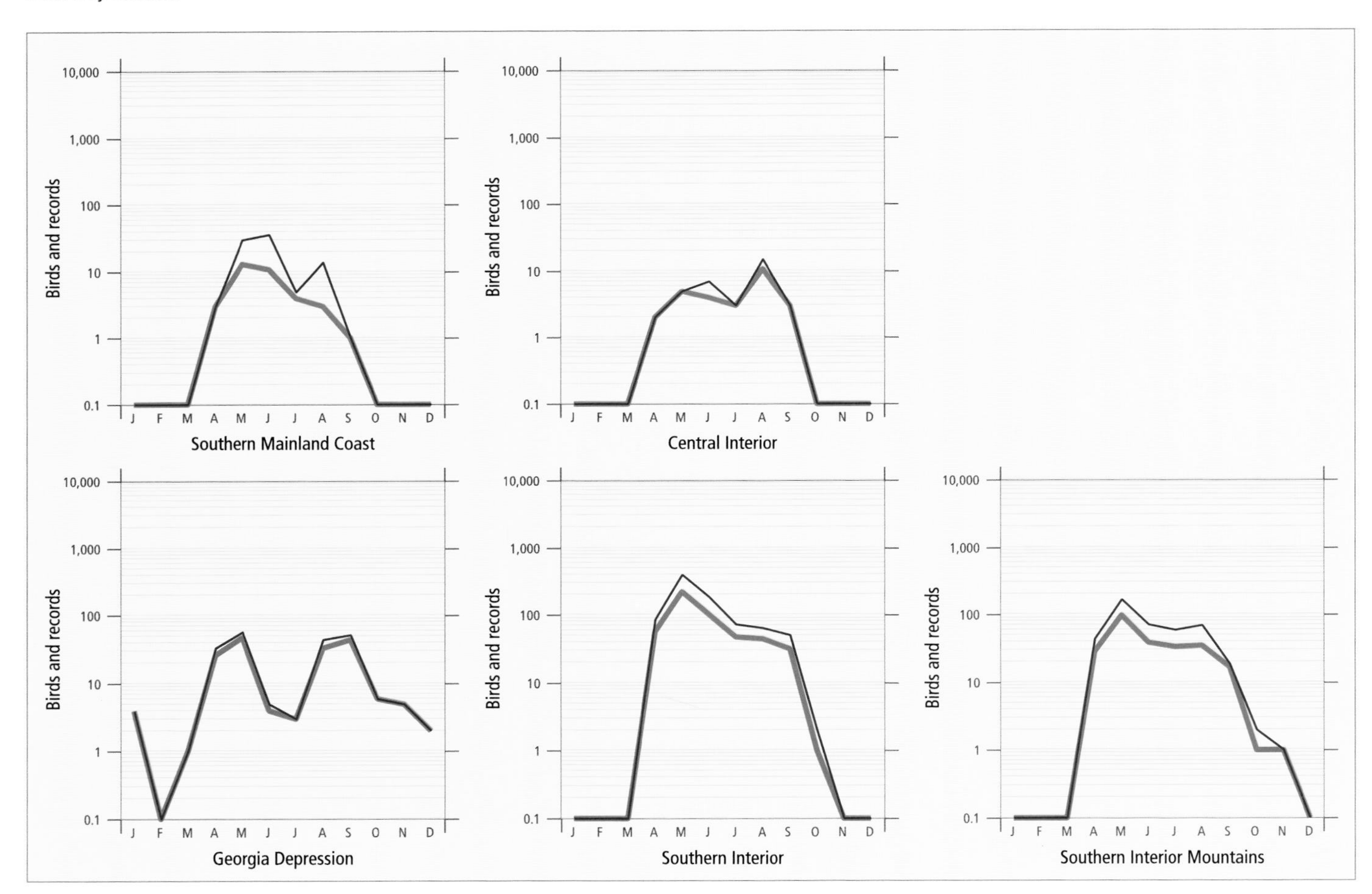

Figure 19. Fluctuations in total number of birds (purple line) and total number of records (green line) for the Nashville Warbler in ecoprovinces of British Columbia. Christmas Bird Counts, Breeding Bird Surveys, and nest record data have been excluded.

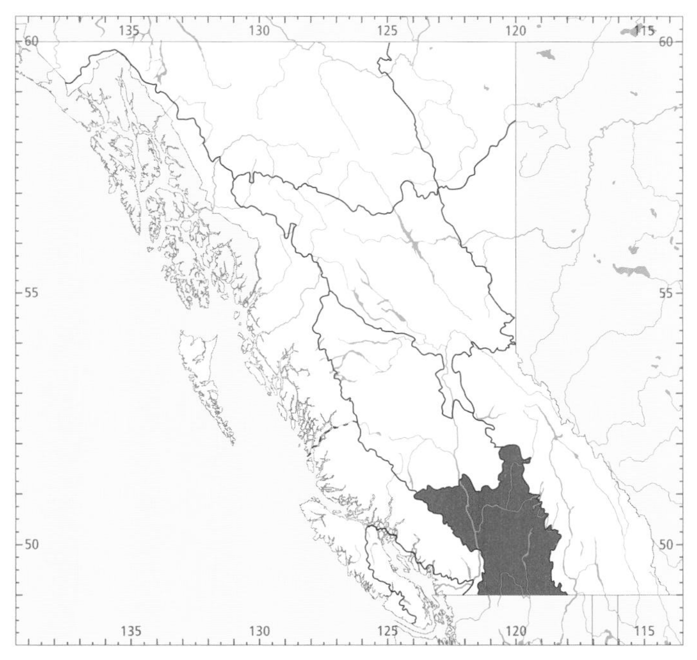

Figure 20. In British Columbia, the highest numbers for the Nashville Warbler in summer occur in the Okanagan valley of the Southern Interior Ecoprovince.

Pemberton and Spuzzum in the Southern Mainland Coast; however, nests with eggs or young have not been found.

The Nashville Warbler reaches its highest numbers in summer in the Southern Interior, from Osoyoos north to Kamloops (Fig. 20). Breeding Bird Surveys for coastal routes for the period 1968 through 1993 contain insufficient data for analysis; the mean number of birds on interior routes increased at an average annual rate of 7% (Fig. 21). Both Canada-wide and continental trends over the period 1980 to 1996 also suggest this warbler's numbers are increasing (Sauer et al. 1997).

In the interior, most nesting records have been close to valley bottom elevations, between 350 and 450 m, but the species probably breeds to about 1,000 m. The characteristic nesting habitat of the Nashville Warbler in British Columbia is the mixed arid forests of ponderosa pine and interior Douglas-fir on the lower, often steep, southerly slopes of the mountains bordering the Okanagan valley (Fig. 22) and small areas of the Columbia valley. There, on dry hillsides, gullies through grassland slopes, and the edges of riparian areas, the warbler frequents an open understorey of mock-orange, snowbrush, saskatoon, choke cherry, common snowberry, tall Oregon-grape, and roses. In the Creston valley, it nests on open shrubby and bushy slopes vegetated with ocean-spray, saskatoon, and mock-orange. Typically, the nesting territory includes tall shrubs and both deciduous and coniferous trees that are used for singing perches and for foraging.

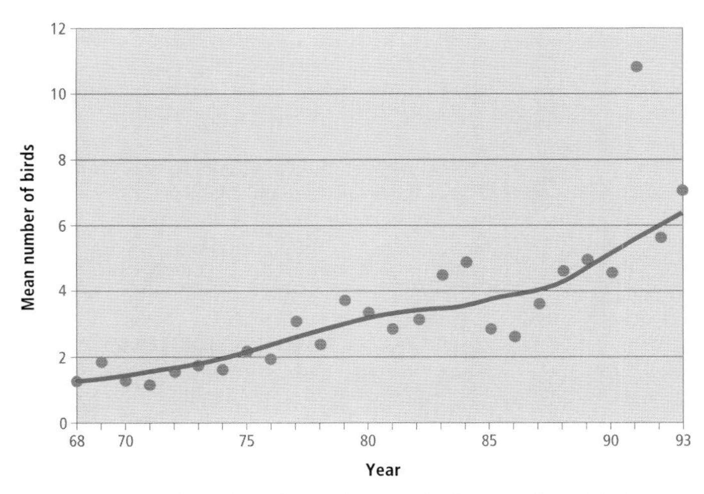

Figure 21. An analysis of Breeding Bird Surveys for the Nashville Warbler in British Columbia shows that the mean number of birds on interior routes increased at an average annual rate of 7% over the period 1968 through 1993 ($P < 0.01$).

In western Washington, the Nashville Warbler nests in the oak-dotted prairies of the Puget Sound region near Tacoma (Bowles 1906; Dawson and Bowles 1909), a habitat more like that on southeastern Vancouver Island (where breeding does not occur) than that of the mainland extension of the Puget Sound region into British Columbia. In eastern Washington, it commonly nests in riparian habitats in ponderosa pine and Douglas-fir zones and sometimes higher, into subalpine fir, as well as in stands of regenerating lodgepole pine and grand fir (Williams 1996).

Breeding on the coast has not been documented. In the interior, the Nashville Warbler has been recorded breeding from 16 May (calculated) to 24 July (Fig. 18).

Nests: Only 9 nests have been described in British Columbia. All nests were situated on the ground, often tucked into a depression on a bank such as a roadcut or under the shelter of ground vegetation. The most frequently used nesting material was grasses (reported in all nests), followed by

Figure 22. In the Okanagan valley of the Southern Interior Ecoprovince, the Nashville Warbler breeds in shrubby draws on dry hillsides adjacent to mixed stands of Douglas-fir and ponderosa pine (Kruger Mountain, 2 June 1998; R. Wayne Campbell).

leaves (in a third of the nests), pine needles, and mosses. Nest linings were of fine grasses.

Eggs: Dates for 10 clutches ranged from 20 May to 28 June. Calculated dates indicate that nests may contain eggs as early as 16 May. Clutch size ranged from 2 to 5 eggs (2E-1, 3E-1, 4E-2, 5E-6). The incubation period is 11 or 12 days (Roth 1977).

Young: Dates for 11 broods ranged from 3 June to 24 July. Brood size ranged from 2 to 5 young (2Y-2, 3Y-4, 4Y-3, 5Y-2). The nestling period is 9 to 10 days (Roth 1977).

Brown-headed Cowbird Parasitism: In British Columbia, 2 of 16 nests found with eggs or young were parasitized by the cowbird. One nest contained 5 cowbird eggs and 2 warbler eggs; another nest had 3 cowbird eggs and 2 warbler eggs. In 2 additional instances, Nashville Warblers were seen tending fledgling cowbirds. Friedmann et al. (1977) report this species as an infrequent host of the Brown-headed Cowbird in the eastern part of its range. At that time, they had no records of parasitism of the western subspecies of Nashville Warbler (see REMARKS) by the Brown-headed Cowbird.

Nest Success: Of 7 nests found with eggs and followed to a known fate, 3 produced at least 1 fledgling.

REMARKS: The Nashville Warbler is unique among the American wood-warblers in having 2 disjunct breeding ranges. The western-nesting population, with its Canadian breeding range confined to British Columbia, has long been recognized as a separate subspecies, *V. r. ridgwayi,* formerly known as the Calaveras Warbler.

A report of this species on the Kwinitsa Breeding Bird Survey lacks sufficient documentation and has been excluded from this account.

For additional information on the life history of the Nashville Warbler, see Knapton (1984) and Williams (1996).

NOTEWORTHY RECORDS

Spring: Coastal – Beaver Lake (Saanich) 29 Apr 1972-1 (Tatum 1973); Fulford Harbour 11 May 1959-1; Surrey 25 Mar 1966-1 at Green Timbers; Cheam Slough 1 May 1976-1; Vancouver 11 Apr 1984-3; Beaver Lake (Vancouver) 24 May 1986-1; Pitt Meadows 2 Apr 1981-2; Harrison Hot Springs 13 Apr 1989-1, 10 May 1974-4; Ralph River (Buttle Lake) 16 May 1996-1 male and 1 female; One Mile Lake (Pemberton) 1 May 1996-1 male; Oxbow Lake 5 May 1996-8 males and 2 females migrating along shrubby road. **Interior** – nr Kilpoola Lake 15 May 1987-2; Oliver 5 Apr 1986-8 banded; West Bench 15 Apr 1985-1; Creston 10 Apr 1984-3; Balfour to Waneta 16 May 1981-37 on survey; Trail 24 May 1902-5; s Castlegar 19 Apr 1981-6, 2 May 1971-3; Okanagan Landing 29 Apr 1907-7; Skihist 3 May 1980-1; Nakusp 24 Apr 1986-1 at sewage lagoon; Enderby 20 May 1941-5 eggs; s Knutsford 18 Apr 1981-1; Lillooet 26 May 1916-1 (RBCM 4113); Paul Creek 27 Apr 1986-2; Salmon Arm 24 Apr 1987-3; Sorrento 28 Apr 1972-2; Celista 20 Apr 1993-1 male; Shuswap Lake Park 24 May 1974-7; Clearwater 9 May 1935-1 (MCZ 284111); Riske Creek 28 Apr 1990-1 at Wineglass Ranch; s Wells Gray Park 5 May 1955-1; Williams Lake 6 May 1977-1; 33 km s Prince George 30 Apr 1972-1 (UBC 13722).

Summer: Coastal – Sea Island (Richmond) 18 Aug 1995-2 banded (Elliott and Gardner 1997); Port Alberni 26 Aug 1984-1 at Shoemaker Bay; Pitt Meadows 1 Jun 1991-1; Gibsons 21 Jul 1988-1; Sechelt 11 May 1997-1; North Bend 5 Sep 1985-1; 6.3 km nw Heriot Bay 27 Jul 1973-1; Alta Lake 18 Aug 1938-1 (UBC 8250); Lillooet Lake 8 Jun 1985-5; Pemberton 22 Jun 1924-12, 30 Jun 1985-2; Spetch Creek (Pemberton) 21 Jul 1996-1 male; Pemberton valley 9 Jun 1981-7; Lakelse Lake 5 Jul 1989-1 singing male; Kitsumkalum Lake 31 Aug 1974-1. **Interior** – Oliver 3 Jun 1992-4 well-feathered young; Kootenay River (Creston) 23 Aug 1947-1 (Munro 1950); Creston 26 Aug 1982-10; Grand Forks 1 Aug 1982-1; Okanagan valley 22 Jun 1922-4 eggs; Vaseux Lake 13 Jul 1980-1 fledgling; White Lake (Okanagan Falls) 10 Jun 1984-5 eggs; Bull River 9 Aug 1970-1; Nelson 24 Jul 1993-5 nestlings with parents, 4 Aug 1994-1 fledgling; Gwillim Lakes 31 Aug 1989-1 male at 2,150 m; nr headwaters Watching Creek 20 Jun 1984-5 eggs; e Falkland 27 Jun 1977-3 young; Shuswap Falls 1 Jun 1927-eggs present; Salmon Arm to Squilax 13 Jun 1973-18 (Sirk et al. 1973); Celista 8 Jun 1993-6; Lillooet 20 Jul 1916-1 (NMC 9763); Bachelor Lake (Kamloops) 25 Aug 1948-1 (UBC 2007); Shuswap Lake Park 2 Jun 1973-3 (Sirk et al. 1973); Celista 16 Jul 1960-6; Adams Lake 9 Jul 1995-2 fledglings; Mount Revelstoke National Park 16 Jun 1982-2; Willowbank Mountain 3 Jun 1996-1 singing male; lower Blaeberry River valley 30 Aug 1997-1 singing male; Dog Creek 13 Jun 1994-1 singing male on territory; Kinbasket Lake 27 Jun 1995-1 singing male; Williams Lake 26 Jun 1977-3; Quesnel 17 Jun 1992-1 singing male; Milburn Lake 26 Aug 1995-1.

Breeding Bird Surveys: Coastal – Recorded from 3 of 27 routes and on 1% of all surveys. Maxima: Chilliwack 26 Jun 1976-2; Pemberton 30 Jun 1985-2. **Interior** – Recorded from 32 of 73 routes and on 34% of all surveys. Maxima: Chu Chua 26 Jun 1993-49; Adams Lake 20 Jun 1994-38; Mabel Lake 20 Jun 1993-30.

Autumn: Interior – Williams Lake 14 Sep 1970-1; Revelstoke 29 Oct 1988-2; Shuswap Lake 6 Sep 1959-3; Tranquille 13 Sep 1986-1; Nakusp 25 Sep 1976-3, 27 Sep 1988-1; Trout Creek Point 26 Sep 1967-1; Vaseux Lake 1 Oct 1975-2. **Coastal** – Port Neville 11 Sep 1975-1; Pacific Rim National Park 4 Sep 1997-1 nr sewage lagoons, 23 Sep 1972-1 at Maltby Slough (Hatler et al. 1978); Gibsons 23 Nov 1985-1; Maplewood Flats 6 Sep 1995-2 (Elliott and Gardner 1997); Surrey 1 Oct 1965-1 at Green Timbers; Island View Beach 12 Oct 1959-1 (Poynter 1960); Victoria 15 Nov 1976-1 adult male; Billings Point (Sooke) 20 to 21 Nov 1983-1 male (Hunn and Mattocks 1984).

Winter: Interior – No records. **Coastal** – Comox 8 Dec 1955-1 (RBCM 13692); Sandhill Creek 31 Dec 1973-1, 3 Jan 1974-1 (Hatler et al. 1978); Saanich 20 Jan 1990-1 on Wascana Street; Victoria 19 Dec 1989 to 31 Jan 1990-1 (Siddle 1990a).

Christmas Bird Counts: Not recorded.

Yellow Warbler

Dendroica petechia (Linnaeus)

YEWA

RANGE: Breeds throughout most of treed North America from northwestern and north-central Alaska, northern Yukon, northwestern and central Mackenzie, northern Saskatchewan, northern Manitoba, northern Ontario, central Quebec, and central Labrador and Newfoundland south to central Alaska, northern Baja California, through mainland Mexico and throughout the eastern United States south into Texas and the central parts of Oklahoma, Arkansas, Mississippi, and Georgia. Winters from southern California, southwestern Arizona and Florida, northern Mexico, and the southern Gulf coast south to the Amazon Basin of Brazil and Peru.

STATUS: On the coast, a *fairly common* to *common* migrant and summer visitant in the Georgia Depression Ecoprovince. In the Coast and Mountains Ecoprovince, a *rare* to *uncommon* spring migrant and summer visitant, *uncommon* autumn migrant, and *accidental* in winter on Western Vancouver Island; *uncommon* to *fairly common* migrant and summer visitant on the Southern Mainland Coast and Northern Mainland Coast; and *very rare* migrant and local summer visitant on the Queen Charlotte Islands.

In the interior, an *uncommon* to *fairly common* migrant and summer visitant in the Southern Interior Ecoprovince; generally *uncommon* to *fairly common*, but locally *very common*, in the Southern Interior Mountains Ecoprovince; *uncommon* to *fairly common* in the Central Interior and Sub-Boreal Interior ecoprovinces; *common* to *very common* in the Boreal Plains Ecoprovince; and *uncommon* in the Northern Boreal Mountains and Taiga Plains ecoprovinces.

Breeds.

Figure 23. Adult female Yellow Warbler incubating in nest in sapling trembling aspen (Beatton Park, 23 June 1996; R. Wayne Campbell).

NONBREEDING: The Yellow Warbler (Fig. 23) is widely distributed throughout the length and breadth of British Columbia from spring through autumn. There are few parts of the province below timberline in which it does not occur. There remain, however, some large areas from which the species has

Figure 24. Deciduous trees already in leaf, including windbreaks bordering marshes and rivers, provide foraging areas for migrating Yellow Warblers in spring (Athalmer, 9 May 1997; R. Wayne Campbell).

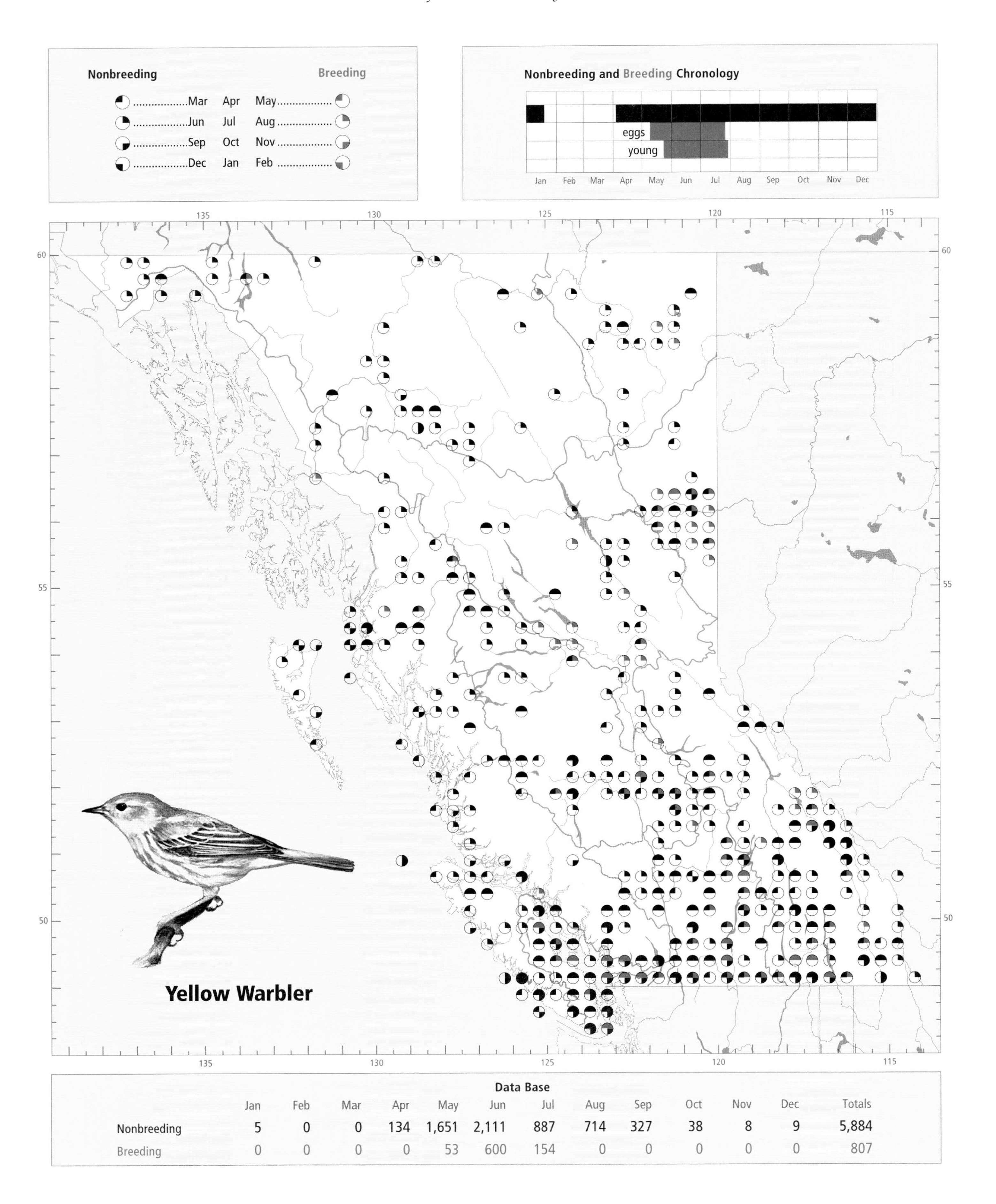

Data Base

	Jan	Feb	Mar	Apr	May	Jun	Jul	Aug	Sep	Oct	Nov	Dec	Totals
Nonbreeding	5	0	0	134	1,651	2,111	887	714	327	38	8	9	5,884
Breeding	0	0	0	0	53	600	154	0	0	0	0	0	807

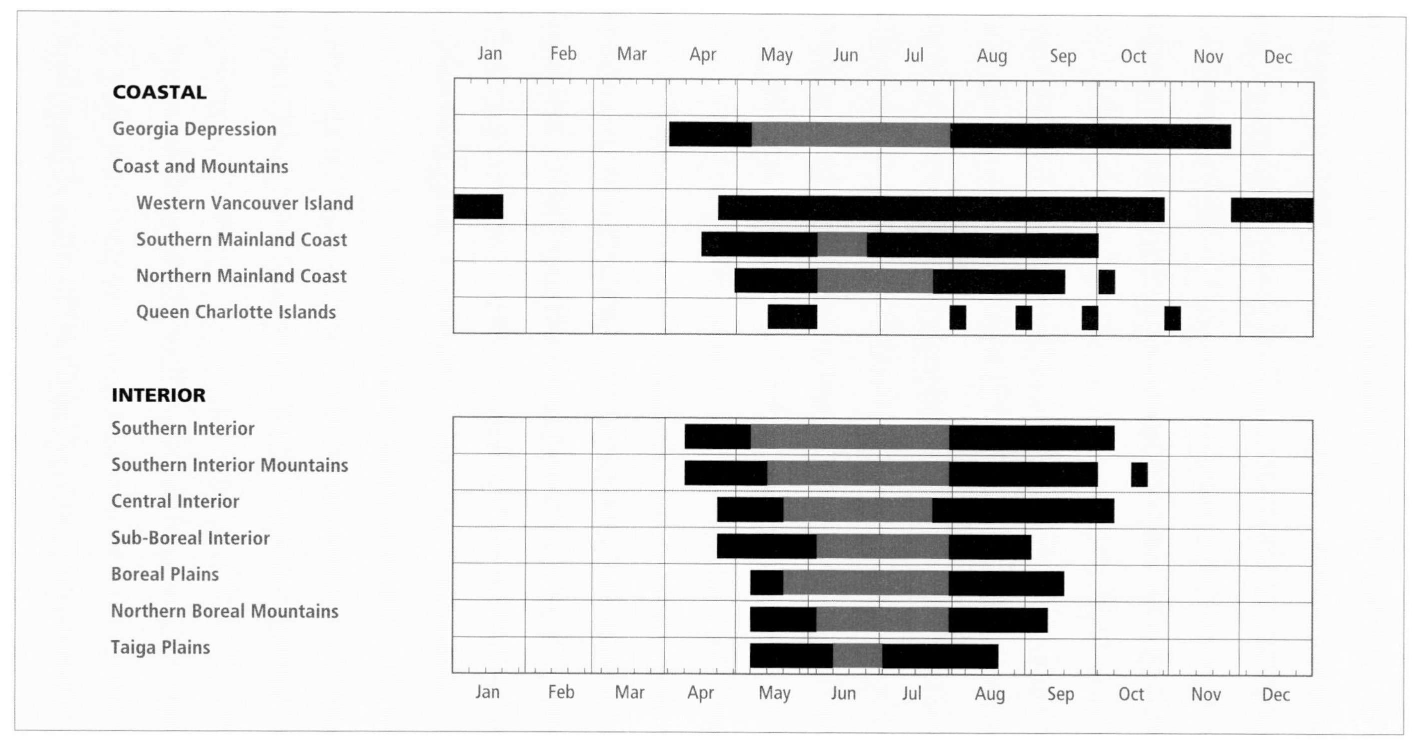

Figure 25. Annual occurrence (black) and breeding chronology (red) for the Yellow Warbler in ecoprovinces of British Columbia. Records are shown for the week in which they occurred.

not been reported, such as parts of the Southern Mainland Coast, the western Chilcotin Plateau, the West Road (Blackwater) River region, the Rocky Mountain Trench from Mackenzie north to Lower Post, and a large area between the Stikine River and Atlin and in the vicinity of Teslin Lake. Most of these gaps are due to inadequate biological exploration, but some probably reflect the scarcity of the deciduous habitats preferred by the Yellow Warbler.

In migration along the coast, the Yellow Warbler has been reported from sea level to about 800 m elevation. In the southern part of the interior, it usually occurs between 280 and 1,200 m, but its presence as a spring migrant depends largely upon the presence of deciduous trees already in leaf (Fig. 24). In the far north, late spring migrants have been recorded to timberline.

The Yellow Warbler is characteristic of the deciduous riparian woodlands throughout the province, both in migration and through the summer. It avoids continuous coniferous forests and occurs mainly on the edges of deciduous and mixed forests, alongside river valleys, wetlands, or estuaries. On steep subalpine slopes, it may occupy the tangled shrub growth of avalanche chutes that characterize so much of the mountainous terrain. In general, migrants use the same habitats that support the breeding population.

On the south coast, the spring migration begins in mid-April with the arrival of a few Yellow Warblers in the Fraser Lowland and on southeastern Vancouver Island (Figs. 25 and 26). It is well established by the end of the month and peaks in early May. Along the conifer-dominated coast of Western Vancouver Island, the migration is not noticeable (Hatler et al. 1973), but there are records from the last week of April to the third week of May. The earliest spring records from the Queen Charlotte Islands are from the third week of May, about a week later than on the adjacent mainland.

In the Southern Interior, the spring migration is not as conspicuous as that of some other warblers, but the first spring arrivals may show up in the Okanagan valley in the second week of April; most normally arrive later in the month or in early May (Figs. 25 and 26). First-of-year records compiled by Cannings et al. (1987) over 31 years ranged from 10 April to 19 May, with a mean of 7 May and a median of 11 May. In the valleys of the Kootenay and Columbia rivers, the earliest arrivals have appeared between the second week of April and the first week of May, as in the Okanagan valley.

The spring migration moves northward rapidly, with first arrivals noted throughout much of the central and northern interior almost simultaneously in the first 10 days of May (Figs. 25 and 26). The spring migration up the eastern side of the Rocky Mountains enters the Peace Lowland as early as the first week of May, but generally between 9 and 16 May. The species reaches the northern boundary of the province by mid-May.

In northern British Columbia, the autumn migration appears to begin shortly after the young are independent. With the cessation of singing, however, it is not easy to distinguish between a decline in numbers resulting from the departure of the summer population and an apparent reduction in numbers due to the absence of song. Peak autumn movements throughout the province occur from mid to late August. In the Peace Lowland, high daily counts of between 6 and 10 birds have been made between 14 and 24 August, and the latest record is 14 September (Figs. 25 and 26).

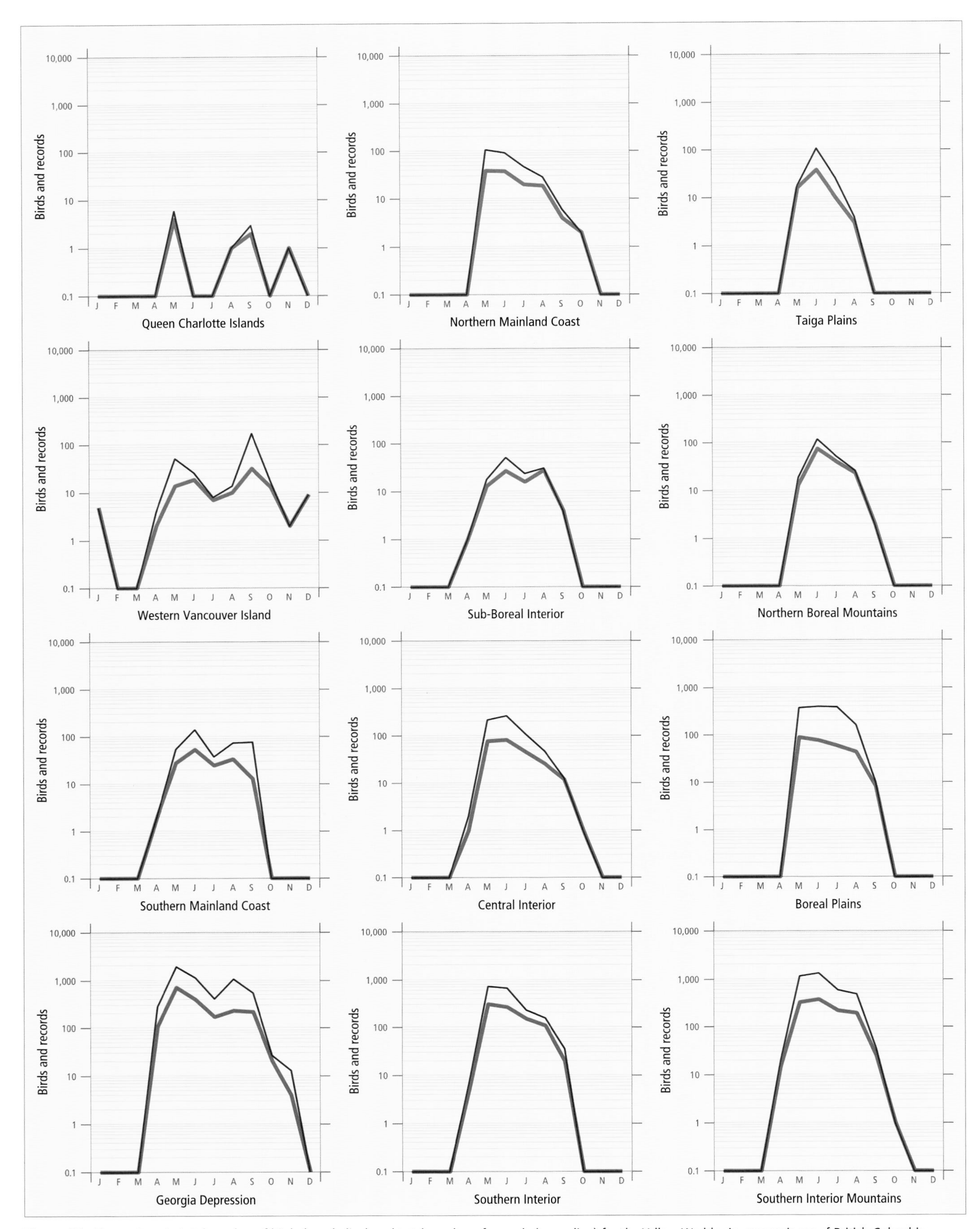

Figure 26. Fluctuations in total number of birds (purple line) and total number of records (green line) for the Yellow Warbler in ecoprovinces of British Columbia. Christmas Bird Counts, Breeding Bird Surveys, and nest record data have been excluded.

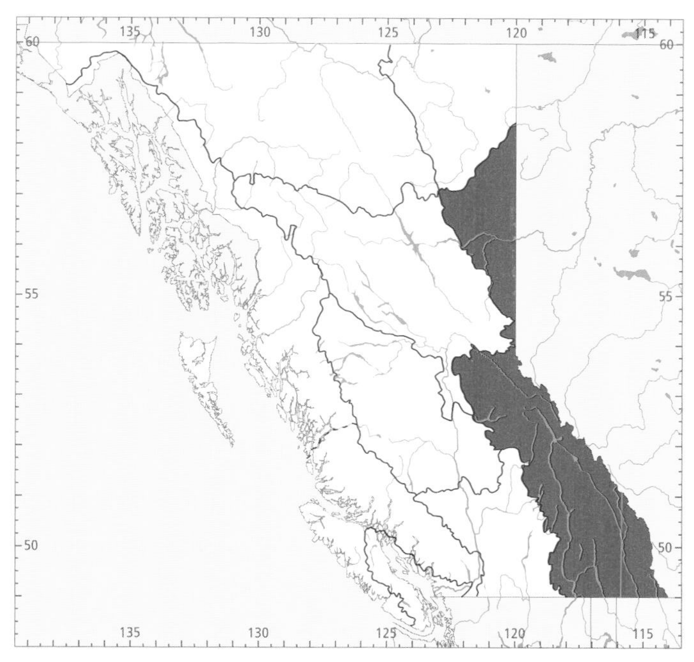

Figure 27. In British Columbia, the highest numbers for the Yellow Warbler in summer occur in the Southern Interior Mountains and Boreal Plains ecoprovinces.

At Mackenzie, in the Sub-Boreal Interior, autumn captures at a banding station have revealed small movements of Yellow Warblers passing through between 13 and 28 August, with the latest bird banded on 11 September. Parallel banding studies at Rocky Point, near Victoria, have recorded Yellow Warblers between 3 August and 2 October, in a banding season extending from 31 July to 23 October; the highest numbers there occur in the last 2 weeks of August. Banding in the Southern Interior, at Vaseux Lake, revealed the same period of highest numbers of migrating Yellow Warblers. Most of the Yellow Warblers have left the northern half of the province by late August, the southern half of the interior by mid-September, and the southern coast by early October.

The Yellow Warbler has been reported in winter only from the south coast, in the vicinity of Tofino (see also REMARKS).

On the coast, the Yellow Warbler has been regularly recorded from 8 April to 24 November; in the interior, it has been recorded from 9 April to 20 October (Fig. 25).

BREEDING: The Yellow Warbler breeds in all the ecoprovinces of British Columbia, from the southern boundary of the province north almost to the northern boundary, and from the ocean shorelines of southeastern Vancouver Island to the Alberta border. Although this warbler is present through the summer along the coast, and undoubtedly breeds, there are no breeding records from Campbell River north to Prince Rupert, including the Queen Charlotte Islands. We do have two breeding records from further north in the Northern Mainland Coast. In the interior, there are only a few records from the vast area of mountains and plateaus extending beyond Mackenzie, north through 5° of latitude, to the Liard River.

This warbler reaches its highest numbers in summer in the valleys of the Kootenay and Columbia rivers, and in the Peace Lowland (Fig. 27). An analysis of Breeding Bird Surveys in British Columbia for the period 1968 through 1993 shows that the mean number of birds on both coastal and interior routes has declined at an average annual rate of 3% and 2%, respectively, (Fig. 28). Breeding Bird Surveys in Canada and across North America from 1966 to 1996 suggest a stable population (Sauer et al. 1997).

On the coast, the Yellow Warbler has been recorded breeding at elevations from sea level to about 900 m; in the interior, it breeds at elevations from 330 to 1,450 m. At Atlin, in the northwestern region of the province, Swarth (1926) describes it as a species of lowland elevations (about 650 m).

The Yellow Warbler breeds in sunlit stands of deciduous vegetation, and has a strong attraction to willow. In 130 recorded observations of the tree or shrub species in which this warbler was observed, 19 different kinds of plants were identified. Fifty-five percent of the observations were in willow, 19% in black cottonwood or poplar, 11% in alder, and 6% in trembling aspen. In general, it prefers shorter trees in dense stands, and shrubbery in riparian habitats along stream

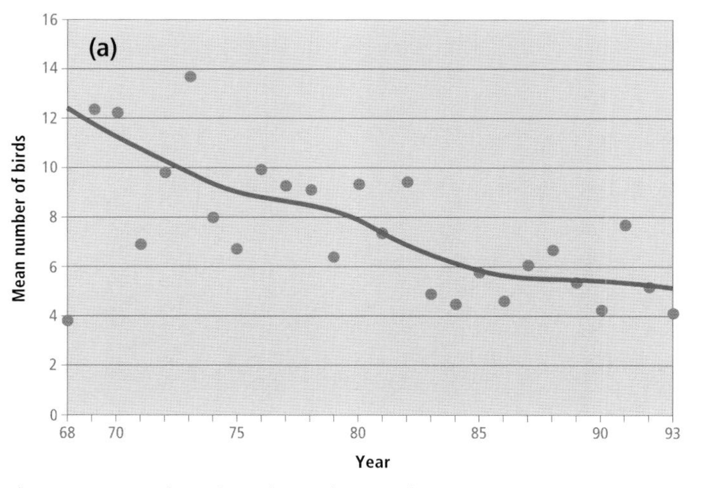

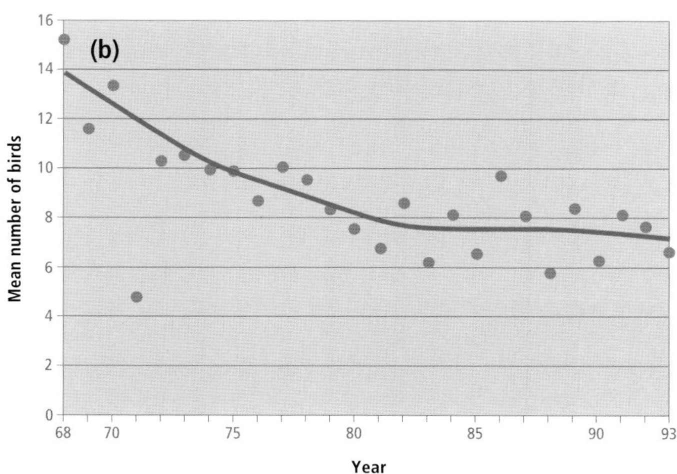

Figure 28. An analysis of Breeding Bird Surveys for the Yellow Warbler in British Columbia shows that the mean number of birds on coastal routes (a) and interior routes (b) decreased at an average annual rate of 3% and 2%, respectively, over the period 1968 through 1993 ($P < 0.01$).

Figure 29. In British Columbia, the Yellow Warbler breeds in dense thickets of wild rose, ocean-spray, willow, and saskatoon bordering wetlands (Creston, 2 July 1997; R. Wayne Campbell).

Figure 30. In the Boreal Plains Ecoprovince of northeastern British Columbia, the Yellow Warbler is consistently one of the most common breeding species found in trembling aspen woodlands, especially where a thick understorey of rose and cow-parsnip exists. Note the nest near the centre of the photograph (Charlie Lake Park, 25 June 1998; R. Wayne Campbell).

courses, on the margins of beaver ponds, sloughs, wet meadows, and, in the northern areas of its distribution, marshes and muskegs. It is also a nesting species along the edges of forest clearings, thickets, or hedgerows of rose, snowberry, spirea, salmonberry, and other shrubs bordering natural or human-made clearings (Fig. 29) such as the cleared rights-of-way of electrical transmission lines, cultivated farmland, fruit or nut orchards, roadsides, well-treed gardens, and suburban parks. In Mount Revelstoke and Glacier national parks, the species was most abundant in red alder–skunk cabbage and red alder–fern plant associations (Van Tighem and Gyug 1983). Both of these are low-elevation plant communities.

The use of trembling aspen forests by this species in British Columbia (Fig. 30) differs remarkably between ecoprovinces. In a study of the birds of the trembling aspen woodlands of the Bulkley valley, in the Central Interior, the Yellow Warbler was one of the less common species in sapling trembling aspen stands and in some plots of mature trembling aspen and mixed conifer-aspen, and was missing from the clearcut areas (Pojar 1995). In the Peace Lowland of the Boreal Plains, however, the Yellow Warbler was reported as "consistently one of the most abundant birds in the trembling aspen study sites." Census counts done in trembling aspen forests, including all stages, found that the only age class avoided by the warblers was the 20- to 40-year-old "pole stage." As second-growth became larger over a 4-year period, the number of warblers appeared to increase, and the proportion seen in old-growth trembling aspen appeared to decrease (Fig. 31). It may be noteworthy that the 2 areas contrasted here are occupied by different subspecies of the Yellow Warbler (see REMARKS).

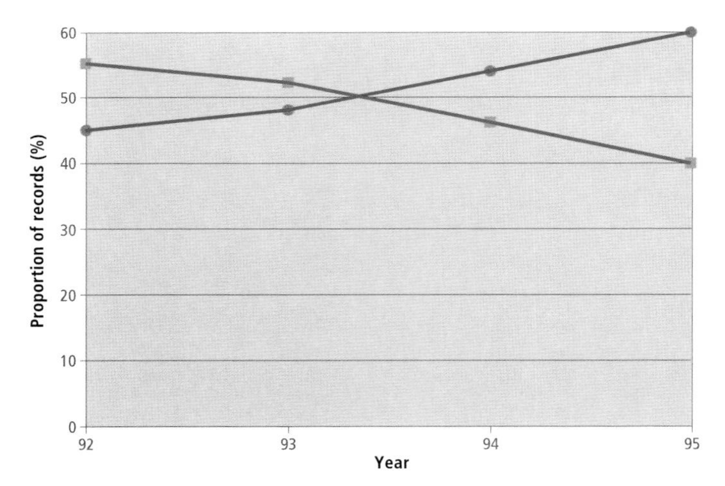

Figure 31. Use of trembling aspen by breeding Yellow Warblers in second-growth (green line) and old-growth (brown line) aspen forests in the Peace Lowland of northeastern British Columbia (adapted from Lance and Phinney 1996).

Most nesting habitats in our sample (85%; n = 385) were described as forested or wooded, of which about one-third were human-influenced. The percentage of nesting habitat in undisturbed forest varied from 93% in the Boreal Plains to 40% in the valleys of the Kootenay and Columbia rivers.

On the coast, the Yellow Warbler has been recorded breeding from 10 May to 25 July; in the interior, it has been recorded breeding from 13 May (calculated) to 29 July (Fig. 25).

Nests: Most nests were found in deciduous trees (56%; n = 391), with willow the most frequently used tree, followed by poplar, trembling aspen, alder, and birch. Most of the remaining nests were in shrubs. Prominent among the native shrubs used were rose, snowberry, and saskatoon. The cup-shaped nest was usually built into a forked branch and included grasses and plant down as the major components. Other frequently used items were animal hair, plant fibres, string, feathers, rootlets, and mosses (Figs. 23 and 32). The lining was usually of fine grasses and plant down. Yellow Warblers nesting at Churchill, Manitoba, at the northern limit of their range, build larger and better-insulated nests than in more moderate climates (Briskie 1995). It is not known whether there is a similar response to altitudinal temperature regimes in British Columbia, but there is no significant difference in clutch size between Yellow Warblers nesting in southern British Columbia and those near the northern boundary, a latitudinal spread of 10°.

The heights of 385 nests ranged from ground level to 14 m, with 61% between 1.2 and 2.8 m.

Eggs: Dates for 450 clutches ranged from 10 May to 21 July, with 52% recorded between 7 June and 23 June. Sizes of 288 clutches ranged from 1 to 6 eggs (1E-39, 2E-26, 3E-33, 4E-125, 5E-61, 6E-4), with 64% having 4 or 5 eggs (Fig. 32).

Across North America, the clutch size of the Yellow Warbler varies with latitude (Briskie 1995). The largest mean clutch size in North America has been recorded in interior and western Alaska (4.9 ± 0.7 eggs, n = 14 completed clutches) (Kessel 1989). In southern Manitoba, mean clutch size was 4.5 ± 0.05 (n = 144 completed clutches) (Goossen and Sealy 1982). Neither in Alaska nor Manitoba were there any completed clutches of fewer than 3 eggs. Our clutch size data are not comparable with the above figures, as the British Columbia statistics include an unknown number of incomplete clutches. In British Columbia, however, there is no evidence of a change in number of eggs per clutch associated with latitude. Both the largest and smallest mean clutch sizes are from the southern parts of the province. The incubation period is 11 days (Curson et al. 1994).

A comparison of the egg period of the nesting season in the southwest coast region, the southern portions of the interior between the Cascade Mountains and the Rocky Mountains, and the Peace Lowland area of the Boreal Plains reveals important differences in the onset of nesting and in

Figure 32. Nest and eggs of the Yellow Warbler (Castlegar, 12 June 1994; Linda M. Van Damme).

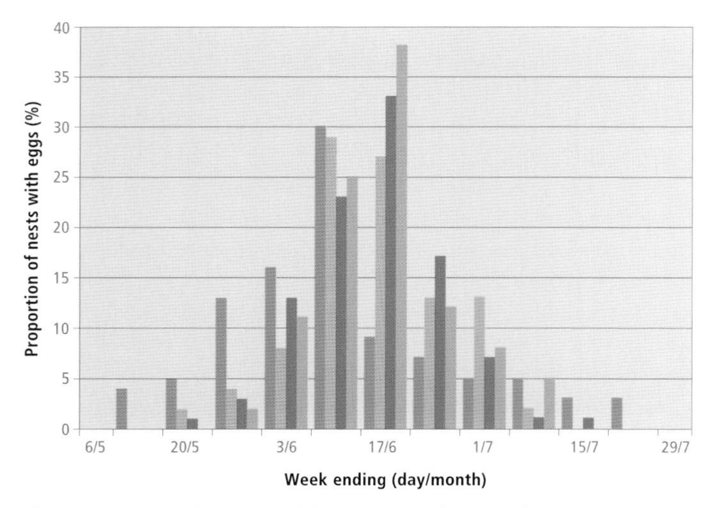

Figure 33. Weekly distribution of the proportion of clutches for the Yellow Warbler in the Georgia Depression Ecoprovince (green bars), Southern Interior Ecoprovince (red bars), Southern Interior Mountains Ecoprovince (yellow bars), and Boreal Plains Ecoprovince (blue bars) in British Columbia. The figure is based on the first week eggs were found in the nest.

the length of the egg phase (Fig. 33). In the Georgia Depression, egg laying begins in the second week of May and eggs are still present into the third week of July. In the southern portions of the interior, egg laying begins in the third week of May and none occurs after the second week of July. In the Boreal Plains, the first eggs appear in the fourth week of May and eggs are present into the first week of July. Thus, the egg period on the coast is 2 weeks longer than in the southern portions of the interior, and nearly a month longer than in the Boreal Plains.

Young: Dates for 274 broods ranged from 24 May to 29 July, with 53% recorded between 20 June and 4 July. Sizes of 182 broods ranged from 1 to 5 young (1Y-36, 2Y-31, 3Y-40, 4Y-59, 5Y-16), with 71% having 2 to 4 young. One brood in British Columbia had a nestling period of 10 days; Curson et al. (1994) report a nestling period of 9 to 12 days.

Brown-headed Cowbird Parasitism: In British Columbia, 15% of 393 nests found with eggs or young were parasitized by the Brown-headed Cowbird (Fig. 34). There were 15 additional records of a Yellow Warbler feeding a cowbird fledgling. On the coast, parasitism was 11% (n = 102); in the interior, it was 16% (n = 291). A single cowbird egg was present in 81% of the parasitized nests (n = 26; Fig. 34), 2 eggs in 15%, and 3 eggs in 1 nest.

In a Manitoba study, the overall incidence of cowbird parasitism was 25% (Goossen and Sealy 1982). There the Yellow Warbler accepted the cowbird eggs in only 52.5% of cases (n = 96). Rejected eggs were either buried under a new nest lining (53.5%) or deserted (46.5%) (Briskie et al. 1990). There is 1 instance in British Columbia of a "three-storey nest": the first nest contained 2 cowbird eggs and 4 of the warbler. They were buried under a new nest floor, and the second nest contained 3 cowbird eggs and 2 warbler eggs, which were also buried under a new nest floor. The third storey of the nest contained 2 warbler eggs.

Friedmann and Kiff (1985) report that the Yellow Warbler is one of the most frequent hosts of the Brown-headed Cowbird, with frequencies of parasitism in parts of the United States as high as 87%.

Nest Success: Of 74 nests found with eggs and followed to a known fate, 42% produced at least 1 fledgling. In Manitoba, the success rate varied between 40% and 64%, with an overall success of 47.9% for all active nests (Goossen and Sealy 1982).

REMARKS: The breeding population of Yellow Warblers in British Columbia is believed to include 3 subspecies and another that passes through in migration (Browning 1994). The nesting subspecies of the coastal region is described as *D. p. rubiginosa;* that of the interior, east to include the Rocky Mountains, is *D. p. morcomi,* while *D. p. amnicola* occurs northeast of the Rocky Mountains in the Boreal Plains and Taiga Plains. In addition, *D. p. banksi* probably migrates through British Columbia on its passages to and from its nesting grounds in the Yukon.

There are many early and late migration dates as well as winter records of the Yellow Warbler for British Columbia (e.g., Mandarte Island 13 Mar 1983-1 [Mattocks and Hunn 1983b]; Swan Lake [Victoria] 12 Dec 1987-1; Westham Island 6 Dec 1969-2; Okanagan Landing 1 Jan 1915-1; Mount Revelstoke

Figure 34. Nest with 3 Yellow Warbler eggs and 1 egg (far right) of the Brown-headed Cowbird (Victoria, 15 June 1995; R. Wayne Campbell). The Yellow Warbler is one of the most frequent hosts of the Brown-headed Cowbird in North America, with parasitism rates reaching 87% in some parts of its range. In British Columbia, only 15% of 393 nests were parasitized by the cowbird.

12 Dec 1982-12; and Vancouver 18 Dec 1976-2 on Christmas Bird Count, reported as the all-time Canadian high count [Anderson 1977]). In all cases, details are incomplete or do not meet our criteria for acceptance. These have all been excluded from this account.

For additional details on the reproductive biology of 2 Canadian populations and parasitism by the Brown-headed Cowbird, see Goossen and Sealy (1982), Briskie et al. (1990), Briskie (1995), or Sealy (1995). Lowther et al. (1999) provide a recent review of the life history of the Yellow Warbler.

NOTEWORTHY RECORDS

Spring: Coastal – Tugwell Lake (Sooke) 19 May 1985-4; Chatham Islands 19 May 1980-4; Victoria 10 May 1907-4 eggs (WFVZ 137); Saanich 8 Apr 1983-2 along power line on Munns Rd; Beaver Lake (Saanich) 8 May 1976-20; Blenkinsop Lake 6 May 1985-10; Bamfield 14 May 1978-6; Galiano Island 18 May 1986-2; Somenos Lake 23 Apr 1977-4; Long Beach 11 May 1997-2; Pacific Rim National Park 23 May 1986-15; McLean Point 27 Apr 1972-2 (Hatler et al. 1978); Barnston Island 28 May 1961-25; Klesilkwa valley 27 Apr 1976-1; Harrison Hot Springs 4 May 1976-1; Quadra Island 2 May 1974-1; Bigsby Inlet 30 May 1990-2 (Siddle 1990b); Queen Charlotte City 20 May 1991-1 (Siddle 1991c); Meziadin Lake 30 May 1978-1 building a nest; Haines Highway 28 May 1979-1. **Interior** – Creston 9 Apr 1984-2; Oliver area 18 May 1969-12; Balfour to Waneta 16 May 1981-104 on survey; Hidden Creek 20 May 1983-35; Erie 27 May 1984-20; Midway 9 May 1905-1 (NMC 3188); Castlegar 11 May 1969-30; Nelson 27 Apr 1980-5, 20 May 1969-2 eggs; Naramata 27 Apr 1965-3; Lytton 8 May 1966-6; ne Enderby 17 May 1981-4 eggs; Kamloops 20 May 1959-5; Tranquille 2 May 1992-1; Adams River 11 May 1997-11 males; Revelstoke 13 Apr 1985-1; Williams Lake 29 Apr 1983-2 males singing on Scout Island, 18 May 1978-5, 25 May 1976-adult incubating; Riske Creek 4 May 1987-1; Anahim Lake 21 May 1948-6; Quesnel 29 May 1963-6; Prince George 26 Apr 1977-1; Nulki Lake 22 May 1945-2 (Munro 1949); Quick 18 May 1978-1; Smithers 31 May 1992-2; Tupper Creek 7 May 1938-1; Swan Lake (Tupper) 21 May 1938-50 (Cowan 1939); s Fort St. John 11 May 1988-1; Charlie Lake 23 May 1977-1 egg; Telegraph Creek (Stikine River) 26 May 1919-3 (Swarth 1922); Fort Nelson 8 May 1980-1 male; Liard Hot Springs 9 May 1975-2 (Reid 1975); Atlin 31 May 1934-1 (CAS 42126).

Summer: Coastal – Rocky Point (Victoria) 28 Aug 1995-42; McGillivray Slough 9 Jun 1963-30; Sumas Prairie 4 Aug 1923-250; Long Beach 6 Jul 1984-1 male; New Westminster 27 Jul 1993-female feeding 1 fledged young; Sumallo River 31 Jul 1984-3 fledglings with adult; Vancouver 10 Aug 1993-92 at Jericho Park (Siddle 1994a); Point Grey 31 Aug 1997-370; Alberni 10 Jun 1975-1; Courtenay 1 Jun 1920-5 nestlings; Pitt Meadows 6 Jul 1972-26 (Campbell et al. 1972a), 21 Jul 1972-2 eggs, female flushed; Mitlenatch Island 24 Jul 1966-2; Triangle Island 29 Jul 1978-2; Kitlope River 5 Jun 1994-15; lower Kateen River 5 Jul 1991-4 eggs (BC Photo 1570); Hazelton 6 Jun 1992-8; Flood Glacier 6 Aug 1919-1 (MVZ 40173). **Interior** – e side Manning Park 2 Aug 1962-female feeding 2 young; Manning Park 16 Jul 1968-4 warbler eggs and 1 Brown-headed Cowbird egg, 20 Jun 1987-27; Six Mile Slough to Duck Lake (Creston) 3 Jul 1997-166 on survey (122 males, 40 females, 4 unsexed); Vaseux Lake 4 Aug 1988-42; Balfour to Waneta 8 Jun 1983-98 on survey; Erie 8 Jun 1983-37; Castlegar 4 Jun 1978-24; Lytton to Skihist 8 Jun 1968-12 on survey; Numa Creek 2 Jun 1970-male carrying nesting material (Poll et al. 1984); Riske Creek 12 Jul 1993-1 warbler and 1 Brown-headed Cowbird nestling; Williams Lake 12 Jun 1976-adult feeding unknown number of young at Scout Island, 6 Aug 1994-8 moving through lake-edge shrubbery; Prince George 5 Jun 1992-4, 24 Aug 1982-1; nr Chetwynd 23 Jul 1975-2; 8 km s Dawson Creek 2 Jul 1993-4 warbler eggs and 1 Brown-headed Cowbird egg, 3 Jul 1984-122; Driftwood River 10 Jun 1941-1; Taylor 11 Jun 1980-5 nestlings; Charlie Lake 24 Aug 1979-8; Fort St. John 27 Jul 1987-5 in flock, 18 Aug 1987-10, 24 Aug 1987-8; Beatton Park 31 Jul 1987-2 fledglings fed by female, 14 Aug 1987-6; Firesteel River 27 Jun 1976-10; Kwadacha River 14 Jul 1997-1 adult male singing; Telegraph Creek (Stikine River) 14 Jun 1922-3 eggs (Swarth 1922); Dease Lake 6 Jun 1962-1 (NMC 50090); Kotcho Lake 22 Jun 1982-5 eggs, incubation well advanced, and 25 Jun 1982-4 nestlings (Campbell and McNall 1982); Aline Lake 26 Jun 1980-3 eggs; Chilkat Pass 30 Jun 1956-3 (Weeden 1960); O'Connor and Tatshenshini rivers 26 Jun 1993-female and 1 fledgling at confluence; Atlin 26 Aug 1924-1 (Swarth 1926).

Breeding Bird Surveys: Coastal – Recorded from 24 of 27 routes and on 77% of all surveys. Maxima: Pemberton 5 Jun 1994-44; Kwinitsa 18 Jun 1978-40; Squamish 10 Jun 1973-35. **Interior** – Recorded from 68 of 73 routes and on 87% of all surveys. Maxima: Tupper 18 Jun 1995-110; Kuskonook 18 Jun 1968-55; Hudson's Hope 7 Jun 1976-53.

Autumn: Interior – McBride (Stikine River) 7 Sep 1977-1; St. John Creek 14 Sep 1985-1; Mackenzie 2 Sep 1994-1, 11 Sep 1996-1; Williams Lake 4 Oct 1982-1; Eagle Lake 13 Sep 1992-1; Lac la Hache 22 Sep 1943-1 (Munro 1945a); Kootenay National Park 20 Oct 1981-1; lower Blaeberry River valley 10 Sep 1995-1; Tranquille 10 Sep 1994-1; Okanagan Landing 3 Sep 1905-3; Naramata 12 Sep 1965-3; West Bench 25 Sep 1980-1; Salmo 21 Sep 1975-4; Vaseux Lake 2 Oct 1995-1 banded. **Coastal** – Rose Spit 1 Sep 1960-2; Masset 1 Nov 1950-1 (RBCM 10513); Sandspit 26 Sep 1991-1 (Siddle 1992a); Triangle Island 1 Sep 1976-1; Harrison Hot Springs 18 Sep 1986-1; Vancouver 4 Sep 1942-20; Iona Island 25 Oct 1981-1; Roberts Bank 24 Nov 1995-1 immature; Reifel Island 1 Nov 1985-1; Tofino 13 Sep 1984-45 of which 20 were in a single small tree, 26 Oct 1984-1, 28 Nov 1995 to 18 Jan 1996-1; Saanich 9 Nov 1991-9 (Siddle 1992a).

Winter: Interior – No records. **Coastal** – Tofino 2 Dec 1995 to 18 Jan 1996-1 at feeder. See REMARKS.

Christmas Bird Counts: Interior – Not recorded. **Coastal** – See REMARKS.

Chestnut-sided Warbler

CSWA

Dendroica pensylvanica (Linnaeus)

RANGE: Breeds from east-central Alberta and south-central Saskatchewan east through southern Manitoba, across central and southern Ontario, southern Quebec, and the Maritime provinces, and southward through the New England states to northwestern New Jersey and southeastern Pennsylvania and along the Appalachian Mountains to northwestern Georgia; disjunct populations south of the primary range are found in Colorado, North Dakota, and Nebraska. Winters from southern Mexico south through Central America, mainly on the Caribbean coast, to Panama and Venezuela. Autumn vagrants regularly occur in California and western Mexico.

STATUS: On the coast, *very rare* in summer and *accidental* in autumn in the Georgia Depression Ecoprovince; in the Coast and Mountains Ecoprovince, *casual* vagrant on the Northern Mainland Coast, *accidental* to Western Vancouver Island, and absent from the Southern Mainland Coast and Queen Charlotte Islands.

In the interior, *casual* vagrant in the Southern Interior and Taiga Plains ecoprovinces; *very rare* summer and *accidental* in spring and autumn in the Southern Interior Mountains Ecoprovince; *accidental* in the Sub-Boreal Interior and Boreal Plains ecoprovinces.

CHANGE IN STATUS: The Chestnut-sided Warbler (Fig. 35) was not known to occur in British Columbia when Munro and Cowan (1947) completed their review of the status of birds in the province. The first record of this "eastern" warbler occurring in British Columbia was a male in full breeding plumage photographed at Red Pass in Mount Robson Park, in the Southern Interior Mountains, on 22 June 1971 (Shepard 1972). Since then, this brightly coloured warbler has been recorded at least 20 times in the province. It has been recorded at least once in all ecoprovinces except the Northern Boreal Mountains and Central Interior, but is still an extremely rare visitor anywhere in the province. Also, see POSTSCRIPT.

OCCURRENCE: The Chestnut-sided Warbler has occurred sporadically in most regions of British Columbia, but there is no ecoprovince with more than 7 records. All breeding season records are of males, some of which appeared to have established territories and were in full song. Near Revelstoke, single males were found in 1988 and 1989. These birds stayed for at least several weeks during the breeding season. Two males may have been present in 1988.

For the Georgia Depression, there are 7 records between 1974 and 1997. A juvenile, probably a bird in first basic plumage (M. Richardson pers. comm.), was recorded at Reifel Island in 1974 and was the first accepted record for the British Columbia coast (Crowell and Nehls 1975a). In 1979, a singing male was observed at Point Grey. In 1986, a single bird was found at Burnaby Lake. In 1990, a single bird was found at Stanley Park, Vancouver, and was noted as the fourth record for Vancouver. The most recent Vancouver records were of a

Figure 35. The Chestnut-sided Warbler was first recorded in British Columbia in 1971 and found breeding in 1998 (see POSTSCRIPT; Anthony Mercieca).

singing male at the University of British Columbia campus in 1995 (Elliott and Gardner 1997) and 1997. The first record for Vancouver Island was of a singing male at Hamilton Marsh, south of Qualicum Beach, in 1982.

For the Coast and Mountains, there are 3 records. Single birds were found at Pondosy Lake (Tweedsmuir Park), Hazelton, and Pacific Rim National Park.

For the Southern Interior, there are 5 records. Three are from the Okanagan: a male caught in a mist net east of Oliver in 1985, a male found near Penticton in 1989, and a late summer migrant at Okanagan Falls in 1989. In 1985, a singing male was found on the Coldwater River, southwest of Merritt, and in 1996 a male was found near Kamloops.

For the Sub-Boreal Interior, the only record is of a bird near Prince George in 1972. For the Boreal Plains, the only record is of a bird near Fort St. John. For the Taiga Plains, single birds have been found twice near Fort Nelson.

Breeding Bird Surveys in British Columbia for the period 1968 through 1993 did not detect this warbler. An analysis of Breeding Bird Surveys across Canada from 1966 to 1996

Figure 36. In British Columbia, the Chestnut-sided Warbler is most often found in dense, early successional stage deciduous thickets of alder and willow, often adjacent to airports and roads (21 June 1997; R. Wayne Campbell).

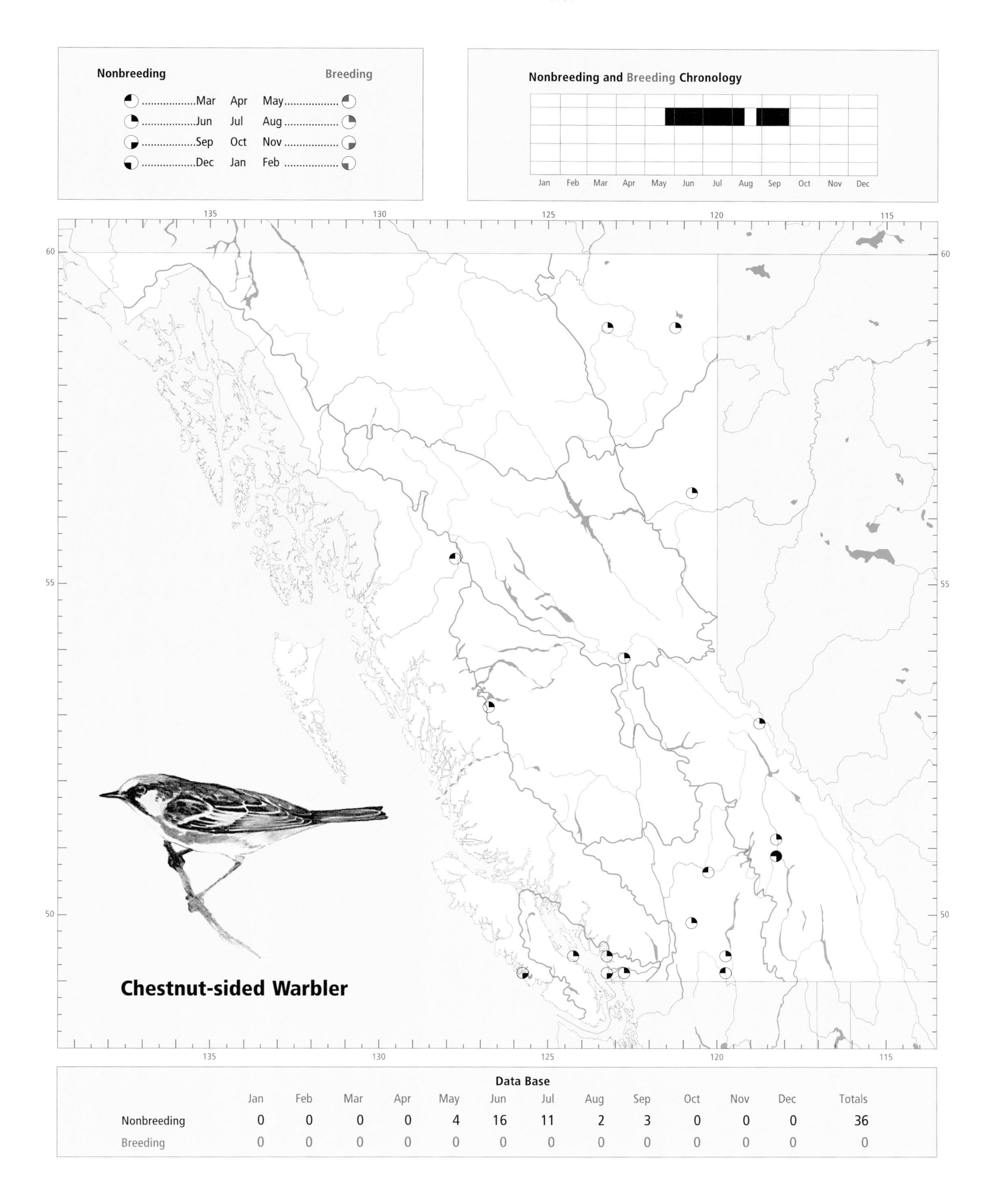

Data Base

	Jan	Feb	Mar	Apr	May	Jun	Jul	Aug	Sep	Oct	Nov	Dec	Totals
Nonbreeding	0	0	0	0	4	16	11	2	3	0	0	0	36
Breeding	0	0	0	0	0	0	0	0	0	0	0	0	0

Figure 37. In the Taiga Plains Ecoprovince of northeastern British Columbia, the Chestnut-sided Warbler has been found in tall stands of willow and regenerating deciduous trees and bushes adjacent to spruce wetlands (near Fort Nelson, 28 June 1998; R. Wayne Campbell).

suggests that populations have remained stable (Downes and Collins 1996). The Chestnut-sided Warbler is thought to have expanded its range and numbers in North America since the early 1800s in response to clearing of mature forests and provision of more shrubby habitat (Richardson and Brauning 1995).

On the coast, the Chestnut-sided Warbler has been reported mainly near sea level. In the interior, it has been recorded at elevations up to about 1,000 m.

Habitat use has been poorly described in British Columbia. In general, this warbler specializes in exploiting early successional stages such as shrubby, deciduous habitat. Most birds observed in British Columbia have been found in shrubby thickets of alder, willow, and black cottonwood along the edges of ponds, streams, or open areas such as airports and roadsides (Figs. 36 and 37). Natural sites reported elsewhere include regenerating burns, blowdowns, flooded streamsides, and riparian thickets (Richardson and Brauning 1995).

In many areas of eastern and central Canada and the United States, local populations have increased after forests were cleared or thinned and shrubs were allowed to regenerate. This warbler does not inhabit urbanized or intensively farmed areas (Richardson and Brauning 1995). The highest densities of Chestnut-sided Warblers in eastern North America tend to occur in areas with more complex vegetation and greater density of shrubs (Freedman et al. 1981; Niemi and Hanowski 1984). Early second-growth deciduous stands, such as those that develop after logging or along transmission lines and roads, are also used.

Migration is not well known in British Columbia, as only a few birds occur in the province each year, but migration patterns are likely similar to those of other "eastern" neotropical migrant warblers. Early migrants arrive in British Columbia in late May, which is similar to the earliest spring records in Alberta (Semenchuk 1992). Spring migrants probably arrive from late May through early June. Most autumn migrants have probably left the province by late August (Figs. 38 and 39).

The Chestnut-sided Warbler occurs sporadically in British Columbia from 22 May to 29 September (Fig. 38).

REMARKS: In Alberta, the first specimen records of the Chestnut-sided Warbler were obtained in 1972 (Salt 1972,

Figure 38. Annual occurrence for the Chestnut-sided Warbler in ecoprovinces of British Columbia. Records are shown for the week in which they occurred.

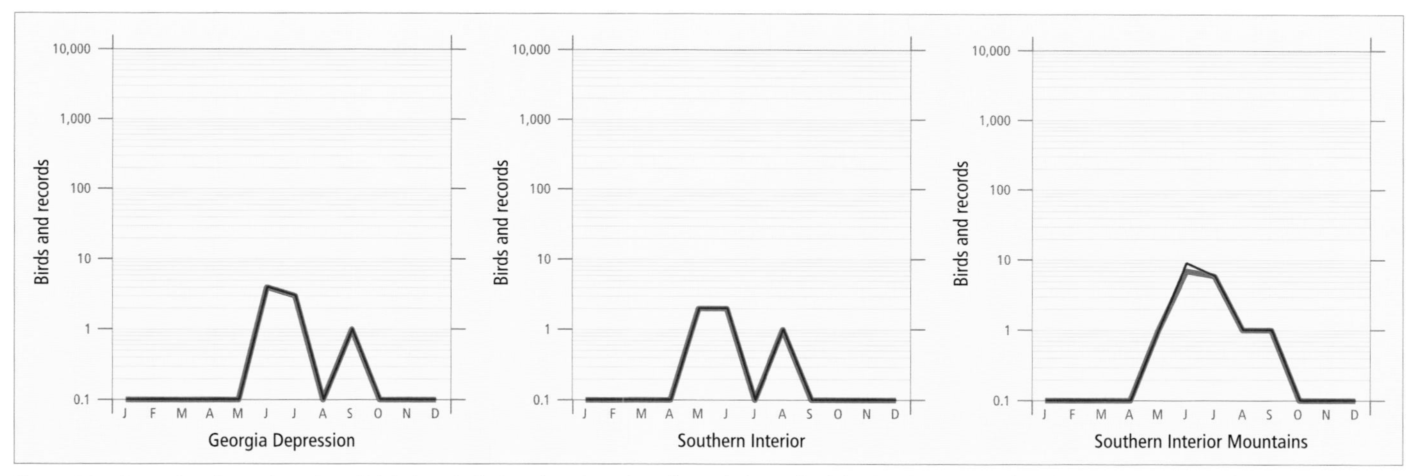

Figure 39. Fluctuations in total number of birds (purple line) and total number of records (green line) for the Chestnut-sided Warbler in ecoprovinces of British Columbia. Data from Breeding Bird Surveys have been excluded.

1973), 1 year after the first documented occurrence in British Columbia. These birds represented considerable range extensions northwestward, although Salt was apparently unaware of the 1971 record from Mount Robson Park in British Columbia. In addition, Salt (1972) uncovered records of breeding from Tupper Creek, 20 km south of Pouce Coupe, in correspondence with collector T.E. Randall. Randall wrote that in 1944 he found the Chestnut-sided Warbler quite plentiful at Tupper Creek and "found two nests with four and five eggs, July 4th." Salt (1972) was unable to confirm these records, as Randall had "lost" his specimens and field notes. That this warbler was plentiful at Tupper Creek in 1944 seems doubtful, considering that none had been found during extensive field work there in 1938 (Cowan 1939) and that there is only 1 record (1990) from the Peace Lowland in the more than 50 years since 1944. Munro and Cowan (1947) certainly discounted Randall's observations, as they used mostly specimen records. While there is an intriguing possibility that a few pairs overshot their normal range and nested in northeastern British Columbia in 1944, breeding remains to be documented in this province.

An additional published record for Vancouver exists: a male found in July 1966 at Point Grey (Orcutt 1967). Documentation was not considered sufficient for acceptance as the first record for British Columbia, and the record has been discounted by all other authors since.

Typically, the Chestnut-sided Warbler nests in the shrubby understorey, placing its nest within 2 m of the ground. Clutch sizes range from 3 to 5 eggs and the incubation period is 11 to 12 days (Richardson and Brauning 1995). The Chestnut-sided Warbler is frequently parasitized (20% to 50% of nests in several studies) by Brown-headed Cowbirds (Friedmann 1963; Friedmann and Kiff 1985).

The breeding ecology of this species is poorly known. Populations in Alberta and Colorado are thought to be isolated from the main range. The Alberta breeding population consists of small numbers of birds near Cold Lake, along the border with Saskatchewan (Richardson and Brauning 1995).

For a thorough review of information on this warbler, see Richardson and Brauning (1995).

POSTSCRIPT: On 19 August 1998, G.S. Davidson found a family of fledged, bobtailed young being fed by an adult at Puntchesakut Lake in the Central Interior. This constitutes the first breeding record for British Columbia (Campbell et al. 2000).

NOTEWORTHY RECORDS

Spring: Coastal – Hazelton 28 May 1992-1 (Bowling 1992). **Interior** – nr Oliver 26 May 1985-1 banded along Camp McKinney Road (Cannings et al. 1987; BC Photo 1027); Kamloops 27 May 1996-1 male (Bowling 1996c); Revelstoke 22 May 1993-1 (Siddle and Bowling 1993).

Summer: Coastal – Burnaby Lake 22 Jun 1986-1 (Mattocks 1986b); Point Grey (Vancouver) 19 to 30 Jun 1979-1 (BC Photo 569); Vancouver 22 Jun 1995-1 adult male (Elliott and Gardner 1997) and 8 Jun 1997-1 adult male, both on the University of British Columbia grounds; Vancouver 27 Jul 1990-1 at Stanley Park, 4th record for area (Siddle 1990c); Qualicum Beach 10 Jul 1982-1 at Hamilton Marsh (Merilees 1982); Pondosy Lake 8 Jul 1978-1. **Interior** – Okanagan Falls 20-27 Aug 1989-1 (Weber and Cannings 1990); Penticton 13 Jun 1989-1 (Weber and Cannings 1990); Coldwater River 16 Jun 1985-1; Revelstoke 19 Jun 1988-2, 2 Jul 1988-1 (BC Photo 1222); Mount Revelstoke 12 Aug 1989-1; Red Pass (Mount Robson Park) 22 Jun 1971-1, first record for British Columbia (Shepard 1972; BC Photo 172); Prince George 21 Jun 1972-1 (BC Photo 273); Beatton Park 6 Jul 1990-1; e Fort Nelson 23 Jun 1982-1 (Campbell and McNall 1982); Fort Nelson 18 Jun 1974-1 (Erskine and Davidson 1976).

Breeding Bird Surveys: Not recorded.

Autumn: Interior – Revelstoke 18 Sep 1988-1. **Coastal** – Reifel Island 29 Sep 1974-1 (Crowell and Nehls 1975a); Pacific Rim National Park 4 Sep 1993-1 (Siddle 1994a).

Winter: No records.

Magnolia Warbler

Dendroica magnolia (Wilson)

MGNW

RANGE: Breeds from northeastern British Columbia and west-central and southern Mackenzie east across the forested parts of the Prairie provinces, central Ontario, south-central and eastern Quebec, and southern Newfoundland, and south into south-central British Columbia, central Alberta and Saskatchewan, and southern Manitoba; in the east, into northern Minnesota, Wisconsin, Michigan, southern Ontario, and northern Pennsylvania, with an extension on the Appalachian highlands south to West Virginia and Virginia. Winters from Mexico to Panama and in the West Indies, occasionally north to southern California, southwestern Arizona, and southern Texas.

STATUS: On the coast, *very rare* vagrant in the Georgia Depression Ecoprovince. In the Coast and Mountains Ecoprovince, *rare* to *uncommon* migrant and summer visitant on the Northern Mainland Coast; absent from Western Vancouver Island, the Southern Mainland Coast, and the Queen Charlotte Islands.

In the interior, a *very rare* vagrant in the Southern Interior Ecoprovince; *rare* to *uncommon* migrant and summer visitant in the Southern Interior Mountains and Central Interior ecoprovinces; *uncommon* to *fairly common* migrant and summer visitant in the Sub-Boreal Interior, Boreal Plains, and Taiga Plains ecoprovinces; *casual* in the Northern Boreal Mountains Ecoprovince.

Breeds.

Figure 40. The Magnolia Warbler breeds in moist coniferous forests throughout its range in British Columbia (Anthony Mercieca).

NONBREEDING: The Magnolia Warbler (Fig. 40) is widely distributed throughout the central and eastern regions of the province from Lakelse Lake and Kleena Kleene in the west to Liard Hot Springs and Kotcho Lake in the northeast and Kootenay National Park in the southeast. In this extensive area, the warbler is found in migration in most wooded habitats, including forests of Engelmann or white spruce; mixed forests of spruce, pine, birch, and trembling aspen (Fig. 41); riparian stands of balsam poplar; or shrubby vegetation, including red-osier dogwood, alder, and willow bordering lakes, ponds, and sloughs. It also occurs irregularly on the south coast in the Georgia Depression, in the Southern Mainland Coast, and in the Southern Interior.

The spring migration probably enters the Boreal Plains and Taiga Plains in the latter half of May (Figs. 42 and 43). It then spreads out across the central portions of the province, reaching the Nechako and Skeena river valleys as early as the end of May but more generally in the first week of June. Semenchuk (1992) and Sadler and Myres (1976) remark that the Magnolia Warbler is not a conspicuous spring migrant through Alberta, and the same must be said for British Columbia. The migration records in the Boreal Plains do not indicate the passage of as many of these warblers as would be expected if this were the primary migration route into the province for the populations of north-central British Columbia. There is no indication of a north-south migration of this species west of the Rocky Mountains in Washington, Idaho, or British Columbia. It should be looked for, however, as there is a small spring and autumn migration through California (Small 1994), and in the autumn of 1995, 3 birds were captured and banded at Sea Island, on the Fraser River estuary (Elliott and Gardner 1997).

Figure 41. During migration in British Columbia, the Magnolia Warbler visits mixed forests of hybrid white spruce, lodgepole pine, paper birch, and trembling aspen in the Sub-Boreal Interior Ecoprovince (Mackenzie, 9 June 1996; R. Wayne Campbell).

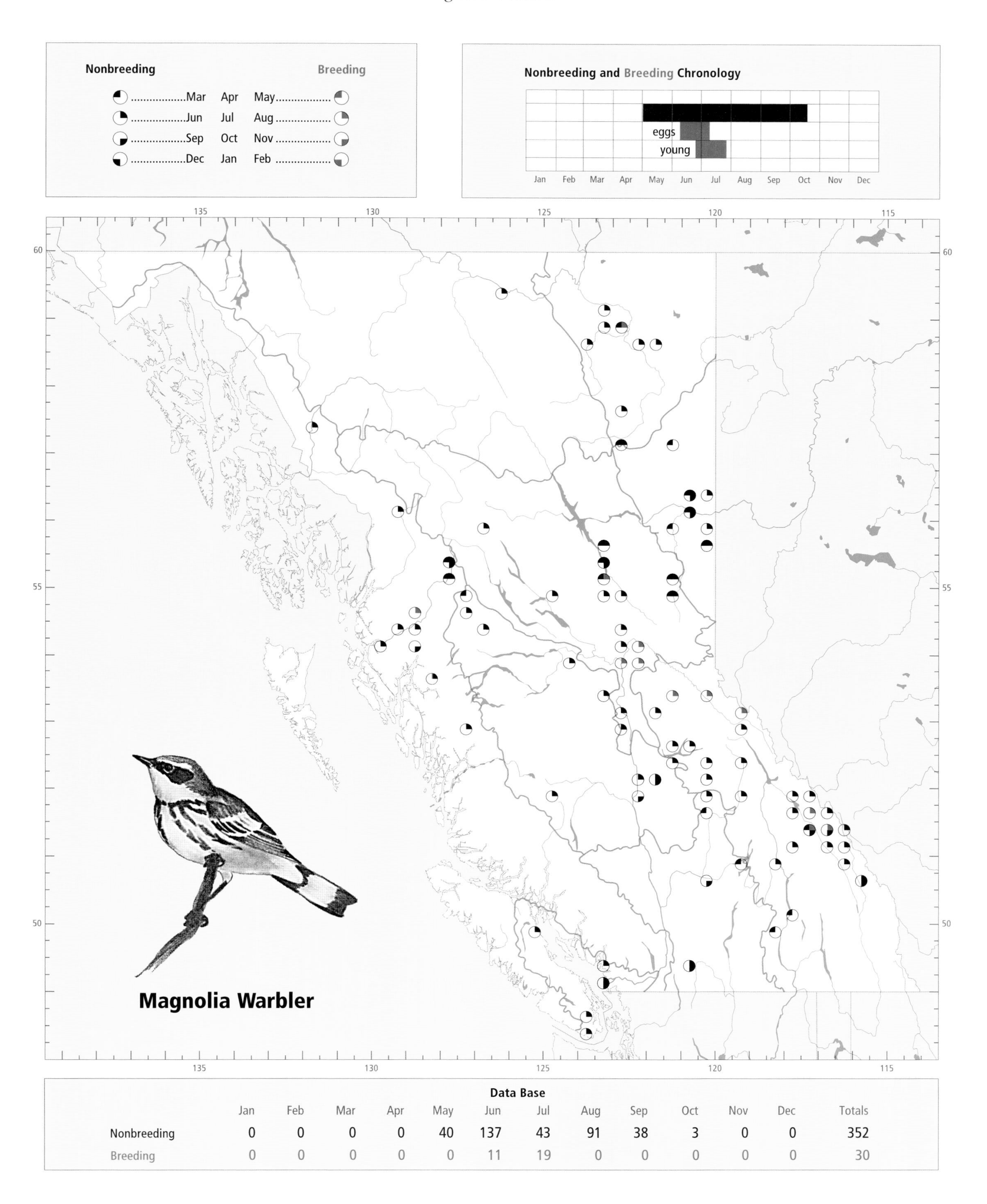

Data Base

	Jan	Feb	Mar	Apr	May	Jun	Jul	Aug	Sep	Oct	Nov	Dec	Totals
Nonbreeding	0	0	0	0	40	137	43	91	38	3	0	0	352
Breeding	0	0	0	0	0	11	19	0	0	0	0	0	30

Figure 42. Annual occurrence (black) and breeding chronology (red) for the Magnolia Warbler in ecoprovinces of British Columbia. Records are shown for the week in which they occurred.

The early June arrival of spring migrants in Kootenay National Park and at Glacier and Mount Revelstoke national parks suggests that they reach the area via the passes through the Rocky Mountains at that latitude. In Alberta, the species is known to extend its range from the foothills into the Rocky Mountains at Banff and Jasper, at the eastern entrances to 2 relatively low passes leading to localities of occurrence in British Columbia.

The peak in numbers in British Columbia occurs in late May and early June, as the last of the spring migrants reach the breeding grounds and males are in full song. Autumn migration in the Boreal Plains and Sub-Boreal Interior begins in late July and proceeds through August and early September (Figs. 42 and 43). Banding stations near Mackenzie, in the Sub-Boreal Interior, showed the Magnolia Warbler to be one of the 7 most abundant passerines between early August and early September (R. Millikin pers. comm.). Most birds have left the province by the end of the first week of September.

On the coast, the Magnolia Warbler has been recorded from 15 May to 18 October; in the interior, it has been recorded from 2 May to 15 October (Fig. 42).

BREEDING: The Magnolia Warbler probably breeds throughout its summer range in the province but few nests have been reported. It is known to nest from Terrace in the west across the interior of the province to Prince George, then both north to Fort Nelson in the Taiga Plains and south to the lower Blaeberry River valley near Golden in the Southern Interior Mountains.

The Magnolia Warbler reaches its highest numbers in summer in the Sub-Boreal Interior (Fig. 44). An analysis of Breeding Bird Surveys for the period 1968 through 1993 could not detect a net change in numbers on interior routes. Across Canada, for the period 1966 through 1996, Magnolia Warbler numbers increased at an average annual rate of 1.4% ($P < 0.05$); throughout their North American range their numbers have increased over the same period at an average annual rate of 1.6% ($P < 0.01$) (Sauer et al. 1997).

In the higher latitudes of British Columbia, the Magnolia Warbler appears to nest over a narrow elevational range between 570 and 810 m; towards the southern extent of its summer range in the Rocky Mountains, it probably nests to about 1,300 m. Nesting habitat is generally associated with the edges of both second-growth and mature coniferous forests and includes mixed stands of Engelmann or white spruce, subalpine fir, trembling aspen, birch, and alder, along with dense associated shrubbery (Fig. 45). In the Boreal Plains, it is found in mixed woods or coniferous forests, particularly around openings and edges where a coniferous shrub layer has developed (Phinney 1998). Salt (1973), describing the nesting habitat in Alberta adjacent to northeastern British Columbia, stated that "sunlit clearings grown to willows and young conifers or flat mountain valleys with stunted alder and black birch" were the summer habitat of this colourful warbler. In British Columbia, three-quarters of all nests were found in mixed forests, and all but 1 of them were in unlogged forest. In a detailed analysis of bird communities in the trembling aspen forests of the Bulkley River valley between 10 May and 10 July, Pojar (1995) examined 5 age categories from clearcut to mature mixed forest and found Magnolia Warblers only in the mature conifer-aspen forests. There they were present at densities of 2.5 to 14.6 singing males per 10 ha. Nesting

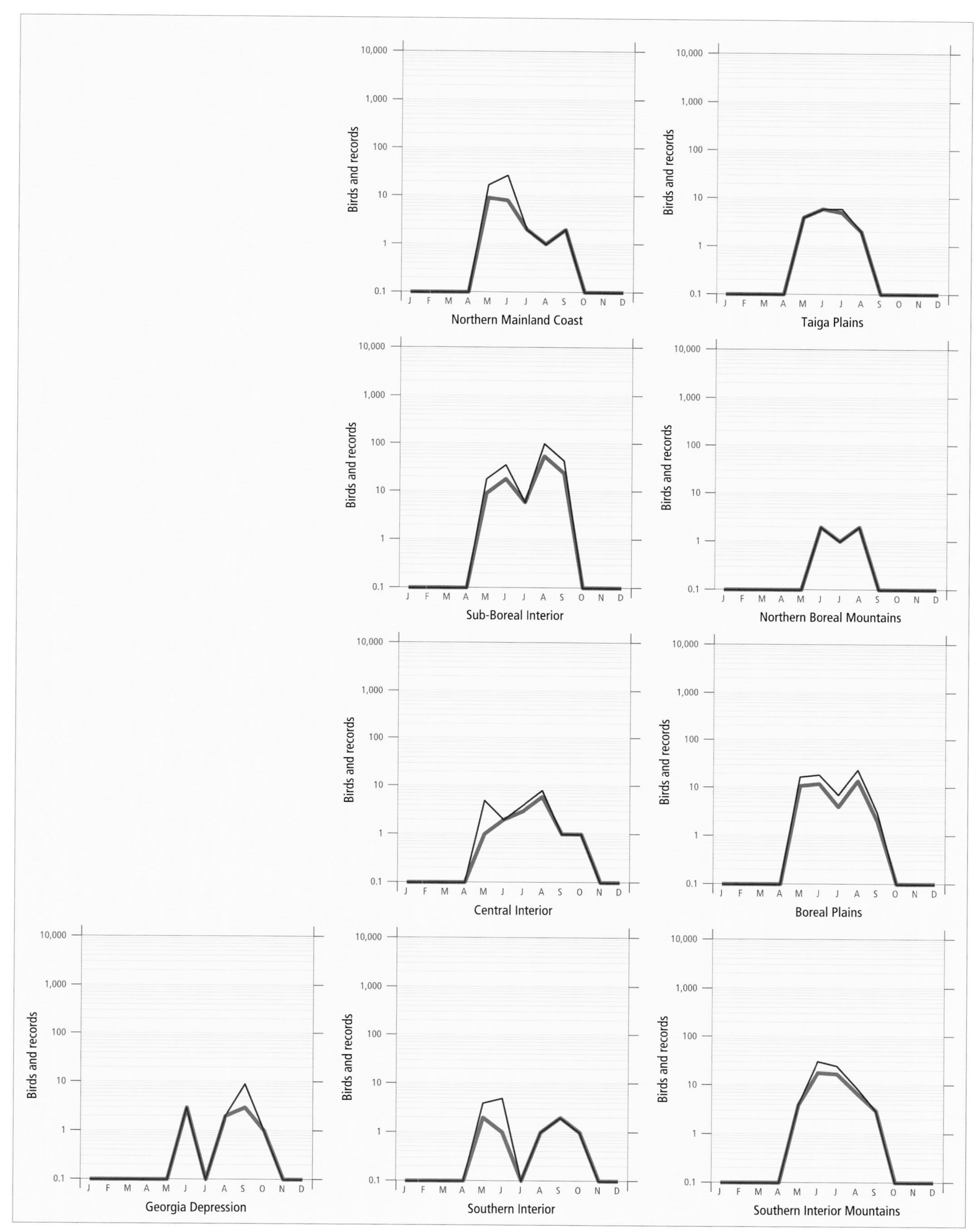

Figure 43. Fluctuations in total number of birds (purple) and total number of records (green) for the Magnolia Warbler in ecoprovinces of British Columbia. Breeding Bird Surveys and nest record data have been excluded.

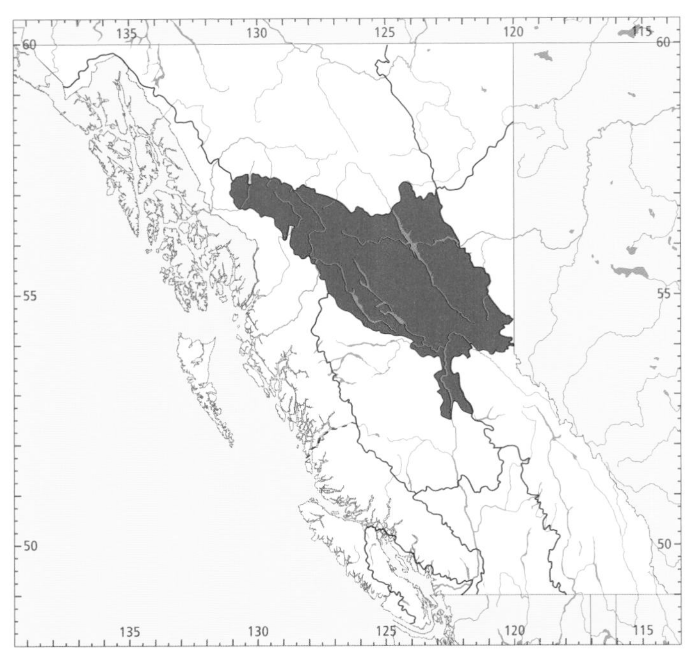

Figure 44. In British Columbia, the highest numbers for the Magnolia Warbler in summer occur in the Sub-Boreal Interior Ecoprovince.

territories in northern British Columbia were frequently close to water.

In British Columbia, the Magnolia Warbler has been recorded breeding from 9 June (calculated) to 26 July (Fig. 42).

Nests: In British Columbia, all nests were elevated in living trees and shrubs, both coniferous and deciduous, a situation similar to that found in Alberta (Salt 1973). In eastern North America, coniferous trees (e.g., spruces, balsam fir, hemlock, and white cedar) are favoured over deciduous trees for nesting (Bent 1953; Peck and James 1987). Nests were usually well hidden and located away from the trunk, often where diverging branches provided secure support. Some were placed where a branch left the trunk. The heights of 7 nests ranged from 0.8 to 1.7 m, with 5 nests between 0.8 and 1.5 m.

Nests were loosely built, small cups of grasses (Fig. 46), plant stems, and rootlets; hair, twigs, conifer needles, spider webs, and plant fibres were used less frequently. Nests were lined with fine black rootlets and fine grasses.

While some Magnolia Warblers do use rootlets that are black, it is likely that the shiny, smooth fibres that look like black rootlets reported in British Columbia are really fungal mycorrhizae in symbiotic relationships with tree rootlets (Ross D. James pers. comm.; McFarland and Rimmer 1996).

Eggs: Dates for 9 clutches ranged from 18 June to 9 July. Calculated dates indicate that eggs can occur as early as 9 June. Clutch size ranged from 1 to 5 eggs (1E-2, 3E-1, 4E-2, 5E-4). The incubation period is 11 to 13 days (Laughlin and Kibbe 1985; Peck and James 1987).

Young: Dates for 15 broods (Fig. 46) ranged from 25 June to 26 July, with 9 recorded between 8 July and 11 July. Sizes of 11 broods ranged from 1 to 5 young (1Y-3, 3Y-1, 4Y-2, 5Y-5). The nestling period is 8 to 10 days (Mouseley 1924; Nice 1926).

Brown-headed Cowbird Parasitism: In British Columbia, 3 of 15 nests found with eggs or nestlings were parasitized by the Brown-headed Cowbird. In addition, 2 cowbird fledglings were seen being tended by Magnolia Warblers. Friedmann (1963) notes that this warbler is seldom parasitized by the Brown-headed Cowbird, citing a survey in Quebec in which 4% of 147 Magnolia Warbler nests had been parasitized. Friedmann et al. (1977) cited another study, in Ontario, where 10% of 60 nests had been parasitized. At that time all known instances of parasitism of the Magnolia Warbler by the Brown-headed Cowbird were in eastern North America. The review by Friedmann and Kiff (1985) had little to add; thus, the only records of this parasitism west of the Great Plains are from British Columbia.

Nest Success: Three nests found with eggs and followed to a known fate were successful in rearing at least 1 fledgling.

REMARKS: The Magnolia Warbler entered the known avifauna of British Columbia relatively late. Prior to ornithological explorations in the Peace River area in 1938 (Cowan 1939), the species had been known only from Revelstoke in 1890, Field in 1892, and the Quesnel area in 1932 and 1936. By the mid-1940s, it was known to be widely distributed in the Sub-Boreal Interior and the northern regions of the Southern Interior

Figure 45. In the northern parts of the Central Interior Ecoprovince of British Columbia, the Magnolia Warbler breeds in mixed stands of spruce with associated dense patches of willows (west of Endako, 20 June 1997; R. Wayne Campbell).

Figure 46. Nest of the Magnolia Warbler with 5 nestlings (35 km northwest of Donald, 10 July 1999; Douglas Leighton).

Mountains ecoprovinces, and to be less common, or transient, in the Peace Lowland (Munro and Cowan 1947). In large part, the scarcity of records was a consequence of a lack of biological exploration in the northern half of the province.

In the past 50 years, the Magnolia Warbler has been confirmed as being widespread, occurring at low densities, and breeding in the forests of the Sub-Boreal Interior, sometimes in fair numbers (see "Breeding Bird Surveys"), as well as in the Boreal Plains and Taiga Plains ecoprovinces and probably up to the subalpine forests of the Rocky Mountains south as far as Kootenay National Park (Poll et al. 1984). In the south, there are summer records as far west as the Columbia River valley at Nakusp. This southern presence appears to be a recent range extension.

There are several reports of the Magnolia Warbler during the nesting season at localities in the southern Rocky Mountain Trench, in the Thompson River valley, and along the Northern Mainland Coast. All of them come from Breeding Bird Surveys and have no supporting documentation. Most are from localities for which there is no other evidence of occurrence, even where a substantial amount of ornithological field work has been done. Because of this, we have not considered these areas to be part of the breeding range of this species.

In addition, confusion regarding the species codes for the Magnolia Warbler and the MacGillivray's Warbler used in Breeding Bird Surveys has contaminated some data to the extent that we were unable to determine the species seen by the observer. In localities where one species is known to be of regular occurrence whereas the other is scarce or unreported, we have not accepted the presence of the scarce one on the basis of Breeding Bird Surveys or of a single banding record alone. In the latter case, a bird in the hand can be identified with certainty, but the species code changes introduce doubt.

On this basis, unsubstantiated summering records involve the following localities: Kitlope Lake, Columbia Lake, Scotch Creek, Adams Lake, and Succour Creek. A Christmas Bird Count record from Victoria on 30 December 1967 has also been excluded.

For a thorough review of the ecology and biology of the Magnolia Warbler in North America, see Hall (1994).

NOTEWORTHY RECORDS

Spring: Coastal – Hazelton 15 May 1992-1. **Interior** – Box Lake 8 May 1981-1 male; Nakusp 15 May 1981-1 male; Fauquier 31 May 1981-1 male; Brouse 16 May 1993-1 male; lower Blaeberry River valley 25 May 1999-1 male singing; Clearwater 27 May 1935-3 (Dickinson 1953); Smithers 31 May 1992-5; Mugaha Creek (Mackenzie) 2 May 1997-1 male singing; confluence of Murray and Flatbed rivers 29 May 1976-1 male; Tupper Creek 31 May 1938-1 (Cowan 1939); Taylor 22 May 1981-1, 24 May 1986-2; Beatton Park 19 May 1987-1, 25 May 1986-3; Sikanni Chief 31 May 1994-1; Fort Nelson 22 May 1987-1.

Summer: Coastal – 16 km n Sooke 26 Jun 1983-1 (Mattocks et al. 1983c); Sea Island 30 Aug 1995-1 (Elliott and Gardner 1997); Mitlenatch Island 24 Aug 1965-1 (Campbell and Kennedy 1965a); Lakelse Lake 18 Aug 1976-1 (Hay 1976); Kwinitsa to Salvers 15 Jul 1971-1; Contact Creek (Stikine River) 11 Jun 1972-1. **Interior** – 11 km n Princeton 30 Aug 1992-1 male with flock Yellow-rumped Warblers; Brouse 16 Jun 1978-1 male, 11 to 14 Jun 1980-1 male, 5 Jul 1978-1 male; Crescent Bay 3 Jun 1989-1 male; Marble Canyon (Kootenay National Park) 3 Jun 1965-1 (Van Tighem 1977); 5 km w Spillimacheen 8 Jun 1996-1 singing male; Revelstoke 1 Jun 1982-1; Glacier 4 Jun 1994-1 banded, 2 Jul 1993-2 banded, 2 Aug 1994-1 banded; Field 27 Aug 1892-2 (Rhoads 1893); Willowbank Mountain 4 Jun 1996-2 singing males; Marl Creek Park 7 Jun 1995-1 singing male; Kinbasket Lake 5 Jul 1995-1 singing male; Bailey's Chute (Wells Gray Park) 15 Jun 1996-1 male; 40 km e Williams Lake 17 Aug 1953-1 (Jobin 1954); Williams Lake 29 Aug 1953-2 (Jobin 1954); Quesnel 5 Jun 1932-1 (UBC 261); 9.6 km w Lazaroff Lake 13 Jul 1967-1 fledgling being fed by adult; nr Barkerville 18 Aug 1961-2 young being fed by female; Mount Terry Fox Park 14 Jul 1987-1 adult male; Mount Robson Park 30 Jul 1984-2 (Rogers 1984b); Robson Motor Village 26 Jul 1971-3 recently fledged young; Indianpoint Lake 28 Jun 1930-3 eggs, 5 Jun to 12 Aug 1933-12 (Dickinson 1953); Nulki Lake 20 Jun 1946-1, 30 Aug 1945-2 (Munro 1949); 15 km e Prince George 22 Jul 1993-1 young being fed by male; Nukko Lake 5 Jun 1944-1 (Munro 1945c); 33.2 km ne Tabor Lake, 8 Jul 1966-5 nestlings; Stuart Lake 1 Aug 1981-1 banded; 145 km n Tudyah Lake 24 Jun 1968-5 eggs; Gagnon Creek 14 Aug 1995-6 banded; Fort St. John 28 Aug 1985-8; Taylor 24 Jul 1982-2 fledglings following female; St. John Creek 31 Aug 1985-1; Trutch 17 Jul 1943-1 (Rand 1944); Mile 335 Alaska Highway 28 Jul 1985-2 fledglings; Sikanni Chief 9 Jun 1993-1; Fort Nelson 12 Aug 1985-1 (Grunberg 1986); Fort Nelson River 8 Jul 1978-1 egg and a large nestling Brown-headed Cowbird, 9 Jul 1978-5 fledglings just from nest; Liard Hot Springs Park 25 Jun 1985-1 (Grunberg 1985d); Liard River 9 Aug 1943-1 (Rand 1944).

Breeding Bird Surveys: Coastal – Recorded from 5 of 27 routes and on 8% of all surveys. Maxima: Kispiox 20 Jun 1993-40; Nass River 21 Jun 1975-12; Meziadin 17 Jun 1975-5. **Interior** – Recorded from 19 of 73 routes and on 9% of all surveys. Maxima: McLeod Lake 29 Jun 1990-43; Ferndale 30 Jun 1990-27; Summit Lake 17 Jun 1992-21.

Autumn: Interior – Charlie Lake 1 Sep 1985-1; Fort St. John 7 Sep 1985-1; Gagnon Creek (Mackenzie) 10 Sep 1995-6 banded, 19 Sep 1995-1 banded; Chilcotin River (Riske Creek) 10 Oct 1994-1 male at Wineglass Ranch; 144 Mile House 9 Sep 1953-1 specimen (Jobin 1954); Marble Canyon (Kootenay National Park) 5 Sep 1965-1 (Van Tighem 1977); lower Blaeberry River valley 12 Sep 1997-1 female; 10 km n Princeton 15 Oct 1991-1 male with flock Yellow-rumped Warblers. **Coastal** – Hazelton 5 Sep 1921-1 (Swarth 1924); Kitimat River 17 Sep 1974-1 (Hay 1976); Vancouver 30 Sep to 7 Oct 1995-1 (Elliott and Gardner 1997); Iona Island 13 Sep 1995-7 banded (Elliott and Gardner 1997); Sea Island 18 Oct 1995-1 (Elliott and Gardner 1997).

Winter: No records.

Christmas Bird Count: See REMARKS.

Cape May Warbler

CMWA

Dendroica tigrina (Gmelin)

RANGE: Breeds from southeastern Yukon, northeastern British Columbia, southwestern Mackenzie, and northern Alberta across Canada in the boreal forests east to central Quebec and southwestern Newfoundland, and south to northern Minnesota, northern Wisconsin, northern Michigan, southern Ontario, northern New York, northern Vermont, northern New Hampshire, central Maine, and Nova Scotia. Winters mainly in the West Indies, but also in Central America, northern Colombia, northern Venezuela, and southern Florida.

STATUS: On the coast, absent.

In the interior, *rare* to locally *uncommon* migrant and summer visitant to the Boreal Plains and Taiga Plains ecoprovinces; *accidental* in the Sub-Boreal Interior and Southern Interior Mountains ecoprovinces.

Breeds.

Figure 47. Cape May Warbler (adult male shown). Since its first occurrence in 1938, this species has become established as a breeding species in northeastern British Columbia (Anthony Mercieca).

CHANGE IN STATUS: Munro and Cowan (1947) included the Cape May Warbler (Fig. 47) in their extralimital list for British Columbia based on a single record from Charlie Lake in 1938 (Cowan 1939). They correctly predicted that it would prove to occur regularly in the Peace River Parklands (Peace Lowland of the Boreal Plains). It was not until the 1970s and 1980s, however, that additional records of this warbler were obtained throughout the northeastern portions of the province. It has now been recorded in the Taiga Plains, mainly near Fort Nelson, and along the Liard River. Breeding has been confirmed in the Boreal Plains and Taiga Plains ecoprovinces. It remains one of the most infrequently observed warblers in British Columbia even though it is a fairly conspicuous bird during migration. It is also very local and may not be observed at the same locality each year during migration.

NONBREEDING: The Cape May Warbler occurs almost exclusively in the Boreal Plains and Taiga Plains. Most records are from a portion of the northern Peace Lowland near Fort St. John and Taylor, from the Fort Nelson Lowland near Fort Nelson and Kledo Creek, and from the Muskwa Plateau along the Sikanni Chief River. This warbler is very scarce in parts of the Kiskatinaw Plateau, south of Dawson Creek (M. Phinney

Figure 48. During spring migration in British Columbia, the Cape May Warbler frequents the lower canopy of mixed forests and often forages in nearby shrubs (west of Fort St. John, 18 June 1996; R. Wayne Campbell).

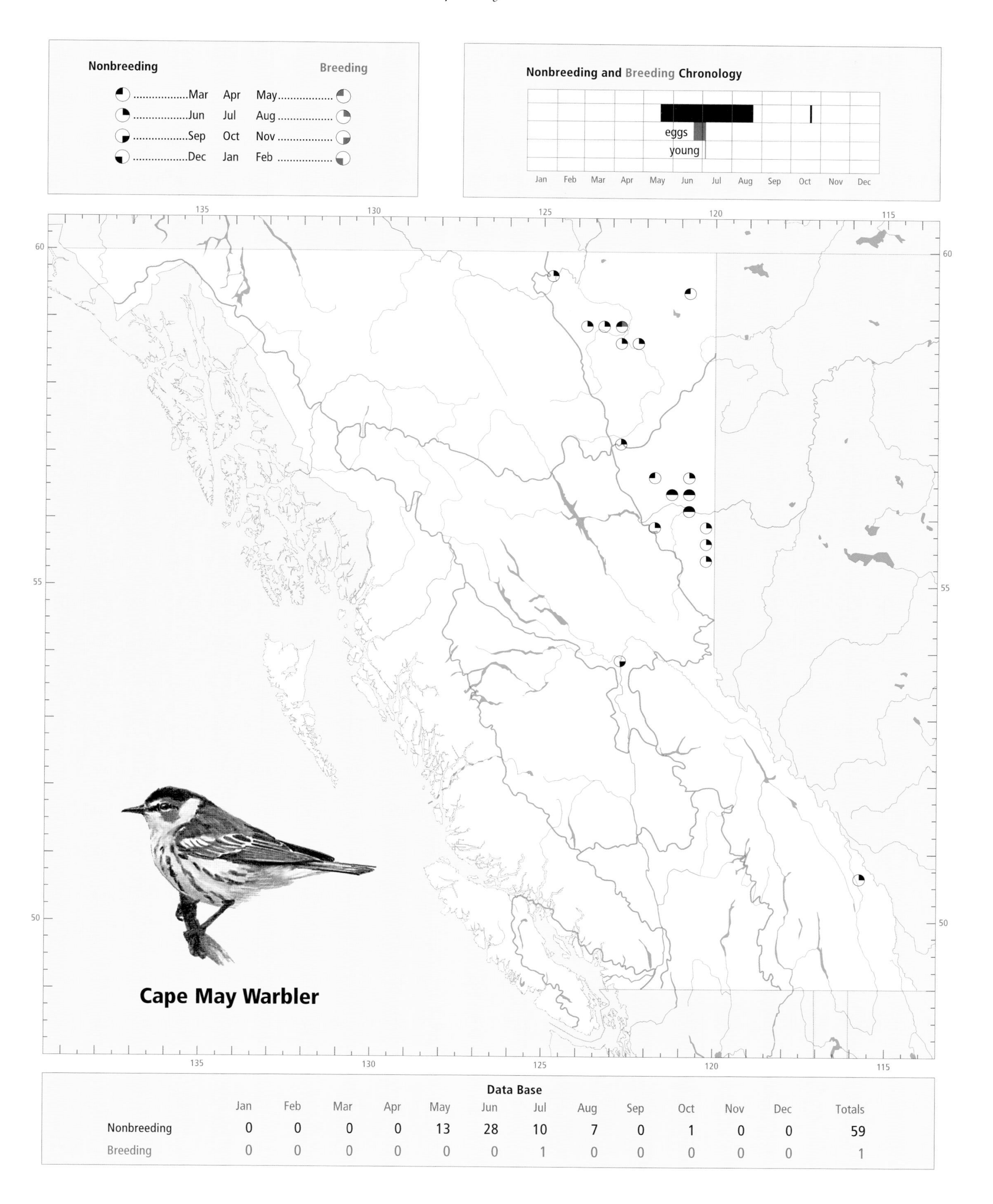

Data Base

	Jan	Feb	Mar	Apr	May	Jun	Jul	Aug	Sep	Oct	Nov	Dec	Totals
Nonbreeding	0	0	0	0	13	28	10	7	0	1	0	0	59
Breeding	0	0	0	0	0	0	1	0	0	0	0	0	1

pers. comm.), but seems to occur regularly to the east in Alberta (C. Helm pers. comm.). There are only 2 records outside the normal range of the Cape May Warbler in British Columbia: a single bird that was photographed at Prince George in October 1971, and a singing male along the edge of the Cross River, just south of Kootenay National Park, in June 1981.

The Cape May Warbler has been reported at relatively low elevations in the northeast, from 230 m in valley bottoms to about 760 m. During spring migration, it frequents the lower forest canopy and also forages in shrubs (Fig. 48). The species is very flexible during migration and is able to exploit varied food supplies when inclement weather hits at stopover areas (S.G. Sealy pers. comm.). Although nonbreeding habitat can be similar to breeding habitat, a wider variety of woodland habitats is used during migration (Dunn and Garrett 1997).

The Cape May Warbler's migration in British Columbia is poorly known. Spring migrants begin to arrive in the northeast in mid-May, with most probably arriving in late May and the first few days of June (Figs. 49 and 50). At Beatton Park, first arrivals in the 1980s occurred between 19 and 25 May (C. Siddle pers. comm.). Migrants enter British Columbia from Alberta, where early migrants appear in the third week of May (Pinel et al. 1993). The autumn movement likely begins in late July, shortly after nesting is completed. Most birds have departed by mid to late August (Figs. 49 and 50).

The Cape May Warbler occurs regularly in northeastern British Columbia from 19 May to 21 August (Fig. 49). The autumn migrant that appeared at Prince George on 22 October did so a full 2 months after most birds had left the breeding areas to the northeast.

BREEDING: The documented breeding distribution of the Cape May Warbler in British Columbia is the Fort Nelson Lowland of the Taiga Plains, in the northeastern corner of the province, although adults feeding fledged young have been reported from the Peace Lowland of the Boreal Plains. Our knowledge of this warbler's breeding distribution is limited because of its scarcity, its tendency to nest high in the canopy, and the lack of observers who visit its habitat. It is likely that the Cape May Warbler breeds in pockets of suitable habitat in scattered areas of the Boreal Plains and Taiga Plains. These areas include mature coniferous forests along the major river drainages and low-elevation plateaus from the Alberta border west to the Rocky Mountain foothills.

The highest numbers for the Cape May Warbler in summer occur in the Taiga Plains (Fig. 51). Breeding Bird Surveys for interior routes for the period 1968 through 1993 contain insufficient data for analysis. An analysis of Breeding Bird Surveys across Canada from 1966 to 1996 suggests that populations are stable (Sauer et al. 1997).

The Cape May Warbler has been reported from elevations between 420 and 660 m. In the "Big Bend" area of the Liard River, most Cape May Warblers were found in the middle to upper slopes above 450 m (Bennett and Enns 1996). Breeding habitat includes fairly dense mature and old-growth stands of white spruce, including stands interspersed with clumps of paper birch, balsam poplar, willow, and alder (Fig. 52) and with a poorly to moderately developed shrub layer (Fig. 52; Enns and Siddle 1996). These stands occur mainly on level or gently sloping river terraces and may have numerous openings.

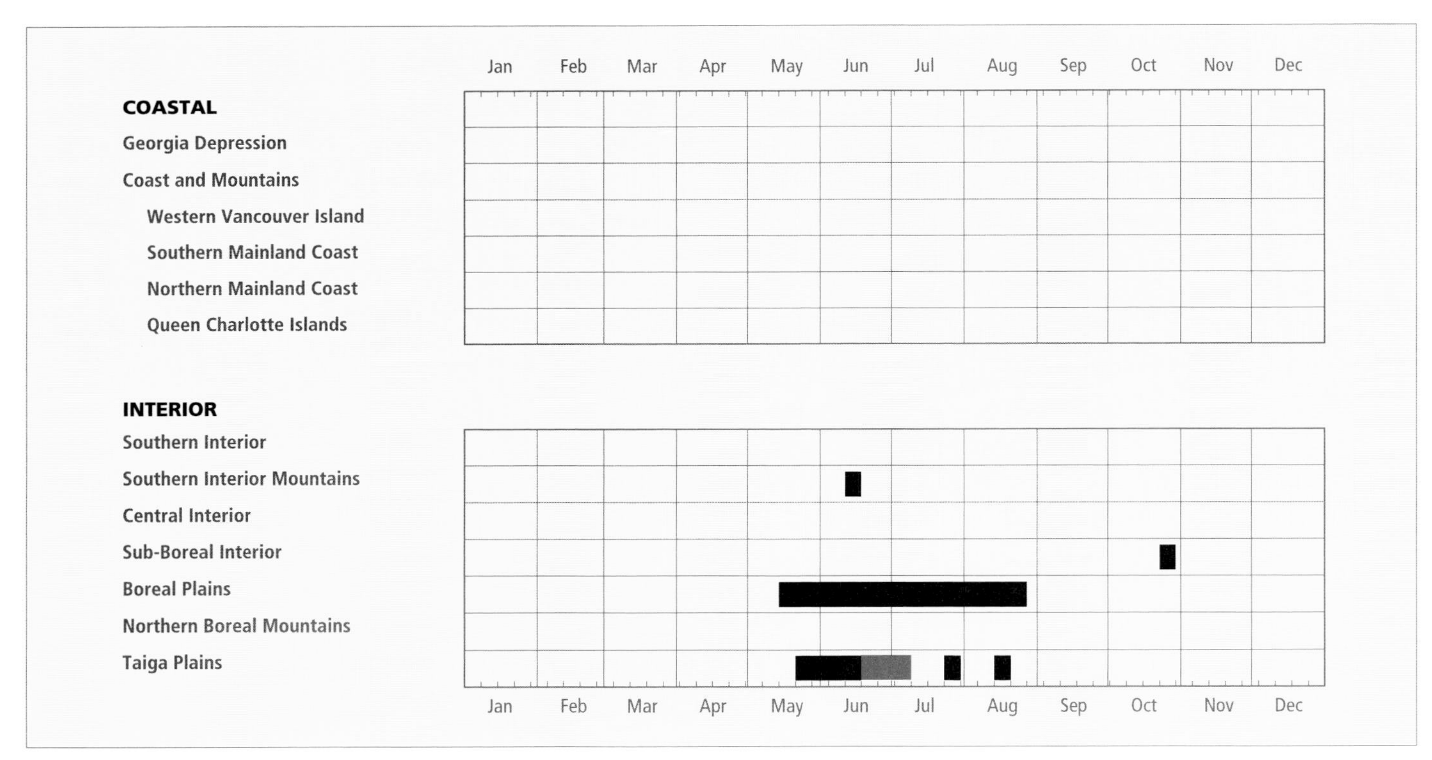

Figure 49. Annual occurrence (black) and breeding chronology (red) for the Cape May Warbler in ecoprovinces of British Columbia. Records are shown for the week in which they occurred.

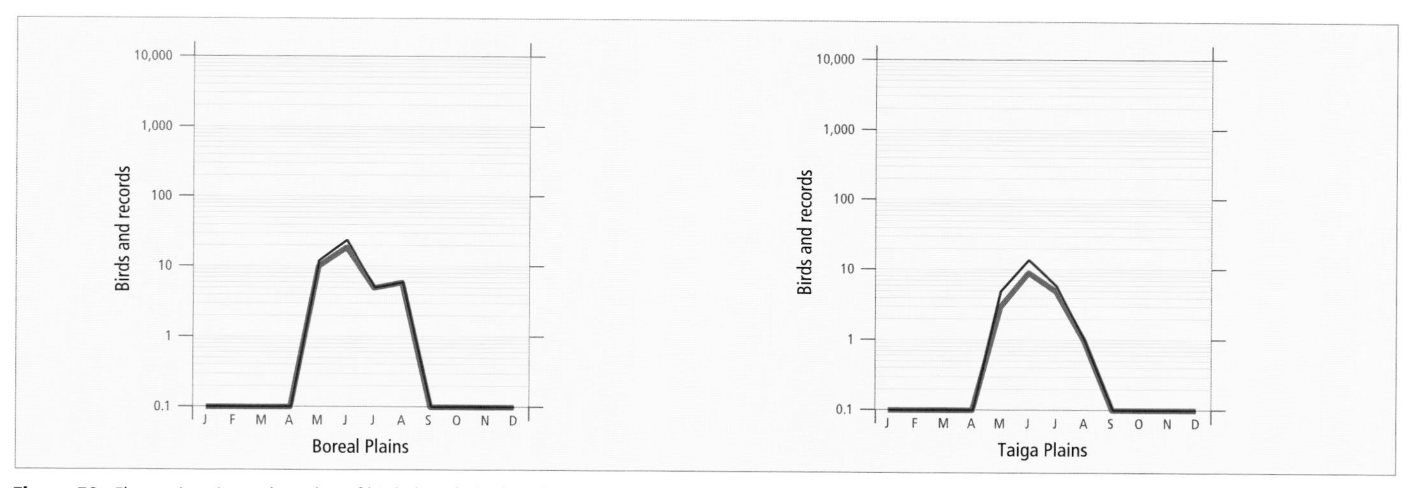

Figure 50. Fluctuations in total number of birds (purple line) and total number of records (green line) for the Cape May Warbler in ecoprovinces of British Columbia. Breeding Bird Surveys and nest record data have been excluded.

Tall spruce trees that extend above the main canopy are used by singing males, and are an important habitat element for breeding territories (Salt 1973). The Cape May Warbler usually occurs in the mid to upper canopies, and frequently ventures down to forage in the lower canopy or shrubs.

Suitable breeding habitat appears to be correlated with the white spruce–currant–horsetail site series in the Boreal White and Black Spruce Biogeoclimatic Zone (Meidinger and Pojar 1991), and probable nesting areas of this warbler can generally be predicted by the occurrence of that habitat (Cooper et al. 1997). In the "Big Bend" reach of the Liard River, the preferred habitat was even-aged stands of white spruce that had developed after hot fires. The stands of white spruce were dense, averaged about 140 years of age and 22 centimetres diameter at breast height, had considerable woody debris on the ground, and had an understorey dominated by mosses and herbs (Bennett and Enns 1996).

A few singing males have been found in selectively logged stands where only the largest spruce have been removed, which indicates that relatively open stands are used.

The Cape May Warbler has been recorded breeding between 22 June (calculated) and 4 July (Fig. 49). It likely breeds from early June to late July.

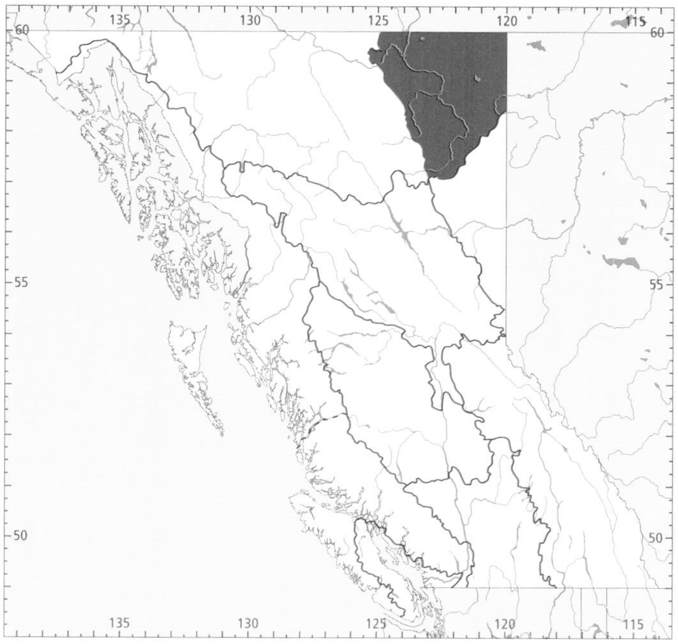

Figure 51. In British Columbia, the highest numbers for the Cape May Warbler in summer occur in the Taiga Plains Ecoprovince.

Figure 52. Mature stands of white spruce, interspersed with a variety of deciduous trees, provide breeding habitat for the Cape May Warbler in northeastern British Columbia (near Fort Nelson, 27 June 1996; R. Wayne Campbell).

Nests: None has been examined in British Columbia. Elsewhere in North America, the Cape May Warbler nest is a bulky cup composed mainly of twigs, grasses, and mosses and lined with hair, feathers, and some rootlets (Baicich and Harrison 1997). The nest is built relatively high in coniferous trees (usually spruce), mainly in the mid to upper canopy between 10 and 20 m (Baltz and Latta 1998).

Eggs: Clutches have not been recorded in British Columbia. One nest found with an unknown number of nestlings on 4 July would have contained eggs no later than 22 June. The incubation period of this warbler is unknown but is likely about 12 days, as in other congeneric warblers.

Clutch size ranges from 4 to 9 eggs, but is usually 5 or 6 eggs (Bent 1953; Morse 1978). The Cape May Warbler, like the Bay-breasted Warbler, has one of the largest clutch sizes of all North American warblers. Clutches of 7 to 9 eggs usually occur during years when food is very abundant (see REMARKS).

Young: Only 1 nest has been documented in British Columbia; it was not directly inspected but contained an unknown number of nestlings on 4 July. An adult feeding 4 fledged young was observed on 5 July. Nestlings can likely occur in northeastern portions of British Columbia from late June through late July. The nestling period is believed to be similar to that of other warblers of the same genus: 9 to 12 days (Baicich and Harrison 1997).

Brown-headed Cowbird Parasitism: Our only record is of an adult Cape May Warbler that was observed feeding a fledgling cowbird on 18 July 1983 (Siddle 1992c). The Cape May Warbler is an infrequent host of the Brown-headed Cowbird (Friedmann 1963), probably because of its preference for extensive stands of mature boreal forest habitat, where few cowbirds occur. This may change because forests in northeastern British Columbia are becoming increasingly fragmented (Cooper et al. 1997) and cowbirds are now quite common in some areas.

Nest Success: Insufficient data.

REMARKS: In eastern North America, the Cape May Warbler is one of a few warblers whose regional abundance and productivity are closely linked to the abundance of their primary food, the eastern spruce budworm (*Choristoneura fumiferana*) (Morse 1989). Clutches of up to 9 eggs, which are unusually large for a warbler species, have been recorded during spruce budworm outbreaks (Morse 1978).

Cooper et al. (1997) suggest that there is some correlation in the distribution of spruce budworm outbreaks and Cape May Warblers in northeastern British Columbia, mainly because pure, mature stands of spruce are more susceptible to budworm outbreaks than younger stands, and mature spruce is the warbler's preferred nesting habitat. Recent inventory field work along the Liard River, in an area that had had a moderate budworm outbreak, found higher concentrations of this warbler than had been previously reported in British Columbia (Bennett and Enns 1996).

There is 1 published record of this warbler for the coast, at the Kitimat River estuary (Bell and Kallman 1976). Supporting documentation is unavailable, however, and we have excluded the record from this account.

Identification of the Cape May Warbler by song alone is difficult because of similarities with the song and habitat use of the Bay-breasted Warbler and Golden-crowned Kinglet.

For additional information on the life history of the Cape May Warbler, see Sealy (1988, 1989), Dunn and Garrett (1997), and Baltz and Latta (1998).

NOTEWORTHY RECORDS

Spring: Coastal – No records. **Interior** – Beatton Park 19 May 1987-1, 19 May 1989-1, 24 May 1985-2, 25 May 1984-2 (Grunberg 1984), 25 May 1986-1 (McEwen and Johnston 1986); Taylor 31 May 1986-1 at Peace Island Park; Fort Nelson 23 May 1987-2; Helmut (Kwokullie Lake) 28 May 1982-1, 30 May 1982-2.

Summer: Coastal – No records. **Interior** – Cross River 17 Jun 1981 (Poll et al. 1984); One Island Lake 8 Jun 1978-1 (Campbell 1978c); Moberly Lake 18 Jun 1988-1; Bear Mountain (Dawson Creek) 26 Jun 1994-1 (Bowling 1994b); Doe Creek (Dawson Creek) 4 Aug 1975-1; Stoddart Creek (Fort St. John) 7 Jul 1975-1, 19 Aug 1987-1; St. John Creek 14 Aug 1987-1; Charlie Lake 17 Jun 1938-1 (Cowan 1939; RBCM 8106), 21 Aug 1987-1; Beatton Park 5 Jun 1984-1 male carrying food, 18 Jul 1983-1 female feeding 1 fledged Brown-headed Cowbird (Siddle 1992c), 11 Aug 1988-1 (Siddle 1988d), 12 Aug 1987-1 (Siddle 1988d); Gutah Creek Jun 1991-1 (Siddle 1992c); Sikanni Chief River 1 Jun 1993-1, 4 Jun 1992-2 (Greenfield 1998); Clarke Lake 17 Jun 1979-1; Fort Nelson 24 Jun 1988-2 pairs, 4 Jul 1990-male feeding unknown number of nestlings; 10 km n Fort Nelson 14 Aug 1991-1 in spruce; Steamboat Creek 2 Jul 1986-1; Kledo Creek 22 Jun 1988-2 pairs (Siddle 1988d), 30 Jun 1987-1, 5 Jul 1985-adult feeding 4 nestlings; Liard River 11 to 22 Jun 1996-21 records during surveys (Bennett and Enns 1996).

Breeding Bird Surveys: Not recorded.

Autumn: Interior – Prince George 22 Oct 1971-1 (BC Photo 268). **Coastal** – No records.

Winter: No records.

Yellow-rumped Warbler — YRWA
Dendroica coronata (Linnaeus)

D. c. auduboni group ("Audubon's" Warbler) — AUWA
D. c. coronata group ("Myrtle" Warbler) — MYWA

Figure 53. Adult male Yellow-rumped ("Myrtle") Warbler, *coronata* group (Revelstoke, 30 April 1997; R. Wayne Campbell). In breeding plumage, the bright yellow patches on the rump, side, and crown and the white eyebrow and throat are the best distinguishing field marks.

Figure 54. Adult male Yellow-rumped ("Audubon's") Warbler, *auduboni* group (Revelstoke, 30 April 1997; R. Wayne Campbell). In breeding plumage, the bright yellow patches on the rump, side, crown, and throat are distinguishing field marks.

RANGE: Breeds from western and central Alaska, northern Yukon, northwestern and central Mackenzie, southwestern Keewatin, northern Manitoba, northern Ontario, northern Quebec, north-central Labrador and Newfoundland south in the west to northern Baja California, southern California, Arizona, New Mexico, the mountains of western Durango and eastern Chiapas in Mexico, western Chihuahua, and western Guatemala, and in the east to northern Minnesota, central Michigan to eastern West Virginia, Pennsylvania, northern New Jersey, western Maryland, Connecticut, and Massachusetts. Winters locally from southwestern British Columbia and the Pacific states south to Arizona, Colorado, and eastern Kansas; east across the central United States to southern Ontario and New England, south through the southern United States and Mexico to eastern Panama and the West Indies.

STATUS: On the coast, a *fairly common* to *very common* migrant and summer visitant in the Georgia Depression Ecoprovince (locally *very common* to *abundant* during both spring and autumn migration), and *uncommon* to locally *fairly common* in winter, particularly on southeastern Vancouver Island south of Nanaimo and on the Fraser River delta; in the Coast and Mountains Ecoprovince, *uncommon* to *fairly common* migrant and summer visitant (locally *very common* in migration) and *casual* in winter in the Southern Mainland Coast and Northern Mainland Coast, *uncommon* to *fairly common* migrant and summer visitant (occasionally *common* in migration) and *very rare* in winter on Western Vancouver Island (locally *common* winter visitant on Stubbs Island), and *very rare* on the Queen Charlotte Islands.

In the interior, a *fairly common* to *common* migrant and summer visitant and *casual* to *very rare* in winter in the Southern Interior, Southern Interior Mountains, and Central Interior ecoprovinces (locally *very common* to *very abundant* during spring and autumn migration), becoming an *uncommon* to *fairly common* migrant and summer visitant (locally *common* to *very common* during migration) in the ecoprovinces further north.

Breeds.

NONBREEDING: The Yellow-rumped Warbler (Figs. 53 and 54) is British Columbia's most widespread and abundant warbler and is distributed throughout much of the province. Along the coast, it occurs from southern Vancouver Island and the Fraser Lowland north throughout most of Vancouver Island and the adjacent southwest mainland coast. Its distribution becomes more irregular further north. There are few observations of this warbler from the Queen Charlotte Islands.

In the interior, the Yellow-rumped Warbler occurs from the international boundary north through the southern and central portions of the interior to the southern portions of the Sub-Boreal Interior and Boreal Plains. Further north, its reported distribution is more patchy, possibly a reflection of a lack of observers in areas of suitable habitat.

There are large areas of the province from which the Yellow-rumped Warbler has not been reported, including much of the mountainous areas of the southwest mainland coast, the McGregor River valley , a large area from the Driftwood River valley northwest to the Stikine River valley, the Rocky Mountain Trench from Mackenzie north to Lower Post, an area between the Stikine River valley and Atlin and Teslin lakes, and an area centred around the junction of the Liard and Fort Nelson rivers. Again, this is probably a result of inadequate field investigations.

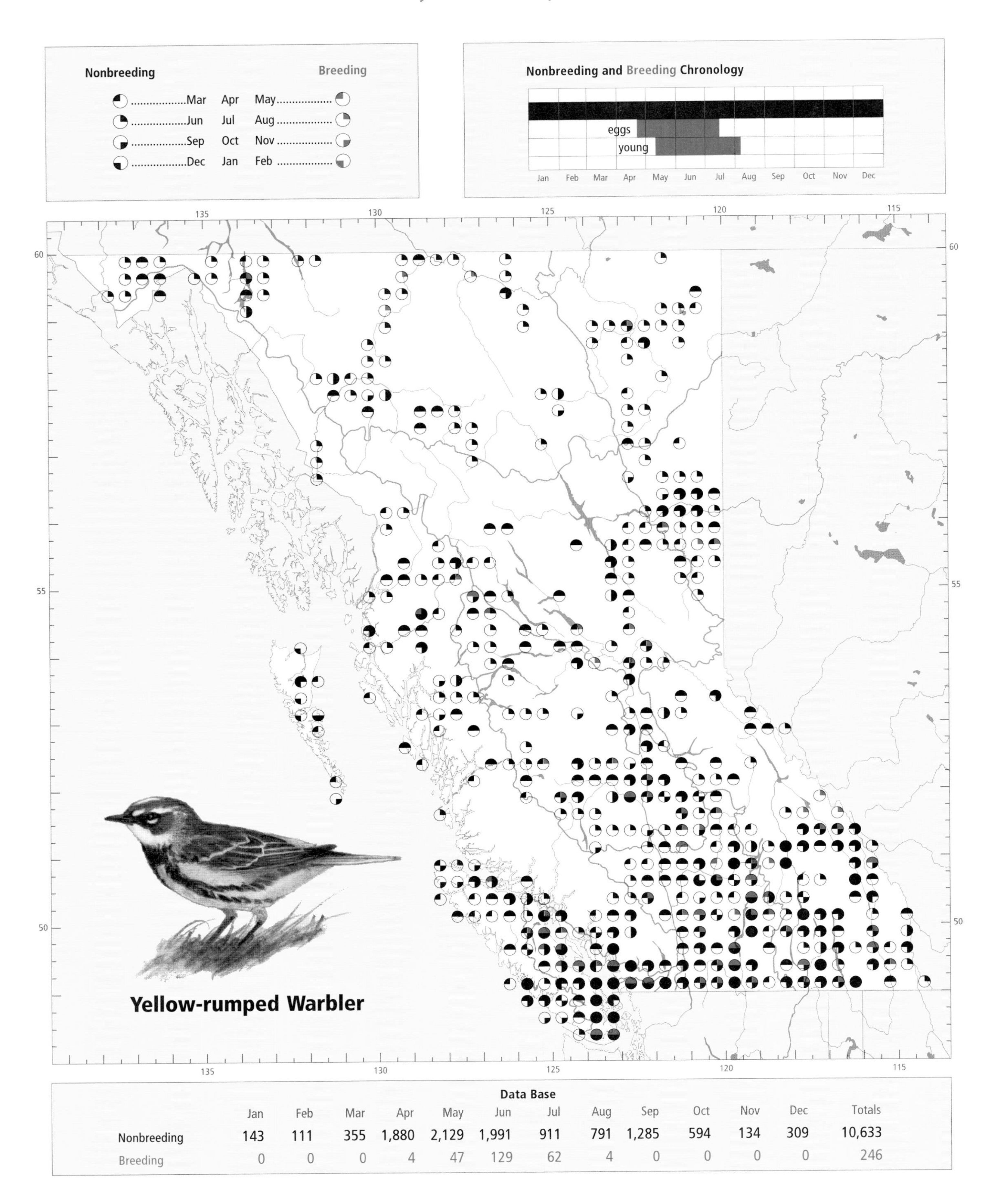

Data Base

	Jan	Feb	Mar	Apr	May	Jun	Jul	Aug	Sep	Oct	Nov	Dec	Totals
Nonbreeding	143	111	355	1,880	2,129	1,991	911	791	1,285	594	134	309	10,633
Breeding	0	0	0	4	47	129	62	4	0	0	0	0	246

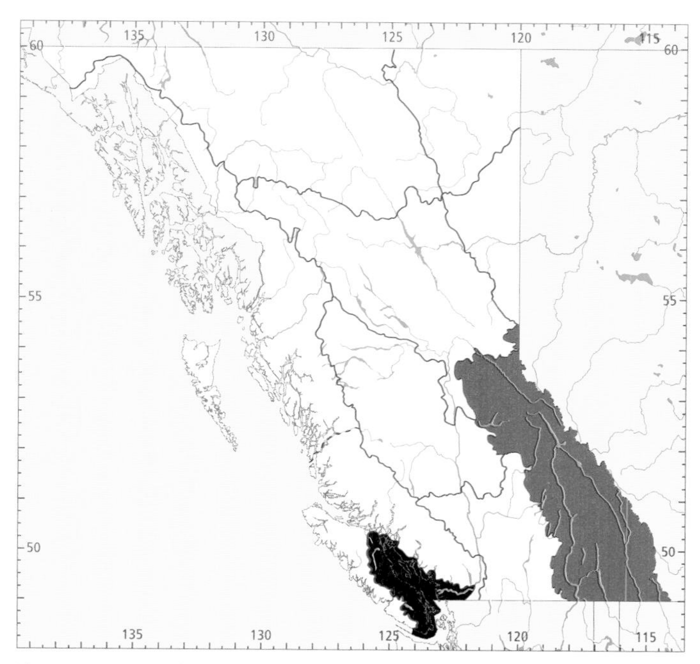

Figure 55. In British Columbia, the highest numbers for the Yellow-rumped Warbler in winter (black) occur in the Georgia Depression Ecoprovince; the highest numbers in summer (red) occur in the Southern Interior Mountains Ecoprovince.

The highest numbers in winter occur in the Nanaimo Lowland and Fraser Lowland of the Georgia Depression (Fig. 55).

During migration, the Yellow-rumped Warbler has been reported from near sea level to 1,830 m elevation on the coast; in the interior, it has been reported from 240 to 2,200 m. It frequents a wide variety of forested and unforested habitats, including subalpine meadows, but in British Columbia it seems to prefer mixed forests or stands that are relatively open in structure. Forest edges and riparian habitats are used extensively.

On the coast, the Yellow-rumped Warbler has been reported from old-growth and second-growth coniferous forests, deciduous forests, and mixed forests, and especially open forests. It also frequents woodland or shrubby riparian habitats such as those adjacent to intertidal areas, marshes, lakes, rivers, and sewage lagoons. The riparian habitats along rivers and lakes are used heavily by migrating flocks in spring. Other habitats have included backyard gardens, treed boulevards (Fig. 56), orchards, and powerline rights-of-way.

In the interior, this warbler has been reported from most forest types (Fig. 61). Other habitats have included grasslands, big sagebrush areas, sphagnum bogs, orchards, and backyards. During spring migration, it is not uncommon to see waves of Yellow-rumped Warblers streaming across open shrublands, hanging like grapes from big sagebrush and other shrubs, a habitat seldom used at any other time.

Little is known of the roosting habitat of this warbler in British Columbia. In the interior, J.M. Cooper found Yellow-rumped Warblers roosting in little pockets and perched on small ledges in a cutbank (Fig. 57). He observed 9 birds flying into the bank during a heavy rain just before sunset. They were still there after dark but had left by dawn.

Figure 56. During migration, the Yellow-rumped Warbler can be found in a wide variety of habitats, including woodlands bordering highways (Abbotsford, 21 April 1997; R. Wayne Campbell).

Figure 57. Yellow-rumped Warblers asleep and roosting on small ledges in a cutbank of a gravel pit (Cache Creek, 10 May 1996; John M. Cooper).

On the coast, the first spring arrivals in the Georgia Depression are masked by overwintering birds, but numbers start to build about the third week of March (Fig. 60). Numbers continue to build until the last week of April, when the peak of migration occurs, and then drop off abruptly in the second week of May. A similar trend occurs in the Southern Mainland Coast. On Western Vancouver Island, arrival is also masked by overwintering birds; there is a small but noticeable movement from the end of March through the first week of May. Birds begin to arrive on the northern portions of the coast about the second week of April and peak in the first week of May.

In the interior, spring migration begins as birds arrive in the southern and central regions as early as the last week of March (Figs. 58, 59, and 60), although the main movement does not begin until at least the first week of April. Numbers build through April and peak in the second week of May in the Southern Interior and Central Interior, and by the third week of May in the Southern Interior Mountains. In the Southern Interior, the numbers of this warbler in passage can be spectacular, with waves of between 500 and 1,500 birds moving through in succession in early May. In the Southern Interior and Central Interior, numbers drop abruptly in the third week of May. In the Southern Interior Mountains, numbers drop by the end of May but remain relatively high through the summer compared with the other ecoprovinces (Fig. 60). Birds arrive in the Sub-Boreal Interior in the second week of April, and numbers peak between the first and third weeks of May.

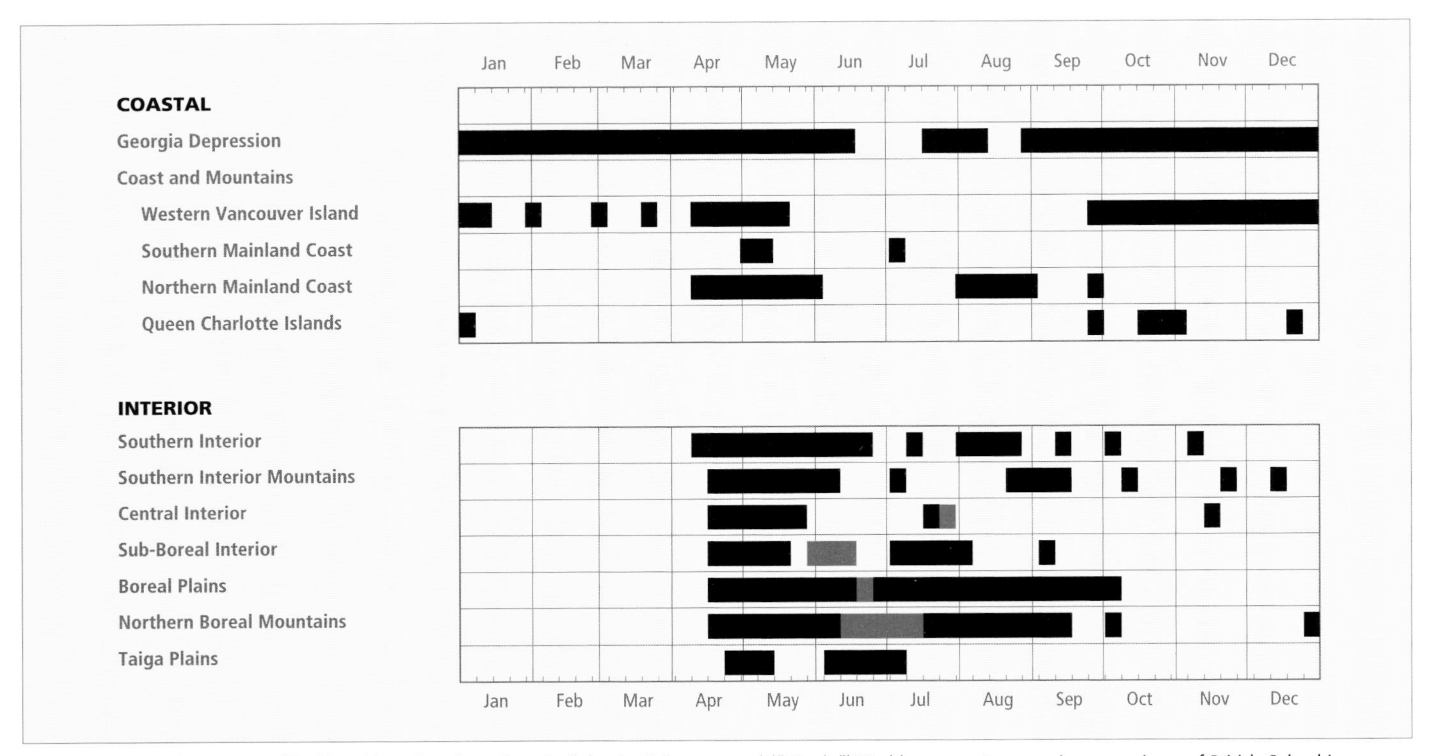

Figure 58. Annual occurrence (black) and breeding chronology (red) for the Yellow-rumped ("Myrtle") Warbler, *coronata* group, in ecoprovinces of British Columbia. Records are shown for the week in which they occurred.

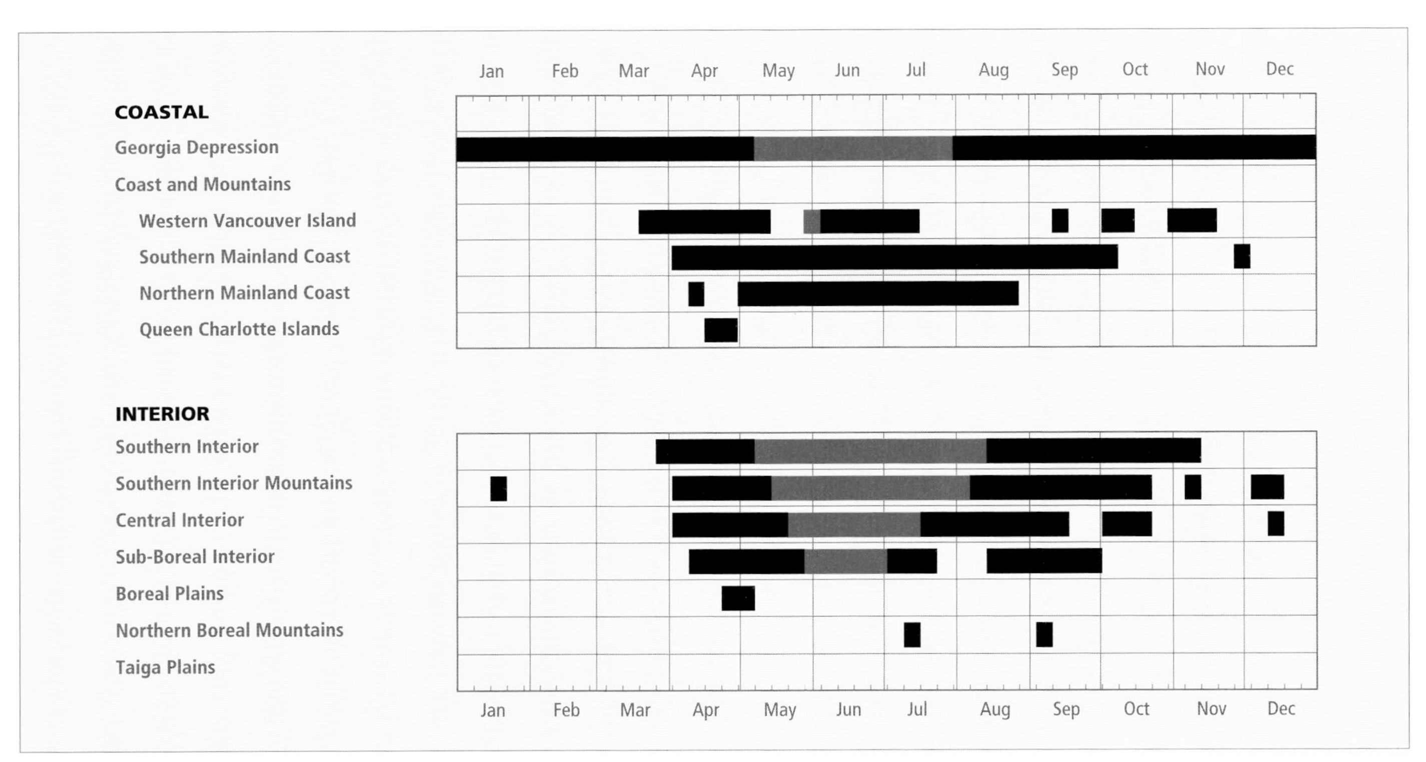

Figure 59. Annual occurrence (black) and breeding chronology (red) for the Yellow-rumped ("Audubon's") Warbler, *auduboni* group, in ecoprovinces of British Columbia. Records are shown for the week in which they occurred.

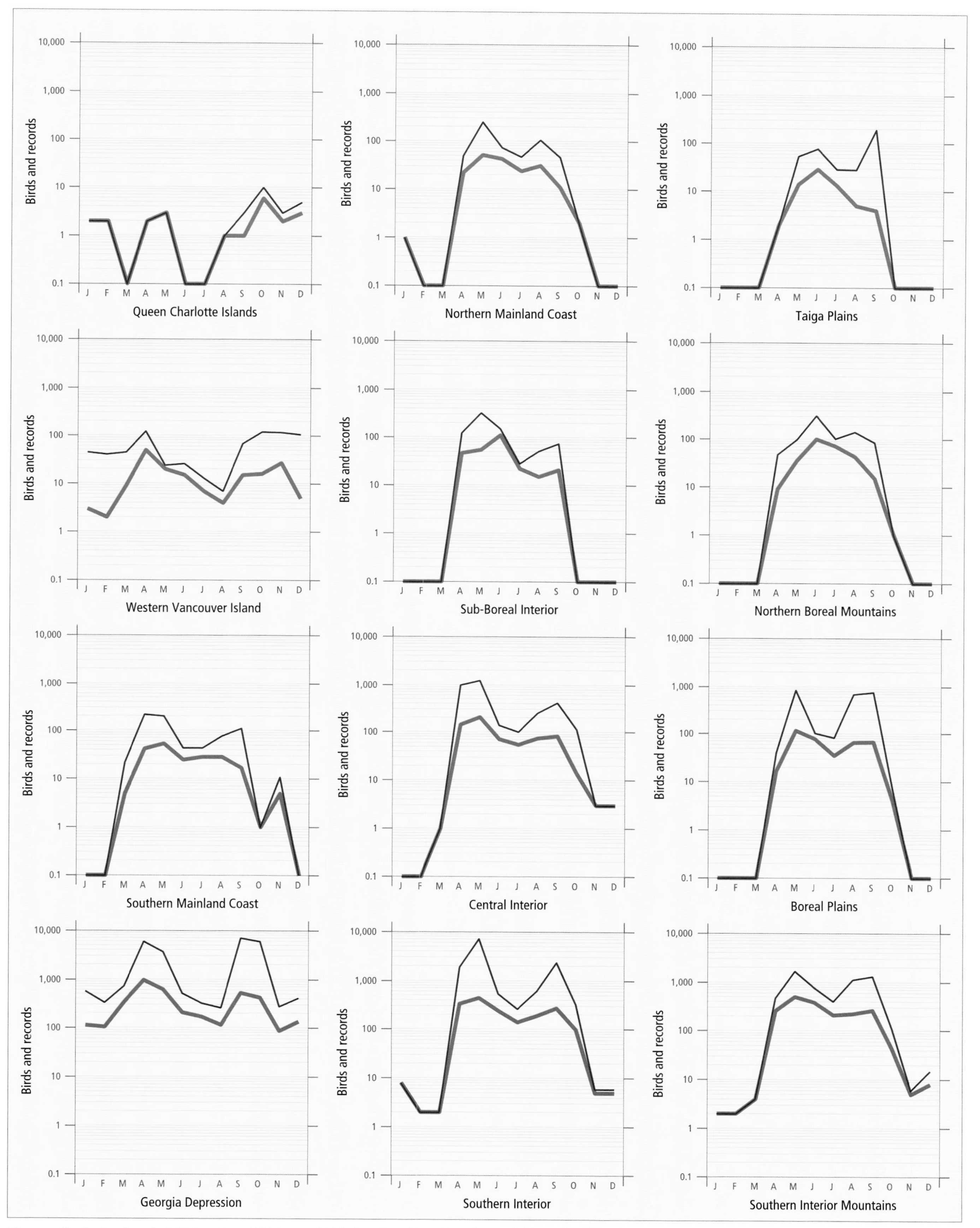

Figure 60. Fluctuations in total number of birds (purple line) and total number of records (green line) for the Yellow-rumped Warbler in ecoprovinces of British Columbia. Christmas Bird Counts, Breeding Bird Surveys, and nest record data have been excluded.

Figure 61. During autumn migration, the Yellow-rumped Warbler uses a wide variety of habitats, including subalpine forests (Copper Mountain, southwest of Nelson, 24 September 1997; R. Wayne Campbell).

Birds may arrive in the Peace Lowland of the Boreal Plains and in the Northern Boreal Mountains as early as the third week of April, but large numbers normally do not appear until the first week of May. Most of these birds likely enter this area from east of the northern Rocky Mountains, and consist mainly of the *coronata* group (see REMARKS). In the Boreal Plains, numbers rise quickly and peak in the second and third weeks of May; most have passed through by the end of that month. A spring movement through the Northern Boreal Mountains and the Taiga Plains cannot be discerned with the limited data we have from these areas.

The autumn migration begins in August in the far north, but because of our limited data it is barely discernible in the 2 northernmost ecoprovinces. In the Atlin region, however, Swarth (1926) noted that the southern movement peaked in the last week of August and first week of September. He reported that flocks of warblers, mostly this species, flitted rapidly through the poplar woods, and there was a constant stream of "Myrtle" Warblers making long flights overhead. Most birds have left that area by mid-September, although stragglers may remain into the third week of that month. Some northern Yellow-rumped Warblers may take a coastal migration route to southern areas. In the Stikine River region, Swarth (1922) noted that this warbler appeared to migrate coastward, suggesting a late summer line of travel down the Stikine River to the coast. He found similar conditions in late summer at the mouth of the Taku River, about 240 km north of the Stikine. This coastwise movement at this latitude likely involves only the *coronata* group (see REMARKS).

Birds moving through the Boreal Plains are most abundant between the second week of August and the second week of September, with no particular peak discernible (C. Siddle pers. comm.). Most birds have gone by the end of the month, although stragglers may remain into the first week of October (Figs. 58, 59, and 60).

In the Sub-Boreal Interior, there are few reports on the autumn migration, which probably occurs from early August to the last week of September. In the Central Interior, birds appear to move through in waves from early August to the last week of September; small numbers may remain until mid-October. A similar pattern occurs in the Southern Interior Mountains. In the Southern Interior, autumn numbers begin to build by the second week of August and peak in the first or second weeks of September. The numbers gradually decline after that, and most birds have left the region by the third week of October.

On the Northern and Southern Mainland coasts, the autumn movement has been almost unreported, but probably occurs between the third week of August and the end of September. On Western Vancouver Island, numbers increase from the first week of September into early October, but migrants soon become difficult to distinguish from the overwintering birds. In the Georgia Depression, Yellow-rumped Warbler numbers begin to increase in the last week of August; they build to a peak in the last week of September and the first week of October. The autumn movement is still obvious into the third week of October, but most migrants have passed through by the end of that month.

The Yellow-rumped Warbler often occurs in mixed-species flocks during migration. In British Columbia, they have been reported with Hammond's and Dusky flycatchers, American Robins, Ruby-crowned and Golden-crowned kinglets, Red-breasted Nuthatches, Orange-crowned, Nashville, Yellow, and Magnolia warblers, Common Yellowthroats, and Dark-eyed Juncos.

The adaptability of this species allows it to undergo a migration that extends well into the autumn, secure in its ability to find food under most of the conditions it would encounter during the season regardless of how far it has to go to reach the wintering grounds (J.P. Hubbard pers. comm.).

Small numbers regularly winter in the Nanaimo and Fraser lowlands of the Georgia Depression and locally on Western Vancouver Island.

The wintering population on Western Vancouver Island is of special interest. On Stubbs Island, just west of Tofino, a small population of 30 to 50 birds spends the winter in a California wax-myrtle (*Myrica californica*) "forest" (Fig. 62), feeding on the wax-myrtle berries. The Yellow-rumped Warbler population that winters there consists entirely of the *coronata* group (A. Dorst pers. comm.). This may be a recent event. Hatler et al. (1978) note the "Myrtle" Warbler only as a migrant in Pacific Rim National Park, and they reported just 1 sighting of the Yellow-rumped Warbler after the end of August.

In British Columbia, California wax-myrtle occurs only in an isolated pocket along the west coast of Vancouver Island between Ucluelet and Tofino. On Stubbs Island, it reaches 5 to 6 m in height and densely covers an estimated 1 ha of land (A. Dorst pers. comm.). This situation is somewhat reminiscent of the wintering habitat of this warbler in eastern North America, where thickets of bayberry (*Myrica pennsylvanica*) and wax-myrtle (*Myrica cerifera*) provide food and shelter for the "Myrtle" Warbler (which bears the latter plant's name), even in areas of heavy snowfall (Wilz and Giampa 1978; Godfrey 1986; Terres 1991). There concentrations can be awesome; hordes flush ahead of the advancing birder, filling the air with hard *chek* calls (Farrand 1983).

Like other winter frugivores, they often flock in that season – a social system linked to the exploitation of patchy, temporarily abundant resources (Morse 1989).

On the coast, the Yellow-rumped Warbler has been recorded regularly throughout the year; in the interior, it has been regularly recorded from 29 March to 15 October (Figs. 58 and 59).

BREEDING: The Yellow-rumped Warbler likely breeds throughout most of its summer range in British Columbia, except on the Queen Charlotte Islands.

Along the coast, it nests in much of the Georgia Depression north to Campbell River. On Western Vancouver Island, in the Coast and Mountains, there is only 1 breeding record: near the northern end of the island, at Sointula. A vast area of mainland coast north to Kitimat lacks any reports of breeding by this species.

In the interior, the Yellow-rumped Warbler has a widespread breeding distribution in suitable habitat south of latitude 53°N. Further north its breeding distribution extends to the northern boundary of the province, although nesting records are few.

This warbler reaches its highest numbers in summer in the Southern Interior Mountains (Fig. 55). An analysis of Breeding Bird Surveys in British Columbia for the period 1968 through 1993 could not detect a net change in numbers on either coast or interior routes. An analysis of Breeding Bird Survey data for 1966 to 1996 yielded similar results for the *auduboni* group across both Canada and North America (Sauer et al. 1997). For the same period, Canadian and North American data for the *coronata* group showed average annual increases of 1.8% ($P < 0.01$) and 1.5% ($P < 0.05$), respectively, although recent trends (1980 to 1996) suggest relatively stable populations (Sauer et al. 1997). Continent-wide Breeding Bird Survey data indicate that British Columbia has some of the highest densities of the *auduboni* group in North America (Sauer et al. 1997).

The Yellow-rumped Warbler has been reported nesting on the coast from near sea level to 490 m elevation; in the interior, it has been reported from 240 m to subalpine habitats (Fig. 63) of at least 2,250 m elevation. Most nests (66%; $n = 197$) were in forested habitats, followed by human-influenced (25%) and riparian habitats.

Of the forested habitats, coniferous forests were selected most frequently (33%; $n = 152$; Fig. 63) and included nearly every coniferous forest type in the province. Mixed woods (16%) and deciduous woods (12%; Fig. 64) were selected less frequently. Riparian habitats comprised 8% of reported nesting habitats, while human-influenced nesting habitats (31%) included suburban, urban, and rural areas, cultivated farmlands, recreational areas, pasture, orchards, and logged areas.

The seral stages used for nesting included young and regenerating forests as well as second-growth, mature, and old-growth forests. Backyard gardens and residential areas were the human-influenced habitats most often reported as nesting sites.

Figure 62. On Western Vancouver Island, a small population of the Yellow-rumped ("Myrtle") Warbler, *coronata* group, winters locally near Tofino, feeding on berries of the California wax-myrtle (*Myrica californica*) (Stubbs Island, May 1999; Adrian Dorst).

Figure 63. The Yellow-rumped ("Myrtle") Warbler, *coronata* group, breeds in open coniferous forests in subalpine areas of the Northern Boreal Mountains Ecoprovince (Stone Mountain, 28 June 1996; R. Wayne Campbell).

Figure 64. In the southern portions of the interior of British Columbia, the Yellow-rumped ("Audubon's") Warbler, *auduboni* group, breeds in a wide variety of forest types, including deciduous woodlands often found along sloughs and rivers (Creston, 11 May 1997; R. Wayne Campbell).

Figure 65. In British Columbia, the Yellow-rumped Warbler nest consists mainly of coarse plant stems and small twigs and is lined with feathers and plant down (Whitehorse, Yukon, 28 June 1999; R. Wayne Campbell). It is placed in both coniferous and deciduous trees.

Although coniferous forests are important habitats for this species throughout much of British Columbia, Bryant et al. (1993) note that this warbler was not found consistently on any transect in clearcut, 15- to 20-year-old, 30- to 35-year-old, 50- to 60-year-old, or old-growth stands of the Coastal Western Hemlock Biogeoclimatic Zone in the Franklin River, Kennedy Lake, and Sproat Lake areas of Western Vancouver Island. B.R. Gates (pers. comm.) reported similar results on Breeding Bird Surveys in the nearby Elsie Lake region on central Vancouver Island.

In the Okanagan valley, the Yellow-rumped Warbler was found in all woodland habitats but was most plentiful above 600 m (Cannings et al. 1987). Schwab (1979), who studied the effect of vegetation structure on breeding bird communities in the Interior Douglas-fir Biogeoclimatic Zone of the east Kootenay region, found the Yellow-rumped Warbler in sites ranging from young conifers under 10 m in height to climax Douglas-fir forests more than 120 years old. Most Yellow-rumped Warblers were found in the mature seral stages of lodgepole pine over 10 m tall (21 to 60 years old), ponderosa pine–Douglas-fir over 10 m (61 to 100 years old), and climax Douglas-fir. In Mount Revelstoke and Glacier national parks, the habitats with the highest densities of Yellow-rumped Warblers were the Engelmann spruce–black cottonwood open forest, the Engelmann spruce–subalpine fir open forest, and the subalpine fir–mountain hemlock open forest (Van Tighem and Gyug 1983).

Further north, in trembling aspen forests near Smithers, Pojar (1993) found the highest densities of singing male Yellow-rumped Warblers in the old trembling aspen and mixed conifer-aspen seral stages. The densities of singing males in all seral stages were as follows: sapling trembling aspen, 0 to 16 per 10 ha (n = 20); mature trembling aspen, 19 to 21 per 10 ha (n = 42); old trembling aspen, 21 to 26 per 10 ha (n = 11); and mixed conifer-aspen, 24 to 26 per 10 ha (n = 20). She also found this warbler in a pure pine stand. Cowan (1939) noted that this warbler is characteristic of the black spruce and white spruce forests of the Peace Lowland. Phinney (1998) found it in coniferous (especially pine) and mixed-wood forests in the Dawson Creek area, and Greenfield (1998) noted it as being common in spruce forests along the Sikanni Chief River.

It is also common in lodgepole pine and trembling aspen forests, white spruce and trembling aspen forests, and the balsam poplar, trembling aspen, and white spruce forests of the Boreal Plains (C. Siddle pers. comm.). In the Stikine and Iskut river valleys of northwestern British Columbia, the Yellow-rumped Warbler was the third most common species recorded on songbird transects in coniferous, deciduous, and mixed forests and open shrublands (Blood et al. 1981).

Morse (1989) notes that the Yellow-rumped Warbler has the largest breeding territories of any warbler.

On the coast, the Yellow-rumped Warbler has been recorded breeding from 22 April to 29 July; in the interior, it has been recorded breeding from 29 April (calculated) to 6 August. Figures 58 and 59 show the known range in breeding dates for both the *coronata* and *auduboni* groups, respectively.

Nests: Most nests were placed in trees (87%; n = 115; Fig. 65), including both coniferous (49%) and deciduous (15%)

Figure 66. In British Columbia, the Yellow-rumped Warbler is a common host for the Brown-headed Cowbird (Creston, 21 June 1997; R. Wayne Campbell).

trees. From the 91 records where the type of tree was noted, the most commonly used trees were Douglas-fir (14%), pines (13%), unidentified firs (12%), willows (11%; Fig. 65), spruce (9%), birch (8%), and trembling aspen (8%); other nest trees included apple, red alder, black cottonwood, cedar, western larch, Rocky Mountain juniper, hemlock, dogwood, bitter cherry, and poplar. Shrubs were used to a lesser degree (13%) and included hazelnut and common snowberry. Nests were placed in the fork or crotch of a branch or were saddled on a branch, often close to the trunk. The heights above ground for 103 nests ranged from 0.3 to 21 m, with 51% between 2.1 and 7.6 m.

Nests consisted of a small, flat cup composed mainly of grasses, twigs, and rootlets. They were lined with feathers, hair, and plant down (Fig. 65).

Eggs: Dates for 98 clutches ranged from 22 April to 15 July, with 52% recorded between 27 May and 20 June. Sizes of 52 clutches ranged from 1 to 5 eggs (1E-6, 2E-3, 3E-10, 4E-24, 5E-9), with 65% having 3 or 4 eggs (Fig. 66). The incubation period is 12 to 13 days (Knight 1905; Harrison 1979).

Young: Dates for 117 broods ranged from 11 May to 6 August, with 51% recorded between 15 June and 7 July. Sizes of 53 broods ranged from 1 to 4 young (1Y-8, 2Y-15, 3Y-16, 4Y-14), with 58% having 2 or 3 young. The nestling period is 10 to 12 days (Harrison 1979).

Brown-headed Cowbird Parasitism: In British Columbia, 21% of 118 nests found with eggs or young were parasitized by the cowbird (Fig. 66). Nest parasitism on the coast was 10% (n = 20); in the interior, it was 22% (n = 118). An additional 43 instances of fledgling cowbirds being fed by Yellow-rumped Warbler foster parents were reported. A comparison of the subspecies shows that both *D. c. auduboni* (20% of 54 nests; 14 fledged cowbirds reported being fed) and *D. c. coronata* (0 of 4 nests; 7 fledged cowbirds being fed) were parasitized by the cowbird. Friedmann and Kiff (1985) note a similar rate of cowbird parasitism at the Sierra National Forest in California (> 22%).

Nest Success: Of 26 nests found with eggs and followed to a known fate, 6 produced at least 1 fledgling, for a success rate of 23%. None of 7 coastal nests was successful; in the interior, nest success was 32% (n = 19).

REMARKS: The *coronata* and *auduboni* groups were formerly regarded as distinct species: the Myrtle Warbler and Audubon's Warbler, respectively (Figs. 53 and 54; American Ornithologists' Union 1957); however, intergradation occurs from southeastern Alaska southeast across central British Columbia to southern Alberta (American Ornithologists' Union 1983).

There are 4 recognized subspecies in British Columbia, 2 in the *D. c. coronata* group and 2 in the *D. c. auduboni* group. Three geographic races have been reported breeding in British Columbia: "Myrtle" Warbler, *D. c. hooveri* (northern British Columbia); "Audubon's" Warbler, *D. c. auduboni* (central and southwestern British Columbia); and *D. c. memorabilis* (southeastern British Columbia) (American Ornithologists' Union 1957). Hubbard (1970) notes, however, that the race *memorabilis* is not distinguishable because of broad intergradation in wing length and plumage characters with northwestern populations. He further states that *memorabilis* does not breed in British Columbia, if in Canada at all. Vagrants no doubt occur, however, in the form of migrants overshooting the breeding ranges in areas such as Idaho and Montana. In addition, *hooveri* appears limited in British Columbia to the northwest, with *coronata* occupying the northeast, as shown by measurements from central Alberta.

Because the *coronata* and *auduboni* groups are distinguishable in the field, we have summarized the available data for the 2 groups below. For the most part, the summaries are based on identifications made using visual characteristics alone. We encourage observers to continue separating both groups in their field notes.

The *coronata* group: The "Myrtle" Warbler's nonbreeding distribution in the province is similar to the species' distribution as a whole (Fig. 67). In the southern portions of the province, however, its numbers are small compared with those of the more abundant *auduboni* group (Table 1). North of latitude 55°N, *coronata* becomes the dominant group. In spring, this group has a widespread distribution throughout the province, but by summer its numbers are concentrated in the northernmost ecoprovinces. Autumn reports are widely scattered, with most coming from the Georgia Depression; most birds have left the northern regions by the end of summer. The Georgia Depression birds may be part of a population that Swarth (1922) noted moving down the river valleys to the coast. Some of these birds stay through the winter in the Georgia Depression and Western Vancouver Island (see NONBREEDING).

We have only 4 nesting reports for the *coronata* group, 2 of which occurred below latitude 55°N: Fort St. James (54°26′) and Stum Lake (52°17′) (Fig. 67). While the observers specifically noted the birds as belonging to the *coronata* group, we

Table 1. Ratios of the *auduboni* group to the *coronata* group of the Yellow-rumped Warbler in ecoprovinces of British Columbia. The ratios are based on records where the observer identified the species group.

Ecoprovince	*auduboni:coronata* ratio	Number of birds
Coastal		
Georgia Depression	3:1	1,790
Western Vancouver Island	1:1	90
Southern Mainland Coast	14:1	61
Northern Mainland Coast	2:1	60
Queen Charlotte Islands	1:4	10
Interior		
Southern Interior	10:1	457
Southern Interior Mountains	7:1	252
Central Interior	5:1	290
Sub-Boreal Interior	2:1	271
Boreal Plains	1:66	269
Northern Boreal Mountains	1:64	129
Taiga Plains	0:27	27

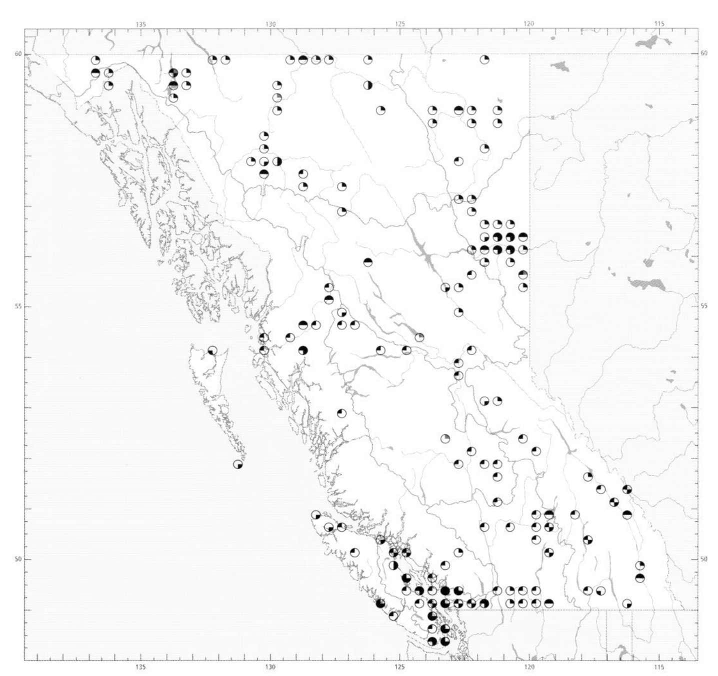

Figure 67. Nonbreeding (black) and breeding (red) distribution of the "Myrtle Warbler" in British Columbia.

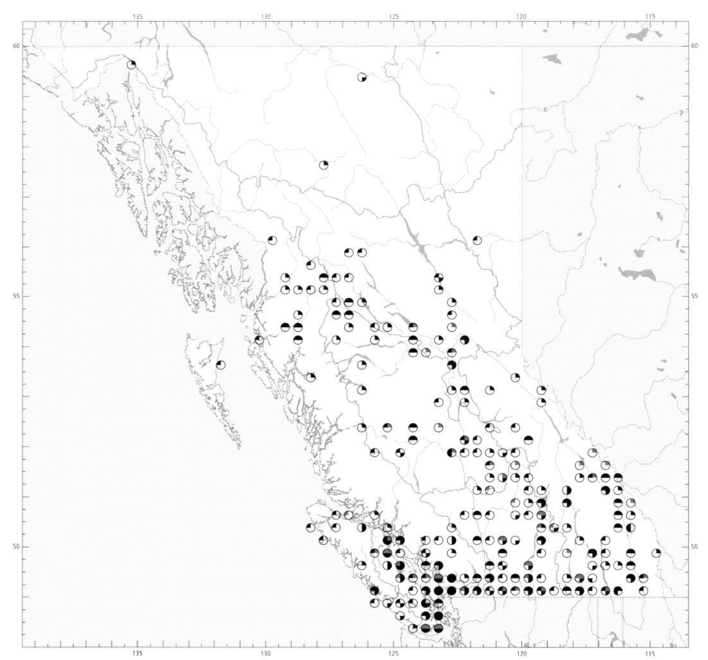

Figure 68. Nonbreeding (black) and breeding (red) distribution of the "Audubon's Warbler" in British Columbia.

could not tell from their reports whether both birds in the pair were seen and identified; thus, the pair in each case may have consisted of members of both groups.

The *auduboni* group: Members of this group have a widespread distribution in British Columbia north to about latitude 56°N (Fig. 68). We have only 7 reports of this group's occurrence further north. South of latitude 56°N, *auduboni* is the dominant form in all seasons (Table 1). At or near the northern limit of this group's range, Stanwell-Fletcher and Stanwell-Fletcher (1943) noted that it was probably the most common warbler at Tetana Lake in the Driftwood River valley; both groups occurred there.

The breeding distribution of the *auduboni* group ranges across southern British Columbia, including Vancouver Island, north to Fort St. James (54°26') and Summit Lake (54°17'), north of Prince George (Fig. 68).

Our data support those of Hubbard (1969), who studied the interbreeding of the *coronata* and *auduboni* groups in southeastern Alaska, British Columbia, and Alberta. He based his analysis primarily on 6 breeding plumage characteristics of males: wing pattern, tail pattern, throat colour, auricular colour, and the presence or absence of a supraloral spot and postocular line. His results clearly demonstrated intergradation between the then-species *D. coronata* and *D. auduboni.* Figure 69 shows the results Hubbard (1969) obtained. Birds displaying characteristics of the *auduboni* group occurred west of the Rocky Mountains and south of about latitude 56°N; birds with characteristics predominantly of the *coronata* group occurred east of the Rocky Mountains, with the exception of those that occurred in the northwestern portion of the province. Intergradation in most characteristics is restricted to a zone of a few hundred kilometres or less in width. Hubbard (1969) noted that the 2 groups interbreed and backcross freely where their ranges meet; away from those areas, however, massive introgression appears to be prevented by the limited area of contact between the 2 groups.

Besides an introgression zone, Hubbard (1969) and Barrowclough (1980) found a number of hybrid zones where

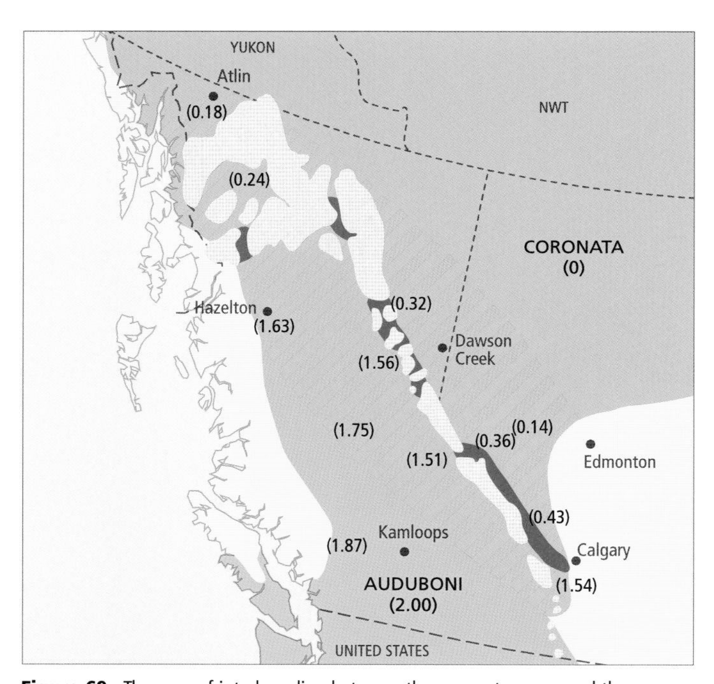

Figure 69. The area of interbreeding between the *coronata* group and the *auduboni* group of the Yellow-rumped Warbler in British Columbia. The introgression zone is cross-hatched and the hybrid zone is red. The white area is primarily unforested habitat. The numbers are selected sample means for plumage characteristics, with pure *coronata* having a value of 0 and pure *auduboni* having a value of 2 (modified from Hubbard 1969).

Figure 70. During the height of autumn migration in southern British Columbia, when dense fog and cloud obscure corridors, Yellow-rumped Warblers are often killed by automobiles as they move along and across major highways (Bridesville, 22 September 1996; R. Wayne Campbell).

the characteristics of the birds were intermediate between the 2 groups. They noted that hybridization probably occurred in the upper reaches of the Iskut and Nass, and perhaps the Skeena and Stikine, river valleys. Hubbard (1969) suggested that other probable areas included the Kechika, Findlay [sic; = Finlay], and Peace rivers (Fig. 69), although based on our available information, the latter area, east of the Rocky Mountains, is likely not a hybridization zone.

Hubbard (1969) proposed that an ancestral population of the Yellow-rumped Warbler inhabited North America during the late Pleistocene, and that the advancing glaciation split it along the Rocky Mountain axis. Part of the population (*D. c. coronata*) adapted to the boreal forest and the other (*D. c. auduboni*) adapted to the montane forests. When the glaciers receded, the 2 separate populations once again came into contact where they could freely interbreed.

During spring and autumn, when dense fog and low clouds often settle in valleys (Fig. 70), migrating Yellow-rumped Warblers are vulnerable to being killed by vehicular traffic along major highways. On 22 September 1996, 11 dead birds were counted along a 100 m stretch of Highway 3 near Bridesville.

An adult male "Audubon's" Warbler banded in Vancouver on 18 April 1954 was recovered on 25 April 1958 in Toronto, some 3,500 km east (Hughes 1959).

For a summary of the systematics and natural history of the Yellow-rumped Warbler in North America, see Hunt and Flaspohler (1998). Morse (1989) provides an ecological and behavioural review of North American warblers, including the Yellow-rumped Warbler.

NOTEWORTHY RECORDS

Spring: Coastal – *D. c. auduboni:* Skirt Mountain 28 May 1973-10; Sooke Harbour 21 May 1945-4 nestlings about 3 days old; Victoria 8 May 1918-4 eggs; Elk Lake 17 Apr 1984-30; Beaver Lake (Saanich) 29 Apr 1972-200 (Tatum 1973); Tofino 24 Mar 1968-1 (Hatler et al. 1978); Westham Island 17 Mar 1970-10 (Campbell et al. 1972a); Reifel Island 4 May 1982-120; Surrey 25 Mar 1979-57, sizeable flock moving through area; Langley 9 Apr 1967-15, in woods at Murray Creek mouth; Langley 22 Apr 1969-25 at airport; Glen Valley 28 Apr 1974-50; Skagit River valley 20 Apr 1971-20; Sechelt 2 May 1981-100; Vancouver 19 Apr 1980-200 at Stanley Park, 25 Apr 1981-75, 15 Apr 1996-214 at John Hendry Park; North Vancouver 13 Apr 1980-50 at Cates Park; Silver Lake (Skagit River valley) 28 Apr 1973-10, 11 May 1974-15; Courtenay 4 Apr 1988-18; Alice Lake 14 May 1961-12; Solander Island 5 May 1975-1; Sunshine Coast 6 Apr 1984-1; Grant Bay 1 May 1969-1 (Richardson 1971); Port Hardy 21 Apr 1939-10; Sointula 28 May 1976-adults feeding nestlings; Tlell 24 Apr 1972-1, singing; Kitimat 29 May 1975-5; Lakelse Lake 13 Apr 1980-2, "Myrtle" subspecies also present; Cedarvale 7 May 1980-4; Meziadin Lake 30 May 1978-2. *D. c. coronata:* Oak Bay 26 Mar 1977-4; Blenkinsop Lake 10 Apr 1990-6; Saanich 16 Apr 1981-16 at Quick's Pond; Beaver Lake (Saanich) 4 Apr 1979-3; Stubbs Island 25 Mar 1984-12, all in winter plumage, 15 Apr 1979-4; Tofino 18 to 20 May 1974-1; Beach Grove 22 Apr 1979-15; Sea Island 29 Apr 1972-200 (Campbell et al. 1974); Reifel Island 1 May 1972-30 (Campbell et al. 1974), 4 May 1982-100; Port Alberni 23 Apr 1988-10; Burnaby Lake 20 Apr 1980-20; Vancouver 13 Apr 1985-65 at Jericho Park, 15 Apr 1974-19 at Lost Lagoon, 20 Apr 1987-100 at Jericho Park, 16 May 1975-40 at Stanley Park; West Vancouver 23 Apr 1972-42 (Campbell et al. 1974); Sechelt 2 May 1981-25; Hope 4 May 1975-1; Alice Lake 3 May 1969-1; Woss Camp 10 May 1980-2; Port Hardy 19 Apr 1936-4; Kimsquit River 22 Apr 1982-1; Kitimat Mission 11 May 1975-8. *Undetermined subspecies:* Mount Douglas 14 May 1993-nestlings; Cowichan Bay 18 Mar 1972-10; Boundary Bay 2 May 1982-120; Reifel Island 14 Apr 1985-15; Langley 5 Apr 1966-30; New Westminster 22 Apr 1939-5 eggs; Surrey 4 Apr 1966-15; Vancouver 14 Apr 1985-25 at Jericho Park, 22 Apr 1988-200 at Stanley Park, 1 May 1969-101 (Weber 1972); West Vancouver 2 May 1982-150; Pitt Meadows 9 Apr 1974-20, 4 May 1929-100; Seabird Island 20 Apr 1984-34; Harrison Hot Springs 22 Mar 1980-1, 4 May 1986-79; Englishman River 22 Mar 1979-1 at estuary (Dawe et al. 1994); Qualicum Beach 12 Apr 1976-14; Squamish 14 Apr 1963-12; McLean Point (Kyuquot) 23 Apr 1972-20; Sedman Creek 23 May 1991-1 at estuary (Dawe 1991); Kitlope Lake 3 to 10 May 1991-62, 21 at estuary, 21 at lower Kitlope, 10 at Lake, 10 at Gardner; Mayer Lake 18 May 1991-1 (Dawe 1991); Kitimat 25 Apr 1980-15; Tergroup 11 Apr 1966-1, first arrival (Crowell and Nehls 1966b); Hazelton 16 May 1992-10. **Interior** – *D. c. auduboni:* Osoyoos Lake 2 May 1974-12; Whipsaw Creek Ecological Reserve (w Princeton) 30 Apr 1974-12; Princeton 6 May 1979-15, 11 May 1975-500, migration through poplar groves; 5 km w Bromley 24 Apr 1971-20; Bromley 30 Apr 1974-30; White Lake 26 May 1975-10; Castlegar 15 May 1976-2 eggs; Robson 21 Apr 1969-6; Kinnaird Park 5 Apr 1971-2; Okanagan Landing 31 Mar 1942-1, first of year, 8 Apr 1927-10, first of year; Dorothy Lake 2 May 1981-4; McQueen Lake 8 May 1973-2 eggs, 11 May 1973-3 nestlings; Adam's River 30 Mar 1997-1; Celista 21 Apr 1948-10, arrival, travelling with "Myrtle" subspecies; Loon Lake (Clinton) 26 May 1963-3 eggs; Leanchoil 15 May 1972-8; North Barriere Lake 7 Apr 1974-10; Tatton Lake 5 May 1942-30 (Munro 1945a); lower Blaeberry River valley 14 Apr 1996-1 male; Lac la Hache 4 Apr 1944-1 (Munro 1945a); Williams Lake 8 Apr 1985-1, arrival; Westwick Lakes 6 May 1955-20; Pelican Lake 25 May 1948-15; Chezacut 2 May 1943-50, 8 May 1943-100, 14 May 1943-10; Prince George 11 Apr 1985-10, second wave came through on 3 May, very few this year, 7 May 1975-20; nr Tabor Creek 4 May 1985-18; Willow River 12 Apr 1985-1, common summer resident; Quick 19 Apr 1981-8 to 10 birds; Mackenzie 17 Apr 1996-1; Chichouyenily Creek 4 May 1997-47; Mugaha Creek 26 Apr 1997-3, 27 Apr 1996-34, 5 May 1996-26; Hudson's Hope 27 Apr 1980-1, 30 Apr 1987-1. *D. c. coronata:* Anarchist Mountain 31 May 1951-1; Princeton 7 May 1979-14; Castlegar 24 Apr 1969-1; Tunkwa Lake 12 May 1968-10; Enderby 14 Apr 1952-1 (UBC 8001); Scotch Creek 16 May 1970-4; Celista 21 Apr 1948-7; Revelstoke 20 Apr 1982-1, arrival; lower Blaeberry River valley 10 May 1996-1 male and 1 female; Kootenay National Park 19 May 1983-30; Riske Creek 18 to 30 Apr 1984-7; Lac la Hache 8 May 1943-12 (Munro 1945a); Williams Lake 12 May 1984-10; Wells Gray Park 14 May 1962-10 (Edwards and Ritcey 1967); Prince George 5 May 1985-2; Francois Lake 26 Apr 1977-1; Ellis Island 23 May 1978-7; Willow River 22 Apr 1980-1, 18 to 20 May 1968-1; Quick 30 Apr 1978-2; Tupper Creek 26 May 1938-1, carrying nest material (Cowan 1939); Tetana Lake 27 Apr 1938-1 (Stanwell-Fletcher and Stanwell-Fletcher 1943); Hudson's Hope 24 Apr 1983-1; Bear Flat to Hudson's Hope 26 Apr 1986-20, all males; Moberly River 10 May 1986-150; e Farrell Creek 11 May 1985-100, large flocks; nr Attachie 21 Apr 1984-1, 20 May 1984-70; Taylor 15 May 1983-8; Bear Flat 4 May 1985-18; Charlie Lake 17 May 1986-20; Beatton Park 3 May 1986-8, 16 May 1986-43, mostly males, 22 May 1986-79; Cecil Lake 24 Apr 1983-2, first spring arrivals; North Pine 6 May 1985-23; Cold Fish Lake 27 May 1976-3; Prophet River Park 13 May 1982-1; Fort Nelson 28 Apr 1974-first noted (Erskine and Davidson 1976); Liard Hot Springs 30 Apr 1975-40 (Reid 1975); Chilkat Pass 14 to 16 May 1977-35 birds a day; Atlin 21 Apr 1934-1 (CAS 42135), 18 May 1977-8. *Undetermined subspecies:* Manning Park 15 Apr 1977-100; nr Oliver 4 Apr 1986-17, 5 Apr 1986-97, 6 Apr 1988-34, at banding station; nr Rosebud Lake 26 Apr 1981-20; Princeton 19 May 1963-112, in flock moving through area; Similkameen River 20 May 1963-450, 1 flock; Osprey Lake (Princeton) 7 May 1979-1,500, both *auduboni* and *coronata* groups, migrating through in large flocks, 7 May 1979-1,200, another huge flock flying through area; Wolfe Lake 3 May 1975-40; Green Lake (Okanagan Falls) 11 Apr 1985-20; Summerland 25 Apr 1947-100; Merritt 11 Apr 1976-340; Kelowna 12 Apr 1974-200; Nakusp 15 Apr 1981-10; s Kamloops 5 May 1984-1,000+, grounded by storm along Highway 5A; Bridge River valley 20 May 1986-500, in tributary watersheds, heavy migratory movement; Celista 23 Apr 1948-62; Revelstoke 27 Mar 1986-1; Kootenay National Park 26 May 1981-2 nestlings, 16 May 1983-25; Glacier National Park 3 May 1982-30, 7 May 1982-25, 14 May 1982-50, 19 May 1982-30; Yoho National Park 4 May 1977-40, grounded in storm; Ottertail River valley 4 May 1977-30, in flock with 8 Orange-crowned Warblers; Yoho National Park 29 May 1979-24; Wapta Lake 28 May 1977-25; Soda Lake 27 Apr 1981-30; 100 Mile House 24 Apr 1962-35; Horse Lake 6 Apr 1976-12; West Lake (Riske Creek) 10 May 1978-42, migrating; nr Riske Creek 10 May 1987-eggs, 1.6 km e Wineglass Ranch; Lac la Hache 30 Apr 1942-250, male:female ratio 10:1 (Munro 1945a); Williams Lake 30 Mar 1979-1, 11 to 19 Apr 1959-40, males suddenly arrived, 20 to 26 Apr 1964-100, flocks feeding in shrub border of lake; Quesnel 6 May 1979-13; Telkwa 7 May 1976-200, 23 May 1994-4 eggs, female near nest; Tumbler Ridge 11 May 1995-20, 25-27 May 1997-20 to 30 at sewage lagoons; Hyland Post 27 May 1976-11; nr Fort Nelson 21 May 1979-12, pond at Mile 17.4 on Clarke Lake Road; Atlin 16 May 1981-19 (Campbell 1981).

Summer: Coastal – *D. c. auduboni:* Jordan River 6 Jul 1976-2; Victoria 29 Jul 1939-nestlings; Reifel Island 4 Aug 1972-20; Aldergrove 23 Aug 1967-14; Maplewood 10 Jul 1981-10, post-breeding wanderers?; Alouette Lake 28 Jul 1963-1; Hope 28 Jul 1990-3 fledglings; Skagit River valley 13 Jul 1982-1; Campbell River 6 Jun 1980-33; Alta Lake 4 Aug to 2 Sep 1941-10 daily; Port Neville 5 Jun 1975-1; Sointula 26 Jul 1976-1 fledgling being fed by adults; Kitimat 24 Aug 1975-2; Lakelse Lake 30 Jun 1974-3, 17 Aug 1977-15, at Park; Nass River (w New Aiyansh) 22 Aug 1977-1. *D. c. coronata:* Alta Lake 6 Jul 1946-1, very common in migration; Terrace 2 Aug 1987-1; New Hazelton 28 Aug 1917-1 (NMC 10967). *Undetermined subspecies:* Parksville 23 Jun 1961-eggs, 4 Jul 1961-nestlings, 6 Jul 1961-nestlings; Mitlenatch Island 12 Jul 1964-2 nestlings; Mayer Lake 8 Aug 1985-1; Lakelse Lake 10 Jul 1974-13; Terrace 15 Jul 1989-5 fledglings. **Interior** – *D. c. auduboni:* Manning Park 12 Jun 1975-106, 6 Aug 1962-3 nestlings being fed by adults; Lightning Lake 26 Aug 1973-40; Manning Park 15 Jul 1968-eggs; Creston 30 Jul 1980-2 nestlings, within 2 or 3 days of fledging; Bull River 13 Jun 1960-4 young, fledged when nest checked; Brookmere 19 Jun 1974-22; Wasa Park 24 and 25 Aug 1971-25 (Dawe 1971); Sorrento 12 Aug 1972-20; Beaverfoot River 31 Aug 1975-100 moving s; Glacier National Park 18 to 21 Jun 1982-17; Lake O'Hara 23 Aug 1975-50; 100 Mile House 10 Jun 1976-3 nestlings; w Riske Creek 15 Jul 1971-3 young; Dog Creek (Alkali Lake [Cariboo]) 13 Jun 1959-5 eggs, incubation started; Anahim Lake 25 Jun 1961-8; Glathelli Lake 10 Aug 1975-6; Prince George 27 Aug 1984-25 with robins and juncos; Cluculz Lake 13 Jun 1970-2 eggs; Pinkut Creek to Burns Lake 11 Jun 1975-19; Summit Lake (Prince George) 3 Jun 1947-nestlings, 30 Jun 1944-nestlings; Adoogacho Creek 15 Jul 1975-1; Chilkat Pass 26 Aug 1974-1. *D. c. coronata:* Windy Joe Mountain 10 Jul 1966-1, at beaver pond; Osoyoos Lake 23 Jun 1957-2; Edgewood 18 Jun 1922-nest collected; Wasa Park 25 Aug 1971-2, only 2 seen in park (Dawe 1971); Nakusp 2 May 1993-1; Celista 3 Aug 1960-2; Scotch Creek 14 Aug 1963-6; Kootenay National Park 4 Jun 1983-1; Succour Summit 1 Jun 1996-1 male; Stum Lake 19 Jul 1971-1 nestling, 27 Jul 1973-4 nestlings (Ryder 1973); Bowron Lake Park 3 Jul 1971-1 (Runyan 1971); Prince George 31 Jul 1982-1, 5 Aug 1982-1; Pine Pass 6 Jun 1976-1, at rest area; 19 km n Tupper Creek 23 Jun 1939-adults feeding young recently out of nest (Cowan 1939); Beryl Prairie 8 Jul 1979-1; Taylor 5 Jun 1982-pair nearly completed nest, 18 Jul 1982-1 fledgling being fed; Beatton Park 12 Jul 1983-11, 2 pairs feeding cowbirds, 12 Aug 1986-42, 14 Aug 1986-25, 27 Aug 1982-40; Charlie Lake 24 Aug 1979-18; North Pine 22 Aug 1984-15; Stoddart Creek (Fort St. John) 30 Aug 1985-30; St. John Creek 31 Aug 1985-29; Spatsizi River 15 Jul 1977-1; Buckinghorse Park 18 Jul 1987-male with fledgling; Tatlatui Lake 14, 15, and 21 Jun 1986-1; Eddontenajon Lake 8 Jun 1976-2; Parker Lake 8 Jul 1978-1; Fort Nelson 12 Jun 1982-10, most common warbler in area at this time; 39 km s and 4 km w Joe Irwin Lake 13 Jun 1988-2, several others singing but not seen; n Boya Lake Park 12 Jul 1978-female with 2 recently fledged young; Lower Liard Crossing 22 Aug 1943-10 (Rand 1944); Atlin 15 Jun 1926-5 eggs, 2 Jul 1980-3, 23 Aug 1934-1 (CAS 42142); Surprise Lake 7 Jun 1975-4. *Undetermined subspecies:* Manning Park 6 Aug 1962-adults feeding 3 nestlings; Ashnola River near Keremeos 28 Jul 1968-4 nestlings; Trail 3 Aug 1982-1 nestling; East Trail 20 Aug 1978-25; Waneta Junction 23 Aug 1980-30; Apex Mountain (Penticton) 12 Aug 1978-50 to 75 birds; Barrett Creek 16 Jun 1984-14; Chute Lake 20 Jul 1975-2 fledglings; Kelowna 18 Aug 1973-50; Wilmer 18 Aug 1977-20; Mount Tom (Gang Ranch) 23 Jul 1993-15, all young of the year; Crowfoot Mountain 24 Aug 1996-100, everywhere through subalpine meadows; Emerald Lake 5 Jun 1976-30 (Wade 1977); Williams Lake 16 Aug 1959-100, stayed to 30 Aug; Nimpo Lake 13 Jul 1971-2 nestlings; Tweedsmuir Park 14 Aug 1982-30, in the Rainbow Range; Prince George 11 Jun 1969-16, 30 Jun 1991-both adults feeding young,

10 Jul 1981-3 fledglings; Fort St. James 1 Jun 1889-4 eggs, 14 Jun 1889-1 egg (MacFarlane and Mair 1908); Summit Lake (Prince George) 25 Jul 1944-adults feeding nestlings; sw Dawson Creek 8 Jun 1993-2 eggs plus 1 Brown-headed Cowbird egg, 17 Jun 1994-2 newly hatched nestlings, 24 Jun 1994-5 eggs, female flushed (Phinney 1998); Beatton Park 23 Jun 1985-6; Charlie Lake 20 Aug 1976-15, migrants, 29 Aug 1978-30; Mason Lake 28 Aug 1982-10; Kinaskan Lake 27 Aug 1979-10; Gladys Lake 22 Jul 1976-10; Fern Lake 13 Aug 1983-10, 18 Aug 1983-20, several groups 4-6 each (Cooper and Cooper 1983); 8 km n Mile 304 Alaska Highway 13 Aug 1985-10; 1.5 km e Fort Nelson 28 Jul 1986-2 recently fledged young; Haines Highway 4 Jun 1983-38, an adult female carrying food, probably for young; Towagh Creek 7 Jun 1983-38; Helmut (Kwokullie Lake) 8 Jun 1982-female building nest; Atlin region 28 Jun 1924-newly hatched nestlings (Swarth 1926); s Lower Post 23 Jun 1983-4 nestlings, when checked all left nest, 24 Jun 1983-5 eggs.

Breeding Bird Surveys: Coastal – Recorded from 23 of 27 routes and on 65% of all surveys. Maxima: Campbell River 30 Jun 1984-40; Alberni 21 Jun 1974-36; Kispiox 20 Jun 1993-35. **Interior** – Recorded from 71 of 73 routes and on 94% of all surveys. Maxima: Christian Valley 26 Jun 1993-73; Adams Lake 28 Jun 1989-59; Ferndale 30 Jun 1990-57.

Autumn: Interior – *D. c. auduboni:* Liard Hot Springs Park 6 Sep 1974-7 in a flock of 45 mixed warblers; Willow River 21 Sep 1964-20; Williams Lake 13 Sep 1982-12, 3 and 4 Oct 1982-12, 18 Oct 1980-1; Canim Lake 5 Sep 1960-18; Bridge Lake 7 Sep 1960-45; Revelstoke 24 Sep 1981-20, 7 Oct 1984-2; Shuswap Lake 6 Sep 1959-50; Okanagan Landing 5 Sep 1929-100, 24 Oct 1944-1, last of year, 8 Nov 1927-1; Monk Park 2 Sep 1995-150+, 7 Sep 1971-45; Kinnaird Park 7 Oct 1969-2, 17 Oct 1969-2, 5 Nov 1969-2; Glade 15 Sep 1968; Vaseux Lake area 1 Sep 1973-14; Osoyoos 10 Oct 1975-4; Richter Pass 21 Sep 1968-19 along Kilpoola Lake Rd; Oliver 10 Sep 1960-80; Manning Park 26 Sep 1970-2. *D. c. coronata:* Atlin 14 Sep 1972-50, 5 Oct 1931-1 last seen (Swarth 1936); Warm Bay Hot Springs 16 Sep 1972-3; Charlie Lake Park 3 Oct 1982-4, last record for autumn; Fort St. John 5 Oct 1986-1, last record for autumn; St. John Creek 12 Sep 1986-45; Stoddart Creek (Fort St. John) 6 Sep 1986-69, highest count for single locality this autumn; Bear Flat 8 Sep 1986-42; Fort St. John 6 Sep 1986-33, 23 Sep 1984-2, 5 Oct 1986-1; Smithers 12 Nov 1985-1; Barkerville 16 Sep 1962-2; Field 9 Oct 1976-2, first 2 weeks of Sep heaviest movement, waves through the area 19 to 24 Sep; Nicholson 15 Sep 1975-1; Enderby 5 Nov 1941-1 (UBC 8003); Nakusp 22 Nov 1986-1; Vernon 3 Oct 1965-3. *Undetermined subspecies:* Atlin 19 Sep 1924-1, last seen (Swarth 1926); Fort Nelson 16 Sep 1986-90 (McEwen and Johnston 1987a); Andy Bailey Lake 22 Sep 1986-3; Williams Lake 14 Oct 1986-40, flocks moving through, 19 Nov 1973-1; Riske Creek 5 Sep 1978-30; Lac la Hache 8 Sep 1942-50 (Munro 1945a); Farwell Canyon 25 Sep 1981-30; Fletcher Lake 22 Sep 1982-50, conspicuous movement; Horse Lake (100 Mile House) 21 Oct 1984-2; lower Blaeberry River valley 18 Sep 1995-111, highest daily autumn migration count, 16 Sep 1997-84+, second highest daily autumn migration count; Golden 25 Oct 1975-1, late sighting; Vernon 28 Nov 1991-1 (Siddle 1992a); Okanagan Lake 3 Oct 1973-55, mainly young of the year; Penticton 21 Sep 1977-50; Vaseux Lake 1 Oct 1975-40; West Bench 7 Oct 1975-6; Columbia Gardens (Trail) 11 Sep 1984-40; Pend-d'Oreille River valley 30 Sep 1982-50; Erie Lake 23 Sep 1978-30; East Trail 12 Sep 1984-50; Oliver 7 Sep 1964-133, on Camp McKinney Rd; Glacier Lake (Cathedral Park) 1 Sep 1980-250; Bridesville 22 Sep 1996-11 dead on highway. **Coastal** – *D. c. auduboni:* Port Neville 10 Sep 1975-8; Egmont 5 Sep 1977-2; Strawberry Island 26 Nov 1972-1; Ross Lake 30 Sep 1971-15; Aldergrove 6 Sep 1967-26; Burnaby Flats 16 Nov 1974-67; Ladner 27 Sep 1981-40, at Harbour Park; Yellowpoint 30 Sep 1987-27, in flock; Cowichan Bay estuary 7 Oct 1987-60; Pachena Point 13 and 15 Nov 1974-1 (Hatler et al. 1978); Island View Beach 15 Sep 1980-53; Saanich 13 Sep 1985-55, just s McIntyre Reservoir; Port Renfrew 5 Oct 1974-35; Victoria 9 Sep 1985-15; Beacon Hill Park 21 Sep 1984-60, 25 Sep 1973-200, 29 Sep 1973-80, 3 Oct 1966-150, 7 Oct 1973-100, 10 Oct 1975-300, 21 Oct 1980-45. *D. c. coronata:* Kitimat 26 Sep 1974-1, at Alcan Smelter site; Cape St. James 26 Sep 1977-3, 16 Oct 1978-3, 29 Oct 1978-1; Quatsino 7 Oct 1935-1; Burnaby Flats 16 Nov 1974-3; Reifel Island 30 Sep 1974-10; Tofino 14 Nov 1982-1; Stubbs Island 1 Nov 1982-10; Long Beach 29 Sep 1984-10, first seen this month, 29 Oct 1982-1; Central Saanich 13 Sep 1985-44, s McIntyre Reservoir, 20 Oct 1986-15; Victoria 3-4 Nov 1983-3; Beacon Hill Park 3 Oct 1966-20 (Crowell and Nehls 1967a). *Undetermined subspecies:* Kitimat River 17 Sep 1974-25, below bridge; Swanson Bay 5 Oct 1935-1 (MCZ 284249); Sandspit 5 Nov 1986-2; Baronet Passage 5 Sep 1986-15; Campbell River 1 Oct 1983-4 (Dawe et al. 1995a); Alta Lake 1 Sep 1941-10; Garibaldi Lake 17 Sep 1982-20, most immature; Baynes Sound nr Courtenay 19 Sep 1981-28 (Dawe et al. 1998); Rolly Lake 30 Nov 1974-6; Debouville Slough 19 Sep 1982-120; North Vancouver 26 Sep 1974-150, at Cates Park; West Vancouver 9 Sep 1977-50, at Whitecliff Park; Burnaby Lake 18 Sep 1983-80; Reifel Island 7 Sep 1983-19, 19 Sep 1971-100, maximum fall numbers, 23 Sep 1983-100, 5 Oct 1983-65; Stubbs Island 14 Oct 1983-30; Beacon Hill Park 6 Oct 1981-100, 14 Oct 1975-150, mostly immatures, 30 Oct 1974-150, mostly immatures; Swan Lake (Saanich) 22 Sep 1985-70; Rocky Point 2 Sep 1995-34, 29 Sep 1995-380, 1 Oct 1995-162; Witty's Lagoon 20 Sep 1958-100, migratory flock.

Winter: Interior – *D. c. auduboni:* Riske Creek 14 Dec 1983-1, at feeder since 19 Oct, last day seen at Wineglass Ranch; Windermere 11 Dec 1982-1, into feeder every day to 7 Jan; Kelowna 31 Dec 1992-1 (Siddle 1992b); Nelson 15 Dec 1993-4 with 1 "Myrtle" subspecies in apple tree, 15 Jan 1994-1; Creston 8 Dec 1993-1. *D. c. coronata:* Nelson 15 Dec 1993-1 with 4 "Audubon's" subspecies in apple tree. *Undetermined subspecies:* Revelstoke 4 Feb 1993-1 (Siddle 1993b); Tranquille 1 Jan to 13 Feb 1994-1 surviving without bird feeders; Vernon 6 Feb 1994-1 (Siddle 1994b); Nakusp 10 Jan 1993-1 (Siddle 1993b), 7 Feb 1994-1 (Siddle 1994b); Creston Dec 1989-5 (Siddle 1990a). **Coastal** – *D. c. auduboni:* Pitt Meadows North 13 Jan 1973-10; Iona Island 11 Jan 1981-12; Westham Island 23 Dec 1972-15; Swan Lake (Saanich) 8 Dec 1986-10. *D. c. coronata:* Masset 3 Jan 1942-1 (Munro and Cowan 1947); Pitt Lake 13 Feb 1984-4; Ambleside Park (West Vancouver) 5 Jan 1975-1, 26 Dec 1974-1 adult male; Stubbs Island 27 Feb 1983-20, 26 Dec 1982-20; Cowichan Bay 23 Dec 1978-2; Saanich 5 Dec 1981-4. *Undetermined subspecies:* Queen Charlotte City 20 Feb 1993-1 (Siddle 1993b); Agassiz 31 Dec 1974-6; Stubbs Island 14 Jan 1983-15, 4 Feb 1983-21, 3 Dec 1987-30 among California wax-myrtle; Reifel Island 1 Jan 1970-4; Duncan 1 Jan 1980-26, near sewage lagoons; Swan Lake (Saanich) 16 Jan 1983-35, 23 Jan 1983-150, 22 Feb 1983-25, 6 Dec 1986-15, 16 Dec 1980-25.

Christmas Bird Counts: Interior – Recorded from 8 of 27 localities and on 5% of all counts. Maxima: Vernon 17 Dec 1989-2; Nakusp 2 Jan 1994-1; Lake Windermere 27 Dec 1981-1; Oliver-Osoyoos 28 Dec 1981-1; Penticton 27 Dec 1983-1; Revelstoke 18 Dec 1993-1; Shuswap Lake Park 21 Dec 1991-1; Vaseux Lake 2 Jan 1994-1. **Coastal** – Recorded from 18 of 33 localities and on 27% of all counts. Maxima: Vancouver 17 Dec 1978-**90**, all-time Canadian high count (Anderson 1979); Ladner 27 Dec 1992-39; Duncan 15 Dec 1990-34.

Black-throated Gray Warbler BTGW
Dendroica nigrescens (Townsend)

RANGE: Breeds from southwestern British Columbia, including Vancouver Island, south through western Washington, western and central Oregon, and California to northern Baja California; east to southern Idaho, Utah, southern and central Wyoming, northwestern and central Colorado; from Oregon south through California, Nevada, western Texas, New Mexico, and southeastern Arizona to northeastern Sonora, Mexico. Winters primarily from Baja California south to central Oaxaca, Mexico.

STATUS: On the coast, a *fairly common* migrant and summer visitant and *casual* in winter in the Georgia Depression Ecoprovince; in the Coast and Mountains Ecoprovince, *uncommon* on the Southern Mainland Coast, locally *rare* on Western Vancouver Island and the Northern Mainland Coast, and absent from the Queen Charlotte Islands.

In the interior, locally *very rare* in the Southern Interior and Central Interior ecoprovinces.

Breeds.

CHANGE IN STATUS: During the past 30 years or so, the Black-throated Gray Warbler (Fig. 71) has become established along the southeast coast of Vancouver Island.

The species was first recorded on Vancouver Island at Comox, where specimens were taken by R.M. Stewart in 1927 and 1929. Unfortunately these remained in his private collection and were unknown to Munro and Cowan (1947). These were the first substantiated records of the species off the mainland.

Neither Fannin (1891, 1898) nor Kermode (1904) included the Black-throated Gray Warbler as being present on Vancouver Island. Swarth and associates, who spent the summer of 1910 collecting birds and mammals on eastern Vancouver Island, did not encounter the species (Swarth 1912). Rhoads (1893) stated that he occasionally heard the song of this species when he was at Victoria and Goldstream in May 1892, but he saw none. Brooks and Swarth (1925) found references by experienced observers to 3 other early observations of this species on the island, at Wellington in 1895 and Cowichan and Nanaimo in 1904, but they cite no specimen records.

Figure 71. Autumn-plumaged Black-throated Gray Warbler. This species reaches the northern limit of its western North American breeding range in British Columbia (Victoria, late August 1989; Tim Zurowski).

On the basis of the available evidence, we suggest that the species was scarce and of irregular occurrence on Vancouver Island during the earlier period of ornithological exploration.

Figure 72. Near the northern limits of its range in British Columbia, the Black-throated Gray Warbler frequents mixed woodlands of black cottonwood, mountain ash, and red alder, with a scattering of Douglas-fir. A thick understorey of wild rose and thimbleberry is often present (Hagensborg, 12 June 1996; Neil K. Dawe).

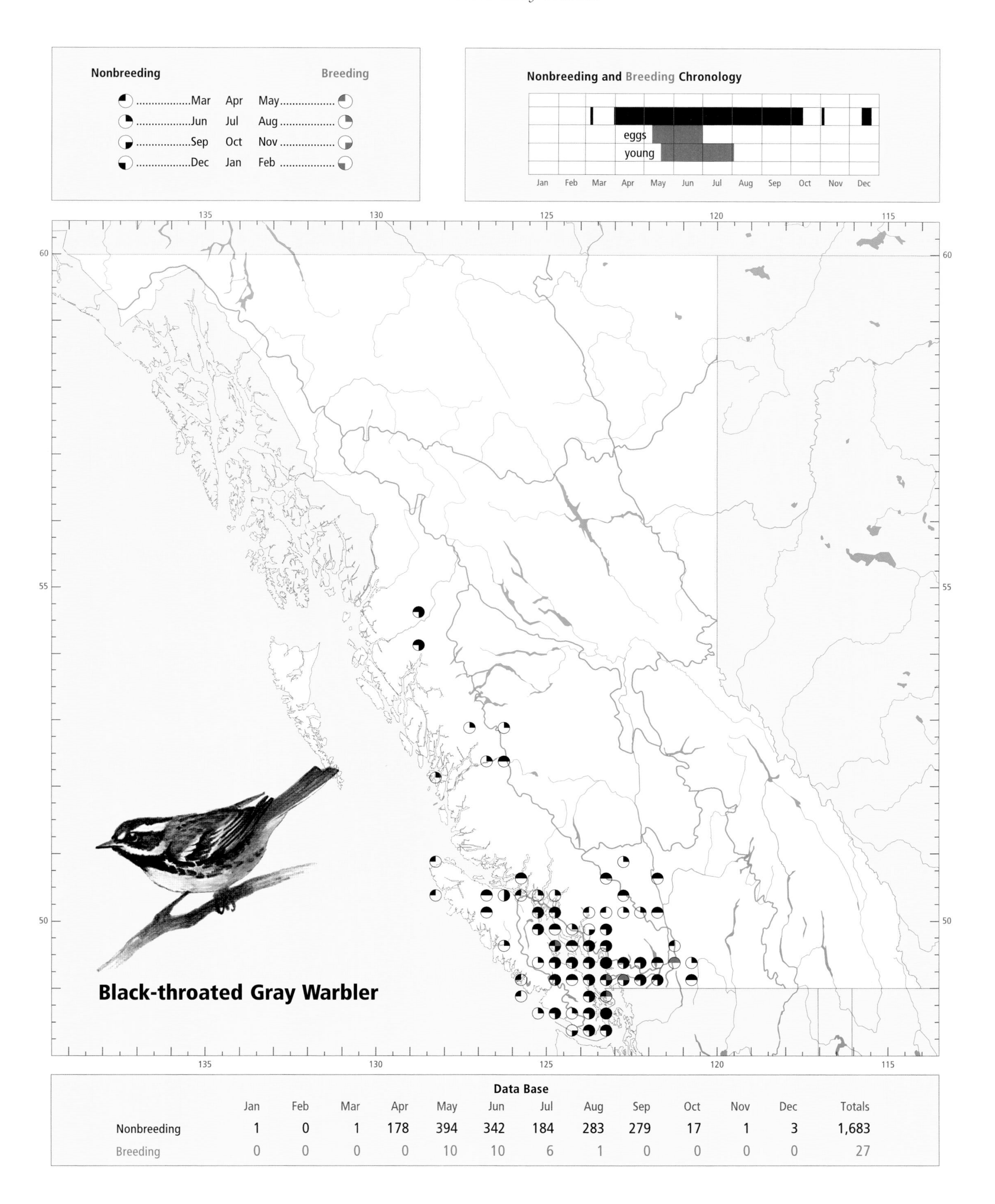

Data Base

	Jan	Feb	Mar	Apr	May	Jun	Jul	Aug	Sep	Oct	Nov	Dec	Totals
Nonbreeding	1	0	1	178	394	342	184	283	279	17	1	3	1,683
Breeding	0	0	0	0	10	10	6	1	0	0	0	0	27

Figure 73. On occasion, the Black-throated Gray Warbler visits bogs with stunted western redcedar, shore pine, Labrador tea, bog laurel, and *Myrica gale* in northern coastal areas of British Columbia (McLoughlin Bay near Bella Bella, 12 June 1996; Neil K. Dawe).

After Stewart's record of 1929, the Black-throated Gray Warbler was next observed on Denman Island in 1930. Subsequently, single birds were recorded at Victoria in 1946, 1951, 1958, and 1961, and on Quadra Island in 1961. This warbler was not seen, however, during extensive summer field work in the Englishman River valley and Parksville areas between 1960 and 1963. In 1964, 2 were observed at Victoria during the autumn migration. There were records from Mitlenatch Island and Departure Bay among the 7 records in 1965. From 1967 the species occurred regularly, with several records almost every year from a steadily increasing number of localities, all of them along the east side of Vancouver Island between Victoria in the south and Cortes Island in the north. Actual numbers of records by year were: 1969-5 records, 1970-7, 1971-5, 1972-6, 1973-18, 1974-34, and 1975-39.

The 5 specimens from Comox in 1927 and 1929 were taken in May, June, and August. Between 1927 and 1970, 28 of 49 records were from the months of August and September, 14 were from June and July, and only 7 were from April and May. Apparently post-breeding dispersal was a major reason that Black-throated Gray Warblers reached Vancouver Island during the early years of the invasion.

The first indication of nesting on Vancouver Island was in 1967, when 2 fledglings with 2 adults were recorded at Miracle Beach (Stirling 1972). Since then, there has been 1 certain record of breeding, a nest with 3 eggs at Courtenay in 1979. A single fledgling was reported at Errington, also in 1979.

NONBREEDING: The Black-throated Gray Warbler has a limited distribution in British Columbia. It occurs primarily in the south coastal region, where it is widely distributed on southeastern Vancouver Island, on the Gulf Islands, and on the mainland close to the coast bordering the Strait of Georgia. Along the Northern Mainland Coast, the species is distributed locally, often near the head of long inlets, with scattered records north as far as Kitimat and Terrace. In the Bella Coola valley, Laing (1942) found it "by no means rare" between Stuie and Hagensborg. In the interior, the easternmost occurrence has been in Manning Park. Other bounding records are from the valley of the middle Fraser River at Lillooet and Pavilion.

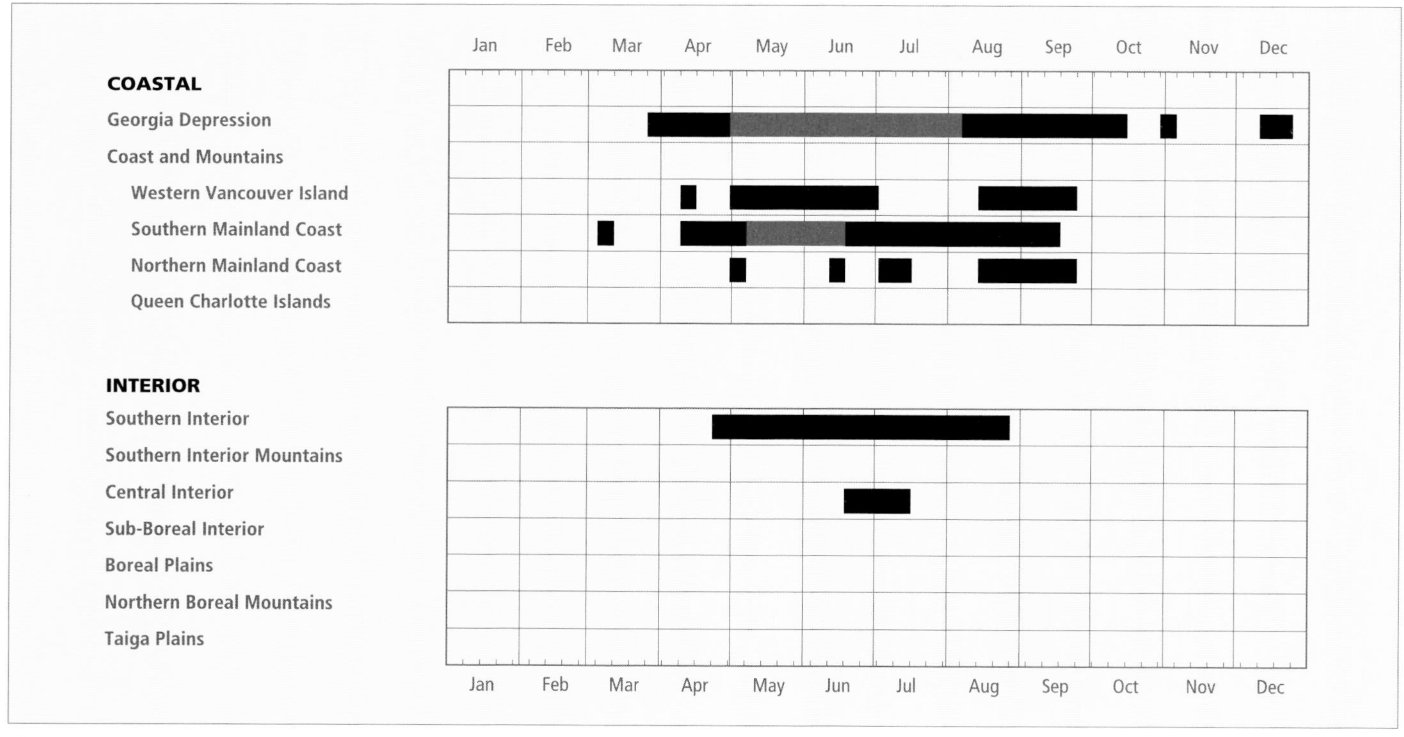

Figure 74. Annual occurrence (black) and breeding chronology (red) for the Black-throated Gray Warbler in ecoprovinces of British Columbia. Records are shown for the week in which they occurred.

The Black-throated Gray Warbler has been recorded from near sea level to 300 m elevation on the coast, and up to 700 m inland, in the vicinity of Alta Lake and Pemberton. It occurs widely in mixed forests and deciduous woodlands at lower elevations, especially along southeastern Vancouver Island and in the Fraser Lowland. Specific habitats chosen there include Douglas-fir–arbutus–Garry oak forests along rocky shorelines; mixed old second-growth forests of Douglas-fir, bigleaf maple, and red alder, with an understorey dominated by ninebark, ocean-spray, and choke cherry; or western hemlock and western redcedar in association with red alder, vine maple, western flowering dogwood, and cascara. Near Bella Coola, this warbler frequents associations of black cottonwood–mountain-ash–red alder with scattered Douglas-firs and western redcedars and an edge understorey of wild rose and thimbleberry (Campbell and Dawe 1992; Fig. 72). It is most frequently seen in forest-edge situations, often near water such as stream sides, lake margins, sewage lagoons, and rocky ocean shores, but also along transmission line rights-of-way, edges of golf courses, wooded gardens, and occasionally bogs (Fig. 73).

On the south coast, at Vancouver and Victoria, spring migrant Black-throated Gray Warblers may arrive during the first week of April, but the height of the migration occurs later in the month and through early May (Figs. 74 and 75). There is only a small decrease in numbers after the migration is over, suggesting that it is largely a movement of birds into the area, with few proceeding to summer habitats along the north coast. The first migrants to reach the north end of Vancouver Island have been recorded between 15 April (Cape Scott) and 16 May (Grant Bay). The earliest of the small number of records from the Northern Mainland Coast was 6 May. In the interior, the earliest of the few spring dates was about a month behind those at the coast, on 29 April.

The onset of the southbound migration is difficult to detect. In the Georgia Depression, a sharp decline in numbers seems to occur in mid-July with the end of the song period, but numbers begin to increase in the first week of August and the migration is at its peak in September (Fig. 75). By mid-October only a few stragglers remain. On the Southern Mainland Coast, the autumn migration appears to take place between mid-August and mid-September.

In winter, the Black-throated Gray Warbler has been reported only from southern Vancouver Island and North Vancouver on the south coast.

On the coast, the Black-throated Gray Warbler has been recorded fairly regularly from 1 April to 12 October; in the interior, it has been recorded from 29 April to 22 August (Fig. 74).

BREEDING: The Black-throated Gray Warbler breeds throughout most of its range in British Columbia; however, we have confirmed nesting from only 7 locations: Saltspring Island, Vancouver, Burnaby, Surrey, Langley, Hope, and Courtenay.

The Black-throated Gray Warbler reaches its highest numbers in summer in the Georgia Depression (Fig. 76). An analysis of Breeding Bird Surveys in British Columbia for

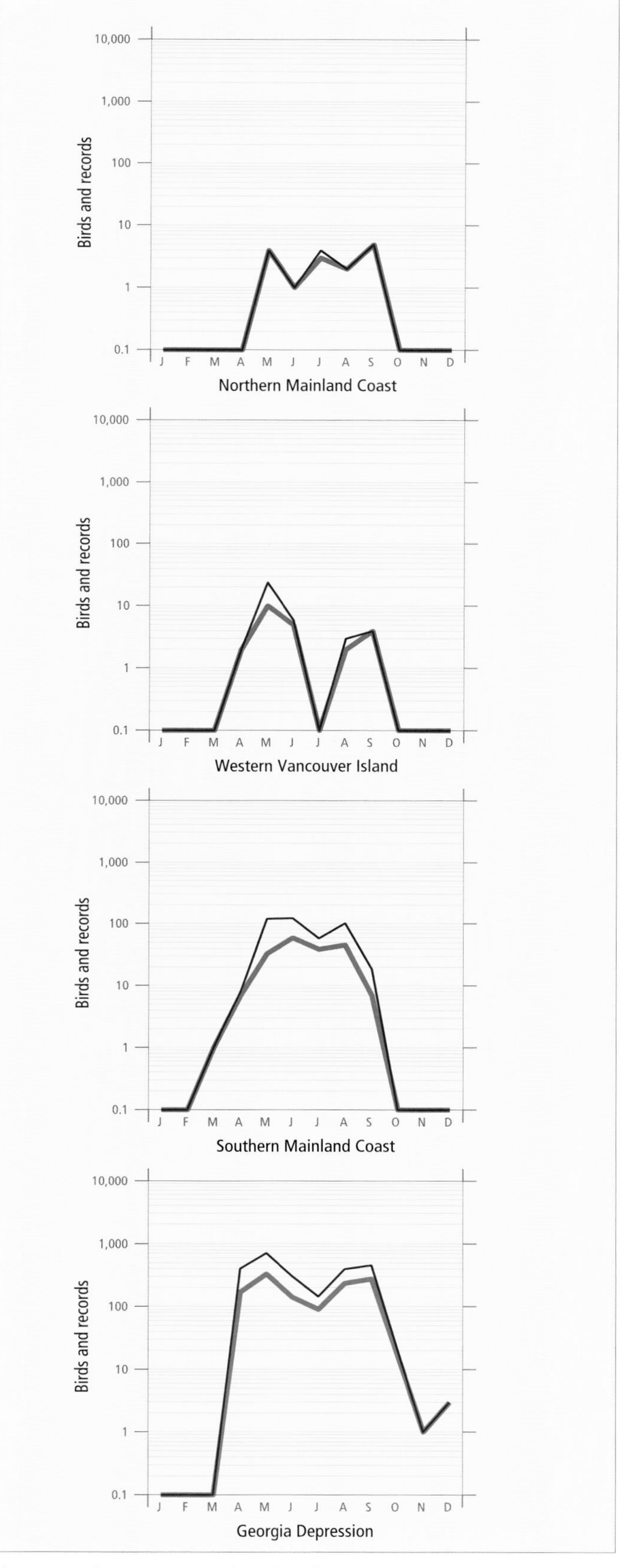

Figure 75. Fluctuations in total number of birds (purple line) and total number of records (green line) for the Black-throated Gray Warbler in ecoprovinces of British Columbia. Breeding Bird Surveys and nest record data have been excluded.

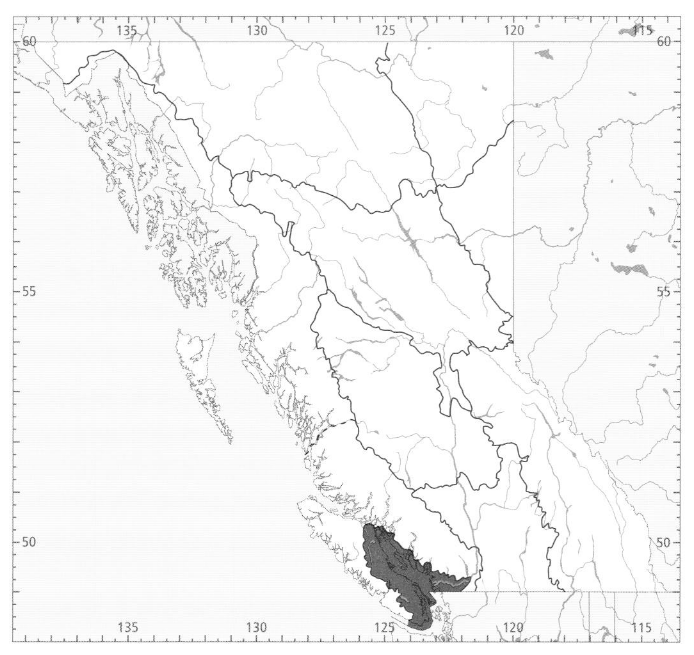

Figure 76. In British Columbia, the highest numbers for the Black-throated Gray Warbler in summer occur in the Georgia Depression Ecoprovince.

the period 1968 through 1993 could not detect a net change in numbers on coastal routes. Throughout its range in western North America, this species showed an average annual rate of increase of 2.2% ($P < 0.05$) between 1966 and 1996 (Sauer et al. 1997).

This warbler breeds at relatively low elevations, and almost all records are from near sea level to 700 m. There are 2 summer records from Manning Park from "lodgepole pine forests at 4,000 feet to the lower edge of alpine meadows at 5,000 feet near Perdue Lake" (Carl et al. 1952). There is little information on the nesting habitat or the nest sites of the Black-throated Gray Warbler in British Columbia. In a study of avian communities in old-growth and managed coniferous forests of Western Vancouver Island, Bryant et al. (1993) found that

Figure 77. On southern Vancouver Island, the Black-throated Gray Warbler breeds in mixed coniferous-deciduous woodlands of Douglas-fir, western redcedar, western hemlock, bigleaf maple, and red alder (near Victoria, 20 July 1997; R. Wayne Campbell).

Figure 78. An important element of the breeding habitat of the Black-throated Gray Warbler in southwestern British Columbia is relatively open but brushy undergrowth (near Victoria, 20 July 1997; R. Wayne Campbell).

in summer this warbler was confined to 50- to 60-year-old stands, rather than old-growth or younger second-growth. In the Georgia Depression, most nesting habitats were recorded as forest. Of these at least 3 were human-influenced; one other was described as wetland, and it too was certainly vegetated with trees or large shrubs. In general, the nesting habitat falls within the categories of vegetation detailed under NON-BREEDING (Figs. 77 and 78).

On the coast, the Black-throated Gray Warbler has been recorded breeding from 6 May (calculated) to 2 August (Fig. 74).

Nests: The Black-throated Gray Warbler uses a variety of trees and tall shrubs for nest sites. Most nests ($n = 8$) were in trees, both coniferous (usually Douglas-fir) and deciduous species (willow, birch, and choke cherry), and 2 were in shrubs. Nests were attached to a forked branch (4), on a branch close to the trunk (2), or secured among the twigs of a branch some distance from the tree trunk (2). From available data, the nesting sites in British Columbia appear to differ from those in the main range of the species in the Pacific states. In Washington they were mainly in Douglas-fir, but in Oregon and California nesting habitat has been described as more varied and included a variety of oaks and chaparral (Small 1994).

Nests were cups woven of grasses, plant fibres and stems, twigs, rootlets, and mosses; all were lined with fine grasses, plant down, and spider webs. Descriptions of the few nests reported in British Columbia do not mention the use of feathers in the lining, a characteristic feature of the species in other areas (Harrison 1984; Bent 1953). The heights of 13 nests ranged from 1.4 to 9 m, with 7 nests between 3 and 5 m. Rathburn (*in* Bent 1953) recorded nests in Washington as high as 12 m.

Eggs: Dates for 7 clutches ranged from 10 May to 30 June. Calculated dates indicate nests may contain eggs as early as 6 May. Clutch sizes ranged from 2 to 5 eggs (2E-1, 3E-2, 4E-2, 5E-2), with 4 having 4 or 5 eggs. The incubation period is unknown (Guzy and Lowther 1997).

Young: Dates for 18 broods ranged from 19 May to 2 August, with 72% recorded between 7 June and 10 July. Sizes of 16 broods ranged from 3 to 5 young (3Y-7, 4Y-7, 5Y-2), with 14 having 3 or 4 young. The nestling period is unknown (Guzy and Lowther 1997).

Brown-headed Cowbird Parasitism: In British Columbia, 1 of 13 nests found with eggs or young was parasitized by the cowbird. In addition, there were 6 instances of fledgling cowbirds being fed by adult Black-throated Gray Warblers. Friedmann and Kiff (1985) found the species to be a regular and not infrequent host of the Brown-headed Cowbird.

Nest Success: Only 1 nest found with eggs was followed to a known fate. It was unsuccessful.

REMARKS: The Black-throated Gray Warbler is regarded as a member of the *Dendroica virens* superspecies group, which includes the Black-throated Green Warbler, Townsend's Warbler, Hermit Warbler, and Golden-cheeked Warbler. The Townsend's and Black-throated Gray warblers are believed to be western species derived from the widely distributed eastern species, the Black-throated Green Warbler, during the prolonged separation resulting from Pleistocene glaciation.

There is a Breeding Bird Survey report of a single Black-throated Gray Warbler at Kuskonook, in the east Kootenay. This observation is far out of the normal range of the species and is not supported by adequate evidence. Reports of this warbler from routes at Princeton and Osprey Lake also lack supporting evidence. These records are not included in this account.

A Christmas Bird Count report of a Black-throated Gray Warbler at White Rock (Cannings 1987) has also been excluded for lack of adequate documentation.

The Black-throated Gray Warbler is the least known of the regularly occurring wood-warblers of British Columbia, and there is a need to know much more precisely its habitat requirements for successful nesting in the province. It is a species of the open coastal mixed forests, an ecosystem that has been subject to great changes that are likely to continue. So far the species appears to have maintained its numbers through the changes, but its ecological requirements and limits of tolerance are unknown.

Morse (1989) and Guzy and Lowther (1997) provide a useful summary of the natural history of the Black-throated Gray Warbler in North America.

NOTEWORTHY RECORDS

Spring: Coastal – Saanich 1 Apr 1983-2, 26 Apr 1987-10; Mayne Island 28 Apr 1974-10; Port Alberni 28 May 1991-24; Mission 12 May 1973-15; Fleetwood (Surrey) 3 Apr 1960-2, 10 May 1960-5 eggs; Surrey 31 May 1962-4 nestlings about 2 days old; Vancouver 28 Apr 1993-75 (Siddle 1993c); Pitt Meadows 26 Apr 1981-10; Tofino 3 May 1974-10; Black Mountain (North Vancouver) 18 May 1974-23; Mount Seymour Park 22 May 1983-8; Gibsons Landing 3 May 1959-12; Comox 14 May 1927-1; Grant Bay 16 May 1969-1 (Richardson 1971); Cape Scott Park 15 Apr 1980-1; Stuie 24 to 27 May 1933-3 (Dickinson 1953; MCZ 284348 to 284350). **Interior** – Manning Park 18 May 1997-1, 31 May 1982-2; Lillooet 7 May 1916-1; Stein River 29 Apr 1978-1.

Summer: Coastal – Sooke 29 Jun 1986-3 fledglings being fed by adults; Victoria 12 Jul 1995-4 fledglings being fed by adults, 11 Aug 1982-10 in Haro woods; Surrey 2 Aug 1961-4 nestlings; Agassiz 6 Jun 1971-12; Ambleside 18 Aug 1974-10 in 1 flock; Carmanah Point 11 Jun 1961-2; Pitt Polder 29 Aug 1976-female feeding a fledgling; Errington 2 Jul 1979-1 fledgling; Halfmoon Bay 16 Jun 1986-3 fledglings being fed by an adult male; Spuzzum 23 Jul 1985-3 fledglings being fed by adults; Gold River 1 Jun 1988-1; Courtenay 30 Jun 1979-3 eggs; Comox 30 Aug 1927-1; Mitlenatch Island 15 Aug 1974-3; Alta Lake 18 Aug 1938-1 (RBCM 8249); Campbell River 19 Jul 1976-3 fledglings being fed by adults; Pemberton 22 Jun 1924-1 (Racey 1948); Stuart Island 25 Jul 1936-2, 6 Aug 1936-1; Fawn Bluff 1 Aug 1938-2; McLoughlin Bay (Bella Bella) 11 Jun 1996-1 singing in bog (Dawe and Buechart 1996; Fig. 73); Bella Coola 29 Jun 1940-1 (FMNH 176239), 1 Jul 1940-1 (FMNH 176240); Hagensborg 26 Jun 1997-3 fledglings being fed by adults, 3 Jul 1938-1 (NMC 28743); Kimsquit 10 Jul 1986-2. **Interior** – Manning Park 22 Aug 1987-1 (Mattocks 1988a); Lillooet 6 Jul 1916-1 (NMC 9774), 2 Aug 1916-1 (NMC 9780); Pavilion 13 Aug 1932-1.

Breeding Bird Surveys: Coastal – Recorded from 18 of 27 routes and on 44% of all surveys. Maxima: Gibsons Landing 23 Jun 1974-31; Squamish 21 Jun 1986-24; Courtenay 18 Jun 1992-19. **Interior** – See REMARKS.

Autumn: Interior – No records. **Coastal** – Terrace 6 Sep 1968-1, 23 Sep 1984-1; Port Neville 18 Sep 1986-1; Hornby Island 7 Sep 1978-1; Port Alberni 3 Nov 1991-1 (Siddle 1992a); Westham Island 3 Oct 1973-1; Victoria 5 Oct 1972 (Tatum 1973); Island View Beach (Central Saanich) 12 Oct 1975-1.

Winter: Interior – No records. **Coastal** – North Vancouver 15 Dec 1987-1, 23 Dec 1987-1; Swan Lake (Saanich) 18 Dec 1994-1 (Bowling 1995b).

Christmas Bird Counts: Coastal – See REMARKS. **Interior** – Not recorded.

Black-throated Green Warbler

Dendroica virens (Gmelin)

BTNW

RANGE: Breeds from northeastern British Columbia, northern Alberta, north-central Saskatchewan, central Manitoba, central Ontario, central Quebec, southern Labrador, and Newfoundland south to central Alberta, central Saskatchewan, southern Manitoba, northern and east-central Minnesota, northern Wisconsin, southern Michigan, Pennsylvania, New Jersey, and southern New England, and south through the Appalachian region through eastern Kentucky, West Virginia, western Maryland, western Virginia, eastern Tennessee, western North Carolina, central Alabama, northern Georgia, and northwestern South Carolina; also in northwestern Arkansas and the coastal plains from southeastern Virginia to eastern South Carolina. Winters mainly from southern Texas and southern Florida south through the Caribbean, southern Mexico, and Central America to Panama; rarely south to northern Colombia.

STATUS: On the coast, *casual* in the Georgia Depression Ecoprovince.

In the interior, *casual* in the Southern Interior Mountains Ecoprovince; *uncommon* migrant and local summer visitant in the Boreal Plains Ecoprovince; *casual* in the Taiga Plains Ecoprovince.

Breeds.

CHANGE IN STATUS: The Black-throated Green Warbler (Fig. 79) was not mentioned, even as a hypothetical species, by Munro and Cowan (1947) in their review of the birds of British Columbia. The first record for the province occurred on 25 June 1965, when a male was collected at Moberly Lake, 15 km north of Chetwynd, by John Hubbard, a graduate student from the University of Michigan (Salt 1966a). Salt also reported that Hubbard heard several singing males in the Chetwynd to Moberly Lake area, which is about 100 km west of the Alberta border. These reports led Godfrey (1966) to speculate that the Black-throated Green Warbler "probably breeds in central-eastern British Columbia." It was still considered accidental in British Columbia as late as 1977 (Campbell 1977). Since then, this warbler has been found regularly in small numbers in the Peace Lowland in northeastern British Columbia, and breeding was first confirmed there in 1981.

NONBREEDING: The Black-throated Green Warbler occurs regularly in British Columbia only in the northeast and almost exclusively in the Peace Lowland of the Boreal Plains. In the northeast, it occurs locally from Hudson's Hope and Chetwynd in the west to Noel Creek in the south, the Alberta border in the east, and the vicinity of Blueberry River in the north. There are only 3 records from the Taiga Plains to the north. Elsewhere, the warbler's occurrence is considered exceptional.

The Black-throated Green Warbler has been reported from a narrow elevational range in northeastern British Columbia, between 650 and 1,100 m. One bird near Revelstoke was recorded at about 1,800 m elevation. On the coast, it has been found near sea level. Nonbreeding habitat in the Boreal Plains is generally similar to breeding habitat (see BREEDING). A bird found near Port Alberni in May was using a mixed riparian stand of red alder and bigleaf maple. During the autumn migration in the Peace Lowland, migrants also use shrubbier habitats such as willow, alder, paper birch, and trembling aspen thickets along the edges of lakes, rivers, and roads.

Figure 79. Autumn-plumaged Black-throated Green Warbler. This species reaches the western limit of its breeding range in North America in northeastern British Columbia (Anthony Mercieca).

Spring migrants begin to arrive in the northeast in the second week of May; most probably arrive by the last week of May and first few days of June (Figs. 80 and 81). Migrants enter British Columbia from the east.

The southward movement likely begins in late July, shortly after nesting is completed, with the main movement in the third and fourth weeks of August. Most birds have left the region by early September. In Alberta, the main autumn movement occurs from late August to early September (Salt 1973).

In northeastern British Columbia, the Black-throated Green Warbler has been recorded from 9 May to 5 September (Fig. 80).

BREEDING: The Black-throated Green Warbler reaches the limits of its western breeding range in northeastern British Columbia (Morse 1993). Breeding populations are small and localized. In Alberta, this warbler breeds in central and northern boreal forests but is scarce even in areas with suitable habitat (Semenchuk 1992), a situation that is apparently similar in British Columbia.

The Black-throated Green Warbler has been recorded breeding in British Columbia only in the Peace Lowland of the Boreal Plains. Records are few, but this local warbler probably breeds in suitable habitat throughout that area. Adults with a fledged young were found in Kiskatinaw Park near

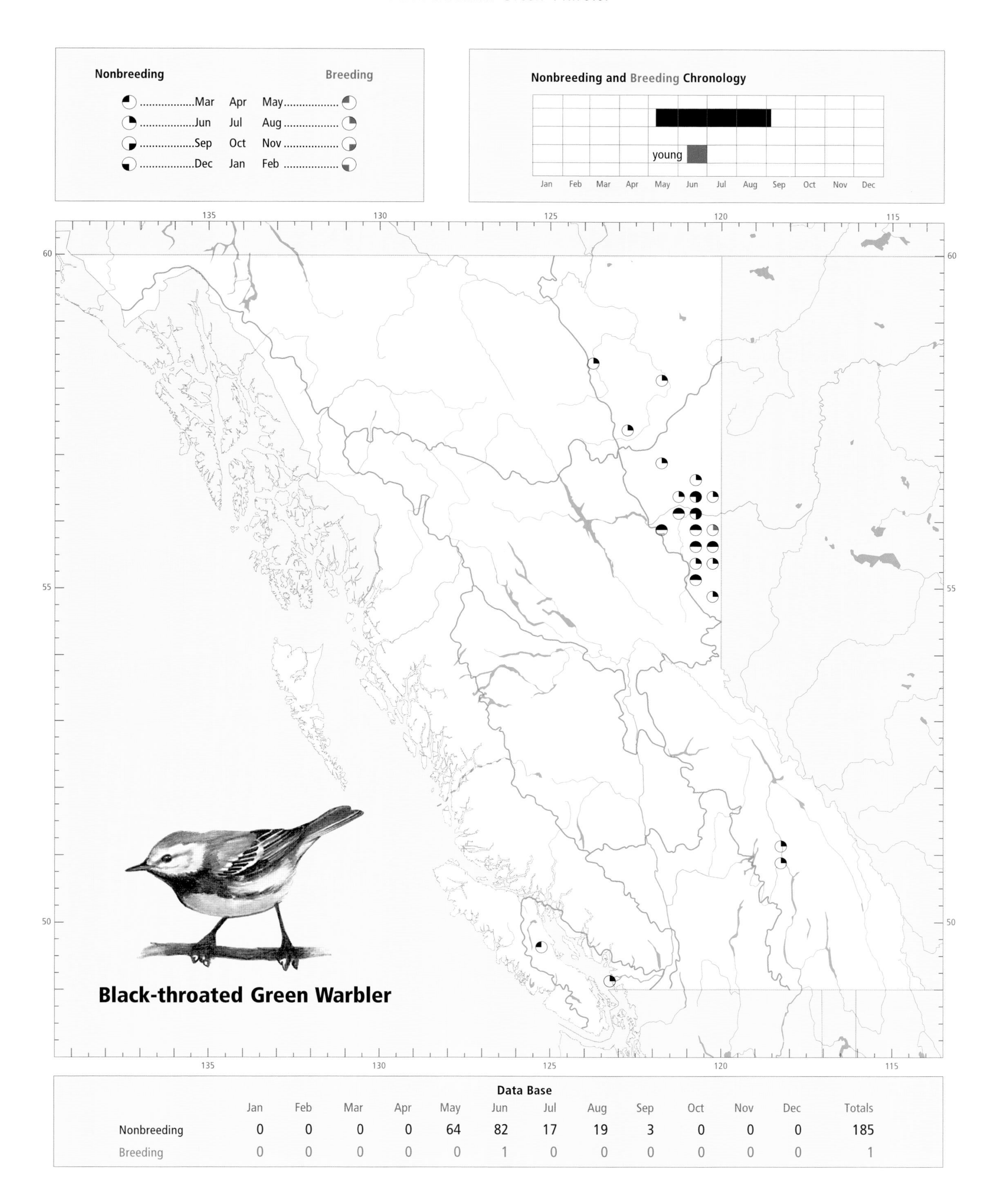

Data Base

	Jan	Feb	Mar	Apr	May	Jun	Jul	Aug	Sep	Oct	Nov	Dec	Totals
Nonbreeding	0	0	0	0	64	82	17	19	3	0	0	0	185
Breeding	0	0	0	0	0	1	0	0	0	0	0	0	1

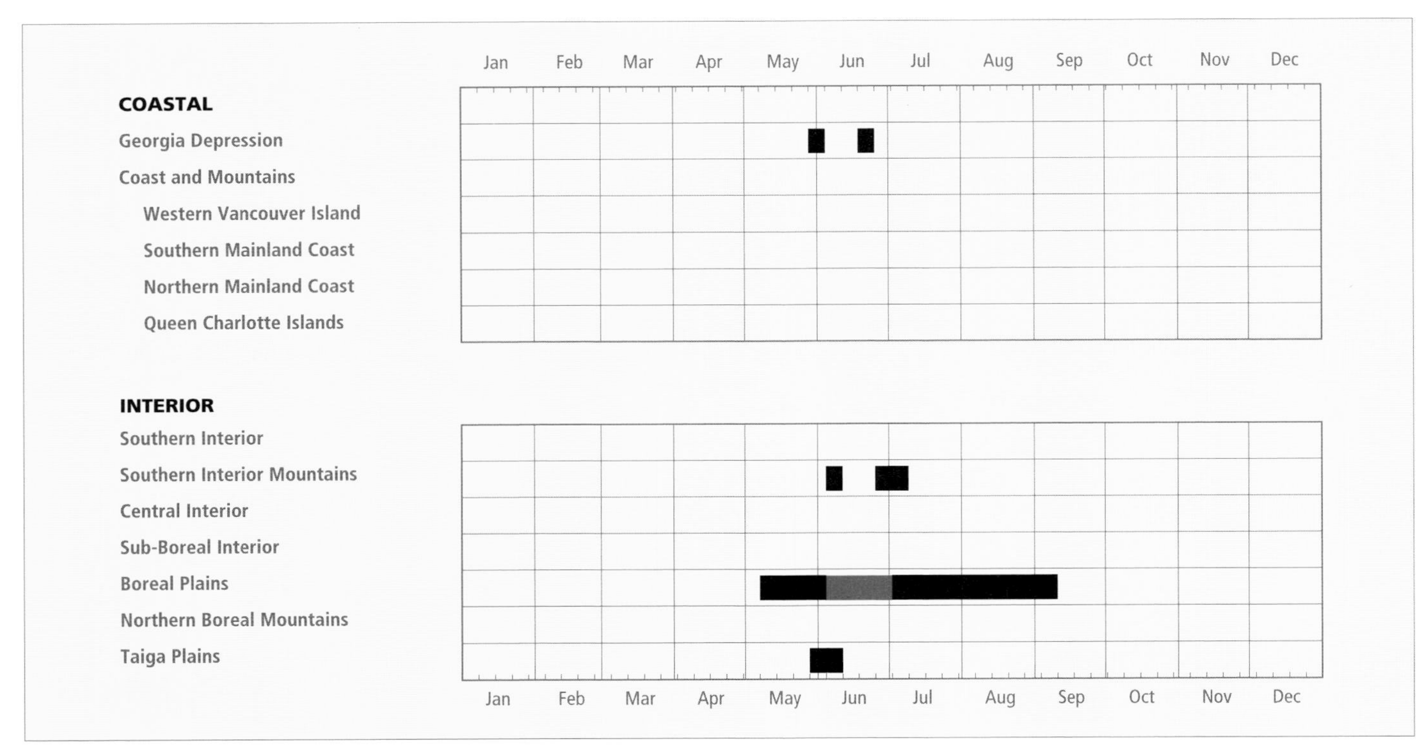

Figure 80. Annual occurrence (black) and breeding chronology (red) for the Black-throated Green Warbler in ecoprovinces of British Columbia. Records are shown for the week in which they occurred.

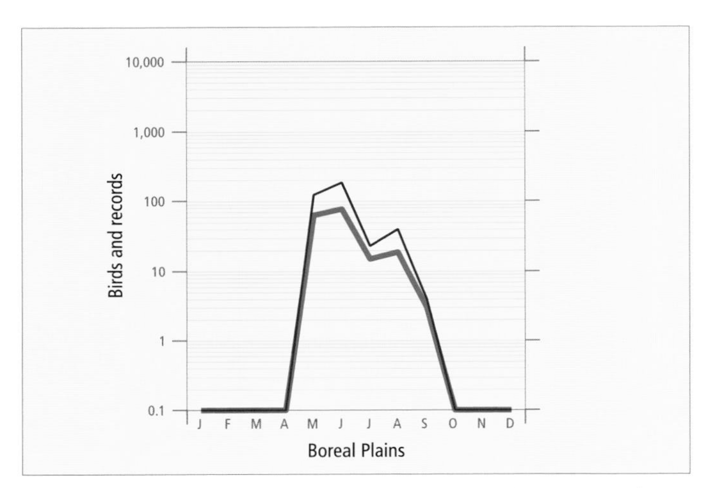

Figure 81. Fluctuations in total number of birds (purple line) and total number of records (green line) for the Black-throated Green Warbler in the Boreal Plains Ecoprovince of British Columbia. Breeding Bird Surveys and nest record data have been excluded.

Dawson Creek, and a male was found feeding a fledged Brown-headed Cowbird on the south bank of the Peace River near Taylor. These 2 locations are less than 20 km apart.

The Black-throated Green Warbler occurs consistently in late spring and early summer near Taylor, Charlie Lake, Stoddart Creek, Kiskatinaw Park, Moberly Lake, Brassey Creek, and Dawson Creek (Siddle 1992d; M. Phinney pers. comm.). It likely breeds west to Chetwynd, south of the Peace River in the Kiskatinaw Plateau, and north to Blueberry River. One summer record from the Tuchodi River suggests that small numbers may breed along some of the major river valleys on the eastern slope of the Rocky Mountains in the Taiga Plains and the Northern Boreal Mountains.

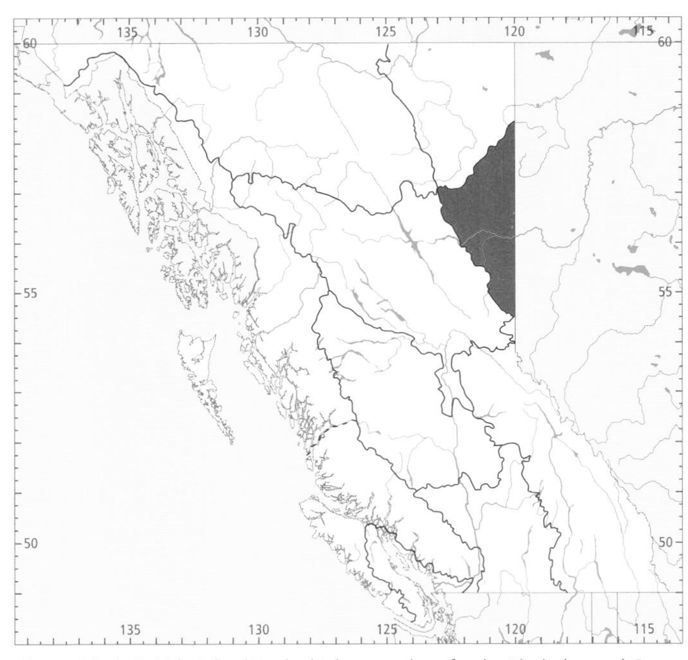

Figure 82. In British Columbia, the highest numbers for the Black-throated Green Warbler in summer occur in the Boreal Plains Ecoprovince.

In British Columbia, the highest numbers for the Black-throated Green Warbler in summer occur in the Boreal Plains (Fig. 82). Breeding Bird Surveys for interior routes for the period 1968 through 1993 contain insufficient data for analysis.

Figure 83. In parts of the Boreal Plains Ecoprovince of British Columbia, the Black-throated Green Warbler breeds in mixed forests of maturing white spruce and trembling aspen (near Dawson Creek, 15 June 1996; R. Wayne Campbell).

An analysis of Breeding Bird Surveys across Canada from 1966 to 1996 indicates that populations are generally stable (Sauer et al. 1997).

The Black-throated Green Warbler has been recorded feeding young just out of the nest at an elevation of 660 m in northeastern British Columbia. In the Boreal Plains, most summer records of this warbler are from riparian stands of white spruce or mixed stands of mature white spruce–trembling aspen–balsam poplar (Fig. 83). River floodplains with stands of larger spruce and deciduous trees are primary habitats. Lower slopes and plateaus are also used. Stands of trembling aspen and balsam poplar are used only if mature white spruce are present. Often, understorey vegetation consists of a combination of willow, prickly rose, baneberry, highbush-cranberry, cow-parsnip, bunchberry, horsetail, fireweed, kinnikinnick, peavine, and American vetch (Fig. 84). This warbler may also be found in riparian stands of old-growth white spruce and balsam poplar along major rivers such as the Liard, Muskwa, Prophet, and Sikanni Chief.

In 1 study in northeastern British Columbia, the habitat used by singing males was associated mainly with tall, mature, riparian mixed white spruce and balsam poplar (Enns and Siddle 1996). These were on moist sites with an understorey consisting of prickly rose, baneberry, highbush-cranberry, bunchberry, fireweed, kinnikinnick, mosses, peavine, and American vetch. Similar habitats are used in Alberta (Salt 1973; Westworth and Telfer 1993). Along Lake Huron, this warbler prefers openings in the canopy of mature forests to places without gaps (Smith and Dallman 1996).

Figure 84. Understorey vegetation in Black-throated Green Warbler breeding habitats may include a variety of shrubs and plants, including cow-parsnip, willow, horsetail, and highbush-cranberry (Moberly Lake, 18 June 1997; R. Wayne Campbell).

Gaps in the canopy permit increased light penetration in the understorey, which results in a greater diversity of vegetation structure and food supply. Stands with small natural disturbances, therefore, may provide more valuable habitat than continuous, unbroken stands.

In British Columbia, the Black-throated Green Warbler probably breeds from early June through late July (Fig. 80).

Nests: Nests have not been found in British Columbia. Elsewhere, the Black-throated Green Warbler usually nests in the understorey of forested areas, mainly 2 to 8 m above ground but sometimes up to 20 m (Bent 1953; Peck and James 1987). Nests are placed mainly in conifers, on a horizontal branch. Nests are typically small cups of grasses, twigs, weed stems, bark strips, and spider webs. They are lined with fine grasses, hair, mosses, and feathers (Baicich and Harrison 1997).

Eggs: Nests with eggs have not been recorded in British Columbia. Eggs can probably occur from early June through early July. Elsewhere, clutch sizes range from 3 to 5 eggs, with 4 being the most common (Bent 1953; MacArthur 1958). The incubation period elsewhere is 12 days (Stanwood 1910; Pitelka 1940).

Young: Nests with young have not been recorded in British Columbia. Brood sizes elsewhere range from 2 to 4 young, with 4 being the most common (Morse 1993). There is 1 record in British Columbia of a recently fledged young as early as 30 June (Siddle 1981). Nestlings can probably occur in the province from about mid-June through mid-July. The nestling period elsewhere is 8 to 10 days (Bent 1953; Nice and Nice 1932).

Brown-headed Cowbird Parasitism: In British Columbia, we have 1 record of an adult male feeding a recently fledged cowbird (Siddle 1981). Nests with cowbird eggs or nestlings have not been found in the province. Friedmann (1963) states that the Black-throated Green Warbler is infrequently parasitized by the Brown-headed Cowbird. In Ontario, however, Peck and James (1987) report that 34% of 32 nests were parasitized.

Nest Success: Insufficient data.

REMARKS: The primary habitat of the Black-throated Green Warbler in northeastern British Columbia consists of riparian stands of mature spruce and mixed spruce–trembling aspen–balsam poplar. These stands have high timber value because of the large size of the trees, and are being harvested rapidly in the Boreal Plains and Taiga Plains. Since the harvested areas tend to be converted to farmland or to pure deciduous or spruce stands with a short harvesting rotation, good habitat may be permanently lost (Cooper et al. 1997c). Further study of this issue is needed in order to guide appropriate habitat management.

In central Saskatchewan, the Black-throated Green Warbler declined significantly from 1972 to 1992 in some study plots in unfragmented, mature forests (Kirk et al. 1997), but in Manitoba no declines were detected in fragmented forests. While Breeding Bird Survey data from British Columbia are insufficient to detect population trends, long-term surveys in areas used by this warbler may be useful in assessing local population changes.

For a thorough review of the biology of the Black-throated Green Warbler in North America, see Morse (1993) and Dunn and Garrett (1997).

NOTEWORTHY RECORDS

Spring: Coastal – Arden Creek (Port Alberni) 28 May 1991-1. **Interior** – Tumbler Ridge 27 May 1997-1; Brassey Creek 25 May 1993-7 on transect (Lance and Phinney 1996); Bear Mountain (Dawson Creek) 28 May 1994-9 on transect (Lance and Phinney 1996); Dawson Creek 11 May 1993-1 male singing from mature mixed-wood forest; Moberly Lake Park 29 May 1990-2 (Siddle 1990c); Taylor 9 May 1987-2 (McEwen and Johnson 1987b), 14 May 1985-1, first spring arrival, 18 May 1985-5 along Peace Island Park Rd; St. John Creek 10 May 1980-1; Beatton Park 14 May 1987-1.

Summer: Coastal – Reifel Island 19 Jun 1982-1, first Vancouver area record. **Interior** – Revelstoke 3 Jul 1989-1; Mount Revelstoke National Park 9 Jun 1990-1 (Siddle 1990c), 1 to 3 Jul 1989-1 (Weber and Cannings 1990); One Island Lake Park 21 Jun 1988-1 (Siddle 1988d); Brassey Creek 11 Jun 1994-11 on transect (Lance and Phinney 1996); Bear Mountain (Dawson Creek) 20 Jun 1993-5 on transect (Lance and Phinney 1996); s Groundbirch 21 Jun 1993-3 males singing in a space of 360 m; Moberly Lake 11 Jun 1983-2, 20 Jun 1965-1 (Salt 1966a; UMMZ 209678); Kiskatinaw Park 30 Jun 1981-2 adults feeding a fledgling; Peace River 2 Jun 1977-4, downstream of confluence with Halfway River; Beatton Park 1 Jul 1990-5 singing males, 11 Aug 1988-7 (Siddle 1989a); Stoddart Creek 7 Jul 1975-1, 22 Aug 1975-4; Boundary Lake (Goodlow) 22 Aug 1984-1; Mile 115 Alaska Highway 23 Jun 1984-1; Sikanni Chief 3 Jun 1987-1 male singing; Dehacho Creek 5 Jun 1997-1; Childers Creek (Tuchodi River) 1 Jun 1994-1 singing from spruce stand in old-growth spruce-poplar floodplain.

Breeding Bird Surveys: Not recorded.

Autumn: Interior –Beatton Park 2 Sep 1982-2; se shore Charlie Lake 5 Sep 1982-1; Taylor 2 Sep 1984-1 on Peace Island Park Rd. **Coastal** – No records.

Winter: No records.

Townsend's Warbler

TOWA

Dendroica townsendi (Townsend)

RANGE: Breeds from east-central Alaska and southern Yukon south through British Columbia and Washington to southern Oregon; east to southwestern Alberta, with disjunct breeding records in southwestern Saskatchewan, northern Idaho, northwestern Montana, and possibly northwestern Wyoming. Winters in 2 distinct geographic areas: one extends from extreme southwestern British Columbia south along the Pacific coast to northern Baja California; the other extends from northern Mexico south to Costa Rica.

STATUS: On the coast, a *fairly common* to *common* migrant and summer visitant in the Georgia Depression Ecoprovince; *very rare* there in winter. In the Coast and Mountains Ecoprovince, *uncommon* to *fairly common* migrant and summer visitant on the Northern Mainland Coast, Southern Mainland Coast, Western Vancouver Island, and the Queen Charlotte Islands; *casual* in winter on Western Vancouver Island and the Queen Charlotte Islands.

In the interior, *fairly common* to *common* migrant and summer visitant in the Southern Interior and Southern Interior Mountains ecoprovinces; generally *uncommon,* but occasionally *fairly common,* in the Central Interior and Sub-Boreal Interior ecoprovinces; *very rare* spring and autumn migrant in the Boreal Plains Ecoprovince and *uncommon* migrant and locally *rare* summer visitant in the Northern Boreal Mountains Ecoprovince; not recorded from the Taiga Plains.

Breeds.

NONBREEDING: The Townsend's Warbler (Figs. 85 and 86) has a widespread distribution throughout the province except in the Taiga Plains, where it has not been recorded. There are large areas, however, where reports are lacking, partly because of the absence of observers and partly because of the distribution of forests unattractive to this warbler. Such areas occur in portions of the Central Interior, the Northern Boreal Mountains, the Boreal Plains, and the Taiga Plains.

Figure 85. Immature Townsend's Warbler. British Columbia is the centre for most of North America's breeding population of this species (Victoria, late August 1989; Tim Zurowski).

The Townsend's Warbler has been reported at elevations from sea level to 1,400 m on the coast, and from 300 to 2,150 m in the interior. In general, the species is most abundant and widespread in forests of Douglas-fir, western redcedar, and grand fir, somewhat less abundant where spruce dominates the forest, and least numerous in forests of lodgepole pine and shore pine.

During migration along the south coast of the province, the Townsend's Warbler occurs mainly throughout many forested areas, moving and feeding in a wide variety of deciduous and coniferous habitats below timberline. Other specific coastal habitats in which the warbler has been found include estuaries, shorelines of the sea coast and of lakes, gulches and gullies, river courses, powerline rights-of-way, gardens, and riparian areas around beaver ponds, sloughs, reservoirs, and sewage lagoons. In the interior, migrating Townsend's

Figure 86. Adult male Townsend's Warbler foraging among bunchgrass and pasture sage in an open grassland during spring migration (Hat Creek, north of Cache Creek, 11 May 1996; John M. Cooper).

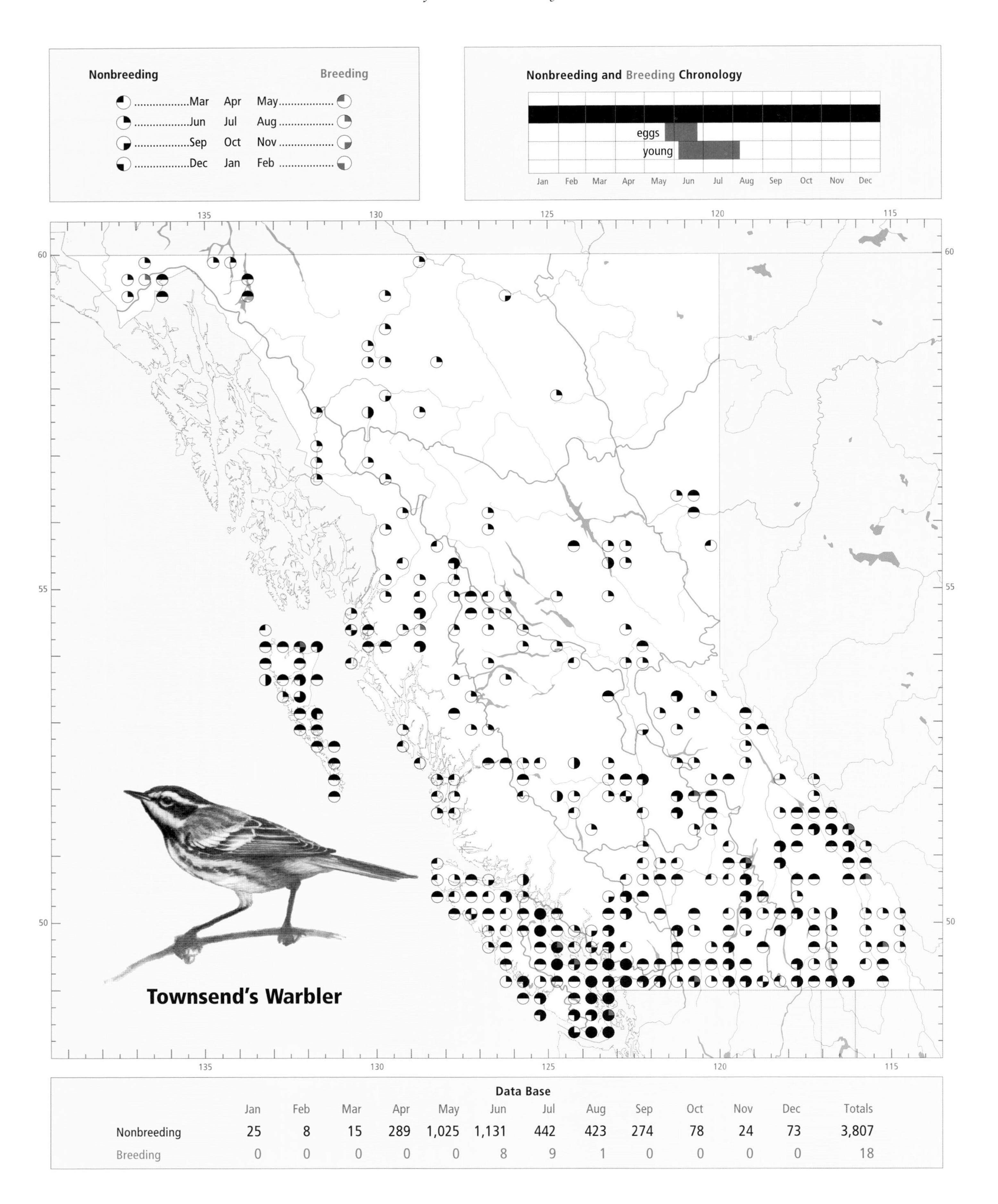

Data Base

	Jan	Feb	Mar	Apr	May	Jun	Jul	Aug	Sep	Oct	Nov	Dec	Totals
Nonbreeding	25	8	15	289	1,025	1,131	442	423	274	78	24	73	3,807
Breeding	0	0	0	0	0	8	9	1	0	0	0	0	18

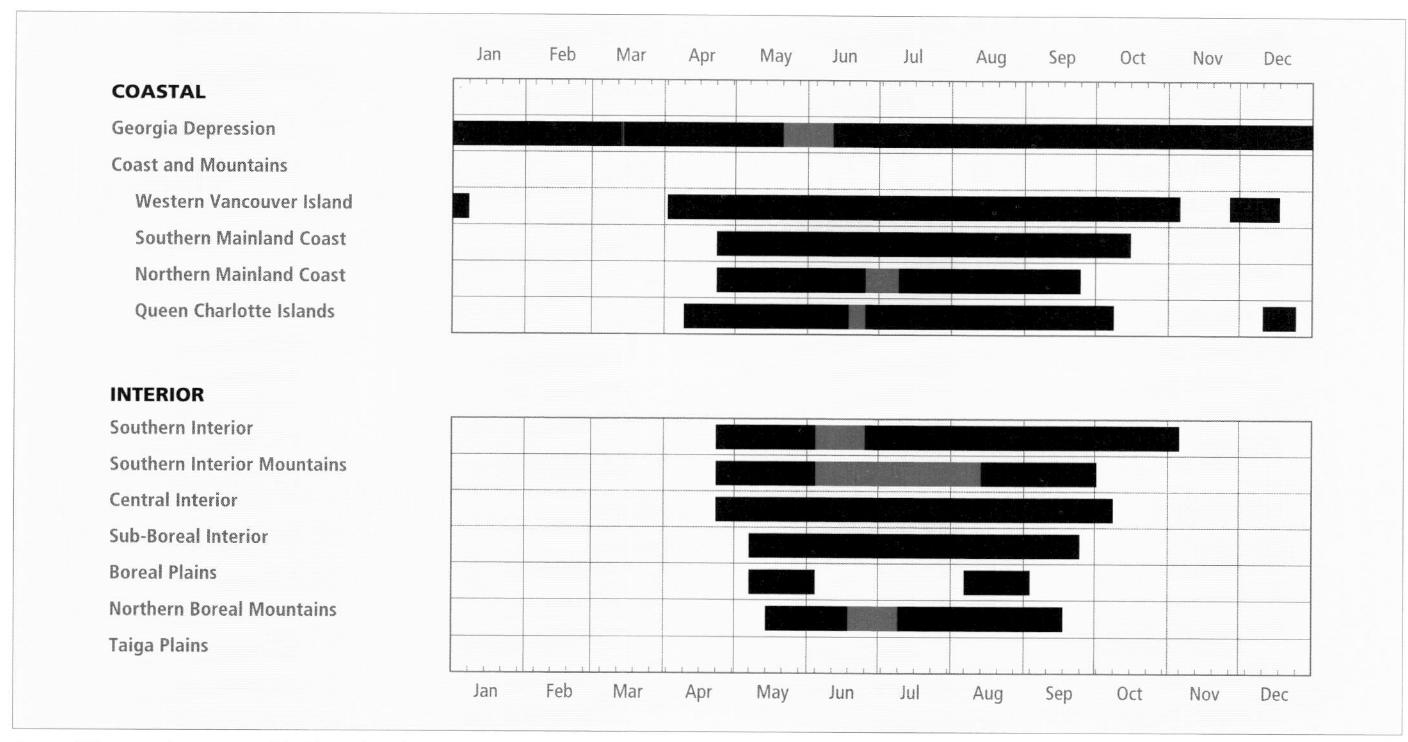

Figure 87. Annual occurrence (black) and breeding chronology (red) for the Townsend's Warbler in ecoprovinces of British Columbia. Records are shown for the week in which they occurred.

Warblers frequent pure and mixed forests of ponderosa pine, lodgepole pine, Engelmann spruce, black cottonwood, western larch, and trembling aspen. It has also been found in patches of willow and saskatoon, and even on alluvial plains, subalpine slopes of grasses and *Dryas* spp., and open grasslands (Fig. 86).

On the coast, the first spring migrants may appear as early as late March. Sightings of newly arriving birds are rare before mid-April, but then increase rapidly to a peak in mid-May (Figs. 87 and 88). On Western Vancouver Island, the first migrants appear in mid to late April, leading to a peak in May. This movement reaches the Queen Charlotte Islands in mid-April and peaks there in the first week of May. Spring arrival on the larger coastal islands is about 2 weeks earlier than on the adjacent mainland (Fig. 87). Reports of migrating Townsend's Warblers along the Northern Mainland Coast are few, with the first arrivals in late April and the largest weekly total in the first week of May. This timetable appears to persist at least as far north as the Chugach Mountains of Alaska (at about latitude 61°N), where the dates of first-arriving males varied between 27 April and 10 May over a period of 3 years. Females arrived later, between 15 and 17 May (Matsuoka et al. 1997a).

In the interior, the first arrivals are often reported in the last week of April in the Southern Interior, Southern Interior Mountains, and Central Interior (Fig. 87). Peak numbers are reached in the third week of May. Further north in the province, the first arrivals have been recorded in mid-May in the Sub-Boreal Interior and at Atlin in the Northern Boreal Mountains (Swarth 1936). In the Boreal Plains, the warbler is a scarce spring migrant usually between the first and third weeks of May (Siddle 1997).

The onset of the southbound migration is difficult to detect. In the region near the northern provincial boundary, records decline in early July, but this is probably a consequence of the end of the song period. The latest records in Atlin are from 1 September. In the Peace Lowland, the Townsend's Warbler reappears as an autumn migrant mainly through the last 2 weeks of August (Fig. 88). In the Prince George area, the decline in numbers is apparent in early August and the latest record is from the second week of September. In the Okanagan valley and in the east and west Kootenay, the migration occurs steadily from about mid-August through September. Most of these warblers have left southern portions of the interior by October. On the coast, the latest records are from the first week of October, except for the Fraser Lowland and southeastern Vancouver Island, where a few stragglers have been recorded in late November and still fewer through the winter.

On the coast, the Townsend's Warbler has been recorded in every month of the year in the Georgia Depression; elsewhere on the coast, it is regularly reported from the last week of April to the third week of September. In the interior, it is regularly reported from 29 April to 30 September (Fig. 87).

BREEDING: The Townsend's Warbler likely breeds throughout most of its range in British Columbia. It is the most abundant nesting warbler in the coastal coniferous forests, and one of the most common warblers in the coniferous forests of the interior, especially at higher elevations.

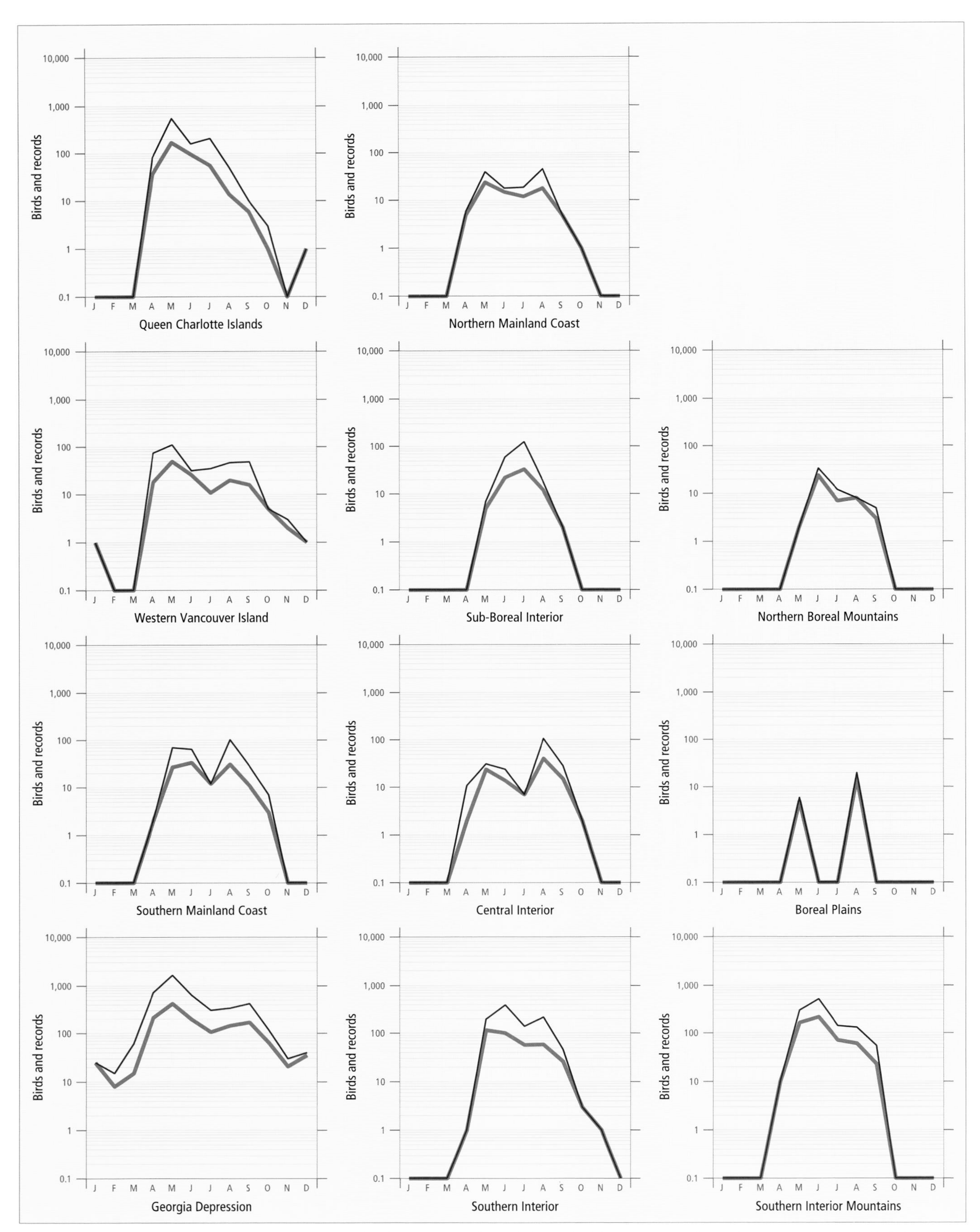

Figure 88. Fluctuations in total number of birds (purple line) and total number of records (green line) for the Townsend's Warbler in ecoprovinces of British Columbia. Christmas Bird Counts, Breeding Bird Surveys, and nest record data have been excluded.

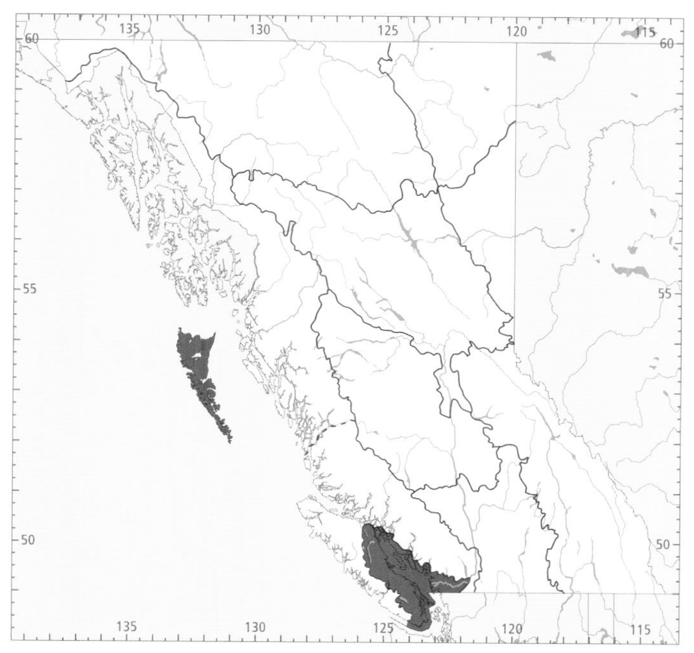

Figure 89. In British Columbia, the highest numbers for the Townsend's Warbler in summer occur in the Georgia Depression Ecoprovince and on the Queen Charlotte Islands in the Coast and Mountains Ecoprovince.

The Townsend's Warbler reaches its highest numbers in summer in the Georgia Depression and on the Queen Charlotte Islands in the Coast and Mountains (Fig. 89), but there are also large numbers in the Southern Interior and in the Kootenays of the Southern Interior Mountains. An analysis of Breeding Bird Surveys for the period 1968 through 1993 could not detect a change in the mean number of birds on either coastal or interior routes. This is also true for the continental population (Sauer et al. 1997). Continent-wide Breeding Bird Survey data indicate that British Columbia has some of the highest densities of the Townsend's Warbler in North America (Sauer et al. 1997).

This warbler has been recorded breeding at elevations from near sea level to 1,200 m on the coast, and from 470 to at least 2,200 m in the interior.

Although this warbler occurs in a variety of forested habitats during the nonbreeding seasons, it is confined to coniferous forests while breeding. On the coast, the conspicuously coloured males are a characteristic element of old second-growth or old-growth forests of Douglas-fir, grand fir, and western redcedar. They are also abundant in the Sitka spruce (Fig. 90), western hemlock, and yellow cedar forests of the central and northern coastal regions. Breeding habitat extends upslope to the subalpine forest, where mountain hemlock, yellow cedar, and amabilis fir dominate the landscape.

Bryant et al. (1993), in their study of avian communities in old-growth and managed forests on Western Vancouver Island, found the Townsend's Warbler to be most numerous in second-growth forests of 30 to 35 years and 50 to 60 years, and somewhat less abundant in old-growth. The warbler's numbers there were positively correlated with crown closure.

In the southern portions of the interior, it is a characteristic nesting species of the subalpine forests of Engelmann spruce and subalpine fir, and of the Douglas-fir, western hemlock, western redcedar (Fig. 91), and western white pine forests of the "wet belt" of the Kootenay and Columbia river valleys. In the southern Rocky Mountains, the Townsend's Warbler is the most abundant summer bird in Kootenay National Park, where it occurs in every watershed (Poll et al. 1984).

On the coast, the Townsend's Warbler has been recorded breeding from 22 May (calculated) to 17 July; in the interior, it has been recorded from 4 June (calculated) to 7 Aug (Fig. 87).

Nests: Only 7 nest locations have been described in British Columbia. An old report of nests found in willows in Vancouver is probably incorrect (Sprunt 1979). Three nests were in unspecified conifer trees, 3 were placed on Douglas-fir limbs, and 1 was placed on a spruce limb. In each case the nest was placed on the upper surface of a branch where it forked. Grasses were present as a major item in 4 nests; twigs, rootlets, plant fibres, mosses, hair, and string were also used. Nest materials in this small sample were markedly different from those in Alaskan nests, which were mainly of spruce and birch twigs and lined with fine grasses and moose hair

Figure 90. On the Queen Charlotte Islands, the Townsend's Warbler prefers to nest in Sitka spruce forests (Queen Charlotte City, 20 May 1996; R. Wayne Campbell).

Figure 91. In parts of the Southern Interior Mountains Ecoprovince, the Townsend's Warbler breeds in mixed coniferous forests of Douglas-fir, western hemlock, and western redcedar (near Christina Lake, 4 July 1997; R. Wayne Campbell).

(Matsuoka et al. 1997a). The heights of 6 nests in British Columbia ranged from 0.9 to 30.0 m. Observations of birds carrying nesting material indicated, however, that many of them were building high in the canopy of giant trees, well out of reach of most people.

Nests found in Washington and Oregon are similar in height to those found in British Columbia. Two nests from the Lake Chelan area of Washington (Bent 1953) are described as being saddled on the limbs of Douglas-firs about 3.6 m from the ground; they consisted largely of western redcedar bark with a few fir twigs and a lining of stems of moss flowers. Mannan et al. (1983) describe 15 nests in the Wallowa Mountains of Oregon. Nine were in grand firs, 5 in Douglas-firs, and 1 in an Engelmann spruce; the heights of the nests were between 4.6 and 20.7 m. In all but 1 case, the nest was placed on a branch where tufts of small branches and needles concealed it from above. The Townsend's Warbler appears to select nest sites in dominant conifers that provide a high density of foliage to conceal nests. In the Chugach Mountains of south-central Alaska, almost all nests were in white spruce (Matsuoka et al. 1997a). They were from 1.8 to 11.8 m (mean = 6.7 m) above the ground, and thus closer to the ground than those in British Columbia.

Eggs: Dates for 4 clutches ranged from 8 to 24 June. Calculated dates indicate that eggs may be present as early as 22 May. Clutch size was 4 or 5 eggs (4E-1, 5E-3). Clutches are larger in Alaska, with a modal clutch size of 6 eggs and a range of 5 to 7 (n = 15 nests) (Matsuoka et al. 1997a). In Alaskan nests, the incubation period, from laying of the last egg to hatching of the last egg, was 11 to 14 days, with a mean of 12.5 days (n = 13 nests) (Matsuoka et al. 1997a).

Young: Dates for 14 broods ranged from 5 June to 7 August, with 57% recorded between 23 June and 17 July. Sizes of 11 broods ranged from 1 to 5 young (1Y-6, 2Y-3, 5Y-2). In Alaska, the nestling period, from hatching of the first egg to departure of the last nestling, ranged from 9 to 10 days, with a mode of 10 days (n = 7 nests) (Matsuoka et al. 1997a). In their study area in Oregon, Mannan et al. (1983) found the nestling period of 1 nest to be 11 days.

Brown-headed Cowbird Parasitism: In British Columbia, none of the 15 nests found with eggs or young was parasitized by the cowbird; however, there have been 7 reports of Townsend's Warblers feeding fledgling cowbirds. Apparently this warbler is an infrequent host of the Brown-headed Cowbird. The single instance reported by Friedmann et al. (1977) was from Vancouver Island (Tatum 1973), and no further occurrences were reported by Friedmann and Kiff (1985).

Nest Success: In British Columbia, 3 nests found with eggs and followed to a known fate were all successful in fledging at least 1 young. Matsuoka et al. (1997a) reported that predation was the primary cause of nest failure in south-central Alaska. The availability of potential nest sites in larger-diameter trees, and the placement of nests higher in trees with lower densities of surrounding woody shrubs, may be important to the reproductive success of Townsend's Warblers throughout Alaska.

REMARKS: The Townsend's Warbler has 2 distinct wintering areas: southwestern British Columbia to California, and southern Mexico through Central America. Although no subspecies are currently recognized, birds wintering in the western United States have shorter wings compared with populations wintering in southern regions (Morrison 1983). The shorter-winged birds, which winter in western Oregon and California, breed on the Queen Charlotte Islands and possibly Vancouver Island. Morrison (1983) also states that "individuals from the two wintering groups can be distinguished from each other with sufficient accuracy to meet accepted subspecies criteria," but so far they have not been recognized as separate subspecies.

Patterns of geographic variation in spring arrival and song in the province appear to support Morrison's (1983) findings. Birds of the Queen Charlotte Islands arrive earlier (see NONBREEDING) and many have different patterns of song display (J. Bowling pers. comm.) than mainland populations. D. Leighton (pers. comm.) suggests that in the interior, Townsend's Warblers singing in the Rocky Mountain spruce forests of the Southern Interior Mountains sound distinctly different from those in the Douglas-fir forests of the Southern Interior. More comparative ecological studies of insular and mainland populations of this species are needed to determine its status in British Columbia (see McNicholl 1980).

In an Alaskan study, 55% of 22 nests with nestlings were infested with the larvae of bird blowflies (Matsuoka et al. 1997a). *Protocalliphora braueri* and *P. spenceri* were both present. In each infested nest, all nestlings had *P. braueri* larvae embedded subcutaneously. Nestling mortality was found in 4 infested nests, but because inclement weather was a confounding factor, mortality could not be unequivocally attributed to parasitism.

There are several offshore records of migrating Townsend's Warblers that landed on ships.

For a description of the nesting ecology of the Townsend's Warbler in Alaska, see Matsuoka et al. (1997a, 1997b). For additional life-history information, see Wright et al. (1998).

NOTEWORTHY RECORDS

Spring: Coastal – Thetis Lake 19 Apr 1990-24; Sumas 24 Apr 1976-33; Surrey 17 Mar 1960-32; Pacific Rim National Park 4 Apr 1981-1; Tofino 22 Apr 1970-45; Harrison Hot Springs 4 May 1984-7; Shadow Lake (Whistler) 29 Apr 1996-1 male singing; Barrier Lake 27 May 1994-2 males singing at 1,360 m elevation, still snow-covered landscape; Campbell River 20 Apr 1981-37; Cape St. James 5 May 1982-200 males and females; Kunga Island 17 May 1977-12; Tlell 13 Apr 1987-4; Awun Lake 18 May 1985-15 (Cooper 1985); Tow Hill 12 Apr 1972-1; Langara Island 11 May 1977-2; km 80 on Haines Highway (Chilkat Pass) 28 May 1979-5. **Interior** – Oliver 1 May 1974-1 at Camp McKinney Rd; West Bench 8 May 1974-8; Revelstoke 15 May 1972-10; Kootenay National Park 19 May 1983-15; 20 km n Golden 16 May 1995-2 males singing; Adams River 3 May 1977-1 at mouth; Tatlayoko Lake 29 Apr 1991-1; Williams Lake 4 May 1986-1; Anahim Lake 14 May 1932-1 (MCZ 284327); Spanish Lake (e Likely) 8 Jun 1992-males singing at 1,500 m; Willow River 11 May 1985-1; Smithers 3 May 1987-1; Mackenzie 4 May 1995-1; Fort St. John 7 May 1989-1; Beatton Park 14 May 1987-1; Atlin 18 May 1930-1 (RBCM 5655).

Summer: Coastal – Rocky Point (Victoria) 7 Aug 1995-12; Saltspring Island 24 Jun 1993-adult male feeding 2 fledglings; e Vargas Island 14 Aug 1970-12 in a flock; Rolley Lake Park 24 Jul 1986-female feeding 4 or 5 fledglings; Grice Bay 7 Jul 1972-adults with fledged young; Hornby Island 12 Aug 1977-14; Falk Lake 10 Jun 1994-2; Masset 17 Apr 1898-1 first arrival (Osgood 1901); Oweegee Lake 24 Aug 1979-6; Great Glacier 14 Aug 1919-1 (MVZ 40196). **Interior** – Manning Park 11 Jun 1979-4 eggs, 26 Jun 1983-100 counted during survey; Oliver 9 Jun 1974-6 at McKinney Rd; Fernie 8 Jun 1958-5 eggs in advanced incubation; Shuswap Lake 20 Jun 1948-5 small nestlings; Kootenay National Park 15 Jun 1983-5 eggs; Yoho National Park 11 Jun 1975-27 along Ice River Rd (Wade 1977), 7 Aug 1977-female feeding 2 recently fledged young; lower Blaeberry River valley 12 Jul 1995-1 female, early autumn migrant; Eagle Lake 28 Aug 1991-1; Chezacut 6 Jun 1975-3 fledglings; nr Bear Lake (Driftwood River) 11 Jul 1938-several (Stanwell-Fletcher and Stanwell-Fletcher 1943); nr Tumbler Ridge 10 Aug 1996-1; Bullmoose Mountain 20 Jun 1996-2 males singing at 1,200 m; Fort St. John 31 Aug 1982-2; 80 km s Nuttlude Lake 28 Jul 1957-2 adults feeding 2 fledglings; Dease Lake 13 Jun 1988-4 at north end.

Breeding Bird Surveys: Coastal – Recorded from 26 of 27 routes and on 62% of all surveys. Maxima: Masset 21 Jun 1994-74; Campbell River 22 Jun 1974-40; Queen Charlotte City 25 Jun 1994-36. **Interior** – Recorded from 48 of 73 routes and on 43% of all surveys. Maxima: Kootenay National Park 14 Jun 1983-44; Illecilewaet 17 Jun 1982-30; Kuskonook 17 Jun 1983-28.

Autumn: Interior – Liard River Hot Springs Park 4 Sep 1969-1 (Erskine and Davidson 1976); Ealue Lake 14 Sep 1977-3; Mackenzie 21 Sep 1995-1; Prince George 8 Sep 1994-1; Indianpoint Lake 25 Sep 1929-1 (MCZ 284321); Quesnel 6 Sep 1900-1 (AMNH 382121); Chezacut 26 Sep 1933-1 (MCZ 284342); Williams Lake 4 Oct 1965-1; 20 km n Golden 7 Sep 1998-1 juvenile; Mount Revelstoke National Park 10 Sep 1982-25; Fauquier 22 Sep 1979-1; Vernon 12 Oct 1994-1; Naramata 3 Nov 1972-1; Conkle Mountain 16 Oct 1988-1. **Coastal** – Masset 28 Sep 1992-1; Port Clements 2 Oct 1971-3; Kyuquot 2 Sep 1984-2; Port Neville 18 Sep 1986-20; Pemberton 10 Sep 1995-1 male; Harrison Hot Springs 27 Oct 1983-1; North Vancouver 15 Sep 1980-35; Cypress Bowl Rd 12 Sep 1982-20; Black Mountain 1 Sep 1985-28, on summit; Mayne Island 13 Oct 1996-6 in flock; Victoria 1 Oct 1954-16 (Flahaut and Schultz 1955a), 13 Nov 1996-1 adult male; Muir Creek 30 Nov 1986-5; Jordan River 30 Nov 1986-1.

Winter: Interior – No records. **Coastal** – Campbell River 3 Jan 1987-1; Thetis Island 26 Dec 1992-1 (Siddle 1993b); North Vancouver 30 Dec 1982-2; Stanley Park (Vancouver) 1 Jan 1960-1; Jericho Park (Vancouver) 18-31 Dec 1995-1 female (Elliott and Gardner 1997); Langley 18 Dec 1986-1 with chickadees and kinglets; Brentwood Bay 5 to 31 Jan 1975-1 at feeder regularly; Triangle Mountain (Colwood) 20 Dec 1991-1 female; Saanich 2 Jan 1982-1, 3 Jan to 27 Feb 1973-1 (Crowell and Nehls 1973b), 18 Feb 1978-1; Swan Lake (Saanich) 31 Dec 1980-1; Jordan River 10 Dec 1983-1, 7 Jan 1984-1; Sooke 16 Feb 1985-1, 16 Dec 1984-1.

Christmas Bird Counts: Interior – Not recorded. **Coastal** – Recorded from 12 of 33 localities and on 7% of all counts. Maxima: Skidegate 16 Dec 1989-**5**, all-time Canadian high count (Monroe 1990a); Pender Island 20 Dec 1976-3; Victoria 20 Dec 1980-3; Sooke 27 Dec 1986-3; Pitt Meadows 15 Dec 1974-2; Comox 16 Dec 1990-2.

Palm Warbler

PAWA

Dendroica palmarum (Gmelin)

RANGE: Breeds from west-central Mackenzie, northern Saskatchewan, northern Manitoba, northern Ontario, south-central Quebec, and southern Labrador and Newfoundland south to northeastern British Columbia, south-central Alberta, south-central Saskatchewan, southeastern Manitoba, northeastern Minnesota, northern Wisconsin, southern-central Ontario, southern Quebec, south-central Maine, and southern Nova Scotia. Winters mainly from coastal Oregon to southern California and from the Gulf states and Delaware south through the Caribbean islands to Yucatan and Honduras.

STATUS: On the coast, *rare* migrant and winter visitant in the Georgia Depression Ecoprovince; in the Coast and Mountains Ecoprovince, locally *very rare* in winter on central Western Vancouver Island and *casual* on the Queen Charlotte Islands.

In the interior, *casual* transient in the Southern Interior, Southern Interior Mountains, Central Interior, Sub-Boreal Interior, and Northern Boreal Mountains ecoprovinces; *uncommon* migrant and summer visitant in the Taiga Plains and Boreal Plains ecoprovinces.

Breeds.

CHANGE IN STATUS: Munro and Cowan (1947) noted that the Palm Warbler (Fig. 92) was a summer visitant and breeding bird only in the extreme northeastern portion of British Columbia. This assessment was made on the basis of 2 records, from Trutch and Muskwa River, provided by Rand (1944), 1 of which included reference to nesting. The known breeding status of the Palm Warbler in British Columbia is essentially the same today, except that we now know that this warbler occurs more frequently in the Taiga Plains than in the Boreal Plains.

Beginning in the 1960s, the Palm Warbler began to appear in winter in the Georgia Depression, a trend that was also noted in Oregon about the same time (Gilligan et al. 1994). The first winter record in British Columbia was of 2 birds at Sidney, Vancouver Island, in 1963. During the early 1970s it was listed as a "casual transient" in the Vancouver area (Campbell et al. 1974) and "accidental" on Vancouver Island (Victoria Natural History Society 1974). During the last 2 decades, autumn migrants and wintering Palm Warblers have begun to occur regularly in the Georgia Depression, and somewhat less regularly on Western Vancouver Island.

NONBREEDING: The Palm Warbler occurs regularly only in northeastern British Columbia east of the Rocky Mountains. There it occurs mainly in the Fort Nelson Lowland and Etsho Plateau of the Taiga Plains, and at Boundary Lake at the northern edge of the Peace Lowland in the Boreal Plains. The only known nonbreeding locality in the Northern Boreal Mountains is in the Liard Plain at Liard Hot Springs Park. It is likely that the Palm Warbler occurs in scattered locations across the plateaus of the Taiga Plains and Boreal Plains west to the eastern slope of the Rocky Mountains, and in the Liard Plain.

Figure 92. Palm Warbler in winter plumage. In winter, this hardy warbler prefers unwooded, open habitats (Esquimalt Lagoon, 12 February 1988; Tim Zurowski).

A few individuals now appear fairly regularly during autumn migration on the coast, where some overwinter. For example, 5 individuals were recorded in the Vancouver area in the autumn and winter of 1995 (Elliott and Gardner 1997). Most records on the south coast are from the Fraser Lowland, especially from Sea, Iona, and Reifel islands, and from Vancouver Island from Sooke to Port Alberni and near Long Beach on the central west coast.

On the Queen Charlotte Islands, there are 5 occurrences, totalling at least 13 individuals, from autumn through early spring. All records are from the vicinity of Masset or Sandspit, 2 of the few places on the Queen Charlotte Islands where open,

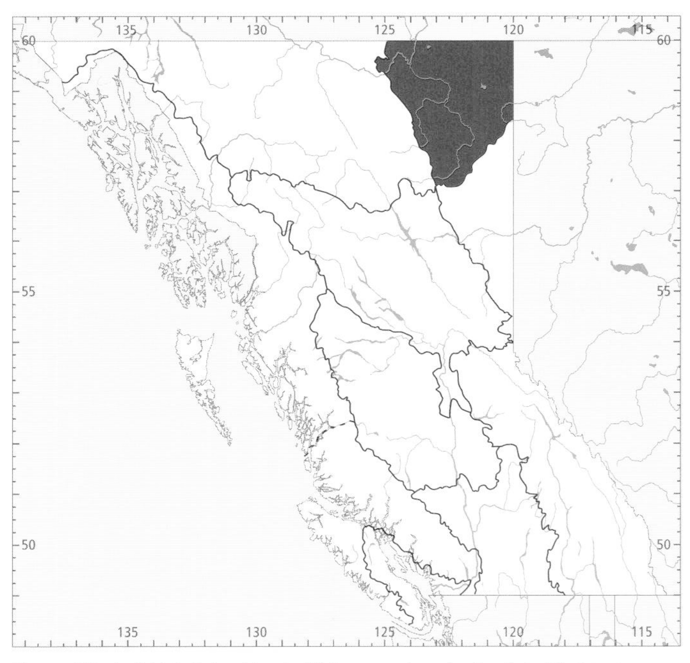

Figure 93. In British Columbia, the highest numbers for the Palm Warbler in summer occur in the Taiga Plains Ecoprovince.

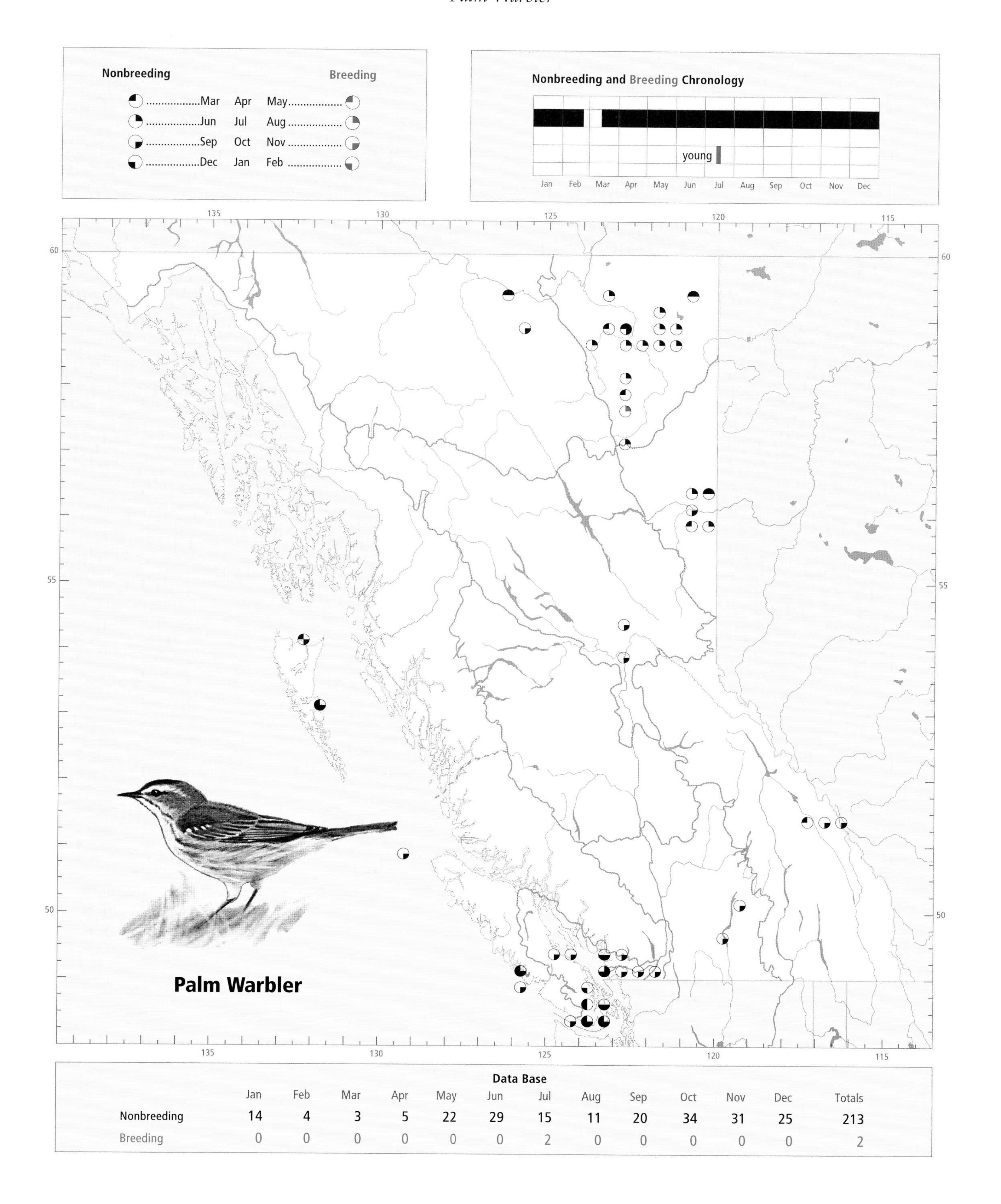

Data Base

	Jan	Feb	Mar	Apr	May	Jun	Jul	Aug	Sep	Oct	Nov	Dec	Totals
Nonbreeding	14	4	3	5	22	29	15	11	20	34	31	25	213
Breeding	0	0	0	0	0	0	2	0	0	0	0	0	2

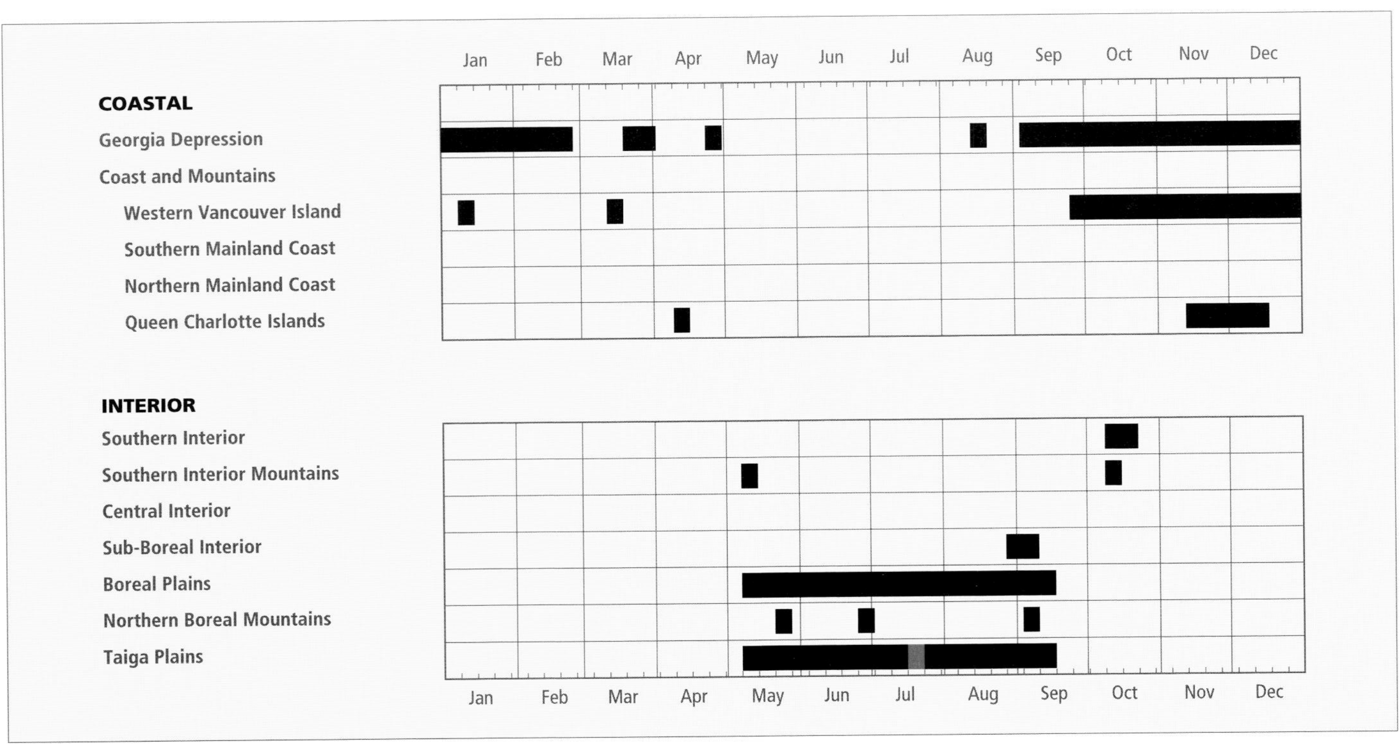

Figure 94. Annual occurrence (black) and breeding chronology (red) for the Palm Warbler in ecoprovinces of British Columbia. Records are shown for the week in which they occurred.

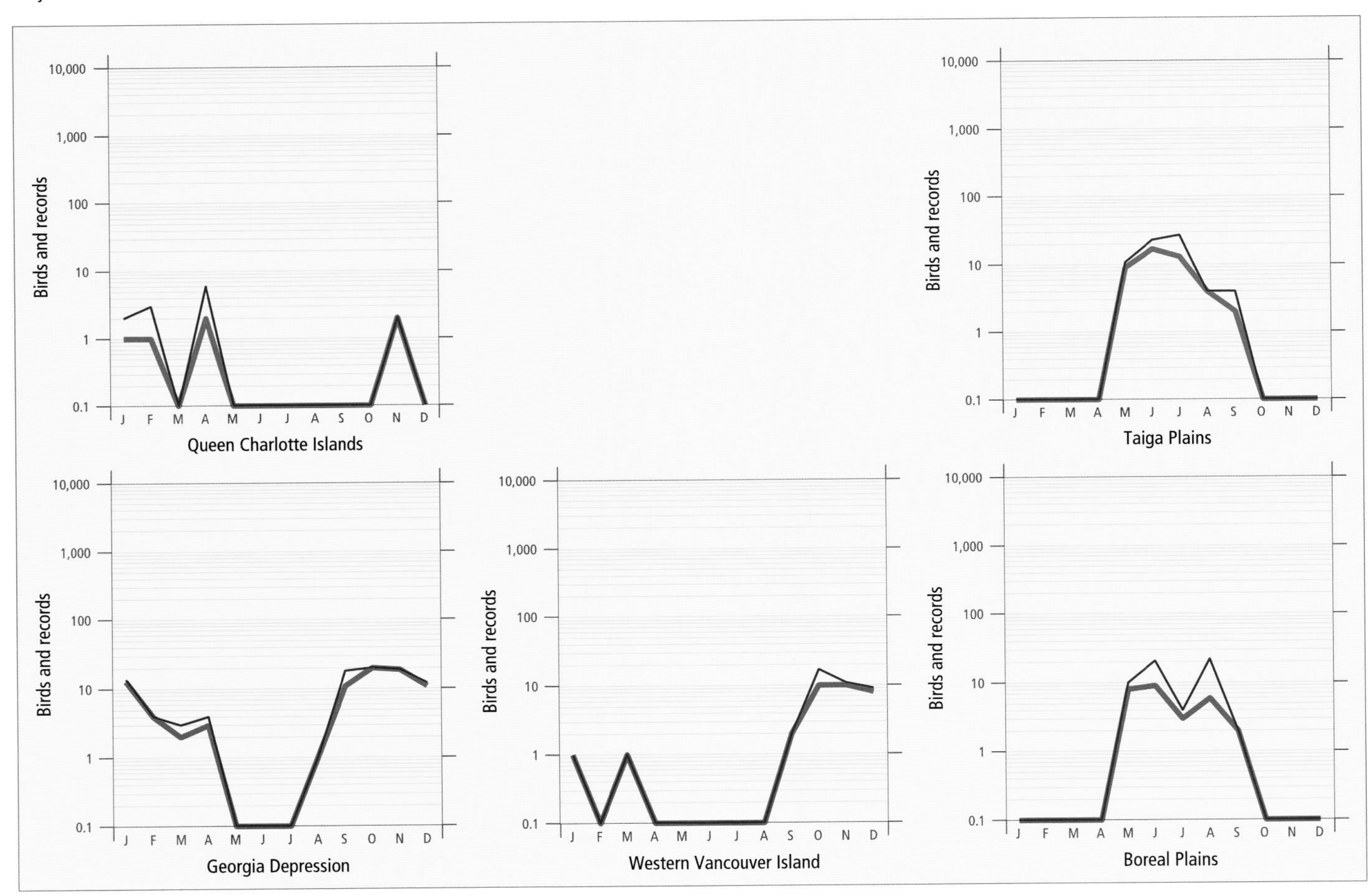

Figure 95. Fluctuations in total number of birds (purple line) and total number of records (green line) for the Palm Warbler in ecoprovinces of British Columbia. Christmas Bird Counts, Breeding Bird Surveys, and nest record data have been excluded.

shrubby habitat occurs. The first record was of 1 bird in November 1985; the next 4 records were from 1991 to 1994. In 1994, at least 2 Palm Warblers overwintered at Masset (Siddle 1994a). These records represent the northernmost wintering locality for the species in North America (Wilson 1996). One autumn migrant has been recorded on Triangle Island, off the northwestern tip of Vancouver Island.

In the southern portions of the interior, the Palm Warbler occurs during migration. In the Southern Interior Mountains, there are spring and autumn records of single birds near Golden. In the Southern Interior, there are 2 records of single birds from the Okanagan valley, and 1 record of 3 birds near Vernon.

The Palm Warbler has been recorded near sea level at the coast and from 300 to 700 m in the interior. It reaches its highest numbers in winter in the Georgia Depression.

During migration and winter on the coast, the Palm Warbler is usually found near sea level, mainly in open, brushy areas. Vegetated spits, causeways, dykes, parks, golf courses, and beaches are typical habitats. In the interior, autumn migrants occur in valley bottoms and riparian thickets along lakeshores, rivers, and sewage lagoons. During migration, the Palm Warbler also frequents suburban gardens, shrubby or weedy roadsides, and sapling willows and trembling aspens at the edge of mature trembling aspen groves.

Migration patterns are not well known, but the Palm Warbler migrates earlier in spring and later in autumn than most other warblers (Wilson 1996). In northeastern British Columbia, spring migrants begin to arrive in the second week of May (Figs. 94 and 95). Most migrants probably arrive in late May. The Palm Warbler enters British Columbia from Alberta, where migrants arrive generally in early to mid-May (Salt 1973). The southward movement begins in mid-August, with the main movement in late August through the first week of September. Migrants move in small flocks; the largest flock recorded in British Columbia consisted of 8 birds. The latest autumn departure dates in the northeast are 16 September in the Taiga Plains and 14 September in the Boreal Plains (Figs. 94 and 95). These dates are relatively late compared with those of other warblers in these parts of the province.

On the coast, autumn migrants begin to arrive in September, but most records are from October and November. The appearance of the Palm Warbler on the west coast of North America during the autumn movement is highly variable; for example, exceptional numbers of about 3 times the norm occurred during 1993 (Wilson 1996). There are few spring records on the coast; the latest is in April.

On the coast, the Palm Warbler occurs irregularly from 6 September to 26 April, occasionally as early as 17 August; in the northeastern interior, it occurs regularly from 9 May to 16 September (Fig. 94).

BREEDING: Although breeding has been confirmed only at Trutch (Rand 1944), the Palm Warbler undoubtedly breeds throughout its summer range in northeastern British Columbia. Singing males have been found at many locations in suitable habitats in the Taiga Plains and Boreal Plains. Nesting likely occurs from the Peace Lowland north to the border with the Yukon and Mackenzie and west to the eastern slope of the Rocky Mountains, and in the Liard Plain of the Northern Boreal Mountains. In the Boreal Plains, the Palm Warbler is more numerous in the north than in the south.

The highest numbers in summer occur in the Taiga Plains (Fig. 93). Breeding Bird Surveys for the period 1968 through 1993 contain insufficient data for analysis. An analysis of Breeding Bird Surveys across Canada from 1966 to 1996 suggests that populations are stable (Sauer et al. 1997).

The Palm Warbler breeds in open to semi-open black spruce bogs or muskegs in northeastern British Columbia (Fig. 96). These habitats are typically wet with a hummocky ground cover of *Sphagnum* moss, scattered stunted black spruce trees in wetter areas, taller spruce along the drier edges of openings, and clumps of willows and birch along the edges. Other common understorey vegetation associated with this habitat in British Columbia includes sedges, horsetails, bog cranberry, and Labrador tea (Fig. 97; Enns and Siddle 1996). Spruce snags and spruce with dead tops are also frequent features of the nesting habitat. Road rights-of-way and oil exploration cutlines, which create more open habitat in forested areas, are also used.

The Palm Warbler is the only warbler in British Columbia that typically uses bogs and muskegs for nesting. An unusually high concentration of nesting Palm Warblers was found in Alberta in a muskeg that had burned several years previously and had a dense growth of regenerating spruce seedlings about 30 cm high (Salt 1973).

During the breeding season, the Palm Warbler generally remains close to the ground when foraging, taking insects from the ground, shrubs, or scattered trees, occasionally foraging higher up in trees or flycatching insects in mid-air (Wilson 1996). In northeastern British Columbia, foraging has been noted in small willows, birch, and sedges along the edges of muskegs. On nesting territories, males sing from the top of tall trees, from bushes (Salt and Salt 1976; Welsh 1971), or from the tip of dead spruce saplings (Enns and Siddle 1996).

In British Columbia, the Palm Warbler probably breeds between early June and late July (Fig. 94).

Nests: Nests have not been described in British Columbia. Elsewhere, nests are usually placed on the ground in *Sphagnum* moss under a sapling or small shrub. In Alberta, nests are often on drier hummocks with tufts of sedges or grasses and a spruce seedling on top; a number of nests have been found a few centimetres off the ground in spruce seedlings (Salt 1973). The nests are small cups of herb stalks, grasses, sedges, bark strips, rootlets, and fern fronds lined with finer grasses, animal hair, and feathers (Wilson 1996).

Eggs: Nests with eggs have not been described in British Columbia. In this province, clutches can likely be found from the second week of June through mid-July. Elsewhere, the clutch size is normally 4 or 5 eggs (Wilson 1996) and the incubation period is 12 days (Knight 1908; F.L. Burns *in* Godfrey 1986).

Young: Nests with young have not been described in British Columbia. Our only breeding record is from near Trutch,

Figure 96. In northeastern British Columbia, the Palm Warbler breeds in open to semi-open black spruce bogs and muskegs (east of Cecil Lake, 24 June 1996; R. Wayne Campbell).

where Palm Warblers were relatively common from 13 to 17 July 1943 and adults were "feeding young" (Rand 1944). Near Boundary Lake, an adult was seen carrying food on 11 July 1984. Nestlings likely occur in British Columbia from late June through late July. The nestling period elsewhere is 12 days (Knight 1908). In Nova Scotia, 4 successful nests produced an average of 4 fledglings each (Welsh 1971).

Brown-headed Cowbird Parasitism: Data are not available for British Columbia. In North America, the Palm Warbler is thought to be rarely parasitized by cowbirds (Friedmann 1963); in northern Alberta, however, cowbird parasitism was reported to be "not uncommon" (Salt 1973).

Nest Success: Insufficient data.

REMARKS: The Palm Warbler is often seen foraging on the ground, characteristically bobbing its tail up and down, much like an American Pipit. In winter, this behaviour is usually the first field mark noticed by birdwatchers.

Two recognizable subspecies of Palm Warbler occur in North America. The "Western" Palm Warbler (*D. p. palmarum*) occurs in the western part of the species' range, including British Columbia. This subspecies typically has a paler plumage than the "Yellow" Palm Warbler (*D. p. hypochrysea*), which occurs from Ontario eastward and has a plumage that is strongly washed with yellow. Their breeding ranges apparently do not overlap, whereas their wintering ranges overlap somewhat (Pittaway 1995; Wilson 1996; Dunn and Garrett 1997).

Figure 97. Hummocks of sphagnum moss with patches of Labrador tea among semi-open stands of black spruce provide nesting sites for the Palm Warbler in British Columbia (west of Goodlow, 19 June 1996; R. Wayne Campbell).

It is unusual for this warbler to overwinter, even in small numbers, in British Columbia. Thirty to 40 birds regularly overwinter in California each year (Small 1994), and a few overwinter in Oregon (Gilligan et al. 1994).

There are few significant conservation concerns for the Palm Warbler in British Columbia. Its preference for peatland or bog nesting habitat and open or logged wintering habitat makes it less susceptible than most other warblers to the impacts of forest clearing.

The Palm Warbler is one of the least studied North American warblers. Our knowledge of its breeding ecology is especially poor in British Columbia, as little ornithological research and inventory have been conducted in its habitat. As Wilson (1996) notes, few Palm Warbler nests have been found anywhere, mainly because the nesting habitat is often difficult for humans to traverse and nests are difficult to find. Naturalists who can record details of habitat use, nest sites, timing of nesting, and clutch sizes will provide useful conservation information. In northeastern British Columbia, it would be especially interesting to document the use of logged areas as possible nesting habitat.

A published record of 8 Palm Warblers at Vernon on 3 October 1990 (Siddle 1991a) contained insufficient detail for inclusion in this account.

NOTEWORTHY RECORDS

Spring: Coastal – Sooke 19 Mar 1990-1 (Siddle 1990b); Swan Lake (Saanich) 23 Apr 1983-2 (Mattocks and Hunn 1983b), 26 Apr 1976-1 (Shepard 1976d); Cowichan Bay 19 Mar 1990-2 (Siddle 1990b); Tofino 14 Mar 1983-1 still in winter plumage; Sandspit 13 Apr 1994-2 (Bowling 1994a); Masset 10 Apr 1994-4 (Bowling 1994a). **Interior** – Moberly Marsh 10 May 1996-1 loosely associated with small flock of Yellow-rumped Warblers (Bowling 1996c); Groundbirch 20 May 1994-1, 25 May 1954-1 (NMC 48216); Cecil Lake 9 May 1981-1; w Boundary Lake (Goodlow) 11 May 1986-1, 12 May 1985-1; Fort Nelson Lowland 22 May 1974-1 (Erskine and Davidson 1976); Parker Lake 13 May 1975-1; Liard Hot Springs 27 May 1981-1; Helmut (Kwokullie Lake) 19 May 1982-2, first spring arrivals, became numerous by 31 May.

Summer: Coastal – Vancouver 17 Aug 1974-1 immature. **Interior** – Cecil Lake 21 Aug 1980-4; German Lake 15 Jun 1980-6 males on territory; Boundary Lake (Goodlow) 22 Aug 1984-4, last autumn record, 26 Aug 1986-8 in flock in shrubs along causeway; Trutch 13 to 17 Jul 1943-relatively common with adults apparently feeding young (Rand 1944); Sikanni Chief 5 Jun 1992-2; Fort Nelson 30 Aug 1985-1 (Grunberg 1986); Fort Nelson area Jun 1974-up to 10 individuals per day (Erskine and Davidson 1976); Esso Resources Road (e Fort Nelson) 1 Jul 1982-5 (Campbell and McNall 1982); Parker Lake 13 Jun 1976-4; Liard Hot Springs 27 Jun 1985-1 (Grunberg 1985d).

Breeding Bird Surveys: Coastal – Not recorded. **Interior** – Recorded from 1 of 73 routes and on less than 1% of all surveys. Maximum: Steamboat 14 Jun 1976-3.

Autumn: Interior – Muncho Lake 5 Sep 1991-1 along highway; Fort Nelson 16 Sep 1986-2 (McEwen and Johnston 1987a); Muskwa River 16 Sep 1943-2 (Rand 1944; NMC 29551); s Fort St. John 6 Sep 1986-1, 14 Sep 1985-1; Summit Lake (Prince George) 9 Sep 1989-1; Prince George 2 Sep 1996-1; Yoho National Park 12 Oct 1975-1; Golden 8 Oct 1977-1; Vernon 24 Sep 1995-3 (Bowling 1996c); Adventure Bay 9 Oct 1963-1; Summerland 19 Oct 1972-1, with Dark-eyed Juncos. **Coastal** – Masset 12 Nov 1985-1 (Hunn and Mattocks 1986); Sandspit 20 to 30 Nov 1991-1 (Siddle 1992a); Triangle Island 5 to 12 Oct 1994-1 (Bowling 1995a); Pitt Lake 9 Nov 1975-1; McCoy Lake 25 Nov 1994-1 (Bowling 1995b); Maplewood Flats (North Vancouver) 30 Sep to 8 Oct 1995-1 immature (Elliott and Gardner 1997); Little Qualicum River 15 Nov 1989-1 on estuary (Weber and Cannings 1990), 15 Nov 1992-1; Nanoose 14 Nov 1992-1; Stubbs Island 7 Nov 1987-1; Tofino 29 Oct 1995-3 (Bowling 1996c), 16 Nov 1993-2; Long Beach 16 Oct 1983-3 (BC Photo 884); Sea Island 28 Nov 1981-1; Huntingdon 16 Sep 1949-7; Ucluelet 10 Oct 1993-6 (Siddle 1994b); Amphitrite Point 27 Sep 1986-1 (Mattocks and Harrington-Tweit 1987a); Sidney Island 16 Sep 1991-1 (Siddle 1992a); Central Saanich 17 Nov 1991-1; Swan Lake (Victoria) 6 Sep 1985-1; Victoria 25 Sep 1989-1; Rocky Point 28 Sep 1995-1; Sooke 26 Nov 1989-1 (Weber and Cannings 1990); Jordan River 10 Oct 1988-1 (Mattocks 1989b).

Winter: Interior – No records. **Coastal** – Masset Jan 1994-2 overwintered (Siddle 1994b); Sandspit Feb 1993-3 (Siddle 1993b); Tofino 26 Dec 1982-2, 13 Jan 1983-1; Delta 22 Feb 1995-1 (Bowling 1995b); Iona Island 21 Dec 1997 to 11 Jan 1998-1, 1 Jan 1978-1; Cowichan Bay 14 Jan 1990-1 (Siddle 1990a); Sidney 12 Jan 1963-2; Central Saanich 4 Jan 1992-1 (Siddle 1992b); Saanich 1 Jan 1989-1 at Quicks Bottom; Esquimalt Lagoon 11 to 13 Feb 1988-1 (BC Photo 1644; Fig. 92).

Christmas Bird Counts: Interior – Not recorded. **Coastal** – Recorded from 4 of 33 localities and on 1% of all counts. Maxima: Skidegate 14 Dec 1991-3; Duncan 27 Dec 1993-2; Vancouver 18 Dec 1977-1; Victoria 19 Dec 1987-1.

Bay-breasted Warbler

Dendroica castanea (Wilson)

BAYW

RANGE: Breeds from southeastern Yukon, southwestern Mackenzie, north-central Alberta, northwestern and central Saskatchewan, central Manitoba, central Ontario, central Quebec, New Brunswick, Prince Edward Island, Nova Scotia, and Newfoundland south to northeastern British Columbia, central Alberta, south-central Saskatchewan, southern Manitoba, northeastern Minnesota, northern Wisconsin, southern Ontario, southern Quebec, northeastern New York, northeastern Vermont, northern New Hampshire, and Maine. Winters mainly from Panama south to Colombia and Venezuela, less regularly north to Costa Rica.

STATUS: On the coast, *casual* on Western Vancouver Island in the Coast and Mountains Ecoprovince.

In the interior, *rare* to *uncommon* migrant and summer visitant to the Boreal Plains and Taiga Plains ecoprovinces; *casual* in the Northern Boreal Mountains, Sub-Boreal Interior, and Southern Interior Mountains ecoprovinces.

Breeds.

Figure 98. The Bay-breasted Warbler (here in autumn plumage) reaches the northwestern limit of its North American breeding range in British Columbia and is the scarcest of our regularly occurring warblers (Anthony Mercieca).

Figure 99. In the Taiga Plains Ecoprovince of British Columbia, the Bay-breasted Warbler inhabits stands of mixed white spruce and trembling aspen with a multilayered canopy, frequent openings, and a dense understorey of highbush-cranberry and paper birch (2 km east of Fort Nelson, 26 June 1996; R. Wayne Campbell).

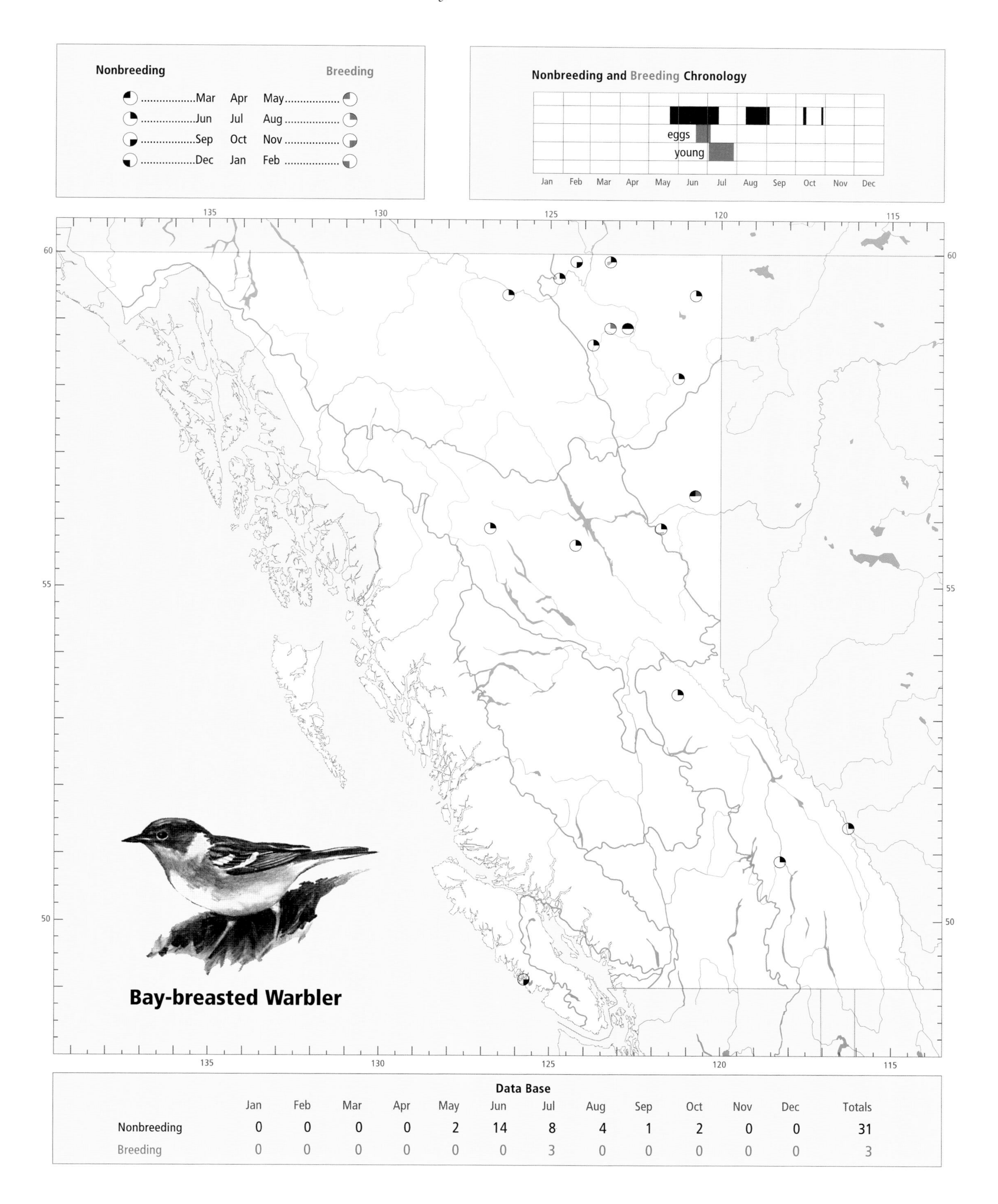

Data Base

	Jan	Feb	Mar	Apr	May	Jun	Jul	Aug	Sep	Oct	Nov	Dec	Totals
Nonbreeding	0	0	0	0	2	14	8	4	1	2	0	0	31
Breeding	0	0	0	0	0	0	3	0	0	0	0	0	3

CHANGE IN STATUS: With only 2 locality records, Munro and Cowan (1947) considered the Bay-breasted Warbler (Fig. 98) to be a scarce summer visitant in the Peace River Parklands (i.e., the Peace Lowland of the Boreal Plains) and areas further northwest. This warbler has now been recorded further north, in several locations in the Taiga Plains. Breeding was assumed, but not confirmed, by Munro and Cowan (1947). Breeding has now been documented in the Boreal Plains and Taiga Plains. The Taiga Plains likely contain the highest density of breeding Bay-breasted Warblers in British Columbia. This species, however, remains the most infrequently observed warbler that occurs regularly in the province.

NONBREEDING: The Bay-breasted Warbler occurs mainly in the Taiga Plains and Boreal Plains. Most records in the Taiga Plains are from the Fort Nelson Lowland near Fort Nelson and Kledo Creek, with a few records from the other areas. In the Boreal Plains, most records are from the Peace Lowland near Fort St. John. Although records are scarce from the Northern Boreal Mountains, it is likely that the Bay-breasted Warbler occurs regularly in the Liard Plain region. In general, its distribution in the northeastern portion of the province is associated with the Boreal White and Black Spruce Biogeoclimatic Zone (Fig. 99).

Single birds have been observed during summer at a few widely scattered localities outside the normal breeding range: Tetana Lake and near Manson River in the Sub-Boreal Interior, and Indianpoint Lake, Field, and Revelstoke in the Southern Interior Mountains.

The Bay-breasted Warbler has been recorded at relatively low elevations in the northeast, from 230 m in valley bottoms to about 760 m. Nonbreeding habitat is similar to breeding habitat (see BREEDING; Fig. 99).

Migration is poorly known in British Columbia. Spring migrants begin to arrive in the northeast in late May, with most probably arriving in the last few days of May and the first week of June (Figs. 100 and 101). This warbler enters British Columbia from Alberta, where early migrants appear in the last week of May (Salt 1973; Pinel et al. 1993). The autumn movement likely begins in late July, shortly after nesting is completed. Most birds have left the province by mid to late August.

In the northeastern interior of British Columbia, the Bay-breasted Warbler has been regularly recorded from 23 May to 2 September (Fig. 100).

BREEDING: The known breeding distribution of the Bay-breasted Warbler in British Columbia is confined to the Peace Lowland and Fort Nelson Lowland in northeastern portions of the province. Our knowledge of this species' breeding distribution is limited because of its scarcity, its habit of using the upper forest canopy, and the lack of observers who visit its habitat. It is likely that the Bay-breasted Warbler breeds locally in most areas of the Taiga Plains and in the Liard Plain ecosection of the Northern Boreal Mountains, wherever mature coniferous forests occur (Fig. 99). These forests are mainly along the major river drainages from the Alberta border west to the foothills of the Rocky Mountains, especially the Fort

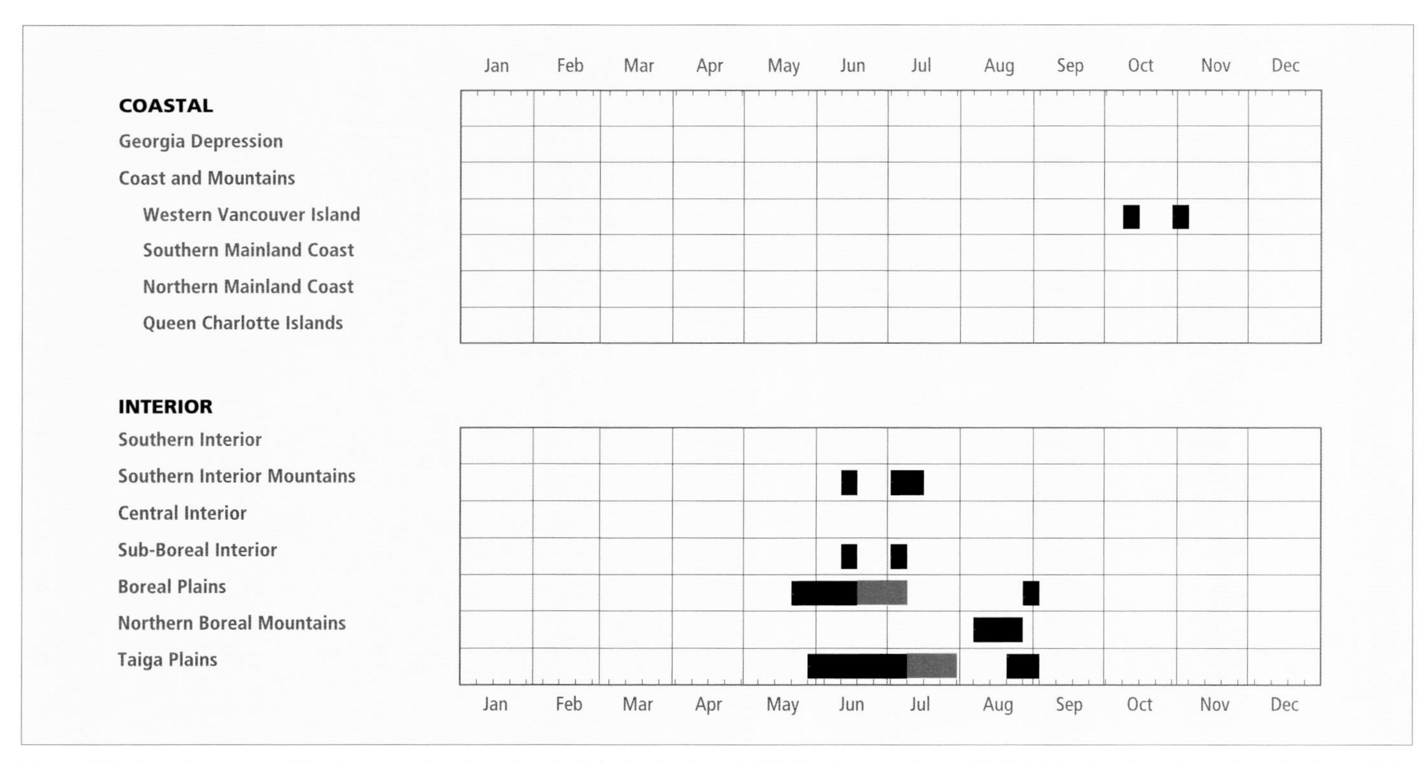

Figure 100. Annual occurrence (black) and breeding chronology (red) for the Bay-breasted Warbler in ecoprovinces of British Columbia. Records are shown for the week in which they occurred.

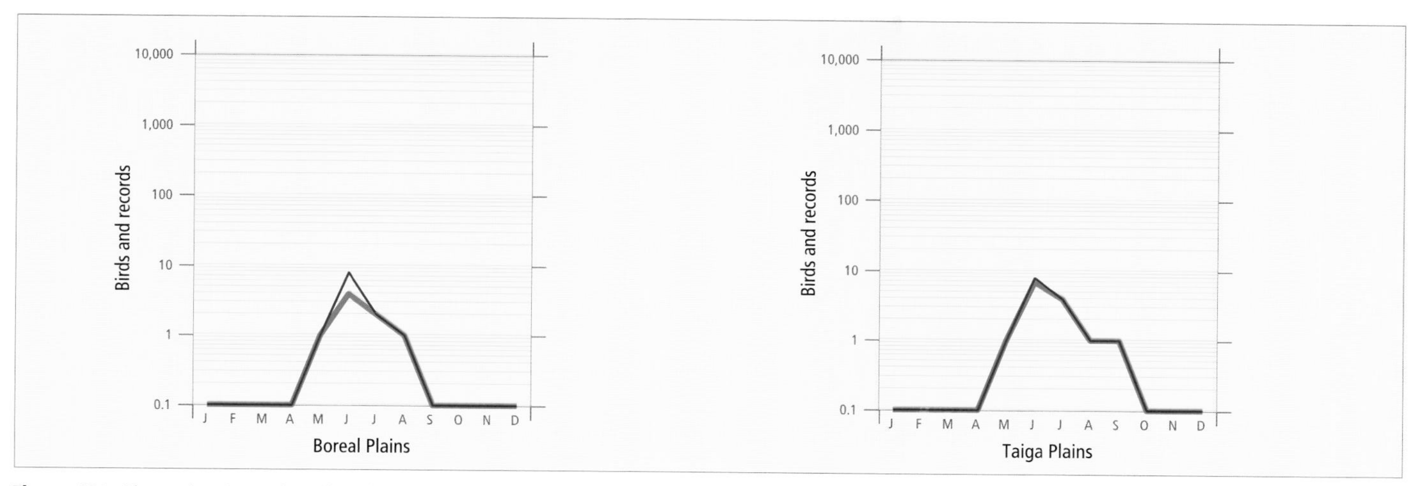

Figure 101. Fluctuations in total number of birds (purple line) and total number of records (green line) for the Bay-breasted Warbler in ecoprovinces of British Columbia. Breeding Bird Surveys and nest record data have been excluded.

Nelson, Muskwa, and Liard river systems. In southern parts of the Boreal Plains, the Bay-breasted Warbler appears to be absent from apparently suitable habitat (M. Phinney pers. comm.).

The highest numbers of the Bay-breasted Warbler in summer occur in the Taiga Plains (Fig. 102). Breeding Bird Surveys for interior routes for the period 1968 through 1993 contain insufficient data for analysis. An analysis of Breeding Bird Surveys across Canada from 1966 to 1996 suggests that populations are declining at an average annual rate of 2.5% ($P < 0.10$) (Sauer et al. 1997).

Breeding habitat in British Columbia includes mature and old-growth stands of white spruce, including stands interspersed with trembling aspen, paper birch, balsam poplar, willow, and alder and a poor to moderately developed shrub layer (Fig. 99; Cowan 1939; Erskine and Davidson 1976; Enns and Siddle 1996). At Moberly Lake, mature trembling aspens at the edge of mature white spruce have been used (C. Siddle pers. comm.).

In the vicinity of the Liard River, most Bay-breasted Warblers were found from the valley bottom to the middle slopes between 250 and 330 m, whereas the ecologically similar Cape May Warbler was found in the middle to upper slopes, mainly above 450 m (Bennett and Enns 1996). Riparian coniferous or mixed forest, a multilayered canopy, and frequent openings were preferred habitat features, whereas the Cape May Warbler preferred a more closed canopy and uniform crown structure. Typical features of valley bottom habitats were trees taller than 20 m, with diameter at breast height greater than 50 cm and a shrubby understorey of highbush-cranberry, prickly rose, and alder. In more upland areas, Bay-breasted Warbler habitat included mature spruce or mixed stands with an understorey of highbush-cranberry and paper birch.

The Bay-breasted Warbler occurs mainly in the mid to upper canopies, and rarely ventures down to the lower canopy or uses shrubs. Salt (1973) notes that males prefer to sing while hidden in the "shade of the intermediate levels."

There is insufficient data to define the extent of the Bay-breasted Warbler's breeding season in British Columbia, but it likely breeds from early June to late July (Fig. 100).

Nests: A nest found in Beatton Park was a small structure composed of small twigs placed on top of a large spruce branch 8 m from the ground. Elsewhere, the Bay-breasted Warbler breeds in small to large coniferous trees, mainly in the mid-canopy between 3.5 and 5.5 m but up to 18 m (Bent 1953; Peck and James 1987). Nests are occasionally found in deciduous trees, close to the trunk or along a main branch (Sealy 1979). Nests are small, bulky cups of fine twigs, grasses, bark strips, and caterpillar webs lined with fine black rootlets and hair (Harrison 1979).

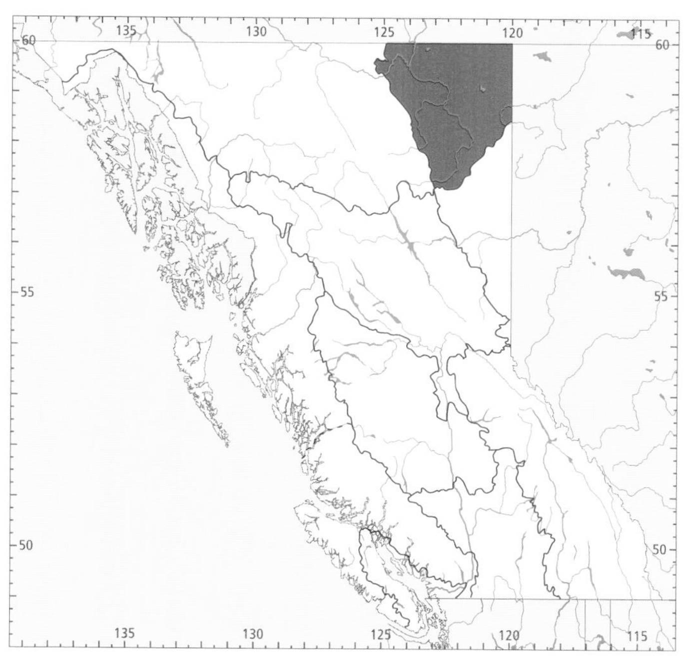

Figure 102. In British Columbia, the highest numbers for the Bay-breasted Warbler in summer occur in the Taiga Plains Ecoprovince.

Eggs: Clutches have not been recorded in British Columbia. Eggs can likely be found between the second week of June and early July. One nest found with nestlings on 3 July would have contained eggs by at least 18 June. The incubation period is 12 to 13 days (Bent 1953; Harrison 1979).

Clutch size ranges from 3 to 7 eggs, but is usually 5 or 6 eggs (Bent 1953; Morse 1978, 1989). Like the Cape May Warbler, the Bay-breasted Warbler has one of the largest clutch sizes of all North American warblers. The larger clutch sizes usually occur during years of high food abundance (see REMARKS).

Young: Only 1 nest with young has been documented in British Columbia: 4 nestlings were observed at Beatton Park being fed by an adult on 3 July. The only other record of breeding is of 2 recently fledged young near Fort Nelson on 28 July. Nestlings likely occur in the northeast from late June through late July. The nestling period is 10 to 12 days (Mendall 1937; Harrison 1984).

Brown-headed Cowbird Parasitism: There are no data on cowbird parasitism in British Columbia. Elsewhere, the Bay-breasted Warbler is an infrequent host of the Brown-headed Cowbird (Friedmann 1963), probably because of its preference for continuous stands of mature boreal forest habitat, where few cowbirds occur. At Delta Marsh, in Manitoba, where cowbirds were common in riparian habitat, Sealy (1979) found that 4 of 6 nests were parasitized. These nests were outside the warbler's normal nesting habitat.

Nest Success: Insufficient data.

REMARKS: Elsewhere, the Bay-breasted Warbler is one of a few warblers whose regional abundance and productivity are closely linked to abundance of their primary food, the eastern spruce budworm (*Choristoneura fumiferana*) (Morse 1989). Clutches averaging 6 or 7 eggs have been recorded during spruce budworm outbreaks (Morse 1978, 1989). Distribution and productivity can be affected by food abundance. For example, Sealy (1979) found this warbler nesting outside typical breeding habitat in Manitoba in areas that were experiencing forest tent caterpillar (*Malacosoma disstria*)outbreaks and for which there was no previous evidence of breeding. This occurred in 1 year only despite the presence of tent caterpillars in the same habitat the following year.

Cooper et al. (1997) suggest that there is an overlap in the distribution of spruce budworm outbreaks and Bay-breasted Warblers in northeastern British Columbia. These areas tend to contain stands of old-growth spruce. A recent inventory along the Liard River found higher concentrations of this warbler than previously reported in British Columbia. The Liard River had experienced a moderate budworm outbreak the previous year, but a minor one during the period of the bird inventory (Bennett and Enns 1996).

Identification of the Bay-breasted Warbler by song alone is difficult because of similarities with the Cape May Warbler in its song and habitat use. Hybridization with Blackpoll and Yellow-rumped warblers has been documented in other regions (Harrison 1984).

For a complete review of the life history of the Bay-breasted Warbler, see Williams (1996a).

NOTEWORTHY RECORDS

Spring: Coastal – No records. **Interior** – Beatton Park 23 May 1981-1; 11 km n Fort Nelson 29 May 1991-1.

Summer: Coastal – No records. **Interior** – Revelstoke 3 Jul 1988-1; Field 11 Jul 1975-1; Indianpoint Lake 12 Jul 1930-1 male (Dickinson 1953); Tetana Lake 6 Jul 1941-1 (RBCM 8950); nr Manson River 16 Jun 1993-1 (Price 1993); Moberly Lake Park Jun 1986-1, 4 Jul 1980-1, 8 Jul 1982-1; Fort St. John 31 Aug 1980-1; Beatton Park 3 Jun 1983-2 males, 30 Jun 1990-2 pairs, 3 Jul 1989-male feeding 4 nestlings small worms, 10 Jul 1980-female carrying food, 31 Aug 1980-1; Charlie Lake 16 Jun 1938-2 (RBCM 8104 and 8105); Kenai Creek 7 and 8 Jun 1995-1; Fort Nelson 28 May 1982-1 (RBCM 17490); w Fort Nelson 28 Jul 1985-2 fledglings, 24 Aug 1988-1; Kledo Creek 15 Jun 1987-1 (Johnston and McEwen 1987b); Steamboat Creek 2 Jul 1986-1; ne Kwokullie Lake 6 Jun 1982-1; lower Liard River 22 Aug 1943-1 (Rand 1944); Liard River 11 to 22 Jun 1996-55 adults recorded on surveys (Bennett and Enns 1996); Maxhamish Lake 19 Jun 1994-1.

Breeding Bird Surveys: Coastal – Not recorded. **Interior** – Recorded from 1 of 73 routes and on less than 1% of all surveys. Maximum: Steamboat 14 Jun 1976-3.

Autumn: Interior – Beaver River (Nelson Forks) 2 Sep 1996-1. **Coastal** – Tofino 29 Oct 1995-1 (Bowling 1996a; BC Photo 1614); Pacific Rim National Park 10 Oct 1993-1.

Winter: No records.

Blackpoll Warbler

Dendroica striata (Forster)

BKPW

RANGE: Breeds from western and north-central Alaska, northern Mackenzie, northern Manitoba, northern Ontario, northern Quebec, northern Labrador, and Newfoundland south to southern Alaska, south-central British Columbia, southwestern and central Alberta, north-central Saskatchewan, central Manitoba, north-central Ontario, and southern Quebec, and through the eastern states to northwestern Massachusetts, central New Hampshire, east-central Maine, and Nova Scotia. Winters in South America from Venezuela, Colombia, and the Guianas south to eastern Peru, northern Argentina, and southern Brazil.

STATUS: On the coast, *very rare* in the Georgia Depression Ecoprovince; in the Coast and Mountains Ecoprovince, *very rare* transient on the Southern Mainland Coast and Northern Mainland Coast, *casual* on Western Vancouver Island and the Queen Charlotte Islands.

In the interior, *accidental* in the Southern Interior Ecoprovince, *uncommon* summer visitant in the Southern Interior Mountains, and *very rare* summer visitant in the Central Interior Ecoprovince; *uncommon* migrant and local summer visitant in the Boreal Plains and Sub-Boreal Interior ecoprovinces, and *fairly common* migrant and summer visitant in the Northern Boreal Mountains Ecoprovince; *rare* in the Taiga Plains Ecoprovince.

Breeds.

Figure 103. Of all the wood-warblers in British Columbia, the Blackpoll Warbler (immature shown) is the most highly migratory (Anthony Mercieca).

NONBREEDING: The Blackpoll Warbler (Fig. 103) occurs throughout the northern two-thirds of the interior of the province. South of latitude 53°N, about the latitude of Williams Lake, occurrences are mainly confined to the Rocky Mountains. Records suggest wide swings in regional abundance. For example, between 1968 and 1978 there appears to have been a rapid increase in numbers of Blackpoll Warblers in the interior ecoprovinces northwest of the Rocky Mountains, followed by an almost symmetrical decrease.

On the coast, the Blackpoll Warbler occurs infrequently during spring and autumn migration, and rarely in summer, in the lower Fraser River valley and on southern Vancouver Island in the Georgia Depression and in widely scattered locations in the Coast and Mountains, including the Queen Charlotte Islands.

The only exceptions to this generalization are summer records from Mosher Creek on the western slope of the Rainbow Range. The fauna of these mountains demonstrates clear northern and interior influences. For example, this is the southernmost locality for the Gray-cheeked Thrush and brown lemming (*Lemmus sibiricus*), both species of boreal distribution; it is also where all 3 species of ptarmigan occur. The presence of the Blackpoll Warbler in summer is a further instance of the occurrence of boreal species in this range.

The Blackpoll Warbler has been recorded from near sea level to about 1,600 m. In migration, it has been observed in a wide variety of forested habitats, including lakeside trembling aspens and willows; riparian stands of black cottonwood and willow; groves of paper birch; muskegs including black spruce, dwarf birch, and willow; dense stands of old-growth lodgepole pine and white spruce; and mixed forests including conifers and deciduous trees bordering marshes, fields, roadsides, and other openings.

The earliest dates that the Blackpoll Warbler has been reported in the province are from the Peace Lowland of the Boreal Plains, where males arrive in early May, with first arrival dates ranging from 9 to 23 May (Siddle 1997). The birds become numerous by the third week of May, and continue to increase in number into early June (Figs. 104 and 105).

The earliest arrival date for the Liard Hot Springs area of the Northern Boreal Mountains is at the end of the second week of May. For the Prince George region and the Tetana Lake area, both in the Sub-Boreal Interior, the dates of earliest arrival are in the third week of May. Further south, in the Central Interior, spring arrival occurs in the fourth week of May. The evidence provided by this sequence of spring arrival dates suggests that a population of this warbler, in the northern and central areas of British Columbia, enters the province from east of the Rocky Mountains in the general region of the Peace River, and then fans out west, north, and south into its summer range. The earliest spring dates in the Rocky Mountains south to Yoho and Kootenay national parks occur in the second and third weeks of May (Figs. 104 and 105), however. Thus, it is likely that the summer population of the southern Canadian Rocky Mountains and adjacent ranges to the west enters the province from Alberta through the several fairly low passes that traverse the mountains at these latitudes. The species was reported as "quite common" in the valleys of Alberta's Jasper National Park in the 1940s (Cowan 1955), and it is a consistent summer visitant in Mount Robson Park, at the western end of Yellowhead Pass.

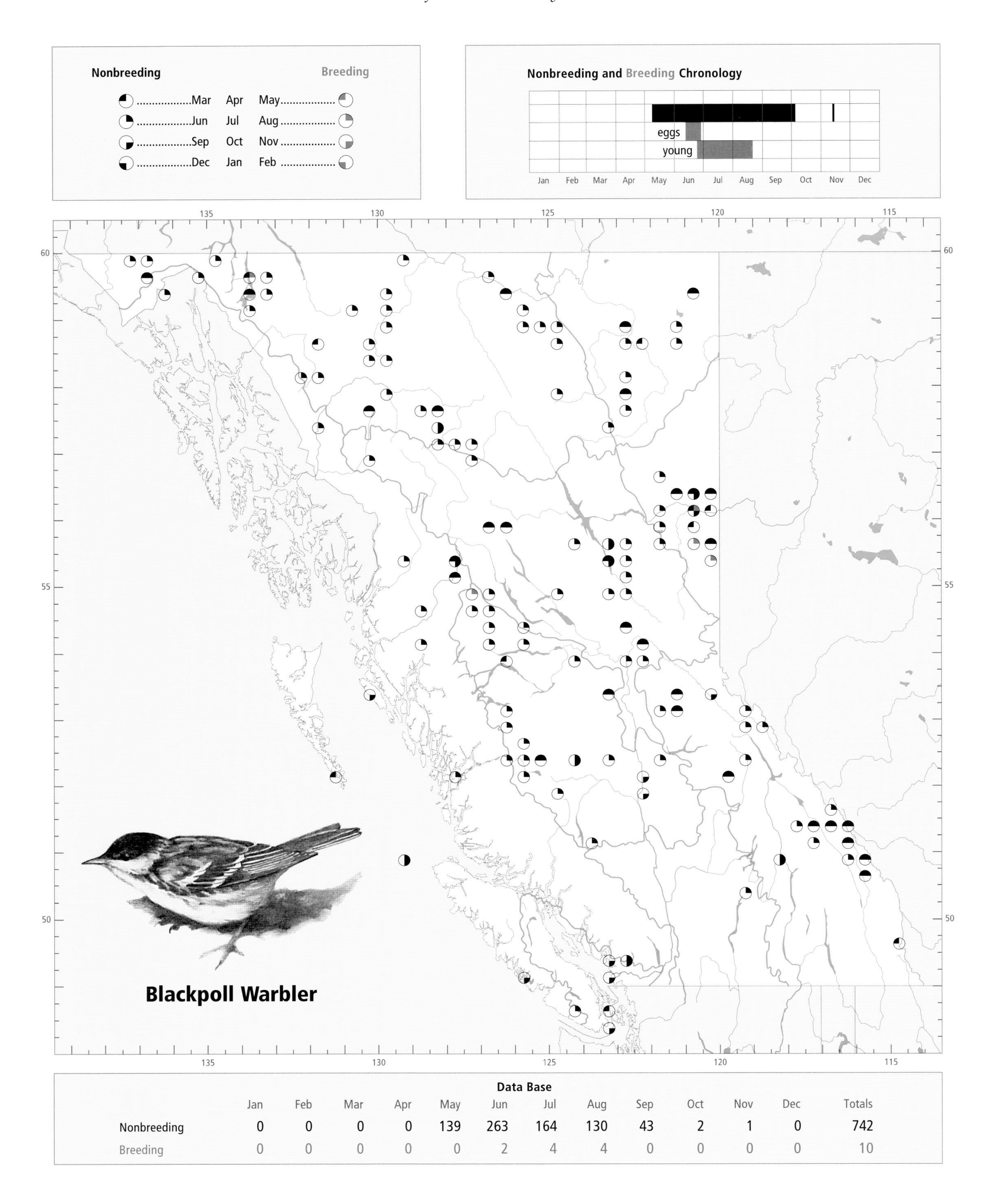

Data Base

	Jan	Feb	Mar	Apr	May	Jun	Jul	Aug	Sep	Oct	Nov	Dec	Totals
Nonbreeding	0	0	0	0	139	263	164	130	43	2	1	0	742
Breeding	0	0	0	0	0	2	4	4	0	0	0	0	10

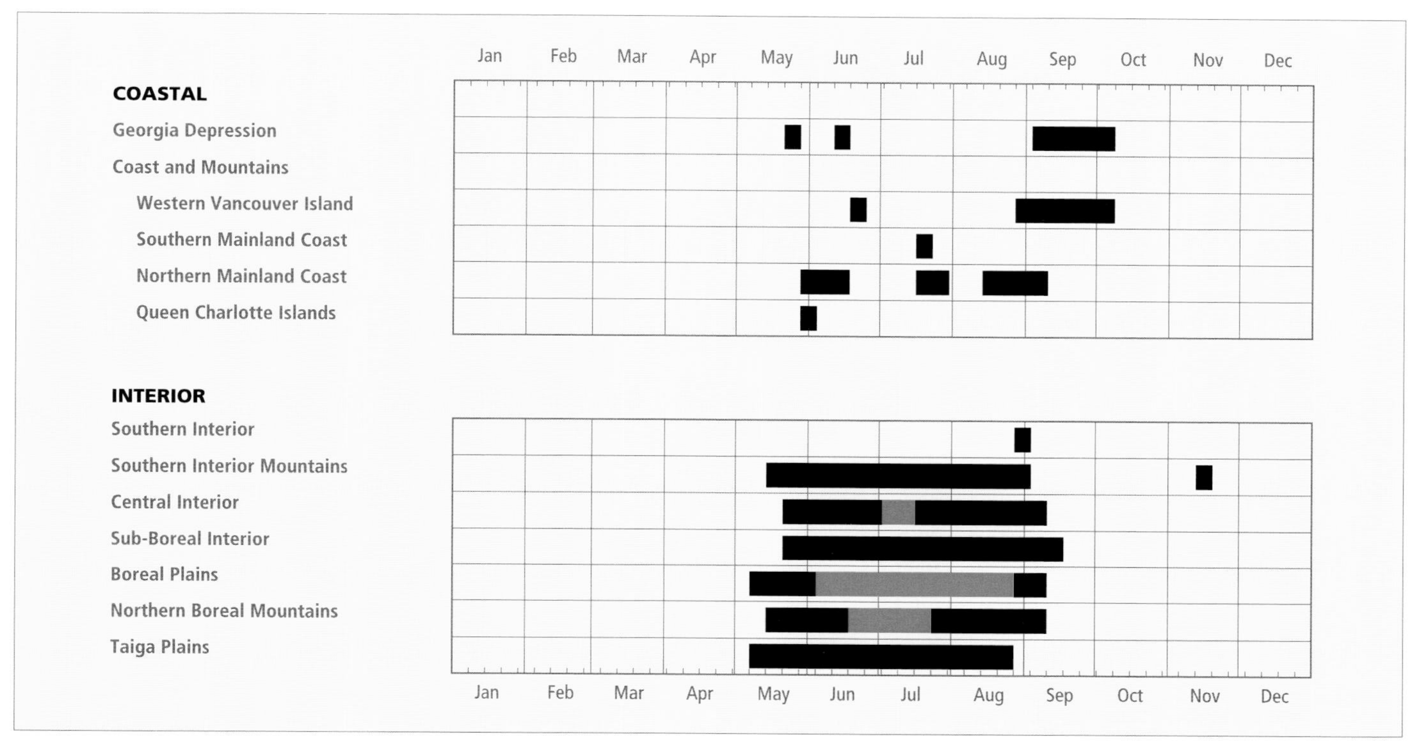

Figure 104. Annual occurrence (black) and breeding chronology (red) for the Blackpoll Warbler in ecoprovinces of British Columbia. Records are shown for the week in which they occurred.

The end of the song period in late June and July makes it difficult to detect the beginning of the autumn migration. By early August, however, there is a noticeable decline in the number of Blackpoll Warblers recorded, and by the first week of September, only stragglers are left in the Northern Boreal Mountains and the Boreal Plains (Figs. 104 and 105). The last migrants have left the interior of the province before the middle of September. There is a single November record from Revelstoke. There are no records to suggest that this species migrates via the interior of the province out of British Columbia across the international boundary into Washington.

On the coast, the Blackpoll Warbler has been recorded between 26 May and 2 October; in the interior, it occurs regularly from 9 May to 9 September, with occasional records as late as 14 November (Fig. 104).

BREEDING: Observations of fledged young suggest that the Blackpoll Warbler breeds at widely scattered localities from Stum Lake, just west of Williams Lake near latitude 52°N, northward throughout much of the Sub-Boreal Interior, Boreal Plains, and Northern Boreal Mountains. Although there are no breeding records for most of northern British Columbia, including the Taiga Plains, the species undoubtedly breeds throughout the area. The presence of the species in summer in the Rainbow Range and Lord River area of the western Chilcotin suggests breeding there, but this has not been confirmed.

The Blackpoll Warbler reaches its highest numbers in summer in the Northern Boreal Mountains (Fig. 106). An analysis of Breeding Birds Surveys for the period 1968 through 1993 could not detect a net change in numbers on interior routes. An analysis of Breeding Bird Surveys across Canada from 1966 to 1996 suggests a small net downward trend, but this arises from an upward trend of 15.0% per year between 1966 and 1979 ($P < 0.01$) followed by a decrease of 10.6% per year between 1979 and 1996 ($P < 0.01$) (Sauer et al. 1997).

The Blackpoll Warbler has been found breeding at elevations between 350 and 1,000 m. Its nesting ecology in its northern range is poorly known. It is a characteristic species of mature white spruce forests and black spruce muskegs in northern Canada and Alaska. It has been found at high elevations, close to the upper limit of trees, establishing territories in the dwarf spruce of those altitudes. During his survey of birds along the Alaska Highway, Rand (1944) recorded the Blackpoll Warbler as "fairly common and apparently breeding in the spruce second-growth" near Trutch. It was found breeding in "fair abundance" on the islands in Atlin Lake (Swarth 1926). In the Driftwood River valley of the Fraser River drainage, 8 pairs occupied nesting territories in the spruce forest bordering mile-long Tetana Lake, where the Blackpoll Warbler selected nesting territory on the edges of spruce forest with deciduous shrubs as well as scattered trembling aspen, black cottonwood, and willow (Stanwell-Fletcher and Stanwell-Fletcher 1943). Throughout its range, edge habitat was provided by river valleys, lakeshores (Fig. 107), beaver ponds, avalanche tracks, and patches of black spruce muskeg, the latter a favoured breeding habitat for the Blackpoll Warbler in the Peace Lowland. During a survey conducted from late June to late July 1992, the habitat classes

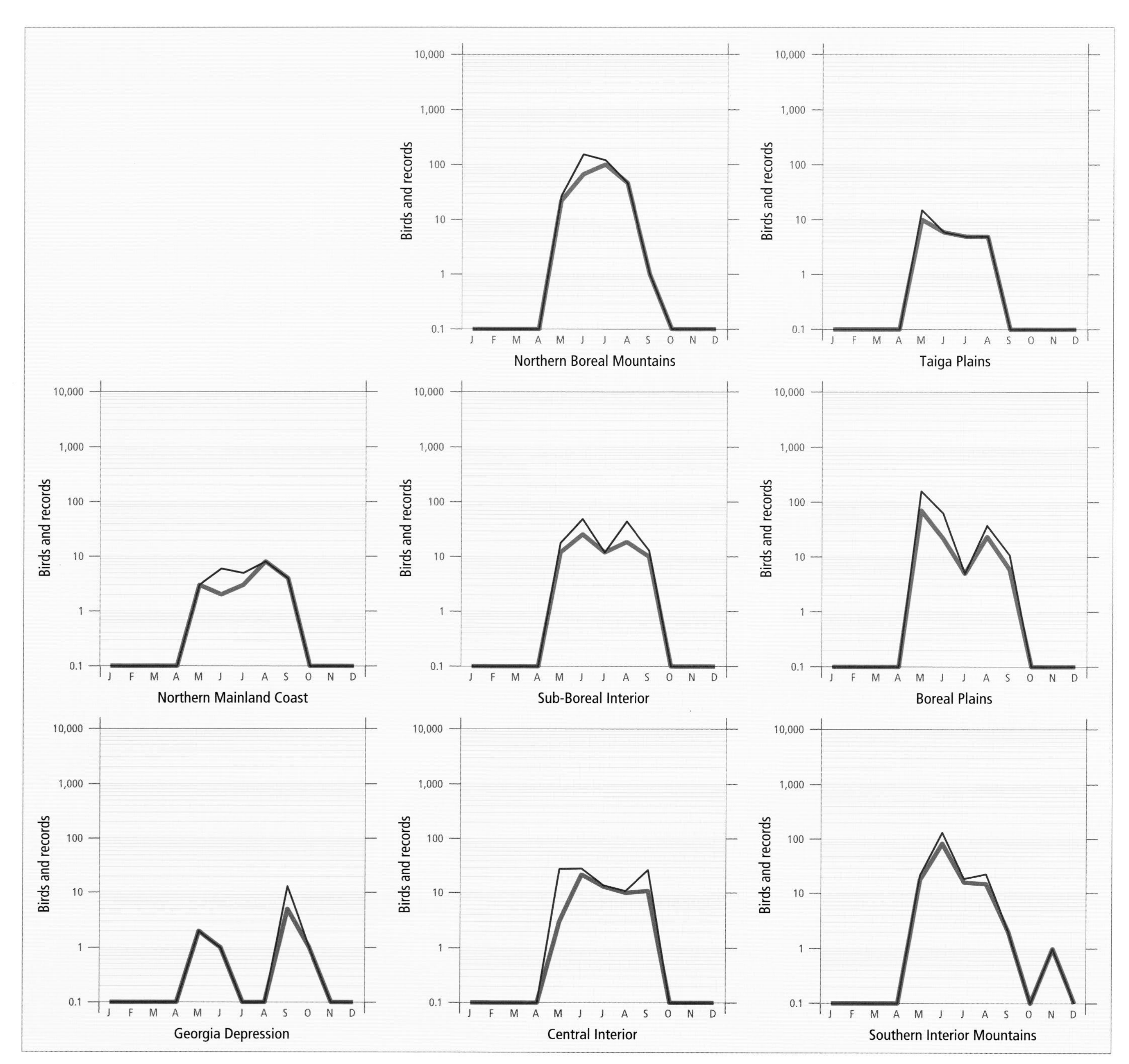

Figure 105. Fluctuations in total number of birds (purple line) and total number of records (green line) for the Blackpoll Warbler in ecoprovinces of British Columbia. Breeding Bird Surveys and nest record data have been excluded.

occupied by Blackpoll Warblers in the Boreal and Taiga plains ecoprovinces were, in descending order of number of warblers observed: black spruce bog, shrub swamp, boreal white spruce–balsam poplar riparian, and montane shrub-grassland. Within these habitat classes, the species occupied wet edge habitat dominated by spruce, or willow-dominated fringes of wet meadows. Associated plant species included scrub birch, Labrador tea, and red honeysuckle (Enns and Siddle 1996).

The Blackpoll Warbler has been recorded breeding in the province from 10 June (calculated) to 21 August (Fig. 104).

Nests: The range of nest sites selected in British Columbia is almost unknown, with just 8 nests found in the province and almost complete lack of detail about them. One nest was saddled on a spruce limb, another on the limb of a deciduous tree; each was between 3 and 3.5 m from the ground. Nest materials in both were mainly grasses, mosses, and twigs.

Eggs: A single nest containing 4 eggs, with incubation "well-started," was found on 28 June. Calculated dates from nestlings of known age indicate that eggs can be found as early as 10 June. Elsewhere, the clutch size has been 4 or

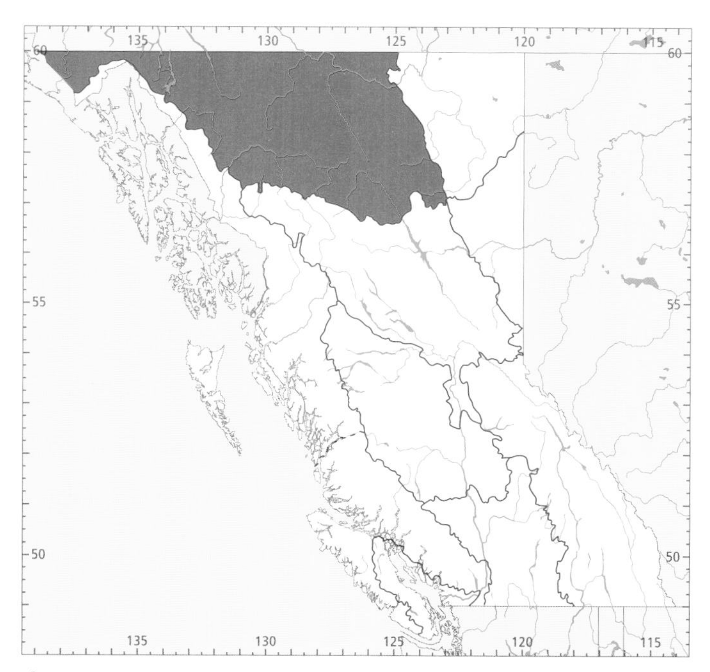

Figure 106. In British Columbia, the highest numbers for the Blackpoll Warbler in summer occur in the Northern Boreal Mountains Ecoprovince.

5 eggs, occasionally 3, and the incubation period is 11 days (Curson et al. 1994).

Young: Dates for 6 broods ranged from 25 June to 21 August, with 3 between 4 July and 18 July. Brood size ranged from 2 to 4 young (2Y-2, 3Y-1, 4Y-3), with 3 having 4 young. The nestling period has been given as 11 or 12 days (Curson et al. 1994).

Brown-headed Cowbird Parasitism: In British Columbia, cowbird parasitism was not found in 6 nests recorded with eggs or young. Elsewhere in North America, there are 2 records of the Blackpoll Warbler as host of the Brown-headed Cowbird (Friedmann et al. 1977).

Nest Success: Insufficient data.

REMARKS: The Blackpoll Warbler population breeding in Alaska, Yukon, and northern British Columbia is a possible source of the birds recorded in small numbers in autumn along the coasts of British Columbia, Washington, Oregon, and California. In British Columbia, the 17 records of this warbler on the southwest coast of the province included 4 individuals seen during spring migration and 12 during autumn migration. This predominance of autumn migrants along the coast has also been noted in Oregon and California (Roberson 1980).

Figure 107. In the Northern Boreal Mountains Ecoprovince, the Blackpoll Warbler breeds along the edges of spruce forests, often near water, where deciduous shrubs such as willow are abundant (Surprise Lake, east of Atlin, 7 July 1996; R. Wayne Campbell).

As of 1953 there were no records of the species from Washington (Jewett et al. 1953), but in recent years there have been a small number; the details are unavailable. From Oregon there have been 9 records, all but 1 of them from the coast and all between 30 September and 10 October (Gilligan et al. 1994). There are many more records from California (about 110 per autumn), 90% of them from the coast (Small 1994).

The small number of observations along the coasts of British Columbia, Washington, and Oregon does not correlate with the larger number of birds recorded from California. This discrepancy could be an artifact caused by the much larger number of observers in California. Also, special environmental features on the coast of California tend to concentrate migrants in a few areas where they are more easily seen. However, on the Atlantic coast, the species is known for its long trans-ocean migration (Nisbett 1970; Murray 1989). Given this propensity, one can postulate a migratory path at sea, nonstop from Alaska to California, with only the occasional individuals appearing on the coasts between.

McCaskie (1970) believed the most likely explanation for the arrival of these birds in California was misorientation of immature migrants. We suggest that the coastal birds may represent the normal migration of a small population occupying a discrete winter range and with its own migratory patterns. The question remains an interesting one for birders and ornithologists to explore.

For a comprehensive summary of current knowledge about this species, see Hunt and Eliason (1999).

NOTEWORTHY RECORDS

Spring: Coastal – Elk Lake (Saanich) 26 to 27 May 1986-1; Seeley Lake 29 May 1962-1; Anthony Island 28 May 1990-1 male in breeding plumage following se storm. **Interior** – nr jct Elk River and Michael Creek 22 May 1984-1; Beaver River valley (Glacier National Park) 14 May 1982-1; Kootenay National Park 16 May 1983-2; 16 km n Golden 20 May 1998-1 male; Anahim Lake 26 May 1948-20; Murtle Lake (Wells Gray Park) 20 May 1959-1 (Edwards and Ritcey 1967); Indianpoint Lake 24 May 1934-1 (MVZ 81473); Punchaw Lake 29 May 1968-1 (Rogers 1968c); Ootsa Lake 31 May 1976-7; Prince George 19 May 1991-1 male; Willow River 25 May 1966-1; Crooked River Park 27 May 1990-4 (Siddle 1990b); Tetana Lake 22 May 1938-1 (Stanwell-Fletcher and Stanwell-Fletcher 1943); Charlie Lake 9 May 1981-1, 11 May 1980-3, 23 May 1983-1, all dates of first arrival; North Pine 18 May 1985-9; Beatton Park 24 May 1985-14, height of migration; s Hyland Post 30 May 1978-2; Parker Lake (Fort Nelson) 13 May 1978-1, 27 May 1975-2; Liard Hot Springs 14 and 16 May 1975-1 (Reid 1975); Helmut (Kwokullie Lake) 21 May 1982-1 (1 or 2 males seen daily, first female recorded 28 May); Chilkat Pass 20 May 1957-1 (Weeden 1960); Atlin 17 May 1981-1 (Campbell 1981), 21 May 1934-1, first arrival (Swarth 1936).

Summer: Coastal – nr Port Renfrew 18 Jun 1974-1; Pitt Meadows 11 Jun 1995-1 adult male along Katziz Marsh Nature Trail; Mosher Creek 16 Jul 1940-2 collected near headwaters at 1,550 m elevation (Laing 1942); Contact Creek 11 Jun 1972-1; Moosehorn Lake (Golden Bear Mine) 18 Jul 1987-1 at access road. **Interior** – Swan Lake (Vernon) 27 Aug 1995-1 immature with migrant Yellow-rumped Warblers; Revelstoke 8 Jun 1986-2; Field 4 Jun 1975-2 (Wade 1977); McIntyre Lake (nr Monashee Pass) 27 Jun 1983-1; Blaeberry 3 Jun 1996-1, 5 Aug 1996-1; Blaeberry River 3 Jun 1996-4 singing males; Lord River 18 Jul 1993-1; Stum Lake 16 Jun 1988-1, 30 Jul 1971-2 fledglings being fed by parent (Ryder 1973); Chezacut Lake 2 Aug 1931-1 (UBC 6433); Anahim Lake 9 Jun 1932-1 (MCZ 284355); Murtle Lake 23 Jul 1950-1 (UBC 2270); Rainbow Range 19 Jun 1932-1 (MCZ 284359); Mount Robson Park 1 to 2 Jun 1996-9, 2 Jun 1973-1; Indianpoint Lake 15 Jun 1930-1 (CZ 284358), 20 Aug 1932-1 (CZ 284364); n Ahbau Lake 17 Jul 1967-1 fledgling fed by male; Nulki Lake 27 Aug 1945-1 (ROM 82163); 37 km n Prince George 12 Jun 1969-1 (NMC 57049); Prince George 27 Jul 1981-1; Driftwood Creek 15 Jul 1979-2 young fed by adults; Pine Pass 24 Aug 1975-3; Gagnon 26 Aug 1995-13, 30 Aug 1995-10, at banding station; One Island Lake 28 Jun 1978-4 eggs; Little Prairie 10 Aug 1930-1 (RBCM 14871); 79 km w Chetwynd 16 to 20 Jun and 18 Jul 1969-23 (Webster 1969a); Brassey and Coldstream creeks 14 Jun 1993-14 on a strip transect (M. Phinney pers. comm.); 23 km s Groundbirch 25 Jun 1994-4 nestlings, eyes not yet open; Tetana Lake 19 Jul 1938-1 (RBCM 8268); Taylor 21 Aug 1975-3 nestlings; Boundary Lake (Goodlow) 23 Jun 1985-1, 20 Aug 1986-5; Firesteel River 27 Jun 1976-25; Stalk Lake 10 Jul 1976-15; Trutch 13 Jul 1943-1 (fairly common to 17 Jul); 28 km s Dease Lake 19 Jun 1962-1 (NMC 50134); Fort Nelson 29 Jul 1967-1; Summit Lake (Stone Mountain Park) 21 Jul 1943-1 (NMC 29550); Muncho Pass 28 Aug 1943-1 (Rand 1944); Liard River 9 Aug 1943-1 (Rand 1944); Atlin 27 Aug 1924-1, latest seen (Swarth 1926); ne Torres Channel (Atlin Lake) 4 Jul 1980-4 nestlings about 1 week old; Lower Macdonald Lake (ne Atlin) 18 Jul 1980-4 recent fledglings; Survey Lake 26 Jun 1980-50; Portage Brule Rapids 11 Jul 1978-2 adults feeding 2 recently fledged young (Campbell 1978a); Mile 42 to 46 Haines Highway Jun 1980-3.

Breeding Bird Surveys: Coastal – Not recorded. **Interior** – Recorded from 14 of 73 routes and 6% of all surveys. Maxima: Wingdam 23 Jun 1968-34; McLeod Lake 14 Jun 1969-20; Pinkut Creek 11 Jun 1975-9.

Autumn: Interior – McBride 1 Sep 1971-1; Hyland Post 3 Sep 1976-1 (Osmond-Jones et al. 1977); Parker Lake (Fort Nelson) 21 Aug 1979-1; Tuaton Lake 3 Sep 1976-1, last record (Osmond-Jones et al. 1977); Beatton Park 7 Sep 1985-3; Cecil Lake 3 Sep 1982-2 at s end; Mackenzie 9 Sep 1996-1, 15 Sep 1995-1; Williams Lake 9 Sep 1953-3 (NMC 48211); Alkali Lake (Cariboo) 7 Sep 1953-3 (Jobin 1954); Revelstoke 2 Sep 1988-1, 14 Nov 1989-1. **Coastal** – Queen Charlotte City 6 Sep 1989-1 immature with small flock of Townsend's Warblers; Triangle Island 4, 5, and 8 Sep 1995-1 immature banded, 2 Oct 1994-1 (Bowling 1995a); Pitt Meadows 29 Sep 1969-1 immature, Menzies Mountain (Pitt Meadows) 9 Sep 1976-1; Vancouver 26 Sep 1962-2 at University of BC campus (Boggs and Boggs 1963a); Iona Island 3 Oct 1997-1 immature; Pacific Rim National Park 15 Sep 1996-1 immature; nr McLean Point (Tofino) 25 Sep 1972-1 (Hatler et al. 1978); Discovery Island 15 Sep 1962-3 (Boggs and Boggs 1963a).

Winter: No records.

Black-and-white Warbler

BAWW

Mniotilta varia (Linnaeus)

RANGE: Breeds from extreme southeastern Yukon, west-central and southwestern Mackenzie, northern Alberta, central Saskatchewan, central Manitoba, northern Ontario, south-central Quebec, Labrador, and Newfoundland south to northeastern British Columbia, central Alberta, southern Saskatchewan, eastern Montana, southwestern South Dakota, western Nebraska, central Kansas, south-central and eastern Texas, central Louisiana, southern Mississippi, southern Alabama, central Georgia, central South Carolina, and southeastern North Carolina. Winters from southern and coastal northern California, southern Arizona, southern Texas, the Gulf coast, and Florida south through the West Indies, Mexico, Central America, and northern South America south to northern Peru.

STATUS: On the coast, *casual* in all seasons in the Georgia Depression Ecoprovince; in the Coast and Mountains Ecoprovince, *casual* on the Southern Mainland Coast and Northern Mainland Coast, *accidental* on Western Vancouver Island, and absent from the Queen Charlotte Islands.

In the interior, *uncommon* migrant and summer visitant to the Boreal Plains and Taiga Plains ecoprovinces; *very rare* vagrant elsewhere in the interior.

Breeds.

CHANGE IN STATUS: Munro and Cowan (1947) state that by the mid-1940s the Black-and-white Warbler (Fig. 108) was known to occur in British Columbia only in the Peace River Parklands. There specimens were obtained and this species was found generally distributed in the Boreal Plains from Tupper and Swan Lake to Fort St. John and Charlie Lake (Cowan 1939). It was not until the 1970s and 1980s that significant additional records of this warbler in the northeast were obtained. The warbler is now known to occur and breed regularly in the Boreal Plains and Taiga Plains.

Figure 108. Although the Black-and-white Warbler (adult male shown) is widespread in northeastern British Columbia, its ecology and specific distribution are poorly known (Anthony Mercieca).

NONBREEDING: The Black-and-white Warbler occurs regularly only in the Boreal Plains and Taiga Plains, where it is distributed throughout the Peace Lowland and the Kiskatinaw Plateau, and throughout the Fort Nelson Lowland in the Taiga Plains. This warbler likely occurs in small numbers throughout much of the Boreal Plains and Taiga Plains wherever deciduous or mixed forests with tall trees occur. Vagrants have appeared in similar habitat in most other ecoprovinces of British Columbia (Fig. 109).

The Black-and-white Warbler has been reported at relatively low elevations in the northeast, from 360 m to about 760 m. Nonbreeding habitat is similar to breeding habitat (see BREEDING).

Spring migrants begin to arrive in the northeast in the second week of May, with most probably arriving in the third and fourth weeks of May (Figs. 110 and 111). Migrants enter

Figure 109. During the nonbreeding seasons, the Black-and-white Warbler has occurred throughout the province as a vagrant, usually in mixed riparian woodlands with dense ground cover, such as this site at Campbell Valley Park, Langley (27 April 1996; R. Wayne Campbell).

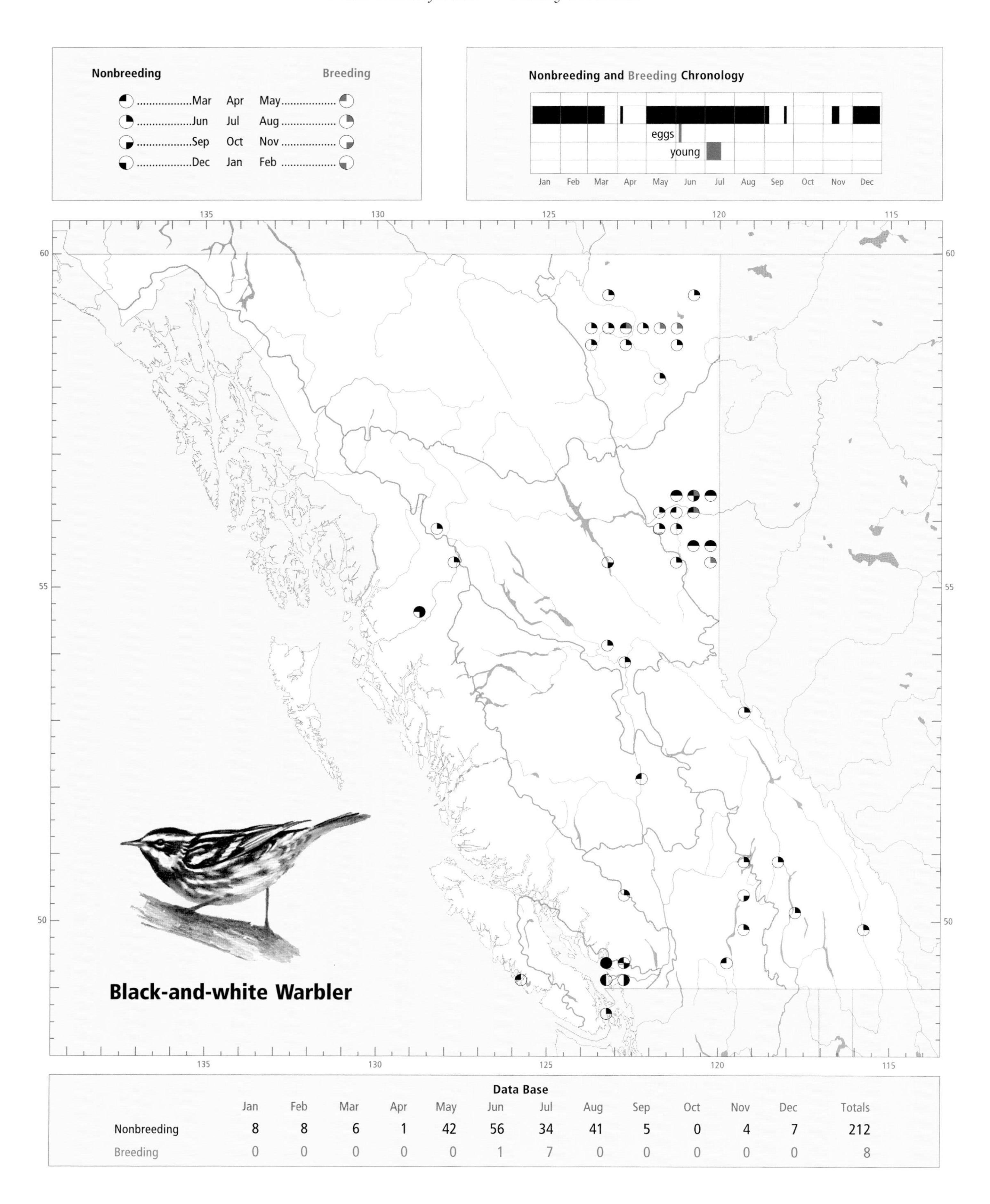

Data Base

	Jan	Feb	Mar	Apr	May	Jun	Jul	Aug	Sep	Oct	Nov	Dec	Totals
Nonbreeding	8	8	6	1	42	56	34	41	5	0	4	7	212
Breeding	0	0	0	0	0	1	7	0	0	0	0	0	8

British Columbia mainly from the east, having passed through Alberta in mid-May (Salt 1973).

The autumn migration is poorly known in British Columbia, but likely begins in late July or early August, shortly after nesting is completed; the main movement is evident in mid-August (Figs. 110 and 111). Almost all birds have left the province by the end of August. The latest autumn departure in the northeast is on 1 September. A single Black-and-white Warbler overwintered at Musqueam Park, in Vancouver, from at least 15 December 1991 through 17 March 1992.

On the coast, the Black-and-white Warbler has been recorded as a vagrant throughout much of the year; in northeastern British Columbia, it occurs regularly from 10 May to 1 September (Fig. 110).

BREEDING: In British Columbia, the breeding distribution of the Black-and-white Warbler includes the Peace Lowland of the Boreal Plains and the Fort Nelson Lowland of the Taiga Plains in northeastern portions of the province. Our knowledge of this warbler's breeding distribution is limited because of the lack of observers who visit its habitat and because of the bird's secretive nesting habits. The Black-and-white Warbler probably breeds in most lower-elevation areas of the Boreal Plains and Taiga Plains wherever deciduous or mixed forests occur.

The Black-and-white Warbler reaches its highest numbers in summer in the Taiga Plains (Fig. 112). Breeding Bird Surveys for interior routes for the period 1968 through 1993 contain insufficient data for analysis. An analysis of Breeding Bird Surveys across Canada from 1966 to 1996 suggests that populations are increasing at an annual rate of 1.8% ($P < 0.01$); North American trends suggest no net change in numbers (Sauer et al. 1997).

The Black-and-white Warbler nests at elevations between 360 and 750 m. Breeding habitat consists mainly of deciduous thickets in wet areas or along the edges of water (Fig. 113). These habitats occur in a wide variety of forest types, including mixed stands of white spruce, trembling aspen, balsam poplar, and paper birch; pure trembling aspen woodlands and balsam poplar stands; and black spruce bogs. It seems to prefer riparian areas along lakes, beaver ponds, streams, and bogs. In 1 study in northeastern British Columbia, breeding habitat was associated with tall willows and alders in the understorey of middle-aged trembling aspen stands (Enns and Siddle 1996). These trembling aspen stands were frequently in moist areas, were usually adjacent to more mature mixed stands, and always had birch present. Along the Sikanni Chief River, alder-willow patches within spruce or mixed spruce–trembling aspen stands in very moist areas were preferred.

In north-central Alberta, similar breeding habitat included fragmented and continuous stands of mixed woods dominated by mature trembling aspen, with high densities of willow in the understorey (Sodhi and Paszkowski 1997). The Black-and-white Warbler has also been characterized as favouring shrub forest edge habitat (Collins et al. 1982). It may use more varied habitats than most other warblers found in northeastern British Columbia.

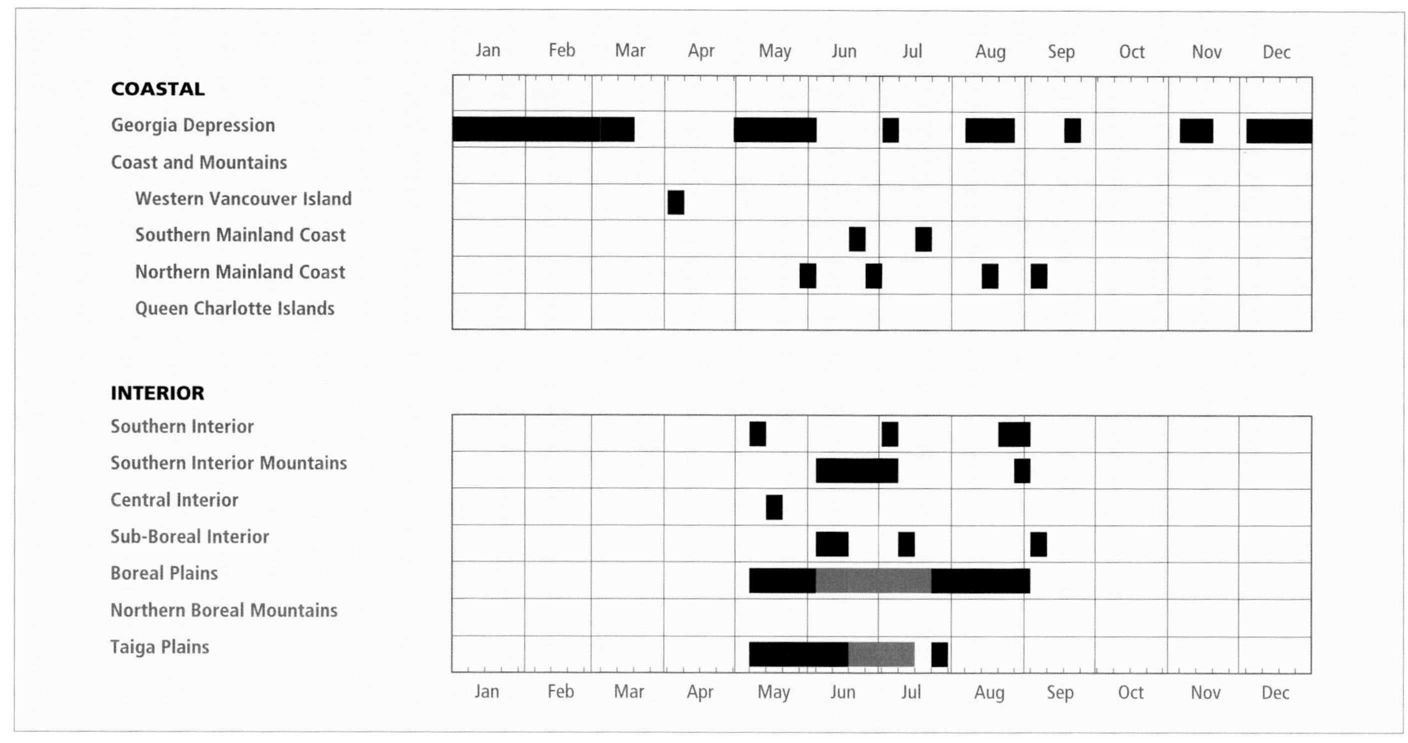

Figure 110. Annual occurrence (black) and breeding chronology (red) for the Black-and-white Warbler in ecoprovinces of British Columbia. Records are shown for the week in which they occurred.

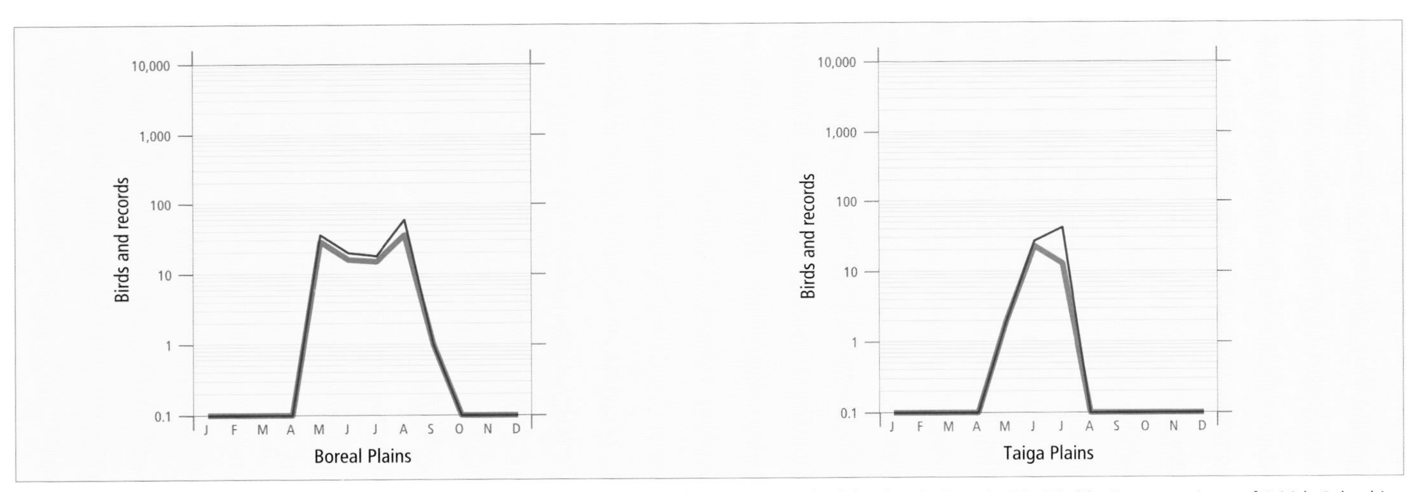

Figure 111. Fluctuations in total number of birds (purple line) and total number of records (green line) for the Black-and-white Warbler in ecoprovinces of British Columbia. Christmas Bird Counts, Breeding Birds Surveys, and nest record data have been excluded.

In British Columbia, the Black-and-white Warbler has been recorded breeding between 4 June (calculated) and 17 July (Fig. 110).

Nests: Only 1 nest has been described in British Columbia. This nest, found near Tupper, was on the ground, well hidden under roots of a willow in "wet woods." It was a cup of shredded bark. The Black-and-white Warbler typically nests on or near the ground against the base of a shrub, among roots, alongside a log, or under an overhanging rock (Morse 1989; Kricher 1995). Nests are typically cups of grasses, bark strips, dead leaves, and rootlets, and are lined with finer material (Kricher 1995).

Eggs: The Tupper nest contained 3 eggs on 6 June. The first egg was estimated to have been laid on 4 June. Eggs can likely be found between the first week of June and mid-July. Clutch size elsewhere is usually 4 or 5 eggs (Baicich and Harrison 1997). The incubation period elsewhere is 10 to 12 days (Kricher 1995).

Young: Nests with young have not been observed in British Columbia. Dates for recently fledged young ranged from 3 to 17 July. Nestlings likely occur from about 16 June through

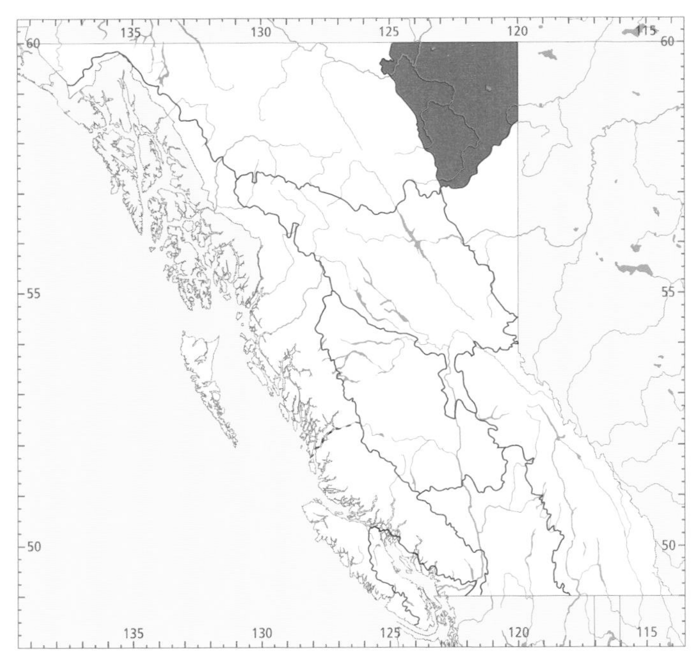

Figure 112. In British Columbia, the highest numbers for the Black-and-white Warbler in summer occur in the Taiga Plains Ecoprovince.

Figure 113. In British Columbia, the Black-and-white Warbler breeds in deciduous thickets, often tall willows, in wet areas adjacent to more open water (near Fort Nelson, 28 June 1998; R. Wayne Campbell).

late July. A late date for fledglings being tended by adults was 4 August. The nestling period elsewhere is 8 to 12 days; this warbler tends to leave the nest relatively early compared with other warblers (Kricher 1995). The sizes of 10 fledged broods in British Columbia ranged from 1 to 3 young.

Brown-headed Cowbird Parasitism: In British Columbia, the single nest we have with eggs also contained a single cowbird egg. Of 11 records of the Black-and-white Warbler feeding fledglings, 1 was of an adult feeding a single cowbird near Toms Lake in the Boreal Plains. Elsewhere, the Black-and-white Warbler is an uncommon host of the Brown-headed Cowbird (Friedmann 1963; Friedmann et al. 1977). In Ontario, however, Peck and James (1987) reported that 20% of nests were parasitized by the cowbird.

Nest Success: Insufficient data.

REMARKS: The Black-and-white Warbler is one of our most distinctive warblers. It is the only member of its genus but is closely related to the *Dendroica* warblers. It has a specialized feeding habit, unique among warblers, of creeping up and down tree trunks and along main branches while foraging, much like nuthatches. It also forages in leafy foliage and branch tips like many other warblers (Dunn and Garrett 1997).

In the northeast, its striking black-and-white plumage could cause it to be confused only with the Blackpoll Warbler, but male and female Black-and-white Warblers have much bolder markings. Observers on the coast should carefully document important field marks, including its "creeping" behaviour, when identifying the Black-and-white Warbler, to avoid any confusion with the Black-throated Gray Warbler. For field marks, see Curson et al. (1994) and Dunn and Garrett (1997).

A published reference to a nest with eggs near Fort St. James in 1899 (Mair and MacFarlane 1908) is erroneous because of a typographical error in the United Kingdom version of the book. Two clutches collected on that expedition, which are now in the United States National Museum, are also incorrectly labelled as being from this species.

For a thorough review of the biology of the Black-and-white Warbler, see Kricher (1995).

NOTEWORTHY RECORDS

Spring: Coastal – Campbell Valley Park (Langley) 28 May 1984-1 adult; Fraser River Park (Vancouver) 23 May 1988-1 (Mattocks 1988); Port Coquitlam 29 May 1973-1 (Jerema 1973); Lennard Island 5 Apr 1977-1 adult male; Terrace 29 May 1976-1 (Hay 1976). **Interior** – West Bench 11 May 1987-1 (Cannings et al. 1987); lower Blaeberry River valley 13 May 1997-1 male; Williams Lake 14 May 1981-1; Fellers Heights 11 May 1993-1; Tupper 20 May 1938-several males arrived (Cowan 1939); Dawson Creek 10 May 1995-1; Taylor 14 May 1988-1, 21 May 1983-3; Charlie Lake 13 May 1981-1; Beatton Park 16 May 1986-1; w side Boundary Lake 14 May 1983-2, first spring arrival; Parker Lake 13 May 1975-1; Alaska Highway 23 May 1974-1, 10 records to early July between Mile 320 and 335 (Erskine and Davidson 1976); Liard Hot Springs Park 15 May 1991-1 adult male singing.

Summer: Coastal – North Saanich 2 Jul 1981-1; Burnaby Lake 9 Aug 1995-1 (Elliott and Gardner 1997); Vancouver 4 Jun 1997-1 at Cecil Green Park; West Vancouver 20 Aug 1988-1 (Mattocks 1989b); Pemberton 17 Jul 1970-1; 8 km n Pemberton Meadows 20 Jun 1981-1; w Hazelton 26 Jun 1975-1. **Interior** – Kelowna 23 Aug 1992-1 (Siddle 1993a); Wasa Park 29 Aug 1971-1 foraging with Red-breasted Nuthatches (Dawe 1971); Nakusp 15 Jun 1991-1 (Siddle 1992c); Revelstoke 5 Jun 1978-1 (first record); Blaeberry River 18 Aug 1995-1 (Ferguson and Halverson 1997); Mount Robson Park 2 Jul 1974-1 (Shepard 1975a); Ness Lake 14 Jul 1968-1 (Rogers 1968d); Prince George 15 Jul 1990-1 (Siddle 1990c); nr Redwillow River 10 Jun 1997-1; Tupper Creek 6 Jun 1941-3 eggs plus 1 Brown-headed Cowbird egg; Sudeten Park 30 Jun 1987-adult male feeding fledged Brown-headed Cowbird; Fellers Heights 12 Jul 1992-2 adults carrying food; e Chetwynd 5 Aug 1975-2; Hudson's Hope 29 Aug 1979-1; nr Taylor 12 Jul 1987-female feeding a downy fledgling; Beatton Park 3 Jul 1997-pair feeding tiny fledgling, 14 Aug 1986-8 (high summer count for a single locality); St. John Creek 17 Aug 1986-6, 31 Aug 1985-1; Beatton Park 2 Jul 1990-male carrying food, 17 Jul 1985-2 begging fledglings following adult female, 4 Aug 1987-pair feeding 2 fledglings, 31 Aug 1982-1; Boundary Lake (Goodlow) 22 Aug 1984-4; Gutah Creek 1 Jun 1991-1 (Siddle 1992c); Sikanni Chief 2 Jun 1996-1; Clark Lake (Fort Nelson) 8 Jul 1978-1 adult and 1 fledgling caught in mist net; Fort Nelson area 2 Jul 1982-6 males at various locations, 4 Jul 1982-1 adult feeding 1 recently fledged young (Campbell and McNall 1982), 9 Jul 1986-2 fledglings; e Fort Nelson 1 Jul 1982-7 males singing at various places along Esso Resources Rd, 4 Jul 1982-2 fledglings with adult female (Campbell and McNall 1982); ne Fort Nelson 6 Jul 1982-adult feeding 3 fledglings nr Mobil Oil camp (Campbell and McNall 1982); Mile 7.5, 317 Rd (ne Fort Nelson) 29 Jul 1979-1; Beaver Lake (ne Fort Nelson) 9 Jul 1978-1 adult female with 2 fledglings; Helmut 2 Jun 1982-1 (RBCM 17483).

Breeding Bird Surveys: Coastal – Recorded from 1 of 27 routes and on less than 1% of all surveys. Maximum: Kispiox 26 Jun 1975-1. **Interior** – Recorded from 4 of 73 routes and on less than 1% of all surveys. Maxima: Tupper 18 Jun 1995-5; Steamboat 25 Jun 1974-1; Fort Nelson 15 Jun 1975-1; Fort St. John 13 Jun 1990-1.

Autumn: Interior – Charlie Lake 1 Sep 1982-1; St. John Creek 1 Sep 1985-1; Mugaha Creek 4 Sep 1997-1; Vernon 2 Sep 1974-2 (Cannings et al. 1987). **Coastal** – Terrace 4 Sep 1968-1 (Hay 1976); Burnaby Lake 11 to 16 Nov 1982-1; Vancouver 22 to 24 Sep 1974-1 at Stanley Park; Saanich 12 and 13 Oct 1993-1.

Winter: Interior – No records. **Coastal** – North Vancouver 3 Dec 1980-1; Vancouver 15 Dec 1991 to 17 Mar 1992-1 at Musqueam Park.

Christmas Bird Counts: Interior – Not recorded. **Coastal** – Recorded from 1 of 33 localities and on less than 1% of all counts. Maximum: Vancouver 15 Dec 1991-**1**, all-time Canadian high count (Monroe 1992a).

American Redstart

AMRE

Setophaga ruticilla (Linnaeus)

RANGE: Breeds from southern Yukon, southeastern Alaska, and southern Mackenzie east across northern Alberta, central Saskatchewan, southern Manitoba, and Ontario to southern Labrador and Newfoundland; in the west, south through British Columbia east of the coastal mountain ranges, into eastern Washington, northern Idaho, and Montana; in the east, south into the states bordering the Great Lakes and through eastern North America to Nova Scotia in the northeast and Georgia in the southeast. Winters from the western coast of northern Baja California and extreme southern Florida south through Central America and the Greater Antilles to Surinam and Peru.

STATUS: On the coast, *rare* migrant and local summer visitant in the Fraser Lowland of the Georgia Depression Ecoprovince; in the Coast and Mountains Ecoprovince, *very rare* migrant and summer visitant on the Southern Mainland Coast, *rare* to *uncommon* migrant and summer visitant on the Northern Mainland Coast, *casual* on Western Vancouver Island, and absent from the Queen Charlotte Islands.

In the interior, an *uncommon* to *fairly common* migrant and summer visitant in the Southern Interior, Southern Interior Mountains, and Central Interior ecoprovinces; *fairly common* to *common* migrant and summer visitant in the Sub-Boreal Interior and Boreal Plains ecoprovinces; *uncommon* migrant and summer visitant in the Northern Boreal Mountains and Taiga Plains ecoprovinces.

Breeds.

NONBREEDING: The American Redstart (Fig. 114) has a widespread distribution throughout most of the province east of the coastal mountain ranges. Along the coast, it appears most years on the southwestern mainland or makes its way to the coast at the heads of deep inlets where major river valleys cut through the mountains, as at Dean, Gardner, and Douglas channels. There are widely scattered records for Vancouver Island and a small population is present each summer in the Pemberton valley.

In the interior, the American Redstart occurs in migration from the eastern foothills of the coastal ranges to the lower-elevation passes through the Rocky Mountains, north in the west to the valley of the Tatshenshini River and in the east to Kwokullie Lake in the Taiga Plains. The small number of locality records in the northern third of the province are mainly concentrated along the roads and highways, an indication that the known distribution in this region is largely a reflection of an absence of observers rather than of birds. Biological inventories now in progress will provide a better picture of the distribution of the American Redstart in the northern regions of the province. The species has a patchy occurrence in the dry interior and is sparsely distributed in the large areas of the northern interior dominated by coniferous forests with few pockets of deciduous trees and shrubs. Such conditions are found in the Rocky Mountain Trench north of Williston Lake, the Cassiar Mountains, and several other ranges and high plateaus within the Northern Boreal Mountains.

Figure 114. Adult female American Redstart incubating eggs (Beatton Park, east of Fort St. John, 23 June 1996; R. Wayne Campbell).

Swarth (1922) reported that at Telegraph Creek, in the Northern Boreal Mountains, from 11 June on "the song of this species was heard everywhere we went in the poplar woods of the lowlands." The same author (1926, 1936) found the species scarce in similar habitat in the Atlin region, and suggested that it was near its northern limit there.

The American Redstart has been reported at elevations from sea level to about 300 m on the coast and from 300 to 1,500 m in the interior. Its occurrence at the coast has been confined to riparian thickets bordering rivers or marshy areas where the habitat featured willow, red alder, red-osier dogwood, Nootka rose, salmonberry, and salal. In the southern interior, it occurs during migration in open forests of ponderosa pine and Douglas-fir; mixed woodlands adjacent to marshes (Fig. 115), lakes, and rivers; stands of western larch;

Figure 115. During spring migration, the American Redstart frequents a wide variety of mixed woodlands, often on south-facing slopes adjacent to water (Illecilewaet, south of Revelstoke, 30 April 1997; R. Wayne Campbell).

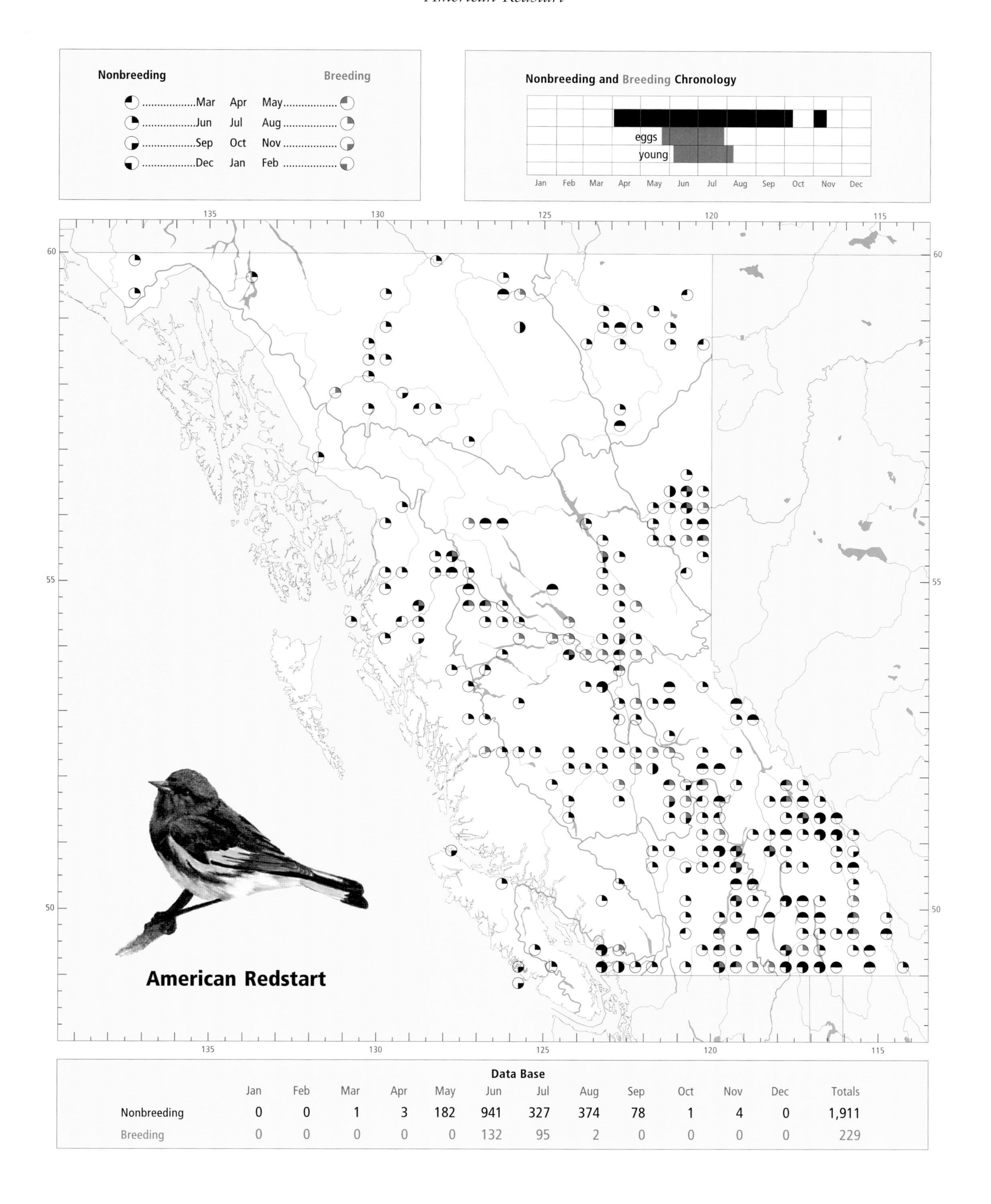

Data Base

	Jan	Feb	Mar	Apr	May	Jun	Jul	Aug	Sep	Oct	Nov	Dec	Totals
Nonbreeding	0	0	1	3	182	941	327	374	78	1	4	0	1,911
Breeding	0	0	0	0	0	132	95	2	0	0	0	0	229

and river floodplains forested with black cottonwood. In the northeast, woodlands of willow, balsam poplar, paper birch, scrub birch, and trembling aspen, as well as riparian vegetation bordering streams, lakes, beaver ponds, and similar wetlands, are preferred. Migrating redstarts are also frequently seen in parks, gardens, and the windbreaks of mixed trees and shrubs that are features of many ranch homes.

In southern regions of the interior, spring migration begins with the appearance of a few birds in April, but consistent arrivals begin in mid-May in both the Southern Interior and Southern Interior Mountains (Figs. 116 and 117). In the Okanagan valley, where there is a long series of records, the first arrivals over a period of 10 years have been noted between 2 May and 5 June, with a mean date of 21 May (Cannings et al. 1987). In both ecoprovinces, the movement does not begin to build until the third week of May; it is at its height in early June both in the Okanagan valley and the valleys of the east and west Kootenay. The front of the migration reaches the Central Interior, the Sub-Boreal Interior, and the Boreal Plains in the last week of May. We suggest, therefore, that the northern half of the province may receive at least part of its population from the migrant stream east of the Rocky Mountains, which is moving to a different timetable. The migration reaches the northern provincial boundary areas during the first week of June (Figs. 116 and 117).

The distinctive differences in plumage colour between the first-spring males and adult male American Redstarts have led to the observation that, elsewhere, most of the adult males arrive on the summer range before the first-spring males. In New York, Francis and Cooke (1986) determined that the first resident second-year males arrived about a week later than the last of the older males. The same migratory timing was noted in the Peace Lowland of British Columbia by Cowan (1939). There the first full-plumaged males reached Tupper Creek on 24 May, while females and first-year males arrived together about a week later. No doubt the sex and age differences in spring migration are characteristic of the species, or even of many species, and are apparent in the American Redstart only because of plumage characteristics.

In the northern regions of the province, the southward migration appears to begin in late July, with the largest numbers moving in the first week of August (Figs. 116 and 117). For example, in the Vanderhoof region most of the resident population of redstarts had left by 30 July, but a further wave of migration through the area, doubtless originating further north, continued through 28 August (Munro 1949). The migration is fast-paced and remarkably synchronous throughout the interior. The latest records in all the interior ecoprovinces except the Northern Boreal Mountains are from late September. Birds moving through coastal localities during the autumn migration are too few to present a pattern, but the latest records are from the third week of September in both the Northern Mainland Coast and, except for a single early November record, on the Fraser River delta.

On the coast, the American Redstart has been recorded from 17 April to 14 November; in the interior, it has been recorded from 4 April to 26 September (Fig. 116).

BREEDING: The American Redstart has a widespread breeding distribution across the province. On the coast, it has been

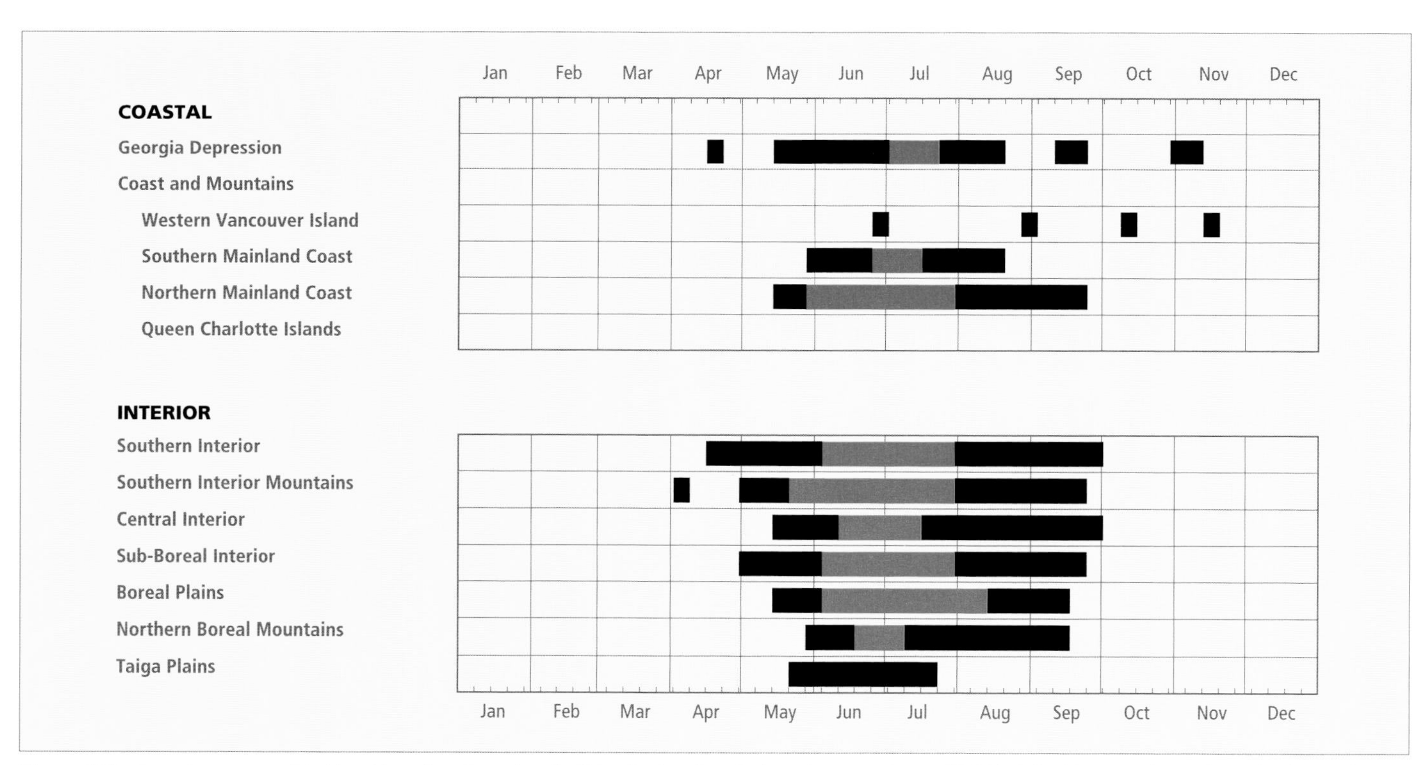

Figure 116. Annual occurrence (black) and breeding chronology (red) for the American Redstart in ecoprovinces of British Columbia. Records are shown for the week in which they occurred.

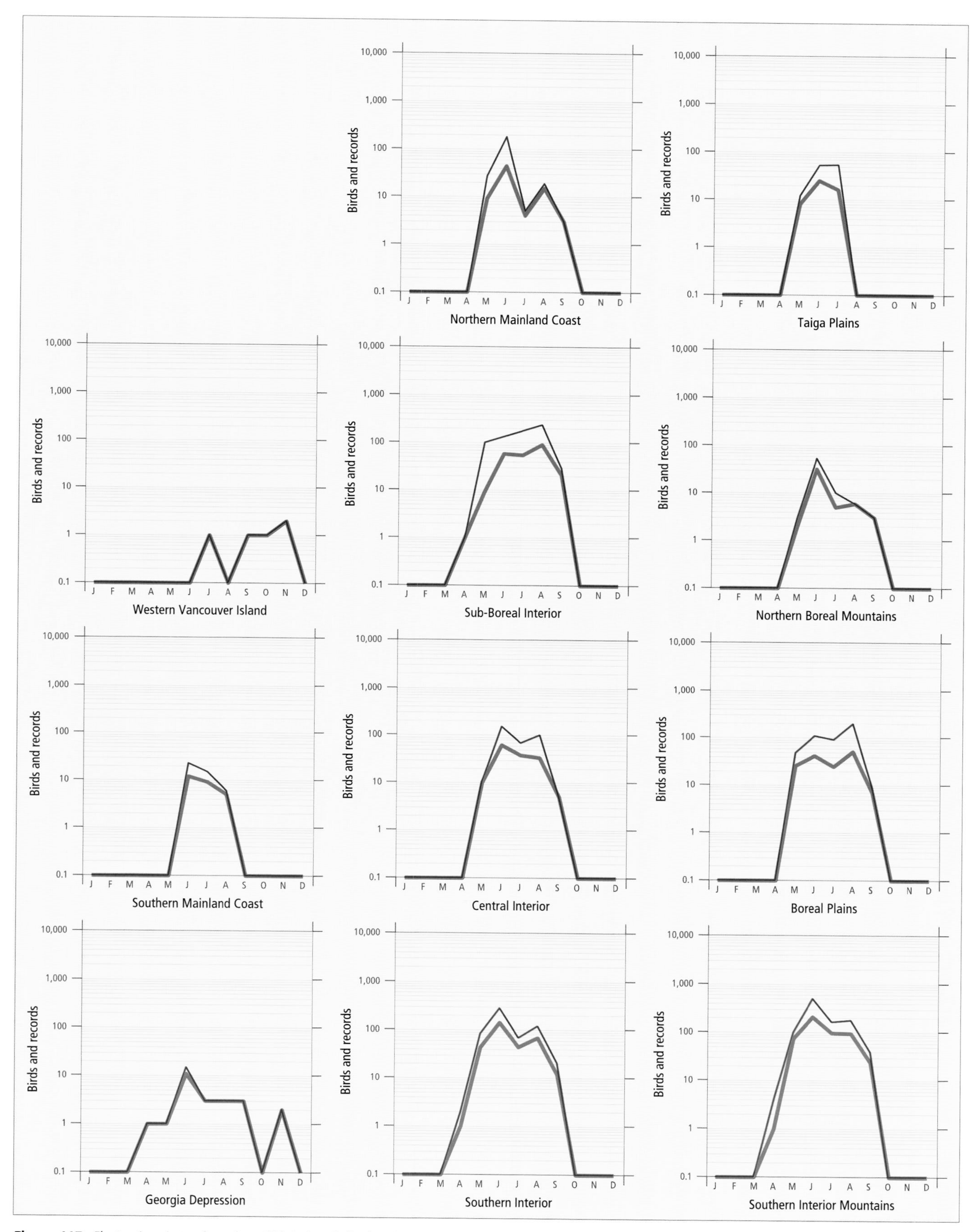

Figure 117. Fluctuations in total number of birds (purple line) and total number of records (green line) for the American Redstart in ecoprovinces of British Columbia. Breeding Bird Surveys and nest record data have been excluded.

reported from Pitt Lake and Hagensborg on the southwest mainland coast, and from further north, in the Skeena River valley near Terrace. Its main breeding distribution lies east of the Coast Mountains from Manning Park east to Cranbrook and north to the Yukon boundary.

In summer, the American Redstart reaches its highest numbers in the valleys of the Sub-Boreal Interior (Fig. 118). Relatively high numbers also occur in parts of the Skeena River valley of the Central Interior, in the Southern Interior Mountains, and in the Peace Lowland, where Cowan (1939) described it as the most abundant nesting warbler.

An analysis of Breeding Bird Surveys in British Columbia for the period 1968 through 1993 could not detect any change in numbers on interior routes; coastal routes provided insufficient data for analysis. Numbers for National and Continental surveys appeared stable for the period 1966 to 1996 (Sauer et al. 1997).

The American Redstart has been reported nesting from near sea level to 200 m elevation on the coast and from 300 to 1,500 m in the interior. In British Columbia, the characteristic nesting habitat of this species features willow shrubs and deciduous trees (Fig. 119). These may occur in a wide variety of forest types and plant associations both adjacent to and far from water (Fig. 120). In the Okanagan valley of the Southern Interior, the redstart prefers "deciduous growth, especially willows and alders bordering streams and lakes" (Cannings et al. 1987). In the Southern Interior Mountains, breeding habitat is varied and includes flooded wetlands of willow; black cottonwood, Sitka and mountain alder, trembling aspen, water birch, hardhack, and thimbleberry; mixed stands of balsam poplar, spruce, and alder with tall willows; and shrubby areas under powerlines composed of willows, wild rose, red-osier dogwood, and black twinberry. In Kootenay National

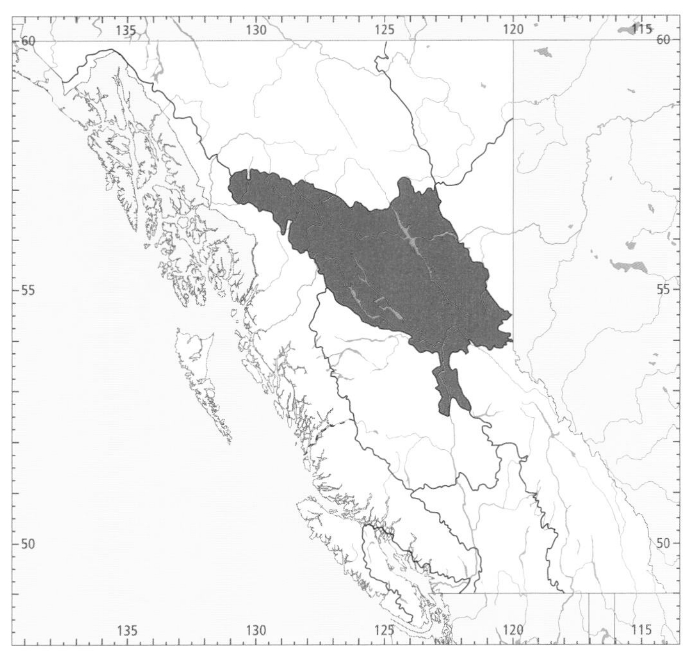

Figure 118. In British Columbia, the highest numbers for the American Redstart in summer occur in the Sub-Boreal Interior Ecoprovince.

Figure 119. Tall willow shrubs and deciduous trees are important components of all breeding habitats for the American Redstart in British Columbia (Hixon, 13 June 1997; R. Wayne Campbell).

Park, the American Redstart inhabits tall shrubbery adjacent to streams and other wetland areas, mixed forests, and alluvial fans dominated by trembling aspen (Poll et al. 1984). North of Golden, in the Rocky Mountain Trench, the redstart can be very common locally in burns or clearcut second-growth (D. Leighton pers. comm.). In the Mackenzie area of the Rocky Mountain Trench, and in the Omineca River valley and Philip Creek area on the Manson Plateau of the Sub-Boreal Interior, breeding American Redstarts are most commonly associated with alders, especially *Alnus incana* and *A. viridis* where they occur as an understorey species with trembling aspen or in mixed aspen-conifer stands (J. Tuck pers. comm.).

In the Boreal Plains of northeastern British Columbia, the American Redstart breeds primarily in the brushy borders of wet deciduous woodlands and shrublands and in tall thickets of willows within dry forest stands. Specific habitats reported by Siddle (1997) include damp willow thickets bordering the edge of mature spruce stands; willow and tamarack saplings in damp sites among young white spruce, balsam poplar, trembling aspen, and water birch forests; shrubby willows and alders along the edges of young balsam poplar and trembling aspen forests; thick tangles of willow, alder, birch, balsam poplar, and trembling aspen, often with red-osier dogwood as a dominant understorey shrub; and extensive willow patches on transmission-line corridors through white spruce, trembling aspen, and balsam poplar forests.

In the Skeena River valley, nesting has occurred in the dense undergrowth of red elderberry, devil's club, willow, and alder, in riparian forests bordering the Skeena River. In the Bulkley River valley, near Smithers, Pojar's (1995) study of the diversity of bird species in trembling aspen forests found that the redstart was absent from clearcut areas but was 1 of the 2 most abundant species in the mature trembling aspen, old trembling aspen, and mixed aspen-conifer stands. In the latter 3 habitats, a census of nesting populations produced estimates of up to 40 singing males per 10 ha and 150 per km^2 (Erskine 1977; Pojar 1995).

Of 138 nesting sites specifically described in British Columbia, 88% were in forest conditions, 9% in wetland areas, and 2% in grasslands. In both of the latter habitats, clumps of trees and shrubs characterize the ecotype selected by the redstarts. More specifically, the nesting territories were in cottonwood floodplain forests; riparian growth involving willow, trembling aspen, and spruce; dense alder thickets; associations including Douglas-fir, birch, and western redcedar; grassy marshland with willows and alders; groves of trembling aspen and willow around grassland ponds; on the edges of cultivated farmland; brushy lakeshores; and even a few in burned areas. Thirty-two percent of nesting areas were close to water.

On the coast, the American Redstart has been recorded breeding from 4 June (calculated) to 28 July; in the interior, it has been recorded breeding from 24 May (calculated) to 7 August (Fig. 116).

Nests: Most nests (55%; *n* = 98) were in deciduous trees (Figs. 114 and 121). Others were placed in shrubs (28%) and in dead trees or snags (8%). More than 94% were in a fork, usually next to the trunk (Fig. 114). The 5 most frequently used nesting materials were: grasses (in 52% of nests), plant down (40%; Fig. 114), plant fibres (32%), string (23%), and animal hair (18%). The heights of 102 nests ranged from ground level to 10 m, with 64% between 1.5 and 3 m.

In rare instances, the American Redstart may use a nest built by another individual. In eastern Canada, Yezerinac (1988) found that after a pair of Yellow Warblers had built a nest and fledged young, redstarts took over the nest and successfully completed incubation to at least the nestling phase.

Eggs: Dates for 123 clutches ranged from 5 June to 28 July, with 55% recorded between 16 June and 29 June. Calculated dates indicate that eggs can occur as early as 24 May. Sizes of 84 clutches ranged from 1 to 5 eggs (1E-6, 2E-13, 3E-14, 4E-36, 5E-15), with 61% having 4 or 5 eggs. The incubation period is 10 to 13 days (Griscom and Sprunt 1957; Peck and James 1987).

Young: Dates for 88 broods ranged from 5 June to 7 August, with 53% recorded between 28 June and 9 July. Sizes of 61 broods ranged from 1 to 5 young (1Y-11, 2Y-10, 3Y-15, 4Y-20, 5Y-5), with 57% having 3 or 4 young. The nestling period is between 7 and 9 days (Sherry and Holmes 1997).

Brown-headed Cowbird Parasitism: In British Columbia, 9% of 131 nests found with eggs or young were parasitized by the cowbird. All parasitized nests were from the interior of the province. There were 6 additional records of an American Redstart feeding a cowbird fledgling. Friedmann (1963)

Figure 120. In the northern parts of the Coast and Mountains Ecoprovince, the American Redstart breeds in a variety of forest types and plant associations adjacent to and far from water (50 km north of Kitwanga, 21 June 1999; R. Wayne Campbell).

Figure 121. In British Columbia, most American Redstart nests are built in the forks of deciduous trees, mainly willows (Beatton Park, 23 June 1996; R. Wayne Campbell).

describes the American Redstart as frequently raising cowbird chicks, but includes the species among those that may bury the foreign egg under a new nest lining. Friedmann and Kiff (1985) refer to studies in Illinois, where the rate of parasitism of the American Redstart had increased from 7% in 1900 to 37% in 1983.

Nest Success: Of 10 nests found with eggs and followed to a known fate, 6 were successful in producing 1 or more young. Lozano et al. (1996) note that "reproductive success in male redstarts is independently affected by both arrival date and age, and that the lower success of subadults may not be a consequence solely of their age and late arrival, but also of their lack of experience in selecting territories early in the breeding season."

REMARKS: A recent study by Hunt (1996) in the regenerating hardwood forests of New Hampshire showed that in earlier successional stages of the forest, there were higher densities of nesting American Redstarts, a higher proportion of older males, smaller territory sizes, and a higher mating success for first spring males than in more mature forests. This combination led to the redstart's higher productivity in younger forest stands compared with older or mature hardwood forests or mixed forests. These findings contradict those of studies in British Columbia (Pojar 1993, 1995) in which nesting densities were higher in the older and mature trembling aspen–birch forests. The recent increase of the hardwood harvest in northeastern British Columbia will provide new opportunities to evaluate this relationship in the context of the boreal hardwood communities.

For additional information on the biology, ecology, and management of the American Redstart, see Morris and Lemon (1988), Lozano et al. (1996), Sherry and Holmes (1997), and Hobson and Villard (1998).

NOTEWORTHY RECORDS

Spring: Coastal – Marpole 17 Apr 1942-1; Vancouver 15 May 1983-1. **Interior** – Fort Steele 21 May 1977-1; Kimberley 24 May 1997-a large flock; Okanagan Landing 2 May 1973-2; White Lake (Shuswap) 17 Apr 1962-2; Barrier Lake 4 Apr 1974-4; Glacier National Park 3 May 1982-2; Lac la Hache 26 May 1959-1; Indianpoint Lake 31 May 1934-1 (MVZ 65671); Mount Tatlow 18 May 1975-1; Prince George to 14 km ene Prince George 29 May 1966-68 on survey (Grant 1966); Chichouyenily Creek 22 May 1994-2 banded; Gagnon Creek 29 May 1994-30 banded; Smithers 18 May 1975-1; Stuart Lake 5 May 1980-1; nr Tumbler Ridge 29 May 1976-1; Tupper Creek 24 May 1938-1; Tetana Lake 20 May 1938-2; Taylor 15 May 1988-1; Sikanni Chief 30 May 1995-1 adult male; Cold Fish Lake 29 May 1976-2; Fort Nelson 23 May 1987-1; Liard Hot Springs Park 28 May 1981-1; Helmut 27 May 1982-1.

Summer: Coastal – Iona Island 19 Aug 1997-1; Seabird Island 7 Aug 1992-1; Sproat Lake 11 Jun 1982-1 (Campbell 1982d); Pitt Lake 20 Jul 1991-3 nestlings (BC Photo 1576), 11 Jun 1995-2, 6 Aug 1995-1 (Elliott and Gardner 1997); Grant Narrows (Pitt Lake) 20 Jul 1991-3 nestlings; Whistler 3 Jul 1997-1 adult male singing; Pemberton 3 Jun 1984-4; Pemberton Meadows 29 Jun 1985-6; Pacific (nr Terrace) 17 Jun 1978-5 eggs; Ferry Island (Terrace) 19 Jul 1979-2 eggs; Spring Creek (Terrace) 16 Jul 1978-4 nestlings; Great Glacier 15 Aug 1919-1 (MVZ 40223). **Interior** – Oliver 25 Jul 1970-nestlings; Grand Forks 15 Jun 1960-4 eggs; 13 km w Hedley 17 Jun 1972-1; Penticton 10 Jun 1913-4 eggs; Nelson 5 Jun 1941-1 egg and 2 nestlings; 2 km n Riondel 13 Jul 1980-1 fledgling; Tamarack Lake 8 Jul 1979-3 nestlings; Shuswap Park 3 Jun 1973-10 (Sirk et al. 1973); Scotch Creek (Shuswap) 7 Jun 1967-2 eggs; Crystal Lake 12 Jul 1961-2 nestlings; Canim Lake 1 Jul 1977-3 young; Hemp Creek 22 Jun 1960-5 eggs; Hotnarko River 9 Jul 1956-1 at mouth (Ritcey 1956); Anahim Lake 7 Jun 1932-3 (MCZ 284526 to 284528); Birch Bay (Horsefly) 19 Jun 1970-4 eggs; Spanish Lake 3 Jun 1992-2 singing males at 1,500 m; Quesnel 31 Aug 1900-2 (RBCM 1948-49); 10 Mile Lake (Quesnel) 8 Jun 1932-eggs present; Mount Robson Park 2 Jun 1973-10 along Kinney Lake Rd; John's Island (Francois Lake) 4 Jun 1977-5; MacLure Lake Park 15 Jun 1978-2 eggs; Fort George 10 Jun 1889-1 egg; Prince George 21 Jul 1981-3 nestlings; Pine Pass 23 Aug 1975-3; Mugaha Creek 6 Aug 1996-30, 10 Aug 1996-24, 15 Aug 1996-13, all on transect counts; 23 km s Groundbirch 25 Jun 1994-4 eggs; Finlay Forks 22 Jun 1930-1 (RBCM 14876); Taylor 8 Jul 1984-26; nr Taylor 7 Jul 1983-1 fledgling just out of nest; Fort St. John 7 Aug 1996-1 young being fed in nest; Charlie Lake 10 Jun 1938-12; Trutch 14 Jul 1943-1 (Rand 1944); Glenora 2 Jul 1919-4 eggs (Swarth 1922); Dease Lake 28 Aug 1996-1; Mile 335 Alaska Highway 19 Jul 1985-1; 1 km e Fort Nelson 9 Jul 1986-2 about 4 days fledged; Muncho Lake 29 Aug 1996-1; Liard Hot Springs Park 19 Jun 1980-5 eggs; Atlin 7 Jun 1934-1 (CAS 42150); Atlin 11 Aug 1924-1 (MVZ 104687); Sediments Creek 21 Jun 1993-12.

Breeding Bird Surveys: Coastal – Recorded from 6 of 27 routes and on 12% of all surveys. Maxima: Kispiox 20 Jun 1993-104; Meziadin 17 Jun 1975-28; Nass River 21 Jun 1975-17. **Interior** – Recorded from 52 of 73 routes and on 48% of all surveys. Maxima: McLeod Lake 25 Jun 1991-71; Telkwa High Rd 27 Jun 1990-55; Ferndale 7 Jun 1969-52.

Autumn: Interior – Muncho Lake 11 Sep 1996-1; McBride River 14 Sep 1977-1; Beatton Park 14 Sep 1985-1; Charlie Lake 2 Sep 1984-1; Gagnon Creek 17 Sep 1995-4; Mugaha Creek 18 Sep 1995-1 banded; Nulki Lake 2 Sep 1953-1 (ROM 89433); 150 Mile House 10 Sep 1953-1 (NMC 48239); Canim Lake 4 Sep 1960-2; 15 km n Golden 17 Sep 1997-3; Golden 4 Sep 1977-5; Yoho National Park 12 Sep 1976-1; Scotch Creek (Shuswap) 22 Sep 1962-1; Savona 26 Sep 1978-1; Nakusp 17 Sep 1975-1. **Coastal** – Pine Island 1 Sep 1976-1 (Shepard 1977); Vancouver 18 Sep 1986-1 at Stanley Park; Westham Island 2 Nov 1972-1 (Crowell and Nehls 1973a); Boundary Bay 5 Nov 1990-1; Long Beach 14 Nov 1973-1 (BC Photo 324).

Winter: No records.

Ovenbird
Seiurus aurocapillus (Linnaeus)

OVEN

RANGE: Breeds from northeastern British Columbia east through southern Mackenzie, northern Alberta, central Saskatchewan, southern Manitoba, central and south-central Ontario, southern Quebec, and the Maritime provinces to southern Newfoundland; south to Colorado in the west and Oklahoma and South Carolina in the east. Winters from the Gulf coast of the southern United States and lowlands in central Mexico south through Central America into northern Colombia and Venezuela.

STATUS: On the coast, *casual* in the Fraser Lowland of the Georgia Depression Ecoprovince; in the Coast and Mountains Ecoprovince, *accidental* on the Northern Mainland Coast and not reported from Vancouver Island, the Southern Mainland Coast, or the Queen Charlotte Islands.

In the interior, *uncommon* to locally *fairly common* migrant and summer visitant in the Boreal Plains and Taiga Plains ecoprovinces; *very rare* in the Sub-Boreal Interior and Southern Interior Mountains ecoprovinces; *casual* in the Southern Interior, Central Interior, and Northern Boreal Mountains ecoprovinces.

Breeds.

CHANGE IN STATUS: Munro and Cowan (1947) described the Ovenbird as an abundant summer visitant to the Peace River Parklands Biotic Area and northwestward. Their reports were only from Tupper in the south to the Fort Nelson and Liard rivers in the north; they had no records from west of the Rocky Mountains. Since that time, the Ovenbird has been reported from every ecoprovince, although only in small numbers and mostly since the 1980s. In the Sub-Boreal Interior and the Southern Interior Mountains, the Ovenbird has been reported more frequently in the last 2 decades than in the previous 7 combined, and there is some evidence that the species may be nesting in the Sub-Boreal Interior. Still, it is an unusual find anywhere west of the Rocky Mountains.

NONBREEDING: The Ovenbird (Fig. 122) has a fairly restricted distribution in British Columbia, with numbers concentrated in the Peace Lowland and the Fort Nelson Lowland, east of the Rocky Mountains. Elsewhere in the interior of the province, its distribution is scattered and sparse. It is seldom reported from the coast. Single Ovenbirds have been found as far west as Kitimat on the Northern Mainland Coast and Kleena Kleene in the Central Interior, and as far south as Richter Pass in the Southern Interior and the Vancouver area of the Georgia Depression; however, the species can be considered of regular occurrence only in the northeastern portions of the province. In neighbouring Washington, it is a rarity confined to the eastern part of the state (Roberson 1980). This regional distribution pattern is typical of species whose migration takes place east of the Rocky Mountains.

The Ovenbird has been reported from near sea level to about 115 m elevation on the coast, and from 285 to 860 m elevation in the interior. It is a secretive species, and by the time it is first reported in the province, it is usually already in its breeding habitat (see BREEDING). A few early and late records for the species suggest that during migration the Ovenbird may frequent floodplains, fields, trembling aspen sapling forests, brushy thickets, and areas of thick undergrowth. An observation of a bird at a landfill is the only recorded use of an industrial habitat by the Ovenbird in British Columbia. Vagrants in the Georgia Depression have been found in suburban habitats and city parks.

Figure 122. The Ovenbird is one of the few warblers that is primarily a ground feeder and ground nester (Anthony Mercieca).

Spring migration is detectable only in the northern portions of the province, where the Ovenbird is found in sufficient numbers. It may first appear in the Boreal Plains by the second week of May, although numbers do not become apparent until the last week of May or the first week of June (Figs. 123 and 124). Early migrants reach the Taiga Plains in the third week of May, with most birds arriving in early June. The Ovenbird may arrive in the Sub-Boreal Interior and Central Interior in the third week in May, but it has not been reported from the Southern Interior Mountains until the last week of May (Figs. 123 and 124).

Autumn migration is difficult to discern because the absence of song makes this secretive species much more difficult to detect. Nowhere in the province do our data allow us to determine a peak for the autumn movement. In the Taiga Plains, birds have not been reported after the third week of July. In the Boreal Plains, the latest reports are from the first few days of September (Figs. 123 and 124). There is some evidence, however, that birds may remain in the province well into September; for example, the last Ovenbird reported from a banding station near Mackenzie in the Sub-Boreal Interior was observed in the third week of September. Vagrants may appear in coastal habitats during the migration period; the November record from Kitimat is unusually late.

The Ovenbird has been regularly recorded in British Columbia from 11 May to 1 September (Fig. 123).

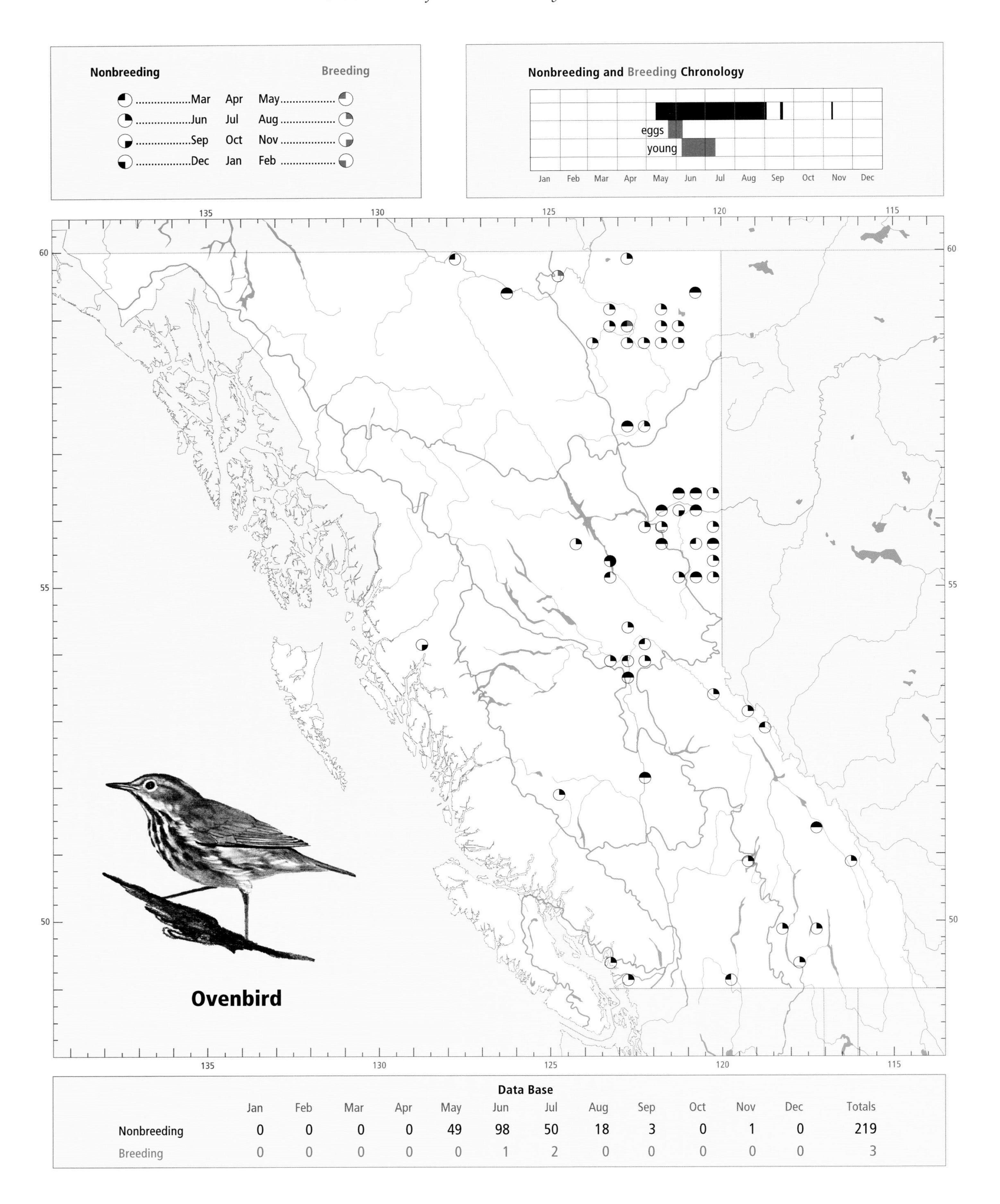

Data Base

	Jan	Feb	Mar	Apr	May	Jun	Jul	Aug	Sep	Oct	Nov	Dec	Totals
Nonbreeding	0	0	0	0	49	98	50	18	3	0	1	0	219
Breeding	0	0	0	0	0	1	2	0	0	0	0	0	3

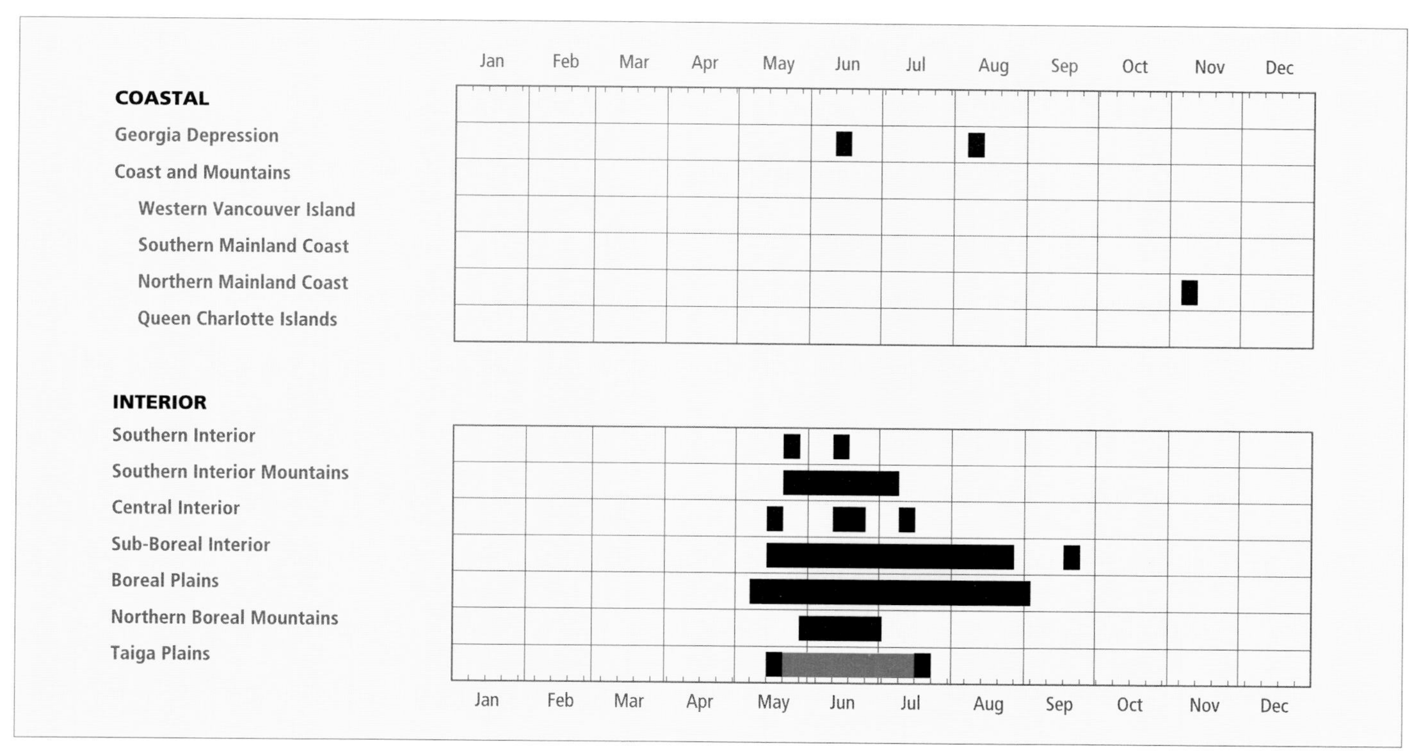

Figure 123. Annual occurrence (black) and breeding chronology (red) for the Ovenbird in ecoprovinces of British Columbia. Records are shown for the week in which they occurred.

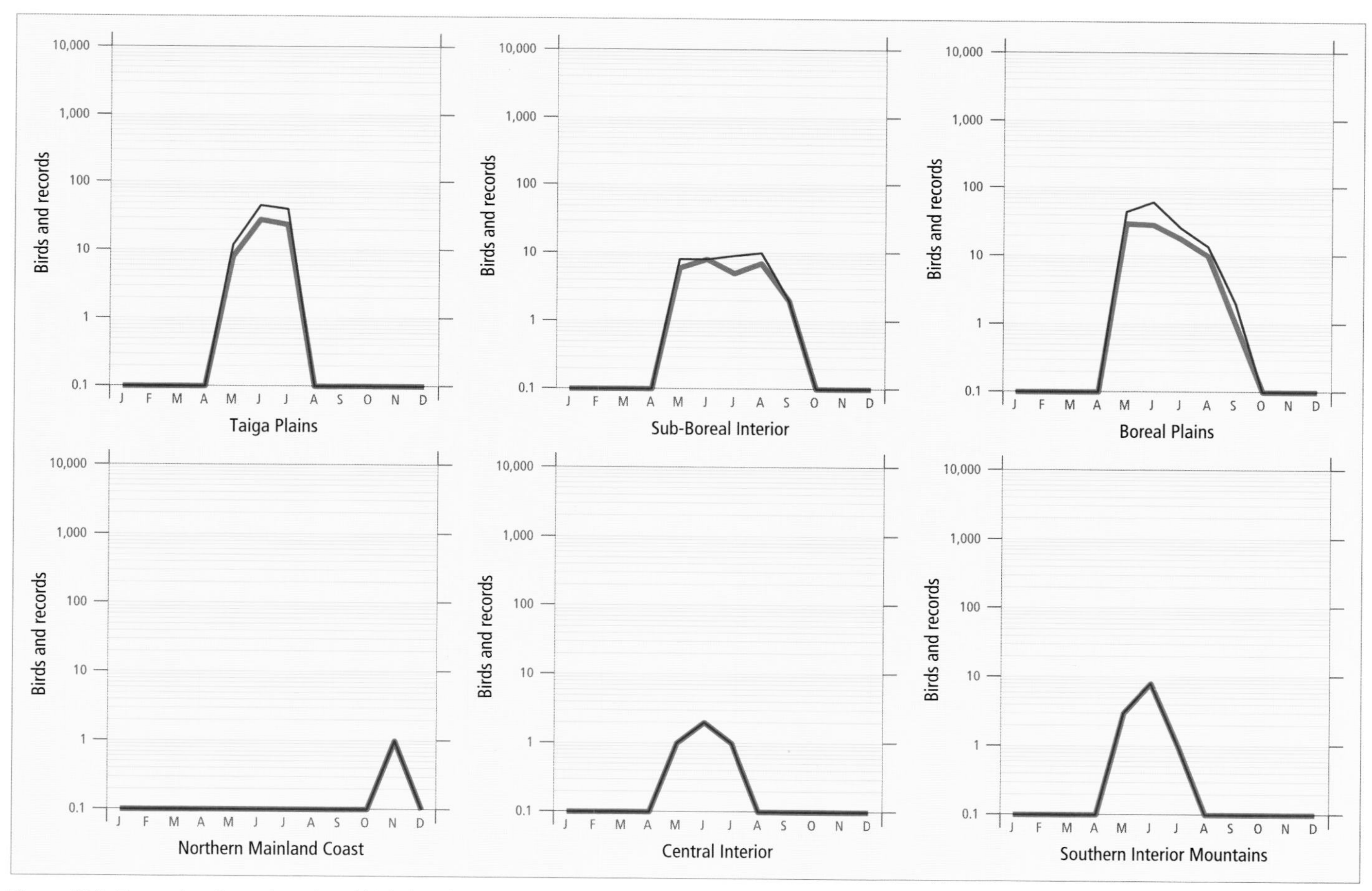

Figure 124. Fluctuations in total number of birds (purple line) and total number of records (green line) for the Ovenbird in ecoprovinces of British Columbia. Breeding Bird Surveys and nest record data have been excluded.

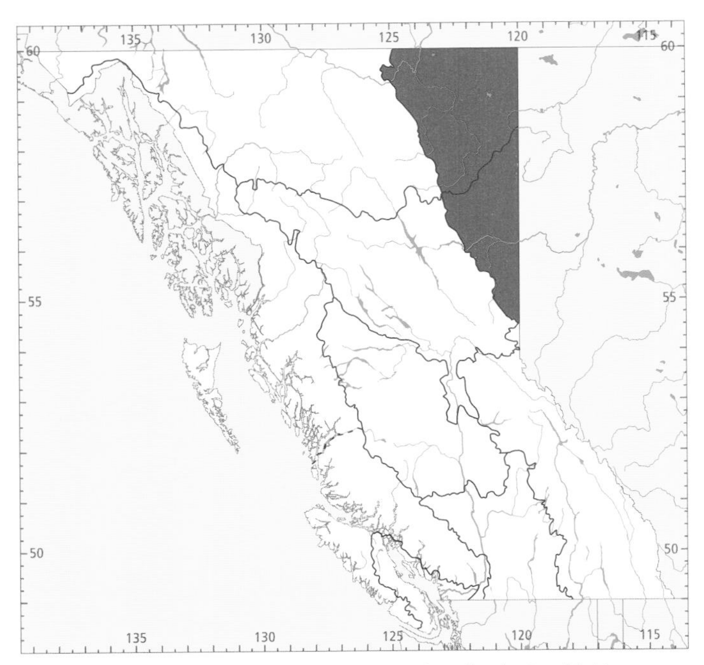

Figure 125. In British Columbia, the highest numbers for the Ovenbird in summer occur in the Boreal Plains and Taiga Plains ecoprovinces.

BREEDING: The Ovenbird likely breeds throughout its summer range east of the Rocky Mountains. To date, however, only 2 nests containing eggs or young have been found, both from the Fort Nelson Lowland. Ovenbird adults accompanied by juveniles or adults carrying food or a fecal sac have been reported from the Peace Lowland of the Boreal Plains, and the Ovenbird is often reported on Breeding Bird Surveys from Tupper and Fort St. John. There are extensive areas of black spruce bogs between their 2 areas of summer concentration, by the Peace and Fort Nelson rivers, which may explain the lack of records from much of this area (Phinney 1992). West of the Rocky Mountains, in the Prince George area of the Sub-Boreal Interior, copulating adults have been seen and nesting birds may occur occasionally (J. Bowling pers. comm.).

The Ovenbird reaches its highest numbers in summer in the Boreal Plains and Taiga Plains (Fig. 125). An analysis of Breeding Bird Surveys in British Columbia for the period 1968 through 1993 could not detect a net change in numbers on interior routes. Sauer et al. (1997) report that on Breeding Bird Survey routes from 1980 to 1996, continental Ovenbird numbers were relatively stable ($P > 0.10$); however, Canada-wide trends for the same period indicate that Ovenbird numbers declined at an average annual rate of 1.1% ($P < 0.01$).

In British Columbia, the Ovenbird has been reported breeding from 285 to 422 m elevation. In the Boreal Plains and Taiga Plains, Enns and Siddle (1996) found it most often in 1 of 3 basic habitat classes: boreal white spruce–trembling aspen (32%; n = 108; Fig. 126), boreal white spruce–balsam poplar riparian (25%), and trembling aspen copse (28%). The preferred sites were productive deciduous stands dominated by paper birch and trembling aspen, with many litter-producing layers of foliage. Understoreys were often messy and fairly dense with either a thick herbaceous or highly productive, woody species composition and high litter production spanning 10 years or more. Plant species common to many breeding sites included highbush-cranberry, Sitka alder, prickly rose, pink pyrola, species of *Erigeron*, tall bluebell, cow-parsnip, meadowrue, horsetail, red-osier dogwood, fireweed, saskatoon, sarsaparilla, creamy peavine, strawberry, American vetch, purple peavine, bluegrass, and kinnikinnick (in open areas with exposed soil and hummocks).

In the Dawson Creek area, Phinney (1992) found the Ovenbird to be most common in pole-stage trembling aspen forest (10 to 20 years) and mature forest (more than 30 years), both forests with a high degree of canopy closure, little or no shrub layer, and abundant leaf litter. Litter depth has been described as an important habitat consideration for territorial males in other parts of the species' range (Van Horn and Donovan 1994; Burke and Nol 1998). Phinney (1992) also found some Ovenbirds in a trembling aspen stand that had undergone a recent mild burn. The stand had retained a complex array of species in its understorey and may have had a temporary abundance of leaf-litter insects as a result of the burn. This suggests that the Ovenbird may also be opportunistic in its choice of habitats.

Near Prince George, the Ovenbird prefers mature trembling aspen stands, often on moister sites with extensive undergrowth that includes black twinberry, red-osier dogwood, ferns, and bunchberry. Van Horn and Donovan (1994) discuss the great structural variation in habitats used by this

Figure 126. In the Taiga Plains Ecoprovince of British Columbia, the Ovenbird breeds in boreal white spruce–trembling aspen associations with a thick undergrowth of Sitka alder and rose (near Fort Nelson, 27 July 1996; R. Wayne Campbell).

Figure 127. In the Boreal Plains Ecoprovince, habitat fragmentation resulting from agriculture activities may have already affected breeding populations of Ovenbirds in northeastern British Columbia (near Progress, 15 June 1996; R. Wayne Campbell).

species across its range, but they note that a partially closed (60% to 90%) canopy, 16 to 22 m above the ground, is repeatedly cited as an important component of territory placement.

Salt and Salt (1976) state that in Alberta the Ovenbird avoids dense coniferous patches and confines itself to extensive stands of trembling aspen and balsam poplar. Van Horn and Donovan (1994) also note that it has a broad tolerance for breeding in different plant communities except sites dominated by conifers. Litter in coniferous forests is frequently devoid of insect life. The scarcity of suitably dry and productive habitat in which deciduous trees can dominate coniferous species may explain the inability of this species to establish itself in coastal habitats.

Throughout its range, the Ovenbird prefers drier areas than the closely related Northern Waterthrush. It appears to prefer uplands and moderately sloped areas but often constructs the nest in or near a break in the forest canopy. Zach and Falls (1975) report that Ovenbirds will use smaller, more tightly packed nesting territories during periods of unusually high food abundance, such as an outbreak of spruce budworm.

The Ovenbird has been recorded breeding in British Columbia from 24 May (calculated) to 11 July (Fig. 123).

Nests: The Ovenbird derives its name from its habit of erecting a light thatch of grass over the nest so that the whole resembles a clay oven similar to the outdoor structures used for baking bread in the early 18th century (Bent 1953). The 2 nests described for British Columbia were not such elaborate constructions; both were simple cups on the ground. Although neither nest was described as domed, 1 was situated beneath a bush and the other was covered with grasses and alder leaves. One was constructed from grass and moss, and the other from grass and leaves.

Eggs: In British Columbia, there have been no observations of nests containing eggs, but calculated dates indicate that eggs may be present as early as 24 May, about 2 weeks after the earliest arrivals are reported. In Michigan, the incubation period ranged from 11 to 14 days, with an average of 12 days and 5.6 hours (Van Horn and Donovan 1994).

Young: Dates for 2 broods ranged from 7 June to 11 July; 1 brood consisted of 3 young and the other of 4 young. In Michigan, the nestling period of 57 birds from 16 nests averaged 7 days and 22.5 hours (Van Horn and Donovan 1994).

Brown-headed Cowbird Parasitism: In British Columbia, cowbird parasitism was not reported in the 2 nests found with young. In eastern North America, Van Horn and Donovan (1994) report cowbird parasitism rates ranging from 10% to 52%. Nests in south-central Ontario suffer relatively low rates of parasitism, e.g., less than 10% (Burke 1998). Friedmann and Kiff (1985) found the high rate of parasitism in North America puzzling because the Brown-headed Cowbird prefers open nests built in open country and seldom bothers forest birds; they noted, however, that cowbird eggs laid in Ovenbird nests are much less likely to fledge than those placed in Song Sparrow or Red-eyed Vireo nests.

Nest Success: Insufficient data.

REMARKS: Of the 3 races of the Ovenbird, only *S. a. aurocapillus* occurs in British Columbia (American Ornithologists' Union 1957; Godfrey 1986).

The Ovenbird has a long history of inappropriate names. It was first considered a wagtail (1766) and then called the Golden-crowned Thrush (1790). In 1827, this thrush-like, ground-dwelling warbler and the waterthrushes were assigned to the genus *Seiurus*, but the common name was not changed until the American Ornithologists' Union meeting of 1886. The current name remains inappropriate because the North American Ovenbird is not a member of the diverse family of neotropical ovenbirds (Furnariidae) among which it winters.

Salvin (1883) mentions an Ovenbird collected at Esquimalt in 1880 by Captain A.H. Markham. This would be the first recorded occurrence of this species on the west coast of northern North America. Brooks and Swarth (1925) decided that the specimen had been collected further south. We reconsidered the validity of this record in light of more recent records from coastal North America, including records from British Columbia and areas south to California (Roberson 1980) and even Skagway, Alaska (31 May 1899; Macoun and Macoun 1909). Two facts weigh heavily against the Esquimalt record. In March 1880, *HMS Triumph* was within the winter range of the Ovenbird at Acapulco, where Captain Markham undoubtedly acquired 2 Golden-winged Warblers (*Vermivora chrysoptera*). The record of those 2 birds follows that of the Ovenbird in his chronological list of acquisitions and is also erroneously attributed to Esquimalt. It seems most likely that this was a clerical error passed on to Salvin (1883).

There are several coastal records for which convincing details are lacking, including a bird that was only heard, not seen, in Point Grey during a Breeding Bird Survey on 16 Jun 1985. All have been excluded from this account.

The Ovenbird can suffer significantly from habitat fragmentation. Studies have found that in some areas the minimum contiguous habitat area required for the Ovenbird to breed successfully ranged from 100 to more than 850 ha, depending on local productivity (Van Horn and Donovan 1994). Small patches may be unsuitable because of factors such as edge effects (increased predation, increased Brown-headed Cowbird access) and the absence of neighbouring members of the same species, or because the lower humidity in smaller tracts contributes to a lower abundance of leaf-litter insects. For example, in southern Ontario Burke and Nol (1998) found that the density and pairing success of territorial males increased significantly with the area of the woodlot core. In Ovenbird territories, leaf litter was deeper in large woodlots than in small woodlots; this influenced prey biomass, which was 10 to 36 times higher in large woodlots than in small woodlots. The researchers also found that Ovenbirds chose territories with significantly higher prey biomass than was found at randomly selected sites within the woodlots. Ovenbird territories in large woodlots were more than 250 m from the forest edge; such distances were not attainable in smaller woodlots (Burke and Nol 1998). Woodlots smaller than 90 ha in core area are now considered sinks, where adult mortality exceeds recruitment (Burke 1998). Kroodsma (1984) also found that Ovenbird territories were significantly denser in the forest interior than near the edge. Habitat fragmentation in the Peace Lowland, resulting from extensive clearing of the mature trembling aspen forests for agriculture, has already removed much of the prime Ovenbird nesting habitat from the Boreal Plains (Fig. 127), and there are increased industrial demands on the remaining mature trembling aspen forests in both the Boreal Plains and Taiga Plains. Continued loss of habitat and fragmentation of remaining habitats could seriously affect populations of the Ovenbird in British Columbia.

A recent review of the biology of the Ovenbird in North America can be found in Van Horn and Donovan (1994).

NOTEWORTHY RECORDS

Spring: Coastal – No records. **Interior** – Richter Pass 22 May 1981-1 along road to Kilpoola Lake; lower Blaeberry River valley (n Golden) 25 May 1996-1 male singing; Williams Lake 15 May 1981-1 found dead beneath window; Mount Robson Park 30 May 1992-1 singing (Bowling 1992); 30 km s Mackenzie 20 May 1996-2; Mackenzie 20 May 1995-1; s Fellers Heights 11 May 1993-1; Tupper Creek 26 May 1938-1 (RBCM 8091), 28 May 1938-1 (RBCM 8093), 31 May 1938-2 (RBCM 8094 and 8095); Taylor 18 May 1985-8 on Peace Island Park road census; Sikanni Chief 22 May 1995-1; Fort Nelson 18 May 1987-1; Liard Hot Springs Park 28 May 1981-1; Iron River at Hyland River 28 May 1981-1.

Summer: Coastal – Surrey 12 Jun 1988-1 (Mattocks 1989a); Vancouver 16 Jun 1985-1 heard at Stanley Park, 1st record for SW mainland coast; North Vancouver 8 Aug 1988-1 struck window. **Interior** – Shuswap Lake Park 15 Jun 1989-1 (Rogers 1989b); Blaeberry River 15 Jun 1996-1 male singing; Kleena Kleene 23 Jun 1965-1 seen and heard singing; Williams Lake 12 Jun 1982-1; Castle Creek (w Dunster) 2 Jul 1968-1; Fort George Canyon 8 Jul 1995-4; 33 km w Prince George 11 Jun 1969-1 male (NMC 57054); nr Bullmoose Creek 10 Jun 1995-1 in trembling aspen forest; Tumbler Ridge area 3 Jul 1990-1; Donna Creek 1 Jun 1993-1; Brassey Creek 17 Jun 1994-23 along 7.6 km × 80 m transect in mainly deciduous forest; Tupper Creek 21 Jun 1938-1 (RBCM 8096); Gething Creek 24 Jun 1976-1 singing; Dawson Creek 3 Jul 1974-2; 1 km sw Hudson's Hope 24 Jun 1992-1 singing; Charlie Lake 10 Jun 1938-2 (RBCM 8097 and 8098); 1 km w Cecil Lake 24 Jun 1992-1 singing nr village; Fort St. John 5 Jun 1922-1 singing (Williams 1933a); Sikanni Chief 10 Jun 1992-1; Fort Nelson 12 Jun 1976-4, 26 Jun 1985-6 among balsam poplar bordering airport, 11 Jul 1967-4 nestlings, 18 Jul 1943-well-grown fledgling accompanying adult; Mile 11.5 on 317 Rd 9 Jun 1979-1; Liard Hot Springs 7 Jun 1980-3 young, 17 Jun 1982-1, 28 Jun 1985-1; Helmut 8 and 10 Jun 1982-3 singing in same area; Nelson Forks 7 Jun 1980-3 nestlings fed by adult; Petitot River Bridge 26 Jun 1985-3.

Breeding Bird Surveys: Coastal – see REMARKS. **Interior** – Recorded from 9 of 73 routes and on 4% of all surveys. Maxima: Fort Nelson 19 Jun 1974-11; Fort St. John 29 Jun 1984-8; Tupper 15 Jun 1974-6.

Autumn: Interior – Cache Creek (Peace River) 1 Sep 1979-2 at jct Hwy 29; Mugaha Creek 18 Sep 1995-1 banded. **Coastal** – Minette Bay 10 Nov 1974-1 at n end (Hay 1976).

Winter: No records.

Northern Waterthrush

Seiurus noveboracensis (Gmelin)

NOWA

RANGE: Breeds from western and north-central Alaska, northern Yukon, southern Mackenzie, northern Alberta, northern Saskatchewan, northern Manitoba, northern Ontario, northern Quebec, northern Labrador and Newfoundland south to southern Alaska, central British Columbia, northeastern Washington (and in an isolated pocket in southwestern Oregon), northern Idaho, and western and central Montana, central Alberta, southern Saskatchewan, and southern Manitoba; and south to New England and northwestern Pennsylvania. Winters from southern Baja California and the coasts of central Mexico south to Ecuador, northern Peru, and northern Brazil, and south from the tip of Florida through the Caribbean islands.

STATUS: On the coast, *very rare* transient and *casual* in winter in the Georgia Depression Ecoprovince; in the Coast and Mountains Ecoprovince, *very rare* transient on the Southern Mainland Coast, *casual* on Western Vancouver Island, *very rare* migrant and summer visitant on the Northern Mainland Coast, and *accidental* on the Queen Charlotte Islands.

In the interior, *uncommon* to locally *common* migrant and summer visitant in the Sub-Boreal Interior, Boreal Plains, and Taiga Plains ecoprovinces; *uncommon* migrant and summer visitant elsewhere in the interior; *accidental* in winter in the Southern Interior Ecoprovince.

Breeds.

NONBREEDING: The Northern Waterthrush (Fig. 128) has a widespread distribution throughout much of interior British Columbia south of latitude 56°N. Further north, it is a fairly regionalized species in both the Boreal Plains and Taiga Plains, but its distribution becomes sparser in the Northern Boreal Mountains. This may be due simply to a lack of observers in those areas. There are large areas of the interior for which we have no information, including the McGregor River Basin, the Rocky Mountain Trench north of Mackenzie, a large area east of the Rocky Mountains north of Fort St. John and south of Kotcho Lake, and much of the Northern Boreal Mountains.

On the coast, it appears irregularly north to the Kispiox region of the Northern Mainland Coast of the Coast and Mountains. Observations are sporadic from Stewart north along the Alaska panhandle border to the Tatshenshini River and the extreme northwestern portions of the province.

The Northern Waterthrush has been recorded from near sea level to 2,600 m elevation. Nonbreeding habitat is similar to breeding habitat (see BREEDING), although during the nonbreeding season, the Northern Waterthrush becomes more flexible in its choice of habitats. During migration, for example, it may occupy damp, shady edges of fast-moving streams and creeks as well as brushy margins of lakes and water control channels. At Creston, Munro (1950) found it in the shrubby mountain slopes, and it has been reported from human-made dugouts and water-filled ditches surrounded by shrubs in

Figure 128. The Northern Waterthrush is a large wood-warbler that nests in wet habitats throughout much of interior British Columbia (Anthony Mercieca).

upland habitats, which it does not frequent during the breeding period (C. Siddle, pers. comm.). During autumn migration on the coast, 2 were captured in Garry oak meadows at Rocky Point near Victoria.

On the southern coast, birds may arrive in the Georgia Depression and Southern Mainland Coast by mid-April, but migrants have not been reported from the Northern Mainland Coast until the third week of May (Figs. 129 and 130). In all coastal regions, our data show little sign of a significant spring movement. In the interior, spring arrivals may appear in the Southern Interior Mountains as early as the first week of April, but migrants do not become numerous until mid-May. They arrive in the Southern Interior and Central Interior by the end of April or the first week of May (Figs. 129 and 130), but significant numbers do not appear until at least the third week of May. In the northern portions of the interior, migrants arrive in the first or second week of May but can be delayed until the end of the month, and significant numbers are not reached until the third week of May or early June (Fig. 130).

Eaton (1995) suggests that populations in British Columbia probably migrate south along the Cascade Mountains. However, birds in the northeastern corner of the province, and perhaps from all the far north, likely move southeast to the Peace Lowland and then into Alberta.

Autumn migration is more protracted than the spring movement and seems to get under way soon after the young are on the wing. In the far north, our data are insufficient to detect any migration; all birds have left the areas by the last week of August. In the Boreal Plains, peak numbers are noted in the last 2 weeks of August and most birds have gone by the third week of September (Figs. 129 and 130). In the Sub-Boreal Interior, the movement is well under way in early August. The highest numbers are reached in the first 3 weeks of the month, although an obvious movement continues into the first

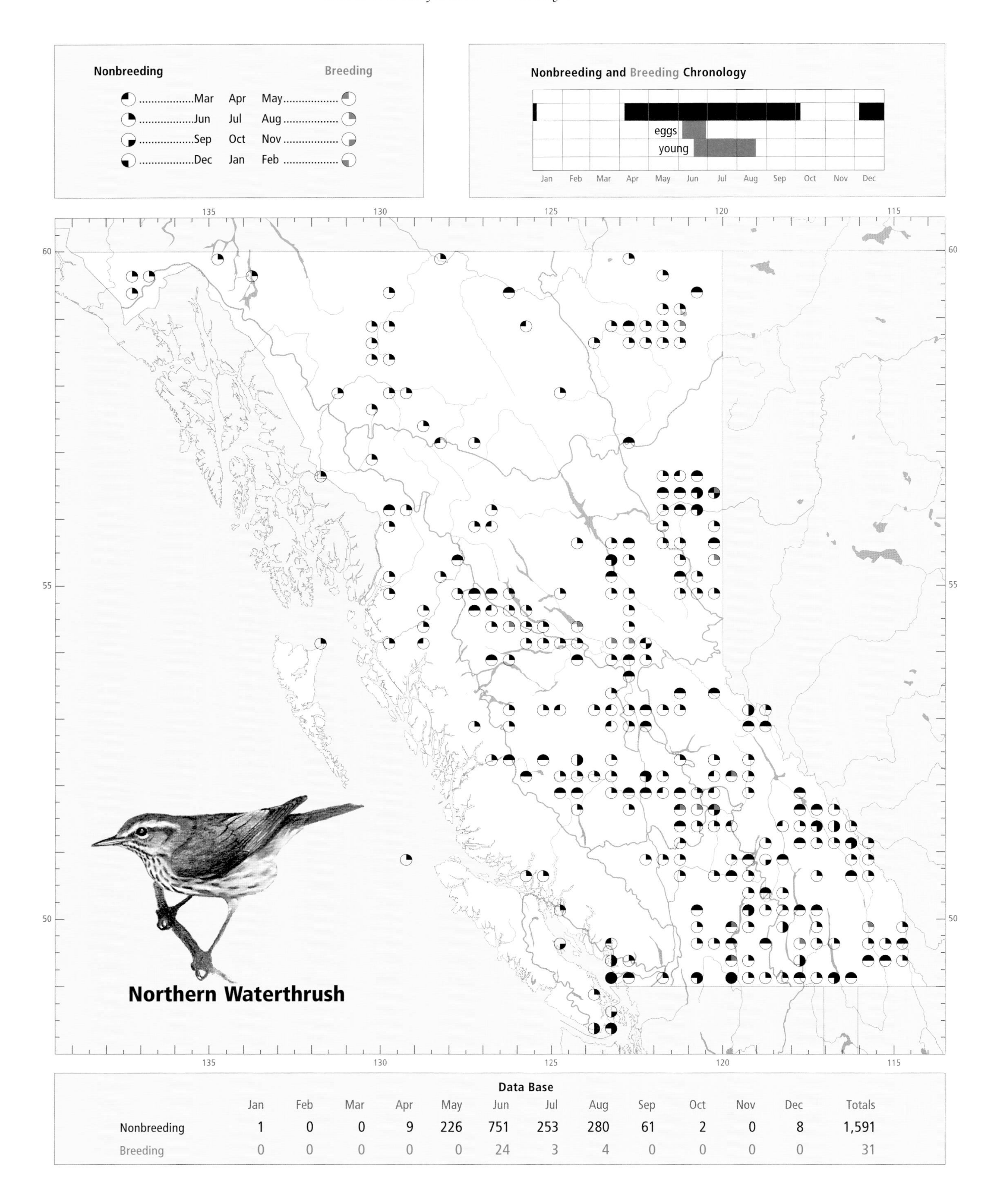

Data Base

	Jan	Feb	Mar	Apr	May	Jun	Jul	Aug	Sep	Oct	Nov	Dec	Totals
Nonbreeding	1	0	0	9	226	751	253	280	61	2	0	8	1,591
Breeding	0	0	0	0	0	24	3	4	0	0	0	0	31

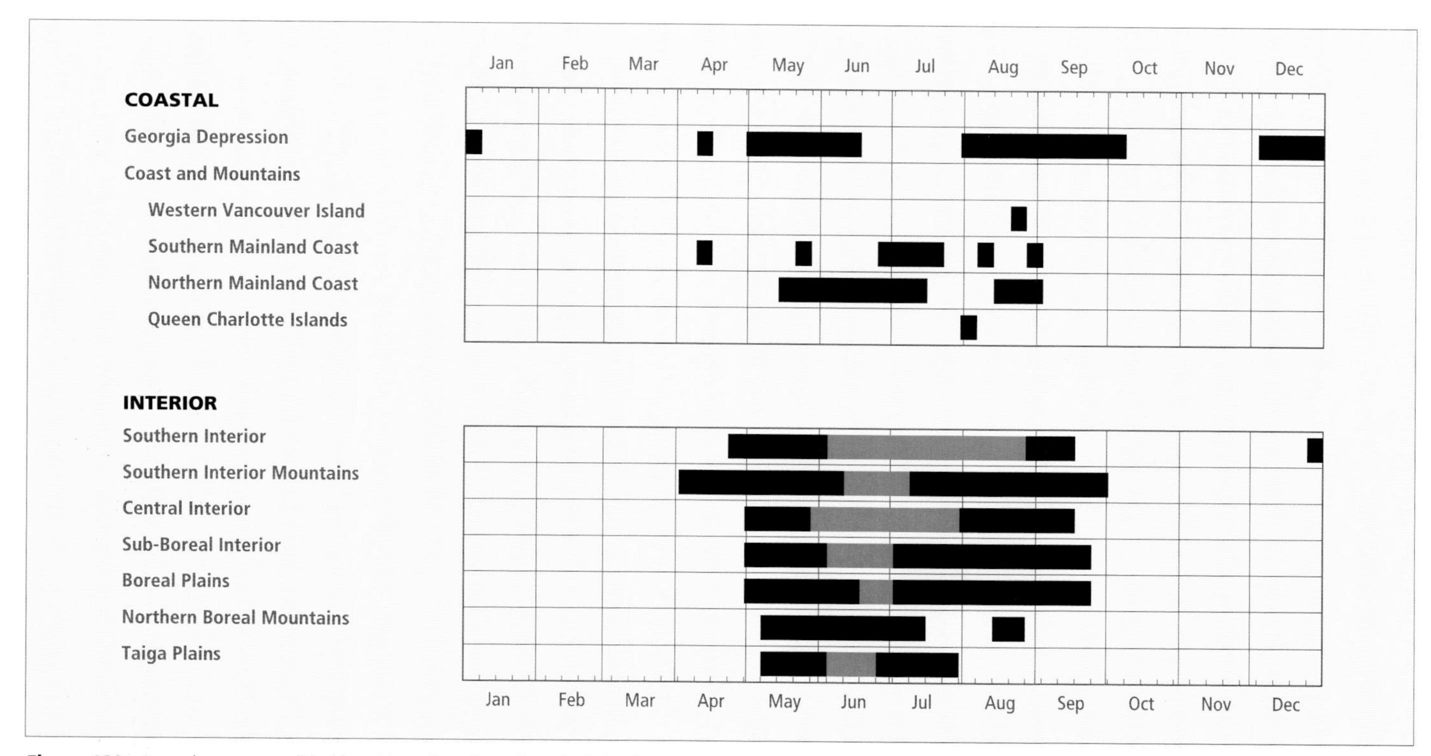

Figure 129. Annual occurrence (black) and breeding chronology (red) for the Northern Waterthrush in ecoprovinces of British Columbia. Records are shown for the week in which they occurred.

week of September. In the Central Interior, a small movement is noticeable the last week of July and the first 2 weeks of August. Munro (1949) also noted a small migration of juveniles moving among the brush along the Nulki Lake shoreline in the second week of August. Most birds have gone from the region by mid-September. In the Southern Interior Mountains, autumn numbers are highest from mid to late August, and all birds have gone by the end of September (Figs. 129 and 130). Near Creston, Munro (1950) noted a local peak of migration on 22 August. In the Southern Interior, our data are insufficient to detect any significant autumn movement; most birds have left by mid-September.

On the coast, small numbers of autumn migrants appear in the Georgia Depression from early August to mid-September, with most birds reported in the last 2 weeks of August. Stragglers may occasionally be found in the first week of October. A few waterthrushes have been caught at a migration monitoring station at Rocky Point, southwest of Victoria (R. Millikin pers. comm.). Two have also been banded on Triangle Island in the last week of August. These records suggest that there may be a small, unnoticed coastal movement, but the birds could also be vagrants from further north.

Rarely, an individual will occur in early winter in wetlands in the southern portions of the province. There are 2 winter records from Oliver in the Southern Interior and 5 from Reifel Island in the Georgia Depression.

On the coast, the Northern Waterthrush has been recorded regularly from 10 April to 2 October and in winter from 7 December to 3 January; in the interior, it has been recorded regularly from 7 April to 24 September and once in winter, on 28 December (Fig. 129).

BREEDING: The Northern Waterthrush has been reported nesting at only a few locations east of the Coast Mountains, including the Okanagan valley in the Southern Interior; the east and west Kootenay regions and Wells Gray Park in the Southern Interior Mountains; near 100 Mile House, Sinkut Lake, and Houston in the Central Interior; Willow River and Fort St. James in the Sub-Boreal Interior; near Boundary Lake in the Boreal Plains; and south of Kotcho Lake in the Taiga Plains. It likely breeds throughout most of its summer range in the interior of the province where there is suitable habitat, although the distribution of our data with any evidence of breeding implies a strong northern component to the population.

On the coast, the Northern Waterthrush likely breeds on the Northern Mainland Coast of the Coast and Mountains. There is 1 report of fledged young from the Kispiox River valley, and singing males are regularly reported on Breeding Bird Surveys in the Port Essington, Terrace, Nass Bay, and Meziadin Lake regions. To date, however, nests with eggs or nestlings or recently fledged young have not been reported.

Even where it is strongly suspected that the Northern Waterthrush breeds, its behaviour and the dense shrubbery in which it nests make finding nests difficult. Recently fledged young are also difficult to observe even though they are flightless, as they tend to hide under the dense vegetation. For example, Cowan (1939) noted that the Northern Waterthrush was an abundant breeding bird about Swan Lake and Charlie Lake; a few pairs nested near Toms Lake and in dogwood thickets bordering the lower 2 miles of Tupper Creek. He did not report any nests containing eggs or young, however. Stanwell-Fletcher and Stanwell-Fletcher (1943) noted that 10

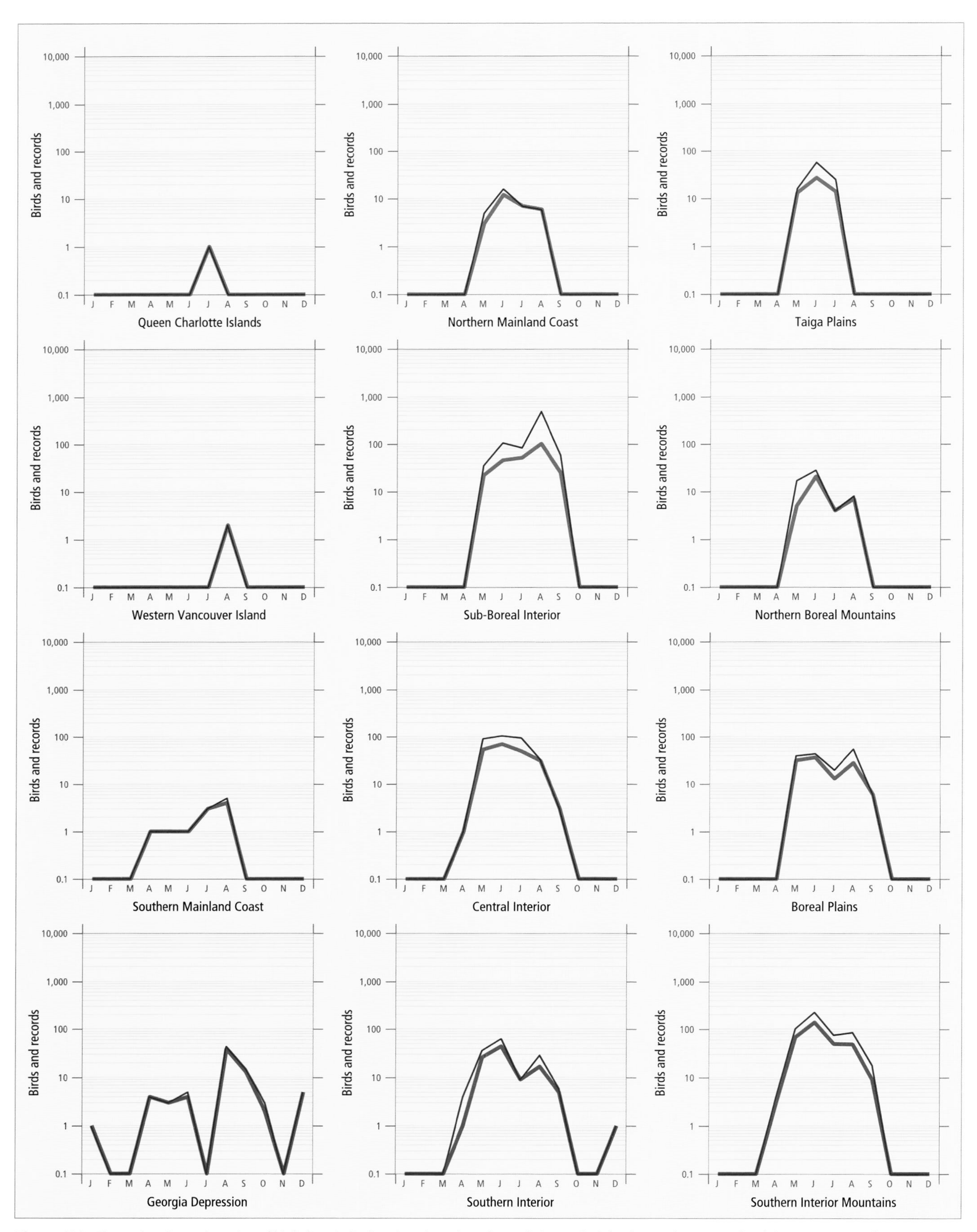

Figure 130. Fluctuations in total number of birds (purple line) and total number of records (green line) for the Northern Waterthrush in ecoprovinces of British Columbia. Christmas Bird Counts, Breeding Bird Surveys, and nest record data have been excluded.

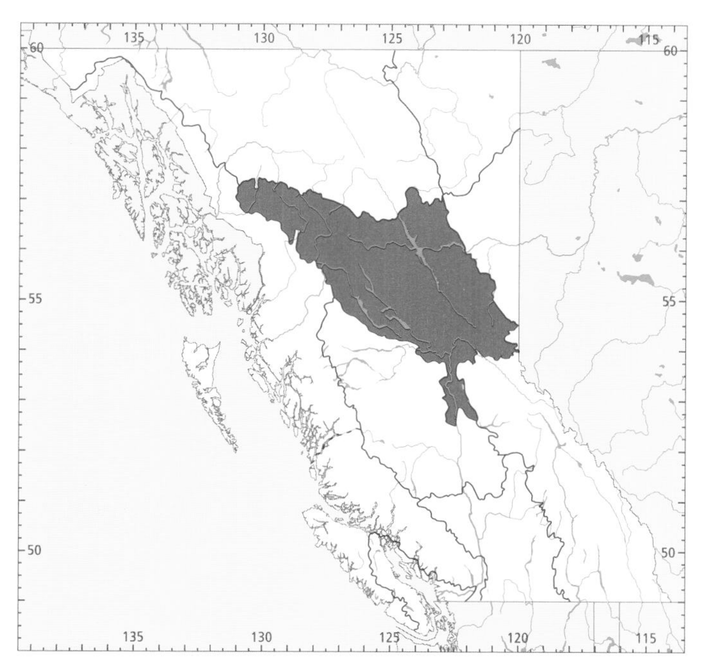

Figure 131. In British Columbia, the highest numbers for the Northern Waterthrush in summer occur in the Sub-Boreal Interior Ecoprovince.

pairs nested along the shore of Tetana Lake in the summer of 1938, but again nests with eggs or young were not reported. J.A. Munro, working in portions of the central interior of the province (1947), the Vanderhoof region (1949), and the Creston area (1950), also failed to report a single nest of the species in areas such as Bouchie and Summit lakes where the waterthrush was "exceedingly common" (Munro 1947).

The Northern Waterthrush reaches its highest numbers in summer in the Sub-Boreal Interior (Fig. 131). An analysis of Breeding Bird Surveys for the period 1968 through 1993 shows that the mean number of birds on interior routes increased at an average annual rate of 4% (Fig. 132). Surveys for coastal routes contain insufficient data for analysis. Sauer et al. (1997) report no significant national or continental trends for this warbler.

The Northern Waterthrush has been found nesting between 405 and 1,350 m elevation. Typically, it establishes and defends by song about 1.0 ha of breeding territory in a wooded swamp or along a willow-bordered stream (Eaton 1995). In British Columbia, it prefers shrub thickets that provide dense cover adjacent to water such as lakes, rivers, creeks, ponds, sloughs, large ditches, and, less frequently, bogs, swamps (Fig. 133), and wet willow shrub-carrs. Flooded vegetation is often used, particularly willow, trembling aspen, and alder mixed with spruce (Fig. 133). Eaton (1995) notes that dense cover near ground level and the presence of water appeared to be the 2 most important habitat requirements throughout the species' breeding range.

In Kootenay National Park, the Northern Waterthrush prefers areas of flooded forest near beaver ponds with a heavy growth of willows, and seepage areas at the base of mountains. Holroyd and Van Tighem (1983) describe preferred habitats as "forests, often white spruce–dominated, with tall deciduous shrubbery, bordering marshes, ponds, lakes, or flood plains." In Mount Revelstoke and Glacier national parks, the highest densities of the Northern Waterthrush were found in the green alder–fern shrubbery (1.13 birds per ha), the spruce–labrador tea–brown moss open forest (0.85 birds per ha), and the alder–skunk cabbage shrubbery (0.57 birds per ha) in the Interior Cedar–Hemlock Biogeoclimatic Zone to lower subalpine sites (1,040 to 1,860 m elevation; Van Tighem and Gyug 1983). At Bouchie Lake, north of Prince George, Munro (1947) found that the birds preferred the willow and alder thickets and revegetated clearings along the lakeshore. In the latter, a pair occupied an area where a marginal strip of willows had been cleared and the area had been overgrown with "a rank vegetation of tall ferns, nettles, and cow parsnip" that concealed much of the debris left by the clearing. At Summit Lake, north of Prince George, Munro (1947) also found that the waterthrush favoured the deciduous growth along the lakeshore.

In her study of the breeding bird communities in trembling aspen forests near Smithers, Pojar (1995) found the highest density of singing waterthrush males in mature trembling aspen (0 to 6.3 singing males per 10 ha), old aspen (0 to 8.0 per 10 ha), mixed conifer-aspen (1.6 to 7.6 per 10 ha), and pure conifer (0 to 8.5 per 10 ha) stands. She also noted a relationship between the presence of ponds or streams in the stands and the presence of the waterthrush; for example, of 88 birds found in the mature or old stands, 79 found near the edge of a pond or a stream. Erskine and Davidson (1976) also found the Northern Waterthrush mainly in alder and willow shrubbery along rivers in the Fort Nelson area.

The Northern Waterthrush has been recorded nesting in British Columbia from 2 June (calculated) to 20 August (Fig. 129).

Nests: Detailed descriptions are available for only 11 of the 44 nests found in British Columbia. Three were nestled among the roots of fallen trees and 1 was under an overhang in the bank of a small stream; the rest were on the ground under logs or other debris (Fig. 134), a clump of sod, or mosses.

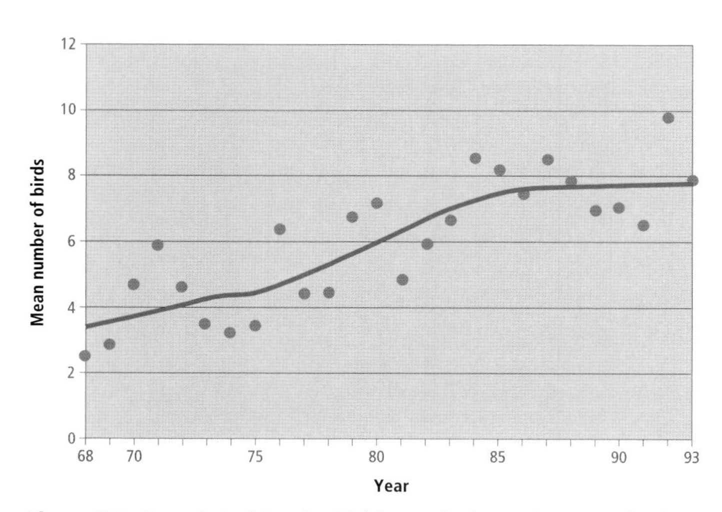

Figure 132. An analysis of Breeding Bird Surveys for the Northern Waterthrush in British Columbia shows that the mean number of birds on interior routes increased at an average annual rate of 4% over the period 1968 through 1993 ($P < 0.01$).

Figure 133. Throughout British Columbia, wooded swamps and flooded forests are favourite breeding habitats for the Northern Waterthrush (west of 70 Mile House, 7 June 1997; R. Wayne Campbell).

The heights of 9 nests ranged from ground level to 0.8 m, with 7 of the nests at ground level.

All the nests were well-concealed cups constructed of mosses, rootlets, and grasses supplemented by twigs, leaves, horsetail, and fragments of rotten western redcedar. The nests were lined with fine grasses, hair, plant down, and bark strips.

Eggs: Dates for 15 clutches ranged from 7 June to 29 June, with 60% recorded between 13 June and 24 June. Calculated dates indicate that eggs can occur as early as 4 June. Sizes of 8 clutches ranged from 4 to 6 eggs (4E-2, 5E-4, 6E-2). The incubation period is 12 days (Eaton 1995).

Young: Dates for 13 broods ranged from 17 June to 20 August, with 54% recorded between 20 June and 7 July. Sizes of 12 broods ranged from 1 to 4 young (1Y-2, 2Y-6, 3Y-2, 4Y-2), with 8 having 2 to 3 young. The nestling period is about 9 or 10 days (Eaton 1995).

Brown-headed Cowbird Parasitism: No cowbird parasitism was found in 17 clutches observed in British Columbia. Friedmann (1963) and Friedmann and Kiff (1985) reported a rate of 62% (n = 37) at 1 site in Wisconsin, 17% (n = 18) in Quebec, and 20% (n = 40) in Ontario. Eaton (1995) concluded that the general low rate compared with other ground-nesting warblers may be due to this species' habit of nesting in wetlands with a thick understorey and to the fact that much of the northern nesting area was beyond the range of the cowbird.

Nest Success: Insufficient data. One nest was apparently raided by a red squirrel.

REMARKS: The Northern Waterthrush has 3 subspecies, 2 of which occur in British Columbia. *S. n. notabilis* nests in northern British Columbia while *S. n. limnaeus* breeds in the central and southeastern regions of the province.

Eaton (1995) notes a number of conservation issues that could affect Northern Waterthrush populations, including concentration of contaminants and toxins in wetlands, reduction of prey biomass by aerial spraying to control the spruce budworm, and effects on waterthrush habitat of drainage or development of wetlands.

A review of the life history and biology of the Northern Waterthrush may be found in Eaton (1995).

Figure 134. In British Columbia, the Northern Waterthrush often constructs its nest on dry ground, in damp situations, among debris and fallen trees (One Island Lake, 14 July 1996; R. Wayne Campbell).

NOTEWORTHY RECORDS

Spring: Coastal – Reifel Island 15 Apr 1970-1 at refuge, first record for the Greater Vancouver area; nr Squamish 10 Apr 1971-1 feeding at edge of Cheakamus River; Stuie 27 May 1892-1 (MCZ 284374); n Kitimat Mission 17 May 1975-3 in boggy area beside main road (Hay 1976); Meziadin Lake 30 May 1978-1 at n end. **Interior** – Manning Park 29 Apr 1979-4 at east end; Champion Lake 24 May 1986-1; Columbia Gardens 12 Apr 1986-2; Okanagan Landing 2 May 1944-1, first of the year; Nakusp 15 May 1976-2; Shuswap Indian Reserve 20 May 1978-5; Eagle River 21 May 1977-10; North Barriere Lake 7 Apr 1974-1 feeding in swamp; 16 km n Golden 17 May 1998-24 singing males along 5 km route; Doc English Gulch 19 May 1978-2; nr Hemp Creek 24 May 1971-1 (Grass 1971); Williams Lake 20 May 1985-4 singing loudly at Scout Island; Anahim Lake 26 May 1948-10; Quesnel 19 May 1966-1, first of year (Rogers 1966c); Prince George 15 May 1981-2; Willow River 3 May 1966-1, early spring date; Round Lake 15 May to 9 Jul 1975-at least 1; Toboggan Lake 30 Apr 1977-1; Smithers 16 May 1985-1 heard; Gagnon Creek 28 May 1994-10 nr migration monitoring station; Tupper Creek 31 May 1938-2 (RCM 8085-6); Swan Lake (Tupper) 3 May 1992-1; Fort St. John 13 May 1981-4; Kluayaz 26 May 1979-12 (Page and Bergerud 1979); Parker Lake 13 May 1975-1; Fort Nelson 19 May 1975-2; Liard Hot Springs 8 May 1975-1 (Reid 1975).

Summer: Coastal – Rocky Point (Victoria) 11 and 20 Aug 1994-1 banded each day; Victoria 23 Aug 1994-1 (Bowling 1995a); Cheam Slough 16 Aug 1975-1; Vancouver 21 Aug 1986-2 at Jericho Park; Golden Ears Park 30 Aug 1982-1 heard singing; Carrington Lagoon 7 Jun 1975-1 (Shepard 1975b); Loughborough Inlet 27 Aug 1943-1 (Rand 1943); Triangle Island 24 and 26 Aug 1995-1 caught and banded both days; Bella Coola 3 Jul 1940-1 (UMMZ 193810); Stuie 12 Aug 1982-1; Kimsquit River 6 Jul 1986-1 at confluence of Cornice Creek; Kumara Lake 31 Jul 1974-1, typical habitat, struck by fact male was singing at this late date; Lakelse Lake Park 17 Aug 1977-1 singing (Shepard 1977); w Greenville 16 Jun 1981-1; Kitwanga 26 Jun 1975-4 east of town; Kispiox River valley 14 Jul 1924-1 young female collected, mostly in juvenile plumage, 27 Aug 1921-1, 37 km n Hazelton (MVZ 42513); 29 km ne Stewart 4 Jul 1977-1 (CAS 69728); Nuttlude Lake 29 Jul 1957-2; Meziadin Lake 26-27 Jun 1991-1; Dokdaon Creek 21 Jul 1919-1 (MVZ 40198). **Interior** – Manning Park 21 Jun 1986-6; Vaseux Lake 4 Aug 1994-5 banded, 15 Aug 1990-1 (UBC 15064); Christina Lake 16 Jun 1984-1; Allison Creek 12 Jul 1997-1 calling in undergrowth along creek; 16 km s Fernie 5 Jul 1991-6, 1 carrying nesting material (Campbell and Dawe 1991); Slocan City 13 Jun 1982-5 eggs, female incubating; s Sparwood 5 Jul 1991-2 (Campbell and Dawe 1991); Elk River 14 Jun 1984-6 at Michel Creek confluence; Peachland 25 Jun 1973-4 nestlings; Davis Lake (Princeton) 19 Jun 1993-1 singing in willow growth around beaver ponds; Tamarack Lake 7 Jul 1979-4 nestlings, soon to leave nest; Marshall Lake (Lillooet) 28 Aug 1968-2 feeding along shoreline; Tappen-Squilax Station 13 Jun 1973-3; Big Bar Lake 9 Jul 1964-1 (NMC 52674); Bridge Lake 19 Jun 1977-8; Yoho National Park 13-27 Aug 1975-4; 100 Mile House 7 Jul 1983-2 downy fledglings; Bridge Lake 17 Jun 1970-3 nestlings, 2 to 3 days old, on northernmost island; Graffunder Lakes 9 Jul 1972-2 fledglings; e Canim Lake 30 Jun 1977-2 fledglings being fed by adults; Williams Lake 25 Jul 1982-15 at Scout Island Nature Centre; Murtle Lake 13 Jun 1987-5 eggs, 3 Jul 1962-2 nestlings; Anahim Lake 25 Jun 1961-6 on east side; Stum Lake 9 Jul 1971-adult female feeding 2 fledglings on beach; Moose Lake (Mount Robson Park) 17 Jun 1972-5; Bowron Lake Park 4 Aug 1975-1; s McBride 10 Jul 1991-3 in swampy area (Campbell and Dawe 1991); nr Sinkut Lake 16 Jul 1945-1 three-quarter-grown fledgling at Bradley's Slough; Shane Lake 17 Jun 1992-2 recently fledged young (Campbell and Dawe 1992); Tabor Lake 17 Jun 1992-2 recently fledged young (Campbell and Dawe 1992); 24 km s Prince George 28 Jun 1986-2 fledglings running on logs by pond with 1 adult; 48 km nw Prince George 7 Jun 1974-6 eggs; Loup Garou Lake 11 Jul 1967-adult feeding 1 fledgling with down tufts still on head; Houston 20 Jun 1968-2 nestlings; Pinkut Creek to Burns Lake 11 Jun 1975-14; Fort St. James 25 Jun 1889-5 eggs (Munro 1949); Topley 28 Jun 1956-7; McLeod Lake 14 Jul 1978-4 fledged young; Mount Babcock 17 Jun 1995-1 in wooded swamp; One Island Lake 28 Jun 1978-4 eggs; Bullmoose Creek area 7 Aug 1994-4; Mugaha Creek 6 Aug 1996-32, 8 Aug 1996-29, 13 Aug 1996-15, 16 Aug 1996-13, 24 Aug 1996-7, all at migration monitoring station; Tetana Lake Jun 1938-20 pairs nested (Stanwell-Fletcher and Stanwell-Fletcher 1943); Moberly Lake 25 Jul 1975-4; Taylor 21 Aug 1975-female feeding 4 fledglings; Charlie Lake 20 and 21 Aug 1975-7; Stoddart Creek (Fort St. John) 30 Aug 1985-3; Boundary Lake (Goodlow) 7 Jun 1986-1, 24 Jun 1978-5 eggs; Fort St. John 15 Aug 1975-female feeding fledgling; Rose Prairie 10 Aug 1975-2; Ningunsaw River 14 Jun 1988-several, between 20 and 50 km n Ningunsaw Pass; Fern Lake 24 Aug 1979-1 (Cooper and Adams 1979); s Dease Lake 18 Jun 1975-3; nw Clarke Lake 17 Jun 1985-1; Raspberry Creek (Fort Nelson) 21 Jun 1953-1, singing; Fort Nelson 12 Jun 1976-7, 28 Jul 1982-1; Fort Nelson River 7 Jul 1978-1; s Kotcho Lake 23 Jun 1982-2 adults with 3 recently fledged young, 30 Jun 1982-23, 1 to 5 called out at every stop along 11.6 km of road, 6 Jul 1982-2 fledglings with 2 adults, young were large, could fly well, and were secretive (Campbell and McNall 1982); Helmut 2-9 Jun 1982-1 to 2 birds almost daily; O'Connor River 26 Jun 1993-1, collecting grubs, likely nesting; km 46 Skagway Rd 25 Jun 1979-1, singing; Hyland River 20 Aug 1978-2 at slough in burn area; Petitot River 1 Jul 1982-1.

Breeding Bird Surveys: Coastal – Recorded from 5 of 27 routes and on 5% of all surveys. Maxima: Meziadin 17 Jun 1975-15; Kispiox 20 Jun 1993-15; Nass River 21 Jun 1975-6; Kitsumkalum 13 Jun 1976-5. **Interior** – Recorded from 55 of 73 routes and on 60% of all surveys. Maxima: McLeod Lake 25 Jun 1991-75; Ferndale 7 Jun 1969-27; Bridge Lake 5 Jun 1976-27; Summit Lake 17 Jun 1992-26.

Autumn: Interior – St. John Creek 21 Sep 1985-1; Mugaha Creek 5 Sep 1996-2, 17 Sep 1996-1; Kinney Lake 2 Sep 1971-10; 16 km w Williams Lake 15 Sep 1954-1 (NMC 48220); nr Clearwater 8 Sep 1959-1 (UKMU 38396); lower Blaeberry River valley 10 Sep 1995-1; Edgewood 24 Sep 1925-1 (Kelso 1926); Manning Park 14 Sep 1973-1 (Belton 1973). **Coastal** – Tsable River 2 Oct 1929-1 at river mouth; Vancouver 12 Sep 1987-2 at Jericho Park.

Winter: Interior – See Christmas Bird Counts. **Coastal** – Reifel Island 23 Dec 1986-1 (Tweit and Mattocks 1987a), 3 Jan 1987-1 (Tweit and Mattocks 1987a), 7 Dec 1970-1 walking on logs and sticks in same habitat as 6 Virginia Rails.

Christmas Bird Counts: Interior – Recorded from 1 of 27 localities and on less than 1% of all counts. Maximum: Oliver-Osoyoos 28 Dec 1987-1. **Coastal** – Recorded from 1 of 33 localities and on less than 1% of all counts. Maximum: Ladner 14 Dec 1974-1.

Connecticut Warbler

Oporornis agilis (Wilson)

COWA

RANGE: Breeds from northeastern British Columbia and central Alberta east in a narrow band across central Canada to western Quebec and south to southern Manitoba, Minnesota, Michigan, and southern Ontario. Winters from northeastern Colombia south to central Brazil, but mainly in the Amazon Basin.

STATUS: On the coast, *accidental* on Western Vancouver Island in the Coast and Mountains Ecoprovince.

In the interior, *rare* migrant and local summer visitant in the Boreal Plains Ecoprovince; locally *very rare* in the Taiga Plains Ecoprovince. Not recorded elsewhere in the interior.

Breeds.

CHANGE IN STATUS: The Connecticut Warbler (Fig. 135) was noted by Munro and Cowan (1947) to be "known only from the Peace River Parklands." This assessment was based on specimens collected near Tupper in June 1938 (Cowan 1939). Since then, the Connecticut Warbler has been recorded at various localities in the Peace Lowland, Kiskatinaw Plateau, and Halfway Plateau, and a small population has been documented near Fort Nelson. Breeding has now been confirmed in the province.

The Connecticut Warbler is generally thought to be rare in British Columbia; Siddle (1992d) and Enns and Siddle (1996) found it to be the third and second least abundant of the scarce warblers, respectively, in the northeast. During the early 1990s, it was recorded regularly but uncommonly north of Chetwynd (Merkins and Booth 1996). However, Lance and Phinney (1994) report it to be the second most "common" warbler, after the Ovenbird, in younger trembling aspen stands in their study area south of Dawson Creek. It may, therefore, be more common locally than generally acknowledged.

NONBREEDING: The Connecticut Warbler has been found in northeastern British Columbia from the Alberta border west to Chetwynd and Moberly Lake, south to the Dawson Creek area and north to Boundary Lake. A small population occurs locally near Fort Nelson.

Most records are from the Peace Lowland and Kiskatinaw Plateau in the Boreal Plains. The Connecticut Warbler appears to be more abundant south of the Peace River, probably because of the presence of more suitable habitat.

In northeastern British Columbia, the Connecticut Warbler has been reported from elevations of 400 to 1,100 m. Nonbreeding habitat is probably similar to breeding habitat (see BREEDING). During autumn migration, the species has been seen in tamarack-spruce bogs with dense shrub growth and brushy willows along creeks.

Spring migrants begin to arrive in northeastern British Columbia in very late May; most birds probably arrive during the first 2 weeks of June (Figs. 136 and 137). This relatively late spring arrival is expected since the Connecticut Warbler returns to Canada from its wintering areas later than most other warblers, except for the Canada, Mourning, and Blackpoll warblers (Francis and Cooke 1986). Migrants enter British Columbia from the east, having arrived in Alberta mainly in late May and early June (Salt 1973). Details of the southward movement in northeastern portions of the province are unknown, but it likely begins in late July or early August, shortly after nesting is completed; the main movement probably occurs in mid-August. The Connecticut Warbler often travels with other warblers. The main autumn movement in Alberta occurs from late August to early September (Salt 1973).

Figure 135. The Connecticut Warbler (adult male shown) is a shy wood-warbler that breeds locally in northeastern British Columbia (©Rob Curtis/VIREO).

In northeastern British Columbia, the Connecticut Warbler has been recorded from 22 May to 21 August (Fig. 136).

BREEDING: The Connecticut Warbler reaches the limit of its northwestern breeding range in the boreal forests of northeastern British Columbia. Breeding populations are small and local, and are largely confined to the Peace Lowland, Kiskatinaw Plateau, and Halfway Plateau in the Boreal Plains and the Fort Nelson Lowland in the Taiga Plains. A few other breeding areas are possible in these ecoprovinces where suitable habitat occurs. In Alberta, the species breeds from the Peace River area through central parts of the province. It is thought to be fairly common but locally distributed (Salt 1973; Flack 1976; Semenchuk 1992).

In British Columbia, the Connecticut Warbler has been recorded breeding on only 2 occasions and at 2 localities, north of Fort St. John in the Peace Lowland of the Boreal Plains and near Fort Nelson. The Peace Lowland record was of a short-tailed fledgling found with an adult female on 2 August 1983. It was found at a site where a singing, territorial male had been recorded on 16 June and where an agitated pair had been present on 20 July. There are several records of adults carrying food from south of Dawson Creek in July, which suggests that breeding occurs there.

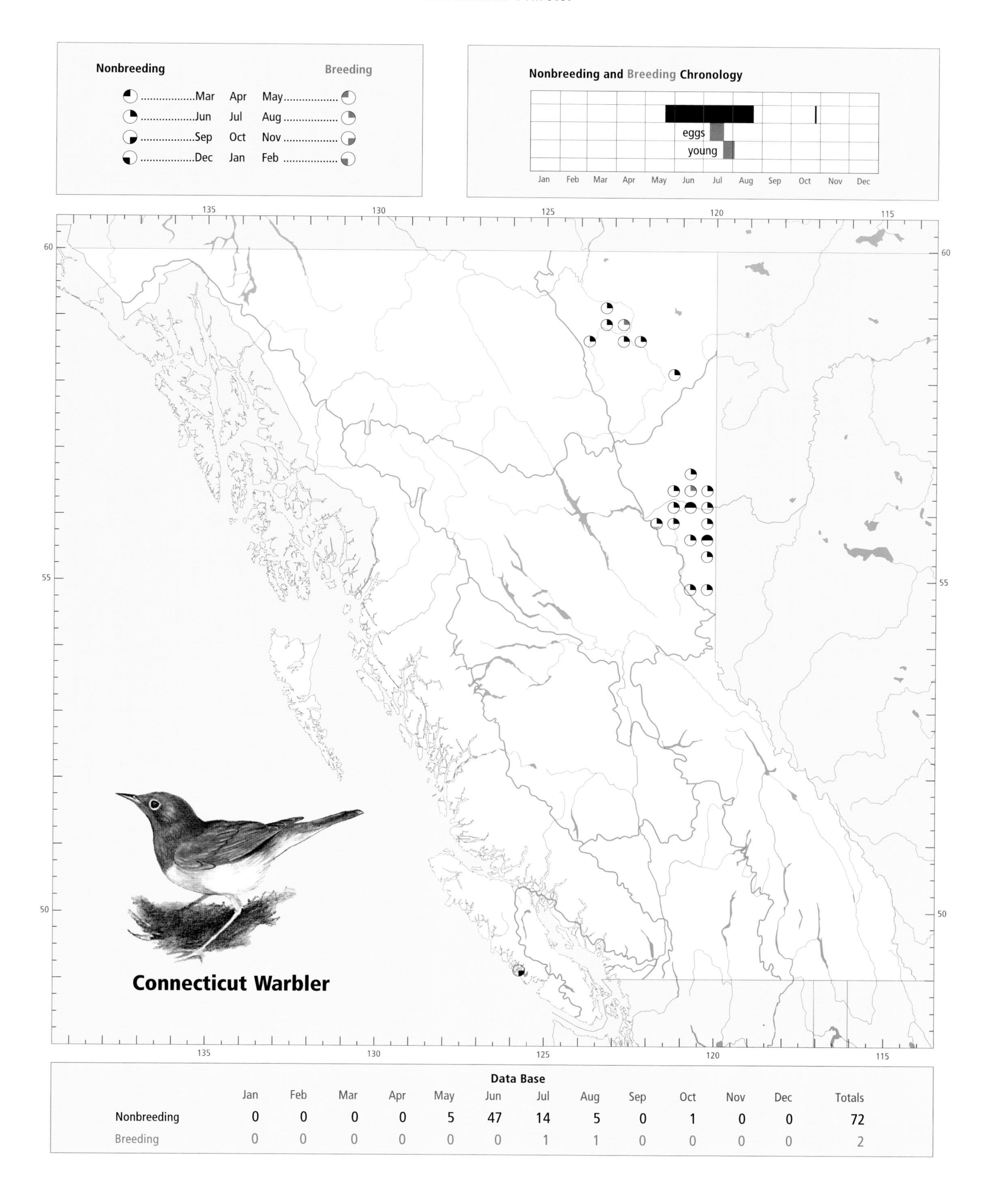

Data Base

	Jan	Feb	Mar	Apr	May	Jun	Jul	Aug	Sep	Oct	Nov	Dec	Totals
Nonbreeding	0	0	0	0	5	47	14	5	0	1	0	0	72
Breeding	0	0	0	0	0	0	1	1	0	0	0	0	2

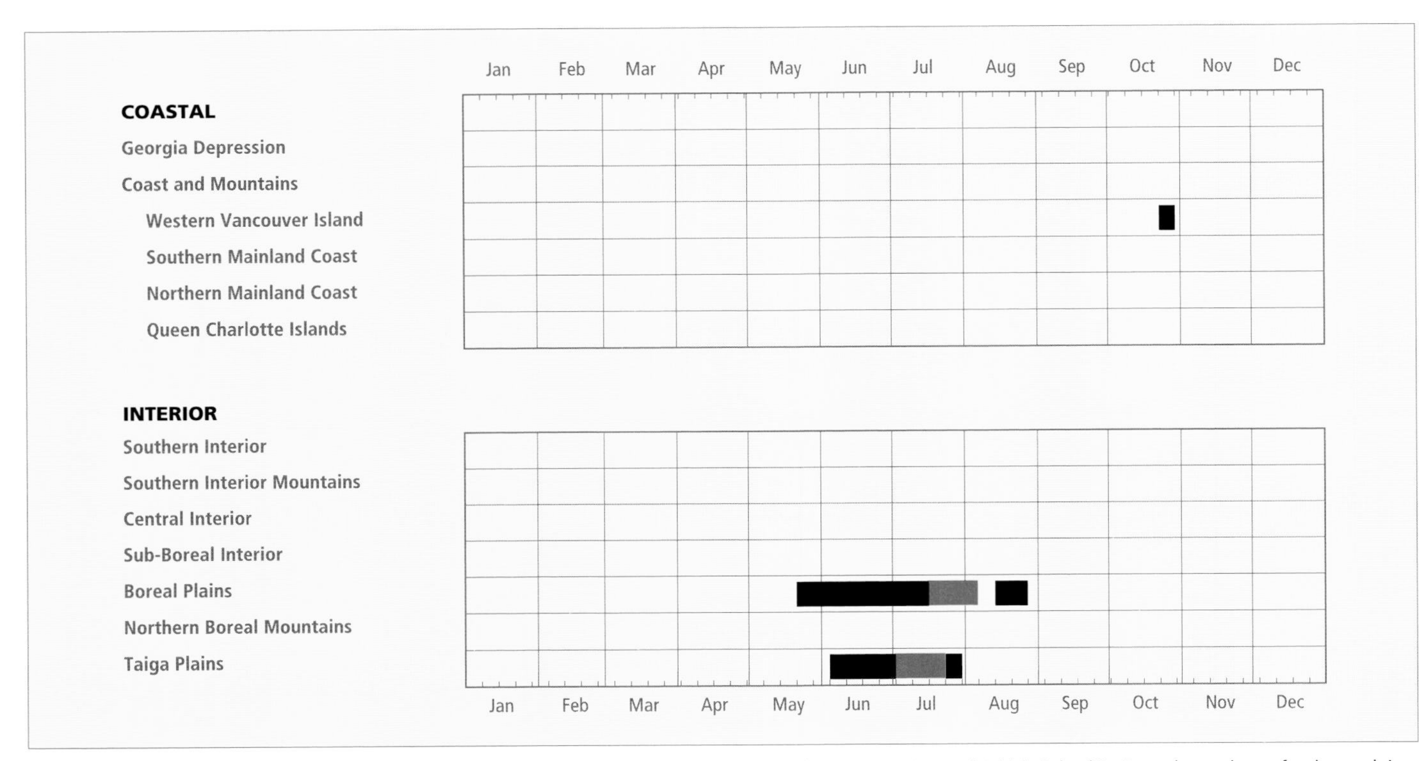

Figure 136. Annual occurrence (black) and breeding chronology (red) for the Connecticut Warbler in ecoprovinces of British Columbia. Records are shown for the week in which they occurred.

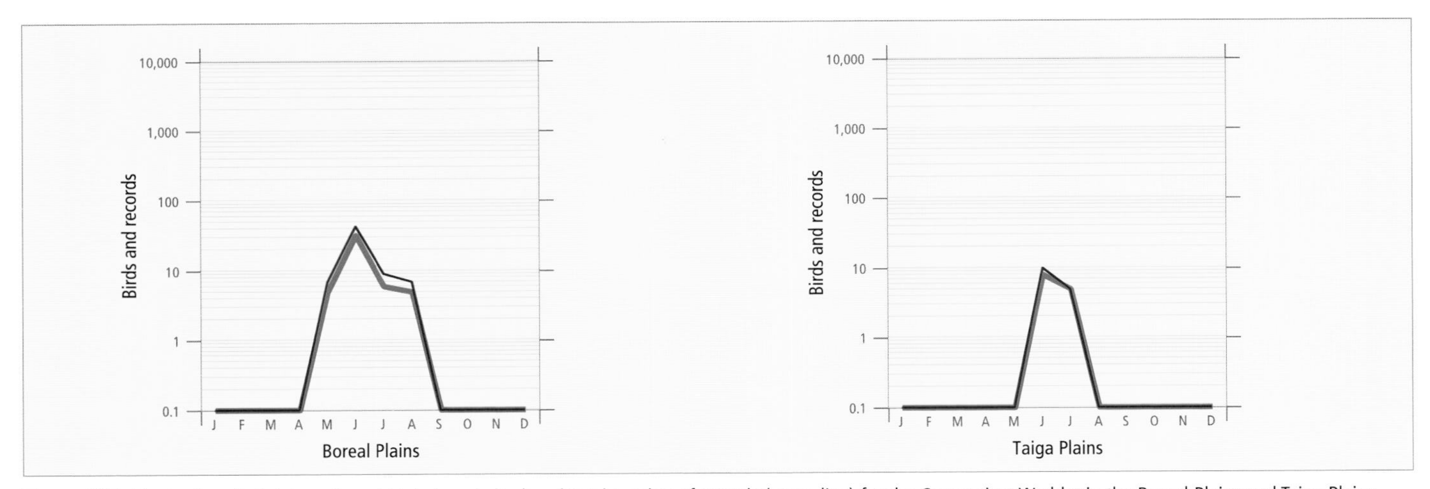

Figure 137. Fluctuations in total number of birds (purple line) and total number of records (green line) for the Connecticut Warbler in the Boreal Plains and Taiga Plains ecoprovinces. Breeding Bird Surveys and nest record data have been excluded.

Siddle (1992d) reports that the Connecticut Warbler occurred consistently at only a few localities in northeastern British Columbia: in an old-growth trembling aspen forest near Gundy (north of Tupper), and in a remnant mature trembling aspen stand east of Cecil Lake. In the northern Peace Lowland, from Taylor north to Charlie Lake, and near Fort Nelson in the Taiga Plains, it occurred in some years but not others (Siddle 1992d), a trend also noted in Alberta (Salt 1973). Lance and Phinney (1996) report it to be regular southwest of Dawson Creek, where they recorded between 20 and 33 individuals each year from 1992 to 1995.

In British Columbia, the highest numbers for the Connecticut Warbler in summer occur in the Boreal Plains (Fig. 138). Breeding Bird Surveys for interior routes for the period 1968 through 1993 contain insufficient data for analysis. An analysis of Breeding Bird Surveys across Canada from 1966 to 1996 suggests that populations are generally stable (Sauer et al. 1997).

The Connecticut Warbler has been recorded breeding at 335 m and 600 m elevations in northeastern British Columbia; summer records, however, range from 300 m to 1,100 m. Most summer records of this warbler were from mature or old trembling aspen forest, with some records from younger stands of mixed willow and trembling aspen, mixed willow and balsam poplar, or stands of smaller trembling aspen on slightly drier ridges. General descriptions of habitat in British Columbia

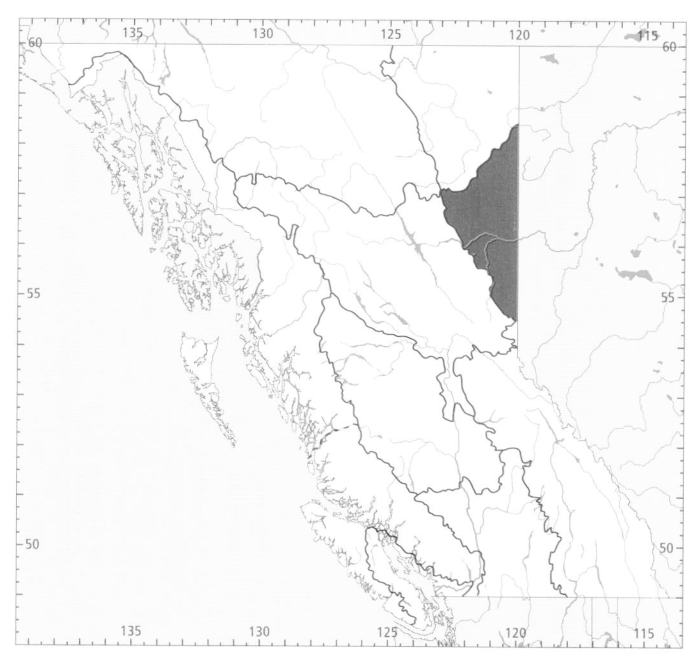

Figure 138. In British Columbia, the highest numbers for the Connecticut Warbler in summer occur in the Boreal Plains Ecoprovince.

include edges of old-growth and younger stands of trembling aspens and white spruce near Swan Lake (Cowan 1939); mixed balsam poplar and white spruce islands in the Peace River (Penner 1976); old-growth trembling aspen stands with white spruce sapling understorey near Tupper, Cecil Lake, and Fort Nelson (Siddle 1992d); and young to old-growth stands of trembling aspen southwest of Dawson Creek (Lance and Phinney 1994). The presence of spruce in the understorey of trembling aspen forests may be a characteristic of good habitat (Fig. 139).

Figure 139. In the Taiga Plains Ecoprovince of British Columbia, the Connecticut Warbler prefers trembling aspen forests with scattered spruce in the understorey (north of Fort Nelson, 28 June 1998; R. Wayne Campbell).

In 1 study in northeastern British Columbia, the habitat used by the Connecticut Warbler was associated with mature or old, large, widely spaced trembling aspens in stands on flat to gently sloping land. White spruce were scattered sporadically through the stand. These stands had a low understorey (< 3 m tall) of shrubs, herbs, and grasses under an even and high canopy. In all cases, there was a noticeable gap between the canopy and shrub levels, with much open space in between. Understorey plants included prickly rose, red-osier dogwood, willow, bunchberry, soopolallie, fireweed, paintbrush, purple peavine, and white geranium (Enns and Siddle 1996). This type of habitat is more common south, rather than north, of the Peace River and may explain why the Connecticut Warbler is more abundant there (C. Siddle pers. comm.).

South of Dawson Creek, this warbler showed a strong preference for pole-stage to mature trembling aspen stands with a well-developed herbaceous layer and sparse shrub layer (Lance and Phinney 1996; Fig. 140). These stands were 25 to 45 years old but had a closed canopy and up to 9,000 stems per ha. This habitat is noticeably different from that described by Enns and Siddle (1996).

In Alberta, the Connecticut Warbler has been found in young to old trembling aspen forests (Westworth and Telfer 1993). In eastern North America, moist deciduous forests and spruce muskegs are said to be its preferred habitat (Pitocchelli et al. 1997), but muskegs do not seem to be used much in British Columbia.

Figure 140. In the Boreal Plains Ecoprovince, the Connecticut Warbler nests in pole-stage to mature trembling aspen woodlands with a well-developed herbaceous layer and sparse shrub layer (near Dawson Creek, 16 July 1995; Mark Phinney).

Males usually sing from the low to middle canopy, occasionally from shrubby undergrowth but rarely from treetops. Most foraging is done in the shrubby understorey or along the ground (Enns and Siddle 1996).

In British Columbia, the Connecticut Warbler has been recorded breeding from 8 July (calculated) to 2 August. It probably breeds from mid-June through early August (Fig. 136).

Nests: Nests have not been found in British Columbia. Elsewhere, the Connecticut Warbler nests on or near the ground in the understorey of forested areas. Nests can be in grass or herbs, in a mossy hummock, at the base of a sapling, or a few centimetres off the ground in the base of a shrub, often wild rose. They are typically well hidden by overhanging vegetation. The nests are compact, deep cups of grasses and rootlets and are lined with fine grasses and hair (Bent 1953; Baicich and Harrison 1997).

Eggs: Nests with eggs have not been recorded in British Columbia. Calculated dates suggest that eggs can be present from at least 8 July to 22 July; however, they probably occur from the latter half of June through late July. Elsewhere, clutch sizes range from 3 to 5 eggs, with 4 eggs being the most common (Bent 1953; Pitocchelli et al. 1997). The incubation period is unknown but is probably 12 or 13 days (Bent 1953).

Young: Nests with young have not been found in British Columbia. Dates for 2 broods (recently fledged young) ranged from 22 July to 2 August. Brood sizes were 1 and 3 young, respectively. Nestlings are likely present in the province from late June through early August. The nestling period is unknown but is probably 8 to 10 days, similar to that of other species in the same genus (Bent 1953).

Brown-headed Cowbird Parasitism: Data are not available for British Columbia. The Connecticut Warbler is not known to be parasitized by cowbirds (Friedmann 1963; Friedmann and Kiff 1985), but this may be due to the lack of nest records rather than absence of parasitism.

Nest Success: Insufficient data.

REMARKS: The Connecticut Warbler is usually inconspicuous unless the male is singing. Because migrants tend to remain well hidden in foliage, spring and autumn migrants, which are usually silent, are very difficult to detect. These traits, which have also been noted in Alberta (Salt 1973), are probably partly responsible for our lack of information on specific migration periods.

In northeastern British Columbia, the primary habitat of the Connecticut Warbler consists of pole-stage to old-growth trembling aspen forest, with or without a shrubby understorey. These stands have high timber value because of the abundance and large size of the trees, and are being harvested rapidly, especially in the Peace Lowland. Because harvested areas tend to be converted to farmland or to pure white spruce or trembling aspen stands with a short rotation, much suitable habitat is being lost (Cooper et al. 1997b) – lost in the long term if converted to farmland or spruce, and useful only in the medium term when converted to trembling aspen after the stand reaches the pole stage but before it is reharvested.

A study in Saskatchewan showed that the Connecticut Warbler is area-sensitive: the larger the trembling aspen stand, the more likely the stand is to contain Connecticut Warblers (Johns 1993). The minimum size for a stand containing this species was 3.5 ha (Johns 1993); south of Dawson Creek, birds were recorded in stands as small as 4 ha (Lance and Phinney 1994). Further study of the impact of logging on this warbler's habitat is needed, but it is obvious that the habitat requirements of the Connecticut Warbler should be incorporated into trembling aspen management plans in northeastern British Columbia.

Two other warblers of this genus occur in northeastern British Columbia: the Mourning Warbler, which is more common than the Connecticut Warbler, and the MacGillivray's Warbler, which is widely distributed in the province but rarely occurs within the range of the Connecticut Warbler. Observers should carefully note field marks and songs when identifying these similar-looking warblers (see Curson et al. 1994).

The Connecticut Warbler is one of the most poorly known songbirds in North America. Pitocchelli et al. (1997) summarize the little that is known about the ecology of this shy wood-warbler.

NOTEWORTHY RECORDS

Spring: Coastal – No records. **Interior** – Brassey Creek 22 May 1994-1; Tomslake 26 May 1994-1; Dawson Creek 24 May 1994-1, 29 May 1993-3; Charlie Lake 29 May 1985-1 (Grunberg 1985c).

Summer: Coastal – No records. **Interior** – s Tumbler Ridge 14 Jun 1998-4 in 4 km along old Monkman hwy; Swan Lake (Tupper) 10 Jun 1978-1, 24 Jun 1938-5 singing males (Cowan 1939); Gundy 12 Jun 1983-3 singing males; Tupper 9 Jun 1976-1, 22 Jun 1938-1 (RBCM 8132); Pouce Coupe 5 Jul 1992-pair carrying food; Fellers Heights 18 Jun 1997-1; 25 km e Chetwynd 5 Aug 1975-2; Moberly Lake sw 3 Jun 1989-1 singing male; Peace River, 65 km w Alberta border 7 Jun 1974-1 (Penner 1976); e Red Creek (Peace River) 8 Jun 1986-1; Stoddart Creek (Fort St. John) 16 Jun 1983-1 male singing, 2 Aug 1983-1 bob-tailed fledgling; St. John Creek 17 Aug 1986-1 (McEwen and Johnston 1987a); Charlie Lake 21 Aug 1997-1 banded; Boundary Lake (Goodlow) 16 Jun 1985-1; 14 km ene Rose Prairie 24 Jul 1992-1 juvenile; Clarke Lake (Fort Nelson) 19 Jun 1974-1 (Erskine and Davidson 1976); Kenai Creek 6 Jun 1994-1 calling, 7 Jun 1993-2 singing; Fort Nelson 14 Jun 1976-1, 22 Jul 1994-adult feeding 3 recently fledged young, 29 Jul 1979-1 at mile 7.5 on 317 Rd.

Breeding Bird Surveys: Coastal – Not recorded. **Interior** – Recorded from 4 of 73 routes and on 2% of all surveys. Maxima: Tupper 20 Jun 1994-6; Steamboat 28 Jun 1980-5; Fort Nelson 27 Jun 1980-5.

Autumn: Interior – No records. **Coastal** – nr Tofino 27 Oct 1994-1 immature male at Comber's Beach.

Winter: No records.

Mourning Warbler MOWA
Oporornis philadelphia (Wilson)

RANGE: Breeds from northeastern British Columbia, southeastern Yukon, northeastern and central Alberta, central Saskatchewan, central Manitoba, central Ontario, southern Quebec, and Newfoundland south to southern Manitoba, north-central and northeastern North Dakota, central Minnesota, central Wisconsin, northeastern Illinois, southern Michigan, and northern Ohio, western Maryland, southeastern New York, and western Massachusetts. Winters from southern Nicaragua south through Costa Rica and Panama to western Colombia, southern Venezuela, and eastern Ecuador.

STATUS: On the coast, *casual* summer and autumn vagrant in the Georgia Depression Ecoprovince.

In the interior, *rare* to locally *uncommon* migrant and summer visitant in the Boreal Plains and Taiga Plains ecoprovinces; *casual* summer vagrant in the Southern Interior Mountains and Northern Boreal Mountains ecoprovinces.

Breeds.

CHANGE IN STATUS: The Mourning Warbler (Fig. 141) was not known to occur in British Columbia during the first half of the 20th century (Munro and Cowan 1947). Neither Cowan (1939) nor Rand (1944) found this eastern warbler during their extensive field work in northeastern British Columbia. The first published records of the Mourning Warbler in British Columbia were from 1974, when it was found to be surprisingly common near Fort Nelson. Erskine and Davidson (1976) found it on 21 different days in June and July, with up to 7 individuals noted on a single day. They also found a bird as far west as Liard Hot Springs and 3 males between Tupper and Tomslake in the southern Boreal Plains. Weber (1976) reports earlier records for British Columbia, involving single males singing near Fort Nelson on 11 July 1967 and Liard Hot Springs on 12 July 1971.

The first record suggesting breeding is from 1978, when a pair was found feeding 1 or more fledglings near Taylor (Campbell and Gibbard 1979). Breeding has been documented in the Peace Lowland of the Boreal Plains and is suggested for the Fort Nelson Lowland of the Taiga Plains. The maps of the Mourning Warbler's breeding range published in Godfrey (1986), which omits British Columbia, and in Pitocchelli (1993), which includes only the Fort Nelson Lowland, are not current.

Pitocchelli (1993) suggests that the opening up of boreal forests through logging, agriculture (Fig. 142), and mining has enabled this warbler to become more widespread in Canada. Large-scale habitat alterations may have enabled it to colonize northeastern British Columbia beginning in the 1950s or 1960s.

Currently, the Mourning Warbler is thought to occur locally in northeastern British Columbia, where it is more numerous than most other eastern North American warblers.

Figure 141. The Mourning Warbler (adult male shown) is a colourful but seldom-seen warbler that inhabits disturbed areas with thick undergrowth (Anthony Mercieca).

For example, Enns and Siddle (1996) found it to be the third most common species of 10 eastern warblers recorded during their surveys in that part of the province in 1992.

NONBREEDING: In British Columbia, the Mourning Warbler occurs regularly only in northeastern portions of the province. Most records are along the Peace River from the British Columbia–Alberta border west to Moberly Lake, around Swan Lake near Tupper, and in the Fort Nelson Lowland. In the Boreal Plains, a few individuals have been found south to Chetwynd and near One Island Lake, west to Hudson's Hope, and north to Cecil Lake. In the Taiga Plains, the Mourning Warbler occurs from southeast of Fort Nelson at Clarke Lake, north to the Petitot River along the British Columbia–Yukon border, northeast of Fort Nelson to near Kotcho Lake, and west to Liard Hot Spring in the Northern Boreal Mountains.

Figure 142. The opening up of boreal forests through agriculture and logging may be a contributing factor to this warbler's more widespread distribution in northeastern British Columbia and across Canada (near Chetwynd, 18 June 1997; R. Wayne Campbell).

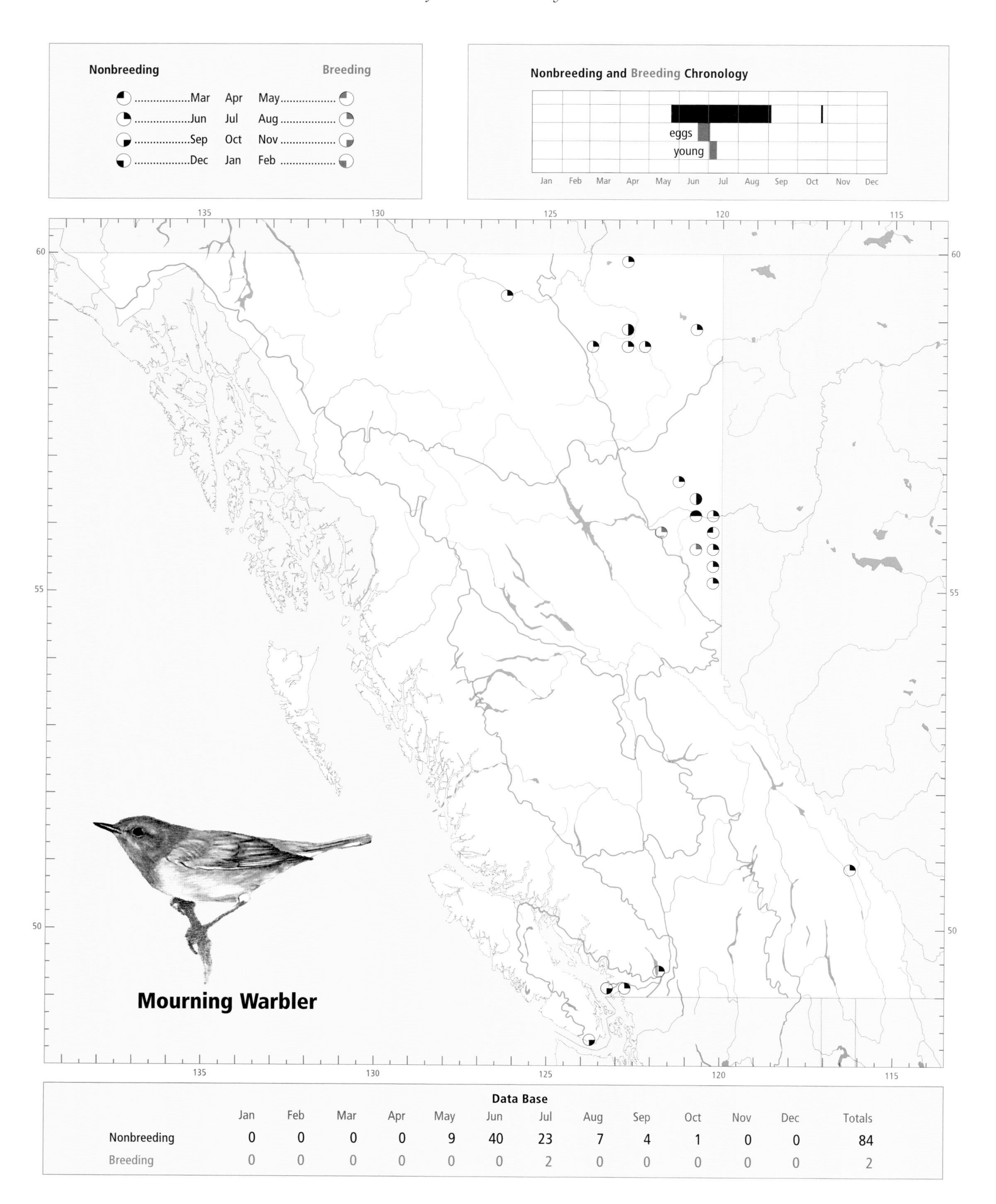

Data Base

	Jan	Feb	Mar	Apr	May	Jun	Jul	Aug	Sep	Oct	Nov	Dec	Totals
Nonbreeding	0	0	0	0	9	40	23	7	4	1	0	0	84
Breeding	0	0	0	0	0	0	2	0	0	0	0	0	2

The Mourning Warbler has been reported from 335 to 1,100 m in northeastern British Columbia. Vagrants on the coast have occurred from near sea level to 125 m. Nonbreeding habitat is poorly described but is likely similar to breeding habitat (see BREEDING). One adult male found on the coast along the Harrison River frequented a marshy backwater vegetated with willows, red alder, and hardhack (Kautesk 1983).

The Mourning Warbler is one of the last warblers to migrate north in the spring. Spring migrants begin to arrive in northeastern British Columbia in very late May in the Boreal Plains and in the first and second weeks of June in the Taiga Plains; most migrants probably arrive during the first 10 days of June (Figs. 143 and 144). Migrants enter British Columbia through Alberta, where they arrive on similar dates (Salt 1973). Details of the southward movement in northeastern portions of the province are unknown, but it likely begins in late July, shortly after nesting is completed, with the main movement probably in mid to late August. The last migrants have gone by the first few days of September (Figs. 143 and 144). In Alberta, the main autumn movement occurs from late August to early September (Salt 1973). Migrants often go unnoticed as they skulk about secretively in dense undergrowth.

In northeastern British Columbia, the Mourning Warbler occurs regularly from 24 May to 2 September (Fig. 143).

BREEDING: The Mourning Warbler reaches the limit of its northwestern breeding range in the Boreal Plains of northeastern British Columbia. Breeding has been confirmed only at Moberly Lake (Peace Lowland) and near Brassey Creek (Kiskatinaw Plateau), both in the Boreal Plains. Because singing males have been noted elsewhere, however, breeding populations likely occur in all regions of the Taiga Plains and Boreal Plains where there is suitable habitat. It is possible that small numbers breed in the Liard Plain of the Northern Boreal Mountains, where singing males have also been found. There are only 2 breeding records in the province, and only 1 nest has been found. In Alberta, the Mourning Warbler breeds west to Peace River and into north-central areas, but has not been documented as far west as the border with British Columbia (Salt and Salt 1976; Semenchuk 1992).

The Mourning Warbler reaches its highest numbers in summer in the Boreal Plains (Fig. 145). Breeding Bird Surveys for interior routes for the period 1968 through 1993 contain insufficient data for analysis. An analysis of Breeding Bird Surveys across Canada from 1966 to 1996 suggests that populations are declining at an annual rate of 1.0% ($P < 0.05$) (Sauer et al. 1997).

Summer records for the Mourning Warbler in its breeding range in northeastern British Columbia have all been from below 1,100 m. Unlike the Connecticut Warbler, which requires mature forests with a shrubby understorey, the Mourning Warbler prefers second-growth or disturbed woodlands (Pitocchelli 1993). Breeding habitat in British Columbia includes mainly overgrown clearings, rose bushes in regenerating clearcuts, dense deciduous thickets along edges of mixed or deciduous forests, and riparian areas along wetlands or streams. It also includes young to older forest stages, often with mixed trembling aspen, white spruce, paper birch, willows, and alders. Mature mixed forests with openings and heavy shrub undergrowth is also used. In floodplains with

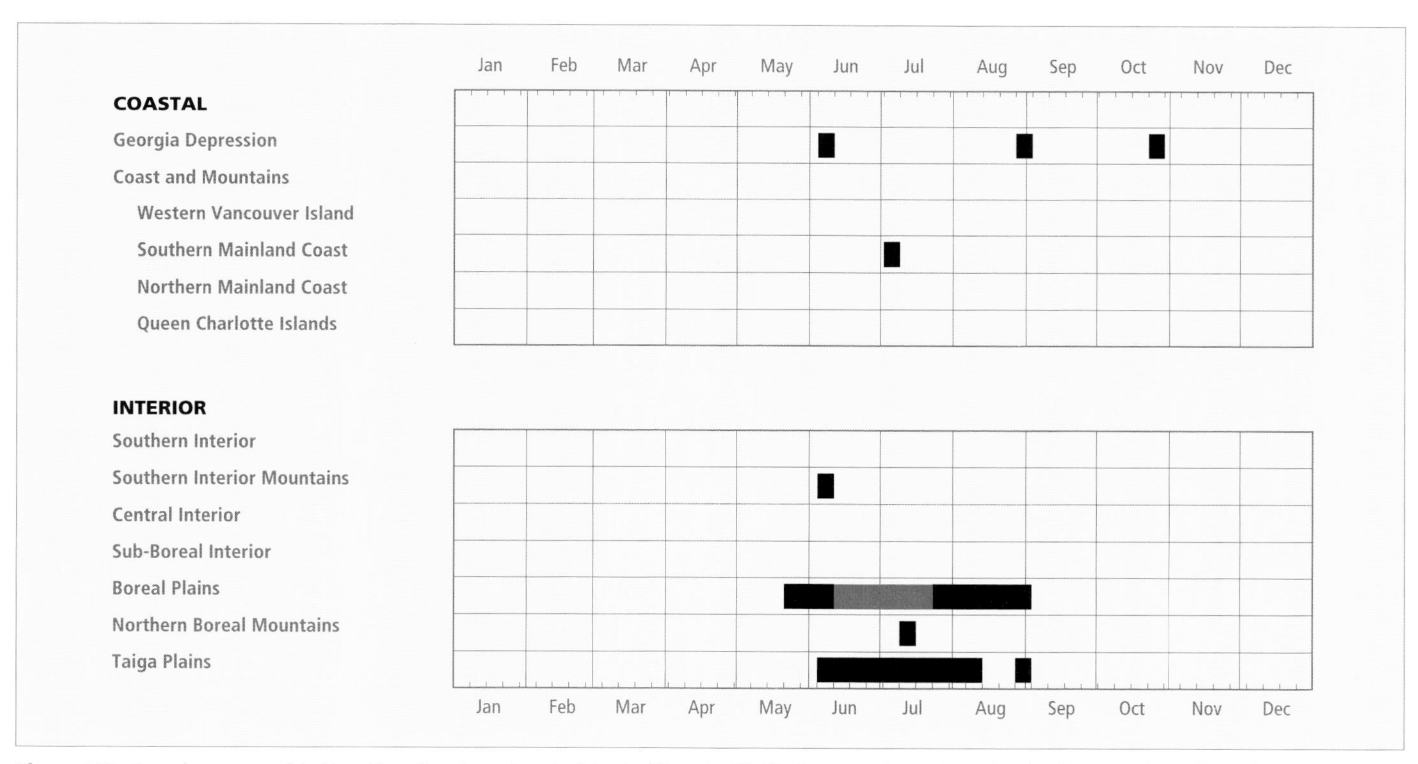

Figure 143. Annual occurrence (black) and breeding chronology (red) for the Mourning Warbler in ecoprovinces of British Columbia. Records are shown for the week in which they occurred.

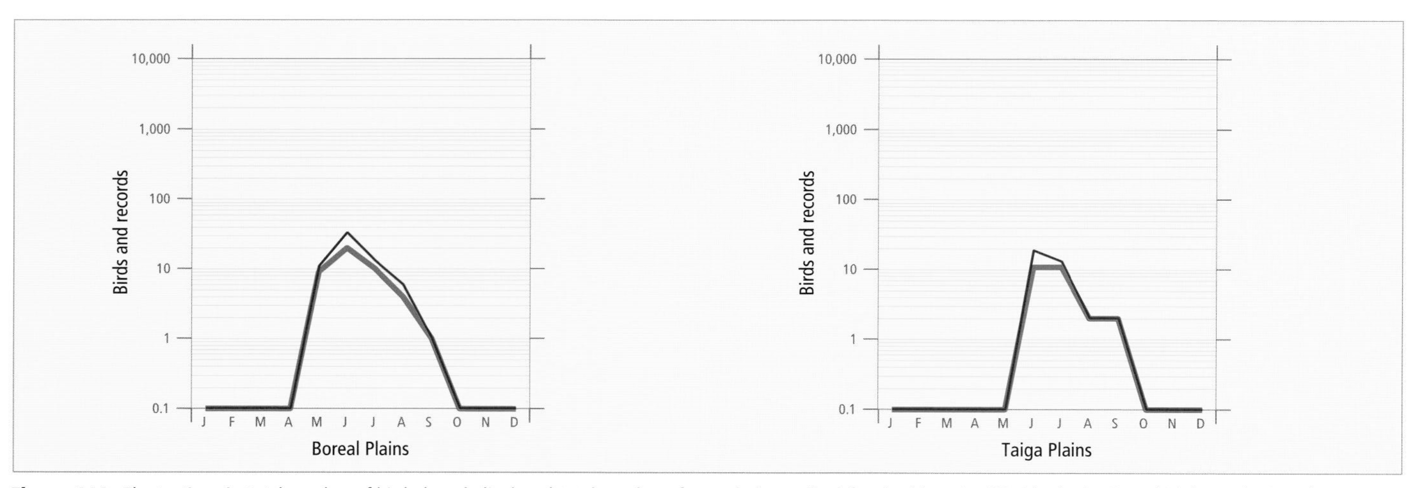

Figure 144. Fluctuations in total number of birds (purple line) and total number of records (green line) for the Mourning Warbler in the Boreal Plains and Taiga Plains ecoprovinces of British Columbia. Breeding Bird Surveys and nest record data have been excluded.

mature balsam poplar, understoreys typically include red-osier dogwood. Human-created habitats such as seismic cutlines, roadsides, logged cutblocks, and ski hills are also used.

A study of forest habitat use in northeastern British Columbia showed that moist mixed stands with openings in the canopy that encouraged heavy shrub development were preferred by this warbler (Enns and Siddle 1996). Understorey shrubs included willows, Sitka alder, rose, and highbush-cranberry. Typical forbs included fireweed, yarrow, Canada goldenrod, thistle, and grasses. In the Big Bend area of the Liard River, a few Mourning Warblers were found in immature deciduous riparian thickets, usually next to standing water (Bennett and Enns 1996); along the Kotcho River, riparian willow thickets were used.

Near Fort Nelson, habitat included alder and willow edges of trembling aspen and mixed conifer-aspen stands along roads, cutlines, and creeks (Erskine and Davidson 1976). In western Alberta, the Mourning Warbler nested in all age classes of trembling aspen stands, including clearcuts and mature stands, but was more than twice as abundant in 30- to 80-year-old stands than in younger stands (Westworth and Telfer 1993).

Males usually sing from isolated young trees surrounded by shrubs. Most foraging is done in the shrubby understorey within 1 to 2 m of the ground.

In British Columbia, the Mourning Warbler has been recorded breeding from 20 June (calculated) to 9 July; it probably breeds from mid-June through late July (Fig. 143).

Nests: One nest has been described in British Columbia. It was built on the ground among cover and consisted of grasses, leaves, and bark strips woven into an oval cup. Elsewhere, the Mourning Warbler builds its nest on or very near the ground among tangles of shrubby deciduous vegetation, among ferns, sedges, and horsetails, or in grass hummocks (Pitocchelli 1993). In Alberta, nests are usually a few inches above the ground, fastened to several stems of bog cranberry, choke cherry, wild rose, or red raspberry bushes (Salt 1973). Nests are usually well hidden by overhanging vegetation. Typical nests are cups of weed stalks, leaves, bark, grasses, and sedges lined with rootlets, fine sedges, grasses, and hair (Cox 1960).

Eggs: Nests with eggs have not been recorded in British Columbia. Eggs can probably occur from mid-June through early to mid-July. Clutch sizes of 3 to 5 eggs are common, but most nests contain 4 eggs (Bent 1953). The incubation period is 12 days (Hofslund 1954).

Young: Only 1 nest with young has been discovered in British Columbia; it contained 4 large nestlings on 2 July. A second record, of a flightless young with its parents on 9 July was probably of a bird that had just left its nest. Nestlings are likely present in British Columbia from late June through late July. The nestling period is 8 or 9 days, which is similar to

Figure 145. In British Columbia, the highest numbers for the Mourning Warbler in summer occur in the Boreal Plains Ecoprovince.

that of other warblers in the same genus (Cox 1960). Young leave the nest several days before they can fly.

Brown-headed Cowbird Parasitism: Nests with cowbird eggs or young have not been found in British Columbia. Of 4 records of adult Mourning Warblers feeding fledglings, however, 2 included cowbirds. Both records were from the vicinity of Taylor. The Mourning Warbler is known to be a fairly common host of the cowbird (Friedmann 1963; Friedmann and Kiff 1985).

Nest Success: Insufficient data.

REMARKS: Mourning and MacGillivray's warblers are very closely related and some researchers have thought that they form a western (MacGillivray's) and eastern (Mourning) subspecies complex, such as those found in flickers and orioles. Recent evidence, however, strongly supports species status (Hall 1979; Pitocchelli 1990, 1992). Hybridization between the 2 species may occur where ranges overlap (Hall 1979), such as in the Rocky Mountain foothills of Alberta (Cox 1973; Salt and Salt 1976). Pitocchelli (1993) notes that he did not find MacGillivray's and Mourning warblers in overlapping areas of northeastern British Columbia in the 1980s. Fifty kilometres southwest of Dawson Creek, however, singing males of both species were found within 50 m of each other in 1993 (Lance and Phinney 1994). Hybridization may explain the presence of broken eye rings in some individuals. Observers are encouraged to carefully note field marks and songs when identifying *Oporornis* warblers in northeastern British Columbia.

Unlike the Connecticut Warbler, which uses mainly mature or old-growth trembling aspen forests for nesting, the Mourning Warbler uses younger forests or shrubby areas. In fact, some human activities that open up boreal forests, such as road building, logging, farming, and mining, may facilitate the expansion of this warbler's breeding range (Fig. 142). Elsewhere in its range, hardwood clearcuts with 70% to 80% second-growth attracted the highest densities of breeding Mourning Warblers. Clearcutting of trembling aspen stands in the northeast may benefit this warbler; as a result, there are few conservation concerns for the Mourning Warbler related to habitat management in British Columbia.

The Mourning Warbler is very inconspicuous unless the male is singing. Migrants and nesting females tend to remain well hidden in foliage near the ground. Migrants are therefore difficult to detect, especially in late summer. In Alberta, Salt (1973) also noted this trait, which is probably responsible in part for our lack of information about autumn migration periods.

Two other warblers of this genus occur in northeastern British Columbia, and both are similar in appearance to the Mourning Warbler. The Connecticut Warbler, which is also restricted to the northeast in British Columbia but is generally rare, has a *complete* white eye ring. The MacGillivray's Warbler, which is relatively common and widely distributed in the province, has a *broken* white eye ring. An obvious field mark that is often used to separate MacGillivray's and Mourning warblers is the absence of any white eye ring in the latter. A small percentage of Mourning Warblers in some but not all populations have a broken white eye ring, however (Pitocchelli 1992). In northeastern British Columbia, Mourning Warblers with eye rings are not known to occur during the breeding season, but autumn migrants may have partial, faint eye rings.

There are several records of Mourning Warblers on coastal Breeding Bird Surveys (e.g., Albion, Campbell River, Kwinitsa, Port Hardy, Port Renfrew, Seabird), but these are erroneous and have been excluded from this account.

For additional life-history information on the Mourning Warbler in North America, see Pitocchelli (1993).

NOTEWORTHY RECORDS

Spring: Coastal – No records. **Interior** – Dawson Creek 24 May 1993-1, 24 May 1994-2; Taylor 24 May 1987-1, 25 May 1985-2, 26 May 1984-1, 26 May 1988-1, 28 May 1983-1, 29 May 1982-3, 31 May 1986-1.

Summer: Coastal – Harrison Mills 3 Jul 1983-1 (Kautesk 1983). **Interior** – Spillimacheen River 8 Jun 1996-1 (Ferguson and Halverson 1997); Redwillow River 30 Jun 1997-1; 50 km s Dawson Creek 20 Jun 1993-3; Brassey Creek 9 Jul 1995-1 flightless young with adults; Tupper 5 Jul 1981-2; Dawson Creek 20 Jun 1993-2; Moberly Lake Park 30 Jun 1990-1 male with food, 2 Jul 1989-4 large nestlings and adult male; Peace River 18 Jun 1974-1 (Penner 1976); Taylor 3 Jun 1983-5 along 7 km of Johnstone Rd, 13 Jun 1982-6 males along 6 km of Johnstone Rd, 15 Jun 1985-5 along Peace Island Park Rd; Peace Island Park (Taylor) 22 Aug 1986-1, 24 Aug 1984-2, 25 Aug 1988-1; Taylor Landing Park 22 Jul 1978-female feeding 1 or more young; Mile 79 Alaska Highway 23 Jun 1984-1; Gutah Creek Jun 1991-1 (Siddle 1992c); Andy Bailey Recreation Area (Fort Nelson) 5 Jul 1990-1 male with food; Kledo Creek 16 Jul 1986-1 fledged young; Fort Nelson 5 Jun 1974-1, Jun and Jul 1974-seen 21 times during surveys (Erskine and Davidson 1976), 11 Jul 1967-1 (Weber 1976); Clarke Lake 6 Jul 1978-3 along short stretch of road; near Kotcho Lake Jul 1992-1; Liard Hot Springs 9 Jul 1974-1 (Reid 1975), 12 Jul 1971-1 (Weber 1976); Liard River Jun 1996-7 records at Big Bend (Bennett and Enns 1996); Petitot River 26 Jun 1985-1 at bridge on Liard Highway.

Breeding Bird Surveys: Coastal – See REMARKS. **Interior** – Recorded from 3 of 73 routes and on 1% of all surveys. Maxima: Fort Nelson 19 Jun 1976-22; Steamboat 14 Jun 1976-9; Tupper 15 Jun 1974-3.

Autumn: Interior – Fort Nelson 1 Sep 1985-1 (Grunberg 1986); Fort Nelson River 2 Sep 1985-1 across from Old Fort Nelson; sw Cecil Lake 2 Sep 1984-1. **Coastal** – Sea Island 26 Oct 1998-1 (Toochin 1999); Rocky Point (Sooke) 1 Sep 1995-1 (Bowling 1996a).

Winter: No records.

MacGillivray's Warbler

MACW

Oporornis tolmiei (Townsend)

RANGE: Breeds from southeastern Alaska, southern Yukon, northwestern British Columbia, southwestern Alberta, and southwestern Saskatchewan south through the western states into southern California and east into South Dakota and New Mexico.

Winters from southern Baja California and the western and central mainland of Mexico, but not the Yucatan Peninsula, south to Honduras and north-central Nicaragua and western Panama.

STATUS: On the coast, *fairly common* to occasionally *common* migrant and summer visitant and *accidental* in winter in the Georgia Depression Ecoprovince; in the Coast and Mountains Ecoprovince, *fairly common* migrant and summer visitant on the Southern Mainland Coast, *uncommon* on Western Vancouver Island and the Northern Mainland Coast, and absent from the Queen Charlotte Islands.

In the interior, a *fairly common* to *common* migrant and summer visitant in the Southern Interior, Southern Interior Mountains, Central Interior, and Sub-Boreal Interior ecoprovinces; *very rare* to *rare* migrant and local summer visitant in the Boreal Plains and Northern Boreal Mountains ecoprovinces; *very rare* in summer in the Taiga Plains Ecoprovince.

Breeds.

NONBREEDING: The MacGillivray's Warbler (Fig. 146) is widely distributed throughout much of the southern two-thirds of the province. It is more local and sparsely distributed in the Boreal Plains, Northern Boreal Mountains, and Taiga Plains. There are also large areas on the coast where the species has not been reported, including much of the exposed coastal areas of Western Vancouver Island and most of the southern and northern regions of the Coast and Mountains between Port Hardy and Prince Rupert. Along the coast north of Vancouver Island, the few records are from the heads of deep inlets where valleys cut through the mountains from the interior (e.g., Kingcome Inlet, Rivers Inlet, and Portland Inlet). The northernmost coastal occurrence is from Stewart (Erskine and Davidson 1976). Despite field work on the central and northern coast, the species has not been found on coastal islands with apparently suitable habitat, such as the Goose Group (Guiguet 1953).

Figure 146. The MacGillivray's Warbler (female shown here) is widely distributed in British Columbia (Victoria, 15 August 1989; Tim Zurowski).

The MacGillivray's Warbler occurs from sea level along the coast to about 1,000 m elevation in the mountains in the interior. During migration it uses a variety of habitats that provide dense cover, including shrub patches, shrubby forest edges, brush piles, willow and red-osier dogwood thickets, shrub-grown wetlands (Fig. 147), open woodlands, moist gullies choked with deciduous shrubs, and early second-growth forests after clearcutting.

Figure 147. During migration in parts of the southern interior of British Columbia, the secretive MacGillivray's Warbler frequents well-shaded habitats of thick shrubs, usually in wet areas (Vaseux Lake, 9 August 1996; R. Wayne Campbell).

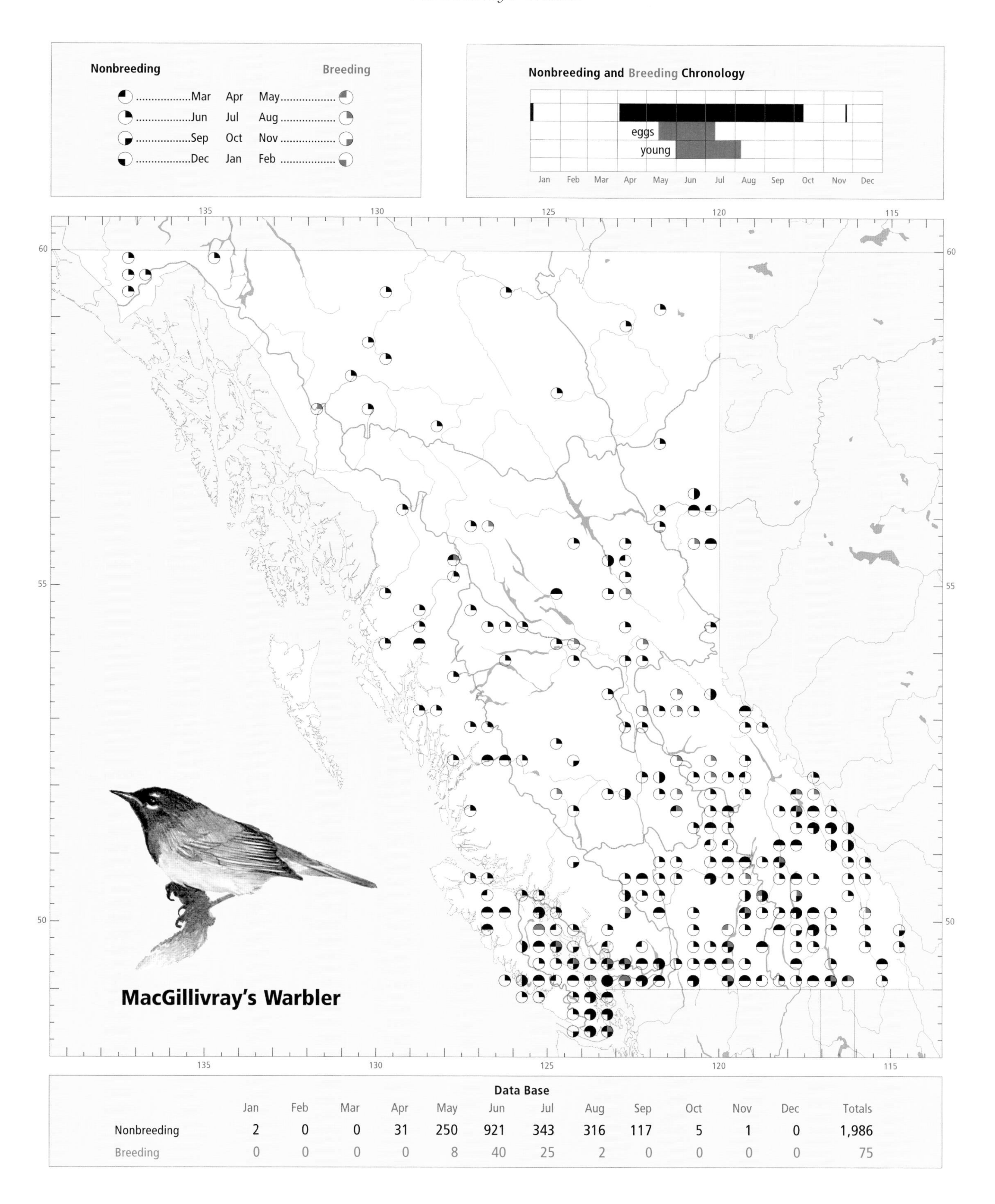

Data Base

	Jan	Feb	Mar	Apr	May	Jun	Jul	Aug	Sep	Oct	Nov	Dec	Totals
Nonbreeding	2	0	0	31	250	921	343	316	117	5	1	0	1,986
Breeding	0	0	0	0	8	40	25	2	0	0	0	0	75

On the southern coast, the spring migration begins with the arrival in the Georgia Depression of occasional birds in early April; most arrive towards the end of the month. Numbers increase steadily, reaching a peak in the second and third weeks of May (Fig. 149). On the Northern Mainland Coast, migrants do not arrive in the vicinity of the lower Skeena River until mid-May. The time lag suggests that MacGillivray's Warblers in this region arrive by a route through the interior of the province rather than up the coast.

In southern regions of the interior, the first arrivals may appear in the first week of May, with large numbers moving through by the third week of that month (Figs. 148 and 149). The greatest numbers pass through the southern parts of the interior in the last week of May. In the Cariboo and Chilcotin areas, the first arrivals appear in early May and the peak movement is evident during the second half of the month. In the Sub-Boreal Interior, migrants appear in small numbers at Prince George in mid-May and at Mackenzie in the third week of May. In both areas, the peak movement occurs later in the month depending upon prevailing edaphic conditions.

The onset of the southbound migration of MacGillivray's Warblers is imperceptible except in the Sub-Boreal Interior, where, after the presence of few birds in late July, numbers increase sharply in the first half of August in what is believed to be the passage of warblers from the north. The last birds leave the northern half of the province during the last week of August in the Northern Boreal Mountains and the second week of September in the Sub-Boreal Interior (Figs. 148 and 149). The migration moves rapidly through the southern half of the interior and, except for a few stragglers, all MacGillivray's Warblers have left the province by the end of September. On the coast, the timing of the southward movement from the Northern Mainland Coast is identical to the movement in the interior at that latitude. In the Georgia Depression, most of these warblers are gone by the end of September, but small numbers linger into early October.

On the British Columbia coast, the MacGillivray's Warbler has been recorded mainly from 4 April to 9 October; in the interior, it has been recorded regularly from 25 April through the third week of September (Fig. 148).

BREEDING: The MacGillivray's Warbler has a widespread breeding distribution from southern and eastern Vancouver Island east across the province to the east Kootenay, north in the west through the interior to Dokdaon Creek (Doch-da-on Creek in Swarth 1922) southwest of Telegraph Creek in the Northern Boreal Mountains. Breeding information is lacking for western and northern Vancouver Island and much of the Sub-Boreal Interior, Boreal Plains, Taiga Plains, and Northern Boreal Mountains. The species probably breeds throughout its summer range in the province, however.

Breeding Bird Surveys in southern parts of the Sub-Boreal Interior have detected large numbers of singing males. Even as long ago as the early 1920s, Swarth (1924) reported this warbler to be one of the commonest nesting birds along the lower Skeena River. He also found it abundant in summer along the Stikine River, from its mouth to near Telegraph Creek (Swarth 1922). Territorial males have been reported along the Tatshenshini River but the species has not been found breeding there.

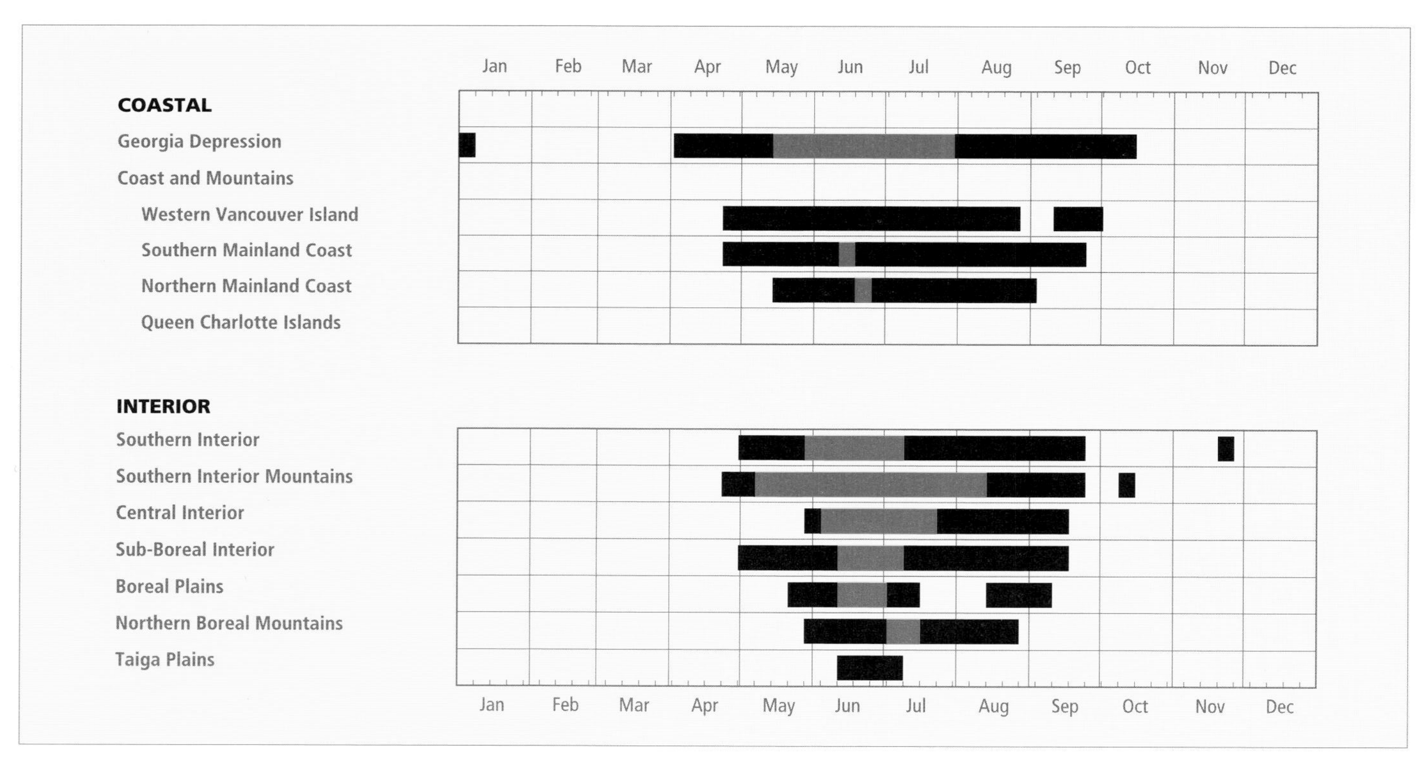

Figure 148. Annual occurrence (black) and breeding chronology (red) for the MacGillivray's Warbler in ecoprovinces of British Columbia. Records are shown for the week in which they occurred.

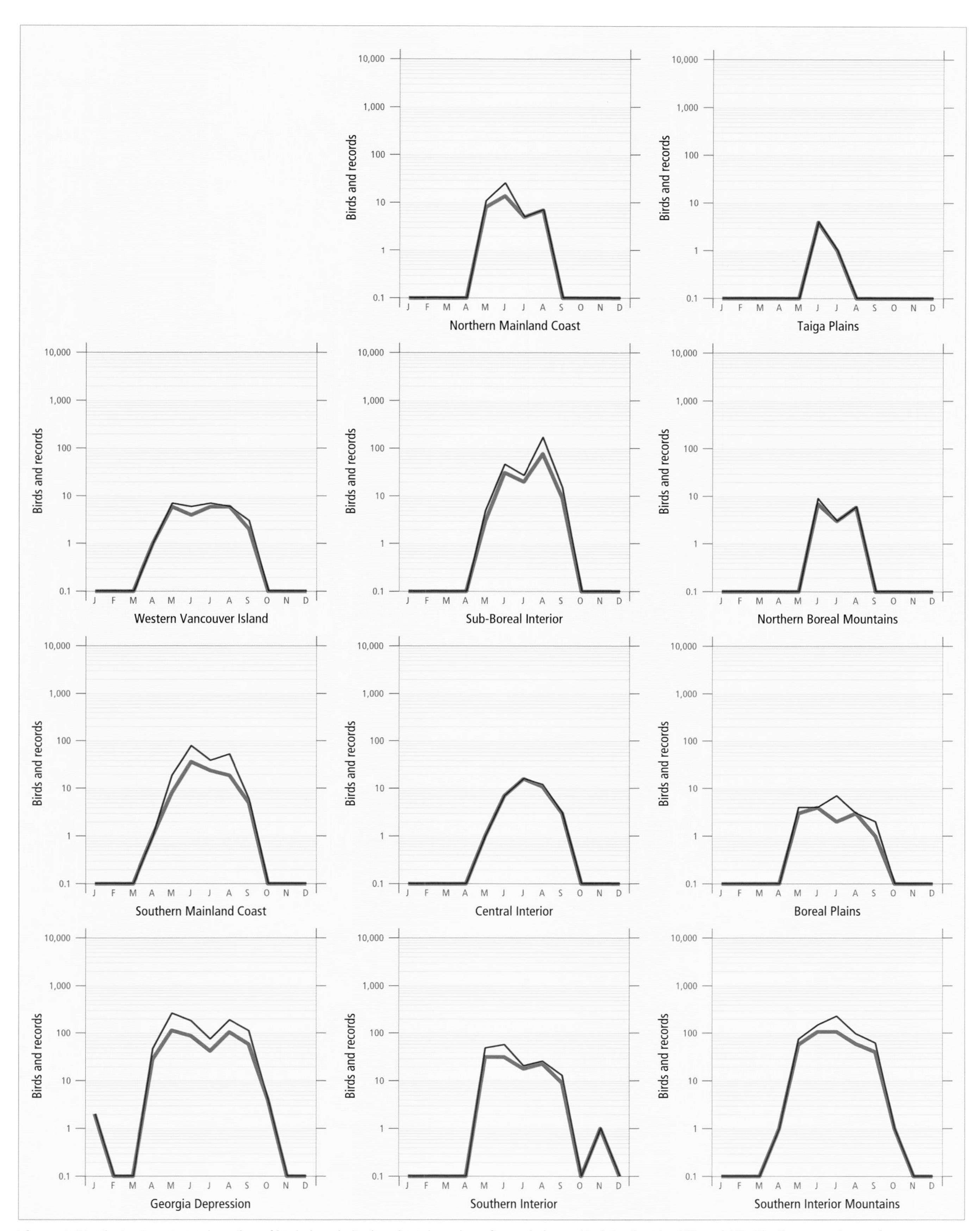

Figure 149. Fluctuations in total number of birds (purple line) and total number of records (green line) for the MacGillivray's Warbler in ecoprovinces of British Columbia. Christmas Bird Counts, Breeding Bird Surveys, and nest record data have been excluded.

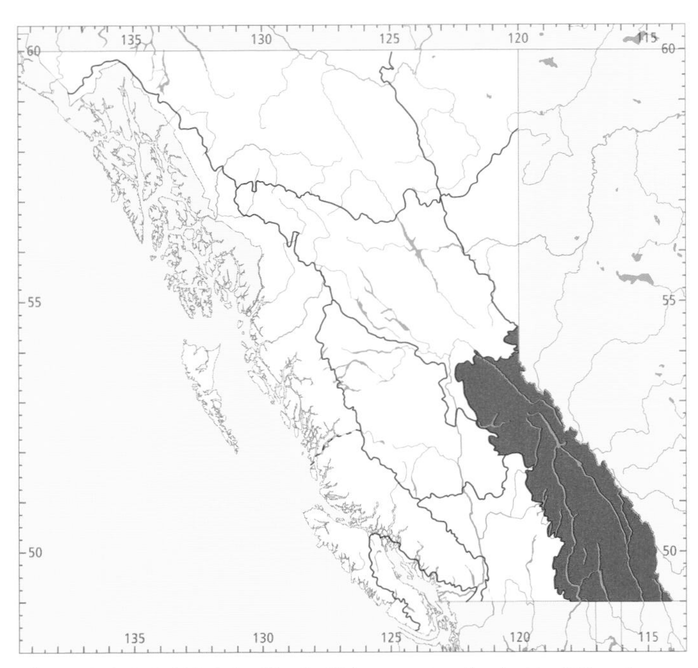

Figure 150. In British Columbia, the highest numbers for the MacGillivray's Warbler in summer occur in the Southern Interior Mountains Ecoprovince.

The highest numbers for the MacGillivray's Warbler in summer occur in the Southern Interior Mountains (Fig. 150), closely followed by southern areas of the Sub-Boreal Interior. An analysis of Breeding Bird Surveys in British Columbia for the period 1968 through 1993 could not detect a net change in numbers on either coastal or interior routes. The same is true for the analysis of Breeding Bird Survey data in North America for the period between 1966 and 1996 (Sauer et al. 1997). Continent-wide Breeding Bird Survey data indicate that British Columbia has some of the highest densities of the MacGillivray's Warbler in North America (Sauer et al. 1997).

The MacGillivray's Warbler has been recorded breeding from sea level to at least 1,500 m (Munro 1947) on the coast, and from 380 to at least 2,000 m in the interior. In Idaho not far south of the international boundary it is known to nest to an altitude of 2,400 m (Burleigh 1972).

In British Columbia, the characteristic breeding habitat of this warbler varies considerably between ecoprovinces but always includes thick, shaded deciduous shrubs, bushes, and thickets in both moist and dry environments (Fig. 151). In the valleys of the Southern Interior, it seeks shaded shrubby areas with a light overstorey of deciduous trees, and finds these in coulees supporting a dense growth of rose, saskatoon, sumac, mock-orange, willow, currant, and other similar shrubs, as well as in riparian stands of black cottonwood with a dense shrubby undergrowth, and in riparian thickets of willow and red-osier dogwood.

Specific plant associations in the Southern Interior Mountains have been described by Campbell and Dawe (1991).

Near Valemount, breeding habitat includes young stands of black cottonwood and trembling aspen with scattered western hemlock and western redcedar accompanied by a dense understorey of shrubs. It also includes roadside patches of thick willow, red-osier dogwood, black twinberry, fireweed, grasses, and mixed deciduous trees up to 7 m in height.

Near Revelstoke, the MacGillivray's Warbler inhabits stands of regenerating lodgepole pine and alder in logged areas, and burns with tangles of thimbleberry, bracken, and fireweed; beaver ponds with shallow pools of young western white pine and alder and dense patches of willow, cow-parsnip, and skunk cabbage; roadside brushy-grassy-weedy habitats with regenerating trembling aspen, black cottonwood, birch, and maple; and wetlands of willow, reed canary grass, fireweed, and thimbleberry bordering black cottonwood and alder woodlands.

In southern areas of the Southern Interior Mountains, near Fernie, the MacGillivray's Warbler breeds in semi-open stands of Douglas-fir and ponderosa pine with saskatoon, willow, young trembling aspen, and cow-parsnip as understorey plants; in associations of subalpine fir and spruce with blue elderberry and other deciduous shrubs; and in shrubby areas with tangles of black twinberry, blue elderberry, thimbleberry, stinging nettle, and small trembling aspens. In the Cariboo and Chilcotin areas, it frequently nests in small patches of rose under a canopy of trembling aspen and scattered conifers.

Figure 151. In portions of the Northern Boreal Mountains Ecoprovince of British Columbia, the MacGillivray's Warbler inhabits mixed woods of white spruce, balsam poplar, and alder with a willow and rose understorey (63 km south of Tatogga Lake, 13 July 1999; R. Wayne Campbell).

In the Georgia Depression, small nesting populations occur in roadside thickets of salmonberry, thimbleberry, black raspberry, currant, gooseberry, salal, mountain-ash, red elderberry, and similar shrubs along the margins of second-growth red alder stands, the borders of forest clearings, farmsteads, and the rights-of-way for country roads and powerlines. In coastal ecoprovinces, the MacGillivray's Warbler has also responded to the opportunities created by the removal of the old-growth coniferous forests. For example, Bryant et al. (1993), in their study of the summer avian communities in coniferous forests of west-central Vancouver Island, found the MacGillivray's Warbler to be consistently present in all age classes of forest except old-growth. It was most numerous in clearcuts and in regenerating forests between 15 and 35 years old.

Studies in the same habitat in Pacific Rim National Park yielded similar results. There the MacGillivray's Warbler was most abundant in a 12-year-old replanted clearcut (Roe 1974). There is no confirmation of nesting in Pacific Rim National Park or elsewhere on Western Vancouver Island (Hatler et al. 1978). It may nest there, but evidence is lacking.

In the lower Skeena River valley, a study of the summer bird communities of the interior trembling aspen forests found this warbler to be 1 of the 16 most abundant bird species (Pojar 1993). It was present in all age classes of forest from clearcut to mature, with its highest densities measured in the mature and old-growth stands with a well-developed understorey. The measured density of these warblers, in terms of singing males per 10 ha, were: clearcut, 0 to 12.74 (n = 60); sapling trembling aspen stage, 0 to 6.37 (n = 60); mature trembling aspen, 4.25 to 12.74 (n = 210); old trembling aspen, 9.5 to 14.33 (n = 22); and mixed conifer-aspen, 0 to 18.20 (n = 60). (Note that n is the number of samples of each ecotype surveyed.)

On the coast, the MacGillivray's Warbler has been recorded breeding from 14 May to 25 July; in the interior, it has been recorded breeding from 13 May to 6 August (Fig. 148).

Nests: Most nests (71%; n = 28; Fig. 152) were well hidden and placed close to the ground in low shrubs; another 15% were in living or dead bracken fern. The nests were cups in which the 3 most frequently appearing materials were grasses (86%), plant stems and fibre (36%), and bark strips (14%).

The heights of 26 nests ranged from ground level to 1.5 m, with 65% between 0.3 and 0.6 m.

Eggs: Dates for 42 clutches ranged from 13 May to 10 July, with 55% recorded between 9 June and 27 June (Fig. 152). Sizes of 30 clutches ranged from 1 to 5 eggs (1E-1, 2E-2, 3E-5, 4E-15, 5E-7), with 50% having 4 eggs. In British Columbia, the incubation period for 5 nests ranged from 12 to 13 days. In Alberta, the incubation period is 11 to 13 days (Semenchuk 1992).

Young: Dates for 33 broods ranged from 31 May to 6 August, with 52% recorded between 25 June and 14 July. Sizes of 20 broods ranged from 1 to 5 young (1Y-8, 2Y-1, 3Y-4, 4Y-5, 5Y-2), with 65% having 1 to 3 young. The nestling period is 8 or 9 days (Pitocchelli 1995).

Brown-headed Cowbird Parasitism: In British Columbia, 15% of 40 nests found with eggs or young were parasitized

Figure 152. Nest and eggs of the MacGillivray's Warbler in a thick patch of rose (Horsefly Lake, 18 June 1970; John K. Cooper).

by the cowbird. Parasitism was 8% (n = 13) on the coast and 19% (n = 27) in the interior. There were 4 additional records of cowbird fledglings being fed by MacGillivray's Warblers. Friedmann (1963) regarded this warbler as a regular but infrequent host for the cowbird, and he referred to just 9 known occasions of parasitism. Fourteen years later, he added 4 more instances (Friedmann et al. 1977). Parasitism of MacGillivray's Warbler by the cowbird appears to occur more frequently in British Columbia than elsewhere in the warbler's range.

Nest Success: Of 5 nests found with eggs and followed to a known fate, 2 produced at least 1 fledgling.

REMARKS: Two subspecies of the MacGillivray's Warbler are recognized by the American Ornithologists' Union (1957) within the species range: *O. t. tolmiei* and *O. t. monticola*. The former occupies the Pacific coastal states and western British Columbia while the latter occurs in eastern portions of the province (Dunn and Garrett 1997).

The MacGillivray's Warbler is considered by some authors (e.g., Pitocchelli 1990) to be conspecific with the Mourning Warbler. The 2 warblers, however, are generally regarded as a species pair that developed after an ancestral common population became climatically separated into eastern and western parts that had no breeding contact. Where this contact has been re-established, as in northeastern British Columbia, the Mourning Warbler appears to be replacing the MacGillivray's Warbler (M. Phinney pers. comm.). A few cases of hybridization between these species have been reported elsewhere (Cox 1973; Patti and Myers 1976).

Observations by Swarth (1922) indicate that in 1919 the MacGillivray's Warbler was an abundant breeding species in the region drained by the Stikine River in the Northern Boreal Mountains Ecoprovince. He describes the warbler as:

> Abundant throughout the whole of the region we explored. First noted at the Junction, June 1, one bird seen and another heard singing. During the next few days they were evidently arriving in abundance, and

thereafter the song was heard nearly everywhere we went. The first young out of the nest was seen July 13, at Doch-da-on Creek. A day or two later they were emerging in numbers, and as we went through the woods fussy parents in attendance followed us about. The species was noted in moderate abundance at each of our subsequent stations – Flood Glacier, Great Glacier and Sergief Island. Last noted, a single bird on Sergief Island September 3.

All March records and all but 1 winter occurrence, including Christmas Bird Counts (e.g., Pearse 1958), have been omitted from this account pending satisfactory documentation.

For additional life-history details and help with distinguishing this warbler from the similar Mourning and Connecticut warblers, see Salt (1973), McNicholl (1980), Curson (1992), Kowalski (1983), Pitocchelli (1995), and Dunn and Garrett (1997).

NOTEWORTHY RECORDS

Spring: Coastal – Triangle Mountain (Colwood) 4 Apr 1984-1; Chatham Islands 20 Apr 1984-1; Maple Ridge 23 May 1959-1 egg; Vancouver 24 Apr 1987-1 on University of BC campus; Harrison Hot Springs 29 Apr 1986-1; Hornby Island 29 Apr 1980-1; Squamish 2 May 1964-1; Golden Ears Park 15 May 1988-1; Qualicum Beach 25 Jul 1999-1 young; Cranberry Lake 26 May 1979-1; 16 km sw Campbell River 14 May 1972-eggs; Tahsis Inlet 24 Apr 1949-10 (Mitchell 1959); Mohun Lake 12 May 1950-1; Nimpkish Lake 8 May 1995-1; Bella Coola 30 May 1932-1 (MCZ 2841385); Stuie 24 May 1932-2 (MCZ 2841387-88); Kitimat 31 May 1975-1 (Hay 1976); Kispiox 16 May 1993-1. **Interior** – Richter Pass 21 May 1954-1 (PMNH 72057); Osoyoos 3 May 1987-7 banded; Creston 16 May 1984-1; Wolfe Lake (Princeton) 4 May 1975-1; Elko 19 May 1904-1 (NMC 4301); Squilax 11 May 1997-1; Shuswap Falls 13 May 1919-3 eggs; New Denver 3 May 1966-1, banded; Lytton 7 May 1966-2; Revelstoke 25 Apr 1981-1, earliest; 16 km n Golden 7 May 1998-2 singing males; Horse Lake (100 Mile House) 29 May 1937-1 (USNM 413383); Prince George 17 May 1997-3, 22 May 1997-4, 27 May 1997-25, 28 May 1997-21; Stuart Lake May 1980-2 banded; Gwillim Lake 5 May 1994-1; Morfee 31 May 1994-2 on transect; Chichouyenily Creek 29 May 1994-5; Brassey Creek 27 May 1994-1; Taylor 24 May 1986-2; Peace River 31 May 1973-1 (Penner 1976).

Summer: Coastal – Rocky Point 24 Aug 1994-8; Victoria 3 Jun 1914-nest and 4 eggs (RBCM E1006); Iona Island 28 Aug 1995-5 (Elliott and Gardner 1997); Hobbs Islet 23 Aug 1967-1; Qualicum Beach 17 Jun 1977-adult female and 2 fledglings; Whonnock 5 Jul 1937-4 eggs; Campbell River 18 Jun 1972-3 or 4 fledglings left nest when approached; Mons Creek (Alta Lake) 16 Jun 1924-3 eggs; Stuart Island 30 Jul 1936-1 (NMC 27315); Khutze Inlet 13 Jun 1936-2 (MCZ 2841409-10); Lakelse Lake 28 Aug 1976-1; 30 Jul 1991-2 fledglings fed by adult; Hazelton 24 Aug 1917-1 (NMC 10940); Kispiox River valley 22 Jun 1921-4 eggs; Stewart 12 Jul 1974-6, 17 km e Stewart (Erskine and Davidson 1976); Nass River 21 Jun 1975-28 on survey. **Interior** – Arrow Creek (Creston) 3 Aug 1989-5 nestlings; Drywash Mountain 6 Aug 1948-brood of young present; Fernie 5 Jul 1991-9 on 30 km survey along Morrissey Rd; Horn Lake (Okanagan) 3 Jun 1908-4 eggs; Kimberley 2 Jun 1974-1; Naramata 27 Jun 1987-nest with 2 eggs and 1 Brown-headed Cowbird egg, 25 Jun 1986-2 young recently out of nest, down tufts on head; Columbia Lake 9 Jun 1948-female incubating 5 eggs; Illecilewaet River 12 Aug 1990-2 fledglings; Emerald Lake 16 Jul 1975-15 (Wade 1977); 111 Mile Creek (0.8 km e Lac la Hache) 10 Jul 1959-4 eggs; Hemp Creek 5 Jun 1959-1; Williams Lake 24 Aug 1976-1 fledgling with female; Itcha Mountain 16 Aug 1931-1 (UBC 6526); Kleena Kleene 2 Jun 1958-1 (Paul 1959); 21.9 km n Valemount 10 Jul 1991-3; Mount Robson Park 6 Jun 1993-6 on bird count; Ten Mile Lake Park (Quesnel) 4 Jul 1936-1; Indianpoint Lake 24 Jun 1935-4 eggs, 6 Jul 1935-4 nestlings; Prince George 5 Jun 1997-16; Kinuseo Falls 19 Jul 1997-2; Vanderhoof 9 Jun 1995-4 eggs; McLeod Lake 14 Jul 1978-3 fledglings with adults; Aleza Lake 3 Jul 1969-4 fledglings just out of nest; Bullmoose Creek 10 Aug 1996-1; Tumbler Ridge 7 Jun 1997-1, 16 Jun 1997-5; Wolverine River 7 Jun 1977-1; Gagnon Creek 10 Aug 1994-10 banded; Tupper Creek 7 Jun 1938-1 (RBCM 8132; Cowan 1939); Moberly Lake 26 Jun 1987-8; Tetana Lake 5 Jul 1938-4 eggs; Charlie Lake 16 Jun 1938-1 (RBCM 8131); Cold Fish Lake 15 Jun 1976-1 (Osmond-Jones et al. 1977); Dokdaon Creek 13 Jul 1919-first young out of nest (Swarth 1922); 10 km ne Telegraph Creek 1 Jun 1919-1; Fort Nelson 12 Jun 1976-1, 8 Jul 1978-1, 9 Jul 1986-1 young about 3 days fledged; Tatshenshini River 9 Jun 1983-1; Liard Hot Springs Park 7 Jun 1985-1 (Grunberg 1985d); Liard Hot Springs 22 Aug 1943-1 (Rand 1944).

Breeding Bird Surveys: Coastal – Recorded from 24 of 27 routes and on 75% of all surveys. Maxima: Kwinitsa 29 Jun 1980-51; Elsie Lake 17 Jun 1970-44; Kitsumkalum 10 Jun 1979-39. **Interior** – Recorded from 63 of 73 routes and on 69% of all surveys. Maxima: Illecilewaet 9 Jun 1983-75; Gosnell 29 Jun 1971-68; Scotch Creek 1 Jul 1993-61.

Autumn: Interior – Gagnon Creek 12 Sep 1994-1 banded, 11 Sep 1995-2 banded; Chezacut 16 Sep 1933-1 (MCZ 2841401); 144 Mile House 8 Sep 1953-1; 16 km n Golden 6 Sep 1998-1 adult male; Revelstoke 8 Oct 1972-1; Paul Lake (Kamloops) 24 Nov 1994-1 (Bowling 1995c); Kamloops 28 Sep 1980-1; Sparwood 13 Sep 1983-1; Naramata 20 Sep 1965-3 (Rogers 1966a); Creston 22 Sep 1984-1; Manning Park 23 Sep 1973-1 (Belton 1973). **Coastal** – Kingcome Inlet 19 Sep 1936-1; Campbell River 27 Sep 1975-1, 9 Oct 1981-1; One Mile Lake (Pemberton) 10 Sep 1995-1; Alta Lake 2 Sep 1941-2; Seabird Island 21 Sep 1991-1 male; Iona Island 1 Oct 1995-1 (Elliott and Gardner 1997); Tofino Inlet 24 Sep 1972-2; Jordan River 16 Sep 1989-1; Rocky Point (Victoria) 3 Sep 1994-2.

Winter: Interior – No records. **Coastal** – Ladner Park 1 and 2 Jan 1992-1.

Common Yellowthroat

Geothlypis trichas (Linnaeus)

COYE

RANGE: Breeds from southeastern Alaska, central Yukon, southwestern Northwest Territories, northern British Columbia, northern Alberta, northern Saskatchewan, northern Manitoba, northern Ontario, central Quebec, and Newfoundland south to northern Baja California, across the Gulf states from Texas to Florida, and in mainland Mexico to Veracruz. Winters in the southern part of its breeding range from northern California, southern Arizona, southern New Mexico, southern Texas, the Gulf states, and South Carolina south through Mexico and Central America to northern South America.

STATUS: On the coast, a *fairly common* to *common* migrant and summer visitant and *very rare* winter visitant to the Georgia Depression Ecoprovince; in the Coast and Mountains Ecoprovince, an *uncommon* to *fairly common* migrant and summer visitant on the Southern and Northern Mainland coasts, locally *uncommon* migrant, *rare* summer visitant, and *casual* in winter on Western Vancouver Island, and *accidental* on the Queen Charlotte Islands.

In the interior, an *uncommon* to *common* migrant and summer visitant and *casual* winter visitant in the Southern Interior Ecoprovince; *common* migrant, locally *very common* summer visitant, and *casual* winter visitant in the Southern Interior Mountains Ecoprovince; locally *fairly common* migrant and summer visitant in the Central Interior and Sub-Boreal Interior ecoprovinces; locally an *uncommon* to *fairly common* migrant and summer visitant in the Boreal Plains, Taiga Plains, and Northern Boreal Mountains ecoprovinces.

Breeds.

NONBREEDING: The Common Yellowthroat (Fig. 153) is distributed across much of the province, usually in close association with wetlands. On the coast, it is widely distributed along eastern Vancouver Island and the lower Fraser River valley, but occurs more locally and less frequently on Western Vancouver Island, and the Southern and Northern Mainland coasts. There is a single autumn record from the Queen Charlotte Islands. The northernmost records on the coastal slope are from the valley of the Tatshenshini River.

In the interior, the Common Yellowthroat is widely distributed in the south along the valleys and plateaus from the eastern slope of the Cascade Mountains to the Rocky Mountains, and from the international boundary in the south to the boundary with the Yukon in the north. It is most numerous and widespread south of the Prince George area. Further north, spruce forests and mountainous terrain predominate, and the marshy habitats with emergent vegetation favoured by this warbler are scarcer. There are concentrations of the species in the Peace and Fort Nelson lowlands, along the upper Stikine River drainage basin, and in the vicinity of Atlin Lake.

In migration along the coast, the Common Yellowthroat has been reported from sea level to an elevation of 700 m. It occurs up to elevations of 1,200 m in the southern portions of the interior but is rarely found above 400 m in the north.

Figure 153. The Common Yellowthroat (adult male shown) is widely distributed in British Columbia in a variety of wet habitats (Anthony Mercieca).

During spring migration, the Common Yellowthroat is usually encountered in the same habitats where it will nest (see BREEDING). It is seldom seen among mixed-species flocks of warblers that move through the treetops during the daylight hours of the migration periods. There are, however, isolated records of yellowthroats in a variety of habitats, including flower gardens, edges of agricultural lands, stagnant ditches with emergent vegetation, patches of broom and gorse, berry bushes, shrubs and low trees, black cottonwood stands, trembling aspen groves, and even dry forests. The use of a wide variety of habitats is much more common during the autumn migration than at other times of the year.

The Common Yellowthroat appears to migrate individually rather than in flocks, and in most ecoprovinces the arrival of the main body of migrating yellowthroats is preceded by scattered occurrences of individual birds from 2 weeks to a month or more before records include several birds. For example, in the Southern Interior, records from mid-April to mid-May involve 1 or 2 birds but from 15 May onward, larger numbers are frequent.

On the coast, all Common Yellowthroats observed between January and mid-March are probably overwintering birds; most observations involve only a few individuals (Fig. 155). After mid-March, records of 3 or more are frequent as the main body of migrants moves into and through the region. The population is close to its maximum in late April and reaches its peak by mid-May (Figs. 154 and 155). Although numbers are small on Western Vancouver Island, spring arrival occurs in late April and the number of birds observed increases slowly through May. On the Northern Mainland Coast, the timing of spring arrival is several weeks behind that on the southwest coast, but peak populations are reached by late May (Fig. 155).

In the interior, the earliest record for the Okanagan valley is 18 April, and from then to early May the few records

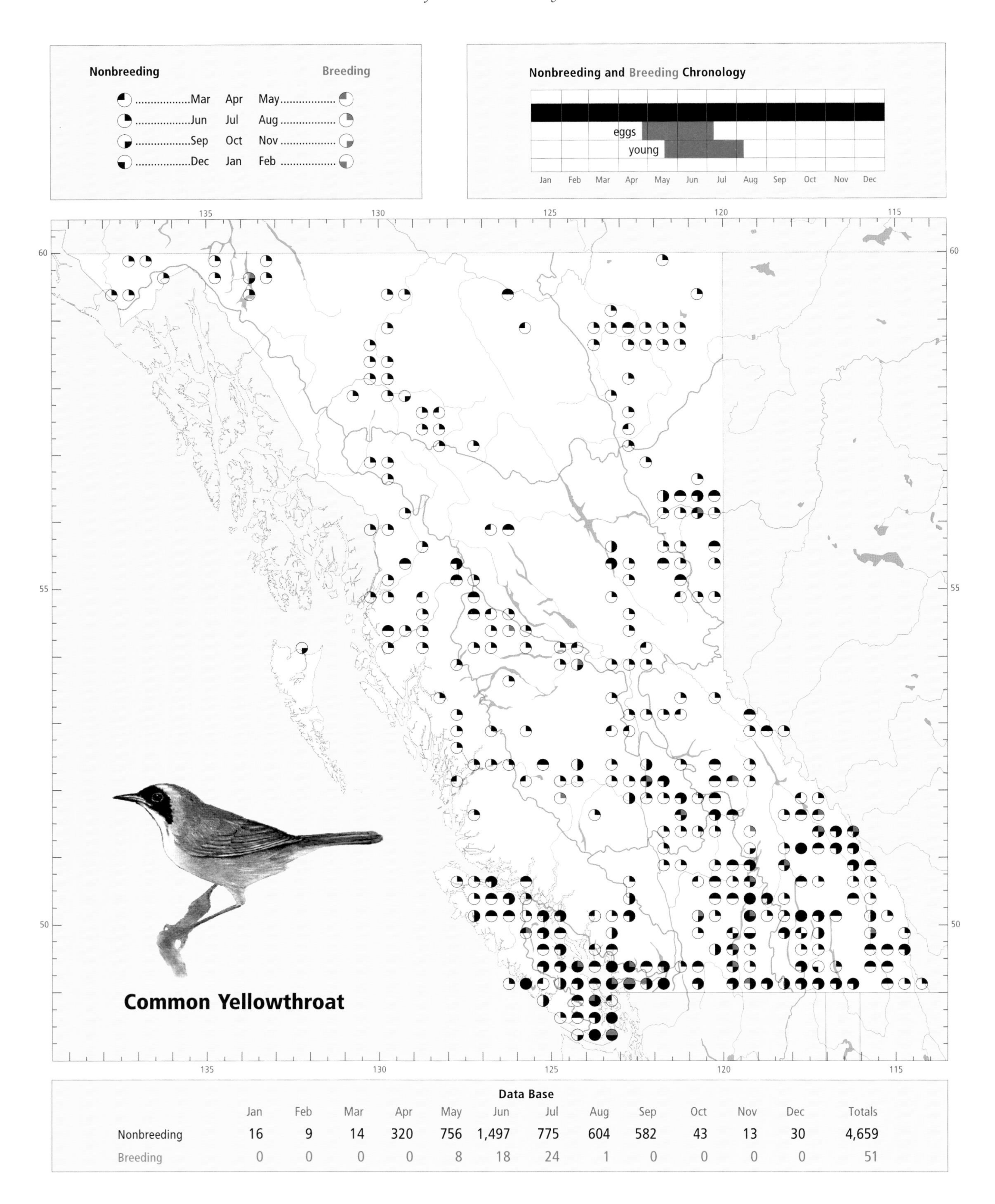

Data Base

	Jan	Feb	Mar	Apr	May	Jun	Jul	Aug	Sep	Oct	Nov	Dec	Totals
Nonbreeding	16	9	14	320	756	1,497	775	604	582	43	13	30	4,659
Breeding	0	0	0	0	8	18	24	1	0	0	0	0	51

are of 1 or 2 birds. The first observation of several birds occurs between the second and third weeks of May as the movement approaches its peak later in the month. In the east and west Kootenay regions of the Southern Interior Mountains, the timing of spring migration is similar but the numbers are greater. In the Sub-Boreal Interior and Boreal Plains, the first arrivals appear in the third week of May (Figs. 154 and 155).

In the northernmost areas of the province, the Common Yellowthroat appears to leave the nesting grounds as soon as the young are independent. In the Fort Nelson Lowland of the Taiga Plains, the latest records are from the end of August, and in the Northern Boreal Mountains most birds have left by the first week of September. In the Peace Lowland, the largest numbers occur between the end of August and mid-September, and all yellowthroats are out of the region by the end of September (Figs. 154 and 155). This is similar to the timing in areas west of the Rocky Mountains, where banding stations near Mackenzie, in the Sub-Boreal Interior, captured small numbers of Common Yellowthroats from the beginning of banding in the first week of August to a final date of 28 September. The period of greatest movement was 26 August to 15 September (V. Lambie pers. comm.).

In the southern parts of the Kootenay, Columbia, and Okanagan valleys, records are continuous until early October, with the largest numbers moving between 5 August and 23 September, and there are scattered observations through December (Figs. 154 and 155). On the coast, the population on the Northern Mainland Coast declines little between June and August, but most of the birds are out of the region by September. The Fraser Lowland and adjacent wetlands of the Southern Mainland Coast reveal a different pattern. There is a gradual reduction in number of records and in number of birds per record from June to August, a slight increase in September, and a drop to a minimum in November with the rapid departure of the summer population.

Migration banding stations at Vaseux Lake in the Okanagan valley and at Rocky Point on the southernmost end of Vancouver Island (R. Milliken pers. comm.) provides another indication of the southbound migration leaving the province. There is a steady movement of small numbers of Common Yellowthroats passing through both regions. At Rocky Point, from 1 to 9 of these warblers were captured for banding daily from 28 July to 25 September, with a concentration between 31 July and 26 August. In the south Okanagan, a thin trickle of birds entered the nets between 3 and 31 August 1995, from 1 to 4 in a 6-hour banding day and a total of 32 for the month (27 banding days). In September, the monthly total was 96 birds, of which 74 were taken between 5 and 20 September. This evidence suggests that the concentrated movement in the Southern Interior was about 3 weeks later than on the coast.

Our data from each of the ecoprovinces do not indicate an increase in numbers as a result of either the addition of young from the breeding effort or the arrival of southbound migrants. The exceptions are the Southern Interior and Sub-Boreal Interior, where between August and September there appear to be slight increases in both number of records and total birds recorded, suggesting an influx of transient migrants from other breeding areas. Thus, the data accumulated over many years suggest that through the late summer and autumn, birds depart from each ecoprovince at somewhat the

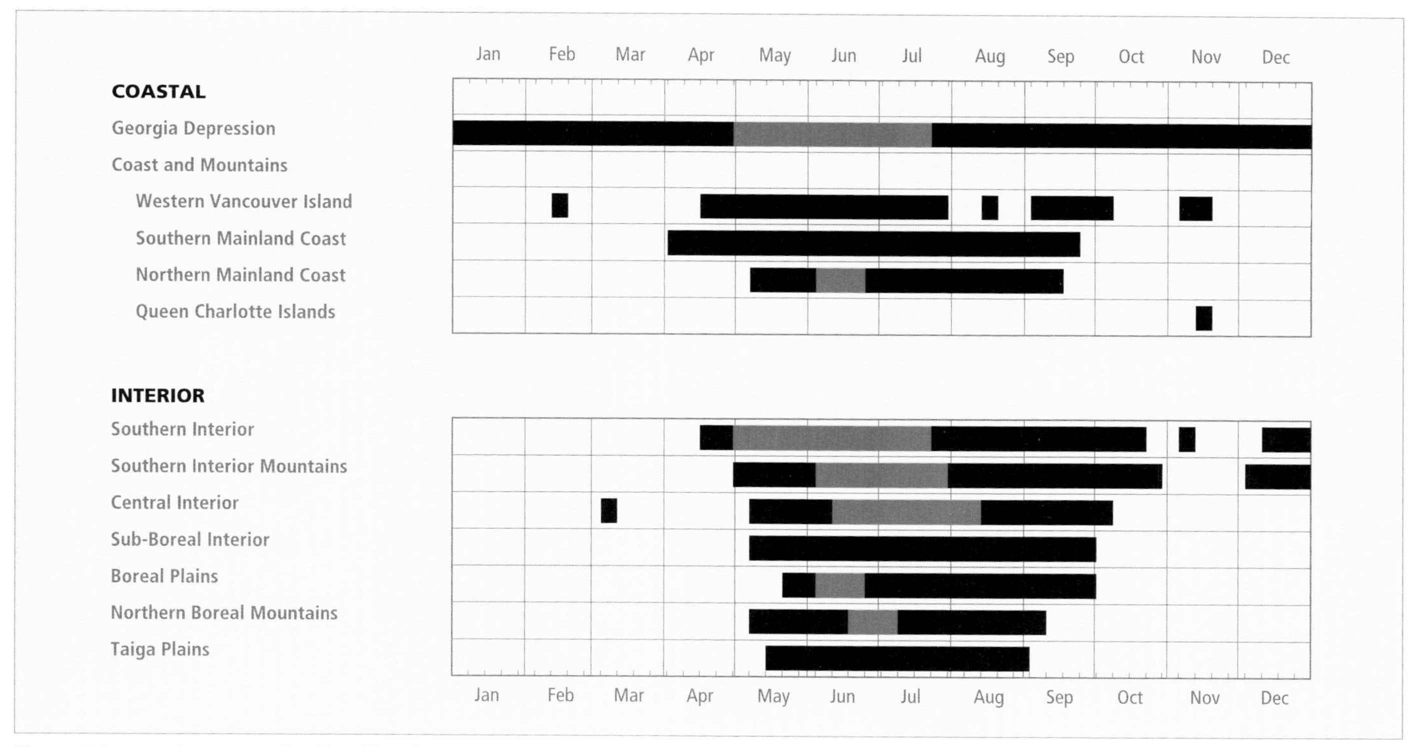

Figure 154. Annual occurrence (black) and breeding chronology (red) for the Common Yellowthroat in ecoprovinces of British Columbia. Records are shown for the week in which they occurred.

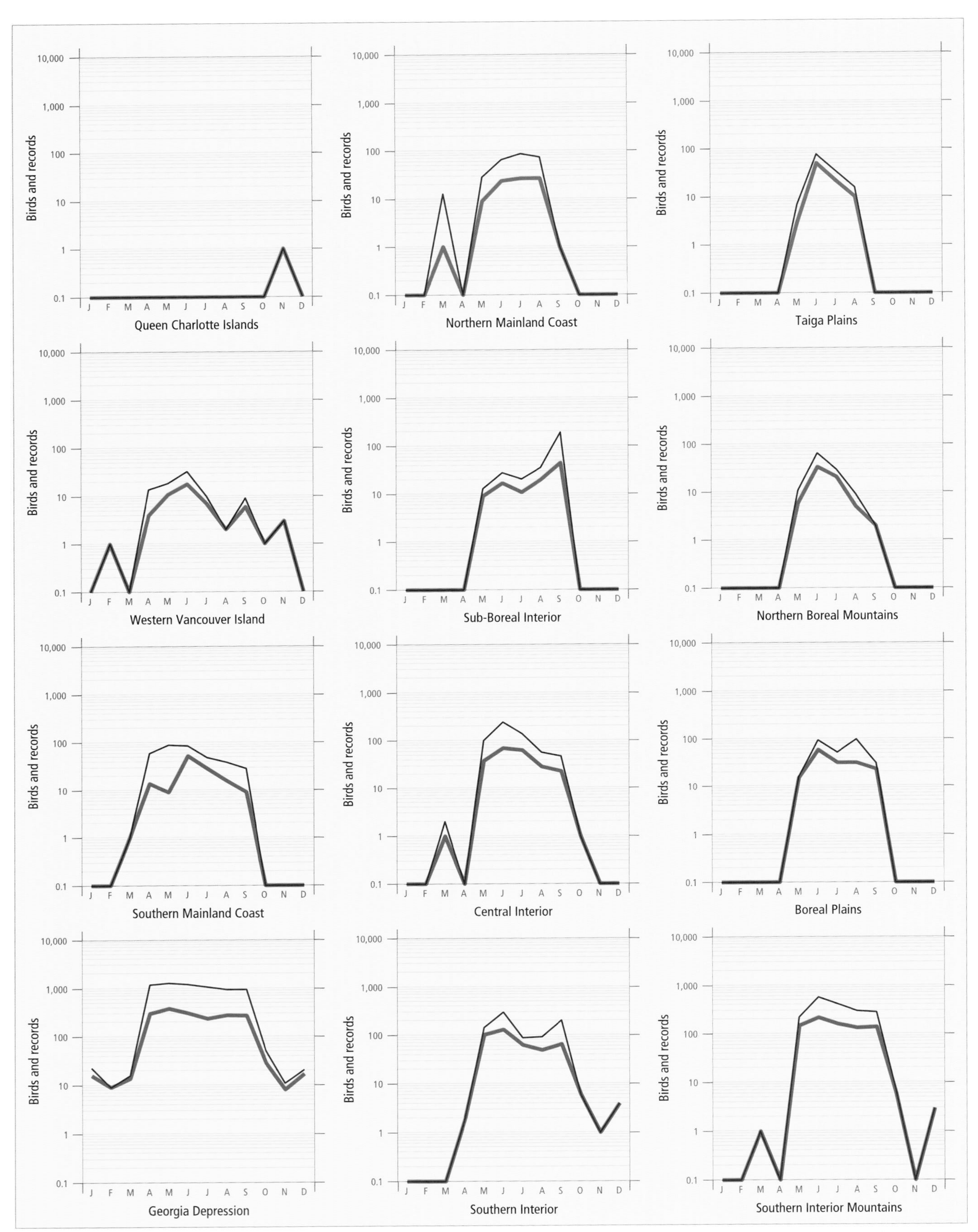

Figure 155. Fluctuations in total number of birds (purple line) and total number of records (green line) for the Common Yellowthroat in ecoprovinces of British Columbia. Christmas Bird Counts, Breeding Bird Surveys, and nest record data have been excluded.

same rate as their numbers are being increased by reproduction and in-migration.

Birds are rarely recorded in winter. On the coast, most have been found in the Fraser Lowland on the southwest mainland and the Nanaimo Lowland on Vancouver Island. In the interior, all are from the Okanagan valley.

On the coast, the Common Yellowthroat occurs regularly from 8 April to 21 October, with occasional records throughout the rest of the year; in the interior, it occurs regularly from 22 April to 7 October, with occasional records to 29 December (Fig. 154).

BREEDING: The Common Yellowthroat undoubtedly breeds throughout its summer range in the province. On the coast, however, there are no records of nesting for northern Vancouver Island, the Queen Charlotte Islands, and much of the northern coast. The species is widely distributed throughout the southern interior but records are sparse in the north, probably because the biological resources of many regions of this very large area have been inadequately explored.

The highest numbers for the Common Yellowthroat in summer occur in the Fraser Lowland of the Georgia Depression and the valleys of the Kootenay and Columbia rivers in the Southern Interior Mountains (Fig. 156). An analysis of Breeding Bird Surveys in the continent for the period 1968 through 1993 could not detect a net change in numbers on either coastal or interior routes. Analyses of all Breeding Bird Survey routes on the continent for the period 1966 through 1996 indicates that the mean number of birds decreased at an average annual rate of 0.3% ($P < 0.02$); in Canada, over the same period, the mean number of birds decreased at an average annual rate of 0.7% ($P = 0.07$) (Sauer et al. 1997).

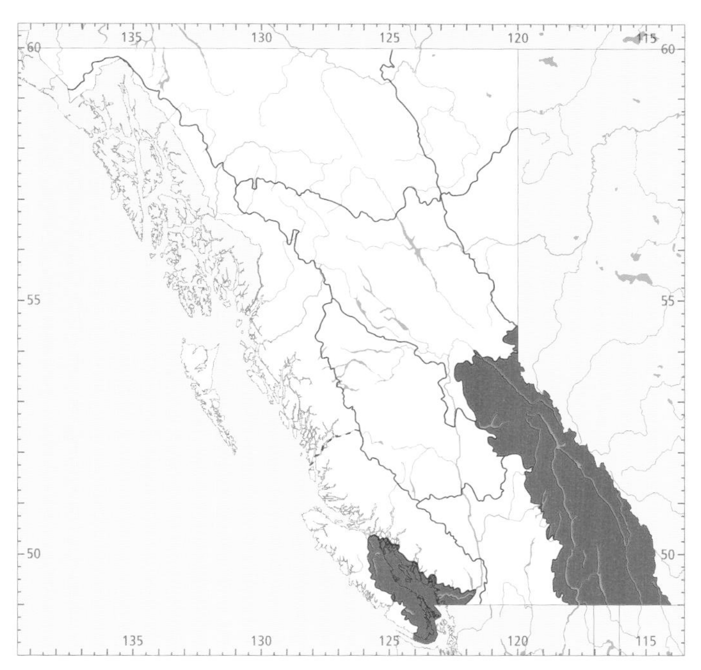

Figure 156. In British Columbia, the highest numbers for the Common Yellowthroat in summer occur in the Georgia Depression and Southern Interior Mountains ecoprovinces.

Figure 157. Throughout its range in North America, the Common Yellowthroat is associated with densely vegetated edges of wetlands, including shallow lakes (Duck Lake, Creston, 2 July 1997; R. Wayne Campbell).

The Common Yellowthroat breeds from sea level to about 125 m elevation on the coast, and from 300 to 1,200 m in the interior.

The Common Yellowthroat is a characteristic nesting species of densely vegetated wetlands, such as shallow lakes (Fig. 157) and ponds and similar habitats dominated by emergent vegetation, including tall sedges, reedbeds, stands of cattail, or a low shrubbery of hardhack, sweet gale, and willows, as well as sedge-grass-bulrush meadows, estuarine salt marshes, marshy borders of rivers, beaver ponds, sewage lagoons, roadsides, ditches, riparian scrub, permanent ponds in fields, wooded swamps (Fig. 158), regenerating trembling aspen clearcuts (Phinney 1998), and abandoned quarries. Cody (1974) and Douglas et al. (1992) determined that within the riparian zone there are distinct floristic regions that result from patterns of moisture. In their studies, the Common Yellowthroat appeared to prefer the low-structured willow vegetation.

The Common Yellowthroat has also been reported in the drier habitats along the edges of agricultural lands, airport runways, or damp roadsides; in tall grass meadows bordered by shrubs, thickets of willow, and young black cottonwood; and even in the undergrowth of a mixed forest of conifer and trembling aspen (Pojar 1993). Such conditions offer the dense cover required for nesting and foraging and the elevated perches used for territorial singing. Provincewide, the Common Yellowthroat nesting habitat ($n = 17$) was most frequently characterized as being within woodland areas, followed by wetland, grassland, and human-altered forest. Within these habitat types, the habitat classes used were mixed vegetation, sedge-bulrush marsh, and 6 other wetland classes represented by single nests.

The Common Yellowthroat has been recorded breeding in coastal British Columbia from 25 April (calculated) to 13 July and in the interior from 3 May to 7 August (Fig. 154).

Nests: Twenty-four percent of nests ($n = 25$) were on the ground; the others were in low vegetation, including sedges, reeds, grasses (Fig. 159), hardhack, rose, and willow. The heights of 23 nests ranged from ground level to 3 m, with 17 nests between ground level and 0.6 m. The nests were neat,

Figure 158. In parts of the Northern Boreal Mountains Ecoprovince of British Columbia, shrub swamps are used by the Common Yellowthroat for nesting (near Hazel Creek [BC-Yukon border], 18 June 1999; R. Wayne Campbell).

Figure 159. The nest of the Common Yellowthroat is a bulky cup of grasses, usually well concealed and built on or near the ground (Davie Hall Lake, 29 June 1983; R. Wayne Campbell).

compact cups constructed mainly of grasses and sedges; leaves, rootlets, plant fibres, and mosses were used in some nests. Linings were typically of fine grasses or sedges, and sometimes included hair.

Eggs: Dates for 29 clutches ranged from 30 April to 7 July, with 55% recorded between 26 May and 4 July. Eggs may be present as early as 25 April (calculated). Sizes of 23 clutches ranged from 1 to 6 eggs (1E-2, 2E-4, 3E-4, 4E-10, 5E-2, 6E-1), with 61% having 3 or 4 eggs. The incubation period of the Common Yellowthroat elsewhere is 12 days (Curson et al. 1994) or, more precisely, 11 days, 3 hours to 12 days, 19.75 hours (Stewart 1953).

Young: Dates for 23 broods ranged from 18 May to 7 August, with 57% recorded between 11 June and 18 July. Sizes of 14 broods ranged from 1 to 5 young (1Y-2, 2Y-5, 3Y-2, 4Y-3, 5Y-2), with 7 having 2 or 3 young (Fig. 159). The nestling period has been reported as 8 to 10 days (Curson et al. 1994) and 8 or 9 days (Stewart 1953). Bent (1953) provides an account of daily changes in the nestlings' plumage that permits their age in days after hatching to be identified.

Brown-headed Cowbird Parasitism: In British Columbia, 18% of 35 nests found with eggs or young were parasitized by the cowbird, similar to 19% of 106 nests in Ontario (Peck and James 1987). All of the parasitized nests were found in the interior, where the rate of parasitism was 32% (n = 19). In addition, another 8 fledgling cowbirds, equally divided between coast and interior, were recorded with Common Yellowthroat foster parents. Friedmann (1963) considers this species a frequent host of the Brown-headed Cowbird.

In a study of a Michigan population of Common Yellowthroats, the average number of yellowthroat fledglings produced by an unparasitized nest was 1.9, compared with 0.1 in nests parasitized by the cowbird (Stewart 1953); the average number of cowbirds fledged per parasitized nest was 0.4 (n = 20). In another study (Hofslund 1957), the average number of cowbird eggs per parasitized yellowthroat nest was 2; only 42.5% of them hatched, and the average number of cowbirds produced per parasitized nest was 0.6.

Nest Success: Of 3 nests found with eggs and followed to a known fate, none was successful.

REMARKS: The Common Yellowthroat is the most geographically variable of the North American warblers, with 15 subspecies widely recognized and as many as 14 additional subspecies proposed in its extensive breeding range from Alaska to Veracruz, Mexico (Dunn and Garrett 1997). Three races occupy the species' range in British Columbia: *G. t. campicola* occurs throughout the province east of the Cascade Mountains and other coastal ranges, *G. t. arizela* inhabits the coastal regions from British Columbia to central California, and *G. t.*

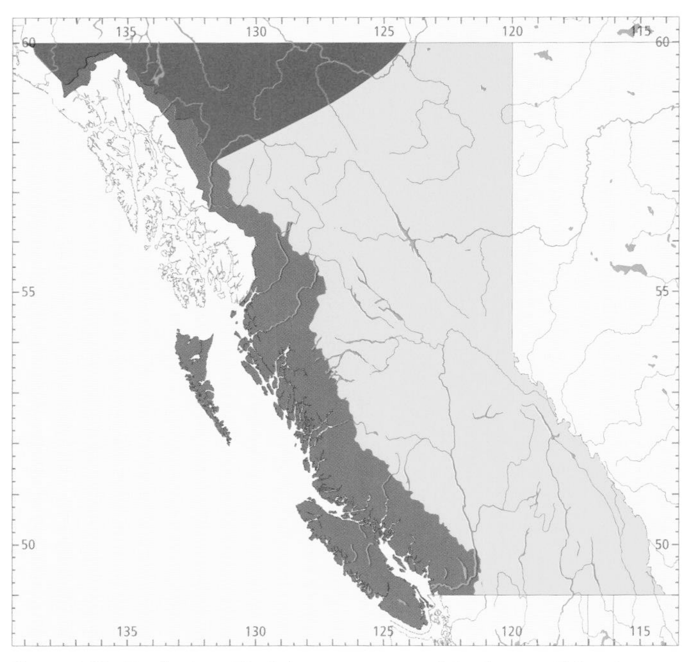

Figure 160. Distribution of 3 of the western races of the Common Yellowthroat – *Geothlypis trichas arizela* (brown), *G. t. campicola* (yellow), and *G. t. yukonicola* (green) – showing their breeding range (modified from Behle 1950, American Ornithologists' Union 1953, and Dunn and Garrett 1997).

yukonicola occupies the extreme northwestern interior of the province. Figure 160 shows the tentative distribution of the recognized subspecies in British Columbia. Behle (1950) identified specimens from the lower Skeena River valley as *G. t. arizela*, but the subspecific identity of the yellowthroats breeding in the Bella Coola valley and the region around Meziadin Lake has not been determined. For details about the characteristics and distribution of the subspecies, see Behle (1950) and Dunn and Garrett (1997).

On Airport Reserve lands on Sea Island, the Common Yellowthroat was the third most abundant breeding songbird after Savannah Sparrow and American Robin. It occurred at densities of 2.6 singing males per 10 ha in mixed woodland/grasslands, 1.3 singing males per 10 ha in hayfields, and 2.3 singing males per km of hedgerow habitat. Construction of the Parallel Runway was estimated to remove habitat for 66% of the estimated 82 singing males recorded on Airport Reserve lands in 1992 (Cooper 1993).

For additional information on life history, see Stewart (1953), Bent (1953), Hofslund (1957), Dunn and Garrett (1997), and Guzy and Ritchison (1999).

NOTEWORTHY RECORDS

Spring: Coastal – Saanich (Rithet's Bog) 23 Apr 1972-14 (Tatum 1972); Somenos Lake 8 Apr 1978-16; Duncan 4 May 1974-4 eggs, 18 May 1974-4 young nestlings; 11 km n Abbotsford 10 May 1944-4 eggs; Seabird Island 27 Apr 1992-8; Pitt Meadows 6 Apr 1986-20 (Mattocks 1986a); Coombs 26 May 1983-female with 1 fledgling at Dudley Marsh; Courtenay 28 Mar 1990-1; Pitt Meadows 14 May 1973-74; Hoomak Lake 17 Apr 1984-5; Alta Creek 14 Apr 1993-10; 35 km s Nimpkish 28 May 1978-3. **Interior** – Osoyoos 3 May 1905-4 eggs; Creston 12 May 1981-1; Rossland 4 May 1973-1; Erie Lake 27 May 1984-12; Vaseux Lake 9 May 1974-3; Swan Lake (Vernon) 4 Apr 1992-1; White Lake (Shuswap) 18 Apr 1962-1; Revelstoke 11 May 1977-1; Wapta Lake 10 May 1977-1; 16 km n Golden 5 May 1998-1 singing male; Williams Lake 5 Mar 1980-2, 23 May 1980-17; McBride 3 May 1981-2; Willow River 31 May 1971-1; Babine Lake 30 Apr 1889-6; Tetana Lake 22 May 1938-1 (Stanwell-Fletcher and Stanwell-Fletcher 1943); Bullmoose Creek 24 May 1994-1; Morfee 22 May 1994-2; Tupper Creek 22 May 1938-1 (Cowan 1939); Dawson Creek 18 May 1991-1 (Phinney 1998); Sikanni Chief 30 May 1995-1; Parker Lake 19 May 1975-5; Liard Hot Springs 13 May 1975-1 (Reid 1975), 27 May 1981-4.

Summer: Coastal – Rocky Point (Victoria) Jul 1994-5, 19 Aug 1994-11; Pitt Meadows 17 Jun 1971-30 (Campbell 1972); Langley 13 Jul 1969-2 nestlings; Pitt Meadows 20 Jun 1973-78; Port Moody 5 Jul 1897-3 eggs; Harrison Hot Springs 2 Aug 1992-9; Morte Lake 28 Aug 1977-30; Quatse Lake 9 Jun 1978-5; Meziadin Lake 20 Jun 1975-3. **Interior** – Manning Park 11 Aug 1978-12; Erie Lake 19 Jul 1984-25 in family groups; 31 km ne Chute Lake 18 Jul 1971-2 or 3 fledglings, abandoned nest found; 16 km ne Merritt 28 Jun 1971-3 eggs plus 1 Brown-headed Cowbird egg; 3 km se Brouse 27 Jul 1978-2 flying fledglings still being fed by parent; Turtle Valley 5 Jun 1977-20; e of Tum Tum River 10 Jun 1991-5 eggs; ne Revelstoke 8 Jul 1977-4 nestlings; Glacier National Park 7 Jun 1982-12; Ottertail (Yoho National Park) 17 Jul 1975-13 counted around sloughs (Wade 1977); 108 Mile House 4 Jul 1980-4 eggs; Chilanko Forks 1 Jul 1984-4; Williams Lake 24 Jun 1977-2 nestlings and 2 Brown-headed Cowbird nestlings in nest, 8 Jun 1978-26; Stum Lake 11 Aug 1971-2 fledglings fed by female; Kleena Kleene 14 Jun 1957-1 (Paul 1959); Moose Lake (Yellowhead Pass) 16 Jun 1972-2 in marsh; Klio Creek (Topley) 10 Jul 1980-2 males, at 1,150 m elevation; Mugaha Creek 2 Jun 1996-6; e Moose Lake (Gwillim Lake Park) 18 Jul 1995-2 pairs; Chetwynd 5 Aug 1975-1; ne Alex Graham Lake 30 Jul 1974-3 fledglings with female; Peace Island Park Road 21 Jun 1988-1 fledgling; Boundary Lake (Goodlow) 25 Jul 1982-10, 26 Aug 1985-21; w Clarke Lake 7 Aug 1946-young; Mile 10.9 (Clarke Lake Rd) 28 Aug 1978-4; Parker Lake 31 Jul 1986-1 fledgling about 14 days fledged; Lower Tatshenshini River valley 5 Jun 1983-6; Crater Creek 5 Jul 1980-4 nestlings.

Breeding Bird Surveys: Coastal – Recorded from 23 of 27 routes and on 69% of all surveys. Maxima: Coquitlam 25 Jun 1993-23; Pitt Meadows 1 Jul 1988-22; Seabird 4 Jul 1993-18. **Interior** – Recorded from 57 of 73 routes and on 62% of all surveys. Maxima: Golden 29 Jun 1980-30; Succour Creek 17 Jun 1994-27; Lac la Hache 27 Jun 1982-20.

Autumn: Interior – Atlin 4 Sep 1929-1 (CAS 32507); Fort St. John 28 Sep 1985-1; Mugaha Creek 15 Sep 1996-6, 28 Sep 1996-1; Riske Creek 2 Oct 1986-1; 15 km n Golden 23 Oct 1995-1 adult male; Salmon Arm 11 Sep 1994-100; Swan Lake (Vernon) 20 Oct 1968-1; Ottertail (Yoho National Park) 8 Oct 1973-1 in horse pasture (Wade 1977); Kelowna 9 Nov 1991-1 at Chichester marsh; Nakusp 3 Oct 1976-2; Okanagan Landing 7 Oct 1944-1; Vaseux Lake 19 Sep 1977-9. **Coastal** – Masset 16 Nov 1995-1 male; Pemberton 10 Sep 1995-3; Alta Creek 23 Sep 1996-2; Pitt Marsh 6 Sep 1972-89 (Campbell et al. 1974); Vancouver 21 Oct 1973-3, 15 Nov 1990 to 26 Jan 1991-2; Tofino 17 Nov 1983-1 (Campbell 1983d); Somenos Lake 3 Sep 1977-19; Rocky Point 18 Oct 1995-2; Jordan River 7 Nov 1987-1.

Winter: Interior – Mount Revelstoke 12 Dec 1982-1; Nakusp 25 Dec 1997-1 male at sewage lagoon; Okanagan Landing 13 Dec 1930-1 (ROM 89482), 27 Dec 1981-1 (Cannings et al. 1987); Kelowna 24 Dec 1994-1, 29 Dec 1991-1, both at Chichester marsh; Duck Lake (Creston) 5 Dec 1997-1. **Coastal** – Jericho Park (Vancouver) 3 Jan 1975-1; Richmond 31 Jan 1992-1 female in winter plumage; Iona Island 19 Dec 1993-1 (Siddle 1994b), 26 Dec 1994 to 2 Jan 1995-1 successfully overwintered (Bowling 1995d); Tofino 15 Feb 1980-1 (BC Photo 540); Reifel Island 21 Jan 1987-1; Ladner 26 Feb 1989-1 at Burns Bog; Somenos Lake 3 Dec 1990-1 (Siddle 1991b); Victoria 28 Dec 1986 to 1 Jan 1987-1, 2 Feb 1990-1 (Siddle 1990a).

Christmas Bird Counts: Interior – Recorded from 1 of 27 localities and on less than 1% of all counts. Maxima: Vernon 27 Dec 1981-1. **Coastal** – Recorded from 3 of 33 localities and on 1% of all counts. Maxima: Ladner 27 Dec 1982-1; Pitt Meadows 29 Dec 1991-1; Vancouver 15 Dec 1991-1.

Wilson's Warbler

Wilsonia pusilla (Wilson)

WIWA

RANGE: Breeds from western and northern Alaska, northern Yukon, northwestern and east-central Mackenzie, northern Saskatchewan, northern Manitoba, northern Ontario, northern Quebec, northern Labrador and Newfoundland south to southern Alaska, central coastal California, northeastern Nevada, south-central Utah, southwestern Colorado, and northern New Mexico; and in the central and southern Canadian provinces south to include Nova Scotia, northeastern Minnesota, northeastern New York, central Maine, and northern Vermont. Winters from coastal California and southern Baja California, the Gulf states, and mainland Mexico south through Central America to western Panama.

STATUS: On the coast, a *fairly common* to *common* migrant, *uncommon* summer visitant, and *casual* in winter in the Georgia Depression Ecoprovince; in the Coast and Mountains Ecoprovince, *uncommon* (occasionally *fairly common*) migrant and summer visitant on Western Vancouver Island, the Southern Mainland Coast, and the Northern Mainland Coast; *uncommon* migrant and summer visitant on the Queen Charlotte Islands.

In the interior, an *uncommon* to *common* spring migrant, *fairly common* summer visitant, *uncommon* autumn migrant, and *casual* in winter in the Southern Interior and Southern Interior Mountains ecoprovinces; *uncommon* to *common* migrant and summer visitant in the Central Interior and Sub-Boreal Interior ecoprovinces; a *fairly common* summer visitant in the Northern Boreal Mountains Ecoprovince; *uncommon* spring and *fairly common* autumn migrant and *rare* summer visitant in the Boreal Plains Ecoprovince; *rare* in the Taiga Plains Ecoprovince.

Breeds.

NONBREEDING: The Wilson's Warbler (Fig. 161) is among the most widely distributed warblers in British Columbia and occurs throughout all ecoprovinces. Although there remain some fairly large areas in which it has not yet been recorded, its apparent absence is almost certainly due to lack of observers. Along the coast, it has been noted on many of the offshore islands, including much of the Queen Charlotte Islands archipelago. In winter, it has been observed most frequently on the south coast in the Nanaimo Lowland and the Fraser Lowland of the Georgia Depression.

Figure 161. The Wilson's Warbler (adult male shown) is among our most widely distributed warblers in British Columbia and breeds from near sea level to timberline. (Victoria, 15 May 1978; Tim Zurowski).

During spring migration along the coast, the Wilson's Warbler has been recorded at elevations from sea level to about 600 m; in the interior, it occurs mostly at valley bottom elevations between 280 and 400 m but has been observed to 1,100 m on snow-free mountain slopes. During autumn migration, elevations up to 2,200 m have been reported.

During migration, the Wilson's Warbler frequents thickets and ribbons of shrubs and low trees that line the edges of lakeshores, beaver ponds, stream banks (Fig. 162), sloughs,

Figure 162. During spring migration, the Wilson's Warbler forages in low shrubs and bushes bordering rivers and streams (Bummers Flats, 47 km north of Cranbrook, 7 May 1997; R. Wayne Campbell).

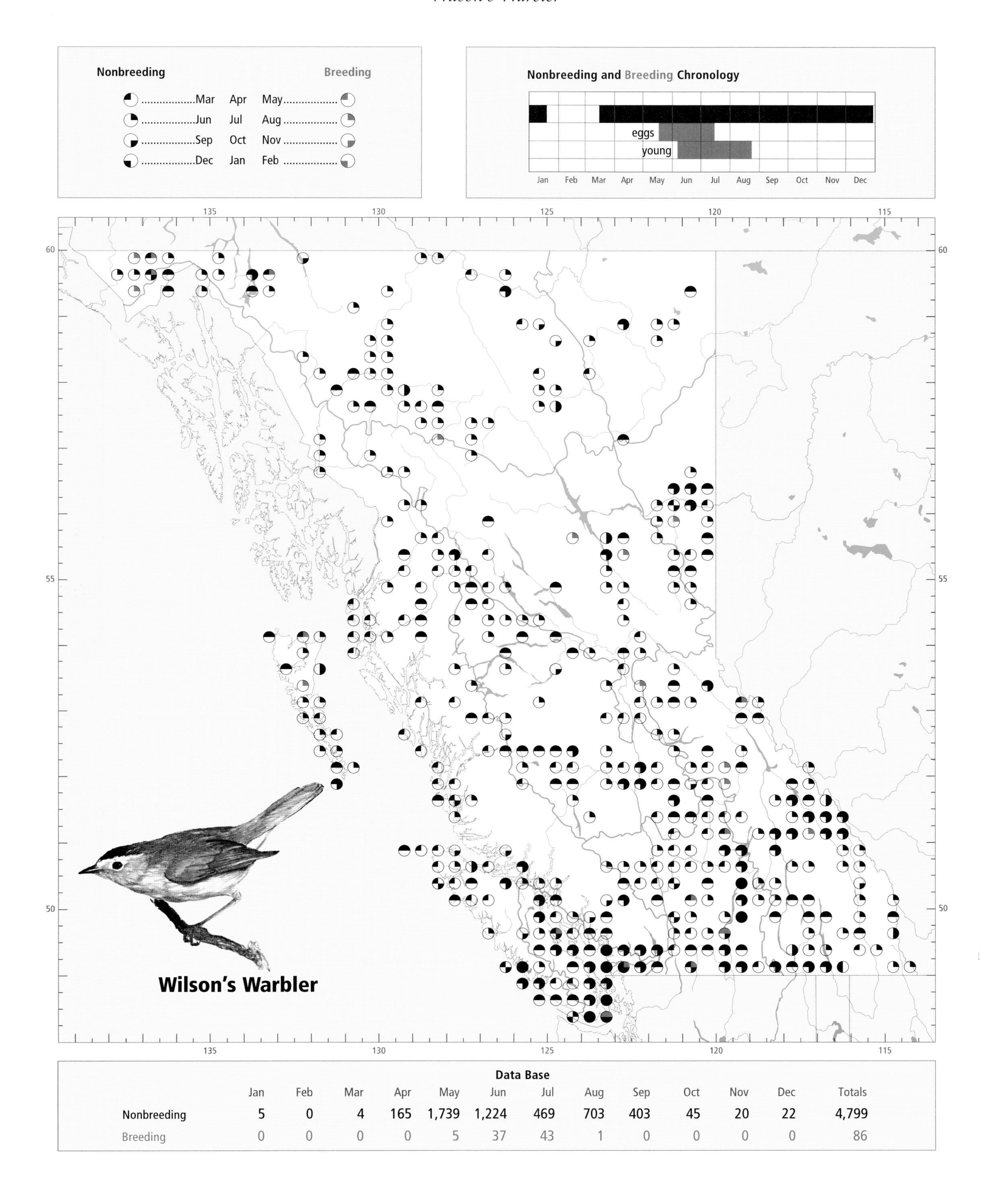

Data Base

	Jan	Feb	Mar	Apr	May	Jun	Jul	Aug	Sep	Oct	Nov	Dec	Totals
Nonbreeding	5	0	4	165	1,739	1,224	469	703	403	45	20	22	4,799
Breeding	0	0	0	0	5	37	43	1	0	0	0	0	86

Figure 163. Annual occurrence (black) and breeding chronology (red) for the Wilson's Warbler in ecoprovinces of British Columbia. Records are shown for the week in which they occurred.

sea beaches, shoreline spits, airports, sewage lagoons, cleared fields, and powerline rights-of-way. In the Boreal Plains, it occurs mostly in mixed woodlands and shrubby edges of willow, birch, and alder in spring, and along muskeg edges, willow hedgerows, and wet areas where willows and other shrubs predominate in autumn (C. Siddle pers. comm.). The Wilson's Warbler is often seen in parks and gardens, especially during the spring migration.

On the south coast, spring migration begins with the arrival of a few warblers in late March and the first week of April (Figs. 163 and 164). The migrating warblers build rapidly to a peak in mid-May, then decline steadily into mid-June as many continue northward or move onto nesting grounds at higher elevations. Passage northward is rapid, with earliest arrivals reaching the mouth of the Skeena River in the last week of April and Atlin as early as the second week of May (Campbell 1981).

In the interior, the migration is almost synchronous over a wide latitude (Figs. 163 and 164), with the first arrivals appearing in the southern portions of the Southern Interior, Southern Interior Mountains, Central Interior, and Sub-Boreal Interior all in the last 2 weeks of April. The first arrivals in the Northern Boreal Mountains and the Boreal Plains occur simultaneously in the first week of May. The front of the migration reaches the Yukon boundary by the first or second week of May.

The autumn migration is remarkably similar throughout the northern half of the province (Fig. 164). We have no observations from the 3 northernmost ecoprovinces after the third week of September. The autumn departure is more prolonged further south, but in the Cariboo and Chilcotin areas, the Thompson and Okanagan valleys, and the east and west Kootenays the bulk of the population has left by the third week of September, although a few birds have been seen as late as December in each of these regions.

On the coast, most of the birds have left the Northern Mainland Coast by mid-September, and the Queen Charlotte Islands and Southern Mainland Coast before mid-October. In the Nanaimo Lowland of Vancouver Island and in the Fraser Lowland, only stragglers remain after mid-October, but there are records for each of the winter months.

On the coast, the Wilson's Warbler occurs regularly from 18 March to 14 October; in the interior, it occurs regularly between 29 April and 23 September (Figs. 163 and 164).

BREEDING: The Wilson's Warbler likely breeds in all ecoprovinces except the Taiga Plains, including offshore islands and ocean shorelines in the west to the interprovincial boundary in the Rocky Mountains in the east, and over the full north-south extent of British Columbia.

This warbler reaches its highest numbers in summer in the Northern Boreal Mountains (Fig. 165). An analysis of Breeding Bird Surveys for the period 1968 through 1993 could not detect a change in the mean number of birds on coastal routes; on interior routes, the mean number of birds decreased at an average annual rate of 4% (Fig. 166). An analysis of summer populations throughout North America reveals no detectable change over the period 1966 to 1996, but in the most recent period (1980 to 1996), the population has declined at an average annual rate of 2.0% ($P < 0.05$) (Sauer et al. 1997).

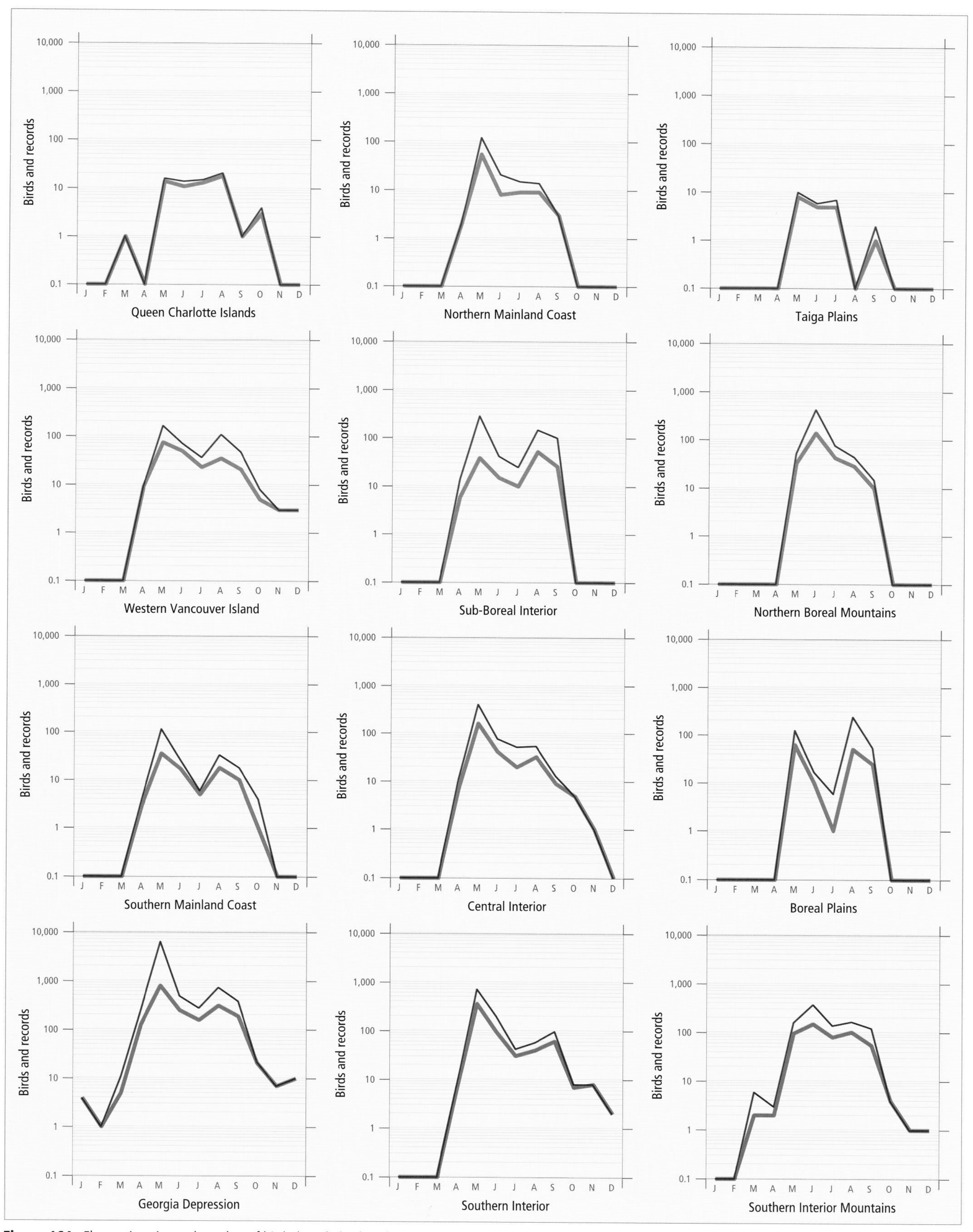

Figure 164. Fluctuations in total number of birds (purple line) and total number of records (green line) for the Wilson's Warbler in ecoprovinces of British Columbia. Christmas Bird Counts, Breeding Bird Surveys, and nest record data have been excluded.

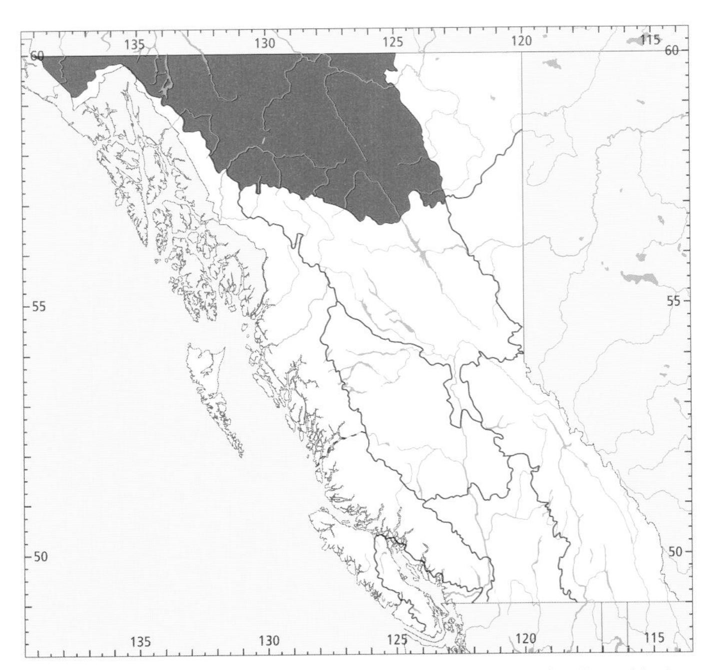

Figure 165. In British Columbia, the highest numbers for the Wilson's Warbler in summer occur in the Northern Boreal Mountains Ecoprovince.

Continent-wide Breeding Bird Survey data indicate that British Columbia has some of the highest densities of the Wilson's Warbler in North America (Sauer et al. 1997).

The Wilson's Warbler breeds from near sea level on Vancouver Island and the Queen Charlotte Islands to timberline elevations in the mountainous regions of the province. In all regions, it selects breeding territories in dense, relatively low vegetation (Figs. 167 and 168). Along the coast, this includes thickets of salmonberry, thimbleberry, black twinberry, false-azalea, and salal. It has also been commonly found in second-growth coastal forests on Vancouver Island, including Douglas-fir, Sitka spruce, and western redcedar, with associated deciduous species such as willow and red alder and a variety of shrubs (Bryant et al. 1993). In the trembling aspen forests of the Bulkley River valley, it was present throughout the summer but nesting was not recorded (Pojar 1995).

In the mountainous regions of the province, the Wilson's Warbler is a characteristic nesting species near timberline habitats (Fig. 168) and the avalanche chutes that transect the forests on steep slopes. In the southern parts of the Rocky Mountains, as represented in Kootenay National Park, the species reaches its highest density in the lower subalpine habitats. The most important vegetation types used are green alder–fern–avalanche shrub, Engelmann spruce–subalpine fir–green alder closed forest, and lodgepole pine–false-azalea–grouseberry closed forest (Poll et al. 1984). The single most characteristic plant species of this warbler's nesting territory is willow.

Douglas et al. (1992) arrived at similar results in their examination of bird distribution patterns within riparian habitat in the Centennial Mountains of Idaho. There 80% of observations of this warbler were associated with willow, with 1 exception the strongest association with any habitat among the 20 bird species studied. The Wilson's Warbler selected the drier sites along the moisture gradient within the willow communities. Pojar (1995), in a study of bird populations in 4 stages of trembling aspen forest (clearcut to old-growth) in the Bulkley River valley, showed, by repetitive sampling between May and July, that the Wilson's Warbler was scarce in clearcuts and most numerous in the mature trembling aspen areas. In terms of presence in census plots, it was recorded in 2 of 6 clearcut plots, 4 of 6 sapling-stage plots, and 6 of 10 plots where

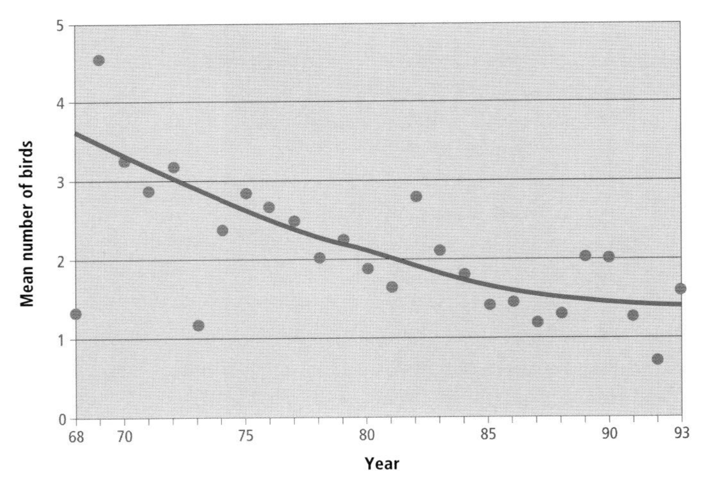

Figure 166. An analysis of Breeding Bird Surveys for the Wilson's Warbler in British Columbia shows that the mean number of birds on interior routes decreased at an average annual rate of 4% over the period 1968 through 1993 ($P < 0.05$).

Figure 167. In portions of the Boreal Plains, the Wilson's Warbler breeds in the shrubby edges of young willow, alder, paper birch, and balsam poplar (Stewart Lake, northwest of Dawson Creek, 1 July 1996; R. Wayne Campbell).

Figure 168. In the mountainous parts of the province, the Wilson's Warbler is a characteristic breeding species near timberline (Nadahini River, 20 June 1983; R. Wayne Campbell).

trembling aspens were 50 to 60 years old. Where the warbler was present in these areas of mature trembling aspen, densities ranged from a low of 1.6 singing males per 10 ha to a high of 16.

The Wilson's Warbler has been recorded breeding in British Columbia from 18 May to 23 August (Fig. 163).

Nests: Most nests in British Columbia were situated in forest lands where the cover was predominantly deciduous and the moisture level encouraged a ground cover of mosses or grasses. Most were on the ground (*n* = 31); other locations included an axe notch in an old stump (1) and rosebushes (2). About one-third of nests were in mossy hummocks, often at the base of a tree or shrub that provided the nest with added shelter. The nests were constructed of dead grasses, used in 82% (*n* = 27), with hair, most frequently moose hair (22%), followed by leaves, sedges, mosses, plant fibres, stems, rootlets, and lichens. The heights of 31 nests ranged from ground level (88%) to 0.9 m.

Eggs: Dates for 56 clutches ranged from 18 May to 15 July, with 55% recorded between 17 June and 4 July. Sizes of 39 clutches ranged from 2 to 6 eggs (2E-2, 3E-6, 4E-9, 5E-20, 6E-2), with 51% having 5 eggs. The incubation period elsewhere is 11 to 13 days (Stewart 1973).

Young: Dates for 30 broods ranged from 7 June to 23 August, with 56% recorded between 5 July and 13 July. Sizes of 19 broods ranged from 1 to 6 young (1Y-2, 2Y-4, 3Y-3, 4Y-5, 5Y-3, 6Y-2), with 11 having 3 to 5 young. The nestling period is reported as 8 to 10 days (Stewart 1973; Curson et al. 1994). In British Columbia, there is no evidence of a second brood after the successful fledging of a first, as has been reported to occur rarely in California (Stewart 1973).

Brown-headed Cowbird Parasitism: In British Columbia, only 1 of 46 nests (2%) found with eggs or young was parasitized by the cowbird. In the parasitized nest, the warbler had covered the egg with a new lining. Four fledgling cowbirds were seen, each with a Wilson's Warbler foster parent. Friedmann (1963) reported only 14 records of Brown-headed Cowbird parasitism of this warbler, 1 in Alberta and the rest in California. Friedmann et al. (1977) extended the list to British Columbia with 3 records of this warbler feeding fledgling cowbirds, all from southern Vancouver Island.

Nest Success: Of 7 nests found with eggs and followed to a known fate, 3 produced at least 1 fledgling.

REMARKS: The Wilson's Warbler is another wood-warbler that shows evidence of separation into eastern and western forms in North America during the most recent glaciation (Morse 1989). The species is now represented by 1 eastern and 2 western geographic races: *W. p. pusilla* in eastern North America west to eastern Alberta (Salt 1973), *W. p. pileolata* from Alaska through most of British Columbia and south into extreme eastern California, and *W. p. chryseola* in southwestern British Columbia (Munro and Cowan 1947) and south along the coast, including most of California (Grinnell and Miller 1944). With the sole exception of the Alberta record, the race *chryseola* has been the only 1 of the 3 to be recorded as a host of the Brown-headed Cowbird.

For many years the western races of this warbler were commonly known as the Black-capped Warbler or Pileolated Warbler.

NOTEWORTHY RECORDS

Spring: Coastal – Rocky Point (Victoria) 9 May 1995-60; Goldstream Park 30 Apr 1963-8 (Edwards 1963); Thetis Lake Park (Victoria) 8 May 1987-15; Colquitz River 18 May 1932-3 eggs; Saltspring Island 6 Apr 1985-6; s Bamfield 13 May 1978-6; Tofino 12 Apr 1971-1; Saturna Island 15 May 1987-10; Reifel Island 25 May 1989-150 at Alaksen National Wildlife Area (Tweit and Heinl 1989); Surrey 17 Mar 1975-6; Alouette River 15 May 1977-75; Burnaby Lake 9 May 1981-100; Seabird Island 27 Apr 1992-2; Pitt Meadows 26 Apr 1981-10, 4 May 1981-50; Pitt Lake 6 May 1983-40; Campbell River 4 Apr 1963-1 (Boggs and Boggs 1963c); Port Neville 18 May 1975-2; Grant Bay 30 Apr 1969-1 (Richardson 1971); Triangle Island 23 May 1978-1; Namu 6 Apr 1981-1; Cape St. James 9 May 1982-1; Bella Coola 3 May 1989-1; Ramsay Island 3 May 1984-1 at n end; Langara Island 10 May 1981-2; Prince Rupert 26 Apr 1983-1; New Aiyansh 28 Apr 1987-1; Pleasant Camp 28 May 1979-6. **Interior** – Oliver to Richter Lake 17 May 1959-30; Creston 17 Mar 1984-5; Vaseux Lake 29 Apr 1931-1; Naramata 21 May 1963-3 along Arawana Rd; Summerland 27 Apr 1949-1 at Research Station; Ashcroft 14 May 1948-10; n side Mission Creek (Westbank) 11 May 1985-26; Kelowna 28 Apr 1974-1; Vernon and Coldstream 16 Apr 1963-2; Revelstoke 27 Apr 1989-2; Moberly Marsh 5 May 1998-1 singing male; 15 km n Golden 14 May 1997-1 male; Lac la Hache 7 May 1943-10 (Munro 1945a); Riske Creek 28 Apr 1987-1; Williams Lake 12 Apr 1961-1 (NMC 48233); 24 km s Prince George 29 Apr 1979-"huge flock"; Smithers 6 May 1992-13; Tumbler Point 11 May 1997-2; Mackenzie 22 Apr 1997-1; Chichouyenily Creek 30 Apr 1994-8; Tupper Creek 12 May 1938-15 (Cowan 1939); Fort St. John 4 May 1982-3, spring arrival; Fort St. John 10 May 1986-2; Sikanni Chief 27 May 1993-1; Hyland Post 27 May 1976-2; w Fort Nelson 19 May 1975-2; Fort Nelson 18 May 1987-1+; Atlin 2 May 1931-1 (RBCM 5667); Chilkat Pass 8 May 1957-1 (Weeden 1960).

Summer: Coastal – Elk Lake 6 Jun 1987-6; Spectacle Lake 7 Jul 1988-8; Surrey 10 Jun 1961-4 eggs; Surrey 9 Aug 1961-2 fledglings; Reifel Island 29 Aug 1986-29; Courtenay 9 Jun 1942-5; Mission 30 Jul 1974-adults feeding 1 Brown-headed Cowbird fledgling; Pitt Meadows 7 Jul 1972-3 fledglings with parents; Port Neville 29 Aug 1975-33; Cape St. James 22 Jul 1979-2; Graham Island 7 Jun 1927-4 nestlings; Hippa Island 27 Jul 1961-2; Great Glacier 13 Aug 1919-1 (MVZ 40213); Three Guardsmen Lake 12 Jun 1979-10. **Interior** – Manning Park 6 Jul 1962-3 nestlings nr Ranger Station; Creston 4 Aug 1988-12; Christina Lake 22 Jul 1977-2 nestlings; Naramata 21 May 1967-3 eggs; Beaver River (Glacier National Park) 19 Jun 1980-4 eggs; 19.5 km s Brunsa Lake 15 Aug 1979-2 fledglings; Murtle Lake (Wells Gray Park) 14 Jul 1956-2 nestlings; Mount Robson 25 Jul 1974-3 fledglings; 1.6 km ne Hay Lake (Naver Creek) 24 Jun 1966-5 eggs; nr Bullmoose Creek 27 Aug 1995-1; nr Gwillim Lake 5 Jun 1994-1; Pine Pass 23 Aug 1975-3 young with both adults; 80 km w Chetwynd 16 Jun 1969-18 (Webster 1969a); e Chetwynd 5 Aug 1975-15 including 2 fledglings with a male; Stoddart Creek (Fort St. John) 20 Aug 1984-42; Fire Flats (Spatsizi Plateau Wilderness Park) 3 Jul 1977-6 eggs; Clear Creek (Haines Highway) 10 Jun 1983-5 eggs; Haines Highway (Mile 75) 25 Jun 1958-6 nestlings, 19 Jul 1957-3 nestlings 3 to 4 days old.

Breeding Bird Surveys: Coastal – Recorded from 27 of 27 routes and on 62% of all surveys. Maxima: Port Renfrew 8 Jun 1986-36; Pemberton 30 Jun 1974-35; Kwinitsa 13 Jun 1971-31. **Interior** – Recorded from 59 of 73 routes and on 48% of all surveys. Maxima: Chilkat Pass 1 Jul 1976-73; Wingdam 23 Jun 1968-71; Gnat Pass 18 Jun 1975-41.

Autumn: Interior – Teslin Lake 12 Sep 1924-1 (Swarth 1926); Atlin area 14 Sep 1972-1; Old Fort Nelson 2 Sep 1985-2; Summit Pass 1 Sep 1943-5; se Charlie Lake 18 Sep 1982-1; Fort St. John 22 Sep 1985-1; Mackenzie 3 Sep 1994-2; Mugaha Creek 23 Sep 1996-1; Williams Lake 14 Oct 1984-1; Riske Creek 31 Oct 1985-1, 4 Nov 1985-1 at Wineglass Ranch; lower Blaeberry River valley 25 Sep 1998-1; Mount Revelstoke 14 Sep 1982-15; Revelstoke 29 Nov 1972-1; Yoho National Park 31 Oct 1976-1 (Wade 1977); Little River (Squilax) 20 Oct 1995-1 adult male; Okanagan Landing 29 Oct 1929-1 (MVZ 104662); Vernon 22 Nov 1981-1; Kelowna 19 Nov 1968-1; Vaseux Lake 16 Sep 1959-12; Kinnaird 9 Oct 1970-1; Grand Forks 1 Nov 1984-1; Oliver 11 Sep 1971-4. **Coastal** – Tow Hill 9 Sep 1921-1 (Patch 1922); Mount Brilliant 28 Sep 1939-1; Cape St. James 11 Oct 1981-1; Port Neville 10 Oct 1974-4; Quadra Island 13 Oct 1964-1 (Boggs and Boggs 1964a); Mitlenatch Island 23 Sep 1965-12 (Campbell and Kennedy 1965); Vancouver 19 Oct 1958-1 (Schultz 1959), 29 Oct 1996-1 female; Victoria 30 Nov 1984-1; Tofino 11 Sep 1971-20, 25 Oct 1989-1, 17 Nov 1983-1.

Winter: Interior – Vernon 26 Dec 1991-1 (Siddle 1992b); Kelowna 9 Dec 1964-1. **Coastal** – Tofino 8 Dec 1982-1; Vancouver 18 Dec 1977-1, 19 and 20 Dec 1991-1; Boundary Bay 29 Dec 1979-1; Saanich 7 Jan 1989-1; Jordan River 3 Dec 1983-1.

Christmas Bird Counts: Coastal – Recorded from 4 of 33 localities and on 2% of all counts. Maxima: Vancouver 29 Dec 1974-1; Surrey 1 Jan 1964-1; Ladner 29 Dec 1979-1; Victoria 17 Dec 1983-1. **Interior** – Not recorded.

Canada Warbler

CAWA

Wilsonia canadensis (Linnaeus)

RANGE: Breeds from extreme southeastern Yukon, northeastern British Columbia, northern Alberta, central Saskatchewan, central Manitoba, northern Ontario, south-central Quebec, New Brunswick, Prince Edward Island, and Nova Scotia south to central Alberta, southern Manitoba, northern Minnesota, southern Wisconsin, northern Illinois, southern Michigan, northern Indiana, and southeastern Ohio, through the Appalachians to eastern Kentucky, eastern Tennessee, northwestern Georgia, western North Carolina, western Virginia, western Maryland, and east-central Pennsylvania, and to northern New Jersey, southeastern New York, and southern New England. Winters in South America from northern Colombia and Venezuela south, east of the Andes, to Peru and northern Brazil.

STATUS: On the coast, *accidental* in the Georgia Depression Ecoprovince; in the Coast and Mountains Ecoprovince, *accidental* on Western Vancouver Island.

In the interior, *uncommon* migrant and local summer visitant in the Boreal Plains and Taiga Plains ecoprovinces; *accidental* in the Sub-Boreal Interior Ecoprovince.

Breeds.

CHANGE IN STATUS: The Canada Warbler (Fig. 169) is not mentioned by Munro and Cowan (1947) in their review of the birds of British Columbia, nor do Godfrey (1986) or the American Ornithologists' Union (1983) mention British Columbia as part of the breeding range of the species. The first record for the province was documented in June 1974 near Mile 320 of the Alaska Highway, about 30 km west of Fort Nelson in the Taiga Plains (Erskine and Davidson 1976). One or 2 birds were seen at this location at least 9 times over the following month. Erskine and Davidson (1976) also recorded an individual bird at another locality, east of the Fort Nelson River. In the Boreal Plains, the Canada Warbler was first recorded near Clayhurst in 1976. Breeding in British Columbia was first verified in 1978 when adults with recently fledged young were found near Fort Nelson.

At about the same time, the Canada Warbler was breeding locally in northeastern and central Alberta, and there was a report from the Alberta–British Columbia boundary (Salt and Salt 1976).

Whether the Canada Warbler became established in British Columbia during the early 1970s or had simply remained undetected because of lack of observers remains open to speculation. By the 1990s, small numbers had been recorded further north, in southeastern Yukon.

NONBREEDING: The Canada Warbler occurs regularly, but locally, in British Columbia in the Boreal Plains and Taiga Plains in the northeastern portions of the province. Most records are from the Peace Lowland, mainly along the Peace River valley, from the Fort Nelson Lowland near Fort Nelson, and from the Sikanni Chief River drainage. There are few records outside these areas.

Figure 169. The Canada Warbler (adult male shown) forages low in the understorey during both migration and nesting seasons; it was discovered in British Columbia in 1974 (Anthony Mercieca).

The Canada Warbler occurs at elevations from 350 to 1,130 m in northeastern British Columbia. Nonbreeding habitat is probably similar to breeding habitat (see BREEDING), except that a wider range of habitats may be used during the autumn migration (Fig. 170). For example, single birds have been found occasionally at Beatton Park, Charlie Lake, in August, but the Canada Warbler does not normally occur in the park during the breeding season (C. Siddle pers. comm.).

The Canada Warbler is one of the last spring migrants to arrive in northeastern British Columbia, with the first birds reaching there during the last few days of May. For the period 1981 through 1989, arrival dates at Taylor ranged from 22 May to 1 June. Most birds probably arrive in northeastern British Columbia in early June (Figs. 171 and 172). This relatively late spring arrival is expected since the Canada Warbler returns to eastern Canada from its wintering areas later than all other warblers except the Mourning and Blackpoll

Figure 170. During migration, especially in the autumn, the Canada Warbler frequents a wider variety of shrub and forest habitats than during the breeding period. These may include willows and alders along quiet streams and beaver ponds, and stretches of tall willows bordering mixed woodlands (Pink Mountain, 25 June 1996; R. Wayne Campbell).

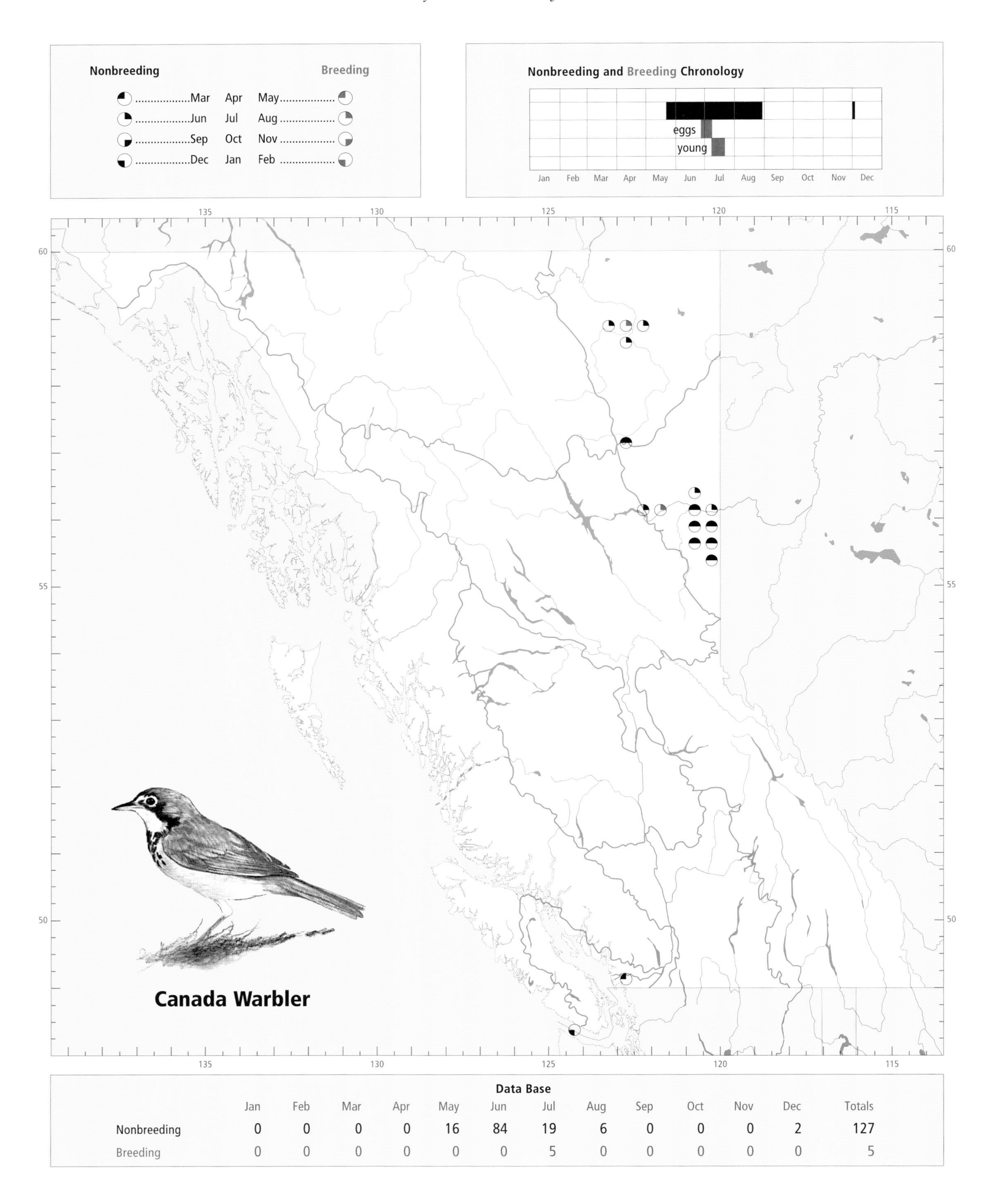

Data Base

	Jan	Feb	Mar	Apr	May	Jun	Jul	Aug	Sep	Oct	Nov	Dec	Totals
Nonbreeding	0	0	0	0	16	84	19	6	0	0	0	2	127
Breeding	0	0	0	0	0	0	5	0	0	0	0	0	5

warblers (Francis and Cooke 1986). In northeastern British Columbia, the Canada Warbler tends to arrive 1 week later than the Blackpoll Warbler. Migrants enter British Columbia from the east, where they arrive in Alberta mainly in the last 2 weeks of May (Salt 1973). The southward movement from northeastern portions of the province likely begins in late July, shortly after nesting is completed, with the main movement from mid-August. The southbound migration appears to be over by late August (Figs. 171 and 172). In Alberta, the main autumn movement occurs from late August to early September (Salt 1973).

In northeastern British Columbia, the Canada Warbler occurs regularly from 25 May to 28 August (Fig. 171).

BREEDING: The Canada Warbler reaches the limit of its northwestern breeding range in northeastern British Columbia. Almost all of our summer records are from the Peace Lowland in the Boreal Plains and from near Fort Nelson in the Fort Nelson Lowland of the Taiga Plains. Breeding populations are probably small and local. Enns and Siddle (1996) report, however, that in 1992 the Canada Warbler was recorded more often than they expected. Lance and Phinney (1994) state that it was fairly common in 1993 in mixed woods southwest of Dawson Creek. These findings suggest that it may be more locally abundant than generally thought.

This uncommon warbler probably breeds in suitable habitat in the Peace Lowland from the Alberta border west to Hudson's Hope, in the Kiskatinaw Plateau south of Dawson Creek, in the Muskwa Plateau, and very locally in the Fort Nelson Lowland near Fort Nelson. In the Taiga Plains, adults with fledged young have been recorded at the Fort Nelson garbage dump, 1 km east of Fort Nelson, and southeast of Fort Nelson, 3 km along the Clarke Lake Road. In the Boreal Plains, a pair with 2 or 3 dependent juveniles and a female with 1 recently fledged young have been found a few kilometres east of Farrell Creek, on the south bank of the Peace River, and an adult with a fledged Brown-headed Cowbird was found near Taylor. In addition, adults carrying food, which suggests nestlings or fledglings nearby, have been recorded near Fort St. John, Taylor, and Dawson Creek.

In neighbouring Alberta, the Canada Warbler is a rare and sparsely distributed summer visitor to northern boreal forests (Semenchuk 1992).

The highest numbers for the Canada Warbler in British Columbia in summer occur in the Boreal Plains (Fig. 173). Breeding Bird Surveys for interior routes for the period 1968 through 1993 contain insufficient data for analysis. An analysis of Breeding Bird Surveys across Canada from 1980 to 1996 suggests that populations declined at an average annual rate of 3.5% ($P < 0.01$) (Sauer et al. 1997).

The Canada Warbler has been recorded breeding at elevations between 300 and 570 m in northeastern British Columbia. Most summer records of this warbler in the Boreal Plains and Taiga Plains are from stands of mixed paper birch, trembling aspen, balsam poplar, and alders with a minor amount of white spruce (Fig. 174). These stands are usually on slopes or hillsides or in gullies. Sites can be either wet or dry, but there is always a rich shrubby understorey, often with red-osier dogwood. Edges between forest and shrubby areas along roads, transmission lines, and ski hills are often used.

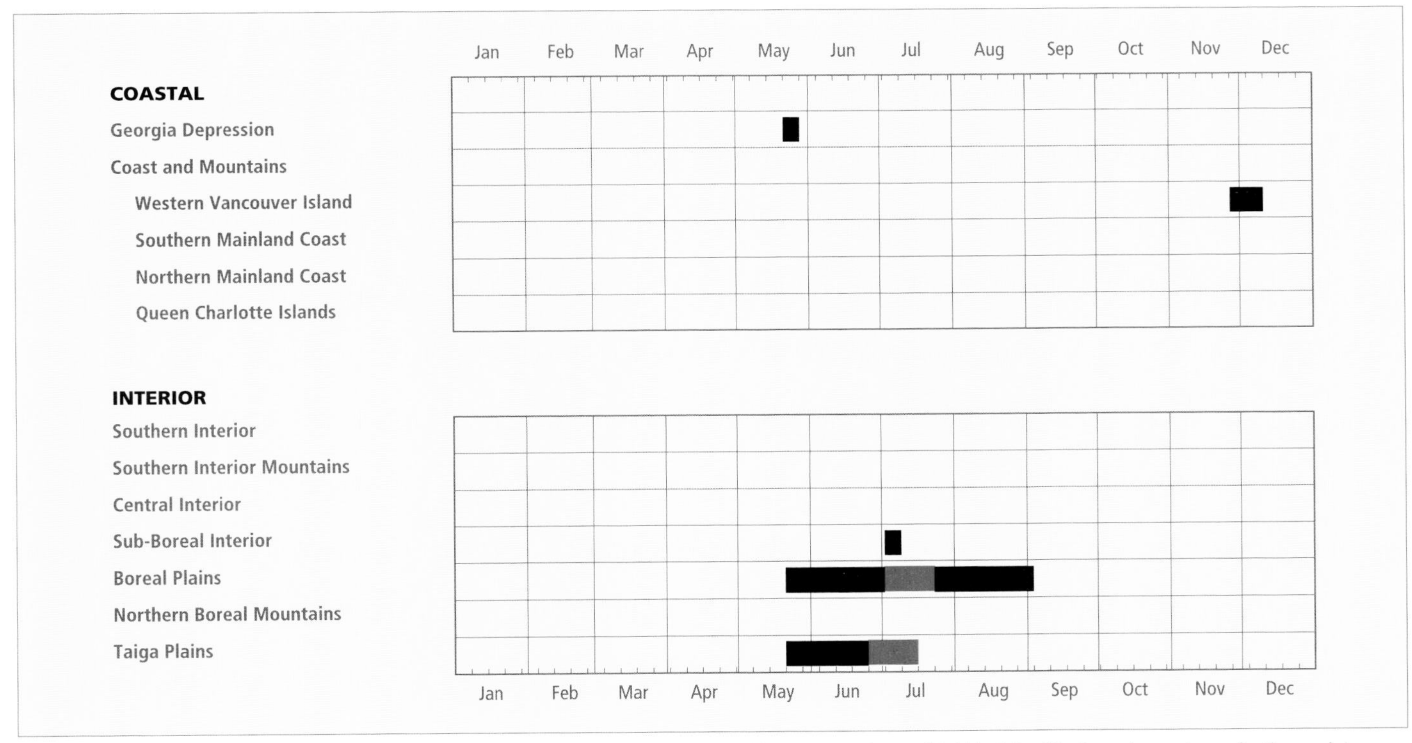

Figure 171. Annual occurrence (black) and breeding chronology (red) for the Canada Warbler in ecoprovinces of British Columbia. Records are shown for the week in which they occurred.

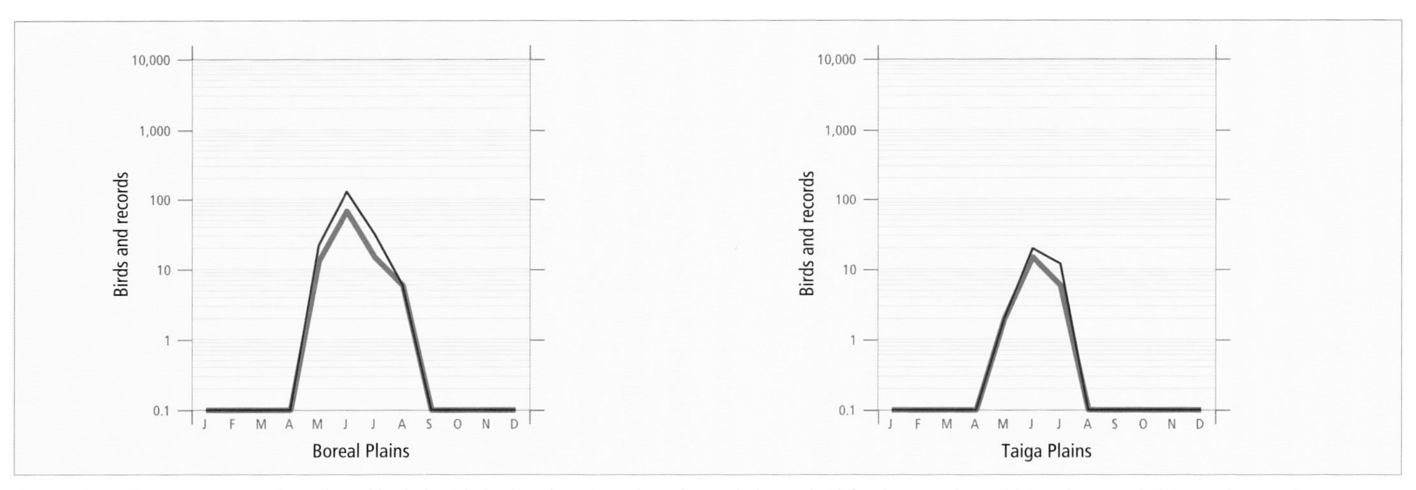

Figure 172. Fluctuations in total number of birds (purple line) and total number of records (green line) for the Canada Warbler in the Boreal Plains and Taiga Plains ecoprovinces of British Columbia. Breeding Bird Surveys and nest record data have been excluded.

Near Fort Nelson and south of Dawson Creek, Canada Warblers were not often found in trembling aspen but were commonly found in moist deciduous or, more often, mixed-wood forests or occasionally balsam poplar stands (Erskine and Davidson 1976; Lance and Phinney 1994). This is consistent with results from 2 studies in central Alberta (Westworth and Telfer 1993; Schieck et al. 1995).

In 1 study in northeastern British Columbia, the habitat used by singing males was associated mainly with mixed stands of paper birch, alder, and white spruce, and occasionally with trembling aspen and balsam poplar (Enns and Siddle 1996). These stands were on moderate to steep slopes with unstable banks and a rich understorey of shrubs, especially soopolallie, prickly rose, red-osier dogwood, alders, and highbush-cranberry. Fallen trees and other woody debris were usually present as well. The Canada Warbler appears to favour sites in mature forests where 1 or 2 fallen trees have left small gaps in the canopy, thereby allowing lush patches of undergrowth to develop. Foraging birds were observed using red-osier dogwood and sapling paper birch up to about 4 m above ground.

In British Columbia, the Canada Warbler has been recorded breeding from 27 June (calculated) to 21 July; it

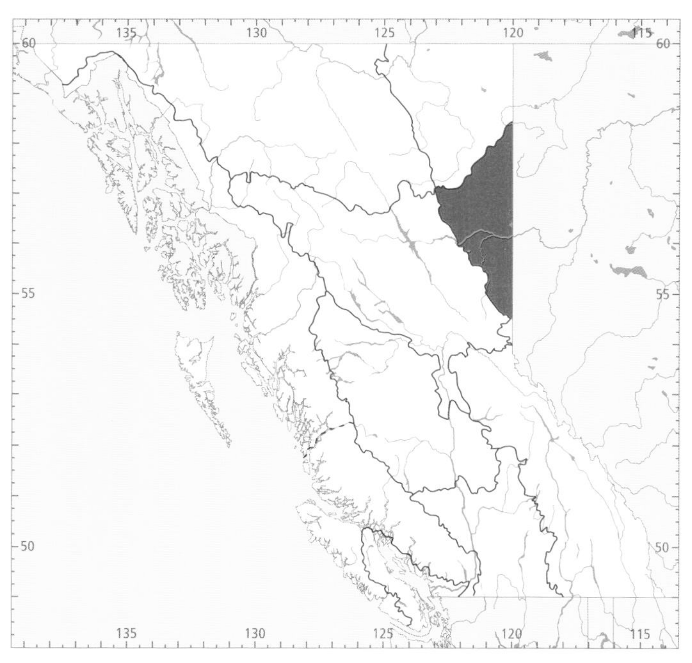

Figure 173. In British Columbia, the highest numbers for the Canada Warbler in summer occur in the Boreal Plains Ecoprovince.

Figure 174. In the Boreal Plains Ecoprovince, the Canada Warbler inhabits mixed-wood forests, usually on slopes, with a dense understorey of shrubs and new tree growth (near Cecil Lake, 26 June 1996; R. Wayne Campbell).

probably breeds from early June through late July (Fig. 171).

Nests: Nests have not been described in British Columbia. Elsewhere, the Canada Warbler nests on or near the ground in moss-covered logs and stumps, under the overhang of streambanks, in grass hummocks, on the sides of moss- or fern-covered rocks, in overturned tree roots, or beside a clump of herbs (Bent 1953; Peck and James 1987). Nests are typically bulky cups of dry leaves, grasses, and plant fibre, lined with fine grass, rootlets, and hair (Bent 1953; Peck and James 1987).

Eggs: Nests with eggs have not been recorded in British Columbia. Based on our records of recently fledged young, nests contained eggs no later than 27 June (calculated), although eggs can probably occur from early June through early July. Elsewhere, clutch sizes range from 3 to 5 eggs, with 4 being the most common (Peck and James 1987; Berger et al. 1991). The incubation period is unknown but is probably similar to that of the Wilson's Warbler: 11 to 13 days (Baicich and Harrison 1997; Conway 1999).

Young: Nests with young have not been observed in British Columbia. Dates for 5 records of recently fledged young ranged from 8 July to 21 July. Nestlings can probably occur in British Columbia from about 18 June through mid-July. The nestling period is unknown but is probably similar to that of the Wilson's Warbler: 10 or 11 days (Baicich and Harrison 1997).

Brown-headed Cowbird Parasitism: Nests with cowbird eggs or nestlings have not been found in British Columbia. There were, however, 2 records of this warbler feeding fledged cowbirds. Elsewhere, the Canada Warbler is infrequently parasitized by cowbirds (Friedmann 1963; Friedmann et al. 1977).

Nest Success: Insufficient data.

REMARKS: The primary habitat of the Canada Warbler in northeastern British Columbia is found on wet, usually unstable slopes with deciduous or mixed deciduous-coniferous forest, well-developed shrub layers, and considerable amounts of woody debris. Mixed woods are being heavily harvested for pulp and oriented strand board in the northeast, especially in the Peace Lowland (L. Pin pers. comm.). Because this warbler uses a variety of seral stages, local logging may not have a negative impact on habitat over the long term.

On the Atlantic coast, the Canada Warbler is thought to be area-sensitive, i.e., it is most likely to occur in contiguous forests larger than 3,000 ha and less likely to occur in smaller tracts (Robbins et al. 1989). In British Columbia, this warbler also occurs in forest habitat well back from edges, but it often occurs along roadsides where forest stands are fragmented. It may be less sensitive to area effects in this province.

The Canada Warbler most often forages on or near the ground in deciduous vegetation (Power 1971). It also hawks flying insects more often than most other warblers, and was once known as the "Canada Flycatching Warbler" (Bent 1953). In Wisconsin, it also forages in scattered small conifers in deciduous stands (Sodhi and Paszkowski 1995).

A report of a Canada Warbler captured during a banding operation at Sea Island, near Vancouver, in September 1959 lacked sufficient detail to be included in this account.

For life-history information for the Canada Warbler in British Columbia and North America, see Cooper et al. (1997), Dunn and Garrett (1997), and Conway (1999).

NOTEWORTHY RECORDS

Spring: Coastal – Pitt Meadows 23 May 1983-1 (Mattocks and Hunn 1983b). **Interior** – s Swan Lake 30 May 1993-1 male singing from dense shrubbery; Brassey Creek 27 May 1994-1 on transect (Lance and Phinney 1996); s Groundbirch 31 May 1994-3 on transect (Lance and Phinney 1996); McQueen Slough 26 May 1990-1; Taylor 22 May 1981-1, 25 May 1985-1, 26 May 1988-1; Fort St. John 27 May 1986-1, 31 May 1982-4 (Grunberg 1982c); Sikanni Chief River 27 May 1994-1.

Summer: Coastal – No records. **Interior** – Bear Mountain (Dawson Creek) 23 Jun 1994-6 on transect (Lance and Phinney 1996); Brassey Creek 19 Jun 1995-3 on transect (Lance and Phinney 1996); Danish Creek (Hudson's Hope) 7 Jul 1997-1; Clayhurst 1 Jun 1989-2 males nr bridge over Peace River, 21 Jun 1976-2; Pine River 17 Jul 1983-1; 10.5 km e Farrell Creek mouth 21 Jul 1992-pair with 2 or 3 fledglings; Taylor 3 Jun 1983-5, 9 Jun 1982-6, 13 Jun 1982-5, 17 Jul 1983-5, 17 Jul 1983-adult with fledgling Brown-headed Cowbird, 24 Jul 1982-3, 8 Aug 1986-1, 18 Aug 1987-1, last of year (Siddle 1988d); Taylor Landing Park 4 Aug 1991-1 female; Beatton Park 14 Aug 1986-1, 28 Aug 1985-1; Gutah Creek Jun 1991-small numbers at confluence of Sikanni Chief River (Siddle 1992c); Alaska Highway (Mile 316.5) 20 Jun 1976-4; Fort Nelson 6 Jun 1985-1 at airport, 7 Jun 1974-1 (10 records through July; Erskine and Davidson 1976), 8 Jul 1986-1 adult feeding a recent fledgling, 10 Jul 1978-5 at garbage dump, 10 Jul 1978-adult female feeding 2 recently fledged young, 10 Jul 1978-adult female with 2 fledglings at km 5 on the Clarke Lake Rd (1st breeding record for the province); Liard Highway (Kilometre 72) 16 Jul 1992-1 (Enns and Siddle 1996).

Breeding Bird Surveys: Coastal – Not recorded. **Interior** – Recorded from 2 of 73 routes and on less than 1% of all surveys. Maxima: Fort Nelson 19 Jun 1976-5; Fort St. John 20 Jun 1982-1.

Autumn: No records.

Winter: Interior – No records. **Coastal** – Jordan River 2 and 3 Dec 1995-1.

Christmas Bird Counts: Not recorded.

Yellow-breasted Chat

YBCH

Icteria virens (Linnaeus)

RANGE: Breeds from southern British Columbia, southern Alberta, southern Saskatchewan, and southern Ontario south in the west to central Baja California and the central Mexican mainland; south in the east from extreme southern Ontario and the Great Lakes states, Vermont, and New Hampshire to Florida and the Gulf coast. Winters from southern Baja California, southern Texas, and Florida south to Panama.

STATUS: On the coast, *very rare* in the Georgia Depression Ecoprovince; in the Coast and Mountains Ecoprovince, *casual* on Western Vancouver Island.

In the interior, *uncommon* migrant and local summer visitant in the Southern Interior Ecoprovince; locally *very rare* in the Southern Interior Mountains and *casual* in the Central Interior and the Sub-Boreal Interior ecoprovinces.

Breeds.

Figure 175. In British Columbia, the Yellow-breasted Chat, our largest wood-warbler, breeds only in the Okanagan and Similkameen valleys (Anthony Mercieca).

NONBREEDING: The Yellow-breasted Chat (Fig. 175), a relatively scarce and local visitor to the province, has occurred across much of southern British Columbia. On the coast, it has been recorded from Ucluelet and Carmanah Point on Western Vancouver Island, the Courtenay-Comox and Cowichan Bay areas of southeastern Vancouver Island, and much of the Lower Mainland to Pitt Meadows. In the interior, it frequents the Okanagan and Similkameen valleys, less frequently the Creston valley, and formerly the Thompson valley between Ashcroft and Shuswap Lake, with vagrants reaching the vicinity of Golden, Williams Lake, and Mackenzie.

On the coast, there have been fewer than 25 records in the approximately 90 years since the first specimen was taken in 1897 (Brooks 1917), and they have been divided almost evenly between summer occurrences and migrants.

The Yellow-breasted Chat occurs from about sea level to 70 m elevation on the coast and between 250 and 800 m elevation in the interior. It occupies the thickest available habitat, primarily in riparian forests of black cottonwood where there is an understorey of willow, mountain alder, Rocky Mountain maple, blue elderberry, and associated shrubs (Fig. 176). It also occupies dense forest-edge thickets in draws above the valley floor, where Columbian hawthorn, trembling aspen, choke cherry, snowberry, prairie rose, and tangles of white clematis, along with other shrubs and forbs, provide the dense undergrowth preferred by the chats. Thickets of

Figure 176. In British Columbia, the Yellow-breasted Chat prefers riparian habitats with dense shrub thickets. A pair of chats nest at this location most years (near the north end of Vaseux Lake, 20 June 1996; Neil K. Dawe).

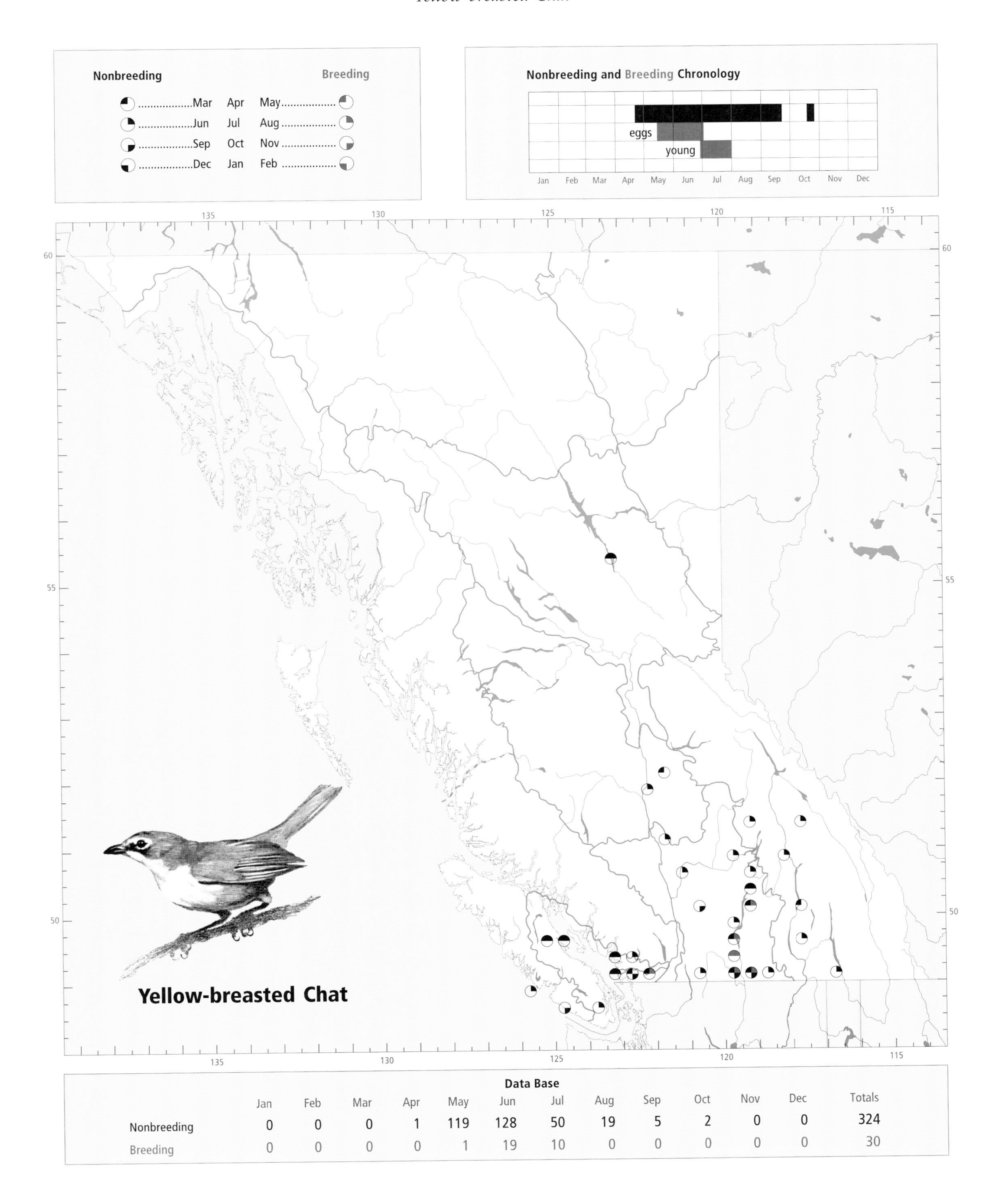

Data Base

	Jan	Feb	Mar	Apr	May	Jun	Jul	Aug	Sep	Oct	Nov	Dec	Totals
Nonbreeding	0	0	0	1	119	128	50	19	5	2	0	0	324
Breeding	0	0	0	0	1	19	10	0	0	0	0	0	30

rose and snowberry, or of Himalayan blackberry, in the uncultivated corners of fields, orchards, and vineyards also serve as habitat for this large warbler. In British Columbia, density of cover appears more important to the chat than the plant species that comprise the shrubbery.

The Yellow-breasted Chat returns to the southern Okanagan valley in May (sometimes as early as late April); the peak of spring migration is evident during the second half of May (Figs. 177 and 178).

Autumn departure dates are difficult to determine because the song period is over in early July and chats are difficult to find. After mid-July, numbers probably decline throughout August, with few birds remaining into October (Figs. 177 and 178). In eastern North America, Dennis (1967) and Thompson and Nolan (1973) suggest that resident chats are present in late July and into August on the midwestern and eastern breeding grounds, and that southward migration probably occurs in late August and September.

On the coast, the Yellow-breasted Chat has been recorded infrequently from 9 May to 26 October; in the interior, it occurs more regularly from 22 April to 20 October (Fig. 177).

BREEDING: The Yellow-breasted Chat breeds only in the southern Okanagan valley, between Osoyoos and Lavington, and in the southern part of the Similkameen valley in Richter Pass. There is also a single breeding record for the south coast, near Mission in the Fraser Lowland.

The highest numbers for the Yellow-breasted Chat in summer occur in the Okanagan valley of the Southern Interior (Fig. 179). Breeding Bird Surveys for interior routes for the period 1968 through 1993 contain insufficient data for analysis. An analysis of Breeding Bird Surveys across North America for the 30-year period from 1966 to 1996 indicate a possible decline throughout the continent. An average annual decline of 3.5% between 1966 and 1979 ($P < 0.01$) was followed by an average annual increase of 1.0% between 1980 and 1996 ($P < 0.01$) (Sauer et al. 1997).

The Yellow-breasted Chat breeds in valley bottoms at elevations between 280 and 600 m. There is 1 record from the coast at 50 m elevation. Both natural forests and human-influenced forests, including mixed deciduous forests, riparian shrublands (Fig. 180), woodlots, orchards, and vineyards, are preferred general breeding habitats. Birds arriving in the spring go directly into the breeding habitat, so the descriptions under NONBREEDING also apply to breeding habitat.

After extensive field study of the species, Gibbard and Gibbard (1992) stated:

> The preferred nesting habitat for the Yellow-breasted Chat is dense to very dense wild rose thickets exhibiting vigorous growth and in close proximity to or containing large shrubs or medium height trees. The size of the rose thicket is variable, from a minimum of approximately 9 square metres to a maximum of approximately 195 square metres.
>
> Rose height averaged 1.25 m. Trees growing within or close to the thicket generally did not exceed 6 m in height, and large shrubs were usually about 3.5 m high. Trembling aspen was the most common tree species, and elderberry, hawthorn and saskatoon the

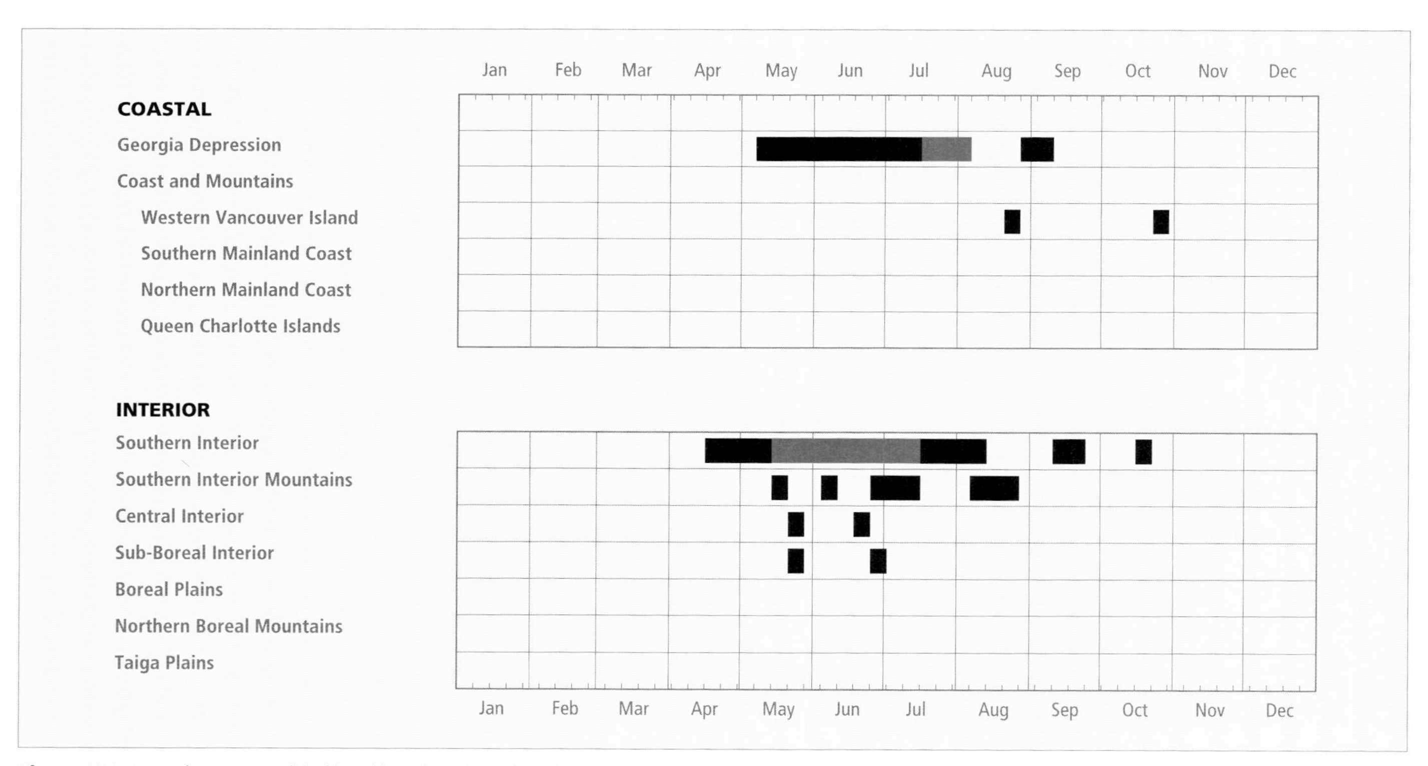

Figure 177. Annual occurrence (black) and breeding chronology (red) for the Yellow-breasted Chat in ecoprovinces of British Columbia. Records are shown for the week in which they occurred.

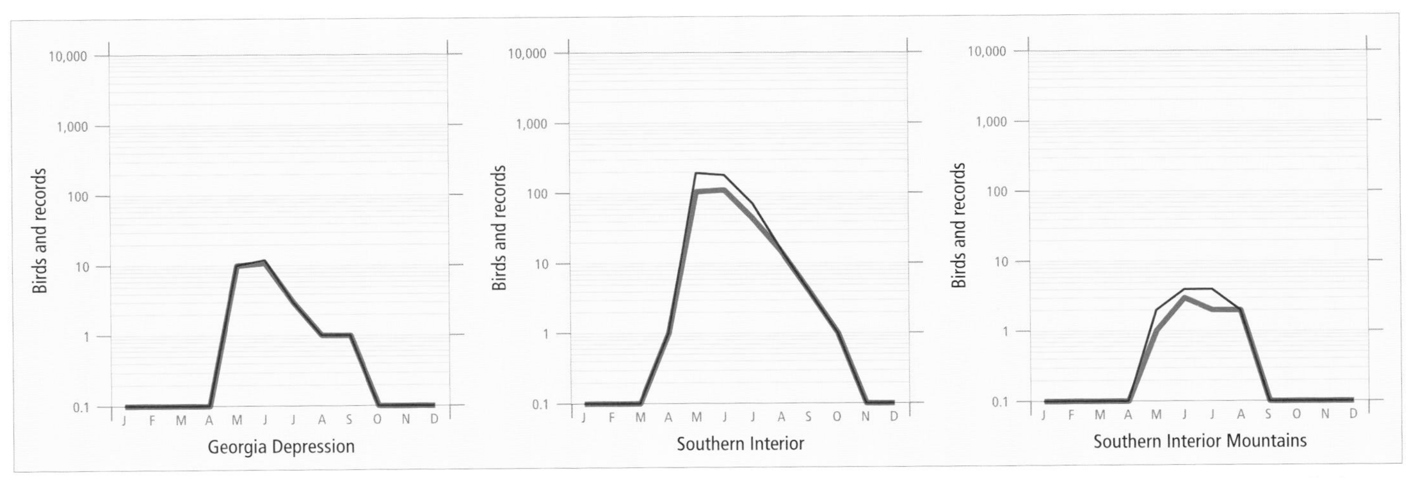

Figure 178. Fluctuations in total number of birds (purple line) and total number of records (green line) for the Yellow-breasted Chat in ecoprovinces of British Columbia. Breeding Bird Surveys and nest record data have been excluded.

> most common shrub species. Birch, alder and willow were less frequently represented.

They also noted that chats were not found where pathways or cattle trails dissected the thickets, nor did they occur where there was an overstorey of large trees. Chats also avoided areas with a high level of traffic noise, even where the habitat appeared otherwise suitable. These necessary conditions for the chat are of localized distribution, and are decreasing locally as a result of land clearing, grazing, and other similar habitat-altering activities in the Okanagan valley.

On the coast, the Yellow-breasted Chat has been recorded breeding from 19 July (calculated) to 31 July; in the interior, it has been recorded breeding from 15 May to 12 July (Fig. 177).

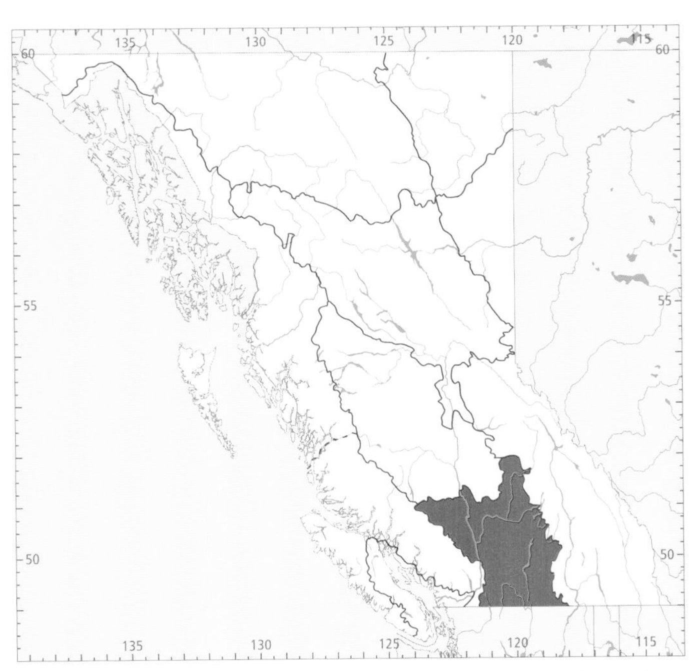

Figure 179. In British Columbia, the highest numbers for the Yellow-breasted Chat in summer occur in the Southern Interior Ecoprovince.

Nests: All nests (n = 6) were placed in thick shrubbery, including rose bushes (3 nests), snowberry (2), and a small willow (1). Nest materials were mainly grasses, along with plant fibres, twigs, leaves, and bark strips. The nest cup was lined with fine grasses. The heights of 9 nests ranged from 0.3 to 1.2 m, with 6 nests between 0.6 and 0.9 m.

Eggs: Dates for 19 clutches ranged from 15 May to 2 July, with 58% recorded between 15 June and 25 June. Sizes of 16 clutches ranged from 1 to 4 eggs (1E-1, 2E-1, 3E-7, 4E-7), with

Figure 180. In the Okanagan valley, the preferred breeding habitat for the Yellow-breasted Chat consists of brushy tangles of riparian growth, especially wild rose thickets (No. 22 Road, north of Osoyoos, 6 July 1997; R. Wayne Campbell).

88% having 3 or 4 eggs. The incubation period elsewhere is 11 or 12 days (Curson et al. 1994).

Young: Dates for 12 broods ranged from 29 June to 31 July. Sizes of 5 broods ranged from 1 to 3 young (1Y-1, 2Y-2, 3Y-2). The nestling period is 8 to 12 days (Curson et al. 1994).

Brown-headed Cowbird Parasitism: In British Columbia, 4 of 23 nests found with eggs or young were parasitized by the cowbird. Friedmann (1963) reports that the Yellow-breasted Chat is a frequent host of the Brown-headed Cowbird in all parts of its range. He also refers to the frequency with which the chat deserts a parasitized nest. Two of the 3 parasitized nests in British Columbia had 3 cowbird eggs each; each nest was deserted prior to hatching. Thompson and Nolan (1973) report a single case of desertion that could likely be attributed to cowbird parasitism.

Nest Success: Insufficient data.

REMARKS: The western race of the Yellow-breasted Chat was formerly known as the Long-tailed Chat.

Two geographic races of the Yellow-breasted Chat are recognized in North America: *I. v. virens* occupies the eastern part of the species' range and *I. v. auricollis* inhabits the western part of the continent from southern British Columbia, Alberta, and Saskatchewan south into central Mexico (American Ornithologists' Union 1957).

As a breeding species, the Yellow-breasted Chat has always had a restricted range in British Columbia. The valley bottom habitats that it prefers occur mainly on lands in demand for human exploitation. These circumstances have led to the designation of the chat as a species of conservation concern in the province. Cannings (1995a) reviewed the data on the species and prepared a status report with recommendations for conservation action. He estimated that not more than 12 pairs were present in the province in any year, and the nesting success appeared to be low. Furthermore, the required habitat had declined, a trend that was forecast to continue. Cannings also recommended that the Yellow-breasted Chat be listed as a threatened species in British Columbia. It is a species at the northern edge of its distribution, however, and the number of pairs that nest in British Columbia probably depends partly on its density in the breeding range to the south of us. Probably the only practical conservation action we can take is to secure the small areas of habitat that we know it finds suitable.

The chat has encountered the same problems elsewhere in Canada and is listed nationally as a threatened species (Cadman and Page 1994).

Thompson and Nolan (1973) provide a detailed life-history study of a population of chats in eastern North America, including mortality rates of eggs and nestlings and causes of nest failure. Burhans and Thompson (1999) discuss the significance of habitat patch size for the nesting success of Yellow-breasted Chats.

NOTEWORTHY RECORDS

Spring: Coastal – Langley 15 May 1987-1 (Mattocks and Harrington-Tweit 1987b); Vancouver 9 May 1927-1 (Turnbull 1929); Comox 31 May 1945-1 (Pearse 1947); nr Courtenay 26 May 1934-1 (Laing 1942). **Interior** – Inkaneep Park 19 May 1963-9; Oliver 22 Apr 1990-1 at entrance to Haynes Ecological Reserve, 4 May 1985-1 banded, 18 May 1968-3; n end Osoyoos Lake 31 May 1969-10; s McIntyre Bluff 26 May 1974-9 (Cannings 1974); White Lake (Okanagan) 15 May 1942-4 eggs; e end Williams Lake 25 May 1985-1 singing at Sugarcane; Chichouyenily 22 May 1994-2.

Summer: Coastal – Cowichan Bay 16 Jun 1984-1 in blackberry tangle; Ucluelet 21 Aug 1966-1; Mission 30 Jul 1966-adults feeding 2 young; Vancouver 31 Aug 1933-1 (RBCM 7037); Sturgeon Slough (Pitt Meadows) 17 Jul 1972-1 (Campbell et al. 1972b); Pitt Meadows 4 Jun 1960-1 (Boggs and Boggs 1960c); Puntledge River (Courtenay) 15 Jun 1940-1 heard singing (Laing 1942); Comox 3 to 6 Jun 1959-1. **Interior** – Osoyoos 15 Jun 1907-4 eggs; Osoyoos Lake 1 Aug 1974-1 at n end (Fig. 180); Richter Pass 16 Jun 1960-3 eggs at Pollock's Ranch, 2 Jul 1960-3 eggs plus 1 Brown-headed Cowbird egg (Campbell and Meugens 1971); Creston 22 Aug 1982-1; Cawston 6 Jun 1948-3 eggs, incubation started, 15 Jun 1950-4 fresh eggs; Similkameen River 9 Jun 1905-2 collected, 27 Jun 1998-4; Duck Lake (Creston) Jun 1974-1, 12 Jul 1984-1 adult and 2 recently fledged young; Vaseux Lake 4 Jun 1929-4 eggs, 6 Jun 1966-5, 6 Aug 1989-2; White Lake (Okanagan Falls) 12 Jun 1963-3; 16 km n South Slocan 7 Aug 1983-1; Naramata 29 Jun 1966-3 nestlings, 9 Jul 1967-1 nestling plus 2 cowbird nestlings; Peachland 1 Jul 1978-6; nr Kalamalka Lake 4 Jul 1958-1 (Bradley 1959); Lavington 25 Jun 1965-3 eggs; Vernon (6 Mile Point) 3 Jun 1979-2; Coldstream 15 Jun 1991-1 at Creekside Park; Enderby 29 Jun 1975-1; Ashcroft 9 Jun 1892-1 singing male (ANS 31411); Chase 17 Jun 1973-2 on Indian Reserve; Squilax 18 Jul 1961-2 (Stirling 1961a); Shuswap Lake 20 Jul 1961-2 nr bridge over river joining Little Shuswap Lake; Revelstoke 6 Jun 1986-2; Glacier National Park 4 Jul 1990-1 at Illecilewaet Camp site area; Alkali Lake (Cariboo) 14 to 21 Jun 1993-1 at China Flats (Roberts and Roberts 1993); Chichouyenily 29 Jun 1994-1.

Breeding Bird Surveys: Interior – Recorded from 2 of 73 routes and on less than 1% of all surveys. Maxima: Oliver 5 Jun 1974-2; Summerland 20 Jun 1987-2. **Coastal** – Not recorded.

Autumn: Interior – Nicola Lake 20 Oct 1976-1; Osoyoos 21 Sep 1928-1 (RBCM 13802). **Coastal** – Surrey 5 Sep 1973-1; Carmanah Point 26 Oct 1950-1 (Irving 1953).

Winter: No records.

Western Tanager

Piranga ludoviciana (Wilson)

WETA

RANGE: Widespread in western North America. Breeds from southeastern Alaska, northern British Columbia, south-central Mackenzie, northern Alberta, and central Saskatchewan south in the western part of the range into Baja California, southern Nevada, southwestern Utah, central and southeastern Arizona, southern New Mexico, and western Texas, and east to eastern Montana, western South Dakota, northwestern Nebraska, central Colorado, and central New Mexico. Winters regularly from southern California and Baja California south through Mexico and Central America to Costa Rica.

STATUS: On the coast, a *fairly common* to *common* migrant, *uncommon* to *fairly common* summer visitant, and *casual* winter visitant in the Georgia Depression Ecoprovince; in the Coast and Mountains Ecoprovince, *uncommon* to *fairly common* migrant and summer visitant on the Southern Mainland Coast, *rare* to *uncommon* migrant and summer visitant on the Northern Mainland Coast, *rare* on Western Vancouver Island, and *casual* on the Queen Charlotte Islands.

In the interior, *fairly common* to *common* spring migrant and summer visitant, *uncommon* autumn migrant, and *casual* in winter in the Southern Interior Ecoprovince; *fairly common* to *common* spring migrant and summer visitant and *uncommon* autumn migrant in the Southern Interior Mountains Ecoprovince; *uncommon* to *fairly common* spring and autumn migrant and summer visitant in the Central Interior, Sub-Boreal Interior, Boreal Plains, and Taiga Plains ecoprovinces; *rare* to locally *uncommon* in the Northern Boreal Mountains Ecoprovince.

Breeds.

NONBREEDING: The Western Tanager (Fig. 181) is widely distributed throughout most of the southern and central portions of the province, including Vancouver Island. The distribution of this tanager in northern areas is more localized and occurrences are more widely scattered. There are few records from the coast between northern Vancouver Island and the mouth of the Skeena River, and only 2 reports from the Queen Charlotte Islands.

Outside the breeding season, the Western Tanager has been reported from near sea level to 250 m elevation on the coast and from 280 to 1,150 m in the interior. During migration, this species uses a wide variety of forested habitats, both deciduous and coniferous, including mature forests, human-influenced forests, flooded wooded lakeshores (Fig. 182), and stands of trees in suburban areas, orchards, parks, and gardens. It occasionally visits garden feeding stations with water sources.

On the south coast, the spring migration begins in mid to late April and reaches its peak a month later (Figs. 183 and 184). The first arrivals reach the Northern Mainland Coast several weeks later, in early May, but identification of a peak is hampered by insufficient data. In the southern portions of the interior, the first arrivals sometimes reach the Okanagan, Kootenay, and Columbia valleys in the second week of April but become regular there by late April and the first week of May (Figs. 183 and 184). In these southern valleys, the number of birds present increases rapidly through May; the greatest numbers are present in late May or early June in the Southern Interior and Southern Interior Mountains.

Figure 181. Adult male Western Tanager (Victoria, 15 July 1996; Tim Zurowski).

The Western Tanager arrives in the Cariboo-Chilcotin areas and the vicinity of Prince George in the latter half of April and in the Peace Lowland in early May. The earliest spring arrival records for the Taiga Plains are from the third week of May.

In the Northern Boreal Mountains, Taiga Plains, and Boreal Plains, the autumn migration appears to unfold gradually, beginning as early as late July. Most tanagers have departed by mid-August (Figs. 183 and 184). In the southern regions of the interior, the migration is usually over by mid-September except for a few stragglers. On the coast, the peak of the southbound migration passes through the Fraser River delta and surrounding mountains from mid-August to mid-September. Only stragglers remain by the first week of October.

In winter, the Western Tanager has been reported at Kamloops in the Southern Interior and from southeastern Vancouver Island and the Fraser Lowland in the Georgia Depression.

On the coast, the Western Tanager has been recorded throughout most of the year, but regularly only from 15 April to 27 September; in the interior, it has been recorded regularly from 7 April to 27 September (Fig. 183).

BREEDING: The Western Tanager has a widespread breeding distribution throughout most of British Columbia from the

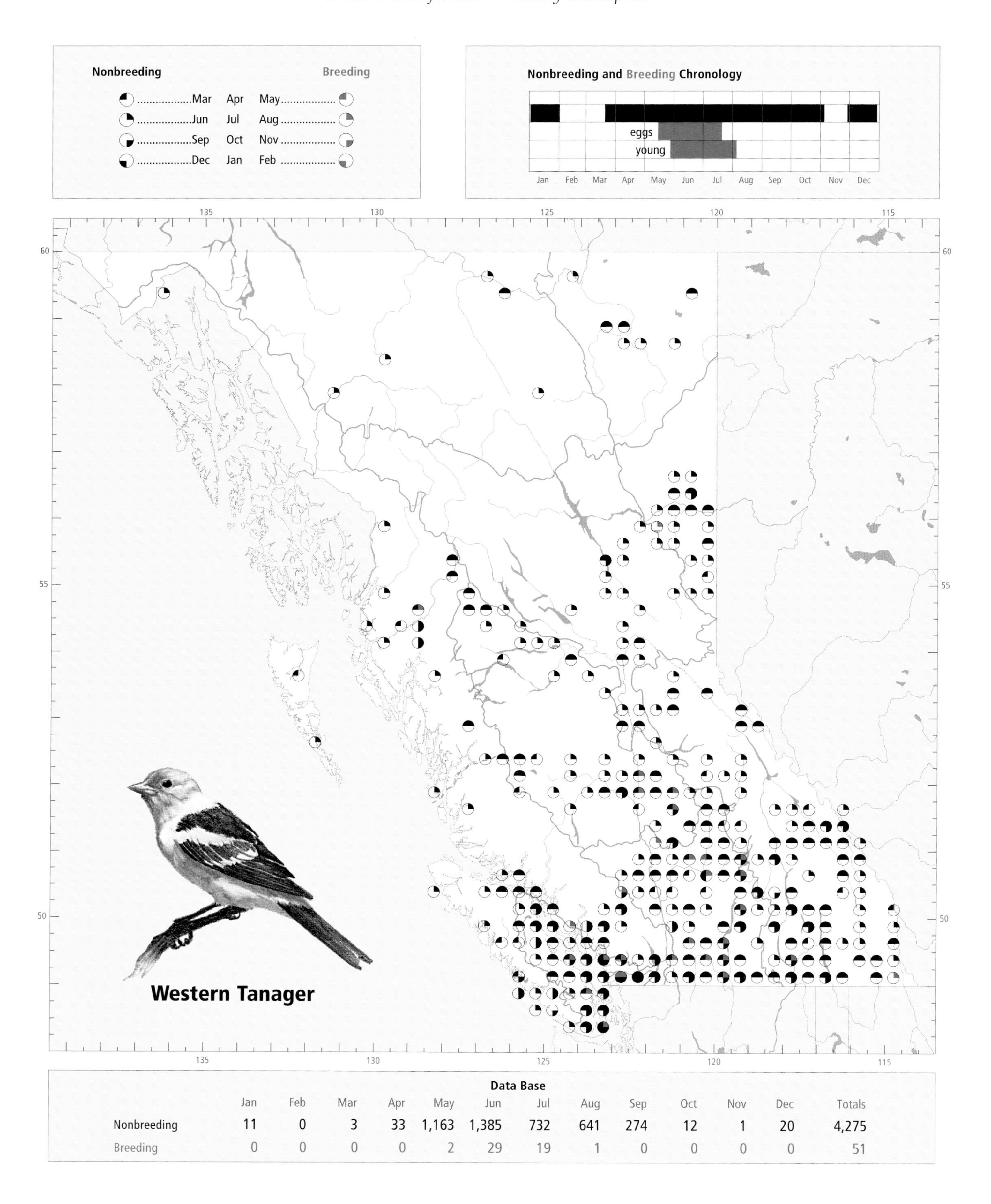

Data Base

	Jan	Feb	Mar	Apr	May	Jun	Jul	Aug	Sep	Oct	Nov	Dec	Totals
Nonbreeding	11	0	3	33	1,163	1,385	732	641	274	12	1	20	4,275
Breeding	0	0	0	0	2	29	19	1	0	0	0	0	51

Peace Lowland southward and including southeastern Vancouver Island. It has not been reported from the Queen Charlotte Islands. Although records are scarce, the species probably breeds locally throughout the northern interior.

This tanager reaches its highest numbers in summer in the Southern Interior (Fig. 185). An analysis of Breeding Bird Surveys in British Columbia for the period 1968 through 1993 could not detect a net change in the mean number of birds on either coastal and interior routes. North American trends for the period 1966 to 1996 were similar (Sauer et al. 1997).

The Western Tanager has been found breeding at elevations from near sea level to 135 m on the coast and from 330 to 1,200 m in the interior. Breeding habitat includes a variety of forest types where it is most numerous in edge or ecotone situations, including a mix of conifers and deciduous trees. These are frequently associated with openings such as beaver ponds, lake margins, rock bluffs, meadows, trembling aspen copses, Douglas-fir and ponderosa pine forests bordering grasslands and shrub-steppes (Fig. 186), and sometimes suburban parks and gardens. In the old-growth and managed forests of western Vancouver Island, consisting of Douglas-fir, western redcedar, western hemlock, Sitka spruce, Pacific yew, and amabilis fir, the Western Tanager occurred only in the oldest age class of the forest (> 200 years; Bryant et al. 1993). In coastal coniferous forests outside British Columbia, a detailed analysis of spring bird communities in the Coast Mountains of Oregon provides comparable data. There Douglas-fir forests of 3 age classes – young forest 40 to 72 years old, mature forest 80 to 120 years old, and old-growth stands 200 to 525 years old – all had Western Tanagers during the

Figure 182. During spring and autumn migration, the Western Tanager forages in a variety of mixed woodlands in British Columbia, including stands of flooded dead trees bordering lakes (Blackwater, northwest of Quesnel, 25 June 1997; R. Wayne Campbell).

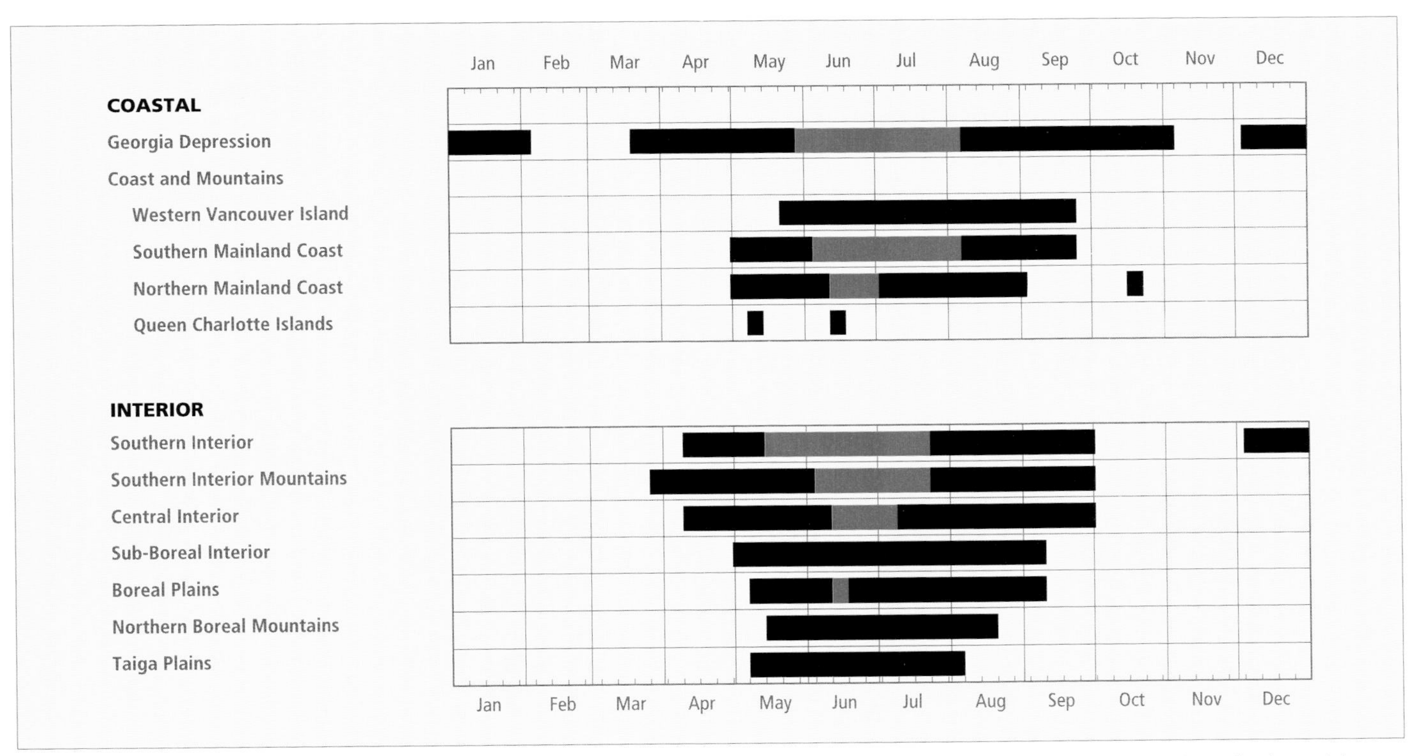

Figure 183. Annual occurrence (black) and breeding chronology (red) for the Western Tanager in ecoprovinces of British Columbia. Records are shown for the week in which they occurred.

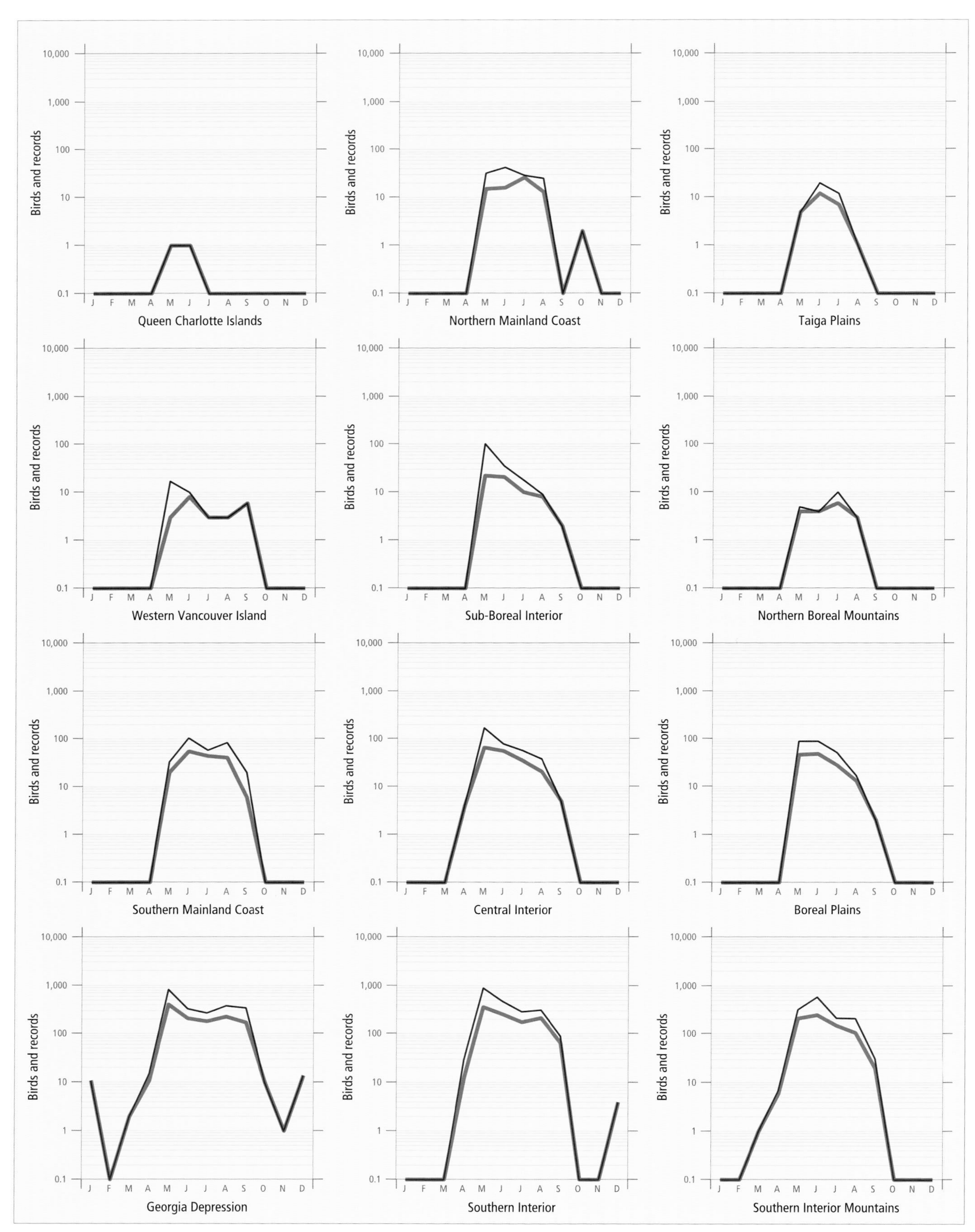

Figure 184. Fluctuations in total number of birds (purple line) and total number of records (green line) for the Western Tanager in ecoprovinces of British Columbia. Christmas Bird Counts, Breeding Bird Surveys, and nest record data have been excluded.

study period from late April through June. Numbers were consistently higher in the young age class (Carey et al. 1991).

Further examples of the variety of forest types in which this tanager breeds come from the Rocky Mountains area, where it was a common summer resident in Kootenay National Park and reached its highest density in the Montane ecoregion, and less frequently in the lower subalpine habitats. Here, it inhabited interior Douglas-fir forests as well as other open mature coniferous or mixed forests. It was most numerous in Douglas-fir–hairy wildrye closed forest, and Douglas-fir–ponderosa pine–wheatgrass open forest (Poll et al. 1984). In the trembling aspen forests of the Bulkley valley, repeated census counts between late May and mid-July revealed that the Western Tanager was generally absent from clearcut stands and from the sapling stage but was generally present in stands of mature trembling aspen (trees 50 to 60 years old) and in stands of 100-year-old trees, but was most numerous in mixed conifer–trembling aspen old-growth, where densities ranged from 3.2 to 9.1 singing males per 10 ha (Pojar 1995). The timing of these counts included both migration and early nesting periods. In the Sub-Boreal Interior, the Western Tanager nests in mature trembling aspen and mixed trembling aspen–conifer forests of the Bulkley River region (Pojar 1995). In the vicinity of Mackenzie, the species prefers mature (more than 80 years old) lodgepole pine or lodgepole pine–trembling aspen mixed forests, although it is occasionally found in white spruce–trembling aspen stands 50 to 100 years old (J. Tuck pers. comm.). Further north, in the vicinity of Dawson Creek in the Boreal Plains, the Western Tanager breeds in mature mixed-wood forests and less so in pure trembling aspen forests (Phinney 1998). Along the Liard River drainage basin, this species occurs during the breeding season mainly in mature stands of white spruce (Erskine and Davidson 1976).

In British Columbia, the Western Tanager generally nests in coniferous forests, especially interior Douglas-fir, but also in mixed forests that include western redcedar, western hemlock, spruce, and associations consisting of such deciduous trees as trembling aspen, black cottonwood, paper birch, red alder, and bigleaf maple. Studies of the species in western Montana and adjacent Idaho, not far south of British Columbia, emphasize the relationship between the Western Tanager and interior Douglas-fir forests (Hejl and Wood 1991). In the association between birds and Douglas-fir forests of 2 age classes – old-growth (200+ years) and "rotation age" (80 to 120 years) – the Western Tanager was among the 4 species primarily associated with old-growth forest.

On the coast, the Western Tanager has been recorded breeding from 30 May to 4 August; in the interior, it has been recorded breeding from 15 May (calculated) to 20 July (Fig. 183).

Nests: All nests (Fig. 187) were placed in trees. Nest trees were predominantly conifers (79%, n = 43), and most nests were found in Douglas-fir (55%; Fig. 187), followed by western hemlock (7%), spruce (7%), and western redcedar (5%). Ponderosa pine, lodgepole pine, and western larch were each used once. Just 21% of nests were in deciduous trees, with

Figure 185. In British Columbia, the highest numbers for the Western Tanager in summer occur in the Southern Interior Ecoprovince.

3 nests each in trembling aspen and willow, and single nests in red alder, domestic apple, and Garry oak.

Nests were mainly constructed of loosely organized twigs and grasses, along with, in descending order of frequency, lichens, mosses, rootlets, animal hair, needles, plant stems, and feathers (Fig. 187).

The heights of 43 nests ranged from 2.4 to 23 m, with 56% between 6.4 and 11 m. Most nests in coniferous trees were placed on top of a branch among the abundant branchlets near the extremity (Fig. 187). The nest location was more variable in deciduous trees, from close to the trunk to the terminal branches.

Eggs: Nests of the Western Tanager are generally placed in such a way that they are difficult to reach or even to see into, so details on clutch size are few. Dates for 22 clutches ranged from 30 May to 20 July, with 59% recorded between 5 June and 27 June. Calculated dates indicate that eggs can occur as early as 15 May. Sizes of 20 clutches ranged from 1 to 5 eggs (1E-3, 2E-1, 3E-3, 4E-11, 5E-2), with 55% having 4 eggs. The incubation period is 13 days (Baicich and Harrison 1997).

Young: Dates for 29 broods ranged from 28 May to 4 August, with 52% recorded between 24 June and 16 July. Sizes of 18 broods ranged from 1 to 4 young (1Y-2, 2Y-3, 3Y-7, 4Y-6), with 72% having 3 or 4 young. The nestling period is about 10 to 13 days (Baicich and Harrison 1997; Ehrlich et al. 1988).

Brown-headed Cowbird Parasitism: In British Columbia, 2 of 39 nests found with eggs or young were parasitized by the cowbird. In addition, there were 13 records of Western Tanagers feeding fledgling Brown-headed Cowbirds; 1 involved 3 cowbird fledglings and another involved 2. This is an unusually high rate of parasitism, as Friedmann (1963) and Friedmann and Kiff (1985) list just 4 instances in North America: 2 from British Columbia and 1 each from Alberta

Figure 186. In parts of the southern portions of the interior of British Columbia, the Western Tanager breeds in mixed forests of ponderosa pine and Douglas-fir bordering grasslands and shrub-steppe habitats (Hat Creek, 7 June 1997; R. Wayne Campbell).

and Montana. Thus, although it is not possible to calculate a rate of parasitism for the province, it appears to be the highest in North America. Our records also include 1 nest deserted by the tanager after a cowbird egg was deposited and 2 others in which both tanager and cowbird nestlings were reared to fledging. Cannings et al. (1987) gives a rate of 33% Brown-headed Cowbird parasitism in the Okanagan valley, but this includes Western Tanagers accompanied by fledgling cowbirds.

Nest Success: Insufficient data. Among the few identified causes of nestling mortality in the province are predation by a Northern Pygmy-Owl and 1 instance of infestation by a blowfly species feeding subcutaneously.

REMARKS: As a nesting species in the province, the Western Tanager is closely associated with interior Douglas-fir forests and is a species of concern to those designing management strategies for such forests. Robbins et al. (1986) refer to declines in the numbers of the Western Tanager in western North America, especially in British Columbia and Montana, between 1968 and 1973. Over the period 1966 to 1979, Breeding Bird Surveys for all of British Columbia did show a decline in the tanager population at an average annual rate of 5.3% ($P < 0.05$); since 1980, however, the population has increased at an average annual rate of 2.1% ($P < 0.10$) (Sauer et al. 1997). For example, the spring of 1996 saw a remarkable migration through parts of the interior of the province that was observed from the Okanagan valley north as far as Prince George (Bowling 1996c). The largest reported flock consisted of 100 birds seen among roadside shrubbery on the east side of Swan Lake, Vernon, on 12 May. The last big flocks were

Figure 187. Many Western Tanager nests in British Columbia are constructed entirely of small twigs and rootlets (Victoria, 7 June 1986; R. Wayne Campbell).

seen in the Vernon area on 20 May, but at Kamloops the migration continued until the end of May. In Prince George, on 19 May, birds were concentrating in trees that still carried the remnants of the previous year's berry crop, as well as foraging on the ground throughout the city. The last flocks reported from the Prince George area were from 25 May. Local reports of the birds' behaviour suggested a food shortage and possibly heavy mortality.

The 1996 spring migration in British Columbia is reminiscent of one that occurred in the San Francisco Bay district of California a century earlier, from 12 to 28 May 1896 (Emerson 1903). Then, the tanagers did serious damage to a cherry crop and large numbers were shot by orchardists.

For a comprehensive summary of current knowledge about this species, see Hudon (1999). Additional information may be found in Bent (1958).

NOTEWORTHY RECORDS

Spring: Coastal – w Sooke 22 Mar 1987-1 male; Saanich 31 Mar 1952-1 (Clay 1953), 6 May 1980-6; Brentwood Bay 11 May 1940-25; Reifel Island 15 Apr 1988-1; Lennard Island 26 May 1978-12; Surrey 30 May 1970-3 eggs; Sea Island 4 May 1995-1 (Elliott and Gardner 1997); Vancouver 18 May 1975-33; North Vancouver 14 May 1984-25; Pitt Meadows 10 May 1976-20; Cheam Slough 27 Apr 1992-1 male; Miracle Beach Park 7 Apr 1962-1 (Westerborg 1962); Gold River 25 May 1974-4; Port Neville 23 May 1975-1; Stuie 22 May 1932-1; Yakoun Lake 13 May 1983-1 (Mattocks and Hunn 1983b). **Interior** – Osoyoos 20 May 1945-40 (90% males); Oliver to Richter Lake 17 May 1959-8 on survey; Balfour to Waneta 27 May 1982-11 on survey; Princeton 18 May 1959-50; Carrs Landing 9 Apr 1967-1; Okanagan Landing 13 May 1937-50; Redstreak 31 Mar 1965-1 male singing (Seel 1965); Ashcroft 15 May 1948-1, (NMC 1852); Peterhope Lake 11 Apr 1976-5; McQueen Lake 28 May 1973-adult feeding nestlings; Celista 28 May 1974-28; Mount Revelstoke National Park 12 Apr 1972-2; North Barriere Lake 7 Apr 1964-1 male; 15 km n Golden 6 May 1996-1 male; Chilcotin River 18 May 1990-39; Williams Lake area 13 Apr 1980-1; Fountain Valley 23 May 1970-100; Willow River 4 May 1966-1, 12 May 1968-25 (mostly males); 25 km s Prince George 9 May 1982-1, 17 May 1975-10; Gagnon Creek 11 May 1997-1; Taylor 10 May 1983-1, 18 May 1985-8; Bear Flat 11 May 1985-1; Beaver Lake (Cassiar) 11 May 1980-1; Fort Nelson 18 May 1987-1; Liard Hot Springs 16 May 1975-1 (Reid 1975).

Summer: Coastal – Sahtlam 30 Jun 1915-nest and 1 egg collected; Goldstream Park 1 Jun 1975-1; Duncan 5 Jun 1971-5; Surrey 20 Jul 1983-4 eggs; Deas Island 20 Jul 1989-5; Alberni 4 Jun 1910-1; Deer Lake (Burnaby) 13 Jun 1971-13; Hope 31 Jul 1993-3 nestlings; Tony Lake 6 Jul 1936-4 nestlings; Campbell River 6 Jun 1980-15; Lost Lake 16 Aug 1946-6; Alta Lake 10 Jun 1945-2; Pemberton 3 Jun 1967-8; Grant Bay 25 Aug 1968-1 (Richardson 1971); Goose Island 1 Aug 1948-1 pair (Guiguet 1953); Bella Coola River 9 Jun 1976-1; De la Beche Inlet 12 Jun 1986-1 male in mixed woods; Terrace 28 Jun 1979-3 nestlings; Haines Highway (Km 35) 18 Jun 1974-1. **Interior** – Creston 20 Jul 1971-young; Kootenay River (Creston) 12 Aug 1948-30 (Munro 1957); 1.6 km ne Big Sand Creek 25 Jun 1974-4 nestlings; Nakusp 7 Aug 1979-3 young; West Bench (Penticton) 26 Aug 1980-8; 5 km ne of Nicola Lake 1 Aug 1980-1 fledgling; Invermere 25 Aug 1983-1 fledgling; Skihist to Lytton 8 Jun 1968-14 on survey; Scotch Creek (Shuswap) 31 Jul 1970-4 nestlings with adults; 15 km n Golden 22 Aug 1997-11, highest count; 100 Mile House 20 Aug 1978-4 young; 32 km n 100 Mile House 12 Jun 1961-2 eggs; Westwick Lakes 30 Jun 1956-1 recently fledged young; Alexis Creek 12 Jun 1976-11 on survey; Bear Lake (Crooked River) 15 Aug 1975-2; Prince George 24 Aug 1982-1; Eaglet Lake 28 Jul 1985-2 fledglings; Youngs Lake 3 Jul 1981-2 fledglings; Pine Pass 17 Aug 1975-1; w Chetwynd 22 Jul 1975-2; Moberly Lake 13 Jun 1984-female incubating; Beatton Park 4 Jul 1981-1 fledgling with parents; Four Mile Creek (Stikine River) 5 Jul 1922-1 female incubating; Clarke Lake 14 Jul 1985-5; 2 km nw Clarke Lake 28 Jul 1985-2 fledglings; Fort Nelson 19 Jun 1976-6, 29 Jul 1968-1; Liard Hot Springs 8 Jul 1992-4 males, 16 Aug 1975-1 (Reid 1975); Coal River 7 Jul 1992-2 males.

Breeding Bird Surveys: Coastal – Recorded from 22 of 27 routes and on 59% of all surveys. Maxima: Squamish 16 Jun 1985-24; Alberni 10 Jun 1969-20; Pemberton 13 Jun 1982-20; Campbell River 14 Jun 1980-15; Kitsumkalum 11 Jun 1978-15; Nass River 21 Jun 1975-15. **Interior** – Recorded from 63 of 73 routes and on 81% of all surveys. Maxima: Beaverdell 20 Jun 1970-63; Canford 4 Jul 1993-46; McLeod Lake 15 Jun 1968-43.

Autumn: Interior – St. John Creek (Fort St. John) 7 Sep 1985-1; Mugaha Creek 9 Sep 1996-1; Riske Creek 11 Sep 1986-1; n side Chilcotin River 27 Sep 1988-1; 15 km n Golden 13 Sep 1997-2; Field 27 Sep 1976-3; Okanagan Landing 30 Sep 1927-1; Revelstoke 11 Sep 1988-1. **Coastal** – Pemberton 10 Sep 1995-4; Harrison Hot Springs 23 Sep 1986-1; Pitt Meadows 27 Sep 1976-1 female; Vancouver 2 Sep 1986-15, 13 Oct 1995 (Elliott and Gardner 1997); Tofino 22 Sep 1983-1; Surrey 28 Sep 1977-50; Victoria 6 Sep 1971-15 (Tatum 1972).

Winter: Interior – Kamloops 28 Dec 1986-1 caught by cat. **Coastal** – Vancouver 3 Dec 1989 to 31 Jan 1990-1 at feeder, 14 Dec 1997 to 13 Jan 1998-1 immature at feeder; Victoria 1 to 31 Dec 1962-1 at feeder all month.

Christmas Bird Counts: Interior – Not recorded. **Coastal** – Recorded from 2 of 33 localities and on less than 1% of all counts. Maxima: Victoria 22 Dec 1962-1; Vancouver 17 Dec 1989-1.

Spotted Towhee

SPTO

Pipilo maculatus (Swainson)

RANGE: Breeds in western North America from southern British Columbia (including Vancouver Island), southern Alberta, and southern Saskatchewan south along the Pacific coast into northwestern Baja California, southern Nevada, southern Arizona, and through the Mexican highlands to central Guatemala, and east to the central Dakotas, central Nebraska, eastern Colorado, eastern New Mexico, and western Texas. Winters from southern British Columbia through the Pacific coast states, Nevada, Utah, and Colorado, south into Baja California and locally along the Mexican mainland. South of this there are discontinuous resident populations as far south as Guatemala (Greenlaw 1996).

STATUS: On the coast, a *fairly common* to *common* resident in the Georgia Depression Ecoprovince; in the Coast and Mountains Ecoprovince, *fairly common* resident in southern areas of the Southern Mainland Coast, *uncommon* resident on Western Vancouver Island, *casual* in the Northern Mainland Coast, and absent from the Queen Charlotte Islands.

In the interior, a *common* summer visitant and *rare* in winter in the Okanagan valley and *fairly common* summer visitant and *very rare* in winter elsewhere in the Southern Interior Ecoprovince; locally *uncommon* summer visitant and *casual* in winter in the southern half of the Southern Interior Mountains Ecoprovince and *very rare* elsewhere; locally *uncommon* summer visitant and *very rare* in winter in the Central Interior Ecoprovince.

Breeds.

NONBREEDING: The Spotted Towhee (Fig. 188) is present throughout the year and is widely distributed on the coast throughout the lowland regions, including the Nanaimo Lowland, Gulf Islands, Georgia Lowland, and Fraser Lowland of the Georgia Depression, and the Southern Mainland Coast from Port Neville and Whistler east to Harrison Hot Springs and Hope. Although scarce on Western Vancouver Island, it occurs the full length of the island. On the mainland coast, the only records north of Port Neville are from the Bella Coola valley and Kitimat, the latter being the northernmost occurrence on the coast.

In the interior, the Spotted Towhee is widely distributed throughout the year in the Okanagan valley and becomes more locally distributed, during migration and in winter, in the South Thompson and Nicola valleys, the Columbia River valley south of Revelstoke in the Southern Interior Mountains, and north as far as Williams Lake and west to Tatlayoko Lake in the Central Interior. It winters regularly only in the Okanagan valley.

The highest numbers in winter occur in the Georgia Depression Ecoprovince (Fig. 189).

In coastal British Columbia, the Spotted Towhee occurs from sea level to 600 m elevation. In the interior, it occurs during migration mostly at valley bottom elevations between 260 and 400 m but has been noted at up to 1,200 m.

Figure 188. The Spotted Towhee (*Pipilo maculatus*) was formerly known as the Rufous-sided Towhee (*P. erythrophthalmus*) (Victoria, 21 January 1991; Tim Zurowski).

On the coast, the Spotted Towhee is a species of the brush-filled ravines, forest edges, open coniferous forests with an undergrowth of salal, and roadside or field-edge tangles of blackberry vines (Fig. 190). It is also a common bird of urban and suburban gardens, where hedges of exotic shrubs, cypress, western redcedar, Douglas-fir, and landscape plantings of broad-leaved evergreens, such as rhododendron and camellia, provide stiff-twigged cover close to the ground. It is a regular visitor at bird feeding stations that provide suitable cover nearby. In the interior, the Spotted Towhee also favours dense shrubbery and thickets, often on south-facing hillsides.

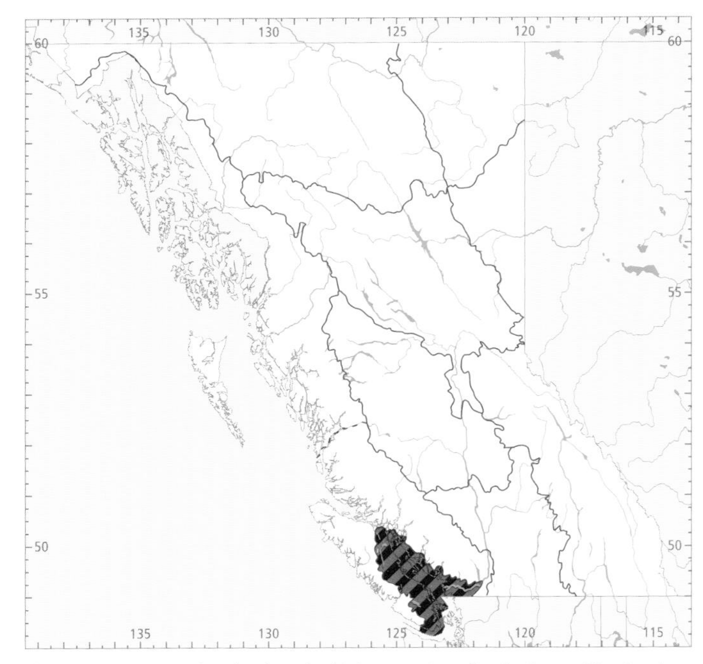

Figure 189. In British Columbia, the highest numbers for the Spotted Towhee in winter (black) and summer (red) occur in the Georgia Depression Ecoprovince.

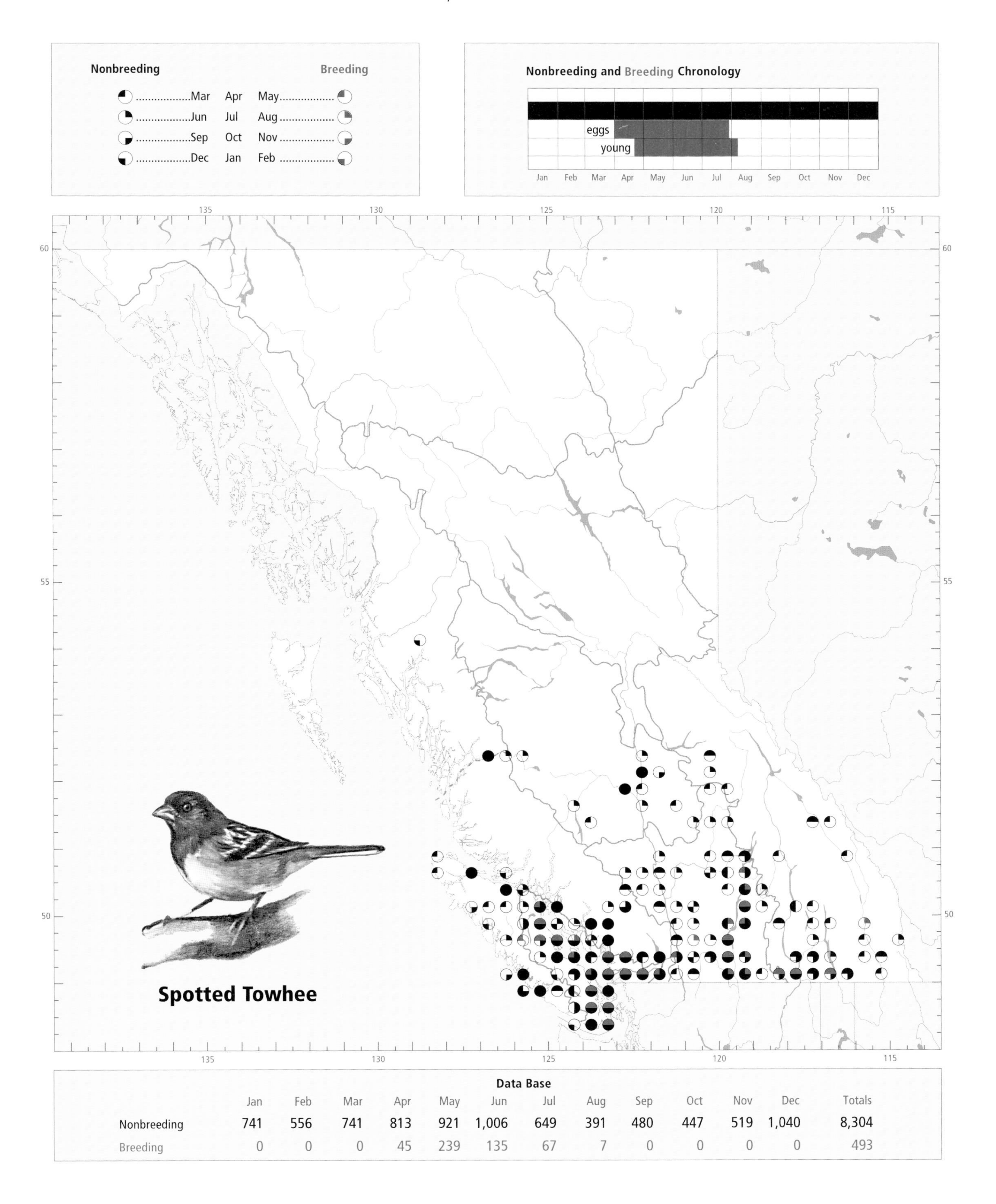

Data Base

	Jan	Feb	Mar	Apr	May	Jun	Jul	Aug	Sep	Oct	Nov	Dec	Totals
Nonbreeding	741	556	741	813	921	1,006	649	391	480	447	519	1,040	8,304
Breeding	0	0	0	45	239	135	67	7	0	0	0	0	493

Figure 190. In winter on the south coast of British Columbia, the Spotted Towhee frequents tangles of blackberry vines bordering agricultural fields and water courses (Aldergrove, 25 February 1996; R. Wayne Campbell).

Apparently most or all of the population is resident on the southern portions of the coast, but there are local movements from higher to lower elevations nearer the coastline. For example, there are no winter records from the valley of the Fraser River east of Harrison Hot Springs. There is also a pronounced shift into the cultivated habitats described above. In the interior, the migration habitat appears to be similar to that used during the breeding season.

In the interior, spring migration is first noticeable in early March, with the arrival of a small number of Spotted Towhees in the Okanagan valley (Figs. 191 and 192). Through March all records are of single individuals. Beginning in early April, several birds can be seen in a day, and the number of birds increases steadily towards a maximum in May (Fig. 192). In the Southern Interior Mountains, the first spring arrivals are about a month behind those in the Okanagan valley but the population still reaches maximum numbers between late April and early May. The few towhees that have been reported from the Central Interior have followed the same migratory timetable in spring as those further south.

In autumn, the last birds have left the Cariboo-Chilcotin areas by the end of September, except for the occasional individual that remains into winter. Throughout the southern interior regions of the province, the autumn departure appears to be in progress by early September; by late September all towhees are gone from the Southern Interior Mountains and the population in the Southern Interior has reached winter numbers (Figs. 191 and 192).

The northernmost winter record in the interior is from Williams Lake.

On the coast, the Spotted Towhee has been recorded every month of the year; in the interior, it has been recorded irregularly throughout the year, mainly between 9 February and 18 October (Fig. 191).

BREEDING: The Spotted Towhee has a widespread breeding distribution across southern British Columbia from Vancouver Island to the Rocky Mountain Trench. It may nest as far north as Riske Creek, where a pair of adults accompanied by a fledgling has been reported.

The Spotted Towhee reaches its highest numbers in summer in the Fraser Lowland and Nanaimo Lowland of the

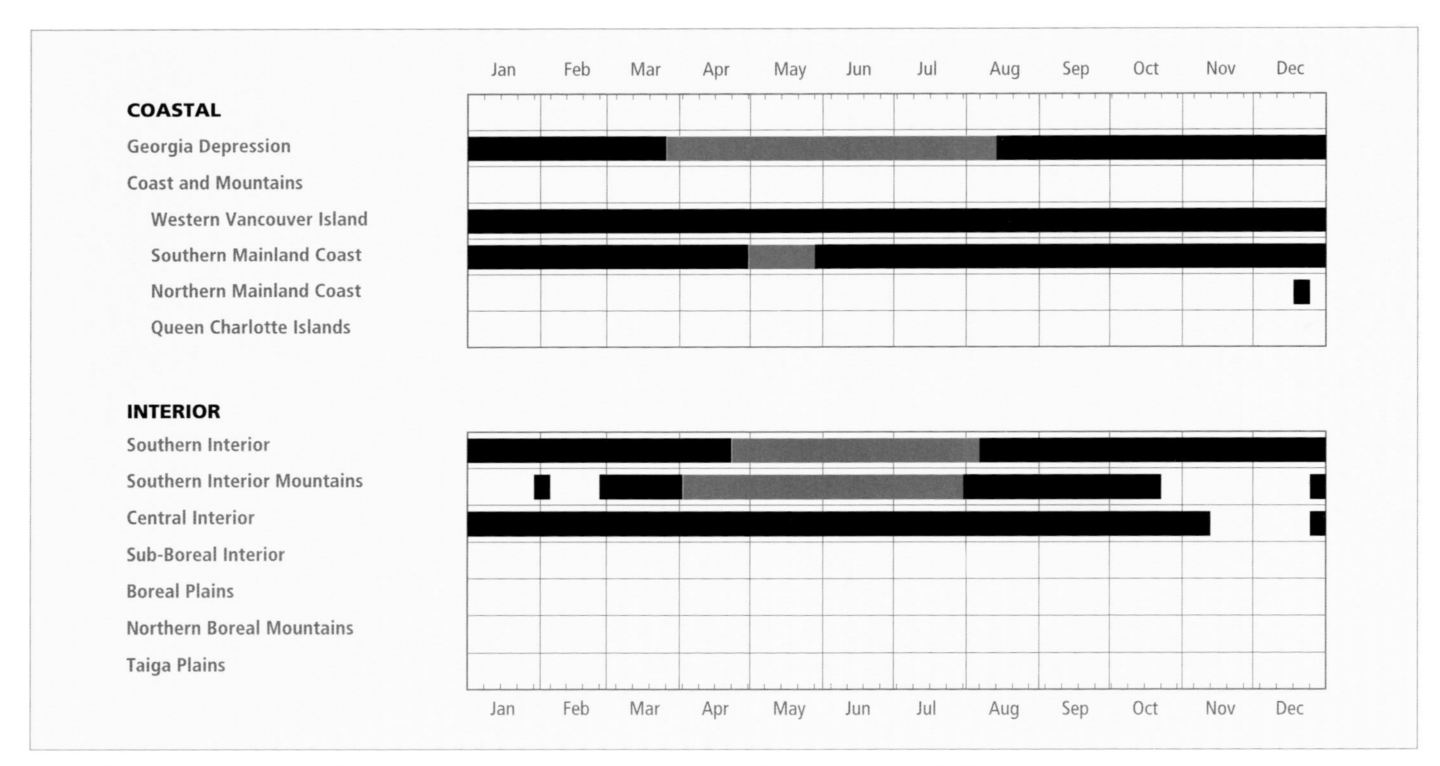

Figure 191. Annual occurrence (black) and breeding chronology (red) for the Spotted Towhee in ecoprovinces of British Columbia. Records are shown for the month in which they occurred.

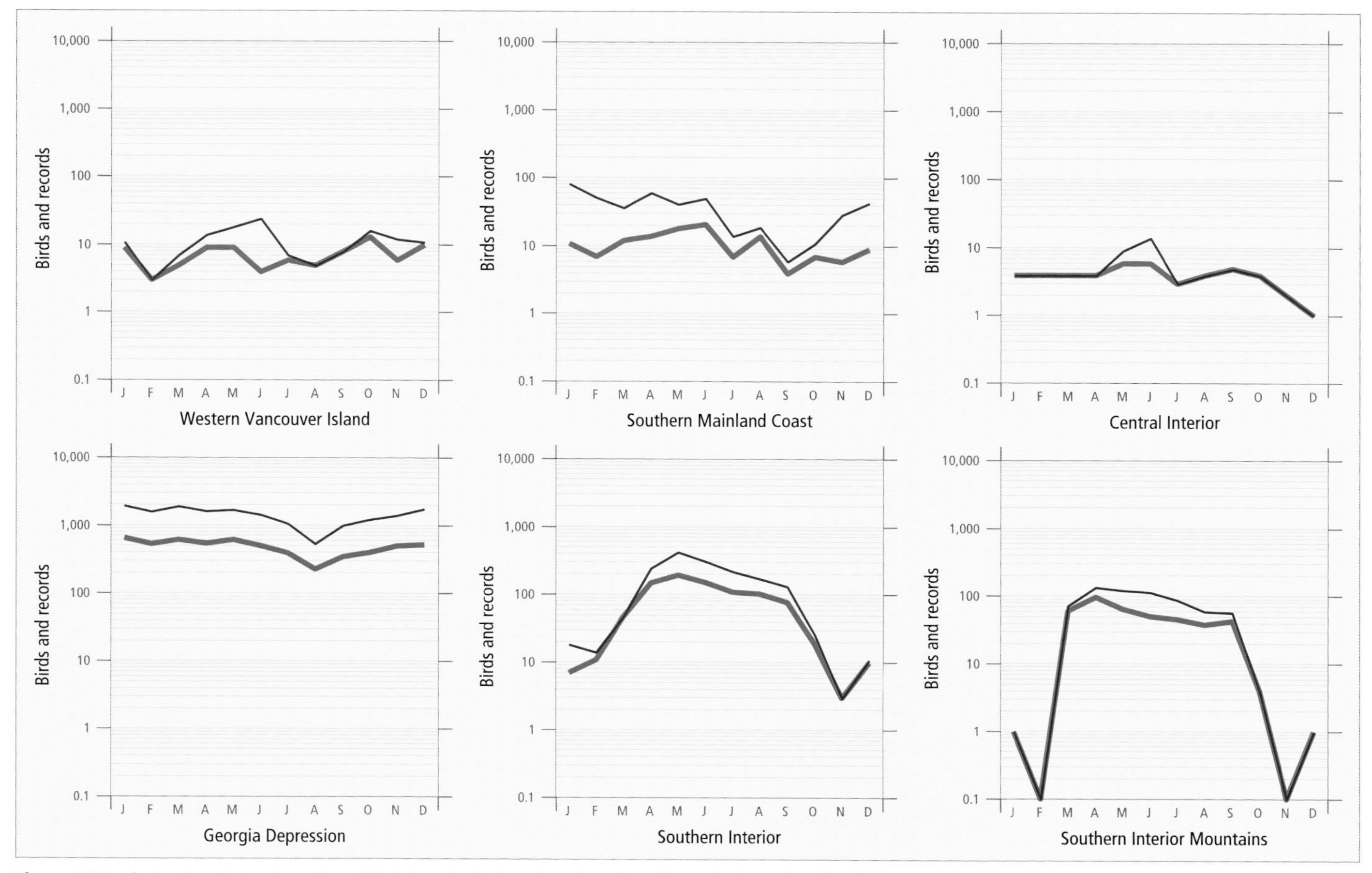

Figure 192. Fluctuations in total number of birds (purple line) and total number of records (green line) for the Spotted Towhee in ecoprovinces of British Columbia. Christmas Bird Counts, Breeding Bird Surveys, and nest record data have been excluded.

Georgia Depression (Fig. 189). An analysis of Breeding Bird Surveys in British Columbia for the period 1968 through 1993 could not detect a net change in numbers on the coastal routes; on interior routes, the mean number of birds increased at an average annual rate of 5% (Fig. 193). An analysis of Breeding Bird Surveys across North America from 1966 to 1996 reveals an increase in mean number of birds at an average annual rate of 0.8% ($P < 0.10$) (Sauer et al. 1997).

Rising (1996), in his review of relative densities of the Spotted Towhee in various parts of the species' range in North America, places the Georgia Depression in the second highest category of nesting density, with averages of 20 to 50 singing males per census route.

The Spotted Towhee breeds from sea level to 975 m elevation on the coast and from 280 to 1,200 m in the interior. Most nesting habitats (82%; n = 33) selected by the towhee were on the edges of forested areas, including areas altered by forest removal (39%). A further 12% of nests were in grasslands and 6% in shrublands. On the coast, the cultivated habitats of urban and suburban areas or farmland were most heavily used, including areas with a well-established vegetation of willow, Nootka rose (Fig. 194), hardhack, and blackberry thickets and tangles. The forested patches used were equally divided between those dominated by deciduous trees (red alder or vine maple) and those dominated by coastal Douglas-fir, along with brushy areas and fallen woody debris. Bryant et al. (1993), in their study of bird communities in the forests of western Vancouver Island, found the Spotted Towhee only in the 15- to 20-year-old regeneration on cutover lands.

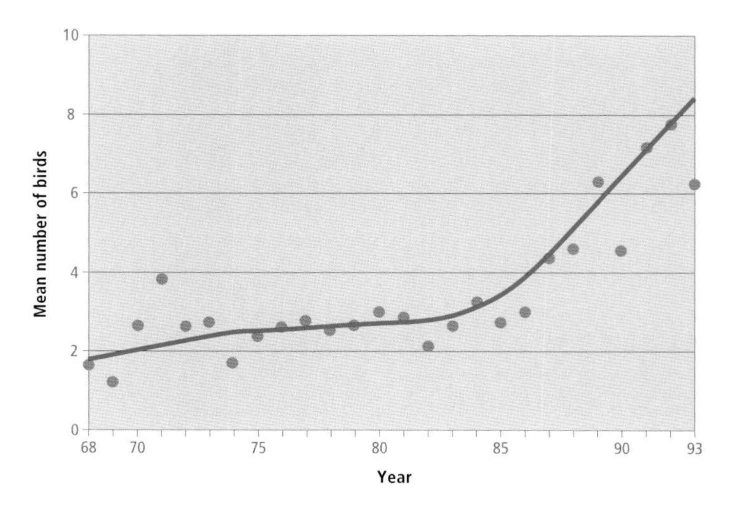

Figure 193. An analysis of Breeding Bird Surveys for the Spotted Towhee in British Columbia shows that the mean number of birds on interior routes increased at an average annual rate of 5% over the period 1968 through 1993 ($P < 0.001$).

Figure 194. On southern Vancouver Island, the Spotted Towhee nests in dense thickets of Nootka rose and Indian-plum with an understorey of forbs and grasses (Island View Beach, Central Saanich, 6 April 1997; R. Wayne Campbell).

In the interior, shrubland habitats are the most frequently used category, followed by woodlands with a tree cover of black cottonwood, trembling aspen (Fig. 195), or interior Douglas-fir and ponderosa pine with an associated shrub understorey, and by dense pockets of shrubs on rangeland.

Figure 195. In parts of the Southern Interior Ecoprovince of British Columbia, the Spotted Towhee breeds in disturbed trembling aspen woodlands with dense patches of shrubs and fallen woody debris as an understorey (Hat Creek, 13 June 1996; R. Wayne Campbell).

Cannings et al. (1987) describe the nesting habitat in the Okanagan valley as the "dense shrubbery of hillside draws, and thickets along watercourses" including "birch groves near water, and trembling aspen copses on rangeland." Less than one-quarter of nesting habitats were urban or suburban. Along the South Thompson River, the towhee selects similar habitat, and includes the shrubby black cottonwood riparian areas bordering the river. At the northern edge of its range, around the junction of the Chilcotin and Fraser rivers, the Spotted Towhee selects nesting areas within the riparian community of the lower grasslands (big sagebrush) areas.

On the coast, the Spotted Towhee has been recorded breeding from 1 April (calculated) to 7 August; in the interior, it has been recorded breeding from 3 April (calculated) to 5 August (Fig. 191).

Nests: Most nest sites (84%; n = 220) were on the ground. Eight percent were in low shrubs, 5% in non-woody vegetation (e.g., forbs, grasses, bracken fern, etc.), and 3% in other locations.

Nests were almost always in thick ground cover where they were sheltered from above (Fig. 196). Such sites included tangles of Himalayan and trailing blackberry vines, thickets of wild roses, common snowberry, or tall Oregon-grape, with a ground cover of forbs and grasses. Nests were placed less frequently in big sagebrush or bunchgrass.

Nests were either set into a depression on the ground or built in a low shrub. They consisted of a base of grasses (79%), leaves (30%), small twigs (19%), plant fibres (16%), bark strips (14%), plant stems (8%), and a variety of other items (Fig. 199). The lining was of fine grasses with occasional feathers, hair, or leaves. The heights of 217 nests ranged from ground level to 1.8 m, with 92% of the nests on the ground.

Eggs: Dates for 300 clutches ranged from 4 April to 29 July, with 51% recorded between 9 May and 7 June. Calculated dates indicate that eggs may be present as early as 1 April. Sizes of 180 clutches ranged from 1 to 6 eggs (1E-8,

Figure 196. Most Spotted Towhee nests in British Columbia were built into the ground and were well concealed by adjacent plants. The nest above (circled) is sheltered by trailing blackberry and common snowberry leaves (View Royal, southern Vancouver Island, 22 July 1996; R. Wayne Campbell).

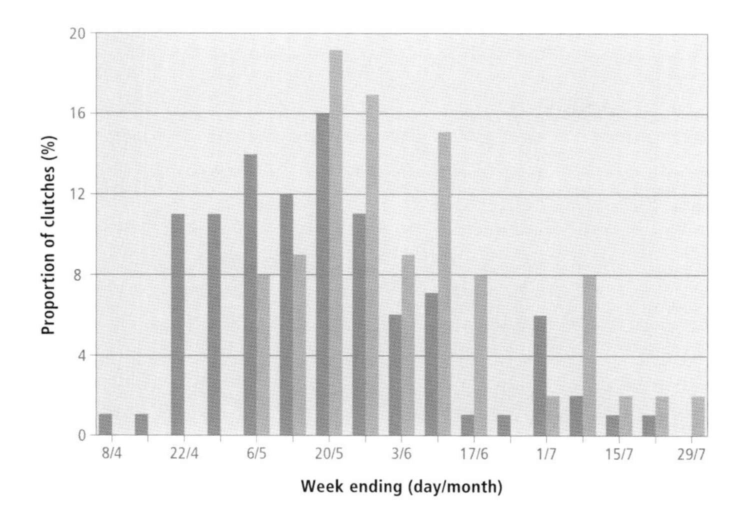

Figure 197. Distribution of nests with eggs of the Spotted Towhee through the egg-laying period in the Georgia Depression Ecoprovince (dark bars; *n* = 128) and in the Southern Interior and Southern Interior Mountains ecoprovinces (light bars; *n* = 53).

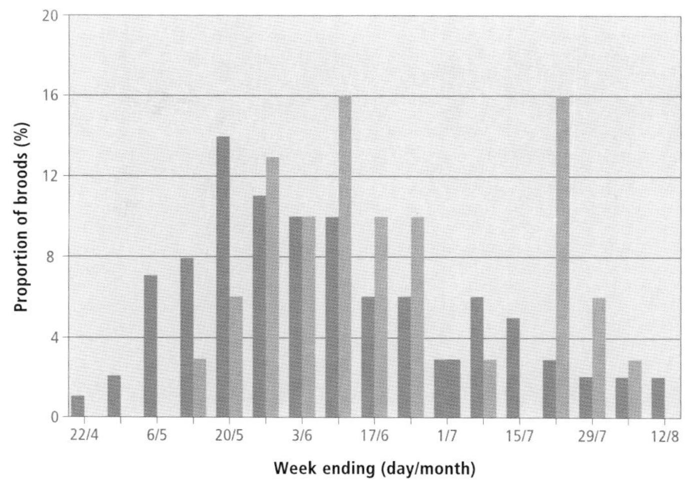

Figure 198. Distribution of nests with young of the Spotted Towhee through the nestling period in the Georgia Depression Ecoprovince (dark bars; *n* = 87) and in the Southern Interior and Southern Interior Mountains ecoprovinces (light bars; *n* = 26).

2E-8, 3E-34, 4E-113, 5E-14, 6E-3), with 63% having 4 eggs. The incubation period is 12 to 14 days (Baumann 1959; Davis 1960).

Young: Dates for 197 broods ranged from 22 April to 7 August, with 52% recorded between 22 May and 29 June. Sizes of 112 broods ranged from 1 to 5 young (1Y-8, 2Y-11, 3Y-42, 4Y-45, 5Y-6), with 78% having 3 or 4 young. The nestling period, as determined by our data and by Baumann (1959), is 9 to 11 days.

As with the egg period, there are differences between the coast and the interior in the period through which nestlings are present (Figs. 197 and 198). On the coast, they occur between 22 April and 7 August, a 17-week period, whereas in the interior they occur between 9 May and 30 July, a 13-week period. Our data on the period during which nestlings are present in the Georgia Depression suggest a remarkably symmetrical distribution, probably reflecting a prolonged period over which renesting takes place. In contrast, the "young" period in the Southern Interior shows a peak in late May followed by a second peak about a month later. This suggests a more concentrated period of renesting in the interior.

Brown-headed Cowbird Parasitism: In British Columbia, 3% of 248 nests found with eggs or young had been parasitized by the cowbird. Four nests each contained 2 cowbird eggs, and another 2 each contained 3 cowbird eggs. There were an additional 16 reports of adult Spotted Towhees feeding fledged cowbirds. Among the fledglings were 12 single young and 4 instances of 2 fledglings being fed by a pair of towhees. An observation at Victoria confirmed that the towhee could fledge both 1 or more of its own young and a cowbird from the same nest.

In British Columbia, the coastal and interior ranges of the Brown-headed Cowbird are occupied by separate subspecies of towhees (see REMARKS). The 2 subspecies appear to experience different levels of parasitism: 1% (*n* = 179) on the coast and 7% (*n* = 69) in the interior. A similar difference is indicated by the numbers of fledgling cowbirds observed in the 2 regions: 4 on the coast and 12 in the interior.

Friedmann (1963) refers to the Rufous-sided Towhee as a very frequent host of the Brown-headed Cowbird, with nearly 300 recorded instances, mainly in the eastern parts of the species' range. At that time the Spotted Towhee and Rufous-sided Towhee were regarded as conspecific. It is significant that Friedmann also comments on the small number of instances reported from California, in the range of the Spotted Towhee.

Friedmann et al. (1977) mention 393 clutches of towhees eggs (from all races) in the collections of the Western Foundation of Vertebrate Zoology in 1976. The series included 297 clutches representing 7 subspecies of the Spotted Towhee, mainly from California. In these, only 2 clutches (1%) were parasitized, whereas the parasitism rate was 16% in 55 clutches

Figure 199. Spotted Towhee nest containing 1 egg and 3 recently hatched young (View Royal, southern Vancouver Island, 22 July 1996; R. Wayne Campbell).

of the Rufous-sided Towhee from eastern North America. Greenlaw (1996), quoting Baumann (1959) and Davis (1960), re-emphasizes the paucity of records of cowbird parasitism of the Spotted Towhee.

Nest Success: Of 44 nests found with eggs and followed to a known fate, 22 produced at least 1 fledgling, for a success rate of 50%. The rate was the same for the coast and the interior.

REMARKS: The Spotted Towhee (*Pipilo maculatus*) was formerly known as the Rufous-sided Towhee (*P. erythrophthalmus*).

Nine subspecies of the Spotted Towhee are recognized in North America north of Mexico (Greenlaw 1996). Two occur in British Columbia: *P. m. oregonus* is the subspecies of the coastal parts of the province north to the northern end of Vancouver Island, whereas *P. m. curtatus* is the subspecies of the interior regions (Munro and Cowan 1947; Godfrey 1986; American Ornithologists' Union 1957). The map, but not the text, presented by Greenlaw (1996) suggests that *P. m. arcticus* extends from the Prairie provinces into British Columbia, but no evidence is given to support this. Recently, J.S. Greenlaw (pers. comm., 31 January 1998) confirmed that *P. m. arcticus* is the subspecies found east of the Rocky Mountains in Alberta. Specimens from the Bella Coola valley have not been examined to determine whether they are *oregonus* or *curtatus*.

Cohen (1899) and Bleitz (1956) refer to the occasional presence of California Quail eggs in the nests of the Spotted Towhee. There is a single instance of this in British Columbia, where a towhee's nest found at Naramata contained 6 towhee eggs and 2 quail eggs. The observer noted that it looked as though the Spotted Towhee had taken over a quail's nest.

For an excellent review of the biology of this species, see Greenlaw (1996).

NOTEWORTHY RECORDS

Spring: Coastal – Victoria 2 May 1964-15; Pender Island 1 May 1977-1; Port Kells 22 Apr 1958-2 eggs hatching; Agassiz 8 Apr 1976-4; Harrison Hot Springs 22 Mar 1980-10; Ponderosa Meadow 28 Apr 1973-3; Sproat Lake Park 8 Apr 1974-1; North Vancouver 4 Apr 1924-4 eggs; Vancouver 6 Apr 1974-23 at Stanley Park on survey; Hope 6 Apr 1969-2, 26 Apr 1976-8, 12 May 1964-4 eggs, 22 May 1964-4 nestlings; 8 km n Boston Bar 5 May 1977-2; Stein River 29 Apr 1978; Quadra Island 24 Mar 1974-5; Cortes Island 29 Mar 1977-1; Cape Scott Park 15 Apr 1980-1; Bella Coola 22 May 1984-1. **Interior** – Trail 1 Mar 1986-1; 6 Apr 1906-4 eggs; Balfour to Waneta 16 May 1981-46 on survey; Creston 10 May 1980-4 eggs, 19 May 1980-4 nestlings; Princeton 15 May 1975-1 on Old Hedley Rd; Wynndel 6 Apr 1984-6; Vaseux Lake 2 May 1974-5; Twin Lakes 10 Apr 1969-6; Castlegar 19 Apr 1981-15; Cranbrook 3 Apr 1937-1; Naramata 2 Mar 1968-1, 2 May 1965-5 eggs, 8 May 1965-nest containing 6 towhee eggs and 2 California Quail eggs, 9 May 1965-4 nestlings; White Lake (Okanagan Falls) 17 Apr 1962-8; Okanagan Landing 6 Mar 1925-1, 30 Apr 1912-4 eggs; Nakusp 10 Apr 1976-1; Botanie Valley 11 May 1969-2; Lillooet 16 May 1916-1 (RBCM 3849); Cherry Creek 2 Mar 1986-1; Sorrento 11 May 1970-2; Revelstoke 2 Mar 1986-1; 15 km n Golden 25 Apr 1996-1; Alkali Lake (Cariboo) 30 May 1994-4; Chilcotin River (Riske Creek) 6 Mar 1987-1 at feeder at Wineglass Ranch all winter, -22°C for several days.

Summer: Coastal – Victoria 20 Jul 1996-4 eggs; nr Esquimalt 22 Jul 1996-3 nestlings; Colwood 4 Jun 1987-18; Tugwell Lake 2 Jul 1983-12; Bamfield 24 Jun 1981-20; Mayne Island 7 Jun 1974-1; Aldergrove 22 Aug 1992-1 unhatched egg; Marion Lake (Maple Ridge) 4 Aug 1975-8; 2.4 km e Deroche 20 Jul 1968-4 nestlings; Surrey 12 Aug 1983-4 nestlings; Black Mountain 20 Jul 1968-3; Alouette Lake 1 Jun 1963-15; Forbidden Plateau 5 Jun 1978-1; Courtenay 13 Jul 1973-3 eggs; Alta Lake 26 Aug 1944-1; Stuie 10 Aug 1982-1. **Interior** – Grand Forks 1 Aug 1982-6; Trail 28 Jul 1982-3 nestlings; Elko 1 Jul 1953-3 (Godfrey 1955); Jaffray 22 Jun 1968-18; Sparwood 20 Jul 1981-11; Westbank 21 Jul 1981-3 eggs, 30 Jul 1981-3 nestlings; Botanie Valley 18 Jul 1964-5; Vernon 23 Jul 1975-22, 15 Aug 1991-adult feeding 3 young; Lillooet 3 Jul 1916-1 (NMC 9718); Barnhartvale 30 Jun 1991-2; Dog Creek 4 Jun 1995-2; confluence of Chilcotin and Fraser rivers 6 Jun 1994-7; Wineglass Ranch (Riske Creek) 7 Jun 1994-1 fledgling with adults.

Breeding Bird Surveys: Coastal – Recorded from 20 of 27 routes and on 78% of all surveys. Maxima: Pitt Meadows 6 Jul 1989-52; Gibsons Landing 4 Jul 1993-52; Albion 20 Jun 1973-51; Coquitlam 30 Jun 1990-49. **Interior** – Recorded from 31 of 73 routes and on 31% of all surveys. Maxima: Syringa Creek 24 Jun 1989-30; Canford 22 Jun 1982-29; Oliver 28 Jun 1992-22.

Autumn: Interior – Williams Lake 9 Nov 1995-1 at feeder; Riske Creek 3 Nov 1986-1; 15 km n Golden 24 Sep 1996-1; Kamloops 13 Oct 1997-1; Summerland 28 Nov 1976-1; Sproule Creek 19 Oct 1978-1; Vaseux Lake 10 Oct 1975-2 at cliffs; Waneta 30 Sep 1982-5. **Coastal** – 9 km e Bella Coola 25 Sep 1977-1; Bella Coola 14 Oct 1989-2 at feeder; Malcolm Island 28 Nov 1975-1 at Pultney Point; Campbell River 25 Oct 1964-10; Hope 26 Nov 1976-8; Little Qualicum River 26 Sep 1977-10 on estuary survey; Harrison Hot Springs 26 Nov 1986-16; Blenkinsop Lake 13 Nov 1982-8.

Winter: Interior – Williams Lake 8 Jan 1971-1; Chilcotin River (Riske Creek) 20 to 31 Dec 1986-1 at Wineglass Ranch, 11 Feb 1987-1; Riske Creek 4 and 15 Jan 1987-1; Tranquille 1 Jan 1989-10, 20 Jan 1989-1; Kamloops 28 Dec 1985-1 at feeder, 18 Dec 1993-1; Nicola 1 Feb 1997-2 at feeder; Lavington 27 Dec 1976-2, 2 Jan 1955-2 at feeder; Summerland 7 Jan 1979-1; Naramata 30 Dec 1977-1; West Trail 24 Dec 1983-1; Sunningdale 29 Jan 1986-1. **Coastal** – Kitimat 19 Dec 1992-1; Bella Coola 2 Dec 1989-2, 31 Dec 1989-1 at feeder; Port Neville 4 Dec 1975-2; Sointula 25 Dec 1988-2; Whistler 18 Dec 1993-5; Brackendale 14 Jan 1916-1 (NMC 9507); Egmont 3 Jan 1975-3; Cranberry Lake 1 Dec 1980-2; Tahsis Inlet 5 Jan to 21 Feb 1949-1 (Mitchell 1959); Porpoise Bay 21 Feb 1981-1; Harrison Hot Springs 1 Jan 1989-12; Qualicum Beach 13 Jan 1976-15; Burnaby Lake 25 Dec 1975-10; Tofino 31 Dec 1980-1; Ucluelet 2 Jan 1972-1; Saltspring Island 11 Feb 1949-40; Harrison Hot Springs 13 Jan 1991-21; Blenkinsop Lake 1 Dec 1985-12.

Christmas Bird Counts: Interior – Recorded from 8 of 27 localities and on 20% of all counts. Maxima: Penticton 27 Dec 1987-6; Vernon 21 Dec 1980-5; Vaseux Lake 28 Dec 1978-4; Kelowna 14 Dec 1991-4. **Coastal** – Recorded from 25 of 33 localities and on 79% of all counts. Maxima: Victoria 18 Dec 1993-718; Vancouver 18 Dec 1983-712; White Rock 2 Jan 1993-579.

American Tree Sparrow

ATSP

Spizella arborea (Wilson)

Figure 200. The American Tree Sparrow winters across southern British Columbia and breeds in the far northern areas of the province (Haines Highway, 23 June 1999; Linda M. Van Damme).

RANGE: Breeds from northern Alaska, Yukon, Mackenzie, Quebec, and Labrador south to southern Alaska, north-central British Columbia, southern Mackenzie, northwestern Saskatchewan, northern Manitoba, northern Ontario, and northern Quebec. Winters from south coastal Alaska (rarely) and southern Canada (British Columbia to the Maritime provinces except southern Manitoba) south to northern California, central Nevada, Arizona, New Mexico, Texas, and Arkansas, and into Tennessee and North Carolina.

STATUS: On the coast, a *rare* migrant and winter visitant on southeastern Vancouver Island and the Fraser Lowland in the Georgia Depression Ecoprovince; in the Coast and Mountains Ecoprovince, *casual* in migration on Western Vancouver Island, *very rare* on the Northern Mainland Coast, and *very rare* in migration and *casual* in winter on the Queen Charlotte Islands and the Southern Mainland Coast.

In the interior, a *rare* migrant and winter visitant in the Southern Interior and Southern Interior Mountains ecoprovinces; a *rare* to *uncommon* migrant, *casual* in summer, and *rare* in winter in the Central Interior and Sub-Boreal Interior ecoprovinces; a *fairly common* to occasionally *very common* migrant and *casual* in summer in the Boreal Plains Ecoprovince; a *common* migrant and summer visitant and *casual* in winter in the Northern Boreal Mountains Ecoprovince; an *uncommon* migrant and summer visitant and *casual* in winter in the Taiga Plains Ecoprovince.

Breeds.

NONBREEDING: The American Tree Sparrow (Fig. 200) is widely distributed across British Columbia. Over most of this range it is scarce enough to be a notable sighting in a birder's field day. Records of its presence are concentrated in the northern third of the province in spring and summer and in the southern third of the province in winter.

On the coast, this sparrow occurs regularly, but infrequently, in the Nanaimo and Fraser lowlands. It is an occasional visitor to Western Vancouver Island. It is unknown along the heavily timbered coastline between the northern tip of Vancouver Island and the mouth of the Skeena River, except for isolated records at Kemano and Bella Coola, each of which lies at the head of a long inlet served by a major east-west river valley passing through the coastal mountains.

In the interior, a few American Tree Sparrows winter in the Okanagan valley and along the valley of the Columbia River south of Revelstoke. Even fewer winter in the Rocky Mountain Trench, all of them south of Harrogate. Further north in the interior, there are a few wintering records from the vicinity of Williams Lake and along the lower Skeena River valley. The highest numbers in winter occur in the Okanagan valley of the Southern Interior (Fig. 201).

In winter, the American Tree Sparrow occurs on the coast at elevations from sea level to about 300 m; in southern portions of the interior, wintering has been recorded at elevations from 280 to 500 m. In the northern regions of the province, this sparrow has been found at elevations between 800 and 1,500 m during the immediate pre- and post-breeding periods.

During migration, tree sparrows frequent a variety of clearings that offer an abundance of dried weed stems, low deciduous shrubs, and open ground for foraging. Such habitat has been found in a variety of locations, including weedy

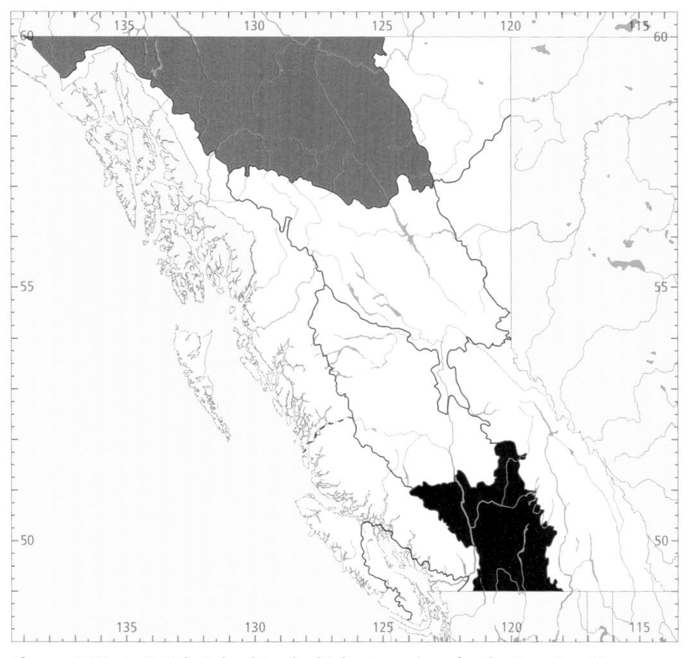

Figure 201. In British Columbia, the highest numbers for the American Tree Sparrow in winter (black) occur in the Southern Interior Ecoprovince; the highest numbers in summer (red) occur in the Northern Boreal Mountains Ecoprovince.

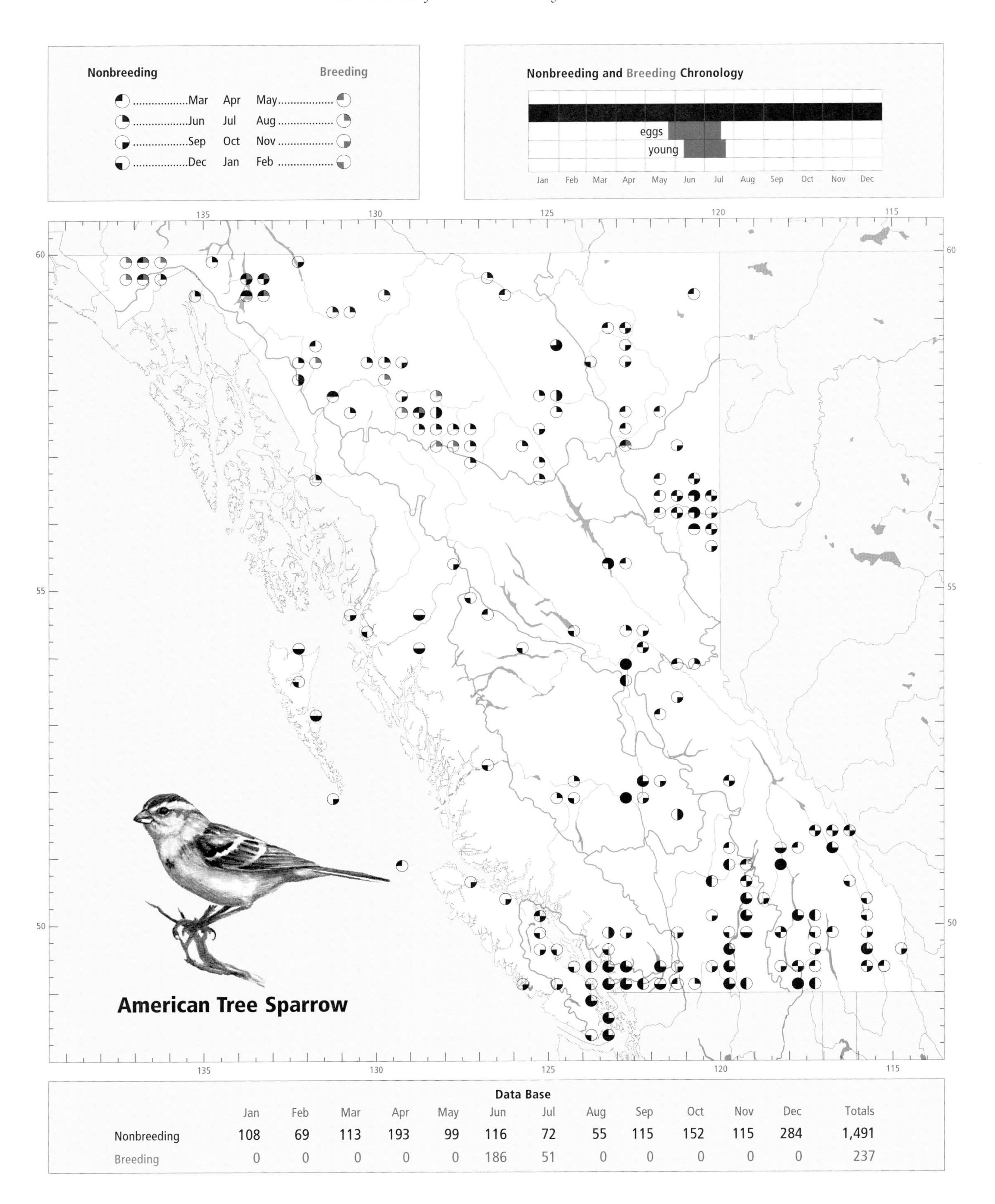

Data Base

	Jan	Feb	Mar	Apr	May	Jun	Jul	Aug	Sep	Oct	Nov	Dec	Totals
Nonbreeding	108	69	113	193	99	116	72	55	115	152	115	284	1,491
Breeding	0	0	0	0	0	186	51	0	0	0	0	0	237

Figure 202. During migration in southern British Columbia, the American Tree Sparrow frequents open grasslands where seeds are available in patches of weedy vegetation (Douglas Lake, 29 September 1996; R. Wayne Campbell).

edges of agricultural fields (Fig. 202), willow thickets, around sewage lagoons, abandoned farmsteads, hedgerows, abandoned roadways, thickets along mixed-forest edges, dykes, recent clearcuts, powerline corridors, field borders, stubble fields, grasslands, road verges, airports, grain elevator yards, and cabin clearings in the northern spruce forest. In winter, shrubby areas, weed patches, overgrowing meadows and clearings, feeders, and cattail marshes (Fig. 205) are frequented.

In the Georgia Depression, a slight increase in the number of birds in late February marks the onset of spring migration. The movement tapers off by mid-March and ends suddenly before the end of the month, with only a few birds remaining into April (Figs. 203 and 204).

The route followed by the relatively small number of migrants that move through the southwestern coast to their northern nesting grounds is unknown. The species appears not to migrate along the coast north of Vancouver Island, and few birds appear in the Cariboo and Chilcotin areas.

In the interior, the spring migration can be detected largely by the departure of the small wintering population (Figs. 203 and 204). In the Southern Interior, this is completed in April, with occasional stragglers in May and June. In the Southern Interior Mountains, there appears to be a small migration entering the province from the south. This occurs largely along the lower Columbia valley in the west Kootenay, beginning in the last week of March and ending in the last week of April. The latest departures of tree sparrows from the Kootenay and Columbia valleys are in the second week of June. In the Cariboo and Chilcotin areas, there is no certain evidence of a spring influx of birds from the south, and the last of the winter records has been in the final days of April. As with the Central Interior, there is little evidence as to the direction of a spring migration into the Sub-Boreal Interior. The first arrivals at Prince George have been recorded in mid-March, with 1 record of 20 birds in the first week of April.

The major spring migration of American Tree Sparrows into British Columbia enters the Peace Lowland, of the Boreal Plains, and the Fort Nelson Lowland, of the Taiga Plains, from Alberta from mid to late April and continues through the first week of May with a few stragglers later in the month (Figs. 203 and 204). The front of this migration moves quickly north and west to reach the Liard River valley the same week it enters the Peace River area, and Atlin by the last week of April.

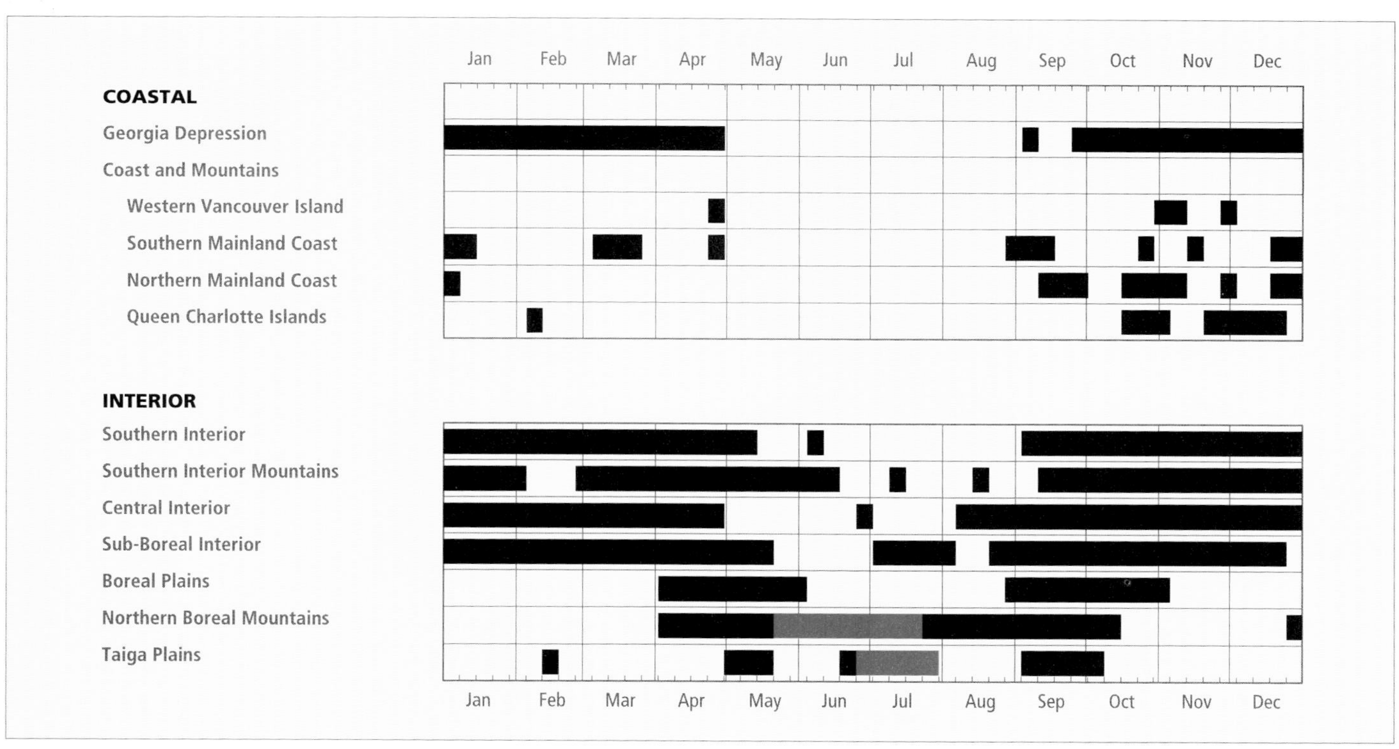

Figure 203. Annual occurrence (black) and breeding chronology (red) for the American Tree Sparrow in ecoprovinces of British Columbia. Records are shown for the week in which they occurred.

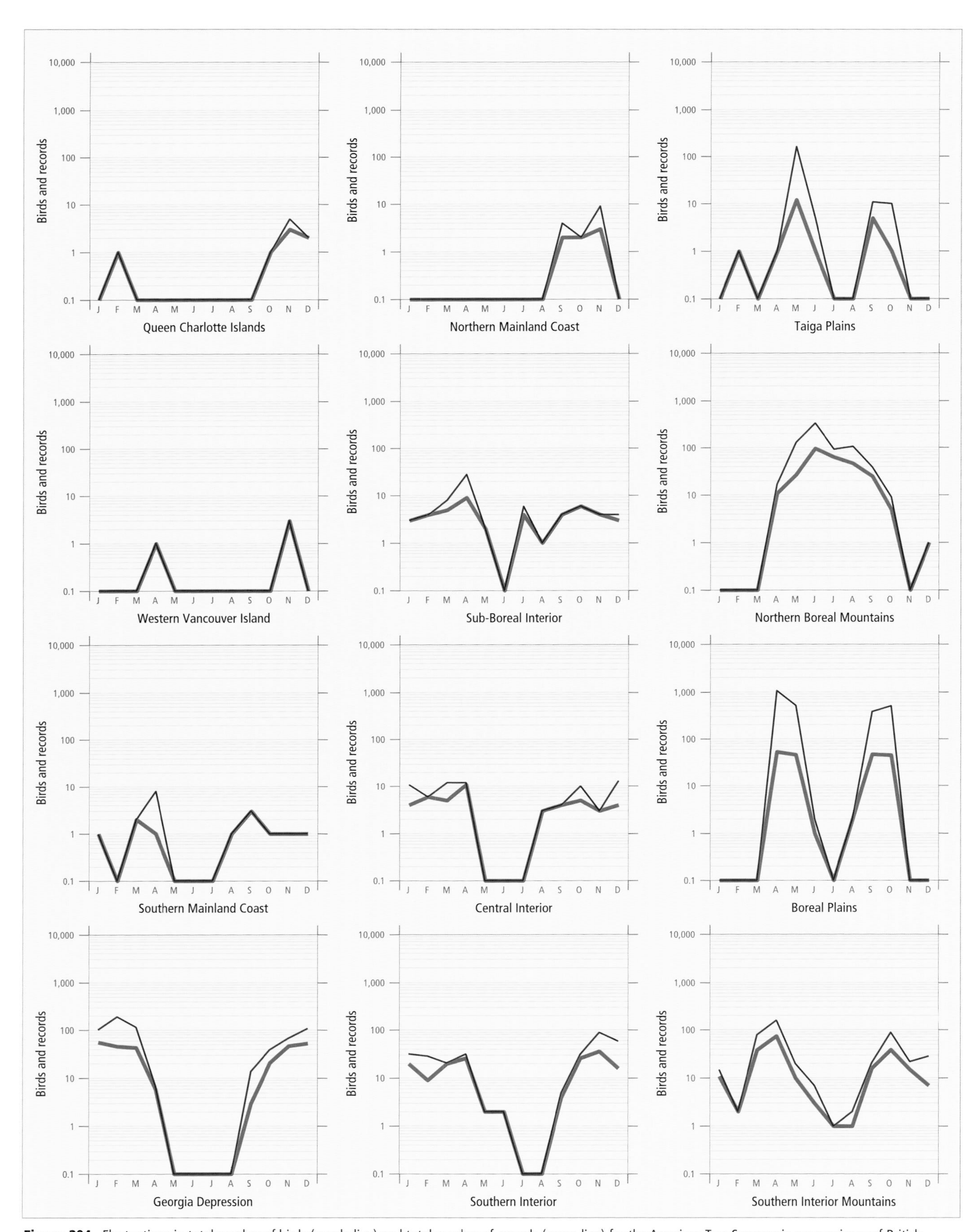

Figure 204. Fluctuations in total number of birds (purple line) and total number of records (green line) for the American Tree Sparrow in ecoprovinces of British Columbia. Christmas Bird Counts, Breeding Bird Surveys, and nest record data have been excluded.

Figure 205. At least 5 American Tree Sparrows spent the winter of 1997-98 in this cattail marsh (Duck Lake, Creston, 15 December 1997; R. Wayne Campbell).

Numbers on the nesting grounds in the Northern Boreal Mountains continue to increase until mid-June.

The autumn migration begins with the appearance of a few birds in the central parts of the province as early as the third week of July (Einar Creek in the Sub-Boreal Interior), but in the Peace Lowland the movement regularly begins in mid-September and peaks there in the first week of October; the last birds have left northern British Columbia by the end of the month (Figs. 203 and 204). In the southern portions of the interior, the first arrivals of the autumn reach the Okanagan, Columbia, and Kootenay valleys by mid-September. Numbers continue to increase until mid- to late November, then decline slightly to wintering numbers as some birds appear to move on southward.

The dates when the American Tree Sparrow is present in Idaho and Washington are consistent with this calendar of migration into southern British Columbia. In Idaho, the earliest autumn record is 13 October and the latest in spring is 13 April (Burleigh 1972). In Washington, the early and late dates given by Jewett et al. (1953) are 7 September and 10 April.

In the interior, the northernmost wintering records are from Summit Lake (Stone Mountain Park) and Flack Creek. On the coast, the northernmost record is from Prince Rupert. The habitat occupied in winter is the same as that used during migration.

On the coast, the American Tree Sparrow is regularly present from 5 October to 26 April. In the southern portions of the interior, it is regularly present from 16 September to 27 April, with occasional observations during summer; in the

Figure 206. In northwestern regions of the Northern Boreal Mountains Ecoprovince of British Columbia, the American Tree Sparrow breeds in willow thickets (Chuck Creek, Haines Highway, 25 June 1983; R. Wayne Campbell).

Figure 207. Most American Tree Sparrow nests in British Columbia were placed in natural situations, nestled deeply in surrounding vegetation (Kusawak Lake, 24 June 1999; Linda M. Van Damme).

Figure 208. The chamber of the American Tree Sparrow nest is generally lined with Willow Ptarmigan feathers (Shini Lakes, 10 July 1983; R. Wayne Campbell).

northern interior it is regularly present from 6 April to 31 October (Fig. 203).

BREEDING: The American Tree Sparrow breeds throughout the Northern Boreal Mountains with concentrations in the far northwest adjacent to the Haines Highway (Fig. 206) and in Spatsizi Plateau Wilderness Park and adjacent areas on the upper Stikine River. The northernmost record is near the Tatshenshini River, while the southernmost is at Mile 148, near Pink Mountain, on the Alaska Highway. The latter record, the only one for the Taiga Plains, is believed to be an exception, for none of the many expeditions into this area, and elsewhere along the Alaska Highway as far northwest as Liard Hot Springs, has encountered the tree sparrow as a nesting species.

The American Tree Sparrow reaches its highest numbers in summer in the Northern Boreal Mountains (Fig. 201). An analysis of Breeding Bird Surveys for the period 1968 through 1993 could not detect a net change in numbers on interior routes.

The American Tree Sparrow has been reported nesting at elevations between 800 and 1,400 m. The characteristic nesting habitat of this sparrow is near the altitudinal limit of trees, where subalpine meadows and frequently wet or marshy wetlands are interspersed with thickets of willows, ranging in height from a few centimetres to 2 m or more (Fig. 206). Spruce, subalpine fir, and scrub birch are frequently present.

The American Tree Sparrow has been recorded breeding in British Columbia from 25 May (calculated) to 23 July (Fig. 203).

Nests: The American Tree Sparrow almost always selects a natural location in which to construct its nest (Fig. 207). Of 117 nests, only 1 was associated with a building. In this sample, 77% were in shrub thickets, predominantly willow; 11% were in other low vegetation (Fig. 207); and 9% were in unspecified locations. Heights for 94 nests ranged from ground level to 1.4 m, with 93% occurring on the ground.

The bulk of the neatly woven cup nests consisted of dried grasses, mosses, and sedges. The lining was of fine grasses and sedges, and generally included feathers (Fig. 208). Feathers were present in 89% of nests and grasses in 84%, followed by sedges (19%), mosses (18%), twigs (3%), leaves (4%), and miscellaneous items (7%). Nest building in Manitoba took 7 days and was followed by a period of 2 to 5 days before egg laying began (Baumgartner 1968).

Eggs: Dates for 152 clutches ranged from 3 June to 18 July, with 53% recorded between 16 June and 26 June. Calculated dates indicate that eggs can occur as early as 25 May. Sizes of 100 clutches ranged from 1 to 6 eggs (1E-2, 2E-2, 3E-10, 4E-41, 5E-44, 6E-1), with 85% having 4 or 5 eggs (Fig. 208). The incubation period elsewhere has been reported as 12 or 13 days (Baumgartner 1968) and 10.5 to 13.5 days (Rees 1973).

Young: Dates for 84 broods ranged from 10 June to 23 July, with 58% recorded between 23 June and 4 July. Sizes of 59 broods ranged from 1 to 5 young (1Y-2, 2Y-1, 3Y-7, 4Y-32, 5Y-17), with 54% having 4 young. The nestling period is 9 to 9.75 days, but young can leave the nest in 8 days if disturbed (Baumgartner 1968).

Brown-headed Cowbird Parasitism: Cowbird parasitism was not found in British Columbia in 236 nests recorded with eggs or young, nor has it been reported elsewhere (Naugler 1993).

Nest Success: Of 17 nests found with eggs and followed to a known fate, 6 produced at least 1 fledging, for a success rate of 35%. At least 7 of the failed nests appeared to have been destroyed by predators. An Arctic ground squirrel, a well-known predator of ground-nesting birds, was the only one identified.

REMARKS: Two subspecies of the American Tree Sparrow are recognized. The one occupying British Columbia is *S. a. ochracea* (American Ornithologists' Union 1957).

It appears that 2 populations of this sparrow occur in British Columbia. One migrates north and south through the province west of the Rocky Mountains and winters from southern British Columbia south through the western states, including Idaho, into northern California and Nevada. The other population enters the northeastern area of the province from Alberta and nests in the northern third of the province, as described above. The migration routes of this population south of about latitude 55°N appear to lie exclusively east of the Rocky Mountains, to the more easterly parts of the known winter range. It is not known whether the 2 populations occupy discrete nesting grounds, and if so, how they are divided.

An American Tree Sparrow banded near Ithaca, New York, on 4 January 1954 was recovered near Delta, British Columbia, on 25 May 1961, some 7 years later.

For a review of the life history of the American Tree Sparrow in North America, see Naugler (1993).

NOTEWORTHY RECORDS

Spring: Coastal – Sea Island 13 Mar 1971-3 (Campbell et al. 1972a); Richmond 24 Mar 1997-3; Pitt Meadows 28 Mar 1972-1 and 11 Apr 1972-1 (Campbell et al. 1974); Campbell River 26 Apr 1975-21; Klesilkwa valley 27 Apr 1971-8; 8 km nw Triangle Island 23 Apr 1987-1 landed on a boat. **Interior** – Balfour to Waneta 16 May 1981-9 on census; Lavington 27 Apr 1970-1; Nakusp 1 Apr 1976-25; New Denver 29 May 1976-1; Kamloops 8 Apr 1990-1; Brennan Creek (Adams Lake) 18 Apr 1981-3; Moberly Marsh (Golden) 30 Mar 1993-6; 15 km n Golden 23 and 24 Apr 1995-1; Williams Lake 23 Apr 1954-1 (NMC 48431); 24 km s Prince George 7 Apr 1976-20; Quick 5 Mar 1982-3 at feeder; Mackenzie 9 Apr 1997-1; Willow River 15 May 1972-1; Taylor 17 May 1986-1 at Peace Island Park; Fort St. John 23 Apr 1988-200 at grain elevators, 11 May 1982-10; Cecil Lake 11 Apr 1986-40, 24 Apr 1983-200; Rose Prairie 17 Apr 1976-200; North Pine 2 May 1983-200 along Road No. 259; 19 km n Sikanni Chief 13 May 1982-50; 58 km n Wonowon 12 May 1982-50; Mile 345 Alaska Highway 3 May 1975-10; Fort Nelson 30 Apr 1978-1; Muncho Lake Park 6 Apr 1965-4; Liard Hot Springs 30 Apr 1975-4 (Reid 1975); Chilkat Pass 14 May 1977-40.

Summer: Coastal – Wilson Creek 20 Jun 1914-1; Garibaldi Park 30 Aug 1974-1 (Thomson 1974); n Stikine River 18 Jun 1975-34; Three Guardsmen Lake 12 Jun 1979-11; Haines Highway 1 Jul 1984-20. **Interior** – Manning Park 10 Jun 1987-1; Balfour to Waneta 8 Jun 1983-4 on census; Needles 15 Jun 1968-1; Revelstoke 13 Jul 1988-1, 19 Aug 1989-2; Tatla Lake 12 Aug 1958-1; Puntzi Lake 13 Aug 1958-1; Einar Creek 20 Jul 1981-3; Mugaha Creek 22 Aug 1995-1 banded; Crooked River 3 Jul 1982-1; Scott Lake 2 Jun 1980-2; nw Fort St. John 30 Aug 1984-1; Tatlatui Lake 27 Jul 1974-1; Mile 147 Alaska Highway 20 Jun 1976-5; Mile 148 Alaska Highway 9 Jul 1967-4 eggs; s end Hotlesklwa Lake 3 Aug 1976-10; Chilkat Pass 10 Jun 1957-4 nestlings; lower Tatshenshini River 12 Jun 1983-53 during 3-hour walk around lake (Campbell et al. 1983), 19 Jun 1983-4 eggs, incubation advanced; Tatshenshini River 19 Jun 1980-26 counted from Goat Creek to Mile 87; Survey Lake 26 Jun 1980-30; Mile 79 Haines Highway 2 Jun 1981-20; Parton River valley 21 Aug 1977-12; Mile 75 Haines Highway 3 Jun 1957-4 eggs; Chilkat Pass, km 128 17 Jul 1984-4 nestlings, 18 Jul 1984-4 eggs.

Breeding Bird Surveys: Coastal – Recorded from 1 of 27 routes and on less than 1% of all surveys. Maximum: Kitsumkalum 12 Jun 1977-1. **Interior** – Recorded from 5 of 73 routes and on 1% of all surveys. Maxima: Gnat Pass 18 Jun 1975-34; Haines Summit 27 Jun 1993-31; Chilkat Pass 10 Jun 1975-27.

Autumn: Interior – Wright Creek 10 Oct 1980-1; Fort Nelson 7 Sep 1987-6, 7 Oct 1986-10; Finbow 16 Sep 1996-1; Andy Bailey Lake 29 Sep 1985-2; Spatsizi Plateau Wilderness Park 8 to 12 Oct 1981-4; Rose Prairie 20 Oct 1984-10; n Fort St. John 18 Sep 1986-38, 31 Oct 1986-1; Fort St. John 22 Sep 1985-45, 28 Sep 1985-75; s Fort St. John 2 Oct 1986-100 in flock foraging in pigweed patch; Willow River 11 Nov 1982-1; Mugaha Creek 20 Oct 1997-1; Riske Creek 12 to 28 Oct 1991-small flocks; 15 km n Golden 5 Oct 1995-3; Nicholson 11 Sep 1976-1; Okanagan Landing 7 Sep 1929-1 (MVZ 83096), 13 Nov 1919-40; Sparwood 26 Oct 1984-6; Richter Lake 13 Sep 1966-2 (Campbell and Meugens 1971). **Coastal** – Bearskin Lake 21 Sep 1986-1; Kispiox River valley 13 Sep 1921-1 (MVZ 42304); Kitimat River 29 Sep 1974-3; Masset 4 Nov 1971-1 with 100 Dark-eyed Juncos; Sandspit 20 Nov 1991-1 (Siddle 1992a); Cape St. James 17 Oct 1981-1; Port Neville 29 Nov 1977-1; Garibaldi Park 12 Sep 1974-1 (Thomson 1974); Campbell River 5 Sep 1975-12; Harrison Hot Springs 30 Oct 1991-1; Hope 18 Nov 1968-1; Vancouver 8 Sep 1987-1 at Jericho Park, 14 Oct 1984-6 (Hunn and Mattocks 1985); Delta 25 Nov 1983-3; Long Beach 3 Nov 1973-1 (BC Photo 678); Duncan 24 Oct 1972-1 (Tatum 1973); Esquimalt 24 Oct 1997-1.

Winter: Interior – Summit Lake (Stone Mountain Park) 26 Dec 1943-1 (NMC 29576); Flack Creek 15 Feb 1985-1; 24 km s Prince George 12 Dec 1982 to 24 Feb 1993-2; Williams Lake 26 Dec 1974-10, 15 Jan to 8 Mar 1974-8 at feeder; Revelstoke 3 Jan 1986-2; Kamloops 1 Jan 1979-1, 28 Dec 1986-17; Vernon 31 Dec 1978-4; Nakusp 1 Jan 1983-1 at feeder most of month; Okanagan Landing 7 Feb 1910-16. **Coastal** – Delkatla Inlet 11 Feb 1973-1; Masset 21 Dec 1986-1; Port Clements 1 Dec 1954-1; Bella Coola 11 Jan 1976-1; Serpentine Fen 1 Jan 1980-3; Boundary Bay 29 Dec 1979-1; Iona Island 6 Dec 1984-6; e end Iona Island 13 Jan 1985-5; Victoria 1 Jan to 28 Feb 1971-1 in Beacon Hill Park (Tatum 1972).

Christmas Bird Counts: Interior – Recorded from 15 of 27 localities and on 28% of all counts. Maxima: Oliver-Osoyoos 28 Dec 1982-65; Kamloops 18 Dec 1993-56; Vernon 21 Dec 1986-51. **Coastal** – Recorded from 17 of 33 localities and on 15% of all counts. Maxima: Pitt Meadows 27 Dec 1977-24; Ladner 27 Dec 1992-21; Vancouver 16 Dec 1984-8.

Chipping Sparrow

Spizella passerina (Bechstein)

CHSP

RANGE: Breeds from east-central and southeastern Alaska, central Yukon, and central Mackenzie, and across Canada from northern Saskatchewan, northern Manitoba, and northern Ontario to central Quebec and southwestern Newfoundland, south in the west through British Columbia and the western states to northern Baja California, southern Nevada, and central and southeastern Arizona; south in the east into parts of Colorado, Kansas, Iowa, Illinois, Michigan, and western New York, and to the highlands of Mexico and northern Central America, as far as Guatemala and Nicaragua. Winters from central California, southern Nevada, central Arizona, central New Mexico, northern Texas, and in the eastern states from Arkansas, Tennessee, Virginia, and Maryland, to the Gulf coast and southern Florida, and south through Mexico and Central America to Nicaragua.

STATUS: On the coast, an *uncommon* to *fairly common* (sometimes locally *common*) migrant and summer visitant and *very rare* in winter in the Georgia Depression Ecoprovince; in the Coast and Mountains Ecoprovince, a *rare* migrant and summer visitant and *very rare* in winter on the Southern Mainland Coast, a *rare* migrant and summer visitant on the Northern Mainland Coast, *very rare* on Western Vancouver Island, and *accidental* on the Queen Charlotte Islands.

In the interior, a *fairly common* to *common* (sometimes *very common*) migrant and summer visitant in the Southern Interior and Southern Interior Mountains ecoprovinces; *very rare* in winter in the Southern Interior and *casual* in the Southern Interior Mountains; *uncommon* to *fairly common* (occasionally *common*) migrant and summer visitant in the Central Interior, Sub-Boreal Interior, Boreal Plains, Northern Boreal Mountains, and Taiga Plains ecoprovinces; *accidental* in winter in the Sub-Boreal Interior.

Breeds.

NONBREEDING: The Chipping Sparrow (Fig. 209) has a widespread distribution throughout the province except in the heavily forested coastal mountains. In some years, a small number of migrants reach offshore islands and occasional birds find suitable habitat at the heads of a few major inlets where rivers cut through the mountains and settlements have created breaks in the forest.

In southern coastal areas of the province, it is not an alpine species, and we have few records from regions dominated by continuous subalpine forests, rock, and snowfields. During northbound migration along the coast, almost all records of the Chipping Sparrow have been from between sea level and about 100 m elevation. In the southern portions of the interior, the species is usually seen in spring at elevations between 280 and 1,200 m. In late summer and early in the southbound migration, it frequently follows subalpine ridges at elevations of 2,000 m or higher (Fig. 211).

Since the Chipping Sparrow is a characteristic summer bird over much of the province, there is much overlap between the habitats it uses during the nonbreeding and breeding periods. On the south coast during spring migration, it is most numerous in shrubby field borders, roadside thickets (Fig. 210), forest edges, shoreline thickets, suburban parks, and similar open areas with shrub cover and open forest. In the interior, it migrates through almost all relatively open terrain that is snow-free, from subalpine forests, shrub patches, fence rows, field borders, pondside thickets, transmission corridors, farmsteads, hedgerows, grain fields, and gardens to open forest of ponderosa pine and Douglas-fir, or western larch. It occurs even more widely in the more northern ecoprovinces, in early and late stages of the trembling aspen forest (Pojar

Figure 209. The Chipping Sparrow has a widespread distribution throughout British Columbia, and occurs from sea level to alpine regions (near Sidley, 7 June 1998; R. Wayne Campbell).

Figure 210. On southeastern Vancouver Island, the Chipping Sparrow frequents shrubby borders and roadside thickets during spring migration (Courtenay, 4 May 1998; R. Wayne Campbell).

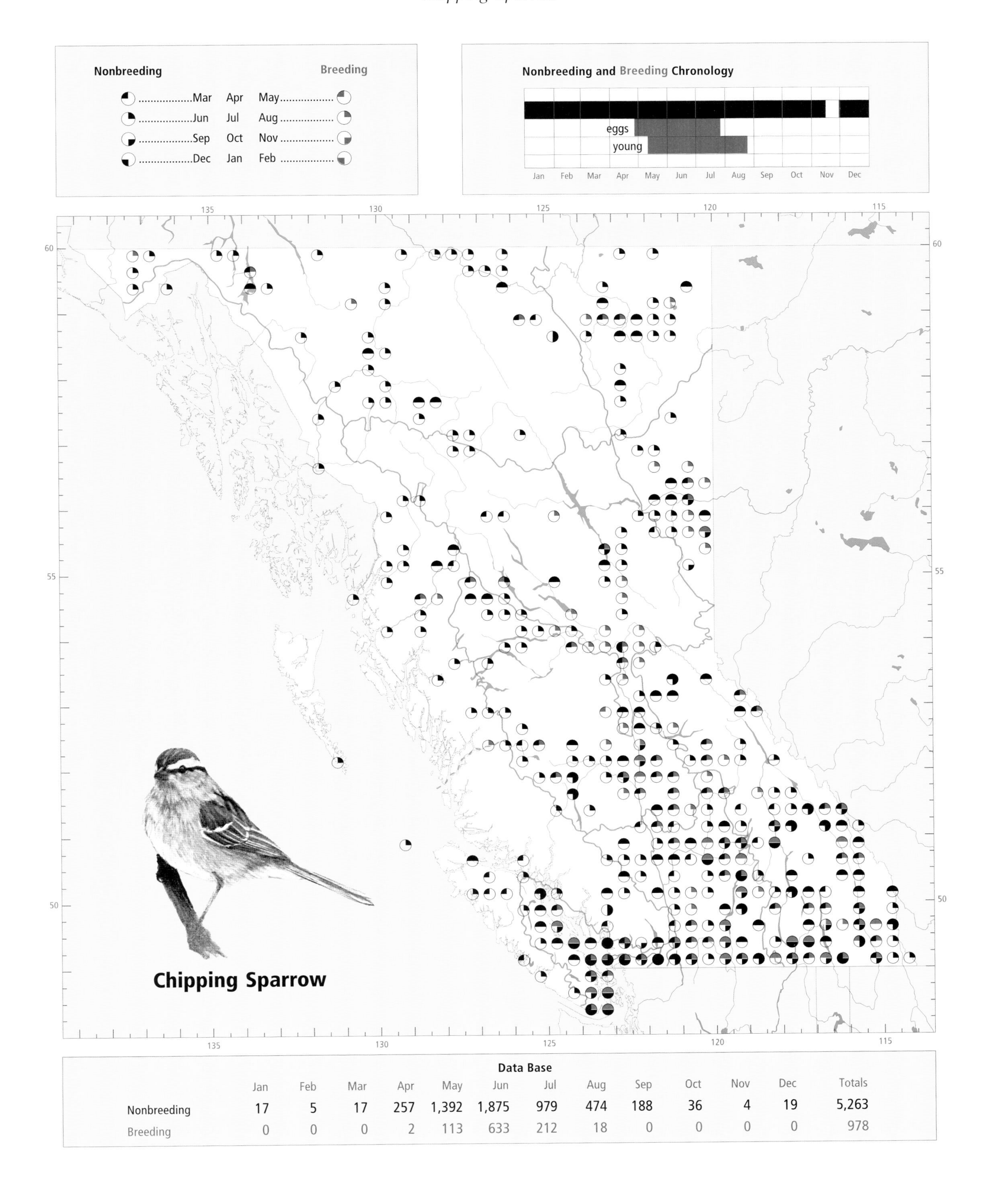

Data Base

	Jan	Feb	Mar	Apr	May	Jun	Jul	Aug	Sep	Oct	Nov	Dec	Totals
Nonbreeding	17	5	17	257	1,392	1,875	979	474	188	36	4	19	5,263
Breeding	0	0	0	2	113	633	212	18	0	0	0	0	978

Figure 211. During autumn migration throughout southern portions of the interior of British Columbia, the Chipping Sparrow often follows sparsely vegetated subalpine ridges (Copper Mountain, southwest of Nelson, 24 September 1997; R. Wayne Campbell).

1995), lodgepole pine forests, subalpine meadows, and, less frequently, in open stands of white spruce or in black spruce muskegs.

On the south coast, spring migration is signalled by the arrival of a small influx of Chipping Sparrows in the first week of April. Numbers increase thereafter to a peak between the second and third weeks of May (Figs. 212 and 213). The northward movement appears unhurried and the first arrivals reach the mouth of the Skeena River in the third week of May, with the migration peaking there in early June.

In the interior, the pattern of spring arrival in the Okanagan valley, of the Southern Interior, and in the valleys of the Kootenay and Columbia rivers, of the Southern Interior Mountains, is similar. The first dates represented by more than a few birds are in the first week of April and the wave of arrivals reaches a peak in the third week of May (Figs. 212 and 213). The front of the migration reaches the Cariboo and Chilcotin areas, the region around Prince George, and the Peace Lowland in early to mid-May. In the latter region, arrival dates for 1977 and for 1980 through 1989 ranged from 6 May to 15 May (n = 11) (C. Siddle, pers. comm.). In these northern regions, the peak in numbers appears to occur in late May. The migrating birds reach the northern boundary of the province in the third week of May.

In northern ecoprovinces, the southward movement begins in early August (Figs. 212 and 213). The last birds have left the Taiga Plains by the end of the second week of August and the Northern Boreal Mountains and Boreal Plains by the second week of September. In the Central Interior, the peak movement occurs in late August, with stragglers noted into late

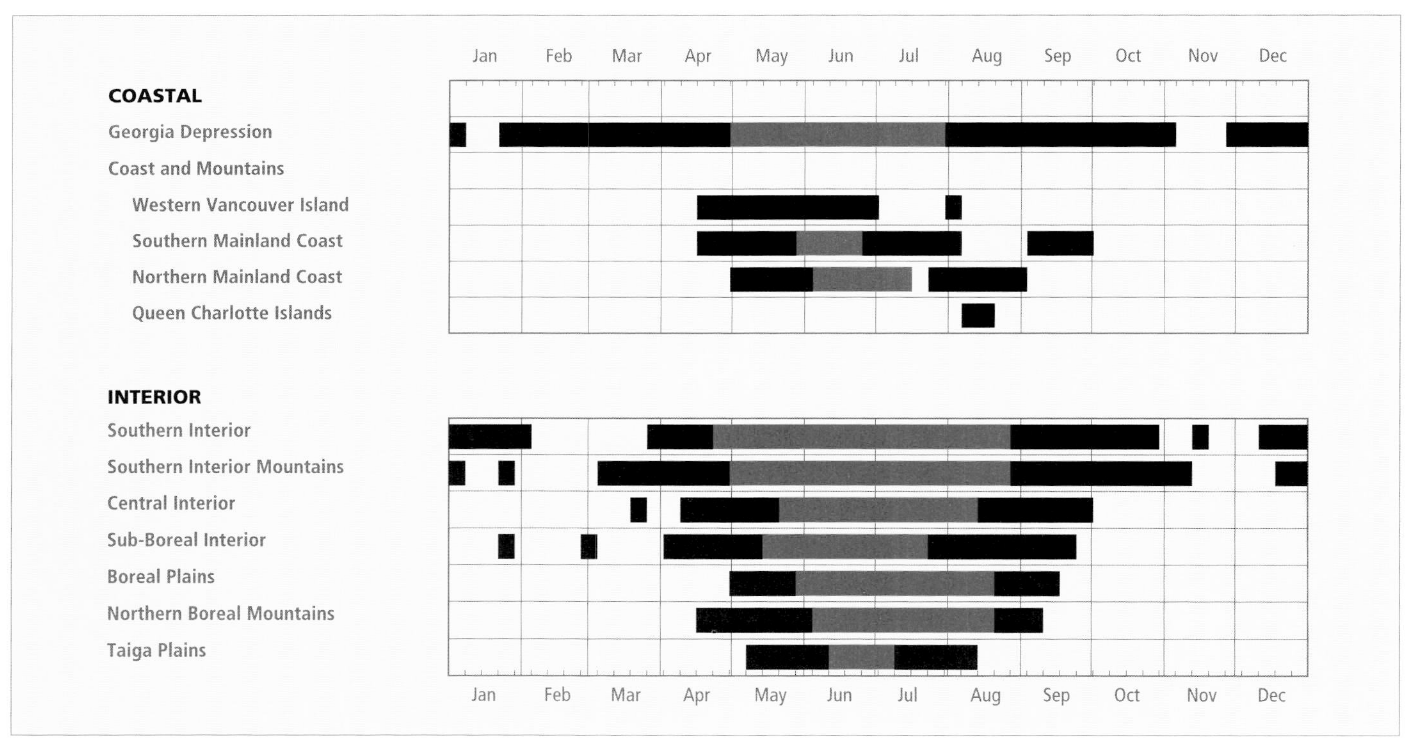

Figure 212. Annual occurrence (black) and breeding chronology (red) for the Chipping Sparrow in ecoprovinces of British Columbia. Records are shown for the week in which they occurred.

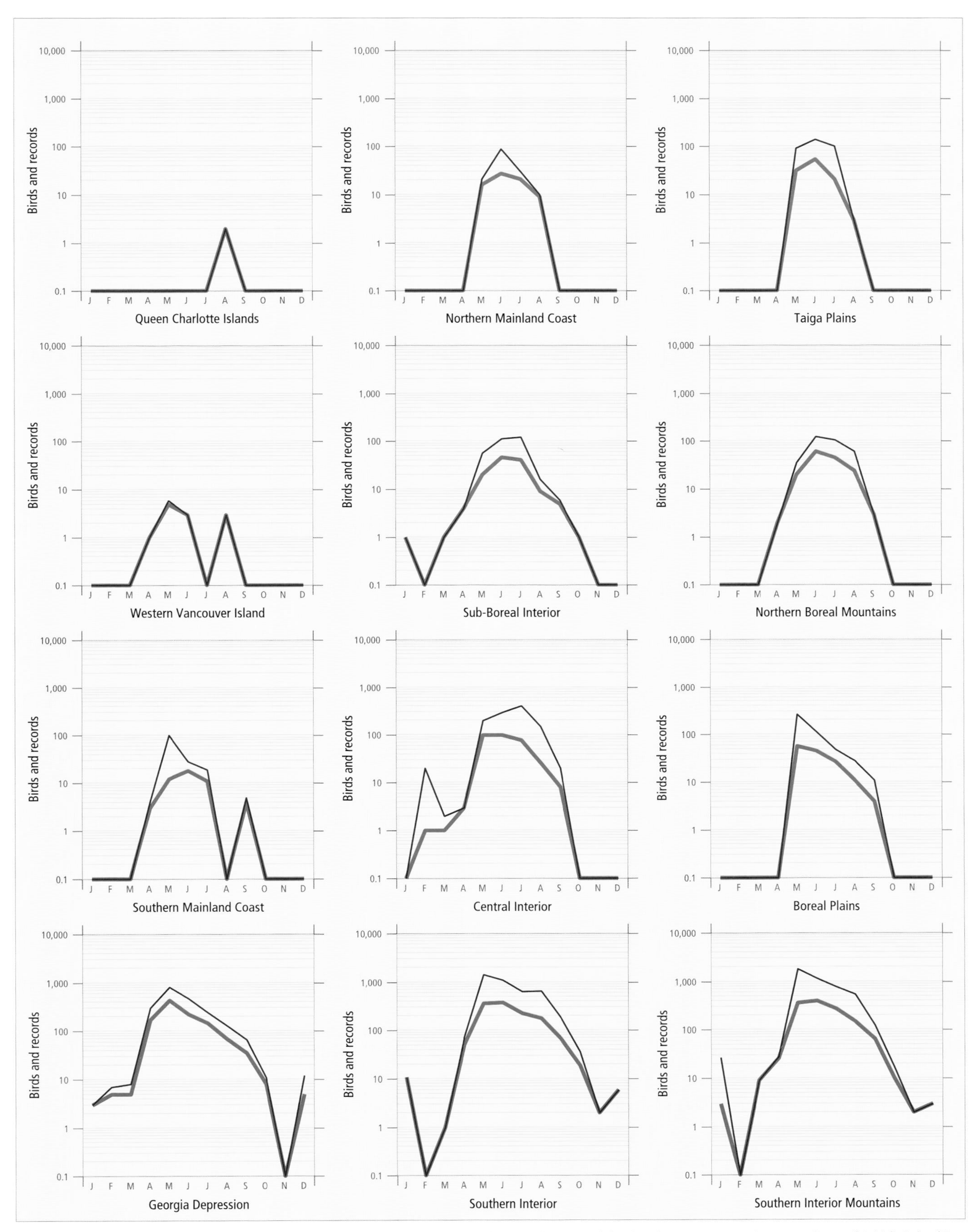

Figure 213. Fluctuations in total number of birds (purple line) and total number of records (green line) for the Chipping Sparrow in ecoprovinces of British Columbia. Christmas Bird Counts, Breeding Bird Surveys, and nest record data have been excluded.

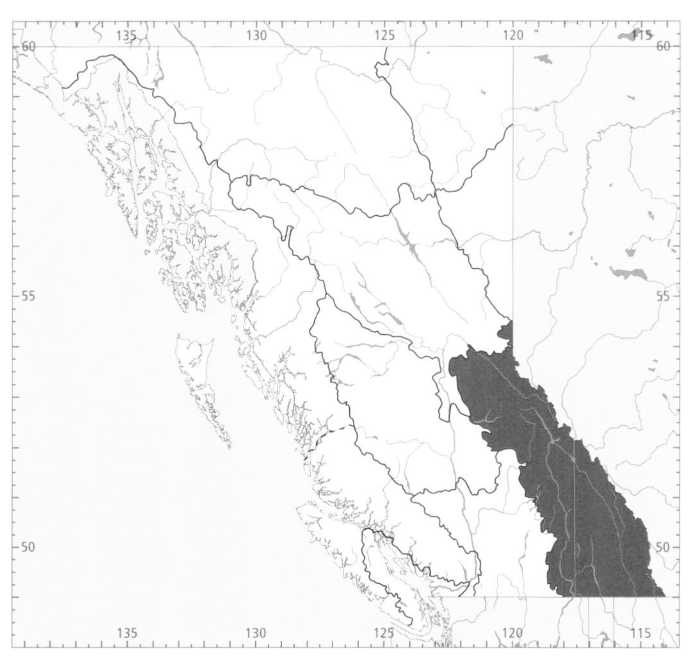

Figure 214. In British Columbia, the highest numbers for the Chipping Sparrow in summer occur in the Southern Interior Mountains Ecoprovince.

September. In the southern regions of the interior, however, there is an increase in numbers of birds observed beginning about the first week of August. All but a few stragglers have left the southern portions of the interior by late October.

On the south coast, autumn migration is evident in the gradual decline in birds observed after the first week of July. The main exodus probably occurs in late August.

Only in the Georgia Depression is there evidence that the Chipping Sparrow occurs throughout the winter months (Fig. 212). While there are a few observations of the species in the southern parts of the interior in December and January, there have been none there in February and the evidence suggests that those sparrows seen in the interior in the early winter either leave or perish.

On the coast, the Chipping Sparrow has been recorded throughout the year, but regularly only from early April to late August; in the interior, it has been recorded throughout most of the year but regularly only from early May to mid-September (Fig. 212).

Figure 215. In British Columbia, the adaptable Chipping Sparrow breeds in a variety of shrubby habitats throughout its summer range, including big sagebrush scrub (Wallachin, east of Cache Creek, 10 June 1998; R. Wayne Campbell).

BREEDING: The Chipping Sparrow undoubtedly breeds throughout its summer range in British Columbia. On the coast, however, there are no records of nesting on Western Vancouver Island or at the few places where it has been recorded in summer along the outer coast between Queen Charlotte Strait and the mouth of the Skeena River, and Portland Canal.

In northern British Columbia, nesting has been documented as far north as Sediments Creek, Atlin, Tagish Lake, Butte Lake, Liard River Hot Springs Park, Trutch, and Kwokullie Lake. Elsewhere in the vast area of northern British Columbia, there are nesting records for several localities along the Stikine River valley, and many records for the Peace Lowland and along the Alaska Highway between Tupper and Fort Nelson. The easternmost records in the south are at Wasa and Kootenay Crossing.

Studies of bird populations in the boreal forests of northern Canada have revealed populations of Chipping Sparrows in coniferous forests at a density of between 3 and 5 singing males per km^2. In the poplar and birch forests along the Alaska Highway, the density was 2 males per km^2 (Erskine 1977).

The Chipping Sparrow reaches its highest numbers in summer in the Southern Interior Mountains (Fig. 214). An analysis of Breeding Bird Surveys in British Columbia for the period 1968 through 1993 could not detect a net change in numbers in either coastal or interior routes.

On the coast, the Chipping Sparrow breeds from near sea level to about 230 m elevation; in the interior, it breeds from valley bottom elevations to about 1,950 m.

The Chipping Sparrow is an adaptable species that has benefited from human alteration of landscapes in British Columbia. It breeds in a wide variety of shrubby habitats, usually in open spaces and often associated with the edges of coniferous and deciduous woodlands, big sagebrush scrub (Fig. 215), wetlands, cutblocks, and agricultural areas. Along the coast, it nests mainly in land modified by human use (72%), of which the largest categories are suburban (27%) and cultivated farms (15%). Open forests of Douglas-fir and Garry oak (6%) are the largest category of wild land, followed by coastal western hemlock–western redcedar forest and shrubland. In the southern portions of the interior, nesting habitat is also dominated by human-use landscapes (44%), but the open forests of interior Douglas-fir or ponderosa pine, or these species in combination, provide another 33% of breeding habitat. Other habitats used in the south include birch woodland, subalpine grassland, subalpine meadow, rangeland, grass-sedge meadow, lodgepole pine stands, Engelmann spruce–subalpine fir, and trembling aspen. Several other categories of semi-open landscapes are used occasionally. Even in the Boreal Plains, about 20% of our records are from human-use landscapes, with mixed forest, white spruce, and trembling aspen woodlands combining to constitute another 50% of

Figure 216. In the Taiga Plains Ecoprovince of British Columbia, the Chipping Sparrow often nests in shrub-clad perimeters of wetlands (near Pink Mountain, 26 June 1998; R. Wayne Campbell).

habitat during the nesting season. In the Taiga Plains, nesting habitat is often associated with wetlands surrounded by patches of willow (Fig. 216), alder, and birch.

On the coast, the Chipping Sparrow has been recorded breeding from 3 May (calculated) to 26 July; in the interior, it has been recorded breeding from 27 April to 26 August (Fig. 212).

Nests: Approximately half of the nests were in coniferous trees (48%; *n* = 500; Fig. 217). Shrubs and brush tangles (38%) and deciduous trees (8%) accounted for almost all the other general nest locations. Nests were attached to or placed on branches, or placed in the fork of a branch (82%). A few nests were placed in a fallen tree, against a rock, or on the ground sheltered beneath a clump of vegetation. Nests are neat cups woven of grasses (89%; *n* = 432; Figs. 219 and 220), hair (59%), rootlets (14%), twigs (10%), stems (7%), and plant fibre (5%), with 13 other classes of material used less frequently. The lining was almost always of fine grasses, hair, and sometimes a few feathers. Materials varied in detail from one ecoprovince to another, but in every ecoprovince grasses and hair (horse, deer, or moose hair) were the 2 materials most frequently used. The heights of 460 nests ranged from ground level to 9.0 m, with 57.4% between 0.8 and 2.2 m.

Eggs: Dates for 588 clutches ranged from 27 April to 26 July, with 54% recorded between 3 and 24 June. Sizes of 353 clutches ranged from 1 to 6 eggs (1E-33, 2E-35, 3E-86, 4E-170, 5E-26, 6E-3), with 73% having 3 or 4 eggs (Fig. 220). The incubation period is 10 to 15 days (Bradley 1940; Walkinshaw 1952; Dawson and Evans 1957; Reynolds and Knapton 1984).

The length of the egg phase of breeding and the total length of the breeding season vary between the ecoprovinces. Both are longest in the Southern Interior and Southern Interior Mountains, where the egg phase is about 14 weeks and the total breeding season, from earliest egg to latest nestling to fledge, can be as long as 18 weeks. In the Georgia Depression, although spring comes earlier, the first eggs are laid about 2 weeks later than in the Southern Interior ecoprovinces. Furthermore, egg laying stops about 3 weeks earlier at the coast, resulting in a breeding season of 11 to 12 weeks. In the central and northern regions of the interior, the egg-laying phase is shortened to about 9 weeks and the total breeding season to about 11 or 12 weeks (Fig. 212).

The pattern of the egg season differs between the coast and the interior. On the coast, egg laying is relatively evenly distributed between the third week of May and mid-June, with 19% of the nests containing eggs in the third week of May. In the southern portions of the interior, the greatest concentration of nests with eggs occurred 2 weeks later, in the first week of June, and involved 22% of the seasonal total. A larger sample of nest data from the coast is required before a more detailed interpretation of the apparent regional difference and its meaning can be attempted. While we have no documented cases of double-brooding, the extended nesting season in the southern portions of the interior permits time for second nesting (Fig. 218).

Young: Dates for 396 broods ranged from 11 May to 24 August, with 51% recorded between 16 June and 7 July. Sizes of 246 broods ranged from 1 to 5 young (1Y-22, 2Y-49, 3Y-89, 4Y-76, 5Y-10), with 56% having 2 or 3 young (Fig. 219). The nestling period is 9 to 12 days (Dawson and Evans 1957; Reynolds and Knapton 1984; Middleton 1998).

Brown-headed Cowbird Parasitism: In British Columbia, 10% of 508 nests found with eggs or young were parasitized by the cowbird (Fig. 220). Parasitism was 3% (*n* = 62) on the coast and 11% (*n* = 446) in the interior. In addition, 13 instances of fledgling cowbirds attended by Chipping Sparrows have been reported. Elsewhere, the rate of parasitism varies greatly from region to region. In Quebec, a 1961 study found a rate of 12% (*n* = 138), similar to ours (Terrill 1961). Friedmann

Figure 217. Although the Chipping Sparrow nests in a wide variety of trees and shrubs, about half of all nests in British Columbia were found in coniferous trees (10 km west of Bridesville, 1 June 1998; R. Wayne Campbell).

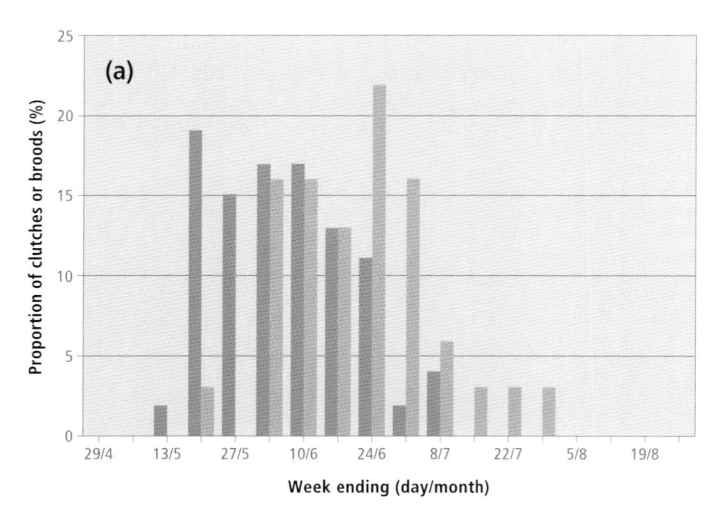

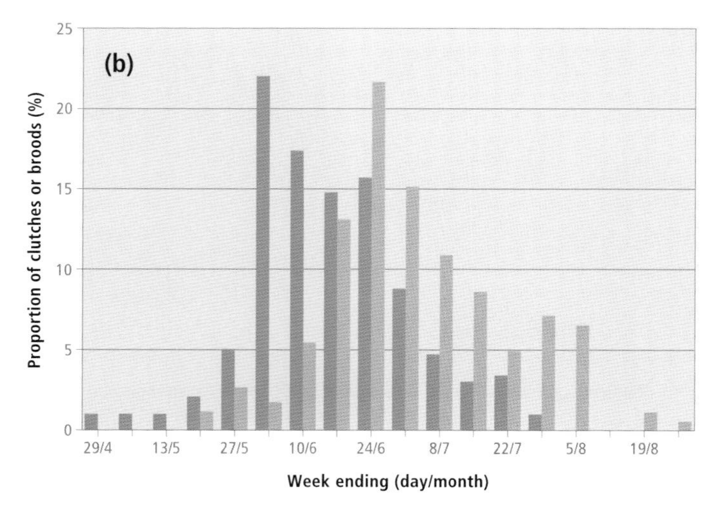

Figure 218. Weekly distribution of the proportion of clutches (dark bars) and broods (light bars) for the Chipping Sparrow in the (a) Georgia Depression Ecoprovince (clutches: $n = 47$; broods: $n = 32$) and (b) Southern Interior and Southern Interior Mountains ecoprovinces combined (clutches: $n = 237$; broods: $n = 185$). Note the difference between the 2 regions in length and pattern of the breeding season.

(1963), however, summarized the information available at that time thus: "Not only is the Chipping Sparrow a very frequent victim – in the total number of known instances of cowbird parasitism – but it seems to be one of the main fosterers in almost every locality." The response of the Chipping Sparrow to cowbird parasitism needs thorough investigation (Middleton 1998), and all data should be interpreted with caution, given the biases inherent in observational studies (Rothstein 1975; Graham 1988).

Two studies have included the Chipping Sparrow in detailed examinations of cowbird parasitism as a biological strategy. A study of host response to the presence of cowbird eggs in their nests used the introduction of either real or artificial cowbird eggs into the nests of wild nesting hosts (Rothstein 1975). The technique consisted of removing 1 of the host eggs and replacing it with an artificial cowbird egg. The Chipping Sparrow deserted 15.8% of the treated nests ($n = 19$).

More recently, in Ontario, the response of 5 different passerine hosts to naturally occurring parasitism by the Brown-headed Cowbird was studied (Graham 1988). This research included 161 Chipping Sparrow nests found while laying of the clutch was in progress. In this random sample, 62% of nests were parasitized. Subsequently, 52% of the parasitized nests were deserted. The number of cowbird eggs in the nest did not alter this response, nor did the number of observational visits made by the researcher. It would appear, therefore, that the desertion of the nest after it had been interfered with by the cowbird resulted more from the invasion by the cowbird than from the presence of the foreign egg. This provides an explanation of the different results obtained by the 2 researchers.

Nest Success: Of 165 nests found with eggs and followed to a known fate, 35% produced at least 1 fledgling. The success rate was 31% ($n = 32$) on the coast and 36% ($n = 133$) in the interior.

Figure 219. Chipping Sparrow nest with 3 nestlings; the nestling period ranges from 8 to 12 days (Castlegar, 23 June 1994; Linda M. Van Damme).

Figure 220. The Chipping Sparrow is a frequent host for the Brown-headed Cowbird in North America. In British Columbia, 10% of 508 nests found with eggs or young were parasitized by the cowbird (Victoria, 10 June 1994; R. Wayne Campbell).

REMARKS: Three subspecies of Chipping Sparrow have been described over its entire range in North America (American Ornithologists' Union 1957). Only *S. p. boreophila* occurs in British Columbia. Recently, Zink and Dittman (1993) suggested, through mitochondrial DNA analysis, that there was no detectable variation in the 3 named subspecies.

For additional information on the species and on the field characteristics that enable the observer to identify the Chipping Sparrow, the Clay-colored Sparrow, and the Brewer's Sparrow in all plumages, see Rising (1996) and Middleton (1998).

NOTEWORTHY RECORDS

Spring: Coastal – Rocky Point (Victoria) 13 May 1995-24; Victoria 10 May 1941-20, 14 May 1954-2 eggs and 2 nestlings; Tofino 21 Apr 1982-1; Long Beach 2 May 1981-2; Ross Lake 11 May 1974-28, 25 May 1976-50 feeding in Ponderosa Meadows; Cultus Lake 31 May 1942-3; Cheam Slough 27 Apr 1982-1; Harrison Hot Springs 4 May 1977-1; Hope 1 May 1958-2; Squamish 20 Apr 1962-1, early arrival (Boggs and Boggs 1962c); Pemberton 25 May 1997-1; Campbell River 12 Mar 1963-2 singing; Pultney Point 12 May 1976-1, only spring sighting; Port Neville Inlet 16 May 1975-1; Hazelton 6 May 1976-1, 18 May 1963-1. **Interior** – Chopaka 12 Apr 1969-5; Osoyoos Lake 4 Apr 1986-1; Osoyoos 28 May 1974-54; Grand Forks 1 May 1983-1; Trail 19 Apr 1939-1; Salmo 7 Mar 1979-1; Penticton 18 May 1959-100 on road beside garbage dump; Castlegar 15 May 1974-4 nestlings; Boston Bar to Lytton 24 May 1964-30 on count; Nakusp 20 May 1976-50; Botanie Creek valley 3 May 1980-9; Vernon 27 Mar 1969-several, first of the year; Invermere 14 Apr 1978-1; Tunkwa Lake 12 May 1968-75; Gold Bridge 20 and 21 May 1986-100; McQueen Lake 11 May 1973-2 eggs and 2 nestlings; Revelstoke 23 Mar 1986-1; nw Lillooet 20 May 1968-12; Leanchoil 10 May 1977-40; McLeod Meadows 16 May 1976-40 with 55 Brown-headed Cowbirds and 2 European Starlings; Horse Lake 22 Mar 1976-2 along road; Tatlayoko Lake 25 Apr 1992-1; Kleena Kleene 10 May 1947-1 (Paul 1959); Lac la Hache 30 Apr 1943-1 (USNM 425354); Riske Creek 7 May 1977-1 at Becher's Prairie (RBCM 18182), 26 May 1972-4 eggs; Puntchesakut Creek 8 May 1944-2 (Munro 1947); McBride 6 May 1968-5; 24 km s Prince George 2 Mar 1985-1; Hudson Bay Mountain 15 May 1975-1; Babine 15 May 1889-3 eggs; Mugaha Creek 29 Apr 1995-1; Tupper Creek 8 May 1938-1, several small flocks up to 20 birds (Cowan 1939); Taylor 18 May 1985-11; e Fort St. John 6 May 1985-1, spring arrival; nr Fort Nelson 7 May 1978-1; Liard Hot Springs 16 May 1975-6 (Reid 1975); Atlin 17 Apr 1930-1 (RBCM 5702).

Summer: Coastal – Sooke 8 Jul 1956-3 eggs; Bamfield 4 Jun 1970-1; Rosewall Creek 27 Jul 1976-12; Little Mountain (Hope) 17 Jun 1962-3 nestlings; Courtenay 26 Jul 1940-young; Upper Campbell Lake 13 Jul 1975-4; Triangle Island 4 Jun 1976-1 feeding in dry seaweed along beach; Rose Harbour 17 Aug 1946-1; Kunghit Island 11 Aug 1946-1; Bella Coola 1 Jun 1968-4 eggs, 3 Jun 1933-1 (MCZ 285193); Hagensborg 13 Jun 1938-1 (NMC 28737); Stuie 21 Jun 1938-2; Kimsquit River 24 Jun 1985-5; 10 km e Kimsquit-1 young barely fledged; Kitimat River 7 Jul 1975-3; Lakelse Lake 19 Aug 1977-1 (Shepard 1977); Green Island 29 Aug 1977-1; Terrace 10 Jun 1975-4 eggs, 23 Jun 1975-3 nestlings; Kitsault 17 to 19 Jun 1980-27; Stewart 17 Jun 1975-1; Meziadin Lake 8 Jun 1976-4; Dokdaon Creek (Stikine River) 20 Jul 1919-1 (MVZ 39948); Pleasant Camp (Haines Highway) 2 Jun 1981-1; Mile 46.5 Haines Highway 18 Jun 1972-2. **Interior** – Flathead 7 Jul 1952-3; 8 km n Grand Forks 1 Aug 1982-10; Apex Mountain (Penticton) 10 Aug 1991-50 (Siddle 1992a); Jaffrey 22 Jun 1968-52; Penticton 26 Aug 1950-4 nestlings, all dead; St. Mary's Alpine Park 14 Aug 1977-3 young; Wasa 17 Aug 1976-35 including immatures; Lightning Lake area 11 Aug 1979-20; Botanie Creek valley 5 Jul 1964-14; Nakusp 20 Jul 1978-10; Zincton 14 Jun 1980-25; 6 km n Enderby 25 Jul 1969-3 eggs; Bridge Lake 4 Jun 1976-2; Emerald Lake 3 Jun 1977-6; 13 km n Clearwater 3 Jun 1935-1 (MCZ 285212); Anahim Lake 4 Jul 1948-20; Mount Robson Park 12 Aug 1973-25; Wistaria 4 Jun 1977-5; Nulki Lake 16 Jun 1945-5 eggs; Cluculz Lake 19 Jul 1986-3 nestlings; 16 km w Prince George 17 Jun 1972-3 nestlings; Francois Lake 4 Aug 1944-50 (Munro 1947); Stellako River 13 Jun 1951-15 in 3.2 km along s side of river; Houston to Topley 26 Jul 1944-200 seen along road between the towns (Munro 1947); Morfee Lake 6 Jul 1996-4 nestlings; sw Dawson Creek 8 Jul 1993-female flushed from 3 eggs; Bear Mountain (Dawson Creek) 11 Jun 1993-4 newly hatched nestlings; Rose Prairie 19 Aug 1975-female feeding 2 fledglings; Tatlatui Lake 1 Jul 1976-2; Dease Lake 3 Jun 1962-2 (NMC 50298-99); Parker Lake area 7 Jul 1978-3 eggs, 12 Aug 1979-1; w Fireside 19 Jun 1981-18; Sediments Creek 21 Jun 1993-4 very small nestlings; Smart River 11 Jun 1978-2; Petitot River 20 Jun 1982-6.

Breeding Bird Surveys: Coastal – Recorded from 22 of 27 routes and on 43% of all surveys. Maxima: Alberni 27 Jun 1979-23; Chemainus 25 Jun 1969-18; Nanaimo River 13 Jun 1976-15. **Interior** – Recorded from 72 of 73 routes and on 98% of all surveys. Maxima: Kuskonook 18 Jun 1968-82; Oliver 21 Jun 1994-67; Williams Lake 6 Jun 1975-64.

Autumn: Interior – Summit Lake Pass 9 to 12 Sep 1972-1, a few also seen nr Summit Lake and Tetsa River (Griffith 1973); 116 km n Fort Nelson 2 Sep 1943-1 (NMC 29581); Fort St. John 11 Sep 1988-7 immatures, last record of autumn; Gagnon Creek 9 Sep 1995-1 banded, 17 Sep 1994-1; Tumbler Ridge 8 Sep 1992-2; 24 km s Prince George 18 Nov 1982-2; Williams Lake 12 Sep 1978-5; Tatlayoko Lake 29 Sep 1991-1; Mount Revelstoke 10 Sep 1982-8; Revelstoke 7 Nov 1986-1; Scotch Creek 2 Oct 1964-6; Okanagan Landing 14 Nov 1929-1; Green Lake (Penticton) 15 Sep 1990-15. **Coastal** – Garibaldi Lake 5 Sep 1987-1; Skagit River valley 29 Sep 1974-2; Langley 8 Oct 1968-4; Somenos Lake 5 Sep 1971-8.

Winter: Interior – Prince George 28 Jan 1986-1 on feeder for several days; Revelstoke 20 Dec 1984-1; Kamloops 15 Dec 1994 to 29 Jan 1995-1 (Bowling 1995d); Vernon 16 Dec 1986-1 (Cannings et al. 1987), 13 Jan 1992-1 (Siddle 1992b), 20 Jan 1986-1 (Cannings et al. 1987); South Slocan 25 Jan 1983-1, present since early December. **Coastal** – Chilliwack 4 Feb 1927-1; South Vancouver 7 Jan 1924-1 (RBCM 7534); Victoria 8 Feb 1969-2 (Crowell and Nehls 1969b); Saanich (Rithet's Bog) 24 Jan 1979-1.

Christmas Bird Counts: Interior – Not recorded. **Coastal** – Recorded from 4 of 33 localities and on less than 1% of all counts. Maxima: North Saanich 31 Dec 1960-1; Nanaimo 31 Dec 1978-1; Ladner 19 Dec 1976-2; Chilliwack 29 Dec 1973-7.

Clay-colored Sparrow

Spizella pallida (Swainson)

CCSP

RANGE: Breeds from west-central and southern Mackenzie, northeastern British Columbia, northern Alberta, northwestern and central Saskatchewan, northern Manitoba, central Ontario, and southwestern Quebec south in the west into Washington and the Rocky Mountain states to Colorado; east to include parts of Kansas, Iowa, Wisconsin, and Michigan, and, sporadically, into New York. Winters from southern Baja California, northern Sonora, and central Texas south through the Mexican highlands to western Guatemala.

STATUS: On the coast, *very rare* spring and autumn vagrant, *casual* in summer, and *accidental* in winter in the Georgia Depression Ecoprovince; in the Coast and Mountains Ecoprovince, *casual* autumn vagrant on Western Vancouver Island and *accidental* spring vagrant on the Northern Mainland Coast.

In the interior, *uncommon* migrant and local summer visitant in the Southern Interior, Southern Interior Mountains, Central Interior, and Sub-Boreal Interior ecoprovinces; *accidental* in winter in the Southern Interior Ecoprovince; *fairly common* migrant and summer visitant in the Boreal Plains Ecoprovince; locally *rare* in the Taiga Plains Ecoprovince.

Breeds.

CHANGE IN STATUS: In the mid-1940s, the Clay-colored Sparrow (Fig. 221) was known to be a regular summer visitant in the Peace Lowland of the Boreal Plains, and to occur irregularly west of the Rocky Mountains at widely scattered locations throughout the southern interior dating back to 1901.

Figure 221. The Clay-colored Sparrow breeds locally throughout southern, central, and northeastern interior regions of British Columbia (Anthony Mercieca).

Prior to 1947, the only nesting reported west of the Rocky Mountains occurred at Trail in the Columbia River valley in 1902 and at Sinkut River in the Nechako River valley in 1946.

Over the past 50 years, the species has steadily expanded its distribution throughout the eastern and southern portions of the province. The expansion has been especially noticeable in the Southern Interior Mountains and in the southern regions of the Sub-Boreal Interior. In the Okanagan valley of the Southern Interior, there were 5 sightings between 1925 and 1946. Since 1958, there have been few years in which the Clay-colored Sparrow has not been sighted, and the first

Figure 222. During autumn migration in British Columbia, the Clay-colored Sparrow often frequents overgrown fields and pastures where forb and grass seeds are abundant (near Fort Steele, 23 September 1996; R. Wayne Campbell).

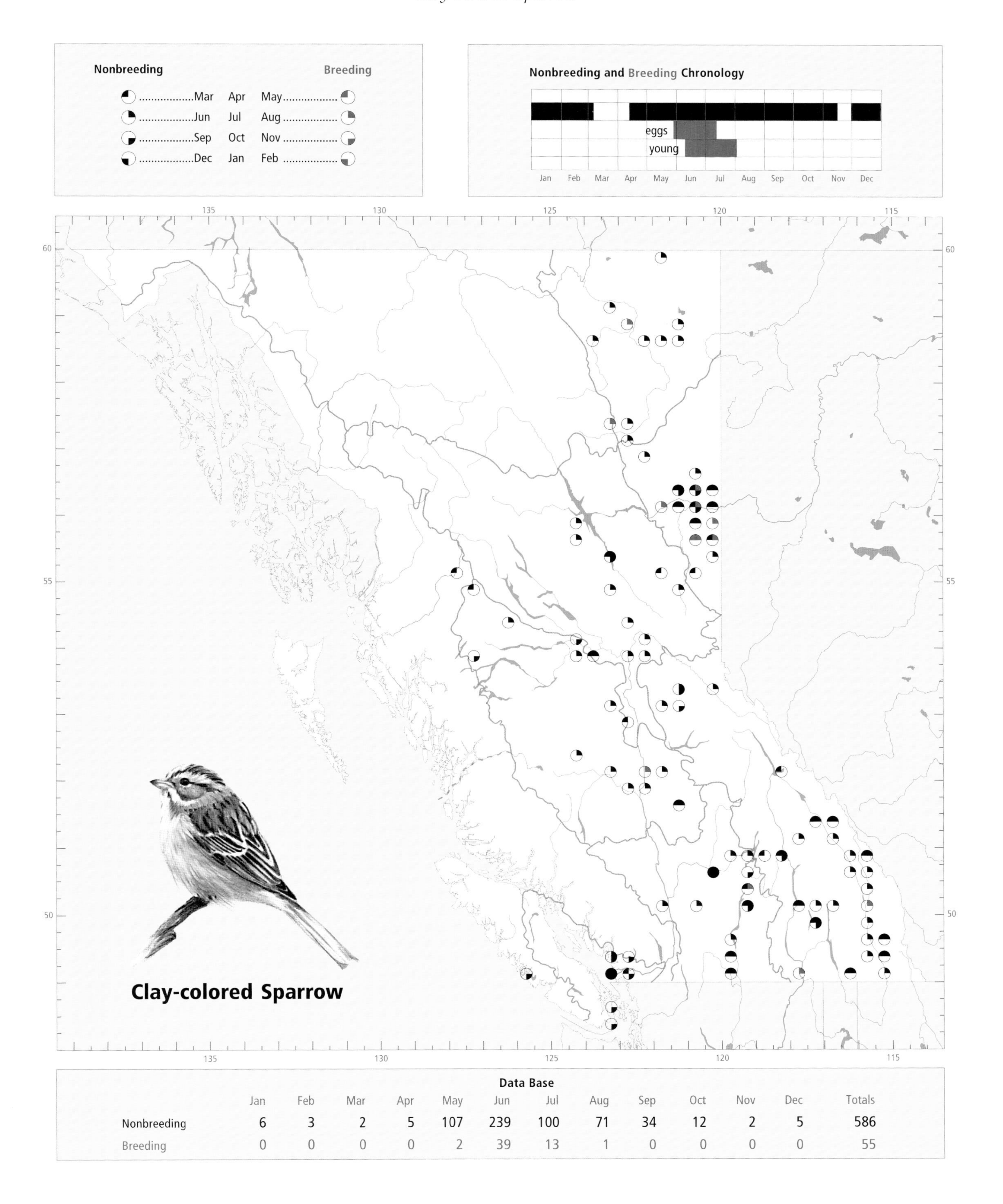

Data Base

	Jan	Feb	Mar	Apr	May	Jun	Jul	Aug	Sep	Oct	Nov	Dec	Totals
Nonbreeding	6	3	2	5	107	239	100	71	34	12	2	5	586
Breeding	0	0	0	0	2	39	13	1	0	0	0	0	55

recorded nesting there was in 1973 (Cannings et al. 1987). At Kamloops, in the Thompson River valley of the same ecoprovince, there have been 48 recorded occurrences up to 1997 (R.R. Howie pers. comm.).

Much the same pattern was repeated in the Rocky Mountain Trench, but the regular presence of the Clay-colored Sparrow did not begin there until 1965. The only recent nesting records in the east and west Kootenay have been at Canal Flats in 1970 and Creston in 1994. In the Cariboo and Chilcotin areas, there are 17 records over 9 years, scattered between 1901 and 1977. Nesting occurred near Williams Lake in 1959 and has been reported annually since 1982. Reports from the Prince George area indicate that the species has occurred regularly in the 1990s. There have been no reports yet of nesting in the Sub-Boreal Interior, but the presence of this sparrow on territory suggests that breeding occurs in the vicinity of Prince George and Mackenzie.

On the coast, the Clay-colored Sparrow was first reported in the Fraser Lowland of the Georgia Depression in 1982. Since then, it has been found near Vancouver, Pitt Meadows, and Chilliwack and on southern and western Vancouver Island.

The extension of the Clay-colored Sparrow's range in British Columbia was probably facilitated in part by forest removal, agricultural land-clearing activities, and transmission corridor clearings, all of which tend to increase the availability of suitable breeding habitat.

NONBREEDING: The Clay-colored Sparrow is widely distributed in northeastern portions of the province and locally distributed in southern regions. It occurs during migration east of the Pacific and Cascade ranges and west of the Rocky Mountains, from the international boundary north to Smithers, Fort St. James, Prince George, and Mackenzie and north throughout the Boreal Plains and Taiga Plains to the Petitot River, near the Northwest Territories boundary.

On the south coast, it is scarce during migration, with most records coming from the Fraser River delta, Greater Vancouver, and southern and west-central Vancouver Island.

In the interior, the Clay-colored Sparrow occurs at elevations between 280 and 700 m during spring migration. No elevations have been reported for the post-breeding autumn period.

During spring migration, the Clay-colored Sparrow frequents a variety of open landscapes, such as open hillsides, clearings, gravel pits, beach edges, dykes, regenerating clearcuts, and lawns where the plant cover is dominated by grasses, dead forbs, or low shrubs. During the autumn migration period, most observations of this sparrow have been made in shrub patches bordering fields, overgrown fields, and pastures (Fig. 222); at the margins of lakes, ponds, and marshes; or in hedgerows.

In the Southern Interior and Southern Interior Mountains, a few early migrants arrive in late April (Figs. 223 and 224). By mid-May, spring migration is evident throughout the entire southern and central interior of the province, including the Sub-Boreal Interior. Further north in the Boreal Plains, especially the Peace Lowland, the first migrants usually arrive between 13 and 17 May (C. Siddle pers. comm.; Figs. 223 and 224).

The beginning of autumn migration is difficult to detect. There are no records after early August in the Taiga Plains

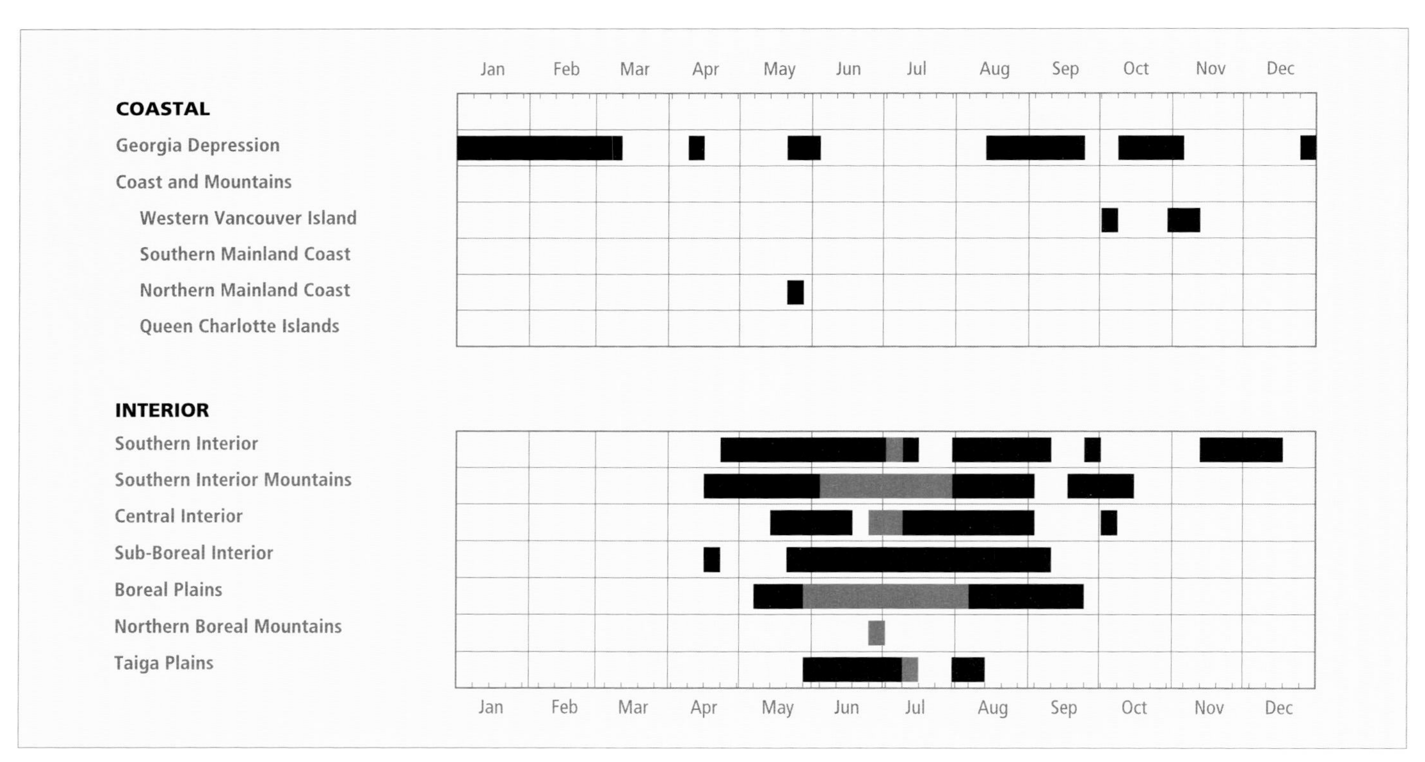

Figure 223. Annual occurrence (black) and breeding chronology (red) for the Clay-colored Sparrow in ecoprovinces of British Columbia. Records are shown for the week in which they occurred.

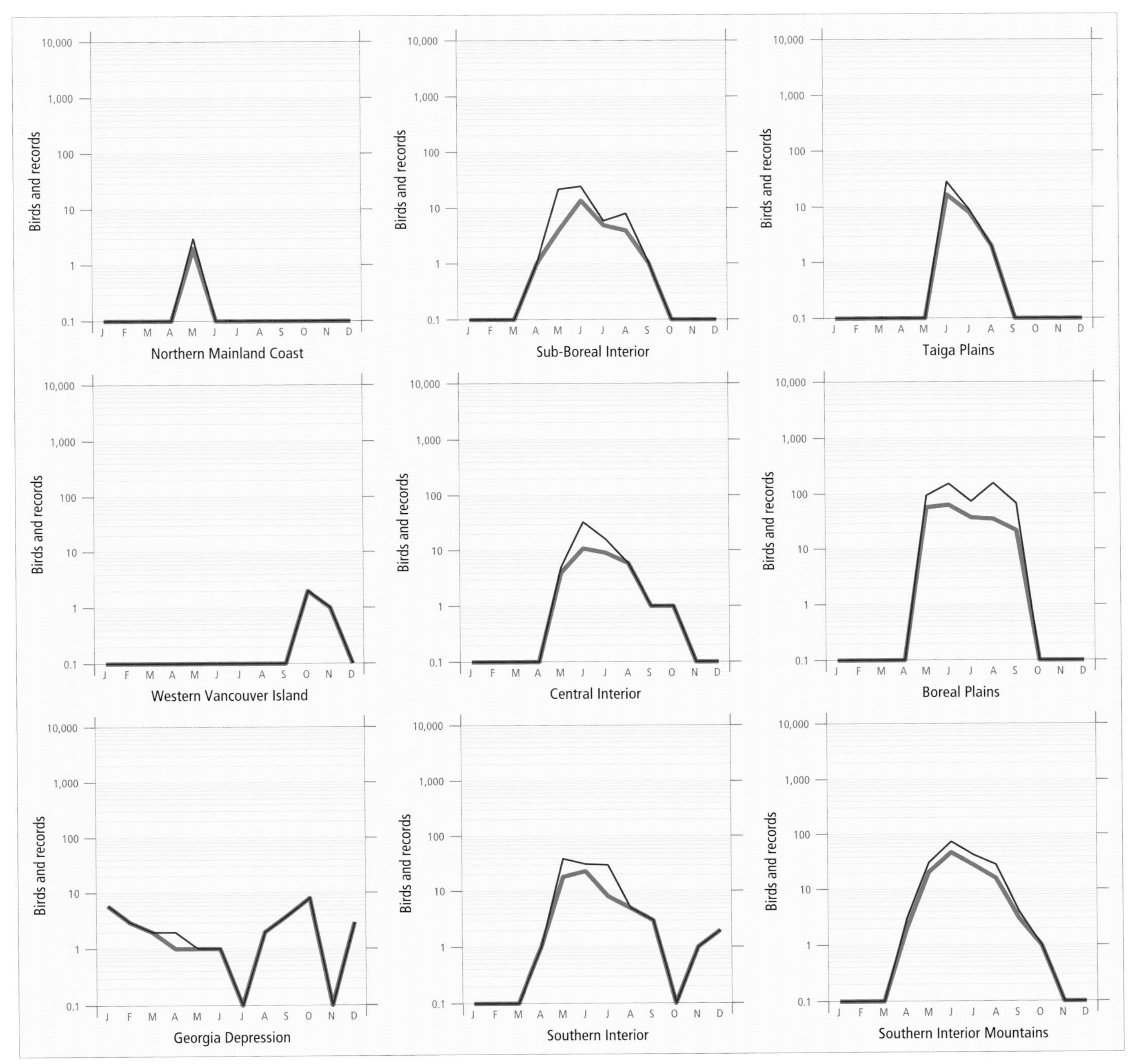

Figure 224. Fluctuations in total number of birds (purple line) and total number of records (green line) for the Clay-colored Sparrow in ecoprovinces of British Columbia. Christmas Bird Counts, Breeding Bird Surveys, and nest record data have been excluded.

and none after early September in the Sub-Boreal Interior. In the Peace Lowland of the Boreal Plains, the number of migrants rises steadily from the first week of August, peaks at the end of August or the first week of September, and then declines sharply to a latest record on 22 September. Except for some stragglers, most Clay-colored Sparrows have left the Boreal Plains by the third week of September; most birds have left the northern regions of the province west of the Rocky Mountains by mid-August, and southern portions of the interior regions by the first week of September (Figs. 223 and 224).

There are 2 winter records for the Clay-colored Sparrow in British Columbia.

On the coast, the Clay-colored Sparrow has been recorded irregularly throughout the year, with autumn having the most records; in the interior, it has been recorded regularly from 28 April to 28 September (Fig. 223).

BREEDING: The Clay-colored Sparrow breeds locally throughout the southern and central interior regions of the province west of the Rocky Mountains, from Richter Pass, Creston, and

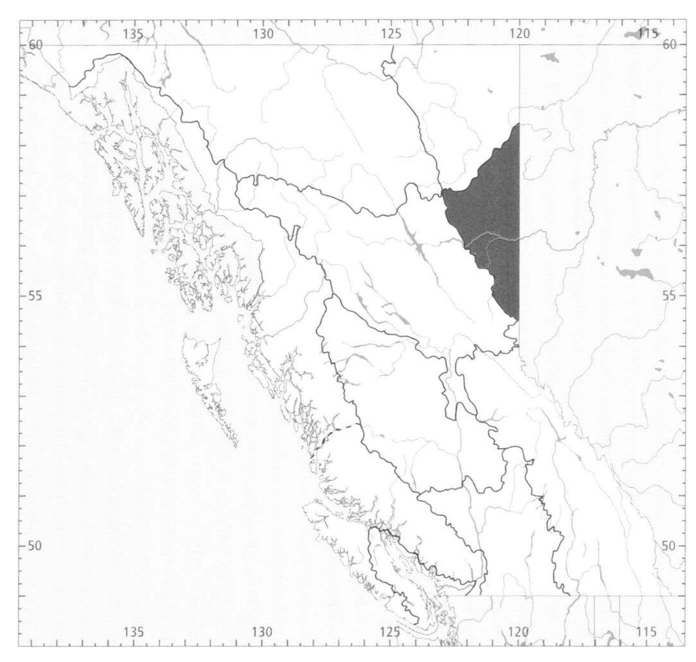

Figure 225. In British Columbia, the highest numbers for the Clay-colored Sparrow in summer occur in the Boreal Plains Ecoprovince.

Canal Flats in the south likely north to the vicinity of the Sinkut River in the Nechako valley in the central interior. It also breeds regularly in northeastern British Columbia from Tupper Creek north to Fort Nelson. Singing males, apparently on territory, have been reported from Quesnel, Prince George, Vanderhoof, the Bulkley River valley, and Mackenzie, which suggests that nesting may occur in these areas, although Munro (1949) comments on the presence of males without apparent mates.

In summer, the Clay-colored Sparrow reaches its highest numbers in the Peace Lowland of the Boreal Plains (Fig. 225). An analysis of Breeding Bird Surveys in British Columbia for the period 1968 through 1993 could not detect a net change in numbers on interior routes. An analysis of Breeding Bird Surveys across Canada between 1966 and 1996 reveals a slow decline in mean number of birds at an average annual rate of 1.2% ($P < 0.01$). Approximately the same rate of decline (1.1%, $P < 0.01$) has been observed over its North American breeding range (Sauer et al. 1997).

The Clay-colored Sparrow has been reported nesting at elevations between 480 and 1,160 m. It inhabits the intermountain valleys where open forests of Douglas-fir and ponderosa pine in the south, and lodgepole pine or spruce further north, occur. Parkland terrain, where groves of trembling aspen, birch, and willow are interspersed with open patches of shrubs, forbs, and grasses, is used (Fig. 226). In portions of the Southern Interior, the Clay-colored Sparrow inhabits grasslands and tends to associate with thickets of rose or extensive patches of sagebrush. Shrub thickets selected by this bird feature snowberry, hawthorn, wild rose, wolf-willow, soopolallie, antelope-brush, and saskatoon. Low shrubbery along fencelines, dense roadside thickets, and patches of low

Figure 226. In the Peace River region of northeastern British Columbia, the Clay-colored Sparrow breeds in parkland habitats where patches of dense shrubs, especially saskatoon and prickly rose, dot the landscape (near Fort St. John, 23 June 1996; R. Wayne Campbell).

Figure 227. In British Columbia, disturbed areas under powerline rights-of-way, with patches of dense shrubs, are frequently used as breeding sites by the Clay-colored Sparrow (near Prince George, 14 June 1997; R. Wayne Campbell).

Figure 228. Nest and eggs of the Clay-colored Sparrow in rose shrub (Kilpoola Lake, 18 June 1993; R. Wayne Campbell).

shrubs under powerline rights-of-way (Fig. 227) are also favoured nesting habitats. In the southern interior Douglas-fir forests of the southern Rocky Mountain Trench near Wasa, Schwab (1979) found a strong correlation between the presence of the Clay-colored Sparrow and shrubby seral stages of forest regeneration.

In the Nulki Lake region of British Columbia, Munro (1949) found a most favoured habitat that was "a tract of slash grown up with a shrubbery chiefly of willow and black twin-berry, and bordering a hay meadow." That area of about 4 ha contained 5 territories.

In the Peace Lowland, habitat classes for reported nests were mainly trembling aspen second-growth and shrubland, followed by cultivated farmland and clearcuts.

The Clay-colored Sparrow has been recorded breeding in British Columbia from 29 May (calculated) to 2 August (Fig. 223).

Nests: Nests were most frequently found in shrubs or low vegetation (64%; $n = 34$), on the ground, often in a grass clump at the base of a shrub or small tree (23%), or in low deciduous bushes (6%; Fig. 228). The heights of 40 nests ranged from ground level to 1.8 m, with 58% between 0.1 and 0.5 m. Grasses were the most frequently used nesting material and were found in all nests ($n = 30$; Fig. 228), followed by rootlets (40%), mammalian hair (37%), and twigs and stems (23%).

Eggs: Dates for 34 clutches ranged from 31 May to 12 July, with 53% recorded between 8 and 28 June. Calculated dates indicate that nests may hold eggs as early as 29 May. Sizes of 25 clutches ranged from 1 to 6 eggs (1E-1, 2E-1, 3E-7, 4E-14, 5E-1, 6E-1), with 56% having 4 eggs (Fig. 228). The incubation period is 10 to 14 days (Knapton 1994).

Young: Dates for 24 broods ranged from 10 June to 2 August, with 58% recorded between 15 June and 8 July. Sizes of 17 broods ranged from 1 to 4 young (1Y-2, 2Y-1, 3Y-6, 4Y-8), with 82% having 3 or 4 young. We have no certain evidence of renesting following a successful first brood, but observation of a pair at McQueen Lake in 1997 strongly suggested such an event (J. Bowling pers. comm.). In southern Manitoba, pairs that successfully raise young do not renest in some years; in other years, as many as half of the pairs that fledge young before 23 June attempt to raise a second brood. Pairs that fledged young after late June did not attempt a second brood (Knapton 1994). The nestling period has been reported as 8 or 9 days (Ehrlich et al. 1988) and 7 to 9 days (Baicich and Harrison 1997).

Brown-headed Cowbird Parasitism: In British Columbia, 2 of 58 nests recorded with eggs or young were parasitized by the cowbird; no cowbird fledglings were noted with sparrow foster parents. Friedmann et al. (1977) report that Clay-colored Sparrows nesting on the Great Plains of North

Dakota and in the provinces of Alberta, Saskatchewan, and Manitoba were heavily parasitized by the Brown-headed Cowbird: 39.4% (n = 33) in North Dakota and 23% (n = 275) on the Canadian prairies. Studies in southwestern Manitoba (Knapton 1979; Friedmann and Kiff 1985) found that 89% of nests (n = 94) were parasitized: 51 contained a single cowbird egg, 26 had 2 eggs, 5 had 3 eggs, and 2 had 4 eggs. In Alberta, 1 Clay-colored Sparrow population rejected or did not incubate the parasitic eggs (Salt 1966). In the Manitoba study, however, 60% (n = 53) accepted the foreign eggs and 40% rejected them by deserting the nest (Hill and Sealy 1994). There appeared to be differences between local populations in this behaviour.

Nest Success: Only 1 nest was found with eggs and followed to a known fate; it was unsuccessful. Knapton (1979) found that of 517 eggs laid, 285 young hatched and 221 young fledged in Manitoba. The average number of fledglings per successful nest was 1.7.

REMARKS: Because the Clay-colored Sparrow is locally distributed throughout the southern interior of British Columbia, significant patches of breeding habitat should be secured and cattle grazing in some areas should be discouraged from May through August. In addition, spring and summer burning should be delayed until autumn.

For a review of the distribution and biology of the Clay-colored Sparrow, see Buech (1980, 1982) and Knapton (1994).

NOTEWORTHY RECORDS

Spring: Coastal – Delta 5 Mar 1983-1, latest date for overwintering bird; Serpentine River 14 Apr 1984-2; Sea Island 26 May 1995-1 (Elliott and Gardner 1997); New Hazelton 21 and 24 May 1993-1 (Siddle and Bowling 1993). **Interior** – Creston 29 May 1993-1 (Siddle and Bowling 1993); nr Wye Lake 6 May 1969-1; White Lake (Okanagan Falls) 15 May 1993-1 (Siddle and Bowling 1993); Bull Mountain 26 May 1995-3 singing males; Deep Lake (Okanagan valley) 4 May 1985-2; New Denver 4 Apr 1978-1, 13 May 1994-1; Nakusp 25 May 1975-1, 16 May 1993-1 (Siddle and Bowling 1993); Vernon 30 May 1979-13; Goose Lake (Vernon) 31 May 1958-1; Kamloops 20 May 1991-4 (Siddle 1991c); Revelstoke 16 Apr 1989-2, 28 Apr 1989-1; Golden 24 May 1993-1 (Siddle and Bowling 1993); 6.5 km se 100 Mile House 17 May 1985-2, first seen in area; Quesnel 31 May 1994-11 singing males; Sinkut Lake 31 May 1945-1; Smithers 27 May 1993-1 (Siddle and Bowling 1993); Mackenzie 18 Apr 1996-1; Mugaha Creek 24 May 1997-6 on survey; 55 km s Dawson Creek (Sunset Hill) 31 May 1994-3 eggs, adult on nest; Dawson Creek 11 May 1993-1; Tumbler Ridge 25 May 1997-2; Fort St. John 13 May 1988-1, spring arrival, 24 May 1986-5 nr sewage lagoons; Boundary Lake (Goodlow) 16 May 1982-1; Goodlow 25 May 1975-1 along Jackfish Rd, 29 May 1983-1.

Summer: Coastal – Beach Grove 16 Aug 1984-1 on dyke path nr sewage pond; Fraser River 26 and 27 Aug 1997-1 on North Arm jetty; Vancouver 2 Jun 1983-1 on Musqueam Indian Reserve; North Vancouver 21 Aug 1982-1 (BC Photo 807). **Interior** – nr Nighthawk 5 Jun 1982-2; Newgate 25 Jun 1997-2 singing males, Jul 1997-fledglings; Creston 5 Jun 1994-4 adults, 22 Jul 1994-2 adults with a recent fledgling; Grasmere 2 Jul 1979-2; Jaffray 22 Jun 1968-3; Fernie 17 Jun 1969-2 at airport (Siddle 1980); Wasa 2 Jul 1979-2; Premier Ridge 7 Jun 1997-1 singing male at 1,160 m; Canal Flats 27 Jul 1970-1 young; Lytton 20 Jun 1987-1 at 250 m elevation; Goose Lake (Vernon) 7 Jul 1973-4 eggs; Okanagan Lake 1 Jul 1979-13 at head of lake along 1.5 km of road on e side of lake; Merritt 20 Jul 1997-4; Otter Lake (Armstrong) 29 Jun 1975-3 fledglings with adults; Kamloops 9 Jun 1990-3; Emerald Lake 5 Jun 1975-1 (Wade 1977); Ottertail 16 Jun 1977-1, at the horse pasture (Wade 1977); 6 km e Otto Creek 22 Aug 1975-1 (Wade 1977); 15 km w Williams Lake 10 Jun 1989-2; Sugarcane (Williams Lake) Jul 1959-4 eggs; Meadow Creek 15 Jun 1980-1; Indianpoint Lake 6 Jun 1934-1 (MVZ 65716); McBride 10 Jul 1991-4; Sinkut River 11 Jun 1946-10 singing males, 7 Aug 1945-4 fledglings, 24 Aug 1946-1 (Munro 1949); Prince George 22 Jun 1969-5; Eaglet Lake 1 Jul 1969-2 at e end; Gagnon Creek 2 Jun 1997-2; Tupper Creek to Pouce Coupe 15 Jun 1974-10; 55 km s Dawson Creek 10 Jun 1994-3 small nestlings; 22 km south of Dawson Creek 5 Jul 1992-3 eggs; 130 km n Fort St. James 17 Jul 1968-1 (Rogers 1968d); s Hudson's Hope 23 Jun 1978-1; Cecil Lake to Fort St. John 14 Jun 1974-20; Montney Creek, at bridge 6 Jul 1976-3; ne Fort St. John 24 Aug 1986-14; Fort St. John 2 Aug 1987-1 adult with 1 young just out of nest; North Pine 30 Aug 1984-22 in a flock; Charlie Lake 10 Jul 1939-4 eggs; Kenai Creek 2 Jun 1996-1 (Greenfield 1998); Mile 128 Alaska Highway 23 Jun 1984-3; Mobil Lake 6 Jul 1982-1; nr Fort Nelson 10 Jul 1967-6 eggs, 12 Jul 1974-1, and 8 Aug 1989-1; Petitot River 20 Jun 1982-1 singing male.

Breeding Bird Surveys: Interior – Recorded from 25 of 73 routes and on 17% of all surveys. Maxima: Tupper 20 Jun 1994-61; Fort St. John 27 Jun 1986-46; Hudson's Hope 7 Jun 1976-33. **Coastal** – Not recorded.

Autumn: Interior – nr Fort St. John 22 Sep 1985-1, last seen this autumn; Mugaha Creek 7 Sep 1997-1; Vanderhoof 2 Sep 1934-1 (MVZ 106244); Kidprice Lake 4-8 Oct 1974-1; Indianpoint Lake 9 Oct 1928-1 (McCabe and McCabe 1929); Revelstoke 28 Sep 1984-2; Kamloops 3-8 Sep 1995-1; Enderby 29 Sep 1943-1 (UBC 10190); Okanagan Landing 24 Nov 1929-1 (A.C. Brooks collection 5156). **Coastal** – Vancouver 31 Oct 1994-1 (Bowling 1995c), 20 to 23 Oct 1990-1; Westham Island 9 Sep 1993-1 at Alaksen National Wildlife Area (Siddle 1994a); Long Beach 6 and 30 Oct 1982-1 (BC Photo 809); Pacific Rim National Park 9 Nov 1986-1; Gordon Head (Saanich) 12 to 14 Oct 1995-1; Sea Island 15 and 20 Sep 1995-1 (Elliott and Gardner 1997); Iona Island 21 to 30 Oct 1995-1 (Elliott and Gardner 1997).

Winter: Interior – Kamloops 1 to 15 Dec 1987-1 at feeder (R.R. Howie pers. comm.). **Coastal** – Ladner 27 Dec 1982 to 5 Mar 1983-1 along Park Canada Rd.

Christmas Bird Counts: Interior – Not recorded. **Coastal** – Recorded from 1 of 33 localities and on less than 1% of all counts. Maximum: Ladner 27 Dec 1982-1 on Park Canada Rd.

Brewer's Sparrow

Spizella breweri (Cassin)

BRSP

RANGE: Breeds from southwestern Yukon, northwestern and southern British Columbia, southern Alberta and Saskatchewan, and southwestern North Dakota south, through the Great Basin region east of the coastal ranges, to southern California and northwestern New Mexico. Winters from southern interior California and southern Nevada southeast to western and central Texas and south to Baja California, the Pacific Lowlands of northern and central Mexico, and the Highlands of west-central Mexico to Guanajuato.

STATUS: On the coast, *casual* vagrant in the Georgia Depression Ecoprovince.

In the interior, *uncommon* to locally *common* migrant and summer visitant in the Southern Interior Ecoprovince; locally, an *uncommon* migrant and summer visitant in the Southern Interior Mountains Ecoprovince; *casual* in the Central Interior and Sub-Boreal Interior ecoprovinces; *uncommon* to locally *fairly common* migrant and summer visitant in the Northern Boreal Mountains Ecoprovince.

Breeds.

Figure 229. Brewer's Sparrow on the big sagebrush shrub-steppe habitat of the Okanagan valley (Round Lake, Richter Pass, 22 June 1983; Mark Nyhof).

NONBREEDING: The Brewer's Sparrow (Fig. 229) is distributed in 2 widely separate regions of the province: the southern portions of the interior, mostly south of latitude 52°N, and the extreme northwestern portions of the Northern Boreal Mountains. In both areas its distribution is rather localized. Elsewhere in the province it is an unusual find.

Two races of the Brewer's Sparrow occur in British Columbia: *S. b. breweri* and *S. b. taverneri* (also known as the "Timberline" Sparrow) (Godfrey 1986; Rising 1996). Although the morphological differences between the 2 subspecies are "subtle and it is not yet clear whether or not they can be recognized in the field" (Rising 1996), the races differ from each other in their distribution and habitat preferences. They probably represent separate species (Sibley and Monroe 1990; Klicka et al. 1999). Because of these apparent differences, we have included a more detailed discussion of the subspecies throughout this account.

The known distribution of *S. b. breweri* in the province includes only the Southern Interior (Fig. 230). There it is known to occur from the international boundary east to Midway, west to Keremeos, and north to at least Ashcroft. Individuals, likely of this subspecies, have been reported from Vernon, Kamloops, Clinton, and perhaps as far north as Dog Creek, south of Williams Lake in the Central Interior. Most of this population is concentrated in the extreme southern Okanagan valley (Fig. 230), from White Lake south to Chopaka.

Most of the Brewer's Sparrow reports from mountainous regions away from the arid big sagebrush shrub-steppe of the Southern Interior are likely of *S. b. taverneri*. This subspecies, however, is known to occur commonly only in 2 widely separate regions of the province. In the Southern Interior Mountains, it has been reported from Elko (Godfrey 1955) north to Thompson Pass. Individuals not identified to subspecies, but likely of the *taverneri* race, have been reported further north, at Berg Lake in Mount Robson Park. *S. b. taverneri* has been reported once from the Williams Lake area in the Central Interior. This race also occurs in numbers in extreme

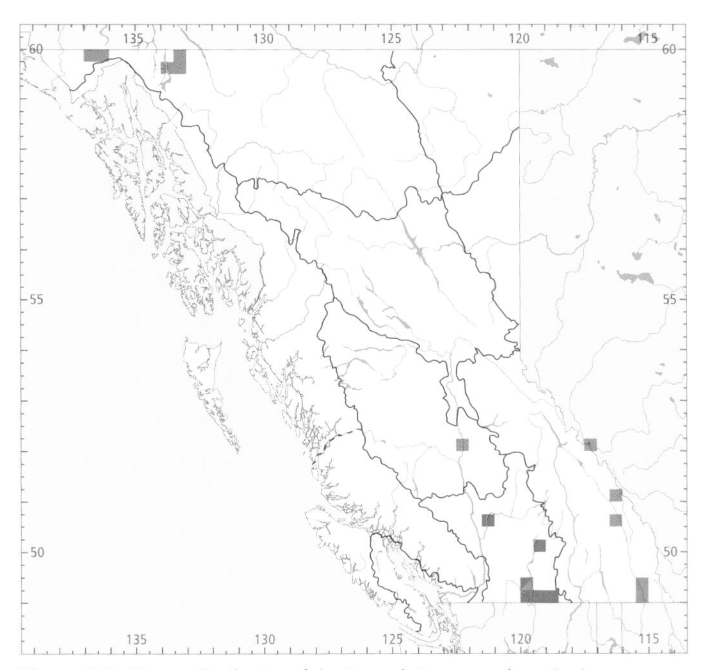

Figure 230. Known distribution of the Brewer's Sparrow subspecies in British Columbia based on specimen data. Red represents "Brewer's" Sparrow (*S. b. breweri*) and pink the "Timberline" Sparrow (*S. b. taverneri*).

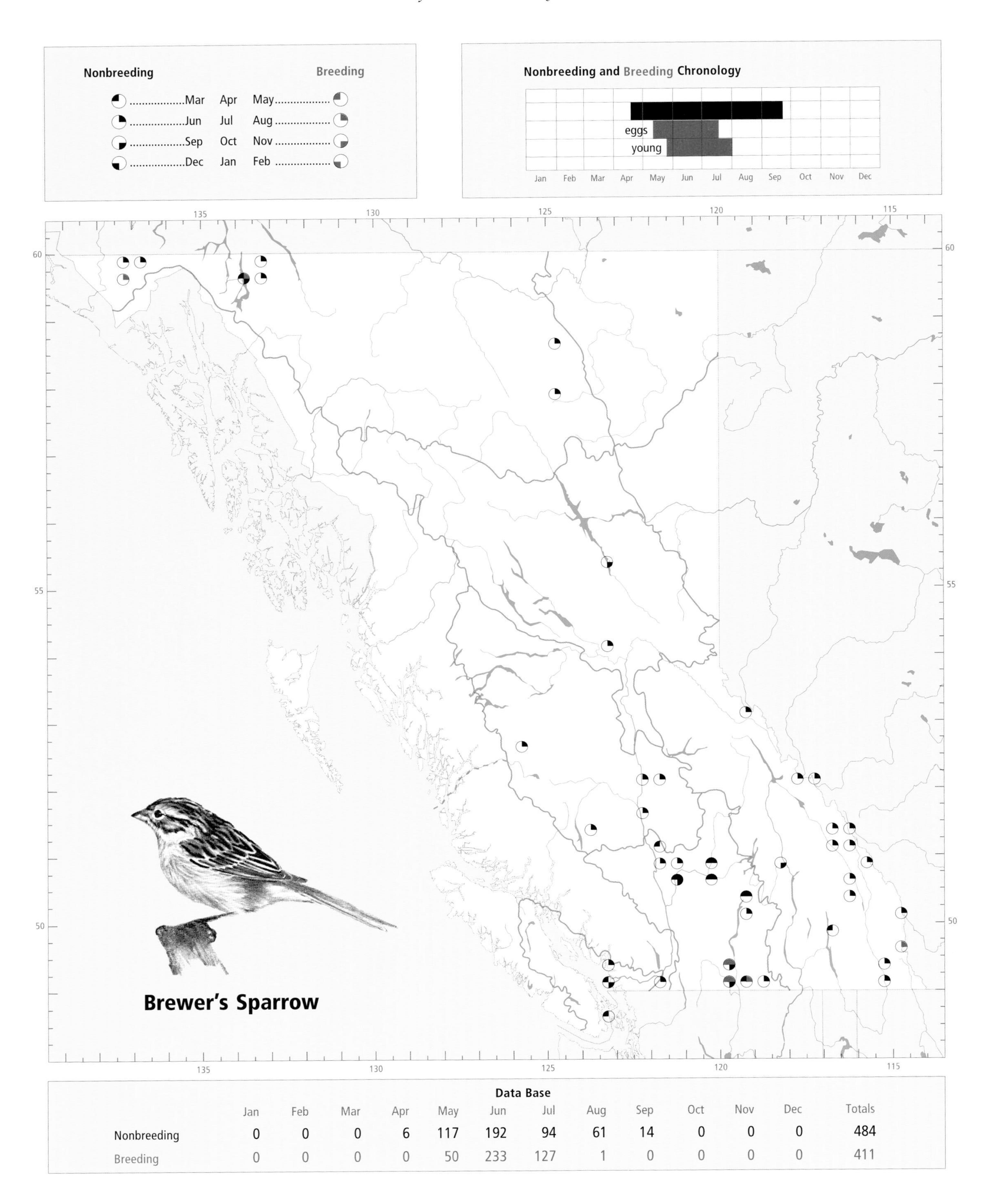

Data Base

	Jan	Feb	Mar	Apr	May	Jun	Jul	Aug	Sep	Oct	Nov	Dec	Totals
Nonbreeding	0	0	0	6	117	192	94	61	14	0	0	0	484
Breeding	0	0	0	0	50	233	127	1	0	0	0	0	411

northwestern British Columbia (Fig. 230), where its primary distribution includes the Atlin area west to the Tatshenshini River. Reports of the Brewer's Sparrow (likely *S. b. taverneri*) have also come from Stone Mountain Park and Kwadacha Wilderness Park (Cooper and Cooper 1983) in the eastern portions of the Northern Boreal Mountains.

In the Central Interior, the Brewer's Sparrow has also been reported from Mount Tatlow in the central Chilcotin Ranges, the Rainbow Range of the Western Chilcotin Upland (Campbell and Dawe 1992), and Hudson Bay Mountain northwest of Smithers. It has been reported once in the Prince George area (Prince George Naturalists Club 1996) and once at a banding station near Mackenzie. The subspecies frequenting all these areas has not been determined but is likely *S. b. taverneri*.

During the nonbreeding season in the regular parts of its range, the Brewer's Sparrow has been reported from 280 to 2,040 m elevation. For a description of its habitat requirements, see BREEDING (Figs. 231 and 232).

The Brewer's Sparrow arrives in the Southern Interior as early as the third week of April (Figs. 233 and 234). Numbers build to a peak around the last week of May, by which time most birds are on their nesting grounds. Munro (1935b) collected 2 specimens of the "Timberline" subspecies at White Lake in late May, and on that basis Cannings et al. (1987) note that the "Timberline" subspecies apparently moves through the Okanagan valley in late May as well. In the Southern Interior Mountains, a noticeable movement has not been documented; the earliest arrivals there are from the third week of May.

In the far north, this sparrow arrives about 6 weeks later than the southern populations, and occurs in numbers from mid-June to early July, when the birds are on territory (Fig. 233).

In autumn, a small movement is apparent in northern populations during the first 2 weeks of August, after which numbers dwindle; stragglers may be found there as late as the third week of September (Figs. 233 and 234).

Little in the way of an autumn movement is discernible in the Southern Interior Mountains; all birds have left that region by mid-September. In the Southern Interior, there is no discernible autumn movement either; reports decrease in number after the second week of July as birds stop vocalizing, and most birds have left the area by the end of August, although a few may remain as late as the third week of September.

In southern British Columbia, the Brewer's Sparrow has been recorded between 19 April and 22 September; in the northern portions of the province, it has been recorded between 29 May and 17 September (Fig. 233).

Figure 231. In the Southern Interior Ecoprovince of British Columbia, the Brewer's Sparrow (*Spizella breweri breweri*) inhabits big sagebrush shrub-steppe communities at low elevations (Richter Pass, 8 August 1996; R. Wayne Campbell).

Figure 232. In northwestern British Columbia, the Brewer's Sparrow (*Spizella breweri taverneri*) is found locally in patches of tall willows near timberline (Three Guardsmen Mountain, 21 June 1983; R. Wayne Campbell).

Figure 233. Annual occurrence (black) and breeding chronology (red) for the Brewer's Sparrow in ecoprovinces of British Columbia. Records are shown for the week in which they occurred.

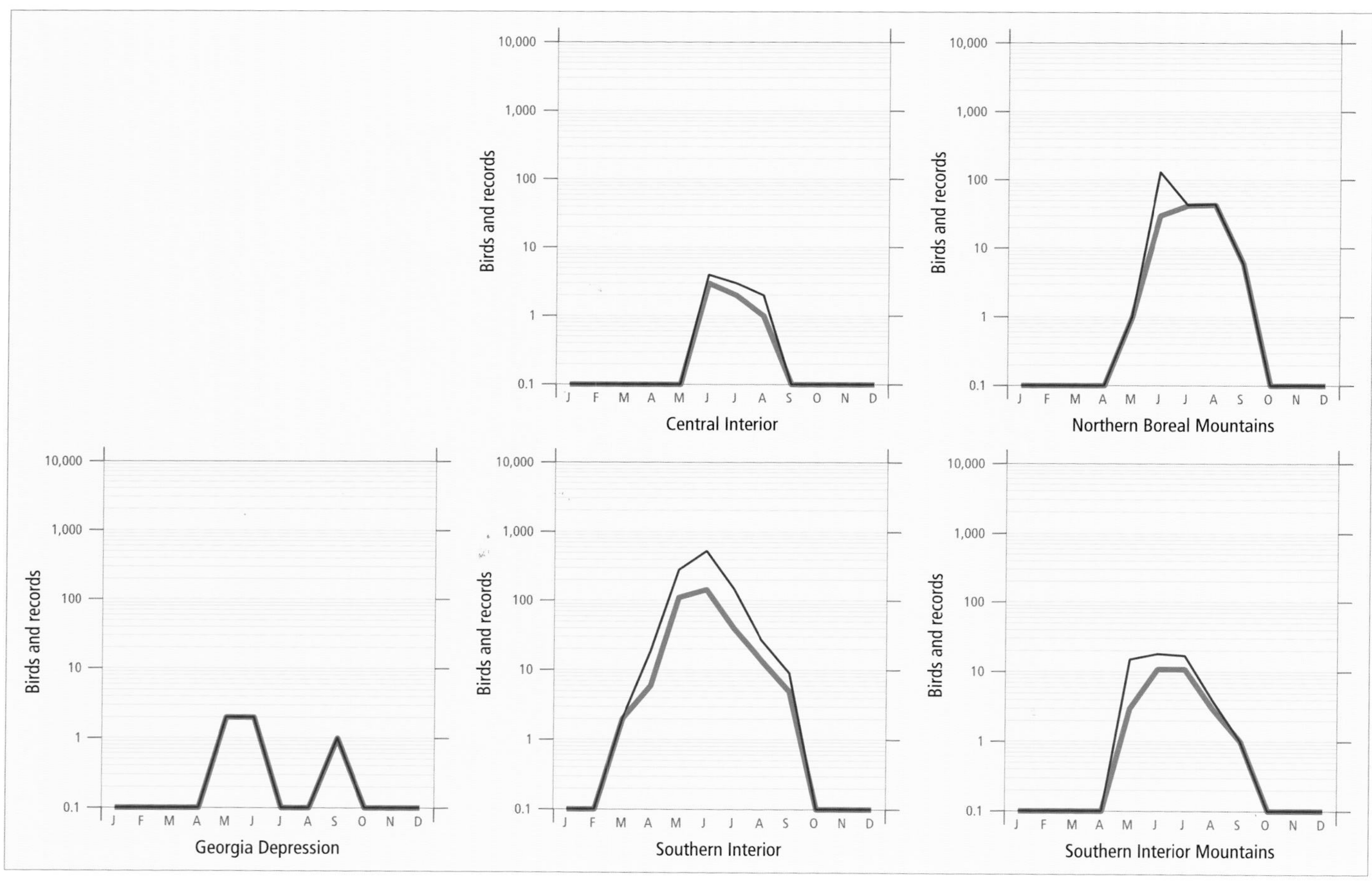

Figure 234. Fluctuations in total number of birds (purple line) and total number of records (green line) for the Brewer's Sparrow in ecoprovinces of British Columbia. Breeding Bird Surveys and nest record data have been excluded.

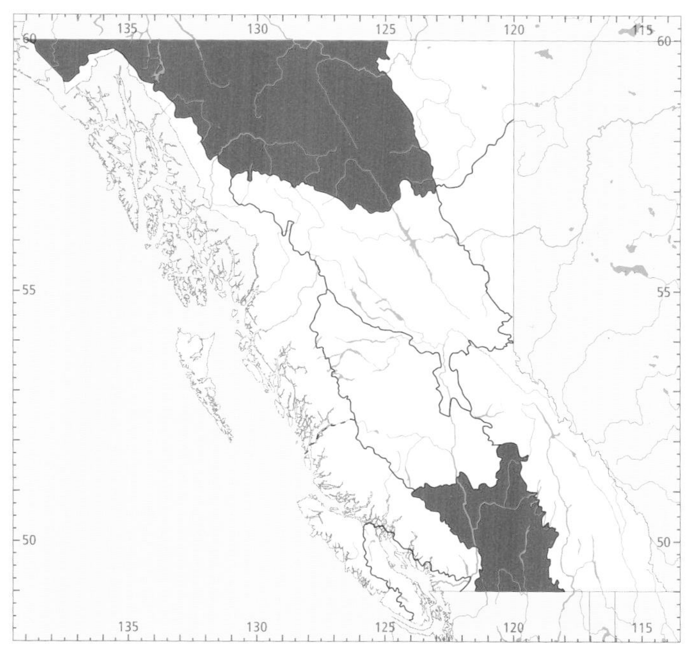

Figure 235. In British Columbia, the highest numbers for the Brewer's Sparrow in summer (red) occur in the Southern Interior and Northern Boreal Mountains ecoprovinces.

BREEDING: The Brewer's Sparrow is known to breed in 3 disjunct areas of British Columbia: the extreme southern portions of the Southern Interior west of the Okanagan River, from the Marron River valley south to Kilpoola Lake near the international boundary; the Sparwood area of the Southern Interior Mountains; and the extreme northwestern corner of the Northern Boreal Mountains, from Atlin westward. It probably also breeds in a number of other areas of the province (see REMARKS).

The Brewer's Sparrow reaches its highest numbers in summer in both the Southern Interior and Northern Boreal Mountains (Fig. 235). Breeding Bird Surveys for interior routes for the period 1968 through 1993 contain insufficient data for analysis.

Over the period 1966 to 1996, the Brewer's Sparrow showed average annual declines of 3.7% ($P < 0.01$) on North American Breeding Bird Surveys; average annual declines of 3.0% ($P < 0.01$) continued to be apparent more recently, over the period 1980 to 1996 (Sauer et al. 1997). The declines presumably relate only to *S. b. breweri.*

The Brewer's Sparrow has been reported nesting in British Columbia between 340 and 2,040 m elevation. Breeding habitat includes arid, subalpine, or alpine shrublands throughout its range in the province (Figs. 231 and 232).

S. b. breweri frequents the big sagebrush shrub-steppe communities of the Southern Interior at lower elevations, between 340 and 750 m. In the same area, however, the Brewer's Sparrow has also been reported at an elevation of 1,860 m on Mount Kobau, where big sagebrush extends to adjoin stands of subalpine fir (Cannings et al. 1987); to our knowledge, the subspecies at these higher elevations has not been determined.

In the southern Okanagan valley, Harvey (1992) found that bird density had a significant negative correlation with shrub density; however, shrub density explained less than 10%

Figure 236. In the Northern Boreal Mountains Ecoprovince of British Columbia, the "Timberline" Sparrow (*Spizella breweri taverneri*) breeds in the clumps of dwarfed shrubs, usually above timberline (Range Creek valley, 10 August 1989; Andrew C. Stewart).

Figure 237. In the southern interior of British Columbia, most Brewer's Sparrow nests were found in big sagebrush shrubs (White Lake, Okanagan Falls, 25 June 1984; Mark Nyhof).

of the total variation in the data set. In Idaho, Petersen and Best (1985) found that sagebrush coverage and shrub height near Brewer's Sparrow nests were significantly different from those in their study area in general. They also found that the sparrow preferred to nest in sagebrush shrubs that were entirely or mostly alive.

S. b. taverneri has been reported during the nesting season from elevations of 870 to 2,040 m. It makes use of shrublands or open grassy habitats interspersed with shrubs or scrubby trees, usually above timberline (Fig. 236). In the south, the habitats of *S. b. taverneri* include the shrubby slopes of avalanche chutes and scrubby clumps of trees, often near water. The birds near Elko were anomalous in that they were found well below timberline, in areas covered mainly by antelope-brush (Godfrey 1955). Although not identified to subspecies, an adult Brewer's Sparrow (likely *S. b. taverneri*) was observed feeding a recently fledged young at 1,600 m elevation near the Erickson Mine southeast of Sparwood. The site was below timberline on an open bedrock slope with scattered willows that had been heavily coppiced by elk browsing. The ground cover there was very low; it was dominated by herbaceous plants and mixed with large areas of bedrock (D.F. Fraser pers. comm.). Schwab (1979) also found the Brewer's Sparrow in habitat similar to that reported by Godfrey (1955), some 90 km northwest of Elko, near Wasa. There it occupied scrubby seral vegetation dominated by deciduous shrubs, primarily antelope-brush and saskatoon, as well as scrubby seral–young conifer transition dominated by ponderosa pine, saskatoon, grasses (*Poa* spp.), and lupines. The subspecies of the Wasa-area birds is not known.

In the far north, *S. b. taverneri* frequents open, grass-covered areas near and above timberline. The damper areas contain clumps of mountain-heather interspersed with scrubby clumps of prostrate subalpine fir (occasionally reaching 4.5 m high) and waist-high tangles of scrub birch and taller willow shrubs (Swarth 1926, 1936). Swarth (1930) notes: "As the sage-brush is to the Brewer [sic] Sparrow ... so is the trailing [scrub] birch to the Timberline Sparrow."

Most nests (99%; n = 134) reported from British Columbia were from the big sagebrush shrub-steppe grasslands of the Southern Interior; 2 were from the subalpine and alpine shrublands of the far north.

In British Columbia, the Brewer's Sparrow has been recorded breeding from 12 May to 1 August (Fig. 233).

Nests: Most nests of *S. b. breweri* (92%; n = 149) were found in shrubs or scrub tangle (Fig. 237); 8% were located among ground vegetation. Shrubs identified to species by observers included big sagebrush (32%), snowberry (1%), and snowbrush (1%). Ground vegetation included unidentified grasses

Figure 238. In British Columbia, as elsewhere in North America, the Brewer's Sparrow is rarely parasitized by the Brown-headed Cowbird (Richter Pass, 18 June 1968; R. Wayne Campbell).

(11 nests), wildrye (9), alfalfa (1), sulphur buckwheat (1), lemonweed (1), and dried plant debris (1). During extensive field work in the southern Okanagan valley in 1997 and 1998, N. Mahony (pers. comm.) located 257 nests and noted the following major nest substrates: big sagebrush (84%), snowberry (6.2%), and wildrye (5.4%). Other shrubs and plants included rabbitbush, mustard spp., lemonweed, diffuse knapweed, rose spp., and bluebunch wheatgrass.

Both nests of *S. b. taverneri* were in shrubs, 1 in a juniper, the other in a scrub birch.

Nests in British Columbia were compact cups of grasses (94%; *n* = 130), plant stems or fibres (13%), and rootlets (2%), and were lined with fine grasses and mammalian hair (porcupine, moose, horse, cow). Nests were built in the fork or crotch of the shrub branches or in hollows on the ground beside clumps of vegetation. The heights of 141 nests ranged from ground level to 1.8 m, with 65 % between 0.2 and 0.4 m.

Eggs: Dates for 251 clutches (all *S. b. breweri*) ranged from 12 May to 18 July, with 52% recorded between 4 June and 30 June. Sizes of 119 clutches ranged from 1 to 5 eggs (1E-9, 2E-14, 3E-32, 4E-62, 5E-2), with 52% having 4 eggs. Only 2 clutches were reported for *S. b. taverneri,* 1 with 3 eggs and 1 with 4; both nests were found between 22 and 29 June. Two nests from southern British Columbia (*S. b. breweri*) indicate an incubation period of between 11 and 12 days; Reynolds (1981) and Rotenberry and Wiens (1989) note a typical period of 11 days.

Young: Dates for 157 broods (all *S. b. breweri*) ranged from 26 May to 1 August, with 56% recorded between 9 June and 8 July. Adults have been reported feeding fledglings as late as 27 August. Sizes of 88 broods ranged from 1 to 4 young (1Y-4, 2Y-14, 3Y-27, 4Y-43), with 80% having 3 or 4 young. We have no reports of nests of *S. b. taverneri* with young. Two *S. b. breweri* nests from British Columbia indicate a nestling period of about 8 days; Rising (1996) notes a period of between 7 and 9 days.

Brown-headed Cowbird Parasitism: In British Columbia, 6 of 154 nests found with eggs or young were parasitized by the cowbird (Fig. 238). All involved *S. b. breweri.* This adds to Friedmann and Kiff's (1985) report that the Brewer's Sparrow has been found acting as a cowbird host only a few times, and in 3 of those instances, this sparrow deserted the nest after parasitism occurred. Brood parasitism varies geographically. In Alberta, Biermann et al. (1987) found 18 instances of cowbird parasitism in 13 of 25 Brewer's Sparrow nests; 9 of those nests had been abandoned by the sparrow. They note that despite the heavy parasitism, the reproductive success of the sparrows was similar to that reported for other members of the genus *Spizella;* cowbird productivity was low, however. In southeastern Washington, 14 of 281 nests (5%) were parasitized by the cowbird (Vander Haegen and Walker, 1999); in central Oregon, none of 110 nests was parasitized (Rotenberry and Wiens 1989).

Nest Success: Of 59 nests found with eggs (all *S. b. breweri*) and followed to a known fate, 36 produced at least 1 fledgling, for a nest success rate of 61%.

REMARKS: The Brewer's Sparrow is shy, wary, and inconspicuous in its nesting range (Terres 1991; Rising 1996). Swarth (1926) notes the secretive nature of the "Timberline" subspecies in northwestern British Columbia:

> The species might easily be overlooked, for besides their habitual wariness the birds are with difficulty

Figure 239. Less than 10% of land in the Okanagan valley of south-central British Columbia is in a relatively undisturbed state (north end of Osoyoos Lake, 16 April 1999; R. Wayne Campbell). The further loss of big sagebrush shrub-steppe plant communities to agricultural and urban developments is a major concern of conservationists.

Figure 240. In some areas of the Southern Interior Ecoprovince, nesting habitat for the Brewer's Sparrow has been seriously impaired by the clearing of big sagebrush to improve domestic livestock range conditions (3 May 1994, near Kilpoola Lake; Ted Lea).

> dislodged from the sheltering cover they frequent. If flushed at a distance from the tops of the balsam thickets on which they often perched when suspicious of danger (and they rarely permitted a near approach), the timber-line sparrow might easily be overlooked amid the tree sparrows, chipping sparrows, and even the *Zonotrichias,* which were in the same surroundings and arising from the bushes near at hand. When flushed they flew long distances, to dive into birch thickets, tangled masses of shrubbery about waist high, and it was rarely that a bird could be dislodged from such a refuge.

Because of the secretive nature of *S. b. taverneri* and the difficult access to much of its mountainous habitat, it may have a wider distribution throughout British Columbia than our data indicate, despite the fact that its song is rather striking and unmistakable. For example, R.J. Cannings (pers. comm.) found 2 Brewer's Sparrows in August among krummholz subalpine fir on Mount Tatlow, and in the second week of July, a Brewer's Sparrow was found singing in willow shrubs adjacent to a small subalpine lake in the Rainbow Range. Further north, a male was found in June singing in an open ski area on Hudson Bay Mountain northwest of Smithers (R. Toochin pers. comm.). All those reports are from the western portions of the Central Interior, far south or west of the sparrow's known breeding range. While the Mount Tatlow birds may have been migrants, the records suggest at least a movement through the area and perhaps breeding.

There is nesting evidence for the Brewer's Sparrow from a number of other areas of the province, but nests with eggs or young or recently fledged young have not yet been found. For example, Willing (1970) discovered a territorial male near Vernon, between 2 Clay-colored Sparrow territories. The Brewer's Sparrow has also been reported throughout the summer near Kamloops, and Howie (1994) notes that it "probably breeds" there.

Although no birds were observed, a number of nests, presumed to be of this species, were found in the Black Canyon area near Ashcroft (Sarell and McGuinness 1996). The nests were small and made of fine grasses and a small amount of horse hair. All were found in big sagebrush at varying heights. The only other vegetation around the nests were sparse, short saskatoon bushes (K.P. McGuinness pers. comm.).

In the east Kootenay, Godfrey (1955) reported 2 small "colonies" of the "Timberline" subspecies near Elko. He noted that "although no actual nests were found, the condition of specimens collected and the behaviour of those not collected left little doubt that they did breed there." We have reviewed the specimens collected by Godfrey (1955) and agree with him that they are all of the *taverneri* race. The small population near Elko is of interest, as it is the only known location where *taverneri* likely nests below timberline in dry habitat similar to that used by *S. b. breweri*. Further north in the Rocky Mountain Trench, near Wasa and Skookumchuck, Schwab (1979) reported 14 observations of this sparrow between early May and late June. More work is required in the Rocky Mountain Trench from the international boundary north to at least Skookumchuck to document the status and subspecies of the Brewer's Sparrow in that area.

At Toby Creek, west of Invermere, an adult *taverneri* was observed feeding a fledged young, and Wade (1977) noted that the Brewer's Sparrow very probably breeds in Yoho National Park, "as they have been recorded throughout the month of July." Cowan (1955) found a nest with eggs of the "Timberline" Sparrow in Elysium Pass, Jasper National Park (Alberta), just east of Mount Robson Park (British Columbia), suggesting that the subspecies may breed in other areas along the provincial border.

The Brewer's Sparrow has been reported from the following Breeding Bird Surveys: Fraser Lake 22 Jun 1969-1; Columbia Lake 26 Jun 1973-1; Pavilion 12 Jun 1977-1; Golden 21 Jun 1992-1. Adequate documentation is lacking for these observations, however, and they have been excluded from this account.

There are many questions to be answered about the breeding distribution of the 2 Brewer's Sparrow subspecies in British Columbia. For example, to which subspecies can the high-elevation birds found on Mount Kobau be assigned? Are the lowland birds in the Wasa area of the Rocky Mountain Trench of the same subspecies as the birds from the lower elevations near Elko? If the "Timberline" subspecies nests at both high and low elevations in strikingly different habitats, are the 2 subspecies separate species, as Sibley and Monroe (1990) and Klicka et al. (1999) suggest? Does the distribution of the "Timberline" race include much of the high-elevation shrub and scrub habitats of the province, and if so, has the sparrow somehow been overlooked because of its secretive nature? While amateur naturalists can contribute significantly to our understanding of the distribution of the "Timberline" race in British Columbia, the answers to questions involving subspecific versus specific determinations probably must involve the collection of specimens and the use of complicated chemical procedures that are now only in the realm of professional ornithologists or their students.

Population estimates for the Brewer's Sparrow in the southern Okanagan valley range from 826 birds, based on the 413 males Harvey (1992) counted during his surveys, to between 800 and 1,000 adults (Sarell and McGuinness 1996). Willing (1970) estimated the density of this sparrow at White Lake, near Okanagan Falls, to be 4 birds per 100 ha. Harvey (1992), in his survey of suitable Brewer's Sparrow habitat in the Okanagan valley south of Penticton, found densities ranging from just under 2 males per 100 ha on the Nighthawk Road to just over 9 males per 100 ha in the Kilpoola Lake West region. His density estimates at White Lake were just under 6 males per 100 ha.

Rising (1996) notes that Brewer's Sparrows "may breed in fairly high densities, and are perhaps loosely colonial." The "loosely colonial" characteristic of the species appears to apply to the southern Okanagan population in British Columbia. There the birds tend to occur more in discrete groups rather than as single pairs scattered across the landscape (N. Mahony pers. comm.). Godfrey (1955) also mentions the colonial nature of the apparent breeding population at Elko.

As with many species of birds in British Columbia, human impacts on the habitat of the Brewer's Sparrow are cause for concern. In the south Okanagan, less than 10% of the area is in a relatively undisturbed state (Redpath 1990). As agricultural (Fig. 239) and urban expansion continues, further loss of the big sagebrush shrub-steppe communities is inevitable unless adequate measures are taken to appropriately steward or otherwise protect these habitats. The impact of allowing livestock grazing on breeding habitat is another cause for concern (Fig. 240).

The 19 August 1898 record of *S. b. taverneri* from Okanagan Landing noted in Cannings et al. (1987) should refer to *S. b. breweri*.

Willing (1970) looked at the role of song in the behaviour and evolution of 3 species of spizellid sparrows in British Columbia, including the Brewer's Sparrow.

For additional life-history and management information on this sparrow, see Rising (1996), Austin (1968), Fraser and Walters (1993), and Rotenberry et al. (1999).

NOTEWORTHY RECORDS

Spring: Coastal – Elk Lake 25 May 1986-1 along Jennings Lane; Sea Island 14 May 1976-1 (Kautesk 1982a). **Interior** – Kilpoola Lake 4 May 1998-6, 14 May 1972-10; Richter Pass 26 May 1974-12; Haines-Lease Ecological Reserve 17 May 1986-2; Twin Lakes Creek (Okanagan) 19 Apr 1978-5, singing; White Lake (Okanagan Falls) 12 May 1946-4 eggs, 22 May 1974-10, 26 May 1991-21, 27 May 1972-5 eggs, 27 May 1931-1 *S. b. taverneri* (Munro 1935a), 30 May 1926-1 *S. b. taverneri* (Munro 1935b); Okanagan Falls 26 May 1973-4 nestlings; sw Wasa 2 May to 29 Jun 1977-a total of 10 observations (Schwab 1979); se Skookumchuck 2 May to 29 Jun 1977-a total of 4 observations (Schwab 1979); Kaslo 21 May 1973-1; Vernon 24 May to 7 Jul 1969-1 unpaired male in typical Clay-colored Sparrow habitat of snowberry and wild rose; Ashcroft 15 May 1948-1; Kamloops 27 May 1991-1 singing male; Mara Hill 20 May 1985-2; Clinton 15 May 1948-1; n Dog Creek (Gang Ranch) 19 Jun 1993-2 males singing in sagebrush grassland; Atlin 29 May 1934-1 (CAS 42073).

Summer: Coastal – Jericho Beach late Jun to early Jul 1991-1 singing in broom shrub; Kilby Park 4 Jun 1984-1. **Interior** – Chopaka 30 Jun 1981-4 nestlings about 1 day old, 30 Jul 1976-25; Midway 12 Jun 1929-1 *S. b. breweri* (NMC 23342); 3 km sw Kilpoola Lake 23 Jul 1997-3 nestlings; Kilpoola Lake 18 Jun 1975-2 nestlings being fed by adults; Mount Kobau 4 Jun 1990-4 eggs, 6 Jun 1991-9; Keremeos 4 Jul 1928-1 *S. b. breweri;* 6 km s Elko 4 to 30 Jun 1953-4, all *S. b. taverneri* (NMC 38826 to 38829); sw Okanagan Falls 17 Jul 1994-3 nestlings with feathers half out of sheaths; White Lake (Okanagan Falls) 10 Jun 1973-25, 25 Jun 1936-50 males singing, 6 Jul 1990-3 eggs, 18 Jul 1997-3 eggs, 18 Aug 1974-11; Blind Creek (Okanagan) 11 Jun 1991-2 nestlings about 7 days old; Yellow Lake (Okanagan Falls) 19 Jun 1983-4 nestlings being fed by adults, 13 Jun 1991-3 nestlings fledged when nest was approached; Erickson 19 Jul 1984-2, feeding newly fledged young; Vernon 7 Jul 1969-1 unpaired male in typical Clay-colored Sparrow habitat of snowberry and rose, from 24 May; Toby Creek (Invermere) 27 Aug 1945-1 female feeding fledgling *S. b. taverneri* (UBC 809); Kamloops 30 Jun 1991-2 in Pruden Pass; Ashcroft 4 Jun 1892-1 *S. b. breweri* (ANS 31112); Mara Hill 15 Jun 1993-1 singing male, 27 Jun 1983-1, 30 Jun 1991-1, 7 Jul 1991-2, 14 Jul 1990-2; Mount Tatlow 20 Aug 1978-2, in krummholz subalpine fir at 2,040 m elevation with Golden-crowned and White-crowned sparrows; Kaufmann Lake 6 Jul 1983-5 on avalanche slope, including 2 singing males; Williams Lake 9 Jun 1951-1 (NMC 48440); 158 Mile House 3 Jul 1901-2 (Brooks 1903); Stanley Glacier 17 Jun 1975-2 singing males in burn; Thompson Pass 28 Jul 1945-1 *S. b. taverneri* (UBC 810); Rainbow Range 12 Jul 1992-1 (Campbell and Dawe 1992); Berg Lake 12 Jun 1973-3; nr Chief Lake 3 Jun 1996-1, km 3 on the 200 Rd; Hudson Bay Mountain Jun 1997-1 male singing on territory; Fern Lake 14 Aug 1983-2 (Cooper and Cooper 1983); Stone Mountain Park 7 Jun 1983-1, Jun 1994-7 birds on territory and recorded on tape, along road to radio tower; Samuel Glacier 27 Jun 1983-9 (Campbell et al. 1983), 29 Jun 1983-3 eggs, adult flushed from nest; Tatshenshini River 8 Jul 1980-1; Monarch Mountain (Atlin) 22 Jun 1929-4 eggs, adult collected (CAS 32406); Spruce Mountain 8 Aug 1924-1 *S. b. taverneri* (MVZ 44855); Carmine Mountain 14 Jun 1983-11 *S. b. taverneri* (RBCM 17786), 17 Jun 1983-6 were seen during 6 hr walk (*S. b. taverneri;* RBCM 17787 and 17788); Sediments Creek 22 Jun 1993-6 males singing between 875 and 1,045 m elevation on steep se-facing mountainside; Haines Highway Mile 85 21 Jul 1944-1 *S. b. taverneri* (ROM 71130).

Breeding Bird Surveys: Coastal – Not recorded. **Interior** – Recorded from 1 of 73 routes and on less than 1% of all surveys. Maximum: Summerland 18 Jun 1994-1.

Autumn: Interior – Atlin 17 Sep 1931-1 *S. b. taverneri* (RBCM 5710); Revelstoke 11 Sep 1984-1; Ashcroft 22 Sep 1938-1 *S. b. breweri* (RBCM 8780); White Lake (Okanagan Falls) 14 Sep 1986-5. **Coastal** – Sea Island 14 Sep 1980-1 (Kautesk 1982a).

Winter: No records.

Vesper Sparrow

VESP

Pooecetes gramineus (Gmelin)

RANGE: Breeds from central and southwestern British Columbia, southern Mackenzie, northern Alberta, northwestern and central Saskatchewan, north-central Manitoba, central Ontario, south-central Quebec, parts of Prince Edward Island, New Brunswick, and Nova Scotia south to western Oregon, northern and eastern California, central Nevada, southwestern Utah, northern and east-central Arizona, south-central New Mexico, Colorado, Missouri, eastern Tennessee, and North Carolina. Winters regularly from central California, southern Nevada, southwestern Utah, southern Arizona, southern New Mexico, Arkansas, Louisiana, and Maryland south to Florida, Texas, continental Mexico as far as Oaxaca and Veracruz, and southern Baja California.

STATUS: On the coast, a *very rare* to *rare* migrant, local summer visitant, and *accidental* in winter in the Georgia Depression Ecoprovince; in the Coast and Mountains Ecoprovince, *very rare* from spring through autumn and *accidental* in winter on the Southern Mainland Coast, *casual* on Western Vancouver Island and the Northern Mainland Coast, and absent from the Queen Charlotte Islands.

In the interior, a *common* to *very common* migrant (sometimes abundant) and summer visitant in the Southern Interior Ecoprovince; *fairly common* to *common* migrant and *uncommon* to *fairly common* summer visitant in the Southern Interior Mountains and Central Interior ecoprovinces; *very rare* migrant and summer visitant in the Sub-Boreal Interior Ecoprovince; *uncommon* migrant and locally *uncommon* to *fairly common* summer visitant in the Boreal Plains Ecoprovince; *casual* in the Northern Boreal Mountains and Taiga Plains ecoprovinces.

Breeds.

NONBREEDING: The Vesper Sparrow (Fig. 241) is widely distributed throughout lower elevations across southern portions of the province from southeastern Vancouver Island east to the Rocky Mountain Trench and north through the Cariboo and Chilcotin areas of the Central Interior. It is less numerous in the Sub-Boreal Interior and scarce anywhere in the Northern Boreal Mountains and Taiga Plains. An exception occurs in the Peace Lowland, in a small area east of the Rocky Mountains, where the species is again widely distributed and occurs regularly.

The Vesper Sparrow occurs at elevations from near sea level on the coast to between 280 and 1,100 m in the interior. During autumn migration, it sometimes uses alpine elevations to 2,000 m.

Since the Vesper Sparrow is a characteristic summer bird of many of the habitats through which it migrates, there is much overlap between the habitats used during migration and when it is on breeding territory. While migrating into and through the province in spring and autumn, it occurs in open, grassy or weedy areas of many types: overgrazed ranges and pastures (Fig. 242), stubble fields, pastures, prairies, airports, openings in trembling aspen parkland, south- and west-facing unforested or sparsely forested hillsides, road verges, clearcuts, regenerating burns, dry agricultural fields, playgrounds, sandspits, and dunes. In autumn it sometimes migrates along alpine and subalpine ridges.

Figure 241. The Vesper Sparrow is a characteristic summer bird of grassland and shrub-steppe habitats across the interior of southern British Columbia (near Chopaka, 13 May 1997; R. Wayne Campbell).

During migration, the Vesper Sparrow reaches its highest numbers in the Okanagan, Similkameen, and Thompson river valleys of the Southern Interior. On the south coast, spring migration begins consistently in the second week of April and reaches its peak in mid-May (Figs. 243 and 244). The dates of first arrivals in the Southern Interior and Southern Interior Mountains are almost identical to those on the coast: early to mid-April. In the Okanagan valley, 34 arrival dates in the Vernon area range from 4 to 28 April, with a median of 13 April (Cannings et al. 1987). Across the southern portions of the interior, the migration reaches its peak in early

Figure 242. In autumn, flocks of Vesper Sparrows are abundant as they migrate over grasslands in parts of the Southern Interior Ecoprovince of British Columbia (near Douglas Lake, 29 September 1996; R. Wayne Campbell).

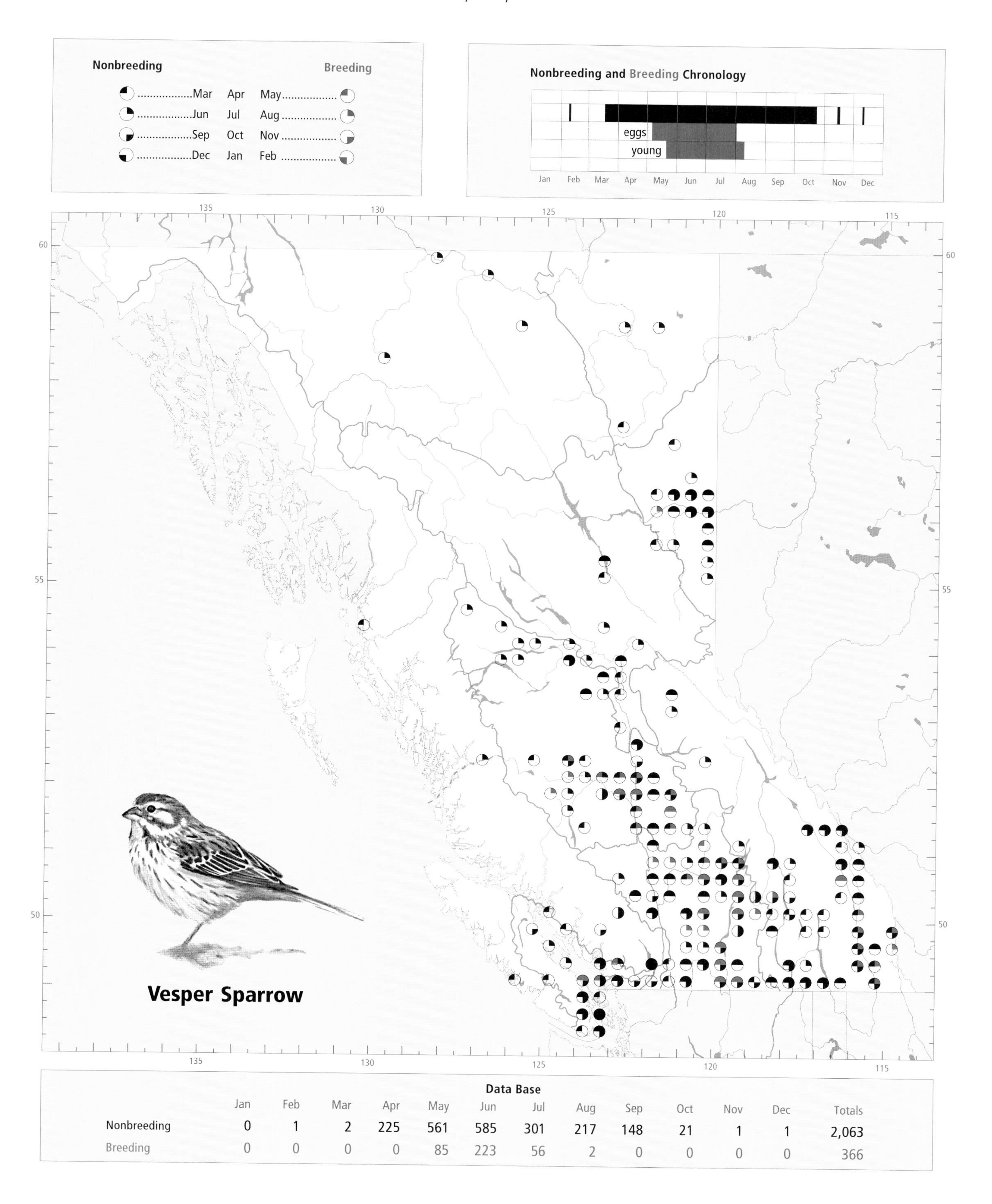

Data Base

	Jan	Feb	Mar	Apr	May	Jun	Jul	Aug	Sep	Oct	Nov	Dec	Totals
Nonbreeding	0	1	2	225	561	585	301	217	148	21	1	1	2,063
Breeding	0	0	0	0	85	223	56	2	0	0	0	0	366

May. The spring return of the Vesper Sparrow reaches the latitude of Prince George in early May. Most records in the Sub-Boreal Interior are from the month of May (Fig. 244). In the Peace Lowland, although the earliest arrival over 11 years (1977 and 1980 to 1989) ranged from 27 April to 9 May (C. Siddle pers. comm.), the movement does not become noticeable until early May and peaks in the third week of May.

The onset of the autumn migration is difficult to detect in our records from the Peace Lowland, where there is a gradual decline in numbers recorded from July through August and a rapid decline through September (Fig. 244). The latest recorded observations are from early October. In the Central Interior, there is a slight decline in numbers of birds from mid-August to early September. The last birds leave the Central Interior in the third week of September. In the southern portions of the interior, the autumn migration is apparent by mid-July and peaks in late August and early September (Fig. 244). Except for a few stragglers, all Vesper Sparrows have left the interior by the end of September. The latest record for the Okanagan valley is 22 October; for the Columbia valley, it is 2 October.

On the coast, small numbers build from August into October, abruptly dropping by the middle of October.

On the south coast, the Vesper Sparrow has been recorded regularly from 10 April to 18 October; in the interior, it has been recorded from 2 April to 22 October (Fig. 243).

BREEDING: The Vesper Sparrow is a characteristic species of the summer avifauna of the relatively arid grassland and shrub steppe of the interior of the province, from the international boundary north through the Cariboo and Chilcotin areas. Its highest breeding densities are in the Okanagan, Similkameen, and Thompson river valleys. It is less abundant but still numerous locally in the lower valleys of the Kootenay and Columbia rivers. It occurs regularly on the extensive rangelands that border the Fraser River north of Lillooet, and in the lower reaches of the Chilcotin River, where it is usually the most abundant breeding grassland species (A. Roberts pers. comm.), and extends to include much of the Cariboo and Chilcotin plateaus. The northernmost breeding records west of the Rocky Mountains are from Chezacut Lake. There is a report of 3 pairs on established territories at Nulki Lake but nests were not found (Munro 1949). The species probably nests regularly in the Peace Lowland, but only a single nest with eggs has been found.

On the coast, the Vesper Sparrow has been recorded breeding in the Fraser Lowland, but no breeding has been documented since 1968. In the Nanaimo Lowland of southeastern Vancouver Island, the species is present through the summer and occasionally breeds locally.

The Vesper Sparrow reaches its highest numbers in summer in the Southern Interior (Fig. 245). An analysis of Breeding Bird Surveys for the period 1968 through 1993 could not detect a net change in numbers on either coastal or interior routes. Survey wide analyses of the Breeding Bird Surveys from 1966 to 1996 show an average annual decrease of 0.8% ($P = 0.01$) (Sauer et al. 1997).

On the coast, the Vesper Sparrow breeds a few metres above sea level; in the interior, this sparrow nests over a wide altitudinal range, from about 280 to 1,500 m.

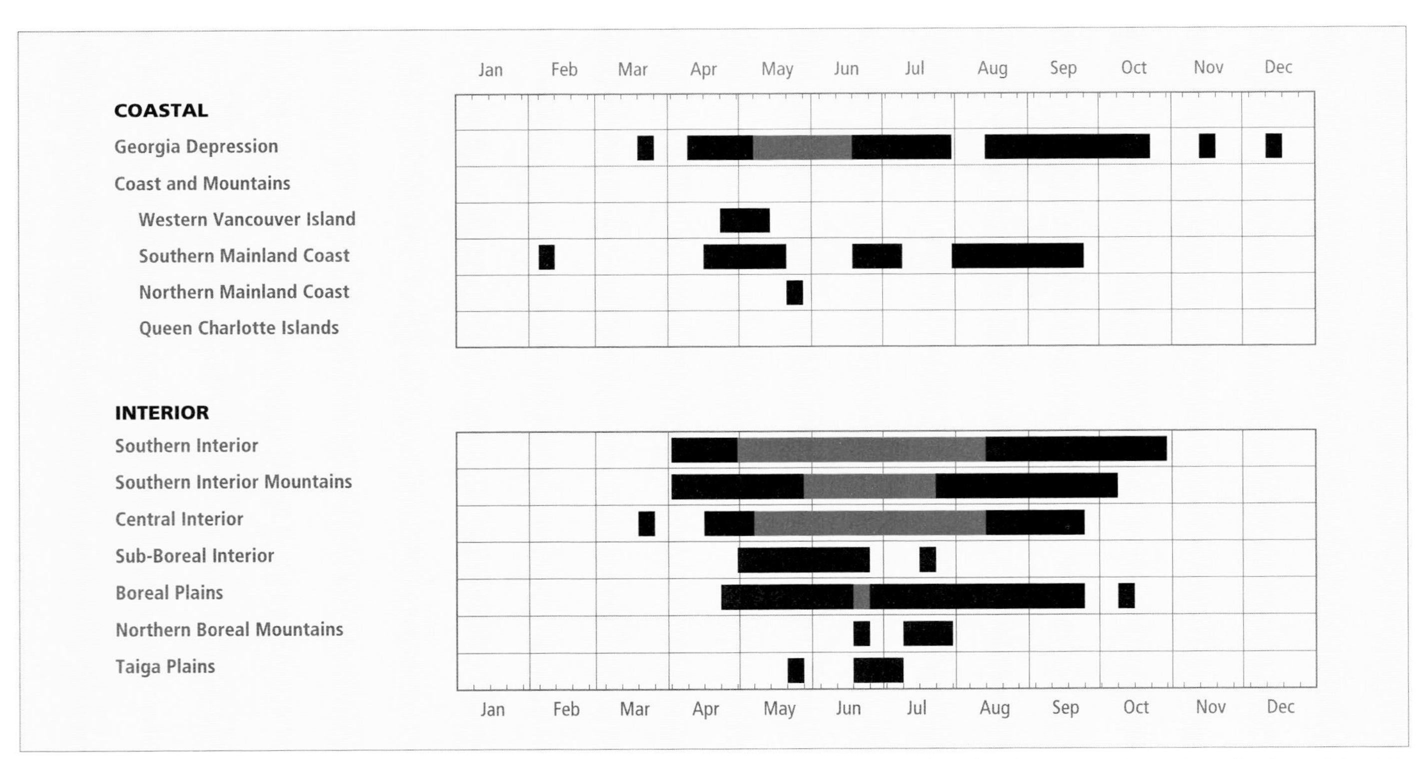

Figure 243. Annual occurrence (black) and breeding chronology (red) for the Vesper Sparrow in ecoprovinces of British Columbia. Records are shown for the week in which they occurred.

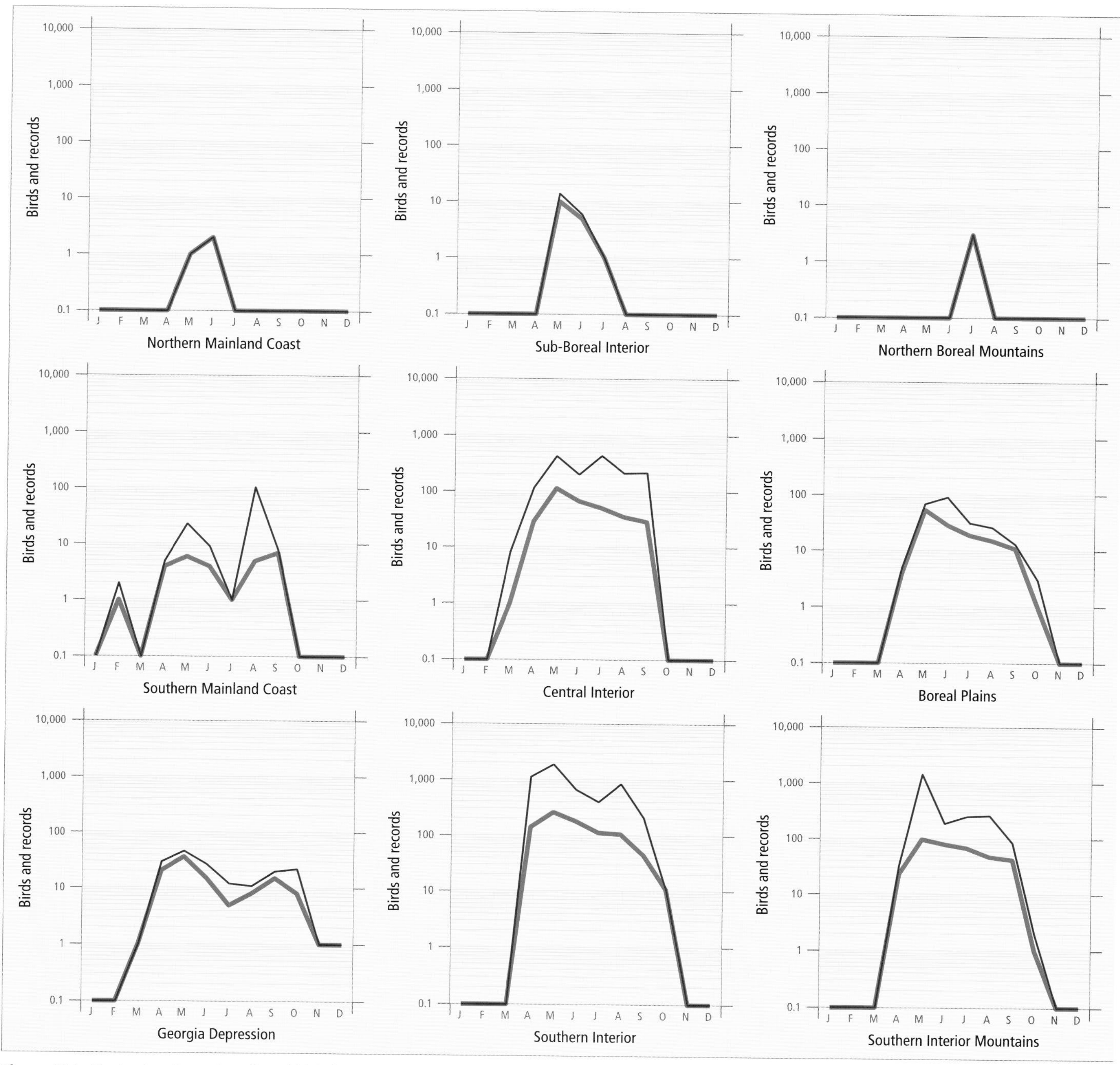

Figure 244. Fluctuations in total number of birds (purple line) and total number of records (green line) for the Vesper Sparrow in ecoprovinces of British Columbia. Christmas Bird Counts, Breeding Bird Surveys, and nest record data have been excluded.

On the coast, the breeding habitat includes pastureland, agricultural fields, and airports with patches of grasses and weeds. In the southern regions of the interior, the Vesper Sparrow breeds in bunchgrass rangelands and sagebrush-steppe (Fig. 246), and in grassy or weedy openings in the parkland forest of ponderosa pine, interior Douglas-fir, or lodgepole pine, often interspersed with scattered patches of trembling aspen. It also moves into recently burned areas where the understorey has been removed, new grass is growing, and charred trees remain standing. In the Peace Lowland, C. Siddle (pers. comm.) describes the summer habitat of this sparrow as the grassy "breaks" found on south- and west-facing slopes along the major river valleys, where the birds favour arid slopes with patches of wolf-willow, roses, and trembling aspens growing between large open grasslands (Fig. 247). It is also found along the edges of dry agricultural fields. In all habitats, the Vesper Sparrow selects areas that carry a cover of short, densely growing grasses or forbs.

In Montana, studies in similar habitat to that used by the birds in British Columbia have shown that the Vesper Sparrow nests "in areas where the vegetation was short and dense, with a relatively high percentage of ground cover, and not in the areas where vegetation was tall and patchy" (Reed 1986). In parts of Washington contiguous to the southern interior of

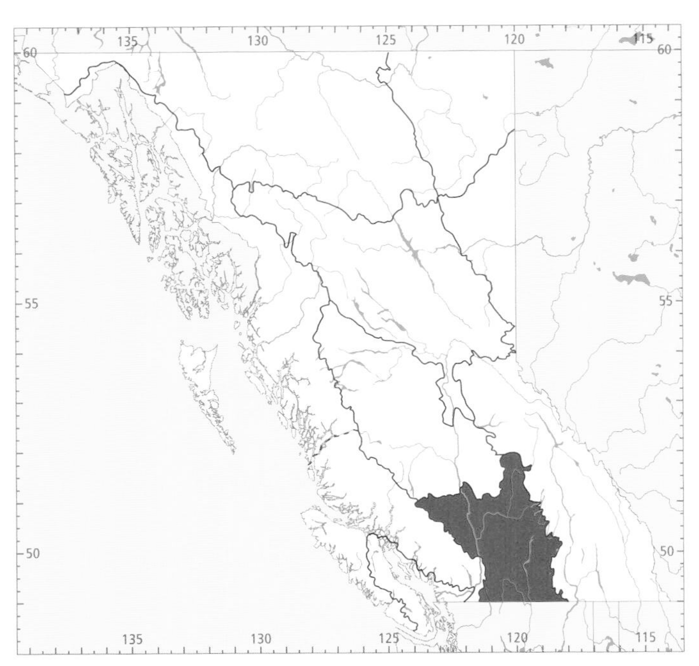

Figure 245. In British Columbia, the highest numbers for the Vesper Sparrow in summer occur in the Southern Interior Ecoprovince.

British Columbia, it is a species of the sagebrush-dominated landscape (Smith et al. 1997). Further north in British Columbia, the species occupies fields and grasslands usually adjacent to trembling aspen groves and frequently close to ponds or lakes. There the vegetation is taller than that on the southern rangelands. In the Peace Lowland, however, this sparrow occurs on open range and overgrazed meadowland, frequently interspersed with trembling aspen groves, and on the edges of grain fields, not unlike areas selected in parts of the Cariboo and Chilcotin areas (Cowan 1939). The species tends to be closely associated with habitat that provides these special characteristics, resulting in its spotty distribution, especially towards the edges of its range in the province.

On the coast, the Vesper Sparrow has been recorded breeding from 7 May (calculated) to 11 June; in the interior, it has been recorded breeding from 6 May (calculated) to 9 August (Fig. 243).

Nests: Most nests (65%; *n* = 92) were in big sagebrush rangeland, followed by grassland with Douglas-fir and ponderosa pine (17%), meadow and pasture (12%), trembling aspen parkland (4%), and cultivated field (2%). Almost all nests (99%; *n* = 176; Figs. 248 and 249) were on the ground, in a depression sheltered by a grass clump or adjacent sagebrush; 2 nests were in shrubs. Grasses formed the main component of all nests; other materials included hair, twigs, coarse plant stems, and plant fibre.

Eggs: Dates for 242 clutches (Fig. 248) ranged from 9 May to 1 August, with 51% recorded between 29 May and 21 June. Calculated dates indicate that nests may contain eggs as early as 6 May. Sizes of 168 clutches ranged from 1 to 5 eggs (1E-8, 2E-8, 3E-53, 4E-89, 5E-10), with 53% having 4 eggs. The incubation period is 11 to 13 days (Rising 1996). The presence of nests with eggs in late July and early August indicates that the species may be double-brooded in British Columbia, as it is reputed to be in other parts of its range (Rising 1996).

Young: Dates for 124 broods ranged from 21 May to 9 August, with 53% recorded between 10 June and 26 June. Sizes of 87 broods ranged from 1 to 6 young (1Y-9, 2Y-11, 3Y-27, 4Y-38, 6Y-2), with 75% having 3 or 4 young (Fig. 249). The nestling period is 7 to 12 days (Rising 1996).

Brown-headed Cowbird Parasitism: In British Columbia, 2 of 215 nests found with eggs or young were parasitized by the cowbird. There were no additional records of Vesper Sparrows feeding cowbird fledglings. On the coast, there were no instances of parasitism (*n* = 3); in the interior, 3 of 212 nests (1%) each contained a single cowbird egg. This low rate of parasitism in British Columbia contrasts sharply with the situation in the Prairie provinces, where 11.8% of 195 nests were parasitized (Friedmann et al. 1977). Friedmann (1963) reported that this sparrow was a fairly frequent host of the Brown-headed Cowbird, and cited instances from throughout the sparrow's range. It has been known to raise cowbird young to fledgling stage.

Nest Success: Of 29 nests found with eggs and followed to a known fate, 20 produced at least 1 fledgling, for a nest success rate of 69%. One nest on the coast was not successful; in the interior, nest success was 71% (*n* = 28).

Elsewhere, it has been shown that the density of ground cover and the amount of litter on the ground strongly influences the success of Vesper Sparrow nests (Wray and

Figure 246. Throughout the Okanagan valley, in the Southern Interior Ecoprovince of British Columbia, the Vesper Sparrow nests among short grasses in big sagebrush shrub-steppe (White Lake, 13 May 1997; R. Wayne Campbell).

Figure 247. In the Boreal Plains Ecoprovince of British Columbia, the Vesper Sparrow inhabits grassy "breaks" on the south- and west-facing slopes along the Peace River (15 km west of Charlie Lake, 18 June 1996; R. Wayne Campbell).

Figure 248. Ground nest containing 3 eggs of the Vesper Sparrow. Overhanging grasses have been parted to expose the nest (Cassidy, 23 May 1984; R. Wayne Campbell).

Whitmore 1979). In their study, 62% of 39 nests were unsuccessful. Successful nests (n = 15), as opposed to unsuccessful nests (n = 24), had a vertical cover density of 15% versus 12%, a percentage of litter cover of 63% versus 54%, and a percentage of bare ground of 33% versus 45% for unsuccessful nests. Predation, largely by crows, accounted for all unsuccessful nests.

REMARKS: Two subspecies are known in British Columbia. The subspecies occupying the province east of the Coast and Cascade mountain ranges is *P. g. confinis.* The coastal plain from British Columbia south to extreme northwestern California is the range of *P. g. affinis* (American Ornithologists' Union 1957).

In British Columbia, the Vesper Sparrow has never been numerous on the southwest coast. Prior to 1947 it was known from 2 specimen records (Chilliwack and Alta Lake), both taken in September 1927. In addition, there were 10 observational records between 1888 and 1947. Five of these were spring records, 2 from the Victoria area and 3 from the Fraser Lowland. There were many naturalists observing birds in the region during these years. Since then, there has been an increase in the number of records and in the consistency of the species' occurrence. For instance, since 1970 the species has been recorded in every year but one. The subspecies *affinis,* however, cannot be distinguished from *confinis* under field conditions, and there has been no systematic study to identify the coastal subspecies present during the nesting season. On the basis of habitat type and geographic continuity with the known distribution of the race in Oregon and Washington, it is assumed that the summer birds in the lowlands of southwestern British Columbia belong to the subspecies *affinis*.

On Vancouver Island, the earliest records for the months of May and June were in 1890, 1891, and 1892. There were no records between then and 1957. From 1971 to 1985, there were a series of reports of Vesper Sparrows in May, June, and July in the area between Duncan, Cobble Hill, and Mill Bay. Most were of single birds, but there is a record of 6 at Cobble Hill on 7 July and of 7 at Cobble Meadows on 17 June. It is reasonable to conclude that the species was nesting in that region. The Nanaimo Lowland, between Nanaimo and Cassidy, has recently (1990 through 1996) provided a series of spring and summer observations, including the only known nesting on Vancouver Island (Fig. 248).

In the Fraser Lowland, the history of the Vesper Sparrow as a summer visitant is somewhat similar to that on Vancouver Island. A.C. Brooks, who lived and did extensive collecting of birds in the Chilliwack region beginning in 1887 (Brooks 1917), did not find the Vesper Sparrow during the summer months. He listed the species as a very scarce migrant. The first spring and summer records were in 1938, and there were sporadic observations of the species in 1944, 1961, 1962, 1968, 1979, 1982, and 1987. The Vesper Sparrow is not mentioned by Campbell et al. (1972a, b) or Butler and Campbell (1987), but the species has been reported more often in recent years (Elliott and Gardner 1997). There are nesting records for the

Figure 249. Vesper Sparrow nest with 4 young. In the interior of British Columbia, broods have been recorded between 26 May and 9 August (Empire Valley, 26 May 1994; R. Wayne Campbell).

Fraser Lowland, at New Westminster in 1938 and Iona Island in 1968, and unconfirmed reports of more recent breeding elsewhere on the Fraser River estuary.

Evidence suggests that under conditions prevailing in the 19th century, the Vesper Sparrow was probably rare but breeding on southern Vancouver Island but not on the Fraser River delta. Subsequently, land clearing and the establishment of large areas planted to grain and other crops, as well as pastureland, provided habitat attractive to this sparrow, and a small population moved in, probably from adjacent Washington. The trend of the population since the late 1960s is unclear but may be reflected in the events in the Pacific states.

Jewett et al. (1953) described the northwestern race of the Vesper Sparrow as quite common in some parts of western Washington north to Dungeness and San Juan Island, east to Yelm and Seattle, and south as far as Camas and Vancouver. Thirty years later, Lewis and Sharpe (1987) stated that "disappearing habitat is causing a gradual abandonment by Vesper Sparrows of former nesting areas in most of western Washington and Vancouver Island." At that time, San Juan Island remained the stronghold for this sparrow in Washington. More recently, *affinis* has been reported to be in danger of extirpation in Washington because of habitat destruction (Smith et al. 1997). In western Oregon, Gilligan et al. (1994), without referring to the subspecies, referred to the Vesper Sparrow as a "locally uncommon summer resident, only from Bandon, Coos County south through Curry County."

Small (1994) refers to the Vesper Sparrow as an extremely rare breeder in its range in California. Fifty years earlier, Grinnell and Miller (1944) had provided no data on breeding records for the state, but described it as a wintering species varying locally from rare to common. It is clear that in the northern Pacific states the western race of the Vesper Sparrow is declining in numbers and in the extent of its breeding range. It remains scarce in coastal British Columbia.

NOTEWORTHY RECORDS

Spring: Coastal – Victoria 3 May 1973-1 at 10 Mile Point; Mount Douglas 9 May 1994-1; 1.8 km wsw Cobble Hill 12 Apr 1978-2; South Pender Island 24 Apr 1992-1 (Siddle 1993c); 11.5 km sw Duncan 10 May 1972-1 (Tatum 1973); Cassidy 30 Apr 1993-2 on fence at airport, 23 May 1984-3 eggs at airport (Fig. 248); Long Beach 25 Apr 1981-1 on sand dunes, 7 May 1969-1 on sand dunes; Port Alberni 21 May 1992-1 (Siddle 1993c); Iona Island 17 May 1968-4 eggs; Serpentine River 10 Apr 1989-1; Pitt Meadows 21 Mar 1997-1; Grant Narrows (Pitt Meadows) 16 to 21 Apr 1995-1 (Elliott and Gardner 1997); Manning Park 22 Apr 1979-2 about 2.6 km nw Skaist River; New Westminster 10 May 1938-4 eggs; e side Pitt Polder 2 May 1975-1; Seabird Island 27 Apr 1992-1; Hope 20 May 1986-15 at airport; Prince Rupert 23 May 1974-1 (Hay 1976). **Interior** – Kilpoola Lake 10 Apr 1977-1, 12 Apr 1969-46; Similkameen River valley 30 May 1922-1; Midway 17 Apr 1905-1 (NMC 3149); Rossland 2 May 1973-1; sw White Lake (Okanagan Falls) 18 Apr 1978-200, 9 May 1990-4 eggs, 26 May 1990-4 young starting to develop pinfeathers, eyes open; Twin Lakes 18 Apr 1978-300 (Cannings et al. 1987); Vaseux Lake 9 May 1992-1; Castlegar 18 Apr 1969-1 nr Selkirk College; Kaslo 2 Apr 1963-1 banded; Okanagan Landing 4 Apr 1942-1, first of the year; Tunkwa Lake 12 May 1968-20; Radium Hot Springs 31 May 1979-adult incubating 4 eggs; Kootenay National Park 12 May 1982-10 in a flock; Lac du Bois area 5 May 1972-38; Revelstoke 19 May 1986-2; Yoho National Park 27 Apr 1977-2 at Ottertail horse pasture; 100 Mile House 19 Mar 1978-1, 1 May 1934-5; Watson Lake (100 Mile House) 9 May 1966-5 (MCZ 285193 through 285197); Alkali Lake (Cariboo) 26 May 1994-2 adults feeding 4 nestlings; Riske Creek 19 Apr 1984-2, 25 Apr 1993-40 estimated, 22 May 1978-3 eggs; Anahim Lake 15 May 1932-1 (MCZ 285075); Chezacut 22 May 1943-12; 24 km s Prince George 5 May 1980-2; Prince George 8 May 1997-1; Nulki Lake 18 May 1945-1 (ROM 88995); Pack River (Mackenzie) 15 May 1994-1; Chetwynd 23 May 1964-3; Cecil Lake 27 Apr 1986-1 singing, spring arrival; Buckinghorse River 23 May 1991-1.

Summer: Coastal – Cobble Hill 17 Jun 1978-7; Cassidy 20 Jun 1990-6 at airport (Siddle 1990c); Englishman River 2 Jul 1981-1 on estuary, 14 Aug 1980-1 (Dawe et al. 1994); Vancouver 10 and 11 Jun 1992-1; Harrison Hot Springs 22 Jun 1975-2; Comox 5 Aug 1973-1 at air force base; Alta Lake 2 Aug 1941-1 (Racey 1948); Bella Coola 19 Jun 1977-4 feeding in backyard. **Interior** – Elko 13 Jul 1970-3 nestlings; Kimberley 28 Jun 1979-12 at airport; Wasa 23 Jul 1971-3 fledglings (Dawe 1971); Stump Lake 31 Aug 1975-150; Vernon 9 Aug 1993-6 nestlings; Mount Wardle (Kootenay National Park) 12 Jun 1979-nest with 4 nestlings (Poll et al. 1984); Tenquille Lake 23 Jul 1960-1 (Bradley 1961); Hat Creek 13 Jul 1996-26; Kamloops 2 Jun 1982-3 eggs, 7 Jun 1986-6; Enderby 15 Jul 1948-2 eggs; Heffley Creek 6 Jun 1975-4 nestlings; Scotch Creek 10 Jul 1973-1; Tatla Lake 19 Jul 1978-10; Bald Mountain (Riske Creek) 9 Aug 1983-50; Farwell Canyon 29 Jul 1993-119 counted along Farwell Rd (Campbell and Stewart 1993); Alkali Lake (Cariboo) 9 Aug 1993-6 young; nr Riske Creek 1 Aug 1960-3 eggs; Nulki Lake 30 Aug 1945, last record (Munro 1949); Prince George 22 Jun 1969-1; Chichouyenily Creek 7 Jun 1997-1; Farrell Creek 22 Jun 1979-5 eggs (BC Photo 1641); Cecil Lake to Fort St. John 14 Jun 1974-23; Muncho Lake 12 Jul 1995-1; Coal River 23 Jul 1977-1 at elevation of 500 m (Campbell 1978d); Liard River 15 Jul 1977-1.

Breeding Bird Surveys: Coastal – Not recorded. **Interior** – Recorded from 43 of 73 routes and on 49% of all surveys. Maxima: Pleasant Valley 22 Jun 1980-59; Summerland 21 Jun 1976-55; Riske Creek 16 Jun 1991-52.

Autumn: Interior – nr Clayhurst 8 Oct 1986-3 along Peace River; Riske Creek 2 Sep 1978-100; Hanceville 20 Sep 1982-4; Salmo 2 Oct 1983-2; Nicola Lake 29 Sep 1996-1; Oliver 22 Oct 1984-1. **Coastal** – Garibaldi Lake 17 Sep 1982-1 at Black Tusk meadows; Harrison Hot Springs 17 Sep 1992-1; Vancouver 4 Sep 1995-1 (Elliott and Gardner 1997); Iona Island 4 Oct 1997-1; Aldergrove 12 Oct 1968-6; Boundary Bay 29 Sep 1960-3; Crescent Beach 18 Oct 1975-1; Victoria 15 Nov 1997-1.

Winter: Interior – No records. **Coastal** – nr Harrison Hot Springs 9 Feb 1984-1; Central Saanich 17 to 26 Dec 1994-1 at Martindale Flats.

Christmas Bird Counts: Not recorded.

Lark Sparrow
Chondestes grammacus (Say)

LASP

RANGE: Breeds from southern interior British Columbia, southeastern Alberta, southern Saskatchewan, and Manitoba south in the west into southern California and Arizona, and in Mexico to Zacatecas and Tamaulipas. In the east, from southern Ontario south through Wisconsin, Minnesota, and Michigan to eastern Texas, Louisiana, western Virginia, and North Carolina. Winters from central California, southern Arizona, eastern Texas, and the Gulf coast south in Mexico to southern Baja California, Chiapas, and Veracruz.

STATUS: On the coast, *very rare* spring and autumn transient and *casual* in summer and winter in the Georgia Depression Ecoprovince; in the Coast and Mountains Ecoprovince, *very rare* vagrant on Western Vancouver Island, *casual* on the Southern Mainland Coast, and absent from the Northern Mainland Coast and Queen Charlotte Islands.

In the interior, an *uncommon* to *fairly common* migrant and local summer visitant in the Southern Interior Ecoprovince; *very rare* migrant and summer visitant in the Southern Interior Mountains Ecoprovince; *very rare* in the Central Interior Ecoprovince; *casual* in the Sub-Boreal Interior Ecoprovince; and *accidental* in the Taiga Plains Ecoprovince.

Breeds.

NONBREEDING: The Lark Sparrow (Fig. 250), although one of the less abundant and most habitat-specific species among the sparrows in British Columbia, is a characteristic summer bird in the southern Okanagan and Similkameen valleys, and occurs regularly as far north as the South Thompson River valley. It is less numerous across the southeastern areas of the interior of the province as far as the Kootenay River valley, and occurs regularly but locally east as far as Nelson and Creston and north into the Cariboo and Chilcotin areas. The easternmost occurrences in the south are from Fernie and Field; the northernmost record west of the Rocky Mountains is from McLeod Lake, north of Prince George. There is a single record from Trutch in the southern Taiga Plains. The species occurs irregularly on the south coast, including southern and eastern Vancouver Island to Tofino and Nimpkish, the Gulf Islands, the lower Fraser River valley to Hope, and the Sunshine Coast to Sechelt. The northernmost coastal occurrence is from Bella Coola.

On the coast, the Lark Sparrow occurs from near sea level to 25 m; in the interior, it has been reported from 280 m in valley bottoms near Osoyoos to about 1,200 m on Anarchist Mountain (Cannings et al. 1987), with most of the population found below 750 m.

The Lark Sparrow frequents open habitats during migration and the post-breeding period. On the coast, these include beaches, trails, brushy edges of fields and fence rows, urban lawns, weedy fields, and feeders. In the interior, habitats are apparently the same as those where nesting occurs; descriptions are found under BREEDING.

Figure 250. Adult Lark Sparrow on territory (near Chopaka, 7 July 1997; R. Wayne Campbell).

The Lark Sparrow is one of the later spring migrants to arrive in the province. In the interior, the earliest arrivals have entered the Southern Interior as early as the second week of April (Figs. 251 and 252), but in the southern Okanagan valley, spring arrival dates over a 15-year period ranged from 6 May to 20 May, with a mean date of 14 May (Cannings et al. 1987). A much smaller number of arrival dates in the Southern Interior Mountains has about the same distribution. In both ecoprovinces, the spring migration peaks in the second half of May. The earliest of the small number of spring arrival dates in the Cariboo and Chilcotin areas are between 13 and 27 May (Fig. 251).

The numbers of Lark Sparrows recorded in the interior decline sharply after mid-July, and most birds have left the province by mid to late August. The latest dates in the southern portions of the interior of the province range from 30 August (Munro and Cowan 1947) to 20 September (Fig. 251).

On the coast, most records during migration are of single birds seen from late April through early June and from September through November. There are few summer and winter records.

On the coast, the Lark Sparrow has been recorded irregularly throughout the year; in the interior, it has been recorded from 12 April to 20 September (Fig. 251).

BREEDING: The Lark Sparrow has a restricted breeding range in British Columbia. It is almost confined to the Okanagan and Similkameen valleys, with 90% of occurrences south of the latitude of Summerland. Savona is likely both the northernmost and westernmost nesting locality. The only breeding record east of the Okanagan valley is from the Trail area. The mention by Brooks (1917) that "in 1899 I saw a breeding pair in the valley," referring to the Chilliwack area, has not been confirmed.

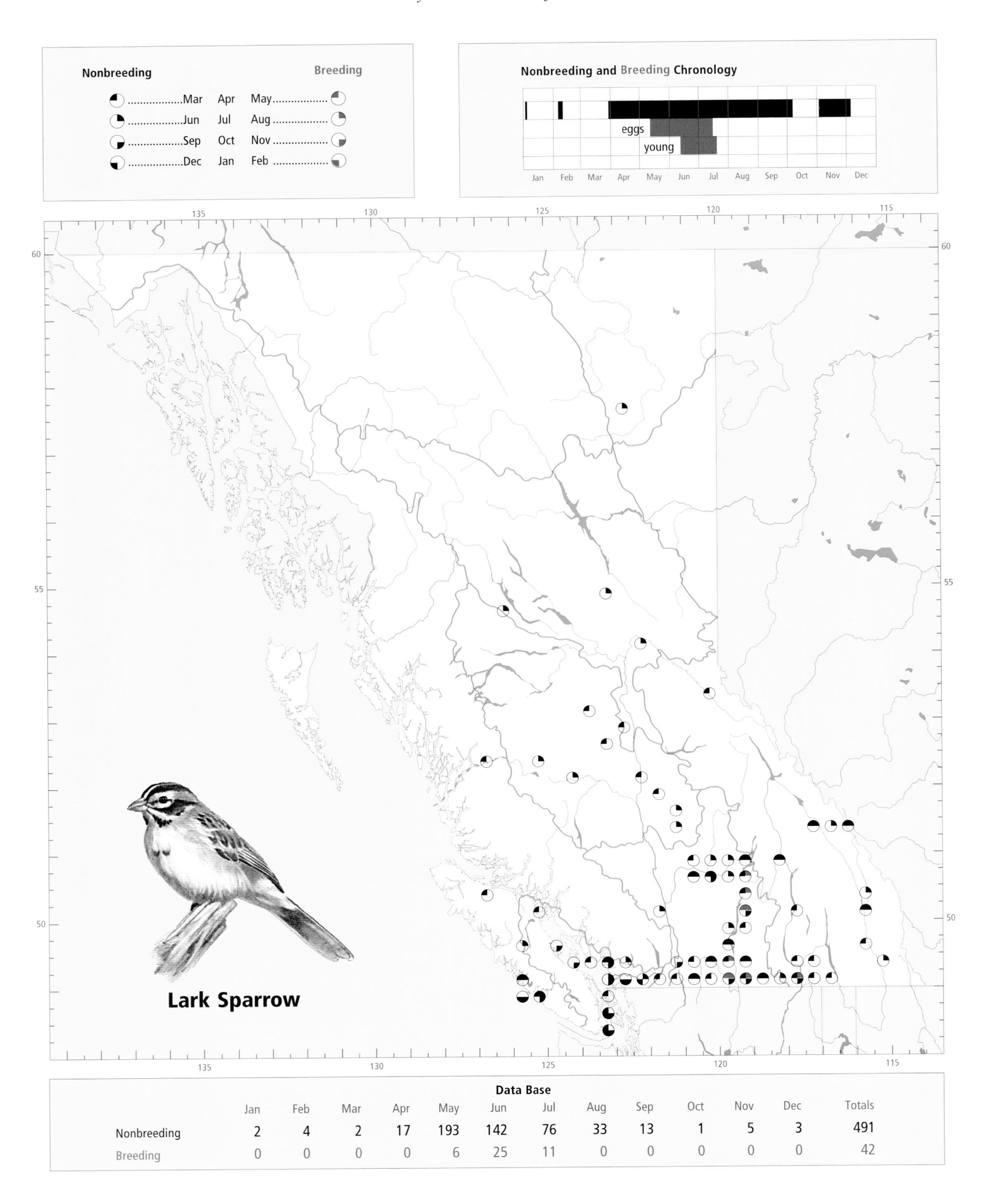

Data Base

	Jan	Feb	Mar	Apr	May	Jun	Jul	Aug	Sep	Oct	Nov	Dec	Totals
Nonbreeding	2	4	2	17	193	142	76	33	13	1	5	3	491
Breeding	0	0	0	0	6	25	11	0	0	0	0	0	42

Figure 251. Annual occurrence (black) and breeding chronology (red) for the Lark Sparrow in ecoprovinces of British Columbia. Records are shown for the week in which they occurred.

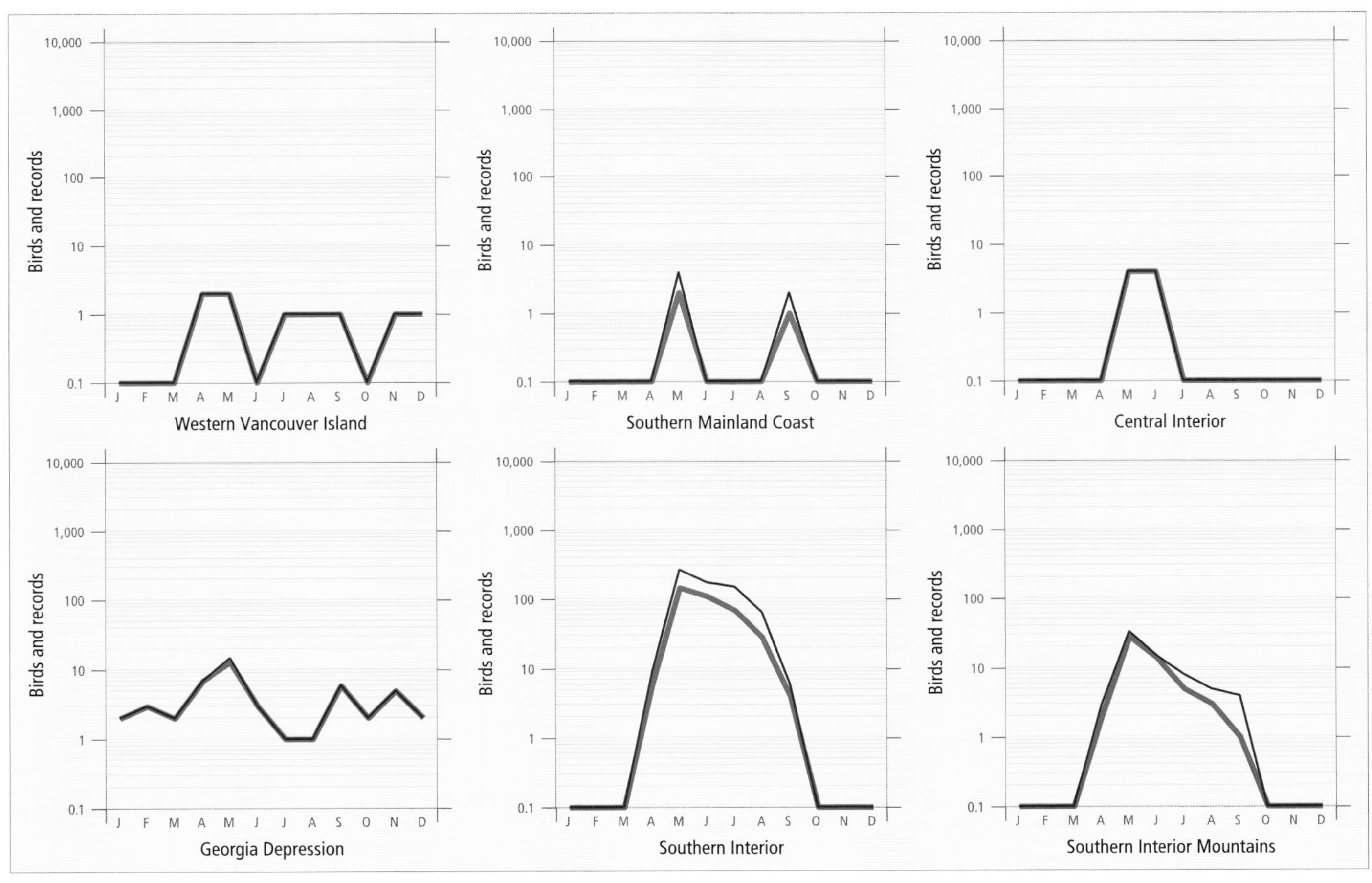

Figure 252. Fluctuations in total number of birds (purple line) and total number of records (green line) for the Lark Sparrow in ecoprovinces of British Columbia. Christmas Bird Counts, Breeding Bird Surveys, and nest record data have been excluded.

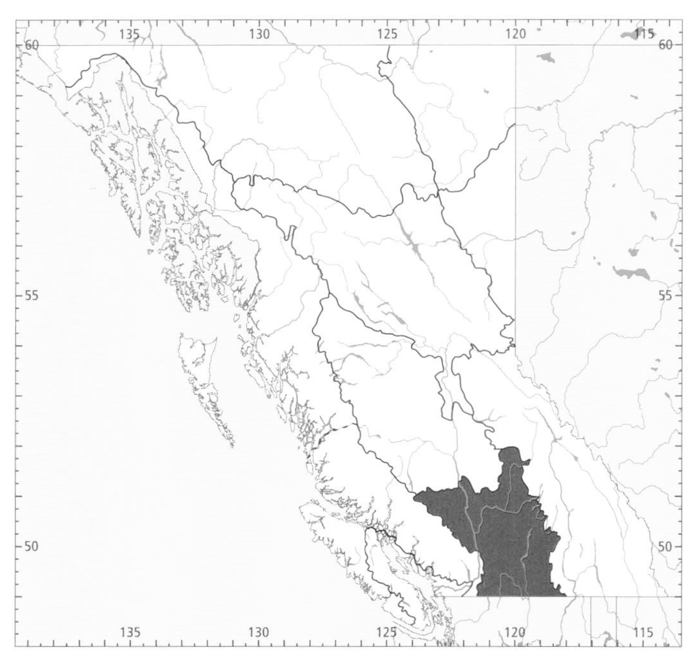

Figure 253. In British Columbia, the highest numbers for the Lark Sparrow in summer occur in the Southern Interior Ecoprovince.

Figure 255. In the southern Okanagan valley of British Columbia, the Lark Sparrow regularly nests in drier habitats with bare ground and sparse big sagebrush cover (Chopaka, 13 May 1997; R. Wayne Campbell).

In summer, the Lark Sparrow reaches its highest numbers in the southern Okanagan valley (Fig. 253). An analysis of Breeding Bird Surveys for the period 1968 through 1993 could not detect a net change in numbers on interior routes. The species does not breed in the areas sampled by the coastal routes. Over the North American range of this sparrow, Breeding Bird Surveys from 1966 to 1996 revealed a decline in numbers by an average annual rate of 3.2% ($P < 0.01$) (Sauer et al. 1997). Like many other species, the Lark Sparrow has experienced large changes in numbers from year to year, although these have not been documented in detail in British Columbia. From 1927 to 1929, it was present in the grasslands around Kamloops in such numbers that 2 or 3 pairs could be encountered in a morning's hike. Several nests were found. It was still present in 1941, and the behaviour of some birds indicated the presence of nests northwest of Kamloops, on the grasslands bordering the road to McQueen Lake. Since then it has become less common, and we have just 1 recent breeding record from the Kamloops area.

The Lark Sparrow has been found nesting at elevations between about 280 and 500 m. It occurs in summer on Anarchist Mountain at elevations as high as 1,200 m, and probably

Figure 254. In the centre of its breeding range in the Southern Interior Ecoprovince of British Columbia, the Lark Sparrow prefers breeding habitats that include antelope-brush with a lower density of big sagebrush and associated perennial grasses (13 km north of Osoyoos, 7 June 1998; R. Wayne Campbell).

Figure 256. All Lark Sparrow nests in British Columbia were placed on the ground, usually beside a low shrub. On the day this photograph was taken, there was a nest under the dead branches in the foreground (13 km north of Osoyoos, 7 June 1998; R. Wayne Campbell).

Figure 257. In British Columbia, Lark Sparrow nests are well concealed by vegetation (13 km north of Osoyoos, 29 May 1998; Mark Nyhof).

nests there occasionally. The summer habitat of the Lark Sparrow is concentrated in the shrub-steppe grasslands of the southern parts of the valleys of the Okanagan and Similkameen rivers.

This sparrow occurs at greatest density in stands of antelope-brush and in somewhat lower density in mixed shrubbery dominated by big sagebrush and associated perennial grasses (Figs. 254 and 255). R. Millikin (pers. comm.) found breeding Lark Sparrows in drier habitats with bare ground and a low big sagebrush cover (Fig. 255). In the northern Okanagan valley and in the Kamloops area, where antelope-brush is absent, it is found in big sagebrush stands and "degraded bunchgrass habitats with saskatoon and other shrubs and adjacent weedy fields" (Cannings et al. 1987). West of Savona, Dawe and Buechert (1996) found singing males in heavily grazed bluebunch wheatgrass habitat with scattered Ponderosa pine, big sagebrush, and some prickly-pear cactus. In Washington state, the preferred habitat is shrub savannah, where shrubs are present but scarce and grasses dominate, especially cheatgrass (Smith et al. 1997). When big sagebrush cover becomes dense, the Lark Sparrow is replaced by the Brewer's Sparrow and the Sage Sparrow (Stepniewski 1994).

The Lark Sparrow has been recorded breeding in British Columbia from 11 May (calculated) to 19 July (Fig. 251).

Nests: All nests were on the ground, usually beside a low shrub or partially concealed by a clump of grasses and other vegetation (Figs. 256 and 257). The nests were cup-shaped and set into depressions. Grasses were the principal component of all nests. Other materials used were hair (17% of nests), stems (11%), plant fibre (11%), and rootlets (6%). All nests were lined with fine grasses.

Eggs: Dates for 26 clutches ranged from 12 May to 15 July, with 54% recorded between 1 June and 4 July. Sizes of 22 clutches ranged from 1 to 5 eggs (1E-1, 2E-1, 3E-1, 4E-11, 5E-8), with 86% having 4 or 5 eggs (Fig. 258). The incubation period elsewhere is 11 to 13 days (Baicich and Harrison 1997). The Lark Sparrow may be double-brooded in British Columbia (Fig. 259).

Young: Dates for 15 broods ranged from 12 June to 19 July, with 9 recorded between 17 June and 29 June. Sizes of 13 broods ranged from 2 to 4 young (2Y-1, 3Y-3, 4Y-9), with 69% having 4 young. The nestling period is 9 to 10 days (Baicich and Harrison 1997).

Brown-headed Cowbird Parasitism: In British Columbia, only 1 of 35 nests found with eggs or young was parasitized by the cowbird. Friedmann (1963) found the Lark Sparrow to be seldom parasitized by the Brown-headed Cowbird. Subsequently, a study in Oklahoma revealed a population of this sparrow enduring a parasitism rate of 46% (Newman 1970).

Figure 258. Complete clutch of 5 eggs of the Lark Sparrow in its typical grass nest (13 km north of Osoyoos, 7 June 1998; R. Wayne Campbell).

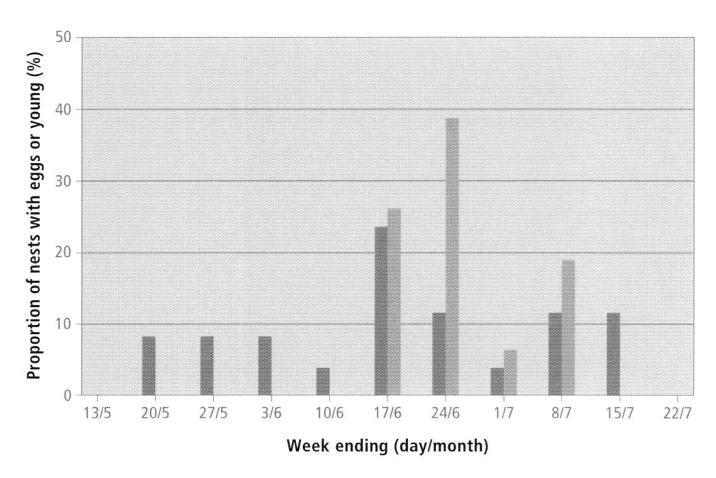

Figure 259. Weekly distribution of the proportion of clutches (dark bars) and broods (light bars) for the Lark Sparrow in the Southern Interior Ecoprovince (clutches: $n = 23$; broods: $n = 14$). The figure suggests that a few pairs may attempt a second brood.

The study also showed that only 9% of the parasitized nests actually fledged young cowbirds, whereas 20% of the sparrow eggs in the parasitized nests and 59% of unparasitized Lark Sparrow eggs resulted in Lark Sparrow fledglings. Friedmann et al. (1977) suggest that the parasitized nests may experience higher losses to predators, perhaps because of the louder, more insistent begging of the cowbird nestlings.

Nest Success: Insufficient data.

REMARKS: Two subspecies of the Lark Sparrow have been described. An eastern subspecies, *C. g. grammacus*, occupies the species range from Minnesota to eastern Texas and eastward, and a western subspecies, *C. g. strigatus*, breeds in the range westward. The latter is the subspecies found in British Columbia (American Ornithologists' Union 1957).

Like many ground-nesting species, the Lark Sparrow is affected by agricultural activities, grazing livestock, and recreational vehicles on their breeding grounds (Kantrud and Kologiski 1982). Some impacts are direct, whereas habitat loss, disturbance, fire control programs, and harvesting activities influence the species indirectly. The effects of human-related activities on ground-nesting birds in British Columbia need to be researched in representative habitats so that appropriate management guidelines can be developed.

Cannings et al. (1987) report an unusual instance of interspecific cooperation: a female Mountain Bluebird was photographed (BC Photo 850) feeding 4 almost fully feathered young Lark Sparrows in their nest.

For additional life-history information for the Lark Sparrow in North America, see Rising (1996).

NOTEWORTHY RECORDS

Spring: Coastal – Mount Tolmie 31 May 1990-1; Saanich 31 Mar to 6 Apr and 13 May 1973-1 (Campbell 1973; Crowell and Nehls 1973b; BC Photo 280); Island View Beach 2 Apr 1956-1 (Lemon 1956); Chilliwack 21 May 1889-1 (Brooks 1917); Carnation Creek (Port Alberni) 15 May 1985-1; Saturna Island 29 May 1990-1 (Siddle 1990b); Long Beach 28 Apr 1975-1 (BC Photo 410); Skagit River valley 23 May 1976-1 feeding on ground in meadow with American Pipits at km 52; Sechelt 27 Apr 1992-1, 21 May 1992-2 (Bowling 1992); Campbell River 14 May 1974-1; Nimpkish River 26 May 1995-1 (McNicholl 1995); Bella Coola 27 May 1991-3 (Siddle 1991c). **Interior** – Cawston 15 May 1958-5 eggs; White Lake (Oliver) 21 May 1967-6; Similkameen 2 May 1941-1; Osoyoos 8 May 1977-6; Oliver to Richter Lake 17 May 1959-12 on count; Trail 19 Apr 1974-1; Waneta Junction 22 May 1980-3; Vaseux Lake 31 May 1978-10 at East Hills; Kimberley 17 May 1984-1 (Rogers 1984); Kamloops 29 May 1985-2; 10 km s Vernon 28 May 1960-4 eggs; Okanagan 12 Apr 1912-1 (FMNH 168272); Yoho National Park 21 May 1976-1 with White-crowned Sparrows (Wade 1977); 130 Mile House 11 May 1946-1 (Munro 1955a); Williams Lake 17 May 1991-1 (Siddle 1991c); Puntchesakut Lake 10 May 1944-1; Quesnel 23 May 1977-1.

Summer: Coastal – Bamfield 16 Aug 1976-1 at Brady's Beach; Long Beach 2 Jul 1987-1 (BC Photo 1186); Sea Island 22 Jun 1995-1 adult (Elliott and Gardner 1997); Mission Flats Park 7 Jun 1906-1; Vancouver 30 Aug 1985-1 (Hunn and Mattocks 1986); Buttle Lake 8 Jun 1997-1. **Interior** – Keremeos 18 Jun 1928-1 (RBCM 14120); Kilpoola Lake 18 Jun 1993-2 adults feeding 2 recently fledged young; Osoyoos 9 Jul 1953-8; Osoyoos to n end Osoyoos Lake 4 Aug 1993-12 on count; Columbia Gardens 19 Jul 1972-2 adults with 1 young, 23 Aug 1980-3; Hedley 4 Jun 1928-1 (NMC 22680); Similkameen River 7 Jun 1917-1; Vaseux Mountain 12 Jun 1980-1; e Vaseux Lake 30 Jun 1985-5 eggs; Okanagan Landing 15 Jul 1911-4 eggs, 30 Aug 1911-1, last seen (Munro and Cowan 1947); Kamloops 7 Jun 1914-1, 11 Jul 1963-10, 6 Aug 1929-2 adults with 3 fledglings; Savona 16 Jun 1996-2 males in full song along Sabiston Creek Rd (Dawe and Buechert 1996); Lac du Bois 10 Jun 1941-4, 3 Aug 1989-3 fledglings; Field 10 Jun 1977-1; Chilanko Forks 28 Jun 1961-1; Anahim Lake 21 Jun 1948-1 (UBC 1825); Willow River 5 Jun 1970-1, only record between 1965 and 1986; Topley 17 Jun 1956-1; nr Trutch 2 Jul 1965-1 flushed from roadside of the Alaska Highway at Mile 190 (Keith 1967).

Breeding Bird Surveys: Coastal – Not recorded. **Interior** – Recorded from 9 of 73 routes and on 5% of all surveys. Maxima: Oliver 18 Jun 1989-8; Grand Forks 2 Jul 1978-3; Summerland 21 Jun 1990-2.

Autumn: Interior – e Kamloops 1 Sep 1960-1; Okanagan Landing 15 Sep 1944-1, last of the year; Oliver 2 Sep 1973-3 along Camp McKinney Rd; Trail 20 Sep 1978-4 at mouth of Beaver Creek. **Coastal** – Comox 15 Sep 1928-1 (MVZ 105980); Hope 23 Sep 1922-2, first record for this locality; Qualicum Beach 7 Sep 1979-1 (BC Photo 588); Boundary Bay 2 Sep 1973-1 (Crowell and Nehls 1974a); Ucluelet 22 Nov 1994-1 (Bowling 1995d); Saanich 27 Nov 1985-1 at feeder; Victoria 4 Nov 1974-1.

Winter: Interior – No records. **Coastal** – Mission 3 Jan 1962-1 banded; Ucluelet 2 Dec 1994-1 (Bowling 1995d); Saanich 5 Dec 1973-1, 7 Feb 1973-1 (Crowell and Nehls 1973b).

Christmas Bird Counts: Not recorded.

Black-throated Sparrow

Amphispiza bilineata (Cassin)

BTSP

RANGE: Breeds from southeastern Washington, south-central and southeastern Oregon, southwestern Idaho, southwestern Wyoming, western and southern Colorado, northwestern Oklahoma, northern Arizona, northwestern New Mexico, and north-central Texas south through eastern California to southern Baja California. Winters from southern Nevada, central and southeastern Arizona, southern New Mexico, and central and southern Texas south through the remainder of the breeding range.

STATUS: On the coast, *very rare* vagrant in the Georgia Depression Ecoprovince.

In the interior, *casual* in the Southern Interior and Southern Interior Mountains ecoprovinces.

CHANGE IN STATUS: The Black-throated Sparrow (Fig. 260) is considered a vagrant anywhere in the north Pacific coast region, and its appearance in British Columbia is best understood in the context of its occurrence in northwestern Oregon and Washington.

Prior to 1959, a year of unusual extralimital occurrence in the Pacific Northwest (Dubois 1959), the Black-throated Sparrow was known from only a single occurrence, near Brooks Lake, Washington, in 1908 (Mattocks et al. 1976). The species was reported infrequently between 1960 and 1996 (e.g., Brown 1960), but more often and in more widespread locations in 1970, 1975, 1977, 1984, 1985, and 1994 (Hunn 1978; see specific references in National Audubon Society Field Notes). Hunn (1978) suggested that severe drought conditions within the breeding range at Malheur National Wildlife Refuge in Oregon and at other nearby locations may have contributed to the vagrancy in those years. The species now breeds regularly in Washington north to at least Vantage and probably Omah Lake (T. Wahl pers. comm.).

Through 1977, nearly 90% of vagrant Black-throated Sparrows in the Pacific Northwest occurred between 6 May and 10 June (Hunn 1978). In northern California, the Black-throated Sparrow arrived at breeding sites in mid-April (McCaskie and DeBenedictis 1966). Hunn (1978) suggests, therefore, that occurrences to the north are of individuals that have "overshot the mark during spring migration and have been wandering for some time."

The Black-throated Sparrow did not occur in the province when Munro and Cowan (1947) reviewed the avifauna, and was first recorded in Canada in June 1959 at Murtle Lake, in the northern portion of the Southern Interior Mountains. Since then it has been reliably reported 18 times between 1960 and 1997.

OCCURRENCE: The Black-throated Sparrow has been reported at widely scattered localities across southern British Columbia, from southern Vancouver Island to Kimberley in the east Kootenay and north to Wells Gray Park.

Figure 260. The Black-throated Sparrow, which prefers arid regions of the southwestern United States, is considered a vagrant anywhere north of Oregon (Adrian Dorst).

It has appeared in a variety of habitats, including clearings in heavily forested areas, open sagebrush flats, low grass areas, open areas adjacent to mixed forests and shrubby environments, and roadsides. It also visits feeders.

In British Columbia, the Black-throated Sparrow has occurred from near sea level to 1,050 m elevation. All records have been of single males.

On the coast, the Black-throated Sparrow has been recorded from 15 May to 19 June; in the interior, it has been recorded from 8 June to 5 July (Fig. 261).

The British Columbia records, updated from Van Damme (1998) and listed in chronological order, are as follows:

(1) Murtle Lake (Wells Gray Park) 8 June 1959-1 collected (NMC 44454) from a small clearing in heavy forest while feeding on the ground at 1,050 m elevation. This constitutes the first record for Canada (Godfrey 1961).
(2) Richter Lake (Richter Pass) 15 June 1977-1 singing near lakeshore (A.L. Pollard pers. comm.).
(3) Near Osoyoos 18 June 1979-1 clearly seen and documented (correspondence from Richard and Margaret James to Charles J. Guiguet; Royal British Columbia Museum).
(4) Spotted Lake (Richter Pass) 27 June 1981-1 singing in open sage (Brunton and Pratt 1986).
(5) Pitt Meadows 20 May 1984-1 well described (Fix 1984).
(6) White Rock 28 May 1984-1 seen by many observers and well documented (Fix 1984).
(7) Surrey 10 June 1984-1 near Barnston Island ferry landing (Harrington-Tweit and Mattocks 1984; BC Photo 970).
(8) Kimberley 12 June 1984-1 studied closely (Rogers 1984).

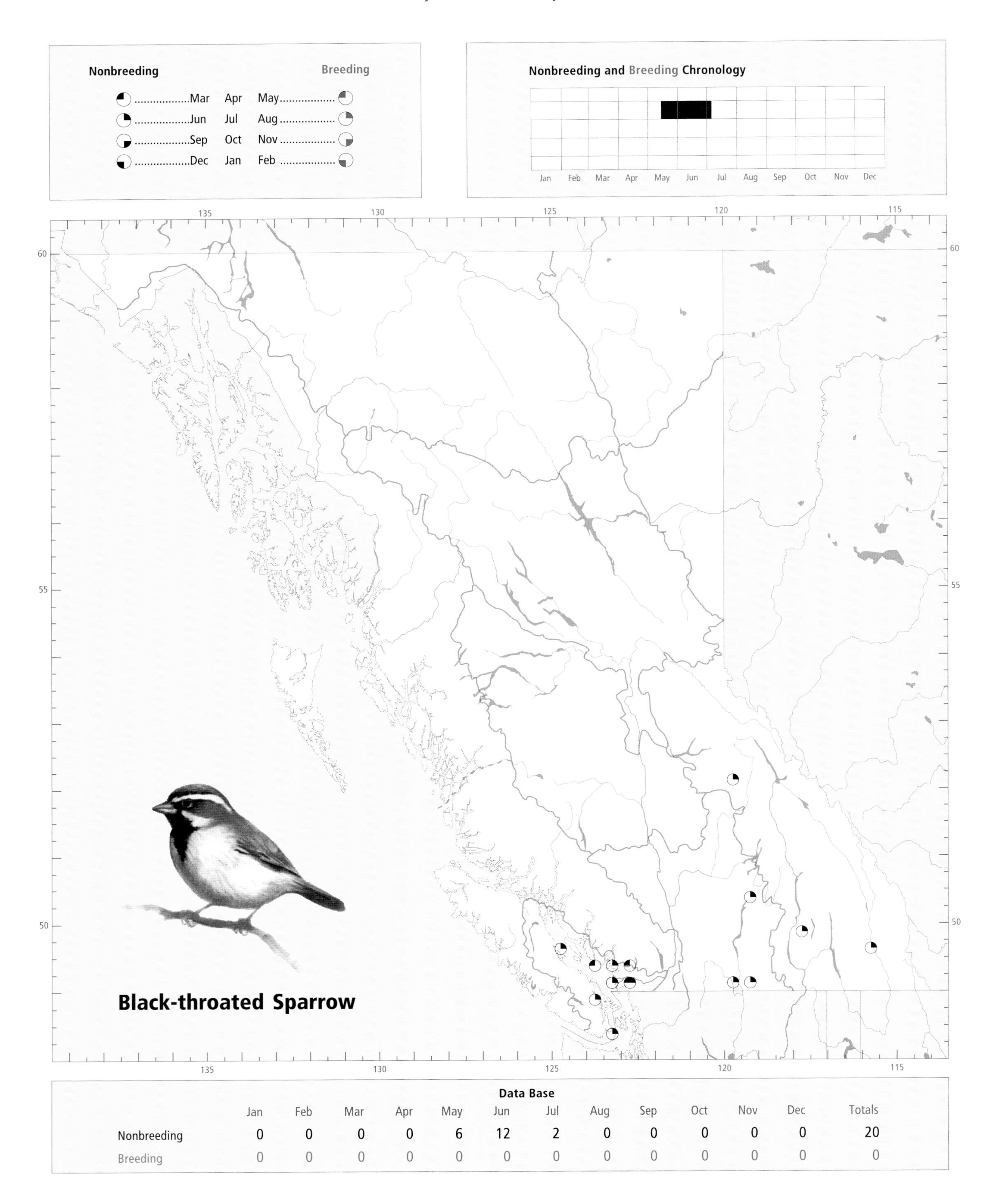

Data Base

	Jan	Feb	Mar	Apr	May	Jun	Jul	Aug	Sep	Oct	Nov	Dec	Totals
Nonbreeding	0	0	0	0	6	12	2	0	0	0	0	0	20
Breeding	0	0	0	0	0	0	0	0	0	0	0	0	0

Figure 261. Annual occurrence for the Black-throated Sparrow in ecoprovinces of British Columbia. Records are shown for the week in which they occurred.

(9) Reifel Island 13 June 1984-1 foraging in low grass at the Alaksen National Wildlife Area (Harrington-Tweit and Mattocks 1984).
(10) Burton 23 June 1984-1 moving among rocks on slope to shrubby lakeshore (G.S. Davidson pers. comm.).
(11) Pitt Meadows 19 May 1985-1 (Mattocks 1985b).
(12) West Vancouver 2 June 1985-1 feeding on grass seeds adjacent to forest edge at Juniper Point (Lighthouse Park) (Campbell 1985d; Harrington-Tweit and Mattocks 1985; BC Photo 1112).
(13) Somenos Lake (Duncan) 18 and 19 June 1992-1 pecking on a gravel road (Marven 1992; BC Photo 1779).
(14) Maple Ridge 15 to 28 May 1994-1 in yard (Bowling 1994a).
(15) Maple Ridge 27 May to 4 June 1994-1 in alternate plumage (J.A. Mackenzie pers. comm.; Escott 1994).
(16) Mount Tolmie (Victoria) 16 June 1994-1 singing (Bowling 1994b).
(17) Vernon 3 July 1994-1 in alternate plumage along Commonage Rd (Bowling 1994b).
(18) Sechelt 17 to 21 May 1996-1 at feeder (Bowling 1996c).
(19) Near Chopaka 5 July 1997-1 singing from fence post.

REMARKS: For a discussion on the conservation and management of grassland species in North America, see Vickery and Heekert (1999).

Lark Bunting

LKBU

Calamospiza melanocorys (Stejneger)

RANGE: Breeds from southern Alberta, southern Saskatchewan, southwestern Manitoba, southeastern North Dakota, and southwestern Minnesota south along the east of the Rocky Mountains to eastern New Mexico, northern Texas, western Oklahoma, eastern Kansas, and northwestern Missouri. Winters from southern California, southern Nevada, central Arizona, southern New Mexico, and north-central Texas south through Baja California, central Texas, and central Mexico.

STATUS: On the southern coast, *very rare* vagrant in summer and autumn in the Georgia Depression Ecoprovince.

In the interior, *casual* vagrant in the Southern Interior, Southern Interior Mountains, and Central Interior ecoprovinces; *accidental* in spring in the Boreal Plains Ecoprovince.

OCCURRENCE: The Lark Bunting (Fig. 262) is a rarity on the west coast of North America north of California, and is considered an irregular straggler to British Columbia (Roberson 1980). On the south coast, there are 7 reports of individual birds, all from the Georgia Depression: 1 from southern Vancouver Island, 3 from the Vancouver area, 2 from the lower Fraser River valley at Pitt Meadows and Chilliwack, and 1 from the Sunshine Coast near Gibsons. In the interior, reliable documentation is limited to reports of 10 individual birds from widely scattered locations: 4 from the Okanagan valley in the Southern Interior, 3 from the Southern Interior Mountains, 2 from the Central Interior, and 1 at Fort St. John in the Boreal Plains, which was also the northernmost occurrence. The Lark Bunting has been recorded in British Columbia at elevations from near sea level on the coast to 800 m in the interior. Its habitat includes big sagebrush flats (Fig. 263), cultivated fields, shrubby hedgerows, weedy roadsides and dykes, and tall grasses surrounding sewage lagoons.

There appears to be no regular movement of the Lark Bunting into British Columbia. It has been recorded on the coast in June, August, September, October, and November, and in the interior from 25 May to 11 July (Fig. 264). The majority of records are from May and June.

In chronological order, records include:

(1) Chilliwack River 1 August 1906-1 adult male collected at Thurston ranch (Macoun and Macoun 1909; NMC 3384).

(2) Okanagan Landing 8 June 1914-1 male collected in a thicket of hawthorns on the shore of Okanagan Lake (Munro 1915; RBCM 3901).

(3) Wistaria 28 May 1939-1 male collected among a newly arrived flock of Western Meadowlarks (Cowan 1940). This record is listed as Ootsa Lake in Munro and Cowan (1947).

(4) Kootenay National Park 26 May 1953-1 male on side of highway near Dolly Varden Creek in the Kootenay River flats. The bird was associated with a flock of Dark-eyed Juncos in a small clearing in a lodgepole pine forest (Banfield 1954; Godfrey 1955; Poll et al. 1984).

(5) Pitt Lake 30 August 1969-1 male in shrubby area along dyke (R.E. Luscher, pers. comm.).

Figure 262. The Lark Bunting is a rarity anywhere in the Pacific Northwest, and is considered an irregular straggler to British Columbia (Anthony Mercieca).

Figure 263. In 1993, a male Lark Bunting was found in big sagebrush flats at Kilpoola Lake, in the Southern Interior Ecoprovince of British Columbia (Kilpoola Lake, 5 June 1997; R. Wayne Campbell).

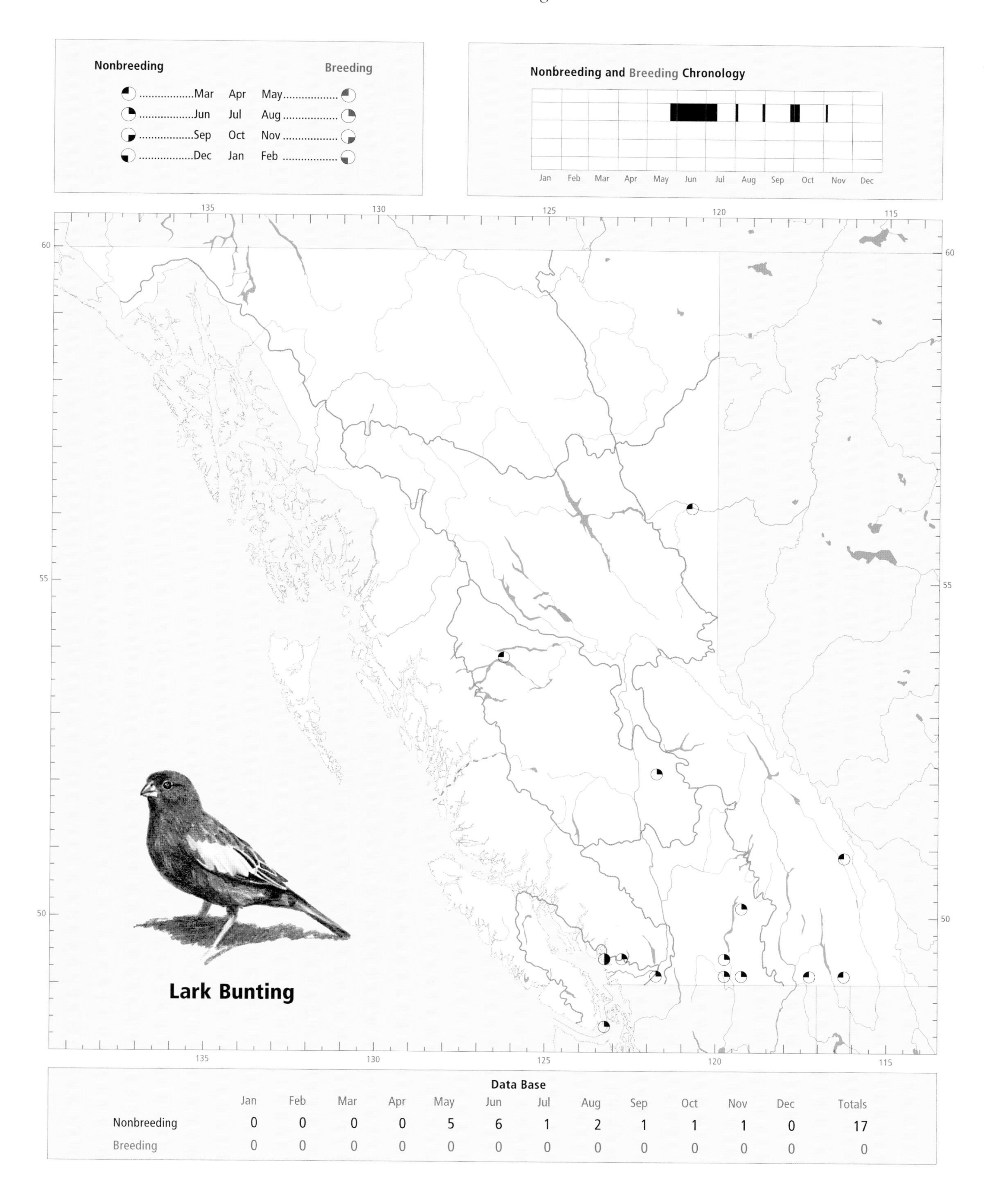

Data Base

	Jan	Feb	Mar	Apr	May	Jun	Jul	Aug	Sep	Oct	Nov	Dec	Totals
Nonbreeding	0	0	0	0	5	6	1	2	1	1	1	0	17
Breeding	0	0	0	0	0	0	0	0	0	0	0	0	0

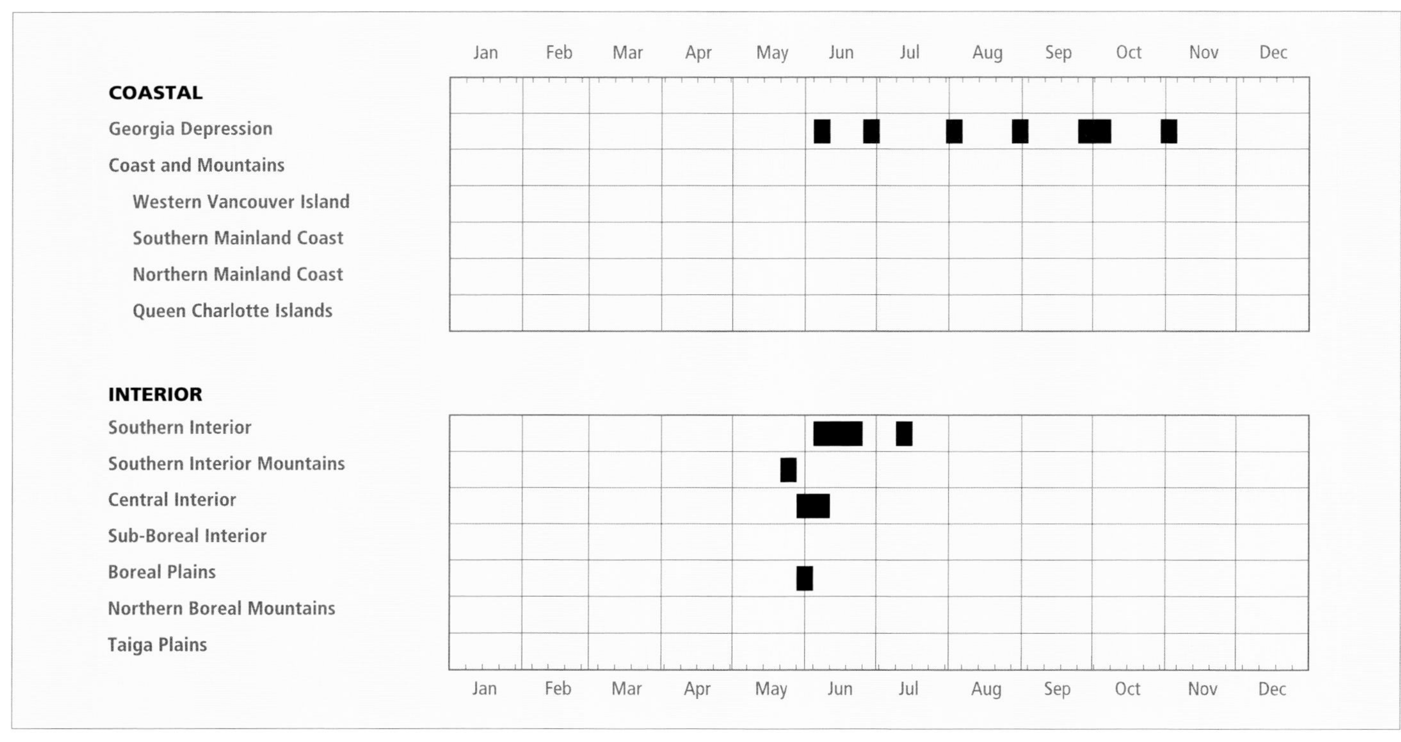

Figure 264. Annual occurrence for the Lark Bunting in ecoprovinces of British Columbia. Records are shown for the week in which they occurred.

(6) West Vancouver 5 October 1975-1 photographed at Ambleside Park (Crowell and Nehls 1976a).
(7) Vaseux Lake 11 July 1978-1 adult male (Williams 1978).
(8) Oak Bay 7 and 8 June 1980-1 male not yet fully adult photographed at feeding station (Guernsey 1980; BC Photo 635; Fig. 265).
(9) North of Osoyoos Lake 13 June 1982-1 male in breeding plumage chased by a male Bobolink near Road 22 (Cannings et al. 1987).
(10) North Vancouver 4 November 1983-1 male at Harbour View Park (Hunn and Mattocks 1984).
(11) Near Creston 25 May 1986-1 adult male perched in willow bush along road to Duck Lake (H.L. Moen pers. comm.).
(12) Vancouver 28 to 30 September 1986-1 juvenile at Jericho Park (BC Photo 1120).
(13) Fort St. John 28 May 1988-1 male at sewage lagoon (Siddle 1988c).
(14) 150 Mile House 10 June 1990-1 adult male singing from fence post bordering agricultural field.
(15) Granthams Landing (near Gibsons) 27 June 1992-1 male (Greenfield 1997).
(16) Kilpoola Lake 24 June 1993-1 adult male singing from sagebrush (G.P. Wilkinson pers. comm.).
(17) 1 km north of Nelway 27 May 1995-1 adult male in alternate plumage (D. Cooper pers. comm.).

REMARKS: The records of 3 birds seen in March and April 1984 north of Creston reported by Butler et al. (1986) are without documentation and have been excluded from this account.

Tate and Tate (1982) listed the Lark Bunting for the first time as a species of "special concern" because summer declines were noted in the key states of North and South Dakota and Texas.

Figure 265. In 1980, a male Lark Bunting visited a feeding station on southern Vancouver Island (Oak Bay, 7 June 1980; Bertha McHaffie-Gow).

Savannah Sparrow
Passerculus sandwichensis (Gmelin)

SAVS

RANGE: Breeds from western and northern Alaska, northern Yukon, the Northwest Territories south of the Arctic Archipelago, northern Ontario, northern Quebec, northern Labrador, and Newfoundland south to southwestern Alaska, coastal regions of west-central California, southern Nevada, southern Utah, east-central Arizona, northern New Mexico, central Colorado, Nebraska, Iowa, Kentucky, eastern Tennessee, western Virginia, central Maryland, and northern New Jersey, and on the Pacific coast of Baja California as far as Magdalena Bay. Also in the interior highlands of Mexico.

Winters from southwestern British Columbia, southern Nevada, southwestern Utah, northern Arizona, central New Mexico, Kansas, Missouri, Tennessee, southern Kentucky, and, east of the Appalachians, Maine south to southern Baja California, southern Texas, the Gulf coast, and southern Florida, and along the highlands of Mexico and Central America south to Guatemala and Honduras.

STATUS: On the coast, *common* to occasionally *very abundant* migrant, locally *fairly common* summer visitant, and *rare* to *uncommon* winter visitant in the Georgia Depression Ecoprovince; in the Coast and Mountains Ecoprovince, an *uncommon* to locally *common* migrant, *very rare* to *rare* summer visitant, and *casual* in winter on Western Vancouver Island and the Northern and Southern Mainland Coast; *uncommon* to locally *fairly common* migrant and *very rare* summer and winter visitant on the Queen Charlotte Islands.

In the interior, *uncommon* to *fairly common*, sometimes locally *common* migrant, *uncommon* to locally *fairly common* summer visitant in the Southern Interior, Southern Interior Mountains, and Central Interior ecoprovinces and *casual* in winter in the Southern Interior; *rare* to *uncommon* migrant and locally *common* summer visitant in the Sub-Boreal Interior Ecoprovince; *uncommon* to *common* migrant and locally *common* summer visitant in the Boreal Plains and Northern Boreal Mountains ecoprovinces; *rare* to *uncommon* migrant and summer visitant in the Taiga Plains Ecoprovince.

Breeds.

Figure 266. Four subspecies of Savannah Sparrow have been recorded in British Columbia, 3 of which breed. The range of each race has not been adequately described for the province. The subspecies vary in body size, colour, bill size, and extent of streaking. On southern Vancouver Island, in the Georgia Depression Ecoprovince, *Passerculus sandwichensis brooksi* (a) adults have a pale body, washed-out yellow lores, a small bill, and limited streaking on the underparts (North Saanich, 16 May 1998; R. Wayne Campbell). In portions of the Southern Interior Mountains Ecoprovince, the adult *P. s. nevadensis* (b) has a rich brown body, bright yellow lores, and more pronounced streaking on the underparts (Elizabeth Lake, Cranbrook, 5 May 1997; R. Wayne Campbell).

NONBREEDING: The Savannah Sparrow (Fig. 266) is the most abundant and widely distributed migrant sparrow in British Columbia. It occurs from the small rocky islets along the Pacific coast east to the boundary with Alberta, and from the international boundary in the south to the boundary with the Yukon in the north. It occurs in every ecoprovince. Within this extensive range, it is a species of open landscapes and is not found in forested areas, especially extensive tracts of dense coniferous forests. The highest numbers in winter occur in the Georgia Depression (Fig. 267).

On the coast, the Savannah Sparrow has been recorded at elevations from sea level to about 500 m during spring migration, and up to about 2,100 m during autumn migration. In the interior, migration takes place at elevations from about 280 to 800 m in spring, and up to about 2,300 m in autumn. The small number of wintering individuals are within a few metres of sea level.

The Savannah Sparrow is a bird of open country. In spring, large numbers move northward along the coast of the province, where they pause to feed in the cultivated fields, farmsteads, airports, log-littered beaches with vegetation (Fig. 268), and rough pastures of the agricultural parts of the Fraser and Nanaimo lowlands, or along the dykes and adjacent tidal wetlands vegetated with grasses, sedges, and a variety of forbs. On the ocean coast, they use the sandspits, beach driftlines, patches of blue wildrye, and forbs between the detritus at beachhead, and also the rocky islets where gullies are well vegetated and small invertebrates abound.

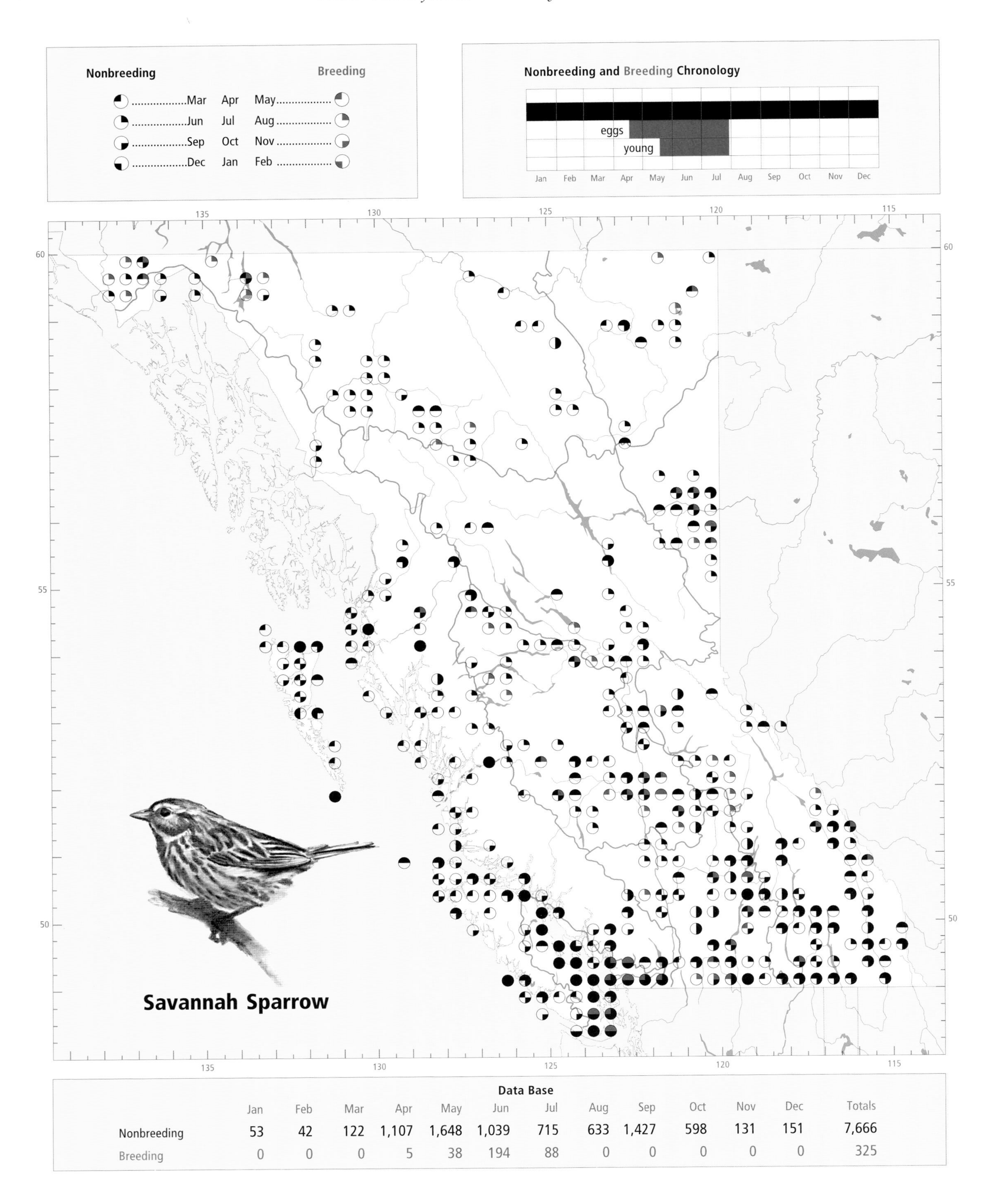

Data Base

	Jan	Feb	Mar	Apr	May	Jun	Jul	Aug	Sep	Oct	Nov	Dec	Totals
Nonbreeding	53	42	122	1,107	1,648	1,039	715	633	1,427	598	131	151	7,666
Breeding	0	0	0	5	38	194	88	0	0	0	0	0	325

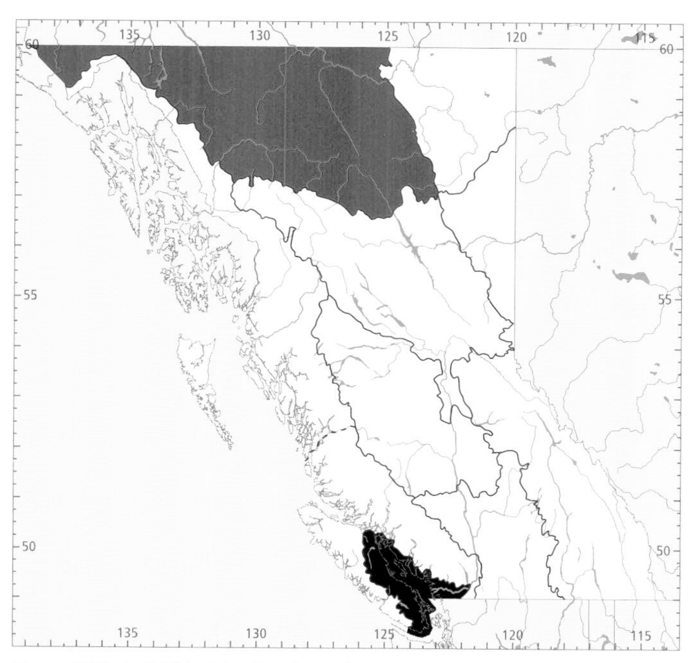

Figure 267. In British Columbia, the highest numbers for the Savannah Sparrow in winter (black) occur in the Georgia Depression Ecoprovince; the highest numbers in summer (red) occur in the Northern Boreal Mountains Ecoprovince.

In the interior, the Savannah Sparrow uses a variety of wild and cultivated landscapes where weeds, grasses, and sedges occur, including meadows and rangelands, cultivated lands (Fig. 269), logging clearcuts, sedge areas adjacent to ponds, roadsides (Fig. 270), railways, airports, and, at higher latitudes, subalpine-tundra, timberline shrubbery of dwarf willows and birches, and the edges of alpine meadows.

Wintering individuals are found mostly in the Fraser and Nanaimo lowlands, mainly on cultivated lands (Fig. 271).

The arrival of the first spring migrants in the Georgia Depression can be observed in the third week of March, but the migration develops slowly until the end of March and then gathers momentum to reach a peak in the last week of April (Fig. 273). It then declines, reaching summer numbers after the third week of May. Elsewhere on the coast, the spring migration passes along Western Vancouver Island and the Southern Mainland Coast from mid-April to the end of May. In the Northern Mainland Coast, a few arrivals have been noted as early as the first week of April, but most pass through during May (Figs. 272 and 273).

In the interior, scattered birds have appeared between late March and mid-April but the main movement, in both the Southern Interior and Southern Interior Mountains, begins in the third week of April and peaks in early May (Figs. 272 and 273). The dates of the first arrivals in the Vernon area over a 19-year period range from 25 March to 28 April (Cannings et al. 1987). Forerunners of the migration reach the Central Interior by the third week of April, and the peak occurs about the second week of May. The exact timing of the spring movement of Savannah Sparrows into the Sub-Boreal Interior is not clear but appears to be about a week later than in the Central interior. Small numbers of these sparrows reach the northern boundary of the Northern Boreal Mountains west of the Rocky Mountains by the third week of April, and numbers peak in mid-June. The pattern of arrival in the Boreal

Figure 268. During spring and autumn migration along the coast, the Savannah Sparrow inhabits open country such as log-littered marine beaches with scattered grasses and forbs (Island View Beach, Central Saanich, 6 April 1997; R. Wayne Campbell).

Plains suggests that the Savannah Sparrow enters from Alberta, the first arrivals in early April and a peak varying by year from the first to the third week of May, at the same time as in the Taiga Plains. Thus the migration moves the length of the province in about a month.

We have few records that clearly identify the start of the autumn migration in the northern interior. In the Northern Boreal Mountains, records indicate a sharp decrease in numbers of birds present after the first week of August (Fig. 273). This timing of migration is reinforced by the observation of a large flock on 18 August in the southeastern part of this ecoprovince (Cooper and Cooper 1983). Most birds have left the region by mid-September, and the latest records are from the first week of October. In the Boreal Plains, an abrupt increase in the number of records and the number of birds per record in the third week of August identifies the migration, which reaches a peak that includes the last week of August and the first or second weeks of September. Numbers decline quickly, and the latest records are in the first week of October. The front of the migration moves rapidly southward, entering the Southern Interior in the third week of August and the Southern Interior Mountains in the first week of September. Intensive banding at Vaseux Lake between early August and late October from 1994 to 1997 yielded the earliest Savannah Sparrow on 14 August and the latest on 13 September, but no substantial migration was detected at this valley bottom banding station. Few birds remain in the southern parts of the interior after the first week of October, and the latest records in the interior of the province are from the first week of November.

In the Georgia Depression, the front of the southbound migration is first noted in the Fraser Lowland in the last week of August. Peak numbers occur between 25 August and 5 September. Numbers are large into early October, and gradually decline into the first week of November as the migration leaves the province (Figs. 272 and 273).

Banding data from the Qualicum National Wildlife Area, near Qualicum Beach, suggest that this sparrow moves through in waves during autumn migration (Fig. 274). In 1983, 556 Savannah Sparrows were banded there between 12 and 29 September. Peak numbers occurred in the third week of September, when nearly 2,000 birds were estimated to be present.

The autumn migration measured by the banding station at Rocky Point, near Victoria, the southernmost point of the province, began on 31 July, when the first Savannah Sparrow was captured. Occasional birds were taken until early September. The peak movements began between 8 and 14 September (1994 to 1997) and ended between 28 September and 10 October. The latest record was from 3 November. Wintering numbers on the southwest mainland are reached by mid-November.

Figure 269. In the Southern Interior Mountains Ecoprovince, migrating Savannah Sparrows frequent cultivated land where grasses afford food and cover (near Athalmer, 9 May 1997; R. Wayne Campbell).

Figure 270. Roadsides with an abundant growth of weeds and grasses are used by migrating Savannah Sparrows in autumn (Duck Lake, Creston, 22 September 1996; R. Wayne Campbell).

Figure 271. In winter, small numbers of Savannah Sparrows inhabit cultivated lands, such as these corn stubble fields in southwestern British Columbia (Central Saanich, 4 February 1999; R. Wayne Campbell).

The autumn migration through south coastal British Columbia is remarkable in that the number of birds recorded per week through the period of sustained migration is about 4 times the number recorded northbound (Fig. 273) (see REMARKS).

Small numbers of Savannah Sparrows winter in the Nanaimo and Fraser lowlands, where cultivated fields, rough pasture, shrub thickets, dykes, and a well-vegetated tidal foreshore provide suitable habitat.

On the coast, the Savannah Sparrow is present throughout the year and has been recorded regularly from mid-March to late October; in the interior, it has been recorded regularly from 25 March to 29 October (Fig. 272).

BREEDING: The breeding distribution of the Savannah Sparrow extends over the entire length of the province but is most concentrated where grasslands, shrublands, weedy fields, alpine meadows, and subalpine tundra are widely distributed.

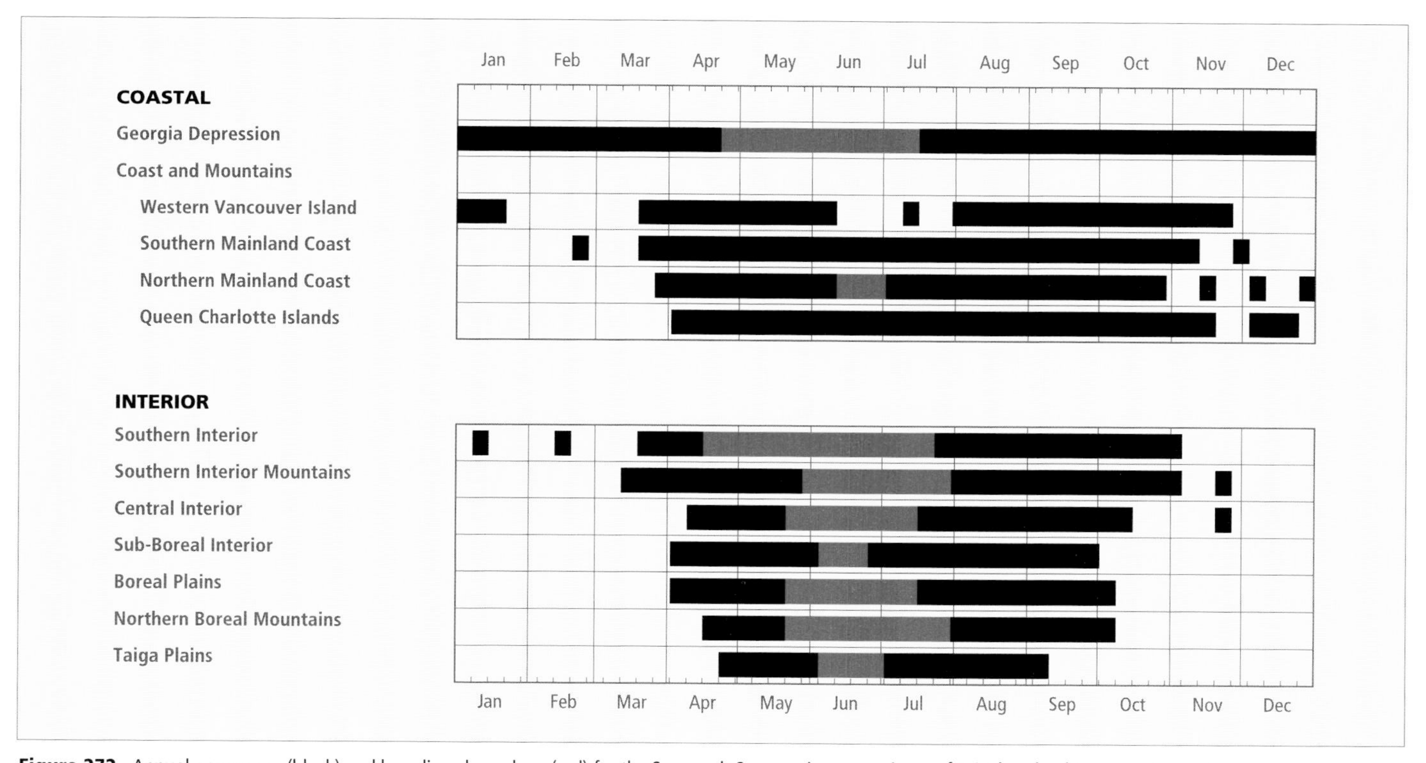

Figure 272. Annual occurrence (black) and breeding chronology (red) for the Savannah Sparrow in ecoprovinces of British Columbia. Records are shown for the week in which they occurred.

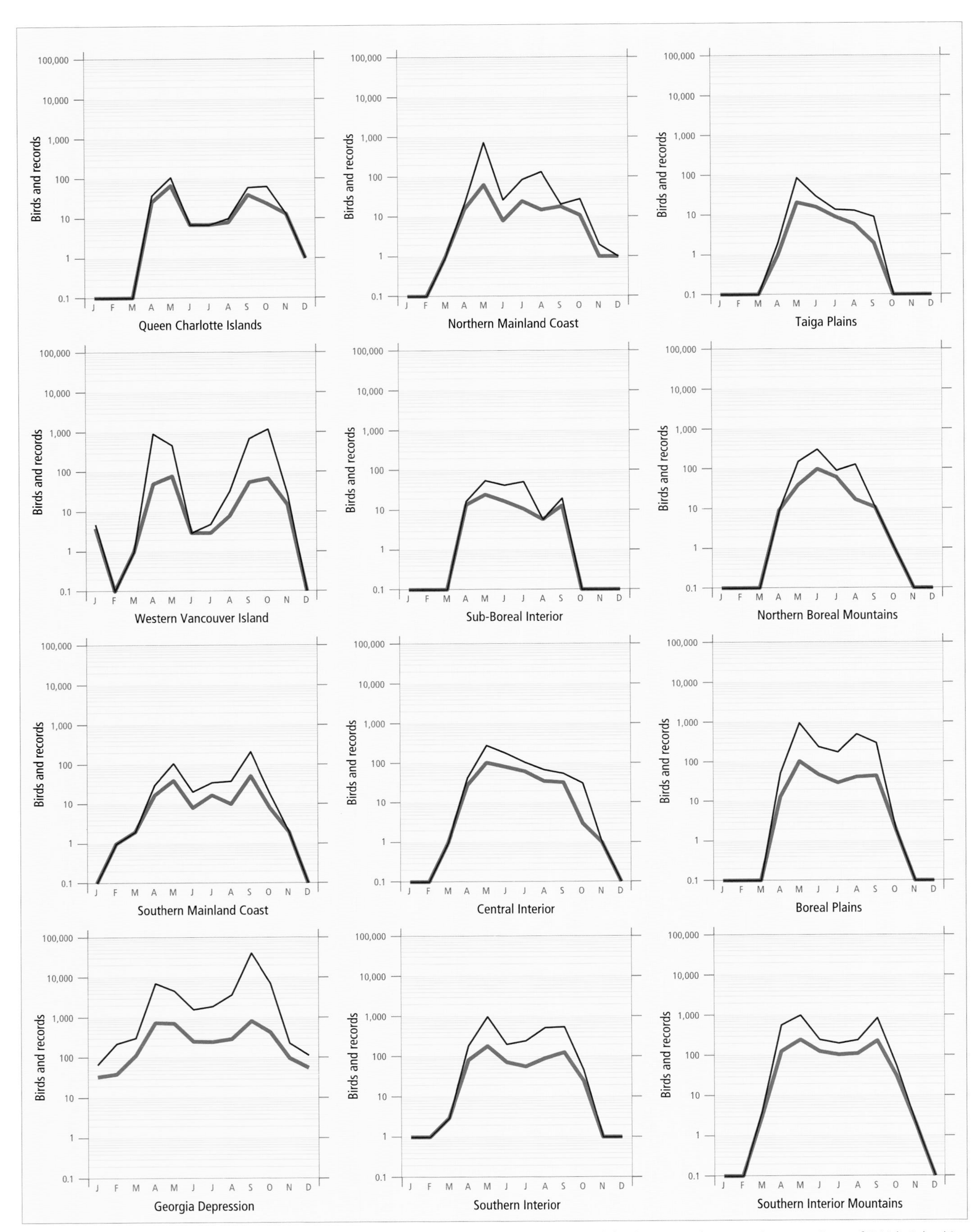

Figure 273. Fluctuations in total number of birds (purple line) and total number of records (green line) for the Savannah Sparrow in ecoprovinces of British Columbia. Christmas Bird Counts, Breeding Bird Surveys, and nest record data have been excluded.

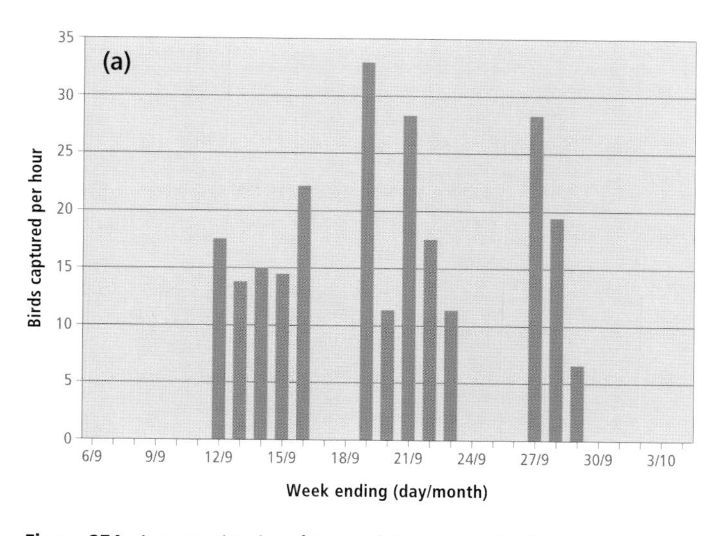

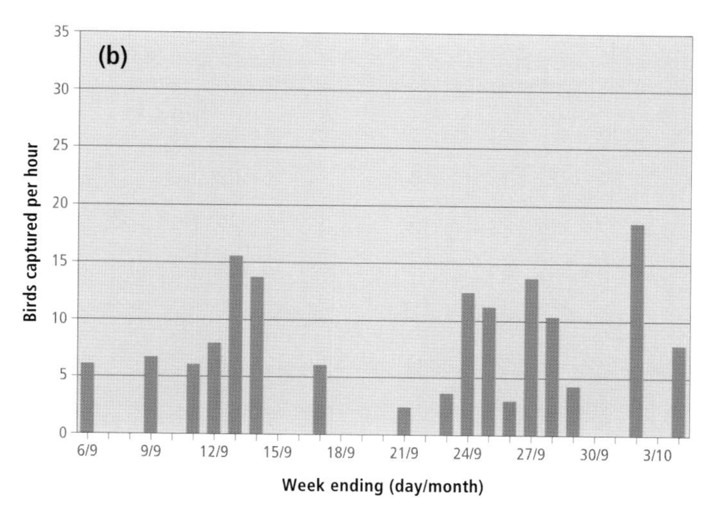

Figure 274. Autumn migration of Savannah Sparrows at Qualicum Beach, British Columbia, in (a) 1983 ($n = 556$) and (b) 1984 ($n = 380$), based on bird captures at a banding station in the Qualicum National Wildlife Area. The figure suggests that the birds moved through in waves. Dates without a bar were not sampled. The larger numbers of birds captured per hour in 1983 were due to upland field manipulation that created an abundance of annual plant seeds, primarily Lamb's-quarters (*Chenopodium album*), which attracted the migrating sparrows.

In the northern half of the province, most reported breeding localities are in the Peace Lowland, along the upper Stikine River, and in the extreme northwestern corner. Along the coast north of the Georgia Depression, there are only 2 nesting records, both at Terrace, but in the Rainbow Range in mid-June, pairs were on territory and 1 of 2 females taken had an egg in the oviduct (Dickinson 1953).

The highest numbers in summer occur in the Northern Boreal Mountains Ecoprovince (Fig. 267). An analysis of the Breeding Bird Surveys in British Columbia for the period 1968 through 1993 could not detect a net change in numbers in either interior or coastal routes. Sauer et al. (1997) report a continent-wide decline of 0.6% ($P < 0.05$) in numbers of the Savannah Sparrow over the period 1966 to 1996.

The Savannah Sparrow has been recorded breeding at elevations from near sea level to 2,010 m. On the coastal lowlands, this sparrow nests primarily along the edges of cultivated crops and meadows, but also along the edges of swamps and small streams, along the shoulders of dykes, and on estuaries (Fig. 276). On the coastal mountains, it occupies meadows of grasses and forbs at and above timberline. In the southern and central parts of the interior, it chooses moist meadows that have been moderately grazed, fields of hay or alfalfa, the grassy and weedy edges of small lakes and ponds, including the upland edges of riparian willow thickets, and recently clearcut or burned areas. In northeastern British Columbia, the Savannah Sparrow may breed in farm fields, hay fields, long-grass fields around wetlands, open shrublands with long grasses, and alpine tundra. Near the northern boundary of the province, as at Haines Pass and Chilkat Pass, it frequently nests in habitat dominated by dwarf willows and birch (Weeden 1960), on grassy flats of low subalpine vegetation (Fig. 275). It is a characteristic summer visitant to many of the alpine regions of the central and southern parts of the province, where it nests in alpine meadows and alpine grasslands. On the Seward Peninsula of Alaska, Kessel (1989) reports densities of up to 12.7 territories per 10 ha in willow-grass meadow habitat.

On the coast, the Savannah Sparrow has been recorded breeding from 24 April (calculated) to 15 July; in the interior, it has been recorded breeding from 17 April (calculated) to 29 July (Fig. 272).

Figure 275. In northwestern British Columbia, the Savannah Sparrow often breeds in grassy flats of open subalpine shrub habitats (near Gladys Lake, 5 June 1996; R. Wayne Campbell).

Figure 276. Exposed nest and eggs of the Savannah Sparrow (Englishman River estuary, 8 July 1993; Neil K. Dawe).

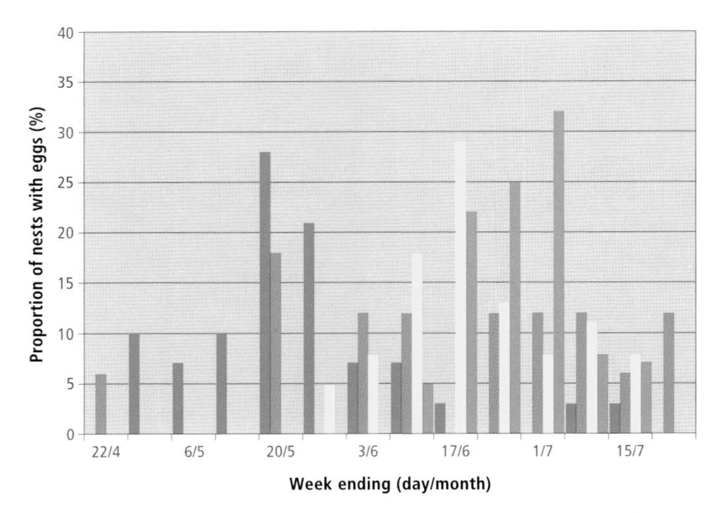

Figure 277. Proportion of Savannah Sparrow nests with clutches by week for the Georgia Depression (green bars; n = 30), Southern Interior (red bars; n = 17), Central Interior (yellow bars; n = 38), and Northern Boreal Mountains (blue bars; n = 59) ecoprovinces. Note the differences among the regions in the length and pattern of the egg-laying season.

Nests: Most nests (n = 134) were in alpine habitat (28%) or in forests altered by logging or fire (31%); other habitats used were grassland (19%), wetland (13%), and shrubland (9%). Within these habitat types, the habitat classes used for most nesting included cultivated farms, mostly in hayfields (24%), grasslands dominated by forbs (17%), alpine grasslands and alpine meadows (26%), marsh and swamp (13%), and subalpine shrub meadows dominated by scrub birch and willows (10%).

Most nests were well concealed and placed on the ground (97%; n = 161); a few were in low vegetation at heights up to 0.5 m. Nests were cup-shaped and placed in depressions in the ground, usually sheltered by overarching vegetation (Fig. 276). The materials used most frequently in nest construction were grasses (90% of nests), sedges (14%), hair (5%), plant fibres (5%), mosses (2%); 7 other items were used in 7% of nests.

Eggs: Dates for 213 clutches ranged from 20 April to 29 July, with 52% recorded between 9 June and 28 June. Calculated dates indicate that nests may hold eggs as early as 17 April. Sizes of 159 clutches ranged from 1 to 6 eggs (1E-2, 2E-4, 3E-11, 4E-83, 5E-55, 6E-4), with 52% having 4 eggs (Fig. 276). The incubation period outside British Columbia varied between 10 days and 13.2 days in 5 regional North American studies (Wheelwright and Rising 1993). In south coastal British Columbia, the Savannah Sparrow is double-brooded (Cooper 1997). In the southern areas of its range outside British Columbia, it is reported to be double- or even triple-brooded (Baicich and Harrison 1997), but in northern areas the nesting season may be too short to permit more than a single brood. The nesting season on the south coast and in the Southern Interior begins in April, with the first eggs laid late in the month. The egg phase peaks in late May and ends in the second to third week of July (Fig. 277). In the Central Interior, egg laying does not begin until late May and is over by mid-July. At the northern extreme of the province, in the Northern Boreal Mountains, egg-laying does not begin until early June and also ends in mid-July.

Young: Dates for 112 broods ranged from 19 May to 29 July, with 51% recorded between 22 June and 3 July. Sizes of 74 broods ranged from 1 to 5 young (1Y-4, 2Y-3, 3Y-10, 4Y-35, 5Y-22), with 77% having 4 or 5 young. The nestling period is 8 to 13 days, usually 10 to 12 days (Rising 1996). In the Georgia Depression, Southern Interior, Southern Interior Mountains, and Northern Boreal Mountains, the ecoprovinces from which we have most nesting data, the last young fledge in the final week of July or the first week of August.

Brown-headed Cowbird Parasitism: In British Columbia, 2 of 203 nests found with eggs or young were parasitized by the Brown-headed Cowbird. There were no additional records of Savannah Sparrows feeding fledgling cowbirds. None of 31 nests on the coast was parasitized. In the interior, 2 of 172 nests (1%) were parasitized; both were near Dawson Creek.

Friedmann (1963) commented that "the Savannah Sparrow is molested infrequently by the Brown-headed Cowbird," with just 28 records continent-wide. By 1985, 26 additional records had been accumulated and it was noted that the incidence of parasitism varied greatly between regions. The highest incidence was found in Wisconsin (Friedmann and Kiff 1985). On Kent Island, New Brunswick, virtually no parasitism was found in 1,150 nests examined between 1987 and 1998 (N.T. Wheelwright pers. comm.), although there were a few reports in the 1960s. In a Michigan study, parasitized nests fledged no sparrows (Potter 1974).

Nest Success: In British Columbia, of 19 nests found with eggs and followed to a known fate, 6 produced at least 1 fledgling. We have found few studies of nest success, but in his Nova Scotia study, Welsh (1975) determined that 14 females were successful in fledging young in 22 nests in 1 summer. A study on Isle Verte, Quebec (LaPointe and Bedard 1986) expresses losses in terms of eggs rather than nests; the results cannot be compared with ours, but only 35.5% (n = 849) of the eggs laid resulted in fledged young.

REMARKS: Of the 17 recognized subspecies of the Savannah Sparrow, 4 have been recorded from British Columbia (American Ornithologists' Union 1953; Wheelwright and Rising 1993). These differ in size, coloration, beak size and shape, relative sizes of such details as wing length and leg length, summer and winter distribution, and migration timing and routes. In general, discrimination between the subspecies can be done with certainty only from a bird in the hand. The subspecies are:

(1) *P. s. sandwichensis,* the subspecies with the largest body, is a migrant along the coast, including Western Vancouver Island (Hatler and Campbell 1975), and a winter visitant to the Georgia Depression. Breeds on Unalaska Island (Swarth 1936a).

(2) *P. s. anthinus* is an abundant migrant through both the south coast and the interior, and the breeding subspecies of the Central Interior and northwestern interior regions of the province from the Cariboo and Chilcotin areas north into the Yukon, and in southwestern Alaska (Kessel 1989).

(3) *P. s. nevadensis,* the breeding subspecies of the southern parts of the interior and the Peace Lowland, also breeds across the Prairie provinces.

(4) *P. s. brooksi,* the subspecies with the smallest body, breeds in the Fraser and Nanaimo lowlands, where it also occasionally winters. It is an early spring migrant and is often nesting by the time the more northerly nesting subspecies arrive in migration (Munro and Cowan 1947; Godfrey 1986).

In a discussion of population regulation, Wheelwright and Rising (1993) remark that in the Savannah Sparrow, population changes are not closely related to events on the breeding ground in the previous year. Population changes appear to result from events during migration or on the wintering grounds. In British Columbia, this is substantiated by a comparison of the total number of birds recorded in the Georgia Depression during the northbound and southbound migration periods. Each of these periods occurs over approximately 12 weeks. The spring period is from the week ending 25 March to that ending on 10 June. The autumn migration appears to extend from the week ending 26 August to that ending 11 November. The Georgia Depression is the ecoprovince with the highest counts of migrating Savannah Sparrows in British Columbia. While the sources of our data do not permit refined comparisons, we believe they are a reasonably random sample of the birds present.

The total number of Savannah Sparrows recorded during the autumn migration through the Georgia Depression is about 48,800; records for the spring migration total 12,900. Records for the other ecoprovinces show that the difference cannot be explained by changes in migration routes used between the 2 seasons. It appears, therefore, that more than two-thirds of the birds moving south out of British Columbia do not return in the spring. Such a large difference in counts made over a period of more than 50 years is unlikely to be a product of chance.

The extensive use of agricultural lands by migrating Savannah Sparrows often exposes them to pesticide residues. An example of this occurred on turnip and radish fields in Richmond, British Columbia, in the autumn of 1986. Large numbers of sparrows were found dead or dying. A sample of 152 Savannah and 5 Lincoln's sparrows was collected for testing. The sample led to an estimate of between 500 and 1,178 dead sparrows in the fields. Carbofuran residues found in the carcasses were identified as the most probable cause of death (Wilson et al. 1995).

Two studies have documented the density of breeding Savannah Sparrows in agricultural lands in the Georgia Depression. In the Koksilah River estuary, on southeastern Vancouver Island, there were an estimated 6 singing males per 10 ha (Cooper 1997). On Vancouver International Airport Reserve lands on Sea Island, there were an estimated 6.6 singing males per 10 ha in 1992 (Cooper 1993). In both studies, the Savannah Sparrow was the most abundant breeding bird.

On Sea Island, there was an estimated population of 229 singing male Savannah Sparrows in 1992. Construction of the Parallel Runway project was slated to develop 338 ha of the 478 ha Airport Reserve lands, resulting in the loss of breeding habitat for 176 (77%) of those birds (Cooper 1993).

For a detailed summary of information on the Savannah Sparrow, see Wheelwright and Rising (1993), Wheelwright and Schultz (1994), and Wheelwright et al. (1997).

NOTEWORTHY RECORDS

Spring: Coastal – Saanich 15 Apr 1984-85; Central Saanich 19 May 1984-nest with young; North Saanich 28 May 1984-3 nestlings; Cowichan Bay 31 May 1980-42; Lulu Island 22 Mar 1924-16, earliest of year; Ladner 28 May 1924-50 breeding birds; Iona Island 20 May 1968-1 nestling; Langley 9 Apr 1967-50, 21 Apr 1969-104; Sumas 28 Apr 1905-5 eggs; Lueyetts Island (Harrison) 24 Mar 1984-1; Qualicum Beach 20 Mar 1973-20 (Dawe 1976); Vancouver 2 May 1982-200 feeding on lawns; Smelt Bay 25 May 1975-6; Whistler 26 Apr 1997-1; Port Neville 22 Mar 1986-1, first seen this spring; Port Hardy 22 Apr 1942-500; Cape Scott Park 15 Apr 1980-25; Safety Cove 20 Apr 1937-1 (MCZ 285028); Goose Island 10 May 1948-18 (Guiguet 1953); Ramsay Island 16 May 1984-8; Langara Island 4 Apr 1971-1; Rose Spit 26 Apr 1979-4; Kitimat Mission 1 May 1975-75, 17 May 1975-120 (Hay 1976); Metlakatla 31 Mar 1902-1, earliest spring arrival (Keen 1910). **Interior** – Osoyoos area 25 Apr 1922-12; Penticton 15 May 1954-5 eggs; nw Princeton 29 May 1987-3 nestlings; Robson 12 Mar 1971-1; Kelowna 20 Apr 1940-4 eggs (RBCM 1337); Vernon 25 Mar 1979-1; Kaslo 18 Mar 1973-1; Nakusp 22 Apr 1982-50, 24 Apr 1980-100; Golden 12 Apr 1994-3; Lac du Bois area 5 May 1972-400; nr Forest Grove 12 Apr 1986-1, first report of year; Williams Lake 2 May 1972-6; Kleena Kleene 25 May 1971-40 (Paul 1964); Hanceville 21 May 1980-4 eggs; Willow River 23 Apr 1972-1, earliest spring date; Giscome 21 Apr 1990-2; Mugaha Creek 15 Apr 1996-2 at marsh; e Tupper 4 Apr 1975-30; Dawson Creek 1 May 1986-7, 27 May 1993-2 eggs plus 1 Brown-headed Cowbird egg; Tetana Lake 7 Apr 1938-1 (Stanwell-Fletcher and Stanwell-Fletcher 1943); Fort St. John 13 May 1983-150-200 along 150 km of road; North Pine 5 May 1985-60; 14 km n Pink Mountain 13 May 1982-25; Fort Nelson 29 Apr 1987-2; 23 km s Atlin 19 Apr 1934-1 (CAS 42035); Chilkat Pass 9 May 1957-1, first seen (Weeden 1960).

Summer: Coastal – Saanich (Rithets Bog) 15 Jul 1986-4 recently hatched young; nr Cowichan Bay 13 Jul 1940-4 eggs; Cowichan Lake 5 Aug 1972-95; Boundary Bay 15 Jul 1987-4 nestlings about 1 week old; Westham Island 10 Jun 1973-1 fledgling; Sea Island 4 Jun 1977-22, 23 Aug 1966-200, 30 Aug 1965-350; Lulu Island 17 Jun 1942-60 along dyke; Squamish River 13 Jul 1997-2 on estuary; Pemberton Meadows 30 Jun 1996-2; Oyster River 13 Jul 1978-4; Campbell River 21 Jul 1968-50; Port Hardy 31 Aug 1936-10; Triangle Island 14 Jul 1978-2; Tlell 21 Jul 1974-1; Terrace 1 Jul 1990-5 nestlings. **Interior** – Osoyoos Lake 3 Jul 1973-3; Manning Park 18 Jul 1968-5 eggs; Naramata 29 Jul 1971-3 fledglings, 8 Aug 1969-2 fledglings being fed by parents; Kimberley 7 Jun 1975-7 at airport; Chase to Kamloops 11 Jul 1963-22; Salmon Arm 24 Aug 1973-30; Kootenay National Park 2 Jun 1981-4 eggs; Wapta Lake 4 Aug 1975-15 (Wade 1977); Riske Creek 15 Jul 1978-5 nestlings, 15 Aug 1978-12; Rainbow Range 17 to 19 Jun 1933-6 males and 2 females collected (Dickinson 1953); Anahim Lake 3 Jun 1948-10 on alkali flats; McBride 10 Jul 1991-13; Goodrich Lake 14 Jul 1963-5 eggs; Sinkut River 13 Jun 1946-5 eggs, female flushed from nest (Munro 1949); Houston 14 Jun 1987-4 nestlings; Fort St. James 12 Jun 1889-4 eggs (Mair and MacFarlane 1908); Topley 7 Jul 1956-25; 4.8 km n Pouce Coupe 12 Jul 1982-4 eggs; n Dawson Creek 8 Jun 1993-5 nestlings, 3 Aug 1975-50; Charlie Lake 25 Jul 1964-1 fledgling being fed by adult; ne Fort St. John 24 Aug 1986-102 at sewage lagoons; Muskwa River 18 Aug 1983-100 (Cooper and Cooper 1983); Kotcho Lake 26 Jun 1982-4 small nestlings (Campbell and McNall 1982); Kwokullie Lake 7 Jun 1982-4 eggs; Atlin 9 Jun 1958-4 eggs, at Wright Creek; Mile 80 Haines Highway 23 Jun 1980-5 eggs, 25 Jul 1980-4 fledglings.

Breeding Bird Surveys: Coastal – Recorded from 17 of 27 routes and on 51% of all surveys. Maxima: Albion 20 Jun 1973-48; Chilliwack 10 Jun 1973-29. **Interior** – Recorded from 60 of 73 routes and on 63% of all surveys. Maxima: Haines Summit 19 Jun 1994-73; Prince George 19 Jun 1994-61; Fort St. John 29 Jun 1987-55.

Autumn: Interior – Kelsall Lake 26 Sep 1972-1 (CAS 68769); O'Donnel River valley 3 Oct 1980-1; Fort Nelson 8 Sep 1986-8; Flatrock 1 Oct 1951-1; ne Fort St. John 4 Oct 1987-1 at sewage lagoons, last record of autumn; Mugaha Creek 20 Sep 1997-1; Quick 24 Nov 1983-1; Willow River 26 Sep 1969-1; Puntchesakut Lake 5 Sep 1944-1 (Munro 1947); Kleena Kleene 10 Oct 1962-30 (Paul 1964); Revelstoke 23 Nov 1985-1; Knutsford 3 Oct 1982-100; Nakusp 8 Sep 1978-30, 29 Oct 1978-1; Okanagan Landing 3 Nov 1927-1; Oliver 22 Oct 1984-10. **Coastal** – Port Neville Inlet 9 Nov 1975-1; Pultney Point 9 Nov 1975-6; 24 Nov 1975-1; Port Hardy 11 Oct 1938-500; Pemberton 14 Sep 1995-1; Squamish River 15 Sep 1996-50+ on estuary; Harrison Hot Springs 27 Nov 1985-1; Sea Island 10 Sep 1965-900, 13 Sep 1965-1,500, both at International Airport; Richmond 16 Sep 1986-500 to 1,178 poisoned by pesticide (Wilson et al. 1995); Long Beach 20 Sep 1980-200, between Sandhill Creek and Wickaninnish; Saanich 17 Nov 1988-6.

Winter: Interior – Vernon 10 Jan 1993-1 (Siddle 1993b); Creston 23 Jan 1998-1; Okanagan Landing 1 Dec 1897 to 31 Jan 1898-1 (Cannings et al. 1987); Osoyoos 13 Feb 1988-1 (Rogers 1988a). **Coastal** – Prince Rupert 9 Dec 1984-1; Bella Coola 21 Feb 1976-1 at feeding station; Cape St. James 7 Dec 1981-1; Vargas Island 10 Jan 1969-1; Kilby 25 Dec 1987-1; Sea Island 28 Feb 1987-55, estimated from daily observations at airport; Delta 10 Dec 1989-10, 29 Dec 1988-4 along railway between 80 St and 72 St; Boundary Bay 18 Feb 1989-40 along dyke; Central Saanich 24 Feb 1988-10; Saanich 7 Jan 1989-12; Jordan River 2 Jan 1984-2; Billings Spit 1 Jan 1984-3.

Christmas Bird Counts: Coastal – Recorded from 18 of 33 localities and on 23% of all counts. Maxima: Vancouver 17 Dec 1978-33; Victoria 18 Dec 1993-21; Ladner 27 Dec 1988-18. **Interior** – Not recorded.

Grasshopper Sparrow

Ammodramus savannarum (Gmelin)

GRSP

RANGE: Breeds from south-central interior British Columbia, eastern Washington, southern Alberta, southern Saskatchewan, southern Manitoba, western and southern Ontario, and southwestern Quebec south in the west through eastern Washington, western Idaho, locally in eastern Oregon, and coastal California; east of the Rocky Mountains, from eastern Montana to southern Maine, south into Texas, Arkansas, Tennessee, Georgia, and Florida. Winters from Arizona, New Mexico, Texas, Arkansas, Tennessee, and South Carolina south through Baja California, mainland Mexico, and Central America to Costa Rica.

STATUS: On the coast, *casual* summer and autumn vagrant in the Georgia Depression Ecoprovince.

In the interior, locally *rare* to *uncommon* migrant and summer visitant in the Southern Interior Ecoprovince; *accidental* in the Central Interior Ecoprovince.

Breeds.

NONBREEDING: The Grasshopper Sparrow (Fig. 278) is locally distributed in the central-southern interior of British Columbia, mainly in the Okanagan valley of the Southern Interior between Osoyoos Lake in the south and Goose Lake north of Vernon. It also occurs irregularly near Spotted Lake and Kilpoola Lake in Richter Pass and near Chopaka in the southern Similkameen valley. It has occurred as a vagrant on Becher's Prairie, west of Williams Lake, in the Central Interior. On the coast, the species has occurred as a vagrant in the Fraser Lowland and near Victoria on southern Vancouver Island.

Other than for a few vagrants, the small number of records of the Grasshopper Sparrow before and after the breeding season have all been from the known summer range (Fig. 279). The important elements of this habitat are described under BREEDING.

Spring migrants arrive on known nesting grounds in the southern Okanagan valley in early May. The species is so secretive that little comment is possible on the build-up of numbers, but observations over the years indicate that it occurs during mid-May (Figs. 280 and 281). The autumn departure appears to occur from September through the first half of October (Cannings et al. 1987).

On the coast, the Grasshopper Sparrow has been recorded as a vagrant between 5 June and 29 November; in the interior, it has been recorded from 1 May to 19 October (Fig. 280).

BREEDING: The few nests that have been found in the province and the behavioural evidence of nesting indicate that the Grasshopper Sparrow breeds mainly in the Okanagan valley and the extreme southern end of the Similkameen River valley. A small breeding population has been reported in the Nicola valley near Chapperon Lake (Cannings 1995b).

The Grasshopper Sparrow reaches its highest numbers in summer in the Okanagan valley of the Southern Interior

Figure 278. The Grasshopper Sparrow is an inconspicuous grassland bird that is distributed locally in the central-southern interior of British Columbia (Anthony Mercieca).

(Fig. 282). Breeding Bird Surveys for interior routes for the period 1968 through 1993 contain insufficient data for analysis. Surveys representing the entire breeding range of this sparrow in North America between 1966 and 1996 reveal that the Grasshopper Sparrow has been declining at an average annual rate of 3.6% ($P < 0.01$). Surveys across Canada east of British Columbia over the same period indicate a decline in numbers between 1980 and 1996 at the average annual rate of 4.2% ($P < 0.05$) (Sauer et al. 1997). Observations over the past 40 years suggest that the presence of this sparrow in the province has become more consistent than in the early years of observation.

In British Columbia, the Grasshopper Sparrow appears to nest in small colonies (up to 6 pairs) or as single pairs, while apparently suitable habitat close to the colonies goes unoccupied (Cannings 1995b). This pattern of breeding distribution has also been commented on by Janes (1983) in Oregon. In British Columbia, most breeding activity occurs at elevations between 300 and 500 m, but nests have been found between 1,000 and 1,160 m. All breeding records in the province are from grassland habitats.

Grasslands with patches of open ground and a generally sparse shrub and grass cover are preferred. In the Okanagan and adjacent valleys, the habitats selected feature bluebunch wheatgrass, Idaho fescue, needle-and-thread grass, the introduced crested wheatgrass, and a variety of forbs, now often

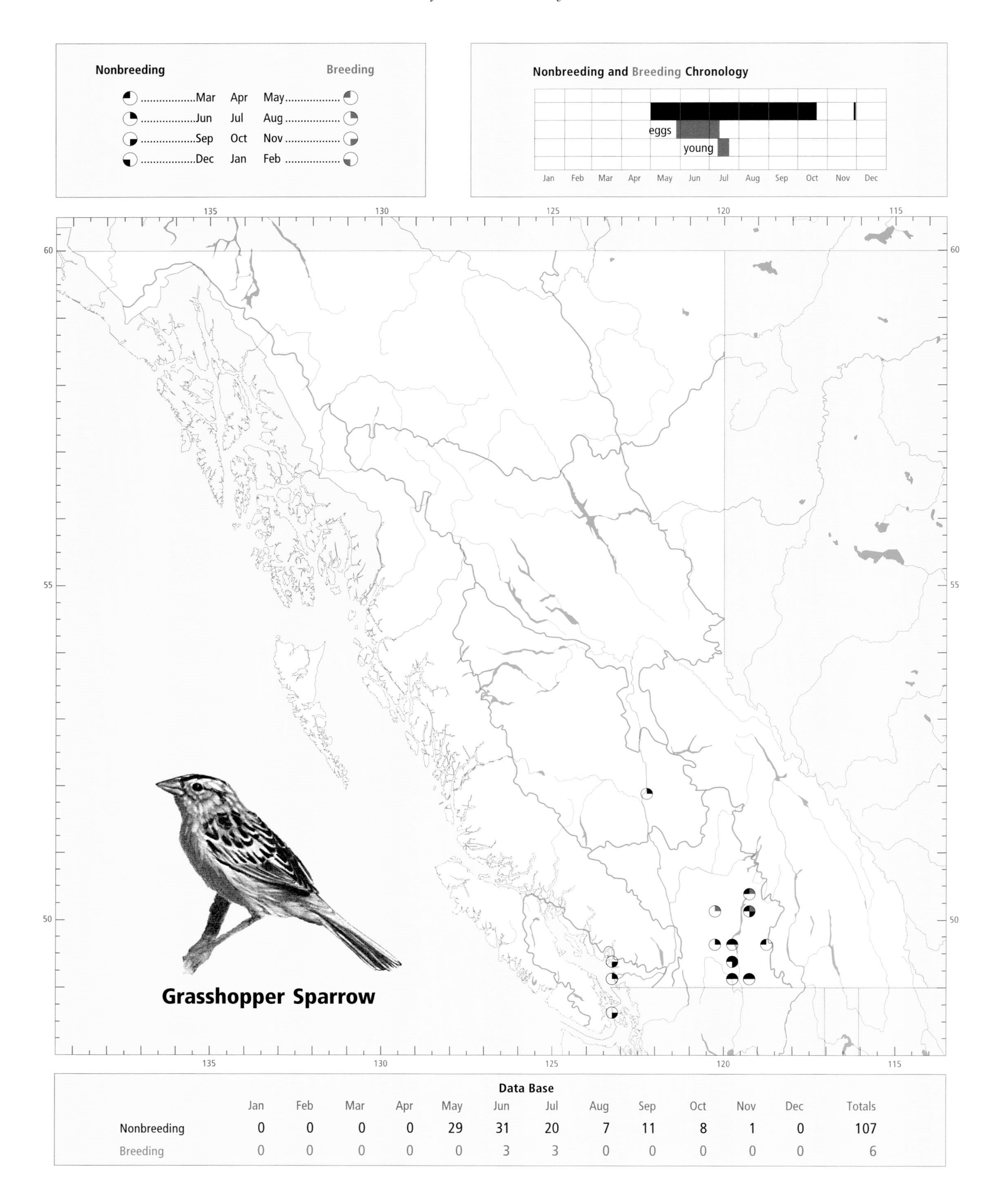

Data Base

	Jan	Feb	Mar	Apr	May	Jun	Jul	Aug	Sep	Oct	Nov	Dec	Totals
Nonbreeding	0	0	0	0	29	31	20	7	11	8	1	0	107
Breeding	0	0	0	0	0	3	3	0	0	0	0	0	6

Figure 279. Before and after the breeding season in British Columbia, the Grasshopper Sparrow may be found in grass-dominated habitats near wetlands (Kilpoola Lake, 8 August 1996; R. Wayne Campbell).

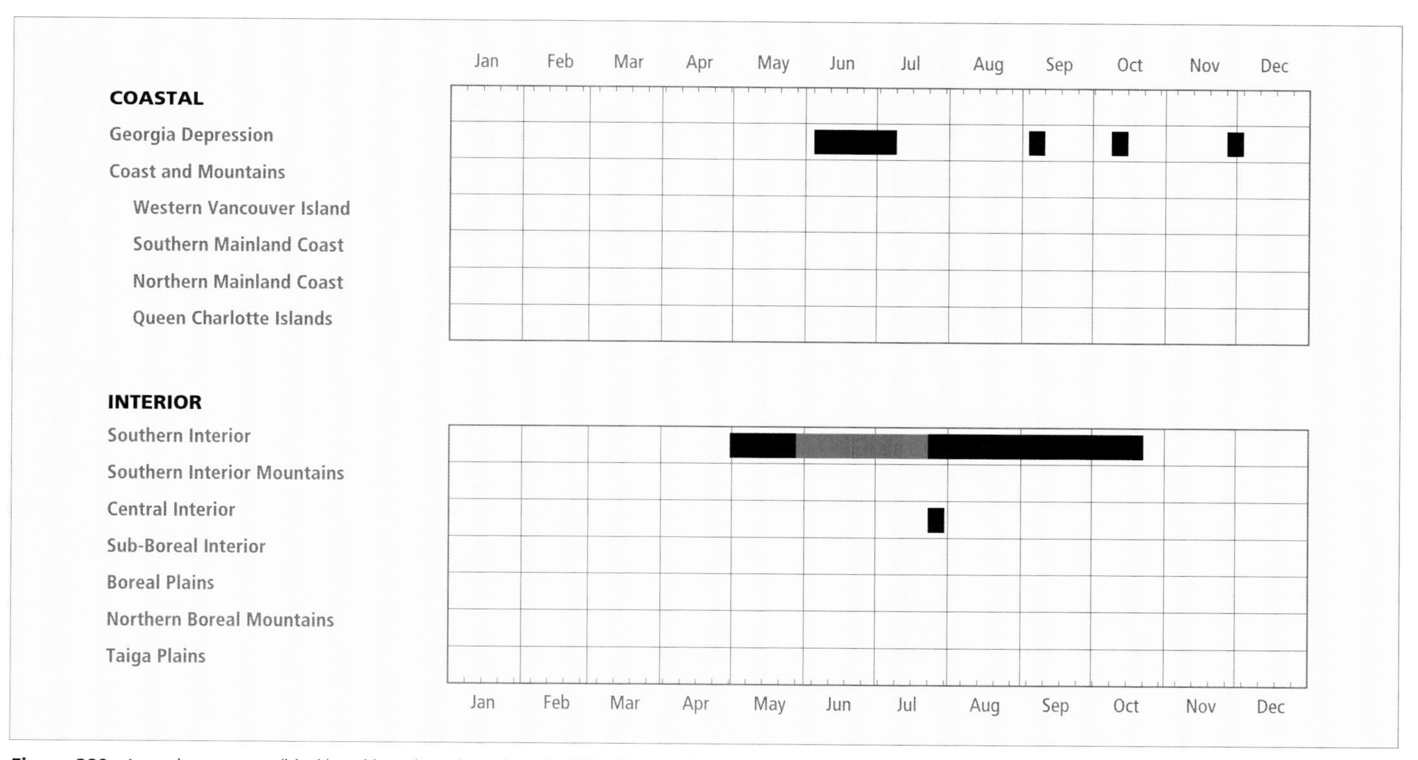

Figure 280. Annual occurrence (black) and breeding chronology (red) for the Grasshopper Sparrow in ecoprovinces of British Columbia. Records are shown for the week in which they occurred.

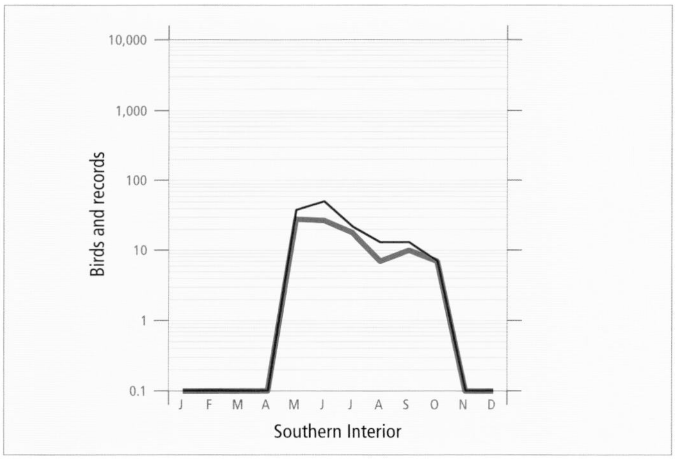

Figure 281. Fluctuations in total number of birds (purple line) and total number of records (green line) for the Grasshopper Sparrow in the Southern Interior Ecoprovince. Breeding Bird Surveys and nest record data have been excluded.

Figure 283. Open grasslands with a variety of native and introduced grasses and forbs are the preferred breeding habitat for the Grasshopper Sparrow in the Okanagan valley in the Southern Interior Ecoprovince of British Columbia (near White Lake, 13 May 1997; R. Wayne Campbell).

dominated by the noxious diffuse knapweed (Fig. 283). On the dry grassy hills, damper areas with clumps of rose provide singing perches used by the males, while the nests are in the adjacent grasses. Other shrubs frequently present include big sagebrush, antelope-brush, and rabbit-brush, but in their presence the sparrows appear to select patches of grass within the shrub stands. The Grasshopper Sparrow avoids tracts of dense sagebrush.

The habitat of this sparrow in British Columbia is a narrow northern extension of the Great Basin Columbia River shrub-steppe that is more extensive in Washington state. In Washington, the Grasshopper Sparrow is uncommon to common below or at the lower limits of the Ponderosa Pine Zone, in shrub-steppe habitats with extensive grass cover (Smith et al. 1997).

In Alberta, Grasshopper Sparrow habitat is described as "a mixture of lush grasses and low, relatively open shrubbery ... and includes overgrown pastures, hay-fields, dry short grass plains ... generally drier than that chosen by the Savannah Sparrow" (Semenchuk 1992).

In British Columbia, the Grasshopper Sparrow has been recorded breeding from 28 May (calculated) to 21 July (Fig. 280).

Nests: Nests are built on the ground, usually in a shallow depression. The nest cup is constructed of grasses and sedges and sometimes includes hair. It is domed, also with grasses and sedges, and has a side opening (Vickery 1996).

Eggs: Dates for 5 clutches ranged from 2 June to 11 July. Clutch size ranged from 1 to 6 eggs (1E-1, 4E-3, 6E-1). Calculated dates indicate that nests may hold eggs as early as 28 May. The incubation period is 11 to 13 days (Nicholson 1936; Smith 1968). The clutch size in North America is 4.30 ± 0.69 (n = 438) (McNair 1987). Outside British Columbia, this sparrow is a frequent renester and may raise 2 broods a year (Smith 1968). Renesting has not been recorded in British Columbia.

Young: One brood was found on 21 July, with 1 young. Fledglings still with their parents were noted on 25 July. The nestling period is 9 days (Baicich and Harrison 1997).

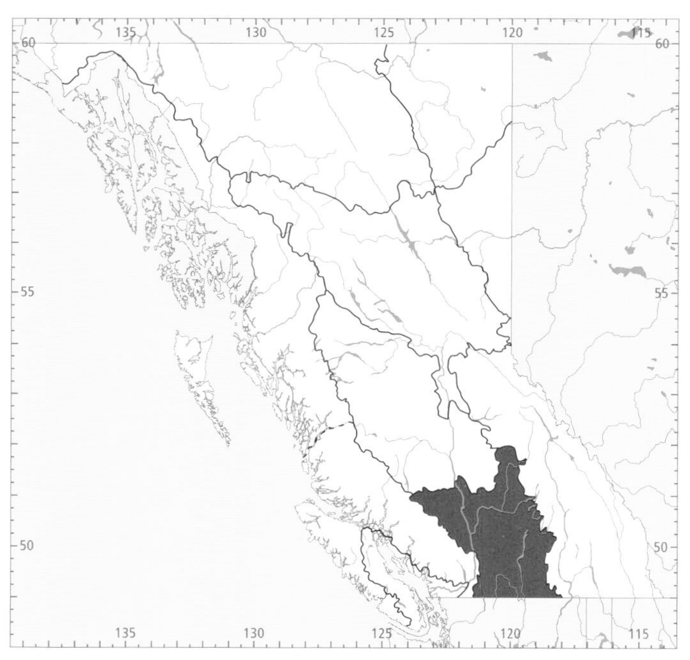

Figure 282. In British Columbia, the highest numbers for the Grasshopper Sparrow in summer occur in the Southern Interior Ecoprovince.

Brown-headed Cowbird Parasitism: In British Columbia, 1 of 6 nests found with eggs or young was parasitized by the cowbird. There were no additional records of Grasshopper Sparrows feeding cowbird fledglings. There are only a small number of records from western North America of the Grasshopper Sparrow as host for the Brown-headed Cowbird (Friedmann and Kiff 1985).

Nest Success: Insufficient data.

REMARKS: In its extensive range north of Mexico, the Grasshopper Sparrow has evolved into 4 subspecies that breed in North America (Vickery 1996), but only the "Western" Grasshopper Sparrow, *A. s. perpallidus*, enters British Columbia. Records of this sparrow in British Columbia suggest that it may be an irregular member of the bird fauna of the province, and may be absent for several years at a time. Since the species was first discovered in British Columbia in 1898, there have been 66 years when the Grasshopper Sparrow was not recorded. The 2 longest periods of apparent absence were 18 years between 1927 and 1945 and 13 years between 1945 and 1958. During these years, some of the most talented bird observers and collectors in the province lived within a few hundred metres of some of the best habitat for the species, and it is unlikely that they would have missed the bird if it had been present.

Since 1958 there have been only 9 years when sparrows were not reported; the longest period of absence was 2 years. With a species as scarce and elusive as the Grasshopper Sparrow, 1 or 2 years without a sighting is probably insignificant. This pattern of occurrence suggests that Grasshopper Sparrows have become more regular in the Okanagan valley despite the changes that a rapidly expanding human population has imposed on the habitat.

In Washington, Grasshopper Sparrows were "more widespread and numerous before the conversion of large tracts of shrubland to agriculture" (Smith et al. 1997). The same authors comment that overgrazing of these arid lands is known to increase shrub cover and result in less favourable conditions for this sparrow. Trampling of nests is also a concern. For additional information on the impact of grazing pressure on the Grasshopper Sparrow and other ground-nesting grassland species, see Blankespoor (1980), Larrison (1981), Kantrud and Kologiski (1982), and Bock and Webb (1984).

The northern end of the Okanagan valley is probably the northernmost part of the summer range for the Grasshopper Sparrow. There it has occurred regularly at 2 sites: Mount Middleton and Goose Lake. Both sites are now adjoined by housing developments and may not provide suitable habitat for the sparrows for much longer (C. Siddle pers. comm.).

Banding records from near McBride, and Breeding Bird Survey records from Adams Lake and Salmon Arm, are well out of the normal range for the species. Because they lack documentation, they have been excluded from this account.

For additional life-history information on the Grasshopper Sparrow, see Kaspari (1991) and Vickery (1996).

NOTEWORTHY RECORDS

Spring: Coastal – No records. **Interior** – Chopaka 9 May 1982-2; Kilpoola Lake 5 May 1998-1; Haynes Lease (n Osoyoos Lake) 9 May 1982-3, first of year; Okanagan Falls 1 May 1913-1 (RBCM 3581); Okanagan Landing 21 May 1906-1, first of year, 17 May 1915-numerous (Munro and Cowan 1947).

Summer: Coastal – Sea Island 5 to 18 Jun 1977-1 singing on territory (Hunn and Mattocks 1977), 8 Jul 1979-1. **Interior** – Chopaka 29 Jun 1982-1; Kilpoola Lake 15 Jul 1990-1 singing male (Preston 1990), 8 Aug 1996-1 foraging in bare ground nr lakeshore; n Osoyoos Lake 9 Jul 1998-1; Vaseux Lake 3 Aug 1980-1 (Cannings 1995b); West Bench 15 Jun 1990-4; Penticton 19 Jun 1924-1 egg (Cannings et al. 1987); Chapperon Lake 22 Jun 1962-4 eggs, incubation fresh; Okanagan 2 Jun 1914-6 eggs; Okanagan Landing 11 Jul 1911-nest with 4 eggs collected, 25 Jul 1916-1 fledging with adults; Mount Middleton 8 Jul 1981-1 singing, 25 Jun 1990-3 singing (Preston 1990); Vernon 14 Jun 1898-1 (RBCM 1871), 11 Jul 1911-4 eggs and a Brown-headed Cowbird egg (Cannings et al. 1987), 26 Aug 1927-1 on Commonage (Cannings et al. 1987); Riske Creek 25 Jul 1985-1.

Breeding Bird Surveys: Coastal – Not recorded. **Interior** – Recorded from 4 of 73 routes and on less than 1% of all surveys. Maxima: Osprey Lake 26 Jun 1977-3; Christian Valley 25 May 1994-1 (see also REMARKS).

Autumn: Interior – Okanagan Centre 11 Oct 1913-1 (MVZ 105883); Okanagan Landing 19 Oct 1927-1. **Coastal** – Vancouver 6 Sep 1976-1 (Crowell and Nehls 1977a); Saanich 8 Oct 1975-1 (BC Photo 447), 29 Nov 1992-1 (Siddle 1993a).

Winter: No records.

Le Conte's Sparrow

LCSP

Ammodramus leconteii (Audubon)

RANGE: Breeds from northeastern and east-central British Columbia, southern Mackenzie, northern Alberta, northern Saskatchewan, northwestern and central Manitoba, north-central Ontario, and west-central Quebec south to southern Alberta and Saskatchewan, northern North Dakota, northwestern and eastern Minnesota, northeastern Wisconsin, and northern Michigan. Winters from west-central Kansas, southern Missouri, southern Illinois, western Tennessee, central Alabama, south-central Georgia, and South Carolina south to the Gulf coast from Florida through Louisiana and southern Texas.

STATUS: On the coast, *casual* in autumn in the Georgia Depression Ecoprovince.

In the interior, *casual* in the Southern Interior, Central Interior, and Sub-Boreal Interior ecoprovinces; *very rare* in the Southern Interior Mountains Ecoprovince; *uncommon* migrant and very local summer visitant in the Boreal Plains and Taiga Plains ecoprovinces.

Breeds.

Figure 284. The Le Conte's Sparrow is an elusive marsh bird that breeds only in extreme northeastern British Columbia (Anthony Mercieca).

Figure 285. In British Columbia, the Le Conte's Sparrow frequents wet habitats, such as this marshy lakeshore (Toms Lake, south of Tupper, 14 June 1996; R. Wayne Campbell).

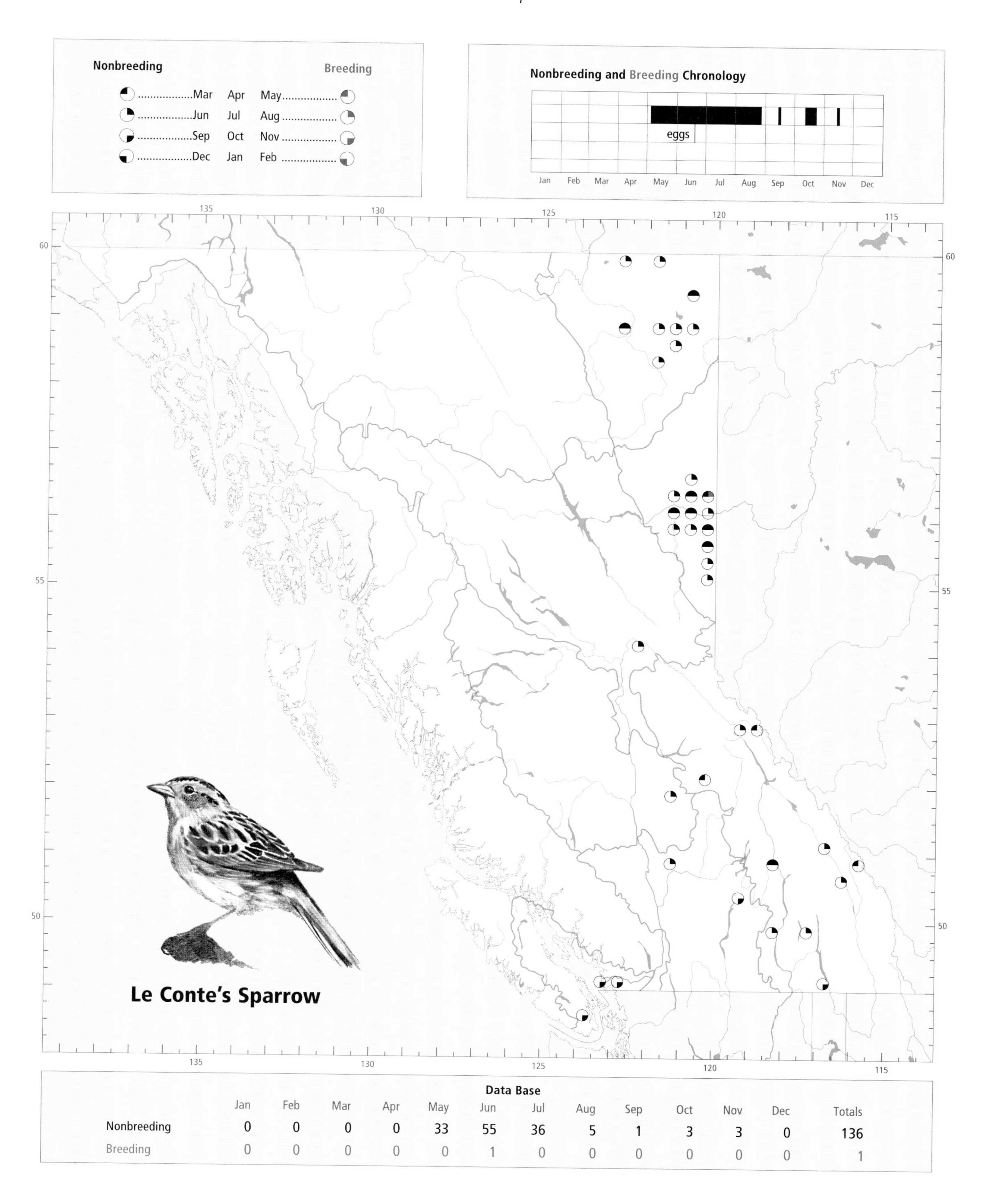

Data Base

	Jan	Feb	Mar	Apr	May	Jun	Jul	Aug	Sep	Oct	Nov	Dec	Totals
Nonbreeding	0	0	0	0	33	55	36	5	1	3	3	0	136
Breeding	0	0	0	0	0	1	0	0	0	0	0	0	1

NONBREEDING: In British Columbia, the Le Conte's Sparrow (Fig. 288) is at the northwestern limit of its range in North America, but its actual distribution in the province is poorly documented. It is known to be very locally distributed in shallow wetlands and fields in the Peace Lowland of the Boreal Plains and in the Fort Nelson Lowland of the Taiga Plains. The species has occurred from time to time in the southern portions of the interior, most often in the valleys of the Kootenay and Columbia rivers. There are only 3 coastal records.

In the interior, vagrants have been recorded from as far south as New Denver, Edgewood, and Creston in the Southern Interior Mountains, and as far north as perhaps the Petitot River in the Taiga Plains. The only coastal records are of autumn vagrants at Cowichan Bay in the Nanaimo Lowland and Boundary Bay and Roberts Bank in the Fraser Lowland.

The Le Conte's Sparrow has been recorded at elevations from near sea level to 1,090 m, but most occurrences in the province are between 400 and 700 m. In general, the Le Conte's Sparrow goes directly to its breeding habitat when it arrives in spring and has seldom been observed during migration in either spring or autumn. The few spring records have been from wet or moist habitats adjacent to marshy lakeshores (Fig. 285), wet meadows, or roadside ditches with dead grasses and short willows, patches of sedges, or open meadowland. Late in the season, it has been recorded from overgrown

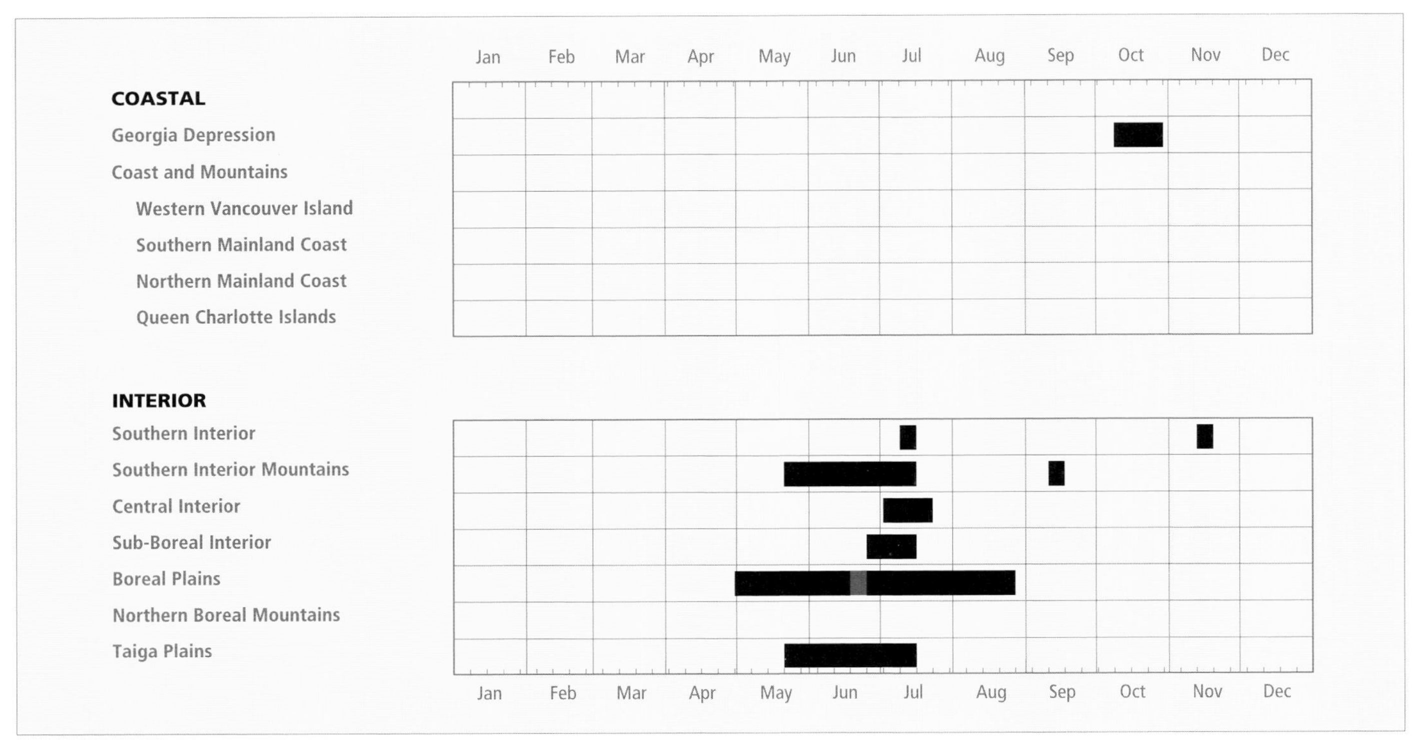

Figure 286. Annual occurrence (black) and breeding chronology (red) for the Le Conte's Sparrow in ecoprovinces of British Columbia. Records are shown for the week in which they occurred.

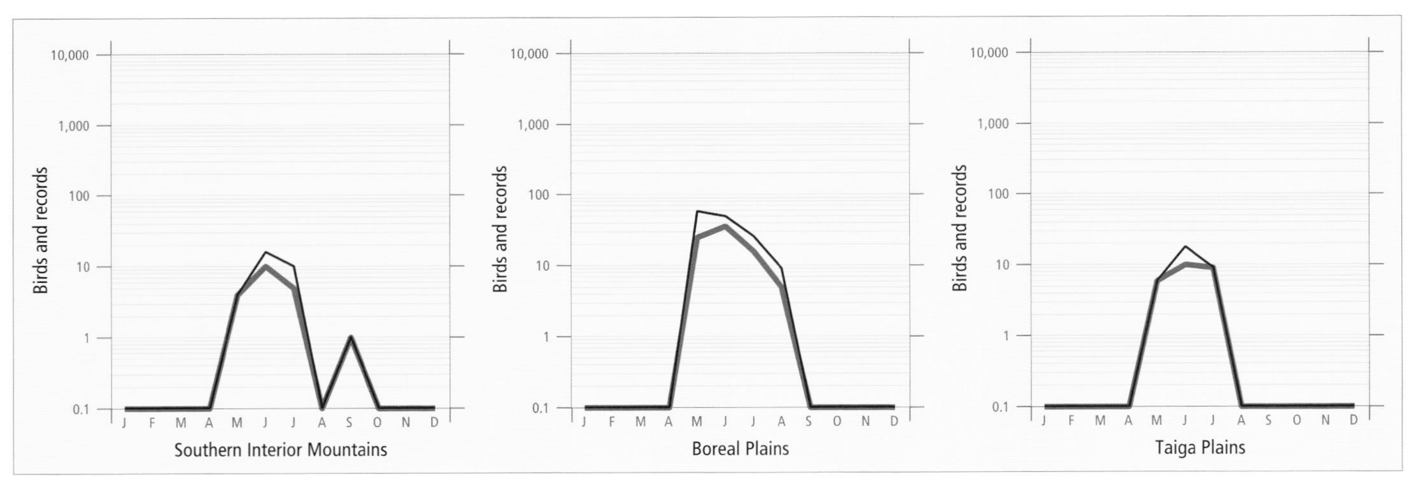

Figure 287. Fluctuations in total number of birds (purple line) and total number of records (green line) for the Le Conte's Sparrow in ecoprovinces of British Columbia. Breeding Bird Surveys and nest record data have been excluded.

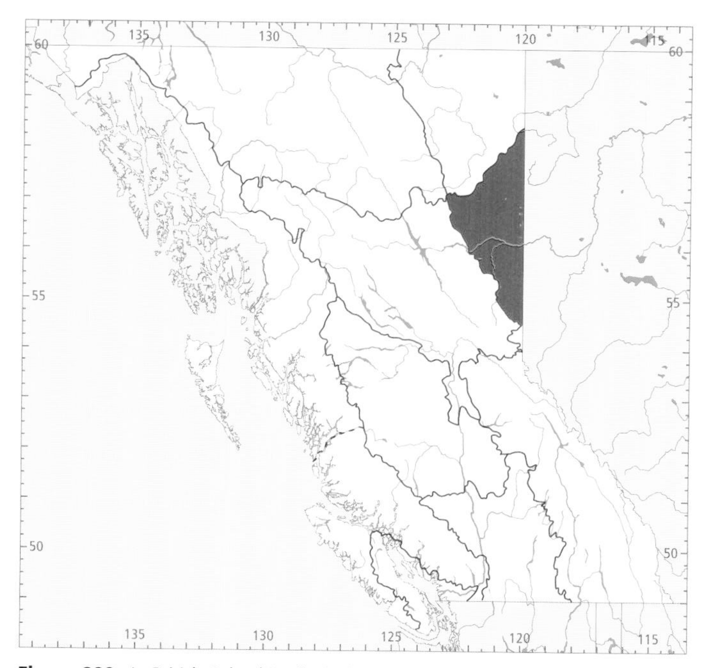

Figure 288. In British Columbia, the highest numbers for the Le Conte's Sparrow in summer occur in the Boreal Plains Ecoprovince.

meadows and tall grasses adjacent to wetlands. For additional information on specific habitat use, see BREEDING.

During migration, the Le Conte's Sparrow takes a route across the Great Plains west of the Mississippi River and east of the Rocky Mountains. The first spring arrivals in British Columbia appear in the Peace Lowland as early as the first week of May, with the height of migration in the last half of May (Figs. 286 and 287). The earliest appearances of the species at localities in the Southern Interior Mountains have been in the last week of May. There are few indications of the timing of the autumn migration. All Le Conte's Sparrows have left the Peace Lowland by the end of August (Figs. 286 and 287). The latest of the few observations of this sparrow in southern regions of the province have been in mid-October and mid-November.

In the western United States, the Le Conte's Sparrow is considered a very rare spring and autumn vagrant, with most records from late May in the spring and mid-October to mid-November in the autumn (LaFave 1965; McCaskie 1975; Roberson 1980; Gilligan et al. 1994). In California, about 1 bird per year has been reported between 1970 and 1990 (Lowther 1996).

On the coast, the Le Conte's Sparrow has been found very irregularly from 13 to 22 October; in the interior, it occurs regularly from 11 May to 26 August, with occasional occurrences as early as 5 May and as late as 16 November (Fig. 286).

BREEDING: The Le Conte's Sparrow likely breeds locally in northeastern British Columbia; our only confirmed breeding record is from the Peace Lowland. It is present in small numbers every summer at several localities in the Peace Lowland and Fort Nelson Lowland. There the males are on territory and in full song. In 1938, searches for nests in the Peace River

Figure 289. The only Le Conte's Sparrow nest discovered in British Columbia was found in a thick clump of sedge just above a moist substrate, similar to this habitat where territorial adults were found (near Boudreau Lake, 1 July 1998; R. Wayne Campbell).

region were unsuccessful but the birds gave every evidence of nesting (Cowan 1939). Only a single nest was discovered there in the next 50 years.

The Le Conte's Sparrow may also breed in the southern interior of British Columbia. Two birds with brood patches, collected near 115 Mile House in 1958, provide evidence of breeding in the vicinity (Erskine and Stein 1964). The species has also been present through the summer along the Columbia River south of Revelstoke.

The Le Conte's Sparrow reaches its highest numbers in summer in the Peace Lowland of the Boreal Plains (Fig. 288). Breeding Bird Surveys in British Columbia for the years 1968 through 1993 contain insufficient data to determine trends in numbers. Across the Canadian range of the species, numbers increased at an average annual rate of 4.0% ($P < 0.01$) between 1980 and 1996 (Sauer et al. 1997).

In the Peace Lowland, the Le Conte's Sparrow occupies the patches of sedges emerging from flooded land with scattered clumps of low-growing willows early in the nesting season. Somewhat drier sites with grasses, sedges, and willows (used as singing perches) are also utilized. In a study of the sparrow's habitat requirements, Enns and Siddle (1996) found that it occupied "damp sites ranging from shrub-carrs to fields in which tall black spruce patches were prevalent. Sedge meadow shrub-carr edges were the most common habitat." They identified 1 location with an abundance of creeping bentgrass close to a shrub-carr complex. The preferred habitat had a plant association that included beaked sedge, horsetail, bog cranberry, scrub birch, Labrador tea, *Sphagnum* moss, palmate coltsfoot, and species of willows (Enns and Siddle 1996). This sparrow also uses damp, short-grass margins of grain fields, wet sedge verges of black spruce–tamarack muskeg, tall grasses at sewage lagoons (Siddle 1990a), and patches of Labrador tea along the shores of muskeg ponds (Campbell and McNall 1982).

The Le Conte's Sparrow is present and breeds in adjacent Alberta (Lowther 1996), but we could find no information on its nesting habitat or nesting biology there.

The Le Conte's Sparrow probably breeds in British Columbia from mid June to mid July (Fig. 286).

Nests: The only Le Conte's Sparrow nest found in British Columbia was a small, compact structure placed in a dense patch of sedges just above moist ground (Fig. 289). It consisted entirely of grasses lined with finer grasses. Elsewhere in its range, nests are built on or close to the ground or water in sites with thick accumulations of dead grasses or in the "drier borders of open marshes beneath tangles of old dead rushes, grasses or sedges" (Walkinshaw 1968).

Eggs: A nest with 4 eggs found on 18 June 1978, at Boundary Lake near Goodlow, is the only record of a clutch size for British Columbia. Elsewhere in North America, the clutch size is 4 to 5 eggs (Lowther 1996). Incubation is by the female alone and ranges from 11 to 13 days (Walkinshaw 1937; Baicich and Harrison 1997). Egg dates in Alberta range from 6 to 24 June (n = 6) (Walkinshaw 1968).

Young: Neither nestlings nor fledglings have been reported from British Columbia. The nestling period is unknown (Lowther 1996).

Brown-headed Cowbird Parasitism: Cowbird parasitism was not reported in British Columbia. Friedmann (1963) and Friedmann and Kiff (1985) report a few instances of this cowbird parasitizing the Le Conte's Sparrow in Alberta, Saskatchewan, Manitoba, and North Dakota.

Nest Success: Insufficient data.

REMARKS: Despite its having so wide a range over relatively accessible areas of the continent, remarkably little information is available on the breeding biology of the Le Conte's Sparrow. Its requirement for nesting habitat associated with the margins of a small variety of wetlands renders it vulnerable to the many changes humans are making to the wetlands of the Great Plains and the agricultural areas along their northern fringe. So far, the Le Conte's Sparrow appears to be maintaining its numbers and distribution. Lowther (1996) comments on the meagre and anecdotal character of the data on this sparrow. He states that "intentional work on Le Conte's Sparrow in the field provides a challenge of observational skills, ingenuity and luck to achieve success."

For additional information on the Le Conte's Sparrow, see Igl and Johnson (1995), Lowther (1996), and Rising (1996).

NOTEWORTHY RECORDS

Spring: Coastal – No records. **Interior** – Revelstoke 27 May 1989-1; Dog Lake (Kootenay National Park) 27 May 1975-1 (Poll et al. 1984); Wells Gray Park 25 May 1962-1 collected at Ray Farm (Edwards and Ritcey 1967); Mount Robson Park 30 May 1990-1 (Siddle 1990a); Tupper Creek 18 May 1938-2 (RBCM 8018 and 8019); Pouce Coupe 23 May 1992-1 (BC Photo 1642); Dawson Creek 17 May 1993-1; n Fort St. John 5 May 1985-1 singing male; Boundary Lake (Goodlow) 11 May 1986-1; Fort Nelson 23 May 1982-1.

Summer: Coastal – No records. **Interior** – Edgewood 14 Jun 1991-2 (Siddle 1992c); New Denver 17 Jun 1990-1 (Siddle 1990c); Revelstoke 14 Jul 1988-1 (BC Photo 1223); Westwick Lakes 15 Jun 1952-1 (Erskine and Stein 1964); nr 115 Mile House 6 Jul 1958-a pair collected with brood patches (Erskine and Stein 1964); nr 111 Mile House 17 Jul 1959-1 male specimen (Erskine and Stein 1964); Giscome 1 and 14 Jul 1989-1 singing (BC Photo 1643); Tupper 23 Jun 1997-4; Swan Lake (Tupper) 28 Jun 1984-1 at s end singing, 5 Jul 1986-2; e Chetwynd 5 Aug 1975-3; Boudreau Lake 1 Jul 1998-3; Dawson Creek 20 Jun 1994-2, 8 Jul 1994-1; Boundary Lake (Goodlow) 18 Jun 1978-4 eggs, 4 Jun 1983-2 at w side, 26 Aug 1986-2 (Johnston and McEwen 1987b); Charlie Lake 12 Jun 1998-1 (BC Photo 1638); Sikanni Chief River 4 and 5 Jun 1997-1; sw Kotcho Lake 7 Jul 1982-1, 9 Jul 1982-1 calling (Campbell and McNall 1982); Kwokullie Lake 1 Jun 1982-4 pairs in 1.6 km of roadside habitat; Petitot River 20 Jun 1982-2.

Breeding Bird Surveys: Coastal – Not recorded. **Interior** – Recorded from 4 of 73 routes and on 1% of all surveys. Maxima: Tupper 5 Jul 1981-5; Fort St. John 29 Jun 1984-2; Needles 14 Jun 1992-2; Zincton 17 Jun 1990-1.

Autumn: Interior – Swan Lake (Vernon) 15 and 16 Nov 1983-1 (BC Photo 1022); Creston 15 Sep 1988-1. **Coastal** – Cowichan Bay 13 Oct 1986-1; Roberts Bank 17 Oct 1992-1; Boundary Bay 22 Oct 1977-1.

Winter: No records.

Nelson's Sharp-tailed Sparrow

Ammodramus nelsoni (Allen)

NSTS

RANGE: Breeds from northeastern British Columbia, south-central Mackenzie, northern Alberta, central Saskatchewan, and central Manitoba south to south-central Alberta, southern Saskatchewan, southern Manitoba, western and southeastern North Dakota, northeastern South Dakota, and northwestern Minnesota; around southwestern Hudson Bay and James Bay in northern Manitoba and Ontario and northwestern Quebec; in southeastern Quebec along the lower St. Lawrence River and along the Atlantic coast from eastern Quebec, Prince Edward Island, and Nova Scotia south to southern Maine. Winters along the mid to south Atlantic coast and the Gulf coast, rarely in coastal California and northwestern Baja California.

Figure 290. In British Columbia, the Nelson's Sharp-tailed Sparrow breeds only in the Peace Lowland of the Boreal Plains Ecoprovince (Anthony Mercieca).

STATUS: On the coast, *accidental* in the Georgia Depression Ecoprovince.

In the interior, locally an *uncommon* summer visitant in the Peace Lowland of the Boreal Plains Ecoprovince.

Breeds.

NONBREEDING: In British Columbia, the Nelson's Sharp-tailed Sparrow (Fig. 290) is at the western limit of its breeding range in North America. It occurs regularly only in the northeastern corner of the province east of the Rocky Mountains, and is locally distributed only from near the Alberta border west to Charlie Lake (northwest of Fort St. John), in the Peace Lowland, south to Swan Lake near Tupper, in the Kiskatinaw Plateau. It has been found at 13 sites, 5 of which are very small wetlands that may be inhabited only by small numbers of sparrows in unusually wet years (Fig. 291).

Small numbers occur regularly only at Swan Lake south of Pouce Coupe, and Boundary Lake east of Fort St. John. Small numbers have also been found irregularly at Cecil Lake, along 184 Road east of Cecil Lake, at North Boundary Lake, and from

Figure 291. During unusually wet summers in the Boreal Plains Ecoprovince of British Columbia, small numbers of Nelson's Sharp-tailed Sparrows may be found in small wetlands scattered throughout the Peace Lowland (near Boundary Lake, Goodlow, 24 June 1996; R. Wayne Campbell).

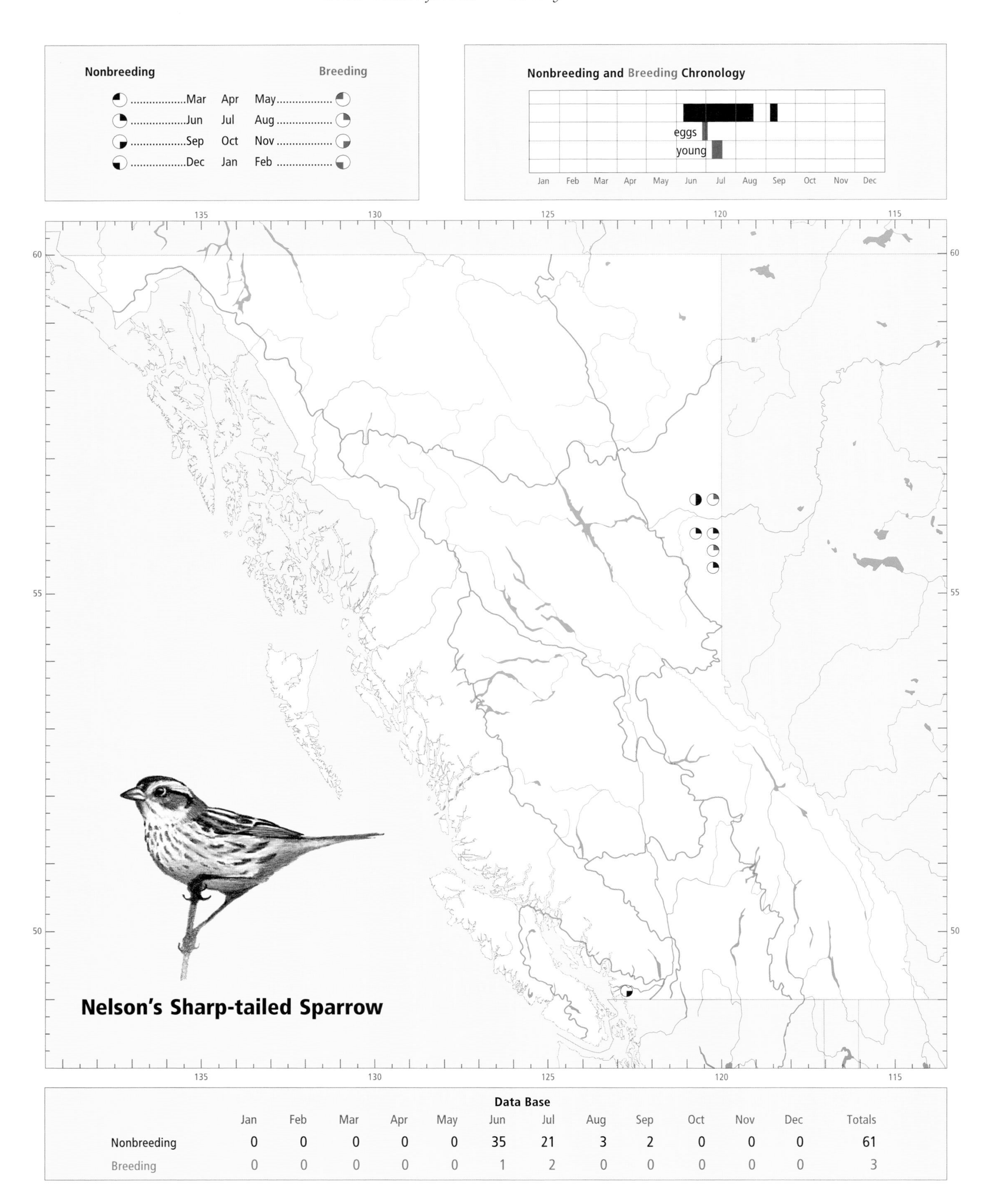

Nelson's Sharp-tailed Sparrow

Data Base

	Jan	Feb	Mar	Apr	May	Jun	Jul	Aug	Sep	Oct	Nov	Dec	Totals
Nonbreeding	0	0	0	0	0	35	21	3	2	0	0	0	61
Breeding	0	0	0	0	0	1	2	0	0	0	0	0	3

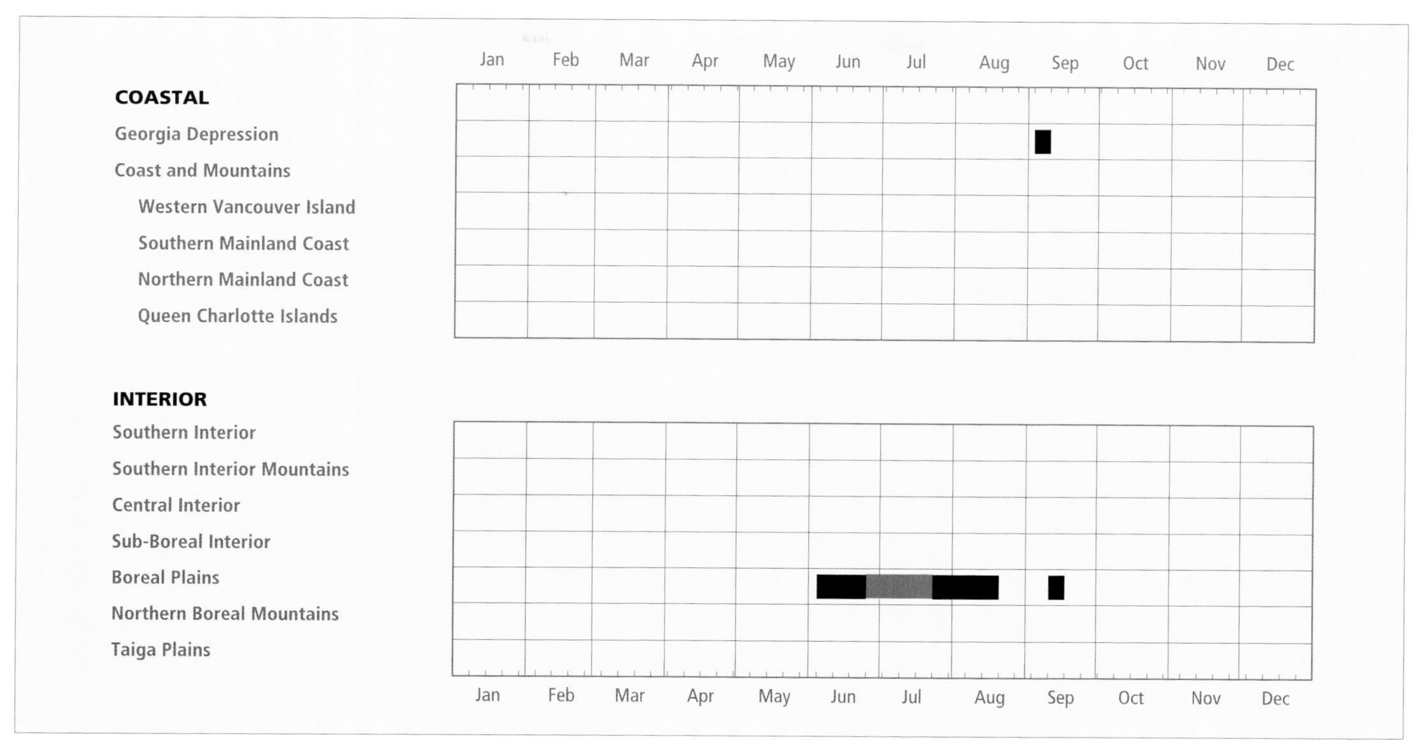

Figure 292. Annual occurrence (black) and breeding chronology (red) for the Nelson's Sharp-tailed Sparrow in ecoprovinces of British Columbia. Records are shown for the week in which they occurred.

smaller wetlands at Alcock Lake, McQueen Slough, near Progress, near Sudeten Park, and the south end of Charlie Lake.

The Nelson's Sharp-tailed Sparrow has been recorded at elevations from 690 to 800 m. Because it moves directly onto its breeding grounds, nonbreeding habitat is similar to breeding habitat (see BREEDING). In the very wet summer of 1996, when water levels were much higher than usual, the Nelson's Sharp-tailed Sparrow was more widely distributed, appearing in flooded areas such as ponds (Fig. 291), densely vegetated wide ditches, and low fields.

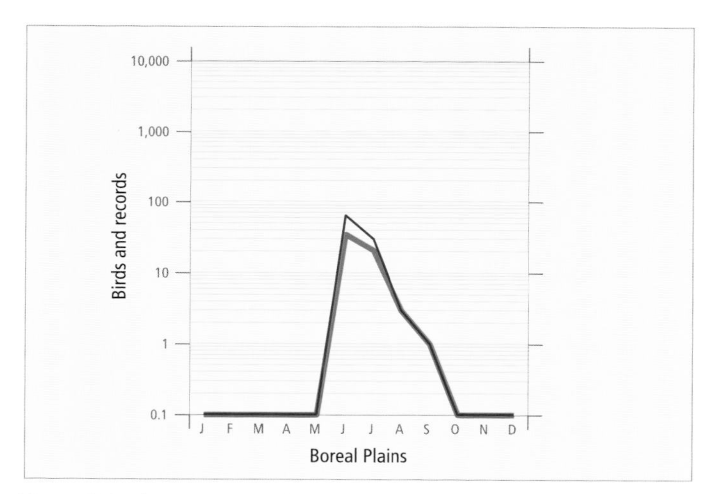

Figure 293. Fluctuations in total number of birds (purple line) and total number of records (green line) for the Nelson's Sharp-tailed Sparrow in the Boreal Plains Ecoprovince of British Columbia. Nest record data have been excluded.

Migration periods are not well known. Spring migrants may appear in the Peace Lowland early in the second week of June, but most probably arrive in the middle of the month (Figs. 292 and 293). Males begin singing upon arrival. In Alberta, the Nelson's Sharp-tailed Sparrow arrives in late May and early June (Semenchuk 1992).

The southward migration probably begins in late July, soon after breeding is completed, and likely extends into late August. Most birds have left the Peace River region by the end of August, although stragglers may be found as late as mid-September (Figs. 292 and 293). In Alberta, autumn records are also scarce but it appears that the last migrants are seen in northern Alberta in the second week of September (Pinel et al. 1993).

In British Columbia, the Nelson's Sharp-tailed Sparrow has been recorded from 8 June to 12 September (Fig. 292).

BREEDING: In British Columbia, the breeding distribution of the Nelson's Sharp-tailed Sparrow is restricted to the Peace Lowland and Kiskatinaw Plateau of the Boreal Plains. We have only 2 nesting records, 1 from the west side of Boundary Lake and the other from the south end of Swan Lake (Fig. 294). The only other evidence of possible breeding is an observation of an adult carrying food into marsh grasses and sedges in late June, at McQueen Slough, northeast of Dawson Creek. The species likely breeds in suitable habitat throughout much of its restricted British Columbia range. It was suspected to be nesting at Charlie Lake in 1938 (Cowan 1939), but no longer occurs there in summer.

The Nelson's Sharp-tailed Sparrow reaches its highest numbers in summer in the Boreal Plains (Fig. 295). Breeding

Figure 294. Wetlands at the south end of Swan Lake in the Boreal Plains Ecoprovince are 1 of 2 sites in British Columbia where the Nelson's Sharp-tailed Sparrow has been recorded nesting (Swan Lake, Tupper, 14 June 1996; R. Wayne Campbell).

Bird Surveys for interior routes for the period 1968 through 1993 did not detect this species. Continent-wide surveys indicate that numbers increased at an average annual rate of 6% ($P < 0.01$) from 1980 to 1996.

The Nelson's Sharp-tailed Sparrow has been recorded breeding at about 725 m. It frequents wetlands, including marshes and wet meadows with stands of emergents and willows. Specifically, this secretive sparrow has been found in damp sedge meadows and marshes bordering woodland lakes, among dead and living willows between creeks and wet grassy meadows, and on willow-covered islets in lakes. In marshes and sedge meadows, clumps of dead and living willows are invariably present, interspersed between large areas of grasses or sedges. In Alberta, this sparrow inhabits marshes in woodland areas, preferring those in which a few clumps of willows are scattered, and in cattails and wet grassy meadows around woodland lakes (Salt and Salt 1976), habitat similar to most of the sites in British Columbia.

The Nelson's Sharp-tailed Sparrow has been recorded breeding in British Columbia from 27 June (calculated) to 17 July (Fig. 292).

Nests: The 2 nests found in the province were placed among clumps of dry grasses. The Boundary Lake nest was built 5 cm above water, neatly concealed among dead grasses next to a willow on a spongy islet in the middle of a sedge marsh. The Swan Lake nest was made of grasses and built on a grassy clump slightly elevated above the substrate; it was also located among long grasses next to a willow patch with water channels on the marshy lakeshore (Fig. 296). Both nests were very compact structures composed of coarse grasses and plant stems and lined with finer dry grasses.

Eggs: The Boundary Lake nest containing 4 "well-incubated" eggs was found on 2 July. Calculated dates indicate that nests can be found with eggs as early as 27 June. The incubation period is 11 or 12 days (Greenlaw and Rising 1994).

Young: A single nest containing 4 nestlings was found at Swan Lake on 17 July. Calculated dates indicate that young

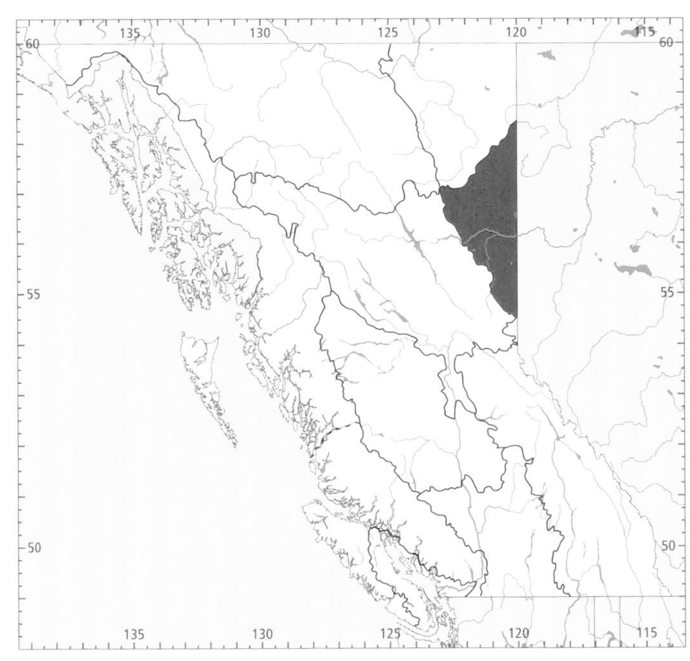

Figure 295. In British Columbia, the highest numbers for the Nelson's Sharp-tailed Sparrow in summer occur in the Boreal Plains Ecoprovince.

may occur as early as 7 July and, assuming that the young fledge, as late as 22 July. The nestling period is 8 to 11 days (Greenlaw and Rising 1994); young become independent 15 to 20 days after fledging (DeRagon 1988).

Brown-headed Cowbird Parasitism: Cowbird parasitism was not found in British Columbia. There is only a single record of Brown-headed Cowbird parasitism elsewhere in North America (Greenlaw and Rising 1994).

Nest Success: Insufficient data.

REMARKS: The Nelson's Sharp-tailed Sparrow was formerly known as Sharp-tailed Sparrow (*Ammodramus caudacutus*) (American Ornithologists' Union 1983). In 1995, the Sharp-tailed Sparrow was split into 2 species: the Saltmarsh Sharp-tailed Sparrow (*A. caudacutus*), which occurs along the northeastern Atlantic coast, and the Nelson's Sharp-tailed Sparrow (*A. nelsoni*), which is widely distributed and occupies 3 disjunct geographical areas (American Ornithologists' Union 1995). Of the 3 subspecies of the Nelson's Sharp-tailed Sparrow, only *A. n. nelsoni* occurs in British Columbia.

Known from only 13 sites in British Columbia, the Nelson's Sharp-tailed Sparrow has one of the most restricted breeding distributions of any passerine in the province. The future of this bird in the province depends upon the protection of its habitat. The wetlands it frequents are at risk from drainage, infilling, water level manipulation, grazing by domestic stock, recreational activities, pesticide and herbicide spraying, and pollution. For example, sharp-tailed sparrows are no longer found at Stoddart Creek, at the south end of Charlie Lake, because of draining and channelling activities (Siddle 1981).

The Nelson's Sharp-tailed Sparrow is one of the least known passerines in the province. First discovered in British Columbia in 1938 at Charlie Lake and Swan Lake (Cowan 1939), it remained uninvestigated until the late 1970s. Field work from 1977 to 1997 found only small numbers of singing males (fewer than 15) at any one site. Shy and inconspicuous, the bird is difficult to locate and may be confused with the related Le Conte's Sparrow (*Ammodramus leconteii*). In addition, its wetland habitat is a challenging one for biologists and birders to work in. Biologists planning to survey breeding populations should consult Greenlaw and Rising (1994) for suggestions on obtaining reliable information on populations through intensive banding activities, mark-and-recapture techniques, and thorough nest searches.

Figure 296. The Nelson's Sharp-tailed Sparrow builds its well-concealed nest among tall grasses adjacent to dense willow stands bordering small lake channels (Swan Lake, Tupper, 14 June 1996; R. Wayne Campbell).

For additional information on the life history, conservation, and identification of the Nelson's Sharp-tailed Sparrow, see Greenlaw (1993), Greenlaw and Rising (1994), Rising (1996), and Sibley (1996).

NOTEWORTHY RECORDS

Spring: No records.

Summer: Coastal – No records. **Interior** – s end Swan Lake (Tupper) 19 Jun 1938-abundant (Cowan 1939), 10 Jun 1983-2 males singing, 14 Jun 1996-11 males singing, 17 Jul 1986-4 nestlings; 2.2 km s Sudeten Park 16 Jun 1996-1 male singing in small marsh; Tom's Lake 17 Jul 1986-1 male singing from tall willows; Alcock Lake 24 Jun 1992-1 male singing (Phinney 1992); McQueen Slough 15 Jul 1986-1 male singing, 8 Jun 1992-2, likely a pair, 14 Jun 1992-1 male singing, 27 Jun 1992-1 adult carrying food; Charlie Lake 12 Jun 1938-2 seen and 1 collected (RBCM 8013), 15 Jun 1938-3 more seen and 1 collected (RBCM 8014); n end Cecil Lake 3 Jul 1978-1 male singing; ne Cecil Lake 16 Jun 1996-3 singing from wet fields and ditches along 184 Rd; Boundary Lake (Goodlow) 10 Jun 1989-2 males singing, 2 Jul 1978-6 birds on census along w side, nest with 4 well-advanced eggs, 8 Jul 1997-4 males singing, 20 Jul 1992-2 males; North Boundary Lake 2 Jul 1978-2 at n end.

Breeding Bird Surveys: Not recorded.

Autumn: Interior – Charlie Lake 12 September 1982-1 in marsh grasses at s end of lake. **Coastal** – Blackie Spit (White Rock) 6 Sep 1974-1.

Winter: No records.

Fox Sparrow

FOSP

Passerella iliaca (Merrem)

RANGE: Breeds from northwestern and interior Alaska, northern Yukon, northwestern Mackenzie, northern Manitoba, northern Ontario, northern Quebec, and northern Labrador south to southwestern Alaska, northern, central interior, and southwestern British Columbia, central Alberta, central Saskatchewan, southern Manitoba, north-central Ontario, southeastern Quebec, and southern Newfoundland, south to northwestern Washington, central Nevada, and central Utah. Winters from southern Alaska and coastal British Columbia south to northern Baja California, and from central Arizona east to southern Wisconsin, southern Ontario and southern Quebec to New Brunswick, Nova Scotia, and southern Newfoundland, and south to southern New Mexico and Texas, and along the Gulf coast to central Florida.

STATUS: On the coast, an *uncommon* resident and locally *fairly common* to *common* during migration and winter in the Georgia Depression Ecoprovince; in the Coast and Mountains Ecoprovince, an *uncommon* resident and locally *fairly common* to *common* during migration and winter on Western Vancouver Island, the Southern and Northern Mainland coasts, and the Queen Charlotte Islands.

In the interior, an *uncommon* migrant and summer visitant to every ecoprovince; in winter, *very rare* in the Southern Interior, Southern Interior Mountains, and Central Interior ecoprovinces; *casual* in the Sub-Boreal Interior and *accidental* in the Boreal Plains ecoprovinces.

Breeds.

NONBREEDING: The Fox Sparrow (Fig. 297) has a widespread distribution in suitable habitat throughout most of the province. On the coast, it occurs from the southern international boundary north throughout Vancouver Island, the Queen Charlotte Islands, and northern portions of the mainland coast. In the interior, its distribution covers the province. There, although it is mainly a bird of higher elevations during migration, it does move into valley bottoms in small numbers and may spend the winter (Cannings et al. 1987). The highest numbers in winter occur in the Georgia Depression (Fig. 298).

The Fox Sparrow has been reported at elevations from near sea level to 2,130 m on the coast and from 280 to 1,830 m in the interior. In British Columbia, it mainly frequents areas of heavy shrub cover along forest and woodland edges, often near water, including estuaries, lagoons, ponds, rivers, lakes, sloughs, and bogs, and along beaches.

On the coast, this sparrow has been reported from shrub-dominated Douglas-fir, Sitka spruce, and subalpine fir and mountain hemlock forests. It also occurs in red alder and black cottonwood stands with dense shrub understoreys, open shrub areas, Garry oak woodlands, and blackberry, salmonberry, hardhack, young alder and scrub willow thickets, especially near water. Clearcuts in the early stages of regrowth are also important habitats for this sparrow. On the west coast

Figure 297. Of the 11 subspecies of Fox Sparrow recognized as occurring in British Columbia, 4 are known to winter on the coast (Reifel Island, Delta, 22 January 1993; Michael I. Preston).

of Vancouver Island, the Fox Sparrow is a conspicuous bird in the wind-pruned forest of Sitka spruce and western hemlock with its understorey of salal, devil's club, and red huckleberry that backs the exposed beaches. The Fox Sparrow can often be seen foraging in the windrows of marine vegetation on the beaches (Hatler et al. 1978). In winter it comes readily to backyard gardens and feeders.

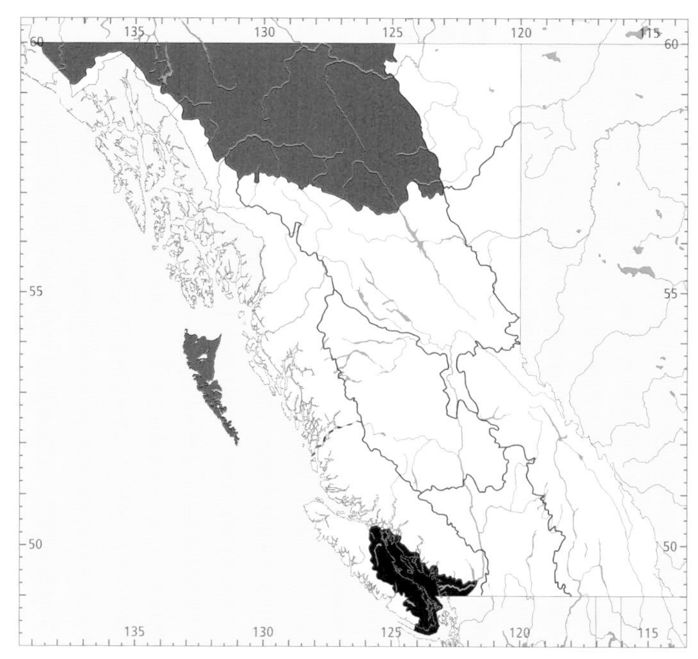

Figure 298. In British Columbia, the highest numbers for the Fox Sparrow in winter (black) occur in the Georgia Depression Ecoprovince; the highest numbers in summer (red) occur on the Queen Charlotte Islands (in the Coast and Mountains Ecoprovince) and in the Northern Boreal Mountains Ecoprovince.

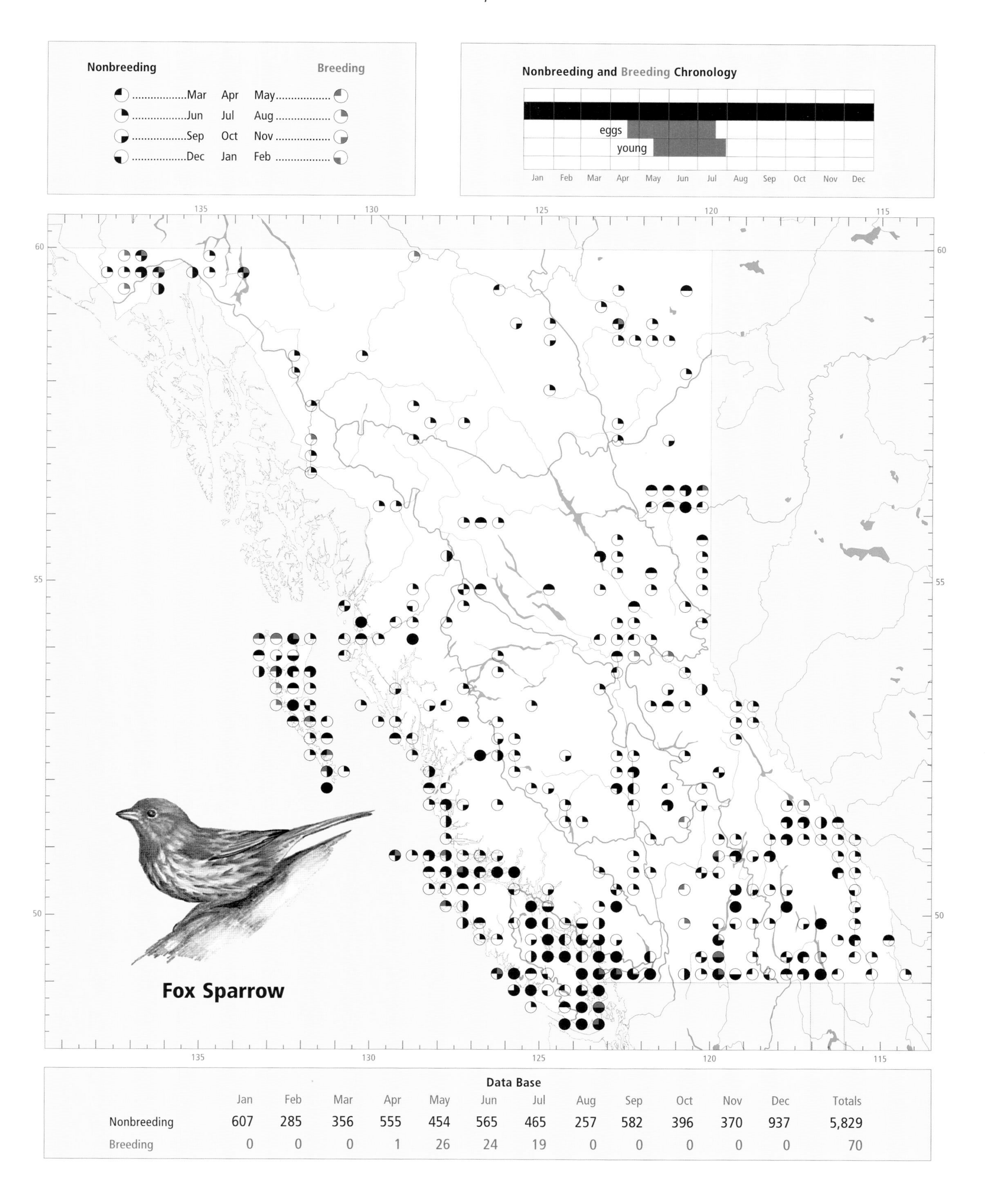

Data Base

	Jan	Feb	Mar	Apr	May	Jun	Jul	Aug	Sep	Oct	Nov	Dec	Totals
Nonbreeding	607	285	356	555	454	565	465	257	582	396	370	937	5,829
Breeding	0	0	0	1	26	24	19	0	0	0	0	0	70

In the interior, the Fox Sparrow frequents shrub thickets, tamarack and spruce muskegs, bogs, floodplain forests, including cottonwood and alder bottomlands, and scrub willow and trembling aspen thickets. Rural and urban gardens are also used.

Spring migration begins in the Georgia Depression with a small rise in numbers through March, peaking in the first or second week of April (Figs. 299 and 300). Numbers remain relatively stable until the second week of May, then decline abruptly as wintering birds leave and migrants pass through on their way to breeding areas at higher elevations or regions further north. On the Southern and Northern Mainland coasts, small movements are discernible between the first week of April and the first week of May. On both Western Vancouver Island and the Queen Charlotte Islands, the pattern is somewhat different as wintering numbers decline through February and into March then begin building again as migrants arrive. Unlike the pattern in the Georgia Depression or the mainland Coast and Mountains, where the spring movement peaks in April, numbers continue to climb on both Western Vancouver Island and the Queen Charlotte Islands and peak in June. This is more likely an indication of an increase in the numbers of observers in those areas at that time of the year coupled with the song period, which makes this sparrow more conspicuous, rather than an indication of the bird's continued spring movement into June.

In the Southern Interior, spring migration begins in mid-April and peaks during the second and third weeks of April. Spring migrants may appear in the Central Interior during the second week of April, with the movement peaking in late April and early May. At Mackenzie, in the Sub-Boreal Interior, spring migrants appear during the last week of April, with peak numbers occurring during the first 2 weeks of May. Munro and Cowan (1947) note that this sparrow travels at high elevations and seldom descends to the low valleys, where most observers are concentrated. Further north, in the Boreal Plains, migrants reach Fort St. John during the second week in April, with the movement peaking over the next 3 weeks.

In the northern portions of the interior, there is little in the way of a discernible autumn movement. Departures appear to begin after the young are off the nest, although an increase in numbers in September suggests that migrants from further north may be passing through. Most birds have left the region by mid-September. In the central and southern portions of the interior, the sparrows begin leaving in early August, with the main period of movement in September; most are gone by the first week of October (Figs. 299 and 300).

Autumn migration on the Northern Mainland Coast begins in August as numbers increase during the last week of that month and through mid-September; most birds have left the area by the end of September. On the Queen Charlotte Islands, a small autumn movement is discernible as numbers increase in the third week of August and through September as birds move further south. Another wave of migrants appears to arrive and winter on the islands. On Western Vancouver Island, a small movement is discernible beginning in August and continuing through September. On the Southern Mainland Coast, numbers decline from mid-September through November. In all these areas, small numbers of

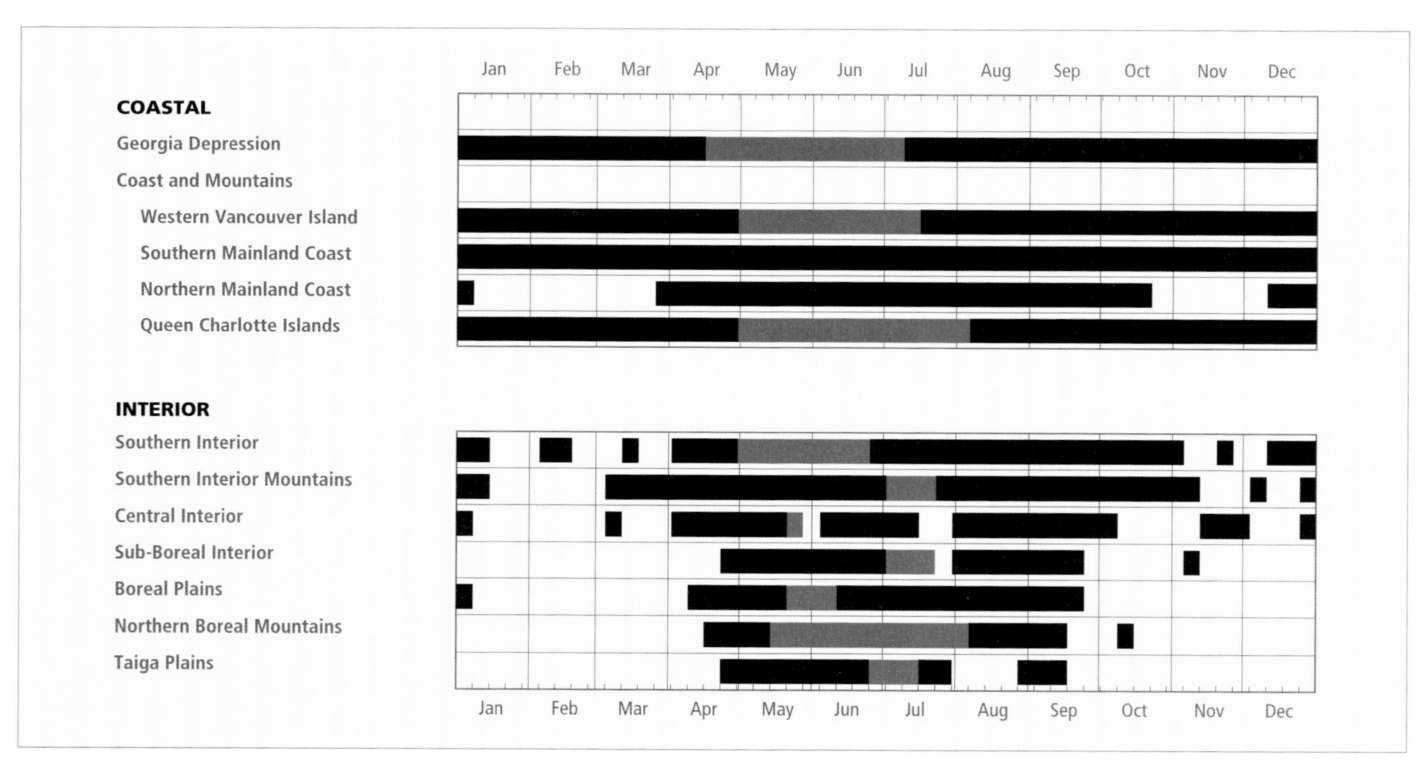

Figure 299. Annual occurrence (black) and breeding chronology (red) for the Fox Sparrow in ecoprovinces of British Columbia. Records are shown for the week in which they occurred.

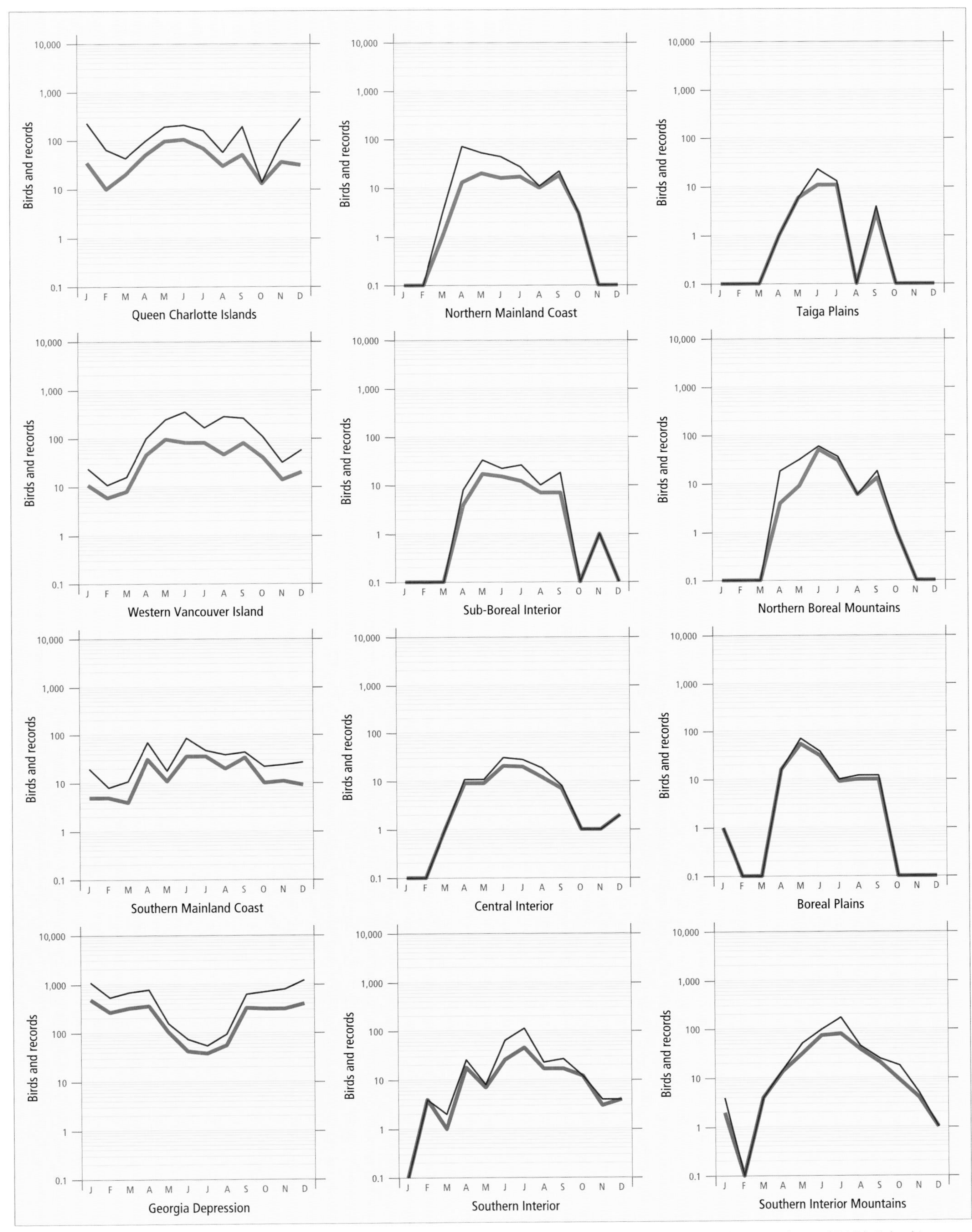

Figure 300. Fluctuations in total number of birds (purple line) and total number of records (green line) for the Fox Sparrow in ecoprovinces of British Columbia. Christmas Bird Counts, Breeding Bird Surveys, and nest record data have been excluded.

Figure 301. On Western Vancouver Island, in the Coast and Mountains Ecoprovince, the Fox Sparrow breeds in red alder bottomlands punctuated with small western hemlocks and an understorey of salmonberry and red elderberry (10.7 km east of Tahsis, 19 June 1997; Neil K. Dawe).

wintering birds may be found. In the Georgia Depression, numbers start to build in early August, and by the first week of September the autumn movement is well under way as birds pass through or stay to overwinter. Numbers there continue to climb into December.

The Fox Sparrow has been reported throughout the year on the coast. It occurs regularly from early April through September in the southern portions of the interior, and from the end of April through mid-September in the northern regions (Fig. 299).

BREEDING: The Fox Sparrow likely breeds throughout most of its summer range in British Columbia, including Vancouver Island and the Queen Charlotte Islands; however, breeding records are lacking from large areas of suitable habitat both along the coast and in the interior.

On the coast, nesting has been documented from the extreme southern portions of the Nanaimo Lowland, the Fraser Lowland, and along the west and north coasts of Vancouver Island, including offshore islands such as Cleland, Dodd, Solander, Triangle, and Pine islands. Further north, the Fox Sparrow has been reported breeding only on the Queen Charlotte Islands, in the Stikine River valley, and north of Pleasant Camp in the Chilkat Pass. Although the Fox Sparrow undoubtedly nests on the Southern and Northern Mainland coasts (see, for example, Brooks 1922; Racey 1948), adequate documentation, in the form of nests with eggs or young, is lacking.

In the interior, the Fox Sparrow has been found breeding west of Keremeos, on Apex Mountain near Penticton, at Aspen Grove and Mamit Lake in the Merritt area, and near

Figure 302. On the Northern Mainland Coast, in the Coast and Mountains Ecoprovince, the Fox Sparrow breeds in sphagnum bogs with western redcedar, shore pine, sweet gale, Labrador tea, bog laurel, sundew, and butterwort (McLoughlin Bay, near Bella Bella, 11 June 1996; Neil K. Dawe).

Figure 303. In parts of British Columbia, the Fox Sparrow breeds in subalpine habitats of spruce, willow, and birch (Stone Mountain Park, 27 June 1996; R. Wayne Campbell).

Chase, all in the Southern Interior; near Bridge Lake in the Central Interior; at Boundary Lake east of Fort St. John in the Boreal Plains; throughout much of the Chilkat Pass area of the extreme northwestern Northern Boreal Mountains; and near Fort Nelson in the Taiga Plains.

The Fox Sparrow reaches its highest numbers in summer on the Queen Charlotte Islands in the Coast and Mountains, and in the Northern Boreal Mountains (Fig. 298). An analysis of Breeding Bird Surveys for the period 1968 through 1993 could not detect a net change in numbers on either coastal or interior routes. Sauer et al. (1997) obtained similar results across both Canada and North America for the period 1966 to 1996.

The Fox Sparrow has been reported nesting on the coast from near sea level to 90 m elevation; there are no documented nesting records from subalpine or alpine areas along the coast, although it is certain that the birds nest there (e.g., McNair Lake). Elliott and Gardner (1997) note that singing males are heard every year in the mountain hemlock clearcuts of Cypress Bowl at 1,000 m elevation (see also Racey 1948). In the interior, the Fox Sparrow nests from 360 to 2,320 m elevation.

The Fox Sparrow's breeding habitat along the coast includes shrub-dominated Sitka spruce forests, red alder bottomlands with thick salmonberry (Fig. 301) or salal understorey, shrub-covered slopes of clearcuts in early stages of regrowth, and boggy areas (Fig. 302). Near Kennedy Lake, Bryant et al. (1993) consistently detected the Fox Sparrow in a clearcut with 5 to 10 years of regrowth and described the sparrow as confined to this age class. On the Sechelt Peninsula, the Fox Sparrow is found in clearcuts at 1,000 m elevation with regenerating fir and western hemlock (T. Greenfield pers. comm.).

In the interior, breeding habitat includes alpine shrublands, subalpine fir and spruce krummholz, willow, and birch (Fig. 303), subalpine meadows, subalpine forests, partially logged and burned forests (Fig. 304), and deciduous thickets such as willow flats and avalanche chutes.

In the Okanagan region, the Fox Sparrow, which normally nested above 1,700 m in subalpine meadows, now breeds

Figure 304. In the Sub-Boreal Interior Ecoprovince, the Fox Sparrow has been recorded breeding at higher elevations where forests have been logged or burned ("Vineyards," 116 km east of Prince George, 17 July 1992; R. Wayne Campbell).

regularly in clearcuts at 1,500 m elevation (R.J. Cannings pers. comm.).

In Mount Revelstoke and Glacier national parks, summer Fox Sparrow densities were highest in the Engelmann spruce–subalpine fir/rhododendron–tall bilberry open forest, the subalpine fir–mountain hemlock/heather-luetkea open forest, and the subalpine fir–whitebark pine (Engelmann spruce)/tall bilberry-heather open forest (Van Tighem and Gyug 1983).

In the Peace Lowland, Cowan (1939) found the Fox Sparrow frequenting dense tangles of willows near lakes and streams as well as in trembling aspen understorey thickets of saskatoon and silverberry. He noted that a requirement for this bird appeared to be damp but not marshy or swampy ground with a litter of dead leaves rather than a heavy growth of weeds. In addition, such conditions had to provide a dense growth of low-growing brush, generally willow, with occasional singing perches elevated above the surrounding brush level. Near Dawson Creek, the Fox Sparrow is regularly seen in 5- to 10-year-old regenerating trembling aspen forests, while along the Peace River near Taylor, it uses tall, old-growth balsam poplar with a heavy shrub understorey of prickly rose and red-osier dogwood.

Breeding birds in the Fort Nelson Lowland were found in willow and alder shrubbery (Erskine and Davidson 1976),

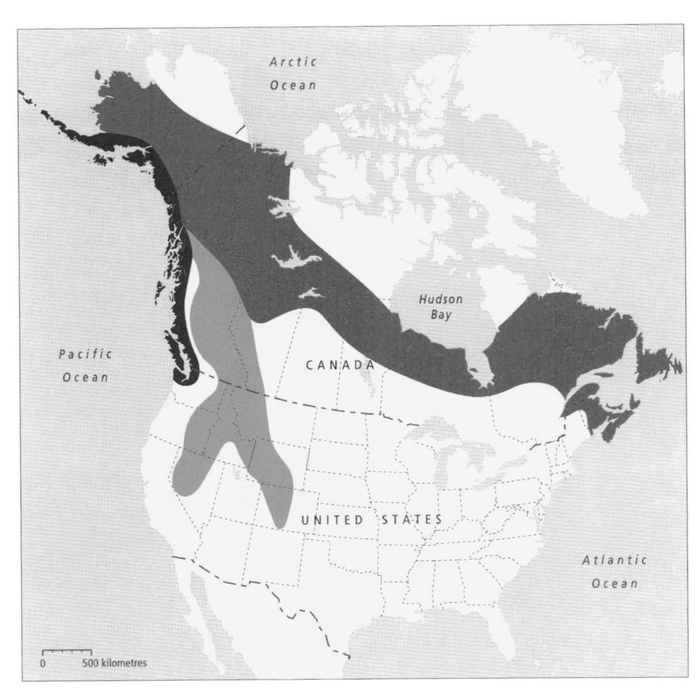

Figure 305. Tentative breeding ranges of the proposed Fox Sparrow complex in British Columbia. The Red Fox Sparrow, *Passerella iliaca* (red), includes *P. i. zaboria;* the Slate-colored Fox Sparrow, *P. schistacea* (orange), includes *P. s. schistacea, P. s. altivagans,* and *P. s. olivacea;* the Sooty Fox Sparrow, *P. unalaschcensis* (purple), includes *P. u. fuliginosa* and *P. u. townsendi* (modified from Zink 1994 and Rising 1995).

while in the Sikanni Chief River area the Fox Sparrow inhabited drowned muskegs with standing dead snags and black spruce (T. Greenfield pers. comm.). In Chilkat Pass, Weeden (1960) found this sparrow breeding in shrubby areas, especially where willow and birch were between 1.5 and 2 m tall.

On the coast, the Fox Sparrow has been recorded breeding from 19 April (calculated) to 30 July; in the interior, it has been recorded breeding from 29 April (calculated) to 30 July (Fig. 299).

Nests: Most nests (40%; *n* = 35) were found in shrubs (including salmonberry, willow, rose, and blackberry), on the ground (34%), and in trees (23%), mainly spruce. Nests in shrubs and trees were attached to a branch or branches, whereas ground nests were built on top of grassy hummocks or hidden among the underbrush or beneath a shrub or tree. Nests were bulky cups constructed of grasses (24 of 26 nests), mosses (5), twigs (2), and stems; they were lined with hair, fine grasses, mosses, and occasionally feathers. The heights of 33 nests ranged from ground level to 3.0 m, with 52% between 0.1 and 0.6 m.

Eggs: Dates for 41 clutches ranged from 22 April to 19 July, with 51% recorded between 19 May and 17 June. Calculated dates indicate that nests may hold eggs as early as 19 April. Sizes of 39 clutches ranged from 1 to 5 eggs (1E-1, 2E-7, 3E-14, 4E-11, 5E-6), with 64% having 3 or 4 eggs. The incubation period ranges from 12 to 14 days (Ehrlich et al. 1988). The incubation period of 1 clutch in Newfoundland was 12 days and 4 hours, with hatching spread over 35 hours (Threlfall and Blacquiere 1982).

Young: Dates for 29 broods ranged from 16 May to 30 July, with 55% recorded between 6 June and 17 July. Sizes of 21 broods ranged from 1 to 4 young (1Y-8, 2Y-4, 3Y-7, 4Y-2), with 57% having 1 to 2 young. The nestling period is 9 to 11 days (Ehrlich et al. 1988). The Fox Sparrow is double-brooded in Alaska, and circumstantial information from Cape St. James on the Queen Charlotte Islands in the summer of 1979 suggests that this may also be the case in parts of coastal British Columbia (J. Bowling pers. comm.).

Brown-headed Cowbird Parasitism: In British Columbia, none of 36 nests was found parasitized by the cowbird, nor have there been any reports of adult sparrows feeding fledged cowbirds. Friedmann (1963) notes the Fox Sparrow is an infrequent victim of the cowbird.

Nest Success: Insufficient data. In Alaska, Fox Sparrow nest success was considered intermediate (45% to 65%) (Rogers 1994).

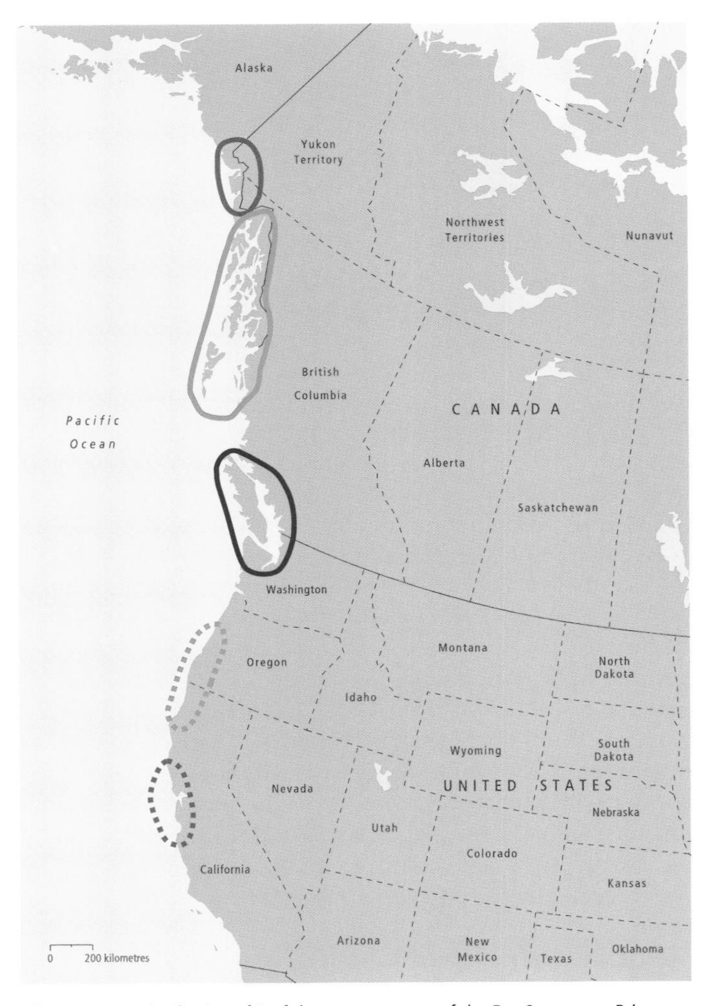

Figure 306. Distribution of 3 of the western races of the Fox Sparrow – *P. i. annectens* (red), *P. i. townsendi* (orange), and *P. i. fuliginosa* (purple) – showing their main breeding (solid line) and wintering (broken line) areas (modified from Bell 1997). Note that the northernmost breeding populations leapfrog over the wintering areas of the southern breeding populations to winter further south (see text for explanation). *P. i. fuliginosa* is mainly local and is an altitudinal migrant.

REMARKS: Of the 18 subspecies of the Fox Sparrow, 10 occur in British Columbia (American Ornithologists' Union 1957; Godfrey 1986). *P. i. zaboria* occurs mainly as a breeding bird in the Peace Lowland and across the northern portions of the province, and has occasionally been reported from the south coastal regions in winter. *P. i. altivagans* breeds in the central and southeastern portions of the interior. *P. i. olivacea* breeds at timberline regions in the southwestern and south-central portions of the province. *P. i. townsendi* breeds mainly on the Queen Charlotte Islands and winters in the Fraser and Nanaimo lowlands. *P. i. fuliginosa* breeds on much of Vancouver Island and the adjacent mainland coast, occasionally north to the Stikine River, excluding the Queen Charlotte Islands; it winters on southeastern Vancouver Island. *P. i. schistacea* breeds at higher elevations in the extreme southern Rocky Mountains. *P. i. annectens, P. i. unalaschcensis, P. i. sinuosa,* and *P. i. insularis* all winter on the south coast of the province.

In his monograph on the Fox Sparrow, Swarth (1924) hinted that the subspecies *P. i. fuliginosa* might require subdivision. On the basis of differences in colour, size, and migration patterns, Webster (1983) suggested a new subspecies, *P. i. chilcatensis,* which breeds from the Chilkat River of southeastern Alaska along the mainland to the Stewart area of British Columbia.

Recent molecular and behavioural evidence suggest that the Fox Sparrow can be divided into a complex of 4 species: (1) the Red Fox Sparrow (*P. iliaca*), which frequents the boreal forests from Newfoundland to Alaska; (2) the Sooty Fox Sparrow (*P. unalaschcensis*) of coastal Alaska and British Columbia; (3) the Slate-colored Fox Sparrow (*P. schistacea*), which occurs through the Rocky Mountains and Great Basin regions; and (4) the Thick-billed Fox Sparrow (*P. megarhyncha*) of the California Mountains (Zink 1994; Rising 1996). Figure 305 shows the tentative distribution of these proposed species in British Columbia.

The migration and winter distribution of the Pacific coast populations of the Fox Sparrow is generally considered to be the pattern that exemplifies leapfrog migration (Bell 1997). Later-breeding, northern populations of coastal British Columbia minimize the cost of spring migration by wintering in California, where food becomes available early in the spring, thereby enhancing the conditions for premigratory fattening. Spring food appears too late for the earlier-breeding southerly populations in British Columbia, which leave their wintering areas before food availability improves anywhere in the viable wintering range; they therefore truncate their migration and winter as far north as possible (Fig. 306) (Bell 1997).

Figure 307. Bird mortality from collisions with windows is a growing concern in North America, where deaths have been estimated to range from 97 to 975 million birds per year (Klem 1990). In the Greater Victoria area, a survey between 1994 and 1996 showed that Fox Sparrows accounted for 2 of 118 birds reported killed by collisions with windows at 1 residence (Victoria, 10 April 1994; R. Wayne Campbell).

Although the Fox Sparrow is known to winter in the interior in small numbers, finding even 1 on an interior Christmas Bird Count would be unusual. Observers and Christmas Bird Count compilers are encouraged to provide adequate documentation for winter occurrences of the Fox Sparrow in the interior.

Klem (1990) has suggested that between 1 and 10 birds are killed annually by each building in the United States when the birds collide with windows. Between 1994 and 1996, an informal survey of window-killed birds in the Greater Victoria area yielded a mortality of 118 birds of 23 species, all killed by windows at 1 residence. Fox Sparrows accounted for only 2% of all kills (Fig. 307).

NOTEWORTHY RECORDS

Spring: Coastal – Port Renfrew 9 May 1983-1; Saanich 10 Mar 1985-18, 27 Mar 1985-15, both at feeder, 22 Apr 1900-4 eggs; Beacon Hill Park (Victoria) 21 Apr 1976-12; Fairy Lake (Port Renfrew) 3 Apr 1974-1 singing; Sidney Island 31 Mar 1974-10 on survey of island; Bamfield 1 Mar 1976-8; White Rock 3 Mar 1982-2; Agassiz 16 Mar 1980-1; Vancouver 2 Apr 1943-53 in 1-hour count, all pale-coloured birds in migration, 8 Apr 1945-16, 26 Apr 1974-15; Upper Campbell Lake 26 May 1936-1 (RBCM 14263); Solander Island 5 May 1976-female flushed from 3 eggs; Port Alice 3 May 1973-20; Cluxewe River 5 May 1991-15 on estuary (Dawe et al. 1995); Port Hardy 10 Apr 1951-12, 17 May 1936-3 nestlings about 1 week old; Cape St. James 25 Apr

1982-5; Anthony Island 21 May 1991-1 (Dawe 1991); Goose Island 10 May 1948-1; Bella Coola 22 Apr 1933-8 (Linsdale and Sumner 1934); Klemtu 5 Apr 1976-20; Low Island 16 May 1977-3 small young; Langara Island 8 May 1936-3 eggs; Cox Island 13 May 1952-1; Masset 21 Mar 1976-1 singing; Skonun Point 14 May 1991-2 (Dawe 1991); Lawyer Island 29 Mar 1979-3, first spring sighting; Kitimat 27 Apr 1975-30 birds singing; Whitesand Island 13 May 1987-1. **Interior** – Erie Lake 7 Mar 1983-1; Arawana 15 Mar 1962-2; Pearson Mountain 1 May 1914-3 eggs; Celista 24 Apr 1948-4; Bridge Lake 22 May 1963-5 eggs; Glacier National Park 3 May 1982-10; 108 Mile House 5 Mar 1986-1; Williams Lake 5 Apr 1978-2; 24 km s Prince George 23 Apr 1983-2; Tupper Creek to Swan Lake May and Jun 1938-common but inconspicuous (Cowan 1939); Taylor 13 Apr 1986-1, spring arrival – silent bird, 14 Apr 1984-1 at km 10 Upper Cache Creek Rd nr Bear Flat, first arrival of spring; Boundary Lake (Goodlow) 6 May 1984-4, 16 May 1982-6, all singing on territories; Fort Nelson 28 Apr 1987-some singing males; Liard Hot Springs 30 Apr 1975-15 (Reid 1975); Chilkat Pass 9 May 1957-1 (Weeden 1960), 14 May 1977-12; Atlin 18 Apr 1977-1.

Summer: Coastal – North Saanich 2 Jul 1976-2 eggs; Pacific Rim National Park 4 Jun 1983-49 between Wickaninnish Inn and Tofino; Reifel Island 1 Jun 1969-4 eggs; Malaspina Lake 19 Jun 1997-3 in full song (Dawe and Dawe 1997); Brooks Peninsula 5 to 14 Aug 1981-at least 11 juveniles recorded, most could fly, some had natal down (Campbell and Summers 1997); Red Mountain (Alta Lake) 26 Jun 1924-13; 16 km w Holberg 17 Jun 1997-1 in full song (Dawe and Dawe 1997); Triangle Island 1 Jul 1977-2 eggs, 13 Jul 1984-3 downy nestlings, 2 Aug 1994-43; Storm Islands 13 Jun 1976-39, total for all islets; Tree Islets 11 Jun 1976-23, total for all islets; Cape St. James 16 Jun 1979-6; Kunghit Island 5 Jun 1986-1; McLoughlin Bay 11 Jun 1996-1 in full song (Dawe and Buechert 1996); Fingal Island 28 Jun 1988-1; McKinney Islands 25 Jun 1976-14 on survey; Hunter Point (Graham Island) 19 Jul 1977-25; Lihou Island 21 Jun 1977-11; Gospel Island 20 Jul 1977-1 recently fledged young with adult; Masset 30 Jul 1989-1 recently fledged young with adult; Langara Island 12 Jul 1968-4 nestlings; Rocher Déboulé Range 20 Jul 1944-1 seen, several heard (Munro 1947); 104 km n Meziadin Junction 7 Jul 1977-1 (CAS 69767); Mile 85 Haines Highway 14 Jul 1944-3 eggs; Mile 46 Haines Highway 22 Jun 1972-nest with small young. **Interior** – Manning Park 26 Jun 1983-15, 6 Jul 1985-15; nr Sage Creek 14 Aug 1983-1; Penticton 21 Jun 1928-1 egg; Apex Mountain (Penticton) 22 Jul 1974-1 fledgling; Mount Scaia 4 Jul 1983-15, 7 Jul 1983-1 fledgling; Meadow Mountain (Argenta) 5 Aug 1992-1 fledgling being fed by adult on ground; McGillivray Creek (Lillooet) 10 to 18 Aug 1916-5, typical *P. i. schistacea* (Taverner 1917); Mount Revelstoke National Park 6 to 12 Aug 1943-several flushed in or near rhododendron thickets (Cowan and Munro 1944-1946); Mount Revelstoke 19 Jul 1978-6; Beaverfoot River 31 Aug 1975-3; Mount Tatlow 6 Jul 1985-2 singing loudly; Wells Gray Park 6 Jul 1978-2 along stream at Helmeken Lodge, singing; Indianpoint Mountain 16 Jul 1927-1 (MCZ 285474); Kiwa Creek 16 Jun 1988-1 on steep slopes above creek; 116 km e Prince George 17 Jul 1992-2 recently fledged young in subalpine habitat (Campbell and Dawe 1992); Seebach Creek 17 Jul 1981-2; n Burnie Lake 23 to 30 Aug 1975-2 (Osmond-Jones et al. 1975); Hudson Bay Mountain 5 Jul 1977-2; Astlais Mountain 1 Jul 1975-12; nr Mackenzie 6 Aug 1994-4 banded; Tupper Creek 15 Jun 1974-1; Boundary Lake (Goodlow) 6 Jun 1982-1 fledgling capable only of fluttering flight; Kluayaz Creek 16 Jul 1976-1, scarce in shrubby areas (Osmond-Jones et al. 1977); Alaska Highway (Mile 147) 20 Jun 1976-8; Fern Lake 13 Aug 1983-1 (Cooper and Cooper 1983); 1.5 km e Fort Nelson 9 Jul 1986-1 fledgling 7 to 8 days out of nest; Atlin 30 Jul 1991-1 young just out of nest; Tatshenshini River 3 Jun 1983-1 egg, 3 Jun 1983-3 nestlings almost full grown (Campbell et al. 1983); Helmut (Kwokullie Lake) 2 to 10 Jun 1982-1; Tats Lake 2 Jul 1983-1, very common; Haines Highway (Mile 80) 24 Jun 1981-4 nestlings about 2 or 3 days old; Haines Highway (Mile 85) 14 Jul 1944-3 eggs; Haines Highway (Mile 45 to 54) 22 Jul 1972-1 nest with small young.

Breeding Bird Surveys: Coastal – Recorded from 8 of 27 routes and on 15% of all surveys. Maxima: Masset 21 Jun 1994-70; Port Renfrew 26 Jun 1974-59; Queen Charlotte City 2 Jun 1989-49. **Interior** – Recorded from 22 of 73 routes and on 8% of all surveys. Maxima: Haines Summit 19 Jun 1994-27; Illecilewaet 9 Jun 1983-18; Chilkat Pass 1 Jul 1976-11.

Autumn: Interior – Atlin 11 Oct 1930-1 (RBCM 5741); Fort Nelson 2 Sep 1985-2 singing constantly, 14 Sep 1985-1 foraging in meadow edge at airport; Nig Creek 15 Sep 1984-1, last record of autumn; Fort St. John 22 Sep 1985-2; Mackenzie 4 Sep 1994-4, 5 Sep 1994-3; Chezacut 4 Oct 1933-1 (MCZ 285518); Williams Lake 14 Nov 1985-1; Kleena Kleene 23 Sep 1961-1 found dead (Paul 1964); Mount Revelstoke 7 Oct 1981-5; Revelstoke 29 Oct 1984-2, 11 Nov 1985-2; Sorrento 3 Nov 1970-2; Kamloops 28 Sep 1980-1, *P. i. altivagans*, 12 Oct 1984-1, *P. i. schistacea*; Okanagan Landing 18 Oct 1913-1 (MVZ 106375); West Bench 30 Sep 1972-4. **Coastal** – Kispiox River valley 14 Sep 1921-5 (Swarth 1924); Green Island 17 Oct 1978-1; Masset 25 Oct 1949-1 (RBCM 10275); Pitt Island 28 Sep 1935-1 (MCZ 285516); Mount Brilliant (MacKenzie Valley) 6 Sep 1938-1 (NMC 18798); Bella Coola 28 Oct 1989-4; Cape St. James 21 Sep 1981-24, 1 to 31 Oct 1981-averaging 6 per day, usually with Song Sparrows and Dark-eyed Juncos, 16 Nov 1981-8; Cape Scott 18 Sep 1935-40; Triangle Island 2 Sep 1994-34 banded; Port Hardy 15 Nov 1986-2; Port Neville 14 Nov 1975-8; Little Qualicum River 27 Oct 1975-16 (Dawe 1976); Pacific Rim National Park 29 Sep 1986-6; Saanich 11 Sep 1983-10; Island View Beach 2 Nov 1986-25; View Royal 12 Nov 1979-24; Rocky Point (Victoria) 4 Sep 1994-14 banded; Jordan River 25 Oct 1987-25.

Winter: Interior – Fort St. John 3 Jan 1988-1, first winter record (Siddle 1988a); Prince George 3 Jan 1988-1, unofficial Christmas Bird Count; Chilcotin River (Riske Creek) 2 to 27 Dec 1994-1 at Wineglass Ranch; Windermere 4 Jan 1983-2; Vernon 1 Dec 1994 to 28 Feb 1995-1 (Bowling 1995b); Okanagan valley 12 Feb 1921-1 (FMNH 306881); nr Kaslo 5 Dec 1965-1 banded; Summerland 11 Feb 1979-1; Taghum 9 Jan 1979-2; Oliver 28 Dec 1988-1 (Rogers 1989). **Coastal** – Drizzle Lake 9 Dec 1977-2; Cape St. James 5 Jan 1982-12, 31 Dec 1981-12; Port Neville Inlet 15 Jan 1975-8, 9 Dec 1977-5 feeding with Song Sparrows, Spotted Towhees, and Dark-eyed Juncos; Quatse River 31 Dec 1990-3 on estuary; Cluxewe 17 Feb 1991-5 on estuary, 23 Dec 1990-3 on estuary (Dawe et al. 1995a); Campbell River 1 Jan 1976-2 on estuary, 26 Dec 1982-2 on estuary (Dawe et al. 1995a); Harrison Hot Springs 17 Feb 1986-3; Fanny Bay 31 Dec 1990-1 (Dawe et al. 1995b); Westham Island 2 Jan 1974-10, 14 Dec 1974-35; Sea and Iona Islands 1 Jan 1977-8; Tofino 3 Jan 1976-6; Saanich 1 Jan 1985-14, 25 Feb 1985-25, 30 Dec 1985-45 coming to feeders during heavy snowfall; Swan Lake (Saanich) 5 Dec 1986-10; Victoria 1 Jan 1975-9 at Beacon Hill Park; Skirt Mountain 8 Jan 1977-6; Witty's Lagoon 15 Jan 1982-13 on beach; Jordan River 10 Dec 1983-35.

Christmas Bird Counts: Interior – See REMARKS. **Coastal** – Recorded from 30 of 33 localities and on 85% of all counts. Maxima: Vancouver 20 Dec 1992-**465**, all-time Canadian high count (Monroe 1993); Victoria 19 Dec 1992-417; Ladner 23 Dec 1984-281.

Song Sparrow
Melospiza melodia (Wilson)

SOSP

RANGE: Breeds from southern Alaska, south-central Yukon, northern British Columbia, south-central Mackenzie, northern Saskatchewan, northern Manitoba, northern Ontario, south-central Quebec, and southwestern Newfoundland south in the west to south-central Baja California and the mainland of Mexico, and from northern New Mexico east to coastal South Carolina. Winters from southern Alaska and coastal and southern British Columbia east across the northern United States, southern Ontario and Quebec, Prince Edward Island, and Nova Scotia, south throughout the breeding range to southern Texas and Florida, including the Gulf coast.

STATUS: On the coast, a *common* resident and winter visitant in the Georgia Depression Ecoprovince; in the Coast and Mountains Ecoprovince, present throughout the year but *uncommon* to *fairly common* (occasionally locally *common*) summer visitant and *uncommon* winter visitant on Western Vancouver Island and on the Southern and Northern Mainland coasts, and *uncommon* to *fairly common* on the Queen Charlotte Islands.

In the interior, present year-round in southern regions but *fairly common* to locally *common* summer visitant and *uncommon* winter visitant in the Southern Interior, Southern Interior Mountains, and Central Interior ecoprovinces; *uncommon* summer visitant and *rare* winter visitant in the Sub-Boreal Interior Ecoprovince; *uncommon* summer visitant in the Boreal Plains Ecoprovince; and *rare* in the Taiga Plains and Northern Boreal Mountains ecoprovinces.

Breeds.

Figure 308. Of the 29 subspecies of Song Sparrow recognized from north of Mexico, 6 occur in British Columbia. The sparrow's plumage is extremely variable throughout its range. In south coastal areas of British Columbia, the resident adult *Melospiza melodia morphna* (a) generally has a darker and redder body than other breeding forms, and also shows a darker gray eyebrow and malar stripe (Reifel Island, Delta, 12 June 1996; Michael I. Preston). In the southern portions of the interior of the province, the adult *M. m. merrilli* (b) is generally lighter, with less heavy streaking, and has a light gray eyebrow line and malar stripe (Meyers Lake, near Rock Creek, 6 June 1997; R. Wayne Campbell).

CHANGE IN STATUS: The Song Sparrow (Fig. 308) has undergone a major change in status in northeastern British Columbia since the mid-1940s (Munro and Cowan 1947). M.Y. Williams (1933a) does not mention the Song Sparrow in his account of birds seen along the Peace River from the town of Peace River to Taylor, or in the regions bordering the Sikanni Chief, Fort Nelson, and Liard rivers. In 1929 and 1930, he undertook geological surveys in parts of what is now the Peace Lowland, and his reference to the Song Sparrow as "apparently not common in wilder parts; common in cultivated areas" is evidently an error of recall or identification. Racey (1930) saw none in the area around Pouce Coupe. The species was not detected during intensive ornithological collecting between Tupper and Charlie Lake in 1938 (Cowan 1939) and between Dawson Creek and Irons Creek in 1943 (Rand 1944). The first specimen records for the Boreal Plains were 2 seen and 1 collected in May 1954 at Swan Lake, near Tupper Creek (Jobin 1955a). Erskine and Davidson (1976) do not include the Song Sparrow in their account of the birds of the Fort Nelson Lowland. Thus, as recently as 1954 the Song Sparrow was rare and local in the Boreal Plains, and in the late 1970s was unknown in the Taiga Plains.

Our records indicate a gradual invasion of northeastern British Columbia by this species since about 1950. There were 5 records from the Boreal Plains from 1950 to 1959, 2 from the 1960s, 34 from the 1970s, and 99 from the 1980s. The actual numbers seen in the 1980s were certainly influenced by the reports received from 1 enthusiastic and experienced observer. From the earliest records of Song Sparrows at Flatrock in October 1951 and at Swan Lake in May 1954, the species has gradually spread as a summer visitant along the course of the Peace River and in adjacent agricultural habitat as far as Hudson's Hope. To the north, there are records from the communities of Pink Mountain and Prophet River. The species was first detected in the Taiga Plains in 1981.

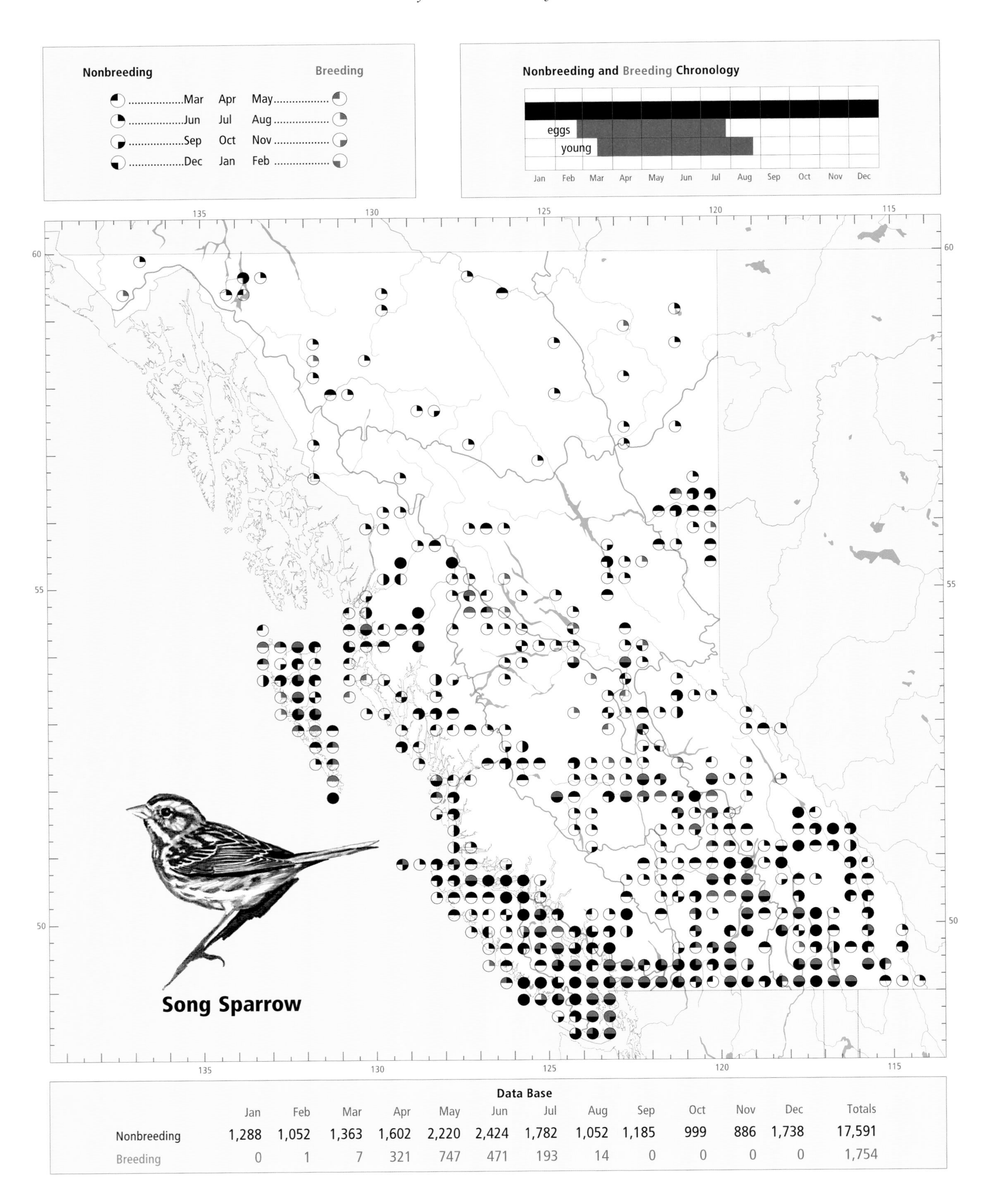

Song Sparrow

Data Base

	Jan	Feb	Mar	Apr	May	Jun	Jul	Aug	Sep	Oct	Nov	Dec	Totals
Nonbreeding	1,288	1,052	1,363	1,602	2,220	2,424	1,782	1,052	1,185	999	886	1,738	17,591
Breeding	0	1	7	321	747	471	193	14	0	0	0	0	1,754

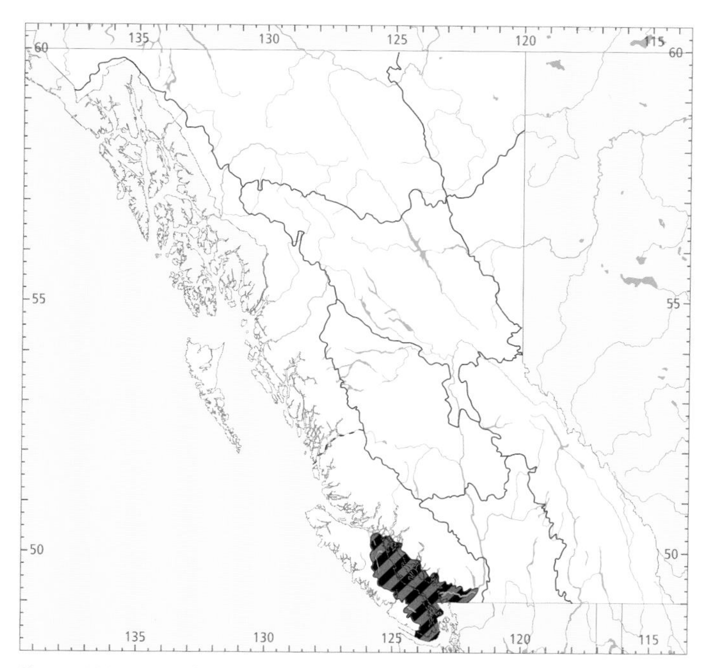

Figure 309. In British Columbia, the highest numbers for the Song Sparrow in both winter (black) and summer (red) occur in the Georgia Depression Ecoprovince.

Figure 310. Throughout its range in British Columbia, the Song Sparrow prefers habitats with a relatively dense, low cover of shrubs, brambles, and tall grasses (Courtenay, 20 April 1998; R. Wayne Campbell).

NONBREEDING: The Song Sparrow is widely distributed throughout the province from spring through autumn, but is absent from most of the northern and central interior during winter.

Along the coast from about latitude 54°N southward, but especially on the Queen Charlotte Islands and in the Georgia Depression, a small part of the local population is resident and is supplemented during migration and winter by large numbers of birds from elsewhere. These winter visitants include subspecies from outside those areas, probably from Alaska and the Yukon Territory (see REMARKS).

In the interior, there is a regular small wintering population in the southern parts of the Southern Interior and Southern Interior Mountains, but wintering numbers decrease to the north until they are infrequent in locations north of about latitude 52°N; they have not been reported north of latitude 55°N. The northernmost interior locality at which wintering has been recorded is Smithers (latitude 54°47′N). On the coast, the northernmost winter record is from Hazelton (latitude 55°15′N). The highest numbers in winter occur in the Fraser Lowland of the Georgia Depression (Fig. 309).

On the coast, the Song Sparrow has been recorded in spring and winter at elevations from sea level to 1,300 m; in the interior, elevations occupied during the nonbreeding seasons extend from 280 to 1,600 m. The highest elevations have been recorded during the autumn migration.

The Song Sparrow is notable for the breadth of the habitats it uses during the nonbreeding parts of the year. Through all seasons, however, it is a bird of relatively dense, low cover (Fig. 310). On the coast, it frequently occurs along the upper edge of sea beaches, where it forages among the driftwood and marine flotsam, and takes shelter in the dense tangles of salal and wind-pruned western hemlock, shore pine, and Sitka spruce that border the edge of the forest. Throughout the province, other habitats include patches of sedges and cattails that border lakes, ponds, and marshes, shrubby tangles bordering deciduous woodlands, patches of fireweed, marshy meadows (Swarth 1922), the shrubby borders of farm clearings, weedy vacant lots, powerline corridors, roadsides, fencelines, and similar openings. It is a regular visitor in suburban gardens, especially in winter, where planted or native shrubs provide the needed cover, and it comes readily to bird feeding stations. It seldom uses grass and grain fields, cropland,

Figure 311. During severe winters in the Southern Interior Mountains Ecoprovince, Song Sparrows roost together among bales of hay or wood piles in human-made shelters (near Creston, 4 February 1997; R. Wayne Campbell).

Figure 312. The Song Sparrows found in the shrubby habitats of Mandarte Island, British Columbia, are probably the most studied population of this North American species. Note the research cabins on this small rocky island (6 July 1995; Andrew C. Stewart).

or dense forests of either coniferous or deciduous trees. During severe winters, the Song Sparrow often forages on beaches and fallen logs and roosts in farm storage buildings (Fig. 311), machinery depots, and private yards, especially with wood piles near bird feeders. The highest densities in winter have been on vegetable farms on the Fraser River delta, in 1 instance, more than 200 birds in a 1 ha field (R.J. Cannings pers. comm.).

The spring migration in the Fraser Lowland is accompanied by a slight increase in the numbers of birds apparent in early March. Other than this, there is little change between January and May, in either total number of birds or number of records. The first returning birds reach the Northern Mainland Coast in the first week of April, about 3 weeks earlier than migrants are noted on the Queen Charlotte Islands (Figs. 313 and 314).

In the interior, the first wave of spring migrants appears along the valleys of the Southern Interior Mountains and in the Central Interior in the second week of March (Fig. 314). The migrants along the east Kootenay valley reach a peak in the third week of March. The timetable for the Southern Interior Mountains is almost a month earlier than for the Southern Interior, where the first surge of migrants appears in the first and second weeks of April. The spring movement reaches the Sub-Boreal Interior in early April and the Boreal Plains in the last week of April, but does not reach the northern boundary of the province until the third week of May. Thus, unlike many of the warblers and northern sparrows, the Song Sparrow appears to have a leisurely migration, taking nearly 2 months to move the length of the province. In part, this may be accounted for by the several subspecies following different timetables (see REMARKS).

The more protracted southbound migration appears to begin in northern British Columbia in late July and continues into September. Stragglers may remain until mid-October. In central and southern regions of the interior, the main autumn movement occurs from late August through mid-September (Fig. 314).

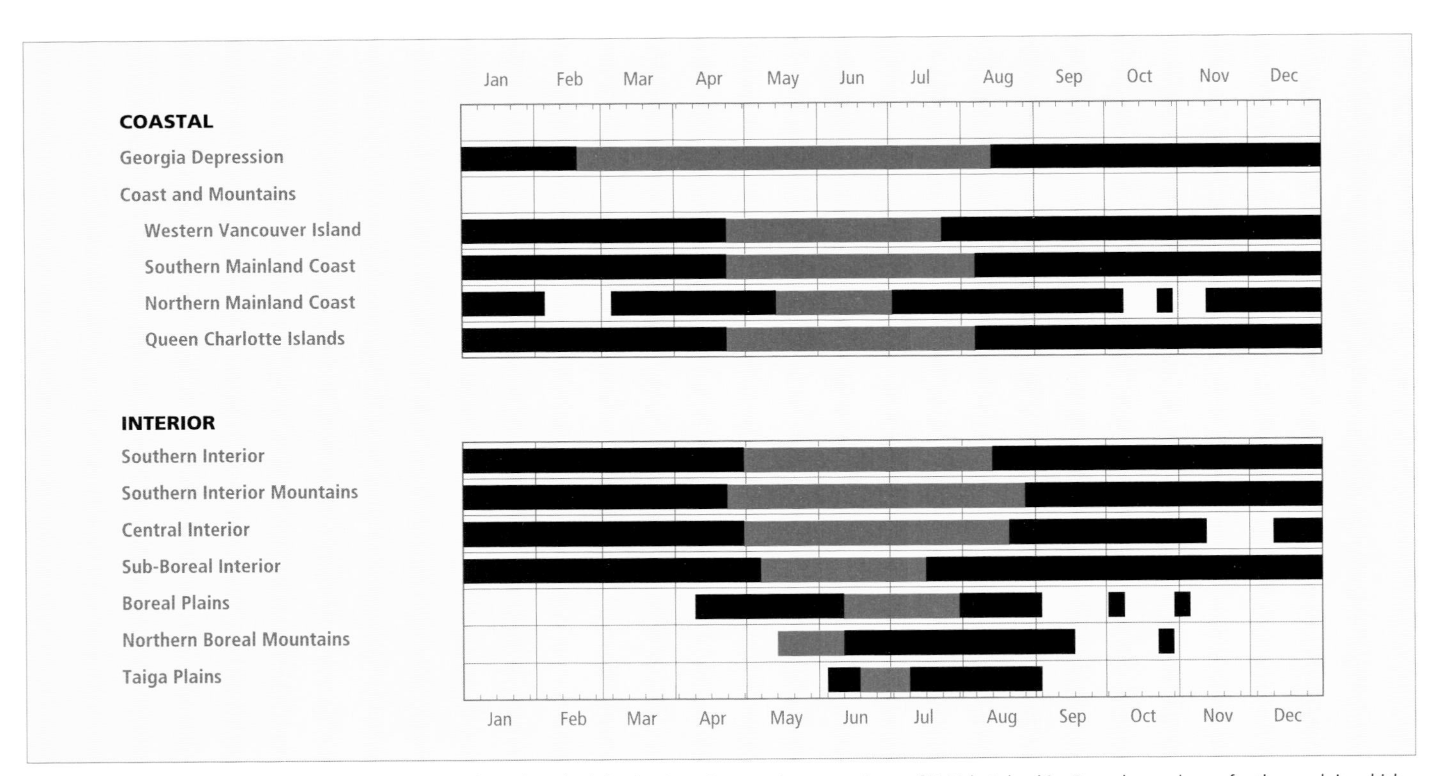

Figure 313. Annual occurrence (black) and breeding chronology (red) for the Song Sparrow in ecoprovinces of British Columbia. Records are shown for the week in which they occurred.

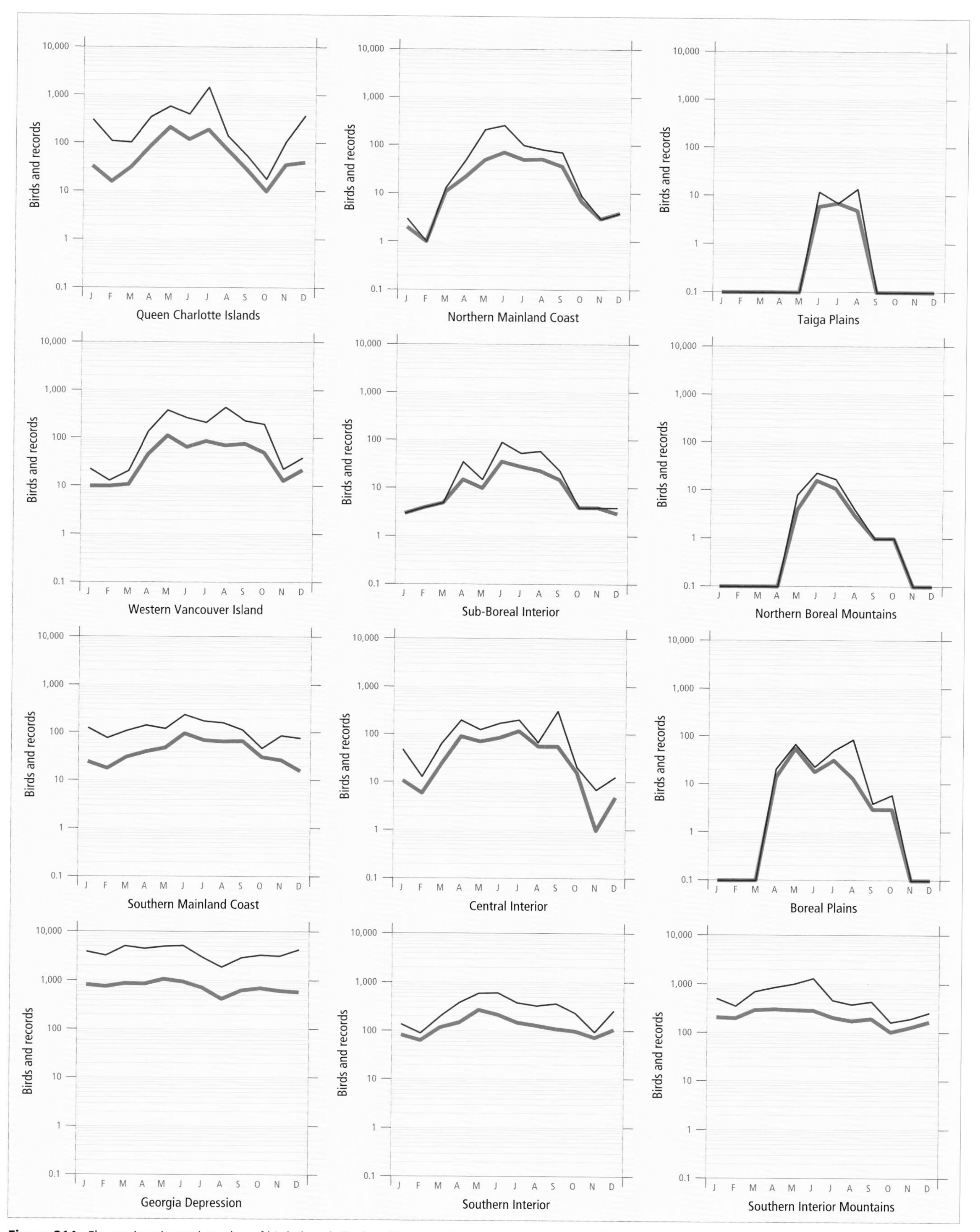

Figure 314. Fluctuations in total number of birds (purple line) and total number of records (green line) for the Song Sparrow in ecoprovinces of British Columbia. Christmas Bird Counts, Breeding Bird Surveys, and nest record data have been excluded.

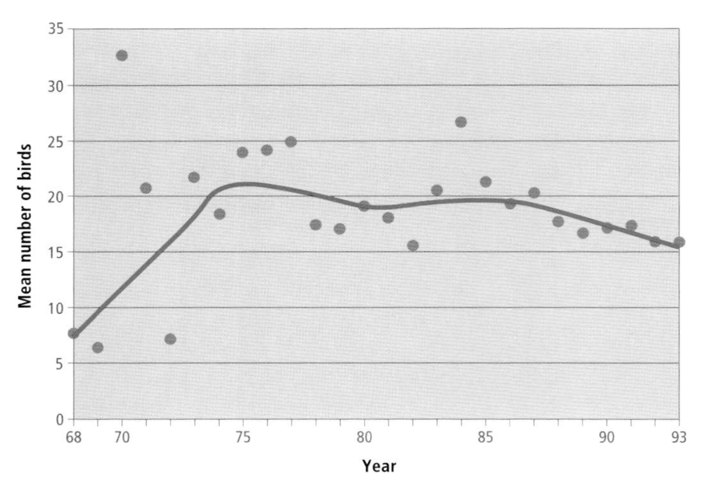

Figure 315. An analysis of Breeding Bird Surveys for the Song Sparrow in British Columbia shows that the numbers of birds on coastal routes decreased at an average annual rate of 1% over the period 1968 through 1993 ($P < 0.05$).

Figure 317. Throughout the interior of British Columbia, the Song Sparrow inhabits wet marshes where dense dry grasses and sedges provide cover for nesting (near Prince George, 14 June 1997; R. Wayne Campbell).

On the coast, most of the summering Song Sparrows on the Queen Charlotte Islands appear to have left the islands by October, when wintering numbers begin to build. On Western Vancouver Island, wintering numbers are reached by the end of October. There appears to be little sequential pattern to the movement of Song Sparrows either northward or southward across British Columbia. It is almost as though there were several discrete populations, each with its own general timetable of response to the migration stimuli.

Figure 316. On the southwest mainland of British Columbia, the Song Sparrow breeds in dense vegetation including that lining broad drainage ditches in farmland (Delta, 31 March 1997; R. Wayne Campbell).

On the coast and in the southern portions of the interior and the Central Interior, the Song Sparrow has been recorded regularly throughout the year; in most of the regions from the Sub-Boreal Interior northward, it has been recorded irregularly between 15 April and 30 October (Fig. 313).

BREEDING: The Song Sparrow breeds throughout the province, including large and small coastal islands (Fig. 312). It becomes much less abundant north of latitude 55°N. This is partly due to the fact that much less effort has been expended on field study of the birds in the northern half of the province, but the experience of field biologists in the region also indicates a much smaller population of Song Sparrows there than in southern British Columbia. The Breeding Bird Surveys also reflect this disparity. The mean annual number of birds per record for each ecoprovince during the period 1968 through 1989 were as follows: 5.8 for the Georgia Depression and Queen Charlotte Islands, 2.6 for the Southern Interior, 3.2 for the Southern Interior Mountains, 2.2 for the Central Interior, 1.8 for the Sub-Boreal Interior, and 1.5 for the Northern Boreal Mountains.

The Song Sparrow reaches its highest numbers in summer in the Georgia Depression (Fig. 309). An analysis of Breeding Bird Surveys for the period 1968 through 1993 could not detect a net change in numbers for interior routes; on coastal routes, the number of birds decreased at an average annual rate of 1% mostly between 1968 and 1993 (Fig. 315). An analysis of Breeding Bird Surveys across Canada for the period 1966 to 1996 reveals a decrease in numbers at an average annual rate of 1.5% ($P < 0.01$) (Sauer et al. 1997); most of this decrease occurred from 1966 to 1979.

The Song Sparrow breeds at elevations from near sea level to 1,200 m on the coast, and from about 300 m to 1,400 m in the interior. Through the extensively varied habitats in which it nests successfully in the province, the Song Sparrow reveals a remarkable adaptability to a variety of climatic and vegetative conditions. The environmental characteristics that appear to be essential are proximity to water, or at least moist habitats; dense but not tall vegetation (Fig. 316), including willow thickets, damp field edges, riverside flats, shrubbery, tall grasses, sedges, rushes, hedges, and brush piles; and exposed ground under the vegetation to which some light can penetrate. This last characteristic may be provided by low, dense shrubbery, such as bramble tangles, salal patches, or patches dominated by willow, dogwood, snowberry, rose, or horsetail; marsh vegetation (Fig. 317); and high grasses or sedges. All of these provide cover along with ease of movement for ground feeding. Dense forest, dry and low grassland, and vegetation that hampers movement on the ground are usually avoided by Song Sparrows (Tompa 1963a).

Two examples illustrate the pattern of summer occurrence of Song Sparrows in forests in British Columbia. In the trembling aspen forests of the Bulkley valley, the Song Sparrow was found in 4 of 6 clearcut sample plots at a mean density of 0.22 singing males per sampling point, on 3 of 6 plots at a mean density of 0.24 in sapling-stage forest, and in 4 of 10 plots at a mean density of only 0.06 males per plot in mature trembling aspen (Pojar 1995). In the coniferous forest of western Vancouver Island, the Song Sparrow was found to be most abundant in clearcuts (32 of 36 stations) and in 15- to 20-year-old-stands (23 of 36 stations), but did not occur in old-growth forest (Bryant et al. 1993). The Song Sparrow's more extensive use of the trembling aspen forest compared with coniferous forest may be a response to the greater extent of shrub understorey in the deciduous forest as it progresses from clearcut to maturity.

On the coast, the Song Sparrow has been recorded breeding from 23 February to 10 August; in the interior, it has been recorded breeding from 21 April (calculated) to 23 August (Fig. 313).

Nests: Nests were situated in a wide variety of sites. Most (30%; n = 635) were placed near the edge of a forest or woodland. Other sites included wetlands (24%), cutover forests (22%), shrublands (21%), grasslands (2%), and miscellaneous (1%). Within these habitat types, 61% of nests (n = 444) were in low vegetation cover and attached to plant stems (Fig. 318), 19% were in tall grasses, and about 9% were in shrubs or low trees. Another 17 types of locations were used in less than 1% of the sample. The body of the nest consisted of grasses, with plant stems, hair, leaves, plant fibres, twigs, rootlets, bark strips, sedges, and mosses as other frequent components (Fig. 318); the lining included fine grasses, mosses, hair, feathers, or plant down.

The heights of 526 nests ranged from ground level to 10 m, with 56% between 0.1 and 0.6 m.

Eggs: Dates for 1,056 clutches (Fig. 319) ranged from 23 February to 26 July, with 50% recorded between 2 May and 9 June. Sizes of 484 clutches throughout the province ranged from 1 to 6 eggs (1E-14, 2E-32, 3E-134, 4E-268, 5E-34, 6E-2). The percentage distribution of clutch size was as follows: 1 egg, 2.9%; 2 eggs, 6.6%; 3 eggs, 27.7%; 4 eggs, 55.4%; 5 eggs, 7.2%; and 6 eggs, 0.4%. These data include incomplete clutches, but are similar to the ratios of 296 complete clutches from Mandarte Island, off southern Vancouver Island. Their

Figure 318. Except in wet marshes, most Song Sparrow nests in British Columbia were built in low vegetation and attached to plant stems such as scouring rush (Victoria, 16 May 1998; R. Wayne Campbell).

Figure 319. Typical nest and eggs of the Song Sparrow in British Columbia (Victoria, 7 May 1973; R. Wayne Campbell).

distribution was as follows: 1 egg, 2.7%; 2 eggs, 10.5%; 3 eggs, 29.7%; 4 eggs, 52.4%; 5 eggs, 4.4%; and 6 eggs, 0.7% (Smith and Roff 1980).

Two studies of the Song Sparrow population of Mandarte Island, from 1960 to 1963 (Tompa 1964) and from 1975 to 1978 (Smith and Roff 1980), provide more detailed information on clutch sizes by laying date for an insular coastal population (Table 2). Clutches begun before 5 May averaged 3.61 eggs (n = 173), those begun between 6 May and 10 June averaged 3.83 eggs (n = 110), and those begun after 10 June averaged 3.10 eggs (n = 19). In still later studies on Mandarte Island it was found that clutch size declines through the breeding season.

A significant source of variation in clutch size is the age of the female. Yearling females lay smaller clutches than older birds, and lay first clutches later in the year (Smith and Roff 1980; Hochachka 1990). There was some seasonal decline within both age groups, and no consistent variation of clutch size between years (Smith and Roff 1980).

The length of the nesting season varies considerably between populations of Song Sparrows nesting in different ecoprovinces (Fig. 320). In the Georgia Depression, the period during which there are nests with eggs begins in the fourth week of February and sometimes continues into the last week of July, a maximum egg period of 23 weeks (Fig. 320). On the Queen Charlotte Islands, egg laying begins in the first week of May and there are nests with eggs into the second week of July, an egg period of 11 weeks. Almost the same pattern prevails in the southern portions of the interior, where in each of the ecoprovinces the egg period is 12 or 13 weeks.

The lengthening of the egg period in the Georgia Depression arises largely as a "tail" at each end of the season, resulting from single clutches being discovered 5 or 6 weeks before and after the more concentrated period of nesting. This tail has not been noted in any of the other ecoprovinces. A smaller cluster beginning the week ending 3 June and continuing until the week ending 8 July suggests a period of second nest attempts.

On Mandarte Island, about one-third of the females nested a second time after a successful first brood; about 6% attempted a third nest (Smith and Roff 1980). There is a single instance of a Song Sparrow successfully raising 4 broods in a year (Smith 1982). In second clutches, hatching success was about 10% lower than in first clutches (Hochachka 1990).

The longer breeding season in the Georgia Depression enables Song Sparrows to increase the total number of young they rear to independence by engaging in multiple nesting (see "Nest Success").

Johnston (1954) reported a cline of increasing clutch size with latitude between California and Alaska. The data derived by J.N.M. Smith and his colleagues at the University of British Columbia reveal some of the complexities of obtaining representative clutch sizes for the Song Sparrow. At this time, the existence of a latitudinal cline is questionable. Certainly there is no increase in clutch size between latitudes 40° and 50°N.

The incubation period in southern British Columbia ranged from 10 to 13 days (n = 5). Mean incubation periods on Mandarte Island were 12.5 days (J.N.M. Smith pers. comm.). Throughout North America, it ranges from 12 to 14 days (Baicich and Harrison 1997).

Young: Dates for 700 broods ranged from 16 March to 23 August, with 52% recorded between 10 May and 22 June. Sizes of 333 broods ranged from 1 to 5 young (1Y-25, 2Y-49, 3Y-117, 4Y-128, 5Y-14), with 74% having 3 or 4 young (Fig. 321). The nestling period of the Song Sparrow in southern British

Table 2. Mean clutch size for Song Sparrows on Mandarte Island, southern British Columbia, by laying date compared with mainland coastal populations (modified from Smith and Roff 1980).

	Clutches begun before 5 May			Clutches begun 6 May to 10 June			Clutches begun after 10 June		
Locality	Mean	n	SE	Mean	n	SE	Mean	n	SE
Mandarte 1960-63	3.77	33	0.105	3.87	23	0.093		–	
Mandarte 1975-78	3.58	140	0.10	3.42	87	0.19	3.10	19	0.37
Mandarte 1960-78	3.61	173	–	3.83	110	–	3.10	19	–
Coastal British Columbia	3.47	57	0.066	3.77	57	0.061		–	

Columbia ranged from 9 to 13 days ($n = 6$). Rising (1996) gives the nestling period as 7 (only when nestlings are disturbed by a predator) to 14 days.

As with the egg period, there are major differences between some ecoprovinces in the period through which young are present in nests (Fig. 320). In the Georgia Depression, the first chicks may be found during the third week of March and the last of them fledge in the second week of August, for a period of 22 weeks. This extended period through which nestlings are present arises from the same circumstances mentioned under "Eggs," and includes renestings. In the Southern Interior, the period with young in the nests spans

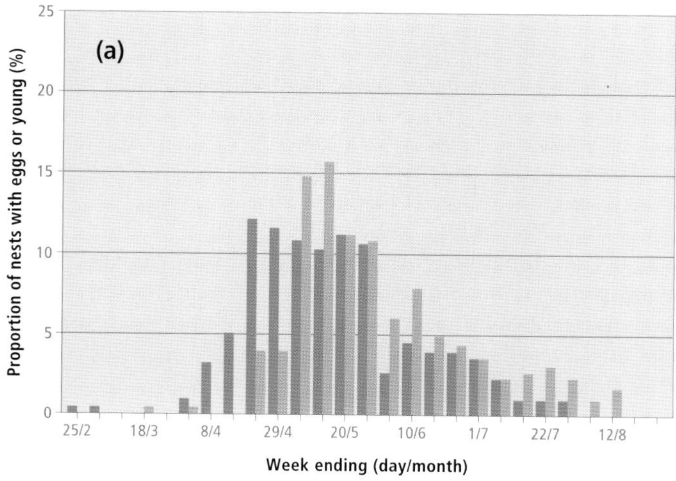

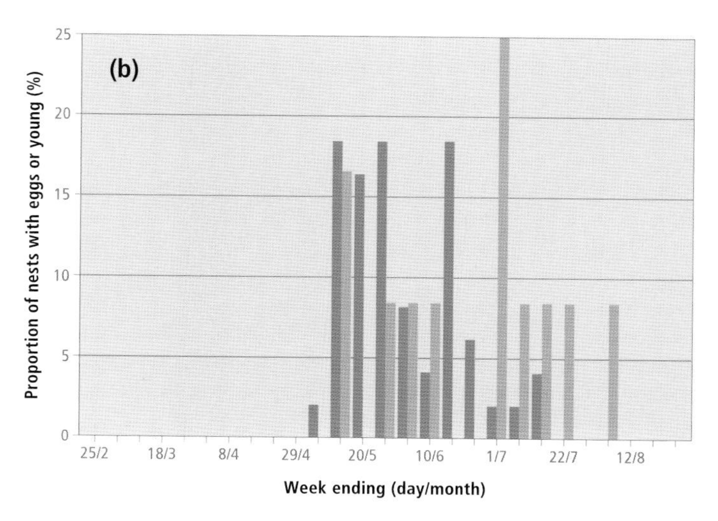

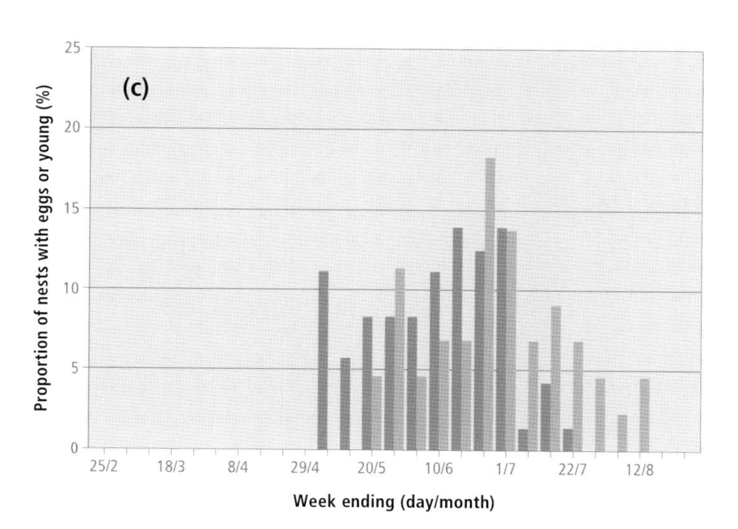

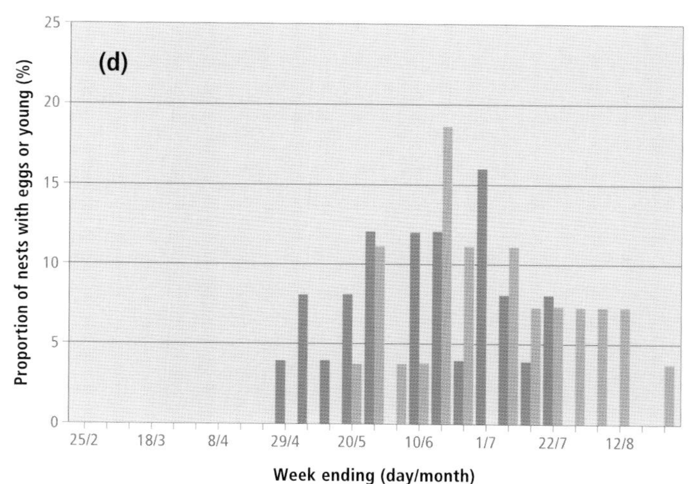

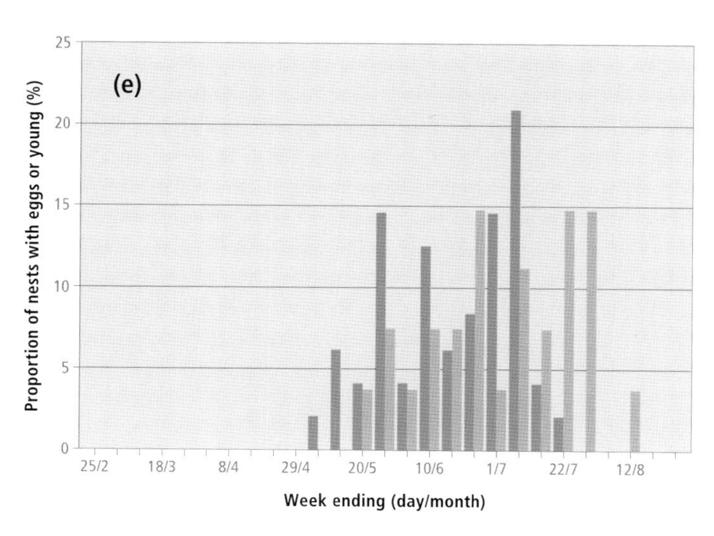

Figure 320. Weekly distribution of the proportion of clutches (dark bars) and broods (light bars) for the Song Sparrow in the (a) Georgia Depression (clutches: $n = 40$; broods: $n = 21$), (b) Coast and Mountains (Queen Charlotte Islands) (clutches: $n = 11$; broods: $n = 6$), (c) Southern Interior (clutches: $n = 11$; broods: $n = 6$), (d) Southern Interior Mountains (clutches: $n = 11$; broods: $n = 6$), and (e) Central Interior (clutches: $n = 11$; broods: $n = 6$) ecoprovinces of British Columbia. The figures are based on the first week eggs or young were found in the nest. On the coast, nests may contain eggs 9 weeks earlier than nests in the interior.

Figure 321. Song Sparrow nest with young; the nestling period of the Song Sparrow in British Columbia ranges from 9 to 13 days (Victoria, 16 May 1998; R. Wayne Campbell).

the 13 weeks between the third week of May and the second week of August. In the Southern Interior Mountains, broods may be found from the third week of May until the fourth week of August. Except for 1 late brood date in the Southern Interior Mountains, the termination of the breeding season with the fledging of the last young occurs in the same week in all ecoprovinces from which we have adequate nesting data. In general, the length of the nesting season is determined by the onset of the summer dry season. Song Sparrows breed later in wet summers, with more pairs breeding into late July and early August (J.N.M. Smith pers. comm.).

Brown-headed Cowbird Parasitism: In British Columbia, 7% of 753 nests found with eggs or young were parasitized by the Brown-headed Cowbird. There were 27 additional records of the Song Sparrow feeding cowbird fledglings. On the coast, 33 of 529 nests (6%) were parasitized; in the interior, 23 of 224 nests (10%) were parasitized. The similarity between Song Sparrow and Brown-headed Cowbird eggs probably causes the frequency of parasitism to be underestimated (Smith and Arcese 1994).

The Mandarte Island population of Song Sparrows (Fig. 312) has been studied both with and without Brown-headed Cowbird parasitism (Smith 1981; Smith and Arcese 1994). The arrival of cowbirds on the island occurred well after the early Song Sparrow clutches had been laid. The first cowbird eggs from year to year ranged from 6 to 19 May. Parasitism ranged from 30% to 80% and averaged 35%. The cowbirds frequently removed an egg from the host nest, causing a reduction of 0.8% in the clutch size of senior females and 0.5% in that of yearling females. The survival of cowbird nestlings was the same as for their Song Sparrow nestmates, but newly fledged cowbirds had a lower survival, mainly because their behaviour made them more vulnerable to crow predation. Female Song Sparrows whose nests were parasitized raised as many young to independence as females of the same age whose nests were not parasitized. It was concluded that, in this population of Song Sparrows, the presence of cowbirds on the island had no influence on the productivity of the breeding population of Song Sparrows.

A study of cowbird parasitism in southern Ontario found that 27% of parasitized Song Sparrow nests were deserted, compared with just 2% of unparasitized nests (Graham 1988). In contrast, on Mandarte Island desertions were slightly higher before the arrival of cowbirds than in their presence (Smith 1981). Nests without apparent cowbird effects can also be deserted when cowbirds remove host eggs (J.N.M. Smith pers. comm.).

Nest Success: Of 205 nests found with eggs and followed to a known fate, 66 produced at least 1 fledgling, for a nest success rate of 32%. Nest success was 34% (n = 154) on the coast and 27% (n = 51) in the interior.

If the ultimate objective of reproduction is to leave independent offspring, clutch size and repetitive nesting have a potential role in influencing reproductive success. Studies of the Mandarte Island population by Smith and Roff (1980) showed that clutches of 4 eggs resulted in slightly more young reaching independence than clutches of 3 eggs, and that females that bred 3 times fledged more young than those nesting only twice.

REMARKS: Thirty-one subspecies of the Song Sparrow have been described throughout its range (American Ornithologists' Union 1957); 6 occur in British Columbia. *M. m. juddi* breeds in the northeastern corner of the province, including the Peace Lowland. *M. m. inexpectata* breeds on the inner islands of southeastern Alaska and in northwestern British Columbia southeast to about latitude 51°N; it winters primarily on the south coast and southern portions of the interior, south into Oregon. *M. m. merrilli* (Fig. 308) breeds from southern portions of the province north to about latitude 51°N, and winters from southern British Columbia south into California and New Mexico. *M. m. caurina* breeds on the coast of southeastern Alaska, north of the southeastern archipelago, and winters along the marine shores of British Columbia and south into California. *M. m. rufina* breeds on the outer islands of southeastern Alaska and along the north coast of British Columbia, including the Queen Charlotte Islands; it winters in its breeding range. *M. m. morphna* (Fig. 308) breeds in the south coastal regions of the province, including Vancouver Island, and as far south as southwestern Oregon. It winters mainly in its breeding range. For details on the 4 breeding subspecies in the province, see Munro and Cowan (1947).

Throughout its nonbreeding and breeding range in British Columbia, the Song Sparrow inhabits areas that are heavily managed by humans, especially around irrigation ditches and other moist to wet environments. Dredging, fires, grazing, pruning, and indiscriminate use of herbicides and pesticides all affect roosting, foraging, and nesting substrates for many species of birds. In some cases, proactive planning of human activities can minimize their impacts on the life cycle of birds.

Rising (1996) provides a summary of life-history details along with an outline of the range occupied by the 27 currently recognized subspecies of Song Sparrow. He properly comments on some of these that may not stand up to closer scrutiny. The research in British Columbia by F. Tompa (1964), R.W. Knapton (1973), and J.N.M. Smith (Smith and Arcese 1994; Smith et al. 1996) and his students (Arcese 1989; Arcese and Smith 1988; Arcese et al. 1992; Rogers et al. 1991, 1997; Hochachka 1990) provides a detailed account of the population biology and reproduction of probably the most intensively studied Song Sparrow population in North America.

NOTEWORTHY RECORDS

Spring: Coastal – San Juan River 3 Apr 1974-3 on estuary; Mandarte Island 23 Feb 1983-1 egg, 16 Mar 1983-1 nestling, 31 Mar 1977-94; Carnation Creek 27 Apr 1981-15 males; Grice Bay 3 May 1974-25 feeding among driftwood; Long Beach 20 to 23 May 1983-23; Westham Island 4 Mar 1973-80; Sea Island 9 Apr 1960-100; n Surrey 4 Mar 1966-1 egg; Klesilkwa valley 25 Mar 1971-8; Spanish Banks (Vancouver) 30 Mar 1974-80; Harrison Hot Springs 13 Apr 1979-20; Kawkawa Lake 6 May 1964-4 eggs, 12 May 1964-4 nestlings; Yale 8 Mar 1964-12; Leiner River 15 Mar 1982-3 on estuary (Dawe and Buechert 1997); Mitlenatch Island 18 to 30 May 1976-30; Solander Island 5 May 1976-24 during trip around island; Port Hardy 12 May 1951-3 fledglings; Triangle Island 4 May 1976-8; Cape St. James 18 Apr 1982-6; Goose Island 20 May 1948-4 pairs (Guiguet 1953); Klemtu 5 Apr 1976-10; Skedans Island 10 May 1983-4 nestlings, eyes closed; Sandspit 14 May 1977-14; Kitlope Lake area 3 to 10 May 1981-21 on estuary; Pitt Island 18 May 1900-5 eggs; Langara Island 10 May 1977-16; Masset 29 Apr 1972-20; Prince Rupert 28 May 1978-3 nestlings, 2 or 3 days old; Kitsault 13 to 17 May 1980-20. **Interior** – Osoyoos Lake 30 Mar 1977-6; Grand Forks 28 Apr 1984-2; Erie Lake 4 Apr 1983-22; Princeton 23 Apr 1977-15; Vaseux Lake 10 Mar 1974-6, 19 May 1976-20; Okanagan River 30 Apr 1911-4 eggs; Nelson 24 Apr 1975-4 eggs; West Bench (Penticton) 2 Mar 1980-5; Tamarack Lake 3 Mar 1979-7; Yoho National Park 2 Apr 1977-6; Williams Lake 26 Mar 1981-26, 5 May 1983-3 eggs; Quesnel 14 Mar 1979-1; 24 km s Prince George 22 Mar 1977-1, 14 Apr 1983-3; n Summit Lake (Crooked River) 30 Apr 1965-2; Smithers 6 Apr 1977-1; McLeod Lake 14 Apr 1996-1; Mackenzie 22 Apr 1996-1; Swan Lake (Tupper Creek) 27 May 1954-1 (Jobin 1955a); n Fort St. John 15 Apr 1984-1, first spring arrival; Telegraph Creek 18 May 1983-5; Atlin 16 May 1930-1 (RBCM 5734).

Summer: Coastal – Mandarte Island 6 Jun 1963-61 banded, 15 Aug 1961-216 (Tompa 1963b); Cleland Island 24 Jul 1967-36 (Campbell and Stirling 1968); Sea Island 7 Jun 1959-40; Langley 10 Aug 1970-2 nestlings plus 1 Brown-headed Cowbird nestling; Deer Lake (Harrison) 16 Jun 1989-4 eggs; Triangle Island 22 Jun 1977-estimated 50 to 100 on island; Atnarko River 22 Aug 1940-3 (FMNH 155227 through 155229); Kawas Islets 10 Jul 1977-59; McKenny Islands 25 Jun 1976-69 on survey; Marble Island 20 Jun 1977-111; Frederick Island 25 Jul 1977-43 along w side; Minette Bay 27 Jun 1991-3 eggs; Kitsault 23 Jul 1980-12; ne Stewart 12 Jul 1974-1; Meziadin Lake 17 Jun 1975-4; Flood Glacier 31 Jul 1919-1 (MVZ 40041). **Interior** – Manning Park 26 Jun 1983-36 on day birding blitz; Erie 20 Jun 1983-54; Creston 23 Aug 1981-3 nestlings; Vaseux Lake area 25 Jun 1976-23; Watch Lake 28 Jun 1974-21; Golden 2 Aug 1976-16; Canim Lake 1 Aug 1978-6; Kleena Kleene 9 Aug 1963-1 fledgling, 2 days out of nest; Stum Lake 9 Jun 1973-7 on all-day count (Ryder 1973); Bouchie Lake 4 Jun 1944-10 pairs (Munro 1947); Ootsa Lake 13 Jul 1936-21; nr Prince George 15 Jul 1973-3 nestlings; Topley 6 Jul 1956-4 eggs; e Chetwynd 5 Aug 1975-10; n Dawson Creek 3 Aug 1975-10; Charlie Lake 25 Jul 1964-1 nestling; n Bell-Irving River 14 Jun 1980-1; Rose Prairie 17 to 19 Aug 1975-25; Ingenika 26 Jul 1980-1 (RBCM 17023); 20 km s Pink Mountain 17 Jun 1981-4; Mason Lake 28 Aug 1982-10; Beatton River 8 Jun 1956-singing males; Level Mountain Range 6 Jun 1978-3 nestlings, half grown; Summit Lake Pass 3 to 11 Aug 1970-2 (Griffith 1973); Fort Nelson 4 Jul 1978-3 nestlings; Cassiar 15 Jun 1980-6, at highway junction; Kotcho Lake 26 Jun 1982-1; 24 km e Atlin 5 Jul 1975-4; Haines Highway 7 Jul 1956-1 at Mile 75.

Breeding Bird Surveys: Coastal – Recorded from 27 of 27 routes and on 98% of all surveys. Maxima: Pitt Meadows 1 Jul 1974-66; Masset 21 Jun 1994-54; Albion 4 Jul 1970-50. **Interior** – Recorded from 64 of 73 routes and on 88% of all surveys. Maxima: Tokay 29 Jun 1973-75; Columbia Lake 26 Jun 1973-40; Grand Forks 26 Jun 1983-27; Golden 19 Jun 1976-27.

Autumn: Interior – Warm Spring Mountain 16 Sep 1972-1 at hot springs; Hyland Post 25 Oct 1980-1; Flatrock 30 Oct 1951-4; nr Mackenzie 2 Sep 1994-6 banded; 24 km s Prince George 30 Nov 1983-1; Chezacut 2 Sep 1941-200; Williams Lake 15 Sep 1986-20, 10 Nov 1982-7; Glacier National Park 5 Oct 1982-13; Mount Revelstoke 17 Nov 1989-2; Scotch Creek 10 Sep 1962-50; Nakusp 21 Nov 1976-5; Penticton 22 Nov 1976-6. **Coastal** – Port Simpson 28 Sep 1969-9 (Crowell and Nehls 1970a); Kitimat 21 Nov 1974-1; Naden Harbour 9 Oct 1974-3; 8 km n Swanson Bay 4 Oct 1935-1 (MCZ 285603); Sandspit 5 Nov 1986-6; Mount Brilliant 18 Sep 1939-1 (NMC 29090); Cape St. James 24 Sep 1981-6; Port Neville 8 Nov 1975-6; Quatse River 18 Nov 1980-1 on estuary (Dawe et al. 1995); Campbell River 25 Oct 1964-80; Egmont 3 Sep 1977-6; Hope 26 Nov 1976-20; Sea Island 3 Oct 1994-144 (Bowling 1995a); Long Beach 2 Oct 1971-10; Port Renfrew 5 Oct 1974-10.

Winter: Interior – Willow River 9 Jan 1966-1, 12 Dec 1968-1; Prince George 21 Jan 1979-1; Quesnel 27 Dec 1983-1; Williams Lake 26 Feb 1988-8, 31 Dec 1978-1; Revelstoke 31 Jan 1997-4 roosting in wood pile, 20 Feb 1989-2; Celista 15 Feb 1948-3; Nakusp 3 Jan 1986-12; Lavington 20 Jan 1973-8; Penticton 1 Jan 1980-7; Castlegar 31 Dec 1978-3; Vaseux Lake to Oliver 31 Dec 1976-44 seen over 8-hour period; Trail 5 Jan 1982-1. **Coastal** – Hazelton 1 Feb 1963-1; Kitsault 10 to 12 Dec 1980-1; Prince Rupert 15 Jan 1984-2; Masset 26 Dec 1972-35; Juskatla 23 Jan 1972-2; Cape St. James 22 Feb 1982-30, 6 Dec 1981-15; Port Neville Inlet 26 Feb 1975-7; Mitlenatch Island 12 to 22 Dec 1965-67 (Campbell and Kennedy 1965); Leiner River 16 Feb 1982-3, 29 Dec 1981-5 on estuary (Dawe and Buechert 1997); Harrison Hot Springs 16 Jan 1983-18; Baynes Sound 10 Jan 1981-21 (Dawe et al. 1998); Skagit River 21 Feb 1971-1; Westham Island 14 Dec 1974-180; Pachena Bay 11 Jan 1976-5; Colwood 28 Feb 1987-5; River Jordan 5 Jan 1974-1; Mandarte Island 23 Feb 1983-1 egg.

Christmas Bird Counts: Interior – Recorded from 22 of 27 localities and on 78% of all counts. Maxima: Oliver-Osoyoos 28 Dec 1988-313; Penticton 27 Dec 1992-231; Vernon 15 Dec 1991-188. **Coastal** – Recorded from 32 of 33 localities and on 96% of all counts. Maxima: Vancouver 17 Dec 1989-**2,692**, all-time Canadian high count (Monroe 1990a); Ladner 23 Dec 1990-1,654; White Rock 2 Jan 1997-1,544.

Lincoln's Sparrow

LISP

Melospiza lincolnii (Audubon)

RANGE: Breeds from central Alaska, Yukon, west-central and southeastern Mackenzie, and northern Saskatchewan east through Ontario, northern Quebec, Labrador and Newfoundland; south through most of British Columbia and in the mountains through southern Alberta to southern California, central Arizona, and northern New Mexico, and from central Alberta east through central-southern Saskatchewan and southern Manitoba to northern New England and northern Nova Scotia. Winters from coastal British Columbia south through western Washington, Oregon, central Arizona, and central Missouri, across to northern Georgia and south to Florida and the Gulf coast, and through Central America except Yucatan and Belize to Honduras and Panama.

STATUS: On the coast, an *uncommon* to *common* migrant, *very rare* in summer, and *rare* to locally *uncommon* winter visitant to the Georgia Depression Ecoprovince; in the Coast and Mountains Ecoprovince, *rare* transient and *casual* in summer and winter on Western Vancouver Island and the Southern Mainland Coast, *uncommon* migrant and summer visitant and *very rare* in winter on the Northern Mainland Coast and the Queen Charlotte Islands.

In the interior, an *uncommon* to *common* migrant and summer visitant in all ecoprovinces; *casual* in winter in the Southern Interior Ecoprovince.

Breeds.

Figure 322. Although widely distributed in British Columbia, the Lincoln's Sparrow is an elusive bird whose natural history is poorly known (Anthony Mercieca).

NONBREEDING: The Lincoln's Sparrow (Fig. 322) has a widespread distribution throughout British Columbia, including Vancouver Island and the Queen Charlotte Islands. There are, however, large areas from which information for this sparrow is lacking, including much of Vancouver Island, the mountainous areas of the Southern Mainland Coast, the Takla Lake and Omenica Mountains regions, and the northern Rocky Mountain Trench between Mackenzie and Lower Post. Biological field investigations of suitable habitat in these areas are needed. The highest numbers in winter occur in the Georgia Depression (Fig. 323).

The Lincoln's Sparrow has been reported at elevations from sea level to 1,370 m on the coast and from the valley bottoms around 280 m to over 2,015 m in the interior. Migration and winter habitats include logging slashes, powerline rights-of-way (Fig. 324), hedgerows and shrub edges adjacent to farmland fields, old field and shrubby pastures, young forests and older forest edges, avalanche chutes, and orchards, as well as riparian areas such as bramble and shrub thickets adjacent to sewage lagoons, lakes, marshes, sloughs, swamps, beaver ponds, estuaries, beaches, grass and sedge meadows, river bottoms, and native bunchgrass prairie.

The Lincoln's Sparrow occurs throughout the year in the Georgia Depression (Figs. 325 and 326). Numbers remain low until birds from further south begin passing through in early April. The main spring movement passes through during the third and last weeks of April, and most birds have left the ecoprovince by the third week of May. A few birds remain through the summer.

On Western Vancouver Island and the Southern Mainland Coast, birds can be found by mid-April, but little in the way of a spring movement has been documented in either region. Further north, on the Northern Mainland Coast and

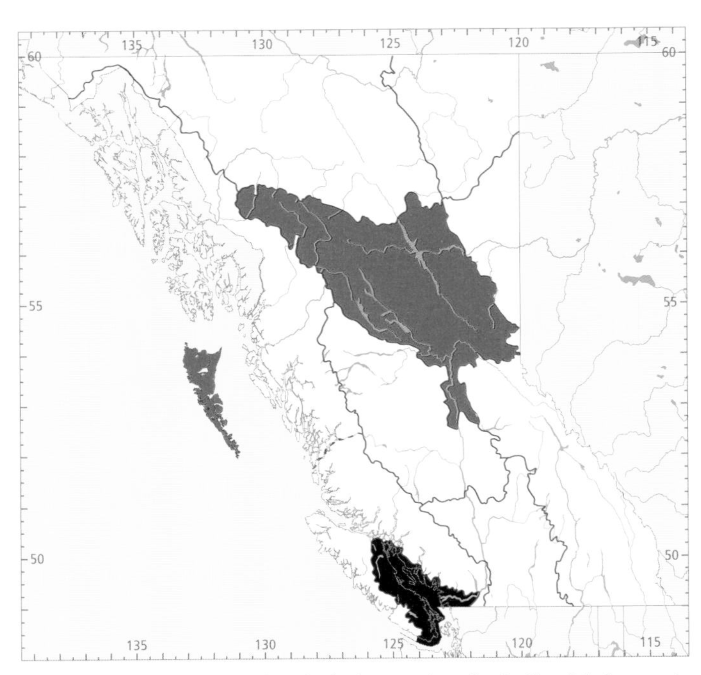

Figure 323. In British Columbia, the highest numbers for the Lincoln's Sparrow in winter (black) occur in the Georgia Depression Ecoprovince; the highest numbers in summer (red) occur on the northern Queen Charlotte Islands in the Coast and Mountains Ecoprovince and in the Sub-Boreal Interior Ecoprovince.

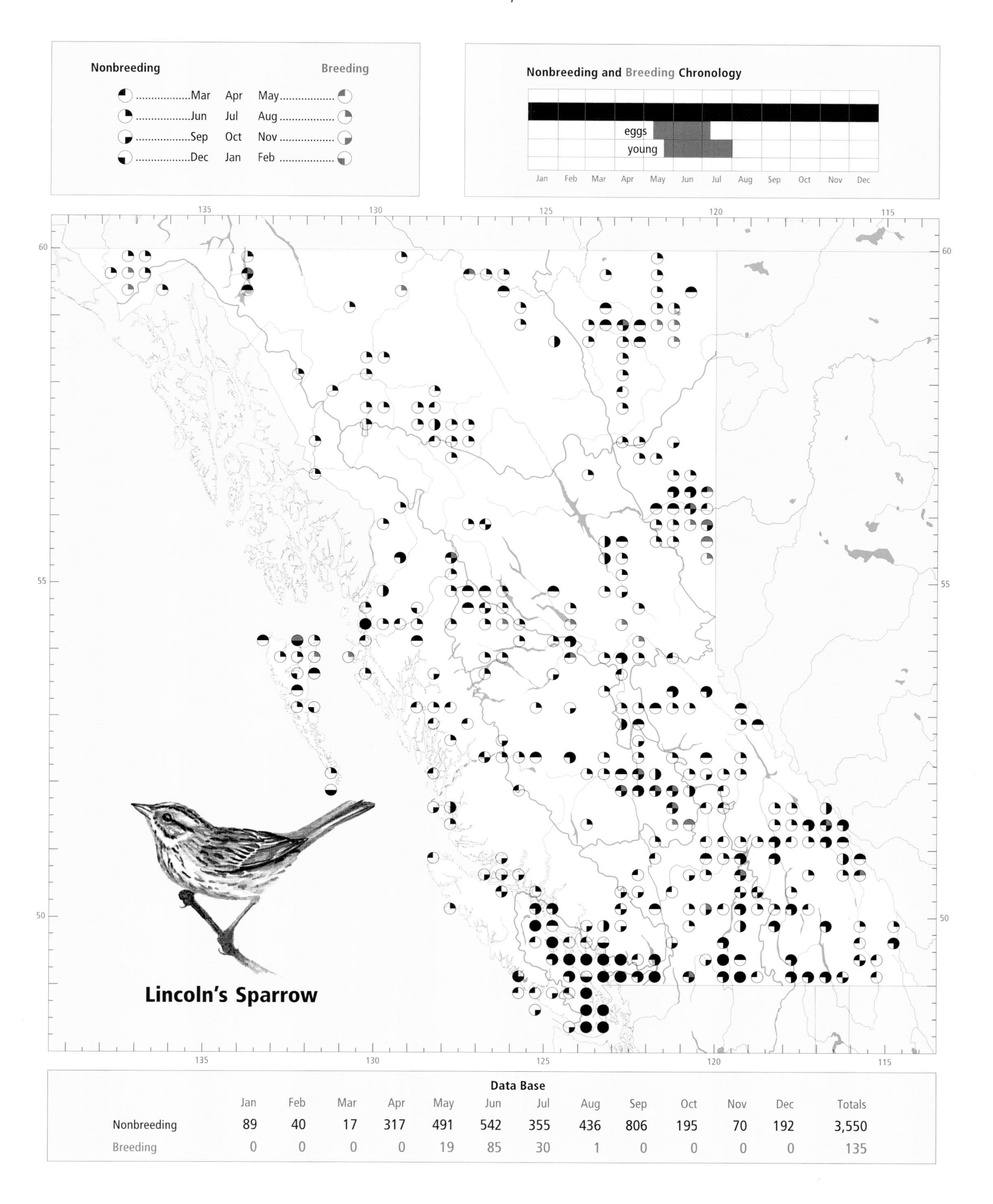

Data Base

	Jan	Feb	Mar	Apr	May	Jun	Jul	Aug	Sep	Oct	Nov	Dec	Totals
Nonbreeding	89	40	17	317	491	542	355	436	806	195	70	192	3,550
Breeding	0	0	0	0	19	85	30	1	0	0	0	0	135

Figure 324. During autumn migration in British Columbia, the Lincoln's Sparrow frequents powerline rights-of-way with brushy areas consisting of bracken, thimbleberry, and regenerating alder and willow (Lost Creek, east of Salmo, 29 September 1997; R. Wayne Campbell).

the Queen Charlotte Islands, birds arrive in the last 2 weeks of April (Fig. 325).

In the Southern Interior, birds may arrive as early as the first week of April. Numbers build thereafter, peaking at the end of April and the first week of May. In the Southern Interior Mountains, the first Lincoln's Sparrows are not reported until the third week of April, and their numbers build through the first week of May. In the Central Interior, birds arrive by mid-April and numbers build shortly thereafter, peaking by the third week of May. Further north, birds arrive in the Sub-Boreal Interior by the third week of April, but there are not enough reports to document a significant spring movement. Munro (1949) also notes the lack of a large spring migration in the Vanderhoof area of the Sub-Boreal Interior. In the Northern Boreal Mountains, birds may arrive in the last week of April (Fig. 325).

East of the Rocky Mountains, in the Boreal Plains, the Lincoln's Sparrow may arrive in very late April, but most appear in early May and numbers continue to build into the summer. Further north, in the Taiga Plains, birds do not appear until the second week of May. A noticeable spring migration through this ecoprovince has not been documented.

Autumn migration in the 2 northernmost ecoprovinces has not been documented; most birds have left those regions by the second week of September. In the Boreal Plains, the autumn movement begins by at least the second week of August and peaks later in the month and in the first week of September. Most birds have left by the third week of September, although stragglers may remain until the first week of October (Figs. 325 and 326). A small migration is noticeable

Figure 325. Annual occurrence (black) and breeding chronology (red) for the Lincoln's Sparrow in ecoprovinces of British Columbia. Records are shown for the week in which they occurred.

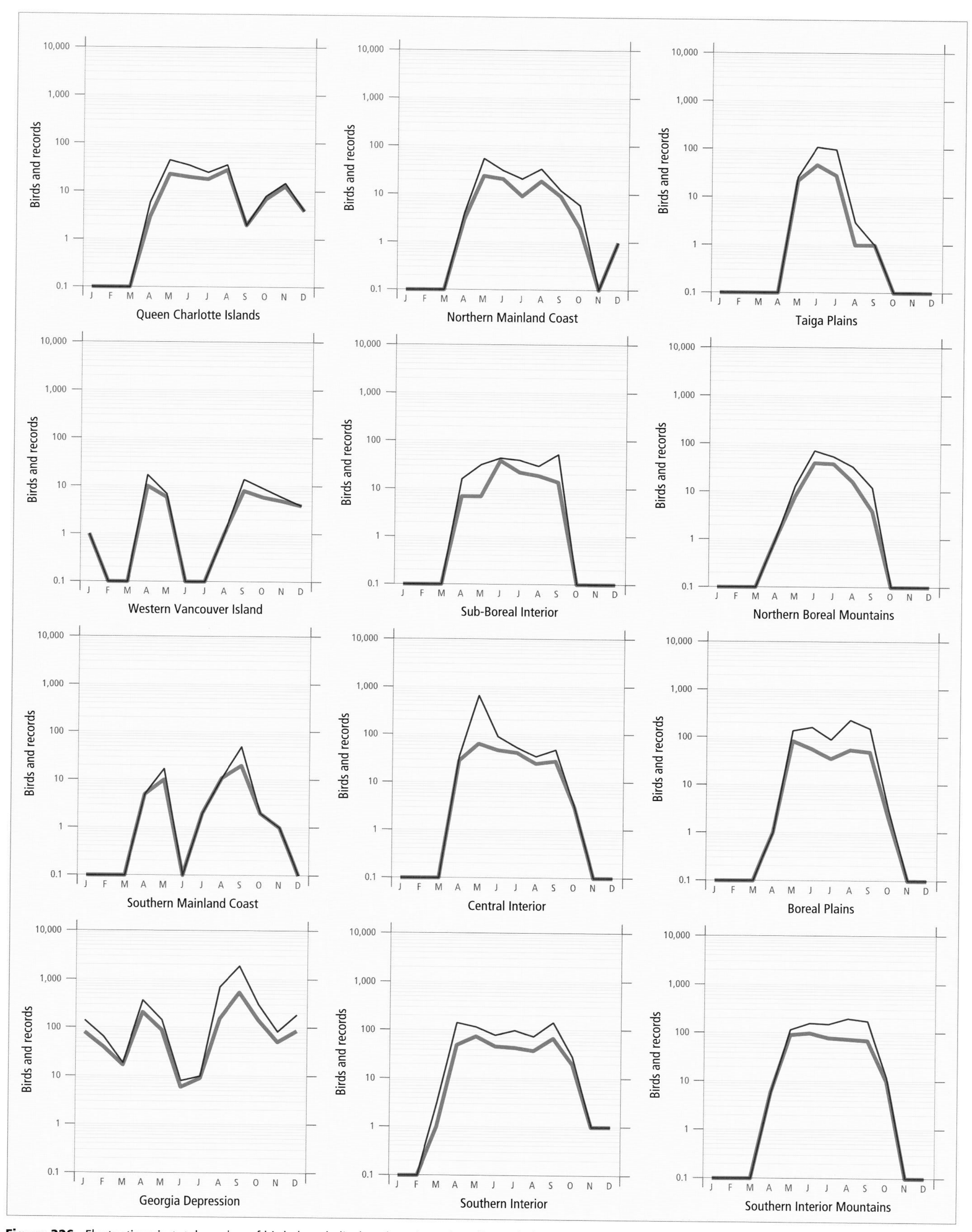

Figure 326. Fluctuations in total number of birds (purple line) and total number of records (green line) for the Lincoln's Sparrow in ecoprovinces of British Columbia. Christmas Bird Counts, Breeding Bird Surveys, and nest record data have been excluded.

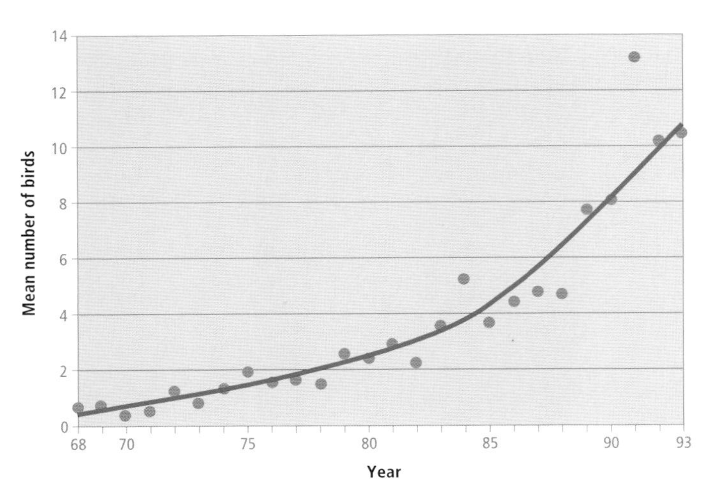

Figure 327. An analysis of Breeding Bird Surveys for the Lincoln's Sparrow in British Columbia shows that the number of birds on interior routes increased at an average annual rate of 14% over the period 1968 through 1993 ($P < 0.001$).

at the end of August and first week of September in the Sub-Boreal Interior, and all birds have left by the end of September.

In the Central Interior, the autumn movement begins in August and is heaviest in the second and third weeks of September; few birds are left after the first week of October. Autumn migration in the Southern Interior Mountains is noticeable from the first week of August through mid-September. Most birds have left by the third week of September, although stragglers may remain until the third week of October. In the Southern Interior, autumn migration is obvious from late August and September, and most birds have passed through by mid-October.

On the northern portions of the coast, a small autumn movement is noticeable through August and most birds have left the region by the end of September, although some winter in these areas occasionally. On the Southern Mainland

Figure 328. On the northern Queen Charlotte Islands, the Lincoln's Sparrow breeds in open, wet habitats where clumps of rushes occur (Delkatla Inlet, near Masset, 1 June 1988; John M. Cooper).

Coast, there is a small increase in autumn numbers by the end of August and the first week of September, and most birds have left the region by early October. On Western Vancouver Island, no noticeable autumn movement has been documented; in rare instances, birds may winter there.

Autumn migration through the Georgia Depression begins in the first week of August. Numbers are high by the third week of August, and remain relatively high through the last week of October as the majority of the migrants pass through. Small numbers remain through the winter (Figs. 325 and 326).

In winter, the few Lincoln's Sparrows remaining in British Columbia are concentrated in the lower Fraser River valley and on southeastern Vancouver Island, but there are a few December records for the Queen Charlotte Islands, Terrace, and the Okanagan valley. There are indications that these birds either move further south in late winter or do not survive (Fig. 325).

On the southern portions of the coast, the Lincoln's Sparrow has been recorded throughout the year; in the northern coastal regions, it has been recorded regularly from 19 April through December; in the interior, it has been recorded regularly from 5 April through 15 October (Fig. 325).

BREEDING: The Lincoln's Sparrow likely breeds throughout its summer range in the interior of the province. On the coast, the only documented nesting locations are at Delkatla Inlet (Fig. 328) and Blue Danube Swamp, both at the north end of Graham Island on the Queen Charlotte Islands, on Porcher Island (Swarth 1924), and in the Kispiox River valley (Swarth 1924); it likely breeds, however, in suitable habitat along the mainland coast north of Vancouver Island. For example, T.T. McCabe (*in* Dickinson 1953) notes that specimens collected at Khutze Inlet in June were "breeding rather commonly on flats in crab[apples] and willows and heavy grass and Siwash Rhubarb. At this time [12 and 13 June] young were out of the nest." Details of nests containing either eggs or young were not reported, however.

Brooks and Swarth (1925) note that the Lincoln's Sparrow is a rare and local breeder on Vancouver Island, and Rising (1996) also includes Vancouver Island in its breeding distribution. Supporting data are lacking, however, and we have found no evidence to suggest that the Lincoln's Sparrow breeds either on Vancouver Island or in coastal regions on the southwest mainland.

The Lincoln's Sparrow reaches its highest numbers in summer on northern Graham Island in the Coast and Mountains and in the Sub-Boreal Interior (Fig. 323). An analysis of Breeding Bird Surveys for the period 1968 through 1993 shows that the number of birds on interior routes increased at an average annual rate of 14% (Fig. 327). Data for coastal routes were insufficient for analysis.

The increase in numbers in the interior agrees with the results of Sauer et al. (1997), who found that over the period 1966 through 1996, the Lincoln's Sparrow showed an average annual increase of 2.8% ($P < 0.01$) across North America and an average annual increase of 2.9% ($P < 0.05$) across Canada.

Figure 329. In portions of the Northern Boreal Mountains Ecoprovince, the Lincoln's Sparrow breeds in wetlands such as this willow swamp (near Cassiar, 1 July 1999; R. Wayne Campbell).

In both cases, the increases occurred primarily over the period 1966 through 1979; an analysis of the surveys for the period 1980 to 1996 could not detect a net change in either continent-wide or Canada-wide numbers. In British Columbia, the large increase is probably a response to clearcut logging (Pojar 1995, Merkins and Booth 1996, 1998).

The Lincoln's Sparrow has been reported nesting at elevations from near sea level on the coast to 1,300 m in the interior, although singing males are now common at 1,700 m in the Okanagan. This sparrow prefers wet habitats with the presence of low, dense shrub growth, particularly willow from 1 to 2.5 m high, and openings where mosses, sedges (Fig. 328), or grasses occur. Nesting habitats include swamps, muskegs, bogs, thickets near beaver ponds, brushy fields, marshy areas with alder or willow thickets (Fig. 329), sedge and rush meadows, streamside willows, black spruce–tamarack bogs, floodplain forests, regenerating burns and logged areas (Fig. 330), trembling aspen groves, avalanche slopes, subalpine parklands, and shrub areas near timberline.

At Delkatla Inlet (Fig. 328), on the northern Queen Charlotte Islands, where the highest density of nesting pairs in the province has been reported, the Lincoln's Sparrow breeds in wetter areas where the rush *Juncus effusus* occurs and along edges where red alder saplings are present (Cooper 1993a).

In the Okanagan valley, the Lincoln's Sparrow frequents willow swamps and open, wet deciduous brush at higher elevations (Cannings et al. 1987). In the Southern Interior Mountains, this sparrow was found in logged areas and burns with regenerating alder, bracken, fireweed, and thimbleberry; in roadside willow, red-osier dogwood, and fireweed, with trees up to 8 m high; and in wetlands of dwarf or scrub water birch, scouring rush, Labrador tea, *Sphagnum* moss, and some open water (Campbell and Dawe 1992). In both Mount Revelstoke and Glacier national parks, the Lincoln's Sparrow was confined to the edges of wet shrub thickets with a dense herbaceous understorey (Van Tighem and Gyug 1983). In the Central Interior, the Lincoln's Sparrow used wetlands of sedge, lodgepole pine, hybrid white spruce, and willow as well as moist areas of scrub birch, willow, and spruce.

In trembling aspen forests of the Sub-Boreal Interior, Pojar (1995) found the Lincoln's Sparrow to be the most abundant species in 4- to 6-year-old clearcuts, with densities ranging from 13 to 29 singing males per 10 ha. Densities from other forest types were as follows: sapling trembling aspen (< 23 years old), 9 to 23 singing males per 10 ha; mature trembling aspen (50 to 60 years old), 0 to 3 singing males per 10 ha; old trembling aspen (> 100 years old), 0 to 4 singing males per 10 ha; and mixed conifer-aspen (50 to 95 years old), 0 to 2 singing males per 10 ha. Stanwell-Fletcher and Stanwell-Fletcher (1943) found the Lincoln's Sparrow to be a common species in the Driftwood River valley, in open meadows, marshes, and willow thickets.

In the Boreal Plains, the Lincoln's Sparrow was a dominant species in all marshy areas where low clumps of willow alternated with mossy clumps and patches of sedges, at edges of beaver ponds, in overgrown clearings, and in transmission corridors in deciduous and mixed woodlands (C.R. Siddle pers. comm.). In the Chilkat Pass region of northwestern British Columbia, it frequents wet, shrubby marshes at timberline.

Most reported nests (39%; n = 36) were found associated with wetland habitats, followed by forested (25%) and human-influenced (19%) habitats.

Of the wetland habitats used, marshes were most often reported (14%), followed by lakes (11%) and by willow swamps and black cottonwood bottomlands (11%). Most of the nests in forested habitats were reported from deciduous and trembling aspen forests (22%). Human-influenced habitats included farmland, pastureland, suburban and logged areas, roadsides, and transportation corridors.

Of the specific habitats reported, 68% of nests (n = 25) were found at the edges of young and mature forests, as well as shrub, sapling, and young tree-dominated (Fig. 331), regenerating clearcuts. Other specific nest habitats included muskeg, beaver ponds, roadsides, farmyards, and airport fields.

The Lincoln's Sparrow has been recorded breeding on the coast from 11 May (calculated) to 1 August, and in the interior from 23 May (calculated) to 27 July (Fig. 325).

Nests: Most nests (92%; n = 38) were found on the ground (Fig. 332); 3 nests were found in shrubs. Ground nests were usually a shallow depression in grasses or mosses, placed beside or in a hummock or clump of taller vegetation that tended to conceal the nest. Nests in shrubs were woven among the branches.

Nests were cups consisting of coarse grasses, sedges, twigs, leaves, and rootlets; they were lined with fine grasses, sedges, or hair (Fig. 332).

The heights of 38 nests ranged from ground level (92%) to 1 m.

Eggs: Dates for 76 clutches ranged from 20 May to 9 July, with 50% recorded between 5 June and 26 June. Calculated dates indicate that eggs can occur as early as 11 May. Sizes of 55 clutches ranged from 1 to 5 eggs (1E-3, 2E-1, 3E-5, 4E-30, 5E-16), with 55% having 4 eggs. The incubation period is 10 to 13 days, with a mean of 11.5 days (Ammon 1995).

Young: Dates for 58 broods ranged from 22 May to 1 August, with 53% recorded between 10 June and 4 July. Sizes of 37 broods ranged from 1 to 5 young (1Y-4, 2Y-8, 3Y-8, 4Y-15, 5Y-2), with 62% having 3 or 4 young. The nestling period is 10 or 11 days (Ammon 1995).

Brown-headed Cowbird Parasitism: In British Columbia, 4 of 66 nests found with eggs or young were parasitized by the cowbird. Nest parasitism was not found on the coast (n = 14); in the interior, 8% of nests were parasitized (n = 52). There were 2 additional records of the Lincoln's Sparrow feeding cowbird fledglings.

Nest Success: Of 6 nests found with eggs and followed to a known fate, 5 produced at least 1 fledgling.

REMARKS: Two of the 3 subspecies of Lincoln's Sparrow are known to occur in British Columbia: *M. l. lincolnii* occurs throughout the interior of the province, and *M. l. gracilis* fre-

Figure 330. Throughout British Columbia, the Lincoln's Sparrow commonly colonizes and breeds in recently disturbed habitats such as logging clearcuts or agricultural clearings (18 km northeast of Hudson's Hope, 18 June 1997; R. Wayne Campbell).

Figure 331. Most Lincoln's Sparrow nests in British Columbia were found on the ground in shrub and young, tree-dominated habitats along the edge of young to mature forests (28 km north of Charlie Lake, in the Boreal Plains Ecoprovince, 18 June 1996; R. Wayne Campbell).

Figure 332. Typical Lincoln's Sparrow ground nest, consisting of coarse grasses and lined with fine grasses, holding an egg and 2 recently hatched chicks (28 km north of Charlie Lake, 18 June 1996; R. Wayne Campbell).

quents the north and central coastal regions, including the Queen Charlotte Islands. In some areas of the province (e.g., Dokdaon Creek), *M. l. gracilis* intergrades with *M. l. lincolnii* (American Ornithologists' Union 1957). W.E. Godfrey's (1986) description of the subspecies distribution of the Lincoln's Sparrow in British Columbia appears to be in error.

Cooper (1993a) has discussed the impacts of humans on a small but dense breeding population of Lincoln's Sparrows resulting from the restoration of marshes at the head of Delkatla Inlet (Fig. 328) on the northern Queen Charlotte Islands.

As with some other riparian species (e.g., the Willow Flycatcher), decreases in Lincoln's Sparrow populations in its breeding habitat have been associated with livestock grazing (Schultz and Leininger 1991). The Lincoln's Sparrow also appears susceptible to disturbance during the nesting season and may react by deserting its nest. Ammon (1995) notes significantly higher nest desertion rates at sites used for picnicking, fishing, and hiking (17.5%) than at sites not used for recreation (4%).

A recent review of the biology of the Lincoln's Sparrow may be found in Ammon (1995).

NOTEWORTHY RECORDS

Spring: Coastal – Oak Bay 23 Apr 1982-14 on golf course; Bamfield 14 May 1978-1; Mesachie Lake 4 May 1986-15; Stubbs Island 14 and 15 Apr 1979-1; Long Beach 26 Apr 1984-3; Cheam Slough 20 Apr 1984-1; Harrison Hot Springs 22 Apr 1986-1; Sandwick 22 Apr 1933-1; Cortes Island 2 May 1977-1 nr Gorge Harbour; Whistler 22 Apr 1985-1; Cape Scott 2 May 1972-1 killed in collision with the lighthouse (Hatler and Campbell 1975); Bella Bella 9 May 1979-4 in song; Bella Coola 27 Apr 1933-1 (MCZ 285548); Kimsquit 24 Apr 1985-2 in salmonberry nr airstrip; Kitlope area 3-10 May 1991-8, 4 at estuary and 4 on lower Kitlope River; Masset 19 Apr 1920-1 (AMNH 405024); Delkatla Creek 20 Apr 1979-1, 27 Apr 1979-4 in song, 13 May 1988-10, 20 May 1988-4 eggs, 22 May 1988-3 nestlings; Kitimat Mission 11 May 1972-7. **Interior** – Osoyoos 15 Apr 1985-1; Oliver 1 Apr 1987-1, 5 Apr 1986-23 banded; nr Midway 26 Apr 1905-1; Balfour to Waneta 16 May 1981-1 on count; Nakusp 20 Apr

1985-1; Seton Lake 14 May 1916-1 (RBCM 3822); Community Creek (Helfey Lake) 21 May 1983-7; Bridge Lake 31 May 1943-5 eggs, incubation advanced; 15 km n Golden 25 Apr 1997-1; Clearwater 6 May 1935-2 (MCZ 285571 and 285572); Riske Creek 13 Apr 1988-1; Williams Lake 22 Apr 1982-6 with flock of White-crowned Sparrows; Redstone 15 May 1943-500; Chezacut 10 May 1943-8, 13 May 1943-20; Quesnel 16 May 1979-1; Mount Robson Park 3 May 1981-5 banded; 24 km s Prince George 23 Apr 1983-10, 1 May 1981-20; Vanderhoof 27 Apr 1997-1; Papoose Lake 14 May 1970-1; Smithers 21 Apr 1989-1; Tupper 7 May 1938-1 (Cowan 1939; RBCM 8043), 25 May 1981-3 eggs; Tetana Lake 17 Apr 1938-1; Fort St. John 30 Apr 1986-1 feeding at waste grain pile; Taylor 18 May 1985-9 on Peace Island Park Rd; Stoddart Creek (Charlie Lake) 3 May 1986-5, 1 in song; Fort Nelson 11 May 1980-2 along Parker Lake Rd; Warm Bay 29 May 1981-4 in song; Helmut 15 to 22 May 1982-1 to 4 birds on territory; Atlin 28 Apr 1934-1 (CAS 42092); Chilkat Pass 1957-arrived in May and seen June and July (Weeden 1960).

Summer: Coastal – Rocky Point (Victoria) 24 Aug 1994-6, 26 Aug 1994-16, 27 Aug 1994-10, 31 Aug 1994-14 banded; Somenos Lake 31 Aug 1980-28; Steveston 4 Jun 1980-2; Englishman River 30 Aug 1979-17 counted on estuary (Dawe et al. 1994); Texada Island 24 Jul 1921-1 nr Blubber Bay; Alta Lake 2 Aug 1941-1 juvenile (Racey 1948); Cape Cook Lagoon 18 Aug 1981-1; Fawn Bluff 16 Aug 1936-1 (NMC 27417); Calvert Island 26 Jul 1937-1 (NMC 28164); Khutze Inlet 12 Jun 1936-3 males (Dickinson 1953); Rose Harbour 16 Aug 1946-1; Blue Danube Swamp 1 Aug 1974-2 recently fledged young being fed by adults; Porcher Island 23 Jun 1921-fledgling just out of nest, collected (Swarth 1924); Masset 28 Jun 1941-1, 24 Jul 1961-5; Delkatla Creek 5 Jun 1984-10, young seem to have fledged; Kitimat 12 Jul 1975-8, 23 Aug 1975-5 on tidal flats; Rocher Déboulé Mountains 20 Jul 1944-1 young being fed by adult (Munro 1947); Kispiox River 23 Jun 1921-1 young bird just out of nest (Swarth 1924); Bear River (Stewart) 17 Jun 1975-1; Meziadin Lake 26 Jun 1991-1; Haines Highway (Mile 46.5) 18 Jun 1972-2. **Interior** – Manning Park 24 Jun 1968-4 nestlings, 11 Jul 1964-15, 17 Jul 1971-4 nestlings; n Sparwood 7 Jun 1984-1 in song (Fraser 1984); Chapperon Lake 15 Jun 1962-4 eggs; 64 km e Ashcroft 29 Aug to 1 Sep 1889-6 (Chapman 1890); 4 km n Enderby 8 Jul 1967-4 eggs; Kootenay National Park 2 Jun 1983-4, 8 Jun 1983-5 eggs; 6 km s Bridge Lake 4 Jun 1976-1 along Eagan Lake Rd; Ottertail River 29 Jun 1975-11 around sloughs, including adults feeding moths to unknown number of young in nest (Wade 1977); 1 km e 111 Mile Creek 5 Jul 1959-4 eggs; Chilcotin River (Riske Creek) 6 Aug 1994-several moving through at Wineglass Ranch; Williams Lake 21 Jun 1969-3 nestlings; 39 km e Anahim Lake 12 Jun 1996-2 (Dawe and Buechert 1996); Chezacut 21 Jun 1943-7; Moose Heights 24 Jul 1974-1 fledgling; nr Barkerville 13 Jul 1930-1 (MCZ 285526); Mount Robson Park 29 Aug 1973-10; Timon and Lightning creeks 26 Jul 1988-2; 21.4 km s McBride 10 Jul 1991-8 (Campbell and Dawe 1991); Nulki Lake 21 Aug 1945-5; Prince George 17 to 30 Jun 1969-6 to 28 birds; Willow River 7 Jun 1965-5 eggs; 2 km e Eaglet Lake 14 Jun 1996-1 (Dawe and Buechert 1996); s Burnie Lake 23 to 30 Aug 1975-1+ (Osmond-Jones et al. 1975); Bulkley Lake 27 Jul 1944-young; Summit Lake (Prince George) 27 Jun 1944-pair feeding young; nr Chetwynd 24 Jul 1975-2; Tupper 28 Jun 1978-6, fairly common along highway; Tupper Creek to Pouce Coupe 15 Jun 1974-18; 15 km s Dawson Creek 8 Jul 1994-2 big nestlings; 7 km s Dawson Creek 9 Jul 1993-4 eggs; Dawson Creek (Pouce Coupe) 8 Jun 1975-34; s Hudson's Hope 17 Aug 1974-20 in flock feeding together; ne Dawson Creek 8 Jun 1993-4 nestlings; nr Two Rivers 5 Jul 1974-10; n Cecil Lake 29 Aug 1984-10+, heavy migration; Cecil Lake 14 Jun 1974-21; Boundary Lake (Goodlow) 26 Aug 1975-15 migrants; Trygve Lake 6 Jul 1976-5; Alaska Highway (Mile 126) 23 Jun 1984-1; Alaska Highway (Mile 128) 23 Jun 1984-1; Firesteel River 27 Jun 1976-4; Hotlesklwa Lake 3 Aug 1976-16; Todagin Creek and Kinaskan Lake 20 Jun 1975-7; Alaska Highway (Mile 393.5) 10 Jul 1978-1; Kledo Creek 16 Jun 1976-1; se Fort Nelson 13 Jun 1976-6 on Parker Lake Rd, 15 Jun 1975-26 on Parker Lake Rd, 26 Jun 1985-12, Fort Nelson River, 21 Jul 1986-1 fledgling out of nest 6 or 7 days, 28 Aug 1978-3 on Clarke Lake Rd at Km 17; O'Connor River 6 Jun 1983-4 eggs (Campbell et al. 1983), 10 Jun 1983-4 small nestlings about 4 days old; Tatshenshini River 5 Jun 1983-2 pairs; ne Kotcho Lake 1 Jun 1982-2 (RBCM 17505 and 17506), 22 Jun 1982-1+, adults bringing food to nest (Campbell and McNall 1982); Tats Lake 7 Jul 1983-1 (RBCM 18279); Rabbit River 8 Jul 1981-5 eggs; n Petitot River 20 Jun 1982-1.

Breeding Bird Surveys: Coastal – Recorded from 6 of 27 routes and on 4% of all surveys. Maxima: Kispiox 20 Jun 1993-11; Masset 21 Jun 1994-7; Nass River 21 Jun 1975-4. **Interior** – Recorded from 54 of 73 routes and on 36% of all surveys. Maxima: Tupper 20 Jun 1994-77; Summit Lake 28 Jun 1991-52; McLeod Lake 3 Jul 1993-42.

Autumn: Interior – Atlin 7 Sep 1931-1 (RBCM 5699); Fort Nelson River 2 Sep 1985-1; Griffith Creek 4 Sep 1976-9; Nig Creek 15 Sep 1984-2; n Fort St. John 7 Sep 1985-29 migrants, 4 Oct 1986-2; Tetana Lake 21 Sep 1938-1 (RBCM 8376); nr Mackenzie 1 Sep 1994-6, 3 Sep 1994-18 banded; Firth Lake 25 Sep 1993-1; Williams Lake 4 Sep 1978-10, 24 Sep 1978-3, 21 Oct 1981-1; Yoho National Park 7 Sep 1976-14; Horse Lake (100 Mile House) 17 Sep 1934-1; Burges and James Gadsden Park 18 Oct 1997-1; Scotch Creek 10 Sep 1962-30; Tranquille 5 Sep 1983-15; Okanagan Landing 6 Sep 1909-7, 23 Oct 1944-1; Kelowna 3 Nov 1991-1 (Siddle 1992a); Laycock Creek (Sparwood) 20 Oct 1984-1; Creston 3 Oct 1977-3. **Coastal** – Kincolith 21 Sep 1980-4; Prince Rupert 18 Oct 1956-3; Mount Brilliant 15 Sep 1940-1; Cape St. James 19 Sep 1978-1; Calvert Island 8 Sep 1935-1 (MCZ 285578); Pemberton 10 Sep 1995-20; Alta Lake 3 Sep 1927-4 (Racey 1948); Brackendale 6 Sep 1916-5 in new, unworn plumage (Taverner 1917); Egmont 6 Oct 1973-1; Yale 1 Sep 1920-1; Squamish River 2 Nov 1986-1 on estuary; Seabird Island 5 Oct 1991-1; Port Coquitlam 11 Sep 1988-35; Jericho Beach (Vancouver) 1 Sep 1975-25, 6 Sep 1987-25; Little Qualicum River 20 Sep 1975-23 on estuary (Dawe 1976), 20 Sep 1976-11 (Dawe 1980); Pitt Polder to Maple Ridge 13 Sep 1967-28; Delta 5 Oct 1983-20 at Brunswick Point; Reifel Island 23 Sep 1983-31; Sea Island 16 Sep 1989-30 (Weber and Cannings 1990), 21 Sep 1980-50; River Jordan 16 Sep 1989-4; Rocky Point (Victoria) 1 Sep 1994-8, 11 Sep 1994-14, 17 Sep 1994-7, 29 Sep 1994-3, 12 Oct 1994-3, 16 Oct 1994-1 banded.

Winter: Interior – See Christmas Bird Counts. **Coastal** – Terrace 22 Dec 1977-1 (BC Photo 754); Cape St. James 26 Dec 1978-1; Serpentine River 10 Dec 1989-10, high count; Iona Island 3 Dec 1974-23; Reifel and Westham islands 27 Dec 1987-3; Sandhill Creek 31 Dec 1973-1, 1 Jan 1974-1; Central Saanich 1 Jan 1989-3; Swan Lake (Saanich) 2 Jan 1989-1.

Christmas Bird Counts: Interior – Recorded from 2 of 27 localities and on less than 1% of all counts. Maxima: Oliver-Osoyoos 28 Dec 1987-1; Penticton 27 Dec 1982-1. **Coastal** – Recorded from 17 of 33 localities and on 22% of all counts. Maxima: Victoria 19 Dec 1992-**30**, all-time Canadian high count (Monroe 1993); Ladner 23 Dec 1991-18; Duncan 2 Jan 1993-13.

Swamp Sparrow
Melospiza georgiana (Latham)

SWSP

RANGE: Breeds from northeastern British Columbia and southwestern Mackenzie east across Canada to northern Ontario and central Labrador, south to east-central British Columbia, central Alberta and Saskatchewan, Nebraska, Ohio, and Maryland. Winters mainly in the southeastern United States, Texas, and central Mexico; small numbers winter on the Pacific coast from southwestern British Columbia to southern California.

STATUS: On the coast, *rare* migrant and winter visitant and *casual* in summer in the Georgia Depression Ecoprovince; in the Coast and Mountains Ecoprovince, *very rare* autumn migrant and winter visitant on Western Vancouver Island, *accidental* in winter on the Queen Charlotte Islands and the Southern Mainland Coast; absent elsewhere.

In the interior, *very rare* migrant and winter visitant in the Southern Interior Mountains and Southern Interior ecoprovinces, *rare* to locally *uncommon* migrant and summer visitant in the northern Central Interior and Sub-Boreal Interior ecoprovinces, *uncommon* to locally *common* migrant and summer visitant in the Boreal Plains Ecoprovince; *fairly common* to locally *common* migrant and summer visitant to the Taiga Plains Ecoprovince; absent from the Northern Boreal Mountains.

Breeds.

Figure 333. In British Columbia, the Swamp Sparrow is an elusive local summer visitant in the Taiga Plains Ecoprovince (©M. Stubblefield/VIREO).

CHANGE IN STATUS: Munro and Cowan (1947) describe the Swamp Sparrow (Fig. 333) as a summer visitant to the Peace River Parklands and the Vanderhoof region. This status was based mainly on research in 1938 that found the Swamp Sparrow to be "locally numerous" in the Peace Lowland (Cowan 1939), and from a few pairs that seemed to be breeding near Vanderhoof (Munro 1949).

Since then, the status of this "eastern" species has changed markedly. The Swamp Sparrow is now known to be a common summer visitant well north of the Peace River in the Taiga Plains, from the Fort Nelson Lowland north to the border with the Northwest Territories. Prior to 1945, only 1 Swamp Sparrow had been documented west of the Rocky Mountains (Vanderhoof) in British Columbia, but in 1945 a few pairs were found near Nulki Lake (Munro 1949). Small numbers now occur regularly in summer in the Sub-Boreal Interior, near Prince George, and breeding is suspected. Its status in the Bulkley Basin, at the northwestern edge of the Central Interior (near Vanderhoof, Nulki Lake, and Topley) remains uncertain, as breeding has yet to be satisfactorily documented and we have no records of birds there since the 1950s.

One significant change in status has been the establishment of small wintering numbers on the southern coast of British Columbia. The first Swamp Sparrow appeared on the coast near Victoria in November 1969. This was followed by birds near Courtenay in August 1972 and near Victoria in December 1974 (McNicholl 1978). The first Swamp Sparrow in the Vancouver area appeared in December 1979 (Weber et

Figure 334. Between 13 December 1989 and 31 January 1990, a Swamp Sparrow frequented a bird feeder daily near Nakusp, in the Southern Interior Mountains Ecoprovince, roosting each evening in overhanging vegetation bordering a small creek (near Nakusp, 29 January 1990; R. Wayne Campbell).

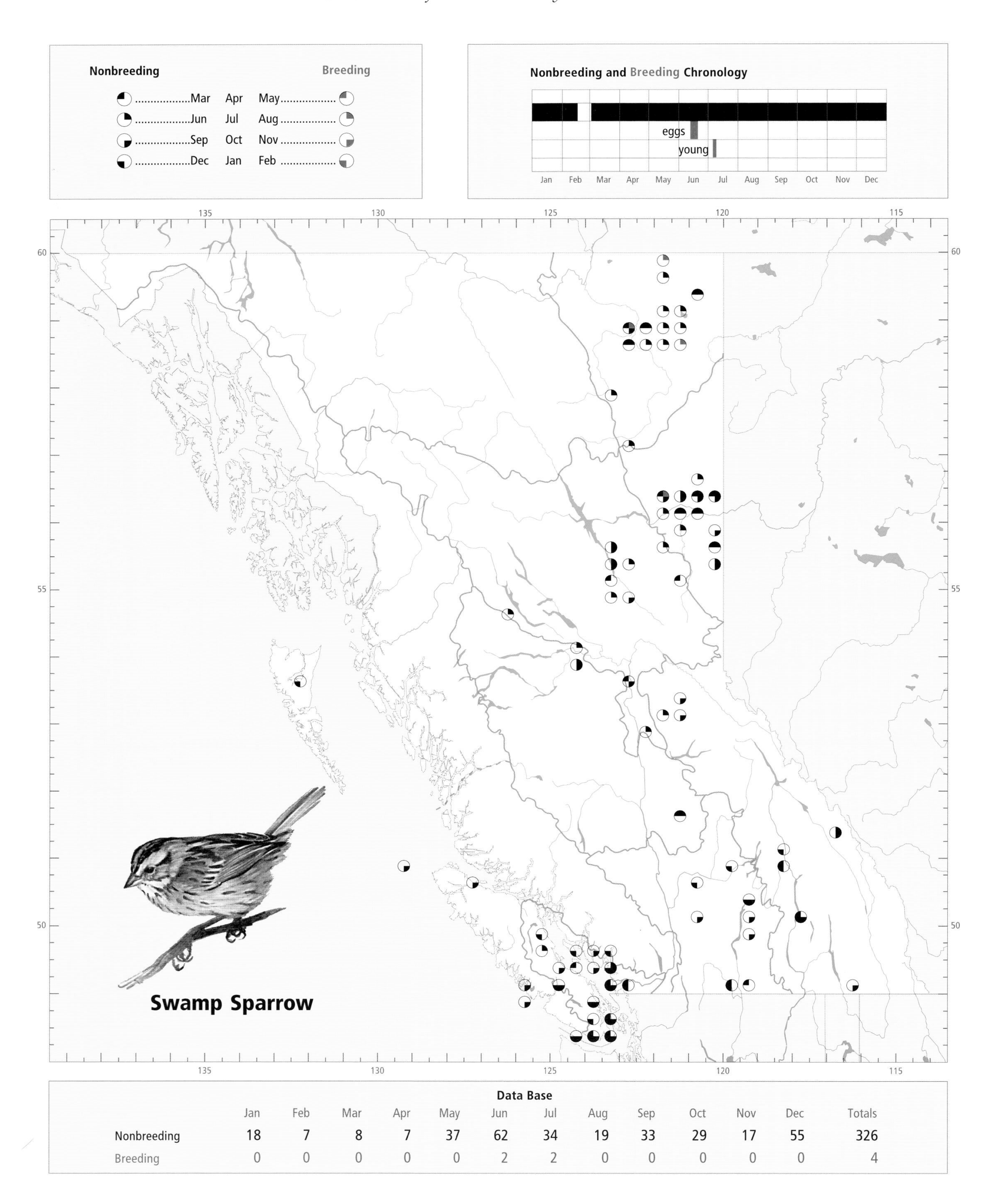

Data Base

	Jan	Feb	Mar	Apr	May	Jun	Jul	Aug	Sep	Oct	Nov	Dec	Totals
Nonbreeding	18	7	8	7	37	62	34	19	33	29	17	55	326
Breeding	0	0	0	0	0	2	2	0	0	0	0	0	4

al. 1980). Since the early 1980s, small numbers of Swamp Sparrows have occurred regularly during Christmas Bird Counts at Victoria and Vancouver. In addition, wintering birds have appeared sporadically on the west coast of Vancouver Island at Jordan River, Long Beach, and Port Alberni, on the east coast north to Campbell River, and along the Sunshine Coast on the mainland.

Small wintering populations have also become established during the last few decades along the Pacific coast from Washington south to southern California (Mowbray 1997). In Oregon, records have increased from 1 or more per year in the 1970s to several per year in the early 1980s, to several dozen per year in the late 1980s, and even more in the 1990s (Gilligan et al. 1994). It is not known whether this trend towards overwintering on the Pacific coast is a true winter range extension or the result of better coverage by observers, but we suspect that the former is more likely.

Another significant change in status is that small numbers have recently begun to overwinter in southern portions of the interior of British Columbia. The Swamp Sparrow was not known to occur in the Okanagan valley before 1987 (Cannings et al. 1987), but there have been at least 1 spring, 3 autumn, and 2 winter occurrences since then. Wintering birds have also been found in recent years at Nakusp, Revelstoke, Kamloops, and Adams River.

NONBREEDING: The Swamp Sparrow occurs regularly throughout most of the northeastern portions of the province, including most of the Taiga Plains and southern parts of the Boreal Plains. Most records are from the Fort Nelson Lowland and the Peace Lowland. The species occurs locally in the Sub-Boreal Interior, in the vicinity of Prince George, and west to Vanderhoof; it also occurs in the northern Southern Interior Mountains near Bowron Lake. Elsewhere in the southern and central interior, it occurs occasionally during migration or winter.

On the coast, the Swamp Sparrow occurs regularly and locally from autumn through spring on Vancouver Island, in the Fraser Lowland, and north to Powell River; it occurs occasionally north to the Queen Charlotte Islands.

During nonbreeding seasons, the Swamp Sparrow uses habitat similar to that used during the breeding season (see BREEDING). On the coast, it winters only near sea level in wet, shrubby areas such as in small brush-rimmed ponds, wet grassy fields edged with Himalayan blackberry and other shrubs, seasonally flooded fields, roadside ditches, and brackish marshes at the mouths of major rivers.

In the interior, it winters only at the lowest elevations in major valleys, usually in brushy or marshy edges of wetlands or creeks (Fig. 334). In the Peace Lowland, migrants also use smaller, damp areas such as tall grass near water, weedy spots along the edges of wetlands, and damp brushy edges of fields. In the Okanagan valley, most autumn migrants use stands of cattails in shallow wetlands.

During migration, any brushy habitat near water can probably be used. For example, although most migrants in the Peace Lowland use willow thickets near water, 1 migrant Swamp Sparrow was found at a puddle on a trail 50 m into an alder and trembling aspen stand. One early autumn arrival on the coast was found in a large revegetated

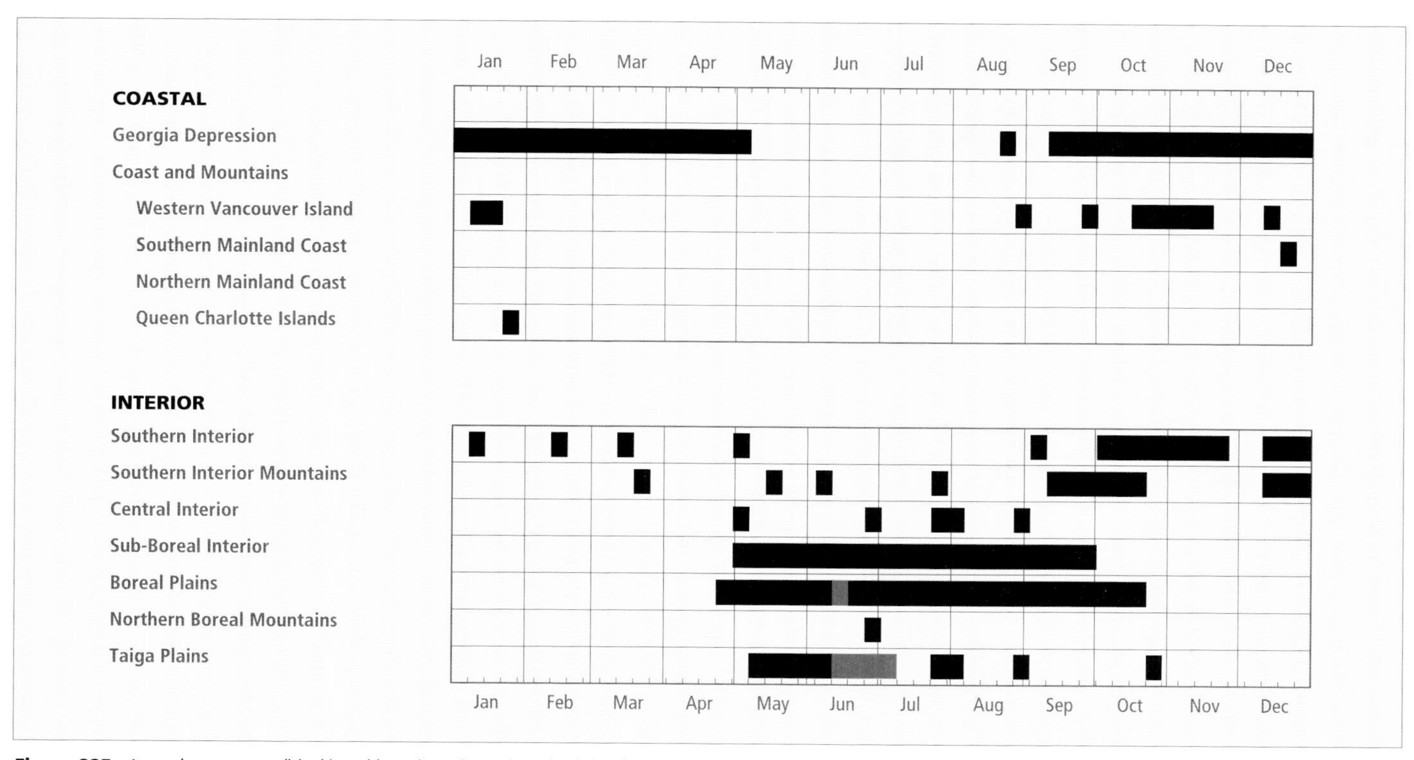

Figure 335. Annual occurrence (black) and breeding chronology (red) for the Swamp Sparrow in ecoprovinces of British Columbia. Records are shown for the week in which they occurred.

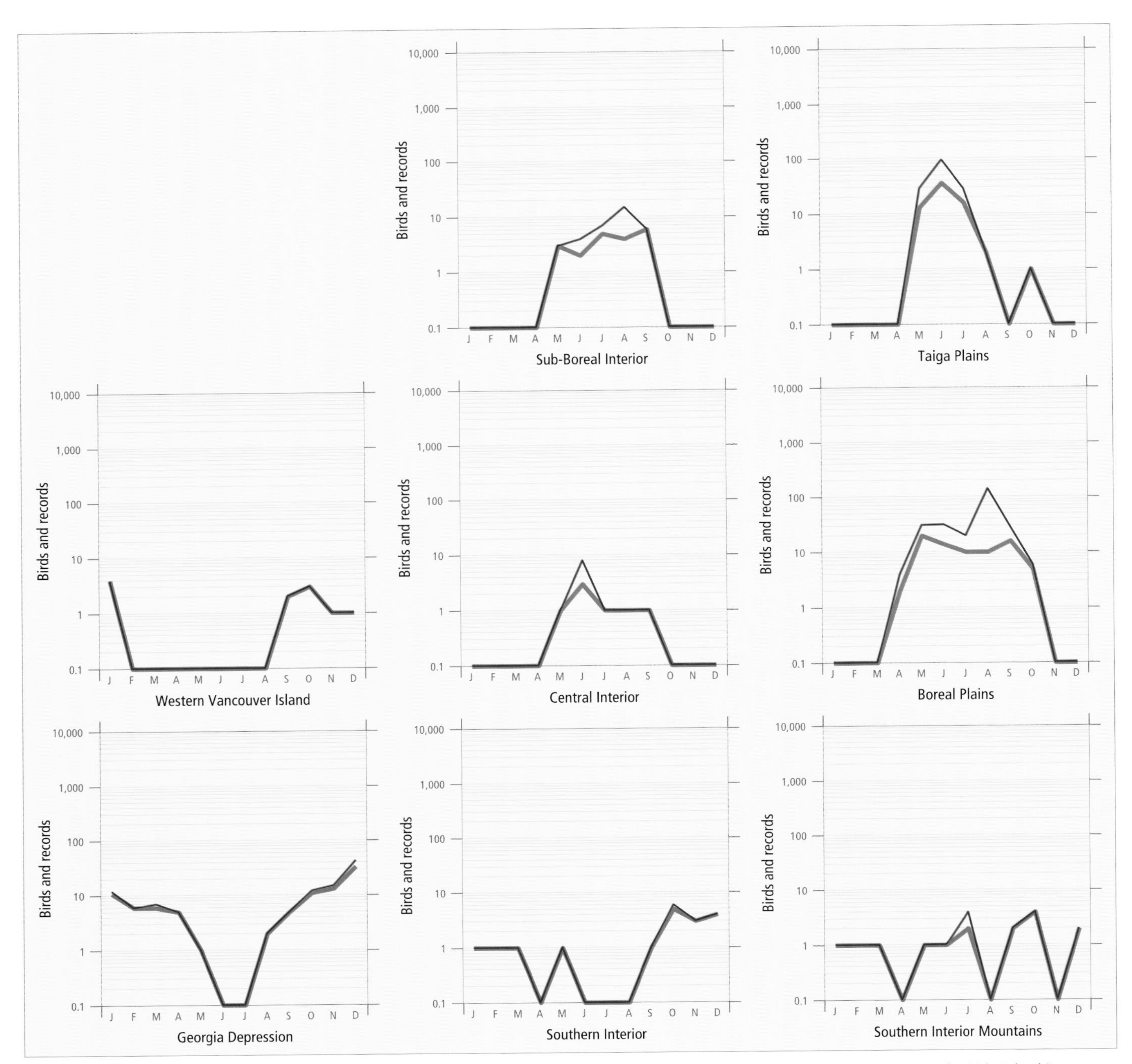

Figure 336. Fluctuations in total number of birds (purple line) and total number of records (green line) for the Swamp Sparrow in ecoprovinces of British Columbia. Christmas Bird Counts, Breeding Bird Surveys, and nest record data have been excluded.

burn near Comox (McNicholl 1978). During the autumn migration, the Swamp Sparrow may pass through habitats at higher elevations than its normal nesting habitat; for example, a dozen birds were found at Pine Pass late one August.

Spring migrants arrive in the Boreal Plains mainly in the last week of April and first week of May (Fig. 335). In some years, early migrants arrive in the last week of April; in other years, the first arrivals can appear as late as mid-May. For example, Cowan (1939) documented the sudden arrival of Swamp Sparrows at Swan Lake during the night of 14 May, when he found his study site full of singing males on the morning of 15 May. In the Taiga Plains, the main movement occurs in mid-May. In the Prince George area, early spring migrants arrive in the first week of May, but the peak movement is in the third week of May. On the south coast, wintering birds appear to be gone by early May.

The autumn movement is more protracted than the spring movement (Fig. 336). In the northeastern interior, the autumn migration usually begins in mid-August but peaks in September, which is earlier than the early October peak autumn movement in Alberta (Semenchuk 1992). Near Fort St. John, autumn migrants begin to appear in habitats not used for

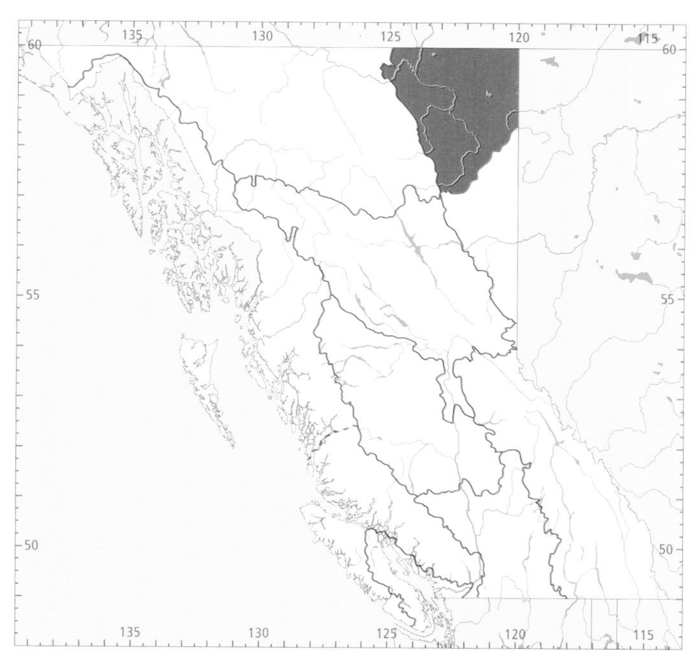

Figure 337. In British Columbia, the highest numbers for the Swamp Sparrow in summer occur in the Taiga Plains Ecoprovince.

nesting as early as the first week of August. The latest departures in the Peace Lowland ranged from 21 September to 16 October between 1982 and 1986. During mild autumns, some birds can remain until late October in the Fort Nelson Lowland and Peace Lowland. In the Prince George area, most birds are gone by early October.

Overwintering Swamp Sparrows begin to arrive on the south coast in mid-September, and the movement probably lasts through October. Most of our coastal winter records are from December, but Swamp Sparrows likely remain at their wintering sites from November through at least March. A few Swamp Sparrows overwinter in southern parts of the interior from Revelstoke south through Nakusp and the Okanagan valley, and west through the Shuswap area to Kamloops. Autumn migrants have appeared in the north Okanagan valley as early as 5 September.

On the coast, the Swamp Sparrow occurs irregularly from late August to 4 May; in the southern portions of the interior, it occurs mainly from October to the end of the year. Further north, it has been reported from 24 April to 26 October (Fig. 335).

BREEDING: The Swamp Sparrow reaches the limit of its western breeding range in northern British Columbia. Breeding has been documented only east of the Rocky Mountains, in the Taiga Plains and Boreal Plains, from the Alberta border west to at least Chetwynd and north to the Northwest Territories border. The Swamp Sparrow probably breeds locally in the Sub-Boreal Interior (see REMARKS), but confirmation is still required there.

In the Taiga Plains, most summer records are from the Fort Nelson Lowland, but the Swamp Sparrow is likely a common breeder in the Etsho Plateau and Petitot Plain. We have no records from the Rocky Mountain foothills in the Muskwa Plateau, but suspect that the Swamp Sparrow occurs locally in suitable habitat there.

In the Boreal Plains, breeding has been confirmed only at Boundary Lake, near Chetwynd, near the Halfway River, and south of Dawson Creek in the Kiskatinaw Plateau. Based on numerous summer records, however, the Swamp Sparrow is apparently a locally fairly common breeder elsewhere in the Peace Lowland (e.g., Swan Lake [Cowan 1939]), the north end of Charlie Lake, and near Fort St. John, and locally in the Halfway Plateau and Clear Hills where suitable habitat exists. There are few summer records from the higher-elevation plateaus north of Fort St. John, and breeding remains to be confirmed there.

In the Sub-Boreal Interior, breeding is suspected to occur in several marshes near Eaglet Lake, Punchaw Lake, Chief Lake, and Vama Vama Creek. On the northern edge of the Central Interior, Munro (1949) reported 3 pairs apparently nesting near Nulki Lake, south of Vanderhoof, based on their behaviour and the presence of a brood patch on the female he collected. Although he did not find active nests or fledged young, this record suggests that breeding probably occurs there.

The highest numbers in summer occur in the Taiga Plains (Fig. 337) in British Columbia. Breeding Bird Surveys for interior routes for the period 1968 through 1993 contain insufficient

Figure 338. In the Taiga Plains Ecoprovince, the Swamp Sparrow breeds in dense beds of sedges along the edges of lakes (Parker Lake, 26 June 1996; R. Wayne Campbell).

data for analysis. An analysis of Breeding Bird Surveys across Canada from 1966 to 1996 suggests that populations are stable (Sauer et al. 1997).

The Swamp Sparrow has been found breeding at elevations between 400 and 900 m in British Columbia. Breeding populations occur generally in lowlands, valley bottoms, or lower-elevation plateaus and are probably absent from elevations above 1,100 m. The Swamp Sparrow breeds along the edges of ponds, lakes, marshes, slow-moving streams, and bogs. Wetlands used for nesting range from secluded woodland ponds to brush-lined dugouts in open agricultural areas. Shallow standing water and dense cover are required for nesting habitat; where these attributes are absent, so is the Swamp Sparrow. Typical nesting habitats are dense beds of sedges (Fig. 338), reedgrasses, or cattails, or where emergent vegetation meets a fringe of willows, Labrador tea, and other shrubs along the edges of wetlands (Fig. 339). Flooded willow swamps and beaver ponds are often used in British Columbia. The Swamp Sparrow forages in the same wet habitats in which it nests, often taking emerging dragonflies and damselflies as they crawl onto plant stems, searching under submerged leaves or probing in mud (Mowbray 1997).

Breeding areas appear to be quite local. In the Peace Lowland, Cowan (1939) estimated that 50 pairs of Swamp Sparrows nested in a localized 2 ha area of alder and willow marshland surrounding a pond at Swan Lake; surprisingly, none was found elsewhere at Swan Lake even where apparently suitable habitat occurred. Cowan (1939) also found a few pairs in a semi-open willow swamp near Tupper, but did not find additional breeding sites in the Boreal Plains during 1938 despite careful searches of apparently suitable habitat. During the 1980s, breeding was suspected to occur at several localities, including Boundary Lake, Lost Lake, German Lake, Cecil Lake, Charlie Lake, Fort St. John, Cache Creek, and along the Upper Cache Road between Hudson's Hope and Fort St. John (C. Siddle, pers. comm.).

In British Columbia, the Swamp Sparrow breeds from 13 June (calculated) through 9 July (calculated) (Fig. 335).

Nests: One nest has been described in British Columbia. It was built 10 cm above the ground, in the branches of a small live willow where dense grass grew (Campbell and McNall 1982). Elsewhere, the Swamp Sparrow builds its nest near the ground, occasionally on the ground, among dense vegetation such as grasses, cattails, or small shrubs. Nests are often placed over water and are generally well concealed by vegetation. They are bulky cups of coarse grasses, sedges, cattails, twigs, and rootlets, and are lined with finer grasses, plant down, and rootlets (Mowbray 1997).

Eggs: Two nests with 4 eggs each have been recorded in British Columbia, 1 on 16 June and the other, with well-incubated eggs, on 20 June (Campbell and McNall 1982). Eggs can probably occur in British Columbia from early June through mid-July. Elsewhere, clutch size ranges from 3 to 5 eggs, but 4 eggs is most common. The incubation period is 12 to 14 days (Ellis 1980).

Young: Dates for 2 recently fledged broods, 1 with 2 fledglings and the other with 4, ranged from 6 July to 9 July; nests with young have not been observed in British Columbia. The nestling period is 9 to 11 days (Mowbray 1997). The young leave the nest 2 or 3 days before they can fly well.

Figure 339. Dense shrub growth along slow-moving streams and water channels is the preferred nesting habitat for the Swamp Sparrow in the Boreal Plains Ecoprovince (Swan Lake, Tupper, 14 June 1996; R. Wayne Campbell).

Brown-headed Cowbird Parasitism: Nests with cowbird eggs or young have not been found in British Columbia. Of 11 records of fledged young that were being cared for by adults, however, 1 was of an adult Swamp Sparrow feeding a fledged cowbird. The Swamp Sparrow is generally thought to be an uncommon host for cowbirds, but it can be heavily parasitized in some local areas (Friedmann 1963; Friedmann et al. 1977).

Nest Success: Insufficient data.

REMARKS: Three subspecies of Swamp Sparrow are recognized but only *M. g. ericrypta* breeds and winters in British Columbia (Mowbray 1997).

A fledged young was thought to have been observed east of Prince George. An acceptable field description is not available, however, and the Prince George Naturalists Club (1996) reports the Swamp Sparrow to be only a probable breeder in central British Columbia. A record of 2 Swamp Sparrows at Columbia Lake during a Breeding Bird Survey on 26 June 1973 lacks sufficient details and has been excluded from this account.

Despite the steady destruction of small wetlands throughout much of its range, the Swamp Sparrow seems to have maintained its numbers (Sauer et al. 1997), or even increased its local range as in British Columbia. Conservation efforts that protect or enhance wetlands, especially the larger ones, are probably the best method of managing habitat for the Swamp Sparrow (Mowbray 1997).

In British Columbia, grazing of shoreline vegetation by cattle is probably the most significant cause of habitat degradation. Local concentrations of Swamp Sparrows should be identified and monitored periodically so that effective management decisions can be made. Naturalists could also help increase our knowledge of breeding distribution by looking for evidence of nesting, including recently fledged young, when birding in wetlands used by Swamp Sparrows.

NOTEWORTHY RECORDS

Spring: Coastal – Langford 4 Mar 1984-2 (Fix 1984); Saanich 7 Apr 1988-1; Serpentine River 29 Apr 1988-1 (Mattocks 1988a); Reifel Island 5 Mar 1984 (Fix 1984), 4 May 1984-1; Delta 1 Apr 1990-1; Little Qualicum River 9 Apr 1994-1. **Interior** – Haynes Point Park 16 Mar 1987-1 (Cannings et al. 1987); Oliver 1 May 1987-1 banded; Nakusp 24 Mar 1990-1 (Siddle 1990b); Watson Lake (100 Mile House) 5 May 1958-1 (Erskine and Stein 1964); Prince George 3 May 1981-1; Gataiga Creek 20 May 1996-1; Bullmoose Creek 11 May 1998-1; Tupper 15 May 1938-1 (Cowan 1939); 30 km w Fort St. John 4 May 1985-1; North Pine 24 Apr 1988-1; Boundary Lake (Goodlow) 27 Apr 1986-3 males singing, 6 May 1984-5; Jackfish Creek (Fort Nelson) 25 May 1975-15; Clarke Lake Rd 19 May 1980-1; Parker Lake (Fort Nelson) 13 May 1975-1, 19 May 1975-3; Petitot River 31 May 1982-3.

Summer: Coastal – Vancouver 26 Aug 1981-1; nr Comox 24 Aug 1972-1 on burn (McNicholl 1978). **Interior** – lower Blaeberry River valley 8 Jun 1997-1; Watson Lake (100 Mile House) 25 Jul 1959-1 (Erskine and Stein 1964); Dragon Lake (Quesnel) 11 Jul 1996-1; confluence Timon and Lightning creeks 26 Jul 1988-2 singing males; Nulki Lake 27 Jun to 4 Jul 1945-at least 3 pairs (Munro 1949), 30 Jun 1945-1 (ROM 83390); Vanderhoof 4 Aug 1919-1 (NMC 13727); Topley 24 Jun 1956-3; Chichouyenily Creek 1 Jun 1996-1; Pine Pass 23 Aug 1975-12; Chetwynd 24 Jul 1975-2; Alcock Lake 4 Jul 1993-1 adult feeding fledgling, 4 Aug 1975-1 adult with 3 fledglings; s Taylor 22 Aug 1975-85 during day-long survey; Halfway River 16 Jun 1990-4 eggs; Boundary Lake (Goodlow) 4 Jun 1983-2, 25 Jul 1982-8, 13 Aug 1986-8, 26 Aug 1985-24; Rose Prairie 18 Aug 1975-5; e Fort Nelson 29 Jun 1982-6 along Esso Resources Rd; Minaker River 28 Jun 1944-1 (Munro and Cowan 1947); Parker Lake (Fort Nelson) 28 Jun 1978-1 recently fledged young, 9 Jul 1978-4 recently fledged young "able to fly somewhat"; Helmut 6 Jun 1982-12 (very numerous), 9 Jun 1982-20; Kotcho Lake 26 Jun 1982-1 (Campbell and McNall 1982); Petitot River 20 Jun 1982-4 eggs, well incubated.

Breeding Bird Surveys: Coastal – Not recorded. **Interior** – Recorded from 8 of 73 routes and on 2% of all surveys. Maxima: Fort Nelson 19 Jun 1976-4; Tupper 20 Jun 1994-3; Punchaw 27 Jun 1991-2.

Autumn: Interior – Fort Nelson 26 Oct 1986-1; ne Fort St. John 21 Sep 1985-1, 3 Oct 1986-1, 14 Oct 1984-1; St. John Creek 12 Sep 1986-6, 4 Oct 1986-1; ne Cecil Lake 16 Oct 1983-1; sw Cecil Lake 3 Sep 1982-3; Charlie Lake 1 Sep 1982-1; s Mugaha Creek 23 Sep 1995-1 banded; Nulki Lake 2 Sep 1945-1 (ROM 83391); 25 km s Prince George 7 Sep 1975-1; Indianpoint Lake 9 Oct 1928-1 (MCZ 285584); lower Blaeberry River valley 23 Sep 1998-1; Swan Lake (Vernon) 5 Sep 1992-1 (Siddle 1993a), 29 Oct 1987-2 (Rogers 1988), 5 Nov 1995-1; Nakusp 3 Oct 1993-1 (Siddle 1994a), 18 Oct 1992-1 (Siddle 1993a); Nicola Lake 19 Nov 1995-1; Kelowna 14 Oct 1991-1 (Siddle 1992a), 15 Oct 1992-1; Yahk 12 Sep 1929-1 (NMC 23378). **Coastal** – Port McNeil 18 Nov 1985-1 (Force and Mattocks 1986); Angus Creek (Sechelt) 5 Oct 1991-1 immature; Port Alberni 11 Nov 1987-1; Long Beach 29 Sep 1983-1; Ucluelet 28 Oct 1994-1; Duncan 16 Sep 1991-1 (Siddle 1992a); Jordan River 29 Oct 1988-1; Rocky Point (Victoria) 25 Sep 1995-1.

Winter: Interior – Swan Lake (Vernon) 20 Dec 1987-1 (Rogers 1988a), 15 Dec 1991-1; Nakusp 13 Dec 1989 through 31 Jan 1990-1 daily at feeder (Siddle 1990a); Tranquille 1 to 10 Jan 1995-1; Adams River 16 Dec 1995-1; Osoyoos Lake 15 Feb 1987-1 (Cannings et al. 1987). **Coastal** – Juskatla 24 Jan 1994-1 (Siddle 1994b); Pender Harbour 26 Dec 1993-2; Langley 2 Jan 1990-1 (Siddle 1990a); Pitt Meadows 6 Dec 1983-1; Reifel Island 27 Dec 1979-1, first Vancouver area record (Weber et al. 1980); Duncan 1 Feb 1991-1 (Siddle 1991b); Ladner 23 Dec 1991-3; Somenos Lake 13 Feb 1990-1; Saanich 22 to 30 Dec 1974-1 (McNicholl 1978); Elk Lake (Victoria) 30 Nov 1969-1 (McNicholl 1978); Goldstream River 7 Dec 1983-2 (Mattocks 1984); Victoria 18 to 21 Dec 1986-1, 1 Jan 1987-2; Jordan River 11 Dec 1983-1, 14 Jan 1989-1.

Christmas Bird Counts: Interior – Recorded from 3 of 27 localities and on 2% of all counts. Maxima: Revelstoke 21 Dec 1991-1; Vernon 15 Dec 1991-1; Nakusp 31 Dec 1989-1; Shuswap Lake Provincial Park Dec 1995-1. **Coastal** – Recorded from 9 of 33 localities and on 5% of all counts. Maxima: Duncan 27 Dec 1993-11; Victoria 18 Dec 1993-6; Port Alberni 2 Jan 1994-6.

White-throated Sparrow

WTSP

Zonotrichia albicollis (Gmelin)

RANGE: Breeds from southeastern Yukon and southwestern Mackenzie east across the forested areas of Canada to Labrador and Newfoundland; south through northeastern and central British Columbia, central Alberta, and central Saskatchewan across to northeastern Minnesota and northern New Jersey. Winters in 2 disjunct areas; in the west, from southwestern British Columbia south along the coastal slope of Washington and Oregon to northern Baja California. The main wintering area occurs in southeastern North America, east of the Great Plains from Michigan through to southern Nova Scotia, south to central Florida, and west through to Texas and southeastern Arizona.

STATUS: On the coast, an *uncommon* migrant and winter visitant to the Georgia Depression Ecoprovince; in the Coast and Mountains Ecoprovince, a *very rare* migrant and winter visitant on Western Vancouver Island and the Southern Mainland Coast, *very rare* on the Northern Mainland Coast, and absent from the Queen Charlotte Islands.

In the interior, a *common* migrant and summer visitant east of the Rocky Mountains in the Taiga Plains and Boreal Plains ecoprovinces, except locally *very common* in the Peace Lowland of the Boreal Plains; *uncommon* to locally *fairly common* migrant and summer visitant in the southern portions of the Sub-Boreal Interior Ecoprovince and northern portions of the Central Interior and Southern Interior Mountains ecoprovinces; *casual* in the northern portion of the Sub-Boreal Interior and the Northern Boreal Mountains ecoprovinces, except locally *uncommon* migrant and summer visitant in the Liard Plain; *rare* migrant in the Southern Interior Ecoprovince and southern portions of the Central Interior and Southern Interior Mountains. In winter, a *rare* visitant to the Southern Interior Mountains and Southern Interior, except locally *uncommon* in the Northern Okanagan Basin; *casual* in the Central Interior and Sub-Boreal Interior; *accidental* in the Boreal Plains.

Breeds.

CHANGE IN STATUS: The White-throated Sparrow (Fig. 340) is generally considered to be a boreal forest species found east of the Rocky Mountains. In British Columbia, however, it was first known from a specimen collected at Saanichton, on Vancouver Island, in 1913. Shortly afterwards, other specimens were collected in the interior, at Grindrod and Vanderhoof in 1919 and in the Kispiox River valley, along the north coast, in 1921. By the mid-1940s, a small breeding population had been discovered in the Vanderhoof–Nulki Lake area (Munro and Cowan 1947). At that time, this sparrow was considered to be a summer visitant only in the Peace River and Vanderhoof areas and casual elsewhere in the province. Munro and Cowan (1947) had overlooked a 1919 winter record from Grindrod and consequently did not record this sparrow as a wintering species. Since the late 1940s, the White-throated Sparrow has become established as an uncommon but regular wintering species on the south coast, and has expanded its breeding distribution in the north-central interior.

(a)

(b)

Figure 340. There are 2 distinct plumage types for the White-throated Sparrow, a white-striped (a) and a tan-striped (b) morph that can be distinguished from each other primarily by the brightness of the crown (illustrations by Mark Nyhof). Both morphs occur in British Columbia.

The second winter record of a White-throated Sparrow in the province came from Vernon, during a Christmas Bird Count on 26 December 1955. From 1960 to 1963, 6 more winter records were documented, 2 from southern Vancouver Island at Victoria and 4 from the north Okanagan at Lavington, Vernon, and Kelowna. Since the mid-1960s the White-throated Sparrow has become a regular and increasingly numerous wintering species in the Georgia Depression (Fig. 341), notably on southeastern Vancouver Island. Although less common in valleys in the southern portions of the interior, this sparrow has become a more frequent winter visitor since about 1985. Wintering birds are now occasionally recorded as far north as Fort St. John in the Boreal Plains and Terrace on the Northern Mainland Coast, but are dependent on bird feeders for survival at those latitudes.

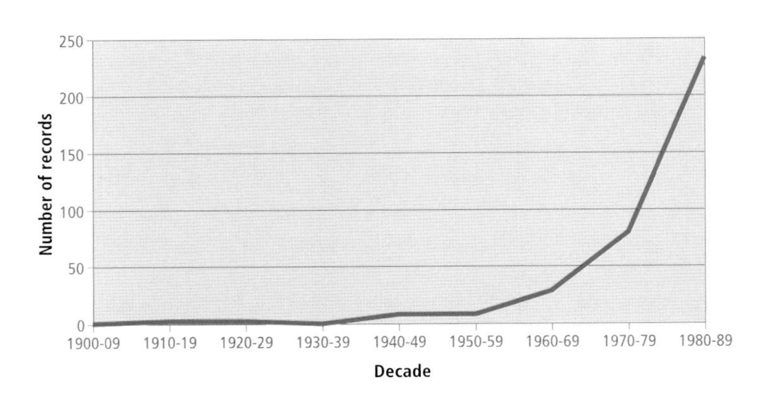

Figure 341. Number of White-throated Sparrow records per decade in the Georgia Depression Ecoprovince, from 1900 to 1989.

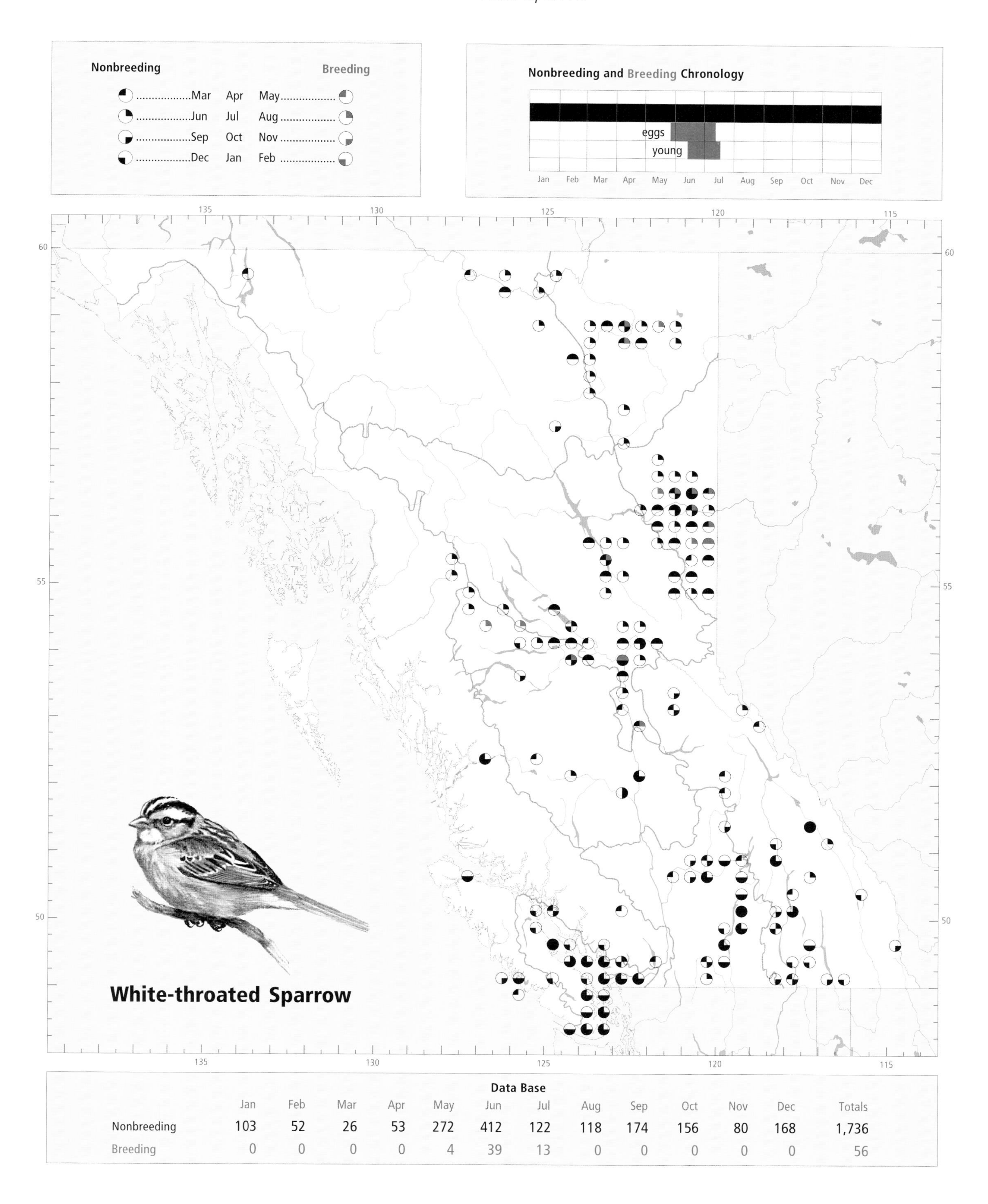

Data Base

	Jan	Feb	Mar	Apr	May	Jun	Jul	Aug	Sep	Oct	Nov	Dec	Totals
Nonbreeding	103	52	26	53	272	412	122	118	174	156	80	168	1,736
Breeding	0	0	0	0	4	39	13	0	0	0	0	0	56

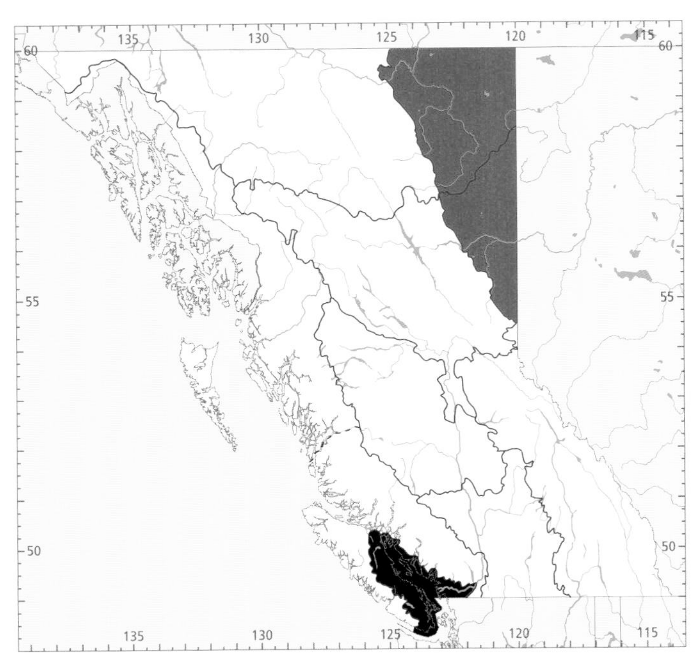

Figure 342. In British Columbia, the highest numbers for the White-throated Sparrow in winter (black) occur in the Nanaimo Lowland and Fraser Lowland of the Georgia Depression Ecoprovince; the highest numbers in summer (red) occur in the Boreal Plains and Taiga Plains ecoprovinces.

Since the mid-1940s, the known breeding distribution of the White-throated Sparrow has increased in northeastern British Columbia to include the Fort Nelson and Liard River areas. It is likely that the sparrow was established in this region long before Rand (1944) and others began to explore the area following the construction of the Alaska Highway. West of the Rocky Mountains, in the Sub-Boreal Interior and Central Interior, the White-throated Sparrow has undergone a significant range expansion that appears unrelated to the increased access and survey effort seen in northeastern British Columbia. For example, the field notes of James A. Munro, an exceptionally observant naturalist and meticulous notetaker, specifically state that in 1944 the White-throated Sparrow was absent from the region about Francois Lake, Bulkley Lake, and places further to the northwest, an area where this sparrow is now generally well established. It also seems unlikely that the numerous museum collecting expeditions conducted in this region between 1919 and 1956 could have missed this conspicuous songster. Since then, the White-throated Sparrow has expanded its range west of Vanderhoof to the Telkwa River and north towards Stuart Lake. From Prince George it is now found as far east as the lower McGregor River valley, south along the Fraser River valley to Quesnel and north to Parsnip Reach on Williston Lake. More recently, this sparrow has become established in the Robson River area of Mount Robson Park.

NONBREEDING: The White-throated Sparrow has a widespread distribution east of the Rocky Mountains throughout the Boreal White and Black Spruce Biogeoclimatic Zone of the Taiga Plains and Boreal Plains and in the Liard Plain of the Northern Boreal Mountains. Elsewhere in the north-central portions of the interior, it is locally distributed in low-elevation deciduous and mixed forests of the Sub-Boreal Spruce Zone from the Telkwa River valley east to the lower McGregor River valley and from the southern arm of Williston Lake south to Quesnel. On the coast, it occurs regularly in the Georgia Depression during the winter and migration periods, but is observed infrequently elsewhere. It has not been recorded on the Queen Charlotte Islands.

The White-throated Sparrow reaches its highest numbers in winter in the Nanaimo Lowland and Fraser Lowland of the Georgia Depression (Fig. 342).

The White-throated Sparrow has been reported from near sea level to 640 m elevation on the coast and from 330 to 2,195 m in the interior. This sparrow rarely occurs above 1,200 m elevation, but there are 2 records from the interior of autumn migrants in subalpine habitats. During spring and autumn migration, this sparrow is most common in deciduous forests with dense shrub undergrowth, but it is also found in mixed deciduous-coniferous forests and in young successional forests such as those found in burns and clearcuts. It especially favours forest edge habitats and natural forest openings as well as the shrubby perimeters of lakes, beaver ponds, wetlands, muskegs, and watercourses. In all cases, it demonstrates a strong affinity for low, dense shrubby vegetation, from which it seldom ventures.

On the coast, the White-throated Sparrow is most frequently reported from backyard feeding stations (Fig. 343), but it is also found in a variety of other human-influenced habitats, including the brushy edges of fields, utility corridors, footpaths, roads, and parking lots, and in low, dense vegetation found in city parks and golf courses. During the winter months in the southern portions of the interior, this sparrow favours backyard gardens with bird feeders, but it is also found in shrub thickets along lakeshores and creeks and the edges of weedy fields and pastures.

On the coast, spring migration begins in the Georgia Depression with a slight increase in numbers during the second

Figure 343. On the south coast, the White-throated Sparrow is most often reported in winter from backyard feeding stations (Victoria, October 1994; Ian McTaggart-Cowan).

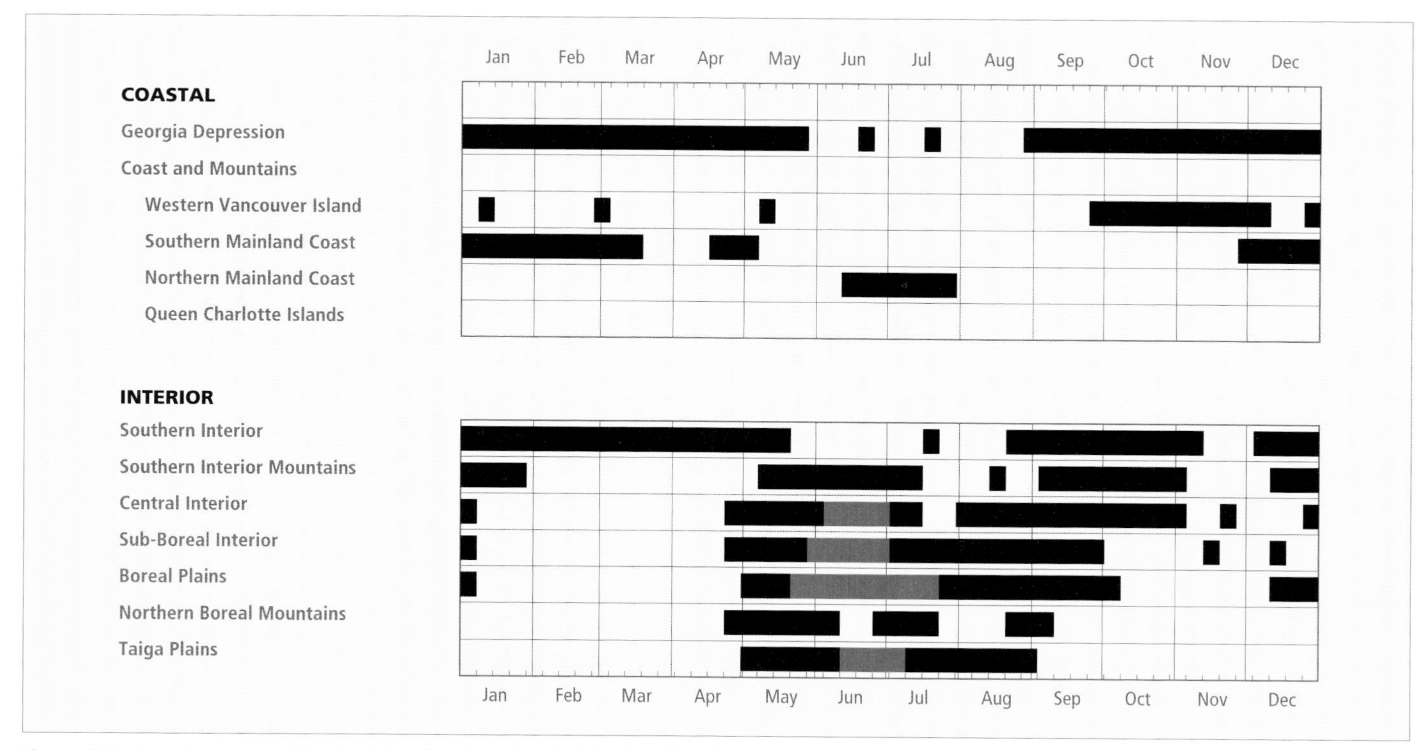

Figure 344. Annual occurrence (black) and breeding chronology (red) for the White-throated Sparrow in ecoprovinces of British Columbia. Records are shown for the week in which they occurred.

week of April, peaking in the last week of the month but dropping sharply after the first week of May (Fig. 345). Only a few stragglers remain after this period. For the rest of the coast, spring migration has not been recorded except at the extreme southern end of the Southern Mainland Coast, where a few birds have been noted during the last week of April and first week of May.

In the interior, a small movement is recorded throughout May in the Southern Interior and Southern Interior Mountains. In the Central Interior and Sub-Boreal Interior, a few early migrants may appear in mid-May, followed by an abrupt peak in numbers later in the month and rapid dispersal onto the breeding grounds.

East of the Rocky Mountains, the White-throated Sparrow moves into the province from Alberta, where the spring migration begins in late April and early May (Semenchuk 1992). In the Boreal Plains and Taiga Plains, the first spring migrants arrive in small numbers during the first week of May. Numbers peak during the third week of May, and birds have dispersed onto breeding grounds by the first week of June (Figs. 344 and 345). In the Liard Plain of the Northern Boreal Mountains, the arrival of spring migrants is slightly later, with birds first reported in the second week of May and peaking in number about the first week of June.

Falls and Kopachena (1994) report that autumn movements of the White-throated Sparrow are more gradual and involve longer stopover periods than in the spring migration. In the northeastern portions of the interior, there is little in the way of a noticeable autumn movement except in the Boreal Plains, where numbers begin to build during the third and fourth weeks of August and peak during early September; most birds have left by the last week of September (Figs. 344 and 345). Sparrows migrating from this region likely move southeast through Alberta, where autumn migrants pass through during the last 2 weeks of September (Salt and Salt 1966). In the Sub-Boreal Interior, a small autumn movement is noticeable from the last week of August to the second week of September. In the Vanderhoof area of the Central Interior, Munro (1946) noted a migratory movement of these sparrows from 24 August to 2 September. In southern portions of the interior, small numbers begin moving through in the first week of September and most have passed through by the second week of October. In the Okanagan valley, small autumn movements have been recorded between 2 September and 4 November (Cannings et al. 1987).

On the coast, the White-throated Sparrow has been recorded in autumn only once north of Vancouver Island. On Western Vancouver Island, it is an infrequent migrant from the last week of September to about mid-October. In the Georgia Depression, the first autumn migrants arrive during early September and peak movements occur during the last week of September and first week of October. Most birds pass through this ecoprovince but a few stay to overwinter. Banding stations at Rocky Point on southern Vancouver Island and at Sea Island on the Fraser River estuary have detected small autumn movements from 22 September to 20 October (R. Millikin pers. comm.). Sparrows remaining on the coast later than October are probably wintering birds.

On the south coast, the White-throated Sparrow has been recorded regularly only in the Georgia Depression from late

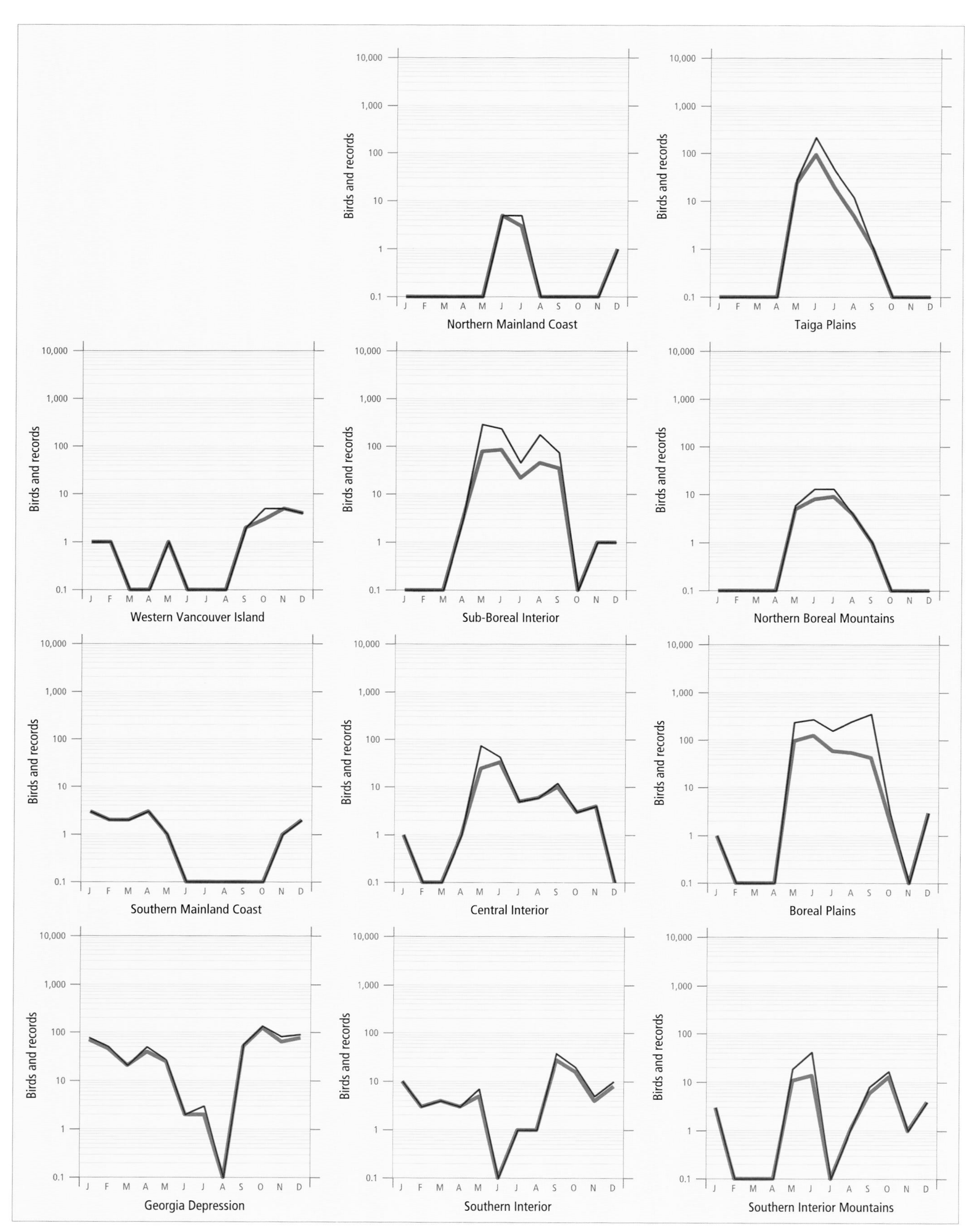

Figure 345. Fluctuations in total number of birds (purple line) and total number of records (green line) for the White-throated Sparrow in ecoprovinces of British Columbia. Christmas Bird Counts, Breeding Bird Surveys, and nest record data have been excluded.

September through early May, and on the north coast during the summer; in the interior, it occurs regularly in southern portions from early September to the third week of May, and in northern portions from mid-May until about the third week of September (Fig. 344).

BREEDING: The White-throated Sparrow breeds in suitable habitats throughout the boreal forest of northeastern British Columbia, from the eastern slope of the Rocky Mountains to the Alberta border, north to the Yukon–Northwest Territories border. It also likely breeds in the Liard Plain of the Northern Boreal Mountains. In the north-central interior, it breeds at low elevations from Mackenzie south to Quesnel and from Miworth, near Prince George, west to Houston. There are no breeding records from the coast, although it likely breeds locally in the Hazelton-Kispiox area, where it was first suggested as a breeding species in 1921 (Swarth 1924).

The White-throated Sparrow reaches its highest numbers in summer in the Boreal Plains and Taiga Plains (Fig. 342). Breeding Bird Surveys for British Columbia for the period 1968 through 1993 contain insufficient data for analysis. An analysis of continent-wide Breeding Bird Surveys for the period 1966 through 1996 shows an average annual decline of 1.1% ($P < 0.01$) (Sauer et al. 1997).

The White-throated Sparrow has been recorded nesting in British Columbia at elevations between 400 and 1,095 m. In the Boreal White and Black Spruce Zone of the northeast, this sparrow uses a variety of shrub-dominated habitats, including mature deciduous, mixed-wood, and floodplain forests with dense shrub undergrowth, recent burns, clearcuts, and forest edge habitats around beaver ponds, wetlands, and bogs. Trembling aspen stands with a dense shrub undergrowth of prickly rose, soopolallie, and highbush-cranberry, seem to be preferred breeding habitat. Elsewhere in Canada's boreal region, the White-throated Sparrow is most common in coniferous forests, mixed forests, bogs, burns, clearcuts, and forest edges (Fig. 346), but uncommon in deciduous broadleaf forests (Erskine 1977).

Bird surveys conducted in predominantly coniferous and mixed forests along the Liard River found the White-throated Sparrow to be most common in mature riparian forests (Bennett and Enns 1996). Deciduous riparian forests had the highest encounter rate – 1.3 sparrows per 100 m of transect – followed by coniferous riparian forest, which had 0.8 birds per 100 m. The White-throated Sparrow was the seventh most abundant of 46 passerine species detected during this inventory.

Songbird surveys conducted from 1992 to 1996 in mature trembling aspen forests south of the Peace River consistently found the White-throated Sparrow, along with the Least Flycatcher, to be the first or second most abundant breeding bird (Jeffrey and Darling 1997). Following logging, it was the third most abundant songbird after the Lincoln's Sparrow and the Clay-colored Sparrow (Darling 1996; Merkins and Booth 1996). It remained at about the same abundance level in trembling aspen plots monitored before and up to 5 years following clearcut logging. In Nova Scotia, a similar study in mixed hardwood forest showed a large increase in White-throated Sparrow numbers following logging (Freedman et al. 1981).

West of the Rocky Mountains in the Sub-Boreal Interior and Central Interior, the White-throated Sparrow also demonstrates an affinity for deciduous forests within the Sub-Boreal Spruce Biogeoclimatic Zone. Most breeding activity in this area was closely associated with trembling aspen forests (Fig. 347). Road surveys using broadcast calls consistently detected this sparrow in trembling aspen stands along the Nechako valley from Fraser Lake to Vanderhoof, at the east end of Francois Lake, at Endako, at Nulki and Tachick lakes, and near Fort St. James (Campbell and Stewart 1994, 1995). Trembling aspen stands in this area typically have a dense shrub undergrowth of highbush-cranberry, black twinberry, prickly rose, and soopolallie. Although similar habitat occurs further west in the Bulkley River valley near Smithers, Pojar (1995) did not record this species in any of the trembling aspen stands she surveyed, suggesting that it has yet to occupy that area.

In the lower McGregor River valley, songbird surveys detected the White-throated Sparrow most commonly in clearcut areas with regenerating vegetation less than 20 years old (Phinney and Lance 1998).

In the Quesnel area, the White-throated Sparrow is found in mixed trembling aspen–white birch stands. Other habitats used in the north-central interior include shrubby wetlands and creek bottoms and black cottonwood floodplains with a thick undergrowth of red-osier dogwood. In 1946, Munro (1949) estimated that 35 breeding pairs occupied a 60 ha bottomland along the Sinkut River.

The White-throated Sparrow has been recorded breeding in British Columbia from 27 May (calculated) to 17 July (Fig. 344).

Nests: All nests ($n = 34$) were found on the ground except for 1 located in a river debris pile. Nests were usually well concealed beneath the shrub layer, in grass clumps, or under

Figure 346. In northeastern British Columbia, seismic lines from gas and oil exploration provide edge habitat where the White-throated Sparrow nests among piles of woody debris and in shrubby areas (One Island Lake, 11 June 1990; R. Wayne Campbell).

Figure 347. The preferred breeding habitat for the White-throated Sparrow in British Columbia includes stands of trembling aspen with a dense shrub growth of highbush-cranberry, black twinberry, prickly rose, and soopolallie (Fort Fraser, 23 May 1994; R. Wayne Campbell).

other low vegetation. Several were located at the base of a deciduous shrub or sapling and 1 was found beneath an upturned root. Most nests were found in, or near, natural forest openings or human-made clearings such as road and powerline corridors, seismic lines (Fig. 346), clearcuts, footpaths, and fields. Several nests were reported among or close to deadfalls and other natural accumulations of woody debris or near human-created brush piles.

Nests were usually constructed of coarse grasses, plant stems, rootlets, and bark strips. Less commonly reported materials included mosses, leaves, and fine twigs. Most nests had a lining of fine grasses; several incorporated mammalian hair, including moose (5 nests), deer (3), and horse (1). The heights of 34 nests ranged from ground level to 1 m.

Eggs: Dates for 36 clutches ranged from 28 May to 12 July, with 53% recorded between 8 and 22 June. Calculated dates suggest that eggs may be found as early as 27 May. Sizes of 24 clutches ranged from 1 to 5 eggs (1E-1, 2E-2, 3E-3, 4E-12, 5E-6), with 50% having 4 eggs. The incubation period is 11 to 14 days (Falls and Kopachena 1994; Baicich and Harrison 1997).

Young: Dates for 19 broods ranged from 14 June to 17 July, with 53% recorded between 27 June and 8 July. Sizes of 13 broods ranged from 1 to 5 young (1Y-2, 2Y-2, 3Y-3, 4Y-5, 5Y-1). The nestling period ranges from 7 to 12 days, but is usually 8 or 9 days (Rising 1996; Baicich and Harrison 1997). It was 9 days for 1 brood in British Columbia, where the young prematurely fledged during a nest inspection (Munro 1949). Fledglings often leave the nest 2 or 3 days before they are capable of sustained flight (Lowther and Falls 1968) and remain with parents for 22 to 24 days (Baicich and Harrison 1997). In eastern North America, the White-throated Sparrow is known to rear a second brood provided the first brood has fledged before the end of June (Falls and Kopachena 1994). This has not been recorded in British Columbia.

Brown-headed Cowbird Parasitism: In British Columbia, 18% of 34 nests found with eggs or young were parasitized by the Brown-headed Cowbird. There were 6 additional records of a White-throated Sparrow feeding a cowbird fledgling. All records of cowbird parasitism come from the Boreal Plains. This rate of parasitism is significantly higher than that reported for other areas in North America, where the White-throated Sparrow is considered to be a rare host of the cowbird (Falls and Kopachena 1994; Rising 1996).

Nest Success: Of 4 nests found with eggs and followed to a known fate, 1 produced at least 1 fledgling. In southern Ontario, predation was the main cause of nest failure in 171 nests found during the incubation period (Falls and Kopachena 1994). Over a 5-year period, the annual whole brood loss to nest predators, such as the red squirrel (*Tamiasciurus hudsonicus*) and eastern chipmunk (*Tamias striatus*), was 44.8% (range: 22% to 70%).

REMARKS: There are no subspecies described for the White-throated Sparrow, but 2 distinct plumage morphs occur in both sexes throughout the range of this species (Lowther 1961). The white-striped and tan-striped morphs (Fig. 340) are distinguished from each other primarily by the brightness of the crown, with the former having more white in the median and superciliary stripes and a blacker crown stripe than its duller-plumaged counterpart. Both plumage types are equally represented in the population, but more males are white-striped and more females are tan-striped. Thus, while the majority of pairs consist of a brightly marked male and a dull female, a large number consist of dull males and bright females (Falls and Kopachena 1994).

The White-throated Sparrow is one of the most distinctive North American sparrows, and its song is one of the most recognizable of all bird songs. Outside the breeding season, however, this sparrow can be surprisingly inconspicuous and difficult to detect. During the spring and autumn migration, most movements occur at night; during daytime it seldom ventures from the low brushy cover it so favours. Nests can also be very difficult to find and are most often discovered accidentally or only through the most determined of searches. In British Columbia, the White-throated Sparrow is seldom seen in flocks of more than 5 or 6 birds, except in the company of more numerous species such as the White-crowned Sparrow, Golden-crowned Sparrow, and Dark-eyed Junco.

Band returns indicate that much of the continental breeding population, from Alberta east to Newfoundland, converges on the main wintering grounds in southeastern North America (Falls and Kopachena 1994). The only significant band return for British Columbia, a sparrow banded at Milwaukee, Wisconsin (7 October 1939), and recovered near Chilliwack (8 June 1941), shows a similar movement pattern. The breeding grounds of sparrows wintering along the west coast of North America remain unknown, however. Wythe (1938) was the first to suggest the possibility that these western sparrows could be coming from a breeding population in British Columbia. Recent observations in the Fraser Lake–Vanderhoof region indicate that the sparrows in the breeding population have a different song pattern from those found east of the Rocky Mountains, suggesting that they could be an isolated population. Further study is required to determine whether this is a distinct western population.

White-throated Sparrow records from Breeding Bird Surveys for Pemberton (6 June 1994) and Meadow Lake (17 June 1970) were determined to be erroneous and have not been included in this account. Three unsupported summer records from the Rocky Mountain Trench, between Brisco and Fairmont Hot Springs (Hennan 1975), were considered doubtful and have also been omitted from this account. The first winter record of a White-throated Sparrow in the Southern Rocky Mountain Trench was at Windermere during the 1995 Christmas Bird Count.

For additional information on the life history of the White-throated Sparrow, see Falls and Kopachena (1994).

NOTEWORTHY RECORDS

Spring: Coastal – Victoria 1 Mar 1960-1 (Briggs 1960), 25 Apr 1970-1 (Crowell and Nehls 1970c); Saanich 11 May 1982-1; Swan Lake (Saanich) 13 Apr 1994-1; Metchosin 15 Apr 1983-4 (Mattocks and Hunn 1983b); Langford 31 Mar 1975-1; Saanichton 21 Mar 1971-1; Duncan 8 May 1974-1; Ucluelet 12 May 1991-1 tan-striped morph singing; Maple Bay 13 Apr 1983-1 singing; Reifel Island 6 May 1996-4 at refuge; Surrey 8 May 1974-2; Vancouver 23 Apr 1985-4 (Mattocks 1985b), 21 May 1962-1 (Bradley 1963); Gibsons 29 Apr 1997-1; Sechelt 4 May 1997-1; Pitt Meadows 27 Mar 1986-1; Harrison Hot Springs 22 Apr 1986-1, 4 May 1988-1; 6 km n Whistler 29 Apr 1996-1 tan-striped morph; Kitimat 13 Mar 1987-1 at feeder. **Interior** – Princeton 5 May 1997-1; Trail 9 May 1990-1 at feeder; Summerland 17 May 1991-1 killed by cat; Blondeaux Creek (Kelowna) 7 May 1978-1; Lavington 26 Mar 1960-1; n Nakusp 31 May 1975-8 on dump road; Barnes Lake (Ashcroft) 20 Apr 1987-1; Kamloops 12 Mar 1990-1 at feeder; nr Community Lake 14 May 1983-1; Celista 20 Apr 1994-1; Revelstoke 27 May 1973-1; Blaeberry River 18 May 1997-1 male singing; Hemp Creek 19 May 1962-1 singing; Murtle Lake 14 May 1959-2 (Edwards and Ritcey 1967); Williams Lake 29 May 1982-1; Anahim Lake 26 Apr 1993-1; Dragon Lake 30 May 1994-1 singing; Yellowhead Lake (Mount Robson Park) 30 May 1992-1 singing; Buchan Creek 24 May 1994-3; Bowron Lake 11 May 1928-1 (McCabe and McCabe 1929); 24 km s Prince George 26 Apr 1977-1; Nulki Lake 23 May 1994-20 along road transect, 30 May 1945-1 (USNM 425357); Sinkut Lake 10 May 1945-2; Fraser Lake 22 May 1994-13, responded to taped calls along road transect; Nautley River 22 May 1994-4; Vanderhoof 12 May 1995-1; Miworth 28 May 1994-1 adult flushed off empty nest; Willow River 24 May 1972-2; Chichouyenily Creek 28 Apr 1996-1; Burden Lake 23 May 1997-1 singing; Tupper Creek 7 May 1938-2 (Cowan 1939); 6 km s Pouce Coupe 24 May 1981-1 singing; Swan Lake Park 28 May 1990-2 eggs; Moberly Lake 28 May 1974-1; Groundbirch 24 May 1954-1 (NMC 48462); Dawson Creek 13 May 1974-1; Worth 9 May 1994-1; Taylor 15 May 1982-10; Fort St. John 1 May 1987-1; Baldonnel 3 May 1980-1; Cecil Lake 14 May 1983-5 singing; Beatton Park 2 May 1987-1; Boundary Lake (Goodlow) 17 May 1986-2; Tuchodi River 31 May 1994-2; Jackfish Creek (Fort Nelson) 12 May 1975-1; Fort Nelson 2 May 1987-1; Liard Hot Springs 16 May 1975-1 (Reid 1975); Helmut 20 May 1982-1; Atlin 27 Apr 1987-1; 14 km w Fireside 28 May 1981-1 singing.

Summer: Coastal – Coquitlam 17 Jul 1995-1 (Elliott and Gardner 1997); Englishman River 22 Jun 1979-1 on estuary (Dawe et al. 1994); Baynes Sound 18 Jul 1981-2 (Dawe et al. 1998); New Hazelton 15 Jul 1993-1 fledgling with adult; 37 km Hazelton 21 Jun 1921-2, 1 of mated pair collected (MVZ 42303); Kispiox 23 Jun 1997-1 singing. **Interior** – Cathedral Park 21 Aug 1983-1 at 2,195 m elevation (treeline) with large flock of Dark-eyed Juncos, White-crowned Sparrows, and Chipping Sparrows; Nakusp 2 Jun 1975-1; Lavington

20 Jul 1957-1 singing; Blaeberry River 24 to 26 Jun 1994-1 male singing; Wineglass Ranch (Chilcotin River) 3 Aug 1986-1; Chilanko Forks 10 Jun 1997-1 singing; Puntzi Lake 10 Jun 1997-3, 1 singing; Quesnel 30 Jun 1995-4 nestlings; Kinney Lake (Mount Robson Park) 19 Aug 1971-1; Mount Robson Park 5 Jun 1994-12 on bird survey; Hixon 12 Jun 1997-1 singing; 24 km s Prince George 1 Jul 1983-3 fledglings; Nulki Lake 11 Jun 1946-4 eggs (Munro 1949); Sinkut River 24 Aug 1945-1; 17 km w Prince George 11 Jul 1991-1 fledgling; Prince George 14 Jun 1969-10; Miworth 10 Jun 1994-3 eggs; Endako 4 Jun 1995-2; Francois Lake 3 Jun 1995-3 singing at east end; Fraser Lake 15 Jun 1987-1; Vanderhoof 30 Jun 1995-2 fledglings with parents, 15 Aug 1919-1 (NMC 13776); Aleza Lake 19 Jun 1963-1 (UBC 11732); Decker Lake 16 Jun 1991-4 eggs; Houston 20 Jun 1997-1 singing, 24 Jun 1984-4 eggs; Topley 21 Jun 1997-1 singing; Fort St. James 4 Jun 1995-5; Stuart Lake 24 Jun 1982-1 banded; Kathlyn Lake 1 Jul 1974-1; Flatbed Creek 10 Jun 1996-1 singing; Bullmoose Creek 9 Jun 1996-1 singing; Parsnip River 30 Jun 1969-1 (NMC 57166); 2 km sw Mackenzie 2 Jun 1996-4 eggs; Mackenzie 8 Aug 1994-3 banded; Tate Creek 28 Jun 1995-1; Dina Lakes 1 Jun 1994-1; Pine Pass 4 Jul 1983-1 singing; Manson River 7 Jul 1993-1 singing (Price 1993); Chetwynd 18 Jun 1997-1 singing; 40 km sw of Dawson Creek 8 Jun 1995-5 eggs; Swan Lake (Tupper) 6 Jun 1962-8; 4 km s Pouce Coupe 28 Jun 1993-1 adult feeding Brown-headed Cowbird fledgling; 8 km s Dawson Creek 24 Jun 1995-1 adult feeding Brown-headed Cowbird fledgling; Bear Mountain (Dawson Creek) 9 Jun 1995-3 eggs; Dawson Creek 8 Jun 1975-38, songbird survey; Doe Creek (Dawson Creek) 4 Aug 1975-6; Fort St. John 21 Jun 1962-5 eggs; Charlie Lake 12 Jun 1939-2 eggs plus 2 Brown-headed Cowbird eggs (Cowan 1939), 10 Jul 1992-1 nestling with 1 Brown-headed Cowbird nestling; Cecil Lake 3 Jul 1978-5 eggs; Goodlow 17 Jul 1996-4 nestlings; Rose Prairie 17 Aug 1975-10, 19 Aug 1975-3 fledglings; Cameron River 14 Jun 1998-4 nestlings, 29 Jul 1986-1; Tuchodi River 8 Jun 1994-3; Trutch 17 Jul 1943-1 (Rand 1944); Minaker River 28 Jun 1944-1 (Munro and Cowan 1947); Beckman Creek 15 Jun 1994-4; Prophet River 18 Jul 1992-1 fledgling; Metlahdoa Creek 1 Jul 1982-16, some singing; Toad River 28 Jun 1996-1; Muskwa River 23 Aug 1997-6; Parker Lake 20 Jun 1980-5 eggs; nr Yoyo 4 Jul 1982-4 nestlings about 5 days old; Yoyo 23 Jun 1982-1; Liard River 22 Jul 1943-1 at bridge crossing (Rand 1944); Fort Nelson River 27 Aug 1997-1 singing; Liard River 22 Aug 1943-1 at Alaska Highway Mile 513 (NMC 29594), 18 Jun 1996-15 on songbird transect (Bennett and Enns 1996).

Breeding Bird Surveys: Coastal – See REMARKS. **Interior** – Recorded from 13 of 73 routes and on 12% of all surveys. Maxima: Summit Lake 25 Jun 1994-43; Tupper 15 Jun 1974-39; Fort Nelson 19 Jun 1974-35.

Autumn: Interior – Fort Nelson River 2 Sep 1985-1; Akie River 7 Sep 1997-1 in subalpine willow thicket at 1,575 m elevation; Fort St. John 2 Oct 1986-2, nr airport; Taylor 2 Sep 1984-30, heavy migration; Mackenzie 10 Sep 1994-7 banded, 22 Sep 1997-1 banded, 13 Nov 1995-1 at feeder; Willow River 25 Sep 1967-2; Prince George 5 Sep 1985-1; Chief Louis Arm (Ootsa Lake) 26 Sep 1997-3; Indianpoint Lake 5 Oct 1929-1 (MCZ 285472); Williams Lake 8 Sep 1965-1, 21 Nov 1987-1 at feeder with Dark-eyed Juncos; Wineglass Ranch (Chilcotin River) 3 Nov 1986-1; Blaeberry River 6 Sep 1998-1 adult; Revelstoke 31 Oct 1989-1; Little Shuswap Lake 30 Sep 1982-6 with flock of Dark-eyed Juncos; Watching Creek 1 Oct 1996-1; lower Tranquille Creek 5 Sep 1993-1; Tranquille 21 Sep 1991-2, 1 adult white-striped morph with juvenile; Brocklehurst (Kamloops) 12 Oct 1997-3; Enderby 30 Sep 1944-1 (UBC 7749); Grindrod 15 Sep 1948-1 (Munro 1953); Kelowna 8 Sep 1945-2; Nakusp 20 Sep 1997-3 at golf course; Lavington 9 Oct 1981-1 at feeder; Okanagan Landing 9 Sep 1931-1 (Munro 1935b); Kelowna 8 Sep 1945-2 with White-crowned and Golden-crowned sparrows, 4 Nov 1979-1; Kokanee Creek Park 18 Oct 1997-1 tan-striped morph; Sparwood 20 Oct 1984-1 (BC Photo 1009); Princeton 5 Oct 1992-1; Creston 12 Oct 1997-3 adult white-striped morphs; Pend-d'Oreille 24 Oct 1982-2; Grand Forks 9 Nov 1989-2 (Walker 1996). **Coastal** – Terrace 19 Oct 1969-1 (Hay 1976); Triangle Island 2 Oct 1994-1 banded; Port Hardy 13 Nov 1986-1; Grant Bay 8 Nov 1968-1 (Richardson 1971); Cortes Island 9 Sep 1985-1; Comox 5 Oct 1928-1 (MVZ 83134); Halfmoon Bay 29 Sep 1995-1; Gibsons 16 Oct 1990-1; Vancouver 2 Sep 1965-1 banded; Pitt Meadows 28 Sep 1969-1 (Crowell and Nehls 1970a); Sea Island 28 Sep 1995-3 banded; Lulu Island (Richmond) 23 Sep 1986-1; Tofino 25 Sep 1985-1 (BC Photo 1041); Chesterman Beach 15 Oct 1981-3; Vargas Island 14 Nov 1968-1; Duncan 28 Oct 1972-2; Brentwood Bay 27 Nov 1995-2 at feeder; Saanichton 6 Oct 1913-1 (RBCM 5050); Sidney 15 Nov 1981-2 at feeder; Cowichan Bay 21 Oct 1973-2; Cadboro Bay (Saanich) 13 Oct 1997-2 at feeder; Ten Mile Point (Saanich) 17 Nov 1987-2 in garden; Rocky Point (Victoria) 1 Oct 1995-3 (Shepard 1995b); Victoria 13 Sep 1970-1; Metchosin 18 Sep 1982-1 at feeder; Jordan River 1 Oct 1988-1.

Winter: Interior – Fort St. John 31 Dec 1997-1 at feeder (BC Photo 1615); Prince George 16 Dec 1990-1 at feeder; Williams Lake 3 Jan 1976-1 wintering with House Sparrows in cattle stockyard; Kamloops 5 Jan 1994-1 at feeder; Adams River 27 Dec 1997-1; Grindrod 11 Jan 1919-1 (UBC 7747); Vernon 26 Dec 1955-1, 26 Dec 1963-1, 21 Dec 1992-3 (Siddle 1993b); Lavington 26 Feb 1960-1 at feeder, 14 Dec 1963-1; Nakusp 10 Dec 1989-1 (Siddle 1990a); Edgewood 21 Jan 1984-1 at feeder; Kelowna 29 Dec 1962-1; Westbank 26 Jan 1991-1 at feeder; Naramata 30 Dec 1989-1 (Siddle 1990a); Castlegar 15 Dec 1977-1 at feeder; Nelson 23 Dec 1997-1 tan-striped morph. **Coastal** – Terrace 19 Dec 1968-1 (Hay 1976); Kitimat 27 Dec 1993-1; Bella Coola 10 Dec 1995-1 at feeder; Port Hardy 28 Feb 1987-1; Willow Point 1 Jan 1975-1; Stories Beach 30 Jan 1975-1 at feeder; Courtenay 4 Jan 1989-1 at feeder; Comox 18 Feb 1996-1 tan-striped morph; Union Bay 26 Feb 1987-1 at feeder; Errington 11 Dec 1976-1 at feeder; Pender Harbour 18 Dec 1996-1; Gibsons 15 Dec 1990-1; Vancouver 30 Dec 1970-1 in Stanley Park; West Vancouver 9 Jan 1987-1; New Westminster 17 Jan 1979-1; Burnaby 12 Dec 1983-1 in blackberry thicket with 6 Golden-crowned Sparrows; White Rock 30 Jan 1982-3 at feeder; Maple Ridge 28 Jan 1975-1 at feeder with White-crowned Sparrows and Dark-eyed Juncos; Tofino 7 Dec 1986-1; Saltspring Island 1 Jan 1993-1 (Siddle 1993b); Duncan 31 Jan 1978-1 at feeder; Cowichan Bay 3 Feb 1990-1; Brentwood Bay 9 Dec 1995-2 at feeder; Central Saanich 18 Jan 1989-1 at feeder; Saanich 1 Dec 1982-2; Cadboro Bay (Saanich) 6 Dec 1993-2, 1 Feb 1982-2; Oak Bay 14 Feb 1987-1; Victoria 4 Feb 1960-1 (Davidson 1960a), 23 Dec 1961-1; Metchosin 15 Jan 1984-3, 28 Feb 1983-4; Sooke 20 Jan 1973-1; Jordan River 3 Dec 1983-1.

Christmas Bird Counts: Interior – Recorded from 11 of 27 localities and on 9% of all counts. Maxima: Vernon 16 Dec 1990-4; Penticton 27 Dec 1993-3; Nakusp 31 Dec 1989-2; Fort St. James 2 Jan 1993-2. **Coastal** – Recorded from 15 of 33 localities and on 10% of all counts. Maxima: Victoria 18 Dec 1993-7; Duncan 21 Dec 1991-2; Ladner 29 Dec 1985-2; Sooke 31 Dec 1988-1; Vancouver 18 Dec 1983-1; White Rock 2 Jan 1984-1; Campbell River 28 Dec 1991-1; Comox 21 Dec 1980-1; Kitimat 27 Dec 1993-1; Skidegate 20 Dec 1986-1; Squamish 22 Dec 1990-1; Masset 21 Dec 1986-1; Sunshine Coast 15 Dec 1990-1; Nanaimo 16 Dec 1990-1; Pitt Meadows 19 Dec 1993-1.

Harris's Sparrow

Zonotrichia querula (Nuttall)

HASP

RANGE: Breeds from northwestern and east-central Mackenzie and southern Keewatin south to northeastern Saskatchewan, northern Manitoba, and northwestern Ontario. Winters primarily from northern Nebraska and central Iowa south to south-central and south coastal Texas, and rarely but regularly north to southeastern Alaska, southern British Columbia, Idaho, Montana, northeastern Saskatchewan, and North Dakota, west to southern California, southern Nevada, southern Utah, southern Arizona, and southern New Mexico, and east to western Tennessee, Arkansas, and northwestern Louisiana.

STATUS: On the coast, *rare* to *uncommon* migrant and winter visitant and *casual* summer vagrant in the Georgia Depression Ecoprovince; in the Coast and Mountains Ecoprovince, *very rare* on Western Vancouver Island and the Southern and Northern Mainland coasts; absent from the Queen Charlotte Islands.

In the interior, *rare* to *uncommon* migrant and winter visitant; *casual* summer vagrant in the Southern Interior Ecoprovince, *accidental* summer vagrant in the Southern Interior Mountains and Central Interior ecoprovinces; *very rare* autumn transient in the Boreal Plains Ecoprovince; *casual* in the Sub-Boreal Interior Ecoprovince and *accidental* in the Taiga Plains Ecoprovince.

OCCURRENCE: The Harris's Sparrow (Figs. 348 and 349) occurs widely across the southern half of the province, including Vancouver Island and the Gulf Islands. There are no records for the Queen Charlotte Islands. It is a scarce species in the Coast and Mountains and throughout the interior generally, outside the Okanagan valley. It has occasionally been recorded in the Peace Lowland in autumn. There are no occurrences in the Northern Boreal Mountains. The Harris's Sparrow is found annually only in the Georgia Depression and portions of the Southern Interior. Most birds are reported as immatures. The highest numbers in winter occur in the Fraser Lowland of the Georgia Depression and in the Okanagan valley in the Southern Interior (Fig. 350).

Figure 348. The Harris's Sparrow wanders into British Columbia during its spring and autumn migration through the central United States and Canada (Anthony Mercieca).

Figure 349. Most records of the Harris's Sparrow in British Columbia are of immatures at feeding stations (©A. Morris/VIREO).

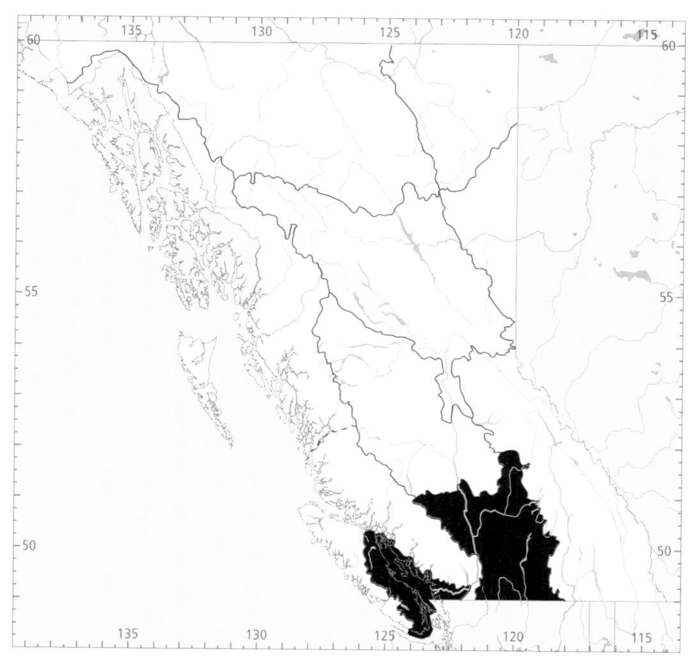

Figure 350. In British Columbia, the highest numbers for the Harris's Sparrow in winter occur in the Georgia Depression and Southern Interior ecoprovinces.

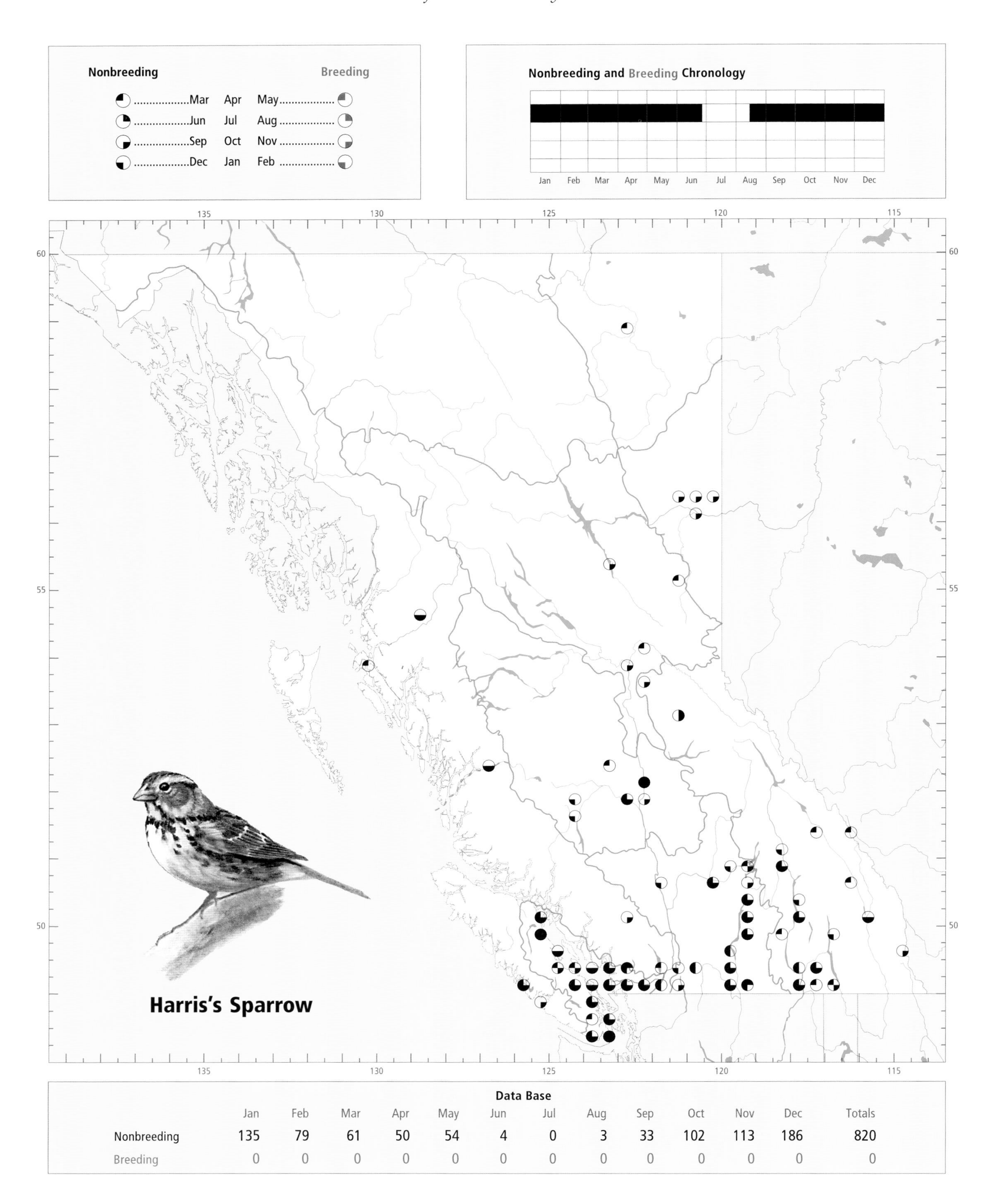

Data Base

	Jan	Feb	Mar	Apr	May	Jun	Jul	Aug	Sep	Oct	Nov	Dec	Totals
Nonbreeding	135	79	61	50	54	4	0	3	33	102	113	186	820
Breeding	0	0	0	0	0	0	0	0	0	0	0	0	0

During winter months, the Harris's Sparrow occurs at fairly low elevations, from sea level to 100 m on the coast and from 280 to 1,150 m in the interior. On the coast, it occupies willow hedgerows, weedy fields, shrubby roadsides, edges of fence rows, blackberry brambles., thistle patches, hawthorn thickets, and urban gardens. In the interior, it frequents shrubby field and woodland edges that support willow, red-osier dogwood, and choke cherry (Fig. 351); patches of weeds; wild rose thickets; shrubby fence rows; weedy agricultural fields with patches of shrubs; ornamental hedges in residential areas; and overgrown clearings near grain elevators (C. Siddle pers. comm.). It is easily attracted to backyard feeding stations. The Harris's Sparrow often forages in mixed-species flocks with Golden-crowned, White-crowned, Song, Fox, Chipping, and American Tree sparrows as well as Dark-eyed Juncos and Spotted Towhees.

The Harris's Sparrow occurs in small numbers both during migration and in winter. Although the primary spring and autumn migration routes are through the central United States and Canada, birds wander widely (Norment and Shackleton 1993). The onset of spring migration on the south coast of British Columbia is revealed by a decline in the number of birds reported through April and early May as wintering birds leave for the north (Fig. 353). Very few birds remain into June. In the Southern Interior, this decline in numbers takes place in March and early April. In general, all Harris's Sparrows have left by early May. In the valleys of the west Kootenay, in the Southern Interior Mountains, and to a lesser extent in the Cariboo and Chilcotin areas, of the Central Interior, there is a slight increase in numbers in March with a high in April, when the migration passes through the region (Fig. 353).

Figure 351. In the southern interior of British Columbia, the Harris's Sparrow often forages along shrubby woodland edges where willow and red-osier dogwood are plentiful (Richter Pass, 22 February 1998; R. Wayne Campbell).

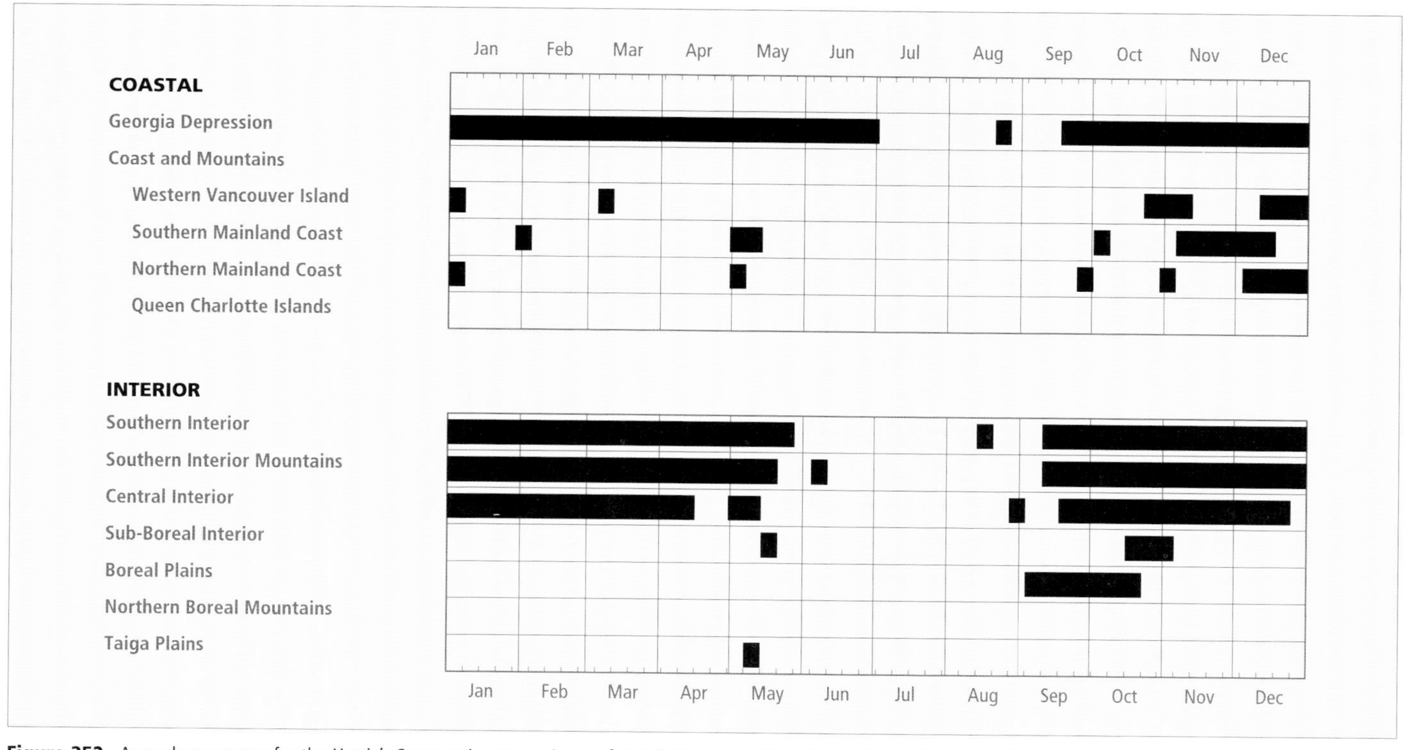

Figure 352. Annual occurrence for the Harris's Sparrow in ecoprovinces of British Columbia. Records are shown for the week in which they occurred.

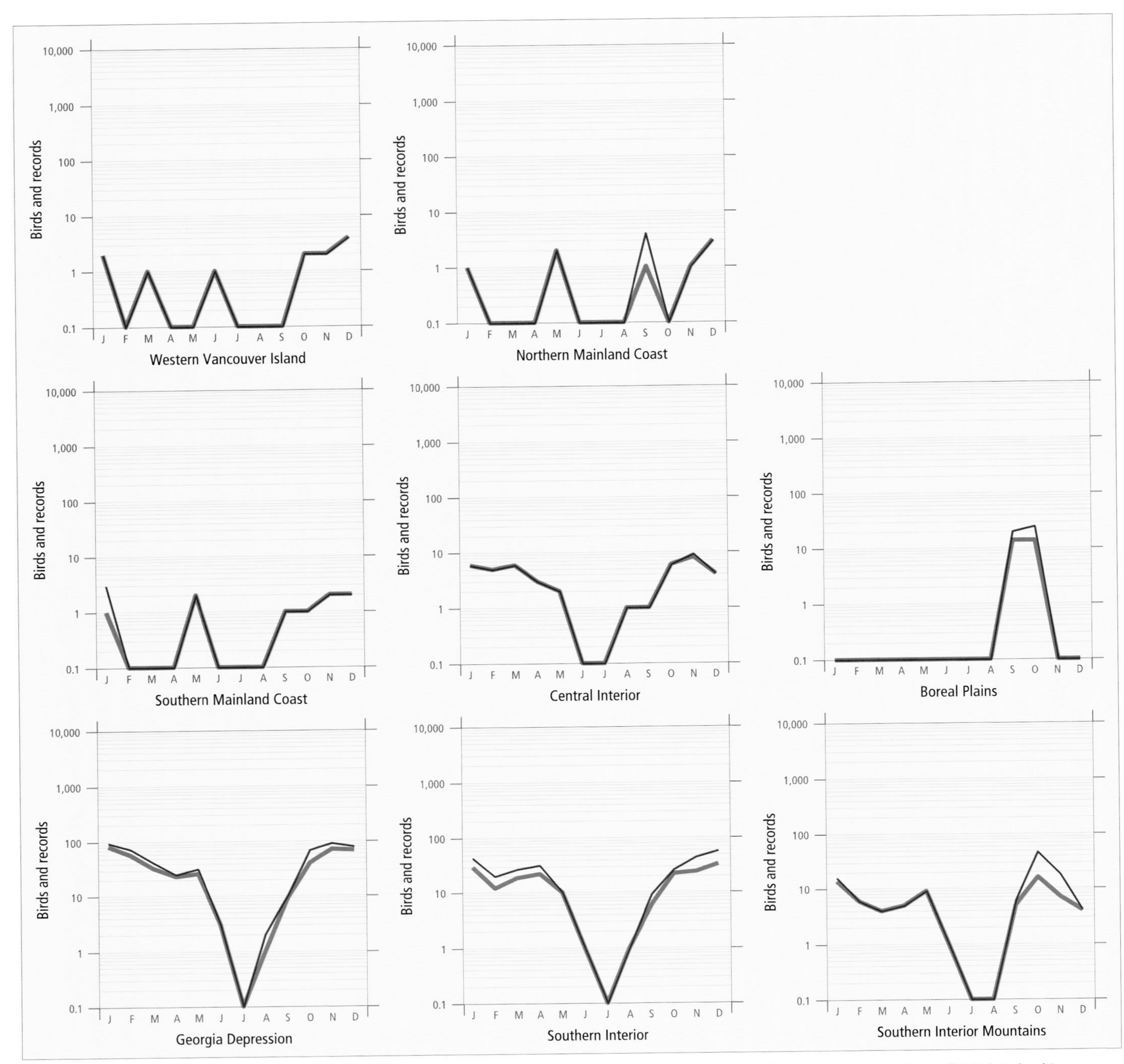

Figure 353. Fluctuations in total number of birds (purple line) and total number of records (green line) for the Harris's Sparrow in ecoprovinces of British Columbia. Christmas Bird Counts have been excluded.

A small autumn movement can be first seen in early September in the Boreal Plains, where it peaks later in the month and into early October (C. Siddle pers. comm.). Elsewhere in the southern portions of the interior, autumn migrants may arrive in late September, but most occur in October and numbers reach winter levels in November and December.

In the Georgia Depression, autumn migrants may arrive in mid-September but the main movement occurs from mid-October through late November.

The autumn migration pattern is similar in Alberta. Migrants are more frequent and migration occurs between early September and November (Salt and Salt 1976; Semenchuk 1992; Pinel et al. 1993).

The Harris's Sparrow has been recorded in British Columbia from 16 August to 25 June. On the coast, it has been recorded regularly from 21 September to 24 May; in the interior, it has been recorded regularly from 12 September to 22 May (Fig. 352).

REMARKS: The Harris's Sparrow was previously known as the Harris' Sparrow.

There are no recognized subspecies of the Harris's Sparrow. The species is the only passerine that breeds solely in Canada.

For a thorough summary of the life history of the Harris's Sparrow in North America, see Norment and Shackleton (1993).

Figure 354. Immature Harris's Sparrow (Bella Coola, 21 February 1999; print taken from 8 mm video by Ron Mayo). Photographing rare and vagrant species is helpful in establishing the status of unusual birds in British Columbia. ▶

NOTEWORTHY RECORDS

Spring: Coastal – Central Saanich 1 Mar 1987-2; North Saanich 6 May 1986-1; Tofino 10 Mar 1986-1; Richmond 10 Apr 1985-3 with small flock of White-crowned Sparrows, 4 May 1986-1; Vancouver 24 May 1984-4 at Jericho Beach (Fix 1984); Harrison Hot Springs 4 May 1988-1, 7 May 1986-1; West Vancouver 31 Mar 1987-2 at Ambleside Park; 10 km w Qualicum 8 May 1985-1 with 5 Golden-crowned Sparrows; Willow Point 10 May 1973-1; Campbell River 14 May 1976-2. **Interior** – Trail 16 May 1977-1; Kaslo 16 May 1973-1; South Slocan 11 May 1973-1; Nelson 22 Apr 1976-1; Summerland 22 May 1990-1 (Siddle 1990b); Kelowna 19 Apr 1990-2; 4 km s Fauquier 5 May 1981-1; Lavington 15 Apr 1970-4 at feeder; Kamloops 26 Apr 1986-1 at feeder; s Nakusp 4 May 1986-1; Nakusp 11 Apr 1976-1, 5 May 1976-1; Blind Bay (Shuswap Lake) 15 May 1970-1 (Schnider et al. 1971); Yoho National Park 6 May 1972-1 male (Wade 1977); Williams Lake 11 May 1986-1; Willow River 18 May 1980-1 with 2 Chipping Sparrows; Tumbler Ridge 16 May 1994-1.

Summer: Coastal – Mitlenatch Island 21 Aug 1965-2 (Campbell and Kennedy 1965); Willow Point 25 Jun 1978-1; Port Hardy Jun 1958-1 (PMNH 72265). **Interior** – Okanagan valley 16 Aug 1969-1; Williams Lake 30 Aug 1973-1 with White-crowned Sparrows; Bowron Lake Park 8 Jun 1975-1; Buckinghorse River 20 Jun 1976-1 at campground.

Breeding Bird Surveys: Not recorded.

Autumn: Interior – 6 km w Boundary Lake (Goodlow) 5 Oct 1986-1; St. John Creek 7 Sep 1985-2 in willows along creek; Charlie Lake 16 Oct 1986-1 at golf course; nr Fort St. John 12 Sep 1985-1, 22 Sep 1986-3, 2 Oct 1986-7; Mount George (Prince George) 16 Oct 1993-1; 24 km s Prince George 2 Nov 1986-1; Indianpoint Lake 24 Sep 1926-2 banded (McCabe and McCabe 1929); 16 km n Williams Lake 18 Sep 1971-1 (Rogers 1972); Williams Lake 1 Oct 1977-1, Nov 1952-2 (Jobin 1953); Riske Creek 16 Oct 1988-1; Revelstoke 16 Sep 1973-1, 12 Oct 1984-2; Edith Lake (Kamloops) 5 Oct 1990-1; Enderby 14 Sep 1974-1 (Rogers 1974); Canal Flats 30 Nov 1985-2 at feeder; Nakusp 25 Sep 1993-1 (Siddle 1994a); Lavington 16 Nov 1968-4; Okanagan Landing 24 Sep 1926-4 specimens (Swenk and Stevens 1929); n Sparwood 6 Oct 1983-7 in weedy field with flock of White-crowned Sparrows; Sparwood 5 Nov 1984-6 at feeder; Nelson 20 Oct 1997-1 banded; Creston 22 Sep 1984-1; Trail 5 Oct 1978-1 at feeder, 6 Nov 1982-2; s Oliver 19 Nov 1990-5 (Siddle 1991a); 2 km e Keremeos 20 Sep 1989-1. **Coastal** – Terrace 27 to 30 Sep 1968-4 (Hay 1976), 3 Nov 1968-1 (Hay 1976); Bella Coola 19 Nov 1989-1 at feeder; Mitlenatch Island 21 Sep 1965-2 in small pines with Chipping Sparrows; Comox 20 Nov 1894-3 (Swenk and Stevens 1929); Sechelt 23 Sep 1985-1; West Vancouver 29 Nov 1986-2 in Ambleside Park; Marpole 28 Oct 1955-7 banded (Hughes 1956); Nanaimo 28 Oct 1984-1 (BC Photo 1137); Tofino 1 Nov 1982-1 (BC Photo 808); Skagit River 4 Oct 1971-1; Iona Island 21 Sep 1986-1; Sea Island 23 Nov 1969-5; Crescent Beach 10 Oct 1960-5, mostly immatures with White-crowned Sparrows; Sumas 10 Oct 1895-2 collected (Brooks 1900); Chalk Island 26 Oct 1974-1 along beach edge; Saanich 21 Sep 1986-1, 10 Nov 1984-4 (Hunn and Mattocks 1985).

Winter: Interior – Williams Lake 4 Dec 1987 to 26 Mar 1988-1 at feeder; Lunch Lake 21 Jan 1995-1 at feeder; Revelstoke 8 Dec 1989-1; Kamloops 4 Jan 1983-1 at feeder; Lillooet 8 Jan 1953-1 (NMC 48441); Vernon 27 Dec 1965-5 (Rogers 1966b); Nakusp 7 Jan 1990-1 (Siddle 1990a), 26 Feb 1977-1; Canal Flats 13 Jan 1985-1 at feeder; Lavington 1 Jan 1969-4, 18 Dec 1972-3 at feeder, 8 Feb 1973-3 at feeder; Coldstream 5 Jan 1966-5 at feeder; Kaslo 20 Jan 1998-1; Nelson 31 Jan 1998-1 at feeder; Castlegar 21 Jan 1975-1 (RBCM 16477); 2 km e Keremeos 13 Dec 1989-2. **Coastal** – Terrace 20 Dec 1968-1 (Hay 1976); Bella Coola 2 Dec 1989-1 at feeder, 21 Feb 1999-1 immature at feeder (Fig. 354); 6 km e Bella Coola 10 Dec 1985-1 (BC Photo 1093); Campbell River 1 Jan 1980-1 (BC Photo 600); Comox 25 Feb 1895-1 (RBCM 862); Hope 29 Jan 1990-3; North Vancouver 6 Jan 1973-2; Vancouver 21 Dec 1986-2; Nanaimo 1 Jan 1986-1 present for several weeks; Tofino 6 Jan 1986-1 at feeder; Chilliwack 9 Jan 1895-3, 2 collected (Brewster 1895); Westham Island 28 Jan 1973-2; Saltspring Island 17 Feb 1993-1 (Siddle 1993b); Central Saanich 27 Dec 1972-3 (Tatum 1973), 12 Feb 1987-2.

Christmas Bird Counts: Interior – Recorded from 10 of 27 localities and on 14% of all counts. Maxima: Vernon 22 Dec 1985-11; Kamloops 15 Dec 1984-3; Penticton 27 Dec 1993-2; Kelowna 15 Dec 1990-2; Revelstoke 29 Dec 1990-2; Oliver-Osoyoos 28 Dec 1993-2. **Coastal** – Recorded from 15 of 33 localities and on 13% of all counts. Maxima: Vancouver 21 Dec 1986-5; Terrace 3 Jan 1971-5; Ladner 23 Dec 1972-5; Victoria 18 Dec 1993-5; Deep Bay 29 Dec 1985-2; Chilliwack 21 Dec 1985-2; Squamish 2 Jan 1989-1; Pitt Meadows 27 Dec 1980-1; Sunshine Coast 23 Dec 1979-1; Campbell River 16 Dec 1979-1; Nanaimo 22 Dec 1985-1.

White-crowned Sparrow

Zonotrichia leucophrys (Forster)

WCSP

Figure 355. Immature White-crowned Sparrow (Alaksen National Wildlife Area, Delta, 20 November 1996; Michael I. Preston).

Figure 356. In southern British Columbia, the spring migration of White-crowned Sparrows (adult shown) peaks in late April and early May (Elizabeth Lake, Cranbrook, 5 May 1997; R. Wayne Campbell).

RANGE: Breeds from northern Alaska east to northern Mackenzie and central Keewatin, and from northern Manitoba east to northern Quebec and Labrador; south to southern Alaska and most of the interior and the extreme southwest coast of British Columbia to southern California, northern Nevada, central Arizona, and northern New Mexico; and from western Alberta and northern Saskatchewan east through northern Manitoba and Ontario to central Quebec, southern Labrador, and Newfoundland. Winters from southern British Columbia, including Vancouver Island, east through Idaho and Kansas to southern West Virginia; south to southern Baja California and northern Mexico, and east to Florida.

STATUS: On the coast, *fairly common* to *common* migrant and summer visitant (locally *very common* during migration) and *fairly common* to locally *common* in winter in the Georgia Depression Ecoprovince; in the Coast and Mountains Ecoprovince, *uncommon* migrant and summer visitant and *casual* in winter on the Southern Mainland Coast; an *uncommon* transient, *rare* in summer, and *very rare* in winter on Western Vancouver Island and the Northern Mainland Coast; *very rare,* mainly in autumn and winter, on the Queen Charlotte Islands.

In the interior, *fairly common* to *common* migrant and summer visitant (locally *very common* to *abundant* during spring migration and locally *very common* autumn migrant) in the Southern Interior, Southern Interior Mountains, and Central Interior ecoprovinces; *uncommon* to locally *fairly common* migrant and summer visitant (occasionally *very common* during spring migration) in the Sub-Boreal Interior and Northern Boreal Mountains ecoprovinces; an *uncommon* transient (locally *fairly common* to *very common* during spring migration) in the Boreal Plains and Taiga Plains ecoprovinces, *rare* there in summer. In winter, *uncommon* to locally *common* in the Southern Interior, *very rare* in the Southern Interior Mountains, *casual* in the Central Interior, and *accidental* in the Sub-Boreal Interior.

Breeds.

NONBREEDING: The White-crowned Sparrow (Figs. 355 and 356) has a widespread distribution in suitable habitat throughout much of British Columbia from spring through autumn. It is absent throughout most of the central and northern interior in winter. Along the coast north of Vancouver Island, its distribution is localized and sparse, including the Queen Charlotte Islands. There are, however, some large regions of the province where records are lacking, including parts of the

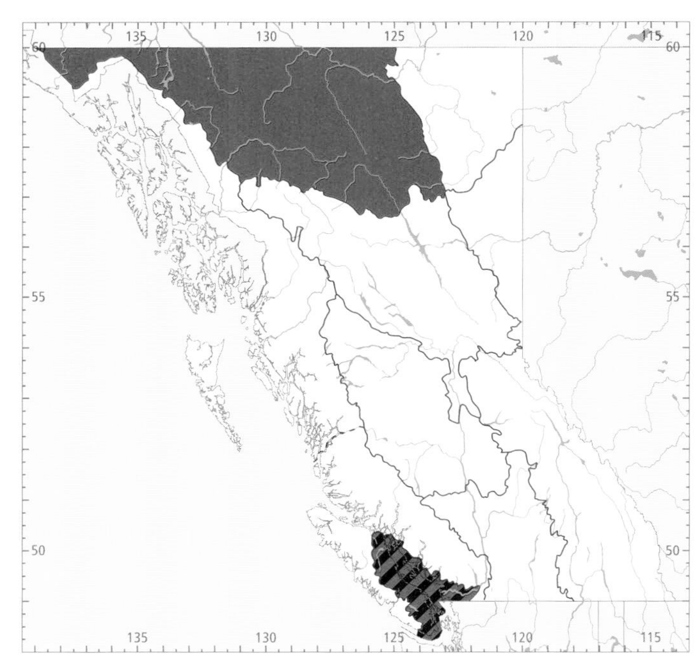

Figure 357. In British Columbia, the highest numbers for the White-crowned Sparrow in winter (black) occur in the Georgia Depression Ecoprovince; the highest numbers in summer (red) occur in the Georgia Depression and the Northern Boreal Mountains ecoprovinces.

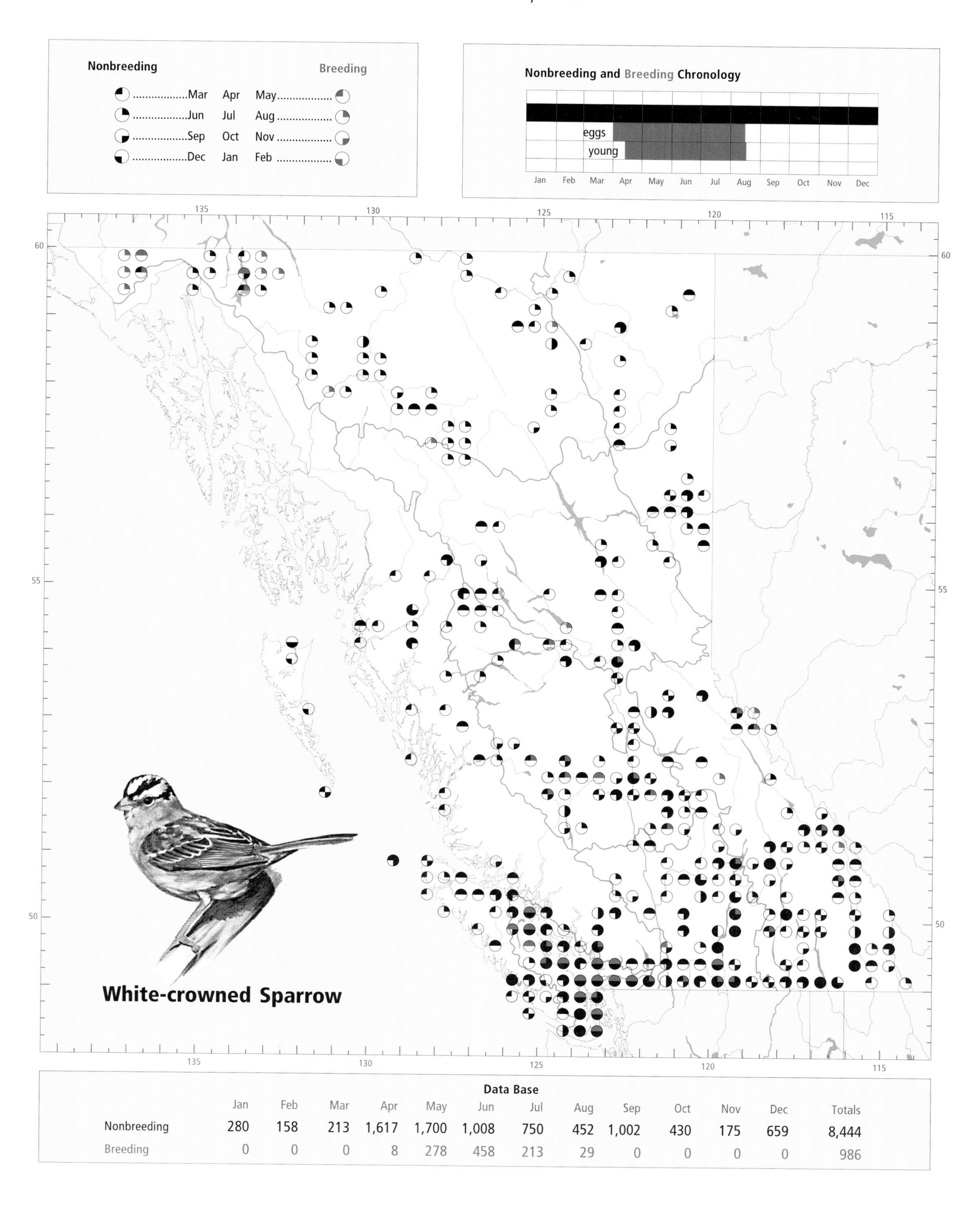

Data Base

	Jan	Feb	Mar	Apr	May	Jun	Jul	Aug	Sep	Oct	Nov	Dec	Totals
Nonbreeding	280	158	213	1,617	1,700	1,008	750	452	1,002	430	175	659	8,444
Breeding	0	0	0	8	278	458	213	29	0	0	0	0	986

Table 3. Christmas Bird Count localities in the Georgia Depression Ecoprovince showing the difference in winter abundance of the White-crowned Sparrow between the Fraser Lowland and Nanaimo Lowland over the period 1974 to 1993.

Ecosection	Location	Number of counts	Mean number of birds	Median number of birds
Nanaimo Lowland	Campbell River	15	7	5
	Comox	17	7	4
	Nanaimo	16	5	4
	Duncan	19	13	10
	Victoria	20	65	47
Fraser Lowland	Chilliwack	18	86	59
	Pitt Meadows	19	21	12
	White Rock	20	80	63
	Ladner	20	331	269
	Vancouver	20	97	66

Southern Mainland Coast, the Nazko Upland and Nechako Plateau, the McGregor River valley, most of the Rocky Mountain Trench between Williston Lake and Lower Post, the northern reaches of the Skeena River and Nass River valleys, and the Boundary Ranges. It reaches its highest numbers in winter in the Georgia Depression (Fig. 357), with larger numbers wintering in the Fraser Lowland than in the Nanaimo Lowland (Table 3).

On the coast, the White-crowned Sparrow has been found at elevations ranging from near sea level to 1,700 m; in the interior, it has been reported from 280 to 2,200 m elevation. On the coast, the White-crowned Sparrow frequents edge habitats (Fig. 358), particularly in areas of heavy shrub growth adjacent to more open areas, including riparian habitat adjacent to beaver ponds, sloughs, sewage lagoons, creek and bog edges, estuaries and other wetlands, agricultural areas and road edges with short grass and weed patches, shrub-dominated logged or burned areas, transmission line and transportation corridors, beaches and spits, golf courses and city parks, airports, and gardens. It also frequents backyard feeding stations. In the interior, it has been reported from habitats similar to those on the coast, as well as from sagebrush grasslands, shrubby and weedy fence rows (Fig. 359), industrial areas such as grain elevators and sheds where spilled grain provides a convenient food source, and avalanche chutes and subalpine habitats during migration.

On the coast, spring migration in the Georgia Depression occurs in early April and peaks about mid-month (Fig. 361). Most migrants have passed through by the end of April. On Western Vancouver Island, numbers build from the second week of April and a small movement is evident from the third week of April through the first week of May. On the Southern Mainland Coast, a small movement is evident from the last week of April to the second week of May. Further north, birds arrive on the Northern Mainland Coast in the first week of April and a small movement is noticeable from the last week of April into the first week of May.

In the interior, the White-crowned Sparrow spring migration occurs very quickly and, although it is rather abrupt in some areas, large numbers of birds move through (Figs. 360 and 361). Numbers in the Southern Interior are augmented

Figure 358. In winter and during migration periods on the south coast, the White-crowned Sparrow frequents edge habitats with heavy shrub growth, including blackberry brambles (Delta, 31 March 1997; R. Wayne Campbell).

Figure 359. In the Southern Interior Ecoprovince of British Columbia, autumn migrant White-crowned Sparrows forage in weedy areas such as along fence rows (near Douglas Lake, 29 September 1996; R. Wayne Campbell).

in the first week of April as birds enter the province from their southern wintering areas. The spring migration usually peaks during late April and the first week of May, and the movement there is over abruptly by the second week of that month. In the Southern Interior Mountains, numbers build in the third week of April and peak in the first week of May. Most birds have passed through by the third week of the month. Birds may arrive in the Central Interior as early as the third week of March, but numbers do not begin to build until the first week of April. The pattern of the movement closely follows that of the Southern Interior Mountains. In the Sub-Boreal Interior, birds arrive somewhat later, in the second week of April, but build quickly and peak in the third week of the month; most have passed through by the second week of May. In the Northern Boreal Mountains, birds may arrive in the first week of April, with an obvious movement through the second week of May. East of the Rocky Mountains, spring migrants in the Boreal Plains may arrive in the third week of April. Numbers build quickly and peak in the first or second weeks of May; the movement is generally over by the third week of that month. In the Taiga Plains, migrants arrive in the last week of April, with the peak of the movement in the second week of May; most have passed through by the third week of May (Figs. 360 and 361).

In British Columbia, the autumn migration of the White-crowned Sparrow is not as conspicuous as the spring migration (Figs. 361 and 362). Most birds have left the Northern Boreal Mountains by the first or second week of September, although stragglers have been reported into the second week of October. In the Boreal Plains, the autumn movement is obvious by the fourth week of August and peaks in the first or second week of September; most birds have left by the last week of September, although a few stragglers may remain until the end of October. A small movement occurs in the Sub-Boreal Interior as numbers build from the end of August through mid-September, after which most birds have left the region. In the Central Interior and Southern Interior Mountains, the autumn migration begins in the last week of August and numbers increase through the third week of September; most birds have left those regions by the first week of October, although stragglers may remain until the end of the month. The pattern of autumn migration in the Southern Interior is similar to that of the Central Interior and the Southern Interior Mountains, except that a noticeable movement may continue until the third week of October, after which most birds have gone (Figs. 360 and 361).

On the northern portions of the coast, there is no noticeable autumn migration and most birds have left the area by mid-September. On the Southern Mainland Coast, a small movement occurs from the last week of August to the third week of September, with most birds gone by the end of the month. In the Georgia Depression, an autumn movement is visible, with large numbers of immatures from mid-September to mid-October.

DeWolfe et al. (1973) found that differences in patterns of movement of the White-crowned Sparrow (*Z. l. gambelii*) suggested that individuals migrate at their own pace instead of moving as members of a tight flock, and that migrating flocks

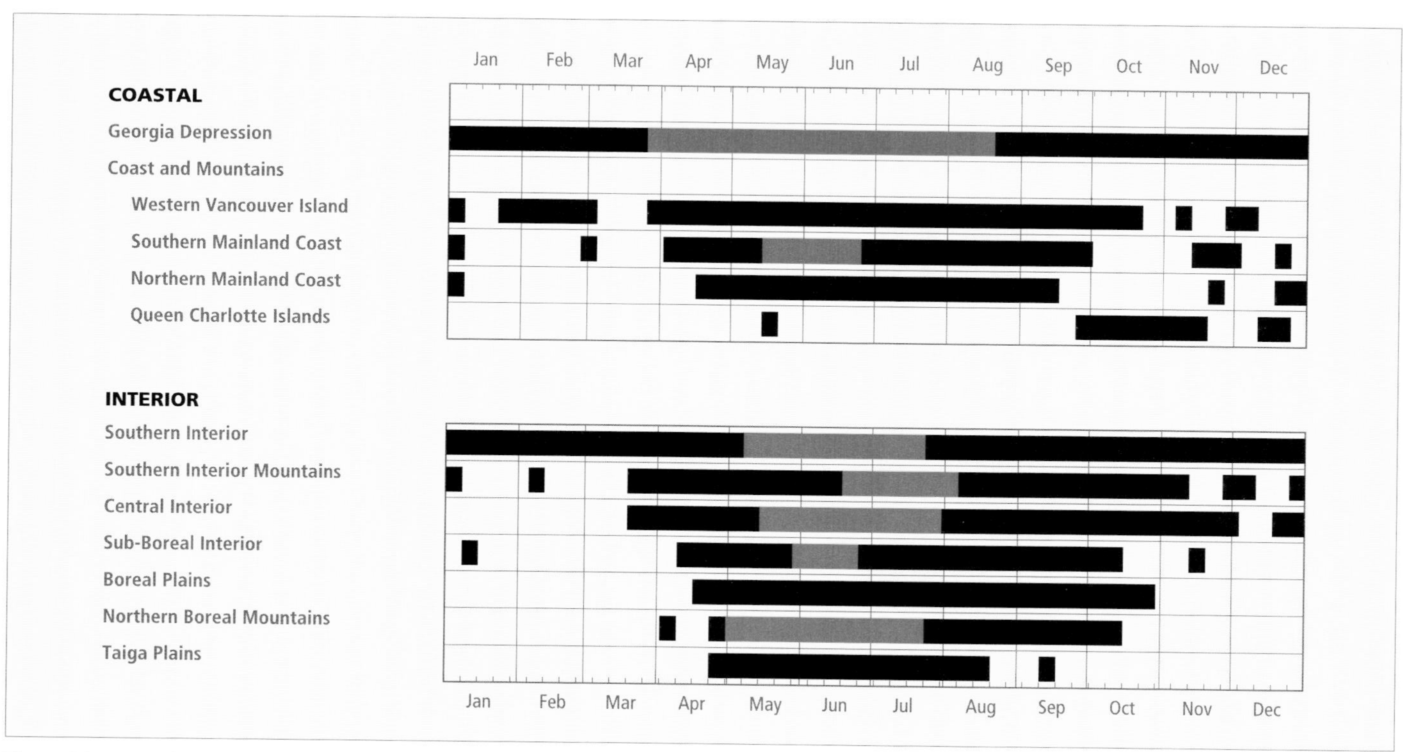

Figure 360. Annual occurrence (black) and breeding chronology (red) for the White-crowned Sparrow in ecoprovinces of British Columbia. Records are shown for the week in which they occurred.

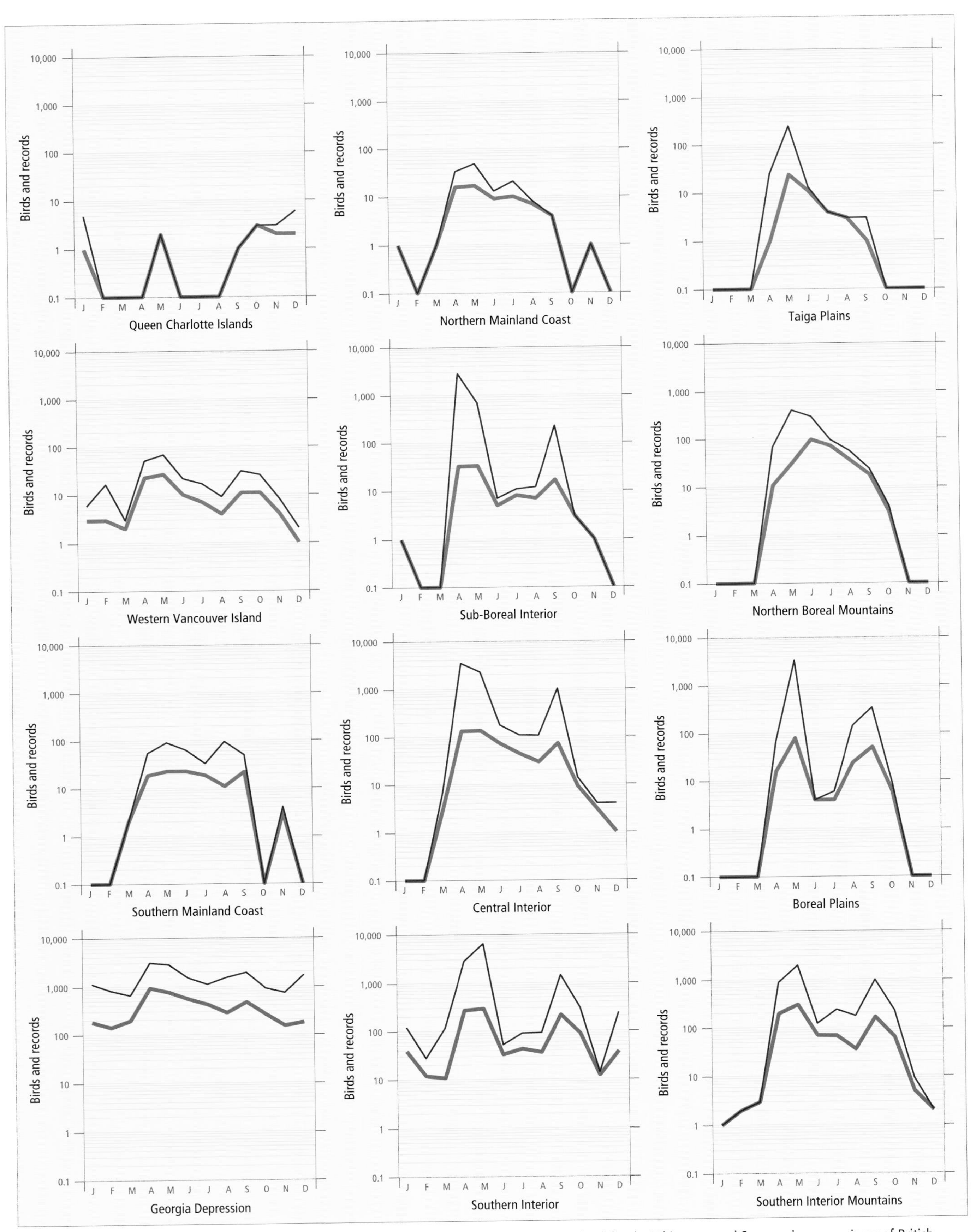

Figure 361. Fluctuations in total number of birds (purple line) and total number of records (green line) for the White-crowned Sparrow in ecoprovinces of British Columbia. Christmas Bird Counts, Breeding Bird Surveys, and nest record data have been excluded.

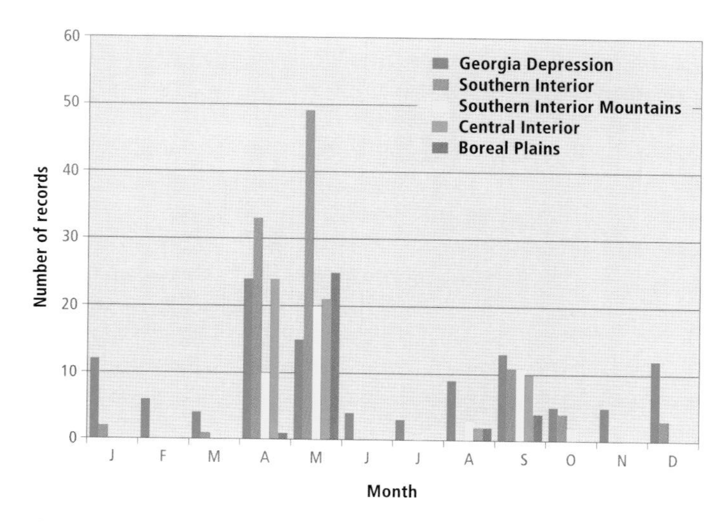

Figure 362. Seasonal distribution of large flocks or groups of flocks (> 20 birds) of the White-crowned Sparrow in ecoprovinces of British Columbia. Throughout the province, the spring migration is far more apparent than the autumn migration.

seldom exceed 8 birds. In British Columbia, however, we have a number of reports of flocks of more than 8 birds (Fig. 362); comments such as "200+ moving by all day," "flocks of 10 to 50 adults," "flock [of 200 birds] stayed until May 10 then left all at once," "hordes of sparrows present a day previously suddenly were not to be found," and "1,400 birds spread out over a sagebrush hillside" are not uncommon.

Some evidence of migratory movements of White-crowned Sparrows banded or recovered in British Columbia is shown in Figure 363, and suggests that the birds migrating through or summering in British Columbia spend the winter in central and southern California and western Arizona. Banding results by DeWolfe et al. (1973) imply that the White-crowned Sparrow (*Z. l. gambelii*) follows 3 general migration routes: coastal, intermountain, and prairie. Rates of travel ranged from 69 to 183 km per day.

There is a differential migration of the sexes, with males arriving on the breeding grounds first, followed by females up to 2 weeks later; the pattern of movement in autumn suggests a later migration of males, both adults and immatures, which could be temporal, geographic, or both (King et al. 1965). There also appears to be a difference in the migration routes taken by adults and young; for example, nearly all birds ($n > 350$) banded in autumn 1995 at Sea Island were young of the year (T. Plath pers. comm.).

Besides the wintering population of White-crowned Sparrows in the Georgia Depression, small wintering numbers have been reported as far north as Masset and Terrace in the northern regions of the Coast and Mountains Ecoprovince. In the interior, most wintering birds are found in the southern Okanagan valley of the Southern Interior, with smaller numbers in the northern Okanagan and Thompson river valleys, in the Southern Interior Mountains (Cranbrook, Revelstoke) and Central Interior (Williams Lake, Smithers), and very rarely in the Sub-Boreal Interior (Prince George).

On the coast, the White-crowned Sparrow occurs throughout the year (Fig. 360). In the southern portions of the interior, it is regularly recorded from early April to mid-October; in the central and northern regions, it regularly occurs from the third week of May to the third week of September.

BREEDING: The White-crowned Sparrow has a widespread breeding distribution throughout most of British Columbia, including eastern Vancouver Island and the Georgia Depression, east across the southern portions of the interior of the province and north to the Yukon boundary. Although Cowan (1939) notes this species as "nesting in small numbers through the woodland and parkland of the eastern part of the [Peace River] district," we could find no evidence that the White-crowned Sparrow nests in either the Boreal Plains or the Taiga Plains. Besides the large areas mentioned under NON-BREEDING for which we have no information, nesting data are lacking for Western Vancouver Island and the mainland coast north of Vancouver Island.

The highest numbers for the White-crowned Sparrow in summer occur in the Northern Boreal Mountains and the Georgia Depression (Fig. 357). An analysis of Breeding Bird Surveys in British Columbia for the period 1968 through 1993 could not detect a net change in numbers on coastal routes; interior routes for the same period contained insufficient data

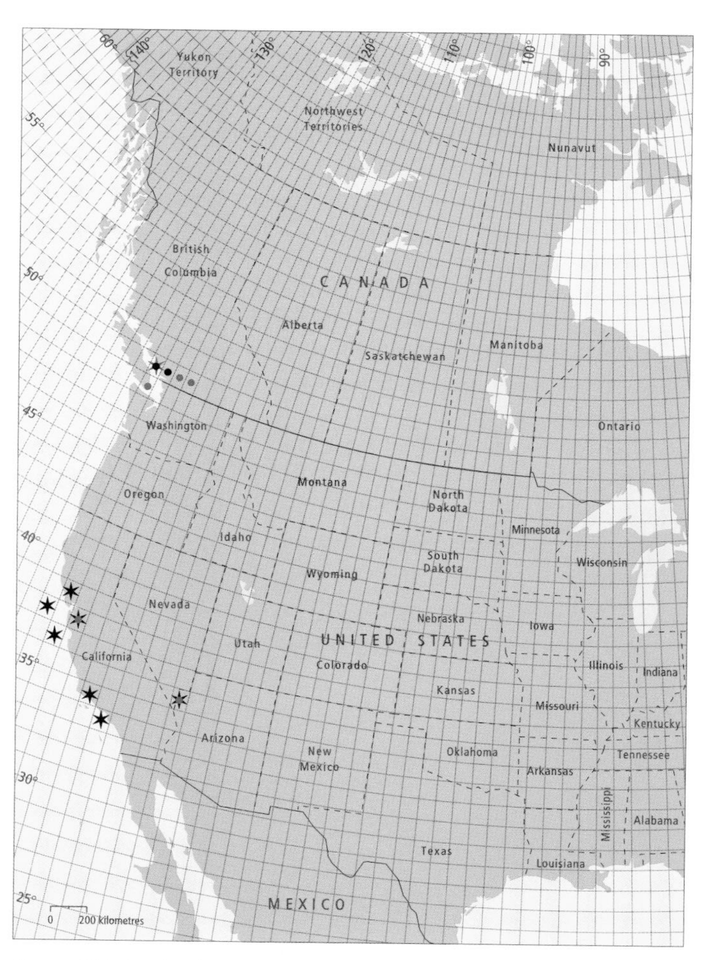

Figure 363. Banding locations (stars) and recovery sites (circles) of White-crowned Sparrows associated with British Columbia. Red indicates birds banded in British Columbia; black indicates birds banded elsewhere.

Figure 364. In central portions of the interior of British Columbia, the White-crowned Sparrow nests along edges of deciduous woodlands in dense vegetation composed of young trembling aspen, willow, prickly rose, and fireweed (Vanderhoof, 20 June 1997; R. Wayne Campbell).

for analysis. Continent-wide surveys for the period 1966 to 1996 indicate that White-crowned Sparrow numbers decreased at an average annual rate of 1.7% ($P < 0.01$), although most of the decline occurred over the period 1966 to 1979 (Sauer et al. 1997). Continent-wide Breeding Bird Survey data indicate that British Columbia has some of the highest densities of the White-crowned Sparrow in North America (Sauer et al. 1997).

On the coast, the White-crowned Sparrow has been reported nesting from near sea level to 1,500 m elevation; in the interior, it has been reported from 330 to 2,200 m elevation. Coastal populations tend to favour lower elevations and low latitudes for nesting (the median elevation of 200 nests was 270 m), whereas interior populations are generally higher-elevation or high-latitude nesters (the median elevation of 94 nests was 900 m). The habitat features important in the breeding territories of all subspecies include grasses, bare ground for foraging, and dense shrubs or small conifers able to provide roosting sites or conceal a nest; standing or running water on or near the territory and tall conifers on the

Figure 365. In the Northern Boreal Mountains Ecoprovince of British Columbia, the White-crowned Sparrow breeds in pockets of dense shrub growth throughout subalpine regions (Stone Mountain Park, 28 June 1996; R. Wayne Campbell).

Figure 366. In British Columbia, most clutches of White-crowned Sparrows contained 4 eggs (Burnaby Lake, 2 June 1969; R. Wayne Campbell).

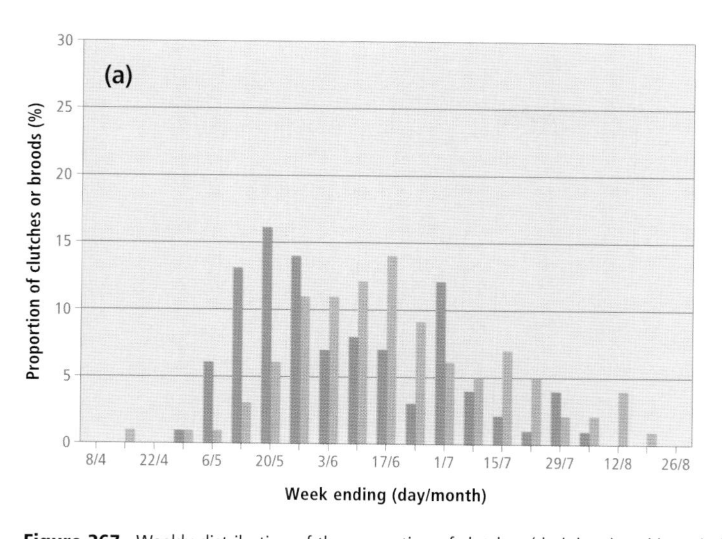

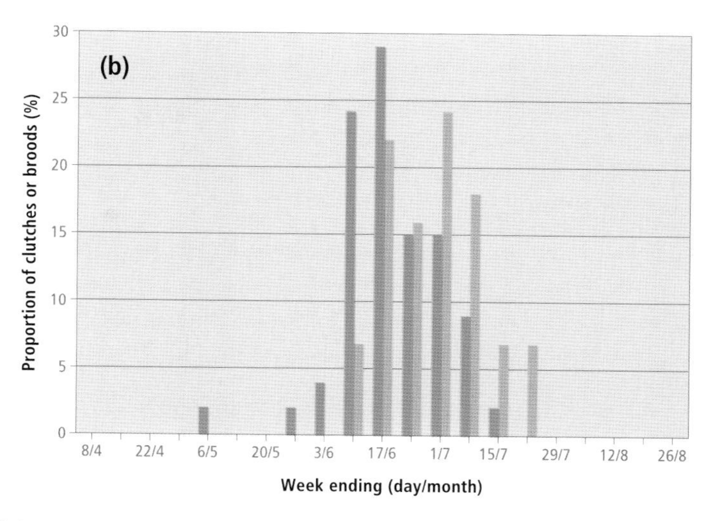

Figure 367. Weekly distribution of the proportion of clutches (dark bars) and broods (light bars) for the White-crowned Sparrow in the (a) Georgia Depression (*Z. l. pugetensis*) (clutches: n = 252; broods: n = 190) and (b) Northern Boreal Mountains (*Z. l. gambelii*) (clutches: n = 55; broods: n = 41) ecoprovinces of British Columbia. The figures are based on the week eggs or nestlings were found in the nest and suggest that southern populations may be multiple-brooded.

periphery of the territory are often present; the latter are used as perches for singing (DeWolfe and DeWolfe 1962).

On central Western Vancouver Island, Bryant et al. (1993) detected the White-crowned Sparrow in all forest age classes except old-growth, but found them consistently only in 15- to 20-year-old and 30- to 35-year-old stands. Mean bird abundance in those forest types ranged from 0 to 0.8 and 0 to 0.7 individuals per sampling station, respectively. In her study of bird communities in trembling aspen forests near Smithers, Pojar (1995) found the White-crowned Sparrow only in clearcuts (mean of 0 to 20 singing males per 10 ha; n = 20) and sapling trembling aspen (mean of 0 to 23 singing males per 10 ha; n = 20).

Most nests (41%; n = 400) were found in habitats associated with humans, followed by forested lands (38%; Fig. 364), shrublands (11%), subalpine and alpine areas (8%; Fig. 365), and grasslands (2%). Habitats associated with humans included suburban, rural, and urban habitats, cultivated farmlands and pastures, recreational areas, transmission corridors, industrial sites, and rangeland. Forested habitats included coniferous forest edges (Douglas-fir, western hemlock, western redcedar, lodgepole pine, Engelmann spruce–subalpine fir, and grand fir), deciduous woods such as black cottonwood riparian and deciduous thickets (willow, birch, and alder), and mixed forests such as lodgepole pine–trembling aspen. Nests were also found in alpine tundra and krummholz, and in subalpine meadows. A number of nests were found adjacent to wetlands.

Specific habitats used by this sparrow included residential backyards and gardens (27%; n = 217), second-growth forest (24%), forest edge (23%), burns (11%), open fields, roadsides, clearcuts and early regenerating forests, riparian habitats, subalpine meadows, beach spits and jetties, orchards, and airports.

On the coast the White-crowned Sparrow has been recorded breeding from 1 April (calculated) to 16 August; in the interior, it has been recorded breeding from 3 May to 5 August (Fig. 360).

Nests: Most nests were found on the ground (66%; n = 388); the remainder were found in shrubs (rose, hardhack, willow thickets, and vines) (20%) and living trees (14%). One nest was found in a garage. Ground nests were usually built beneath a clump of vegetation or among ground cover (72%; n = 215), at the base of trees or among tree roots (27%), and beside a rock or log. Nests in shrubs or trees were found in the fork or the crotch of a branch (n = 40). The heights of 383 nests ranged from ground level to 4.5 m. Kern (1984) discusses racial differences in the nests of White-crowned Sparrows.

Eggs: Dates for 565 clutches (Fig. 366) ranged from 13 April to 15 August, with 53% recorded between 23 May and 28 June. Calculated dates indicate that eggs can occur as early as 1 April. In the central Sierra Nevada, 1 egg is laid per day, beginning an average of 2.5 days after completion of the nest (*Z. l. oriantha*) (Morton et al. 1972). In British Columbia, nests in the south may contain eggs up to 5 weeks earlier than nests of northern populations. Some southern pairs appear to have multiple broods while those nesting entirely north of 56°N latitude, in the Northern Boreal Mountains, show no indication of double brooding (Fig. 367). This supports Morton's (1976) contention that the northern limit of multiple brooding is about latitude 55°N. Sizes of 334 clutches ranged from 1 to 6 eggs (1E-19, 2E-16, 3E-56, 4E-190, 5E-52, 6E-1), with 57% having 4 eggs (Fig. 366). Clutch sizes of southern populations have a mode of 4, whereas clutch sizes of northern populations are bimodal at 4 and 5 (Fig. 368). King and Hubbard (1981) also found a difference in the mode of clutch sizes, which ranged from 3 in the south (latitude 40°03′N) to 5 in the north (latitude 64°48′N).

The incubation period of 3 nests in British Columbia ranged from 11 to 12 days. King and Hubbard (1981) found that the mean incubation period was geographically consistent at about 12.2 days, although they noted that slight deviations were correlated with cold temperatures. Morton et al. (1972), however, found that the mean incubation period of *Z. l. oriantha* in California varied from 10.9 to 13.4 days between years, a significant difference.

Young: Dates for 422 broods ranged from 13 April to 16 August, with 50% recorded between 4 June and 5 July. Nests in the south may contain broods up to 8 weeks earlier than those in the north (Fig. 368). Sizes of 260 broods ranged from 1 to 5 young (1Y-23, 2Y-46, 3Y-70, 4Y-108, 5Y-13), with 68% having 3 or 4 young. In British Columbia, the nestling period of 3 nests ranged from 9 to 11 days; 1 of the nests had a nestling period of 9 days, 5 hours, and 10 minutes. This is similar to the nestling period of between 8 and 10 days reported by King and Hubbard (1981).

Brown-headed Cowbird Parasitism: In British Columbia, 2% of 469 nests found with eggs or young were parasitized by the cowbird. There were 5 additional records of a White-crowned Sparrow feeding a cowbird fledgling. On the coast, 7 of 338 nests (2%) were parasitized; in the interior, 4 of 131 nests (3%) were parasitized. Friedmann and Kiff (1985) note that the Brown-headed Cowbird has greatly extended its area of sympatry with the White-crowned Sparrow, making the latter a frequent host choice of the parasite. Trail and Baptista (1993) provide a very thorough discussion of Brown-headed Cowbird parasitism on populations of the "Nuttall's" White-crowned Sparrow.

Nest Success: Of 81 nests found with eggs and followed to a known fate, 42 produced at least 1 fledgling, for a nest success rate of 52%. Nest success was 53% (n = 72) on the coast and 44% (n = 9) in the interior.

REMARKS: Five subspecies of the White-crowned Sparrow are recognized, and 3 are known to occur in British Columbia.

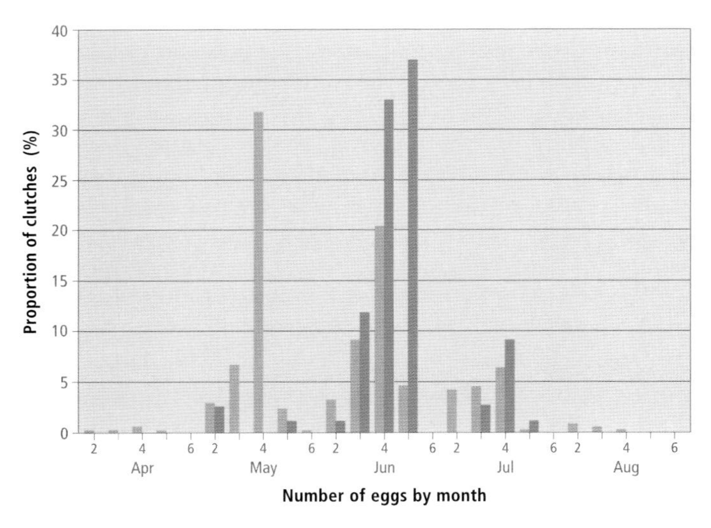

Figure 368. Monthly distribution of clutch sizes of the White-crowned Sparrow between April and August for southern populations (below latitude 54°N) (light bars; n = 262) and northern populations (above latitude 54°N) (dark bars; n = 54). Clutch sizes of southern populations have a mode of 4 eggs, whereas clutch sizes of northern populations are bimodal at 4 and 5 eggs.

Table 4. Field marks helpful in distinguishing among the three subspecies of adult White-crowned Sparrows that occur in British Columbia (adapted from Pyle et al. 1987 and Rising 1996).

Subspecies	Breast	Lores	Bill
Z. l. oriantha	Gray	Black	Dark pink
Z. l. gambelii	Gray to brown	White	Yellow-orange or orange-pink
Z. l. pugetensis	Brown-gray to darkish brown	White	Yellow to orange-yellow

Z. l. gambelii breeds from northwestern to central-southern British Columbia, where it intergrades with *Z. l. oriantha* from Jasper National Park southward (American Ornithologists' Union 1957), although Godfrey (1986) notes it intergrading with *Z. l. oriantha* only at Elko, in the southeastern corner of the province. In British Columbia, *Z. l. gambelii* winters primarily in the southern portions of the province. *Z. l. pugetensis* breeds on the southwest coast of British Columbia, including Vancouver Island, and may occasionally winter (American Ornithologists' Union 1957). Most wintering birds on the coast, however, appear to be *Z. l. gambelii*.

In the field, *Z. l. pugetensis* is easily distinguishable from the other subspecies found in the province. It differs from them in bill colour and in aspects of plumage colour that are distinctive and readily seen when the birds are near (Table 4). Its song is also distinctive (Chilton et al. 1995), and its migration and nesting periods (Fig. 367a) differ from those of the other subspecies. The vast majority also breed within 2 km of salt water (G. Chilton pers. comm.). Observers are encouraged to distinguish between the subspecies when possible.

In the Georgia Depression, *pugetensis* arrives on its nesting grounds in southwestern British Columbia 2 to 3 weeks before the large numbers of migrating *gambelii* pass through the region. The dates of first songs of *pugetensis* in Victoria over 19 years between 1936 and 1963 range from 21 March to 16 April, with a median of 1 April. The arrival dates of *gambelii* over a period of 10 years between 1944 and 1963 range between 23 April and 7 May, with a median of 26 to 27 April.

Z. l. leucophrys and *Z. l. nuttalli* at one time applied to White-crowned Sparrow populations in British Columbia but are now restricted to populations that do not enter the province. Both Brooks and Swarth (1925) and Munro and Cowan (1947) list the former subspecies and a number of specimens from Comox in the collection of the Royal British Columbia Museum (e.g., RBCM 14227 to 14230) are identified as the latter race.

Our knowledge of the status and distribution of the subspecies of the White-crowned Sparrow in British Columbia is inadequate.

For life-history information and aids to identification of the subspecies of the White-crowned Sparrow, see DeWolfe (1968), Pyle et al. (1987), Chilton et al. (1995), Dunn et al. (1995), and Rising (1996).

NOTEWORTHY RECORDS

Spring: Coastal – Victoria 2 May 1964-40, 10 May 1895-6 eggs; Fairy Lake (Port Renfrew) 1 May 1974-1; Cowichan Bay 15 May 1970-142, 21 to 31 May 1970-198; Lennard Island 17 Apr 1980-6; Vancouver 4 May 1975-80 in Queen Elizabeth Park; Surrey 10 May 1969-3 eggs plus 1 Brown-headed Cowbird egg; Matsqui 19 Apr 1968-33; Aldergrove 13 Apr 1978-2 eggs, 13 Apr 1978-1 nestling; Huntingdon 28 Apr 1946-5 eggs; Cheam Slough 23 Apr 1977-75; Harrison Hot Springs area 11 May 1984-44; Squamish 14 Apr 1963-10; Emory Creek 18 May 1970-4 eggs; Leiner River 10 Apr 1982-first arrival, 29 Apr 1982-7 at estuary (Dawe and Buechert 1997); Green Lake (Whistler) 3 Mar 1975-1 at feeder, 30 Apr 1969-10; Port Alice 3 May 1973-20; Sayward 26 Apr 1985-1; Triangle Island 20 May 1976-1 feeding among beached seaweed with Savannah Sparrows; Cape Scott Park 1 May 1975-a few individuals migrating with flocks of Golden-crowned Sparrows; Cape St. James 18 May 1982-1; McInnes Island 29 Apr 1964-1+; Bella Coola 11 May 1933-earliest arrival (Blanchard and Erickson 1949); Lawyer Islands 21 Apr 1979-1, first spring sighting; Terrace 20 Apr 1966-first arrival, most have gone after 17 May (Crowell and Nehls 1966b); Terrace 30 Apr 1987-12. **Interior** – Keremeos to Osoyoos 6 May 1978-239 counted through Richter Pass, a noticeable movement everywhere; Oliver 12 Mar 1977-100, 6 Apr 1988-134, 18 Apr 1985-100; Anarchist Mountain 2 May 1974-300; Osoyoos 24 May 1937-3 eggs; Salmo 26 May 1984-10 at feeder; Vaseux Lake 30 Apr 1974-220; White Lake (Okanagan Falls) 29 Apr 1978-130, 2 May 1974-200, 3 May 1974-600, subspecies *Z. l. gambelii;* southern Okanagan valley 9 May 1981-small numbers still passing through; Okanagan Landing 24 Apr 1941-150, 4 May 1996-1,400 spread out over a sage-brush hillside; Skaha Lake 25 May 1963-3 nestlings; Vernon 12 May 1965-a few small flocks; Nakusp 17 Apr 1984-10, 19 Apr 1980-28, 20 Apr 1985-38, 24 Apr 1985-50, 3 May 1976-100, 5 May 1976-200, 6 May 1985-75; Columbia Lake 23 Apr 1938-19 (Johnstone 1949); Vernon 23 Apr 1991-migrant flocks (Siddle 1991c); Kamloops 2 May 1992-1,500, subspecies *Z. l. gambelii* (Bowling 1992); Chase 4 May 1974-200; Mount Revelstoke National Park 28 Apr 1982-20; Leanchoil 13 May 1975-20 (Wade 1977); 6 km se 100 Mile House 28 Apr 1984-150, subspecies *Z. l. gambelii;* Riske Creek 15 to 30 Apr 1984-47, 1 to 18 May 1984-85; Lac la Hache 30 Apr 1943-200 (Munro and Cowan 1947); s Wells Gray Park 6 May 1955-5; Alexis Creek 19 May 1969-4 eggs; Williams Lake 30 Apr to 5 May 1968-37 banded, 21 Apr 1978-16, 23 Apr 1982-100, 28 Apr 1965-650, 30 Apr 1968-1,000, moving through all day, singing loudly, and next day valley covered with feeding White-crowned Sparrows, 2 to 10 May 1972-200; Chezacut 8 May 1943-200, 13 May 1943-300, 28 May 1941-10; 5 km n Horsefly 25 Apr 1979-30; Quesnel to McLeod Lake 25 Apr 1962-100, flocks all along road, 6 May 1979-47, singing and feeding in small flocks; Indianpoint Lake 25 Apr 1925-earliest arrival (Blanchard and Erickson 1949); nr McBride 2 May 1968-128 banded; 24 km s Prince George 16 Apr 1984-200, 22 Apr 1976-200, 29 Apr 1985-2,000+ left by 15 May; Prince George 3 May 1992-300, subspecies *Z. l. gambelii* (Bowling 1992); Giscome 14 Apr 1988-1; Willow River 11 May 1969-flocks of 200 to 300; Quick 26 Apr 1977-75; Smithers 21 Mar 1989-1, 25 Apr 1992-hundreds, subspecies *Z. l. gambelii* (Bowling 1992); Mackenzie 16 Apr 1996-1; 26 km n Azouzetta Lake 11 May 1982-100; 3 km n Azouzetta Lake 11 May 1982-20; Lynx Creek (Hudson's Hope) 27 Apr 1980-40; Hudson's Hope 5 May 1979-1,000, in flocks of 10 to 50 adults, many singing; w Cache Creek (Bear Flat) 21 Apr 1984-1, spring arrival; Fort St. John 1 May 1987-100, 7 May 1987-hordes of sparrows present a day previously suddenly were not to be found (C.R. Siddle pers. comm.), 11 May 1982-50; Grandhaven 8 May 1983-20, 15 May 1984-100, in small flocks scattered along road, possibly grounded by heavy, cold rain; Highway 29 nr Wilder Creek 4 May 1985-200, 4 km w Alaska Highway; 7 km n Pink Mountain 13 May 1982-45; 14 km n Pink Mountain 13 May 1982-30; Buckinghorse River 13 May 1982-50; Hyland Post 10 May 1982-250; 20 km n Trutch 13 May 1982-20; Fort Nelson 29 Apr 1987-25; nr Liard River 6 Apr 1965-30 banded; Liard Hot Springs 29 Apr 1965-31 (DeWolfe et al. 1973); Helmut (Kwokullie Lake) 19 to 31 May 1982-1, surprisingly scarce; Atlin 26 Apr 1931-1 (Swarth 1936), 3 May 1926-2 eggs; Chilkat Pass 8 May 1957-1 (Weeden 1960).

Summer: Coastal – Jordan Meadows 5 Jul 1953-2; Sooke Hills 3 Jun 1984-20; Spectacle Lake (Victoria) 22 Jun 1975-50; Genoa Bay 3 Aug 1974-128; Stubbs Island 13 Aug 1960-adult with fledgling; Barnston Island 14 Jun 1959-25; Langley 21 Jul 1968-50, 16 Aug 1959-2 nestlings nr fledging; Skagit River 19 Jul 1986-1; Gold Creek (Garibaldi Park) 9 Jun 1963-pair giving alarm notes (Dow 1963); Harrison Hot Springs 31 Aug 1975-75; Courtenay 15 Aug 1972-2 eggs; Point Holmes 15 Aug 1950-30; Anvil Island 21 Jun 1959-4 eggs; Campbell River 30 Aug 1973-150 moving through; Nimpkish Lake 9 Jul 1988-1, nesting (Mattocks 1989a); Adam and Eve River area 12 and 13 Jul 1983-8, common along all logging slash roads; lower Adam River area 21 Jun 1984-11; Mount Currie 3 Jun 1967-8; Fulmore Lake 5 Jun 1976-2; Port Neville Inlet 17 Jul 1976-4, 11 km n head of inlet; Kimsquit River valley 3 Jul 1985-10, singing; Kemano 3 Jul 1975-1 singing on territory in townsite; Kaien Island 5 Jun 1983-3, 1 singing on top of small cedar; Stikine 3 Aug 1975-adult feeding fledgling; Haines Highway 25 Jul 1949-1 (NMC 35424); White Pass 13 Jul 1985-1 with insects in bill. **Interior** – Blackwall Peak 25 Aug 1973-10; Three Brothers Mountain 10 Aug 1974-12; Manning Park 18 Jul 1977-2 eggs; Flathead River valley 9 Aug 1956-1 (NMC 40598); nr Apex Mountain (Penticton) 28 Aug 1988-10; s Fernie 5 Jul 1991-2, 25.7 km along Lodgepole Creek Road (Campbell and Dawe 1991); n Sparwood 22 Jul 1983-4 newly fledged birds being fed by an adult; Mount Scaia 6 Jul 1983-1 fledgling able to flutter away but not able to fly; Lac La Jeune 5 Jul 1986-1 singing at 1,300 m; Tenquille Lake 23 to 31 Jul 1960-1+ (Bradley 1961); 61 km n Revelstoke 2 Jul 1991-11, including 3 fledglings unable to fly (Campbell and Dawe 1991); China Head Mountain 10 Jul 1979-7; Sandbar Island 21 Jul 1976-3 eggs; Mount Tatlow 6 Jul 1985-4, singing with Chipping and Lincoln's sparrows; Green Lake (70 Mile House) 17 Jun 1970-adults feeding 2 White-crowned Sparrow and 2 Brown-headed Cowbird nestlings; Emerald Lake 5 Aug 1944-nestlings present; Lorin Lake 16 Jul 1985-1; Kleena Kleene 24 Jun 1956-3, chiefly a passage migrant (Paul 1959), 25 Jul 1967-1 fledgling, 2 days from nest with parent; Alexis Creek 5 Jun 1969-4 young, 18 Jun 1940-numerous young; Nimpo Lake 16 Jul 1972-feeding 3 nearly grown Brown-headed Cowbird nestlings; Red Pass 22 Jun 1971-1 or more fledglings with adults; Mount Robson Park 20-22 Jul 1970-rather local at low elevations, common at timberline (Stirling 1970), 29 Aug 1973-50; McBride 10 Jul 1991-1 in song (Campbell and Dawe 1991); Cottonwood Island (Prince George) 3 Jul 1990-adult feeding fledgling; 38 km n Prince George 24 Jun 1969-2, females carrying grubs; Willow River 28 Aug 1980-small flocks, earliest fall migration date; s Burnie Lake 23 to 30 Aug 1975-1+ (Osmond-Jones et al. 1975); Fort St. James 14 Jun 1889-4 eggs (MacFarlane 1908); Carp Lake area 25 Aug 1975-3; Tupper Creek 29 Jun 1938-1 (RBCM 8060); Driftwood Range 2 Jul 1941-1; Sunset Prairie 26 Jul 1975-3; n Dawson Creek 3 Aug 1975-

5; Halfway River 2 Jul 1930-1 (Williams 1933a); Fort St. John 30 Aug 1984-41, a common migrant; ne Cecil Lake 29 Aug 1984-21; Stoddart Creek (Fort St. John) 28 Aug 1984-10; Trygve Lake 5 Jul 1976-2; Tatlatui Lake 20 Jun 1976-1, 12 Jul 1976-1 bird with nest material (Osmond-Jones et al. 1977); Beatton River 8 Jun 1956-small flocks are common; Dawson River 31 Aug 1976-8; Fern Lake 17 Aug 1983-1 fledgling (Cooper and Cooper 1983); Gnat Lake 18 Jun 1962-1 (NMC 50323); Summit Pass 3 to 11 Aug 1970-adults only, fairly common (Griffith 1973); Cabin Lake (Kwokullie Lake) 20 Jun 1982-2, 1 with food in bill; confluence of Toad and Liard Rivers 2 Aug 1977-1; Liard River 6 Aug 1977-1, subspecies *Z. l. leucophrys;* Carmine Mountain 14 Jun 1983-9, on an 8-hour hike; Tatshenshini River 12 Jun 1983-6; Haines Highway (Mile 79) 2 Jun 1981-15; Haines Highway (Mile 87.5) 18 Jun 1980-12; s Gladys Lake 19 Jul 1980-adult with 4 recently fledged young feeding at roadside.

Breeding Bird Surveys: Coastal – Recorded from 20 of 27 routes and on 75% of all surveys. Maxima: Chemainus 25 Jun 1969-27; Squamish 10 Jun 1973-23; Pemberton 14 Jun 1992-23; Pitt Meadows 6 Jul 1989-21. **Interior** – Recorded from 38 of 73 routes and on 22% of all surveys. Maxima: Haines Summit 27 Jun 1993-93; Gnat Pass 18 Jun 1975-13; Pennington 30 Jun 1991-13; Succour Creek 17 Jun 1993-12; Nicola 6 Jul 1974-12.

Autumn: Interior – Atlin 15 Sep 1972-2, 10 Oct 1930-1 (RBCM 6718); Fort Nelson 15 Sep 1986-3 (McEwen and Johnston 1987a); Summit Pass 2 to 7 Sep 1943-6 (Rand 1944); Finbow 2 Sep 1997-1, in brush alongside airstrip; s end Charlie Lake 27 Oct 1984-1, last record this autumn; Fort St. John 7 Sep 1985-47, 14 Sep 1985-20, 22 Sep 1985-12; Willow River 11 Oct 1972-1, late autumn migration date; Mackenzie 6 Oct 1995-1 at feeder; Giscome 13 Nov 1994-1 at feeder; 24 km s Prince George 2 to 7 Sep 1985-20, 10 Sep 1984-100, 16 to 22 Sep 1983-60; nr Barkerville 6 Sep 1932-1; Chezacut 18 Sep 1941-500; Williams Lake 5 Sep 1978-10, 28 Sep 1978-50, mostly immatures, 4 Oct 1978-6, 31 Oct 1982-1; Canim Lake 4 Sep 1960-24 around campsite; Chilcotin River (Riske Creek) 22 Nov 1985-seen in small flocks from mid-September along n side river at Wineglass Ranch; Riske Creek 22 Nov 1985-1; Fletcher Lake 11 Sep 1986-40, flock of mixed adult and juvenile birds stopped for 3 days; Tatla Lake 28 Sep 1960-25 (Paul 1964); Yoho National Park 10 Sep 1975-85; Emerald Lake (Yoho National Park) 13 Sep 1975-60, many immatures (Wade 1977); Bridge Lake 7 Sep 1960-24, at campsite; Shuswap Lake 6 Sep 1959-50; Kamloops 7 Sep 1985-300; Nicola Lake 10 Sep 1978-100, 20 Oct 1976-30; Sparwood 2 Nov 1984-4; n Sparwood 24 Sep 1983-50, only 2 adults in flock, 2 Oct 1983-35; Madeline Lake (Penticton) 4 Oct 1973-25; Okanagan Lake 21 Sep 1977-67, feeding on saskatoon berries with Cassin's Finches; White Lake (Okanagan Falls) 7 Oct 1979-20; Vaseux Lake 16 Sep 1959-40; Baldy Mountain 11 Sep 1960-14; Oliver to Richter Lake 18 Sep 1959-50; Manning Park 2 Oct 1973-12. **Coastal** – Terrace 24 Nov 1986-1; Masset 9 Oct 1990-1 (Siddle 1991a); Mount Brilliant 14 Sep 1939-1 (NMC 29087), 15 Sep 1939-1 (NMC 29088); Cape St. James 1 Nov 1978-1; Port Mellon 29 Nov 1987-1; Skagit River valley 30 Sep 1971-many in migration; Pitt Meadows to Maple Ridge 13 Sep 1967-58; Blackie Spit 12 Sep 1971-70, autumn migration; Boundary Bay 12 Nov 1962-122; Delta 26 Nov 1988-70; Pachena Point 7 Oct 1974-6, 11 Oct 1974-4, at lighthouse; Saanich 30 Oct 1985-15, 6 Nov 1985-25; River Jordan 16 Sep 1989-15.

Winter: Interior – Prince George 10 Jan 1993-1 (Siddle 1993b); Williams Lake 26 Dec 1974-4; Revelstoke 1 Dec 1994-1 (Bowling 1995b); Kamloops 3 Jan 1983-40, 31 Dec 1989-25; Lavington 8 Jan 1973-4; Kelowna 26 Dec 1964-3; Cranbrook 5 Jan 1937-1; Penticton 20 Jan 1979-20; Canyon 7 Feb 1980-1; Oliver 31 Dec 1976-110, nr foot of Seacrest Hill; 6 km n Oliver 5 Feb 1988-11; 1 km sw Richter Lake 31 Jan 1975-12. **Coastal** – Masset Jan 1994-5 (Siddle 1994b); Sewell 20 Dec 1980-1 with Dark-eyed Juncos; Boundary Bay 29 Dec 1979-637, all singles; Tsawwassen 9 Dec 1973-125; Delta 8 Feb 1975-75, in various locations; Westham Island 14 Jan 1979-30, 7 Dec 1974-40,14 Dec 1974-86; Sea and Iona islands 24 Jan 1970-33; Lulu Island 19 Jan 1974-200; Tofino 2 Jan 1986-3, 15 Feb 1980-2, 7 Dec 1982-2; Saanich 9 Jan 1988-40.

Christmas Bird Counts: Interior – Recorded from 11 of 27 localities and on 35% of all counts. Maxima: Oliver-Osoyoos 28 Dec 1988-198; Vaseux Lake 29 Dec 1988-159; Penticton 30 Dec 1989-152. **Coastal** – Recorded from 26 of 33 localities and on 64% of all counts. Maxima: Ladner 21 Dec 1975-**774**, all-time Canadian high count (Anderson 1976b) (Anderson [1980] incorrectly notes the Ladner Christmas Bird count of 763 White-crowned Sparrows on 29 Dec 1979 as the all-time Canadian high count); Vancouver 18 Dec 1977-432; Chilliwack 15 Dec 1984-290.

Golden-crowned Sparrow
Zonotrichia atricapilla (Gmelin)

GCSP

Figure 369. Golden-crowned Sparrow in winter plumage (Victoria, 26 September 1998; R. Wayne Campbell).

Figure 370. Golden-crowned Sparrow in breeding plumage (Victoria, 28 March 1983; Tim Zurowski).

RANGE: Breeds from western and north-central Alaska, central Yukon, and southwestern Mackenzie south through the mountainous regions of British Columbia and southwestern Alberta to the northern border of western Washington. Occasionally breeds near sea level on the coast in extreme southwestern British Columbia. Winters mainly from southwestern British Columbia south along the coastal plain through western Washington and Oregon to northern Baja California.

STATUS: On the coast, a *common* to *very common* migrant and winter visitant to the Georgia Depression Ecoprovince, locally *rare* in summer on southeastern Vancouver Island and the Fraser Lowland; an *uncommon* to *very common* migrant (occasionally *abundant* during spring migration) and *rare* winter visitant to the Coast and Mountains Ecoprovince, except *uncommon* during winter on Western Vancouver Island. In summer, locally *fairly common* visitant to high-elevation areas on the mainland coast, *rare* elsewhere in the Coast and Mountains, including Western Vancouver Island and the Queen Charlotte Islands.

In the interior, an *uncommon* to *fairly common* migrant, except *very rare* in the Boreal Plains and Taiga Plains ecoprovinces; *uncommon* to locally *common* summer visitant to high-elevation areas of the interior, except *very rare* in the eastern portions of the Southern Interior Ecoprovince and southern portions of the Southern Interior Mountains Ecoprovince. In winter, locally *very rare* visitant to the Central Interior and Southern Interior ecoprovinces; *casual* in the Southern Interior Mountains.

Breeds.

CHANGE IN STATUS: In the mid-1940s, Munro and Cowan (1947) reported that a "few" Golden-crowned Sparrows (Figs. 369 and 370) wintered in the Puget Sound Lowlands and the Gulf Islands within the area now known as the Georgia Depression. Since that time, this sparrow has become a regular wintering bird here, particularly on southeastern Vancouver Island (Fig. 371).

NONBREEDING: From spring through autumn, the Golden-crowned Sparrow has a widespread distribution throughout most mountainous regions of British Columbia, where it is a characteristic timberline species. There are few records, however, for the mountains east of the Cascade Range and the Fraser River south of latitude 51°N. In winter, the species is absent from the northern two-thirds of the interior, but from Williams Lake south it has a sparse, localized distribution in low-elevation valleys of the Central Interior and Southern Interior. A few birds winter in the Southern Interior Mountains. On the coast, this sparrow is a widespread migrant but during winter is found mainly in the Georgia Depression. The Golden-crowned Sparrow reaches its highest numbers in winter on southeastern Vancouver Island and in the Fraser Lowland of the Georgia Depression (Fig. 372).

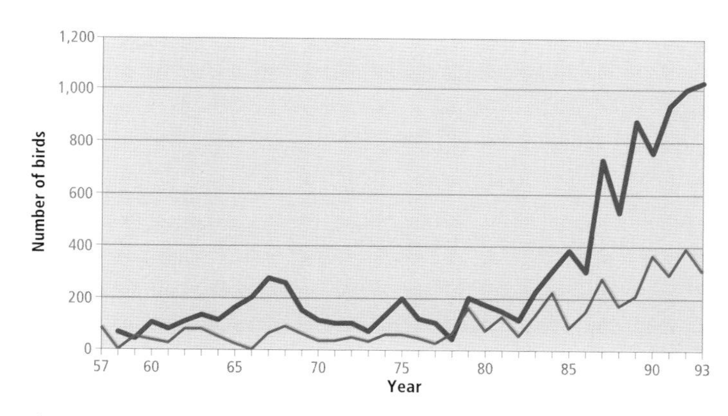

Figure 371. Fluctuations in Golden-crowned Sparrow numbers from Christmas Bird Counts, 1957 through 1993, for Victoria (purple line) and for Ladner and Vancouver counts combined (green line). There was a dramatic increase in numbers from the late 1970s to the early 1990s.

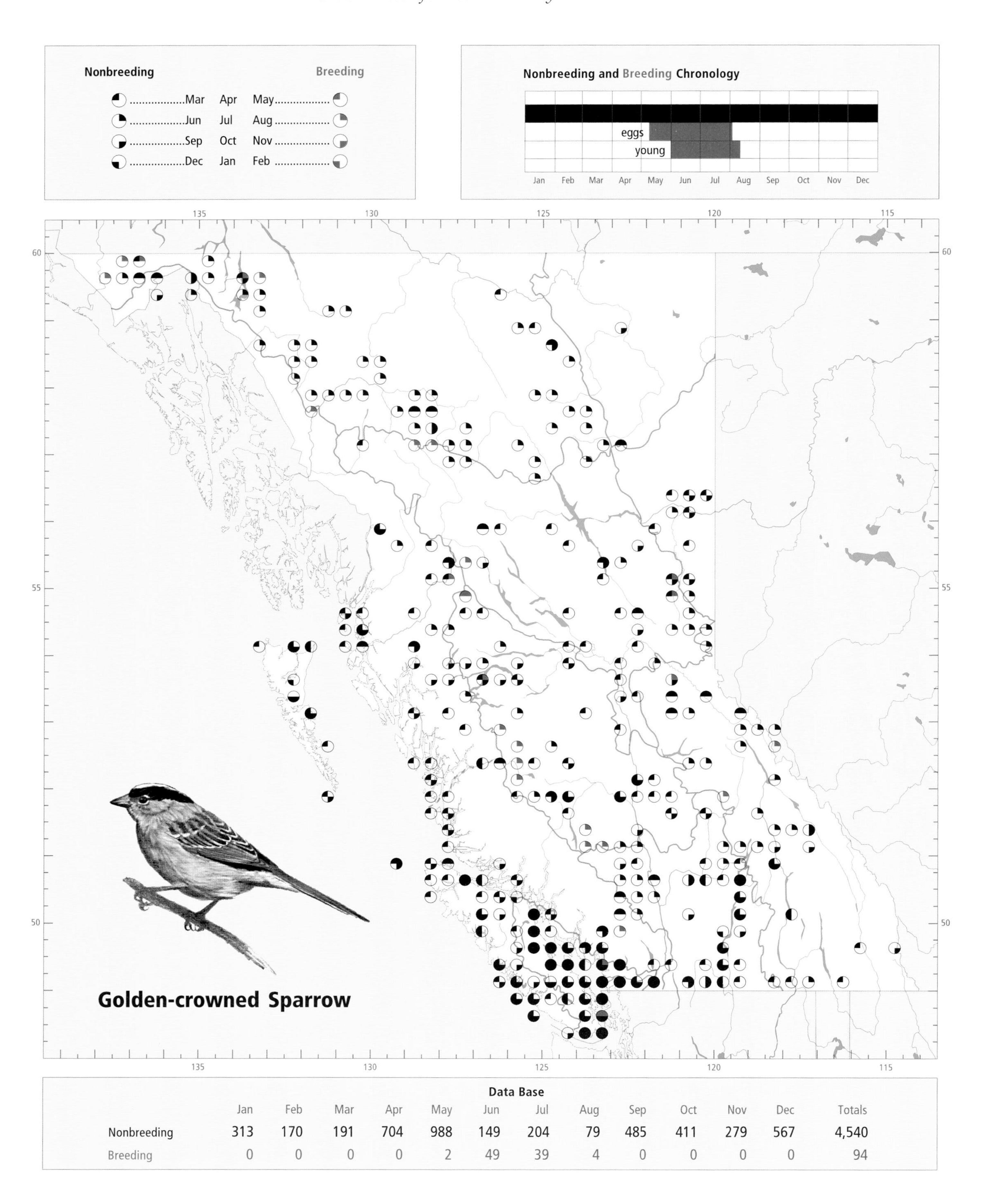

Data Base

	Jan	Feb	Mar	Apr	May	Jun	Jul	Aug	Sep	Oct	Nov	Dec	Totals
Nonbreeding	313	170	191	704	988	149	204	79	485	411	279	567	4,540
Breeding	0	0	0	0	2	49	39	4	0	0	0	0	94

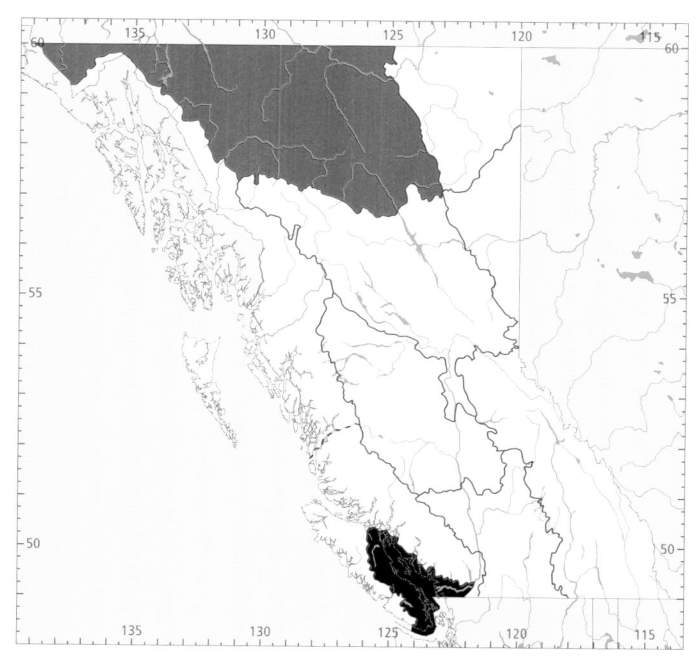

Figure 372. In British Columbia, the highest numbers for the Golden-crowned Sparrow in winter (black) occur in the Georgia Depression Ecoprovince; the highest numbers in summer (red) occur in the Northern Boreal Mountains Ecoprovince.

Figure 373. During spring migration in British Columbia, the Golden-crowned Sparrow frequents a variety of shrubby habitats, including the perimeters of wetlands (Rathtrevor Beach Park, Vancouver Island, 5 April 1998; R. Wayne Campbell).

The Golden-crowned Sparrow has been recorded during the nonbreeding season at elevations ranging from sea level to 2,286 m on the coast and from 280 to 2,438 m in the interior. Throughout the winter and early spring migration period, this sparrow is found at low elevations on the coast and in valley bottoms in the interior. During autumn, it can be found in suitable habitats from sea level to alpine.

Like many of the sparrows, it shows a strong preference for open shrubby habitats and avoids areas of dense forest or open, sparsely vegetated areas away from cover. During winter and spring migration on the coast, the Golden-crowned Sparrow prefers shrubby edge habitats adjacent to more open areas, including the shrub perimeters of wetlands (Fig. 373) and lakes, shrubby riparian areas, recent burns, shore pine bogs, beaches, and estuaries. This sparrow also uses a wide variety of human-modified habitats such as recent clearcuts, utility line corridors, road edges, dykes, weedy fields and pastures, fencelines, and hedgerows, as well as a diversity of urban habitats such as gardens, hedges, city parks, golf courses, and airports. It is a common visitor to bird feeders. During autumn migration, this sparrow also uses a variety of high-elevation habitats in subalpine and alpine areas (Fig. 374) (see BREEDING). In the interior, it favours habitats similar to those used on the coast.

On the coast, the first noticeable spring movement occurs in the Georgia Depression, where migrants begin arriving from the main wintering grounds in California during the second week of April (Fig. 376). These early migrants are often distinguishable from local wintering birds by their more advanced stage of breeding plumage. Spring numbers build steadily through the month, reaching a peak in the first week of May. By early June, only a small number remain in this area, most likely breeding birds (see BREEDING).

On Western Vancouver Island, the spring migration does not become noticeable until the third week of April and normally peaks during the first week of May; most birds have

Figure 374. In autumn, the Golden-crowned Sparrow migrates in higher-elevation habitats, including subalpine areas on Vancouver Island (Mount Service, near Cowichan Lake, 16 August 1995; R. Wayne Campbell).

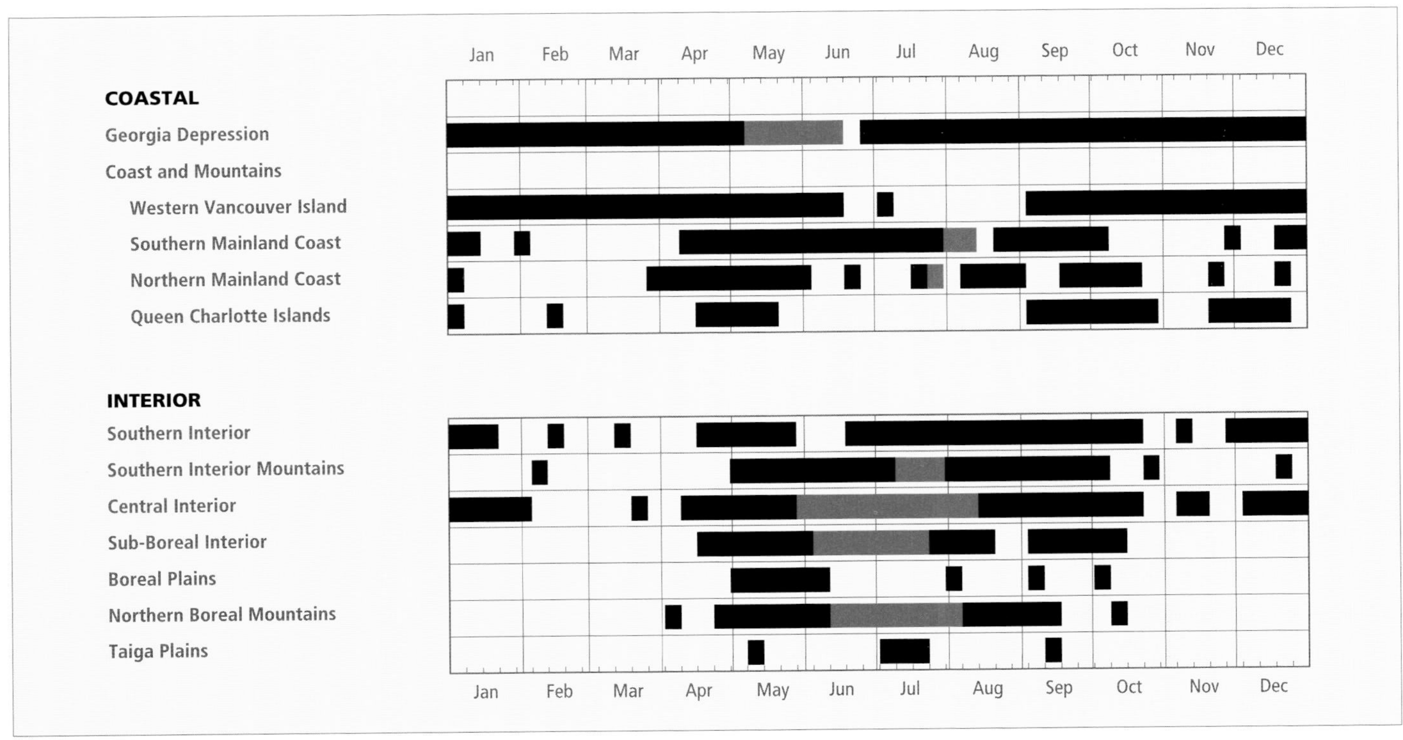

Figure 375. Annual occurrence (black) and breeding chronology (red) for the Golden-crowned Sparrow in ecoprovinces of British Columbia. Records are shown for the week in which they occurred.

passed through by the third week. During late April 1993, a large movement of spring migrants was observed along the shoreline of Pacific Rim National Park (Siddle and Bowling 1993). Very large spring movements have also been recorded on the outer coast of Washington (Kelly 1968), suggesting that many migrants skirt the innermost coast during spring migration.

On the Southern and Northern Mainland coasts, spring migration begins during the third week of April and peaks during the first week of May. The migration front has passed through southern areas by the third week of May and northern areas by the end of May. A small passage of spring migrants also moves through the Queen Charlotte Islands in mid-May.

In the interior, the spring migration is substantially smaller, and consequently much less noticeable, than that recorded along the coast (Fig. 376). During this time, Golden-crowned Sparrows are frequently reported in mixed flocks with the more abundant and closely related White-crowned Sparrow. The earliest spring migrants arrive in small numbers in the Southern Interior during the last week of April, and numbers peak during the first 2 weeks of May. In the Okanagan valley, Cannings et al. (1987) also note a small spring movement during the first 2 weeks of May. Our data indicate that only a few stragglers are left by the end of May. A similar movement occurs in the Central Interior. Edwards and Ritcey (1967) report that the first arrival of this sparrow in Wells Gray Park, in the Southern Interior Mountains, occurs around mid-May. The Stanwell-Fletchers (1943) also report first arrival dates in mid-May in the Driftwood River valley of the Sub-Boreal Interior. In the latter 2 ecoprovinces, a small movement occurs throughout May, with no obvious peak in numbers. In the Northern Boreal Mountains, spring migrants arrive and disperse onto their breeding grounds in late May and early June. A few spring migrants occasionally stray into the Boreal Plains and Taiga Plains in May (Figs. 375 and 376).

The timing of dispersal onto the subalpine and alpine breeding grounds is influenced by spring weather and snow melt. Above-average mountain snowpack levels or cool spring weather can delay this upward movement by as much as 2 or 3 weeks (Norment et al. 1998). In the Central Interior, Norment et al. (1998) recorded the arrival dates of singing males during 2 years on Hudson Bay Mountain; first arrival dates varied by 14 days between years, which the authors attributed to differences in spring snow melt.

In British Columbia, the migration in autumn is more protracted and much less conspicuous than in spring (Figs. 376 and 377), and likely occurs primarily at higher elevations through the mountains (Munro and Cowan 1947). In the northern portions of the interior, the southward migration probably begins shortly after the main nesting season, in mid to late July. In the Stikine River watershed of the Northern Boreal Mountains, Swarth (1922) observes that this sparrow is "noticeably fewer in numbers" in its subalpine breeding habitat by the third week of July. A few migrants occasionally stray into the Taiga Plains and Boreal Plains from September to early October. There are few September records from the Sub-Boreal Interior, suggesting that the autumn migration skirts through the mountainous regions of this ecoprovince,

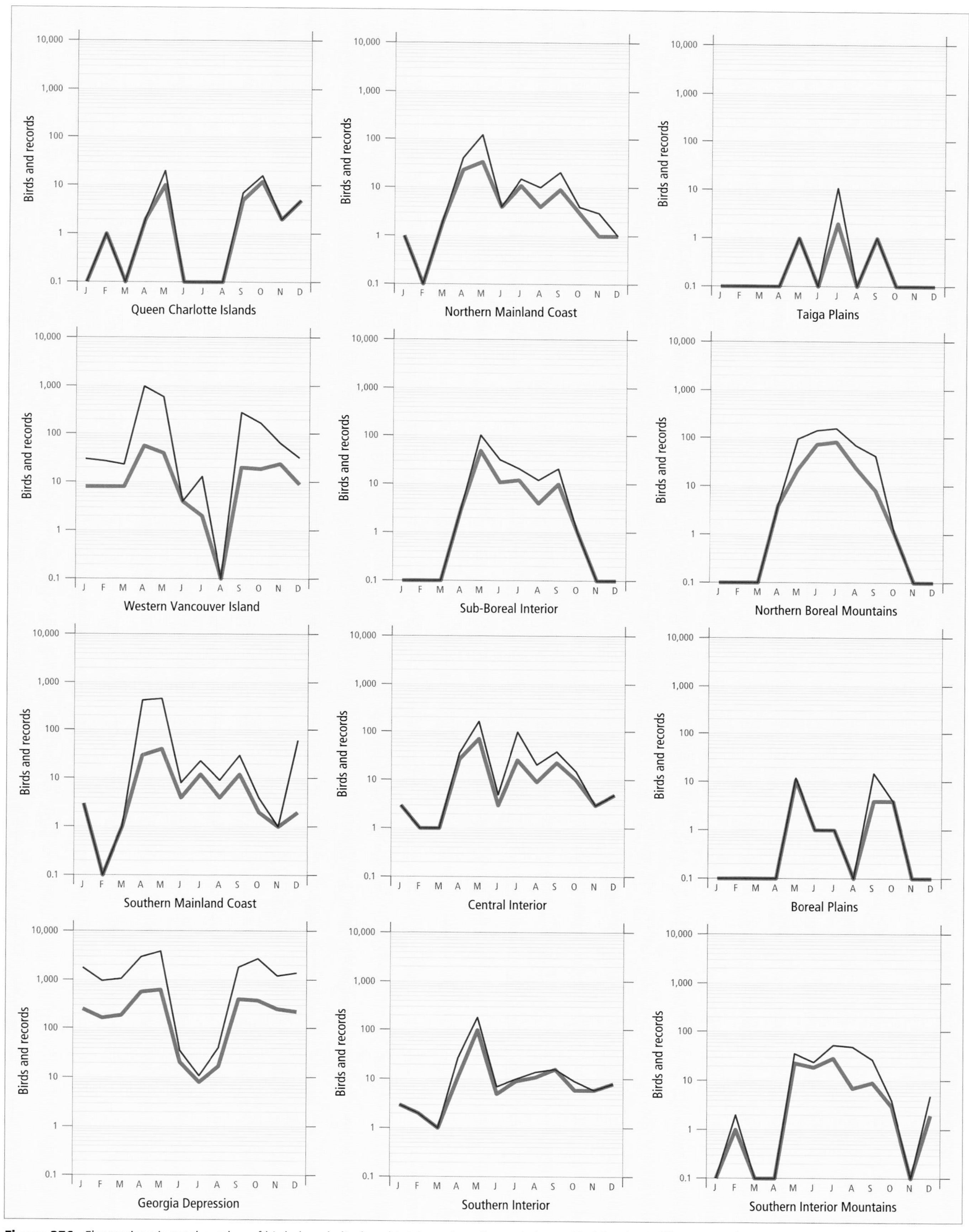

Figure 376. Fluctuations in total number of birds (purple line) and total number of records (green line) for the Golden-crowned Sparrow in ecoprovinces of British Columbia. Christmas Bird Counts, Breeding Bird Surveys, and nest record data have been excluded.

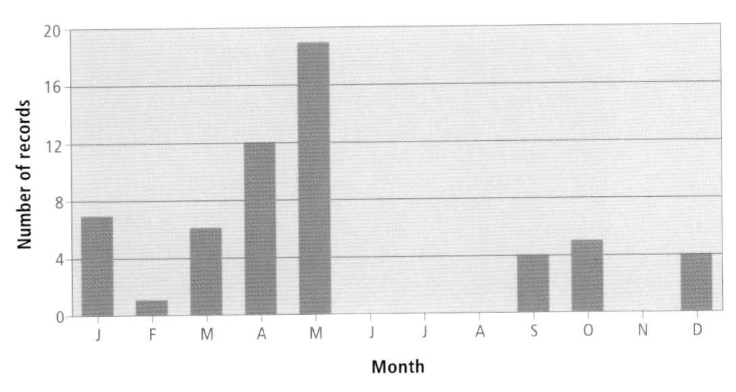

Figure 377. Seasonal distribution of large flocks (≥ 50 birds) of the Golden-crowned Sparrow in the Georgia Depression and the Coast and Mountains ecoprovinces combined. As is the case elsewhere in British Columbia, the spring migration is far more pronounced than the autumn movement.

where there are few observers. In the central and southern ecoprovinces, a small autumn movement begins about the third week of August and continues through September. Most birds have moved through the region by the second week of October (Figs. 375 and 376).

On both the Northern and Southern Mainland coasts, there is little in the way of a discernible autumn movement, but most birds have left this area by the end of September. On the Queen Charlotte Islands, a small movement is evident with the arrival of a few birds in the second week of September; it continues to the second week of October. In the Georgia Depression, early migrants arrive during the last 2 weeks of August and the main movement occurs during the second week of September; most birds have passed through this ecoprovince by the second week of October. The timing of the autumn movement on Western Vancouver Island is similar to that in the Georgia Depression. Ogle and Martin (1997) report a small movement of these sparrows through the upper subalpine areas on Vancouver Island from 8 September to 17 October.

In Washington, the main autumn movement of Golden-crowned Sparrows occurs along the Cascade Range, where they frequent moist subalpine meadows (Kelly 1968). In British Columbia, most autumn records are from low elevations, but this is likely a reflection of where most observer effort occurs. The extent to which this sparrow migrates at high elevations in the province has yet to be adequately documented.

Migratory movements of the Golden-crowned Sparrow along the Pacific coast of North America are apparent from band recoveries (Fig. 378). Most are of wintering birds banded in California and recovered during the spring migration in British Columbia, or autumn migrants banded in British Columbia and recovered during the winter period in California. One sparrow banded during early spring in southeastern Alaska was recovered in Victoria during winter.

On the coast, the Golden-crowned Sparrow has been reported throughout the year in the Georgia Depression; elsewhere, it occurs primarily from mid-April to late October. In the interior, it occurs regularly from late April to early October (Fig. 375).

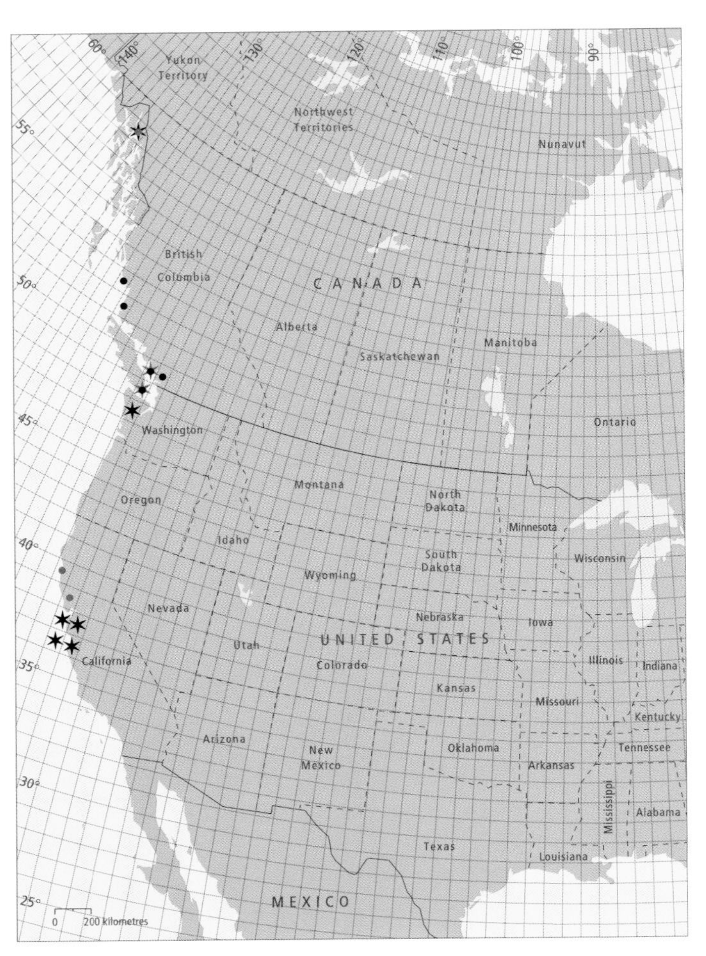

Figure 378. Banding locations (stars) and recovery sites (circles) of Golden-crowned Sparrows associated with British Columbia. Red indicates birds banded in British Columbia; black indicates birds banded elsewhere.

BREEDING: The Golden-crowned Sparrow is widely distributed throughout most mountainous regions of British Columbia during the breeding season. There are, however, few summer occurrences and no breeding records for the southern portion of the Southern Interior Mountains, including the Monashee, Selkirk, and Purcell mountains, and for the Rocky Mountains south of Mount Robson Park. There is also no evidence of nesting in the insular mountains of Vancouver Island and the Queen Charlotte Islands.

Most breeding records are from along, or just east of, the Coast Mountains from Mount Garibaldi in the south to Carmine Mountain in the far northwest. Nesting has also been documented in the Shuswap Highland of Wells Gray Park, in the Cariboo Mountains near Everett Creek, and in the central Rocky Mountains from Mount Robson Park north to the Wolverine River near Tumbler Ridge. In British Columbia, nesting likely occurs throughout the summer range of this sparrow (Fig. 379), but records are sparse in suitable habitats of the Sub-Boreal Interior and the northern portion of the Southern Interior Mountains, and completely lacking through the eastern portions of the Northern Boreal Mountains.

The Golden-crowned Sparrow reaches its highest numbers in summer in the Northern Boreal Mountains (Fig. 372).

Figure 379. In the Northern Boreal Mountains Ecoprovince, the Golden-crowned Sparrow typically breeds near timberline in low shrubby habitats composed of scrub birch and willows with scattered clumps of white spruce and subalpine fir. At this latitude and elevation, snowfall can occur at any time during the breeding season (headwaters of the Prophet River, 15 July 1998; Andrew C. Stewart).

Breeding Bird Surveys for the period 1968 through 1993 contain insufficient data for analysis. Sauer et al. (1997) also provide no analysis for this species due to the inadequacies of continuous Breeding Bird Survey routes in its habitats.

The Golden-crowned Sparrow has been reported nesting from near sea level to 1,585 m elevation on the coast, and from 855 to 2,100 m elevation in the interior. All nests, except for 2 anomalous breeding records on the coast, were found in mountainous areas at mid to high elevations. Breeding typically occurs at upper elevations of the Mountain Hemlock, Engelmann Spruce–Subalpine Fir, and Spruce-Willow-Birch biogeoclimatic zones, or at lower elevations of the Alpine Tundra Zone. In southern regions, nesting occurs within a narrow elevational band near timberline, between areas of closed subalpine forest and sparsely vegetated alpine areas. In the Central Interior, near Smithers, Norment et al. (1998) found that this sparrow occupied an area from timberline (1,500 m) to about 150 m downslope. In the Spruce-Willow-Birch Biogeoclimatic Zone of the Northern Boreal Mountains, where deciduous shrub habitats dominate the mid to upper elevations (Pojar and Stewart 1991), nesting often occurs over a broader elevational range (Fig. 380). Breeding areas include a variety of low shrub and dwarf conifer habitats, including krummholz, scrub birch parkland, subalpine basins, alpine shrublands, shrubby subalpine burns, avalanche chutes, cold air drainage shrubfields, and subalpine scrub birch and willow riparian areas. Areas of taller shrubs or krummholz are frequently used by males as singing perches.

Two nontypical, low-elevation nests were found in human-modified habitats on the southwest coast (Campbell 1975). One nest was found on the ground in an abandoned field with a cover of grasses and scattered shrubs (Fig. 381); the other was in a coniferous hedge in a suburban garden. On the west coast of Alaska, the Golden-crowned Sparrow commonly nests in naturally occurring shrub thickets near sea level (Kelly 1968; Hendricks 1987).

Figure 380. In parts of the Northern Boreal Mountains Ecoprovince, the Golden-crowned Sparrow nests in subalpine deciduous shrub habitats that dominate the landscape (Stone Mountain Park, 28 June 1996; R. Wayne Campbell).

Edwards and Ritcey (1967) describe the breeding habitat of the Golden-crowned Sparrow in Wells Gray Park, in the Southern Interior Mountains, as shrubby slopes occurring "at or above timberline." Stepniewski (1975) notes that, in the Chilcotin Ranges of the Central Interior, this sparrow is characteristic of "all subalpine parkland vegetation types" and also occurs "in willow/scrub birch thickets" in high-elevation valleys. Stanwell-Fletcher and Stanwell-Fletcher (1943) characterize nesting habitat in the Omineca Mountains of the Sub-Boreal Interior as "scattered clumps of scrub subalpine [balsam] fir above timberline." Swarth (1922) describes the breeding habitat near Telegraph Creek, in the Northern Boreal Mountains, as "dense thickets of prostrate subalpine [mountain balsam] fir" at the extreme upper limit of timber at an altitude of between about 900 and 1,200 m. Further north, in the Atlin region, Swarth (1926) found this sparrow breeding above timberline on rolling, grass-covered mountaintops interspersed with thickets of prostrate subalpine fir and trailing scrub birch.

In Alberta, the Golden-crowned Sparrow has been recorded breeding in the Rocky Mountains as far south as Banff National Park (Holroyd and Van Tighem 1983). In Washington, nesting has not been confirmed, although breeding pairs have been reported in the Cascade Range near the international boundary (Farner and Buss 1957) and in the Olympic Mountains (Smith et al. 1997).

On the coast, the Golden-crowned Sparrow has been recorded breeding from 10 May (calculated) to 3 August; in the interior, breeding has been recorded from 29 May to 11 August (Fig. 375; Norment et al. 1998). The unusually early start date for the coast is a result of low-elevation nesting. In Alaska, the start of nesting is apparently governed by the degree of spring snow melt (Kelly 1968).

Nests: Most nests were found on the ground (80%; n = 44), with the remainder in low deciduous shrubs (11%; Fig. 382) or small conifers (9%). One nest was found in a suburban garden hedge. Ground nests were generally well concealed under clumps of grasses and herbaceous vegetation or under low mats of prostrate shrubs. Above-ground nests were bulky structures placed in the crotch, or on the branches, of a low deciduous shrub, including scrub birch (4 nests) and willow (1), or on the sprawling limb of a low conifer, most commonly subalpine fir (3). Nests were thick cups lined with fine grasses and, in several cases, ptarmigan feathers or mammalian hair. Grasses, sedges, mosses, fine twigs, rootlets, and plant stems were the most frequently used nest materials.

The heights for 44 nests ranged from ground level to 0.8 m, with 80% of the nests reported to be on the ground. Norment et al. (1998) observed that 52% of nests (n = 29) found near Smithers were on the ground; the remainder were above-ground nests at an average height of 82 cm. Rising (1996) and Baicich and Harrison (1997) note that this sparrow rarely nests off the ground, but our data and those reported by Norment et al. (1998) indicate that above-ground nests are not uncommon in British Columbia.

In the Northern Boreal Mountains, active nests have been found within 37 m of each other. Norment et al. (1998) found nests as close as 45 m from each other on Hudson Bay Mountain in the Central Interior, and estimated a breeding density of approximately 100 pairs/km^2.

Figure 381. Site of the unusual, low-elevation Golden-crowned Sparrow nest found on southern Vancouver Island. Note that the vegetation at this site has closed in substantially since used by this sparrow (North Saanich, 16 May 1998; R. Wayne Campbell).

Eggs: Dates for 59 clutches ranged from 29 May to 3 August, with 53% recorded between 18 June and 15 July. Calculated dates indicate that eggs may be found as early as 10 May for low-elevation nests on the coast. Sizes of 31 clutches ranged from 1 to 5 eggs (1E-1, 2E-1, 3E-8, 4E-18, 5E-3), with 58% having 4 eggs. One nest found in British Columbia with 6 nestlings indicates that 6-egg clutches also occur (see also "Young"). The usual clutch size ranges from 3 to 5 eggs (Kelly 1968; Rising 1996; Baicich and Harrison 1997), but clutches as large as 7 eggs have been reported from Alaska (Swarth 1926).

The incubation period for 2 nests in British Columbia was 12 days (Norment et al. 1998); Baicich and Harrison (1997) report that the incubation period ranges from 11 to 13 days. In Alaska, the incubation period for 1 nest was between 12 and 12.5 days (Hendricks 1987).

Young: Dates for 35 broods ranged from 1 June to 11 August, with 54% recorded between 21 June and 10 July. Sizes of 20 broods ranged from 1 to 6 young (1Y-1, 2Y-3, 3Y-5, 4Y-9, 5Y-1, 6Y-1), with 70% having 3 or 4 young. Hendricks (1987) reports that nestlings open their eyes at 3 days and fledge at 9 days of age. They are capable of flying short distances 4 days after fledging. Norment et al. (1998) give the nestling period for 3 nests in British Columbia as 9.75 days.

Double-brooding in the Golden-crowned Sparrow has been reported from Alaska (Loof 1939). Although renesting following nest failure has been reported in the Central Interior (Norment et al. 1998), there is no evidence that this sparrow produces more than 1 brood per year in British Columbia.

Brown-headed Cowbird Parasitism: In British Columbia, cowbird parasitism was not found in 46 nests recorded with eggs or young, and there were no records of this sparrow feeding a cowbird fledgling (see also REMARKS). The Golden-crowned Sparrow has not been recorded as a host of the cowbird anywhere in North America (Friedmann and Kiff 1985; Rising 1996).

Nest Success: Of 3 nests found with eggs and followed to a known fate, 1 produced at least 1 fledgling. There were no recorded causes of nest failure in British Columbia. Given the harsh environment in which this species breeds, however, weather conditions are likely one of the main factors influencing nesting success. In Jasper National Park, Alberta, a nest with 4 recently hatched nestlings was lost to predation by a Columbian ground squirrel (I. McTaggart-Cowan *in* Kelly 1968).

Figure 382. The Golden-crowned Sparrow nest is a bulky structure usually placed on the ground but occasionally placed on the branches of low deciduous shrubs (Taseko Mountain, 20 July 1993; R. Wayne Campbell).

REMARKS: The Golden-crowned Sparrow is the least studied of the "crowned" sparrows in North America, and little is known of its breeding biology or ecology. In British Columbia, the breeding distribution and high-elevation movements of this sparrow are poorly documented. The extent and frequency with which it nests at low elevations on the coast also requires clarification.

The nest of the Golden-crowned Sparrow was one of the last North American songbird nests to be described. Early nest records from the mid-1800s in California were later proven erroneous (Swarth 1926). The first published description was of a nest found along the gold rush trail through the White Pass, British Columbia, in 1899 (Bishop 1900), although clutches had been collected in Alaska in 1895 (Norment et al. 1998).

Avian poisoning from natural plant toxins has been reported in British Columbia. On 23 September 1971, between 200 and 500 songbirds were found poisoned at Prince Rupert; based on post-mortem analysis, cyanogenic glucoside was determined to be the probable cause of death. In its normal state, cyanogenic glucoside is a nontoxic form of cyanide, but it is converted to the active acid by enzymes in the bird's crop and gizzard. In a subsample of 45 birds in this instance, 10 were Golden-crowned Sparrows (Wilson et al. 1995).

The Golden-crowned Sparrow is a nocturnal migrant, and collisions with lighthouses have been reported during the spring migration along the coast of Washington (Kelly 1968). Lighthouse-caused mortalities have not been reported in British Columbia but may also occur. In the urban environments of southern British Columbia, human-related sources of mortality include predation by domestic cats and collisions with windows.

In British Columbia, most Golden-crowned Sparrows nest in subalpine or alpine environments, well outside the breeding range of the Brown-headed Cowbird. Low-elevation nesters, however, such as those recorded in the Georgia Depression, would likely be susceptible to cowbird parasitism. A 12 June record of a juvenile cowbird following an adult Golden-crowned Sparrow near Victoria, on southeastern Vancouver Island, lacked sufficient documentation to be considered evidence of parasitism.

A Golden-crowned Sparrow was banded near a feeder in Victoria on 28 February 1993 and was subsequently recaptured at the same location on 5 February 1994 and 8 March 1997, suggesting a winter-site fidelity (A.C. Stewart, unpubl. data). In California, Ralph and Mewaldt (1975) found that 17% of adult Golden-crowned and White-crowned sparrows experimentally translocated from wintering sites had returned to them by the following year.

Although the Golden-crowned Sparrow is strictly a bird of western North America, transients have been recorded in a wide range of locations throughout the continent, including Saskatchewan, Ontario, and Nova Scotia (Godfrey 1986), Utah (Balph 1976), Iowa (Burns 1977), Minnesota (Johnson and Johnson 1987), Wisconsin (Sontag 1993), Michigan (Wykoff 1979), New York (Bull 1988), Massachusetts (Gove and Gordon 1990), and Florida (Hoffman et al. 1991). Off the continent, it has been recorded on Wrangel Island and on the Chukchi Peninsula in Russia, and in Japan (Rising 1996).

A clutch of 4 eggs collected on Burnaby Mountain (UBC E1529), in the Fraser Lowland, by Walter S. Maguire on 23 June 1938 has been excluded from this account. This clutch was not positively identified as belonging to the Golden-crowned Sparrow, and the nest description and habitat appear to be more characteristic of the White-crowned Sparrow.

For additional information on the life history of this sparrow, see Kelly (1968), Rising (1996), and Norment et al. (1998).

NOTEWORTHY RECORDS

Spring: Coastal – Metchosin 10 Apr 1980-35; Victoria 7 May 1955-175 (Davidson 1972); Sidney 30 Apr 1961-22; Cowichan River 3 May 1975-100 in salmonberry thickets; Tofino 26 Apr 1969-50; Grice Bay 3 May 1974-250 feeding along beach; Lennard Island 22 Apr 1977-12; Pacific Rim National Park 29 Apr 1993-300, large spring fallout (Siddle and Bowling 1993); Delta 2 May 1971-200 (Campbell et al. 1972); Vancouver 4 May 1975-80 in Queen Elizabeth Park; Aldergrove 7 May 1970-51, moving and feeding along edge of woods; Qualicum Beach 25 Apr 1980-42; Gibsons Landing 3 May 1959-6; Vancouver 4 May 1894-100 at Jericho Beach, 8 May 1984-170 at Lost Lagoon; Harrison Hot Springs 22 Apr 1986-3, 11 May 1984-9; Alta Creek 3 May 1992-20; Tahsis Inlet 1 May 1949-20 (Erskine 1959); Campbell River 27 Apr 1974-10; Cortes Island 23 Apr 1982-8; Winter Harbour 15 Apr 1969-1 (Richardson 1971); lower Adam River 6 May 1982-100 in logging slash; Cluxewe River 17 Mar 1991-2 on estuary (Dawe et al. 1995); Port Hardy 27 Apr 1938-100; Quatse River 27 Apr 1951-20; Port Neville Inlet 23 Apr 1974-5; Cape Scott 1 May 1975-100; Triangle Island 20 May 1978-1 at feeder; Pine Island 19 Mar 1977-2; Calvert Island 16 May 1933-3 (Linsdale and Sumner 1934); Cape St. James 6 May 1982-1; Goose Island 12 May 1948-70 foraging on beach (Guiguet 1953); Namu 1 May 1981-200 foraging in tidal area; Bella Coola 20 Apr 1933-3, 27 Apr 1998-200 feeding on spilled grain, 5 May 1933-1 (MCZ 285462); Lyell Island 28 Apr 1982-1; Kimsquit River 30 Apr 1982-1; Kitlope Lake 3 May 1991-21; Lord Bight (Langara Island) 10 May 1981-5; Masset 19 Apr 1990-1 (Siddle 1990b), 18 May 1920 (MZV 83131); Rose Spit 5 May 1979-1; Prince Rupert 2 May 1976-12; Metlakatla 24 Apr 1910-1 (Keen 1910); Green Island 29 Mar 1966-1; Gnarled Islands 17 May 1987-10 (Rodway and Lemon 1991); Port Simpson 4 May 1886-1; Kitwanga 30 Apr 1978-1; Hazelton 1 May 1963-1; Stonehouse Creek 19 May 1978-1 (Blood and Chutter 1978). **Interior** – Manning Park 29 Apr 1979-6; Grand Forks 29 Apr 1995-1 (Walker 1996); Creston 13 May 1984-1; Vaseux Lake 10 May 1973-2; West Bench (Penticton) 30 Apr 1968-1; Penticton 6 May 1974-8 at Mac's Lake (Cannings 1974); Okanagan Landing 26 Apr 1912-1; Lavington 27 May 1962-1; Vernon 3 May 1921-5; Nakusp 4 May 1976-1, 10 May 1986-1; Lac du Bois Grassland Park 5 May 1972-6; Kamloops 18 May 1984-1; Sorrento 5 May 1970-2; Revelstoke 3 May 1989-1; Watson Lake (100 Mile House) 9 May 1959-15; Kleena Kleene 26 Apr 1951-1 (Paul 1959); Cochin Lake 14 Apr 1989-1; Westwick Lakes 7 May 1955-2; Chilcotin River (Riske Creek) 17 Apr 1984-1 in shrub thicket at Wineglass Ranch; Williams Lake 8 May 1970-6, 22 May 1982-1; Riske Creek 1 May 1987-3; Anahim Lake 29 Apr 1992-1; Puntchesakut Lake 9 May 1944-1 (ROM 85991); Blackwater River 25 May 1983-2 nr bridge crossing; Bowron Lake Park 17 May 1975-3 (O'Brien and Bell 1975); 24 km s Prince George 25 May 1976-1; Ootsa Lake 10 May 1997-15; Nulki Lake 6 May 1946-2; Prince George 26 Apr 1992-1 with flock of White-crowned Sparrows; Vanderhoof 23 Apr 1983-1; Willow River 4 May 1968-3; Hudson Bay Mountain (Smithers) 29 May 1995-earliest date for eggs in nest (Norment et al. 1998); Tyhee Lake 1 May 1987-3; Quick 4 May 1985-20; Mackenzie 5 May 1997-1; Tetana Lake 14 May 1938-1 (RBCM 8294); Munro Creek (Manson River) 16 May 1993-1 singing (Price 1993); Moberly Lake 4 May 1998-1; Taylor 31 May 1986-1; n Beatton Park 1 May 1985-1 with 50 White-crowned Sparrows; e Cecil Lake 6 May 1988-2; 14 km n Pink Mountain 13 May 1982-1; Hyland Post 27 May 1976-1; Summit Lake (Stone Mountain Park) 18 May 1975-1, 26 May 1997-9 feeding on ground after 20 cm snowfall; Chilkat Pass 9 May 1957-1 (Weeden 1960), 14 May 1977-20; Atlin 9 May 1930-1 (RBCM 5727).

Summer: Coastal – Sooke 10 Jun 1970-1 (Tatum 1971); Victoria 20 Jul 1888-1 (MCZ 225782); Saanich 23 Aug 1981-3; North Saanich 1 Jun 1973-3 nestlings (Campbell 1975); Englishman River 29 Jun 1980-2 on estuary (Dawe et al. 1994); Qualicum Beach 11 Aug 1975-3 (Dawe 1976); Vancouver 6 Jul 1972-4; North Vancouver 15 Jun 1967-3 nestlings; Surrey 11 Jun 1964-1; Forbidden Plateau 16 Jul 1995-1; Mitlenatch Island 15 Jul 1966-1 banded (Campbell and Kennedy 1965); Garibaldi Park 3 Aug 1959-4 eggs; Campbell River 9 Aug 1975-2 on estuary; Whistler Mountain 26 Jun 1924-1 (UBC 5922); Triangle Island 4 Jun 1974-1 in scrub crab apple nr top of island; Tree Islets 11 Jun 1976-1; nr Stuie 16 Jul 1940-1 fledgling; Lawyer Islands 23 Jun 1979-1 adult in shoreline brush; Mount Thornhill 8 Aug 1977-1 in alpine; Nine Mile Mountain 21 Jul 1921-5 (Swarth 1924); Kinskuch Lake 23 Jun 1956-1 at treeline with beak full of insects; Moosehorn Lake 18 Jul 1987-11; Three Guardsmen Lake 12 Jun 1979-12; Tina Creek 6 Jun 1978-1 (Blood and Chutter 1978); Summit Lake (White Pass) 12 Jun 1899-2 adults constructing nearly completed nest (Bishop 1900). **Interior** – Cathedral Lakes 7 Jul 1951-2; Stagleap Park 20 Jul 1980-1 singing; Botanie Creek 18 Jul 1964-1; Tranquille Lake 17 Aug 1996-2, early migrants; Tenquille Lake 23 Jul 1960-1 (Bradley 1961); Mount McLean 15 Jul 1916-1 (RBCM 3619); Enderby 6 Aug 1941-1 (UBC 7820); Yalakom Mountain 28 Jul 1969-3 fledglings; 3 km ne Mount Pinkerton 18 Jul 1993-4 eggs; Powell Creek 17 Jul 1993-8, 7 adults and 1 recent fledgling; China Head Mountain 10 Jul 1979-1; Adams Plateau 28 Aug 1966-1; Crowfoot Mountain 16 Jul 1960-8; Mount Tatlow 9 Jul 1989-4 eggs, advanced; Avalanche Mountain 26 Jun 1984-1; Cherry Creek 8 Jul 1981-6 in subalpine, 1 carrying food to nest; Hemp Creek 2 Jun 1950-1; Herman Valley (Wells Gray Park) 13 Jul 1960-3 eggs; Eureka Peak 2 Jul 1988-9 singing; Rainbow Range 13 Jun 1956-4 eggs, 22 Jun 1992-5 nestlings; Itcha Mountain 14 Aug 1931-1 (UBC 5923); Tonquin Pass 12 Jul 1931-nest with young (Munro and Cowan 1947); Tonquin Ridge 30 Aug 1974-1 at 2,438 m elevation (Cannings et al. 1974); Moose Pass 30 Aug 1973-20; Two Sisters Mountain 3 Aug 1971-2 (Runyan 1971); Indianpoint Lake 12 Jun 1924-1 (MCZ 285427); Nekik Mountain 8 Jul 1977-1 in krummholz; Hook Lake 17 Jun 1977-5 in krummholz; Mount Babcock 17 Jun 1995-5 eggs; Mast Creek 20 Jul 1976-3 eggs, nr headwaters; Hudson Bay Mountain 11 Aug 1996-latest date for nestlings (Norment et al. 1998); ne Nine Mile Mountain 5 Jun 1958-4 eggs; Driftwood Range 2 Jul 1941-1 (Stanwell-Fletcher and Stanwell-Fletcher 1943); Kitchener Lake 20 Jun 1976-1 adult building nest (Osmond-Jones et al. 1977); Stalk Lake 11 Jul 1976-6; Fire Flats 11 Jun 1977-4 eggs; nr Akie River 15 Aug 1997-1 fledgling with adult; Pink Mountain 21 Jul 1982-4 (Grunberg 1982d); Didene Creek 15 Jul 1977-1; Mount Kirk 16 Jul 1919-2 bobtailed fledglings (Swarth 1922); Dokdaon Creek 16 Jul 1919-2 (Swarth 1922); Eaglenest Range (Spatsizi Plateau Wilderness Park) 7 Aug 1959-3 fledglings; Ross River 3 Jun 1976-2; 19 km n Telegraph Creek 4 Jun 1919-1 (MVZ 39932); Mount Barrington 19 Jul 1982-2; Haworth Lake 3 Aug 1976-30 in willow thickets; Prophet River 15 Jul 1998-9, including 3 fledged young; Fern Lake area 20 Aug 1983-10 in willows (Cooper and Cooper 1983); Gnat Lake 11 Jun 1962-5 (NMC 50331 to 50335); Tatsamenie Lake 18 Jul 1987-1; Kakuchuya Creek 29 Jun 1978-6; Tanzilla Butte 8 Jun 1962-2 (NMC 50328 to 50329); Dudidontu River 2 Jun 1978-2; Summit Pass 30 Jun 1987-2 adults with nearly completed nest; Muncho Lake 28 Aug 1943-5 (Rand 1944); Wilson Creek (Atlin) 30 Jun 1914-1 (RBCM 3613); Kahan Creek 8 Jul 1977-1 recent fledgling; Mount

Prudence 30 Jul 1981-2 fledglings; nr Tats Lake 2 Jul 1983-6 nestlings, eyes closed; Shallow Lake 24 Jun 1979-1 singing; Monarch Mountain 19 Jun 1924-5 eggs (Swarth 1926); Spruce Mountain 3 Aug 1924-1 (MVZ 106340); nr Wright Creek (Atlin) 16 Jun 1975-2 in alpine scrub; s Carmine Mountain 19 Jun 1983-4 nestlings, eyes closed; Datlasaka Range 10 Jul 1980-4 half-grown nestlings; Chilkat Pass 14 Jun 1957-4 nestlings, 4 or 5 days old.

Breeding Bird Surveys: Coastal – Recorded from 2 of 27 routes and on 1% of all surveys. Maxima: Pemberton 30 Jun 1974-2; Seabird 10 Jul 1977-2. **Interior** – Recorded from 8 of 73 routes and on 2% of all surveys. Maxima: Haines Summit 27 Jun 1993-32; Chilkat Pass 1 Jul 1976-31; Summit Lake 18 Jun 1972-2.

Autumn: Interior – White Pass 10 Sep 1924-1 (MVZ 106341); Atlin 12 Sep 1913-12 (Kermode and Anderson 1914); Haines Highway 9 Sep 1972-3; Fort Nelson 13 Sep 1987-1; Summit Pass 1 Sep 1943-1 (Rand 1944); upper Griffith Creek 4 Sep 1976-13; Flatrock 1 Oct 1951-1 at edge of large marsh; Fort St. John 7 Sep 1980-1, 3 Oct 1986-1 at sewage lagoon; Pyramis Peak 13 Oct 1998-1 (Sherrington 1998); Gagnon Creek 16 Sep 1997-1; Nilkitkwa Lake 16 Sep 1980-1 in burn; Sinkut River 20 Sep 1952-3 (Munro 1955a); Kidprice Lake 4 Oct 1974-1; Ootsa Lake 4 Oct 1997-5 along brushy fenceline; Chezacut 16 Sep 1933-1 (MCZ 285463); Everett Creek 12 Sep 1992-12 in alpine meadow; Williams Lake 24 Sep 1978-2 on beach, 8 Oct 1954-1 (NMC 48454) ; Chilcotin River (Riske Creek) 18 Sep 1983-1 killed by cat at Wineglass Ranch; nr Horse Lake 28 Sep 1975-1; Revelstoke 19 Sep 1986-1; Bralorne 22 Sep 1933-1 (ROM 68576); Tranquille 18 Oct 1987-1; Enderby 14 Oct 1941-1 (UBC 7821); Grindrod 13 Sep 1947-1 (UBC 7814); Vernon 21 Sep 1965-1 (Rogers 1966a); Lavington 19 Sep 1970-1; Kelowna 10 Sep 1991-1 (Siddle 1992a); Okanagan Landing 18 Oct 1933-1 (USNM 425360); Corbett Lake 1 Oct 1979-1; Sparwood 1 Oct 1983-1 (Fraser 1984); Chute Lake 3 Sep 1942-1; Arawana 12 Sep 1965-1; West Bench (Penticton) 29 Sep 1973-1; Manning Park 4 Oct 1969-1. **Coastal** – 37 km n Hazelton 1 Sep 1921-1, first migrant (Swarth 1924); Green Island 16 Oct 1978-2 at lightstation; Prince Rupert 20 Oct 1981-1 killed by cat; Kitimat Arm 17 Sep 1974-5 (Hay 1976); Masset 4 Sep 1937-1 (RBCM 10288); Nanika River 4 Oct 1974-2; Queen Charlotte City 30 Sep 1971-1; Cape St. James 12 Oct 1981-3; Safety Cove (Calvert Island) 11 Sep 1934-1 (MCZ 285432); Triangle Island 6 Sep 1994-1 banded; Port Neville 14 Sep 1977-16 feeding on beach; Port Hardy 25 Sep 1936-20; lower Adam River 18 Sep 1984-6 in logging slash; Campbell River 23 Sep 1979-1 on estuary; Garibaldi Lake 3 Oct 1965-2; Port Mellon 29 Nov 1987-1; Pender Harbour 18 Sep 1980-1; Comox 27 Sep 1989-2; Union Bay 18 Sep 1988-12 in blackberry thicket; Forbidden Plateau 15 Sep 1962-1 (Boggs and Boggs 1963a); Cypress Park 27 Sep 1987-80; Hollyburn Mountain 3 Oct 1982-8 migrating (1,219 m); Jericho Park 9 Sep 1974-15; Qualicum Beach 10 Nov 1974-25 (Dawe 1976); Parksville 4 Sep 1983-1 banded; Fanny Bay 17 Sep 1990-1 (Dawe et al. 1995a); Chilliwack 13 Sep 1888-1 (Cooke 1913); Pitt Polder 13 Sep 1967-10 with White-crowned Sparrows; Vancouver 18 Sep 1965-30 at airport; Nanaimo River 10 Sep 1989-1 on estuary; Clayoquot 17 Sep 1907-1 (NMC 4182); Tofino 9 Sep 1996-100; Lennard Island 13 Oct 1976-100 in migration; Long Beach (Pacific Rim National Park) 11 Sep 1970-8; Mount Arrowsmith 10 Sep 1938-2; Central Saanich 19 Sep 1970-25; Mill Bay 4 Sep 1924-95 (Sprot 1925); Pachena Point 16 Oct 1974-10; Victoria 11 Oct 1953-200 (Flahaut and Schultz 1954a); Saanich 1 Sep 1979-6; Blenkinsop Lake (Saanich) 21 Nov 1987-26; Goldstream River 22 Sep 1984-14 on estuary; Colwood 27 Sep 1986-20; Mount Finlayson 12 Oct 1985-23 at summit; East Sooke Park 13 Sep 1984-12; River Jordan 5 Oct 1986-6.

Winter: Interior – Williams Lake 8 Dec 1984-1 at feeder with Harris's Sparrow, 31 Dec 1988-1; Eagle Lake 3 Feb 1992-1; Chilcotin River (Riske Creek) 21 Jan 1989-1 at feeder at Wineglass Ranch; Revelstoke 22 Dec 1992-1 (Siddle 1992b); Grindrod 3 Jan 1973-1; Enderby 6 Dec 1972-1 at feeder (Rogers 1973b); Vernon 15 Feb 1973-1 at feeder; Okanagan Landing 21 Jan 1930-1; Penticton 26 Dec 1967-1 (Rogers 1968b); Osoyoos Lake 28 Dec 1973-1. **Coastal** – Stewart 22 Dec 1991-1 at feeder; Kaien Island 3 Jan 1982-1 in shrubbery; Masset 12 Feb 1941-1 (RBCM 10285); Queen Charlotte City 1 Dec 1971-1 (Crowell and Nehls 1972b); Skidegate Inlet 20 Dec 1986-1; Bella Coola 24 Dec 1989-2 at feeder; Sointula 12 Feb 1989-3; Willow Point (Campbell River) 30 Jan 1975-2 at feeder; Woss 18 Jan 1976-4; Mitlenatch Island 21 Dec 1965-2 (Campbell 1965); Squamish River 29 Jan 1980-1 on estuary; Comox 21 Jan 1989-20 in blackberry tangle; Union Bay 11 Feb 1994-7 at feeder; Trent River 19 Dec 1987-19 on estuary; Qualicum Beach 7 Dec 1974-20 (Dawe 1976); Delta 2 Dec 1995-60 (Elliott and Gardner 1997); Vancouver 19 Feb 1895-4; Matsqui 4 Feb 1968-4; Boundary Bay 29 Dec 1979-15; Marpole 26 Dec 1971-3 at feeder; Westham Island 16 Jan 1966-10; Steveston 21 Dec 1963-15; Ahousat 24 Feb 1991-9 at feeder; Tofino 2 Dec 1979-5; Saltspring Island 9 Dec 1978-6; Pender Island 29 Dec 1967-2 (Stevens 1968); Somenos Lake 30 Jan 1988-5; Cowichan Bay 3 Dec 1977-18; Sidney 1 Dec 1978-10; Central Saanich 8 Jan 1956-60 (Davidson 1972); Pachena Point 15 Feb 1975-1 at lightstation; Saanich 27 Dec 1982-15; Cordova Spit 13 Feb 1983-16; Victoria 11 Jan 1953-30 (Davidson 1953); Metchosin 11 Feb 1978-20.

Christmas Bird Counts: Interior – Recorded from 4 of 27 localities and on 3% of all counts. Maxima: Penticton 27 Dec 1980-6; Oliver-Osoyoos 28 Dec 1993-1; Vaseux Lake 23 Dec 1982-1; Vernon 21 Dec 1986-1. **Coastal** – Recorded from 26 of 33 localities and on 66% of all counts. Maxima: Victoria 18 Dec 1993-**1,017** all-time Canadian high count (R.J. Cannings, pers. comm.); Duncan 2 Jan 1993-335; Ladner 23 Dec 1990-267.

Dark-eyed Junco

DEJU

Junco hyemalis (Linnaeus)

J. h. hyemalis group ("Slate-colored" Junco) SCJU

J. h. oreganus group ("Oregon" Junco) ORJU

RANGE: Breeds from western and northern Alaska and northern Yukon east to the northern regions of Manitoba, Ontario, Quebec, Labrador, and Newfoundland; south to southern California and in the mountains to northern Baja California, east to northeastern Arizona and northwestern New Mexico, central Colorado, Wyoming, and Montana, to southern Manitoba, northern and east-central Minnesota, Wisconsin, and Michigan, to southern New England states and south in the Appalachians to Georgia and northwestern South Carolina. Winters from south coastal Alaska and coastal and southern British Columbia east across southern Canada to Newfoundland, south to northern Baja California and northern Mexico, east to Texas and along the Gulf coast to northern Florida.

STATUS: On the coast, *uncommon* to *very common* resident in the Georgia Depression, locally *abundant* during migration and in winter. Elsewhere on the coast, including all of the Coast and Mountains Ecoprovince, *fairly common* to locally *common* resident, occasionally *very common* to *abundant* during migration or in winter.

In the interior, a *common* to *very common* resident (occasionally *abundant* during migration or in winter) in the Southern Interior, Southern Interior Mountains, and Central Interior ecoprovinces; *fairly common* to locally *common* migrant and summer visitant in the Sub-Boreal Interior, Boreal Plains, and Northern Boreal Mountains ecoprovinces; *uncommon* to *fairly common* migrant and summer visitant (occasionally *common* during migration) to the Taiga Plains Ecoprovince. In winter, *uncommon* to *fairly common* in the Sub-Boreal Interior, *accidental* in the Boreal Plains, Northern Boreal Mountains, and Taiga Plains ecoprovinces.

Breeds.

Figure 383. Adult Dark-eyed Junco, *hyemalis* group (Michael I. Preston). The gray head, back, and breast with the white belly are the best distinguishing field marks.

Figure 384. Adult Dark-eyed Junco, *oreganus* group (Victoria, 27 September 1998; R. Wayne Campbell). The black hood, brown back, and pinkish sides are the best distinguishing field marks.

NONBREEDING: The Dark-eyed Junco (Figs. 383 and 384) is the most widespread and abundant of the sparrows in British Columbia, and occurs throughout the province from spring through autumn. In winter, it is absent from most of the interior north of Fort St. James (latitude 54°30'N). The highest numbers in winter occur in the Georgia Depression (Fig. 385). In coastal regions, the largest numbers are due to late autumn or wintering birds, whereas in the interior, large numbers are more common during migration periods, with spring migration more evident than the autumn movement (Fig. 386).

Three groups of the Dark-eyed Junco have been reported from British Columbia. Two occur regularly in the province: the *hyemalis* group or "Slate-colored" Junco (Fig. 383), and the *oreganus* group or "Oregon" Junco (Fig. 384). A third, the *caniceps* group or "Gray-headed" Junco, has been reported only once (Dawe 1976a) (see also REMARKS). Because the 2 common groups are usually separable in the field and differ from each other in their numbers and summer distribution, we have included a more detailed discussion of them throughout this account.

During the nonbreeding seasons in spring and autumn, both the "Oregon" Junco and the "Slate-colored" Junco have a widespread distribution across southern and central British Columbia, including Vancouver Island and the Queen Charlotte Islands, although the "Oregon" Junco is far more plentiful. North of latitude 56°N and east of the Rocky Mountains, the "Slate-colored" Junco is the more abundant of the two. In winter, the "Oregon" Junco is the dominant group in British Columbia; the "Slate-colored" Junco occurs only in small numbers.

On the coast, the Dark-eyed Junco has been reported from near sea level to more than 1,800 m elevation; in the interior, it has been reported from valley bottoms to krummholz at

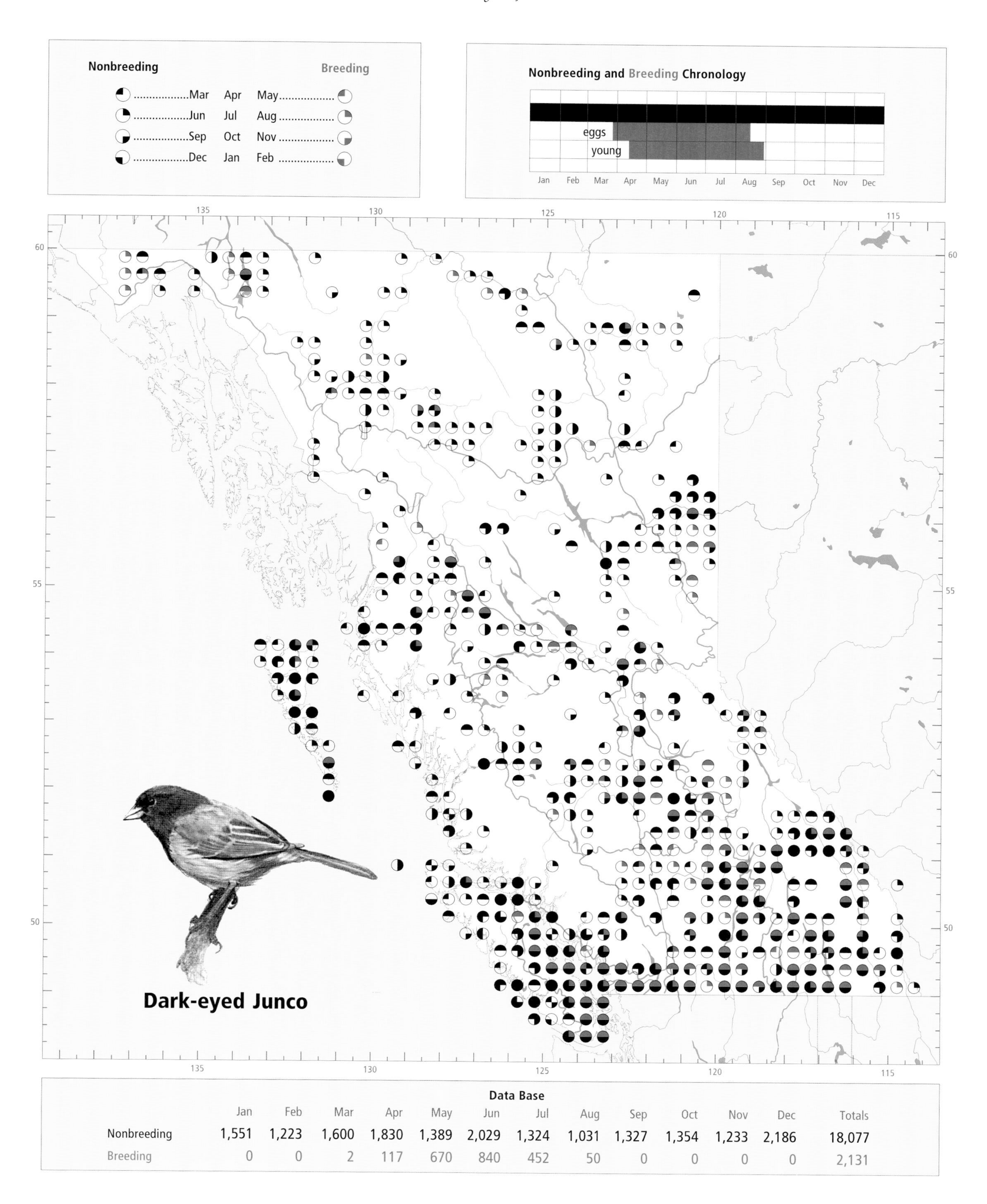

Data Base

	Jan	Feb	Mar	Apr	May	Jun	Jul	Aug	Sep	Oct	Nov	Dec	Totals
Nonbreeding	1,551	1,223	1,600	1,830	1,389	2,029	1,324	1,031	1,327	1,354	1,233	2,186	18,077
Breeding	0	0	2	117	670	840	452	50	0	0	0	0	2,131

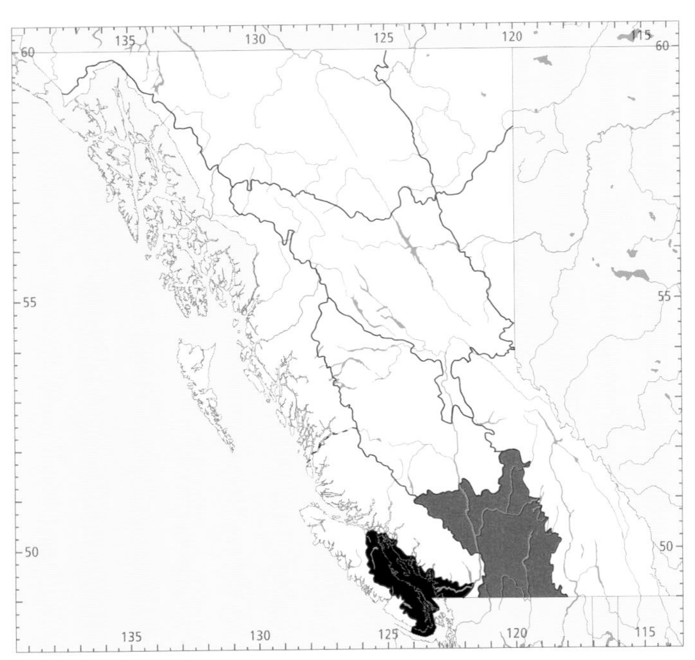

Figure 385. In British Columbia, the highest numbers for the Dark-eyed Junco in winter (black) occur in the Georgia Depression Ecoprovince; the highest numbers in summer (red) occur in the Southern Interior Ecoprovince.

2,100 m. During the nonbreeding season, it frequents primarily the edges of open fields, brushy areas, and forest edges. In British Columbia, it has been reported from open farmlands; pastures; flooded and fallow fields; beach edges and spits (Fig. 387); weed patches along roadsides; meadows; clearcuts; open burns; urban and rural yards and gardens (especially those with bird feeders); powerline rights-of-way; riparian habitat adjacent to estuaries, sloughs, sewage ponds, lagoons, and beaver ponds; sphagnum bogs, lakeshores, rivers, and marshes; holly farms and nurseries; shrub areas such as willow thickets, woodlots, and coniferous, deciduous, and mixed forest edges.

On the south coast, overwintering numbers in the Georgia Depression mask an overall spring migration through the ecoprovince, although undoubtedly a movement arrives or passes through from areas further south (Fig. 389; see also REMARKS). Winter numbers start to decline by February and the decline continues to the end of April as birds leave the lowland areas and move further north or to higher elevations. This pattern is common in the lowlands on the south coast (see, for example, Campbell et al. 1972b; Tatum 1973; Dawe 1976a; Dawe et al. 1998). On the Southern Mainland Coast and on Western Vancouver Island, our data show little in the way of a spring movement. Further north in the Northern Mainland Coast, numbers begin to arrive in the last week of March and build through the first week of May. On the Queen Charlotte Islands, wintering numbers begin leaving by February and appear to be replaced by birds moving into the islands from southern areas.

In the Southern Interior, wintering numbers tend to cloud an overall spring movement, although away from urban

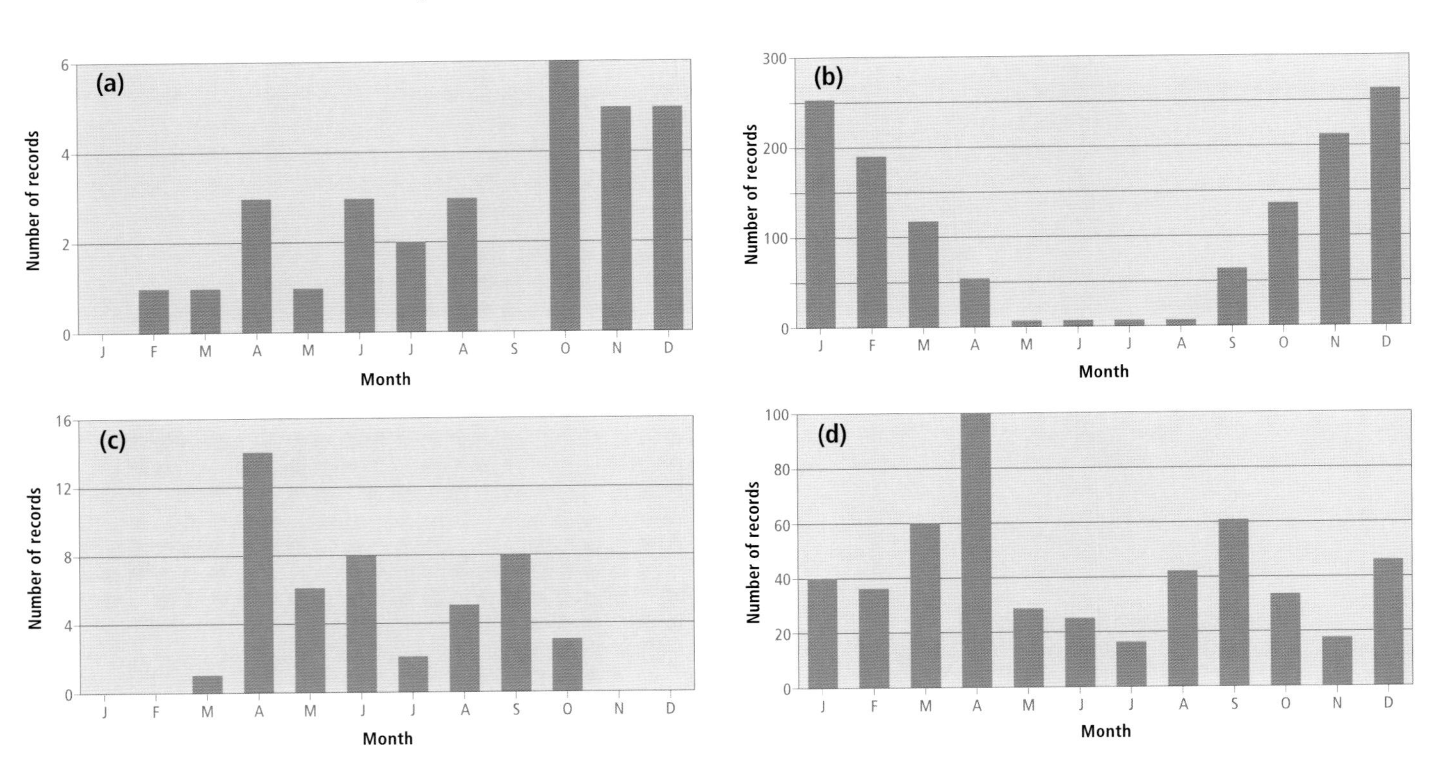

Figure 386. Seasonal distribution of large flocks or groups of flocks (> 20 birds) of the Dark-eyed Junco in the (a) north coast (Northern Mainland Coast, Queen Charlotte Islands), (b) south coast (Georgia Depression, Western Vancouver Island, Southern Mainland Coast), (c) northern interior (Sub-Boreal Interior, Boreal Plains, Northern Boreal Mountains, Taiga Plains), and (d) southern interior (Southern Interior, Southern Interior Mountains, Central Interior) of British Columbia. In coastal regions, the largest numbers are due to late autumn or wintering birds; in the interior, large numbers are more common during migration periods, with spring migration more evident than the autumn migration. Note the differences in scale of the *y*-axes.

areas, where wintering birds tend to concentrate, a spring movement is easier to detect. There birds may arrive in late March and early April. Generally, numbers build through April (Fig. 389) then drop in May as birds move out of the valley bottoms to higher elevations or to areas further north. In the Okanagan valley, birds leave the valley bottoms in March and April and move to the higher Douglas-fir and subalpine forests of the area (Cannings et al. 1987). Spring migration in the Southern Interior Mountains is a little more obvious as numbers begin to build in the third week of March, peak in the first week of April, and continue into the first week of May. As in the Southern Interior, wintering birds move out of the valley bottoms in March and April to forested areas at higher elevations. For example, Wade (1977) notes that in Yoho National Park, the Dark-eyed Junco arrives in numbers in early April and is abundant from May to August; by mid-October it has become relatively scarce. In the Central Interior, numbers first appear about the second week of March and climb through the first and second weeks of April, dropping slowly thereafter as birds move further north or to higher elevations. Munro (1947) notes a small movement at Puntchesakut Lake between 8 and 23 May.

In the Sub-Boreal Interior, small numbers of overwintering birds mask the early arrivals, but numbers begin to build about the third week of March and continue through April, peaking in the last 2 weeks of the month. Both groups move through the Mackenzie area, but the "Oregon" group is more abundant. There the "Oregon" Junco migration begins as early as mid-March, whereas the "Slate-colored" Junco does not appear until the second week of April (W.J. Tuck pers. comm.).

Figure 387. On the coast of British Columbia, the Dark-eyed Junco is commonly found in migration along beach edges and spits (Rose Spit, Queen Charlotte Islands, 21 May 1996; R. Wayne Campbell).

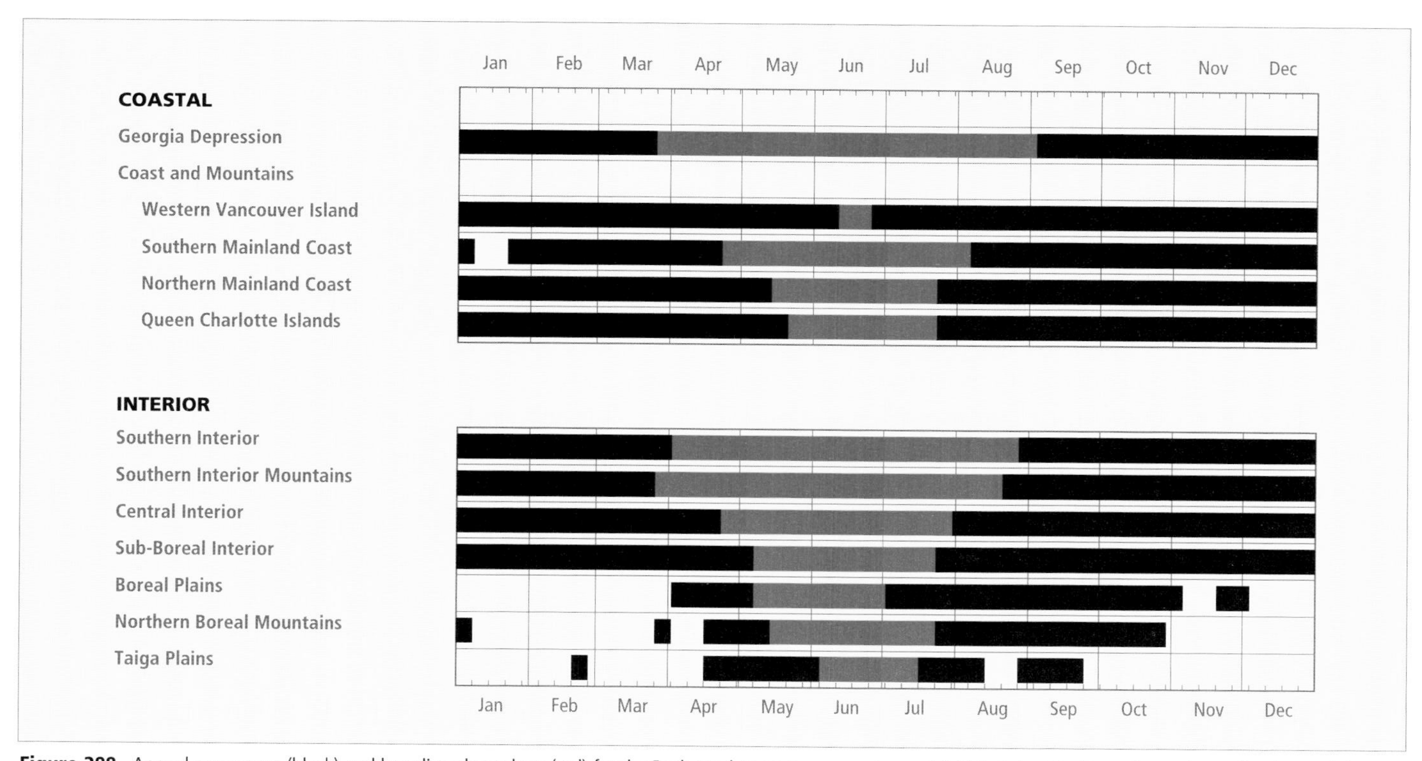

Figure 388. Annual occurrence (black) and breeding chronology (red) for the Dark-eyed Junco in ecoprovinces of British Columbia. Records are shown for the week in which they occurred.

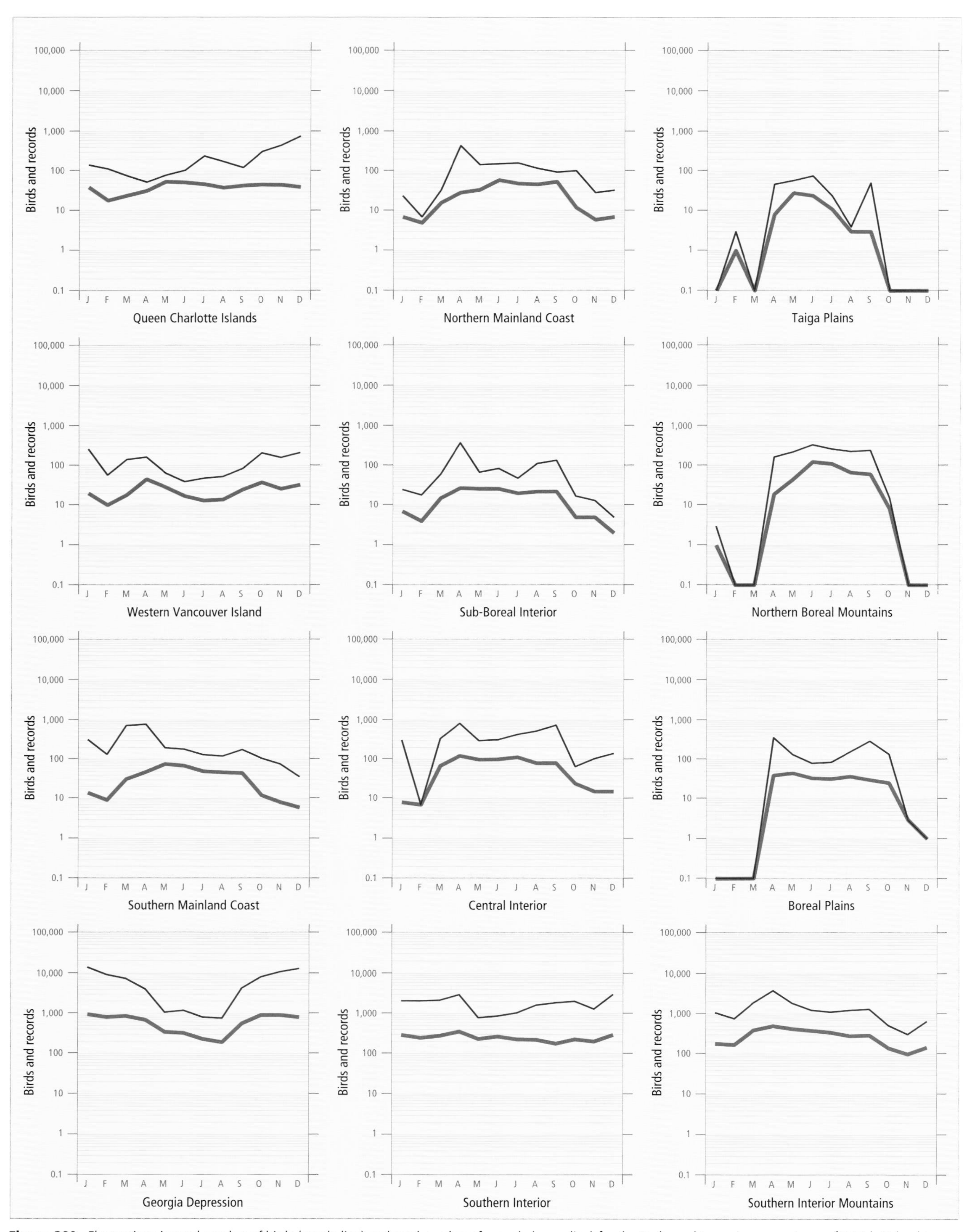

Figure 389. Fluctuations in total number of birds (purple line) and total number of records (green line) for the Dark-eyed Junco in ecoprovinces of British Columbia. Christmas Bird Counts, Breeding Bird Surveys, and nest record data have been excluded.

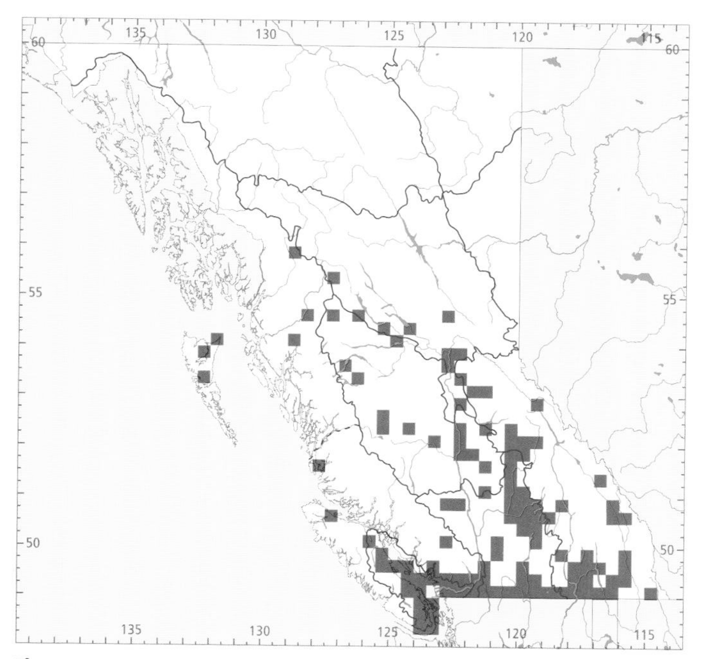

Figure 390. Known breeding distribution of the "Oregon" Junco in British Columbia.

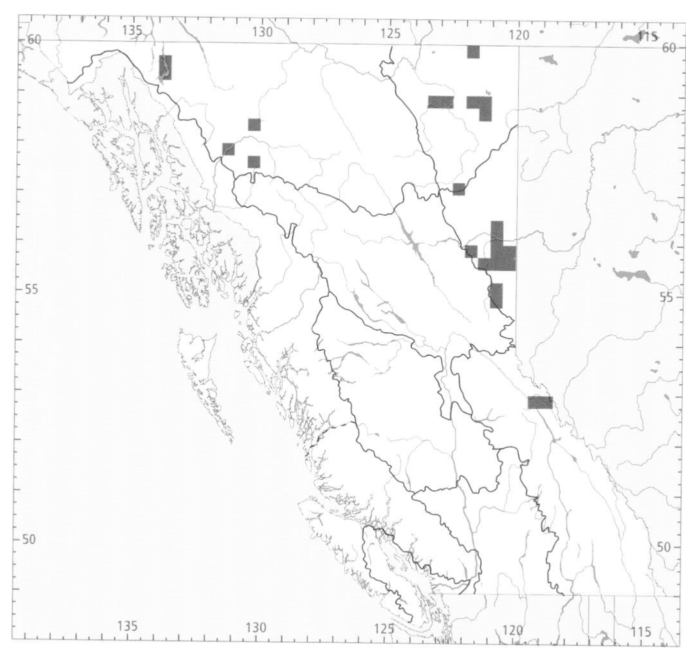

Figure 391. Known breeding distribution of the "Slate-colored" Junco in British Columbia.

Stanwell-Fletcher and Stanwell-Fletcher (1943) also note the "Oregon" Junco as a "very common spring and autumn migrant, and a fairly common summer resident" at Tetana Lake in the Sub-Boreal Interior, while the "Slate-colored" Junco is only a "fairly common spring and autumn migrant" there.

In the Northern Boreal Mountains, birds may arrive in the first week of April (Fig. 388); numbers are not seen until the third week of April, and they continue increasing into May. East of the Rocky Mountains in the Boreal Plains, birds may arrive in the first week of April, with numbers moving through to the end of the month. Our earliest reports from the Taiga Plains are in the third week of April, and numbers there peak in the last week of April or first week of May.

The autumn movement begins by the third week of August in the Northern Boreal Mountains. Numbers continue to move through September and most birds have gone by the end of that month; a few stragglers may be found in the area until the end of October. East of the Rocky Mountains, most birds have also left the Taiga Plains by the end of September. In the Boreal Plains, the autumn movement begins in August, with the main movement occurring in the last 2 weeks of that month and continuing to mid-October. Nearly all birds have left the region by the end of October (Figs. 388 and 389).

In the Sub-Boreal Interior, a small autumn movement is noticeable in the last week of August and the first 2 weeks of September, and most birds have left by mid-October. In the Central Interior, the autumn migration begins in early August (Munro 1947) and is rather protracted, with the main movement continuing to the end of September and smaller numbers continuing into October. Most birds have left the region by the second week of October, leaving small numbers to overwinter. In both the Southern Interior and Southern Interior Mountains, a movement is noticeable in late August and early September, particularly in the valley bottoms, as birds move down from the higher elevations or from areas further north. The autumn movement there continues through September and into the third week of October (Fig. 389).

On the Northern Mainland Coast, our data reveal little in the way of an autumn movement, but most birds have left the area by the end of October. On the Queen Charlotte Islands, an influx is noted in September and numbers climb through the autumn into December as the wintering population builds. On the southern portions of the coast, numbers begin to build around the end of August and continue climbing through the autumn to peak in December and January with the wintering population.

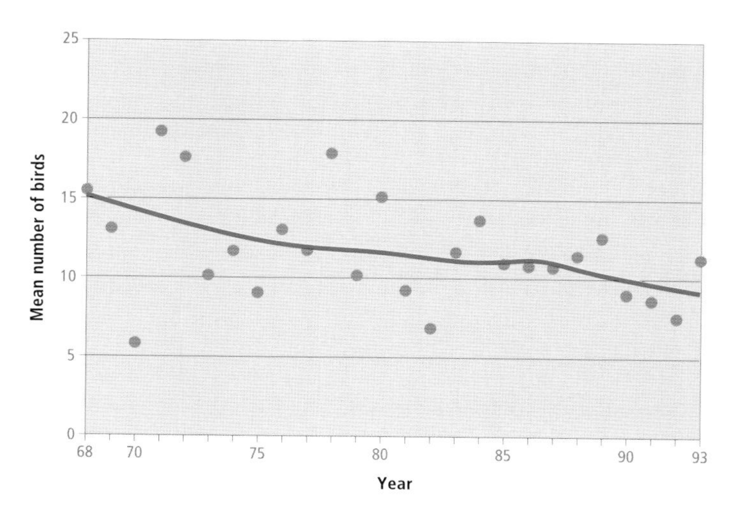

Figure 392. An analysis of Breeding Bird Surveys for the Dark-eyed Junco in British Columbia shows that the number of birds on coastal routes decreased at an average annual rate of 1% over the period 1968 through 1993 ($P < 0.10$).

Figure 393. The Dark-eyed Junco breeds in a wide variety of forested habitats throughout British Columbia. In the Southern Interior Ecoprovince, a few breed in shaded stands of mixed deciduous trees (Richter Pass, 10 October 1998; R. Wayne Campbell).

Figure 394. In parts of the Sub-Boreal Interior Ecoprovince, the Dark-eyed Junco breeds at the edges of trembling aspen forests (Germansen Landing, 3 July 1998; R. Wayne Campbell).

The Dark-eyed Junco regularly winters in large numbers along the coast, including Vancouver Island and the Queen Charlotte Islands. It also occurs regularly in large numbers in the Southern Interior. Smaller numbers winter in the southern valley bottoms of the Southern Interior Mountains and the Central Interior, and the species occurs regularly although locally in winter in the Sub-Boreal Interior. It has been reported in winter once each from the Taiga Plains, Northern Boreal Mountains, and Boreal Plains.

The Dark-eyed Junco occurs throughout the year along coastal British Columbia and in the central and southern portions of the interior of the province. In the northern portions of the interior, it occurs regularly from mid-April to late September (Figs. 388 and 389).

BREEDING: The Dark-eyed Junco has a widespread breeding distribution throughout most of British Columbia. There are few breeding records for Western Vancouver Island, much of the mainland coast north of Vancouver Island, and some areas of the far northern interior, but the Dark-eyed Junco undoubtedly breeds throughout its summer range in British Columbia.

The "Oregon" Junco has a widespread breeding distribution across the southern half of the province, including Vancouver Island and the Queen Charlotte Islands. It has been reported nesting on the coast as far north as Swan Lake, 83 km northwest of Kitwanga, and in the interior, to Redrocky Lake, some 80 km north of Prince George (Fig. 390); it likely nests north to at least Tetana Lake in the Driftwood River valley, as it is a "common summer resident" there (Stanwell-Fletcher and Stanwell-Fletcher 1943). Nesting records for the "Slate-colored" Junco were mainly from east of the Rocky Mountains and in the far northern portions of the province; it has also been reported nesting in the Mount Robson area of the Southern Interior Mountains (Fig. 391). The "Slate-colored" Junco is a common breeder in the boreal forest of northern British Columbia. In the Dawson Creek and Fort St. John areas, it is the common breeding junco; the "Oregon" Junco is occasionally seen there only during migration (Phinney 1998; C. Siddle pers. comm.).

During the nesting season, Swarth (1922) found both the "Oregon" Junco and the "Slate-colored" Junco at Flood Glacier, some 80 km upstream of the mouth of the Stikine River, with the "Oregon" Junco being dominant. He noted that 64 km upstream from Flood Glacier, nothing but the "Slate-colored" Junco was seen, whereas 48 km downstream from Flood Glacier only the "Oregon" Junco was found.

The Dark-eyed Junco reaches its highest numbers in summer in the Southern Interior (Fig. 385). An analysis of Breeding Bird Surveys in British Columbia for the period 1968 through 1993 shows that the mean number of birds on coastal routes decreased at an average annual rate of 1% (Fig. 392); an analysis of interior routes could not detect a net change in numbers.

Sauer et al. (1997) found that, based on Canada-wide surveys, the mean number of "Oregon" Juncos declined at an average annual rate of 2.5% ($P < 0.001$) from 1980 to 1996. Across North America, the numbers of both the "Oregon" and "Slate-colored" juncos appear to be declining as well. From 1966 to 1996, the numbers of the "Oregon" Junco declined at an average annual rate of 1.2% ($P < 0.001$), with most of the decline occurring from 1980 to 1996 (1.09%; $P < 0.001$). Similar trends were found for the "Slate-colored" Junco: its numbers declined at an average annual rate of 1.5% ($P < 0.001$) from 1966 to 1996, with most of the decline also occurring from 1980 to 1996 (1.6%; $P < 0.001$) (Sauer et al. 1997). Continent-wide Breeding Bird Survey data indicate that British Columbia has some of the highest densities of the "Oregon" Junco in North America (Sauer et al. 1997).

The Dark-eyed Junco has been reported breeding from near sea level to 1,800 m elevation on the coast, and from 200 to 2,400 m elevation in the interior.

Some habitats used during the breeding season (e.g., urban and suburban) are similar to those used during the nonbreeding season (see NONBREEDING), although the junco frequents a variety of forested habitats at higher elevation, particularly forest openings and edges, to a greater degree and is often found deeper in the forest when nesting (Figs. 393 and 394). It also makes greater use of subalpine and alpine meadows during the nesting season, and in the north it uses spruce-tamarack bogs and muskeg as nesting habitat.

On the west coast of Vancouver Island, Bryant et al. (1993) found the Dark-eyed Junco abundant and widespread in all age classes of forests except for 50- to 60-year-old stands. The juncos were most abundant in clearcuts and 15- to 20-year-old stands; their abundance was positively correlated with elevation and negatively correlated with canopy closure.

In Schwab's (1979) study of the effect of vegetation structure on breeding bird communities in the Wasa area of the east Kootenay, the Dark-eyed Junco was most common in the climax Douglas-fir forests dominated by conifers taller than 10 m with an open understorey of scattered common snowberry and grasses. This was followed by young coniferous forests dominated by conifers less than 10 m in height, with some common snowberry and kinnikinnick, and the young conifer/mature seral conifer transition dominated by conifers less than 10 m in height, with some up to 24 m in height and an open understorey of saskatoon and grasses.

In Mount Revelstoke and Glacier national parks, the highest densities of Dark-eyed Juncos were found in the subalpine fir–mountain hemlock open forest interspersed with heath tundra or herb meadow, the Engelmann spruce–subalpine fir open forest, and the mixed open forests of Engelmann spruce and black cottonwood with yellow dryas–willowherb mat (Van Tighem and Gyug 1983). The most important vegetation types used by the juncos included recent burns, well-drained, shrubby avalanche chutes, and Engelmann spruce–subalpine fir/rhododendron–tall bilberry open forest.

Table 5. Dark-eyed Junco density in various seral stages of trembling aspen forests near Smithers, British Columbia (1991-92) (adapted from Pojar 1995).

Seral stage	Density range (singing males per 10 ha)	*n*
Clearcut	17.8 – 29.7	6
Sapling trembling aspen	9.6 – 13.9	6
Mature trembling aspen	10.4 – 19.1	10
Old trembling aspen	12.7 – 21.2	4
Mixed conifer-aspen	14.6 – 17.5	3
Conifer	6.4 – 8.5	3

Pojar (1995) found the Dark-eyed Junco in all seral stages of the trembling aspen forests she studied in the Smithers area, although the highest junco densities were found in clearcuts and old trembling aspen forest (Table 5). Merkins and Booth (1998) noted an immediate colonization of clearcuts by the Dark-eyed Junco following clearcutting of mature (60- to 100-year-old) trembling aspen stands in the Peace River region. The presence of slash piles was thought to be an important habitat component of the clearcuts for the juncos. Junco abundance was significantly higher in younger than in older clearcuts.

In the Boreal Plains, spruce woodland edges, mixed woodland edges of white spruce with various mixtures of trembling aspen, balsam poplar, and paper birch, relatively pure trembling aspen woodlands, mature balsam poplar floodplain forest, black spruce muskeg, and riparian wooded creek edges are used primarily by the "Slate-colored" Junco during the nesting season (C. Siddle, pers. comm.).

In the Stikine River area, near Telegraph Creek and Glenora, Swarth (1922) noted that the "Slate-colored" Junco was mainly a bird of the valleys, occurring primarily in open woods. He also observed it "well up in the mountains, not in the dense spruce forests of the middle heights, but at the

Figure 395. Human-influenced habitats, such as early regenerating clearcuts and burns, are favourite nesting sites for the Dark-eyed Junco in British Columbia (Grizzly Lake, near Prince George, 14 June 1997; R. Wayne Campbell).

Figure 396. Most Dark-eyed Junco nests in British Columbia are built on the ground, often adjacent to a rock or log (15 km east of Nelson, 1 July 1994; Linda M. Van Damme).

upper edge of the timber where the trees were more scattered and of smaller size."

Most nests in British Columbia (69%; n = 833) were reported from forested habitats (Figs. 393 and 394), followed by human-influenced sites (22%; Fig. 395), alpine areas (4%), and riparian habitats (3%). Forested sites included coniferous (34%; n = 578), mixed (17%), and deciduous (6%) forests. Human-influenced nest sites included rural (10%) and suburban (7%) areas; cultivated farmlands, orchards, and vineyards (6%); regenerating clearcuts and burns (4%; Fig. 395); and urban and recreational areas. Riparian nest sites were associated with swamps, marshes, sloughs, lakes and ponds, bogs, muskeg, and bottomlands.

On the coast, the Dark-eyed Junco has been recorded breeding from 1 April (calculated) to 29 August; in the interior, it has been recorded breeding from 27 March to 22 August (Fig. 388).

Nests: Most nests were found on relatively flat ground (66%; n = 683), although a number were reported from dirt or clay banks, hillsides, or cliffs (23%), followed by shrubs, stumps, or trees (6%), and buildings (2%). Most ground nests were located among vegetation such as tall grasses, bracken, heather, and ferns, or at the base of a tree or beside a rock or a log (Fig. 396). Nests in banks or on hillsides usually had an overhang of grasses that concealed the nest. Nests found in trees or shrubs were usually located in the fork or crotch of a branch. A few nests were reported in hanging flower baskets (Fig. 397), at least 1 of which was "watered every day," and 1 bird nested in a greenhouse in a flat of marigolds. Nests were usually built in cuplike depressions and were constructed of grasses (100%; n = 593), mosses and lichens (8%), rootlets (5%), and plant fibres such as stems, twigs, bark strips, and leaves. Nest lining included fine grasses, hair (moose, deer, cattle, horse, black bear, caribou, porcupine quills), feathers (ptarmigan), and plant down. The heights of 672 Dark-eyed Junco nests ranged from ground level to 10.0 m, with 86% found at ground level. Of these, 360 were "Oregon" Junco nests ranging in height from ground level to 3.0 m, with 85% at ground level. All nests attributed to the "Slate-colored" Junco (n = 15) were found on the ground.

Figure 397. In 1997, a pair of juncos used this hanging flower pot as their nest site. After fledging the first brood, they built another nest in the same pot (the nest rims touched each other) and fledged a second brood. In 1998, a pair of juncos, possibly the same pair, built their first nest in the same pot and raised a brood (North Saanich, Vancouver Island, 3 June 1997; John M. Cooper).

Figure 398. Clutch of Dark-eyed Junco eggs (left) with 2 Brown-headed Cowbird eggs (right) (Victoria, 6 May 1993; R. Wayne Campbell). Only 3% of 840 Dark-eyed Junco nests in British Columbia were parasitized by the Brown-headed Cowbird.

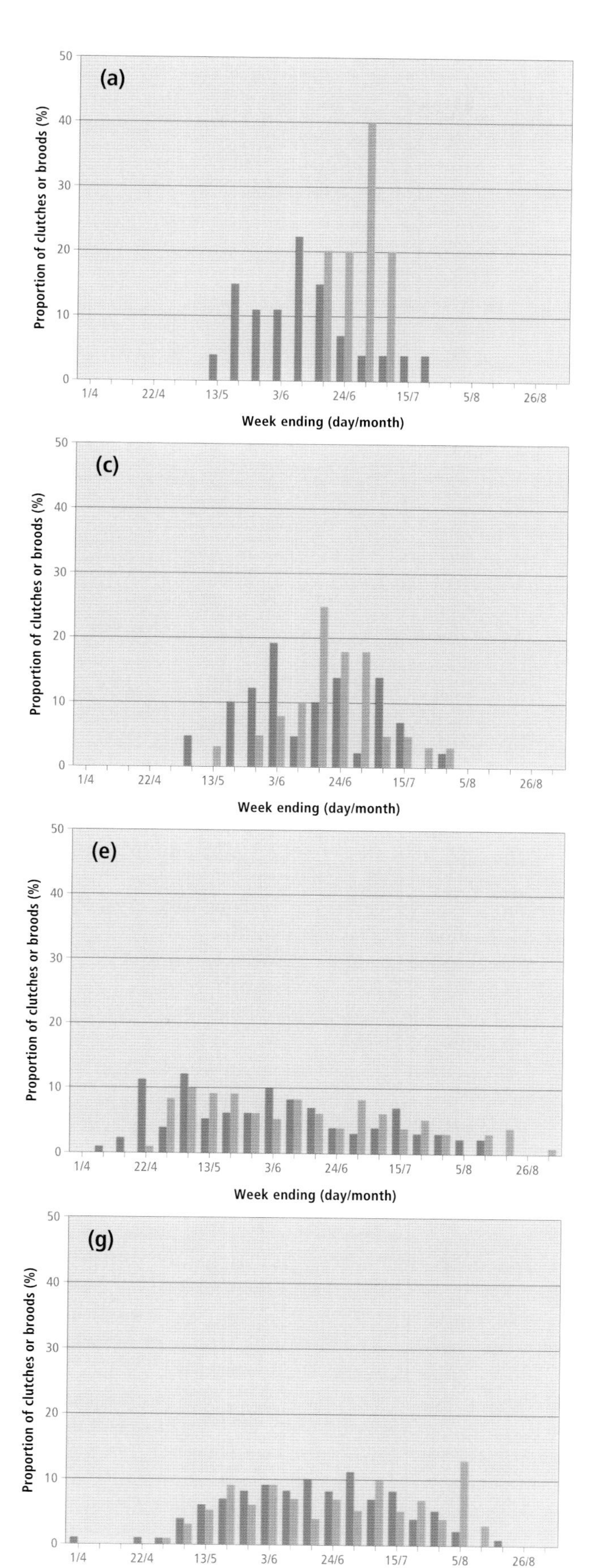

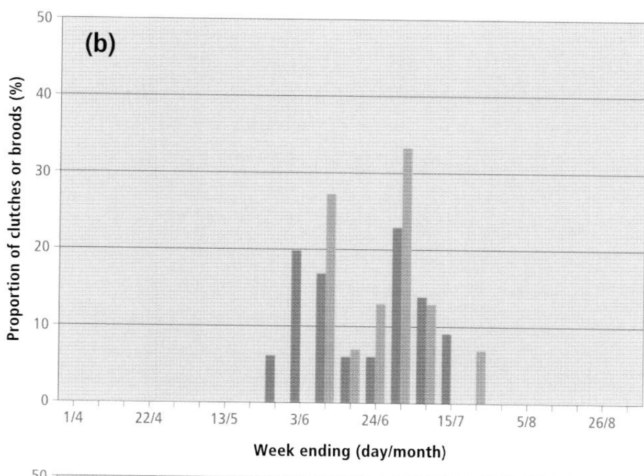

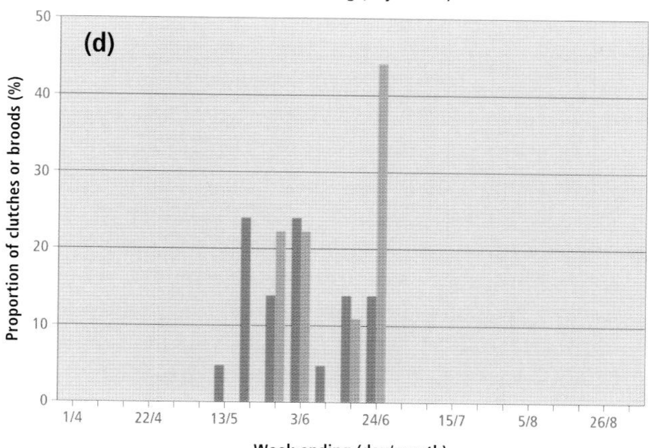

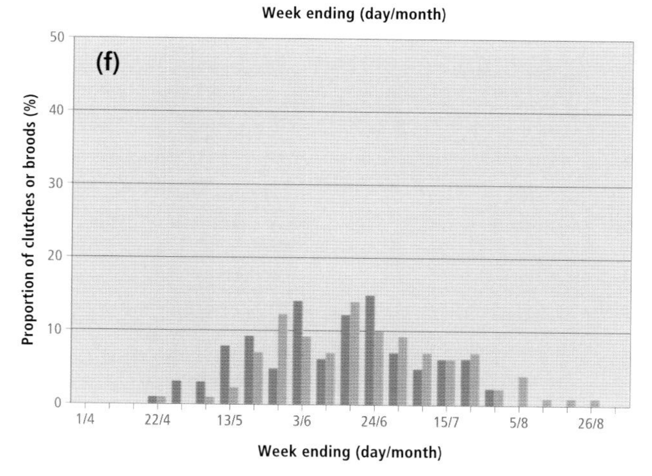

Figure 399. Weekly distribution of the proportion of clutches (dark bars) and broods (light bars) for the Dark-eyed Junco in the (a) Sub-Boreal Interior (clutches: $n = 27$; broods: $n = 5$), (b) Northern Boreal Mountains (clutches: $n = 35$; broods: $n = 15$), (c) Central Interior (clutches: $n = 42$; broods: $n = 40$), (d) Boreal Plains (clutches: $n = 32$; broods: $n = 7$), (e) Georgia Depression (clutches: $n = 117$; broods: $n = 78$), (f) Southern Interior (clutches: $n = 190$; broods: $n = 129$), and (g) Southern Interior Mountains (clutches: $n = 168$; broods: $n = 94$) ecoprovinces of British Columbia. The figures are based on the week eggs or young were found in the nest. Nesting in the southern portions of the province may begin up to 5 weeks earlier than in the central regions of the province and up to 8 weeks earlier than in the northern regions. The figures suggest that the junco is double-brooded in the far north, while in the southern portions of the province it may have 3 and perhaps as many as 4 broods.

Eggs: Dates for 1,244 clutches (Fig. 398) ranged from 27 March to 15 August, with 53% recorded between 20 May and 29 June. Nests in the southern portions of the province may contain eggs up to 5 weeks earlier than nests in the central portions of the province and up to 8 weeks earlier than nests in the northern regions (Fig. 399). Sizes of 615 clutches ranged from 1 to 5 eggs (1E-25, 2E-31, 3E-84, 4E-365, 5E-110), with 59% having 4 eggs. Three nests found with 6 young indicate 6-egg clutches occur. In British Columbia, the clutches of southern populations have a mode of 4 eggs while those of northern populations appear to have a mode of 4 eggs early in the season and 5 eggs later in the season (Fig. 400). Most 3-egg clutches from northern Utah represented renesting following either a successful first clutch or predation (Smith and Anderson 1982; Smith 1988), and our data support that hypothesis: a higher proportion of 3-egg clutches in British Columbia appear later in the season than either 4- or 5-egg clutches (Table 6). Smith and Anderson (1982) suggest that 5-egg clutches are usually associated with delayed breeding due to late-lying snowpack; this might explain the bimodal clutch size in the northern portions of the province, although latitudinal distribution likely played a role as well. The Dark-eyed Junco appears to be double-brooded in the far north; in the southern portions of the province, it may have 3 and perhaps as many as 4 broods (Fig. 399). The incubation period of 4 nests in British Columbia ranged from 12 to 13 days, with a median of 12 days.

Of the 1,244 clutches mentioned above, 749 clutches were attributed to the "Oregon" Junco. These clutches occurred between 8 April (calculated) and 7 August, with 52% recorded between 20 May and 30 June. Sizes of 298 "Oregon" clutches ranged from 1 to 5 eggs (1E-11, 2E-9, 3E-55, 4E-167, 5E-56), with 56% having 4 eggs.

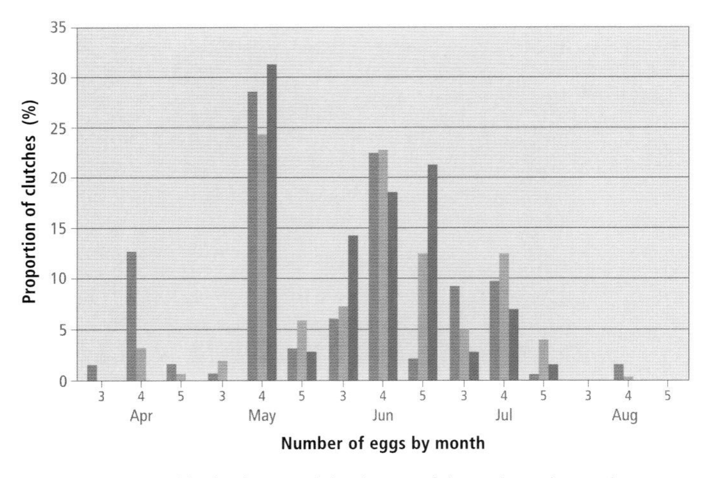

Figure 400. Monthly distribution of clutch sizes of the Dark-eyed Junco between April and August for southern populations on the coast (dark green bars; n = 133), southern populations in the interior (light green bars; n = 359), and northern populations in the interior (purple bars; n = 70). Clutch sizes of southern populations have a mode of 4 eggs while those of northern populations appear to have a mode of 4 eggs early in the season and a mode of 5 eggs later in the season.

Table 6. Proportion of various clutch sizes of the Dark-eyed Junco found in British Columbia early (before 15 June) and late (after 15 June) in the nesting season.

	Proportion (%)		
Nesting season	3 eggs (n = 88)	4 eggs (n = 367)	5 eggs (n = 113)
Early	41	66	54
Late	59	34	46

Dates for 38 clutches of the "Slate-colored" Junco ranged from 13 May to 15 July, with 55% recorded between 30 May and 27 June. Sizes of 32 clutches ranged from 1 to 5 eggs (1E-4, 2E-3, 3E-6, 4E-15, 5E-4), with 66% having 3 or 4 eggs.

Young: Dates for 887 broods (Fig. 401) ranged from 13 April to 29 August, with 51% recorded between 26 May and 30 June. Sizes of 370 broods ranged from 1 to 6 young (1Y-29, 2Y-40, 3Y-107, 4Y-163, 5Y-28, 6Y-3), with 73% having 3 or 4 young. The nestling period for 3 nests in British Columbia ranged from 9 to 11 days.

Of the 887 broods mentioned above, 537 were identified as "Oregon" Junco broods. They were found between 13 April and 22 August, with 51% recorded between 24 May and 30 June. Sizes of 185 "Oregon" Junco broods ranged from 1 to 6 young (1Y-11, 2Y-21, 3Y-51, 4Y-90, 5Y-11, 6Y-1), with 76% having 3 or 4 young.

Dates for 14 "Slate-colored" Junco broods ranged from 26 May to 10 July, with 64% recorded between 1 June and 23 June. Brood size ranged from 1 to 5 young (1Y-2, 2Y-1, 3Y-8, 4Y-2, 5Y-1), with 57% having 3 young.

Brown-headed Cowbird Parasitism: In British Columbia, 3% of 840 Dark-eyed Junco nests found with eggs or young were parasitized by the cowbird (Fig. 398). There were 41 additional records of a Dark-eyed Junco feeding a cowbird fledgling. On the coast, 3 of 197 nests (2%) were parasitized; in the interior, 23 of 643 nests (4%) were parasitized.

Of the 840 nests mentioned above, 399 nests were attributed to the "Oregon" Junco, 3% of which were parasitized by the cowbird. There were 4 additional records of an "Oregon" Junco feeding a cowbird fledgling. On the coast, 1 of 121 nests (1%) was parasitized; in the interior, 11 of 278 nests (4%) were parasitized. Friedmann (1963) noted that the "Oregon" Junco is an infrequent victim of the cowbird; 3 of his 4 records of cowbird parasitism were from British Columbia. Later, Friedmann and Kiff (1985) reported a few more instances of this junco acting as host to the cowbird.

Of 43 nests attributed to the "Slate-colored" Junco that were found with eggs or young, 16% were parasitized by the cowbird. There were 6 additional records of a "Slate-colored" Junco feeding a cowbird fledgling. Friedmann (1963) noted that this junco is an "infrequently reported host; probably it is molested very slightly by the brown-headed cowbird."

Nest Success: Of 161 Dark-eyed Junco nests found with eggs and followed to a known fate, 58 produced at least

Figure 401. Recently fledged Dark-eyed Junco young (Rithets Bog, Saanich, southern Vancouver Island, 22 April 1995; R. Wayne Campbell).

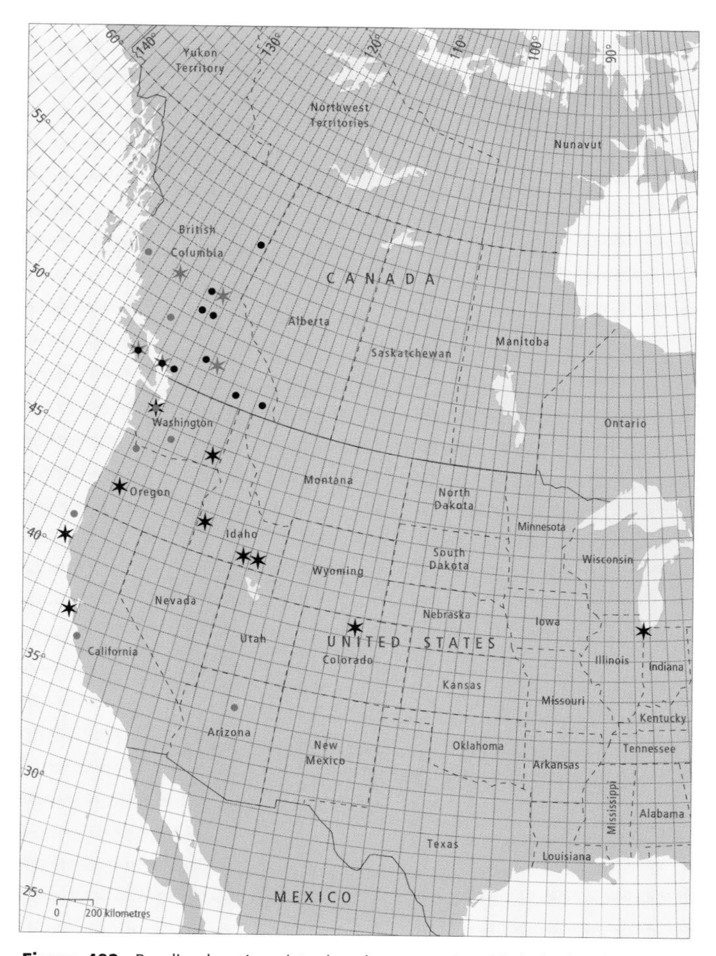

Figure 402. Banding locations (stars) and recovery sites (circles) of Dark-eyed Juncos associated with British Columbia. Red indicates birds banded in British Columbia; black indicates birds banded elsewhere.

1 fledgling, for a nest success rate of 36%. Nest success on the coast was 39% (n = 36); in the interior, it was 35% (n = 125).

Of the 161 nests, 93 were attributed to the "Oregon" Junco. Of these, 36 produced at least 1 fledgling, for a nest success rate of 39%. Nest success was 37% (n = 27) on the coast and 39% (n = 66) in the interior.

Of 8 nests of the "Slate-colored" Junco found with eggs and followed to a known fate, 3 produced at least 1 fledgling.

REMARKS: Until 1973, the various groups of the Dark-eyed Junco were treated as separate species: Slate-colored Junco (*Junco hyemalis*), Oregon Junco (*J. oreganus*), White-winged Junco (*J. aikeni*), Gray-headed Junco (*J. caniceps*), and Guadalupe Junco (*J. insularis*) (American Ornithologists' Union 1973). The Guadalupe Junco has recently been given species status again by a number of authors (Byers et al. 1995).

Four subspecies of the Dark-eyed Junco are known to occur regularly in British Columbia. *J. h. cismontanus* ("Slate-colored" Junco) occurs in the northern and central parts of the interior of the province. The following 3 are members of the "Oregon" Junco group: *J. h. oreganus* frequents the northern portions of the coast south to Calvert Island, including the Queen Charlotte Islands; *J. h. shufeldti* occurs primarily in the Georgia Depression; and *J. h. montanus* has been reported from central and southern British Columbia. There is some intergradation between *oreganus* and *shufeldti* on Vancouver Island (American Ornithologists' Union 1957; Godfrey 1986). A fifth subspecies, *J. h. caniceps* ("Gray-headed" Junco), has been reported only once, from near Qualicum Beach.

We have summer records for the "Oregon" Junco from Kluayaz Creek, Refuge Lake, Eddontenajon Lake, and Telegraph Creek, which suggests that breeding may occur; adequate documentation is lacking, however.

In west-central Illinois, Gottfried and Franks (1975) studied the winter habitat use of the Dark-eyed Junco. They found that the juncos spent most of the diurnal hours feeding in open areas with trees, shrubs, or tall forbs as components, including short-grass pastures and agricultural fields with wooded or shrubby boundaries. Up to 95% of their feeding activity occurred on the ground. Their foraging areas were usually at one end of their home range (17 to 33 ha), up to 1 km from their night roosts at the other end of the home range. Night-roost sites consisted of areas with dense vegetation. There the birds usually roosted in coniferous trees at locations where the foliage was densest. Although the temperature inside the dense foliage was only 1°C warmer

Table 7. Winter sex ratios of the Dark-eyed Junco in the southern ecoprovinces of British Columbia.

Ecoprovince	*n*	Males	Females	Proportion of females (%)
Georgia Depression	100	406	255	39
Southern Interior	26	60	30	33
Southern Interior Mountains	16	271	156	37

than outside the roost site, the wind velocity was considerably reduced: when the wind speed was 29 kph in the open, it was less than 1 kph at the roost site. There are no descriptions of roost sites for the Dark-eyed Junco from British Columbia. Field observers are encouraged to provide information on this important habitat component.

Ketterson (1976) found that the sex ratio of the Dark-eyed Junco wintering in the eastern and central regions of North America varied clinally along a latitudinal gradient. The proportion of females ranged from around 70% in the south to 20% in the north. The winter sex ratios of the Dark-eyed Junco from the 3 southern ecoprovinces in British Columbia have a preponderance of males (Table 7), which agrees with Ketterson's findings that females generally tend to winter further south than males. The British Columbia figures seem to be appreciably higher for their latitude than Ketterson's figures, which may be due to the fact that winters in southern British Columbia are generally milder than those at similar latitudes east of the Continental Divide.

Banding returns suggest that juncos from British Columbia may winter as far south as central California and northern Arizona (Fig. 402). Of the juncos trapped in British Columbia, the one originating furthest to the east was banded near Chicago on 13 April 1989 and recovered near Fort St. John on 15 September the same year. A long-lived junco was banded near Barkerville, British Columbia, on 28 April 1927 and recovered near Kitimat, British Columbia, nearly 7 years later, in the second week of April 1934.

NOTEWORTHY RECORDS

Spring: Coastal – Witty's Lagoon (Metchosin) 6 Mar 1976-50; Saanich 13 Apr 1995-5 young approximately 3 to 5 days old, 23 Apr 1984-1 egg plus 1 Brown-headed Cowbird egg; Mill Bay 24 Mar 1925-200 to 300 birds; Sidney Island 31 Mar 1974-75 found on all the island; Carnation Creek 28 Apr 1981-20 in the watershed (*J. h. oreganus*); Maple Bay 21 Apr 1973-115, 1 nest with 4 eggs; Reifel Island 5 Mar 1974-25 (*J. h. oreganus*); Fleetwood 4 Mar 1961-32 feeding; Ridgedale 11 Apr 1968-60; Mission 6 Apr 1994-2 eggs, adult on nest; Vancouver 18 Apr 1959-60 at Stanley Park; Qualicum 22 Apr 1975-3, 20 Apr 1976-3, 4 Apr 1977-1, all latest spring departures (Dawe and Buechert 1995); Black Mountain (Vancouver) 18 May 1974-30; Harrison Hot Springs 1 Mar 1975-250, 13 Apr 1979-75; Hope 29 Apr 1962-3 eggs, 19 May 1964-4 nestlings; Strathcona Park 14 Mar 1988-2; Egmont area 11 May 1979-17; Sayward 26 Apr 1985-6, at feeder; Port Hardy 9 Apr 1951-35; Port Neville Inlet 12 Mar 1974-3, 30 Mar 1976-200, 5 Apr 1977-200, all along roadsides; Cape Scott Park 15-19 Apr 1980-10; Goose Island 10 May 1948-10, on Islet 4 (Guiguet 1953); Andrew Point 21 May 1977-2 fledglings being fed by adults; Klemtu 5 Apr 1976-40; Kitlope Lake 3-10 May 1991-8, 4 at estuary, 4 at lower Kitlope; nr Juskatla 22 Mar 1972-25; Langara Island 13 May 1981-2, nr Lord Bight; Masset 21 Apr 1972-10; Rose Point 24 May 1960-4 eggs; Baron Island 12-19 May 1987-1; Prince Rupert 10 Mar 1979-6, first for 1979; Exchamsiks River 20 Apr 1981-4, at confluence of Skeena; Kitsault 14 May 1980-5 eggs; nr Skunsnat Lake 18 May 1981-10; Three Guardsmen Pass 28 May 1979-2, at timberline. **Interior** – Anarchist Mountain 4 May 1975-15; Pend-d'Oreille River 26 Mar 1980-50; Copper Mountain (Nelson) 8 Apr 1976-165; White Lake (Okanagan Falls) 19 Apr 1978-152 in big sagebrush; nr White Lake (Okanagan Falls) 17 Apr 1983-4 eggs, female incubating; West Bench (Penticton) 13 Apr 1968-56; 4 km s Castlegar 27 Mar 1986-1 egg, female flew off nest; Castlegar 23 May 1972-1 egg plus 1 Brown-headed Cowbird egg; nr Osprey Lake 4 Apr 1979-150, feeding along the roadside, 16 Apr 1976-120 in 2 flocks of 70 and 50; Glen Lake (Okanagan) 20 Apr 1944-4 nestlings, newly hatched; Kimberley 11 Apr 1982-100, scattered in small groups along roadside; Merritt 11 Apr 1976-50; Edgewood 3 Apr 1985-100; Premier Lake 18 Apr 1976-25; Wasa 6 Apr 1982-12; Nakusp 24 Mar 1985-20, 31 Mar 1976-100; Brouse 25 Mar 1978-50; New Denver 3-24 Apr 1976-50; w side Windermere Lake 1 Mar 1983-11; Falkland 31 Mar 1977-100, in garden at about 1,060 m elevation; Lac du Bois Grassland Park 5 May 1972-100; Little Shuswap Lake 31 Mar 1962-60; Sorrento 18 Mar 1972-62; Revelstoke 21 Mar 1985-24; Mount Revelstoke National Park 10 May 1982-20; Clinton to 70 Mile House 3 May 1978-1; Glacier National Park 7 May 1982-50; Yoho National Park 2 Apr 1977-20; 100 Mile House 13 Apr 1981-25; Walker Valley 2 Apr 1976-100; nr Horse Lake 22 Mar 1978-40 along road; Donneley 3 Apr 1985-100; Kleena Kleene 7 Apr 1958-2 (Paul 1959), 4 May 1947-eggs present; Westwick Lakes 11 May 1958-5 nestlings being fed by adults; Williams Lake 10 Mar 1950-1, 11 Mar 1979-4, first ones seen here this spring, 15 Mar 1985-20, first flock to arrive at feeder this winter, snow still covering ground; Chezacut 18 Mar 1941-20, 9 Apr 1941-30, 24 Apr 1941-12, 8 May 1943-20; Indianpoint Lake 7-13 Apr 1927-129, 14-20 Apr 1927-167, 21-27 Apr 1927-110 (McCabe and McCabe 1927); 24 km s Prince George 10 Mar 1979-2, 26 Mar 1978-10, 5 Apr 1985-20, 10 Apr 1986-50, 22 Apr 1982-100, all on farmlands 8 km e Fraser River (all *J. h. oreganus*); 14 km ene Prince George 29 May to 10 Jul 1966-16 (Grant 1966); Prince George 10 May 1983-5 eggs; Willow River 25 Apr 1966-150, largest flock seen; Smithers 20 Mar 1977-1 (*J. h. hyemalis*); Flatbed River 30 May 1976-4 eggs, "Slate-colored" group; Pine Pass 9 May 1968-16; 26 km n Azouzetta Lake 11 May 1982-5; 15 km sw Dawson Creek 13 May 1991-4 eggs (*J. h. hyemalis*); 5 km s Pouce Coupe 23 May 1994-4 eggs, all Brown-headed Cowbird (*J. h. hyemalis*); Driftwood River valley 28 Mar 1941-1, early arrival (Stanwell-Fletcher and Stanwell-Fletcher 1943); McIntyre Lake 25 Mar 1979-20; Hudson's Hope 28 May 1974-11; e Farrell Creek 14 Apr 1984-1 (*J. h. hyemalis*); Taylor 13 Apr 1986-70, 14 May 1986-10, 24 May 1986-4 nestlings; Fort St. John 20 Apr 1986-18; e Fort St. John 7 Apr 1988-2, spring arrival; nr Bear Flat 20 Apr

1986-10; Charlie Lake 1 May 1985-25; Cecil Lake 11 Apr 1986-20, with 5 American Tree Sparrows, 27 Apr 1986-20, 1 May 1985-25; Montney 7 Apr 1984-1; Rose Prairie 12 Apr 1986-50; 14 km n Pink Mountain 13 May 1982-4; Nig Creek 25 May 1922-1 *J. h. hyemalis* (NMC 17656); Spatsizi Plateau Wilderness Park 27 May 1977-4 eggs; 6 km n Telegraph Creek 26 May 1919-1 *J. h. hyemalis* (MVZ 39949), 27 May 1919-1 *J. h. hyemalis* (MVZ 39953); 40 Mile Flats 1 Apr 1981-hundreds on bare ground; Kluachon Lake 22 Apr 1981-5; Beggerlay Creek 1 Apr 1981-5; Ealue Lake 21 Apr 1981-5; Jackfish Creek (Fort Nelson) 25 May 1975-1; Fort Nelson 19 Apr 1987-2, 30 Apr 1975-10, 6 May 1975-15; Liard Hot Springs 28 Apr 1975-12 (Reid 1975), 30 Apr 1975-20 (Reid 1975), 27 May 1981-3; Helmut 17 to 31 May 1982-1, recorded almost daily but in very low numbers; Atlin 16 May 1981-24, 21 May 1981-42; Chilkat Pass 9 May to 6 Sep 1957-small numbers (Weeden 1960).

Summer: Coastal – North Saanich 1-31 Aug 1986-10, daily at feeder; Mesachie Lake 6 Jun 1974-1 fledgling; Vesuvius Bay 10 Aug 1990-4 eggs; Marion Lake 4 Aug 1975-14 (*J. h. oreganus*); Matsqui 17 Aug 1982-4 nestlings, left nest and flew successfully to nearby bushes; Aldergrove 27 Jun 1993-2 fledglings; Cultus Lake 21 Jun 1959-100; Cheam Ridge 17 Aug 1983-22; Mount Liumchin 18 Aug 1928-fledglings just out of nest; Pitt Meadows 31 Jul 1974-10; Alouette Lake 1 Jun 1963-6 (*J. h. oreganus*); Golden Ears Park 21 Aug 1974-1 fledgling; Hope 29 Jul 1990-4 eggs; Oyster River 31 Aug 1975-100, in alpine areas; Black Tusk (Garibaldi Park) 17 Jun 1972-25; Helm Lake 3 Aug 1986-4 nestlings, 3 days old; Brooks Peninsula 14 Aug 1981-1 fledgling (Campbell and Summers 1997); Adam to Eve rivers area 21 Jun 1984-13, mostly males seen or heard; Stuart Island 29 Jul 1936-20; Port Hardy 15 Jun 1934-4 eggs; Cape St. James 15 Aug 1981-20; Shearwater 29 Jul 1986-5; upper Kitlope River 3 Jun 1994-3; Chaatl Island 12 Jun 1959-24, on high ridges; Yakoun Lake 17 Jul 1974-14; Graham Island 27 Jun 1910-3 eggs; Tlell 26 Jul 1974-37; nr Delkatla Inlet 17 Jul 1988-3 nestlings being fed; Prince Rupert 20 Jul 1958-50; e Terrace 22 Jul 1981-3 eggs; nr Terrace 5 Aug 1991-1 fledgling; w Greenville 16 Jun 1981-10; Kitwanga to Kispiox 26 Jun 1975-25; Kitsault 17-19 Jun 1980-27; Bear River (Stewart) 17 Jun 1975-1; Meziadin Lake 8 Jun 1976-2; Mount Pearson 27 Aug 1919-1, at 2,300 m elevation (CMNH P102222). **Interior** – Frosty Mountain 11 Aug 1974-65 (*J. h. oreganus*); Ripple Mountain 29 Jul 1980-20, at 2,300 m elevation, several immatures; Manning Park 2 Aug 1969-60, 22 Aug 1968-2 nestlings; Three Brothers Mountain 10 Aug 1974-55 (*J. h. oreganus*); Flathead River valley 11 Jun 1978-2; Castlegar 5 Jun 1972-1 nestling, 29 Jul 1977-1 egg, 15 Aug 1977-2 eggs; Apex Mountain (Penticton) 22 Jul 1974-14; around Jaffrey 22 Jun 1968-22 (*J. h. oreganus*); Summerland 28 Jul 1991-1 adult feeding fledgling Brown-headed Cowbird; Mount Brent 20 Jul 1977-21, 5 in alpine meadows; Chute Lake 2 Jun 1970-2 eggs with 2 Brown-headed Cowbird eggs; Brookmere 19 Jun 1974-20; Whatshan Lake 3 Jun 1984-13; Wasa Park 11 Aug 1971-1 adult feeding a Brown-headed Cowbird fledgling (Dawe 1971); Mount Scaia 21 Jul 1981-25, at 2,050 m elevation; Idaho Mountain 30 Jul 1977-10; Lytton to Skihist 11 Jun 1966-20; Canford 18 Jun 1974-12; Glimpse Lake 22 Aug 1978-1 fledgling; Lake Bonnie Gem 3 Aug 1980-22, at 2,130 m elevation; Ewin Creek 20 Jun 1982-25; Tenquille Lake 23 to 31 Jul 1960-a number seen (Bradley 1961); Kamloops 29 Aug 1960-100; Salmon Arm 14 Jul 1984-1 nestling; Shuswap Lake Park 14 Jun 1982-10; Celista 16 Jul 1960-36; Crowfoot Mountain 21 Jul 1973-20 (*J. h. oreganus*); Beaverfoot River 30 Aug 1975-60 (*J. h. oreganus*); Emerald Lake (Yoho National Park) 16 Jul 1975-20 (Wade 1977); Amiskwi River valley 1 Jul 1976-68; Horse Lake (100 Mile House) 19 Jul 1978-50, on road to lake; Kleena Kleene 25 Aug 1967-2 fledglings; Lac la Hache area 28 Jun 1970-9 (*J. h. oreganus*); Hemp Creek 6 Jul 1978-2, at Helmeken Lodge; Murtle Lake (Wells Gray Park) 18 Aug 1990-60, in 4 flocks; Riske Creek 12 Jun 1978-10; Chezacut 31 Aug 1941-30; Swiftcurrent Creek 15 Aug 1973-33; Titetown Lake Jul 1983-10 (*J. h. oreganus*), 3 adults feeding Brown-headed Cowbird fledglings; Kiwa Creek 16 Jun 1988-1, on steep slopes above creek in second-growth subalpine fir and spruce; w McBride 10 Jul 1991-8, on Bell Mountain Road to summit of Little Bell Mountain (Campbell and Dawe 1991); Little Bell Mountain 10 Jul 1991-19, including 3 fledglings at summit near treeline (Campbell and Dawe 1991); Goodrich Lake 27 Jul 1963-4 eggs, in sw corner of lake; 24 km s Prince George 27 Aug 1984-50, with American Robins and warblers; Ootsa Lake 13 Jul 1936-13; Nulki Lake 15 Aug 1945-20, 21 Aug 1945-100; Prince George area 14 Jun 1994-1 adult feeding fledged Brown-headed Cowbird, 21 Aug 1972-3 fledglings; e Prince George 21 Jul 1991-2 eggs; 2 km e Giscome 26 Jul 1991-group of fledglings accompanied by Brown-headed Cowbird; Burnie Lakes 23 to 30 Aug 1975-some (Osmond-Jones et al. 1975); Pinkut Creek to Burns Lake 11 Jun 1975-23; Bear Lake (nr Prince George) 15 and 16 Aug 1975-18; Telkwa 8 Jul 1992-1 adult feeding fledged Brown-headed Cowbird, 21 Aug 1993-adult pair feeding 2 Brown-headed Cowbird fledglings; Topley 3 Jul 1956-20 (*J. h. oreganus*); Bullmoose Creek 2 Jun 1977-3, 3 km up the Wolverine River valley from confluence of creek at about 800 m (*J. h. oreganus*); 77 km w Chetwynd 16 to 20 Jun 1969-20 (Webster 1969b); e Chetwynd 5 Aug 1975-15; Buffalo Lake 29 Jun 1968-10; nr Brassey Creek 30 Jun 1993-2 eggs; n Dawson Creek 3 Aug 1975-10 (*J. h. hyemalis*); Taylor 29 Jul 1982-1 fledgling, 21 and 22 Aug 1975-34; Charlie Lake 21 Aug 1975-14; Beatton River 26 Jul 1982-2 fledglings; Beatton Park 10 Jul 1987-1 fledgling; Ingenika 29 Jul 1980-1; w Hudson's Hope 3 Jul 1974-1, W.A.C. Bennett Dam; Tatlatui Lake 3 Jul 1976-6; Stalk Lake 9 Jul 1976-8; Buckinghorse Creek 16 Jul 1976-22, in groups of adults and young; Kinaskan Lake 1 Jun 1978-2 (*J. h. hyemalis*); Kinaskan Lake 27 Aug 1979-20; Todagin Lake 11 Jul 1963-10; Dawson River 31 Aug 1976-25; Eddontenajon Lake 10 Jun 1976-6; Spatsizi Plateau Wilderness Park 12 Aug 1976-2 families with 6 fledglings; n end Haworth Lake 4 Aug 1976-50; 14 km s Dease Lake 28 Jun 1962-1 (NMC 50275); Tetsa River 15 Jun 1980-2, at Mile 373 on Alaska Highway; Alaska Highway (Mile 370) 1 Jun 1975-2, on Tetsa Rd; Evie Lake 2 Aug 1980-3 fledglings; Fort Nelson 12 Jun 1976-14 (*J. h. hyemalis*); Parker Lake (Fort Nelson) 10 Jul 1978-2 nestlings being fed by adult; e Fort Nelson 6 Jun 1982-2 fledglings; Alsek River area 4 Jun 1983-4 nestlings; s Basement Creek 9 Jun 1983-1 excited pair on creek gravel bars; Cassiar 13 Jun 1972-15 (*J. h. hyemalis*); Cassiar area 19 Jun 1975-24 *J. h. oreganus*, 2 *J. h. hyemalis;* Survey Lake 26 Jun 1980-30; Gladys Lake (Atlin) 19 Jul 1980-23 along road to lake.

Breeding Bird Surveys: Coastal – Recorded from 27 of 27 routes and on 77% of all surveys. Maxima: Kitsumkalum 14 Jun 1981-80; Kispiox 20 Jun 1993-68; Chemainus 24 Jun 1973-41. **Interior** – Recorded from 73 of 73 routes and on 99% of all surveys. Maxima: Christian Valley 25 May 1994-88; Spillimacheen 26 Jun 1976-77; Telkwa High Rd 27 Jun 1990-69.

Autumn: Interior – Atlin 19 Sep 1924-4 (Swarth 1926), 8 Oct 1931-1 *J. h. hyemalis* (CAS 34097); Liard River 23 Oct 1980-2; Fort Nelson 18 Sep 1985-10, nr airport (*J. h. hyemalis*); Little Tahltan River Sep 1986-21; n Hyland Post 10 Sep 1976-50, in burn; Hyland Post 8 to 12 Oct 1981-5 (*J. h. hyemalis*), 25 Oct 1980-3; Cold Fish Lake 21 Sep 1976-4; Buckinghorse River 7 Sep 1974-40; Flatrock 30 Oct 1951-25, in open burn; Fort St. John 14 Oct 1997-20, several in flock were *J. h. oreganus*, balance were *J. h. hyemalis;* Cecil Lake 5 Oct 1986-30; Beatton Park 26 Sep 1982-32; Stoddart Creek (Fort St. John) 5 and 6 Sep 1986-90, 17 Sep 1986-17; Taylor 20 Nov 1983-1, very late autumn record; 24 km s Prince George 10 Sep 1982-100 (*J. h. oreganus*), 18 Oct 1982-2 (*J. h. hyemalis*), 29 Nov 1983-4 (*J. h. oreganus*); Williams Lake Oct 1982-22, in last week, 12 Nov 1987-20, at feeding station; Canim Lake 5 Sep 1960-40 (*J. h. oreganus*); Riske Creek 6 Sep 1978-16, 10 Sep 1978-61, 21 Sep 1978-50, 25 Sep 1978-30; Golden 2 Sep 1977-50; Bridge Lake 7 Sep 1960-60 (*J. h. oreganus*); 28 km e Revelstoke 1 Sep 1990-3 fledglings; Magna Bay 14 Sep 1975-50; Scotch Creek to Celista 19 Oct 1964-214, 1 flock of 160 birds; Scotch Creek and Shuswap area 10 Sep 1962-250; Kamloops 5 Oct 1965-32, in small flocks of 10, 14, and 8 along highway edge; Coldstream 20 Nov 1955-30, at feeder; Lavington 18 Nov 1972-20, at feeder; e Elk River 24 Oct 1983-25; n Sparwood 19 Sep 1983-43; Vaseux Lake 14 Sep 1960-50, 26 Nov 1975-41; John's Creek (Hedley) 13 Oct 1985-100, heavy migration through mountains at around 600 m; Anarchist Mountain 15 Sep 1960-60; Baldy Mountain (Oliver) 11 Sep 1960-55; Blackwall Peak 27 Sep 1975-30 (*J. h. oreganus*). **Coastal** – Kitimat 4 Oct 1974-25 at golf course; Masset 10 Oct 1971-100; Delkatla Inlet Oct 1971-25, 10 Nov 1981-150; Queen Charlotte City 30 Sep 1971-15; Takakia Lake 21 Sep 1976-15 (*J. h. oreganus*); Cluxewe River 11 Nov 1990-40, on estuary (Dawe et al. 1995); Grant Bay 1 Oct 1968-30; Campbell River 9 Nov 1975-100, nr airport; Trent River 21 Nov 1987-130, small groups everywhere on estuary; Union Bay 5 Nov 1987-40 at feeder; Strathcona Park Oct 1977-134, at lodge; Little Qualicum River 8 to 10 Nov 1975-1 (*J. h. caniceps*) nr estuary (Dawe 1976a, 1976b); Maple Ridge 10 Sep 1979-30 in a flock, 21 Nov 1964-62, various places in flocks of 10, 16, 24, and 12; Mount Seymour Park 13 Sep 1975-40 (*J. h. oreganus*); North Vancouver 22 Oct 1972-100; Cypress Park 19 Sep 1981-200 (*J. h. oreganus*); Little Qualicum River 15 Nov 1983-150 on estuary; Englishman River 16 Nov 1979-37, on estuary (Dawe et al. 1994); Huntingdon 23 Sep 1948-158, first autumn flocks of 75, 30, 8, 20, 15, and 5; Reifel Island 24 Sep 1974-50 (*J. h. oreganus*), 12 Oct 1983-125 in the Alaksen National Wildlife Area; Gabriola Island 4 Sep 1977-25; Beaver Point 22 Nov 1977-46; Mount Tzuhalem 21 Nov 1983-130, on nw side; Island View Beach (North Saanich) 2 Nov 1986-50; Saanich 7 Oct 1975-60; s Sooke Lake 10 Sep 1977-25; Metchosin 3 Sep 1978-36; River Jordan 5 Oct 1986-20.

Winter: Interior – Atlin 1 Jan 1981-3 at feeder (*J. h. hyemalis*); Fort Nelson 24 Feb 1995-1 (*J. h. hyemalis*) at landfill; Fort St. John 1 to 20 Dec 1984-1 (Grunberg 1985b); Quick 22 Jan 1986-2, 8 Dec 1985-1; Willow River 19 Dec 1983-1; Prince George 15 Dec 1982-4, in backyard; Quesnel 21 Feb 1979-6; Sugarcane 27 Dec 1987-1; Lac la Hache 28 Feb 1986-1; Riske Creek 11 Dec 1983-2 (*J. h. hyemalis*); Riske Creek 1 Jan to early Feb 1985-1 (*J. h. hyemalis*); Revelstoke 20 Dec 1984-2; Celista 24 Feb 1948-42 (*J. h. oreganus*); Lillooet 1 Jan 1963-110; Swansea Mountain 25 Dec 1983-4, on Big Horn Ranch; Nakusp 19 Jan 1978-50, 30 Dec 1984-7; Lavington 1 Jan 1969-50 at feeder; Kelowna 31 Dec 1975-26; Summerland 2 Jan 1974-100, 3 Dec 1969-50, 20 Dec 1967-300, all at research station; Trout Creek Point (Okanagan Lake) 5 Dec 1973-62; Penticton 24 Dec 1978-75 at Macs Lake; Nelson 4 Jan 1981-22 at feeder, 3 Dec 1978-40 at feeder; s Castlegar 18 Feb 1979-15, 10 Dec 1978-10 at feeder; West Bench (Penticton) to Okanagan River 26 Dec 1974-259; Trail 1 Jan 1981-1 (*J. h. hyemalis*), 26 Dec 1982-3, at feeder; Creston area 6 Jan 1982-45; Salmo 1 Jan 1975-20; Fruitvale 12 Jan 1980-15. **Coastal** – Kitimat 4 Jan 1975-8, 16 Feb 1975-3 (*J. h. hyemalis*), 22 Dec 1974-16; Masset 11 Dec 1971-185; Delkatla Inlet 11 Feb 1973-9; Naden Harbour 1 Dec 1975-100; nr Tlell 16 Feb 1972-67 at feeders, 5 Dec 1971-150, around since 15 Nov and population increased until 15 Dec; Bella Coola Jan 1982-10 at feeder, 2 Dec 1989-20 at feeder; Cape St. James 4 Jan 1982-6; Campbell River 1 Jan 1963-100; Harrison Hot Springs 3 Jan 1976-60, 16 Feb 1976-90; Baynes Sound 10 Jan 1981-47 (Dawe et al. 1998), 17 Jan 1981-125 (Dawe et al. 1998); Qualicum Beach 4 Jan 1984-204, 12 Dec 1983-253 in fields; Vancouver 13 Jan 1979-107 in Stanley Park; Skagit River valley 1 Jan 1971-1 (*J. h. hyemalis*); Agassiz 3 Jan 1975-50, 19 Feb 1984-100, 22 Dec 1974-100; Westham Island 14 Dec 1974-146 (*J. h. oreganus*); Tofino 18 Dec 1969-40; Vargas Island 1 Jan 1971-40; Mount Tzuhalem 19 Jan 1974-200; Saanich 1 Jan 1987-150; 2 Dec 1984-100 at airport; Isabella Point (Saltspring Island) 9 Dec 1978-123; Mount Tolmie (Victoria) Jan 1925-numerous flocks (Preece 1925); Ten Mile Point (Saanich) 30 Dec 1973-165 (*J. h. oreganus*); Sooke 31 Dec 1983-89 (*J. h. oreganus*); Langford 24 Dec 1975-215 (*J. h. oreganus*); Langford 10 Dec 1974-200; River Jordan 5 Jan 1974-2.

Christmas Bird Counts: Interior – Recorded from 23 of 27 localities and on 84% of all counts. Maxima: Penticton 27 Dec 1993-1,481; Oliver-Osoyoos 28 Dec 1993-1,139; Kelowna 18 Dec 1993-1,060. **Coastal** – Recorded from 32 of 33 localities and on 99% of all counts. Maxima: Victoria 18 Dec 1993-**5,930**, all-time Canadian high count (Ortega 1995); Vancouver 22 Dec 1985-3,052; Duncan 27 Dec 1993-2,125.

Lapland Longspur
Calcarius lapponicus (Linnaeus)

LALO

RANGE: Holarctic. In North America, breeds from western and northern Alaska, northern Yukon, Banks, Prince Patrick, Melville, and northern Ellesmere islands south to the Aleutians, south-coastal Alaska, northern Mackenzie, southern Keewatin, northeastern Manitoba, northern Ontario, northern Quebec, and northern Labrador. Winters from southwestern British Columbia, southern Alberta, and southern Saskatchewan east to southern Ontario and Nova Scotia, south to northern California, northern Utah, Colorado, Oklahoma, northwestern Texas, Arkansas, Tennessee, and Maryland. Also occurs in Greenland and the Palearctic.

STATUS: On the coast, an *uncommon* to occasionally *fairly common* migrant and winter visitant and *very rare* in summer in the Georgia Depression Ecoprovince, particularly in the Fraser Lowland; in the Coast and Mountains Ecoprovince, generally a *rare* spring and autumn transient on Western Vancouver Island, *casual* there in summer; *rare* spring and autumn transient, *casual* in winter on the Northern Mainland Coast; *casual* spring and *rare* autumn transient on the Queen Charlotte Islands, *casual* there in winter; and *casual* in spring and autumn on the Southern Mainland Coast.

In the interior, *very common* to *abundant* (occasionally *very abundant*) transient in the Boreal Plains Ecoprovince; *common* to locally *very common* in the Taiga Plains and Northern Boreal Mountains ecoprovinces; *uncommon* to locally *fairly common* transient in the Southern Interior and Central Interior ecoprovinces; *rare* spring and *common* autumn transient in the Southern Interior Mountains Ecoprovince; *uncommon* in the Sub-Boreal Interior Ecoprovince. In winter, *casual* in the Southern Interior and Southern Interior Mountains; *accidental* in the Sub-Boreal Interior and Boreal Plains.

OCCURRENCE: The Lapland Longspur (Figs. 403 and 404) is a widespread but sparsely distributed transient throughout much of British Columbia. It is a bird of open, treeless habitats ranging from rocky offshore islets on the coast to low-elevation grasslands and alpine tundra in the interior. This longspur is one of the earliest spring migrants and begins its northward movement through the province at about the same time as the American Pipit and well over a month before the much scarcer Smith's Longspur. A spectacularly large spring and autumn movement occurs east of the Rocky Mountains through the Boreal and Taiga plains and along the British Columbia/Yukon border area of the Northern Boreal Mountains.

In British Columbia, the Lapland Longspur has been recorded at elevations ranging from sea level to 2,660 m on the coast and from 280 to 2,440 m in the interior. During early spring, flocks of Lapland Longspurs begin sweeping north, seeking out the first snow-free areas, including low-elevation grasslands (Fig. 405), agricultural fields, and roadsides in the interior, and agricultural fields, estuaries, and beaches along the coast (Fig. 406). In northern portions of the province, where natural forest openings are less common, it makes greater use of human-created habitats, including fields, road edges, gravel pits and roadside clearings, sewage lagoon dykes, airstrips, gas and petroleum well sites, utility corridors, recent clearcuts, and settlement clearings. During the autumn migration, longspurs use habitats similar to those in spring, but also make use of the alpine areas. In winter, the Lapland Longspur is found primarily on the south coast in open grassy or sparsely vegetated habitats (Fig. 407), including cultivated fields, grazed pastures, roadsides, estuaries, rocky shorelines and islets, beaches, lagoons, jetties, causeways, sedge wetlands, golf courses, playing fields, and airports. On rare occasions it goes to bird feeders.

Figure 403. Male Lapland Longspur in breeding plumage (© Brian E. Small/VIREO). The Lapland Longspur migrates through British Columbia to and from its tundra breeding grounds in the far north.

Figure 404. Lapland Longspur in winter plumage (© S. Young/VIREO). A small number of Lapland Longspurs winter most years in extreme southwestern British Columbia.

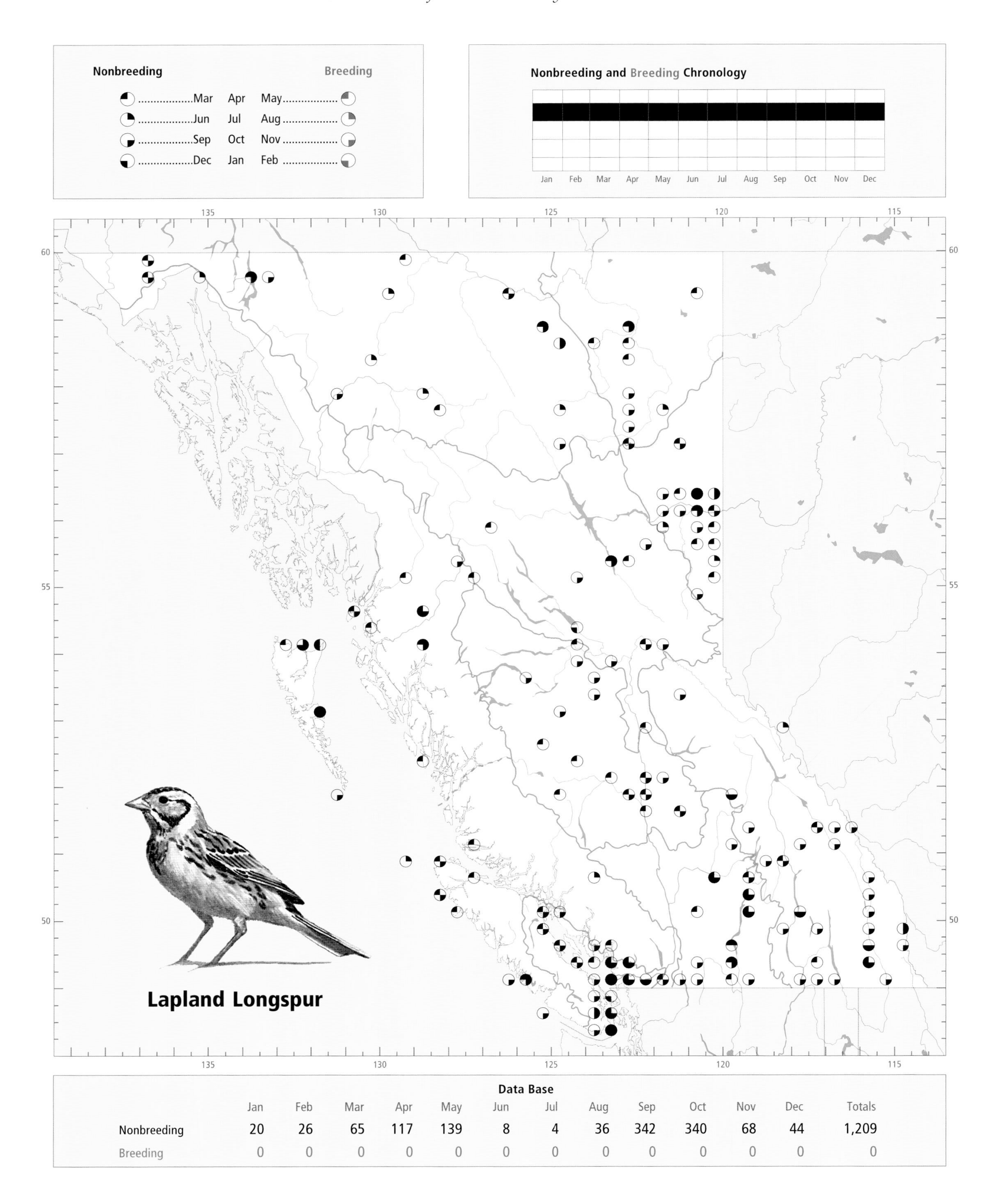

Data Base

	Jan	Feb	Mar	Apr	May	Jun	Jul	Aug	Sep	Oct	Nov	Dec	Totals
Nonbreeding	20	26	65	117	139	8	4	36	342	340	68	44	1,209
Breeding	0	0	0	0	0	0	0	0	0	0	0	0	0

Figure 405. During spring and autumn migration in the southern interior of British Columbia, the Lapland Longspur frequents low-elevation grasslands (Hat Creek, 4 November 1996; R. Wayne Campbell).

Figure 406. Along the coast of British Columbia, the Lapland Longspur is often found foraging on log-littered beaches during migration (Tlell, 20 May 1996; R. Wayne Campbell).

Figure 407. In winter, small numbers of Lapland Longspurs are most often found in open grassy or sparsely vegetated regions of the Fraser Lowland in the Georgia Depression Ecoprovince (Delta, 31 March 1997; R. Wayne Campbell).

In northeastern British Columbia, the Lapland Longspur is usually reported in large single-species flocks. Elsewhere in the province, it is more often observed in mixed flocks with 1 or more other species, usually Horned Larks, American Pipits, and Snow Buntings. It is also occasionally seen with Western Meadowlarks; Gray-crowned Rosy-Finches; Savannah, White-crowned, Chipping, and American Tree sparrows; and Smith's Longspurs.

The spring migration of the Lapland Longspur along the coast is small. In the Georgia Depression, it is difficult to distinguish migrants from wintering numbers, but migration likely begins in early March; migrants are reported in this ecoprovince throughout April, with a few stragglers recorded into late May (Figs. 408 and 409). Elsewhere on the coast, a very small movement occurs along the Southern Mainland Coast from the first week of March to the third week of April; the timing of arrival and departure is about 1 week later on the Northern Mainland Coast. The first spring migrants do not appear on Western Vancouver Island until the second week of April, and arrive 1 week later on the Queen Charlotte Islands; the last birds have left both areas by the third week of April. The majority of spring records on the coast are of single birds, with the largest flocks reported from the Georgia Depression.

In the interior, the first spring migrants begin to arrive in the Southern Interior during the second week of March and reach the Central Interior about 1 week later. Small numbers move through both areas until about the third week of April, but a few stragglers are reported into late May (Figs. 408 and 409). In the Southern Interior Mountains and Sub-Boreal Interior, single birds and small flocks begin to arrive in mid-March and have passed through by early May.

Further north, the first longspurs begin to arrive in the Boreal Plains in early April, heralding the beginning of a large migratory movement from the prairie and parkland regions of Alberta. From late April through early May, flocks of up to 1,000 birds or more quickly pass through this area; only a few small flocks remain by the third week of May. In the Peace Lowland, where most migrants are recorded, the Lapland Longspur is found in agricultural areas and shows a preference for grain stubble fields (Phinney 1998). Longspurs are also attracted to fields with standing water, small ponds, or water-filled ditches, where they gather to drink. Further north, in the Taiga Plains, the spring migration is less well documented but appears to begin in early May and is largely over by the third week of the month. In the Northern Boreal Mountains, the first migrants generally begin to arrive during the last week of April; by the third week of May the main movement has passed through.

The earliest spring migrants arriving at the Atlin Lake and Chilkat Pass areas of northwestern British Columbia and along the Alaska Highway in southwestern Yukon appear to precede the main "prairie" migration through northeastern British Columbia (Figs. 408 and 409). West et al. (1968) note that the arrival of these first migrants precludes the possibility that they are using the prairie route; they must therefore use alternative routes along the coast and through interior valleys of British Columbia. Although a spring movement is recorded along both routes, our data suggest that they are very minor in extent and do not fully account for the large

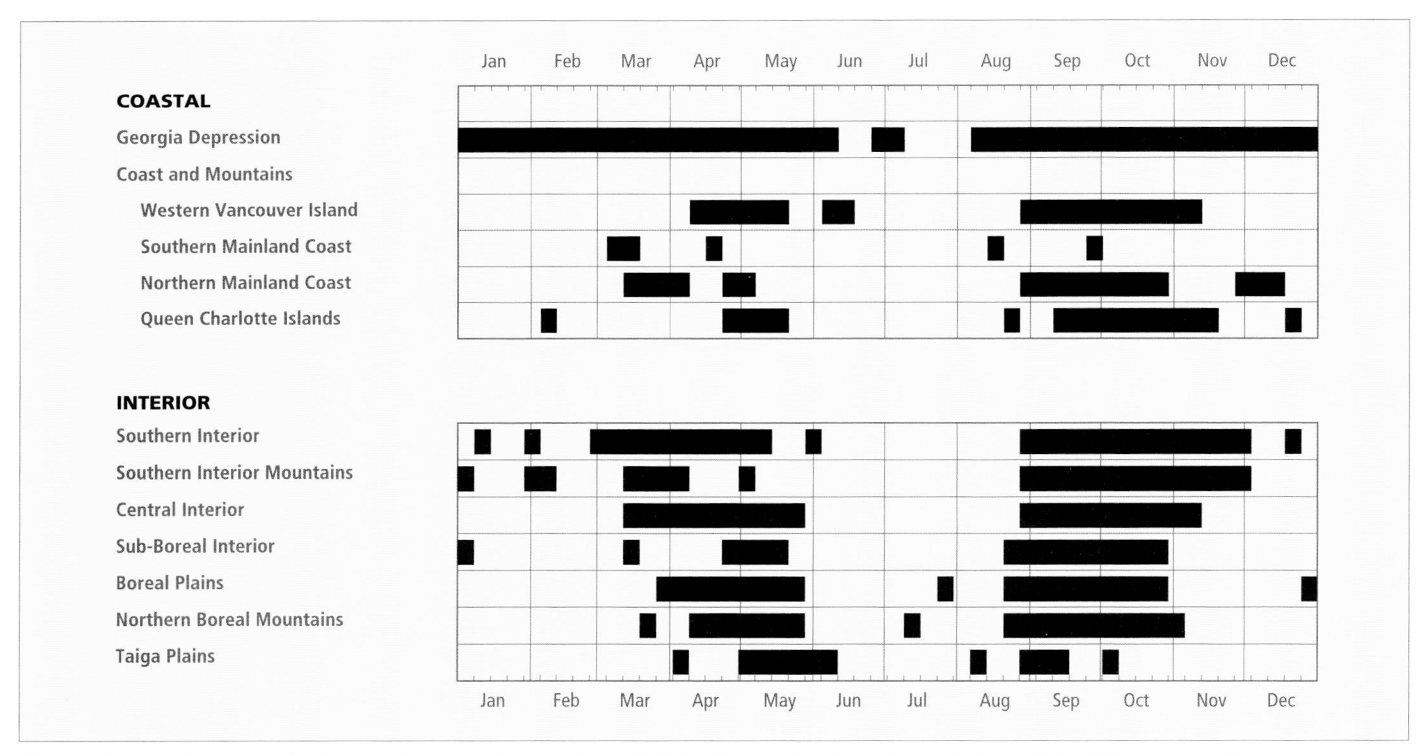

Figure 408. Annual occurrence for the Lapland Longspur in ecoprovinces of British Columbia. Records are shown for the week in which they occurred.

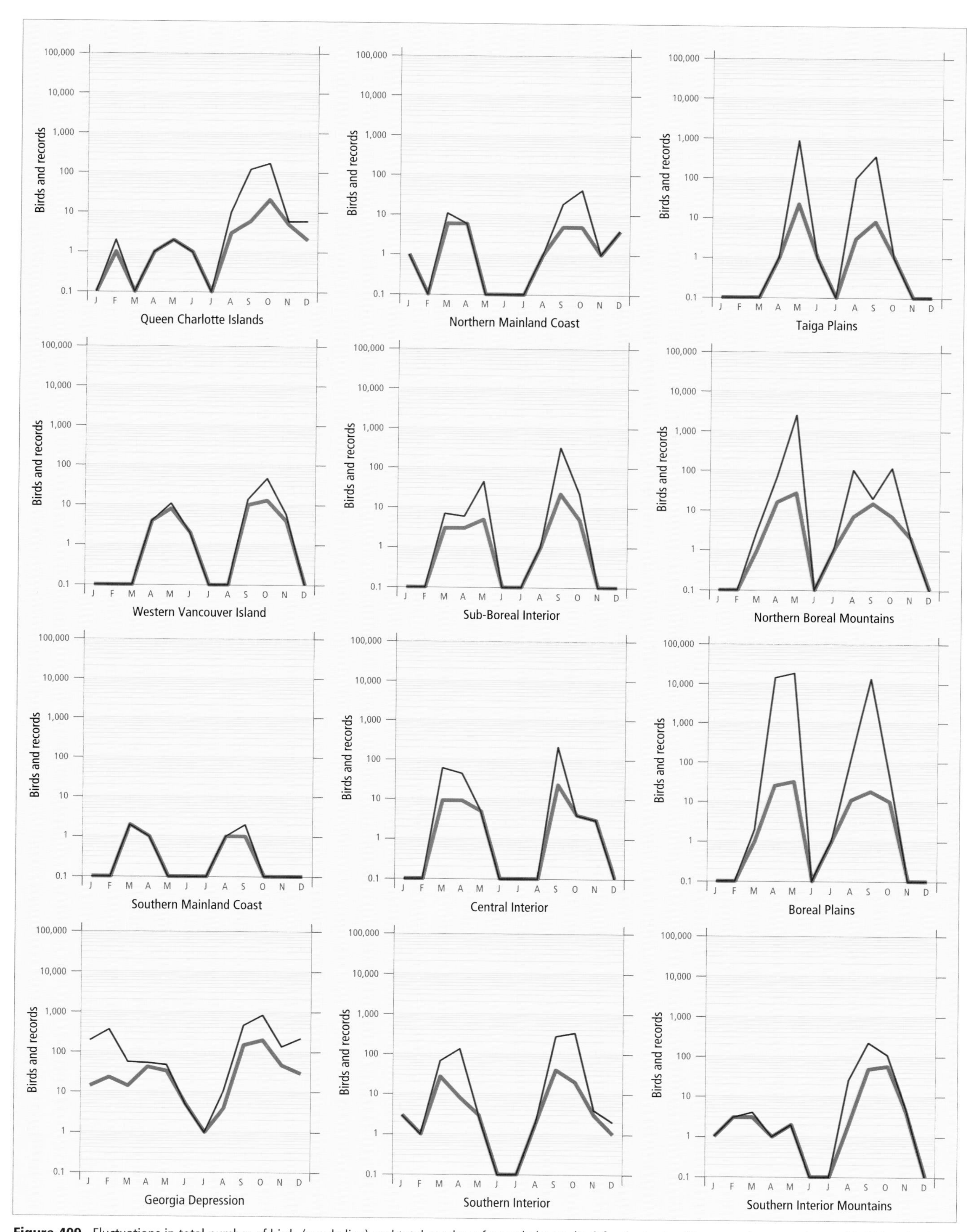

Figure 409. Fluctuations in total number of birds (purple line) and total number of records (green line) for the Lapland Longspur in ecoprovinces of British Columbia. Christmas Bird Count data have been excluded.

numbers that first appear in the extreme northwestern corner of British Columbia. The Lapland Longspur is a strong flier (Rising 1996), however, and may be capable of migrating quickly over these lightly populated areas, thereby avoiding detection. In northern Alberta, longspurs have been noted passing rapidly over forested areas during migration (Salt and Salt 1966).

During autumn, the main movement of longspurs retraces the spring route east of the Rocky Mountains through the Taiga and Boreal plains and into Alberta. A distinct southward movement is also noted along the coast and throughout the southern portions of the interior, where significantly more records and larger flock sizes than in the spring are reported (Fig. 409). There are several autumn records of longspurs using alpine habitats, mostly along the Rocky Mountains. On the coast, there is a single record from the alpine areas in the southern Coast Mountains, but this species was not detected during autumn surveys of alpine areas on Vancouver Island (Ogle and Martin 1997).

On the coast, the first migrants begin arriving in small numbers in mid-August, with the main movement occurring from late September through early October. Only a few stragglers remain after the end of October, except in the Georgia Depression, where some birds overwinter.

In the interior, the first southward movement is noted in most ecoprovinces when a few early migrants appear in late August (Figs. 408 and 409). In northern areas, peak movements probably occur during the first 2 weeks of September, but this is not well documented. In the Boreal Plains, the autumn migration is comparable in magnitude to the spring migration (Fig. 409), but flock sizes tend to be smaller, with most consisting of fewer than 50 birds (C. Siddle pers. comm.). Large aggregations occasionally occur, however, such as a huge gathering of approximately 12,000 longspurs recorded near Prespatou, north of Fort St. John. By mid-September most longspurs have passed through the north; a few stragglers remain into late October. In the southern ecoprovinces, the autumn movement peaks during late September except in the Southern Interior, where large flocks are occasionally reported into mid-October. Most migrants have passed through by the end of October; stragglers are occasionally reported into late November.

The Lapland Longspur is an irregular wintering species in British Columbia. Most winter records come from the Georgia Depression, especially the western portion of the Fraser Lowland (Fig. 407), including Iona, Sea, Lulu, and Westham islands, and the Ladner-Delta areas. A smaller number of winter records come from Greater Victoria on southern Vancouver Island. Elsewhere on the coast, there are a few records of from 1 to 4 birds on the Queen Charlotte Islands, at Terrace on the Northern Mainland Coast, and at the Squamish River estuary on the Southern Mainland Coast. In the interior, there are a few scattered records mostly of single birds near Kamloops and in the north Okanagan valley in the Southern Interior, and near Cranbrook, Nakusp, and Wells Gray Park in the Southern Interior Mountains. There is a single record of 6 longspurs wintering in the Peace Lowland of the Boreal Plains, and a single longspur in winter at Fort St. James in the Sub-Boreal Interior.

On the coast, the Lapland Longspur has been recorded through most of the year but mainly from mid-September through mid-May; in the interior, it is reported regularly from late August to early November and from mid-March to late May (Fig. 408). There are several June and July records in British Columbia, mostly of single birds.

REMARKS: Known in Old World literature as the Lapland Bunting. Of the 3 subspecies found in North America, only *C. l. alascensis* occurs in British Columbia (American Ornithologists' Union 1957). Once known as the Alaska Longspur, this subspecies breeds in the Arctic tundra of western Canada and Alaska, including the islands in the Bering Sea (Rising 1996).

As with many passerines, the initial spring flocks of Lapland Longspurs consist predominantly of males. West et al. (1968) report that the first flocks of longspurs migrating through northern British Columbia and the Yukon into Alaska consist of 90% males, but as the migration progresses, the proportion drops to 25% males. Although the onset of spring migration can vary from year to year, the ratio of males to females provides an easy method for determining the stage of migration. Many records in our data base note the preponderance of males in these early spring flocks.

Phinney (1998) reports that the appearance of the Merlin in the Boreal Plains near Dawson Creek coincides with the spring arrival of the Lapland Longspur. Our data show several instances of Merlin following and preying upon longspurs in this region during both spring and autumn migration.

The Lapland Longspur's habit of foraging along roadsides and flushing at close range makes it vulnerable to vehicle collisions. Irving (1961) reported that while driving to Atlin, he found 15 longspurs that had been killed or injured by a vehicle travelling ahead of him. Similar incidents have been observed near Dawson Creek (M. Phinney pers. comm.) and along the British Columbia portion of the Alaska Highway.

The cryptic plumage of the Lapland Longspur during the nonbreeding season and its habit of mixing in flocks of more abundant species make it an easy species to overlook. Even during spring, the distinctively marked males are seldom in full breeding plumage when they migrate through the province. These factors, combined with its relative scarcity in populated areas of southern British Columbia, have likely resulted in this species being underreported. With experience, observers can readily identify the Lapland Longspur by its distinctive flight calls.

NOTEWORTHY RECORDS

Spring: Coastal – Esquimalt Lagoon (Metchosin) 24 Apr 1985-1; Clover Point (Victoria) 24 May 1971-1 in breeding plumage (Tatum 1973); Central Saanich 5 Apr 1986-2 in field; Tofino 19 Apr 1996-1 on golf course; Delta 9 Mar 1975-22 flying over in 1 large flock; Sea Island 2 May 1965-8 at Vancouver airport; Iona Island 31 May 1971-1; Chilliwack 31 Mar 1888-1 (Cooke 1910); Delta 30 Mar 1980-5 at Brunswick Point; Campbell River 27 Apr 1975-2 on estuary; Cluxewe River 20 Apr 1991-1 on estuary (Dawe et al. 1995); Guise Bay 19 May 1974-2; Point Holmes 21 Apr 1952-1 (RBCM 10072); McInnes Island 25 Mar 1964-1; Sandspit 6 May 1979-1 at airport; Langara Island 23 Apr 1927-1 in muskeg (Darcus 1930); Kitimat River 15 Mar 1980-5 in short grass on estuary; Terrace 23 Apr 1976-1 in schoolyard (Hay 1976); New Aiyansh 30 Apr 1987-1. **Interior** – Christina Lake 20 Apr 1971-1 on dyke at s end (Walker 1996); Okanagan valley 31 May 1914-1; nr Quilchena 1 Mar 1997-1; Knutsford 4 Apr 1982-125 in ploughed field, 17 Apr 1983-1 in grassland; Dog Creek 17 Mar 1991-10, all males with Snow Buntings and Horned Larks (Stewart 1991); Chilcotin (Riske Creek) 13 Mar 1984-1 male at Wineglass Ranch; Farwell Canyon 16 Mar 1985-25 in grassland with Horned Larks; Deer Park (Chilcotin) 10 May 1991-1; Quesnel 15 Mar 1979-1; Giscome 15 Mar 1992-5; Moricetown 13 Mar 1979-1; Tetana Lake 12 May 1938-1 (RBCM 8995); Sunset Hill 27 May 1993-8 in recent clearcut; Dawson Creek 15 Apr 1990-1,000 (Siddle and Bowling 1993), 22 Apr 1994-10,000 (Bowling 1994a); Fort St. John 10 May 1982-2,500 in ploughed field; nw Fort St. John 3 May 1985-2,500, about 85% males; sw North Pine 27 Mar 1982-2 feeding in stubble field, 18 Apr 1988-600; Charlie Lake 21 May 1983-1; Sikanni Chief River 7 May 1992-40 in grassy field; Buckinghorse River 13 May 1982-10; Hyland Post 10 May 1982-10; Steamboat 3 May 1985-100 in mixed flock with White-crowned Sparrows and Gray-crowned Rosy-Finches; Fort Nelson 3 Apr 1987-1, 1 May 1987-250 at airport, 22 May 1982-1; Helmut 15 May 1982-50; Liard River 7 May 1975-3 (Reid 1975); s Cormier Creek 29 Apr 1965-30 along Highway 37 (West et al. 1968); Atlin 21 Mar 1981-3; Atlin Lake 2 May 1958-2,000 between Atlin and BC-Yukon border (Irving 1961); Chilkat Pass 7 May 1957-1 (Weeden 1960); Kusawak Lake 21 May 1977-1 feeding on ground with Snow Buntings, American Tree Sparrows, and 2 Smith's Longspurs.

Summer: Coastal – Esquimalt Lagoon 24 Aug 1980-2 with 6 Horned Larks; Saanich 20 Aug 1979-2; Sidney Island 29 Aug 1987-1; Tofino 10 Jun 1982-1 at airport in full breeding plumage; Iona Island 4 Jun 1990-1, 2 Jul 1985-1; Tsawwassen 7 Aug 1960-6 on beach (Boggs and Boggs 1961a); Meager Mountain 14 Aug 1932-1 in alpine area nr summit (Carter 1933); Triangle Island 13 Jun 1976-1 in breeding plumage; Sandspit 23 Aug 1991-1 (Siddle 1992a); Masset 26 Aug 1957-8 on gravel beach (Mills 1960); Chilkoot Pass 12 Jul 1975-10 feeding on snow patches with Snow Buntings. **Interior** – Naramata 27 Aug 1942-1; Harmer Ridge (Sparwood) 30 Aug 1984-16 on high-elevation, reclaimed mine spoil; Tonquin Ridge 30 Aug 1974-10 in alpine area at 2,440 m elevation (Cannings et al. 1974); Kelly Lake 30 Aug 1981-10 with 25 American Pipits; e Moberly Lake 25 Jul 1930-1 (Williams 1933b); Baldonnel 29 Aug 1982-100; Beatton Park 20 Aug 1988-1; Boundary Lake (Goodlow) 22 Aug 1984-3; nr Ipec Lake 28 Aug 1979-50 (Cooper and Adams 1979); Conroy Creek 9 Jun 1922-1 (Williams 1933a); Spatsizi Plateau Wilderness Park 15 Jul 1959-1; Summit Pass 30 Aug 1943-7 (Rand 1944); Fort Nelson 30 Aug 1987-10 on golf course; Cassiar 25 Aug 1962-44.

Breeding Bird Surveys: Not recorded.

Autumn: Interior – Haines Highway 23 Oct 1981-100; Atlin 1 Nov 1932-1 (RBCM 5760); Summit Lake (Stone Mountain Park) 7 Sep 1943-1 (NMC 29607); Fort Nelson 6 Sep 1986-80 (McEwen and Johnston 1987a); Stikine River 27 Sep 1971-1; 11.8 km n Buckinghorse River 12 Sep 1997-75, on edge of Alaska Highway; Nig Creek (n Prespatou) 15 Sep 1984-12,000 in large fields; Pink Mountain 4 Oct 1981-several small flocks in alpine area; Pesika River 4 Sep 1997-4 in alpine area on s side of valley; ne Fort St. John 5 Sep 1986-500 at sewage lagoons, 23 Oct 1982-1 at sewage lagoons with flock of Snow Buntings; Cache Creek (Peace River) 1 Sep 1985-34, n of Highway 29; Two Rivers 2 Oct 1982-1; Roman Mountain 19 Sep 1998-200 in alpine area; 32 km s Nation River 20 Oct 1973-10 along roadside; Willow River 7 Sep 1983-40; Beverly Lake (Prince George) 23 Sep 1957-20; Giscome 22 Oct 1994-1; Nulki Lake 29 Sep 1951-100 (Munro 1955a); n Ootsa Lake 1 Sep 1997-1 in well-grazed pasture; Indianpoint Lake 23 Sep 1929-1 (MCZ 285803); Mount Davidson 15 Sep 1982-20; Riske Creek 8 Sep 1978-20 with Horned Larks; Williams Lake 1 Nov 1950-1 (Jobin 1952); Buffalo Lake 24 Sep 1933-2 (RBCM 11439-40); lower Blaeberry River valley 7 Oct 1997-7 in cattle pasture; Revelstoke 30 Nov 1988-2; Knutsford 10 Oct 1983-25, 18 Nov 1984-2; Enderby 28 Sep 1954-1 (UBC 7777); Okanagan Landing 16 Oct 1934-100; Nakusp 22 Sep 1980-15; McDonald Creek (Nakusp) 2 Sep 1985-1; Harmer Ridge 17 Sep 1983-40 on reclaimed mine spoil; West Bench (Penticton) 12 Oct 1970-100; Princeton 6 Sep 1890-1 (MCZ 44382); Cranbrook 1 Oct 1940-20 (Johnstone 1949); Newgate 16 Oct 1930-1 (NMC 24639); Duck Lake (Creston) 15 Sep 1980-13; Manning Park 27 Sep 1972-2. **Coastal** – 37 km n Hazelton 8 Sep 1921-1 (MVZ 42269); Terrace 30 Nov 1968-1; Green Island (Dundas Island) 7 Oct 1977-23 at lightstation; Masset 7 Oct 1939-2 (RBCM 10352 to 10353), 4 Nov 1971-2; Sandspit 6 Oct 1991-35 (Siddle 1992a), 8 Nov 1986-1 at airport; Cape St. James 27 Sep 1981-85, 12 Nov 1978-8; Cape Scott 16 Sep 1935-1; Grant Bay 14 Oct 1968-5; Angus Creek (Sechelt) 15 Oct 1992-1; Mission Point (Sechelt) 15 Sep 1992-1; Stubbs Island (Tofino) 17 Oct 1982-14 feeding on grassy beach dune, 9 Nov 1982-1; Chesterman Beach (Tofino) 2 Sep 1983-1; Pitt Meadows 11 Oct 1981-25; Skagit River 29 Sep 1974-2; Iona Island 3 Oct 1981-100; Duncan 8 Sep 1972-10; Pachena Point 23 Sep 1974-4 on gravelly, weedy areas; Oak Bay 17 Sep 1977-17 on golf course.

Winter: Interior – Hemp Creek (Wells Gray Park) 3 Feb 1953-1 in field, collected (Edwards and Ritcey 1967); Knutsford (Kamloops) 2 Dec 1984-2; Vernon 11 Jan 1992-1 with Horned Larks (Siddle 1992b); Okanagan Landing 3 Feb 1922-unknown number with Snow Buntings and Horned Larks; Cranbrook 5 Jan 1937-1 (RBCM 10911). **Coastal** – Terrace 1 Dec 1968-1 (Crowell and Nehls 1969b); Rose Spit 18 Dec 1986-3; Delkatla Inlet 11 Feb 1973-2; Pitt Meadows 28 Jan 1973-8; Iona Island 15 Dec 1992-20 (Siddle 1992b); Lulu Island 1 Feb 1949-30 feeding in field; Westham Island 25 Feb 1985-30; Delta 8 Feb 1975-130 in 2 flocks, with Western Meadowlarks; Chain Islets 23 Dec 1976-22; Victoria 10 Jan 1971-3 (Crowell and Nehls 1971b).

Christmas Bird Counts: Interior – Recorded from 4 of 27 localities and on 2% of all counts. Maxima: North Pine 27 Dec 1986-6; Fort St. James 2 Jan 1994-1 (BC Photo 1713); Vernon 22 Dec 1985-1; Nakusp 3 Jan 1986-1. **Coastal** – Recorded from 6 of 33 localities and on 3% of all counts. Maxima: Ladner 21 Dec 1963-102; Victoria 18 Dec 1976-22; Vancouver 15 Dec 1991-20.

Smith's Longspur
Calcarius pictus (Swainson)

SMLO

RANGE: Breeds from east-central Alaska and adjacent northwestern British Columbia and from northern Alaska east across northern Yukon and northern and east-central Mackenzie to southern Keewatin, northeastern Manitoba, and extreme northern Ontario. Winters from Kansas and central Iowa south to Oklahoma, central Texas, and northwestern Louisiana, and east to Arkansas, Mississippi, Tennessee, and Alabama.

STATUS: On the coast, *casual* in the Georgia Depression Ecoprovince; in the Coast and Mountains Ecoprovince, *casual* on Western Vancouver Island and the Queen Charlotte Islands; *accidental* on the Northern Mainland Coast.

In the interior, locally *uncommon* migrant and summer visitant in the Northern Boreal Mountains Ecoprovince; *casual* in the Taiga Plains, and *accidental* in the Southern Interior, Southern Interior Mountains, Central Interior, Sub-Boreal Interior, and Boreal Plains ecoprovinces.

Breeds.

Figure 410. Adult male Smith's Longspur (© T. Vezo/VIREO). This species breeds only in tundra-like habitats in extreme northwestern British Columbia.

Figure 411. Male Smith's Longspur in nonbreeding plumage (Mount Tolmie, southern Vancouver Island, 16 September 1990; Tim Zurowski). During the spring and autumn migration periods, the Smith's Longspur has been reported as a vagrant from widely scattered locations throughout British Columbia.

CHANGE IN STATUS: Munro and Cowan (1947) include the Smith's Longspur (Figs. 410 and 411) in the avifauna of British Columbia based on 3 records: "Boundary Pass" (now North Kootenay Pass) in 1858 (Blakiston 1861-62), the Kispiox valley in 1921, and Swan Lake (near Tupper) in 1938 (Cowan 1939). The species was reported infrequently in British Columbia between 1958 and 1970 (2 or 3 birds per decade), mostly in the regions of Kelsall Lake and Spatsizi Plateau Wilderness Park. Then, in the 1970s, there were 36 records totalling 152 birds, followed by 19 records including 26 birds in the 1980s. Almost all the records between 1970 and 1989 were from the mountainous headwaters of the Tatshenshini, Kelsall, Alsek, and Stikine rivers. It cannot be concluded from these data that there was an increase in the numbers of Smith's Longspurs during those decades. Rather, a surge of biological exploration that accompanied the creation of Mount Edziza Park, Tatlatui Park, and Spatsizi Plateau Wilderness Park was largely responsible for the discovery of the Smith's Longspur as a summer visitant in those regions of the province (Osmond-Jones et al. 1977; Bergerud and Butler 1978). On the other hand, the Chilkat pass area was intensively studied by R.B. Weeden from 1957 to 1959, with a total of 22 weeks of field work spanning the period 7 May to 10 September (Weeden 1960). During that time, he saw 2 Smith's Longspurs. In the same region, in June and July 1979 and 1980, 6 nests with attendant adults were found and other birds were seen (Martin et al. 1981). Also, 8 of the 14 vagrant records are from the 1980s. We suggest that there was an increase in the numbers of this species in the province during the 1970s and 1980s.

The first indication of possible breeding in the province came when a pair was observed and the female collected, on 17 July 1944, near Kelsall Lake, Chilkat Pass (Clarke 1945; Weeden 1960). This female had a large brood patch. The first confirmed breeding records for the province were in 1979 near Kelsall Lake, Chilkat Pass (Martin et al. 1981). It is likely that the Smith's Longspur has been present as a nesting species for many years, undiscovered in the vastness of its remote habitat.

NONBREEDING: The Smith's Longspur occurs regularly in British Columbia only in the northwestern portion of the province, where it is distributed locally in the Northern Boreal Mountains from Chilkat Pass southeast to Spatsizi Plateau Wilderness Park. Elsewhere in the interior, it occurs as a vagrant during the migration periods in the Southern Interior, Central Interior, Sub-Boreal Interior, Boreal Plains, and Taiga Plains. On the coast, it has occurred during the migration periods as a vagrant from Sooke, on southern Vancouver Island, north to Hazelton, in the Northern Mainland Coast of the Coast and Mountains.

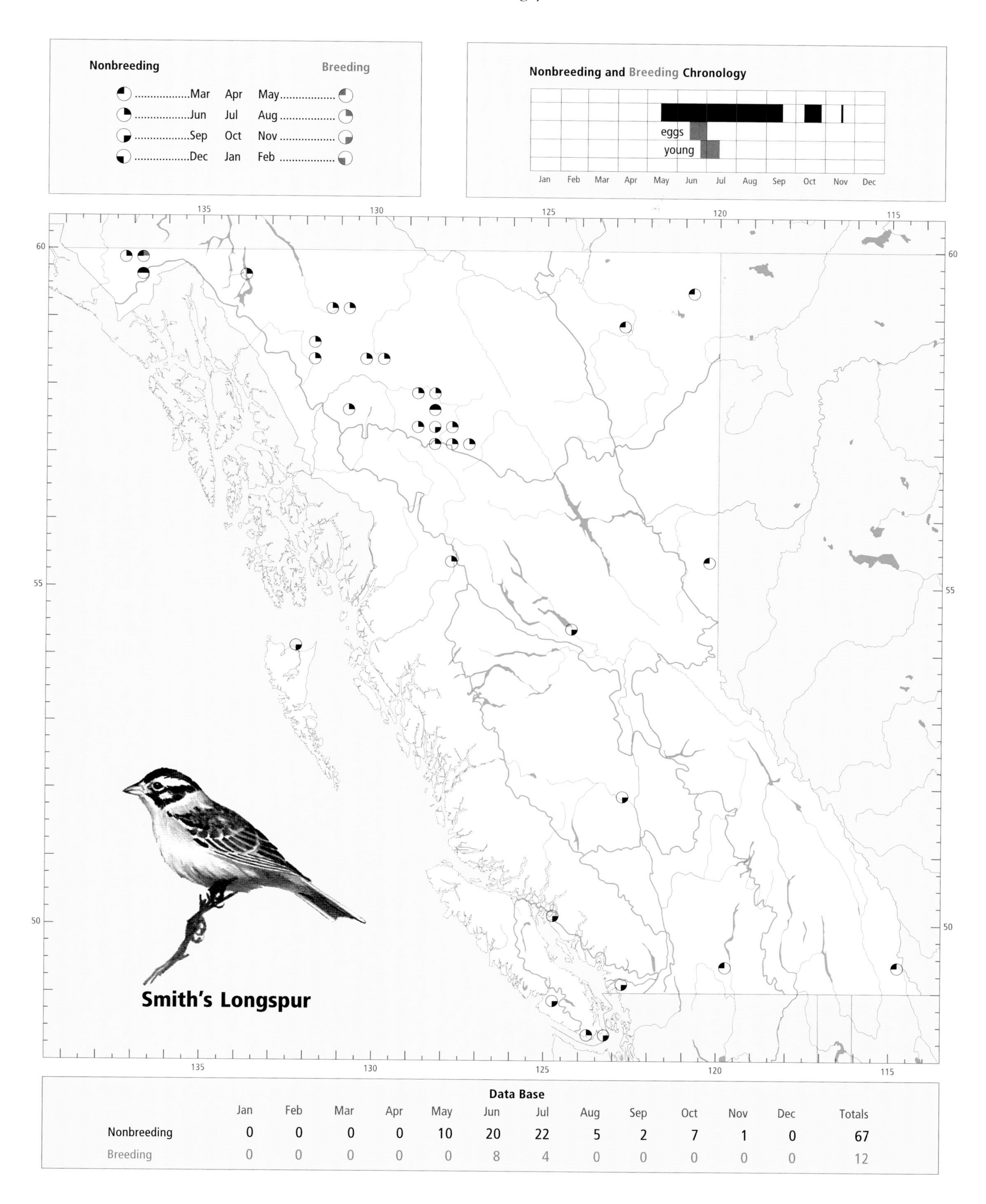

Data Base

	Jan	Feb	Mar	Apr	May	Jun	Jul	Aug	Sep	Oct	Nov	Dec	Totals
Nonbreeding	0	0	0	0	10	20	22	5	2	7	1	0	67
Breeding	0	0	0	0	0	8	4	0	0	0	0	0	12

Figure 412. When the Smith's Longspur reaches its breeding grounds in extreme northwestern British Columbia each spring, it forages for a brief time in vegetated flats with dwarf birches, forbs, and grasses (near Kelsall Lake, 25 June 1999; R. Wayne Campbell).

The Smith's Longspur has been recorded at elevations from sea level to 1,400 m. There is little information on habitat use in the nonbreeding season. The few vagrants on the south coast were observed in patches of dry grasses and weeds in logged land, alongside gravel roads, or in field verges, and were occasionally noted to perch in shrubs. In the interior, it has been observed feeding in clearcut areas, weedy areas near willows on the margin of a cattail swamp, and grassy meadows bordering a lake. Near its breeding grounds in late May, this species feeds in large flocks on open grass-herb slopes or flats vegetated with dwarf birches, forbs, grasses, and sedges at subalpine sites (Fig. 412). During the spring and autumn migrations across the Great Plains, the species uses pastures, stubble fields, airports, and grassy flats around large lakes (Briskie 1993).

We have little information on the spring migration into the province (Figs. 413 and 414). The species does not occur west of the Rocky Mountains, nor does it appear to reach the province directly by way of Alberta. There are few records of the Smith's Longspur in western Alberta (Sadler and Myres 1976; Salt and Salt 1976), and its scarcity there appears to be confirmed by the paucity of migration records in the Boreal Plains and Taiga Plains of British Columbia. The 3 spring records from northeastern British Columbia are from 18 to 27 May, and 1 from the Okanagan valley is from 27 May.

In the Northern Boreal Mountains, there have been few observers in the field in the spring months and the earliest record available is from 18 May. There are a number of records between 27 May and 9 June. Large flocks were seen at an elevation of 1,000 m near Hyland Post on 27 May. Thus, the species apparently arrives on its nesting grounds in British Columbia in late May and early June (Figs. 413 and 414). The origin and migration routes remain unknown, but it is probably safe to assume that the British Columbia–bound birds travel a route across Saskatchewan with birds bound for the Wrangel Mountains of south-central Alaska and the St. Elias Mountains of the Yukon (Godfrey 1986), then continue south into the contiguous ranges of the Northern Boreal Mountains of British Columbia. Males reach the breeding grounds several days before females (Briskie 1993).

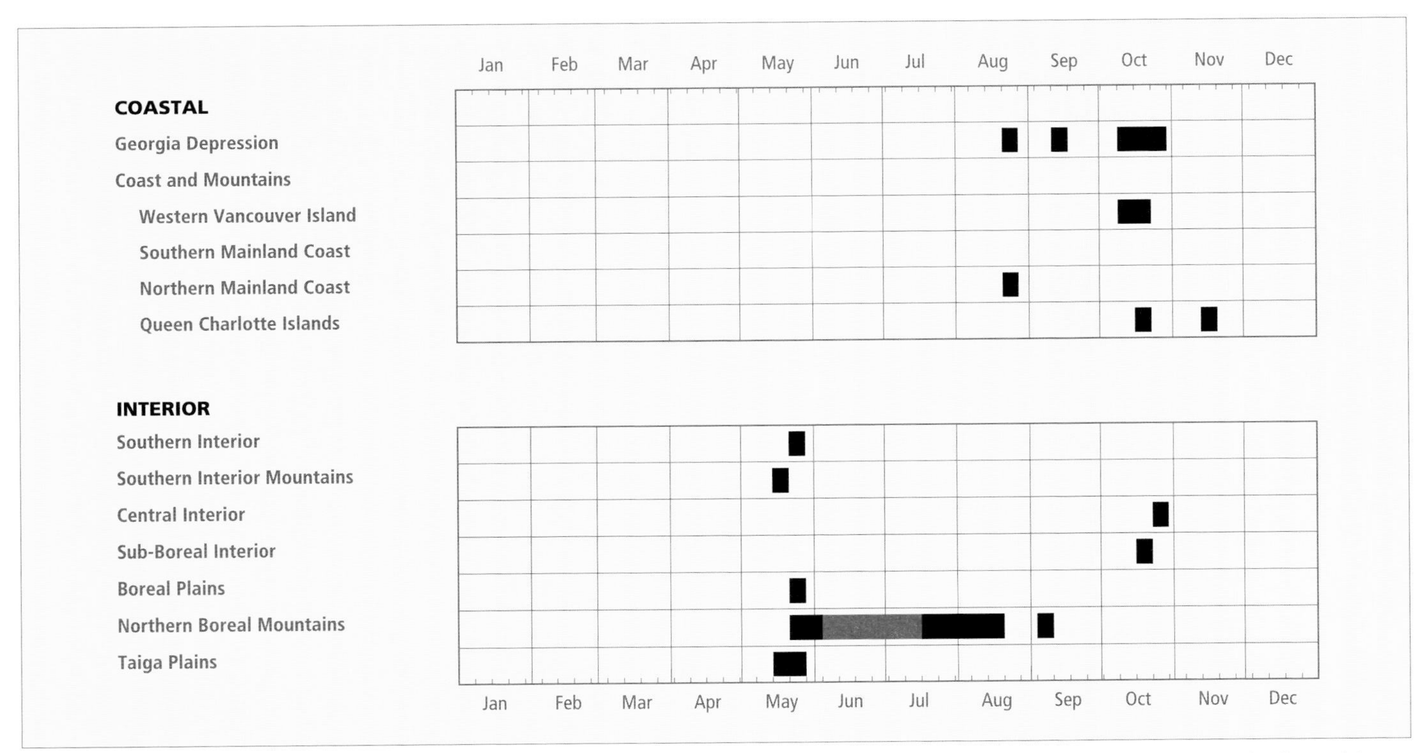

Figure 413. Annual occurrence (black) and breeding chronology (red) for the Smith's Longspur in ecoprovinces of British Columbia. Records are shown for the week in which they occurred.

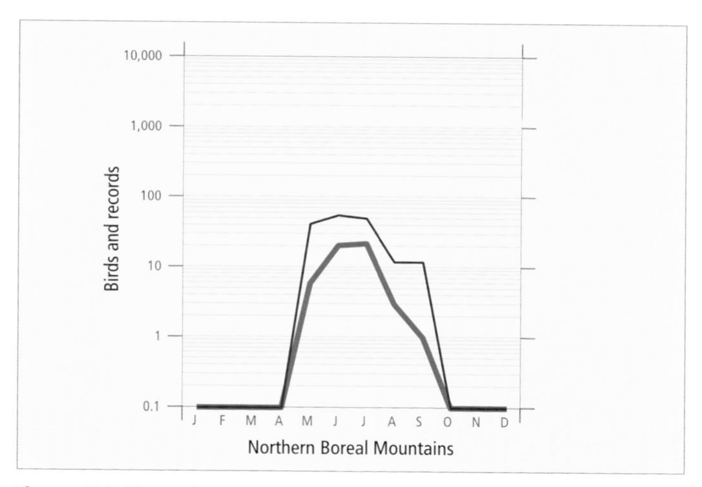

Figure 414. Fluctuations in total number of birds (purple line) and total number of records (green line) for the Smith's Longspur in the Northern Boreal Mountains Ecoprovince of British Columbia. Nest record data have been excluded.

Autumn migrants begin leaving the breeding grounds by mid-August; the latest date recorded for the Northern Boreal Mountains is 3 September. Flocks of 10 and 12 are on record for Spatsizi Plateau Wilderness Park on 12 August and 3 September. Although the vagrants in central and southern British Columbia are off course, they nevertheless help indicate the timing of migration. There are records of 9 vagrants between 19 and 26 October in the interior and between 26 August and 11 November on the coast. In Alberta, the Smith's Longspur is an uncommon migrant that is seen more often in spring than in autumn. The autumn migration occurs in the first half of September, mainly in the eastern parts of Alberta (Semenchuk 1992) and Saskatchewan (Salt and Salt 1976). Migrants frequently travel in flocks of as many as 200 birds (Rising 1996). In British Columbia, the largest autumn flock recorded consisted of 30 birds. There are no winter records of the Smith's Longspur.

The Smith's Longspur has been recorded in the province from 15 May to 18 November (Fig. 413).

BREEDING: The Smith's Longspur breeds in the extreme northwestern portion of British Columbia only in the vicinity of Kelsall Lake, near Chilkat Pass (Martin et al. 1981). Breeding numbers appear to be small, but field workers in the region note that most sightings on the summer range involve pairs of birds behaving as though on territory. There are records of fledglings in the late summer in the Mount Edziza Park, Spatsizi Plateau Park, and Tatlatui Park areas (Osmond-Jones et al. 1977). These observations suggest that breeding is more widely distributed over the Northern Boreal Mountains than has been confirmed. A careful search for nesting Smith's Longspurs in the Spatsizi area has yet to be made.

The Smith's Longspur reaches its highest numbers in summer in the Northern Boreal Mountains (Fig. 415). In British Columbia, breeding Bird Surveys conducted from 1968 to 1993 did not have a survey route in the area inhabited by this species. Rising (1996) mentions that Smith's Longspur populations have been known to fluctuate from year to year. Failure to locate the species on its breeding grounds in British Columbia in some years may be due in part to these fluctuations experienced elsewhere, but it is probably due more to a lack of observers in the remote habitat of the species.

Nesting habitat in the Chilkat Pass area was in subalpine tundra characterized by a shrub cover of willow, scrub birch, and shrubby cinquefoil, and a ground cover of sedges, lichens, mosses, and perennial forbs (Martin et al. 1981; Fig. 416). In the Wrangel Mountains of east-central Alaska, the Smith's Longspur was nesting either on an alpine plateau at 2,500 to 2,700 m elevation in well-drained areas dominated by *Dryas* hummocks with scattered *Rhododendron lapponicum,* or in *Carex* tussocks with associated birch and willow shrubs (Sage 1976). These habitats are similar in structure to those adjacent to Kelsall Lake, but different from those sought by the species on the arctic slope of Alaska and the Yukon, where it prefers "deep tussock meadows on the floor of wide alpine valleys" (Kessel and Gibson 1978).

The Smith's Longspur has been found breeding in British Columbia from 14 June (calculated) to 14 July (calculated) (Fig. 413).

Nests: Seven nests found in the province were in depressions in the ground on flat exposed sites or in the sides of hummocks. They were constructed of grasslike material and lined with sedges, grasses, hair, or feathers, frequently those of the Willow Ptarmigan. In Ontario, nests (n = 14) were similarly situated and constructed, but sometimes in mosses and lichens near the nest. One nest was in a clump of rhododendron, another under a small dwarf birch; a third was hidden in tall, dead grass clumps (Peck and James 1987).

Eggs: Dates for 5 clutches ranged from 18 June to 1 July. Calculated dates indicate that eggs can be found as early as 14 June. Clutch size ranged from 4 to 6 eggs. In 3 of the 5

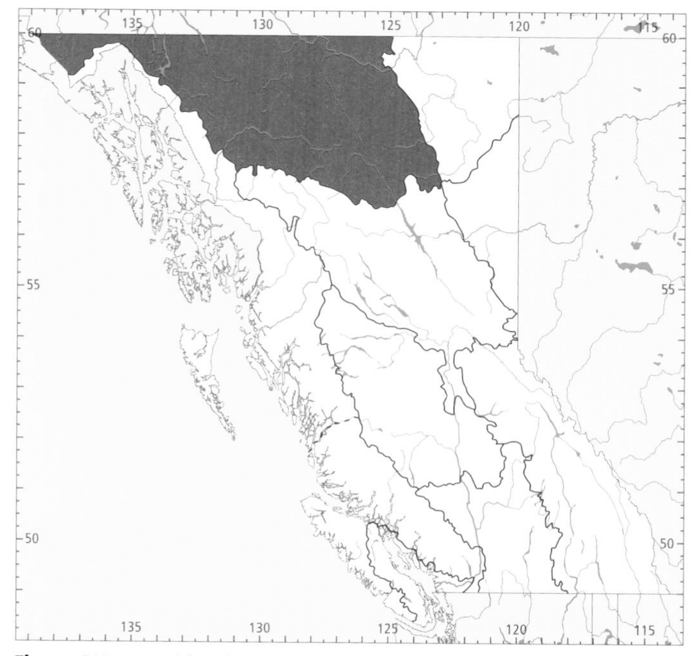

Figure 415. In British Columbia, the highest numbers for the Smith's Longspur in summer occur in the Northern Boreal Mountains Ecoprovince.

Figure 416. In British Columbia, breeding habitat for the Smith's Longspur includes subalpine tundra with a shrub cover of willows and dwarf birches and a ground cover of lichens and mosses (near Kelsall Lake, 30 June 1999; R. Wayne Campbell).

nests, hatching occurred from 25 to 29 June. In northern Ontario, the clutch size also ranges from 4 to 6 (usually 4) eggs (Rising 1996). The incubation period is 11 to 13 days (Jehl 1968; Briskie 1993). Renesting after the loss of a clutch is uncommon.

Young: Dates for 5 broods ranged from 25 June to 3 July. Calculated dates indicate that young can be found as late as 14 July. Brood size ranged from 2 to 6 young. The nestling periods for 2 nests in British Columbia were 9 and 14 days; elsewhere, it is reported to be 7 to 9 days (Briskie 1993). A single brood is reared per season.

Brown-headed Cowbird Parasitism: Not recorded. The nesting distribution of the 2 species do not coincide anywhere. There are no records of the Smith's Longspur being parasitized by the Brown-headed Cowbird (Friedmann and Kiff 1985).

Nest Success: In British Columbia, the single nest found with eggs and followed to a known fate was successful. During a 6-year study at Churchill, Manitoba, nest success ranged from a low of 33% in 1990 to a high of 91% in 1992 (mean = 65%, n = 79 nests) (Briskie 1993).

REMARKS: Kemsies (1961) suggests that 3 subspecies be recognized, but the supposed differences between the proposed subspecies were subsequently found to be associated with seasonal wear and fading of breeding plumage (Jehl 1968).

Populations occur at low densities throughout the species' breeding range. The total world population is unlikely to be more than about 75,000 birds (Briskie 1993).

Like Snow Buntings in parts of the same alpine wilderness, Smith's Longspurs nesting in British Columbia offer attractive biological and aesthetic challenges to students of birds. Their habitat differs in many ways from that used in the breeding grounds of arctic Canada and Alaska, and their nesting biology still holds many secrets.

For additional information on the ecology and life history of the Smith's Longspur, see Briskie (1993).

NOTEWORTHY RECORDS

Spring: Coastal – No records. **Interior** – White Lake (Okanagan Falls) 27 May 1966-1 (Cannings et al. 1987); North Kootenay Pass 15 May 1858-1 (Blakiston 1863); Swan Lake (Tupper) 27 May 1938-1 (RBCM 8004) (Cowan 1939); Fire Flats May 1977-4 (Bergerud and Butler 1978); Hyland Post 29 May 1976-30; Fort Nelson 18 May 1987-1; Kwokullie Lake 24 May 1982-1 (RBCM 17500); Kusawak Lake 21 May 1977-2.

Summer: Coastal – Muir Point (Sooke) 26 Aug 1980-1; Kispiox River valley 25 Aug 1921-1 (MVZ 42271). **Interior** – Kitchener Lake 6 Jul 1976-5; Kluayetz Creek 16 Jul 1976-2; Buckinghorse Lake 20 Jun 1976-1 (Osmond-Jones et al. 1977); Mount Edziza Park 14 Aug 1986-1; Spatsizi Plateau Park 12 Aug 1976-10 adults with fledged young; Tanzilla Butte 8 Jun 1962-1 (NMC 50355); Big Gnat Lake 8 Jun 1962-13; Dease Lake 29 Jun 1978-2; Wright Creek (e Atlin) 9 Jun 1958-1 (UBC 8580); Dudidontu River valley 2 Jun 1978-6; Haines Highway 17 Jul 1944-1 (NMC 31674); Kelsall Lake 25 Jun 1958-2 (Weeden 1960), 18 Jun 1979-4 eggs (first nesting for province), 2 Jul 1980-2 nestlings, 25 Jun 1980-6 eggs hatching (Martin et al. 1981).

Breeding Bird Surveys: Not recorded.

Autumn: Interior – Griffith Creek 3 Sep 1976-12; Fort St. James 19 Oct 1978-1; Chilcotin River (Riske Creek) 26 Oct 1988-1 at Wineglass Ranch. **Coastal** – Masset 15 Oct 1994-1; Delkatla Inlet 18 Nov 1983-1; Cortes Island 25 Oct 1976-1; nr Bamfield 18 Oct 1980-1; Pitt Meadows 11 Oct 1981-1; Mount Tolmie (Victoria) 16 Sep 1990-1 (Fig. 411).

Winter: No records.

Snow Bunting

SNBU

Plectrophenax nivalis (Linnaeus)

RANGE: Holarctic. In North America, breeds from northern Alaska, northern Yukon, northwestern Mackenzie, and Banks and northern Ellesmere islands south to southern Alaska (including the Aleutian Islands), northwestern British Columbia, southwestern and central Yukon, east-central Mackenzie, central and southwestern Keewatin, Southampton and Belcher islands, northern Quebec, and northern Labrador. In the Palearctic, breeds from Greenland and Spitsbergen across Siberia to Kamchatka and south into northern Britain, central Siberia, and the Commander Islands. Winters in North America from west-central and southern Alaska and southern British Columbia east across southern Canada to Newfoundland; south in the west to California, Utah, and Colorado. East of the Rocky Mountains, it winters south into Kansas, Kentucky, and North Carolina, and casually to Texas, Tennessee, Mississippi, Georgia, and Florida. In the Palearctic, it winters from the breeding range south into central Europe, southern Russia, Manchuria, Sakhalin, and the Kurile Islands.

STATUS: On the coast, a regular but *uncommon* to occasionally *fairly common* migrant and winter visitant in the Georgia Depression Ecoprovince; in the Coast and Mountains Ecoprovince, a *very rare* autumn transient and *casual* in winter on Western Vancouver Island, *casual* on the Southern Mainland Coast, *very rare* transient on the Northern Mainland Coast, *casual* there in winter; and *very rare* autumn transient and winter visitant on the Queen Charlotte Islands, *accidental* there in summer.

In the interior, an *uncommon* to sometimes locally *very common* migrant and winter visitant in the Southern Interior, Southern Interior Mountains, Central Interior, Boreal Plains, and Taiga Plains ecoprovinces; *very rare* during migration and winter in the Sub-Boreal Interior. In the Northern Boreal Mountains, a *common* migrant and summer visitant and a *very rare* winter visitant.

Breeds.

CHANGE IN STATUS: In the mid-1940s, the Snow Bunting (Figs. 417 and 418) was considered a transient throughout the interior and on the southern coast of British Columbia, and an irregular winter visitant across southern regions of the province (Munro and Cowan 1947). Today, its nonbreeding status has remained relatively unchanged but the Snow Bunting has now been found breeding in the northwestern corner of the province, in the Tatshenshini River drainage (Campbell and Van Der Raay 1985). A number of recent summer records suggest a more widespread breeding distribution in the Northern Boreal Mountains that may extend east to near Lower Post and south to Mount Edziza and Spatsizi Plateau wilderness parks. Lyon and Montgomerie (1995) suggest that there has been a recent expansion of the breeding range of the Snow Bunting southward on the southern coast of Hudson Bay and

Figure 417. In British Columbia, the Snow Bunting breeds only in alpine areas in the Northern Boreal Mountains Ecoprovince (© T. Vezo/VIREO).

Figure 418. The Snow Bunting, here in winter plumage, winters in small flocks in open country over much of the province (Victoria, 28 October 1984; Tim Zurowski).

in Scotland. Perhaps the same has occurred in British Columbia, but the nature of biological exploration in extreme northwestern British Columbia leaves the possibility that the species had been present there for many years and went undiscovered.

NONBREEDING: The Snow Bunting is widely distributed throughout the interior of British Columbia and sparsely distributed in widely scattered areas on the south coast, including Vancouver Island. It occurs irregularly and locally on the Northern Mainland Coast and the Queen Charlotte Islands.

It reaches its highest numbers in winter in the Boreal Plains (Fig. 419), where more than 700 birds have been reported on Christmas Bird Counts.

Outside the breeding season, most records are from lower elevations where open, sparsely vegetated habitat occurs. The Snow Bunting has been found from sea level to 1,780 m

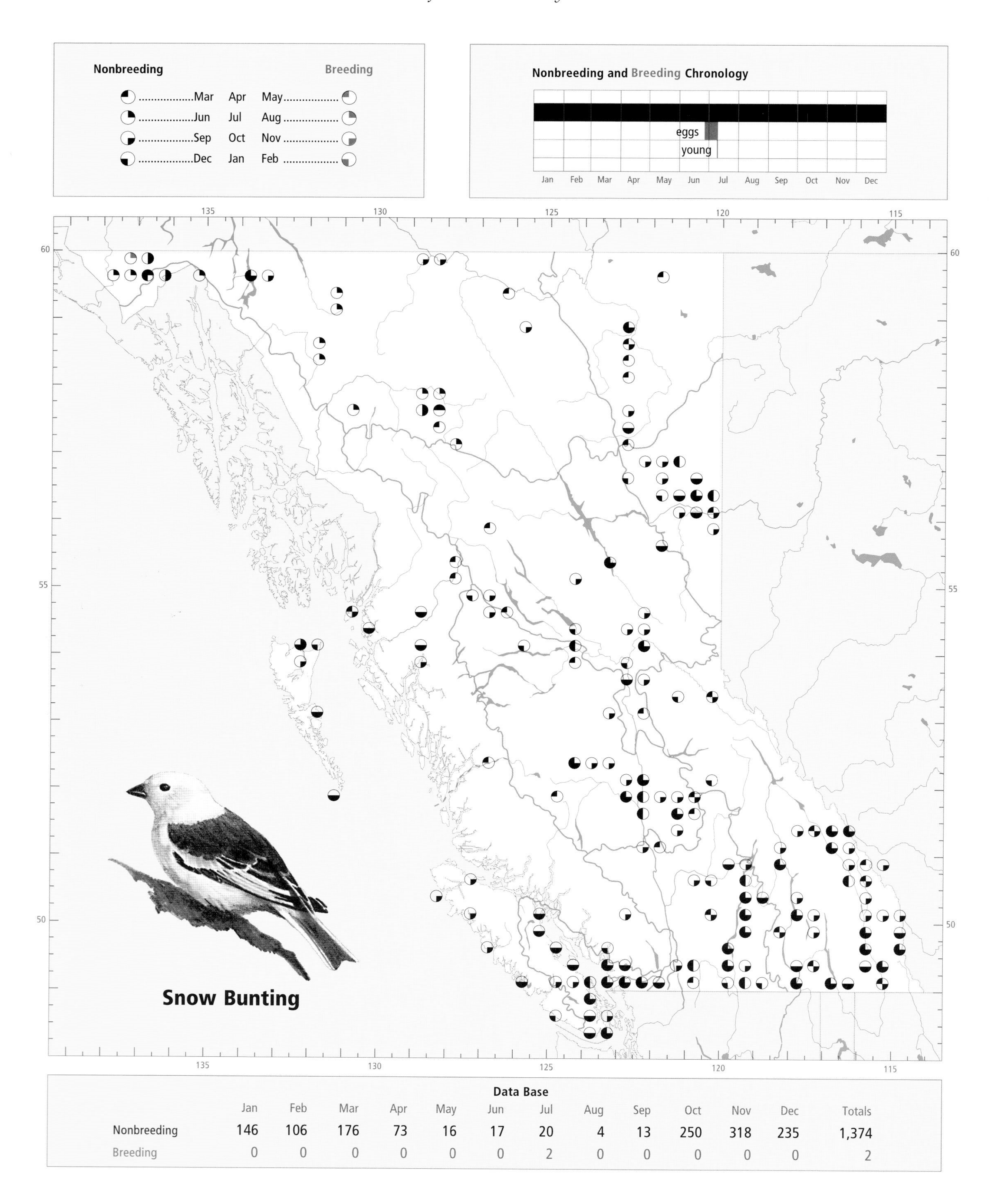

Data Base

	Jan	Feb	Mar	Apr	May	Jun	Jul	Aug	Sep	Oct	Nov	Dec	Totals
Nonbreeding	146	106	176	73	16	17	20	4	13	250	318	235	1,374
Breeding	0	0	0	0	0	0	2	0	0	0	0	0	2

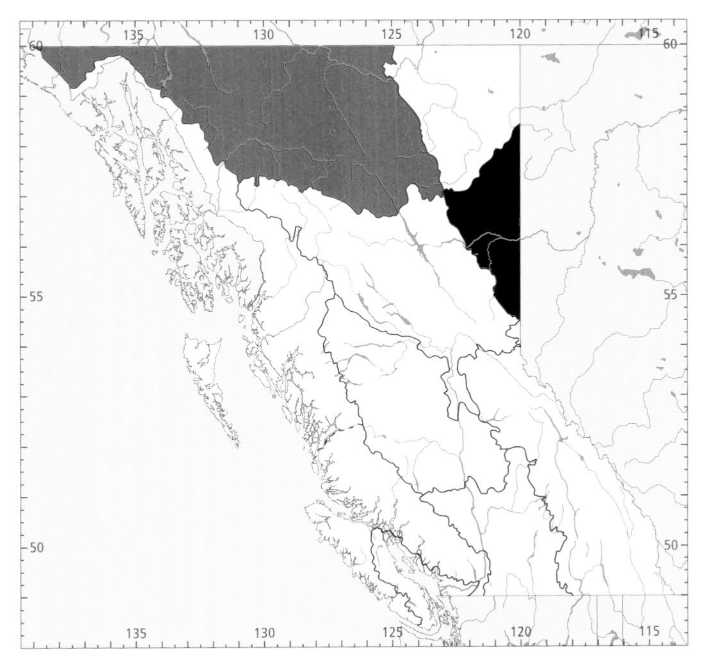

Figure 419. In British Columbia, the highest numbers for the Snow Bunting in winter (black) occur in the Boreal Plains Ecoprovince; the highest numbers in summer (red) occur in the Northern Boreal Mountains Ecoprovince.

elevation on the coast, and from about 270 to 2,150 m elevation in the interior.

During the nonbreeding parts of the year, the Snow Bunting inhabits almost any open, exposed environment such as grasslands (Fig. 420), windswept fields (Fig. 423), sandspits, extensive lakeshores, and road verges that are widespread in the interior of the province, but will use airports, roadsides, rock-strewn islets, cattle feedlots, stockyards, extensive areas of mine tailings, clearings associated with sewage ponds, golf courses, and jetties. In the north, early in the autumn migration, small flocks gather on the upper alplands and migrate along them, but very few Snow Buntings have been seen in alpine habitats further south. Along the forested coast, the few records are generally confined to major sandspits and to clearings associated with human activity.

Figure 420. In autumn, flocks of Snow Buntings are conspicuous in grasslands and along road edges in the southern interior of British Columbia (Douglas Lake, 3 November 1996; R. Wayne Campbell).

In British Columbia, the Snow Bunting has been most frequently recorded on fields or grasslands (29%; n = 231; Figs.

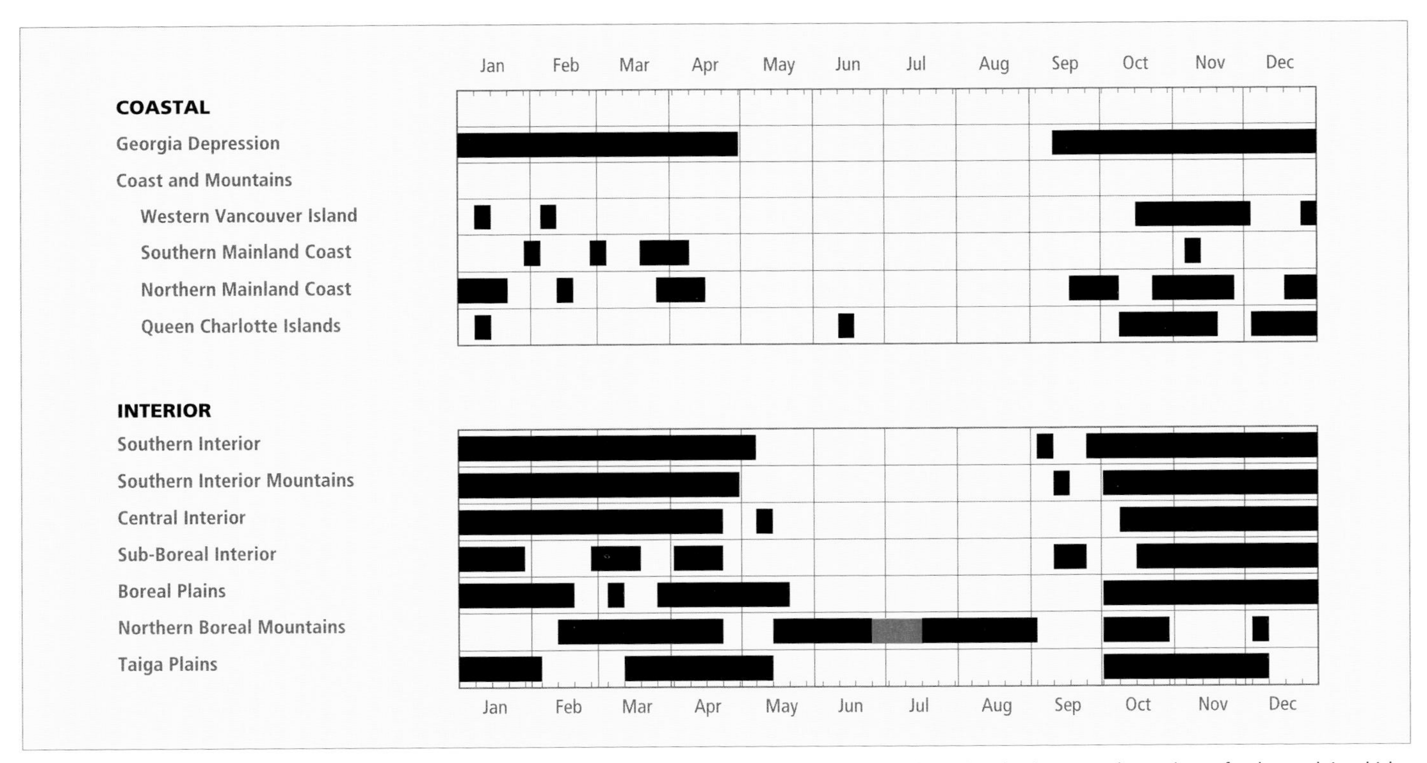

Figure 421. Annual occurrence (black) and breeding chronology (red) for the Snow Bunting in ecoprovinces of British Columbia. Records are shown for the week in which they occurred.

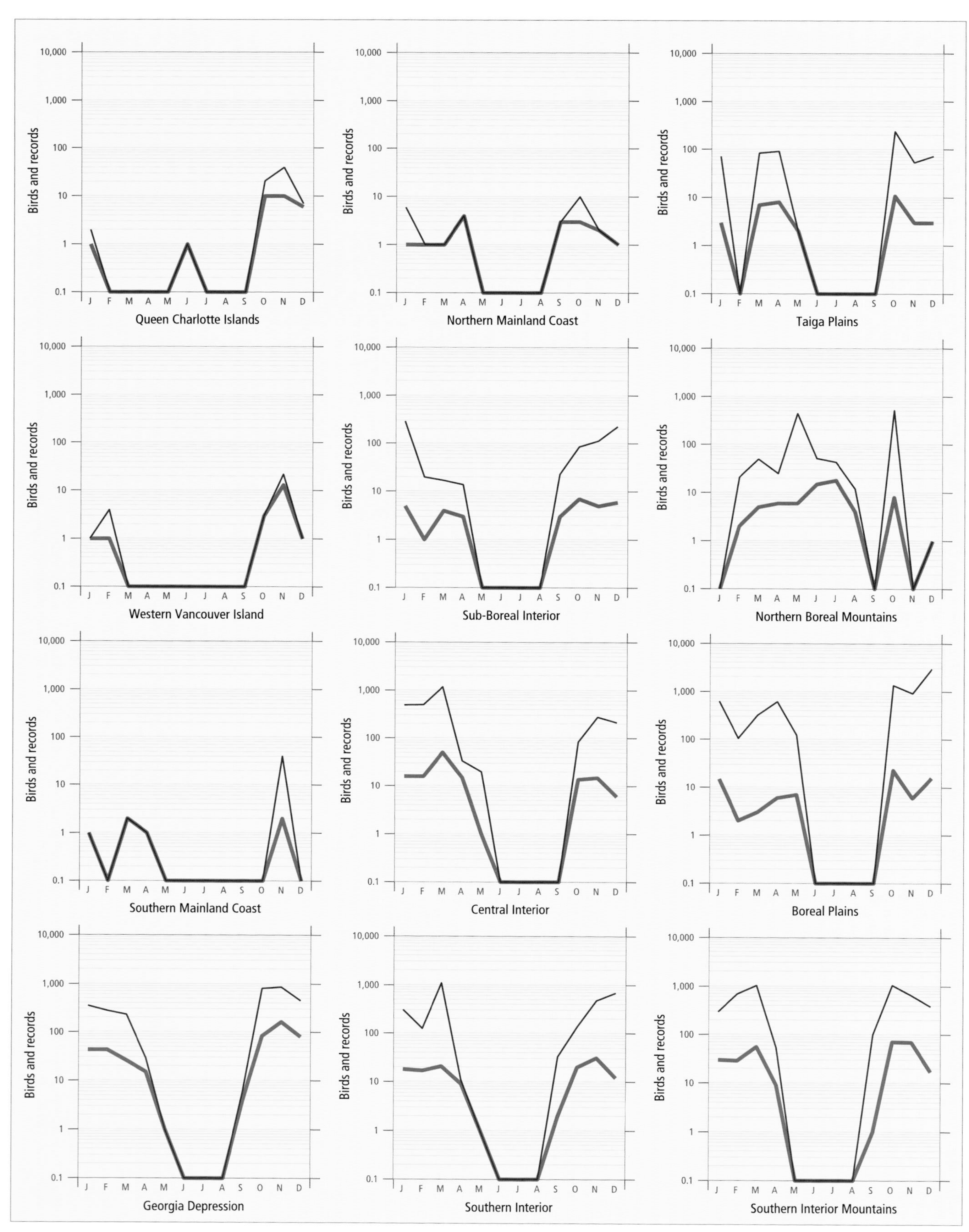

Figure 422. Fluctuations in total number of birds (purple line) and total number of records (green line) for the Snow Bunting in ecoprovinces of British Columbia. Christmas Bird Counts and nest record data have been excluded.

Figure 423. Snow-covered grassy and grain stubble fields in the Boreal Plains Ecoprovince are preferred foraging habitats for the Snow Bunting in winter (Dawson Creek, 10 November 1996; R. Wayne Campbell).

Figure 424. In British Columbia, the Snow Bunting breeds only in alpine areas in the Northern Boreal Mountains Ecoprovince similar to those shown here (Kusawak Lake, 20 June 1999; R. Wayne Campbell).

420 and 423), at airports (14%), or along roads (14%; Fig. 420) and coastal jetties (10%), and less frequently on reclaimed mine sites, sewage works, lawns, and expanses of pavement. It is frequently seen in the company of Horned Larks, Gray-crowned Rosy-Finches, and Lapland Longspurs – other occupants of the same alpine and arctic habitats. The Snow Bunting has the unusual habit of searching out patches of powder-dry snow in which it goes through the motions of dustbathing (Austin 1968). Lyon and Montgomerie (1995) suggest that the Snow Bunting uses this behaviour in spring and late winter to wear away black feather edges on its plumage.

In the Georgia Depression, there is little indication of the onset of spring migration (Fig. 422). Weekly totals remain approximately the same from early December to late March. By mid-April there are only scattered records ending in the last week of the month. Further north, on the Northern Mainland Coast, there is little sign of a movement. In the Southern Interior, winter numbers decline through January and early February, followed by an increase in March (Figs. 421 and 422). This sequence suggests a departure of many wintering birds in January and February that is followed in March by the passage of birds from winter ranges further south. Throughout the interior west of the Rocky Mountains, the last Snow Buntings have usually left by mid-April. East of the Rocky Mountains, in the Boreal Plains, the migration peaks between mid-April and early May and continues until about mid-May. In the Northern Boreal Mountains, the few records suggest a small wintering population that moves out by early April to be replaced by larger numbers in early May. This is the only ecoprovince with regular records of the species through June, July, and August (Figs. 421 and 422).

The start of the spring migration in January and February at British Columbia latitudes is consistent with the arrival of the species in the low Arctic regions of North America in mid-March (Lyon and Montgomerie 1995).

In the Northern Boreal Mountains, the birds have begun their autumn movement by mid-August and most have left that area by the third week of October. In the Boreal Plains and Taiga Plains, numbers begin arriving in October and wintering numbers build into December. The first autumn appearance of Snow Buntings in southern British Columbia may occur during the second and third weeks of September, although numbers do not appear until early October (Figs. 421 and 422). In the Southern Interior Mountains, the migration peaks in mid to late October and then drops slightly as some birds move further south. Wintering bird numbers build into December or January, and remain relatively constant in all areas. This pattern of migration suggests that there is more than a single stream of migrants, each probably originating in a different source population. On the coast, birds begin to arrive in the Georgia Depression by mid-September and the peak can occur between late October and mid-November. Numbers decline slightly into December as some birds move further south, leaving the wintering population behind.

The Snow Bunting occurs in British Columbia throughout the year. On the coast, it has been recorded regularly from mid-October to mid-April; in the interior, it has been found throughout the year (Fig. 421).

BREEDING: In British Columbia, the Snow Bunting has a restricted breeding distribution, confined to the Northern Boreal Mountains. There it is known to breed in alpine areas west of the Haines Highway, including the Datlasaka Range (Fig. 424) (Campbell and Van Der Raay 1985). Evidence of more widespread breeding includes observations of apparent nest building at Kakuchuya Creek and a male specimen from Tats Lake with testes of breeding size (Watson 1986). There are also summer records with details suggesting that the breeding distribution in the province may extend as far south and east as Mount Edziza and Spatsizi Plateau wilderness parks (Campbell and Van Der Raay 1985), including Mink Creek in the latter area (D.F. Hatler pers. comm.).

The highest numbers in summer occur in the Northern Boreal Mountains (Fig. 419); indeed, it appears to be confined to that ecoprovince in summer. Breeding Bird Surveys

Figure 425. In British Columbia, boulder fields are used by the Snow Bunting during the breeding season (Datlasaka Range, Haines Highway, 19 July 1980; R. Wayne Campbell).

for interior routes for the period 1968 through 1993 contain insufficient data for analysis.

Almost all records in June and early July in the province have been from above timberline, in moraine above tundra and in other rocky habitats, frequently close to cliff faces and snow patches.

Elevations occupied by the Snow Bunting during the breeding season range from 1,500 to 2,200 m. Nesting habitat on the Seward Peninsula of Alaska is described as cliff and block-fields juxtaposed with dry low dwarf shrub mat tundra (Kessel 1989). In other parts of the species' range, breeding habitat is described as barren tundra, rocky terrain, boulder fields (Fig. 425), scree slopes, cracks in large rock faces, and cliffs (Lyon and Montgomerie 1995; Byers et al. 1995). As in a Greenland population (Lyon and Montgomerie 1995), there is no evidence of nonbreeders away from the breeding grounds in the summer.

The Snow Bunting has been recorded breeding in the province from 26 June (calculated) to 10 July (Fig. 421).

Nest: The only nest found in British Columbia was situated in a vertical crevice in an alpine cliff face. The nest and nest cup were composed largely of grasses with an abundance of feathers. Elsewhere, nest locations are described as "well back in a narrow fissure, usually not visible from the cavity entrance" (Lyon and Montgomerie 1995).

Eggs: Nests with eggs have not been found in British Columbia, but the presence of nestlings on 10 July indicates that eggs were present on 26 June. Observations made in various parts of the Nearctic indicate that laying usually begins about 10 June but varies from year to year. On the Cumberland Peninsula of Baffin Island in the Northwest Territories, clutches have been initiated from 4 June to 7 July, with a mean of 20 June (Lyon and Montgomerie 1995). The female usually initiates incubation upon laying egg 3 or 4, leading to some asynchronous hatching of the chicks in the larger broods. In North America, clutch size varies from 3 to 9 eggs (Rising 1996), but is usually 5 or 6, with a mean of 5.8 (n = 46) (Nethersole-Thompson 1966). The incubation period is 10.5 to 13.6 days (n = 17) (Lyon and Montgomerie 1995). Byers et al. (1995) report that a few pairs may attempt 2 broods, but Lyon and Montgomerie (1995) suggest that apparent instances were probably pairs that had lost earlier clutches before they hatched.

Young: Dates for 2 broods were 10 July; the broods consisted of 4 and 5 young. The nestling period is 10 to 14 days (Baicich and Harrison 1997).

Brown-headed Cowbird Parasitism: Cowbird parasitism was not found in British Columbia, nor has it been reported elsewhere in North America (Friedmann et al. 1977; Friedmann and Kiff 1985).

Nest Success: Insufficient data. In the Northwest Territories, 41.8% of females in 4 samples (n = 126) successfully fledged at least 1 young (Lyon and Montgomerie 1995).

REMARKS: The Snow Bunting shares with the Hoary Redpoll the distinction of being the most northerly nesting passerine in North America. Both have been recorded nesting at Alert on northern Ellesmere Island in the Northwest Territories (Godfrey 1986).

At the southern edge of its breeding range, in British Columbia, there has been little serious ornithological exploration in the high alpine rocky habitat to which the Snow Bunting is closely restricted. We suggest that the species will be found to occupy a wider range than is now known.

NOTEWORTHY RECORDS

Spring: Coastal – Gabriola Island 14 Apr 1968-1 (Crowell and Nehls 1968c); Iona Island 16 Mar 1974-50 on sewer jetty; Tsawwassen 28 Apr 1975-2 on ferry jetty; Vancouver 22 Mar 1990-20; Bella Coola 4 Mar 1942-1 (RBCM 10021); Kispiox 10 Apr 1975-1 at 18 Mile along Kispiox Rd. **Interior** – Three Brothers Mountain 25 Apr 1982-2; Okanagan Falls 1 May 1913-1 found dead (Anderson 1914); Nelson 9 Apr 1972-4 on lakefront; Wardner 21 Mar 1937-300 (Johnstone 1949); Chapperon Lake 4 Mar 1986-1,000 (Rogers 1986a); 10 km e Bridge Lake 9 May 1984-20 feeding in grassy fields; Kleena Kleene 28 Mar 1954-40; s Riske Creek 12 Mar 1985-150 along road to Farwell Canyon; Chezacut 7 Mar 1940-100 (Paul 1959); Ten Mile Lake (Quesnel) 21 Apr 1968-1 (Rogers 1968c); Vanderhoof 19 Mar 1994-300 (Bowling 1995c); Tetana Lake 18 Mar 1941-common, 8 Apr 1938-12 (Stanwell-Fletcher and Stanwell-Fletcher 1943); Boundary Lake (Goodlow) 27 Apr 1986-350; 44 km n Prophet River 13 May 1982-1; Alaska Highway (Mile 255) 28 Mar 1975-25; Chilkat Pass 7 to 25 May 1958-common (Weeden 1960), 14 May 1977-200 in many small flocks; Atlin 9 Mar 1981-12 in town, first seen this spring, 17 Apr 1930-1, latest seen in spring.

Summer: Coastal – Masset 17 Jun 1920-1 (MVZ 106696), injury to head had prevented migration. **Interior** – Mink Creek (Spatsizi Plateau Wilderness Park) 13 Jun 1983-1 adult flying from a cliff face; Cocoa Crater (Mount Edziza Park) 11 Jul 1981-11 above 1,650 m; Mount Edziza Park 30 Jul 1977-4 females; Kakuchuya Creek 5 Jun 1978-1 adult carrying nesting material; Chilkat Pass (Km 105), 4 Jul 1958-1 female with 3 large follicles (Weeden 1960); Datlasaka Range 10 Jul 1980-5 nestlings, 10 Jul 1980-4 small fledglings with 2 adults; Haines Highway (4 km se of Km 150) 2 Jul 1980-2 adults feeding 2 fledglings; s Tats Lake 15 Jul 1983-10 both sexes, no young.

Breeding Bird Surveys: Not recorded.

Autumn: Interior – Chilkat Pass 3 to 6 Sep 1958-common (Weeden 1960); Haines Highway 20 Oct 1981-500; Atlin 17 Oct 1933, earliest in autumn (Swarth 1936); Atlin Park 16 Sep 1986-30 on nunatak in middle of Llewellyn Glacier; Fort Nelson 9 Oct 1984-20 at airport, first autumn flock seen; ne Cecil Lake 16 Oct 1983-100 in stubble field; ne Fort St. John 6 Oct 1985-2, first of autumn; 22 km ne Dawson Creek 19 Nov 1978-500; nr Parsnip River 21 Sep 1983-15 on logging road; Summit Lake (Prince George) 16 Sep 1973-4 (Rogers 1974); Riske Creek 11 Oct 1959-25; Tunkwa Lake 6 Nov 1962-40; Okanagan Landing 5 Oct 1925-1; nr Minnie Lake 24 Sep 1983-30 on grasslands; Harmer Ridge (Elk Valley) 16 Sep 1983-100 using reclaimed mining sites; Kimberley 2 Nov 1991-60 (Siddle 1992a); Apex Mountain (Penticton) 20 Oct 1973-50 at 2,150 m (Rogers 1974). **Coastal** – North Beach 23 Oct 1971-8; Kitimat Arm 17 Sep 1974-1 (Hay 1976); 16 km s Masset 18 Nov 1977-15; Sandspit 21 Oct 1991-1 (Siddle 1992a); Port Hardy 1 Nov 1950-2; Grant Bay 29 Oct 1968-2 (Richardson 1971); Mount Elphinstone 4 Nov 1979-3 on gravel road through clearcut at 900 m elevation; Hope 5 Nov 1924-40; Iona Island 14 Sep 1980-3, 1 Nov 1981-25; Tsawwassen 24 Nov 1972-25 (Crowell and Nehls 1973a); Central Saanich 17 Oct 1984-500, 17 Nov 1984-100; Chesterman Beach 13 Oct 1972-1 (Hatler et al. 1978); Ferrer Point (Pacific Rim National Park) 8 Nov 1969-1 (Hatler et al. 1978); Cowichan Bay 25 Sep 1971-1, first record this month for southern Vancouver Island (Tatum 1972); Coburg Peninsula 3 Nov 1984-4; Clover Point (Victoria) 1 Nov 1972-6 (Tatum 1973); Otter Point (w Sooke) 11 Nov 1985-2.

Winter: Interior – Atlin 7 Dec 1930-1 (RBCM 5754), 24 Feb 1946-1 (UBC 6801); Chilkat Pass 16 Feb 1985-20 (Grunberg 1985b); 6 km ne Fort Nelson 4 Jan 1985-15; Fort Nelson 6 Dec 1984-70 at airport; Charlie Lake 29 Dec 1987-500; e Bear Flat 1 Jan 1983-120 in a flock; Cecil Lake 12 Feb 1984-100; Chetwynd 22 Jan 1982-2; Prince George 2 Jan 1972-140; Farwell Canyon 27 Dec 1975-200; Springhouse 9 Jan 1959-100 (Erskine and Stein 1964); Revelstoke 1 Dec 1976-100; Kamloops 28 Dec 1986-285 (Cannings 1987); Armstrong 6 Jan 1962-100; n Sparwood 3 Jan 1984-30; Kimberley 19 Feb 1967-200. **Coastal** – Kitimat River 19 Jan 1980-6 on estuary; Cape St. James 27 Dec 1978-1; Vancouver 8 Dec 1963-60 at airport; Tsawwassen 20 Jan 1973-4 on jetty; Boundary Bay 5 Jan 1983-35 at airport; Green Point 8 Feb 1949-4 on beach; Greater Chain Islet 26 Dec 1978-1; Cattle Point (Victoria) 8 Jan 1988-1; Whiffen Spit (Sooke) 1 Dec 1985-10.

Christmas Bird Counts: Interior – Recorded from 19 of 27 localities and on 22% of all counts. Maxima: North Pine 19 Dec 1992-705; Kamloops 15 Dec 1984-620; Fort St. John 26 Dec 1977-284. **Coastal** – Recorded from 14 of 33 localities and on 11% of all counts. Maxima: Ladner 22 Dec 1973-44; Vancouver 26 Dec 1963-40; Prince Rupert 18 Dec 1993-30; Kitimat 28 Dec 1985-30.

Rose-breasted Grosbeak

RBGR

Pheucticus ludovicianus (Linnaeus)

RANGE: Breeds from northeastern British Columbia, southwestern and south-central Mackenzie, northern Alberta, central Saskatchewan, southern Manitoba, western and southern Ontario, southwestern Quebec, New Brunswick, Prince Edward Island, and Nova Scotia south to central and eastern Alberta, north-central North Dakota, eastern South Dakota, eastern Nebraska, Kansas, central Oklahoma, southern Missouri, and parts of Illinois, Indiana, Ohio, and New Jersey, and through the Appalachian states to Georgia. Winters from central Mexico south into Venezuela, Colombia, Ecuador, and Peru.

STATUS: On the coast, *very rare* in the Georgia Depression Ecoprovince; in the Coast and Mountains Ecoprovince, *accidental* on Western Vancouver Island and absent elsewhere.

In the interior, *very rare* in the Southern Interior, Southern Interior Mountains, Sub-Boreal Interior, and Northern Boreal Mountains ecoprovinces; locally *fairly common* migrant and summer visitant in the Boreal Plains Ecoprovince and locally *uncommon* in the Taiga Plains Ecoprovince.

Breeds.

CHANGE IN STATUS: The first records of the Rose-breasted Grosbeak (Fig. 426) in British Columbia were by Williams (1933a), who reported that he heard it "commonly along Sikanni Chief, Fort Nelson, and Liard rivers" in the Taiga Plains in June 1922. In the early 1930s, the species was also "fairly common" in similar habitat in Alberta, along the 145 km of highway between Edmonton and Athabasca Landing (Randall 1933).

The 1938 I. McTaggart-Cowan and P.W. Martin expedition to the Peace River region obtained the first specimen records for the province and elaborated on the migration pattern and habitat use of the Rose-breasted Grosbeak there (Cowan 1939). Rand (1944), however, spent 71 field days in 1943 studying birds and collecting specimens for the Canadian Museum of Nature along the Alaska Highway between Dawson Creek and Irons Creek, a distance of 510 km to the north and west, but made no mention of the Rose-breasted Grosbeak. Thus, in 1947 the Rose-breasted Grosbeak was known only as a summer visitor along the route of the Alaska Highway north of the Peace Lowland, as far as the valley of the Liard River north of Fort Nelson (Munro and Cowan 1947).

Rand (1944) reported a 1944 sighting of the species at Liard Hot Springs, but there were no further published reports of the species until 1976. This hiatus was probably the result of few observers and the local distribution of the Rose-breasted Grosbeak. From 1973 to 1975, field work in the same regions studied by Rand brought to light several observations of the species that had been made between 1953 and 1968 (Erskine and Davidson 1976). In addition, Erskine saw Rose-breasted Grosbeaks on 9 of 47 field days during his field work between 14 May and 10 July 1974.

Figure 426. Over the past 50 years, the Rose-breasted Grosbeak (adult male shown) has expanded westward and southward from its traditional range in northeastern British Columbia (Anthony Mercieca).

Since then, the species has been recorded in the Boreal Plains and Taiga Plains in varying numbers almost every year. Records from the Taiga Plains and Northern Boreal Mountains, combined with those from the centre of the species' distribution in the Peace Lowland, indicate that there has been some increase in the numbers and distribution of the Rose-breasted Grosbeak in northern British Columbia during the past 40 years.

A second aspect of the changing status of this species in British Columbia can be seen in the appearance of small numbers on an irregular basis in southern portions of the province. Many competent bird observers were active in the southern regions of the interior and in the southern portions of coastal British Columbia in the early years of the 20th century. During the first half of the century, they collectively logged many thousands of field days, but except for a nesting pair observed at Celista in 1948, none of them encountered a Rose-breasted Grosbeak before 1955 on the coast and 1975 in the interior. Since then a total of 32 of these grosbeaks

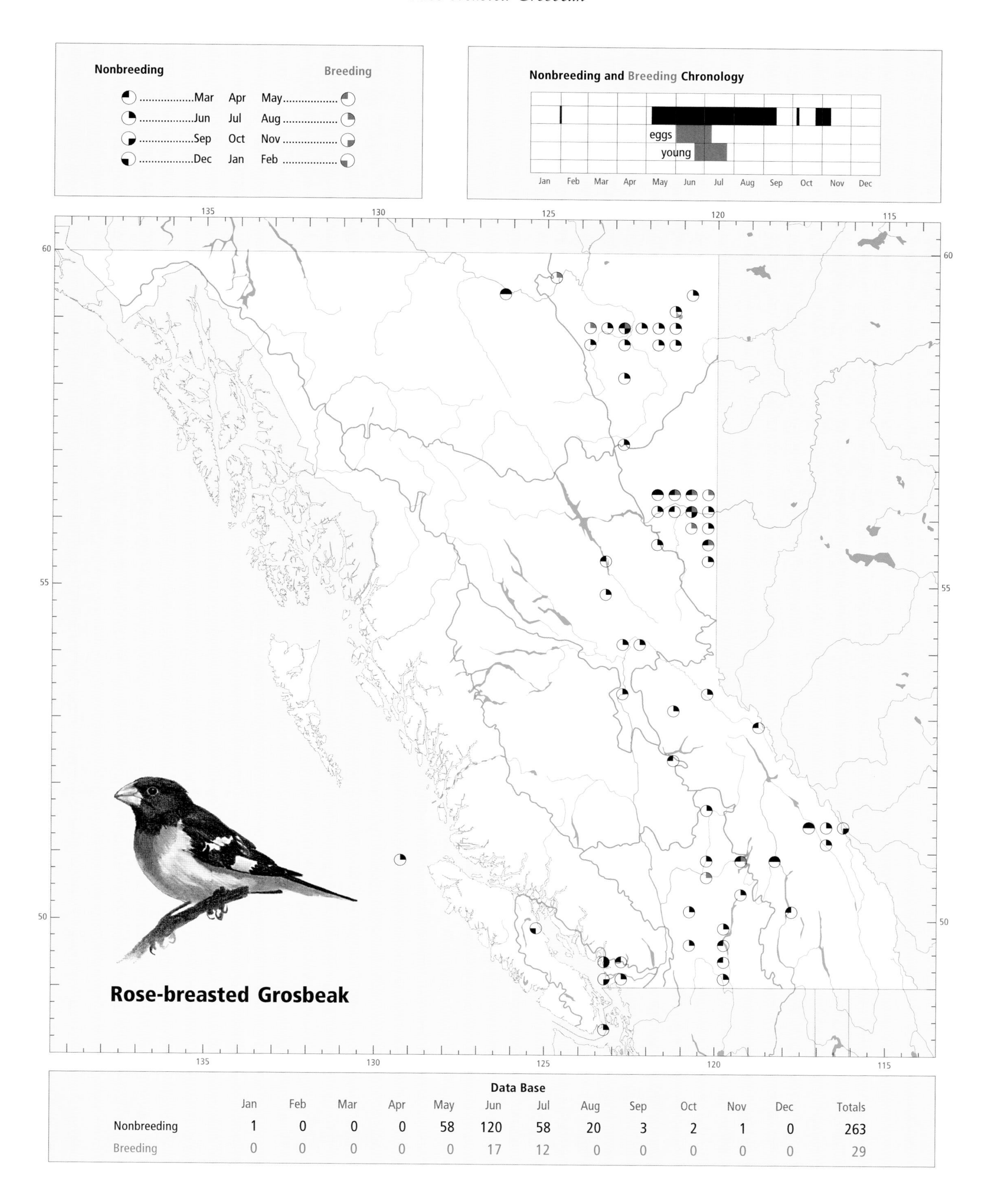

Data Base

	Jan	Feb	Mar	Apr	May	Jun	Jul	Aug	Sep	Oct	Nov	Dec	Totals
Nonbreeding	1	0	0	0	58	120	58	20	3	2	1	0	263
Breeding	0	0	0	0	0	17	12	0	0	0	0	0	29

has been recorded in coastal and interior British Columbia south of the Cariboo and Chilcotin areas: 8 in May, 14 in June, 4 in July, 5 from August to October, and 1 in January.

The northern population is obviously an extension from the occupied range of the species in Alberta (Semenchuk 1992). A few of the grosbeaks reported in the southern parts of the province were on the western slope of the Rocky Mountains and may have filtered through the mountains from the Alberta population. Most interior and coastal occurrences, however, appear to represent an element in the population that moves between summer and winter ranges by way of the Pacific slope of the continent.

This Pacific slope component can be identified in California, where prior to 1944 there were just 5 records (Grinnell and Miller 1944); more recently, the average has been 15 records of spring migrants annually, between late May and early July, and additional records during the autumn migration (Small 1994). The movement of a small number of these grosbeaks along the Pacific slope of the continent can also be seen in Oregon (first noted in 1972 and now about 50 records [Gilligan et al. 1994]) and Washington (first recorded in 1956 [Roberson 1980]).

We suggest that most of the Rose-breasted Grosbeaks that have been recorded in southern British Columbia and along the Pacific states are not accidental strays from the population that migrates up the Mississippi valley into Alberta and northeastern British Columbia. Rather, they derive from a small number of migrants that follow a western route north from their wintering areas in Middle America. This may be the early stage in an extension of the species range westward into the range of the Black-headed Grosbeak.

NONBREEDING: The Rose-breasted Grosbeak occurs regularly in the province only in the northeast but extending northwestward, in a narrow area along the Liard River drainage as far as Liard Hot Springs Park, and north at least as far as Kwokullie Lake. It is a sporadic migrant in the southern portions of the interior of the province, and its occurrence is unusual on the coast.

Outside the nesting season, the Rose-breasted Grosbeak has been reported from a wide variety of forested habitats, including river floodplains, islands largely treed with white spruce, black spruce at the edge of muskegs, and tamarack bogs, but more usually in trembling aspen forests, even before the trees are leafed out.

During migration, the southern occurrences have been recorded at elevations between sea level and 800 m. In the northeast, elevations during migration do not differ from those occupied during the nesting season.

On the south coast, the first arrivals have been recorded as early as 30 May and 6 June, but most of the records are between 6 and 27 June (Figs. 427 and 428). In the southern portions of the interior, the earliest record is 16 May, but most have been between 24 May and 19 June. In the Peace Lowland, the earliest arrivals have been recorded on 8 May, but the peak of the migration appears to occur between mid-May and the first week of June (Figs. 427 and 428). The dates of first arrivals in the Fort St. John region between 1980 and 1988

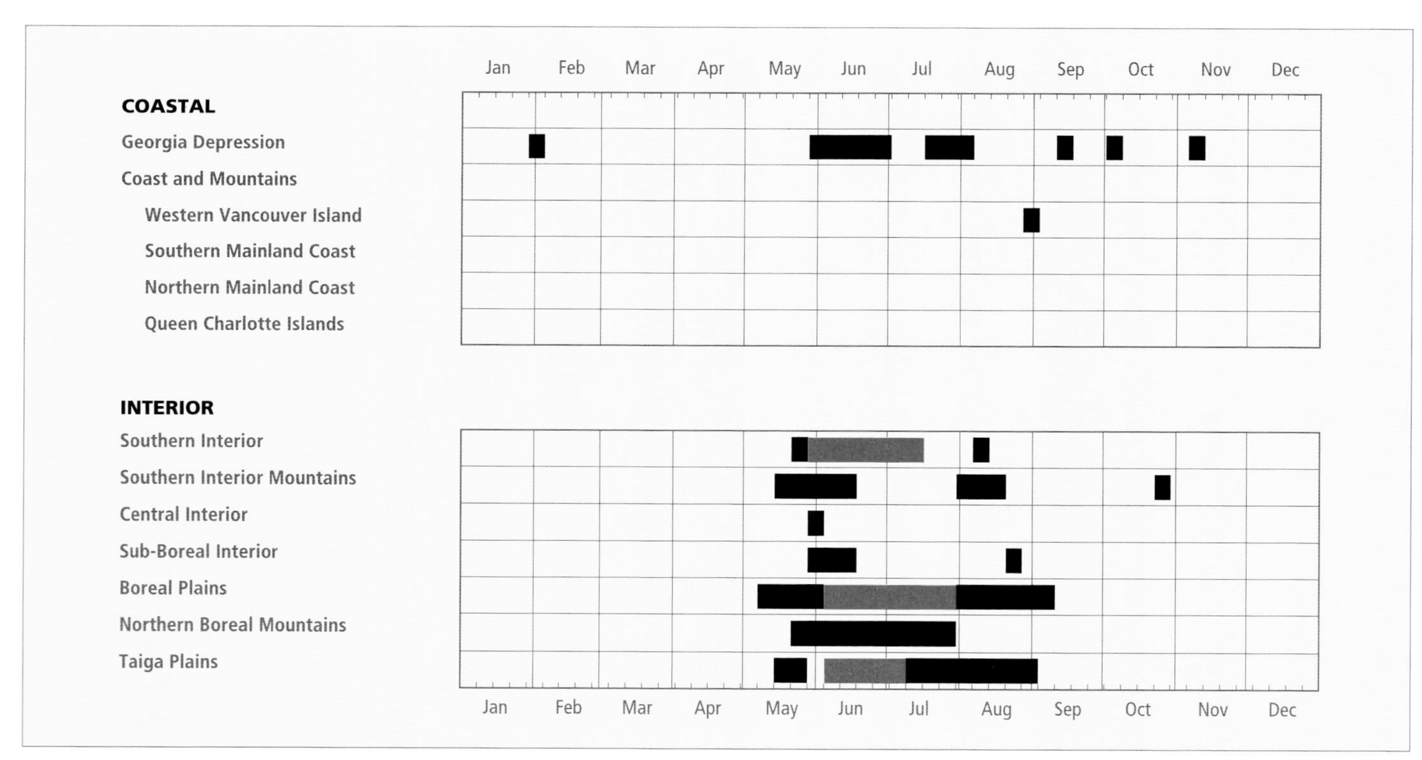

Figure 427. Annual occurrence (black) and breeding chronology (red) for the Rose-breasted Grosbeak in ecoprovinces of British Columbia. Records are shown for the week in which they occurred.

were 8 May to 22 May, with 7 records between 17 and 22 May (C. Siddle pers. comm.).

The southbound migration in the north is concentrated in the latter half of August, with no records later than the first week of September.

On the south coast, the Rose-breasted Grosbeak has been recorded irregularly from 31 January to 9 November; in the interior, it has been recorded regularly from 14 May to 25 August (Fig. 427).

BREEDING: The Rose-breasted Grosbeak breeds regularly in the Boreal Plains and Taiga Plains, and probably at Liard Hot Springs in the Northern Boreal Mountains of northeastern British Columbia. In the southern regions of the province, the species has been recorded nesting only at Celista, on the shores of Shuswap Lake, and at Barnhartvale, Kamloops.

The Rose-breasted Grosbeak reaches its highest numbers in the Boreal Plains (Fig. 429). Breeding Bird Surveys for British Columbia for the period 1968 through 1993 contain

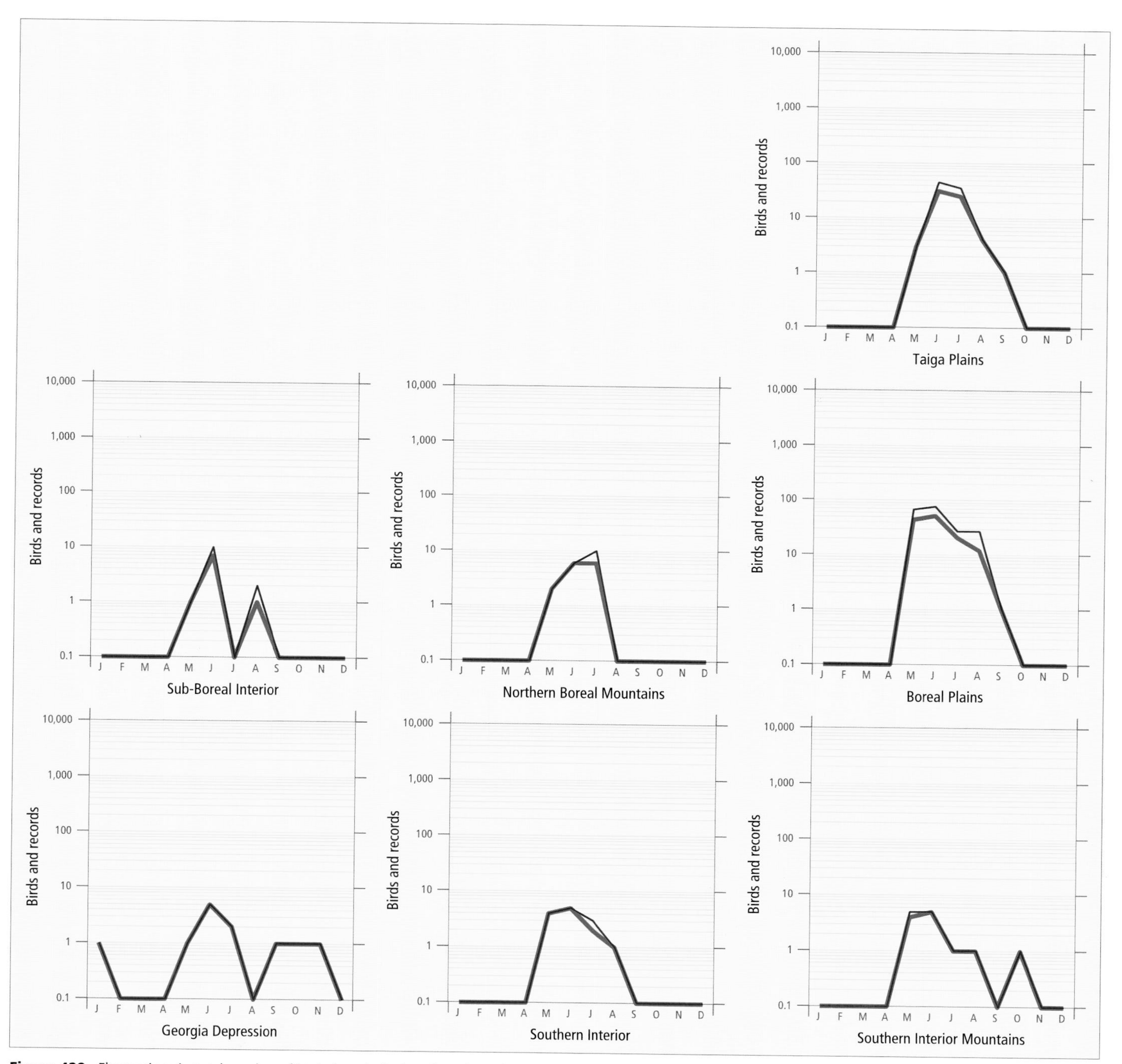

Figure 428. Fluctuations in total number of birds (purple line) and total number of records (green line) for the Rose-breasted Grosbeak in ecoprovinces of British Columbia. Breeding Bird Surveys and nest record data have been excluded.

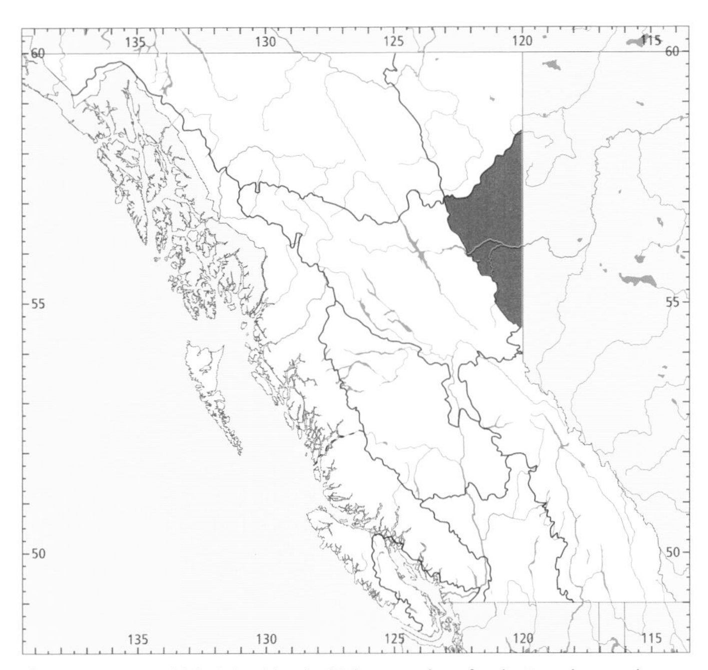

Figure 429. In British Columbia, the highest numbers for the Rose-breasted Grosbeak in summer occur in the Boreal Plains Ecoprovince.

insufficient data for analysis. An analysis of these surveys across Canada, however, indicates an overall decline at an average annual rate of 1.3% ($P < 0.10$); the population increased from 1966 to 1979 but experienced a severe decline (4.7%; $P < 0.01$) between 1980 and 1996 (Sauer et al. 1997).

Breeding has been documented at valley bottom elevations between 285 and 750 m in the Peace Lowland, Fort Nelson Lowland, and Muskwa Plateau. In northeastern British Columbia, the Rose-breasted Grosbeak establishes its breeding territories in mature and second-growth deciduous or mixed woodlands in wet as well as dry areas. These habitats often feature beaver ponds, muskegs, swampy areas and streamsides, and drier hillsides and seismic line rights-of-way (Fig. 434). The moist but well-drained areas feature willow and balsam poplar mixed with scattered white spruce; the poorly drained boggy spots support black spruce or tamarack; and the drier sites support a tree cover of trembling aspen with a varied understorey of shrubs and forbs (Fig. 430). Many of the shrubs in all habitat types produce berries in late summer (e.g., highbush-cranberry, blueberry, soopolallie, choke cherry, honeysuckle, mountain-ash, cloudberry, and raspberry).

Figure 430. In the Boreal Plains Ecoprovince of British Columbia, the Rose-breasted Grosbeak inhabits relatively open trembling aspen woodlands with a varied understorey of shrubs and forbs (Beatton Park, 21 June 1996; R. Wayne Campbell).

Figure 431. In British Columbia, the Rose-breasted Grosbeak constructs its nest in forks and crotches of deciduous trees, often saplings, and always closer to the top of the tree than to the bottom. Note the bulky nest in the upper right corner (Beatton Park, Fort St. John, 21 June 1996; R. Wayne Campbell).

Figure 432. All Rose-breasted Grosbeak nests in British Columbia were composed of twigs, rootlets, and coarse plant stems (Beatton Park, 23 June 1996; R. Wayne Campbell).

Figure 433. Nest and eggs of the Rose-breasted Grosbeak (Beatton Park, 23 June 1996; R. Wayne Campbell).

The Rose-breasted Grosbeak has been recorded breeding in British Columbia from 2 June (calculated) to 24 July (Fig. 427).

Nests: Thirteen of the nests in our sample of 15 were in living deciduous trees, mostly trembling aspen and willow (Fig. 431). The others were in deciduous shrubs. Small or sapling trees were more often selected than large trees. In general, the nests were in forks or crotches of upright stems; they were closer to the top of the tree or bush than the bottom, and were usually located closer to the main trunk than away from it (Figs. 431 and 432).

Nest construction varied. Most were substantial but roughly built structures consisting largely of twigs and rootlets with the occasional use of coarse grasses, plant stems, and plant fibres (Fig. 433). Most were lined with fine rootlets but some had no lining. The heights of 15 nests ranged from 1.2 to 9.1 m, with 8 nests between 2.2 and 4 m.

Eggs: Dates for 8 clutches (Fig. 433) ranged from 2 June to 8 July, with 11 recorded between 15 and 29 June. Sizes of 11 clutches ranged from 3 to 5 eggs (3E-2, 4E-8, 5E-1). The incubation period is 11 to 14 days (Peck and James 1987).

Young: Dates for 4 broods ranged from 21 June to 24 July, with 10 recorded between 23 June and 7 July. Sizes of 13 broods ranged from 1 to 4 young (1Y-6, 2Y-2, 3Y-4, 4Y-1). The nestling period is 9 to 12 days (Baicich and Harrison 1997). The Rose-breasted Grosbeak is generally single-brooded throughout its range, but there are records of second nestings (Austin 1968; Rothstein 1973).

Brown-headed Cowbird Parasitism: Eighteen Rose-breasted Grosbeak nests were found in British Columbia with eggs or young; none was parasitized. Friedmann (1963) reports that the Rose-breasted Grosbeak is a frequent victim of the cowbird and has been known to rear cowbird young. In Ontario, Peck and James (1987) report that 7.5% of 275 grosbeak nests were parasitized by the Brown-headed Cowbird. The absence of any observation of cowbird parasitism in British Columbia may be an artifact of small sample size. The Brown-headed Cowbird is a prevalent species in the Boreal Plains, and there is ample opportunity for parasitism to occur.

Nest Success: Only a single nest was found with eggs or young and followed to a known fate. It was unsuccessful.

REMARKS: The Rose-breasted and Black-headed grosbeaks are regarded as a superspecies (American Ornithologists' Union 1983), and there are a few instances of hybridization where their summer ranges overlap (Austin 1968; Rising 1983).

Populations of many songbirds are influenced by local forest structure, which affects microclimate, availability of foraging and nesting sites, and risk of predation. Some researchers have suggested that clearing of forests has created forest edges, which not only increase the density and diversity of birds but allow some species to expand their ranges into new territory (e.g., Johnston 1947; Anderson et al. 1977; Strelke and Dickson 1980). Kroodsma (1984) provides insightful comments and concerns about the effect of edge on breeding forest bird species.

Figure 434. During the past few decades, the Rose-breasted Grosbeak may have expanded its breeding range in northeastern British Columbia by following the edges of rights-of-way created by oil and gas exploration activities (36 km north of Prophet River, 28 June 1998; R. Wayne Campbell).

NOTEWORTHY RECORDS

Spring: Coastal – Mike Lake 30 May 1988-1. **Interior** – Shuttleworth Creek 28 May 1986-1; Naramata 29 May 1977-1; Nakusp 16 May 1981-1; Celista 22 May 1948-1 male singing from trembling aspen; Revelstoke 24 May 1990-1 (Siddle 1990b); lower Blaeberry River 29 May 1996-1; Mount Robson Park 25 May 1992-2, first record for region (Bowling 1992); Mackenzie 31 May 1996-1; Tupper Creek 25 May 1938-2 (Cowan 1939); Pouce Coupe 25 May 1994-adult female on nest; Dawson Creek 19 May 1992-6; Taylor 17 May 1986-1, 28 May 1983-5; Beatton Park 8 May 1982-1, unusually early; Fort Nelson 16 May 1996-1 male in trembling aspen, 25 May 1974-1 (Erskine and Davidson 1976); Liard Hot Springs 27 May 1981-1.

Summer: Coastal – Victoria 6 Jun 1955-1 at Braefoot, 27 Jun 1972-1 (Tatum 1973); Beaver and Elk Lake (Victoria) 5 to 12 Jun 1994-1 male (Bowling 1995b); Burnaby 31 Jul 1993-1 (Siddle 1993c); Triangle Island 28 Aug 1995-1 banded. **Interior** – Allison Lake 1 Jun 1997-1 male at feeder (BC Photo 1627); Inkaneep Park 7 Aug 1978-1; Westbank Park 10 Jun 1993-1 (Siddle 1993c); Monck Lake (Merritt) 6 Jun 1990-1; Vernon 19 Jun 1993-1 (Siddle 1993c); Barnhartvale 10 Jul 1977-4, female feeding nestlings, male nearby; McQueen Lake 12 Jul 1997-1 adult male; Celista 5 Jun 1948-2 adults and 4 eggs; Revelstoke 31 Jul 1993-1 (Siddle 1993c); Leanchoil 1 Jun 1975-1 (Wade 1977); Ottertail River (Yoho National Park) 6 Jun 1976-1 at viewpoint (Wade 1977); lower Blaeberry River 5 Jun 1996-1 (Bowling 1996d); Clearwater mid-Jun 1997-1; Bowron Lake Park 16 Jun 1990-1 male photographed; Hixon 12 Jun 1997-1 male singing; Holmes River 15 Aug 1992-1 (Siddle 1993a); McBride 2 Jun 1991-1 male in trembling aspen; Shelley 8 Jun 1994-1 male killed at window (Bowling 1994b); Willow River 1 Jun 1971-1; 13 km w Little Prairie 8 Jun 1954-1 (NMC 48288); 4 km s Pouce Coupe 15 Jun 1994-4 eggs, male on nest, 8 Jul 1994-3 nestlings in a different nest; 8 km south of Dawson 2 Jul 1993-4 nestlings; Dawson Creek 3 Aug 1975-4 young out of nest; w Dawson Creek 3 Aug 1975-1; n Dawson Creek 3 Aug 1975-adults feeding 4 fledglings; Hudson's Hope 1 Jun 1954-1 (RBCM 15855); Beatton Park 31 Aug 1982-1; Boundary Lake (Goodlow) 23 Jun 1985-adult pair and 1 recently fledged young; Sikanni Chief River 5 to 25 Jun 1922-heard "commonly along Sikanni Chief, Fort Nelson and Liard rivers" (Williams 1933a); Parker Creek 27 Jul 1967-family group; 31 km s Fort Nelson 2 Aug 1968-4 fledglings begging for food; Raspberry Creek (Fort Nelson) 23 Jun 1953-adult male on nest with 3 newly hatched nestlings and 1 egg; Fort Nelson 2 Aug 1968-1 family group (Erskine and Davidson 1976); 50 km w Kledo Creek 25 Jul 1967-3 fledglings; Lower Liard River Crossing 2 Jun 1944-1 (Rand 1944); Liard Hot Springs 1 Jul 1975-1 fledgling, 6 Jul 1993-3 (Siddle 1994a); Helmut 6 Jun 1982-2, 1 collected (RBCM 17468); Liard River, 4 Jul 1980-adults feeding 3 nestlings; Kotcho Lake 18 Jun 1982-2, 25 Aug 1975-2.

Breeding Bird Surveys: Coastal – Not recorded. **Interior** – Recorded from 5 of 73 routes and on 2% of all surveys. Maxima: Tupper 20 Jun 1994-16; Fort Nelson 19 Jun 1974-7; Fort St. John 20 Jun 1982-4.

Autumn: Interior – Fort Nelson 2 Sep 1996-1 male at feeder; Taylor 4 Sep 1987-1 at Peace Island Park; Field 26 Oct 1975-1 immature male (Wade 1977). **Coastal** – Vancouver 9 Oct 1992-1 (Siddle 1993a), 9 Nov 1986-1 at Lost Lagoon; Little Mountain (Vancouver) 12 Sep 1978-1.

Winter: Interior – No records. **Coastal** – Oyster River 31 Jan 1972-1 male.

Christmas Bird Counts: Not recorded.

Black-headed Grosbeak

Pheucticus melanocephalus (Swainson)

BHGR

RANGE: Breeds from southern British Columbia, southern Alberta, and southwestern Saskatchewan south in the west to northern Baja California, and in the east from Nebraska, Kansas, New Mexico, and Texas through the Mexican highlands to Oaxaca. Winters rarely in coastal California, southern Texas, and Louisiana, but mostly in southern Baja California and south in mainland Mexico to Oaxaca and Veracruz.

STATUS: On the coast, an *uncommon* to *common* migrant and summer visitant in the Georgia Depression Ecoprovince; in the Coast and Mountains Ecoprovince, an *uncommon* to *fairly common* summer visitant on the Southern Mainland Coast, *very rare* on Western Vancouver Island and the Northern Mainland Coast, and *casual* on the Queen Charlotte Islands.

In the interior, *uncommon* to *fairly common* migrant and summer visitant in the Southern Interior and Southern Interior Mountains ecoprovinces; *casual* in the Central Interior and the Sub-Boreal Interior ecoprovinces.

Breeds.

Figure 435. The Black-headed Grosbeak (adult female shown) breeds across southern British Columbia wherever edges of deciduous forests and shrub thickets provide adequate cover for nests (Vaseux Lake, 27 June 1982; Mark Nyhof).

CHANGE IN STATUS: An analysis of changes in distribution of the Black-headed Grosbeak (Fig. 435) in the province reveals a steady expansion of its range over the past 50 years. Prior to 1946, there had been 1 report of the species on western Vancouver Island (Munro and Cowan 1947). Since then, there have been 12 additional reports distributed between River Jordan in the south and Port Hardy in the north, most of them in late spring and early summer. As of 1946, the species was unknown on the Northern Mainland Coast and the Queen Charlotte Islands, despite many years of ornithological research there. It is now known at Terrace and Kitimat in early summer, and there are at least 3 observations from the Queen Charlotte Islands, all during the autumn migration.

In the interior, the occupied area of the Southern Interior has been extended westward, up the Similkameen valley as far as Princeton, and northward to include the Shuswap and Adams lakes areas. In the Southern Interior Mountains, observations recorded before 1946 were confined to the Creston valley and the west Kootenay from Trail in the south to Revelstoke in the north. There were no records from the Rocky Mountain Trench despite extensive museum collecting and field research there. The first Black-headed Grosbeak was reported from the Rocky Mountain Trench in 1953, and the species now occupies the valley from Newgate to Golden, with 1 record as far north as Mount Robson Park.

While the extension of the range of the Black-headed Grosbeak in the province has been obvious, the change in its nesting status has been more conservative. The confirmed breeding distribution has changed little on the coast, with nests on Vancouver Island confined to the Greater Victoria area, except for a single nesting at Courtenay. On the mainland, nesting has been concentrated in the Fraser Lowland between Marpole and Sumas Prairie, and likely north to Pemberton in the Southern Mainland Coast.

In the Southern Interior, there were only 2 known nestings in the Okanagan valley before 1946, and none in the Similkameen valley. Beginning in 1967, records of breeding in the Okanagan valley have steadily increased, and the species has been found nesting throughout the valley since the early 1980s, with a centre of abundance between Okanagan Falls and Osoyoos Lake (Cannings et al. 1987). There have been sporadic nestings in the southern Similkameen River valley.

In the Kootenay valley, breeding was unknown until adults with fledglings were seen, suggesting nesting, near Creston in 1949. In recent years, the Black-headed Grosbeak has become a regular breeding species, mainly in the region between Castlegar and Creston. The increases documented during Breeding Bird Surveys support these observations.

These expansions in numbers and distribution are probably responses to changing habitat. The large-scale deforestation that took place as the heavily forested delta of the Fraser River was converted into the most densely populated part of British Columbia increased the habitat preferred by this grosbeak. Comparable habitat changes have taken place in the interior valleys of southern British Columbia.

The changes we have observed occurred almost entirely in the densely populated areas of the province, where there have been active ornithologists since the beginning of the century. Furthermore, the timing of the dramatic increase in number of birds reported, and in number of birds per record, supports the conclusion that it is not an artifact. The increase began on the coast in the 1960s and in the interior in the 1970s (Fig. 436). We know of no changes in number or activity of observers that would explain this regional difference.

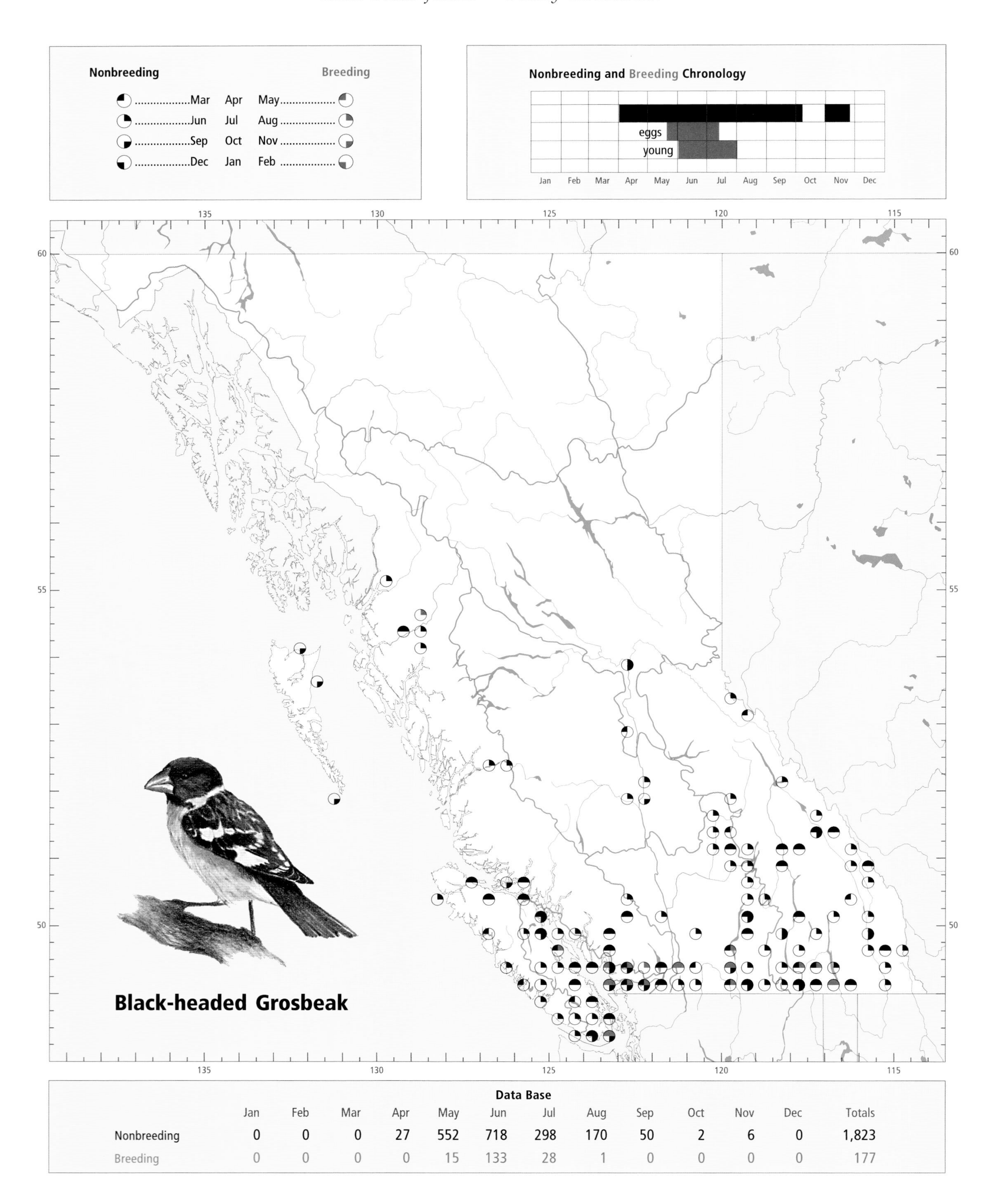

Data Base

	Jan	Feb	Mar	Apr	May	Jun	Jul	Aug	Sep	Oct	Nov	Dec	Totals
Nonbreeding	0	0	0	27	552	718	298	170	50	2	6	0	1,823
Breeding	0	0	0	0	15	133	28	1	0	0	0	0	177

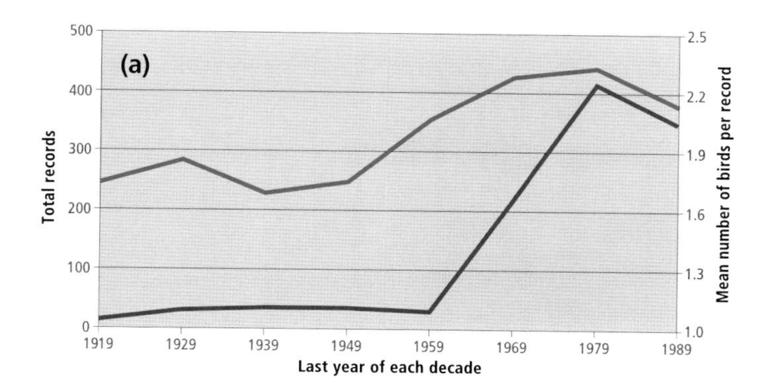

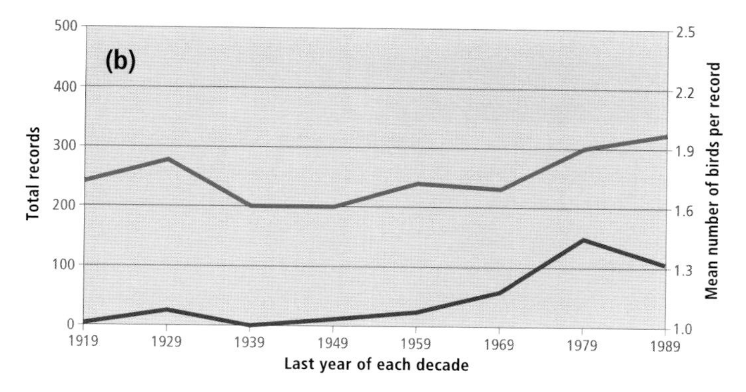

Figure 436. Changes in number of records per decade (green line) and number of birds per record by decade (purple line) of the Black-headed Grosbeak in (a) coastal and (b) interior regions of British Columbia.

NONBREEDING: The Black-headed Grosbeak is a characteristic spring and summer bird of lower elevations across southern British Columbia from Vancouver Island east to the BC-Alberta boundary and north to about latitude 51°N in the interior. Further north, occurrence is sporadic and infrequent. On the mainland coast, all records north of Vancouver Island are at the heads of deep fiords where rivers cut through the coastal mountains. Thus, the species has occurred in summer in the Bella Coola valley, on the north coast near the mouth of the Skeena River; the northernmost coastal record is from the mouth of the Nass River.

In the interior, the northernmost occurrence is from the Prince George region, and the easternmost records are from Sparwood and from Marble Canyon, in Kootenay National Park. Throughout the interior, this grosbeak is strongly associated with riparian habitats and the dense shrubby vegetation on the edges of wetlands. During migration, however, it is frequently more widely distributed and can be found in a variety of mixed coniferous and deciduous forest types (Fig. 437), as well as in urban and suburban parks, gardens, vineyards, and orchards.

On the coast, the Black-headed Grosbeak migrates at elevations from sea level to about 700 m; in the interior, it has been reported at elevations as high as 1,900 m.

On the south coast, birds may arrive in early April, but the main migration generally does not begin until early May (Figs. 438 and 439). In the Fraser Lowland and along the southeast coast of Vancouver Island, the migration of the Black-headed Grosbeak is at its peak in late May, by which time most birds appear to have selected nesting habitat. On the Northern Mainland Coast, the earliest arrivals appear in the first week of May, but the population is small and occurrence irregular, with just 6 records from this geographic region since 1972.

In the interior, the Black-headed Grosbeak usually arrives between early and mid-May (Figs. 438 and 439). Over a period of 70 years, the earliest and latest first-of-year records in the Okanagan valley were from 9 and 24 May (Cannings et al. 1987). In the valleys of the Kootenay and Columbia rivers, the pattern of spring arrival is similar to that in the Southern Interior, with numbers building steadily from early May to the second or third week of June (Fig. 439). The earliest arrival dates in Revelstoke have been 21 and 22 May, and there is no evidence of a migration reaching north of that region. In the North Thompson River valley, the northern extreme of the spring migration appears to be the region of Hemp Creek and Mabel Lake, in Wells Gray Park, where grosbeaks arrive in early June.

Figure 437. During migration in the Southern Interior Mountains Ecoprovince of British Columbia, the Black-headed Grosbeak moves through diverse habitats, including mixed stands of black cottonwood, willow, and red-osier dogwood (Creston, 25 September 1997; R. Wayne Campbell).

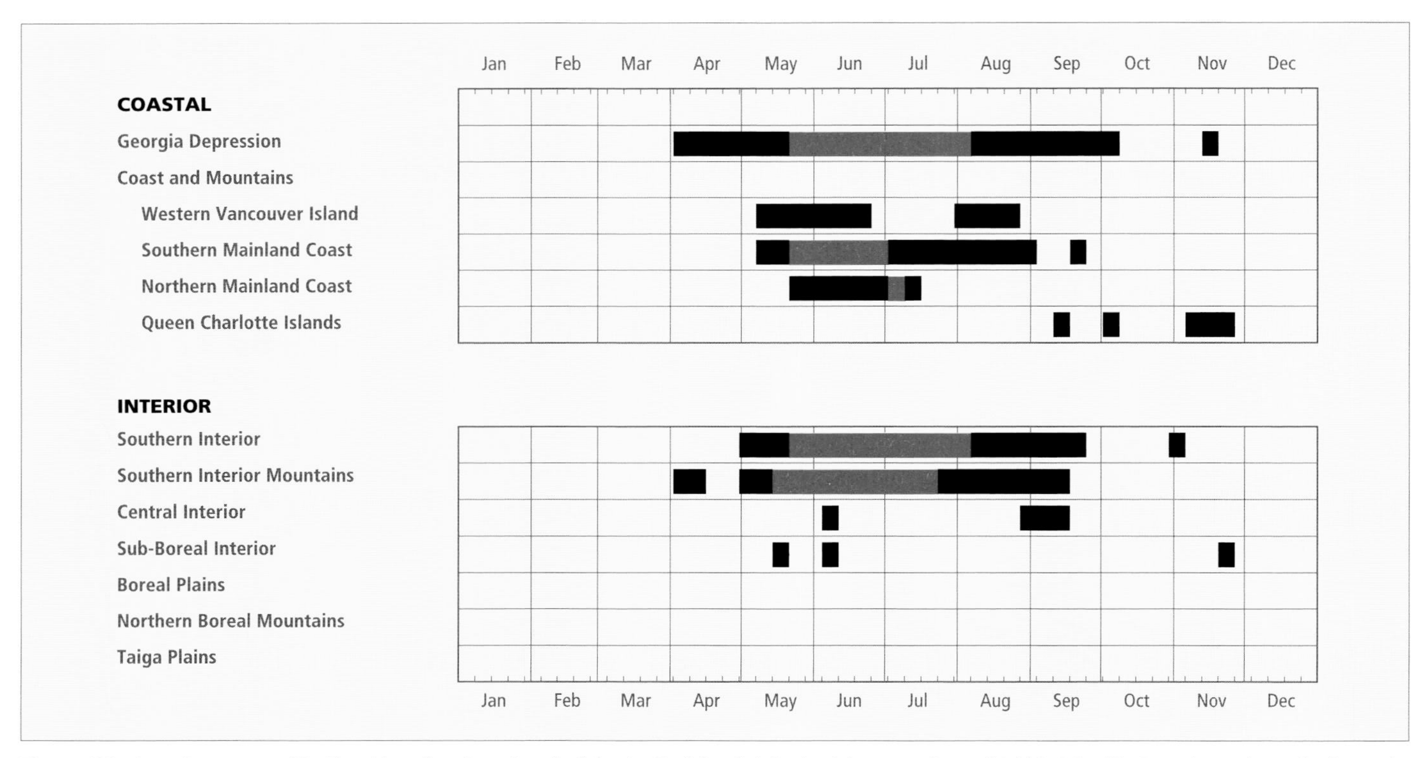

Figure 438. Annual occurrence (black) and breeding chronology (red) for the Black-headed Grosbeak in ecoprovinces of British Columbia. Records are shown for the week in which they occurred.

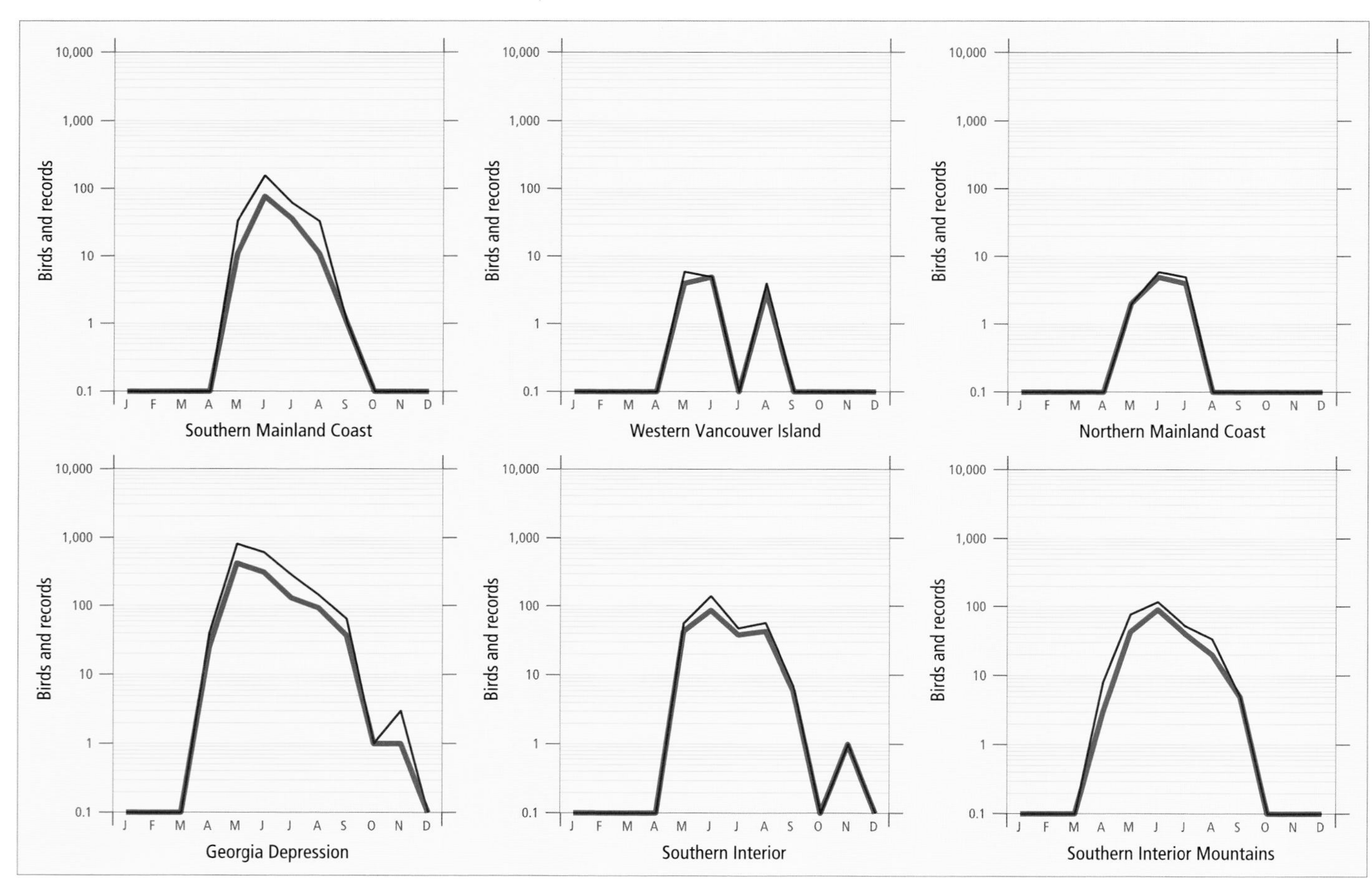

Figure 439. Fluctuations in total number of birds (purple line) and total number of records (green line) for the Black-headed Grosbeak in ecoprovinces of British Columbia. Christmas Bird Counts, Breeding Bird Surveys, and nest record data have been excluded.

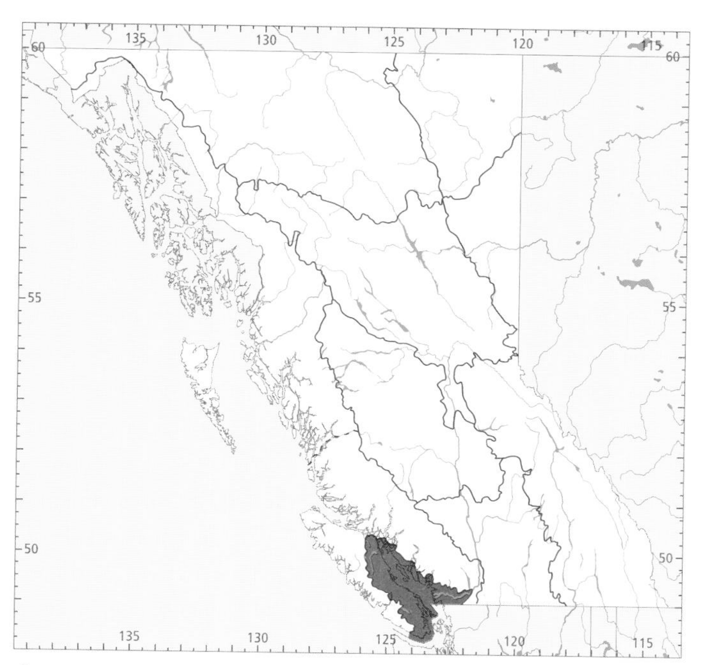

Figure 440. In British Columbia, the highest numbers for the Black-headed Grosbeak in summer occur in the Fraser and Nanaimo lowlands of the Georgia Depression Ecoprovince.

The autumn migration appears to begin in late July, almost as soon as the young are independent. In the interior, the latest autumn records from the northern areas are from Prince George on 25 November and Riske Creek on 14 September. The latest records for the Southern Interior Mountains are from mid-September. In the Southern Interior, records decline sharply during August, with most birds having departed by mid-September.

All records for the Queen Charlotte Islands are from autumn, between mid-September and late November; on Western Vancouver Island, migration observations include 9 August and 24 October.

Figure 441. In the Southern Mainland Coast of the Coast and Mountains Ecoprovince of British Columbia, the Black-headed Grosbeak breeds in mixed deciduous woodlands of black cottonwood and bigleaf maple (Laidlaw, west of Hope, 8 August 1996; R. Wayne Campbell).

On the coast, the Black-headed Grosbeak occurs regularly from 2 April to 2 October; in the interior, it has been recorded regularly from 1 May to 15 September (Fig. 438).

BREEDING: The Black-headed Grosbeak breeds primarily on the Nanaimo Lowland of southeastern Vancouver Island, in the Fraser Lowland, on the Sunshine Coast, in the Okanagan valley, in the west and east Kootenays, and north in the interior, likely to the vicinity of Shuswap Lake and Golden.

On the coast, there are no breeding records from Western Vancouver Island or from the entire coast north of Vancouver Island except for a single occurrence in the lower Skeena River valley.

The Black-headed Grosbeak reaches its highest numbers in summer in the Fraser and Nanaimo lowlands of the Georgia Depression (Fig. 440). The highest reported local density of breeding birds, however, is from the riparian lands where the Kootenay River enters Kootenay Lake in the Southern Interior Mountains. An analysis of Breeding Bird Surveys in

Figure 442. Black-headed Grosbeak nest in dense thicket of red-osier dogwood, a frequently used nest site in British Columbia (Creston, 5 June 1994; R. Wayne Campbell).

British Columbia for the period 1968 through 1993 could not detect a change in numbers on either coastal or interior routes. Throughout the North American survey area, Black-headed Grosbeak numbers for the period 1980 to 1996 increased at an average annual rate of 1.8% ($P < 0.01$) (Sauer et al. 1997).

The Black-headed Grosbeak has been recorded breeding at elevations from sea level to about 700 m on the coast, and from 300 to 600 m in the interior.

This grosbeak establishes its nesting territories primarily in the deciduous woodland (Fig. 441) and shrub thickets associated with edges, especially those bordering streams, sloughs, marshes, ponds, or small lakes. One of the highest local nesting densities for this species occurs on the extensive delta of the Kootenay River where it enters Kootenay Lake. There ribbons of black cottonwood trees parallel the rivers and dykes that support a lush and diverse shrub layer of red-osier dogwood, blue elderberry, saskatoon, choke cherry, wild cherry, black hawthorn, and willows.

The species also finds suitable habitat in suburban areas with roadsides and fencelines bordered with shrubs and small trees. Gardens, suburban parks, orchards, and vineyards sometimes provide breeding territory. Nests in British Columbia were in woodland (76%), in human-influenced woodland (10%), and in shrubbery (10%). Within the woodland category, 55% of nesting territories were in mixed woodland, 20% in black cottonwood, and 10% in an urban setting; single nests were found in ponderosa pine stands, shrubland, and marsh. In terms of habitat specifics, 85% were near water, 12% on edges, and 3% in unspecified bottomland.

On the coast, the Black-headed Grosbeak has been recorded breeding from 26 May to 1 August; in the interior, it has been recorded from 20 May (calculated) to 30 July (Fig. 438).

Nests: Nest were found almost equally in shrubs or trees ($n = 95$). Eleven different types of trees or shrubs were named in the records; the following were the most frequently reported: red-osier dogwood (22%; Fig. 442), elderberry (22%), willow (20%), apple (13%), vine maple (11%), and birch (6%)

Figure 443. Nearly 80% of all Black-headed Grosbeak clutches reported from British Columbia contained 3 or 4 eggs (Six Mile Slough, Creston, 5 June 1994; R. Wayne Campbell).

Figure 444. Black-headed Grosbeak nest with 4 nestlings. The nestling period is between 10 and 14 days (Creston, 17 June 1997; Linda M. Van Damme).

($n = 64$). In general, nests were placed in forked branches near the tops of the trees or shrubs. Nests were cup-shaped or saucer-shaped, composed of twigs (57% of nests; Fig. 443), grasses (20%), stems (15%), and rootlets (15%); needles, leaves, hair, and plant down were used less frequently. Saucer-shaped nests had an open construction, sometimes permitting the eggs to be seen from below. The heights of 87 nests ranged from 0.8 to 8 m, with 60% between 1.8 and 3 m.

Eggs: Dates for 109 clutches ranged from 23 May to 13 July, with 53% recorded between 3 and 19 June. Calculated dates indicate that eggs can occur as early as 20 May. Sizes of 71 clutches ranged from 1 to 5 eggs (1E-2, 2E-9, 3E-27, 4E-29, 5E-4), with 79% having 3 or 4 eggs (Fig. 443). Hill (1995), in a review of the clutch size over the full nesting range of the species, found a median and modal clutch size of 3 eggs. The incubation period is 12 days (Austin 1968) or between 12 and 14 days (Hill 1995). Referring to observations in California, Austin (1968) notes that the eggs begin hatching on the 12th day of incubation and may hatch within a few hours of one another or over a period of 2 days.

Young: Dates for 69 broods (Figs. 435 and 444) ranged from 2 June to 1 August, with 52% recorded between 19 June and 5 July. Sizes of 52 broods ranged from 1 to 4 young (1Y-11, 2Y-12, 3Y-16, 4Y-13), with 56% having 3 or 4 young. There were 2 records of adults feeding 5 fledged young. In 1 nest, the 4 young hatched over a period of 3 days and the nestling period was at least 12 days. Young can leave the nest as early as day 9 if disturbed, but the nestling period is usually between 10 and 14 days (Ritchison 1983).

Brown-headed Cowbird Parasitism: In British Columbia, 6% of 107 nests found with eggs or young were parasitized by the cowbird. On the coast, none of 50 nests was parasitized by the cowbird. In the interior, 6 of 57 nests (11%) were parasitized. No instances have been recorded of this grosbeak feeding cowbird fledglings in British Columbia.

The Black-headed Grosbeak is an infrequent host of the Brown-headed Cowbird over many parts of its range

(Friedmann 1963, 1971; Friedmann et al. 1977). Friedmann and Kiff (1985) do not list it among the species known to have successfully reared Brown-headed Cowbird chicks.

Nest Success: Of 10 nests found with eggs and followed to a known fate, only 1 produced at least 1 fledgling. Predation by Black-billed Magpie and Steller's Jay was noted. In an intensive study of a population of 25 marked pairs of Black-headed Grosbeaks nesting in the Sandia Mountains of New Mexico, jay predation was the most important cause of nest failure, and varied with the habitat structure. In habitat heavily used by jays, nest success was about 20%, whereas grosbeaks nesting in habitat unattractive to jays had 79% success (Hill 1988). It was also found that the nest sites with the highest success were those occupied by males 3 years and older and their mates.

Failed nest attempts early in the season are always followed by a second attempt (Hill 1995).

REMARKS: Two subspecies are described from British Columbia: *P. m. melanocephalus* is the subspecies of the interior of the province, while *P. m. maculatus* occurs in coastal regions.

The November photographic record for Prince George (see NOTEWORTHY RECORDS) was first identified as a female Rose-breasted Grosbeak, but re-examination revealed that it was a Black-headed Grosbeak – the latest record for the province.

Many winter records for British Columbia (e.g., Agassiz 3 January 1912, Fleetwood 2 February 1966, and New Denver 3 February 1972) as well as a bird reported on the Chilliwack Christmas Bird Count of 30 December 1972 (Sargeant 1973) are undocumented and have therefore been excluded from this account.

For additional information on the natural history and conservation of the Black-headed Grosbeak, see Hill (1995).

NOTEWORTHY RECORDS

Spring: Coastal – Victoria 29 May 1916-4 eggs; Cowichan Lake 31 May 1924-3; Barnston Island 19 May 1973-12; Surrey 3 Apr 1966-1 pair, 7 Apr 1965-1 pair; Burnaby Lake 9 May 1981-10; Port Kells 4 May 1964-5; Chilliwack 31 May 1888-1 (MCZ 193005); North Vancouver 31 May 1924-3 eggs; Golden Ears Park 29 May 1982-12; Campbell River 25 May 1980-6; Alta Creek 21 May 1997-1 adult male; Port Hardy 11 May 1980-1; Kaskis River 26 May 1985-1. **Interior** – Osoyoos 5 May 1985-2 along oxbows; French Slough (Creston) 23 May 1981-4 eggs; Creston 18 May 1985-1; Balfour to Waneta 16 May 1981-28 on census; Similkameen River (s of Keremeos) 20 May 1963-4; Fernie 31 May 1983-1 (Fraser 1984); Invermere 5 Apr 1978-1; North Barriere Lake 7 Apr 1974-1; Mount Revelstoke National Park 22 May 1981-1; Golden 31 May 1991-1 (Siddle 1991c); Moberly Marsh 22 May 1995-1 male singing; Quesnel 19 May 1989-1.

Summer: Coastal – Rocky Point (Victoria) 21 Aug 1995-1; Victoria 9 Jun 1960-2 nestlings; Langford 18 Jul 1985-4; Carmanah Point 24 Aug 1951-2 (Irving 1953); Cowichan Bay 7 Jun 1987-1 on estuary; Ladner 14 Jun 1973-3 nestlings; Huntingdon 14 Aug 1959-6; Nicomen Island (Deroche) 15 Jun 1972-13; Burnaby Lake 30 Jul 1992-adult feeding 5 young just out of nest, 20 Aug 1994-3 fledglings; Langley 13 Jul 1969-3 eggs; Surrey 3 Jun 1963-3 eggs plus 1 Brown-headed Cowbird egg (RBCM 1925); Pitt Meadows 20 Jul 1972-3 nestlings, 25 Jul 1971-40 (Campbell et al. 1972b); Alberni valley 7 Jun 1910-1 (MVZ 16292); Beaver Creek (Alberni) 19 Jun 1931-1; Coquitlam 30 Jun 1990-21 on survey; Cypress Park 23 Jun 1968-1; Hollyburn Mountain 18 Jun 1972-2; Hope 24 Jun 1905-1; Sandford Island 9 Aug 1967-1; Kennedy Lake 20 Jun 1974-1; Courtenay 1 Jun 1945-1; Brackendale 13 Jun 1906-1 (NMC 9512); Horseshoe Lake (n Stillwater) 24 Jun 1936-1 (NMC 27285); Whistler 18 Jun 1993-1; Pemberton 22 Jun 1924-1 (UBC 5670), 22 Jun 1993-1; Fawn Bluff 1 Aug 1936-a family group (Laing 1942); Knight Inlet 9 Jun 1976-1; Bella Coola 6 Jul 1940-1 (FMNH 177317); Stuie 2 Aug 1938-1 (Laing 1942); Kitimat 10 Jul 1975-2; Lakelse Lake 3 Jul 1983-1; Terrace 5 Jul 1977-male incubating eggs (BC Photo 775); Kitsumkalum 8 Jun 1980-2; Greenville 29 Jun 1993-1. **Interior** – n Newgate 30 Jun 1953-3 (Godfrey 1955); Cawston 6 Jul 1958-3 (Campbell and Meugens 1971); Grand Forks 18 Jun 1929-1; Creston 5 Jun 1994-female flushed from 4 eggs, 1 Jul 1997-female feeding 2 fledglings, 7 Jul 1993-2 eggs; Fruitvale 23 Aug 1981-10 at meadows; w side Christina Lake 12 Jul 1979-1; Keremeos 8 Jul 1981-adults feeding 5 young just out of nest; Vaseux Lake 3 Jun 1983-3 eggs; Waldo 10 Jun 1953-2 (Godfrey 1955); Naramata 14 Jun 1988-4 downy nestlings, 30 Jul 1989-3 young; s Kimberley 21 Jul 1977-a pair with fledged young; Sparwood 3 Jun 1977-1; Lytton 19 Jun 1976-1; Sorrento 9 Jun 1971-1; Celista 22 Jul 1960-1; Barriere 24 Jun 1995-1; Golden 26 May 1994-1 at Moberly Flats, 15 km n Golden 5 Jul 1998-3 naked young in nest; Hemp Creek Jun 1955-1; Kinbasket Lake 20 Jun 1996-3 males singing; Riske Creek 31 Aug 1989-1; Williams Lake 10 Jun 1996-1 at feeder; Prince George 7 Jun 1964-1 (Weber 1964).

Breeding Bird Surveys: Coastal – Recorded from 18 of 27 routes and on 51% of all surveys. Maxima: Seabird 2 Jul 1989-27; Chilliwack 25 Jun 1992-23; Pitt Meadows 30 Jun 1985-21; Coquitlam 30 Jun 1990-21. **Interior** – Recorded from 26 of 73 routes and on 12% of all surveys. Maxima: Syringa Creek 19 Jun 1993-7; Mabel Lake 12 Jun 1994-6; Salmo 15 Jun 1969-3; Summerland 20 Jun 1987-3; Oliver 18 Jun 1989-3; Creston 18 Jun 1984-3.

Autumn: Interior – Prince George 25 Nov 1973-1 (BC Photo 346); Riske Creek 14 Sep 1989-1 (Weber and Cannings 1990); Glacier National Park 11 Sep 1982-1; Okanagan Landing 23 Sep 1929-1 (MVZ 105137); Skaha Lake 1 Nov 1970-1; Trail 15 Sep 1986-1. **Coastal** – Masset 16 Sep 1953-1 (RBCM 10349); Tlell 6 Oct 1990-1 (Siddle 1991c); Cape St. James 19 Nov 1981-1; Port Neville 17 Sep 1986-1; Miracle Beach Park 1 Sep 1960-1 (Stirling 1961); West Vancouver 2 Oct 1973-1; Huntingdon 13 Sep 1956-9; Boundary Bay 12 Nov 1962-3; Sea Island 12 Sep 1995-1 male banded (Elliott and Gardner 1997); Victoria 29 Sep 1986-1.

Winter: See REMARKS.

Christmas Bird Counts: See REMARKS.

Lazuli Bunting

LZBU

Passerina amoena (Say)

RANGE: Breeds from southern British Columbia, southern Alberta, and southern Saskatchewan south through parts of western North Dakota, western South Dakota, western Nebraska, east-central Colorado, and southwestern Kansas; south to western Oklahoma, northern New Mexico, northern Arizona, southern Nevada, southern California, and northwestern Baja California; and west to the Cascade Mountains of western Oregon and Washington. Winters from southeastern Arizona south to western Mexico.

STATUS: On the coast, locally *rare* to *uncommon* summer visitant in the Georgia Depression Ecoprovince; in the Coast and Mountains Ecoprovince, *casual* on Western Vancouver Island, locally *rare* to *uncommon* on the Southern Mainland Coast, and *accidental* on the Northern Mainland Coast. Not recorded from the Queen Charlotte Islands.

In the interior, *fairly common* migrant and summer visitant in the Southern Interior and Southern Interior Mountains ecoprovinces; locally *rare* to *uncommon* summer visitant in the Central Interior Ecoprovince; *casual* in the Sub-Boreal Interior Ecoprovince; *accidental* in the Boreal Plains and Northern Boreal Mountains ecoprovinces.

Breeds.

Figure 445. Over the past several decades, the Lazuli Bunting has increased its range northward and westward from the southern interior of British Columbia (Anthony Mercieca).

Figure 446. Over most of its distribution in British Columbia, the Lazuli Bunting arrives directly on its breeding grounds, which includes brushy habitats on bushy hillsides (8 km east of Castlegar, 4 June 1997; R. Wayne Campbell).

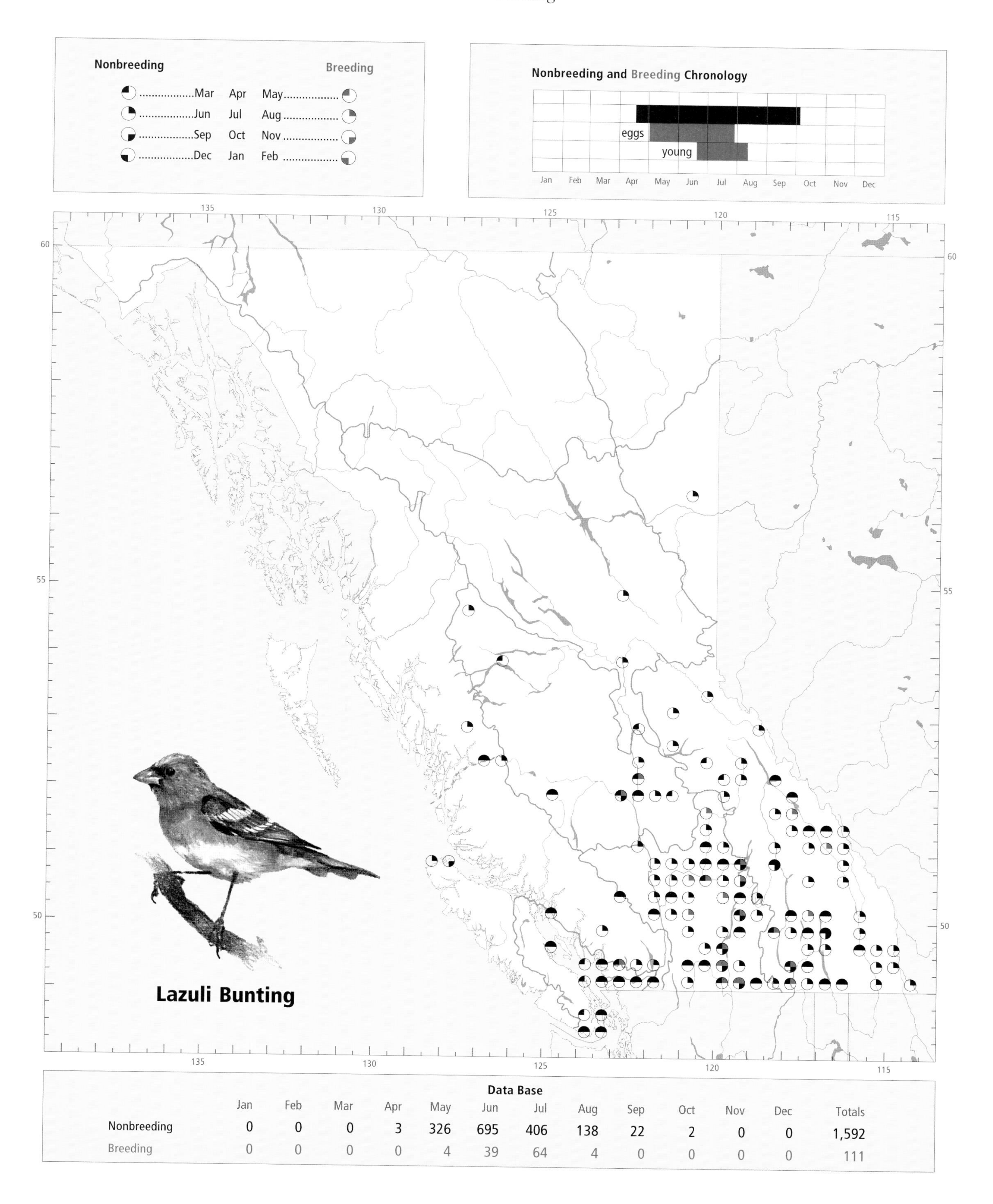

Data Base

	Jan	Feb	Mar	Apr	May	Jun	Jul	Aug	Sep	Oct	Nov	Dec	Totals
Nonbreeding	0	0	0	3	326	695	406	138	22	2	0	0	1,592
Breeding	0	0	0	0	4	39	64	4	0	0	0	0	111

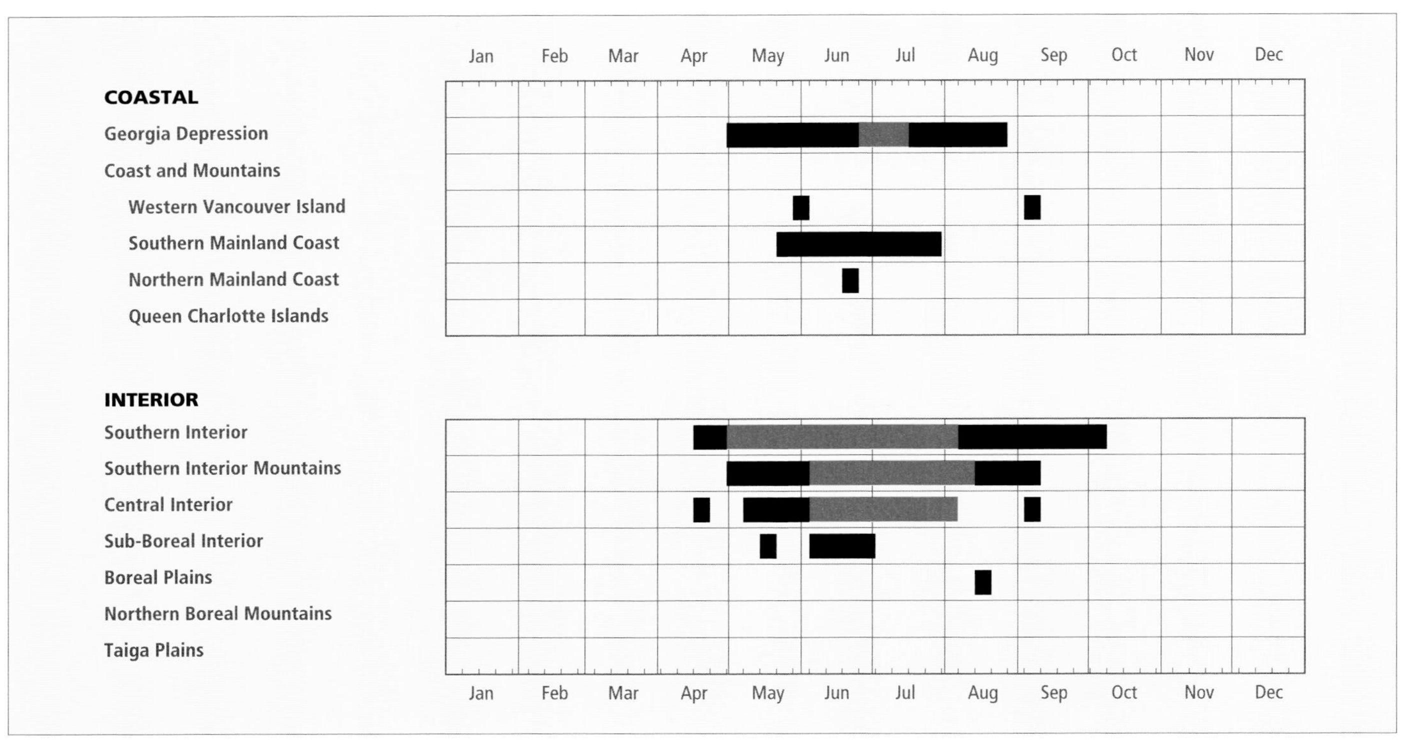

Figure 447. Annual occurrence (black) and breeding chronology (red) for the Lazuli Bunting in ecoprovinces of British Columbia. Records are shown for the week in which they occurred.

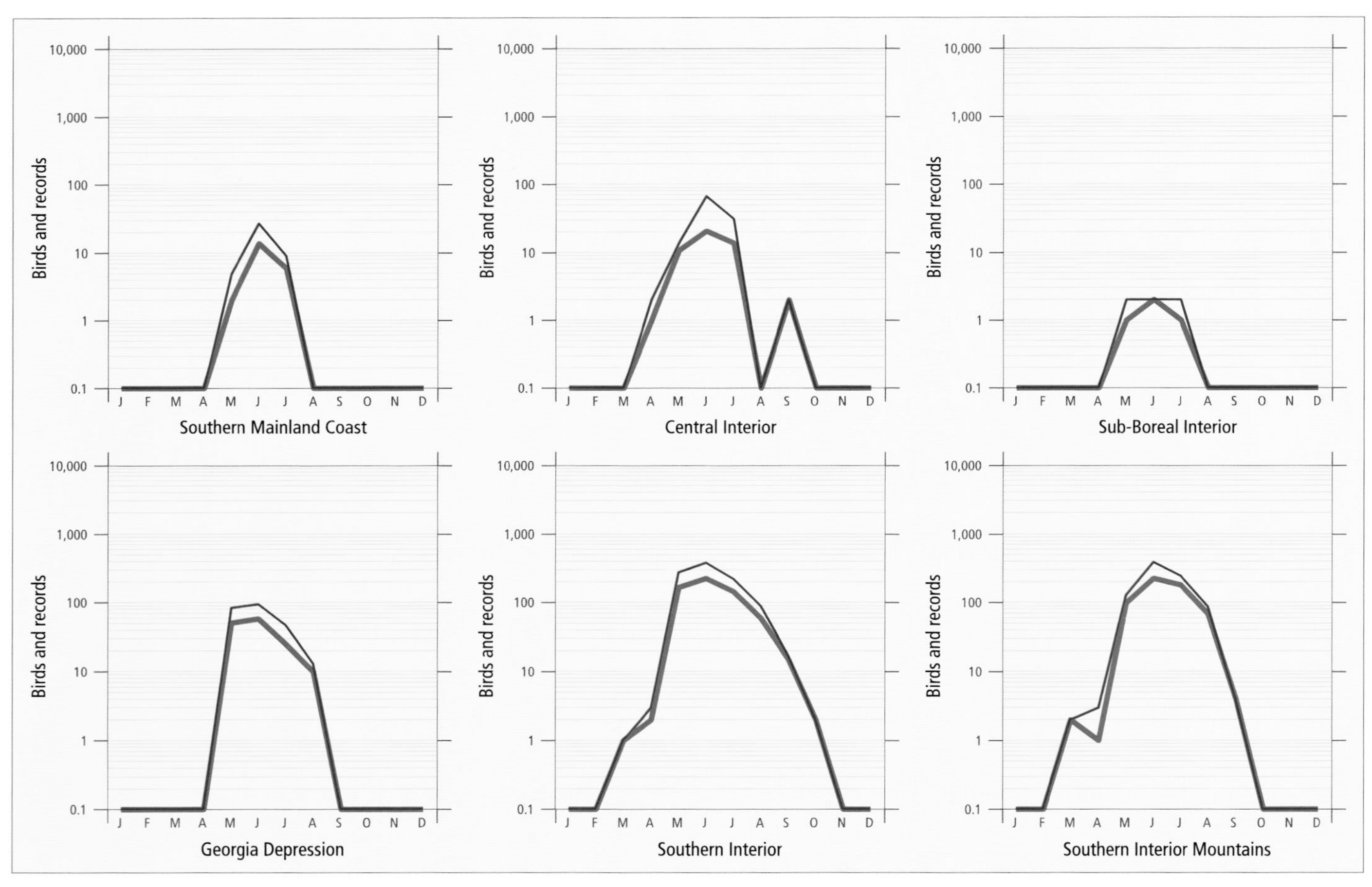

Figure 448. Fluctuations in total number of birds (purple line) and total number of records (green line) for the Lazuli Bunting in ecoprovinces of British Columbia. Breeding Bird Surveys and nest record data have been excluded.

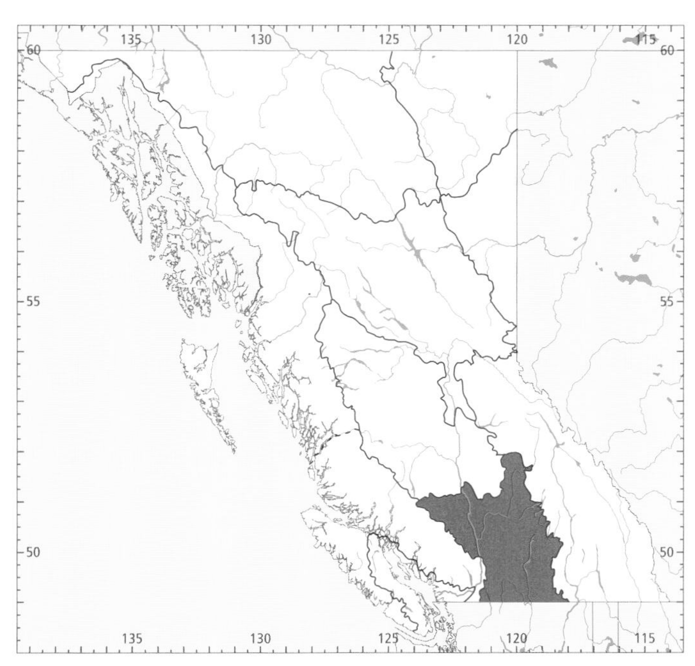

Figure 449. In British Columbia, the highest numbers for the Lazuli Bunting in summer occur in the Southern Interior Ecoprovince.

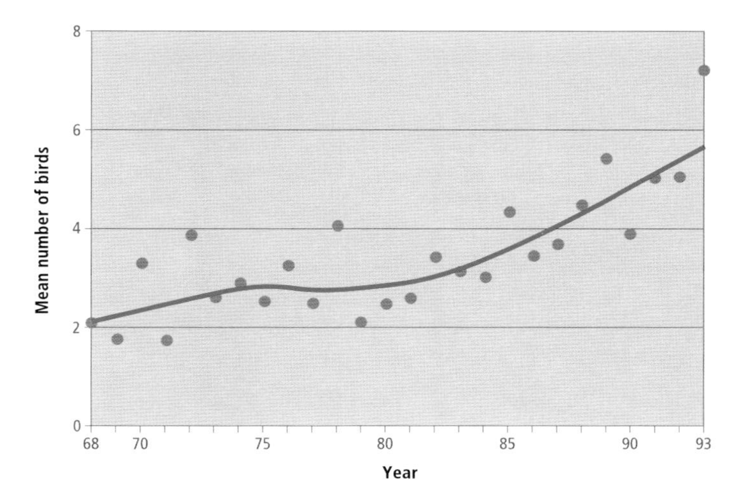

Figure 450. An analysis of Breeding Bird Surveys for the Lazuli Bunting in British Columbia over the period 1968 through 1993 shows that the mean number of birds on interior routes increased at an average annual rate of 4% ($P < 0.10$).

CHANGE IN STATUS: In the mid-1940s, the Lazuli Bunting (Fig. 445) was known only from the Southern Interior and Southern Interior Mountains ecoprovinces of southern British Columbia. It ranged from Botanie Creek valley northeast of Lytton in the north, south through the Okanagan valley and east through Newgate, Creston, and Trail near the international boundary, to Cranbrook and Canal Flats (Munro and Cowan 1947).

In the succeeding 50 years, this species has slowly increased its range northward and westward into the Central Interior. There a specimen was taken at Wistaria in May 1939, a record not reported by Munro and Cowan (1947). It was next detected near Williams Lake in 1951. In 1958, the Lazuli Bunting was recorded west of Williams Lake, near the Chilcotin River. From then to the present, this bunting has been recorded in the northeastern Cariboo and Chilcotin areas almost every year, with a maximum of 18 birds reported in any year. Most of these occurrences have been within a triangle from Lac la Hache in the south, Soda Creek in the north, and Farwell Canyon in the west, about 30 km in each direction. In the Central Interior, the westernmost record is from Kleena Kleene and the northernmost from Telkwa. In 1991 it appeared in the vicinity of Prince George. The northernmost records in the province west of the Rocky Mountains have been from McLeod Lake, north of Prince George (1991 and 1997), and east of the Rocky Mountains, near Fort St. John (1979).

The expansion of this bunting's range into the Central Interior appears to be part of a general increase in numbers in the province and an expansion within the southern portions of the interior in general. Notes made by the early ornithologists and naturalists in British Columbia indicate that the Lazuli Bunting was an uncommon species. In the years since 1967, however, it has been reported in new locations in the interior annually and in increasing numbers. On the coast, where it was scarce and where several years would elapse between reported sightings, the Lazuli Bunting has been recorded every year since at least 1989. It appears that this species has maintained or increased its numbers at the centre of its range in British Columbia, and has gradually increased its distribution to occupy regions where human alteration of the landscape has created a more favourable environment for it.

NONBREEDING: The Lazuli Bunting is widely distributed in suitable habitat across the southern third of the interior of the province, north to about the vicinity of Quesnel, in the southern Sub-Boreal Interior. There is 1 isolated record for Fort St. John in the Boreal Plains. On the coast, it has been reported on Vancouver Island and rarely on the central and northern mainland coast. The northernmost records there are from Bella Coola and Kimsquit, where river valleys cut through the

Figure 451. Semi-open habitats with thick clumps of ocean-spray and saskatoon are used for nesting by Lazuli Buntings in parts of the Southern Interior Mountains Ecoprovince of British Columbia (Creston, 2 June 1997; R. Wayne Campbell).

Figure 452. Mixed-shrub habitats of Oregon-grape, ocean-spray, and saskatoon in semi-open environments are frequently used by nesting Lazuli Buntings in British Columbia (near Christina Lake, 12 May 1997; R. Wayne Campbell).

mountains from the interior and provide access to estuarine habitat similar to that in parts of the Fraser Lowland.

On the coast, the Lazuli Bunting is confined to elevations below 100 m. In the interior, it has been found mostly from valley bottom elevations of 280 to 500 m, but there is a single occurrence at 1,400 m.

In the interior, this is a characteristic species of the brush patches (Fig. 446), shrub-choked gullies and hillsides, margins of riverside black cottonwood groves, trembling aspen stands bordering ponds or lakes, and similar edges where shrubs mark the transition from woodland to dry grassland, or clearings such as fields, orchards, roadways, sewage lagoons, railroad tracks, or the scrub tangles among the heavy talus below cliff faces. The species of shrubs are less important than the structure of the habitat, but include wild rose, mock-orange, ocean-spray (Fig. 451), saskatoon, snowberry, choke cherry, hawthorn, and sumac. There are records of singing males in small clumps of young trembling aspen on avalanche slopes in the heart of the Rocky Mountains (Poll et al. 1984).

On the coast, the habitat used is similar in structure but differs in species, and includes thimbleberry, hardhack, red elderberry, Nootka rose, Scotch broom, and Himalayan blackberry thickets. In both the coast and interior, the Lazuli Bunting sometimes invades logging clearcuts.

On the coast, the earliest spring migrants appear in the first week of May, and the peak movement is from the second week of May to mid-June (Figs. 447 and 448). In the Southern Interior, there are 2 late April records for the southern Okanagan valley, but from early May on observations are recorded continuously. First-of-year dates for Vernon over 24 years range from 4 May to 29 May, with a median date of 19 May (Cannings et al. 1987). The earliest dates and the course of the migration are the same in the valleys of the Kootenay and Columbia rivers of the Southern Interior Mountains. The migration does not reach the Central Interior until the third week of May, where it rises rapidly to a peak in mid-June.

The Lazuli Bunting begins its autumn migration in July and the movement is at its height in August (Figs. 447 and 448). In the Central Interior, there are no August records, but there are 2 for early September. In the valleys of the southern portions of the interior, there are many records for August, but there are few for September; the latest observation is from the Columbia valley on 9 September. In the Okanagan valley, a trickle of birds persists in some years through September to early October.

On the coast, the few records indicate a similar timing, with most of the autumn migrants recorded in August and just 1 record for September.

On the coast, the Lazuli Bunting has been recorded regularly from 11 May to 26 August, with occasional records as early as 2 May and as late as 9 September; in the interior, it has been recorded regularly from 21 April to 25 September, with occasional records as late as 4 October (Fig. 447).

BREEDING: The Lazuli Bunting breeds locally on the southwest coast in the Fraser Lowland, and throughout the southern portions of the interior north to Revelstoke and Kinbasket Lake in the Southern Interior Mountains, locally in the Central Interior, and likely near Quesnel in the Sub-Boreal Interior.

In British Columbia, the Lazuli Bunting reaches its highest numbers in summer in the Okanagan valley of the Southern Interior (Fig. 449). An analysis of Breeding Bird Surveys for the period 1968 through 1993 shows that the mean number of birds on interior routes increased at an average annual rate of 4% (Fig. 450); coastal routes for the same period contained insufficient data for analysis. Across North America, analysis of Breeding Bird Surveys over the period 1966 to 1996 suggests that Lazuli Bunting numbers are relatively stable or increasing slightly (Sauer et al. 1997).

The Lazuli Bunting nests near sea level on the coast and breeds at elevations between 280 and 950 m in the interior. There are records of males in the breeding season at elevations of 1,100 m in the Golden area of the Southern Interior Mountains. Breeding is concentrated in the Southern Interior, where this bunting occupies shrubby gullies in grassland and trembling aspen parkland habitats, within the Interior Douglas-fir and Ponderosa Pine biogeoclimatic zones. It also occurs in similar habitats in the west Kootenay and in the Cariboo and Chilcotin areas. Because the Lazuli Bunting moves directly onto its breeding habitat over most of its distribution in the province, much of the detailed description given under NONBREEDING applies equally here. Breeding usually takes place in open and semi-open country (Fig. 452), including ranchland, the edges of cultivated farms and orchards, woodlots, brushy draws, bottomlands, and the margins of lakes and ponds, roadside thickets, powerline rights-of-way, logging debris on deforested land, and sometimes suburban gardens. The advance of land clearing in the Pacific states has resulted in the Lazuli Bunting moving into irrigated lands and some extensive lumbered areas at lower elevations (Austin 1968). While the same changes have been noted in British Columbia, most nest territories have still been in forest edge or shrublands rather than in human-influenced areas. The plant species associated with breeding distribution are the same as those noted for the nonbreeding season.

The Lazuli Bunting has been recorded breeding in the province from 3 May (calculated) to 11 August (Fig. 447).

Nests: Most nests (85%; $n = 46$) were in shrubs (Fig. 453). Most shrubs were unidentified as to species, but 19% of nests were in rose briars and 2% in elders. Fifteen percent were in trees, and just 1 nest was in a conifer. The heights of 42 nests ranged from ground level to 2.2 m, with 78% recorded between 0.6 m and 1.2 m.

Nests were cups of loosely assembled material, with grasses occurring in 92% (Fig. 454), plant fibres in 32%, mammalian hair in 10%, plant down in 16%, bark strips in 11%, and stems, twigs, spider webs, and feathers collectively in 19% of nests. The nest lining was mainly of fine grasses, plant fibres, and mammalian hair.

Eggs: Dates for 65 clutches ranged from 3 May to 28 July, with 54% recorded between 19 June and 11 July. Sizes of 33 clutches ranged from 1 to 6 eggs (1E-2, 2E-4, 3E-14, 4E-11, 5E-1, 6E-1), with 76% having 3 or 4 eggs (Fig. 454). The incubation period determined for 1 nest in British Columbia was 11 days. Elsewhere, it is given as 11 to 14 days, with an average of 12 days (Greene et al. 1996). The average clutch size of 3.18 eggs ($n = 33$) in our sample is smaller than that of Alberta (3.44 eggs per nest; $n = 16$) but similar to clutch sizes in Montana (3.13; $n = 60$) (Greene et al. 1996).

Young: Dates for 48 broods ranged from 20 June to 11 August, with 52% recorded between 5 July and 28 July. Sizes of 34 broods ranged from 1 to 5 young (1Y-4, 2Y-8, 3Y-14, 4Y-7, 5Y-1), with 65% having 2 or 3 young. Two of the 4 broods of single young had been parasitized by the Brown-headed Cowbird, but the foreign eggs had been removed by observers. The nestling period is 9 to 11 days, but chicks will leave the nest at 8 days if disturbed and may stay for 13 days if undisturbed (Greene et al. 1996). There is no indication that the Lazuli Bunting in British Columbia attempts to raise 2 broods in a summer.

Figure 453. Most Lazuli Bunting nests in British Columbia were constructed in shrubs like the colourful mock-orange to the right of the Douglas-fir stump (Mount Kruger, Richter Pass, 6 July 1997; R. Wayne Campbell).

Figure 454. Nearly three-quarters of all complete Lazuli Bunting clutches in British Columbia contained 3 or 4 eggs (Mount Kruger, Richter Pass, 6 July 1997; R. Wayne Campbell).

Brown-headed Cowbird Parasitism: In British Columbia, 18% of 55 nests found with eggs or young were parasitized by the Brown-headed Cowbird. Friedmann (1963) recorded a small number of instances of cowbird parasitism of this species from California to Washington, and in 1977 referred to a rate of 7% in 107 nests. Among our records, there are 5 instances of parasitism involving single cowbird eggs. A deserted nest found in November contained 4 cowbird eggs and 3 bunting eggs. Cannings et al. (1987) refer to 5 records of adult Lazuli Buntings feeding fledgling cowbirds.

Nest Success: Of 9 nests found with eggs and followed to a known fate, 1 produced at least 1 fledgling.

REMARKS: The Lazuli Bunting and Indigo Bunting are regarded by the American Ornithologists' Union (1983) as a superspecies. They occupy separate ranges over most of North America, the Lazuli Bunting in the western parts of the continent and the Indigo Bunting in the east. Their ranges do overlap in some areas, however, and hybridization has been recorded (Emlen et al. 1975; Baker 1991, 1994). In British Columbia, a male Lazuli-Indigo Bunting hybrid was photographed at Brouse, in the Southern Interior Mountains, on 22 July 1990 (G.S. Davidson pers. comm.).

We have a record of 5 Lazuli Buntings on a Breeding Bird Survey route at Gnat Pass, in the Northern Boreal Mountains, and Mason (1951) observed a bird in February near Kawkawa Lake (Hope) that she identified as a Lazuli Bunting. Both records lack adequate documentation and have been excluded from this account.

For additional details on the life history of this species, see Greene et al. (1996).

NOTEWORTHY RECORDS

Spring: Coastal – Mount Tolmie (Victoria) 17 May 1992-1 (Bowling 1992), 28 May 1959-6 (Davidson 1966); Saanich 20 and 21 May 1973-4; Mount Douglas (Victoria) 22 May 1993-4 (Siddle 1993c); Agassiz 27 May 1991-1; Chilliwack 11 May 1890-1 (MVZ 105178); Gabriola Island 27 May 1976-1; North Burnaby 2 May 1967-1; Sechelt 27 May 1980-1; Comox 20 May 1997-1; Mansons Landing 27 May 1975-1; Pemberton 25 May 1997-4. **Interior** – Oliver to Richter Lake 17 May 1959-12; Creston 5 May 1987-1; Kaslo 3 May 1987-1; Kilpoola Lake 21 Apr 1977-2; 4.8 km e Cawston 22 May 1971-2; Osoyoos 7 May 1907-5 eggs; Trail 3 May 1959-1; Balfour to Waneta 16 May 1981-15 on survey; Okanagan Falls 29 May 1913-nest with 6 eggs collected (RBCM E1066); Okanagan Landing 23 May 1938-1; Kamloops 24 May 1986-5 at Paul Creek; Magna Bay 14 May 1981-1 singing male; Revelstoke 16 May 1981-1; Moberly Marsh 22 May 1995-1 singing male; 15 km n Golden 25 May 1996-1 singing male; Yoho National Park 24 May 1973-1; Kleena Kleene 18 Apr 1964-2 (Paul 1964a); Chilcotin River 9 May 1987-1 at Wineglass Ranch; Williams Lake 26 May 1978-2; Murtle Lake 28 May 1959-1 (Edwards and Ritcey 1967); Quesnel 16 May 1979-2; Wistaria 28 May 1939-1 (RBCM 8782).

Summer: Coastal – Rocky Point (Victoria) 26 Aug 1995-1; Mount Douglas (Saanich) 4 Jun 1959-3; Iona Island 23 Jun 1990-3; Huntingdon 19 Jul 1958-7; Abbotsford 10 Jun 1992-1 adult carrying food; Coquitlam 9 Jul 1994-adult feeding 2 young in nest; Harrison Hot Springs 10 Jun 1984-2, 16 Aug 1992-1; Sechelt 17 Jul 1994-nest with 4 nestlings at airport; Comox 23 Jun 1893-1 (NMC 1701), 2 Jul 1997-2 singing males; Pemberton Meadows 29 Jun 1985-5, 21 Jul 1996-1 male; Cape Scott 1 Jun 1976-1; Bella Coola 14 Jun 1981-1. **Interior** – Grand Forks 3 Jun 1995-2 singing males and female carrying nesting material; Creston 2 Jul 1977-10 males seen in a day; Trail 6 Jun 1981-6, 12 Jul 1975-2 nestlings; Castlegar 11 Aug 1978-3 nestlings; Elko 3 Jun 1953-8; Hedley 28 Jun 1973-1; Nelson 10 Jun 1982-3; Trout Creek Point (Summerland) 25 Jul 1957-4 eggs; Naramata 5 Aug 1965-2 fully feathered nestlings; Zincton 12 Jun 1960-3 eggs (RBCM E1855); Canal Flats 5 Jun 1983-5; Botanie Creek valley 6 Jul 1929-1 (RBCM 7519); Lillooet 18 Jul 1916-2 (NMC 9727-28); 19 km n Lillooet 23 Jun 1932-2 (MCZ 284695-6); Fountain Valley 19 Jun 1971-1; Ashcroft 3 Jun 1892-1 (ANS 31192); Heffley Creek 19 Jul 1981-5, 2 adults and 3 fledglings; Hat Creek 3 Jul 1990-1; Squilax 6 Jun 1973-5 on natural slide; Ochre Creek 1 Jul 1975-2 males, 20 Jul 1977-1 male; Adams Lake 1 Jul 1963-3; Yoho National Park 21 Jun 1975-1 male at Wapta marsh; Glacier National Park 14 Jun 1983-1; Bencoth Hill 20 Jun 1985-3 nestlings; Kinbasket Lake 5 Jul 1995-7 singing males; Chilcotin River (Riske Creek) 12 Jun 1991-4 eggs and 1 Brown-headed Cowbird egg at Wineglass Ranch, 29 Jun 1991-3 nestlings, 28 Aug 1991-4 recent fledglings; Chimney Creek (Williams Lake) 10 Jun 1989-13; Williams Lake 16 Jun 1985-14 singing males along Highway 20 between Williams Lake and the Fraser River; Quesnel 30 Jun 1975-1 fledgling; Prince George 8 Jun 1991-1 found dead; nr McLeod Lake (Crooked River) 25 Jun 1991-1 male; 4 km s McLeod Lake (Crooked River) 23 Jun 1997-1 male; Montney 19 Aug 1975-1.

Breeding Bird Surveys: Coastal – Recorded from 4 of 27 routes and on 9% of all surveys. Maxima: Seabird 6 Jul 1991-11; Pemberton 14 Jun 1992-5; Albion 4 Jul 1970-3; Chilliwack 26 Jun 1993-3. **Interior** – Recorded from 41 of 73 routes and on 46% of all surveys. Maxima: Canoe Point 19 Jun 1994-25; Salmon Arm 1 Jul 1968-21; Mabel Lake 30 Jun 1973-21; Needles 14 Jun 1992-19 (see also REMARKS).

Autumn: Interior – Riske Creek 5 Sep 1986-1; Revelstoke 3 Sep 1988-1; West Bench (Penticton) 25 Sep 1967-1; Okanagan Landing 4 Oct 1941-1 (MVZ 105177); Castlegar 9 Sep 1969-1; Osoyoos 1 Sep 1973-3. **Coastal** – Shushartie 9 Sep 1935-1 (NMC 26251).

Winter: Interior – No records. **Coastal** – No records (see REMARKS).

Indigo Bunting

INBU

Passerina cyanea (Linnaeus)

RANGE: Breeds from southeastern Saskatchewan, southern Manitoba, northern Minnesota, western and southern Ontario, southwestern Quebec, southern Maine, and southern New Brunswick south to southern New Mexico, central and southeastern Texas, the Gulf coast, and central Florida, and west to eastern Wyoming, eastern Colorado, western Kansas, and central New Mexico. Winters rarely from southern Texas, the Gulf coast, and southern Florida south mainly through Middle America, the Greater Antilles and the Bahamas to Panama and northwestern Colombia.

STATUS: On the coast, *very rare* in the Georgia Depression Ecoprovince including southern Vancouver Island; in the Coast and Mountains Ecoprovince, *accidental* on Western Vancouver Island.

In the interior, *very rare* migrant and local summer visitant in the Southern Interior and Southern Interior Mountains ecoprovinces.

Breeds.

Figure 455. The Indigo Bunting is a migratory songbird that has recently expanded its breeding range from eastern North America into the southwestern United States and southern British Columbia (Anthony Mercieca).

Figure 456. The Indigo Bunting frequents patches of brushy and weedy habitats scattered in open rangelands in southern British Columbia (near Grand Forks, 12 May 1997; R. Wayne Campbell).

Figure 457. On southern Vancouver Island, the Indigo Bunting has been found in brushy habitats bordering second-growth woodlands (near Mount Tolmie, Victoria, 9 July 1997; R. Wayne Campbell).

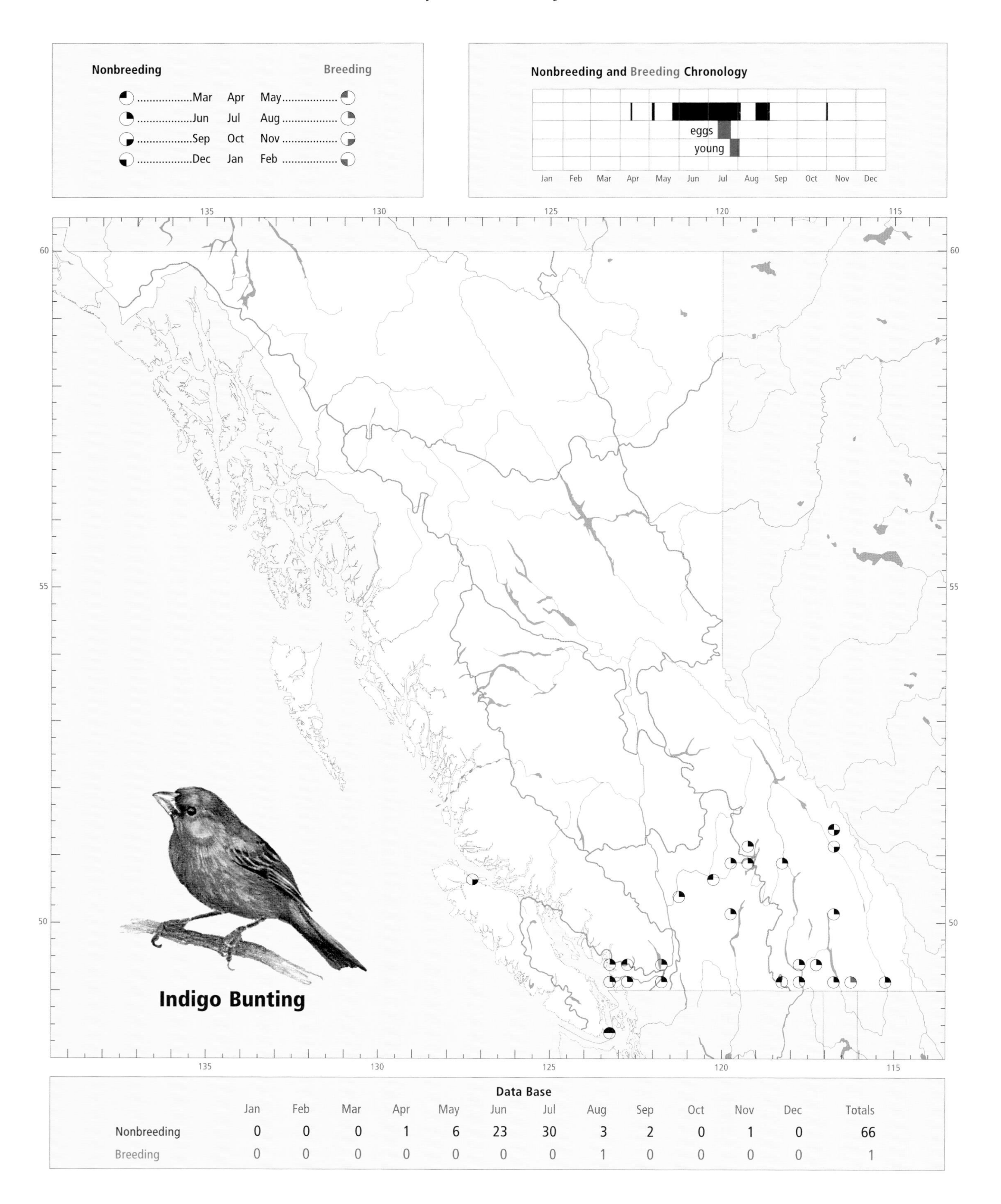

Data Base

	Jan	Feb	Mar	Apr	May	Jun	Jul	Aug	Sep	Oct	Nov	Dec	Totals
Nonbreeding	0	0	0	1	6	23	30	3	2	0	1	0	66
Breeding	0	0	0	0	0	0	0	1	0	0	0	0	1

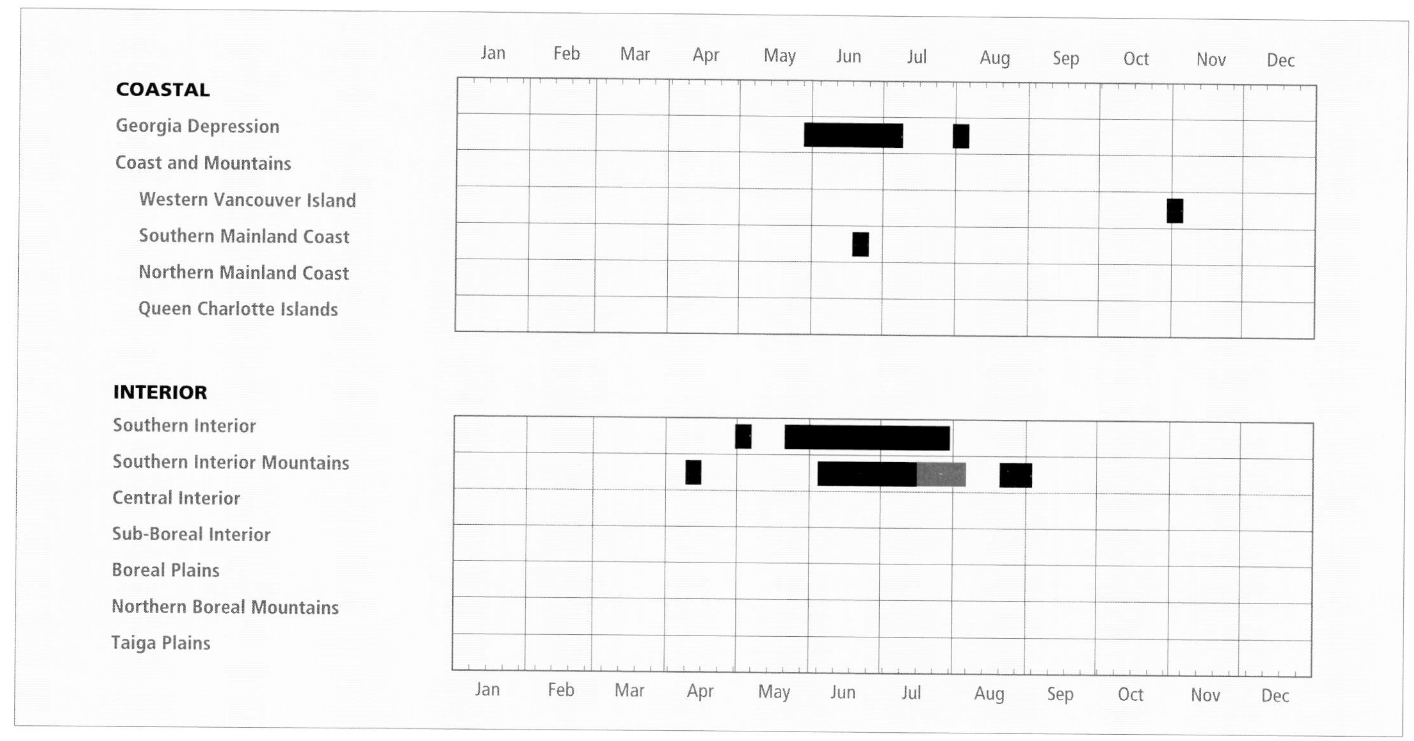

Figure 458. Annual occurrence (black) and breeding chronology (red) for the Indigo Bunting in ecoprovinces of British Columbia. Records are shown for the week in which they occurred.

Figure 459. Breeding habitat for the Indigo Bunting family of bobtailed young found along Highway 21 south of Creston included roadside patches of ocean-spray shrubs (3 July 1997; R. Wayne Campbell).

CHANGE IN STATUS: The Indigo Bunting (Fig. 455) was first reported in British Columbia in 1948, at Shuswap Lake, on the basis of a well-described sighting by D. and G.R. Ryder. The occurrence of this bunting was subsequently confirmed in the province in 1953 when a bird was collected at Port Hardy, on northern Vancouver Island (Young 1989). Another individual was found at Trail in 1958. Since the 1970s, 1 or 2 birds have appeared in 14 of the 29 years ending in 1997.

NONBREEDING: The Indigo Bunting is a rarity north of California and is considered an irregular straggler to southern British Columbia (Roberson 1980). On the south coast, reliable documentation has come from Port Hardy and Victoria on Vancouver Island, the Greater Vancouver area, and Agassiz in the Fraser River valley. In the interior, the Indigo Bunting has been found across southern portions of the province from Spences Bridge east to Roosville in the east Kootenay and north to Shuswap Lake and Golden.

The Indigo Bunting has been recorded in British Columbia from near sea level to 800 m in the interior. It has occurred in brushy and weedy habitats along the edges of cultivated lands, rangelands (Fig. 456), roads, landfills, riparian woods, old fields, second-growth woodlands (Fig. 457), and bird feeders. It has most often been found in associations of willow, birch, and saskatoon shrubs.

There is no regular movement of the Indigo Bunting into British Columbia. Most records are from June and July (Fig. 458). In Oregon, most records (n = 28) are from the last week in May and the first week of June (Gilligan et al. 1994), while in California most occurrences are of spring vagrants.

Most observations have been of singing males for only 1 or 2 days. The longest stay was of a singing male at South Slocan over a period of 27 days between 30 June and 26 July 1972 (Street and Merilees 1974).

On the coast, the Indigo Bunting has been recorded in most weeks mainly from 30 May to 2 July; in the interior, it has been recorded in most weeks mainly from 26 May to 20 August (Fig. 458).

BREEDING: The single known breeding record for the province was from near Creston in the Southern Interior Mountains. The breeding record on 2 August was of a brood of 3 bobtailed young being fed regularly by an adult female (and occasionally by the male) in an ocean-spray shrub (Fig. 459) along Highway 21, south of Creston.

The incubation period of the Indigo Bunting is 12 or 13 days (Holcomb 1966; Taber and Johnston 1968) and the nestling period is 9 to 12 days (Payne 1992). In British Columbia, calculated dates indicate that breeding may occur from at least 11 July to 2 August (Fig. 458).

Brown-headed Cowbird Parasitism: A single instance of cowbird parasitism has been noted in British Columbia, when a female was seen feeding a juvenile cowbird near Revelstoke (see NOTEWORTHY RECORDS). The Indigo Bunting is a common host of the Brown-headed Cowbird and is parasitized throughout its range in North America (Payne 1992).

REMARKS: The Indigo Bunting interbreeds with the Lazuli Bunting where ranges overlap in the Great Plains and in parts of North Dakota and Montana (Sibley and Short 1959; Kroodsma 1975). In British Columbia, there was 1 instance of a male Indigo × Lazuli Bunting seen with a female Lazuli Bunting and juveniles near Nakusp, in the Southern Interior Mountains (Siddle 1990c).

For a summary of the life history of the Indigo Bunting in North America, see Payne (1992).

POSTSCRIPT: On 2 and 3 August 1998, an adult male Indigo Bunting was present on Lochside Trail, north of Martindale Road in Central Saanich, on southern Vancouver Island (B. Begg pers. comm.).

NOTEWORTHY RECORDS

Spring: Coastal – Mount Tolmie (Victoria) 30 May 1992-1 male singing; Pitt Meadows 31 May 1966-1 male singing. **Interior** – Kamloops 26 May 1975-1 male (BC Photo 443); Golden 12 Apr 1995-1 male (BC Photo 1634).

Summer: Coastal – Seabird Island (Agassiz) 23 Jun 1968-1 male singing (Cannings 1974); Vancouver 4 Jun 1985-1 male (Harrington-Tweit and Mattocks 1985), 2 Jul 1977-1 adult along Marine Drive; North Vancouver 13 Jun to 2 Jul 1993-1 adult male at Premier Street landfill; Lions Bay (West Vancouver) 1 to 3 Jun 1973-1 male at feeder (BC Photo 312). **Interior** – Creston 20 Aug 1973-1 pair eating timothy seeds (Butler et al. 1986), 2 Aug 1993-pair feeding 3 bobtailed young; Summit Creek (Creston) 6 and 7 Jul 1984-1 male singing; Roosville 3 Jul 1977-1 male singing; Trail 15 Jun 1958-1 male singing; South Slocan 30 Jun to 26 Jul 1972-1 male singing (Street and Merilees 1974); Spences Bridge 20 Jun 1973-1 male (Cannings 1974); Shuswap Lake 4 Jun 1948-1 male singing from willow and saskatoon, 22 Jun 1948-1 male singing from willow and birch shrubs; Cooper Creek (Argenta) 1 Jul 1980-1 adult male singing; Magna Bay 7 Jul 1973-1 male singing (Cannings 1974; BC Photo 313); Scotch Creek 25 Jul 1989-a pair (Weber and Cannings 1990); Revelstoke 29 Jun 1997-adult female feeding juvenile Brown-headed Cowbird in willows, 16 Jul 1988-a pair.

Breeding Bird Surveys: Not recorded.

Autumn: Interior – Golden 1 Sep 1977-1 male (Rogers 1978). **Coastal** – Port Hardy 1 Nov 1953-1 male in first basic plumage (PMNH 72124; Young 1989).

Winter: No records.

Bobolink

Dolichonyx oryzivorus (Linnaeus)

BOBO

RANGE: Breeds from central and southern British Columbia, southern Alberta, southern Saskatchewan, southern Manitoba, central Ontario, southern Quebec, and the Maritime provinces south to eastern and south-central Washington, eastern Oregon, northeastern Nevada, northern Utah, eastern Colorado, Kansas, northern Missouri, central Illinois, central Indiana, southern Ohio, southern Pennsylvania, and central New Jersey to north-central Kentucky, northeastern Tennessee, western North Carolina, Virginia, and Maryland. Winters in southern South America (mostly east of the Andes) in eastern Bolivia, central Brazil, Paraguay, and northern Argentina.

STATUS: On the coast, a *very rare* to *rare* vagrant in the Georgia Depression Ecoprovince.

In the interior, *uncommon* to locally *fairly common* spring migrant and summer visitant to the Southern Interior and Southern Interior Mountains ecoprovinces; *uncommon* migrant and summer visitant to the Central Interior Ecoprovince; *rare* late spring migrant and local summer visitant in the Sub-Boreal Interior Ecoprovince; *casual* in the Boreal Plains Ecoprovince.

Breeds.

CHANGE IN STATUS: Munro and Cowan (1947) list the Bobolink (Fig. 460) as a "summer visitant, of irregular appearance, to the meadows in the Osoyoos-Arid and Dry Forest biotic areas [southern Okanagan valley], casual in the Puget Sound [Fraser] Lowlands." Since the 1940s, the Bobolink has increased in abundance and is more widely distributed throughout the southern interior of the province. Range expansion occurred with the clearing and dyking of land for agriculture, especially hay crops of timothy and alfalfa.

The Bobolink was first reported in British Columbia near Penticton in 1913 (Anderson 1914). Since then, the species has bred regularly in the Okanagan valley in "the few large hayfields along the valley bottom and occasionally in the outlying valleys" (Cannings et al. 1987). A survey conducted in 1994, however, did not find birds at traditional sites near Vaseux Lake, Swan Lake, or northern regions of the Okanagan valley (Van Damme 1999). In the east Kootenay, Johnstone (1949) reported the Bobolink as a scarce summer visitant as far north as Ta Ta Creek. Jobin (1952) first reported the Bobolink as far north as the Cariboo and Chilcotin areas, and by the late 1950s the species had extended its breeding range to 115 Mile House and Alkali Lake (Cariboo) in the Central Interior (Erskine and Stein 1964). Breeding birds were found at Horsefly by the mid-1970s. Today the northernmost limit of its breeding range is in the Quesnel area in the Sub-Boreal Interior. There is an isolated breeding record from Giscome, northeast of Prince George, in 1989. Although northward expansion has taken place into the Central Interior and southern Sub-Boreal Interior, the species still remains a rare and local summer visitant in those two ecoprovinces.

Figure 460. Each year, small numbers of Bobolinks make a round-trip of about 20,000 km to breed in southern portions of interior British Columbia. Males (a) arrive in the breeding areas first, followed about a week later by the females (b) (Anthony Mercieca).

NONBREEDING: The Bobolink occurs regularly in British Columbia only in the southern and central interior of the province, where it reaches the northernmost limit of its range in western North America. Vagrants, however, have been found in northern Alaska (Kessel and Gibson 1978). It is locally but widely distributed across the southern portions of the interior of the province from Princeton east to Sparwood in the east Kootenay. It is sparsely distributed north to Williams Lake in the Chilcotin-Cariboo Basin and to Giscome in the Nechako Lowland. There are several occurrence records for the Fort St. John area and Toms Lake in the Boreal Plains. On the coast, it is a vagrant in the lower Fraser River valley and southern Vancouver Island.

The Bobolink is a bird of lower elevations, occupying habitat within the valley bottoms; it has been recorded from

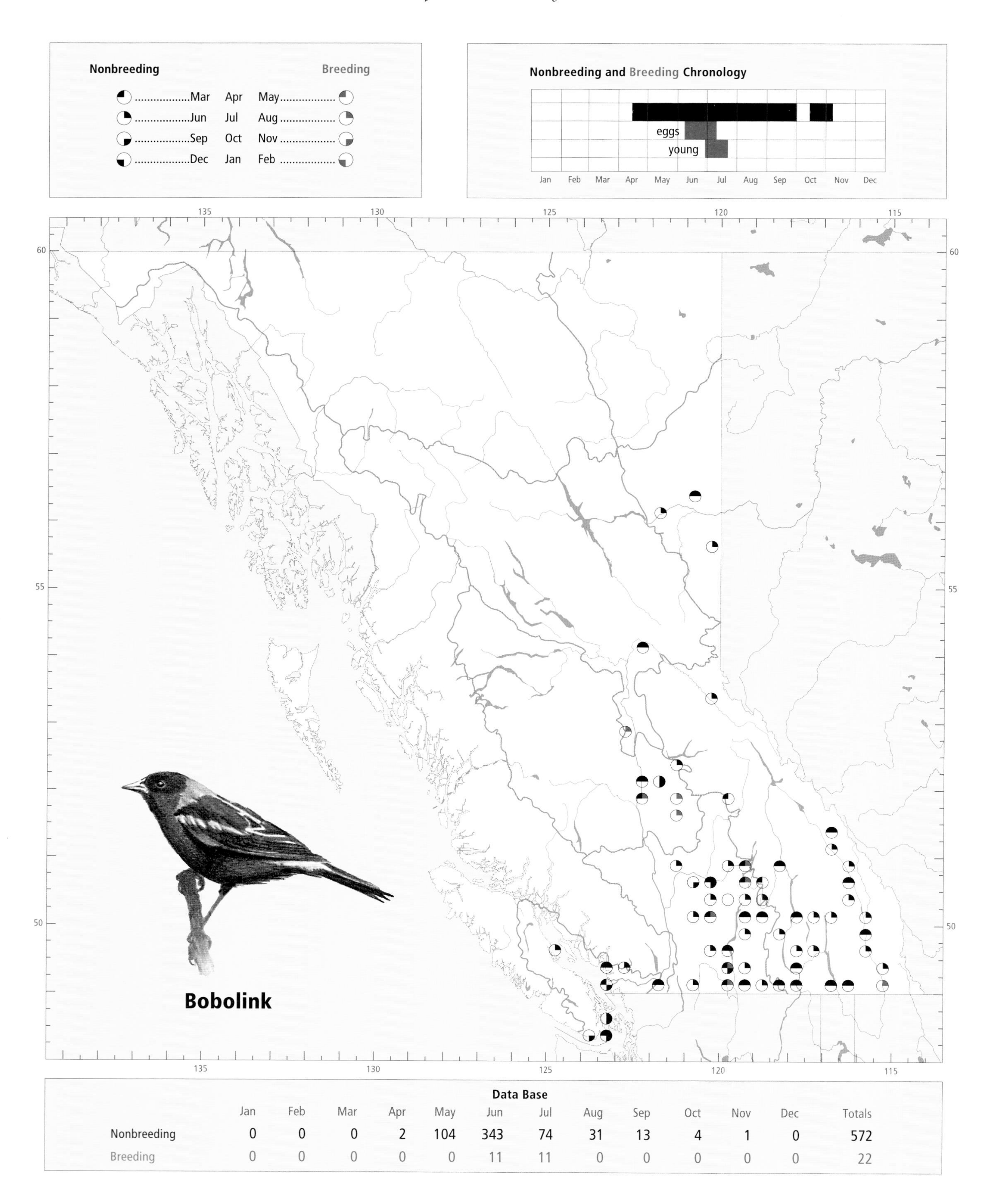

Data Base

	Jan	Feb	Mar	Apr	May	Jun	Jul	Aug	Sep	Oct	Nov	Dec	Totals
Nonbreeding	0	0	0	2	104	343	74	31	13	4	1	0	572
Breeding	0	0	0	0	0	11	11	0	0	0	0	0	22

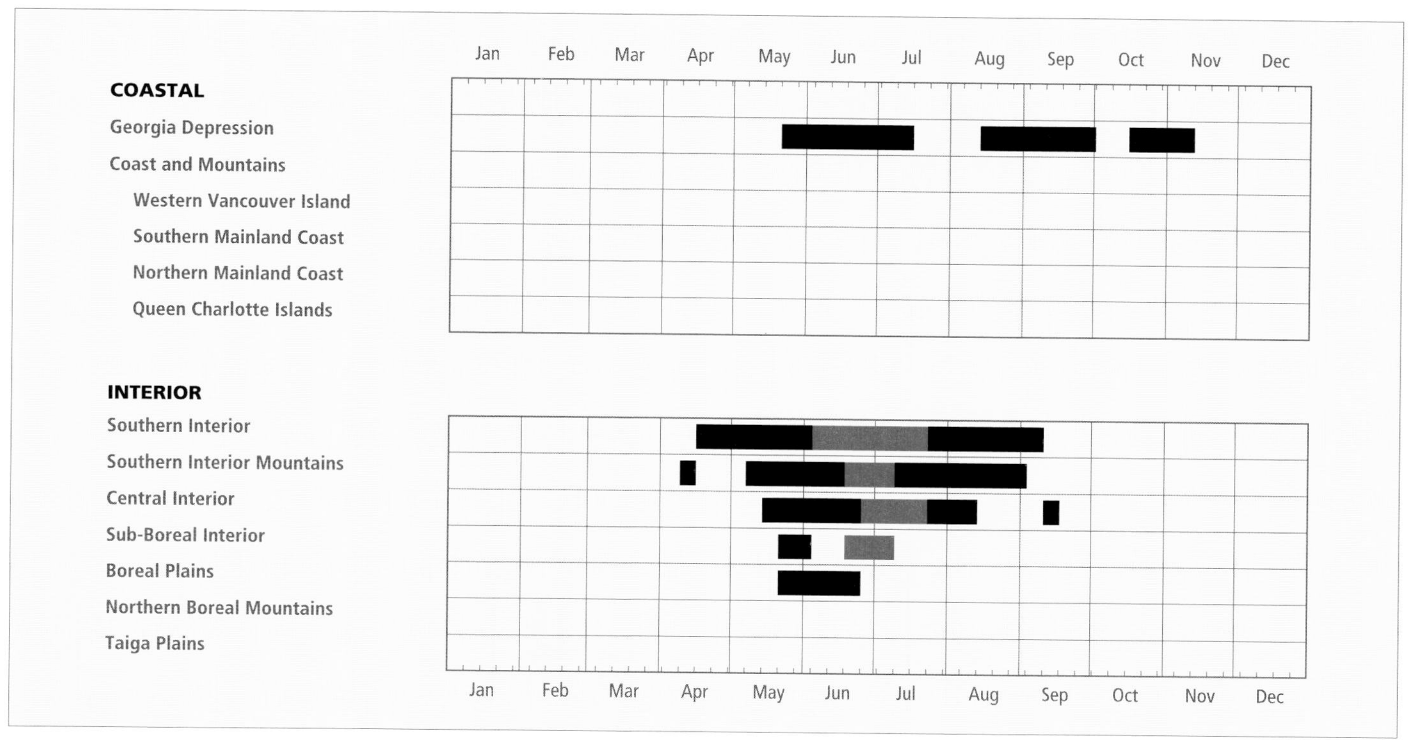

Figure 461. Annual occurrence (black) and breeding chronology (red) for the Bobolink in ecoprovinces of British Columbia. Records are shown for the week in which they occurred.

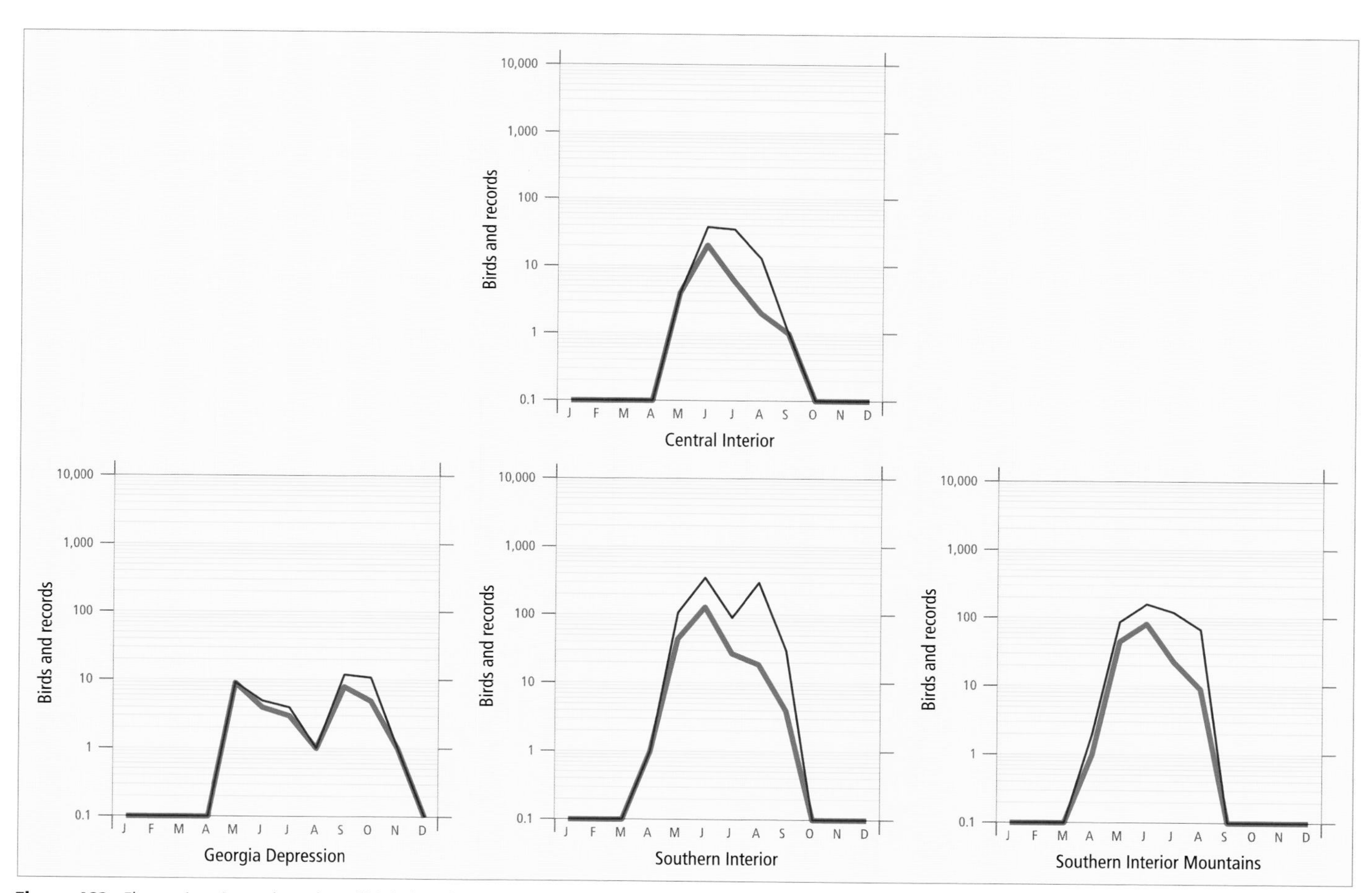

Figure 462. Fluctuations in total number of birds (purple line) and total number of records (green line) for the Bobolink in ecoprovinces of British Columbia. Breeding Bird Surveys and nest record data have been excluded.

near sea level to 120 m at the coast and between 280 and 1,000 m in the interior. It occurs mainly in human-influenced landscapes. See BREEDING for details on habitat.

On the coast, the Bobolink has occurred infrequently during the last 2 weeks of May on extreme southern Vancouver Island, in the Fraser River delta, and in scattered locations in the lower Fraser River valley of the Georgia Depression.

In the southern interior, spring migrants can arrive as early as mid-April, but the main movement occurs during the last 2 weeks of May and the first week of June (Figs. 461 and 462). In the southern Okanagan valley, the first spring migrants appear in the latter half of May, with males arriving up to a week before females (Cannings et al. 1987). This movement, during mid to late May and early June, occurs throughout the southern areas of the breeding range in the Kootenay, Creston, and Columbia valleys of the Southern Interior Mountains. Migrants arrive in the Thompson valley of the Southern Interior in the latter half of May through mid-June, and continue a northward movement into the Cariboo and Chilcotin areas of the Central Interior. It is probable that migrants occurring in the Peace Lowland arrive from east of the Rocky Mountains, from the Peace River area of Alberta (Semenchuk 1992).

In mid to late July, adults and fledged young congregate in loose post-breeding flocks, and by early August autumn migration is initiated (Figs. 461 and 462). In the central and southern regions of the interior, autumn migration takes place rapidly in August and few birds remain through September. On southern Vancouver Island, there are scattered records, mostly of single birds, from early September to late October, and a single late record for early November. Bobolink populations in western North America migrate eastward in late

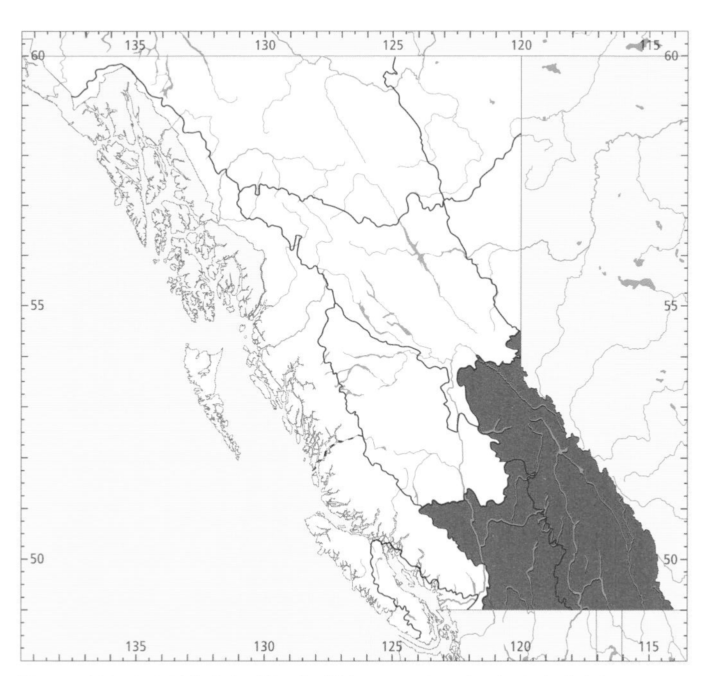

Figure 463. In British Columbia, the highest numbers for the Bobolink in summer occur in the Southern Interior and Southern Interior Mountains ecoprovinces.

Figure 464. In the Southern Interior Ecoprovince of British Columbia, the Bobolink breeds in hayfields (north end of Osoyoos Lake at Road 22, 6 July 1997; R. Wayne Campbell).

summer rather than southward; from August until mid-September, they form large concentrations along the Atlantic Ocean from New Jersey to Florida (Howell 1932; Stone 1937) before beginning their arduous journey to wintering grounds in South America. The Bobolink's annual migratory cycle covers approximately 20,000 km, which surpasses the distance for all other Western Hemisphere passerines (Hamilton 1962).

On the coast, the Bobolink has been recorded infrequently from 21 May to 7 November; in the interior, it has been recorded regularly from mid-May to late August (Fig. 461).

BREEDING: The Bobolink has a localized breeding distribution in suitable open habitat in the southern one-third of the interior. It breeds from Osoyoos Lake and Waldo Meadows in the south, north through the central and northern Okanagan to Williams Lake. Further north, there are 2 known nesting locations: Dragon Lake near Quesnel and likely at Giscome northeast of Prince George in the southern region of the Sub-Boreal Interior. The locations of major known and suspected breeding sites for the Bobolink in British Columbia are listed in Table 8.

This species reaches its highest numbers in summer in the Okanagan valley in the Southern Interior and in the Kootenay valley in the Southern Interior Mountains (Fig. 463). An analysis of Breeding Bird Surveys in British Columbia for the period 1968 through 1993 could not detect a net change in numbers. Breeding Bird Surveys in Canada show a decline of 1.7% ($P < 0.01$) from 1966 to 1996, and a decline of 5.5% ($P < 0.01$) from 1980 to 1996. Surveys from across North

Table 8. Total Bobolink counts and sex ratios in the major known and suspected breeding sites surveyed in British Columbia in 1994 (modified from Van Damme 1999).

Location	Total birds	Sex ratio (male:female)[1]
Southern Interior		
North end of Osoyoos Lake (Fig. 571)	48	2.0:1
Grand Forks	62	5.2:1
Southern Interior Mountains		
Creston	59	3.9:1
Castlegar	9	3.5:1
Brouse Loop (Fig. 573)	19	5.3:1
Edgewood	25	3.2:1
Central Interior		
150 Mile House (Fig. 572)	6	6.0:0
Total	228	3.4:1

[1] Female Bobolinks are more difficult to see than the conspicuous males, particularly when the latter are singing or displaying. In most well-studied populations, sex ratios are equal or slightly skewed towards females (see Martin and Gavin 1995).

America show declines of 1.6% ($P < 0.01$) and 3.8% ($P < 0.01$) over the same time periods (Sauer et al. 1997).

Breeding occurs at elevations from 280 to 950 m. Throughout its breeding range, the Bobolink is closely associated with open, contiguous habitats (Figs. 464, 465, and 466), primarily fields where forage crops of timothy and alfalfa are cultivated. Other habitats include pastures, weedy fields, and moist meadows, but the key habitat component is a mixture of grasses and broad-leaved forbs (Fig. 464). In a population survey conducted in 1994, a mixture of plant species such as sedges, horsetails, tall grasses, buttercups, daisies, red and white clovers, thistles, common dandelions, American vetch, and either an alfalfa or timothy hay crop were common to all sites (Van Damme 1999).

In the interior, the Bobolink has been recorded breeding from 8 June to 22 July (Fig. 461).

Nests: Nests were located on the ground (100%; $n = 11$) in hayfields, meadows, and open tall-grass fields. Two nests were each located in a ground depression or hollow. A single nest was reported to consist of dry grasses. In Oregon, Wittenburger (1978) describes the nest as a loosely constructed, well-concealed cup in a ground hollow, frequently at the base of tall weedy plants.

Eggs: Dates for 11 clutches ranged from 8 June to 10 July. Sizes of 11 clutches ranged from 2 to 6 eggs (2E-1, 4E-5, 5E-2, 6E-3). The incubation period is 11 to 13 days (Martin 1971; Baicich and Harrison 1997).

Young: Dates for 5 broods ranged from 29 June to 22 July. Calculated dates indicate that young can be found as early as 20 June. Brood size ranged from 2 to 6 young. The nestling period is 10 to 14 days, but young may leave the nest before they are able to fly (Baicich and Harrison 1997).

Figure 465. In the Central Interior Ecoprovince of British Columbia, the Bobolink prefers pasture fields with a mixture of grasses and broad-leaved forbs (150 Mile House, 12 July 1996; R. Wayne Campbell).

Figure 466. Open, contiguous habitats that have been altered by humans are the preferred breeding habitat of the Bobolink in the Southern Interior Mountains Ecoprovince (Brouse, north of Nakusp, 10 August 1996; R. Wayne Campbell).

Brown-headed Cowbird Parasitism: In British Columbia, there were no cases of cowbird parasitism in 11 nests found with eggs or young. The Bobolink is an infrequent host, with only 33 cases scattered throughout its North American range (Friedmann 1963). Bobolinks routinely chase or attack Brown-headed Cowbirds entering their territory (Martin and Gavin 1995).

Nest Success: Insufficient data.

Figure 467. A decline in Bobolink populations in North America has been associated with intensively farmed areas. In British Columbia, the time of hay cutting can seriously affect the success of fledging (north end of Osoyoos Lake at Road 22, 11 July 1998; R. Wayne Campbell).

REMARKS: A record for Triangle Island on 22 June 1977 lacks sufficient documentation and has been excluded from this account.

Other potential breeding sites in British Columbia where Bobolinks have been observed on territory include Meadow Creek, Castlegar, Edgewood, Ta Ta Creek, Cherryville, Lavington, Douglas Lake, Nicola Lake, Nicola River, Kamloops, Parson, and 150 Mile House.

The arrival dates of Bobolinks in British Columbia are earlier than in other areas in North America. T.A. Gavin (pers. comm.) suggests that Bobolinks in the province may not be moving north via the same route as birds in the eastern United States. There is a "population" of Bobolinks that winters west of the Andes Mountains in South America (Martin and Gavin 1995). Almost nothing is known about this, but the early spring arrival dates for British Columbia suggest that Bobolinks are coming from a different area of South America than eastern birds, perhaps west of the Andes.

There has been a significant decline in Bobolink populations in North America, with the greatest declines occurring in the central part of the range and apparently associated with the most densely populated (by humans) and intensively farmed regions (S. Droege pers. comm.). The decline may not have occurred in British Columbia. Major threats to Bobolinks in the province are habitat loss and changing agricultural practices. In the southern Okanagan valley, grassland habitats have disappeared because of increasing pressure from residential and agricultural developments. More than 90% of the land in the south Okanagan and lower Similkameen valleys has been altered from its original state (Redpath 1990). The shift to alfalfa-dominated fields (with faster regrowth) and the mechanization of modern hay-cropping equipment have resulted in earlier and repeated mowing, with a negative impact on the reproductive success of the Bobolink (Bollinger 1988). The hot, dry climate of the south Okanagan is conducive to this early mowing, and during the 1994 survey some hayfields were cut as early as 3 June. With cooler, wetter climates in other regions of the southern interior, the first hay cutting is delayed until late June and early July, which still poses a threat to young that have not yet fledged (Fig. 467).

For population monitoring and management recommendations, see Bollinger et al. (1988, 1990), Dale (1993), and Johnson and Temple (1990). For additional information on the breeding behaviour and life history of the Bobolink, see Gavin and Bollinger (1985), Bollinger and Gavin (1989, 1992), Martin and Gavin (1995), and Wittenburger (1978).

NOTEWORTHY RECORDS

Spring: Coastal – Victoria 21 May 1958-1; Vancouver 24 May 1982-1. **Interior** – Osoyoos Lake 27 Apr 1977-1; Oliver 21 Apr 1977-1; White Lake (Okanagan Falls) 6 May 1978-1; Trail 15 Apr 1986-2 at Columbia Gardens; Creston 18 May 1985-1; Douglas Lake 29 May 1979-1; Wasa 21 May 1977-1; Edgewood 21 May 1982-10; Kamloops 18 May 1992-3; Brouse 26 May 1985-10; Yoho National Park 31 May 1976-4 (Wade 1977); Williams Lake 24 May 1979-1 at Scout Island; Giscome 27 May 1995-3; Willow River 22 May 1976-2; 8 km s North Pine 23 May 1976-2.

Summer: Coastal – Kent 21 Jun 1976-2; Courtenay 15 Jul 1990-2 (Siddle 1990c). **Interior** – Osoyoos Lake 19 Aug 1978-100 at n end of lake; Keremeos 5 Jul 1973-50, all males; Okanagan River 8 Jun 1958-4 eggs; Nicks Island (Creston) 15 Jul 1981-20, mostly males; Waldo 6 Jul 1941-4 nestlings; Ta Ta Creek 3 Jun 1948-4 (Johnstone 1949); Monck Park 7 Jun 1982-10 along road into park; Nicola Lake 2 Jun 1975-4, at n end of lake; Penticton 26 Jun 1913-10 collected (Anderson 1914); Okanagan Landing 27 Aug 1943-7; Brouse 12 Jun 1977-7, 11 Aug 1989-40 (Weber and Cannings 1990); Nicola Lake 1 Jun 1981-1 male; Meadow Creek 27 Jun 1983-8; Lumby 1 Jun 1972-6, singing; Enderby 25 Jun 1953-6 eggs; Celista 19 Jun 1948-4 eggs, 22 Jul 1960-2 nestlings; Parson 15 Jun 1974-1 male in field; Revelstoke 28 Aug 1986-1 at airport; 115 Mile House 12 Jul 1958-3 recently fledged young fed by adults (Erskine and Stein 1964); Alkali Lake (Cariboo) 2 Jun 1981-3 in grasslands, 12 Jul 1972-3 nestlings; Lac la Hache 16 Jul 1958-3 nestlings; Williams Lake 10 Jul 1988-20 in meadows at e end of lake; Giscome 3 Jun 1996-6, 1 Jul 1989-3 in hay field; Toms Lake 18 Jun 1995-1 male singing (Phinney 1998); 5 km s North Pine 17 Jun 1979-1 in hay field.

Breeding Bird Surveys: Coastal – Not recorded. **Interior** – Recorded from 17 of 73 routes and on 18% of all surveys. Maxima: Grand Forks 28 Jun 1987-30; Mabel Lake 30 Jun 1973-13; Spillimacheen 21 Jun 1977-7.

Autumn: Interior – 150 Mile House 10 Sep 1953-1 (RBCM 15840); Tranquille 5 Sep 1983-25 in hay fields; e Kamloops 1 Sep 1960-3; Kamloops 4 Sep 1986-1 (Rogers 1987); Vaseux Lake 7 Sep 1971-1 at s end of lake. **Coastal** – Delta 10 Sep 1994-1; Saanich 17 Sep 1983-1, 18 Sep 1974-4 in transition plumage, 23 Sep 1979-1, 16 Oct 1979-7 feeding in weedy field, 20 Oct 1982-1, 28 Oct 1967-1, 7 Nov 1981-1 killed and dropped by Northern Shrike (RBCM 23198); Rocky Point (Victoria) 22 Sep 1995-1.

Winter: No records.

Red-winged Blackbird
Agelaius phoeniceus (Linnaeus)

RWBL

Figure 468. In wetlands of southern British Columbia, adult male Red-winged Blackbirds establish territories in late February and March (Myers Lake, 5 June 1997; R. Wayne Campbell).

Figure 469. Adult female Red-winged Blackbird (Myers Lake, 5 June 1997; R. Wayne Campbell).

RANGE: Breeds from southern Alaska, southern Yukon, west-central and southern Mackenzie, northern Saskatchewan, central Manitoba, northern Ontario, and southern Quebec, and east to include New Brunswick, Prince Edward Island, Nova Scotia, and southwestern Newfoundland, south in the west to northern Baja California, and along both coasts of Mexico and Central America as far as Nicaragua and Costa Rica; in eastern North America south to the Gulf coast in Texas and Florida. Winters from southern British Columbia, Idaho, Montana, North Dakota, southern Minnesota, the southern Great Lakes region, southern Ontario, and New England south throughout the remainder of the breeding range.

STATUS: On the coast, a *very common* to *very abundant* resident in the Georgia Depression Ecoprovince; in the Coast and Mountains Ecoprovince, a *rare* migrant and *very rare* summer and winter visitant on Western Vancouver Island and the Southern Mainland Coast; *very rare* locally throughout the year on the Northern Mainland Coast and Queen Charlotte Islands.

In the interior, a *common* to *very common* migrant and summer visitant and irregular, but locally *uncommon* to *common*, in winter in the Southern Interior Ecoprovince; *common* to *very common* migrant and summer visitant and *uncommon* locally in winter in the Southern Interior Mountains and Central Interior ecoprovinces; *uncommon* migrant and summer visitant and *very rare* in winter in the Sub-Boreal Interior Ecoprovince; *uncommon* to *fairly common* migrant and summer visitant in the Boreal Plains Ecoprovince; and *rare* to *uncommon* migrant and local summer visitant in the Taiga Plains and Northern Boreal Mountains ecoprovinces.

Breeds.

CHANGE IN STATUS: By the mid-1940s, there had been only 2 occurrences of the Red-winged Blackbird (Figs. 468 and 469) north of the Boreal Plains: a specimen from Atlin (Swarth 1936) and another from the Fort Nelson River (Williams 1933b). Since then, the species has extended its range further into the Taiga Plains (Campbell and McNall 1982) and west along the Alaska Highway, where there are nesting records for Coal River and the lower Tatshenshini River valley. Elsewhere, it has slowly become established locally in the Sub-Boreal Interior and Northern Boreal Mountains, where small numbers have been recorded at 35 different locations. The largest numbers in any record have been 20 birds at Mile 419 of the Alaska Highway, 12 at Sawmill Lake (near Telegraph Creek), and 9 at the lower Tatshenshini River valley. The latter small colony appeared to be nesting there. The Red-winged Blackbird, however, has not found much suitable habitat for colony foundation in the largely mountainous terrain existing in the far north of the province.

The appearance of this species in the northern one-third of British Columbia west of the Rocky Mountains seems to have occurred recently. About 94% of our records for the region have been reported since 1972, with a concentration between 1971 and 1981. In a part of the province where biological exploration has been sparse and infrequent, it is impossible to separate in our data the part played by the tempo of exploration from evidence of birds entering an area not previously occupied by them. With few exceptions, the new regional records have been obtained by summer field parties rather than resident observers. The absence of the species before the 1940s, however, appears to be well established by the extensive studies made by Swarth (1922) along the Stikine River

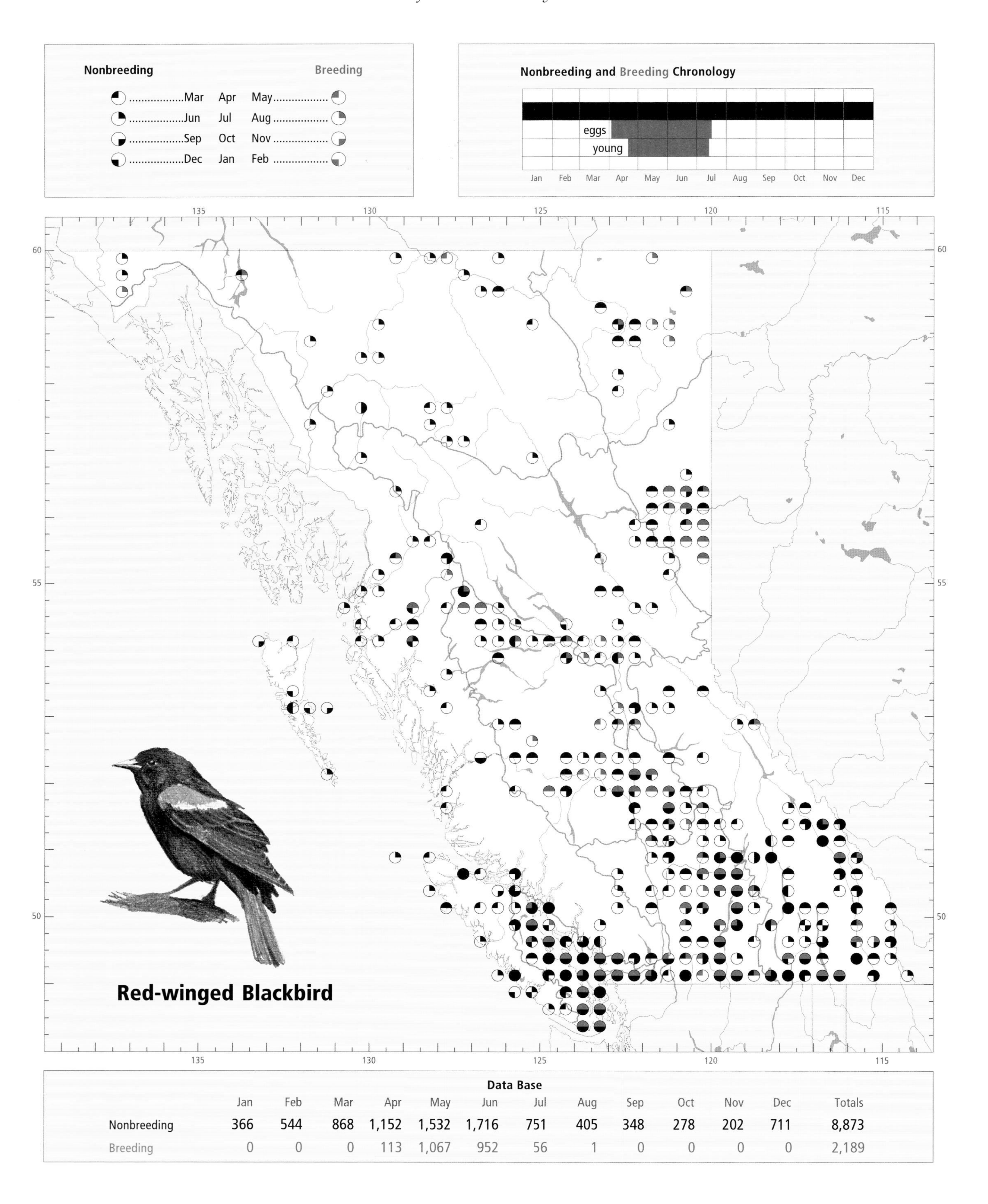

Data Base

	Jan	Feb	Mar	Apr	May	Jun	Jul	Aug	Sep	Oct	Nov	Dec	Totals
Nonbreeding	366	544	868	1,152	1,532	1,716	751	405	348	278	202	711	8,873
Breeding	0	0	0	113	1,067	952	56	1	0	0	0	0	2,189

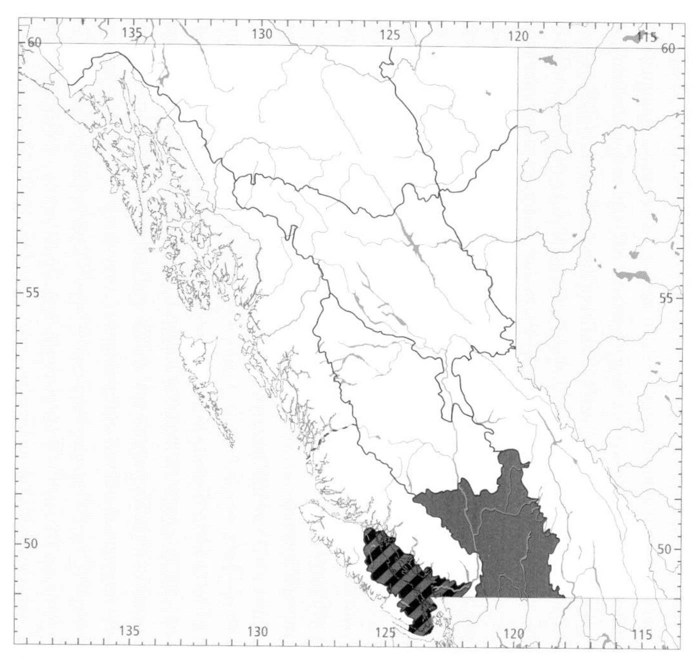

Figure 470. In British Columbia, the highest numbers for the Red-winged Blackbird in winter (black) occur in the Georgia Depression Ecoprovince; the highest numbers in summer (red) occur in the Georgia Depression and Southern Interior ecoprovinces.

valley and Rand (1944) along the Alaska Highway between Dawson Creek and Watson Lake in the Yukon. The Rusty Blackbird was the only icterid either of them encountered. Erskine and Davidson (1976) provide a brief summary of other bird work in the Alaska Highway area from the early 1950s to the early 1970s. They refer to the presence of the Red-winged Blackbird "mainly at Parker Lake," near Fort Nelson.

NONBREEDING: The Red-winged Blackbird is distributed across much of British Columbia, usually in close association with low-elevation wetlands. In the southern half of the province, it occurs from Vancouver Island east to the Rocky Mountains and regularly north through the central interior to the Peace Lowland. Further north, it is widespread but its distribution is more local and sparse. It is a rare sight on the Queen Charlotte Islands. The Red-winged Blackbird is one of a select group of species that has been recorded in all 9 terrestrial ecoprovinces. The highest numbers in winter occur in the Fraser Lowland of the Georgia Depression (Fig. 470).

The Red-winged Blackbird is found mainly at lower elevations but has been reported from sea level on the coast to 1,200 m elevation in the interior. It avoids heavily timbered regions. It is most abundant around small wetlands supporting an emergent vegetation of rushes, sedges, cattails, shrubs,

Figure 471. On the Fraser River delta of coastal British Columbia, large flocks of Red-winged Blackbirds, often mixed with European Starlings and Brewer's Blackbirds, forage in agricultural fields during the winter (Reifel Island, 2 December 1995; R. Wayne Campbell).

Figure 472. Throughout autumn and winter on the south coast, flocks of Red-winged Blackbirds gather in favourite pre-roost sites in the late afternoon before departing to cattail marshes to spend the evening (Westham Island, 7 November 1992; R. Wayne Campbell).

Figure 473. In late February, many male Red-winged Blackbirds are already on territory at their breeding grounds in the Okanagan valley (14 km south of Oliver, 23 February 1998; R. Wayne Campbell).

and tall grasses on a substrate rich in organic materials. It is also a familiar sight around human habitations (Fig. 472). On southeastern Vancouver Island and the Fraser River estuary, this blackbird frequents the shrubby borders of ponds, sloughs, lagoons, dykes, agricultural fields (Figs. 471), patches of dead cattail, and small trees adjacent to cattle feedlots, farmsteads, and rough undeveloped land. It also forages in parks and gardens, stubble fields, upper beaches, and small islands, and frequently visits bird feeding stations. In winter, the blackbird's habitat is strongly influenced by human presence and activity, but spring sees a gradual movement away from such surroundings to more natural environments (Fig. 473). During spring on the coast, most blackbirds are in the vegetation surrounding freshwater and brackish marshes, sloughs, lagoons, ponds, backwaters, and dykes.

In the southern and central portions of the interior of the province, the small number of wintering birds generally survive by using such anthropogenic resources as cattle feed lots, garbage dumps, sewage lagoons, stubble fields, unharvested grain fields, and bird feeding stations. With the advance of spring, the flocks arriving in the Okanagan, Thompson, Columbia, and Kootenay river valleys and the Cariboo and Chilcotin areas move through field edges, shrubs and trees around sewage lagoons, suburban areas of mixed vegetation, and stream courses lined with riparian willows and black cottonwoods.

Further north, in the Peace Lowland, the Red-winged Blackbird utilizes wet fields, pipeline rights-of-way, airfields, roadside sloughs, and the more restricted habitat provided by tamarack and spruce ponds, muskegs, and beaver ponds. The autumn migrants make more use of stubble fields, irrigation ponds, and grasslands bordering the host of small lakes, ponds, and sloughs that had served a summer population.

The Red-winged Blackbird is a resident in southern coastal areas of the province (Figs. 474 and 475). There is little indication of a spring migration into the Georgia Depression. Wintering numbers there, however, begin to decline in February and March as some of the population likely moves north. Elsewhere on the coast, small numbers of blackbirds arrive in early March on the Southern Mainland Coast and, except for those occupying the lowland near Hope, the Pitt River, and the Pemberton valley, they appear to be birds in transit, as numbers decline from then into early summer. On Western Vancouver Island and the Northern Mainland Coast, there is only slight evidence of a spring migration (Fig. 474).

In southern portions of the Southern Interior and Southern Interior Mountains, there is a gradual increase in the numbers of Red-winged Blackbirds through February and March, with the largest numbers recorded in the last week of March and the first week of April (Fig. 475). In the Central Interior, the spring movement is rapid, although somewhat obscured by the presence of wintering birds. It is first apparent in early

March and peaks in late March. Numbers decline through April into May as some birds move further north and local breeding begins. The first birds to appear in the Prince George region of the Sub-Boreal Interior sometimes reach there as early as late February and early March, and the small numbers entering the southern parts of that ecoprovince appear to build from late March through April (Fig. 475). In contrast to the relatively subdued tempo of the spring migration west of the Rocky Mountains, the arrival of spring migrants in the Peace Lowland of the Boreal Plains is dramatic. On a date in late April or early May, dawn finds large flocks of newly arrived blackbirds, entirely of males, noisily occupying the leafless trees bordering the still ice-bound lakes and ponds. A week or so later the females arrive, also in same-sex flocks, and activity becomes frenzied. By mid-May the population is at breeding density. The spring arrival in the Taiga Plains and the Northern Boreal Mountains is similar to that in the Boreal Plains but roughly 1 to 2 weeks later (Figs. 474 and 475).

The autumn migration in the far northern areas of British Columbia occurs mainly in late July and early August, with a few stragglers in the Northern Boreal Mountains, Taiga Plains, Boreal Plains, and even the Nechako Lowland adjacent to Prince George in late August and early September (Figs. 474 and 475). In the Central Interior, the number of records declines steadily beginning in June, but total numbers rise with the appearance of the post-breeding flocks with their young of the year. There is a sharp decline in numbers from August through November as most birds depart, leaving the wintering birds. Red-winged Blackbird numbers in the east and west Kootenays and in the Okanagan and Thompson river valleys also experience gradual declines from a high in August to only the wintering population in November.

On the coast, scattered individuals have been recorded at Bella Coola, on the Queen Charlotte Islands, and along the central and southern coastal regions, but there is no obvious coastal migration. Most birds have left the north coast by the end of August; occasionally a few birds overwinter. In the Georgia Depression, a gradual increase in numbers from July to October may reflect the input of young raised locally but may also include an influx of birds from outside. In all areas, the tendency of the post-breeding blackbirds to aggregate in flocks can lead to an apparent increase in numbers locally. Numbers begin to decline by the end of October and reach wintering levels by the end of November.

The Red-winged Blackbird is present in southern coastal areas and southern and central portions of the interior throughout the year. In the Boreal Plains, it occurs regularly from mid-April through August; further north, it occurs regularly from May through July (Fig. 474).

BREEDING: The Red-winged Blackbird has a widespread breeding distribution across the southern half of the province, from the coastal lowlands of southeastern Vancouver Island to the Flathead valley in the Rocky Mountains. The species has been recorded breeding in every ecoprovince.

Along the coast, the breeding range is almost confined to the Georgia Depression and a few small areas of the Southern

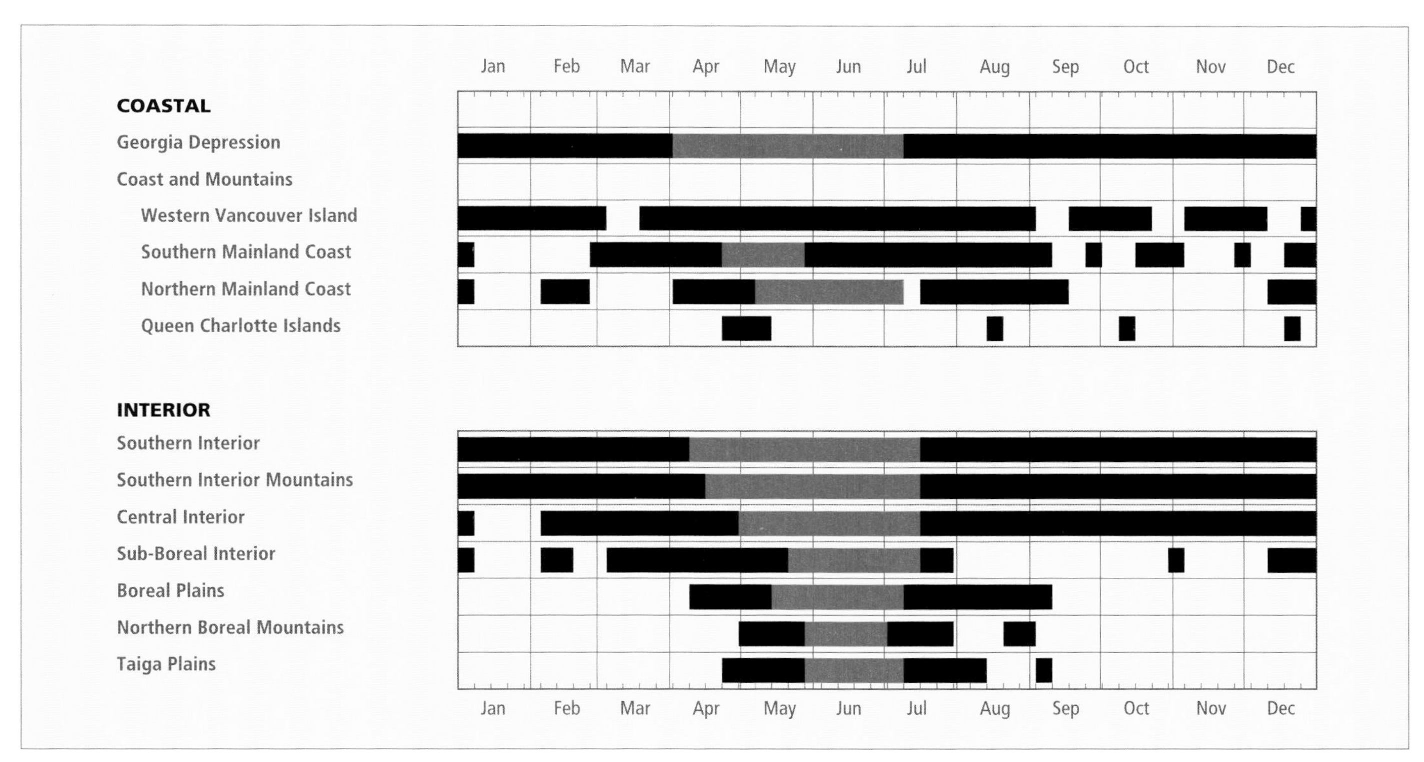

Figure 474. Annual occurrence (black) and breeding chronology (red) for the Red-winged Blackbird in ecoprovinces of British Columbia. Records are shown for the week in which they occurred.

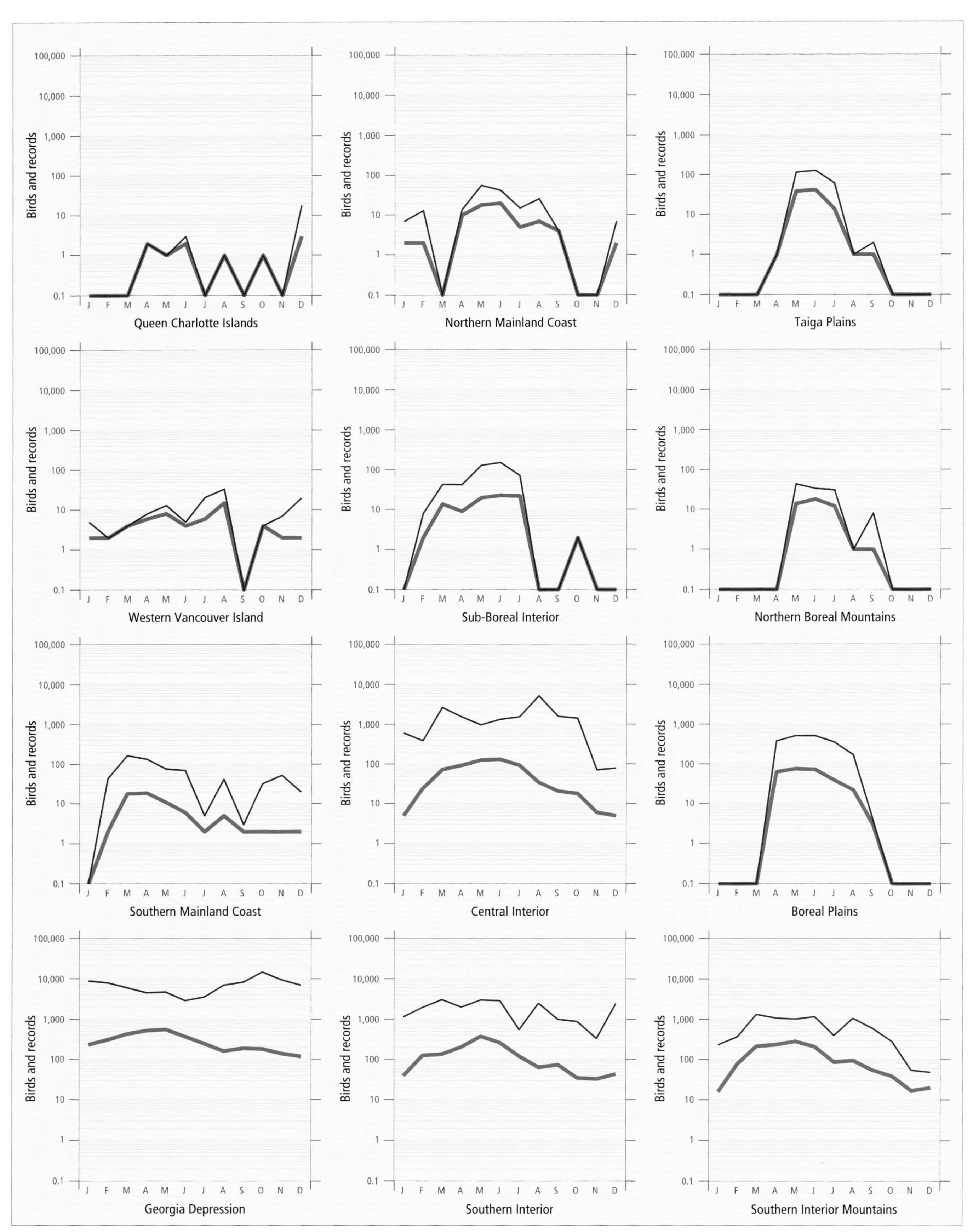

Figure 475. Fluctuations in total number of birds (purple line) and total number of records (green line) for the Red-winged Blackbird in ecoprovinces of British Columbia. Christmas Bird Counts, Breeding Bird Surveys, and nest record data have been excluded.

Figure 476. Throughout its breeding range in British Columbia, the Red-winged Blackbird prefers to build its nest in dense stands of cattails (Richter Pass, 1 May 1994; R. Wayne Campbell).

Mainland Coast contiguous to the Fraser Lowland. There are no breeding records from Western Vancouver Island and the Queen Charlotte Islands.

East of the Coast Mountains, the Red-winged Blackbird breeds from Manning Park in the west to Kootenay National Park in the east, and north through the lowland valleys of the Southern Interior and the extensive parklands of the Central Interior to the valleys of the Nechako and Skeena rivers. The northernmost nesting localities in the Southern Interior Mountains are in the vicinity of Yellowhead Pass. All known nesting areas in the Sub-Boreal Interior are in a small area close to Prince George. West of the Rocky Mountains, there is just 1 record of the species nesting between latitudes 55° and 59°N: at the mouth of the Kitsault River. This is a vast area with few observers reporting. In addition, there are few known nesting localities in the Northern Boreal Mountains: 2 in the Liard River valley, 1 in the lower Tatshenshini River valley, and 1 near Atlin. In northeastern British Columbia, the Red-winged Blackbird is a regular breeding species in the Peace Lowland but is less frequent in the Fort Nelson Lowland, where it occurs north to Parker Lake near Fort Nelson, and Kwokullie Lake.

The Red-winged Blackbird reaches its highest numbers in summer in the Georgia Depression and Southern Interior (Fig. 470). An analysis of Breeding Bird Surveys in British Columbia for the period 1968 through 1993 could not detect a net change in numbers on either coastal or interior routes. Throughout North America, Breeding Bird Surveys between 1966 and 1996 show a decline in population at an average annual rate of 1% ($P < 0.01$); across Canada during the same period, the decline was 0.9% per year ($P < 0.01$) (Sauer et al. 1997).

Figure 477. Many important nesting wetlands for the Red-winged Blackbird in the Central Interior Ecoprovince of British Columbia have been protected as ecological reserves, parks, and wildlife management areas, or locally as nature sanctuaries (north end of Williams Lake, 27 June 1997; R. Wayne Campbell).

Figure 478. Smaller ponds with emergent vegetation such as cattails are used by Red-winged Blackbirds for nesting throughout the Boreal Plains Ecoprovince of British Columbia (near Swan Lake, Tupper, 12 June 1996; R. Wayne Campbell).

On the coast, the Red-winged Blackbird nests from intertidal shores at sea level to 160 m elevation; in the interior, it breeds at elevations from 280 to 1,000 m.

In British Columbia, most Red-winged Blackbirds (92%; *n* = 1,794) breed in wetlands (Fig. 476) such as marshes (Fig. 477), sloughs, lake edges, swamps, and beaver ponds where cattails (Fig. 476), bulrushes, sedges, or tall grasses grow in shallow water. A variety of perennial plants and shrubs was also present. By midsummer there is usually an abundance of secondary aquatic plants in the nesting wetlands – pondweed, water-milfoil, arrowhead, buckbean, patches of water-smartweed – with attendant invertebrates that provide foraging opportunities for the blackbirds. Bordering shrubs are usually dominated by willow but often include hardhack and sweet gale. In all ecoprovinces, a few blackbirds (4%; *n* = 1,794) use upland habitat, including shrubbery and deciduous trees along forest edges; another 2% select their nesting sites in human-influenced habitat, including ditches, dykes, field borders, and orchards.

Towards the northern edge of the species' range, it uses smaller ponds (Fig. 478) and even muskeg habitats in the predominantly spruce-dominated forests of the Boreal White and Black Spruce Zone, where the emergent cattails and rushes may be absent.

Throughout its range, the Red-winged Blackbird nests in groups and loose colonies (see "Colonies").

In the southern interior of the province, the Red-winged Blackbird often has to compete for nesting territory with the larger Yellow-headed Blackbird. There are many documented cases of the later-arriving but larger Yellow-headed Blackbird displacing the Red-winged Blackbird from nesting territory.

Studies in the Cariboo and Chilcotin areas of British Columbia and in eastern Washington (Orians 1985) have revealed that where marshes are surrounded by open grassland, the Yellow-headed Blackbirds prevail, but where woodlots border the marshes, the Red-winged Blackbirds were able to retain the nesting territories in the parts of the marsh closest to the trees. Orians (1985) found that "where the angle that the tree tops projected to the marshes was less than 30° the Yellow-headed Blackbirds were present, but if trees projected an angle greater than 30°, the Yellow-headed Blackbird was absent" and the Red-winged Blackbird nested. This factor, among others, frequently leads to the Red-winged Blackbird nesting in the cattail stands growing in the shallower water nearer the shoreline trees (Campbell 2000), while the Yellow-headed Blackbird occupies the bulrush stands growing in the deeper water, further from shore. This is not a plant-oriented separation; in some marshes, the Red-winged Blackbird was confined to cattails but the Yellow-headed Blackbird used both species.

The Marsh Wren is another species that uses many of the same wetlands sought by the Red-winged and Yellow-headed blackbirds. It is an important – often the most important – predator on the eggs and young of the Red-winged Blackbird (Picman 1980a, 1980b). Tension between the 2 species sometimes results in the blackbirds nesting in areas of the marsh with more widely spaced clumps of cattail, where defence against the wren is more effectual, and the wren nesting in the denser vegetation, where it can better escape the attacks of the blackbirds (Picman et al. 1988).

On the coast, the Red-winged Blackbird has been recorded breeding from 3 April to 6 July; in the interior, it has been recorded breeding from 15 April (calculated) to 13 July (Fig. 474).

Colonies: Throughout its nesting range in North America, the Red-winged Blackbird almost always nests in colonies (Nero 1984; Yasukawa and Searcy 1995). Orians (1985) and Yasukawa and Searcy (1995) regard the aggregation of nests to be more a consequence of the distribution of suitable habitat than of an innate gregariousness, however. The colonial nesting of the Red-winged Blackbird involves 2 interrelated behavioural characteristics. The first is the innate attraction to marshy wetlands as nesting habitat; the second is aggregation of nesting females on the territory defended by some males and not others. Some researchers propose that this polygyny provides some defence against predatory Marsh Wrens (Picman 1980a, 1980b; Picman et al. 1988). Other than this, there has been no demonstrated advantage for the success of the nesting effort that can be attributed to the aggregation of male territories on the nesting grounds or the clustering of female nesters in the areas defended by males, the 2 obvious aspects of coloniality.

It is difficult to determine the number of nesting birds involved in a "colony" of Red-winged Blackbirds. The traditional technique of counting the more conspicuous of the 2 genders and regarding each as representing a nesting pair is not applicable because of the species' polygyny (Helms et al. 1994; Yasukawa and Searcy 1995; Campbell 2000). An additional complication of this approach is the number of floating males that are frequently present within a colony. The only reliable technique for ascertaining colony size is to count active nests. There is often a sequence of females nesting within a male territory, however, so no single count of active nests will reveal the population (Dolbeer 1976). The wetland habitat makes nest counting difficult to accomplish accurately, and makes it likely to cause considerable disturbance in the colony.

While our data include many references to colonies, few of them state that serious attempts were made to count all active nests within the colony. Under these circumstances, we have not given a table of colony sizes. In British Columbia, the estimates of numbers of "nests" in a "colony" ranged from 2 to 400. All the colonies of more than 50 "nests" estimated were in the Okanagan and west Kootenay valleys, where there are some extensive marshy areas. In all other regions, colony sizes did not exceed an estimated 50 nests.

Nests: Provincewide, most nests (67%; n = 1,317) were built in emergent vegetation, including cattails (34%; Fig. 479), bulrushes (19%), and reeds (14%). Other nest sites included shrubs (19%), deciduous trees (5%), and sedges and grasses (4%).

Where the nests are in marshlands, nest construction begins with a platform of fine vegetation, frequently dead sedges, woven around several vertical stems. On this the cup-shaped nest of wet marsh vegetation is built. An inner cup of mud is usually part of the structure, and is lined with fine grasses. The heights of 1,077 nests ranged from water or ground level to 8.0 m, with 70% between 0.3 and 0.9 m. Early nests are frequently flooded by rising water. In the Nulki Lake area, Munro (1949) reported that sedge clumps were the only available nest sites early in the season, and almost all nests were flooded or lost stability as young increased in weight. On the coast, some nests with eggs have been found attached to the tall ecotype of *Carex lyngbyei* in estuaries which normally gets inundated by the rising tides; all were eventually abandoned.

Eggs: Dates for 1,424 clutches (Fig. 479) ranged from 3 April to 16 July, with 52% recorded between 15 May and 8 June. Calculated dates indicate that eggs can occur as early as 3 April. Sizes of 1,160 clutches ranged from 1 to 7 eggs (1E-129, 2E-128, 3E-253, 4E-522, 5E-113, 6E-13, 7E-2), with 67% having 3 or 4 eggs. The incubation period in British Columbia is 10 to 11 days; elsewhere it has been reported as 11 to 13 days (Yasukawa and Searcy 1995).

Figure 479. An incomplete clutch of Red-winged Blackbird eggs laid in a typical cup-shaped nest built of grasses and attached to the stems of cattails (near Fort St. John, 3 June 1997; R. Wayne Campbell).

Figure 480. Nestling Red-winged Blackbirds leave their nests between 11 and 16 days (west of Charlie Lake, 22 June 1996; R. Wayne Campbell).

Young: Dates for 706 broods (Fig. 480) ranged from 23 April to 13 July, with 55% recorded between 23 May and 16 June. Sizes of 607 broods ranged from 1 to 9 young (1Y-52, 2Y-101, 3Y-212, 4Y-221, 5Y-16, 6Y-1, 7Y-1, 8Y-2, 9Y-1), with 71% having 3 or 4 young. The nestling period was 11 to 16 days; in an Alaskan study, fledging occurred on days 9 and 10 (n = 66) (McGuyire 1986).

The length of the nesting season changes between ecoregions and is dependent on the time nest building and egg laying are initiated. In all ecoprovinces from which we have adequate data, the last nestlings fledged between 6 and 10 July. The proximate factors that trigger nest building in the Red-winged Blackbird are probably the seasonal pattern of food availability (Caccamise 1978). The sequence of events that terminates the nesting season is poorly understood. The length of the nesting season was 94 days in the Georgia Depression, 75 days in the Southern Interior, 72 days in the Southern Interior Mountains, and 53 days in the Boreal Plains (Fig. 474).

Brown-headed Cowbird Parasitism: In British Columbia, 2% of 1,416 nests found with eggs or young were parasitized by the Brown-headed Cowbird. There was 1 additional record of a Red-winged Blackbird feeding a cowbird fledgling. Parasitism on the coast was 1% (n = 478); in the interior, it was 3% (n = 938). Cannings et al. (1987) mention 29 parasitized nests (n = 258), for a rate of 11.2% parasitism in their Okanagan valley sample. We have no explanation for the discrepancy other than sampling differences. Friedmann (1963) and Robertson and Norman (1976) have called attention to a low rate of cowbird parasitism in marsh nests, as opposed to upland nests, of the Red-winged Blackbird. Ontario data cited by Robertson and Norman (1976) revealed a parasitism rate of 2.4% (n = 2,039), similar to the rate for British Columbia. A parasitism rate of 7.7% has been reported for Washington (Freeman et al. 1990), but the report does not refer to the nest ecology.

Ward et al. (1996) have shown experimentally that the Red-winged Blackbird in British Columbia does not reject Brown-headed Cowbird eggs, and even incubates clutches containing none of its own eggs. The authors reason that evolutionary lag is the best explanation for this acceptance of cowbird eggs by the blackbird hosts. In Ontario, about a third of parasitized nests were deserted (Robertson and Norman 1976).

Nest Success: Of 275 nests found with eggs and followed to a known fate, 67 produced at least 1 fledgling, for a nest success rate of 24%. Nest success on the coast was 22% (n = 153); in the interior, it was 27% (n = 122).

Picman et al. (1988) provide a summary of their extensive inquiry into nesting success and nesting mortality of Red-winged Blackbirds on the delta of the Fraser River. The success rate between 1976 and 1982 ranged from 18.9% to 45.1% (Table 9). In 1976 and 1977 combined (n = 399 nests), 23.5% of nests were successful (Picman 1980a, 1980b). Predation accounted for the failure of 45% of all nests. In camera-monitored dummy nests (n = 47), the Marsh Wren accounted for 77% of the predation (Picman et al. 1988).

REMARKS: The Red-winged Blackbird is probably the most abundant land bird in North America (Nero 1984). Meanley and Royall (1976) estimate the winter population in the United States to be about 190 million birds.

Yasukawa and Searcy (1995) list 14 subspecies of this blackbird in North America, of which 3 breed in British Columbia. These are *A. p. caurinus* in the Georgia Depression, *A. p. nevadensis* in the Southern Interior and Southern Interior Mountains, and *A. p. arctolegus* in the northern parts of the province, south to include the Central Interior. Of these *caurinus* is largely resident, *nevadensis* winters in small numbers in the Okanagan valley, and *arctolegus* leaves the province in winter except for the occasional winter straggler in

Table 9. Success rates and sources of failure of Red-winged Blackbird nests at Delta, British Columbia (from Picman et al. 1988).

Nesting history	1976	1977	1978	1979	1980	1981	1982	Combined 1976-82
% successful nests	18.9	28.7	38.0	45.1	35.7	45.0	38.2	35.1
% depredated nests	40.8	50.5	42.9	36.9	51.7	43.7	45.8	44.2
% of nests destroyed by high tides	18.5	0.0	3.1	1.0	2.1	4.0	2.1	4.8
% of nests failed for unknown reason	21.8	20.8	16.0	17.0	10.5	7.3	13.9	15.9
Total number of nests	211	188	163	206	143	151	144	1,206

the summer range that may be of this subspecies. The distribution of the subspecies in southern British Columbia during migration has not been studied.

In parts of North America, especially where wintering birds gather in enormous roosting aggregations, the Red-winged Blackbird is a serious agricultural pest, especially of cereal crops and corn. In British Columbia, however, it is not regarded as such.

For a summary of the life history of the Red-winged Blackbird in North America, see Yasukawa and Searcy (1995).

NOTEWORTHY RECORDS

Spring: Coastal – Victoria 1 May 1981-29 counted in Beacon Hill Park; Delta 5 May 1922-40; Burnaby Lake 5 Apr 1970-3 eggs, 21 Apr 1970-3 nestlings; Agassiz 7 Mar 1976-300; Ross Lake 30 Mar 1971-20 on mudflats; Pitt Meadows 3 Apr 1975-104; Hatzic Lake 5 May 1964-4 eggs, 24 May 1964-4 nestlings, 27 May 1962-10 nests; Campbell River 3 Mar 1963-100, 16 Apr 1983-1,098 foraging on estuary (Dawe et al. 1995a); Port Neville Inlet 9 Mar 1976-1; Tweedsmuir Park 29 May 1975-2 eggs (Osmond-Jones et al. 1977); Kitlope Lake 3 May 1991-10 on estuary; Delkatla Inlet (Masset) 27 Apr 1979-1; Kitimat 9 May 1980-1 egg; Terrace 20 May 1979-4 eggs. **Interior** – Oliver 25 Apr 1947-4 eggs; Grand Forks 20 May 1962-11 nests, 4 eggs collected; Erie Lake (Salmo) 14 Mar 1983-34; Duck Creek (Creston) 1 May 1956-2 eggs, nest abandoned; 23 km w Nelson 12 May 1980-4 nestlings about 2 days old; 16 km s Hedley 7 May 1960-3 nests; Trout Creek Point (Summerland) 13 Mar 1973-200; Alleyne Lake 28 May 1993-3 eggs; Kelowna 25 Apr 1994-4 young nestlings at Chichester Bird Sanctuary; Enderby 16 May 1974-11 nests; Kamloops 1 May 1973-4 nestlings; South Thompson River between Chase and Monte Creek 6 May 1952-4 nests; Parson 26 Mar 1977-55; 100 Mile House 13 May 1980-3 eggs; Dutch Lake (Clearwater) 22 May 1960-4 nestlings; Chezacut 7 Mar 1943-2; Moose Lake (Mount Robson Park) 31 May 1972-30; nr Nulki Lake 19 May 1945-12 nests at Bradley's Slough; 24 km s Prince George 7 Mar 1986-7; Tyhee Lake (Telkwa) 27 May 1984-4 eggs; Quick 29 Mar 1987-20; McLeod Lake (Crooked River) 12 Apr 1997-20 males; Mackenzie 16 Mar 1996-2 males; e Chetwynd 1 May 1965-12; Swan Lake (Tupper) 14 May 1938-100 (Cowan 1939); Dawson Creek area 17 Apr 1991-1 early arrival (Phinney 1998); Toms Lake 31 May 1994-3 nestlings; Tetana Lake 16 May 1938-2 (Stanwell-Fletcher and Stanwell-Fletcher 1943); Fort St. John 11 Apr 1986-3; s end Cecil Lake 22 Apr 1984-30, 17 May 1975-2 eggs; Hyland Post 8 May 1984-1; Alaska Highway (Mile 419) 18 May 1975-20; Fort Nelson 24 Apr 1987-1; Liard Hot Springs 4 May 1975-1 (Reid 1975); Helmut 14 May 1982-1; Coal River 31 May 1980-5 eggs.

Summer: Coastal – nr Sooke 5 Jun 1984-7; Vancouver 9 Aug 1965-300 at airport; West Vancouver 15 Jul 1982-2 nestlings at Ambleside Park; Qualicum Beach 6 Jul 1983-3 nestlings; Comox 17 Aug 1952-1,000 in flock (Flahaut 1953a); Quadra Island 12 Jun 1980-4; Stuie 29 Jun 1979-1; Kunghit Island 18 Aug 1946-1 (UBC 1128); Terrace 4 Jul 1979-4 nestlings. **Interior** – nr Rock Creek 3 to 4 Jun 1997-69 nests at Meyers wetlands (Campbell 2000); Creston 16 Aug 1982-600; South Slocan 10 Jun 1971-4 eggs; Fernie 11 Jun 1981-4 nestlings; Chapperon Lake 24 Jun 1974-5 eggs; Vernon 31 Aug 1973-1,050 at O'Keefe's pond, 5 Aug 1892-3 fledglings (PAS 30971 to 30973); Swan Lake (Vernon) 15 Jun 1963-400 pairs estimated; Kootenay National Park 8 Jul 1981-5 eggs at Daer Pond; Ottertail River (Yoho National Park) 3 Jun 1975-2 eggs at sloughs nr river mouth; Kleena Kleene 2 Jun 1965-20 nests; Riske Creek 8 Jul 1983-600 in flock; Westwick Lakes 14 Jun 1978-8 nests; Meldrum Creek 14 Aug 1995-4,000 feeding in freshly cut grain crop; Stum Lake 10 Jul 1973-3 eggs (Ryder 1973); Red Pass 8 Jul 1972-3 fledglings; 10 Mile Lake (Quesnel) 16 Jul 1970-nestlings; Moose Pass 1 Jun 1991-4 eggs; Bowron Lake Park 27 Jul 1971-2 fledglings; Sinkut Lake 14 Jun 1945-12 nests at Collin's Marsh (Munro 1949); Prince George 9 Jul 1973-3 nestlings; Topley 25 Jul 1956-1; Smithers 1 Jun 1985-4 nestlings; One Island Lake 28 Jun 1976-12 nests; Kitsault River 18 Jun 1980-nest with 4 nestlings on estuary; 15 km s Dawson Creek 4 Jul 1993-3 eggs; n Fort St. John 24 Jun 1978-11 nests at Nurnberger's Marsh; North Pine 20 Aug 1984-37; sw Cecil Lake 23 Jul 1983-1 fledgling; Boundary Lake (Goodlow) 16 Jun 1985-4 nestlings; Sawmill Lake (Stikine River) 22 Jul 1987-12; e Fort Nelson 5 Jul 1982-5 eggs; Camp Yoyo 6 Jul 1982-4 nestlings along Esso Resources Rd; Parker Lake 2 Aug 1985-1 fledgling; lower Tatshenshini River valley 5 Jun 1983-2 eggs; Kwokullie Lake 6 Jun 1982-5 eggs; Atlin 20 Jun 1982-4 eggs, 20 Aug 1934-1 (RBCM 5866).

Breeding Bird Surveys: Coastal – Recorded from 19 of 27 routes and on 59% of all surveys. Maxima: Pemberton 2 Jul 1977-72; Victoria 5 Jul 1980-67; Seabird 6 Jun 1971-58. **Interior** – Recorded from 55 of 73 routes and on 75% of all surveys. Maxima: Grand Forks 1 Jul 1991-381; Salmon Arm 30 Jun 1993-223; Wasa 30 Jun 1975-111.

Autumn: Interior – Parker Lake 7 Sep 1985-2; ne Fort St. John 9 Sep 1984-2, last record of autumn; 25 km s Prince George 29 Oct 1984-1, stayed until mid-November; Williams Lake 20 Sep 1980-500; Glacier National Park 13 Sep 1982-2 fledglings; Cranbrook 11 Sep 1944-200 (Johnstone 1949). **Coastal** – Hope 26 Nov 1976-42; Pitt Meadows 27 Oct 1974-1,200 in mixed flock of Red-winged and Brewer's blackbirds and Brown-headed Cowbirds; Ladner 6 Nov 1922-3,700, 3 large flocks, 3,000, 500, and 200 mixed blackbird species; Boundary Bay 17 Oct 1979-690; Island View Beach (North Saanich) 12 Oct 1958-5,000, mixed flock, mostly Brewer's Blackbirds.

Winter: Interior – Williams Lake 2 Jan 1977-200, flock wintering nr city; Revelstoke 19 Dec 1986-26, 1 Jan 1988-1; Salmon Arm 18 Feb 1971-2; Nakusp 29 Dec 1983-3; Kelowna 8 Jan 1972-400; Creston 24 Jan 1980-100; Oliver 31 Dec 1975-1,500. **Coastal** – Terrace 21 Feb 1970-12, had wintered there (Crowell and Nehls 1970b); Kitimat 31 Dec 1974-5, 4 Jan 1975-4; Queen Charlotte City 20 Dec 1986-6; Bella Coola 28 Dec 1989-11; Campbell River 10 Jan 1963-1; Squamish River 26 Feb 1980-30; Langley 1 Jan 1968-2,000 in flock with European Starlings and Brewer's Blackbirds; Pitt Meadows 6 Feb 1971-400; Ladner 22 Dec 1976-1,500; Somenos Lake (Duncan) 7 Jan 1978-160; Saanich 5 Jan 1981-60; Metchosin 31 Dec 1974-150.

Christmas Bird Counts: Interior – Recorded from 17 of 27 localities and on 55% of all counts. Maxima: Vaseux Lake 31 Dec 1975-2,250; Vernon 15 Dec 1991-1,534; Oliver-Osoyoos 29 Dec 1989-1,205. **Coastal** – Recorded from 26 of 33 localities and on 66% of all counts. Maxima: Pitt Meadows 30 Dec 1979-5,090; Ladner 30 Dec 1961-3,075; Victoria 14 Dec 1991-2,160.

Western Meadowlark

WEME

Sturnella neglecta (Audubon)

RANGE: Breeds from central and northeastern British Columbia, north-central Alberta, central Saskatchewan, southern Manitoba, western Ontario, northeastern Minnesota, northern Wisconsin, northern Michigan, southern Ontario, and northwestern Ohio south to northwestern Baja California, southern California, northwestern Sonora, western and central Arizona, in the Mexican highlands and to west-central Texas, northwestern Louisiana, southwestern Tennessee, southern Illinois, northern Indiana, central Ohio, and western New York. Winters from southern British Columbia, central Idaho, central Colorado, southern South Dakota, southern Wisconsin, and northern Indiana south to southern Baja California, Michoacán, the state of Mexico, Veracruz, and the Gulf coast east to northwestern Florida, occurring east to central Kentucky, eastern Tennessee, and eastern Alabama.

STATUS: On the coast, *uncommon* migrant and summer visitant and locally *uncommon* to *fairly common* in winter in the Georgia Depression Ecoprovince; in the Coast and Mountains Ecoprovince, *rare* migrant and *very rare* summer and winter visitant on Western Vancouver Island, *rare* migrant and *very rare* summer and winter visitant on the Southern Mainland Coast, *casual* on the Northern Mainland Coast, and *very rare* autumn migrant on the Queen Charlotte Islands.

In the interior, *uncommon* to *common* migrant and summer visitant and locally *uncommon* to *fairly common* winter visitant in the Southern Interior Ecoprovince; *uncommon* migrant and summer visitant and *very rare* in winter in the Southern Interior Mountains Ecoprovince; *uncommon* migrant and summer visitant and *casual* in winter in the Central Interior Ecoprovince; *rare* spring migrant and locally *very rare* summer visitant, and *casual* in autumn in the Sub-Boreal Interior Ecoprovince; *rare* spring migrant and summer visitant and *casual* in autumn in the Boreal Plains Ecoprovince; *casual* in the Taiga Plains Ecoprovince; accidental in the Northern Boreal Mountains Ecoprovince.

Breeds.

CHANGE IN STATUS: In the mid-1940s, the Western Meadowlark (Fig. 481) was resident in extreme southwestern British Columbia, including Vancouver Island, and throughout the Okanagan valley. Elsewhere, it occurred as a regular summer visitant from southern portions of the province north through the Cariboo and Chilcotin areas to Quesnel. It was considered casual in the Peace Lowland of the Boreal Plains (Munro and Cowan 1947).

Over the next 5 decades, the species slowly extended its range northward in the interior to the Prince George area, where it now occurs locally each year. In the Boreal Plains of northeastern British Columbia, the Western Meadowlark is now a regular summer visitant and local breeder, probably the result of an expanding population in northwestern Alberta from 1892 to the early 1990s (Pinel et al. 1993).

Figure 481. The Western Meadowlark, one of our most recognized grassland birds, is a common sight in summer across the interior of southern British Columbia (Brisco, 8 May 1997; R. Wayne Campbell).

In south coastal areas of the province, including southeastern Vancouver Island, the Gulf Islands, and the lower Fraser River valley, the Western Meadowlark was a regular breeding species in the early 1900s (Brooks and Swarth 1925), but by the early 1940s it was becoming less common and more restricted as habitat was being lost to urbanization. In a story related to Bruce Whittington (pers. comm.), an old-timer recalled the meadowlarks singing so noisily during school days on the Saanich Peninsula that his teacher had to close the windows so the children could concentrate on their studies. Today, the song of the meadowlark is seldom heard there and, although the species still occurs in small numbers on Vancouver Island, breeding has not been reported since 1977; on the Lower Mainland, the meadowlark has not been reported breeding since 1986 (Butler and Campbell 1987).

Christmas Bird Count data from select areas on the south coast between 1958 and 1997 indicate a steady decline in numbers (Fig. 482). A significant decline occurred between 1960 and 1964, followed by a gradual decline from the mid-1960s to the 1990s. Rapid urban expansion and the subsequent loss of habitat during the 1960s followed by steady urban growth and agricultural changes are the likely reasons for the declining wintering populations in the fastest-growing metropolitan region of the province.

Declines in the numbers of meadowlarks wintering in the southern Okanagan valley were also noted by Cannings et al. (1987), especially between the late 1950s and mid-1980s. Slower urban growth in the northern portion of the valley coincided with a less marked decline in meadowlark numbers during the same period. Urban expansion in the Okanagan is second only to that in the Fraser River delta and southeastern Vancouver Island. The replacement of some agricultural practices that favoured the Western Meadowlark by other land uses (such as hothouses, ginseng operations)

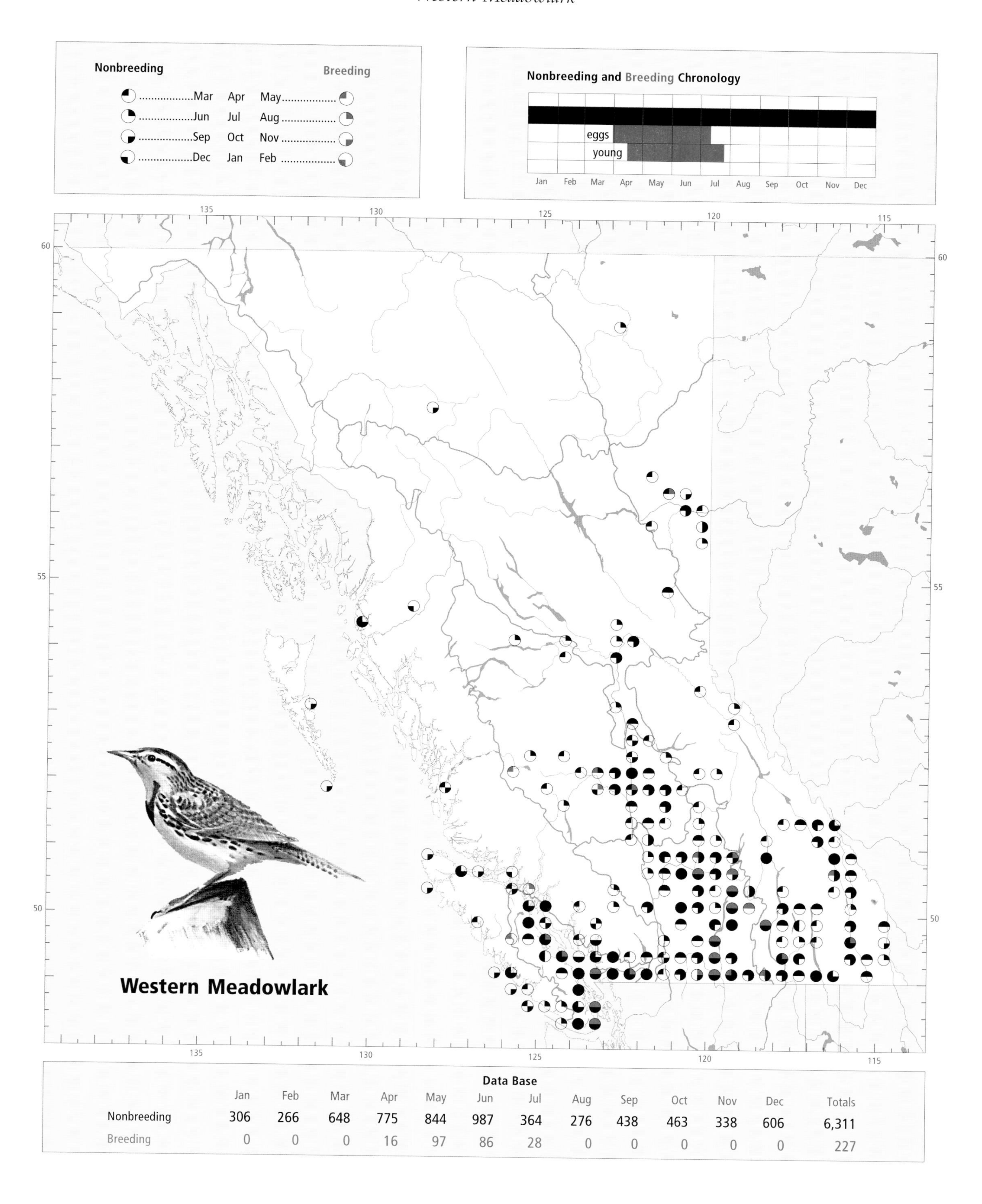

Data Base

	Jan	Feb	Mar	Apr	May	Jun	Jul	Aug	Sep	Oct	Nov	Dec	Totals
Nonbreeding	306	266	648	775	844	987	364	276	438	463	338	606	6,311
Breeding	0	0	0	16	97	86	28	0	0	0	0	0	227

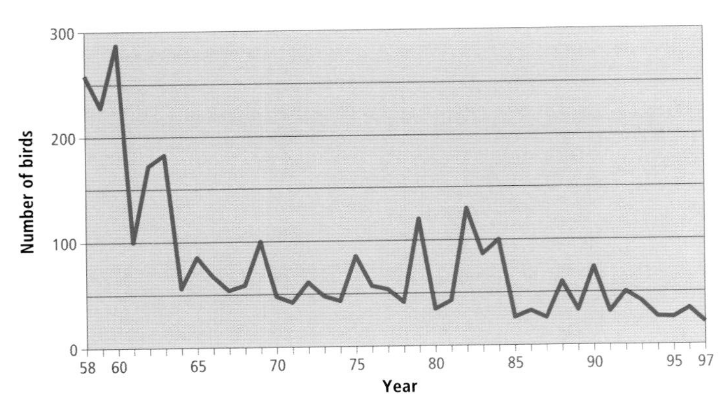

Figure 482. Fluctuations in the numbers of Western Meadowlarks reported on Christmas Bird Counts for the Georgia Depression Ecoprovince over the period 1958 through 1997. Totals are summed from the Vancouver, Ladner, and Victoria counts.

may have contributed to the declines. In the central and northeastern portions of the province, the clearing of forests for agricultural purposes was likely the major contributing factor to the expansion of the meadowlark into these regions.

NONBREEDING: The Western Meadowlark is widely distributed at low to mid elevations across the southern regions of the province. On the coast, it occurs regularly, and recently more locally, on southeastern Vancouver Island, from Victoria to Campbell River, and in the Fraser Lowland on the mainland. It occurs infrequently on the rest of Vancouver Island and north along the coast to Prince Rupert and New Aiyansh, as well as on the Queen Charlotte Islands.

In the interior, the Western Meadowlark occurs regularly north to the Cariboo and Chilcotin areas, and irregularly and locally in the Nechako and Bulkley river valleys. There is a disjunct population distributed locally throughout the Peace Lowland. Further north, it is sparsely distributed.

The Western Meadowlark reaches its highest numbers in winter in the Fraser Lowland of the Georgia Depression (Fig. 483).

On the south coast, the Western Meadowlark occurs throughout the coastal lowlands from sea level to 100 m elevation; in the interior, it ranges from valley bottoms at 280 m elevation to 1,850 m elevation, with most records occurring below 1,000 m. It generally prefers open country with grasslands and pastures. On the coast, the Western Meadowlark frequents the upper edges of estuarine marshes, agricultural fields, and grassy fields. In the interior, lower-elevation grasslands, shrub grasslands, pastures, abandoned fields, and grassy openings in ponderosa pine parkland are the main nonbreeding habitats. In the Peace Lowland, the Western Meadowlark occupies agricultural lands; along the north side of the Peace River, it occupies native grasslands. In winter, many meadowlarks associate with cattle-feeding operations or other agricultural sites in the interior where spilled grain is available.

On southern portions of the mainland coast, spring migration is evident as the wintering birds begin to leave the

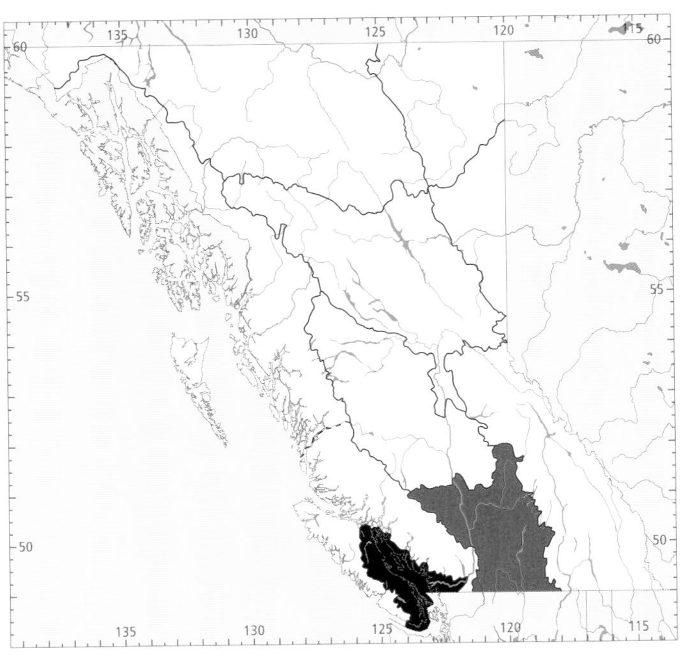

Figure 483. In British Columbia, the highest numbers for the Western Meadowlark in winter (black) occur in the Georgia Depression Ecoprovince; the highest numbers in summer (red) occur in the Southern Interior Ecoprovince.

area in early March and a steady decline occurs until early June (Figs. 484 and 485). Most wintering birds leave for other breeding locations and the small coastal breeding population does not appear to be augmented by any noticeable influx of spring migrants from the south.

In the Southern Interior, spring migration is evident most years in late February and early March, when singing males establish territories as snow is leaving the ground at lower elevations. Numbers increase through April and peak in early May (Figs. 484 and 485). In the Okanagan valley, arrival dates over 18 years ranged from 4 February to 18 March, with a mean of 28 February (Cannings et al. 1987). Early migrants may arrive in the Southern Interior Mountains during the third week of March, and numbers increase during April and peak in early May. The southern Rocky Mountain Trench and the valleys in the west Kootenay do not appear to be major spring migration corridors for the Western Meadowlark, as incoming summer birds arrive, breed, and depart with no significant swelling in numbers. In the Central Interior, the first spring migrants may arrive as early as the last week of February but the main movement does not begin until the last week of March; numbers increase through early May. The Western Meadowlark arrives in the Peace Lowland during the latter half of April.

In northeastern regions of the province, the autumn migration likely occurs during late August and early September; a few birds may linger into November. Elsewhere, the Western Meadowlark may assemble in post-breeding flocks of up to 75 birds before moving south. In the Central Interior and Southern Interior Mountains, autumn migration peaks in the second week of September, with the last birds leaving by the end of October. In the Southern Interior, the southern movement occurs from the end of August through September, and most birds have gone by mid-October (Fig. 485). On the south coast, autumn migrants begin appearing during the latter half of September and rapidly increase to wintering numbers by the second week of October (Fig. 485).

There are no winter records for Western Meadowlarks in the northern areas of British Columbia and only 2 for the Central Interior; the northernmost winter records are from Tranquille, north of Kamloops, and Williams Lake. The winter population in the Southern Interior and Southern Interior Mountains has probably peaked by early December, with an increase in reports after November perhaps being due to the arrival of late stragglers or to aggregations of birds into more visible areas. Historically, winter numbers in the Okanagan were higher, with Christmas Bird Counts showing a more significant decline in the southern part of the valley than in the north (Cannings 1987).

On the coast, the northernmost winter records are from Quatse River on Vancouver Island (Dawe et al. 1995) and Terrace on the mainland.

The Western Meadowlark is present in south coastal areas and southern portions of the interior throughout the year (Figs. 484 and 485). In the Central Interior, the meadowlark regularly occurs from early March through the end of September; in the Boreal Plains it occurs regularly from the first week of April to the first week of August.

BREEDING: The Western Meadowlark has a widespread breeding distribution across the province east of the Coast Mountains from Princeton to Cranbrook and Wardner and north through the Columbia, Okanagan, and Nicola river valleys to Williams Lake and Quesnel. Formerly it nested on the south coast, including southeastern Vancouver Island. A disjunct population breeds locally in the Peace Lowland in northeastern British Columbia and may be an extension of populations breeding in adjacent Alberta.

The Western Meadowlark reaches its highest numbers in summer in the Southern Interior (Fig. 483). An analysis of Breeding Bird Surveys in British Columbia for the period 1968 through 1993 could not detect a net change in numbers on interior routes; coastal routes for the same period contain insufficient data for analysis. Throughout its Canadian range, the Western Meadowlark declined at an average annual rate of nearly 2% ($P < 0.01$) from 1966 to 1996; continent-wide surveys showed an average annual decline of 0.6 % ($P = 0.01$) for the same period (Sauer et al. 1997).

On the coast, the Western Meadowlark breeds in coastal lowlands at elevations below 100 m; in the interior, it has been recorded breeding from 280 to 1,200 m. The characteristic nesting habitat of this species in British Columbia features open spaces, including natural grasslands (Fig. 486), agricultural grasslands (Fig. 487), and other habitats such as alfalfa fields, pastures, and abandoned fields with forb and grass cover. A few nests were found in open forest and parkland areas where grasses were the main understorey component, or in grassy shrubland habitats. Single nests were situated in a wetland and on a sand dune. Grass cover in British Columbia, as elsewhere in the bird's North American range, is an

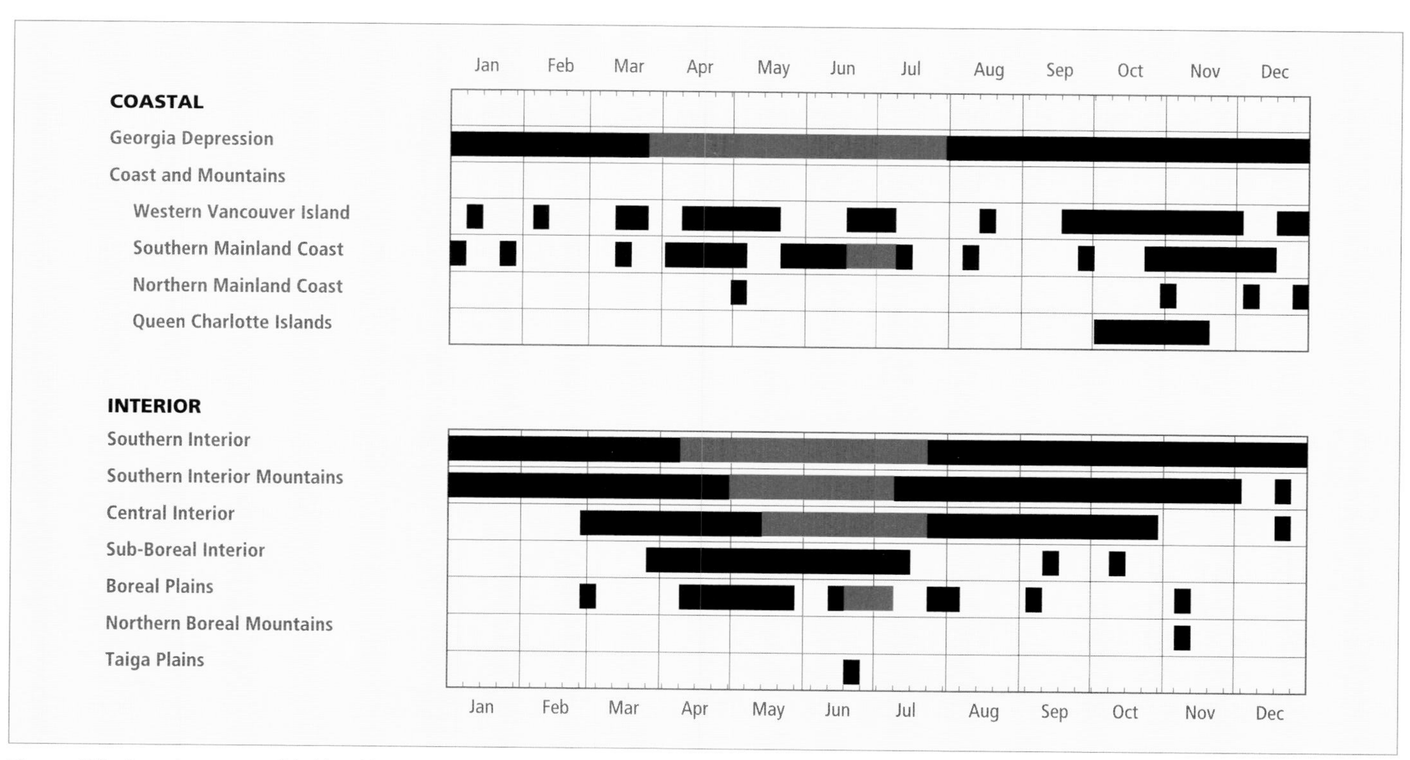

Figure 484. Annual occurrence (black) and breeding chronology (red) for the Western Meadowlark in ecoprovinces of British Columbia. Records are shown for the week in which they occurred.

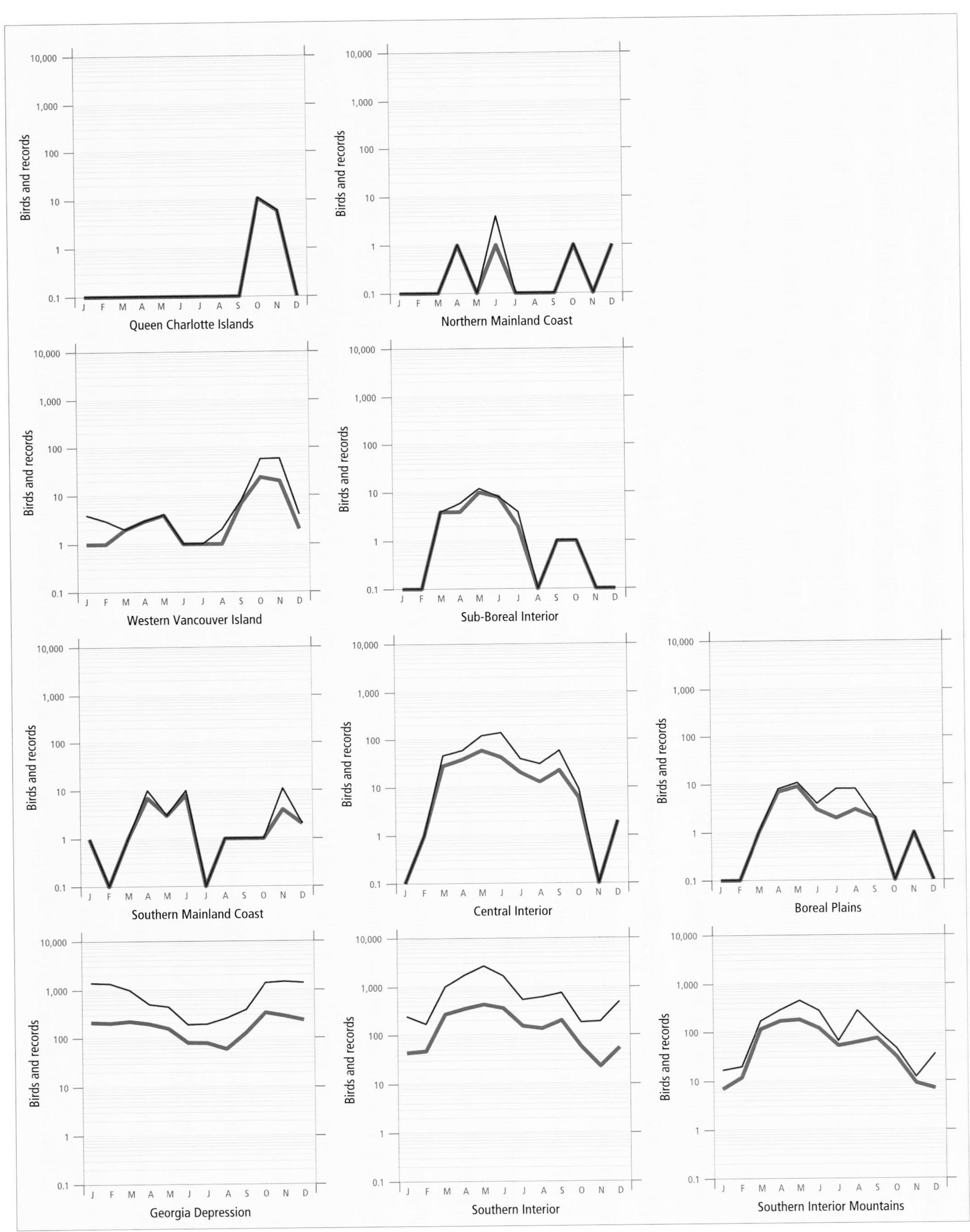

Figure 485. Fluctuations in total number of birds (purple line) and total number of records (green line) for the Western Meadowlark in ecoprovinces of British Columbia. Christmas Bird Counts, Breeding Bird Surveys, and nest record data have been excluded.

Figure 486. In the Southern Interior Ecoprovince of British Columbia, the Western Meadowlark nests in native grasslands where scattered big sagebrush provides singing perches and cover (near Kilpoola Lake, 8 August 1990; R. Wayne Campbell).

Figure 487. In the Boreal Plains Ecoprovince of northeastern British Columbia, the Western Meadowlark breeds locally in agricultural grasslands (near Fort St. John, 18 Jun 1966; R. Wayne Campbell).

Figure 488. Western Meadowlark nests in British Columbia are built on the ground and are usually well concealed by grasses and other vegetation (Richter Pass, 26 May 1979; R. Wayne Campbell).

important component of all nesting habitats. Lanyon (1994) indicates that the meadowlark is most common during the breeding season in native grasslands and pastures, but also uses hay fields, weedy cropland borders, orchards, and other open areas.

On the coast, the Western Meadowlark has been recorded breeding from 31 March (calculated) to 25 July; in the interior, it has been recorded breeding from 14 April (calculated) to 20 July (Fig. 484).

Nests: All nests (n = 104) were situated on the ground (Fig. 488) and were hidden within or beneath vegetation. More than 50% were located beside a clump of vegetation. Others were simply listed as well-concealed depressions in tall grasses; 1 was found at the base of a tree and 1 beside a rock. Most nests (n = 83) were constructed of coarse, dried grasses and lined with finer grasses; 1 was constructed entirely of knapweed. Eight nests were described as being cup-shaped, while 30 had roofed coverings simply described as being "covered," "domed," or "arched" (Fig. 488). A few nests were listed as "grass balls," and several had short tunnels leading to the nest cup.

Eggs: Dates for 143 clutches (Fig. 488) ranged from 10 April to 11 July, with 52% recorded between 10 May and 8 June. Calculated dates indicate that eggs can occur as early as 31 March. Sizes of 107 clutches ranged from 1 to 7 eggs (1E-11, 2E-5, 3E-5, 4E-19, 5E-58, 6E-8, 7E-1), with 54% having 5 eggs. The incubation period is usually 13 or 14 days, rarely 15 or 16 days (Lanyon 1994).

Young: Dates for 84 broods ranged from 15 April to 25 July, with 51% recorded between 30 May and 30 June. Sizes of 54 broods ranged from 1 to 5 young (1Y-7, 2Y-7, 3Y-6, 4Y-19, 5Y-15), with 63% having 4 or 5 young. The nestling period is 10 to 12 days (Lanyon 1994).

Brown-headed Cowbird Parasitism: In British Columbia, 2 of 139 nests found with eggs or young were parasitized by the cowbird. Both parasitized nests were from the interior (n = 116). There were no instances of a Western Meadowlark feeding a fledged cowbird. Friedmann (1963) and Friedmann and Kiff (1985) list 36 records of nest parasitism in North America and conclude that the Western Meadowlark is an infrequent host for the cowbird.

Nest Success: Of 16 nests found with eggs and followed to a known fate, 5 produced at least 1 fledgling, for a nest success rate of 31%. All but 1 nest was from the interior. Nestlings leave the nest and hide in tall grasses before they can fly, making data on nesting success difficult to collect.

REMARKS: Lanyon (1994) states that the availability of suitable breeding habitat controls the size of meadowlark breeding populations. He notes that agricultural activities such as grazing and trampling by livestock, mowing, the use of pesticides (Griffin 1959), and surface tillage for weed control are detrimental to nests and flightless young. In British Columbia, it is apparent that the conversion of agricultural lands and estuarine backshore habitats to urbanized landscapes is largely responsible for the rapid decline of this species on the south coast. Likewise, urban growth in the Okanagan valley has resulted in permanent habitat loss.

The 22 June 1993 Breeding Bird Survey for Gnat Pass during which 17 Western Meadowlarks were recorded was considered unlikely and has been excluded from this account.

For a thorough review of the ecology and natural history of the Western Meadowlark in North America, see Lanyon (1994).

NOTEWORTHY RECORDS

Spring: Coastal – Victoria 12 Mar 1966-20, 10 Apr 1908-eggs and young collected, 15 Apr 1958-3 nestlings, 23 May 1892-1 fledgling; Tofino 13 Mar 1973-1; Sea Island 12 May 1974-29; Alouette River 5 Mar 1977-40; Huntingdon 5 May 1921-20; Pitt Meadows 2 Mar 1980-32, 24 May 1968-30; Port Hardy 2 May 1939-1; Namu 29 Apr 1983-1; New Aiyansh 30 Apr 1987-1. **Interior** – Richter Pass 4 Apr 1979-34; Oliver to Richter Lake 17 May 1959-100; Osoyoos 18 Apr 1908-5 eggs; Creston 24 May 1981-1 fledgling; White Lake (Okanagan Falls) 23 Mar 1986-25, 12 May 1977-44; Castlegar 24 Mar 1970-15; Douglas Lake 27 Apr 1980-2 nestlings; Nicola River 16 Mar 1997-30; Vernon 29 Apr 1991-3 fledglings; Nakusp 13 Mar 1976-2; Kamloops 1 Mar 1986-1, 20 Mar 1988-200; Invermere 14 May 1976-2 nestlings, three-quarters full grown; Lac du Bois 5 May 1972-60; nr Golden 1 Apr 1998-1; Lonesome Lake 31 May 1939-young present; Alexis Creek 31 May 1970-5 eggs; Williams Lake 14 Mar 1960-6; Prince George 15 Mar 1992-2; Nulki Lake (Vanderhoof) 22 Mar 1999-2; Willow River 28 Mar 1967-1; Cameron River (Wonowon) 4 to 6 Mar 1965-1; Peace River 13 Apr 1988-1.

Summer: Coastal – Carmanah Point 5 Jul 1979-1; Saanich 25 Jul 1947-1 large young in nest; Vancouver 20 Jul 1964-12; Tsawwassen 25 Jul 1986-2 fledglings; Englishman River 6 Jul 1977-5 eggs. **Interior** – White Lake (Okanagan Falls) 28 Aug 1962-75; Penticton 10 Aug 1974-42; Madeline Lake 7 Jul 1975-4 nestlings; Kimberley 30 Aug 1977-175; Cranbrook 17 Aug 1976-13; Edgewood 7 Jul 1931-5 eggs; nr Douglas Lake 1 Jul 1978-30; Kalamalka Lake Park 11 Jul 1958-5 eggs; Vernon 21 Jul 1993-2 nestlings; Spences Bridge to Chase 15 Aug 1966-78; Revelstoke 3 Aug 1989-3 fledglings; Kinbasket Lake 5 Jul 1995-3; Mount Robson Park 4 Jun 1994-1; Tatla Lake 1 Jul 1969-5 fledglings; Chilco Ranch (Alexis Creek) 12 Jun 1983-1 fledgling; Riske Creek 6 Jul 1983-3 eggs, 20 Jul 1983-3 nestlings; Quesnel 8 Jun 1988-3 fledglings; Barbour Creek 28 Jun 1997-1; Dawson Creek 26 Jun 1993-1; 40 km nw Fort St. John 25 Jun 1978-4 eggs; Fort Nelson 21 Jun 1988-1.

Breeding Bird Surveys: Coastal – Recorded from 5 of 27 routes and on 8% of all surveys. Maxima: Nanaimo River 8 Jun 1975-8; Pemberton 22 Jun 1986-5; Seabird 6 Jun 1970-3. **Interior** – Recorded from 41 of 73 routes and on 61% of all surveys. Maxima: Pleasant Valley 15 Jun 1984-109; Lavington 14 Jun 1975-95; Oliver 28 Jun 1981-81.

Autumn: Interior – Hyland Post 6 Nov 1983-1; Fort St. John 9 Sep 1984-1; Willow River 10 Sep 1985-1; Dawson Creek 11 Nov 1999-1; Prince George 15 Oct 1997-1; Williams Lake 24 Oct 1978-1; Deer Park (Riske Creek) 22 Sep 1992-22; Roundup Lake 5 Sep 1978-12; Kamloops 5 Oct 1965-23, 11 Nov 1986-1; Vernon 9 Sep 1977-3 juveniles; Okanagan Landing 7 Nov 1931-50; Nakusp 28 Sep 1981-1; Bridesville 9 Sep 1977-24. **Coastal** – Prince Rupert 31 Oct 1982-1 nr grain elevator; Sandspit 14 Nov 1994-1; Cape St. James 6 Oct 1984-1; Cape Scott 20 to 29 Sep 1935-1; Fort Rupert 1 Nov 1950-12; Port Hardy 20 Oct 1939-6; Cranberry Lake 2 Nov 1984-24; Boundary Bay 22 Oct 1921-40; Vancouver 23 Sep 1964-17; Reifel Island 17 Sep 1965-16; Central Saanich 22 Nov 1983-50.

Winter: Interior – Williams Lake 28 Feb 1985-1; Chilcotin River (Riske Creek) winter 1986-1 at Wineglass Ranch; Golden 26 Feb 1977-1; Tranquille 17 Jan 1988-4, 15 Feb 1988-4; Kamloops 10 Feb 1980-30; Vernon 31 Dec 1952-120; Penticton 1 Jan 1968-6; Nelson 9 Jan 1999-1; Creston 22 Dec 1980-15. **Coastal** – nr Prince Rupert 9 Dec 1983-1; Port Neville 2 Dec 1975-1; Port McNeil 14 Jan 1982-4; Quatse River 31 Dec 1980-2 on estuary (Dawe et al. 1995); Squamish River 22 Jan 1978-1; Brackendale 15 Dec 1975-1; Comox 11 Dec 1948-50; Harrison Bay 3 Jan 1986-31; Sumas Prairie 2 Feb 1974-60; Boundary Bay 29 Dec 1979-59; White Rock 24 Jan 1982-150; Saanich 1 Jan 1988- 40.

Christmas Bird Counts: Interior – Recorded from 6 of 27 localities and on 12% of all counts. Maxima: Vernon 21 Dec 1980-58; Penticton 27 Dec 1987-17; Oliver-Osoyoos 28 Dec 1987-12. **Coastal** – Recorded from 18 of 33 localities and on 39% of all counts. Maxima: Ladner 2 Jan 1961-153; Victoria 27 Dec 1958-126; White Rock 30 Dec 1973-109.

Yellow-headed Blackbird
Xanthocephalus xanthocephalus (Bonaparte)

YHBL

RANGE: Breeds from western Oregon, central Washington, south coastal and northeastern British Columbia, northern Alberta, north-central Saskatchewan, central Manitoba, extreme western and southern Ontario, northern Minnesota, and northern Michigan south to southern California, northeastern Baja California, southwestern and east-central Arizona, New Mexico, northern Texas, western Oklahoma, central Kansas, western and northern Missouri, central Illinois, northwestern Indiana, and northwestern Ohio. Winters from Washington, Oregon, California, central Arizona, Colorado, and southeastern Texas south to southern Baja California, northern Guerrero, Chiapas, Puebla, and central Veracruz.

STATUS: On the coast, *uncommon* migrant, *fairly common* local summer visitant, and *very rare* in winter in the Georgia Depression Ecoprovince; in the Coast and Mountains Ecoprovince, *very rare* spring vagrant and *casual* autumn vagrant on Western Vancouver Island; *very rare* spring vagrant and *casual* in summer on the Southern Mainland Coast, *casual* on the Northern Mainland Coast, and absent from the Queen Charlotte Islands.

In the interior, *fairly common* to *common* migrant, *common* to *abundant* summer visitant, and *very rare* in winter in the Southern Interior Ecoprovince; *uncommon* spring migrant, *fairly common* to *common* local summer visitant, *very common* to *abundant* autumn migrant, and *casual* in winter in the Southern Interior Mountains Ecoprovince; *fairly common* to *common* spring migrant, *common* to *abundant* local summer visitant, *fairly common* autumn migrant, and *casual* in winter in the Central Interior Ecoprovince; *rare* migrant and *uncommon* local summer visitant in the Sub-Boreal Interior Ecoprovince; *uncommon* migrant and *fairly common* local summer visitant in the Boreal Plains Ecoprovince; *casual* vagrant in the Taiga Plains and Northern Boreal Mountains ecoprovinces.

Breeds.

Figure 489. The male Yellow-headed Blackbird is conspicuous in many interior wetlands (Myers Lake, 6 June 1997; R. Wayne Campbell).

Figure 490. In late May and early June, female Yellow-headed Blackbirds are busy feeding the first young to hatch in southern British Columbia (Myers Lake, 6 June 1997; R. Wayne Campbell).

CHANGE IN STATUS: The Yellow-headed Blackbird (Figs. 489 and 490) has significantly expanded its breeding range in British Columbia during the 1900s. Munro and Cowan (1947) reported it to be a locally abundant breeder in the dry interior from the Okanagan valley north to the Cariboo Basin, but becoming much less common north through the northern Central Interior to Vanderhoof and west through the Chilcotin Plateau. By the mid-1940s, the Okanagan valley, Nicola valley, and southern Cariboo from about 100 Mile House to Springhouse appeared to contain the most significant colonies in the province, and Tachick and Nulki lakes, near Vanderhoof, were the northernmost breeding locations.

The Yellow-headed Blackbird was not found in the Boreal Plains prior to the mid-1940s (Cowan 1939; Munro and Cowan 1947). The first birds recorded there were 9 counted on Cecil Lake in 1976. Small breeding colonies have since been established at a few localities in the Peace Lowland. In the Peace River region of Alberta, the Yellow-headed Blackbird was noted as scarce (Salt and Wilk 1958; Semenchuk 1992), but the birds nesting in the Peace Lowland of British Columbia likely originated from Alberta. Pinel et al. (1993) note increases in numbers of Yellow-headed Blackbirds in southern Alberta over the last few decades.

In the Sub-Boreal Interior, the Yellow-headed Blackbird was first recorded in the 1970s near Prince George, where it now occurs regularly in small numbers. Although breeding has yet to be confirmed near Prince George, it occurs to the west near Vanderhoof; all these colonies are within the northern edge of the Central Interior. Further south, Munro (1947)

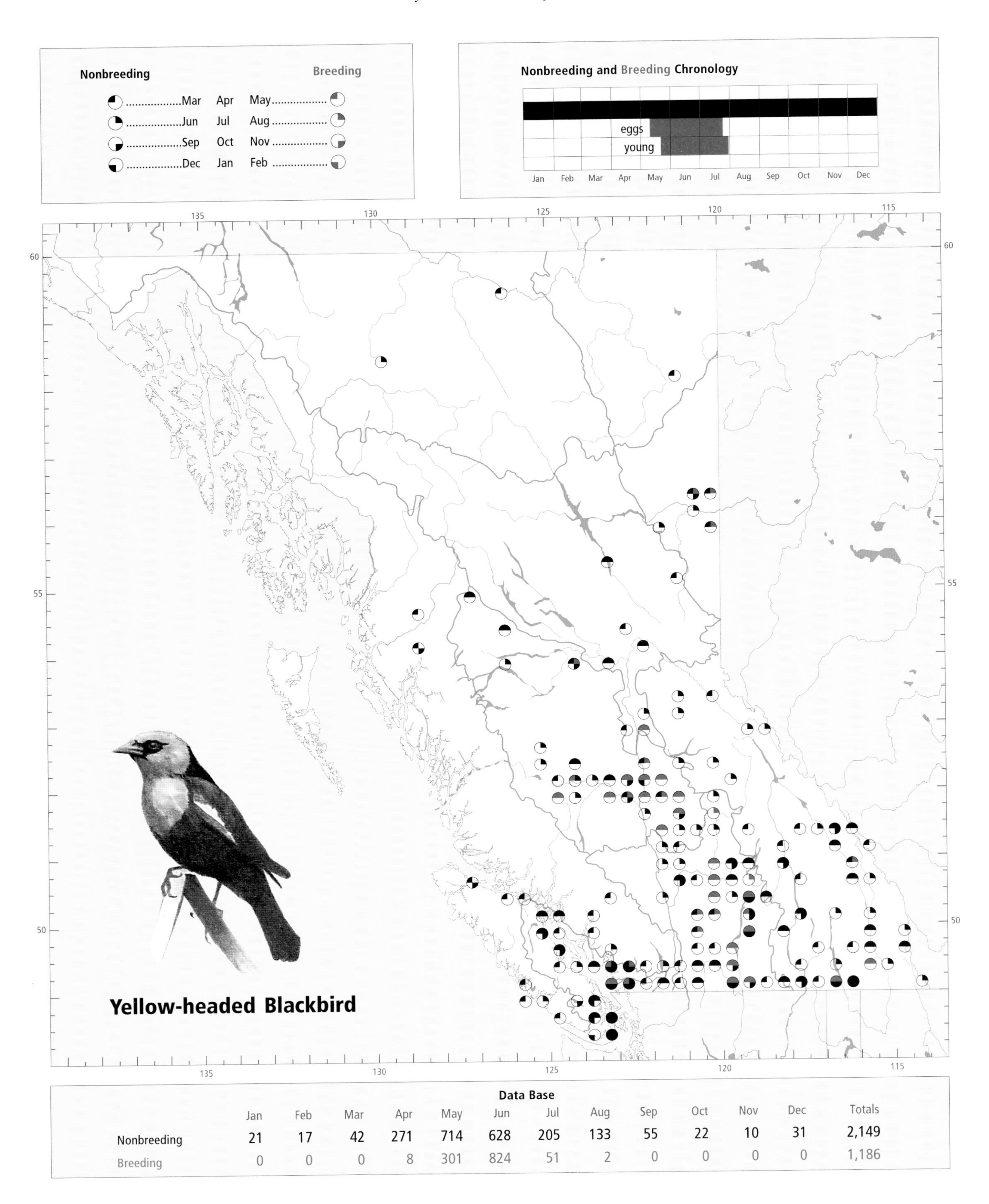

Data Base

	Jan	Feb	Mar	Apr	May	Jun	Jul	Aug	Sep	Oct	Nov	Dec	Totals
Nonbreeding	21	17	42	271	714	628	205	133	55	22	10	31	2,149
Breeding	0	0	0	8	301	824	51	2	0	0	0	0	1,186

Figure 491. During spring migration in the southern portions of the interior of British Columbia, the Yellow-headed Blackbird forages among loose piles of hay around barnyards (Creston, 28 April 1997; R. Wayne Campbell).

reported the Yellow-headed Blackbird to be only a casual spring vagrant in the Quesnel area. Breeding colonies are now known to occur near Quesnel, at Dragon Lake and Dale Lake.

In the Chilcotin Plateau, the Yellow-headed Blackbird remained relatively scarce through the 1940s and 1950s. Munro and Cowan (1947) cited a noteworthy record of 4 birds at Alexis Creek in 1946, and Paul (1959) reported only migrants near Kleena Kleene from 1947 to 1958. Paul (1959) did note that F. Shillaker had found 4 birds near Chezacut in 1939 and that breeding was suspected at "Maxwell's Lake." Several years later, Shillaker found them breeding in small numbers near Redstone and Chezacut.

Numbers were obviously low during this period, as Paul (1959) did not find any colonies up to 1958. Paul (1964) then found a colony of about 30 pairs in 1961 and 1962 at a wetland on a neighbouring ranch near Kleena Kleene. Yellow-headed Blackbirds did not nest on his ranch in those years, but he subsequently recorded 50 pairs nesting on his land in 1963, and 40 to 50 pairs each year thereafter until 1967. The Yellow-headed Blackbird is currently a locally common breeder on the Chilcotin Plateau.

In the Okanagan valley, the abundance of Yellow-headed Blackbirds fluctuated markedly during the 20th century. Munro (1915) noted that it was unusually abundant in 1914, whereas it was normally scarce. Allan Brooks noted it to be common in 1925 at the north end of Okanagan Lake, where it had formerly been rare (Cannings et al. 1987). At Swan Lake, near Vernon, hundreds were thought to be nesting in the 1930s, more than 900 were nesting in 1950, and more than 1,200 in 1963. Through the 1970s, however, numbers declined and fewer than 100 were found at Swan Lake in 1978 (Cannings et al. 1987).

Similar fluctuations occurred at Richter Pass, near Osoyoos, but declines were likely due to low water levels and periodic draining of some key wetlands. Population declines at Richter Pass occurred at the same time that populations were increasing at the north end of Osoyoos Lake (Campbell and Meugens 1971), which was likely the natural result of birds relocating to more stable wetlands.

One of the best Yellow-headed Blackbird wetlands in the Okanagan valley was lost in the 1980s when Alki Lake was filled by an expanding garbage dump near Kelowna. The loss of this wetland, plus significantly lower numbers at Swan Lake and elsewhere, suggest that the species' relative abundance in the Okanagan valley is lower now than during the middle part of the 20th century.

In the Southern Interior Mountains, the Yellow-headed Blackbird was thought to occur at Creston only during migration, and was quite rare in some years. Nesting was not noted in 1947 or 1948, and only a single migrant was noted in 1948 (Munro 1950). By 1956, small numbers were nesting at Duck Lake in the Creston valley (Munro 1957). In the 1980s, the species was noted as a common and widespread breeder in cattail marshes in the Creston valley (Butler et al. 1986) and remained so through the 1990s.

In the east Kootenay, breeding was thought to occur only at Elizabeth Lake, Cranbrook, where 12 to 15 pairs nested every summer (Johnstone 1949). Yellow-headed Blackbirds are now known to nest in the Kootenay River floodplain in suitable habitat north to at least Windermere Lake.

Figure 492. Cattle feedlots, with spilled grain and associated insects, are important feeding sites for migrant Yellow-headed Blackbirds in southern British Columbia (Myers Lake, 13 April 1999; R. Wayne Campbell).

On the coast, Munro and Cowan (1947) noted the Yellow-headed Blackbird only as a casual visitor to the Fraser Lowland and the Gulf Islands. In 1968, a breeding colony was discovered on Sea Island, Richmond. This colony represented the first breeding range expansion to the Pacific coast of North America; indeed, it was located less than 200 m from tidal marshes in the Strait of Georgia. It reached a peak in 1970, when 36 nests were recorded, then declined for various reasons (see REMARKS). A small colony now exists on nearby Iona Island. In the Fraser Lowland, pairs have nested occasionally at a few other locations during the last 3 decades, with the most consistent location being Pitt Marsh.

Overwintering in British Columbia was unknown during the first half the 20th century (Munro and Cowan 1947), but it is not unexpected for 1 or 2 birds to remain during at least part of the winter on the south coast or at cattle feedlots in the Okanagan valley.

NONBREEDING: The Yellow-headed Blackbird is widely distributed from spring through autumn at low elevations throughout most of the southern and central portions of interior British Columbia. It is most abundant from the Southern Thompson Upland east to the Okanagan valley, and north through the Cariboo Basin and Chilcotin Plateau. In the Southern Interior Mountains, most birds occur in the southern valleys; few birds occur in mountainous areas. Numbers are lower in the west and east Kootenay, but the largest flocks of autumn migrants that have been recorded in British Columbia

Figure 493. During late summer and early autumn in the southern interior of British Columbia, the Yellow-headed Blackbird forages in large mixed flocks with European Starlings, Red-winged Blackbirds, and Brown-headed Cowbirds (Creston, 22 September 1996; R. Wayne Campbell).

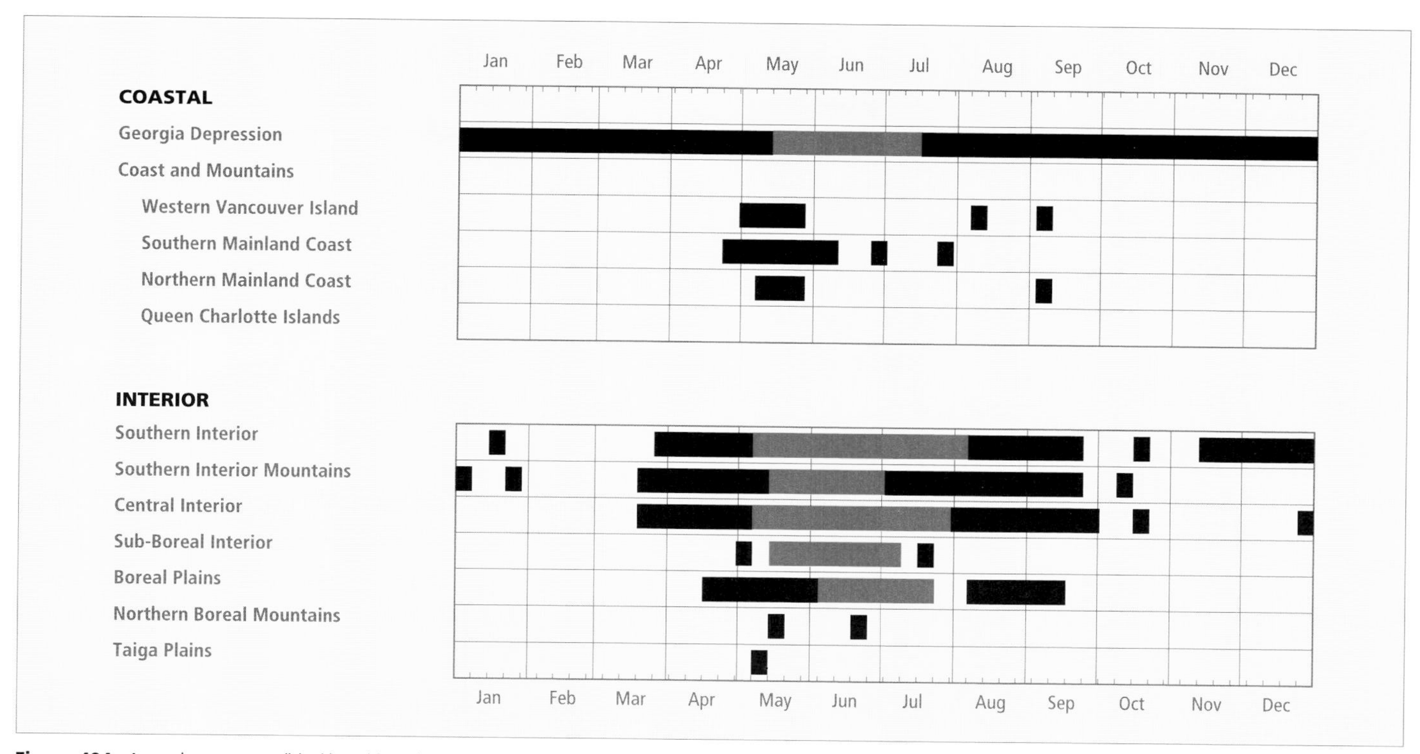

Figure 494. Annual occurrence (black) and breeding chronology (red) for the Yellow-headed Blackbird in ecoprovinces of British Columbia. Records are shown for the week in which they occurred.

occur in some years near Creston. The Yellow-headed Blackbird is a rare spring and summer visitor to Kootenay National Park (Wade 1977) and Yoho National Park (Poll et al. 1984). In Yoho National Park, most birds appear at Leanchoil or Ottertail, but do not stay because there is no suitable breeding habitat.

In the Central Interior, the Yellow-headed Blackbird is relatively abundant from 70 Mile House north to Williams Lake, west to Kleena Kleene, and east to Bridge Lake. North of Williams Lake, populations become much sparser and more locally distributed through to the Prince George, Vanderhoof, and Smithers areas. In the Peace Lowland of the Boreal Plains, the Yellow-headed Blackbird occurs very locally but in larger numbers than in the Sub-Boreal Interior. Vagrants have occurred as far north as Fort Nelson, Liard Hot Springs, and Gnat Pass in the Northern Boreal Mountains.

On the southwest mainland coast, the Yellow-headed Blackbird occurs mainly in the Fraser Lowland of the Georgia Depression. On Vancouver Island, it occurs infrequently along coastal areas from Victoria north to Port Hardy on the east coast, and north to Long Beach on the west coast. Only vagrants occur in the mountain valleys of the Southern Mainland Coast. In the Northern Mainland Coast, a few vagrants have appeared at Kitimat and as far north as Terrace. The Yellow-headed Blackbird has not been recorded on the Queen Charlotte Islands.

On the coast, the Yellow-headed Blackbird occurs from sea level to only a few metres above sea level. In the interior, it occurs from about 280 to about 1,250 m.

In all regions, wetlands with emergent vegetation, grassy uplands, and farmlands are preferred habitats. In early spring, barnyards and feedlots are favoured because insects and grain are relatively abundant there compared with more natural habitats (Figs. 491 and 492). During migration, the Yellow-headed Blackbird may feed on spilled grain along railways. In the Creston valley, large flocks of migrants forage with European Starlings, Red-winged Blackbirds, and Brown-headed Cowbirds in cattle feedlots, and autumn migrants make use of the harvested and unharvested agricultural fields (Fig. 493). Other habitats include landfills, alluvial floodplains (in rugged mountainous areas), and sewage lagoons.

On the coast, the Yellow-headed Blackbird has been observed in open chicken runs, manure piles, estuaries, marine beaches, lawns, and sewage ponds. In autumn, it also frequents corn stubble fields and may appear at backyard bird feeders.

On the coast, spring migrants begin to arrive in the Georgia Depression in the second week of April, with a peak movement in the last week of April or early May (Fig. 495). Elsewhere on the coast, migrants are extremely rare. On Western Vancouver Island, spring vagrants may appear from early to late May. In the interior, early spring migrants begin to arrive in the Southern Interior, Southern Interior Mountains, and Central Interior in the fourth week of March (Figs. 494 and 495). Peak spring movements occur in these ecoprovinces in late April and early May, and breeding numbers build through mid-May. In the Sub-Boreal Interior, the first spring migrants arrive in early May, with the peak movement in the third week of May. In the Boreal Plains, spring migrants can arrive as early as the third week of April but the peak movement occurs in mid to late May. Males arrive about 2 weeks earlier than females in the Okanagan valley (Cannings et al.

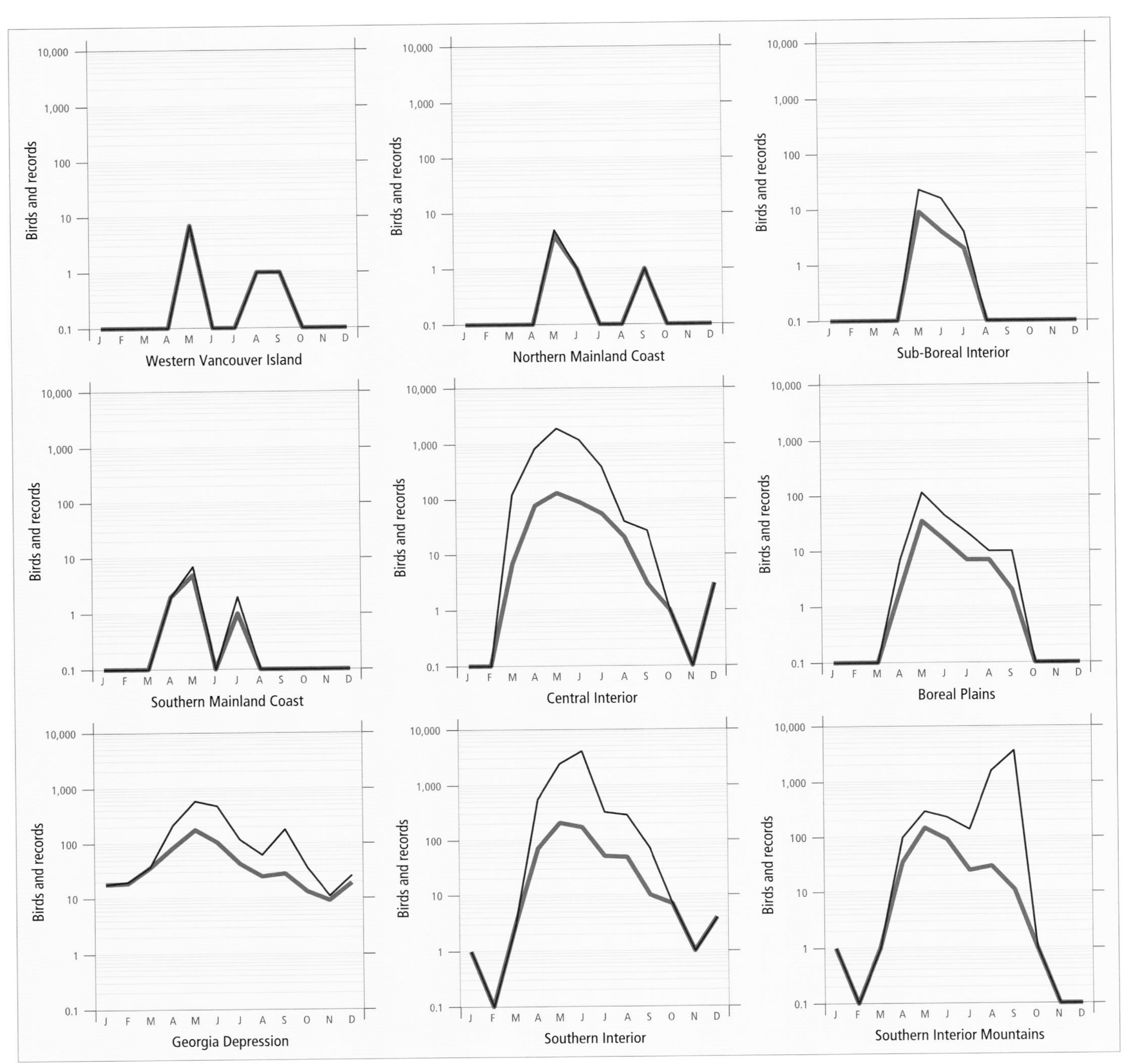

Figure 495. Fluctuations in total number of birds (purple line) and total number of records (green line) for the Yellow-headed Blackbird in ecoprovinces of British Columbia. Christmas Bird Counts, Breeding Bird Surveys, and nest record data have been excluded.

1987), and 3 or 4 days earlier in the east Kootenay (Johnstone 1949).

In late July, the Yellow-headed Blackbird drifts away from its nesting marshes to join mixed, post-breeding flocks of premigrant Red-winged and Brewer's blackbirds in agricultural areas where farms and feedlots provide food (Fig. 493). The autumn migration is poorly defined in most regions, as few large flocks have been noted. In the Boreal Plains, all birds have left by the second week of September; most probably leave in August. In the Central Interior, all have left by the end of September. In the Southern Interior Mountains, flocks of up to 3,000 birds have been noted from mid-August to early September. These records, plus several records of about 200 birds as late as 20 September, indicate that the main autumn migration occurs from late August to mid-September (Figs. 494 and 495). All birds have left this ecoprovince by mid-October. In the Southern Interior, most migrants have left by the third week of September, with a few stragglers remaining into October in some years.

On the coast, autumn migrants leave the Fraser Lowland mainly in September, with the odd straggler remaining through the first week of October. This blackbird does not occur regularly in winter in British Columbia. During Christmas Bird Counts, however, Yellow-headed Blackbirds are occasionally recorded on the coast at Ladner and Victoria, and in the interior at Oliver-Osoyoos, Creston, Kelowna, and Vernon.

On the coast, the Yellow-headed Blackbird occurs in the Georgia Depression throughout the year but only from 24 April to 4 September elsewhere; in the interior, it occurs regularly from late March to late September (Fig. 494).

BREEDING: The Yellow-headed Blackbird is a locally distributed breeding bird in marshy wetlands in the valleys and lower plateaus of the southern portions of the interior of British Columbia. It breeds mainly from the Southern Thompson Upland east to the East Kootenay Trench, and from the international boundary north to near Vanderhoof. Its centre of breeding abundance is in the Southern Interior and Central Interior. In the latter ecoprovince, it breeds in wetlands associated with the grasslands and parkland areas of the Cariboo Basin and the eastern part of the Chilcotin Plateau.

The Yellow-headed Blackbird is also a relatively common breeding bird in the Southern Interior, especially in the Southern Thompson Upland, Northern Okanagan Basin, Southern Okanagan Basin, and Okanagan Highland. In the Southern Interior Mountains, it breeds locally, with significant colonies located only in the Creston Valley Wildlife Management Area, at Elizabeth Lake in marshes along the Kootenay River floodplain. Summer records in the East Kootenay Trench suggest that breeding may occur as far north as Golden. The Yellow-headed Blackbird also breeds locally in the Peace Lowland of the Boreal Plains.

In southern parts of interior British Columbia, the breeding distribution of the Yellow-headed Blackbird is linked mainly to the distribution of the Bunchgrass, Interior Douglas-fir, and Ponderosa Pine biogeoclimatic zones, except in the western Southern Interior Mountains, where it is linked to the Interior Cedar–Hemlock Biogeoclimatic Zone.

On the coast, the Yellow-headed Blackbird is a scarce and local breeding bird in the Fraser Lowland of the Georgia Depression. It does not breed on Vancouver Island, on other coastal islands, or on the mainland coast north of Vancouver. The only regularly occupied colony extant is at Iona Island, near the mouth of the north arm of the Fraser River, where a few pairs nest each year in a rehabilitated dredge spoil area (see REMARKS). Since the 1970s, lone pairs have nested occasionally at Jericho Beach Park (Vancouver), at Serpentine Fen (Surrey), and near Chilliwack. At Pitt Marsh, Pitt Meadows, 1 or more pairs may nest in most years. Up to 60 birds were counted there in the 1960s, but numbers were much lower in the 1990s.

The Yellow-headed Blackbird reaches its highest numbers in summer in the Southern Interior and Central Interior (Fig. 496). An analysis of Breeding Bird Surveys for the period 1968 through 1993 could not detect a net change in numbers on interior routes; on the coast, the data were insufficient for analysis. Breeding Bird Surveys reveal that during the period 1966 to 1979, the Canadian population increased at an average annual rate of 7.7% ($P < 0.01$) (Sauer et al. 1997).

On the coast, the Yellow-headed Blackbird breeds at elevations from just above sea level to about 20 m elevation. In the interior, nests have been found at elevations from 280 to about 1,200 m. In general, nesting occurs in valley bottoms or on plateaus.

The Yellow-headed Blackbird is a characteristic breeding bird of marsh wetlands in dry grasslands or open farmlands. In general, suitable nesting habitat has 3 components: (1) robust stands of emergent vegetation over standing water for placement of nests, (2) an abundance of aquatic insects for food, and (3) sufficient edge along stands of emergent vegetation to allow foraging on emerging aquatic insects. In British Columbia, nesting habitats ($n = 750$) are always associated with wetlands that have emergent vegetation. Virtually all wetlands used for nesting ($n = 558$) have dense stands of bulrushes or cattails (97%). Rare exceptions are wetlands that do not have bulrushes or cattails but instead have flooded willows or willows growing over water (3%). One study in the Cariboo and Chilcotin areas showed that the Yellow-headed Blackbird tends to occur in more productive wetlands than the Red-winged Blackbird (Orians 1966).

Colonies usually occur in small wetlands such as sloughs, ponds (Fig. 497), marshes, or small lakes (Figs. 498 and 499) ranging in size from 0.5 to 5 ha. In larger, deeper lakes, such as Stump Lake, Williams Lake (see Fig. 477), Okanagan Lake, or Osoyoos Lake, nesting is often localized at one end of the lake or in shallower bays where there is dense emergent vegetation. In larger, shallower lakes that have widespread stands of dense emergent vegetation, such as Duck Lake (Creston) or Swan Lake (Vernon; Fig. 498), colonies can be spread out

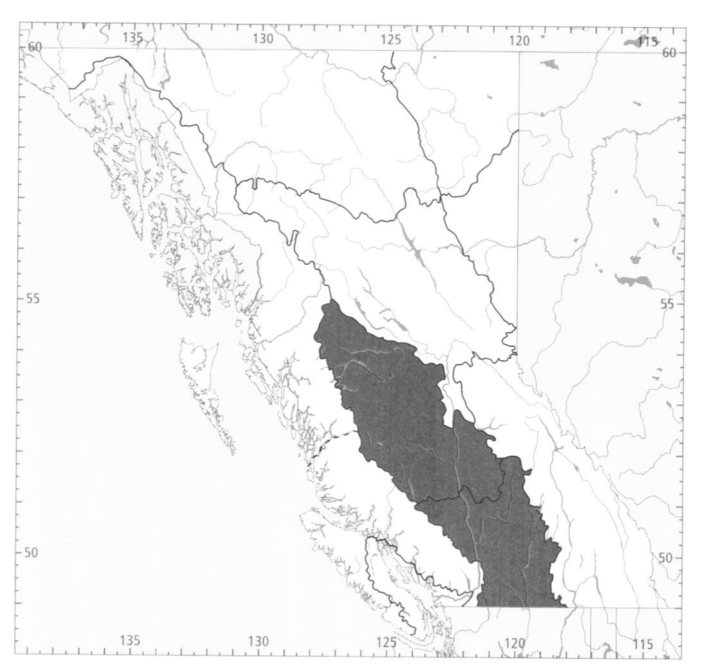

Figure 496. In British Columbia, the highest numbers for the Yellow-headed Blackbird in summer occur in the Southern Interior and Central Interior ecoprovinces.

Figure 497. There is only 1 colony of Yellow-headed Blackbirds that is regularly occupied each year on the south coast of British Columbia. It is situated in a cattail marsh in a rehabilitated dredge spoil area (Iona Island, Richmond, 24 August 1998; R. Wayne Campbell).

over much of the marshy area. Since water depth largely controls the distribution of emergents, most nesting habitats begin a few metres out from the shoreline and extend towards deeper water until emergents begin to thin out. Occasionally, shallow areas in the middle of wetlands may have islands of emergents, and the Yellow-headed Blackbird will readily nest there.

Fluctuating water levels affect the quality and quantity of stands of emergents from year to year, and therefore the distribution of nesting Yellow-headed Blackbirds. For example, more than 100 pairs of Yellow-headed Blackbirds may nest at the south end of Stump Lake. Relatively low water levels in 1991 and 1992 resulted in most of the emergent vegetation being in very shallow water, which is unsuitable for this blackbird. Nests were restricted to the outer edges of the cattail stands, and blackbird numbers were lower than normal. In 1998, very high water submerged the emergents and they were not available when the blackbirds returned. By the time the plants had grown sufficiently, the blackbirds had nested or moved elsewhere (R.R. Howie pers. comm.).

On the coast the Yellow-headed Blackbird has been recorded breeding from 13 May (calculated) to 9 July; in the interior, it has been recorded from 11 May (calculated) to 30 July (Fig. 494).

Colonies: The Yellow-headed Blackbird almost always nests in colonies. In British Columbia, lone pairs rarely occur and are usually found only in marginal habitats or in areas at the extremes of their breeding range. The sizes of 60 colonies where active nests were counted ranged from 2 to 600 nests (Table 10). The 2 largest colonies with documented active nests were at Stump Lake (300 nests) and Swan Lake (Vernon) (200 nests). Both lakes are wetlands with extensive bulrush beds. Historically, the largest colony in British Columbia probably occurred at Swan Lake. Swan Lake has a relatively large wetland with extensive beds of cattails along the eastern shore. It contained about 900 adult birds in 1950, more than 1,200 adult birds in 1963, and 504 males in 1965 (Cannings et al. 1987). These numbers suggest that there were likely more than 600 nests in some years in the 1960s (see REMARKS).

Nests: The Yellow-headed Blackbird builds its nest over water in relatively thick stands of emergent vegetation. In British Columbia, 96% of 558 nests were placed in stands of "rushes," mainly bulrushes (48%; Fig. 500), "reeds" (bulrush or cattail; 35%), or cattails (10%). Live cattails or bulrushes are used only very late in the nesting season. Occasionally, willow shrubs growing in water or hanging over water were used (3%).

Table 10. Locations and sizes of major colonies of the Yellow-headed Blackbird in British Columbia.

Location	Date	Nests	Pairs	Nest location	Source[1]
COASTAL: Colonies with ≥ 5 nests or pairs					
Georgia Depression					
Burnaby Lake	29 May 1970	36	36	–	1
Iona Island	31 May 1995	10	–	–	
INTERIOR: Colonies with ≥ 30 nests or pairs					
Southern Interior					
Glimpse Lake (Nicola valley)	29 July 1980	–	30	Cattail	
Myers Lake (5 km sw Rock Creek)	26 May 1970	15	75	Reeds	
Okanagan Lake	28 June 1980	–	72	–	
Spotted Lake	19 May 1963	47	68	Reeds	2
Armstrong	10 June 1980	17	60	Cattail	
Stump Lake	28 May 1995	300	–	–	
Swan Lake (Vernon)	2 June 1995	200	–	Cattail	
Southern Interior Mountains					
Elizabeth Lake	14 July 1984	50	50	Cattail	
Central Interior					
123 Mile House	15 June 1963	38	45	Reeds	
Hanceville (Jones Lake)	19 June 1980	26	30	Cattail	
Kleena Kleene	12 July 1967	–	40	Reeds	
Riske Creek (nr Roundup Lake)	14 June 1978	58	80	Bulrush	
Westwick Lakes	14 June 1963	95	115	–	
Westwick Lakes	14 June 1978	123	147	Bulrush	
Williams Lake	20 May 1963	8	50	Reeds	

[1] All data are estimates and are from the British Columbia Nest Records Scheme unless otherwise noted (1, Campbell et al. 1972a; 2, Campbell and Meugens 1971).

The heights above the water level of 476 nests ranged from 0.0 to 3.0 m, with 68% between 0.3 and 0.6 m. The highest nests were in willows. Nests recorded at or near the water level had probably been built higher, but rising waters likely reduced the original height above water. Nests in reeds are woven into several vertical stalks; those in shrubs are woven into several twigs or small branches. In bulrush and cattail stands, dead stalks from the previous year are usually used.

Nests (n = 489) were bulky cups of grasses (62%; Figs. 500 and 501), rushes (55%), plant fibre (34%), sedges (8%), mosses (3%), leaves (2%), and feathers (2%). Grass was reported more frequently from the Central Interior (87% of 222 nests) compared with the Southern Interior (58% of 154 nests), Georgia Depression (14% of 97 nests), and Southern Interior Mountains (1 of 8 nests). In general, nests are built mainly of bulrush in bulrush marshes and of cattails in cattail marshes.

Figure 498. Wetlands used for nesting by the Yellow-headed Blackbird in the Okanagan valley of British Columbia typically have extensive beds of cattails (Swan Lake, Vernon, 2 June 1995; R. Wayne Campbell).

Figure 499. In the Central Interior Ecoprovince of British Columbia, many Yellow-headed Blackbird colonies are found in dense patches of bulrushes (Rock Lake, near Riske Creek, 8 June 1995; R. Wayne Campbell).

Figure 500. Yellow-Headed Blackbird nests are bulky cups of rushes and coarse grasses, often lined with finer grasses and placed in thick stands of emergent vegetation (Myers Lake, 3 June 1997; R. Wayne Campbell).

Figure 501. Most Yellow-headed Blackbird clutches in British Columbia contained 3 or 4 eggs (Stump Lake, 20 May 1992; John M. Cooper).

Figure 502. Many young Yellow-headed Blackbirds leave the nest before they can fly, and hide in dense emergent vegetation, where they are fed (Swan Lake, Vernon, 2 June 1995; R. Wayne Campbell).

Nests are typically lined with grasses; where cattails are present, cattail "down" is often used. Many nests do not have a finer lining material, however.

Eggs: Dates for 692 clutches (Fig. 501) ranged from 13 May to 24 July, with 55% recorded between 27 May and 14 June. Calculated dates indicate that eggs can occur as early as 11 May. Egg laying begins about a week earlier in the Southern Interior than in the Georgia Depression or the Central Interior, and about 4 weeks earlier than in the Boreal Plains (Fig. 503).

Birds in colonies at higher elevations likely lay eggs later than those in colonies at lower elevations. For example, in 1992 egg laying began at Stump Lake (elevation 745 m) on 10 May, whereas in a small colony east of Quilchena (elevation 1,150 m), birds were just beginning to build nests on that date.

Sizes of 821 clutches ranged from 1 to 6 eggs (1E-127, 2E-125, 3E-189, 4E-353, 5E-26, 6E-1), with 66% having 3 or 4 eggs (Fig. 501). The modal clutch size in all ecoprovinces is 4 eggs. The incubation period for 3 clutches ranged from 11 to 13 days. Elsewhere, the incubation period is 12 or 13 days, with the 12-day period occurring more commonly early in the nesting season (Twedt and Crawford 1995).

Unlike the Red-winged and Brewer's blackbirds, the Yellow-headed Blackbird raises only a single brood each year. Replacement clutches are laid if the first clutch is lost, but usually contain fewer eggs (Twedt and Crawford 1995). In the Okanagan valley, the latest date for the beginning of a replacement clutch is 22 June (Cannings et al. 1987).

Young: Dates for 359 broods ranged from 22 May to 30 July, with 56% recorded between 8 June and 21 June. Sizes of 430 broods ranged from 1 to 5 young (1Y-53, 2Y-114, 3Y-145, 4Y-109, 5Y-9), with 86% having 2 to 4 young.

The nestling period for 1 brood in British Columbia was 12 days. The nestling period reported elsewhere is 9 to 12 days (Twedt and Crawford 1995). Young leave the nest several days before they can fly, and hide in vegetation near the nest (Fig. 502). They remain dependent on their parents for food during this preflight phase. They also pursue the female and vocally beg for food for several days after fledging.

Brown-headed Cowbird Parasitism: In British Columbia, less than 1% of 537 nests found with eggs or young were parasitized by the cowbird. Parasitism occurred in 1 of 440 nests in the interior and in 1 of 97 nests on the coast. Parasitized nests contained either 1 or 2 cowbird eggs. Cowbird fledglings being fed by the Yellow-headed Blackbird were not recorded. Elsewhere, the Yellow-headed Blackbird is parasitized only incidentally by cowbirds (Willson 1966; Dufty 1994).

Nest Success: Of 84 nests found with eggs and followed to a known fate, 38% produced at least 1 fledgling. Nest success was 29% (n = 63) in the interior and 67% (n = 21) on the coast.

Nest predation is common in the Yellow-headed Blackbird. In 1 British Columbia study, 15 km southeast of Hanceville, 25 of 65 (38%) active nests lost their contents or were preyed upon, 18 of them during the egg-laying period including 10 when only a single egg was present (Lyon et al. 1992). The predators in this study were not identified. In Manitoba, between 45% and 62% of nests fledged at least 1 young (Lightbody and Weatherhead 1987).

Fluctuating water levels may reduce nest success. During the spring and early summer of 1996, a heavy snow pack and wet spring resulted in rising water levels through the nesting period in the southern and central portions of interior British Columbia. Many Yellow-headed Blackbird colonies were affected. For example, at 108 Mile Lake, on 12 June 1996, 2 nests with 3 young each and 1 nest with 2 eggs were under water, and a fourth nest with eggs was almost under water; some young had already fledged from this colony. Dropping water levels can also expose nests to increased predation by mammals.

Marsh Wrens occasionally destroy Yellow-headed Blackbird eggs in areas cohabited by the 2 species (Bump 1986), but the territorial centres of the 2 species are usually spatially separated by the behavioural dominance of the blackbirds (Leonard and Picman 1986). Three clutches in our sample were noted to have been pecked by Marsh Wrens.

Starvation is the major cause of nestling mortality (Willson 1966), and may be linked to cold, rainy weather (Orians 1966). In the Cariboo, nestling survival was found to be higher in more productive wetlands (Orians 1966). In British Columbia, 8 of 231 nests recorded with young contained at least 1 dead nestling.

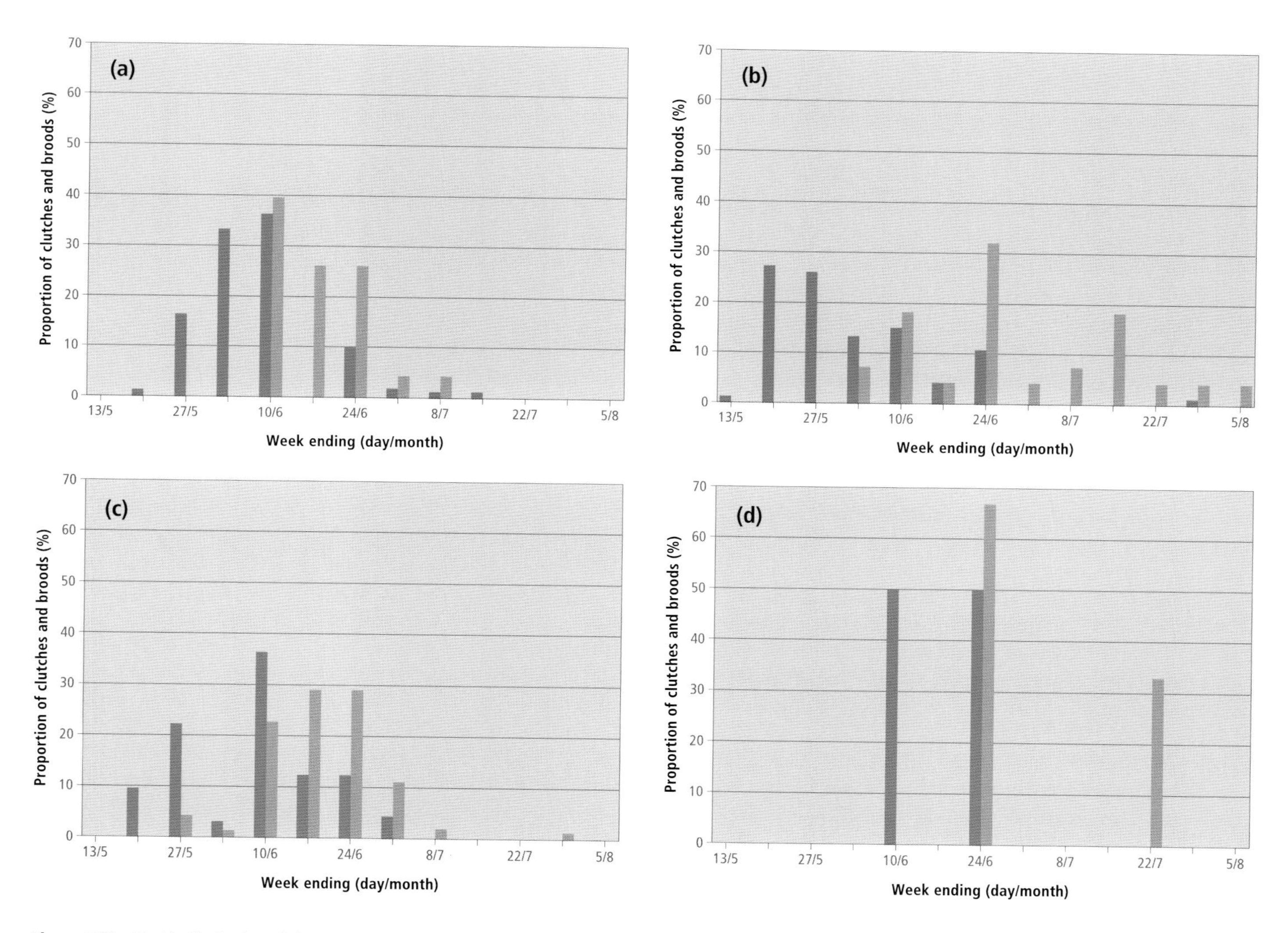

Figure 503. Weekly distribution of the proportion of clutches (dark bars) and broods (light bars) for the Yellow-headed Blackbird found in the (a) Georgia Depression (clutches: n = 83; broods: n = 69), (b) Southern Interior (clutches: n = 134; broods: n = 28), (c) Central Interior (clutches: n = 236; broods: n = 134), and (d) Boreal Plains (clutches: n = 2; broods: n = 3) ecoprovinces of British Columbia. Egg laying begins about a week earlier in the Southern Interior than in the Georgia Depression or the Central Interior, and about 4 weeks earlier than in the Boreal Plains.

Figure 504. Throughout British Columbia, small and large wetlands supporting populations of birds restricted to breeding in marshes are being filled in or drained at an alarming rate for local industrial, residential, and agricultural purposes (15 km northwest of Fort St. John, 23 June 1996; R. Wayne Campbell).

REMARKS: The Yellow-headed Blackbird is polygynous, with most territorial males having more than 1 mate. In a study in Washington, the number of females paired with each territorial male ranged from 0 to 8, with an average of 2 to 4 on different wetlands (Orians 1980). Because of this, observers cannot estimate the number of nests or "pairs" from a count of males alone. Counts of males on a nesting marsh should be reported as numbers of males. Accurate estimates can be made only by counting active nests, which contain eggs or young. Empty nests should be counted separately, as a significant percentage of nests do not receive eggs (Miller 1968).

North American populations are considered stable. Local fluctuations in population size and distribution can be expected to occur depending on wetland conditions. In general, local populations increase as water levels rise and decrease as water levels decline (Lederer et al. 1975). Sometimes these fluctuations can be dramatic. For example, North Dakota has the highest density of breeding Yellow-headed Blackbirds in North America. In 1981 and 1982, the Missouri Coteau physiographic region contained the highest densities in the state, but in 1990 no Yellow-headed Blackbirds occurred there at all. This decline was attributed to severe drought conditions (Twedt and Crawford 1995). In Alberta, this blackbird is more common in the central and northern parts of the province during dry years, probably as a result of more birds searching for suitable nesting habitat outside of their normal breeding range (Semenchuk 1992).

Besides range expansions in British Columbia, expansions in eastern parts of the species' range have likely occurred (Twedt and Crawford 1995). Breeding colonies were not found in Ontario until 1961 (Peck and James 1987) or in Michigan until 1956 (Brewer et al. 1991). In Washington, the Yellow-headed Blackbird is expanding its breeding distribution in the western parts of the state, although it still occurs very locally and uncommonly (Smith et al. 1997).

Threats to habitat are mainly related to draining of wetlands (Fig. 504), but this threat is minimized because the Yellow-headed Blackbird tends to nest in deeper and larger wetlands that are less likely to be drained for human development (Twedt and Crawford 1995). In British Columbia, however, there are several examples of local population decline or extirpation due to draining of wetlands. Near Kelowna, Alki Lake was converted to an extension of the local landfill, which destroyed one of the best marshes for the Yellow-headed Blackbird in the Okanagan valley (Cannings et al. 1987).

In Richter Pass, near Osoyoos, upper and lower Richter lakes fluctuate periodically from dry to full, from a combination of draining for irrigation and medium-term weather cycles. Dry periods may last several years, and the Yellow-headed Blackbird is absent then. After the lakes refill and emergent growth becomes re-established, the Yellow-headed Blackbird nests there once again. On the coast, a colony on Sea Island, Richmond, that contained 36 nests in 1970 was abandoned the following year because the wetland was drained (Butler and Campbell 1987). By 1985, small numbers were again nesting at this site, and continued to do so until 1992 (Cooper 1993). The wetland was filled in 1993 as the Vancouver International Airport was expanded.

Populations can fluctuate from year to year even in seemingly stable wetlands. For example, at Swan Lake near Vernon, there were about 1,200 breeding Yellow-headed Blackbirds in 1965, but numbers dwindled through the 1970s to fewer than 100 in 1978 (Cannings et al. 1987). An apparent influx of breeding populations in the Chilcotin Plateau coincided with these peak counts at Swan Lake in the early and mid-1960s (Cannings et al. 1987), and roughly correlates with the discovery of expanding populations in Ontario and Michigan. Currently, the Yellow-headed Blackbird is a widespread breeder in the Chilcotin Plateau.

In response to plans for expanding airport facilities on Sea Island, which included filling of the remaining marsh used by nesting Yellow-headed Blackbirds, Transport Canada funded an attempt to establish a nesting population at Jericho Park in Vancouver. From 1990 through 1994, Yellow-headed Blackbird eggs were transplanted from the British Columbia interior into Red-winged Blackbird nests at Jericho Park. Several young were fledged each year, but none returned to nest there and the attempt to establish a nesting colony was unsuccessful (Searing and Cooper 1992; Searing and Schieck 1995). The Yellow-headed Blackbird colony on Sea Island moved to a marsh adjacent to Iona Island that had been rehabilitated by the Greater Vancouver Regional District.

Yellow-headed Blackbirds and Red-winged Blackbirds may nest in the same wetland, but male Yellow-headed Blackbirds are dominant and will displace territorial male Red-winged Blackbirds. The result is that Yellow-headed Blackbirds nest over deeper waters and Red-winged Blackbirds are relegated to shallower waters (Orians and Willson 1964; Willson 1966).

Nestlings are fed a diet consisting mainly of aquatic insects. A study in the Cariboo found that the primary insects fed to nestlings were damselflies at Rush Lake, dipterans at Westwick Lakes, and sawflies at 150 Mile House (Orians 1966).

The same study showed that adults tend to feed on aquatic insects during mid-day but feed more on terrestrial insects in upland areas during the morning and late afternoon (Orians 1980).

For a summary of life-history information on the Yellow-headed Blackbird throughout its worldwide range, see Twedt and Crawford (1995).

NOTEWORTHY RECORDS

Spring: Coastal – Carmanah Point 16 May 1951-1 (Irving 1953); Ucluelet 6 May 1972-1 (Hatler et al. 1978), 16 May 1992-1 (Siddle and Bowling 1993); Reifel Island 24 Apr 1972-16 at refuge; Skagit River 24 Apr 1971-1 at Ponderosa Meadows; Vedder River 4 May 1992-1 male on territory; Tofino 10 May 1975-1 (Hatler et al. 1978); Sea Island 17 Apr 1985-5, 29 Apr 1972-22, 19 May 1971-35 including a flock of 25 females and immatures (Campbell et al. 1972b), 27 May 1969-4 eggs, 29 May 1970-4 nestlings; Iona Island 31 May 1995-11 (Elliott and Gardner 1997); Burnaby Lake 24 May 1970-4 eggs; Vancouver 4 Mar 1988-1, 14 May 1988-2 eggs; Campbell Lake 26 May 1994-4 eggs; Hope 20 May 1986-2 at airport; Courtenay 1 May 1986-1 (BC Photo 1151); Gibsons 15 Apr 1987-1; Sechelt 22 May 1990-2 (Siddle 1990b); Port Coquitlam 20 Mar 1973-1 (Jerema 1973); Pitt Meadows 28 Apr 1972-13, 28 May 1976-15; Salmo River (Kelsey Bay) 17 May 1979-1 on estuary; Cranberry Lake (Powell River) 7 May 1985-1; Miracle Beach Park 10 May 1989-1; Quadra Island 2 May 1974-4; Cortes Island 29 May 1975-1 nr Mansons Landing; Whistler 3 May 1997-1 adult male at landfill; Port Neville Inlet 3 May 1977-1; Pine Island 15 to 18 May 1976-1; Kitimat River 9 May 1980-1 pair on estuary; Terrace 22 May 1976-1 (Hay 1976); Kispiox 20 May 1982-1. **Interior** – Richter Lake 4 May 1973-80; Meyers Lake (Rock Creek) 16 May 1976-4 eggs; Leach Lake 25 Mar 1987-1; Duck Lake (Creston) 8 Apr 1990-38, 20 May 1981-4 eggs; August Lake (Princeton) 5 May 1998-1 male; Penticton 22 May 1969-4 nestlings; Wasa 21 Apr 1980-1; Nakusp 11 May 1985-3; Lac du Bois 5 May 1972-20; Okanagan Landing 26 Mar 1951-1, first of year (Cannings et al. 1987), 2 May 1937-200, 1st of year; Okanagan Lake 27 Apr 1978-100 tallied in North Arm; Vernon 3 Apr 1977-12, 30 Apr 1985-100; Swan Lake (Vernon) 13 May 1968-200, 29 May 1965-504 (Cannings et al. 1987); Adams Lake 24 May 1984-2 at Lammer's Island; Revelstoke 17 Apr 1977-1; Rogers Pass 27 May 1984-1; Field 15 May 1972-1 (Wade 1977); 103 Mile Lake 18 May 1963-4 eggs; 122 Mile House 23 May 1961-4 nestlings; Riske Creek 27 May 1978-125 nr Roundup Lake; Williams Lake 23 Mar 1978-2, 25 Mar 1979-1, 14 Apr 1980-50; Kleena Kleene 22 May 1963-100; McBride 16 Apr 1989-1; nr Nulki Lake 27 Apr 1997-11 males; Prince George 15 May 1968-1 (Rogers 1968c); Willow River 5 May 1983-1; Giscome Canyon 18 May 1986-11; Old Man Lake 24 May 1979-100; Mackenzie 15 May 1996-2; Crooked River Park (n Prince George) 4 May 1981-1; Tumbler Ridge 1 May 1993-1 male; McQueen Slough (Dawson Creek) 17 Apr 1990-6; Cecil Lake 27 May 1976-9; Boundary Lake (Goodlow) 25 Apr 1987-1, 16 May 1982-20, 17 May 1986-1; Kenai Creek 9 May 1995-1; 6 km e Liard Hot Springs 18 May 1981-1.

Summer: Coastal – Central Saanich 17 Jul 1983-1; Bamfield 12 Aug 1965-1 (Hatler et al. 1978); Sea Island 8 Jul 1970-2 nestlings; Iona Island 18 Jul 1993-2 adults and 2 juveniles; Richmond 9 Jul 1968-2 eggs; Vancouver 21 Jun 1989-2, 17 Jul 1993-2; Bowen Island 8 Jul 1987-1; Little Qualicum River 29 Aug 1978-1 on estuary (Dawe 1980); Burnaby Mountain 7 Aug 1986-1, found dead against window; Pitt Lake 5 Jun 1976-34, 13 Jun 1964-60; Wilson Creek (Sechelt) 1 Jul 1987-1; Courtenay 16 Jun 1957-10; Squamish River 24 Jul 1983-2. **Interior** – n end Osoyoos Lake 21 Jun 1974-40; Richter Pass 7 Jun 1962-3 newly fledged young, 29 Jun 1973-200; Duck Lake (Creston) 16 Aug 1982-1,500; Kimberley 21 Jul 1976-4 fledglings at Adrian Lake; Alleyne Lake 24 Jul 1974-1 nestling and 1 egg; Nakusp 3 Aug 1977-1 recent fledgling; Illecilewaet River 18 Aug 1990-2; Brisco 7 Jul 1991-72, at least 12 nests; Swan Lake (Vernon) Jun 1963-1,200, 7 Jun 1950-900 (Cannings et al. 1987), 12 Jun 1937-662 (Munro and Cowan 1947); Mahood Lake 2 Jun 1975-1; Westwick Lakes 15 Jun 1978-200, 80 nests checked; Westwick Lakes 21 Jul 1956-4 eggs, 27 Jul 1956-3 nestlings; Mount Robson Park 1 Jun 1973-1; Dragon Lake 3 Jul 1995-2 nestlings; Ten Mile Lake (Quesnel) 9 Jun 1978-12; Eaglet Lake 17 Jun 1990-2 (Siddle 1990c); Mugaha Creek 1 Jun 1997-1 in marsh; McQueen Slough (Dawson Creek) 18 Jun 1993-3 nestlings, 18 Jun 1993-4 eggs; Taylor 29 Aug 1982-1, last autumn record; Boundary Lake (Goodlow) 16 Jul 1983-4 fledglings.

Breeding Bird Surveys: Coastal – Recorded from 2 of 27 routes and on 1% of all surveys. Maxima: Point Grey 12 Jun 1988-1; Seabird 6 Jun 1971-1. **Interior** – Recorded from 25 of 73 routes and on 27% of all surveys. Maxima: Mabel Lake 9 Jun 1991-103; Salmon Arm 30 Jun 1993-101; Grand Forks 26 Jun 1982-60.

Autumn: Interior – Fort St. John 9 Sep 1984-9, last autumn record, 11 Sep 1983-1, last autumn record; Nulki Lake 16 Oct 1994-1; Riske Creek 5 Sep 1978-6, 16 Sep 1985-1; 100 Mile House 30 Sep 1984-20; Swan Lake (Vernon) 21 Oct 1949-5 (ROM 76954); Golden 9 Sep 1976-1; Revelstoke 14 Sep 1986-1; Brouse 11 Sep 1977-7; Trail 15 Sep 1981-1; Wynndel 14 Oct 1988-1; Creston 9 Sep 1989-3,000 (Weber and Cannings 1990), 14 Sep 1985-200, 20 Sep 1978-200; Osoyoos Lake 16 Sep 1990-3. **Coastal** – Kitimat River 5 Sep 1979-1; Malcolm Island 4 Sep 1976-1; Oyster River 16 Sep 1979-1; Royston 13 Nov 1989-1 with 50 Brewer's Blackbirds; Sea Island 30 Nov 1983-2; Reifel Island 1 Sep 1968-100, 6 Oct 1974-20; Lake Cowichan 17 Oct 1986-1; Central Saanich 15 Oct 1988-2.

Winter: Interior – Williams Lake 26 Dec 1969-1; Westwick Lakes Dec 1952-a few wintered (Erskine and Stein 1964); Vernon 19 Jan 1973-1; Kelowna 15 Dec 1990-1, 21 Dec 1957-1 (NMC 48256); Creston 2 Jan 1988-1, 25 Jan 1993-1 (Siddle 1993b). **Coastal** – Pitt Meadows 6 Jan 1980-1, 26 Feb 1989-1; Vancouver 5 Jan 1985-1 at Jericho Beach, 10 Feb 1990-2; Cowichan Bay 6 Jan 1973-1; Sea Island 5 Feb 1984-1; Reifel Island 4 Jan 1990-1; Vancouver 30 Dec 1989-1; Sidney 16 Feb 1976-1; Saanich 23 Jan 1979-1, 24 Feb 1987-1.

Christmas Bird Counts: Interior – Recorded from 4 of 27 localities and on 2% of all counts. Maxima: Oliver-Osoyoos 28 Dec 1982-1; Creston 2 Jan 1988-1; Kelowna 16 Dec 1989-1; Vernon 16 Dec 1990-1. **Coastal** – Recorded from 2 of 33 localities and on less than 1% of all counts. Maxima: Ladner 28 Dec 1980-2; Victoria 19 Dec 1981-2.

Rusty Blackbird

RUBL

Euphagus carolinus (Müller)

RANGE: Breeds from western and north-central Alaska, northern Yukon, northwestern and central Mackenzie, southern Keewatin, northern Manitoba, northern Ontario, northern Quebec, north-central Labrador, and Newfoundland south to southwestern and south-coastal Alaska, northwestern and south-central British Columbia, central Alberta, central Saskatchewan, central Manitoba, central Ontario, southern Quebec, northern New York, western Massachusetts, central New Hampshire, central Maine, and Nova Scotia. Mostly winters in the southeastern United States, but also found locally from south-coastal Alaska, central British Columbia, central Alberta, southern Saskatchewan, southern Manitoba, southern Ontario, and the northern United States (east of the Rocky Mountains) south to central and southeastern Texas, the Gulf coast, and northern Florida, and west to Montana, central Colorado, and eastern New Mexico; rarely in southern Arizona and coastal California.

Figure 505. The Rusty Blackbird is primarily a bird of wooded wetlands in the central and northern interior of British Columbia. The males (shown) arrive in the breeding areas first, followed by the females (© A. Morris/VIREO).

STATUS: On the coast, a *rare* migrant and *very rare* vagrant in winter in the Georgia Depression Ecoprovince; in the Coast and Mountains Ecoprovince, *very rare* vagrant on Western Vancouver Island, *casual* vagrant on the Southern Mainland Coast, *rare* migrant and summer visitant and *very rare* vagrant in winter on the Northern Mainland Coast, and *accidental* on the Queen Charlotte Islands.

In the interior, an *uncommon* migrant and summer visitant in the Southern Interior, Southern Interior Mountains, Central Interior, Sub-Boreal Interior, Northern Boreal Mountains, and Taiga Plains ecoprovinces; *fairly common* to occasionally *very common* migrant and *uncommon* summer visitant in the Boreal Plains Ecoprovince. In winter, *very rare* vagrant in the Southern Interior, Southern Interior Mountains, Central Interior, and Sub-Boreal Interior.

Breeds.

CHANGE IN STATUS: The winter status and known breeding distribution of the Rusty Blackbird (Fig. 505) in British Columbia has changed considerably since Munro and Cowan (1947). In the mid-1940s, this blackbird was considered to be a summer visitant in the northern interior and a migrant or transient through the southern portions of the interior and along the coast. During this period there was only a single breeding record (at Bulkley Lake in the Central Interior) and a single winter record (from the Okanagan valley in the Southern Interior). Godfrey (1986) made no significant changes to this assessment and depicted the breeding distribution as lying north of approximately latitude 53°N (at about Quesnel).

Today the Rusty Blackbird nests in all ecoprovinces of the interior and several valleys on the Northern Mainland Coast. The southernmost nest record is from the Southern Interior, within 40 km of the international boundary. In winter, it occurs regularly in low numbers in the southern half of British Columbia as far north as Hazelton on the coast and Smithers in the interior.

The breeding distribution of the Rusty Blackbird has probably not changed significantly during the past 50 years, but because of a vastly improved road network into formerly inaccessible forest wetland areas, it is now much better understood. The modest increase in winter records, however, is almost certainly the result of more birds wintering in British Columbia. Similar increases in numbers of wintering Rusty Blackbirds have also been observed in western Alberta (Pinel et al. 1993).

NONBREEDING: From early spring through autumn, the Rusty Blackbird is widely but locally distributed throughout much of the interior and northern coast of British Columbia, where it inhabits wooded wetlands, bogs, and swampy lake shorelines. On the coast, it frequently winters in small numbers in agricultural areas of the lower Fraser valley and on southeastern Vancouver Island. It is occasionally recorded during winter at Tofino on Western Vancouver Island and along the Skeena River and Kitimat valley areas of the Northern Mainland Coast. There is a single winter record from the Queen Charlotte Islands. In the interior, it has been recorded during winter at irregular locations throughout the southern and central portions of the province, as far north as Smithers at latitude 55°N. Because of the sparse, irregular winter distribution of the Rusty Blackbird, no ecoprovince can be identified for highest winter numbers in British Columbia.

The Rusty Blackbird has been reported from near sea level to 1,430 m elevation on the coast and from 300 to 1,500 m in the interior. In spring and autumn, the Rusty Blackbird is most often associated with water bodies, including the edges of lakes, beaver ponds, rivers, and creeks. It forages along flooded shorelines, frequently wading into shallow water in search of insect prey (Skutch 1996). Following the nesting season, it gathers in multifamily groups and is often seen in human-created habitats, including stubble fields (Fig. 506),

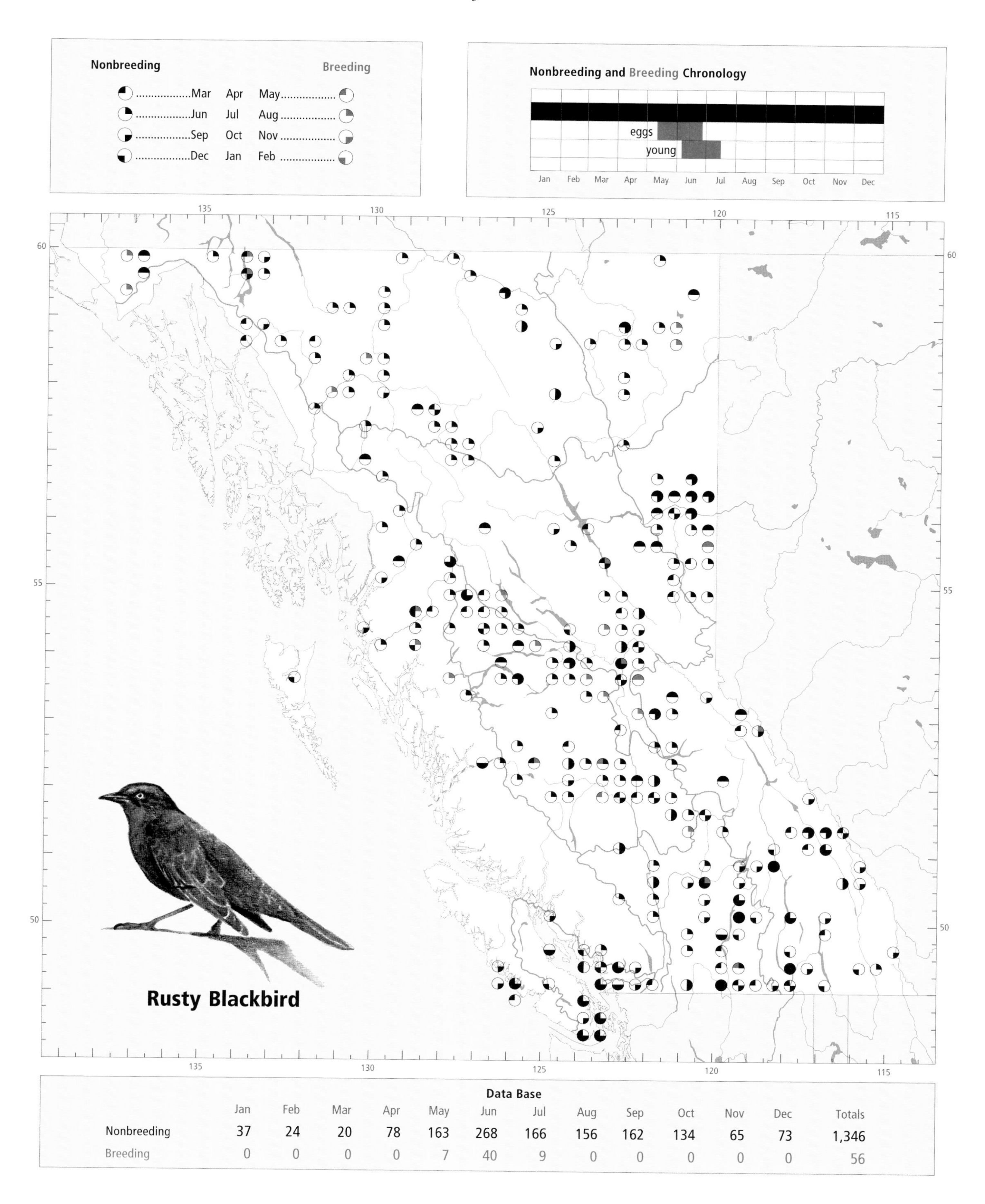

Data Base

	Jan	Feb	Mar	Apr	May	Jun	Jul	Aug	Sep	Oct	Nov	Dec	Totals
Nonbreeding	37	24	20	78	163	268	166	156	162	134	65	73	1,346
Breeding	0	0	0	0	7	40	9	0	0	0	0	0	56

Figure 506. During the nonbreeding seasons on the southern mainland coast of British Columbia, the Rusty Blackbird forages in stubble fields, often with flocks of European Starlings and Red-winged and Brewer's blackbirds (Pitt Meadows, 26 February 1996; R. Wayne Campbell).

pastures, and ploughed fields often in the company of other blackbird species. During the migration and winter periods, it uses agricultural fields, livestock feedlots, and sewage lagoons, and occasionally comes to bird feeders. The Rusty Blackbird avoids very open areas in winter or during cold periods in the spring, and prefers to roost in wooded areas (Bent 1958).

There is no evidence of a north-south movement along the entire coast, and Rusty Blackbirds probably enter and exit this region from the east along valleys from the interior. In the northern interior, Brooks and Swarth (1925) noted that most of these blackbirds entered and left this region from the east and few moved along southern routes. On the southern coast, the first indication of a spring movement occurs in mid-March, when the small numbers of wintering birds begin to leave; only a few stragglers remain here after early April. On the Northern Mainland Coast, the first migrants begin to arrive in early April. There are no perceptible peak movements on the coast in spring.

In the southern portions of the interior, the first spring migrants begin arriving in early to mid-March, augmenting the small number of wintering birds (Fig. 508). During mild, early springs, they occasionally reach the central portion of the interior near Prince George by the second week of March, but typically do not arrive at this latitude until mid-April. Further north, in the Boreal Plains, spring migrants first arrive from Alberta in mid-April and quickly peak in number during the last week of the month (Figs. 507 and 508). The highest spring numbers occur in the Boreal Plains (Fig. 508), with flocks of 100 birds or more occasionally being reported. In the Taiga Plains, spring migration is poorly documented but probably begins and peaks about 1 week later than in the Boreal Plains. Many birds moving through these ecoprovinces probably continue west into the Sub-Boreal Interior and Northern Boreal Mountains or move northwest into the Yukon and Alaska.

In the Northern Boreal Mountains, the spring migration of the Rusty Blackbird is also poorly documented, but apparently begins in mid to late April. At Atlin, Swarth (1936) noted

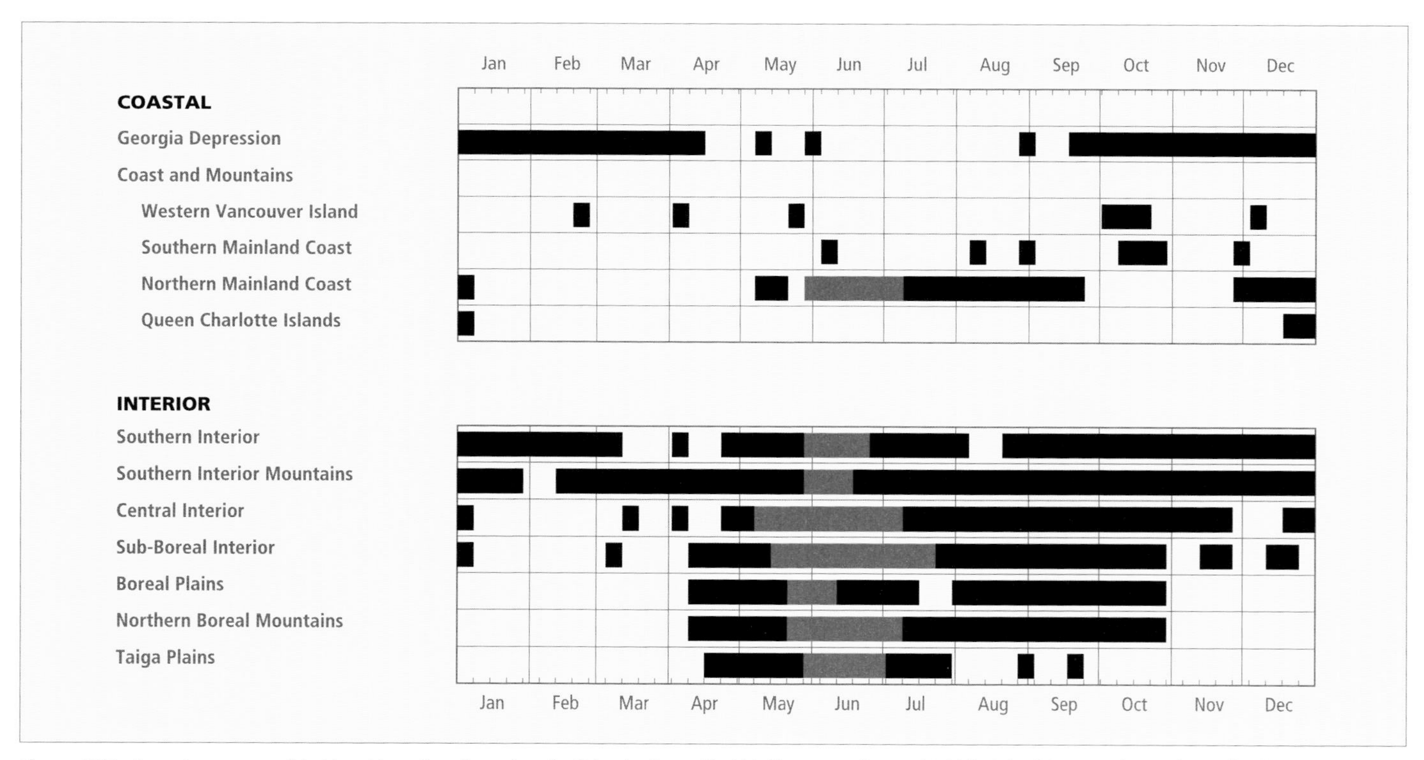

Figure 507. Annual occurrence (black) and breeding chronology (red) for the Rusty Blackbird in ecoprovinces of British Columbia. Records are shown for the week in which they occurred.

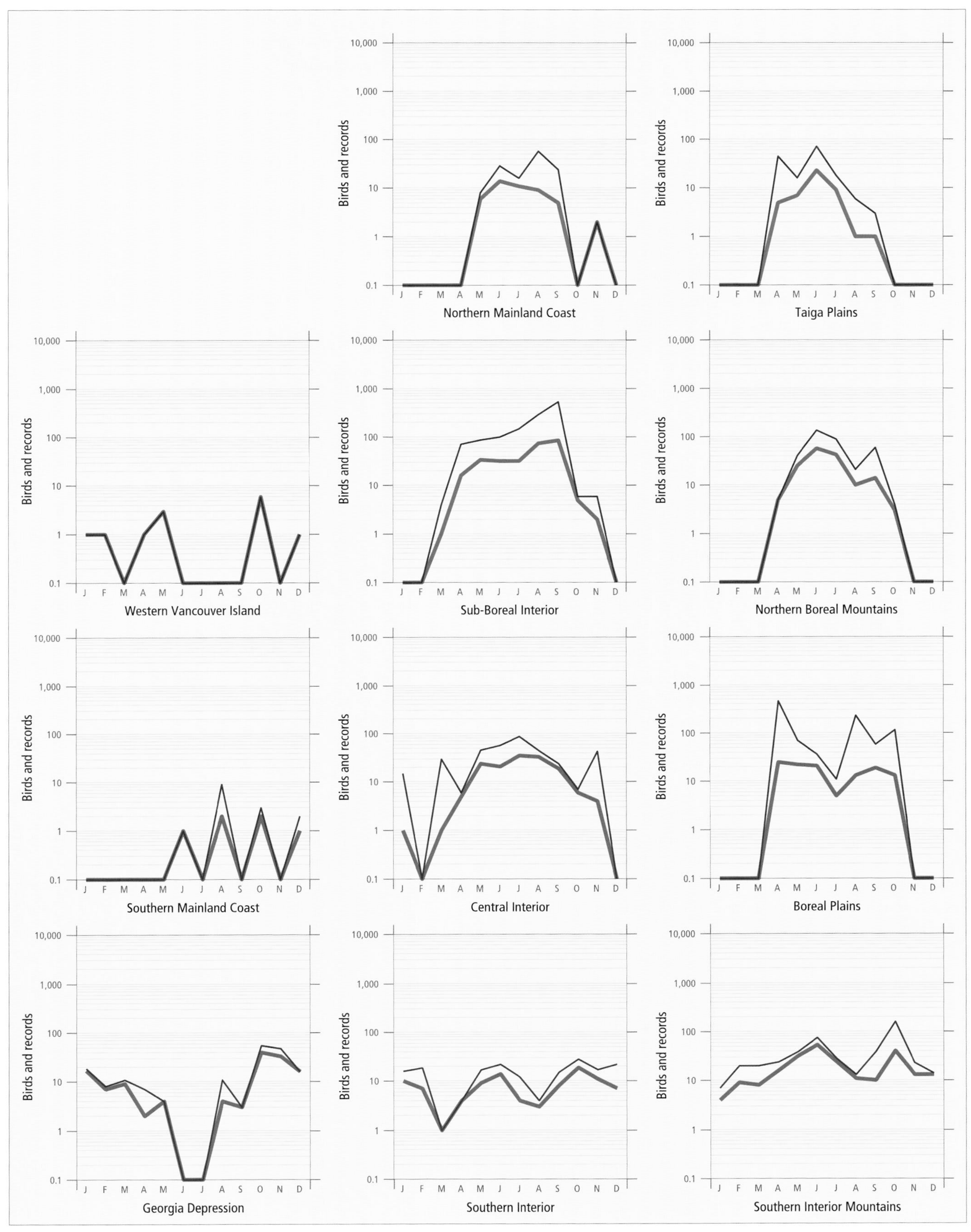

Figure 508. Fluctuations in total number of birds (purple line) and total number of records (green line) for the Rusty Blackbird in ecoprovinces of British Columbia. Christmas Bird Counts, Breeding Bird Surveys, and nest record data have been excluded.

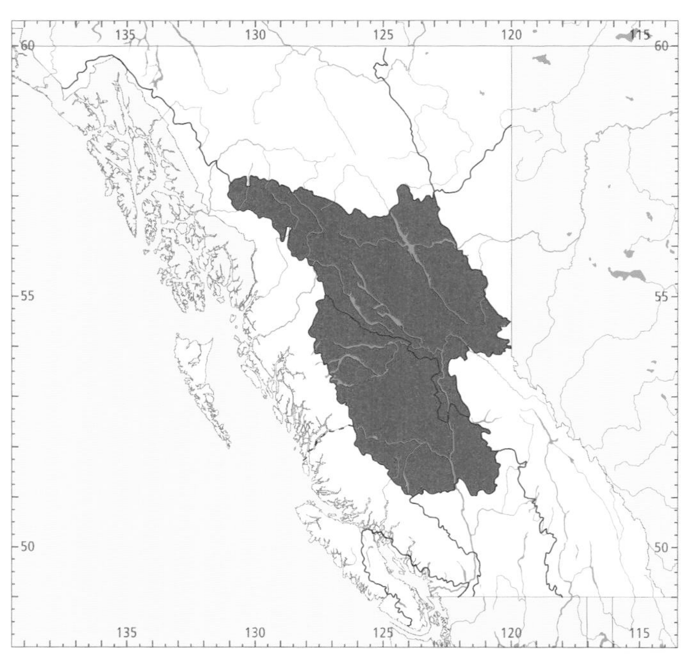

Figure 509. In British Columbia, the highest numbers for the Rusty Blackbird in summer occur in the Central Interior and Sub-Boreal Interior ecoprovinces.

that this blackbird's earliest spring arrival varied from 14 to 30 April. Peak movements in this ecoprovince probably occur during late April to early May and are followed by a rapid dispersal onto the breeding grounds.

The autumn migration is more gradual than the spring movement and begins with the formation of small post-breeding flocks in late July and early August. At this time these secretive wetland nesters once again become noticeable and begin local movements along lake shorelines and into agricultural areas. In the northern ecoprovinces, the autumn movement is under way by mid-August but a few stragglers remain into the last week of October. The main movement through British Columbia occurs east of the Rocky Mountains and, as in spring, the largest numbers are recorded in the Boreal Plains. These blackbirds move into Alberta, where most continue eastward through the central portion of the province (Pinel et al. 1993). Peak numbers in the Boreal Plains are reached by late August, with the occasional large flock occurring as late as the first week of October. In the Sub-Boreal Interior, migration occurs between mid-August and the end of November. In the Southern Interior Mountains, the first migrants appear in late August, with small flocks occurring as late as early November. In the Southern Interior, the migration is not obvious because of the small numbers involved.

On the coast, autumn migration is signalled by local increases in numbers on the Northern Mainland Coast in late August; most birds have left this region by the third week of September, although occasionally a few remain to winter. On the south coast, the first migrants begin to arrive in late August and birds remaining after early November probably stay to winter.

The Rusty Blackbird winters regularly in very small numbers in the Georgia Depression. It occurs occasionally in winter on Western Vancouver Island, and at Kitimat, Terrace, and Hazelton on the Northern Mainland Coast. There is 1 record of a single bird at Port Clements, on the Queen Charlotte Islands. The Rusty Blackbird also winters in small numbers in southern and central portions of the interior. It has been recorded in the interior during winter as far north as Smithers, Barrett (near Houston), Fort St. James, and Prince George; in these areas, it is found in human-created habitats

Figure 510. In some areas of its breeding range in British Columbia, the Rusty Blackbird nests in shrubby wetlands, often in coniferous forest environments (25 km east of Vanderhoof, 15 July 1997; R. Wayne Campbell).

Figure 511. Many Rusty Blackbird nests in British Columbia were found in flooded woody vegetation, often placed in trees or shrubs overhanging water (near Endako, 20 June 1997; R. Wayne Campbell).

Figure 512. Fallen trees and flooded snags along shorelines of wetlands are important foraging sites for the Rusty Blackbird during the breeding season (Cleswuncut Lake, 13 June 1998; R. Wayne Campbell).

with artificial food sources, especially livestock feedlots. Curiously, the highest winter numbers (up to 60 birds) have been reported from Smithers, which is near the northernmost wintering location for this species in British Columbia.

On the coast, particularly in autumn and winter, the Rusty Blackbird occurs frequently in mixed flocks with Brewer's Blackbirds, but will also accompany Red-winged Blackbirds, European Starlings, and Brown-headed Cowbirds. During spring, late summer, and autumn in the interior, it is often found with Brewer's and Red-winged blackbirds, European Starlings, and occasionally Yellow-headed Blackbirds and Brown-headed Cowbirds.

The Rusty Blackbird may occur throughout the year in both the coastal and interior portions of the province. It is recorded regularly from mid-October to early March on the south coast and from mid-May to mid-September on the north coast; in the interior, the Rusty Blackbird is recorded regularly from mid-April to the end of October (Fig. 507).

BREEDING: The Rusty Blackbird breeds primarily throughout the forested plateau landscapes of the Central Interior, Sub-Boreal Interior, Boreal Plains, and Taiga Plains, and in the valley bottoms and plateaus of the Northern Boreal Mountains. It is regularly observed in the wet valley bottoms of the Southern Interior Mountains, from Revelstoke north, but breeding has been confirmed only in Mount Robson Park. There are 2 mid-elevation breeding records from the Southern Thompson Upland and Okanagan Highland ecosections of the Southern Interior. On the coast, breeding probably occurs in most major valleys of the Northern Mainland Coast, but has been confirmed only around Terrace and at the head of Kitimat Inlet.

The Rusty Blackbird reaches its highest numbers in summer in the Central Interior and Sub-Boreal Interior (Fig. 509). Breeding Bird Surveys in British Columbia contain insufficient data for analysis. Sauer et al. (1997) report that continent-wide surveys for the period 1966 through 1996 show an annual average decline of 8.4% ($P < 0.10$). Much of the breeding range of the Rusty Blackbird lies outside the coverage of these surveys, however, and Avery (1995) points out the inadequacies of these data for monitoring this species.

The Rusty Blackbird has been found nesting from 80 to 915 m elevation on the coast and from 540 to 1,465 m elevation in the interior. There appear to be suitable breeding habitats across the northern portion of the province. The paucity of breeding records from large areas of northern British Columbia is mostly a reflection of poor access and a scarcity of observers. The highly secretive nature of this blackbird during the breeding season and the inhospitable nature of its wetland habitats also contribute to the general lack of summer records.

In British Columbia, nesting habitats include beaver ponds, swampy lakeshores, shrubby wetlands (Fig. 510), black spruce and tamarack bogs, muskegs, lake islands, meandering creeks, and river oxbows and back channels. Nearly all breeding sites are closely associated with water and most are in coniferous forest environments. The flooded woodland perimeters of beaver ponds are especially favoured by this

Figure 513. In British Columbia, the nest of the Rusty Blackbird is a bulky structure composed of coniferous and deciduous twigs, rootlets, plant stems, and coarse grasses, and lined with fine grasses (Bridge Lake, 1 July 1962; R. Wayne Campbell).

blackbird for both nesting and foraging during the breeding season. One unusual breeding site was found in a regenerating clearcut approximately 100 m from a creek. The use of forest openings created by logging has also been reported in eastern North America (Avery 1995).

Breeding has been confirmed in most low- to mid-elevation, forested biogeoclimatic zones in British Columbia, except for the Coastal Douglas-fir Zone and the Ponderosa Pine Zone. Most breeding records, however, come from the Sub-Boreal Spruce, Boreal White and Black Spruce, and Sub-Boreal Pine–Spruce zones. Summer observations suggest that the Rusty Blackbird may also nest at the lower elevations of the Engelmann Spruce–Subalpine Fir and Spruce–Willow–Birch zones.

Avery (1995) describes the breeding habitat of the Rusty Blackbird across North America as including fens, alder-willow bogs, muskegs, beaver ponds, and swampy forest openings along lakes and streams. In the Canadian boreal forest, which includes much of northern British Columbia, this blackbird is a characteristic species of coniferous forest bogs, swamps, and floodplain forests (Erskine 1977).

Orians (1985) describes the Rusty Blackbird as being loosely colonial during the breeding season. In British Columbia, most breeding records are of solitary pairs, but groupings of up to 5 nesting pairs have been reported for some wetlands. Avery (1995) finds evidence of both solitary and colonial nesting elsewhere in North America, but concludes that further study is required to clarify whether or not this species is a truly colonial breeder.

Figure 514. Most Rusty Blackbird clutches in British Columbia contained 4 or 5 eggs (Atlin, 4 June 1980; R. Wayne Campbell).

On the coast, the Rusty Blackbird has been recorded breeding from 30 May (calculated) to 6 July; in the interior, it has been recorded from 12 May (calculated) to 16 July (Fig. 507).

Nests: Most nests (59%; n = 32) were found in small to medium-sized coniferous trees, usually black spruce, followed by deciduous shrubs (31%) and small deciduous trees (6%). Many nests (18) were located over standing water, usually among flooded woody vegetation, or in trees or shrubs overhanging water; 11 were reported in dead or dying trees and shrubs (Figs. 511 and 512). Two nests were located on a lake island and 1 was built in a natural cavity in a lodgepole pine. Nests were typically placed in thick shrub clumps or among horizontal branches, usually near the trunk but occasionally along a limb up to 60 cm from the main stem. The heights of 31 nests in British Columbia ranged from 0.2 to 6.1 m, with 58% between 0.6 and 1.5 m. The lowest nests were typically located in dead, flooded trees or shrubs; the higher nests were generally on drier ground, well concealed in live coniferous trees.

Nests were typically bulky cups constructed of spruce twigs and grasses, with smaller amounts of rootlets, plant stems, and lichens (Fig. 513). Several were constructed without twigs and consisted of grasses, plant stems, mosses, and lichens; 1 nest also contained pine needles. All were lined with fine grasses; 4 also used feathers, 1 incorporated deer hair, and another contained a small amount of *Sphagnum* moss.

The Rusty Blackbird characteristically constructs the inner shell of its nest with wet, rotted vegetation, which dries to form a hard surface similar to papier-mâché (Skutch 1996). This vegetative material is easily mistaken for mud (Bent 1958) and was reported as such in 9 British Columbia nests. Nests are not reused by the Rusty Blackbird but may be used in subsequent years by other wetland species, notably the Solitary Sandpiper (Avery 1995).

Eggs: Dates for 19 clutches ranged from 17 May to 27 June, with 53% recorded between 27 May and 20 June. Calculated dates suggest that eggs may occur as early as 12 May. Sizes of 15 clutches ranged from 3 to 6 eggs (3E-3, 4E-5, 5E-4, 6E-3), with 60% having 4 or 5 eggs (Fig. 514). The incubation period is 14 days and incubation is carried out by the female (Bent 1958; Baicich and Harrison 1997).

Young: Dates for 21 broods ranged from 6 June to 16 July, with 50% recorded between 13 and 27 June. Calculated dates indicate that young can occur as early as 31 May. Sizes of 13 broods ranged from 3 to 6 young (3Y-4, 4Y-6, 5Y-1, 6Y-2), with 83% having 3 or 4 young. The nestling period ranges from 11 to 13 days (Avery 1995; Baicich and Harrison 1997). Young often leave the nest several days before they are capable of sustained flight.

Brown-headed Cowbird Parasitism: In British Columbia, only 1 of 27 nests found with eggs or young was parasitized by the Brown-headed Cowbird. This nest contained a clutch of 7 eggs, 3 of which were cowbird eggs. There were no records of this blackbird feeding a fledged cowbird. Elsewhere in North America, there are only 2 confirmed records of cowbird parasitism, both from Alberta, but the Rusty Blackbird has not been known to successfully rear cowbird young (Friedmann et al. 1977; Friedmann and Kiff 1985; Avery 1995). Given this blackbird's northern breeding distribution and preference for swampy coniferous forests, it seems unlikely that it would be a frequent cowbird host.

Nest success: Of 4 nests found with eggs and followed to a known fate, 1 produced at least 1 fledgling. One nest placed on a low branch overhanging a lakeshore was destroyed by wave action resulting from a windstorm. Avery (1995) reports that the agitated behaviour of nesting Rusty Blackbirds towards Gray Jays suggests that the latter are potential nest predators. This was supported in British Columbia by the observation of a Gray Jay eating a nestling from a nest containing 3 Rusty Blackbird young. Our data also indicate that these blackbirds behave aggressively towards the Northern Harrier and pine marten during the breeding season, suggesting that they may also be potential nest predators.

REMARKS: Two subspecies of the Rusty Blackbird are recognized by the American Ornithologists' Union (1957), but only *E. c. carolinus* occurs in British Columbia.

The relatively large movement of Rusty Blackbirds through northeastern British Columbia suggests that breeding may be more common east of the Rocky Mountains than our data currently indicate. There are extensive areas of black spruce and tamarack bogs in the Clear Hills Ecosection of the Boreal Plains and in the Hay River Lowland and Northern Alberta Upland ecoregions of the Taiga Plains. Beaver ponds, small lakes, and meandering creeks abound in this region, providing extensive areas of potential breeding habitat for the Rusty Blackbird. Although road access into these muskeg regions is extremely limited, future surveys may show that they support higher breeding numbers than found elsewhere in British Columbia.

The Rusty Blackbird is an early spring migrant and occasionally encounters late snowfalls and bouts of cold weather. During such weather conditions, it has been known to kill and eat small birds, presumably because normal foods are scarce (Campbell 1974a; Skutch 1996). This behaviour has not been recorded in British Columbia.

For additional information on the life history of the Rusty Blackbird in North America, see Avery (1995).

NOTEWORTHY RECORDS

Spring: Coastal – Metchosin 9 Mar 1974-1 (Shepard 1974); Central Saanich 3 Mar 1972-3 (Tatum 1973); Duncan 20 Mar 1974-1 at sewage ponds; Somenos Lake (Duncan) 1 Apr 1972-6; MacKenzie Beach (Tofino) 22 May 1987-1; Tofino 2 Apr 1989-1; Reifel Island 9 Apr 1974-1 at refuge; Pitt Meadows 31 Mar 1983-1 (Mattocks and Hunn 1983b); Agassiz 10 May 1889-1 (NMC 717); Chilliwack River 31 May 1906-1 (NMC 3369); Sechelt 30 Mar 1989-1; 1.6 km s Terrace 28 May 1977-1; Kitsault 14 May 1980-2; Tulsequah River 12 May 1996-2 in floodplain slough. **Interior** – 10 km e Grand Forks 8 Mar 1987-2 (Walker 1996), Summerland 1 May 1965-1; Venner Meadows 26 May 1985-3; Midway 8 Apr 1905-1 (NMC 3125); 16 km ne Okanagan Falls 30 May 1986-4 eggs, adults close to nest; Nakusp 13 Mar 1976-8, 27 Mar 1976-1; Kamloops 2 Apr 1994-1, 23 May 1889-1 (NMC 577); Revelstoke 21 Mar 1986-2, 26 May 1986-2, 31 May 1985-1; Nicholson 3 Apr 1976-4; Leanchoil 8 Apr 1976-1 (Wade 1977); Glacier National Park 26 Apr 1982-3; Golden 18 May 1986-1; Clearwater 30 May 1935-1 (MCZ 284611); nr Hanceville 17 May 1981-6 eggs, 2 males in attendance at nest with female; 29 km s Williams Lake 2 Apr 1977-1 in cattle feedlot with Red-winged Blackbirds and European Starlings; Williams Lake 5 Apr 1975-1; Anahim Lake 28 Apr 1932-2 (MCZ 284599-600); Stum Lake 22 May 1978-5 eggs, female incubating; Puntchesakut Lake 30 May 1944-2 (Munro 1947); Kinney Lake (Mount Robson Park) 28 May 1974-1; Indianpoint Lake 29 Apr 1927-2 banded (Robertson 1928); nr Ootsa Lake 13 May 1997-2 in flooded willows; 24 km s Prince George 20 Apr 1980-20; Ste. Marie Lake 18 May 1978-6 eggs; Nulki Lake 20 May 1945-4; Prince George 27 Apr 1982-1 at sewage pond; Willow River 9 Mar 1986-4; Smithers 18 Mar 1975-30; Bullmoose Creek 11 May 1998-1; Muskeg Lake 18 May 1997-10; Mugaha Creek (Mackenzie) 25 Apr 1998-15 recorded on transect; Mount Lemoray 9 May 1968-4; Chetwynd 9 May 1968-2; Tupper Creek 29 May 1938-5 (RBCM 7894 to 7898); Swan Lake Park (Tupper) 27 May 1992-4 eggs; Weston Creek (Williston Lake) 29 Apr 1998-5 feeding along edge of ice-covered pond; Driftwood River 15 Apr 1941-1 (Stanwell-Fletcher and Stanwell-Fletcher 1943); Moberly River 30 Apr 1983-6 at beaver pond; nr Cache Creek (Peace River) 14 Apr 1984-25, 20 Apr 1985-40, some singing; North Pine 24 Apr 1987-50 at sewage lagoon; Boundary Lake (Goodlow) 25 Apr 1987-110; Ningunsaw River 27 May 1980-1; Hyland Post 10 May 1982-3; Hatin Lake 31 May 1978-1; Fort Nelson 18 Apr 1986-1, 24 Apr 1986-32; Liard Hot Springs 1 May 1975-4 (Reid 1975); Helmut 14 May 1982-2; Atlin 14 Apr 1931-1, earliest spring arrival (Swarth 1936); Chilkat Pass 8 May 1957-1 (Weeden 1960).

Summer: Coastal – Vancouver 28 Aug 1977-3 at Lost Lagoon; Squamish 8 Jun 1961-1 (Boggs and Boggs 1961d); Pemberton 30 Aug 1984-8; w Chikamin Mountain 20 Jul 1975-1 in alpine (Hazelwood 1976); Nanika Lake 13 Jun 1979-3 nestlings, only a few days old; Kitimat 13 Jun 1980-3 eggs; Lakelse Lake 30 Jun 1974-15, including fledged young (Crowell and Nehls 1974d), 21 Aug 1977-40; Kitsault 23 Jul 1980-1; Hazelton 4 Jun 1921-1 (MVZ 42237); Stewart 17 Jun 1975-1; Bell-Irving River 17 Jun 1975-1; Snowbank Creek 20 Jun 1975-1. **Interior** – Oliver 5 Aug 1970-2 at pond; Venner Meadows 11 Jun 1989-2; Coquihalla Lakes 9 Jun 1992-2 (Siddle 1992c); 29 km n Gemmal Lake 27 Jun 1971-1 fledgling fed by adult; Kane Valley 3 Jul 1992-6 (Siddle 1995); Lytton 4 Jun 1964-1 (NMC 52696); Oyama 25 Jul 1990-4 (Siddle 1990c); Lost Lake (Westbank) 20 Jun 1988-6 on island; Lillooet 16 Jun 1964-1 (NMC 56714); White Lake (Kamloops) 20 Jun 1976-2 recently fledged young, within 3 m of nest; Knutsford (Kamloops) 25 Jun 1953-1 (PMNH 72092); Invermere 19 Aug 1982-1 in large mixed flock with Brewer's and Red-winged blackbirds; Revelstoke 1 Jun 1986-2 at hydro substation; Bridge Lake 8 Jun 1959-4 eggs plus 3 Brown-headed Cowbird eggs; North Barriere Lake 4 Aug 1974-1 in wetland shrub thicket; Williams Lake 15 Jun 1970-3 fledged young; Bond Lake 9 Jun 1974-4; Dewar Lake 26 Jun 1977-2; Stillwater (Wells Gray Park) 2 Aug 1951-1 (Edwards and Ritcey 1967); Anahim Lake 25 Jun 1994-1 recent fledgling with adults; Chezacut Lake 6 Aug 1931-1 (UBC 6932); Moose Lake (Mount Robson Park) 14 Jun 1972-4 nestlings, nearly fledged; Ten Mile Lake (Quesnel) 6 Jun 1978-1 catching insects at pond; Bowron Lake Park 28 Aug 1971-3, in autumn plumage (Runyan 1971); Cinema 16 Jul 1967-2 nestlings; Opatcho Lake 11 Jun 1967-6 nestlings; Ootsa Lake 13 Jul 1936-16; Nadina River 15 Jun 1997-3 in wetland at river mouth; 15 km sw Prince George 27 Jun 1972-5 eggs, 1 of 5 nests; Prince George 13 Jun 1969-15; Parrot Lakes 18 Jun 1981-2; nr Endako 20 Jun 1997-5; Burnie Lakes 23 Aug 1975-1 in subalpine habitat (Osmond-Jones et al. 1975); Bulkley Lake 27 Jul 1944-3 in field with 30 Brewer's Blackbirds; Decker Creek 11 Jun 1975-2; Salmon River (n Prince George) 7 Jun 1970-4 nestlings; Bear Lake (Crooked River) 15 Aug 1975-30; Topley 8 Jul 1956-24; Mugaha Creek (Mackenzie) 7 Jun 1998-2 recently fledged young nr nest; Smokehouse Creek 6 Jun 1976-1 at beaver pond in black spruce bog; Mount Lemoray 22 Jun 1978-5; Tetana Lake 25 Aug 1938-1 (RBCM 8374); Taylor 22 Aug 1975-132; Bear Flat 17 Aug 1986-10; Fort St. John 11 Jun 1982-5 in swampy area; Charlie Lake 20 Aug 1975-12; German Lake 21 Jun 1986-3; Tatlatui Lake 21 Jun 1986-2; Ingenika 24 Jul 1980-1; Buckley Lake 27 Jul 1982-6; Telegraph Creek 27 Aug 1977-5; Level Mountain 6 Jun 1978-2 at small lake (1,310 m); Dease Lake 22 Jun 1962-3 eggs; nr Clarke Lake 28 Aug 1978-6 in swampy area with pond; nr Metlahdoa Creek 14 Jun 1982-3 nestlings, 1 being eaten by Gray Jay, 24 Jun 1982-3 recently fledged young with adults; Parker Lake 13 Jun 1976-6; Basement Creek (Tatshenshini River) 4 Jun 1983-5 eggs, incubation well advanced; Blue River (Cassiar) 2 Jun 1978-2; Liard Hot Springs 3 Jun 1975-1 (Reid 1975); Davie Hall Lake 2 Jul 1980-11 along lakeshore; nr Surprise Lake (Atlin) 10 Jun 1975-2; Survey Lake 26 Jun 1980-30, 27 Jun 1980-1 fledgling just out of nest; Goat Creek (Haines Highway) 18 Jun 1980-5; Log Cabin 15 Jun 1899-2 (Bishop 1900).

Breeding Bird Surveys: Coastal – Recorded from 3 of 27 routes and on 1% of all surveys. Maxima: Kitsumkalum 13 Jun 1976-1; Kwinitsa 24 Jun 1979-1; Kispiox 20 Jun 1993-1. **Interior** – Recorded from 18 of 73 routes and on 5% of all surveys. Maxima: Prince George 23 Jun 1973-24; Golden 21 Jun 1973-7; McLeod Lake 11 Jun 1970-5; Wingdam 10 Jun 1970-5; Brookmere 4 Jul 1980-5.

Autumn: Interior – Atlin 8 Oct 1931-1, latest autumn record (Swarth 1936); lower Liard River 25 Oct 1980-2; Liard Hot Springs 6 Sep 1974-20, mostly males in full autumn plumage; Taku River 6 Sep 1996-1 in floodplain back channel; Summit Pass 1 Sep 1943-25 (Rand 1944); Fern Lake 8 Sep 1979-2 (Cooper and Adams 1979); Dawson River 6 Sep 1976-2; Finbow 14 Sep 1996-1 male in autumn plumage; Rose Prairie 11 Oct 1986-7; North Pine 20 Sep 1986-7 at sewage lagoon; w Boundary Lake (Goodlow) 5 Oct 1986-60; e Cecil Lake 21 Oct 1986-8; ne Fort St. John 23 Oct 1982-1 at sewage lagoon; Fort St. John 22 Sep 1985-23 in willow swamp, 2 males singing; Moberly River 8 Sep 1986-1; Mugaha Creek (Mackenzie) 12 Sep 1996-12; Smithers 23 Nov 1981-14 in town with Brewer's and Red-winged blackbirds; Parsnip River 22 Nov 1964-5 (Rogers 1965a); Giscome 13 Oct 1965-1 collected; Eaglet Lake 26 Oct 1996-1; Tabor Lake 17 Sep 1998-17; 24 km s Prince George 13 Nov 1987-1 eating livestock feed grain; nr Ootsa Lake 18 Sep 1997-1 at feeder; Moose Lake (Mount Robson Park) 2 Sep 1971-3 in marsh; Chezacut 14 Sep 1933-3 (MCZ 284607-9); Riske Creek 28 Sep 1985-1, 30 Oct 1984-2 feeding in a cattail marsh; Chilcotin River (Riske Creek) 3 Nov 1984-1 feeding in cattail marsh at Wineglass Ranch; Field 20 Oct 1975-17; Revelstoke 1 Oct 1983-10 on shrubby island in Columbia River; Scotch Creek 9 Nov 1962-3 along fenceline with Brewer's Blackbirds; Dog Lake (Kootenay National Park) 20 Oct 1981-1 in flooded spruce along shore (Poll et al. 1984); Invermere 11 Sep 1977-15; Grindrod 17 Oct 1940-2 (Munro 1953); Tranquille (Kamloops) 24 Oct 1987-1 (Rogers 1988), 26 Nov 1988-2; Swan Lake (Vernon) 23 Oct 1950-2 (Munro 1953); Okanagan Landing 26 Nov 1919-5, 1 collected (MVZ 104950); Nakusp 5 Oct 1978-8; Argenta 31 Oct 1998-3 foraging along edge of marshy area; Sparwood 30 Oct 1984-1 (BC Photo 1011); Nelson 17 Nov 1968-2; 1.6 km n Trail 14 Nov 1982-1; Grand Forks 25 Sep 1989-2 (Weber and Cannings 1990); Oliver 5 Sep 1985-4. **Coastal** – 37 km n Hazelton 15 Sep 1921-1 (MVZ 42245); Greenville 21 Sep 1980-20; Metlakatla 26 Nov 1901-1 (Kermode 1904; RBCM 1838); Bella Coola 28 Oct 1989-2 at feeder; Pitt Lake 14 Oct 1968-1 male in reeds at s end; Megin River (Clayoquot Sound) 4 Oct 1985-1 (RBCM 1042); Pitt Meadows 22 Nov 1970-6 (Campbell et al. 1972b); Sea and Iona islands 13 Oct 1979-4; Iona Island 19 Sep 1983-1 with Brewer's Blackbirds; Tofino 1 Oct 1976-1 with 2 Brewer's Blackbirds; Central Saanich 24 Sep 1983-1 in corn field with 1,500 European Starlings and Red-winged Blackbirds, 28 Oct 1972-3 (Tatum 1973).

Winter: Interior – Barrett 6 Jan 1993-15 (Siddle 1993b); Nicholson 20 Dec 1977-1; Revelstoke 7 Jan 1986-2 at city landfill; Kamloops 14 Jan 1996-2 at cattle feed lot; Otter Lake (Vernon) 10 Feb 1980-6; Nakusp 29 Feb 1976-10 feeding under bird feeder; Lumby 8 Dec 1968-12; Okanagan Landing 5 Dec 1918-3 along lakeshore, 2 collected (Munro 1922a); Ellison 21 Jan 1985-1 nr small slough; Cranbrook 26 Feb 1921-1 collected; Kinnaird 14 Dec 1968-1; nr Creston 25 Jan 1993-1 (Siddle 1993b); Oliver 28 Dec 1988-1. **Coastal** – Port Clements 6 Jan 1995-1 (Bowling 1995b); Bella Coola 2 Dec 1989-2 at feeder; Pitt Meadows 23 Jan 1971-2 (Campbell et al. 1972b; BC Photo 108); Maple Ridge 8 Jan 1981-2; Sea Island 23 Dec 1981-1; Port Alberni 13 Jan 1995-1 (Bowling 1995b); Tofino 8 Dec 1990-1 (Siddle 1991b); Duncan 30 Dec 1976-1 with 30 Brewer's Blackbirds; Central Saanich 23 Feb 1972-2 (Tatum 1973).

Christmas Bird Counts: Interior – Recorded from 7 of 27 localities and on 9% of all counts. Maxima: Smithers 27 Dec 1993-60; Revelstoke 29 Dec 1990-6; Fort St. James 2 Jan 1993-3. **Coastal** – Recorded from 7 of 33 localities and on 2% of all counts. Maxima: Terrace 3 Jan 1971-20; Pitt Meadows 27 Dec 1980-4; Kitimat 15 Dec 1979-3.

Brewer's Blackbird

Euphagus cyanocephalus (Wagner)

BRBL

RANGE: Breeds from northeastern British Columbia and southwestern Mackenzie, north-central Alberta, central Saskatchewan, southern Manitoba, central Ontario, and northern New York south to northwestern Baja California, southern California, southern Nevada, central Arizona, southern New Mexico, Texas, Oklahoma, Colorado, northwestern Nebraska, southern Minnesota, southern Wisconsin, northeastern Illinois, southwestern Indiana, and southern Michigan. Winters from southern British Columbia, central Alberta, Montana, Kansas, Oklahoma, Missouri, and the northern Gulf states south to Baja California and central Mexico.

STATUS: On the coast, *common* resident becoming *abundant* locally in late summer and *very abundant* locally in winter in the Georgia Depression Ecoprovince; in the Coast and Mountains Ecoprovince, *rare* local migrant and winter visitant on Western Vancouver Island, *fairly common* resident, becoming locally *common* in winter, in southern parts of the Southern Mainland Coast, *uncommon* local resident in the Northern Mainland Coast, and *very rare* vagrant on the Queen Charlotte Islands.

In the interior, *fairly common* spring migrant, *common* summer visitant, *common* to *abundant* autumn migrant, and locally *uncommon* in winter in the Southern Interior, Southern Interior Mountains, and Central Interior ecoprovinces; *fairly common* migrant, *uncommon* summer visitant, and *very rare* in winter in the Sub-Boreal Interior Ecoprovince; *fairly common* to locally *very common* migrant and summer visitant in the Boreal Plains Ecoprovince; *rare* migrant and summer visitant in the Taiga Plains Ecoprovince, *accidental* in winter; and *very rare* migrant and summer visitant in the Northern Boreal Mountains Ecoprovince. In the Southern Interior, Southern Interior Mountains, and Central Interior ecoprovinces, local wintering populations may be *very common* in some years.

Breeds.

Figure 514. Adult male Brewer's Blackbird. In British Columbia Brewer's Blackbird breeding populations appear to be declining; throughout much of the rest of North America they are stable (Creston, 25 April 1997; R. Wayne Campbell).

Figure 515. Adult female Brewer's Blackbird (Sawmill Lake, Okanagan valley, 30 May 1998; R. Wayne Campbell).

CHANGE IN STATUS: The Brewer's Blackbird (Figs. 514 and 515) was formerly confined to central and southern British Columbia. Munro and Cowan (1947) listed Smithers and Tetana Lake (latitude about 56°N) as the northernmost limit of its distribution in the province. Since then, the range of the Brewer's Blackbird has expanded northward. The first record from the Northern Boreal Mountains was from Nuttlude Lake, where a female with 1 immature was observed on 30 July 1957. These birds may or may not have originated there, but the first confirmed breeding in that ecoprovince was at Liard Hot Springs in June 1979. In the Boreal Plains, Cowan (1939) did not record a single Brewer's Blackbird during extensive field work in the Peace Lowland in 1937, although Racey (1930) listed the species as "common" in the Peace River Block in June 1930. During summer field work in the region in 1923, 1929, and 1930, Williams (1933a, 1933b) did not locate Brewer's Blackbirds, although he stated that it was possible that the species was present. It has now become a locally common breeding bird in the Boreal Plains. In the Taiga Plains, the first Brewer's Blackbirds were noted at Fort Nelson in July 1969. It is now an uncommon breeding bird near Fort Nelson and a local breeder elsewhere in the Taiga Plains north to the Yukon border.

Populations in the north appear to have become established and continue to increase as clearing of forests for farmland has increased. Highway corridors and their associated human developments may have acted as pipelines for the range expansion, as northern records are mainly from areas modified by humans and records from wilderness areas are extremely rare. In Alberta, the range of the Brewer's Blackbird has expanded considerably as it has colonized roadways and developed areas in northern parts of the province (Semenchuk 1992).

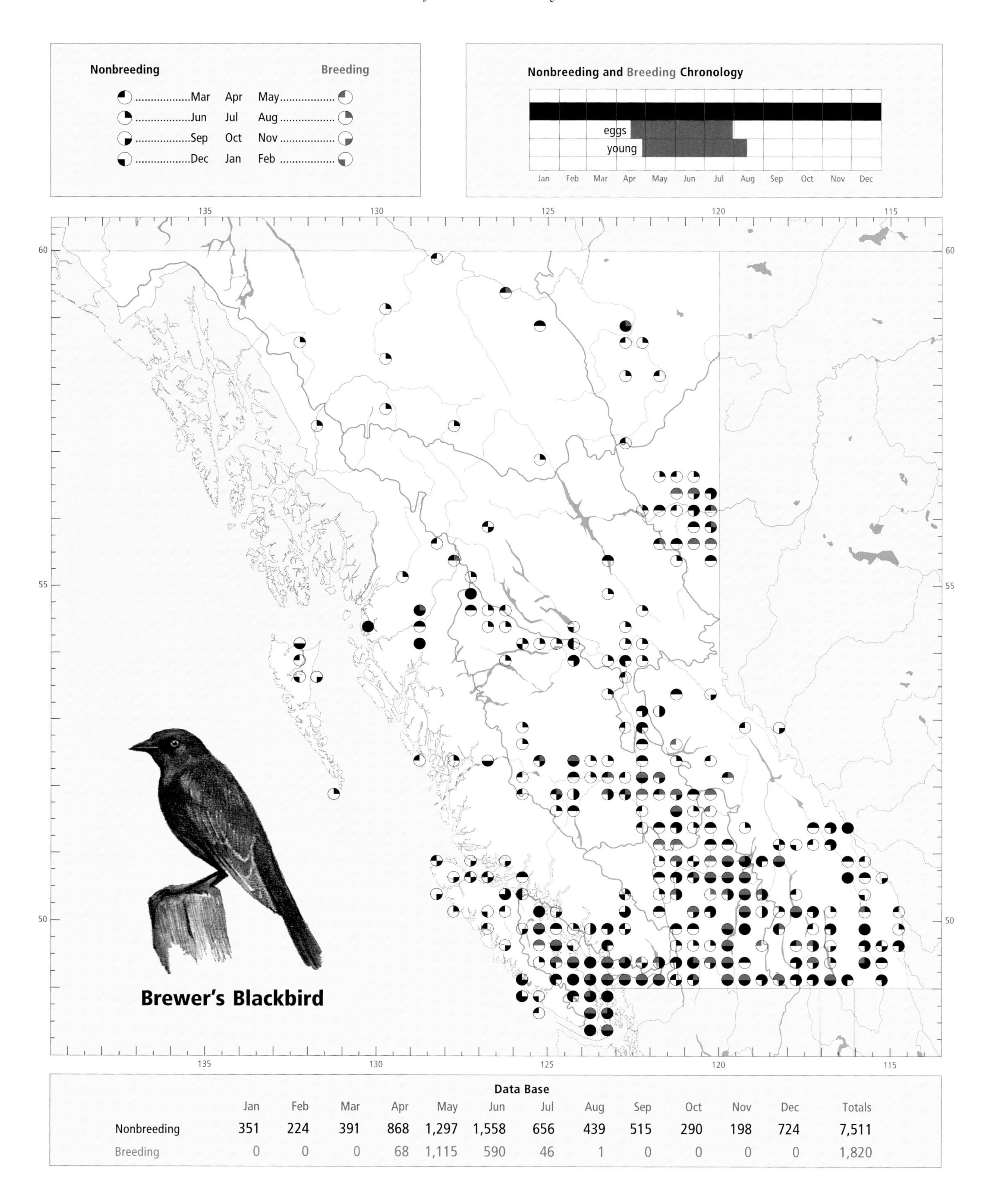

Data Base

	Jan	Feb	Mar	Apr	May	Jun	Jul	Aug	Sep	Oct	Nov	Dec	Totals
Nonbreeding	351	224	391	868	1,297	1,558	656	439	515	290	198	724	7,511
Breeding	0	0	0	68	1,115	590	46	1	0	0	0	0	1,820

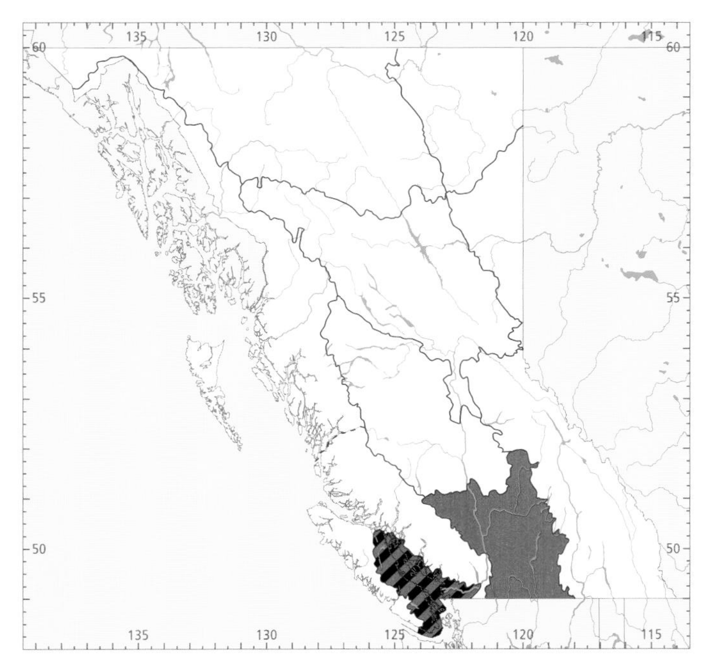

Figure 516. In British Columbia, the highest numbers for the Brewer's Blackbird in winter (black) occur in the Georgia Depression Ecoprovince; the highest numbers in summer (red) occur in the Georgia Depression and Southern Interior ecoprovinces.

Munro and Cowan (1947) also referred to the Okanagan valley as the only wintering area in the interior. Significant wintering populations now occur locally along the Yellowhead Highway corridor from Prince George to Smithers. Small numbers also overwinter in the west and east Kootenays, and near Williams Lake.

NONBREEDING: The Brewer's Blackbird is widely distributed at low elevations throughout most of British Columbia from spring through autumn. It is most abundant in the south, becoming progressively less abundant further north, except that moderately higher numbers occur in the Boreal Plains. The Brewer's Blackbird is extremely scarce in the northwestern portion of the province, and is absent from the Atlin or Tatshenshini River areas. It is scarce along the Northern Mainland Coast south of the Skeena River, and occurs only occasionally on the Queen Charlotte Islands. On Western Vancouver Island, it is more common than further north on the coast, but occurs in very small numbers and very locally. In the Taiga Plains, it occurs along the Alaska Highway corridor and associated human settlements. Elsewhere, the Brewer's Blackbird is one of the more common birds in open areas.

The Brewer's Blackbird occurs most abundantly in the Georgia Depression during nonbreeding seasons and in the Southern Interior during late summer and early autumn. The highest numbers in winter occur in the Georgia Depression (Fig. 516).

On the coast, the Brewer's Blackbird occurs from sea level up to a few hundred metres elevation, but most birds frequent relatively low elevations. In the interior, it occurs from the valley bottoms up to about 1,200 m. Occasionally, autumn migrants may pass through higher elevations; for example, 2 October records were from 1,450 m (Palliser River) and 1,800 m (Blackwall Mountain).

Figure 517. Throughout British Columbia, the Brewer's Blackbird is a common sight in agricultural areas with recently ploughed fields (Saanich, 6 April 1997; R. Wayne Campbell).

Figure 518. Across southern British Columbia, the Brewer's Blackbird frequents farmlands with livestock, feeding troughs, and grain storage containers throughout the year (Pitt Meadows, 25 February 1996; R. Wayne Campbell).

In all regions, agricultural lands and rural areas are preferred habitats, as they are during the breeding season (see BREEDING). These habitats include pastures, ploughed fields (Fig. 517), stubble fields, vegetable fields, berry patches, and nurseries. Other human-modified habitats include barnyards (Fig. 518), sewage lagoons, garbage dumps, vacant lots, grassy road rights-of-way, dykes, golf courses, railway lines, campgrounds, and lawns. In winter, flooded fields, pastures with livestock (Fig. 519), and backyards with bird feeders are used. In heavily forested areas, only open areas created by humans are used.

Natural habitats include marshes, estuaries, beaches, mud flats, lake shorelines, sedge meadows, riparian thickets, and deciduous bottomlands. In the interior, late migrants and overwintering birds frequent exposed shorelines around ponds, feedlots, and urban and rural areas.

Figure 519. In winter, especially in the south-central interior of British Columbia, flocks of Brewer's Blackbirds associate with cattle where the ground has been bared during feeding (12 km north of Osoyoos, 11 December 1992; R. Wayne Campbell).

In the Central Interior, Munro (1949) noted that at Nulki Lake birds fed along narrow beaches in the early spring before water levels rose, then again in the late summer as lake levels dropped. He also noted that summer flocks fed on waste grain and weed seeds, as no grasshoppers were present in the summer of 1948. Elsewhere in the Central Interior, Munro (1947) observed early spring migrants foraging on manure piles heaped up in barnyards and summer birds feeding on grasshoppers in open fields. At Creston, Munro (1950) observed that in late summer, birds foraged along the margins of sloughs, often wading belly-deep in the shallow water. They also foraged on exposed sandbars covered with flotsam, then roosted in nearby willow thickets.

In urban areas, birds use parks, cemeteries, schoolyards, buildings, and lawns. On the University of British Columbia campus in Vancouver, Verbeek (1962, 1964) noted that on October mornings, flocks of Brewer's Blackbirds foraged in parking lots for food scraps left by students the day before. Later in the autumn, however, cars filled the parking areas in the morning before birds arrived after daylight, which precluded foraging on open gravel or pavement. The flocks then foraged in barnyards and stock pens on campus.

Large flocks can form late in the day, then move en masse to nocturnal roosts. After dawn, flocks move to foraging areas. Roosts are usually situated along the edges of open habitat, such as on ridges, along the edges of fields, along the edge of water, or along railway lines, roads, or dykes. Flocks roost mainly in heavy upland cover, including thick coniferous hedges, large coniferous trees, and riparian thickets; in wetlands they roost in cattails, bulrushes, or willows. When not foraging or roosting, flocks often perch on roadside utility wires.

On the coast, wintering birds begin to leave the Georgia Depression in late March or early April, about the same time that resident birds move to nesting areas (Fig. 521). Migrants are scarce elsewhere on the coast. In the interior, spring migrants begin to arrive in the Southern Interior in the third week of March and in the Southern Interior Mountains in the last week of March and first week of April. The peak movement occurs in mid to late April in both ecoprovinces (Fig. 521). In the Central Interior, spring migrants begin to arrive

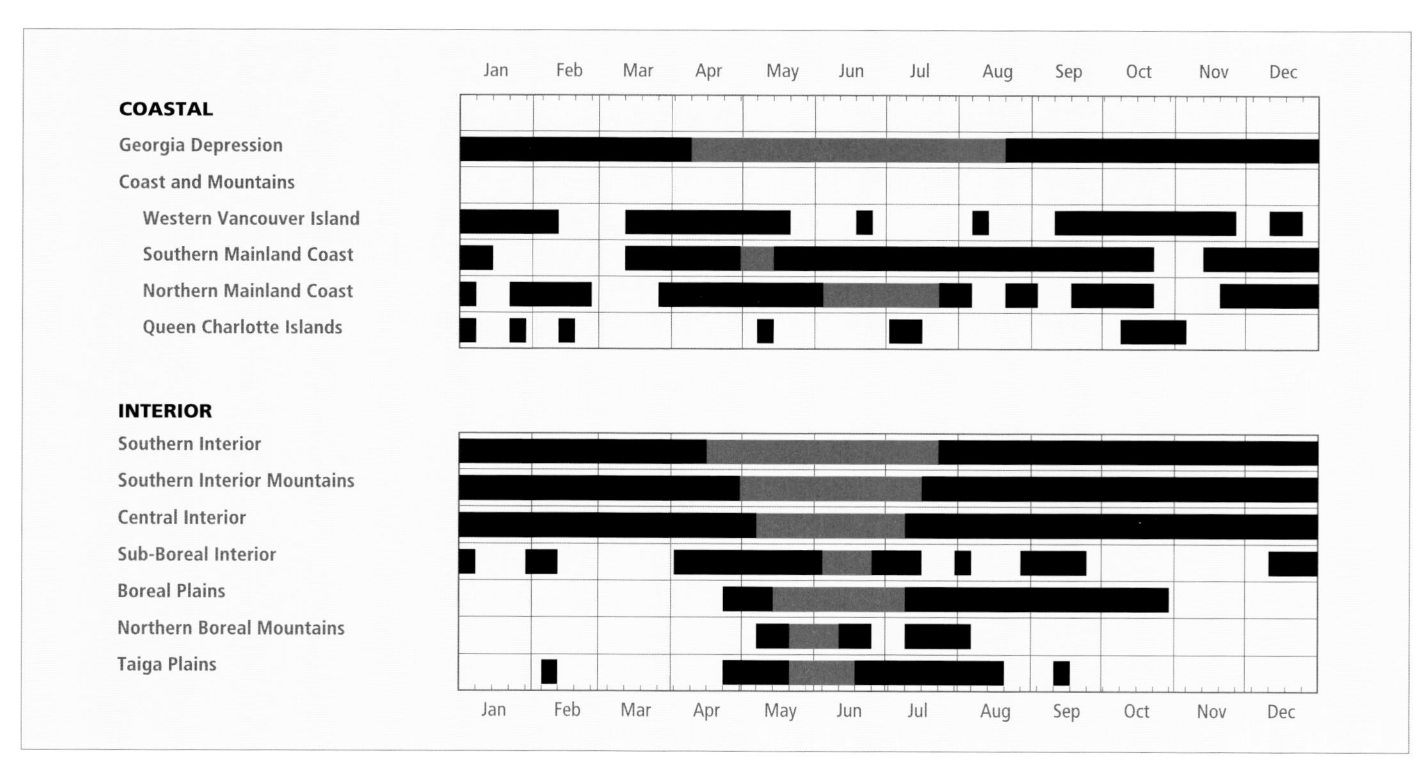

Figure 520. Annual occurrence (black) and breeding chronology (red) for the Brewer's Blackbird in ecoprovinces of British Columbia. Records are shown for the week in which they occurred.

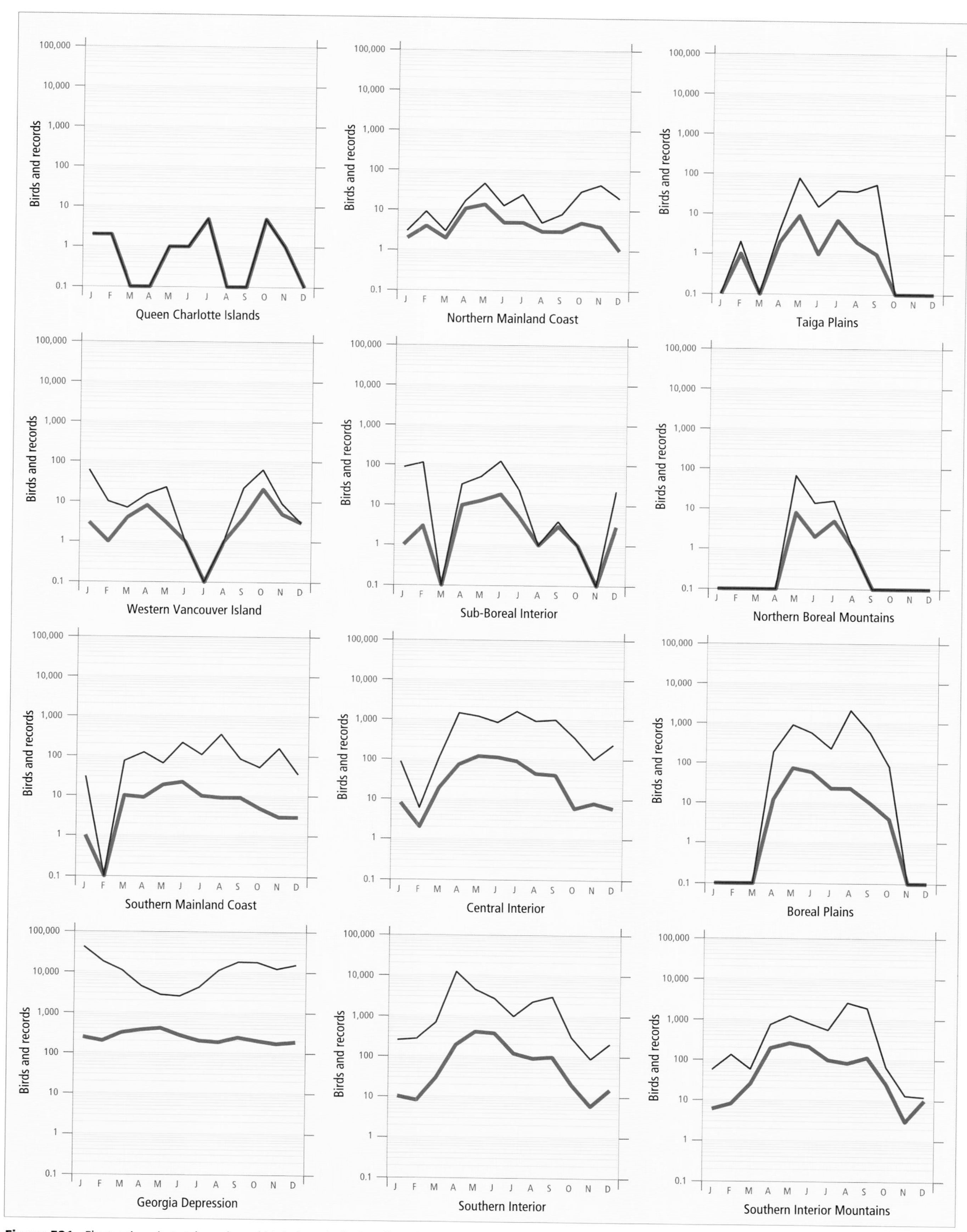

Figure 521. Fluctuations in total number of birds (purple line) and total number of records (green line) for the Brewer's Blackbird in ecoprovinces of British Columbia. Christmas Bird Counts, Breeding Bird Surveys, and nest record data have been excluded.

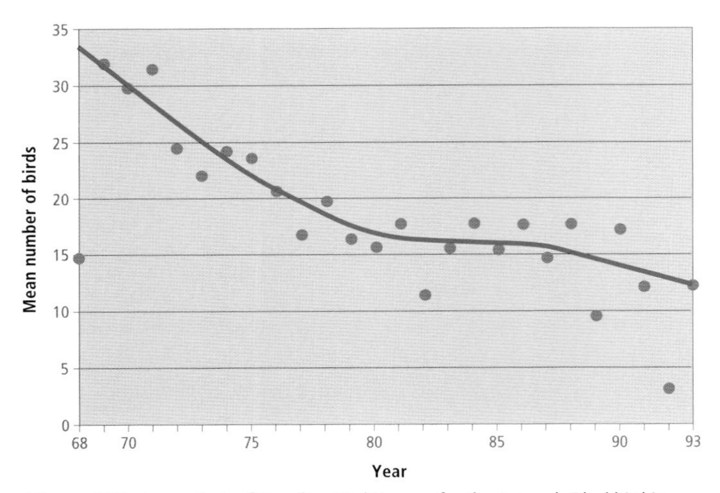

Figure 522. An analysis of Breeding Bird Surveys for the Brewer's Blackbird in British Columbia shows that the mean number of birds on interior routes decreased at an average annual rate of 5% over the period 1968 through 1993 ($P < 0.05$).

in the second week of March; numbers increase in the last week of March and peak in mid-April. In the Boreal Plains and Taiga Plains, spring migrants arrive from late April through mid-May, peaking a few days earlier in the Boreal Plains (Fig. 521). All migrants move quickly to nesting areas after arriving.

After nesting is completed, the Brewer's Blackbird congregates in good foraging areas, which are mainly in agricultural sites. These flocks occur provincewide from August through mid-September. At this time, the Brewer's Blackbird can forage in mixed flocks with Red-winged, Yellow-headed, and Rusty blackbirds, Brown-headed Cowbirds, Common Grackles (in the Boreal Plains), and American and Northwestern crows.

In the interior, the autumn migration occurs mainly in September, but continues into early October in the south. Only stragglers remain in the northeastern interior after September, and after mid-October further south (Figs. 520 and 521). On the coast, numbers build rapidly from September through October as birds move from the interior into their wintering areas in the Georgia Depression.

The farmlands of the Fraser River delta in the Georgia Depression are home to the largest wintering population of Brewer's Blackbirds in the province. Many thousands can occur in a single field. The Ladner Christmas Bird Count consistently reports the most Brewer's Blackbirds of any Christmas Bird Count in Canada. A few overwinter on Western Vancouver Island and stragglers have occurred in winter on the Queen Charlotte Islands and the Northern Mainland Coast. In the interior, wintering numbers are very low, but small numbers overwinter regularly in local areas from Prince George south. In some years, local wintering populations in the interior may number several hundred birds. Records of overwintering in the interior have become more frequent in recent years.

On the coast, the Brewer's Blackbird occurs regularly throughout the year. It also occurs throughout the year in the southern and central portions of the interior, and regularly from late April to mid-September in the northern interior (Figs. 520 and 521).

BREEDING: The Brewer's Blackbird has a widespread breeding distribution over much of British Columbia. Its centre of breeding abundance is in the Georgia Depression and the Southern Interior, where it is one of the most common breeding birds in the Okanagan valley (Cannings et al. 1987).

In the Central Interior and Southern Interior Mountains, the Brewer's Blackbird is a less numerous breeder than in the Southern Interior, but it is still widespread. Breeding populations are more local in the Sub-Boreal Interior, being restricted mainly to human-modified habitats such as farms and roadsides. The species is more abundant and widespread in the Boreal Plains; it is a local breeder in the Taiga Plains.

On the coast, the Brewer's Blackbird is an infrequent breeder outside of the Georgia Depression. Very small numbers nest locally on western and northern Vancouver Island, but the Brewer's Blackbird is a regular breeding bird on farmlands on the east side of the island. Along the Skeena River drainage, breeding likely occurs in most areas where farming takes place, but our only breeding records are from Terrace and near Kispiox. The Brewer's Blackbird likely breeds very locally along other major river valleys on the mainland coast, but breeding has yet to be confirmed on the Southern Mainland Coast. The Brewer's Blackbird does not nest on the Queen Charlotte Islands or on other offshore islands.

The Brewer's Blackbird reaches its highest numbers in summer in the Georgia Depression and Southern Interior (Fig. 516). An analysis of Breeding Bird Surveys for the period 1968 through 1993 shows that numbers on interior routes declined at an average annual rate of 5% (Fig. 522); on the coast, no net change in numbers could be detected. Breeding Bird Surveys across Canada and North America from 1966 to 1996 suggest that populations have remained stable (Sauer et al. 1997).

The Brewer's Blackbird breeds at elevations from near sea level to 450 m on the coast, but most nests are found below

Figure 523. In suburban areas across southern British Columbia, the Brewer's Blackbird nests solitarily or in small colonies in evergreen trees and hedges (near Grand Forks, 24 April 1997; R. Wayne Campbell).

Figure 524. Tangles of vegetation close to wetlands are used by nesting Brewer's Blackbirds in the Okanagan valley of British Columbia (12 km north of Osoyoos, 7 June 1998; R. Wayne Campbell).

Figure 525. In the Boreal Plains Ecoprovince of British Columbia, the Brewer's Blackbird usually nests on the ground in thick cover (14 km northeast of Hudson's Hope, 23 June 1998; R. Wayne Campbell).

200 m. In the interior, nests have been found as high as 1,140 m in the Central Interior and 930 m in the Southern Interior. In general, nesting occurs in valley bottoms or at lower elevations.

In British Columbia, nesting habitats are mainly associated with human-modified habitats (59%; n = 602), followed by forested lands (14%), shrublands (10%), wetlands (7%), and grasslands (6%). Modified habitats include mainly farmland (41%; n = 343), rural areas (28%), and suburban areas (23%; Fig. 523), but also urban areas (4%), orchards and vineyards (3%), and clearcuts (1%). In more natural habitats, shrublands (45%; n = 184), wetlands and marshes (28%), and forest edges (20%) were most frequently used.

Throughout the province, nesting areas are often close to wetlands or wet, low-lying areas that are used for foraging. In farmlands and rural areas, nesting habitats usually include hedgerows, tangles of vegetation (Fig. 524), or ditches with sloped banks. In grasslands and shrublands, nesting habitat includes copses of shrubs, thick ground cover (Fig. 525), or the edges of clumps of trees. In wetlands, willow thickets, emergent vegetation such as cattails, or hummocky ground are used. In suburban areas, spacious gardens with ornamental trees, especially conifers, are often used. Other nesting habitats include riparian thickets and open ponderosa pine woodlands.

There are some regional differences in nesting habitats used. Farmlands and rural areas are widely used throughout the province, but are most important in the Georgia Depression, Southern Interior, Sub-Boreal Interior, and Boreal Plains. Shrublands are used extensively in 2 ecoprovinces, mainly big sagebrush in the Southern Interior, and ocean-spray, Himalayan blackberry, or Scotch broom in the Georgia Depression. In the Southern Interior Mountains, most breeding records are from marshes but antelope-brush shrublands are also used. In the Taiga Plains, the Brewer's Blackbird nests in brushy and open areas along roads (Erskine and Davidson 1976). In the Boreal Plains, nests are often found in open burns (Symons 1967) or in abandoned fields with a sparse shrub cover. In the Georgia Depression, upland areas of marine beaches or spits, especially in estuaries, are used if shrubs or trees are present. In mountainous areas, this blackbird breeds only in sites where major transportation corridors or farming activity have opened up forested lands, or where wetlands occur.

Table 11. Location, size, and nest substrate of major (≥ 10 nests) Brewer's Blackbird colonies in British Columbia.[1]

Location	Survey year	Total nests	Nest substrate
South Coast			
Deroche	1962	11	shrubs
Extension	1958	13	shrubs
Pitt Meadows	1976	10	willows and ground
Rosewall Creek	1977	12	oceanspray
Comox	1988	12	shrubs
Interior			
Osoyoos	1983	14	wild rose
Oliver	1983	13	wild rose
Sawmill Lake (Falkland)	1998	14	shrubs
16 km e Hedley	1960	22	shrubs
Nicola River	1963	12	unknown
S Thompson River	1963	16	shrubs
Wells Gray Park	1962	16	deciduous trees
Westwick Lakes	1955	21	ground

[1] All data are from the British Columbia Nest Records Scheme.

Figure 526. A colony of 12 pairs of Brewer's Blackbirds nested in thickets of ocean-spray at Rosewall Creek, Vancouver Island, in 1977. The colony was abandoned in the early 1990s (Rosewall Creek, 9 May 1997; Neil K. Dawe).

In dry parts of eastern Washington characterized by big sagebrush grasslands and pothole wetlands, nesting colonies were always located in big sagebrush near water, with a high perch nearby where males could watch for danger (Horn 1970). This description would apply in the Okanagan valley of the Southern Interior as well.

On the coast, the Brewer's Blackbird has been recorded breeding from 17 April (calculated) to 14 August; in the interior, it has been recorded breeding from 23 April (calculated) to 19 July (Fig. 520). The long breeding season in the Georgia Depression spans 4 months.

Colonies: The Brewer's Blackbird may nest in small colonies or loose aggregations of pairs in British Columbia (Table 11). Colony sizes recorded in British Columbia where active nests were counted ($n = 48$) ranged up to 22 nests, with a mean of 8 nests. The largest colony on record was located in shrubs along the Similkameen River about 16 km east of Hedley, where 22 pairs nested in 1960. Other large colonies have been reported from Westwick Lakes (21 nests) in the Central Interior, the South Thompson River (16 nests) in the Southern Interior, and Wells Gray Park (16 nests) in the Southern Interior Mountains. Colonies as large as 50 nests have been estimated in the Georgia Depression.

Colonies tend to occur in the same places from year to year. A colony at Rosewall Creek, Vancouver Island, that contained 12 nests in 1977 was abandoned in the early 1990s. This colony was in a thicket of ocean-spray that grew along a beach spit (Fig. 526). The thicket remains in place and the cause of the abandonment is unknown.

Nests: In British Columbia, most nests were placed in deciduous and coniferous shrubs (38%; $n = 831$), in trees (29%), or on the ground (17%). Other general nest locations include on beams or sides of buildings (2.2%), on posts or poles (1.9%), in snags or stumps (1.7%), in vines (1.6%), or in reeds or cattails (1.3%). Nests in live trees were mainly in coniferous trees (65%; $n = 243$). Single nests have been found in an abandoned car, in a haystack, and among the roots of a fallen tree. One unusual nest, at Quinsam Lake on Vancouver Island, was in a burned-out woodpecker cavity, where the fire had created a level platform 20 cm deep.

There are some geographic differences in nest placement in British Columbia (Table 12). In the Georgia Depression, shrubs and trees were recorded about equally. In the Southern Interior, shrubs were recorded most frequently, while in the Southern Interior Mountains, trees were dominant nest locations and ground nests were rare. In the Central Interior and Boreal Plains, ground nests were the predominant location.

Table 12. General nest locations for the Brewer's Blackbird in ecoprovinces of British Columbia.

Ecoprovince	n	Tree (%)	Shrub (%)	Ground (%)	Building (%)	Other (%)
Georgia Depression	285	38.9	40.0	12.6	1.8	6.7
Southern Interior	386	22.0	46.9	14.8	2.6	13.5
Southern Interior Mountains	47	60	13	4	0	23
Central Interior	59	3	2	71	3	20
Boreal Plains	41	34	10	54	0	2

The heights of 793 nests ranged from ground level to 10.0 m, with 55% between 0.5 and 2.2 m. Nests on the ground were placed under or next to a clump of vegetation, among thick ground cover, or at the base of a sapling. Nests in shrubs were usually placed among tangles of branches or stalks. Nests in trees were usually saddled on a branch next to the trunk, rarely at the end of a branch. In 2 cases, old nests were reused.

Most nests were bulky cups of grasses, small twigs, and plant stems, rarely with mosses (2.5%; *n* = 714; Figs. 527 and 529), leaves (1.7%), and feathers (4 nests). Human-made products such as nylon rope (1.4%) and paper (2 nests) were also used. In nests above ground level, mud was often used to bind twigs, grasses, and plant stems in a central core. Nests on the ground generally did not contain twigs. The nest cup was almost always lined with fine grass and rootlets. On farmlands, nest cups were often lined with horse or cow hair. The nest in the burned-out woodpecker cavity consisted only of black-tailed deer hair, moss, and fine rootlets.

Eggs: Dates for 1,098 clutches (Fig. 529) ranged from 17 April to 15 July, with 55% recorded between 9 May and 1 June. Calculated dates indicate that eggs can occur as early as 17 April and as late as 30 July. Egg laying begins about 1 to 2 weeks earlier on the coast than in the southern portions of the interior, and 3 or 4 weeks earlier than in the central and northern parts of the species' range (Fig. 528).

Sizes of 653 clutches ranged from 1 to 7 eggs (1E-53, 2E-45, 3E-67, 4E-138, 5E-239, 6E-107, 7E-4), with 58% having 4 or 5 eggs. A nest found with 8 nestlings indicates that clutches can have as many as 8 eggs. At least in the Southern Interior, clutches in May tend to have 5 eggs but clutches in June are more likely to contain 4 eggs. Elsewhere, Baicich and Harrison (1997) report a range of 3 to 7 eggs, but usually 5 or 6 eggs. The median incubation period in British Columbia was 12 days (*n* = 9), with a range of 11 to 13 days.

The Brewer's Blackbird is double-brooded throughout much of its range in British Columbia (Fig. 528). In the Okanagan valley, most pairs are thought to raise 2 broods each year, beginning the first clutches in late April or early May and second clutches in early June (Cannings et al. 1987). At Creston, first clutches are begun about 7 May and second clutches about 7 June (Butler et al. 1986). At Nulki Lake, which is at the extreme northern edge of the Central Interior, careful study indicated that only 1 brood was produced each year (Munro 1949). In the Sub-Boreal Interior and Taiga Plains, only 1 brood is produced each year.

Young: Dates for 706 broods ranged from 28 April to 14 August, with 53% recorded between 21 May and 13 June. Sizes of 437 broods ranged from 1 to 8 young (1Y-21, 2Y-47, 3Y-85, 4Y-130, 5Y-118, 6Y-33, 7Y-2, 8Y-1), with 57% having 4 or 5 young (Fig. 527).

The median nestling period in British Columbia was 13 days (*n* = 9), with a range of 12 to 15 days, the same as that reported elsewhere (Baicich and Harrison 1997). Fledglings can remain with their parents for an additional 12 or 13 days.

Brown-headed Cowbird Parasitism: In British Columbia, 3% of 797 nests found with eggs or young were parasitized by

Figure 527. Nest containing 5 young Brewer's Blackbirds (Salmon Arm, 29 May 1995; R. Wayne Campbell).

cowbirds (Fig. 529). Parasitism in the interior was 4% of 531 nests; parasitism on the coast was 1% of 266 nests. There were 5 reports of Brewer's Blackbirds feeding fledgling cowbirds, all from the interior. Parasitism in the Okanagan valley was reported to be somewhat higher (9%; n = 161; Cannings et al. 1987) than in the interior in general. Parasitism has been recorded only in the Georgia Depression, Southern Interior, Southern Interior Mountains, and Central Interior. Parasitized nests contained 1 cowbird egg (14 nests), 2 eggs (8 nests), or 3 eggs (2 nests).

The Brewer's Blackbird is one of the most heavily parasitized songbirds, and may be the most heavily parasitized bird in the Canadian Prairies (Friedmann et al. 1977). Data up to the mid-1970s showed a parasitism rate of 23% for Alberta,

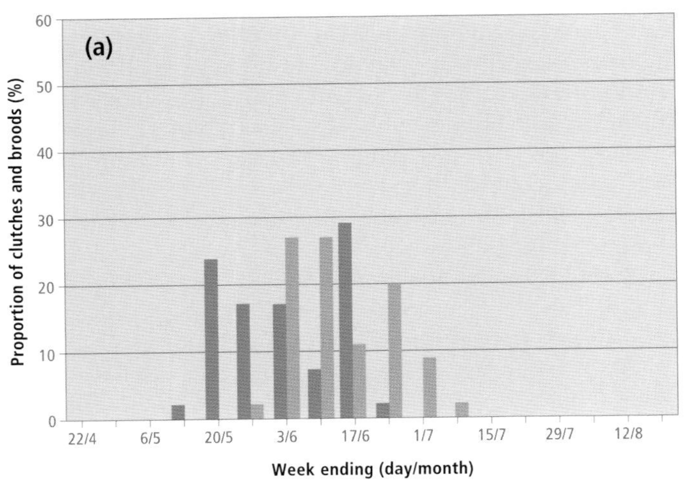

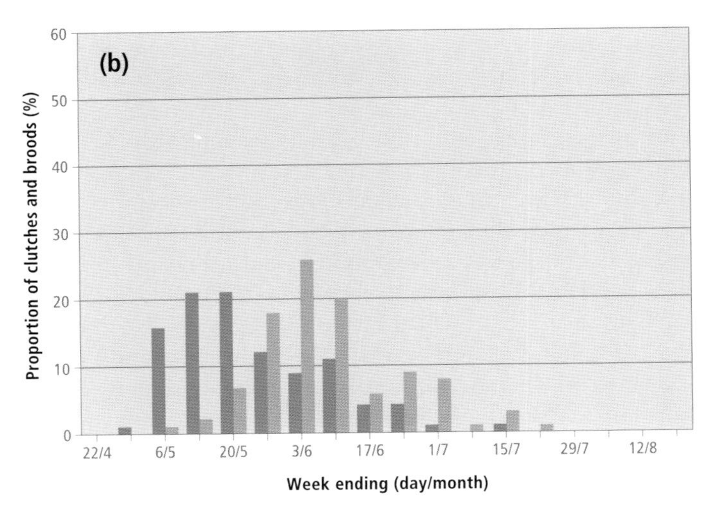

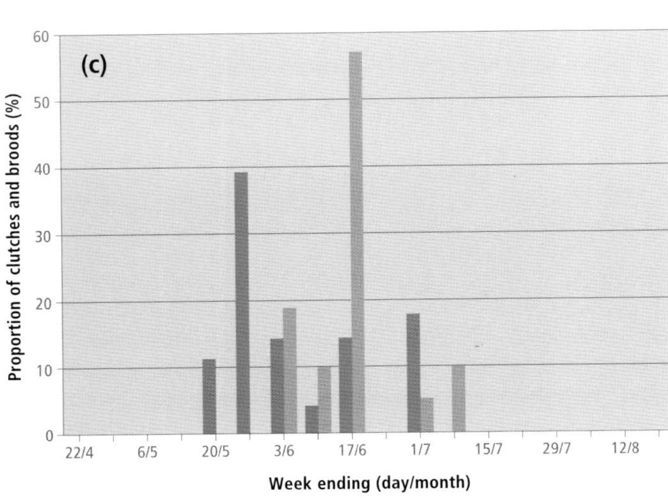

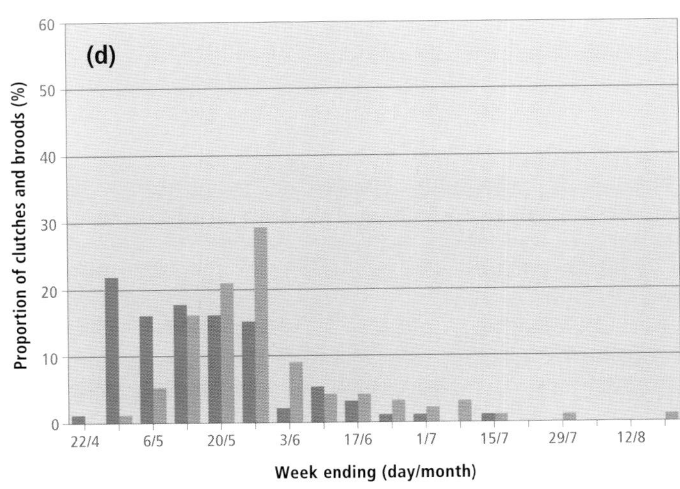

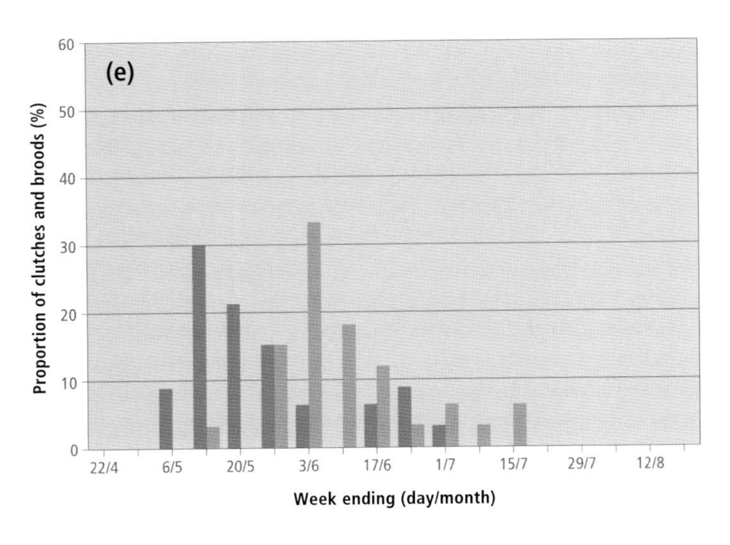

Figure 528. Weekly distribution of the proportion of clutches (dark bars) and broods (light bars) for the Brewer's Blackbird in the (a) Central Interior (clutches: n = 41; broods: n = 44), (b) Southern Interior (clutches: n = 302; broods: n = 199), (c) Boreal Plains (clutches: n = 28; broods: n = 21), (d) Georgia Depression (clutches: n = 186; broods: n = 180), and (e) Southern Interior Mountains (clutches: n = 33; broods: n = 33) ecoprovinces of British Columbia. The figures are based on the week eggs or young were found in the nest. Nesting on the coast may begin 1 to 2 weeks before nesting in the southern portions of the interior and 3 to 4 weeks before nesting in the central and northern regions. Over its range, the Brewer's Blackbird experiences a notable shift in the length of its nesting season, ranging from 14 to 15 weeks in the Georgia Depression and Southern Interior ecoprovinces to 8 weeks in the Boreal Plains ecoprovince. The figures also suggest that this blackbird is double-brooded throughout much of its range in the province.

Figure 529. Of nearly 800 Brewer's Blackbird nests found in British Columbia with eggs or young, only 3% were parasitized by the Brown-headed Cowbird (near Penask Lake, 19 May 1994; R. Wayne Campbell).

Saskatchewan, and Manitoba, which is significantly higher than the rate in British Columbia.

Nest Success: Of 136 nests found with eggs and followed to a known fate, 57% successfully fledged at least 1 young. Interior nest success was 60% (n = 104); coastal success was 47% (n = 32). The main causes of nest failure were predation (42%; n = 77), desertion (35%), and flooding (4%). Other causes of nest failure included human disturbance (Fig. 530), interference by the Brown-headed Cowbird, fire, wind, and death of the female; 1 clutch was pecked by a wren. In nests that were preyed upon, the known predators were mainly crows (18%) and squirrels (12%). Single nests were preyed upon by Sharp-shinned Hawk, Black-billed Magpie, Steller's Jay, weasel, and domestic cat. The Sharp-shinned Hawk was observed taking nestlings out of their nest. The weasel attacked a nest that was built in a wood pile.

Of 415 nests with a known number of nestlings, 6.5% contained at least 1 dead nestling at one time or another.

REMARKS: Large-scale range expansions by the Brewer's Blackbird in eastern North America have also been noted. In the 19th century, the Brewer's Blackbird was a western bird, typically occurring west of the Mississippi River. By 1953, breeding populations had spread to western Ontario and wintering populations had reached the Gulf coast (Walkinshaw and Zimmerman 1961). The winter range and population expansion in the southeastern United States continued through the 1970s and was linked to changing agricultural practices, which provided winter food for the Brewer's Blackbird (Stepney 1975).

Grazing by cattle has been found to severely impact the nesting success of ground-nesting colonies in eastern Washington (Furrer 1975).

Figure 530. Removing cover from ground-nesting Brewer's Blackbirds in British Columbia can result in the death of eggs and young by exposure to sun and rain or predators. This photograph shows a protective layer of vegetation that was cut and deposited over a nest containing large young (15 km north of Chetwynd, 1 July 1998; R. Wayne Campbell).

NOTEWORTHY RECORDS

Spring: Coastal – Sidney 22 Apr 1973-6 eggs; Aldergrove 30 Apr 1990-adult feeding nestlings; Burnaby 21 May 1962-30 to 50 nests estimated in conifers in Forest Lawn Cemetery; Langley 15 May 1966-fledglings; Tofino Inlet 15 May 1970-20 (Campbell and Shepard 1971); Tofino 14 Mar 1983-3; Agassiz 3 May 1975-6 eggs; Port Neville Inlet 19 Mar 1975-15; Tahsis Inlet 3 Apr 1949-1 (Mitchell 1959); Sewall 13 May 1983-1; Hazelton 19 May 1978-12. **Interior** – Manning Park 19 Apr 1979-300; Osoyoos 16 Mar 1977-200; Creston 7 May 1980-7 eggs, 22 May 1980-6 nestlings, about 2 days old; Grand Forks 20 Mar 1980-300; Trail 13 May 1984-50 at Columbia Gardens; e Princeton 16 Apr 1976-10,000, including some Red-winged Blackbirds; nr Hedley 7 May 1960-22 nests in colony; Penticton 1 May 1934-6 eggs; Nicola River 20 May 1963-estimated 22 pairs in colony; Brouse 19 Apr 1979-40, 30 Apr 1983-6 eggs; Coldstream 11 May 1958-4 fledglings (Cannings et al. 1987); Lytton to Spences Bridge 23 May 1964-100; Falkland 17 Apr 1977-250; Invermere 17 Mar 1978-26; Monte Creek 19 May 1983-5 fledglings; Clinton 5 May 1994-5 nestlings; Golden 16 Apr 1977-35; 100 Mile House 3 Mar 1984-27; Chilcotin River (Riske Creek) 15 May 1983-5 eggs at Wineglass Ranch; Westwick Lakes 26 May 1958-5 nestlings, nearly fledged; Williams Lake 1 Apr 1980-85; Chezacut 14 Apr 1941-90; Alcock Lake 20 May 1994-3 eggs; Briar Ridge 24 May 1993-5 eggs; Mugaha Creek 26 Apr 1997-1; ne Attachie 18 May 1980-200; Baldonnel 24 Apr 1980-6; Montney 5 May 1979-200; nr Charlie Lake 26 May 1984-8 nests in colony; Boundary Lake (Goodlow) 27 Apr 1986-32; Alaska Highway (Mile 419) 18 May 1975-50; Fort Nelson 6 May 1975-50 at dump; Hyland River 11 May 1992-1 (Bowling 1992).

Summer: Coastal – Aldergrove 14 Aug 1991-4 nestlings; Surrey 15 Jul 1965-4 eggs; Sea Island 11 Jun 1975-125; Seabird Island 29 Aug 1983-200; Courtenay 25 Jul 1990-3 nestlings, 28 Aug 1949-1,500; Comox 17 Jun 1988-12 nests in colony, 31 Aug 1949-1,500 (Flahaut 1950); Campbell River 26 Aug 1982-1,000; Brooks Peninsula 6 Aug 1981-1 (Campbell and Summers 1997); Woss 12 Jul 1995-adult carrying food to young away from nest; Pemberton 4 Jun 1967-100; Cape St. James 8 Jul 1979-1; Terrace 18 Jun 1982-1 recent fledgling, 5 Jul 1977-5 eggs; New Aiyansh 20 Jun 1993-4 (Siddle 1993c); Contact Creek (Stikine River) 10 Jun 1972-1. **Interior** – Creston 2 Jun 1980-3 fledglings; Nelson 9 Jun 1974-5 fledglings; St. Mary River 12 Aug 1976-100; Sparwood 23 Aug 1983-100; Boston Bar 21 Jun 1995-4 nests with young in colony; nr Douglas Lake 16 Aug 1980-500; Vernon 27 Jun 1975-8 nestlings; Enderby 19 Jul 1968-3 young; Revelstoke 22 Jun 1990-7 fledglings; Clinton 10 Aug 1994-4 fledglings; Westwick Lakes 15 Jun 1955-21 nests in colony, 29 Jun 1955-4 nestlings; Riske Creek 27 Jul 1978-300; Nulki Lake 8 Jul 1945-175 (Munro 1949); Mosley Creek 30 Jun 1991-1; Mackenzie 29 Aug 1997-1; Chetwynd 22 Jun 1978-49 on Jackfish Rd; Briar Ridge 1 Jun 1993-5 small nestlings; Sunset Hill 8 Jul 1993-5 eggs; Alaska Highway (Mile 41) 29 Aug 1982-500; Fort St. John 28 Aug 1979-300; Charlie Lake 11 Jun 1976-3 fledglings; Fort Nelson 3 Jun 1974-5 eggs, 26 Jun 1985-5 fledglings, 18 Aug 1985-25; Liard Hot Springs 7 Jun 1979-5 nestlings, 10 Jun 1979-4 nests in colony.

Breeding Bird Surveys: Coastal – Recorded from 19 of 27 routes and on 53% of all surveys. Maxima: Seabird 22 Jun 1972-92; Chilliwack 26 Jun 1976-73; Victoria 30 Jun 1973-69. **Interior** – Recorded from 54 of 73 routes and on 70% of all surveys. Maxima: Salmon Arm 16 Jun 1976-152; Mabel Lake 8 Jun 1976-107; Columbia Lake 26 Jun 1973-102.

Autumn: Interior – Fort Nelson 16 Sep 1986-55; ne Cecil Lake 23 Oct 1983-1; Beatton River 22 Sep 1984-200 in canyon; Dawson Creek 3 Oct 1982-50; Smithers 17 Nov 1981-68; Kleena Kleene 28 Oct 1992-1; Chezacut 6 Sep 1941-400; Lac la Hache 2 Oct 1982-330 with 80 Red-winged Blackbirds; Cache Creek to Wallachin 5 Sep 1978-300; Merritt 5 Oct 1975-85; Keremeos to Cawston 11 Oct 1969-80; Creston 14 Sep 1985-400; Grand Forks 25 Sep 1989-100 (Weber and Cannings 1990). **Coastal** – Prince Rupert 22 Nov 1981-2; Masset 15 Oct 1954-1 (RBCM 10369); Tlell 3 Nov 1942-1 (Cooke 1947); Bella Coola 12 Nov 1989-9; Cape Scott 30 Sep 1935-19; Black Tusk (Garibaldi Park) 11 Oct 1971-1; Egmont 6 Oct 1973-1; Hemlock Valley 24 Sep 1972-12; Pitt Meadows 2 Sep 1973-832, 24 Sep 1967-3,000, 27 Oct 1974-1,200; Hope 26 Nov 1976-150; Iona Island 6 Oct 1974-1,100; Delta 4 Nov 1978-3,000; Tofino 15 Nov 1972-2; Langley 13 Sep 1962-1,000; Mud Bay 10 Oct 1979-1,000; Ucluelet 22 Oct 1985-20; Island View Beach (North Saanich) 12 Oct 1958-5,000 in mixed flock, mostly Brewer's Blackbirds; Saanich 15 Nov 1958-1,800; Oak Bay 9 Sep 1979-300.

Winter: Interior – Fort Nelson 9 Feb 1975-2; Smithers 16 Dec 1978-200; Fort St. James 17 Dec 1988-1; Vanderhoof 21 Jan 1979-36; Prince George 2 Jan 1972-88 (Rogers 1972b); Quesnel 9 Feb 1979-84; Williams Lake 3 Jan 1976-21; Revelstoke 1 Jan 1993-2 (Siddle 1993b); Sicamous 28 Dec 1976-1; Windermere 24 Feb 1983-2; Nakusp 30 Dec 1991-1; Wasa 25 Dec 1982-1. **Coastal** – Port Clements 6 Jan 1995-1 (Bowling 1995b); Bella Coola 2 Dec 1989-10, 29 Dec 1989-14; Port Neville Inlet 17 Dec 1975-1; 16 km se Campbell River 11 Feb 1973-200; Reifel Island 1 Jan 1970-20,000; Westham Island 18 Feb 1972-770; Cloverdale 10 Feb 1968-600; Surrey 3 Feb 1982-1,000; Langley 1 Jan 1968-2,000; Bamfield 2 Jan 1994-1; Central Saanich 1 Dec 1974-1,500, 24 Jan 1987-400.

Christmas Bird Counts: Interior – Recorded from 18 of 27 localities and on 40% of all counts. Maxima: Vernon 27 Dec 1981-370; Oliver-Osoyoos 29 Dec 1991-340; Smithers 30 Dec 1989-316. **Coastal** – Recorded from 26 of 33 localities and on 71% of all counts. Maxima: Ladner 2 Jan 1961-**6,217**, all-time Canadian high count (Anderson 1976b); Vancouver 26 Dec 1958-4,796; White Rock 22 Dec 1974-2,287.

Common Grackle

COGR

Quiscalus quiscula (Linnaeus)

RANGE: Breeds from northeastern British Columbia, south-central Mackenzie, northern Alberta, northwestern and central Saskatchewan, central and northeastern Manitoba, central Ontario, southern Quebec, New Brunswick, Prince Edward Island, Nova Scotia, and southwestern Newfoundland south to central and southern Texas, the Gulf coast, and southern Florida; and west to eastern Wyoming, central Colorado, and central and southeastern New Mexico. Winters from southern Minnesota, Wisconsin, Michigan, Ontario, New York, Vermont, New Hampshire, and Maine south through central portions of Nebraska, Kansas, and Oklahoma to northern Texas, southeastern New Mexico, central and southern Texas, and southern Florida.

STATUS: On the coast, a *very rare* vagrant in the Georgia Depression Ecoprovince; absent from the Coast and Mountains Ecoprovince.

In the interior, *casual* vagrant in the Southern Interior, Central Interior, Sub-Boreal Interior, and Northern Boreal Mountains ecoprovinces; *very rare* vagrant in the Southern Interior Mountains Ecoprovince; *uncommon* to locally *fairly common* migrant and summer visitant in the Boreal Plains Ecoprovince; and locally *uncommon* migrant and summer visitant in the Taiga Plains Ecoprovince.

Breeds.

CHANGE IN STATUS: Munro and Cowan (1947) describe the Common Grackle (Fig. 531) as a rare transient in the Peace Lowland, where at least 4 birds were collected during field work in May 1938 (Cowan 1939). The birds' occurrence in the Peace Lowland was not surprising as the species had been reported by Preble (1908) in similar habitat at several localities in northern Alberta, including Lesser Slave Lake, and in the Northwest Territories, on the upper Mackenzie River at Fort Smith and Fort Simpson.

The Common Grackle was not recorded again in northeastern British Columbia until 1968, when several were reported from the Fort Nelson area (Erskine and Davidson 1976). In 1974 it was again reported from Fort Nelson, and also from the Boreal Plains. The expansion into northern British Columbia continued, with records from the Sub-Boreal Interior in 1975 and the Northern Boreal Mountains in 1976. The Common Grackle now occurs regularly in summer in the Boreal Plains and Taiga Plains.

In southern portions of the province, the earliest record was of a group of 4 vagrant grackles that appeared in the Shuswap Lake area of the Southern Interior in 1948. Single vagrants were also recorded there in 1976 and 1991. In the upper Columbia valley, the first record was at Columbia Lake in 1969. Single birds were subsequently recorded in the east Kootenay trench in 1976, 1982, 1996, and 1997, and there were multiple sightings of single birds at Revelstoke in 1984, 1985, 1986, 1989, and 1990. All but 3 of these occurrences were at localities on transportation routes through the Rocky Mountains from Alberta. Most of these records are from the Southern Interior Mountains, which adjoins the BC-Alberta boundary.

Figure 531. Adult male Common Grackle (near Swan Lake, Tupper, 16 June 1996; R. Wayne Campbell).

A possible source of grackles occurring in south-central British Columbia may be found in Idaho; a single bird was reported there in 1950 but by the early 1970s the Common Grackle had become an established breeding species (Taylor and Trost 1985). All coastal records have been from the Georgia Depression, where the Common Grackle first appeared in 1968 (Campbell and Sirk 1969), and there have been scattered records there since 1981.

It seems likely that the expansion of the Common Grackle's breeding range into British Columbia is a response to habitat change accompanying agriculture. Orians (1985) commented on the population increase and range expansion of several species of blackbirds, which he explained as a response to the increased availability of winter food supplies, in the form of spilled grain, in eastern and central North America. This is unlikely to be the case with the Common Grackle in British Columbia, as Breeding Bird Surveys, both continent-wide and in Canada, reveal a steady decline in populations of this grackle from 1966 to 1996 (Sauer et al. 1997).

NONBREEDING: The Common Grackle is locally distributed in the regions of the province east of the Rocky Mountains, where it is confined to lower elevations and is generally associated with wetlands and areas of human occupation. There its numbers are highest in the Peace Lowland. West of the Rocky Mountains, the Common Grackle is a scarce vagrant, and single birds and flocks of up to 14 grackles have been reported from widely scattered locations. Most observations have been during the spring migration. It is logical to assume that in northern regions of British Columbia wanderers were part of the population coming through northern Alberta. The habitat favoured by these vagrants includes human-made clearings in otherwise treed habitats, wetlands, backyards with

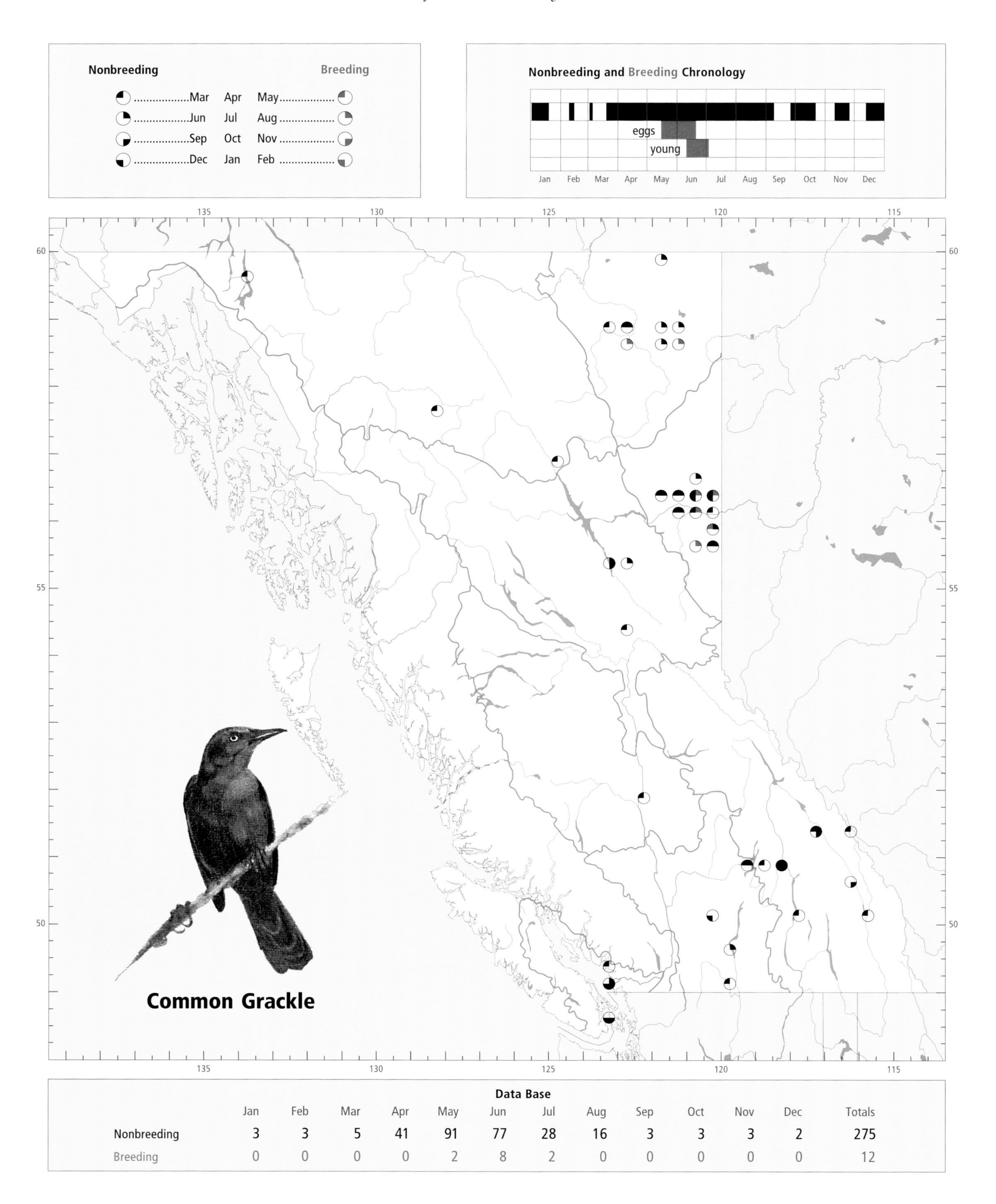

Data Base

	Jan	Feb	Mar	Apr	May	Jun	Jul	Aug	Sep	Oct	Nov	Dec	Totals
Nonbreeding	3	3	5	41	91	77	28	16	3	3	3	2	275
Breeding	0	0	0	0	2	8	2	0	0	0	0	0	12

Figure 532. During migration, the Common Grackle may stop briefly to feed on spilled grain along railway tracks in southeastern British Columbia (Revelstoke, 30 April 1997; R. Wayne Campbell).

bird feeders, homesteads, ponds on abandoned farms, airstrips, and public campgrounds.

In the southern half of the province, Common Grackle records from the Georgia Depression, Southern Interior, and Southern Interior Mountains were clustered along the transcontinental corridor of the Canadian Pacific Railway and the Trans-Canada Highway. Again, the habitat used was land largely altered by human activity: farmyards, sewage lagoons, spilled grain along railway tracks (Fig. 532), a horse corral, a park warden station, service yards at a hydroelectric station, orchards, a harvested cornfield, backyard gardens with bird feeders. Riparian black cottonwood–alder–willow woodlands, islets, and Douglas-firs beside a stream were also used. In the north, spring migrants foraged in damp fields and farmyards.

In the Boreal and Taiga plains, where the Common Grackle is a breeding species, migrants arrive onto the nesting grounds (see BREEDING).

On the coast, the Common Grackle has been found on lowlands, less than 100 m above sea level; in the interior, it has been recorded from elevations of 300 to 860 m.

Migration is most evident in northeastern portions of the province (Figs. 533 and 534). In the Boreal Plains, the first spring migrants may arrive in the second week of April but most arrive later in the month. In the Taiga Plains, the first migrants arrive in early May (Erskine and Davidson 1976), but numbers do not peak until the second week of June. Numbers decline in the third week of July and migrants leave in small groups throughout August. The last birds leave the Taiga Plains in the second week of August and the Boreal Plains by the end of August. There are 2 January records, probably of

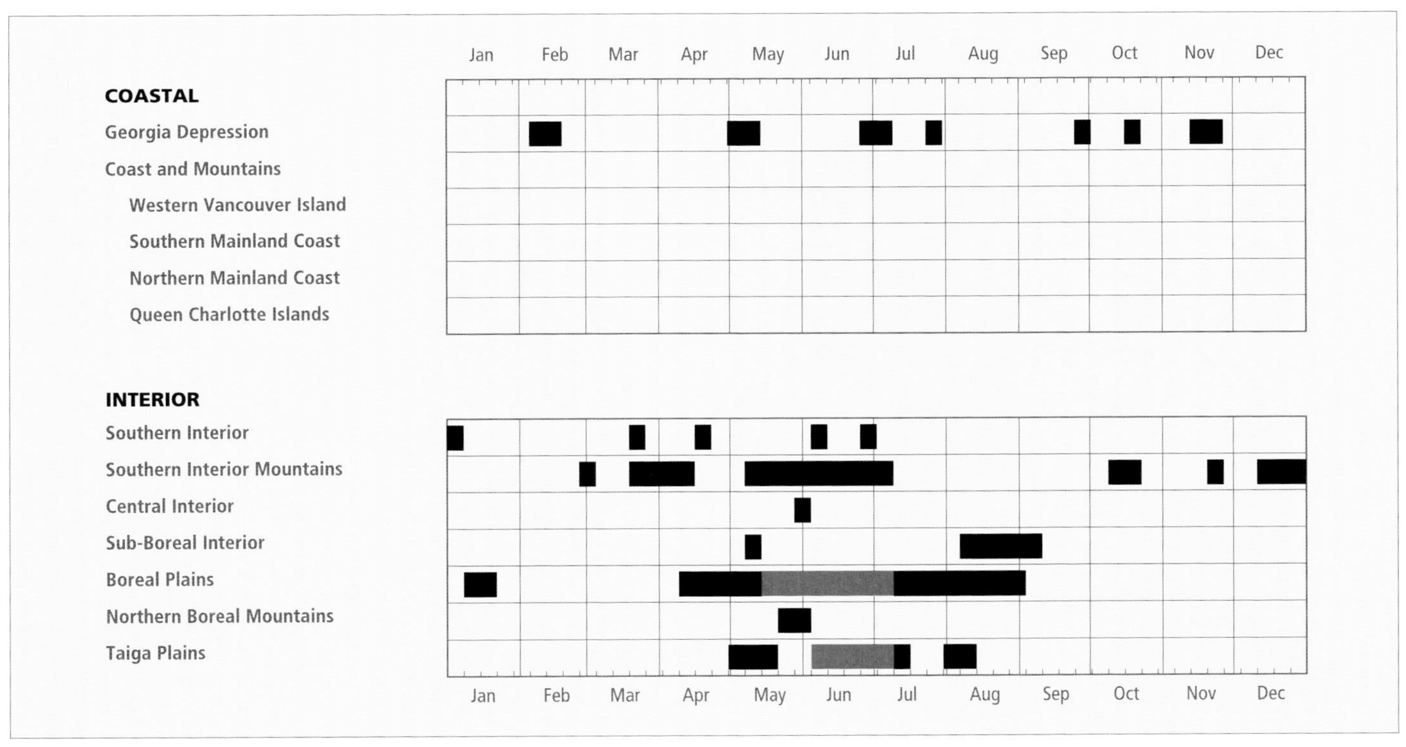

Figure 533. Annual occurrence (black) and breeding chronology (red) for the Common Grackle in ecoprovinces of British Columbia. Records are shown for the week in which they occurred.

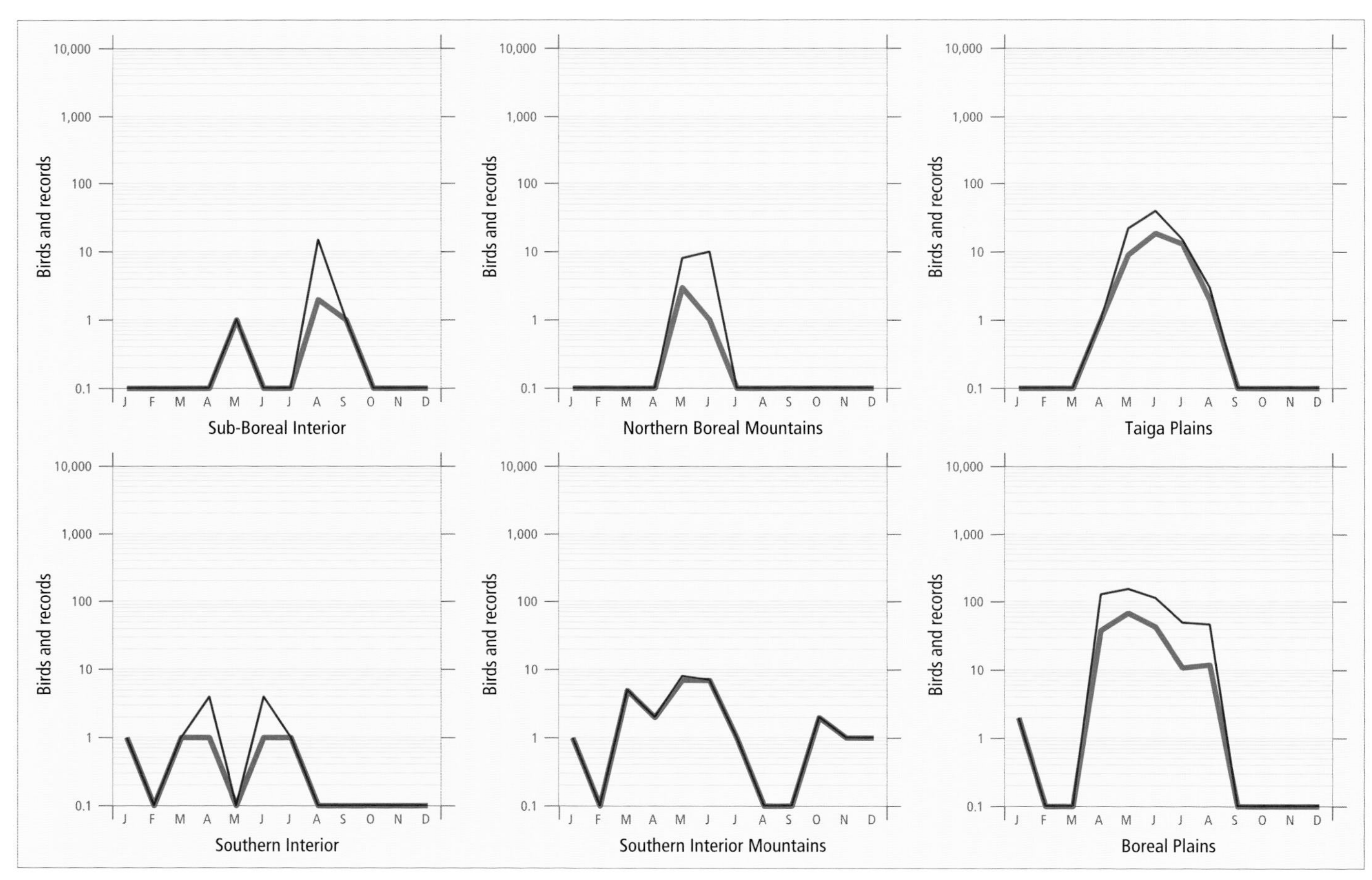

Figure 534. Fluctuations in total number of birds (purple line) and total number of records (green line) for the Common Grackle in ecoprovinces of British Columbia. Christmas Bird Counts, Breeding Bird Surveys, and nest record data have been excluded.

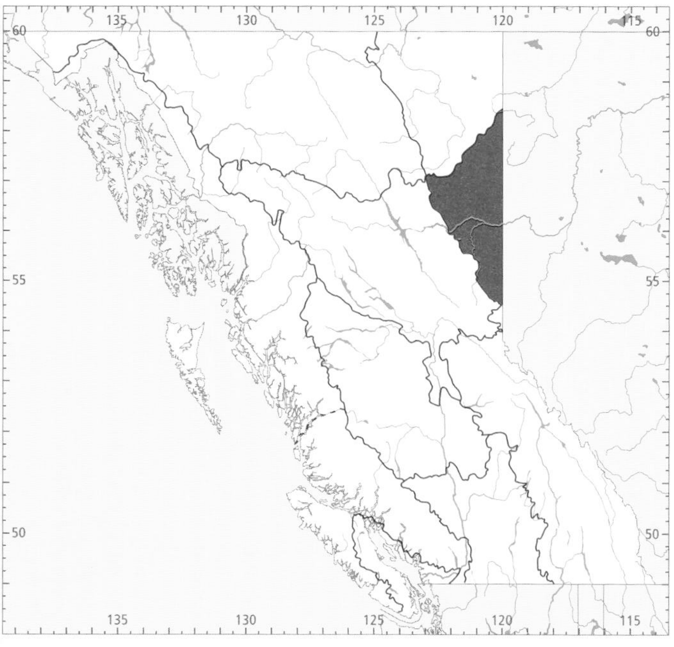

Figure 535. In British Columbia, the highest numbers for the Common Grackle in summer occur in the Boreal Plains Ecoprovince.

the same bird, at Fort St. John in 1987, but no indication that it successfully wintered there.

On the coast, the Common Grackle has occurred sporadically between 11 February and 23 November; in the interior, there are scattered records of solitary birds throughout the year, but it occurs regularly in the Boreal Plains from 14 April to 29 August (Fig. 533).

BREEDING: In British Columbia, the Common Grackle is at the western limit of its breeding range and breeds east of the Rocky Mountains only in the Taiga Plains and Boreal Plains. Many of the records are associated with wetlands near grazing areas or farmland.

The Common Grackle reaches its highest numbers in summer in the Boreal Plains (Fig. 535).

An analysis of Breeding Bird Surveys in British Columbia for the period 1968 through 1993 did not detect a significant change in numbers on the 2 interior routes reporting grackles. Sauer et al. (1997) found an average annual decline of 1.6% ($P < 0.01$) in Canada and across North America for the period 1966 through 1996.

The Common Grackle has been recorded breeding in the interior at elevations from 350 to 800 m.

In British Columbia, this species usually nests in cattails or willows over or near shallow water (Figs. 536 and 537), but

Figure 536. In the Boreal Plains Ecoprovince of British Columbia, many Common Grackle nests in dense patches of dead flooded willows along the shores of lakes and ponds (Swan Lake, Tupper, 14 June 1996; R. Wayne Campbell).

Figure 537. Small patches of cattails in wetlands throughout the Boreal Plains of British Columbia may support small colonies of breeding Common Grackles (west of Charlie Lake, 22 June 1996; R. Wayne Campbell).

nests have been recorded in swampy areas in black spruce forests and on the edge of trembling aspen groves, always associated with water.

The Common Grackle has been recorded breeding in British Columbia from 16 May (calculated) to 2 July (Fig. 533).

Colonies: Although present at low density in the Boreal Plains, the Common Grackle frequently aggregates in small, loose groups of nesting birds. Sizes of 6 such aggregations ranged from 2 to 13 nests (2 nests-1, 4 nests-1, 5 nests-2, 7 nests-1, 13 nests-1).

Nests: Nests of the Common Grackle in British Columbia have been situated in marshy or swampy wetlands, near the edges of ponds or small lakes surrounded by trembling aspen or black spruce, or associated with oxbows. Occasionally,

Figure 538. Common Grackle nest in a cavity in a mature balsam poplar stump (Swan Lake, Tupper, 13 June 1996; R. Wayne Campbell).

Figure 539. Common Grackle nest in a stove pipe in an abandoned building (near Swan Lake, Tupper, 12 June 1996; R. Wayne Campbell).

Figure 540. Full clutch of 4 eggs of the Common Grackle (west of Charlie Lake, 20 June 1996; R. Wayne Campbell).

Figure 541. Common Grackle nestlings in their typical coarse grass and cattail nest (west of Charlie Lake, 22 June 1996; R. Wayne Campbell).

trees included those killed by flooding in beaver ponds and shallow parts of lakes. The nests were either built in emergent vegetation, such as cattails (Fig. 537) and bulrushes, and resembled large Red-winged Blackbird nests, or were built in willow thickets along the shoreline or in shallow water. Nests have also been recorded in a black spruce snag, in a cavity in a dead balsam poplar (Fig. 538), and in a building (Fig. 539). Elsewhere in its range, the Common Grackle places its nest in a wide variety of open or partially open habitats, including woodlands or the edges of deciduous or coniferous forests, clumps of trees adjacent to farms and ranches, cattail marshes (Snelling 1968), flooded alder swamps and the girders of railway bridges (Erskine 1971), and tall conifers in urban settings (Peer and Bollinger 1997a). The heights of 24 nests ranged from ground level to 16.0 m, with 15 nests between 0.5 and 1.2 m.

In British Columbia, nests were constructed of coarse grasses, cattails, plant stems, twigs, and in 1 case, mud. These cups were lined with finer materials such as feathers, fine grasses, and strawberry runners. When built in cattails, nest cups were placed between 0.3 and 1.5 m above the surface of the water.

Eggs: Dates for 9 clutches ranged from 27 May to 20 June, with 6 recorded between 29 May and 12 June. Calculated dates indicate that eggs can be present as early as 16 May. Sizes of 4 clutches (Fig. 540) ranged from 1 to 6 eggs (1E-1, 4E-1, 5E-1, 6E-1). Elsewhere, clutch sizes range from 1 to 7 eggs; the incubation period averages 13.5 days and ranges from 11.5 to 15 days (n = 94) (Peer and Bollinger 1997a).

Young: Dates for 7 broods ranged from 11 June to 2 July, with 5 recorded between 12 June and 2 July. Sizes of 5 broods (Figs. 541 and 542) ranged from 1 to 4 young (1Y-1, 2Y-1, 4Y-3), with 3 having 4 young. The nestling period is typically 12 to 15 days, but can range from 10 to 17 days (Peer and Bollinger 1997a).

Brown-headed Cowbird Parasitism: Cowbird parasitism was not found in British Columbia in 10 nests recorded with eggs or young. The Common Grackle is an infrequent host of the Brown-headed Cowbird, with no instances of parasitism found in 401 Common Grackle nests studied in Illinois, and only 20 cases reported throughout the continent (Peer and Bollinger 1997b). This grackle is not known to have fledged cowbirds (Friedmann and Kiff 1985). Experimental studies of several aspects of the parasitic interaction between these

2 species reveal that the Common Grackle is more aggressive towards Brown-headed Cowbird effigies than towards those of another passerine of similar size. It is a rejector species, and young cowbirds that hatch in grackle nests have a low survival rate (Peer and Bollinger 1997b).

Nest Success: Insufficient data.

REMARKS: The Common Grackle is also known as the Bronzed Grackle or Purple Grackle. It shows a high degree of fidelity to both breeding and wintering sites. Thus, if the species continues to find satisfactory breeding conditions in the areas of British Columbia that it has colonized, the local population is likely to increase. So far, there have been no large post-breeding feeding aggregations such as those that occur in the southern parts of its range, where it has become a serious pest of grain and fruit farmers (Peer and Bollinger 1997a).

For a detailed review of the biology of the Common Grackle, see Peer and Bollinger (1997a).

Figure 542. After 10 to 12 days in the nest, some young leave and hide among cattails and bushes, and are fed later by the adults (west of Charlie Lake, 22 June 1996; R. Wayne Campbell).

NOTEWORTHY RECORDS

Spring: Coastal – Vancouver 6 May 1968-1 with 7 Brewer's Blackbirds (Campbell and Sirk 1969); Quadra Island 26 Mar 1982-1 (Campbell 1982c). **Interior** – Oliver 21 Mar 1976-1 with American Robins; Nakusp 20 May 1986-1 in farmyard; Columbia Lake 20 May 1969-1 (Wilson et al. 1972); Celista 18 Apr 1948-4; Revelstoke 4 Mar 1989-1, 27 May 1984-1 (Rogers 1984b); lower Blaeberry River valley 20 May 1996-2 males; Yoho National Park 7 May 1976-1 with 2 Brewer's Blackbirds; nr Riske Creek 28 May 1982-1 male; Bear Lake (96 km n Prince George) 13 May 1977-1; Tupper Creek 7 May 1938-3 (Cowan 1939), 29 May 1938-1 (RBCM 7891); Baldonnel 7 May 1974-1; McQueen Slough (Dawson Creek) 27 May 1994-6 eggs; Fort St. John 14 Apr 1984-2 at airport, early arrivals; Cache Creek (Peace River) 23 Apr 1983-1; s end Charlie Lake 24 Apr 1984-30 in newly seeded playing field; Ingenika Point 25 May 1986-1 at feeder; se Hyland Post 30 May 1976-6 at small pond; Fort Nelson Apr 1974-1 (Erskine and Davidson 1976), 1 May 1974-1 (BC Photo 368); Parker Lake 5 May 1975-3; Atlin 29 May 1981-1 in Tarahne Park.

Summer: Coastal – Reifel Island 25 Jun to 26 Jul 1989-1. **Interior** – ne Shuswap Lake 1 Jul 1991-1; Moberly Marsh (n Golden) 16 Jun 1997-1 male; Riske Creek 30 May 1994-1 male foraging edge wetland; Pine Pass 23 and 24 Aug 1975-14; Mugaha Creek 12 Aug 1998-1; Arras 20 Jun 1993-3 eggs; 15 km s Dawson Creek 4 Jul 1993-2 full-grown fledglings with adult female; McQueen Slough (Dawson Creek) 1 Jun 1994-6 eggs, 12 Jun 1996-small nestlings, 23 Jun 1996-fledglings close to nest; Fort St. John 1 Jun 1986-11, 22 Jun 1996-colony of 13 nests; Alaska Highway (Mile 47) 29 Aug 1985-20, last autumn record; 8 km n St. John Creek 7 Jun 1977-2 young just out of nest; Charlie Lake 16 Jun 1996-colony of 5 nests; Boundary Lake (e Goodlow) 11 Jun 1989-nestlings, 2 Jul 1978-18, including 4 or 5 recently fledged young, in willow marsh; Rose Prairie 18 Aug 1975-5; Hyland Post Jun 1976-10 (Osmond-Jones et al. 1977); Oxbow Lake 4 Jun 1993-2 pairs carrying food; 1.5 km ne Clarke Lake 17 Jun 1985-2 fledglings; 2.5 km nw Clarke Lake 17 Jun 1985-5 in beaver swamp with willows; nr Fort Nelson 8 Aug 1985-1; Parker Lake 30 Jul 1968-2 fledglings; s Kotcho Lake 9 Jul 1982-1; nr Petitot River 25 Jun 1982-4 young.

Breeding Bird Surveys: Coastal – Not recorded. **Interior** – Recorded from 2 of 73 routes and on less than 1% of all surveys. Maxima: Fort St. John 28 Jun 1985-8; Tupper 20 Jun 1994-4.

Autumn: Interior – Gagnon Creek 7 Sep 1995-1; Beaver River (Glacier National Park) 9 Oct 1981-1; Invermere 20 Oct 1992-1 (Siddle 1993a). **Coastal** – Lulu Island 18 Oct 1970-1; Saanich 27 Sep 1986-1 (BC Photo 1141); Central Saanich 30 Sep 1986-1 with Brewer's and Red-winged blackbirds, European Starlings, and Northwestern Crows, 12 to 23 Nov 1988-1.

Winter: Interior – Fort St. John 8 Jan 1987-1; Revelstoke Jan 1990-1 (Siddle 1990a); Douglas Lake 2 Jan 1991-1. **Coastal** – Beach Grove 11 Feb 1989-1; North Saanich 11 to 13 Feb 1989-1.

Christmas Bird Counts: Interior – Recorded from 1 of 27 localities and on less than 1% of all counts. Maximum: Revelstoke 30 Dec 1989-1. **Coastal** – Not recorded.

Brown-headed Cowbird

Molothrus ater (Boddaert)

BHCO

Figure 543. Over the past 100 years, the Brown-headed Cowbird has expanded its range northwestward in North America to include much of British Columbia. The species' mating system is complex, and the sexual and social behaviour of males (a) and females (b) is being investigated (Victoria, 8 August 1999; R. Wayne Campbell). In late summer, juveniles (c) wander widely and are often tame around humans (Kyuquot Sound, 16 June 1980; Brent Matsuda).

RANGE: Breeds from central and eastern Alaska, southwestern Yukon, northern British Columbia, and southern Mackenzie east to north-central Saskatchewan, southern Manitoba, central Ontario, and southern Quebec; south to Baja California and Veracruz, and east across the Gulf coast to southern Florida. Winters from southwestern British Columbia and the western portions of the Pacific coast states east to southern Utah and northern New Mexico, across to the southern portions of Ontario, New Brunswick, and Nova Scotia, south to the southern limits of the breeding range.

STATUS: On the coast, a *fairly common* to locally *very common* migrant and summer visitant to the Georgia Depression Ecoprovince. Elsewhere on the coast, an *uncommon* to locally *common* migrant and summer visitant, with the exception of the Queen Charlotte Islands, where it is *rare*. In winter, *fairly common* in the Fraser and Nanaimo lowlands of the Georgia Depression, *casual* on the Northern Mainland Coast of the Coast and Mountains Ecoprovince, and *accidental* on Western Vancouver Island.

In the interior, a *fairly common* to locally *very common* migrant and summer visitant in the Southern Interior and Southern Interior Mountains ecoprovinces. Further north in the Central Interior, Sub-Boreal Interior, and Boreal Plains ecoprovinces, it is *uncommon* to locally *very common*, and in the Northern Boreal Mountains and Taiga Plains, it is *uncommon* to locally *common*. In winter, *very rare* in the Southern Interior; *casual* in both the Southern Interior Mountains and the Central Interior.

Breeds.

CHANGE IN STATUS: The exact date of the appearance of the Brown-headed Cowbird (Fig. 543) in British Columbia is not known. It was first reported, from the Revelstoke area, in May 1890 (Macoun and Macoun 1909), not 1939 as reported by Lemon (1969). From 1890 to 1910, it was reported in the Okanagan valley from Penticton north to Martin Prairie near Vernon (Fig. 544a). Brooks considered the cowbird common in the Okanagan when he arrived there in 1898 (Ward and Smith 1998). The first probable nesting record in the province was documented in 1892 near Vernon when a juvenile was collected; the host species was not reported. On the coast, probable vagrants were first recorded in Victoria in 1893 and later at Sumas in 1897 (Brooks 1917), both in the Georgia Depression.

Through the 1920s, the Brown-headed Cowbird was reported only once more from the coast, at Hope, in the Southern Mainland Coast, in 1923. By then it was also being reported from the Southern Interior Mountains – Cranbrook in 1915, Field in 1922 (Ulke 1923), Indianpoint Lake in 1929 – as well as the Boreal Plains – Fort St. John and Rolla in 1922 (Williams 1933a, 1933b) and Tupper Creek in 1928 (Fig. 544b). Brooks and Swarth (1925) noted the Brown-headed Cowbird as a summer visitant in southern British Columbia, "tolerably common east of, and a straggler west of the Cascade Mountains."

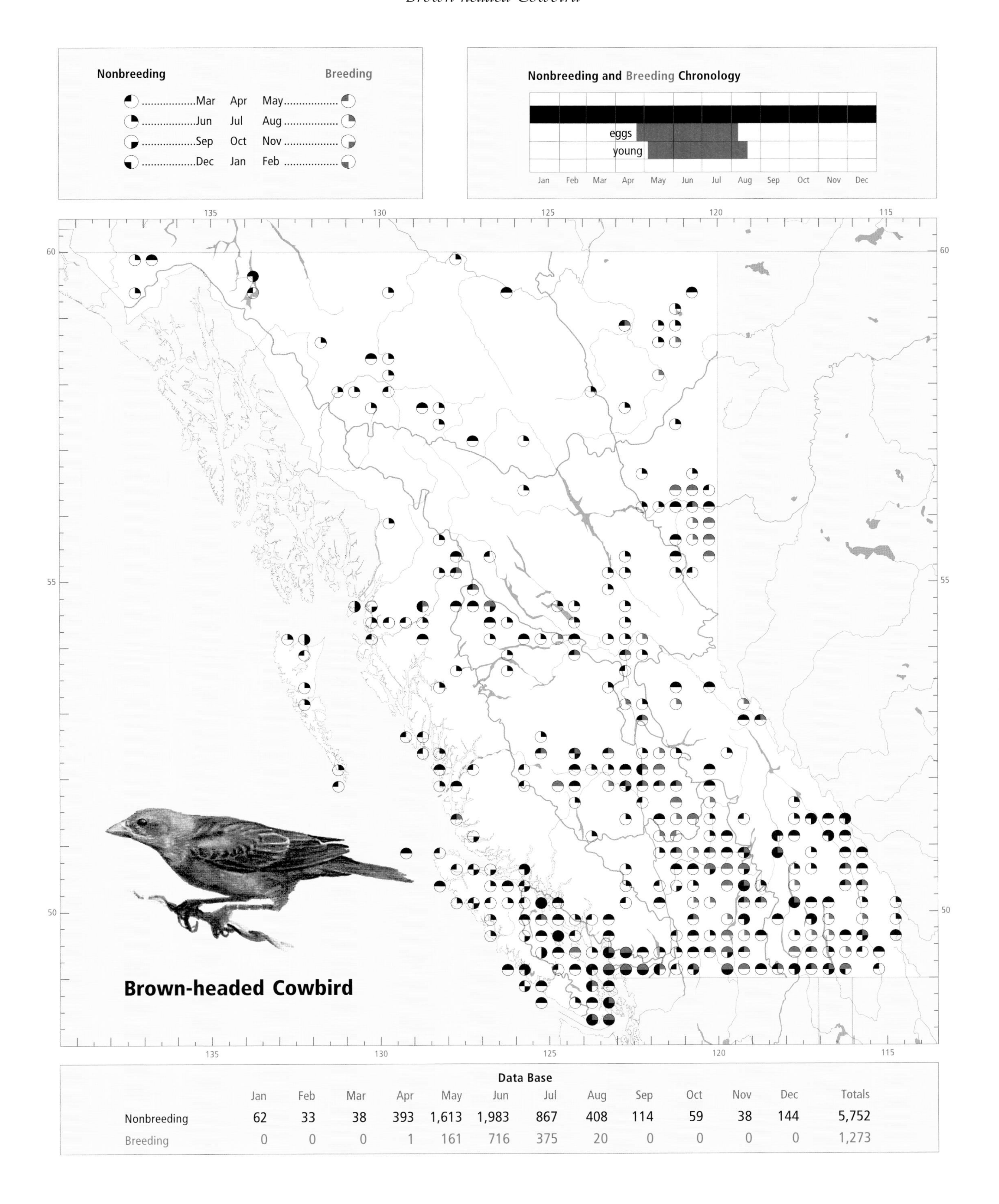

Data Base

	Jan	Feb	Mar	Apr	May	Jun	Jul	Aug	Sep	Oct	Nov	Dec	Totals
Nonbreeding	62	33	38	393	1,613	1,983	867	408	114	59	38	144	5,752
Breeding	0	0	0	1	161	716	375	20	0	0	0	0	1,273

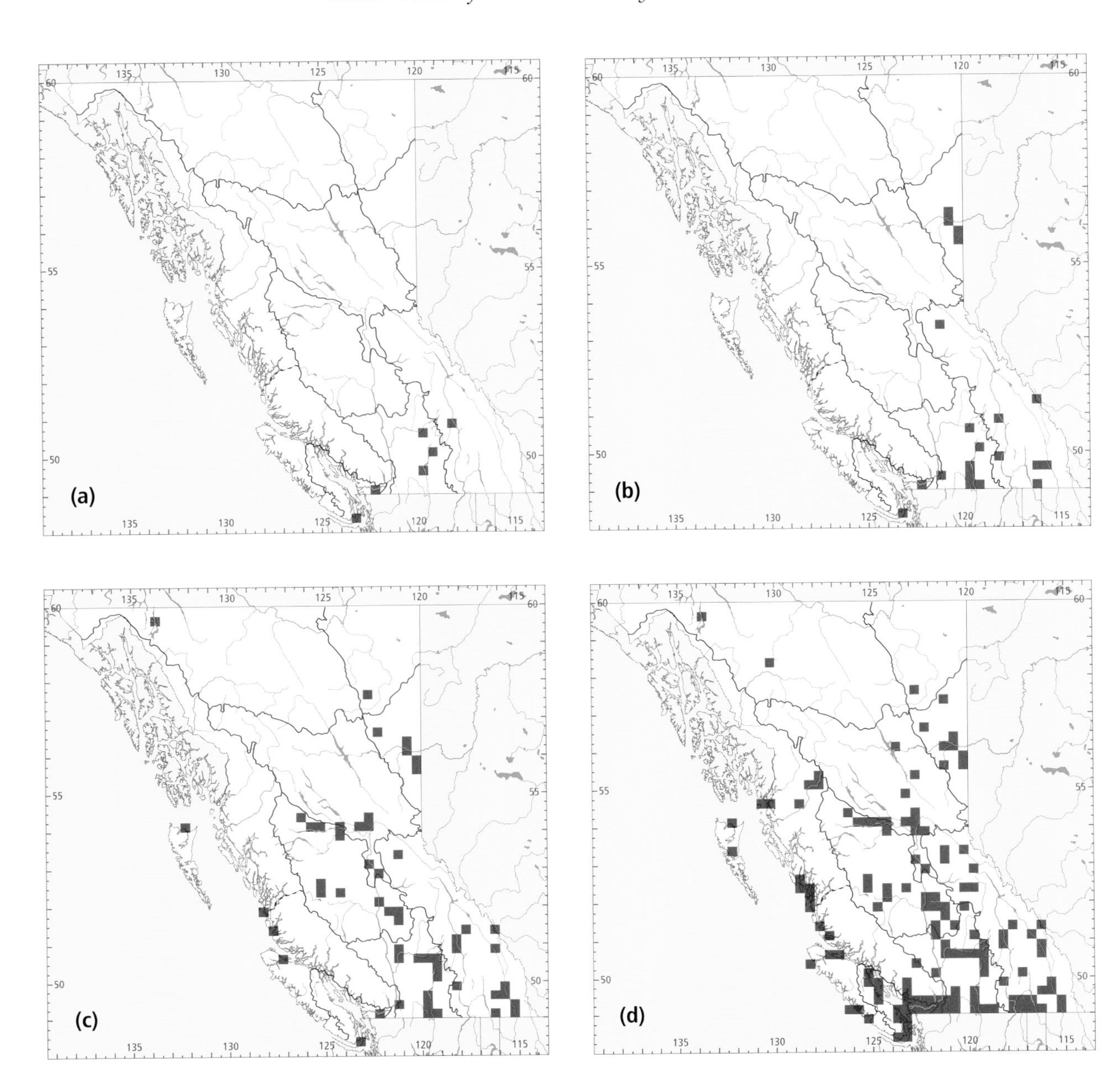

(e)

Figure 544. Range expansion of the Brown-headed Cowbird in British Columbia from 1890 to 1990 by National Topographic Series quadrangles. Cowbirds were first reported from the southern portions of the province (a), and even by 1930 (b) most reports were from the southern interior regions. By 1950 (c), the cowbird had been reported in the Central Interior and Sub-Boreal Interior ecoprovinces, but was still a rare species along the coast. That changed in the late 1950s and early 1960s, and by 1970 (d) the species could be found in suitable habitat the length of the coast, including much of Vancouver Island. At the same time, it was reported from more locations and probably increased in abundance in the interior south of latitude 56°N. By 1990 (e), the Brown-headed Cowbird could be found throughout the province.

By the late 1940s, the cowbird was considered an abundant summer visitant in southern portions of interior British Columbia and the Peace Lowland, but less common in the Cariboo and Chilcotin areas and in the central parts of the province (Munro and Cowan 1947). The cowbird was present at Prince George and Vanderhoof by 1944 and west to Nimpo Lake by 1948 (Fig. 544c). Johnstone (1949) noted that Brown-headed Cowbirds were very numerous in the east Kootenay in the summer of 1937 but were "conspicuous by their absence" in the summers of 1938 to 1941. In the summers of 1944 and 1948, they were again "quite numerous." Munro (1949) considered the cowbird "fairly common" in the Vanderhoof area in the summers of 1945 and 1946. Elsewhere in the province, it was considered casual in the boreal forest: there were at least 2 records from Atlin in the Northern Boreal Mountains in 1931 and 1934 (Swarth 1936), and a specimen was collected near Trutch in the Taiga Plains in 1947.

The cowbird was also considered casual along the coast during this time. In his review of the birds of the Vancouver district, Cumming (1932) does not mention the cowbird. There were, however, a number of reports from further north, including Port Hardy in 1934, Masset in 1942, and Goose Group in 1948. The question that arises is: how did the cowbird find its way to Port Hardy (reported there at least over the period 1934 through 1939), Masset, and some coastal islands in between when the bird was virtually absent from the south coastal region? Perhaps those birds arrived from the northeast as cowbirds from the Peace Lowland population spread westward. The first documented instance of cowbird parasitism on the coast came from the 1930s: in 1937, a young cowbird "scarcely of independent means" was collected on Calvert

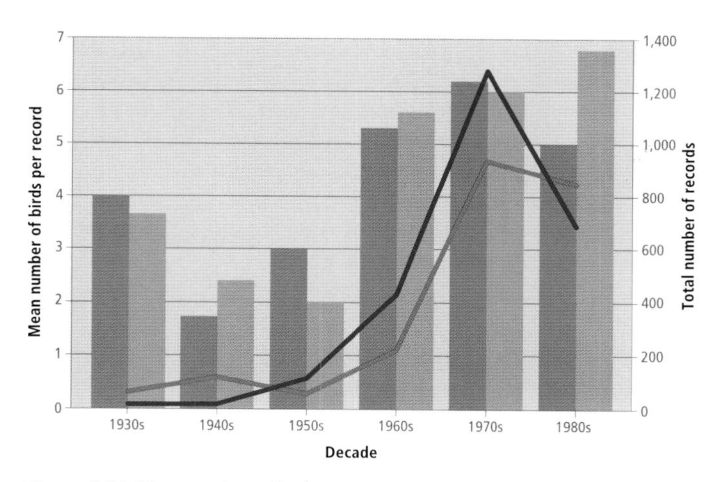

Figure 545. Mean number of birds per record (dark green bars) and total number of records (dark purple line) on the coast and mean number of birds per record (light green bars) and total number of records (light purple line) in the interior by decade for the Brown-headed Cowbird in British Columbia. Although the number of records is a function of the numbers of observers, it does suggest that cowbird observations in the interior were relatively stable from the 1930s through the 1950s, whereas cowbird observations on the coast (where there were proportionately more observers) were low through the 1930s and 1940s but began to increase in the 1950s as the birds began to colonize the area. Large increases in the number of Brown-headed Cowbird observations and the number of birds seen with each observation came in the 1960s; in the 1980s, such increases continued in the interior but declined on the coast.

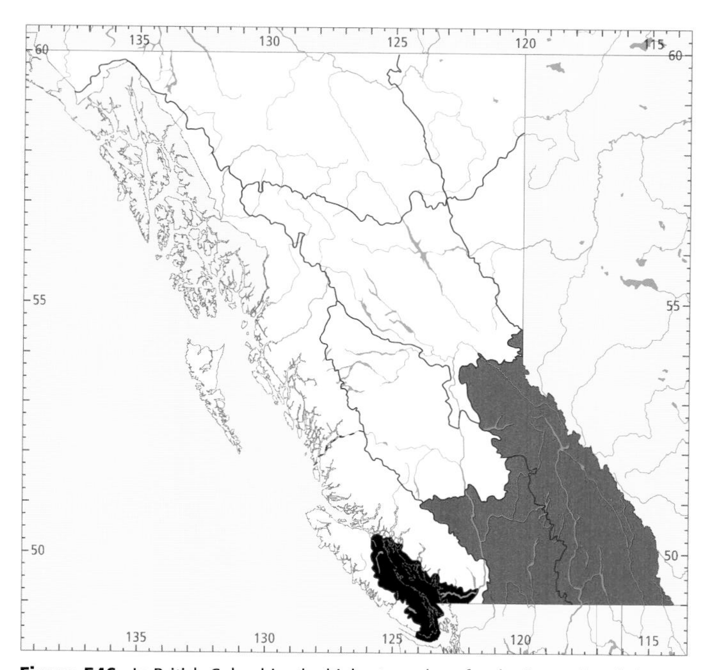

Figure 546. In British Columbia, the highest numbers for the Brown-headed Cowbird in winter (black) occur in the Georgia Depression Ecoprovince; the highest numbers in summer (red) occur in the Southern Interior and Southern Interior Mountains ecoprovinces.

Island, north of Vancouver Island (Laing 1942), and in 1939, a cowbird egg in a Yellow Warbler nest, along with 3 host eggs, was collected on Sea Island, near Vancouver (Meugens 1956).

The cowbird began appearing with greater frequency in the Georgia Depression in the 1950s (Fig. 545). It was reported from Saanichton and the Comox area in 1953, and it was regularly reported from both Victoria and Vancouver by 1955 (Holdom 1960), Huntingdon in 1957 (Racey 1958), and Sidney in 1958. By 1959 the cowbird was being reported from many areas in the Fraser and Nanaimo lowlands (Holdom 1960). Brown-headed Cowbird numbers appear to have increased dramatically both at the coast and in the interior beginning in the 1960s (Fig. 545). By then, it was found along the coast north to Kitwanga (1960) and Hazelton (1963), including Western Vancouver Island (Blunder Island in 1961, Browning Inlet in 1968) (Fig. 544d; Richardson 1971). In the interior, it could be found north to Finlay Forks in the Sub-Boreal Interior by 1966, although it was then an uncommon migrant and summer visitant in the Prince George area of that ecoprovince. Two specimens were collected at Dease Lake in the Northern Boreal Mountains in 1962, bringing the number of records on file to 4 from the Northern Boreal Mountains and 1 from the Taiga Plains.

By the 1990s, Brown-headed Cowbirds were being reported throughout the province, including the northernmost reaches of the Northern Boreal Mountains and Taiga Plains: Cassiar Junction in 1972, Fort Nelson and Fireside in 1974, Survey Lake in 1980, and Helmut (Kwokullie Lake) in 1982 (Fig. 544e).

Because of poor and often restricted coverage by early collectors and naturalists, it is difficult to provide a convincing

Figure 547. In British Columbia, Brown-headed Cowbirds prefer open habitats, especially rangeland and pastures with livestock, throughout the year (Brisco, 27 September 1996; R. Wayne Campbell).

Figure 548. On the southwest mainland coast of British Columbia, flocks of Brown-headed Cowbirds forage at feedlots with European Starlings and Brewer's and Red-winged blackbirds in winter (Delta, 26 February 1996; R. Wayne Campbell).

statement about subspecific changes in populations and ranges across southern British Columbia. Brown-headed Cowbirds that occur in the Okanagan valley are morphologically distinct from birds on the south coast. It seems likely that we have had a colonizing subspecies on the coast (*M. a. obscurus*) with another subspecies (*M. a. artemisiae*) that may have always been present in the interior. Also, because Brown-headed Cowbirds are unusually mobile compared with other songbirds, it is hard to interpret extralimital observations (J.N.M. Smith pers. comm.).

NONBREEDING: The Brown-headed Cowbird has a widespread distribution from spring through autumn throughout most of British Columbia, including Vancouver Island and the Queen Charlotte Islands.

It reaches its highest numbers in winter in the Georgia Depression, where it is usually found in numbers only in the western portions of the Fraser Lowland and in the Nanaimo Lowland of eastern Vancouver Island, from Victoria north to Comox (Fig. 546).

The Brown-headed Cowbird is usually found at lower elevations both on the coast and in the interior. On the coast, most reports are from the lowlands of the Georgia Depression, although cowbirds have been reported accompanying Roosevelt elk (*Cervus elaphus roosevelti*) at elevations as high as 1,370 m (Green Mountain, near Nanaimo). In late summer, young cowbirds use alpine ridges for foraging. In the interior, the cowbird has been found at elevations between 460 and 2,440 m (south of Tatlayoko Lake), and fledged juveniles are now frequently reported in alpine meadows by hikers who experience their "tameness" around camps. The cowbird's use of higher elevations in British Columbia may be a recent development. Rothstein et al. (1980) document a recent altitudinal range expansion by the Brown-headed Cowbird in the Sierra Nevada of California, and Hanka (1985) notes a similar range expansion in Colorado. Both range expansions were attributed to habitat alteration caused by heavy cattle grazing.

The Brown-headed Cowbird frequents primarily open habitats in both urban and rural settings. Farmlands, especially pastures, rangeland, and feedlots with horses or cattle (Figs. 547 and 548), are heavily used, as well as fields with agricultural crops, nursery lands, urban parks, backyard feeders and gardens, powerline rights-of-way, alpine meadows,

Figure 549. Open habitats, including clearcuts, are used by the Brown-headed Cowbird in spring and autumn throughout British Columbia (Powell River, 8 May 1999; R. Wayne Campbell).

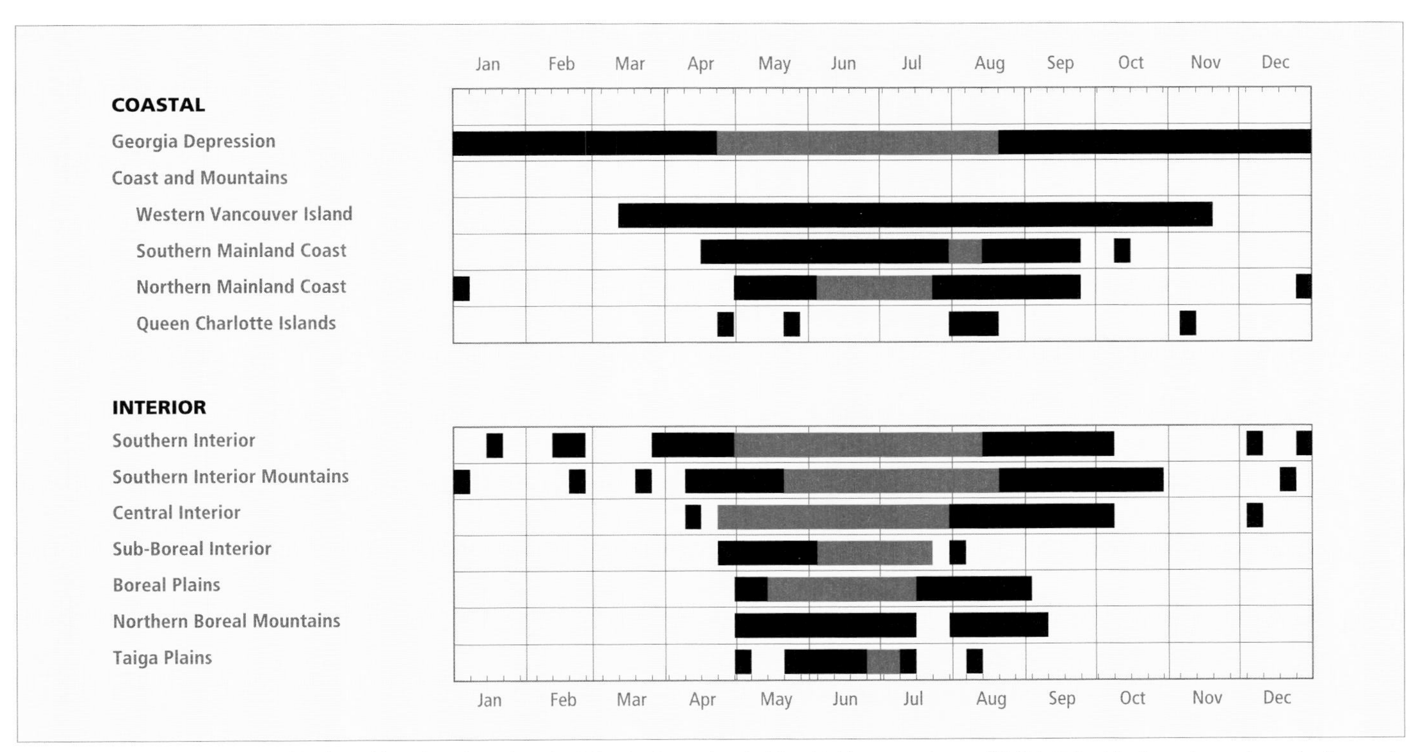

Figure 550. Annual occurrence (black) and breeding chronology (red) for the Brown-headed Cowbird in ecoprovinces of British Columbia. Records are shown for the week in which they occurred.

clearcuts (Fig. 549), garbage dumps, and forest or wooded edges near water, including lagoons, creeks, marshes, reservoirs, estuaries, ponds, and bogs.

Early spring arrivals in the Georgia Depression are difficult to detect because of the overwintering birds. About the first week of April, however, numbers start to swell and climb into the first week of May, with the movement continuing until the end of the month (Fig. 551). On Western Vancouver Island, birds may arrive in the third week of March, with higher numbers appearing by the third week of April (Figs. 550 and 551). The movement there peaks about the first week of May and is over by the end of that month. A similar pattern holds for the Southern Mainland Coast. Further north, in the Northern Mainland Coast, birds arrive by the first week of May and a small movement is discernible throughout the month.

In the Southern Interior, migrants may arrive as early as the last week of March, but numbers do not appear until the first week of May. The movement there peaks around the third week of May and is over by the first week of June. In the Southern Interior Mountains and the Central Interior, the first birds arrive by mid-April, with numbers appearing by the end of the month and continuing into early June. Further north, in the Sub-Boreal Interior, little in the way of a movement is discernible; birds arrive there by the end of April. Birds arrive by the first week of May in the Northern Boreal Mountains, with a noticeable movement into early June. East of the Rocky Mountains, birds arrive in the Boreal Plains by the first week of May and numbers build through the end of the month. In the Taiga Plains, birds also arrive by the first week of May but a movement is not discernible.

The autumn migration appears to consist of 2 separate movements. The first occurs as the early-nesting adults leave the breeding areas. Most cowbird eggs (95%; n = 776 nests with eggs) have been laid by mid-July and the numbers of adult cowbirds in all the ecoprovinces have declined noticeably by that time (Fig. 552). Johnstone (1949) comments on the fact that he had not recorded adult cowbirds later than 12 July in the east Kootenay. Adrian Paul, at Kleena Kleene in the Central Interior, also notes:

> The egg-laying is over by the end of June and early in July the breeders flock up and go elsewhere. A few days later their places with the horses will be taken by the young birds. Casual observers perhaps think there has not been any change but after mid-July a bird seen perching on a horse's back or head is little more than a month old (Satterfield 1966).

Rothstein et al. (1980) note a similar decline in the numbers of adult cowbirds in the Sierra Nevada and suggest that the cowbirds leave the mountains at that time. Bent (1965) also comments on the disappearance of cowbirds from the breeding areas and suggests that it is for concealment during the moulting season. S.G. Sealy (pers. comm.) suggests that the cowbirds probably leave to seek more appropriate resources to support the moult. Whatever the reason, most adults appear to leave British Columbia abruptly, for there is little in the way of an obvious movement anywhere in the province. A more protracted movement consists of the few late-nesting adults and the young as they leave the province,

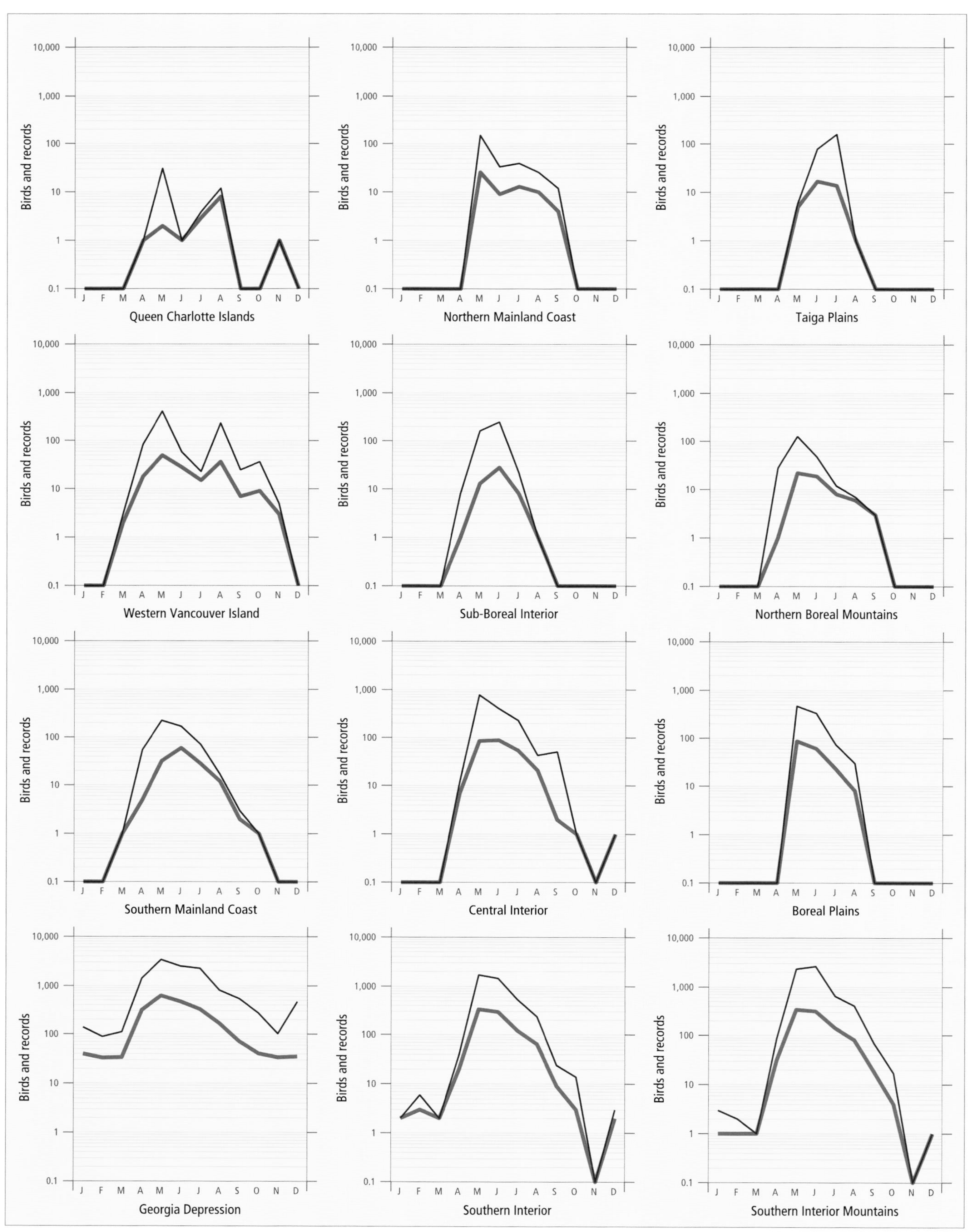

Figure 551. Fluctuations in total number of birds (purple line) and total number of records (green line) for the Brown-headed Cowbird in ecoprovinces of British Columbia. Christmas Bird Counts, Breeding Bird Surveys, and nest record data have been excluded.

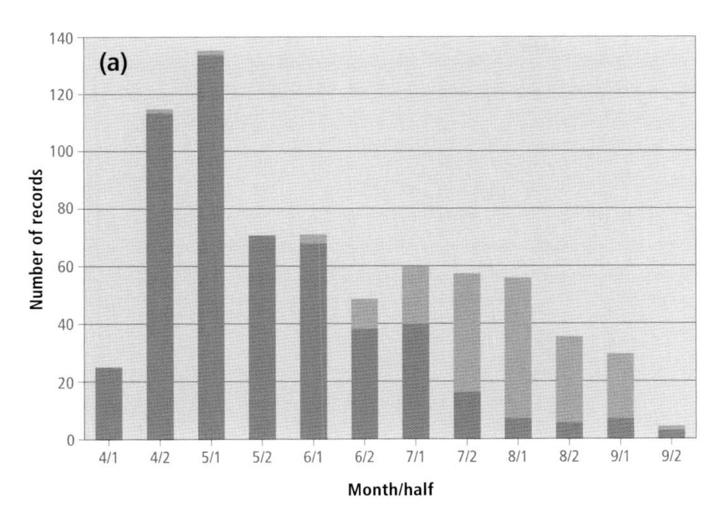

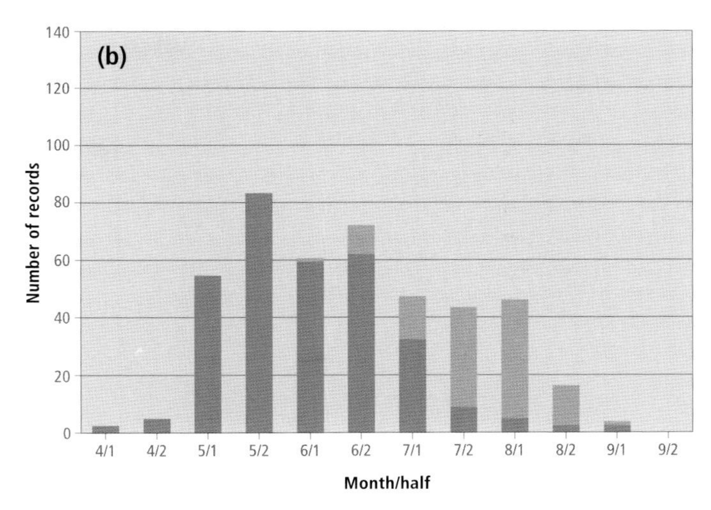

Figure 552. Semi-monthly distribution of records reporting adult (dark bars) and immature (light bars) Brown-headed Cowbirds on the coast (a) and in the interior (b) from April to September in British Columbia. Throughout the province, most adult cowbirds have left the breeding areas by mid-July.

primarily through the month of August. At that time, young birds are often found in flocks conspicuously devoid of adults.

In the far north, birds start leaving the region by early July and most have left by the first week in September (Figs. 550 and 551). In the Boreal Plains and Sub-Boreal Interior, most birds have left by late July, although stragglers may be found into September and October, respectively.

In the Central Interior, numbers drop dramatically by the end of July, although stragglers may be found until the end of September. Small numbers have occasionally wintered there.

In the Southern Interior Mountains and Southern Interior, cowbird numbers decline from mid-July to the end of the month, and most birds have left those regions by the end of August. A few may remain until early October. Small numbers may overwinter occasionally.

On the Northern Mainland Coast, numbers have declined by mid-July and birds have left the region by the third week of September. Only rarely do birds spend the winter there. A similar pattern applies to the Southern Mainland Coast, although a few birds remain into the second week of October; we have no reports of wintering cowbirds from this region.

On Western Vancouver Island, a small autumn movement is discernible in the first 3 weeks of August. Most birds have left by the third week of October, with stragglers remaining until the third week of November.

Cowbirds begin leaving the Georgia Depression by the end of July or the first week of August; most migrants have left the ecoprovince by the end of September, leaving behind a small overwintering population (Figs. 550 and 551).

The few banding returns suggest that some of the cowbirds that summer in coastal British Columbia winter in northern California and southwestern Arizona (Fig. 553).

The Brown-headed Cowbird may be found throughout the year in much of the Georgia Depression. Elsewhere on the coast, with the exception of the Queen Charlotte Islands, it has been reported regularly between the end of April and the third week of September. In the southern and central portions of the interior, the cowbird has been recorded regularly from mid-April through the first week of October; further north, it occurs from early May to early September (Fig. 550).

BREEDING: The known breeding distribution of the Brown-headed Cowbird extends across much of the interior of the province west of the Rocky Mountains and south of latitude 56°N, although independent, free-flying fledglings have been found north to Liard Hot Springs. East of the Rocky Mountains, it is known to breed north to the Fort Nelson area. On the coast, except for the Georgia Depression, where it is a common breeder, documented breeding observations are scarce. For example, there is no evidence of nesting on Western Vancouver Island, although fledged young have been found there, and there are only a few nesting records for the mainland coastal regions north of Vancouver Island. The cowbird now likely breeds regularly throughout its summer range in the province, however, except perhaps for the Queen Charlotte Islands, where only a single instance of parasitism has been found although fledged young have been recorded recently (S.G. Sealy pers. comm.).

The Brown-headed Cowbird reaches its highest numbers in summer in the valleys of the Southern Interior and Southern Interior Mountains (Fig. 546). An analysis of Breeding Bird Surveys in British Columbia for the period 1968 through 1993 shows that the number of birds on coastal routes declined at an average annual rate of 4% (Fig. 554); an analysis of interior routes for the same period could not detect a net change in numbers. Across Canada, Breeding Bird Surveys indicate a drop in cowbird numbers as well; from 1980 to 1996, numbers declined at an average annual rate of 2% ($P < 0.01$). Throughout the bird's North American range, surveys also suggest that Brown-headed Cowbird numbers declined at an average annual rate of 0.8% ($P < 0.01$) over the same period (Sauer et al. 1997).

On the coast, the Brown-headed Cowbird has been reported nesting from near sea level to 335 m elevation; in the interior, it has been reported from 280 to 1,600 m elevation

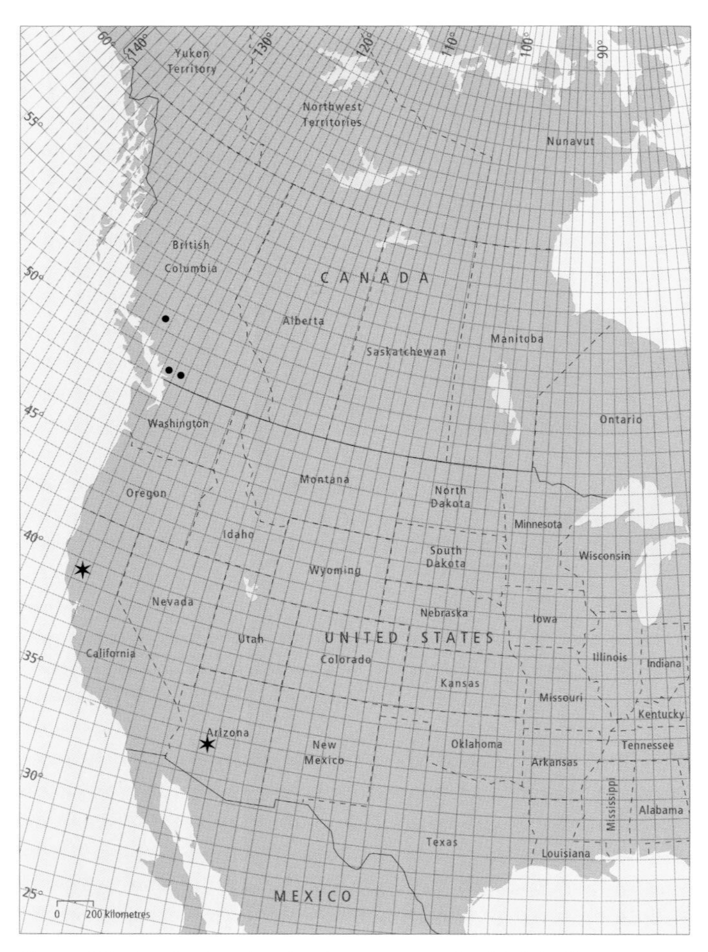

Figure 553. Banding locations (stars) and recovery sites (circles) of Brown-headed Cowbirds associated with British Columbia.

Table 13. Common hosts of the Brown-headed Cowbird in British Columbia. Species were included only if their frequency of cowbird parasitism was at least 10% and if 15 or more nests had been monitored. There is immense site-to-site variation in host selection by cowbirds, depending mostly on the ratio of cowbirds to suitable host individuals locally.

Host species	Parasitism rate (%)	Total number of nests (parasitized and unparasitized)
Warbling Vireo	49	98
Willow Flycatcher	36	210
Red-eyed Vireo	27	56
Veery	24	74
American Goldfinch[1]	22	146
Yellow-rumped Warbler	21	118
Dusky Flycatcher	20	93
Magnolia Warbler	20	15
Least Flycatcher	19	37
Common Yellowthroat	18	35
Lazuli Bunting	18	55
White-throated Sparrow	18	34
Yellow Warbler	15	393
MacGillivray's Warbler	15	40
Golden-crowned Kinglet	14	22
Ruby-crowned Kinglet	14	21
Nashville Warbler	13	16
Yellow-breasted Chat	17	23
Cassin's Finch[1]	11	35
Alder Flycatcher	10	19
Hammond's Flycatcher	10	41
Chipping Sparrow	10	508
Pine Siskin[1]	10	62

[1] Unsuitable host of the cowbird: does not normally rear cowbird young (see REMARKS).

(Fonnesbeck 1998). In all ecoprovinces, it is more abundant at the lower elevations. The Brown-headed Cowbird prefers open habitats with low or scattered trees, including wooded edges (Fig. 555), pastures, orchards, and residential areas (Lowther 1993). It prefers wood-field ecotones rather than extensive woods or extensive fields. Extensive forest is generally avoided.

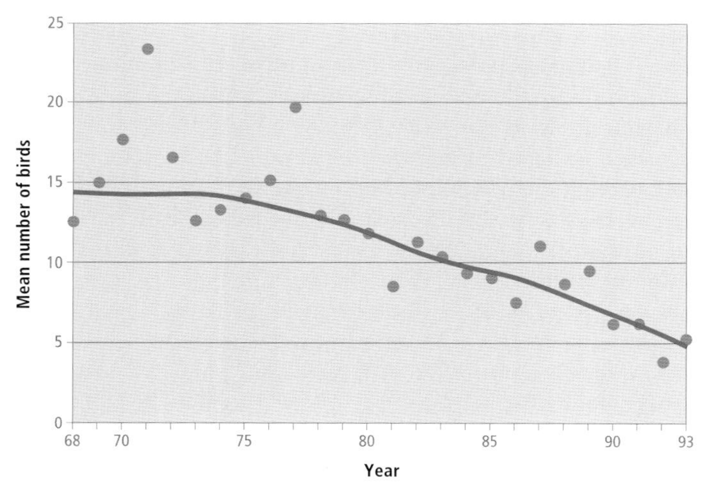

Figure 554. An analysis of Breeding Bird Surveys for the Brown-headed Cowbird in British Columbia for the period 1968 through 1993 shows that the number of birds on coastal routes has declined at an average annual rate of 4% ($P < 0.05$).

During their field work through southern portions of the interior, Dawe and Buechert (1996) found the Brown-headed Cowbird frequenting willow-dominated riparian habitat; willow and red-osier dogwood riparian habitat with black cottonwood and some cattail; pasturelands with cattle; open grass and forb-dominated fields and hillsides with scattered wild rose (Fig. 556), mock-orange, and red-osier dogwood shrubs; and big sagebrush habitat with antelope-bush, common rabbit-brush, wild rose, saskatoon, and soopolallie. In all these habitats, many of the common host species (see Table 13) were also noted. In the southern Okanagan valley, Ward and Smith (2000) found that the Brown-headed Cowbird preferred, in order, low-elevation riparian woodlands with antelope-bush–ponderosa pine savannah, low-elevation ponderosa pine forests with brushy draws, mixed elevations with fir-aspen stands, and high elevations with lodgepole pine–spruce stands. The least preferred habitat was low-elevation sage grasslands.

In Mount Revelstoke and Glacier national parks, the highest Brown-headed Cowbird densities occurred in wet shrub thickets and sedge fen interspersed with strips of hemlock-cedar forest as well as areas dominated by vegetation indicative of frequent snow avalanching (Van Tighem and Gyug 1983).

Pojar (1995) found the Brown-headed Cowbird in all her trembling aspen stands in Smithers, except clearcuts; cowbird densities were highest in mixed conifer-aspen stands followed by mature and old trembling aspen stands (Table 14). Most of the variation in Brown-headed Cowbird numbers was explained by shrub height, suggesting that the birds preferred areas with tall shrubs. In the Chetwynd area, Merkins and Booth (1998) found that the Brown-headed Cowbird was more abundant in mature trembling aspen forests and subsequent 6-year-old regenerating clearcuts than in younger clearcuts; detections were highest in the 6-year-old clearcuts. In the Dawson Creek area, Phinney (1998) notes that the cowbird also frequents agricultural areas but that they are "particularly common in mature aspen forest, especially around openings and edges."

Table 14. Densities of Brown-headed Cowbirds (number of males detected in 10 ha plots) in trembling aspen forests near Smithers, British Columbia, in 1991 and 1992 (from Pojar 1995). The median density was calculated for all plots in each habitat type for both years (n = number of plots).

Forest type	Forest age	Median density	n
Clearcut	Under 7 years	0	6
Sapling trembling aspen	7-23 years	11.2	6
Mature trembling aspen	50-60 years	12.7	10
Old trembling aspen	100 years	12.7	4
Mixed aspen-conifer	50-60 and 95 years	15.0	6

In British Columbia, most nests (34%; n = 580) parasitized by the cowbird were associated with rural habitats, including cultivated farmland (12%), farmyards (11%), orchards (7%), and pasturelands (4%); followed by forested habitats (30%), including deciduous (13%), coniferous (9%), and mixed (8%) forests. In the Okanagan valley, Fonnesbeck (1998) found brood parasitism to be significantly lower in higher-elevation forests than in valleys. Riparian sites (16%), suburban and urban areas (11%), and shrublands (6%) were also important habitats to the cowbird. Specific habitats included backyard and farmyard gardens and residential areas (33%; n = 419), mature and second-growth forests and regenerating clearcuts, forest edges and road edges through forests (29%), and riparian habitats (25%), including forested shores of lakes and streams, cottonwood bottomlands, and riparian willow thickets. Orchards, urban parklands, big sagebrush shrublands, and

Figure 555. In British Columbia, breeding habitat for the Brown-headed Cowbird includes edges with scattered trees and shrubs bordering woodlands, pastures, and agricultural environments (near Hat Creek, 13 July 1996; R. Wayne Campbell).

Figure 556. In the Southern Interior Ecoprovince of British Columbia, the Brown-headed Cowbird favours open habitats with a variety of shrubs, including wild rose (near Oliver, 8 June 1998; R. Wayne Campbell).

Table 15. Species parasitized by the Brown-headed Cowbird in British Columbia. Details of host nests found with cowbird eggs or young and assumed host species reported with fledged cowbirds are shown.

	Cowbird eggs			Cowbird nestlings			Cowbird fledglings		
Host species	Chronology	Mean (*n*)	Max. no. in nest	Chronology	Mean (*n*)	Max. no. in nest	Chronology	Mean (*n*)	Max. no.
Spotted Sandpiper[1]	2 Jun	1 (1)	1	–	–	–	–	–	–
Olive-sided Flycatcher	15-26 Jun	1 (2)	1	21 Jun	1 (1)	1	–	–	–
Western Wood-Pewee	29 May – 15 Jul	1.3 (7)	3	22 Jun – 5 Jul	1 (1)	1	12 Jul – 18 Aug	1.5 (4)	3
Alder Flycatcher	29 Jun	1 (1)	1	13 Jul	1 (1)	1	8 Aug	1 (1)	1
Willow Flycatcher	31 May – 25 Jul	1.3(77)	4	1-23 Jul	1 (8)	1	7 Jul – 8 Aug	1 (4)	1
Least Flycatcher	9 Jun – 8 Jul	1.2 (5)	2	3-15 Jul	1 (3)	1	–	–	–
Hammond's Flycatcher	20 Jun – 13 Jul	1 (3)	1	4-18 Jul	1.3 (3)	2	–	–	–
Dusky Flycatcher	28 May – 14 Jul	1.3(19)	3	19-30 Jun	1 (5)	1	30 Jun	1 (1)	1
Gray Flycatcher	19 Jun – 29 Jul	1 (2)	1	–	–	–	–	–	–
Pacific-slope Flycatcher	6 Jun – 16 Jul	1.4(11)	3	–	–	–	–	–	–
Say's Phoebe	14 Jul	1 (1)	1	–	–	–	–	–	–
Eastern Kingbird[1, 3]	–	–	–	14 Jun	1 (1)	1	5 Aug	1 (1)	1
Cassin's Vireo	20 May – 9 Jul	1.3 (4)	2	9-10 Jun	1 (2)	1	23 Jun – 22 Jul	1 (2)	1
Hutton's Vireo	20 May	2 (1)	2				15 Jun	1 (2)	1
Warbling Vireo	30 May – 15 Jul	1.7(43)	5	15 Jun – 25 Jul	1.1 (14)	2	25 Jun – 30 Jul	1 (9)	1
Red-eyed Vireo	7 Jun – 7 Jul	1 (13)	1	30 Jun – 12 Jul	1.2 (5)	2	9-31 Aug	1 (4)	1
Cliff Swallow[1]	12 Jun	1 (1)	1	–	–	–	–	–	–
Barn Swallow[1, 3]	16 Jul	1 (1)	1	–	–	–	–	–	–
Bushtit[1, 2]	10-13 Jun	1 (2)	1	–	–	–	27 Jul	1 (1)	1
Bewick's Wren	12-16 Jun	1 (3)	1	30 Jun – 2 Jul	1 (1)	1	–	–	–
House Wren[1, 3]	–	–	–	31 Jul	1 (1)	1	–	–	–
Marsh Wren[1, 2]	–	–	–	23 Jun	1 (1)	1	–	–	–
Golden-crowned Kinglet	14 May – 10 Jun	1 (2)	1	10 Jun	1 (1)	1	30 Jun – 28 Jul	1 (6)	1
Ruby-crowned Kinglet	8-10 Jun	1 (1)	1	1-10 Aug	1 (2)	1	25 Jun – 3 Aug	1 (10)	1
Western Bluebird[1, 2]	22-28 Jun	2 (1)	2	–	–	–	–	–	–
Mountain Bluebird[1]	23 May – 27 Jun	1 (3)	1	–	–	–	–	–	–
Townsend's Solitaire	18 Jun – 6 Jul	1.9(10)	4	7 Jul	2 (1)	2	–	–	–
Veery	8 Jun – 26 Jul	1.6(19)	5	–	–	–	–	–	–
Swainson's Thrush	26 May – 5 Aug	1.1(35)	2	24 Jun – 5 Aug	1.2 (9)	2	6 Jul – 11 Aug	1.4 (5)	2
Hermit Thrush	22 May – 23 Jun	1.7(10)	8	30 Jun – 9 Jul	1.3 (3)	2	5 Aug	1 (1)	1
American Robin[1]	–	–	–	25 Jun	2 (1)	2	–	–	–
Varied Thrush[1, 2]	1 Jun	1 (1)	1	–	–	–	–	–	–
Gray Catbird[1]	7-30 Jun	1.2 (9)	2	10 Jul	1 (1)	1	–	–	–
Cedar Waxwing[1]	30 May – 22 Jul	1.4(20)	2	30 Jun – 5 Jul	1 (2)	1	4 Aug	2 (1)	2
Tennessee Warbler	12-17 Jun	2 (1)	2	–	–	–	–	–	–
Orange-crowned Warbler	3 Jun – 6 Jul	1.4 (5)	2	14-17 Jun	2 (1)	2	23 Jun – 31 Jul	1.2 (7)	2
Nashville Warbler	1 Jun – 3 Jul	4 (2)	5				28 Jun – 20 Jul	1 (2)	1
Yellow Warbler	16 May – 20 Jul	1.2(56)	3	10 Jun – 20 Jul	1.1 (18)	2	27 Jun – 26 Aug	1.1 (15)	2
Magnolia Warbler	20 Jun	1 (1)	1	8-14 Jul	1 (3)	1	7-25 Jul	1 (2)	1
Cape May Warbler	–	–	–	–	–	–	18 Jul	1 (1)	1
Yellow-rumped Warbler	23 May – 1 Jul	1 (11)	1	14 Jun – 20 Jul	1 (15)	1	8 Jun – 30 Aug	1.1 (43)	2
Black-throated Gray Warbler	25 Jun	1 (1)	1	–	–	–	30 Jun – 3 Aug	1 (6)	1
Black-throated Green Warbler	–	–	–	–	–	–	1 Jul	1 (1)	1
Townsend's Warbler	–	–	–	–	–	–	30 Jun – 5 Aug	1 (7)	1
Black-and-white Warbler	6 Jun	1 (1)	1	–	–	–	30 Jun	1 (1)	1
American Redstart	5 Jun – 11 Jul	1.3(12)	3	13-21 Jul	1 (2)	1	28 Jun – 27 Jul	1 (6)	1
Mourning Warbler	–	–	–	–	–	–	21-24 Jul	1 (2)	1
MacGillivray's Warbler	20 Jun – 9 Jul	1.2 (6)	2	–	–	–	13 Jun – 18 Jul	1 (4)	1
Common Yellowthroat	10 Jun – 30 Jul	1.1 (7)	2	24 Jun – 12 Jul	1.3 (3)	2	26 Jun – 1 Aug	1 (8)	1
Wilson's Warbler	29 Jun	1 (1)	–	–	–	–	5-30 Jul	1 (4)	1
Canada Warbler	–	–	–	–	–	–	18 Jul	1 (2)	1
Yellow-breasted Chat	5 Jun – 2 Jul	1.5 (4)	3	9 Jul	2 (1)	2	–	–	–
Western Tanager	20 Jun – 2 Jul	1.5 (2)	2	15 Jul	1 (1)	1	2 Jun – 15 Aug	1.3 (13)	3
Spotted Towhee	18 May – 14 Jul	2.2 (6)	3	7 Jul	2 (1)	2	6 Jul – 21 Aug	1.3 (16)	2
Chipping Sparrow	23 May – 7 Aug	1.3(51)	4	4 Jun – 9 Aug	1.1 (14)	2	24 Jun – 9 Aug	1 (13)	1
Clay-colored Sparrow	10 Jun	2 (1)	2	10 Jun	2 (1)	2	–	–	–
Brewer's Sparrow	28 May – 8 Jul	1.2 (6)	2	–	–	–	–	–	–
Vesper Sparrow	30 May – 9 Jun	1.5 (2)	2	–	–	–	–	–	–

▶

◀ **Table 15**

Host species	Cowbird eggs			Cowbird nestlings			Cowbird fledglings		
	Chronology	Mean (*n*)	Max. no. in nest	Chronology	Mean (*n*)	Max. no. in nest	Chronology	Mean (*n*)	Max. no.
Lark Sparrow	6 Jul	4 (1)	4	–	–	–	–	–	–
Savannah Sparrow	27 May – 18 Jun	1 (1)	1	27 Jun – 2 Jul	1 (1)	1	–	–	–
Grasshopper Sparrow	11 Jul	1 (1)	1	–	–	–	–	–	–
Song Sparrow	5 May – 26 Jul	1.7(42)	5	12 May – 17 Aug	1 (27)	3	26 May – 27 Aug	1.1 (27)	2
Lincoln's Sparrow	8 Jun – 8 Jul	1.8 (4)	3	–	–	–	28 Jun – 12 Jul	1 (2)	1
Swamp Sparrow	–	–	–	–	–	–	23 Jun	1 (1)	1
White-throated Sparrow	31 May – 29 Jun	1.4 (5)	2	10 Jul	1 (1)	1	24 Jun – 19 Aug	1 (6)	1
White-crowned Sparrow	10 May – 30 Jun	1.2 (9)	2	17 Jun – 16 Jul	2.5 (2)	3	12 Jun – 31 Jul	1 (5)	1
Dark-eyed Junco	23 Apr – 29 Jul	2 (23)	8	5 Jun – 11 Aug	1.4(19)	3	13 May – 21 Aug	1.2 (41)	2
Black-headed Grosbeak	29 May – 26 Jun	1.8 (6)	4	3-20 Jun	2.5 (2)	4	–	–	–
Lazuli Bunting	12 Jun – 15 Jul	1.8 (9)	4	1 Jul	1 (1)	1	8 Aug	1 (1)	1
Indigo Bunting	–	–	–	–	–	–	29 Jun	1 (1)	1
Red-winged Blackbird	10 May – 16 Jul	1.2(28)	2	26 May – 15 Jun	1 (4)	1	17 Jun	1.5 (1)	1
Western Meadowlark	13 May – 1 Jun	1.5 (2)	2	–	–	–	–	–	–
Yellow-headed Blackbird	24 May – 1 Jul	1.5 (2)	2	–	–	–	–	–	–
Rusty Blackbird	8 Jun	3 (1)	3	–	–	–	–	–	–
Brewer's Blackbird	4 May – 26 Jun	1.6(24)	3	31 May – 28 Jun	1.3 (3)	2	10 Jun – 27 Jul	1 (5)	1
Bullock's Oriole	–	–	–	–	–	–	1 Jul	1 (2)	1
Purple Finch	–	–	–	–	–	–	6 Jun – 6 Jul	1.7 (3)	2
Cassin's Finch[1]	28 May – 14 Jun	1.8 (4)	3	–	–	–	–	–	–
House Finch[1]	5 May – 24 Jul	1.7(21)	3	8-17 Jun	1 (2)	1	7 Jun – 15 Jul	1 (2)	1
Pine Siskin[1]	17 May – 29 Jul	1.2 (5)	2	27-29 Jul	1 (1)	1	21 Jun – 20 Jul	1 (9)	1
American Goldfinch[1]	26 May – 26 Jul	1.2(36)	3	30 Jun – 16 Jul	1 (2)	1	6-15 Jul	1 (2)	1
Evening Grosbeak	–	–	–	–	–	–	4-7 Aug	1 (2)	1
House Sparrow[1]	–	–	–	5 May	1 (1)	1	17 -20 Aug	1 (1)	1

[1] Unsuitable host of the cowbird: does not normally rear cowbird young (see REMARKS).
[2] First report of this species as a cowbird host in North America.
[3] New information since the publication of Volume 3.

transportation corridors were reported to a lesser degree. These habitats reflect the distribution of nest recorders more than the choices of the cowbirds. For a discussion of surveying the nesting activity of cowbirds without bias in the southern Okanagan valley of British Columbia, see Ward and Smith (2000).

On the coast, the Brown-headed Cowbird has been recorded breeding from 23 April to 17 August; in the interior, it has been recorded breeding from 4 May to 13 August (Fig. 550).

Nests: The Brown-headed Cowbird is a generalist, obligate brood parasite that lays its eggs in the nests of other species and leaves the host species to raise their young. Eighty-three species of birds in British Columbia have been reported as cowbird hosts (Table 15); of these, 23 would be considered common hosts (Table 13). The heights of 696 nests containing cowbird eggs ranged from ground level to 15.0 m, with 54% between 0.9 and 2.1 m.

Eggs: Dates for 996 host nests containing cowbird eggs ranged from 23 April to 7 August, with 53% recorded between 7 June and 30 June. The number of cowbird eggs laid per nest in 775 host nests ranged from 1 to 8 (1E-538, 2E-174, 3E-47, 4E-11, 5E-4, 8E-1) (Fig. 559), with 69% of nests having 1 egg.

Both the northern and southern populations of the Brown-headed Cowbird in British Columbia lay a modal "clutch" of 1 egg per host nest (Fig. 558). Host nests on the coast may receive eggs up to 2 weeks earlier than those in the southern and central portions of the interior, and up to 3 weeks earlier

Figure 557. In British Columbia, the Warbling Vireo has the highest rate of Brown-headed Cowbird parasitism (49%) (Richter Pass, 12 June 1969; R. Wayne Campbell).

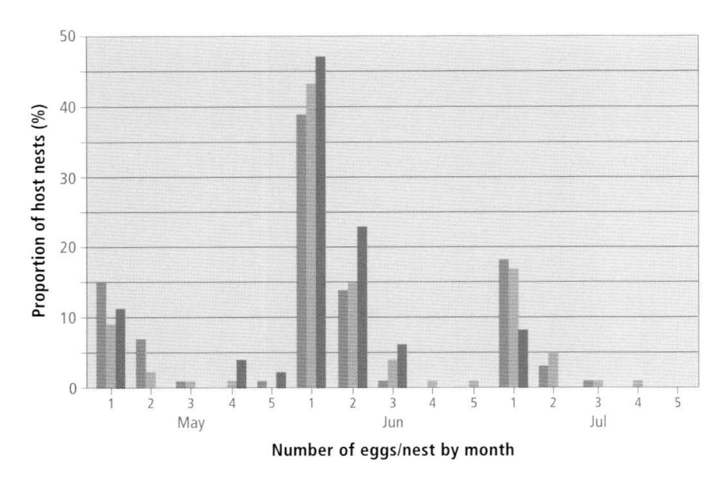

Figure 558. Monthly distribution of numbers of Brown-headed Cowbird eggs per host nest between May and July for southern populations on the coast (dark green bars), southern populations in the interior (light green bars), and northern populations in the interior (purple bars). All populations appear to have a modal "clutch" size of 1 egg.

than nests in the Boreal Plains (Fig. 560). The incubation period has been variously reported to be between 10 and 13 days (Wetherbee and Wetherbee 1961; Briskie and Sealy 1990). Normally the cowbird eggs hatch a day or so before the eggs of most hosts (Sealy et al. 1995; Smith 1995). Lowther (1993) notes that the term *clutch*, in the usual sense, does not apply to the cowbird; rather, cowbirds lay eggs daily in "sequences" of 1 to 7 eggs (mean of 4.0 to 4.6) punctuated by pauses of 1 or more days through the nesting season.

Young: Dates for 278 host nests containing cowbird young ranged from 5 May to 17 August, with 51% recorded between 21 June and 15 July. The number of cowbird young per nest found in 196 host nests ranged from 1 to 4 (1Y-165, 2Y-27, 3Y-3, 4Y-1), with 84% of nests having 1 young. The nestling period is between 8 and 13 days (Lowther 1993).

Nest Success: Data for British Columbia are insufficient. In a large number of instances, observers interfered with cowbird success by removing the cowbird eggs from the nest.

Figure 559. Hermit Thrush nest with 8 Brown-headed Cowbird eggs and 2 Hermit Thrush eggs (Charlie Lake, 29 May 1970; Andrew C. Stewart).

The nest success of the Brown-headed Cowbird varies greatly with host species (where success is almost zero in nests of ejectors and varies greatly among nests of other hosts), host size (generally better in larger hosts), cowbird population density relative to host density, and the local level of nest predation. Other studies have shown that the overall survival of the Brown-headed Cowbird from egg to adulthood is 3% (Lowther 1993).

REMARKS: It has been presumed that only one of the 3 subspecies of the Brown-headed Cowbird, *M. a. artemisiae*, occurs in British Columbia, but see CHANGE IN STATUS.

The Brown-headed Cowbird has experienced the greatest expansion in distribution of any North American bird species (Graham 1988; Rothstein 1994). Before the influx of European settlers, it was a bird of the Great Plains; today its distribution includes most of North America south of the Arctic. This range expansion has been attributed to a number of factors, including the opening up of forested lands across North America and the introduction of cattle by the European settlers as they subdued new lands (Graham 1988). Forests were cut to create large areas for agriculture. Grain left in the fields after harvesting, along with the availability of feeding stations across the continent, gave the cowbird a winter food supply. In addition, where the forests were not opened completely, their fragmentation provided increased edge by which the cowbird gained access to new landscapes. These factors, coupled with the cowbird's brood parasitic behaviour, which lets it migrate between widely separated feeding and breeding areas, enabled the cowbird to rapidly colonize new territory and accounts for the significant increase in the number of host species that it now parasitizes (Rothstein 1994).

In the late 1920s, before it extended its distribution across North America, the Brown-headed Cowbird was known to have parasitized 157 species (Friedmann 1929); the number now exceeds 220 species (Lowther 1993).

Recently, cowbird parasitism has actually contributed to the precarious position of some taxa; the survival of the Kirtland's Warbler (*Dendroica kirtlandii*), "Least" Bell's Vireo (*Vireo bellii pusillus*), and "Southwestern" Willow Flycatcher (*Empidonax trailii extimus*) now appears to depend on wildlife managers removing cowbirds from the host's breeding habitat (Mayfield 1977; Rothstein 1994). Rothstein (1994) points out, however, that the latter 2 host species "would probably have survived coexistence with cowbirds had not most of their riparian habitat also been destroyed. Habitat restoration, not cowbird control, holds the most promise for the long term management of these hosts." The populations of some host species that are heavily parasitized by the cowbird may represent "sinks" that are maintained only through immigration from other populations that suffer less intense cowbird parasitism (Marvil and Cruz 1989; Rogers et al. 1997).

In British Columbia, species of concern that may be threatened by the Brown-headed Cowbird include the Willow Flycatcher and the Warbling Vireo. The vireo, once common in the south Okanagan valley, is now virtually absent from the valley floor; its decline appears to have been due to the combination of habitat fragmentation and cowbird parasitism, factors that are now threatening the viability of other songbird populations in that region (Smith et al. 1998).

For some songbirds, however, particularly the most frequent of the cowbird's victims, which are widespread and common species, there is little evidence of any harmful effects on their populations. For example, in a Song Sparrow population on Mandarte Island, British Columbia, female sparrows that were parasitized at least once during a breeding season raised as many young to independence as females that were not parasitized (Smith 1981). Colonization of Mandarte Island by 1 or 2 Brown-headed Cowbirds had little effect on Song Sparrow numbers because of such factors as

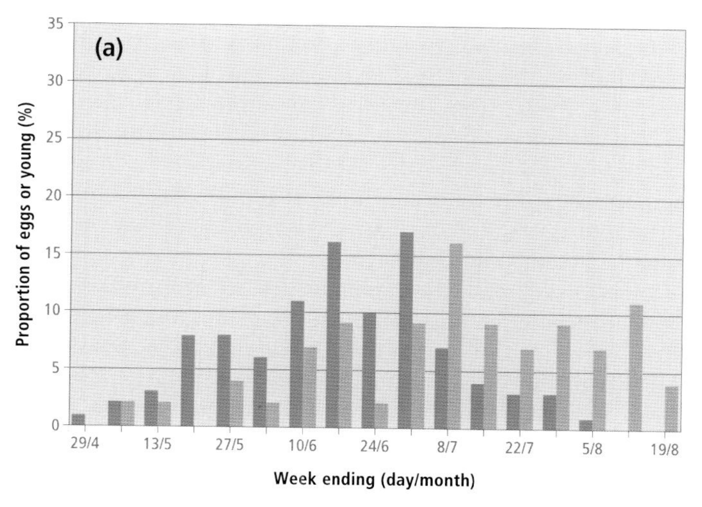

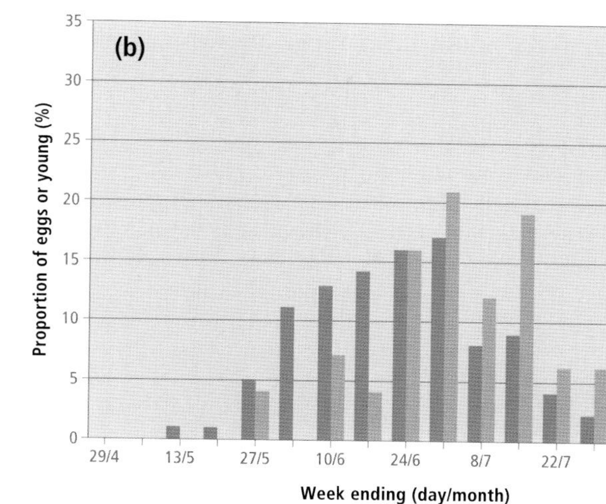

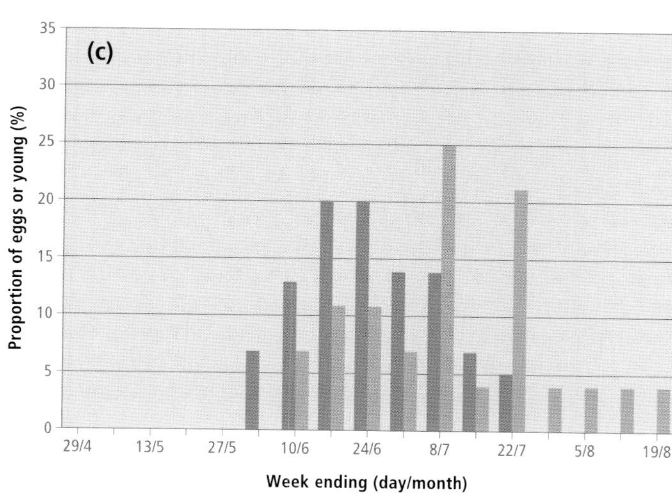

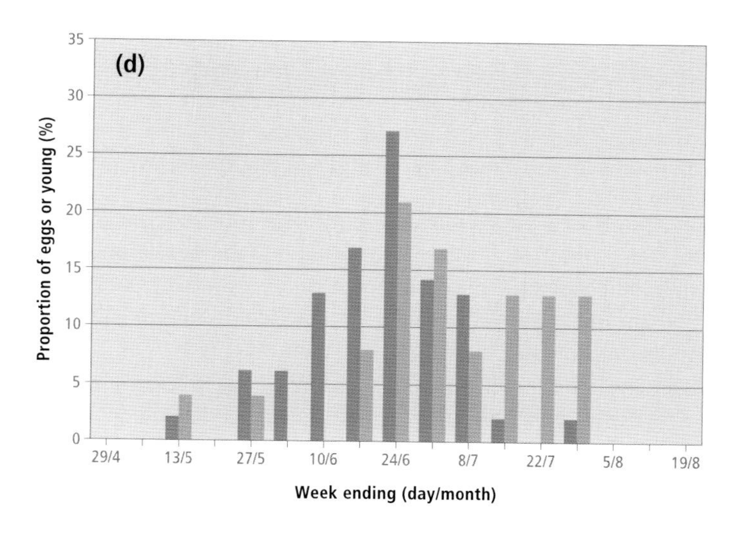

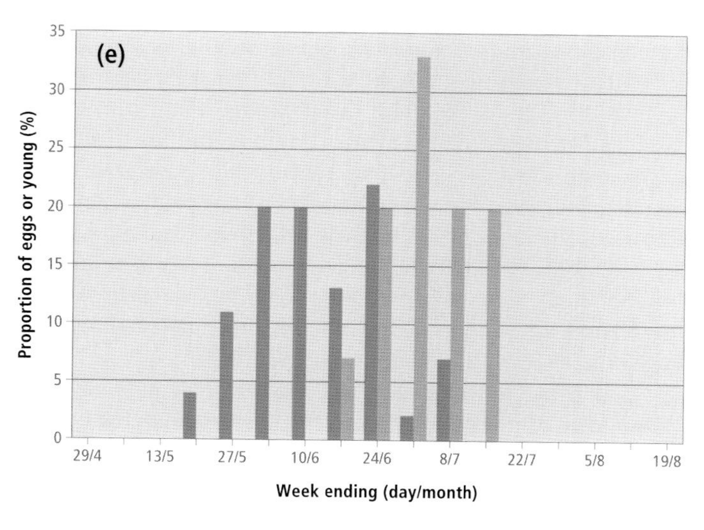

Figure 560. Weekly distribution of host nests containing Brown-headed Cowbird eggs (dark bars) and cowbird nestlings (light bars) in the (a) Georgia Depression (nests with eggs: $n = 135$; nests with young: $n = 45$), (b) Southern Interior (nests with eggs: $n = 466$; nests with young: $n = 68$), (c) Southern Interior Mountains (nests with eggs: $n = 56$; nests with young: $n = 28$), (d) Central Interior (nests with eggs: $n = 64$; nests with young: $n = 24$), and (e) Boreal Plains (nests with eggs: $n = 45$; nests with young: $n = 15$) ecoprovinces of British Columbia. The figures are based on the week the eggs or young were found in the nest. Nesting on the coast may begin up to 2 weeks before nesting in the southern and central interior regions, and up to 3 weeks before nesting in the northern regions of the province.

incomplete overlap in laying seasons, multiple broods in sparrows, the sparrows' ability to rear their young with young cowbirds, and density-dependent host reproductive success (Smith and Arcese 1994). See also Paulson (1992: 91)

In addition, not all species successfully rear cowbird young. Of more than 220 species of birds known to have been parasitized by the Brown-headed Cowbird (= cowbird victims), only 144 species have been reported to have actually reared cowbird young (= cowbird hosts) (Lowther 1993). Some species, such as the Spotted Sandpiper, are unsuitable hosts because their young are precocious. Others, such as the Tree Swallow (Mills 1988), Bushtit, or Marsh Wren (Picman 1986), may be unsuitable because they are cavity-nesters or construct cavity-like nests, or because of their behaviour or size. Tree Swallows, for example, frequently leave their eggs for long periods during colder weather, and the cowbird embryos may be less tolerant to the cold (Mills 1988); Marsh Wrens associate with Red-winged Blackbirds in their nesting areas, a species that may provide protection from cowbird parasitism through group defence against the cowbirds (Picman 1986; Freeman et al. 1990).

Some passerines, such as the Brewer's Sparrow, may abandon nests that have been parasitized by the cowbird (Biermann et al. 1987). Others, such as the American Robin, Gray Catbird, Sage Thrasher, Eastern Kingbird, Baltimore and Bullock's oriole, and Cedar Waxwing, are known as *ejectors* because they recognize cowbird eggs and remove them from the nest. The Yellow Warbler, while not directly ejecting the cowbird egg, often buries it by constructing a new nest on top of the cowbird egg and sometimes eggs of its own (Sealy 1995).

The Cedar Waxwing's partially frugivorous diet makes it unlikely that the cowbird young would survive, even if the cowbird eggs were accepted. Similarly, passerines that feed their young mainly on seeds, such as the House Finch and American Goldfinch, are also unsuitable hosts for the young cowbird, which requires a high-protein diet of arthropods. Kozlovic et al. (1996) found that in 99 House Finch nests parasitized by the cowbird, most of the cowbird young had retarded growth and died in the nest in just over 3 days; only 1 fledged. While all these species can be victims of the cowbird, few are suitable hosts.

In some instances, the cowbird may also lay its eggs in inactive nests, where there is no chance of the eggs hatching. In their study of Red-winged Blackbird–Brown-headed Cowbird interactions, Freeman et al. (1990) found that nearly 22% of 130 instances of parasitism occurred in inactive nests.

Observers should be aware that the sight of an adult of any species feeding a fledgling Brown-headed Cowbird does not necessarily mean that the species was host to the cowbird. Sealy and Lorenzana (1997) summarize 10 instances with evidence of either (1) a species other than the host feeding a cowbird young, (2) at least 2 different species feeding a cowbird young, (3) 2 pairs of a species feeding 3 cowbird fledglings although neither pair had raised all 3 in the nest, (4) a species feeding 3 cowbird fledglings although it had raised only 1 in the nest, or (5) a species feeding a fledged cowbird, fully grown, while also feeding its own recently fledged young.

There are also 5 reports of provisioning of young by the Brown-headed Cowbird (Lorenzana and Sealy 1998), suggesting that the propensity to provision young has not been lost by some individuals of this brood parasite. The reports document both male and female cowbirds feeding cowbird nestlings and fledglings.

Habitat fragmentation now affects more than one-third of British Columbia, roughly coinciding with the normal range of the Brown-headed Cowbird in the province (Harding 1994). Protecting riparian habitats and restricting the conversion of large tracts of forested lands into the fragmented habitat that favours the Brown-headed Cowbird are perhaps the 2 most important short-term means of managing the effects of cowbird parasitism on songbird populations in British Columbia.

Cowbird control programs may be an appropriate management technique only where a species of limited distribution is severely threatened by the cowbird, as in the case of the Kirtland's Warbler. Had the warbler's habitat not been severely compromised by human activities, however, the cowbird may have had little effect on its population. Smith (1995) cautions against initiating cowbird control programs and notes the need for better monitoring of the effects of cowbird parasitism on populations in the province, particularly those of rarer host species.

In studies of forest structure and nest site selection by Song Sparrows and Brown-headed Cowbirds, Larison et al. (1998) found that habitat structure affected the probability of parasitism at 2 scales: (1) nests with abundant lateral cover at ground level were less likely to be parasitized, perhaps because host movement near the nest was concealed, and (2) nests with abundant foliage cover at a height of 2 to 3 m were more likely to be parasitized, suggesting that the foliage cover above the nest may have provided perches close to the nest from which the female cowbird could watch host activities and locate nests. Further studies of these relationships may lead to methods of reducing cowbird parasitism through habitat management and restoration efforts.

The cowbird appears to possess a number of adaptations for parasitism. For example, it lays its eggs quickly, typically spending less than 1 minute in host nests compared with the 21 to 104 minutes taken by nonparasitic passerine species (Sealy et al. 1995). The rapid laying also reduces the cowbird's chances of being attacked by hosts. Once laid, the larger cowbird eggs usually hatch before the eggs of their hosts because cowbird embryos develop more rapidly than host embryos. This adaptation is complemented by what appears to be an ability of the cowbird embryo to respond to stimuli from the host eggs and hatch early, while the presence of the cowbird egg(s) in the nest prolongs the incubation of the smaller host eggs (McMaster and Sealy 1998).

"Infanticidal" behaviour on the part of cowbirds may be a means of increasing the reserve of usable nests for the cowbird (Smith and Arcese 1994; Arcese et al. 1992). The hypothesis is that the female cowbird preys on the young in unusable host nests in an apparent attempt to cause the host to renest. Rogers et al. (1997) note that cowbirds are a greater conservation threat than has been recognized to date. Their results support the hypothesis that cowbirds can threaten host populations by acting as predators. Elliott (1999) has documented on video a female cowbird killing 6 Blue-winged Warbler (*Vermivora pinus*) nestlings in an unparasitized nest, and is investigating the relative probability that such "infanticidal" behaviour is adaptive.

Woodward (1983) found that perching heights and home ranges of fledgling cowbirds in Maryland and Virginia reflect their hosts' foraging heights and territories, respectively. He also noted that hosts fed fledgling cowbirds more than they did an equivalent weight of their own young, probably because of the loud, persistent calling of the fledgling cowbirds.

For a summary of current research on the Brown-headed Cowbird, see Morrison et al. (1999).

NOTEWORTHY RECORDS

Spring: Coastal – Victoria 30 May 1893-1 (RBCM 1829), 20 May 1961-150 at roost site; Metchosin 22 Apr 1980-30 in garden on ground; Triangle Mountain (Colwood) 24 Apr 1985-25; Saanich 10 Apr 1985-7, 20 Apr 1978-23, 23 Apr 1984-1 egg in Dark-eyed Junco nest; 29 Apr 1985-25 at feeder; Chatham Islands 4 May 1985-10; all Saltspring Island 29 May 1974-10; Somenos Lake (Duncan) 20 Apr 1974-12; Ucluelet 6 May 1972-200; Cowichan 1 May 1966-28; Lennard Island 12 Mar 1977-1, first of season, 2 Apr 1977-10, first large group, stayed on lawns for 2 weeks; Tofino 2 May 1971-8; Nanaimo 1 May 1982-15; Westham Island 22 Apr 1974-60; Richmond 5 May 1963-1 large nestling in House Sparrow nest; Reifel Island 30 May 1972-3 nestlings in Song Sparrow nest, 31 May 1972-3 eggs in Song Sparrow nest; Ambleside Park 1 May 1977-75 in playing fields, 5 May 1983-75, feeding on golf course in single flock; Maple Ridge 6 May 1969-26; Sumas 26 May 1897-1 (Brooks 1917); Cheam Slough 21 Mar 1976-10, 1 May 1976-50; Agassiz 16 Apr 1976-100; Chilliwack 28 Apr 1980-1; Skagit River valley 17 to 26 Apr 1971-1; Ross Lake 25 May 1976-120, feeding on dry lake bottom; Little Qualicum River 25 Apr 1977-6 on estuary, 27 Apr 1979-27 (Dawe and Buechert 1995); Sechelt 28 Mar 1996-1; Oyster River 12 May 1976-25; Mitlenatch 30 Apr 1977-30; Stories Beach 4 May 1977-25; Egmont 10 May 1977-10, feeding in yard; Quadra Island 24 May 1976-20; Cape Scott 29 Apr 1976-16, 25 May 1975-12, gradually building over past 3 weeks, 27 May 1982-1; Namu 29 Apr 1983-50; Calvert Island 10 May 1937-1 (NMC 28177); Kwatna River 17 May 1995-4; Masset Inlet 26 Apr 1984-1, 27 May 1984-30, in meadow; Lawyer Island 5 May 1979-2, first of spring; Terrace 19 May 1969-34; Hazelton 6 May 1963-1. **Interior** – Manning Park 31 May 1982-17; Keremeos 11 Apr 1969-6; Oliver 27 Mar 1951-1 (RBCM 10052), 24 May 1982-10; Richter Pass 4 May 1974-5; Osoyoos 4 May 1983-1 egg in Brewer's Blackbird nest; Trail 20 Apr 1983-2, 1 May 1982-15; Salmo 26 Apr 1983-6, at feeder with Brewer's and Red-winged blackbirds; Kootenay River (Creston) 17 May 1948-12; Princeton 23 Apr 1977-1; White Lake (Okanagan Falls) 12 May 1977-100, 26 May 1975-1 nestling in Red-winged Blackbird nest; Penticton 20 May 1988-50, at farm; Mahoney Lake 22 May 1974-25; Castlegar 23 May 1972-1 egg in Dark-eyed Junco nest; Sproule Creek 25 Mar 1979-1; Cranbrook 1 May 1942-2; Summerland 13 May 1977-25 at research station; Brouse 6 May 1978-10; Tunkwa Lake 12 May 1968-150; Enderby 8 May 1953-4 cowbird eggs in unknown host nest; Martin Prairie May 1895-1 (RBCM 1830); Sorrento to Salmon Arm 29 May 1970-25 on survey; Sorrento 15 May 1971-50; Radium Hot Springs 6 May 1981-1; Lac du Bois 5 May 1972-10; Shuswap Lake Park 26 Mar 1974-1; Revelstoke 25 May 1890-1 (Macoun and Macoun 1909; Cowan and Munro 1944-46), Apr 1989-1, 9 May 1985-50; Parson 23 May 1976-250; McLeod Meadows 16 May 1976-55 feeding on roadside; nr Golden 30 Apr 1998-1 male; Bridge Lake 17 May 1958-1, 22 May 1963-3 eggs in Song Sparrow nest; Yoho National Park 1 May 1977-1; Ottertail 1 May 1976-3; Field 30 Apr 1977-1; Wapta Lake (Yoho National Park) 19 May 1975-40; Field 14 May 1976-80; 100 Mile House 14 Apr 1985-1, feeding with Red-winged, Yellow-headed, and Brewer's blackbirds, 24 May 1982-100, at edge of ploughed field with Yellow-headed Blackbirds; Watson Lake (Lac la Hache) 12 May 1975-20; Riske Creek 15 May 1978-21; 111 Mile House 29 Apr 1981-4, feeding with Yellow-headed and Brewer's blackbirds; Williams Lake 23 Apr 1986-1+; 4 km s 150 Mile House 10 May 1983-2 nestlings in Yellow-headed Blackbird nest; Nimpo Lake 27 May 1948-1 (UBC 1817); Chezacut 26 May 1977-60, here for 2 weeks and increasing yearly; 24 km s Prince George 29 Apr 1987-8, 13 May 1982-20, feeding on farm grains and berries; Prince George 13 May 1975-10; Dee Lake 12 May 1991-60 (36 females; Siddle 1991c); Houston 17 May 1989-35; Telkwa 15 May 1975-1, 23 May 1985-13; Quick 23 May 1984-15; e Smithers 24 May 1977-2; nr confluence of Flatbed and Murray rivers 28 May 1976-1; Babine Lake 22 May 1978-14; Mackenzie 26 Apr 1998-1 male at feeder; Tupper Creek 12 May 1928-2 (RBCM 8668 and 8669); 5 km s Pouce Coupe 23 May 1994-4 eggs in Dark-eyed Junco nest; Rolla 12 May 1922-1 (Williams 1933b); Attachie 9 May 1968-60; Bear Flat 20 May 1984-70; Fort St. John to Blueberry River 12 to 18 May 1922-males in song (Williams 1933a); Charlie Lake 9 May 1985-13, 29 May 1970-8 eggs in Hermit Thrush nest (Fig. 559); sw corner Cecil Lake 3 May 1980-1, first record this spring, 10 May 1982-10; n Fort St. John 3 May 1986-1; Beatton Park 11 May 1983-15, 24 May 1985-12; Tupper Creek 20 May 1938-1 egg in Dark-eyed Junco nest; Sikanni Chief River valley 30 Apr 1984-28 in horse corrals; Coldfish Lake 28 May 1976-3; Fort Nelson 5 May 1987-1; Liard Hot Springs 10 May 1975-2; Lower Liard Crossing 11 May 1975-20; Helmut (Kwokullie Lake) 24 May 1982-2; Atlin 4 May 1981-1, first of season, 20 May 1981-21; Surprise Lake (Atlin) 18 May 1977-16.

Summer: Coastal – Clover Point (Victoria) 18 Jul 1985-96; Swan Lake (Saanich) 10 Jul 1975-159 feeding in newly cut hay field; Blenkinsop Lake (Saanich) 25 Aug 1981-11 immatures; Sidney Island 8 Aug 1988-12 immatures; Cowichan Bay 9 Jun 1987-30 in 1 flock at Koksilah Marsh; Bamfield Inlet 26 Aug 1976-35 immatures; Brady Beach 16 Aug 1976-13 immatures; Cleland Island 3 Aug 1970-10, 17 Aug 1974-27 in 1 flock; Tofino 15 Jul 1973-1 fledgling being fed by both Wilson's Warbler adults; Clayoquot 2 Aug 1961-1 (RBCM 10741); Bartlee Island 11 Aug 1961-10 immatures; Green Mountain (Nanaimo) 27 Jul 1986-1 at 1,480 m elevation accompanying elk; Canoe Pass (Delta) 17 Aug 1963-55; Westham Island 28 Aug 1975-50; Sea Island 2 Jul 1939-1 egg in Yellow Warbler nest; Pitt Meadows 3 Jun 1973-16, 26 Jun 1973-200; Maple Ridge 5 Jun 1974-20 in flock of Red-winged Blackbirds; McGillivray Slough (Chilliwack) 9 Jun 1963-40; Huntingdon 15 Aug 1958-50; Mission 31 Aug 1974-1 nestling being fed by Red-eyed Vireo, 7 Aug 1965-1 fledgling being fed by Song Sparrow which was also feeding 2 Wilson's Warbler fledglings, 17 Aug 1965-1 nestling being fed by Song Sparrow; Deroche 5 Aug 1971-1 egg in Swainson's Thrush nest; Cheam Slough

5 Jun 1984-14; Ross Lake 3 Jul 1970-1 fledgling being fed by Song Sparrow; 8.5 km ne Haney 2 Jul 1963-1 fledgling being fed by Black-throated Gray Warbler; Mountain Slough 26 Jun 1974-10; Harrison Hot Springs 10 Jun 1984-15, 31 Aug 1975-3; Seabird Island 4 Jun 1984-11; Hope 27 and 28 Jul 1923-1, first time seen in 18 years; Mitlenatch Island 3 Jun 1974-16; Nasparti River 6 and 7 Aug 1981-16 immatures, feeding nr mouth; Nimpkish 10 Jun 1978-4 at camp; Mount Currie 3 Jun 1967-10; Winter Harbour 21-31 Jul 1968-5 per day (Richardson 1971); Triangle Island 8 Jun 1976-1; Calvert Island 10 Aug 1937-1 (NMC 53349); Cape St. James Jun 1979-1; Anthony Island 10 Aug 1979-1 (RBCM 732); Bella Bella 21 Jun 1976-5, 7 Aug 1977-16 immatures; Salmon Point 17 Jul 1976-13 immatures; Sandspit 3 Aug 1991-2 (Siddle 1992a); Queen Charlotte City 17 Aug 1957-1 feeding on ground nr cattle; Langara Island 31 Jul 1971-1 (UMMZ 218035); Masset 8 Aug 1942-5 (1 collected, RBCM 10372); Lakelse Lake Park 29 Aug 1976-1; Terrace 14 Jun 1978-1 egg in Yellow Warbler nest; Kitwanga 1 Jun 1960-8; Hazelton 6 May 1963-1; New Hazelton 15 Jun 1993-1 nestling in Warbling Vireo nest; Carnaby 26 Jun 1975-10 at Skeena River nr town; Stewart 8 Jun 1991-3 (Siddle 1992c). **Interior** – Manning Park 1 Jun 1976-35; Sawmill Lake 13 Jun 1974-17; Keremeos 15 Jun 1969-14; Trail 15 Jul 1984-2; s Duck Lake (Creston) 20 Jun 1980-550 foraging in short grass with Yellow-headed and Red-winged blackbirds and starlings, 20 Jun 1981-4 eggs in Dark-eyed Junco nest; Wynndel 16 Aug 1982-100 on survey; Meyer's Flat (Rock Creek) 13 Jul 1980-30; Vaseux Lake 29 Jul 1974-40 at s end of lake; Castlegar 5 Jun 1972-1 egg in Dark-eyed Junco nest, 29 Jul 1977-1 egg in Dark-eyed Junco nest; White Lake (Okanagan Falls) 3 Jun 1972-50; Penticton 7 Aug 1950-1 egg in Chipping Sparrow nest; Okanagan Lake 4 Jul 1897-1 (MCZ 247891); Cranbrook 2 Jul 1915-1 (RBCM 3254); Naramata 30 Aug 1968-1 fledgling being fed by Yellow-rumped Warbler; n Sparwood 1 Jun 1984-50, 9 Jun 1984-50, 13 Jun 1984-60, 22 Jun 1984-160, all feeding in seeded field; Alleyne Lake 2 Aug 1987-1 nestling being fed by Song Sparrow; Wasa 8 Jun 1977-20; Elk River 20 Jun 1982-1; Okanagan Landing 2 Jul 1965-8 eggs in Dark-eyed Junco nest; Vernon 1 Aug 1892-1 juvenile (ANS 30951); Skihist Mountain 1 Aug 1964-25; Birkenhead Lake 14 Jul 1976-1; Hat Creek 5 Jul 1990-11 at ranch; Savona 13 Aug 1975-10; Cache Creek 16 Aug 1944-1 flock; Celista 22 Jul 1960-12 around cattle; Minton Creek 18 Jun 1978-16 all following herd of cattle; Field 26 Aug 1922-1 (Ulke 1923); 14 km n Lac la Hache 5 Jul 1931-1 (MCZ 284668); 25 km sw Tatlayoko Lake 1 Aug 1981-1 at 2,440 m elevation (BC Photo 766); Chezacut Lake 13 Aug 1931-1 (UBC 6832); Bald Mountain (Riske Creek) 18 Jun 1980-3 eggs in Yellow-rumped Warbler nest; Phililoo Lake 11 Jul 1975-15; Wells Gray Park 6 Jul 1978-1 feeding off horses; Williams Lake 1 Jun 1963-50; Anahim Lake 21 Jul 1961-20; Stum Lake 9 Jun 1973-10, 26 Jul 1971-1 egg in Song Sparrow nest, 26 Jul 1971-1 nestling being fed by Song Sparrow; Murtle River 26 Jun 1951-30; Mount Robson Park 6 Jun 1973-4, 18 Jun 1973-20, 17 Jul 1974-2 nestlings in Dark-eyed Junco nest; Dragon Lake 9 Jun 1944-1; Quesnel 9 Jun 1994-1 egg in Orange-crowned Warbler nest; Milburn Lake 8 Jun 1991-1 egg in Veery nest; Bowron Lake Park 8 Aug 1961-5 close to a moose, 1 on back of moose, 13 Aug 1975-1 nestling, host unknown; Indianpoint Lake 27 Jul 1930-1 (MCZ 284667); McBride 10 Jul 1991-2; Nulki Lake 1 Jul 1974-50; 16 km w Prince George 9 Jun 1976-2 eggs in Warbling Vireo nest, 14 Jun 1994-1 fledgling being fed by Dark-eyed Junco, 16 Jun 1969-45, 26 Jun 1972-1 egg in Chipping Sparrow nest, 1 Jul 1972-1 nestling in American Redstart nest, 28 Jul 1972-2 fledglings being fed by Swainson's Thrush; Telkwa 21 Aug 1993-2 fledglings being fed by Dark-eyed Juncos; Pinchi Lake 6 Jun 1977-84 at resort; Wolverine River 20 Jul 1976-1; sw Fellers Heights 14 Jun 1992-2 nestlings in Orange-crowned Warbler nest; Dawson Creek 8 Jun 1975-6, 3 Jul 1974-20; Tupper Creek 28 Jun 1938-1 (RBCM 8212); n Arras 8 Jul 1993-1 egg in Lincoln's Sparrow nest; 8 km se Farmington 26 Jun 1978-20; n Dawson Creek 3 Aug 1975-5; Beryl Prairie 8 Jul 1979-7; Cache Creek (Peace River) 9 Jun 1985-60 by cattle ranch at km 6 along Upper Cache Rd; Taylor 15 Jun 1980-10, mostly males; Alaska Highway (Mile 41) 29 Aug 1982-1 with 500+ Brewer's Blackbirds feeding in newly broken field, last record of autumn; Aiken Lake 2 Aug 1979-1; Charlie Lake 15 Jul 1977-1 nestling in Least Flycatcher nest; 19 Aug 1997-1 fledgling being fed by White-throated Sparrow; Beatton Park 8 Jun 1982-10 n Fort St. John, 17 Aug 1986-5, with 30 Brewer's Blackbirds, last record of autumn; Halfway River 12 Jul 1930-1 (RBCM 14890); 1 km w Trutch 20 Jun 1947-1 (UKMU 00024739); Kitchener Lake 2 and 3 Aug 1979-1; Tuaton Lake 9 Aug 1984-1 fledgling walked about camp feeding on insects; Beatton River 8 Jun 1956-1; Dease Lake 6 Jun 1962-1 (NMC 50191); Buckley Lake 19 Aug 1986-several observed; n Gnat Creek 14 Jun 1980-10 by horses on Highway 37; sw Kotcho Lake 29 Jun 1982-1 egg in Alder Flycatcher nest; Andy Bailey Park 17 Jul 1992-1 fledgling being fed by Yellow-rumped Warbler; nr Fort Nelson 2 Jul 1982-21, foraging in drying marsh; Fort Nelson 15 Jun 1975-4, 6 Aug 1985-1 foraging on lawns, likely migrant from areas outside Fort Nelson Lowland; Parker Lake 18 Jun 1976-1 (RBCM 15339), 9 Jul 1978-1; Fort Nelson River 8 Jul 1978-1 nestling ready to fledge in Magnolia Warbler nest; Kotcho Lake 26 Jun 1982-1, on large island; 8 km s Cassiar Junction 12 Jun 1972-1; Atlin 4 Jul 1934-1 (Munro and Cowan 1947); 60 km wnw Fireside 20 Jun 1974-1 (DMNH 40846); Tatshenshini River 7 Jun 1983-2; Liard Hot Springs 7 Jul 1992-1 fledgling with adult Magnolia Warbler.

Breeding Bird Surveys: Coastal – Recorded from 21 of 27 routes and on 67% of all surveys. Maxima: Albion 4 Jul 1970-56; Victoria 30 Jun 1973-40; Seabird 6 Jun 1971-36. **Interior** – Recorded from 58 of 73 routes and on 83% of all surveys. Maxima: Lavington 3 Jul 1972-96; Golden 19 Jun 1976-65; Salmo 14 Jun 1975-35; Syringa Creek 22 Jun 1991-35; Creston 18 Jun 1984-35.

Autumn: Interior – Atlin 28 Aug 1931-several with horses in pasture, 3 Sep 1934-1 (Swarth 1936; RBCM 5867); Mackenzie 10 Oct 1994-1; Quick 29 Sep 1990-50; Riske Creek 3 Oct 1989-1; Glacier National Park 5 Oct 1982-13; Revelstoke 28 Oct 1985-1; Kamloops 26 Sep 1965-9, 5 Oct 1965-9 feeding along edge of highway; Spences Bridge 17 Sep 1978-6; Kimberley 1 Sep 1977-12. **Coastal** – Port Simpson 14 Sep 1969-5; nr Kwinamass River 17 Sep 1985-1; Masset 11 Nov 1952-1 (RBCM 10373); Strachan Bay 5 Sep 1967-2; Port Neville Inlet 9 Oct 1976-1; Port Hardy 14 Oct 1934-2, 16 Oct 1939-10, 15 Nov 1950-1; Sechelt 20 Oct 1995-1; Maple Ridge 24 Sep 1967-60; Sea Island 5 Sep 1971-100, latest concentration; Tofino 11 Oct 1980-1.

Winter: Interior – Williams Lake 5 Dec 1987-1 with flock of Red-winged Blackbirds at feeder; Revelstoke 20 Dec 1984-1; Vernon 15 Feb 1985-4, 15 Feb 1989-1 (Rogers 1989), 31 Dec 1987-2 (Rogers 1988); Nakusp 2 Jan 1978-3; Creston 24 Feb 1993-2 (Siddle 1993b); Oliver 15 Jan 1989-1 (Rogers 1989a). **Coastal** – Agassiz 22 Feb 1975-7; Langley 10 Feb 1968-12; Pitt Meadows 6 Jan 1980-6, 28 Dec 1986-250; Beach Grove 1 Jan 1981-10 at feeder; Surrey 9 Jan 1984-15 nr Mud Bay; Reifel Island 14 Dec 1974-17; Vancouver 4 Dec 1984-10 flying to roost nr Lost Lagoon; Duncan 31 Dec 1977-2 at feeder; Central Saanich 1 Jan 1983-5, 22 Dec 1973-2; Saanich 1 Dec 1973-15.

Christmas Bird Counts: Interior – Recorded from 2 of 27 localities and on 2% of all counts. Maxima: Oliver-Osoyoos 28 Dec 1986-6; Revelstoke 20 Dec 1984-1. **Coastal** – Recorded from 10 of 33 localities and on 24% of all counts. Maxima: Pitt Meadows 28 Dec 1986-250; Ladner 21 Dec 1963-200; White Rock 2 Jan 1994-127.

Baltimore Oriole
Icterus galbula (Linnaeus)

BAOR

RANGE: Breeds from northeastern British Columbia, northwestern Alberta, central Saskatchewan, southern Manitoba, southern Ontario, extreme southern Quebec, central New Brunswick, and central Nova Scotia south through the eastern and central states to Georgia and Louisiana, and west to the western edge of the Great Plains. Winters mainly from southern Mexico south to northern Colombia, northern Venezuela, and Trinidad.

STATUS: On the coast, *casual* in the Georgia Depression Ecoprovince.

In the interior, an *uncommon* to locally *fairly common* migrant and summer visitant to the Boreal Plains Ecoprovince; *casual* in the Taiga Plains, Northern Boreal Mountains, and Sub-Boreal Interior ecoprovinces; *accidental* in the Southern Interior and Southern Interior Mountains ecoprovinces.

Breeds.

CHANGE IN STATUS: The Baltimore Oriole (Fig. 561) is a recent addition to the avifauna of British Columbia. In the 1940s, Munro and Cowan (1947) listed it as an extralimital species based on a single specimen collected at Chilliwack in 1927. This oriole was not encountered again until 1955, when an adult male was observed at Coldstream in the Okanagan valley (Munro 1955b). In 1960, it was discovered nesting east of the Rocky Mountains at Pouce Coupe in the Boreal Plains (Weber 1976). Since that time it has become a regular, but local, breeding species in that ecoprovince.

Early ornithological explorations of the Boreal Plains during the 1930s by Racey (1930), Williams (1933b), and Cowan (1939) failed to detect the Baltimore Oriole. Symons (1973) also made no mention of this oriole during the period he homesteaded in this region, from the late 1940s to the mid-1950s. Many of the areas covered by these early surveyors, including the Peace River valley, Pouce Coupe, Tupper, Dawson Creek, Fort St. John, and Charlie Lake, would later become the centre of this bird's distribution in British Columbia. Sometime during the late 1950s, this oriole probably expanded its range west into British Columbia from the Grand Prairie area of northwestern Alberta, where it has been established since at least 1944 (Soper 1949). From 1960 to 1962, the Baltimore Oriole was recorded at Pouce Coupe, Swan Lake, Tupper, Charlie Lake, and Hudson's Hope (Weber 1976). In 1962, it was also recorded on the Cameron River, south of Wonowon. By the late 1960s or early 1970s, the Baltimore Oriole had become an established, but sparsely distributed, breeding bird of the open deciduous woodlands in the Boreal Plains. In his latest treatise on Canadian birds, however, Godfrey (1986) did not show the breeding distribution of this oriole extending into British Columbia.

NONBREEDING: In British Columbia, the Baltimore Oriole is at the extreme western limit of its summer distribution in North America. It occurs regularly only east of the Rocky Mountains, in the deciduous woodlands of the Boreal Plains. Further north, there are a few records from the Fort Nelson Lowland in the Taiga Plains and from Liard Hot Springs in the Northern Boreal Mountains. West of the Rocky Mountains, this oriole has been recorded near the Crooked River in the Sub-Boreal Interior, south of Nicholson in the Southern Interior Mountains, and at Coldstream in the Southern Interior. There are also several records from the Georgia Depression on the south coast.

Figure 561. The Baltimore Oriole (*Icterus galbula*), now a distinct species, was formerly known as the Northern Oriole and was considered conspecific with the Bullock's Oriole (*I. bullockii*). In British Columbia, males (a) arrive before females (b) and quickly establish breeding territories (Anthony Mercieca).

The Baltimore Oriole has been recorded near sea level on the coast; in the interior, it has been recorded at elevations from 305 to 900 m. This oriole arrives directly onto its breeding grounds in the Boreal Plains, where nonbreeding and breeding habitats are similar (see BREEDING). A lack of records from other regions of British Columbia indicates that it moves into and out of the province through northwestern Alberta.

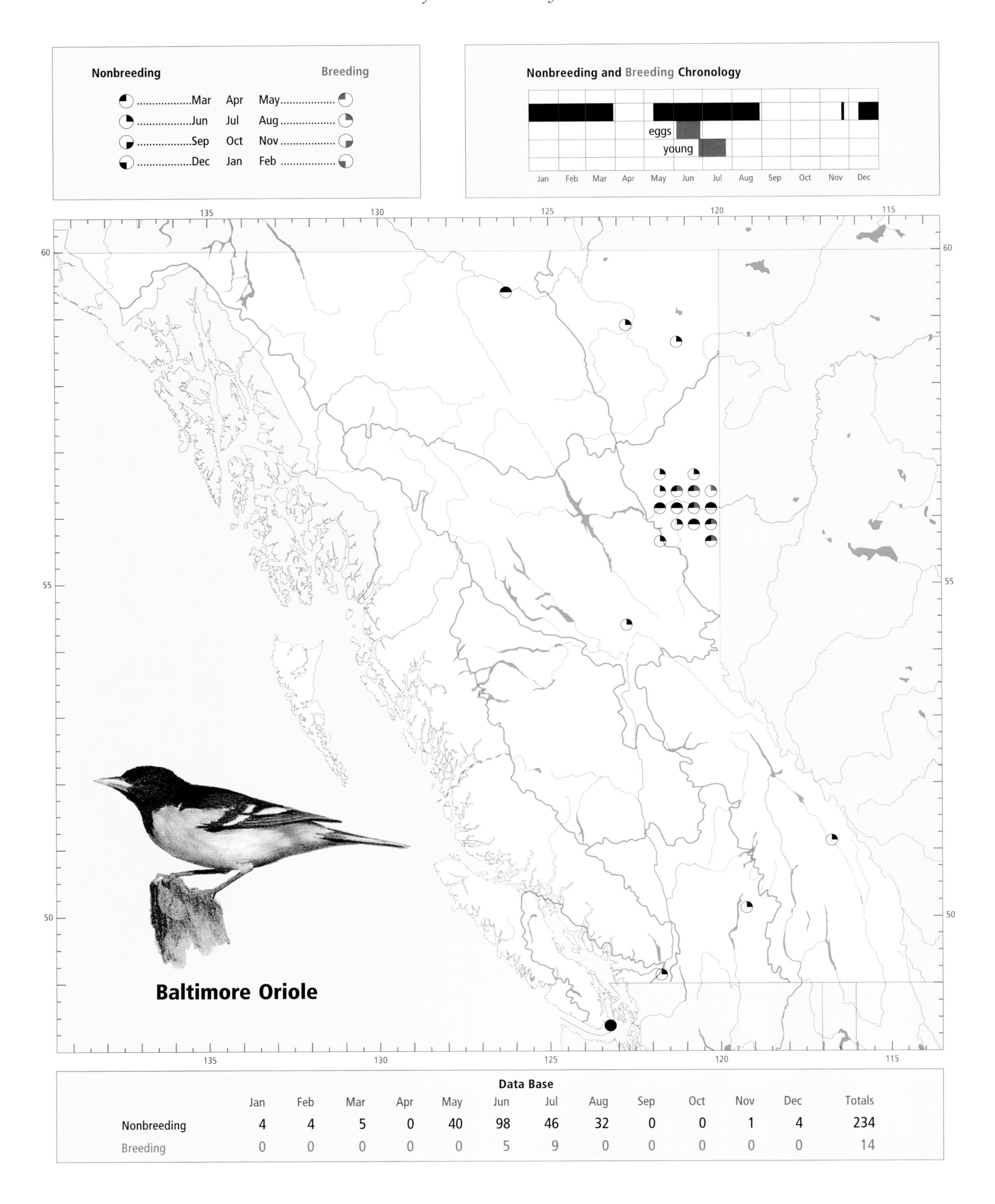

Data Base

	Jan	Feb	Mar	Apr	May	Jun	Jul	Aug	Sep	Oct	Nov	Dec	Totals
Nonbreeding	4	4	5	0	40	98	46	32	0	0	1	4	234
Breeding	0	0	0	0	0	5	9	0	0	0	0	0	14

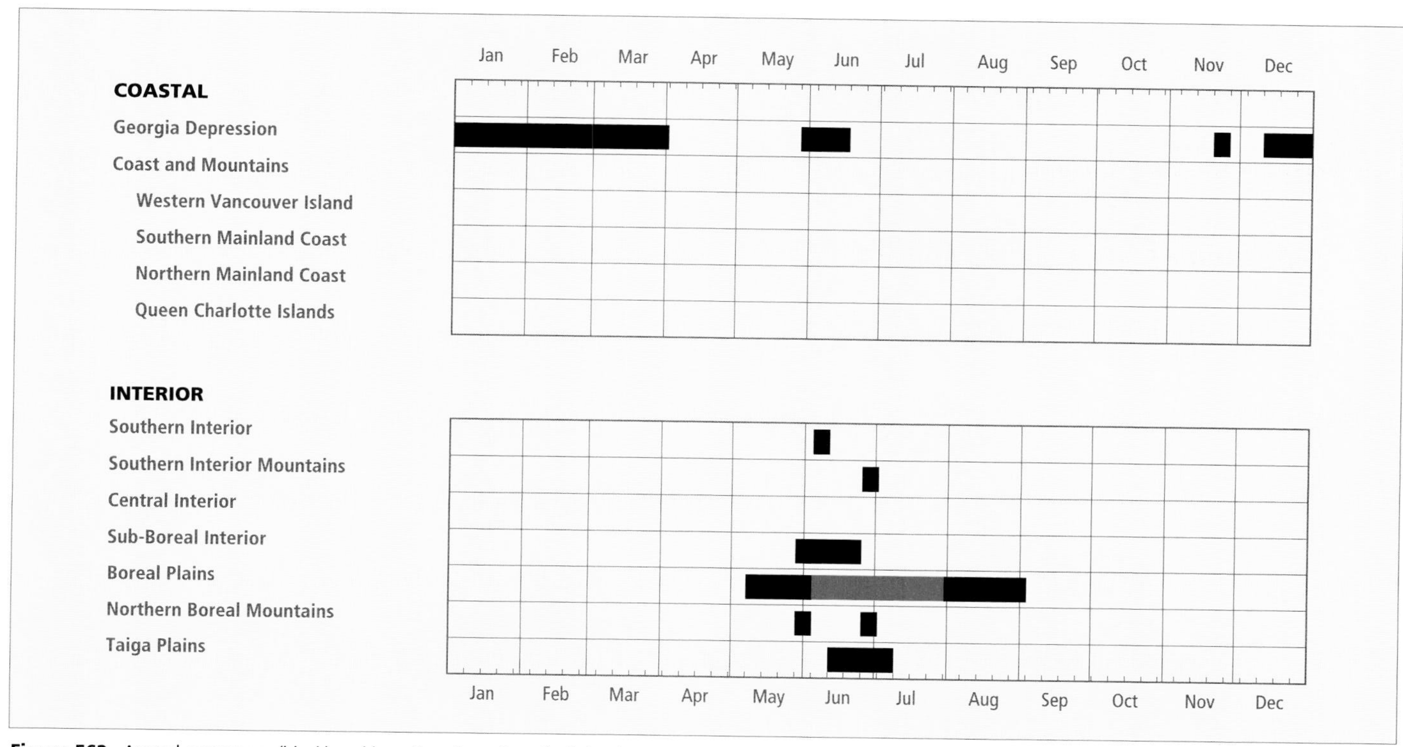

Figure 562. Annual occurrence (black) and breeding chronology (red) for the Baltimore Oriole in ecoprovinces of British Columbia. Records are shown for the week in which they occurred.

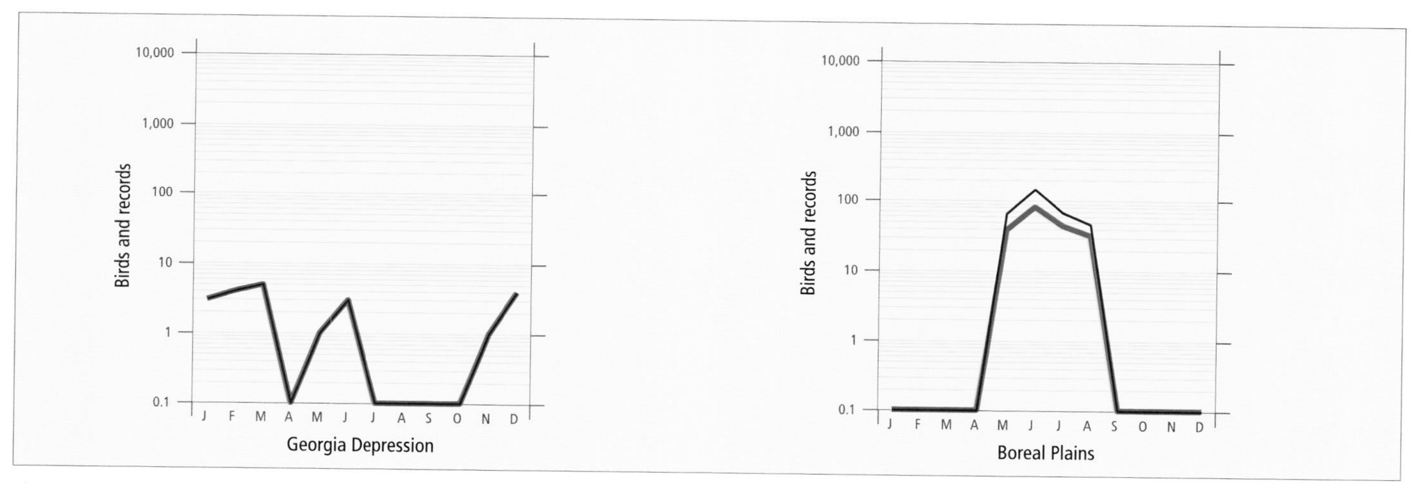

Figure 563. Fluctuations in total number of birds (purple line) and total number of records (green line) for the Baltimore Oriole in ecoprovinces of British Columbia. Breeding Bird Surveys and nest record data have been excluded.

The Baltimore Oriole first arrives in the Boreal Plains in mid-May (Fig. 562). Early spring migrants recorded in the Fort St. John area over an 8-year period (1980 to 1988) provide a median first arrival date of 18 May (range: 14 to 27 May). The first birds to arrive are typically males that precede the main influx, consisting of both sexes, by about a week. There are no obvious peaks in this spring movement, and numbers quickly build to breeding levels by the last week of May or first week of June (Fig. 563). In Alberta, orioles also arrive in mid-May (Pinel et al. 1993), suggesting a rapid movement onto these westernmost breeding grounds.

The Baltimore Oriole is an early-autumn migrant, and most have left the continent by late summer (Rising and Flood 1998). This oriole has left British Columbia for its neotropical wintering grounds by the last week of August; there are no autumn records from the interior and only a single November record from the south coast. In the Boreal Plains, orioles probably begin to leave shortly after the young become independent, and their numbers decline by late July. Most have left by mid-August, with only a few stragglers remaining until the last week of that month (Fig. 563). There is no detectable movement elsewhere in the province.

Tatum (1972) reports an overwintering bird that regularly visited a feeder in Victoria from 15 December 1971 through 28 March 1972. Our data also contain a 12 December record, which is probably of the same bird.

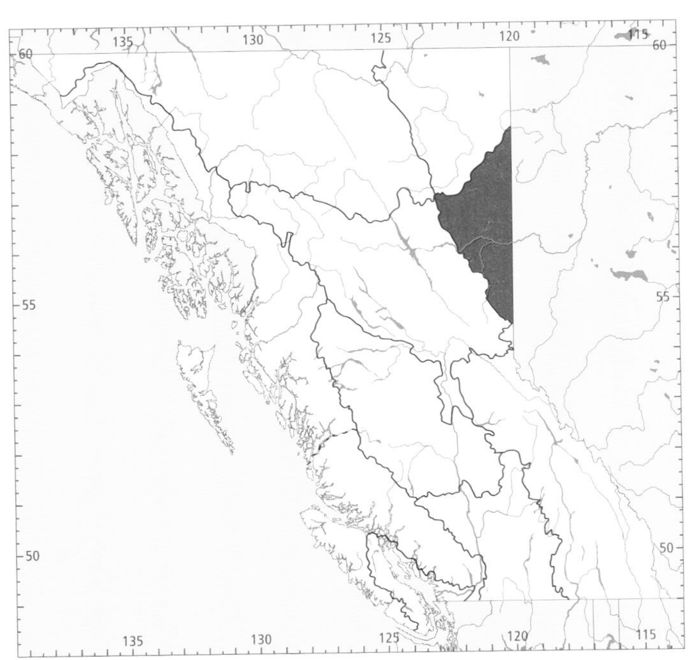

Figure 564. In British Columbia, the highest numbers for the Baltimore Oriole in summer occur in the Boreal Plains Ecoprovince.

The Baltimore Oriole has been recorded in the Boreal Plains from 12 May (Phinney 1998) to 28 August; it has been recorded irregularly elsewhere. A single bird was recorded in the Georgia Depression from 12 December to 28 March (Fig. 562).

BREEDING: The Baltimore Oriole breeds in the Boreal Plains of northeastern British Columbia. Active nests have been located near Pouce Coupe, Briar Ridge, Valley View, and Charlie Lake. Observations of recently fledged young recorded at Bear Flat, Taylor, Fort St. John, Cecil Lake, and Boundary Lake suggest a widespread but sparse breeding distribution in the Peace Lowland of the Boreal Plains. There is a single extralimital record of this oriole "feeding young" near Nicholson in the Southern Interior Mountains. Breeding is unconfirmed for the Taiga Plains, but is suggested by a record of an adult carrying food east of Fort Nelson.

The Baltimore Oriole reaches its highest numbers in summer in the Peace Lowland of the Boreal Plains (Fig. 564). Breeding Bird Surveys for interior routes for the period 1968 through 1993 contain insufficient data for analysis. Breeding Bird Surveys across Canada from 1980 to 1996 indicate that numbers declined at an average annual rate of 2.7% ($P < 0.01$); continent-wide, numbers declined at an average annual rate of 1.7% ($P < 0.01$) over the same period (Sauer et al. 1997).

Songbird surveys conducted near Worth, about 60 km northeast of Chetwynd, detected the Baltimore Oriole in 4 of 6 years (Merkins and Booth 1998). There, this oriole was found in very low numbers in mature stands of trembling aspen ranging from 60 to 100 years of age.

The Baltimore Oriole has been reported nesting at elevations from 610 to 750 m. Nesting was recorded in open, deciduous woodlands, typically near natural or human-created forest openings such as lakeshores, clearings, or road edges. All nests were reported in young (pole-sapling stage) to mature woodlands dominated by trembling aspen (Fig. 565). Observations of recently fledged broods suggest that nesting also occurs in balsam poplar floodplain forests.

The Baltimore Oriole has been recorded breeding in the province from 4 June to 25 July (Fig. 562).

Nests: Only 6 nests have been reported in British Columbia; all were found in trembling aspen trees. Most nests were placed near the end of a branch in the upper tree canopy. Nests were described as a deep pouch suspended from a forked branch and constructed mainly of fine woven grasses and plant fibres. One closely inspected nest was constructed of fine grasses, plant down, bark strips, and moose hair. The heights for 6 nests ranged from 4.5 to 15.0 m, with 4 nests between 6.0 and 14.0 m.

Eggs: Dates for 3 clutches ranged from 4 June to 15 June. Only 1 nest was checked for contents and it contained 4 eggs on 15 June. Calculated dates suggest that nests could contain eggs as late as 25 June. Elsewhere, clutch sizes are usually 4 or 5 eggs, but can range from 3 to 7 (Rising and Flood 1998). The incubation period is 11 to 14 days.

Figure 565. In the Boreal Plains Ecoprovince of British Columbia, the Baltimore Oriole prefers to nest in stands of mature trembling aspen (Beatton Park, Charlie Lake, 23 June 1996; R. Wayne Campbell).

Young: Dates for 11 broods ranged from 27 June to 25 July, with 7 recorded between 1 July and 7 July. Nests were not checked to determine brood size. Observations of recently fledged young indicate that broods with up to 6 young may occur. The nestling period is 11 to 14 days, with most young leaving the nest after 12 or 13 days from hatching (Rising and Flood 1998).

Brown-headed Cowbird Parasitism: Only 1 nest with eggs or young was inspected closely and it was not parasitized by the cowbird. There were also no records of this oriole feeding Brown-headed Cowbird fledglings.

The Baltimore Oriole is an ejector of cowbird eggs which are typically removed by this host shortly after parasitism occurs. In Manitoba, Sealy et al. (1995) experimentally parasitized 16 oriole nests and recorded a 100% rejection rate; in 14 of the nests, orioles removed or otherwise destroyed cowbird eggs within 1 hour after they were introduced. Rising and Flood (1998) report a parasitism rate ranging from 1.4 to 5.9% across North America, and indicate few cases where cowbirds were successfully fledged.

Nest Success: Insufficient data.

REMARKS: No subspecies are recognized for the Baltimore Oriole (Rising and Flood 1998). The Baltimore Oriole and the Bullock's Oriole (*Icterus bullockii*) were formerly known as the Northern Oriole (*I. galbula*) (American Ornithologists' Union 1973, 1983); both are now recognized as separate species, as they had been before 1973 (American Ornithologists' Union 1957, 1995).

Baltimore and Bullock's orioles hybridize where their ranges overlap in the Great Plains region of North America (Sibley and Short 1964). Hybridization has also been noted in the Milk River valley of southern Alberta (Salt and Salt 1976). In British Columbia, Baltimore and Bullock's orioles may occur sympatrically in the Rocky Mountain Trench near Nicholson. Brooks (1942a) observed that a Baltimore Oriole specimen collected at Chilliwack on 11 June 1927 was a "nearly typical *galbula*," except for "a faint yellow supercilium" suggesting a "trace of *bullockii* blood."

Baltimore Oriole nesting densities have been observed to increase in response to outbreaks of the forest tent caterpillar (*Malacosoma disstria*). In Manitoba, Sealy (1980) recorded an increase of from 5.4 to 10.5 pairs/ha during the second year of a 2-year tent caterpillar outbreak. This forest defoliator is found throughout the deciduous forests of northeastern British Columbia and periodically causes severe defoliation of trembling aspen stands (Humphreys 1995).

For additional information on the natural history and identification of the Baltimore Oriole, see Rising (1973, 1983), Sealy (1980), Butcher (1991), Lee and Birch (1998), and Rising and Flood (1998).

NOTEWORTHY RECORDS

Spring: Coastal – Victoria 28 Mar 1972-1, last observation of overwintering bird (Tatum 1972), 30 May 1981-1 adult male (Hunn and Mattocks 1981a). **Interior** – Sudeten Park (Tupper) 25 May 1981-1; Dawson Creek 12 May 1993-1 (Phinney 1998); Kiskatinaw Park 20 May 1988-2; McQueen Slough 26 May 1990-1 in trembling aspen trees; w Hudson's Hope 27 May 1988-1 at Alwin Holland Park, 31 May 1973-2 males singing in open, mature trembling aspen forest (Thormin 1973); Old Fort (Peace River) 14 May 1988-1, early arrival; Taylor 27 May 1984-8 along 3.5 km road transect; Clayhurst 17 May 1981-1 at ferry crossing; Cecil Lake 28 May 1980-2 singing in trembling aspens; Beatton River 19 May 1986-1, early arrival; Charlie Lake 25 May 1985-6; Liard Hot Springs 28 May 1981-2 singing in balsam poplars.

Summer: Coastal – Swan Lake (Victoria) 2 Jun 1981-1; Chilliwack 11 Jun 1927-1 specimen collected (Brooks 1942a). **Interior** – Coldstream 5 Jun 1955-1 adult male (Munro 1955b); 18 km s Nicholson 26 Jun 1976-adults feeding young; Bear Lake (Crooked River) 2 and 4 Jun 1981-1; e Chetwynd 24 Jun 1970-1 in bog area; Pouce Coupe 2 Jul 1960-adults feeding nestlings (Weber 1976); Sudeten 4 Jul 1974-2; around Dawson Creek 7 Jun 1975-8; 3.2 km n Rolla 26 Jun 1978-1; e Briar Ridge 4 Jun 1992-female incubating, 15 Jun 1992-4 eggs (Phinney 1998); Valley View 27 Jun 1995-female attending young in nest; Clayhurst 21 Jun 1976-8 at ecological reserve; Taylor 5 Jul 1998-1 bobtailed fledgling with adult at Peace Island Park, 22 Aug 1975-4; Worth 2 Jun 1993-3 detected on songbird survey; Hudson's Hope 23 Jun 1978-1; Bear Flat 5 Jun 1988-2, 5 Jul 1997-1 bobtailed fledgling with adult; Charlie Lake 7 Jun 1977-female incubating unknown number of eggs, 13 Jul 1978-6 fledged young being fed by female; Beatton Park 12 Jul 1983-2 fledglings, barely able to fly; Fort St. John 7 Jul 1991-3 recently fledged young; Montney Creek 6 Jul 1976-1 nr bridge crossing; Cecil Lake 3 Jul 1978-3, 9 Jul 1981-1 fledgling fed by adult; St. John Creek 28 Aug 1986-1, late sighting; Boundary Lake (Goodlow) 25 Jul 1982-1 recent fledgling fed by male; Cameron River 14 Jun 1962-1 at Northfork Ranch s Wonowon; Metlahdoa Creek (e Fort Nelson) 25 Jun 1982-1 adult male carrying food; Fort Nelson 12 Jun 1982-1 flew into balsam poplar; nr Fort Nelson 5 Jul 1978-1 adult male in trembling aspens at Mile 313 of Alaska Highway; Liard Hot Springs 28 Jun 1985-1.

Breeding Bird Surveys: Coastal – Not recorded. **Interior** – Recorded from 3 of 73 routes and on 2% of all surveys. Maxima: Fort St. John 27 Jun 1986-20; Tupper 20 Jun 1994-18; Hudson's Hope 7 Jun 1976-4.

Autumn: Interior – No records. **Coastal** – Victoria 24 to 26 Nov 1984-1 male (Hunn and Mattocks 1985).

Winter: Interior – No records. **Coastal** – Victoria 12 Dec 1971, 15 Dec 1971 to 28 Mar 1972-1 (Tatum 1972; BC Photo 236).

Christmas Bird Counts: Interior – Not recorded. **Coastal** – Recorded from 1 of 36 counts and on less than 1% of all counts. Maximum: Victoria 1 Jan 1972-1.

Bullock's Oriole

BUOR

Icterus bullockii (Swainson)

Figure 566. The Bullock's Oriole continues to expand its breeding range in British Columbia north along the south coast and into portions of the central interior. The male (a) may feed the female (b) in the nest (Anthony Mercieca). She carries out all incubation duties.

RANGE: Breeds from southern British Columbia, southern Alberta, southwestern Saskatchewan, eastern Montana, southwestern North Dakota, and central South Dakota south, east of the coastal areas of Washington and Oregon, to northern Mexico and Texas, and east to western Nebraska, western Kansas, and western Oklahoma. Winters from coastal California south to Guatemala and in small numbers on the Gulf coast.

STATUS: On the coast, an *uncommon* migrant and local summer visitant and *casual* in winter in the Georgia Depression Ecoprovince; in the Coast and Mountains Ecoprovince, *very rare* vagrant on Western Vancouver Island, *rare* summer visitant on the Southern Mainland Coast, and *accidental* on the Northern Mainland Coast; absent from the Queen Charlotte Islands.

In the interior, *uncommon* to *fairly common* migrant and summer visitant and *casual* in winter in the Southern Interior Ecoprovince; *uncommon* migrant and local summer visitant to the Southern Interior Mountains and Central Interior ecoprovinces.

Breeds.

CHANGE IN STATUS: In the mid-1940s, the Bullock's Oriole (Fig. 566) was considered a regular summer visitant throughout the valleys of the Similkameen and Okanagan rivers north to Clinton and Lytton in the dry southern interior of British Columbia. It also occurred, less commonly and more locally, on the south coast in the vicinity of Chilliwack and Sumas in the eastern Fraser River valley (Munro and Cowan 1947). The species may have nested as far north as Alkali Lake (Cariboo), south of Williams Lake, where an unoccupied nest was found in July 1941 (Munro and Cowan 1947).

During the late 1950s and 1960s, the Bullock's Oriole spread westward on the southern mainland coast to the Fraser River delta, where it was found nesting in 1970 (Campbell et al. 1972a). It now occurs regularly and breeds locally throughout the Lower Mainland region. During the 1980s and 1990s, it became an irregular summer visitant to the Sunshine Coast, north of Vancouver, as well as in the Pemberton area in the southern Coast and Mountains.

On Vancouver Island, the Bullock's Oriole was first reported at Carmanah Point, on the southwest coast, in 1952 (Bruce 1953). Over the next 20 years, the species was seen only infrequently. It was suspected of having bred in 1971 when an old nest was located in Beacon Hill Park, Victoria, in April 1972 (Tatum 1973). The first active nest was found in 1972 in Courtenay, 220 km north of Victoria, along the east side of Vancouver Island. The species continued to expand its range throughout southeastern Vancouver Island during the 1980s and early 1990s, reaching Campbell River in the north and Port Alberni in the west, where it is suspected of nesting. Elsewhere on the island, it is considered a vagrant.

Figure 567. During the nonbreeding seasons in British Columbia, the Bullock's Oriole frequents human-influenced habitats, such as this poplar windbreak (Pitt Meadows, 6 May 1990; R. Wayne Campbell).

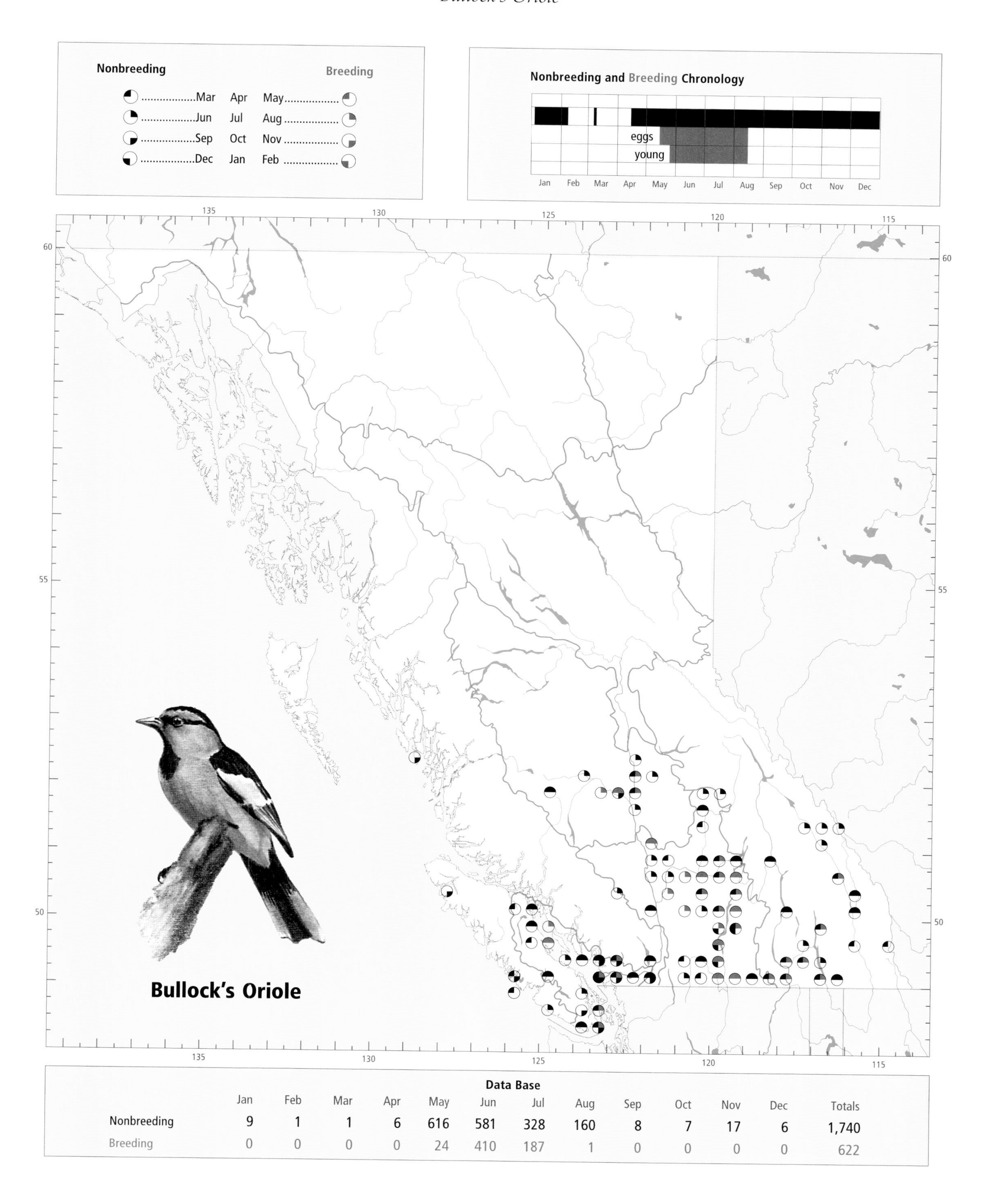

Data Base

	Jan	Feb	Mar	Apr	May	Jun	Jul	Aug	Sep	Oct	Nov	Dec	Totals
Nonbreeding	9	1	1	6	616	581	328	160	8	7	17	6	1,740
Breeding	0	0	0	0	24	410	187	1	0	0	0	0	622

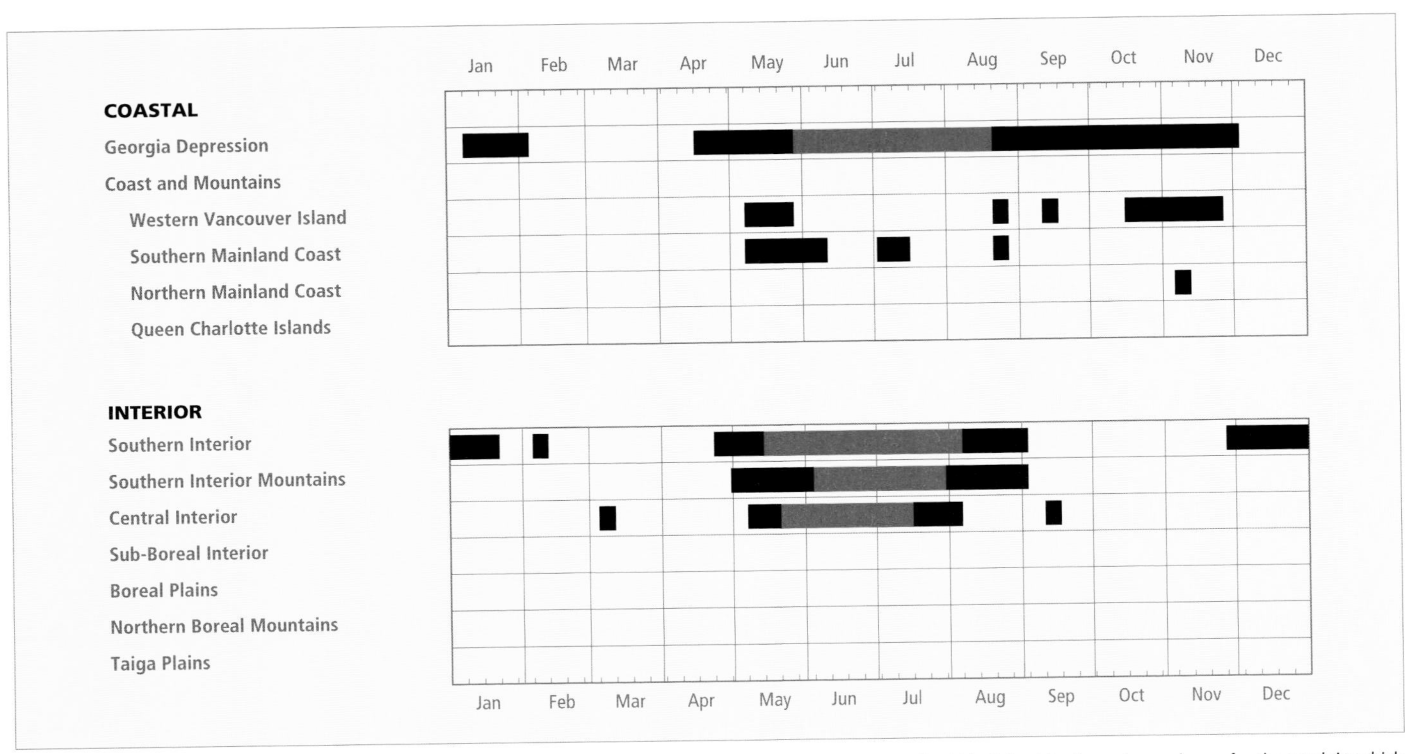

Figure 568. Annual occurrence (black) and breeding chronology (red) for the Bullock's Oriole in ecoprovinces of British Columbia. Records are shown for the week in which they occurred.

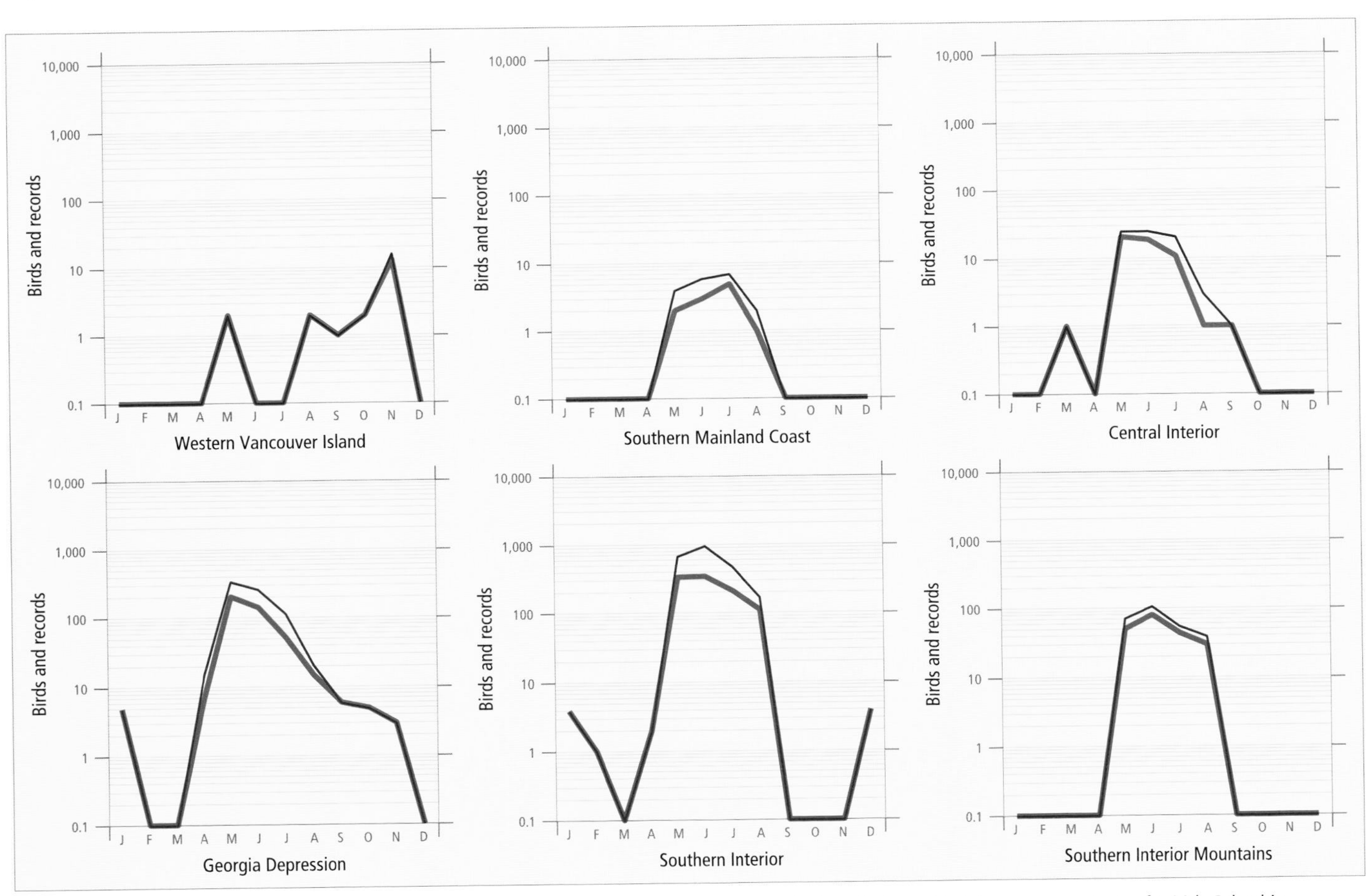

Figure 569. Fluctuations in total number of birds (purple line) and total number of records (green line) for the Bullock's Oriole in ecoprovinces of British Columbia. Christmas Bird Counts, Breeding Bird Surveys, and nest record data have been excluded.

The Bullock's Oriole also expanded its range in the interior. During the 1970s through the 1990s, it was found breeding north to Soda Creek in the Central Interior and east to Radium in the Southern Interior Mountains.

In Washington state, the Bullock's Oriole has shown a similar range expansion westward from the interior, along river valleys to the coast, and eastward in the interior (Smith et al. 1997).

NONBREEDING: The Bullock's Oriole is widely distributed across southern portions of the province from Vancouver Island east to the British Columbia–Alberta border and north through the interior to southern portions of the Central Interior. It is fairly widespread in the Okanagan, Thompson, and southern Kootenay valleys, but becomes scarcer with increasing latitude. It is sparsely distributed in the Cariboo and Chilcotin areas and absent from the Sub-Boreal Interior. On the coast, the species occurs regularly throughout the lower Fraser River valley east to Hope, locally in the Pemberton area, and on Vancouver Island from Victoria north to Campbell River.

The Bullock's Oriole occurs at relatively low elevations, from near sea level to 125 m on the coast and up to 1,150 m in the interior. Throughout the province, it generally frequents the edges of deciduous woodlands, usually along valley bottoms, but it also occurs in woodlands fringing agricultural areas (Fig. 567), parks, golf courses, and grasslands. In the interior, it prefers rangeland copses, and shrub-steppe and grassland habitats where black cottonwood, water birch, trembling aspen, saskatoon, hawthorn, choke cherry, and big sagebrush are dominant features of the landscape. On the coast, the Bullock's Oriole prefers stands of mature broadleaf trees, particularly black cottonwood, Lombardy poplar, paper birch,

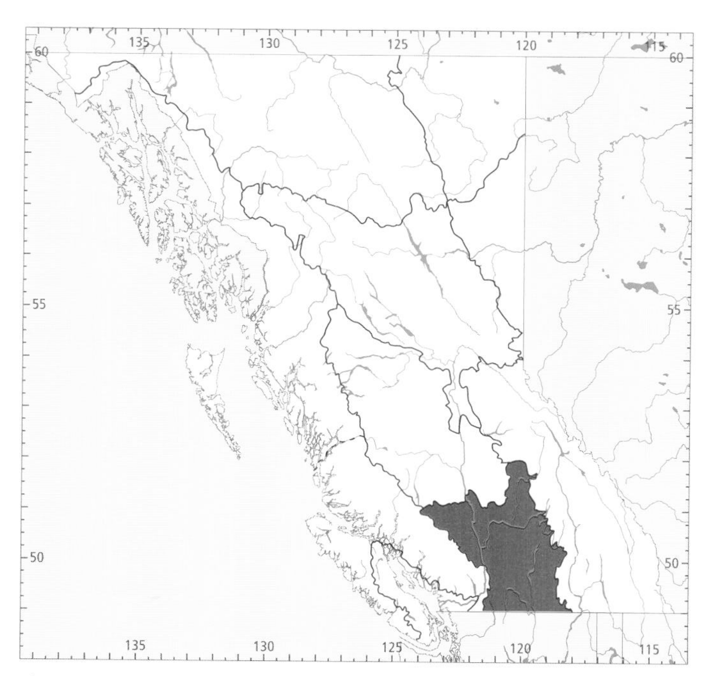

Figure 570. In British Columbia, the highest numbers for the Bullock's Oriole in summer occur in the Southern Interior Ecoprovince.

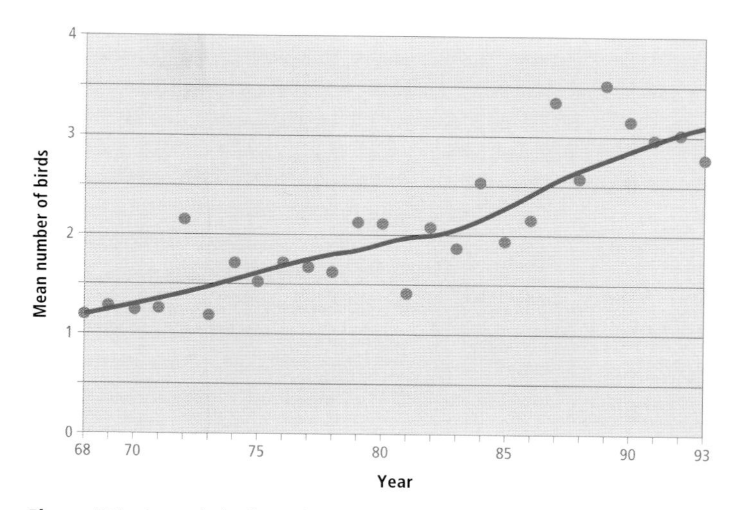

Figure 571. An analysis of Breeding Bird Surveys for the Bullock's Oriole in British Columbia shows that the mean number of birds on interior routes increased at an average annual rate of 3.9% over the period 1968 through 1993 ($P < 0.05$).

and, on Vancouver Island, bigleaf maple, red alder, and mixed woodlands with Douglas-fir. It also shows a strong preference for shade trees near water such as rivers, marshes, sloughs, farm ponds, and sewage lagoons. It commonly feeds in thickets and bushes around such areas. The oriole also visits suburban gardens and backyards with large trees and bird feeders.

Across southern British Columbia, early spring migrants may appear in late April but most birds arrive during the first half of May. On the coast, early migrants may appear in late April, with the peak movement evident during the second and third weeks of May (Figs. 568 and 569). In the Southern Interior, the first migrants can appear in late April, but most birds arrive in the second week of May, with peak numbers occurring in early June (Fig. 569). The mean date of first arrival for Penticton (n = 19 years) and Vernon (n = 35 years), both in the Okanagan valley, was 12 May (Cannings et al. 1987). In the Southern Interior Mountains and Central Interior, the earliest migrants may occur during the first or second weeks of May, respectively, with peak numbers evident later in the month or in early June.

The southward migration in British Columbia occurs mainly during the first 3 weeks of August, after which the Bullock's Oriole becomes scarce (Figs. 568 and 569). Stragglers may linger into November on the coast and September in the interior. Infrequently, single birds may overwinter. In the Okanagan valley, the latest departure date over an 8-year period in Vernon ranged from 8 to 26 August, with a mean of 18 August (Cannings et al. 1987).

On the coast, the Bullock's Oriole has been recorded regularly between early May and early August; in the interior, it has been recorded regularly between early May and mid-August (Fig. 568).

BREEDING: The Bullock's Oriole breeds across southern portions of the province from southeastern Vancouver Island east to Radium in the Rocky Mountain Trench, and from the

Figure 572. Throughout most of its range in British Columbia, the Bullock's Oriole prefers to nest in riparian woodlands (near Hat Creek, 11 June 1998; R. Wayne Campbell).

Figure 573. In the Okanagan valley, the Bullock's Oriole prefers to nest in trees bordering lakes, rivers, and water channels. Note the old nest suspended near the tip of a branch (north of Osoyoos, 17 April 1999; R. Wayne Campbell).

Figure 574. A male Bullock's Oriole at the nest, which hangs like a pensile pouch and is usually attached to the branch from the top and sides (Creston, 23 June 1993; R. Wayne Campbell).

international boundary north through the Okanagan and Thompson valleys to at least Hanceville and Soda Creek in the Central Interior. On the coast, it breeds only in the Georgia Depression from Sea Island east throughout the Fraser Lowland to Chilliwack, and locally in the Nanaimo Lowland, on southeastern Vancouver Island, from Victoria north to Campbell River. The presence of singing males in the Alberni valley, on southwestern Vancouver Island, suggests that the species may also nest there.

The Bullock's Oriole reaches its highest numbers in summer in the Okanagan valley of the Southern Interior (Fig. 570). An analysis of Breeding Bird Surveys for the Bullock's Oriole in British Columbia shows that the number of birds on interior routes increased at an average annual rate of 3.9% over the period 1968 through 1993 (Fig. 571). Coastal routes contain insufficient data for analysis. Nationally, surveys show an average annual increase of 4.1% over the period 1966 to 1996 (P = 0.03); numbers in continent-wide surveys appear stable for the same period (Sauer et al. 1997).

This species has been recorded breeding at elevations ranging from near sea level to 125 m on the coast and from 280 to 1,120 m in the interior.

In British Columbia, the Bullock's Oriole nests primarily along the edges of riparian woodlands (Fig. 572) or in groves of deciduous trees. Most records (68%; n = 427) were found in human-influenced habitats, including wooded farmlands (67%; n = 289), orchards, gardens, campgrounds, and golf courses. In the Okanagan valley, this oriole nests most commonly in deciduous bottomlands, along treed lakeshores and other waterways (Fig. 573), in deciduous groves, in gardens, and in rangeland copses below about 600 m (Cannings et al. 1987).

On the coast the Bullock's Oriole has been recorded breeding from 30 May (calculated) to 15 August; in the interior, it has been recorded breeding from 14 May (calculated) to 31 July (Fig. 568).

Nests: Most nests in British Columbia were placed in trees (96%; *n* = 463), with some found in tall shrubs and small saplings (4%). Deciduous trees and shrubs accounted for 91% of nest sites. Of these, *Populus* spp. (39%; typically black cottonwood, trembling aspen, or ornamental poplars), *Betula* spp. (22%; water birch, paper birch, or ornamental birches), *Salix* spp. (6%; willows), and *Acer* spp. (6%; maples) were preferred. The remaining 9% of nests were located in coniferous species such as Douglas-fir, ponderosa pine, and western larch.

The Bullock's Oriole builds a pendent, pouch-shaped nest, suspended from a twig or branch fork and fastened at the sides as well as at the top (Fig. 574). The nest is usually situated near the end of a branch and often near the top of a tree or tall shrub. Occasionally nests overhang water. Rarely, the oriole builds its nest in a horizontal position (Fig. 575). A new nest is usually built each year, but occasionally a nest from a previous year will be modified and reused.

Most nests were woven from mammalian hair or wool (57%; *n* = 272) and often included grasses and plant fibres (38%) as well as miscellaneous materials such as string and twine. Wool, string, twine, and horse hair were the most commonly reported nest linings. The heights of 441 nests ranged from 0.4 to 30 m, with 68% between 4.5 and 9.0 m.

Eggs: Dates for 148 clutches ranged from 17 May to 12 July, with 55% recorded between 4 June and 21 June. Calculated dates indicate that eggs can occur as early as 14 May. Sizes of 46 clutches ranged from 1 to 6 eggs (1E-1, 2E-2, 3E-7, 4E-19, 5E-14, 6E-3), with 72% having 4 or 5 eggs (Fig. 576). The incubation period is 11 days (Rising and Williams 1999).

Young: Dates for 476 broods ranged from 26 May to 15 August, with 51% recorded between 23 June and 3 July. Sizes of 103 broods ranged from 1 to 6 young (1Y-2, 2Y-19, 3Y-53, 4Y-23, 5Y-4, 6Y-2), with 51% having 3 young. The nestling period is 14 days (Rising and Williams 1999). Young leaving the nest in the Okanagan valley after about 15 July probably represent replacement clutches (Cannings et al. 1987).

Brown-headed Cowbird Parasitism: Cowbird parasitism was not found in British Columbia in 125 nests recorded with eggs or young. There were, however, 2 records of this oriole feeding fledgling cowbirds.

The Bullock's Oriole is a known ejector of Brown-headed Cowbird eggs (Rothstein 1975). Friedmann and Kiff (1985) note only 4 occurrences of cowbird parasitism for this species, all from the western United States. Rising and Williams (1999), however, reported that 12 of 90 Bullock's Oriole nests checked in western Washington were parasitized by the Brown-headed Cowbird.

Nest Success: On the coast, the only nest that was found with eggs and followed to a known fate was successful in fledging at least 1 young. In the interior, no nests met these criteria but several nests were observed to lose young or completely fail due to nest predation. Observed or suspected nest predators included red squirrel (6 nests), American Kestrel (2), Steller's Jay (1), and American Crow (1).

REMARKS: The Bullock's Oriole was considered conspecific with the Baltimore Oriole by the American Ornithologists' Union (1973, 1983), but recent studies by Corbin et al. (1979) and Rohwer and Manning (1990) show that the 2 groups are

Figure 575. Occasionally the Bullock's Oriole builds its nest in a horizontal position straddling a branch (Oliver, July 1991; Mark Nyhof).

Figure 576. Nest and eggs of the Bullock's Oriole (Richter Pass, 12 June 1992: R. Wayne Campbell).

distinct species. Two subspecies are recognized; *I. b. bullockii,* the nominate race, occurs in British Columbia.

Early breeding records for Vancouver Island in 1964 (Taylor 1984), Revelstoke in 1971 (E. Callin pers. comm.), and Hope in 1973 (R. Ferguson pers. comm.) lacked substantiating information and were not included in this account.

For additional information on the life history, conservation, and identification of the Bullock's Oriole, see Bent (1958), Rohwer and Johnson (1992), Bock et al. (1993), Lee and Birch (1998), and Rising and Williams (1999).

NOTEWORTHY RECORDS

Spring: Coastal – Victoria 28 Apr 1993-4, 14 May 1983-12; Saanich 19 May 1984-nest completed; Florencia Bay (Pacific Rim National Park) 12 May 1984-1; Port Alberni 31 May 1992-1; Ladner 11 May 1980-3, 30 May 1972-female building nest; Sumas Mountain 13 May 1982-3; Yarrow 31 May 1981-nesting pair; Reifel Island 5 May 1990-1; Lulu Island 1 May 1990-2 males, spring arrival; Port Alberni 27 May 1996-1 (Bowling 1996c); Burnaby Lake 11 May 1980-1 adult male; Seabird Island 11 May 1984-1; Harrison Hot Springs 27 May 1986-3; Pitt Meadows 21 May 1995-10; Sechelt 27 May 1997-1 (Bowling 1997c); Courtenay 15 May 1988-1; Campbell River 16 Apr 1980-2; Fry Lake 18 May 1981-1. **Interior** – Osoyoos 28 Apr 1962-1 male (UBC 12002), 2 May 1985-1, spring arrival, 17 May 1968-incubating eggs; Goat River (Creston) 22 May 1947-1 (Munro 1957); Trail 22 May 1969-1; Christina Lake 21 May 1980-9; Keremeos 13 May 1975-2 males; Duck Lake (Creston) 8 May 1981-1 along Kootenay River; Vaseux Lake 7 May 1980-4 males and 2 females, 13 May 1976-13 on count; Princeton 13 May 1975-2; South Slocan 6 May 1973-1, spring arrival; Naramata 4 May 1966-1, spring arrival, 30 May 1975-adult chasing Brown-headed Cowbird from nest; Summerland 23 Apr 1991-1, very early spring arrival; Hosmer 31 May 1984-1 male at feeder; Balfour to Waneta 16 May 1981-6 on count; Kimberley 18 May 1981-1; Okanagan Landing 24 May 1922-latest spring arrival in 16 years (Munro and Cowan 1947); 1.6 km w of Coldstream 24 May 1958-5 eggs; Columbia Lake 25 May 1946-2 (Johnstone 1949); Lytton to Spences Bridge 23 May 1965-3 on count; Windermere Lake 29 May 1983-1 pair; Kamloops 15 May 1982-3; Cache Creek 31 May 1975-4; Shuswap Lake Park 26 May 1980-male; Revelstoke 19 May 1973-1, spring arrival; Little Fort 29 May 1971-1; Darfield 25 May 1985-1 male; Kleena Kleene 8 Mar 1948-1 (Paul 1959); Williams Lake 9 May 1987-1 at feeder; Chilcotin River (Riske Creek) 19 May 1985-1, 23 May 1984-4 at Wineglass Ranch; Williams Lake 9 May 1987-1, spring arrival.

Summer: Coastal – Victoria 3 Jun 1972-1 (Tatum 1972); Saanich 5 Jul 1981-adults feeding nestlings; Mount Newton 24 Aug 1985-3; Carmanah Point 26 Aug 1952-1 (Irving 1953; RBCM 10057); Sidney 11 Jun 1991-2 nestlings; Cowichan Bay 24 Jun 1998-adults feeding nestlings; Ladner 4 Jun 1972-3 pairs, at least 2 nests (BC Photo 261); Richmond 1 Jun 1984-eggs, 12 Jun 1987-adult attending nestlings; White Rock 5 Jul 1984-1 fledgling; Cloverdale 7 Aug 1971-3 nestlings; Surrey 18 Jun 1963-4 eggs; Matsqui 28 Jul 1984-male feeding 3 fledglings; Chilliwack 11 Jun 1928-1 (RBCM 6595); Somass River 6 Jun 1993-1; Qualicum Beach 30 Jun 1980-1 (BC Photo 549); Courtenay 30 Jun 1978-3 nestlings, 15 Aug 1972-young left nest (Davidson 1972a); Cranberry Lake 7 Jul 1979-3 nestlings, nearly fledged; Campbell River 7 Jun 1975-1; Pemberton 12 Jul 1980-1; Pemberton Meadows 5 Jun 1983-1. **Interior** – Richter Pass 7 Jun 1975-6 eggs; Chopaka 14 Jul 1969-adults feeding nestlings; Osoyoos 1 Jun 1969-6 eggs; Bridesville 20 Jun 1989-pair at nest; Creston 27 Jun 1983-nestlings; Oliver 27 Jun 1970-adults feeding nestlings; Duck Lake (Creston) 6 Jun 1981-nest under construction; Vaseux Lake 23 Jun 1968-nestlings, 26 Aug 1975-1 adult female; Okanagan Landing 26 Aug 1916-1 (Munro and Cowan 1947); Castlegar 24 Jun 1993-pair feeding nestlings; Okanagan Falls 26 Jun 1981-unspecified number of young in nest; South Slocan 20 Jun 1976-female on nest; Penticton 25 Jun 1968-nest raided by American Kestrel, 4 Jul 1965-2 nestlings leaving nest, 15 Aug 1976-adult feeding fledged Brown-headed Cowbird; Naramata 5 Jul 1963-4 young in nest, 31 Jul 1967-2 young; Merritt 1 Jun 1986-fledglings leaving nest; Nicola 21 Jun 1980-female at nest; Kalamalka Lake 13 Jul 1978-last fledgling leaving nest; Vernon 1 Jul 1983-adult feeding fledged Brown-headed Cowbird at Commonage; Grindrod 1 Jun 1949-1 (UBC 7013); Radium 26 Jun 1998-adults feeding nestlings; Kamloops 15 Jul 1986-fledglings leaving nest, 30 Aug 1986-1; Ashcroft 9 Jun 1892-1 (ANS 30997); Chase 15 Jul 1970-3 fledglings; Kamloops 11 Jul 1963-6 nestlings; n Clinton 8 Jul 1990-3 eggs, 20 Jul 1990-3 nestlings; Lillooet 1 Aug 1916-1 (NMC 9654); McMurdo 20 Jun 1993-1 male; Moberly Marsh (Golden) 12 Jun 1995-1 male; Clearwater 16 Jun 1971-1; Alkali Lake (Cariboo) 10 Jun 1973-female building nest; Hanceville 1 Jul 1980-adults feeding nestlings; Chilcotin River (Riske Creek) 27 Jun 1985-nestlings at Wineglass Ranch; Dog Creek 16 Jun 1968-2; Kleena Kleene 5 Jun 1949-2; Williams Lake 3 Jun 1972-female on nest; 27 Jun 1978-male feeding young in nest; Soda Creek 23 Jun 1977-adults feeding nestlings; 24 km s Prince George 8 Jul 1983-3 fledged young with adults.

Breeding Bird Surveys: Coastal – Recorded from 8 of 27 routes and on 14% of all surveys. Maxima: Chilliwack 25 Jun 1992-15; Albion 5 Jul 1979-9; Seabird 6 Jul 1991-4. **Interior** – Recorded from 27 of 73 routes and on 27% of all surveys. Maxima: Lavington 15 Jun 1981-19; Oliver 30 Jun 1980-13; Mabel Lake 27 Jun 1990-9.

Autumn: Interior – Riske Creek 11 Sep 1986-1. **Coastal** – McInnes Island 5 Nov 1963-1; Mahatta River 12 Nov 1974-1; North Alouette River 2 Oct 1971-1; Vancouver 15 Oct 1983-1 male; Chilliwack 2 Nov 1985-1 (BC Photo 1059); Vancouver 7 Sep 1984-1 at Jericho Park; Sea Island 30 Sep 1978-1; Tofino 16 to 23 Oct 1997-1 adult female, 9 Nov 1978-1, 17 Nov 1983-1, 18 Nov 1982-2; Chesterman Beach (Tofino) 10 Sep 1982-1; Langley 14 Sep 1982-1; Shawnigan Lake 17 Oct 1978-1 at feeder; Victoria 16 Nov 1976-1 (Shepard 1977), 24 to 26 Nov 1984-1 male.

Winter: Interior – Kelowna 14 Dec 1991 to 18 Jan 1992-1; Westside 16 Dec 1995-1 (Bowling 1996b); Okanagan Mission 6 Feb 1968-1 male (Cannings et al. 1987); Naramata 1 Dec 1973 to 4 Jan 1974-1 (Shepard 1974; BC Photo 331). **Coastal** – Ladner 14-29 Jan 1978-1 (BC Photo 519).

Christmas Bird Counts: Interior – Recorded from 1 of 27 localities and on less than 1% of all counts. Maximum: Kelowna 14 Dec 1991-1. **Coastal** – Not recorded.

Brambling

BRAM

Fringilla montifringilla (Linnaeus)

RANGE: In North America, breeds in the western Aleutian Islands, Alaska (Sykes and Sonneborn 1998). Casual in migration and winter from the western Aleutian Islands, northern and southeastern Alaska, southern British Columbia, central Alberta, Saskatchewan, Manitoba, and North Dakota south to northwestern California, Nevada, Utah, and Colorado and along the west coast of Washington and Oregon, and east to Minnesota, Michigan, Ontario, Quebec, Pennsylvania, New York, Massachusetts, New Jersey, and Nova Scotia.

Native to northern Eurasia. Breeds from northern Scandinavia, northern Russia, and northern Siberia south to southern Scandinavia, central Russia, Transbaicalia, northern Amurland, Anadyrland, Kamchatka, and the Sea of Okhotsk. Winters from the British Isles and southern portions of the breeding range south to the Mediterranean region, northern Africa, the Near East, Iran, northwestern India, Tibet, China, Taiwan, and Japan, and casually to the Faroe Islands, Iceland, Madeira, and the Philippines.

Figure 577. The Brambling is now a regular spring and autumn migrant in southeastern Alaska. In the future, British Columbia birdwatchers can expect more records of this European finch at their feeders (© D. Tipling/VIREO).

STATUS: On the coast, *very rare* autumn and winter vagrant in the Georgia Depression Ecoprovince; in the Coast and Mountains Ecoprovince, *casual* spring vagrant on Western Vancouver Island; *casual* spring vagrant, *accidental* in summer, and *very rare* autumn and winter vagrant on the Queen Charlotte Islands; *accidental* on the Northern Mainland Coast and absent from the Southern Mainland Coast.

In the interior, *very rare* vagrant in the Southern Interior and Southern Interior Mountains ecoprovinces; *casual* in the

Figure 578. In British Columbia, the Brambling is most often noticed at feeders but is also found with sparrows, juncos, and Snow Buntings foraging in weedy and grassy areas in fields, along fence rows, and along margins of wetlands and lakes (Myers Lake, 12 November 1998; R. Wayne Campbell).

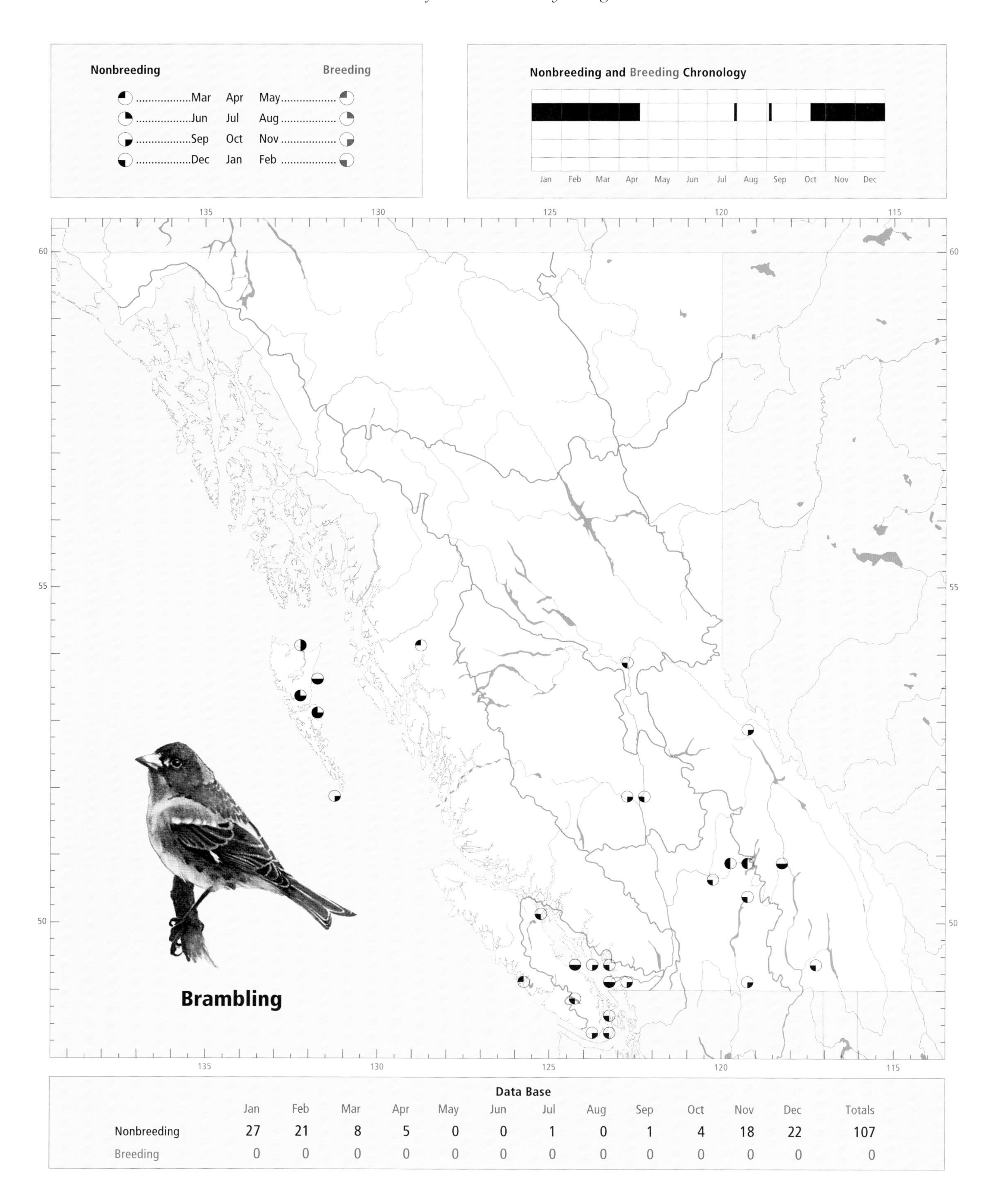

Data Base

	Jan	Feb	Mar	Apr	May	Jun	Jul	Aug	Sep	Oct	Nov	Dec	Totals
Nonbreeding	27	21	8	5	0	0	1	0	1	4	18	22	107
Breeding	0	0	0	0	0	0	0	0	0	0	0	0	0

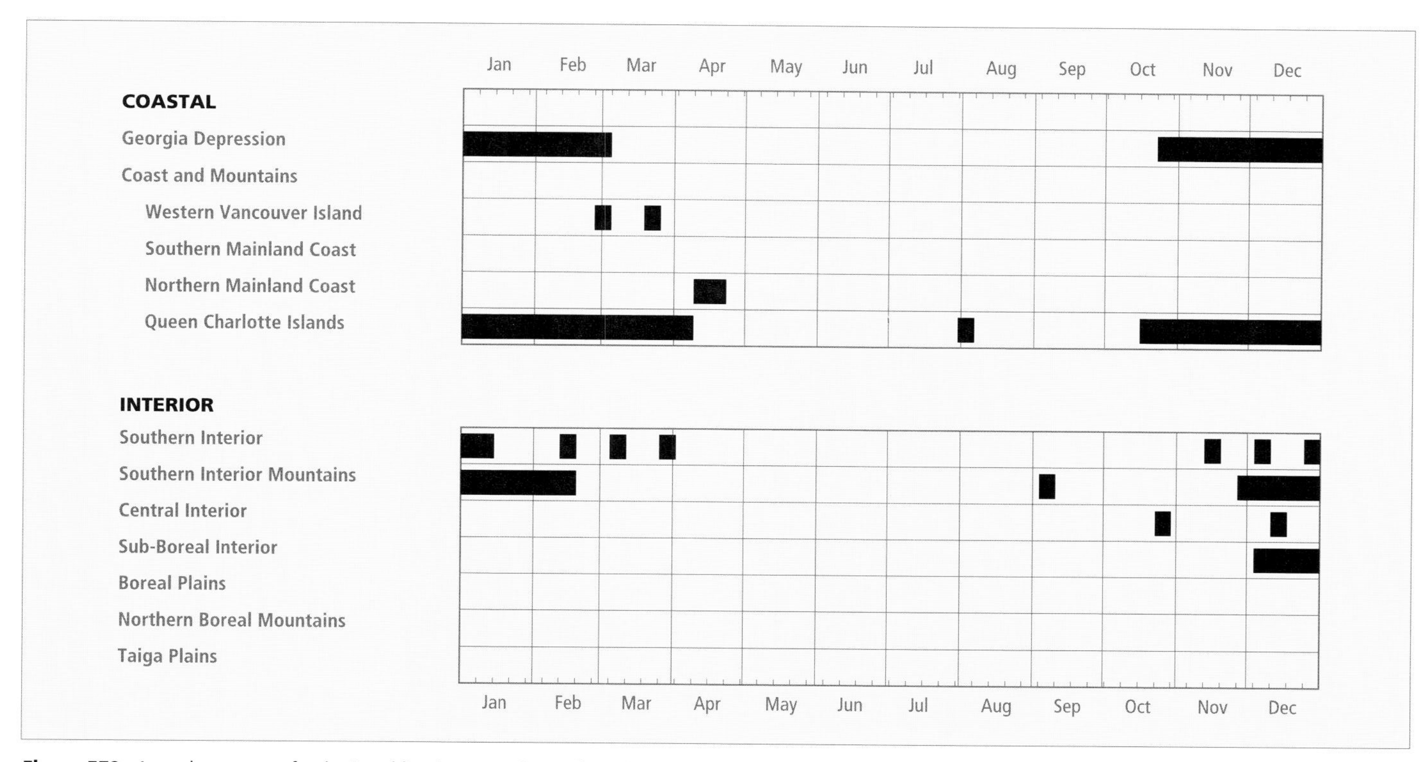

Figure 579. Annual occurrence for the Brambling in ecoprovinces of British Columbia. Records are shown for the week in which they occurred.

Central Interior Ecoprovince; *accidental* in the Sub-Boreal Interior Ecoprovince; not recorded elsewhere in the province.

CHANGE IN STATUS: The Brambling (Fig. 577) was first reported in the province in February 1971, when a single bird was found at a feeder in Tlell, on the Queen Charlotte Islands (Crowell and Nehls 1972b; Campbell 1974). Nearly a year later, another bird visited Tlell, and since then the Brambling has been found on 23 occasions on the Queen Charlotte Islands through 1997.

The Brambling was reported annually in British Columbia between 1980 and 1995, but there are no records for 1996 and 1997.

The casual occurrence of the Brambling in western North America was first noted in the western Aleutian Islands in 1914 (Hanna 1916), where this species is now considered a regular spring and autumn migrant with 150 to 200 birds some years (Gibson 1981; Sykes and Sonneborn 1998). The first Brambling south of Alaska was found in Oregon in 1968 (Banks 1970), and it continues to occur there casually from October through February (Gilligan et al. 1994). In Washington state, there are only 4 records through 1989 (Mattocks and Hunn 1982a, 1982b; Tweit 1989). Since 1984, Bramblings have wandered as far south as California from November through February. Although the Brambling has expanded its range and

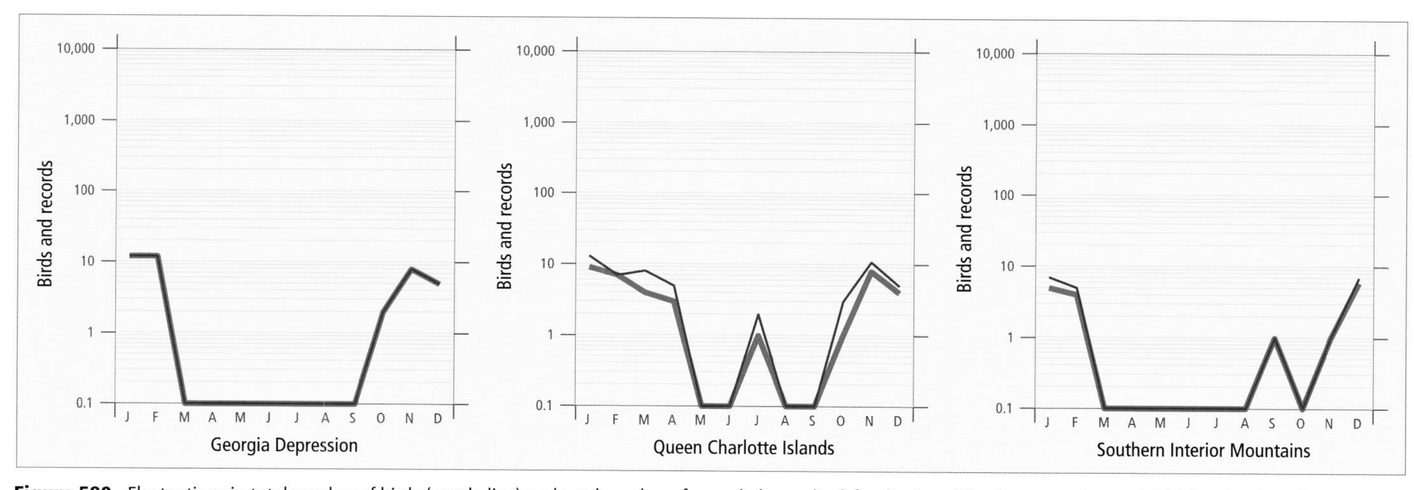

Figure 580. Fluctuations in total number of birds (purple line) and total number of records (green line) for the Brambling in ecoprovinces of British Columbia. Christmas Bird Counts have been excluded.

occurrence south into western North America, it still remains a vagrant there.

OCCURRENCE: The Brambling is a rarity in western North America outside of Alaska. In British Columbia, it has been found on Vancouver Island from Campbell River south to Tofino and Sooke, and on the Queen Charlotte Islands from Masset to Cape St. James. On the southern mainland coast, it has been recorded from the Fraser River delta and Vancouver north to Langdale on the Sunshine Coast.

In the interior, the Brambling has been recorded from Myers Lake to Kamloops and east to Shuswap Lake in the Southern Interior, from Nelson and Revelstoke north to Valemount in the Southern Interior Mountains, and in the Central Interior at Williams Lake and Alkali Lake (Cariboo). There is 1 record from Prince George in the Sub-Boreal Interior.

The Brambling has been recorded from near sea level to 800 m elevation. On the coast, it is most often seen at bird-feeding stations associating with Dark-eyed Juncos, House Finches, and House Sparrows. In the interior, it also frequents feeding stations as well as willow shrubs and weedy fields, shrubby and weedy lake margins (Fig. 578), and weedy patches in agricultural areas. One bird was found in a mixed deciduous woodland adjacent to a beaver pond. In other parts of North America, the Brambling frequents mixed deciduous-coniferous forests, forest edges, birch and willow scrub, and weedy fields, where it forages on the ground (American Ornithologists' Union 1983).

There is no regular movement of the Brambling into British Columbia. Most occurrences (85%; n = 78) have been recorded between 3 November and 26 February (Figs. 579 and 580). There is 1 summer record from the northern Queen Charlotte Islands.

REMARKS: There are no recognized subspecies of the Brambling.

For a summary of the nonbreeding ecology of the Brambling in northern Europe, see Eriksson (1970).

NOTEWORTHY RECORDS

Spring: Coastal – Tofino 1 Mar 1984-1 (BC Photo 903); Chesterman Beach 22 to 25 Mar 1984-1 in winter plumage feeding with Dark-eyed Juncos; Sandspit 10 Mar 1992-5 in red alders, 2 Apr 1994-3 (Bowling 1994a), 3 Apr 1986-1; Queen Charlotte City 5 Mar 1990-1 (Siddle 1990b); Kitimat 10 to 22 Apr 1992-1. **Interior** – Chase 7 Mar 1987-1 (Rogers 1987); Scotch Creek 30 Mar 1987-1 (BC Photo 1188).

Summer: Coastal – Masset 30 Jul 1994-2 at feeder (Bowling 1994b). **Interior** – No records.

Breeding Bird Surveys: Not recorded.

Autumn: Interior – Valemount 4 Sep 1994-1 (Bowling 1995a); Chilcotin River (Riske Creek) 25 Oct to 6 Nov 1991-1 female at Wineglass Ranch; Revelstoke 26 Nov 1991-1 (Siddle 1992a); Myers Lake 12 Nov 1998-1 with sparrows and juncos foraging along lakeshore (Fig. 578). **Coastal** – Delkatla Inlet (Masset) 17 Oct 1991-3 (Siddle 1992a); Masset 20 Nov 1995-1 with flock of juncos, Tlell 9 Nov 1972-1 photographed; Queen Charlotte City 20 to 30 Nov 1983-1 or 2 at feeder (Mattocks 1984; Campbell 1984b), 5 Nov 1989 to 5 Mar-1; Sandspit 3 Nov 1993-3 (Siddle 1994a); Cape St. James 8 Nov 1981-1 moulting male killed by cat (Campbell 1982b); Parksville 28 Nov 1985-1 feeding with Dark-eyed Juncos; Langdale 29 Oct to 4 Nov 1984-1 (Hunn and Mattocks 1985); Reifel Island 7 and 8 Nov 1971-1 (Campbell et al. 1972b; BC Photo 192); Sooke 20 Nov 1983-1 (Campbell 1984b; BC Photo 909).

Winter: Interior – Prince George 3 to 28 Dec 1998-1; nr Alkali Lake (Cariboo) 14 Dec 1992-1 male foraging with Common Redpolls; Kamloops 4 to 6 Dec 1983-1 (Rogers 1984); Scotch Creek 26 Dec 1985-1 (Johnston and McEwen 1986), 10 Jan 1986-1 (Rogers 1986); Shuswap Lake 7 Jan 1986-1 (BC Photo 1101); Revelstoke 17 Dec 1995-1 male (Bowling 1996b), 27 Dec 1989-1 at feeder, 1 Jan 1990-2 stayed all winter (Siddle 1990a); Vernon 12 Feb 1989-1 (Rogers 1989); Nelson 12 Dec 1991-1, 7 Jan 1992-1 male (Siddle 1992b), 18 Feb 1992 found dead below window. **Coastal** – Tlell 5 Feb 1972-1, 7 Feb to early Mar 1971-1 feeding with Dark-eyed Juncos (Campbell 1974; BC Photo 219); Queen Charlotte City 1 to 31 Dec 1983-2 at feeder, 1 to 7 Jan 1984-1 coming to feeder (Mattocks 1984), Jan 1990-1 stayed all winter (Siddle 1990a); Campbell River 8 Feb 1984-1 (Campbell 1984b); Parksville 19 Dec 1985-1 feeding with Dark-eyed Juncos; 10 km w Port Alberni 16 Jan 1998-1 along Beaver Creek Rd (O'Shaughnessy 1998); Vancouver 6 Jan 1991-1 (Siddle 1991b), 19 Jan to 25 Feb 1984-1 on 45th Ave, 25 Feb 1984-1; Sea Island 10 Dec 1990-1; Westham Island 4 Feb 1992-1 (Siddle 1993b), 31 Dec 1992-1 (Siddle 1993b); North Saanich 14 Dec 1990-1, 10 Feb 1991-1 female feeding on ground with Dark-eyed Juncos; Colwood 15 Dec 1990 to 26 Feb 1991-1 at feeder (Siddle 1991b).

Christmas Bird Counts: Interior – Not recorded. **Coastal** – Skidegate 16 Dec 1989-1.

Gray-crowned Rosy-Finch

GCRF

Leucosticte tephrocotis (Swainson)

RANGE: Breeds from western and north-central Alaska and central Yukon south through southern Alaska, British Columbia, and southwestern Alberta to eastern Oregon, central California, New Mexico, and Colorado. Winters from the Aleutian Islands and southern Alaska throughout the rest of its breeding range, casually east to Manitoba and Iowa.

STATUS: On the coast, an irregular and *very rare* local summer visitant and *uncommon* to *fairly common* (sometimes *very common*) migrant and winter visitant in the Georgia Depression Ecoprovince; in the Coast and Mountains Ecoprovince, locally *common* migrant, irregular *uncommon* summer visitant and *fairly common* winter visitant on the Southern Mainland Coast, *casual* vagrant on Western Vancouver Island and the Queen Charlotte Islands, and local *rare* resident on the Northern Mainland Coast.

In the interior, *very rare* summer visitant and locally *rare* to *common* (sometimes *abundant*) migrant and winter visitant in the Southern Interior Ecoprovince; in the Southern Interior Mountains Ecoprovince, locally an *uncommon* summer visitant and *very rare* migrant and winter visitant in the west Kootenay and *very common* spring migrant, locally *uncommon* summer visitant, and *common* autumn migrant and winter visitant in the northern east Kootenay; in the Central Interior Ecoprovince, *fairly common* summer visitant and *common* (sometimes locally *abundant*) migrant and winter visitant; in the Sub-Boreal Interior Ecoprovince, *very rare* summer visitant and locally *rare* to *uncommon* migrant and winter visitant; in the Northern Boreal Mountains Ecoprovince, *common* migrant and *uncommon* to *fairly common* summer visitant; *casual* transient in the Boreal Plains Ecoprovince and *accidental* in the Taiga Plains Ecoprovince.

Breeds.

NONBREEDING: The Gray-crowned Rosy-Finch (Fig. 581) is widely distributed during some nonbreeding seasons in interior British Columbia, especially during late summer and autumn. In most parts of the interior, however, it occurs irregularly at lower elevations and may be locally scarce or abundant during different years. The Gray-crowned Rosy-Finch occurs most commonly in the southeastern portions of the province from the international boundary north to about Kinbasket Lake and west to the Okanagan valley.

Migrants and wintering populations occur north through southern parts of the Cariboo and Chilcotin areas, but wintering birds are scarce north of Williams Lake. It is apparent that northern breeding populations move to southern parts of the province for the winter, as the northernmost winter record from the interior is from Smithers. In the Boreal Plains, the Gray-crowned Rosy-Finch occurs only as an occasional transient. In the Northern Boreal Mountains, it occurs during nonbreeding seasons as a late spring and late summer migrant.

Figure 581. Most records for the Gray-crowned Rosy-Finch in British Columbia that suggest breeding are of fledged young or juveniles (shown) in high alpine habitats (Cathedral Lakes, 7 August 1989; Tim Zurowski).

On the coast, migrant Gray-crowned Rosy-Finches occur mainly in the Fraser Lowland of the Georgia Depression and on Southern Mainland Coast of the Coast and Mountains, less regularly in the Nanaimo Lowland on southeastern Vancouver Island and on the Northern Mainland Coast, and rarely on Western Vancouver Island. It has been reported only twice from the Queen Charlotte Islands. There are numerous winter records of small flocks in the Georgia Depression, but elsewhere on the coast wintering birds are very rare. The northernmost coastal winter records are from near Hazelton.

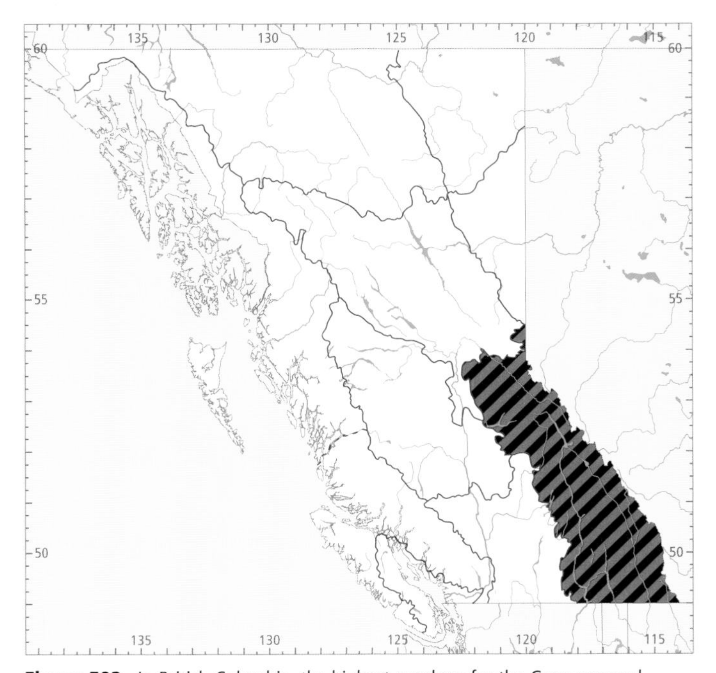

Figure 582. In British Columbia, the highest numbers for the Gray-crowned Rosy-Finch in both winter (black) and summer (red) occur in the Southern Interior Mountains Ecoprovince.

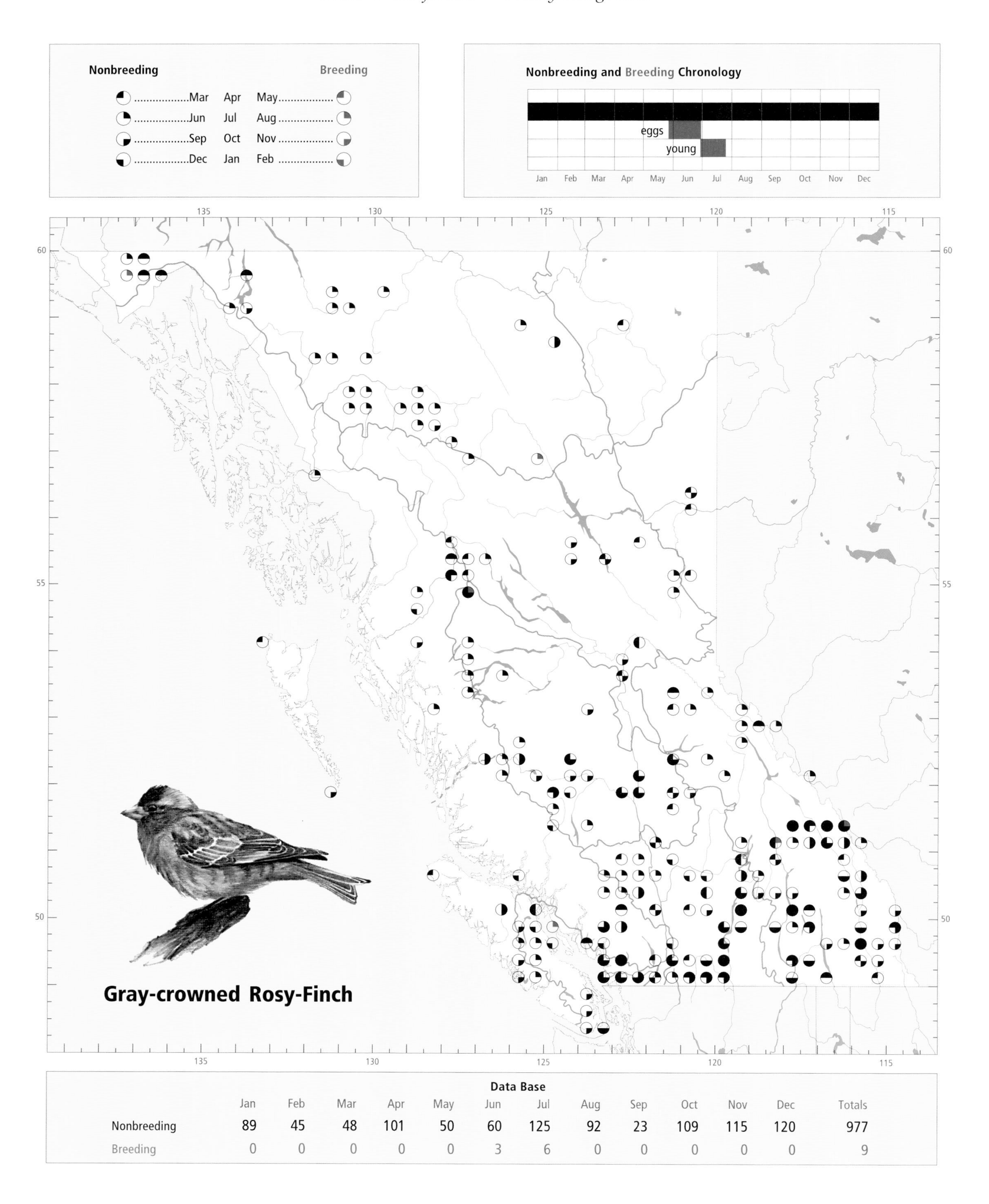

Data Base

	Jan	Feb	Mar	Apr	May	Jun	Jul	Aug	Sep	Oct	Nov	Dec	Totals
Nonbreeding	89	45	48	101	50	60	125	92	23	109	115	120	977
Breeding	0	0	0	0	0	3	6	0	0	0	0	0	9

Figure 583. In autumn, the weedy edges of roads provide suitable foraging areas for the Gray-crowned Rosy-Finch throughout the interior of British Columbia (Hat Creek, 4 November 1996; R. Wayne Campbell).

Figure 584. In the Okanagan valley of British Columbia, where the Gray-crowned Rosy-Finch is an irregular visitor during the nonbreeding seasons, shrub-steppe habitats with windswept slopes provide foraging areas during winter (White Lake, Okanagan Falls, 12 January 1996; R. Wayne Campbell).

The Gray-crowned Rosy-Finch has been reported at elevations from near sea level to 1,524 m on the coast and from the valley bottoms at 280 m to over 2,500 m in the interior. The species reaches its highest numbers in winter in the Southern Interior Mountains (Fig. 582).

Nonbreeding habitat includes a wide range of open plant communities, as this finch is usually found on the ground. It frequents alpine and subalpine areas after nesting, where large flocks gather and roam the mountaintops before snowstorms force them to lower elevations. In late summer and early autumn, the Gray-crowned Rosy-Finch frequents tundra, scree slopes, rocky mountain slopes, alpine wetlands, subalpine meadows, snow fields, and avalanche chutes – habitats similar to those used during the nesting season.

In the interior, it occurs later in the autumn at lower elevations, along roadsides (Fig. 583) and lakeshores and in clearcuts, especially during storms. In winter, flocks congregate mainly in the valley bottoms, where they frequent weedy fields, stubble fields, grass fields, roadsides, railway rights-of-way, gravel pits, and even garbage dumps. In drier areas of the southern portions of the interior, the Gray-crowned Rosy-Finch occurs in natural grasslands during migration and in winter, including shrub-steppe habitats with big sagebrush (Fig. 584). Cannings et al. (1987) note that "flocks wheel about constantly, landing on windswept slopes where seeds are plentiful." In autumn, large concentrations sometimes occur in alpine areas (Fig. 585) and along railways, where birds feed on spilled grain, an event most often reported from near Field and Revelstoke in the Southern Interior Mountains. It also congregates in livestock feedlots. In forested areas of the southern Rocky Mountains, mine slag piles, which are open and sparsely vegetated, attract some flocks. This finch rarely appears at backyard feeders.

On the coast, the Gray-crowned Rosy-Finch occurs in winter on rocky beaches and beaches with drift logs, on dirt banks, along causeways and dykes, in open weedy gardens, and on grassy and gravel playing fields. At higher elevations on Vancouver Island during August and September, Ogle and Martin (1997) found it along the edges of glaciers and rock faces in alpine bowls but not in other alpine or subalpine habitats.

This finch rarely perches in trees, although it has been reported perching briefly in trembling aspens, alders, birches, and spruce near the edges of open areas. The Gray-crowned Rosy-Finch is known to roost in caves, mine shafts, buildings, and even Cliff Swallow nests (Leffingwell and Leffingwell 1931). In British Columbia, it has been observed entering crevices and caves in limestone cliffs along the Fraser River west of Williams Lake in winter (A. Roberts pers. comm.).

After the young have fledged, family groups form flocks of up to a few hundred birds and roam through the high country (Figs. 585 and 586). Migratory movements are not well known, but northern populations, at least, shift southward, mainly in September. Migrants follow alpine routes along mountain ranges (Cooper 1994). Although some birds move to lower elevations by early August, most stay in alpine areas until snowfalls begin. Near Field, an immature bird was recorded at low elevation as early as 3 August (Wade 1977). In the Southern Interior and Southern Interior Mountains, the main movements to lower elevations occur from late October to early November. In the Central Interior, the major autumn movement occurs from mid-October to mid-November.

On the coast, as in the southern portions of the interior, autumn migrants arrive at lower elevations mainly between late October and early November (Figs. 587 and 588). Recent information shows that migrants occur on alpine peaks in the Leeward Island Mountains of Vancouver Island in August and stay through at least early October. The origin of these birds is unknown; since breeding has not been documented on Vancouver Island, these birds likely come from mainland coastal mountains.

Throughout the British Columbia wintering range, large flocks remain together through early spring (Fig. 586), then most birds move to higher elevations in April and early May (Figs. 587 and 588). Surprisingly, the largest flocks we have documented (up to 3,000 birds near Field, in Yoho National

Figure 585. Part of a premigratory flock of 500 Gray-crowned Rosy-Finches swirling over a pass at 2,470 m in the Cayoosh Range in the Southern Interior Ecoprovince of British Columbia (headwaters of Melvin Creek, 25 September 1992; John M. Cooper).

Park) occur from late April to late June, a period when most birds should be on the nesting grounds. These large flocks were recorded in years when late spring snowstorms drove large numbers of prenesting birds down from higher elevations.

Spring migrants leave low elevations in the Southern Interior Mountains in April, with the occasional straggler remaining through early May (Figs. 587 and 588). In the Southern Interior, spring migrants occur regularly at lower elevations through May, and even into June if the weather is poor.

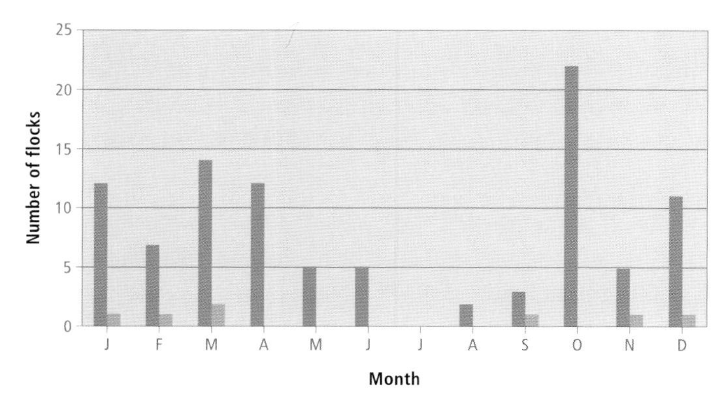

Figure 586. Combined monthly distribution of flocks of 100+ Gray-crowned Rosy-Finches for coastal (light bar) and interior (dark bar) British Columbia.

In the Central Interior, spring migrants have left the lower elevations by the first week of May. Near Kleena Kleene, Paul (1964b) noted an unusually large spring migration in the second week of April, when several thousand birds moved through. In the Boreal Plains, spring transients have appeared from mid-April to mid-May.

Although this finch does not undergo irruptions like some other finches such as crossbills, it may be unusually abundant in some years compared with others. Exceptionally high numbers in winter occurred in the Southern Interior in 1963, 1969, and 1981, and in the Southern Interior Mountains in 1975, 1977, 1983, and 1984.

The Gray-crowned Rosy-Finch seldom occurs in mixed-species flocks, but a few may move with groups of Horned Larks, Common Redpolls, American Pipits, Snow Buntings, Savannah Sparrows, or White-crowned Sparrows.

On the coast and in the southern portion of the interior, the Gray-crowned Rosy-Finch has been recorded irregularly throughout the year (Fig. 587).

BREEDING: The breeding distribution of the Gray-crowned Rosy-Finch in British Columbia is poorly known. Few nests have been found because of its remote alpine habitat and use of rock cliffs for nesting. The timing of breeding may be heavily influenced by weather, as this finch nests in habitats with typically severe climatic conditions. The breeding habitat per se may also vary somewhat from the summer range. This species forages over a very large area and is often seen foraging at locations where it would never nest.

The Gray-crowned Rosy-Finch likely has a widespread breeding distribution in alpine areas of interior British Columbia. Breeding has been documented in the southern Rocky Mountains north to Mount Robson Park, in Cathedral Park in the Okanagan Range, near Tenquille Lake in the Leeward Pacific Ranges, at Mount Revelstoke and Glacier national parks in the Rocky Mountains, near Barkerville in the Quesnel Highland, in the southern Cassiar Mountains, at Spatsizi Plateau Wilderness Park, and near the Tatshenshini River.

The Gray-crowned Rosy-Finch likely breeds in most mountainous regions in the interior where there are substantial amounts of alpine and rocky habitat. For example, although it breeds in the mountains above the Ashnola River, west of Keremeos, it is not known to breed in the smaller alpine areas above the Okanagan valley (Cannings et al. 1987).

On the coast, the Gray-crowned Rosy-Finch breeds locally in mainland coastal mountains. The only confirmed records are from Garibaldi Park northeast of Squamish, near Beta Lake east of Powell River, and from the Rocher Déboulé Range near Smithers. The Gray-crowned Rosy-Finch likely breeds locally from Garibaldi Park and Golden Ears Park north along the Coast Mountains to the Rocher Déboulé Range, then probably in higher numbers through the Boundary Ranges to the British Columbia–Yukon border. Summer records also suggest the possibility of breeding on mountains east of the Sunshine Coast and near Hope. Although there are numerous records for central Vancouver Island from June through autumn, breeding has not been documented. The species is not known to occur in summer on any other coastal islands.

In British Columbia, the Gray-crowned Rosy-Finch reaches its highest numbers in summer in the Southern Interior Mountains (Fig. 582). Within this ecoprovince, it has been described as a common summer resident in Kootenay National Park (Poll et al. 1984). Breeding Bird Surveys for the period 1966 through 1996 in British Columbia, in Canada, and across North America contain insufficient data on this species for analysis, as survey routes seldom enter its breeding habitat (this study; Sauer et al. 1997).

The Gray-crowned Rosy-Finch nests at higher elevations than any other songbird in British Columbia. Very few nests have been found. Nests have been reported between 1,500 and 2,010 m elevation on the coast and from 1,500 to 2,550 m in the interior, but the Gray-crowned Rosy-Finch probably nests up to 2,500 m in the southern portions of the interior. In general, breeding probably occurs in any alpine areas that have rock faces with adjacent tundra, or that have sparse vegetation. Typical summer habitat includes snow patches, alpine meadows, tundra, scree slopes, rocky alpine outcroppings, avalanche chutes through subalpine coniferous stands, shorelines of alpine lakes, and streambanks flowing through tundra (Fig. 589).

At Chilkat Pass, the Gray-crowned Rosy-Finch is thought to breed above 1,200 m where rocky ledges are common (Weeden 1960). Elsewhere, in the Tatshenshini River, it was found only high on mountains, among rock piles where vegetation was very sparse. In Kootenay National Park,

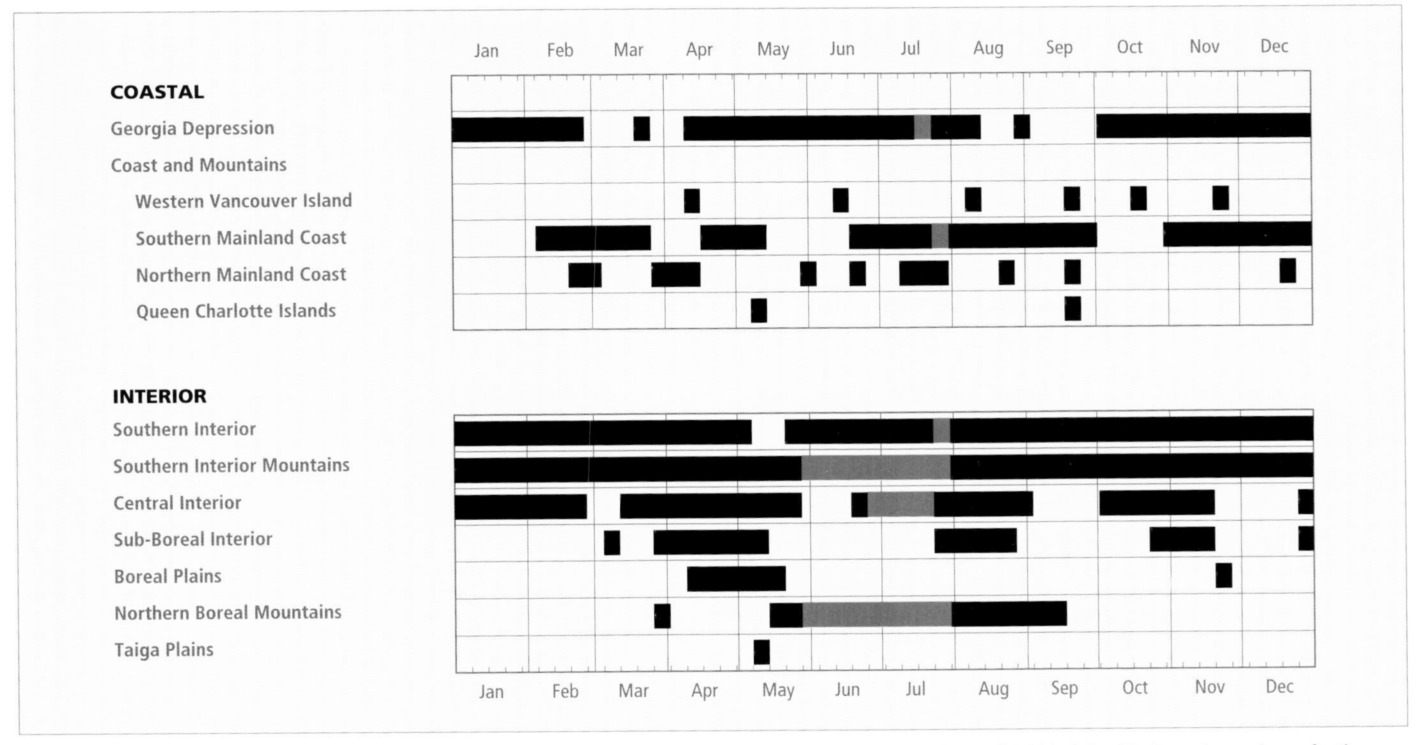

Figure 587. Annual occurrence (black) and breeding chronology (red) for the Gray-crowned Rosy-Finch in ecoprovinces of British Columbia. Records are shown for the week in which they occurred.

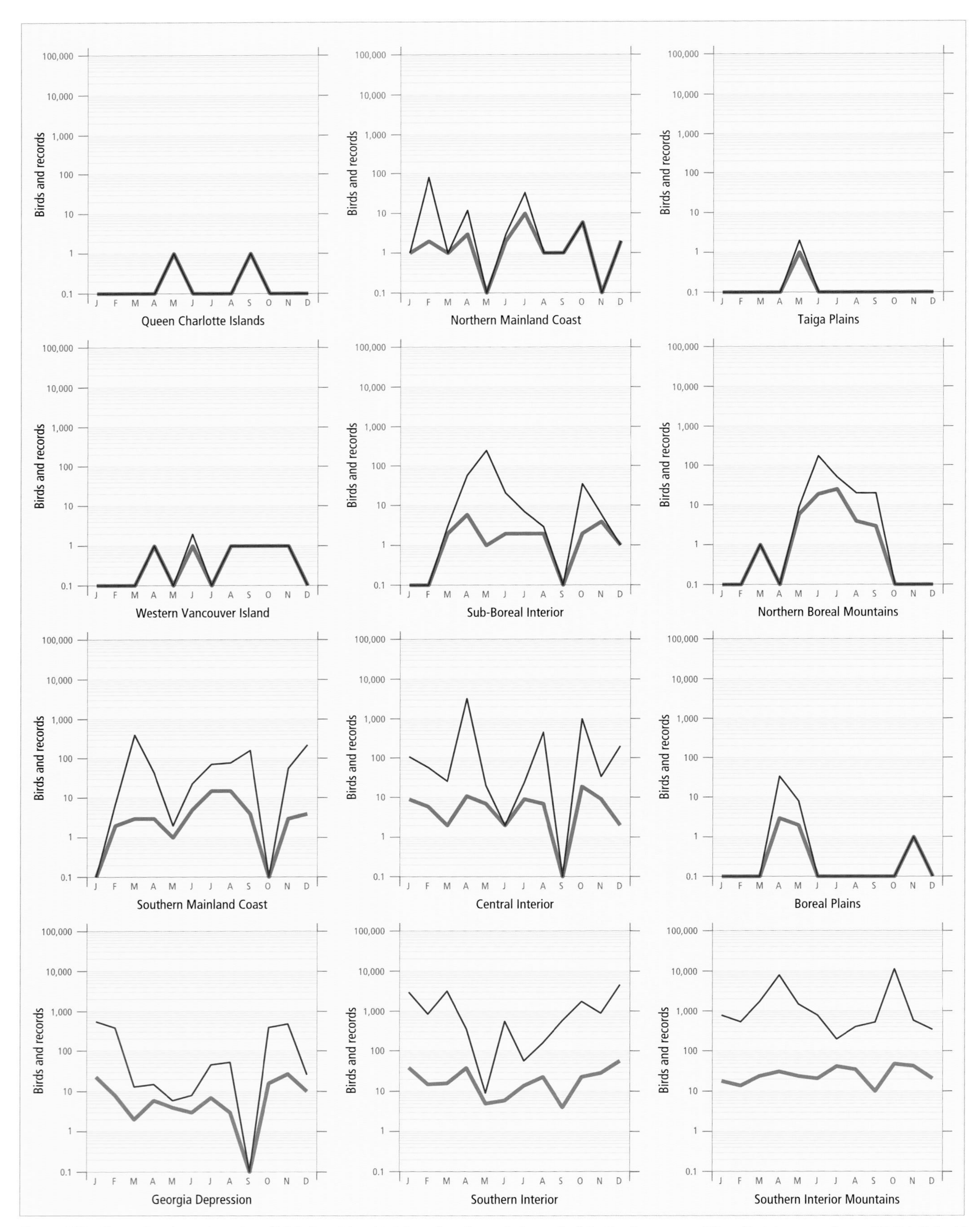

Figure 588. Fluctuations in total number of birds (purple) and total number of records (green line) for the Gray-crowned Rosy-Finch in ecoprovinces of British Columbia. Christmas Bird Counts and nest record data have been excluded.

Figure 589. Nesting habitat for the Gray-crowned Rosy-Finch in British Columbia includes sparsely vegetated alpine areas with snow patches and rocky outcroppings (headwaters of Powell Creek, 21 July 1993; R. Wayne Campbell).

breeding takes place where rock faces occur in heath tundra, although some birds forage in open areas near the subalpine larch and subalpine fir timberline (Poll et al. 1984).

Nesting adults often forage along cliff faces, working ledges for insects and seeds. At Mount Revelstoke National Park, nesting birds took caterpillars, beetle larvae, and various seeds (Cowan and Munro 1944-46). The Gray-crowned Rosy-Finch forages in habitats that may seem unusual for a songbird. For example, it will search ice-covered lakes in July for insects, and probe bare earth at the edge of snow patches (Wade 1977). Many observers have watched it foraging for insects on alpine snow banks.

On the coast the Gray-crowned Rosy-Finch has been recorded breeding from 10 July (calculated) to 22 July; in the interior, it has been recorded breeding from 27 May (calculated) to 25 July (Fig. 587).

Nests: Three nests have been described in British Columbia; all were in rocky, alpine areas. A nest at Beta Lake, east of Powell River, was described as being among rocks. One near Smithers was built in a deep crack in a rock face, just below a glacier. A nest near the Chilkat Pass was built in a vertical rock crevice and consisted of alpine vegetation and feathers. In general, nests appear to be placed in crevices on rocky cliffs or outcroppings in alpine areas (Fig. 590).

Nests are bulky cups of grasses, rootlets, plant down,

Figure 590. In British Columbia, the Gray-crowned Rosy-Finch builds its nest on rocky cliffs or outcroppings in high alpine areas (RCAF Mountain, 22 July 1993; R. Wayne Campbell).

mosses, and feathers. In other areas, nests have been used for 2 or more successive years, with additional material being added each year (Johnson 1983).

Eggs: Only 1 nest with eggs has been observed in British Columbia; it was found on 29 June, but the nest was too deep in a crevice for the eggs to be counted accurately. Calculated dates from records of recent fledglings suggest that eggs can be present as early as 27 May. The early date appears significant because the breeding season has been described as running from mid-June through August, although the Aleutian subspecies can lay eggs by late April (Johnson 1983; MacDougall-Shackleton, in press). Clutches elsewhere usually contain 4 to 6 eggs; the incubation period is about 14 days and ranges from 12 to 16 days (Johnson 1983).

Young: Dates for 8 broods ranged from 29 June to 25 July. Sizes of 4 broods ranged from 1 to 4 young (1Y-2, 2Y-1, 4Y-1). The nestling period is 15 to 22 days. After fledging, the young remain dependent on their parents for about 2 weeks (Johnson 1983).

Fledged young (Fig. 581) have been recorded in British Columbia as early as 29 June (Tatshenshini River) and 30 June (near Barkerville). These dates appear to be rather early, given the relatively harsh conditions at high alpine elevations, and may be related to relatively early springs in those years. Recently fledged young have been observed through late July, and family groups have been reported into at least mid-August. Apparently the family group can remain together through the autumn.

Brown-headed Cowbird Parasitism: There are no records of cowbird parasitism in British Columbia or elsewhere in North America. The Gray-crowned Rosy-Finch's breeding areas are generally not capable of supporting cowbirds because of their harsh conditions.

Nest Success: Insufficient data.

REMARKS: The taxonomy of rosy finches has given rise to much discussion for many years. The American Ornithologists' Union (1957) listed 3 species in North America: Gray-crowned

Rosy-Finch, Black Rosy-Finch (*L. atrata*), and Brown-capped Rosy-Finch (*L. australis*). These 3 species were then merged, along with the Asian Rosy Finch (*L. arctoa*) into 1 species, the Rosy Finch (*L. arctoa*) (American Ornithologists' Union 1983). Sibley and Monroe (1990) considered the Asian and North American rosy finches to be separate species; the American Ornithologists' Union (1998) followed suit, and went further by again splitting the North American Rosy Finch into the former 3 species and hyphenating the name Rosy-Finch.

There are 6 subspecies of Gray-crowned Rosy-Finch in North America (American Ornithologists' Union 1957), although the boundaries of distribution are uncertain in some. Two subspecies occur in British Columbia and are distinguishable in the field: *L. t. tephrocotis* has a gray head and brown throat and cheeks; *L. t. littoralis*, previously known as Hepburn's Rosy-Finch and currently known as the subspecies "Gray-cheeked" Rosy-Finch, has a gray head and cheek but black throat. *L. t. tephrocotis* breeds from northern Alaska, central Yukon, and western Alberta south through interior British Columbia to southeastern British Columbia and northwestern Montana. It winters from southern British Columbia, central Alberta, and southern Saskatchewan south to California, Nevada, Utah, New Mexico, Nebraska, and South Dakota. *L. t. littoralis* breeds from central Alaska and southwestern Yukon south through the mountains of southeastern Alaska and the Coast Mountains of British Columbia to the Cascade Range in Washington, Oregon, and northern California.

Munro and Cowan (1947) describe the distribution of the 2 subspecies in British Columbia as follows: *tephrocotis* "occupies the eastern portions of the species' range, and nests in the alplands. In winter, it occurs in small numbers among flocks of *littoralis* in the western valleys." The subspecies *littoralis* "occupies the western portion of the species' range. It nests in the alplands and winters irregularly in the valleys, more commonly in the interior than on the coast."

Mixed flocks of both subspecies occur in most regions during nonbreeding seasons. In the Southern Interior Mountains, both subspecies occur commonly in autumn and winter, but the eastern subspecies (*tephrocotis*) usually vastly outnumbers the western subspecies (*littoralis*); sometimes the reverse is true, however. On the coast, *tephrocotis* is relatively uncommon. We have only 2 breeding records that identify the subspecies: *tephrocotis* at Mount Revelstoke National Park (Cowan and Munro 1944-46) and *littoralis* at Tenquille Lake, near Pemberton. The breeding distribution of each subspecies in British Columbia is poorly documented.

Road kills occasionally occur because flocks tend to use ditches, banks, and shoulders of highways and other roads during winter and fly up when vehicles approach. Near Vernon, 22 dead rosy-finches were found on a roadside in January 1976; both subspecies were present.

Naturalists can add to our knowledge of this colourful finch in British Columbia by recording the subspecies observed and the proportions of each subspecies in mixed migrant and wintering flocks, and by looking for evidence of nesting in alpine areas (especially on Vancouver Island and the mainland coast).

This species was previously known as Rosy Finch and Gray-crowned Rosy Finch.

For a review of the migration, nesting biology, and genetic relationships of the Gray-crowned Rosy-Finch in North America, see King and Wales (1965), Johnson (1983), Marten and Johnson (1986).

NOTEWORTHY RECORDS

Spring: Coastal – Mount Strachan 22 May 1977-1; Little Mountain (Hope) 4 Mar 1922-200 (Thacker and Thacker 1923); West Vancouver 18 Apr 1993-8 (Siddle 1993a); Pitt Polder 23 Mar 1976-9; s Campbell River 1 May 1982-1; Cape Scott Park 15 Apr 1980-1; Kispiox River 14 Apr 1970-1; Langara Island 8 May 1981-1; Skeena Crossing 3 Apr 1975-10; 25 km n Hazelton 3 Apr 1990-40; Stonehouse Creek 19 May 1978-small flock (Blood and Chutter 1978). **Interior** – Kilpoola Lake 28 May 1973-many; Kimberley 13 Apr 1945-180 (Johnstone 1949), 26 Apr 1974-75; Eager 10 Mar 1947-800 (Johnstone 1949); Harmer Ridge (Sparwood) 17 May 1984-470 at 1,900 m, another 300 at 2,000 m (Fraser 1984); Nakusp 2 Apr 1978-200; Kalamalka Lake 6 Mar 1952-1,000; Revelstoke 7 Apr 1992-3,000 (Siddle 1993b); Nicholson 6 Mar 1977-250; Glacier National Park 15 Mar 1983-100, 28 Apr 1982-300; Kicking Horse River 17 Apr 1977-500 in 3 flocks; Field 16 Apr 1977-3,000, 27 Apr 1975-400, 17 May 1976-300, 24 May 1975-300 (Wade 1977); Revelstoke Dam 27 Mar 1983-100; Lac la Hache 29 Apr 1945-50 (Munro 1955b); Farwell Canyon 31 Mar 1980-25; Kleena Kleene 13 Apr 1958-3,000 (Paul 1964b); Moose Lake (Cariboo) 25 May 1972-75; nr Moricetown 30 Mar 1975-2; Mount Lemoray 7 May 1968-250; Mackenzie 14 Apr 1996-14; Tumbler Ridge 12 Apr 1996-4 at feeder, 14 May 1996-6 at feeder; Fort St. John 22 Apr 1988-15, 1 and 2 May 1987-2; Fort Nelson 13 May 1986-2 at airport after 30 cm snowfall; Chilkat Pass 21 May 1977-2; Atlin 29 Mar 1933-1 (RBCM 5755).

Summer: Coastal – w Sutton Pass 14 Jun 1981-2; Sumas 14 Jun 1987-3 (MCZ 247882 to 247884); Sumallo River 6 Jul 1974-23 on roadside; Mount Cheam 8 Aug 1981-50; Golden Ears Park 1 Jul 1975-2; Mount Albert Edward 12 Jul 1981-4; Mount Steele (Sechelt) 20 Jul 1988-21; Panther Peak 20 Jul 1987-15; Beta Lake 22 Jul 1981-nest among rocks; Sentinel Glacier 28 Jul 1984-adults feeding 1 fledgling; Panorama Ridge (Garibaldi Park) 3 Aug 1986-1 fledgling begging from adult, 21 Aug 1982-40, with many young; Black Tusk (Garibaldi Park) 1 Aug 1994-female with 2 immatures at 2,310 m elevation; Whistler Mountain 24 Jun 1924-19; ne Bella Coola 25 Jun 1934-1 (MCZ 284744); Caribou Mountain (Stuie) 17 Aug 1938-2 (NMC 28805 and 28806); Yule Lake 24 Jun 1936-2 (MCZ 284745 to 284746); Sibola Peak 17 Jul

1976-5; Emerald Glacier 12 Aug 1956-15; Terrace Mountain (Terrace) 10 Jun 1979-6 (Page and Bergerud 1979); Hazelton 16 Jul 1917-5 (Taverner 1919); Stikine River 23 Jul 1919-20 on mountain above Dokdaon Creek (MVZ 39889 to 39908); Llewellyn Glacier 31 Aug 1975-5; White Pass 13 Jun 1899-1 (FMNH 165687); Three Guardsmen Mountain 3 Jul 1984-4. **Interior** – Cathedral Park 27 Jul 1985-1 or more fledglings, 1 Aug 1976-adults feeding fledglings, 19 Aug 1983-60; Melvin Creek 22 Jun 1992-400; Gwillim Lakes 12 Aug 1989-40, including juveniles; Tenquille Lake 31 Jul 1930-adults and fledglings about snow fields (MVZ 105415); Eva Lake (Revelstoke) 12 Jul 1937-3 pairs feeding nestlings; Goodsir Pass 26 Jul 1982-2 families with 2 fledglings (Poll et al. 1984); Glacier National Park 29 Jun 1983-15; Mount Tatlow 8 Jul 1989-5; Amiskwi River 14 Jun 1975-100 (Wade 1977); Ottertail River (Yoho National Park) 3 Jun 1976-200 (Wade 1977); Lake O'Hara 25 Jul 1984-adults feeding fledglings; Field 21 Jun 1976-300, 28 Aug 1975-100 (Wade 1977); Mount Assiniboine Park 7 Aug 1963-1 fledgling with adults; Niut Mountain 18 Aug 1995-450 at 1,950 m elevation; Perkins Peak 6 Jul 1980-10, 15 Jul 1992-1 fledged young with adult; Barkerville 30 Jun 1903-fledged young above timberline (Brooks 1903); Robson Meadow 12 Jun 1972-35; Berg Lake (Mount Robson Park) 15 Jul 1971-adults feeding fledglings; w McBride 10 Jul 1991-23; Gosnell Creek 30 Jun 1975-1; 8 km w Smithers 20 Jul 1974-1 fledgling capable of short fluttering flights; 16 km w Smithers 29 Jun 1975-nest too far in crevice to accurately count eggs; Nilkitkwa Lake 15 Aug 1986-2; Mount Spieker 27 Jul 1994-6 at 1,800 m elevation; 15 km n Pelly Lake 25 Jul 1980-1 fledgling with adults; Nuttlude Lake 16 Jul 1963-9; Mount Edziza Park 11 Jul 1981-1; Spatsizi Plateau Wilderness Park 7 Aug 1959-many pairs with fledglings; Stone Mountain Park 6 Jun 1983-60; Mount McDame 19 Jun 1956-10; Samuel Glacier 28 Jun 1983-12, 29 Jun 1983-2 recent fledglings (Campbell et al. 1983); Datlasaka Range 10 Jul 1980-4 nestlings; Nadahini Creek 4 Jul 1984-4.

Breeding Bird Surveys: Not recorded.

Autumn: Interior – Summit Pass 10 Sep 1971-small flocks of 6 to 9 birds (Griffith 1973); Griffith Creek (Spatsizi Plateau Wilderness Park) 3 Sep 1976-4; Manson Lakes 27 Oct 1972-25; Fort St. John 22 Nov 1986-1; Mackenzie 15 Nov 1996-2; 24 km s Prince George 18 Nov 1978-1; West Lake Park 3 Nov 1985-2; Kleena Kleene 27 Oct 1960-500 in 10 flocks (Paul 1964b); Field 5 Oct 1973-1,000, 21 Oct 1975-1,000 (Wade 1977); Lake O'Hara 17 Sep 1976-40 (Wade 1977); Hungabee Lake 20 Sep 1975-150 (Wade 1977); Enderby 15 Nov 1945-1 (UBC 7176); Melvin Creek 22 Sep 1992-500 (Cooper 1994); e Canal Flats 8 Nov 1983-62 in 4 flocks along Kootenay River; nr Peachland 2 Nov 1963-150; Harmer Ridge (Sparwood) 21 Oct 1983-600 (Fraser 1984); Edgewood 30 Oct 1924-1,000 (Kelso 1926); Mount Erickson 1 Nov 1984-30 (Fraser 1984); Sparwood 18 Oct 1984-1,200, 22 Oct 1984-1,500; Bull River 9 Nov 1983-40; Alice Siding Nov 1984-2 (Butler et al. 1986); Richter Pass 7 Nov 1977-100; Blackwall Mountain 18 Oct 1964-1,500, 2 Nov 1974-195. **Coastal** – Port Simpson Oct 1886-1 (RBCM 779); Kitimat 17 Sep 1974-1 (Hay 1976); Bella Coola 15 Nov 1975-1; Cape St. James 17 Sep 1978-1; Garibaldi Park 7 Sep 1986-130 at Black Tusk; Puzzle Mountain 1 Oct 1977-4; Mount Seymour Park 27 Oct 1991-60 (Siddle 1992a), 30 Oct 1989-80, 1 Nov 1991-250 (Siddle 1992a); Pitt Lake 8 Nov 1975-34 on dyke; Hope 3 Nov 1922-30 (Thacker and Thacker 1923); Schooner Cove (Pacific Rim National Park) 21 Nov 1973-1; Surrey 10 Oct 1960-60; Oak Bay 17 Nov 1973-28; Crofton 21 Oct 1974-7, 1st record for Cowichan valley; Sooke 25 Nov 1958-1, 24 Nov 1996-1.

Winter: Interior – nr Babine Lake 22 Jan 1979-4; Willow River 25 Dec 1958-1; Barkerville Dec 1900-present (Brooks 1901); w Williams Lake 8 Feb 1992-150 roosting in limestone cliff on ranch; Williams Lake 30 Dec 1989-200, 7 Feb 1982-40; Sheep Creek (e Riske Creek) 29 Jan 1978-80; Yoho National Park 26 Jan 1977-175; Golden 16 Jan 1977-300; Glacier National Park 1 Feb 1983-100; Mount Hunter 13 Feb 1977-200 (Wade 1977); Edge Hills (w Clinton) 13 Dec 1981-flocks of 200, 350, and 1,000; Sorrento 11 Jan 1972-200; nr Cache Creek 30 Dec 1977-200; Vernon 1 Jan 1976-300; Winfield 19 Feb 1969-200; nr Savona 20 Jan 1985-120; Summerland 7 Dec 1968-200; Eager 22 Jan 1941-200 (Johnstone 1949); West Bench (Penticton) 14 Dec 1963-1,000, 27 Jan 1954-1,000, 30 Jan 1974-300, 25 Feb 1969-400; Apex Mountain (Penticton) 27 Jan 1979-400; Vaseux Lake 25 Jan 1969-150. **Coastal** – Hazelton 27 Feb 1968-80 (Crowell and Nehls 1968b); Terrace 23 Dec 1977-1; Port Neville 2 Dec 1977-4; s Yale 24 Dec 1977-150; Port Mellon 16 Dec 1989-60; Cypress Park 30 Jan 1989-150; North Vancouver 25 Jan 1973-30; Surrey 1 Jan 1979-82; Fort Langley 21 Feb 1963-24; White Rock 8 Jan 1983-43, 10 Jan 1973-47; Sumas 2 Feb 1974-350.

Christmas Bird Counts: Interior – Recorded from 13 of 27 localities and on 16% of all counts. Maxima: Penticton 27 Dec 1986-600; Vaseux Lake 31 Dec 1977-395; Oliver--Osoyoos 29 Dec 1989-347. **Coastal** – Recorded from 8 of 33 localities and on 2% of all counts. Maxima: Sunshine Coast 16 Dec 1989-60; Terrace 26 Dec 1977-35; Pitt Meadows 29 Dec 1973-35.

Pine Grosbeak PIGR

Pinicola enucleator (Linnaeus)

Figure 591. The Pine Grosbeak is a common bird at feeders during the nonbreeding seasons in northern regions of British Columbia. At this time, both males (a) and females (b) can be heard giving soft warbling calls as they forage at feeders or fruit-bearing trees or shrubs (R. Wayne Campbell).

RANGE: Holarctic. In North America, breeds from western and central Alaska, northern Yukon, northwestern and central Mackenzie, northern Manitoba, northern Ontario, northern Quebec, northern Labrador, and Newfoundland south to southern Alaska, British Columbia, central California, extreme west-central Nevada, northern and east-central Arizona, northern New Mexico, and, east of the Rocky Mountains, to northern Alberta, northern Saskatchewan, central Manitoba, southern Ontario, northern Michigan, southern Quebec, northern New Hampshire, northern Vermont, central Maine, and Nova Scotia. Winters from western and central Alaska, southern Yukon, southern Mackenzie, and southern Canada south through the breeding range, casually or sporadically as far as central New Mexico, northern and north-central Texas, northwestern Oklahoma, north-central Arkansas, Missouri, Kentucky, Virginia, and the Carolinas.

Also occurs in the Palearctic.

STATUS: On the coast, occasionally *fairly common* during winter at lower elevations, but of infrequent, irregular occurrence in the south; *very rare* migrant and visitant in the Georgia Depression Ecoprovince; in the Coast and Mountains Ecoprovince, *uncommon* resident on the Queen Charlotte Islands, the Southern Mainland Coast, and the Northern Mainland Coast, and *casual* on Western Vancouver Island.

In the interior, *fairly common* to occasionally *common* during winter at lower elevations throughout the interior, but of more erratic occurrence in the southern ecoprovinces; an *uncommon* to *fairly common* resident at higher elevations of the Southern Interior, Southern Interior Mountains, Central Interior, Sub-Boreal Interior, and Northern Boreal Mountains ecoprovinces; an *uncommon* to *fairly common* migrant and winter visitant, but *very rare* summer visitant in the Taiga Plains and Boreal Plains ecoprovinces.

Breeds.

NONBREEDING: The Pine Grosbeak (Fig. 591) is a widely but sparsely distributed resident of most subalpine coniferous forests in British Columbia. To the casual observer, however, this species is most commonly associated with lower-elevation habitats, including human-modified landscapes, which it often frequents from late autumn through early spring. On the Queen Charlotte Islands and probably much of the mainland coast, it is a year-round inhabitant of lower-elevation coniferous forests. In the northern interior, north of

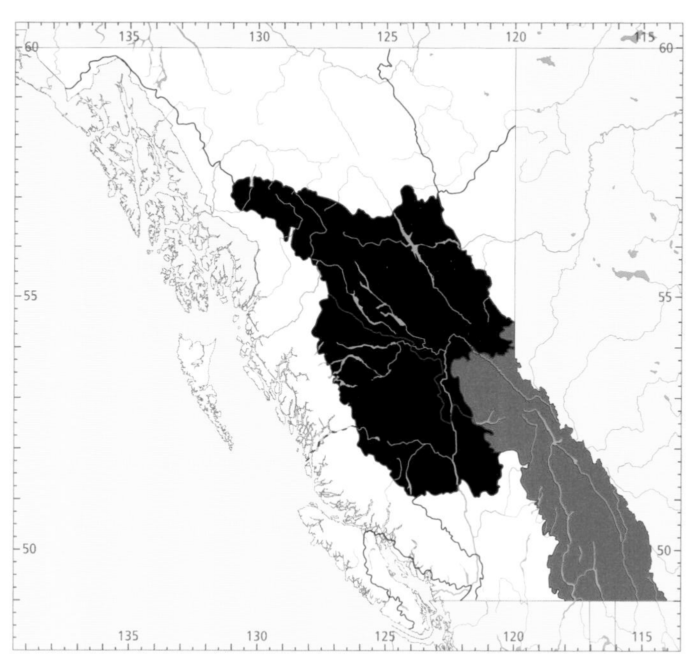

Figure 592. In British Columbia, the highest numbers for the Pine Grosbeak in winter (black) occur in the Sub-Boreal Interior and Central Interior ecoprovinces; the highest numbers in summer (red) occur in the Southern Interior Mountains Ecoprovince.

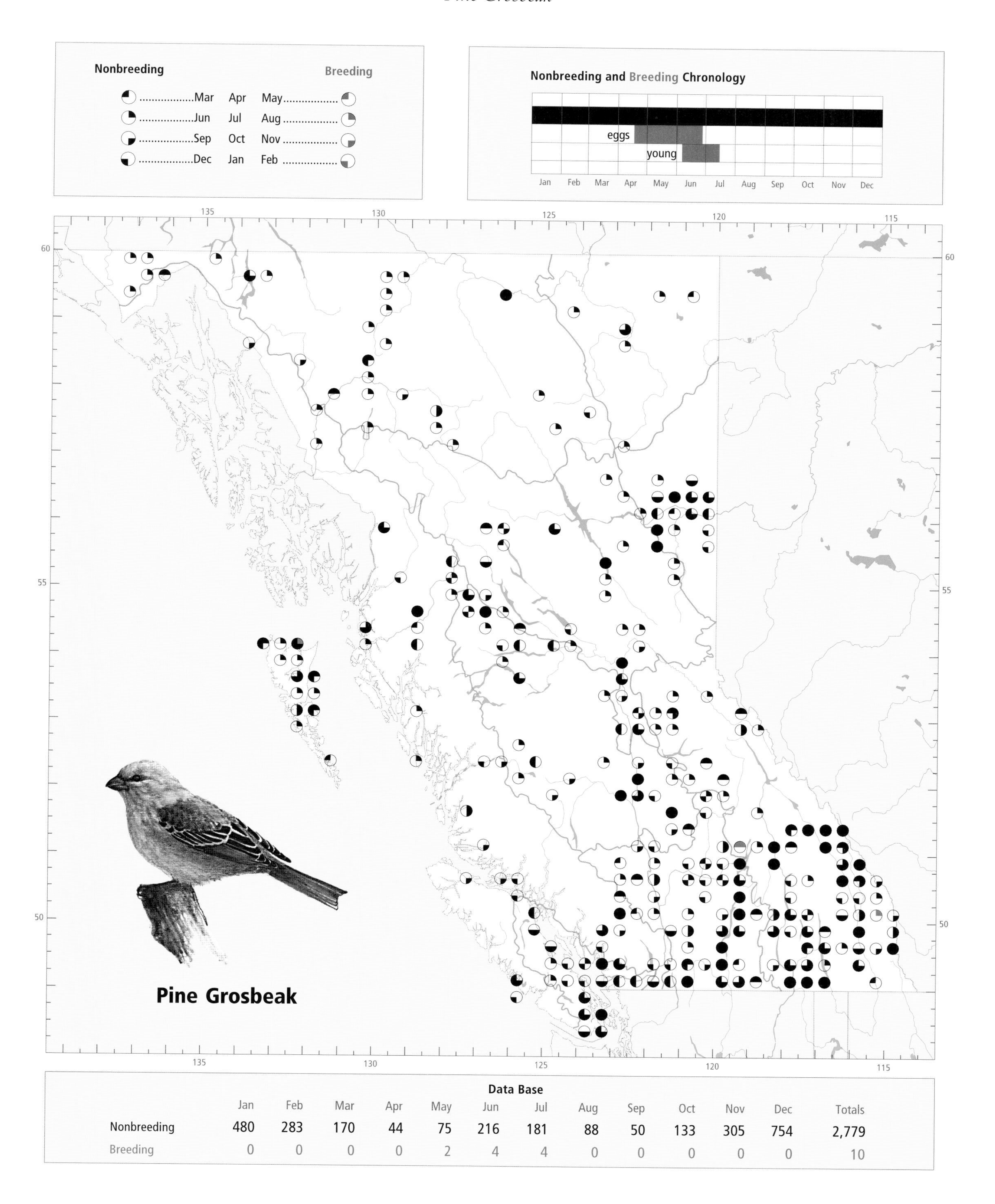

Data Base

	Jan	Feb	Mar	Apr	May	Jun	Jul	Aug	Sep	Oct	Nov	Dec	Totals
Nonbreeding	480	283	170	44	75	216	181	88	50	133	305	754	2,779
Breeding	0	0	0	0	2	4	4	0	0	0	0	0	10

Figure 593. In late autumn and winter, small flocks of Pine Grosbeaks can be seen foraging on weed seeds and salt along highways in British Columbia. Fruit-bearing bushes are often nearby (near Tumbler Ridge, 13 November 1996; R. Wayne Campbell).

latitude 54°N, it is probably a resident of upper-elevation boreal and sub-boreal forests. Much remains to be learned, however, about the distribution and life history of this enigmatic species in British Columbia.

The highest numbers in winter occur in the sub-boreal forests of the northern portions of the Central Interior and southern portions of the Sub-Boreal Interior (Fig. 592).

The Pine Grosbeak has been recorded at elevations ranging from sea level to 1,980 m on the coast and from valley floors to 2,135 m in the interior. During the nonbreeding period, it is found in a wide variety of forest and shrub habitats, ranging from timberline krummholz down to riparian thickets in interior grasslands and the shrubland fringes of estuaries on the coast. During winter, it regularly descends from its subalpine breeding habitats (see BREEDING) to lower elevations, where it frequents a variety of open-forested and non-forested habitats in search of berries, seeds, and buds. It is also a frequent visitor to bird feeding stations.

During winter, the Pine Grosbeak occurs in higher numbers in the northern half of the province than in the southern half, and more regularly at low elevations in the interior than on the coast. In the south, it is also more common at the lower elevations in mountainous regions than in the coastal lowlands or interior plateaus. This latter pattern has also been observed in western Alberta; there wintering grosbeaks were most abundant where the Alberta Plateau abuts the Rocky Mountains (Adkisson 1981).

The Pine Grosbeak eats the buds and fruits of a wide selection of woody plants throughout its range in the northern hemisphere (Newton 1972). In British Columbia, it is probably a lack of these food reserves in the subalpine areas that initiates their wandering, downward movements following the breeding period. A. Brooks (*in* Dawson and Bowles 1909) noted that the periodic influxes of this bird in the southern valleys of the interior appeared unrelated to the severity of the winter.

In the interior, the Pine Grosbeak frequents a wide variety of low-elevation coniferous, mixed, and deciduous forests during winter. It is also attracted to forest edges and openings, where it feeds on the fruits of various shrubs, especially rose hips and common snowberry (Dawson and Bowles 1909). It also frequents human-modified landscapes, including weedy field edges, road verges (Fig. 593), orchards, and urban areas. In settled areas, it is readily attracted to the fruit of domestic crab apple and mountain-ash, as well as to bird feeders. An analysis of Project FeederWatch data for the period 1988 to 1998 shows that this grosbeak was among the 10 most common bird species visiting winter feeding stations in the northern interior.

Pine Grosbeak movements in British Columbia, like those of the Steller's Jay, are primarily altitudinal in nature. Following the breeding period, the earliest nesters occasionally move to the lower elevations in small family groups as early as mid-May. In most years, however, grosbeaks do not appear in numbers at low elevations until late October or early November (Figs. 596 and 597). In the Okanagan valley, they typically arrive in November and leave again for higher elevations in March (Cannings et al. 1987). In the valleys of the Southern Interior, Christmas Bird Counts indicate that winter numbers peak every second or third year (Fig. 594).

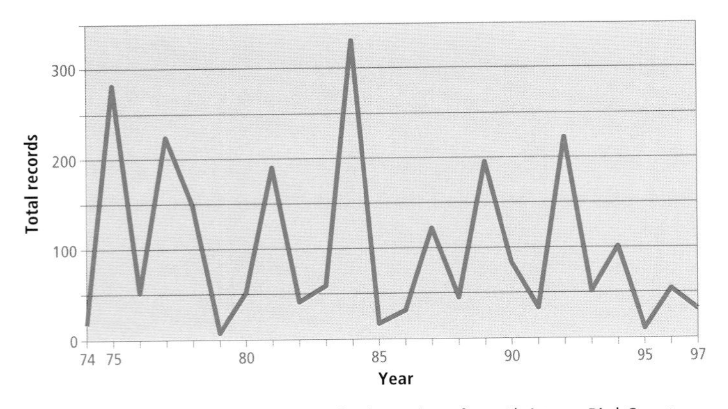

Figure 594. Fluctuations in Pine Grosbeak numbers from Christmas Bird Counts for the Southern Interior Ecoprovince (Penticton, Vernon, and Shuswap Lake) for the period 1974 through 1997. Influxes to the valley bottoms occur every second or third winter.

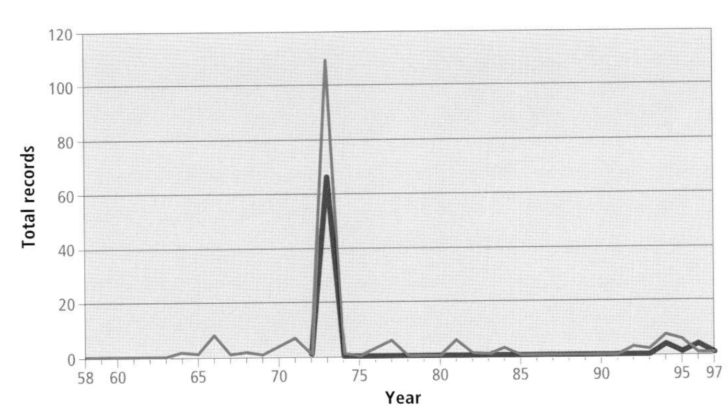

Figure 595. Fluctuations in Pine Grosbeak numbers from Christmas Bird Counts in the Georgia Depression Ecoprovince for Vancouver and Victoria (green line) for the period 1958 through 1997 and for Chilliwack, Pitt Meadows, and White Rock (purple line) for the period 1972 through 1997.

During years of adequate food resources, however, most birds probably remain in the subalpine areas through the winter.

The Pine Grosbeak also makes irruptive, southward movements in North America, in response to the natural fluctuations of northern seed and berry crops (Bock and Lepthien 1976). These latitudinal movements are very pronounced and synchronous in boreal forests east of the Rocky Mountains, including the Taiga Plains and Boreal Plains of British Columbia. In the mountainous regions of western North America, including British Columbia, these southward movements are less frequent and are often difficult to distinguish from local altitudinal movements. During the winter of 1921 to 1922, however, Brooks (1922a) recorded 1 such invasion into the southern portions of the interior of British Columbia. This movement consisted of birds of the northern race, *P. e. alascensis,* which were described as notably larger and paler in coloration than the local race, *P. e. montana.* Earlier influxes into this region were recorded during the winters of 1901 to 1902, 1906 to 1907 (Dawson and Bowles 1909), and 1916 to 1917, but the geographic race of these birds was not noted.

The Pine Grosbeak is notably scarce on the south coast, but conspicuous winter invasions are occasionally recorded. Notable influxes were observed during the winters of 1894 to 1895, 1922 to 1923, 1940 to 1941 (Laing 1942), and 1952 to 1953 (Holdom 1953). A more recent widespread invasion was recorded in the Georgia Depression during the winter of 1973 to 1974. During this period, record high numbers were reported in many Christmas Bird Count areas in this ecoprovince, including Chilliwack, Pitt Meadows, Vancouver, Victoria, and White Rock (Fig. 595).

The Pine Grosbeak generally occurs in small flocks of 10 or fewer birds during the nonbreeding period, although groups numbering 30 or more are occasionally reported. In Kootenay National Park, winter flocks typically range from 5 to 10 birds, with the largest flock numbering 30 birds (Poll et al. 1984). In the Peace Lowland of the Boreal Plains, most flocks (80%; n = 127) recorded during an 8-year period consisted of 8 or fewer birds (C.R. Siddle pers. comm.). Throughout British Columbia, the Pine Grosbeak seldom occurs in mixed flocks, except when near bird feeders or other areas of common food resources, when it loosely associates with species such as the Pine Siskin, Common Redpoll, Purple Finch, House Finch, Evening Grosbeak, Red Crossbill, and Bohemian Waxwing. During years when different Pine Grosbeak subspecies winter in the same geographic area, the respective races tend to remain in separate flocks (Adkisson 1981).

The Pine Grosbeak has been recorded throughout the year in subalpine areas of British Columbia; it generally occurs at lower elevations from late October through mid-March (Fig. 596).

BREEDING: The breeding distribution of the Pine Grosbeak in British Columbia is poorly known. A nest with young reported from the Queen Charlotte Islands and 2 nests from the Southern Interior Mountains represent the only documented breeding records. Observations of paired adults, nest building, adults with fledged young, and specimens collected in breeding condition suggest that breeding is widespread throughout the province, however. Most nesting is believed to occur in subalpine areas, but on the Queen Charlotte Islands it also

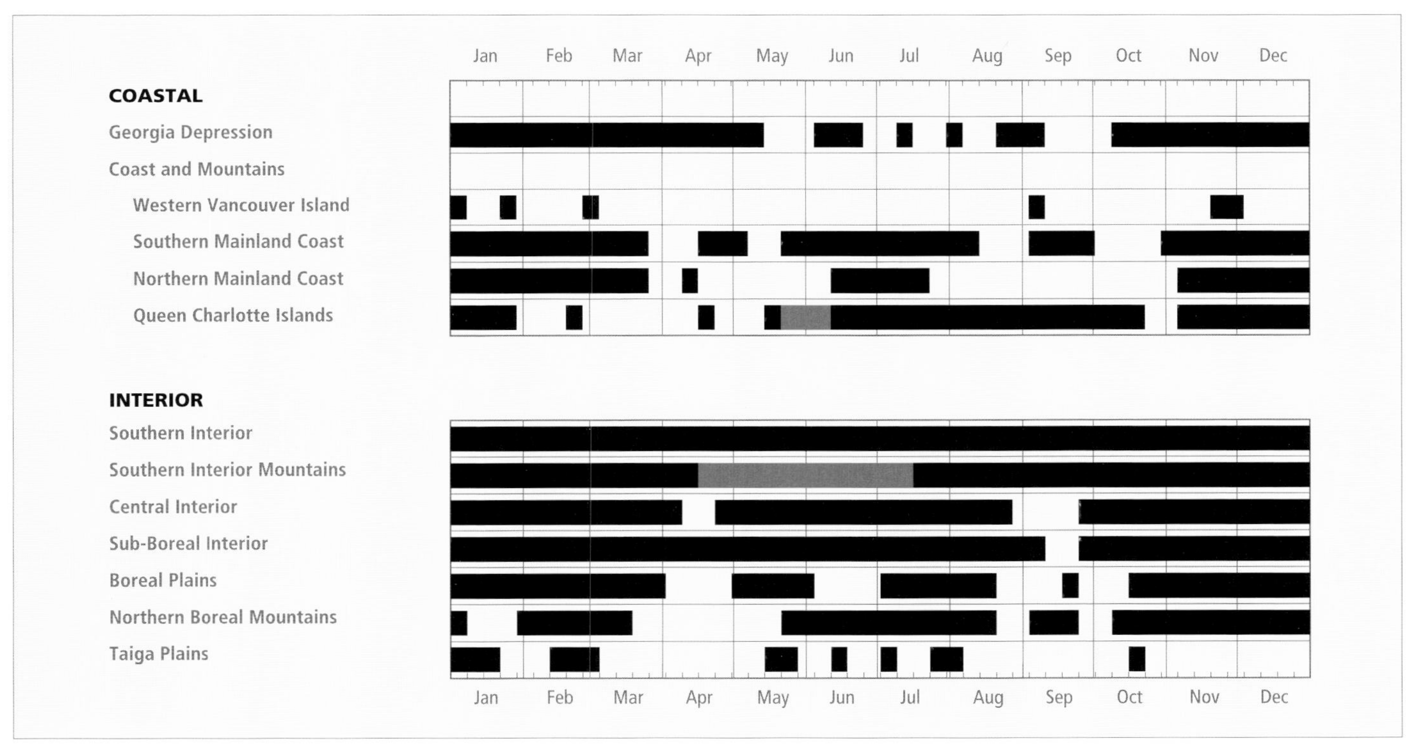

Figure 596. Annual occurrence (black) and breeding chronology (red) for the Pine Grosbeak in ecoprovinces of British Columbia. Records are shown for the week in which they occurred.

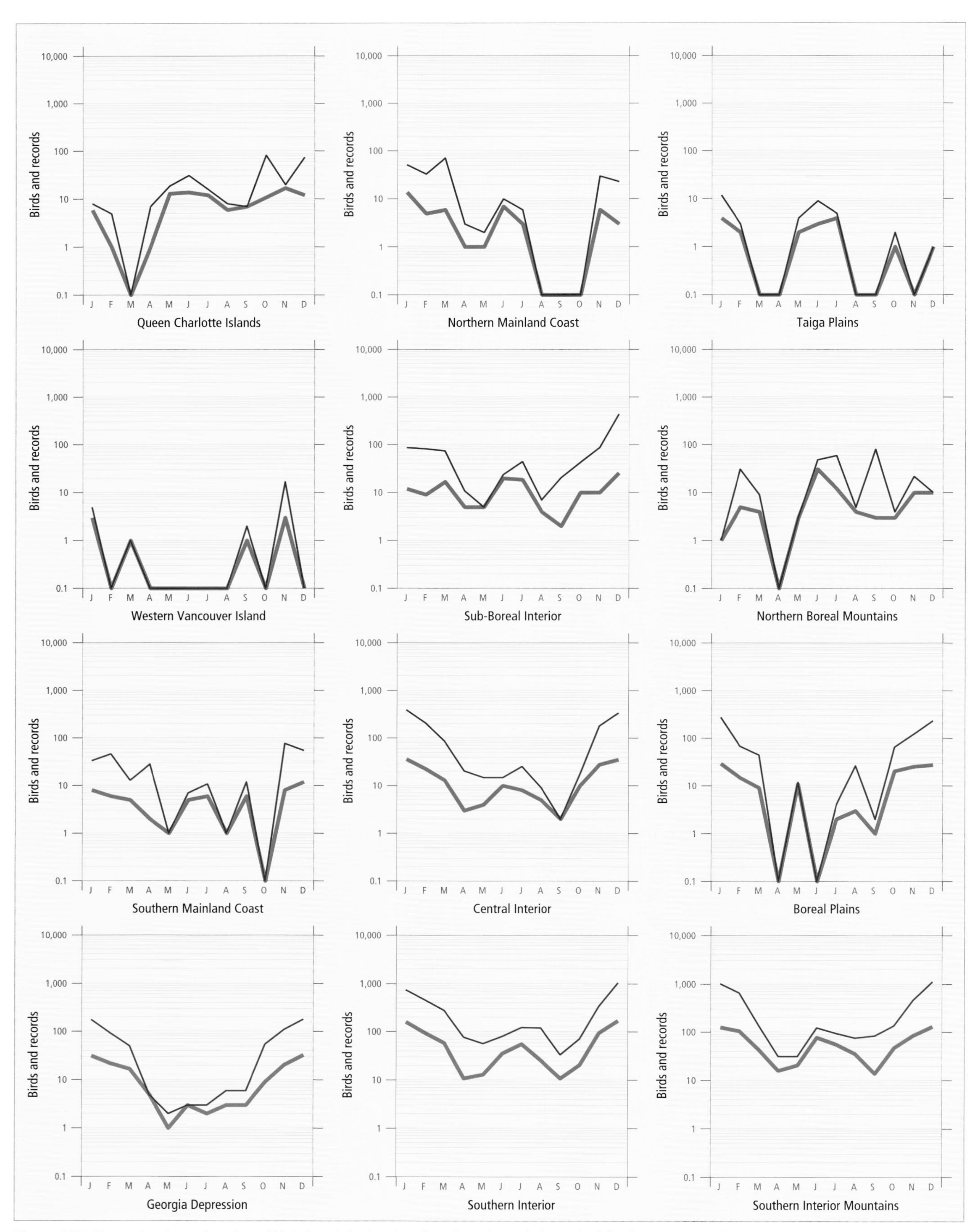

Figure 597. Fluctuations in total number of birds (purple line) and total number of records (green line) for the Pine Grosbeak in ecoprovinces of British Columbia. Christmas Bird Counts, Breeding Bird Surveys, and nest record data have been excluded.

occurs in lower-elevation forests of western hemlock and Sitka spruce. Although not yet documented, breeding likely occurs at mid to low elevations on the Northern Mainland Coast and at the upper elevations of the Boreal White and Black Spruce and Sub-Boreal Spruce zones of the northern interior.

On the coast, breeding has been confirmed only on Graham Island in the Queen Charlotte Islands (Darcus 1930). A. Brooks (*in* Dawson and Bowles 1909) reported adults feeding fledged young near the international boundary on the Southern Mainland Coast. This observation, along with other scattered summer records (Brooks 1923; Laing 1942), suggests that the species is a sparse breeder along the mainland portion of the Coast and Mountains. Contrary to other sources (Brooks 1939; Munro and Cowan 1947; Godfrey 1986; American Ornithologists' Union 1957), however, breeding on Vancouver Island remains unsubstantiated (see REMARKS).

Two nests found in the Southern Interior Mountains, near Celista and Nitta Creek, are the only known breeding records for the interior of British Columbia. Other related observations, however, including recent songbird studies, suggest a widespread breeding distribution throughout the Engelmann Spruce–Subalpine Fir Zone of the southern two-thirds of the interior. The northernmost record that suggests breeding is of adults feeding fledged young near Germansen Landing in the Sub-Boreal Interior (Fig. 598). Although records are lacking in the Northern Boreal Mountains, breeding is believed to occur throughout the Spruce-Willow-Birch Zone and at higher elevations of the Boreal White and Black Spruce Zone. In the Taiga Plains and Boreal Plains, there are few summer records to suggest breeding in this region of the province, although nesting may occur in upper-elevation boreal forests. Erskine (1977) did not list the Pine Grosbeak as a breeding species in spruce forests of the Fort Nelson Lowland. In northern Alberta, where similar boreal habitats are found, the breeding status of this species is also ambiguous (Salt and Salt 1966; Pinel et al. 1993).

The Pine Grosbeak is a representative but sparsely distributed breeding species of subalpine forests in the interior of British Columbia (Fig. 599). In the Manson River area of the Sub-Boreal Mountains, Gyug (1997) considered it to be among the 30 most common breeding birds found in the Engelmann Spruce–Subalpine Fir Zone. At Sicamous Creek in the Southern Interior Mountains, Dickinson and Leupin (1997) found the Pine Grosbeak to be the least common of the 10 "core" breeding bird species found in this subalpine zone. Other songbird studies in the Southern Interior Mountains also regularly detected this bird in low numbers in the subalpine forests (e.g., Poll et al. 1984; Davis et al. 1999).

The Pine Grosbeak uses a wide range of coniferous forest habitats during the breeding season, and appears to be a habitat generalist. Erskine (1977) could find no clear habitat preferences during his study of boreal forest communities in Canada. Adkisson (1999) reports that the species prefers open coniferous forest, including habitats disturbed by partial logging. Although a similar broad pattern of habitat use occurs in British Columbia, it has frequently been recorded in young or open subalpine coniferous forests during the breeding period. Habitats include older burns or clearcuts, selectively logged areas, forest edges along meadows and avalanche chutes, beaver ponds and lakeshores, and subalpine parkland or krummholz. The burns were estimated to be from 15 to 60 years of age, and several had a deciduous component of willows.

Figure 598. In the Sub-Boreal Interior Ecoprovince of British Columbia, the Pine Grosbeak has been observed feeding fledged young in shrubs and trees around wetlands (45 km north Germansen Landing, 4 July 1998; R. Wayne Campbell).

This apparent predilection for young seral forest, edge habitats, and open subalpine areas is also supported by several breeding songbird studies. At Kootenay National Park in the Southern Interior Mountains, Poll et al. (1984) found the Pine Grosbeak most frequently in open subalpine coniferous forests. In the Quesnel Highland east of Quesnel, Davis et al. (1999) also detected this grosbeak during a study of "seral" subalpine forests. There it was found in all 3 forest age classes examined, ranging from logged sites from 11 to 29 years of age to old burns estimated to be 80 to 275 years old. Climax forests were not sampled during the study of Davis et al. (1999), preventing an assessment of the relative importance of mature subalpine forests to the birds.

Other studies in the Engelmann Spruce–Subalpine Fir Zone that monitored bird communities following logging indicate that tree cover is important for breeding grosbeaks. Gyug (1997) found that they avoided recent clearcuts (≤ 5 years old) and small remnant forest patches (≤ 0.40 ha) within

Figure 599. The Pine Grosbeak is a sparsely distributed breeding species in interior subalpine forests of British Columbia (Pesika River, 3 September 1997; Andrew C. Stewart).

recent clearcuts. Dickinson and Leupin (1997) could detect no significant changes in breeding abundance following selective logging where the forest was opened up but not completely removed.

In the central interior region of the province, mid- to low-elevation coniferous forests are also used by the Pine Grosbeak. For example, in the Sub-Boreal Interior northeast of Prince George, Phinney and Lance (1998) found it in both mature and shrub-herb stage forests during June breeding surveys. Singing males were detected in the Engelmann Spruce–Subalpine Fir Zone and in the Sub-Boreal Spruce Zone, above 870 m elevation. In the Bulkley valley area of the Central Interior, Erskine (1977) detected it in mature spruce forests within the Sub-Boreal Spruce Zone. There he estimated a breeding density of 1 male/km^2.

There have been no comparable songbird studies in subalpine forests on the coast, but this species probably breeds in a similar range of forest habitats found in the Mountain Hemlock Zone. In the Coastal Western Hemlock Zone of the Queen Charlotte Islands, the Pine Grosbeak was found during the breeding period in small, open clearcuts or selectively logged patches, but was not encountered in dense mature forest (Adkisson 1999).

The highest numbers in summer occur in the Engelmann Spruce–Subalpine Fir Zone of the Southern Interior Mountains (Fig. 592). The Pine Grosbeak has been recorded nesting from near sea level on the coast to between 1,370 and 1,570 m in the interior.

Breeding Bird Surveys in British Columbia, for the period 1968 through 1993 for both coastal and interior routes contain insufficient data for analysis. Canada-wide surveys indicate an average annual decline of 9.3% ($P = 0.03$) for the period 1966 to 1996 (Sauer et al. 1997). Most continental survey routes lie well south of the breeding distribution of the Pine Grosbeak, diminishing the usefulness of these data for detecting meaningful changes. In addition, few routes in British Columbia traverse the subalpine forests where most breeding occurs.

On the coast, the Pine Grosbeak has been recorded breeding from 25 May (calculated) to 7 June; in the interior, it has been recorded breeding from 18 April (calculated) to 15 July (Fig. 596).

Nests: Only 2 nests have been described in British Columbia. The first nest, found near Celista, was placed near the top of a thick, stunted Douglas-fir tree, at a height of 1.2 m; it was constructed of twigs and grasses, and lined with rootlets. The second nest, found near Nitta Creek, was placed at a height of 2.1 m on the limb of a subalpine fir tree and was constructed primarily of twigs. Elsewhere in North America, nests are usually placed from 2 to 4.5 m above ground in a coniferous tree (Godfrey 1986; Adkisson 1999).

Eggs: Dates for 3 clutches ranged from 11 May to 27 June. Two clutches had 4 eggs each, the only clutch sizes reported for British Columbia. Observations of fledged young being fed by adults suggest that nests could contain eggs as early as 18 April. This is significantly earlier than the late May to early June egg-laying period reported elsewhere for North America (Adkisson 1999) and Europe (Newton 1972). Clutch sizes in North America are usually 3 or 4 eggs; rarely, 2 or 5 eggs (Adkisson 1999). The incubation period is 13 or 14 days.

Young: Only 3 nests containing young have been reported in British Columbia. Dates ranged from 7 June to 15 July. Two nests contained 4 young each. Observations of fledged young with adults indicate that broods containing as many as 5 young can occur in British Columbia. Baicich and Harrison (1997) report that the nestling period is about 14 days, but data from British Columbia indicate that it can be as long as 20 days. There is no evidence of double-brooding in this species (Newton 1972; Adkisson 1999).

Brown-headed Cowbird Parasitism: None of the 3 nests reported with eggs or young in British Columbia was parasitized by the cowbird. There were also no reports of cowbird fledglings being fed by the Pine Grosbeak. Brown-headed Cowbird parasitism has not been reported for this species elsewhere on the continent (Friedmann and Kiff 1985; Adkisson 1999).

Nest success: The only nest found with eggs and followed to a known fate was successful in fledging young.

REMARKS: Of 8 subspecies recognized in North America, 4 are believed to breed in British Columbia (American Ornithologists' Union 1957; Cannings 1998). *P. e. alascensis* occurs east of the Rocky Mountains in northeastern British Columbia (see BREEDING); *P. e. flammula* breeds in the northwestern portion of the province, south to about Tetana Lake, in the Driftwood River valley; *P. e. carlottae* breeds on the Queen Charlotte Islands and possibly along the mainland coast and on Vancouver Island (see BREEDING); and *P. e. montana* breeds from central through southern British Columbia. Adkisson (1977, 1999) describes the 3 northern subspecies – *P. e. alascensis, eschatosus,* and *leucurus* – as a single, continuously distributed, clinally variable race, which he proposes be named *leucurus.*

The taxonomic status of the Pine Grosbeak in British Columbia, especially along the coast, requires clarification. Brooks (1922a) proposed recognizing the subspecies *P. e. carlottae* for the Queen Charlotte Islands based on a type specimen collected at Masset. After examining a July specimen collected on Vancouver Island by Swarth (1912) plus 2 winter specimens, Brooks (1939) later expanded the breeding range of this subspecies to include Vancouver Island and the mainland coast. Rand (1943) reviewed coastal specimens in the National Museum collection and concurred that they were likely all *P. e. carlottae.* Later, Adkisson (1977) analyzed morphological variations of Pine Grosbeaks across North America. He determined that specimens collected along the mainland coast of British Columbia, within the range described for *P. e. carlottae,* were indistinguishable from *P. e. flammula* specimens. Significantly, he noted that specimens from the Queen Charlotte Islands, with a single exception, were morphologically distinct from 3 specimens collected from the adjacent mainland coast (n = 3). Furthermore, Adkisson (1981) found that the vocalizations of the Queen Charlotte Islands birds were distinct from those from other geographic locations, suggesting a long period of isolation. The latter observations strongly suggest that *P. e. carlottae* is endemic to the Queen Charlotte Islands and warrants further study. The apparent confusion in taxonomy will probably be resolved only through mitochondrial DNA analysis.

Although human access into subalpine environments has increased dramatically over the past several decades in British Columbia, only a single Pine Grosbeak nest has been reported in the last 50 years. Biologists and naturalists in all regions are encouraged to record and report all sightings of this species during the main breeding period from late May through June. Observations during this period are important for determining the breeding distribution and habitat preferences of this secretive species. Observations on Vancouver Island and in the Taiga Plains and Boreal Plains are especially significant, since it remains unclear whether this species breeds in these areas.

The Pine Grosbeak is exceedingly tame and often easily approached (Newton 1972). This lack of wariness combined with its attraction to road verges in search of road salts, grit, and weed seeds makes it vulnerable to vehicle collisions. Our data contain numerous references to road mortalities, although usually of single birds. Other human-related causes of mortality include window collisions in settled areas and a single instance of hip chain entanglement. This species is a new addition to the growing list of forest birds that needlessly become caught in discarded hip chain string in North America (Loegering 1997).

Munro and Cowan (1947) report a Pine Grosbeak specimen collected on 15 June 1910 on Mount Douglas. This record should read Douglas Peak (south of Port Alberni) collected on 15 July 1910 (see Swarth 1912). This particular specimen, that of a juvenile male, was the primary source used by Brooks (1939) to infer breeding on Vancouver Island. Since mid-July is within the period of breeding dispersal, however, there can be no certainty that it came from a nest on Vancouver Island. We have no subsequent information to suggest that breeding occurs there.

On 5 January 1997, a new provincial high Christmas Bird Count of 610 Pine Grosbeaks was established at Prince George.

For further information on vocalizations, morphology, and life history of the Pine Grosbeak in North America, see Adkisson (1977, 1981, 1999).

NOTEWORTHY RECORDS

Spring: Coastal – Central Saanich 16 Mar 1974-6; Crofton 4 Mar 1978-6; Schooner Cove (Tofino) 4 Mar 1972-1 (Hatler et al. 1978); Surrey 21 Mar 1960-12; Sumas 19 Mar 1897-1 (MCZ 247365); Skagit River 16 Apr 1972-22; Klesilkwa River 25 Mar 1971-1, singing; Campbell River 20 Mar 1966-6, in willows; Nest Islets 27 May 1977-2; Sandspit 18 Apr 1987-7; Tlell 18 May 1936-1, collected; Cox Island 21 May 1977-2; Porcher Island 31 May 1921-a few breeding pairs seen (Brooks 1923); Kitimat 4 Mar 1975-15 feeding on mountain-ash berries; Lakelse Lake 18 Mar 1975-10; Pleasant Camp 28 May 1979-2. **Interior** – Lightning Lake (Manning Park) 4 Mar 1981-15; Kerr Creek (Midway) 10 May 1919-1 (MVZ 105388); Rossland 4 Mar 1973-2; Cranbrook 3 Apr 1948-11 (Johnstone 1949); Summerland 2 Mar 1972-12; Naramata 16 May 1979-3 fledglings fed by adults; Penticton 1 Mar 1970-30, nr Mac's Lake; Kelowna 1 Mar 1978-12 feeding on crab apples; Wasa 3 Mar 1941-12 (Johnstone 1949); Monashee Pass 5 Mar 1976-15; Nakusp 11 Mar 1976-9; Invermere 14 Mar 1982-15; Community Lake 19 May 1984-2; nr Celista 9 Apr 1948-22, 15 May 1948-4 eggs; Revelstoke 20 Mar 1972-9; 6.4 km se 100 Mile House 28 May 1982-6 in lodgepole pines; Murtle Lake (Wells Gray Park) 24 May 1950-2; Anahim Lake 28 Apr 1932-2 (MCZ 284740 and 284741); Puntchesakut Lake 30 May 1944-1 (Munro 1947); Quick 13 Mar 1983-5 in mixed flock with Pine Siskins and Common Redpolls; Hudson Bay Mountain (Smithers) 26 May 1980-1; Moberly Lake 2 Mar 1999-22, coming to feeder; Bulkley House 3 Mar 1938-1 (RBCM 7855); Tetana Lake 24 Mar 1938-1 (Stanwell-Fletcher and Stanwell-Fletcher 1943); Farrell Creek 31 May 1939-1 (RBCM 14906); Worth 10 May 1993-1; Fort St. John 16 Mar 1982-10, feeding in fruit tree, 14 May 1981-2; 6.4 km n Telegraph Creek 31 May 1919-1 (MVZ 39868); Tanzilla River 2 Mar 1978-5; Dease Lake 2 Mar 1978-5; Liard Hot Springs 10 Mar 1975-2 (Reid 1975); Helmut 24 May 1982-2 (RBCM 17462 and 17463); Atlin 12 Mar 1914-1 (RBCM 3400).

Summer: Coastal – Pender Island 4 Aug 1978-3; Douglas Peak (s Port Alberni) 15 Jul 1910-1 (MVZ 15868; Swarth 1912); Vancouver 20 Aug 1978-2; Mount Davis (Hope) 26 Jul 1987-1 fledgling fed by adult male; Gentian Ridge 24 Jul 1983-3, at timberline; nr Alta Lake 26 Jun 1924-1 (2,130 m); Pemberton Meadows 9 Jun 1985-2; Rivers Inlet 2 Jul 1937-2 (NMC 28156 and 28157); King Island Jun 1939-uncommon (Laing 1942); Lawn Hill (Queen Charlotte

City) 20 Jun 1973-3 (UMMZ 223450 to 223452); Masset Inlet 7 Jun 1980-2; Beresford Bay 3 Jul 1986-3; Masset 7 Jul 1988-2; Balance Rock 7 Aug 1985-2; Graham Island 7 Jun 1927-nest containing well-grown young (Darcus 1930); Masset 2 Jun 1920-1, type specimen (Brooks 1922a; MVZ 100803); Rocher Déboulé Range 19 Jul 1944-1 (Munro 1947); Flood Glacier 2 Aug 1919-1 male, singing in treetop (Swarth 1922); Rainy Hollow 24 Jun 1980-1. **Interior** – Lightning Lake (Manning Park) 11 Aug 1979-50; Red Mountain (Rossland) 3 Aug 1980-4; Rossland 15 Jul 1929-2 (Rand 1943); Apex Mountain (Penticton) 27 Jun 1999-2 adults, nest building; Naramata 15 Jul 1984-3 fledglings fed by adults; Redding Creek 14 Jun 1975-2, breeding pair in old burn (1,220 m); Nitta Creek 27 Jun 1999-4 eggs, 15 Jul 1999-3 or 4 nestlings, 1 fledged when nest inspected; Colin Creek (Fairmont Hot Springs) 18 Jun 1998-1 dead female entangled in hip chain line; Silver Star Mountain (Vernon) 9 Jun 1980-1; Chuwels Mountain 2 Jun 1996-1; nr Celista 20 Jun 1948-4 nestlings, ready to leave nest; Adams Plateau 28 Aug 1966-8; Crowfoot Mountain 16 Jul 1960-9, nr timberline; Revelstoke National Park 11 Aug 1943-8, all post-breeding adult males (Cowan and Munro 1944-46); Mount Revelstoke National Park 5 Jul 1937-6, mated pairs; Vermilion Pass 20 Jul 1983-2 fledged young; nr Amiskwi Lake (Yoho National Park) 27 Aug 1975-8 (Wade 1977); 6 km se 100 Mile House 2 Jul 1984-3; Kerbyville Creek 22 Jul 1997-1; Chilcotin River (Riske Creek) 5 Aug 1983-2 at Wineglass Ranch; Lac la Hache 20 Jun 1975-5 fledglings fed by both adults; nr Philip Lake (Wells Gray Park) 26 Jun 1971-6 (Grass 1971); Phyllis Lake 9 Jun 1977-2; Red Pass (Mount Robson Park) 8 Jun 1973-2; Bowron Lake Park 3 Jun 1971-1, road kill (Runyan 1971); Indianpoint Lake 10 Jun 1934-1 (MVZ 65682); Goodrich Lake 23 Jul 1963-1 fledgling with adult female; Vanderhoof 12 Jun 1951-3; Summit Lake (Prince George) 2 Aug 1975-3; Tatlow 6 Jul 1975-1; Topley 6 Jul 1956-5; Mast Creek 13 Jun 1977-2, breeding pair in burn (1,060 m); w Chetwynd 22 Jul 1975-7; 6.4 km s Germansen Landing 7 Jul 1967-1 fledgling with adults; Moberly Lake 18 Jul 1930-1 (Williams 1933a); e Chetwynd 5 Aug 1975-12; Stoddart Creek (Fort St. John) 6 Jul 1975-3; w Hudson's Hope 23 Aug 1975-2; Wonowon 19 Aug 1975-3; Kitchener Lake (Tatlatui Park) 25 Jun 1976-2; Klahowya Creek 21 Jul 1976-3; Akie River 1 Aug 1997-1, in old subalpine burn; Dokdaon Creek 17 Jul 1919-7 or 8 (Swarth 1922); Hyland Post 7 Jun 1976-1; Haworth Lake 5 Aug 1976-2, beside avalanche chute; Dease Lake 5 Jun 1962-3 (NMC 50197 to 50199); Fort Nelson 15 Jun 1975-6; s Basement Creek 9 Jun 1983-2; Cottonwood River (Cassiar) 12 Jun 1962-2 (NMC 50202 and 50203); Liard Hot Springs 15 Jul 1975-40 (Reid 1975); Kelsall Lake 18 Jun 1972-1 (CAS 68738); Survey Lake 26 Jun 1980-2.

Breeding Bird Surveys: Coastal – Recorded from 3 of 27 routes and on 2% of all surveys. Maxima: Queen Charlotte City 25 Jun 1994-9; Masset 25 Jun 1991-3; Kitsumkalum 16 Jun 1991-1. **Interior** – Recorded from 25 of 73 routes and on 8% of all surveys. Maxima: Mount Morice 16 Jun 1971-14; Wingdam 17 Jun 1969-11; Punchaw 16 Jun 1974-8.

Autumn: Interior – Atlin 16 Nov 1980-3; Liard River 23 Oct 1980-1; Fort Nelson 15 Oct 1986-2; Samotua River 18 Sep 1986-40; Hyland Post 4 Nov 1983-2; North Pine 10 Nov 1985-30; Fort St. John 19 Sep 1985-2, at feeder; 9.7 km n Germansen Landing 24 Oct 1972-10; Nilkitkwa River 14 Nov 1978-30, in burn; Mugaha Creek 6 Sep 1999-20; Aleza Lake 10 Oct 1964-1 (UBC 12029); s Prince George 28 Nov 1983-6; s Cheslatta 12 Nov 1997-17 feeding on willow galls; n Valemount 25 Nov 1977-8; Williams Lake 10 Nov 1984-20; Kleena Kleene 30 Nov 1950-12 (Paul 1959); Buffalo Lake 2 Nov 1962-4; Bridge Creek 28 Nov 1977-20; Yoho National Park 13 Sep 1975-30; China Head Mountain 9 Nov 1974-20 (2,135 m); s Knouff Lake 21 Oct 1984-10; Celista 4 Oct 1964-7, in orchard; Nakusp 30 Nov 1977-3; Okanagan Landing 21 Nov 1929-20; Baldy Mountain (n Sparwood) 15 Nov 1983-30; Wildhorse River 17 Oct 1982-3; Summerland 27 Nov 1973-20; Kinnaird 29 Oct 1969-6; Granby River 28 Nov 1958-20; Salmo 12 Nov 1979-10. **Coastal** – Stuhini Creek (Taku River) 17 Sep 1988-1, at creek mouth; Prince Rupert 12 Nov 1983-2; Masset 2 Oct 1971-20; Atnarko River (nr Stuie) 25 Sep 1940-2 (FMNH 177567 and 177568); Port Hardy 9 Sep 1934-2; Port Neville 21 Nov 1977-1; Courtenay 5 Sep 1954-2; Kennedy Lake (Ucluelet) 29 Nov 1971-15 (Hatler et al. 1978); Vancouver 10 Nov 1973-35 (Crowell and Nehls 1974a); Mount Tuam (Saltspring Island) 2 Nov 1986-1 at top of mountain; Bear Hill (Saanich) 13 Nov 1973-18; Colwood 24 Oct 1975-20.

Winter: Interior – Atlin 1 Dec 1930-1 (MVZ 105362); Liard Hot Springs 9 Feb 1975-5 (Reid 1975); Fort Nelson 11 Jan 1975-8, at airport; 14.5 km s Dease Lake 4 Jan 1963-1 (NMC 50210); Richards Creek (Prophet River) 23 Feb 1995-20, at headwaters; North Pine 20 Jan 1985-28; Flatrock 9 Jan 1983-6; e Halfway River 16 Jan 1983-33; Fort St. John 16 Jan 1982-35; Tetana Lake 19 Feb 1938-1 (RBCM 7867); 1.6 km e Dawson Creek 31 Dec 1978-30, perched in trembling aspen beside road; Moberly Lake 26 Feb 1999-1, killed nr feeder by Northern Shrike; Driftwood Creek (Smithers) 19 Jan 1975-30; Francois Lake 26 Dec 1987-25; Prince George 30 Dec 1972-195 (Rogers 1973b); 24 km s Prince George 28 Dec 1984-40; s Cheslatta 1 Dec 1997-8 on road; Quesnel 21 Feb 1979-15; Williams Lake 8 Feb 1959-100 feeding on weed seeds; Kleena Kleene Jan 1956-12 (Paul 1959); Yoho National Park 18 Dec 1977-34; Golden 26 Dec 1977-42; Clinton 24 Jan 1895-1 (RBCM 750); Revelstoke 20 Dec 1984-20; Celista 29 Jan 1948-22; Tranquille 9 Feb 1985-30; nr Lac le Jeune 15 Jan 1987-50; Edgewater 27 Feb 1978-15; Invermere 4 Jan 1978-22; Kamloops 2 Jan 1976-20; Glenbank 20 Jan 1985-30; Okanagan Landing 24 Dec 1925-27 (Munro 1926); Nakusp 2 Jan 1978-67; Oyama 5 Feb 1970-6; Wasa 15 Feb 1938-15 (Johnstone 1949); New Denver 23 Jan 1982-35; Kimberley 11 Dec 1973-20; Summerland 26 Dec 1973-43; Creston 1 Feb 1982-14; Porcupine Creek (Nelson) 12 Feb 1984-33; Salmo 12 Jan 1975-30; Manning Park 21 Jan 1983-7. **Coastal** – New Hazelton 17 Feb 1979-18; Greenville 4 Jan 1982-6; Terrace 4 Jan 1982-10; Mount Hays (Prince Rupert) 12 Feb 1984-2; Masset 1 Dec 1971-12; Tlell 3 Dec 1971-12; Bella Coola River 8 Jan 1976-10, feeding on wild currant bushes; Mount Washington 1 Dec 1981-20; Comox 1 Jan 1923-23; Squamish River 7 Dec 1980-7 nr estuary; nr Sunshine Valley (Hope) 10 Dec 1972-20; Harrison Hot Springs 11 Feb 1974-15; Burrard Inlet (Vancouver) 16 Jan 1895-12; North Vancouver 16 Dec 1973-24; Vancouver 7 Jan 1895-10; Reifel Island 27 Jan 1974-20 in crab apple tree; Long Beach 22 Jan 1974-2 (Hatler et al. 1978); Quamichan Lake 22 Dec 1973-12; Prevost Hill (Saanich) 8 Jan 1974-14.

Christmas Bird Counts: Interior – Recorded from 27 of 27 localities and on 72% of all counts. Maxima: Smithers 29 Dec 1985-450; Revelstoke 19 Dec 1992-244; Fort St. James 2 Jan 1992-237. **Coastal** – Recorded from 21 of 33 localities and on 13% of all counts. Maxima: Terrace 27 Dec 1981-124; Vancouver 26 Dec 1973-81; Campbell River 2 Jan 1984-52.

Purple Finch

Carpodacus purpureus (Gmelin)

PUFI

RANGE: Breeds from central and northeastern British Columbia, southern Yukon, southwestern Mackenzie, northern and central Alberta, central Saskatchewan, south-central Manitoba, northern Ontario, southern Quebec, and Newfoundland south along the Pacific coast to southern California, southeastern British Columbia, central Alberta, southeastern Saskatchewan, north-central North Dakota, Minnesota, central Wisconsin, central Michigan, northern Ohio, West Virginia, central Pennsylvania, and southeastern New York. Winters on the Pacific coast from southwestern British Columbia south through western Washington, central and western Oregon, and California to northern Baja California, and from central parts of the eastern range south to the Gulf coast of the United States.

STATUS: On the coast, *uncommon* to *fairly common* resident in the Georgia Depression Ecoprovince; in the Coast and Mountains Ecoprovince, *rare* to *uncommon* resident in the Southern Mainland Coast; *very rare* migrant and summer visitant to Western Vancouver Island and the Northern Mainland Coast, *casual* in both areas in winter; *accidental* in winter on the Queen Charlotte Islands.

In the interior, *rare* spring migrant and summer visitant, *casual* in autumn and winter in the Southern Interior Ecoprovince; *rare* spring migrant and summer visitant, *very rare* in autumn and *casual* in winter in the Southern Interior Mountains Ecoprovince; *uncommon* spring migrant and summer visitant, and *very* rare autumn migrant in the Central Interior, *casual* there in winter; *uncommon* migrant and summer visitant in the Sub-Boreal Interior Ecoprovince, *casual* there in autumn and winter; *uncommon* to locally *fairly common* migrant and summer visitant in the Boreal Plains Ecoprovince; *uncommon* migrant and summer visitant in the Taiga Plains and Northern Boreal Mountains ecoprovinces.

Breeds.

Figure 600. Adult male (a) and female (b) Purple Finch (Victoria, 26 April 1991; Tim Zurowski).

NONBREEDING: The Purple Finch (Fig. 600) is widely but sparsely distributed throughout much of British Columbia. On the coast, it occurs regularly in good numbers throughout the Georgia Depression, but becomes quite rare and local on Western Vancouver Island and on the Southern Mainland Coast. On the Northern Mainland Coast, it occurs rarely along some of the major river systems such as the Dean, Kitimat, and Skeena rivers, probably because of more open habitats created by river floodplains and human development. The Purple Finch appears to be absent from much of the outer coastal mainland and the outer coastal islands, except for a single winter record from the Queen Charlotte Islands. It reaches its highest numbers in winter in the Georgia Depression (Fig. 601).

In the interior, the Purple Finch occurs regularly in the wetter parts of south-central British Columbia, roughly corresponding to the distribution of the northern Interior Cedar–Hemlock and Sub-Boreal Spruce biogeoclimatic zones. It occurs regularly in eastern parts of the Central Interior, throughout the Sub-Boreal Interior, and in the southern and central parts of the Boreal Plains. Further north, it occurs locally and in small numbers.

In drier regions, such as most of the Southern Interior south of the Thompson River and in the East Kootenay Trench of the Southern Interior Mountains, the Purple Finch occurs only rarely. Cannings et al. (1987) note that the Purple Finch occurs sporadically only in winter in the Okanagan valley. In the Southern Interior Mountains, the species is quite rare across the southern portions, with only a few records from wetter parts of the west Kootenay and from the Rocky Mountains. It is not mentioned by Butler et al. (1986) or Munro (1950) at Creston, or by Johnstone (1949) in the east Kootenay. It becomes more regular north of Invermere.

The Purple Finch appears to be absent from much of the Chilcotin Plateau in the western Central Interior. Paul (1959, 1964a) did not mention it in his reviews of birds near Kleena

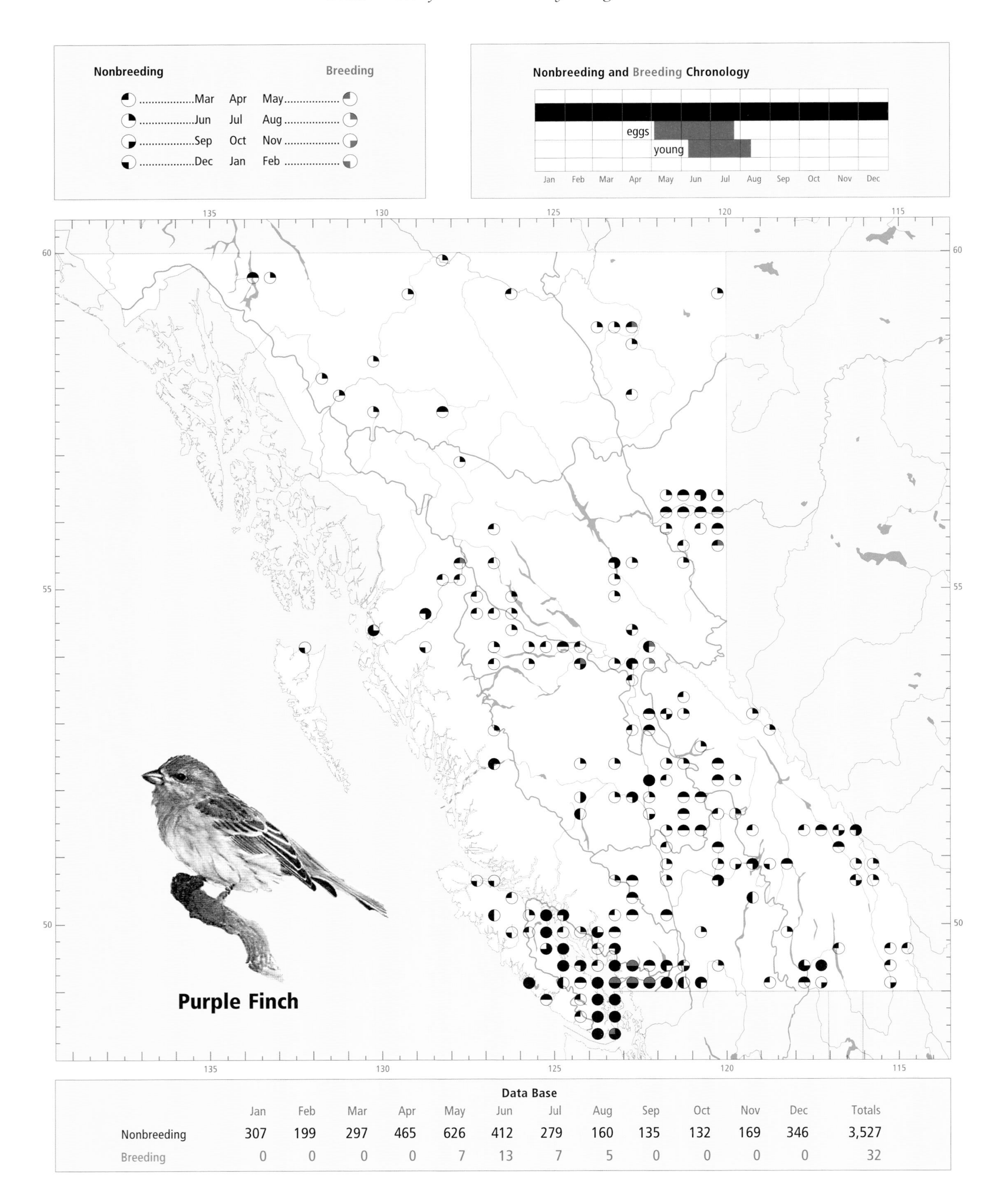

Data Base

	Jan	Feb	Mar	Apr	May	Jun	Jul	Aug	Sep	Oct	Nov	Dec	Totals
Nonbreeding	307	199	297	465	626	412	279	160	135	132	169	346	3,527
Breeding	0	0	0	0	7	13	7	5	0	0	0	0	32

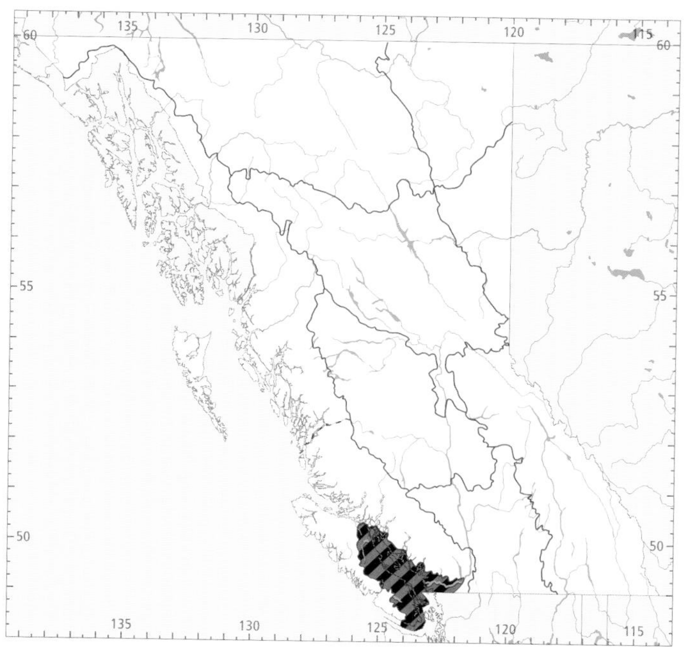

Figure 601. In British Columbia, the highest numbers for the Purple Finch in winter (black) and summer (red) occur in the Georgia Depression Ecoprovince.

Kleene, nor do we have any records west of Chilcotin Lake. Further east, it is uncommon but regular from about Riske Creek east to the Quesnel Highland.

On the coast, the Purple Finch occurs from near sea level to at least a few hundred metres; in the interior, it occurs mainly at elevations from about 300 to 900 m, but occasionally up to 1,700 m. The extent of this finch's use of higher-elevation habitats is not known. Migrants frequent habitats similar to those used during the breeding season (see BREEDING), except that riparian and edge habitats (Fig. 602) and bird feeders are used more extensively during migration. Unlike many other finches, the Purple Finch feeds on a wide range of seeds, buds, and insects (Wootton 1996), so it prefers mixed and open forests with a rich, brushy understorey that provides a diversity of food. It avoids dense and dark forests or open grasslands. In the Boreal Plains, most migrant Purple Finches are found along the edges of beaver ponds, muskegs, lakeshores, and streambanks, in woodlots or hedgerows bordering farms, and in stands of trembling aspen. In the Dawson Creek area, this species shifts from using trembling aspen in early spring to using more mixed forests in the nesting season (Phinney 1998). In the interior, the few wintering birds were always associated with backyard bird feeders.

On the coast, migrants and wintering birds frequent suburban and rural areas, riparian red alder, bigleaf maple, and black cottonwood forest mixed with conifers along rivers and wetlands, the edges of natural and human-made clearings, stands of Douglas-fir and western hemlock mixed with deciduous trees, treed estuaries, edges of marine coast, brushy areas, and fields with scattered clumps of trees. Backyard feeders are used extensively from autumn through early spring. The Purple Finch will even take nectar from hummingbird feeders, although not as often as House Finches.

Figure 602. In migration, the Purple Finch uses riparian and edge habitats throughout its range in British Columbia (Victoria, 9 April 1999; R. Wayne Campbell).

On the coast, there appears to be a spring movement in the Georgia Depression as small numbers pass through in April, about the same time that resident birds are beginning to establish breeding territories. The presence of resident birds masks the early arrival of migrants. The peak movement is almost a month earlier than in the interior (Figs. 603 and 604). The Purple Finch disappears from most of the coast in late summer, except in the Georgia Depression, where a small movement may pass through from mid-September to early October. Numbers climb into December as the wintering population builds.

In the southern portions of the interior, spring migrants may arrive in early April but small numbers peak in the first 2 weeks of May. A peak movement in the Central Interior occurs from the last week of April to the second week of May. In the Sub-Boreal Interior, early spring migrants can arrive as early as the first week of April, but the peak movement occurs quickly during the end of April and beginning of May. In the Boreal Plains, early arrivals appear by the third week of April and the peak occurs in the first 2 weeks of May. The peak movement in the Taiga Plains and Northern Boreal Mountains is in early to mid-May.

Autumn migration occurs in late summer through most of the interior, with the last flocks moving through the Boreal Plains in late August and only a few stragglers remaining until the third week of September. Throughout the rest of the interior, numbers decline noticeably after July, although small peaks in September in the Central Interior and Southern Interior Mountains may indicate small movements there (Figs. 603 and 604).

Salt (1952) and Wootton (1996) suggest that British Columbia's Purple Finch populations nesting west of the Rocky Mountains (both *C. p. californicus* and *C. p. purpureus*) winter largely in California and Arizona, whereas those nesting east of the Rocky Mountains winter in the eastern United States. Although limited, banding data suggest otherwise. One bird banded in April 1975 in Minnesota was recovered in June 1976 at War Lake, north of Prince George. Another bird, banded in Kansas in January 1984, was recovered at Telkwa in May 1987. These band recoveries suggest that Purple Finches breeding in interior British Columbia winter in the central or eastern United States. A bird banded at Tacoma, Washington, in March 1985 was recovered on Saltspring Island in June 1987, which is consistent with a coastal migration route.

Purple Finch populations do not irrupt to the same degree as those of other cardueline finches, such as Red and White-winged crossbills and Pine Siskins, that move en masse in response to changing cone and seed crops. Although eastern North American populations do have quasi-cyclic mass movements on the winter range, western populations apparently do not (Wootton 1996). Unusually high numbers (465 to 746) were recorded on the Vancouver Christmas Bird Counts in 1949, 1950, 1956, and 1957, which suggests that influxes occur during some winters. During winter, however, the Purple Finch is more abundant in the Georgia Depression than during the breeding season.

In coastal British Columbia, the Purple Finch occurs regularly throughout the year. In the southern and central portions of the interior, it occurs regularly from early April to the end of July and irregularly throughout the year. In the northern interior, it occurs regularly from late April or early May to the end of July, except in the Boreal Plains, where it can be regularly found through August (Fig. 603).

BREEDING: The breeding distribution of the Purple Finch in British Columbia is poorly known. On the coast, breeding has been confirmed in the Georgia Depression in the Nanaimo Lowland on southeastern Vancouver Island, on the Gulf Islands, and in the Fraser Lowland. Elsewhere on the coast, the Purple Finch has been reported breeding only from the Hazelton area on the Northern Mainland Coast. Breeding probably occurs on Vancouver Island locally west to Pacific Rim National Park; we have no evidence that it occurs on northwestern Vancouver Island. On the mainland, breeding probably occurs north of the Fraser Lowland to the Georgia Lowland and Southern Pacific Ranges, and on the Northern Mainland Coast along the Skeena and Bella Coola rivers, and possibly other major rivers. Breeding has not been recorded on the outer coastal islands, including the Queen Charlotte Islands.

In the interior, breeding has been reported only from Nulki Lake in the Central Interior, near Prince George and Aleza Lake in the Sub-Boreal Interior, Dawson Creek and Bissette Creek in the Boreal Plains, and Fort Nelson in the

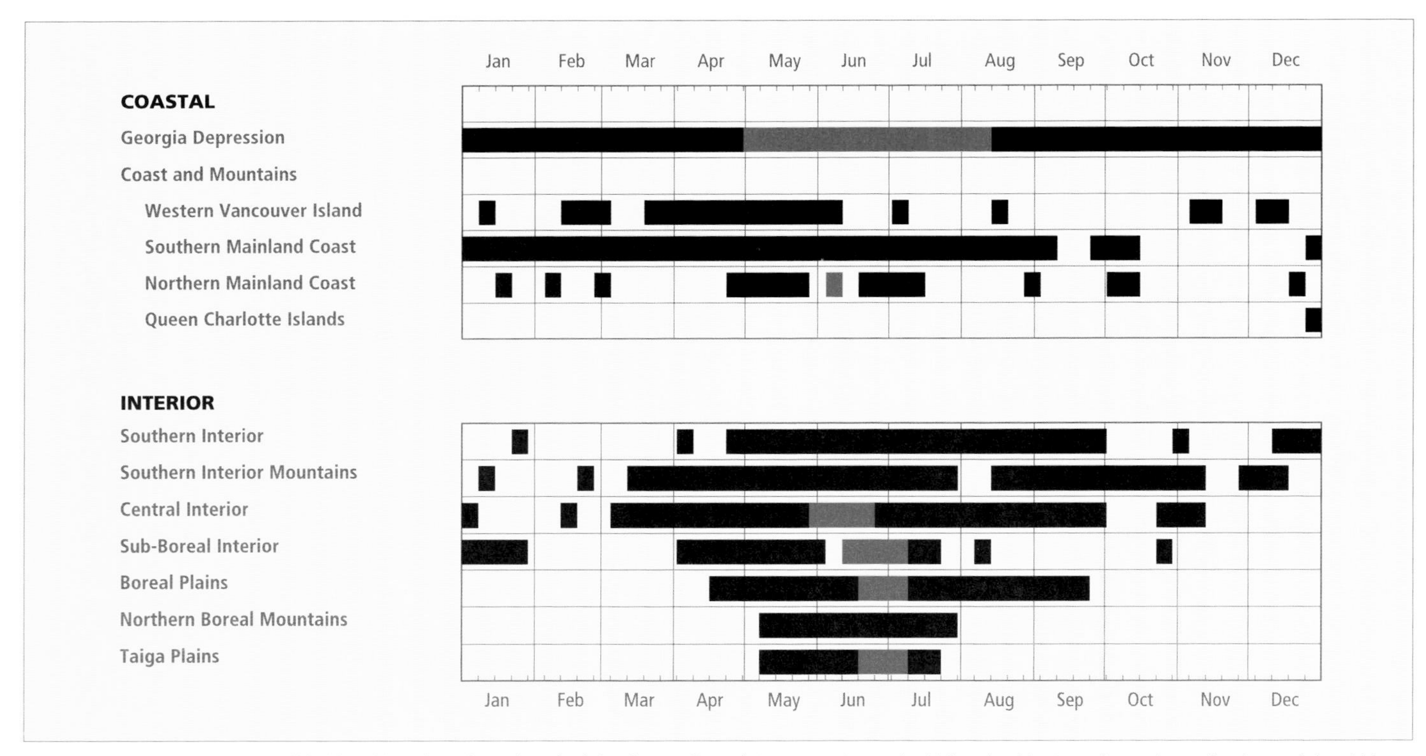

Figure 603. Annual occurrence (black) and breeding chronology (red) for the Purple Finch in ecoprovinces of British Columbia. Records are shown for the week in which they occurred.

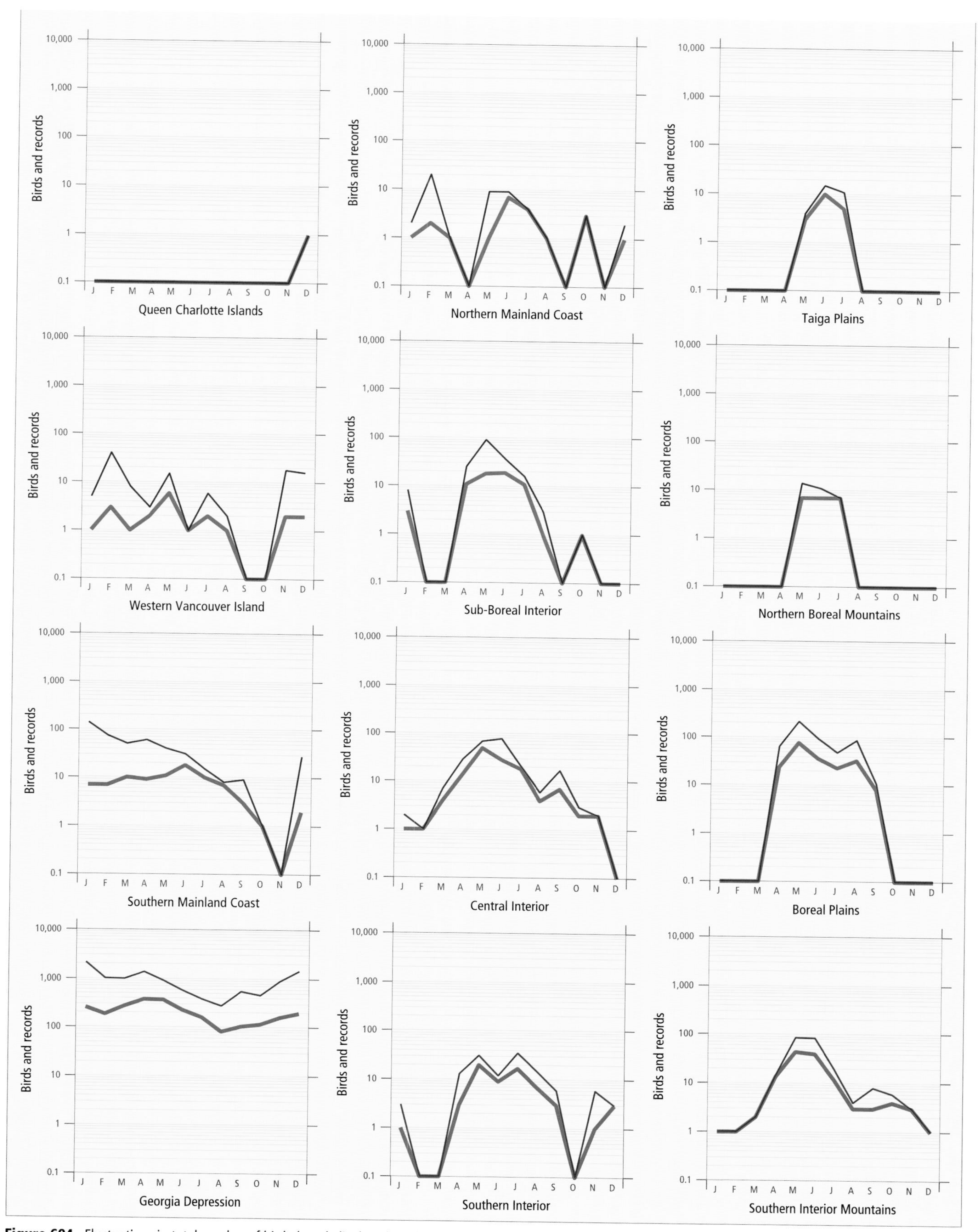

Figure 604. Fluctuations in total number of birds (purple line) and total number of records (green line) for the Purple Finch in ecoprovinces of British Columbia. Christmas Bird Counts, Breeding Bird Surveys, and nest record data have been excluded.

Taiga Plains. There is evidence of breeding in the form of fledged young from Kootenay National Park, Invermere, and Bowron Lake Park in the Southern Interior Mountains; 100 Mile House, Williams Lake, Riske Creek, and Bulkley Lake in the Central Interior; and Taylor in the Boreal Plains. It is likely, however, that the Purple Finch breeds throughout much of its summer range in the province.

Godfrey (1986) and Wootton (1996) include all of northern British Columbia within the breeding range of this finch, but records of nests with eggs or young, or recently fledged young, have yet to be obtained for the Northern Boreal Mountains.

In British Columbia, the Purple Finch reaches its highest numbers in summer in the Georgia Depression (Fig. 601). An analysis of Breeding Bird Surveys for the period 1968 through 1993 could not detect a net change in numbers on interior or coastal routes. An analysis of Breeding Bird Surveys across Canada from 1966 to 1996 suggests that populations declined at an average annual rate of 2.3% ($P < 0.01$); across North America, they declined at an average annual rate of 1.8% ($P < 0.01$) (Sauer et al. 1997).

Nests have been found at elevations from near sea level to 200 m on the coast, and from 490 to 900 m in the interior. In the interior, nesting likely occurs at considerably higher elevations than are currently documented. For example, in Kootenay National Park the Purple Finch is most common in the Montane Spruce Zone (Poll et al. 1984), which extends up to about 1,700 m.

Figure 605. In the Sub-Boreal Interior Ecoprovince of British Columbia, the Purple Finch breeds in mixed forests of paper birch, hybrid white spruce, and lodgepole pine (near Prince George, 13 June 1997; R. Wayne Campbell).

The Purple Finch generally breeds in moist coniferous forests at relatively moderate elevations (Figs. 605 and 606). On the coast, it breeds mainly in coniferous or mixed forests, but openings of some kind are usually present. It avoids the dense, dark interior of forests and seeks out more open habitats such as along edges of clearings, rivers, creeks, ponds, and rock outcroppings, and also human-modified habitat along roads, powerlines, farms, and suburbs. On southeastern Vancouver Island and the Gulf Islands, open stands of Douglas-fir, often mixed with arbutus and bigleaf maple, are used for breeding. On the lower slopes of mountainous areas, breeding habitats include mixed or pure stands of western hemlock and Douglas-fir. In very wet habitats, such as on the west coast of Vancouver Island, the Purple Finch frequents edge habitats, often in red alder stands, where it likely breeds (Hatler et al. 1973). In the interior, it avoids the dry, low-elevation ponderosa pine and Douglas-fir forests of the south; for example, it does not occur at all during the breeding season in the Okanagan valley (Cannings et al. 1987) or near Creston (Van Damme 1996). In the east Kootenay, it occurs along riparian river corridors in spring, where it likely breeds, but is absent from the dry valley bottoms. In Kootenay National Park, it occurs in spring and summer in subalpine fir and Engelmann spruce stands, on drier slopes with trembling aspen, along the edges of avalanche paths, and in valley bottom lodgepole pine and spruce forests (Poll et al. 1984). In Wells Gray Park, the Purple Finch occupies lower valley areas of mixed conifers and trembling aspens as well as small trembling aspen copses. In the Peace Lowland, it breeds in mixed or deciduous woodlands such as trembling aspen–balsam poplar–white spruce forests and white spruce–tamarack–trembling aspen–balsam poplar–birch forests (C. Siddle pers. comm.). In 1 study along the Liard River, the few Purple Finches encountered in June were found in immature deciduous stands on slopes above the river (Bennett and Enns 1996).

General nesting habitat in British Columbia is mainly human-influenced (63%; n = 30) or natural forest (37%). Human-influenced habitat includes well-treed rural and suburban areas (92%; n = 13), with 1 nest in an orchard. In more natural areas, stands of Douglas-fir, spruce (Figs. 605 and 606), and black cottonwood were used. Stands of trembling aspen were used only in the Boreal Plains. At Burnaby Lake, in the Georgia Depression, the forested edge of a wetland was used for nesting. The finch tends to leave urban and rural areas in mid-spring for more treed nesting areas, but a few birds apparently remain behind to nest in garden habitats.

On the coast, the Purple Finch has been recorded breeding from 1 May (calculated) to 11 August; in the interior, it has been recorded breeding from 3 June (calculated) to 7 July (Fig. 603).

Nests: Most nests were built in live conifers (71%; n = 21), mainly Douglas-fir, spruce, and subalpine fir. Coniferous

Figure 606. In the Northern Boreal Mountains Ecoprovince, the Purple Finch likely breeds in wet mixed stands of white spruce and trembling aspen (5.5 km north of Atlin, 11 June 1999; R. Wayne Campbell).

ornamental shrubs were used occasionally. Two nests were in bigleaf maples and 1 nest was in a trembling aspen. The most unusual nest was one near Nulki Lake, which was on the ground in the centre of a willow clump at the edge of a trembling aspen stand (Munro 1949). This appears to be the only record of a nest on the ground (Wootton 1996).

The heights of 24 nests ranged from ground level to 18 m, with 50% between 1.8 and 4.5 m. Several observations of birds carrying nesting material into the upper parts of mature conifers suggest that many nests are built high above the ground and are therefore not easily detected. This may explain why relatively few nests have been reported. Most nests were built saddled on a branch away from the trunk, 2 nests were in the crown of a tree, 1 was in a crotch of a main branch, and 1 was in a scrape on the ground. Nests are cups of dried grasses, twigs, and rootlets, occasionally with mosses, lichens, and spider's web. Nest cups are lined with fine grass and hair, and occasionally plant down.

Eggs: Dates for 18 clutches ranged from 4 May to 7 July, with 56% recorded between 21 May and 18 June. Calculated dates suggest that eggs can be found as early as 1 May and as late as 25 July. The range of egg dates, which spans a period of more than 2.5 months, suggests that at least some females in southern parts of British Columbia nest twice per year, which is expected as elsewhere the Purple Finch typically raises 2 broods annually (Yunick 1983). Egg dates on the coast are more than 4 weeks earlier than those in the interior (Fig. 603).

Sizes of 13 clutches ranged from 1 to 6 eggs (1E-2, 2E-1, 4E-2, 5E-7, 6E-1), with 54% having 5 eggs. The median clutch size of 5 eggs in British Columbia is larger than the mean clutch size of 4 eggs obtained from a sample of 258 clutches from the United States and eastern Canada (Wootton 1996). Because clutch size increases with latitude, however, and larger clutches (4.5 eggs per clutch) were found in Canada, the larger clutch sizes in British Columbia follow the established pattern. The incubation period is 12 or 13 days (Wootton 1996).

Young: Dates for 15 broods ranged from 8 June to 11 August, with 60% recorded between 18 June and 7 August. Sizes of 10 broods ranged from 1 to 5 young (1Y-4, 3Y-2, 4Y-2, 5Y-2). The nestling period is 13 to 16 days (Wootton 1996).

Brown-headed Cowbird Parasitism: There were no cases of parasitism by cowbirds in 15 nests found with eggs or young. There were, however, 3 cases (1 in the Boreal Plains and 2 in the Georgia Depression) of fledged cowbirds being fed by Purple Finches, so some parasitism does occur. The Purple Finch is an infrequent victim of cowbird parasitism; Wootton (1996) reported that 5.9% of 372 nests were parasitized, but none of these was from western North America.

Nest Success: Of 2 nests found with eggs and followed to a known fate, 1 produced fledglings. The failed nest was a ground nest that contained 3 dead nestlings and 1 egg 10 days after it was initially discovered with 4 eggs (Munro 1949). The female Purple Finch performs most of the incubation (Wootton 1996), yet Munro (1949) flushed only the male from the ground nest and never did observe a female in the vicinity.

REMARKS: There are 2 subspecies of Purple Finch, both of which occur in British Columbia. *C. p. purpureus* breeds from northern British Columbia south to about Lac la Hache in the Central Interior and Wasa Park in the east Kootenay; a few occur west to the central mainland coast. *C. p. californicus* occurs along the Pacific coast of North America; in British Columbia, it breeds on Vancouver Island, the Gulf Islands, and the southern mainland coast east to the Coast Mountains. Cumming (1933) reported *C. p. californicus* at Pavilion, which is within an area of possible overlap between the subspecies, but specimens have not been found to confirm the record. Godfrey (1986) referred to *C. p. taverneri* as a subspecies that occurs in British Columbia, but this subspecies is no longer considered valid (Wootton 1996).

Observers are encouraged to make identifications carefully where the range of the Purple Finch overlaps with its close relatives the Cassin's and House finch. Female and juvenile Cassin's and Purple finches are especially difficult to separate, and males can be as well (Kaufman 1990). In the Okanagan, Thompson, and Creston valleys and in the East Kootenay Trench, it is particularly important to document well any Purple Finch sightings, especially nesting and wintering birds, as there are few records from these drier habitats where the Cassin's Finch is present. On the coast, it is possible to confuse the very common House Finch, which routinely nests in suburban gardens, with the much less common Purple Finch, which very rarely nests in garden habitats.

We have excluded from this account several unsubstantiated records of Purple Finch from the southern portions of the interior, especially winter records. For example, 9 of 27 interior Christmas Bird Count localities (Smithers, Burns Lake–Francois Lake, Lake Windermere, Quesnel, Salmon Arm, Revelstoke, Vernon, Cranbrook, Creston) have reported the Purple Finch, but none of these records is satisfactorily documented. Confusion with the Cassin's Finch is a strong possibility in southern localities. Two published records that we have also excluded involve 8 birds on the 1926 Christmas Bird Count at Okanagan Landing (Munro 1927), and 4 birds on the 1948 Christmas Bird Count at Summerland (Tait 1949). Cannings et al. (1987) also exclude these 2 records based on the lack of documentation and the possibility of confusion with the Cassin's Finch.

There are few conservation concerns for this widely distributed songbird. Since the Purple Finch prefers open woodlands, selective logging can improve its habitat in stands of dense trees. Extensive clearcuts, on the other hand, probably reduce populations by eliminating habitat (Wootton 1996).

Common predators of the Purple Finch are probably small forest hawks such as Sharp-shinned and Cooper's hawks, and small falcons such as the Merlin and American Kestrel (Wootton 1996). One unusual record was of the remains of a Purple Finch found in Barn Owl pellets taken at Alaksen National Wildlife Area in Delta (Dawe et al. 1978).

Additional information on the ecology of the Purple Finch in North America can be found in Wootton (1996).

NOTEWORTHY RECORDS

Spring: Coastal – San Juan River 15 May 1974-2; Sidney Island 18 May 1984-60; 2 km e Deroche 4 May 1969-4 eggs; Haney 21 May 1963-6 eggs; Burnaby 28 May 1993-4 fledglings; Tofino Inlet 3 May 1974-8; Vancouver 22 Apr 1971-1 egg, 1 May 1955-3 eggs; Harrison Hot Springs 8 Mar 1986-25; Hope 14 Apr 1922-30 (Thacker and Thacker 1923); Pemberton 15 Apr 1996-1; Campbell River 21 May 1976-2 fledglings; Cortes Island 18 May 1975-12; Port Neville 24 Mar 1987-8; Bella Coola 21 Apr 1933-1 (MCZ 284709); Terrace 16 May 1976-9. **Interior** – Rock Creek 27 May 1984-1; Elk River 15 May 1984-3; Skihist Park 10 May 1969-4; Seton Creek 13 May 1916-1 (RBCM 3403); Invermere 18 Mar 1983-1; Revelstoke 5 Apr 1989-1; Falkland 4 Apr 1977-10; Kamloops 24 May 1986-1 male; Green Lake (70 Mile House) 22 May 1959-1; Bridge Lake 9 May 1958-1; Riske Creek 29 Mar 1987-1, earliest arrival date; nr Tatlayoko Lake 1 Apr 1991-1; Quesnel 9 May 1966-1 (Rogers 1966c); Prince George 3 Apr 1987-6, increased to about 6 or 7 pairs at a feeder in a week; Vanderhoof 20 Mar 1983-4; Smithers 15 May 1975-1; Mackenzie 9 Apr 1998-1, 15 May 1996-43; 5 km s Hudson's Hope 30 May 1976-1 (ROM 126459); Cache Creek (Peace River) 17 Apr 1986-1 nr bridge; Fort St. John 27 Apr 1985-14, 12 May 1986-15; Hyland Post 29 May 1976-5; 5 km s Prophet River Park 26 May 1981-1; Prophet River Park 13 May 1982-1; Fort Nelson 25 Apr 1987-1, 24 May 1975-2; Liard Hot Springs 10 May 1975-1 (Reid 1975); Atlin 29 May 1981-1.

Summer: Coastal – Deroche 11 Aug 1969-3 young; Vancouver 18 Jul 1971-4 nestlings; Haney 8 Jun 1965-5 nestlings; Pitt Meadows 21 Jun 1972-5 nestlings; Oxbow Lake (Pemberton) 7 Jun 1997-2; Diana Island 5 Jul 1973-4; Kimsquit 7 Jul 1939-1 (NMC 29058); Hazelton 4 Jun 1921-female incubating. **Interior** – Invermere 14 Jul 1981-3 fledglings; Kootenay National Park 2 Aug 1965-4 fledglings; Pavilion 13 Aug 1932-1 (Cumming 1933); Chilcotin River (Riske Creek) 26 Jul 1992-2 fledglings at Wineglass Ranch; Eagle Lake 1 Jun 1991-1; Bowron Lake Park 9 Jun 1975-3 fledglings; Nulki Lake 6 Jun 1945-4 eggs, 16 Jul 1945-family with fledged young (Munro 1949); 30 km e Prince George 27 Jun 1968-5 eggs; Prince George 30 Jun 1983-4 fledglings; Stellako River to Francois Lake 13 Jun 1951-25; Vanderhoof 6 Jun 1949-4 eggs; Aleza Lake 3 Jul 1969-4 nestlings; Trapper Creek (Chetwynd) 25 Jun 1969-1 (NMC 57111); Mackenzie 5 Jul 1994-1; Mugaha Creek 11 Aug 1996-3 banded in 1 morning, only ones all summer; nr Dawson Creek 6 Jul 1986-15, 7 Jul 1994-nest with 1 egg and 1 newly hatched chick (Phinney 1998); Bissette Creek 7 Jul 1994-1 egg; Cheslatta 5 Jun 1976-1; Tupper Creek 15 Jun 1974-1; Alces River 22 Aug 1984-1; Bear Flat 12 Jun 1986-30 in 1 flock; Taylor 21 Aug 1975-4 fledglings; Cache Creek (Peace River) 17 Jun 1984-2; Mount Gordon (Stikine River) 21 Jul 1949-2 (RBCM 9855-6); Telegraph Creek 5 Jul 1919-1 (MVZ 39874); Kledo Creek 21 Jun 1984-1, 19 Jul 1985-5; Hyland River 29 Jul 1981-1.

Breeding Bird Surveys: Coastal – Recorded from 22 of 27 routes and on 55% of all surveys. Maxima: Campbell River 23 Jun 1973-18; Chemainus 30 Jun 1988-16; Courtenay 18 Jun 1992-14. **Interior** – Recorded from 32 of 73 routes and on 17% of all surveys. Maxima: Prince George 15 Jun 1980-14; Nicola 17 Jun 1973-13; Telkwa High Rd 23 Jun 1974-12.

Autumn: Interior – St. John Creek (Fort St. John) 7 Sep 1985-2, 22 Sep 1986-1, last autumn record; Mackenzie 23 Oct 1996-1; Eagle Lake 27 Sep 1992-1; Nulki Lake 2 Sep 1945-1 (Munro 1949); Wistaria Nov 1939-1 (RBCM 8852); Barkerville 16 Sep 1962-4; Williams Lake 23 Oct 1970-2 nr creek; Riske Creek 15 Sep 1978-4, 22 Oct 1983-1; Chilcotin River 5 Nov 1990-1; Golden 1 Sep 1977-3; Field 23 Sep 1976-1 (Wade 1977); Boulder Creek 1 Oct 1976-3 (Wade 1977); Scotch Creek 2 Nov 1962-6; Adams River 15 Sep 1962-3; Kamloops 2 Sep 1989-1 at feeder; Invermere 23 Oct 1976-1; Salmo area 5 Nov 1983-1; Castlegar 1 Nov 1997-1 male at feeder; Nelson 27 Nov 1996-1 male; Newgate 15 Oct 1930-1 (NMC 24638). **Coastal** – Terrace 9 Oct 1969-1 (Hay 1976); Prince Rupert 5 Oct 1983-1; Egmont 3 Sep 1977-3; Squamish 25 Sep 1971-4; Hope 11 Oct 1922-1 (Thacker and Thacker 1923); Tofino 6 Nov 1982-9; Saanich 1 Oct 1986-68, 9 Nov 1988-40.

Winter: Interior – 43 km n Prince George 1 Jan 1983-1; Newlands 28 Jan 1989-2; Prince George 17 Jan 1989-5; nr Tatlayoko Lake 13 Feb 1991-1; Williams Lake 3 Jan 1982-1, 4 Jan 1981-2; Vernon 12 Nov to 30 Dec 1981-1 (BC Photo 743), 17 Dec 1995-1; Nelson 15 Dec 1996-1 male in flock of House Finches. **Coastal** – Prince Rupert 16 Jan 1984-2, 26 Feb 1984-18; Masset 26 Dec 1971-1 (Crowell and Nehls 1972b); Kitimat 18 Dec 1982-1 (Horwood 1983); Sointula 12 Feb 1989-2; Telegraph Cove 12 Dec 1940-6 (Munro and Cowan 1947); Waring Peak 11 Jan 1984-5; Pitt River 22 Jan 1983-45; Courtenay 30 Jan 1942-200; Union Bay 23 Feb 1989-46; Harrison Hot Springs 26 Dec 1988-14, 3 Feb 1986-24, 29 Feb 1980-15; North Vancouver 27 Jan 1976-200; Tofino 7 Dec 1982-10; Manning Park 23 Jan 1971-50; Skagit River 1 Jan 1971-30; Saltspring Island 11 Feb 1949-34.

Christmas Bird Counts: Interior – Recorded from 1 of 27 localities and on 1% of all counts. Maxima: Vernon 16 Dec 1990-1. **Coastal** – Recorded from 25 of 33 localities and on 75% of all counts. Maxima: Vancouver 26 Dec 1957-632; Victoria 18 Dec 1993-227; Duncan 2 Jan 1993-224.

Cassin's Finch

Carpodacus cassinii (Baird)

CAFI

Figure 607. In Canada, the Cassin's Finch breeds only in south-central British Columbia and extreme southwestern Alberta. Males (a) establish a territory only when females (b) begin to search for a nest site (Anthony Mercieca).

RANGE: Breeds from south-central British Columbia, southwestern Alberta, western Montana, Wyoming, and South Dakota south along the eastern slope of the Cascade Mountains and Sierra Nevada to California, and along the eastern slope of the Rocky Mountains to northern New Mexico. Also breeds locally in coastal California and Baja California. Winters from southern interior British Columbia through its breeding range, and south along mountainous areas through Arizona and New Mexico to northern Mexico.

STATUS: On the coast, *casual* spring to autumn vagrant and *accidental* in winter in the Georgia Depression Ecoprovince; in the Coast and Mountains Ecoprovince, *casual* vagrant on the Southern and Northern Mainland coasts and *accidental* on the Queen Charlotte Islands; absent from Western Vancouver Island.

In the interior, *uncommon* to locally *common* migrant and summer visitant and *uncommon* to *fairly common* winter visitant to the Southern Interior and Southern Interior Mountains ecoprovinces; *uncommon* migrant and summer visitant and *casual* winter visitant in the Central Interior Ecoprovince; *casual* in the Sub-Boreal Interior Ecoprovince.

Breeds.

CHANGE IN STATUS: The Cassin's Finch (Fig. 607) was previously known as a common summer visitant in the Southern Interior and parts of the Southern Interior Mountains, where it was also known to winter, but was only a casual visitant in the Cariboo and Chilcotin areas of the Central Interior (Munro and Cowan 1947). It now breeds locally in southern parts of the Central Interior and is a casual summer visitant in Quesnel and the upper Fraser River area near Valemount and McBride. It is now also a very rare vagrant on the coast.

Hahn (1996) speculates that the Cassin's Finch is gradually expanding its range to the northwest, based on summer records from the Yukon and Alaska. Records from Kitimat, Hazelton, and the Queen Charlotte Islands, which are documented here, support this contention.

NONBREEDING: The Cassin's Finch is widely distributed at low and intermediate elevations in southern portions of interior British Columbia. It occurs most abundantly in the Southern Interior, becoming more locally distributed in southern parts of the Southern Interior Mountains. In the Central Interior, it occurs locally north in the Cariboo Basin and Fraser River Basin to Williams Lake and Riske Creek, and casually west in

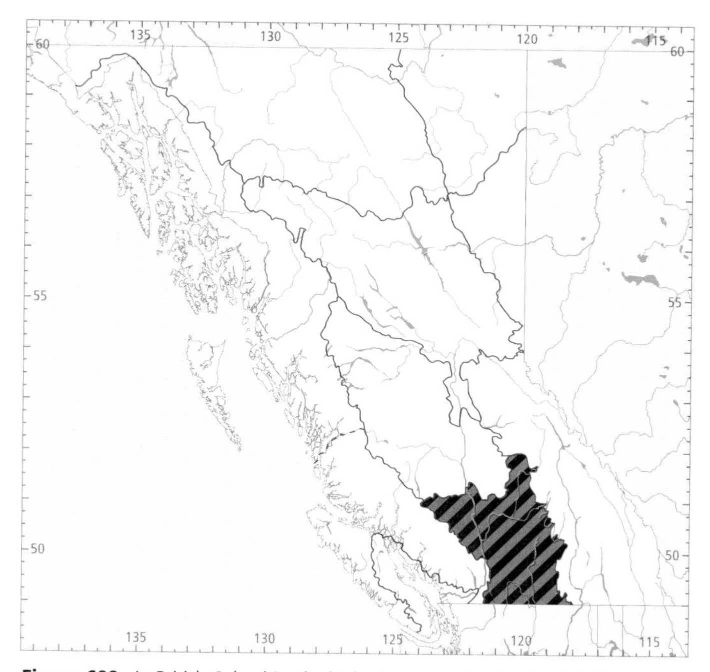

Figure 608. In British Columbia, the highest numbers for the Cassin's Finch, both in winter (black) and summer (red), occur in the Southern Interior Ecoprovince.

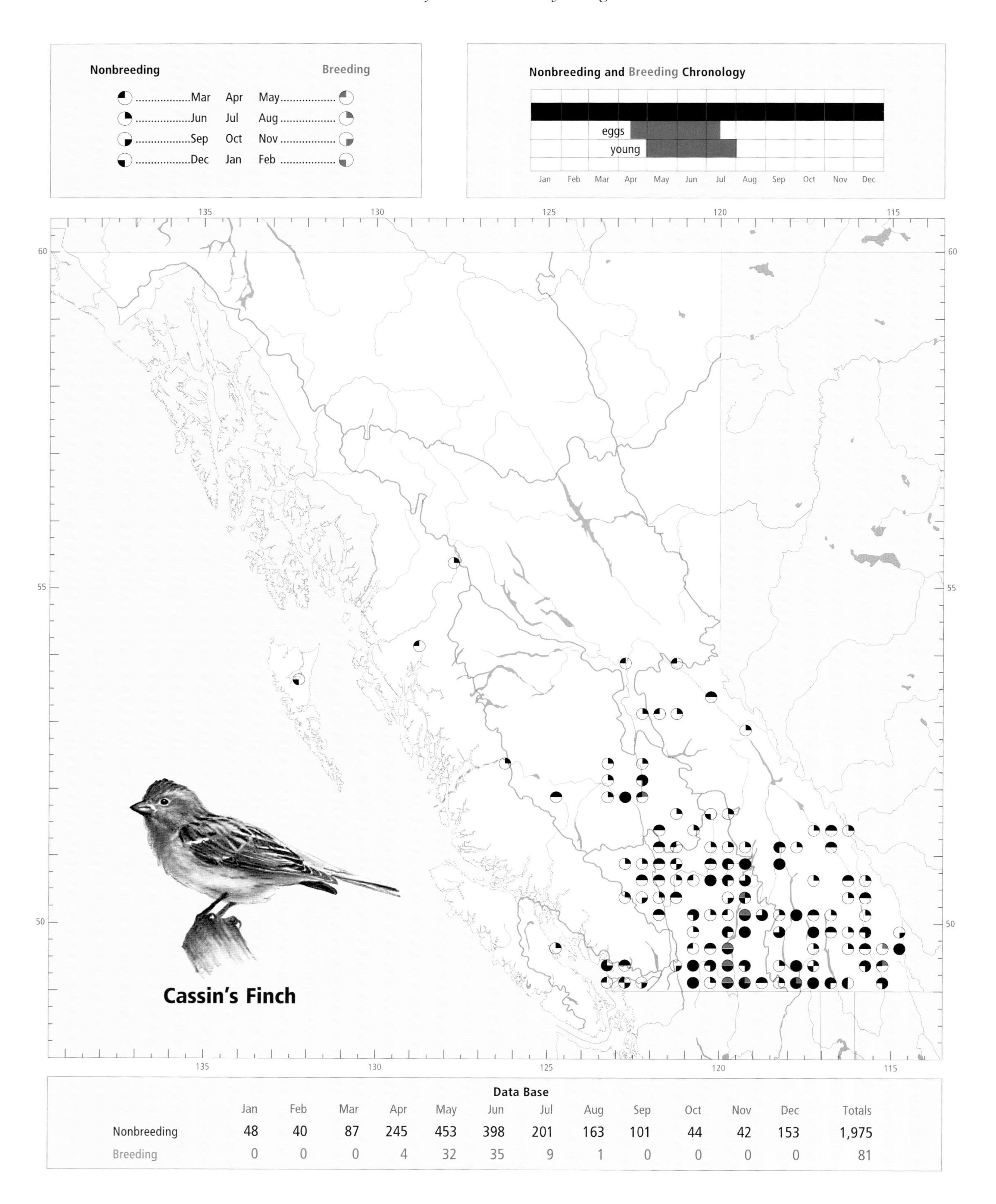

Data Base

	Jan	Feb	Mar	Apr	May	Jun	Jul	Aug	Sep	Oct	Nov	Dec	Totals
Nonbreeding	48	40	87	245	453	398	201	163	101	44	42	153	1,975
Breeding	0	0	0	4	32	35	9	1	0	0	0	0	81

Figure 609. In British Columbia, the Cassin's Finch is most commonly found in ponderosa pine or mixed ponderosa pine–Douglas-fir forests throughout the year (White Lake, Okanagan Falls, 19 November 1995; R. Wayne Campbell).

the Chilcotin Plateau to Stum Lake and Kleena Kleene. In the Southern Interior Mountains, it occasionally occurs north to McBride and Papoose Lake. In winter, the highest numbers of the Cassin's Finch in British Columbia occur in the Southern Interior (Fig. 608).

On the coast, the Cassin's Finch is a very rare visitor. There are a handful of records from the Fraser Lowland in the Fraser River delta and from near Hope. Further north, there are single records at Hazelton and south of Port Clements, at Juskatla on the Queen Charlotte Islands.

In the interior, the Cassin's Finch occurs regularly in drier parts of southern British Columbia, roughly corresponding to the distribution of the Bunchgrass, Ponderosa Pine, and Interior Douglas-fir biogeoclimatic zones. In all areas, the Cassin's Finch is most abundant in ponderosa pine or mixed ponderosa pine and Douglas-fir forests (Figs. 609 and 613). Nonbreeding habitat is similar to breeding habitat (see BREEDING) and consists mainly of open ponderosa pine, Douglas-fir, and mixed riparian stands, but the Cassin's Finch also uses brushy riparian habitat along rivers, gullies, and lakeshores, as well as pastures, weedy roadsides, and burns, for foraging. In winter, backyard feeders and gardens are used more often than at other times of the year.

Small numbers of birds occur in subalpine forests and may occur up to timberline (e.g., 1,900 m in Cathedral Park and eastern Manning Park and 2,100 m on Scaia Mountain). In subalpine areas, the species frequents the forested edges of meadows and the high canopy of forested areas. High-elevation records tend to be from mid to late summer, suggesting that some birds move briefly to higher elevations after nesting is completed, a behaviour noted elsewhere (Orr 1968). In the west Kootenay, the Cassin's Finch occurs in the Interior Cedar–Hemlock Zone where valleys have been opened up through logging or clearing of forests for farmland. In the Central Interior, it occurs at lower elevations along valley bottoms. In all interior areas, it can often be found on dirt and gravel roads and roadsides, where it seeks out salt-enriched material. On the coast, all records are from suburban or rural areas, usually at backyard bird feeders.

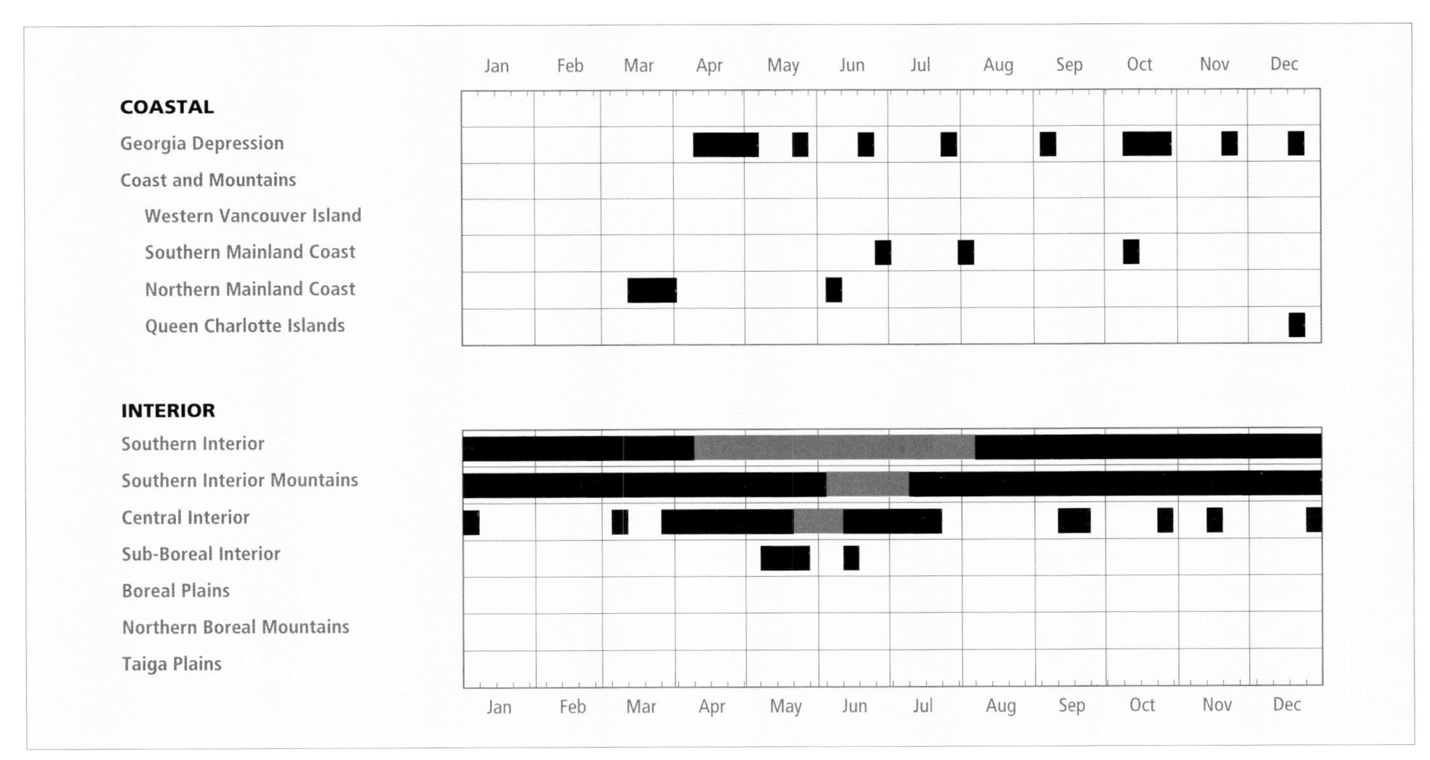

Figure 610. Annual occurrence (black) and breeding chronology (red) for the Cassin's Finch in ecoprovinces of British Columbia. Records are shown for the week in which they occurred.

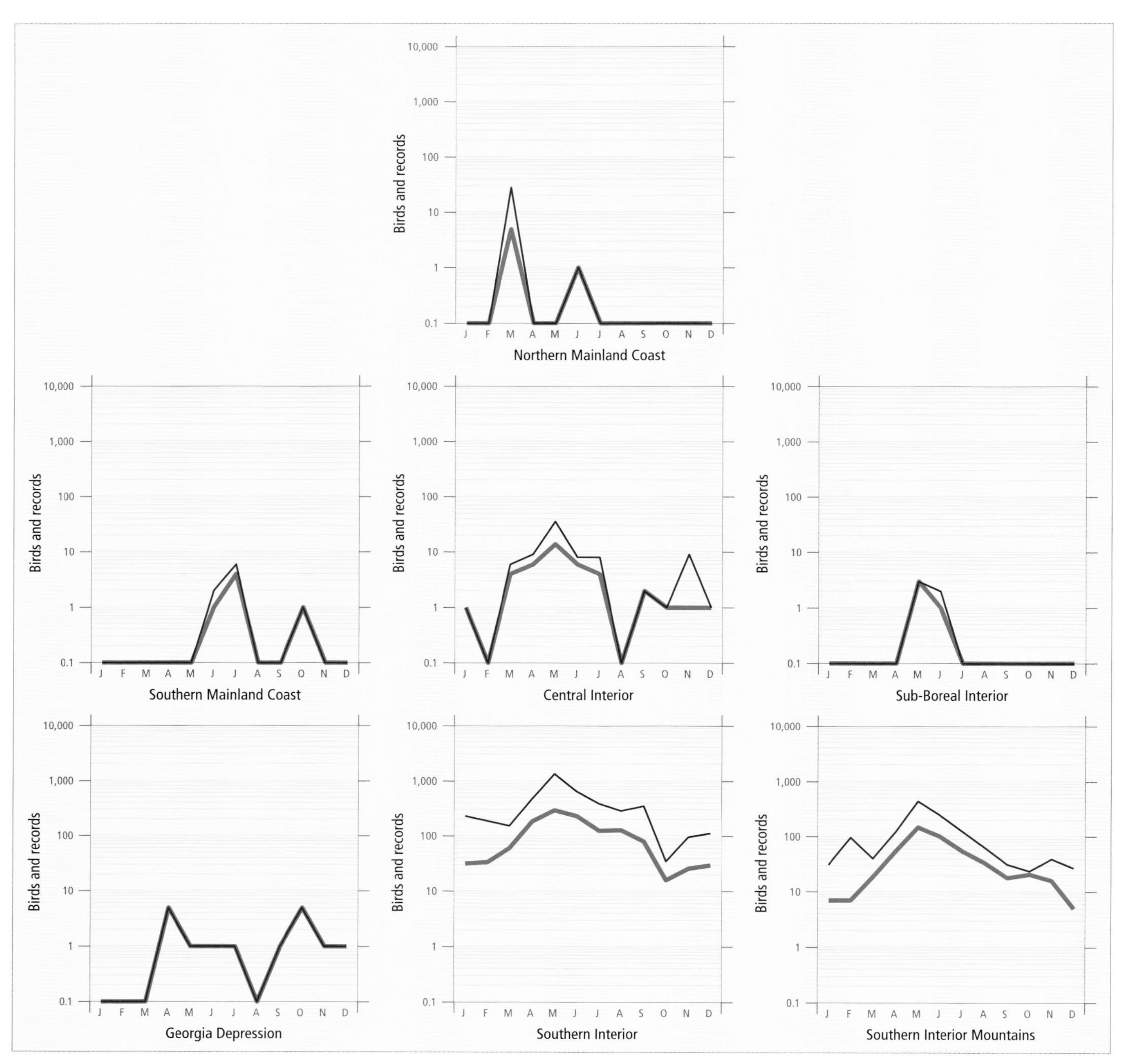

Figure 611. Fluctuations in total number of birds (purple line) and total number of records (green line) for the Cassin's Finch in ecoprovinces of British Columbia. Christmas Bird Counts, Breeding Bird Surveys, and nest record data have been excluded.

In the interior, spring migrants may arrive in early March, but the first large influx occurs in the last week of March or the first week of April. The main spring migration occurs in April and May (Figs. 610 and 611).

Premigration flocks form in early August and some may leave the region by the end of the month, although the peak movement is in early to mid-September (Fig. 611). Most migrants have left the southern portions of the interior by late September, leaving a small population to spend the winter.

Very small numbers overwinter at feeders in most years, but occasionally large flocks overwinter in ponderosa pine forests when there is a large cone crop. In British Columbia, the Cassin's Finch reaches its highest numbers in winter in the Southern Interior (Fig. 608); wintering populations are highest in southern parts of the Okanagan valley (Cannings et al. 1987).

The Cassin's Finch does not irrupt in the same way as some other cardueline finches, such as crossbills and Pine Siskins, which move en masse in response to changing cone and seed crops. This is reasonable since its diet does not depend entirely on cone crops but also on insects, berries, herbaceous seeds, and vegetative buds of conifers (Hahn 1996).

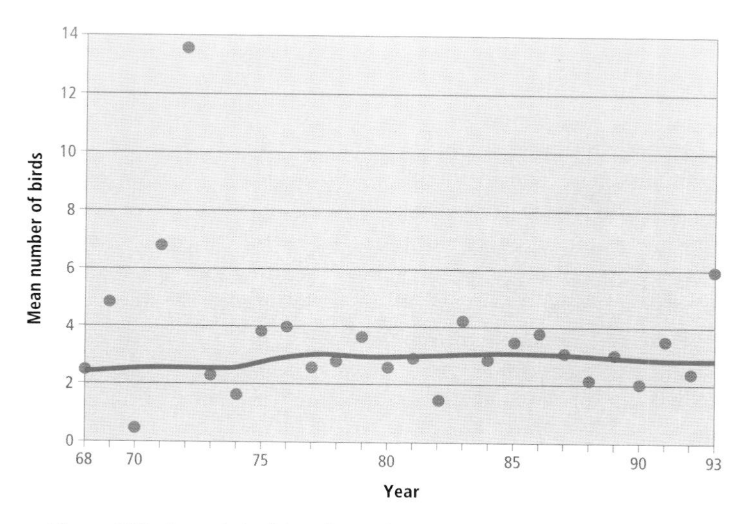

Figure 612. An analysis of Breeding Bird Surveys for the Cassin's Finch in British Columbia shows that the mean number of birds on interior routes has increased at an average annual rate of 2% over the period 1968 through 1993 ($P < 0.10$).

It is likely, however, that wintering populations in British Columbia rely on ponderosa pine cone crops. Presumably, large numbers of Cassin's Finches are present during winters when cone crops are bountiful and scarce when cone crops are poor.

In coastal British Columbia, the Cassin's Finch occurs sporadically from April through October, and can be found occasionally into December. In the Southern Interior and Southern Interior Mountains, the Cassin's Finch occurs year-round; in the Central Interior, it occurs regularly from late March to mid-July, and casually the rest of year (Fig. 610).

BREEDING: The Cassin's Finch breeds in the southern and central portions of interior British Columbia. Breeding has been documented from the Coldwater and Nicola rivers west and southwest of Merritt, east through the lower Similkameen and Okanagan valleys, to the west Kootenay near Trail and Rossland, and the East Kootenay Trench and Kishenina Creek; then north in the east Kootenay to Wasa and throughout the Okanagan valley north to the South Thompson River and Shuswap Lake. The furthest northern breeding localities are near Riske Creek and at Williams Lake in the Central Interior.

Summer records of birds suggest that breeding probably occurs west of the Fraser River to Pemberton, Carpenter Lake, and Kleena Kleene. Fledged young have been found in the Chilcotin as far west as Mosher Creek in Tweedsmuir Park. There are large areas of the Chilcotin Plateau and Chilcotin Ranges with apparently suitable habitat (grassland and Douglas-fir) for which we have no records. The presence of fledglings at Mosher Creek suggests that there may be local populations in some parts of the western Cariboo and Chilcotin areas.

In the east Kootenay, breeding is known to occur north to Wasa, but it also likely occurs north to about Golden, locally near Revelstoke, and in most of the major river valleys running north from the international boundary. The Cassin's Finch may breed in Yoho National Park, but it does not occur there every year (Wade 1977).

Breeding is poorly documented in the Thompson Basin, but likely occurs throughout the southern interior west to Pavilion and Lillooet. Breeding has yet to be documented at Creston, but undoubtedly occurs there. Breeding has not been documented on the coast.

In British Columbia, the Cassin's Finch reaches its highest numbers in summer in the Southern Interior (Fig. 608). An analysis of Breeding Bird Surveys for the period 1968 through 1993 shows that numbers on interior routes increased at an average annual rate of 2% (Fig. 612); data for coastal routes were insufficient for analysis. An analysis of Breeding Bird Surveys across North America from 1966 to 1996 could not detect a net change in numbers (Sauer et al. 1997).

The Cassin's Finch breeds mainly in dry and open coniferous forests, or the ecotone between grasslands and forests, at low to moderate elevations (Fig. 613). Nesting has been documented at elevations from 280 to 1,890 m, but only 9% of nest records are from above 1,000 m. The species' breeding distribution correlates roughly with the distribution of the Bunchgrass, Ponderosa Pine, and Interior Douglas-fir biogeoclimatic zones. It is most abundant in forests of ponderosa pine or mixed ponderosa pine and Douglas-fir, but also nests along riparian strips of black cottonwood and trembling aspen along wetlands and rivers.

Figure 613. The Cassin's Finch breeds mainly in dry and open coniferous forests throughout its summer range in British Columbia (11 km east of Grand Forks, 4 July 1997; R. Wayne Campbell).

Figure 614. Most Cassin's Finch nests in British Columbia contained 4 eggs (Richter Pass, 27 May 1962; R. Wayne Campbell).

In a study in the East Kootenay Trench, the Cassin's Finch was most abundant during the breeding season in mature mixed ponderosa pine and Douglas-fir stands (Schwab 1979). These stands were relatively open and had an understorey mainly of soopolallie. Only small numbers of Cassin's Finches were found in mature lodgepole pine or climax Douglas-fir stands; none were found in grassland, scrub, or young coniferous forest habitat.

Dry rocky mountainsides with scattered conifers are also used for nesting; these dry hillsides can extend the finch's use of habitat into higher elevations than in adjacent, denser forest. The Cassin's Finch rarely breeds in the higher-elevation subalpine forests of spruce and subalpine fir.

In wetter regions, such as the west Kootenay, the Cassin's Finch has likely colonized once dense forests that were harvested and turned into farmlands. Summer records from Williams Lake and the upper Fraser River near Valemount suggest that farmlands or other altered habitats are being used almost exclusively there. West of Williams Lake, the Cassin's Finch frequents riparian forests and the edges of lodgepole pine stands and grasslands.

The general nesting habitat in British Columbia was described mainly as human-influenced (56%; n = 110) or natural forest (39%), but shrublands (3%) and riparian thickets were also used. Human-influenced habitats included mainly rural farms and pastures (43%; n = 73). In more natural habitats, ponderosa pine woodlands (18%) and mixed Douglas-fir and ponderosa pine woodlands (12%) were used most often; but also trembling aspen copses, big sagebrush shrublands with scattered ponderosa pines, mixed subalpine fir and hybrid white spruce forest, lodgepole pine forest, and black cottonwood riparian forest. Unlike the House Finch, the Cassin's Finch tends to avoid nesting in ornamental trees in suburban or garden habitats.

The Cassin's Finch has been found breeding in British Columbia from 14 April (calculated) to 22 August (Fig. 610).

Nests: Most nests were built in live conifers (79%; n = 67); deciduous trees (16%) and shrubs (3%) were also used. Nest trees were mainly ponderosa pine (58%; n = 60), Douglas-fir (22%), and trembling aspen (12%), but there were single records of a nest in lodgepole pine, western redcedar, juniper, birch, and saskatoon. Nest trees ranged in age from young to mature trees.

The Cassin's Finch nests semi-colonially (Samson 1976), an unusual breeding strategy for a forest songbird. Examples of semi-colonial nesting in British Columbia are rare, however. Near Celista, a group of 4 nests was found in a single trembling aspen tree, and near Pritchard 7 active nests were found close together in an unknown number of trembling aspens.

The height above ground of 64 nests ranged from 1.2 to 24 m, with 53% between 4.6 and 11 m. Most nests were built saddled near the tip of a horizontal branch (33%; n = 51) or on a branch away from the trunk (31%). Others were positioned in the tip of a tree (18%), in the fork of a branch (14%), or on a branch against the tree trunk (4%). Nests were shallow cups of grasses, hair, twigs, and rootlets; occasionally lichens, herb stems, mosses, and bark strips were also used. Nest cups were lined with fine grasses and hair, usually cow or horse hair; occasionally lichens, plant down, and mosses were used.

Eggs: Dates for 44 clutches ranged from 18 April to 15 July, with 52% recorded between 20 May and 7 June. Calculated dates suggest that eggs can be found as early as 14 April. The range of egg dates, which spans a period of 2 months, suggests that at least some females nest twice per year. Cannings et al. (1987) reported 1 record, at Naramata on 4 July, of a male feeding fledglings while its mate was gathering nest material. This record and several late egg dates suggest that the Cassin's Finch is double-brooded in British Columbia; elsewhere, the Cassin's Finch is thought to be single-brooded (Mewaldt and King 1985; Hahn 1996).

Sizes of 25 clutches ranged from 1 to 5 eggs (1E-1, 2E-1, 3E-7, 4E-9, 5E-7), with 64% having 4 or 5 eggs (Fig. 614). The incubation period is probably about 12 days (Orr 1968).

Young: Dates for 37 broods ranged from 30 April to 22 August, with 51% recorded between 25 May and 28 June. Sizes of 18 broods ranged from 1 to 5 young (1Y-4, 2Y-4, 3Y-3, 4Y-4, 5Y-3), with 61% having 2 to 4 young. Fledged young accompanying parents have been noted into early September. The nestling period for a single nest in British Columbia was 13 or 14 days; elsewhere, the nestling period is unknown (Hahn 1996).

Brown-headed Cowbird Parasitism: In British Columbia, 11% of 35 nests found with eggs or young were parasitized by cowbirds. Two parasitized nests contained 1 cowbird egg; 1 contained 2 eggs, and 1 contained 3 eggs. Three of the parasitized nests were found in Richter Pass in the vicinity of Kilpoola Lake; the other was at Naramata. Nest parasitism by cowbirds had not been reported in major reviews of cowbird parasitism (Friedmann 1963; Friedmann et al. 1977) or of the Cassin's Finch (Hahn 1996), but was noted by Cannings et al. (1987) in the Okanagan valley. There were no records of the Cassin's Finch with cowbird nestlings or with fledged cowbirds.

Nest Success: Of 9 nests found with eggs and followed to a known fate, 5 produced fledglings. Thirteen of 15 nests that failed had been preyed upon and 2 fell from the tree. At Naramata, between 1969 and 1982, 7 nests with eggs or young were preyed upon by the Clark's Nutcracker, 2 by red squirrel (Cannings et al. 1987), and 1 by a Common Raven. In several cases, adult Clark's Nutcrackers fed Cassin's Finch nestlings to their fledglings. The Gray Jay was another reported nest predator in British Columbia (Kingery and Kingery 1995).

REMARKS: Published winter references to the Cassin's Finch at Kitimat (Horwood 1992) and from several coastal Christmas Bird Counts were not included in this account because of unsatisfactory documentation. For example, a record of 16 Cassin's Finches on the Christmas Bird Count at Port Alberni on 4 Jan 1976 undoubtedly should refer to Purple Finches.

Where the range of the Cassin's Finch overlaps with those of its close relatives, the Purple Finch and House Finch, observers are encouraged to make identifications carefully. Female and juvenile Cassin's and Purple finches are especially difficult to separate, and males can be as well (Kaufman 1990).

There are no significant conservation concerns for the habitat of the Cassin's Finch. Since it prefers open forests, selective logging and small clearcuts may even improve habitat in some areas where forests are dense (Hahn 1996). In the east Kootenay, however, Schwab (1979) found that the Cassin's Finch did not occur in stands where merchantable timber had been removed and only young trees were left, with an occasional very large veteran.

The Cassin's Finch is particularly vulnerable to salmonellosis at feeders because of its foraging characteristics. In Kamloops, 10 birds died at 1 location between 15 and 25 May 1995 (R.R. Howie pers. comm.).

Additional information on the ecology of the Cassin's Finch can be found in Hahn (1996).

NOTEWORTHY RECORDS

Spring: Coastal – Vancouver Apr 1967-1 (UBC 13334); North Vancouver 21 Apr 1984-1 (RBCM Photo 1051), 30 Apr 1987-1 (Mattocks and Harrington-Tweit 1987b); Pitt Meadows 10 Apr 1985-1; Kitimat 13 Mar 1991-2. **Interior** – Oliver 24 Apr 1971-40; 19 km w Osoyoos 17 May 1969-4 eggs; Keremeos 9 Apr 1969-9; Balfour to Waneta 16 May 1981-40 on songbird survey; Trail 3 Apr 1983-8; Vaseux Lake 14 May 1972-25; Okanagan Falls 17 May 1969-30; Skaha Lake 16 May 1962-5 nestlings; White Lake (Okanagan Falls) 12 May 1978-20, 21 May 1962-5 eggs; West Bench 10 May 1972-large flocks (Cannings et al. 1987), 26 May 1969-4 eggs; Princeton 8 May 1979-32; Okanagan valley 9 May 1981-100, small flocks everywhere; Sparwood 20 May 1983-3; Naramata 31 Mar 1965-nest building, 14 May 1972-adults feeding 2 fledglings (Cannings et al. 1987); se Skookumchuck May 1977-27 on survey (Schwab 1979); Fintry 22 May 1982-35; Quilchena Creek 12 May 1992-2; Brouse 9 May 1982-10; Dutch Creek 23 May 1977-16; Invermere 3 Apr 1983-3, 28 Apr 1979-10; Coldstream 18 Apr 1961-5 eggs, 30 Apr 1961-3 nestlings and 2 eggs, 7 May 1961-5 nestlings; Cache Creek 13 May 1983-1; Hat Creek 30 May 1990-15; Celista 22 May 1948-4 nests in aspen tree; Sorrento 25 Mar 1972-10; Pritchard 26 May 1962-7 pairs nesting close together; Revelstoke 17 May 1985-1; Riske Creek 1 Apr 1987-1; Doc English Gulch 27 Apr 1979-1; Farwell Canyon 23 May 1985-6; Williams Lake 7 Mar 1984-2, 28 May 1989-6; Kleena Kleene 28 May 1954-4 (Paul 1959); Papoose Lake 5 May 1970-3 (RBCM 14925 to 14927); Prince George 7 to 21 May 1995-1 at feeder.

Summer: Coastal – Comox 23 Jun 1940-1 (RBCM 13973); Maple Ridge 23 Jul 1973-1 (ROM 119605); Mosher Creek (Stuie) 31 Jul 1940-2 (RBCM 13969 and 13970; Munro and Cowan 1947); Hazelton 10 Jun 1992-1. **Interior** – Richter Pass 4 Jul 1975-4 eggs; Creston 21 Aug 1947-1, only record for year (Munro 1950); Kishenina Creek 8 Jul 1988-2 fledglings with adults picking at gravel; Trail 4 Jul 1982-3 nestlings; nr Hedley 11 Jun 1972-17; White Lake (Okanagan Falls) 4 Jun 1967-5 eggs; Wardner 20 Jun 1939-pair feeding young (Johnstone 1949); Naramata 1 Aug 1982-red squirrel sitting by nest eating nestlings, 22 Aug 1971-1 fledgling just able to fly; Needles 25 Jun 1983-15; Wasa Park 1 Jun 1977-1 fledgling; Skookumchuck 29 Aug 1976-12; Scaia Mountain 7 Jul 1983-3; Coldstream 15 Jul 1963-3 eggs; Panorama Mountain 7 Jul 1991-9; Glacier National Park 9 Jun 1983-3; Graffunder Lakes 9 Jul 1972-1; Barriere 7 Jun 1934-2 (CMNH P115738 and P115739); 87 Mile House (Cariboo) 14 Jun 1940-1 fledgling with male; Riske Creek 15 Jul 1991-1 fledgling with adults; Chilcotin River 21 Jul 1987-1 fledgling with adult; Alkali Lake (Cariboo) 17 Jul 1978-1; Williams Lake 24 Jun 1977-2 fledglings with adults; Alexis Creek 17 Jun 1979-1; Stum Lake 9 Jun 1973-1 (Ryder 1973); Cottonwood River Park 15 Jun 1977-2; Bowron Lake Park 26 Jul 1971-1 (Runyan 1971).

Breeding Bird Surveys: Coastal – Recorded from 1 of 27 routes and on less than 1% of all surveys. Maximum: Pemberton 30 Jun 1985-2. **Interior** – Recorded from 32 of 73 routes and on 30% of all surveys. Maxima: Summerland 17 Jun 1989-23; Canford 14 Jun 1984-19; Oliver 24 Jun 1993-17.

Autumn: Interior – Williams Lake 14 Sep 1981-1, 12 Nov 1987-9; Riske Creek 23 Sep 1988-1, 27 Oct 1984-1; Revelstoke 29 Nov 1988-2; Scottie Creek 21 Sep 1974-4; Melvin Creek 23 Sep 1992-3; Kamloops 11 Sep 1983-15; Blowdown Creek 12 Nov 1993-10; Nakusp 25 Nov 1979-1; Wasa Park 19 Sep 1976-12; Summerland 28 Nov 1976-10; Stemwinder Park 9 Sep 1979-50, 21 Sep 1984-15; Trail 30 Nov 1980-12; Baldy Mountain (Osoyoos) 11 Sep 1960-17. **Coastal** – Hope 11 Oct 1922-1 (Thacker and Thacker 1923); West Vancouver 17 and 18 Oct 1984-1 (Hunn and Mattocks 1985), 22 Oct 1987-1 (Mattocks 1988); Vancouver 4 Sep 1965-1; Blackie Spit (Surrey) 14 to 20 Oct 1984-1 (Hunn and Mattocks 1985); Abbotsford 21 Nov 1996-1 male.

Winter: Interior – Chilcotin River (Riske Creek) 31 Dec 1988 to 1 Jan 1989-1, first winter record at Wineglass Ranch; Kamloops 19 Dec 1980-2; Lumby 11 Jan 1986-20; New Denver 5 Dec 1982-10; Okanagan Landing 6 Jan 1927-30 (Munro and Cowan 1947); Edgewood 11 Feb 1979-10; Nelson 5 Jan 1986-8; Creston 1 Feb 1982-75; Manning Park 2 Feb 1983-12, along road to ski area. **Coastal** – North Vancouver 17 Dec 1986-1.

Christmas Bird Counts: Interior – Recorded from 14 of 27 localities and on 36% of all counts. Maxima: Vaseux Lake 29 Dec 1988-312; Penticton 27 Dec 1988-235; Vernon 16 Dec 1990-97. **Coastal** – Recorded from 1 of 33 localities and on less than 1% of all counts. Maximum: Port Clements 20 Dec 1994-1 (Bowling 1995b) (see also REMARKS).

House Finch
HOFI

Carpodacus mexicanus (Müller)

RANGE: Breeds from southern British Columbia (including Vancouver Island), northern Idaho, western Montana, north-central and southeastern Wyoming, western Nebraska, and west-central Kansas south to Baja California, Sonora to northern Sinaloa, Oaxaca, interior Chiapas and San Cristobal, and south-central Texas. Distribution in southern Mexico is confused by the release of cage birds. Also, birds introduced on Long Island, New York, in 1940 are now established in eastern North America from southern Minnesota, northern Michigan, southern Ontario, southern Quebec, and southern Maine south to northern Florida and northeastern Louisiana and west to central Oklahoma and central North Dakota. Winters throughout the breeding range and east to southern Texas.

STATUS: On the coast, a *common* to *very common* resident in the Georgia Depression Ecoprovince; in the Coast and Mountains Ecoprovince, locally *uncommon* to *common* in the southern portions of the Southern Mainland Coast and on Western Vancouver Island, *casual* on the Northern Mainland Coast, and absent from the Queen Charlotte Islands.

In the interior, a *fairly common* to locally *common* resident in the Southern Interior Ecoprovince, except *very common* in the Okanagan valley; *uncommon* to locally *fairly common* in the southern portions of the Southern Interior Mountains and Central Interior ecoprovinces, becoming *very rare* in the northern portions; *casual* in the Sub-Boreal Interior, except locally *uncommon* resident at Quesnel, *rare* visitant in Prince George, and *very rare* visitant in Mackenzie; *casual* in the Northern Boreal Mountains Ecoprovince; *accidental* in the Boreal Plains Ecoprovince; absent from the Taiga Plains Ecoprovince.

Breeds.

CHANGE IN STATUS: The House Finch (Fig. 615) is a recent addition to the British Columbia avifauna. It was unknown in the province until 1935, when a pair of House Finches were first reported nesting at Penticton in the Okanagan valley (Cowan 1937; Brooks 1942b). By 1937, it had also arrived on the coast, where it was observed nesting at Victoria on southern Vancouver Island (Cowan 1937). From these 2 small pioneering populations (Fig. 616a), the House Finch launched its rapid range expansion into British Columbia (Edwards and Stirling 1961).

In the mid-1940s, Munro and Cowan (1947) considered the House Finch to be a local summer visitant in the Victoria area and in the Okanagan valley as far north as Vernon. By 1950, it had been seen at Celista on Shuswap Lake in the Southern Interior, and several specimens had been collected at Williams Lake in the Central Interior (Fig. 616b; Jobin 1952). The House Finch was migratory during the decade following its first appearance in British Columbia. In the Okanagan valley, it did not begin to overwinter until at least 1945 (Cannings et al. 1987), and it was not reported as a wintering bird on the coast until 1951 (Clay 1952). By the mid-1950s, it was wintering regularly, both on the coast and in the interior.

Figure 615. Since its arrival in the Okanagan valley of British Columbia in 1935, the House Finch has expanded its range across southern portions of the province and into the central interior. On the south coast, the male (a) and female (b) House Finch may start nest building in early March (Victoria, 19 October 1997; R. Wayne Campbell).

Through the 1950s, the House Finch rapidly expanded its range on the coast, and by 1960 (Fig. 616c) it was well established on southern Vancouver Island and the Fraser River delta of the mainland coast (Edwards and Stirling 1961). It had also been reported at Carmanah Point (Irving 1953) on the west coast of Vancouver Island, at Beach Grove near Campbell River, and further north, at Sonora Island, where several were banded. In the Southern Interior, it had become well established throughout the Okanagan valley and had also been reported at Kamloops in the Thompson River valley (Edwards and Stirling 1961) and at Spences Bridge. Further north in the Central Interior, it had been reported in the area between Dog Creek and Williams Lake.

By 1970 (Fig. 616d), the House Finch had broadened its range in the Georgia Depression and the Southern Interior. It had also moved east into the Southern Interior Mountains, where it was reported in the lower Columbia River valley

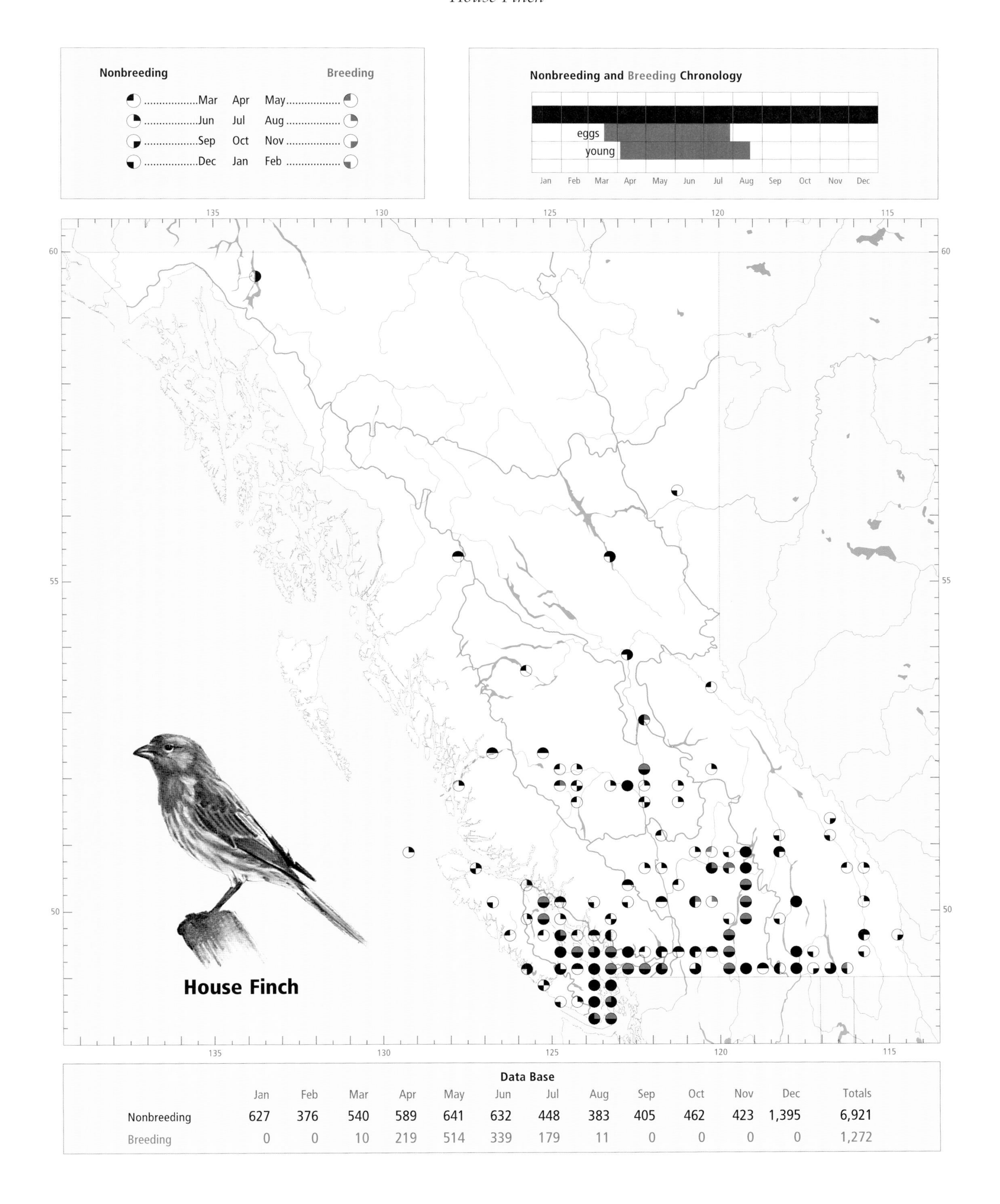

Data Base

	Jan	Feb	Mar	Apr	May	Jun	Jul	Aug	Sep	Oct	Nov	Dec	Totals
Nonbreeding	627	376	540	589	641	632	448	383	405	462	423	1,395	6,921
Breeding	0	0	10	219	514	339	179	11	0	0	0	0	1,272

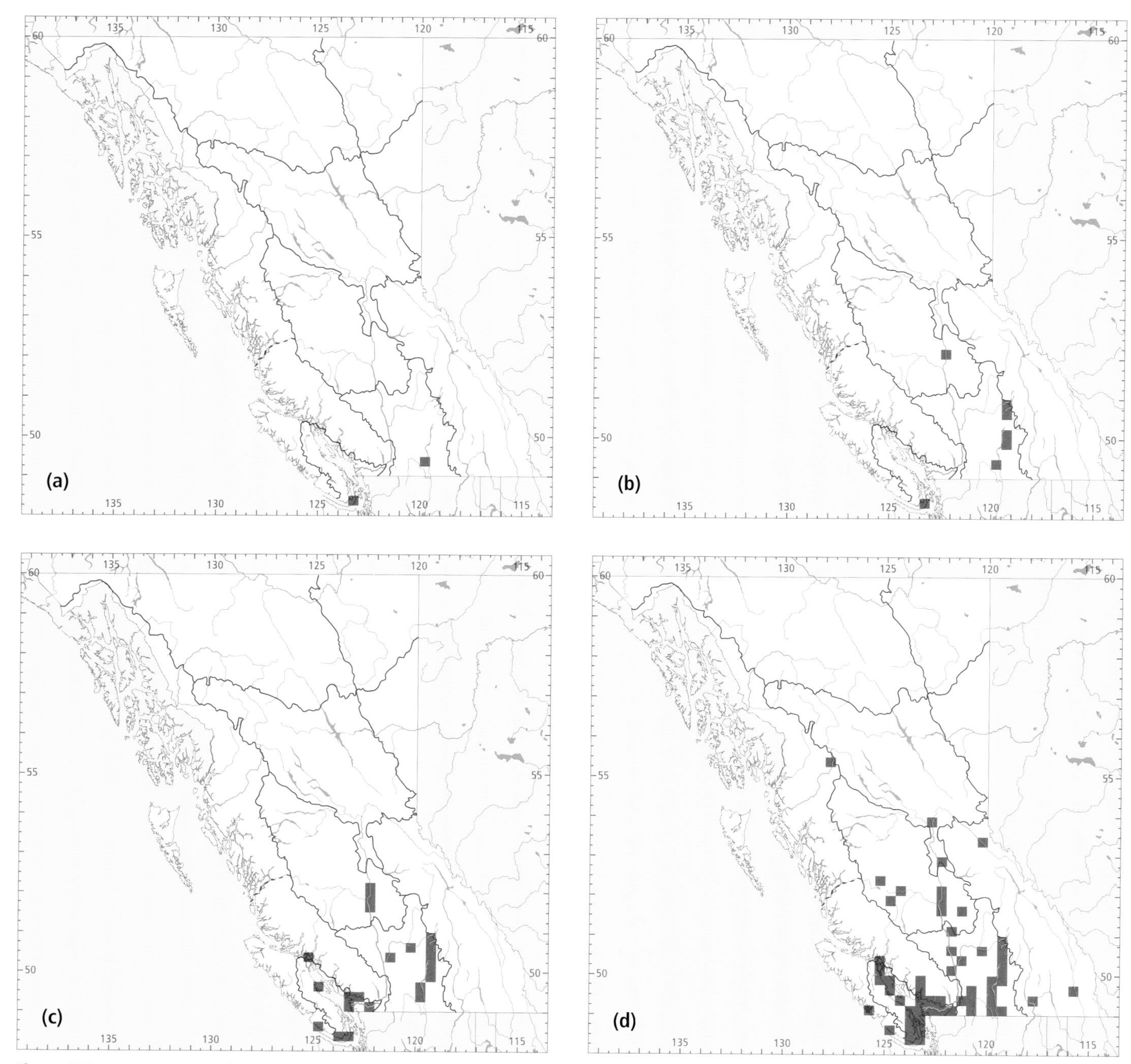

Figure 616. Range expansion of the House Finch in British Columbia from 1935 to 1970 by National Topographic Series quadrangles. Pioneering birds were first found in the late 1930s (a) in the Okanagan valley near Penticton and in the Nanaimo Lowland of the Georgia Depression near Victoria. By the 1950s (b), the House Finch occurred north in the Okanagan valley to Vernon and had been reported from the Shuswap Lake area and Williams Lake. By the 1960s (c), it had become well established on southern Vancouver Island and the Fraser River delta on the coast, and in the Okanagan valley in the interior. By the 1970s (d), it had broadened its range in the Georgia Depression, Southern Interior, and Central Interior and had been reported from the Southern Interior Mountains, the Sub-Boreal Interior, and the Northern Mainland Coast.

near Trail, in the East Kootenay Trench near Cranbrook, and at McBride in the Upper Fraser Trench. Elsewhere, it had moved west towards Chilanko Forks, Kleena Kleene (Paul 1964a, 1964b), and Anahim Lake, and north to Quesnel and Prince George in the Sub-Boreal Interior (Weber 1968). It had also been reported at Hazelton, on the Northern Mainland Coast.

Edwards and Stirling (1961) review the arrival and early stages of range expansion of this finch in British Columbia. Through published records, they trace its northward expansion into the Okanagan valley through the dry interior valleys of Washington. They were unable, however, to find evidence of a similar movement through either Oregon or Washington to explain its arrival on the coast of British Columbia. Lacking evidence of a coastal route, they conclude that this finch was most likely introduced to Victoria through the release of captive birds. Although this possibility cannot be entirely ruled out, we believe that this finch most probably arrived through natural means. The House Finch is clearly capable of moving over extensive areas of unsuitable habitat, as was documented by Ian McTaggart-Cowan on 30 May 1944, when he collected an adult male House Finch at Topaz Lake in Jasper Park, Alberta (Cowan 1955). Recent observations at Atlin and Fort St. John are also testimony to the House Finch's ability to arrive at locations far removed from established populations. Furthermore, the species was clearly undergoing a rapid northward expansion during the period when it arrived on the south coast of British Columbia.

NONBREEDING: The House Finch is a widespread but locally distributed resident across the southern portion of British Columbia, where it is closely associated with human-created habitats. With the possible exceptions of the introduced European Starling and House Sparrow, this finch is more reliant on human-created habitats than any other passerine species in the province. In the interior, it is sparsely distributed north of latitude 51°N, and is seldom observed north of Prince George (latitude 54°N). On the coast, it is absent from the Queen Charlotte Islands and much of the mainland coast north of Powell River. The highest numbers in winter occur in the Georgia Depression (Fig. 617).

The House Finch has been reported from sea level to about 350 m elevation on the coast; in the interior, it occurs from 280 to 1,200 m elevation. On the coast, this finch is found year-round in most urban and agricultural landscapes of the Georgia Depression. It is most abundant around areas of human habitation, particularly in suburban areas, but also uses a wide variety of other anthropogenic habitats, ranging from city core areas to golf courses, semi-open parks, cemeteries, nurseries, airports, and agricultural fields. In autumn, the House Finch is attracted to weedy areas such as roadsides, dykes, utility corridors, fallow fields, and grassy backshore areas along beaches and estuaries.

In the city of Vancouver, Lancaster and Rees (1979) found that trees and shrubs were important habitat elements for the House Finch, which favoured wooded residential areas and avoided poorly vegetated industrial and commercial areas. Elsewhere on the coast, the species has a very localized distribution near human settlements on Western Vancouver Island and the Southern Mainland Coast. North of Powell River on the mainland coast, there are only single spring records from Hagensborg and Hazelton. The main distribution of this finch on the south coast corresponds closely to the Coastal Douglas-fir Biogeoclimatic Zone and the drier subzones of the Coastal Western Hemlock Zone.

In the interior, the House Finch prefers human-modified habitats similar to those found on the coast, but also uses irrigated agricultural areas, especially orchards and vineyards. Natural shrub thickets in gullies and around wetland perimeters are favoured in open grassland areas. In the Okanagan valley, this finch is also closely associated with humans (Edwards and Stirling 1961; Cannings et al. 1987). The best year-round habitats occur within the Bunchgrass, Ponderosa Pine, and Interior Douglas-fir biogeoclimatic zones, and in the driest subzones of the Interior Cedar–Hemlock Zone. Throughout its range in British Columbia, this finch avoids closed forested habitats, and is rarely observed in wilderness areas. There are no records from subalpine or alpine regions.

In most regions of the province, there is no significant difference in seasonal habitat selection. In the city of Vancouver, Lancaster (1976) found that this species was most abundant in lightly wooded apartment and residential areas (30% shrub and tree cover), but noted a shift to more open residential areas (7% shrub and tree cover) during winter. He attributed this seasonal shift in habitat preference to the prevalence of bird feeders in the latter areas.

The House Finch is generally nonmigratory, and in most parts of the province numbers remain relatively stable throughout the year. The greatest numbers and the greatest fluctuations in abundance occur on the coast, in the Georgia Depression. There numbers begin to decline from the high in mid-December and the decline continues through early March, when courtship and nest building start for some early breeders (Fig. 619). These low numbers continue until the first wave of post-breeding flocks begins to appear in the second week of July, reaching an annual peak in numbers in late October. During this period, autumn flocks, consisting mostly of juveniles and numbering up to several hundred birds, gather in areas of abundant food. Large flocks of this nature are regularly recorded in the Fraser Lowland at Sea Island, Westham Island, Canoe Passage, and Boundary Bay, and on the Saanich Peninsula in the Nanaimo Lowland.

During winter, the House Finch is readily attracted to bird feeding stations and shows a strong preference for sunflower seeds. In many areas of the province, particularly in regions of regular winter snow cover, this finch probably depends on human-supplied food for survival. During severe winters in more northerly locations, House Finches may migrate south

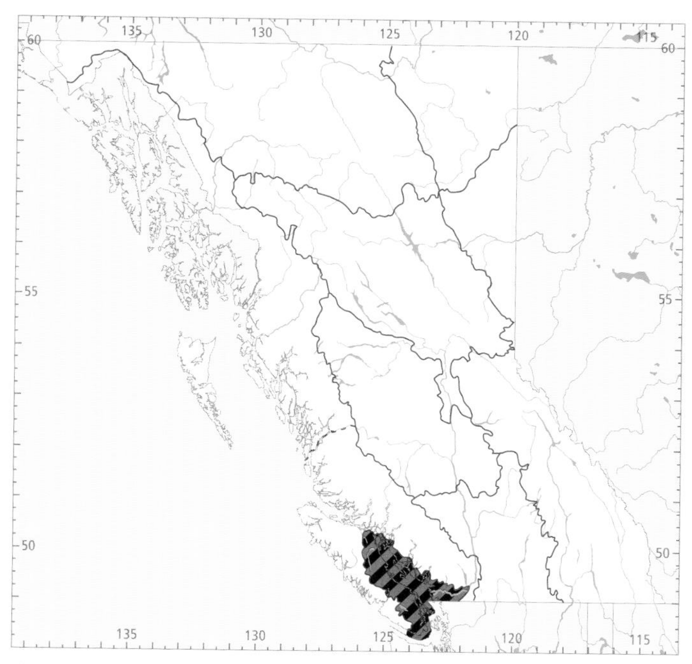

Figure 617. In British Columbia, the highest numbers for the House Finch in winter (black) and summer (red) occur in the Georgia Depression Ecoprovince.

or suffer winter mortality, as was observed in the northern Okanagan valley during the winter of 1968 to 1969 (Cannings et al. 1987). Conversely, a northward trend in wintering appears to occur during mild winters. For example, 2 of these finches wintered far north of their normal range, in Fort St. John, during the mild winter of 1997 to 1998.

In eastern North America, Sprenkle and Blem (1984) found that the House Finch was poorly suited to cold winter temperatures but was able to partially compensate for this limitation by eating high-quality foods provided at bird feeders. Dawson et al. (1983) also found these eastern finches to be significantly better acclimatized to cold winter temperatures than California finches, the original source of this introduced population. In Alberta, the House Finch has been recorded at bird feeders in winter as far north as Edmonton (Bancroft and Parsons 1991), suggesting that it may have some ability to survive in harsh winter environments. In British Columbia, it remains to be seen how much further north this finch can extend its regular wintering distribution.

Christmas Bird Counts in the Georgia Depression and Southern Interior show an increasing trend in House Finch numbers for both ecoprovinces (Fig. 620). In the city of Vancouver, Weber (1972) and Lancaster (1976) found winter House Finch densities ranging from 1.9 to 7.9 birds per 10 ha and 0.2 to 8.1 birds per 10 ha, respectively. Both researchers observed that the distribution of this finch in the city was significantly influenced by the location of bird feeding stations.

The House Finch rarely strays far from trees or other structures on which it can perch or roost (Hill 1993). In British Columbia, small flocks regularly roost in dense hedges, ornamental evergreens, and vines covering the walls of buildings. Larger roosts, ranging from 200 to 1,000 birds, have been reported in ivy-covered buildings in the downtown areas of Victoria (Tatum 1971, 1972) and Vancouver (Merilees 1994). One of the few evergreens it apparently avoids roosting in is the native western redcedar (Bell 1954; Tatum 1972).

The House Finch occurs throughout the year in southern British Columbia (Fig. 618). In the Central Interior, it has been regularly reported from 18 March to 26 August.

BREEDING: The House Finch is distributed across southern British Columbia during the breeding season and likely breeds throughout most of its summer range; 98% of all our breeding records (n = 449), however, come from the Georgia Depression and the Okanagan valley of the Southern Interior. In many areas of recent range expansion, evidence of breeding has yet to be established. For example, in the Southern Interior Mountains, adults with fledged young have been reported at Trail, Nelson, and Invermere, but nesting has been confirmed only at Creston. In the Central Interior, breeding has been recorded only at Williams Lake and near Tatla Lake. The northernmost breeding record comes from Quesnel, in the Sub-Boreal Interior. In all regions, nesting nearly always occurs in close association with people in urban, rural, and agricultural environments.

The House Finch reaches its highest numbers in summer in the Fraser and Nanaimo lowlands of the Georgia Depression (Fig. 617). An analysis of Breeding Bird Surveys for the period 1968 through 1993 shows that the mean number of birds increased on both coastal and interior routes at an

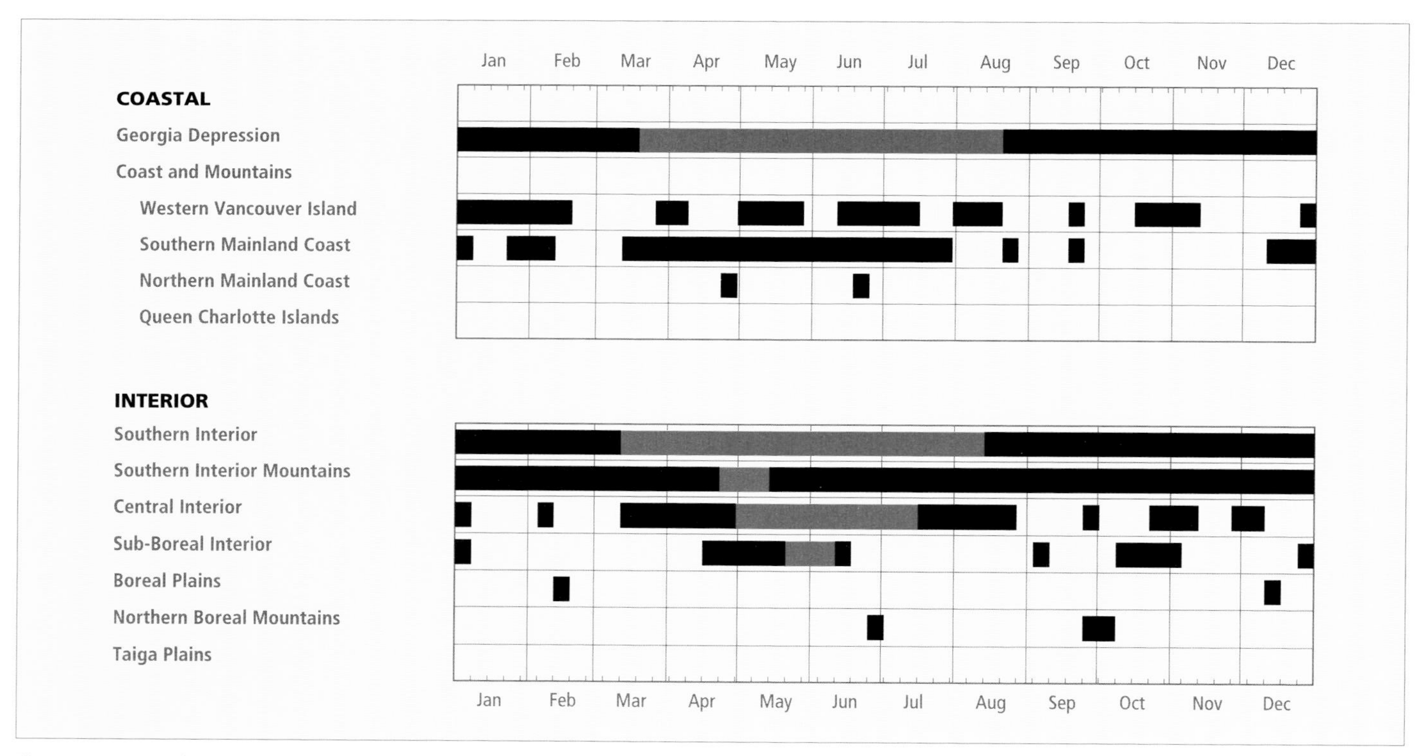

Figure 618. Annual occurrence (black) and breeding chronology (red) for the House Finch in ecoprovinces of British Columbia. Records are shown for the week in which they occurred.

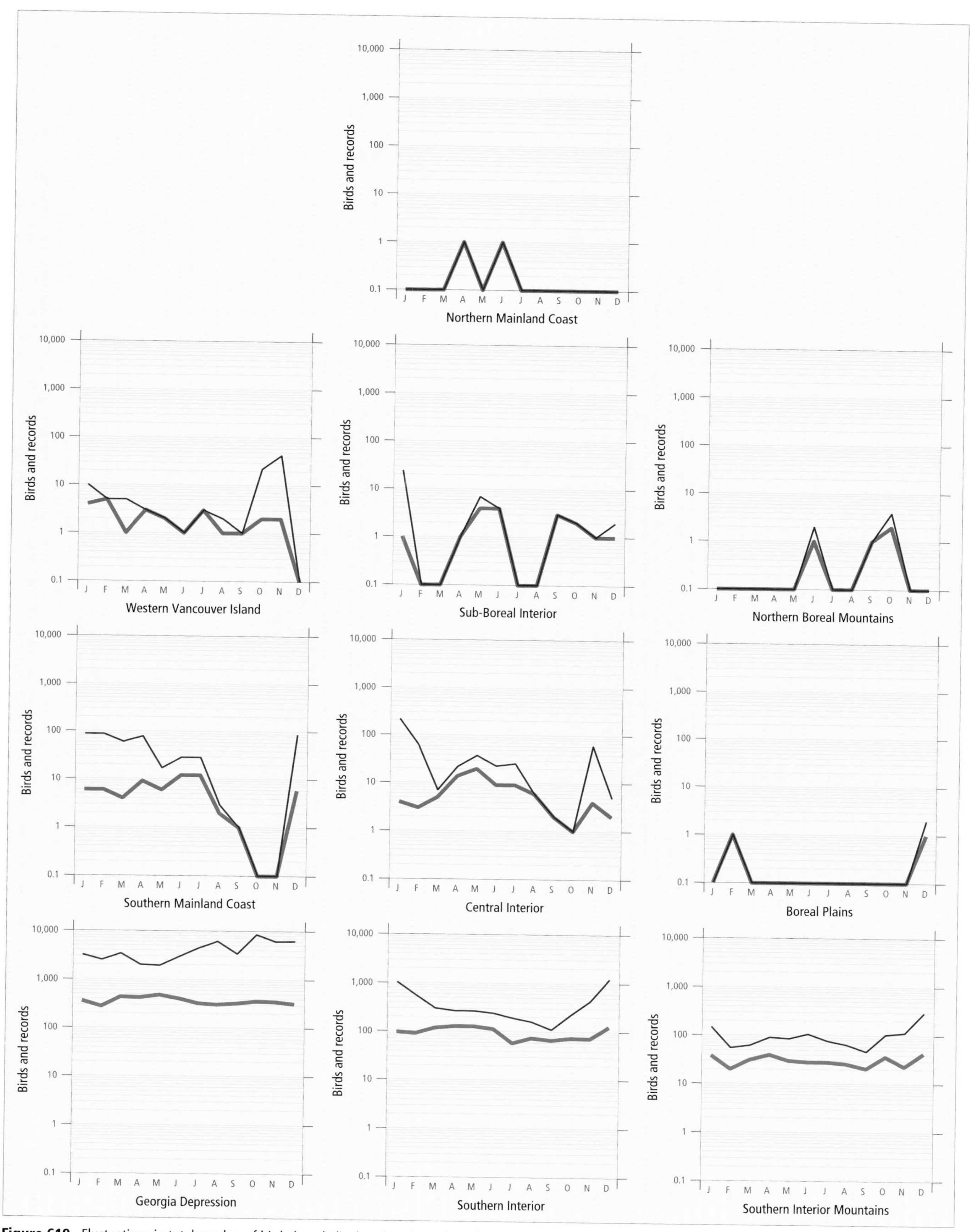

Figure 619. Fluctuations in total number of birds (purple line) and total number of records (green line) for the House Finch in ecoprovinces of British Columbia. Christmas Bird Counts, Breeding Bird Surveys, and nest record data have been excluded.

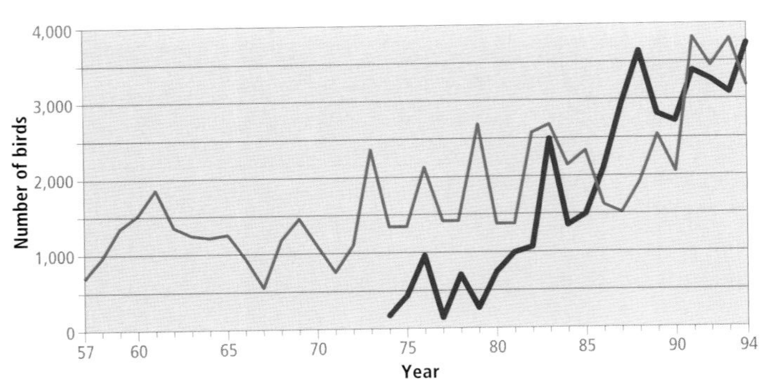

Figure 620. Fluctuations in total number of House Finches from Christmas Bird Counts for Ladner, Vancouver, and Victoria in the Georgia Depression Ecoprovince (green line) for the period 1957 through 1994, and for Salmon Arm, Penticton, and Vaseux Lake in the Southern Interior Ecoprovince (purple line) for the period 1974 through 1994. Counts in both regions indicate an increasing trend in the total number of wintering birds.

average annual rate of 6% (Fig. 621). Nationally, the numbers increased at an average annual rate of 8.8 % ($P < 0.01$) over the period 1966 to 1996; continentally, they increased at an average annual rate of 1.1% ($P < 0.06$) over the same period (Sauer et al. 1997). In the city of Vancouver in the 1970s, Weber (1972) and Lancaster (1976) reported House Finch breeding season densities of 1.9 to 9.7 birds per 10 ha and 0.6 to 12.8 birds per 10 ha, respectively.

The House Finch breeds at elevations near sea level to 335 m on the coast, and from 280 to 1,005 m in the interior. Most nesting habitats on the coast and in the interior were described as human-influenced (97%; n = 619), with the remaining 3% characterized as forest, wetland, or grassland habitats. Human-influenced habitats included suburban (46%; n = 595; Fig. 622), cultivated farmland (26%), urban (10%), rural (7%), and orchard (6%). In the Okanagan valley of the Southern Interior, Cannings et al. (1987) report that this finch almost always nests near areas of human habitation.

On the coast, the House Finch has been recorded breeding from 26 March (calculated) to 18 August; in the interior, it has been recorded breeding from 18 March (calculated) to 12 August (Fig. 618).

Nests: House Finch nests were found in a diverse range of sites. Most were placed in trees and shrubs (62%; n = 471; Fig. 623). Evergreens (53%) were used most often and included a wide variety of ornamental trees and shrubs, such as cypress and cedars (14%), spruce (4%), and junipers (3%). Nine percent of nests were built in vines or ivies, usually growing on buildings. Deciduous trees and shrubs were used for only 8% of nests. Where sufficient information was recorded, 83% of 277 nests were noted as occurring in "exotics" or "ornamentals" and only 17% in native plant species. Of the native species, Douglas-fir (36 nests) was the most commonly used in both interior and coastal areas.

Eleven percent of all nests were built in or on human-created structures, including shed rafters, beams or shelves under sundecks and carports, window shades, and wall vents. Included in this category were some unusual nest locations: in a covered street lamp, on a porch light fixture, on top of a sliding garage door, in a Stevenson screen (weather station), on top of an orchard prop pole, in an opening in a balcony railing, in a bird feeder, and in a coil of wire on a radio telescope array pole. Twenty-seven nests were also found in hanging flower baskets and 6 in nest boxes. A single nest was built on the ground beneath some roadside shrubbery.

In an Ontario study, the House Finch preferentially selected blue spruce trees and evergreen vines for nesting (n = 24), and avoided deciduous trees (Graham 1988). In other areas of North America, the blue spruce is also a favoured nest tree (Woods 1968); in British Columbia, spruce trees were used in 44 instances; 19 were identified as this ornamental conifer.

The heights of 451 nests ranged from ground level to 45 m, with 59% between 1.8 and 3.0 m. The highest nest was on a 15th floor balcony of a high-rise apartment building. This finch appears to look for any convenient structure or platform on which to construct its nest, most of which were described as among branches in shrubs or trees, near the trunk,

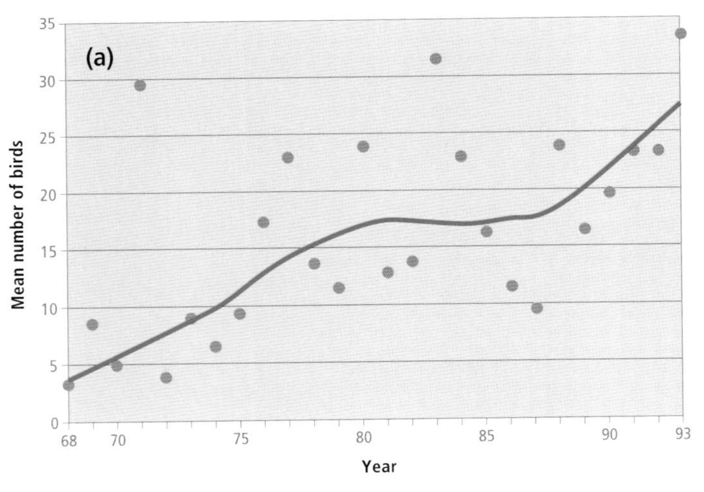

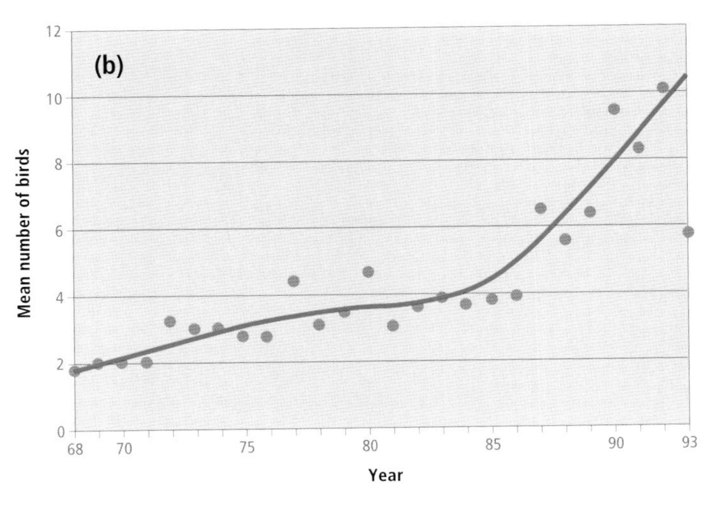

Figure 621. An analysis of Breeding Bird Surveys for the House Finch in British Columbia shows that the mean number of birds on coastal routes (a) increased at an average annual rate of 6.0% over the period 1968 through 1993 ($P < 0.001$); the mean number of birds on interior routes (b) also increased at an average annual rate of 6.0% over the same period ($P < 0.05$).

or part way out on a limb. Ten nests were constructed near the end of a conifer limb. Two nests were built in abandoned nests of other species: a Barn Swallow nest (Cannings et al. 1987) and an American Robin nest.

Some regional differences were noted between nests on the coast (n = 237) and those in the interior (n = 226; Fig. 624). On the coast, this finch was more likely to nest in evergreens and less likely to nest in deciduous trees than in the interior. In the interior, human-created structures and hanging baskets were used more frequently for nesting than on the coast.

Nests were typically constructed of grasses and fine plant stems, and less frequently with rootlets and fine twigs. Human-made products were reported in 28% of nests, and included materials such as string, thread, rope fibres, paper, cigarette filters, cotton batten, pieces of plastic or cloth, audio tape, tinsel, knitting yarn, lint, and upholstery stuffing. Nests were usually lined with fine grasses and mammalian hair (horse, domestic sheep, dog, and cat) and less often with plant down or feathers.

The House Finch has been known to reuse nests for a second or even third brood (Cannings et al. 1987; Hill 1993). It will also nest in close proximity to other House Finch nests, occasionally less than 1 m apart (Woods 1968). In British Columbia, there are 2 records of active nests within 2.5 m of each other, and 3 other records of finch nests occurring in the same tree. There are also 2 records of 5 nests located within a confined area, including in a hedge bordering a driveway and in landscape shrubbery along the side of building.

Figure 622. On the south coast of British Columbia, the House Finch breeds mainly in suburban habitats often associated with shrubby backyards and ornamental gardens (Victoria, 6 June 1998; R. Wayne Campbell).

Figure 623. In the Okanagan valley of British Columbia, small colonies of House Finches can often be found breeding in ornamental evergreens lining suburban yards (Oliver, 13 May 1997; R. Wayne Campbell).

The House Finch is capable of constructing its nest in as little as 2 days. Nest building is more protracted early in the breeding season, however, and can take as long as 3 weeks (Evenden 1957; Cannings et al. 1987).

Eggs: Dates for 688 clutches ranged from 21 March to 28 July, with 51% recorded between 30 April and 9 June. Calculated dates indicate that nests can contain eggs as early as 18 March. Sizes of 316 clutches ranged from 1 to 7 eggs (1E-27, 2E-25, 3E-28, 4E-110, 5E-114, 6E-10, 7E-2), with 71% having 4 or 5 eggs.

Six breeding records suggested that this finch was attempting a second clutch. There were also 2 records of 3 clutches occurring in the same nest during a breeding season.

The incubation period in British Columbia ranged from 11 to 18 days (n = 13), with a median of 12 days. Several observers noted that incubation began before the last egg was laid. Hill (1993) reports that incubation is normally 13 or 14 days, but may be as long as 17 days when interrupted by cold weather.

Young: Dates for 575 broods ranged from 4 April to 18 August, with 51% recorded between 14 May and 27 June. Sizes of 245 broods ranged from 1 to 7 young (1Y-8, 2Y-20, 3Y-56, 4Y-89, 5Y-69, 6Y-2, 7Y-1), with 64% having 4 or 5 young. The nestling period in British Columbia ranged from 12 to 19 days (n = 12), with a median of 15 days. This is similar to the range of 11 to 19 days reported for other areas in North America (Hill 1993). Populations in the Georgia Depression and Southern Interior appear to produce 2, and occasionally 3, broods a year (Fig. 625). In the Okanagan valley, Cannings et al. (1987) report an instance of 3 broods produced from a single nest. Elsewhere in North America, most females nest more than once during a breeding season but have not been known to produce more than 3 broods (Hill 1993).

Females occasionally begin a second nest before the first nestlings have fledged, leaving most care of the first brood to the male (Evenden 1957). Our data contain more than 20 records of solitary adult males feeding recently fledged young, suggesting that "overlap nesting" may also occur in British Columbia.

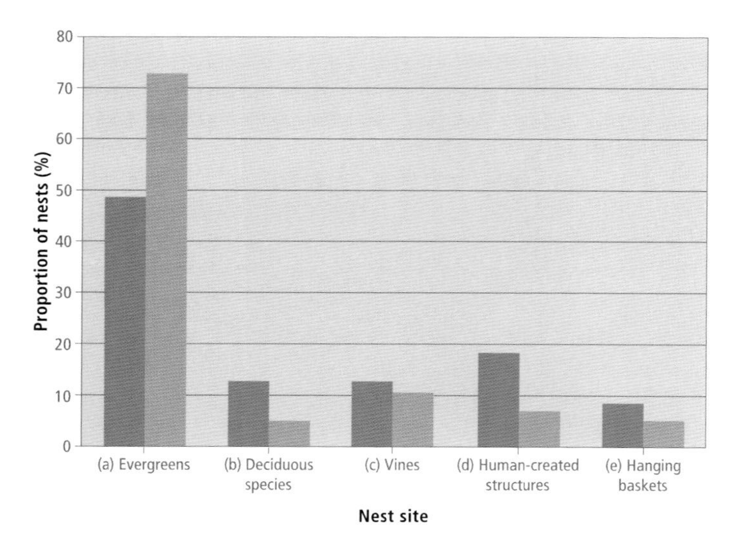

Figure 624. In British Columbia, the House Finch uses a broad range of nest sites on the coast (light bars) and in the interior (dark bars). Most nests were placed in trees and shrubs, including a wide variety of evergreens (a) and deciduous species (b), as well as in vines (c), in or on human-created structures (d), and in hanging baskets (e). Evergreens were used more commonly on the coast than in the interior, whereas deciduous trees and shrubs, human-made structures, and hanging baskets were selected more frequently in the interior than on the coast. Vines were used in similar proportions in both regions.

Brown-headed Cowbird Parasitism: In British Columbia, 6% of 404 nests found with eggs or young were parasitized by the cowbird. On the coast, 8 nests were parasitized by the cowbird (4%; n = 202); in the interior, 17 nests were parasitized (8%; n = 202).

The House Finch is an unsuitable host for the Brown-headed Cowbird. In an Ontario study, Brown-headed Cowbirds parasitized 99 (24.4%) of 406 House Finch nests (Kozlovic et al. 1996). Although 88.4% of cowbird eggs hatched, nestlings survived very poorly on the vegetative diet provided by these finches and, on average, lived only 3.2 days; only a single cowbird reached fledging age. A comparison of our data from British Columbia with the above study is not possible because cowbird eggs were routinely removed by observers from most parasitized House Finch nests. For 1 nest in which no interference occurred, however, a cowbird nestling survived for at least 6 days before observations ended. There were also 2 records of adult House Finches feeding a cowbird fledgling.

Nest Success: Of 134 nests found with eggs and followed to a known fate, 83 produced at least 1 fledgling, for a nest success rate of 62%. Nest success on the coast was 61% (n = 54); in the interior, it was 63% (n = 80).

Nest predation (n = 33) was reported as the main cause of nest failure. Observed or suspected nest predators included red squirrel (12 nests), chipmunk (4), domestic cat (3), Clark's Nutcracker (3), Northwestern Crow (2), Steller's Jay (2), Common Raven (1), gopher snake (1), and yellow-jacket wasp (1). Weather events, including windstorms, heavy rainfall, and cold temperatures were implicated in 17 nest failures. One nest failed when the incubating female fell prey to a Sharp-shinned Hawk. Two of 13 nests were believed to have been abandoned after egg laying had begun because of human disturbance, and a third one because of cowbird parasitism.

In the Okanagan valley, single instances were reported of nests being destroyed or usurped by the House Wren, House Sparrow, and American Robin (Cannings et al. 1987). In a California study, House Finch nests were frequently usurped by House Sparrows during the early part of the breeding season; the main cause of nestling mortality there was predation by domestic cats (Evenden 1957).

REMARKS: The House Finch was formerly known as the Common House Finch and has also been referred to as the "Linnet." The latter name is appropriate only to the Eurasian species *Carduelis cannabina* (Clement et al. 1993). The American Ornithologists' Union (1957) recognizes 3 subspecies of the House Finch, but only *C. m. frontalis* occurs in British Columbia.

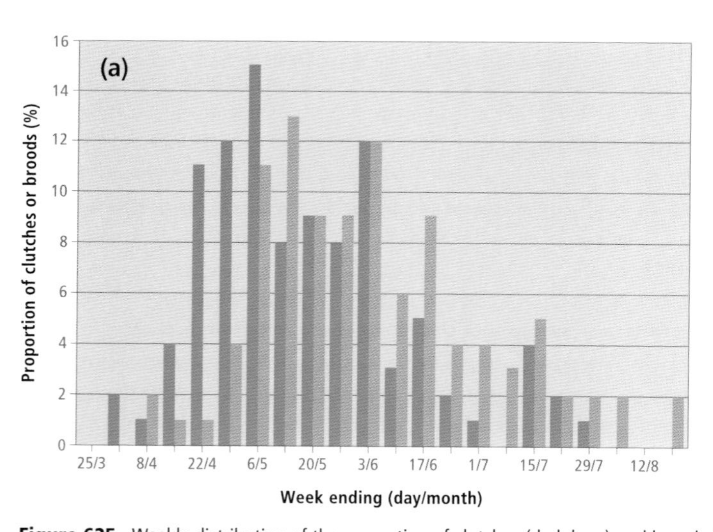

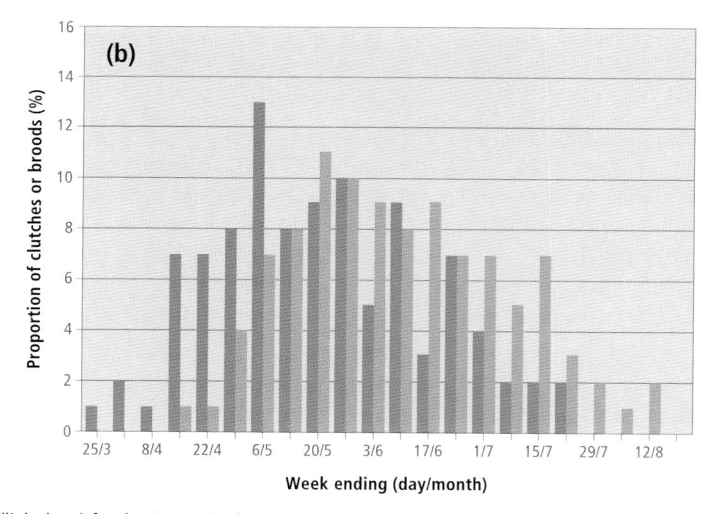

Figure 625. Weekly distribution of the proportion of clutches (dark bars) and broods (light bars) for the House Finch in the (a) Georgia Depression (clutches: n = 169; broods: n = 127) and the (b) Southern Interior (clutches: n = 167; broods: n = 159) ecoprovinces of British Columbia. The figures are based on the week eggs or young were found in the nest. The figures suggest that this finch is double-brooded and may occasionally have 3 broods a year.

Lesions attributed to avian-pox (Fig. 626) have been observed in House Finches throughout much of western North America. In California, McLure (1989) found that avian-pox affected up to 23.5% of the House Finches he examined. These infections were more common in winter than at other seasons and more prevalent in urban than in rural environments. Symptoms of avian-pox vary from loss of toes and damaged mandibles to blindness, the latter usually resulting in death through starvation or predation. Avian-pox is transmitted through biting insects, such as mosquitoes, and through close contact, such as at bird feeders (Zahn and Rothstein 1999). It has also been linked to recent changes in plumage coloration (see below). This disease also appears in House Finches in British Columbia.

The House Finch damages agricultural crops such as tree fruits, figs, grapes, berries, and seed crops, and is considered one of the most significant native bird pests in western North America (Palmer 1970a; Marzluff et al. 1994). In the late 1960s, more than 50,000 of these finches were destroyed, through poison baits and trapping, at a single California fig orchard (Palmer 1970b). In British Columbia, it occasionally causes damage to blueberry and wine grape crops, but is generally considered to be a minor agricultural pest (Weber 1989).

Beal (1907) analyzed the stomach contents of 1,206 birds in California and found that weed seed comprised 86% of the total diet, followed by fruit (11%), animal materials (2%), and miscellaneous items (1%). In Vancouver, British Columbia, Weber (1972) observed that 60% of the seasonal foraging activities of this finch occurred on the ground, and nearly half of ground foraging occurred in sparse grass (weed patches) habitats, which made up less than 2% of his urban study plots. The most important dietary items recorded in this study were seeds from birch trees, bird feeders, and weeds. Less important food items included broom seeds, mountain-ash berries, saskatoon berries, and apples. Elsewhere in British Columbia, the House Finch has been observed feeding on seeds from the dandelion, yellow salsify, and goldenrod, and on berries from the Pacific crab apple, black hawthorn, Himalayan blackberry, and mountain-ash. At Stubbs Island near Tofino, it has been observed eating California wax-myrtle berries. This finch also feeds on sucrose solutions (Avery et al. 1999) and is readily drawn to hummingbird feeders. During spring, it frequently eats the buds of fruit trees.

Female House Finches show a preference for the reddest males during the breeding season. Throughout its range, the plumage of the male House Finch varies from pale yellow to bright red (Hill 1993). This coloration is derived from 3 types of carotenoid pigments found in its diet. Experimental feeding of captive males demonstrates that all individuals have the potential for the full range of plumage coloration, depending on the availability of these dietary pigments during the moult period (Brush and Power 1976; Hill 1992). Zahn and Rothstein (1999) found that these pigments are not limiting in the natural environment and that the recent increase in the numbers of yellow males may be related to population stress. They document a significant decline in the number of males with red plumage in western North America since 1960 and propose that it has resulted from the stress effects of avian-pox.

Figure 626. In British Columbia, as in most of western North America, the House Finch is affected by avian-pox, a disease transmitted by biting insects or through close contact around feeders. Symptoms range from damaged mandibles to blindness; note the unusually large growth covering the right eye (Victoria, 25 March 1999; R. Wayne Campbell).

An analysis of House Finch band returns from British Columbia, Washington, Idaho, and Oregon shows that only 4% of birds moved as far as 200 km from their banding location and suggests a nonmigratory population (Aldrich and Weske 1978). In British Columbia, most finches were recovered near their point of banding. The only distant movement was of a finch banded near Langley on 20 June 1965 and recovered in December of the same year near Tacoma, Washington.

Six House Finch specimens (MCZ 258944 through 258949) from the Museum of Comparative Zoology in Cambridge, Massachusetts, have been excluded from this account. All were reportedly collected in the Okanagan valley by J.A. Munro between 15 July 1914 and 10 August 1918. Munro (1939, 1953) makes no reference to these early records and we presume that these specimens were incorrectly catalogued or are possibly misidentified Cassin's Finches. An early record of 4 House Finches from Deer Lake in Burnaby on 1 May 1924 has also been excluded from this account for lack of documentation.

For an early account of the range expansion of the House Finch into British Columbia, see Edwards and Stirling (1961).

NOTEWORTHY RECORDS

Spring: Coastal – Victoria 28 May 1937-2 adults feeding nestlings (Cowan 1937); Discovery Island (Victoria) 14 Apr 1962-1 female building nest; Oak Bay 28 Mar 1984-3 eggs; Somenos Lake (Duncan) 1 Apr 1984-12; Tofino Inlet 31 Mar 1962-5; Ladner 10 Mar 1974-150; Vancouver 4 Apr 1973-100; Sea Island 12 Apr 1959-120; Surrey 15 Apr 1992-5 eggs, 25 May 1965-3 nestlings in ground nest; White Rock 10 Apr 1979-5 eggs; Vancouver 14 Mar 1979-36 at feeder, 18 Apr 1958-5 eggs, on University of BC campus; Pitt Meadows 29 May 1966-4; Harrison Hot Springs 25 Mar 1985-14; Gold River 25 May 1974-1; Squamish 20 Apr 1962-1; Campbell River 21 Mar 1975-50, 2 May 1973-6 eggs; Hagensborg 11 Apr 1985-1; Hazelton 28 Apr 1963-number unknown. **Interior** – Osoyoos 6 Mar 1974-20; Grand Forks 28 Apr 1984-1; Trail 18 Apr 1982-7; Creston 11 May 1991-3 nestlings; Vaseux Lake 31 Mar 1990-5 eggs; Penticton 27 May 1936-1 addled egg remaining in nest (Brooks 1942b); Kinnaird 30 Apr 1969-1; Castlegar 27 May 1979-2, feeding in weeds; Naramata 21 Mar 1992-4 eggs, 4 Apr 1992-4 nestlings; Kimberley 27 May 1974-1 at feeder; Kalamalka Lake 31 Mar 1962-4 eggs; Celista 24 Apr 1948-10; Rayleigh (Kamloops) 20 Apr 1993-4 recently hatched nestlings; Revelstoke 12 May 1983-1 singing; Clinton 13 Apr 1962-1; Kleena Kleene 19 May 1963-2 (Paul 1964); Chilcotin River (Riske Creek) 12 Mar 1994-2, 22 May 1987-2 at feeder at Wineglass Ranch; Riske Creek 30 Mar 1987-2; Williams Lake 22 Mar 1978-1, 5 May 1970-4 eggs; Quesnel 29 May 1963-4; McBride 2 May 1968-1 banded; Prince George 16 Apr 1964-1 (Weber 1968).

Summer: Coastal – Saanich 31 Aug 1996-3 recent fledglings fed by adult in bird feeder; Swan Lake (Saanich) 3 Jul 1981-80, many were immatures; Central Saanich 25 Jul 1974-300; Bamfield 17 Aug 1977-2; Chemainus 12 Aug 1992-4 fledglings in bird feeder; Tofino 30 Jul 1983-1 singing from top of tree; Sea Island 10 Jul 1964-800 at airport; Delta 6 Aug 1979-200 at Serpentine Fen; Surrey 28 Jul 1976-3 eggs and 1 recently hatched nestling; White Rock 18 Aug 1976-4 nestlings, nr fledging; Baynes Sound 22 Aug 1981-18 (Dawe et al. 1998); Pitt Meadows 20 Jun 1965-10; Harrison Hot Springs 4 Jul 1988-3; Little Mountain (Hope) 24 Jul 1962-13 at Thacker Ecological Reserve (Kelleher 1963); Upper Campbell Lake 13 Jul 1975-4; Port McNeill 11 Jul 1988-several adults with begging juveniles (Mattocks 1989a); Triangle Island 21 Jun 1976-1; Namu 19 Jun 1976-2 feeding on grasses nr intertidal zone; Bella Coola 17 Jul 1978-1 singing; Kispiox 23 Jun 1993-1. **Interior** – Osoyoos 31 Jul 1976-10; Rock Creek 1 Jun 1998-2 at feeder; Trail 3 Jun 1984-8 at feeder; Creston 15 Jun 1983-1; Oliver 17 Jul 1968-2 eggs plus 3 Brown-headed Cowbird eggs; Skaha Lake 5 Jun 1977-3; Okanagan Falls 19 Jun 1976-4; Nelson 3 Jul 1983-2 fledglings with adults at feeder; Naramata 7 Jun 1976-7 eggs, 20 Jun 1976-7 nestlings; Summerland 27 Jun 1976-10; Jimsmith Lake Park (Cranbrook) 20 Aug 1964-10; Kelowna 12 Aug 1993-2 nestlings; Quilchena 5 Jul 1984-3 nestlings; Merritt 8 Jun 1981-4 eggs; Botanie Creek (Lytton) 18 Jul 1964-8; 29 km s Lillooet 10 Jun 1964-1 (NMC 52745); Vernon 4 Jul 1943-15 in thickets; Monte Creek 10 Jun 1988-3 nestlings; Kamloops 3 Jun 1959-1, 8 Jun 1986-5 eggs; Invermere 27 Jun 1983-2 fledglings with adult male; Walhachin 10 Jun 1998-1; Buffalo Lake 29 Jun 1968-6; 100 Mile House 20 Jul 1985-3; Kleena Kleene 15 May to 20 Jun 1963-2, nest built and abandoned (Paul 1964); 9.6 km w Tatla Lake 9 Jul 1992-4 recent fledglings being fed by adult; Chilcotin River (Riske Creek) 5 Jun 1983-2 perched on powerline at Wineglass Ranch; Alkali Lake (Cariboo) 27 Jun 1970-2, nesting in garden; Dempsey Lake 14 Aug 1986-1, being chased by Sharp-shinned Hawk; Puntzi Lake 11 Jun 1997-1 on lawn; Chilanko Forks 28 Jun 1961-4; Williams Lake 6 Jun 1976-4 nestlings, 26 Aug 1949-1 (NMC 48318); Wells Gray Park 20 Jul 1975-1 singing at Ray Farm (Goward 1976); Anahim Lake 22 Jun 1961-1; Quesnel 6 Jun 1977-3 nestlings; Prince George 1 Jun 1968-1 (Rogers 1968d); Mackenzie 17 Jun 1995-1 at feeder; Atlin 26 Jun 1983-2.

Breeding Bird Surveys: Coastal – Recorded from 18 of 27 routes and on 53% of all surveys. Maxima: Victoria 30 Jun 1974-122; Coquitlam 25 Jun 1993-59; Pitt Meadows 4 Jul 1982-50. **Interior** – Recorded from 18 of 73 routes and on 14% of all surveys. Maxima: Lavington 18 Jun 1988-54; Kamloops 14 Jun 1992-49; Oliver 24 Jun 1990-25.

Autumn: Interior – Atlin 4 Oct 1980-3, females or juveniles feeding on weeds; Williams Lake 8 Nov 1987-30, coming to feeder; Riske Creek 27 Oct 1989-1; nr Dog Creek 30 Sep 1953-1 (NMC 48310); e Kamloops 1 Sep 1960-4; Sorrento to Salmon Arm 2 Sep 1970-2; Nakusp 14 Oct 1990-2 (Siddle 1991a); Lavington 12 Nov 1962-8 at feeder; n Sparwood 21 Oct 1984-1 at feeder; Madeline Lake (Penticton) 19 Sep 1982-2; Summerland 30 Nov 1965-30; Penticton 28 Nov 1977-120 along road to Mac's Lake; Nelson 13 Sep 1997-1 fledgling being fed by adult; Trail 28 Nov 1982-37; Osoyoos 19 Oct 1962-10. **Coastal** – Campbell River 5 Sep 1975-20 at estuary; Alice Lake (Cheekye) 20 Sep 1975-1; Oyster River 17 Sep 1978-4; Trent River 22 Oct 1988-2 perched in shrub at estuary; Comox Harbour 2 Oct 1974-18; Pitt Meadows 7 Oct 1962-6; Fanny Bay 30 Sep 1990-11 (Dawe et al. 1995b); Baynes Sound 19 Sep 1981-18 (Dawe et al. 1998); Aldergrove 30 Sep 1968-5, 12 Oct 1968-21; Crescent Beach 30 Oct 1965-100, feeding on goldenrod seeds; Surrey 12 Sep 1995-1 recent fledgling fed by adult male; Sea Island 5 Nov 1965-600 at airport, 18 Nov 1979-800; Lulu Island 20 Oct 1976-350 along New Westminster Highway, 7 Nov 1973-250 along west dyke; Stubbs Island (Tofino) 10 Nov 1996-35 feeding on California wax-myrtle berries; Central Saanich 12 Sep 1987-250 in corn stubble, 16 Oct 1965-300 (Stirling 1966); Cowichan Bay 25 Sep 1981-6 in high shrubs along road; Victoria 3 Oct 1970-200 in ivy-covered walls of City Hall (Tatum 1971); Oak Bay 26 Sep 1982-60; Rocky Point (Victoria) 22 Sep 1995-16 banded; Whiffin Spit (Sooke) 1 Nov 1980-20.

Winter: Interior – nr Fort St. John 16 Dec 1997-2 at bird feeder (BC Photo 1712); Quesnel 30 Dec 1978-2 at feeder; Williams Lake 6 Jan 1984-50, 1 captured by Northern Shrike; Wineglass Ranch (Chilcotin River) 2 Dec 1988-1 at feeder with Evening Grosbeaks; Revelstoke 27 Dec 1989-2; Sorrento 20 Feb 1971-1; Salmon Arm 14 Feb 1980-1 (PMA 280120441); Kamloops 7 Jan 1984-75; Swan Lake (Vernon) 17 Dec 1960-40; Vernon 9 Jan 1966-40; Oyama 5 Feb 1970-3; Merritt 3 Jan 1998-150; Naramata 26 Dec 1973-170; Summerland 7 Dec 1968-100; Penticton 27 Dec 1976-180; Nelson 11 Dec 1997-14 at feeder; Blewett 10 Jan 1991-14; Kinnaird 19 Jan 1971-1, captured by Northern Pygmy-Owl; Creston 13 Dec 1997-38 eating mountain-ash berries; Lister 20 Jan 1987-14 at feeder; Trail 19 Dec 1982-24 at feeder; Osoyoos 10 Feb 1974-12 in marsh. **Coastal** – Sonora Island Jan 1955-6 banded; nr Woss Lake 31 Jan 1976-5; Union Bay 24 Jan 1988-20 at bird feeder; Courtenay 14 Jan 1987-7; Kye Bay 23 Jan 1988-3; Harrison Hot Springs 22 Jan 1986-25; Burnaby 13 Feb 1971-1 killed by Sharp-shinned Hawk; Parksville 1 Dec 1976-36 in flock; Agassiz 12 Dec 1975-100; Aldergrove 11 Feb 1968-15; Boundary Bay 29 Dec 1979-250; Ladner 17 Jan 1971-100 (Campbell et al. 1972b); Delta 9 Dec 1973-285; Buttertubs Marsh (Nanaimo) 2 Dec 1976-35; St. Mary Lake (Saltspring Island) 9 Dec 1978-12; Duncan 30 Jan 1988-12 at sewage lagoons; Island View Beach 26 Dec 1986-50; Pender Island 5 Feb 1979-12; Saanich 16 Feb 1974-150; Carmanah Point 13 Jan 1952-1 (Irving 1953); Victoria Dec 1951-4 (Clay 1952), 11 Dec 1971-1,000 roosting in ivy on building (Tatum 1972); Whiffin Spit (Sooke) 31 Dec 1983-3.

Christmas Bird Counts: Interior – Recorded from 16 of 27 localities and on 50% of all counts. Maxima: Penticton 27 Dec 1988-2,584; Kelowna 19 Dec 1987-2,013; Vernon 18 Dec 1983-1,510. **Coastal** – Recorded from 22 of 33 localities and on 73% of all counts. Maxima: Vancouver 19 Dec 1993-1,983; Ladner 29 Dec 1979-1,383; White Rock 4 Jan 1987-1,264.

Red Crossbill

RECR

Loxia curvirostra (Linnaeus)

RANGE: Holarctic. Resident in Canada and the United States from southwestern Alaska, southern Yukon, southwestern Mackenzie, northern Saskatchewan, central Manitoba, southern Ontario and Quebec, southern New Brunswick, and Newfoundland south through forested and mountainous regions to northern California, New Mexico, Arizona, southwestern Alberta, south-central Saskatchewan, southern Manitoba, northern Minnesota, Wisconsin, New York, and Maine. Disjunct populations breed in the Cypress Hills of southeastern Alberta and southwestern Saskatchewan and in the Appalachian Mountains; also breeds occasionally south to Mississippi and northern Baja California and through central Mexico to north-central Nicaragua.

Winters throughout its normal range, but may winter south to the southern United States in some years.

Also found in Eurasia and North Africa.

STATUS: May be absent or very numerous in many portions of the province at any given season.

On the coast, a *fairly common*, at times *very common*, resident in the Georgia Depression Ecoprovince; in the Coast and Mountains Ecoprovince, a *fairly common*, at times *very common*, resident on Western Vancouver Island and the Queen Charlotte Islands, a *fairly common*, at times *common*, resident on the Southern Mainland Coast, and an *uncommon*, at times *common*, resident on the Northern Mainland Coast.

In the interior, a *common*, at times *very common*, resident in the Southern Interior, Southern Interior Mountains, Central Interior, and Sub-Boreal Interior ecoprovinces; a *very rare* migrant and summer visitant in the Boreal Plains and Northern Boreal Mountains ecoprovinces, *accidental* in winter in the latter ecoprovince; and *casual* in the Taiga Plains Ecoprovince.

Breeds.

NONBREEDING: The Red Crossbill (Fig. 627) is a highly nomadic species that may occur at any locality at almost any time of year. Distribution is highly dependent on the yield of coniferous cone crops and the stage of cone ripening. Throughout most of its North American range, the Red Crossbill will vacate large geographic areas if cone crops are poor and irrupt into areas where cones are plentiful (Adkisson 1996).

During nonbreeding seasons, it occurs in all forested regions of British Columbia but is less abundant in the northern one-third of the province than elsewhere in the interior and on the coast. Although Adkisson (1996) shows the Red Crossbill to be resident in all of British Columbia, there are few winter records from most of the northern third of the province. We are therefore less certain of the species' year-round status in the far north. It is possible that most northern Red Crossbills shift southward during the coldest months. The bird tends to be more common than its close relative the White-winged Crossbill in the southern and central portions of the interior, but is less common in northern areas.

Figure 627. The Red Crossbill may be absent or relatively common at any time of the year throughout much of British Columbia. Flocks of males (a) and females (b) are vocal when flying but silent when feeding on cones of firs and pines. Juveniles (c) can be found in British Columbia throughout most of the year (all R. Wayne Campbell).

The Red Crossbill is apparently quite scarce in the Taiga Plains. Erskine and Davidson (1976) did not see any during 2 summers of field work near Fort Nelson, nor did Greenfield (1998) during 6 years of field work in the vicinity of the Sikanni Chief River. We also have few records for this crossbill from the Northern Boreal Mountains. Swarth (1926) noted the White-winged Crossbill to be common in Atlin but made no mention of the Red Crossbill. The same situation prevailed along the Tatshenshini River (Campbell et al. 1983). The species is very rare in the trembling aspen parkland areas of the Boreal Plains; Cowan (1939) did not record even a single bird during his work there. It occurs locally each year in forested areas of the Boreal Plains.

On the coast, the Red Crossbill occurs along the length of the mainland coast, on Vancouver Island and the Queen Charlotte Islands, and on most offshore islands. It is reported much more often on the coast than the White-winged Crossbill. In winter, the highest numbers of the Red Crossbill occur in the

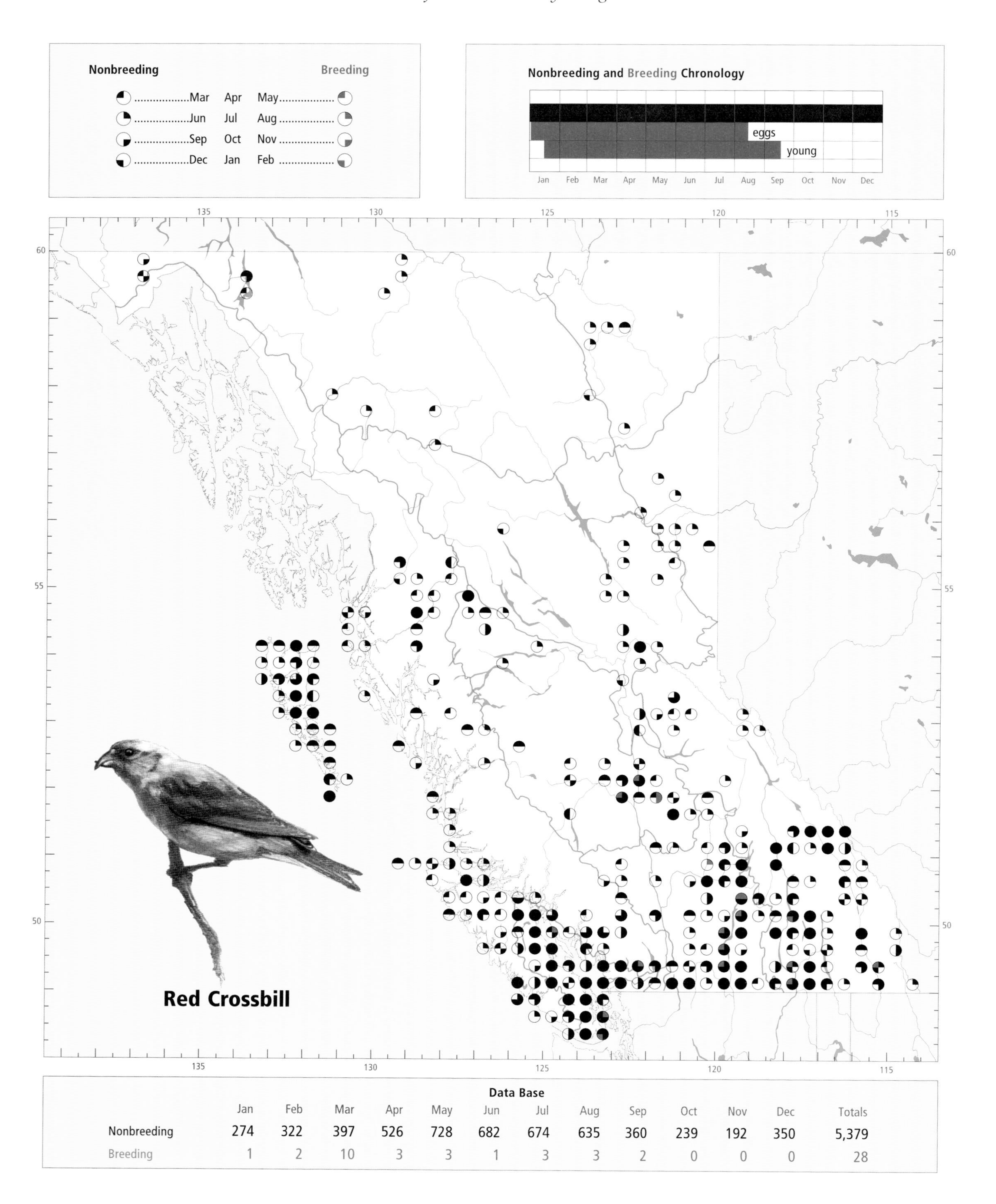

Data Base

	Jan	Feb	Mar	Apr	May	Jun	Jul	Aug	Sep	Oct	Nov	Dec	Totals
Nonbreeding	274	322	397	526	728	682	674	635	360	239	192	350	5,379
Breeding	1	2	10	3	3	1	3	3	2	0	0	0	28

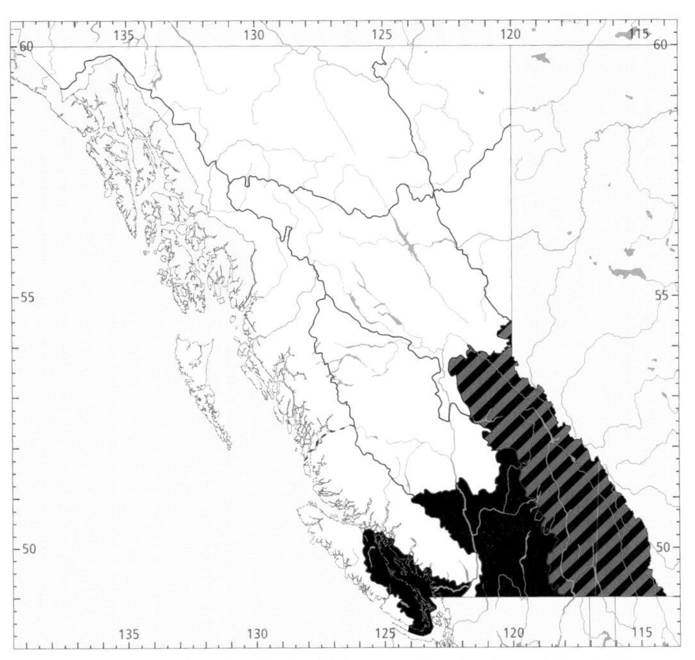

Figure 628. In British Columbia, the highest numbers for the Red Crossbill in winter (black) occur in the Georgia Depression, Southern Interior, and Southern Interior Mountains ecoprovinces; the highest numbers in summer (red) occur in the Southern Interior Mountains Ecoprovince.

Georgia Depression, Southern Interior, and Southern Interior Mountains (Fig. 628). On the coast, the Red Crossbill occurs at elevations from near sea level to at least 1,620 m; in the interior, it occurs at elevations from 280 m to at least 2,130 m. It may occur at all elevations up to timberline at any time of year but is more often found at lower and middle elevations. On the coast, nonbreeding habitat includes all forested areas from shore pine at sea level up to high-altitude mountain hemlock and subalpine fir forests at timberline. On treeless Triangle Island, Red Crossbills were observed feeding on cow-parsnip (Vermeer et al. 1976). In autumn and winter, it frequents treed suburbs, where it feeds on berries of ornamental and native trees and occasionally takes suet and seed from feeders.

In the interior, nonbreeding habitat includes ponderosa pine, Douglas-fir, and spruce forests. In addition, larch, lodgepole pine, and mixed-wood riparian forests, and occasionally deciduous stands and subalpine meadows, are used. In the Okanagan, it is most common in or near Douglas-fir forests. Near Okanagan Lake, a flock of adults and young fed on unripe choke cherries and insect larvae on the undersides of black cottonwood leaves (Cannings et al. 1987). In Yoho National Park, it is most common in spruce and pine forests with plentiful cones (Wade 1977). It very rarely occurs in non-treed habitats. In winter, the Red Crossbill may be seen along the edges of highways eating salt-encrusted gravel. Such behaviour is most often seen in the southern portions of the interior.

Migrations of the nomadic Red Crossbill include both local and long distance movements between areas with abundant cone crops. These movements are often irregular and can occur at any time of year. As with White-winged Crossbills, there may be 1 to 3 movements per year, the timing of which depends on the stage of cone ripening (Adkisson 1996) (see REMARKS).

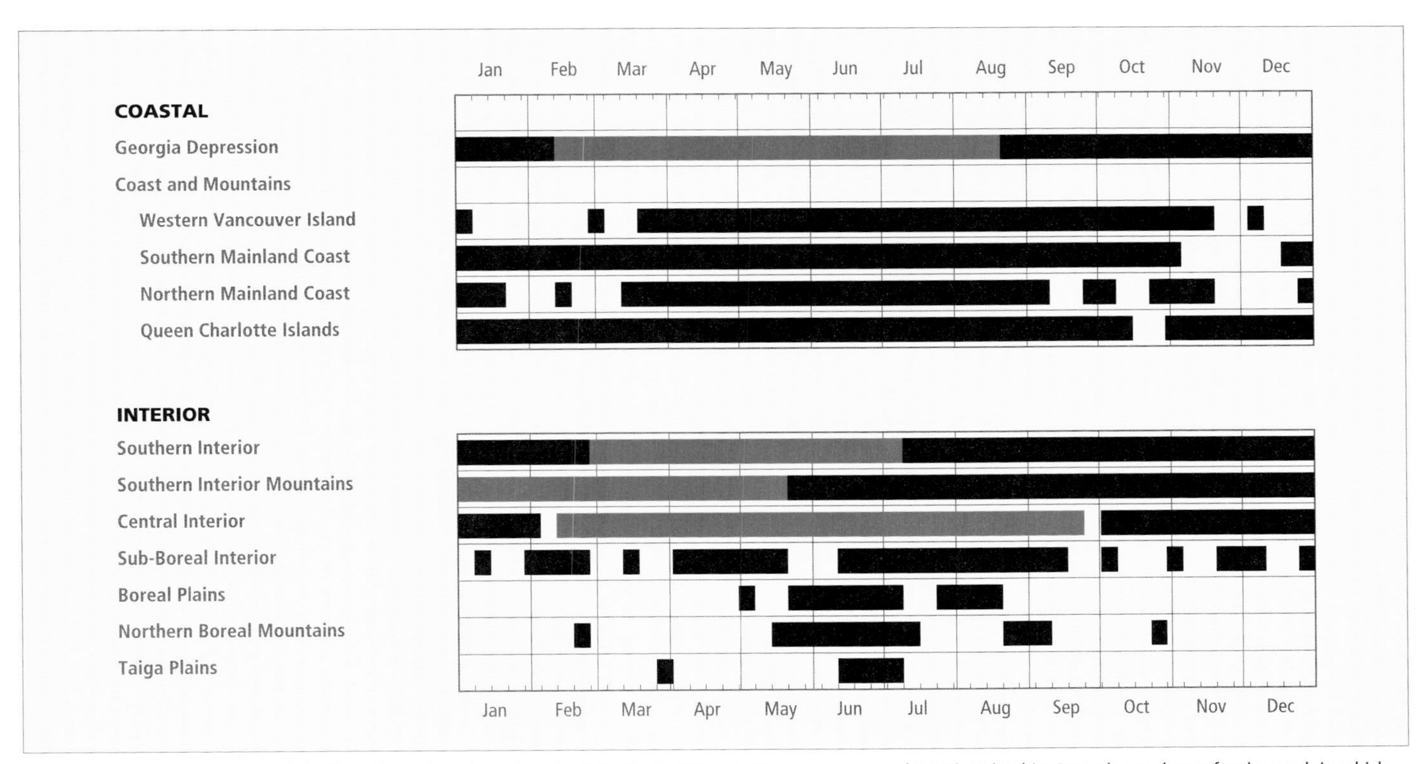

Figure 629. Annual occurrence (black) and breeding chronology (red) for the Red Crossbill in ecoprovinces of British Columbia. Records are shown for the week in which they occurred.

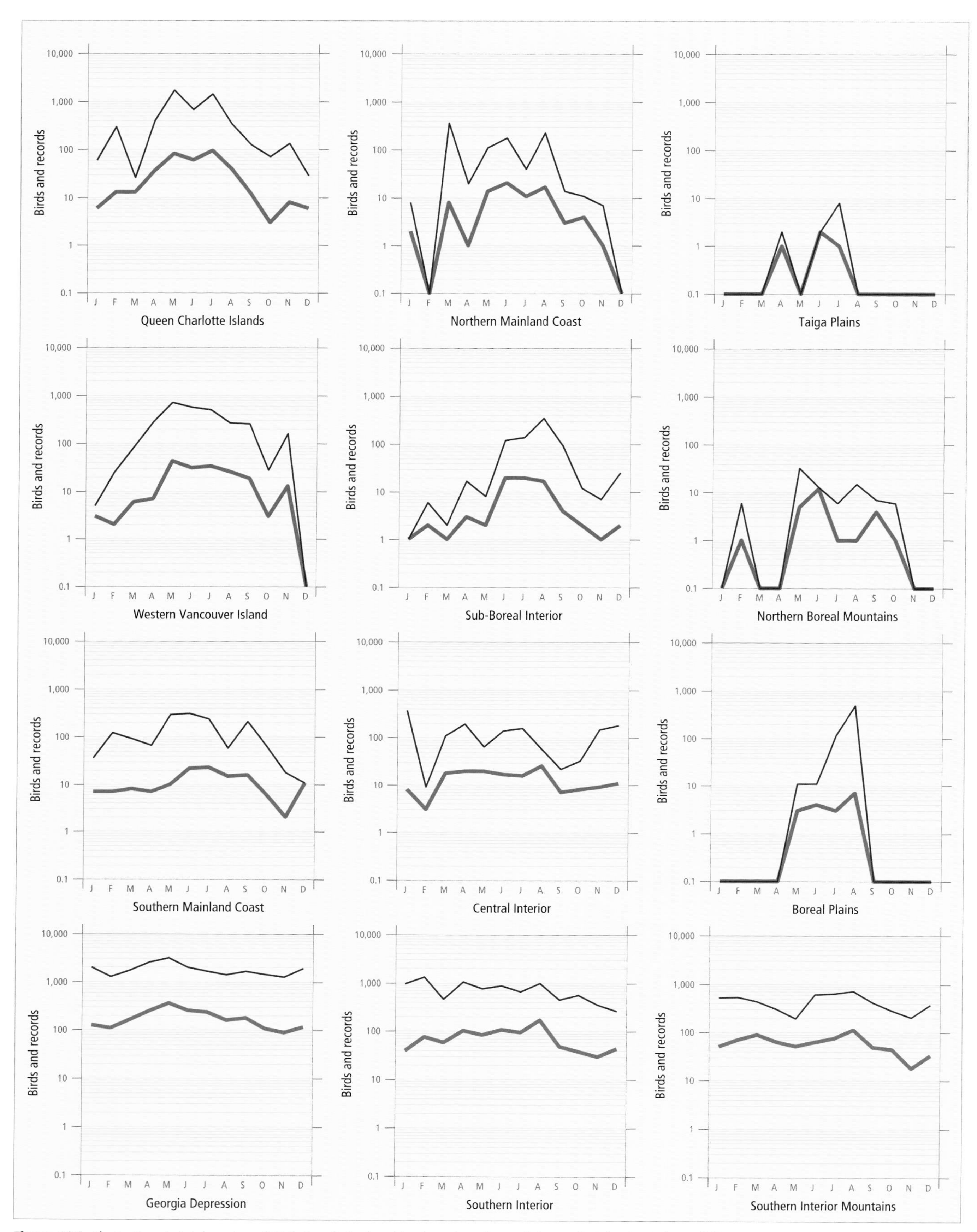

Figure 630. Fluctuations in total number of birds (purple line) and total number of records (green line) for the Red Crossbill in ecoprovinces of British Columbia. Christmas Bird Counts, Breeding Bird Surveys, and nest record data have been excluded.

Figure 631. On the Queen Charlotte Islands of British Columbia, the Red Crossbill is found throughout the year in Sitka spruce forests (Collison Inlet, 28 May 1996; R. Wayne Campbell).

Across North America, major movements usually occur from May to June and from September to October (Austin 1968). Data for parts of British Columbia support this trend. Throughout the coast, records of 40+ birds tend to occur mainly from April through June (46%; n = 171), with May (20%) being the peak month; they also occur from September through October (12%). The largest flocks tend to occur from May through July on the Queen Charlotte Islands and from April through June on Western Vancouver Island. Throughout the interior, the largest flocks tend to occur from June through August (44%; n = 63). In the Southern Interior, many large flocks occur in June (25%; n = 28); in the Boreal Plains, 6 of 7 large flocks documented occurred in August.

Irruptive movements of Red Crossbills, in which very large numbers of birds move in or out of large geographic areas, occur in response to food supply, as populations exploit cone crops (Adkisson 1996). In their review of the birds of British Columbia, Munro and Cowan (1947) noted this phenomenon, stating that the Red Crossbill is resident, off and on, in all forested areas, but "violently cyclic in numbers."

Our data show that major irruptions occurred in 1976 and 1977 in the Southern Interior Mountains, and in 1975 in the Sub-Boreal Interior and Boreal Plains. Notable local irruptions have also been documented. In the late winter of 1918, one occurred in the Okanagan valley prior to nesting. Munro (1919b) had noted the first Red Crossbill of the year on 9 January but estimated that the density reached about 600 birds in 5 km^2 by 15 February. Another irruption occurred in the Okanagan valley in the winter of 1978 to 1979, when record numbers were recorded on Christmas Bird Counts. The species had been virtually absent in the previous few winters and was absent again in the following winter, of 1979 to 1980 (Cannings et al. 1987).

During the 12 years from 1937 to 1948, the Red Crossbill was abundant in the east Kootenay in 1937, 1944, and 1945; scarce in 1938, 1941, 1942, 1943, 1946, and 1947; and absent in 1939, 1940, and 1948 (Johnstone 1949). In Mount Revelstoke National Park, a few Red Crossbills occurred in 1942, but none was noted in 1943 (Cowan and Munro 1944-46).

Munro (1947) noted the absence of Red Crossbills from vast areas of the Sub-Boreal Interior during the summer of 1944, and also mentioned their scarcity throughout southern parts of the province in the early 1940s. Munro (1949) found them absent from the Vanderhoof area in the summers of 1946 and 1947, until early August, when numerous small flocks appeared in spruce stands with heavy cone crops.

On the coast, Guiguet (1953) found numerous small flocks on Goose Island from mid-May to 28 June 1948, then none until 20 July, when numbers returned. Major irruptions occurred throughout Vancouver Island in 1973 and 1974. The 1973 irruption was also detected by Hatler et al. (1978) in Pacific Rim National Park. In the Nimpkish River valley, McNicholl (1995) did not find the Red Crossbill in May 1995, but large numbers appeared suddenly in June. In subalpine habitats on several mountains on Vancouver Island, the Red Crossbill was the most abundant bird in September 1997, with hundreds noted daily (Ogle and Martin 1997). A major irruption to the Queen Charlotte Islands occurred in 1977.

Red Crossbills sometimes flock with Pine Siskins, and less commonly with Mountain Chickadees, Red-breasted Nuthatches, and Cassin's Finches. Red and White-winged crossbills join in foraging flocks in trees but do not associate in flight. Both crossbill species shift their diets among seeds of different coniferous species, depending on ripening phenology (Benkman 1987).

In coastal British Columbia, the Red Crossbill occurs regularly year-round; in the interior, it occurs year-round in the southern and central portions of the interior and mainly from May to August in the Boreal Plains and Northern Boreal Mountains (Figs. 629 and 630).

BREEDING: The known breeding range of the Red Crossbill in British Columbia is scattered throughout the southern portions of the province. Breeding on the coast has been documented only at Victoria and Saanich on southern Vancouver Island, Powell River on the Sunshine Coast, and Golden Ears Park on the Southern Mainland Coast, although there are reports of fledged young from the Gulf Islands and from Sooke north to Port Alberni and Qualicum Beach on Vancouver Island. On the Northern Mainland Coast, the Red Crossbill has been reported breeding on Porcher Island, although the presence of eggs or nestlings has not been documented. There are no breeding records from the Fraser Lowland in the Georgia Depression or from Western Vancouver Island or the Queen

Charlotte Islands, but the Red Crossbill likely breeds throughout its summer distribution along coastal British Columbia.

In the interior, the Red Crossbill has been confirmed breeding in the Southern Interior from Summerland north to Okanagan Landing in the Okanagan valley and at Heffley Lake and Westsyde, north of Kamloops; in the Southern Interior Mountains, breeding has been confirmed only in the southern parts of the west and east Kootenay at Nakusp and near Slocan, Rossland, and Lake Koocanusa, south of Fernie; and in the Central Interior at Lac la Hache, Bald Mountain near Riske Creek, and Williams Lake. There are also a number of records of fledged young from each of these ecoprovinces, but no breeding records or records of fledged young from the 4 northern ecoprovinces. The Red Crossbill likely breeds in suitable habitat throughout its summer range in the interior.

In British Columbia, the Red Crossbill reaches its highest numbers in summer in the Southern Interior Mountains (Fig. 628). Breeding Bird Surveys for the period 1968 through 1993 show that interior and coastal populations are stable. Breeding Bird Surveys across Canada from 1966 to 1996 suggest that populations are also stable, but across North America, populations increased an average annual rate of 2.3% over the same period ($P < 0.01$) (Sauer et al. 1997). Continent-wide Breeding Bird Survey data indicate that British Columbia has some of the highest densities of the Red Crossbill in North America (Sauer et al. 1997).

Nests have been found at elevations from near sea level to 600 m on the coast, and from valley bottoms at 300 m up to 1,800 m in the interior. Breeding habitat consists of mature coniferous forests with large cone crops (Figs. 631 and 632). In the Okanagan valley, the preferred breeding habitat is found in Douglas-fir forests.

Throughout the species' range, nesting occurs only when roaming flocks find good crops of cones (Adkisson 1996). As with the White-winged Crossbill, females may lay more than 1 clutch in rapid succession if food is plentiful. The breeding chronology is well described by Munro (1919b) for birds near Okanagan Landing: large numbers of birds settled in the area in January and early February, paired birds and courtship behaviour were noted in mid-February, nest building began in early March, eggs were present by late March, and young with adults were flocking together by June.

Nesting habitat is mainly coniferous forest (58%; $n = 50$; Figs. 631 and 632) and treed suburbs or rural areas (38%). One nesting area was in a treed wetland, and 1 was on a treed estuary. Males tend to sing from treetops within 60 m of nest trees (Munro 1919b; Adkisson 1996).

On the coast the Red Crossbill has been recorded breeding from 16 February to 14 August; in the interior, it has been recorded from 2 January (calculated) to 17 September (Fig. 629).

Nests: As in other parts of the bird's range, nest building can begin as early as December in British Columbia. In the Okanagan, nest building has been observed from 28 December to 20 July, although nests built in winter may not lead to nesting (Cannings et al. 1987). In Yoho National Park, a Red Crossbill was observed carrying nesting material on 18 February 1977 (Wade 1977). On the coast, nest building has been noted in early February.

Figure 632. In the south-central interior of British Columbia, the Red Crossbill breeds in Douglas-fir forests (Johnstone Creek Park, 15 April 1999; R. Wayne Campbell).

Most nests (94%; $n = 18$) were built in conifers, usually ponderosa pine or Douglas-fir. Nests were usually well concealed in foliage and were saddled on top of a branch away from the trunk. The heights above ground ranged from 1.8 m to 20 m, with 56% between 6 and 10.5 m. Nests are bulky cups of twigs, grasses, mosses, and needles lined with plant down, fine grasses, and hair.

Munro (1919b) described 1 Okanagan nest as an inner cup of black tree moss, lichens, dry grasses, and weed stalks surrounded by fine fir twigs, with those used on the rim having little tufts of bright green lichen. The nest was lined with fine grasses and 1 breast feather from a Red-tailed Hawk.

Eggs: Dates for 10 clutches ranged from 16 February to 14 August with 6 recorded between 8 March and 21 March. Calculated dates indicate that nests could have eggs as early as 2 January. Sizes of 4 clutches ranged from 1 to 3 eggs (1E-1, 2E-1, 3E-2). In the Okanagan, 6 of 7 clutches contained 4 eggs (Cannings et al. 1987), which may indicate that clutches of 4 eggs are more common than was thought. In addition, several broods of 4 nestlings have been reported, confirming that some clutches have a minimum of 4 eggs. Clutch sizes elsewhere range from 2 to 4 eggs, but 3 eggs is normal (Adkisson 1996).

Incubation may begin with the laying of the first egg during cold weather (Adkisson 1996), a trait that was noted in the Okanagan by Munro (1919a). The incubation period is usually 14 days (Adkisson 1996).

Young: Dates for 19 broods ranged from 29 January to 17 September, with 58% recorded between 18 March and 30 July. Calculated dates suggest that nestlings can be found as early as 16 January. Sizes of 15 broods ranged from 1 to 4 young (1Y-6, 2Y-3, 3Y-3, 4Y-3). Fledged broods recorded in British Columbia contained 1 to 4 young, with an average of 2.0 young per brood (n = 40). Four of these broods contained 4 young. The nestling period is 15 to 25 days (Adkisson 1996).

In the Okanagan, fledged young have been noted as early as 21 March (Cannings et al. 1987). At Indianpoint Lake, McCabe and McCabe (1933) presumed that nestlings were present in March 1931. Up to 4 broods may be reared annually (Adkisson 1996).

Brown-headed Cowbird Parasitism: There are no records of cowbird parasitism in British Columbia. There is only 1 record of parasitism elsewhere, near London, Ontario (Friedmann 1963).

Nest Success: Insufficient data.

REMARKS: The taxonomy of the Red Crossbill has been confusing for some time. Unlike the White-winged Crossbill, the Red Crossbill varies considerably in bill and body size and shape, and as many as 8 North American subspecies have been recognized (American Ornithologists' Union 1957). These subspecies were thought to have discrete breeding distributions but to wander widely in nonbreeding seasons; 2 subspecies were thought to breed in British Columbia. Munro and Cowan (1947) listed *L. c. sitkensis* as the subspecies occurring mainly on the coast, although they noted a few specimens from the southern portions of the interior. *L. c. bendirei* was found mainly in the interior, but there were specimens for the coast from Vancouver Island and even the Queen Charlotte Islands.

The existence of 2 different subspecies in both coastal and interior British Columbia during spring and summer was puzzling in the 1940s because subspecies are usually geographically isolated during the breeding season. In the Red Crossbill, the subspecies seemed to occur simultaneously in the same areas.

Recent research has discovered that there are 8 different "subspecies," or *call types*, of Red Crossbill north of Mexico, each with its own distinctive flight calls, which help maintain reproductive isolation (Groth 1993a, 1993b). Most of these "subspecies" are highly nomadic and may roam over much of the species' North American range, mixing with other "subspecies." Six of the 8 call types have been documented in British Columbia (Adkisson 1996), so up to 6 "subspecies" of Red Crossbill may occur here. Some of them have calls that can be distinguished from each other by human hearing.

Individuals may forage in a variety of conifers, but each "subspecies" has a strong association with different key coniferous species, especially during breeding. The 6 "subspecies" that may occur in British Columbia specialize in foraging in the following key conifers: Type 1, Sitka spruce and western hemlock (coastal); Type 2, ponderosa pine, Engelmann spruce, and lodgepole pine (interior and coastal); Type 3, western hemlock and Sitka spruce (coastal); Type 4, Douglas-fir (interior and coastal), Type 5, lodgepole pine and Engelmann spruce (interior), and Type 7, Engelmann spruce and lodgepole pine (interior) (Benkman 1987; Groth 1993a, 1993b). It is likely that 2 or more subspecies may breed in the same general area and at the same time, but may use different ecological niches. There is a strong correlation between bill morphology, call type, and the key conifer; thus, each subspecies is specialized for extracting seeds from specific conifers.

The timing of breeding in the Red Crossbill is the most erratic of any North American bird (Tordoff and Dawson 1965). Worldwide, it may be found breeding somewhere in every month of the year. In North America, there are egg dates from mid-December to early September. Recent evidence indicates, however, that breeding ceases in the autumn when day length becomes shorter than 12 hours; after the annual moult, it can resume in late December at day lengths of 10.5 hours (Adkisson 1996). In British Columbia, there are many reports of courtship behaviour and nest building during December and January, but it is believed that most of these early nests are abandoned before eggs are laid. Nonbreeding pairs can occur at all times of the year (Adkisson 1996).

McCabe and McCabe (1933) conducted research on Red Crossbills at Indianpoint Lake north of Bowron Lake Park in British Columbia and contributed a thoughtful paper on the breeding ecology of this enigmatic species. Their thoughts on timing of nesting, age of first breeding, and moult patterns stimulated early discussion that helped focus future worldwide research.

Because of mineral deficiencies in plant foods, many northern finches, including crossbills, are attracted to salt-impregnated and calcium-rich seepage areas. In British Columbia, these can occur on disturbed hillsides, on coastal mudflats, along the shores of marshes and sloughs, and along roadsides.

There may be some conservation concerns for the Red Crossbill, even if the bird seems relatively common. The subspecies endemic to Newfoundland has possibly been extirpated because of competition for cones with introduced red squirrels and because of accelerated logging of old-growth forests. Logging has also caused declines in Europe and the northeastern United States, probably by reducing cone crops (Adkisson 1996). Shorter rotations, which occur in British Columbia, may reduce overall food availability because conifers produce their largest cone crops after about 60 years of age (Benkman 1993a).

Winter records are scarce from northern parts of British Columbia, and breeding records are scarce for all regions. Naturalists can help improve our understanding of the distribution of the Red Crossbill by looking for this species in winter, carefully distinguishing between Red and White-winged crossbills, and noting evidence of breeding such as courtship behaviour, nests with eggs or young, or recent fledglings.

Additional information on the fascinating ecology and taxonomy of the Red Crossbill can be found in Adkisson (1996), Benkman (1987, 1989, 1993b), Groth (1993a), and Holimon et al. (1998).

NOTEWORTHY RECORDS

Spring: Coastal – Lizard Lake (Port Renfrew) 18 Apr 1974-76; Oak Bay 2 Mar 1970-200 (Sherman 1970), 15 May 1978-100; Victoria 4 Mar 1970-150; Elk Lake (Saanich) 20 May 1974-106; Observatory Hill (Saanich) 15 May 1994-6; Prevost Hill 30 Apr 1978-100; Chilliwack River 5 Apr 1969-58; Hope 8 May 1975-50; Golden Ears Park 12 Mar 1988-100; Gold River 25 May 1974-80; Miracle Beach Park 27 Apr 1974-200; Triangle Island 28 May 1974-80 (Vermeer et al. 1976); Goose Island (Goose Group) 10 May 1948-40, 26 May 1948-120, 6 flocks of 18 to 30 each (Guiguet 1953); Lyell Island 10 May 1982-180; Kunga Island 22 May 1982-105; Skidegate 18 May 1960-40; Hiellen River 2 May 1979-180; Porcher Island summer 1921-breeding; Legate Creek 30 Mar 1977-230. **Interior** – Manning Park 8 May 1975-90; Penticton 21 Mar 1910-fledglings (Cannings et al. 1987); Rossland 20 May 1972-2 young not able to fly well; South Slocan 6 Mar 1994-female incubating; Lake Koocanusa 18 May 1983-3 nestlings; nr Westsyde (Kamloops) 8 Mar 1981-nest with 3 eggs; Monashee Mountain 4 Mar 1977-50; Okanagan Landing 20 Mar 1918-nest with 3 eggs, female with 1 more egg in oviduct collected (Munro 1919b); Okanagan Landing 18 Mar 1918-4 nestlings (Munro 1919b); Celista 10 Apr 1948-100; Bald Mountain (Riske Creek) 1 Mar 1988-1 fledgling with traces of down; Quesnel 6 May 1979-6; Wells Gray Park 29 May 1971-6; Alexis Creek 8 May 1977-11; Indianpoint Lake Mar 1931-nestlings present (McCabe and McCabe 1933); Ross and Spatsizi rivers 28 May 1976-20; nr Hyland Post 30 May 1976-7; n Fort Nelson 1 Apr 1986-2; Chilkat Pass 15 May 1977-2.

Summer: Coastal – Jordan Meadows 21 Jun 1946-25; Central Saanich 6 Jun 1993-2; Surrey 5 Aug 1987-125; Cowichan Bay 18 Aug 1973-168; Keeha Bay 11 Jun 1970-35; Mount Steele 27 Jul 1987-1 at 1,620 m elevation; Qualicum 18 Aug 1976-3 fledglings; Harrison Lake 21 Jul 1987-60; Cranberry Lake (Powell River) 19 Jul 1976-4 fledglings 1 to 2 days out of nest; Miracle Beach Park 1 Jul 1968-200 (Crowell and Nehls 1968d); Cortes Island 10 Jun 1978-50; Triangle Island 5 Jun 1976-50 (Vermeer et al. 1976), 14 Jul 1984-50; Cape St. James Jun 1979-35; Goose Island 5 Jun 1948-100 (Guiguet 1953); Bella Coola 26 Aug 1942-2 (RBCM 10004 and 10005); Reef Island 11 Jul 1977-138; Kimsquit 20 Jun 1939-1 (NMC 29031); Tlell River 9 Jun 1985-70; Frederick Island 25 Jul 1977-38; Kwinamass River 21 Aug 1985-150; Telegraph Creek 22 Jun 1919-fledglings (Munro and Cowan 1947; UBC 872). **Interior** – Nicks Island (Creston) 6 Jun 1983-50; Flathead River 14 Aug 1983-15; Bonanza Pass 31 Aug 1982-12; Naramata 5 Aug 1983-3 fledglings with parents; Pruden Pass 30 Jun 1991-100; Tenquille Lake 23 Jul 1960-present (Bradley 1961); Vernon 5 Jul 1993-nest with 2 nestlings; Revelstoke to Downie Creek Aug 1955-200 dead on highway (Sugden 1956); Chilco Lake 21 Jul 1993-1 fledgling with adults; Farwell Canyon 22 Jun 1968-80; Mount Robson Park 27 Jul 1973-60; Carp Lake Park 25 Aug 1975-85; Pine Pass 17 Aug 1975-50; e Tupper 4 Aug 1975-125; w Dawson Creek 3 Aug 1975-58; Chetwynd to Hudson's Hope 17 Aug 1975-59; Wonowon 19 Aug 1975-130; Fire Flats (Spatsizi Plateau Wilderness Park) 11 Jul 1975-common at Fire Flats; Alaska Highway (Mile 176) 4 Jul 1978-8; Kinaskan Lake 10 Jun 1991-2; French and Dease rivers 25 Aug 1977-15.

Breeding Bird Surveys: Coastal – Recorded from 23 of 27 routes and on 46% of all surveys. Maxima: Kitsumkalum 28 Jun 1987-118; Albion 12 Jun 1988-105; Port Renfrew 14 Jun 1980-65. **Interior** – Recorded from 52 of 73 routes and on 49% of all surveys. Maxima: Oliver 20 Jun 1986-114; Kamloops 24 Jun 1984-97; Canford 5 Jul 1990-63.

Autumn: Interior – Haines Highway 26 Oct 1981-6; Babine Lake Rd 21 Nov 1981-7; Smithers 2 Nov 1977-18; Morice River 11 Nov 1979-8; Willow River 5 Sep 1983-70; Barkerville 16 Sep 1962-15; Williams Lake 18 Nov 1976-100; Lac la Hache 17 Sep 1993-1 young just out of nest; Mount Revelstoke Park 10 Nov 1983-50; Vernon 10 Sep 1978-50; Peter Hope Lake 20 Oct 1976-50; McNulty Creek 14 Oct 1985-several flocks of 10 to 20 birds during the day. **Coastal** – Masset 10 Oct 1971-50; McInnes Island 12 Nov 1963-7; Campbell River 3 Oct 1973-100; Diamond Head 10 Sep 1972-50; Buttle Lake 5 Sep 1988-15; Skagit River 29 Sep 1974-50; Port Renfrew 29 Sep 1985-flocks of 25 and 30; Muir Creek 30 Nov 1986-100; Witty's Lagoon (Metchosin) 11 Oct 1988-30.

Winter: Interior – Crehan Creek 23 Feb 1995-8; nr Babine Lake 22 Jan 1979-3 along Babine Lake Rd; 25 km s Prince George 14 Jan 1984-1; Williams Lake 29 Dec 1976-80; Yoho National Park 18 Feb 1977-1 carrying nesting material (Wade 1977), 18 Dec 1976-65; Nakusp 29 Jan 1984-3 recently fledged young; Okanagan Landing 15 Feb 1918-600 in 3 square miles (Munro 1919a); Nakusp to Shelter Bay 26 Jan 1977-83 in small flocks along roadside. **Coastal** – Greenville 4 Jan 1982-4; Masset 29 Jan 1983-9; Juskatla 31 Jan 1972-25; Port Clements to Tlell 12 Feb 1973-185 in numerous small flocks; Anthony Island 1 Jan 1983-1; Cape St. James 25 Dec 1978-1; Mount Seymour Park 28 Jan 1975-100; Cowichan Bay 1 Jan 1977-100; Port Alberni 27 Jan 1973-100; Alouette Lake 24 Feb 1988-60; Manning Park 23 Jan 1971-200; Saanich 28 Feb 1967-nestlings heard; Victoria 16 Feb 1958-female incubating eggs.

Christmas Bird Counts: Interior – Recorded from 19 of 27 localities and on 53% of all counts. Maxima: Oliver-Osoyoos 28 Dec 1993-381; Vaseux Lake 23 Dec 1983-374; Penticton 26 Dec 1978-343. **Coastal** – Recorded from 28 of 33 localities and on 54% of all counts. Maxima: Victoria 18 Dec 1993-1,830; Duncan 16 Dec 1973-452; Sooke 27 Dec 1986-437.

White-winged Crossbill

Loxia leucoptera (Gmelin)

WWCR

RANGE: Holarctic. Breeds from Alaska to Newfoundland, south in the Cascade Range to central Oregon and in the Rocky Mountains to Colorado and New Mexico; in the east, south to Nova Scotia, New Brunswick, Maine, and New York. Winter range is the same as the breeding range except that the species occasionally wanders south to California, Nevada, Texas, and Florida.

Also in Old World.

STATUS: May be absent or very numerous in many portions of the province at any given season.

On the coast, *very rare* to locally *uncommon* at times throughout the year in the Georgia Depression Ecoprovince; *very rare* throughout the year in the Coast and Mountains Ecoprovince.

In the interior, *uncommon* to locally *very common* at times throughout the year in the Southern Interior, Southern Interior Mountains, and Central Interior; *uncommon* to locally *common* in the Sub-Boreal Interior and Boreal Plains ecoprovinces; and *uncommon* to locally *common* from May to September in the Northern Boreal Mountains and Taiga Plains ecoprovinces.

Breeds.

NONBREEDING: During nonbreeding periods, the White-winged Crossbill (Fig. 633) has a widespread distribution throughout interior British Columbia, although winter records in the north are scarce. The nonbreeding range is generally the same as the breeding range. This crossbill occurs in all ecoprovinces, but seems to be more abundant in mountainous regions or where spruce and larch occur. It is much less abundant in drier areas where lodgepole pine or ponderosa pine forests occur, such as in the Central Interior and Southern Interior, but is present in higher-altitude forests there. Nonbreeding periods are not predictable as this crossbill may breed throughout the year (Godfrey 1986) and may be simultaneously breeding and not breeding in different parts of the province depending on the abundance of coniferous cones (see BREEDING and REMARKS).

The White-winged Crossbill occurs irregularly on the coastal mainland from the mountains of North and West Vancouver north to Garibaldi Park and east through mountainous areas of the Pacific and Cascade ranges to Manning Park. We have no records from the mainland coast between Garibaldi Park and Kitimat. On the Queen Charlotte Islands, numbers appear irregularly. Wandering birds occasionally show up on offshore islands. In winter, the highest numbers for the White-winged Crossbill in British Columbia occur in the Southern Interior Mountains (Fig. 634).

On the coast, the White-winged Crossbill occurs from near sea level to at least 1,500 m elevation; in the interior, it occurs from 280 m to at least 2,100 m. In the interior plateau of southern portions of the province, it tends to occur at lower elevations during the winter, when spruce cone crops are less

Figure 633. The White-winged Crossbill has an atypical migration pattern compared to that of other songbirds. Linked with food supply, migrations can occur at any time of the year (© B. Henry/VIREO).

productive. In the interior, nonbreeding habitat includes spruce (Fig. 635), larch, lodgepole pine, Douglas-fir, and western hemlock forests, treed muskegs, birch woodlands, and mixed-wood riparian forests. It very rarely occurs in non-treed habitats. It also forages in black cottonwood riparian habitats, eats small buds off willows and birches and berries off shrubs, and eats salted gravel on roadways. In winter, it also occurs in suburban gardens and at bird feeders. On the coast, nonbreeding habitat includes a wide range of coniferous forests ranging from clumps of Sitka spruce on estuaries

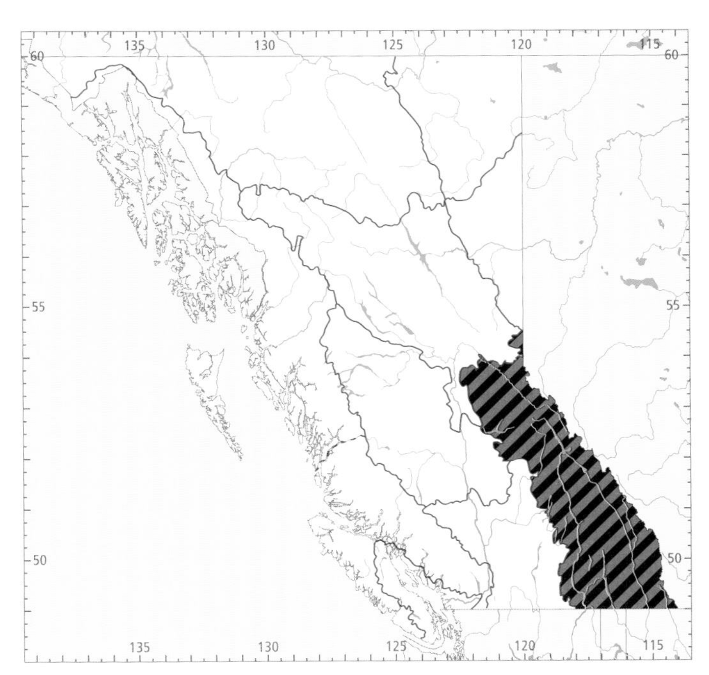

Figure 634. In British Columbia, the highest numbers for the White-winged Crossbill both in winter (black) and summer (red) occur in the Southern Interior Mountains Ecoprovince.

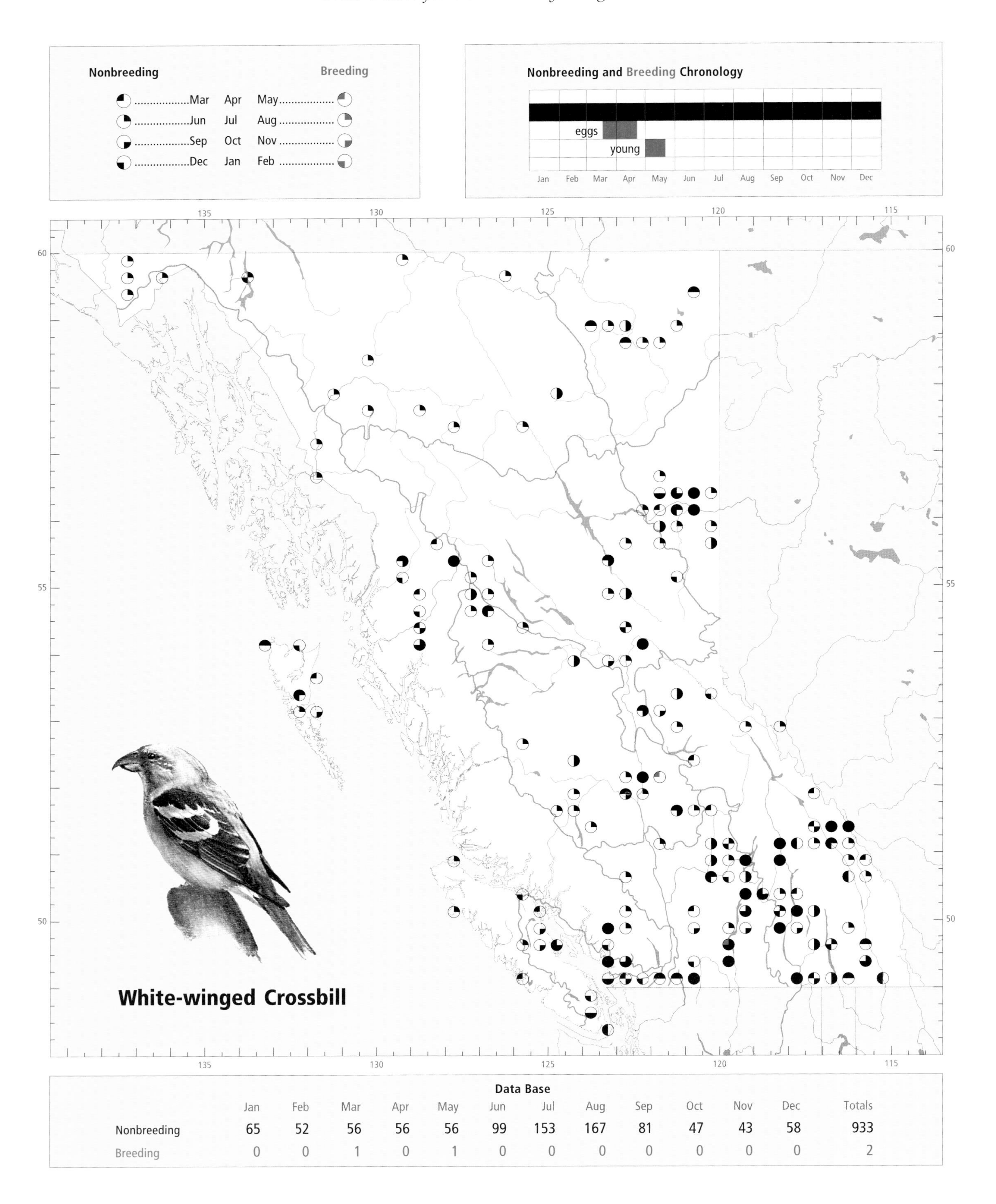

Data Base

	Jan	Feb	Mar	Apr	May	Jun	Jul	Aug	Sep	Oct	Nov	Dec	Totals
Nonbreeding	65	52	56	56	56	99	153	167	81	47	43	58	933
Breeding	0	0	1	0	1	0	0	0	0	0	0	0	2

Figure 635. During the nonbreeding seasons in British Columbia, the White-winged Crossbill prefers spruce forests, where it can be present in small flocks at any time of the year (Stagleap Park, 29 September 1997; R. Wayne Campbell).

to high-altitude mountain hemlock, probably up to timberline. On treeless Triangle Island, a few White-winged Crossbills have been observed feeding on the seeds of giant cow-parsnip (Vermeer et al. 1976). In autumn and winter, this crossbill also feeds on berries of ornamental and native trees and takes suet and seed from feeders.

White-winged Crossbill migration does not follow the typical British Columbia songbird pattern of birds moving northward in spring and southward in autumn, as crossbills are nomadic (Figs. 636 and 637). Migrations may be local or long distance movements linked with food supply; they are irregular and can occur at any time of year (see REMARKS). Near Kamloops, peak spring movements usually occur in March and peak autumn movements in September. However, at 1 study site northwest of Kamloops in 1997, no White-winged Crossbills were observed in June, whereas flocks were abundant during early August and fairly common in late September. At Atlin, Swarth (1926) noted that numbers increased through mid-June, then flocks split off into pairs, apparently preparing to nest.

Irruptive movements, where large numbers of birds can move in or out of a large geographic area, occur in response to food supply (Benkman 1992). These movements occur frequently as populations follow coniferous cone crops, but few have been documented in British Columbia. Notable irruptions into the Okanagan valley occurred during the summer of 1918 and winter of 1985 (Cannings et al. 1987). In Mount Revelstoke National Park, the White-winged Crossbill was fairly numerous in 1937, but none was observed in 1942 or 1943 (Cowan and Munro 1944-46). At Tofino, flocks were recorded in May 1976, May 1991, and May 1996, the only records for that area. In Manning Park, very large numbers appeared during the summer and autumn of 1985; as many as 500 were noted at Lightning Lake in August and similar large flocks were seen elsewhere in the park. In other years there, typical records were from 1 to 25 birds. Based on our records, many irruptions occurred over much of the province in 1985. Conversely, during the winter of 1990-1991, the White-winged Crossbill was absent or uncommon throughout British Columbia (Siddle 1991b).

Figure 636. Annual occurrence (black) and breeding chronology (red) for the White-winged Crossbill in ecoprovinces of British Columbia. Records are shown for the week in which they occurred.

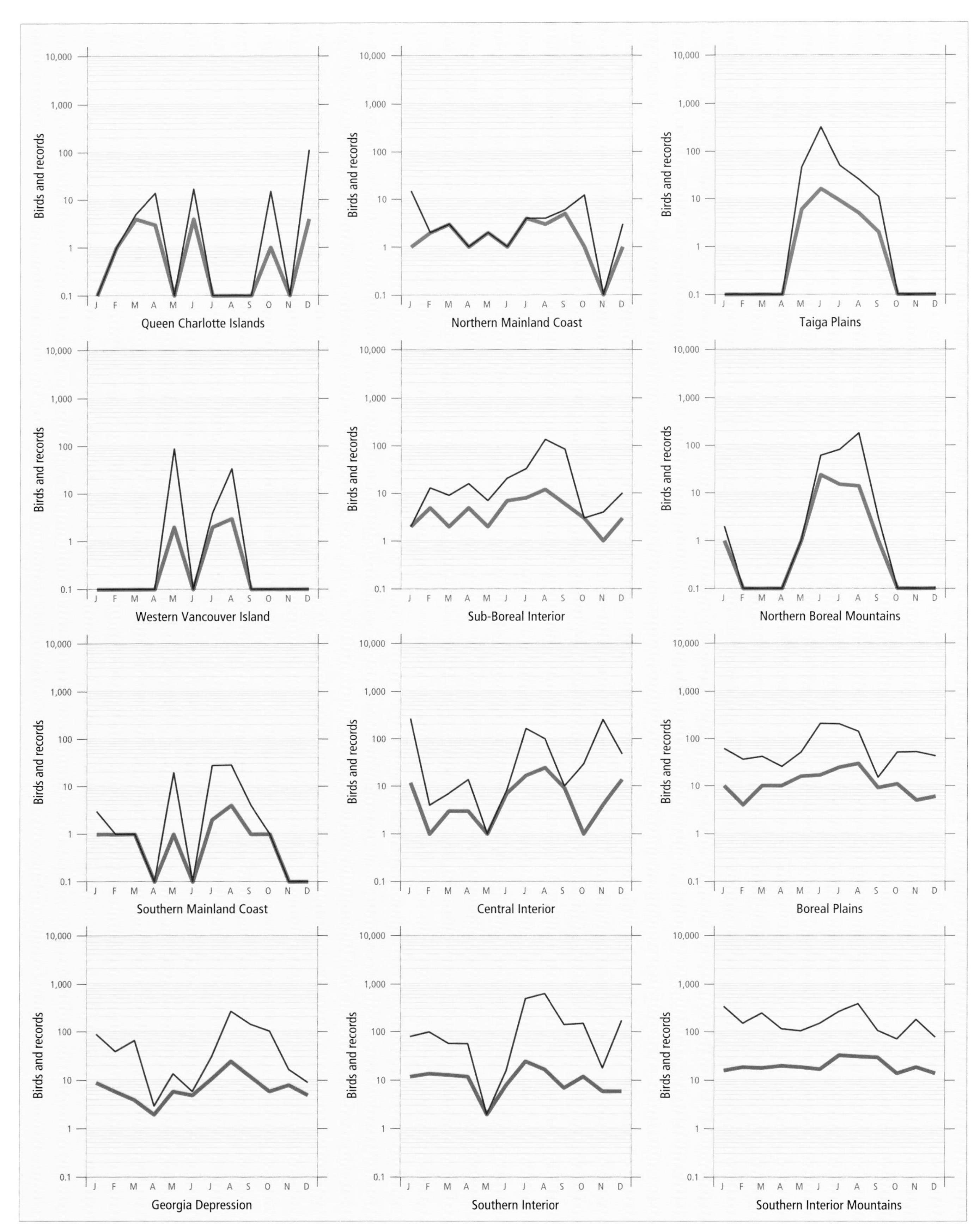

Figure 637. Fluctuations in total number of birds (purple line) and total number of records (green line) for the White-winged Crossbill in ecoprovinces of British Columbia. Christmas Bird Counts, Breeding Bird Surveys, and nest record data have been excluded.

Provincewide, the largest flocks tend to occur from June through March. Of 27 records of 40+ birds, none was from April and only 1 from May. On the coast, most records of the White-winged Crossbill are from the late summer through winter; but there was 1 flock of 40+ birds in each of the months of May, August, and September.

During nonbreeding seasons in British Columbia, the White-winged Crossbill often flocks with the Red Crossbill, Common Redpoll, and Pine Siskin.

In the Southern Interior Mountains and Boreal Plains, the White-winged Crossbill occurs regularly throughout the year. Throughout much of the rest of southern and central British Columbia, it occurs irregularly year-round. In the far north, it can be found regularly from June through August (Fig. 636).

BREEDING: The breeding distribution of the White-winged Crossbill is an enigma in British Columbia. Our breeding data include an active nest from Summerland in the Southern Interior and recently fledged young reported from Spokin Lake in the Central Interior. There are a few reports of adults feeding fledged young at the summit of the Salmo-Creston Highway and in Mount Revelstoke National Park in the Southern Interior Mountains, at Bridge Lake in the Central Interior, near Prince George in the Sub-Boreal Interior, near Fort St. John in the Boreal Plains, and northeast of Fort Nelson in the Taiga Plains. We have no indication that the White-winged Crossbill breeds along coastal British Columbia, other than a few reports of singing males (Forbidden Plateau, Strathcona Park on Vancouver Island; Mount Seymour Park, Gold Creek, and Cheam Peak in the extreme southern portions of the Southern Mainland Coast; and Alice Arm on the Northern Mainland Coast) and a number of specimens with enlarged testes taken on the lower Stikine River (Swarth 1922).

Godfrey (1986) showed all of mainland British Columbia within this crossbill's breeding range but questioned whether Vancouver Island and the Queen Charlotte Islands should be included. More recently, Benkman (1992) included all of mainland British Columbia within the breeding range, except the western slope of the Coast Mountains. Based on sightings of individuals, pairs, or small flocks, many authors have suggested that the White-winged Crossbill probably breeds in British Columbia in areas where it occurs regularly.

As noted earlier, however, breeding on the coast has yet to be documented despite the fact that the White-winged Crossbill occurs sporadically over much of the coast. Since coastal birds may belong to wandering flocks, breeding cannot be assumed, but considering the abundance of coniferous cone crops, especially spruce, it is likely that this crossbill also nests locally on the coast.

In British Columbia, the White-winged Crossbill reaches its highest numbers in summer in the Southern Interior Mountains (Fig. 634). Breeding Bird Surveys in British Columbia for the period 1968 through 1993 contain insufficient data for analysis. Breeding Bird Surveys across Canada and North America from 1966 to 1996 suggest that populations are stable (Sauer et al. 1997).

During the breeding season, the White-winged Crossbill has been found at elevations between 450 and 2,100 m in British Columbia. In southern areas, nesting occurs mainly in high-elevation, subalpine forests; further north, it breeds from valley bottoms up to timberline (Fig. 638). Adequate food supply is the primary habitat element required for breeding to occur (Benkman 1990).

In the Okanagan valley, the White-winged Crossbill occurs mainly above 1,000 m in spruce forests (Cannings et al. 1987). In Kootenay National Park, it nests in high-elevation Engelmann spruce, subalpine fir, and alpine larch stands, and is less common in disturbed sites than the Red Crossbill (Poll et al. 1984). In the "Big Bend" area of the Liard River, it prefers riparian coniferous stands (Bennett and Enns 1996). In the Tatshenshini River Basin, White-winged Crossbills were found up to timberline (Campbell et al. 1983), but none was found above timberline in nearby Chilkat Pass (Weeden 1960).

The timing of breeding probably varies from year to year as the White-winged Crossbill is known to breed throughout the year (Benkman 1992). For example, courtship behaviour was noted in the middle of winter near Vernon (Cannings et al. 1987) and near Fort St. John, which suggests that nesting may have begun soon after.

In the interior, the White-winged Crossbill has been recorded nesting from 18 March (calculated) to 21 May (Fig. 636), but likely breeds from late winter through autumn (see REMARKS).

Nests: One nest has been described in British Columbia. It was built 7.5 m above the ground at the end of a live pine branch. The nest was constructed of stems, rootlets, and grasses, and was lined with fine rootlets and 2 feathers. Elsewhere, the White-winged Crossbill builds its nest in a coniferous tree, usually spruce, from 1 to 20 m above the ground. Nests are often well concealed by foliage and are constructed of twigs, grasses, lichens, rootlets, and bark, and lined with rootlets, mosses, lichens, hair, and insect cocoons (Benkman 1992).

Like those of other cardueline finches, White-winged Crossbill nests are often clustered in areas having favourable conditions instead of throughout a landscape (Benkman 1992). To date, this habit has not been documented in British Columbia.

Eggs: Only 1 nest with eggs has been recorded in British Columbia. The nest contained 4 eggs on 21 March. Calculated dates indicate that eggs can be present from 18 March to 22 April. Eggs likely occur from late winter through autumn, however. Elsewhere, most clutches of this crossbill contain 2 to 4 eggs. The incubation period is unknown, but is probably 12 to 14 days (Benkman 1992). Up to 4 clutches per year may be laid.

Young: Nests with young have not been observed in British Columbia, but there is 1 report of 2 recently fledged young on 21 May. Dates calculated from limited data indicate that nestlings can be present in British Columbia as early as 1 May, but young likely occur from early spring through autumn. Fledged broods of up to 4 young have been observed in British Columbia. The nestling period is unknown but may be 15

Figure 638. In years when spruce forests have large cone crops, the White-winged Crossbill probably breeds throughout the Taiga Plains Ecoprovince of British Columbia (10 km north of Prophet River, 28 June 1998; R. Wayne Campbell).

to 25 days, as in the Red Crossbill (Adkisson 1996). Up to 4 broods per year may be produced.

Brown-headed Cowbird Parasitism: There are no records of cowbird parasitism in British Columbia. The White-winged Crossbill is not known to be parasitized by cowbirds (Benkman 1992).

Nest Success: Insufficient data.

REMARKS: Three subspecies of White-winged Crossbill are recognized, but only *L. l. leucoptera* occurs in North America (Benkman 1992).

It is only in recent years that the incredible life history of crossbills has become better understood. The White-winged Crossbill wanders nomadically across boreal forests in North America; some evidence suggests that home ranges may encompass the width of North America from Alaska to Newfoundland (Benkman 1987). The entire population may concentrate in 1 or several large geographic regions of the continent in any one year in response to food supplies.

Crossbill migrations are quite different from those of most other northern hemisphere songbirds, and do not follow the typical north-to-south pattern. All movements appear to be linked to food availability, and 1 to 3 movements per year are usual. The first movement occurs in mid to late May as birds search for developing cone crops (Engelmann and other spruces in British Columbia) sufficiently large to support breeding. A second major movement occurs in October and early November, coinciding with shedding of coniferous seeds; crossbills may remain in an area if seed abundance remains high and leave when seed availability declines. Large irruptions occur occasionally when cone crops fail, and flocks can move south of their normal breeding range (Benkman 1992).

The timing and extent of breeding are controlled by food availability. This crossbill has been recorded nesting in every month of the year in various regions of Canada (Godfrey 1986), but winter nesting is infrequent (Benkman 1992). There are 3 seasonal nesting periods. The earliest begins in January and February and runs through June, but depends on large spruce cone crops where seeds are held through winter; this is the least consistent nesting period. A second period begins in March and April, runs through June, and coincides with the opening of black spruce cones; after nesting, birds search for newly developing cone crops. The main nesting season occurs from early July through late November, coinciding with the maturation of most coniferous cones; multiple nesting is most likely to occur during this period if cone crops are large (Benkman 1992).

When multiple nestings occur, they usually do so in rapid succession. Females leave their mates, who remain to feed the fledglings, and then they renest. Young can begin to breed at less than 5 months old, so spring fledglings may nest the following autumn (Benkman 1992).

Crossbills eating salt along highways suffer significant mortality, as thousands likely collide with vehicles each year. A long-term threat to populations is the logging of boreal spruce forests and the shortening of forest harvest rotations. This reduces cone crops because maximum crops occur only after 60 years (Benkman 1992).

In Washington, the White-winged Crossbill is thought to breed in some years in the northern Cascades and Okanogan Highlands in the northern part of the state, but there are no confirmed records of nesting. In 1995, this crossbill was the most common singing bird in parts of Okanogan County, Washington, and was undoubtedly nesting (Smith et al. 1997).

For more information, see Benkman (1992).

NOTEWORTHY RECORDS

Spring: Coastal – Skagit River 28 Mar 1975-5; Tofino 1 May 1976-55; Cypress Park 14 Mar 1992-1, 31 Mar 1982-20; Campbell River 26 Apr 1975-1; Queen Charlotte Islands 20 Apr 1987-5; Kiusta 28 Mar 1978-2; Kispiox River 19 May 1979-1. **Interior** – Creston 25 Mar 1979-7; Kimberley 3 May 1979-50; Summerland 21 Mar 1910-4 eggs; Rose Hill (Kamloops) 11 Mar 1979-35 in Douglas-fir; Celista 4 Apr 1962-14; Ross Lake (Yoho National Park) 13 Mar 1977-100 (Wade 1977); Spokin Lake 21 May 1983-adults feeding 2 newly fledged young; 24 km s Prince George 15 Apr 1984-adults feeding 4 newly fledged young; Gagnon Creek 25 May 1997-1; Taylor 29 Apr 1994-2; 30 km w Fort St. John 20 May 1984-2 well-feathered fledglings begging food from male; Taylor 26 May 1984-12; Spatsizi Plateau Wilderness Park May 1977-several flocks (Osmond-Jones et al. 1977); Clarke Lake 5 May 1980-6; Helmut (Kwokullie Lake) 22 May 1982-25.

Summer: Coastal – Skagit River 22 Aug 1985-20; West Vancouver 8 Aug 1991-2 (Siddle 1992c); Strathcona Park 17 Aug 1985-140; Gold Creek 8 Aug 1985-30; Garibaldi Park 13 Jul 1983-20; Pemberton 30 Aug 1984-6; Brooks Peninsula 6 Aug 1981-30 (Campbell and Summers 1997); Herbert Island 12 Jul 1975-3; Cape St. James Jun 1979-13; Alliford Bay 6 Jun 1987-2; Langara Island 13 Jun 1970-1 (UMMZ 218041); Moricetown 22 Jun 1975-1; Kitimat 29 Aug 1978-2; Alice Arm 4 Aug 1957-1; Dokdaon Creek 21 Jul 1919-1 (MVZ 39882); Flood Glacier 4 Aug 1919-1 (MVZ 39888); Rainy Hollow 27 Jun 1991-10. **Interior** – Lightning Lake (Manning Park) 18 Jul 1985-150, 30 Jul 1985-250, 6 Aug 1985-500, Aug 1991-hundreds (Siddle 1992c); Champion Lakes Park 11 Aug 1978-143; Kimberley 28 Jul 1977-70; Michel Creek Jun 1974-1; Kootenay National Park 4 Jun 1983-50; Silver Star Mountain (Vernon) 8 Jun 1975-6; Whitewood Lake 7 Aug 1997-small flocks everywhere; Mount Tatlow 21 Aug 1978-6; Bridge Creek (100 Mile House) 25 Aug 1985-female feeding fledgling; Lorin Lake 16 Jul 1985-4; Eagle Lake 16 Jul 1978-30; Sinkut River 29 Aug 1945-6 (Munro 1949); Carp Lake Park 25 Aug 1975-12; Mugaha Creek 30 Aug 1997-5; n Pine Pass 17 Aug 1975-50; Tupper 29 Jun 1938-6 (Cowan 1939); Kiskatinaw River 3 Jul 1974-100 at campground; Hudson's Hope 22 Aug 1975-30 at W.A.C. Bennett Dam; s Taylor 5 Jun 1982-25; Taylor 22 Jun 1985-70; Boundary Lake (Goodlow) 21 Jun 1986-30; Wonowon 19 Aug 1975-34; Laslui Lake 29 Jul 1976-25; Nuttlude Lake 30 Jul 1957-40; Kinaskan Lake 27 Aug 1979-25; Fern Lake (Kwadacha Wilderness Park) 12 Aug 1983-20 (Cooper and Cooper 1983); e Fort Nelson 24 Jun 1982-2 (Campbell and McNall 1982); nr Clarke Lake 29 Jul 1979-19; Kledo Creek Park 26 Jun 1985-212; Alaska Highway (Mile 340) 10 Jul 1978-5; Atlin mid-Jun 1924-flocks of 50 to 60 (Swarth 1926); Tatshenshini River 3 Jun 1983-5 (Campbell et al. 1983); Helmut (Kwokullie Lake) 9 Jun 1982-adults feeding actively begging fledgling; Carmine Mountain 17 Jun 1983-12 singles during hike along foothills (Campbell et al. 1983).

Breeding Bird Surveys: Coastal – Not recorded. **Interior** – Recorded from 18 of 73 routes and on 5% of all surveys. Maxima: Fort Nelson 19 Jun 1976-18; Fort St. John 28 Jun 1985-18; Steamboat 14 Jun 1976-13.

Autumn: Interior – Muskwa River 17 Sep 1943-8 (Rand 1944); Fern Lake (Kwadacha Wilderness Park) 7 Sep 1979-3 (Cooper and Cooper 1983); Grindrod 14 Nov 1918-1 (UBC 7825); nr Halfway River (Peace River) 5 Nov 1983-9; Taylor 20 Nov 1983-26; Mugaha Creek 7 Sep 1997-3; Glacier Gulch 21 Nov 1981-250; Willow River 5 Sep 1983-70; Barkerville 16 Sep 1962-6; Emerald Lake 27 Sep 1975-35; Hills 24 Nov 1985-50; Mount Revelstoke National Park 1 Sep 1991-3 immatures with adults; Celista 10 Sep 1962-102; Apex Mountain (Penticton) 20 Oct 1976-29; Stagleap Park 11 Sep 1991-adult feeding immature; Lightning Lake (Manning Park) 29 Oct 1985-60. **Coastal** – 60 km n Hazelton 12 Sep 1921-1 (MVZ 42260); Alice Arm 23 Oct 1956-12; Garibaldi Lake 3 Oct 1965-1; Mount Washington 22 Sep 1988-15; Comox 15 Nov 1933-7 (RBCM 14053-9); Cypress Park 1 Sep 1985-20, 29 Sep 1985-40; Mount Seymour Park 28 Oct 1995-35 (Elliott and Gardner 1997), 29 Oct 1991-60 (Siddle 1992a); Surrey 10 Oct 1995-2 (Elliott and Gardner 1997).

Winter: Interior – Atlin 17 Jan 1981-2; Stoddart Creek (Fort St. John) 2 Dec 1979-23, 1 Feb 1987-30; Smithers Dec 1943-1 (Munro 1947); Willow River 14 Dec 1968-8; McBride 16 Dec 1984-1; Williams Lake 26 Dec 1974-20; 3 Jan 1988-261 in 1 flock; Glacier National Park 6 Dec 1977-25; Goose Lake (Kamloops) 29 Dec 1985-51 in Douglas-firs; Falkland 7 Dec 1977-22; Wilson Lake (Nakusp) 17 Jan 1981-150; Whatshan Lake 12 Jan 1986-40; Summerland 21 Dec 1930-75 (Tate 1931); Cranbrook 28 Feb 1947-14 (Johnstone 1949); Manning Park 26 Jan 1981-35. **Coastal** – Greenville 4 Jan 1982-15; Masset 17 Dec 1989-110; Graham Island 20 Dec 1916-1 (MVZ 105629); Black Tusk (Garibaldi Park) 11 Jan 1975-3; Cypress Park 26 Feb 1992-12 (Siddle 1992b); Baldy Mountain (Shawnigan Lake) 16 and 17 Jan 1992-80 (Siddle 1992b); Prevost Hill 8 Dec 1973-1.

Christmas Bird Counts: Interior – Recorded from 20 of 27 localities and on 29% of all counts. Maxima: Vaseux Lake 27 Dec 1985-239; Revelstoke 21 Dec 1991-200; Oliver-Osoyoos 29 Dec 1985-193. **Coastal** – Recorded from 13 of 33 localities and on 4% of all counts. Maxima: Masset 17 Dec 1989-110; Pender Harbour 22 Dec 1991-72; Squamish 21 Dec 1991-66.

Common Redpoll

Carduelis flammea (Linnaeus)

CORE

RANGE: Holarctic. Breeds from northern and western Alaska east through the subarctic to northern Labrador, south to northwestern British Columbia, central Alberta and Saskatchewan, northern Ontario, southeastern Quebec, and Newfoundland. Also breeds in Eurasia.

Winters from southern and central parts of the breeding range south to the northern United States. Also winters in Eurasia. Introduced populations have become established in the Southern Hemisphere in New Zealand and on Lord Howe Island.

STATUS: May be absent or very numerous in most of the interior portions of the province during the nonbreeding seasons.

On the coast, *rare* to *uncommon* migrant and winter visitant (at times *fairly common* to *common* locally) in the Georgia Depression Ecoprovince; in the Coast and Mountains Ecoprovince, *accidental* in winter on Western Vancouver Island, *very rare* in winter and *casual* in summer on the Southern Mainland Coast, *rare* migrant and winter visitant on the Northern Mainland Coast, and *casual* in winter on the Queen Charlotte Islands.

In the interior, *fairly common,* at times *very common,* migrant and winter visitant in the Southern Interior and Southern Interior Mountains ecoprovinces, *accidental* in summer in the latter; a *common,* at times *very common* to *abundant,* migrant and winter visitant and *casual* in summer in the Central Interior, Sub-Boreal Interior, Boreal Plains, and Taiga Plains ecoprovinces. Our data suggest that the Common Redpoll is an *uncommon* to *fairly common* migrant and summer visitant and *accidental* in winter in the Northern Boreal Mountains Ecoprovince, but observer effort is low in winter and the redpoll probably occurs there in good numbers throughout the year.

Breeds.

NONBREEDING: The Common Redpoll (Fig. 639) has a widespread distribution throughout interior British Columbia during the nonbreeding seasons, generally from October through April. It likely winters throughout the interior, although records from the Northern Boreal Mountains are scarce in winter. In the central and southern portions of the interior, and at southern portions of the coast, numbers fluctuate widely between years, the redpoll being absent in some and very abundant in others.

The species reaches its highest numbers in winter in the Sub-Boreal Interior, where more than 3,700 birds have been reported on a single Christmas Bird Count (Fig. 640). It is also quite numerous and widespread in the Southern Interior Mountains. The Common Redpoll is scarce in winter in Kootenay National Park, but more numerous in the Columbia valley (Poll et al. 1984), which suggests that it is more numerous at lower elevations. It is also a fairly numerous winter visitant at Yoho National Park (Wade 1977) and less common in January and February at Creston (Van Damme 1996).

Figure 639. In the northern interior of British Columbia, the Common Redpoll is a widespread bird at feeders with niger seeds (4 April 1994; Michael I. Preston).

On the coast, the Common Redpoll occurs irregularly on southern Vancouver Island, in the Nanaimo Lowland, and in the Fraser Lowland on the adjacent mainland, north through the Squamish River valley, the Sunshine Coast, Bella Coola, Terrace, and Greenville. It has also been reported at Masset and Aliford Bay on the Queen Charlotte Islands, and there is a single winter record from Tofino, on Western Vancouver Island.

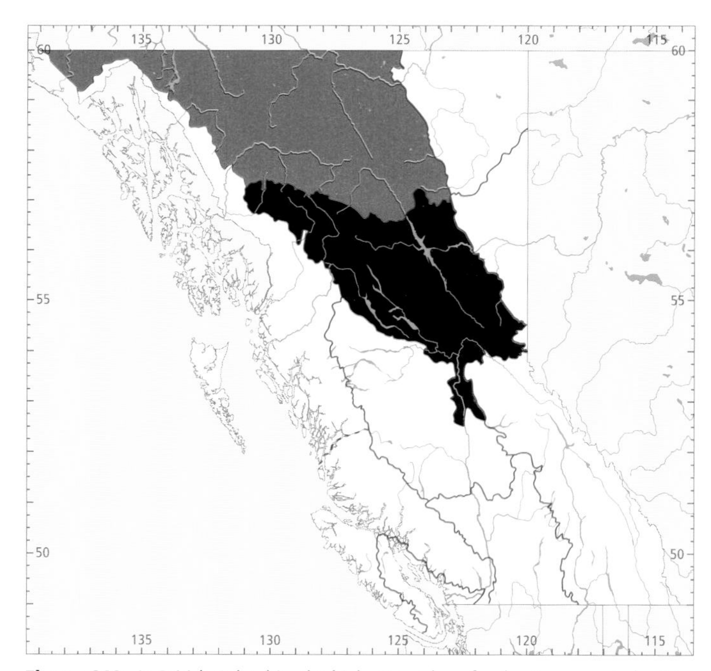

Figure 640. In British Columbia, the highest numbers for the Common Redpoll in winter (black) occur in the Sub-Boreal Interior Ecoprovince; the highest numbers in summer (red) occur in the Northern Boreal Mountains Ecoprovince.

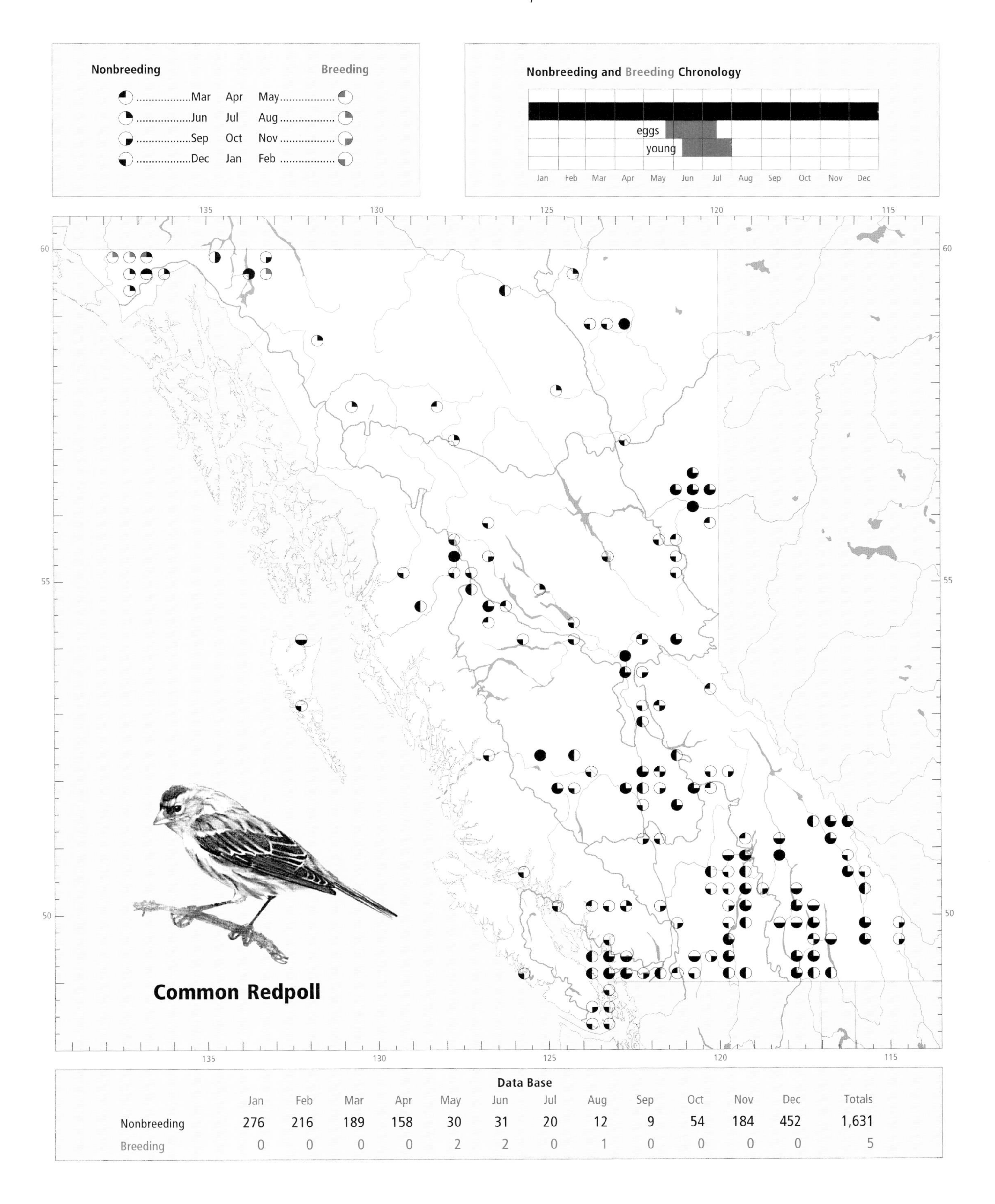

Data Base

	Jan	Feb	Mar	Apr	May	Jun	Jul	Aug	Sep	Oct	Nov	Dec	Totals
Nonbreeding	276	216	189	158	30	31	20	12	9	54	184	452	1,631
Breeding	0	0	0	0	2	2	0	1	0	0	0	0	5

Figure 641. In the Sub-Boreal Interior Ecoprovince of British Columbia, Common Redpolls frequent edge habitat and deciduous vegetation along waterways during the nonbreeding season (Crooked River, 15 November 1996; R. Wayne Campbell).

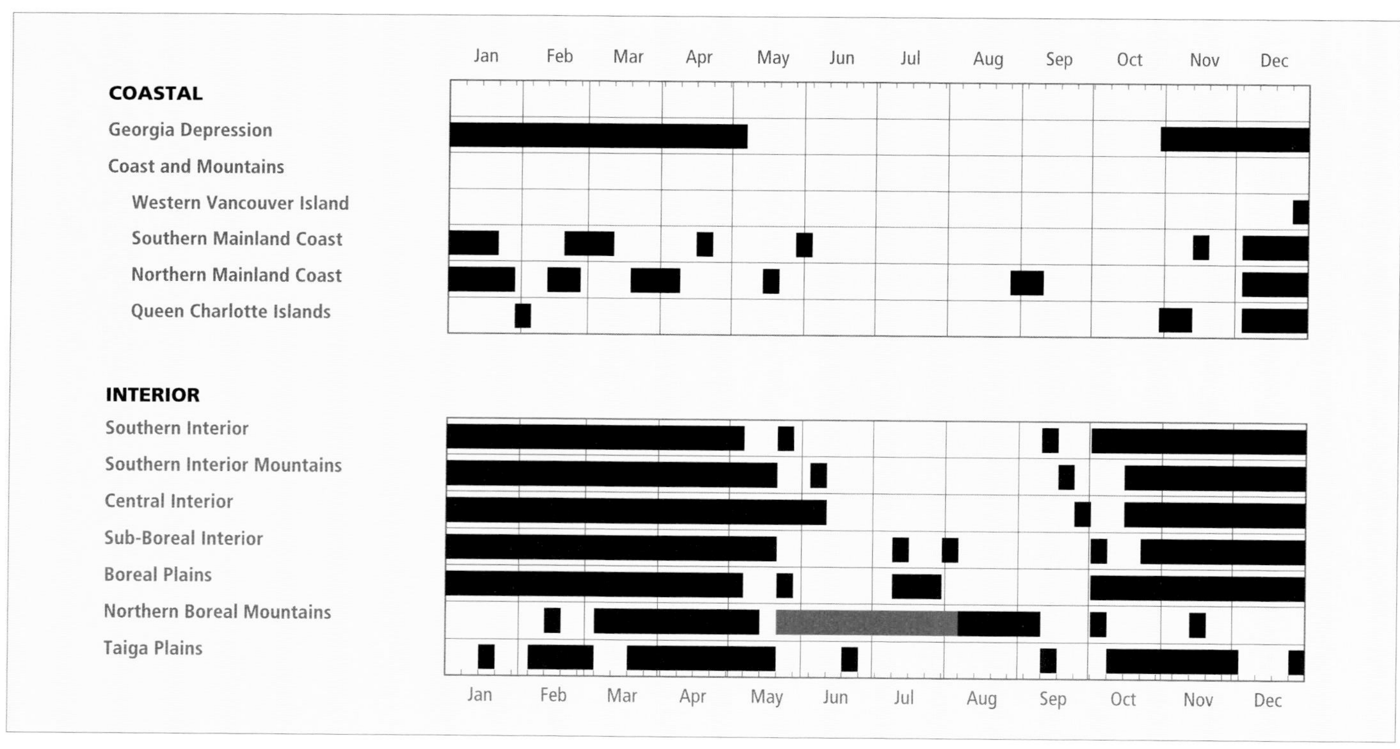

Figure 642. Annual occurrence (black) and breeding chronology (red) for the Common Redpoll in ecoprovinces of British Columbia. Records are shown for the week in which they occurred.

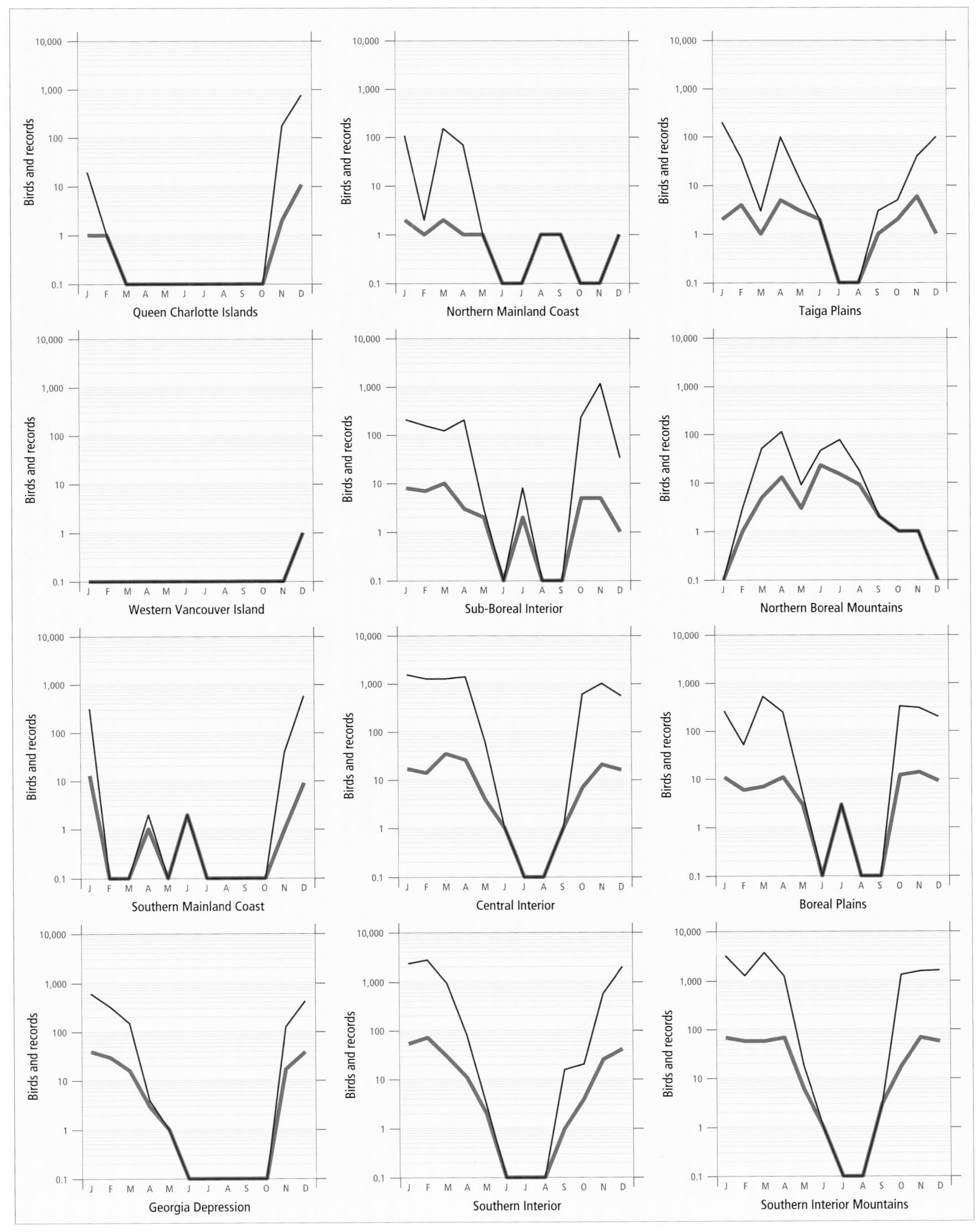

Figure 643. Fluctuations in total number of birds (purple line) and total number of records (green line) for the Common Redpoll in ecoprovinces of British Columbia. Christmas Bird Counts, Breeding Bird Surveys, and nest record data have been excluded.

During nonbreeding seasons, the Common Redpoll occurs at elevations from near sea level to 800 m on the coast, and from 275 to 1,600 m in the interior. In the Chilkat Pass area, Weeden (1960) notes that the Common Redpoll uses higher elevations (alpine areas) after the breeding season. Few redpolls occur in unbroken coniferous forest; most congregate in valleys or along waterways, where edge habitat, deciduous vegetation, and weedy patches are more abundant (Fig. 641). In all areas, birch and alders are the preferred trees for foraging, as their catkins provide abundant seeds (Fig. 644). Wintering redpolls are often seen along roads, fences, and lake edges where these trees, shrubs, and seed-bearing plants are common (Fig. 645). The Common Redpoll also uses willow thickets, weedy grasslands, weedy fields, roadsides, shrub-filled gullies, mixed forests, suburban and urban gardens, and bird feeders. In agricultural areas, it is often seen in grain fields or along hedgerows. The seeds of lamb's-quarters, common in disturbed sites, and shrubby cinquefoil have been identified as a food source in several regions. Redpolls are frequently attracted to salt-encrusted gravel on roads and road edges. They also forage for insects in coniferous trees, and in the Fort St. John area, they feed on spilled grain in open storage sheds.

Weeden (1960) described the Common Redpoll as an "abundant migrant" but only a "common breeder" at Chilkat Pass. He noted that flocks of adults and immatures began to form after late June. In Spatsizi Plateau Wilderness Park, migrant flocks have been recorded in August (Osmond-Jones et al. 1977). From these and other observations, it appears that the Common Redpoll congregates in flocks shortly after nesting is completed and roams around the breeding range, heading south only in mid to late autumn (Figs. 642 and 643). In the Taiga Plains and Boreal Plains, autumn migrants begin to arrive as early as mid-September but large numbers do not appear until the first week of October, peaking towards the end of the month and early November (Fig. 643). Redpoll numbers decline from December to February or March. In the Sub-Boreal Interior, birds begin arriving in numbers in early October and peak in November; some birds continue their movement further south, leaving a small wintering population. In the central and southern portions of the interior, this redpoll may arrive in mid to late September but not in any numbers until the end of October or early November; numbers generally peak in December (Figs. 642 and 643). These flocks often arrive abruptly, and areas that contain no redpolls one day may have large numbers the next. The timing of autumn arrivals varies widely from year to year. In some years, the birds do not appear at all.

On the coast, birds may arrive by early November but large numbers do not appear until mid-November, peaking in December and January.

The distribution of wintering redpolls varies widely from year to year. As with many other finches, redpolls can be virtually absent from traditional wintering areas in some years and very numerous in others. In the Okanagan valley, for example, Christmas Bird Count totals varied from zero to several hundred Common Redpolls during the 1960s to 1980s (Cannings et al. 1987). In the east Kootenay, redpolls were absent during the winters of 1943 to 1947 (Johnstone 1949). During the winter of 1992-1993, Siddle (1993b) reported that the Common Redpoll occurred in higher numbers than usual in the east Kootenay, none occurred at Grand Forks, normal numbers occurred in the north Okanagan, and numerous small flocks occurred at Prince George and Vanderhoof.

The irregular and local distribution of wintering redpolls can probably be explained mainly by 2 factors: the tendency to form large flocks, so that most individuals from an area gather in relatively few groups at any one time, and the redpoll's dependence on deciduous tree seed crops, which leads to concentrations in areas with high seed abundance.

Notable irruptions occurred in the Okanagan valley in the winters of 1908-1909 (Venables 1910), 1929-1930 (Cannings et al. 1987), and 1972-1973; in the Southern Interior Mountains in 1982-1983, 1984, and 1990; in the Central Interior in 1968-1969, 1979, 1982, and 1985; in the Sub-Boreal Interior in

Figure 644. In British Columbia, the catkins of paper birch provide abundant seeds for flocks of foraging Common Redpolls (Delta, 2 May 1970; R. Wayne Campbell).

Figure 645. In the Boreal Plains Ecoprovince of British Columbia, nonbreeding Common Redpolls frequently forage in weedy areas along roads and fencelines (Rose Prairie, 28 October 1984; Chris Siddle).

Figure 646. Although the Common Redpoll is known to breed in subalpine willow thickets in the Atlin area of northwestern British Columbia, its breeding range is poorly documented (Ruby Creek, 7 July 1996; R. Wayne Campbell).

1985, 1990, and 1993-1994; and in the Boreal Plains in 1986. Hochachka et al. (1999) analyzed the irruptive movements of Common Redpolls in North America during the winter of 1993-94 and suggested that the irregular irruptive migration of redpolls may be more allied to conventional winter migration than to nomadism.

On the coast, an unusual irruption occurred on the Queen Charlotte Islands in December 1971 and birds remained common throughout the month, with as many as 150 redpolls being observed at Masset. In most years even single birds would be notable on the Queen Charlotte Islands.

In spring, wintering redpolls on the coast and in the southern and central interior, north to the Sub-Boreal Interior, disappear, often suddenly. Most birds have left by the last week of March on the coast and by the third week of April in the interior. In both regions, stragglers may remain through May (Fig. 642). In the Boreal Plains and Taiga Plains, most birds have left by the end of April.

Throughout most of British Columbia, the Common Redpoll often flocks with small numbers of its close relative the Hoary Redpoll. In the Boreal Plains, however, the Hoary Redpoll may often outnumber the Common Redpoll, especially in winter. In the southern portions of the interior, Common Redpolls also flock with Pine Siskins and American Goldfinches, and occasionally with Red and White-winged crossbills and Pine Grosbeaks.

On the coast, the Common Redpoll occurs irregularly from early November to late April. In the far northern interior, west

Figure 647. Throughout extreme northwestern British Columbia, the Common Redpoll probably breeds in deciduous thickets in floodplains and along the edges of streams and rivers (near Tatshenshini River, 3 October 1987; Lance Goodwin).

of the Rocky Mountains, it probably occurs regularly throughout the year. In the northeastern interior, it occurs regularly from early October to early May, and in the southern and central portions of the interior from the end of October to early May (Fig. 642).

BREEDING: The Common Redpoll has been documented breeding only in the far northwestern corner of interior British Columbia, in the vicinity of Chilkat Pass and Alsek River, and near Atlin. Weeden (1960) noted it as a common breeder in Chilkat Pass, yet he found only a single nest. Near Carmine Mountain, about 30 km northwest of Chilkat Pass, Campbell et al. (1983) found that the Common Redpoll occurred sparsely during the nesting season; it was apparently less abundant in 1983 than during the late 1950s. Swarth (1930, 1936) found the Common Redpoll "nesting in fair abundance in some localities" near Atlin, but collected only a single fledged young on 2 August to document nesting. It likely breeds throughout the St. Elias Mountains and Haines Highway areas east to at least the Teslin Plateau.

This main breeding range is similar to that reported by Godfrey (1986). In Alberta, the Common Redpoll may breed locally following long, cold winters (Salt and Salt 1976). It has not been documented breeding in the vast mountainous areas with apparently suitable habitat in central-northern portions of British Columbia. It seems reasonable to expect breeding to occur as far south as Spatsizi Plateau.

In British Columbia, the Common Redpoll reaches its highest numbers in summer in the Northern Boreal Mountains (Fig. 640). Breeding Bird Survey data for British Columbia, Canada, and North America as a whole are insufficient for trend analysis (this study; Sauer et al. 1997).

Information on breeding in British Columbia is limited, but nesting habitat appears to occur in subalpine areas (Figs. 646 and 647). The Common Redpoll has been found nesting at elevations from about 360 m on the Alsek River to 1,000 m at Chilkat Pass. Early summer records in the Tatshenshini River area suggest that breeding habitat comprises mainly subalpine willow thickets, mixed spruce and deciduous thickets at edges of openings, and edge habitat along streams and rivers (Figs. 647 and 648). Nesting habitat is often associated with wetlands. A nest found along the Alsek River was on a floodplain with scrubby thickets of alders, willows, and balsam poplars between channels.

In British Columbia, the Common Redpoll has been recorded breeding from 24 May (calculated) to 2 August (Fig. 642).

Figure 648. One of only 3 nests of the Common Redpoll found in British Columbia was neatly placed in the shelter of an overhang on a river bank (Turnback Canyon, Alsek River, 30 June 1992; William J. Merilees).

Figure 649. Thousands of redpolls are killed each year along highways in the interior of British Columbia, where salting and road maintenance activities provide attractive foraging areas (near Golden, 30 January 1997; R. Wayne Campbell).

Nests: One nest was on the ground, concealed "junco fashion," under the overhanging lip of a river flood-channel bank (Fig. 648; Merilees 1995). Two nests, 0.8 and 1.7 m above ground, were placed in willow shrubs. Elsewhere, nests are usually situated 1 to 2 m above ground in a tree or shrub; occasionally they are found on the ground or high in trees. Nests have been described as scruffy cups of fine twigs, grasses, and plant stems, lined with plant down, feathers, and hair (Baicich and Harrison 1997).

Eggs: Dates for 4 clutches ranged from 26 May to 30 June. Calculated dates indicate that eggs can occur as early as 24 May and as late as mid-July. Egg laying in western Alaska also begins in the last week of May (Troy and Shields 1979). Sizes of 3 clutches ranged from 1 to 6 eggs (1E-1, 4E-1, 6E-1). Clutches usually contain 4 or 5 eggs, but may range from 3 to 7 eggs; the incubation period ranges from 10 to 13 days (Baicich and Harrison 1997).

Young: We have no records of nests with nestlings in British Columbia. A fledgling estimated to be 2 days out of the nest was collected on 2 August near Atlin (Swarth 1930). Calculated dates suggest that nestlings can occur from 10 June to the end of July. Weeden (1960) noted that flying young could be found in Chilkat Pass after late June. The nestling period is 11 to 14 days (Baicich and Harrison 1997).

Brown-headed Cowbird Parasitism: We have no records of cowbird parasitism of the Common Redpoll in British Columbia. This northern finch is very rarely parasitized; only 2 cases, both from the Canadian Prairies, were reported by Friedmann (1963) and Friedmann et al. (1977).

Nest Success: The 1 nest found with eggs on 27 May was found empty on 13 June and was presumed to have been predated.

REMARKS: The Common and Hoary redpolls are considered to form a superspecies (Troy 1985; American Ornithologists' Union 1983). Recent work near Churchill, Manitoba, by Seutin et al. (1992) found that the 2 species coexist in the same habitat.

The Common Redpoll is known in Great Britain as the Redpoll.

The breeding ecology of redpolls has not been carefully studied. Female redpolls lay 2 clutches each breeding season (Troy and Shields 1979). Some female Common Redpolls, at least, are thought to mate with 2 different males, building a nest and mating with a second male while simultaneously feeding large nestlings in their first nest with their first mate (Seutin et al. 1991). There is even circumstantial evidence that female redpolls may raise broods hundreds of kilometres apart during a single breeding season (Seutin et al. 1991). Fledglings are usually tended by males alone (Seutin et al. 1991).

During the winter months, many Common Redpolls are killed by vehicular traffic as they pick up salt-encrusted grit from roads (Fig. 649). Carcasses should be examined for bands and the rarer Hoary Redpoll should be searched for among them.

Identification of Common and Hoary redpolls can be difficult in the field because they are similar in plumage (Czaplak 1995). Subtle differences between the species in redness of the breast and poll and in darkness or lightness of other parts are further complicated by the fact that older birds are brighter than younger birds (Seutin et al. 1992). In addition, all redpolls become darker as the year progresses because of feather wear after the autumn moult (Knox 1988). Most observers presume – correctly or incorrectly – that redpolls with dark rump patches and red breasts are Common Redpolls and those with light rumps and pink breasts are Hoary Redpolls. In the field, only individuals with the lightest rumps are usually identified as Hoary Redpolls; the rest are assumed to be Common Redpolls.

For a thorough discussion of the irruptive migration pattern of Common Redpolls, see Bock and Lepthien (1976) and Hochachka et al. (1999).

NOTEWORTHY RECORDS

Spring: Coastal – Skagit River 16 Apr 1972-2; Vancouver 4 Mar 1982-35; Sechelt 2 May 1990-1 (Siddle 1990b); Terrace 20 Mar 1966-150 (Crowell and Nehls 1966a), 4 Apr 1970-70 (Crowell and Nehls 1970c), 15 May 1970-1 (Crowell and Nehls 1970c); New Hazelton 27 Mar 1971-1. **Interior** – Kilpoola Lake 12 Apr 1969-9; Fort Steele 20 Mar 1945-65 (Johnstone 1949); Kelowna 6 Mar 1976-150; Okanagan Landing 2 May 1907-last departure (Cannings et al. 1987); Nakusp 27 Mar 1976-100, 22 Apr 1982-100; Grindrod 2 Apr 1954-1 (UBC 7904); Kamloops 12 Mar 1926-208; Sorrento 3 May 1970-2; Revelstoke Mar 1990-2,000 (Siddle 1990b), 17 May 1982-2; Mount Revelstoke National Park 27 May 1982-1; Yoho National Park 13 Apr 1973-latest spring departure (Wade 1977); Field 17 Mar 1977-100; Horse Lake (100 Mile House) 18 Mar 175-10; Williams Lake 28 Mar 1982-500, 4 Apr 1979-200, 6 Apr 1958-160 (Erskine and Stein 1964), 25 Apr 1982-wintering flock at feeder left on this date, 2 May 1972-60, 15 May 1954-1 (NMC 48342); Horsefly Lake Apr 1979-400 (largest flock in 5 years); Quesnel 14 Apr 1979-200; McBride 1 Apr 1990-1; Papoose Lake 14 May 1970-1 (RBCM 14929); Willow River 23 Apr 1972-1, latest date from 1965 to 1986; Prince George 2 May 1991-1 (Siddle 1991c), 17 May 1992-2; Houston 4 Apr 1975-1 (RBCM 18183); Quick (Telkwa) 26 Apr 1985-350; Smithers 7 Mar 1975-35; Mackenzie 4 Apr 1996-100+, 20 May 1996-1; 15 km s Pouce Coupe 23 Apr 1979-60; North Pine 1 Mar 1987-50; n Fort St. John 27 Mar 1982-40; Beatton Park 17 Apr 1983-100, 25 May 1986-2; Flatrock 29 Mar 1981-1, 11 Apr 1982-350; Rose Prairie 12 Mar 1983-5; Hyland Post 1 Apr 1976-3; Parker Lake (Fort Nelson) 13 Apr 1975-50; Liard Hot Springs Park 14 Mar 1975-20, 15 Apr 1975-50, 9 May 1975-6 (Reid 1975); Atlin 25 Mar 1935-1 (RBCM 5769); Chilkat Pass 26 May 1957-3 eggs (Weeden 1960), 28 May 1979-1.

Summer: Coastal – Alta Lake 2 Jun 1932-1; Kispiox River 29 Aug 1921-1 (Swarth 1924). **Interior** – Revelstoke 4 Jun 1989-1; Prince George 31 Jul 1982-1; Burns Lake 14 Jul 1956-7; Fort St. John 11 Jul 1987-1; Hotlesklwa Lake 3 Aug 1976-10; Fern Lake (Kwadacha Wilderness Park) 2 Aug 1983-1 (Cooper and Cooper 1983); Parker Lake (Fort Nelson) 18 Jun 1976-1; Atlin Jun 1928-1 recent fledgling (Swarth 1936), 2 Aug 1929-1 fledgling out of nest for only a day or 2 (Swarth 1930); Tatshenshini River 7 Jun 1983-1 (Campbell et al. 1983); Liard River 20 Jun 1922-1 (Williams 1933a); Three Guardsmen Lake 12 Jun 1979-2; Chilkat Pass 9 Jun 1978-2, 22 Jun 1980-1 egg; Turnback Canyon (Alsek River) 30 Jun 1992-6 eggs (Merilees 1995; BC Photo 1777; Fig. 648); Kwatini Creek 18 Jun 1980-6; w Kelsall Lake 1 Jul 1984-40.

Breeding Bird Surveys: Coastal – Not recorded. **Interior** – Recorded from 3 of 73 routes and on less than 1% of all surveys. Maxima: Chilkat Pass 1 Jul 1976-24; Pennington 30 Jun 1991-11; Haines Summit 27 Jun 1993-1.

Autumn: Interior – Laidlaw Lake 14 Nov 1974-1; Log Cabin 1 Sep 1924-1 (MVZ 105460); Atlin 5 Oct 1931-1 (CAS 34036); Fort Nelson 16 Sep 1986-3, 11 Oct 1987-3; ne Cecil Lake 3 Nov 1985-200; Beatton Park 2 Oct 1988-9 (Siddle 1989a); Nilkitkwa Lake 14 Nov 1978-250; nr Morfee Lakes 27 Nov 1994-165; Quick (Telkwa) 30 Oct 1985-200; Walcott 27 Oct 1985-200, 13 Nov 1988-500; nr Buth Creek 20 Oct 1997-15+; Prince George 5 Oct 1983-70, 26 Nov 1990-700 along 20 km of road (Siddle 1991a); 25 km s Prince George 3 Oct 1983-50, 8 Nov 1984-150; Williams Lake 25 Oct 1980-50; Canim Lake 11 Nov 1986-50; Watson Lake (100 Mile House) 28 Sep 1974-1 (RBCM 11893); Vernon 5 Nov 1980-80; Yalakom River 13 Nov 1993-8; Yoho National Park 2 Nov 1975-earliest autumn arrival (Wade 1977); Revelstoke 19 Sep 1986-1, 23 Nov 1982-90; Kootenay National Park 24 Nov 1965-30 (Seel 1965); Scotch Creek 12 Sep 1962-16, 5 Oct 1964-6; Okanagan Landing 24 Oct 1908-11, 1 Nov 1925-75; Kuskanax Creek 28 Nov 1982-100; Columbia Lake 26 Oct 1948-31 (Johnstone 1949); Sparwood 23 Sep 1983-1, 27 Oct 1984-300, 30 Oct 1984-650 (Fraser 1984); Trail 17 Nov 1968-70. **Coastal** – Kispiox River 7 Sep 1921-1 (Swarth 1924); Masset 4 Nov 1971-150; North Bend (Boston Bar) 13 Nov 1896-40; Mount Seymour Park 4 Nov 1989-3; Reifel Island 4 Nov 1986-5 (Mattocks and Harrington-Tweit 1987a); Richmond 24 Nov 1985-40 (Force and Mattocks 1986).

Winter: Interior – Kledo Creek Park 18 Jan 1975-100 in 3 flocks (Erskine and Davidson 1976); Alaska Highway (Mile 150) 29 Dec 1989-100; Fort St. John 27 Jan 1985-100 (Grunberg 1985b); Tumbler Ridge 3 Feb 1996-106; Quick 16 Dec 1984-150; Vanderhoof 21 Feb 1998-1,000; Williams Lake 28 Jan 1964-200; Glacier National Park 3 Dec 1981-200; Nicholson 13 Jan 1977-100; w Clinton 13 Dec 1981-150; Shuswap Lake 1 Jan 1977-60; Vernon 14 Jan 1973-thousands (Cannings et al. 1987); Coldstream 6 Feb 1958-300 (Cannings et al. 1987); Invermere 6 Feb 1982-135; Nakusp 29 Feb 1976-50; Skaha Lake 9 Jan 1972-100; Creston 6 Jan 1982-700, 1 Feb 1982-351; Cawston 23 Feb 1974-500 (Shepard 1974). **Coastal** – Greenville 4 Jan 1982-20; Terrace 17 Feb 1968-2 (Crowell and Nehls 1968b); Masset 19 Dec 1971-150, 23 Dec 1971-85, 24 Dec 1972-125, 27 Dec 1992-14 (Siddle 1993b), 4 Feb 1994-1 (Siddle 1994c); Alliford Bay 30 Dec 1941-2; Bella Coola 24 Feb 1976-large flock; Cortes Island 29 Jan 1975-7; Alta Lake 25 Dec 1945-250; Whistler 18 Jan 1992-300; Sechelt 16 Dec 1989-30 (Siddle 1990a); Vancouver 7 Jan 1970-61 (Weber 1972); Cheam Slough (Harrison Hot Springs) 3 Feb 1982-15; Sea Island 18 Dec 1977-200; Westham Island 23 Jan 1986-100; Lulu Island 12 Feb 1990-50; Tofino 27 Dec 1989-1, first in 11 yrs; Victoria 25 Jan 1993-1 (Siddle 1993b).

Christmas Bird Counts: Interior – Recorded from 26 of 27 localities and on 73% of all counts. Maxima: Prince George 19 Dec 1993-3,762; Smithers 29 Dec 1985-1,383; Nakusp 3 Jan 1987-991. **Coastal** – Recorded from 13 of 33 localities and on 11% of all counts. Maxima: Terrace 23 Dec 1984-184; Vancouver 18 Dec 1977-150; Ladner 29 Dec 1985-129.

Hoary Redpoll
Carduelis hornemanni (Holboll)

HORE

RANGE: Holarctic. Breeds in North America from western and northern Alaska, northern Yukon, northern and east-central Mackenzie, and southern Victoria Island east to northeastern Manitoba and to Ellesmere and northern Baffin islands. Winters irregularly from southern and central parts of the breeding range south across southern Canada and the northern United States. Also breeds across northern Eurasia.

STATUS: On the coast, a *casual* vagrant in the Fraser Lowland of the Georgia Depression Ecoprovince and on the Southern Mainland Coast and Queen Charlotte Islands in the Coast and Mountains Ecoprovince. Not reported from Vancouver Island or the Northern Mainland Coast.

In the interior, an *uncommon* to occasionally *very common* migrant and winter visitant in the Boreal Plains Ecoprovince; *very rare* in the Southern Interior, Southern Interior Mountains, Central Interior, and Sub-Boreal Interior ecoprovinces; *casual* in the Northern Boreal Mountains and Taiga Plains ecoprovinces.

CHANGE IN STATUS: Munro and Cowan (1947) describe the Hoary Redpoll as a scarce winter visitant to widely scattered locations in the interior of the province west of the Rocky Mountains. While it still remains an irregular and scarce winter visitant there, field observations between 1975 and 1989 have shown that the Hoary Redpoll is a regular migrant and winter visitant east of the Rocky Mountains in the Boreal Plains. The status of this species remains unclear elsewhere in northern British Columbia.

OCCURRENCE: The Hoary Redpoll (Fig. 650) has a sparse and scattered distribution throughout the interior of British Columbia west of the Rocky Mountains. It occurs most regularly in the Peace Lowland of the Boreal Plains, where it can be found in variable numbers most winters. On the coast, it has been reported only from the Queen Charlotte Islands, Bella Coola, Green Lake, Sumas, and Pitt Meadows; we have no records from Vancouver Island. The highest numbers in winter occur in the Peace Lowland of the Boreal Plains (Fig. 651).

The Hoary Redpoll has been recorded from near sea level on the coast and between 400 and 900 m elevation in the interior. In the Peace Lowland, it frequents agricultural fields (Fig. 652) and brushy areas, including hedgerows adjacent to fields; trembling aspen copses, Sitka alder, red-osier dogwood, and willow along streams and gullies; weedy field edges; and grain fields at various stages of harvest. It readily goes to waste grain piles in fields, along railway tracks, and near grain elevators. In 1 instance, a flock of 100 birds frequented a farmyard near Fort St. John and fed on grain in an open storage shed. The Hoary Redpoll also occurs in birch and alder woodlands. Elsewhere in the interior, it has been reported among flocks of Common Redpolls foraging in birch or apple trees, in abandoned landfill sites and open fields, along road edges, and at backyard feeders. On the coast, all birds were found with Common Redpolls; roadside shrubs and dyke vegetation were the only habitats reported.

Figure 650. Although the Hoary Redpoll occurs regularly in northeastern British Columbia each winter, its status in other northern areas of the province is unknown (Fort St. John, 8 March 1986; Chris Siddle).

Autumn migrants appear in the Peace Lowland as early as the second week of October and their numbers build into December and January (Figs. 653 and 654). The first arrival dates for Hoary Redpolls near Fort St. John ranged from 12 October to 3 December, with a median arrival date of 1 November over 7 years. This redpoll may appear in pure flocks

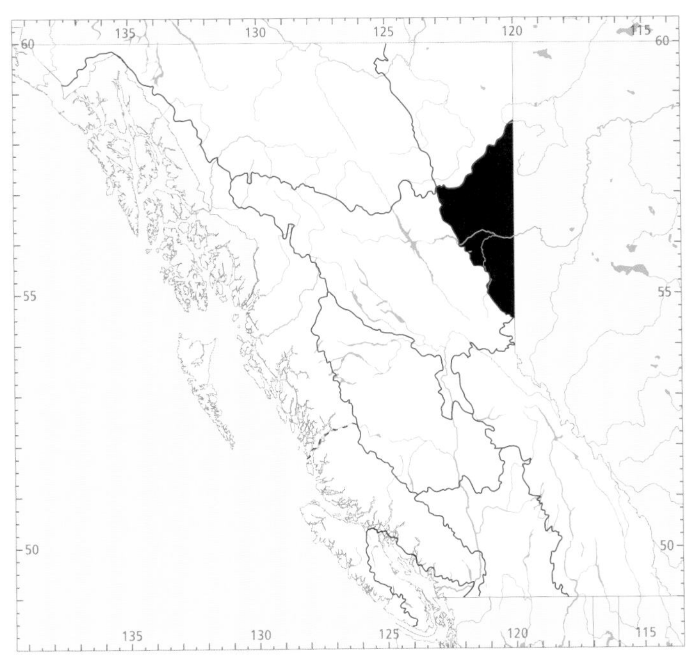

Figure 651. In British Columbia, the highest numbers for the Hoary Redpoll in winter occur in the Peace Lowland of the Boreal Plains Ecoprovince.

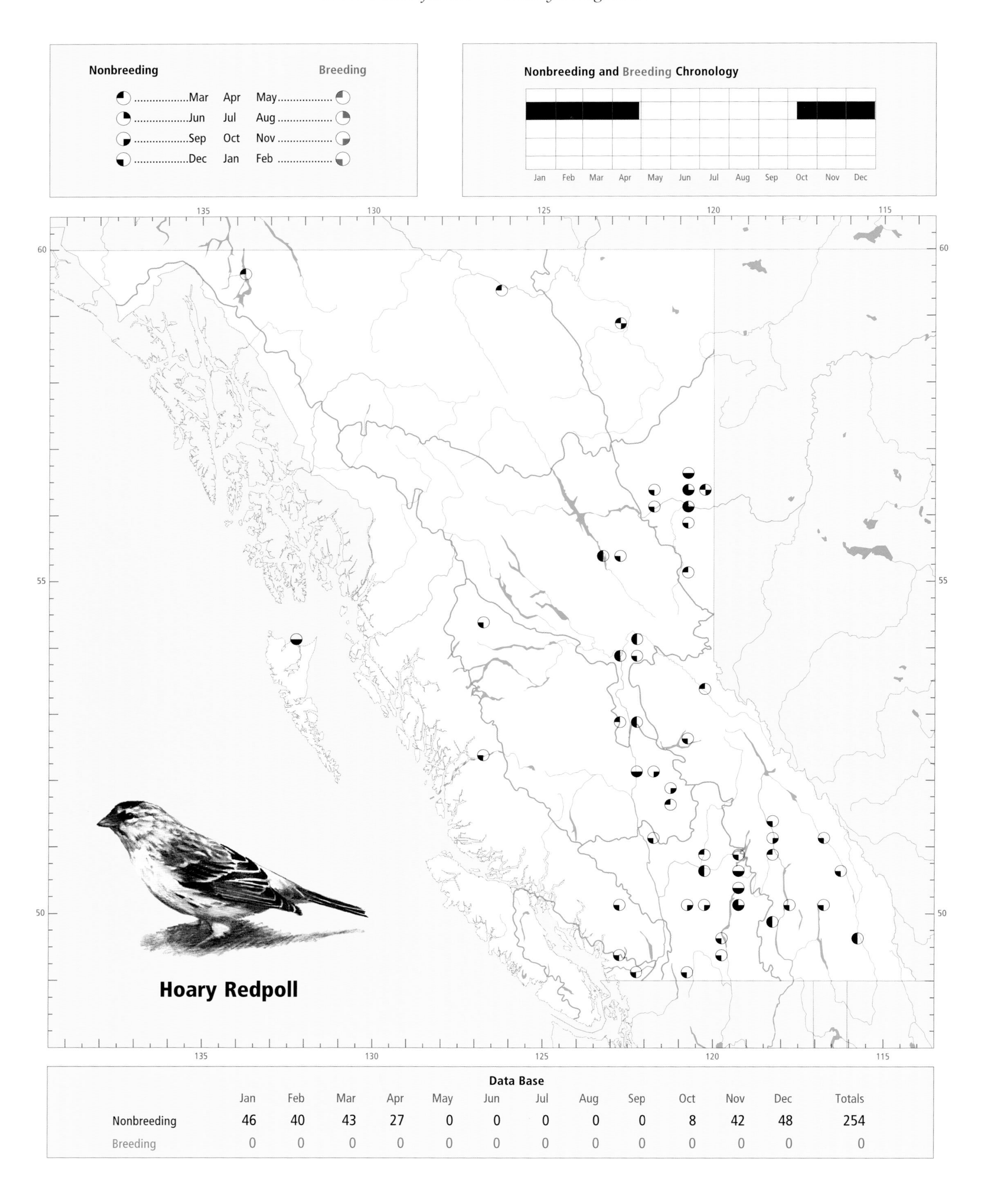

Data Base

	Jan	Feb	Mar	Apr	May	Jun	Jul	Aug	Sep	Oct	Nov	Dec	Totals
Nonbreeding	46	40	43	27	0	0	0	0	0	8	42	48	254
Breeding	0	0	0	0	0	0	0	0	0	0	0	0	0

Figure 652. In British Columbia, the Hoary Redpoll frequents extensive patches of grasses and weeds in agricultural fields (Montney, 11 November 1996; R. Wayne Campbell).

of up to 50 birds or in mixed flocks with the Common Redpoll, which arrives 2 to 4 weeks earlier. In the Sub-Boreal Interior, Central Interior, Southern Interior, and Southern Interior Mountains, the Hoary Redpoll arrives in November or December (Fig. 653).

During winters of heavy redpoll visitation, the Hoary Redpoll can become very numerous in the Peace Lowland. For example, during the winter of 1984 to 1985, an estimated 1,200 birds were present in the North Pine–Montney–Rose Prairie area north of Fort St. John. During the winter of 1987 to 1988, however, both redpoll species were scarce and only a total of 7 Hoary Redpolls was recorded in the same area. In the 1980s the Hoary Redpoll was recorded every winter except the winter of 1982 to 1983.

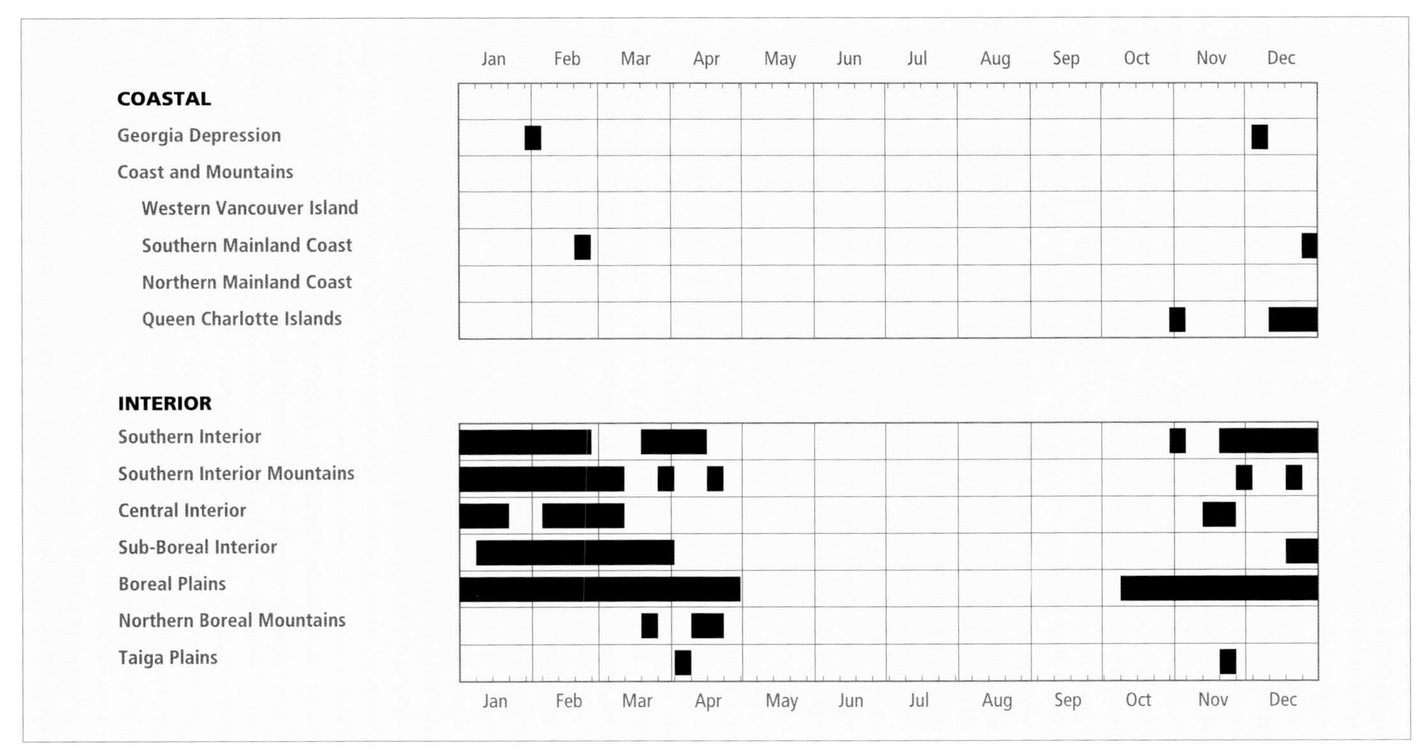

Figure 653. Annual occurrence for the Hoary Redpoll in ecoprovinces of British Columbia. Records are shown for the week in which they occurred.

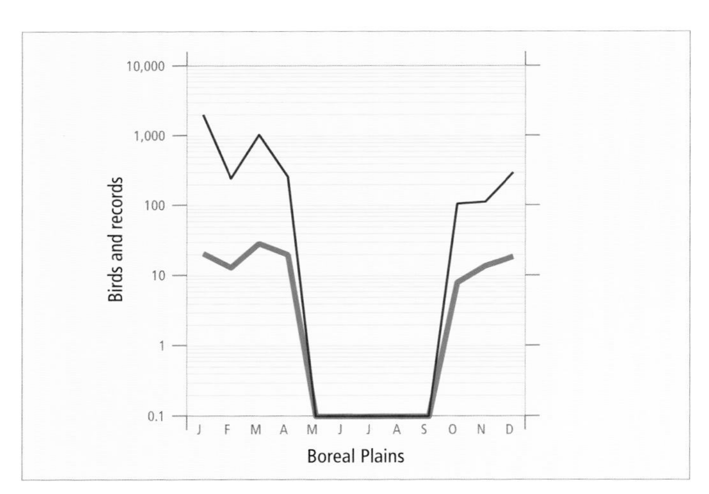

Figure 654. Fluctuations in total number of birds (purple line) and total number of records (green line) for the Hoary Redpoll in the Boreal Plains Ecoprovince of British Columbia. Christmas Bird Counts have been excluded.

In the Peace Lowland, the Hoary Redpoll can remain fairly common throughout March, with an increase in flock size in March some years that suggests, perhaps, a influx from the southeast. By mid-April it becomes scarce and by the end of the month all birds have left for the breeding grounds to the north (Figs. 653 and 654). In the central and southern portions of the interior, the latest spring departure dates range from the second week of March to the third week of April.

On the coast, the Hoary Redpoll has been reported sporadically between early November and late February; in the interior, it has been recorded from 12 October to 27 April (Fig. 653).

REMARKS: Two subspecies of the Hoary Redpoll are recognized in North America; the race found in British Columbia is *C. h. exilipes* (American Ornithologists' Union 1957). The Hoary Redpoll is also known as the Arctic Redpoll.

Separating the Hoary and Common redpolls in the field can be difficult, and has probably resulted in the former species being underreported. Observers are encouraged to look for the following distinguishing plumage characteristics of the Hoary Redpoll: (1) an unmarked white rump; (2) white undertail coverts with at most a single thin, dark streak; (3) minimal side streaking, in which streaks never appear heavy or well marked; and (4) predominantly gray and white, not brownish, upper parts.

A mid-May record of the Hoary Redpoll occurring in the Driftwood River valley reported by Stanwell-Fletcher and Stanwell-Fletcher (1943) has been omitted from this account due to inadequate documentation.

For additional life-history and identification information, see Clement et al. (1993) and Czaplak (1995).

NOTEWORTHY RECORDS

Spring: Coastal – No records. **Interior** – Mount Heinz (Trail) 27 Mar 1949-1; Vernon 9 Apr 1936-1 (MVZ 82943); Knutsford (Kamloops) 30 Mar 1985-2; Rayleigh (Kamloops) 21 Mar 1982-2; 100 Mile House 5 Mar 1982-1; Quesnel 1 Mar 1896-1 (RBCM 786), 3 Mar 1901-1 (MVZ 105440), 26 Mar 1896-1 (MCZ 247327); Prince George 9 Mar 1969 (Rogers 1969); Tumbler Ridge 10 Apr 1995-4; Mackenzie 21 Mar 1995-3 with 40 Common Redpolls, 29 Mar 1997-3; Flatrock 11 Apr 1982-4; Fort St. John 29 Mar 1987-20, 13 Apr 1986-60, 19 Apr 1987-1; Goodlow 29 Mar 1986-1; Cecil Lake 17 Apr 1982-2; Charlie Lake 27 Apr 1985-1 at south end; North Pine 2 Mar 1985-200; nr Montney 7 Mar 1982-100, 27 Apr 1985-30; Fort Nelson 6 Apr 1975-5; Liard Hot Springs 21 Mar 1975-1 and 9 Apr 1975-1 (Reid 1975).

Summer: No records.

Autumn: Interior – Fort Nelson 19 Nov 1986-1; Rose Prairie 20 Oct 1984-30, 11 Nov 1984-30; Montney 28 Oct 1984-40, 6 Nov 1988-2; Boundary Lake (Goodlow) 1 Nov 1981-2; Cecil Lake 11 Nov 1987-2, 26 Nov 1987-32; Beatton Park 12 Oct 1985-2; Fort St. John 27 Oct 1986-1; 21 km w Williams Lake 14 Nov 1950-1 (NMC 48346); Lac la Hache 15 Nov 1950-1 (RBCM 15880); Mount Revelstoke National Park 27 Nov 1976-1; Nicola Lake 25 Nov 1984-1; Grindrod 22 Nov 1950 to 7 Jan 1951-many in winter (Munro 1953); Vernon 1 Nov 1985-6 (Rogers 1986a); Nicola 25 Nov 1984-2. **Coastal** – Masset 4 Nov 1971-2.

Winter: Interior – Rose Prairie 27 Jan 1985-400 with 100 Common Redpolls (Grunberg 1985b); North Pine 6 Jan 1985-10, 27 Jan 1985-400, 8 Feb 1986-38, 23 Dec 1984-60; Cecil Lake 20 Jan 1985-300 (Grunberg 1985b); Beatton Park 5 Feb 1989-15, 23 Dec 1985-15; nr Halfway River 28 Jan 1984-3 along Upper Cache Rd; Fort St. John 14 Feb 1988-35, 21 Feb 1987-10, 27 Dec 1987-25; Baldonnel 29 Jan 1982-12, Feb 1982-3 to 20 per day; Farrell Creek (Peace River) 26 Dec 1985-5; Taylor 11 December-3 along Johnstone Rd; Tower Lake 9 Feb 1986-1; Mackenzie 13 Jan 1996-1; Bear Lake (Driftwood River) early January 1938-10 to 12 (Stanwell-Fletcher and Stanwell-Fletcher 1943); Chetwynd 23 Jan 1982-4; Emerson Creek 11 Feb 1990-25 (Siddle 1990a), 14 Feb 1990-50 (Siddle 1990a); Prince George 21 Jan 1979-2, 10 Feb 1974-1 (BC Photo 140), 23 Feb 1987-2; Quesnel 17 Jan 1896-1 (MCZ 247366), Jan 1899-1 (MVZ 105438), 3 Feb 1896-1 (RBCM 785); Williams Lake 17 Jan 1951-1 (NMC 48352), 20 Jan 1951-1 (NMC 48353); Clinton 12 Feb 1894-1 (RBCM 784); Revelstoke 20 Dec 1982-2; Kamloops 15 Dec 1984-3, 16 Dec 1990-2; Grindrod 7 Jan 1951-1 male identified as *C. h. exilipes* (ROM 83015); Vernon 20 Jan 1973-11, 8 and 9 Feb 1992-1 (Siddle 1992b), 22 Dec 1957-10 (Grant 1958); Lavington 24 Jan 1914-1 (MVZ 82942); Okanagan Landing 17 Feb 1966-6; Nakusp 25 Feb 1990-1 (Siddle 1990b); Edgewood 27 Feb 1917-1; Cooper Creek (Lardeau) 15 Jan 1998-1; Summerland 26 Dec 1977-6; Manning Park 15 Feb 1981-12. **Coastal** – Masset 23 Dec 1971-5, 27 Dec 1991-1 (Siddle 1992b; BC Photo 1646); Bella Coola 24 Feb 1976-flock; Green Lake (Whistler) 28 Dec 1969-2 with 20 Common Redpolls; Pitt Meadows 3 Feb 1965-1 with 12 Common Redpolls; Sumas 5 Dec 1895-1 (MCZ 247326).

Christmas Bird Counts: Interior – Recorded from 5 of 27 localities, and on 6% of all counts. Maxima: North Pine 23 Dec 1984-90; Fort St. John 3 Jan 1976-50; Lake Windermere 30 Dec 1984-40. **Coastal** – Not recorded.

Pine Siskin

Carduelis pinus (Wilson)

PISI

RANGE: Breeds from central and south-coastal Alaska, central Yukon, southwestern Mackenzie, northwestern and east-central Saskatchewan, central Manitoba, northern Ontario, central Quebec, southern Labrador, and Newfoundland south to south-central California, southern Nevada, northern and southeastern Arizona, southern New Mexico, western Texas, west-central Mexico, southwestern Oklahoma, Kansas, Missouri, central Illinois, central Indiana, southwestern and northern Ohio, Pennsylvania, and southern New Jersey. Winters from south-coastal Alaska and coastal and southern British Columbia east across southern Canada and the northern United States. Also winters in northern Mexico. During major irruption years, winters south to the Gulf coast and northern Florida.

STATUS: May be scarce or very numerous in most regions of the province at any given season.

On the coast, a *fairly common*, at times *abundant*, migrant, an *uncommon*, at times *very common*, summer visitant, and *fairly common*, at times *very abundant*, winter visitant in the Georgia Depression Ecoprovince; in the Coast and Mountains Ecoprovince, an *uncommon*, at times *very common*, migrant, an *uncommon*, at times *common*, summer visitant, and an *uncommon*, at times *very common*, winter visitant on Western Vancouver Island, the Southern Mainland Coast, the Northern Mainland Coast, and the Queen Charlotte Islands.

In the interior, a *fairly common*, at times *abundant*, migrant, an *uncommon*, at times *very common*, summer visitant, and an *uncommon*, at times *very common*, winter visitant in the Southern Interior and Southern Interior Mountains ecoprovinces; an *uncommon*, at times *common*, migrant and summer visitant, and *very rare*, at times *fairly common*, in winter in the Central Interior and Sub-Boreal Interior ecoprovinces; a *very rare*, at times *common*, migrant and summer visitant in the Boreal Plains, Northern Boreal Mountains, and Taiga Plains ecoprovinces; *accidental* in the Boreal Plains in winter.

Breeds.

NONBREEDING: The Pine Siskin (Fig. 655) is the most widespread and abundant of the cardueline finches in British Columbia. It is widely distributed throughout the province from spring through autumn, including Vancouver Island and the Queen Charlotte Islands. Our lack of information from large areas of the province is for the most part due to inadequate coverage by field observers. During winter, the Pine Siskin is found mainly at low elevations on the south coast. It also winters regularly elsewhere on the coast as far north as the Queen Charlotte Islands and Stewart at the head of the Portland Canal; except for the Bella Coola valley, however, we have no records for the mainland coast between Pender Harbour and Kitimat. In the interior, it is recorded regularly in winter as far north as Williams Lake (latitude 52°N). Elsewhere in the interior, it occurs erratically as far north as Smithers in the Central Interior and Mackenzie in the Sub-Boreal Interior.

Figure 655. The Pine Siskin is an erratic species in British Columbia, often very numerous in one year and scarce the next (Nelson, 2 December 1997; Linda M. Van Damme).

There is a single winter record from the Boreal Plains. The highest numbers in winter occur in the Georgia Depression (Fig. 656).

On the coast, the Pine Siskin is found from near sea level to 1,850 m elevation; in the interior, it is found from the valley bottoms around 280 m up to 2,285 m elevation. It is reported primarily from coniferous or mixed coniferous-deciduous forests (Fig. 657). It also occurs in deciduous woodlands, especially where red alder or paper birch provide an abundant source of seeds. It also makes use of utility rights-of-way, road

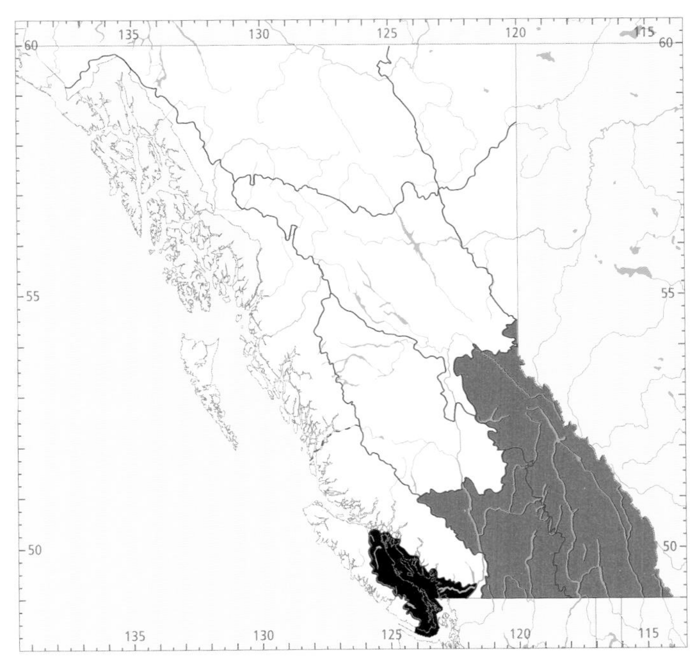

Figure 656. In British Columbia, the highest numbers for the Pine Siskin in winter (black) occur in the Georgia Depression Ecoprovince; the highest numbers in summer (red) occur in the Southern Interior and Southern Interior Mountains ecoprovinces.

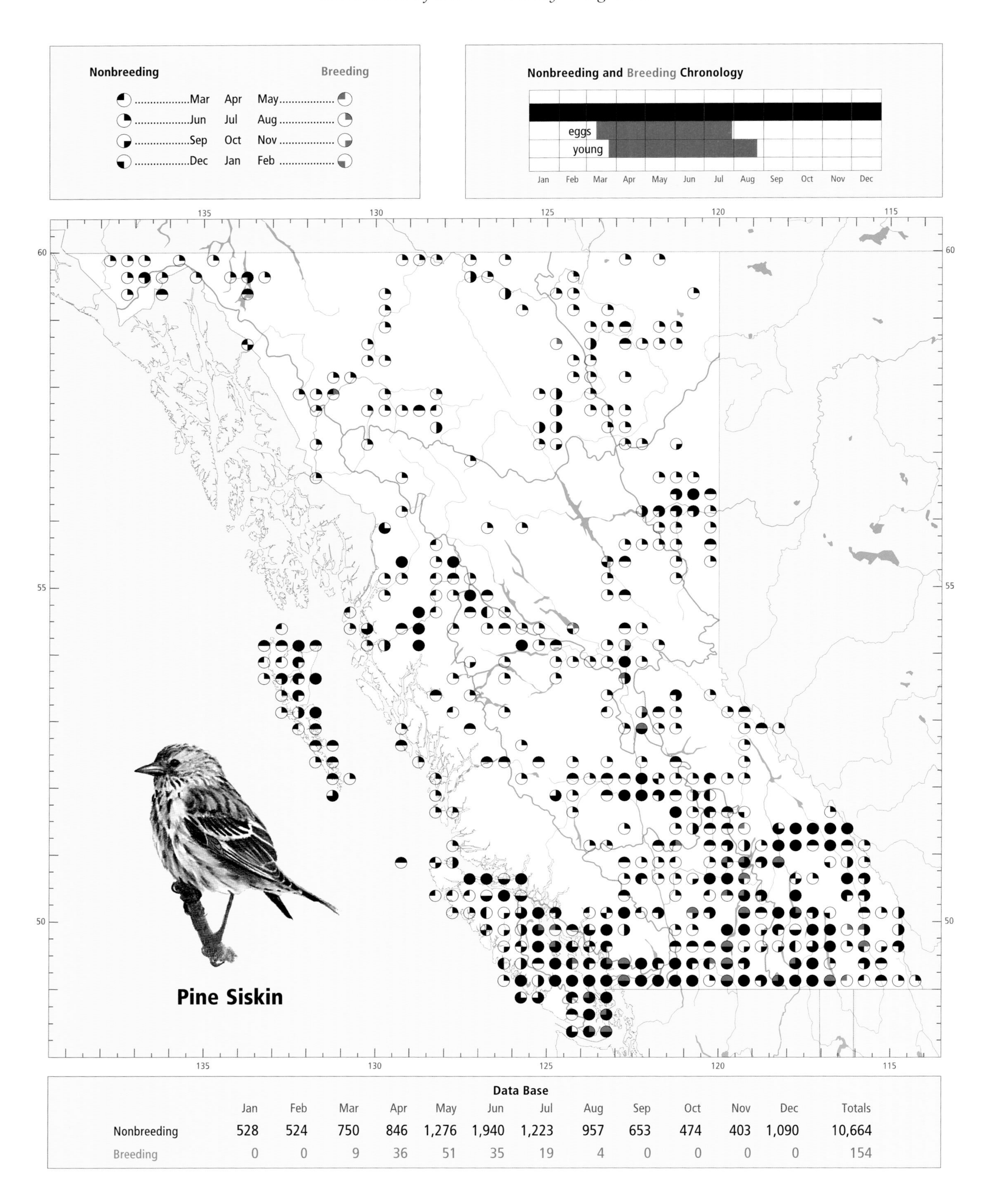

Pine Siskin

Data Base

	Jan	Feb	Mar	Apr	May	Jun	Jul	Aug	Sep	Oct	Nov	Dec	Totals
Nonbreeding	528	524	750	846	1,276	1,940	1,223	957	653	474	403	1,090	10,664
Breeding	0	0	9	36	51	35	19	4	0	0	0	0	154

Figure 657. In the Southern Interior Ecoprovince of British Columbia, the Pine Siskin frequents ponderosa pine and Douglas-fir forests throughout the year (Johnstone Creek Park, 15 April 1999; R. Wayne Campbell).

Figure 658. Throughout British Columbia, flocks of Pine Siskins are commonly found in weedy fields, where they forage on seeds such as hairy cat's ear, thistle, and dandelion (Stikine River, 10 July 1996; R. Wayne Campbell).

verges, weedy fields, recent clearcuts, and other disturbed areas where weeds such as dandelion, hairy cat's ear, and thistles are plentiful (Fig. 658). Human-influenced habitats such as orchards and backyard gardens, especially those with feeders containing niger and sunflower seeds, are commonly used by the siskin (Fig. 661).

Movements of the Pine Siskin are difficult to determine, and in some areas wanderings have often been interpreted as migratory movements (Dawson 1997). On the coast, siskin numbers slowly decline from their winter highs in the Georgia Depression as the birds likely move further north or to higher elevations (Fig. 660). A small movement then appears from the third week of March through the end of May as numbers rise there; it peaks in the last week of April and first 2 weeks of May (Fig. 660). A similar but more erratic pattern occurs on Western Vancouver Island. On the Southern and Northern mainland coasts, there is no noticeable movement although siskin numbers do fluctuate throughout the year. In general, higher numbers and more observations are reported through the spring and summer months in the Northern Mainland Coast, while highest numbers generally occur in winter in the Southern Mainland Coast, suggesting a small south-to-north migration. On the Queen Charlotte Islands, there is a noticeable influx through May as birds move in from the south.

In the interior, spring movements are quite protracted in the Southern Interior and Southern Interior Mountains, with no obvious peaks in numbers. In the Southern Interior, there is a gradual increase in winter numbers about the second week of February, continuing into May. In the Okanagan valley, spring movements usually occur in April and May (Cannings et al. 1987). Pine Siskin numbers in the Southern Interior Mountains gradually increase into March and then decline into the summer months. In the Central Interior, a gradual rise in numbers into May also occurs, but no obvious peak is discernible.

Further north, in the Sub-Boreal Interior, numbers occasionally winter but first reports are usually not made until the second week of March, with a gradual rise in numbers through May; there are no obvious peaks. In the Northern Boreal Mountains, siskins are first reported about the third week of May and numbers rise into June.

East of the Rocky Mountains, in the Boreal Plains, siskins are first reported in the last week of February, although not many birds are present until mid-May. In the Taiga Plains, the first siskins are reported in the last week of May.

In August, following the main breeding period, siskins often wander in search of the ripening seed crops of coniferous and deciduous trees and various composite plants. During this period, siskin flocks are commonly observed foraging in open or sparsely forested habitats, including weedy fields, orchards, recent clearcuts, and alpine meadows near timberline (Fig. 662). The onset of the autumn migration is difficult to distinguish from these post-breeding wandering movements.

In the northern interior, the main autumn movement begins from about the third week of August to the second week of September. The latest autumn record for the Taiga Plains is from the third week of September, but stragglers probably remain into the third or fourth week of October, as in the Northern Boreal Mountains and Sub-Boreal Interior. In the Boreal Plains, the last siskins are recorded during the first week of November. In the Sub-Boreal Interior, numbers peak during the last week of August and decline through September; a few birds occasionally remain into winter. In the Central Interior, numbers fluctuate through August and September, showing a small peak in the first week of October. In the Southern Interior and Southern Interior Mountains, autumn numbers decline slowly through September and October after showing small peaks in August. In the Okanagan valley, siskin numbers have dropped sharply by November (Cannings et al. 1987).

On the north coast, numbers decline through August and September and reach winter levels by October. On the south coast, autumn migrants begin to arrive in early September and build in numbers through October. They reach winter numbers by the first week of November.

Across North America, the Pine Siskin has an erratic wintering pattern, often being plentiful one year and scarce the next (Dawson 1997). The siskin is also known for its irruptive southward movements, especially into the southeastern United States, when northern seed crops become scarce (Bock and Lepthien 1976). In British Columbia, winter numbers are partly regulated by coniferous cone crops. Most species of conifers produce an abundance of cones at intervals of 2 to 6 years (Eremko et al. 1989).

During the winter of 1985-86 in southwestern Washington, a large influx of siskins and crossbills was attributed to an abundant Douglas-fir cone crop (Manuwal and Huff 1987). In British Columbia, long-term cone crop data are scarce, but a comparison of coastal Douglas-fir cone estimates with Christmas Bird Count data for Vancouver suggests a relationship in some winters (Fig. 663).

Christmas Bird Counts for the Georgia Depression and Southern Interior show large and generally synchronous fluctuations (Fig. 664). Peak numbers in both ecoprovinces were recorded in 1983, 1993, and 1996; low counts were recorded in 1977, 1986, and 1987. Major peaks recorded in the Southern Interior in 1978 and in the Georgia Depression in 1981, 1985, and 1989 were not reflected in the corresponding ecoprovince; the large peak recorded in 1989 on the south coast coincided with a major, near continent-wide irruption (Yunick 1997). Analyses of Christmas Bird Counts for the continent indicates that the south coast of British Columbia is a primary wintering ground for the Pine Siskin in North America (Root 1988; Sauer et al. 1996) (see also, Fig. 742d on p. 652).

The Pine Siskin readily uses bird feeding stations (Fig. 661). An analysis of Project FeederWatch data for British Columbia for the period 1988 to 1998 shows that the Pine Siskin was the second or third most abundant bird visiting winter feeding stations along the coast; in the interior, it was among

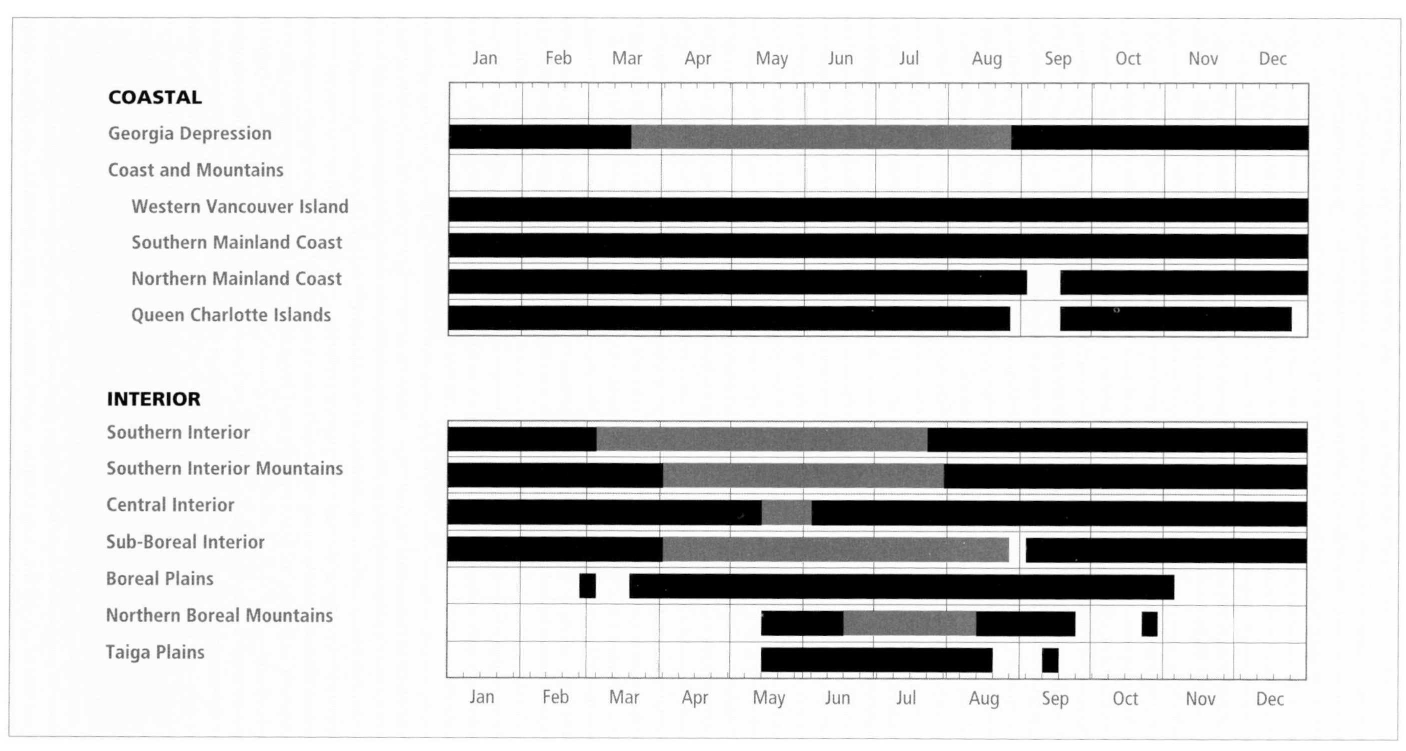

Figure 659. Annual occurrence (black) and breeding chronology (red) for the Pine Siskin in ecoprovinces of British Columbia. Records are shown for the week in which they occurred.

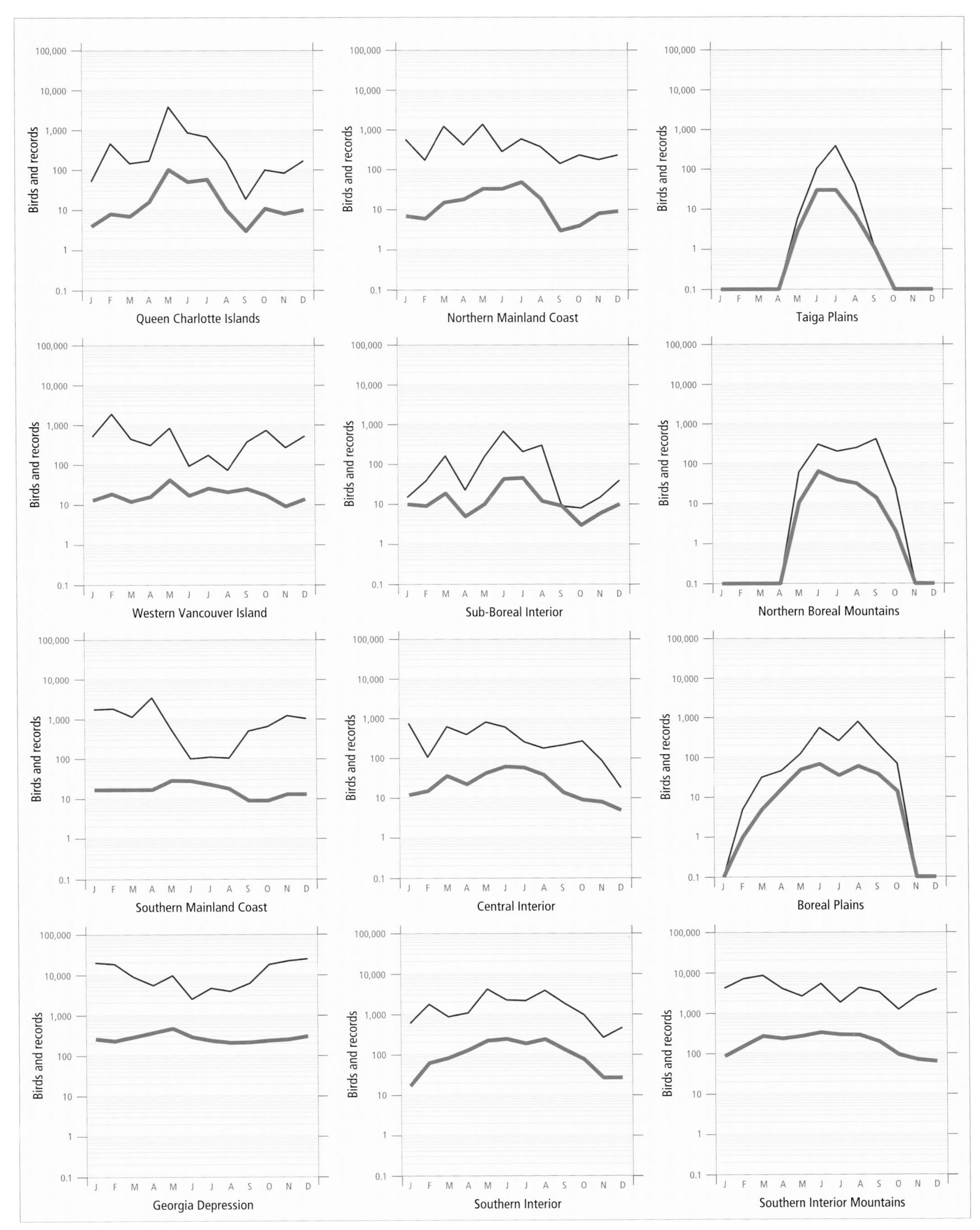

Figure 660. Fluctuations in total number of birds (purple line) and total number of records (green line) for the Pine Siskin in ecoprovinces of British Columbia. Christmas Bird Counts, Breeding Bird Surveys, and nest record data have been excluded.

Figure 661. Feeders filled with niger and sunflower seeds are attractive to Pine Siskins throughout the year in British Columbia (Saanich, southern Vancouver Island, 13 December 1998; R. Wayne Campbell).

the top 10 species using feeders in the Southern Interior, Southern Interior Mountains, and Central Interior. The northernmost winter FeederWatch record for the interior was from Burns Lake in the Central Interior.

The highly gregarious Pine Siskin occurs in flocks ranging in size from a few birds to occasionally several thousand; flock sizes of 50 to 200 birds are common (Palmer 1968). In British Columbia, large flocks were most commonly observed during the nonbreeding periods in the Georgia Depression and during the late spring and early autumn migration periods in the Southern Interior (Fig. 665). In the Georgia Depression, spring and autumn flocks as large as 5,000 birds have been reported (Pearse 1939; Tatum 1970). Siskins were usually reported in single-species flocks but occasionally occurred in mixed flocks, most notably with the Common Redpoll and American Goldfinch. The Pine Siskin also associates loosely with other cardueline finches, such as the Red Crossbill, White-winged Crossbill, and Evening Grosbeak, especially in areas of common food resources or where mineral salts have accumulated (see REMARKS). Elsewhere, it is commonly observed at bird feeders with the Purple Finch, Cassin's Finch, and Evening Grosbeak (Balph 1979; Popp 1987).

On the coast and in the southern portions of the interior, the Pine Siskin has been recorded regularly throughout the year; in the north-central portions of the province, including the Boreal Plains, Sub-Boreal Interior, and northern half of the Central Interior, it occurs regularly from 25 March to 4

Figure 662. In late summer, Pine Siskin flocks wander widely in subalpine parkland and alpine habitats near timberline in search of seeds from conifers and composite plants (6 km west of Mount Milton, near Valemount, 9 August 1997; Andrew C. Stewart).

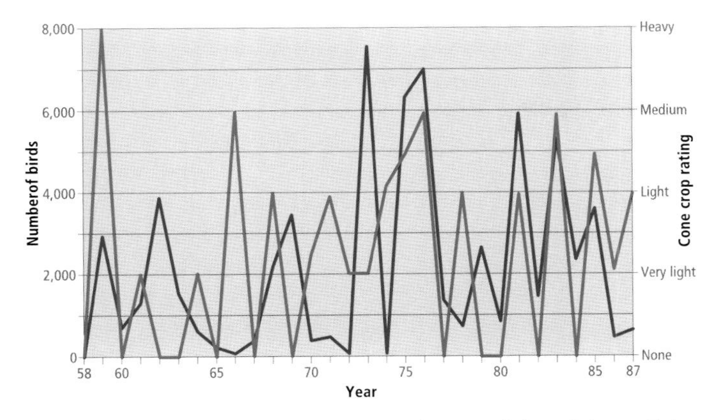

Figure 663. Comparison of numbers of Pine Siskins recorded on Christmas Bird Counts for Vancouver (purple line) and cone crop abundance estimates for coastal Douglas-fir (green line) for the period 1958 through 1987 (adapted from Eremko et al. 1989). Fluctuations appear to be unrelated for much of the early period, but are synchronous for the years 1980 to 1987.

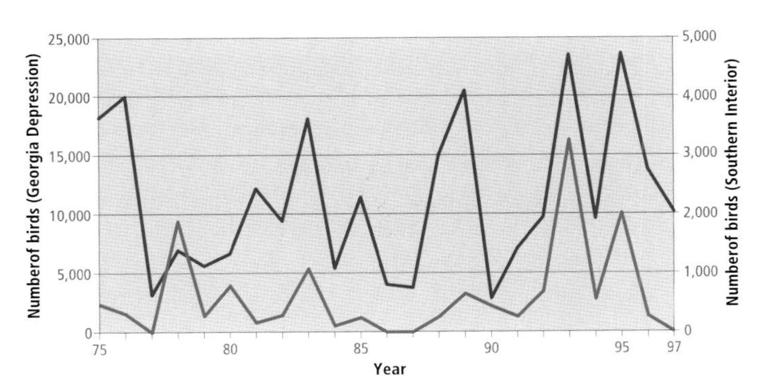

Figure 664. Fluctuations in winter numbers of the Pine Siskin (1975-97) based on combined Christmas Bird Counts in the Georgia Depression Ecoprovince (dark line, left axis) (Campbell River, Chilliwack, Comox, Deep Bay, Duncan, Ladner, Nanaimo, Pender Islands, Pitt Meadows, Vancouver, Victoria, White Rock) and the Southern Interior Ecoprovince (light line, right axis) (Penticton, Shuswap Lake Park, Vaseux Lake, Vernon).

November. Further north, it occurs regularly from the third week of May to mid-September in the Northern Boreal Mountains and mid-August in the Taiga Plains (Fig. 659).

BREEDING: The Pine Siskin is widely distributed throughout British Columbia during the breeding period, and probably breeds throughout its range. Breeding records are lacking for large areas of the northern interior and much of the coast, however. On the coast, breeding has been confirmed only for southeastern Vancouver Island and the Fraser Lowland of the Georgia Depression. In the interior, most breeding records are from the Southern Interior and the southern half of the Southern Interior Mountains. Other scattered records are from Williams Lake, Quesnel, in and around Prince George, and Fort St. James; the northernmost records are from Telegraph Creek and the Tetsa River in the Northern Boreal Mountains. Like other cardueline finches, the siskin is an erratic breeder, frequently ranging from plentiful one year to scarce or absent the next (Godfrey 1986). In British Columbia, most breeding is likely associated with coniferous cone crop cycles.

The Pine Siskin reaches its highest numbers in summer in the Southern Interior and Southern Interior Mountains (Fig. 656). An analysis of Breeding Bird Surveys for the period 1968 through 1993 shows that the mean number of birds on coastal routes decreased at an average annual rate of 9% (Fig. 666); an analysis of interior routes for the same period could not detect a net change in numbers. Numbers appear to be stable in both Canada-wide and continent-wide surveys for the period 1966 through 1996 (Sauer et al. 1997).

Continent-wide Breeding Bird Survey data indicate that the southeastern portion of British Columbia supports the highest breeding density of siskins in North America (Sauer et al. 1997) (see also, Fig. 740d on p. 641). The wettest portions of this region support the greatest diversity of tree species in British Columbia, including 17 of the 22 coniferous species

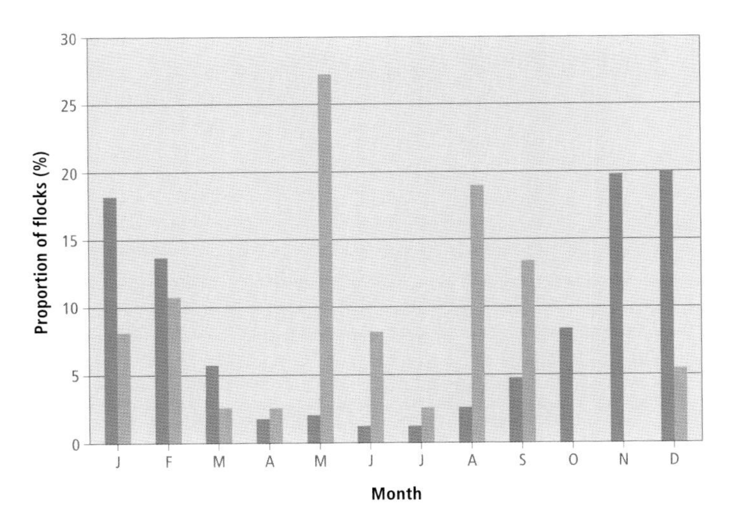

Figure 665. Monthly distribution of large flocks (≥ 100 birds) of Pine Siskins in the Georgia Depression (dark bars; $n = 329$) and Southern Interior (light bars; $n = 37$) ecoprovinces. In the Georgia Depression, large flocks dominate the autumn to early spring period; in the Southern Interior, most large flocks occur during the spring and autumn migration periods.

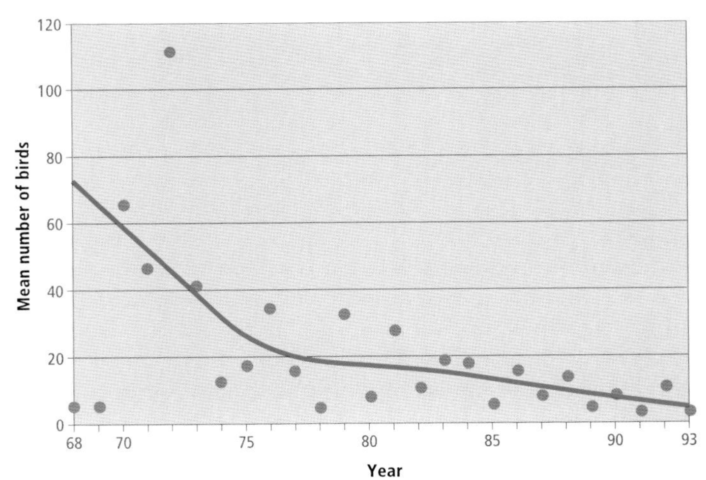

Figure 666. An analysis of Breeding Bird Surveys for the Pine Siskin in British Columbia shows that the number of birds on coastal routes decreased at an average annual rate of 9% over the period 1968 through 1993 ($P < 0.01$).

Figure 667. The range and breeding habitats used by the Pine Siskin in British Columbia are poorly documented. In the Northern Boreal Mountains Ecoprovince, the species has been recorded building nests in moose-browsed willows and alders on a river floodplain (Tatshenshini River, 15 June 1993; John M. Cooper).

Figure 668. In northern portions of the Central Interior Ecoprovince of British Columbia, the Pine Siskin breeds in mixed forests of trembling aspen and hybrid white spruce (Sunset Lake, 22 June 1997; R. Wayne Campbell).

found here (Ketcheson et al. 1991; Parish et al. 1996); the resulting range of cone crops is probably the factor that contributes the most to the high breeding densities.

Brooks and Swarth (1925) described the Pine Siskin as "abundant and of general distribution"; "occurs at all altitudes, in the interior and on the coast"; and "breeds throughout the province." Their description is as true today as it was in the 1920s, and the siskin remains among the most widespread and abundant breeding birds in British Columbia.

Bennett and Enns (1996) consider the Pine Siskin to be an "abundant" species in the Boreal White and Black Spruce Zone in the Liard River area of the Taiga Plains. There it was recorded in 6 of the 7 forest habitats examined, with the highest densities occurring in mature forests of mixed paper birch, trembling aspen, and white spruce. In the Interior Douglas-fir Zone of the Central Interior, Waterhouse and Dawson (1998) found the Pine Siskin to be "locally abundant" in Douglas-fir forests; they attributed its abundance to a "huge cone crop" from the preceding year.

The Pine Siskin has been reported breeding from near sea level on the coast to 1,380 m in the interior. Most breeding records were from low-elevation human-influenced (68%; $n = 74$) or forest (28%) habitats. Of the former, suburban habitats were most commonly reported (30%; $n = 63$), followed by cultivated farmland (24%) and rural (16%). These data are not fully representative of the range of breeding habitats used by the siskin. For example, although only 2 nests in our data were from the Engelmann Spruce–Subalpine Fir Zone, numerous breeding bird studies in British Columbia suggest that the Pine Siskin is a widespread breeder in subalpine forests (Van Tighem and Gyug 1983; Poll et al. 1984; Catt 1991; Dickinson and Leupin 1997; Gyug 1997; Phinney and Lance 1998; Davis et al. 1999). Two of these studies compared the subalpine to lower-elevation forests and found higher siskin

Table 16. Pine Siskin densities (singing males per 10 ha) in trembling aspen forests near Smithers, British Columbia, 1991 and 1992 (from Pojar 1995). Median density was calculated for all plots in each habitat type for both years (n = number of plots).

Forest type	Forest age	Median density	n
Sapling aspen	7-23 years	0	6
Mature aspen	50-60 years	1.3	10
Old aspen	100-year-old veterans	2.2	4
Mixed aspen-conifer	50-60 and 95 years	8.3	6

densities in the former (Poll et al. 1984; Phinney and Lance 1998). Furthermore, a nest reported near timberline in Washington (Jewett et al. 1953) suggests that breeding occurs up to the highest elevations of the subalpine habitats.

Keller and Anderson (1992) compared breeding bird communities in fragmented and unfragmented old-growth subalpine forests in Wyoming. They found that Pine Siskins were distributed throughout both forest classes, but were significantly more abundant in stands altered by old, strip-cut logging. In a similar study in north-central British Columbia, Gyug (1997) examined the effects of experimental logging on breeding birds in the Engelmann Spruce–Subalpine Fir Zone. There the Pine Siskin was among the most abundant songbirds, but no significant difference in habitat use could be detected between clearcuts, remnant forest patches within clearcuts, or old-growth forest. A comparable study of old logging (25 to 29 years) in subalpine and lower-elevation forests near Grand Forks yielded a similar result (Gyug and Bennett 1995).

Breeding bird studies in trembling aspen–dominated forests indicate that conifers are an important habitat component for siskins. In the Central Interior near Smithers, Pojar (1995) detected the Pine Siskin in most trembling aspen types, except for the sapling stage, but it was most abundant in mixed forests of aspen and hybrid white spruce (Table 16; Fig. 668). She classified the siskin as a conifer-related species. Further north, in the homogeneous trembling aspen forests of the Boreal Plains, siskins were detected in very low numbers during all years of a 6-year study (Merkins and Booth 1998).

The Pine Siskin has a long breeding period in British Columbia; nest-building activities have been observed as early as 19 February in the interior and 25 February on the coast. It has been recorded breeding from 22 March (calculated) to 20 August on the coast, and from 11 March (calculated) to 24 August in the interior (Fig. 659).

Nests: Most nests were found in trees (85%; n = 80), the majority of which were coniferous species, notably Douglas-fir (36%; Fig. 669), ponderosa pine (20%), and hybrid white spruce (14%). Other conifers included lodgepole pine (2 nests), western white pine, Engelmann spruce, Sitka spruce, and subalpine fir (1 nest each). In settled areas, a variety of ornamental conifers were used (10 nests), as were domestic apple trees (3). A few were reported in deciduous shrubs, including willow (2 nests) and ocean-spray (2). Nests were usually saddled on a horizontal limb, most often placed well out from the trunk (44%; n = 62), or near the end of the branch (40%). A few were placed near the main tree stem (5 nests) or near the top of a shrub (2 nests; Fig. 667).

Figure 669. Nest and eggs of the Pine Siskin (Mount Douglas Park, Saanich, 30 April 1973; R. Wayne Campbell).

Nests were typically neat cups of grasses and fine twigs lined with hair. Less commonly used nest materials included plant stems, mosses, rootlets (Fig. 669), lichens, and string. Mammalian hair (horse and dog) was the most commonly identified lining; other materials included fine grasses, plant down, lichens, and feathers. The heights of 79 nests ranged from 0.6 to 20 m, with 61% between 2.2 and 8 m. This height range is broader than the 0.9 to 15.2 m range reported for other areas in North America by Dawson (1997), but nests as high as 30 m have been reported in Washington (Dawson and Bowles 1909). The Pine Siskin usually nests in loose colonies (Dawson 1997). In Washington, Dawson and Bowles (1909) indicate that as many as 6 nests can occur in a single tree. In British Columbia, we have a single record of 4 nests occurring close to each other, but none were in the same tree.

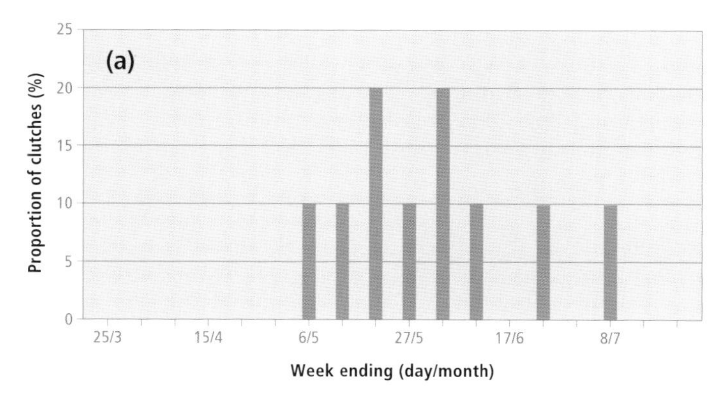

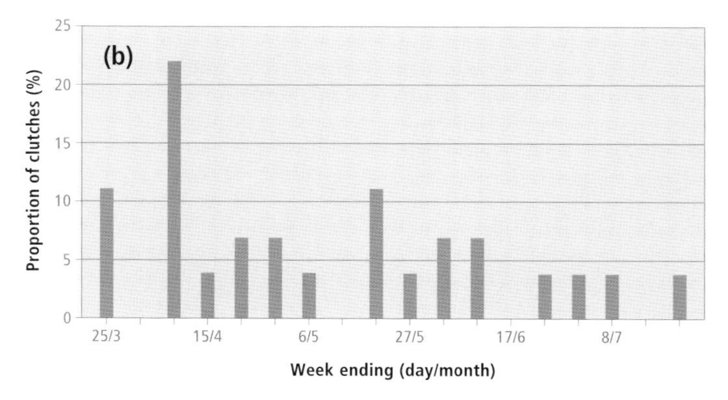

Figure 670. Weekly distribution of the proportion of clutches for the Pine Siskin in the (a) Georgia Depression (n = 10) and (b) Southern Interior (n = 27) ecoprovinces of British Columbia. The figures are based on the week eggs were found in the nest. In the Southern Interior, the second peak in clutches, 6 weeks after the first, suggests that double-brooding may occur.

Eggs: Dates for 71 clutches ranged from 19 March to 29 July, with 54% recorded between 25 April and 4 June. Calculated dates indicate that eggs can occur as early as 11 March. Sizes of 39 clutches ranged from 1 to 5 eggs (1E-3, 2E-3, 3E-11, 4E-17, 5E-5), with 72% having 3 or 4 eggs (Fig. 669). Clutch sizes across North America range from 1 to 6 eggs, but are usually 3 or 4 eggs (Dawson 1997).

The Pine Siskin usually produces a single brood per year, but at times may be double-brooded (Dawson 1997). Clutch dates for the Southern Interior indicate 2 peaks, suggesting that double-brooding occurs in this ecoprovince. The first peak occurs in the first week of April and a second, smaller peak occurs in the third week of May (Fig. 670b). Double-brooding in the Georgia Depression is not apparent (Fig. 670a).

For 1 nest in British Columbia, the incubation period was 13 days, the same as that reported for a nest in New Hampshire (Weaver and West 1943).

Young: Dates for 81 broods ranged from 24 March to 24 August, with 51% recorded between 23 April and 15 June. Sizes of 35 broods ranged from 1 to 5 young (1Y-4, 2Y-2, 3Y-15, 4Y-13, 5Y-1), with 80% having 3 or 4 young. The nestling period for 2 nests was 15 days. Elsewhere, the nestling period is generally 14 or 15 days, but can be as short as 9 or 10 days if the nest is disturbed (Baicich and Harrison 1997; Dawson 1997).

Brown-headed Cowbird Parasitism: In British Columbia, 10% of 62 nests found with eggs or young were parasitized by the cowbird. Parasitism was found in 13% ($n = 47$) of interior nests but was not detected in nests from the coast ($n = 15$). There were also 9 records of a Pine Siskin feeding a cowbird fledgling; 5 of these were from the coast. This relatively high rate of parasitism is probably biased by the disproportionate number of low-elevation nests in our data base. Nests occurring in the earliest part of the breeding season (i.e., March and April) or in closed coniferous forest habitats, especially in the subalpine environment, would likely escape this nest parasite.

Figure 671. Listless behaviour, fluffed appearance, and birds found in sleeping postures during the day are indications that Pine Siskins have likely contracted salmonellosis, a disease that may kill thousands of siskins in some years (Saanich, southern Vancouver Island, 13 December 1998; R. Wayne Campbell).

Figure 672. During some winters in British Columbia, large numbers of Pine Siskins are attracted to road salt and are killed by vehicular traffic (Highway 1 near Woolsey Creek, Mount Revelstoke National Park, 1 February 1991; Mas Matsushita, ©Parks Canada).

Most cardueline finches are poor candidates for parasitism because cowbird young typically languish and die on the seed diet provided to nestlings (see **House Finch**). Pine Siskins are apparently no exception, and parasitized nests are only rarely successful (Dawson 1997). Data from British Columbia, however, suggest that some Brown-headed Cowbirds do survive and fledge.

Nest Success: Of 15 nests found with eggs and followed to a known fate, 5 produced at least 1 fledgling, for a nest success rate of 33%. Nest predation was observed or suspected as the cause of 8 failures. Nest predators included Clark's Nutcracker (2 nests), American Crow (2), Steller's Jay, red squirrel, racoon, and domestic cat (1 nest each). Also, domestic cats were observed to prey on recently fledged young on 2 occasions. Of 6 nests found with Brown-headed Cowbird eggs, 4 were left alone and all failed; the other 2 nests had the parasite eggs removed by the observer and both were successful.

Pine Siskin nests are vulnerable to windstorms (Palmer 1968) and our data show 2 instances of active nests being blown out of trees. One resulted in the loss of 3 nestlings, but in the second case, 3 large young survived the fall and were placed back in the tree by the observer using a cottage cheese container as a substitute nest; the parents found the young and resumed feeding them. In a related example of parental tenacity, a nest containing 4 eggs that had been accidentally cut down during pruning was placed on another tree limb; the female resumed incubation and the nest eventually fledged 2 young.

REMARKS: Three subspecies of the Pine Siskin have been recognized (American Ornithologists' Union 1957; Dawson 1997); *C. p. pinus* is the widespread northern race and breeds throughout British Columbia (Cannings 1998). The Pine Siskin is sometimes placed in the genus *Spinus*.

The Pine Siskin is very susceptible to salmonellosis, especially when congregating at bird feeding stations (Fig. 661) (Dawson 1997). During the winter of 1992 to 1993, a wide-

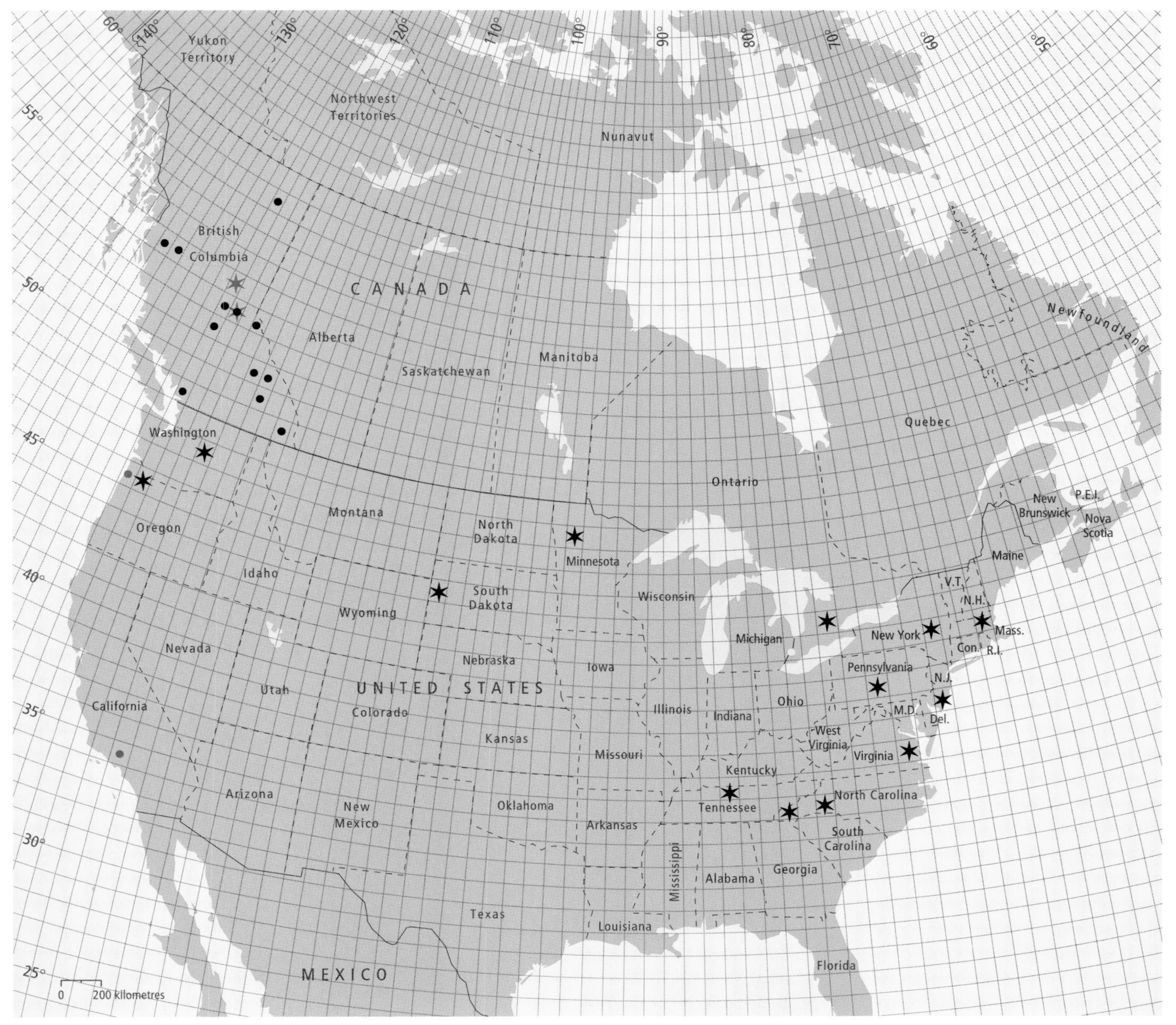

Figure 673. Banding locations (stars) and recovery sites (circles) of Pine Siskins associated with British Columbia. Red indicates birds banded in British Columbia; black indicates birds banded elsewhere.

spread outbreak killed thousands of siskins in Oregon, Washington, and British Columbia. An estimated 1,000 siskins succumbed to this disease in the Fraser Lowland and on Vancouver Island during that winter; *Salmonella typhimurium* was isolated from a sample of these birds (Wilson et al. 1995). Infected birds can be identified by their listless behaviour, fluffed up appearance, dull eyes, diarrhea, and feeble flight capabilities (Fig. 671). Once infected, most probably succumb to this disease or fall prey to predators such as the Sharp-shinned Hawk, Cooper's Hawk, Northern Pygmy-Owl, Northern Shrike, or domestic cat. Siskins can be discouraged from using infected feeding stations by eliminating the use of sunflower and niger seeds.

Like many cardueline finches, the Pine Siskin is strongly attracted to materials containing salt (Bennetts and Hutto 1984). In British Columbia, Swarth (1922) observed siskins eating the plaster chinking from log cabins at Glenora, along the Stikine River. Likewise at Indianpoint Lake near Barkerville, siskins were attracted to stove ashes, freshly set portland cement, chimney mortar, and a blue clay spoil pile from a cellar hole (McCabe and McCabe 1928). Unfortunately, siskins are also attracted to road salt used along transportation routes and, in some winters, are killed in large numbers by vehicular traffic (Fig. 672). Highways traversing regions of high snowfall, particularly through mountain passes, appear to have the greatest incidence of collision mortality in British Columbia. Known trouble spots include Highway 1 from Three Valley Lake to Rogers Pass (Fig. 672), Highway 23 from Galena Bay to Nakusp (J.G. Woods pers. comm.), and Highway 3 through Manning Park.

In Mount Revelstoke and Glacier national parks, more than 1,000 siskins were killed by vehicular traffic along Highway 1 during the winter of 1982-83 (Van Tighem and Gyug 1983). In another winter, 104 dead siskins were counted along a 2 km length of Highway 1 east of Revelstoke (Yunick 1997). Further south, approximately 120 siskins were killed on 25 December 1983 along a 30 m stretch of Deep Creek Road near Enderby (Wilson et al. 1995). The most compelling figures, however, resulted from single collision incidents in which 2 separate vehicles killed 129 and 208 birds, respectively, in Mount Revelstoke and Glacier national parks (Van Tighem and Gyug 1983). Most road mortalities occur during winter or early spring but are occasionally recorded at other times of the year. In early July, for example, a small number of siskins were killed along with many Evening Grosbeaks along Highway 3 through Manning Park (Smith 1981; Wilson 1981; Wilson et al. 1995). These birds had apparently been attracted to residual salts from earlier salting activities along the highway.

Mortality from pesticide poisoning has been suspected in British Columbia. For example, on 12 September 1970, 54 siskins were found dead or dying in a residential yard in Sardis (Wilson et al. 1995). Many were observed to fall from the trees, convulse, and die. A broccoli crop in a nearby field was being sprayed with pesticide at the same time; unfortunately, it was not possible to perform a lab analysis to confirm the cause.

Banding data for British Columbia illustrate the erratic, continental scale of Pine Siskin movements in North America (Fig. 673). Most birds had been banded in eastern North America and recovered in British Columbia. Banding locations included Washington, Oregon, South Dakota, Minnesota, southern Ontario, Tennessee, North Carolina, Virginia, New Jersey, Pennsylvania, New York, and Massachusetts; the average distance from banding site to point of recovery was 2,800 km (range: 360 to 4,300 km). Two birds banded in British Columbia were recovered in northwestern Oregon and southern California, respectively.

Yunick (1983) reviewed Pine Siskin banding data from eastern North America. He determined that siskins have low winter site fidelity and show wide geographic variation between years. Banding data for British Columbia show that continent-wide movements can occur between winters. For example, a bird banded on 28 December 1989 at Nashville, Tennessee, was recovered the following winter at Terrace on 16 December 1990, a distance of approximately 3,800 km.

A Pine Siskin nest with 3 eggs collected on 22 June 1889 at Fort St. James (MacFarlane and Mair 1908) was reported incorrectly as 27 June by Munro and Cowan (1947) and Munro (1949).

NOTEWORTHY RECORDS

Spring: Coastal – Saanich 4 May 1982-250; Cowichan Bay 19 May 1982-125; Sidney Island 16 Apr 1988-250; Wickaninnish Bay (Pacific Rim National Park) 19 May 1988-100 feeding on willow catkins with American Goldfinches; Duncan 5 Apr 1969-4 nestlings; Tofino 18 May 1974-250; Ladner 15 Mar 1974-200; Surrey 24 May 1961-250; Burnaby Lake 12 Apr 1970-150; Langley 19 May 1949-5 eggs (UBC 1891); Skagit River 3 Mar 1986-50, killed along Highway 3, 23 Mar 1971-200; Ross Lake 25 May 1976-200; Sproat Lake Park 8 Apr 1974-110; West Vancouver 4 Mar 1979-350; Vancouver 24 Apr 1974-100 at Stanley Park; Harrison Hot Springs 30 Apr 1984-100; Union Bay 12 May 1988-60, feeding on dandelion heads; Courtenay 7 Apr 1936-1 collecting nesting material, 8 May 1939-5,000 (Pearse 1939); Hornby Island 11 Mar 1979-200, in red alder grove; Cranberry Lake (Powell River) 12 May 1977-4 nestlings, about 10 days old; Campbell River 2 Mar 1981-600; Callaghan Creek 26 Apr 1986-3,000 (Mattocks 1986a); Triangle Island 27 May 1974-30; Stuie 20 Mar 1926-2 (RBCM 14067-8); Sandspit 14 May 1977-1,200, feeding on dandelion seeds nr airport; Kitlope River 3 May 1991-21, at estuary; Lord Bight (Langara Island) 30 May 1981-300, large flocks swarming through trees along shore; Kitimat 22 Mar 1980-1, collecting small twigs for nest; Lakelse Lake 13 Apr 1980-100; Terrace to Smithers 20 Mar 1977-800, feeding along gravel shoulders of Highway 16, some killed by traffic; Legate Creek 30 Mar 1977-180, feeding with Red Crossbills; Stewart 21 Mar 1994-8 at feeder; Taku River 15 May 1995-20; Pleasant Camp 28 May 1979-1 along Haines Highway. **Interior** – Manning Park 25 May 1975-550; Grand Forks 23 Mar 1979-4 eggs; Balfour to Waneta 16 May 1981-524, on count; Creston 21 Apr 1984-3 nestlings; Ainsworth Hot Springs 15 Mar 1979-40, some gathering nesting material; Lytton 8 May 1966-500; 6.4 km ne of Nicola 1 May 1975-5 eggs; Okanagan Landing 18 Mar 1918-2, constructing nest in ponderosa pine; Lavington 19 Mar 1979-3 eggs; Nakusp 1 Mar 1981-200; D'Arcy 3 Mar 1997-60, at feeder; Shuswap Park 23 May 1974-500; Bridge Creek (Revelstoke) 19 May 1949-5 eggs; Three Valley Lake 10 Mar 1984-229; Rogers Pass 7 Mar 1984-1,093 along Highway 1; 100 Mile House 25 Mar 1976-150; Lac la Hache 21 Apr 1962-150; Williams Lake 18 Mar 1983-2, building nest, 8 May 1976-1 recently fledged young, 30 May 1963-200; Quesnel 18 May 1988-4 eggs, 29 May 1963-100; Summit Lake (Crooked River) 12 Mar 1983-30; Tate Creek 18 May 1993-1; Taylor 4 May 1984-4 fledglings being fed by adults; Fort St. John 20 Mar 1984-18, some singing, 20 Apr 1984-10, some collecting nesting material; Hyland Post 27 May 1976-12; nr Clarke Lake 19 May-1; Fort Nelson 28 May 1987-4; Chilkat Pass (Haines Highway) 14 May 1977-10; Atlin 20 May 1981-12 (Campbell 1981).

Summer: Coastal – Central Saanich 9 Jul 1983-2,000; North Saanich 13 Jun 1986-1 adult feeding Brown-headed Cowbird fledgling; Long Beach 6 Jun 1968-1 constructing nest in Sitka spruce (Hatler et al. 1978); Surrey 26 Aug 1960-350; Qualicum 20 Aug 1976-3 nestlings, left nest this date (Dawe 1980); Courtenay 26 Jun 1933-200; Garibaldi Lake 29 Aug 1982-10; Brooks Peninsula 6 Aug 1981-10; 3.2 km s Campbell River 25 Aug 1973-500 eating dandelion and thistle seeds with American Goldfinches; Port Neville 29 Jun 1975-21 feeding on red alder seeds; Pemberton 3 Jun 1967-10; Stubbs Island (Tofino) 9 Jul 1975-30; Triangle Island 4 Jun 1974-6; Calvert Island 1 Jul 1988-1 (Rodway and Lemon 1991); Cape St. James 18 Jul 1979-30; Anthony Island 30 Jun 1977-100; Bella Coola 18 Jul 1978-3; Reef Island 11 Jul 1977-150; Kimsquit River 3 Jul 1985-3; Marble

Island 8 Jun 1977-45; Queen Charlotte City 4 Aug 1979-50; Tlell River 16 Jun 1977-70; Kitimat 21 Jul 1975-275; McDonnell Lake 27 Jun 1980-50; w Greenville 16 Jun 1981-60; Kitwanga Lake 28 Aug 1979-25; Meziadin Lake 8 Jun 1976-20; Flood Glacier 3 Aug 1919-1 (MVZ 39916); Barrington River 20 Jul 1982-1, at headwaters in subalpine forest; Haines Highway 21 Jun 1972-60, at Mile 46. **Interior** – Three Brothers Mountain (Manning Park) 10 Jul 1974-750; Erie 12 Aug 1983-164; Ripple Mountain (Newgate) 29 Jul 1980-29 at 2,285 m elevation; Kootenay River (Creston) 29 Aug 1947-300, feeding in weed patch (Munro 1957); Creston 6 Jun 1982-280; Oliver 25 Aug 1964-100; White Lake (Penticton) 9 Jun 1974-45; Summerland 21 Jun 1969-2 adults feeding Brown-headed Cowbird fledgling, 2 Jul 1968-3 eggs plus 2 Brown-headed Cowbird eggs; Sparwood 14 Jun 1984-60, feeding on dandelions; Lytton to Skihist 11 Jun 1966-200, along road; Okanagan Centre 26 Jun 1892-5 eggs; Whatshan Lake 3 Jun 1984-120; Brouse 2 Jun 1984-200; Radium 27 Aug 1971-175; Chase 18 Jul 1980-66; Meadow Creek 18 Aug 1973-400 (Sirk at al. 1973); White Lake (Shuswap Lake) 15 Aug 1980-300; Loon Lake (Clinton) 1 Jun 1975-3 nestlings; Field 21 Jun 1975-250, feeding on dandelion seeds (Wade 1977); Kleena Kleene 9 Jun 1966-100 (Paul 1968); Williams Lake 1 Jun 1963-100; Stum Lake 6 Jul 1973-20 (Ryder 1973); 6 km w Mount Milton 9 Aug 1997-20, in alpine meadows (2,100 m; Fig. 652); Moose Lake (Mount Robson Park) 17 Jun 1972-100; Cinema 24 Aug 1966-3 nestlings; 24 km s Prince George 5 Jun 1983-250, feeding on dandelion seeds; Prince George 10 Jul 1993-1 adult feeding fledged Brown-headed Cowbird; Francois Lake 4 Jun 1977-78; Morice River 29 Jun 1975-30; Fort St. James 22 Jun 1889-3 eggs (MacFarlane and Mair 1908); Carp Lake Park 25 Aug 1975-45; Swan Lake (Tupper) 7 Jul 1984-30; e Chetwynd 5 Aug 1975-50; Moberly Lake to Hudson's Hope 6 Jun 1976-250, along Highway 29; Dawson Creek 3 Jul 1974-20; Taylor 21 Aug 1975-134; Fort St. John 22 Jul 1983-30; Wonowon 19 Aug 1975-60; Tatlatui Lake 1 Jul 1976-25; Pink Mountain 3 Jul 1978-8; Nuttlude Lake 28 Jul 1957-40; nr Telegraph Creek 20 Jun 1919-3 eggs, collected (Swarth 1922); Tetsa River 3 Aug 1970-adults with recently fledged young; Jackfish Creek 16 Jun 1982-5; Kledo Creek 19 Jul 1985-60; s Fort Nelson 13 Jul 1981-30, flock chased by a Merlin; Denedin River 25 Jul 1997-20, feeding in spruce and balsam poplars; Liard River 4 Aug 1943-30, at bridge crossing (Rand 1944); Liard Hot Springs 16 Jun 1982-15; Helmut 7 Jun 1982-2; Atlin 23 Aug 1974-100; Tatshenshini River 15 Jun 1993-nest under construction (Fig. 667); Range Creek (Alsek River) 10 Aug 1989-10; Survey Lake 26 Jun 1980-100; Petitot River 26 Jun 1985-3, at Liard Highway bridge crossing.

Breeding Bird Surveys: Coastal – Recorded from 27 of 27 routes and on 80% of all surveys. Maxima: Kitsumkalum 13 Jun 1976-356; Pemberton 2 Jul 1976-250; Chilliwack 22 Jun 1975-86. **Interior** – Recorded from 72 of 73 routes and on 94% of all surveys. Maxima: Prince George 18 Jun 1984-299; Wingdam 17 Jun 1969-254; Spillimacheen 13 Jun 1981-213.

Autumn: Interior – Atlin 23 Sep 1924-1 (Swarth 1926); Liard River 25 Oct 1980-20; Steamboat Mountain (Muskwa River) 14 Sep 1943-1 (Rand 1944); Fern Lake 6 Sep 1979-300 (Cooper and Adams 1979); Pesika River 4 Sep 1997-10, in subalpine forest; Griffith Creek 5 Sep 1976-30; Fort St. John 6 Sep 1986-32, 29 Oct 1987-1, latest sighting; Hudson's Hope 14 Oct 1979-14; Glacier Gulch 18 Nov 1981-40, feeding in conifers; Indianpoint Lake 2 Sep 1927-16, all banded (McCabe and McCabe 1928); Ten Mile Lake (Quesnel) 5 Oct 1986-6; Lac la Hache 7 Oct 1946-250 (Munro and Cowan 1947); Kleena Kleene 12 Sep 1962-60 (Paul 1964); Field 21 Oct 1975-150, eating birch seeds (Wade 1977); Golden 4 Sep 1976-100; Mount Revelstoke National Park 16 Oct 1982-200; Scotch Creek 10 Sep 1962-250; Turtle Valley 16 Sep 1980-200; Nakusp 10 Nov 1985-250; Okanagan Landing 22 Sep 1963-120; Vaseux Lake 14 Oct 1973-80; w Creston 22 Sep 1984-500. **Coastal** – Stuhini Creek (Taku River) 17 Sep 1988-1, at creek mouth; Kispiox River valley 1 Sep 1921-100 (Swarth 1924); Terrace 14 Oct 1974-200; Prince Rupert 12 Nov 1983-50; Masset 29 Oct 1971-50; Port Hardy 24 Sep 1938-200; Port Neville 11 Nov 1975-200; Alta Lake 10 Sep 1944-375; Campbell River 14 Nov 1981-500; Stories Beach (s Campbell River) 20 Nov 1976-1,200; Courtenay 13 Nov 1987-362; North Vancouver 7 Oct 1983-500; Mount Douglas (Saanich) 13 Oct 1975-500; Saanich Oct 1969-5,000, in a flock (Tatum 1970), 20 Nov 1998-1, caught nr feeder by Northern Shrike; Saseenos (Sooke) 24 Nov 1975-1,000; w Sooke 25 Nov 1984-1,800, feeding in red alders.

Winter: Interior – Stoddart Creek (Fort St. John) 26 Feb 1984-5, several in song flights; Smithers 19 Jan 1975-40; Williams Lake 19 Feb 1983-2, collecting grass and hair for nest; Tatla Lake 27 Jan 1989-1; 100 Mile House 20 Feb 1983-1, collecting lichens from Douglas-fir branches for nest; Glacier National Park 20 Feb 1983-227; Mount Revelstoke National Park 8 Dec 1983-1,000 in flock, many being killed on Highway 1; Revelstoke 26 Feb 1984-1,000; Sorrento 13 Dec 1970-50; Celista 5 Feb 1948-300; Falkland 12 Jan 1979-50 feeding on abundant Douglas-fir cones with Red Crossbills; Enderby 25 Dec 1983-120, killed along Deep Creek Rd (Wilson et al. 1995); Invermere 18 Feb 1981-100; Nakusp 2 Jan 1984-350, in 3 separate flocks; New Denver 18 Feb 1979-350; Okanagan Landing 24 Feb 1918-150 (Munro and Cowan 1947); Princeton to Hedley 21 Jan 1984-125, 25 to 40 killed on Highway 3; Salmo Feb 1972-1,000; Creston 6 Jan 1982-280. **Coastal** – Stewart 9 Dec 1991-3 at feeder; Prince Rupert 9 Dec 1984-60; Kitimat River 30 Jan 1975-300; Juskatla 18 Dec 1971-50; Port Clements 12 Feb 1973-350; Skidegate 12 Feb 1973-60; Cape St. James 7 Dec 1981-1; Bella Coola 6 Dec 1998-50 at feeder; Port Hardy 15 Feb 1938-200; Ida Lake (Woss) 26 Feb 1983-200; Alta Lake 25 Dec 1945-200; Elk Falls Park 1 Jan 1976-2,000; Campbell River 10 Jan 1976-1,000 on estuary; Harrison Hot Springs 4 Dec 1975-160; Pitt Meadows 13 Dec 1975-1,500; North Saanich 22 Feb 1969-1,000 (Tatum 1970); Metchosin 25 Feb 1983-1 building nest; Jordan River 30 Dec 1977-120.

Christmas Bird Counts: Interior – Recorded from 22 of 27 localities and on 66% of all counts. Maxima: Revelstoke 21 Dec 1983-3,277; Nakusp 2 Jan 1984-2,060; Shuswap Lake Provincial Park 2 Jan 1994-1,490. **Coastal** – Recorded from 30 of 33 localities and on 88% of all counts. Maxima: Victoria 18 Dec 1993-7,998; Vancouver 26 Dec 1973-7,527; Squamish 2 Jan 1989-5,498.

American Goldfinch AMGO

Carduelis tristis (Linnaeus)

Figure 674. American Goldfinch in (a) breeding plumage and (b) in winter plumage. This species is widely distributed across southern British Columbia, but in winter it becomes more local (Creston, 20 April 1995 and Nelson, 2 December 1997; Linda M. Van Damme).

RANGE: Breeds from southern British Columbia, north-central Alberta, central Saskatchewan, west-central and southern Manitoba, central Ontario, southern Quebec, New Brunswick, Prince Edward Island, Nova Scotia, and southwestern Newfoundland south to southern California, northern Baja California, eastern Oregon, central Nevada, southern Colorado, northern New Mexico, central Oklahoma, extreme northeastern Texas, northern Louisiana, northern Mississippi, central Alabama, central Georgia, and South Carolina. Winters from southern British Columbia, the northern United States, southern Manitoba, southern Ontario, New Brunswick, and Nova Scotia south to northern Baja California, northern Sonora, southern New Mexico, western and southern Texas, northern Coahuila, Nuevo Leon, Tamaulipas, Veracruz, the Gulf coast, and southern Florida.

STATUS: On the coast, *fairly common* to *very common* migrant and summer visitant in the Georgia Depression Ecoprovince, becoming locally *abundant* during autumn migration, generally *uncommon* but occasionally *common* or locally *very common* in winter; in the Coast and Mountains Ecoprovince, an *uncommon* migrant and summer visitant on Western Vancouver Island and the Southern Mainland Coast, *very rare* on the Northern Mainland Coast, *casual* on the Queen Charlotte Islands.

In the interior, *fairly common* to *very common* migrant and summer visitant in the Southern Interior and Southern Interior Mountains ecoprovinces, *uncommon* but occasionally *common* or locally *very common* there in winter; *uncommon* migrant and summer visitant in the Central Interior Ecoprovince, *casual* there in winter. *Accidental* in the Sub-Boreal Interior (see REMARKS) and Boreal Plains ecoprovinces. Not known to occur in the Northern Boreal Mountains or Taiga Plains ecoprovinces.

Breeds.

NONBREEDING: The American Goldfinch (Fig. 674) is widely distributed at lower elevations across southern British Columbia south of latitude 51°N, including Vancouver Island. On the coast, its distribution becomes scattered and sparse away from the Georgia Depression and even more so north of Vancouver Island; occasionally it occurs as far north as the Queen Charlotte Islands and Prince Rupert.

In the interior, the American Goldfinch occurs regularly throughout most of the southern valleys and lowlands north to the Williams Lake area, but it is far more common in the Okanagan, Thompson, and west Kootenay valleys than in the

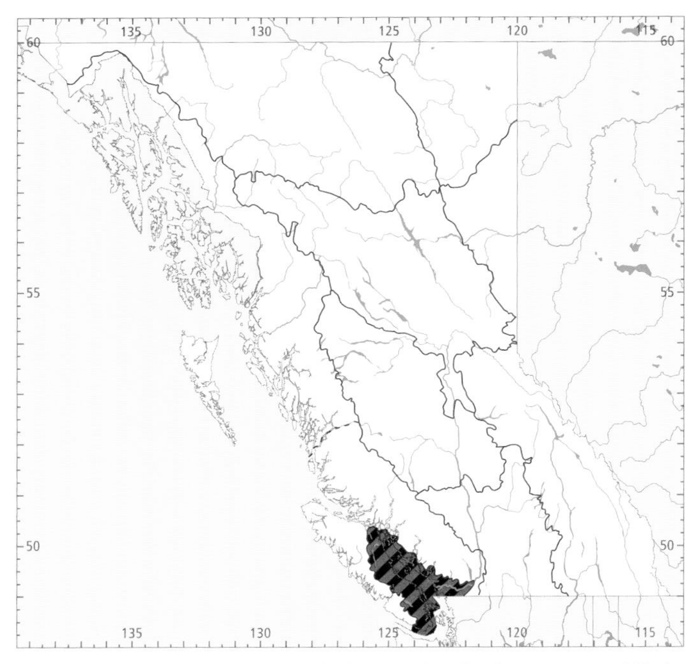

Figure 675. In British Columbia, the highest numbers for the American Goldfinch in winter (black) and summer (red) occur in the Georgia Depression Ecoprovince.

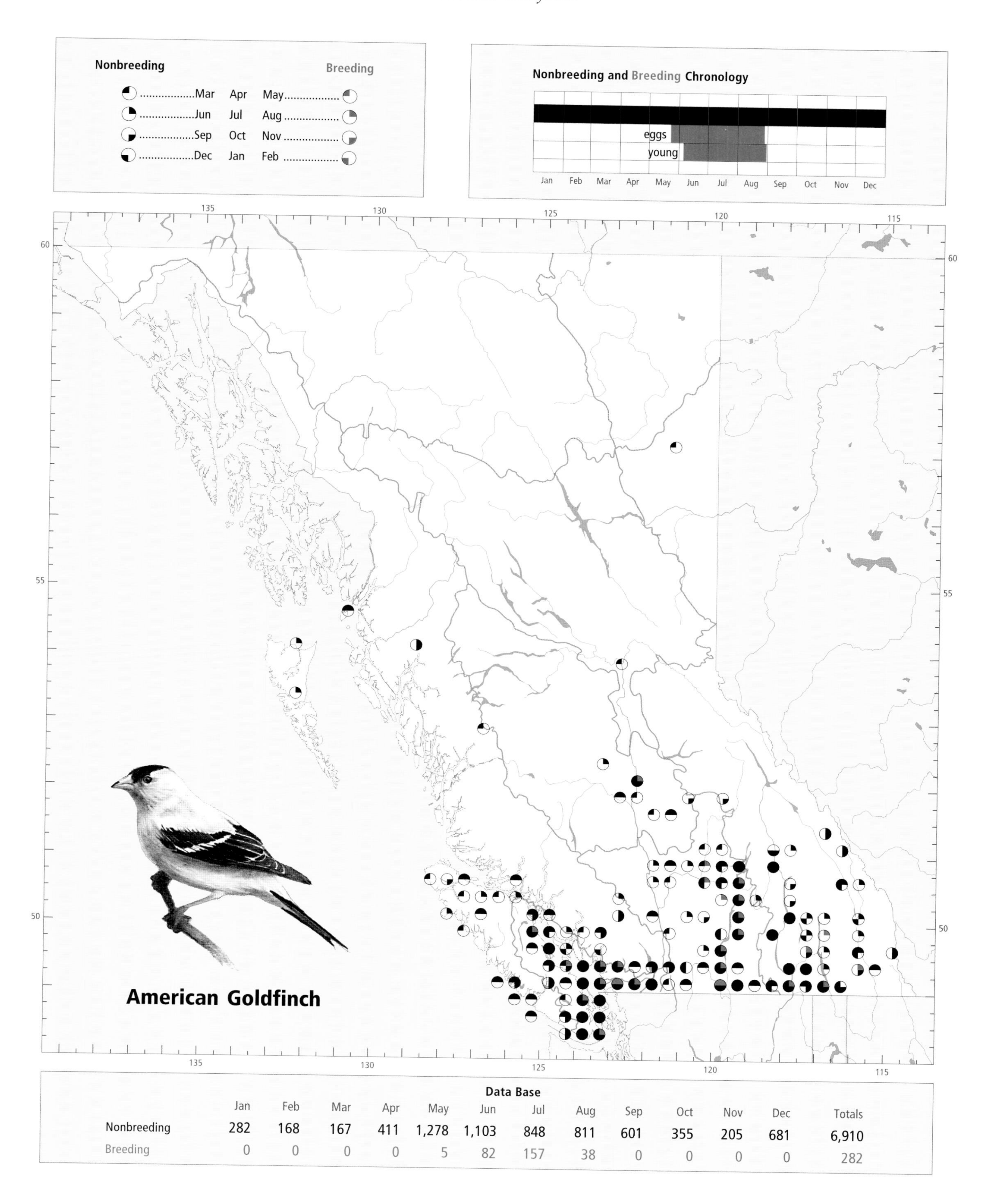

Data Base

	Jan	Feb	Mar	Apr	May	Jun	Jul	Aug	Sep	Oct	Nov	Dec	Totals
Nonbreeding	282	168	167	411	1,278	1,103	848	811	601	355	205	681	6,910
Breeding	0	0	0	0	5	82	157	38	0	0	0	0	282

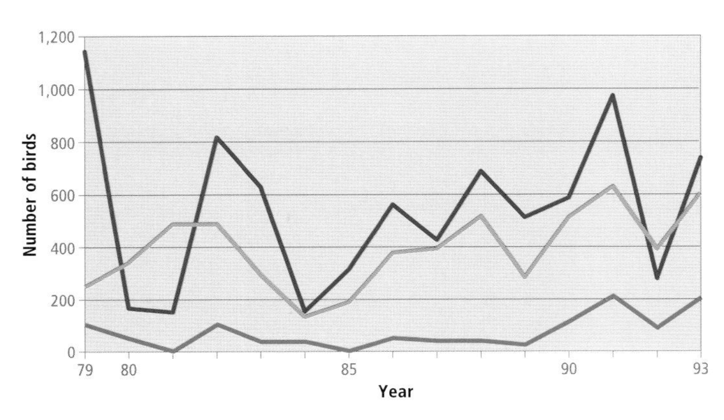

Figure 676. Fluctuations in winter numbers of the American Goldfinch (1979-93) based on Christmas Bird Counts in the Fraser Lowland (purple line) (Chilliwack, Ladner, Pitt Meadows, Vancouver, White Rock) and the Nanaimo Lowland (green line) (Campbell River, Comox, Duncan, Nanaimo, Victoria) of the Georgia Depression, and in the Southern Interior (orange line) (Oliver-Osoyoos, Penticton, Shuswap Lake Park, Vaseux Lake, Vernon).

southern Rocky Mountain Trench or the Williams Lake area. North of Williams Lake, it has been reported only twice, near Prince George in the Sub-Boreal Interior and at Nig Creek in the Boreal Plains (see also REMARKS).

In winter, the American Goldfinch concentrates in the Georgia Depression and the southern portions of the Southern Interior and Southern Interior Mountains. A few birds have occasionally wintered as far north as Williams Lake in the Central Interior. It reaches its highest numbers in winter in the Fraser Lowland of the Georgia Depression (Figs. 675 and 679). While numbers may fluctuate considerably from one year to the next, Christmas Bird Count data for the Georgia Depression and Southern Interior suggest that wintering numbers over the 15-year period 1979 to 1993 were relatively stable (Fig. 676).

Figure 677. During the nonbreeding season, the American Goldfinch frequents powerline rights-of-way in portions of the southern interior of British Columbia (near Dodge Creek, 27 September 1997; R. Wayne Campbell).

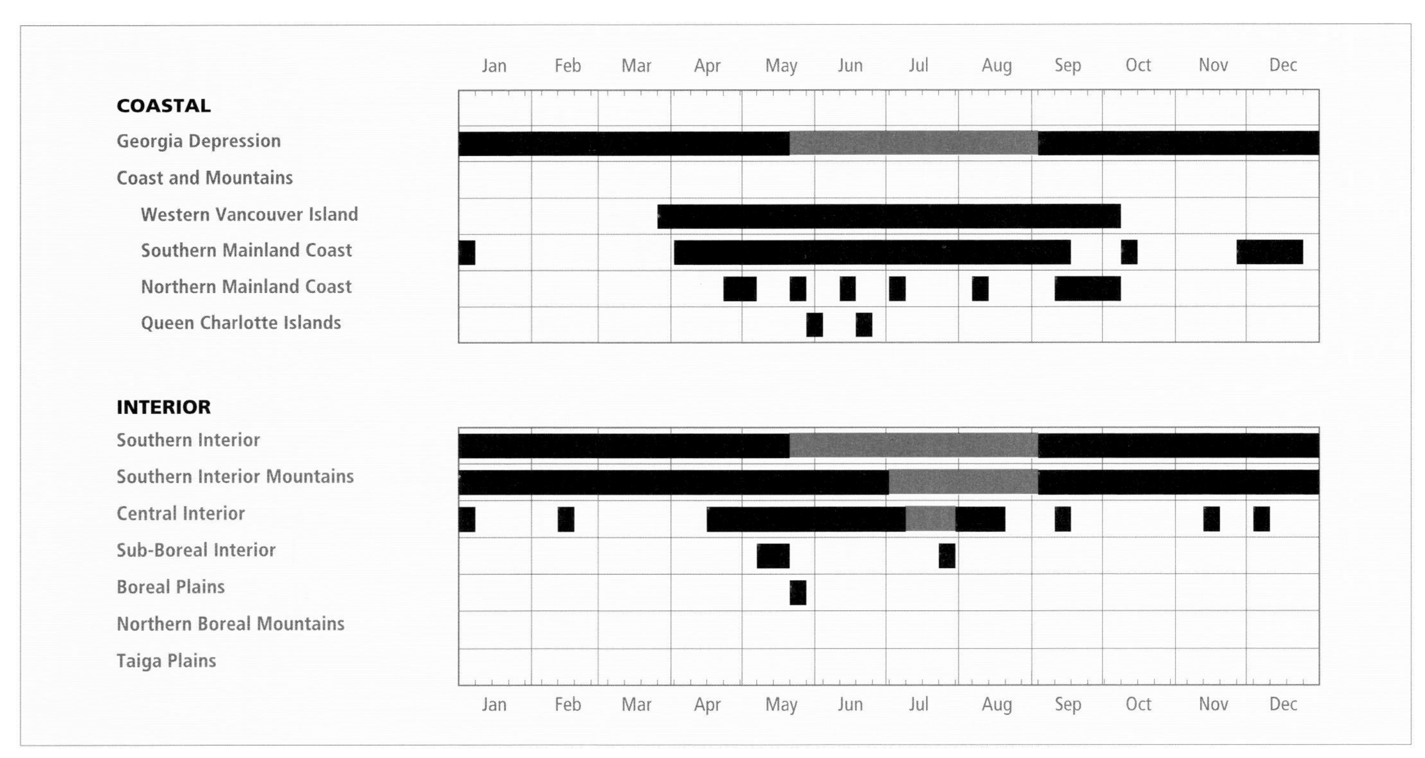

Figure 678. Annual occurrence (black) and breeding chronology (red) for the American Goldfinch in ecoprovinces of British Columbia. Records are shown for the week in which they occurred.

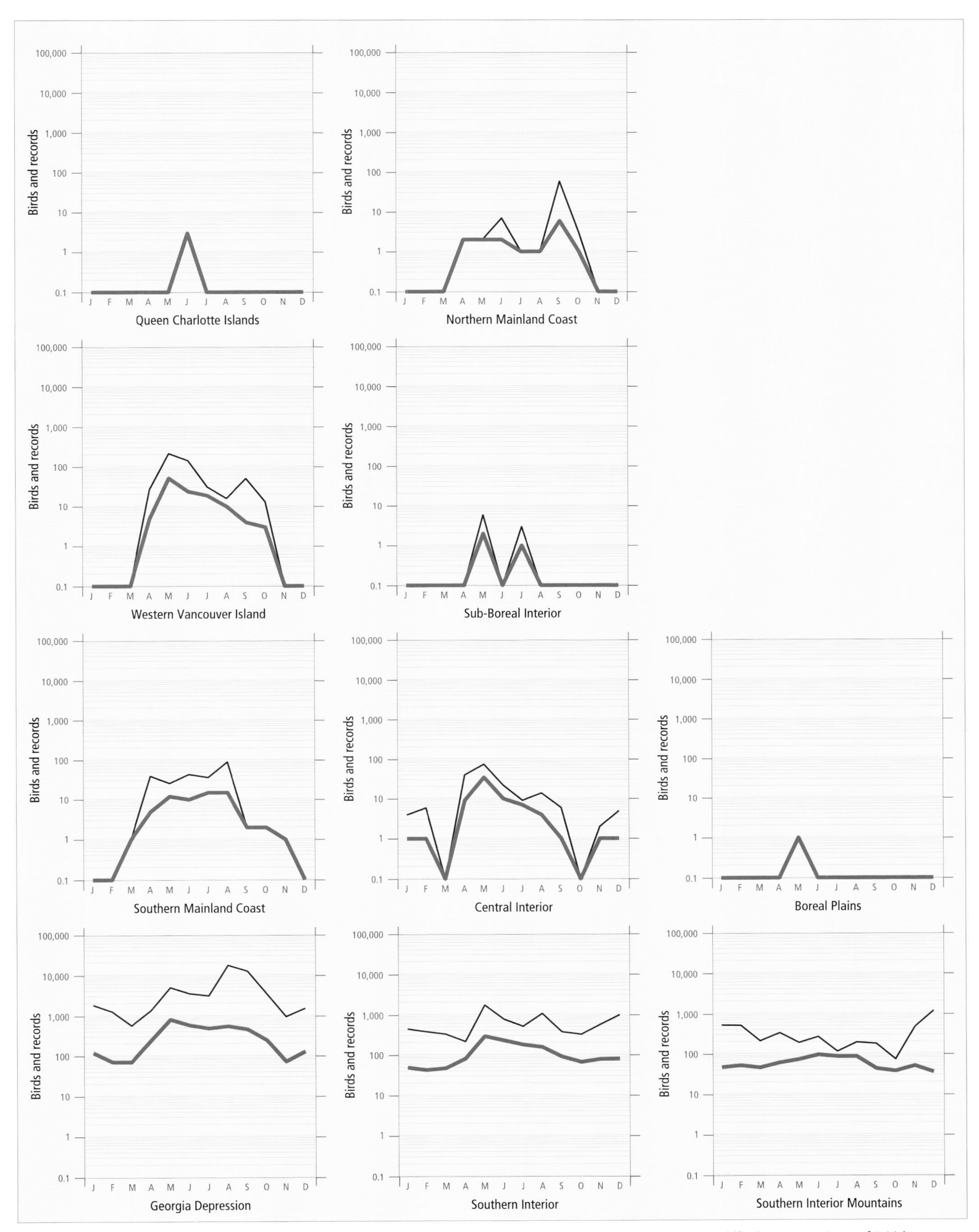

Figure 679. Fluctuations in total number of birds (purple line) and total number of records (green line) for the American Goldfinch in ecoprovinces of British Columbia. Christmas Bird Counts, Breeding Bird Surveys, and nest record data have been excluded.

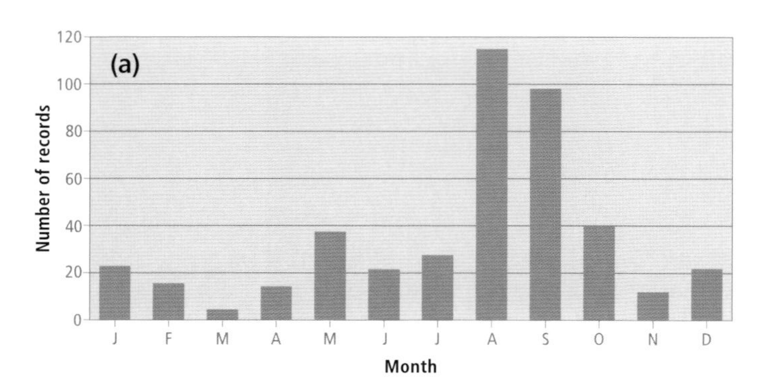

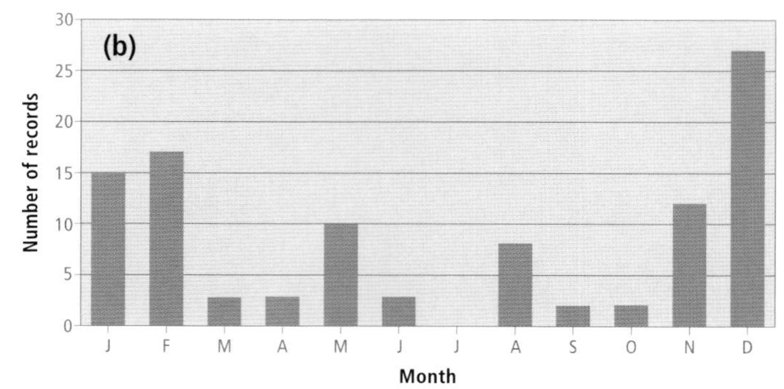

Figure 680. Monthly distribution of large flocks (≥ 25 birds) of American Goldfinches in the (a) Georgia Depression and (b) southern portions of the interior (Southern Interior and Southern Interior Mountains ecoprovinces). In the Georgia Depression, large flocks dominate the autumn migration, whereas in the interior large flocks occur primarily in the winter months. Note the differences in scale.

The American Goldfinch is found mainly at lower elevations throughout its nonbreeding range in the province, from near sea level to 240 m on the coast and from 280 to 1,140 m in the interior.

It tends to remain in typical breeding habitat until lack of food, low temperatures, or snow force it to move (Middleton 1993). In autumn, large flocks may be found at farmlands and open fields wherever extensive weed patches such as thistle occur. On the coast, dyked upland adjacent to estuaries is often used. These open areas often have an abundance of thistles (Dawe et al. 1994).

The American Goldfinch also frequents deciduous and riparian woodlands, especially in winter, where trees heavy with edible seeds such as birch, red alder, and western redcedar provide food. The winter flocks are nomadic and are closely tied to these food sources (Middleton 1993). Other habitats used by the goldfinch in British Columbia during the nonbreeding season included powerline rights-of-way (Fig. 677), city parks and residential gardens, backyard feeding stations, orchards, and nurseries.

In winter, the American Goldfinch often joins other birds that commonly forage in rural and suburban woodlands, such as Pine Siskins and, less frequently, Common Redpolls. It also forages in flocks with Brown Creepers, Black-capped Chickadees, Dark-eyed Juncos, House Finches, Red Crossbills, Red-breasted Nuthatches, Song Sparrows, or White-crowned Sparrows as they move through open woodlands and small patches of mixed forest.

In the spring, the wintering numbers in the Georgia Depression make it difficult to discern any movements, but there appears to be a departure of at least part of the coastal wintering population in the second and third weeks of March. It is not known whether these birds simply disperse to other areas within the Georgia Depression or move west or north along the coast or into the interior. Numbers start to build around the third week of April and peak in the second or third week of May (Figs. 678 and 679). On Western Vancouver Island, goldfinches may arrive as early as the first week of April, but most birds do not appear until mid-May.

In the Southern Interior, wintering birds also make it difficult to determine the arrival dates of the spring migrants. There is an obvious decline in winter numbers around mid-March, but migrants are not evident until the first week of May, reaching a peak in the third week of the month. In the Southern Interior Mountains, wintering residents again cloud the spring arrival. Wintering birds begin declining about the second week of March and migrants may arrive in the first week of April, building to a peak in the third week of the month. In the Central Interior, migrants arrive in the third week of April, and numbers are apparent until the end of May.

In much of the interior, autumn migration begins with the appearance of family groups in the last week of July or the first week of August. Most birds have left the Central Interior by the end of July and there is no discernible autumn influx (Figs. 678 and 679). In the Southern Interior Mountains, there is also little in the way of autumn migration, and most breeders and young have left the region by the third week of September. In the Southern Interior, an autumn movement is apparent as numbers begin to build in the last week of July, peak around the third week of August, and continue into mid-September, by which time most of the breeders and their young have left the province. Goldfinch numbers increase in all southern areas of the interior in November as flocks gather

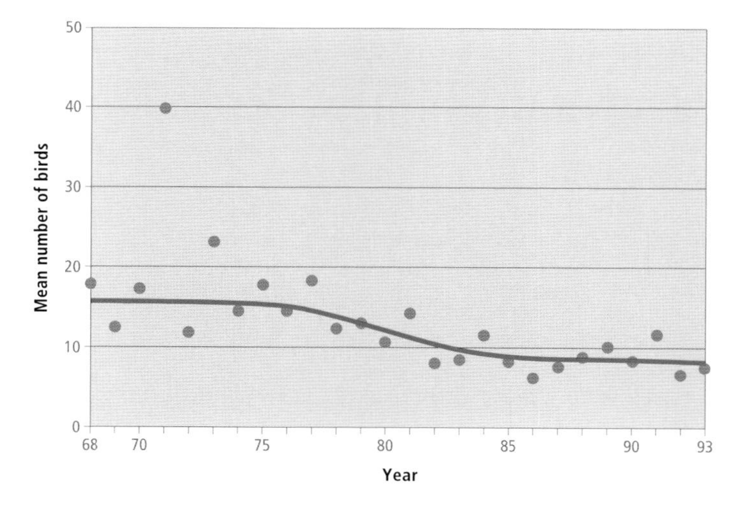

Figure 681. An analysis of Breeding Bird Surveys for the American Goldfinch in British Columbia shows that the number of birds on coastal routes decreased at an average annual rate of 4% over the period 1968 through 1993 ($P < 0.05$).

Figure 682. On the southwestern mainland coast of British Columbia, the American Goldfinch inhabits agricultural areas with shrubs for nests and patches of dandelions and thistles for foraging (Blackie Spit, 7 August 1996; R. Wayne Campbell).

Figure 683. In the southern Okanagan valley of British Columbia, the American Goldfinch breeds in open riparian shrublands adjacent to weedy areas (near Osoyoos, 11 July 1998; R. Wayne Campbell).

(Fig. 680), and then the numbers dwindle somewhat with the colder weather in January (Fig. 679).

Little in the way of an autumn movement is apparent on Western Vancouver Island, and most birds have left the area by the first week of October. A small movement is discernible in the Southern Mainland Coast from the second to the fourth week of August, with most birds gone by September. In the Georgia Depression, arrivals begin in early August and peak in the last week of August or the first week of September as large flocks of goldfinches move through the region (Fig. 680). Most birds have left the region by mid-October, although numbers increase again by mid-December and remain fairly constant through the winter until the departures in March (Fig. 679).

In the Georgia Depression and southern portions of the Southern Interior and Southern Interior Mountains, the American Goldfinch has been reported throughout the year (Fig. 678). On Western Vancouver Island and the Southern Mainland Coast, it occurs regularly from early April to mid-September; in the Central Interior, it has been regularly reported from late April through early August (Fig. 678).

BREEDING: The American Goldfinch has a widespread breeding distribution in extreme southern coastal British Columbia, from the Nanaimo Lowland of the Georgia Depression east across much of the Fraser Lowland. Breeding records are lacking elsewhere on the coast.

In the interior, it is known to breed in the Southern Interior from the international boundary north to Rayleigh, just north of Kamloops; in the Southern Interior Mountains from the international boundary north to Nakusp in the west Kootenay and Cranbrook in the east Kootenay (although fledged young have been reported as far north as Revelstoke in this ecoprovince); and locally in the Williams Lake area of the Central Interior. There are no breeding records from the Fraser River valley between Hope and Williams Lake, or the Rocky Mountain Trench north of Cranbrook. It is likely, however, that the American Goldfinch breeds throughout its summer range in the province south of the Williams Lake area.

In summer, the American Goldfinch reaches its highest numbers in the Georgia Depression (Fig. 675). An analysis of Breeding Bird Surveys for the period 1968 through 1993 shows that the number of birds on coastal routes decreased at an average annual rate of 4% (Fig. 681); an analysis of data for interior routes for the same period could not detect a net change in numbers. Sauer et al. (1997) note a Canada-wide decline of 4.3% between 1966 and 1979 ($P < 0.01$), but there is no apparent trend for the period 1980 through 1996. Continent-wide, goldfinch numbers declined at an average annual rate of 4.7% between 1966 and 1979 ($P < 0.01$), but since 1980 they have increased at an average annual rate of 0.4% ($P < 0.05$) (Sauer et al. 1997).

The American Goldfinch has been reported nesting at elevations from near sea level to 185 m on the coast and from 305 to 920 m in the interior. During the nesting season, it prefers weedy fields and areas of early successional growth as well as cultivated field and road edges, orchards, and gardens.

In British Columbia, it uses a wide range of habitats, from sites that have a largely natural character (31%; $n = 197$; Figs. 682 and 683) to those that are almost completely modified by human activity (69%). The natural habitats included remnant stands of deciduous or coniferous trees (20%), riparian habitat (6%; Fig. 683), shrublands (3%), and grasslands (2%). Most nests associated with human-modified habitat were reported from orchards (30%; $n = 130$), followed by agricultural sites

(25%; Fig. 682), backyard gardens (rural 9%, urban 8%, and suburban 6%), woodland edges (9%), riparian habitats (9%), and meadows (4%).

The American Goldfinch is one of the latest-nesting passerines in North America, and there is a close relationship between onset of nesting and thistle blooms (Middleton 1993). In the east, the goldfinch usually does not nest until late June or early July, but our records indicate that in British Columbia most goldfinches have completed their egg laying by mid-July (97% of clutches on the coast, 78% in the interior; Fig. 678).

On the coast, the American Goldfinch has been recorded breeding from 24 May to 28 August; in the interior, it has been recorded breeding from 24 May to 30 August (Fig. 678).

Nests: Most nests (76%; *n* = 158) were built in trees (deciduous, 66%; coniferous, 6%); the rest were placed in shrubs (24%). Deciduous nest trees identified to type (*n* = 81) included orchard fruit trees (57%; apple, pear, peach, plum, cherry), maple (27%; vine, bigleaf), Pacific crab apple (6%), alder, birch, cottonwood, Lombardy poplar, walnut, and horse-chestnut. Coniferous trees included various fir, spruce, and pine species. The 2 most often used shrubs included wild rose (5) and elderberry (4), followed by spirea, saskatoon, big sagebrush, ocean-spray, red-osier dogwood, salmonberry, hazelnut, lilac, and raspberry canes.

Within trees and shrubs, the nest was usually placed in the fork of a branch (57%; *n* = 87) or was saddled on a larger branch (31%). Seven percent were described as being near the end of a branch and 5% were near the top of the tree or shrub.

Figure 684. Most American Goldfinch nests in British Columbia were placed in the fork of a branch and consisted of plant down and spider webbing (Creston, 27 July 1998; Linda M. Van Damme).

In British Columbia, nests (Fig. 684) were small, compact, densely woven cups constructed primarily of grasses and lined with fine grasses, mammalian hair (horse, dog, sheep wool), plant down (thistle, cottonwood, milkweed, dandelion, willow), and human artifacts such as string and mattress stuffing. Many nests contained a variety of other materials, including mosses, plant stems, spider webbing, lichens, leaves, twigs, and artificial fibres.

The heights of 155 nests ranged from ground level to 12.0 m, with 62% between 1.8 and 4.0 m.

Eggs: Dates for 169 clutches ranged from 24 May to 28 August, with 55% recorded between 23 June and 12 July. Sizes of 107 clutches ranged from 1 to 6 eggs (1E-13, 2E-10, 3E-11, 4E-21, 5E-28, 6E-24), with 68% having 4 to 6 eggs (Fig. 685). The incubation period is normally between 12 and 14 days (Middleton 1993).

The egg-laying period was longer in the interior than on the coast (Fig. 686). This may have some significance for the problem of declining goldfinch numbers on the coast but not in the interior. By mid-July on the coast, most of the egg laying appears to have taken place, which means that most of the goldfinch nests are potential victims of the cowbird. In the interior, however, the goldfinch continues laying beyond mid-July, thereby ensuring that about 20% of its clutches are laid when few adult cowbirds are around to parasitize the goldfinch nests (see also "Brown-headed Cowbird Parasitism"). Note also that the 2 regions are populated by separate subspecies.

Young: Dates for 113 broods ranged from 6 June to 30 August, with 52% recorded between 9 July and 2 August. Sizes of 70 broods ranged from 1 to 6 young (1Y-7, 2Y-17, 3Y-16, 4Y-18, 5Y-8, 6Y-4), with 73% having 2 to 4 young. Middleton (1993) questions the value of the nestling period calculations because the timing of nest departure by the nestlings is determined by many factors. Nestlings may leave undisturbed nests anytime after day 12.

Nests in the interior tended to hold young later in the season than coastal nests (Fig. 686), although a nest with 3 eggs collected at Sea Island near Vancouver on 28 August suggests that nests at the coast can contain young as late as 8 September.

Brown-headed Cowbird Parasitism: In British Columbia, 22% of 146 nests found with eggs or young were parasitized by the cowbird. Parasitism on the coast was 23% (*n* = 65); in the interior, it was 21% (*n* = 81). These rates are high; the incidence of parasitism in other studies ranges from 0% to just over 9% (Middleton 1993). There were also 2 cases of American Goldfinch feeding fledged cowbirds in British Columbia.

Middleton (1993) reports that early nests may be commonly parasitized by cowbirds in some areas but that later nests (e.g., after 16 July in Ontario) may be immune because the cowbird's egg-laying season has already ended. The situation is probably similar in British Columbia, as most cowbirds have completed their egg laying by mid-July (see

Figure 685. Clutch of 4 American Goldfinch eggs on lining of plant down and dried plant stems (Salmon Arm, 31 July 1993; R. Wayne Campbell).

Brown-headed Cowbird). Unlike in the east, however, where "the breeding seasons of these species barely overlap" (Middleton 1993), they appear to overlap significantly in British Columbia, especially on the coast (see also "Nest Success").

The American Goldfinch seldom rears a cowbird to independence. While the cowbird usually hatches, the mainly granivorous diet of the goldfinch appears to retard the growth of the cowbird, and most cowbird nestlings die before they fledge, usually within 4 days of hatching (Middleton 1991). Nevertheless, cowbird parasitism results in the loss of some goldfinch eggs and nestlings that are deprived of food, which can lead to local nest failure and affect the reproductive success of the goldfinch. While most of the American Goldfinch population appears to be unaffected by cowbird parasitism (Middleton 1993), this may not be the case for the coastal population in British Columbia, and further study is warranted.

Nest Success: Of 37 nests found with eggs and followed to a known fate, 15 produced at least 1 fledgling, for a nest success rate of 41%. Nest success on the coast was 31% (n = 13); in the interior, it was 46% (n = 24). For a possible explanation of the lower nest success rate on the coast, see "Eggs" and "Brown-headed Cowbird Parasitism" above.

REMARKS: Four subspecies of the American Goldfinch have been described, 2 of which occur in British Columbia: *C. t. pallidus,* which is distributed throughout the interior of the province, and *C. t. jewetti,* which occurs along the coast (American Ornithologists' Union 1957; Middleton 1993).

As Middleton (1993) notes, the long-term future of this species would benefit from the conservation of riparian habitats along river bottoms and floodplains and early successional habitats. In addition, appropriate landscaping could enhance the urban environment for the goldfinch, as would less rigorous maintenance of "weed-free" areas.

Mitchell (1959) had 3 observations of the American Goldfinch from Tahsis, on the west coast of Vancouver Island, but he noted them as hypothetical because a positive identification was not made. We have excluded these records from this account. Laing (1942) notes that the goldfinch "is known to some residents of the Bella Coola valley," but details are lacking.

In the interior, we have observation records on file from McBride in the Southern Interior Mountains and from Crooked River and Sukunka River in the Sub-Boreal Interior; these records lack details, however, and have been excluded from this account.

Two American Goldfinches banded near Sutter, in northern California, and near Corvallis, Oregon, respectively, were recovered in the lower Fraser River valley.

For a summary of life-history information for the American Goldfinch in North America, see Middleton (1993).

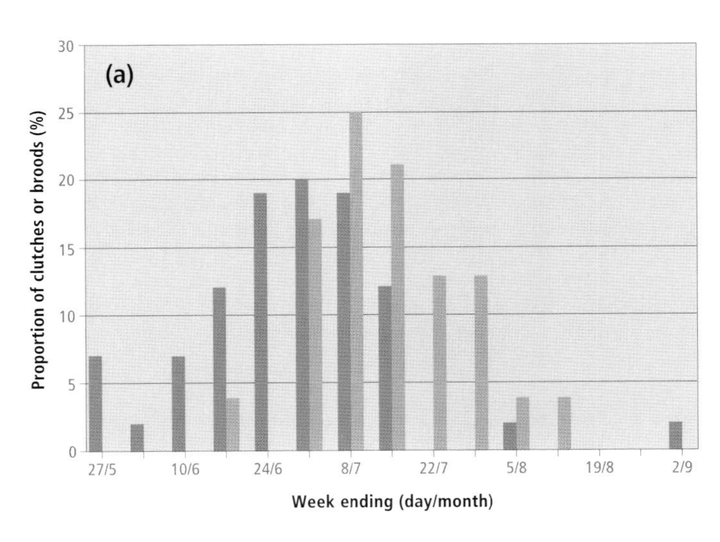

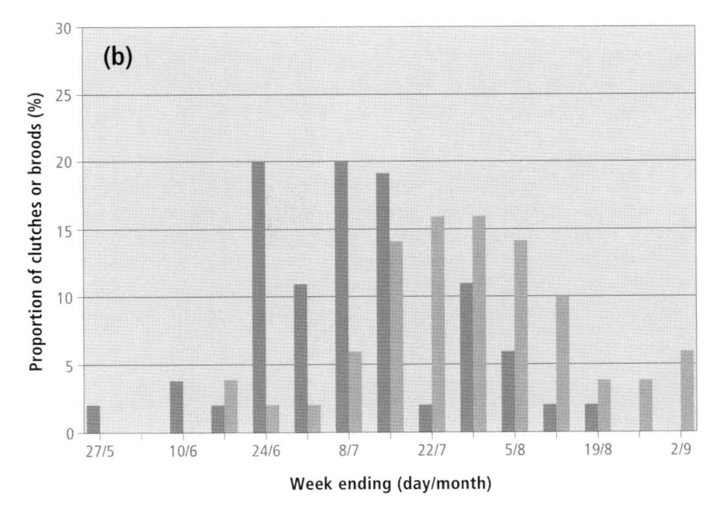

Figure 686. Weekly distribution of the proportion of clutches (dark bars) and broods (light bars) for the American Goldfinch in the (a) Georgia Depression and (b) Southern Interior ecoprovinces of British Columbia. The figures are based on the week eggs or young were found in the nest. Although nesting begins about the same time at both the coast and the interior, nesting at the coast is not as protracted as nesting in the interior; on the coast, 97% of the egg laying takes place before mid-July, compared with 78% in the interior.

NOTEWORTHY RECORDS

Spring: Coastal – Saltspring Island 17 Mar 1973-30; Ucluelet 15 May 1972-15; Lennard Island 1 Apr 1977-5, first of year, 3 Apr 1977-11, most stayed for 1 week, possible nesters, 12 May 1980-50, flying around garden; Sea Island 15 May 1984-130 on airport fence; Burnaby 24 May 1965-4 eggs; New Westminster 25 May 1944-6 eggs, incubation fresh; Sumas 25 Apr 1976-100; Skagit River valley 24 Apr 1971-8; False Creek 8 Apr 1865-50; Burnaby Mountain 11 May 1975-100; Hope 6 Apr 1929-1, 19 Apr 1926-24, half in mature plumage (Thacker 1944); Comox 7 May 1937-100, considerable migration (Pearse 1946); nr Port Neville 22 May 1975-6; Port Hardy 6 May 1938-4, 10 May 1940-12, spring arrival; Guise Bay 15-19 Apr 1980-4; Green Island (Dundas Island) 24 and 26 Apr 1978-1, first spring sighting. **Interior** – Keremeos 24 May 1961-5 eggs; Okanagan Falls 16 May 1964-100; Balfour to Waneta 16 May 1981-64 on bird count; Brouse 20 Mar 1983-25; Creston 11 Apr 1984-50; Richter Pass 12 Apr 1969-6; Kamloops 21 May 1959-4; Cache Creek to Clinton 2 May 1944-10; Barriere 27 May 1984-2; 103 Mile Lake 16 Apr 1941-8; 105 Mile Lake 17 Apr 1941-4; Williams Lake 28 Apr 1980-20 feeding on birch catkins; Riske Creek 28 Apr 1987-1, early arrival date; Kimsquit River 28 Apr 1985-1; 3 km se McBride 12 May 1985-1; Crooked River Park 7-11May 1981-5; nr Prince George 16 May 1997-1 at feeder (BC Photo 1760); Nig Creek 27 May 1922-1 in song (Williams 1933a).

Summer: Coastal – River Jordan 21 Jun 1982-2, on wires at e end; Victoria area 31 Jul 1972-1 egg; Saanich 15 Jun 1981-150; Central Saanich 28 Aug 1973-150; Long Beach (Pacific Rim National Park) 2 Jun 1968-100, birds are everywhere along edge of sand and thickets; Sea Island 15 Jun 1966-80, 20 Jun 1966-200, 3 Aug 1963-500, 15 Aug 1966-1,200, 24 Aug 1965-1,000, all at Vancouver International Airport, 28 Aug 1951-3 eggs (UBC 1904); Tsawwassen Beach 27 Aug 1960-200; Crescent Beach 16 Jun 1966-2 nestlings; around Agassiz 6 Jun 1971-39; Pitt Meadows 12 Jul 1970-100; Harrison Hot Springs 6 Aug 1990-25; Little Qualicum River 6 Aug 1982-3 nestlings on estuary; Sayward 15 Aug 1935-50; Cluxewe River 18 Aug 1991-1 on estuary (Dawe et al. 1994); Loughborough Inlet 17 Aug 1936-2; Queen Charlotte City 2 Jun 1990-1 (Siddle 1990c); Masset 18 Jun 1938-1 (RBCM 10215); Kitimat 17 Jun 1974-6 (Hay 1976); Green Island (Dundas Island) 7 to 10 Aug 1978-1. **Interior** – Keremeos 6 Jun 1961-4 nestlings; Oliver 6 Aug 1989-110; Richter Pass 14 Jul 1969-1; Midway 6 Jun 1929-1 (NMC 23338); nr Creston 28 Aug 1980-3 nestlings and 1 fledgling sitting on branch beside nest; Naramata 30 Aug 1972-2 nestlings, adults feeding young; nw confluence of Summit Creek and Kootenay River (Creston) 8 Jul 1989-6 eggs, female incubating; n Sparwood 27 Aug 1983-2; Kaslo 15 Jul 1975-1 Brown-headed Cowbird fledgling about 3 times larger than American Goldfinch host; Coldstream 18 Aug 1963-3 eggs; Adams Lake 16 Jul 1980-15 at s end; Chase 17 Jun 1973-10; Onyx Creek 29 Jul 1973-10 (Cannings, S.G. 1973); Revelstoke 12 Jun 1994-1 fledgling, 20 Jun 1980-80 feeding with Black-capped Chickadees, Brown Creepers, and Red-breasted Nuthatches; Chilcotin River (Riske Creek) 16 Jul 1988-1 fledgling being fed by adult at Wineglass Ranch; Williams Lake 7 Aug 1972-many fledglings being fed by adults in garden, 18 Aug 1970-1; Sugarcane (Williams Lake) 27 Jul 1959-2 eggs.

Breeding Bird Surveys: Coastal – Recorded from 20 of 27 routes and on 64% of all surveys. Maxima: Albion 26 Jun 1977-44; Pitt Meadows 9 Jul 1977-44; Chilliwack 25 Jun 1992-44; Seabird 6 Jun 1971-39; Nanaimo River 15 Jun 1974-34. **Interior** – Recorded from 30 of 73 routes and on 32% of all surveys. Maxima: Lavington 20 Jun 1976-34; Grand Forks 28 Jun 1992-32; Canford 22 Jun 1982-22.

Autumn: Interior – Williams Lake 10 Sep 1984-6 nestlings being fed thistle seeds by adults, 12 Nov 1987-2 at Scout Island; Canim Lake 4 Sep 1960-2 at campsite; Golden 1 Sep 1975-12; Salmon Arm 8 Sep 1985-10; Enderby 3 Sep 1960-12; Glenbank 10 Nov 1979-100; Castlegar 15 Nov 1981-50; Vaseux Lake 19 Nov 1975-30; West Bench (Penticton) 6 Nov 1976-30; Pend-d'Oreille River 26 Oct 1983-3 nr Four Mile Creek; Nakusp 13 Nov 1983-75; Richter Pass 19 Oct 1962-35. **Coastal** – Kitimat River 10-26 Sep 1974-10 (Hay 1976); Kitimat 4 Oct 1974-3 on golf course; Alta Lake 12 Sep 1941-1; Sunshine Coast 2 Nov 1986-5; Campbell River 20 Sep 1963-3 fledglings with adult; Hope 11 Oct 1934-late record (Thacker 1944); Laidlaw 30 Nov 1974-1; Pitt Meadows 15 Sep 1967-325; e Delta 12 Sep 1983-300, by Boundary Bay; Sea Island 8 Sep 1964-900 at Vancouver International Airport; Witty's Lagoon (Metchosin) 24 Sep 1975-2.

Winter: Interior – Williams Lake 2 Jan 1989-4, 18 Feb 1984-6, 4 Dec 1987-5 at water outflow from pumps; Revelstoke 2 Feb 1989-2 in downtown area, 28 Dec 1989-1; Nakusp 27 Jan 1984-30, 18 Dec 1982-200; Lavington 2 Jan 1955-30, 9 Dec 1973-120 in mixed flock with Common Redpolls; Okanagan Landing 3 Dec 1933-200; New Denver 5 Dec 1982-100; Nelson 5 Jan 1986-30, 14 Dec 1980-300 in trees with Common Redpolls and Pine Siskins; nr Nelson 22 Feb 1984-60 in orchard; Castlegar 5 Jan 1975-95 feeding on cedar seeds and female catkins of paper birch, 31 Dec 1978-10 feeding in birch trees with Common Redpolls; Fruitvale 30 Dec 1980-30; 5 km e Grand Forks 12 Jan 1980-120 feeding on weed seeds. **Coastal** – Vancouver 13 Jan 1979-130 at Stanley Park; Boundary Bay 29 Dec 1979-125; Burnaby 14 Feb 1971-200; Lulu Island 27 Jan 1971-200; Duncan 30 Jan 1971-150 at sewage lagoon; Saanich 26 Dec 1966-105.

Christmas Bird Counts: Interior – Recorded from 16 of 27 localities and on 58% of all counts. Maxima: Salmon Arm 19 Dec 1992-539; Oliver-Osoyoos 28 Dec 1988-378; Shuswap Lake Provincial Park 21 Dec 1982-300 (Howie 1983). **Coastal** – Recorded from 16 of 33 localities and on 37% of all counts. Maxima: Vancouver 16 Dec 1979-722; Ladner 29 Dec 1979-400; White Rock 29 Dec 1991-296.

Evening Grosbeak

Coccothraustes vespertinus (Cooper)

EVGR

Figure 687. The Evening Grosbeak has a complex status in British Columbia. It is considered an irruptive migrant whose abundance is often associated with wild and ornamental seed and fruit crops. The male (a) is more brightly coloured than the less yellow female (b) (Victoria, 22 December 1989; Tim Zurowski).

RANGE: Breeds from northeastern British Columbia, northern Alberta, central Saskatchewan, southern Manitoba, central Ontario, southern Quebec, New Brunswick, Prince Edward Island, Nova Scotia, and Newfoundland south in the mountains to central California, west-central and eastern Nevada, central and southeastern Arizona, and southern New Mexico; in the Mexican highlands to Michoacán, Mexico, Puebla, and west-central Veracruz; and, east of the Rocky Mountains, to southwestern South Dakota, north-central and northern Minnesota, northern Wisconsin, central Michigan, southern Ontario, northern New York, and central Massachusetts. Winters throughout the breeding range and south, sporadically, to southern California, southern Arizona, Oaxaca, western and central Texas, the northern portions of the Gulf states, Georgia, and South Carolina, and casually to the Gulf coast and central Florida.

STATUS: May be scarce or very numerous in many portions of the province at any given season.

On the coast, an *uncommon* to *very common* migrant and winter visitant and *uncommon* to *common* summer visitant in the Georgia Depression Ecoprovince; in the Coast and Mountains Ecoprovince, a *common* to *abundant* migrant and summer visitant in southern portions of the Southern Mainland Coast, *uncommon* to *very common* there in winter; *rare* to *uncommon* migrant and visitant on Western Vancouver Island (*casual* in winter), northern portions of the Southern Mainland Coast, and the Northern Mainland Coast; *accidental* on the Queen Charlotte Islands.

In the interior, a *common* to *abundant* summer and winter visitant and *common* to *very common* migrant in the Southern Interior Ecoprovince; *uncommon* to *abundant* migrant and winter visitant and *uncommon* to *fairly common* summer visitant in the Southern Interior Mountains and Central Interior ecoprovinces; *uncommon* to *fairly common* migrant and summer visitant and *uncommon* to locally *very common* winter visitant in the Boreal Plains and southern portions of the Sub-Boreal Interior ecoprovinces; *rare* migrant and summer visitant and *very rare* in winter in the Taiga Plains Ecoprovince; *very rare* in summer in the Northern Boreal Mountains Ecoprovince and northern portions of the Sub-Boreal Interior.

Breeds.

NONBREEDING: The Evening Grosbeak (Fig. 687) occurs across much of the province, where it is associated with a wide variety of forested habitats and human-modified landscapes. It is

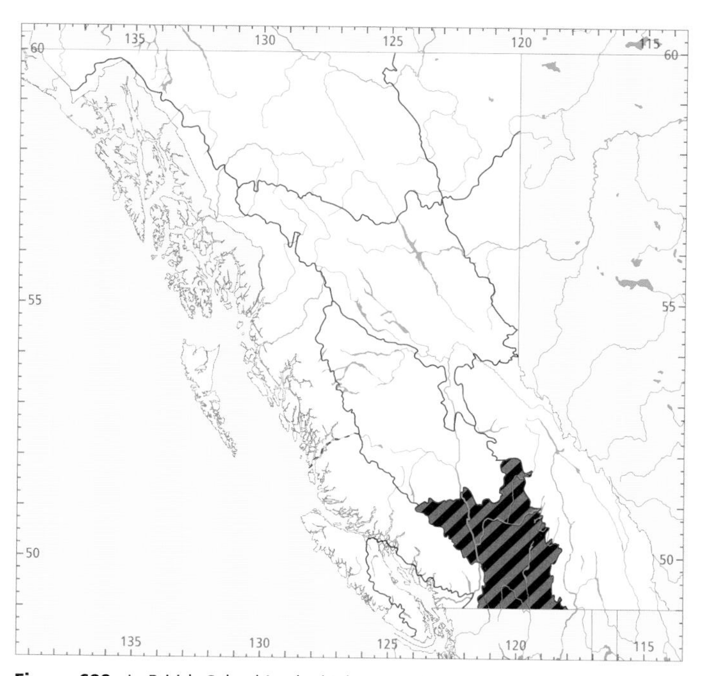

Figure 688. In British Columbia, the highest numbers for the Evening Grosbeak in winter (black) and summer (red) occur in the Southern Interior Ecoprovince.

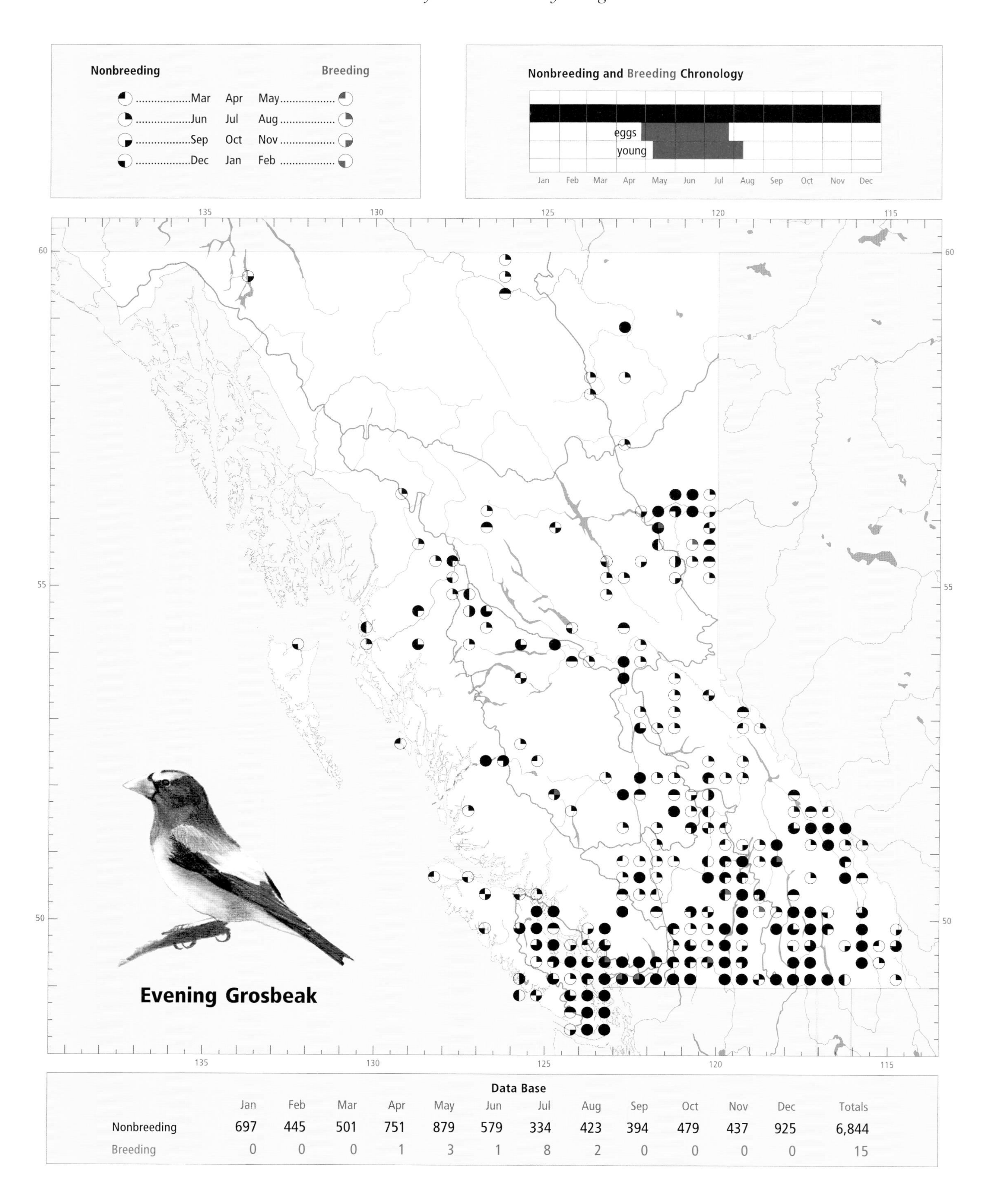

Data Base

	Jan	Feb	Mar	Apr	May	Jun	Jul	Aug	Sep	Oct	Nov	Dec	Totals
Nonbreeding	697	445	501	751	879	579	334	423	394	479	437	925	6,844
Breeding	0	0	0	1	3	1	8	2	0	0	0	0	15

Figure 689. In spring, the Evening Grosbeak is associated with areas of native bigleaf maple on the coast and introduced Manitoba maple in the interior, which provide large winged seeds and new leaves as food (Victoria, 10 April 1996; R. Wayne Campbell).

Figure 690. In autumn, berries of many tree species, including arbutus on the south coast, provide food for migrating Evening Grosbeaks (Mount Tolmie, Saanich, 10 October 1996; R. Wayne Campbell).

widespread throughout the southern and central portions of the interior. North of latitude 56°N, it occurs mainly east of the Rocky Mountains and is very scarce in the mountainous northwestern region of the province. It has been reported only once at Atlin. On the coast, it is found chiefly on southeastern Vancouver Island and south of Bute Inlet along the mainland coast. It is sparsely distributed on western and northern Vancouver Island and along much of the central and northern portions of the mainland coast, except for the main valleys of the Bella Coola and Skeena rivers. It has been reported once on the Queen Charlotte Islands. The highest numbers in winter occur in the Southern Interior (Fig. 688).

The Evening Grosbeak occurs from sea level to 1,400 m elevation on the coast and from 280 to 1,830 m elevation in the interior. Its habitat preferences during the nonbreeding period are poorly known, but its distribution appears to be closely associated with areas of berry and seed production. On the coast, this grosbeak frequents areas with Douglas-fir, bigleaf maple (Fig. 689), paper birch, arbutus (Fig. 690), red-osier dogwood, vine maple, black hawthorn, and Pacific crab apple, where it forages for seeds and berries. Around areas of human settlement, it is attracted to introduced trees that produce large, winged seeds, especially Manitoba maple (*Acer negundo*), Chinese elm (*Ulmus paravifolia*), and American elm (*U. americana*), as well as a variety of berry-producing ornamental plants. It is also a frequent visitor to bird feeding stations, especially those stocked with sunflower seeds. In the interior, it frequents forests of Douglas-fir, ponderosa pine, white, Engelmann, or hybrid spruce, trembling aspen, black cottonwood, or balsam poplar, and understoreys, or forest openings, with Douglas maple, choke cherry, pin cherry, saskatoon, or red-osier dogwood. In urban areas, it is readily drawn to bird feeding stations, and seed- or berry-producing ornamental plants. It also frequents gardens and orchards, where it forages on domestic fruits, including apples, crab apples, and cherries. In the Okanagan valley, it favours the large-seeded berries of the introduced Russian olive (Cannings et al. 1987).

Like many of the cardueline finches, the Evening Grosbeak is highly nomadic and erratic in occurrence throughout its North American range. In British Columbia, numbers often fluctuate greatly between seasons and from year to year. The reasons for these fluctuations are poorly understood, but are probably related to food availability and recent breeding patterns (see BREEDING).

The onset of spring migration in British Columbia is variable and difficult to ascertain because of the presence of overwintering birds (Figs. 691 and 692). In the Georgia Depression, spring movements appear to begin in late March or early April, and reach a small peak in numbers by the last week of April; numbers decline through May and drop sharply by the end of that month. Elsewhere on the coast, spring movements are small and poorly documented, but likely occur in April and May. In southern portions of the interior, spring migration is first noted with the gradual departure of wintering birds from the valleys and the arrival of migrants from other regions. In the Okanagan valley, in the Southern Interior, wintering birds start to leave in mid-April, about the same time that migrant flocks begin arriving from the south (Cannings et al. 1987). By early June, most grosbeaks have left the valley floors of the Southern Interior and Southern Interior Mountains and moved to higher-elevation breeding areas. In the Central Interior, winter numbers are augmented by the arrival of spring migrants in late April, with peak numbers occurring in early May (Fig. 692). In the Sub-Boreal Interior, spring movements are also poorly documented but probably occur mostly throughout May (Fig. 692). East of the Rocky Mountains, spring migrants begin to arrive in the Boreal Plains in late April, and further north they arrive in the Taiga Plains in early May. Peak numbers occur in both ecoprovinces during mid-May, followed by a breeding dispersal in early June. Migrants arriving in the northeast probably move into the province from Alberta, which experiences a spring movement from mid-March to mid-May (Pinel et al. 1993).

Following the breeding season, small flocks of adults with fledglings become evident in most regions by late July or early August. In the Okanagan valley, grosbeaks begin returning to the valley bottoms in early August, with increasing numbers and flock sizes building into September (Cannings et al. 1987). Autumn movements are poorly documented, but probably occur in the northern regions in September and continue into October in southern regions (Figs. 691 and 692). Birds remaining in the province after late October probably overwinter, provided that food reserves remain adequate.

In winter, the Evening Grosbeak has a widespread distribution at low elevations throughout much of its range in British Columbia. It occurs regularly in valleys of the Southern Interior and in southern valleys of the Southern Interior Mountains. In the central and northeastern portions of the interior, its distribution appears to be highly localized around human settlements, especially near bird feeding stations. West of the Rocky Mountains, it has been recorded in winter as far north as Mackenzie in the Sub-Boreal Interior. East of the Rocky Mountains, it frequently winters in the Peace Lowland of the Boreal Plains. The northernmost winter occurrences are from Fort Nelson in the Taiga Plains. On the coast, most wintering occurs in the Georgia Depression and along southern portions of the Southern Mainland Coast. Elsewhere on the coast, there are scattered winter records from Bamfield, Tahsis, Sayward, and Port Hardy on Vancouver Island and from Bella Coola, Kitimat, Prince Rupert, Terrace, and Hazelton in the central and northern portions of the mainland coast. There is a single winter record from Masset, on the Queen Charlotte Islands.

Winter populations of grosbeaks in British Columbia show large annual fluctuations and often demonstrate a biennial rhythm pattern (Fig. 693). Bock and Lepthien (1976) analyzed Christmas Bird Count data for 10 boreal seed-eating

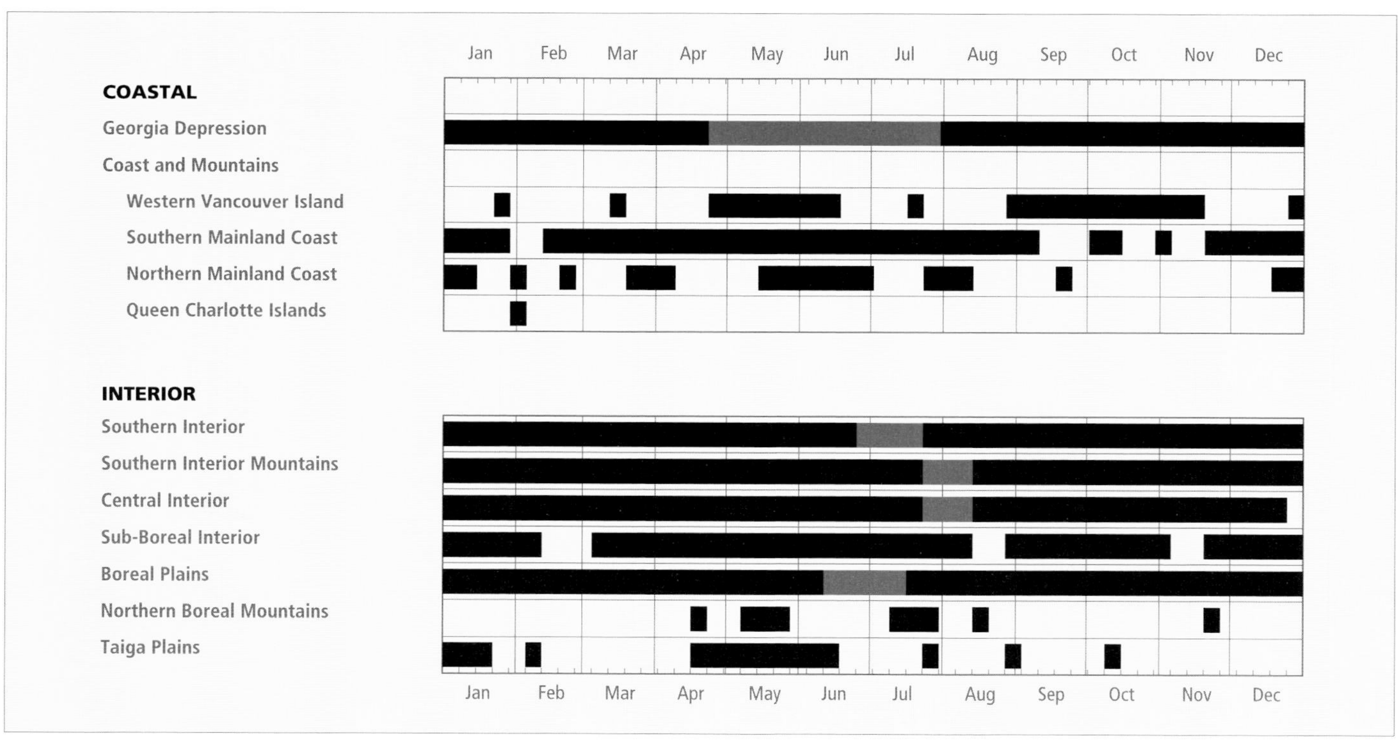

Figure 691. Annual occurrence (black) and breeding chronology (red) for the Evening Grosbeak in ecoprovinces of British Columbia. Records are shown for the week in which they occurred.

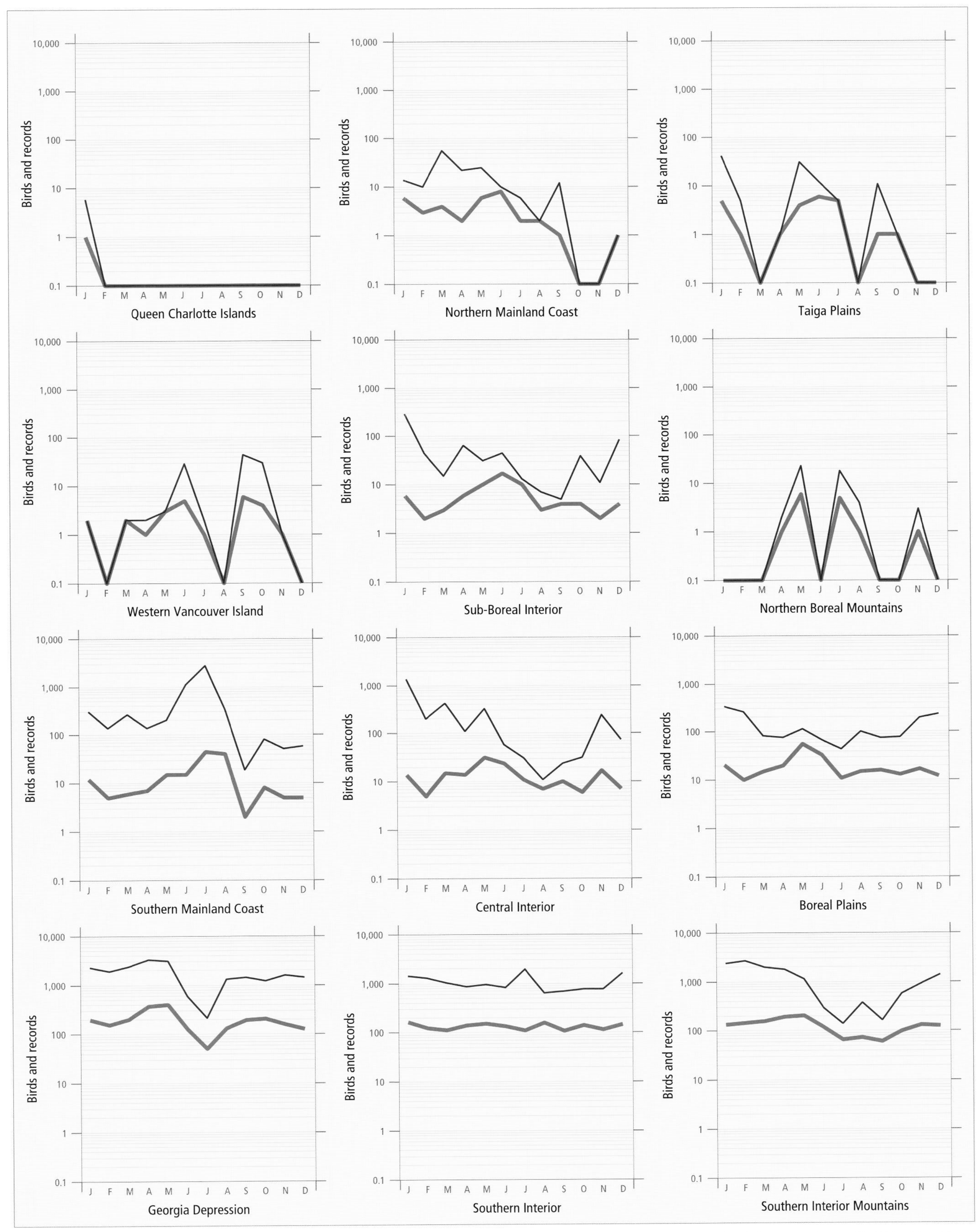

Figure 692. Fluctuation in total number of birds (purple line) and total number of records (green line) for the Evening Grosbeak in ecoprovinces of British Columbia. Christmas Bird Counts, Breeding Bird Surveys, and nest record data have been excluded.

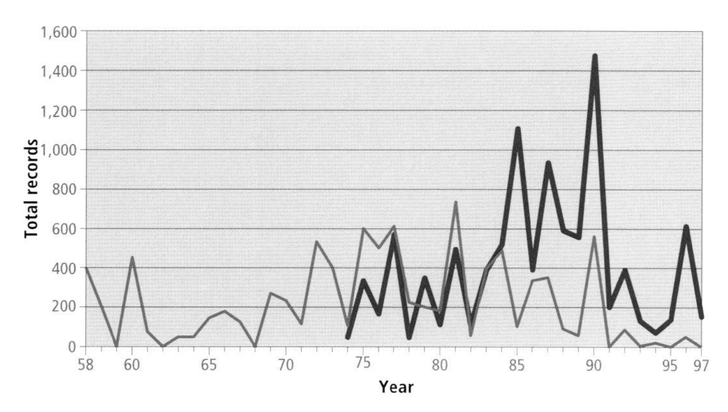

Figure 693. Fluctuations in Evening Grosbeak numbers from Christmas Bird Counts for the Georgia Depression Ecoprovince (green line) (Ladner, Vancouver, and Victoria) for the period 1958 through 1997 and the Southern Interior Ecoprovince (purple line) (Penticton, Vernon, and Shuswap Lake) for the period 1974 through 1997. Fluctuations on the coast and in the interior for the period 1974 through 1997 are frequently synchronous, suggesting large regional population shifts into and out of southern British Columbia. The general biennial rhythm of these fluctuations is characteristic of many boreal seed-eating birds. The sharp drop in numbers after 1990 coincides with the decline of a long-term infestation of western spruce budworm in the Southern Interior Ecoprovince.

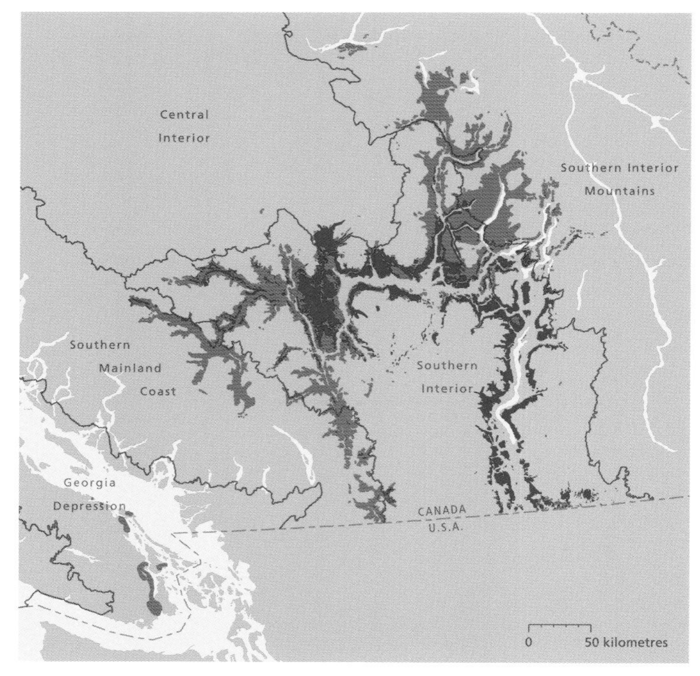

Figure 694. Defoliation history of the western spruce budworm in British Columbia from 1909 to 1992 (brown = 1 to 9 occurrences; red = 10 to 19 occurrences) (adapted from Parfett et al. 1994). More than 65% of all Douglas-fir forests attacked by this insect fall within the Southern Interior Ecoprovince, where the highest summer densities of the Evening Grosbeak occur. Areas with the greatest frequency of repeat infestation (red) occur mainly along narrow elevational bands of the Okanagan and Thompson valleys. Smaller, less frequent outbreaks have occurred on the south coast, including the Fraser Canyon area, the Pemberton and Skagit valleys, and southern Vancouver Island. In the Southern Interior Mountains Ecoprovince, budworm outbreaks have occurred in the vicinity of Adams Lake, Wells Gray Park, and Quesnel Lake.

passerines in North America and found that 8 species, including the Evening Grosbeak, showed regular synchronous fluctuations. These fluctuations are believed to be related to the biennial cycles of various seed crops.

The Evening Grosbeak is present throughout the year in the interior and in southern portions of the coast (Fig. 691).

BREEDING: The breeding distribution of the Evening Grosbeak in British Columbia is poorly known. Only 6 nests have been documented in the province, but many observations of fledglings with attending adults suggest that this grosbeak is a widespread breeder. It is known to breed in the Boreal Plains of the northeast, in the Central Interior and Southern Interior Mountains south of latitude 52°N, and in the Southern Interior. On the south coast, breeding has been confirmed for the mainland portion of the Georgia Depression but likely also occurs on southeastern Vancouver Island.

The Evening Grosbeak reaches its highest numbers in the summer in the Southern Interior (Fig. 688). An analysis of Breeding Bird Surveys in British Columbia for the period 1968 through 1993 could not detect a net change in numbers on either coastal or interior routes. Nationally, the numbers increased at an average annual rate of 2.3% during the period 1966 to 1979 ($P < 0.5$) but have declined at an average annual rate of 5.2% ($P < 0.01$) since 1980; continentally, the numbers appear stable (Sauer et al. 1997). Considering the erratic, wide-ranging breeding pattern of this species, Breeding Bird Survey data should be interpreted cautiously, especially for small geographic areas or over short time frames. Erskine (1977) observes that it is difficult to determine their breeding status during these surveys.

Breeding season numbers in British Columbia show considerable annual variation, particularly in the northern half of the bird's breeding range. In the Central Interior, Pojar (1995) detected Evening Grosbeaks in only 1 of 2 survey years. In the second year of her study, grosbeaks were found in mature and "old" trembling aspen forests or in mixed conifer-trembling aspen forests; grosbeak densities in these habitats ranged from 1.2 to 7.6 birds per 10 ha. During a 6-year study of trembling aspen forests in the Boreal Plains, Evening Grosbeaks were also detected irregularly and in small numbers

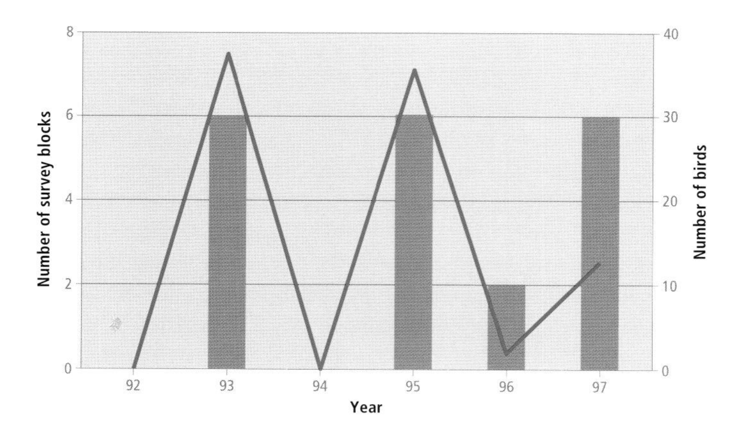

Figure 695. Number of Evening Grosbeaks observed (line) and number of survey blocks (bars) in which grosbeaks were detected (bars) in trembling aspen forests in the Boreal Plains Ecoprovince from 1992 to 1997 (adapted from Merkins and Booth 1998).

Figure 696. The Evening Grosbeak is a habitat generalist in British Columbia and uses a variety of coniferous forests in the Central Interior Ecoprovince (near Gang Ranch, 8 June 1997; R. Wayne Campbell).

(Merkins and Booth 1998). From 1992 to 1997, grosbeaks were consistently recorded in the odd years and were usually undetected in the even years, possibly reflecting natural cycles in seed crops or insect abundance (Fig. 695).

Throughout its North American range, breeding appears to be closely associated with outbreaks of forest-defoliating insects, especially the eastern spruce budworm (*Choristoneura fumiferana*) and western spruce budworm (*C. occidentalis*). In Idaho, Montana, Oregon, and Washington, Langelier (1983) found that grosbeak numbers varied greatly between similarly structured forests, but that its breeding abundance was closely linked to western spruce budworm densities. This grosbeak was most abundant in forests where budworms were plentiful and rare where budworm densities were low. In southern British Columbia, western spruce budworm distribution also appears to be an important predictor of Evening Grosbeak breeding patterns and densities (Fig. 694). In northern British Columbia, the eastern spruce budworm occurs in the Boreal Plains and Taiga Plains, and in the Liard Plain of the Northern Boreal Mountains (Humphreys 1995). Major budworm outbreaks have occurred in the Smith, Liard, and Fort Nelson river valleys, and probably account for the periodic summer occurrence of the Evening Grosbeak in this far northern region of the province.

In Colorado, Evening Grosbeak nesting has also been linked to outbreaks of the large aspen tortrix, *Choristoneura conflictana* (Scott and Bekoff 1991); this forest defoliator is common in trembling aspen stands of central and northeastern British Columbia (Humphreys 1995). Between 1946 and 1994, major infestations periodically occurred in aspen woodlands from about Prince George north to the Liard River.

On the coast, a single Evening Grosbeak nest was found at approximately 200 m elevation. In the interior, breeding has been reported at elevations from 490 to 915 m. The Evening Grosbeak is a habitat generalist and uses a broad range of forest types during the breeding season (Langelier 1983). In British Columbia, it apparently nests in a wide variety of coniferous (Fig. 696), mixed-wood, and deciduous forests. On the south coast, its breeding range approximates the distribution of Douglas-fir and bigleaf maple, within the Coastal Douglas-fir Biogeoclimatic Zone and the drier subzones of the Coastal Western Hemlock Zone. In the southern portion of the interior, it appears to breed in mid- to low-elevation forests within the Ponderosa Pine, Interior Douglas-fir, Montane Spruce, and Interior Cedar–Hemlock zones, wherever Douglas-fir prevails. Breeding bird studies by Catt (1991) and Davis et al. (1999) in the Engelmann Spruce–Subalpine Fir Zone of the Southern Interior Mountains suggest that this grosbeak may also breed at higher elevations. In Kootenay National Park, Catt (1991) estimated that densities ranged between 0.08 and 0.4 birds per 10 ha in mature forests of Engelmann spruce and subalpine fir. Further north in the Sub-Boreal Spruce Zone of the Central Interior and Sub-Boreal Interior, breeding likely occurs wherever Douglas-fir is found or in mixed forests of hybrid white spruce and trembling aspen. East of the Rocky Mountains, it probably breeds throughout the Boreal White and Black Spruce Zone, in forests of white spruce, trembling aspen, and balsam poplar (Fig. 697). In

Manitoba and Alberta, nesting has been recorded in urban environments (Speirs 1968a; Hoffmann and Pletz 1988); this has not yet been observed in British Columbia.

The Evening Grosbeak apparently makes little use of the black spruce–tamarack bogs found throughout the Taiga Plains and Boreal Plains, nor does it appear to utilize the expansive lodgepole pine forests of the Sub-Boreal Pine-Spruce and Montane Spruce zones in the Central Interior.

On the coast, the Evening Grosbeak has been recorded breeding from 27 April (calculated) to 26 July; in the interior, it has been recorded breeding from 16 June (calculated) to 10 August (calculated) (Fig. 691).

Nests: Only 6 nests have been reported in British Columbia. All were found in trees, evenly divided between coniferous and deciduous species. Nest tree species included Douglas-fir (2 nests), lodgepole pine, red alder, trembling aspen, and balsam poplar. Five nests were saddled on horizontal limbs and placed 1.8 to 3.0 m from the main tree stem. One was described as being positioned near "the top" of the nest tree (Speirs 1968b).

Nests were typically flattened cups with an outer shell constructed mainly of fine twigs. The lining for 1 nest consisted chiefly of fine rootlets; another was lined with grasses and smaller amounts of lichens, rootlets, and berry stems. The heights of 6 nests ranged from 5.5 to 16.0 m, with 4 nests occurring between 6.0 and 15.2 m.

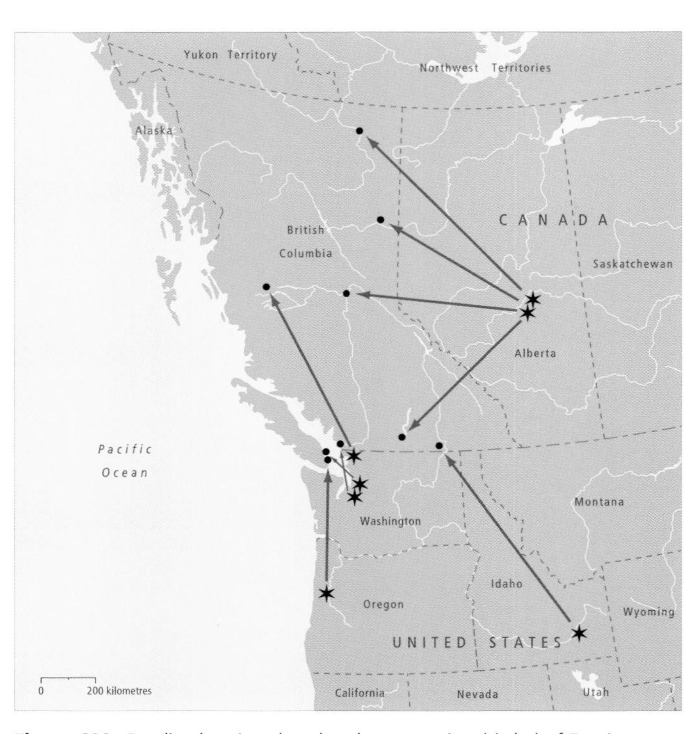

Figure 698. Banding locations (stars) and recovery sites (circles) of Evening Grosbeaks associated with British Columbia.

Figure 697. East of the Rocky Mountains, in northeastern British Columbia, the Evening Grosbeak breeds in mixed-wood forests of white spruce and trembling aspen with scattered balsam poplars (near Pink Mountain, 27 June 1998; R. Wayne Campbell).

Four Evening Grosbeak nests have been reported in Alberta, all in deciduous trees, including paper birch, balsam poplar, and *Populus* spp. (Speirs 1968a; Hoffmann and Pletz 1988). In the Pacific Northwest states of Oregon, Idaho, and Washington, Langelier (1983) reported finding 5 nests in Douglas-fir and 1 in a ponderosa pine; the average nest height was 12.8 m. In Colorado, the only region where the nesting ecology of this species has been closely studied, all nests (n = 64) were found in coniferous tree species, including ponderosa pine, Douglas-fir, lodgepole pine, and Engelmann spruce (Bekoff et al. 1987). In this last study, ponderosa pine was selected more frequently than expected based on availability; nest heights for all 4 tree species averaged 9.3 m.

Eggs: A nest with 5 eggs found on 8 July is the only record of a clutch size for British Columbia. Observations of recently fledged young suggest that nests may contain eggs as early as 27 April or as late as 26 July. Breeding records from late April through early May are unusually early for this species, which typically begins nesting from late May through June. This grosbeak lays from 2 to 5 eggs, with 3 or 4 being the most frequent clutch size (Speirs 1968a). The incubation period ranges from 12 to 14 days (Baicich and Harrison 1997). There is no evidence that this species is double-brooded, but second clutches have been recorded following nest failure (Scott and Bekoff 1991).

Young: Dates for 3 broods ranged from 8 July to 26 July. Only 1 nest was inspected closely enough to determine the number of young, and it contained 2 nestlings. A second nest

Figure 699. Evening Grosbeaks are attracted to residual road salts along the edges of highways in British Columbia. In some years, large numbers are killed by vehicular traffic. Note the carcasses lining the gravel shoulder on the highway (Manning Park, 15 July 1980; Douglas J. Wilson).

was estimated to contain between 3 and 4 young, based on nestling vocalizations (Speirs 1968b). Observations of recently fledged young suggest that nestlings can be found as early as 9 May or as late as 10 August in British Columbia.

Baicich and Harrison (1997) give the nestling period as 13 or 14 days. Young remain in the nesting area from 2 to 5 days after fledging (Scott and Bekoff 1991). The dependency period for fledged young is unknown, but several October records from British Columbia of young being fed by adults suggest that it is lengthy.

Brown-headed Cowbird Parasitism: In British Columbia, only 2 occupied nests were closely inspected, and neither contained cowbird eggs or young. In the Southern Interior, however, there were 2 records of fledged cowbirds being fed by adult Evening Grosbeaks. Hoffmann and Pletz (1988) reported this grosbeak feeding 2 cowbird nestlings in Alberta; Friedmann and Kiff (1985) also reported 2 instances of parasitism from eastern North America. The Evening Grosbeak switches from its normal seed diet to insects during the breeding season (Speirs 1968a), making it a potentially more suitable host for the cowbird than most finches.

Nest Success: Insufficient data for British Columbia. In Colorado, nest success was reported to be 54.7%; nest failures were attributed to severe weather conditions such as cold temperatures and wind, abandonment, or predation (Bekoff et al. 1989). Most nest failures in this Colorado study occurred during the incubation period.

REMARKS: The Evening Grosbeak is often placed in the genus *Hesperiphona*. The American Ornithologists' Union (1957) identifies *C. v. brooksi* as the only Evening Grosbeak subspecies breeding in British Columbia; *C. v. vespertinus* is recorded as breeding as far west as northeastern Alberta. Other authorities, however, suggest that the latter subspecies ranges as far west as the eastern slope of the Rocky Mountains, including northeastern British Columbia (e.g., Godfrey 1986; Prescott 1994; Cannings 1998). Band returns also suggest that *C. v. vespertinus* breeds in northeastern British Columbia (Fig. 698).

Evening Grosbeaks banded in Washington, Oregon, and Idaho have been recovered in southern and central British Columbia (Fig. 698). Five grosbeaks banded during the nonbreeding period in central Alberta (presumably *C. v. vespertinus*) were recovered in British Columbia, 3 during the breeding period near Fort St. John and Fort Nelson, and single birds during the nonbreeding period at Prince George and Penticton. The latter 2 returns suggest that populations of *C. v. vespertinus* and *C. v. brooksi* intermingle during the nonbreeding period in British Columbia.

Banding studies indicate that Evening Grosbeaks are geographically wide-ranging in North America but may be more philopatric in the western third of their range. Yunick (1983) reviewed 18 years of banding data for eastern North America (n = 2,637) and found that grosbeaks showed little tendency to return to the same wintering sites. In northern Utah, Balph and Lindahl (1978) reported a re-encounter of 14% of banded grosbeaks (n = 107) over 2 winters.

The small number of Evening Grosbeak nests found in British Columbia is typical of most regions in North America. Although highly conspicuous during most of the year, the Evening Grosbeak becomes very secretive during the breeding season (Speirs 1968b; Bekoff et al. 1989). This covert

Figure 700. The Evening Grosbeak is a primary avian predator of the western spruce budworm and breeds in areas where this Douglas-fir defoliator has become abundant. During severe budworm outbreaks, forest canopies have a reddish appearance resulting from the damage these insects cause to the foliage (Fraser Canyon near Gilt Creek, July 1977; Al Dawson).

behaviour, combined with the fact that nests are typically well concealed and placed high in trees, makes finding nests very challenging. These difficulties are further compounded by the erratic breeding pattern of this species. Scott and Bekoff (1991) caution against climbing nest-trees during the incubation period, as they found that this grosbeak has a low tolerance to human disturbance during that time.

In Montana, Bennetts and Hutto (1984) found that several cardueline finches, including the Evening Grosbeak, demonstrate a strong attraction to soils containing salts. This affinity for salt has also been observed in British Columbia. For example, at Birkenhead Lake Park in the Southern Interior, small flocks of Evening Grosbeaks were attracted to a site that had been spread with a salt 3 years earlier. Our data also indicate that these grosbeaks are attracted to gravel road surfaces that have been treated with calcium chloride for dust suppression. Unfortunately, this attraction to salts occasionally has tragic results. During late June and early July 1980, for example, an estimated 1,000 to over 2,000 Evening Grosbeaks were killed along Highway 3 through Manning Park (Smith 1981; Wilson 1981; Fig. 699). This mortality resulted when large numbers of grosbeaks were attracted to residual road salts found along the edge of the highway, exposing them to collisions with vehicular traffic. The unusually large concentration of grosbeaks in this general area was the result of high western spruce budworm infestations; grit and spruce budworm larvae were identified in the crops and gizzards of these grosbeaks (Wilson 1981; Wilson et al. 1995). Our data indicate that an earlier kill of approximately 500 grosbeaks had occurred along this same stretch of Highway 3, from about Snass Creek to Manning Park Lodge, in early July 1974. Wood (1982) confirmed that a major western spruce budworm outbreak had occurred in the upper Skagit River area from 1974 through 1980. Road mortalities of this nature have also been experienced in other regions of North America (Speirs 1968a).

Evening Grosbeak mortality due to salmonellosis has been reported in British Columbia (Wilson et al. 1995). In November 1992, large numbers of birds from central Vancouver Island and the lower Fraser valley, including an estimated 50 Evening Grosbeaks, succumbed to a *Salmonella* outbreak, which probably resulted from increased use of bird feeders on the west coast. The previous summer, Evening Grosbeaks had also been reported dying at feeders in Prince George and Dawson Creek, but an analysis of 11 birds was inconclusive in determining the cause of this mortality.

Across North America, Evening Grosbeaks have been observed to converge on areas where spruce budworm outbreaks occur. Torgersen and Campbell (1982) reported that this grosbeak was a primary predator of the western spruce budworm in Washington, significantly reducing the numbers of both larvae and pupae. Erskine (1977) observed that Evening Grosbeaks and Tennessee Warblers were excellent biological indicators of the eastern spruce budworm and frequently signalled this insect's presence before tree damage became apparent. In Colorado, Bekoff et al. (1987) found this grosbeak nesting in areas where western spruce budworm (Fig. 700) as well as large aspen tortrix (aspen leaf-roller) were abundant. All of these forest insects occur in British Columbia and periodically reach epidemic levels (Harris et al. 1985; Humphreys 1995). On southeastern Vancouver Island, Evening Grosbeak flocks have been observed feeding on winter moth larvae, an introduced insect pest threatening the native Garry oak. Other forest defoliators, such as the 2-year-cycle budworm (*Choristoneura biennis*), western blackheaded budworm (*Acleris gloverana*), western hemlock looper (*Lambdina fiscellaria lugubrosa*), and Bruce spanworm (*Operophtera bruceata*), also reach epidemic levels in British Columbia, but the significance of these forest insects for breeding Evening Grosbeaks is unknown.

NOTEWORTHY RECORDS

Spring: Coastal – Victoria 6 May 1970-100; Bamfield 29 Apr 1976-2; Maple Bay Mar 1987-30; Ganges Harbour (Saltspring Island) 25 Mar 1977-30; Port Alberni 3 Mar 1996-106 (Bowling 1996c); Vancouver 24 May 1980-100 at Queen Elizabeth Park; Langley 25 May 1984-4 recent fledglings fed by adult; Aldergrove 2 Apr 1967-50; Surrey 20 May 1977-1 recent fledgling; Mission 6 Apr 1968-70; Chilliwack River 5 Apr 1969-32 feeding in bigleaf maples; North Vancouver 4 Mar 1975-42 at feeder; Harrison Hot Springs 1 Mar 1988-45; Alice Lake Park 3 May 1969-75; Campbell River area 12 Mar 1973-100, Nimpkish Lake 8 May 1995-1 (McNicholl 1995); Mons 30 Apr 1969-40; San Josef Bay 22 May 1974-1; Bella Coola River 7 Mar 1976-20, feeding on maple seeds; Moore Islands 17 May 1988-1 (Rodway and Lemon 1991); Prince Rupert 8 Apr 1981-12; Kitimat 24 Mar 1966-12 (Hay 1976); Terrace 2 Apr 1966-10 (Hay 1976); Hazelton 16 May 1992-8. **Interior** – Manning Park 18 May 1963-150; Oliver 31 May 1990-100; Grand Forks 28 Apr 1984-6; Trail 21 Apr 1982-86; Balfour to Waneta 27 May 1982-67; Princeton 7 May 1979-25, foraging on ground under ponderosa pine tree; Penticton 26 Mar 1976-50; Vaseux Lake 13 May 1979-30; Cranbrook 4 Mar 1940-5 (Johnstone 1949); New Denver 5 Mar 1982-50; Wasa Lake 30 Apr 1975-10; Vernon 31 May 1993-2 fledged young with adults; ne Vernon 22 May 1972-60, in trembling aspen copse; Kamloops 12 Mar 1926-30; Invermere 13 Apr 1981-15 at feeder; Carpenter Lake 28 Mar 1975-4; Sorrento 3 Mar 1972-20; Celista 1 Apr 1948-32; Golden 26 Mar 1977-30; Kleena Kleene 4 Mar 1968-1, first of year; Williams Lake May 1985-200; 3.2 km se McBride 6 Mar 1986-1; s Cheslatta 7 May 1997-11 at feeder; 24 km s Prince George 10 Apr 1982-14 at feeder; Quick (Telkwa) 13 Mar 1983-100; Smithers 6 May 1992-40; Tupper Creek 20 May 1938-2 (Cowan 1939); Chetwynd 1 May 1965-2; Tetana Lake 17 Apr 1938-1 (Stanwell-Fletcher and Stanwell-Fletcher 1943); Dawson Creek 13 May 1974-1; Hudson's Hope 2 Mar 1984-1; Worth 12 May 1993-3, on songbird survey; Fort St. John 26 Apr 1982-22; Fort Nelson 11 May 1975-20; Liard Hot Springs 15 May 1975-11 (Reid 1975).

Summer: Coastal – Metchosin 28 Jun 1983-3 fledglings fed by adults; Saanich 20 May 1985-1 female carrying nesting material, 3 Jun 1984-48; Duncan 3 Aug 1974-20; Tofino 3 Jun 1996-9; Vancouver 18 Aug 1979-50 at Queen Elizabeth Park; Surrey 29 Aug 1978-100; Skagit River 10 Jun 1979-1,000 to 1,500 along 6.4 km length of Highway 3, 3 Jul 1974-500, killed along Highway 3, 15 Jul 1980-600 dead along Highway 3 (Wilson 1981); Qualicum 22 Aug 1981-37; West Vancouver 25 Jul 1938-3 or 4 nestlings (Speirs 1968b); North Vancouver 31 Aug 1970-3 fledged young at feeder with adults, 28 Aug 1971-100; Pitt Meadows 17 Aug 1969-52; Harrison Hot Springs 31 Aug 1975-55; Hope to Ross Lake 12 Jul 1980-1,400; Comox 8 Jun 1995-1 fledgling at feeder with adults; Cortes Island 26 Jul 1979-40; Hagensborg 1 Jul 1938-1 (RBCM 4783); Prudhomme Lake 30 Jun 1976-2 (Crowell and Nehls 1974d); Kitwanga Lake 25 Jul 1982-1; Hazelton 5 Jun 1921-1 (MVZ 42249); Cranberry Junction 23 Jul 1981-5; n Hanna Creek 14 Jun 1980-2, along Highway 37. **Interior** – 22 km e of Osoyoos 27 Jun 1981-2 adults, female building nest; Grand Forks 2 Jun 1980-1 fledgling with adults; Trail 25 Aug 1984-60; w Creston 14 Aug 1984-1 fledged young; nr Wardner 4 Jun 1938-1 female collected, large egg in oviduct (Johnstone 1949); 11 km e Princeton 12 Jul 1997-1 bob-tailed fledgling being fed by adult, 7 Aug 1992-1 adult male feeding fledged Brown-headed Cowbird; Vaseux Lake 28 Aug 1971-40 eating saskatoon berries; Schoonover Mountain 12 Jun 1913-3 at 1,070 m elevation (Anderson 1914); South Slocan 1 Aug 1983-30 in orchard; Naramata 4 Aug 1988-1 adult female feeding fledgling Brown-headed Cowbird and fledgling Evening Grosbeak; Okanagan Lake Park 2 Jun 1976-50 feeding on gravel campsite pads; Michel Creek (Sparwood) 22 Jun 1983-7 at creek mouth (Fraser 1984); Kelowna 18 Aug 1973-25; New Denver 16 Jul 1978-5; Jackass Mountain 29 Jun 1975-4 along Highway 1; 4.8 km sw Lumby 22 Jul 1973-4 recent fledglings fed by adults; 6.4 km se Brouse 5 Aug 1979-6 fledglings with 2 adult females; Nakusp 4 Jun 1982-10; 14 km sw Westwold 8 Jul 1991-5 eggs; Yalakom River 2 Aug 1969-4 fledglings with adults; Scotch Creek 20 Aug 1973-25; Revelstoke 7 Aug 1973-4 recent fledglings fed by adult; Hawk Creek 8 Jul 1970-6 (Christman 1970); Clearwater 8 Jun 1935-1 (MCZ 284707); Kleena Kleene 10 Aug 1966-4 recent fledglings with adult; Wineglass Ranch (Chilcotin River) 7 Aug 1995-4 fledged young; Lac la Hache 1 Jun 1959-2; Williams Lake 10 Aug 1989-5 fledglings with adult male at feeder; Wells Gray Park 20 Jul 1975-18, at Ray Farm; Rainbow Range 14 Aug 1982-1; nr Quesnel 6 Jul 1995-3 fledglings with adults; Indianpoint Lake 23 Jul 1929-1 (MCZ 284697); Nulki Lake 29 Aug 1945-5 (Munro and Cowan 1947); Prince George 27 Jun 1969-8, 6 Aug 1968-3 fledglings fed seeds from saskatoon berries by adult; Tatlow 1 Jun 1975-2; Blackhawk Lake 20 Jul 1997-1 fledgling fed by adult female; s of Swan Lake Park 11 Jun 1995-1 female on nest; Sunset Hill 14 Jun 1993-1, female building nest, 8 Jul 1993-2 nestlings; sw Fellers Heights 9 Jun 1993-1, female gathering twigs and flying to nest; Tetana Lake 18 Jun 1938-1 (RBCM 8257); Moberly Lake 27 Jun 1993-3 bob-tailed fledglings fed by 2 adults; Hudson's Hope 29 Aug 1979-30; Taylor 17 Jul 1983-15; e Red Creek (Peace River) 8 Jun 1986-4 observed on edge of road; North Pine 19 Aug 1984-20; Sikanni Chief River 15 Jun 1982-2, along Alaska Highway; Muskwa River 14 Jul 1998-2 downstream from Crehan Creek; Beckman Creek 13 Jun 1994-1, heard during bird survey; Fort Nelson 7 Jun 1986-5 (Johnston and McEwen 1986); Liard Hot Springs 27 Jul 1983-7 (Grunberg 1983); Smith River 25 Jul 1966-4 (Erskine and Davidson 1976).

Breeding Bird Surveys: Coastal – Recorded from 17 of 27 routes and on 22% of all surveys. Maxima: Pemberton 22 Jun 1975-197; Alberni 10 Jun 1969-20; Albion 10 Jun 1984-16. **Interior** – Recorded from 51 of 73 routes and on 46% of all surveys. Maxima: Adams Lake 28 Jun 1989-70; Summerland 22 Jun 1988-60; Grand Forks 27 Jun 1993-56.

Autumn: Interior – Atlin 20 Nov 1986-3; Fort Nelson 2 Sep 1985-11; n Fort St. John 28 Nov 1982-50 at feeder; Taylor 23 Sep 1984-34 feeding on red-osier dogwood berries; Bullhead Mountain (Hudson's Hope) 14 Oct 1979-4; Dawson Creek 19 Nov 1978-3 at feeder; Germansen Landing 21 Oct 1972-10; Mount Lemoray 13 Oct 1998-5 along Highway 97 (Sherrington 1998); Tumbler Ridge 19 Nov 1997-10; Coffin Lake 7 Nov 1979-18; Telkwa River 2 Nov 1979-3; 24 km s Prince George 31 Oct 1981-20 at feeder; s Cheslatta Lake 13 Nov 1997-40 at feeder; Williams Lake 17 Nov 1982-14; Kleena Kleene 13 Oct 1959-6 (Paul 1964); nr Bridge Lake 8 Sep 1960-5; Revelstoke 29 Oct 1984-40; Scotch Creek 10 Sep 1962-72; Invermere 19 Nov 1977-25; Vernon 28 Nov 1981-50; Brouse 1 Sep 1984-28; Okanagan Landing 5 Sep 1929-70; Wasa 25 Oct 1982-25; Sparwood 23 Oct 1984-6; Penticton 2 Nov 1973-30; Cranbrook 22 Oct 1938-14 (Johnstone 1949); South Slocan 27 Nov 1987-80; West Bench (Penticton) 21 Sep 1975-40; Apex Mountain (Penticton) 20 Oct 1973-1 (1,830 m elevation); Pend-d'Oreille River 11 Oct 1969-10; Trail 30 Nov 1980-25; Flash Lake (Manning Park) 8 Oct 1978-80. **Coastal** – Kitimat 21 Sep 1974-12 (Hay 1976); Bella Coola 25 Nov 1989-3 at feeder; Cortes Island 12 Sep 1977-12; n Campbell River 28 Sep 1979-25; Union Bay 7 Nov 1987-40; Coquitlam 15 Sep 1979-28; North Vancouver 3 Sep 1980-45; Vancouver 16 Nov 1942-60; Ross Lake 4 Oct 1971-30; Chilliwack 29 Nov 1889-1 (MVZ 105262); Mission 25 Sep 1972-38; Surrey 30 Sep 1965-50; Blackie Spit (Surrey) 3 Sep 1977-80; Reifel Island 10 Sep 1980-25 eating crab apples; Tofino 14 Sep 1984-21; Duncan 20 Sep 1971-15; Bamfield 10 Sep 1977-2; Victoria 13 Oct 1973-44; Colwood 14 Sep 1978-20; Goldstream Park 13 Nov 1987-110; Sheringham Point 30 Nov 1986-42; River Jordan 5 Oct 1986-13.

Winter: Interior – Fort Nelson 18 Jan 1975-30; North Pine 23 Dec 1984-50 at feeder; Montney 27 Dec 1986-60 at feeder (Johnston and McEwen 1987a); Quick (Telkwa) 17 Feb 1987-40 at feeder; 24 km s Prince George Dec 1986-80, mostly males; Quesnel 7 Feb 1979-37, at feeder eating sunflower seeds; Williams Lake 2 Jan 1989-638, winter bird count; Riske Creek 10 Jan 1986-42; Golden 27 Feb 1977-30; Little Fort 22 Feb 1998-23 at feeder; Castledale 27 Feb 1977-250; Sorrento 23 Jan 1973-108; Windermere 10 Feb 1983-37, feeding in crab apple tree; Nakusp 30 Dec 1984-80; Okanagan Landing 18 Dec 1924-1, killed by Northern Shrike, 21 Dec 1944-250 (Munro 1945c); Hosmer 1 Jan 1984-200, eating sunflower seeds at feeder; nr Fort Steele 16 Dec 1997-63, eating maple seeds; Penticton 27 Dec 1976-82; Cranbrook 7 Feb 1937-6 (Johnstone 1949); South Slocan 8 Feb 1984-120; Creston 1 Feb 1982-116, winter bird survey; Salmo 26 Feb 1975-50; Fruitvale 3 Jan 1980-50 at feeder; Trail 21 Feb 1982-75; Rock Creek 17 Dec 1997-12 at feeder. **Coastal** – 32 km n Terrace 17 Dec 1975-1; Terrace 13 Jan 1976-9 (Hay 1976); Prince Rupert 1 Feb 1981-8, feeding on berries; Masset 31 Jan 1970-6; Bella Coola 2 Dec 1989-3 at feeder; Campbell River area 28 Feb 1974-75; Courtenay 4 Feb 1956-50; Harrison Hot Springs 19 Feb 1976-75; North Vancouver 18 Jan 1979-100; Skagit River 1 Jan 1971-8; White Rock 11 Feb 1911-100 (Campbell et al. 1972b); Surrey 29 Jan 1962-84; Duncan Dec 1972-151 (Tatum 1973); Ucluelet Inlet 24 Jan 1974-1, at head of inlet; North Saanich 31 Jan 1974-45; Langford 17 Feb 1974-54.

Christmas Bird Counts: Interior – Recorded from 26 of 27 localities and on 84% of all counts. Maxima: Vernon 16 Dec 1990-1,150; Smithers 31 Dec 1988-995; Prince George 19 Dec 1993-510. **Coastal** – Recorded from 23 of 33 localities and on 58% of all counts. Maxima: Vancouver 20 Dec 1981-696; White Rock 30 Dec 1989-545; Squamish 22 Dec 1990-245.

House Sparrow HOSP

Passer domesticus (Linnaeus)

Figure 701. Often considered a nuisance species and agricultural pest, the introduced House Sparrow is now well established in human settlements across southern British Columbia. The actual nesting season can last 6 months but males (a) can begin breeding displays in autumn and winter before mating with females (b) occurs the following spring (Saanich, 24 November 1996; R. Wayne Campbell).

RANGE: Introduced to North America (New York) in 1850, now resident from central and northeastern British Columbia, southwestern Mackenzie, northwestern and central Saskatchewan, northern Manitoba, central Ontario, southern Quebec, and the Maritime provinces south throughout the continental United States and most of Mexico. Native to Europe, northern Russia, and Siberia south to northern Africa, Arabia, India, and Myanmar (Burma). Also introduced to southern Africa, Australia, New Zealand, and South America.

STATUS: On the coast, a *common* to *very common* resident in the Georgia Depression Ecoprovince; in the Coast and Mountains Ecoprovince, locally *rare* to *uncommon* resident, except *accidental* on the Queen Charlotte Islands.

In the interior, locally *common* to *very common* resident in the Southern Interior and Southern Interior Mountains ecoprovinces; locally *common* in the Central Interior, Sub-Boreal Interior, and Boreal Plains ecoprovinces; locally *rare* in the Taiga Plains Ecoprovince; *casual* in the Northern Boreal Mountains Ecoprovince.

Breeds.

NONBREEDING: The House Sparrow (Fig. 701) is a widespread but locally distributed resident of most human settlement and farmland areas in British Columbia. It is most plentiful in the human-populated areas of the south coast and in the dry southern valleys of the interior. It is essentially absent from the Queen Charlotte Islands and the sparsely settled mainland coast between Powell River and Kitimat. In the interior, it is found in most areas of permanent human settlement except for the northwestern portion; there, in the Northern Boreal Mountains, it has yet to become established in the villages of Atlin, Dease Lake, Telegraph Creek, Iskut, Lower Post, Fireside, and Fort Ware. The northernmost population is at Fort Nelson, in the Taiga Plains. The highest numbers in winter occur in the Georgia Depression (Fig. 702).

The House Sparrow has been reported from sea level on the coast to about 1,220 m elevation in the interior. Throughout the province, this Old World sparrow is closely associated with human-occupied buildings and anthropogenic food sources. It is found in a variety of human-modified landscapes, ranging from highly urbanized commercial and industrial sites (Fig. 703) to sparsely settled rural and farmland areas. It

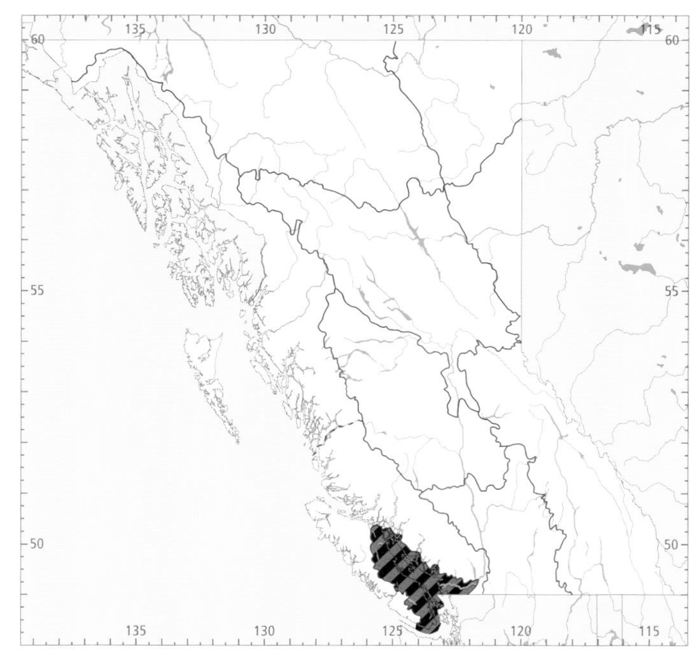

Figure 702. In British Columbia, the highest numbers for the House Sparrow in winter (black) and in summer (red) occur in the Georgia Depression Ecoprovince.

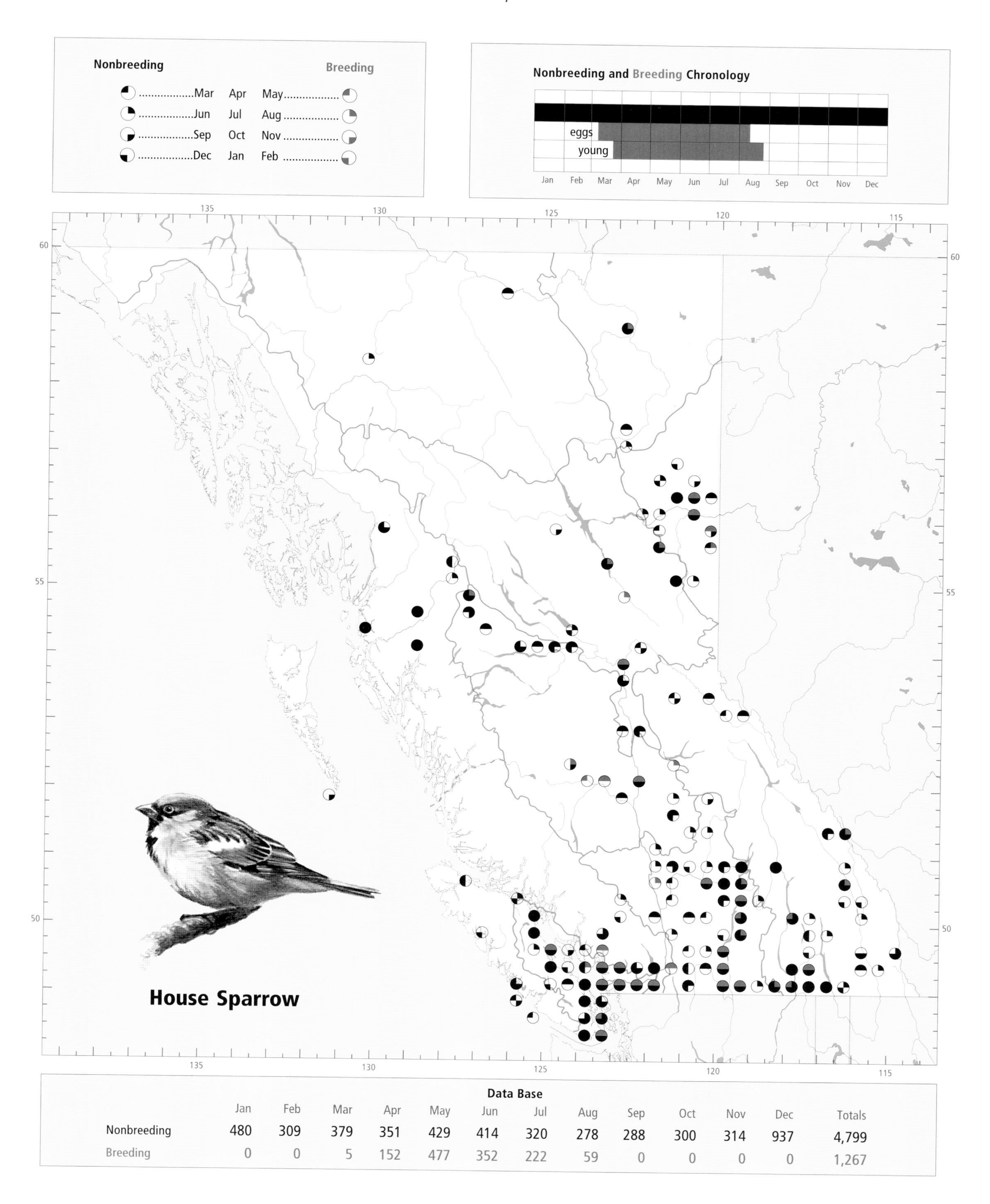

Data Base

	Jan	Feb	Mar	Apr	May	Jun	Jul	Aug	Sep	Oct	Nov	Dec	Totals
Nonbreeding	480	309	379	351	429	414	320	278	288	300	314	937	4,799
Breeding	0	0	5	152	477	352	222	59	0	0	0	0	1,267

Figure 703. Throughout its range in British Columbia, the House Sparrow prefers habitats modified by humans, including highly urbanized environments (Vancouver, 7 August 1996; R. Wayne Campbell).

regularly inhabits areas around railway yards, grain elevators, grain-shipping facilities, farms, livestock barns and feedlots, poultry pens, and garbage landfills, where it forages for spilled grain, animal feeds, or discarded human foods. In cities and towns it is often seen near fast-food restaurants, sidewalk cafes, bakeries, concession stands, and parks, where it scrounges for or solicits bread crumbs, seeds, and other human handouts. It is especially numerous in residential areas, particularly those with an abundance of bird feeding stations.

Weber (1972) and Lancaster and Rees (1979) studied urban bird communities in the city of Vancouver and found that the House Sparrow, along with the European Starling, were the 2 most abundant species. The latter researchers found that these 2 species, along with the Rock Dove, accounted for up to 95% of the bird fauna in industrial and commercial areas. Both studies recorded the highest House Sparrow densities in residential areas and noted the importance of feeding stations in influencing this bird's distribution.

The House Sparrow is a bird of open or lightly wooded landscapes and is not found in closed-forest habitats. In Vancouver, for example, Lancaster (1976) found that sparrow numbers dropped sharply when total vegetative cover reached 60%, and the birds were absent from woodland areas (Table 17). Throughout its distribution in British Columbia, the House Sparrow generally occurs at elevations below 1,000 m. It has not been recorded in subalpine or alpine habitats, and it is recorded only rarely in wilderness areas.

Although movements suggesting migration have been reported in New Jersey (Broun 1972), the House Sparrow is considered nonmigratory throughout its range in North America (Lowther and Cink 1992). In British Columbia, sparrow numbers show little seasonal fluctuation and indicate a nonmigratory population (Figs. 704 and 705), but local movements related to food supply or nesting sites probably occur. From spring to autumn, individuals or small flocks are occasionally observed in areas far outside their established distribution. Such occurrences have been recorded at Liard River (Reid 1975), Dease Lake, Germansen Landing, Indianpoint Lake, Porcher Island (Brooks 1923), and Cape St. James.

During winter in the interior, this resourceful sparrow survives periods of extreme cold by roosting in heated or partially heated building structures such as attics, soffits, livestock barns, grain elevators, and outbuildings, or in electrical structures such as light fixtures, street lights, and electric signs (Fig. 706). Nevertheless, it periodically suffers significant winter mortality, especially in regions with high snowfall or during lengthy periods of extreme cold. During the winter of 1942 to 1943, for example, Munro (1945a) noted that "only a few survived a period of exceptionally low temperature" around the Williams Lake area of the Central Interior.

Table 17. Mean density of House Sparrows (per 10 ha) by urban habitat class and season in Vancouver, British Columbia (modified from Lancaster 1974).

Urban class	Spring	Summer	Autumn	Winter	Mean (all seasons)
Commercial	12.9	13.0	10.0	13.0	12.2
Industrial	3.0	3.4	2.2	2.1	2.7
Apartment residential	20.9	28.3	30.4	27.5	26.8
Typical residential 1	10.0	10.5	7.5	10.4	9.6
Typical residential 2	34.1	41.9	25.9	30.3	33.1
Older residential	11.0	12.0	13.0	12.0	12.0
Woody residential	7.0	11.4	10.4	3.5	8.1
Mixed woodland	0.0	0.0	0.0	0.0	0.0

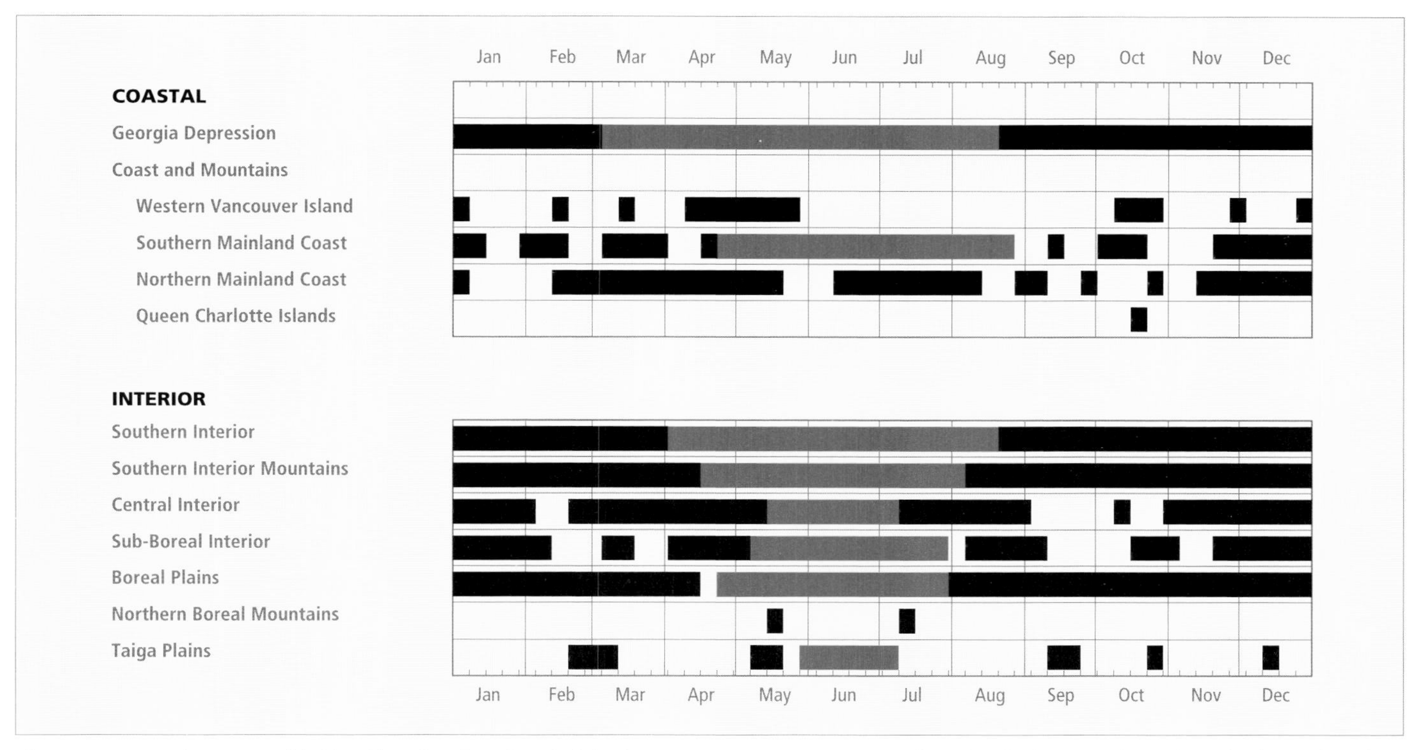

Figure 704. Annual occurrence (black) and breeding chronology (red) for the House Sparrow in ecoprovinces of British Columbia. Records are shown for the week in which they occurred.

In the Okanagan valley of the Southern Interior, Christmas Bird Count data show that large fluctuations occur during some years and indicate a marked decline in the mid-1980s (Fig. 707). Long-term Christmas Bird Counts in the Georgia Depression suggest a stable or increasing trend in sparrow numbers (Fig. 708; see also BREEDING).

The House Sparrow is a common visitor to bird feeding stations in British Columbia. Project FeederWatch data for the period 1988 through 1998 indicate that it was among the 10 most abundant birds using winter feeding stations in the Georgia Depression, Southern Interior, Central Interior, and Boreal Plains. In Vancouver, Lancaster (1976) found it to be the most commonly reported visitor to bird feeders in his study.

The House Sparrow is present throughout the year in most human settlements and farmland areas on the coast and in the interior (Fig. 704).

BREEDING: The House Sparrow likely breeds throughout its established range in British Columbia, but records are lacking for many areas. On the coast, nesting has been recorded in the Georgia Depression and in southern portions of the Southern Mainland Coast, but has yet to be documented on the mainland coast north of Gibsons or on Western Vancouver Island. In the interior, breeding has been recorded in all ecoprovinces except the Northern Boreal Mountains, but we lack documented evidence of nesting for many communities. For example, breeding is suspected but unconfirmed in Princeton, Merritt, Greenwood, Kimberley, Cranbrook, Fernie, Sparwood, Canal Flats, Invermere, Golden, Revelstoke, Cache Creek, Clinton, Lac la Hache, 100 Mile House, Quesnel, McBride, Fraser Lake, Burns Lake, Houston, Fort St. James, and Tumbler Ridge. The northernmost breeding record in the province is from Fort Nelson in the Taiga Plains.

The House Sparrow reaches its highest numbers in summer in the Fraser and Nanaimo lowlands of the Georgia Depression (Fig. 702). An analysis of Breeding Bird Surveys for the period 1968 through 1993 shows that the mean number of birds on interior routes decreased at an average annual rate of 8% (Fig. 709); analysis of coastal routes for the same period could not detect a net change in numbers. In Canada, the House Sparrow declined at an average annual rate of 2.5% for the period 1966 to 1996 ($P < 0.01$); continent-wide, it declined at an average annual rate of 2.2% during the same period ($P < 0.01$) (Sauer et al. 1997). Changes in agricultural practices since the 1960s are suspected to be the cause of this continental decline (Lowther and Cink 1992).

The House Sparrow has been recorded breeding from sea level on the coast to 1,005 m in the interior. It breeds in the same habitat that it occupies at other times of the year, and nesting almost always occurs in close proximity to buildings. Kelleher (1963) examined cavity-nesting avifauna in the town of Hope and in nearby natural forest habitats. He found that this sparrow was a common breeder in town but entirely absent from areas away from human habitation. In Tennessee, Pitts (1979) noted that in rural areas these sparrows rarely used nest boxes more than 300 m from human dwellings. In British Columbia too, nest boxes are rarely used along Bluebird Nestbox Trails in the interior (Pollock 1990, 1992), except when they are near buildings occupied by humans or livestock. In the Central Interior, the House Sparrow has never

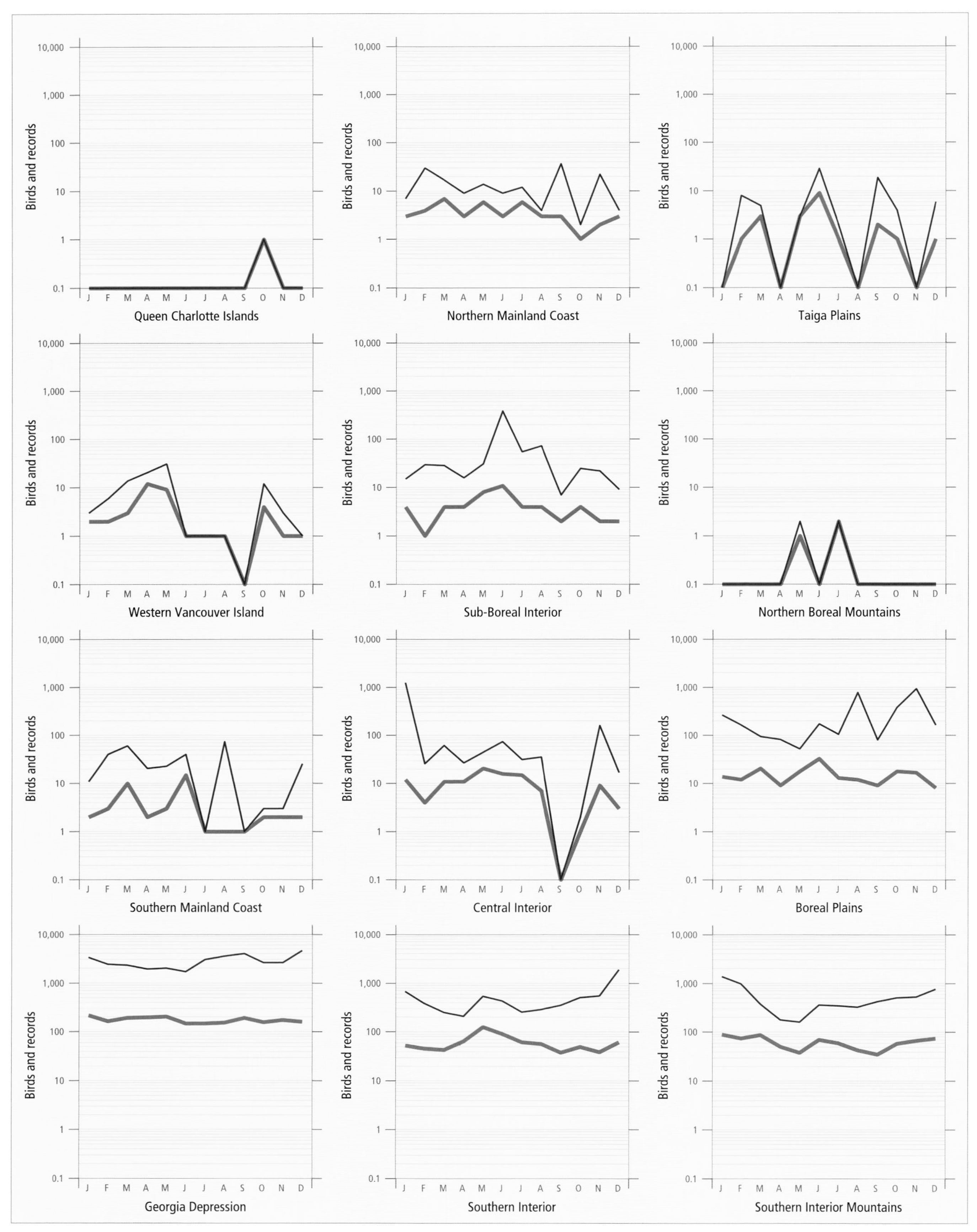

Figure 705. Fluctuations in total number of birds (purple line) and total number of records (green line) for the House Sparrow in ecoprovinces of British Columbia. Christmas Bird Counts, Breeding Bird Surveys, and nest record data have been excluded.

Figure 706. In many cities and towns in the interior of British Columbia, House Sparrows roost each evening in buildings and electrical structures, such as storefront lights, to keep warm during winter (Trail, 26 January 1990; R. Wayne Campbell).

been recorded nesting in any of the approximately 1,000 Mountain Bluebird nest boxes placed mostly in open forest or grassland habitats around Williams Lake (A. Roberts pers. comm.).

In Vancouver, Weber (1972) reported breeding densities ranging from 4.0 to 12.2 breeding pairs per 10 ha in his residential study plots. Also in Vancouver, Lancaster (1976) examined urban avifauna in areas he classified as industrial, commercial, residential, and woodland. He recorded this sparrow in all habitats except woodland, and found the highest summer densities in residential areas (Table 17).

On the coast, the House Sparrow has been recorded nesting from 8 March to 25 August; in the interior, it has been recorded nesting from 3 April (calculated) to 19 August (Fig. 704).

Nests: Nests were found in nest boxes (46%; n = 620), building structures (35%; Figs. 710) such as houses), garages, sheds, barns, roof shelters, commercial or industrial buildings,

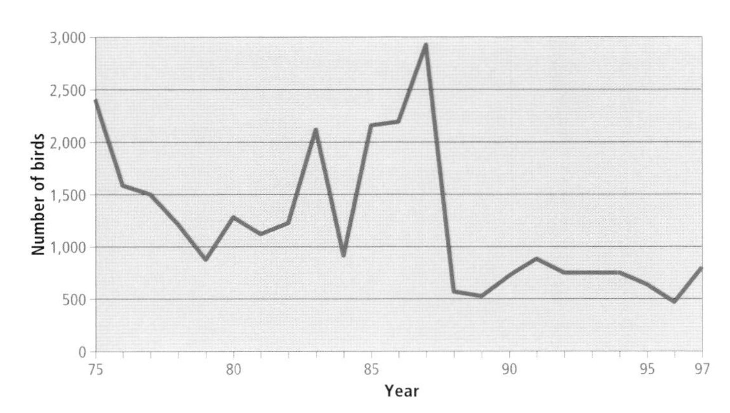

Figure 707. Fluctuations in winter numbers of the House Sparrow based on Christmas Bird Counts for Vernon, Penticton, and Vaseux Lake in the Southern Interior Ecoprovince for the period 1975 through 1997. Note the sharp decline in numbers after 1987.

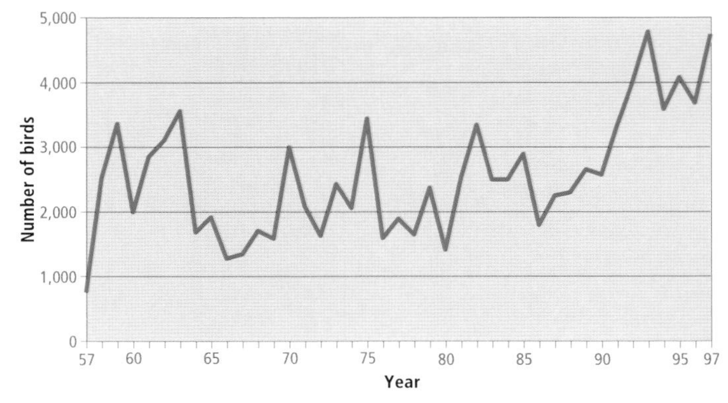

Figure 708. Fluctuations in winter numbers of the House Sparrow based on Christmas Bird Counts for Ladner, Vancouver, and Victoria in the Georgia Depression Ecoprovince for the period 1957 through 1997.

and grain elevators, and in other human-made structures (6%) such as utility poles, street lights, signs, traffic lights, marine pilings, and bridges. Most nests (93%; n = 620) were placed within cavities or crevices, or on partially enclosed ledges; the remaining 7% were non-enclosed, globe-shaped nests placed in trees, vines, hedges, or shrubs. Only 4 nests were found in natural or excavated tree cavities.

In the town of Hope, Kelleher (1963) found 86% of 44 House Sparrow nests in building cavities and 11% in nest boxes. A similar result was obtained in Vancouver, where Weber (1975) reported 91% of 56 sparrow nests in buildings and only 5% in nest boxes. The higher level of nest box use reflected in our data (for all regions) is probably exaggerated by a higher reporting rate compared with other nest types. It may also be a reflection of the greater availability of nest boxes today, since these 2 studies were conducted in the 1960s.

The House Sparrow occasionally builds its nest in or on nests constructed by other species (Bent 1958; Lowther and Cink 1992). In British Columbia, it has been recorded using the nests of the Cliff Swallow (26 nests), American Robin (3), Bank Swallow (2) (Campbell and Meugens 1971), European Starling, House Wren, Downy Woodpecker, and Bald Eagle (1 nest each). The latter nest site held the sparrow nest within the bulky stick structure; the eagle nest was active and contained 2 eaglets. The House Sparrow has also been recorded usurping the active nests of Violet-green (Wootton 1965), Tree, and Cliff swallows, as well as of the American Robin, Chestnut-backed Chickadee, and House Finch.

Nests were generally untidy, bulky structures constructed of grasses (93%; n = 416), feathers (50%), and cultural debris (21%) such as pieces of string, paper, plastic, fabric, and cigarette butts. Less commonly reported materials included leaves, plant stems, rootlets, bark strips, mammalian hair, mosses, and twigs. Most nests were lined with fine grasses and feathers.

Cavity, crevice, and ledge type nests varied in size and bulk depending on available space. For example, only a minimum of new materials were added to the interior of Cliff Swallow nests, whereas considerably more materials were used in more sizeable spaces. Nest box nests usually filled the box to the roof, often creating a domed-over nest structure. Non-cavity type nests are typically globular structures (30 to 40 cm in diameter) with a side entrance (Lowther and Cink 1992). In British Columbia, most non-cavity type nests were either concealed in thick ornamental spruce trees or placed in wall vines sheltered by a roof overhang.

The House Sparrow is a gregarious species and often breeds in small colonies (Lowther and Cink 1992). Most of our records were of individual nests, but some had a clumped distribution that could be described as a colonial nesting. The best example of a colony in British Columbia was 5 nests found together in a wall vine that were close enough to touch each other. On an Alberta ranch, a large colony of 110 nests was found in a row of 34 blue spruce trees (McGillivray 1980).

The heights of 555 nests ranged from 0.9 to 15.0 m, with 59% between 2.4 and 5.0 m. Weber (1975) reported an average nest height of 5.1 m (n = 56) for this sparrow in his Vancouver study area. In metropolitan Toronto, House Sparrow nest heights were influenced by building height; they were higher in areas with 2-storey as opposed to single-storey buildings (Savard and Falls 1981).

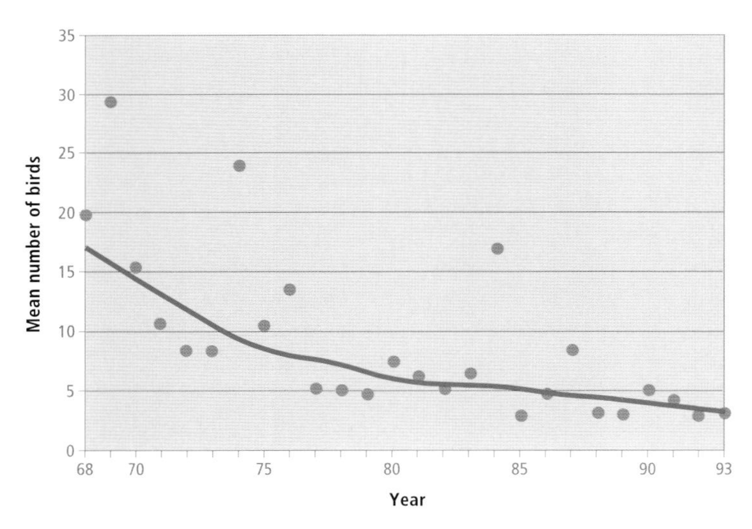

Figure 709. An analysis of Breeding Bird Surveys for the House Sparrow in British Columbia shows that the number of birds on interior routes decreased at an average annual rate of 8% over the period 1968 through 1993 ($P < 0.001$).

Figure 710. Almost any crevice or cavity is used by the House Sparrow for nesting throughout its range in British Columbia. Note the worn entrance above the nest behind the siding on this house (Parksville, 5 April 1998; R. Wayne Campbell).

Figure 711. In British Columbia, the average clutch size of the House Sparrow is 4 or 5 eggs (Saanich, 5 May 1999; R. Wayne Campbell).

Nest-building activities have been recorded as early as 27 February in the Okanagan valley of the Southern Interior (Cannings et al. 1987) and 28 February in the Georgia Depression. Nest construction during this early part of the breeding season is often leisurely and protracted. For example, a nest in the Georgia Depression took 42 days to complete before the first egg was laid. During the peak of the breeding season, however, nests can be constructed in a matter of a few days (Lowther and Cink 1992).

Eggs: Dates for 536 clutches ranged from 8 March to 11 August, with 50% recorded between 3 May and 6 June. Sizes of 263 clutches ranged from 1 to 7 eggs (1E-12, 2E-19, 3E-35, 4E-58, 5E-101, 6E-36, 7E-2), with 60% having 4 or 5 eggs (Fig. 711). The incubation period for 1 closely observed nest was 13 days. Elsewhere in North America, the average incubation period is 11 days but can range from 10 to 14 days (Lowther and Cink 1992).

Young: Dates for 734 broods ranged from 24 March to 25 August, with 51% recorded between 21 May and 6 July. Sizes of 219 broods ranged from 1 to 7 young (1Y-8, 2Y-33, 3Y-62, 4Y-61, 5Y-49, 6Y-5, 7Y-1), with 56% having 3 or 4 young. The nestling period for 1 nest was 18 days. Elsewhere it is usually 14 days, but young may fledge as early as 10 days of age if the nest is disturbed (Lowther and Cink 1992). Fledged young become independent after 7 to 10 days.

In British Columbia, the House Sparrow is usually multi-brooded. Data for the Georgia Depression Ecoprovince indicate that as many as 3 broods were produced annually (Fig. 712). At Hope on the Southern Mainland Coast, Kelleher (1963) found that most nests produced 2 broods (44%; n = 34), with 38% producing a single brood and the remaining 18% producing 3 broods. In the Okanagan valley of the Southern Interior, 2 broods are usual, with some nests occasionally producing 3 broods (Cannings et al. 1987). Double-brooding has been recorded as far north as Fort St. John in the Boreal Plains; early fledging dates (i.e., 14 May), on par with the Okanagan valley, suggest that 3 broods may be possible there.

Brown-headed Cowbird Parasitism: In British Columbia, only 1 of 359 nests found with eggs or young was parasitized by the cowbird. There was also a single record of a House Sparrow feeding a fledgling cowbird. Both records were from the Georgia Depression. The House Sparrow is rarely parasitized elsewhere in North America, largely because of differences in habitat preference and inaccessibility of sparrow nests to cowbirds (Lowther and Cink 1992). Conspecific brood parasitism has also been reported in the House Sparrow (Kendra et al. 1988).

Nest Success: Of 73 nests found with eggs and followed to a known fate, 34 produced at least 1 fledgling, for a nest success rate of 47%. Nest success on the coast was 53% (n = 49); in the interior, it was 33% (n = 24). Not included in this analysis were an additional 89 nests that were intentionally destroyed by humans or where the clutch was collected for scientific purposes.

Predation was identified as one of the main causes of nest failure. Observed or suspected predators included European Starling (2 nests), Crested Myna, American Kestrel, Steller's Jay, red squirrel, and striped skunk (1 nest each). In a related case, dead House Sparrow nestlings were scavenged from a nest box by an ermine (short-tailed weasel), suggesting that it is also a potential nest predator.

Nest collapse was determined to be the cause of 8 nest failures; 5 of these were in old Cliff Swallow nests. Weather conditions were the probable cause of 2 failures. In 1 case the clutch was abandoned when heavy rains penetrated and soaked the nest box; in the other, solar overheating apparently killed a brood located inside an aluminum window awning. There were also 8 cases where whole broods were found dead in the nest; abandonment due to wet, cool weather or parental death were the suspected causes of mortality.

REMARKS: The House Sparrow, formerly known as the English Sparrow, was introduced to North America from English and German stock belonging to the nominate race, *P. d. domesticus* (American Ornithologists' Union 1957; Cannings 1998). Its release in 1850 was one of the most successful avian introductions on the continent but is a testament to the folly of introducing alien species. For much of the 20th century, it

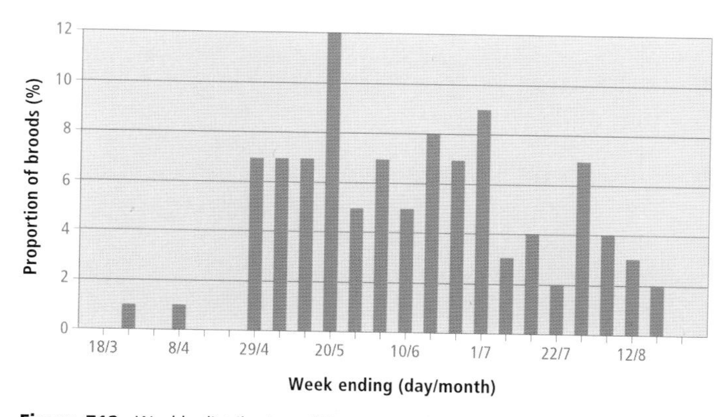

Figure 712. Weekly distribution of House Sparrow nests with young in the Georgia Depression Ecoprovince. The peaks in the weeks ending 20 May, 1 July, and 29 July suggest that triple-brooding occurs in this ecoprovince.

Figure 713. Unsightly droppings on buildings and messy nests under eaves have contributed to the status of the House Sparrow as a pest species throughout its range in North America (Victoria, 19 July 1997; R. Wayne Campbell).

has largely been considered a nuisance species and agricultural pest (Lowther and Cink 1992). In British Columbia, it is a minor pest around cattle feedlots (Weber 1989) and at some fast-food outlets in cities, where its droppings may accumulate (Fig. 713). Because of its non-native status in North America, the House Sparrow is one of the few passerines not protected by international, federal, or provincial regulations.

The arrival of the House Sparrow in British Columbia is poorly documented. Brooks and Swarth (1925) report that it arrived in Vancouver about 1890, but provide no substantiating records. Carl and Guiguet (1957) surmised that it probably first arrived in the province "within a few years" of 1886, shortly after it was known to have colonized California and northern Utah. Records from northern Washington and southern Alberta suggest that it probably arrived in the province later than 1890. For example, Dawson and Bowles (1909) indicated that it had arrived in Spokane, Washington, about 1895 and had first reached Seattle in 1897. Further to the northeast, in Alberta, it was not recorded until the summer of 1898, when several were seen in the town of Red Deer (Houston 1978). Based on these records, it appears likely that this sparrow began its invasion of British Columbia sometime between 1895 and 1898. It had colonized the south coast by the turn of the 19th century, as Kermode (1904) considered it to be "quite common" in coastal cities by 1904. It had probably also arrived in the interior about this time, but was not reported until 6 May 1909, when it was first recorded at Okanagan Landing in the Okanagan valley (Cannings et al. 1987). By 1919 it had

Figure 714. Throughout urbanized landscapes of the Georgia Depression Ecoprovince of British Columbia, the House Sparrow is a major prey item in the diet of urban-nesting Cooper's Hawks (Beacon Hill Park, Victoria, 30 July 1998; Andrew C. Stewart).

Figure 715. The House Sparrow is a very adaptable and opportunistic species. In many places in British Columbia, it has been seen picking insects from the grills of cars as they stop in parking lots and gas stations (15 June 1999; R. Wayne Campbell).

been recorded in the Central Interior at Telkwa (Brooks and Swarth 1925), and by 1921 had reached Porcher Island on the Northern Mainland Coast (Brooks 1923). Further north, in the Peace River region of the Boreal Plains, Williams (1933a) noted that it was absent in 1922 but was "common" at Taylor and Fort St. John by 1930. Racey (1930) also considered it to be a "common" species during his 1930 explorations south of the Peace River in the vicinity of Dawson Creek and Pouce Coupe. By at least 1943, shortly after the completion of the Alaska Highway, the sparrow had reached as far north as Fort Nelson in the Taiga Plains (Rand 1944).

Except for some minor range extensions into areas of recent human settlement, such as Mackenzie and Tumbler Ridge in the Sub-Boreal Interior, the distribution of the House Sparrow appears to have changed little since it was described by Munro and Cowan (1947). Curiously, it has yet to move into the settled areas of west-central British Columbia. Although reported as far west as Tatla Lake (Paul 1959), it has apparently not become established there (R. Travers pers. comm.) or elsewhere in the Cariboo and Chilcotin areas west of Redstone and Chezacut. It has also failed to colonize the Bella Coola valley (R. Mayo pers. comm.) or other isolated human settlements on the central coast. It is also scarce at Clearwater and Blue River in the North Thompson valley (T. Goward pers. comm.).

The House Sparrow is an early breeder and typically establishes possession of nest sites before migratory cavity-nesters arrive. It is also known for forcibly evicting other species from their nests, even occasionally destroying their eggs or young (see Wootton 1965). These characteristics have not endeared it to humans and have contributed to its reputation as a pest, especially it competes for nest boxes erected for other birds. Restricting the entrance size or putting up additional nest boxes remain the best techniques for ensuring that nest sites will be available for other urban cavity-nesters. Although Kelleher (1963) found that cavities with perches were preferred as nest sites over those lacking perches, removing perches from nest boxes is an ineffective method of discouraging their use by this sparrow.

The House Sparrow is an important prey species for some predatory birds in urban environments. Our data contain records of this sparrow being captured, frequently near bird feeders, by the Sharp-shinned Hawk, Cooper's Hawk, Merlin, Northern Pygmy-Owl, and Northern Shrike. In the Georgia Depression, this sparrow is regularly taken by the Barn Owl (Campbell et al. 1987) and by urban-nesting Cooper's Hawks (Fig. 714; A.C. Stewart, unpubl. data). In Alberta and Saskatchewan, it is the most important prey species for urban-nesting Merlins (James and Smith 1987; Sodhi and Oliphant 1993).

On 19 December 1997, a new provincial high Christmas Bird Count of 2,736 House Sparrows was established at Victoria.

The American Ornithologists' Union (1983, 1998) incorrectly describes the distribution of the House Sparrow in British Columbia as "central and southeastern"; this should read "southern, central, and northeastern" British Columbia.

For information on the early history and range expansion of the House Sparrow in North America and in Canada, see Barrows (1889), Weaver (1939), and Houston (1978). For a summary of the ecology and life history of the House Sparrow in North America, see Lowther and Cink (1992).

NOTEWORTHY RECORDS

Spring: Coastal – Victoria 28 May 1988-60; Cadboro Bay (Saanich) 30 Apr 1999-6 nestlings, recently hatched; Pachena Point 12 Apr 1975-1 at lightstation; Ucluelet 24 May 1986-5; Tofino 10 May 1980-7, 1st record; Westham Island 22 Mar 1971-200 (Campbell et al. 1972b); Richmond 5 May 1963-1 Brown-headed Cowbird nestling; Surrey 8 Mar 1961-2 eggs; Crescent Beach 7 Apr 1958-4 eggs; New Westminster 21 Apr 1961-6 eggs; Haney 25 May 1962-7 eggs; White Rock 23 Apr 1975-5 nestlings; Crescent Beach 17 Apr 1958-3 nestlings; North Surrey 2 Apr 1961-2 nestlings; Crescent Beach 24 Apr 1958-3 fledglings; Chilliwack 24 Mar 1969-1 adult feeding nestlings; Cultus Lake 11 Apr 1959-50; Vancouver 16 Apr 1974-300, at grain elevators; Harrison Hot Springs 1 Apr 1974-20; Courtenay 12 Apr 1917-3, first time observed; Port Hardy 30 Apr 1939-3; Kitimat 4 May 1975-3; Prince Rupert 17 Apr 1983-4; Stewart 2 Mar 1996-6. **Interior** – Manning Park 18 May 1963-2; Osoyoos 9 May 1983-4 fledglings; 3.2 km n Osoyoos 30 Apr 1958-5 nestlings, just hatched; Trail 4 Mar 1984-20, 17 May 1980-4 fully feathered nestlings; ne Skaha Lake 17 May 1996-4 eggs; West Bench (Penticton) 30 Mar 1974-40; Cranbrook 2 Apr 1977-20; Naramata 4 Apr 1992-2 eggs; Summerland 10 May 1969-4 fledglings; Skihist Park 27 May 1967-10; nr Douglas Lake 28 Mar 1981-17; Okanagan Landing 6 May 1909-earliest record in Okanagan valley (Cannings et al. 1987); Vernon 18 Apr 1994-7 eggs, 25 Apr 1994-7 nestlings; Kamloops 27 Apr 1975-20; Invermere 4 Mar 1979-12; Revelstoke 14 Apr 1977-12; Field 14 Mar 1977-20 (Wade 1977); Wineglass Ranch (Chilcotin River) 9 May 1987-1 at feeder; Williams Lake 30 May 1963-8; Quesnel 29 May 1963-6; Red Pass (Mount Robson Park) 30 May 1974-1; Indianpoint Lake 15 May 1929-1 (MCZ 284540); Willow River 9 Apr 1976-1; Prince George 12 Mar 1974-15, 14 May 1995-5 eggs; Burns Lake 11 Mar 1997-20 at feeder; Endako 23 May 1978-2 pairs, nesting in Cliff Swallow nests; Fort Fraser 24 Mar 1977-3; Vanderhoof 24 Mar 1977-6; Houston 25 Mar 1977-5; Telkwa 15 May 1975-1; Tumbler Ridge 14 Mar 1998-10; Chetwynd 2 Apr 1993-10 at feeder; Tupper Creek 7 May 1938-2 collected (Cowan 1939); Moberly Lake 23 Mar 1997-1 at feeder; Fort St. John 26 Mar 1983-40 at grain elevator, 14 May 1987-1 fledgling with adults, 31 May 1975-4 eggs; Buckinghorse River 14 May 1974-1 (Erskine and Davidson 1976); Fort Nelson 1 Mar 1975-3; Liard River 15 May 1975-2 (Reid 1975).

Summer: Coastal – Ogden Point (Victoria) 24 Aug 1984-105; Mandarte Island 9 Jul 1960-5; Canoe Passage 18 Jul 1959-50; Crescent Beach 25 Jun 1962-2 adults feeding nestlings in marine piling; Surrey 17 Aug 1990-1 adult feeding Brown-headed Cowbird fledgling; Burnaby 21 Jul 1962-3 eggs with 4 Cliff Swallow eggs (RBCM 795); Vancouver 28 Jul 1969-113 (Weber 1972); Harrison Hot Springs 25 Aug 1976-75; Seabird Island 3 Jun 1984-3; Hope 25 Aug 1961-1 adult feeding nestlings; Boston Bar 7 Jul 1992-1; Campbell River 27 Aug 1979-25; Sayward 4 Jul 1999-7; Kitimat 18 Jul 1975-1; Prince Rupert 30 Jun 1974-6; Terrace 10 Aug 1975-2; New Hazelton 23 Jun 1997-1. **Interior** – Manning Park 11 Jul 1964-4 nr buildings; Oliver 2 Jun 1976-47; Osoyoos 29 Jun 1969-30; Trail 10 Jun 1984-10; Penticton 8 Aug 1993-3 nestlings; White Lake (Okanagan Falls) 24 Jun 1974-1; Okanagan Falls 19 Jun 1976-20; ne Skaha Lake 15 Aug 1995-5 nestlings; Galloway 27 Jun 1953-6; Summerland 18 Jun 1985-6 nestlings; Naramata 20 Jun 1992-5 nestlings; Sparwood 18 Aug 1984-12 at feeder; Lytton 6 Jun 1985-3; Nakusp 15 Jun 1981-2; Kamloops 2 Jun 1973-15; Enderby 3 Jun 1976-10; Salmon Arm 5 Jun 1977-10; Radium 4 Aug 1981-1; Cache Creek 29 Aug 1980-10; Revelstoke 15 Jul 1976-30; Little Fort 27 Jul 1973-5; Golden 8 Jul 1991-2; Field 10 Jun 1975-40 (Wade 1977); 100 Mile House 1 Aug 1978-6; Buffalo Lake 12 Jun 1965-20; Williams Lake 5 Jun 1985-3 nestlings; Chezacut 20 Jun 1966-2 entering nest under building eves; Quesnel 25 Jun 1997-2; Red Pass (Mount Robson Park) 24 Jun 1972-2; McBride 10 Jul 1991-37 nr train station; Prince George 5 Jun 1995-3 nestlings, well feathered, 18 Jun 1973-5 eggs; Endako 5 Jun 1977-2 pairs entering Cliff Swallow nests; Francois Lake 4 Aug 1944-12 (Munro 1947); Fort Fraser 18 Jun 1998-3 picking insects from truck radiator; Vanderhoof 19 Jun 1997-2 recently fledged young; Houston 21 Jun 1997-2; Fort St. James 7 Jun 1977-2; Telkwa Aug 1919-1, early record (Brooks and Swarth 1925); Smithers 5 Jul 1974-2 nestlings; Mackenzie 8 Jun 1996-6, 17 Jun 1997-5 nestlings; Chetwynd 9 Jun 1996-2 adults feeding nestlings; Swan Lake (Tupper) 6 Jun 1962-20; Tupper Creek 7 Jun 1962-10; Pouce Coupe 12 Jun 1996-2; Dawson Creek 28 Jun 1978-2 nestlings, 3 Jul 1974-20; Beryl Prairie 8 Jul 1979-2 on farm; Hudson's Hope 18 Jun 1997-4 picking insects from vehicle grill (see Fig. 715); Fort St. John 4 Jul 1974-5 nestlings, 2nd brood, 29 Aug 1984-400 at grain elevator; Wonowon 25 Jun 1996-2; Trutch 20 Jun 1976-1; Buckinghorse River 20 Jun 1976-4 at campground; Dease Lake 9 Jul 1996-1 adult male; Fort Nelson 11 Jun 1982-4 nestlings, 12 Jun 1976-5; Liard River 11 Jul 1978-1.

Breeding Bird Surveys: Coastal – Recorded from 16 of 27 routes and on 46% of all surveys. Maxima: Victoria 6 Jul 1976-159; Chilliwack 26 Jun 1976-100; Point Grey 16 Jun 1974-70. **Interior** – Recorded from 25 of 73 routes and on 25% of all surveys. Maxima: Kamloops 14 Jun 1987-54; Oliver 2 Jun 1976-47; Lavington 20 Jun 1976-45.

Autumn: Interior – Fort Nelson 16 Sep 1943-12 (Rand 1944); Rose Prairie 27 Nov 1983-100, flying over farmyard; Montney 11 Nov 1982-10; Fort St. John 28 Nov 1982-200 at grain elevator; Germansen Landing 21 Oct 1972-10; Chetwynd 28 Nov 1993-20 at feeder; Smithers 3 Nov 1977-22; Telkwa 18 Nov 1981-11 perched in tree; Willow River 30 Oct 1966-1; Burns Lake 18 Nov 1997-20 at feeder; Indianpoint Lake 6 Oct 1929-1 (MCZ 284538); Wells Gray Park 20 Oct 1952-4, first record in park (Ritcey 1953); Cache Creek 5 Sep 1978-15; Invermere 24 Sep 1977-15; Salmon Arm 6 Nov 1970-100; Monte Creek 26 Nov 1994-26 at feeder; Kamloops 8 Sep 1980-8; Sparwood 5 Oct 1983-4; Trail 6 Nov 1983-42 at feeder; Grand Forks 23 Nov 1979-30 feeding along highway; Osoyoos 11 Oct 1971-60; Kelowna 10 Oct 1944-50; Richter Pass 21 Sep 1968-21. **Coastal** – Stewart 21 Nov 1992-2 at feeder; Terrace 12 Nov 1974-20; Prince Rupert 4 Sep 1983-35 in rail yard; Refuge Bay (Porcher Island) Sep 1921-1 in chicken yard (Brooks 1923); Cape St. James 18 Oct 1981-1 adult female; Campbell River 2 Oct 1981-30; Garibaldi Lake 3 Oct 1965-1; Harrison Hot Springs 19 Oct 1974-2; Maple Ridge 26 Nov 1989-2 at feeder; Vancouver 4 Sep 1968-116 (Weber 1972); Bowser 28 Nov 1989-2 at feeder; Aldergrove 6 Nov 1968-10 feeding on grain in duck pen; Sea Island 29 Sep 1969-200; Steveston 28 Sep 1963-100; Nanaimo 24 Nov 1990-6 at feeder; Tofino 30 Nov 1980-3; Saturna Island 8 Nov 1973-1; Ganges Harbour 22 Nov 1977-8; Duncan 22 Sep 1979-5; Ucluelet 22 Oct 1985-9; Cadboro Bay 22 Nov 1998-1 taken at feeder by Northern Shrike; Victoria 23 Sep 1986-100.

Winter: Interior – Fort Nelson 24 Feb 1995-8 at bird feeder; Wonowon 27 Feb 1995-2; Fort St. John 30 Jan 1983-100; Chetwynd 6 Feb 1993-30 at feeder; Fort Fraser 2 Jan 1996-1 at feeder; Burns Lake 28 Feb 1998-20 at feeder; Quesnel 17 Jan 1989-1 at feeder; Williams Lake 2 Jan 1977-390 on bird survey; 100 Mile House 15 Dec 1985-15 in horse corral with Brewer's Blackbirds; Field 18 Dec 1977-73; Golden 1 Jan 1978-20; Revelstoke 2 Dec 1972-50; Sorrento 18 Dec 1970-150; Windermere 7 Feb 1981-40; Invermere 8 Feb 1981-50; Vernon 23 Dec 1977-150; Nelson 21 Jan 1977-230 at landfill; Penticton 27 Dec 1976-308; Kaleden 21 Jan 1979-175 at game farm; nr Creston 14 Dec 1997-94; Trail 4 Jan 1983-100; Rossland 17 Feb 1997-5 at feeder; Grand Forks 29 Dec 1979-750 feeding on spilled grain. **Coastal** – Stewart 28 Feb 1995-5 at feeder; Prince Rupert 15 Feb 1984-20 at feeder; Campbell River 7 Dec 1979-20; Tahsis Inlet Dec 1948-1 (Mitchell 1959); Harrison Hot Springs 3 Dec 1975-25; Maple Ridge 8 Jan 1989-10 at feeder; Bowser 8 Dec 1991-16 at feeder; Burnaby Lake 20 Jan 1974-58; Westham Island 16 Jan 1966-250; Steveston to Deas Island Tunnel 21 Dec 1963-1,500; Tofino 7 Jan 1981-2 at feeder; Duncan 3 Dec 1983-30 at sewage lagoons; Victoria 18 Jan 1986-60 at Beacon Hill Park; Saanich 21 Feb 1985-75.

Christmas Bird Counts: Interior – Recorded from 25 of 27 localities and on 77% of all counts. Maxima: Oliver-Osoyoos 28 Dec 1987-1,991; Vernon 22 Dec 1985-1,732; Penticton 27 Dec 1987-888. **Coastal** – Recorded from 25 of 33 localities and on 68% of all counts. Maxima: Vancouver 19 Dec 1993-2,339; Victoria 18 Dec 1993-2,039; Ladner 21 Dec 1963-1,874.

Casual, Accidental, Extirpated, and Extinct Species

Northern Parula

NOPA

Parula americana (Linnaeus)

RANGE: Breeds from extreme southeastern Manitoba, central Ontario, southern Quebec, New Brunswick, Prince Edward Island, and Nova Scotia south to northeastern Minnesota, northern Wisconsin, northern Michigan, and the middle Atlantic states, south to south-central and southern Texas, the Gulf coast and southern Florida, and west to the eastern edge of the Plains states. Winters primarily from Veracruz and Oaxaca in Mexico south to Guatemala and Belize; and from central Florida and the Bahama Islands south throughout the West Indies to Tobago.

STATUS: On the coast, *casual* in the Georgia Depression Ecoprovince.

OCCURRENCE: Two records. On 24 June 1994, a male sang from Lombardy poplars bordering the Quilchena Golf Course in Richmond, south of Vancouver (Bowling 1994b).

On 9 and 10 November 1996, an immature male was seen in Port Alberni, on southern Vancouver Island (Bowling 1997a).

The Northern Parula is a regular spring transient in coastal California. It is rare, but occurs almost annually in Oregon and has recently been found in Washington (Small 1994; Gilligan et al. 1994; T.R. Wahl pers. comm.).

REMARKS: For a thorough summary of the life history of the Northern Parula in North America, see Moldenhauer and Regelski (1996).

Black-throated Blue Warbler

BTBW

Dendroica caerulescens (Gmelin)

RANGE: Breeds from southern Ontario, southern Quebec, New Brunswick, Prince Edward Island, and Nova Scotia south to northeastern Minnesota, northern Wisconsin, central Michigan, northern Ohio, in the Appalachians through West Virginia, western Maryland, eastern Kentucky, western Virginia, eastern Tennessee, and western South and North Carolina to northeastern Pennsylvania, northern New Jersey, southern New York, and southern New England. Winters from southern Florida and the Bahama Islands south through the Greater Antilles and adjacent mainland coasts.

STATUS: On the coast, *casual* in the Georgia Depression Ecoprovince.

In the interior, *accidental* in the Southern Interior Ecoprovince; *casual* in the Southern Interior Mountains Ecoprovince.

OCCURRENCE: Six records, all males. The Black-throated Blue Warbler has been recorded in spring, autumn, and winter in southern British Columbia.

On 22 September 1984, an immature Black-throated Blue Warbler was photographed (BC Photo 937) near the wildlife centre in the Creston Valley Wildlife Management Area. The warbler appeared tired and wanted to sleep despite repeated attempts by a Song Sparrow and a Lincoln's Sparrow to drive it away (Davidson and Van Damme 1987). The following year, on 1 October, another bird was found in the same general location associating with Black-capped Chickadees foraging among low shrubs and small trees bordering a small wetland. On 25 and 27 November 1987, a single bird was carefully observed flitting and foraging among Douglas-firs in a backyard in Maple Bay, on southeastern Vancouver Island. On 2 October 1988, a Black-throated Blue Warbler was spotted among a flock of migrating Yellow-rumped ("Audubon's") and Townsend's warblers near Mission Creek, Kelowna, in the central Okanagan valley.

Figure 716. From 12 December 1993 to 16 January 1994, a Black-throated Blue Warbler frequented a suet feeder in Nakusp in the Southern Interior Mountains Ecoprovince of British Columbia (22 December 1993; Gary S. Davidson).

Five years later, on 12 December 1993, a first-year bird was observed at a feeding station in Nakusp (G.S. Davidson pers. comm.). It remained in the area for just over 1 month, feeding on suet mixed with dried fish food (Fig. 716) and apples whose tough skin had been opened by a Northern Flicker. Unfortunately, on 16 January 1994, it flew into a plate glass window and died. The carcass was preserved as a museum specimen (UBC 15172). The sixth and final record occurred on 20 April 1994, when a bird was reliably heard and seen in deciduous trees in the University of British Columbia

Endowment Lands on the north side of West 16th Avenue in Vancouver.

The Black-throated Blue Warbler is considered a rarity anywhere on the west coast of North America (Roberson 1980). The occurrences in British Columbia follow the general pattern for this western vagrant, which is considered an extremely rare spring transient (and summer visitant) and more regular autumn transient, with most records occurring between late September and early November (Roberson 1980; Gilligan et al. 1994; Small 1994). There are several winter records on the west coast of North America (Roberson 1980).

REMARKS: For additional life-history information on the Black-throated Blue Warbler, see Holmes (1994).

Hermit Warbler

Dendroica occidentalis (Townsend)

HEWA

RANGE: Breeds from southwestern Washington south through the Coast Ranges in the Cascade Mountains and Sierra Nevada to southern California and west-central Nevada. Winters mainly in mountainous regions of Central America from Mexico to north-central Nicaragua.

STATUS: On the coast, *casual* in the Georgia Depression Ecoprovince.

OCCURRENCE: Three records. An adult male Hermit Warbler was present at Mount Douglas Park, Saanich, from 18 to 29 April 1992 (Bowling 1992), in loose association with Yellow-rumped Warblers in mixed coniferous-deciduous forests. A hybrid Hermit/Townsend's warbler was also seen in the area from 29 April to at least 8 May (Bowling 1994a). An adult male was seen on the adjacent mainland, at Queen Elizabeth Park in Vancouver, on 25 April 1992. The third record, a pure adult male Hermit Warbler, was found singing near Sproat Lake, west of Port Alberni, in a mixed coniferous forest of Douglas-fir, western redcedar, and western hemlock, on 17 May 1993; it was described thoroughly. It had been in the area since 14 May.

REMARKS: Hermit and Townsend's warblers hybridize extensively along narrow zones in Oregon and Washington (Morrison and Hardy 1983; Rohwer and Wood 1998). Because hybrids are highly variable and can resemble either parental species (Pearson 1997), we have not included sightings of adults without thorough documentation that included either specimens, photographs, taped songs, or detailed field notes. Songs and sightings submitted from the following places were therefore not included in this account: Sooke, River Jordan, Victoria (Bowling 1992, 1994a), Swan (Lost) Lake in Saanich (Godfrey 1986), Thetis Lake (Taylor 1984), Duncan, Cowichan Lake, Cowichan Bay, Langley, Ucluelet, Long Beach, Chilliwack, Nanaimo, Tofino, Port Alberni, Pitt Meadows, Burrard Inlet, Vancouver (Bowling 1992), Brackendale (Spreadborough 1916), Coombs, near Alberni Valley (Innes and Innes 1995), Qualicum Beach, Campbell River, Gold River, and Manning Park.

Pearson (1997) notes that hybrids can be distinguished from pure individuals by several characteristics (Table 18).

For a recent summary of the life history of the Hermit Warbler in North America, see Dunn and Garrett (1997) and Pearson (1997).

Table 18. Characteristics of pure Hermit Warbler and Townsend's Warbler individuals that allow hybrids to be distinguished (after Pearson 1997).

Characteristic	Hermit Warbler	Townsend's Warbler
Crown	entirely yellow	entirely black
Extent and intensity of yellow on breast	absent	extensive and intensive
Streaks on mid and lower flanks	absent	extensive
Back colour	medium gray	olive green
Bib corner size	weak	strong and pronounced

Blackburnian Warbler

BBNW

Dendroica fusca (Muller)

RANGE: Breeds from central Alberta (locally), central Saskatchewan, central Manitoba, central Ontario, southern Quebec, and the Maritime provinces south along the eastern seaboard to Massachusetts and New York, and south along the Appalachian Mountains to Georgia. Winters from Costa Rica, Panama, Colombia, and northern Venezuela south in the Andes through Ecuador to central Peru and Bolivia.

STATUS: On the coast, *accidental* in the Georgia Depression Ecoprovince.

In the interior, *accidental* in the Southern Interior, Northern Boreal Mountains, and Boreal Plains ecoprovinces, and *casual* in the Taiga Plains Ecoprovince.

OCCURRENCE: Six records:

(1) Adult male seen at Thorson's Landing, south of Fort St. John in the Boreal Plains, on 9 July 1930 (Williams 1933b).
(2) Adult male heard singing and observed foraging on a western redcedar limb in mixed woods in Fleetwood (North Surrey) in the Georgia Depression on 25 August 1960 (G.R. Ryder pers. comm.).
(3) Adult male singing from spruce trees near Toad River in the Northern Boreal Mountains Ecoprovince, on 11 July 1978.
(4) Adult male singing near the Fort Nelson airport on 4 July 1990 (R. Toochin pers. comm.).
(5) Adult male in alternate plumage, near Steamboat, 60 km west of Fort Nelson in the Taiga Plains, on 5 June 1995 (Spitman 1996).
(6) Immature carefully observed in Vernon in the Southern Interior Ecoprovince, on 29 August 1995 (C. Siddle pers. comm.).

REMARKS: We have excluded 8 reports that lacked sufficient documentation; these were from Duncan, Mission, Osoyoos, Golden, Prince George, Stuart Lake, Fort Nelson (heard only), and Liard Hot Springs Park.

For additional life-history information on the Blackburnian Warbler, see Morse (1994).

Yellow-throated Warbler

YTWA

Dendroica dominica (Linnaeus)

RANGE: Breeds from east-central Oklahoma, southeastern Kansas, northern Missouri, southeastern Iowa, central Illinois, central Indiana, central Ohio, central Pennsylvania, and central New Jersey south to south-central and eastern Texas, the Gulf coast, and central Florida. Also breeds in the northern Bahama Islands. Winters from coastal South Carolina and Georgia south throughout the Florida peninsula and west along the Gulf coast to eastern Louisiana; also from coastal southern Texas south through northern Central America and the Bahama Islands to Costa Rica.

STATUS: On the coast, *accidental* in the Georgia Depression Ecoprovince.

OCCURRENCE: A male was present at a suet and seed feeder on Gabriola Island, in the Gulf Islands of British Columbia, from 3 to 25 January 1998 (S. Baillie pers. comm.; O'Shaughnessy 1998). It was photographed on 16 January (BC Photo 1613; Fig. 717). The subspecies was identified as *albilora* (K.L. Garrett pers. comm.).

The Yellow-throated Warbler is expanding its range northward to at least extreme southeastern Ontario in eastern North America and to Colorado and northwestern New Mexico in the west (Hall 1996). There is 1 record of unsuccessful nesting in California (Small 1994).

Figure 717. A male Yellow-throated Warbler visited a suet and seed feeder on Gabriola Island, in the Georgia Depression Ecoprovince of British Columbia, for most of January 1998. This is the first provincial record of the species (BC Photo 1613; 16 January 1998; Marie O'Shaughnessy).

REMARKS: For additional life-history information on the Yellow-throated Warbler, see Hall (1996).

Prairie Warbler

PRWA

Dendroica discolor (Vieillot)

RANGE: Breeds from eastern Nebraska, eastern Kansas, central Missouri, northern Illinois, central Wisconsin, northern Michigan, southern Ontario, southern Pennsylvania, southeastern New York, Massachusetts, and southern New Hampshire south to eastern Oklahoma, eastern Texas, the Gulf coast, and southern Florida. Winters from central Florida, the Gulf coast, and the Bahama Islands south throughout the West Indies to islands off the coast of northern Central America.

STATUS: On the coast, *casual* in the Coast and Mountains Ecoprovince.

OCCURRENCE: Two records. A single bird, probably an adult, was present for 39 days, from 18 December 1993 to 25 January 1994, near Masset on the Queen Charlotte Islands (Siddle 1994b). On 8 September 1995, an immature male was captured in banding nets on Triangle Island, off northwestern Vancouver Island (Bowling 1996a).

In California, the Prairie Warbler has occasionally been recorded in spring and summer, but regularly in September and October (Roberson 1980; Small 1994). Further north, in Oregon, it has been documented on only 4 occasions, between 24 August and 28 September (Gilligan et al. 1994).

REMARKS: A well-described adult male Prairie Warbler, seen on 17 June 1977, on Mount Kobau was put on the hypothetical list for the Okanagan valley because it was seen by a single observer only (Cannings et al. 1987).

For a summary of the life history of the Prairie Warbler, see Nolan et al. (1999).

Hooded Warbler

HOWA

Dendroica citrina (Boddaert)

RANGE: Breeds from southeastern Nebraska, central and northeastern Iowa, central Illinois, southern Michigan, southern Ontario, northwestern Pennsylvania, central and southeastern New York, southern Connecticut, and Rhode Island south to eastern Texas, the Gulf coast, and northern Florida, and west to eastern Oklahoma. Winters from coastal California and along the Pacific slope from Durango and Nuevo Leon to Guatemala, Honduras, and north-central Nicaragua.

STATUS: On the coast, *casual* in the Georgia Depression and Coast and Mountains ecoprovinces.

OCCURRENCE: Two records. A single bird was present in Tofino, on Western Vancouver Island, from 14 to 16 December 1989 (Siddle 1990a). The second bird was an adult male recorded in Cecil Green Park at the University of British Columbia campus in Vancouver on 4 June 1997 (Bain and Holder 1997).

The Hooded Warbler is very rare anywhere on the west coast of North America (Roberson 1980). It is still considered a rare spring and autumn transient, and very rare in winter, in California, and recently it has been found breeding there (Small 1994). There are 8 records in Oregon between 20 May and 8 November, through the early 1990s (Gilligan et al. 1994). Roberson (1980) lists only 1 record for Washington, a bird present in Seattle from 31 December 1975 through 4 April 1976.

REMARKS: For recent summaries of the identification and life history of the Hooded Warbler in North America, see Evans Ogden and Stutchberry (1994) and Dunn and Garrett (1997).

POSTSCRIPT: A female was present at the airport in Sechelt on the Sunshine Coast from 12 to 17 December 1999 (T. Greenfield pers. comm.).

Painted Redstart

PARE

Myioborus pictus (Swainson)

RANGE: Breeds from northwestern and central Arizona (casually in southeastern California), southwestern New Mexico, and locally in the Trans-Pecos Region of western Texas south through the highlands to north-central Nicaragua. Winters mainly from northwestern Mexico south through the remainder of the breeding range.

STATUS: On the coast, *accidental* in the Georgia Depression Ecoprovince.

OCCURRENCE: A Painted Redstart was sighted in Ambleside Park, West Vancouver, on 4 November 1973 (Weber and Kautesk 1975), in association with a small flock of Black-capped and Chestnut-backed chickadees, Golden-crowned Kinglets, and Dark-eyed Juncos moving through a mixed forest of western hemlock, western redcedar, red alder, and bigleaf maple.

There are no records for Washington (Wahl and Paulson 1991) or Oregon (Schmidt 1989; Gilligan et al. 1994), but further south, in southern California, the Painted Redstart has been recorded in every month of the year and there is a breeding record (Unitt 1974; Roberson 1980; Small 1994).

In northeastern North America, where the Painted Redstart is considered a vagrant, most sightings were in October and November, including a bird captured at Pickering, Ontario, on 4 November 1971, exactly 2 years before the West Vancouver bird (Speirs and Pegg 1972; Weber and Kautesk 1975).

REMARKS: The Painted Redstart was formerly placed in the genus *Setophaga.*

Scarlet Tanager

SCTA

Piranga olivacea (Gmelin)

RANGE: Breeds from eastern North Dakota, southern Manitoba, southern Ontario, and southern New Brunswick south to central Nebraska, western Kansas, Oklahoma, central Arkansas, west-central Tennessee, northern Alabama, northern Georgia, northwestern South Carolina, eastern North Carolina, central Virginia, and Maryland. Winters in Panama, Colombia, Ecuador, and northwestern Bolivia.

STATUS: On the coast, *accidental* in the Georgia Depression Ecoprovince.

OCCURRENCE: A Scarlet Tanager was collected at Comox, on the southeast coast of Vancouver Island, on 17 November 1926 (Laing 1932). The specimen (RBCM 13912; Fig. 718) was originally identified as a Western Tanager (*P. ludoviciana*), but was later recognized as an immature male Scarlet Tanager.

This species is a rarity anywhere on the west coast of North America. In California, it is considered a very rare spring transient from early May to the end of June and a rare but regular autumn transient from late August to late November (Small 1994). Further north, there are 4 records for Oregon and 2 for Alaska, all in either May or June (Roberson 1980; Gilligan et al. 1994).

REMARKS: There are 3 other records for British Columbia between 1981 and 1996, all without supporting documentation. At least 1 of these was known to be a caged bird that was released near Abbotsford in 1985.

Figure 718. Immature male Scarlet Tanager collected at Comox, Vancouver Island, on 17 November 1926 (RBCM 13912; Michael C.E. McNall).

Green-tailed Towhee

Pipilo chlorurus (Audubon)

GTTO

RANGE: Breeds from southwestern and central Oregon, southeastern Washington, southern Idaho, southwestern Montana, northwestern and southeastern Wyoming, and north-central Colorado south to southern California and extreme northern Baja California, southern Nevada, central Arizona, southern New Mexico, and western Texas. Winters from southern Arizona, southern New Mexico, and western and southern Texas south to southern Baja California and central Mexico.

STATUS: On the coast, *casual* in the Georgia Depression Ecoprovince.

In the interior, *accidental* in the Southern Interior Mountains Ecoprovince.

OCCURRENCE: Five records. Our first documented record was of an adult that spent 5 days between 9 and 13 December 1974, foraging in a patch of pigweed at Fauquier, on the east side of Arrow Lakes. The bird was seen by several observers and was well described (L. Bumpus pers. comm.). In late 1984 and early 1985, an adult frequented a feeder at Comox, on southeastern Vancouver Island, from 10 November to 28 February (Hunn and Mattocks 1985; Mattocks 1985a). It was photographed on 17 November 1984 (BC Photo 996; Fig. 719). On 21 and 22 May 1993, a male was seen and heard singing at Queen Elizabeth Park, Vancouver (G. Ansell pers. comm.; BC Photo 1527). On 26 May 1996, an adult male "in brilliant spring plumage" was seen and heard singing at Wilson Creek, on the Sunshine Coast (Bowling 1996a). The final record was of an adult that spent 3 days foraging below a feeder in Glen Valley, near Abbotsford, from 4 to 6 June 1997 (BC Photo 1782; Fig. 720).

Figure 719. In late 1984 and early 1985, an adult Green-tailed Towhee regularly visited a feeder on southeastern Vancouver Island, in the Georgia Depression Ecoprovince of British Columbia (Comox, 17 November 1984; Tim Zurowski).

In southeastern Oregon, where the Green-tailed Towhee is a locally fairly common breeder, spring migrants arrive in late April and early May and autumn departure peaks in late August and early September (Gilligan et al. 1994). The British Columbia records in May are likely of spring overshoots; the winter occurrences are more difficult to explain.

REMARKS: For a summary of the life history of this poorly known species, see Dobbs et al. (1998).

Figure 720. In early June 1997, an adult Green-tailed Towhee (foreground) visited a feeder in the lower Fraser River valley (Glen Valley, 5 June 1997; Nanny Mulder ten Kate).

Sage Sparrow

SASP

Amphispiza belli (Cassin)

RANGE: Breeds from eastern Washington, eastern Oregon, southern Idaho, southwestern Wyoming, and northwestern Colorado south to southern California, central Baja California, southern Nevada, southwestern Utah, northeastern Arizona, and northwestern New Mexico. Winters in southerly portions of its breeding range. Resident along the California coast.

STATUS: On the coast, *accidental* in the Georgia Depression Ecoprovince.

In the interior, *casual* in the Southern Interior Ecoprovince.

OCCURRENCE: Six records. The first Canadian record was of a female of undetermined age collected on Lulu Island, south of Vancouver, on 2 October 1930 (Cumming 1932; Godfrey 1966; RBCM 6913). Four decades later, an adult was seen singing from the top of a big sagebrush at the summit of Richter Pass on 2 and 3 May 1970 (Cannings et al. 1987). On 11 June 1980 and 24 September 1981, an adult was seen and heard singing on sagebrush near White Lake, in the Okanagan valley (Cannings et al. 1987). The next record was from the Southern Mainland Coast, where an adult male was found at Katzie Marsh in Pitt Meadows on 27 April 1982 (Weber 1985). On 2 March 1996, a Sage Sparrow was located along the Grey Canal Road in Vernon (Bowling 1996a).

The Sage Sparrow breeds in California, eastern Oregon, and eastern Washington as close to British Columbia as Okanogan County. It is considered a vagrant west of the Cascade Mountains (Roberson 1980). In eastern Oregon, spring migrants usually first arrive in late February, with the main passage occurring in the middle of March; autumn migration peaks in mid-September (Littlefield 1990).

Figure 721. A Sage Sparrow on southern Vancouver Island, in the Georgia Depression Ecoprovince of British Columbia (along Dallas Road, Victoria, 19 February 1998; Jo Ann MacKenzie).

REMARKS: At least another 5 sightings (Comox, Anarchist Mountain, Cawston, Chopaka, and White Lake) have been examined, including sight records reported by S.J. Darcus in the summers of 1933 and 1934 in the Okanagan (see Parham 1937). They have been excluded from the account because of lack of documentation.

For details on the life history of the Sage Sparrow in North America, see Martin and Carlson (1998).

Postscript: A single bird was present along Dallas Road in Victoria from 16 to 19 February 1998 (Marie O'Shaughnessy, Laurie J. Savard, and Jo Ann MacKenzie pers. comm.; BC Photo 1783; Fig. 721).

The species' status on the coast has been elevated to *casual* in the Georgia Depression Ecoprovince.

Baird's Sparrow

BASP

Ammodramus bairdii (Audubon)

RANGE: Breeds from southeastern Alberta, southern Saskatchewan, and southwestern Manitoba south to central and eastern Montana, southern South Dakota, southeastern North Dakota, and west-central Minnesota. Winters from southeastern Arizona, southern New Mexico, and western and central Texas south to north-central Mexico.

STATUS: On the coast, *casual* in the Georgia Depression Ecoprovince.

OCCURRENCE: Three records, all from the Southern Mainland Coast of British Columbia. The first occurrence, overlooked by Brooks and Swarth (1925) and Munro and Cowan (1947), was a male of undetermined age collected at Vancouver in June 1889 (Stone 1899; ANSP 47715).

On 30 April 1978, a single Baird's Sparrow was found in Stanley Park, Vancouver, associating with about 10 Savannah Sparrows and up to 15 American Pipits (Kautesk 1982; Mattocks and Hunn 1978b). On 3 June 1981, another bird was discovered in partly dried mud patches with scattered weeds on the northwestern corner of Sea Island, south of Vancouver. Two Baird's Sparrows were seen at this location on 4 June 1981. We suspect that as many as 3 birds were present in the area until 16 June, during which time they were seen by many observers (Kautesk 1982; Weber 1981).

REMARKS: Other sightings, all reported in May and June, lack adequate documentation. They were from Comox, Ladner, Okanagan Landing, Prince George, Saanich Peninsula, Surrey, and Victoria.

McCown's Longspur
Calcarius mccownii (Lawrence)

MCLO

RANGE: Breeds from southwestern Alberta, southern Saskatchewan, and western North Dakota south through Montana and central and southeastern Wyoming to north-central Colorado and west-central Kansas. Winters from western Oklahoma to southeastern Arizona, south into northern Mexico.

STATUS: On the coast, *casual* in the Georgia Depression Ecoprovince.

In the interior, *accidental* in the Southern Interior Mountains Ecoprovince.

OCCURRENCE: Three records. On 1 June 1887, an adult male in full breeding plumage was collected on a little ridge of natural prairie at Chilliwack (Brooks 1917; MVZ 106635). Three years later, on 20 June 1890, 2 females were collected in the exact same location (Brooks 1917; MCZ 106636 and 244679).

The third record is from Tobacco Plains, across the Kootenay River from Newgate, where a female was collected on 29 May 1930 (Rand 1943; NMC 24633).

Roberson (1980) considers these records as spring overshoots west of the Alberta breeding range.

REMARKS: A single McCown's Longspur reportedly seen by J.A. Munro at Okanagan Landing on 18 August 1926 lacks details and has been excluded from this account. It is considered hypothetical by Cannings et al. (1987).

For additional information on the life history of this North American species, see With (1994).

Chestnut-collared Longspur
Calcarius ornatus (Townsend)

CCLO

RANGE: Breeds from southeastern Alberta, southern Saskatchewan, and southwestern Manitoba south, east of the Rocky Mountains, to northeastern Colorado, western Kansas, north-central Nebraska, and western Minnesota. Winters mainly from northern Arizona east to Oklahoma and Texas and south from Chihuahua and Coahuila to Zacatecas and Aguascalientes in central Mexico.

STATUS: On the coast, *casual* in the Georgia Depression and Coast and Mountains ecoprovinces.

In the interior, *casual* in the Southern Interior Mountains Ecoprovince.

OCCURRENCE: Six records. The first record of the Chestnut-collared Longspur in the province was of a female in worn summer plumage collected in the Kispiox River valley, north of Hazelton, on 8 July 1921 (Swarth 1924). Since then there have been 3 additional coastal occurrences: a male was photographed on Faber Islets, Barkley Sound, Vancouver Island, on 18 June 1972 (Hatler 1973; BC Photo 224), a male was observed foraging among grasses and rocks on the Victoria Golf Course in Oak Bay from 1 to 3 May 1977 (Crowell and Nehls 1977c), and a male in breeding plumage was seen in a grassy field with scattered clumps of small red alders at Maplewood Flats in North Vancouver from 8 to 10 July 1981 (Weber et al. 1982).

There are 2 acceptable records for the interior: a female was collected on the northern Tobacco Plains, across the Kootenay River from Newgate, on 28 May 1930 (Anderson 1932; NMC 24629); in 1984, 1 and 2 males in full breeding plumage were observed on 29 and 30 May, respectively, foraging in a strip of weedy, rocky land along the shore of Arrow Lakes at Nakusp. Rogers (1984a) incorrectly lists 2 birds for each day.

REMARKS: For a summary of life-history information on the Chestnut-collared Longspur in North America, see Hill and Gould (1997).

Rustic Bunting

RUBU

Emberiza rustica (Pallas)

RANGE: Occurs annually in North America in the western Aleutian Islands of Alaska and less regularly further east in the Aleutians and on the Alaskan islands in the Bering Sea. Native to northern Europe and Asia, where it breeds from northern Scandinavia, northern Russia, and northern Siberia southeast to southeastern Siberia, northern Sakhalin, the Sea of Okhotsk, and Kamchatka. Winters in eastern China, Japan, and, rarely, the Commander Islands.

STATUS: On the coast, *accidental* in the Georgia Depression Ecoprovince and *casual* in the Coast and Mountains Ecoprovince.

OCCURRENCE: Three records. The first Canadian record of the Rustic Bunting was from the Queen Charlotte Islands, where 2 birds were observed near Queen Charlotte City on 26 October 1971 (Crowell and Nehls 1972a; Godfrey 1986). The second record was of a bird in winter plumage that was present for 88 days at Jordan River, 80 km west of Victoria, from 25 November 1983 to 20 February 1984 (Campbell 1984b; Hunn and Mattocks 1984; Mattocks 1984). It regularly visited a feeder and usually associated with a large flock of Dark-eyed Juncos. It was photographed on 3 December 1983 (BC Photo 883; Fig. 722). On 8 December 1990, a single bird arrived in a backyard in Tofino, on the west coast of Vancouver Island, and remained in the vicinity of a feeder until at least 12 April 1991, when it was last reported (Siddle 1991b; Fig. 723).

The Rustic Bunting occurs annually in small numbers as migrants in the Aleutian Islands and irregularly elsewhere in south coastal Alaska (Roberson 1980; West and Bailey 1986). Recently, the species has been seen with increasing frequency along the Pacific coast of North America from British Columbia to northern California (Gilligan et al. 1994; Small 1994).

Figure 722. Rustic Bunting in winter plumage on southwestern Vancouver Island, in the Coast and Mountains Ecoprovince of British Columbia (Jordan River, 80 km west of Victoria, 3 December 1983; Tim Zurowski).

REMARKS: Two other sightings have been reported. A single bird was found in Coquitlam, east of Vancouver, on 12 December 1988, and another was seen briefly near Port Hardy, on northern Vancouver Island, on 4 January 1991. Both lack adequate details for inclusion here.

Figure 723. During the winter of 1990-91, a Rustic Bunting (centre) visited a feeder almost daily in Tofino, on the west coast of Vancouver Island, in the Coast and Mountains Ecoprovince of British Columbia (Tofino, January 1991; Aurora M. Paterson).

McKay's Bunting

Plectrophenax hyperboreus (Ridgway)

MCBU

RANGE: Breeds in Alaska on islands in the Bering Sea. Winters on the coast of western Alaska, rarely in the Aleutian Islands and south coastal Alaska (Sealy 1972).

STATUS: On the coast, *accidental* in the Coast and Mountains Ecoprovince.

OCCURRENCE: A McKay's Bunting was photographed near Wickaninnish Beach in Pacific Rim National Park, on the central west coast of Vancouver Island, on 12 February 1980 (Godfrey 1986).

The McKay's Bunting is one of the rarest of North American birds and is usually restricted to the vicinity of the Bering Sea where Snow Buntings do not breed. Roberson (1980) lists only 3 occurrences of this species for the west coast of North America outside Alaska. Single birds were present at Ocean Shores, Washington, from 16 December 1978 to 2 March 1979 and in Oregon, at the South Jetty of the Columbia River, from 23 February to mid-March 1980 (see Gilligan et al. 1994).

REMARKS: For a summary of the natural history of this little-studied species, see Lyon and Montgomerie (1995).

Blue Grosbeak

Guiraca caerulea (Linnaeus)

BLGB

RANGE: Breeds from central California, southern Nevada, southern and eastern Utah, southern Colorado, central South Dakota, western Iowa, central Illinois, southern Ohio, northern Kentucky, and southern Tennessee to southeastern Pennsylvania and extreme southern New York south to Texas, the Gulf coast, and central Florida to Costa Rica. Winters from Baja California and northern Mexico south to Costa Rica, and in the Bahamas and Cuba.

STATUS: In the interior, *casual* in the Southern Interior Mountains Ecoprovince.

OCCURRENCE: Two records. The first well-documented record for British Columbia was a single bird observed on 22 May 1982 on brushy slopes above the Goat River in Erickson, in the Creston valley (Butler et al. 1986). The second record was of an adult female Blue Grosbeak clearly seen and documented on 2 July 1991 near Downie Creek, about 41 km north of Revelstoke (Campbell and Dawe 1991). The bird was found among thimbleberry and fireweed in a regenerating burn known locally as the Rhizina Root Rot Test Area, just south of Downie Creek.

On the west coast of North America, the Blue Grosbeak is a very rare vagrant anywhere north of California. Other than the 2 British Columbia records, it has been recorded only twice, both in winter and both in Oregon (Gilligan et al. 1994), and not in Washington through 1990 (Wahl and Paulson 1991). Elsewhere in Canada, the Blue Grosbeak has been recorded as far west as Saskatoon, Saskatchewan (Godfrey 1986).

The northward expansion of the species' breeding range in the eastern United States during the 20th century was possibly due to the cutting of forests (Robbins et al. 1986).

REMARKS: For a summary of the distribution and life history of the Blue Grosbeak in North America, see Ingold (1993).

Dickcissel

DICK

Spiza americana (Gmelin)

Figure 724. Dickcissel at ground feeder on southern Vancouver Island, in the Georgia Depression Ecoprovince of British Columbia (Saanich, 26 September 1982; Tim Zurowski).

RANGE: Breeds from eastern Montana, southeastern Saskatchewan, southern Manitoba, northwestern and central Minnesota, northern Wisconsin, central Michigan, southern Ontario, central New York, and Massachusetts south to central Colorado, eastern New Mexico, western and southern Texas, southern Louisiana, central Mississippi, central Georgia, and South Carolina. Winters mainly from Mexico to northern South America; locally in small numbers in coastal lowlands from southern New England south to Florida and west to southern Texas.

STATUS: On the coast, *casual* in the Georgia Depression Ecoprovince; in the Coast and Mountains Ecoprovince, *casual* on Western Vancouver Island.

In the interior, *casual* in the Southern Interior Ecoprovince.

OCCURRENCE: Six records. A pair of Dickcissels, including a singing male on territory, was discovered in a newly mown alfalfa field on the east side of Vaseux Lake, in the Okanagan valley, on 12 June 1922. The male was collected (NMC 17883). This was the first Canadian record for this species west of the Rocky Mountains (Brooks 1923a). An adult female was present in Victoria from 14 November 1960 to 14 January 1961 (Davidson 1960, 1966; Boggs and Boggs 1961a, 1961b; Stirling 1961c). It was trapped and banded near a feeder on 16 November (C.J. Guiguet pers. comm.). The third record was of an adult male flushed from long grasses in sand dunes on Stubbs Island, near Tofino, on 3 June 1976 (Crowell and Nehls 1976d; BC Photo 442). On 3 August 1980, a female was observed in a mixed flock of migrant sparrows foraging in big sagebrush just south of Vaseux Lake (Cannings et al. 1987). The fifth record was of a bird photographed at a feeder in Saanich, on southern Vancouver Island, on 26 September 1982 (Hunn and Mattocks 1983; BC Photo 803; Fig. 724). In 1993, a male, likely the same bird, was present at feeders in Ucluelet from 23 to 24 May (Siddle 1993c; BC Photo 1812; Fig. 725) and in Tofino from 2 to 7 June, both on the west coast of Vancouver Island.

This species was found in, or adjacent to, open country, including pastures and cultivated fields, weedy and shrubby edges, grassy sand dunes, and sagebrush flats, as well as backyard feeders.

The Dickcissel is a rarity on the west coast of North America and is considered an irregular straggler to British Columbia (Roberson 1980; Gilligan et al. 1994).

REMARKS: There are 7 other records that appear to have been reliably reported but for which we are awaiting documentation. These are from Pitt Lake, Redstone, Williams Lake, Chilcotin River, Osoyoos, near Princeton, and Mackenzie.

We have excluded 2 coastal and 3 interior reports that lacked sufficient documentation: Marpole, Queen Charlotte City, Creston, Kamloops, and Cache Creek.

Figure 725. Male Dickcissel in breeding plumage at feeder on the west coast of Vancouver Island, in the Coast and Mountains Ecoprovince of British Columbia (Ucluelet, 23 May 1993; Donald G. Cecile).

Great-tailed Grackle

GTGR

Quiscalus mexicanus (Gmelin)

RANGE: Resident from southern interior California, southern Nevada, southern Utah, northern New Mexico, southeastern Colorado, Kansas, southern Nebraska, southwestern Missouri, southwestern Arkansas, and southwestern Louisiana south through Central America, to Costa Rica and Panama and along both coasts of South America.

STATUS: On the coast, *accidental* in the Coast and Mountains Ecoprovince.

In the interior, *accidental* in the Southern Interior Ecoprovince.

OCCURRENCE: Two records. A single bird was present at Cape St. James, on the southern tip of the Queen Charlotte Islands, from 6 to 9 May 1979. It was photographed on 7 May (BC Photo 630; Fig. 726). A female appeared in Vernon, in the Okanagan valley, on 4 December 1993 (Siddle 1994b) and remained over the winter until 11 April 1994 (Bowling 1994a).

The breeding range of the Great-tailed Grackle expanded greatly during the early part of the 20th century (American Ornithologists' Union 1983) and continued to do so during the 1990s. The species' recent range expansion in the western United States continues. Although still considered a vagrant in Oregon, the Great-tailed Grackle occurs almost annually at Malheur National Wildlife Refuge (Gilligan et al. 1994). It is, however, a species of infrequent occurrence in Washington (Wahl and Paulson 1991).

Figure 726. The first record of the Great-tailed Grackle in British Columbia was from Cape St. James, on the southern tip of the Queen Charlotte Islands, in the Coast and Mountains Ecoprovince of British Columbia (7 May 1979; Jack Bowling).

REMARKS: There are 3 recognized subspecies of the Great-tailed Grackle in North America. The British Columbia birds were well photographed but the races have not yet been identified. It is likely that birds occurring in British Columbia belong to the subspecies *Q. m. monsoni* Phillips, as this race has been rapidly expanding its range in the southern intermontane area (American Ornithologists' Union 1998).

Orchard Oriole

OROR

Icterus spurius (Linnaeus)

RANGE: Breeds from southeastern Saskatchewan, southwestern Manitoba, central Minnesota, central Wisconsin, central Michigan, southern Ontario, central New York, southern Vermont, and northern Massachusetts south to eastern Chihuahua, Coahuila, southern Texas, the Gulf coast, and central Florida; west to eastern Montana, eastern Wyoming, eastern Colorado, and southeastern New Mexico; and on the Mexican plateau. Winters from central Mexico and northern Yucatan Peninsula south through Central America to Panama, northern Colombia, and northwestern Venezuela.

STATUS: On the coast, *accidental* in the Georgia Depression Ecoprovince.

OCCURRENCE: On 22 May 1997, an immature male Orchard Oriole was well observed in mixed woodlands and at an orchard on northern Saltspring Island (R.B. Weeden pers. comm.). Eight days later, on 30 May, probably the same bird was seen at a hummingbird feeder in Sidney, on southern Vancouver Island (W. Fletcher pers. comm.).

During the past 30 years, the Orchard Oriole has slowly expanded its breeding range in North America northward into southern Manitoba and westward from Colorado towards the Pacific coast (Scharf and Kren 1996). In western North America, it is considered a rare spring transient from late March to early April, and a regular autumn transient from mid-August to late October, in California (Small 1994). Further north, in Oregon, vagrants have been recorded in May, June, September, November, and December (Gilligan et al. 1994).

REMARKS: For a detailed summary of the life history of the Orchard Oriole, see Scharf and Kren (1996).

Hooded Oriole

HOOR

Icterus cucullatus (Swainson)

RANGE: Breeds from northern coastal and central California, southern Nevada, southwestern Utah, central Arizona, southern New Mexico, and western and southern Texas south to Baja California, and through much of Mexico. Winters from northern Mexico (rarely southern California, southern Arizona, and southern Texas) south through the breeding range, and on the Pacific coast of Mexico from Nayarit to Oaxaca.

STATUS: On the coast, *casual* in the Georgia Depression Ecoprovince; in the Coast and Mountains Ecoprovince, *accidental* on the Northern Mainland Coast.

OCCURRENCE: Six records. The first occurrence for the province was only recently located in files at the Royal British Columbia Museum. In correspondence to C.J. Guiguet, Curator of Vertebrates, dated 24 May 1978, Mr. and Mrs. W. Williams adequately described a male Hooded Oriole that fed at their hummingbird feeder in Oak Bay, southern Vancouver Island, on 20 May 1978. Several neighbours also watched the bird. It was not seen the following day.

Nineteen years later, on 6 and 7 May 1996, a male visited another hummingbird feeder in Sidney, also on southern Vancouver Island (Bowling 1996c). Eighteen days later, on 25 May, another male was seen in Coquitlam, east of Vancouver, on the southwest mainland coast (Bowling 1996c). It seems unlikely that the same bird was involved in both sightings.

The next 3 records were all from 1997. At Terrace, east of Prince Rupert on the Northern Mainland Coast, an overwintering male visited a feeder almost daily for 135 days from 19 November 1997 to 2 April 1998 (BC Photo 1788; Fig. 727). On 19 July, a male visited a feeder near Esquimalt Lagoon, on southern Vancouver Island (B.R. Gates pers. comm.), and an adult male was seen at Blackie Spit, on the adjacent mainland coast, on 19 September (J.E. Williams pers. comm.).

In recent decades, following the planting of Washington fan palms, which are the preferred nesting tree for the Hooded Oriole, the species has expanded its range northward (Roberson 1980; Small 1994). It was first recorded in Oregon on 15 May 1963 (Browning 1966), and over the next 3 decades, was recorded on at least another 22 occasions (Gilligan et al. 1994). Almost half of these records were of spring birds that had probably overshot their normal summer range.

Figure 727. Male Hooded Oriole on the Northern Mainland Coast of the Coast and Mountains Ecoprovince of British Columbia (Terrace, 1 February 1998; Jo Ann MacKenzie).

Lesser Goldfinch

LEGO

Carduelis psaltria (Say)

RANGE: Resident from southwestern Washington, western Oregon, northern California, northern Utah, northern Colorado, northwestern Oklahoma, and central and southern Texas south to southern Baja California, through Pacific Middle America to northwestern South America.

STATUS: On the coast, *casual* in the Georgia Depression Ecoprovince.

In the interior, *accidental* in the Southern Interior Mountains Ecoprovince.

OCCURRENCE: Five records. On 9 June 1931, a male Lesser Goldfinch (*C. p. hesperophila*) was collected at Indianpoint Lake, near Bowron Lake Park, in the east-central portion of the province (MCZ 281803). T.T. McCabe (*in* Dickinson 1953) noted: "One lone [male] suddenly appeared below the dining room window and lit on a dandelion. I heard its plaintive note at the same time as I rushed for the gun ... crop filled with dandelion seeds ... moderately fat ... Looked and listened several times in course of day but [could] find no mate. A south wind has been blowing but should not suppose it likely to have blown this bird so far out of its range as it is not an especially strong wind."

On 17 May 1958, a female Lesser Goldfinch, also *C. p. hesperophila*, was collected during a heavy spring migration period at Huntingdon, on the extreme southern mainland coast of the province (Racey 1958; PMNH 72201). A single immature Lesser Goldfinch was carefully studied and well documented in Jericho Park, Vancouver, on 15 and 16 September 1983. A month later, on 22 and 23 October, possibly the same bird was present in Sechelt, about 40 km northwest of Vancouver (T. Greenfield pers. comm.; Hunn and Mattocks 1983). On 6 September 1993, a single bird was observed in Jericho Park, Vancouver (Siddle 1994a).

REMARKS: Formerly in the genus *Spinus*, the Lesser Goldfinch was also known as Dark-backed, Green-backed, or Arkansas goldfinch (American Ornithologists' Union 1983; Godfrey 1986).

There are 4 other unpublished records for the province, all without convincing details. Three of these are from the Lower Mainland (Vancouver, Surrey, and Langley); the other is from the Okanagan valley.

For a summary of the life history of the Lesser Goldfinch in North and Central America, see Watt and Willoughby (1999).

POSTSCRIPT: An adult male was present at a niger seed feeder in Princeton from 11 to 16 May 1999 (R.J. Herzig pers. comm.; BC Photo 1789; Fig. 728).

Figure 728. Adult male Lesser Goldfinch in the Southern Interior Ecoprovince of British Columbia (Princeton, 12 May 1999; R. Jerry Herzig).

Additions to the Avifauna of British Columbia, 1987 through 1999

ADDITIONS TO THE AVIFAUNA OF BRITISH COLUMBIA, 1987 THROUGH 1999

The following species are either new additions to the avifauna of British Columbia or have been recorded as new breeding species since the publication of Volumes 1 and 2 (1990) and Volume 3 (1997).

Clark's Grebe CLGR

Aechmophorus clarkii (Lawrence)

CHANGE IN STATUS: Breeds.

BREEDING: Since the late 1980s, there have been several nesting reports of paired Western and Clark's grebes from Duck Lake, near Creston, and Shuswap Lake, near Salmon Arm (S. Forbes and F. Kime pers. comm.). On 6 June 1996, a nest containing 4 eggs, attended by a pair of Clark's Grebes, was found at "Christmas" Island in Shuswap Lake, representing the first confirmed breeding in the province. Later that summer, on 25 July, 2 chicks were seen swimming with the adults among a Western Grebe colony on the island (M. Ward pers. comm.).

The nearest active breeding colony is located in northern Grant County, Washington, about 400 km southeast of Salmon Arm (Smith et al. 1997).

REMARKS: See Storer and Nuechterlein (1992) for information on the breeding biology of the Clark's Grebe in North America. See also, Volume 1, p. 374.

Northern Fulmar NOFU

Fulmarus glacialis (Linnaeus)

CHANGE IN STATUS: Breeds.

BREEDING: Between 1974 and 1977, up to 6 pairs of the Northern Fulmar have been suspected of breeding on Triangle Island, off the northwestern tip of Vancouver Island (Campbell et al. 1990a). The nests were inaccessible and could not be checked for contents. On 14 June 1979, a fisherman from Port Hardy, anchored at Triangle Island to photograph seabirds, managed to scale the precipitous cliffs on "Puffin Rock" and discovered 3 Northern Fulmars in incubating positions (W. Johnson pers. comm.). One bird was flushed from a single, large white egg (BC Photo 1855).

Between 1982 and 1989, mock courtship behaviour, feeding activities at "nest" sites, and Northern Fulmars "sitting" in Pelagic Cormorant nests have been observed, but conclusive proof of additional nesting was not obtained (Rodway et al. 1990).

REMARKS: Most Northern Fulmars observed on and around Triangle Island have been described as "double dark" in plumage characteristics. See Fisher (1952) for a discussion of extremes of colour variation in this pelagic species. See also, Volume 1, pp. 186-187.

Murphy's Petrel MUPE

Pterodroma ultima (Murphy)

RANGE: Breeds in the south-central Pacific Ocean in the Austral, Tuamotu, and Pitcairn islands. Ranges at sea in the northern and central Pacific Ocean north to the Hawaiian Islands, to the Pacific coast off California and Mexico, and casually off Washington and Oregon.

STATUS: *Casual* offshore.

OCCURRENCE: Five records. During pelagic surveys off the British Columbia coast from 1997 to 1999, Michael D. Bentley observed single Murphy's Petrels at 49°12', 133°47' (16 February 1997), 49°14', 134°02' (23 February), 51°40', 136°30' (6 July 1998), and 53°18', 135°39' (8 July 1998), 50°19', 132°08' (16 June 1999), (K. Morgan pers. comm.).

The first documented record for North America was a bird found dead near Newport, Oregon, on 15 June 1981 (Gilligan et al. 1993). Prior to 1990, the Murphy's Petrel was considered accidental in California (30 April 1989; Erickson et al. 1989). Recently, as exploration of offshore waters continues, the species appears to be a fairly common visitor to pelagic waters off California during the northern hemisphere spring from early April to mid-June (Small 1994).

REMARKS: See Stallcup (1990) for additional details on the distribution and identification of the Murphy's Petrel in the nearshore Pacific Ocean.

Manx Shearwater MASH

Puffinus puffinus (Brünnich)

RANGE: Breeds in the North Atlantic on islands off Newfoundland and Massachusetts, and from Iceland and the Faroe and Shetland islands south around most of the British Isles to western France, and in Madeira and the Azores. Ranges at sea to the western Atlantic along the coast of North America from Newfoundland south to Florida and Bermuda, and to the east coast of South America from Trinidad to Argentina. Casual in Washington, California, and the Gulf coast of North America.

STATUS: *Casual* in summer offshore.

OCCURRENCE: Two records. A single bird was well observed at sea off the northern tip of Vancouver Island (50°45′, 129°34′) by Michael Bentley on 18 May 1998 (K. Morgan pers. comm.). About 3 weeks later, on 6 June 1998, a single bird was carefully studied near the southeast edge of Laskeet Bank, southwest of Hecate Strait off the Queen Charlotte Islands (M. Force pers. comm.).

The Manx Shearwater is an extremely rare vagrant to the Pacific Ocean off North America (Lee and Haney 1996). There are several summer and autumn records for California (Small 1994) and 3 for south coastal Alaska where the species is considered a "casual summer visitant" (Kessel and Gibson 1978; Roberson 1980).

REMARKS: See Lee and Haney (1996) for details on the identification, distribution, taxonomy, and life history of the Manx Shearwater in North America.

Red-tailed Tropicbird RTTR

Phaethon rubricauda (Boddaert)

RANGE: Breeds on islands in the Pacific Ocean from the Hawaiian Islands and Bonin and Volcano islands south to northeastern Australia and Lord Howe, Kermadec, Tuamotu, and Pitcairn islands; and in the Indian Ocean near Mauritius, in the Cocos-Keeling Islands, and off the northwestern coast of Australia. Ranges at sea from Japan, Taiwan, and the Hawaiian Islands south throughout the breeding range and east to waters well off California and islands off Mexico.

STATUS: *Accidental* in the Georgia Depression Ecoprovince.

OCCURRENCE: Parts of a carcass of a Red-tailed Tropicbird were discovered along the Lupin Falls trail in Buttle Lake Park on central Vancouver Island on 5 June 1992 (B. Brooks pers. comm.; RBCM 23668).

The Red-tailed Tropicbird ranges widely in the tropics of the Pacific and Indian oceans and is recorded with some regularity only off Baja California in North America (Roberson 1980). The British Columbia occurrence coincides with higher than average ocean temperatures in the eastern Pacific Ocean.

REMARKS: See Schreiber and Schreiber (1993) for a detailed summary of the life history of the Red-tailed Tropicbird in the world.

Black Vulture

Coragyps atratus (Bechstein)

Figure 729. Sketch from original field notes of a Black Vulture seen near Keremeos, in the southern Okanagan valley, on 3 May 1975 (illustration by Glenn R. Ryder).

RANGE: Resident from southern Arizona, Chihuahua, western Texas, eastern Oklahoma, southern Missouri, southern Illinois, southern Indiana, central Ohio, south-central Pennsylvania, and New Jersey south to the Gulf coast and southern Florida, and throughout Middle America and south America to central Chile and central Argentina. Recorded in summer (and possibly breeding) north to southeastern New York and Maine. Wanders casually north to southwestern Yukon, Saskatchewan, North Dakota, southern Ontario, southern Quebec, New Brunswick, Prince Edward Island, and Nova Scotia.

STATUS: *Casual* in the Southern Interior Ecoprovince.

Figure 730. The all-black body, short squarish tail, and white at the base of the primary feathers identify this bird as a Black Vulture (north of Osoyoos, 16 June 1982; B. Murphy).

OCCURRENCE: Three records, all from the Okanagan valley. The Black Vulture was previously listed by Campbell et al. (1990b, p. 487) as "hypothetical" pending additional information on the status of captive birds in the Pacific Northwest. It now appears that the records are valid as they correspond with recent range expansions in North America (Buckley 1999).

The first record, an adult, was seen northeast of Keremeos on 3 May 1975 (G.R. Ryder pers. comm.; Fig. 729). It was being attacked and chased by two adult Golden Eagles. The second occurrence was a single bird observed soaring high overhead along a cliff edge about 1 km south of Okanagan Falls Park on 25 June 1981 (Brunton and Pratt 1986). The third record, an adult, was seen and photographed on 16 June 1982 (Fig. 730; BC Photo 1860) while soaring over Road 22, south of Oliver (B. Murphy pers. comm.).

REMARKS: See Buckley (1999) for a thorough summary of the life history of the Black Vulture in the Americas.

Whooper Swan WHSW

Cygnus cygnus (Linnaeus)

RANGE: Breeds from Greenland, Iceland, the Faroe Islands, Scotland, Scandinavia, and northern Russia east to Anadyrland and Kamchatka, and south to Poland, the Caspian Sea, Turkestan, and Ussuriland. One breeding record on Attu, Aleutian Islands, Alaska. Winters south to central Europe, the eastern Mediterranean, Black, and Caspian seas; and from Korea and Japan south to eastern China and east to the central Aleutian Islands.

STATUS: *Accidental* in the Georgia Depression and Coast and Mountains ecoprovinces; *accidental* in the Southern Interior Ecoprovince.

OCCURRENCE: There are 3 records, 1 in summer and 2 in the autumn.

The first record was discovered only recently in the historical field notes of Theed Pearse who lived at Comox on Vancouver Island. On 11 November 1977, knowledgeable birdwatchers (e.g., Tex Lyon) discovered an adult Whooper Swan among a flock of 26 adult and juvenile Trumpeter Swans in a bay near Port Hardy on northern Vancouver Island. The flock had departed early the following day. This sighting is noteworthy because on 23 October 1977, 2 Whooper Swans were found among a flock of 26 Trumpeter Swans at Cordova in southcoastal Alaska (Kessel and Gibson 1978).

A lone adult Whooper Swan was present in Seal Bay, in the Comox-Courtenay area of Vancouver Island, from 25 to 27 July 1996 (Innes 1997). It was photographed on 26 July (BC Photo 1509) and later confirmed by Daniel D. Gibson (pers. comm.).

In the Southern Interior, an adult and juvenile were present on Mamit Lake, 16 km south of Logan Lake, from 7 to 17 November 1999 (W.C. Weber pers. comm.). Thorough searches of areas in southern British Columbia where swans are known to winter (e.g., South Thompson River; Fraser River delta, Nicomen Slough, and Harrison River) failed to locate the 2 whoopers.

The Whooper Swan is considered an uncommon local winter visitant in the western and central Aleutian Islands of Alaska from early November through mid-April (Kessel and Gibson 1978) where it often occurs in family groups or small flocks.

REMARKS: See Roberson (1980) for tips on identifying the Whooper Swan in the field.

Falcated Duck FADU

Anas falcata (Georgi)

RANGE: Breeds in eastern Siberia from the Yenisei River east to the Sea of Okhotsk and Kamchatka, and south to Lake Baikal, Mongolia, Amurland, Sakhalin, and Japan. Winters from Japan south to Korea and eastern China, less frequently to Iran, India, Myanmar (formerly Burma), Vietnam, and southeastern China.

STATUS: *Casual* in Western Vancouver Island of the Coast and Mountains Ecoprovince.

OCCURRENCE: In 1932, a male Falcated Duck was seen near Vernon, in the Okanagan valley (Brooks 1932), but was not included in the official list of British Columbia birds because of the strong possibility that it had escaped from captivity (Campbell et al. 1990a, p. 391).

In 1995 and 1996, a male Falcated Duck, likely the same individual in both years, visited mudflats in the Tofino area on Western Vancouver Island (A. Dorst pers. comm.). In 1995 it was seen from 20 January to 19 February, and in 1996 from 27 January through 1 February.

The Falcated Duck is considered a very rare spring migrant and casual in other seasons in the western and central Aleutian Islands and Pribilof Islands in Alaska (Kessel and Gibson 1978).

REMARKS: The Falcated Duck is also known as the Falcated Teal.

Spectacled Eider SPEI

Somateria fischeri (Brandt)

RANGE: Breeds on the Arctic coast of Alaska from Point Barrow south to St. Lawrence Island and the mouth of the Kuskokwim River, and along the Arctic coast of Siberia from the Yana Delta east to the Chukotski Peninsula. Winters in openings in the Bering Sea icepack.

STATUS: *Accidental* in the Georgia Depression and Coast and Mountains ecoprovinces.

OCCURRENCE: Two records. On 22 September 1962, Charles J. Guiguet observed a male "in full or near full plumage" near James Island off southern Vancouver Island. Godfrey (1986) and Campbell et al. (1990a, p. 391) considered this record hypothetical. Recently, field notes were discovered in archives in the Royal British Columbia Museum, which were not submitted to Earl Godfrey for examination. Guiguet's typed notes clearly identify the species and also indicate that the bird was seen by a fishing companion familiar with bird identification. On 15 June 1988, a male was seen roosting with 3 Harlequin Ducks on intertidal rocks in Alliford Bay, Queen Charlotte Islands, by several experienced birdwatchers.

These records apparently constitute the first occurrences of the Spectacled Eider south of its breeding range in Alaska (American Ornithologists' Union 1998).

REMARKS: The Spectacled Eider is the least known North American waterfowl species. See Johnson (1989) for a summary of its biology in the Bering Sea.

White-tailed Kite WTKI

Elanus leucurus (Vieillot)

RANGE: Resident locally from southwestern Washington south to Baja California, in southern Arizona, Sinaloa, Nayarit, and Chihuahua, in peninsular Florida, from northeastern Kansas, southern Oklahoma, southern Louisiana, southwestern Mississippi, east-central and southern Texas, Tamaulipas, and southeastern Oaxaca south through Middle America to eastern Panama south to Bolivia, Brazil, central Argentina, and central Chile in South America.

STATUS: *Casual* in the Georgia Depression Ecoprovince.

OCCURRENCE: Three records. An adult was located on Reifel Island, Delta, on 26 April 1990 (Siddle 1990b). The second record was an adult observed in Pitt Meadows, 30 km east of Vancouver, on 18 and 19 April 1995 (T. Plath pers. comm.). The final record, an adult, appeared in North Saanich on 21 April 1992 and 5 days later an adult was also found in Colwood, 24 km away (Begg 1992). The last 2 records probably involved the same individual. There are additional records for April, May, June, and September for which we have not received full details.

In Oregon, the first White-tailed Kite was found in the mid-1920s and since then their numbers have increased and decreased with fluctuations in California. The species is considered a "rare to locally fairly common fall and winter visitant" on the coast and in the interior valleys of western Oregon from late August through April (Gilligan et al. 1994). In Washington state, it was first reported in 1975 and infrequently since then on both the coast and in the interior (Roberson 1980).

REMARKS: The White-tailed Kite was formerly known as the Black-shouldered Kite.

Crested Caracara CRCA

Caracara plancus (Miller)

RANGE: Resident in central and southern Florida and Cuba (including the Isle of Youth), and from northern Baja California, southern Arizona, Sonora, Sinaloa, Zacatecas, Nuevo León, central and southern Texas, and southwestern Louisiana south locally through Middle America, and throughout most of South America south to northern and central Peru and northern Brazil.

STATUS: *Accidental* in the Coast and Mountains Ecoprovince.

OCCURRENCE: A single bird was present from 4 May to at least 3 June 1998 at Drury Inlet due west of Port Hardy on northern Vancouver Island (Plath 1998; BC Photo 1854). The caracara foraged with Northwestern Crows along the beach at low tides and regularly fed on food scraps thrown out daily by workers at the logging camp.

The Crested Caracara is a rare straggler north of its range in the southern United States. On the Pacific coast, it has been recorded only 3 times in California (Small 1994) and once in Oregon (Morrison 1996).

REMARKS: See Morrison (1996) for a summary of the life history of the Caracara in North America.

Yellow Rail YERA

Coturnicops noveboracensis (Gmelin)

RANGE: Breeds locally from northwestern Alberta, southern Mackenzie, central Saskatchewan, northern Manitoba, northern Ontario, central Quebec, New Brunswick, eastern Maine south to south-central Oregon, southern Alberta, southern Saskatchewan, North Dakota, central Minnesota, central Wisconsin, northern Michigan, southern Ontario, and southern Quebec, and to northern Illinois and central Ohio. Winters from coastal North Carolina south to southern Florida, west through the central and southern Gulf states to central and southeastern Texas and Arkansas, and from Oregon south to southern California.

STATUS: *Casual* in the Southern Interior Mountains and Boreal Plains ecoprovinces.

OCCURRENCE: Four records. On 10 June 1989, a bird was heard calling in a sedge meadow at Boundary Lake, east of Fort St. John, in the Peace Lowland. The bird was not seen but the "ticking" call of the species was "unmistakable" (Weber and Cannings 1990). The second record was of a bird flushed on 10 June 1994 in a sedge meadow on the floodplain of the Kickinghorse River in Yoho National Park (Sherrington 1994). The third record, also from the Southern Interior Mountains, was photographed (BC Photo 1776) while foraging on the edge of cattail and bulrushes at "Cranberry Marsh" near Valemount. The fourth record involved 2 or 3 birds that were regularly heard in a marsh at Chetwynd, in the Boreal Plains, from 3 June to 13 July 1997 (Bowling 1997d).

The Yellow Rail is today a very local summer and winter visitant in Oregon and California where land clearing, cattle grazing, and the disappearance of extensive coastal marshes and wet meadows have impacted its historical populations (Remsen 1977; Gilligan et al. 1994; Small 1994).

REMARKS: The Yellow Rail was considered hypothetical by Campbell et al. (1990b, p. 488). See Roberson (1980) for notes on field identification of the Yellow Rail.

Pacific Golden-Plover PGPL
Pluvialis fulva (Gmelin)

RANGE: Breeds along the Bering coast of Alaska, and in Eurasia from the Arctic coast of Siberia south to the Stanovoi and Koryak mountains and the Gulf of Anadyr. Winters in the Hawaiian Islands, and in the Old World from northeastern Africa, the Red Sea, India, southern China, Taiwan, and islands of Polynesia south to the Malay Peninsula, Australia, Tasmania, New Zealand, and the Tonga and Tuamotu islands.

STATUS: On the coast, *very rare* to *rare* spring and autumn migrant in the Georgia Depression Ecoprovince and on Western Vancouver Island in the Coast and Mountains Ecoprovince.

In the interior, *very rare* autumn migrant in the Southern Interior Ecoprovince; *accidental* in the Sub-Boreal Interior and Boreal Plains ecoprovinces.

OCCURRENCE: The current status of this species is not well known as biologists and birdwatchers have only recently started to distinguish between American and Pacific golden-plovers in the field.

Figure 731. The status of the Pacific Golden-Plover in British Columbia is not well known because it was recently elevated from being a subspecies of the Lesser Golden-Plover, *Pluvialis dominica*, to full species status (near Long Beach, Vancouver Island, 7 May 1998; Adrian Dorst).

Paulson (1993) indicates that the Pacific Golden-Plover (Fig. 731) is largely restricted to the Pacific coast of North America. In British Columbia, small numbers migrate each year through south coastal portions of the province in spring and autumn (Gates and Taylor 1994). A. Dorst (pers. comm.) suggests that all golden-plovers on the central west coast of Vancouver Island in spring are migrating *fulva*. Since 1997, he has recorded up to 5 birds there each spring between 30 April and 25 May. The species is found less frequently there in autumn, mainly from 11 September to 25 October.

In the interior of the province, the Pacific Golden-Plover is not recorded every year. In the Southern Interior it has been found only in autumn (Cannings et al. 1997; Howie 1994). Elsewhere there are single occurrences in spring from the Prince George area in the Sub-Boreal Interior (Prince George Naturalists Club 1996) and Dawson Creek in the Boreal Plains (Bowling 1995c).

REMARKS: In the early 1980s, Connors (1983) demonstrated that *Pluvialis fulva* was a valid species and should not be considered a subspecies of *P. dominica*. A decade later, *fulva* was elevated to full species status (American Ornithologists' Union 1993). See also, Volume 2, p. 114.

Paulson (1993) reviews the status of the American and Pacific Golden-Plover in the Pacific Northwest and has detailed notes on identifying the 2 species.

Mongolian Plover MGPL
Charadrius mongolus (Pallas)

RANGE: Breeds in central and northeastern Eurasia from the Pamir Mountains east to the Chukotski Peninsula, Kamchatka, and the Commander Islands, and south to western China and Tibet; has bred in North America in northern and western

Alaska. Winters in the Old World from the Red Sea, Iran, India, Southeast Asia, southeastern China, the Philippines, and the Mariana and Caroline islands south to southern Africa, the Seychelles, Sri Lanka, Andaman Islands, Indonesia, New Guinea, and Australia.

STATUS: *Accidental* in the Georgia Depression and Coast and Mountains ecoprovinces.

OCCURRENCE: Two records. In 1993, an adult male in breeding plumage was carefully identified on Rose Spit, Queen Charlotte Islands, on 22 June and an adult male was present at Esquimalt Lagoon, southern Vancouver Island, from 26 to 29 July (Davidson 1994).

The Mongolian Plover is a rare spring migrant (mid-May through early June) and very rare autumn migrant (late July through early September) in the western Aleutian Islands. South of Alaska, the species has been recorded between 11 July and 21 October, mainly in Oregon and California (Paulson 1993).

REMARKS: For additional information on the identification, behaviour, and habitat of the Mongolian Plover in North America refer to Paulson (1993).

Mountain Plover — MOPL

Charadrius montanus (Townsend)

RANGE: Breeds locally from southern Alberta, southwestern Saskatchewan, northern Montana, and southwestern North Dakota south through Wyoming, western Nebraska, Colorado, and western Kansas to northeastern Utah, northwestern, central, and southeastern New Mexico, western Texas, and western Oklahoma. Winters locally from central California, Nevada, southern Arizona, and central and coastal Texas south to southern Baja California and northern Mexico.

STATUS: *Accidental* in the Georgia Depression Ecoprovince.

OCCURRENCE: A single bird, associating with American Golden-Plovers and Black-bellied Plovers, was carefully studied on Westham Island, Delta, on 28 September 1986 (R. Toochin pers. comm.).

The Mountain Plover is considered a vagrant in the Pacific Northwest with 5 records from the coast (a winter visitor, 19 November to 10 March) and 6 records from interior locations (a migrant, 6 to 29 May and 19 July to 3 November) in Washington, Oregon, and Idaho (Paulson 1993).

REMARKS: See Paulson (1993) for a thorough summary of the status of the Mountain Plover in the Pacific Northwest and helpful hints on identification of various plumages.

Wood Sandpiper — WOSA

Tringa glareola (Linnaeus)

RANGE: Breeds in North America in the western and central Aleutian Islands; and in Eurasia from Scandinavia, northern Russia, and northern Siberia south to southern Europe, Turkestan, northern Mongolia, Kamchatka, the Kurile and Commander islands, and the Chukotski Peninsula. Winters from the Mediterranean region, Iran, India, northern Thailand, and southern China south to southern Africa, Sri Lanka, the Malay Peninsula, East Indies, and Australia.

STATUS: *Casual* in the Coast and Mountains Ecoprovince.

OCCURRENCE: Two records. An adult was present at Masset, on the northern Queen Charlotte Islands, from 3 to 9 November 1994 (Bowling 1995a). The following year a single bird was discovered in Pacific Rim National Park on 21 July 1995 (Bowling 1995d).

The Wood Sandpiper is an Old World species that is a fairly common shorebird in spring and autumn in the western Aleutian Islands (Paulson 1993). It is also a very rare breeder in the western and central Aleutian Islands (White et al. 1974).

There are unconfirmed sight records for California in August (Bailey et al. 1986) and Washington in November (Paulson 1993).

REMARKS: See Tree and Kieser (1982) for helpful tips on separating the Lesser Yellowlegs and Wood Sandpiper in the field.

Upland Sandpiper — UPSA

Bartramia longicauda (Bechstein)

CHANGE IN STATUS: Breeds.

BREEDING: The Upland Sandpiper was suspected of breeding in the province since the early 1900s (Brooks 1920). During the next 75 years, there were frequent reports of territorial adults calling from fence posts and wires, and agitated adults feigning injury in agricultural fields. The most often reported sites included local sites in the Cariboo and Chilcotin areas, near Rossland, and throughout the Peace River region (Godfrey 1986; Campbell et al. 1990b, pp. 154-155).

On 11 June 1997, an adult was flushed from a nest containing 4 eggs near Progress in the Peace Lowland of the Boreal Plains Ecoprovince (H. Fraser pers. comm.). The nest was located on the ground, in a slight depression, among tall grasses in an overgrown fallow field. The nest hollow was lined with dry grasses, a few leaves, and some coarse plant stems.

REMARKS: See Paulson (1993) for a summary of the present and historical status of the Upland Sandpiper in the Pacific Northwest. Buss (1951) and Kirsch and Higgins (1976) provide information on the breeding biology and management of this red-listed grassland bird (Fraser et al. 1999).

Great Knot GRKN

Calidris tenuirostris (Horsfield)

RANGE: Breeds in the mountains of northeastern Siberia from the lower Kolyma to Anadyrland. Winters from the Persian Gulf, India, and Malaysia, east and south to the Philippines, East Indies, New Guinea, and Australia.

STATUS: *Casual* in the Georgia Depression Ecoprovince.

OCCURRENCE: Two records. A first-year bird was watched foraging on intertidal mud flats at Boundary Bay, 20 km south of Vancouver, on 13 May 1987. The second record, also from the southwest mainland coast, involved a single bird located at Iona Island, Richmond, on the unusual date of 12 January 1998 (T. Plath pers. comm.).

There are a few late-spring records in North America, mostly from around the Bering Sea (Paulson 1993). The only vagrant records from the Pacific Northwest are in autumn from coastal Oregon (Gilligan et al. 1994).

REMARKS: See Paulson (1993) for details on the identification and life history of the Great Knot in the Pacific Northwest.

Baird's Sandpiper BASA

Calidris bairdii (Coues)

Figure 732. Baird's Sandpiper chick (Haines Highway, 2 July 1996; Linda M. Van Damme).

CHANGE IN STATUS: Breeds.

BREEDING: A recently hatched chick was discovered in muskeg habitat along the Haines Highway, in extreme northwestern British Columbia, on 2 July 1996 (L.M. Van Damme pers. comm.; Fig. 732). Although the Baird's Sandpiper has been suspected of breeding in this portion of the province since the late 1980s, the chick found in 1996 represents the first breeding record for British Columbia.

This record extends the breeding range of the Baird's Sandpiper about 900 km southeast from the species' known breeding range in central Alaska (American Ornithologists' Union 1998).

REMARKS: See also, Volume 2, pp. 190-191. Drury (1961) and Norton (1972) provide details on the breeding biology of this high-arctic nesting species.

Laughing Gull LAGU

Larus atricilla (Linnaeus)

RANGE: Breeds on the Pacific coast of western Mexico from the head of the Gulf of California and northwestern Sonora south to Colima; and in the Atlantic-Gulf-Caribbean region from southern New Brunswick and southern Nova Scotia south locally along the coast of Florida and west to southern Texas, through the West Indies to islands off the north coast of Venezuela and to French Guiana, and on the islands off Campeche and Yucatán. Winters along the Pacific coast from Mexico south to northern Peru; and from the Gulf coast and North Carolina south throughout the Gulf-Caribbean region to the coast of South America.

STATUS: *Accidental* in the Georgia Depression Ecoprovince.

OCCURRENCE: A bird in first summer plumage was seen and carefully documented at Iona Island, Richmond, on 28 July 1998. (T. Plath pers. comm.).

The Laughing Gull is extremely rare outside the Salton Sea, California, where post-breeding individuals visit annually in summer and early autumn (Roberson 1980). North of California, it has been recorded at Lower Klamath National Wildlife Refuge, Oregon, on 24 April 1983 (Gilligan et al. 1994) and at the north jetty of the Columbia River, Washington, on 1 September 1975 (Roberson 1980).

REMARKS: See Grant (1982) for a thorough discussion on field identification of medium-sized gulls.

Franklin's Gull FRGU

Larus pipixcan (Wagler)

CHANGE IN STATUS: Breeds.

BREEDING: In the late 1970s, rumours persisted that Franklin's Gulls were breeding on a large island in Kotcho Lake in the Taiga Plains of northeastern British Columbia. On 26 June 1982, the island was carefully searched but no evidence of nesting was found, although a flock of 80 birds was scared up by an aircraft from a marshy area of the island the day earlier (Campbell and McNall 1982).

On 19 June 1984, 3 nests were found in a marshy section of the island in Kotcho Lake (H. Fraser pers. comm.). Two nests contained 2 eggs, the other a single egg. The nests were built over shallow water and composed of a variety of dry and wet aquatic vegetation. The nest saucer was lined with dry grasses and reeds.

The Franklin's Gull breeds commonly in southern and central Alberta east of the Rocky Mountains (American Ornithologists' Union 1998). Since the 1950s and 1960s, it has been expanding its range in areas of northern Alberta (Semenchuk, 1992), which may account for the species' peripheral nesting in British Columbia.

REMARKS: See also, Volume 2, pp. 228-229. Guay (1968) reports additional information on the breeding biology of the Franklin's Gull in Alberta.

Black-legged Kittiwake BLKI

Rissa tridactyla (Linnaeus)

CHANGE IN STATUS: Breeds.

BREEDING: On 23 June 1997, 3 nests were discovered on Holland Rock in Chatham Sound, south of Prince Rupert, on the Northern Mainland Coast. One nest contained 1 egg and 2 nests were empty. On July 22, there was 1 chick in 1 nest and 2 nests contained a single egg (J. Bowling pers. comm.).

The Black-legged Kittiwake has been extending its breeding range in southeast Alaska. In the 1970s, about 1,600 birds were known to breed at 2 colonies along the outside coast of southeast Alaska (Sowls et al. 1978). By the early 1980s, over 3,100 birds were found breeding at 8 colonies and the species' range had expanded south along the Pacific coast to Glacier Bay (Sowls et al. 1982; American Ornithologists' Union 1983).

REMARKS: See also, Volume 2, pp. 274-275. Springer et al. (1982) summarize the breeding biology of the Black-legged Kittiwake in the northeast Pacific Ocean.

Red-legged Kittiwake RLKI

Rissa brevirostris (Bruch)

Figure 733. Adult Red-legged Kittiwake, in winter plumage, at Sandspit, Queen Charlotte Islands, British Columbia, 6 December 1992 (Michael D. Bentley).

RANGE: Breeds in Alaska in the Pribilof, Aleutian, and Commander islands. Winters in extreme northern Pacific Ocean, occurring east to the Gulf of Alaska.

STATUS: *Accidental* in the Coast and Mountains Ecoprovince.

OCCURRENCE: A winter adult was photographed at Little Spit Point in Sandspit on Moresby Island, Queen Charlotte Islands, on 6 December 1992 (Fig. 733; BC Photo 1602). The bird was first observed in a weakened condition and was found dead, and partially eaten, the following day (M.D. Bentley pers. comm.).

The Red-legged Kittiwake is a rarity anywhere south of Alaska (Roberson 1980). There are 7 vagrant occurrences for Oregon, 6 of which have been found dead on beaches in January, March, and December (Gilligan et al. 1994).

REMARKS: See Byrd and Williams (1993) for a summary of the distribution and life history of this pelagic gull.

Least Tern LETE

Sterna antillarum (Lesson)

RANGE: Breeds along the Pacific coast from central California south to southern Baja California and Chiapas; locally in the interior of North America; along the Atlantic-Gulf coast from Maine south to Florida and west to Tamaulipas. Winters in South America along the Pacific coast south to Peru and along the Atlantic coast from Colombia east to eastern Brazil.

STATUS: *Accidental* in the Southern Interior Ecoprovince.

OCCURRENCE: A Least Tern was discovered flying along a water channel at the north end of Osoyoos Lake on 6 June 1998 (E. Lemon and M. Smith pers. comm.). The following day it was photographed (BC Photo 1635; J.A. MacKenzie pers. comm.).

The Least Tern is a regular, but local, summer visitant along the west coast of California north to San Francisco Bay (American Ornithologists' Union 1998). It is casual in spring and summer on the Pacific coast north to Washington (Roberson 1980; Gilligan et al. 1994).

REMARKS: See Thompson et al. (1997) for a thorough summary of the status and biology of this extensively studied species.

Oriental Turtle-Dove — OTDO

Streptopelia orientalis (Latham)

RANGE: Breeds in Asia from the Ural Mountains east to the Sea of Okhotsk and Japan, and south to southern Asia. Winters in the southern part of the breeding range.

STATUS: *Accidental* in Western Vancouver Island of the Coast and Mountains Ecoprovince.

OCCURRENCE: A single bird was present at a feeder in Tofino, on Western Vancouver Island, from 15 to 25 August 1992 (Davidson 1993). It was photographed on 18 August (BC Photo 1848; Fig. 734).

REMARKS: The Oriental Turtle-Dove is also known as Rufous Turtle-Dove.

Figure 734. A skittish, unmarked Oriental Turtle-Dove regularly visited a feeder on the central west coast of Vancouver Island, British Columbia, for a few days in late summer 1992 (Tofino, 18 August 1992; Adrian Dorst).

Xantus's Hummingbird — XAHU

Hylocharis xantusii (Lawrence)

Figure 735. Xantu's Hummingbird at Gibsons, British Columbia (25 November 1997; Ernie Carlson).

RANGE: Resident in southern Baja California, casually north to latitude 29°N, including islands in the Gulf of California north to Isla San José.

STATUS: *Accidental* in the Georgia Depression Ecoprovince.

OCCURRENCE: A male was present at a sugar feeder in Gibsons, 30 km northwest of Vancouver, for 310 days from 16 November 1997 through 21 September 1998 (Bowling 1998a, 1998b, 1998c; Shepard 1998a; 1998b; T. Greenfield pers. comm.; Fig. 735).

The Xantus's Hummingbird is exceedingly rare in California (Small 1994). There are several records in winter and spring, and in 1988, 2 unsuccessful nesting attempts were monitored (Hainebach 1992).

Broad-tailed Hummingbird — BTHU

Selasphorus platycercus (Swainson)

STATUS: *Accidental* in the Southern Interior and Southern Interior Mountains ecoprovinces.

OCCURRENCE: Two records. On 8 July 1995, an adult male was well described as it visited a feeder near Cranbrook (W. Spiers pers. comm.). An adult male was carefully observed at a feeder on the Old Hedley Road, 11 km east of Princeton, on 10 and 11 July 1997 (R.J. Herzig pers. comm.).

The Broad-tailed Hummingbird is an occasional visitant and local summer resident, primarily in southeastern Oregon (Gilligan et al. 1994) and may breed in north-central Idaho (Burleigh 1972).

RANGE: Breeds from northern Wyoming to southeastern California, western Texas and south to Guatemala. Winters from northern Mexico south to Guatemala.

REMARKS: See Calder and Calder (1992) for a current summary of the distribution and biology of the Broad-tailed Hummingbird in North and Central America.

Acorn Woodpecker ACWO

Melanerpes formicivorus (Swainson)

RANGE: Resident from central and southern Washington, northwestern Oregon south through California to southern Baja California; from southern Utah, northern Arizona, northern New Mexico, western and central Texas, Nuevo León, and southwestern Tamaulipas south through the highlands of Middle America to extreme western Panama; and in South America in the Andes of Colombia.

STATUS: *Accidental* in the Georgia Depression Ecoprovince.

OCCURRENCE: A single bird was discovered in a backyard in Maple Ridge, 40 km east of Vancouver, on 15 June 1996. It was photographed the following day (Bastaja 1996; BC Photo 1607).

The Acorn Woodpecker is considered a rarity north of Oregon where it is a common permanent resident of the interior valleys and hills of southwestern portions of the state (Roberson 1980; Gilligan et al. 1994).

Synopsis:
The Birds of British Columbia into the 21st Century

SYNOPSIS: THE BIRDS OF BRITISH COLUMBIA INTO THE 21ST CENTURY

In the four volumes of *The Birds of British Columbia,* we have assembled and analyzed data on the patterns of geographic and seasonal occurrence and habitat preferences of all species of birds known to occur in British Columbia. In interpreting the data, we have focused on changes in distribution and abundance with time for individual species and on the basic details of the reproductive performance of all breeding species. For the breeding species we have also looked at trends in their numbers over the years represented by our data.

We are concerned about the unevenness of regional coverage, especially the paucity of data from large parts of northern British Columbia. However, only the efforts of upcoming ornithologists and field-naturalists can remedy that. Our study reveals the areas and types of data most urgently needed. We emphasize the importance of seeking data selectively and to a purpose, and we doubt that gathering thousands of records of common species in areas of abundance is going to advance the sensitivity of our understanding or the cause of bird conservation.

Today there is as much or more focus on ecosystems and their populations than on individual species. But, as a prerequisite to sound conclusions, there is still no substitute for a sound knowledge of the species that compose the populations. Thus, with the species accounts completed, it is now important to search for patterns within the data that will add to our knowledge. Patterns to look for include: the species associated with each ecoregion, ecosection, or biogeoclimatic zone; concentration areas that may indicate sites of special importance for many species; species richness and species density in specific habitats within each ecoregion, ecosection, or biogeoclimatic zone; major migration routes and staging areas of individual species; species with records of consistent decrease or increase; species with expanding or contracting ranges; and especially, changes in species richness and density that accompany human alterations of the environment.

In these final chapters of the work, we examine data from all four volumes and synthesize information on the above and related matters as a contribution to the understanding of bird populations and behaviour. We end with what we believe are some of the "New Philosophies, Concerns, and Conservation Challenges" that will be required in this new millennium to maintain the diversity and viability of the bird populations dependent on the extensive and biologically rich area that is British Columbia.

AVIAN BIODIVERSITY, ECOLOGICAL DISTRIBUTION, AND PATTERNS OF CHANGE

Ian McTaggart-Cowan, Neil K. Dawe, R. Wayne Campbell, and Andrew C. Stewart

SPECIES AND CATEGORIES OF OCCURRENCE: CHANGES SINCE 1947

In 1947, Munro and Cowan (1947) listed 367 species of birds known to have occurred in British Columbia. Most were represented by specimen evidence. As of 1999, the avifauna known from the province totalled 483[1] species of living birds (Fig. 736) (Campbell 1999) plus the extinct Passenger Pigeon (Campbell et al. 1990b). The total includes several changes that have resulted from new insights into the nature of bird species, rather than the arrival of new species from elsewhere. One native species, the Sage Grouse, and 2 introduced species, the Mountain Quail and Northern Bobwhite, have been extirpated and are maintained for the record, as is the Passenger Pigeon.

Before the 1947 and 1999 lists can be compared for numbers of species involved, adjustments must be made to the 1947 list. Some changes have arisen from new evidence on the systematics of a few species. Other changes are the result of the acceptance of species on the basis of confirmed sightings, rather than on the more rigorous specimen evidence required by Munro and Cowan (1947).

In the first category are 4 changes of systematic status from species to subspecies. In 1947 there were 2 species of flickers recognized in western Canada – Red-shafted (*Colaptes cafer*) and Yellow-shafted (*C. auratus*) – and the Myrtle Warbler (*Dendroica coronata*) and Audubon's Warbler (*D. auduboni*) were regarded as separate species. In both genera, the 2 one-time species are now regarded as conspecific. The same applies to the Lesser Canada Goose (*Branta leucopareia*) and the Cackling Goose (*B. minima*), both included in 1947 as species distinct from the Canada Goose (*Branta canadensis*) (Munro and Cowan 1947). Now all are included in a single species complex. A similar fate overtook Harlan's Hawk (*Buteo harlani*), now regarded as a variant of the Red-tailed Hawk (*B. jamaicensis*). These changes result in the removal of 4 species from the Munro and Cowan (1947) list.

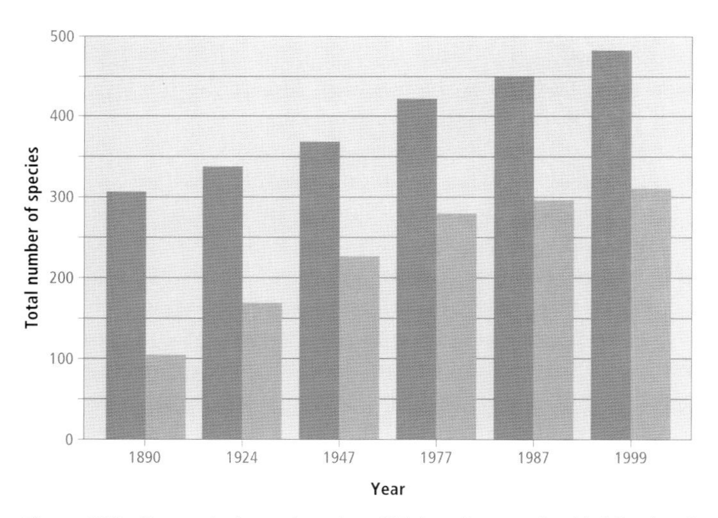

Figure 736. Changes in the total number of bird species occurring (dark bars) and breeding (light bars) in British Columbia between 1890 and 1999 (modified from Campbell et al. 2000).

On the other hand, recent research on the Traill's Flycatcher, at that time known as *Empidonax traillii*, led to the conclusion that 2 species were included. These are now recognized as the Alder Flycatcher (*E. alnorum*) and the Willow Flycatcher (*E. traillii*) (American Ornithologists' Union 1973, 1983). In the same way, the Western Flycatcher (*Empidonax difficilis*) was split and now includes the Pacific-slope Flycatcher (*E. difficilis*) and the Cordilleran Flycatcher (*E. occidentalis*) (Johnson 1980). Similar intensive study of the Solitary Vireo led to the conclusion that the 2 subspecies of 1947 were indeed true species (Johnson et al. 1988). The names Blue-headed Vireo (*Vireo solitarius*) and Cassin's Vireo (*V. cassinii*) have been applied. These "splitting" actions added 3 species to the provincial bird list without altering the biological situation. Thus the net result of changes in the systematics of the birds was the reduction of the 1947 species list by 1. Later re-examination of museum collections also brought to light single specimens of Eurasian Kestrel and Acadian Flycatcher that had escaped identification.

At that time also there were sight records of 8 species that were not represented by specimen evidence and were therefore not included in Munro and Cowan (1947). These include: Snowy Egret, Ferruginous Hawk, Willet, Caspian Tern, Crested Auklet, Horned Puffin, Red-headed Woodpecker, and Ash-throated Flycatcher. They have since been confirmed and are now considered valid records. On the basis of these modifications, the list of bird species known from the province in 1947 included about 376 species, which suggests that in the 52 years between 1947 and 1999, 108 species have been added to the catalogue of birds known to have occurred in British Columbia.

Much remains to be done in the application of modern biochemical and cytological techniques to the examination of avian systematics and taxonomy. It is certain that the next half-century will see more changes in the understanding of speciation among birds, and in how best to express these relationships in nomenclature. The result will be further changes in the scientific names applied to some of the birds of British Columbia.

[1] We have excluded the Timberline Sparrow noted in Campbell's (1999) list as it has not yet been formally accepted by the American Ornithologists' Union (1998) as a species distinct from the Brewer's Sparrow.

Table 19. Birds new to the British Columbia list of breeding species since the publication of Munro and Cowan (1947).

Pacific Loon	Least Sandpiper	White-throated Swift	Cape May Warbler
Clark's Grebe	Baird's Sandpiper	Black-chinned Hummingbird	Black-throated Green Warbler
Northern Fulmar	Upland Sandpiper	Anna's Hummingbird	Palm Warbler
Brandt's Cormorant	Hudsonian Godwit	White-headed Woodpecker	Bay-breasted Warbler
Green Heron	Short-billed Dowitcher	Yellow-bellied Flycatcher	Blackpoll Warbler
Black-crowned Night-Heron	Franklin's Gull	Gray Flycatcher	Black-and-white Warbler
Mute Swan	Ring-billed Gull	Philadelphia Vireo	Ovenbird
American Black Duck	California Gull	Blue Jay	Connecticut Warbler
Surf Scoter	Black-legged Kittiwake	Sky Lark	Mourning Warbler
Broad-winged Hawk	Caspian Tern	Purple Martin	Canada Warbler
Ferruginous Hawk	Forster's Tern	Boreal Chickadee	Lark Sparrow
Gyrfalcon	Thick-billed Murre	White-breasted Nuthatch	Le Conte's Sparrow
Chukar	Marbled Murrelet	Rock Wren	Nelson's Sharp-tailed Sparrow
Ring-necked Pheasant	Horned Puffin	Canyon Wren	Smith's Longspur
Rock Ptarmigan	Rock Dove	Northern Mockingbird	Snow Bunting
Wild Turkey	Flammulated Owl	European Starling	Rose-breasted Grosbeak
Northern Bobwhite	Spotted Owl	Crested Myna	Indigo Bunting
American Golden-Plover	Great Gray Owl	Sprague's Pipit	Common Grackle
American Avocet	Boreal Owl	Chestnut-sided Warbler	Baltimore Oriole
Lesser Yellowlegs	Black Swift	Magnolia Warbler	White-winged Crossbill
Solitary Sandpiper			

The bird fauna of British Columbia, as with that of any large geographic area of the North Temperate Zone, can be organized into several different categories based on residence status (see full status definitions in Volume 1, pp. 148-149). They include breeding species, including summer *migrants* and *visitants* as well as *residents;* winter *visitants; transients* that regularly move through the province during spring and autumn but do not breed here and do not usually remain during the winter; a small group of oceanic trans-equatorial summer *visitants;* and *vagrants*, including a variety of vagrant species that are not present in most years. The latter category includes species referred to in our volumes as *accidental* (a single occurrence) and *casual* (2 to 6 records), along with species known from over 6 records but of very infrequent occurrence, which we note as *very rare*.

Figure 737. Prior to the mid-1940s, the Caspian Tern was known as a vagrant summer visitant to British Columbia. Recently, the species has been found breeding at 3 locations in the province (Shuswap Lake, 19 Jun 1998; R. Wayne Campbell).

In all categories of occurrence there is constant change and few hard boundaries. New discoveries in the field will move species from one category to another. Some species are expanding their ranges in adjacent states, provinces, or territories and will appear within British Columbia. Others that usually leave the province during the winter will find conditions that permit some of their number to remain.

Of the 484 species recorded from the province, 309 have been reported nesting here. Eighty-one of the known breeding species result from studies since 1947 (Table 19). Some of the newly discovered nesting species, such as the Caspian Tern (Fig. 737), had previously been recorded as summer visitants and were suspected of breeding in the province, but evidence was not obtained until recently. Several more species still wait to have their breeding status confirmed. A number of species have been added to the list through continuing exploration of the northern third of the province, especially the Boreal Plains and Taiga Plains ecoprovinces.

The current breeding list includes 182 species (60% of the total breeding species) that occur in the province year-round (14 of them are introduced species), 126 that are summer visitants only, and 2 extirpated species (Table 20). The distinction between the resident breeding species and those that are summer visitants only is of some conservation

Table 20. Primary status and number of species in the avifauna of British Columbia.

Primary Status	Number
Resident[1]	182
Resident nonbreeder[2]	1
Summer visitants (breeders)	126
Summer visitants (nonbreeders)[3]	6
Winter visitants[4]	11
Transient/wintering[5]	14
Transients[6]	16
Irregular occurrence[7]	124
Extirpated or extinct[8]	4
Total	484

1 Excluded are 2 extirpated species known to have bred in the province (Northern Bobwhite and Mountain Quail).
2 Breeding for the Western Gull has yet to be documented.
3 Oceanic visitors from the Southern Hemisphere include Black-footed Albatross, Pink-footed Shearwater, Flesh-footed Shearwater, Buller's Shearwater, Sooty Shearwater, and Short-tailed Shearwater.
4 Species that come to British Columbia primarily in winter: Yellow-billed Loon, Cattle Egret, Snow Goose, Eurasian Wigeon, Black Scoter, Rough-legged Hawk, Thayer's Gull, Glaucous Gull, Snowy Owl, Harris's Sparrow, and Hoary Redpoll.
5 Species that move through British Columbia primarily as transients, but also winter: Greater White-fronted Goose, Brant, Tundra Swan, Black-bellied Plover, Whimbrel, Ruddy Turnstone, Black Turnstone, Surfbird, Sanderling, Western Sandpiper, Rock Sandpiper, Dunlin, Long-billed Dowitcher, and Lapland Longspur.
6 Species that move through the province primarily as transients that do not breed and seldom winter here: Pacific Golden-Plover, Marbled Godwit, Red Knot, Semipalmated Sandpiper, White-rumped Sandpiper, Pectoral Sandpiper, Sharp-tailed Sandpiper, Stilt Sandpiper, Buff-breasted Sandpiper, Long-billed Dowitcher, Red Phalarope, Pomerine Jaeger, Parasitic Jaeger, Long-tailed Jaeger, Heermann's Gull, Sabine's Gull, and Common Tern.
7 See Table 21.
8 Includes Sage Grouse, Mountain Quail, Northern Bobwhite, and Passenger Pigeon; Mountain Quail and Northern Bobwhite were known to have bred.

importance as the latter are influenced primarily by the conditions prevailing during the spring to autumn periods, whereas the residents must find survival conditions throughout the entire year, and winter is often the testing time.

A small group of 6 species, mainly oceanic breeders from the Southern Hemisphere, regularly move to British Columbia waters to summer: Black-footed Albatross, Pink-footed Shearwater, Flesh-footed Shearwater, Buller's Shearwater, Sooty Shearwater, and Short-tailed Shearwater.

Eleven species are primarily winter visitants. All but 1, the Cattle Egret, nest in the arctic regions of east Asia and North America and are often present in the province from autumn to spring. Included are: Yellow-billed Loon, Cattle Egret, Snow Goose, Eurasian Wigeon, Black Scoter, Rough-legged Hawk, Thayer's Gull, Glaucous Gull, Snowy Owl, Harris's Sparrow, and Hoary Redpoll.

A further 14 species also winter in numbers in the province but are primarily transients, passing through to their breeding or wintering areas: Greater White-fronted Goose, Brant, Tundra Swan, Black-bellied Plover, Whimbrel, Ruddy Turnstone, Black Turnstone, Surfbird, Sanderling, Western Sandpiper, Rock Sandpiper, Dunlin, Long-billed Dowitcher, and Lapland Longspur.

Sixteen species are primarily transients, using the habitats of the province only during their spring and autumn migrations: Pacific Golden-Plover, Marbled Godwit, Red Knot, Semipalmated Sandpiper, White-rumped Sandpiper, Pectoral Sandpiper, Sharp-tailed Sandpiper, Stilt Sandpiper, Buff-breasted Sandpiper, Red Phalarope, Pomerine Jaeger, Parasitic Jaeger, Long-tailed Jaeger, Heermann's Gull, Sabine's Gull, and Common Tern.

Another 124 species are considered vagrants to the province (Table 21), and many of their occurrences have little or no biological meaning. It is obvious that the status of *casual*, with a maximum of 6 documented occurrences, is an arbitrary limit without biological significance. Indeed many other species among those now included as *casual* or *accidental* will sooner or later "graduate" to one of the other categories as over time more observers encounter the species and add to their number of records.

Birds have a prodigious ability to travel, and it is not surprising that a few individuals stray from the routes and destinations characteristic of their species. Miraculously, some encounter an observer that recognizes them and reports their occurrence. Examples are the Wood Stork that appeared at Telegraph Creek or the Red-faced Cormorant off Haida in Masset Sound. Most of them, however, pay for their mistake unnoticed. Species of irregular occurrence and their probable geographic sources are listed in Table 21.

Many species from the fauna of eastern Siberia, which normally migrate along the "lead line" of the western shoreline of the North Pacific Ocean, can easily be displaced to the eastern side of the Bering Strait and then follow the eastern shores southward to be seen in British Columbia or the American states to the south. There are some 25 species of irregular occurrence that arrived in British Columbia by this mischance (Table 21). Examples include the Spotted Redshank, Spoonbill Sandpiper, Dusky Thrush, Siberian Accentor, and Eurasian Kestrel.

Others from the east Asian avifauna extend their breeding range into western Alaska but retain an inherited migration route and winter range in Asia or beyond. For them it is even easier to follow the wrong route south when the urge to leave the breeding grounds takes hold. Species such as the Yellow Wagtail, Red-throated Pipit, Bristle-thighed Curlew, and Northern Wheatear are examples.

Another major group of species, about 35 strong on our list, consists of birds that normally reach their northern limit in the states south of British Columbia. A few individuals sometimes overfly their normal target. The Little Blue Heron, Costa's and Broad-tailed hummingbirds, Acorn Woodpecker, Ash-throated Flycatcher, Western Scrub-Jay, and Lesser Goldfinch are examples of this group. Several appear to be expanding their ranges northward and are likely to be found nesting within the provincial boundaries in the next few decades.

Table 21. Vagrant species in British Columbia and their probable source areas.

Laysan Albatross[1]	Yellow Rail[4]	Ivory Gull[6]	Brown Thrasher[10, 4]
Short-tailed Albatross[2]	Common Moorhen[3]	Elegant Tern[3]	Siberian Accentor[7]
Mottled Petrel[2]	Whooping Crane[4]	Aleutian Tern[1]	Yellow Wagtail[1, 7]
Murphy's Petrel[2]	Mongolian Plover[7]	Kittlitz's Murrelet[1]	Black-backed Wagtail[7, 1]
Black-vented Shearwater[2]	Snowy Plover[3]	Xantus's Murrelet[3]	Red-throated Pipit[1]
Red-tailed Tropicbird[2]	Mountain Plover[3, 10]	Crested Auklet[1]	Northern Parula[4]
Brown Pelican[2]	Black-necked Stilt[3]	Parakeet Auklet[1]	Black-throated Blue Warbler[4]
Red-faced Cormorant[1]	Spotted Redshank[7]	Oriental Turtle-Dove[3]	Hermit Warbler[3]
Magnificent Frigatebird[2]	Wood Sandpiper[7]	White-winged Dove[3]	Blackburnian Warbler[4]
Least Bittern[3]	Willet[5]	Black-billed Cuckoo[10, 3]	Yellow-throated Warbler[4]
Great Egret[3]	Terek Sandpiper[7]	Yellow-billed Cuckoo[3, 10]	Prairie Warbler[4]
Snowy Egret[3]	Bristle-thighed Curlew[1]	Xantus's Hummingbird[3]	Hooded Warbler[4]
Little Blue Heron[4]	Far Eastern Curlew[7]	Ruby-throated Hummingbird[4]	Painted Redstart[3]
White-faced Ibis[3]	Bar-tailed Godwit[7, 1]	Costa's Hummingbird[3]	Scarlet Tanager[4]
Wood Stork[4]	Great Knot[7]	Broad-tailed Hummingbird[3]	Green-tailed Towhee[3]
Black Vulture[3]	Red-necked Stint[7, 1]	Red-headed Woodpecker[4, 10]	Black-throated Sparrow[3]
Fulvous Whistling-Duck[5]	Little Stint[7]	Tropical Kingbird[3]	Sage Sparrow[3]
Emperor Goose[6]	Temminck's Stint[7]	Acorn Woodpecker[3]	Lark Bunting[10]
Ross's Goose[6]	Curlew Sandpiper[7]	Acadian Flycatcher[4]	Baird's Sparrow[10]
Whooper Swan[7]	Spoonbill Sandpiper[7]	Cordilleran Flycatcher[3]	McCown's Longspur[10]
Falcated Duck[7]	Ruff[7]	Black Phoebe[3]	Chestnut-collared Longspur[10]
Garganey[7]	South Polar Skua[8]	Ash-throated Flycatcher[3]	Rustic Bunting[7]
Baikal Teal[7]	Laughing Gull[4]	Great Crested Flycatcher[4]	McKay's Bunting[1]
Tufted Duck[7]	Little Gull[9]	Thick-billed Kingbird[3]	Blue Grosbeak[3]
Steller's Eider[1]	Black-headed Gull[9]	Gray Kingbird[4]	Painted Bunting[4]
Spectacled Eider[1]	Iceland Gull[9]	Scissor-tailed Flycatcher[5]	Dickcissel[10]
King Eider[6]	Lesser Black-backed Gull[9]	Loggerhead Shrike[3]	Great-tailed Grackle[5]
Common Eider[6]	Slaty-backed Gull[1]	Western Scrub-Jay[3]	Orchard Oriole[4]
Smew[7]	Great Black-backed Gull[9, 4]	Blue-gray Gnatcatcher[3, 4, 5]	Hooded Oriole[3]
White-tailed Kite[3]	Red-legged Kittiwake[1]	Northern Wheatear[1]	Brambling[7]
Eurasian Kestrel[7]	Ross's Gull[6]	Dusky Thrush[7]	Lesser Goldfinch[3]

1 Probable source area: North Pacific Ocean, Bering Sea
2 Probable source area: Pacific Ocean south of latitude 30°N
3 Probable source area: southwestern North America and Mexico
4 Probable source area: eastern North America, eastern Mexico, and Central America
5 Probable source area: south-central North America
6 Probable source area: nearctic or holarctic regions
7 Probable source area: northeast Asia
8 Probable source area: Antarctic seas
9 Probable source area: North Atlantic Ocean
10 Probable source area: Great Plains

For many among the large fraternity of birdwatchers, the finding of a species new to the list of the birds of British Columbia, especially if it is a bird from a faraway place, arouses great interest and adds zest to an always-fascinating hobby. However, from a biological standpoint, most vagrant birds are unfortunate creatures that have malfunctioned somehow and "gotten lost." As such they are of little biological importance. Their removal from the population is part of the constant process of natural selection. The majority of the species new to the list fall into this category of waifs and strays, but among them are probably a few species that will be recognized later as pioneers – the first arrivals of a species on the move and expanding its range. The recolonization of British Columbia by birds since the devastation caused by the last glacial advances is still proceeding. Even in the span of a lifetime it is possible to observe several species of both birds and mammals undergo changes in distribution that bring them across the artificial boundary lines that we have erected to keep our territories a bit tidy. The now plentiful House Finch is one such, with the first arrivals noted in Penticton and Victoria in 1935 and 1937, respectively (Cowan 1937).

Many of the additions since Munro and Cowan (1947) published their review are not accidental arrivals. These species probably have been present in the province for many years but occur in areas that had received little biological attention. The provincial population has increased by some 2.9 million people, and roads and railroads have penetrated into previously remote areas. Resource extraction has become a ubiquitous force in changing landscapes. New communities have developed, each a focal point of contact between people and birds. Bird-wise observers now live year-round in regions previously unexplored. It is not surprising that new discoveries have been made. For example, prior to 1947, few ornithologists had visited the Boreal Plains and Taiga Plains ecoprovinces. Thus, it is almost certain that many species, apparently new to the avifauna of the province since 1947, had been present in the region for many years and were simply not encountered.

Of special interest are the 18 species, 12 of them marine or shorebirds, that have expanded their breeding ranges to include parts of British Columbia. Five of these – Northern Fulmar, Black-legged Kittiwake, Thick-billed Murre, Horned

Figure 738. Ring-billed Gull chick (Shuswap Lake, 19 June 1998; R. Wayne Campbell). In 1968, the Ring-billed Gull extended its breeding range into British Columbia from colonies in the United States. Today, the species is well established at 4 colony sites and its numbers are increasing each year.

Puffin, and Least Sandpiper – expanded their ranges southward from their northern centres of abundance. They have moved into areas that had been well studied and their prior absence confirmed. Twelve species – Brandt's Cormorant, Green-backed Heron, Black-crowned Night-Heron, American Avocet, California Gull, Ring-billed Gull (Fig. 738), Forster's Tern, Caspian Tern, White-headed Woodpecker, Gray Flycatcher, and European Starling – are expansions from the south. There, the breeding birds of the areas they entered were also well known. Several species have entered British Columbia from the east. Notable arrivals during the last 50 years have been the Broad-winged Hawk, Common Grackle, and Baltimore Oriole in the Peace Lowland, and the Barred Owl, Blue Jay, Sprague's Pipit, and Indigo Bunting, which appear to have moved across the Rocky Mountains onto the western slope of the continent. The gulls, terns, and owl seem to have established themselves, but the presence of the other species in this list of origins appears tentative. During the same period, 2 species, the Mute Swan and American Black Duck, escaped from captivity and have become feral.

A recent important development has been the growing concern about the impact on wildlife species and their habitats of activities destructive to the original environment, such as urbanization, deforestation, mineral extraction, water impoundment, and drainage of wetlands. This has led to laws and regulations designed to mitigate the impacts of some natural-resource extraction on wildlife and its habitat. It is normal procedure today to undertake an environmental impact assessment as part of the approval process for developments that may significantly alter the environment. The consequence has been the appearance of a cadre of well-informed biologists exploring remote areas and gathering data for environmental impact reports. The reports of such studies have made important contributions to the data base on the birds of the province (see e.g., Bryant et al. 1993; Pojar 1995; Bennett 1998; Phinney 1998). These and similar intensive regional studies of parts of northern British Columbia by Weeden (1960), Reid (1975), Erskine and Davidson (1976), and the Royal British Columbia Museum (e.g., Campbell and McNall 1982; Campbell et al. 1983) added important details to what was known of the migration of "eastern" birds into northeastern and northern British Columbia.

The many changes to the birds of British Columbia that have been noted during this relatively short period of years emphasize the dynamic nature of bird distribution in the province. Climatic warming adds a new dimension and will probably increase the rate of change.

THE BIODIVERSITY OF BIRDS IN BRITISH COLUMBIA

The maintenance of biological diversity has become a central philosophy of biological conservation and is well entrenched in the everyday lexicon of North Americans. From its early focus on species and habitats, the concept of biodiversity has evolved to include a scope not appreciated by most users of the term. Heywood and Baste (1993:5-18) state:

> Biodiversity is defined as the total diversity and variability of living things and of the systems of which they are a part. This covers the total range of variation and variability among systems and organisms, at the bioregional, landscape, ecosystem and habitat levels, at the various organismal levels down to species, populations and individuals, and at the level of the populations and genes. It also covers the complex sets of structural and functional relationships within and between these different levels of organisation, including human action, and their origin and evolution in space and time.

This is a comprehensive definition, but one beyond human capacity to attain, and difficult to translate into operational goals and realizable objectives. In an attempt to begin such a translation, Bisby (1993:25-82) offers:

> The biodiversity within an area can be characterized by measures of *species richness, species diversity, taxic diversity*, and *functional diversity* ...
> (a) *Species richness* measures the number of species within an area, giving equal weight to each species.
> (b) *Species diversity* measures the species in an area adjusting for both sampling effect and species abundance.

(c) *Taxic diversity* measures the taxonomic dispersion of species, thus emphasizing evolutionarily isolated species that contribute greatly to the assemblage of features and options.
(d) *Functional diversity* assesses the richness of functional features and interrelations in an area, identifying food webs along with keystone species and guilds.

A most important difference between these two definitions is the role of the species. Bisby (1993) uses the species as the smallest level of concern, whereas Heywood and Baste (1993:5-18) extend concern to "species, populations, and individuals." Both rest on the generally accepted definition of the biological species as a group of interbreeding natural populations that is reproductively isolated from other such groups. To be pragmatic, we can also recognize subspecies in higher vertebrates, and where this is possible we should use the subspecies distinction.

Most who study birds or who find in birds a captivating avocation are comfortable with the idea of species as distinguishable entities and can identify many of them. It is obvious that "biological species are not an arbitrary construct of the human mind" (Mayr and Ashlock 1991), but it is important to appreciate that the way in which we conceive the species is a creation of the human mind, a product of the scientific mission to introduce order into an apparently disorderly world. It has been aptly noted that "without taxonomy to give shape to the bricks, and systematics to tell us how to put them together, the house of biological science is a meaningless jumble" (May 1990).

Most of us seek order, and a degree of finality also, in the names applied to the creatures we wish to know and in the way they are arranged in books, on the Internet, and in other sources we depend on. However, our concepts of the species and arrangement of them in a classification are hypotheses, and as such they will evolve. The avian taxonomist (the identifier and classifier of birds) and the systematist (the arranger of birds into an order reflecting relatedness) review the most recently available data and reach decisions that respect this information. As new data arise from continuing research, the hypotheses may require change.

The last 20 years have seen the adaptation of powerful new research tools to the fields of biological taxonomy and systematics. Many of these have provided new evidence that have added to our understanding of biological species. In particular, the application of data from biochemical and genetic research has permitted taxonomists and systematists to re-examine many problem taxa and to make changes that better reflect their relationships within the great assemblage of birds (Sibley and Monroe 1990). The task is far from complete. Those who seek more information on these topics will find it in Mayr and Ashlock (1991) or in Bisby (1993).

A major study of birds or of ecological regions within a larger geographical area, such as that covered in *The Birds of British Columbia*, makes its contribution to the understanding of biological diversity at the alpha level, species richness (Magurran 1988). Our volumes, as with most of the literature on birds, use the "Biological Species Concept" (BSC) as proposed and developed by Ernst Mayr (1942, 1969, 1982) and Mayr and Ashlock (1991). It is the basis of the American Ornithologists' Union Check-list of North American Birds and the host of other checklists, field guides, and other bird literature worldwide.

Mayr and Ashlock (1991:475) define a biological species "as a group of interbreeding natural populations that is reproductively isolated from other such groups." Earlier, Mayr (1969) explained that, in this species concept, members of a species constitute (1) a reproductive community, (2) an ecological unit, and (3) a large intercommunicating gene pool (Meffe and Carroll 1997). The BSC applies well to birds and mammals and has stood the test of nearly 60 years of use. Most ornithologists, amateur or professional, are comfortable with the concept and use the schemes derived from it to organize their information.

In practice, the reproductive isolation criterion has proven difficult to apply, especially where geographic isolation effectively prevents interbreeding. Also, the occurrence of interbreeding is not an all-or-none biological criterion, and there is no logical point to guide conclusions as to how much interbreeding is tolerable within the definition. Furthermore the BSC concept is not easily applicable to fossil species evolving over time, nor to parthenogenetic species, plants, and many invertebrate phyla (Scudder 1974).

Several other species concepts have been described, largely by systematists working with groups other than birds and mammals. These include: *Ecological Species Concept* – a species is a lineage (or closely related set of lineages) that occupies an adaptive zone minimally different from that of any other lineage in its range and that evolves separately from all lineages outside its range (Van Valen 1976); *Evolutionary Species Concept* – a species is a single lineage of ancestral descendant populations or organisms that maintains its identity from other such lineages and has its own evolutionary tendencies and historical fate (Wiley 1978); *Recognition Species Concept* – species are the most inclusive population of individual biparental organisms that share a common fertilization system (Paterson 1985); *Phylogenetic Species Concept* – a phylogenetic species is an irreducible (basal) cluster of organisms that is diagnosably distinct from other such clusters, and within which there is a parental pattern of ancestry and descent (Cracraft 1989); *Cohesion Species Concept* – a species is the most inclusive population of individuals having the potential for phenotypic cohesion through intrinsic cohesion mechanisms (Templeton 1989); *Internodal Species Concept* – individual organisms are conspecific by virtue of their common membership of a part of the genealogical network between two permanent-splitting events or between a permanent split and an extinction event (Kornet 1993); and *Genotypic Cluster Species Concept* – species are groups that remain recognizable in sympatry because of the morphological and genotypic gaps between them (Mallet 1995).

One of these, the Phylogenetic Species Concept (PSC), also defined as "the smallest diagnosable cluster of individual

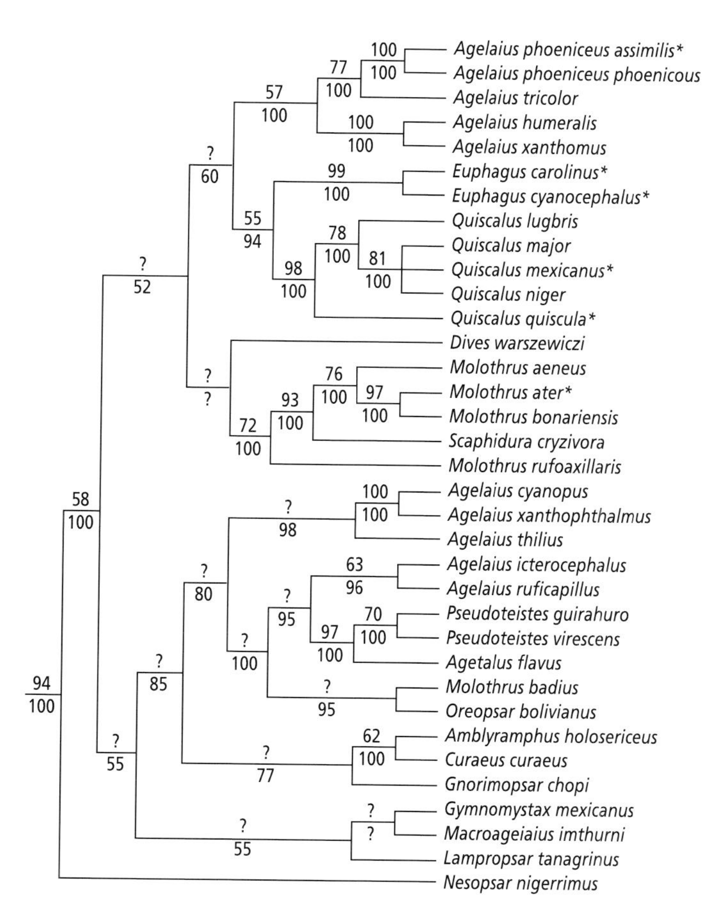

Figure 739. In the Phylogenetic Species Concept, the focus of classification is on the branching through time as organsims evolve. Here, the expression of the relationships among taxa of the grackles and allies is shown through a diagram of genealogical relatedness, or a cladogram. Statistical support for lineages based in 2× weighting is indicated by bootstrap values (above branches) and jacknifed taxon values (below branches). Nodes supported by fewer than 50% of the pseudoreplicates are indicated with a question mark; taxa that occur in British Columbia are indicated with an asterisk (modified from Meffe and Carroll 1997).

organisms within which there is a parental pattern of ancestry and descent" (Meffe and Carroll 1997), has been discussed for at least 40 years (Henig 1950, 1965). In this concept, the focus of classification is on the branching through time as organisms evolved, and the expression of the relationships among taxa is through diagrams of genealogical relatedness, or cladograms (Fig. 739). Although it has found some usefulness in the presentation of zoological information, it shows no sign of replacing the Biological Species Concept in general usefulness.

The matter of the species concept is not a trivial one. In many countries, including Canada, species have a legal validity, especially in laws and regulations concerned with conservation. Changing from a biological to a phylogenetic species concept could have interesting and important [legal] consequences (Meffe and Carroll 1997:66-69) (see also "Rare, Threatened, or Endangered Species," below).

Scudder (1974) concludes his examination of species concepts, as did Dobzhanski (1972), saying that within the world of organisms "what is actually found is a remarkable variety of different kinds of species." For the purposes of recognizing and classifying birds the Biological Species Concept remains the most generally useful. Notwithstanding some problems with the application of the Biological Species Concept, Sibley and Monroe (1990:1111), in their monumental work on the birds of the world, use it throughout, while they base the arrangement of species on "evidence of phylogenetic relationships from comparisons of their DNAs."

Bird Species Richness in the Ecosystems of British Columbia

The Province of British Columbia covers an area of over 900,000 km^2 and spans a north-south distance of about 11° of latitude, extending on the mainland from the 49th to the 60th parallel of latitude. The most southerly tip of Vancouver Island extends south to 48°14′. Throughout this distance, the province encloses a cross-section of the North American cordilleran system from the Pacific Ocean to the crest of the Rocky Mountains (Chapman et al. 1956). In terms of geology, physiography, climate, and biological variety, it has the greatest regional diversity in Canada (Chapman et al. 1956; Farley 1979).

Since early in the twentieth century the region has been a challenge to those whose interests required classification of this diversity. Geologists, physiographers, pedologists, and biologists have each described and mapped the features of the province from their special perspective (Farley 1979). Meidinger and Pojar (1991) provide a useful summary of the evolution of ecological classification in the province.

Research, largely on the forest flora of the province, has led to the description of 14 distinctive "biogeoclimatic zones" (Anonymous 1988; Meidinger and Pojar 1991). This biogeoclimatic framework is widely used in the province as a basis of forestland management. In Meidinger and Pojar (1991), 23 authors have combined to produce a useful treatise summarizing the most important details of each biogeoclimatic zone, including flora, wildlife habitats, and a small selection of the vertebrate species.

Biologists primarily concerned with the birds and mammals of the province, seeking a better "fit" between their distributional data and the major vegetation zones, proposed the recognition of 1 marine and 9 terrestrial *ecoprovinces* as best describing the distribution of the mammals and birds of the province (Demarchi et al. 1990). These replace the system of 13 "Biotic Areas" defined by Munro and Cowan (1947). The ecoprovinces are described and illustrated in Volume 1, pp. 57-144, and have been used throughout these volumes (see also endpapers of this volume).

Among the ecoprovinces, the Coast and Mountains differs from the others in that it is a narrow north-south strip extending about 11° of latitude for the full length of the province, nowhere wider than about 2° of longitude (Demarchi et al. 1990). The influence of the bordering ocean has dominated its biological history and moulded the present flora and fauna. From the Juan de Fuca Strait to Portland Canal, a distance of about 900 km, the ecoprovince consists of a belt of mountain peaks rising from the sea with little or no attached coastal shelf, but deeply incised by narrow steep-sided fiords. What was a shelf is largely drowned and appears as a complex archipelago that is dominated by two major islands. These are Vancouver Island to the south, which shares most of its avifauna with the mainland close by, and the Queen Charlotte Islands, now frequently referred to as Haida Gwaii, off the northern coast of the province and consisting of two large islands and several hundred small ones. Both island systems are far enough off the mainland shore that they had their own icecap during the Wisconsin glaciation, and almost certainly had some ice-free refugia that held a biota (Heuser 1989; Schofield 1989; Ogilvie 1989). The resident and nesting birds of the Queen Charlottes bespeak their long history of isolation. Included are at least 4 resident, endemic subspecies of birds (Cowan 1989) (see also "Insular Endemic Taxa" below).

The ecoregions within the Coast and Mountains Ecoprovince (Demarchi et al. 1990), however, bear little relation to the distribution of birds within this coastal strip. Accordingly we have divided this ecoprovince into 4 sectors. From south to north, they are Western Vancouver Island, Southern Mainland Coast, Northern Mainland Coast, and the Queen Charlotte Islands. For simplicity, we have treated these subdivisions of the Coast and Mountains Ecoprovince as if they were ecoprovinces.

The ecoprovinces differ greatly in area from the smallest, the Georgia Depression with an area of about 27,726 km^2 between latitudes 48°14′N and 50°24′N, to the most extensive, the Coast and Mountains Ecoprovince with an area of over 220,000 km^2 and a north-south extent of about 11°. The ecoprovinces differ also in the variety of avian habitats represented. The most diverse ecoprovinces, the Southern Interior, Southern Interior Mountains, and Central Interior, include representations of 10, 8, and 10 biogeoclimatic zones, respectively (Table 22), and have the greatest diversity of breeding birds (Table 23).

The Distribution of Breeding and Summering Species

Most published statements of regional species richness of birds refer only to the breeding species (Orians 1997). In Table 23 we list for each of the terrestrial ecoprovinces all the species of birds known to have nested or summered in the ecoprovince, including each region of the Coast and Mountains Ecoprovince. For a number of species, British Columbia supports some of the highest summering densities in North America (Fig. 740).

Most of the published comparisons of species richness of breeding birds have been made across widely separated regions supporting biome-level differences in biota. In contrast, our data are from an area spanning about 11° of latitude with more modest ecosystem diversity. However, even here some well-known relationships are apparent. A well-documented

Table 22. Areal extent (km^2) of the terrestrial components of biogeoclimatic zones and select geographical features by ecoprovinces in British Columbia.

	Biogeoclimatic Zone[1]																	
Ecoprovince[2]	Coastal Douglas-fir	Coastal Western Hemlock	Mountain Hemlock	Interior Cedar–Hemlock	Ponderosa Pine	Bunchgrass	Interior Douglas-fir	Engelmann Spruce–Subalpine Fir	Montane Spruce	Sub-Boreal Pine-Spruce	Sub-Boreal Spruce	Boreal White and Black Spruce	Spruce-Willow-Birch	Alpine Tundra	Glaciers	Lakes	Ocean	**Totals**
Georgia Depression (GD)	2,567	13,533	1,269											373	17	697	9,270	27,726
Western Vancouver Island (WVI)		17,389	2,163											274	8	414	5,549	25,797
Southern Mainland Coast (SMC)		26,614	11,945				429	608						13,553	6,742	1,244	14,753	75,888
Northern Mainland Coast (NMC)		36,553	18,576	7,714				5,363			16			14,057	12,499	2,304	10,062	107,144
Queen Charlotte Islands (QCI)		9,322	524											99		122	3,274	13,341
Southern Interior (SI)		424		2,505	2,474	2,045	21,937	10,136	11,833	94	830			4,417	294	1,336		58,325
Southern Interior Mountains (SIM)				35,216	693		3,525	50,961	4,637		4,980	38		29,790	3,288	3,634		136,762
Central Interior (CI)		31	7	56		803	16,447	12,031	9,923	23,276	35,708			6,875	662	4,045		109,864
Sub-Boreal Interior (SBI)				3,567				48,060			51,679	4,848	428	21,817	1,925	4,713		137,037
Boreal Plains (BP)								668				38,510		3		312		39,493
Northern Boreal Mountains (NBM)								7,701			2,717	38,257	70,947	62,078	4,473	3,639		189,812
Taiga Plains (TP)												68,402	418	4		878		69,702
Totals	2,567	103,866	34,484	49,058	3,167	2,848	42,338	135,528	26,393	23,370	95,930	150,055	71,793	153,340	29,908	23,338	42,908	990,891

1 Based on 1995 Biogeoclimatic Classification.
2 Based on 1995 Ecoprovince Classification.

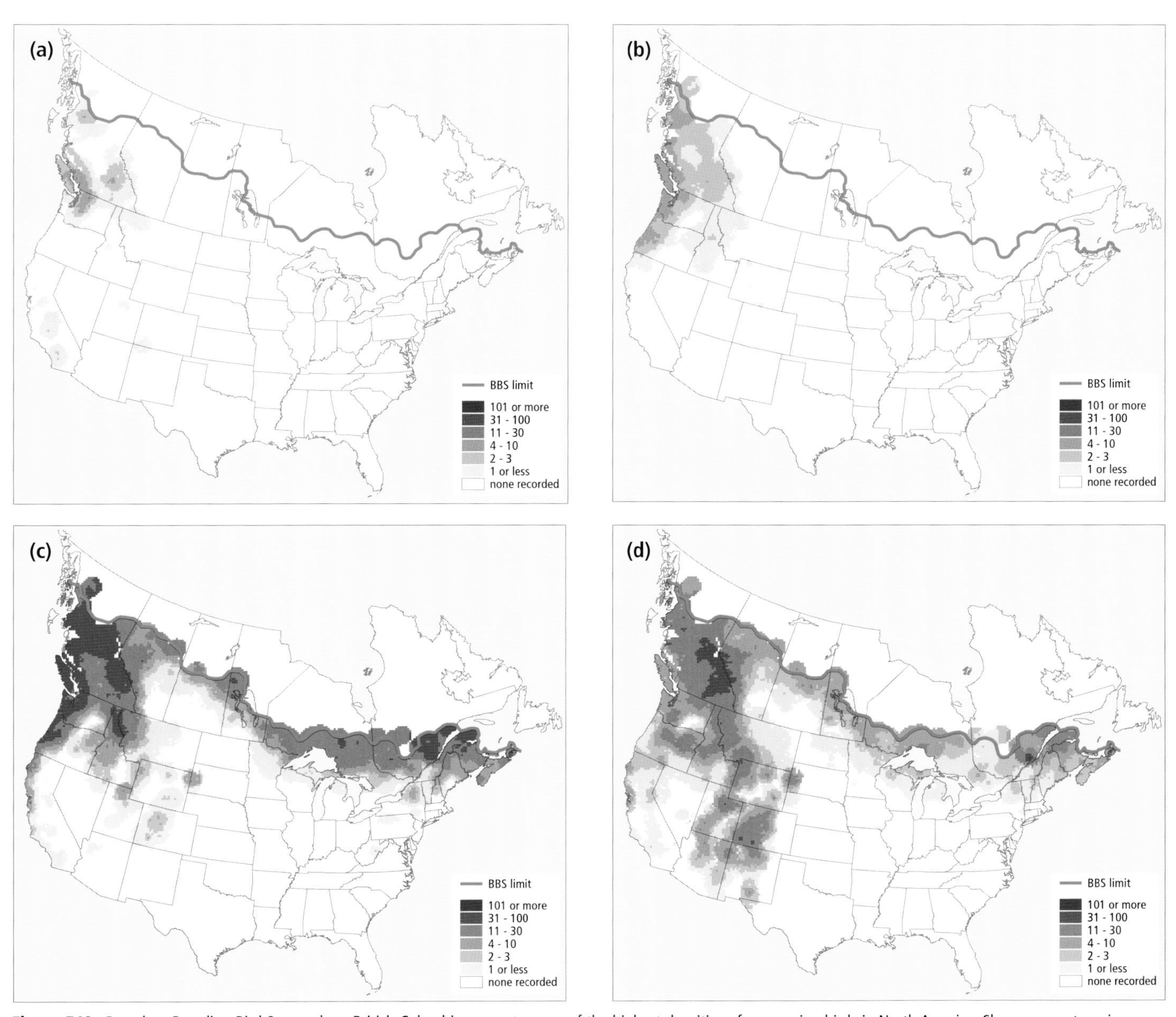

Figure 740. Based on Breeding Bird Survey data, British Columbia supports some of the highest densities of summering birds in North America. Shown are categories of relative abundance based on the number of birds seen on survey routes (Sauer et al. 1997). These summer density maps are for (a) Black Swift, (b) Rufous Hummingbird, (c) Swainson's Thrush, and (d) Pine Siskin.

negative relationship between higher latitudes in the Northern Hemisphere and species richness in breeding birds (Orians 1997) is clearly apparent in the interior of British Columbia (Table 23).

The most diverse breeding avifauna (213 species) is found in the Southern Interior Ecoprovince (Fig. 741), followed by the Central Interior and the Southern Interior Mountains (199 and 198 species). The arrangement of the ecoprovinces in Table 23, both for the coast and the interior, places them in approximate order of the average latitude between 48°14'N and 60°N and reveals a steady decline in the number of species of breeding birds in the interior of the province from south to north, with the smallest roster in the Taiga Plains (112 species). Decline in numbers with advancing latitude is greater in British Columbia than it was over the larger area reviewed by Briggs (1996). It is also evident that coastal forces have led to a different set of circumstances in the 2 coastal ecoprovinces than have operated in the interior of the province. For example, the Georgia Depression has the highest number of species reported of any ecoprovince (415 species) followed by the Coast and Mountains (375), both considerably higher than those with the highest number of species in the interior – the Southern Interior (338) and Southern Interior Mountains (322) (Table 24). The mild coastal winters, moderated by the Pacific Ocean, account for much of the difference.

Another well-known relationship is the increase in number of species with increasing size of the study area (Briggs 1996). This is probably valid when sampling ecologically similar areas. In British Columbia, where the comparison is between ecoprovinces differing in latitude and often in altitude,

Table 23. Breeding and summering birds in British Columbia by ecoprovince. The table is ordered first by those species breeding in all 9 ecoprovinces, then in 8, then 7, and so on. Within those categories, species are listed taxonomically. For each species, ● = breeds; ○ = summers. Lightly shaded columns indicate the subdivisions of the Coast and Mountains Ecoprovince used in these volumes and have not been included in the row totals.

	Ecoprovince													Total Ecoprovinces	
Species	GD	CaM	*WVI*	*SMC*	*NMC*	*QCI*	SI	SIM	CI	SBI	BP	NBM	TP	Breeding	Summering
Common Loon	●	●	●	●	●	●	●	●	●	●	●	●	●	9	9
Red-necked Grebe	●	●	○	○	●	○	●	●	●	●	●	●	●	9	9
Canada Goose	●	●	●	●	●	●	●	●	●	●	●	●	●	9	9
Mallard	●	●	●	●	●	●	●	●	●	●	●	●	●	9	9
Northern Pintail	●	●	○	○	○	●	●	●	●	●	●	●	●	9	9
Green-winged Teal	●	●	●	●	●	●	●	●	●	●	●	●	●	9	9
Bufflehead	●	●	○	○	●	○	●	●	●	●	●	●	●	9	9
Common Goldeneye	●	●	○	●	●	○	●	●	●	●	●	●	●	9	9
Bald Eagle	●	●	●	●	●	●	●	●	●	●	●	●	●	9	9
Northern Goshawk	●	●	●	●	●	●	●	●	●	●	●	●	●	9	9
Red-tailed Hawk	●	●	●	●	●	●	●	●	●	●	●	●	●	9	9
American Kestrel	●	●	○	●	●	○	●	●	●	●	●	●	●	9	9
Ruffed Grouse	●	●	●	●	●		●	●	●	●	●	●	●	9	9
Blue Grouse	●	●	●	●	●	●	●	●	●	●	●	●	●	9	9
American Coot	●	●		●		○	●	●	●	●	●	●	●	9	9
Killdeer	●	●	●	●	●	●	●	●	●	●	●	●	●	9	9
Spotted Sandpiper	●	●	●	●	●	●	●	●	●	●	●	●	●	9	9
Common Snipe	●	●	○	○	●	●	●	●	●	●	●	●	●	9	9
Rock Dove[1]	●	●	○		●	○	●	●	●	●	●	●	●	9	9
Great Horned Owl	●	●	●	●	●		●	●	●	●	●	●	●	9	9
Common Nighthawk	●	●	●	●	●		●	●	●	●	●	●	●	9	9
Belted Kingfisher	●	●	●	●	●	●	●	●	●	●	●	●	●	9	9
Hairy Woodpecker	●	●	●	●	●	●	●	●	●	●	●	●	●	9	9
Northern Flicker	●	●	●	●	●	●	●	●	●	●	●	●	●	9	9
Eastern Kingbird	●	●	○	●	○	○	●	●	●	●	●	●	●	9	9
Gray Jay	●	●	●	●	●		●	●	●	●	●	●	●	9	9
Common Raven	●	●	●	●	●	●	●	●	●	●	●	●	●	9	9
Tree Swallow	●	●	●	●	●	●	●	●	●	●	●	●	●	9	9
Cliff Swallow	●	●	●	●	●		●	●	●	●	●	●	●	9	9
Barn Swallow	●	●	●	●	●	●	●	●	●	●	●	●	●	9	9
Black-capped Chickadee	●	●		●	●		●	●	●	●	●	●	●	9	9
Swainson's Thrush	●	●	●	●	●	●	●	●	●	●	●	●	●	9	9
Hermit Thrush	●	●	●	●	●	●	●	●	●	●	●	●	●	9	9
American Robin	●	●	●	●	●	●	●	●	●	●	●	●	●	9	9
European Starling[1]	●	●	●	●	●	●	●	●	●	●	●	●	●	9	9
Orange-crowned Warbler	●	●	○	○	○	●	●	●	●	●	●	●	●	9	9
Yellow Warbler	●	●	○	●	●	○	●	●	●	●	●	●	●	9	9
Yellow-rumped Warbler	●	●	●	○	●	○	●	●	●	●	●	●	●	9	9
Chipping Sparrow	●	●	○	●	●	○	●	●	●	●	●	●	●	9	9
Savannah Sparrow	●	●	○	○	●	●	●	●	●	●	●	●	●	9	9
Fox Sparrow	●	●	●	○	○	●	●	●	●	●	●	●	●	9	9
Song Sparrow	●	●	●	●	●	●	●	●	●	●	●	●	●	9	9
Dark-eyed Junco	●	●	●	●	●	●	●	●	●	●	●	●	●	9	9
Red-winged Blackbird	●	●	○	●	●	○	●	●	●	●	●	●	●	9	9
Brewer's Blackbird	●	●	○	●	●	○	●	●	●	●	●	●	●	9	9
Pied-billed Grebe	●	●	●	●	●	●	●	●	●	●	●		●	8	8
American Wigeon	●	○	○	○	○	○	●	●	●	●	●	●	●	8	9
Blue-winged Teal	●	●	○	●	○	●	●	●	●	●	●	●	○	8	9
Northern Shoveler	●	○	○	○	○	○	●	●	●	●	●	●	●	8	9
Ring-necked Duck	●	○			○		●	●	●	●	●	●	●	8	9
Barrow's Goldeneye	○	●	○	●	●	○	●	●	●	●	●	●	●	8	9
Common Merganser	●	●	●	●	●	●	●	●	●	●	●	●	○	8	9
Osprey	●	●	●	●	●	○	●	●	●	●	●	●	○	8	9
Sharp-shinned Hawk	●	○	○	○	○	○	●	●	●	●	●	●	●	8	9
Merlin	●	●	●	●	●		●	●	●	●	●	●	○	8	9
Spruce Grouse		●		●	●		●	●	●	●	●	●	●	8	8

▶

◄ ***Table 23***

Species	Ecoprovince													Total Ecoprovinces	
	GD	CaM	*WVI*	*SMC*	*NMC*	*QCI*	SI	SIM	CI	SBI	BP	NBM	TP	Breeding	Summering
White-tailed Ptarmigan	●	●	●	●	●		●	●	●	●		●	●	8	8
Sora	●	●		●	○	○	●	●	●	●	●	●	○	8	9
Red-breasted Sapsucker	●	●	●	●	●	●	●	●	●	●	●	●		8	8
Downy Woodpecker	●	●	●	●	●	○	●	●	●	●	●	●	○	8	9
Three-toed Woodpecker	●	●			●		●	●	●	●	○	●	●	8	9
Red-eyed Vireo	●	●	○	●	●		●	●	●	●	●	○	●	8	9
Horned Lark	●	●	○	●	●		●	●	●	●	○	●	●	8	9
Violet-green Swallow	●	●	●	●	●	○	●	●	●	●	●	●	○	8	9
Northern Rough-winged Swallow	●	●	●	●	●	○	●	●	●	●	●	●		8	8
Bank Swallow	●	○		○	○		●	●	●	●	●	●	●	8	9
Red-breasted Nuthatch	●	●	●	●	●	●	●	●	●	●	●	●	○	8	9
House Wren	●	●	○	●	○		●	●	●	●	●		●	8	8
Mountain Bluebird	●	●		●	●		●	●	●	●	●	●	○	8	9
Varied Thrush	●	●	○	●	●	●	●	●	●	●	●	●	○	8	9
American Redstart	●	●		●	●		●	●	●	●	●	●	○	8	9
MacGillivray's Warbler	●	●	○	●	●		●	●	●	●	●	●	○	8	9
Wilson's Warbler	●	●	○	○	○	●	●	●	●	●	●	●	○	8	9
Lincoln's Sparrow	○	●	○	○	●	●	●	●	●	●	●	●	●	8	9
Rusty Blackbird		●		○	●		●	●	●	●	●	●	●	8	8
Brown-headed Cowbird	●	●	○	●	●	○	●	●	●	●	●	○	●	8	9
House Sparrow[1]	●	●	○	●	○		●	●	●	●	●	○	●	8	9
Horned Grebe	○	○	○	○	○	○	●	●	●	●	●	●	●	7	9
Gadwall	●	○	○		○	○	●	●	●	●	●	○	●	7	9
Lesser Scaup	○	○		○			●	●	●	●	●	●	●	7	9
Harlequin Duck	●	●	●	●	●	●	●	●	●	●	○	●		7	8
White-winged Scoter	○	○	○	○	○	○	●	●	●	●	●	●	●	7	9
Ruddy Duck	●						●	●	●	●	●	○	●	7	8
Northern Harrier	●	○	○	○			●	●	●	●	●	●	○	7	9
Golden Eagle	●	○	○	○	○		●	●	●	●	●	●	○	7	9
Black Tern	●	○	○				●	●	●	●	●		●	7	8
Great Gray Owl							●	●	●	●	●	●	●	7	7
Northern Saw-whet Owl	●	●		●	●	●	●	●	●	●	●	○		7	8
Rufous Hummingbird	●	●	●	●	●	●	●	●	●	●	○	●	○	7	9
Black-backed Woodpecker		●			●		●	●	●	●	●	○	●	7	8
Western Wood-Pewee	●	●	○	●	●		●	●	●	●	●	○	○	7	9
Hammond's Flycatcher	●	●	○	●	●	○	●	●	●	●	○	○	●	7	9
Cassin's Vireo	●	●	○	●	○		●	●	●	●		●	○	7	8
Warbling Vireo	●	●	○	●	●		●	●	●	●	●	○	○	7	9
Black-billed Magpie	●						●	●	●	●	●	●	○	7	8
Boreal Chickadee	○	○			○		●	●	●	●	●	●	●	7	9
Brown Creeper	●	●	●	●	○	●	●	●	●	●	●	○	○	7	9
Marsh Wren	●	○	○	○			●	●	●	●	●	○	●	7	9
Golden-crowned Kinglet	●	●	●	●	●	●	●	●	●	●	○	●	○	7	9
Ruby-crowned Kinglet	●	●	○	●	●	○	●	●	●	●	○	●	○	7	9
American Pipit	○	●	○	●	●	○	●	●	●	●	○	●	●	7	9
Bohemian Waxwing		●		●	○	○	○	●	●	●	●	●	●	7	8
Cedar Waxwing	●	●	●	●	●	○	●	●	●	●	●	○	○	7	9
Tennessee Warbler		●			●		●	●	○	●	●	●	●	7	8
Common Yellowthroat	●	●	○	○	●		●	●	●	●	○	●	○	7	9
Western Tanager	●	●	○	●	●	○	●	●	●	●	●	○	○	7	9
Clay-colored Sparrow	○						●	●	●	●	●	●	●	7	8
White-crowned Sparrow	●	●	○	●	○		●	●	●	●	○	●	○	7	9
Western Meadowlark	●	●	○	●	○		●	●	●	●	●	○	○	7	9
Purple Finch	●	●	○	○	●		○	●	●	●	●	○	●	7	9
American Bittern	●	○		○			●	●	●	●	●		○	6	8
Great Blue Heron	●	●	●	●	●	●	●	●	●	●	○	○		6	8
Wood Duck	●	●	●	●			●	●	●	●	○			6	7
Cinnamon Teal	●	●		●		○	●	●	●	●	○			6	7
Canvasback	○	○			○	○	●	●	●	●	●	●		6	8
Hooded Merganser	●	●	●	●	●	●	●	●	●	●	○	○	○	6	9

►

◀ ***Table 23***

Species	GD	CaM	*WVI*	*SMC*	*NMC*	*QCI*	SI	SIM	CI	SBI	BP	NBM	TP	Total Ecoprovinces: Breeding	Total Ecoprovinces: Summering
Cooper's Hawk	●	●	○	●			●	●	●	●				6	6
Sharp-tailed Grouse							●	●	●	○	●	●	●	6	7
Sandhill Crane	●	●	○	●	●	●	●	○	●	●	○	○	●	6	9
Solitary Sandpiper	○	○		○		○	●	●	●	●	○	●	●	6	9
Wilson's Phalarope	●						●	●	●	●	●	○		6	7
Herring Gull	○	●	○	○	●	○	●	○	●	●	○	●	●	6	9
Mourning Dove	●	●	○	●	○	○	●	●	●	●	○	○	○	6	9
Northern Hawk Owl							●	●	●	●	●	●	○	6	7
Northern Pygmy-Owl	●	●	●	●	○		●	●	●	●	○	○		6	8
Barred Owl	●	●		●	●		●	●	●	●	○	○	○	6	9
Long-eared Owl	●						●	●	●	●	○	●		6	7
Boreal Owl		●					●	●	●	●	○	○	●	6	8
Yellow-bellied Sapsucker							●	●		●	●	●	●	6	6
Pileated Woodpecker	●	●	○	●	○		●	●	●	●	○	○	○	6	9
Olive-sided Flycatcher	●	○	○	○	○		●	●	●	●	○	●	○	6	9
Least Flycatcher	●	○		○	○		○	●	●	●	●	○	●	6	9
Dusky Flycatcher		●		●	●		●	●	●	●	○	○	●	6	8
Pacific-slope Flycatcher	●	●	●	○	○	●	●	●	●	●	○	○		6	8
Western Kingbird	●	●	○	●	○		●	●	●	●	○			6	7
Steller's Jay	●	●	●	●	●	●	●	●	●	●	○	○		6	8
American Crow		●			●		●	●	●	●	●	○	○	6	8
Chestnut-backed Chickadee	●	●	●	●	●	●	●	●	●	●		○		6	7
Winter Wren	●	●	●	●	●	●	●	●	●	●	○	○	○	6	9
American Dipper	●	●	●	●	●	●	●	●	●	●	○	○	○	6	9
Townsend's Solitaire	●	○	○	○	○		●	●	●	●	○	●	○	6	9
Townsend's Warbler	●	●	○	○	●	●	●	●	●	○	○	●		6	8
Northern Waterthrush	○	○	○	○	○	○	●	●	●	●	●	○	●	6	9
Vesper Sparrow	●	○		○	○		●	●	●	●	●	○	○	6	9
Golden-crowned Sparrow	●	●		●	●		○	●	●	●	○	●	○	6	9
Yellow-headed Blackbird	●	○	○	○	○		●	●	●	●	●	○		6	8
Pine Siskin	●	○	○	○	○	○	●	●	●	●	○	●	○	6	9
Eared Grebe	○	○		○			●	●	●	●	●	○	○	5	9
Trumpeter Swan	○	●	○		●	○	○	○	●		●	●	●	5	8
Redhead	○	○				○	●	●	●	●	●	○	○	5	9
Peregrine Falcon	●	●	●	●	●	●	●	○	●	○	●			5	7
Rock Ptarmigan	●	●		●	●				●	○		●	●	5	6
Virginia Rail	●	●	○	●			●	●	●	○		○		5	7
Bonaparte's Gull	○	○	○	○	○	○	○	○	●	●	●	●	●	5	9
Short-eared Owl	●						●	●	●	●	○	○		5	7
Vaux's Swift	●	●	○	●	●	●	●	●	○	●	○			5	7
Calliope Hummingbird	●						●	●	●	●	○			5	6
Alder Flycatcher	○	○		○	○		○	○	●	●	●	●	●	5	9
Say's Phoebe	○	○	○		○		●	●	●	●	○	●	○	5	9
Mountain Chickadee	○	●		●	○		●	●	●	●	○	○		5	8
Veery	○	●		○	●		●	●	●	●	○			5	7
Magnolia Warbler	○	●			●		○	●	●	●	○	○	●	5	9
House Finch	●	○	○	○	○		●	●	●	●		○		5	7
Red Crossbill	●	●	○	●	○	○	●	●	●	○	○	○	○	5	9
Evening Grosbeak	●						●	●	●	○	●			5	6
Western Grebe	●	○	○	○	○	○	●	●	●	○	○			4	7
Turkey Vulture	●	●	○	●			●	●	○	○				4	6
Ring-necked Pheasant[1]	●	○				○	●	●	●					4	5
Willow Ptarmigan		●		●	●				●	●		●		4	4
Semipalmated Plover	●	●	○	○	○	●	○	○	●	○	○	●	○	4	9
Long-billed Curlew	○	○		○	○		●	●	●	●				4	6
Mew Gull	●	●	●	●	●	●	○	○	○	○	○	●	●	4	9
Lewis's Woodpecker	●	○		○			●	●	●	○				4	6
Red-naped Sapsucker	○						●	●	●	●				4	5
Willow Flycatcher	●	○	○	○	○		●	●	●	○				4	6
Clark's Nutcracker	○	●		●	○		●	●	●	○	○	○	○	4	9

▶

◀ ***Table 23***

Species	Ecoprovince													Total Ecoprovinces	
	GD	CaM	*WVI*	*SMC*	*NMC*	*QCI*	SI	SIM	CI	SBI	BP	NBM	TP	Breeding	Summering
Rock Wren	●	○		○			●	●	●	○			○	4	7
Western Bluebird	●	●		◉			●	●		○				4	5
Gray Catbird	●	○		○	○		●	●	●	○	○			4	7
Blackpoll Warbler	○	○	○		○		○	○	●	●	●	●	○	4	9
Spotted Towhee	●	●	○	◉	○		●	●	○					4	5
White-throated Sparrow	○	○			○		○	○	●	●	●	○	●	4	9
Black-headed Grosbeak	●	●	○	◉	◉		●	●	○	○				4	6
Lazuli Bunting	●	○	○	○			●	●	●	○	○	○		4	8
Bobolink	○						●	●	●	●	○			4	6
Bullock's Oriole	●	○	○	○			●	●	●					4	5
Gray-crowned Rosy-Finch	○	○		○	○		○	●	●	●		●		4	7
Pine Grosbeak	○	●		○	○	◉	●	●	●	○	○	○	○	4	9
American Goldfinch	●	○	○	○	○	○	●	●	●	○				4	6
Red-throated Loon	●	●	◉	◉	◉	◉		○			○	●		3	5
Wild Turkey[1]	●						●	●						3	3
California Quail[1]	●						●	●						3	3
American Avocet	●						●	●	○		○			3	5
Greater Yellowlegs	○	○	○	○	○	○	●	○	●	●	○	○	○	3	9
Lesser Yellowlegs	○	○	○	○	○	○	○	○	○	○	●	●	●	3	9
Ring-billed Gull	○	○	○	○	○	○	●	○	●	●	○	○		3	8
Glaucous-winged Gull	●	●	◉	◉	◉	◉	●			○	○			3	5
Caspian Tern	●	○	○	○	○	○	●	○	●	○	○			3	7
Barn Owl	●	●		◉			●	○	○					3	5
Flammulated Owl							●	●	●					3	3
Western Screech-Owl	●	●	○	◉	◉		●	○						3	4
White-throated Swift	○						●	●	●					3	4
Yellow-bellied Flycatcher					○			●		●	○	○	●	3	5
Eastern Phoebe		○			○			●			●		●	3	4
Blue Jay	●	○		○				●			●		○	3	5
White-breasted Nuthatch	○						●	●	○	○	●			3	6
Nashville Warbler	○	○	○	○	○	○	●	●	●	○	○	○	○	3	9
Brewer's Sparrow	○						●	●	○	○		●		3	6
Rose-breasted Grosbeak	○	○	○				●	○	○	○	●	○	●	3	9
Cassin's Finch	○	○			○		●	●	●					3	5
Pacific Loon	○	●	○	○	◉	○	○	○	○	○	○	●	○	2	9
Clark's Grebe							●	●	○					2	3
Brandt's Cormorant	●	●	◉	○	○	○								2	2
Double-crested Cormorant	●	○	○	○	○	○	○	○	●	○	○			2	7
Pelagic Cormorant	●	●	◉	◉	◉	◉								2	2
Green Heron	●	●	○	◉			○	○						2	4
Mute Swan[1]	●						●							2	2
Surf Scoter	○	○	○	○	○	○	○	○	○	○	●	○	●	2	9
Red-breasted Merganser	○	●	○	○		◉	○	○		○		●	○	2	7
Swainson's Hawk	○						●	○	●	○		○		2	6
Prairie Falcon	○						●	○	●					2	4
Chukar[1]							●		●					2	2
Gray Partridge[1]	●						●							2	2
Black Oystercatcher	●	●	◉	◉	◉	◉								2	2
Upland Sandpiper	○	○	○			○	○	○	●	○	●	○	○	2	9
Least Sandpiper	○	●	○	○	○	◉	○	○	○	○	○	●	○	2	9
Short-billed Dowitcher	○	●	○	○	○	◉	○	○	○	○	○	●		2	8
California Gull	○	○	○	○	○	○	●	○	●	○	○	○		2	8
Arctic Tern	○	●	○	○	◉	○	○	○	○	○	○	●		2	8
Pigeon Guillemot	●	●	◉	◉	◉	◉								2	2
Marbled Murrelet	●	●	◉	○	◉	◉								2	2
Rhinoceros Auklet	●	●	◉	◉	◉	◉								2	2
Tufted Puffin	●	●	◉	○	◉	◉								2	2
Band-tailed Pigeon	●	●	◉	◉	○	○	○	○	○	○				2	6
Burrowing Owl	●						●							2	2
Common Poorwill							●	○	●					2	3

▶

◄ ***Table 23***

Species	GD	CaM	*WVI*	*SMC*	*NMC*	*QCI*	SI	SIM	CI	SBI	BP	NBM	TP	Breeding	Summering
	Ecoprovince													Total Ecoprovinces	
Black-chinned Hummingbird		○		○			●	●	○					2	4
Williamson's Sapsucker	○						●	●						2	3
Hutton's Vireo	●	●	*●*	*●*										2	2
Philadelphia Vireo											●		●	2	2
Northwestern Crow	●	●	*●*	*●*	*●*	*●*								2	2
Purple Martin	●	○	○					●			○		○	2	5
Bushtit	●	●	○	*●*			○							2	3
Canyon Wren							●	●						2	2
Gray-cheeked Thrush										○	○	●	●	2	4
Black-throated Gray Warbler	●	●	○	*●*	○		○		○					2	4
Bay-breasted Warbler								○		○	●	○	●	2	5
Black-and-white Warbler	○	○		○	○		○	○		○	●	○	●	2	8
Connecticut Warbler											●		●	2	2
Canada Warbler											●		●	2	2
Yellow-breasted Chat	○						●	●	○	○				2	5
American Tree Sparrow		○		○	○		○	○	○	○	○	●	●	2	8
Lark Sparrow	○	○	○				●	●	○	○			○	2	7
Swamp Sparrow	○							○	○	○	●		●	2	6
Common Grackle	○						○	○	○	○	●	○	●	2	8
White-winged Crossbill	○	○	○	○	○	○	●	○	●	○	○	○	○	2	9
Northern Fulmar	○	●	*●*	○	○	○								1	2
Fork-tailed Storm-Petrel	○	●	*●*	*●*	*●*	*●*								1	2
Leach's Storm-Petrel	○	●	*●*	*●*	*●*	*●*								1	2
American White Pelican	○						○	○	●	○		○	○	1	7
Black-crowned Night-Heron	●						○	○	○					1	4
American Black Duck	●													1	1
Greater Scaup	○	○	○	○	○	○	○	○		○	○	●	○	1	8
Long-tailed Duck	○	○	○		○	○		○	○	○	○	●	○	1	8
Broad-winged Hawk							○			○	●		○	1	4
Ferruginous Hawk							●	○						1	2
Gyrfalcon												●		1	1
Mountain Quail (extirpated?)[1]	●													1	1
American Golden-Plover	○	○	○	○	○	○	○	○		○	○	●	○	1	8
Wandering Tattler	○	○	○	○	○	○	○					●		1	4
Hudsonian Godwit	○	○			○			○			○	●		1	5
Baird's Sandpiper	○	○	○	○	○	○	○	○	○	○	○	●	○	1	9
Red-necked Phalarope	○	○	○	○	○	○	○	○	○	○	○	●	○	1	9
Franklin's Gull	○	○	○			○	○	○	○	○	○	○	●	1	9
Black-legged Kittiwake	○	●	○	○	*●*	○								1	2
Forster's Tern	○						○	●		○	○			1	5
Common Murre	○	●	*●*	○	○	*●*								1	2
Thick-billed Murre		●	*●*			○								1	1
Ancient Murrelet	○	●	○	○	*●*	*●*	○							1	3
Cassin's Auklet	○	●	*●*	*●*	*●*	*●*								1	2
Horned Puffin	○	●	*●*		○	*●*								1	2
Spotted Owl		●		*●*										1	1
Black Swift	○	○	○	○	○		●	○	○	○		○		1	7
Anna's Hummingbird	●	○	○	○	○		○		○	○				1	5
White-headed Woodpecker							●							1	1
Gray Flycatcher							●							1	1
Northern Shrike									○		○	●		1	3
Blue-headed Vireo											●			1	1
Sky Lark	●													1	1
Pygmy Nuthatch							●	○	○					1	3
Bewick's Wren	●	○	○	○										1	2
Northern Mockingbird	○	○	○		○	○	●	○	○	○	○	○		1	8
Sage Thrasher							●	○						1	2
Crested Myna[1]	●													1	1
Sprague's Pipit									●					1	1
Chestnut-sided Warbler	○	○	○		○		○	○	●	○	○		○	1	8
Cape May Warbler								○			○		●	1	3

►

◄ ***Table 23***

Species	Ecoprovince													Total Ecoprovinces	
	GD	CaM	*WVI*	*SMC*	*NMC*	*QCI*	SI	SIM	CI	SBI	BP	NBM	TP	Breeding	Summering
Black-throated Green Warbler	○							○			●		○	1	4
Palm Warbler	○										○	○	●	1	4
Ovenbird	○							○	○	○	○	○	●	1	7
Mourning Warbler	○	○		○				○			●	○	○	1	6
Grasshopper Sparrow	○						●		○					1	3
Le Conte's Sparrow							○	○	○	○	●		○	1	6
Nelson's Sharp-tailed Sparrow											●			1	1
Smith's Longspur	○	○			○							●		1	3
Snow Bunting		○				○						●		1	2
Indigo Bunting	○						○	●	○					1	4
Baltimore Oriole	○							○			●		○	1	4
Common Redpoll		○		○			○	○	○	○	○	●	○	1	8
Sanderling	○	○	○	○	○	○	○	○	○	○	○	○	○	0	9
Semipalmated Sandpiper	○	○	○		○	○	○	○	○	○	○	○	○	0	9
Western Sandpiper	○	○	○	○	○	○	○	○	○	○	○	○	○	0	9
Pectoral Sandpiper	○	○	○	○	○	○	○	○	○	○	○	○	○	0	9
Long-billed Dowitcher	○	○	○	○	○	○	○	○	○	○	○	○	○	0	9
Yellow-billed Loon	○	○	○	○	○	○	○	○	○	○		○	○	0	8
Stilt Sandpiper	○	○	○				○	○	○	○	○		○	0	8
Common Tern	○	○	○				○	○	○	○	○		○	0	8
Lapland Longspur	○	○	○	○	○	○	○	○		○	○	○	○	0	8
Rough-legged Hawk	○						○	○		○	○	○	○	0	7
Greater White-fronted Goose	○	○	○	○	○	○	○	○			○	○		0	6
Whimbrel	○	○	○	○	○	○		○			○	○	○	0	6
Parasitic Jaeger	○	○	○	○	○	○	○	○		○	○			0	6
Long-tailed Jaeger	○	○	○	○	○	○	○	○	○			○		0	6
Great Egret	○						○	○	○	○				0	5
Tundra Swan	○			○	○	○	○	○			○	○		0	5
Black Scoter	○	○	○	○	○	○	○	○				○		0	5
Black-bellied Plover	○	○	○	○	○	○		○		○	○			0	5
Ruddy Turnstone	○	○	○	○	○	○		○		○	○			0	5
Buff-breasted Sandpiper	○	○	○		○	○	○			○	○			0	5
Sabine's Gull	○	○	○	○	○	○	○	○			○			0	5
Scissor-tailed Flycatcher	○	○		○			○			○		○		0	5
Harris's Sparrow	○	○		○			○	○	○					0	5
Cattle Egret	○						○	○	○					0	4
Brant	○	○	○	○	○	○	○		○					0	4
Dunlin	○	○	○	○	○	○					○		○	0	4
Thayer's Gull	○	○	○	○	○	○					○	○		0	4
Glaucous Gull	○	○	○	○		○					○	○		0	4
Blackburnian Warbler	○						○			○			○	0	4
Lark Bunting	○						○	○	○					0	4
Pacific Golden-Plover	○	○	○				○							0	3
Willet	○	○	○							○				0	3
Marbled Godwit	○	○	○		○			○						0	3
Black Turnstone	○	○	○	○	○	○				○				0	3
Red Knot	○	○	○			○				○				0	3
White-rumped Sandpiper	○						○				○			0	3
Red Phalarope	○	○	○		○	○		○						0	3
Pomarine Jaeger	○	○	○	○	○	○		○						0	3
Black-billed Cuckoo							○	○		○				0	3
Ash-throated Flycatcher	○	○	○	○				○						0	3
Brown Thrasher	○							○			○			0	3
Black-throated Sparrow	○						○	○						0	3
Dickcissel		○	○				○		○					0	3
Laysan Albatross	○	○	○			○								0	2
Black-footed Albatross[2]	○	○	○	○	○	○								0	2
Pink-footed Shearwater[2]	○	○	○	○	○	○								0	2
Sooty Shearwater[2]	○	○	○	○	○	○								0	2
Brown Pelican	○	○	○		○									0	2
Least Bittern	○						○							0	2

►

◄ *Table 23*

Species	GD	CaM	*WVI*	*SMC*	*NMC*	*QCI*	SI	SIM	CI	SBI	BP	NBM	TP	Breeding	Summering
	Ecoprovince													Total Ecoprovinces	
Snowy Egret	o						o							0	2
White-faced Ibis	o							o						0	2
Ross's Goose							o	o						0	2
Eurasian Wigeon	o								o					0	2
Snowy Plover	o	o	o			o								0	2
Black-necked Stilt	o								o					0	2
Surfbird	o	o	o	o	o	o								0	2
Sharp-tailed Sandpiper	o	o	o			o								0	2
Ruff	o						o							0	2
Little Gull	o										o			0	2
Heermann's Gull	o	o	o											0	2
Western Gull	o	o	o											0	2
Yellow-billed Cuckoo	o						o							0	2
Ruby-throated Hummingbird	o										o			0	2
Chestnut-collared Longspur	o	o	o		o									0	2
Short-tailed Albatross	o													0	1
Mottled Petrel		o	o											0	1
Flesh-footed Shearweater[2]		o	o	o	o	o								0	1
Buller's Shearwater[2]		o	o			o								0	1
Short-tailed Shearwater[2]		o	o	o	o	o								0	1
Black-vented Shearwater		o	o		o									0	1
Red-tailed Tropicbird	o													0	1
Magnificent Frigatebird		o				o								0	1
Black Vulture							o							0	1
Emperor Goose	o													0	1
Garganey	o													0	1
Tufted Duck	o													0	1
Steller's Eider	o													0	1
King Eider	o													0	1
Common Eider	o													0	1
White-tailed Kite	o													0	1
Yellow Rail											o			0	1
Common Moorhen	o													0	1
Whooping Crane										o				0	1
Mongolian Plover	o													0	1
Terek Sandpiper	o													0	1
Bar-tailed Godwit	o													0	1
Red-necked Stint	o													0	1
Little Stint	o													0	1
Rock Sandpiper		o	o		o									0	1
Curlew Sandpiper		o				o								0	1
Spoonbill Sandpiper	o													0	1
South Polar Skua		o	o		o	o								0	1
Black-headed Gull	o													0	1
Elegant Tern	o													0	1
Aleutian Tern		o				o								0	1
White-winged Dove	o													0	1
Xantus's Hummingbird	o													0	1
Costa's Hummingbird	o													0	1
Red-headed Woodpecker							o							0	1
Acorn Woodpecker	o													0	1
Acadian Flycatcher							o							0	1
Cordilleran Flycatcher								o						0	1
Black Phoebe	o													0	1
Western Scrub-Jay	o													0	1
Blue-gray Gnatcatcher	o													0	1
Red-throated Pipit	o													0	1
Northern Parula	o													0	1
Hermit Warbler	o													0	1
Hooded Warbler	o													0	1
Green-tailed Towhee	o													0	1
Sage Sparrow							o							0	1

►

◄ *Table 23*

Species	GD	CaM	*WVI*	*SMC*	*NMC*	*QCI*	SI	SIM	CI	SBI	BP	NBM	TP	Total Ecoprovinces Breeding	Total Ecoprovinces Summering
	Ecoprovince														
Baird's Sparrow	○													0	1
McCown's Longspur	○													0	1
Blue Grosbeak								○						0	1
Hooded Oriole	○													0	1
Brambling		○				○								0	1
Lesser Goldfinch								○						0	1
Regular breeding species	181	164	74	123	116	72	213	198	199	167	139	135	112		
Percent of total breeding species	59	53	24	40	38	23	69	64	64	54	45	44	36		
Total summering species	265	243	184	210	205	143	258	258	240	231	222	203	186		
Percent of total summering species	62	57	43	49	48	34	61	61	56	54	52	48	44		
Total number of breeding and summering species														309	426
Proportion of provincial area (%)	3	22					6	14	11	14	4	19	7		

1 Introduced.
2 Regular oceanic visitants from the Southern Hemisphere.

as well as in size and plant associations, the generalization is inappropriate. For example, the Southern Interior, with only 6% of the provincial area, has the highest breeding species count, 214, and the Georgia Depression, with just 3% of the provincial area has 181 breeding species, fourth highest of the 9 ecoprovinces. The largest ecoprovinces, the Coast and Mountains and Northern Boreal Mountains with 22% and 19%, respectively, of the provincial area, have 164 and 135 breeding species. Briggs (1996) graphs the declining numbers of breeding species in North and Central America with increasing latitude. For the northern parts of British Columbia the decline in our numbers of breeding species is a bit more rapid than that shown by Briggs (1996).

An examination of the species of birds nesting in each ecoprovince reveals that 45 species, about 15% of the breeding species, occur in all 9 ecoprovinces; another 32 occur in all but 1 ecoprovince (Table 23). These 77 species, one-quarter of all the nesting species, are the real generalists. They have a wide capacity for adaptation to such circumstances as predominant cover type, food plants or food organisms, temperature regimes, day length, and length of the potential nesting season. It is also true, as pointed out by Dobkin (1994), that no species occurs throughout an entire ecoprovince but only where their special ecological conditions prevail. These generalists are for the most part tolerant of some human disturbance of habitat and are unlikely candidates for special conservation measures.

At the other end of the spectrum are those that occur as breeding species in only a single ecoprovince. There are 53 such species in the province (Table 25). This group of species includes several that, in British Columbia, are at the northern or southern limit of their breeding range. They may be restricted by their inability to adapt to opportunities in other ecoprovinces. In no case do we know what the limiting factor or factors are. These species, because of where they occur, are especially vulnerable to human-induced environmental changes, including continued disturbance, and accidents such as spills of oil and toxic chemicals, and global climate change.

On the coast, the Georgia Depression has 7 species that have bred nowhere else in the province (Table 25). Three are introduced and 1 has been extirpated. Of the remaining 3 species, only the Black-crowned Night-Heron does not live in close association with humans.

The Coast and Mountains Ecoprovince provides another example of specialist birds concentrating in a single ecoprovince (Table 25). There, 10 species that breed nowhere else in British Columbia are open-ocean oriented. Their nest sites

Table 24. Total number of bird species reported from each ecoprovince in British Columbia.

Coastal		
Georgia Depression		415
Coast and Mountains		375
Western Vancouver Island	297	
Southern Mainland Coast	300	
Northern Mainland Coast	281	
Queen Charlotte Islands	230	
Interior		
Southern Interior		338
Southern Interior Mountains		322
Central Interior		292
Sub-Boreal Interior		263
Boreal Plains		270
Northern Boreal Mountains		239
Taiga Plains		216

Figure 741. The most diverse breeding avifauna in British Columbia is found in the Southern Interior Ecoprovince (Osoyoos, 5 June 1997; R. Wayne Campbell).

Table 25. Breeding species unique to each of the 9 ecoprovinces in British Columbia.

Ecoprovince	Number of unique species	Species
Georgia Depression	7	Black-crowned Night-Heron, American Black Duck (introduced), Mountain Quail (extirpated), Anna's Hummingbird, Sky Lark (introduced), Bewick's Wren, Crested Myna (introduced).
Coast and Mountains	10	Northern Fulmar, Fork-tailed Storm-Petrel, Leach's Storm-Petrel, Black-legged Kittiwake, Common Murre, Thick-billed Murre, Ancient Murrelet, Cassin's Auklet, Horned Puffin, Spotted Owl.
Southern Interior	8	Ferruginous Hawk, Black Swift, White-headed Woodpecker, Gray Flycatcher, Pygmy Nuthatch, Northern Mockingbird, Sage Thrasher, Grasshopper Sparrow.
Southern Interior Mountains	2	Forster's Tern, Indigo Bunting.
Central Interior	3	American White Pelican, Sprague's Pipit, Chestnut-sided Warbler.
Sub-Boreal Interior	0	
Boreal Plains	7	Broad-winged Hawk, Blue-headed Vireo, Black-throated Green Warbler, Mourning Warbler, Le Conte's Sparrow, Nelson's Sharp-tailed Sparrow, Baltimore Oriole.
Northern Boreal Mountains	12	Greater Scaup, Long-tailed Duck, Gyrfalcon, American Golden-Plover, Wandering Tattler, Hudsonian Godwit, Baird's Sandpiper, Red-necked Phalarope, Northern Shrike, Smith's Longspur, Snow Bunting, Common Redpoll.
Taiga Plains	4	Franklin's Gull, Cape May Warbler, Palm Warbler, Ovenbird.

must provide access to ocean food sources close to the protection from terrestrial predators offered by islands. Several species also require cover sufficient to conceal nest and contents from aerial predators. The 4 smallest of the species breeding only in the Coast and Mountains – the Fork-tailed Storm Petrel, Leach's Storm-Petrel, Cassin's Auklet, and Ancient Murrelet – supplement the physical cover of dense vegetation with that of nocturnal behaviour on the nesting grounds. All these ocean birds are highly vulnerable to spills of petroleum products and other oils as well as direct disturbance of the colonies.

In the interior, the 12 species that, in British Columbia, nest only in the Northern Boreal Mountains, find in this ecoprovince an extension of the arctic and alpine tundra habitats that characterize their primary breeding environments further north in Alaska and Yukon. Without exception, they are at the southern extremity of their breeding ranges. It is safe to say that no two of them are there because of precisely the same combination of environmental circumstances. The Snow Bunting nests in a different habitat complex than the Smith's Longspur. The same is true for the Wandering Tattler, the Red-necked Phalarope, and the others. None of these species nests widely over the ecoprovince, and several are known from just one or a few locations. Thus, each is vulnerable to point-source alteration of the precise set of circumstances they are using. The (mis)placement of an airstrip, mine, wilderness campsite, or similar artefact could be disastrous to them.

Such human modifications of the habitat are under our control. Not so are the long-term swings in climate that over the centuries have washed like a tide over the continents and carried the birds with them like flotsam. Under such circumstances a species either adapts, moves, or vanishes.

Political boundaries are of no biological importance; the normal ebb and flow of bird numbers from year to year may mean that some species nesting regularly in adjacent states, provinces, or territories will nest in British Columbia some years and not others.

It is at the margins of their distribution that the selection pressure exerted on the species by the environment is probably at its greatest. There the small differences in inherited capabilities enable some individuals to survive where others fail. The small populations that adapt to conditions near the edge of tolerance are probably where evolution is most active (Mayr 1967, 1982). They may also hold great promise for conserving endangered species and biological diversity in general (Scudder 1989, 1992; Brooks 2000; Channell and Lomolino 2000). Thus, some geographically marginal populations must be recognized as having significant importance to the survival of a species (see also data on rarity, below).

The Southern Interior Ecoprovince, with the highest species richness in the province, has 8 species that breed nowhere else in the province. This is because the Southern Interior has the highest number of biogeoclimatic zones within its boundaries (Table 22). Twelve additional native bird species nest mainly in the Southern Interior but have small breeding populations in the Southern Interior Mountains and Central Interior, likely due to the fact that the 3 ecoprovinces share at least 2 of 3 southernmost interior biogeoclimatic zones. The Interior Douglas-fir zone extends into both the Southern Interior Mountains and Central Interior. In addition, the Ponderosa Pine zone extends into the Southern Interior Mountains and the Bunchgrass zone into the Central Interior. The latter 2 ecoprovinces have 2 and 3 species respectively that breed nowhere else in the province (Table 25). Two species, the Sprague's Pipit and Indigo Bunting, are only incidental breeders to British Columbia, and the Forster's Tern and American White Pelican are both species at the northern limits of their ranges, west of the Great Divide. The pelican is significant in that it is geographically isolated from other breeding populations.

The Boreal Plains Ecoprovince, with its roster of a hawk, a vireo, 2 warblers, 2 sparrows, and an oriole as unique nesting species, owes its fauna to its position on the east side of the Rocky Mountains where it is reached by a migration of largely neotropical wintering species that nest primarily in the northern boreal forests. Both the Boreal Plains and Taiga Plains are dominated by the Boreal White and Black Spruce Biogeoclimatic Zone, and the 2 ecoprovinces hold 11 species of birds that so far are not known to nest elsewhere in British Columbia.

The Sub-Boreal Interior ecoprovince is the only ecoprovince known to support no unique nesting species.

The Distribution of Wintering Species in British Columbia

While the focus of attention in the discussion of species richness is on breeding species, it is informative to also consider the species that regularly winter in each of the ecoprovinces. The breeding and the wintering species each have an important dependence upon the provincial ecosystems that is likely to be associated with differences in behaviour and adaptation. For a number of species, British Columbia supports some of the highest wintering densities in North America (Fig. 742).

A total of 293 species has been reported in winter at least once in British Columbia; of that figure, 209 are known to regularly winter and their distribution by ecoprovince is shown in Table 26. Change in species richness with latitude is dramatically illustrated by the wintering species. From December through February, only 27 species have been reported regularly from the Taiga Plains and 26 from the Northern Boreal Mountains; the increase with descending latitude is large. The greatest number of regularly wintering species in an interior ecoprovince has been recorded from the Southern Interior (133 species), but the greatest total of wintering species in the province is in the Georgia Depression (180 species), where the wintering community is enriched by some 40 species of marine birds, 25 species of non-seagoing anseriformes, all the North American species of loons, and 5 of the 6 grebes. In both number of species and number of individuals, the Georgia Depression has the greatest concentration of wintering birds in Canada. The significance of the Georgia Depression for wintering birds is further indicated by the 34 regularly wintering species unique to that ecoprovince (Table 27).

Only 3 other ecoprovinces have unique wintering species: the Southern Interior (9 unique species), Coast and

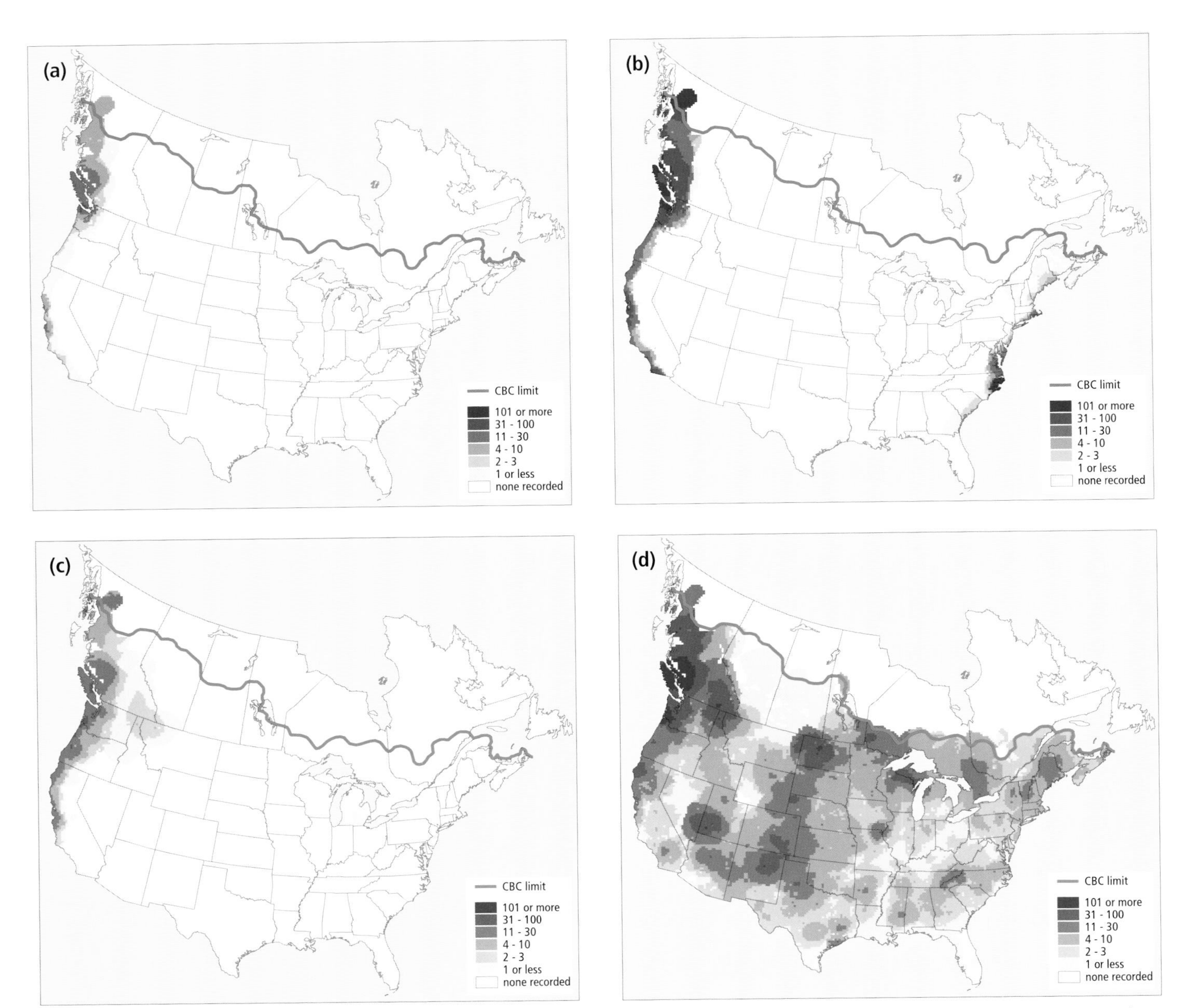

Figure 742. Based on Christmas Bird Count data, British Columbia supports some of the highest densities of wintering birds in North America. Shown are categories of relative abundance based on the number of birds seen within count areas (Sauer et al. 1996). The winter density maps are for (a) Pacific Loon, (b) Surf Scoter, (c) Chestnut-backed Chickadee, and (d) Pine Siskin.

Mountains (3), and Boreal Plains (1). In winter, only 10 species regularly occur in all 9 of the ecoprovinces, with another 9 species occurring in all but 1 of the ecoprovinces (Table 26).

Locations in British Columbia with Greatest Concentrations of Bird Species: The "Hot Spots"

In seeking to identify the regions of the province that have the greatest numbers of different species, we emphasize that areas north of latitude 54°N have had inadequate coverage. Figures 743, 744, and 745 illustrate the patterns of distribution of total species, breeding species, and wintering species in the National Topographic Series (NTS) quadrangles within British Columbia. They also reveal the many gaps in the data on distribution, especially for the winter season. Often, throughout the northern regions of the province, our knowledge of the concentrations of species represents the field work of just a few observers.

In the southern half of the province, Figures 743, 744, and 745 indicate that from 26 to 30 NTS quadrangles (about 5%-7% of the province) have the greatest concentrations of birds in total, as breeding species and as wintering species.

The Georgia Depression and the Southern Interior ecoprovinces dominate with the highest species richness. The Southern Interior, largely the Okanagan valley, has the greatest variety of breeding species (Fig. 744), but the Georgia Depression dominates in winter occurrence and total species recorded (Figs. 743 and 745).

Table 26. Regular and total wintering birds in British Columbia by ecoprovince. The table is ordered first by those species regularly wintering in all 9 ecoprovinces, then 8, then 7, and so on. Within those categories, species are listed taxonomically. For each species ● = regularly winters; ○ = occasionally winters. Lightly shaded columns indicate the subdivisions of the Coast and Mountains Ecoprovince used in these volumes and have not been included in the row totals.

	Ecoprovince													Total Ecoprovinces	
Species	GD	CaM	*WVI*	*SMC*	*NMC*	*QCI*	SI	SIM	CI	SBI	BP	NBM	TP	Regular Wintering	Total Wintering
Northern Goshawk	●	●	●	●	●	●	●	●	●	●	●	●	●	9	9
Ruffed Grouse	●	●	●	●	●		●	●	●	●	●	●	●	9	9
Great Horned Owl	●	●	●	●	●		●	●	●	●	●	●	●	9	9
Hairy Woodpecker	●	●	●	●	●	●	●	●	●	●	●	●	●	9	9
Northern Shrike	●	●	●	●	●		●	●	●	●	●	●	●	9	9
Gray Jay	●	●	○	●	●		●	●	●	●	●	●	●	9	9
Common Raven	●	●	●	●	●	●	●	●	●	●	●	●	●	9	9
Black-capped Chickadee	●	●		●	●		●	●	●	●	●	●	●	9	9
Pine Grosbeak	●	●	○	●	●	●	●	●	●	●	●	●	●	9	9
Common Redpoll	●	●	○	○	●	○	●	●	●	●	●	●	●	9	9
Mallard	●	●	●	●	●	●	●	●	●	●	●	●		8	8
White-tailed Ptarmigan	●	●		●			●	●	●	●		●	●	8	8
Downy Woodpecker	●	●	●	●	●		●	●	●	●	●		●	8	8
Three-toed Woodpecker	●	○		○	○		●	●	●	●	●	●	●	8	9
Pileated Woodpecker	●	●	●	●	●		●	●	●	●	●		●	8	8
Steller's Jay	●	●	●	●	●	●	●	●	●	●	●		●	8	8
American Dipper	●	●	●	●	●	●	●	●	●	●	○	●	●	8	9
Snow Bunting	●	○	○	○	○	○	●	●	●	●	●	●	●	8	9
House Sparrow[1]	●	●	○	●	●		●	●	●	●	●		●	8	8
Common Goldeneye	●	●	●	●	●	●	●	●	●	●	●			7	7
Bald Eagle	●	●	○	●	●	●	●	●	●	●	●	○		7	8
Spruce Grouse		○			○		●	●	●	●	●	●	●	7	8
Blue Grouse	●	●	●	●	●	●	●	●	●			●	●	7	7
Rock Dove[1]	●	●	○	●	●	●	●	●	●	●	●	○	○	7	9
Boreal Chickadee	○	○		○	○		●	●	●	●	●	●	●	7	9
European Starling[1]	●	●	●	●	●	●	●	●	●	●	●		○	7	8
Bohemian Waxwing	○	●	○	○	●	○	●	●	●	●	●	●		7	8
Evening Grosbeak	●	●	○	●	●	○	●	●	●	●	●		○	7	8
Common Merganser	●	●	●	●	●	●	●	●	●		●	○		6	7
Merlin	●	●	●	●	●		●	●	●	●	○			6	7
Northern Pygmy-Owl	●	●	●	●	●		●	●	●	●				6	6
Barred Owl	●	●	○		●		●	●		●	●			6	6
Great Gray Owl	○	○	○	○			●	●	●	●	●		●	6	8
Northern Flicker	●	●	●	●	●	●	●	●	●	●				6	6
Black-billed Magpie	○	○		○	○		●	●	●	●	●	●		6	8
Mountain Chickadee	●	●		●	○		●	●	●	●	○			6	7
Red-breasted Nuthatch	●	●	●	●	●	●	●	●	●	●	○			6	7
American Tree Sparrow	●	○		○	○	○	●	●	●	●		●	○	6	8
Dark-eyed Junco	●	●	●	●	●	●	●	●	●	●	○	○	○	6	9
Red-winged Blackbird	●	●	○	●	●	○	●	●	●	●				6	6
Brewer's Blackbird	●	●	○	●	●	○	●	●	●	●			○	6	7
Red Crossbill	●	●	●	●	●	●	●	●	●	●		○		6	7
Pied-billed Grebe	●	●	●	●	○	○	●	●	●					5	5
Great Blue Heron	●	●	●	●	●	●	●	●	●	○				5	6
Trumpeter Swan	●	●	●	●	●	●	●	○	●	●		○		5	7
American Wigeon	●	●	●	●	●	●	●	●	●					5	5
Ring-necked Duck	●	●	●	●	●	●	●	●	●					5	5
Lesser Scaup	●	●		●	●	●	●	●	●		○	○		5	7
Bufflehead	●	●	●	●	●	●	●	●	●					5	5
Barrow's Goldeneye	●	●	○	●	●	●	●	●	●	○				5	6
Hooded Merganser	●	●	●	●	●	●	●	●	●					5	5
Northern Harrier	●	●	●			●	●	●	●					5	5
Sharp-shinned Hawk	●	●	●	●	●	●	●	●	●					5	5
Red-tailed Hawk	●	●	●	●	●	●	●	●	●					5	5

►

◀ **Table 26**

Species	GD	CaM	*WVI*	*SMC*	*NMC*	*QCI*	SI	SIM	CI	SBI	BP	NBM	TP	Total Ecoprovinces: Regular Wintering	Total Ecoprovinces: Total Wintering
	Ecoprovince														
Golden Eagle	●	○	○	○			●	●	●	○	●	○		5	8
American Coot	●	●	○	○	◍	◍	●	●	●					5	5
Killdeer	●	●	◍	◍	◍	◍	●	○	●	●				5	6
Common Snipe	●	●	◍	◍	◍	◍	●	●	●					5	5
Boreal Owl	○	○		○			●	●	●	○	○	●	●	5	9
Northern Saw-whet Owl	●	●	◍	◍	◍	◍	●	●	●	○	○	○		5	8
Belted Kingfisher	●	●	◍	◍	◍	◍	●	●	●					5	5
Chestnut-backed Chickadee	●	●	◍	◍	◍	◍	●	●	●					5	5
Brown Creeper	●	●	◍	◍	◍	◍	●	●	●		○			5	6
Winter Wren	●	●	◍	◍	◍	◍	●	●	●	○				5	6
Golden-crowned Kinglet	●	●	◍	◍	◍	◍	●	●	●	○				5	6
Townsend's Solitaire	●	●		◍		○	●	●	●	○				5	6
American Robin	●	●	◍	◍	◍	◍	●	●	●	○			○	5	7
Varied Thrush	●	●	◍	◍	◍	◍	●	●	●	○				5	6
Song Sparrow	●	●	◍	◍	◍	◍	●	●	●	○				5	6
Gray-crowned Rosy-Finch	●	○		○	○		●	●	●	●				5	6
House Finch	●	●	○	◍			●	●	●	○	○			5	7
White-winged Crossbill	●	○		○	○	○	●	●	●	●	○	○		5	8
Pine Siskin	●	●	◍	◍	◍	◍	●	●	●	○	○			5	7
American Goldfinch	●	●		◍			●	●	●					5	5
Common Loon	●	●	◍	◍	◍	◍	●	●	○	○		○		4	7
Horned Grebe	●	●	◍	◍	◍	◍	●	●						4	4
Red-necked Grebe	●	●	◍	○	◍	◍	●	○	●					4	5
Western Grebe	●	●	◍	◍	◍	◍	●	●	○					4	5
Canada Goose	●	●	◍	◍	◍	◍	●	●	○	○	○			4	7
Tundra Swan	●	○	○	○	○	○	●	●	●					4	5
Gadwall	●	●	○	◍		○	●	●						4	4
Green-winged Teal	●	●	◍	◍	◍	◍	●	●			○	○		4	6
Greater Scaup	●	●	◍	◍	◍	◍	●	●		○				4	5
Rough-legged Hawk	●	○			○		●	●	●	○	○			4	7
Ring-necked Pheasant[1]	●	○			○		●	●	●					4	5
Rock Ptarmigan	●								●			●	●	4	4
Sharp-tailed Grouse							●	○	●	○	●		●	4	6
Herring Gull	●	●	◍	◍	◍	◍	●	●	○			○		4	6
Mourning Dove	●	○		○			●	●	●	○				4	6
Western Screech-Owl	●	●	◍	◍	○		●	●						4	4
Long-eared Owl	●	○		○	○		●	●	●	○				4	6
Short-eared Owl	●	○	○	○			●	●	●	○	○			4	7
American Crow							●	●	●	●	○			4	5
Northern Shoveler	●	●	○	○		◍	●	○						3	4
Northern Pintail	●	●	◍	◍	◍	◍	●	○	○			○		3	6
Canvasback	●	●	○	◍	◍	◍	●	○						3	4
Redhead	●	○		○		○	●	●						3	4
White-winged Scoter	●	●	◍	◍	◍	◍	●	○						3	4
Long-tailed Duck	●	●	◍	○	◍	◍	○	○	●					3	5
Cooper's Hawk	●	○	○	○	○		●	●						3	4
American Kestrel	●	○	○	○	○	○	●	○	●	○	○			3	7
Gyrfalcon	●						●		○	○		●		3	5
Peregrine Falcon	●	●	◍		◍	◍	●			○				3	4
Willow Ptarmigan		●			◍				●		○	●		3	4
Ring-billed Gull	●	○	○		○	○	●	●						3	4
Thayer's Gull	●	●	◍	◍	◍	◍	●							3	3
Glaucous-winged Gull	●	●	◍	◍	◍	◍	●	○						3	4
Northern Hawk Owl	○	○			○		○	○	○	○	●	●	●	3	9
Spotted Owl	●	●		◍			●							3	3
Black-backed Woodpecker		○			○		●	●	●					3	4
Clark's Nutcracker	○	○		○	○		●	●	●	○				3	6
White-breasted Nuthatch	○						●	○	●	○	●			3	6
Spotted Towhee	●	●	○	◍	○		●	○	○					3	5
Purple Finch	●	●	◍	◍	◍	○	○	○	●					3	5

▶

◀ *Table 26*

Species	Ecoprovince													Total Ecoprovinces	
	GD	CaM	*WVI*	*SMC*	*NMC*	*QCI*	SI	SIM	CI	SBI	BP	NBM	TP	Regular Wintering	Total Wintering
Cassin's Finch	○	○				○	●	●	●					3	5
Red-throated Loon	●	●	●	○	●	●	○							2	3
Pacific Loon	●	●	●	○	○	●	○	○						2	4
Yellow-billed Loon	●	●	○	○	○	●	○							2	3
Brandt's Cormorant	●	●	●	○	○	○								2	2
Double-crested Cormorant	●	●	●	○	●	●	○							2	3
Pelagic Cormorant	●	●	●	○	●	●								2	2
Brant	●	●				●								2	2
Harlequin Duck	●	●	●	●	●	●	○	○						2	4
Surf Scoter	●	●	●	●	●	●	○							2	3
Black Scoter	●	●	●	○	●	●								2	2
Red-breasted Merganser	●	●	●	○	●	●	○	○						2	4
Ruddy Duck	●	○	○	○		○	●	○	○		○			2	6
Wild Turkey[1]	●						○	●						2	3
California Quail[1]	●						●	○						2	3
Virginia Rail	●	○	○	○		○	●	○						2	4
Black Oystercatcher	●	●	●	○	●	●								2	2
Black Turnstone	●	●	●	●	●	●								2	2
Surfbird	●	●	●	●	●	●								2	2
Sanderling	●	●	●		●	●	○							2	3
Rock Sandpiper	●	●	○		●	●								2	2
Dunlin	●	●	●	●	●	●	○			○				2	4
Mew Gull	●	●	●	●	●	●	○							2	3
California Gull	●	○	○				●	○						2	4
Western Gull	●	●	●			○								2	2
Glaucous Gull	●	○	○	○	○	○	●	○						2	4
Common Murre	●	●	●	●	●	●								2	2
Pigeon Guillemot	●	●	●	○	●	●								2	2
Marbled Murrelet	●	●	●	○	●		○							2	3
Ancient Murrelet	●	●	○		●	●		○						2	3
Barn Owl	●	○		○			●							2	3
Snowy Owl	●	○	○		○	○	○	○	○	○	●			2	7
Red-breasted Sapsucker	●	●	●	○	●	●	○		○					2	4
Hutton's Vireo	●	●	●	○										2	2
Northwestern Crow	●	●	●	●	●	●								2	2
Canyon Wren							●	●						2	2
Marsh Wren	●	○	○	○			●	○	○					2	5
Ruby-crowned Kinglet	●	○	○	○	○		●	○	○					2	5
Cedar Waxwing	●	○	○	○	○	○	●	○	○	○				2	6
Yellow-rumped Warbler	●	●	●	○	○	○	○	○	○					2	5
Fox Sparrow	●	●	●	●	●	●	○	○	○		○			2	6
White-throated Sparrow	●	○	○	○			●	○	○	○	○			2	7
Harris's Sparrow	●	○	○	○	○		●	○	○					2	5
White-crowned Sparrow	●	○	○	○	○	○	●	○	○	○				2	6
Golden-crowned Sparrow	●	●	○	●	●	○	○	○	○					2	5
Western Meadowlark	●	○	○	○	○	○	●	○	○	○				2	6
Brown-headed Cowbird	●	○			○		○	●	○					2	5
Hoary Redpoll	○	○	○	○	○	○	○	○	○	○	●	○	●	2	9
Eared Grebe	●	○	○	○			○	○						1	4
Laysan Albatross		●	●											1	1
Northern Fulmar	○	●	●	●	●									1	2
American Bittern	●	○				○								1	2
Cattle Egret	●	○	○	○	○	○	○	○						1	4
Green Heron	●													1	1
Black-crowned Night-Heron	●													1	1
Greater White-fronted Goose	●	○			○	○	○	○	○					1	5
Snow Goose	●													1	1
Mute Swan[1]	●						○	○						1	3
Wood Duck	●	○	○	○		○	○							1	3
Eurasian Wigeon	●	○	○	○		○	○	○						1	4

▶

◀ ***Table 26***

	Ecoprovince													Total Ecoprovinces	
Species	GD	CaM	*WVI*	*SMC*	*NMC*	*QCI*	SI	SIM	CI	SBI	BP	NBM	TP	Regular Wintering	Total Wintering
American Black Duck	●													1	1
Prairie Falcon	○						●	○						1	3
Chukar[1]	○						●	○	○					1	4
Gray Partridge[1]	○						●		○					1	3
Mountain Quail[1]	●													1	1
Black-bellied Plover	●	○	○			○								1	2
Greater Yellowlegs	●	○		○		○								1	2
Spotted Sandpiper	●	○		○			○							1	3
Whimbrel	●	○	○											1	2
Ruddy Turnstone	●	○	○			○								1	2
Western Sandpiper	●	○		○	○									1	2
Long-billed Dowitcher	●	○		○		○								1	2
Red Phalarope	●	○	○											1	2
Bonaparte's Gull	●	○	○	○	○		○							1	3
Black-legged Kittiwake	○	●	○	●	●	○								1	2
Cassin's Auklet	●	○	○	○		○								1	2
Rhinoceros Auklet	●	○	○	○		○								1	2
Band-tailed Pigeon	●	○	○		○									1	2
Anna's Hummingbird	●	○	○	○	○		○	○	○					1	5
Lewis's Woodpecker	○						●							1	2
White-headed Woodpecker							●	○						1	2
Blue Jay	○						○	○	○	○	●			1	6
Sky Lark[1]	●													1	1
Horned Lark	○	○			○		●	○	○		○			1	6
Bushtit	●	○	○	○			○							1	3
Pygmy Nuthatch							●	○						1	2
Bewick's Wren	●	○	○	○										1	2
Western Bluebird	○						●							1	2
Hermit Thrush	●	○	○			○		○						1	3
Crested Myna[1]	●													1	1
American Pipit	●	○	○		○	○	○	○						1	4
Savannah Sparrow	●	○	○	○	○	○	○							1	3
Lincoln's Sparrow	●	○	○	○	○	○	○							1	3
Lapland Longspur	●	○		○	○	○	○	○		○	○			1	6
Rusty Blackbird	○	○	○	○	○	○	●	○	○	○				1	6
Brambling	○	○				○	○	○	○	○				0	6
Mountain Bluebird	○	○	○			○	○	○	○					0	5
Orange-crowned Warbler	○	○	○			○	○	○	○					0	5
Osprey	○	○			○		○	○						0	4
Say's Phoebe	○						○	○	○					0	4
Violet-green Swallow	○	○		○			○	○						0	4
Brown Thrasher	○	○			○		○	○						0	4
Common Yellowthroat	○	○	○				○	○						0	4
Wilson's Warbler	○	○	○				○	○						0	4
Chipping Sparrow	○						○	○		○				0	4
Swamp Sparrow	○	○	○	○		○	○	○						0	4
Yellow-headed Blackbird	○						○	○	○					0	4
Common Grackle	○						○	○			○			0	4
Fork-tailed Storm-Petrel	○	○	○		○	○			○					0	3
American White Pelican	○						○	○						0	3
Turkey Vulture	○	○		○			○							0	3
Ross's Goose	○						○	○						0	3
Blue-winged Teal	○						○	○						0	3
Sora	○	○				○	○							0	3
Burrowing Owl	○						○	○						0	3
Loggerhead Shrike	○	○			○		○							0	3
Northern Mockingbird	○						○	○						0	3
Sooty Shearwater	○	○	○											0	2
Short-tailed Shearwater	○	○	○		○	○								0	2

▶

◄ *Table 26*

Species	Ecoprovince													Total Ecoprovinces	
	GD	CaM	*WVI*	*SMC*	*NMC*	*QCI*	SI	SIM	CI	SBI	BP	NBM	TP	Regular Wintering	Total Wintering
Leach's Storm-Petrel	○	○	○											0	2
Brown Pelican	○	○	○											0	2
Cinnamon Teal	○						○							0	2
King Eider	○	○			○	○								0	2
Sandhill Crane	○	○				○								0	2
Semipalmated Plover	○	○	○											0	2
Least Sandpiper	○	○				○								0	2
Heermann's Gull	○	○		○										0	2
Sabine's Gull	○	○				○								0	2
Tufted Puffin	○	○	○											0	2
Red-naped Sapsucker	○							○						0	2
Tree Swallow	○						○							0	2
Rock Wren	○						○							0	2
Swainson's Thrush	○						○							0	2
Gray Catbird							○	○						0	2
Nashville Warbler	○	○	○											0	2
Townsend's Warbler	○	●	○			○								0	2
Palm Warbler	○	●	○			○								0	2
Northern Waterthrush	○						○							0	2
Western Tanager	○						○							0	2
Clay-colored Sparrow	○						○							0	2
Vesper Sparrow	○	○		○										0	2
Lark Sparrow	○	○	○											0	2
Bullock's Oriole	○						○							0	2
Black-footed Albatross		○	○											0	1
Black-vented Shearwater	○													0	1
Great Egret	○													0	1
Emperor Goose	○													0	1
Tufted Duck	○													0	1
Ferruginous Hawk							○							0	1
American Golden-Plover	○													0	1
American Avocet	○													0	1
Lesser Yellowlegs	○													0	1
Willet	○													0	1
Wandering Tattler	○													0	1
Long-billed Curlew	○													0	1
Marbled Godwit	○													0	1
Red Knot	○													0	1
Pectoral Sandpiper	○													0	1
Sharp-tailed Sandpiper		○				○								0	1
Pomarine Jaeger								○						0	1
Parasitic Jaeger	○													0	1
Franklin's Gull	○													0	1
Little Gull	○													0	1
Thick-billed Murre		○				○								0	1
Horned Puffin		○	○			○								0	1
Calliope Hummingbird							○							0	1
Tropical Kingbird	○													0	1
Western Kingbird	○													0	1
Northern Rough-winged Swallow								○						0	1
Cliff Swallow	○													0	1
Barn Swallow	○													0	1
Yellow Warbler		○	○											0	1
Black-throated Gray Warbler	○													0	1
Black-and-white Warbler	○													0	1
MacGillivray's Warbler	○													0	1
Canada Warbler		○	○											0	1
Rose-breasted Grosbeak	○													0	1
Baltimore Oriole	○													0	1
Bullock's Oriole							○							0	1

►

◄ *Table 26*	Ecoprovince													Total Ecoprovinces	
Species	GD	CaM	*WVI*	*SMC*	*NMC*	*QCI*	SI	SIM	CI	SBI	BP	NBM	TP	Regular Wintering	Total Wintering
Regular wintering species	180	119	85	87	97	80	133	97	91	45	36	26	27		
Percent of regular wintering species	86	57	41	42	46	38	64	46	44	22	17	12	13		
Total wintering species	272	220	169	160	150	143	202	171	127	84	61	41	34		
Percent of total wintering species	93	75	58	55	51	49	69	59	43	29	21	14	12		
Number of regular and total wintering species														209	293

[1] Introduced.

Localities represented with the highest total species richness in British Columbia are the Fraser Lowland from Mission and Chilliwack to the Strait of Georgia, the Nanaimo Lowland of southeastern Vancouver Island from Victoria and Sidney north to Campbell River, the Tofino area of Western Vancouver Island, the Okanagan valley from Osoyoos north to Vernon, the South Thompson River valley near Kamloops, the Columbia River confluence with Kootenay Lake at Creston, the Williams Lake area in the Cariboo Basin, and the Fort St. John area of the Peace Lowland.

Bird habitat in the Fraser and Nanaimo lowlands of the Georgia Depression Ecoprovince and the Okanagan valley in the Southern Interior Ecoprovince is under the greatest threat in the province from urbanization and agricultural development (see also the next chapter, "What Lies in Store for the Birds of British Columbia?").

Table 27. Regularly wintering species unique to each of the 9 ecoprovinces in British Columbia.

Ecoprovince	Number of unique species	Species
Georgia Depression	34	Eared Grebe, American Bittern, Cattle Egret, Green Heron, Black-crowned Night-Heron, Greater White-fronted Goose, Snow Goose, Mute Swan (introduced), Wood Duck, Eurasian Wigeon, American Black Duck (introduced), Mountain Quail (extirpated), Black-bellied Plover, Greater Yellowlegs, Spotted Sandpiper, Whimbrel, Ruddy Turnstone, Western Sandpiper, Long-billed Dowitcher, Red Phalarope, Bonaparte's Gull, Cassin's Auklet, Rhinoceros Auklet, Band-tailed Pigeon, Anna's Hummingbird, Sky Lark (introduced), Bushtit, Bewick's Wren, Hermit Thrush, Crested Myna (introduced), American Pipit, Savannah Sparrow, Lincoln's Sparrow, Lapland Longspur.
Coast and Mountains	3	Laysan Albatross, Northern Fulmar, Black-legged Kittiwake.
Southern Interior	9	Prairie Falcon, Chukar (introduced), Gray Partridge (introduced), Lewis's Woodpecker, White-headed Woodpecker, Horned Lark, Pygmy Nuthatch, Western Bluebird, Rusty Blackbird.
Southern Interior Mountains	0	
Central Interior	0	
Sub-Boreal Interior	0	
Boreal Plains	1	Blue Jay.
Northern Boreal Mountains	0	
Taiga Plains	0	

Except for the coastal areas, where the marine presence is a dominant influence on the bird species present, each of these "hot spots" features an abundance of riparian habitat. Each is also attractive for human development.

Among the NTS quadrangles with greatest numbers of breeding species, the Southern Interior dominates with 8 quadrangles having between 109 and 158 species (Fig. 744); the Vaseux Lake area has the greatest number of breeding species in the province, with 3 other quadrangles in the Okanagan valley having more breeding species than any other quadrangles in British Columbia. The Salmon Arm and Kamloops areas also have more than 100 breeding species. Creston with 107 breeding species has the highest count in the east and west Kootenay regions. Further north, only the Williams Lake area in the Central Interior has over 100 breeding species.

On the coast, only 4 quadrangles held more than 100 breeding species (102 to 119 species), with the Langley and Surrey areas reporting the highest numbers.

The winter distribution of species concentrations again emphasizes the importance of the Okanagan valley and the Georgia Depression to the birds of the province (Fig. 745). The Georgia Depression has 22 quadrangles holding more than 100 wintering species, with 4 quadrangles having more than 200 wintering species. Coastal concentrations also occur at Tofino Inlet on Western Vancouver Island and in Masset Inlet on the Queen Charlotte Islands. The Southern Interior has 8 quadrangles with 100 species or more, half of which have more than 150 wintering species. Three other areas in the interior have more than 100 wintering species: the Kamloops and Shuswap areas of the Southern Interior and the Williams Lake area of the Central Interior.

These data represent the analysis of species concentrations as determined by a coarse filter. Although the quadrangles change in area with longitude, each represents between 744 and 1,035 km². Obviously, detailed local inventory

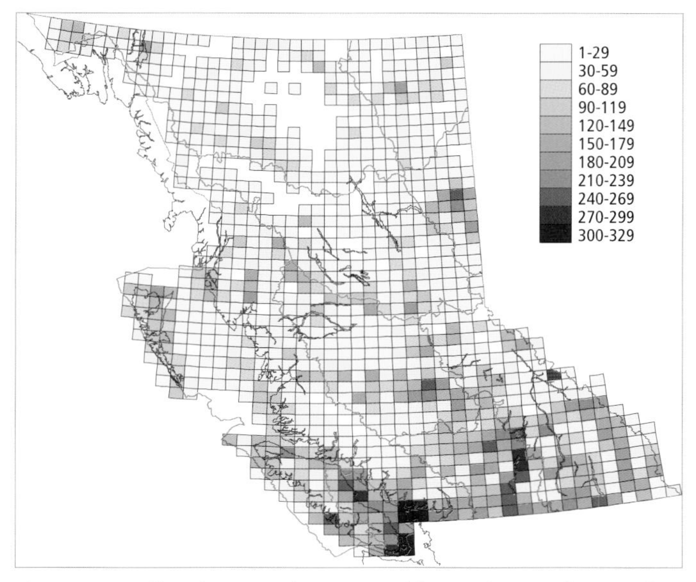

Figure 743. Avifaunal species richness reported from each National Topographic Series quadrangle in British Columbia. The "hot spots" in the province – those areas with one or more quadrangles reporting over 210 species – include the Fraser and Nanaimo lowlands of the Georgia Depression; the Tofino area of Western Vancouver Island; the Okanagan valley, Shuswap, and Kamloops areas of the Southern Interior; the Creston, Trail, and Revelstoke areas of the Southern Interior Mountains; the Williams Lake area of the Central Interior; and the Peace Lowland of the Boreal Plains.

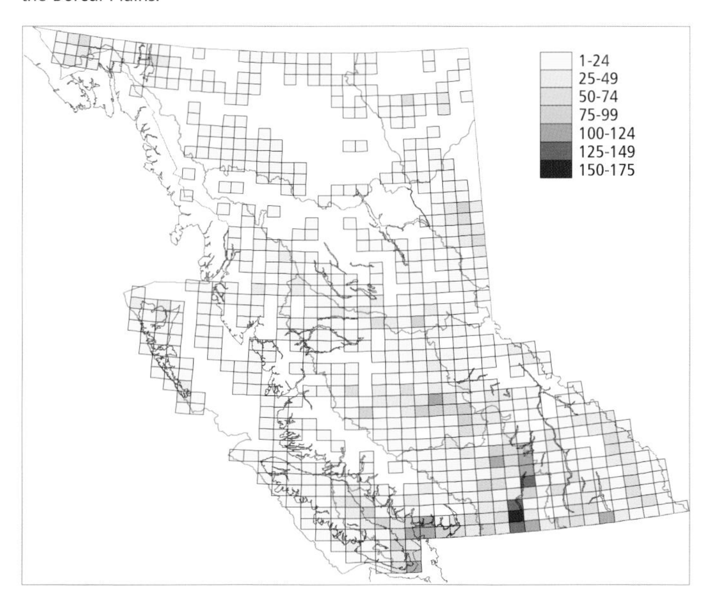

Figure 744. The species richness of breeding birds reported from each National Topographic Series quadrangle in British Columbia. Areas with one or more quadrangles reporting over 100 breeding species include the southern Nanaimo Lowland and the Fraser Lowland of the Georgia Depression; the southern Okanagan valley, Vernon, and Kamloops areas of the Southern Interior; the Creston area of the Southern Interior Mountains; and the Williams Lake area of the Central Interior.

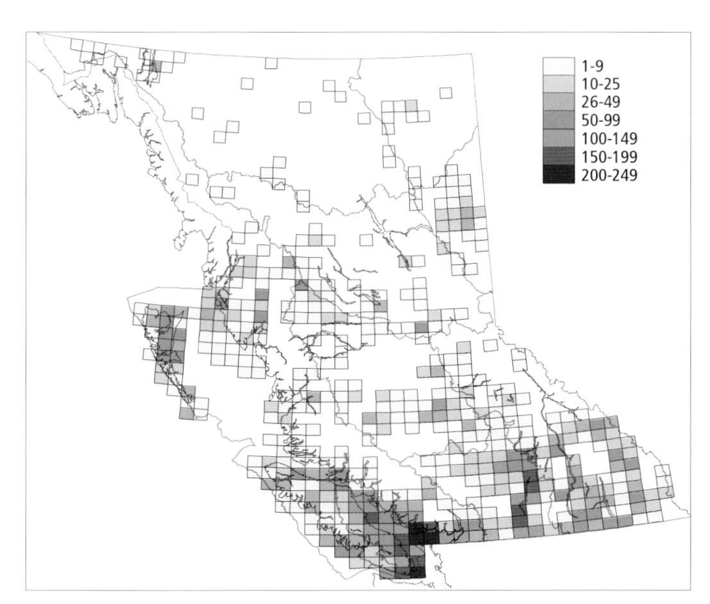

Figure 745. The species richness of wintering birds reported from each National Topographic Series quadrangle in British Columbia. Areas with one or more quadrangles reporting 100 or more wintering species include most of the Nanaimo Lowland and the Fraser Lowland of the Georgia Depression; the Tofino area on Western Vancouver Island; the Masset area on the Queen Charlotte Islands; the Okanagan valley, Shuswap, and Kamloops areas of the Southern Interior; and the vicinity of Williams Lake in the Central Interior.

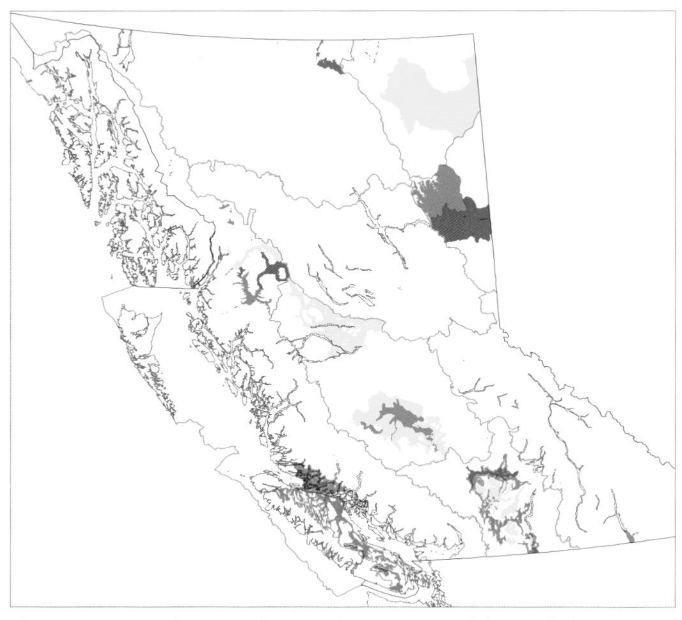

Figure 746. Major known and potential staging areas of the Sandhill Crane through British Columbia during the spring and autumn migration. The map is based on the largest flock sizes reported from various biogeoclimatic units and extrapolated to show the potential habitat distribution of these migrants. Sandhill Crane flocks totalling 100 to 200 birds (green), > 200 to 1,500 birds (yellow), and >1,500 birds (red) have been recorded on 3 major migration routes through the province: along the coast, through the southern and central portions of the interior, and along the east side of the Rocky Mountains.

of species richness and density is still required to provide data for gap analysis and other approaches to decision-making for special conservation needs. Much careful field study, applying both quantitative techniques and wider coverage, also remains to be done to determine the special areas in the northern half of the province and to refine present data in the south.

Major Known Migrations through British Columbia

British Columbia lies on 2 major migration routes: the Pacific Flyway and the western edge of the Central Flyway that runs through northeastern British Columbia. Within the province there are 5 major pathways that birds use: coastal routes, including both the near coast, heavily used by passerines, waterfowl, and shorebirds, and an important route some 40 km offshore where, for example, thousands of Pacific and Common loons, Greater White-fronted Geese, Northern Phalarope, and Bonaparte's Gulls move north along the coast. The coastal areas are of conservation concern because of the potential damage to birds caused by ocean pollution (e.g., ships pumping out their bilge tanks or oil spills). In the interior, the intermountain trenches, especially the Okanagan valley, north through the central interior, and the west Kootenay valley with its major stopping area at Creston, are significant, as is the northeast corner of the province. Some species (e.g., Barrow's Goldeneye, Harlequin Duck) have an east-west migration of short distance from the nesting areas in the interior to their coastal wintering grounds.

There are, however, a few species that are known to concentrate in huge numbers in relatively small staging areas and thus are especially vulnerable to the effects of such anthropogenic factors as oil spills and loss of their staging habitats through activities such as urbanization or industrial development. These important staging areas are significant habitats for various populations as they make their way to northern breeding grounds or southern wintering areas. For example, Brant are dependent on a few staging areas in the Strait of Georgia (e.g., Boundary Bay, Parksville–Qualicum Beach, Comox Harbour). Many of the birds are now known to make a direct flight from the Parksville–Qualicum Beach area to Izembek Lagoon in the Alaskan Peninsula, some 2,000 km distant (W.S. Boyd, pers. comm.). Were their habitat to be degraded in some way, it could affect the Brant's ability to acquire the necessary fat reserves for the next leg of their journey, resulting in poor reproductive success on the breeding grounds or not making the journey successfully.

Other such species dependent on staging areas in the province include, for example, the Sandhill Crane (White Lake [Okanagan Falls], Becher's Prairie, Kispiox valley, Nig Creek, Liard Hot Springs), Western Sandpiper (Fraser River delta, Tofino), and Surf Scoter (Big Bay). The 3 routes used by the crane (Fig. 746; Vol. 2, p. 108) are really a template of the major routes from the south across the province used by a great many species.

While all migration routes are of significant conservation concern, the staging areas, in particular, are locations where their identification and protection are warranted.

Endemism in the Birds of British Columbia

The Wisconsin period of glaciation saw the region that is now British Columbia almost totally ice-covered. There may have been scattered alpine nunataks on the mainland of the province, but there is no evidence that these held any surviving biota. On the contrary, there is evidence that parts of the Queen Charlotte Islands experienced a different glacial history (Prest 1984). In the words of Clague et al. (1982), "recent field studies, supported by radiocarbon dates, suggest that the Late Wisconsin ice cover on the Queen Charlotte Islands was rather restricted." Biological evidence suggests that parts of the islands did not experience total destruction of their flora and fauna during the most recent glaciation (McCabe and Cowan 1945; Heuser 1989; Cowan 1989).

There are no species of birds endemic to the province, but the re-examination of such strongly characterized subspecies from the endemic avifauna of the Queen Charlotte Islands as the Northern Saw-whet Owl (*Aegolius acadicus brooksi*) and the Hairy Woodpecker (*Picoides villosus picoideus*) may lead to changes in their systematic status.

The avifauna of the Queen Charlotte Islands includes 72 breeding species, only about 62% of the species found in equivalent habitat on the adjoining mainland. Obviously the 80 km water barrier has been significant in shaping the avifauna of the Queen Charlotte Islands. Among the breeding birds are 4 endemic subspecies: the Hairy Woodpecker and Northern Saw-whet Owl already mentioned, a Steller's Jay (*Cyanocitta stelleri carlottae*), and a Pine Grosbeak (*Pinicola enucleator carlottae*) (Cowan 1989). All of these are resident on the islands, and it is reasonable to infer *in situ* evolution at the subspecies level. Among the migrant species that breed on the Queen Charlotte Islands, the Townsend's Warbler appears to have distinctive features that would justify subspecific recognition. Distinctions include differences in wing size, breeding chronology, song, and the use of a winter range separate from that of the main body of the species (Morrison 1983).

Another 6 species of passerine birds sampled from populations on the Queen Charlotte Islands have measurable differences from those inhabiting the mainland coast (Cowan 1989). These are the Chestnut-backed Chickadee, Brown Creeper, Dark-eyed Junco, Fox Sparrow, Song Sparrow, and Pine Grosbeak. The size differences in body parts suggest that this nascent endemism may include genetic changes involving physiology and behaviour.

Geological evidence also suggests that small parts of the western coast of Vancouver Island were incompletely glaciated (Hebda and Haggarty 1997), but it is unclear whether these refugia were adequate in size or plant cover to support a vertebrate fauna. Vancouver Island is separated from the nearest point of the mainland by only about 2 km of open water, not much of a barrier to birds. Even so, while the mainland coast supports 164 species of breeding birds, Vancouver Island has 157 species, including 10 marine species that do not nest on the mainland coast. Thus the barrier effect of insularity has excluded about 7% of available species of birds. The effect has been much greater on the mammals, with only 60% of the non-volant species of the adjacent mainland reaching the island.

The avifauna of Vancouver Island includes at least 4 endemic subspecies: a White-tailed Ptarmigan (*Lagopus leucurus saxatilis*), Western Screech-Owl (*Otus kennicottii saturatus*), Northern Pygmy-Owl (*Glaucidium gnoma swarthi*), and Hutton's Vireo (*Vireo huttoni insularis*). Two of the island's endemic vertebrates, the ptarmigan and the endangered Vancouver Island Marmot (*Marmota vancouverensis*), are alpine-adapted taxa that might have survived the most recent glaciation of the island.

All the endemic subspecies of the Queen Charlotte Islands and Vancouver Island are considered of special management concern. The populations of each of them are small, resident, and restricted, and occupy habitats that are subject to the pressures of human presence.

On the mainland of British Columbia, there is only 1 endemic bird subspecies, the Gray Jay (*Perisoreus canadensis pacificus*) described from the Rainbow Mountains of the western Chilcotin.

Species Density in the Breeding Birds of British Columbia

While the macrodistribution of the birds of the province, i.e., the species richness, is now fairly well documented, only incidental progress has been made in developing the next phase of biodiversity: determining more precisely the habitats chosen by each species for nesting, and measuring the species density, i.e., the numbers of each species per unit area in the different habitats. However, it is easy to understand why little progress has been made in determining the species' densities in the variety of habitats throughout British Columbia. To illustrate, we discuss the problems associated with determining the species densities of birds in British Columbia forests.

A.J. Erskine (1977), of the Canadian Wildlife Service, pioneered this aspect of bird study in Canada. In 1968 he began the establishment of a series of bird census plots in the northern coniferous forest biome across Canada. At that time there were few descriptions of the birds of this vast area and no quantitative data that could be used to guide conservation action. Erskine's study plots included parts of what are now known as the Boreal White and Black Spruce and the Sub-Boreal Spruce biogeoclimatic zones in the Taiga Plains and Central Interior ecoprovinces of British Columbia (Erskine 1971, 1972, 1976). All are integral parts of the great northern coniferous forest of North America. *Birds in Boreal Canada* (Erskine 1977) was a pioneering attempt to make these data available as a guide to bird conservation in Canada.

To be meaningful, census figures must be related to a measurable area of relatively uniform habitat. They are almost meaningless when given only in context of areas as large and as regionally diverse as ecoprovinces. For example, Kessel (1998), in her study of the habitat used by birds in the taiga of central Alaska, identified 12 habitats distinguished by their plant cover. Each fulfilled a different role in the bird community. Her study area could be closely matched within the Spruce-Willow-Birch Biogeoclimatic Zone in the Northern Boreal Mountains Ecoprovince of British Columbia.

The British Columbia Forest Service has identified and described 14 biogeoclimatic zones, each characterized by distinctive biogeoclimatic features. Within these zones, 99 subzones have been found useful in forest management, along with 105 variants (Anonymous 1995). It is uncertain how many of these classification levels serve as useful descriptors of distinctive habitats for assemblages of bird species. However, a beginning has been made in measuring the number of birds of each species present on samples of some of the biogeoclimatic units.

A bird seeking a nest site certainly responds to different aspects of the environment than we use in our forest classification system, but it is probable that many of the biogeoclimatic units offer different attractions to the breeding bird fauna.

Since the 1980s growing public concern over destructive forestry practices has led to an important philosophical change in how forestry is practised in the province. The forests, with all the forest-dwelling creatures, are now recognized as important elements of a national heritage; forest ecosystems are understood to have many values beyond the mere production of fibre as a source of national wealth through employment (Anonymous 1995). This new ecological sensitivity has stimulated a series of bird population studies of some forest ecosystems and the changes imposed on them by different schemes of forest removal. In general, the studies have followed a pattern set by Erskine (1977).

Counting Methods

Bird numbers are difficult to determine and the species differ in their detectability. Each species has its characteristic period of song, song frequency, and other characteristics that make it difficult to plan a census that will provide the greatest likelihood of detecting all the birds present. At the same time, observers differ in their knowledge of the birds of the study region according to their experience, training of eye and ear, sensory abilities, and dedication to the task at hand. In addition, most census studies of birds are confined to 1 or 2 nesting seasons and do not allow for changes in abundance between years. Species such as the crossbills are attracted to forest areas that have an abundant cone crop; several species of warblers not only aggregate to areas where there are outbreaks of forest insects but increase their productivity there. These are among the sources of bias that require caution in the conclusions that can be drawn in comparing the density of birds in different habitats.

Several different census techniques have been developed for use with different kinds of birds and different habitats. These are well covered in a variety of publications, including Williams (1936), the International Bird Census Committee (1970), Wetmore et al. (1985), Krebs (1989), and Stork and Samways (1995).

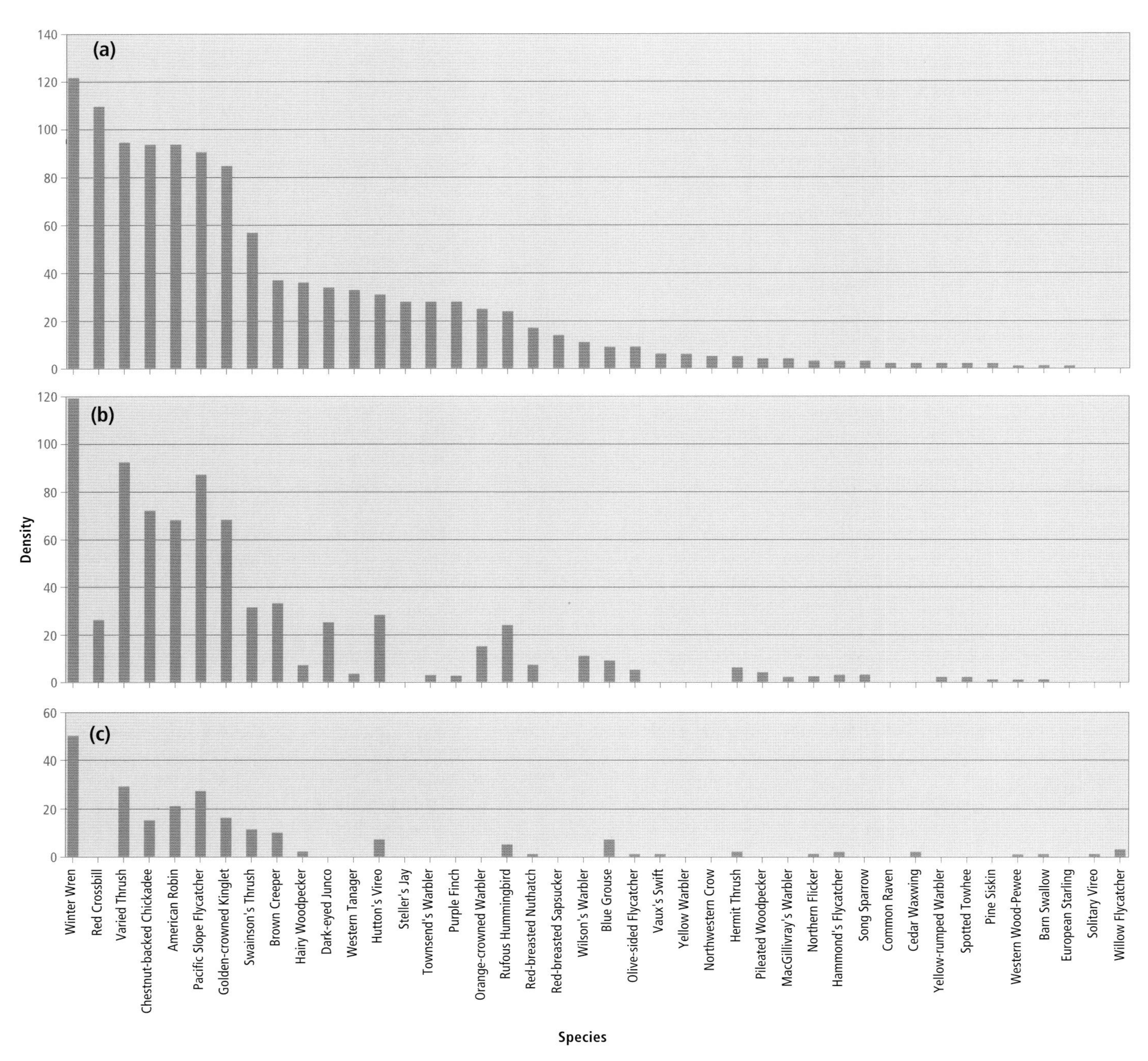

Figure 747. Density of birds per 100 ha in old-growth forest in the Coastal Western Hemlock Biogeoclimatic Zone of Western Vancouver Island in the Coast and Mountains Ecoprovince. Data were analyzed for the same sites using 3 methods: (a) all detections, (b) maximum number of "singing" males per station, and (c) average number of "singing" males per station (modified from Bryant et al. 1993). See Table 29, **L**, for a description of the forest type.

Erskine (1977) took the advice of the International Bird Census Committee and used a spot mapping technique. Kessel and Gibson (1994) used an adaptation of the same approach. Most studies in the province in recent years have used fixed point counts on circular plots of known radius (Williams 1936). Wetmore et al. (1985) proposed a single visit modification of the fixed point count technique and applied it to sample areas in British Columbia. This technique had obvious cost advantages, but it has not gained wide use. The repetitive sampling techniques provide the advantage of wider contact with the less numerous species in a census area.

Kessel (1998), in her work in Alaska, emphasizes the role of the precise habitat choices made by most species in influencing the results of census counts. She states:

> I conclude that most bird species are quite specific in aspects of their structural habitat requirements throughout their geographic range. Much of the geographic habitat variation that has been reported in North America appears to be the result of examining only macrohabitats or of measuring factors ... that may not be affecting the ecology of a given species. When defining avian habitats, a priori knowledge of a species' general biology and natural history characteristics is vital to selecting biologically meaningful habitat variables to examine. I concur with the statement of Knopf and Sedgwick (1992:742). "Published information on passerine habitats and habitat selection contains a large component of statistically valid but biologically meaningless relationships."

Recent initiatives in British Columbia by the Resources Inventory Committee include a standard methodological approach to songbird inventory in the province. This involves 3 basic levels of inventory ranging from simple presence/absence inventories to absolute abundance surveys (Resources Inventory Committee 1999). All songbird inventories conducted or funded by provincial government agencies are to follow these standard methodologies and should help facilitate future comparisons between survey areas.

Most accounts available to us express results as birds per 10 hectares or per 100 hectares, but some provide the data as mean numbers per species per census plot. We have converted them all to birds per 100 ha. Pojar (1995) while using the fixed point count methodology also did a limited trial of the spot mapping technique on the same study area. She reports that the total density of birds determined by the point count method was about 4 times greater than that obtained by spot mapping. She also found that although the numbers obtained by the 2 methods were different, the relative abundance of the species remained much the same.

Those involved professionally with the task of censusing birds will be aware of the many pitfalls that go with the field. Our review of experience in British Columbia shows that the numbers arrived at by the point count technique are highly sensitive to the choices made as the field counts are translated into estimates of the numbers using each sample plot.

Table 28. Differences in species richness and density resulting from choices of methods for recording numbers of birds per station.

	All detections	Maximum "singing" males per station	Average "singing" males per station
Number of species[1]	40	31	23
Total density (birds/100 ha)	1,161	738	210

[1] Data are from the same sites (see also Figure 747) using the 3 different methods (from Bryant et al. 1993).

Most authors have used the average number of individuals detected during the repeated counts at each point. A few authors (e.g., Bryant et al. 1993) provide estimated densities based on each of the average, median, and maximum numbers of birds recorded at each point during the repetitive counts.

In Figure 747 we give an example of the differences in species richness and density resulting from technical choices of how best to record numbers per station. Total density figures derived from the maximum detections of "singing" males per station are more than 3.5 times the number estimated using the average number of "singing" males per station, and 8 more species were detected. Density figures reported from all detections had 9 more species than the maximum number of "singing" males per station and over 1.5 times the density (Table 28).

The highest total densities derived by all the studies in the province examined by us are those reported by Pojar (1995) (Table 29). Her estimates are based on the maximum counts from each station (1995:4), which has been shown to increase both numbers of species and density (Bryant et al. 1993). There is also the possibility that some exaggeration could have resulted from accidental underestimation of the distance to the singing bird (R. Pojar pers. comm.). The differences may also be real, however, reflecting an unusually rich local habitat.

The presently accepted techniques of bird census provide data primarily on the more abundant and widely distributed species. Population sizes of rare species, colonial nesters such as marine and marsh birds, many waterfowl, birds of prey, nocturnal birds, or others with very precise habitat requirements can be determined only through studies focused on the species of concern (Dunn et al. 1997). Thus the studies proceeding in British Columbia are unlikely to provide reasonably reliable data on many species that for one reason or another are sparsely distributed.

Examples of Bird Density Studies in Some Biogeoclimatic Zones in British Columbia

The studies of bird density in the province are few in number compared with the complexity of the habitat. Thus, few

Table 29. A summary of bird species richness and density based on select studies in old-growth coniferous and broad-leaf forests of British Columbia. Forest types are ordered by total density.

Dominant species	Biogeoclimatic zone	Ecoprovince or region	Number of species	Total density[1]	Top 4 species as percent of total density
Coniferous forest					
A Lodgepole pine[2]	Sub-Boreal Pine-Spruce and Montane Spruce	Central Interior	17	204	70
B Western hemlock[3]	Coastal Western Hemlock	Western Vancouver Island	23	210	61
C Western hemlock[3]	Coastal Western Hemlock	Western Vancouver Island	24	214	62
D Engelmann spruce, subalpine fir clearcut[8]	Engelmann Spruce-Subalpine Fir	Sub-Boreal Interior	20	230	79
E White spruce[4]	Boreal White and Black Spruce	Taiga Plains	37	252	42
F Engelmann spruce, lodgepole pine, subalpine fir[5]	Engelmann Spruce–Subalpine Fir	Southern Interior	29	471	53
G Douglas-fir, western larch[6]	Interior Cedar-Hemlock	Southern Interior Mountains	17	475	73
H Douglas-fir, lodgepole pine[7]	Interior Douglas-fir	Central Interior	21	480	54
I Douglas-fir, lodgepole pine[7]	Interior Douglas-fir	Central Interior	36	488	58
J Engelmann spruce, subalpine fir[8]	Engelmann Spruce–Subalpine Fir	Sub-Boreal Interior	37	491	56
K Douglas-fir[9]	Coastal Western Hemlock	Southern Mainland Coast	39	673	33
L Douglas-fir, western redcedar, Sitka spruce[10]	Coastal Western Hemlock	Western Vancouver Island	31	738	50
M Engelmann spruce, subalpine fir[11]	Engelmann Spruce–Subalpine Fir	Southern Interior Mountains	40	745	51
N Western hemlock, mountain hemlock, amabilis fir[5]	Mountain Hemlock	Southern Mainland Coast	22	1,511	79
Broad-leaf forest					
O Mature trembling aspen[12]	Boreal White and Black Spruce	Taiga Plains	21	178	49
P Mature trembling aspen[13]	Boreal White and Black Spruce	Boreal Plains	23	396	51
Q Mature trembling aspen[12]	Sub-Boreal Spruce	Central Interior	17	464	72
R Clearcut trembling aspen[14]	Sub-Boreal Spruce	Central Interior	19	1,155	53
S Mature trembling aspen[14]	Sub-Boreal Spruce	Central Interior	35	2,929	34

1 Unless otherwise indicated, total density is the number of "singing" males per 100 ha.
2 Waterhouse (1995); composite totals from 2 biogeoclimatic zones; all detections per 100 ha.
3 Buckner et al. (1975).
4 Bennett (1998).
5 Wetmore et al. (1985).
6 Gyug and Bennett (1995).
7 Waterhouse and Dawson (1998).
8 Gyug (1997).
9 Horvath (1963).
10 Bryant et al. (1993).
11 Davis et al. (1999); all detections per 100 ha.
12 Erskine (1977).
13 Merkins and Booth (1998); all detections per 100 ha.
14 Pojar (1995).

generalizations of biological significance can be made. However, we have attempted to develop the data from several of these studies to provide comparable information on bird density in a number of forest communities. Most of the data are taken from unpublished reports that show a range of circumstances in 9 of the 14 biogeoclimatic zones in British Columbia. With few exceptions the studies had been done to examine the consequences to birds of forest removal either by clearcut or in selective patterns. Studies we have drawn on include: Bennett (1998), Bryant et al. (1993), Buckner et al. (1975), Davis et al. (1999), Erskine (1977), Gyug (1997), Gyug and Bennett (1995), Horvath (1963), Merkins and Booth (1998), Pojar (1995), Waterhouse (1995), Waterhouse and Dawson (1998), and Wetmore et al. (1985). The results of these studies have been summarized in Table 29; Appendix 4 presents the results of the studies in graphic form.

At this time only a few of the biogeoclimatic subzones that are major bird habitats, and even fewer of the important variants, have been given even preliminary study. The absence of data from other habitats limits conclusions that can be drawn and leaves us ignorant of the precise habitat needs of the majority of the breeding birds of the province. Only about 20% of the breeding species found in British Columbia have appeared on the census counts we have studied. We hope that these results will stimulate increased effort to determine the most important bird habitats, and especially to develop greater focus on the thinly distributed species. None of the publications or reports we have examined indicates that statistical techniques were used to determine the sampling required to achieve a designated level of detection.

We have analyzed the species density in 14 examples of coniferous forest and 5 broad-leaf forest habitats in British

Table 30. The 10 most abundant breeding bird species reported from each of 14 samples of 8 different coniferous forest types in British Columbia based on species densities. For more details of the forest type see Table 29. The species are listed in order of overall dominance and then in taxonomic order within those with tied dominance values. Numbers under forest type are the abundance ranks for each species in that forest type.

	Forest type[1]														Number of studies where species was recorded
Species	A	B	C	D	E	F	G	H	I	J	K	L	M	N	
Yellow-rumped Warbler	1			7	5	5	2	1	2	2	2		7		10
Winter Wren		1	2			8				9	9	1	6	3	8
Golden-crowned Kinglet		5	5			1	4			4		5	4	1	8
Swainson's Thrush		7	4		1					5	5	8	10		7
American Robin	9	4		8					9		1	6		10	7
Dark-eyed Junco	3			3		9	3	5		7			3		7
Red-breasted Nuthatch	8						6	4	3	10	6				6
Varied Thrush		2	6									2	8	4	5
Townsend's Warbler						3	1			1			2	2	5
Red Crossbill	2							3	1			10	9		5
Pine Siskin	5				6			2	10				1		5
Chestnut-backed Chickadee		6	3									4		5	4
Brown Creeper		8					8					7		6	4
Hermit Thrush						4		6	6					8	4
Wilson's Warbler			8	4						6			5		4
Pacific-slope Flycatcher		3	1									3			3
Cassin's Vireo							7		4		4				3
Warbling Vireo					7		9								3
Ruby-crowned Kinglet						2		10		8					3
Chipping Sparrow				5			10	9							2
Blue Grouse		9	7												2
Rufous Hummingbird			10								8				2
Yellow-bellied Sapsucker					9										2
Least Flycatcher					4										2
Hutton's Vireo		10										9			2
Gray Jay	4					7									2
Mountain Chickadee						6		8							2
Townsend's Solitaire	6							7							2
MacGillivray's Warbler				6							10				2
Western Tanager							5		8						2
Brown-headed Cowbird											7				1
Ruffed Grouse									7						1
Spruce Grouse	10														1
Northern Flicker														7	1
Alder Flycatcher				9											1
Say's Phoebe														9	1
Steller's Jay			9												1
Boreal Chickadee	7														1
Tennessee Warbler					3										1
Magnolia Warbler					2										1
Cape May Warbler					10										1
Black-throated Gray Warbler											3				1
Blackpoll Warbler										3					1
American Redstart					8										1
Savannah Sparrow				2											1
Lincoln's Sparrow				1											1
Golden-crowned Sparrow				10											1
Purple Finch						10									1
White-winged Crossbill									5						1

[1] A - lodgepole pine; B, C - western hemlock; D - Engelmann spruce, subalpine fir clearcut; E - white spruce; F - Engelmann spruce, lodgepole pine, subalpine fir; G - Douglas-fir, western larch; H, I - Douglas-fir, lodgepole pine; J, M - Engelmann spruce, subalpine fir; K - Douglas-fir; L - Douglas-fir, western redcedar, Sitka spruce; N - western hemlock, mountain hemlock, amabilis fir.

Columbia. A numerical summary of the results is given in Table 29. As mentioned elsewhere, the total densities are susceptible to even small deviations in the gathering of data in the field, so caution should be used in making detailed comparisons between the total densities given here. Some authors have concentrated on obtaining statistically useful data and, as a result, have not included species that occurred too infrequently to yield reliable figures of density. The absence of these species makes little difference to the total density, but it reduces the usefulness of the study for calculating species diversity or for identifying species of special conservation concern.

While it is important to know major breeding concentrations of regularly occurring species, it is also important to identify even sparse populations of the scarcer species. Among these will be the species with the most particular habitat requirements, which are likely to be most sensitive to imposed changes.

Among the coniferous forests sampled, birds are least numerous in the Central Interior stands of lodgepole pine and in the western hemlock forests of Western Vancouver Island. They are most numerous in our only census of a Mountain Hemlock zone forest (Wetmore et al. 1985), and in an Engelmann Spruce–Subalpine Fir zone forest (Davis et al. 1999) (Table 29). In general, the distribution of our data is too sparse to reveal geographic trends overall, or trends in species densities.

Dominance is a widely recognized feature of bird populations and is most evident in our analyses. We have used the percentage of total density contributed by the 4 most abundant species at each site as indication of the degree of species dominance. In general, these studies indicate that usually 4 species at each site contribute 50% or more of the total density (Table 29). The most extreme situation encountered was in the single Mountain Hemlock zone study, where 2 species, the Golden-crowned Kinglet and Townsend's Warbler, contributed 65% of the total density (Appendix 4 and Tables 29 and 30).

The composition of the 10 most abundant breeding species in the 8 coniferous forest zones for which we have data reveals the diversity in the groups of species that select each habitat (Table 30, Fig. 747, and Appendix 4). Fifty-four species appear in the combined top 10 groups of the various surveys of coniferous forests. Six species, the Yellow-rumped Warbler, Winter Wren, Golden-crowned Kinglet, Swainson's Thrush, American Robin, and Dark-eyed Junco, occurred in at least 7 of the 14 studies represented, 70% occurred in less than 4, and 46% occurred once only. It appears also that the intensity of dominance is negatively correlated with the richness of the avifauna, which is a well-known relationship in many groups.

In the broad-leaved forests surveyed (Table 31; Appendix 4), 24 species appeared in the combined top 10 groups. There, 9 species – Least Flycatcher, American Robin, Dusky Flycatcher, Swainson's Thrush, Orange-crowned Warbler, Yellow-rumped Warbler, American Redstart, MacGillivray's Warbler, and Dark-eyed Junco – occurred in at least 3 of the 5 types sampled. Four of these species were also dominant in the coniferous forests surveyed.

It is impossible to know all the species that are missed in the sampling, but several authors listed the species they saw outside the boundaries of their census areas, or those that occurred too infrequently in their samples to yield a useful statistic. We emphasize that the scarce species may well provide as much or more conservation-oriented data as the abundant species; they should be included in all studies.

The 2 areas for which we have pre- and post-clearcutting figures reveal a decline in species richness of about 46%, and in bird density of between 53% and 61%, resulting from the removal of the forest trees (Table 29).

As the human population continues to expand, increasing the demands it makes on wild land, it is urgent that a program be established to determine, in comparable terms, the species and their densities in all the distinctive habitats for birds in the province. As Newton (1998:27) observes: "One of the most urgent requirements in avian population research is for more work in extensive areas of natural habitat before such areas are gone forever. Only then can we tell the full

Table 31. The 10 most abundant breeding bird species reported from some deciduous forest types in British Columbia based on species densities. The key to the letters for forest type may be found in Table 29. The species are listed in order of overall dominance and then in taxonomic order within those with tied dominance values. Numbers under forest type are the abundance ranks for each species in that forest type.

Species	Broad-leaf forest type[1]					Number of studies where species was recorded
	O	P	Q	R	S	
Least Flycatcher	9	1	5		4	4
American Robin	10		6	3	8	4
Dusky Flycatcher			4	5	2	3
Swainson's Thrush	2	3	3			3
Orange-crowned Warbler			10	7	6	3
Yellow-rumped Warbler		6	8		3	3
American Redstart	5		1		1	3
MacGillivray's Warbler			7	9	9	3
Dark-eyed Junco	7			2	7	3
Chipping Sparrow	6			8		2
Yellow-bellied Sapsucker	8	10				2
Warbling Vireo		9	2			2
Red-eyed Vireo	4	7				2
Ovenbird	1	5				2
Brown-headed Cowbird		8	9			2
Rufous Hummingbird				6		1
Cassin's Vireo					5	1
Black-capped Chickadee					10	1
Yellow Warbler		4				1
Magnolia Warbler	3					1
Savannah Sparrow				10		1
Lincoln's Sparrow				1		1
White-throated Sparrow		2				1
White-crowned Sparrow				4		1

[1] O, P, Q, S - mature trembling aspen; R - clearcut trembling aspen.

extent of human influence on population phenomena, and view our findings in a more appropriate perspective."

Even the very modest beginning that has been made in the study of the numerical distribution of birds in British Columbia adds a new dimension to our concepts of how birds use the complex of habitats available. The single-plane maps of bird distribution that accompany each species account provide a most incomplete impression of the real situations, but it is the best we can do at the present state of knowledge. As the understanding of patterns of bird density increases it will be possible to provide 3-dimensional maps that come much closer to depicting what actually occurs. On a much larger scale the North American Breeding Bird Survey (Sauer et al. 1997) is already producing maps of the regional density of some species and groups of species (see, e.g., Figs. 740 and 742). However, the lack of data on the population density of the scarcer birds is a most serious deficiency. To the extent that we are using density figures to guide decisions on the destruction of local forest habitats, we are lacking much of the most essential information.

Future Needs

Data on species richness are relatively easy to obtain, those on species density much less so. A considerable effort by the province has been devoted to measuring the numbers and species of birds present at scattered points in the forested parts of British Columbia. This level of field study, were it coordinated and focused, could in a few years produce a valuable body of data on the birds of British Columbia. We now know how they are distributed at the ecoprovince level. We know also, in a general way, the ecological conditions they require as we have described them in the species accounts. Now we are in a position to move on, to concentrate on measuring in detail how populations of species at all levels of abundance are tied to the great diversity of habitats. This objective can be reached most expeditiously by redirecting some of the resources to identifying regions in all ecoprovinces that support especially rich bird communities. The task requires quantitative data on population densities of species at all levels of abundance. Hearsay information is no longer satisfactory. Solid data are essential to ensure wise use of the industrial resources while maintaining the wealth of our natural heritage.

Future management should focus on the ecosystem and emphasize the maintenance of habitat conditions that encourage the survival of all natural assemblages of birds and other wildlife components while mitigating disturbances caused by human activities. It is therefore important, and pressing as we have said, that we gather sufficient quantitative data on population densities from unmodified habitats throughout the province.

This brief summary gives some idea of where we are at this time, and it points the way ahead. It lists the birds and their densities, in samples of each of the biogeoclimatic zones for which we have that information. We have found no studies of the density of birds in the Coastal Douglas-fir Zone, the dry, unique coastal zone of the Georgia Depression; the Coastal Western Hemlock Zone of the Queen Charlotte Islands; the Ponderosa Pine Zone in any of its 4 variants; the Bunchgrass Zone; the Spruce-Willow-Birch Zone, and the Alpine Tundra Zone. There is only one rather brief study of the very extensive Mountain Hemlock Zone (Wetmore et al. 1985). Evidence suggests that each of these biogeoclimatic zones has areas of special bird assemblages.

The freshwater wetlands of the interior of the province support a rich and diversified avifauna. Since 1987, a Co-operative Waterfowl Survey has been conducted annually. It consists of 6 replicated counts on each of roughly 400 wetlands in British Columbia. Its purpose is to determine numbers and abundance of breeding and migrating waterbirds in the interior of British Columbia in May, based on their distribution and behaviour (Breault and Watts 2000). Participants in 2000 included the Canadian Wildlife Service, Ducks Unlimited Canada, the B.C. Wildlife Branch, the Kamloops Naturalist Club, the Canadian Parks Service, and biological consultants.

Table 32 presents the data for duck breeding pair use on trend wetlands, a subset of 290 wetlands surveyed consistently since 1988. These data have been analyzed to show departures from the long-term average as well as from the immediately preceding year. Analysis for long-term trends would provide additional useful information; however, such analysis would have to take into account water levels on wetlands, winter and spring weather conditions, migration and nesting chronology, and the management and research programs taken on some species – a complicated undertaking (A. Breault pers. comm.).

For example, in the latter case consider Barrow's Goldeneye trends. This species has undergone a number of research and management initiatives over periods of the survey ranging from the addition of over 500 artificial nest boxes at some of the wetland sites, to the collection of breeding adult females for scientific studies. Today, the provincial index for this goldeneye is now less than 200 breeding pairs. The extent to which this index reflects enhancement or collection programs is unknown but the Barrow's Goldeneye is now likely returning to its pre-"enhancement" levels (A. Breault pers. comm.).

It is noteworthy that in 8 of the 13 years, there were pairs of Eurasian Wigeon on the nesting grounds (Table 32) during the breeding season, the first instance in 1988. There are no confirmed breeding records of the species for North America, but the presence of indicated pairs on the British Columbia breeding grounds suggests that the species might be nesting there.

There has also been some intensive study of the rich marine bird communities (Vermeer and Butler 1989; Vermeer et al. 1992; Butler and Vermeer 1994; Vermeer and Morgan 1997).

CHANGES IN BIRD POPULATIONS IN BRITISH COLUMBIA

All populations of living organisms experience changes in the number of individuals present from day to day throughout the year. Young are born or hatched, and each year-class gradually or suddenly succumbs to one of the many destructive events that are part of every life. Rates of production of young and mortality rates differ from species to species and

Table 32. Mean number of breeding pairs of ducks seen per survey in the British Columbia Interior Cooperative Wetland Survey, 1987-2000.[1] Numbers below means are standard deviations.

Species	Year														'88-'99 mean	2000, % change from '99	2000, % change from '88-'99 mean
	'87	'88	'89	'90	'91	'92	'93	'94	'95	'96	'97	'98	'99	'00			
American Wigeon	138.9	137.2	185.0	163.7	235.0	240.3	232.3	249.7	243.8	127.7	118.2	144.8	103.7	107.8	181.8	4.0	–40.7
	114.7	21.4	21.6	11.7	37.4	32.2	39.6	27.1	21.0	26.7	32.3	17.3	12.1	33.6	55.6		
Blue-winged Teal	91.6	76.7	106.3	39.7	80.5	127.3	124.3	69.3	72.5	14.8	21.2	121.8	30.0	108.0	73.7	260.0	46.5
	61.7	59.6	58.0	34.7	76.5	72.8	72.6	54.3	51.0	10.8	20.6	66.8	25.4	73.9	40.7		
Cinnamon Teal	117.4	100.7	64.7	68.0	71.5	144.0	68.0	76.3	40.0	17.0	25.3	40.7	24.2	41.2	61.7	70.3	–33.3
	75.1	30.1	20.4	16.9	28.0	47.4	23.0	11.0	10.4	5.0	11.5	11.5	16.8	15.3	36.1		
Eurasian Wigeon	0.0	0.2	0.0	0.3	0.0	0.5	0.2	0.7	0.0	0.3	0.2	0.3	0.0	0.0	0.2	0.0	–100.0
	0.0	0.4	0.0	0.5	0.0	0.8	0.4	0.5	0.0	0.5	0.4	0.5	0.0	0.0	0.2		
Gadwall	44.1	21.5	42.5	31.3	61.8	97.7	98.5	98.2	114.7	55.8	67.3	97.3	50.0	70.7	69.7	41.3	1.4
	35.0	7.3	14.5	13.2	24.4	34.2	16.2	21.0	33.8	21.5	33.0	24.9	25.6	27.9	30.7		
Green-winged Teal	78.7	148.3	84.2	88.2	162.0	151.2	183.0	158.7	147.3	94.5	77.5	126.2	98.5	69.7	126.6	–29.3	–45.0
	73.4	63.5	9.4	24.1	19.6	29.5	56.5	28.7	32.4	26.5	11.7	23.2	18.4	6.7	36.3		
Mallard	191.3	285.3	304.5	308.2	403.0	405.0	395.8	423.2	537.3	346.5	335.5	449.0	325.8	363.3	376.6	11.5	–3.5
	132.1	60.0	62.6	18.8	50.9	64.1	34.7	64.4	67.8	38.5	53.1	41.2	42.5	106.2	72.9		
Northern Pintail	24.4	66.2	43.3	22.7	35.0	30.8	43.2	28.7	43.5	16.5	11.2	13.5	11.0	10.0	30.5	–9.1	–67.2
	23.8	25.8	10.0	4.5	7.2	7.4	24.7	9.7	14.0	6.2	5.3	4.6	3.8	3.3	16.8		
Northern Shoveler	75.6	77.5	69.7	62.3	75.8	100.5	69.5	95.8	80.5	57.2	61.8	98.0	51.3	91.5	75.0	78.2	22.0
	61.0	10.1	26.4	14.8	29.2	27.7	18.0	10.3	14.1	18.2	22.7	23.5	18.5	32.5	16.3		
Wood Duck	0.0	0.2	0.8	6.3	9.2	21.7	15.5	15.0	10.2	2.3	1.8	7.2	2.8	2.2	7.8	–23.5	–72.1
	0.0	0.4	0.8	2.2	3.3	3.4	4.4	5.0	2.6	1.5	1.7	4.4	2.5	1.2	6.8		
Total dabblers	762.6	913.7	903.0	790.8	1,134.0	1,319.0	1,230.7	1,216.0	1,289.8	732.6	720.0	1,098.8	697.3	864.3	1,003.8	23.9	–13.9
	549.6	81.4	197.4	71.5	232.2	231.7	64.8	136.2	129.5	117.3	158.3	143.1	127.8	261.9	236.5		
Barrow's Goldeneye	246.7	276.3	328.5	287.8	353.0	278.0	299.0	317.5	262.0	188.3	178.2	175.8	175.0	188.5	260.0	7.7	–27.5
	196.2	44.7	34.1	11.4	18.4	51.2	32.1	14.9	31.1	31.6	6.7	23.0	25.5	17.4	64.4		
Bufflehead	145.7	267.0	224.5	247.8	271.0	239.2	286.7	293.7	315.5	236.2	225.2	203.0	199.5	203.5	250.8	2.0	–18.9
	111.7	49.5	38.3	72.6	27.7	19.0	51.8	25.8	36.9	48.3	22.5	12.6	18.5	10.7	36.4		
Canvasback	11.9	19.8	10.2	19.2	18.0	9.2	14.2	11.7	16.5	12.8	9.3	6.5	5.8	4.5	12.8	–22.9	–64.8
	7.5	14.3	4.5	3.2	6.5	3.8	3.1	4.2	3.3	5.3	3.7	2.3	3.8	2.8	4.8		
Common Goldeneye	0.3	4.3	6.3	23.3	28.2	26.7	50.3	41.5	48.2	26.8	10.0	9.2	11.2	9.7	23.8	–13.4	–59.4
	0.5	2.7	3.0	8.1	4.4	8.9	19.1	8.0	12.0	5.8	3.8	4.5	1.9	4.2	16.2		
Hooded Merganser	0.6	1.3	2.3	13.2	13.2	12.0	21.8	27.2	22.3	17.8	7.7	14.2	15.0	10.7	14.0	–28.9	–23.8
	0.8	1.4	0.5	6.4	4.1	6.0	6.4	8.2	6.1	3.0	4.2	4.4	4.3	2.7	7.8		
Redhead	52.3	49.3	43.8	49.7	54.3	55.7	49.3	67.5	70.3	41.7	35.3	53.0	39.8	49.8	50.8	25.1	–1.9
	42.2	8.0	8.9	11.2	15.7	22.3	10.7	13.8	14.5	11.9	16.9	9.3	10.6	22.2	10.4		
Ring-necked Duck	30.0	73.2	57.5	89.5	107.7	99.8	78.8	105.5	107.2	66.2	48.0	66.0	54.2	59.0	79.5	8.9	–25.8
	28.7	17.5	15.4	18.6	26.9	11.5	15.7	12.9	18.7	13.2	11.4	13.8	17.7	10.5	21.9		
unid. Scaup	133.3	189.5	114.2	95.0	135.8	242.7	142.8	148.2	143.5	109.0	110.8	82.3	77.5	89.5	132.6	15.5	–32.5
	93.4	82.6	22.8	22.8	30.0	258.4	26.6	44.6	33.1	69.8	73.7	14.2	19.0	30.1	46.9		
Total divers	625.4	884.2	793.9	830.2	986.3	964.0	949.5	1,014.4	986.0	703.3	626.5	610.0	578.0	624.8	827.2	8.1	–24.5
	506.7	123.4	31.7	62.6	67.6	293.9	72.1	72.0	52.5	151.0	142.2	21.6	85.5	80.7	162.0		
Total ducks	1,388.0	1,797.8	1,696.9	1,621.0	2,120.3	2,283.0	2,180.2	2,230.4	2,275.8	1,436.0	1,346.5	1,708.8	1,275.2	1,489.2	1,831.0	16.8	–18.7
	1051.4	174.9	220.0	56.5	263.7	419.6	98.9	159.4	155.5	261.7	206.9	150.9	208.2	331.8	374.8		

1 From Breault and Watts (2000).

with changing circumstances. The interaction of intrinsic and environmental factors results in populations that are seldom the same from one year to the next. In general, however, over time a relatively stable pattern of relationships develops between the numbers of birds and the resources of the environment.

At our latitudes (48°14′ to 60°N) a basic core of species can be expected to appear on about the same breeding or wintering grounds, in about the same numbers, and close to the same dates year after year. For thousands of birdwatchers one of the highlights of spring is anticipating the dates upon which old friends return. Departure from the normal path of ups and downs in the population often signals important changes in the environment.

Long-term changes in populations of birds are the basic statistics used to guide conservation action. Short-term increases or decreases in the numbers of a species are usually unimportant, but statistically significant changes in a breeding population, especially declines continuing over a decade or longer, may be grounds for concern.

While this predictability is the norm, it is well known to most naturalists that some species behave quite differently. Several species show cyclic changes in numbers over periods of about six years. In British Columbia, the Ruffed Grouse, Sharp-tailed Grouse, and Willow Ptarmigan are notable species whose numbers fluctuate cyclically. Other species are specialist feeders and require food sources that are highly variable in availability. The crossbills roam vast areas of coniferous forests in search of abundant cone crops. Such an event often serves to stimulate nesting. Although it is not as nomadic as a crossbill, the Bohemian Waxwing responds similarly to an abundant crop of wild fruits and berries. The Black-backed Woodpecker is drawn to the scorched remnants of forests ravaged by fire, and a number of the warblers also respond numerically to outbreaks of forest insects. Each of these is an example of species for which trends of change in abundance can be detected only by censuses taken over areas as large as a continent and over many years.

The only source of continent-wide information on the breeding populations of North American birds is the Breeding Bird Surveys (BBS) program. This is administered cooperatively by the United States Geological Survey, the Patuxent Wildlife Research Center, and the Canadian Wildlife Service. For a description of the census process, see Downes and Collins (1996).

In 1968, the program began in Canada and has gradually extended its coverage throughout the country; it now has collected over 30 years of continuous data on bird populations over much of southern Canada. The BBS is most successful in monitoring the moderately abundant to abundant species of passerine birds. By virtue of the number of "hits" required for statistical significance, it is less successful with the scarce species and has accounted for only about 34 of the 157 species of nonpasserines that nest in the province. Among the nonpasserine species, the BBS techniques are not applicable to the marine species, most of the raptors, the waterfowl, or the varied group of subalpine and alpine nesters. In contrast, BBS has detected 58% of the total roster of passerine species known to nest in the province. This could be improved if the pattern of BBS routes was expanded to include the alpine and subalpine habitats throughout the province, and the northern half of the province, where there is not yet a single BBS route that has operated continuously for more than 10 years and only 2 beyond 5 years. This means that changes in numbers of the species comprising the distinctive bird fauna of the Boreal and Taiga plains, the Northern Boreal Mountains, and 90% of the Sub-Boreal Interior ecoprovinces remain undetected. The same is true of the Coast and Mountains Ecoprovince north of latitude 50°, except for the Queen Charlotte Islands.

Thus, there are serious limitations to our one, continent-wide source of reasonably reliable data on breeding populations of birds. We need such data to provide early warning of serious disturbances in the environment. Perhaps recent proposals from the Canadian Wildlife Service (1994) will result in the gathering of data in at least a few localities where the potential for human impact on bird populations is greatest.

As we note elsewhere, British Columbia has the greatest habitat diversity of any of the provinces and consequently requires more detailed monitoring. In the absence of volunteers willing and able to run at least a basic skeleton of routes, Erskine et al. (1992) suggest that official participation might be required to assure that some of the most important remote routes are staffed.

Species Experiencing Long-Term Changes in Numbers in British Columbia

When analyzing the bird populations of the province we have separated them into coastal and interior regions. This has been done to facilitate identification of possible regional differences in any population changes. Certainly the maritime climate creates distinctively different habitats along the Pacific coast than any occurring to the east of the coastal mountain ranges.

Only 234 of the 309 bird species known to have bred in the province have so far been detected by the Breeding Bird Surveys. Among them, 60 species (26%) showed statistically significant changes in populations through the period 1968 to 1993 in either or both of the coast or interior regions (Tables 33 and 34). Of these, 28 have been decreasing and 33 increasing. Ten of the 60 species showed significant changes at the provincial but not the regional level.

On the coast there are 19 decreasing species and 8 increasing, while in the interior a reverse relationship exists with 10 decreasing and 19 increasing. In relation to the total list of species, these results show that in the interior there are about twice as many increasing as decreasing species, while on the coast there are about twice as many decreasing as increasing species. Keeping in mind that almost all the BBS routes on the coast are in the most densely populated region of the province, the Georgia Depression, it is tempting to conclude that the differences between the coast and interior region arise from some aspects of habitat change caused by human activity.

Table 33. Species showing a decrease ($P < 0.10$) in numbers on coastal or interior Breeding Bird Survey routes summarized from British Columbia. Canadian and North American values are provided for comparison.

	Average annual change (%)			
Species	BC Coast[1]	BC Interior[1]	Canada[2]	Continent[2]
Great Blue Heron	–6.2	i.d.[3]	n.c.[4]	+2.1
Blue Grouse	–5.3	i.d.	–3.6	–3.6
Killdeer	–4.8	n.c.	–2.5	–0.4
Common Snipe	–7.7	n.c.	n.c.	n.c.
Band-tailed Pigeon	–5.8	i.d.	n.c.	–2.7
Mourning Dove	i.d.	–2.7	+1.7	–0.3
Rufous Hummingbird	–5.0	n.c.	n.c.	–2.7
Olive-sided Flycatcher	–5.1	–4.8	–3.7	–3.9
Western Wood-Pewee	n.c.	–1.4	n.c.	–1.5
Willow Flycatcher	–3.7	n.c.	n.c.	–1.2
Pacific-slope Flycatcher	+8.3	–7.3	+4.4	n.c.
Red-eyed Vireo	n.c.	–2.1	+1.0	+1.1
Barn Swallow	–3.7	n.c.	–1.6	–0.3
Black-capped Chickadee	–2.4	n.c.	+1.1	+1.4
Golden-crowned Kinglet	–2.0	n.c.	n.c.	n.c.
Ruby-crowned Kinglet	–0.9	n.c.	–1.2	n.c.
European Starling	–2.5	–4.1	–2.0	–1.1
Yellow Warbler	–2.7	–2.1	n.c.	+0.6
Wilson's Warbler	n.c.	–3.5	n.c.	n.c.
Song Sparrow	–0.9	n.c.	–1.5	–0.6
Dark-eyed Junco	–1.4	n.c.	n.c.	–1.2
Brewer's Blackbird	n.c.	–4.6	n.c.	n.c.
Brown-headed Cowbird	–4.3	n.c.	–1.7	–1.1
Pine Siskin	–8.8	n.c.	n.c.	n.c.
American Goldfinch	–4.1	n.c.	n.c.	–0.8
House Sparrow	n.c.	–7.7	–2.5	–2.2
	Province[5]			
Ruffed Grouse	–7.4		n.c.	n.c.
Black Swift	–9.8		–10.0	n.c.
Purple Finch	–1.5		–2.3	–1.8

1 This study (1968-93).
2 Sauer et al. 1997 (1966-96).
3 Insufficient data.
4 Could not detect a change.
5 Anonymous (1999).

Species Showing Declining Population Trends in British Columbia

Of the 28 species declining in the province (Table 33), the Great Blue Heron (Fig. 748) is certainly the one of greatest concern. The decline in numbers of the Great Blue Heron on its British Columbia range has been discussed by Butler (1997). It is not declining across Canada, and the continental surveys reveal an increase throughout much of its North American range (Fig. 749). The species is resident in British Columbia and there is no doubt that the decline is the result of a local situation. Evidence is strong for predation on nestlings by an increasing population of Bald Eagles as part of the story, but there seem to be contributing environmental problems as well (R.W. Butler, pers. comm.). Loss of old-field habitat and ditches through urbanization is likely playing a role as well.

Ten species – Ruffed Grouse, Common Snipe, Pacific-slope Flycatcher, Red-eyed Vireo, Black-capped Chickadee, Golden-crowned Kinglet, Yellow Warbler, Wilson's Warbler, Brewer's Blackbird, and Pine Siskin – appear to be declining in British Columbia only, which is cause for further monitoring. The Red-eyed Vireo winters in tropical South America and it is always possible that the part of the population that summers in British Columbia has actually encountered problems on the winter range.

Declines for 4 widely distributed species appear to be restricted primarily to western North America and include the Mourning Dove, Willow Flycatcher, Dark-eyed Junco, and American Goldfinch. For these, the cause of the steady decline should also be sought.

Five western species, including the Blue Grouse, Band-tailed Pigeon, Black Swift, Rufous Hummingbird, and Western Wood-Pewee, show declines in British Columbia, an area which constitutes a significant portion of their breeding ranges. The Black Swift, which winters in Mexico and Central America, breeds primarily in British Columbia and shows the greatest decline of all birds detected by the BBS in this province.

Figure 748. The Great Blue Heron, a resident species across southern portions of the province, is 1 of 27 indigenous species of birds in British Columbia showing long-term declines in numbers in either or both of their coastal and interior populations (Vancouver, 1 December 1995; R. Wayne Campbell).

The Blue Grouse is a resident in British Columbia and is declining significantly within the range of the coastal subspecies. The declining entry for all Canada arises from the same data as those for British Columbia. The species is still common locally and is known to increase in the early stages of forest regeneration and decrease as closed canopy situations return. The bird is also open to hunting. Though remedial action is probably within our hands, the continent-wide decline indicates that there are problems in populations in the western American states that may also apply in British Columbia.

In Canada, the Band-tailed Pigeon nests only in southwestern British Columbia. It winters in southwestern United States and adjacent Mexico, so the decline noted for coastal British Columbia and the continent arise from the same regional population (Fig. 750). The cause and location of the decline are unknown. This species was hunted throughout much of its range but is no longer exposed to hunting in British Columbia or Washington, although a limited hunt is still allowed in Oregon and California.

The Ruby-crowned Kinglet is the only species showing declines both in the province and Canada-wide. The remaining 8 species – Killdeer, Olive-sided Flycatcher, Barn Swallow, European Starling, Song Sparrow, Brown-headed Cowbird, Purple Finch, and House Sparrow – all show a downward trend in their numbers continent-wide (Sauer et al. 1997). The causes for the declines may be complex.

The Barn Swallow is a species that has adapted to live in close proximity to humans and it is difficult to imagine environmental sources of impact occurring on its nesting grounds.

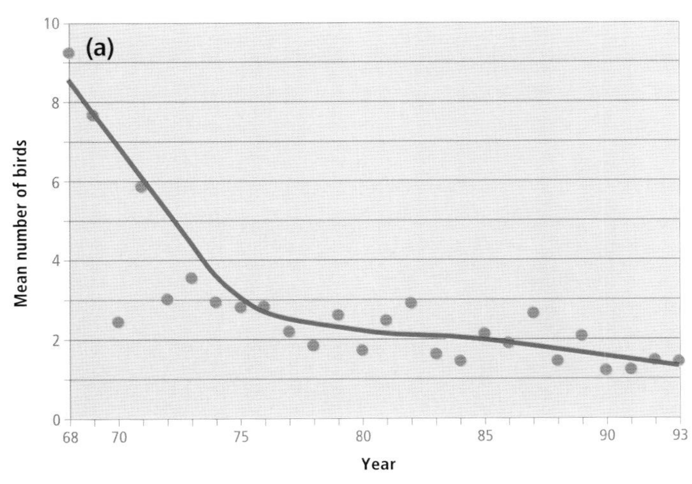

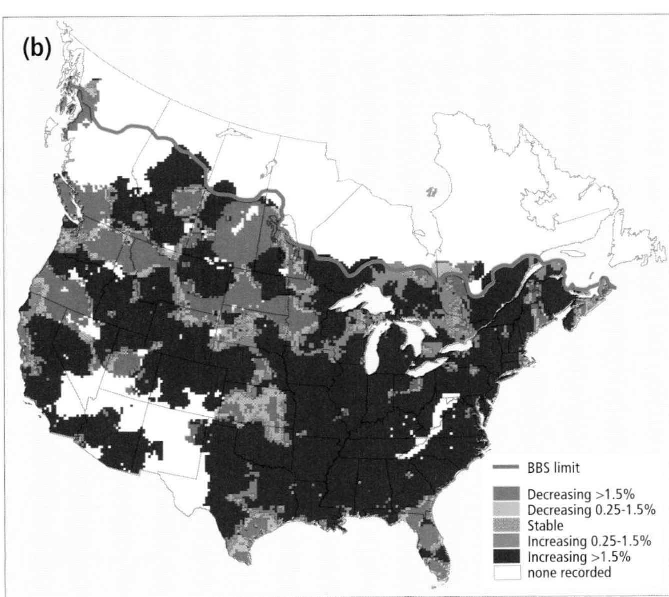

Figure 749. An analysis of Breeding Bird Surveys for the Great Blue Heron in British Columbia shows: (a) The number of birds on coastal routes decreased at an average annual rate of 6.2% over the period 1968 through 1993 ($P < 0.01$); interior routes had insufficient data for analysis (this study). (b) Throughout much of its North American range, however, the Great Blue Heron has exhibited increasing trends (Sauer et al. 1997).

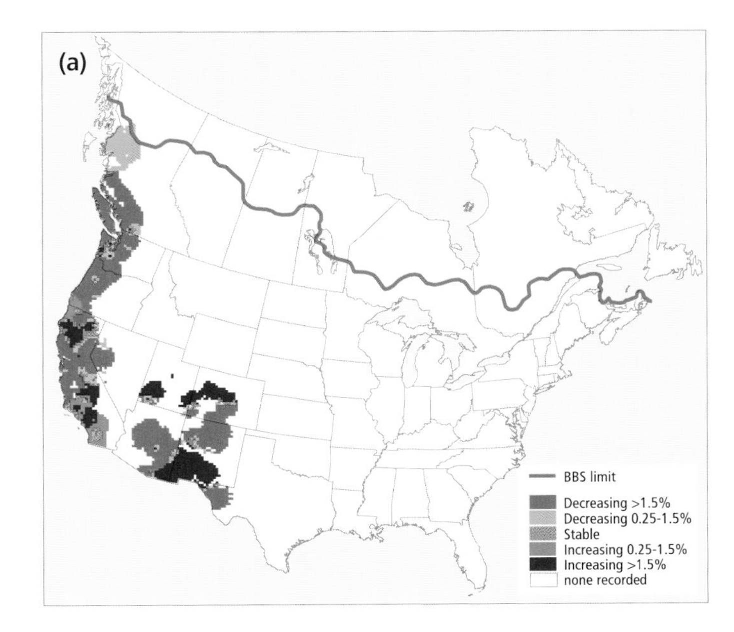

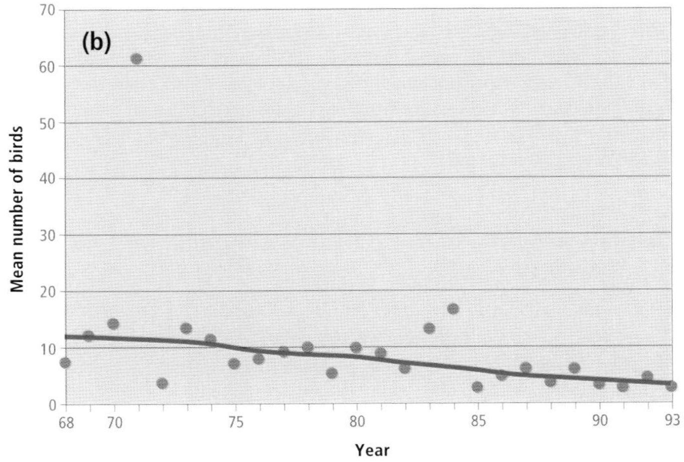

Figure 750. (a) Based on Breeding Bird Survey data, the Band-tailed Pigeon is showing significant declines throughout much of its North American range (Sauer et al. 1997). (b) An analysis of Breeding Bird Surveys for the Band-tailed Pigeon in British Columbia shows that the number of birds on coastal routes decreased at an average annual rate of 5.8% over the period 1968 through 1993 ($P < 0.01$); interior routes had insufficient data for analysis (this study).

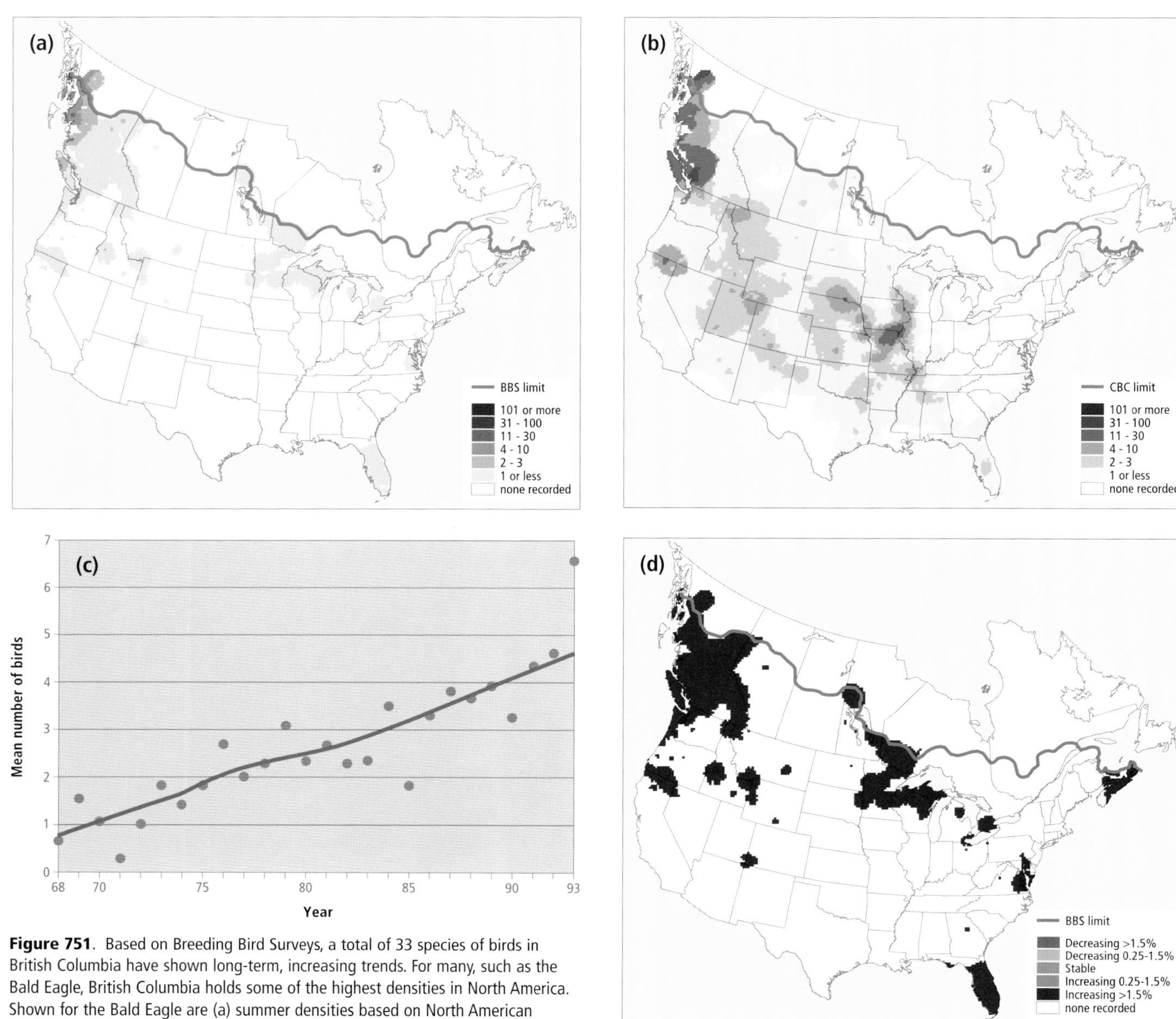

Figure 751. Based on Breeding Bird Surveys, a total of 33 species of birds in British Columbia have shown long-term, increasing trends. For many, such as the Bald Eagle, British Columbia holds some of the highest densities in North America. Shown for the Bald Eagle are (a) summer densities based on North American Breeding Bird Survey (BBS) data (Sauer et al. 1997), (b) winter densities, based on North American Christmas Bird Count data (Sauer et al. 1996), (c) long-term BBS trends (1968 through 1993) for coastal British Columbia showing an average annual increase of 7.6% ($P < 0.001$) (this study), and (d) BBS North American trend data (Sauer et al. 1997).

Dramatic declines of Barn Swallow populations in Alaska since the early 1900s have also occurred, the causes of which are unknown (Kessel and Gibson 1994). Few tears, however, will be shed for the plight of the European Starling, Brown-headed Cowbird, or House Sparrow; however, it would be useful to know whether their declines arise from changes that may also have an impact on some of the scarcer species outside the BBS detection.

For many of these species it is unlikely that there is anything we can do to decrease or reverse the trend each is following. Their nesting populations are widely dispersed across the province in a variety of habitats, and none of them is rare, so seeking to reserve habitat specifically for their maintenance is not warranted. Appropriate action would be to monitor the populations, make further searches for answers, and wait.

Species Showing Increasing Population Trends in British Columbia

Among the 33 increasing species identified by the Breeding Bird Surveys in British Columbia, 13 appear to be part of a continent-wide trend (Table 34). These include the Common Loon, Bald Eagle (Fig. 751), Red-tailed Hawk, Hairy Woodpecker, Pileated Woodpecker, Warbling Vireo, Solitary Vireo, Common Raven, Red-breasted Nuthatch, House Wren, Spotted Towhee, Lincoln's Sparrow, and House Finch. Unfortunately we are no wiser as to the reasons for many of these population increases than we are for the decreases. Some, such as the Lincoln's Sparrow, appear to have taken advantage of the shrubby regrowth of large areas of the province that have been progressively clearcut. Pojar (1995), Gyug (1997), and Merkins and Booth (1998), for example, all found the Lincoln's Sparrow to be the dominant bird species in recently logged sites in the north interior. The increasing numbers of House Finches in British Columbia, and perhaps the rest of the continent, has almost certainly resulted from the increased popularity of bird feeding.

Table 34. Species showing an increase ($P < 0.10$) in numbers on coastal or interior Breeding Bird Survey routes summarized from British Columbia. Canadian and North American values are provided for comparison.

Species	Average annual change (%)			
	BC Coast[1]	BC Interior[1]	Canada[2]	Continent[2]
Common Loon	i.d.[3]	+4.3	+3.1	+2.8
Bald Eagle	+7.6	i.d.	+8.5	+9.2
Herring Gull	i.d.	+3.8	−3.0	−2.6
Black Tern	i.d.	+13.8	−3.5	−3.1
Alder Flycatcher	n.r.[4]	+9.8	n.c.[5]	+0.6
Hammond's Flycatcher	+7.6	n.c.	n.c.	n.c.
Pacific-slope Flycatcher	+8.3	-7.3	+4.4	n.c.
Warbling Vireo	n.c.	+2.7	+2.3	+1.2
Solitary Vireo[7]	n.c.	+4.7	+3.0	+3.0
Steller's Jay	+2.7	i.d.	n.c.	n.c.
Northwestern Crow	+1.8	n.r.	+1.4	+1.2
Common Raven	+5.2	n.c.	+3.3	+3.2
Mountain Chickadee	n.r.	+2.6	n.c.	n.c.
Bushtit	+9.2	n.r.	n.r.	n.c.
Red-breasted Nuthatch	n.c.[5]	+5.8	+4.6	+2.8
House Wren	i.d.	+4.0	+1.9	+1.4
Varied Thrush	n.c.	+3.8	n.c.	n.c.
Orange-crowned Warbler	n.c.	+6.9	n.c.	n.c.
Nashville Warbler	i.d.	+7.1	n.c.	+1.2
Northern Waterthrush	i.d.	+3.9	n.c.	n.c.
Spotted Towhee	n.c.	+5.0	+2.2	+0.8
Lincoln's Sparrow	i.d.	+14.3	+2.9	+2.8
Lazuli Bunting	i.d.	+3.9	+4.8	n.c.
Bullock's Oriole	i.d.	+3.9	+4.1	n.c.
Cassin's Finch	n.r.	+1.6	n.c.	n.c.
House Finch	+6.1	+6.1	+8.8	+1.9
	Province[6]			
Red-tailed Hawk	+5.1		+3.0	+3.2
Sora	+5.4		n.c.	n.c.
Hairy Woodpecker	+3.0		+3.9	+1.5
Pileated Woodpecker	+4.1		+4.7	+1.6
Common Yellowthroat	+2.3		−0.7	−0.3
Black-headed Grosbeak	+2.7		+2.7	n.c.
Evening Grosbeak	+6.2		n.c.	n.c.

1 This study (1968-93).
2 Sauer et al. (1997) (1966-96).
3 Insufficient data.
4 Not reported.
5 Could not detect a change.
6 Anonymous (1999).
7 Since split into Cassin's and Blue-headed Vireo.

Of more immediate interest to provincial naturalists are the 18 species that appear to be increasing in British Columbia but not in the rest of Canada or in their continental range. These include the Sora, Herring Gull, Black Tern, Hammond's Flycatcher, Pacific-slope Flycatcher (coastal only; declining in the interior), Steller's Jay, Mountain Chickadee, Bushtit, Varied Thrush, Orange-crowned Warbler, Nashville Warbler, Northern Waterthrush, Common Yellowthroat, Lazuli Bunting, Bullock's Oriole, Black-headed Grosbeak, Cassin's Finch, and Evening Grosbeak. There seems to be no common thread of habitat use or habitat change to link any of these to observed conditions. Two of the species – Herring Gull and Black Tern – are anomalies, as both appear to be decreasing elsewhere throughout their range.

Among the 26 increasing species for which we have regional analysis, 19 are increasing in the interior region and 8 in the coastal region. Only 1 species, the House Finch, is increasing in both regions, while the Pacific-slope Flycatcher is increasing on the coast but decreasing in the interior. These observations suggest that local circumstances may be implicated.

RARE, THREATENED, OR ENDANGERED SPECIES IN BRITISH COLUMBIA

In the status accounts of all species we have used standard terms (Bull 1974) to describe the approximate abundance of the bird in each ecoprovince and by season. These designations are based on the number of individuals seen per locality per day, per year, or for all time (Campbell et al. 1990, Vol. 1:148). *Rare* is the lowest category among the species of regular occurrence and is defined as 1 to 6 individuals per season. Inasmuch as 4 seasons are defined, a rare species present throughout the year could be represented by records of up to 24 individuals per year. Rarity, as used in the conservation literature, is seldom so quantitatively defined.

No matter how defined, rarity in living organisms can have several origins (Rabinowitz et al. 1986). One of these is marginality. It is usual for the population of birds in any habitat to be dominated by a small number of species, and as much as 20% of the species may occur in densities of less than 1 per 100 ha. Many of these species are nowhere in the dominating group. It is usual also for populations of a species to fluctuate in numbers and for these changes to result in successive expansions into less favourable habitats followed by contractions back into the more favourable habitats.

In British Columbia, a high proportion of the species of birds reach the margins of the habitat within the province and are, to that extent, in marginal situations along the edges of their normal distribution here. We have recorded breeding range extensions of many species within the past 50 years (e.g., Least Sandpiper, Barred Owl, Least Flycatcher, Magnolia Warbler, House Finch). Areas of less favourable habitat, once occupied by marginal populations of these and other species are now incorporated into the regularly occupied favourable habitats. Are these the result of climate change or of evolutionary modification of the species resulting in increased adaptive ability? We pay only passing attention to the changes in species distribution within the province. When the margin happens to be close to the provincial boundary, especially the international boundary, the ebb and flow of a species attracts much more attention. It may even lead to a status assignment that provides special attention under law (e.g., Sage Thrasher).

The scientific task of gathering and analyzing the data on the fauna and on the status of individual taxa is but the first phase of designing and implementing sound conservation measures. In many instances an appropriate response can be made under the management authority of the responsible government agency. However, often competing interests requiring the attention of more than one agency are involved. Under such circumstances effective wildlife conservation, including arranging for the survival of endangered species, requires legislative action to establish priorities. Under authority of the Migratory Birds Conventions Act, the final responsibility for the welfare of most bird species rests with the Government of Canada. On 11 April 2000, Canada tabled legislation to protect endangered Canadian wildlife from extinction and help those at risk of extinction recover to sustainable levels. The Species at Risk Act (SARA), when enacted, will legally recognize the Committee on the Status of Endangered Wildlife in Canada (COSEWIC) for the first time. The Act provides the authority to prohibit the destruction of endangered or threatened species and their critical habitat on all lands in Canada. In May 2000, COSEWIC listed 3 species as endangered in British Columbia – Burrowing Owl, Spotted Owl, and Sage Thrasher – and 4 species as threatened – "Anatum" Peregrine Falcon, Marbled Murrelet, White-headed Woodpecker, and Yellow-breasted Chat. Two species, the Burrowing Owl and American White Pelican, have legal protection under the provincial Wildlife Act.

A recent judicial determination under the Endangered Species Act of the United States requires the United States Fish and Wildlife Service, when bringing forward a species for designation under the act, to identify and designate the critical habitat required by the species (McCabe 2000). We suggest that this provision should be considered for the Canadian wildlife protection legislation.

Provincial Species of Concern

There is widespread public concern in British Columbia that we should conduct ourselves in such a manner that the extraordinary diversity of living organisms and associations for which we are trustees can be passed on to our successors with all options intact.

We are the custodians of the richest biota in Canada. Fortunately, large parts of it have proven to be highly adaptable and resilient to the changes that human populations have imposed. For many species, however, the present conditions are unsustainable. These include species of birds that have declined to the point of endangerment or are judged vulnerable to extirpation. There are several additional species that have declined in numbers steadily through the past 30 years but are not yet threatened; 7 or 8 insular endemics of extraordinary significance; a group of species for which British Columbia provides a significant part of the world population, and a further group of species that, in Canada, are found only in this province. All these, because of their special circumstances, should be given high priority within our general responsibility to build a sustainable relationship with this rich and fascinating land and its resources.

Obviously, it is preferable to detect a species or other taxon vulnerable to a forced decline in numbers or distribution well before the decline becomes a serious threat to its survival. Unfortunately, even in such well-known and popular groups as the birds and mammals, the level of monitoring required to make this possible is seldom available. Then, too, the human species appears to be crisis-oriented and disinclined to respond effectively to early warnings. The result is that we are increasingly confronted with threatened taxa that are already in deep trouble and may be beyond help.

On the world scene, the Species Survival Commission of the International Union for the Conservation of Nature has been striving for 50 years to improve the situation by providing leadership, along with technical and scientific counsel, to biodiversity conservation projects worldwide (Klemens and Howes 1999). International goals and priorities are not always identical with those of nations or regions within nations. In recognition of this, the Commission in recent years has been improving the applicability of criteria and terminology to national and regional circumstances (Klemens and Howes 1999). There has been widespread success in this endeavour.

Often the main factor in the decline of a species is destruction of its habitat. This leads to the inevitable decision that the most effective action to save these species is within national jurisdictions. Avery et al. (1994), have argued that national conservation priorities may have important differences from global priorities and that a "system is needed which identifies national priorities but puts them in a global context." The same thinking can be applicable to local (e.g., provincial) lists.

British Columbia is now fully integrated into the global system of identifying and categorizing the rare taxa of birds within its boundaries. Fraser et al. (1999) present review accounts of each of the 78 taxa (species and subspecies) of birds that are believed to be threatened with extirpation in British Columbia (for an update, see Conservation Data Centre 2000). An additional species, the Sage Grouse, was extirpated before 1900 (Munro and Cowan 1947). The province maintains regularly updated lists of the taxa that are endangered, threatened, or vulnerable. These categories are defined as:

Endangered: Any indigenous species that, on the basis of the best available scientific evidence, is indicated to be threatened with extirpation or extinction throughout all or a significant part of its British Columbia range.

Threatened: Any indigenous species that, on the basis of the best available scientific evidence, is indicated to be experiencing a definite, noncyclical decline throughout all or a major part of its British Columbia range, or any species with an extremely restricted distribution in a habitat with a high probability of environmental degradation. These species are likely to become endangered if the factors affecting population viability are not changed.

Vulnerable: Any indigenous species that is particularly at risk in British Columbia because of low or possibly declining population (Fraser at al. 1999).

Currently, the provincial Conservation Data Centre's "Red List" contains the names of 38 species of birds at the highest level of risk, and a "Blue List" names 42 species or subspecies vulnerable to extirpation because of special sensitivity to "human activities or natural events" (Conservation Data Centre 2000). Similar lists for Canada are maintained by COSEWIC (1999). Some of the taxa listed for British Columbia are absent from the Canada-wide list because the geographic focus is national rather than provincial.

Most species of land birds have a "heartland" of their distribution. This special area of their habitat is where the special geographic and ecological conditions they require are consistently available. In these "soft" areas the species reaches its highest density. Toward the periphery of their range, the physical and environmental attributes become less favourable or discontinuous and the species occurs in smaller numbers and less consistently. Several authors, including Shumaker and Babbel (1980) and Soulé (1973) describe these geographically marginal populations as peripheral. Frequently they may be separated spatially from the central regions of the species range.

The Peripheral Species Argument

Returning to the British Columbia Red and Blue lists, about 25 of the Red-listed taxa and 19 of the Blue-listed taxa (57% of the total), though rare and declining within the province, are still reasonably numerous in some other parts of the species' range. This observation has led some ornithologists (e.g., Atwood 1994) to question the wisdom of directing special conservation effort and funding to the preservation of populations that are at the ecological edge of a species' range. In many instances these marginal populations cross state or provincial boundaries where their lack of consistency becomes a matter of concern to those charged with maintaining regional biodiversity. Atwood (1994) and some others maintain that it is unwise to focus conservation effort and funding on such populations. To do so serves to skew the allocation of always scarce funds available for conservation. They argue that the parochial decisions frequently lack a valid ecological context and lead to a confusion of the issues and could threaten public credibility. Furthermore they reason that money and resources spent on maintaining a so-called peripheral population in one province or state will not be available for truly threatened populations.

Hunter and Hutchinson (1994) have examined these concerns and suggest that there are four reasons why the locally directed conservation action may be appropriate:

Genetic diversity: Maintaining genetic diversity is a primary objective in the protection of animal taxa throughout their ranges, and evidence is mounting that populations at the edge of the geographic range of a taxon differ genetically from those at the centre of its distribution. (See later discussion.)

Ecological roles: It is generally agreed among ecologists that an ecosystem can lose its integrity by the gradual attrition of its component species. It is not necessarily the most abundant species that occupy the most important roles in such systems. As the action takes place at a local level it is important to maintain biological diversity at a local scale.

Local values: People appreciate the species that surround them where they live. They derive special pleasure from rare species and learn from the vicissitudes of threatened species in their vicinity. "The affinity that people feel for things that are close to them, things that they are able to experience directly, can be a great motivating force. It can persuade some people to become conservationists who are not moved by distant, abstract events" (Hunter and Hutchinson 1994:1163-1165).

Umbrella species: In large parts of our province we know quite a lot about the species of birds and mammals of the different ecosystems but little or nothing about the many invertebrates that characterize these ecoysystems. Efforts made to conserve an ecosystem because of its vertebrates then become the medium for conserving the many other species there.

Hunter and Hutchinson (1994) conclude that "the arguments against a parochial approach are simple and compelling." "Nevertheless, making globally endangered species the highest priority does not negate the fact that the locally endangered species still merit our concern and action."

Marginality and Genetic Change

Perhaps the most important contribution of geographically isolated, peripheral populations is genetic change that accumulates as a consequence of isolation and the selection imposed on the segregated population by its marginal habitat. Lessica and Allendorf (1995) refer to empirical evidence in plants of genetic change occurring under these conditions.

The ecologically marginal habitats "are characteristically spatially diverse and temporally unstable" and "selection in them is for colonisation ability and adaptation to a diverse array of density independent factors" (Scudder 1989:180-185). Though little or no work has been done on bird species, studies on insects suggest that after an expansion of range, any subsequent population contraction promotes gene flow from these marginal populations that serves to increase the genetic diversity of the central population. "It is now generally recognised that small, marginal populations are the mainspring of all ongoing evolutionary change" (Mayr 1967, 1982); "large, central populations, in which selection usually favours density-dependent factors, are not the arena of ongoing evolution" (Scudder 1989).

In British Columbia, because of the great north-south extent of the province across more than 11° of latitude, 9 ecoprovinces, and 14 biogeoclimatic zones, a high proportion of the species of birds reach the margins of their most-favoured habitat within the province and are to that extent in ecologically marginal situations along the edges of their central distribution. Then, too, for many species, suitable habitat is discontinuous, and pockets of suitability may be discovered by pioneering individuals that establish colonies, i.e., small populations geographically isolated from the central habitat. Examples of this in British Columbia include the American White

Figure 751. In British Columbia, the American White Pelican is an example of a geographically marginal species in the province (Okanagan valley, 25 July 1999; R. Wayne Campbell).

Pelican (Fig. 751), Western Grebe, Eared Grebe, Bobolink, and the "Timberline" race of the Brewer's Sparrow, likely soon to be classified as a distinct species (Klicka et al. 1999).

Since 1947, we have recorded range extensions of many species (e.g., Black-crowned Night-Heron, Green Heron, American Avocet, Least Sandpiper, Caspian Tern, Forster's Tern, Barred Owl, Blue Jay, and Sprague's Pipit). Most have established disjunct populations well removed from any others. Are these changes in distribution the result of a factor such as climate change or of evolutionary modification of the species, resulting in increased adaptive ability?

It has been almost an axiom of conservation theory that habitat at the centre of a species range represents the most favourable conditions and that as the margin of the range is reached the conditions will be less favourable and discontinuous and will support populations that are lower in density and in stability. It has also been assumed that a species in decline will withdraw first from the marginal locations and concentrate densities in the core area of high suitability.

Recently, Channell and Lomolino (2000:84-86) examined the evidence for these assumptions. They analyzed the pattern of range contraction in 245 species drawn from all vertebrate classes along with arthropods, molluscs, and plants. They found that 98% of the species examined maintained populations in at least some marginal areas of their former range. Indeed, 68% (167 species) maintained "a greater than expected portion of their range in the periphery and not the core, and 37% (91) of the declining species survived only as remnant populations on the periphery of the historical range." These findings confirm the importance of careful review of the facts before decisions are made as to where best to focus conservation attention on declining species.

The conservation response to threatened populations of such marginal (peripheral) species breeding in British Columbia, and the reasoning behind the policy, are well expressed by Fraser et al. (1999).

Declining Populations of Species Not Yet Rare

In our treatment of the passeriform species in volumes 3 and 4 we have given analyses of the Breeding Bird Surveys from both the coast and the interior of the province. The data for several species, extending over the period 1968 through 1993, show significant declines in their numbers. Table 33 lists 27 indigenous species – 9 nonpasserines and 18 passerines – that have been losing breeding numbers at average annual declines of from 1% to nearly 10%.

Atwood (1994), in his look at endangered small land birds in the western United States, found that not 1 of at least 27 declining species was recognized by federal or state wildlife agencies despite the fact that the declines were based on data, often from 2 different sources. In British Columbia, a similar situation occurs. Long-term surveys of 27 indigenous species (Table 33), such as the Blue Grouse (Fig. 752), suggest that their numbers are declining on the coast and/or in the interior; however, only one of those species, the Great Blue Heron, appears on the provincial Red or Blue lists. None has yet reached the point of rarity, but it would seem wise to include all of them in some category for special monitoring. It may be easier to reverse a decline before a species reaches rarity than to leave it until it achieves the Red List classification. As Scott (1994) notes, "the best time to save a species is when it is still common."

Table 35. Species for which British Columbia has a high conservation responsibility. Numbers are the estimated percentages of the global population that breed in British Columbia.

Barrow's Goldeneye	60-90%
Bald Eagle	20-35%
Blue Grouse	40-60%
White-tailed Ptarmigan	50-70%
Black Oystercatcher	30-35%
Marbled Murrelet	20-30%
Ancient Murrelet	74%
Cassin's Auklet	80%
Rhinoceros Auklet	57%
Rufous Hummingbird	40-60%
Hammond's Flycatcher	30-40%
Northwestern Crow	50-90%
Chestnut-backed Chickadee	25-40%
Townsend's Solitaire	10-20%
Varied Thrush	30-40%
American Dipper	15-30%
Townsend's Warbler	40-60%
MacGillivray's Warbler	30-40%
Western Tanager	10-20%
Golden-crowned Sparrow	40-60%
Gray-crowned Rosy-Finch	20-40%

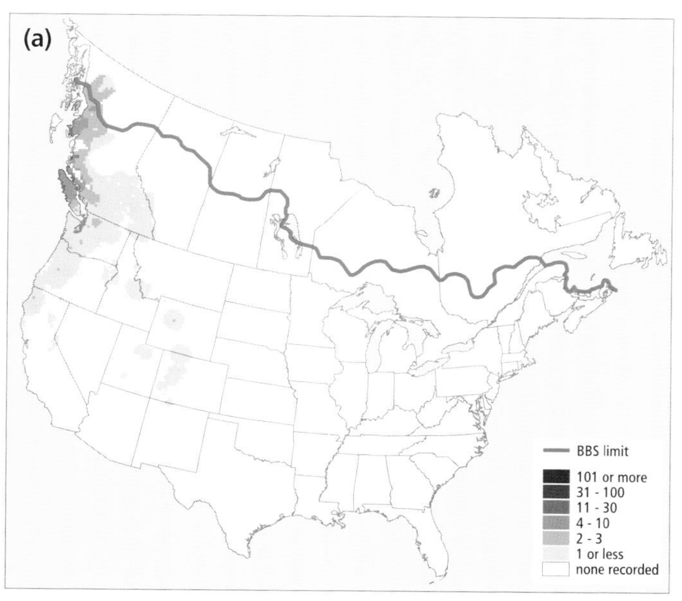

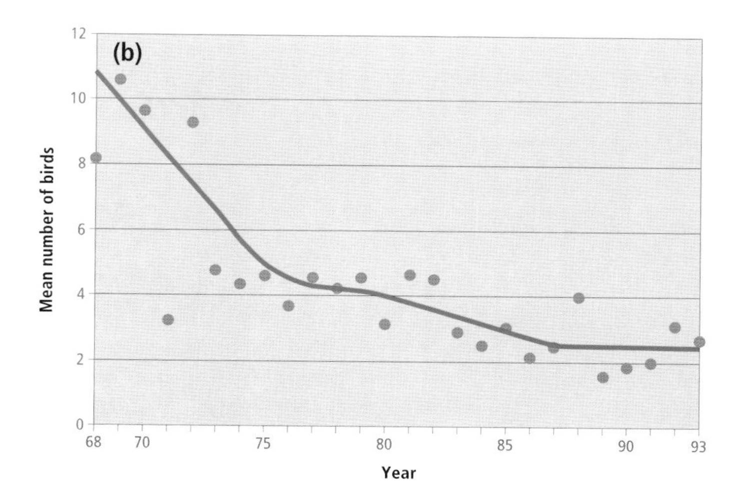

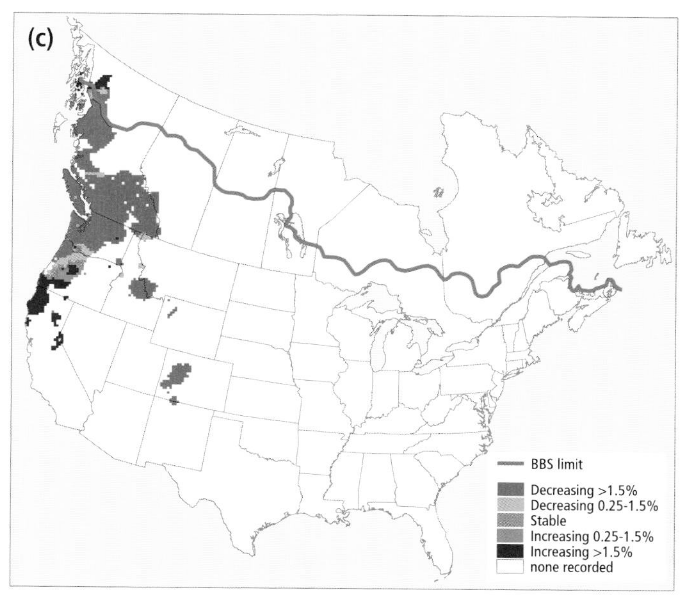

Figure 752. Based on Breeding Bird Survey (BBS) data, British Columbia supports (a) the highest densities of the Blue Grouse in North America (Sauer et al. 1997). It is also one of 27 indigenous species showing long-term declines in British Columbia. For example, an analysis of BBS data from coastal surveys over the period 1968 through 1993 (b) shows that the mean number of Blue Grouse decreased at an average annual rate of 5.3% ($P < 0.05$); interior surveys had insufficient data for analysis (this study). Throughout much of its North American range, BBS trend data also suggest that the species is declining (c) (Sauer et al. 1997). Apparently neither the Blue Grouse, nor 25 of the other 26 species showing long-term declines in British Columbia meet the criteria of the provincial Red and Blue lists.

Species of Global Significance among the Breeding Birds of British Columbia

On the basis of the best available information, British Columbia provides the breeding requirements for 20% or more of the global populations of at least 21 species. Table 35 summarizes these from volumes 1 (p. 4) and 3 (p. 11).

For each of these species we have a special responsibility to the international community for adequate conservation measures. Research to determine their ecological requirements, their populations, and the state of their habitats within the province should be a priority. At this time the plight of the two murrelets is receiving priority attention. The Marbled Murrelet is threatened by the steady removal of the old-growth forests it depends on for much of its nesting habitat. The Ancient Murrelet, with all of its nesting habitat on the Queen Charlotte Islands, is threatened by inadvertently introduced Norway rats and by raccoons, which were introduced to Graham Island more than 50 years ago without proper concern for their potential impact on ground-nesting birds. Raccoons are now widespread there and are a serious predator.

Species of National Significance

At the national level, British Columbia supports the only nesting populations in Canada of about 34 indigenous species of birds – 23 nonpasserines and 11 passerine species (Table 36)

Table 36. Canadian breeding bird species known to occur only in British Columbia.

Fork-tailed Storm-Petrel	White-throated Swift
Brandt's Cormorant	Black-chinned Hummingbird
Pelagic Cormorant	Anna's Hummingbird
Black Oystercatcher	Red-breasted Sapsucker
Glaucous-winged Gull	Williamson's Sapsucker
Pigeon Guillemot	White-headed Woodpecker
Marbled Murrelet	Gray Flycatcher
Ancient Murrelet	Pacific-slope Flycatcher
Cassin's Auklet	Northwestern Crow
Rhinoceros Auklet	Chestnut-backed Chickadee
Tufted Puffin	Bushtit
Horned Puffin	Pygmy Nuthatch
Band-tailed Pigeon	Canyon Wren
Flammulated Owl	Bewick's Wren
Western Screech-Owl	Western Bluebird
Spotted Owl	Hutton's Vireo
Vaux's Swift	Black-throated Gray Warbler

(Campbell et al. 1989, 1997). These species are also logical candidates for special monitoring. Several marine species in the global and national lists are especially vulnerable to marine pollution (e.g., oil spills) because of their aggregative behaviour associated with the breeding season.

Insular Endemic Taxa

The 7 insular endemic taxa, 3 confined to Vancouver Island and 4 to the Queen Charlotte Islands, have been described earlier (p. 660). Each of the Queen Charlotte Islands subspecies is on the provincial Blue List, largely because of lack of accurate knowledge of the precise status. Most are forest birds and likely to have been impacted by the widespread clearcutting of the forest of the islands. We suggest that an early priority should be to obtain accurate, habitat-specific density figures and trend assessments for each of them. At the same time, there is need for a detailed appraisal of the systematic status of each, using benign techniques.

Birds as Fellow Passengers

The birds of British Columbia enrich the lives of vast numbers of us, both residents and visitors. No one can ever measure the sensual and intellectual delights of watching throughout our lifetimes the nearly 500 different species, often legions of them, go about their lives around us. They bring vitality to every landscape, and for those who study their life histories, they provide an endless feast for the imagination.

Bird feeders are ubiquitous and the sale of bird books, bird food, and the accessories of the avocation is a billion-dollar industry. Visiting tourists join us in expeditions to experience the many marine and shorebirds of our coastlines or teeming in our marshlands, or to witness the almost inaudible presence of the unique birds of our ancient forests. A small assortment of species that make up the ducks, geese, and galliformes also provide a carefully controlled recreational harvest and source of food.

In earlier years, much was made of the economic gains birds offered by eating the insects and other pests of our gardens and agricultural lands. But surely our society has grown beyond the necessity of arguing in dollars for the lives of these vibrant, fascinating, and often exquisitely beautiful creatures that share our passage through life. We all need to acknowledge a responsibility to manage our lives, our resource extraction, and our use and disposal of hazardous products so as to not harm these or any other creatures. We also need to be more aware of creating new hazards that cause the accidental death of birds, such as erecting certain types of windows, buildings, and power lines. It is hoped that we will also learn to manage our own numbers and our demands on the biosphere, so as not to devastate the landscapes and habitats that birds need to survive. Undoubtedly, this will be our biggest challenge, as we shall see.

WHAT LIES IN STORE FOR THE BIRDS OF BRITISH COLUMBIA? NEW PHILOSOPHIES, CONCERNS, AND CONSERVATION CHALLENGES

Neil K. Dawe, Ian McTaggart-Cowan, R. Wayne Campbell, and Andrew C. Stewart

> We end, I think, at what might be called the standard paradox of the twentieth century: our tools are better than we are, and grow better faster than we do. They suffice to crack the atom, to command the tides. But they do not suffice for the oldest task in human history: to live on a piece of land without spoiling it.
>
> – Aldo Leopold

This final section may seem to be a paradox in a book about birds, because here we spend more time on the subject of people than we do on these feathered creatures. However, we hope to show that the future of birds in British Columbia is significantly tied to the way we live, to the way we build our homes and our communities, to how we manage ecosystems and the organisms dependent on them, and to our economic activities, both here and around the globe. As a result of virtually all our actions, we are modifying the ecosystems birds use as never before. By making some changes in the way we live on this planet – in the way we treat the ecosystems that support all life, and perhaps more importantly, in the way we think about our relationship with the Earth – we may be able to ensure that the abundance and diversity of birds we have described in these volumes will be here for future generations to marvel at, much as we have over the past quarter-century that these volumes have been in the making.

Figure 753. Adult female Spruce Grouse (near Pack River, 2 July 1998; R. Wayne Campbell). Traditionally, natural resource agencies managed specifically for commodities, such as trees, or harvestable wildlife species, such as waterfowl and grouse. Today, there are growing conflicts between people who value the land and its resources as commodities and those who value the land and its resources for things other than their direct utilitarian use.

OF RESOURCE MANAGEMENT, COMMODITIES, BIODIVERSITY, THE ECONOMY, AND WILD BIRDS

Traditionally, management of the land by natural resource agencies has been restricted to narrowly defined objectives. Fish and wildlife agencies often managed specifically for game or commercial species, such as grouse (Fig. 753), waterfowl, salmon, and furbearers; and forestry agencies managed the trees for their value as commodities. Out of this traditional approach, conflicts arose between users of the land – between the people who value the land and its resources as commodities and those who value the land for reasons other than its direct utilitarian use (Knight 1996).

This latter group, the environmentalists, used the land for their recreation; they appreciated chickadees, alligator lizards, and *Calypso* orchids – as well as the commercial and game species – for their intrinsic values, and considered commodity users as the enemy. They formed environmental organizations to conserve the land and its wildlife values and did not consider their own activities on the land as anything but benign. On the other hand, the commodity users defended their actions, citing the importance of growth of the economy, and often asking the question, "Do you want jobs or the environment?"

Today, some of that has changed. Ecosystem management has now garnered the attention of resource agencies (Christensen et al. 1996). It argues for the preservation of sustainable ecosystems, not primarily for goods and services for humans but for the ecosystem structures and processes, such as the maintenance of biological diversity, that are necessary to deliver those goods and services. Nevertheless, ecosystem management is not without its critics (e.g., Stanley 1995).

Commodity users also are now aware that many of their past activities, such as clearcut logging and its attendant impacts on biodiversity, community water quality, and fish-bearing streams, need to be modified, though changes are occurring slowly, as we shall see. Even the contention that the "non-consumptive" users of the landscape have benign impacts has been seriously questioned (Wilkes 1977; Knight 1996), and in some cases shown to be false (Searle 1998).

Add to these problems the modifications we as a society are making to ecosystems arising out of urban sprawl and from our agricultural and industrial needs, coupled with the attendant pollution, and it appears, as Leopold said, that we have yet to learn how to live on the land without spoiling it.

Living on the land without spoiling it means taking what we need for our well-being in the form of commodities and recreation while maintaining the functioning biodiversity of the land, of which birds are a part. This implies sustainable use of our renewable resources and wiser use of those resources that are not renewable. It also implies that throughput of resources to our economy takes place in a sustainable fashion so that the biosphere can assimilate the waste products of the economy. Today we are in the midst of massive global changes, including atmospheric climate change, topsoil loss, loss of aquifers, desertification, stratospheric ozone depletion, and loss of biodiversity, to name but a few. The continued growth of non-renewable energy use and resource throughput in our global economy appears unsustainable (Goodland 1991; Daly and Cobb 1994; Douthwaite 1999).

One major question that needs to be considered is the role of current neoclassical economics, which has no connectivity to the biosphere (Christensen 1991); as a result, biodiversity has little value in our economy. A fervent belief that the marketplace's high prices for scarce resources will lead to a search for substitutes, and thus act as an agent for conservation of those resources, has prompted one Nobel laureate economist to remark, "If it is very easy to substitute other factors for natural resources, then ... the world can, in effect, get along without natural resources" (Solow 1974). On the other hand, as Clark (1973) points out, the most profitable strategy for harvesting natural resources – in his example, whales – is to harvest them all immediately and put the proceeds from the marketplace sale of the whales in the bank, unless, of course, the whales grow at a rate faster than an investment account. But are natural resources worth more to us than just their value as commodities?

We, along with other organisms, are a part of the biodiversity of the planet and depend on the ecosystems of the Earth for our very survival, much as birds and other forms of wildlife do. And there is much more to biodiversity than simply commodities and recreational values.

The biodiversity of ecosystems cleans our air and water, replenishes our aquifers, prevents erosion, creates our topsoil, moderates our weather, assimilates our wastes, cycles nutrients, sequesters atmospheric carbon, provides genetic diversity, and pollinates our flora. And ecosystems do all that at no cost to us. Where we have had to assume the responsibility for some of these ecosystem functions because we have overtaxed the biosphere (e.g., waste disposal, ozone depletion, global warming, degraded habitats), the costs can be extreme.

An example of this arose in New York in 1996. New York City's water comes from the Catskill Mountains. It is purified by the watershed – by root systems, soil organisms, and filtration and sedimentation as it flows through the soil. However, human actions in the watershed (e.g., sewage, fertilizer, pesticides) had reduced the efficiency of the natural processes to the point where the water no longer met government standards. The city had two options. They could build a filtration plant at a cost of between $6 billion and $8 billion, with ongoing operating costs of $300 million per year. Or they could restore the integrity of the watershed, which meant buying land in and around the watershed and subsidizing better sewage treatment plants at the local level. The total cost of the latter option, which is the option the city chose, was estimated at between $1 billion and $1.5 billion. Thus investing in the natural capital produced savings of at least $6 billion over 10 years and gave an internal rate of return of at least 90% in a payback of 4 to 7 years (Chichilnisky and Heal 1998).

Ecologists and economists have calculated the value of such ecosystem services to humanity at US$33 trillion annually ("almost certainly an underestimate") (Costanza et al. 1997), nearly twice the global gross national product (GNP). And birds contribute to these ecological services to humanity through such activities as suppression of rodents and insects harmful to agriculture, pollination of plants, and seed dispersal. Economics aside, however, it is folly to reduce the functionality of the very systems that allow life to continue on this planet, including our life and the lives of the myriad other organisms that share the planet with us.

All living things, including humans, have the same basic ecological needs: food, clean air and water, shelter from the elements, and a place free from disturbance to breed, rear their young where necessary, rest, and maintain their cleanliness and health. Birds are no different (Fig. 754). They depend on the biodiversity of ecosystems for their survival; and like us, they often rely on a number of widely separate ecosystems. For example, the Townsend's Warbler winters as far south as Costa Rica and nests in the mature coniferous forests of British Columbia and Alaska at the northern limits of its range.

Figure 754. Birds, such as the Osprey, need clean water and air, shelter from the elements, and places free from disturbance to feed, breed, rest, and maintain themselves (Creston, 27 April 1997; R. Wayne Campbell). Like humans, they rely on the biodiversity of ecosystems to provide them with their needs.

So what befalls a Townsend's Warbler that survives the winter and returns to its breeding habitat in British Columbia, only to find a sprawling subdivision or a clearcut where a mature forest once stood?

It is comforting to think that the warbler might simply move to the trees on the other side of the clearcut and, in fact, such crowding into the remaining habitat is often noted in the year following the creation of the clearcut (Schmiegelow et al. 1997). While we know of no studies that have marked individual birds in a forested area and then followed them to determine their fate after their habitat was altered, it is likely that such a territorial species will suffer at least a local decline.

Adult birds would arrive on the breeding grounds first and establish their territories in the forested habitat bordering the clearcut. Assuming that a proportional number of adults – those that had bred at the now clearcut site and those from the adjacent forest – survived to return, it would likely mean that some of the adults, and most of the yearlings that returned later, would either be forced into marginal habitat to attempt to breed or simply remain in the area as nonbreeders. Thus the post-breeding warbler population for the year following the clearcut would be less than the previous year's population, given a relatively stable population and other things being equal. Eventually, the Townsend's Warbler population would stabilize at a level roughly in proportion to the amount of mature forest that remains, but in lower numbers than it was before the clearcut. Now that may not be a significant problem if there is an abundance of mature forest remaining and the population viability of the warbler can be ensured; however, if you leave, say, only 5% of the habitat, you could end up with only 5% of the birds, perhaps a non-viable remainder.

THE HUMAN FACTOR GLOBALLY

On 12 October 1999, the human population reached 6 billion (United Nations 2000) and the annual rate of population growth, which had climbed to 2.2% by 1963 and 1964 (Worldwatch Institute 1997), had dropped to 1.3%. Many have been encouraged by the reduction in the population growth rate. The present rate, however, still adds around 78 million people each year to the Earth and should it continue, the human population will double in just over 50 years. Closely associated with population growth is growth in resource consumption – all 6 billion human beings, like other creatures, use resources of the Earth to survive.

How much do we, as one species, use? In 1986, when the population was about 5 billion, ecologists estimated that the human species was appropriating, in some form or another, nearly 40% of potential terrestrial net primary productivity[1] (NPP) (Vitousek et al. 1986). That figure includes 4% of current terrestrial NPP consumed directly by humans and domestic animals, 26.7% consumed by wholly human-dominated systems (e.g., cities, industrial sites, croplands, plantations, etc.), and 11.7% of potential terrestrial NPP lost as a result of human activity (e.g., the potential terrestrial NPP lost through conversion of natural ecosystems, with their high productivity, to croplands, with their relatively low productivity). In addition, we consumed or otherwise diverted 25% of the marine NPP.

We have changed the face of the planet in many ways, including transforming or degrading between one-third and one-half of the land on the Earth; increasing the CO_2 concentration in the atmosphere by 20%; appropriating half of the surface freshwater; exploiting, overexploiting, or depleting over 60% of the major marine fisheries; and directly driving to extinction as much as one-quarter of the Earth's bird species over the past 2,000 years (Vitousek et al. 1997). As the human population continues to increase, it is clear that this appropriation will continue to increase, leaving fewer wild places for birds and suggesting that, in the near term, only those species that can adapt to the human changes of the landscape will likely survive. Rock Doves, European Starlings, House Sparrows, and American Robins are examples of species that can live closely with humans; Golden Eagles, Sandhill Cranes, and Canyon Wrens may not.

THE HUMAN FACTOR IN BRITISH COLUMBIA

Many of us tend to think that population growth and the problems associated with it are something only the Third World must deal with; however, even here in British Columbia, problems associated with population growth and our consumptive demands are becoming more and more apparent.

Some human activities, although occurring far from British Columbia's borders, create problems that are finding their way into the province's environment. For example, toxic chemicals used around the world, including PCBs, have been found in large concentrations in the snow-covered peaks of British Columbia's mountains. The airborne chemicals precipitate over cold surfaces and, since British Columbia has a high proportion of mountainous terrain, the province is essentially a dumping ground for the global economy (Anonymous 1999b). When the snow melts, these chemicals find their way into the surface water. Although the chemicals are not soluble in water, they can be absorbed in the body fat of animals that drink the water. There they accumulate and are concentrated as they are passed up the food chain. Fish in alpine lakes may hold the chemicals at such levels that they become toxic to eat in numbers and could affect the health or breeding success of loons and other piscivorous birds that feed on the fish.

Another factor has been the commodity approach to resource management, which, coupled with our reliance on technology (Holling 1994) and our demand for economic growth, has caused us to harvest resources in an unsustainable manner.

[1] Net primary productivity (NPP) is the amount of the sun's energy captured by plants through photosynthesis that is available for use by other organisms.

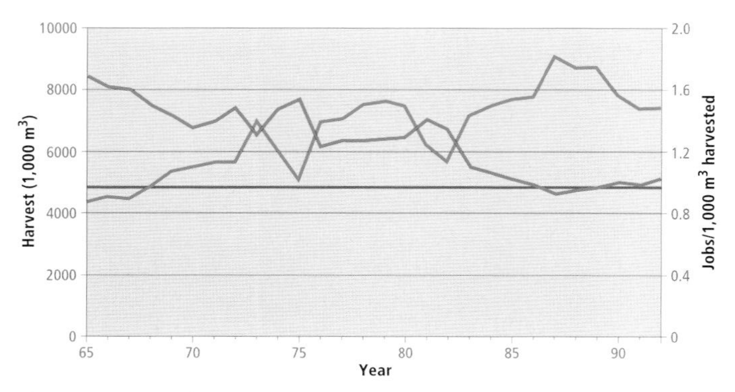

Figure 755. Annual harvest in British Columbia forests, 1965-92 (green line), and forest jobs per 1,000 m^3 harvested (red line); the purple line indicates long-term sustainable yield (British Columbia Ministry of Forests 2000 and Global Forest Watch Canada 2000).

This ultimately has impacts on the avifauna of the province. Forestry in British Columbia provides but one example.[2]

The Forests

For most of the period between 1969 and 1992, British Columbia's forests were harvested above the long-run sustained yield (Fig. 755; see also Harding 1994). The Ministry of Forests estimates that existing cutting rights held by forest companies exceed the projected future timber supplies by 20% (Marchak et al. 1999). At the start of this new millennium, 90% of managed forestlands in British Columbia are still being logged above long-term sustainable levels (Global Forest Watch Canada 2000; Ryan 2000), suggesting that little has changed since 1992. The argument "jobs or the environment" doesn't seem to hold either, for the number of jobs per 1,000 m^3 of timber harvested has continually declined despite the increasing harvest rate over the years (Fig. 755).

This rapid conversion of forested lands, from mature and biologically diverse forests to what essentially are second-growth plantations has affected the biodiversity of the ecosystems and the birds and other wildlife using them. For example, Matsuoka et al. (1997b) suggest that changes to the structure of old-growth and mature forests in the Pacific Northwest, particularly the loss of large conifers, are likely to result in decreased reproductive success of the Townsend's Warbler. In British Columbia, forest practices have already invoked "at risk" status for 28 taxa of terrestrial vertebrates (Bunnell 1998).

Clearcutting (Fig. 756), because it removes all living and dead trees older than the cutting cycle, takes away the habitat of all those species of birds dependent on older trees and snags for their survival. In the 1997-98 period, 92% of cutblocks in approved plans in British Columbia were clearcut, despite government and industry claims that clearcutting is no longer

Figure 756. Clearcutting removes the habitat for those species dependent on older trees and snags for their survival (Strathcona Park, Vancouver Island, 15 August 1997; R. Wayne Campbell).

[2] Neither here nor elsewhere is it our intent to single out any one commodity sector; the problems we discuss are society-based, not commodity sector–based. Our examples were chosen simply from available data that included birds and their habitats. Other examples of non-sustainable resource use could certainly be given.

acceptable (Sierra Legal Defence Fund 1998). While some believe that this situation can be ameliorated through proper management – for example, large live trees, snags, and downed wood can be provided in even-aged stands, including clearcuts, by retaining living trees and snags (Bunnell et al. 1997) – little retention of old-growth characteristics in cutblocks has been practised in British Columbia. However, even with tree and snag retention, unless all the habitat components required by birds remain, there will still likely be declines in bird numbers, for it is the ecosystem – of which trees are only a part – that matters.

Another factor affecting the biodiversity of the avifauna of our forests may be fragmentation. One species at least, the Marbled Murrelet, is thought to be suffering some effects of fragmentation (Ralph 1994). However, in western forests there are few isolated patches surrounded by non-forested habitats. Rather, they are usually interconnected with other stands of different age and species composition (Hejl 1994). Thus fragmentation may not be as serious a problem here as in the east, where it is caused by urbanization or agricultural development. Bunnell et al. (1997) note that the little data available on fragmentation in western forests suggest only "weak support for the concept," although they caution that fragmentation needs more study and should not be ignored. The effect on bird species of having 63% of British Columbia's forests fragmented by roads or transportation routes and of having only 15% of the province's forestlands in contiguous blocks larger than 200 km^2 (Global Forest Watch Canada 2000) is still open to question.

Nevertheless, recent studies have shown that logging practices generally decrease the abundance of almost all permanent resident birds and half the migrants associated with old-growth forests; they also show that the density of cavity-nesting species is directly related to snag density (Hejl 1994). Pre- and post-clearcutting figures for two areas in British Columbia reveal a decline in species richness of about 46% and in avian density of between 53% and 61% resulting from the removal of the forest trees (see Table 29; Gyug 1997 and Pojar 1995). In the Pacific Northwest, including British Columbia, loss of cavity sites is thought to be the most common threat to forest-dwelling vertebrates (Bunnell 1998). As a result, those bird species associated with old-growth forests or snags are less abundant than they were 100 years ago.

Based on current bird population data from 3 seral stages and estimates of past and current forest area, Raphael et al. (1988) hypothesized that many species associated with Douglas-fir forests in northern California have suffered considerable declines in their populations since pre-settlement times resulting from modifications of the forest landscape (Table 37). Although similar data have not been analyzed for British Columbia, there is no reason to believe that the results would be much different. Hejl (1994) predicts that if current forest-use practices continue, species associated with old-growth forests, snags, and burns will continue to decline.

The reduction of Marbled Murrelet nesting areas coupled with high predation at the nest (perhaps due to the fragmented landscape, which allows easier access by avian predators) appear to be two of a number of factors that explain the decline of murrelet populations throughout its range (Ralph 1994). The same holds true for the Spotted Owl, which has declined throughout its range over the past century primarily due to habitat loss through logging, although urbanization and competition with the Barred Owl (Fig. 757) have also contributed (Gutierrez 1994). Based on a habitat model, Northern Goshawk populations in coastal British Columbia and Alaska may have declined by as much as 40% due to loss of old-growth forest (Crocker-Bedford 1990). And in the interior, harvesting of older Douglas-fir and ponderosa pine forests can have a serious impact on Flammulated Owl breeding habitat by affecting cavity and prey availability, while the harvesting of mature western larch and Douglas-fir stands is a primary threat to Williamson's Sapsuckers in the Okanagan Highland (Fraser et al. 1999).

Table 37. Estimated population changes for birds in Douglas-fir forests in northwestern California from pre-settlement times to the 1980s resulting from forest landscape modifications.[1]

Species	Population change (%)
Winter Wren	–47
Golden-crowned Kinglet	–45
Hammond's Flycatcher	–43
Common Raven	–42
Pacific-slope Flycatcher	–39
Pileated Woodpecker	–38
Red-breasted Nuthatch	–37
Red Crossbill	–37
Hairy Woodpecker	–36
Chestnut-backed Chickadee	–36
Townsend's Solitaire	–36
Brown Creeper	–35
Blue Grouse	–34
Hermit Thrush	–32
Western Tanager	–28
Cassin's Vireo	–26
Band-tailed Pigeon	–25
American Robin	–22
Cooper's Hawk	–19
Warbling Vireo	–19
Evening Grosbeak	–17
Red-breasted Sapsucker	–15
Steller's Jay	–5
Northern Flicker	–4
Olive-sided Flycatcher	2
Pine Siskin	2
Brown-headed Cowbird	5
Dark-eyed Junco	18
Yellow-rumped Warbler	41
Purple Finch	76
Western Wood-Pewee	78
Black-throated Gray Warbler	87
MacGillivray's Warbler	162
Orange-crowned Warbler	190
Chipping Sparrow	241
Fox Sparrow	243

1 Modified from Raphael et al. (1988).

Figure 757. Altering habitats for species with narrow ecological requirements like the Spotted Owl often results in their displacement by similar species, like the Barred Owl shown here, whose ecological niches are much broader (Victoria, 21 May 1999; R. Wayne Campbell).

Table 38. Comparative population growth rates of Canada, British Columbia, and the three fastest growing regions of the province (1991-96).

Region	Growth rate (%)[1]
Canada	**5.7**
British Columbia	**13.5**
Okanagan-Similkameen	15
Eastern Vancouver Island[2]	19
Fraser valley/Lower Mainland	22

1 Data from BC Stats.
2 Excluding the Capital Regional District.

Urbanization and Agriculture

It is, however, the growth of communities across the province with their rapid urbanization and attendant agricultural and industrial development that is having the biggest impact on birds, other wildlife, and wild lands. The mountainous geography of British Columbia has created a situation where land preferable for human settlement, land with high agricultural capabilities, and land with high wildlife values all occur in the same location – the valley bottoms. These areas comprise only about 10% of the province's total land area (Moore 1990). Thus, there were – and still are – regular conflicts in the use of these lands, conflicts that, more often than not, result in threatened species and the loss of bird habitat and bird diversity.

Three areas will serve as examples: the Nanaimo Lowland of Vancouver Island and the adjacent Gulf Islands, the Fraser Lowland, and the Okanagan valley.

Nanaimo Lowland and the Gulf Islands

The Nanaimo Lowland and the Gulf Islands, ranging from Sooke in the south to Campbell River in the north, cover an area that holds about 86% of Vancouver Island's population. Based on recent census figures, much of this area has had the highest growth rate in British Columbia (Table 38). Of the top 5 cities in the province showing the highest growth rates from 1991 to 1996, 4 were from the east coast of Vancouver Island: Courtenay (48.2% increase in population), Ladysmith (32.4%), Qualicum Beach (31.0%), and Parksville (28.3%).

This concentration of people has created increased development pressures on wild lands. A recent Sensitive Ecosystem Inventory (SEI) (Ward et al. 1998) of all the ecosystem landscapes that once occurred extensively on the east coast of Vancouver Island showed that less than 8% of the 4,121 km^2 study area was unmodified – that is, the former wild landscapes that comprised this area are now either urban or rural landscapes or consist of forests younger than 100 years. When some modified landscapes with remaining wildlife values are factored in, such as older, second-growth forests and seasonally flooded agricultural fields, the total increases to just over 19%. Thus, an area of over 4,000 km^2, composed of a variety of unmodified ecosystems – habitat for migrant and resident bird populations – has been reduced to about 800 km^2 in little more than a century and a half. Some of the ecosystems in the Sensitive Ecosystems Inventory include the following:

Older forests: Less than 3% of the SEI study area consists of forests older than 100 years and those remaining are highly

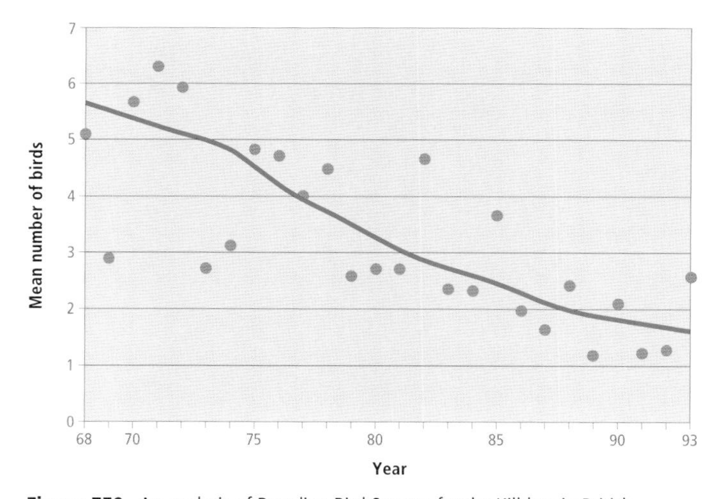

Figure 758. An analysis of Breeding Bird Surveys for the Killdeer in British Columbia shows that the mean number of birds on coastal routes decreased at an average annual rate of 4.8% over the period 1968 through 1993 ($P < 0.01$).

fragmented by urbanization. Nearly three-quarters of the Coastal Douglas-fir Biogeoclimatic Zone that occurs in British Columbia lies on the east coast of Vancouver Island and the Gulf Islands; today only remnants of this rare, productive ecosystem exist, with only 15 of the 609 SEI polygons (map units) identified as older forest exceeding 100 ha.

Sparsely vegetated ecosystems: These include naturally rare areas, such as sand and gravel spits. Unfortunately, many of the spits on the east coast of Vancouver Island have been considered as prime building locations and today have extensive development on them ranging from light industrial (Campbell River) to housing (Little Qualicum River) to recreational vehicle resorts, where people live the year around (Englishman River). Campbell River is in the fortunate position of being able to consider moving much of the light industry from Tyee Spit and creating open space; however, few municipalities have demonstrated such foresight and even when they want to, it is often too difficult and costly to completely remove an existing subdivision. Today only 34 of these sites remain unmodified and cover less than 0.1% of the SEI study area. Once ideal nesting habitat for Killdeer, the loss of these habitats may be playing a role in the decline of Killdeer numbers in coastal British Columbia (Fig. 758).

Estuarine and freshwater wetlands: These ecosystems today cover only 1.7% of the SEI study area. By the turn of the 20th century, one-third of the estuarine wetlands along the southeast coast of Vancouver Island had been modified, primarily through dyking (Prentice and Boyd 1988). With the exception of the Campbell River estuary, which gained estuarine marsh habitat through a marsh creation project (Brownlee et al. 1984; Dawe et al., in press), all other major estuaries on the east coast of Vancouver Island have lost habitat. For two – the Nanaimo and Cowichan estuaries – the extent of the loss has exceeded 50% (Levings and Thom 1994). Estuaries are among the most productive of habitats based on net primary production (Krebs 1994); in terms of the ecosystem services they provide, they have the highest value per hectare per year of any ecosystem on the Earth (Costanza et al. 1997). Along the east coast of Vancouver Island, estuaries are critical habitats to hundreds of thousands of water birds that winter or stage along this coastline (see, e.g., Butler and Vermeer 1994; Dawe and Buechert 1995; Dawe et al. 1998). We know of no studies that document the loss of freshwater wetlands on the east coast of the island, but that loss probably follows a pattern similar to the loss of estuarine wetlands.

While some species have been able to adapt to the modified landscape, the impacts on a majority of the ecosystems of the Nanaimo Lowland have placed at least 13 bird taxa at risk on eastern Vancouver Island and the Gulf Islands, including the Double-crested Cormorant, Brandt's Cormorant, American Bittern, Great Blue Heron, Green Heron, Northern

Figure 759. Richmond, British Columbia, in 1930 (left) and 1995 (right), showing the effects of urbanization over a 65-year period on some of the most fertile agricultural lands in British Columbia. The value of the lands to migratory birds has been much reduced as well. Vancouver International Airport is at the top of the 1995 photograph (photographs courtesy of National Air Photo Library A2238-36 [1930] and Triathlon Mapping Corporation [1995]).

Goshawk (*laingi* subspecies), Peregrine Falcon (*anatum* subspecies), Marbled Murrelet, Barn Owl, Western Screech-Owl (*saturatus* subspecies), Northern Pygmy-Owl (*swarthi* subspecies), Hutton's Vireo, and Vesper Sparrow (*affinis* subspecies) (Fraser et al. 1999). Other species as well, such as the Sky Lark and Western Meadowlark, have probably declined locally due to urbanization.

Fraser Lowland

As on the southeast coast of Vancouver Island, significant changes in the ecosystems of the Fraser Lowland have taken place since European settlement of the region began. Around the turn of the century, the lower Fraser River wetland complexes were reduced by nearly 70% through conversion to agricultural lands (Fraser River Estuary Study Steering Committee 1978). Sumas Lake, which once covered an area of some 3,600 ha, was dyked and drained by 1924 (Siemens 1966) and is now called Sumas Prairie. In addition to the loss of the water volume from this area, an additional 8,000 ha of marshland and sloughs were eventually reduced to remnants and along with them, significant populations of ducks and geese disappeared (Leach 1982). Over half of the 64,500 ha of wet meadows and riparian habitats present in the 1890s have been lost (Levings and Thom 1994).

Figure 760. Greenhouses are now beginning to cover the Fraser delta. In the process, the fertile agricultural soils are removed from direct food production and the wildlife values of the agricultural lands, important to millions of migratory birds, are severely reduced (Delta, midwinter 1997-98; W. Sean Boyd).

Conversion of the remaining wetlands is still occurring. Between 1967 and 1982, wetlands in the southwestern Fraser Lowland were converted to other uses at a rate of 109 ha per year (Pilon and Kerr 1984). Impacts on wetlands and their associated riparian edges in the valley have affected a number of species, including the American Bittern, Yellow-billed Cuckoo, and Great Blue Heron (Fraser et al. 1999). Butler (1997), considering the projected growth of the human population on the south coast, concluded that this coastal heron's future "does not look very rosy."

The lower Fraser River valley is the most densely populated region of the province, with an estimated 2.2 million people (55% of the provincial population) (British Columbia Stats 1999). Between 1980 and 1987, about 4,350 ha of once rural land were urbanized – an annual conversion rate equal to the creation of a community about one and one-half times the land area of White Rock every year (Moore 1990). Within that time period (1981 to 1986), the growth rate in the valley was just over 9%. While we could not find any recent land conversion statistics, a growth rate for the valley of over 22% during the period 1991-96 suggests that urbanization continues unabated.

Although agricultural development has played a role in negatively affecting a number of species of birds on the Fraser River delta, such as the Horned Lark (*strigata* subspecies) and Short-eared Owl, others have benefited, such as the waterfowl, some raptors, and shorebirds. This estuarine delta supports up to 1.4 million birds during migration, many of which have now come to depend to some extent on the agricultural lands (Butler and Campbell 1987).

The lower Fraser River valley is also an area of nationally significant agricultural production and contains some of the province's most productive farmlands. However, urbanization is reducing the areal extent of these lands in significant ways. Between 1980 and 1987, agricultural land was second only to undisturbed land for urban development, and any new areas that were created for agricultural production were more likely to be of lower agricultural productivity and capability than the farmland lost to urbanization (Moore 1990).

Thus, aside from covering our most agriculturally productive lands with asphalt, houses, golf courses, and industries, rather than letting the lands produce food for humans, urbanization of these agricultural lands also reduces the habitat for birds. The effects of urbanization on both agriculture and bird habitat is graphically shown in Fig. 759.

Recently, the construction of massive greenhouses, which make use of hydroponics to grow foods, has begun to cover much of the productive soils (Fig. 760). One of the largest operations (including greenhouse, parking, and driveways) covers over 25 ha (Corporation of Delta 2000), with even larger greenhouse complexes being planned. No longer is urbanization the only threat to agricultural lands; industrial agricultural practices themselves now affect the wildlife values of these lands.

These and other impacts on the ecosystems of the lower Fraser River valley have placed at least 10 bird taxa at risk, including the American Bittern, Great Blue Heron, Green Heron,

Table 39. Mean and cumulative numbers of bird species in an urban environment in Vancouver, B.C., in relation to total vegetation cover and habitat diversity.

Urban class	Total[1] vegetative cover (%)	Habitat diversity	Mean no. species (seasonal range)	Cumulative no. species (seasonal range)
Commercial	2.6	0.14	4.08 – 5.25	5 – 7
Industrial	4.3	0.21	3.38 – 4.36	7 – 8
Apartment residential	34.6	0.65	6.50 – 7.67	10 – 12
Typical residential 1	52.0	0.70	5.90 – 7.80	12 – 18
Typical residential 2	57.3	0.69	6.25 – 8.17	13 – 20
Older residential	69.5	1.02	6.31 – 8.83	13 – 21
Woody residential	84.2	1.04	9.09 – 12.17	17 – 25
Mixed woodland	171.3	1.49	8.00 – 12.00	17 – 23

[1] Total vegetative cover may exceed 100% due to cumulative results of multiple vegetation layers.
Source: Modified from Lancaster and Rees (1979).

Peregrine Falcon (*anatum* subspecies), Sandhill Crane, Yellow-billed Cuckoo, Barn Owl, Short-eared Owl, Hutton's Vireo, and Horned Lark (*strigata* subspecies) (Fraser et al. 1999).

Another aspect of urbanization is loss of avian biodiversity. Lancaster and Rees (1979), in their study of bird communities in urban habitats of Vancouver, found that the mean and cumulative numbers of species in all seasons were lowest in their commercial and industrial plots and highest in their lightly developed residential and woodland sites (Table 39). In addition, they found that introduced species – the European Starling, House Sparrow, or Rock Dove – dominated the avifauna of all but their wooded residential site and their woodland site. Migrant numbers were highly correlated with components of vegetation cover and foliage height diversity.

Clergeau et al. (1998), in their comparative study of bird abundance and diversity in Quebec City (Canada) and Rennes (France), also found that bird species diversity decreases with urbanization. In both those cities as well, 3 species – House Sparrow, European Starling, and Rock Dove – dominated the avifauna in most of the study plots. The theme of decreasing diversity with urbanization recurs worldwide.

Figure 761. More species of birds depend on habitats in the Okanagan valley for nesting than in any other area of British Columbia. Rapid agricultural development, including ginseng and vineyard operations, such as this vineyard near Vaseux Lake, threaten an already diminished land base (2 September 1998; Richard J. Cannings).

Okanagan Valley
More species of plants and animals depend on habitats in the Okanagan valley than in most other areas of British Columbia or Canada. Today many of these organisms are threatened – again, primarily by agricultural and urban development. By the 1990s, only 9% of the Okanagan valley remained in a relatively undisturbed state (Redpath 1990). Over 60% of the grasslands and shrub habitats in the southern portions of the valley have been modified and about 85% of wetland and riparian habitats have been destroyed (British Columbia Ministry of Environment, Lands and Parks 1998).

Ecosystem modification and habitat loss has threatened at least 20 bird taxa in the Okanagan valley to the point where their populations are now considered at risk or have been extirpated. Included are the American Bittern, Great Blue Heron, Prairie Falcon, Sage Grouse (extirpated), Sharp-tailed Grouse (extirpated throughout much of the valley), Sandhill Crane, Long-billed Curlew, Flammulated Owl, Western Screech-Owl (*macfarlanei* subspecies), Burrowing Owl (would likely have been extirpated, but small numbers are maintained through reintroduction efforts), Short-eared Owl, Lewis's Woodpecker, White-headed Woodpecker, Canyon Wren, Sage Thrasher, Yellow-breasted Chat, Brewer's Sparrow (*breweri* subspecies), Lark Sparrow, Grasshopper Sparrow, and Bobolink (Fraser et al. 1999). In addition, species such as the Mourning Dove, Yellow Warbler, and Wilson's Warbler have shown declines on interior Breeding Bird Surveys and their numbers are likely being affected in the Okanagan valley.

Examples of ecosystem modification and habitat loss in the Okanagan valley include:

Grasslands: More than half of the Red- and Blue-listed vertebrate species in the southern Okanagan valley are associated with grasslands, and 2 extirpated birds, the Sage Grouse and Sharp-tailed Grouse, are grassland species. These limited ecosystems are under tremendous development pressure today from both an expanding human population and a growing wine industry (Fig. 761).

In Colorado grasslands, Bock et al. (1999) found that grassland-nesting songbirds were nearly twice as abundant in their "interior" study plots at least 200 m away from suburban housing developments as they were in their "edge" study plots adjacent to such developments, despite the fact that the grassland habitat of the interior plots did not differ from that of the edge plots. Species significantly more abundant in plots away from the suburban edge included Vesper Sparrow, Savannah Sparrow, Grasshopper Sparrow, Bobolink, and Western Meadowlark. Breeding Bird Survey data indicate that all those species have been declining on a continental scale (Sauer et al. 1997), and all occur in the Okanagan valley.

Grasslands also suffer from livestock grazing, fragmentation, shrub removal and recreational activities such as off-road vehicular access (British Columbia Ministry of Environment, Lands and Parks 1998). A number of species that breed in the shrub-steppe habitats of the Okanagan valley are thought to respond negatively to livestock grazing (Table 40) (Bock et al. 1993); however, there are few long-term data from the Okanagan that might allow us to determine any changes in the population levels of these species.

Table 40. Species breeding in shrub-steppe habitats of the Okanagan valley that probably respond negatively to grazing.[1]

Northern Harrrier	Brewer's Sparrow
Swainson's Hawk	Vesper Sparrow
Red-tailed Hawk	Lark Sparrow
Ferruginous Hawk	Savannah Sparrow
Long-billed Curlew	Grasshopper Sparrow
Burrowing Owl	White-crowned Sparrow
Short-eared Owl	Western Meadowlark
Sage Thrasher	

[1] Modified from Bock et al. (1993).

Wetlands and riparian habitats: These are used by wildlife species disproportionately more than any other habitat type. Once, these ecosystems covered much of the southern Okanagan valley bottomlands; however, humans have reduced the wetlands to less than 4% of the land area through draining and filling (Fig. 762). In addition, wetlands have been modified through poor game management practices, such as poisoning the coarse fish and amphibian populations in order to plant game fish, water consumption for irrigation and urban areas, livestock trampling of riparian vegetation, stormwater runoff, and excessive agricultural pesticide and fertilizer use.

Perhaps the most significant adverse impact on riparian habitats in western riparian ecosystems has been improper grazing practices (Krueper 1993). A number of species that breed in riparian habitats in the Okanagan valley have been found to respond negatively to livestock grazing elsewhere (Table 41) (Bock et al. 1993).

Low-elevation forests: The low-elevation ponderosa pine and Douglas-fir forests in the southern Okanagan valley are relatively dry and parkland-like in nature. These drier forests tend to support more shrubs, and thus shrub-nesting species, than other forest types. Cattlegrazing is also encouraged in these areas and cattle graze on and trample shrubs, thereby reducing the available habitat for shrub-nesters. Data suggest that there are fewer shrub-nesters in the drier forests than there once were (Bunnell 1998). Other impacts come from forest practices that remove snags and old-growth trees. Today, with the valley bottoms fully developed, urbanization is moving up the valley walls to the low-elevation forests, and there is now concern that these forests may disappear (Ministry of Environment, Lands and Parks 1998).

THE ROAD WE'VE TAKEN

It is clear that our activities on the landscape have affected birds in British Columbia in many ways. From inappropriate forest practices to direct loss of bird habitat through urbanization and agricultural activities, bird diversity and abundance is threatened across much of the province. There are obvious problems to solve and many will not be easy to overcome. Nevertheless, what choice do we have if we want to maintain

Figure 762. The last marsh in Penticton being filled in (1996, Stephen R. Cannings). Once, marshes such as this covered many hectares in Penticton; they have since been relegated to a few cattails in ditches.

the rich diversity of birds that adds so much to our quality of life, a diversity that depends on healthy ecosystems in British Columbia?

And it's not just here in British Columbia where concern lies. Throughout the world, ecologists and biologists are voicing their opinions – supported by our best science – that humans are changing the biosphere in ways that impose severe environmental constraints on the planet, its ecosystems, and their inhabitants (see, e.g., Union of Concerned Scientists 1992, 1997; Karr 1993; Royal Society and United States National Academy of Sciences 1992; Ludwig et al. 1993; Tickell 1993; Vitousek 1994; Mangel et al. 1996). Many of these changes, such as ozone depletion, global warming, and loss of biodiversity, are already apparent and, if present trends continue, they could ultimately lead to a greater loss of our wild birds and ultimately affect our well-being. We are now at a crossroads.

Our past activities to ensure healthy, diverse, and sustainable bird populations, whether we are ornithologists, biologists, naturalists, or environmentalists, have focused on research and management; protection of habitat through legislation, regulations, and, in some cases, civil disobedience; and where economically feasible, purchase of habitats as protected areas. But it has not been enough. In spite of all the efforts of ecologists and environmentalists around the world, ecosystems – and the birds and other wildlife they support – are in worse shape today than they were 30 years ago. What we're doing is not working. Three current reports will serve to illustrate.

The World Wide Fund for Nature (formerly World Wildlife Fund) recently released the *Living Planet Report* on the state of the world's natural environment (Loh et al. 1999). They calculate what they call the Living Planet Index, a measure of the changes in the natural wealth of the Earth over time. The main factors considered include the area of the world's forests and the populations of different marine and freshwater species. From 1970 to 1995, the Living Planet Index declined by 30% (Fig. 763), suggesting that 30% of the earth's natural capital has been lost within one human generation. Both

Table 41. Species breeding in riparian habitats in the Okanagan valley that usually respond negatively to grazing.[1]

American Kestrel	Chipping Sparrow
Calliope Hummingbird	Lincoln's Sparrow
Willow Flycatcher	White-crowned Sparrow
Cedar Waxwing	Dark-eyed Junco
Yellow-rumped Warbler	Lazuli Bunting
MacGillivray's Warbler	Red-winged Blackbird
Common Yellowthroat	Bullock's Oriole
Wilson's Warbler	Cassin's Finch
Yellow-breasted Chat	American Goldfinch

[1] Modified from Bock et al. (1993).

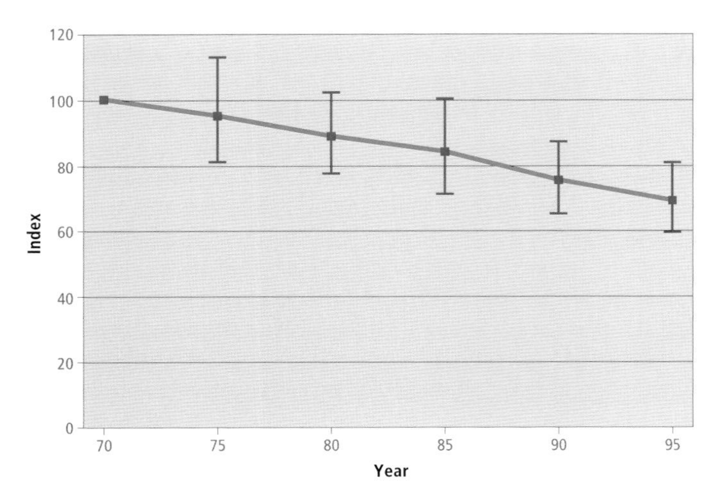

Figure 763. The Living Planet Index is a measure of the health of the world's natural ecosystems. This index suggests that about 30% of the world's natural wealth was lost within the past 25 years (Loh et al. 1999).

consumption of resources and pollution have been growing at around 2% each year since 1970. The main reasons behind this loss of natural wealth are expanded economic activities and the increasing human population (Loh et al. 1999).

A second report, issued by the United Nations in 1999, involves the work of scientists from more than 100 countries and notes the following outlook for the 21st century:

> The beginning of a new millennium finds the planet Earth poised between two conflicting trends. A wasteful and invasive consumer society, coupled with continued population growth, is threatening to destroy the resources on which human life is based. At the same time, society is locked in a struggle against time to reverse these trends and introduce sustainable practices that will ensure the welfare of future generations ... assessment shows both that time is running out on some issues and also that new problems are surfacing to compound an already difficult situation. (United Nations Environment Programme 1999: Chapter 5)

Finally, a report on the state of the world's resources and ecosystems was released in April 2000 through the collaborative efforts of the United Nations Development Programme, United Nations Environment Programme, World Bank, and World Resources Institute (World Resources Institute 2000). Their report notes that

> if we choose to continue our current patterns of [resource] use, we face almost certain declines in the ability of ecosystems to yield their broad spectrum of benefits – from clean water to stable climate, fuelwood to food crops, timber to wildlife habitat. We can use another option, however. It requires reorienting how we see ecosystems, so that we learn to view their sustainability as essential to our own.

The study found that nearly every measure that was used to assess the health of ecosystems revealed that the systems that support all life on this planet are being utilized more than ever and they are being degraded at an accelerating rate (World Resources Institute 2000).

While these reports suggest that we must change the way we treat the world, their shortcoming lies in the fact that they don't seem to have a clue about how to do it. The significance of the reports becomes clear, we believe, when one considers the collaborators, such as the United Nations and the World Bank, and the fact that they are now at least admitting that our present course is not sustainable.

WHERE DO WE GO FROM HERE?

Much as the Ptolemaic view of an earth-centred universe had to change, so too does our view of a human-centred earth, including views that all the Earth's bounty is here solely for us; that our economy (making use of that bounty) can grow forever; and that we can have an ever increasing human population. Ehrenfeld (1981:7-8) calls it the "doctrine of final causes" and notes that "the idea of using a Nature created for us, the idea of control, and the idea of human superiority became associated early in our history." There is little doubt that the Biblical charge "Be fruitful, and multiply, and replenish the earth, and subdue it: and have dominion over the fish of the sea, and over the fowl of the air, and over every living thing that moveth upon the Earth" (Genesis 1:28) has been "terribly misinterpreted" but religiously followed.

In addition, the conservation and management approaches we have practised in the past must change; if continued, they will ultimately fail because they only address the symptoms, not the problem (see Meffe 1992). As Mangel et al. (1996:338-339) point out:

> Managers of wild living resources have not seen the human population issue as something directly part of their profession and activities. In the past, the concepts and practices of wild-living-resource conservation proceeded as if the human population problem can be ignored in day-to-day planning and actions. This is no longer feasible, and population growth must be recognized as a critical conservation problem, both in training and actions of resource managers.

But addressing the human population problem alone will not be enough, for the consumptive nature of the population must also be considered, the measure of which is reflected in economic growth. This ultimately has an effect on wildlife populations and their conservation, an effect that is now being seriously discussed by wildlife professionals. For example, economic growth was recently addressed through a "Special Coverage" section of *The Wildlife Society Bulletin* (Leopold 2000). In it, Czech (2000) argues that economic growth and wildlife conservation are conflicting goals. He notes that for virtually all imperilled species, economic growth eliminates their welfare factors through the principle of competitive exclusion. As a result, economic growth may be considered a limiting factor to their conservation.

The ecological concept *limiting factor* defines any factor that when changed produces a change in the average or equilibrium density of a species (Krebs 1994). So, for example, the availability of large nesting trees and a sparse shrub layer would be a limiting factor for a population of Townsend's Warblers if their abundance were lower when such nest sites are absent. Czech (2000) argues that the concept may be applied to wildlife in the aggregate, suggesting that economic growth is a limiting factor to wildlife conservation. This is a significant concept for conservationists and wildlife managers to consider, for if most of their resources to date have been invested in nonlimiting factors for conservation, it means their work has been relatively futile (Czech 2000).

Given this situation, where do we go from here? We offer three areas where immediate conservation measures could be taken. If we are to maintain viable bird populations in the province at the diversity and population levels we have today, we believe they should at least be considered as part of a new conservation challenge.

(1) Move from Resource or Species Management to Ecosystem Management

Ecosystem management (Christensen et al. 1996) and its integral component, adaptive management (Holling 1978; Walters 1986), are recent concepts slowly starting to be implemented by resources agencies in British Columbia (most notably with forestry practices in Clayoquot Sound) and across North America. As with any new concept, ecosystem management has a number of difficulties associated with its gaining acceptance. These include its lack of definition as a concept, its changing goals (resource extraction to protection of the ecosystem), and its challenge of old assumptions (the Earth is a resource for humans, control instead of adaptation, all problems are soluble). It also asks that we understand management problems not only in biological but in larger contexts, including political, social, and economic contexts (Grumbine 1997).

Nevertheless, the concept is a simple one: rather than manage for a particular species or commodity, as we have done in the past, manage for a sustainable source of the species or commodity, that is, the ecosystem that supports it. And since our knowledge of these natural systems is rather limited, the *default* condition of ecosystem management should simply be to *retain the natural state* rather than manipulate the system dynamics or species within the system, unless direct intervention has clearly been shown to be necessary (Holling and Meffe 1996). Where some intervention is deemed necessary, adaptive management should be used, rather than the command-and-control prescriptions of the past.

Adaptive management, a structured, continual process of "learning by doing," is an integral component of ecosystem management (Holling 1978; Walters 1986; Christensen et al. 1996). Management (policy) objectives are clearly established based on the best available science, and the results are monitored to provide timely feedback to managers, who may then modify their management activities accordingly. The approach "probably never converges to a state of blissful equilibrium involving full knowledge and optimum productivity" (Walters 1986:8). "Adaptive management is grounded in the admission that humans do not know enough to manage ecosystems" (Lee 1999), a concept refreshingly free from the hubris of previous management theory.

In the past, we have tried to manage natural systems and their components, with all their complexities and ranges in natural variation, but we have been focusing on areas beyond our abilities and knowledge. Ironically, where we do have the ability and knowledge to manage, we often deny that management is needed. As Maser (1997) points out, good management entails knowing what we can manage and what we can't, and having the wisdom to leave the latter to manage itself. Holling and Meffe (1996), too, note the irony of our attempts to command and control ecosystems that are complex, poorly understood, and nonlinear, while ignoring the basic source of our environmental problems – human population growth and consumption – "where control is viable, reasonable and could be effective."

What needs to occur in natural resource management in British Columbia is a paradigm shift, not only by resource managers and environmentalists but by policymakers, communities, and the citizens of our communities, for we are all stakeholders. We must align ourselves to ensure that the base for our economy, our social well-being, and wildlife populations – the ecosystems of the province – are managed sustainably.

(2) Move from Ever-Growing Communities to More Sustainable Communities

Throughout many areas of British Columbia, urbanization is occurring at a rapid rate with little forethought as to the consequences of this growth. Small communities in distinct regions of the province that could collectively be fairly self-reliant are quickly trying to become larger cities. And the larger cities, in turn, become dependent on a wide array of subsidies, technologies, and far-off ecosystems to maintain their viability.

We have seen what happens to wild lands and the bird populations dependent on them as a result of our continual quest for growth. Even food, which we all require in order to survive, is taken for granted, as communities swallow up their prime agricultural lands when they expand.

This dependence on subsidies, technologies, and other people in faraway places also puts communities at risk, for when technologies or distant crops fail, the impacts on our local well-being could be severe. As Berry (1993:21) notes, "the only sustainable city … is a city in balance with its countryside: a city … that would live off the *net* ecological income of its supporting region, paying as it goes all its ecological and human debts." Unfortunately, most, if not all, communities in British Columbia are, in a word, unsustainable.

John F. Kennedy once said that "the great enemy of the truth is very often not the lie … but the myth – persistent, persuasive, and unrealistic." So it is with growth: "You can't stop

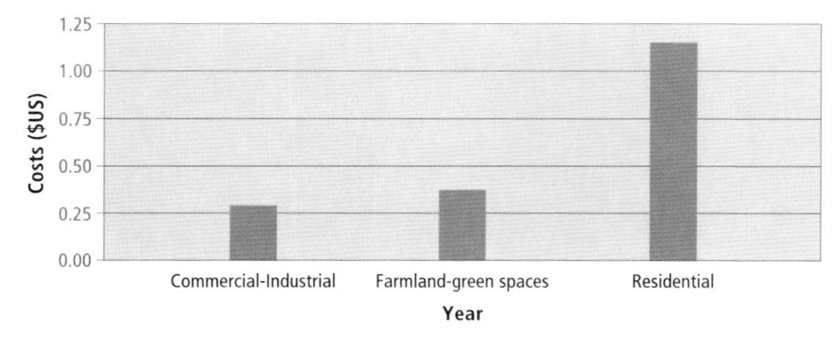

Figure 764. The median cost per dollar of revenue raised to provide public services to different land uses. The figures are based on a summary of 69 studies that reviewed the costs of community services in 18 states in the United States (American Farmland Trust 2000).

growth. Growth benefits the community by increasing the tax base. Our community will die if we stop growth." And the myths are perpetuated because most of us have little knowledge of sustainable communities, and few of our decision-makers understand even the basic concepts of the natural laws of ecology or physics.[3] As a result, decisions are made every day that affect the health of ecosystems with little understanding of the consequences of those decisions. Many have already described how our communities can become more sustainable (see, e.g., Alexander et al. 1977; Roseland 1992; Berry 1993; Orr 1994; Douthwaite 1996; Maser 1997;). Yet the myths persist and, in turn, nothing is done by decision-makers at most local levels in British Columbia to move our communities closer towards sustainability.

Does an increased tax base benefit communities? Many studies have shown that when a community grows, instead of benefiting the community with increased tax revenue, the taxes realized from the new residential developments do not cover the ongoing costs of servicing the developments (e.g., costs of garbage collection, sewers, water, street lighting, policing, fire protection, ambulance service, schools, burgeoning bureaucracies, etc.), even after levies such as development cost charges are made (Scerbinski et al. 1994; American Farmland Trust 2000; see also the review by Fodor 1999). These servicing costs have been shown to be highest on residential property and lowest on commercial and industrial properties, farmland, and naturescapes, such as forested lands, open space, parks, and nature reserves (Fig. 764).

What might community taxes have to do with birds and their habitat? Ehrenfeld (1972) recounts a story of a community where a group of citizens wanted to acquire 80 acres of land to be set aside for limited recreational purposes at a cost of around $500,000. Some residents wondered if the town might not be better off by developing the land for residential purposes; however, calculating the real costs of such development showed that if the community bought the land for green space, parks, or wild lands – habitat for birds – the land would be paid for in 10 years at about the same annual costs were a residential development to be built, but with no subsequent expenses except for nominal maintenance costs.

In addition to increased taxes, there are hidden costs to growth and sprawl. For example, how would one calculate the real costs to British Columbians of the loss of our agricultural lands to urbanization, particularly if there are agricultural shortages in other areas of the globe, areas that we have come to depend upon for our food? The issue of food security is worthy of increased attention.

Could this have been avoided? We could have reserved most of the valley bottomlands primarily for agriculture and green space, with housing and industry relegated to valley walls and uplands (McHarg 1969; Alexander et al. 1977). Even today, bottomlands not needed for cultivation could be kept as green space and wild lands as an investment for future needs, where this is still feasible. This would also serve to remove many flood-prone lands from development.

And what of the question of whether our communities have to grow or die? Davis and Baxter (2000) looked at the consequences of the absence of population growth on the economy of the Greater Vancouver Regional District and found that

> there is no compelling reason to conclude that a stable population will in itself necessarily have a deleterious effect on a region's per-capita income. Economic well-being of a region is substantially less dependent on population growth than it is on the manner in which the region utilizes its natural capital, the quality and extent of the education and training of its workforce, the activities in which it specializes, the flexibility and adaptability of its institutions, and the attitudes and entrepreneurship of its residents. Consequently, the possibility of approximating regional demographic stability should not be excluded from consideration by either policy makers or the general public. *We find no acceptable rationale for the position that a region must choose to either grow or die* [emphasis added].

Political pressures and demands from society to extract and harvest resources will only increase with further growth of the human population. Thus, without major changes, it will get harder, not easier, to protect the ecosystems upon which humans, birds, and all wildlife depend (Stanley 1995). We must work, at the community, regional, and national levels, to align our population with sustainable use of natural resources and the waste assimilation and regenerative capabilities of the Earth's ecosystems. Douthwaite (1996:59) puts it this way:

> A sustainable world will not be one dominated by large companies and run according to the conditions

[3] Ecology, taught as a core course from at least Kindergarten to Grade 12 would be a consequential step to take here; see Ornstein and Ehrlich (1989), Orr (1994), Dawe and Dunn (1995).

necessary for maintaining international competitiveness and speeding economic growth. It will be one of small communities that run their own affairs and that, rather than trading across the globe, meet or make most of their requirements from their local resources. For it is only if communities develop cultures that enable them to live indefinitely within the limits of their own places that humankind as a whole will be able to live sustainably within the limits of the natural world.

(3) Move from an Ever-Growing Economy to a Sustainable, Steady-State Economy

As we have mentioned, you cannot consider the problems that affect the birds of British Columbia – problems that have been created by an ever-expanding human population – without also considering the consumptive nature of that population. Why? Because in a finite world, each person's share of the natural capital must decrease as the population increases. Because 6 billion people living as Mother Teresa did would have nowhere near the impact on the biosphere that 6 billion people would have living as we do in the Western Hemisphere. In fact, if everyone on the Earth today consumed material goods at the rate British Columbians enjoy, we would need at least two more planet Earths to assimilate the wastes and supply the resources that we use (Wackernagle and Rees 1996). If we are going to maximize our population, we will have to minimize everything we do over and above our basic survival needs (Hardin 1968).

Conventional indicators of a successful economy, such as the GNP, include only commodity production. The health of the base for that production – the ecosystem – plays no role in the indicator. The economy, exemplified by the current neoclassical economic model (Fig. 765), can grow forever in physical throughput of natural capital because it is not bounded by limits (Daly 1996). And grow it does, accompanied by the attendant loss or degradation of ecosystems and overharvest of the natural capital, affecting the birds and, indeed, all organisms dependent on the ecosystems. This is nothing new; throughout our history we have overexploited resources to the point of collapse and often to extinction (Ludwig et al. 1993).

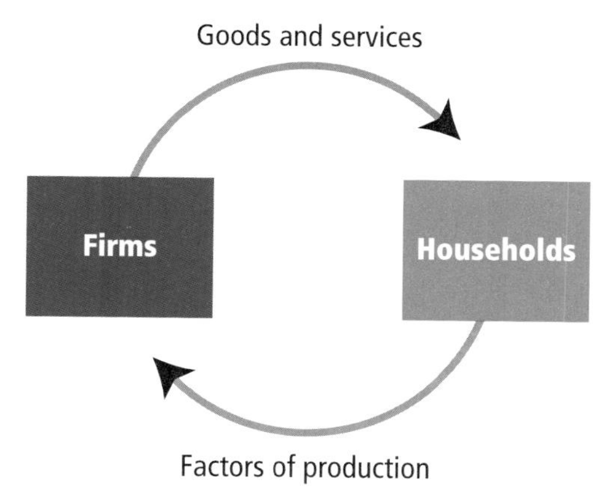

Figure 765. Modern economic theory shows the macroeconomy as an isolated system. The circular flow is simply an abstract exchange value and as such has no dependence on the ecosystems of the biosphere. Since the flow has no dependence on the biosphere, one need not be concerned about unsustainable use of resources, pollution, or the environmental services the ecosystems provide. But where is the natural capital from which the goods are made and where do the industrial and household wastes go? (Modified from Daly 1996.)

However, we do know that the natural capital is basically finite. Nonrenewable resources, such as fossil fuels and copper, certainly are, and renewable resources, such as salmon and forests, are finite to the extent that when we use them, they don't grow back instantly to replace what we have just taken. (A sustainable harvest of renewable resources can, however, make them seem infinite for our purposes.) Therefore, there will come a point when continuous growth of the economy will exceed the limits of the available natural capital and its regenerative capabilities, as well as the waste-assimilating capacity of the biosphere.

Today, we have reached a critical stage where additional growth beyond our present excesses will unquestionably increase costs rather than increase benefits (Daly and Cobb 1994). There is already abundant evidence to suggest that our present economic growth model is not working as economists suggest it should. Douthwaite (1999) discusses how economic growth has increased chronic illness, damaged family and community life, decreased education, enriched the few, impoverished the many, and damaged the ecosystems of Britain. He could find few aspects of British life that improved in quality recently, as a result of growth.

Even in Canada, it is clear that our standard of living does not necessarily equate with quality of life. For example, a report by Human Resources Development Canada found that, since 1975, the GDP has continued to grow, while the Index of Social Health based on issues such as infant mortality, teen suicides, child and seniors poverty, and average weekly earnings has fallen steadily (deGroot-Maggetti 2000). Statistics Canada figures show that the average household income in 1997 was slightly less than 1978 levels, despite the fact that in the 1970s there were far fewer dual-income families and we worked fewer hours on average (Cordon 1999). So after nearly 20 years of perhaps the highest economic growth we have ever seen, more Canadians per household are working longer hours for less money. As one economist put it, "That is an incredible condemnation of the way our economy is working."

Some believe that globalization of the economy is another factor exacerbating economic problems, and thus the problems of the biosphere. They argue that globalization weakens national autonomy and the power of national, regional, and local communities, while strengthening the transnational corporations whose job it is to grow for their shareholders. They suggest that we should instead be seeking to develop domestic production for local, regional, or national consumption first, and trading internationally only with the excess (Keynes 1933; Daly 1996). As Wendell Berry (1993:19-20) observes:

> Global thinking is not possible ... Global thinkers have been and will be dangerous people ... Unless one is

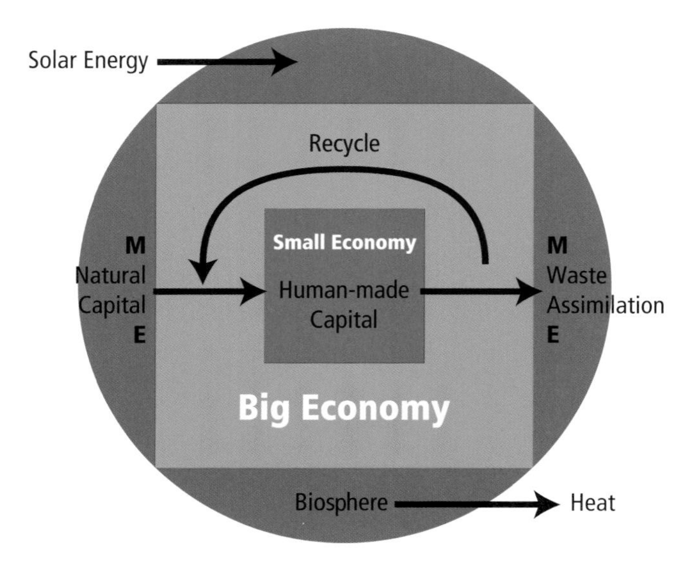

Figure 766. A new vision of the macroeconomy sees it as an open subsystem bounded by the biosphere. The economy takes matter (M) and energy (E) from the natural capital of the biosphere and rearranges it to produce human-made capital. Waste products from the economy are returned to the biosphere to be assimilated or recycled. The model shows that a small economy (green) can grow relative to the biosphere; however, once it becomes a big economy (red) and reaches the bounds of the biosphere, it cannot get any larger without consequences; it is limited by the available natural capital and the regeneration and waste assimilation capabilities of the biosphere (modified from Daly 1996).

> willing to be destructive on a large scale, one cannot do something except locally, in a small place ... If we could think locally, we would take far better care of things than we do now. The right local questions and answers will be the right global ones. The Amish question "What will this do to our community?" tends toward the right answer for the world.

Perhaps the time has come to seriously consider sustainable, preferably local, economies. There is little point, however, in trying to fine-tune a faulty economic model. Rather, a new model needs to be widely accepted that recognizes the limits to economic growth (Fig. 766) (Daly 1996). And since there are limits in this new economic model, it then follows that an economy based on low-entropy resources that are changed into commodities, eventually to become high-entropy waste, cannot grow forever. The economy must move towards a steady state, where quantitative growth is replaced with qualitative growth. This concept, known as sustainable development, has the economy developing or improving in quality, but not growing, much as the Earth itself can develop without growing (Daly 1996).

Ecosystems are good models for our sustainable economies. They are dynamic, not static systems; they have full employment, produce their own commodities, deal with their own wastes, and are continually developing, but grow little in areal extent. Our human economy must do the same.

MAGIC, TABOOS, AND WORLDVIEWS

Ludwig (1993), in his discussion of environmental sustainability and resource management, suggests that the practice of resource management has been based on the theories of *magic*, dictated by the association of ideas under the influence of desire. As long as we have desires, "we shall invent magical theories in an attempt to reconcile the irreconcilable." The same, perhaps, can be said for society as a whole. As long as we cling to our magical beliefs that hoping for something will make it true, we can continue to ignore the realities before our eyes, realities dictated by reason and experience. In this magical world, discussion of anything that dashes these hopes and desires is considered taboo. Talk of stabilizing or reducing the human population or moving from perpetual economic growth to a sustainable economy is seldom discussed openly. "Social, economic and religious forces in our society have imposed a taboo on discussion and implementation of appropriate action. It is time to end this taboo" (Ludwig 1993).

Today, there are two prevalent worldviews (Costanza 2000) (Table 42). The first one – the *technological optimist* view – holds that resources are unlimited, particularly energy; that humans can control nature; and that human ingenuity can solve any problem that may arise through technology. Therefore, the more people we have on the Earth, the greater chance that one of the billions will become the right problem-solver at the right time. It is basically the arrogance of humanism that Ehrenfeld (1981) so eloquently discusses.

The other – the *technological skeptic* view – assumes that resources and technological progress are limited and that the carrying capacity of ecosystems must be respected, and therefore there is a limit to the number of people the Earth can support.

So the question is, which worldview will take us safely into this new millennium, along with the birds and other organisms of the province? Certainly, neither scientific realism nor available evidence supports the view of the technological

Table 42. Some characteristics of the basic worldviews.[1]

Technological optimist	Technological skeptic
technological progress can deal with any future challenge	technological progress is limited and ecological carrying capacity must be preserved
competition	cooperation
linear systems with no discontinuities or irreversibilities	complex, nonlinear systems with discontinuities and irreversibilities
humans dominant over nature	humans in partnership with nature
everybody for themselves	partnership with others
market as guiding principle	market as servant of larger goals

1 After Costanza (2000).

optimists. And if we apply the Precautionary Principle,[4] there is no question that the technological skeptic's view is the only one that does not close any options. If it turns out that the skeptic is wrong, we can still choose to resume growth and deplete resources secure in the knowledge that humans, through technology and innovation, will solve any problems that may arise. But what if the technological optimist is wrong? It takes little imagination to conjure up a planet with myriad people and no time or resources left for change.

Paradoxically, the popular worldview today is that of the technological optimist. So how does one answer the question: Where will human society be some time in the future? It depends. If we rid ourselves of our taboos about discussing population and economic growth and clear our minds of our magical hopes, beliefs, and desires, we may be able to accept the present reality that we have not been treating the Earth very well. Then we can begin serious discussions and actions in an attempt to ameliorate the problems of "too many people, too much consumption" (Dawe and Dunn 1995).

If we can't do that, we simply have to look at what's happening to ecosystems throughout British Columbia today and extrapolate. And what of the birds described in these volumes? Those nesting in wetlands and in the majestic Douglas-firs of old-growth forests (Fig. 767), staging along coastal shores and interior valleys by the millions, or wintering on our productive estuaries and seasonally flooded farmlands – what will be their fate?

Figure 767. The future for most birds, such as this Pileated Woodpecker, will depend on humans accepting the present reality that we have not been treating the Earth very well. Managing ecosystems for all their important functions, including being a sustainable source of our commodities and recreational needs, and bringing our population and resource consumption in line with the carrying capacity of the biosphere and the needs of other organisms are significant components of a new conservation challenge that may help ensure abundant bird populations both in the near term and in the future (Victoria, 17 July 1999; R. Wayne Campbell).

---

[4] Basically, the Precautionary Principle holds that the burden of proof of safety should be borne by the proponent of a new technology, not by the public; and second, that, where there are threats of serious or irreversible damage, lack of scientific certainty should not be used as an excuse for postponing measures to prevent environmental degradation (Freestone and Hey 1996). In other words, when unsure, assume the worst and choose the path with the least risk.

Appendices

To reduce costs and the size of this final volume, we have placed the Appendices on the World Wide Web site of the Ministry of Environment, Lands and Parks, Wildlife Branch. Interested readers can access them at <http://www.elp.gov.bc.ca/wld>, where they can be downloaded as PDF files. Included are the following:

APPENDIX 1: Migration Chronology

APPENDIX 2: Summary of Christmas Bird Counts in British Columbia, 1957 through 1993

APPENDIX 3: Summary of Breeding Bird Surveys in British Columbia, 1968 through 1993

APPENDIX 4: Examples of Bird Species Richness and Density in Old-growth Coniferous and Deciduous Forests of British Columbia

REFERENCES CITED

As with the nonpasserine components of *The Birds of British Columbia* (see Campbell et al. 1990a, 1990b), the passerines are covered in self-contained volumes. Each volume has its own list of references, even though this means that repetition occurs.

Citations of most unpublished material contain a reference to the 2-volume *A Bibliography of British Columbia Ornithology* (Campbell et al. 1979b, 1988a). Copies of the papers cited in that work are on file at the Royal British Columbia Museum in Victoria, and are also found in the personal library of R.W. Campbell, now housed in the provincial WBT Wildlife Data Centre in Victoria.

Adkisson, C.S. 1977. Morphological variation in North American Pine Grosbeaks. Wilson Bulletin 89:380-395.

________. 1981. Geographic variation in vocalizations and evolution of North American Pine Grosbeaks. Condor 83:277-288.

________. 1996. Red Crossbill (*Loxia curvirostra*). *In* A. Poole and F. Gill (editors). The birds of North America, No. 256. The Academy of Natural Sciences, Philadelphia, Pennsylvania; The American Ornithologists' Union, Washington, D.C. 24 pp.

________. 1999. Pine Grosbeak (*Pinicola enucleator*). *In* A. Poole and F. Gill (editors). The birds of North America. The Birds of North America, Inc., Philadelphia, Pennsylvania. 20 pp.

Aldrich, J.W. and J.S. Weske. 1978. Origin and evolution of the eastern House Finch population. Auk 95:528-536.

Alexander, C., S. Ishikawa, and M. Silverstein with M. Jacobson, I. Fiksdahl-King, and S. Angel. 1977. A Pattern Language. Oxford University Press, New York.

American Farmland Trust. 2000. Cost of community services studies. Fact Sheet. Washington, D.C. <http://www.farmlandinfo.org/fic/tas/tafs/tafs-cocs.html>

American Ornithologists' Union. 1957. Check-list of North American birds, 5th edition. Lord Baltimore Press, Maryland. 691 pp.

________. 1973. Corrections and additions to the "thirty-second supplement to the check-list of North American birds." Auk 90:887.

________. 1983. Check-list of North American birds, 6th edition. Allen Press, Lawrence, Kansas. 877 pp.

________. 1995. Fortieth supplement to the American Ornithologists' Union check-list of North American birds. Auk 112:819-830.

________. 1998. Check-list of North American birds: the species of birds of North America from the Arctic through Panama, including the West Indies and Hawaiian Islands, 7th edition. Allen Press, Lawrence, Kansas. 829 pp.

Ammon, E.M. 1995. Lincoln's Sparrow (*Melospiza lincolnii*). *In* A. Poole and F. Gill (editors). The birds of North America, No. 191. The Academy of Natural Sciences, Philadelphia, Pennsylvania; The American Ornithologists' Union, Washington, D.C. 20 pp.

Anderson, E.M. 1914. Report on birds collected and observed during April, May and June, in the Okanagan Valley, from Okanagan Landing south to Osoyoos. Pages 7-16 *in* Report of the Provincial Museum of Natural History for the year 1913, Victoria, British Columbia.

Anderson, R.M. 1932. Annual report of the National Museum of Canada 1930. National Museum of Canada Division of Biology, Ottawa, Ontario. Pages 13-22.

Anderson, R.R. 1976a. Summary of highest counts of individuals for Canada. American Birds 30:643-644.

________. 1976b. Summary of all-time highest counts of individuals for Canada. American Birds 30:645-648.

________. 1977. Summary of highest counts of individuals for Canada. American Birds 31:916-918.

________. 1979. Summary of highest counts of individuals for Canada. American Birds 33:708-709.

________. 1980. Summary of highest counts of individuals for Canada. American Birds 34:708-710.

Anderson, S.H., K. Mann, and H.H. Shugart. 1977. The effect of transmission-line corridors on bird populations. American Midland Naturalist 97:216-221.

Anonymous. 1988. Biogeoclimatic zones of British Columbia – an annotated full colour map. British Columbia Forest Service Research Branch, Victoria.

________. 1994. Canadian Landbird Monitoring Strategy. Canadian Wildlife Service, Hull, Quebec. 44 pp.

________. 1995. Forest practices code of British Columbia – biodiversity guidebook. British Columbia Ministry of Environment, Lands and Parks, Victoria. 108 pp.

________. 1999. British Columbia trend summary. Time period 1966 to 1996. <http://www.mbr.nbs.gov/cgi-bin/guildsel.pl>. 2 pages.

________. 1999b. Scientists find toxic brew in B.C. mountains' snow. Victoria Times-Colonist, 3 October 1999, Section A3.

Arcese, P. 1989. Intrasexual competition, mating system and natal dispersal in Song Sparrows. Animal Behaviour 38:958-979.

Arcese, P. and J.N.M. Smith. 1988. Effects of population density and supplemental food on reproduction in Song Sparrows. Journal of Animal Ecology 57 (1988):119-136.

Arcese, P., J.N.M. Smith, W.H. Hochachka, C.M. Rogers, and D. Ludwig. 1992. Stability, regulation, and the determination of abundance in an insular Song Sparrow population. Ecology 73:805-822.

Atwood, J.L. 1994. Endangered small landbirds of the western United States. Studies in Avian Biology 15:328-339.

Austin, O.L. 1968. Life histories of North American cardinals, grosbeaks, buntings, towhees, finches, sparrows, and their allies – Parts 1 to 3. United States National Museum Bulletin No. 237, Washington, D.C. 1889 pp.

Avery, M., D.W. Gibbons, R. Porter, T. Tew, B. Tucker, and G. Williams. 1994. Revising the British Red Data List for birds: the biological basis for British conservation priorities. Ibis S232-S239.

Avery, M.L. 1995. Rusty Blackbird (*Euphagus carolinus*). *In* A. Poole and F. Gill (editors). The birds of North America, No. 200. The Academy of Natural Sciences, Philadelphia, Pennsylvania; The American Ornithologists' Union, Washington, D.C. 16 pp.

Avery, M.L., C.L. Schreiber, and D.G. Decker. 1999. Fruit sugar preferences of House Finches. Wilson Bulletin 111:84-88.

Baicich, P.J. and C.J.O. Harrison. 1997. A guide to the nests, eggs, and nestlings of North American birds. Academic Press, New York. 347 pp.

Bain, M. and M. Holder. 1997. Cross Canada Round-up, June-July 1997, British Columbia. Birders Journal 6(4):159-176.

Baker, M.C. 1991. Response of male Indigo and Lazuli buntings and their hybrids to song playback in allopatric and sympatric populations. Behaviour 1119:225-242.

________. 1994. Does exposure to heterospecific males affect sexual preferences of female buntings (*Passerina*)? Animal Behaviour 48:1349-1355.

Balph, D.F. 1979. Behavioral flexibility of Pine Siskins in mixed species foraging groups. Condor 81:211-212.

Balph, M.H. 1976. Golden-crowned Sparrows in northern Utah. North American Bird Bander 1:67.

Baltz, M.E. and S.C. Latta. 1998. Cape May Warbler (*Dendroica tigrina*). *In* A. Poole and F. Gill (editors). The birds of North America, No. 332. The Birds of North America, Inc., Philadelphia, Pennsylvania. 20 pp.

Bancroft, J. and R.J. Parsons. 1991. Range expansion of the House Finch into the prairie provinces. Blue Jay 49:128-136.

Banfield, A.W. 1954. A Lark Bunting in Kootenay National Park, British Columbia. Canadian Field-Naturalist 68:135.

Banks, R.C. 1970. Records of the Brambling in North America. Auk 87:165-167.

Barrowclough, G.F. 1980. Genetic and phenotypic differentiation in a wood warbler (Genus *Dendroica*) hybrid zone. Auk 97:655-668.

Barrows, W.B. 1889. The English Sparrow in North America, especially in its relations to agriculture. United States Department of Agriculture Division of Economic Ornithology and Mammalogy Bulletin No. 1, Washington, D.C. 405 pp.

Baumann, S.A. 1959. The breeding cycle of the Rufous-sided Towhee, *Pipilo erythrophthalmus* (Linnaeus), in central California. Wasserman Journal of Biology 17:161-220.

Baumgartner, A.M. 1968. *Spizella arborea* (Wilson) Tree Sparrow. Pages 1137-1165 *in* O.L. Austin (editor). Life histories of North American cardinals, grosbeaks, buntings, towhees, finches, sparrows, and allies – Order Passeriformes: Family Fringillidae (Part 2 – Genera *Pipilo* (part) through *Spizella*). Dover Publications, New York.

Beal, F.E.L. 1907. Birds of California in relation to the fruit industry. Part 1. United States Department of Agriculture and Biological Survey Bulletin 30:13-17.

Begg, B. 1992. A Black-shouldered Kite sight record for Vancouver Island. British Columbia Birds 2:19-20.

Behle, W.H. 1950. Clines in the Yellow-throats of western North America. Condor 52:193-219.
Bekoff, M., A.C. Scott, and D.C. Conner. 1987. Nonrandom nest-site selection in Evening Grosbeaks. Condor 89:819-829.
________, ________, and ________. 1989. Ecological analysis of nesting success in Evening Grosbeaks. Oecologia 81:67-74.
Bell, C.P. 1997. Leap-frog migration in the Fox Sparrow: minimizing the cost of spring migration. Condor 99:470-477.
Bell, G.M. 1954. The Linnet or House Finch. Victoria Naturalist 10:104-106.
Bell, L.M. and R.J. Kallman. 1976. The Kitimat River estuary, status of environmental knowledge to 1976. Special Estuary Series No. 6. Pacific Environmental Institute, West Vancouver, B.C.
Belton, D. 1973. Park naturalist's report, Manning Provincial Park. British Columbia Parks Branch Unpublished Report, Victoria. 33 pp. (Bibliography 2223).
Benkman, C.W. 1987. Food profitability and the foraging ecology of crossbills. Ecological Monographs 57:251-267.
________. 1989. On the evolution and ecology of island populations of crossbills. Evolution 43:1324-1330.
________. 1990. Foraging rates and the timing of crossbill reproduction. Auk 107:376-386.
________. 1992. White-winged Crossbill (*Loxia leucoptera*). *In* A. Poole, P. Stettenheim, and F. Gill (editors). The birds of North America, No. 27. The Academy of Natural Sciences, Philadelphia, Pennsylvania; The American Ornithologists' Union, Washington, D.C. 20 pp.
________. 1993a. Logging, conifers, and the conservation of crossbills. Conservation Biology 7:473-479.
________. 1993b. Adaptation to single resources and the evolution of crossbill (*Loxia*) diversity. Ecological Monographs 63:305-325.
Bennett, S. 1998. Fort Nelson forest bird inventory, 1988: Smith and Dunedin drainages. British Columbia Ministry of Environment, Lands and Parks Unpublished Report, Fort St. John. 47 pp.
Bennett, S. and K.A. Enns. 1996. A bird inventory of the Boreal White and Black Spruce biogeoclimatic zone near the "Big-Bend" of the Liard River. British Columbia Ministry of Environment, Lands and Parks, Habitat Protection Branch Unpublished Report, Fort St. John. 45 pp.
Bennetts, R.E. and R.L. Hutto. 1984. Attraction of social fringillids to mineral salts: an experimental study. Journal of Field Ornithology 56:187-189.
Bent, A.C. 1953. Life histories of North American wood warblers (Part 1). United States National Museum Bulletin No. 203, Washington, D.C. 367 pp.
________. 1958. Life histories of North American blackbirds, orioles, tanagers and their allies. United States National Museum Bulletin No. 211, Washington, D.C. 549 pp.
________. 1965. Life histories of North American blackbirds, orioles, tanagers, and allies. United States National Museum Bulletin 211, Washington, D.C. 531 pp.
Berger, R., G.A. McPeek, and R.J. Adams. 1991. The atlas of breeding birds of Michigan. Michigan State University Press, Lansing, Michigan.
Bergerud, A.T. and H.E. Butler. 1978. Life history studies of caribou in Spatsizi Wilderness Park 1977-78. British Columbia Ministry of Recreation and Conservation, Parks Branch Report, Victoria. 156 pp.
Berry, W. 1993. Sex, economy, freedom and community. Pantheon Books, New York.
Biermann, G.C., W.B. McGillivray, and K.E. Nordin. 1987. The effect of cowbird parasitism on Brewer's Sparrow productivity in Alberta. Journal of Field Ornithology 58:350-354.
Bisby, F.A. (editor). 1993. Characterization of biodiversity. Pages 27-82 *in* Global Biodiversity Assessment. Cambridge University, London. 1140 pp.
Bishop, L.B. 1900. Birds of the Yukon region, with notes on other species. North American Fauna Bulletin No. 19, Washington, D.C. Pages 47-135.
Blakiston, T.W. 1863. On the birds of the interior of British North America. Ibis 39:39-87, 121-155.
Blanchard, B.D. and M.M. Erickson. 1949. The cycle in the Gambel Sparrow. University of California Publications in Zoology 47:225-318.
Blankespoor, G.W. 1980. Prairie restoration: effects on nongame birds. Journal of Wildlife Management 44:667-672.
Bleitz, D. 1956. Eggs of a California Quail in the nest of a Spotted Towhee. Condor 58:77-78.1.
Blood, D.A. and M. Chutter. 1978. Raptor nesting survey for Shakwak Highway improvement project. Don Blood and Associates Unpublished Report, Lantzville, British Columbia. 70 pp.
Blood, D.A., M. Chutter, and G. Anweiller. 1981. An annotated list of the birds of the Stikine region, B.C. Donald A. Blood and Associates Unpublished Report, Lantzville, British Columbia. 32 pp. (Bibliography 2278).
Bock, C.E. and L.W. Lepthien. 1976. Synchronous eruptions of boreal seed-eating birds. American Naturalist 110:559-571.
Bock, C.E. and B. Webb. 1984. Birds as grazing indicator species in southeastern Arizona. Journal of Wildlife Management 48:1045-1049.
Bock, C.E., J.H. Bock, and B.C. Bennett. 1999. Songbird abundance in grasslands at a suburban interface on the Colorado high plains. Studies in Avian Biology 19:131-136.
Bock, C.E., V.A. Saab, T.D. Rich, and D.S. Dobkin. 1993. Effects of livestock grazing on neotropical landbirds in western North America. Pages 296-309 *in* D.M. Finch and P.W. Stangel (editors). Status and management of neotropical migratory birds. United States Department of Agriculture, Forest Service General Technical Report RM-229, Estes Park, Colorado.
Boggs, B. and E. Boggs. 1960a. The fall migration – northern Pacific coast region. Audubon Field Notes 14:65-67.
________. and ________. 1960b. The winter season – northern Pacific coast region. Audubon Field Notes 14:334-336.
________. and ________. 1960c. The spring migration – northern Pacific coast region. Audubon Field Notes 14:414-416.
________. and ________. 1961a. The fall migration – northern Pacific coast region. Audubon Field Notes 15:68-70.
________. and ________. 1961b. The winter season – northern Pacific coast region. Audubon Field Notes 15:352-353.
________. and ________. 1961c. The spring migration – northern Pacific coast region. Audubon Field Notes 15:433-434.
________. and ________. 1961d. The nesting season – northern Pacific coast region. Audubon Field Notes 15:487-489.
________. and ________. 1962a. The fall migration – northern Pacific coast region. Audubon Field Notes 16:67-69.
________. and ________. 1962b. The winter season – northern Pacific coast region. Audubon Field Notes 16:357-359.
________. and ________. 1962c. The spring migration – northern Pacific coast region. Audubon Field Notes 16: 440-442.
________. and ________. 1963a. The fall migration – northern Pacific coast region. Audubon Field Notes 17:58-61.
________. and ________. 1963b. The winter season – northern Pacific coast region. Audubon Field Notes 17:351-353.
________. and ________. 1963c. The spring migration – northern Pacific coast region. Audubon Field Notes 17:427-429.
________. and ________. 1964a. The fall migration – northern Pacific coast region. Audubon Field Notes 18:65-68.
Bollinger, E.K. 1988. Breeding dispersion and reproductive success of Bobolinks in an agricultural landscape. Ph.D. Thesis, Cornell University, Ithaca, New York.
Bollinger, E.K. and T.A. Gavin. 1989. The effects of site quality on breeding-site fidelity in Bobolinks. Auk 106:584-594.
________. and ________. 1992. Eastern Bobolink populations: ecology and conservation in an agricultural landscape. Pages 497-506 *in* J.M. Hagan and D.W. Johnston (editors). Ecology and conservation of neotropical migrant landbirds. Smithsonian Institution Press, Washington, D.C.
Bollinger, E.K., T.A. Gavin, and D.C. McIntyre. 1988. Comparison of transects and circular-plots for estimating Bobolink densities. Journal of Wildlife Management 52:777-786.
Bollinger, E.K., P.B. Bollinger, and T.A. Gavin. 1990. Effects of hay-cropping on eastern populations of the Bobolink. Wildlife Society Bulletin 18:142-150.
Bowles, J.H. 1906. A list of the birds of Tacoma, Washington, and vicinity. Auk 23:138-148.
Bowling, J. 1992. The spring season: March 1 – May 31, 1992 – British Columbia/Yukon region. American Birds 46:465-469.
________. 1994a. Spring season: March 1 – May 31, 1994 – British Columbia/Yukon region. American Birds 48:332-334.
________. 1994b. Summer season: June 1 – July 31, 1994 – British Columbia/Yukon region. National Audubon Society Field Notes 48:979-981.
________. 1995a. Autumn migration: August 1 – November 30, 1994 – British Columbia/Yukon region. National Audubon Society Field Notes 49:87-92.
________. 1995b. Winter season: December 1, 1994 – February 28, 1995 – British Columbia/Yukon region. National Audubon Society Field Notes 48:184-189.

________. 1995c. Spring season: March 1 – May 31, 1995 – British Columbia/Yukon region. National Audubon Society Field Notes 48:294-300.

________. 1995d. Summer season: June 1 – July 31, 1995 – British Columbia/Yukon region. National Audubon Society Field Notes 49:966-969.

________. 1996a. Autumn migration: August 1 – November 30, 1995 – British Columbia/Yukon region. National Audubon Society Field Notes 50:99-105.

________. 1996b. Winter season: 1995-1996: December 1, 1995 – February 29, 1996 – British Columbia/Yukon region. National Audubon Society Field Notes 50:208-213.

________. 1996c. Spring season: March 1 – May 31, 1996 – British Columbia/Yukon region. National Audubon Society Field Notes 50:321-326.

________. 1996d. Summer season: June 1 – July 31, 1996 – British Columbia/Yukon region. National Audubon Society Field Notes 50:986-988.

________. 1997a. Autumn migration: August 1 – November 30, 1996 – British Columbia/Yukon region. National Audubon Society Field Notes 51:105-108.

________. 1997b. Winter season: December 1, 1996 – February 28, 1997 – British Columbia/Yukon region. National Audubon Society Field Notes 51:789-792.

________. 1997c. Spring migration: March 1 – May 31, 1997 – British Columbia/Yukon region. National Audubon Society Field Notes 51:913-917.

Bradley, D.M. 1961. Birds and mammals seen at the Vancouver Natural History Society camp, Tenquille Lake area, July 23 to 31st, 1960. Bulletin of the Vancouver Natural History Society 113:3-5.

________. 1963. A spring record of the White-throated Sparrow at Vancouver, British Columbia. Canadian Field-Naturalist 77:232.

Bradley, H.L. 1940. A few observations on the nesting of the Eastern Chipping Sparrow. Jack-Pine Warbler 18:35-46.

Bradshaw, C. 1992. Mystery photographs. British Birds 85:647-649.

Breault, A. and P. Watts. 2000. 2000 cooperative wetland survey results for the interior of British Columbia – preliminary analyses of waterfowl use of trend wetlands. Canadian Wildlife Service Report, Delta, British Columbia.

Brewer, R., G.A. McPeek, and R.J. Adams. 1991. The atlas of breeding birds of Michigan. Michigan State University Press, East Lansing, Michigan.

Briggs, J.C. 1996. Global biogeography. Elsevier, Amsterdam.

Briggs, T. 1960. *Zonotrichia albicollis*. Victoria Naturalist 16:114.

Briskie, J.V. 1993. Smith's Longspur (*Calcarius pictus*). *In* A. Poole, P. Stettenheim, and F. Gill (editors). The birds of North America, No. 34. The Academy of Natural Sciences, Philadelphia, Penn-sylvania; The American Ornithologists' Union, Washington, D.C. 16 pp.

________. 1995. Nesting biology of the Yellow Warbler at the northern limit of its range. Journal of Field Ornithology 60:531-543.

Briskie, J.V. and S.G. Sealy. 1990. Evolution of short incubation periods in the parasitic cowbirds, *Molothrus* spp. Auk 107:789-794.

Briskie, J.V., S.G. Sealy, and K.A. Hobson. 1990. Differential parasitism of Least Flycatchers and Yellow Warblers by the Brown-headed Cowbird. Behavioral Ecology and Sociobiology 27:403-410.

British Columbia Ministry of Environment, Lands and Parks. 1998. Habitat atlas for wildlife at risk: south Okanagan and lower Similkameen. Ministry of Environment, Lands and Parks, Victoria. 124 pp.

British Columbia Ministry of Forests. 2000. Employment in the forest industry in B.C. <http://www.for.gov.bc.ca/hfp/forsite/jtfacts/3-2-can-employ.htm>

British Columbia Stats. 1999. British Columbia municipal and regional district population estimates. British Columbia Ministry of Finance and Corporate Relations, Population Section, Victoria.

Brooks, A. 1900. Notes on some of the birds of British Columbia. Auk 17:104-107.

________. 1901. Notes on the winter birds of the Cariboo District, B.C. Ottawa Naturalist 15:152-154.

________. 1903. Notes on the birds of the Cariboo District, British Columbia. Auk 20:277-284.

________. 1917. Birds of the Chilliwack district. Auk 34:28-50.

________. 1922. Notes on the mammals and birds of the Hudsonian Zone, Cascade Mts., 49th parallel. Murrelet 3:8-9.

________. 1922a. Notes on the American Pine Grosbeaks with the description of a new subspecies. Condor 24:86-88.

________. 1923. Notes on the birds of Porcher Island. Auk 40:217-224.

________. 1923a. Some recent records for British Columbia. Auk 40: 700-701.

________. 1932. The occurrence of the Falcated duck (*Eunetta falcata*) in Okanagan, British Columbia. Murrelet 13:92.

________. 1939. The Pine Grosbeak of southwestern British Columbia. Murrelet 20:57-59.

________. 1942a. Additions to the distributional list of birds of British Columbia. Condor 44:33-34.

________. 1942b. The extension of range of two northwestern finches. Murrelet 23:78-79.

Brooks, A. and H.S. Swarth. 1925. A distributional list of the birds of British Columbia. Pacific Coast Avifauna No. 17, Berkeley, California. 158 pp.

Brooks, T. 2000. Living on the edge. Nature 403:26-28.

Broun, M. 1972. Apparent migratory behavior in the House Sparrow. Auk 89:187-189.

Brown, R.M. 1960. Black-throated Sparrows in south-central Oregon. Condor 62:220-221.

Browning, M.R. 1966. Additional records on the birds of southwestern Oregon. Murrelet 47:76.

________. 1994. A taxonomic review of *Dendroica petechia* (Yellow Warbler) (Aves: Parulinae). Proceedings of the Biological Society of Washington 104:24-51.

Brownlee, M.J., E.R. Mattice, and C.D. Levings. 1984. The Campbell River Estuary: a report on the design, construction and preliminary follow-up study findings of intertidal marsh islands created for purposes of estuarine rehabilitation. Can. Ms. Rpt. of Fish. and Aquatic Sciences No. 1789.

Bruce, I.E. 1953. Birds at Carmanah Point. Victoria Naturalist 10:19-22; 10:28-31.

Brunton, D.F. and T. Pratt. 1986. Sightings of a Black-throated Sparrow, *Amphispiza bilineata*, and a Black Vulture, *Coragyps atratus*, in British Columbia. Canadian Field-Naturalist 100:256-257.

Brush, A.H. and D.M. Power. 1976. House Finch pigmentation: carotenoid metabolism and the effect of diet. Auk 93:729-731.

Bryant, A.A., J.-P.L. Savard, and R.T. McLaughlin. 1993. Avian communities in old-growth and managed forests of western Vancouver Island, British Columbia. Canadian Wildlife Service Technical Report Series No. 167, Delta, British Columbia. 115 pp.

Buckley, N.J. 1999. Black Vulture (*Coragyps atratus*). *In* The Birds of North America, No. 411 (A. Poole and F. Gill, editors). The Birds of North America, Inc., Philadelphia, PA. 24 pp.

Buckner, C.H., A.J. Erskine, R. Lidstone, B.B. McLeod, and M. Ward. 1975. The breeding bird community of coast forest stands on northern Vancouver Island. Murrelet 56:6-11.

Buech, R.R. 1980. Christmas tree plantations – for the birds? American Christmas Tree Journal 24:22-24.

________. 1982. Nesting ecology and cowbird parasitism of Clay-colored, Chipping, and Field sparrows in a Christmas tree plantation. Journal of Field Ornithology 53:363-369.

Bull, J. 1974. Birds of New York state. Doubleday/Natural History Press, Garden City, New York. 655 pp.

________. 1988. A first substantiated record of Golden-crowned Sparrow from New York state. Kingbird 38:8.

Bump, S.R. 1986. Yellow-headed Blackbird nest defense: aggressive responses to Marsh Wrens. Condor 86:328-335.

Bunnell, F.L. 1998. Next time try data: a plea for variety in forest practices. Paper presented at Interior Forestry Conference – Investing Today in Tomorrow's Forests. University College of the Cariboo, Kamloops, British Columbia, 24-25 March 1998.

Bunnell, F.L, L.L. Kremsater, and R.W. Wells. 1997. Likely consequences of forest management on terrestrial, forest-dwelling vertebrates in Oregon. Report M-7 of the Centre for Applied Conservation Biology, University of British Columbia. 130 pp.

Burhans, D.E. and F.R. Thompson. 1999. Habitat patch size and nesting success of Yellow-breasted Chats. Wilson Bulletin 111:210-215.

Burke, D.M. 1998. The effect of forest fragmentation on food abundance, nest site habitat, and reproductive success of forest breeding birds: a study in the Peterborough region of Ontario. Ph.D. Thesis, Trent University, Peterborough, Ontario.

Burke, D.M. and E. Nol. 1998. Influence of food abundance, nest-site habitat, and forest fragmentation on breeding Ovenbirds. Auk 115:96-104.

Burleigh, T.D. 1972. Birds of Idaho. Caxton Press, Coldwell, Idaho. 407 pp.

Burns, G. 1977. Golden-crowned Sparrow (first Iowa sighting). Iowa Bird Life 47:63.
Butcher, G.S. 1991. Mate choice in female Northern Orioles with a consideration of the role of the black male coloration in female choice. Condor 93:82-88.
Butler, R.W. 1997. The Great Blue Heron. University of British Columbia Press. Vancouver, British Columbia. 167 pp.
________. 1999. Winter abundance and distribution of shorebirds and songbirds on farmlands on the Fraser River delta, British Columbia, 1989-1991. Canadian Field-Naturalist 113:390-395.
Butler, R.W. and R.W. Campbell. 1987. The birds of the Fraser River delta: populations, ecology, and international significance. Canadian Wildlife Service Occasional Paper No. 65, Ottawa, Ontario. 73 pp.
Butler, R.W. and K. Vermeer (editors). 1994. The abundance and distribution of estuarine birds in the Strait of Georgia, British Columbia. Canadian Wildlife Service Occasional Paper No. 83, Delta, British Columbia. 78 pp.
Butler, R.W., B.G. Stushnoff, and E. McMackin. 1986. The birds of the Creston valley and southeastern British Columbia. Canadian Wildlife Service Occasional Paper No. 58, Ottawa, Ontario. 37 pp.
Byers, C.J., J. Curson, and U. Olsson. 1995. Sparrows and buntings. Houghton Mifflin Company, Boston, Massachusetts. 334 pp.
Byrd, G.V. and J.C. Williams. 1993. Red-legged Kittiwake (*Rissa brevirostris*). *In* A. Poole and F. Gill (editors). The birds of North America, No. 60. The Academy of Natural Sciences, Philadelphia, Pennsylvania; The American Ornithologists' Union, Washington, D.C. 12 pp.
Caccamise, D.F. 1978. Seasonal patterns of nesting mortality in the Red-winged Blackbird. Condor 80:290-294.
Cadman, M.D. and A.M. Page. 1994. Status of the Yellow-breasted Chat (*Icteria virens*) in Canada. Committee on the Status of Endangered Wildlife in Canada Report, Canadian Wildlife Service, Ottawa, Ontario. 39 pp.
Calder, W.A. and L.L. Calder. 1992. Broad-tailed Hummingbird (*Selas phorus platycercus*). *In* The Birds of North America, No. 16 (A. Poole, P. Stettenheim, and F. Gill, editors). Philadelphia: The Academy of Natural Sciences; Washington, D.C.: The American Ornithologists' Union. 16 pp.
Campbell, R.W. 1965. Mitlenatch Island Nature Park, December 19-22, 1965. British Columbia Ministry of Recreation and Conservation, Parks Branch Unpublished Report, Victoria. 8 pp. (Bibliography 2501).
________. 1972. Checklist of Vancouver birds, 1972 edition. University of British Columbia, Department of Zoology, Vancouver. Leaflet.
________. 1973. Lark Sparrow on Vancouver Island. Canadian Field-Naturalist 87:471-472.
________. 1974. First records of the Brambling for British Columbia. Canadian Field-Naturalist 88:486-487.
________. 1974a. Rusty Blackbird preys on sparrows. Wilson Bulletin 86:291-293.
________. 1975. Golden-crowned Sparrow breeding on Vancouver Island. Canadian Field-Naturalist 89:175-176.
________. 1977. Checklist of British Columbia birds (to June 1977). British Columbia Provincial Museum, Victoria. Leaflet.
________. 1978a. Birds observed at the Coal and Liard rivers and Portage Brule Rapids Ecological Reserve. British Columbia Provincial Museum Unpublished Report, Victoria. 3 pp. (Bibliography 2442).
________. 1978b. Census of waterbirds at Cecil Lake, northeastern British Columbia. British Columbia Provincial Museum Unpublished Report, Victoria. 10 pp. (Bibliography 2441).
________. 1978c. Breeding birds of One Island Lake, northeastern British Columbia. British Columbia Provincial Museum Unpublished Report, Victoria. 4 pp. (Bibliography 2439).
________. 1978d. Birds observed at the Coal and Liard rivers and Portage Brulé rapids ecological reserve. British Columbia Provincial Museum Unpublished Report, Victoria. 3pp.
________. 1981. Spring migrants observed 16-21 May 1981 at Atlin, British Columbia. British Columbia Provincial Museum Unpublished Report, Victoria. 7 pp. (Bibliography 2447).
________. 1982a. Wildlife atlases progress report. B.C. Naturalist 20(1):8-9.
________. 1982b. Wildlife atlases progress report. B.C. Naturalist 20(2): 8-10.
________. 1982c. Wildlife atlases progress report – spring 1982. B.C. Naturalist 20(3):6-8.
________. 1982d. Wildlife atlases progress report – summer 1982. B.C. Naturalist 20(4):5-7.
________. 1983a. Wildlife atlases progress report. B.C. Naturalist 21(1):4-5.
________. 1983b. Wildlife atlases progress report – winter 1982-83. B.C. Naturalist 21(2):4-5.
________. 1983c. Wildlife atlases progress report – spring/summer 1983. B.C. Naturalist 21(3):4-6.
________. 1983d. Wildlife atlases progress report – fall 1983. B.C. Naturalist 21(4):4-6.
________. 1984a. Wildlife atlases progress report – winter 1983. B.C. Naturalist 22(1):4-5.
________. 1984b. Wildlife atlases progress report – winter 1983-84. B.C. Naturalist 22(2):6-7.
________. 1985a. Wildlife atlases progress report – winter 1984-85. B.C. Naturalist 23(1):6-7.
________. 1985b. Wildlife atlases progress report – spring 1985. B.C. Naturalist 23(2):6-7.
________. 1985c. Wildlife atlases progress report – summer 1985. B.C. Naturalist 23(3):6-7.
________. 1985d. Wildlife atlases progress report – fall 1985. B.C. Naturalist 23(4):6-7, 9.
________. 1999. Pocket checklist of British Columbia birds. WBT Wild Bird Trust of British Columbia Special Publication No.1, West Vancouver, British Columbia. Leaflet.
Campbell, R.W. and N.K. Dawe. 1991. Field report of the joint Royal British Columbia Museum and Canadian Wildlife Service expedition across southern British Columbia, June 29 to July 13, 1991. Royal British Columbia Museum, Victoria. 287 pp.
________. and ________. 1992. Central British Columbia ornithology field trip (Victoria–Williams Lake–Bella Coola–Prince George–Victoria), 6-19 July 1992. British Columbia Ministry of Environment Unpublished Report, Victoria. 244 pp.
Campbell, R.W. and V. Gibbard. 1979. British Columbia Nest Records Scheme: 23rd annual report. B.C. Naturalist 17:1-3.
Campbell, R.W. and K. Kennedy. 1965. Mitlenatch Island Nature Park summer report, June 23 – August 27, 1965. British Columbia Ministry of Recreation and Conservation, Parks Branch Unpublished Report, Victoria. 101 pp. (Bibliography 2492).
________. and ________. 1965a. An afternoon of birding on Mitlenatch Island. Blue Jay 23:158-160.
Campbell, R.W. and M.C.E. McNall. 1982. Field report of the Provincial Museum expedition in the vicinity of Kotcho Lake, northeastern British Columbia, June 11 to July 9, 1982. British Columbia Provincial Museum Unpublished Report, Victoria. 307 pp. (Bibliography 2495).
Campbell, R.W. and A.L. Meugens. 1971. The summer birds of Richter Pass, British Columbia. Syesis 4:93-123.
Campbell, R.W. and M.G. Shepard. 1971. Summary of spring and fall pelagic birding trips from Tofino, British Columbia. Vancouver Natural History Society Discovery 150:13-16.
Campbell, R.W. and G.P. Sirk. 1969. Common Grackle sighted at Vancouver, British Columbia. Murrelet 50:38.
Campbell, R.W. and A.C. Stewart. 1993. British Columbia ornithology field trip (Okanagan–Shuswap–Cariboo–Nechako), May 20-29, 1993. British Columbia Ministry of Environment, Lands and Parks Unpublished Report, Victoria. 256 pp.
________. and ________. 1994. British Columbia ornithology field trip (Okanagan–Williams Lake–Quesnel–Vanderhoof–Fraser Lake), 18 May to 7 June 1994. British Columbia Ministry of Environment, Lands and Parks Unpublished Report, Victoria. 77 pp.
________. and ________. 1995. British Columbia ornithology field trip (Okanagan–Creston–Kamloops–Williams Lake–Fraser Lake–Fort St. James), 26 May to 9 June 1995. British Columbia Ministry of Environment, Lands and Parks Unpublished Report, Victoria. 71 pp.
Campbell, R.W. and D. Stirling. 1968. Notes on the natural history of Cleland Island, British Columbia, with emphasis on the breeding bird fauna. Pages HH25-HH43 *in* Report of the Provincial Museum of Natural History and Anthropology for the year 1967, Victoria, British Columbia.
Campbell, R.W. and K.R. Summers. 1997. Vertebrates of Brooks Peninsula. Pages 12.1-12.39 *in* Brooks Peninsula: an ice age refugium on Vancouver Island. British Columbia Ministry of Environment, Lands and Parks Occasional Paper No. 5, Victoria.
Campbell, R.W. and B.M. Van Der Raay. 1985. First breeding record of the Snow Bunting for British Columbia. Wilson Bulletin 97:128-129.
Campbell, R.W., M.G. Shepard, and R.H. Drent. 1972a. Status of birds in the Vancouver area in 1970. Syesis 5:180-220.
Campbell, R.W., M.G. Shepard, and W.C. Weber. 1972b. Vancouver birds in 1971. Vancouver Natural History Society Special Publication, Vancouver, British Columbia. 88 pp.

Campbell. R.W., M.G. Shepard, B.A. MacDonald, and W.C. Weber. 1974. Vancouver birds in 1972. Vancouver Natural History Society Special Publication, Vancouver, British Columbia. 96 pp.
Campbell, R.W., H.R. Carter, C.D. Shepard, and C.J. Guiguet. 1979b. A bibliography of British Columbia ornithology – Volume 1. British Columbia Provincial Museum Heritage Record No. 7, Victoria. 185 pp.
Campbell, R.W., J.M. Cooper, and M.C.E. McNall. 1983. Field report of the Provincial Museum expedition in the vicinity of Haines Triangle, northwestern British Columbia, May 27 to July 4, 1983. British Columbia Provincial Museum Unpublished Report, Victoria. 351 pp. (Bibliography 2477).
Campbell, R.W., D.A. Manuwal, and A.S. Harestad. 1987. Food habits of the Common Barn-Owl in British Columbia. Canadian Journal of Zoology 65:578-586.
Campbell, R.W., N.K. Dawe, and T.D. Hooper. 1988a. A bibliography of British Columbia ornithology – volume 2. Royal British Columbia Museum Heritage Record No. 19, Victoria. 591 pp.
Campbell, R.W., N.K. Dawe, I. McT.-Cowan, J.M. Cooper, G.W. Kaiser, and M.C.E. McNall. 1990a. The birds of British Columbia – Volume 1: Nonpasserines (Introduction, loons through waterfowl). Royal British Columbia Museum and Canadian Wildlife Service, Victoria. 535 pp.
________, ________, ________, ________, ________, and ________. 1990b. The birds of British Columbia – Volume 2: Nonpasserines (Diurnal birds of prey through woodpeckers). Royal British Columbia Museum and Canadian Wildlife Service, Victoria. 632 pp.
Campbell, R.W., N.K. Dawe, I. McT.-Cowan, J.M. Cooper, G.W. Kaiser, M.C.E. McNall, and G.E.J. Smith. 1997. The birds of British Columbia – Volume 3: Passerines (Flycatchers through vireos). British Columbia Ministry of Environment, Lands and Parks (Wildlife Branch) and Environment Canada (Canadian Wildlife Service), Victoria. 693 pp.
Campbell, R.W., M.L. Funk, and L. Davis. 1999. British Columbia Nest Records Scheme: 43rd annual report – 1997 nesting season. WBT Wild Bird Trust of British Columbia Wildlife Report No. 3, West Vancouver. 22 pp.
Campbell, R.W., N.K. Dawe, L.M. Van Damme, and S.R. Johnson. 2000. British Columbia birds – A complete checklist. WBT Wild Bird Trust of British Columbia Special Publication No. 3. West Vancouver, 42 pp.
Campbell, T. 1997. Substrate segregation by nesting blackbirds at Myers wetlands, British Columbia. University of Victoria, Department of Biology Unpublished Research Paper, Victoria. 29 pp.
________. 2000. Observations on nesting Red-winged and Yellow-headed blackbirds at Myers wetlands, British Columbia. Blue Jay 5877-80.
Canadian Wildlife Service. 1994. Canadian bird trends. <http://199.212.18.79/birds.birdinfo.cfm.id=6420>
Cannings, R.A., R.J. Cannings, and S.G. Cannings. 1987. Birds of the Okanagan valley, British Columbia. Royal British Columbia Museum, Victoria. 420 pp.
Cannings, R.J. 1973. Shuswap birds 1973 – annotated list. British Columbia Parks Branch Unpublished Report, Victoria. 15 pp. (Bibliography 2528).
________. 1974. The Indigo Bunting in British Columbia. Canadian Field-Naturalist 88:232-233.
________. 1987. The 87th Christmas Bird Count – western and northern Canada. American Birds 41:533-535.
________. 1995a. Status of the Yellow-breasted Chat in British Columbia. British Columbia Ministry of Environment, Lands and Parks Wildlife Bulletin No. B-18, Victoria. 10 pp.
________. 1995b. Status of the Grasshopper Sparrow in British Columbia. British Columbia Ministry of Environment, Lands and Parks Wildlife Bulletin No. B-77. 11 pp.
________. 1998. The birds of British Columbia: a taxonomic catalogue. British Columbia Ministry of Environment, Lands and Parks Wildlife Bulletin No. B-86, Victoria. 243 pp.
Cannings, R.J., B. Frederick, and A. Stepniewski. 1974. Year-end hike: 1974 – Tonquin area – Mount Robson Park. British Columbia Parks Branch Unpublished Report, Victoria. 5 pp. (Bibliography 2542).
Cannings, S.G. 1973. Mount Robson vertebrate report, 1973. British Columbia Parks Branch Unpublished Report, Victoria. 18 pp. (Bibliography 2558).
Carey, A.B., M.M. Hardt, S.P. Horton, and B.L. Birwell. 1991. Spring bird communities in the Oregon Coast Range. Pages 123-142 *in* Wildlife and vegetation of unmanaged Douglas-fir forests. United States Department of Commerce Pacific Northwest Research Station, Portland, Oregon.
Carl, G.C. and C.J. Guiguet. 1957. Alien animals in British Columbia. British Columbia Provincial Museum Handbook No. 14, Victoria. 94 pp.
Carl, G.C., C.J. Guiguet, and G.A. Hardy. 1952. A natural history survey of the Manning Park area, British Columbia. British Columbia Provincial Museum Occasional Paper No. 9, Victoria. 130 pp.
Carter, N.M. 1933. Exploration in the Lillooet River watershed. Canadian Alpine Journal 21:8-18.
Caswell, F.D. and K.M. Morse. 1993. Evaluating the status of waterfowl populations in Canada. Pages 8-15 *in* E.H. Dunn, M.D. Cadman, and B. Falls (editors). Monitoring bird populations: the Canadian experience. Canadian Wildlife Service Occasional Paper No. 95. Ottawa, Ontario. 62 pp.
Catt, D.A. 1991. Bird communities and forest succession in the subalpine zone of Kootenay National Park, British Columbia. M.Sc. Thesis, Simon Fraser University, Burnaby, British Columbia. 99 pp.
Channell, R. and M.V. Lomolino. 2000. Dynamic biogeography and conservation of endangered species. Nature 403:84-86.
Chapman, J.D., D.B. Turner, A.L. Farley, and R.I. Ruggles. 1956. British Columbia atlas of resources. 92 pp.
Chichilnisky, G. and G. Heal. 1998. Economic returns from the biosphere. Nature 391:629-630.
Chilton, G., M.C. Baker, C.D. Barrentine, and M.A. Cunningham. 1995. White-crowned Sparrow (*Zonotrichia leucophrys*). *In* A. Poole and F. Gill (editors). The birds of North America, No. 183. The Academy of Natural Sciences, Philadelphia, Pennsylvania; The American Ornithologists' Union, Washington, D.C. 28 pp.
Christensen, N.L., A.M. Bartuska, J.H. Brown, S. Carpenter, C. D'Antonio, R. Francis, J.F. Franklin, J.A. MacMahon, R.F. Noss, D.J. Parsons, C.H. Peterson, M.G. Turner, R.G. Woodmansee. 1996. The report of the Ecological Society of America Committee on the Scientific Basis for Ecosystem Management. Ecological Applications 6:665-691.
Christensen, P. 1991. Driving forces, increasing returns, and ecological sustainability. *In* R. Costanza (editor). Ecological economics: the science and management of sustainability. Columbia University Press, New York.
Christman, M.J. 1970. Occurrence of birds in the Vermilion Pass burn, Kootenay National Park. Canadian Parks Service Unpublished Report, Radium Hot Springs, British Columbia. 51 pp. (Bibliography 2592).
Clague, J.J., R.W. Matthews, and B. Warner. 1982. Late Quaternary Geology of eastern Graham Island, Queen Charlotte Islands, British Columbia. Canadian Journal of Earth Sciences 19:1786-1795.
Clark, C. 1973. Profit maximization and the extinction of animal species. Journal of Political Economy 81:950-961.
Clarke, C.H.D. 1945. Biological reconnaissance of the Alaska Military Highway with particular reference to the Yukon Territory and to various proposed national park areas. Department of Mines and Resources, Lands, Parks and Forests Branch Unpublished Report, Ottawa, Ontario. 118 pp.
Clay, J.O. 1952. House-Finch or Linnet (*Carpodacus mexicanus*). Victoria Naturalist 9:32-33.
________. 1953. Report of the chairman of the ornithology group. Victoria Naturalist 10:3-4.
Clement, P., A. Harris, and J. Davis. 1993. Finches and sparrows. Princeton University Press, Princeton, New Jersey. 500 pp.
Clergeau, P., J-P.L. Savard, G. Mennechez, and G. Falardeau. 1998. Bird abundance and diversity along an urban-rural gradient: a comparative study between two cities on different continents. Condor 100:413-425.
Cody, M.L. 1974. Competition and the structure of bird communities. Monographs in Population Biology No. 7, Princeton University Press, New Jersey.
Cohen, D.A. 1899. Nesting and other habits of the Oregon Towhee. Condor 1:61-63.
Collins, S.L., F.C. James, and P.G. Risser. 1982. Habitat relationships of wood-warblers (Parulidae) in northern central Minnesota. Oikos 39:50-58.
Committee on the Status of Endangered Wildlife in Canada. 1999. List of wildlife. <http://www.cosewic.gc.ca>
Conservation Data Centre. 2000. Provincial list status and CDC ranks. <http://www.elp.gov.bc.ca/rib/wis/cdc/list.htm>
Conway, C.J. 1999. Canada Warbler (*Wilsonia canadensis*). *In* A. Poole and F. Gill (editors). The birds of North America, No. 421. The Birds of North America, Inc., Philadelphia, Pennsylvania. 24 pp.

Cooke, F.S. 1947. Notes on some fall and winter birds of the Queen Charlotte Islands, British Columbia. Canadian Field-Naturalist 61:131-133.

Cooke, W.W. 1910. The migration of North American sparrows, seventh paper. Bird-Lore 12:240-242.

________. 1913. The migration of North American sparrows. Bird-Lore 15:301-303.

Cooper, J.M. 1985. Notes on the vertebrates of McClinton Bay and Masset, Queen Charlotte Islands, 15 May – 14 June 1985. Royal British Columbia Museum Unpublished Report, Victoria. 124 pp.

________. 1993. Breeding bird surveys of the airport reserve on Sea Island, Richmond, British Columbia. LGL Ltd. Report for the Vancouver International Airport Authority, Vancouver, British Columbia. 34 pp.

________. 1993a. Preliminary impact assessment on migratory birds of the proposed causeway breaching at Delkatla Inlet, Masset, British Columbia. Sirius Environmental Research Report, Sidney, British Columbia. 25 pp.

________. 1994. Mobbing of a Common Raven by migrant Rosy Finches. British Columbia Birds 4:13-14.

________. 1997. Breeding bird surveys at the Koksilah River estuary, 1997. Sirius Environmental Research Unpublished Report for the Habitat Conservation Trust Fund, Victoria, British Columbia. 18 pp.

Cooper, J.M. and M. Adams. 1979. The birds and mammals of Kwadacha Wilderness Park (August 13 – September 8, 1979). Royal British Columbia Museum Unpublished Report, Victoria. 10 pp. (Bibliography 2687).

Cooper, J.M. and D.L.P. Cooper. 1983. A second report on the summer vertebrates of the Fern Lake area of Kwadacha Wilderness Park. British Columbia Provincial Museum Unpublished Report, Victoria. 7 pp. (Bibliography 2688).

Cooper, J.M., K.A. Enns, and M.G. Shepard. 1997. Status of the Bay-breasted Warbler in British Columbia. British Columbia Ministry of Environment, Lands and Parks Wildlife Working Report No. 79, Victoria. 24 pp.

________, ________, and ________. 1997a. Status of the Cape May Warbler in British Columbia. British Columbia Ministry of Environment, Lands and Parks Wildlife Working Report No. WR-82, Victoria. 23 pp.

________, ________, and ________. 1997b. Status of the Connecticut Warbler in British Columbia. British Columbia Ministry of Environment, Lands and Parks Wildlife Working Report No. WR-83, Victoria. 32 pp.

________, ________, and ________. 1997c. Status of the Black-throated Green Warbler in British Columbia. British Columbia Ministry of Environment, Lands and Parks Wildlife Working Report No. WR-80, Victoria. 24 pp.

Corbin, K.W., C.G. Sibley, and A. Ferguson. 1979. Genetic changes associated with the establishment of sympatry in orioles of the genus *Icterus.* Evolution 33:624-633.

Cordon, S. 1999. Getting ahead is getting tougher. Victoria Times-Colonist, 5 April, page A1.

Corporation of Delta. 2000. Status of greenhouse building permit applications. Municipal Report. Delta, British Columbia.

Costanza, R. 2000. Visions of alternative (unpredictable) futures and their use in policy analysis. Conservation Ecology 4:5. <http://www.consecol.org/vol4/iss1/art5>

Costanza, R., R. d'Arge, R. de Groot, S. Farber, M. Grasso, B. Hannon, K. Limburg, S. Naeem, R.V. O'Neill, J. Paruelo, R.G. Raskin, P. Sutton, and M. van den Belt. 1997. The value of the world's ecosystem services and natural capital. Nature 387:253-260.

Cowan, I. McT. 1937. The House Finch at Victoria, British Columbia. Condor 39:225.

________. 1939. The vertebrate fauna of the Peace River district of British Columbia. British Columbia Provincial Museum Occasional Paper No. 1, Victoria. 102 pp.

________. 1940. Bird records from British Columbia. Murrelet 21:69-70.

________. 1955. Birds of Jasper National Park, Alberta, Canada. Canadian Wildlife Service Wildlife Management Bulletin Series 2, No. 8, Ottawa, Ontario. 66 pp.

________. 1989. Birds and mammals on the Queen Charlotte Islands. Pages 175-186 *in* G.G.E. Scudder and N. Gessler (editors). The Outer Shores. Queen Charlotte Islands Museum Press, Queen Charlotte City. 127 pp.

Cowan, I. McT. and J.A. Munro. 1944-1946. Birds and mammals of Mount Revelstoke National Park. Canadian Alpine Journal 100-121; 237-256.

Cox, G.W. 1960. A life history of the Mourning Warbler. Wilson Bulletin 72:5-28.

________. 1973. Hybridization between Mourning and MacGillivray's warblers. Auk 90:190-191.

Cracraft, J. 1983. Species concepts and species analysis. Current Ornithology 1:159-187.

________. 1989. Speciation and its ontogeny: the empirical consequences of alternate species concepts for understanding patterns and processes of differentiation. Pages 28-59 *in* J. Otte and J.A. Endler (editors). Speciation and its consequences. Sinauer Associates, Sunderland, Massachusetts.

Crocker-Bedford, D.C. 1990. Status of the Queen Charlotte Goshawk in southeast Alaska. United States Forest Service, Ketchikan, Alaska. 16 pp.

Crowell, J.B. and H.B. Nehls. 1966a. The fall migration – northern Pacific coast region. Audubon Field Notes 20:449-453.

________. and ________. 1966b. The spring migration – northern Pacific coast region. Audubon Field Notes 20:539-542.

________. and ________. 1967a. The fall migration – northern Pacific coast region. Audubon Field Notes 21:67-72.

________. and ________. 1967b. The winter season – northern Pacific coast region. Audubon Field Notes 21:448-452.

________. and ________. 1967c. The spring migration – northern Pacific coast region. Audubon Field Notes 21:532-535.

________. and ________. 1967d. The nesting season – northern Pacific coast region. Audubon Field Notes 21:596-600.

________. and ________. 1968a. The fall migration – northern Pacific coast region. Audubon Field Notes 22:78-83.

________. and ________. 1968b. The winter season – northern Pacific coast region. Audubon Field Notes 22:468-472.

________. and ________. 1968c. The spring migration – northern Pacific coast region. Audubon Field Notes 22:567-571.

________. and ________. 1968d. The nesting season – northern Pacific coast region. Audubon Field Notes 22:638-642.

________. and ________. 1969a. The fall migration – northern Pacific coast region. Audubon Field Notes 23:94-99.

________. and ________. 1969b. The winter season – northern Pacific coast region. Audubon Field Notes 23:508-513.

________. and ________. 1970a. The fall migration – northern Pacific coast region. Audubon Field Notes 24:82-88.

________. and ________. 1970b. The winter season – northern Pacific coast region. Audubon Field Notes 24:530-533.

________. and ________. 1970c. The spring migration – northern Pacific coast region. Audubon Field Notes 24:635-638.

________. and ________. 1971a. The fall migration – northern Pacific coast region. American Birds 25:94-100.

________. and ________. 1971b. The winter season – northern Pacific coast region. American Birds 25:614-619.

________. and ________. 1972a. The fall migration – northern Pacific coast region. American Birds 26:107-111.

________. and ________. 1972b. The winter season – northern Pacific coast region. American Birds 26:644-648.

________. and ________. 1973a. The fall migration – northern Pacific coast region. American Birds 27:105-110.

________. and ________. 1973b. The winter season – northern Pacific coast region. American Birds 27:652-656.

________. and ________. 1974a. The fall migration – northern Pacific coast region. American Birds 28:93-98.

________. and ________. 1974b. The winter season – northern Pacific coast region. American Birds 28:679-684.

________. and ________. 1974c. The spring migration – northern Pacific coast region. American Birds 28:840-845.

________. and ________. 1974d. The nesting season – northern Pacific coast region. American Birds 28:938-943.

________. and ________. 1975a. The fall migration – northern Pacific coast region. American Birds 29:105-112.

________. and ________. 1975b. The winter season – northern Pacific coast region. American Birds 29:730-735.

________. and ________. 1975c. The spring migration – northern Pacific coast region. American Birds 29:897-902.

________. and ________. 1976a. The fall migration – northern Pacific coast region. American Birds 30:112-117.

________. and ________. 1976b. The winter season – northern Pacific coast region. American Birds 30:755-760.

________. and ________. 1976c. The spring migration – northern Pacific coast region. American Birds 30:878-882.

________. and ________. 1976d. The nesting season – northern Pacific coast region. American Birds 30:992-996.
________. and ________. 1977a. The fall migration – northern Pacific coast region. American Birds 31:212-216.
________. and ________. 1977b. The winter season – northern Pacific coast region. American Birds 31:364-367.
________. and ________. 1977c. The spring migration – northern Pacific coast region. American Birds 31:1037-1041.
Cumming, R.A. 1932. Birds of the Vancouver district, British Columbia. Murrelet 13:1-15.
________. 1933. California Purple Finch and Nuttall's Poorwill at Pavilion, B.C. Murrelet 14:78.
Curson, J. 1992. Identification of Connecticut, Mourning and MacGillivray's warbler in female and immature plumages. Birders Journal 1:275-278.
Curson, J., D. Quinn, and D. Beadle. 1994. Warblers of the Americas. Houghton Mifflin, New York. 252 pp.
Czaplak, D. 1995. Identifying Common and Hoary redpolls in winter. Birding 26:447-457.
Czech, B. 2000. Economic growth as the limiting factor for wildlife conservation. Wildlife Society Bulletin 28:4-15.
Dale, B. 1993. Haying and grassland birds: summation of discussion. Page 33 *in* G.L. Holroyd, H.L. Dickson, M. Regnier, and H.C. Smith (editors). Proceedings of the Third Prairie Conservation and Endangered Species Workshop. Provincial Museum of Alberta Natural History Occasional Paper No. 19, Edmonton.
Daly, H.E. 1996. Beyond growth: the economics of sustainable development. Beacon Press, Boston.
Daly, H.E. and J.B. Cobb. 1994. For the common good: redirecting the economy toward community, the environment, and a sustainable future, 2nd edition. Beacon Press, Boston.
Darcus, S.J. 1930. Notes on birds of the northern part of the Queen Charlotte Islands. Canadian Field-Naturalist 44:45-49.
Darling, L. 1996. Aspen wildlife diversity and stand parameters executive summary: 1992-1996. British Columbia Ministry of Environment, Lands and Parks, Wildlife Branch Unpublished Report, Victoria. 15 pp.
Davidson, A.R. 1953. Winter birds in Victoria. Victoria Naturalist 9:94.
________. 1960. A Dickcissel in Victoria. Victoria Naturalist 17:58 (Bibliography 1705).
________. 1960a. Bird notes. Victoria Naturalist 16:90.
________. 1966. Annotated list of birds of southern Vancouver Island. Victoria Natural History Society Mimeographed Report, Victoria, British Columbia. 23 pp.
________. 1972. Then and now. Victoria Naturalist 29:42.
________. 1972a. The Bullock's Oriole. Victoria Naturalist 29:5.
Davidson, G. and L. Van Damme. 1987. First British Columbia record of the Black-throated Blue Warbler, *Dendroica caerulescens*. Canadian Field-Naturalist 101:104-105.
Davis, C. and D. Baxter. 2000. Grow or die: prescription for population change in urban regions? Unpublished Report.
Davis, J. 1960. Nesting behavior of the Rufous-sided Towhee in coastal California. Condor 62:434-456.
Davis, L.R., M.J. Waterhouse, and H.M. Armleder. 1999. A comparison of the breeding bird communities in seral stages of the Engelmann spruce–subalpine fir zone in east-central British Columbia. British Columbia Ministry of Forests Working Paper No. 39, Victoria. 45 pp.
Dawe, N.K. 1971. Nature interpretation report on Wasa, Moyie, and Jimsmith provincial parks, British Columbia. British Columbia Parks Branch Unpublished Report, Victoria. 97 pp. (Bibliography 1172).
________. 1976. Flora and fauna of the Marshall-Stevenson unit, Qualicum National Wildlife Area, August 1976. Canadian Wildlife Service Unpublished Report, Delta, British Columbia. 201 pp. (Bibliography 1862).
________. 1976a. A Gray-headed Junco in British Columbia. Murrelet 57:51.
________. 1980. Flora and fauna of the Marshall-Stevenson unit, Qualicum National Wildlife Area (update to June 1979). Canadian Wildlife Service Unpublished Report, Delta, British Columbia. 149 pp. (Bibliography 2788).
________. 1991. A report on the Canadian Nature Tours trip to the Queen Charlotte Islands, 16 to 27 May 1991. Canadian Nature Tours Unpublished Report, Toronto, Ontario. 23 pp.
Dawe, N.K. and R. Buechert. 1995. Bird use of the Little Qualicum River estuary, Vancouver Island, British Columbia, 1975-1979. Technical Report Series No. 240. Canadian Wildlife Service, Pacific and Yukon Region, British Columbia.
________. and ________. 1996. Central and southern interior ornithological field trip – 10 to 21 June 1996. Canadian Wildlife Service Unpublished Report, Qualicum Beach, British Columbia. 36 pp.
________. and ________. 1997. Bird use of the Tahsis and Leiner river estuaries, Vancouver Island, British Columbia, 1981-1982. Canadian Wildlife Service Technical Report Series No. 289, Delta, British Columbia. 34 pp.
Dawe, N.K. and J.T. Dawe. 1997. Northern and Central Vancouver Island field trip, 16 to 29 June 1997. Canadian Wildlife Service Unpublished Report, Qualicum Beach, British Columbia. 16 pp.
Dawe, N.K. and M. Dunn. 1995. Wildlife management and human implications – the biologist's challenge. Bioline 13:1-7.
Dawe, N.K., T. Martin, and D.E.C. Trethewey. 1994. Bird use of the Englishman River estuary, Vancouver Island, British Columbia, 1979-1980 and 1988-1989. Canadian Wildlife Service Technical Report Series No. 208, Delta, British Columbia. 143 pp.
Dawe, N.K., T. Martin, T. Barnard, D. Howard. 1995. Bird use of the Quatse River estuary. Vancouver Island, British Columbia, 1990-1991. Canadian Wildlife Service Technical Report Series No. 212, Delta, British Columbia. 88 pp.
Dawe, N.K., T. Martin, T. Barnard, and A. Koch. 1995. Bird use of the Cluxewe estuary, Vancouver Island, British Columbia, 1990-1991. Canadian Wildlife Service Technical Report Series No. 209, Delta, British Columbia. 56 pp.
Dawe, N.K., R. Buechert, and D.E.C. Trethewey. 1995a. Bird use of the Campbell River estuary, Vancouver Island, British Columbia, 1982-1984. Canadian Wildlife Service Technical Report Series No. 233, Delta, British Columbia. 68 pp.
Dawe, N.K., R. Buechert, T. Barnard, and C. Cook. 1995b. Bird use of the Fanny Bay–Little Bay wetlands, Vancouver Island, British Columbia. Canadian Wildlife Service Technical Report Series No. 228, Delta, British Columbia. 118 pp.
Dawe, N.K., R. Buechert, and D.E.C. Trethewey. 1998. Bird use of Baynes Sound–Comox harbour, Vancouver Island, British Columbia, 1980-1981. Technical Report Series No. 286. Canadian Wildlife Service, Pacific and Yukon Region, British Columbia.
Dawe, N.K., G. Bradfield, W.S. Boyd, D.E.C. Trethewey, and A.N. Zolbrod. In press. Marsh creation in a northern Pacific estuary: is thirteen years of monitoring vegetation dynamics enough? Conservation Ecology.
Dawson, W.L. and J.H. Bowles. 1909. The birds of Washington – Volume 1. Occidental Publishing Company, Seattle, Washington. 458 pp.
Dawson, W.R. 1997. Pine Siskin (*Carduelis pinus*). *In* A. Poole and F. Gill (editors). The birds of North America, No. 280. The Academy of Natural Sciences, Philadelphia, Pennsylvania; The American Ornithologists' Union, Washington, D.C. 24 pp.
Dawson, W.R. and F.C. Evans. 1957. Relation of growth and development to temperature regulation in nestling Field and Chipping sparrow. Physiology and Zoology 30:315-327.
DeBenedictis, P.A. 1995. Gleanings from the technical literature. Sharp-tailed Sparrow: a parable of field ornithology. Birding 27:312-314.
deGroot-Maggetti, G. 2000. GDP climbs yet quality of life falls. Victoria Times-Colonist, 25 February, page A13.
Demarchi, D.A., R.D. Marsh, A.P. Harcombe, and E.C. Lea. 1990. The environment. Pages 54-144 *in* R.W. Campbell, N.K. Dawe, I. McTaggart-Cowan, J.M. Cooper, G.W. Kaiser, and M.C.E. McNall. The birds of British Columbia – Volumes 1 and 2 – Nonpasserines (loons through woodpeckers). Royal British Columbia Museum, Victoria.
Dennis, J.V. 1967. Fall departure of the Yellow-breasted Chat (*Icteria virens*) in eastern North America. Bird-Banding 38:130-135.
DeRagon, W.R. 1988. Breeding ecology of Seaside and Sharp-tailed sparrows in Rhode Island salt marshes. M.S. Thesis, University of Rhode Island, Kingston.
DeWolfe, B.B. 1968. *Zonotrichia leucophrys gambelli* (Nuttall), Gambel's White-crowned Sparrow, pages 1324-1338. *Zonotrichia leucophrys pugetensis* (Grinnell), Puget Sound White-crowned Sparrow, pages 1344-1352. *In* O.L. Austin (editor). Life histories of North American cardinals, grosbeaks, buntings, finches, sparrows, and allies (Part 3). United States National Museum Bulletin 237, Washington, D.C.
DeWolfe, B.B. and R.H. DeWolfe. 1962. Mountain White-crowned Sparrows in California. Condor 64:378-389.
DeWolfe, B.B., G.C. West, and L.J. Peyton. 1973. The spring migration of Gambel's Sparrows through southern Yukon Territory. Condor 75:43-59.

Dickinson, J.C. 1953. Report on the McCabe collection of British Columbia birds. Bulletin of the Museum of Comparative Zoology 109(2):123-211.

Dickinson, T. and E. Leupin. 1997. The effects of alternative silvicultural practices on the songbird communities at Sicamous Creek and other Engelmann spruce–subalpine fir forests. Pages 254-271 *in* C. Hollstedt and A. Vyse (editors). Sicamous Creek silvicultural systems project: workshop proceedings. 24-25 April 1996. British Columbia Ministry of Forests Working Paper No. 24, Victoria.

Dobbs, R.C., P.R. Martin, and T.E. Martin. 1998. Green-tailed Towhee (*Pipilo chlorurus*). *In* A. Poole and F. Gill (editors). The birds of North America, No. 386. The Birds of North America, Inc., Philadelphia, Pennsylvania. 24 pp.

Dobkin, D.S. 1994. Conservation and management of neotropical migrant landbirds in the Northern Rockies and Great Plains. University of Idaho Press, Moscow. 220 pp.

Dobzhanski, T.H. 1970. Genetics of the evolutionary process. Columbia University Press, New York.

________.1972. Species of *Drosophila*. New excitement in an old field. Science 177:604-669.

Dolbeer, R.A. 1976. Reproductive rate and temporal spacing in nesting of Red-winged Blackbirds in upland habitat. Auk 93:343-355.

Douglas, D.C., J.T. Ratti, R.A. Black, and J.R. Alldredge. 1992. Avian habitat associations in riparian zones of Idaho's Centennial Mountains. Wilson Bulletin 104:485-500.

Douthwaite, R. 1996. Short circuit: strengthening local economies for security in an unstable world. Lilliput Press, Dublin; Green Books, Totnes.

________. 1999. The growth illusion: how economic growth has enriched the few, impoverished the many and endangered the planet. New Society Publishers, Gabriola Island, British Columbia.

Dow, D.D. 1963. A natural history of the southern portion of Garibaldi Provincial Park. British Columbia Parks Branch Unpublished Report, Victoria. 95 pp. (Bibliography 2827).

Downes, C.M. and B.T. Collins. 1996. The Canadian Breeding Bird Survey, 1966-1994. Canadian Wildlife Service Progress Notes No. 210, Ottawa, Ontario. 36 pp.

Drent, R. 1969. Sixty-ninth Christmas Bird Count – Ladner, B.C. Audubon Field Notes 23:134.

Drury, W.H. 1961. The breeding biology of shorebirds on Bylot Island, Northwest Territories, Canada. Auk 78:176-219.

Dubois, H.M. 1959. Black-throated Sparrows in northwestern Oregon. Condor 61:435.

Dufty, A.M. 1994. Rejection of foreign eggs by Yellow-headed Blackbirds. Condor 96:799-801.

Dunn, E.H., M.D. Cadman, and B. Falls (editors). 1997. Monitoring bird populations: the Canadian experience. Canadian Wildlife Service Occasional Paper No. 95, Ottawa, Ontario. 62 pp.

Dunn, J.L. and K.L. Garrett. 1997. A field guide to warblers of North America. Houghton Mifflin, New York. 656 pp.

Dunn, J.L., K.L. Garrett, and J.K. Alderfer. 1995. White-crowned Sparrow subspecies: identification and distribution. Birding 27:182-200.

Eaton, S.W. 1995. Northern Waterthrush (*Seiurus noveboracensis*). *In* A. Poole and F. Gill (editors). The birds of North America, No.182. The Academy of Natural Sciences, Philadelphia, Pennsylvania; The American Ornithologists' Union, Washington, D.C. 20 pp.

Edwards, R.Y. 1963. Twenty-seventh breeding bird census – disturbed Douglas-fir coast forest. Audubon Field Notes 17:499-500.

Edwards, R.Y. and R.W. Ritcey. 1967. The birds of Wells Gray Park, British Columbia. British Columbia Parks Branch, Victoria. 37 pp. (Bibliography 1025).

Edwards, R.Y. and D. Stirling. 1961. Range expansion of the House Finch into British Columbia. Murrelet 42:38-42.

Ehrenfeld, D.W. 1972. Conserving life on earth. Oxford University Press, New York.

________. 1981. The arrogance of humanism. Oxford University Press, New York.

Ehrlich, P.R., D.S. Dobkin, and D. Wheye. 1988. The birder's handbook: a field guide to the natural history of North American birds. Simon and Schuster, New York. 785 pp.

Elliott, K. and W. Gardner. 1997. Vancouver birds in 1995. Vancouver Natural History Society, British Columbia. 92 pp.

Elliott, P.F. 1999. Killing of host nestlings by the Brown-headed Cowbird. Journal of Field Ornithology 70:55-57.

Ellis, H.K. 1980. Ecology and breeding biology of the Swamp Sparrow in a southern Rhode Island peatland. M.S. Thesis, University of Rhode Island, Kingston.

Emerson, W.O. 1903. A remarkable flight of Louisiana Tanagers. Condor 5:250-252.

Emlen, J.T., J.D. Rising, and W.L. Thompson. 1975. A behavioral and morphological study of sympatry in Indigo and Lazuli buntings of the Great Plains. Wilson Bulletin 87:145-179.

Enns, K. and C. Siddle. 1996. The distribution, abundance and habitat requirements of selected passerine birds of the Boreal and Taiga plains of British Columbia. British Columbia Ministry of Environment, Lands and Parks, Wildlife Branch Wildlife Working Report No. WR-76, Victoria. 44 pp.

Eremko, R.D., D.G.W. Edwards, and D. Wallinger. 1989. A guide to collecting cones of British Columbia conifers. Forestry Canada and British Columbia Ministry of Forests, Victoria. 114 pp.

Eriksson, E. 1970. Wintering and autumn migration ecology of the Brambling, *Fringilla montifringilla*. Sterna 9:77-90.

Erskine, A.J. 1959. Christmas bird census, 1958 – Ladner, British Columbia. Canadian Field-Naturalist 73:44.

________. 1971. Some new perspectives on the breeding ecology of Common Grackles. Wilson Bulletin 83:352-370.

________. 1972. The co-operative breeding bird survey in Canada, 1971. Canadian Wildlife Service Progress Notes No. 26, Ottawa, Ontario. 31 pp.

________. 1976. A preliminary catalogue of bird census plot studies in Canada, Part 3. Canadian Wildlife Service Progress Notes No.59, Ottawa, Ontario. 24 pp.

________. 1977. Birds in boreal Canada: communities, densities and adaptations. Canadian Wildlife Service Report Series No. 41, Ottawa, Ontario. 73 pp.

________. 1992. The cooperative Breeding Bird Survey in Canada, 1989-91. Canadian Wildlife Service Progress Notes No. 199. Environment Canada, Ottawa. 14 pp.

Erskine, A.J. and G.S. Davidson. 1976. Birds in the Fort Nelson lowlands of northeastern British Columbia. Syesis 9:1-11.

Erskine, A.J. and R.C. Stein. 1964. A re-evaluation of the avifauna of the Cariboo parklands. Pages 18-35 *in* Provincial Museum of Natural History and Anthropology report for the year 1963, Victoria, British Columbia.

Erskine, A.J., B.T. Collins, E. Hayakawa, and C. Downes. 1992. The cooperative breeding bird survey in Canada, 1989-1991. Canadian Wildlife Service Progress Notes No. 199, Ottawa, Ontario. 14 pp.

Escott, N.G. 1994. Ontario's first Black-throated Sparrow, with a review of extralimital records in Canada and eastern North America. Birders Journal 3:289-293.

Evans Ogden, L.J. and B.J. Stutchbury. 1994. Hooded Warbler (*Wilsonia citrina*). *In* A. Poole and F. Gill (editors). The birds of North America, No. 110. The Academy of Natural Sciences, Philadelphia, Pennsylvania; The American Ornithologists' Union, Washington, D.C. 20 pp.

Evenden, F.G. 1957. Observations on nesting behavior of the House Finch. Condor 59:112-117.

Falls, J.B. and J.G. Kopachena. 1994. White-throated Sparrow (*Zonotrichia albicollis*). *In* A. Poole and F. Gill (editors). The birds of North America, No. 128. The Academy of Natural Sciences, Philadelphia, Pennsylvania; The American Ornithologists' Union, Washington, D.C. 32 pp.

Fannin, J. 1891. Check list of British Columbia birds. Queen's Printer, Victoria, British Columbia. 49 pp.

________. 1898. Check list of British Columbia birds. Pages 15-57 *in* A preliminary catalogue of the collections of natural history and ethnology in the Provincial Museum. Queen's Printer, Victoria, British Columbia.

Farley, A.L. 1979. Atlas of British Columbia. University of British Columbia Press, Vancouver. 136 pp.

Farner, D.S. and I.O. Buss. 1957. Summer records of the Golden-crowned Sparrow in Okanogan County, Washington. Condor 59:141.

Farrand, J. 1983. The Audubon Society master guide to birding – Volume 3 – warblers to sparrows. Alfred A. Knopf, New York. 399 pp.

Ferguson, R.S. and L. Halverson. 1997. Checklist of birds of the Upper Columbia River valley, British Columbia. Matrix Resource Services, Golden, British Columbia. Leaflet.

Fisher, J. 1952. The Fulmar. Collins, St. James Place, London. 496 pp.

Fix, D. 1984. The spring migration – northern Pacific coast region. American Birds 38:948-952.

Flack, J.A.D. 1976. Bird populations of aspen forests in western North America. Ornithological Monograph 19, Washington, D.C. 68 pp.

Flahaut, M.R. 1950. The fall migration – northern Pacific coast region. Audubon Field Notes 4:30-32.

________. 1953a. The fall migration – north Pacific coast region. Audubon Field Notes 7:31-33.

Flahaut, M.R. and Z.M. Schultz. 1954a. The fall migration – north Pacific coast region. Audubon Field Notes 8:36-38.

________. and ________. 1955a. The fall migration – north Pacific coast region. Audubon Field Notes 9:47-50.

Fodor, E. 1999. Better not bigger: how to take control of urban growth and improve your community. New Society Publishers, Gabriola Island, British Columbia.

Fonnesbeck, C.J. 1998. The abundance, distribution and brood parasitism of upland-breeding Warbling Vireos in a fragmented forest landscape. M.Sc. Thesis, University of British Columbia, Vancouver. 73 pp.

Force, M.P. and P.W. Mattocks. 1986. The fall migration – northern Pacific coast region. American Birds 40:316-324.

Francis, C.M. and F. Cooke. 1986. Differential timing of spring migration in wood warblers (Parulinae). Auk 103:548-556.

Fraser, D.F. 1984. An annotated list of the birds of the Elk and Flathead drainages. Westar Mining Environmental Service Unpublished Report, Sparwood, British Columbia. 110 pp.

Fraser, D.F. and E. Walters. 1993. Preliminary species management plan for Brewer's Sparrow, subspecies *breweri*, in British Columbia. British Columbia Ministry of Environment, Lands and Parks Report, Victoria. 12 pp.

Fraser, D.F., W.L. Harper, S.G. Cannings, and J.M. Cooper. 1999. Rare birds of British Columbia. British Columbia Ministry of Environment, Lands, and Parks, Victoria. 236 pp.

Fraser River Estuary Study Steering Committee. 1978. Fraser River estuary study – summary. Government of Canada and Province of British Columbia, Victoria.

Freedman, B.C., L.A. Beauchamp, I.A. McLaren, and S.I. Tingley. 1981. Forestry management practices and populations of breeding birds in a hardwood forest in Nova Scotia. Canadian Field-Naturalist 95:307-311.

Freeman, S., D.F. Gori, and S. Rohwer. 1990. Red-winged Blackbirds and Brown-headed Cowbirds: some aspects of a host-parasite relationship. Condor 92:336-340.

Freestone, D. and E. Hey. 1996. Origins and development of the precautionary principle. *In* D. Freestone and E. Hey (editors). The precautionary principle and international law. Kluwer Law International, The Hague, London, and Boston.

Friedmann, H. 1929. The cowbirds. A study in the biology of social parasitism. C.C. Thomas, Springfield, Illinois. 421 pp.

________. 1963. Host relations of the parasitic cowbirds. United States National Museum Bulletin 233, Washington, D.C. 276 pp.

Friedmann, H. and L.F. Kiff. 1985. The parasitic cowbirds and their hosts. Proceedings of the Western Foundation of Vertebrate Zoology 2:226-303.

Friedmann, H., L.F. Kiff, and S.I. Rothstein. 1977. A further contribution to knowledge of the host relations of the parasitic cowbirds. Smithsonian Contributions to Zoology No. 235, Washington, D.C. 75 pp.

Furrer, R.K. 1975. Breeding success and nest site stereotypy in a population of Brewer's Blackbirds (*Euphagus cyanocephalus*). Oecologia 20:339-350.

Gavin, T.A. and E.K. Bollinger. 1985. Multiple paternity in a territorial passerine: the Bobolink. Auk 102:550-555.

Gibbard, R.T. and V. Gibbard. 1992. Yellow-breasted Chat survey – final report. British Columbia Ministry of Environment, Lands and Parks Unpublished Report, Penticton, British Columbia. 14 pp.

Gibson, D.D. 1981. Migrant birds at Shemya Island, Aleutian Islands, Alaska. Condor 83:65-77.

Gilligan, J., M. Smith, D. Rogers, and A. Contreras. 1994. Birds of Oregon – status and distribution. Cinclus Publications, McMinnville, Oregon. 330 pp.

Global Forest Watch Canada. 2000. Canada's forests at a crossroads: an assessment in the year 2000. World Resources Institute, Washington, D.C. 114 pp.

Godfrey, W.E. 1955. Additional notes on birds of the East Kootenay, British Columbia. Pages 85-94 *in* National Museum of Canada Bulletin No. 136, Ottawa, Ontario.

________. 1961. First Canadian record of the Black-throated Sparrow. Canadian Field-Naturalist 75:162.

________. 1966. The birds of Canada. National Museum of Canada Bulletin No. 203, Ottawa, Ontario. 428 pp.

________. 1986. The birds of Canada, revised edition. National Museum of Canada, Ottawa, Ontario. 595 pp.

Goodland, R. 1991. The case that the world has reached limits. *In* R. Goodland, H. Daly, S. El Serafy, and B. Von Droste (editors). Environmentally sustainable economic development: building on Bruntland. Paris: UNESCO.

Goossen, J.P. and S.G. Sealy. 1982. Production of young in a dense nesting population of Yellow Warblers, *Dendroica petechia*, in Manitoba. Canadian Field-Naturalist 96:189-199.

Gottfried, B.M. and E.C. Franks. 1975. Habitat use and flock activity of Dark-eyed Juncos in winter. Wilson Bulletin 87:374-383.

Gove, G.W. and J.B. Gordon. 1990. Sightings of a Golden-crowned Sparrow at Dartmouth, Massachusetts. Bird Observer 18:163.

Goward, T. 1976. Addendum to the birds of Wells Gray Park – an annotated list. British Columbia Parks Branch Unpublished Report, Victoria. 8 pp. (Bibliography 3050).

Graham, D.S. 1988. Responses of five host species to cowbird parasitism. Condor 90:588-591.

Grant, J. 1958. Christmas bird census, 1957 – Vernon, British Columbia. Canadian Field-Naturalist 72:45-46.

________. 1966. Thirtieth breeding-bird census – northern subalpine forest. Audubon Field Notes 20:630-631.

Grass, A. 1971. Wells Gray Park naturalist's report, 1971. British Columbia Parks Branch Unpublished Report, Victoria. 45 pp. (Bibliography 1175).

Greene, E., V.R. Muether, and W. Davison. 1996. Lazuli Bunting (*Passerina amoena*). *In* A. Poole and F. Gill (editors). The birds of North America, No. 232. The Academy of Natural Sciences, Philadelphia, Pennsylvania; The American Ornithologists' Union, Washington, D.C. 24 pp.

Greenfield, T. 1997. Birds of the Sunshine Coast – a checklist. Sunshine Coast Natural History Society, Sechelt, British Columbia. Leaflet.

________. 1998. Birds observed along the Sikanni Chief River, northeastern British Columbia, 1992-1997. British Columbia Birds 8:3-18.

Greenlaw, J.S. 1993. Behavioral and morphological diversification in Sharp-tailed Sparrows (*Ammodramus caudacutus*) of the Atlantic Coast. Auk 110:286-303.

________. 1996. Spotted Towhee (*Pipilo maculatus*). *In* A. Poole and F. Gill (editors). The birds of North America, No. 263. The Academy of Natural Sciences, Philadelphia, Pennsylvania; The American Ornithologists' Union, Washington, D.C. 32 pp.

Greenlaw, J.S. and J.D. Rising. 1994. Sharp-tailed Sparrow (*Ammodramus caudacutus*). *In* A. Poole and F. Gill (editors). The birds of North America, No. 112. The Academy of Natural Sciences, Philadelphia, Pennsylvania; The American Ornithologists' Union, Washington, D.C. 28 pp.

Griffin, D.N. 1959. The poisoning of meadowlarks with insecticides. Wilson Bulletin 71:193.

Griffith, D.E. 1973. Notes on the birds of Summit Lake Pass, British Columbia. Vancouver Natural History Society Discovery 2:45-51.

Grinnell, J. and A.H. Miller. 1944. The distribution of the birds of California. Pacific Coast Avifauna 27. 608 pp.

Griscom, L. and A. Sprunt. 1957. The warblers of America. Devin-Adair, New York. 356 pp.

Groth, J.G. 1993a. Evolutionary differentiation in morphology, vocalizations, and allozymes among nomadic sibling species in the North American Red Crossbill (*Loxia curvirostra*) complex. California Publications in Zoology 127:1-143.

________. 1993b. Call matching and positive assortative mating in Red Crossbills. Auk 110:398-401.

Grumbine, R.E. 1997. Reflections on "What is Ecosystem Management?" Conservation Biology 11:41-47.

Grunberg, H. 1982a. The autumn migration – northwest Canada region. American Birds 36:197-198.

________. 1982b. The winter season – northwest Canada region. American Birds 36:311-312.

________. 1982c. The spring migration – northwest Canada region. American Birds 36:873-874.

________. 1982d. The nesting season – northwest Canada region. American Birds 36:996-997.

________. 1983. The spring migration – northwestern Canada region. American Birds 37:890-891.

________. 1984. The nesting season – northwest Canada region. American Birds 38:935-936.

________. 1985a. The autumn migration – northwest Canada region. American Birds 39:77-78.
________. 1985b. The winter season – northwest Canada region. American Birds 39:187-189.
________. 1985c. The spring migration – northwest Canada region. American Birds 39:326-327.
________. 1985d. The nesting season – northwestern Canada region. American Birds 39:937-938.
________. 1986. The autumn migration – northwestern Canada region. American Birds 40:141-142.
Guay, J.W. 1968. The breeding biology of the Franklin's Gull (*Larus pipixcan*). Ph.D. Thesis, University of Alberta, Edmonton. 129 pp.
Guernsey, V. 1980. A rare bird. Victoria Naturalist 37:30.
Guiguet, C.J. 1953. An ecological study of Goose Island, British Columbia, with special reference to mammals and birds. British Columbia Provincial Museum Occasional Paper No. 10, Victoria. 78 pp.
Gutierrez, R.J. 1994. Changes in the distribution and abundance of Spotted Owls during the past century. Studies in Avian Biology 15:293-300.
Guzy, M.J. and P.E. Lowther. 1997. Black-throated Gray Warbler (*Dendroica nigrescens*). *In* A. Poole and F. Gill (editors). The birds of North America, No. 319. The Academy of Natural Sciences, Philadelphia, Pennsylvania; The American Ornithologists' Union, Washington, D.C. 20 pp.
Guzy, M.J. and G. Ritchison. 1999. Common Yellowthroat (*Geothlypis trichas*). *In* A. Poole and F. Gill (editors). The birds of North America, No. 448. The Birds of North America, Inc., Philadelphia, Pennsylvania. 24 pp.
Gyug, L.W. 1997. Donna Creek forestry/biodiversity project: breeding bird point counts 1995 and 1996. Peace/Williston Fish and Wildlife Compensation Program Unpublished Report, Prince George, British Columbia. 46 pp.
Gyug, L.W. and S.N. Bennett. 1995. Bird use of wildlife tree patches 25 years after clearcutting. British Columbia Ministry of Environment, Lands and Parks, Penticton. 30 pp.
Hahn, T.P. 1996. Cassin's Finch (*Carpodacus cassinii*). *In* A. Poole and F. Gill (editors). The birds of North America, No. 240. The Academy of Natural Sciences, Philadelphia, Pennsylvania; The American Ornithologists' Union, Washington, D.C. 20 pp.
Haineback, K. 1992. First records of Xantus' Hummingbird in California. Western Birds 23:133-136.
Hall, G. 1979. Hybridization between Mourning and MacGillivray's warblers. Bird-Banding 50:101-107.
Hall, G.A. 1994. Magnolia Warbler (*Dendroica magnolia*). *In* A. Poole and F. Gill (editors). The birds of North America, No. 136. The Academy of Natural Sciences, Philadelphia, Pennsylvania; The American Ornithologists' Union, Washington, D.C. 16 pp.
________. 1996. Yellow-throated Warbler (*Dendroica dominica*). *In* A. Poole and F. Gill (editors). The birds of North America, No. 223. The Academy of Natural Sciences, Philadelphia, Pennsylvania; The American Ornithologists' Union, Washington, D.C. 16 pp.
Hamilton, T.R. and E.M. Novis. 1994. The migratory pattern of House Finches in eastern North America. North American Bird Bander 19:45-48.
Hamilton, W.J. 1962. Bobolink migratory pathways and their experimental analysis under night skies. Auk 79:208-233.
Hanka, L.R. 1985. Recent altitudinal range expansion by the Brown-headed Cowbird in Colorado. Western Birds 16:183-184.
Hanna, G.D. 1916. Records of birds new to the Pribilof Islands including two new to North America. Auk 33:400-403.
Hardin, G. 1968. The tragedy of the commons. Science 162:1243-1262.
Harding, L.E. 1994. Songbirds in decline. Pages 319-322 *in* L.E. Harding and E. McCullum (editors). Biodiversity in British Columbia: our changing environment. Environment Canada, Canadian Wildlife Service, Pacific and Yukon Region, Delta, British Columbia.
Harper, B., S. Cannings, D. Fraser, and W.T. Munro. 1994. Provincial lists of species at risk. Pages 16-23 *in* L.E. Harding and E. McCallum (editors). Biodiversity in British Columbia: Our changing environment. Canadian Wildlife Service, Delta. 426 pp.
Harrington-Tweit, B. and P.W. Mattocks. 1984. The nesting season – northern Pacific coast region. American Birds 38:1054-1056.
________. and ________. 1985. The nesting season – northern Pacific coast region. American Birds 39:953-956.
________. and ________. 1987. The nesting season – northern Pacific coast region. American Birds 41:318-321.
Harris, J.W.E., R.I. Alfaro, A.F. Dawson, and R.G. Brown. 1985. The western budworm in British Columbia 1909-1983. Pacific Forest Research Centre Information Report BC-X-257, Victoria. 32 pp.
Harrison, H.H. 1979. A field guide to western birds' nests. Houghton Mifflin, Boston, Massachusetts. 279 pp.
________. 1984. Wood warblers world. Simon and Schuster, New York. 334 pp.
Harvey, D.H. 1992. The distribution, density and habitat of Brewer's Sparrows, *Spizella breweri,* in the Okanagan valley of British Columbia. British Columbia Ministry of Environment, Lands and Parks, Victoria. 35 pp.
Hatler, D.F. 1973. Chestnut-collared Longspur in British Columbia. Canadian Field-Naturalist 87:66.
Hatler, D.F. and R.W. Campbell. 1975. Notes on spring migration, including sex segregation, of some western Savannah Sparrows. Syesis 8:401-402.
Hatler, D.F., R.W. Campbell, and A. Dorst. 1973. Birds of Pacific Rim National Park, British Columbia. Canadian Wildlife Service Unpublished Report, Edmonton, Alberta. 383 pp. (Bibliography 1147).
________, ________, and ________. 1978. Birds of Pacific Rim National Park, British Columbia. British Columbia Provincial Museum Paper No. 20, Victoria. 194 pp.
Hay, R.B. 1976. An environmental study on the Kitimat region with special reference to the Kitimat River estuary. Canadian Wildlife Service Unpublished Report, Delta, British Columbia. 85 pp. (Bibliography 1204).
Hebda, R.J. and J.C. Haggarty (editors). 1997. Brooks Peninsula: an ice age refugium on Vancouver Island. British Columbia Ministry of Environment, Lands and Parks Occasional Paper No. 5, Victoria.
Hejl, S.J. 1994. Human-induced changes in bird populations in coniferous forests in western North America during the past 100 years. Studies in Avian Biology 15:232-246.
Hejl, S.J. and R.E. Wood. 1991. Bird assemblages in old-growth and rotation-aged Douglas-fir/ponderosa pine stands in the northern Rocky Mountains: a preliminary assessment. Pages 93-100 *in* D.M. Baumgartner and J.E. Lotan (editors). Symposium proceedings of interior Douglas-fir: the species and its management. Washington State University, Pullman.
Helms, C.O., W.J. Bleier, D. Otis, and G.M. Linz. 1994. Population estimates of breeding blackbirds in North Dakota, 1967, 1981, 1982, and 1990. American Midland Naturalist 132:256-263.
Hendricks, P. 1987. Breeding biology and nestling development of Golden-crowned Sparrows in Alaska. Wilson Bulletin 99:693-696.
Hennan, E.G. 1975. Columbia River marshes; British Columbia waterfowl habitat assessment. Ducks Unlimited (Canada) Unpublished Special Report No. 7104, Kamloops, British Columbia. 48 pp. (Bibliography 3208).
Hennig, W. 1950. Grundzuge einer theorie der phylogenischen systematik. Berlin. Deutscher Zentralverlag.
________. 1965. Phylogenetic Systematics. Annual Review of Entomology 10:97-116.
Heuser, C.J. 1989. North Pacific coastal refugia. Pages 91-106 *in* G.G.E. Scudder and N. Gessler (editors). The outer shores. Queen Charlotte Islands Museum Press, Queen Charlotte City. 327 pp.
Heywood, V.H. and I. Baste (editors). 1993. Introduction in global biodiversity assessment. Cambridge University Press, Cambridge, United Kingdom. 1140 pp.
Hill, D.P. and L.K. Gould. 1997. Chestnut-collared Longspur (*Calcarius ornatus*). *In* A. Poole and F. Gill (editors). The birds of North America, No. 288. The Academy of Natural Sciences, Philadelphia, Pennsylvania; The American Ornithologists' Union, Washington, D.C. 20 pp.
Hill, G.E. 1988. Age, plumage brightness, territory quality, and reproductive success in the Black-headed Grosbeak. Condor 90:379-388.
________. 1992. Proximate basis of variation in carotenoid pigmentation in male House Finches. Auk 109:1-12.
________. 1993. House Finch (*Carpodacus mexicanus*). *In* A. Poole and F. Gill (editors). The birds of North America, No. 46. The Academy of Natural Sciences, Philadelphia, Pennsylvania; The American Ornithologists' Union, Washington, D.C. 24 pp.
________. 1995. Black-headed Grosbeak (*Pheucticus melanocephalus*). *In* A. Poole and F. Gill (editors). The birds of North America, No. 143. The Academy of Natural Sciences, Philadelphia, Pennsylvania; The American Ornithologists' Union, Washington, D.C. 20 pp.
Hill, D.P. and S.G. Sealy. 1994. Desertion of nests parasitized by cowbirds: have Clay-colored Sparrows evolved an anti-parasite defence? Animal Behavior 48:1063-1070.

Hobson, K.A. and M.-A. Villard. 1998. Forest fragmentation affects the behavioral response of American Redstarts to the threat of cowbird parasitism. Condor 100:389-394.
Hochachka, W. 1990. Seasonal declines in reproductive performance of Song Sparrows. Ecology 71:1279-1288.
Hochachka, W.M., J.V. Wells, K.V. Rosenberg, D.L. Tessaglia-Hymes, and A.A. Dhondt. 1999. Irruptive migration of Common Redpolls. Condor 101:195-204.
Hoffman, W. and H. Pletz. 1988. 1988 nest record for an Evening Grosbeak near Elk Island Park. Alberta Naturalist 18:101-103.
Hoffman, W., R. Sawicki, C. Thompson, and M. Carrington. 1991. Golden-crowned Sparrow appears in Florida. Florida Field Naturalist 19:19-21.
Hofslund, P.B. 1954. Incubation period of the Mourning Warbler. Wilson Bulletin 66:71.
________. 1957. Cowbird parasitism on the Northern Yellowthroat. Auk 74:42-48.
Holcomb, L.C. 1953. Pine Grosbeaks in Surrey, B.C. Murrelet 34:48.
________. 1966. The development of grasping and balancing coordination in nestlings of seven species of altricial birds. Wilson Bulletin 88:57-63.
Holdom, M.W. 1960. Brown-headed Cowbirds in southwestern British Columbia. Canadian Field-Naturalist 74:176.
Holimon, W.C., C.W. Benkman, and M.F. Willson. 1998. The importance of mature conifers to Red Crossbills in southeast Alaska. Forest Ecology and Management 102:167-172.
Holling, C.S. (editor). 1978. Adaptive environmental assessment and management. John Wiley and Sons, New York.
________. 1994. An ecologist's view of the Malthusian conflict. Pages 79-103 *in* K. Lindahl-Kiessling and H. Landberg (editors). Population, economic development, and the environment. Oxford University Press, New York.
Holling, C.S. and G.K. Meffe. 1996. Command and control and the pathology of natural resource management. Conservation Biology 10:328-337.
Holmes, R.T. 1994. Black-throated Blue Warbler (*Dendroica caerulescens*). *In* A. Poole and F. Gill (editors). The birds of North America, No. 87. The Academy of Natural Sciences, Philadelphia, Pennsylvania; The American Ornithologists' Union, Washington, D.C. 24 pp.
Holroyd, G.L. and K.J. Van Tighem. 1983. The ecological (biophysical) land classification of Banff and Jasper national parks – Volume 3: the wildlife inventory. Canadian Wildlife Service Report, Edmonton, Alberta. 444 pp.
Horn, H.S. 1970. Social behavior of nesting Brewer's Blackbirds. Condor 72:15-23.
Horvath, O. 1963. Contribution to nesting ecology of forest birds. M.S.F. Thesis, University of British Columbia, Vancouver. 181 pp.
Horwood, D. 1983. The eighty-third Audubon Christmas Bird Count – Kitimat, B.C. American Birds 37:440.
________. 1992. Birds of the Kitimat Valley. Kitimat Centennial Museum, British Columbia. 178 pp.
Houston, C.S. 1978. Arrival of the House Sparrow on the prairies. Blue Jay 36:99-102.
Howell, A.H. 1932. Florida bird life. Florida Department of Game and Fresh Water Fishes, Gainesville. 579 pp.
Howie, R.R. 1983. The eighty-third Audubon Christmas Bird Count – Shuswap Lake Provincial Park. American Birds 37:444.
________. 1994. Birds of Kamloops – a checklist. Issued by the author, Kamloops, British Columbia. Leaflet.
Hubbard, J.P. 1969. The relationships and evolution of the *Dendroica coronata* complex. Auk 86:393-432.
________. 1970. Geographic variation in the *Dendroica coronata* complex. Wilson Bulletin 82:335-369.
Hudon, J. 1999. Western Tanager (*Piranga ludoviciana*). *In* A. Poole and F. Gill (editors). The birds of North America, No. 432. The Birds of North America, Inc., Philadelphia, Pennsylvania. 28 pp.
Hughes, W.M. 1956. Observations of our less common birds. Vancouver Natural History Society News 105:4-6.
________. 1959. Audubon Warbler (*Dendroica auduboni*). News from the Bird Banders 34(4):45.
Humphreys, N. 1995. History of population fluctuations and infestations of important forest insects in the Prince George forest region 1924-1994. Canadian Forestry Service, Pacific Forest Research Centre FIDS Report 95-9, Victoria. 68 pp.
Hunn, E.S. 1978. Black-throated Sparrow vagrants in the Pacific Northwest. Western Birds 9:85-89.
Hunn, E.S. and P.W. Mattocks. 1977. The nesting season – northern Pacific coast region. American Birds 31:1178-1183.
________. and ________. 1981. The autumn migration – northern Pacific coast region. American Birds 35:216-219.
________. and ________. 1981a. The spring migration – northern Pacific coast region. American Birds 35:854-856.
________. and ________. 1983. The autumn migration – northern Pacific coast region. American Birds 37:214-218.
________. and ________. 1984. The autumn migration – northern Pacific coast region. American Birds 38:236-240.
________. and ________. 1985. The autumn migration – northern Pacific coast region. American Birds 39:92-96.
________. and ________. 1986. The autumn migration – northern Pacific coast region. American Birds 40:321-324.
Hunt, P.D. 1996. Habitat selection by American Redstarts along a successional gradient in northern hardwoods forest: evaluation of habitat quality. Auk 113:875-888.
Hunt, P.D. and B.C. Eliason. 1999. Blackpoll Warbler (*Dendroica striata*). *In* A. Poole and F. Gill (editors). The birds of North America, No. 431. The Birds of North America, Inc., Philadelphia, Pennsylvania. 24 pp.
Hunt, P.D. and D.J. Flaspohler. 1998. Yellow-rumped Warbler (*Dendroica coronata*). *In* A. Poole and F. Gill (editors). The birds of North America, No. 376. The Birds of North America, Inc., Philadelphia, Pennsylvania. 28 pp.
Hunter, M.L. and A. Hutchinson. 1994. The virtues and shortcomings of parochialism: conserving species that are locally rare, but globally common. Conservation Biology 8:1163-1165.
Igl, L.D. and D.H. Johnson. 1995. Dramatic increase in Le Conte's Sparrow in conservation reserve program fields in the northern Great Plains. Prairie Naturalist 27:89-94.
Ingold, J.L. 1993. Blue Grosbeak (*Guiraca caerulea*). *In* A. Poole and F. Gill (editors). The birds of North America, No. 79. The Academy of Natural Sciences, Philadelphia, Pennsylvania; The American Ornithologists' Union, Washington, D.C. 16 pp.
Innes, D. 1997. A swan for British Columbia. B.C. Naturalist 35(3):5-6.
Innes, D.W. and M. Innes. 1995. Plumage and songs of a putative hybrid between Hermit and Townsend's Warblers found in the Alberni Valley, Vancouver Island. British Columbia Birds 5:12-14.
International Bird Census Committee. 1970. Recommendation for an international standard for a mapping method in bird census work. Audubon Field Notes 24:727-736.
Irving, E.B. 1953. Birds at Carmanah Point. Victoria Naturalist 10:19-22; 28-31.
Irving, L. 1961. The migration of Lapland Longspurs to Alaska. Auk 78:327-342.
IUCN Species Survival Service Commission. 1994. IUCN Red List Categories. <http://www.iucn.org/themes/ss/redlist/criteria.htm>
James, P.C. and A.R. Smith. 1987. Food habits of urban-nesting Merlins, *Falco columbarius,* in Edmonton and Fort Saskatchewan. Canadian Field-Naturalist 101:592-594.
Janes, S.W. 1983. Status, distribution, and habitat selection of the Grasshopper Sparrow in Morrow County, Oregon. Murrelet 64:51-54.
Jeffrey, M. and L. Darling. 1997. Wildlife diversity of unharvested aspen stands in the Dawson Creek forest district. British Columbia Ministry of Environment, Lands and Parks, Wildlife Branch Unpublished Report, Victoria. 27 pp.
Jehl, J.R. 1968. Geographic and seasonal variation in Smith's Longspur, *Calcarius pictus*. Transactions of the San Diego Society of Natural History 15:1-5.
Jewett, S.A., W.A. Taylor, W.T. Shaw, and J.W. Aldrich. 1953. Birds of Washington state. University of Washington Press, Seattle. 767 pp.
Jobin, L. 1952. Some uncommon birds collected in the Cariboo parkland biotic area of British Columbia. Murrelet 33:9.
________. 1953. Notes on some unusual birds in the Cariboo district of British Columbia. Murrelet 34:31.
________. 1954. Additional bird records for the Cariboo parklands, British Columbia. Condor 56:223.
________. 1955. Interesting records of birds collected in the Peace River parklands, British Columbia. Canadian Field-Naturalist 69:65-66.
________. 1955a. First record of the Dusky Song Sparrow, *Melospiza melodia juddi* Bishop, for British Columbia. Canadian Field-Naturalist 69:66.
Johns, B.W. 1993. The influence of grove size on bird species richness in aspen parklands. Wilson Bulletin 105:256-264.
Johnson, E.J. 1983. Nesting biology of the Rosy Finch on the Aleutian Islands, Alaska. Condor 85:447-452.

Johnson, G. and R. Johnson. 1987. Golden-crowned Sparrow in Chippewa County. Loon 59:101-102.

Johnson, N.K. 1980. Character variation and evolution of sibling species in the *Empidonax difficilis – flavescens* complex (Aves: Tyrannidae). University of California Press, Berkeley. 151 pp.

Johnson, N.K. and J.A. Marten. 1988. Evolutionary genetics of flycatchers II. Differentiation in the *Empidonax difficilis* complex. Auk 105:177-191.

Johnson, N.K., R.M. Zink, and J.A. Marten. 1988. Genetic evidence for relationships in the avian family Vireonidae. Condor 90:428-445.

Johnson, R.G. and S.A. Temple. 1990. Nest predation and brood parasitism of tallgrass prairie birds. Journal of Wildlife Management 54:106-111.

Johnston, R.F. 1954. Variation in breeding season and clutch size in Song Sparrows of the Pacific coast. Condor 56:268-273.

Johnston, V.R. 1947. Breeding birds of the forest edge in Illinois. Condor 49:45-53.

Johnston, W.G. and C. McEwen. 1986. The winter season – northwestern Canada region. American Birds 40:303-304.

________. and ________. 1987a. The winter season – northwestern Canada region. American Birds 41:303-305.

________. and ________. 1987b. The nesting season – northwestern Canada region. American Birds 41:1462-1464.

Johnstone, W.B. 1949. An annotated list of the birds of the East Kootenay, British Columbia. British Columbia Provincial Museum Occasional Paper No. 7, Victoria. 87 pp.

Kantrud, H.A. and R.L. Kologiski. 1982. Effects of soils and grazing on breeding birds of uncultivated upland grasslands of the northern Great Plains. United States Department of the Interior, Fish and Wildlife Service Wildlife Research Report No. 15. Washington, D.C. 33 pp.

Karr, J.R. 1993. Protecting ecological integrity: an urgent societal goal. Yale Journal of International Law 18:297-306.

Kaspari, M. 1991. Central place foraging in Grasshopper Sparrows: opportunism or optimal foraging in a variable environment? Oikos 60:307-312.

Kaufman, K. 1990. A field guide to advanced birding. Houghton Mifflin, Boston, Massachusetts. 299 pp.

Kautesk, B. 1982. Baird's Sparrow at Vancouver: first records for British Columbia. Discovery 11:61-63.

________. 1983. Mourning Warbler near Harrison Mills: first record for southwestern B.C. Discovery 12:107.

Keith, S. 1967. New bird records from Alaska and the Alaska Highway. Canadian Field-Naturalist 81:196-200.

Keen, J.H. 1910. Bird migration in northern British Columbia. Ottawa Naturalist 24:116-117.

Kelleher, K.E. 1963. A study of the hole-nesting avifauna of southwestern British Columbia. M.Sc. Thesis, University of British Columbia, Vancouver. 168 pp.

Keller, M.E. and S.H. Anderson. 1992. Avian use of habitat configurations created by forest cutting in southeastern Wyoming. Condor 94:55-65.

Kelly, J.W. 1968. Golden-crowned Sparrow. Pages 1352-1364 *in* O.L. Austin (editor). Life histories of North American cardinals, grosbeaks, buntings, towhees, finches, sparrows, and allies (Part 3). United States National Museum Bulletin No. 237, Washington, D.C.

Kelso, J.E.H. 1926. Birds of the Arrow Lakes, West Kootenay, British Columbia. Ibis 2:689-723.

Kemsies, E. 1961. Subspeciation in the Smith's Longspur, *Calcarius pictus*. Canadian Field-Naturalist 75:143-149.

Kendra, P.E., R.R. Roland, and D.W. Tallamy. 1988. Conspecific brood parasitism in the House Sparrow. Wilson Bulletin 100:80-90.

Kermode, F. 1904. Catalogue of British Columbia birds. British Columbia Provincial Museum, Victoria. 92 pp.

Kermode, F. and E.M. Anderson. 1914. Report on birds collected and observed during September, 1913, on Atlin Lake, from Atlin to south end of the lake. Pages 19-20 *in* Report of the Provincial Museum of Natural History for the year 1913, Victoria, British Columbia.

Kern, M.D. 1984. Racial differences in nests of White-crowned Sparrows. Condor 86:455-466.

Kessel, B. 1989. Birds of the Seward Peninsula, Alaska. University of Alaska Press, Fairbanks. 330 pp.

________. 1998. Habitat characteristics of some passerine birds in western North American taiga. University of Alaska Press, Fairbanks. 117 pp.

Kessel, B. and D.D. Gibson. 1978. Status and distribution of Alaska birds. Studies in Avian Biology No. 1, Allen Press, Lawrence, Kansas. 100 pp.

________. 1994. A century of avifaunal change in Alaska. Studies in Avian Biology 15:4-13.

Ketcheson, M.V., T.F. Braumandl, D. Meidinger, G. Utzig, D.A. Demarchi, and B.M. Wikeem. 1991. Interior cedar-hemlock zone. Pages 161-181 *in* D. Meidinger and J. Pojar (editors). Ecosystems of British Columbia. British Columbia Ministry of Forests Research Branch, Victoria. 330 pp.

Ketterson, E.D. 1976. Geographic variation and its climatic correlates in the sex ratio of eastern-wintering Dark-eyed Juncos (*Junco hyemalis hyemalis*). Ecology 57:679-693.

Keynes, J.M. 1933. "National self-sufficiency." *In* D. Moggeridge (editor). The collected writings of John Maynard Keynes, Volume 21. London: Macmillan Cambridge University Press for the Royal Economic Society.

King, J.R. and J.D. Hubbard. 1981. Comparative patterns of nestling growth in White-crowned Sparrows. Condor 83:362-369.

King, J.R. and E.E. Wales. 1965. Photoperiodic regulation of testicular metamorphosis and fat deposition in three taxa of Rosy Finches. Physiology and Zoology 38:49-68.

King, J.R., D.S. Farner, and L.R. Mewaldt. 1965. Seasonal sex and age ratios in populations of the White-crowned Sparrows of the race *gambelii*. Condor 67:489-504.

Kingery, H.E. and V.C. Kingery. 1995. Gray Jay as predator on Cassin's Finch nestlings. Colorado Field Ornithologists Journal 29:17.

Kirk, D.A., A.W. Diamond, A.R. Smith, G.E. Holland, and P. Chytyk. 1997. Population changes in boreal forest birds in Saskatchewan and Manitoba. Wilson Bulletin 109:1-27.

Klem, D. 1990. Collisions between birds and windows: mortality and prevention. Journal of Field Ornithology 61:120-128.

Klemens, M. and C. Howes (editors). 1999. Draft guidelines for the application of IUCN Red List criteria at the national and regional levels. Pages 58-70 *in* Newsletter of the Species Survival Commission IUCN – The World Conservation Union. No. 31-32. 118 pp.

Klicka, J., R.M. Zink, J.C. Barlow, W.B. McGillivray, and T.J. Doyle. 1999. Evidence supporting the recent origin and species status of the Timberline Sparrow. Condor 101:577-588.

Knapton, R.W. 1973. Some ecological effects of social behaviour in the Song Sparrow, *Melospiza melodia*. B.Sc. Thesis, University of British Columbia, Vancouver. 133 pp.

________. 1979. Breeding ecology of the Clay-colored Sparrow. Living Bird 17:137-158.

________. 1984. Parental feeding of nestling Nashville Warblers: the effects of food type, brood-size, nestling age, and time of day. Wilson Bulletin 96:594-602.

________. 1994. Clay-colored Sparrow (*Spizella pallida*). *In* A. Poole and F. Gill (editors). The birds of North America, No. 120. The Academy of Natural Sciences, Philadelphia, Pennsylvania; The American Ornithologists' Union, Washington, D.C. 16 pp.

Knight, O.W. 1905. Notes on the warblers found in Maine. Journal of the Maine Ornithological Society 7:71-76.

________. 1908. Birds of Maine. Charles H. Glass, Bangor, Maine.

Knight, R.L. 1996. Aldo Leopold, the land ethic, and ecosystem management. Journal of Wildlife Management 60:471-474.

Knox, A.G. 1988. The taxonomy of redpolls. Ardea 76:1-26.

Kornet, D.J. 1993. Permanent splits as speciation events: formal reconstruction of the internodal species concept. Journal of Theoretical Biology 164:407-435.

Kowalski, M.P. 1983. Identifying Mourning and MacGillivray's warblers: geographic variation in the MacGillivray's Warbler as a source of error. North American Bird Bander 8:56-57.

Kozlovic, D.R., R.W. Knapton, and J.C. Barlow. 1996. Unsuitability of the House Finch as a host of the Brown-headed Cowbird. Condor 98:253-258.

Krebs, C.J. 1989. Ecological methodology. Harper and Rowe, New York.

________. 1994. Ecology: the experimental analysis of distribution and abundance. HarperCollins College Publishers, New York. 801 pp.

Kricher, J.C. 1995. Black-and-white Warbler (*Mniotilta varia*). *In* A. Poole and F. Gill (editors). The birds of North America, No. 158. The Academy of Natural Sciences, Philadelphia, Pennsylvania; The American Ornithologists' Union, Washington, D.C. 20 pp.

Kroodsma, R.L. 1975. Hybridization in buntings (*Passerina*) in North Dakota and eastern Montana. Auk 92:66-80.

________. 1984. Effect of forest edge on breeding forest bird species. Wilson Bulletin 96:426-436.

Krueper, D.J. 1993. Effects of land use practices on western riparian ecosystems. *In* D.M. Finch and P.W. Stangel (editors). Status and management of neotropical migratory birds; 1992 September 21-25; Estes

Park, CO. United States Department of Agriculture, Forest Service, Rocky Mountain Forest and Range Experiment Station, General Technical Report RM-229, Fort Collins, Colorado. 422 pp.
LaFave, L.D. 1965. The Le Conte's Sparrow, a new bird for the state of Washington. Murrelet 46:26.
Laing, H.M. 1932. An unusual record of Scarlet Tanager. Canadian Field-Naturalist 46:192.
———. 1942. Birds of the coast of central British Columbia. Condor 44:175-181.
Lancaster, R.K. 1976. Bird communities in relation to the structure of urban habitats. M.Sc. Thesis, University of British Columbia, Vancouver. 119 pp.
Lancaster, R.K. and W.E. Rees. 1979. Bird communities and the structure of urban habitats. Canadian Journal of Zoology 57:2358-2368.
Lance, A.N. and M. Phinney. 1994. Bird diversity and abundance following aspen clearcutting in the Boreal White and Black Spruce Biogeoclimatic Zone. British Columbia Ministry of Forests Report, Victoria. 40 pp.
———. and ———. 1996. Bird species composition and abundance in mixedwood landscapes in the Boreal White and Black Spruce Biogeoclimatic Zone. Industrial Forestry Service Ltd. Report to Canadian Forest Service, Prince George, British Columbia.
Langelier, L.A. 1983. Habitat selection of common breeding bird species in western spruce budworm outbreak areas. M.Sc. Thesis, University of Idaho, Moscow. 119 pp.
Lanyon, W.E. 1994. Western Meadowlark (*Sturnella neglecta*). *In* A. Poole and F. Gill (editors). The birds of North America, No. 104. The Academy of Natural Sciences, Philadelphia, Pennsylvania; The American Ornithologists' Union, Washington, D.C. 20 pp.
LaPointe, G. and J. Bedard. 1986. Savannah Sparrow, *Passerculus sandwichensis,* reproductive success. Canadian Field-Naturalist 100:264-267.
Larison, B., S.A. Laymon, P.L. Williams, and T.B. Smith. 1998. Song Sparrows vs. cowbird brood parasites: impacts of forest structure and nest-site selection. Condor 100:93-101.
Larrison, E.J. 1981. Birds of the Pacific Northwest. University of Idaho Press, Moscow. 337 pp.
Laughlin, S.B. and D.P. Kibbe. 1985. The atlas of breeding birds of Vermont. University Press of New England, Hanover, New Hampshire.
Leach, B.A. 1982. Waterfowl on a Pacific estuary: a natural history of man and waterfowl on the lower Fraser River. British Columbia Provincial Museum Special Publication No. 5, Victoria. 211 pp.
LeBaron, G.S. 1994. The ninety-fourth Christmas Bird Count. National Audubon Society Field Notes 48:435.
Lederer, R.J., W.S. Mazen, and P.J. Metropulos. 1975. Population fluctuation in a Yellow-headed Blackbird marsh. Western Birds 6:1-6.
Lee, C.-T. and A. Birch. 1998. Field identification of female and immature Bullock's and Baltimore Orioles. Birding 30:282-283.
Lee, D.S. and J.C. Haney. 1996. Manx Shearwater (*Puffinus puffinus*). *In* A. Poole and F. Gill (editors). The birds of North America, No. 257. The Academy of Natural Sciences, Philadelphia, Pennsylvania; The American Ornithologists' Union, Washington, D.C. 28 pp.
Lee, K.N. 1999. Appraising adaptive management. Conservation Ecology 3:3. <http://www.consecol.org/vol3/iss2/art3>
Leffingwell, D.J. and A.M. Leffingwell. 1931. Winter habits of the Hepburn Rosy Finch at Clarkston, Washington. Condor 33:140-147.
Lemon, E.K. 1956. Sight record of the Lark Sparrow. Victoria Naturalist 13:19.
———. 1969. Bewick Wren host to Brown-headed Cowbird. Canadian Field-Naturalist 83:395-396.
Leonard, M.L. and J. Picman. 1986. Why are nesting Marsh Wrens and Yellow-headed Blackbirds spatially segregated? Auk 103:135-140.
Leopold, B.D. 2000. Dollars versus sense. Wildlife Society Bulletin 28:1.
Lessica, P. and F.W. Allendorf. 1995. When are peripheral populations valuable for conservation? Conservation Biology 9:753-760.
Levings, C. and R.M. Thom. 1994. Habitat changes in Georgia Basin: implications for resource management and restoration. *In* R.H.C. Wilson, R.J. Beamish, F. Aitkens, and J. Bell (editors). Review of the marine environment and biota of Strait of Georgia, Puget Sound and Juan de Fuca Strait. Canadian Technical Report of Fisheries and Aquatic Sciences 1948.
Lewis, M.G. and F.A. Sharpe. 1987. Birding in the San Juan Islands. Mountaineers Books, Seattle, Washington. 219 pp.
Lightbody, J.P. and P.J. Weatherhead. 1987. Polygyny in the Yellow-headed Blackbird: female choice versus male competition. Animal Behavior 35:1670-1684.
Linsdale, J.M. and E.L. Sumner. 1934. Winter weights of Golden-crowned and Fox sparrows. Condor 36:107-112.
Littlefield, C. 1990. Birds of Malheur National Wildlife Refuge. Oregon State University Press, Corvallis.
Loegering, J.P. 1997. Wildlife mortality and entanglement by discarded hip chain string. Wilson Bulletin 109:353-355.
Loh, J., J. Randers, A. MacGillivray, V. Kapos, M. Jenkins, B. Groombridge, N. Cox, and B. Warren. 1999. Living planet report. World Wide Fund for Nature, Gland, Switzerland.
Looff, H.B. 1939. Returns of Golden-crowned and Kodiak Fox sparrows to breeding grounds on Kodiak Island, Alaska. Bird-Banding 10:85-88.
Lorenzana, J.C. and S.G. Sealy. 1998. Adult brood parasites feeding nestlings and fledglings of their own species: a review. Journal of Field Ornithology 69:364-375.
Lowther, J.K. 1961. Polymorphism in the White-throated Sparrow, *Zonotrichia albicollis* (Gmelin). Canadian Journal of Zoology 39:281-292.
Lowther, J.K. and J.B. Falls. 1968. White-throated Sparrow. Pages 1364-1392 *in* O.L. Austin (editor). Life histories of North American cardinals, grosbeaks, buntings, towhees, finches, sparrows, and allies (Part 3). United States National Museum Bulletin No. 237, Washington, D.C.
Lowther, P.E. 1993. Brown-headed Cowbird (*Molothrus ater*). *In* A. Poole and F. Gill (editors). The birds of North America, No. 47. The Academy of Natural Sciences, Philadelphia, Pennsylvania; The American Ornithologists' Union, Washington, D.C. 24 pp.
———. 1996. Le Conte's Sparrow (*Ammodramus leconteii*). *In* A. Poole and F. Gill (editors). The birds of North America, No. 224. The Academy of Natural Sciences, Philadelphia, Pennsylvania; The American Ornithologists' Union, Washington, D.C. 16 pp.
Lowther, P.E. and C.L. Cink. 1992. House Sparrow (*Passer domesticus*). *In* A. Poole, P. Stettenheim, and F. Gill (editors). The birds of North America, No. 12. The Academy of Natural Sciences, Philadelphia, Pennsylvania; The American Ornithologists' Union, Washington, D.C. 20 pp.
Lowther, P.E., C. Celada, N.K. Klein, C.C. Rimer, and D.A. Spector. 1999. Yellow Warbler (*Dendroica petechia*). *In* A. Poole and F. Gill (editors). The birds of North America, No. 454. The Birds of North America, Inc., Philadelphia, Pennsylvania. 32 pp.
Lozano, G.A., S. Perreault, and R.E. Lemon. 1996. Age, arrival date and reproductive success of male American Redstarts *Setophaga ruticilla*. Journal of Avian Biology 27:164-170.
Ludwig, D. 1993. Environmental sustainability: magic, science, and religion in natural resource management. Ecological Applications 3:555-558.
Ludwig, D., R. Hilborn, and C. Walters. 1993. Uncertainty, resource exploitation, and conservation: lessons from history. Science 260:17-18.
Lyon, B. and R. Montgomerie. 1995. Snow Bunting and McKay's Bunting (*Plectrophenax nivalis* and *Plectrophenax hyperboreus*). *In* A. Poole and F. Gill (editors). The birds of North America, Nos. 198-199. The Academy of Natural Sciences, Philadelphia, Pennsylvania; The American Ornithologists' Union, Washington, D.C. 28 pp.
Lyon, B.E., L.D. Hamilton, and M. McGrath. 1992. The frequency of conspecific brood-parasitism and the pattern of laying determinancy in Yellow-headed Blackbirds. Condor 94:590-597.
MacArthur, R.H. 1958. Population ecology of some warblers of northeastern coniferous forests. Ecology 39:599-619.
MacFarlane, R. 1908. Notes on the mammals and birds of northern Canada. *In* C. Mair (editor). Through the Mackenzie Basin. A narrative of the Athabasca and Peace River Treaty expedition of 1899. William Briggs, Toronto, Ontario.
MacFarlane, R. and C. Mair. 1908. List of birds and eggs observed and collected in North-West Territories of Canada, between 1888 and 1894. Pages 284-470 *in* C. Mair and R. MacFarlane (editors). Through the MacKenzie Basin. Willliam Briggs, Toronto, Ontario. 494 pp.
Macoun, J. and J.M. Macoun. 1909. Catalogue of Canadian birds. Department of Mines, Geological Survey Branch, Ottawa, Ontario. 761 pp.
Magurran, A.E. 1988. Ecological diversity and its measurement. Princeton University Press, Princeton, New Jersey. 179 pp.
Mair, C. and R. MacFarlane. 1908. Through the Mackenzie Basin – a narrative of the Athabasca and Peace River treaty expedition of 1899. William Briggs, Toronto, Ontario. 494 pp.
Mallett, J. 1995. A species definition for modern synthesis. Trends in Ecology and Evolution 10:294-299.
Mangel, M., L.M. Talbot, G.K. Meffe, M.T. Agardy, D.L. Alverson, J. Barlow, D.B. Botkin, G. Budowski, T. Clark, J. Cooke, R.H. Crozier,

P.K. Dayton, D.L. Elder, C.W. Fowler, S. Funtowicz, J. Giske, R.J. Hofman, S.J. Holt, S.R. Kellert, L.A. Kimball, D. Ludwig, K.N. Magnusson, B.S. Malayang III, C. Mann, E.A. Norse, S.P. Northridge, W.F. Perrin, C. Perrings, R.M. Peterman, G.B. Rabb, H.A. Regier, J.E. Reynolds III, K. Shernian, M.P. Sissenwine, T.D. Smith, A. Starfield, R.J. Taylor, M.F. Tillman, K. Toft, J.R. Twiss, Jr., J. Wilen, and T.P. Young. 1996. Principles for the conservation of wild living resources. Ecological Applications 6:338-362.

Mannan, R.W., B.S. White, and M.L. Morrison. 1983. Observations of nesting Townsend's Warblers in northern Oregon. Murrelet 64:23-25.

Manuwal, D.A. and M.H. Huff. 1987. Spring and winter bird populations in a Douglas-fir sere. Journal of Wildlife Management 51:586-595.

Marchak, M.P., S.L. Aycock, and D.M. Herbert. 1999. Falldown: forest policy in British Columbia. The David Suzuki Foundation and Ecotrust Canada, Vancouver, British Columbia.

Marten, J.A. and N.K. Johnson. 1986. Genetic relationships of North American cardueline finches. Condor 88:409-420.

Martin, J.W. and B.A. Carlson. 1998. Sage Sparrow (*Amphispiza belli*). *In* A. Poole and F. Gill (editors). The birds of North America, No. 326. The Birds of North America, Inc., Philadelphia, Pennsylvania. 20 pp.

Martin, K., S. Hannon, and R. Moses. 1981. Nesting of Smith's Longspurs in British Columbia. Canadian Field-Naturalist 95:469-470.

Martin, S.G. 1971. Polygyny in the Bobolink: habitat quality and adaptive complex. Ph.D. Thesis, Oregon State University, Corvallis. 163 pp.

Martin, S.G. and T.A. Gavin. 1995. Bobolink (*Dolichonyx oryzivorus*). *In* A. Poole and F. Gill (editors). The birds of North America, No. 176. The Academy of Natural Sciences, Philadelphia, Pennsylvania; The American Ornithologists' Union, Washington, D.C. 24 pp.

Marven, D. 1992. Black-throated Sparrow: new to Vancouver Island, B.C. Birders Journal 1:290-291.

Marvil, R.E. and A. Cruz. 1989. Impact of Brown-headed Cowbird parasitism on the reproductive success of the Solitary Vireo. Auk 106:476-480.

Marzluff, J.M., R.B. Boone, and G.W. Cox. 1994. Historical changes in populations and perceptions of native pest birds in the west. Pages 202-220 *in* J.R. Jehl and N.K. Johnson (editors). A century of avifaunal change in western North America. Studies in Avian Biology No. 15, Lawrence, Kansas.

Maser, C. 1997. Sustainable community development: principles and concepts. St. Lucie Press, Delray Beach, Florida.

Mason, R. 1951. Winter observation of the Lazuli Bunting near Hope, British Columbia. Murrelet 32:27.

Matsuoka, S.M., C.M. Handel, and D.D. Roby. 1997a. Nesting ecology of Townsend's Warblers in relation to habitat characteristics in a mature boreal forest. Condor 99:271-281.

Matsuoka, S.M., C.M. Handel, D.D. Roby, and D.L. Thomas. 1997b. The relative importance of nesting and foraging sites in selection of breeding territories by Townsend's Warblers. Auk 114:657-667.

Mattocks, P.W. 1984. The winter season – northern Pacific coast region. American Birds 38:349-351.

________. 1985a. The winter season – northern Pacific coast region. American Birds 39:201-204.

________. 1985b. The spring season – northern Pacific coast region. American Birds 39:340-344.

________. 1986a. The spring season – northern Pacific coast region. American Birds 40:514-518.

________. 1986b. The nesting season – northern Pacific coast region. American Birds 40:1244-1248.

________. 1988. The autumn season – northern Pacific coast region. American Birds 42:121-126.

________. 1988a. The spring season – northern Pacific coast region. American Birds 42:1331-1335.

________. 1989a. The summer season – northern Pacific coast region. American Birds 43:154-156.

________. 1989b. The autumn migration – northern Pacific coast region. American Birds 43:157-161.

Mattocks, P.W. and B. Harrington-Tweit. 1987a. The fall migration – northern Pacific coast region. American Birds 41:132-136.

________. and ________. 1987b. The spring migration – northern Pacific coast region. American Birds 41:478-482.

Mattocks, P.W. and E.S. Hunn. 1978a. The autumn migration – northern Pacific coast region. American Birds 32:245-250.

________. and ________. 1978b. The spring migration – northern Pacific coast region. American Birds 32:1045-1049.

________. and ________. 1982a. The winter season – northern Pacific coast region. American Birds 36:323-325.

________. and ________. 1982b. The spring migration – northern Pacific coast region. American Birds 36:886-888.

________. and ________. 1983a. The winter season – northern Pacific coast region. American Birds 37:329-332.

________. and ________. 1983b. The spring migration – northern Pacific coast region. American Birds 37:903-906.

Mattocks, P.W., E.S. Hunn, and T.R. Wahl. 1976. A checklist of the birds of Washington state, with recent changes annotated. Western Birds 7:1-24.

Mattocks, P.W., B. Harrington-Tweit, and E.S. Hunn. 1983. The summer season – northern Pacific coast region. American Birds 37:1019-1022.

May, R.M. 1990. Taxonomy as destiny. Nature 347:129-130.

Mayfield, H. 1977. Brown-headed Cowbird: agent of extermination? American Birds 31:107-113.

Mayr, E. 1942. Systematics and the origin of species from the viewpoint of a zoologist. Reprinted by Dover Publications, New York. 334 pp.

________. 1967. Population size and evolutionary parameters. Pages 47-58 *in* P.S. Moorehead and M.M. Kaplon (editors). Mathematical challenges to the Neo-Darwinian interpretation of evolution. Wistar Institute Symposium Monograph No. 5.

________. 1969. Principles of systematic zoology. McGraw-Hill, New York. 428 pp.

________. 1982. The growth of biological thought: diversity, evolution and inheritance. Harvard University Press, Cambridge, Massachusetts.

Mayr, E. and P.D. Ashlock. 1991. Principles of systematic zoology, 2nd edition. McGraw-Hill, New York. 473 pp.

McCabe, R.E. 2000. Views on endangered species. Outdoor News 54:2. Wildlife Management Inst. Washington, D.C.

McCabe, T.T. and E.B. McCabe. 1927. Analysis of sexes in a junco migration. Condor 29:272-273.

________. and ________. 1928. The plumage of the Pine Siskin. Condor 30:221-227.

________. and ________. 1929. British Columbia records of certain unusual sparrows. Condor 31:35.

________. and ________. 1933. Notes on the anatomy and breeding habits of crossbills. Condor 35:136-147.

McCabe, T.T. and I. McTaggart-Cowan. 1945. *Peromyscus maniculatus macrorhinus* and the problem of insularity. Transactions of the Royal Canadian Institute (1945):117-215.

McCaskie, R.G. 1970. The Blackpoll Warbler in California. California Birds 3:95-104.

________. 1975. Le Conte's Sparrow in California and the western United States. Western Birds 6:65-66.

McCaskie, R.G. and P. DeBenedictis. 1966. Birds of northern California. Golden Gate Audubon Society, Berkeley, California.

McEwen, C.A. and W.G. Johnston. 1986. The spring migration – northwestern Canada region. American Birds 40:497-498.

________. and ________. 1987a. The autumn migration – northwestern Canada region. American Birds 41:116-118.

________. and ________. 1987b. The spring migration – northwest Canada region. American Birds 41:461-463.

McFarland, K.P. and C.C. Rimmer. 1996. Horsehair fungus, *Marasmius androsaceus*, used as nest lining by birds of the subalpine spruce-fir community in the northeastern United States. Canadian Field-Naturalist 110:541-543.

McGillivray, W.B. 1980. Communal nesting in the House Sparrow. Journal of Field Ornithology 51: 371-372.

McGuyire, A.D. 1986. Some aspects of the breeding biology of Red-winged Blackbirds in Alaska. Wilson Bulletin 98:257-266.

McHarg, I. 1969. Design with nature. Natural History Press, New York.

McLure, H.E. 1989. Epizootic lessons of House Finches in Ventura County, California. Journal of Field Ornithology 60:421-430.

McMaster, D.G. and S.G. Sealy. 1998. Short incubation periods of Brown-headed Cowbirds: how do cowbird eggs hatch before Yellow Warbler eggs? Condor 100:101-102.

McNair, D.B. 1987. Egg data slips – are they useful for information on egg-laying dates and clutch size? Condor 89:369-376.

McNicholl, M.K. 1978. Sight records of Veery and Swamp Sparrow on Vancouver Island, British Columbia. Murrelet 59:102-104.

________. 1980. Songs of MacGillivray's and Townsend's warblers in coastal British Columbia. Western Birds 11:157-159.

________. 1995. Bird censuses in the Nimpkish Valley, British Columbia, 1995. BioFor Resource Consulting Ltd. Unpublished Report for Canadian Forest Products Ltd., Woss, British Columbia. 105 pp.

Meanley, B. and W.C. Royal. 1876. Nationwide estimates of blackbirds and starlings. Pages 39-40 *in* Proceedings of the Seventh Bird Control Seminar, Bowling Green State University, Bowling Green, Ohio.

Meffe, G.K. 1992. Techno-arrogance and halfway technologies: salmon hatcheries on the Pacific coast of North America. Conservation Biology 6:350-354.

Meffe, G.K. and C.R. Carroll. 1997. The species in conservation. Pages 57-86 *in* G.K. Meffe and C.R. Carroll (editors). Principles of conservation biology. Sinauer Associates, Sunderland, Massachusetts. 729 pp.

Meidinger, D. and J. Pojar. 1991. Ecosystems of British Columbia. British Columbia Ministry of Forests, Victoria. 330 pp.

Mendall, H.L. 1937. Nesting of the Bay-breasted Warbler. Auk 54:429-439.

Merilees, W.J. 1982. Chestnut-sided Warbler on Vancouver Island. Thrush 2:7. (Bibliography 3586).

________. 1994. The House Finch "colony" at Price Waterhouse Centre. Discovery 23:112-113.

________. 1995. A bank-nesting Common Redpoll. Discovery 24:68-69.

Merkins, M. and B. Booth. 1996. Monitoring changes in wildlife diversity during operational hardwood harvesting: aspen clear-cutting in the Dawson Creek forest district. British Columbia Ministry of Environment, Lands and Parks Unpublished Progress Report, Victoria. 20 pp.

________. and ________. 1998. Monitoring changes in wildlife diversity during operational hardwood harvesting: aspen clearcutting in the Dawson Creek Forest District (1992-1997). British Columbia Ministry of Environment, Lands and Parks Report, Victoria. 78 pp.

Meugens, A.L. 1956. Western extension of range of cowbird. Victoria Naturalist 12:103.

Mewaldt, L.R. and J.R. King. 1985. Breeding site faithfulness, reproductive biology, and adult survivorship in an isolated population of Cassin's Finches. Condor 87:494-510.

Middleton, A.L. 1991. Failure of Brown-headed Cowbird parasitism in nests of the American Goldfinch. Journal of Field Ornithology 62:200-203.

________. 1993. American Goldfinch (*Carduelis tristis*). *In* A. Poole and F. Gill (editors). The birds of North America, No. 80. The Academy of Natural Sciences, Philadelphia, Pennsylvania; The American Ornithologists' Union, Washington, D.C. 24 pp.

________. 1998. Chipping Sparrow (*Spizella passerina*). *In* A. Poole and F. Gill (editors). The birds of North America, No. 334. The Birds of North America, Inc., Philadelphia, Pennsylvania. 32 pp.

Miller, R.S. 1968. Conditions of competition between Red-winged and Yellow-headed blackbirds. Journal of Animal Ecology 37:43-62.

Mills, A.M. 1988. Unsuitability of Tree Swallows as hosts to Brown-headed Cowbirds. Journal of Field Ornithology 59:331-333.

Mills, E.L. 1960. Bird observations in the Queen Charlotte Islands, British Columbia. Canadian Field-Naturalist 74:156-158.

Mitchell, G.J. 1959. Bird observations at Tahsis Inlet, Vancouver Island, British Columbia. Canadian Field-Naturalist 73:6-13.

Moldenhauer, R.R. and D.J. Regelski. 1996. Northern Parula (*Parula americana*). *In* A. Poole and F. Gill (editors). The birds of North America, No. 215. The Academy of Natural Sciences, Philadelphia, Pennsylvania; The American Ornithologists' Union, Washington, D.C. 24 pp.

Monroe, B.L. 1990. Summary of highest counts of individuals for Canada. American Birds 44:1044-1048.

________. 1990a. Summary of highest counts of individuals for Canada and the United States. American Birds 44:1038-1043.

________. 1993. Summary of highest counts of individuals for Canada. American Birds 47:1025-1026.

Moore, K.E. 1990. Urbanization in the Lower Fraser Valley, 1980-1987. Technical Report Series No. 120. Canadian Wildlife Service, Pacific and Yukon Region, British Columbia.

Morris, M.M.J. and R.E. Lemon. 1988. American Redstart nest placement in southwestern New Brunswick. Canadian Journal of Zoology 66:212-216.

Morrison, J.L. 1996. Crested Caracara (*Caracara plancus*). *In* A. Poole and F. Gill (editors). The birds of North America, No. 249. The Academy of Natural Sciences, Philadelphia, Pennsylvania; The American Ornithologists' Union, Washington, D.C. 28 pp.

Morrison, M.L. 1981. The structure of western warbler assemblages: analysis of foraging behavior and habitat selection in Oregon. Auk 98:578-588.

________. 1983. Analysis of geographic variation in Townsend's Warbler. Condor 85:385-391.

Morrison, M.L. and J.W. Hardy. 1983. Hybridization between Hermit and Townsend's Warblers. Murrelet 64:65-72.

Morrison, M.L., L.S. Hall, S.K. Robinson, S.I. Rothstein, D.C. Hahn, and T.R. Rich. 1999. Research and management of the Brown-headed Cowbird in western landscapes. Studies in Avian Biology No. 18, Allen Press, Lawrence, Kansas. 312 pp.

Morse, D.H. 1978. Populations of Bay-breasted and Cape May warblers during an outbreak of the spruce budworm. Wilson Bulletin 90:404-413.

________. 1989. American warblers, an ecological and behavioral perspective. Harvard University Press, Cambridge, Massachusetts. 406 pp.

________. 1993. Black-throated Green Warbler (*Dendroica virens*). *In* A. Poole and F. Gill (editors). The birds of North America, No. 55. The Academy of Natural Sciences, Philadelphia, Pennsylvania; The American Ornithologists' Union, Washington, D.C. 20 pp.

________. 1994. Blackburnian Warbler (*Dendroica fusca*). *In* A. Poole and F. Gill (editors). The birds of North America, No. 102. The Academy of Natural Sciences, Philadelphia, Pennsylvania; The American Ornithologists' Union, Washington, D.C. 20 pp.

Morton, M.L. 1976. Adaptive strategies of *Zonotrichia* breeding at high latitude or high altitude. Acta XVI International Congress of Ornithology 1976:322-336.

Morton, M.L., J.L. Horstmann, and J.M Osborn. 1972. Reproductive cycle and nesting success of the mountain White-crowned Sparrow (*Zonotrichia leucophrys oriantha*) in the central Sierra Nevada. Condor 74:152-163.

Mousley, H.M. 1924. A study of the home life of the Northern Parula and other warblers at Hatley, Stanstead County, Quebec. Auk 41:263-288.

Mowbray, T.B. 1997. Swamp Sparrow (*Melospiza georgiana*). *In* A. Poole and F. Gill (editors). The birds of North America, No. 279. The Academy of Natural Sciences, Philadelphia, Pennsylvania; The American Ornithologists' Union, Washington, D.C. 24 pp.

Munro, J.A. 1915. Two new records for British Columbia. Auk 32:107-108.

________. 1919a. Notes on some birds in the Okanagan valley, British Columbia. Auk 36:64-74.

________. 1919b. Notes on the breeding habits of the Red Crossbill in the Okanagan valley, British Columbia. Condor 21:57-60.

________. 1922a. Additional notes on the wintering birds of the Okanagan valley, British Columbia. Canadian Field-Naturalist 36:12-13.

________. 1926. Christmas bird censuses, 1925 – Okanagan Landing, British Columbia. Canadian Field-Naturalist 40:11.

________. 1927. Christmas bird census returns, 1926 – Okanagan Landing, British Columbia. Canadian Field-Naturalist 41:15.

________. 1935a. Bird life at Horse Lake, British Columbia. Condor 37:185-193.

________. 1935b. Recent records from British Columbia. Condor 37:178-179.

________. 1945a. The birds of the Cariboo parklands, British Columbia. Canadian Journal of Research 23:17-103.

________. 1945b. Preliminary report on the birds and mammals of Glacier National Park, British Columbia. Canadian Field-Naturalist 59:175-190.

________. 1945c. Christmas bird census, 1944 – Okanagan Landing, Vernon, British Columbia. Canadian Field-Naturalist 59:37.

________. 1946. Birds and mammals of the Vanderhoof region, British Columbia, with comments on other resources. Canadian Wildlife Service Unpublished Report, Ottawa, Ontario. 161 pp.

________. 1947. Observations of birds and mammals in central British Columbia. British Columbia Provincial Museum Occasional Paper No. 6, Victoria. 165 pp.

________. 1949. The birds and mammals of the Vanderhoof region, British Columbia. American Midland Naturalist 41:1-138.

________. 1950. The birds and mammals of the Creston region, British Columbia. British Columbia Provincial Museum Occasional Paper No. 8, Victoria. 90 pp.

________. 1953. Bird records from the southern interior of British Columbia. Murrelet 34:15-17.

________. 1955a. Additional observations of birds and mammals in the Vanderhoof region, British Columbia. American Midland Naturalist 53:56-60.

________. 1955b. A sight record of a Baltimore Oriole in British Columbia. Murrelet 36:43.

Munro, J.A. and I. McTaggart-Cowan. 1947. A review of the bird fauna of British Columbia. British Columbia Provincial Museum Special Publication No. 2, Victoria. 285 pp.

Murray, B.G. 1989. A critical review of the transocean migration of the Blackpoll Warbler. Auk 106:8-17.

Naugler, C.T. 1993. American Tree Sparrow (*Spizella arborea*). *In* A. Poole, P. Stettenheim, and F. Gill (editors). The birds of North America, No. 37. The Academy of Natural Sciences, Philadelphia, Pennsylvania; The American Ornithologists' Union, Washington, D.C. 12 pp.

Nero, R.W. 1984. Redwings. Smithsonian Institution Press, Washington, D.C. 160 pp.

Nethersole-Thompson, D. 1966. The Snow Bunting. Oliver and Boyd, Edinburgh and London, United Kingdom. 316 pp.

Newman, G.A. 1970. Cowbird parasitism and nesting success of Lark Sparrows in southern Oklahoma. Wilson Bulletin 82:304-309.

Newton, I. 1972. Finches. Collins, London, United Kingdom. 288 pp.

________. 1998. Population limitation in birds. Academic Press, San Diego, California. 597 pp.

Nice, M.M. 1926. A study of nesting Magnolia Warblers (*Dendroica magnolia*). Wilson Bulletin 33:185-199.

Nice, M.M. and L.B. Nice. 1932. A study of two nests of the Black-throated Green Warbler. Bird-Banding 3:95-105, 157-172.

Nicholson, W.H. 1936. Notes on the habits of the Florida Grasshopper Sparrow. Auk 53:318-319.

Niemi, G.J. and J.M. Hanowski. 1984. Relationships of breeding birds to habitat characteristics in logged areas. Journal of Wildlife Management 48:438-443.

Nisbett, I.C.T. 1970. Autumn migration of the Blackpoll Warbler: evidence for long flight provided by regional survey. Bird-Banding 41:207-240.

Nolan, V., E.D. Ketterson, and C.A. Buerkle. 1999. Prairie Warbler (*Dendroica discolor*). *In* A. Poole and F. Gill (editors). The birds of North America, No. 455. The birds of North America, Inc., Philadelphia, Pennsylvania. 28 pp.

Norment, C.J. and S.A. Shackleton. 1993. Harris' Sparrow (*Zonotrichia querula*). *In* A. Poole and F. Gill (editors). The birds of North America, No. 64. The Academy of Natural Sciences, Philadelphia, Pennsylvania; The American Ornithologists' Union, Washington, D.C. 20 pp.

Norment, C.J., P. Hendricks, and R. Santonocito. 1998. Golden-crowned Sparrow (*Zonotrichia atricapilla*). *In* A. Poole and F. Gill (editors). The birds of North America, No. 352. The Birds of North America, Inc., Philadelphia, Pennsylvania. 20 pp.

Norton, D.W. 1972. Incubation schedules of four species of calidrine sandpipers at Barrow, Alaska. Condor 74:164-176.

O'Brien, D. and K. Bell. 1975. The avifauna of Bowron Lake Provincial Park – report for summer 1975. British Columbia Parks Branch Unpublished Report, Victoria. 22 pp. (Bibliography 3791).

Ogilvie, R.T. 1989. Disjunct vascular flora of northwestern Vancouver Island in relation to Queen Charlotte Islands' endemism and Pacific coastal refugia. Pages 127-130 *in* G.G.E. Scudder and N. Gessler (editors). The outer shores. Queen Charlotte Islands Museum Press, Queen Charlotte City. 327 pp.

Ogle, S. and K. Martin. 1997. The use of alpine habitats by fall migrating birds on Vancouver Island, 1996-1997. Canadian Wildlife Service Unpublished Report, Delta, British Columbia. 19 pp.

Orcutt, L. 1967. Intermediate naturalists are busy birders. Vancouver Natural History Society News 134:5-8 (Bibliography 1730).

Orians, G.H. 1966. Food of nestling Yellow-headed Blackbirds, Cariboo Parklands, British Columbia. Condor 68:321-337.

________. 1980. Some adaptations of marsh-nesting blackbirds. Princeton University Press, Princeton, New Jersey. 295 pp.

________. 1985. Blackbirds of the Americas. University of Washington Press, Seattle. 163 pp.

________. 1997. Global biodiversity I. Patterns and processes. Pages 87-122 *in* G.K. Meffe and C.R. Carroll (editors). Principles of conservation biology. Sinauer Associates, Sunderland, Massachusetts. 729 pp.

Orians, G.H. and M.F. Willson. 1964. Interspecific territories of birds. Ecology 45:736-744.

Ornstein, R. and P. Ehrlich. 1989. New world, new mind: moving toward conscious evolution. Simon and Schuster, New York.

Orr, D.W. 1994. Earth in mind: on education, environment, and the human prospect. Island Press, Washington, D.C.

Orr, R.T. 1968. Cassin's Finch. Pages 280-289 *in* O.L. Austin (editor). Life histories of North American cardinals, grosbeaks, buntings, towhees, finches, sparrows, and allies (Part 1). United States National Museum Bulletin No. 237, Washington, D.C.

Ortega, B. 1995. Brief summary of the 94th Christmas Bird Count high counts. American Birds 49:886.

Osgood, W.H. 1901. Natural history of the Queen Charlotte Islands, British Columbia – natural history of the Cook Inlet region, Alaska. United States Department of Agriculture Division of Biological Survey North American Fauna No. 21:1-87.

O'Shaughnessy, M. 1998. Two lifers in one day on Vancouver Island. Victoria Naturalist 54(5):18.

Osmond-Jones, E.J., L. Bonner, M. Sather, B. Smith, and M. MacColl. 1975. A fisheries and wildlife survey of the Burnie Lake Park proposal. British Columbia Parks Branch Unpublished Report, Victoria. 44 pp. (Bibliography 3802).

Osmond-Jones, E.J., M. Sather, W.G. Hazelwood, and B. Ford. 1977. Spatsizi and Tatlatui wilderness parks: an inventory of wildlife, fisheries, and recreational values in a northern wilderness park. British Columbia Parks Branch, Victoria. 292 pp.

Page, R. and A. Bergerud. 1979. The caribou calf mortality study 1979: a progress report. University of Victoria Department of Biology Unpublished Report, Victoria. 65 pp. (Bibliography 3808).

Palmer, R.S. 1968. *Spinus pinus* (Wilson), Pine Siskin. Pages 424-447 *in* O.L. Austin (editor). Life histories of North American cardinals, grosbeaks, buntings, towhees, finches, sparrows, and allies (Part 1). United States National Museum Bulletin 237. Washington, D.C.

Palmer, T.K. 1970a. The House Finch and starling in relation to California's agriculture. General Meeting Working Group Granivorous Birds, International Biological Program, Section Productivity of Terrestrial Communities, Arnhem/The Hague – The Netherlands, 6-8 September 1970. Pages 1-25.

________. 1970b. House Finch (Linnet) control in California. Pages 173-178 *in* R.H. Dana (editor). Proceedings Fourth Vertebrate Pest Conference, University of California, Davis.

Parham, H.J. 1937. A nature lover in British Columbia. Witherby, London. 1081 pp.

Parish, R., R. Coupe, and D. Lloyd (editors). 1996. Plants of southern interior British Columbia. Lone Pine Publishing, Vancouver. 463 pp.

Patch, C. 1922. A biological reconnaissance on Graham Island of the Queen Charlotte group. Canadian Field-Naturalist 36:101-105.

Paterson, H.E.H. 1985. The recognition concept of species. Transvaal Museum Monograph 4:21-29.

Patti, S.T. and M.L. Myers. 1976. A probable Mourning × MacGillivray's Warbler hybrid. Wilson Bulletin 88:490-491.

Paul, A. 1964. More range expansion of the House Finch. Murrelet 45:11.

________. 1968. Pine Siskins in British Columbia. Murrelet 49:24.

Paul, W.A.B. 1959. The birds of Kleena Kleene, Chilcotin district, British Columbia, 1947-1958. Canadian Field-Naturalist 73:83-93.

________. 1964a. Birds of Kleena Kleene, Chilcotin District, B.C., 1959-1962. Canadian Field-Naturalist 78:13-16.

________. 1964b. Notes on *Leucosticte tephrocotis littoralis* Hepburn's Rosy Finch. Murrelet 45:28.

Paulson, D.R. 1992. Northwest bird diversity: From extravagant past and changing present to precarious future. Northwest Environmental Journal 8:71-118.

________. 1993. Shorebirds of the Pacific Northwest. University of Washington Press, Seattle. 406 pp.

Payne, R.B. 1992. Indigo Bunting (*Passerina cyanea*). *In* A. Poole, P. Stettenheim, and F. Gill (editors). The birds of North America, No. 4. The Academy of Natural Sciences, Philadelphia, Pennsylvania; The American Ornithologists' Union, Washington, D.C. 24 pp.

Pearse, T. 1939. Siskins numerous on Vancouver Island, B.C. Murrelet 20:43.

________. 1946. Notes on changes in bird populations in the vicinity of Comox, Vancouver Island – 1917. Murrelet 27:4-9.

________. 1947. Long-tailed Chat on Vancouver Island. Canadian Field-Naturalist 61:22.

________. 1958. Christmas bird census 1957 – Comox, British Columbia. Canadian Field-Naturalist 72:44-45.

Pearson, S.F. 1997. Hermit Warbler (*Dendroica occidentalis*). *In* A. Poole and F. Gill (editors). The birds of North America, No. 303. The Academy of Natural Sciences, Philadelphia, Pennsylvania; The American Ornithologists' Union, Washington, D.C. 20 pp.

Peck, G.K. and R.D. James. 1987. Breeding birds of Ontario: nidiology and distribution – Volume 2: passerines. Royal Ontario Museum Life Sciences Miscellaneous Publication, Toronto, Ontario. 387 pp.

Peer, B.D. and E.K. Bollinger. 1997a. Common Grackle (*Quiscalus quiscula*). *In* A. Poole and F. Gill (editors). The birds of North America, No. 271. The Academy of Natural Sciences, Philadelphia, Pennsylvania; The American Ornithologists' Union, Washington, D.C. 20 pp.

———. and ———. 1997b. Explanations for the infrequent cowbird parasitism on Common Grackles. Condor 99:151-161.
Penner, D.F. 1976. Peace River sites C and E environmental impact studies. Renewable Resources Consulting Services Ltd., Edmonton, Alberta. 307 pp. (Bibliography 3833).
Petersen, K.L. and L.B. Best. 1985. Brewer's Sparrow nest-site characteristics in a sagebrush community. Journal of Field Ornithology 56:23-27.
Phinney, M. 1992. Birds of Dawson Creek area: Peace River specialties. Appendix to report for British Columbia Ministry of Forests Project B28, Prince George.
———. 1998. Spring and summer birds of Dawson Creek, 1991-1995. WBT Wild Bird Trust of British Columbia Wildlife Report No. 4, West Vancouver. 60 pp.
Phinney, M. and A.N. Lance. 1998. Analysis of field data from the 1997 forest and grassland songbird inventory of the McGregor model forest. McGregor Model Forest Association Unpublished Report, Prince George, British Columbia. 29 pp.
Picman, J. 1980a. Impact of Marsh Wrens on reproductive strategy of Red-winged Blackbirds. Canadian Journal of Zoology 58:337-350.
———. 1980b. Responses of Red-winged Blackbirds to nests of Long-billed Marsh Wrens. Canadian Journal of Zoology 58:1821-1827.
———. 1986. Attempted nest parasitism of the Marsh Wren by a Brown-headed Cowbird. Condor 88:381-382.
Picman, J., M. Leonard, and A. Horn. 1988. Antipredation role of clumped nesting by marsh-nesting Red-winged Blackbirds. Behavioral Ecology and Sociobiology 22:9-15.
Pilon, P. and M.A. Kerr. 1984. Land use monitoring on wetlands in the southwestern Fraser Lowland, British Columbia. Environment Canada, Lands Directorate, Pacific and Yukon Region, Vancouver.
Pinel, H.W., W.W. Smith, and C.R. Wershler. 1993. Alberta birds, 1971-1980 – Volume 2: passerines. Provincial Museum of Alberta, Edmonton. 238 pp.
Pitelka, F.A. 1940. Breeding behavior of the Black-throated Green Warbler. Wilson Bulletin 52:3-18.
Pitocchelli, J. 1990. Plumage, morphometric and song variation in Mourning (*Oporornis philadelphia*) and MacGillivray's (*O. tolmiei*) warblers. Auk 107:161-171.
———. 1992. Plumage and size variation in the Mourning Warbler (*Oporornis philadelphia*). Condor 94:198-209.
———. 1993. Mourning Warbler (*Oporornis philadelphia*). *In* A. Poole and F. Gill (editors). The birds of North America, No. 72. The Academy of Natural Sciences, Philadelphia, Pennsylvania; The American Ornithologists' Union, Washington, D.C. 16 pp.
———. 1995. MacGillivray's Warbler (*Oporornis tolmiei*). *In* A. Poole and F. Gill (editors). The birds of North America, No. 159. The Academy of Natural Sciences, Philadelphia, Pennsylvania; The American Ornithologists' Union, Washington, D.C. 16 pp.
Pitocchelli, J., J. Bouchie, and D. Jones. 1997. Connecticut Warbler (*Oporornis agilis*). *In* A. Poole and F. Gill (editors). The birds of North America, No. 320. The Academy of Natural Sciences, Philadelphia, Pennsylvania; The American Ornithologists' Union, Washington, D.C. 16 pp.
Pittaway, R. 1995. Recognizable forms – subspecies of the Palm Warbler. Ontario Birds 13:23-27.
Pitts, T.D. 1979. Nesting habitats of rural and suburban House Sparrows in northwest Tennessee. Journal of the Tennessee Academy of Sciences 54:145-148.
Pojar, J. and A.C. Stewart. 1991. Spruce-willow-birch zone. Pages 251-262 *in* D. Meidinger and J. Pojar (editors). Ecosystems of British Columbia. British Columbia Ministry of Forests, Research Branch, Victoria. 330 pp.
Pojar, R.A. 1993. The diversity of bird communities in interior aspen forests in the western end of the dry cool subzone of the sub-boreal spruce (SSBS dk) in the Prince Rupert forest region – baseline studies. British Columbia Forest Service Unpublished Report, Victoria. 74 pp.
———. 1995. Breeding bird communities in aspen forests of the sub-boreal spruce (dk subzone) in the Prince Rupert Forest region. British Columbia Ministry of Forests Land Management Handbook No. 33, Victoria. 59 pp.
Poll, D.M., M.M. Porter, G.L. Holroyd, R.M. Wershler, and L.W. Gyug. 1984. Ecological land classification of Kootenay National Park, British Columbia. Volume 2: wildlife resource. Alberta Institute of Pedology, University of Alberta, Edmonton. 260 pp.
Pollock, H. 1990. B.C. bluebird nesting survey 1990. Issued by the author, Victoria, British Columbia. 3 pp.
———. 1992. B.C. bluebird nesting survey 1992. Issued by the author, Victoria, British Columbia. 3 pp.
Popp, J.W. 1987. Scanning behaviour of finches in intermixed groups. Condor 90:512-513.
Potter, P.E. 1974. Breeding behavior of Savannah Sparrows in southeastern Michigan. Jack-Pine Warbler 52:50-63.
Power, D.M. 1971. Warbler ecology: diversity, similarity, and seasonal differences in habitat. Ecology 52:434-443.
Poynter, G.A. 1960. A report on the birds of the lower Vancouver Island region for 1959. Victoria Natural History Society, Victoria, British Columbia. 27 pp. (Bibliography 308).
Preble, E.A. 1908. North American Fauna No. 27. United States Department of Agriculture, Bureau of Biological Survey, Washington, D.C. 574 pp.
Preece, W.H.A. 1925. January bird notes from Mount Tolmie, Victoria, British Columbia. Canadian Field-Naturalist 39:175-176.
Prentice, A. and W.S. Boyd. 1988. Intertidal and adjacent upland habitat in estuaries located on the east coast of Vancouver Island – a pilot assessment of their historical changes. Canadian Wildlife Technical Report Series No. 38, Delta, British Columbia.
Prescott, D.R.C. 1994. Intraspecific and geographic trends in body size of a differential migrant, the Evening Grosbeak. Auk 111:693-702.
Prest, V.K. 1984. The Late Wisconsin glacier complex. Pages 22-36 *in* R.J. Fulton (editor). Quaternary stratigraphy of Canada – a Canadian contribution to IGCP project 24. 210 pp.
Preston, A. 1990. Canyon Wren, Sage Thrasher, White-headed Woodpecker, Gray Flycatcher and Grasshopper Sparrow in the south Okanagan. British Columbia Ministry of Environment, Lands and Parks Unpublished Report, Penticton.
Price, M. 1993. Birds of the Donna Creek area: an annotated checklist for the period 5/14/93 – 7/13/93. Unpublished Report by the author, Vancouver, British Columbia. 21 pp.
Prince George Naturalists Club. 1996. Checklist of north-central B.C. birds. Issued by the Prince George Naturalists Club, British Columbia. Leaflet.
Pyle, P., S.N.G. Howell, R.P. Yunick, and D.F. DeSante. 1987. Identification guide to North American passerines. Slate Creek Press, Bolinas, California. 278 pp.
Rabinowitz, D., S. Cairns, and T. Dillon. 1986. Seven forms of rarity and their frequency in the flora of the British Isles. Pages 182-204 *in* M.E. Soulé (editor). Conservation biology: the science of scarcity and diversity. Sinauer Associates, Sunderland, Massachusetts.
Racey, K. 1930. Peace River notes. Murrelet 11:70-71.
———. 1948. Birds of Alta Lake region, British Columbia. Auk 65:383-401.
———. 1958. Bird notes from Huntingdon, B.C. Murrelet 39:25.
Ralph, C.J. 1994. Evidence of changes in populations of the Marbled Murrelet in the Pacific Northwest. Studies in Avian Biology 15:286-292.
Ralph, C.J. and L.R. Mewaldt. 1975. Timing of site fixation upon the wintering grounds of sparrows. Auk 92:698-705.
Rand, A.L. 1943. On some British Columbia birds. Canadian Field-Naturalist 57:60-63.
———. 1944. Birds of the Alaska Highway in British Columbia. Canadian Field-Naturalist 58:111-125.
Randall, T.E. 1933. A list of the breeding birds of the Athabasca district, Alberta. Canadian Field-Naturalist 47:1-6.
Raphael, M.G., K.V. Rosenberg, and B.G. Marcot. 1988. Large scale changes in bird populations of Douglas-fir forests northwestern California. Bird Conservation 3:63-83.
Redpath, K. 1990. Identification of relatively undisturbed areas in the south Okanagan and Similkameen valleys, British Columbia. Canadian Wildlife Service Technical Report Series No. 108, Delta, British Columbia. 9 pp.
Reed, J.M. 1986. Vegetation structure and Vesper Sparrow territory location. Wilson Bulletin 98:144-147.
Reid, T.C. 1975. Liard River Hot-Springs Park natural history observations (autumn, 1974 and winter-spring-summer, 1975). British Columbia Parks Branch Unpublished Report, Victoria. 117 pp. (Bibliography 3893).
Resources Inventory Committee. 1999. Inventory methods for forest and grassland songbirds: standards for components of British Columbia's biodiversity no. 15 version 2. Ministry of Environment, Lands and Parks, Resources Inventory Branch, Victoria. 37 pp.

Reynolds, J.D. and R.W. Knapton. 1984. Nest-site selection and breeding biology of the Chipping Sparrow. Wilson Bulletin 96:488-493.
Reynolds, T.D. 1981. Nesting of the Sage Thrasher, Sage Sparrow, and Brewer's Sparrow in southeastern Idaho. Condor 83:61-64.
Rhoads, S.N. 1893. The birds observed in British Columbia and Washington during spring and summer, 1892. Proceedings of the Academy Natural Sciences of Philadelphia 1892:21-65.
Richardson, F. 1971. Birds of Gibson Bay and Browning Inlet, northwest Vancouver Island, British Columbia. Murrelet 52:29-40.
Richardson, M. and D.W. Brauning. 1995. Chestnut-sided Warbler (*Dendroica pensylvanica*). *In* A. Poole and F. Gill (editors). The birds of North America, No. 190. The Academy of Natural Sciences, Philadelphia, Pennsylvania; The American Ornithologists' Union, Washington, D.C. 20 pp.
Rimmer, C.C. and K.P. McFarland. 1998. Tennessee Warbler (*Vermivora peregrina*). *In* A. Poole and F. Gill (editors). The birds of North America, No. 350. The Birds of North America, Inc., Philadelphia, Pennsylvania. 24 pp.
Rising, J.D. 1973. Morphological variation and status of the orioles, *Icterus galbula, I. bullockii,* and *I. abeillei,* in the northern Great Plains and Durango. Canadian Journal of Zoology 51:1267-1273.
________. 1983. The Great Plains hybrid zones. Pages 131-157 *in* R.F. Johnston (editor). Current ornithology. Volume 1. Plenum Press, New York.
________. 1996. A guide to the identification and natural history of the sparrows of the United States and Canada. Academic Press, London, United Kingdom. 365 pp.
Rising, J.D. and N.J. Flood. 1998. Baltimore Oriole (*Icterus galbula*). *In* A. Poole and F. Gill (editors). The birds of North America, No. 384. The Birds of North America, Inc., Philadelphia, Pennsylvania. 32 pp.
Rising, J.D. and P.L. Williams. 1999. Bullock's Oriole (*Icterus bullockii*). *In* A. Poole and F. Gill (editors). The birds of North America, No. 416. The Birds of North America, Inc., Philadelphia, Pennsylvania. 20 pp.
Ritcey, R.W. 1953. Winter wildlife report Wells Gray Park, 1952-1953. British Columbia Forest Service Unpublished Report, Victoria. 56 pp. (Bibliography 3919).
Ritchison, G. 1983. Breeding biology of the Black-headed Grosbeak in northern Utah. Western Birds 14:159-167.
Robbins, C.S., D. Bystrak, and P.H. Geissler. 1986. The Breeding Bird Survey: its first fifteen years, 1965-1979. United States Department of the Interior, Fish and Wildlife Service Resource Publication 157, Washington, D.C. 196 pp.
Robbins, C.S., D.K. Dawson, and B.A. Dowell. 1989. Habitat area requirements of breeding forest birds of the middle Atlantic states. Wildlife Monographs 103:1-34.
Roberson, D. 1980. Rare birds of the west coast of North America. Woodcock Publications, Pacific Grove, California. 496 pp.
Roberts, G. and A. Roberts. 1993. Biodiversity in the Cariboo-Chilcotin grasslands. British Columbia Ministry of Environment, Lands and Parks Unpublished Report, Williams Lake. 143 pp.
Robertson, J.M. 1928. Echoes from the annual report file. News from the Bird-Banders 3:15-16.
Robertson, R.J. and R.F. Norman. 1976. Behavioral defenses to brood parasitism by potential hosts of the Brown-headed Cowbird. Condor 78:166-173.
Rodway, M.S. and M.J.F. Lemon. 1991. British Columbia seabird colony inventory: report no. 7: northern mainland coast. Canadian Wildlife Service Technical Report Series No. 121, Delta, British Columbia. 182 pp.
Rodway, M.S., M.J.F. Lemon, and K.R. Summers. 1990. British Columbia seabird colony inventory: report #4 – Scott Islands – census results from 1982 to 1989 with reference to the Nestucca oil spill. Canadian Wildlife Service Technical Report Series No. 86, Delta, British Columbia. 109 pp.
Roe, N.A. 1974. Birds and disturbed forest succession after logging in Pacific Rim National Park, Vancouver Island, British Columbia and a contribution towards the development of an interpretive plan for logged areas. M.Sc. Thesis, University of Calgary, Calgary, Alberta. 212 pp.
Rogers, C.M., J.N.M. Smith, W.M. Hochachka, A.L.E.V. Cassidy, M.J. Taitt, P. Arcese, and D. Schluter. 1991. Spatial variation in winter survival of Song Sparrows *Melospiza melodia.* Ornis Scandinavica 22:387-395.
Rogers, C.M., M.J. Taitt, J.N.M. Smith, and G. Jongejan. 1997. Nest predation and cowbird parasitism create a demographic sink in wetland-breeding Song Sparrows. Condor 99:622-633.
Rogers T.H. 1965a. The fall migration – northern Rocky Mountain–Intermountain region. Audubon Field Notes 19:60-63.
________. 1966a. The fall migration – northern Rocky Mountain–Intermountain region. Audubon Field Notes 20:72-76.
________. 1966b. The winter season – northern Rocky Mountain–Intermountain region. Audubon Field Notes 20:442-445.
________. 1966c. The spring migration – northern Rocky Mountain–Intermountain region. Audubon Field Notes 20:532-535.
________. 1968a. The fall migration – northern Rocky Mountain–Intermountain region. Audubon Field Notes 22:69-72.
________. 1968b. The winter season – northern Rocky Mountain–Intermountain region. Audubon Field Notes 22:460-463.
________. 1968c. The spring migration – northern Rocky Mountain–Intermountain region. Audubon Field Notes 22:557-560.
________. 1968d. The nesting season – northern Rocky Mountain–Intermountain region. Audubon Field Notes 22:628-632.
________. 1969. The spring migration – northern Rocky Mountain – Intermountain region. Audubon Field Notes 23:607-611.
________. 1972. The fall migration – northern Rocky Mountain – Intermountain region. American Birds 26:88-92.
________. 1973a. The fall migration – northern Rocky Mountain–Intermountain region. American Birds 27:85-91.
________. 1973b. The winter season – northern Rocky Mountain–Intermountain region. American Birds 27:639-643.
________. 1974. The fall migration – northern Rocky Mountain – Intermountain region. American Birds 28:78-83.
________. 1984. The winter season – northern Rocky Mountain–Intermountain region. American Birds 38:337-340.
________. 1984a. The spring migration – northern Rocky Mountain–Intermountain region. American Birds 38:936-939.
________. 1984b. The nesting season – northern Rocky Mountain–Intermountain region. American Birds 38:1041-1044.
________. 1985. The nesting season – northern Rocky Mountain–Intermountain region. American Birds 39:327-329.
________. 1986. The winter season – northern Rocky Mountain–Intermountain region. American Birds 40:304-306.
________. 1986a. The fall migration – northern Rocky Mountain–Intermountain region. American Birds 40:498-502.
________. 1987. The autumn migration – northern Rocky Mountain–Intermountain region. American Birds 41:118-121.
________. 1988. The autumn migration – northern Rocky Mountain–Intermountain region. American Birds 42:104-108.
________. 1988a. The winter season – northern Rocky Mountain–Intermountain region. American Birds 42:296-299.
________. 1989. The winter season – northern Rocky Mountain–Intermountain region. American Birds 43:356-360.
________. 1989a. The spring season – northern Rocky Mountain–Intermountain region. American Birds 43:511-513.
________. 1989b. The nesting season – northern Rocky Mountain–Intermountain region. American Birds 43:1342-1345.
Rohwer, S. and M.S. Johnson. 1992. Scheduling differences of molt and migration for Baltimore and Bullock's orioles persist in a common environment. Condor 94:992-994.
Rohwer, S. and J. Manning. 1990. Differences in timing and number of molts for Baltimore and Bullock's orioles: implications to hybrid fitness and theories of delayed plumage maturation. Condor 92:125-140.
Rohwer, S. and C. Wood. 1998. Three hybrid zones between Hermit and Townsend's warblers in Washington and Oregon. Auk 115:284-310.
Root, T. 1988. Atlas of wintering North American birds: an analysis of Christmas Bird Count data. University of Chicago Press, Chicago, Illinois. 312 pp.
Roseland, M. 1992. Toward sustainable communities: a resource book for municipal and local governments. National Roundtable on the Environment and the Economy. Ottawa, Ontario.
Rotenberry, J.T. and J.A. Wiens. 1989. Reproductive biology of shrubsteppe passerine birds: geographical and temporal variation in clutch size, brood size, and fledging success. Condor 91:1-14.
Rotenberry, J.T., M.A. Patten, and K.L. Preston. 1999. Brewer's Sparrow (*Spizella breweri*). *In* A. Poole and F. Gill (editors). The birds of North America, No. 390. The Birds of North America, Inc., Philadelphia, Pennsylvania. 24 pp.
Roth, J.L. 1977. Breeding biology of the Nashville Warbler in northern Michigan. Jack-Pine Warbler 55:129-141.
Rothstein, S.I. 1973. Extreme overlap between first and second nestings in the Rose-breasted Grosbeak. Wilson Bulletin 83:242-243.

________. 1975. An experimental and teleonomic investigation of avian brood parasitism. Condor 77:250-271.
________. 1994. The cowbird's invasion of the Far West: history, causes and consequences experienced by host species. Studies in Avian Biology 15:301-315.
Rothstein, S.I., J. Verner, and E. Stevens. 1980. Range expansion and diurnal changes in dispersion of the Brown-headed Cowbird in the Sierra Nevada. Auk 97:253-267.
Royal Society and United States National Academy of Sciences. 1992. Joint statement on population growth and sustainability. <http://www.spiritone.com/~orsierra/rogue/popco/warn/warn01.htm>
Runyan, B. 1971. Bowron Lake Park natural history report, 1971. British Columbia Parks Branch Unpublished Report, Victoria. 51 pp. (Bibliography 4043).
Ryan, J.C. 2000. State of the northwest: revised 2000 edition. Report 9. Northwest Environment Watch, Seattle, Washington.
Ryder, G.R. 1973. Report on White Pelican Provincial Park, 1973. British Columbia Parks Branch Unpublished Report, Victoria. 122 pp. (Bibliography 1134).
Sadler, T.S. and M.T. Myres. 1976. Alberta birds 1961-1970. Provincial Museum of Alberta, Natural History Section Occasional Paper No. 1, Edmonton. 314 pp.
Sage, B.L. 1976. The breeding distribution of Smith's Longspur in Alaska. Condor 78:116-117.
Salt, G.W. 1952. The relation of metabolism to climate and distribution in three finches of the genus *Carpodacus*. Ecological Monographs 22:121-152.
Salt, W.R. 1966. A nesting study of *Spizella pallida*. Auk 83:274-281.
________. 1966a. Some unusual bird records from the Peace River district. Canadian Field-Naturalist 80:114-115.
________. 1972. Western records of the Chestnut-sided Warbler. Canadian Field-Naturalist 86:390-391.
________. 1973. Alberta vireos and wood warblers. Provincial Museum and Archives of Alberta Publication No. 3, Edmonton. 141 pp.
Salt, W.R. and J.R. Salt. 1966. The birds of Alberta, 2nd edition. Queen's Printer, Edmonton, Alberta. 511 pp.
________. and ________. 1976. The birds of Alberta, 3rd edition. Hurtig Publishers, Edmonton, Alberta. 498 pp.
Salt, W.R. and A.L. Wilk. 1958. The birds of Alberta. Queen's Printer, Edmonton, Alberta. 511 pp.
Samson, F.B. 1976. Territory, breeding density, and fall departure in Cassin's Finch. Auk 93:477-497.
Sarell, M.J. and K.P. McGuinness. 1996. Status of the Brewer's Sparrow in British Columbia. British Columbia Ministry of Environment, Lands and Parks Wildlife Working Report WR-77, Victoria. 12 pp.
Sargeant, J. 1973. The seventy-third Christmas Bird Count – Chilliwack, B.C. American Birds 27:179.
Satterfield, R. 1966. Cowbirds. Victoria Naturalist 22:96-97.
Sauer, J.R., S. Schwartz, and B. Hoover. 1996. The Christmas Bird Count Home Page. Version 95.1. Patuxent Wildlife Research Center, Laurel, Maryland. <http://www.mbr.hbs.gov/bbs/cbc.html>
Sauer, J.R., J.E. Hines, G. Gough, I. Thomas, and B.G. Peterjohn. 1997. The North American Breeding Bird Survey results and analysis – version 96.3. United States Department of the Interior, Patuxent Wildlife Research Center, Laurel, Maryland.
Savard, J.-P. L. and J.B. Falls. 1981. Influence of habitat structure on nesting height of birds in urban areas. Canadian Journal of Zoology 59:924-932.
Scerbinski, V.S., S.C. Christoffersen, and E.H. Granitz. 1994. New neighbours, new taxes? The escalation of property taxes due to population growth. Real Estates Issues 19:41-44.
Scharf, W.C. and J. Kren. 1996. Orchard Oriole (*Icterus spurius*). *In* A. Poole and F. Gill (editors). The birds of North America, No. 255. The Academy of Natural Sciences, Philadelphia, Pennsylvania; The American Ornithologists' Union, Washington, D.C. 24 pp.
Schieck, J., M. Nietfeld, and J.B. Stelfox. 1995. Differences in bird species richness and abundance among three successional stages of aspen-dominated boreal forests. Canadian Journal of Zoology 73:1417-1431.
Schmidt, O. (editor). 1989. Rare birds of Oregon. Oregon Field Ornithologists Special Publication No. 5, Eugene. 190 pp.
Schmiegelow, F.K.A., C.S. Machtans, and S.J. Hannon. 1997. Are boreal birds resilient to forest fragmentation? An experimental study of short-term community responses. Ecology 78:1914-1932.
Schnider, B., D. Beacham, and T. Stevens. 1971. Shuswap Lake Nature House annual report (1971). British Columbia Parks Branch Unpublished Report, Victoria. 15 pp. (Bibliography 4089).
Schofield, W.B. 1989. Structure and affinities of the Bryoflora of the Queen Charlotte Islands. Pages 109-120 *in* G.G.E. Scudder and N. Gessler (editors). The outer shores. Queen Charlotte Islands Museum Press, Queen Charlotte City. 327 pp.
Schreiber, E.A. and R.W. Schreiber. 1993. Red-tailed Tropicbird (*Phaethon rubricauda*). *In* The Birds of North America, No. 43 (A. Poole and F. Gill, editors). Philadelphia: The Academy of Natural Sciences; Washington, D.C.: The American Ornithologists' Union. 24 pp.
Schultz, T.T. and W.C. Leininger. 1991. Nongame wildlife communities in grazed and ungrazed montane riparian sites. Great Basin Naturalist 51:286-292.
Schultz, Z.M. 1959. The fall migration – north Pacific coast region. Audubon Field Notes 13:57-58.
Schwab, F.E. 1979. Effect of vegetation structure on breeding bird communities in the dry zone Douglas-fir forest of southeastern British Columbia. M.Sc. Thesis, University of British Columbia, Vancouver. 135 pp. (Bibliography 4113).
Scott, A.C. and M. Bekoff. 1991. Breeding behavior of Evening Grosbeaks. Condor 93:71-81.
Scott, D.M. 1977. Cowbird parasitism on the Gray Catbird at London, Ontario. Auk 94:18-27.
Scudder, G.G.E. 1974. Species concepts and speciation. Canadian Journal of Zoology 52:1121-1134.
________. 1989. The adaptive significance of marginal populations: a general perspective. Pages 180-185 *in* C.D. Levings, L.B. Hottby, and M.A. Henderson (editors). Proceedings of the national workshop on effects of habitat alteration on salmonid stocks. Canadian Special Publication, Fisheries and Aquatic Sciences.
________. 1992. Marginal populations and environmental change. Pages 374-378 *in* L. Berg and P.W. Delaney (editors). Proceedings of the Coho Workshop. Nanaimo, B.C. May 26-28, 1992.
Sealy, S.G. 1972. Additional winter records of the McKay's Bunting. Canadian Field-Naturalist 86:386-388.
________. 1979. Extralimital nesting of Bay-breasted Warblers: response to forest tent caterpillars. Auk 96:600-603.
________. 1980. Reproductive responses of Northern Orioles to a changing food supply. Canadian Journal of Zoology 58:221-227.
________. 1988. Aggressiveness in migrating Cape May Warblers: defense of an aquatic food source. Condor 90:271-274.
________. 1989. Defense of nectar resources by migrating Cape May Warblers. Journal of Field Ornithology 60:89-93.
________. 1995. Burial of cowbird eggs by parasitized Yellow Warblers: an empirical and experimental study. Animal Behavior 49:877-889.
Sealy, S.G. and J.C. Lorenzana. 1997. Feeding of nestling and fledgling brood parasites by individuals other than the foster parents; a review. Canadian Journal of Zoology 75:1739-1752.
Sealy, S.G., D.L. Neudorf, and D.P. Hill. 1995. Male Northern Orioles eject cowbird eggs: implications for the evolution of rejection behavior. Condor 97:369-375.
Searing, G.F. and J.M. Cooper. 1992. Establishment of a Yellow-headed Blackbird colony in the Vancouver area. LGL Ltd. 1991 Annual Report for Transport Canada, Richmond, British Columbia. 13 pp.
Searing. G.F. and J. Schieck. 1995. Yellow-headed Blackbird transplant program, Vancouver, British Columbia. LGL Ltd. Report for the Vancouver International Airport Authority, Richmond, British Columbia. 14 pp.
Searle, R. 1998. Forever wild? Not our national parks. Victoria Times-Colonist, 2 April, p. A11.
Seel, K.E. 1965. The birds of Kootenay National Park, first report, 1965, of field studies. Parks Canada Unpublished Report, Radium Hot Springs, British Columbia. 41 pp. (Bibliography 4123).
Semenchuk, G.P. (editor). 1992. The atlas of breeding birds of Alberta. Federation of Alberta Naturalists, Edmonton. 391 pp.
Seutin, G., P.T. Boag, and L.M. Ratcliffe. 1992. Plumage variability in redpolls from Churchill, Manitoba. Auk 109:771-785.
Seutin, G., P.T. Boag, B.N. White, and L.M. Ratcliffe. 1991. Sequential polyandry in the Common Redpoll (*Carduelis flammea*). Auk 108:166-170.
Shepard, M.G. 1972. First occurrence of the Chestnut-sided Warbler in British Columbia. Discovery 1:5-6.
________. 1974. British Columbia birds – winter season, 1973-1974. Vancouver Natural History Society Discovery, New Series 3(1):4-11.
________. 1975a. British Columbia birds – spring and summer, 1974. Vancouver Natural History Society Discovery, New Series 3(2):32-38.
________. 1975b. British Columbia birds – spring 1975. Vancouver Natural History Society Discovery, New Series 4(2):41-44.

________. 1976a. British Columbia birds – July to September, 1975. Vancouver Natural History Society Discovery, New Series 4(3): 67-69.

________. 1976b. British Columbia birds – October to December, 1975. Vancouver Natural History Society Discovery, New Series 5(1): 10-13.

________. 1976c. British Columbia birds – January to March 1976. Vancouver Natural History Society Discovery, New Series 5(2): 30-32.

________. 1976d. British Columbia birds – April to June 1976. Vancouver Natural History Society Discovery 5(3):48-50.

________. 1977. British Columbia birds – October to December 1976. Vancouver Natural History Society Discovery, New Series 6(1): 18-20.

Sherman, K. 1970. Crossbills in Victoria. Victoria Naturalist 26:98.

Sherrington, P. 1994. Yellow Rail in Yoho National Park. British Columbia Birds 4:15-16.

________. 1998. Bird highlights from the Pine Pass area, northeastern BC, Oct 3-15, 1998. Unpublished Report by the author, Cochrane, Alberta. 6 pp.

Sherry, T.W. and R.T. Holmes. 1997. American Redstart (*Setophaga ruticilla*). *In* A. Poole and F. Gill (editors). The birds of North America, No. 277. The Academy of Natural Sciences, Philadelphia, Pennsylvania; The American Ornithologists' Union, Washington, D.C. 32 pp.

Shumaker, K.M. and G.R. Babbel. 1980. Patterns of allozyme similarity in ecologically central and marginal populations of *Hordeum jubatum* in Utah. Evolution 34:110-116.

Sibley, C.G. and L.L. Short. 1959. Hybridization in the buntings (*Passerina*) of the Great Plains. Auk 76:443-463.

________. and ________. 1964. Hybridization in the orioles of the Great Plains. Condor 66:130-150.

Sibley, C.G. and B.L. Monroe. 1990. Distribution and taxonomy of birds of the world. Yale University Press, New Haven, Connecticut. 1, 111 pp.

Sibley, D. 1996. Field identification of the Sharp-tailed Sparrow complex. Birding 28:197-208.

Siddle, C. 1980. Notes on the birds of Fernie, British Columbia – spring/ summer 1969. Royal British Columbia Museum Unpublished Report, Victoria. 11 pp. (Bibliography 4157).

________. 1981. Potential effects of the activities of the Scurry-Rainbow oil company on the avifauna of the south end of Charlie Lake. British Columbia Fish and Game Branch Unpublished Report, Fort St. John. 12 pp.

________. 1988a. The winter season – northwestern Canada region. American Birds 42:292-293.

________. 1988b. The autumn migration – northwestern Canada region. American Birds 42:293-295.

________. 1988c. The spring season – northwestern Canada region. American Birds 42:462-463.

________. 1988d. The nesting season – northwestern Canada region. American Birds 42:1316-1317.

________. 1989a. The autumn migration – northwestern Canada region. American Birds 43:134-136.

________. 1990a. The winter season – British Columbia and the Yukon region. American Birds 44:312-317.

________. 1990b. The spring season – British Columbia and the Yukon region. American Birds 44:482-486.

________. 1990c. The summer season – British Columbia/Yukon region. American Birds 44:1173-1176.

________. 1991a. The autumn migration – British Columbia/Yukon region. American Birds 45:142-145.

________. 1991b. The winter season – British Columbia/Yukon region. American Birds 45:306-308.

________. 1991c. The spring migration – British Columbia/Yukon region. American Birds 45:486-489.

________. 1992a. The autumn migration – British Columbia/Yukon region. American Birds 46:139-142.

________. 1992b. The winter season – British Columbia/Yukon region. American Birds 46:303-306.

________. 1992c. The summer season – British Columbia/Yukon region. American Birds 46:1167-1170.

________. 1992d. The declining population of warblers in northeastern British Columbia. Pages 143-151 *in* S. Rautio (editor). Community action for endangered species. Federation of British Columbia Naturalists and Northwest Wildlife Preservation Society, Vancouver, 28 and 29 September 1991.

________. 1993a. The autumn migration – British Columbia region. American Birds 47:136-139.

________. 1993b. The winter season – British Columbia/Yukon. American Birds 47:290-293.

________. 1993c. The summer season – British Columbia/Yukon region. American Birds 47:1141-1143.

________. 1994a. Autumn migration – British Columbia/Yukon region. American Birds 48:142-144.

________. 1994b. Winter season – British Columbia/Yukon Region. National Audubon Society Field Notes 48:240-242.

________. 1995. Noteworthy bird sightings – 1992. British Columbia Birds 3:38-42.

________. 1997. The warblers of the North Peace area of British Columbia (status, distribution, habitats, and breeding chronology). WBT Wild Bird Trust of British Columbia Manuscript, West Vancouver. 44 pp.

Siddle, C. and J. Bowling. 1993. The spring season – British Columbia/ Yukon. American Birds 47:445-447.

Siemens, A.H. 1966. The process of settlement in the Lower Fraser Valley – in its provincial context. Pages 27-49 *in* A.H. Siemens (editor). Lower Fraser Valley – evolution of a cultural landscape. British Columbia Geographical Series No. 9. Tantalus Research Ltd., Vancouver.

Sierra Legal Defence Fund. 1998. British Columbia Forestry Report Card: 1997-1998. Sierra Legal Defence Fund, Vancouver. <http://www.sierraleg.org>

Sirk, G., R.J. Cannings, and M.G. Shepard. 1973. Shuswap Lake Park annual report – 1973. British Columbia Parks Branch Unpublished Report, Victoria. 66 pp. (Bibliography 4169).

Skutch, A.F. 1996. Orioles, blackbirds and their kin: a natural history. University of Arizona Press, Tucson. 291 pp.

Small, A. 1994. California birds: their status and distribution. Ibis Publishing, Vista, California. 342 pp.

Smith, D.W. and J. Ireland. 1992. First record of the Black-shouldered Kite for Canada. Western Birds 23:177-178.

Smith, J.N.M. 1981. Cowbird parasitism, host fitness, and age of the host female in an island Song Sparrow population. Condor 83:152-161.

________. 1982. Song Sparrow pair raises four broods in one year. Wilson Bulletin 94:584-585.

________. 1995. Cowbird parasitism: is it a threat to B.C.'s songbirds? B.C. Field Ornithologist (September):4-6.

Smith, J.N.M. and P. Arcese. 1994. Brown-headed Cowbirds and an island population of Song Sparrows: a 16-year study. Condor 96:916-934.

Smith, J.N.M. and D.A. Roff. 1980. Temporal spacing of broods, brood size, and parental care in Song Sparrows (*Melospiza melodia*). Canadian Journal of Zoology 58:1007-1015.

Smith, J.N.M., M.J. Taitt, C.M. Rogers, P. Arcese, L.F. Keller, A.L.E.V. Cassidy, and W.M. Hochachka. 1996. A metapopulation approach to the population biology of the Song Sparrow *Melospiza melodia*. Ibis 138:120-128.

Smith, J.N.M., D. Ward, and C.J. Fonnesbeck. 1998. Community impacts of cowbird parasitism in the south Okanagan valley. Canadian Society of Zoologists Bulletin 29:98.

Smith, K.G. and D.C. Andersen. 1982. Food predation and reproductive ecology of the Dark-eyed Junco in northern Utah. Auk 99:650-661.

Smith, M.R., P.W. Mattocks, and K.M. Cassidy. 1997. Breeding birds of Washington state – location data and predicted distributions. Volume 4 *in* K.M. Cassidy, C.E. Grue, M.R. Smith, and K.M. Dvornich (editors). Washington state gap analysis – final report. Seattle Audubon Society Publications in Zoology No. 1, Seattle, Washington. 538 pp.

Smith, R. and M. Dallman. 1996. Forest gap use by breeding Black-throated Green Warblers. Wilson Bulletin 108:588-591.

Smith, R.L. 1968. Grasshopper Sparrow (*Ammodramus savannarum*). Pages 725-745 *in* O.L. Austin (editor). Life histories of North American cardinals, grosbeaks, buntings, towhees, finches, sparrows, and allies. United States National Museum Bulletin 237, Washington, D.C.

Snelling, J.C. 1968. Overlap in feeding habits of Red-winged Blackbirds and Common Grackles nesting in a cattail marsh. Auk 85:560-585.

Sodhi, N.S. and L.W. Oliphant. 1993. Prey selection by urban-breeding Merlins. Auk 110:727-735.

Sodhi, N.S. and C.A. Paszkowski. 1995. Habitat use and foraging behavior of four parulid warblers in a second-growth forest. Journal of Field Ornithology 66:277-288.

———. and ———. 1997. The pairing success of male Black-and-white Warblers, *Mniotilta varia,* in forest fragments and a continuous forest. Canadian Field-Naturalist 111:457-458.

Sogge, M.K., W.M. Gilbert, and C.V. Riper. 1994. Orange-crowned Warbler (*Vermivora celata*). *In* A. Poole and F. Gill (editors). The birds of North America, No. 101. The Academy of Natural Sciences, Philadelphia, Pennsylvania; The American Ornithologists' Union, Washington, D.C. 20 pp.

Solow, R. 1974. The economics of resources or the resources of economics. American Economics Review 64:1-14.

Sontag, C. 1993. Golden-crowned Sparrow (*Zonotrichia atricapilla*): 17 January 1993, Sheboygan Co. Passenger Pigeon 55:275-276.

Soper, J.D. 1949. Birds observed in Grande Prairie–Peace River region of northwestern Alberta. Auk 66:233-257.

Soulé, M. 1973. The epistasis cycle: a theory of marginal populations. Annual Review of Ecology and Systematics 4:165-187.

Sowls, A.L., S.A. Hatch, and C.J. Lensink. 1978. Catalog of Alaskan seabird colonies. United States Fish and Wildlife Service, Biological Services Program FWS/OBS-78/78. Washington, D.C.

Sowls, A.L., D.R. Nysewander, J.L. Trapp, and J.W. Nelson. 1982. Marine bird and mammal survey of the outer coast of southeast Alaska. United States Fish and Wildlife Service Unpublished Report, Anchorage, Alaska. 162 pp.

Speirs, D.H. 1968a. *Hesperiphona vespertina vespertina* Grinnell, Eastern Evening Grosbeak. Pages 206-237 *in* O.L. Austin (editor). Life histories of North American cardinals, grosbeaks, buntings, towhees, finches, sparrows, and allies (Part 1). United States National Museum Bulletin No. 237, Washington, D.C.

———. 1968b. *Hesperiphona vespertina brooksi* Grinnell, Western Evening Grosbeak. Pages 237-251 *in* O.L. Austin (editor). Life histories of North American cardinals, grosbeaks, buntings, towhees, finches, sparrows, and allies (Part 1). United States National Museum Bulletin No. 237, Washington, D.C.

Speirs, J.M. and E. Pegg. 1972. First record of Painted Redstart (*Setophaga picta*) for Canada. Auk 89:898.

Spitman, P. 1996. Blackburnian Warbler sighting – Fort Nelson area, British Columbia. Birders Journal 5(1):24-26.

Spreadborough, W. 1916. Birds of Lillooet and Brackendale, B.C. July 1 – August 24, 1916. National Museum of Canada Unpublished Report, Ottawa, Ontario. 119 pp. (Bibliography 4240).

Sprenkle, J.M. and C.R. Blem. 1984. Metabolism and food selection of eastern House Finches. Wilson Bulletin 96:184-195.

Springer, A.M., E.C. Murphy, D.G. Roseneau, and M.I. Springer. 1982. Population status, reproductive ecology and trophic relationships of seabirds in northwestern Alaska. Pages 127-242 *in* Environmental assessment of the Alaskan continental shelf, final report of principal investigators, volume 30. United States Bureau of Land Management and National Oceanic and Atmospheric Administration, Outer Continental Shelf Environmental Assessment Program, Boulder, Colorado.

Sprot, G.D. 1925. Bird banding report – September, 1924 to March, 1925 "Mavisburn" banding station. Mill Bay, Vancouver Island, British Columbia. Canadian Field-Naturalist 39:179-180.

Sprunt, A. 1979. Townsend's Warbler. Pages 102-104 *in* L. Griscom and A. Sprunt (editors). The warblers of America. Revised and updated by E.E. Riley. Doubleday, Garden City, New York.

Stallcup, R. 1990. Ocean birds of the nearshore Pacific: a guide for the sea-going naturalist. Point Reyes Bird Observatory, Stinson Beach, California. 214 pp.

Stanley, T.R. 1995. Ecosystem management and the arrogance of humanism. Conservation Biology 9:255-262.

Stanwell-Fletcher, J.F. and T.C. Stanwell-Fletcher. 1943. Some accounts of the flora and fauna of the Driftwood valley region of north central British Columbia. British Columbia Provincial Museum Occasional Paper No. 4, Victoria. 97 pp.

Stanwood, C.J. 1910. The Black-throated Green Warbler. Auk 27:289-294.

Stepney, P.H.R. 1975. Wintering distribution of Brewer's Blackbird: historical aspect, recent changes, and fluctuations. Bird-Banding 46:106-125.

Stepniewski, A.M. 1975. Summary of birds observed in the Chilcotin Range–Leeward Coast Mountains district: July-August. Unpublished personal report, Williams Lake, British Columbia. 19 pp.

———. 1994. Birds of Wahluke Slope (Saddle Mountain NWR/ Wahluke Slope Wildlife Area: Hanford site biodiversity inventory). Contract report to the Nature Conservancy No. WA FO-022094. Seattle, Washington. 69 pp.

Stevens, G. 1968. Sixty-eighth Christmas Bird Count – Pender Islands, B.C. Audubon Field Notes 22:112.

Stewart, A.C. 1991. Observations of spring migrants and other wildlife in the central Cariboo – March through June, 1991. British Columbia Ministry of Environment, Lands and Parks Unpublished Report, Williams Lake. 20 pp.

Stewart, R.E. 1953. A life history study of the Yellow-throat. Wilson Bulletin 65:99-115.

Stewart, R.M. 1973. Breeding behavior and life history of Wilson's Warbler. Wilson Bulletin 85:21-30.

Stirling, D. 1961a. A report on the flora and fauna of Shuswap Park. British Columbia Parks Branch Unpublished Report, Victoria. 25 pp.

———. 1961b. Sight records of unusual birds in the Victoria area for 1959. Murrelet 41:10-11.

———. 1961c. Sixty-first Christmas Bird Count – Victoria, B.C. Audubon Field Notes 15:102.

———. 1966. Bird report (Victoria) number four – 1965. Victoria Natural History Society Mimeo, Victoria, British Columbia. 6 pp. (Bibliography 310).

———. 1970. An annotated list of birds of Mount Robson Provincial Park. British Columbia Parks Branch Unpublished Report, Victoria. 15 pp. (Bibliography 1198).

———. 1972. Birds of Vancouver Island for birdwatchers. Published by the author, Victoria, British Columbia. 27 pp.

Storer, R.W. and G.L. Nuechterlin. 1992. Western and Clark's Grebe. *In* A. Poole, P. Stettenheim, and F. Gill (editors). The birds of North America, No. 26. The Academy of Natural Sciences, Philadelphia, Pennsylvania; The American Ornithologists Union, Washington, D.C.

Stork, N.E. and M.J. Samways. 1995. Inventorying and monitoring. Pages 453-543 *in* V.H. Heywood (editor). Global Biodiversity Assessment. United Nations Environment Programme. Cambridge, UK. 1, 144 pp.

Street, L. and W. Merilees. 1974. A second sight record of the Indigo Bunting for British Columbia. Canadian Field-Naturalist 88:84.

Strelke, W.K. and J.G. Dickson. 1980. Effect of forest clear-cut edge on breeding birds in east Texas. Journal of Wildlife Management 44:559-567.

Sugden, B.A. 1956. A Big Bend tragedy. North Okanagan Naturalists Club Newsletter for February, Vernon, British Columbia.

Swarth, H.S. 1912. Report on a collection of birds and mammals from Vancouver Island. University of California Publications in Zoology 10:1-124.

———. 1922. Birds and mammals of the Stikine River region of northern British Columbia and southeastern Alaska. University of California Publications in Zoology 24:125-314.

———. 1924. Birds and mammals of the Skeena River region of northern British Columbia. University of California Publications in Zoology 24:315-394.

———. 1926. Report on a collection of birds and mammals from the Atlin region, northern British Columbia. University of California Publications in Zoology 30:51-162.

———. 1930. Notes on the avifauna of the Atlin region, British Columbia. Condor 32:216-217.

———. 1936. A list of the birds of the Atlin region, British Columbia. Proceedings of the California Academy of Sciences 23:35-58.

———. 1936a. Savannah Sparrow migration routes in the northwest. Condor 38:30-32.

Swenk, M.H. and O.A. Stevens. 1929. Harris's Sparrow and the study of it by trapping. Wilson Bulletin 41:129-177.

Sykes, P.W. and D.W. Sonneborn. 1998. First breeding record of Whooper Swan and Brambling in North America at Attu Island, Alaska. Condor 100:162-164.

Symons, R.D. 1973. Hours and the birds – a Saskatchewan record. University of Toronto Press, Toronto, Ontario.

Taber, W. and D.W. Johnston. 1968. *Passerina cyanea* (Linnaeus), Indigo Bunting. Pages 80-111 *in* O. L. Austin (editor). Life histories of North American cardinals, grosbeaks, buntings, towhees, finches, sparrows, and allies (Part 1). United States National Museum Bulletin 237, Washington, D.C.

Tait, E. 1949. Christmas bird census, 1948 – Summerland, British Columbia. Canadian Field-Naturalist 62(2):65.

Tate, E.M. 1931. Christmas bird census, 1930 – Summerland, Okanagan Lake, British Columbia. Canadian Field-Naturalist 45:36.

Tate, J. and D.J. Tate. 1982. The blue list for 1982. American Birds 36:126-135.

Tatum, J.B. 1970. Annual bird report for southern Vancouver Island – 1969. Victoria Natural History Society Mimeo, Victoria, British Columbia. 34 pp.

________. 1971. Bird report for southern Vancouver Island (1970). Victoria Natural History Society, Victoria, British Columbia. 64 pp. (Bibliography 312).

________. 1972. Annual bird report (1971) for southern Vancouver Island. Victoria Natural History Society, Victoria, British Columbia. 66 pp.

________. 1973. Annual bird report (1972) for southern Vancouver Island. Victoria Natural History Society, Victoria, British Columbia. 80 pp.

Taverner, P.A. 1917. Preliminary list of specimens taken by C.H. Young and Wm. Spreadborough at Brackendale, Lillooet, and McGillivray Creek, British Columbia, between June 11 and September 12, 1916. Pages 359-371 *in* Summer Report of the Geological Survey, Department of Mines, 1916, Ottawa, Ontario.

________. 1919. The summer birds of Hazelton, British Columbia. Condor 21:80-86.

Taylor, D.M. and C.H. Trost. 1985. The Common Grackle in Idaho. American Birds 39:217-218.

Taylor, K. 1984. A birders guide to Vancouver Island. Published by the author, Victoria, British Columbia. 106 pp.

Templeton, A.R. 1989. The meaning of species and speciation: a genetic perspective. Pages 3-27 *in* D. Otte and J.A. Endler (editors). Speciation and its consequences. Sinauer Associates, Sunderland, Massachusetts.

Terres, J.K. 1991. The Audubon encyclopedia of North American birds. Wings Books, New York. 1,109 pp.

Terrill, L.M. 1961. Cowbird hosts in southern Quebec. Canadian Field-Naturalist 75:2-11.

Thacker, B.M. and T.L. Thacker. 1923. Extracts from note-book for Little Mountain, Hope, B.C. – year 1922. Migrant 1:20-21. (Bibliography 1929).

Thacker, T.L. 1944. Willow Goldfinch in western British Columbia. Murrelet 25:9.

Thompson, B.C., J.A. Jackson, J. Burger, L.A. Hill, E.M. Kirsch, and J.L. Atwood. 1997. Least Tern (*Sterna antillarum*). *In* The birds of North America, No. 290 (A. Poole and F. Gill, editors). The Academy of Natural Sciences, Philadelphia, PA, and The American Ornithologists' Union, Washington, D.C. 32 pp.

Thompson, C.F. and V. Nolan. 1973. Population biology of the Yellow-breasted Chat (*Icteria virens* L.) in southern Indiana. Ecological Monographs 43:145-171.

Thomson, D. 1974. Naturalist program at Black Tusk 1974. British Columbia Parks Branch Unpublished Report, Victoria. 17 pp. (Bibliography 4360).

Thormin, T.W. 1973. Recent status of birds in the Peace River region of British Columbia. Renewable Resources Consulting Service Ltd., Unpublished Report, Edmonton, Alberta. 13 pp. (Bibliography 4361).

Threlfall, W. and J.R. Blacquiere. 1982. Breeding biology of the Fox Sparrow in Newfoundland. Journal of Field Ornithology 53:235-239.

Tickell, C. 1993. The human species: a suicidal success? The Geographical Journal 159:219-226.

Tompa, F.S. 1963a. Factors determining the number of Song Sparrows on Mandarte Island, B.C. Ph.D. Thesis, University of British Columbia, Vancouver. 124 pp.

________. 1963b. Behavioural response of Song Sparrows to different environmental conditions. Pages 729-739 *in* Proceedings XIII International Ornithological Congress, Helsinki, Finland.

________. 1964. Factors determining the number of Song Sparrows, *Melospiza melodia* (Wilson), on Mandarte Island, B.C., Canada. Acta Zoologica Fennica 1109:1-73.

Toochin, R. 1999. A Mourning Warbler on Sea Island: a first Vancouver record. Discovery 28:18-21.

Tordoff, H.B. and W.R. Dawson. 1965. The influence of day length on reproductive timing in the Red Crossbill. Condor 67:416-422.

Torgersen, T.R. and R.W. Campbell. 1982. Some effects of avian predators on the western spruce budworm in north-central Washington. Environmental Entomology 11:429-431.

Trail, P.W. and L.F. Baptista. 1993. The impact of Brown-headed Cowbird parasitism on populations of the Nuttall's White-crowned Sparrow. Conservation Biology 7:309-315.

Tree, A.J. and J.A. Kieser. 1982. Field separation of Lesser Yellowlegs and Wood Sandpiper. Honeyguide 110:40-41.

Troy, D.M. and G.F. Shields. 1979. Multiple nesting attempts by Alaskan redpolls. Condor 81:96-97.

Turnbull, J.D. 1929. Some Vancouver, B.C. records. Auk 46:122-123.

Twedt, D.J. and R.D. Crawford. 1995. Yellow-headed Blackbird (*Xanthocephalus xanthocephalus*). *In* A. Poole and F. Gill (editors). The birds of North America, No. 192. The Academy of Natural Sciences, Philadelphia, Pennsylvania; The American Ornithologists' Union, Washington, D.C. 28 pp.

Tweit, B. 1989. The winter season – northern Pacific coast region. American Birds 43:356-360.

Tweit, B. and S. Heinl. 1989. The spring season – northern Pacific coast region. American Birds 43:527-530.

Tweit, B. and P.W. Mattocks. 1987a. The winter season – northern Pacific coast region. American Birds 41:318-321.

Ulke, T. 1923. Birds observed at Yoho National Park, British Columbia, 1917 and 1918. Canadian Field-Naturalist 37:54-55.

Union of Concerned Scientists. 1992. Scientist's warning to humanity. <http://www.ucsusa.org/about/warning.html>

________. 1997. World scientists' call for action at the Kyoto Climate Summit. <http://www.ucsusa.org/warming/gw.worldsci.html>

United Nations. 2000. Charting the progress of populations. Department of Economic and Social Affairs, Population Division. New York.

United Nations Environment Programme. 1999. GEO-2000: Global environment outlook. Earthscan Publications, London, United Kingdom.

Unitt, P. 1974. Painted Redstarts attempt to breed in California. Western Birds 5:94-96.

Van Damme, L.M. 1996. Creston Valley – checklist of birds. Nelson Naturalists, Nelson, British Columbia.

________. 1998. The status of the Black-throated Sparrow in British Columbia. B.C. Naturalist 36:5; 36:7.

________. 1999. Status of the Bobolink in British Columbia. British Columbia Ministry of Environment Wildlife Working Report No. WR-93, Victoria. 11 pp.

Vander Haegan, M. and B. Walker. 1999. Parasitism by Brown-headed Cowbirds in shrub-steppe of eastern Washington. pp. 34-40. *In* M.L. Morrison, L.S. Hall, S.K. Robinson, S.I. Rothstein, D.C. Hahn, and T.D. Rich (eds.). Research and management of the Brown-headed Cowbird in western landscapes. Studies in Avian Biology 18, Lawrence, Kansas.

Van Horn, M.A. and T. Donovan. 1994. Ovenbird (*Seiurus aurocapillus*). *In* A. Poole and F. Gill (editors). The birds of North America, No. 88. The Academy of Natural Sciences, Philadelphia, Pennsylvania; The American Ornithologists' Union, Washington, D.C. 24 pp.

Van Tighem, K.J. 1977. The avifauna of Kootenay National Park, British Columbia. Parks Canada Unpublished Report, Radium Hot Springs, British Columbia. 151 pp.

Van Tighem, K.J. and L.W. Gyug. 1983. Ecological land classification of Mount Revelstoke and Glacier national parks, British Columbia. Volume II: wildlife resource. Parks Canada, Calgary, Alberta. 254 pp. (Bibliography 4432).

Van Valen, L. 1976. Ecological species, multi-species, and oaks. Taxonomy. 25: 233-239.

Venables, E.P. 1910. A few winter bird notes. Ottawa Naturalist 24:20-21.

Verbeek, N.A.M. 1962. An ecological study of the Brewer's Blackbird during the winter. B.Sc. Thesis, University of British Columbia, Vancouver. 55 pp.

________. 1964. A time and energy budget study of the Brewer's Blackbird. Condor 66:70-74.

Vermeer, K., and R.W. Butler (editors). 1989. The ecology and status of marine and shoreline birds in the Strait of Georgia, British Columbia. Canadian Wildlife Service Special Publication. Ottawa, Ontario. 186 pp.

Vermeer, K. and K.H. Morgan (editors). 1997. The ecology, status, and conservation of marine and shoreline birds of the Queen Charlotte Islands. Canadian Wildlife Service Occasional Paper No. 93. Ottawa, Ontario. 150 pp.

Vermeer, K., R.W. Butler, and K.H. Morgan (editors). 1992. The ecology, status, and conservation of marine and shoreline birds on the west coast of Vancouver Island. Canadian Wildlife Service Occasional Paper No. 75. Ottawa, Ontario. 132 pp.

Vermeer, K., K.R. Summers, and D.S. Bingham. 1976. Birds observed at Triangle Island, British Columbia. Murrelet 57:35-42.

Vickery, P.D. 1996. Grasshopper Sparrow (*Ammodramus savannarum*). *In* The A. Poole and F. Gill (editors). The birds of North America, No. 239. The Academy of Natural Sciences, Philadelphia, Pennsylvania; The American Ornithologists' Union, Washington, D.C. 24 pp.

Vickery, P.D. and J.R. Herkert (editors). 1999. Ecology and conservation of grassland birds in the western hemisphere. Studies in Avian Biology No. 19:1-299.

Victoria Natural History Society. 1974. Victoria bird check list. Victoria Natural History Society, Victoria, British Columbia. Leaflet.

Vitousek, P., P. Ehrlich, A. Ehrlich, and P. Matson. 1986. Human appropriation of the products of photosynthesis. BioScience 36:368-373.

Vitousek, P., H.A. Mooney, J. Lubchenco, and J.M. Melillo. 1997. Human domination of the Earth's ecosystems. Science 277:494-499.

Vitousek, P.M. 1994. Beyond global warming: ecology and global change. Ecology 75:1861-1876.

Wackernagel, M. and W. Rees. 1996. Our ecological footprint: reducing human impact on the earth. New Society Publishers, Gabriola Island, British Columbia.

Wade, C. 1977. The birds of Yoho National Park. Parks Canada Unpublished Report, Ottawa, Ontario. 799 pp. (Bibliography 1994).

Wahl, T.R. and D.R. Paulson. 1991. A guide to bird finding in Washington. Print Shop, Lynden, Washington. 178 pp.

Walker, R. 1996. Birds of the Boundary District (Grand Forks–Christina Lake) – frequency, numbers, and migration. Unpublished report by the author, Christina Lake, British Columbia. 52 pp.

Walkinshaw, L.H. 1937. Leconte's Sparrow breeding in Michigan and South Dakota. Auk 54:309-320.

________. 1952. Chipping Sparrow notes. Bird-Banding 23:101-108.

________. 1968. Le Conte's Sparrow (*Passerherbulus caudacutus* Latham). Pages 765-776 *in* O.L. Austin (editor). Life histories of North American cardinals, grosbeaks, buntings, towhees, finches, sparrows, and allies (Part 2). United States National Museum Bulletin No. 237, Washington, D.C.

Walkinshaw, L.H. and D.A. Zimmerman. 1961. Range expansion of the Brewer's Blackbird in eastern North America. Condor 63:162-177.

Walters, C. 1986. Adaptive management of renewable resources. Macmillan, New York.

Ward, D. and J.N.M. Smith. 1998. Morphological differentiation of Brown-headed Cowbirds in the Okanagan valley, British Columbia. Condor 100:1-7.

________. and ________. 2000. Interhabitat differences in parasitism frequencies by Brown-headed Cowbirds in the Okanagan valley, British Columbia. Pages 210-219 *in* J.N.M. Smith, T.L. Cook, S.I. Rothstein, S.K. Robinson, and S.G. Sealy (editors). Ecology and management of cowbirds and their hosts. University of Texas Press, Austin.

Ward, P., G. Radcliffe, J. Kirkby, J. Illingworth, and C. Cadrin. 1998. Sensitive ecosystems inventory: East Vancouver Island and Gulf Islands, 1993-1997. Volume 1: methodology, ecological descriptions and results. Technical Report Series No. 320, Canadian Wildlife Service, Pacific and Yukon Region, British Columbia.

Waterhouse, M.J. 1995. Breeding bird communities in riparian habitats of the MSxv and SBPSxc in the Cariboo forest region. British Columbia Ministry of Forests Pilot Project Report, 27 October 1995, Williams Lake. 21 pp.

Waterhouse, M.J. and R. Dawson. 1998. Bird communities in interior Douglas-fir forests in the Cariboo forest region. Pages 90-112 *in* A. Vyse, C. Hollstedt, and D. Huggard (editors). Managing the dry Douglas-fir forests of the southern interior: workshop proceedings, April 29-30, 1997. British Columbia Ministry of Forests, Research Branch Working Paper 34/1998, Kamloops. 199 pp.

Watson, A. 1986. Notes on the sexual cycle of some Baffin Island birds. Canadian Field-Naturalist 100:396-397.

Watt, D.J. and E.J. Willoughby. 1999. Lesser Goldfinch (*Carduelis psaltria*). *In* A. Poole and F. Gill (editors). The birds of North America, No. 392. The Birds of North America, Inc., Philadelphia, Pennsylvania. 24 pp.

Weaver, R.L. 1939. The northern distribution and status of the English Sparrow in Canada. Canadian Field-Naturalist 53:95-99.

Weaver, R.L. and F.H. West. 1943. Notes on the breeding of the Pine Siskin. Auk 60:492-504.

Weber, W.C. 1964. Notes on the birds of the Prince George area, British Columbia. University of British Columbia, Department of Zoology Unpublished Manuscript, Vancouver. 10 pp.

________. 1968. Notes on the birds of the Prince George area, British Columbia. University of British Columbia, Department of Zoology Report, Vancouver. 13 pp. (Bibliography 4502).

________. 1972. Birds in cities: a study of populations, foraging, ecology, and nest-sites of urban birds. M.Sc. Thesis, University of British Columbia, Vancouver. 269 pp. (Bibliography 1156).

________. 1975. Nest-sites of birds in residential areas of Vancouver, British Columbia. Canadian Field-Naturalist 89:457-460.

________. 1976. Mourning Warbler and Northern Oriole in northeastern British Columbia. Murrelet 57:68-69.

________. 1981. Unusual bird sightings: April to July, 1981. B.C. Naturalist 19(3):4.

________. 1985. Vancouver bird records committee report for 1982. Discovery 14:51-55.

________. 1989. Pest birds of agriculture in British Columbia. Pages 13-33 *in* I.E.P. Taylor and R.I. Alfaro (editors). Issues in B.C. vertebrate pest management. Association of Professional Biologists of British Columbia and Professional Pest Management Association of British Columbia, Vancouver.

Weber, W.C. and R.J. Cannings. 1990. The autumn migration – British Columbia and Yukon region. American Birds 44:144-149.

Weber, W.C. and B. Kautesk. 1975. A sight record of the Painted Redstart near Vancouver, British Columbia. Western Birds 6:67.

Weber, W.C., D. Kragh, and D.M. Mark. 1980. First Swamp Sparrow record for the Vancouver area. Discovery 9:28-29.

Weber, W.C., M.K. Daly, and B.M. Kautesk. 1982. First Vancouver record of the Chestnut-collared Longspur. Discovery 10:94-95.

Webster, J.D. 1969a. Thirty-third breeding bird census – climax coastal forest. Audubon Field Notes 23:716-717.

________. 1969b. Thirty-third breeding bird census – white spruce forest. Audubon Field Notes 23:717.

________. 1983. A new subspecies of Fox Sparrow from Alaska. Proceedings of the Biological Society of Washington 96:664-668.

Weeden, R.B. 1960. The birds of Chilkat Pass, British Columbia. Canadian Field-Naturalist 74:119-129.

Welsh, D.A. 1971. Breeding and territoriality of the Palm Warbler in a Nova Scotia bog. Canadian Field-Naturalist 85:31-37.

________. 1975. Savannah Sparrow breeding and territoriality on a Nova Scotia dune beach. Auk 92:235-251.

West, G.C. and E.P. Bailey. 1986. Rustic Bunting, Purple Finch, and Cassin's Finch in south coastal Alaska. Murrelet 67:32.

West, G.C., L.J. Peyton, and L. Erving. 1968. Analysis of spring migration of Lapland Longspurs to Alaska. Auk 85:639-653.

Westerborg, B. 1962. Twenty-sixth breeding bird census – disturbed second-growth coast forest. Audubon Field Notes 16:524-525.

Westerborg, B.J. and D. Stirling. 1963. Twenty-seventh breeding bird census – disturbed second-growth coast forest. Audubon Field Notes 17:500-501.

Westworth, D.A. and E.S. Telfer. 1993. Summer and winter bird populations associated with five age-classes of aspen forest in Alberta. Canadian Journal of Forestry 23:1830-1836.

Wetherbee, D.K. and N.S. Wetherbee. 1961. Artificial incubation of eggs of various bird species and some attributes of neonates. Bird-Banding 32:141-159.

Wetmore, S.P., R.A. Keller, and G.E.J. Smith. 1985. Effects of logging on bird populations in British Columbia as determined by a modified point-count method. Canadian Field-Naturalist 99:224-233.

Wheelwright, N.T. and J.D. Rising. 1993. Savannah Sparrow (*Passerculus sandwichensis*). *In* A. Poole and F. Gill (editors). The birds of North America, No. 45. The Academy of Natural Sciences, Philadelphia, Pennsylvania; The American Ornithologists' Union, Washington, D.C. 28 pp.

Wheelwright, N.T. and C.B. Schultz. 1994. Age and reproduction in Savannah Sparrows and Tree Swallows. Journal of Animal Ecology 63:686-702.

Wheelwright, N.T., J.J. Lawler, and J.H. Weinstein. 1997. Nest-site selection in Savannah Sparrows: using gulls as scarecrows? Animal Behavior 53:197-208.

White, C.M., F.S.L. Williamson, and W.B. Emison. 1974. *Tringa glareola* – a new breeding species for North America. Auk 91:175-177.

Wiley, E.O. 1978. The evolutionary species concept reconsidered. Systematic Zoology 27:17.

Wilkes, B. 1977. The myth of the non-consumptive user. Canadian Field-Naturalist 91:343-349.

Williams, A.B. 1936. The composition and dynamics of a beech-maple climax community. Ecological Monographs 6:317-408.

Williams, J.M. 1996. Nashville Warbler (*Vermivora ruficapilla*). *In* A. Poole and F. Gill (editors). The birds of North America, No. 205. The Academy of Natural Sciences, Philadelphia, Pennsylvania; The American Ornithologists' Union, Washington, D.C. 20 pp.

________. 1996a. Bay-breasted Warbler (*Dendroica castanea*). *In* A. Poole and F. Gill (editors). The birds of North America, No. 206. The Academy of Natural Sciences, Philadelphia, Pennsylvania; The American Ornithologists' Union, Washington, D.C. 20 pp.

Williams, M. 1978. Bird surveys in the south Okanagan valley between Okanagan Falls and Osoyoos, May 29, 1978 to September 1, 1978.

Canadian Wildlife Service Unpublished Report, Delta, British Columbia. 121 pp.
Williams, M.Y. 1933a. Biological notes, covering parts of the Peace, Liard, Mackenzie and Great Bear River basins. Canadian Field-Naturalist 47:23-31.
________. 1933b. Fauna of the former Dominion Peace River block, British Columbia. Pages 14-22 *in* Provincial Museum of Natural History and Anthropology Report for the year 1932, Victoria.
Willing, R.M. 1970. The role of song in the behaviour and evolution of three species of spizellid sparrow: sonograph and field play-back studies. M.Sc. Thesis, Simon Fraser University, Burnaby, British Columbia. 115 pp.
Willson, M.F. 1966. Breeding ecology of the Yellow-headed Blackbird. Ecological Monographs 36:51-77.
Wilson, D.J. 1981. Traffic takes heavy grosbeak toll. B.C. Naturalist 19: 18-19.
Wilson, L.K., I.E. Moul, K.M. Langelier, and J.E. Elliott. 1995. Summary of bird mortalities in British Columbia and Yukon, 1963-1994. Canadian Wildlife Service Technical Report Series No. 249, Delta, British Columbia. 205 pp.
Wilson, M.C., E. Wilson, and L. Wilson. 1972. Ornithological notes from Columbia Lake, British Columbia. Syesis 5:63-65.
Wilson, W.H. 1996. Palm Warbler (*Dendroica palmarum*). *In* A. Poole and F. Gill (editors). The birds of North America, No. 238. The Academy of Natural Sciences, Philadelphia, Pennsylvania; The American Ornithologists' Union, Washington, D.C. 20 pp.
Wilz, K.J. and V. Giampa. 1978. Habitat use by Yellow-rumped Warblers at the northern extremities of their winter range. Wilson Bulletin 90:566-574.
With, K.A. 1994. McCown's Longspur (*Calcarius mccownii*). *In* A. Poole and F. Gill (editors). The birds of North America, No. 96. The Academy of Natural Sciences, Philadelphia, Pennsylvania; The American Ornithologists' Union, Washington, D.C. 24 pp.
Wittenburger, J.F. 1978. The breeding biology of an isolated Bobolink population in Oregon. Condor 80:355-371.
Wood, R.O. 1982. History of population fluctuations and infestations of important forest insects in the Vancouver forest region. Canadian Forestry Service, Pacific Forest Research Centre, Victoria. 43 pp.
Woods, R.S. 1968. *Carpodacus mexicanus frontalis* (Say), House Finch. Pages 290-314 *in* O.L. Austin (editor). Life histories of North American cardinals, grosbeaks, buntings, towhees, finches, sparrows, and allies. United States National Museum Bulletin No. 237, Washington, D.C.
Woodward, P.W. 1983. Behavioral ecology of fledgling Brown-headed Cowbirds and their hosts. Condor 85:151-163.
Wootton, A. 1965. House Sparrows versus Violet-green Swallows. Blue Jay 23:191.
Wootton, J.T. 1996. Purple Finch (*Carpodacus purpureus*). *In* A. Poole and F. Gill (editors). The birds of North America, No. 208. The Academy of Natural Sciences, Philadelphia, Pennsylvania; The American Ornithologists' Union, Washington, D.C. 20 pp.
World Resources Institute. 2000. A guide to world resources 2000-2001: people and ecosystems: the fraying web of life. World Resources Institute, Washington, D.C.
Worldwatch Institute. 1997. Worldwatch Institute, Washington, D.C.
Wray, T. and R.C. Whitmore. 1979. Effects of vegetation on nesting success of Vesper Sparrows. Auk 96:802-805.
Wright, A.L., G.D. Hayward, S.M. Matsuoka, and P.H. Hayward. 1998. Townsend's Warbler (*Dendroica townsendi*). *In* A. Poole and F. Gill (editors). The birds of North America, No. 333. The Birds of North America, Inc., Philadelphia, Pennsylvania. 20 pp.
Wykoff, J.N. 1979. First record of Golden-crowned Sparrow in Michigan. Jack-Pine Warbler 57:109,
Wythe, M.W. 1938. The White-throated Sparrow in western North America. Condor 40:110-117.
Yasukawa, K. and W.A. Searcy. 1995. Red-winged Blackbird (*Agelaius phoeniceus*). *In* A. Poole and F. Gill (editors). The birds of North America, No. 184. The Academy of Natural Sciences, Philadelphia, Pennsylvania; The American Ornithologists' Union, Washington, D.C. 28 pp.
Yezerinac, S.M. 1988. American Redstarts using Yellow Warbler's nest. Wilson Bulletin 105:529-531.
Young, B.E. 1989. First specimen record of the Indigo Bunting, *Passerina cyanea*, in British Columbia. Canadian Field-Naturalist 103:415.
Yunick, R.P. 1983. Age and sex determination of Purple Finches during the breeding season. North American Bird Bander 8:48-51.
________. 1997. Geographical distribution of re-encountered Pine Siskins captured in upstate, eastern New York during the 1989-1990 irruption. North American Bird Bander 22:10-15.
Zach, R. and J.B. Falls. 1975. Response of the Ovenbird (Aves: Parulidae) to an outbreak of spruce budworm. Canadian Journal of Zoology 53:1669-1672.
Zahn, S.N. and S.I. Rothstein. 1999. Recent increase in male House Finch plumage variation and its possible relationship to avian pox. Auk 116:35-44.
Zar, J.H. 1974. Biostatistic analysis. Prentice-Hall, Englewood Cliffs, New Jersey. 620 pp.
Zink, R.M. 1994. The geography of mitochondrial DNA variation, population structure, hybridization, and species limits of the Fox Sparrow *Passerella iliaca*. Evolution 48:96-111.
Zink, R.M. and D.L. Dittmann. 1993. Population structure and gene flow in the Chipping Sparrow and a hypothesis for evolution in the genus *Spizella*. Wilson Bulletin 105:399-413.

CONTRIBUTORS

The following 3,629 people contributed information, mainly personal observations of birds, to this volume.

Abbey, E.
Abbey, J.
Abbott, Elizabeth
Abelson, B.
Ableson, D.
Ackerman, Andy
Acres, Peter
Adams, B.E.
Adams, Bob
Adams, Ellen
Adams, Ian
Adams, John
Adams, Kaye
Adams, Nadine
Adams, W.
Adamson, Anne
Adkisson, Curtis S.
Ainley, Marika
Ainsley, Jim
Aitchison, Cathy
Alcorn, Gordon D.
Aldcroft, David S.
Alderman, E.S.
Alderwood, Damion
Alessandro, D.
Alexander, A.M.
Alexander, Daphne
Alexander, Hector
Alexander, Ken C.
Alexander, Linda
Alexandrovich, Arlene
Alford, C.E.
Algar, David
Allan, Ian V.F.
Allard, Marlene
Allen, Andrea
Allen, Dorothy
Allen, Marty
Allen, N.
Allies, Kelly
Allies, Wilson F.
Allison, David
Allison, Marnee
Alton, G.M.
Alton, J.
Alton, Rudy
Ambedian, B.
Ambedian, Dody
Ambedian, Doug
Ambridge, David
Amulifan, A.
Amundsen, Adeline
Anderess, Alice
Anderson, Alex
Anderson, Astrid
Anderson, Bob
Anderson, C.
Anderson, C.G.
Anderson, D.
Anderson, E .
Anderson, Elton A.
Anderson, Errol M.
Anderson, Gladys
Anderson, Harvey
Anderson, Jerry
Anderson, John
Anderson, Morgan
Anderson, Nancy
Anderson, R.M.
Anderson, T.M.
Anderson, Vicky
Anderson, W.B.
Anderson, W.M.
Anderson, William J.
Anderson, Y.
Andresen, Andy
Andress, Alice
Andrew, J.
Andrews, Art
Andrews, Dianne
Angle, Neil
Anker, Bill
Anse, Rick
Ansell, Gerry
Ansell, Margarite
Ansell, R.
Ansell, S.
Ansell, Wendy
Antifeau, Ted
Antoniazzi, Cathy
Antoniazzi, Helen
Anweiler, Gary
Arcese, Peter
Archibald, Chris
Armstrong, Elizabeth
Arndt, Janice
Arnold, G.
Arnold, John
Arvey, Dale
Arvey, W.D.
Asfeldt, Henrik
Ashby, Dennis
Asher, Dale
Asher, Richard C.
Askeldy, R.
Askevold, Ingolf
Askevold, R.
Askey, Frances
Askey, Jim
Asplin, Heather
Assaly, Robin
Atherton, D.
Atkins, Vicky
Atkinson, Chris
Atkinson, G.
Atkinson, Pat
Atkinson, R.N.
Auden, K.F.
Austin, Madeline
Authier, M.C.
Axhorn, Peter
Bacon, T.
Badger, G.B.
Bailey, G.
Bailey, Steve
Bailiff, B.
Bailik, J.
Baillie, Steve
Bain, A.S.
Baird, Robin W.
Baker, Betty
Baker, Bob
Baker, Brian
Baker, Eric
Baker, G.R.
Baker, Gerry
Baker, Margaret
Baker, Mary
Baker, R.E.
Baker, Stanley W.F.
Balding, James
Baldwin, James
Ball, Sheila
Bambauer, Luke
Bambauer, R.
Bambauer, Sheila
Bamfferman, Dean
Ban, J.
Bandy, P.J.
Bangs, E.A.
Bangs, O.
Banks, Mardy
Banks, R.
Banks, Richard C.
Banwell, June
Barclay, D.A.
Barclay, Grace
Barker, Ruth
Barkley, William D.
Barlow, Jon C.
Barnard, Anthony E.
Barnard, Geoff
Barnes, Jack
Barnes, M.
Barnes, T. Barry
Barnett, Henry
Barnett, Ian M.
Barnett, Jennifer M.
Baron, Toni
Barr, Andrew
Barr, R.B.
Barreca, Jeanette
Barrick, Mary
Barss, A.F.
Barstow, Vicki
Bartels, Avery
Bartels, Letty
Bartholomew, K.
Bartkow, Virginia G.
Bartle, Joe
Bartlett, Colin
Barton, A.S.
Barvis, Margaret
Bassetti, Paolo
Bastaja, Daniel S.
Bastien, C.H.
Batchelder, C.F.
Bateman, Robert
Bates, Erica
Bath, Fermine
Bath, H.R.
Bath, T.
Baumbrough, John
Baumbrough, June
Bavin, Helen
Bavin, Ryan
Bavington, B.
Baxter, Susan
Bazett, Stephen J.
Beacham, E. Derek
Beacher, Marc-André
Beal, Ron
Beals, Alice
Beam, B.
Bear, Alan
Beattie, F. Norman
Beauchesne, E.A.
Beauchesne, Suzanne
Beaumont, Art
Beaumont, Barbara
Becevel, Marion
Beck, Joan
Beck, Ruth
Beck, Sarah
Becker, Bruno
Becker, Phyllis
Beckett, J.
Beckett, Kenneth Ray
Beebe, Frank L.
Beech, Brent
Beere, Mark
Beetstra, R. Dorothy
Begg, Barbara
Begg, J.R.
Beggs, Andrea
Beggs, Karen
Bekhuvs, Timothy J.
Belither, Ann
Bell, A.
Bell, E.E.
Bell, F.C.
Bell, Grace M.
Bell, H.M.S.
Bell, Kevin M.
Bell, Marcella E.
Bell, Marius
Bell, William
Belsom, Betty
Belsom, Sid
Belt, C.
Belton, Desmond E.J.
Bely, Carl
Bely, Ethel
Bendell, James F.
Bender, Fred
Beninger, P.
Bennett, Bruce
Bennett, S.P.
Bennett, Stephen
Bennett, Stewart
Bennie, Fred
Bennie, P.M.
Bennie, Winnifred M.
Benson, D.A.
Benson, Margaret
Benson, W.A.
Bentley, Michael D.
Benton, Marion
Berg, Gail
Berger, Allan E.
Bergerud, A. Tom
Bernadet, Nathan
Bertram, Douglas F.
Best, Alan
Best, Robin
Betuzzi, Sue
Bewick, Mary
Beynon, D. Edwin
Bichon, Ryan
Bidnon, Ryan
Bigg, Michael
Biggar, James
Billings, Ray
Bindernagel, John
Birck, Nancy
Birkel, Anne
Biro, Mike
Birt, Jackie
Bishop, James
Bishop, Jean
Bittner, A.
Bittner, B.
Bittner, D.E.
Bjornson, Don
Blackbourne, Sylvia
Blackmore, Ann K.
Blanchard, B.
Blanchard, D.
Blanchard, George
Blanshard, Claude
Blanshard, Hazel
Blanshard, John
Bloem, Gerard
Blogg, Dennis
Blokker, Peter F.
Blomgrem, Bemgt
Blood, Donald A.
Bloom, F.C.
Bloom, Gerard
Blow, Barbara
Blusson, S.
Boal, P.
Boardman, R.
Boas, F.
Boby, J.
Bodman, Geoff F.
Boer, Steve
Boggs, Bob
Boggs, Elsie
Boggs, O.D.
Bolton, F.D.
Bomford, Ted
Bonar, Norma
Bonar, Richard L.
Bond, P.G.
Bond, Talbot H.
Bond, Tracy
Bonner, Lynne
Bonner, Vera
Bonsall, K.
Bonsel, W.
Boone, John A.
Boone, T.
Boorman, Gladys
Boot, Leslie
Booth, Barry
Booth, E.J.
Booth, I.
Bopp, Victor
Borden, F.
Bosomworth, Myrna
Bostock, David
Boston, R.E.
Botting, Evelyn
Bouchard, Lorraine
Bouman, Dan
Bousfield, E.G.
Bowe, Larry
Bowen, Colin R.
Bowen, Dan
Bowen, Edna
Bowling, G.J.
Bowling, Jack
Bowling, James
Bowman, Robert I.
Boxall, Peter C.
Boyce, J.
Boyce, Kenneth C.
Boyce, Wayne C.
Brabandt, A.
Bracewell, Vince
Braden, B.
Bradford, Sherry
Bradley, Dorothy M.
Bradley, Robert
Bradley, S.F.
Bradshaw, Jan
Bradshaw, Jim
Brahandt, A.
Brandell, Bruce R.
Brandon, Jean
Brandt, Charles A.E.
Brandt, Herbert W.
Braun, Abe
Breadon, James
Brebart, Kevan
Brehart, Josanne
Breitkreutz, A.R.
Brenchley, Anne
Brenton, W.D.
Brewster, A.
Brewster, Norman
Brewster, William
Bricknell, Peter
Bridge, Irene
Bridge, Richard
Briggs, Gwen
Briggs, J.
Briggs, Tom R.
Brighouse, Tom
Brin, Neville
Broadland, Ken R.
Broadley, Tom
Brodie, E.
Brook, C.
Brook, G.
Brookes, Stu
Brooks, Allan
Brooks, Allan C.
Brooks, Betty
Brooks, C.
Brooks, D.
Brooks, E.J.
Brooks, E.S.
Brooks, G.
Brooks, H.S.
Broughton, Colleen
Broughton, L.
Brown, Andrew
Brown, Bonita J.
Brown, Douglas
Brown, E.
Brown, George
Brown, Gordon
Brown, Ken
Brown, Murray
Brown, P.
Brown, Q.
Brown, Rory
Brown, T.
Brown, Valerie
Brown, W.W.
Brownlee, P.
Brownlie, Denise
Brownlie, E.
Brownlow, Harry
Bruce, Carol
Bruce, Marie
Brucker, Maury
Brun, Brenda
Brunton, Daniel F.
Bryant, Andrew
Bryant, George
Bryant, Margaret
Bryant, W.S.
Bryden, Coleen

Bryden, Colleen
Buchanan, Daphne
Buchanan, Melda
Buchanan-Simpson, M.
Buck, Nancy
Buckley, P.A.
Buechert, Ron
Buelow, Markus
Buffam, Frank
Buis, Christopher A.
Buller, R.L.
Bullock, J.C.C.
Bumpus, Ruth
Bunnell, Fred L.
Burbidge, M. (Don)
Burbride, J.
Burger, Alan
Burgess, Tom E.
Burke, Claudia
Burke, Thomas W.
Burleigh, T.D.
Burn, Jim
Burnett, Bill
Burnett, I.
Burnett, Ona
Burnett, Stella
Burnett, Terry
Burnett, W.J.
Burns, James G.
Burnton, Dan
Burr, H.
Burton, Clyde H.
Burton, J.
Burton, Walter
Bush, David
Butcher, Beverly H.
Butcher, George
Butler, E.T.
Butler, H.E.
Butler, Robert W.
Butler, Sharon
Butt, Christy
Butt, Colin
Buttimer, Andreas
Byatt, Archie
Byers, Don
Calderwood, Damon
Caldwell, Barbara
Caldwell, Jim
Caldwell, Ken
Caldwell, Lola
Caldwell, Margaret
Calef, George
Calif, Neil
Callin, Ed
Callin, Elmer
Calvert, Gilbert
Calvert, Joy
Cameron, C.
Cameron, E.
Cameron, I.
Cameron, Mabel
Camfferman, Dean
Camilletti, Monica
Camilletti, Orlando
Camp, Frank E.
Campbell, B.
Campbell, Carol
Campbell, Caryl
Campbell, D. Sean
Campbell, Eileen C.
Campbell, J.W.
Campbell, Lucile M.
Campbell, Mildred W.
Campbell, R. Wayne
Campbell, Robert L.
Campbell, Tessa N.
Cannings, Jean
Cannings, Richard J.
Cannings, Robert A.
Cannings, Stephen R.
Cannings, Sydney G.
Cantwell, G.G.
Capes, Phil
Carder, G.W.
Carefoot, Thomas H.
Carhart, Ernest
Carl, G. Clifford
Carl, Ruth
Carlisle, Susan
Carlson, Gary R.
Carlson, J.S.
Carmichael, Agnes
Carmichael, Bruce
Carmichael, Nicola
Carpentier, G.
Carr, Melissa
Carruthers, Donald
Carson, Irene
Carson, J.
Carter, Harry R.
Cartes, Leslie
Cartwright, Joan
Caspell, W.
Casperson, R.
Casselman, L.
Cassidy, Alice
Catchpole, Rob
Cathcart, Margaret
Cavers, Beth
Cawley, P.J.
Cecile, Donald G.
Cerenzia, Bob
Cerenzia, Joan
Chaddock, Arnold
Chambers, James
Chambers, Mike
Chambers, W. Jim
Chandler, J.
Chan-Sing, E.
Chaplin, S.
Chapman, Barb
Chapman, David P.
Chapman, Erika
Chapman, Fred
Chapman, Ross
Chapman, Sharon
Chappell, Eileen
Charbonneau, Alan R.
Charles, Walter
Charlesworth, Chris
Chase, J.D.
Chatwin, S.
Chatwin, Trudy
Chaundy, Rosalind
Chesterfield, Norman
Child, Ken N.
Chohan, Tonya
Christensen, James
Christmas, M.J.
Chruszcz, Bryan
Chungranes, Vi
Chutter, Myke J.
Chynell, T.L.
Chytyk, Paul
Claire, Stephanie
Clance, J.
Clapham, Paul
Clapham, Rosemary
Clapman, Fred
Clark, A.H.
Clark, Colin
Clark, J.
Clark, Mary E.
Clarke, Gerry
Clarke, Iris
Clarke, Jenny
Clarke, Lois
Clarke, W.H.
Clarkson, Norman
Clavir, M.
Clay, J.O.
Clayton, Gerry E.
Clayton, Margaret
Clayton, O.
Clements, Paul
Clow, Stewart
Clowater, James S.
Clulow, George
Coale, H.K.
Coates, B.
Cober, W.A.
Cochette, A.
Cochran, J.D.
Cochrane, June
Cochrane, Win
Code, D.
Code, I.
Coffin, Ed
Colby, Gwen
Colby, Norris A.
Coles, P.
Collard, Jennifer
Collard, Kelly
Collie, M.
Colling, B.
Collins, Jack
Collins, Mary
Colls, D.
Colonel, Cyril
Colt, Will C.
Comer, John
Comfort, David
Comfort, Scott
Commons, Michelle
Conant, Jim
Condrashoff, S.F.
Coneybeare, Howard
Connelly, Derek
Connor, C.F.
Constabel, Fred
Conway, Jack
Conway, Kay
Cook, Ann
Cook, Chris
Cook, F. Stanley
Cook, James D.
Cook, K.
Cook, M.
Cook, Peter
Cooke, E.A.
Cooke, Fred
Coombes, D.
Coon, Mae
Cooper, A.D.
Cooper, Aziza
Cooper, Bunny Adele
Cooper, Carrie
Cooper, Dianne L.P.
Cooper, Dick
Cooper, Doug
Cooper, Gillian
Cooper, J.L.
Cooper, John K.
Cooper, John M.
Cooper, L.M.
Cooper, Louise V.
Cooper, T.K.
Cooper, T.M.
Cooper, V.
Copley, Claudia R.
Copley, Darren R.
Copping, R.P.
Corbett, Andrew
Corbould, Fraser
Corder, L.
Cormack, F.
Cormann, J.
Corry, Michael
Cortez, Adeline
Cortez, Pascal
Cossentine, Carol
Costanzo, Brenda
Coste, D.
Cott, J.
Coulson, Evi
Coulson, Jane
Coulson, Mel
Counsell, Eric M.
Counsell, Jean
Cousens, Bruce
Cousens, H.
Cousens, J.
Cousineau, Cecile
Cousineau, L.
Cousineau, Vic
Coutts, H.
Coventry, L.
Coventry, R.
Cowan, Bruce
Cowan, Larry
Cox, Terry
Coyle, Tom
Crabbe, Joan
Crack, David T.
Craig, Jean
Cramer, Howard
Crandall, C.W.
Craven, Dale
Craven, Harold
Craven, Ruth
Crawford, Dosie
Crawford, M.
Crawford, Neil
Crawford, S.
Crins, Bill
Crins, William J.
Crompton, Dr. E.
Cross, Erv
Cross, Ruth
Crossman, E.
Cross-Tallman, Linnea
Crowell, John B.
Crowther, Guy
Crowther, Marjorie
Cruckshank, Kinder
Csizmadia, L.
Csizmadia, M.
Cull, G.
Cullen, Jennifer
Cullen, Leo
Cullen, R.
Cuming, Jean
Cuming, Tom
Cumming, A.
Cumming, R.H.
Cumming, Sue
Cummings, Bruce
Cummings, J.
Cummings, Neva
Cummings, R.P.
Cunningham, Beryl J.
Cunningham, John
Cunningham, John
Cunningham, M.
Cunningham, Norm
Curry, Gwen
Curwen, M.A.
Cushing, John
Cuthbert, Calvin
Cuthrie, D.
Cuthrie, P.
Cynthia, J.
Czaijkowski, Chris
D'Alessandro, C.
D'Andrea, John
Dafoe, Eric
Dafoe, Peggy
Dahan, J.
Dahlke, Anna M.
Dahlke, G.M.
Dale, Bill
Dale, Chris A.
Dale, Elthea
Dalsam, Bob
Dalsam, Irene
Daly, Carol
Daly, Mark K.
Danlock, Tye
Danzen, Pat
Darby, John
Darling, Laura
Darney, Mike
Dates, P.
Daubin, A.
Dave, Bill
Daves, N.L.
Davidson, Albert R.
Davidson, Carol
Davidson, Douglas H.
Davidson, Eleanor
Davidson, Gary S.
Davidson, Harry
Davidson, Marie
Davies, Brian
Davies, Dorothy
Davies, Edith
Davies, Jennifer
Davies, John
Davis, Alice
Davis, Ann
Davis, Henry
Davis, Larry
Davis, Lyndis
Davis, N.L.
Dawe, Helen
Dawe, Jordan T.
Dawe, Karen E.
Dawe, Neil K.
Dawe, Peter
Dawson, Rick
Day, Bev
Day, Clifford A.
de Boon, Frank
de Burgh, B.
Deagle, George
Deal, John
Dean, Anna
Deanfide, Deborah
DeAnna, Mel
Deaton, James
DeCamp, Gwen
Deegan, Catheryn
Deegan, Rick
Deemerey, Albert
Dejager, Anne
Delaise, R.
Delter, J.A.
Demarchi, Dennis A.
Demarchi, Raymond A.
DeMarco, John
Demartini, Al
Demers, Chris
Denham, Joe
Denis, David
Denison, R.A.
Dennison, Barry
Denny, Mary L.
Derkson, Lydia
Deshield, Ursula
Despard, E.
DeWolfe, Barbara
Dhondt, A.A.
Diakow, Brent
Diakow, Wayne
Dick, J.A.
Dick, J.H.
Dick, Pauline
Dick, T.A.
Dicken, F.
Dickens, Harry B.
Dickinson, Tom
Dicks, K.
Dickson, Rita
Diduck, Andrew
Diduck, Dorothy
Diggle, Paul
Dilabio, Bruce
Dillabough, Eileen
Dillard, M.
Dillon, Jenny
Dimock, Doris
Din, Nasar A.
Dirkson, Bob
Dirkson, Irene
Disney, John
Disney, Sebastian
Dittrich, Dieter
Dixon, Chris
Dixon, Joseph
Dizzey, A.
Dobovicnik, G.
Dobozy, T.
Dobson, David
Dobson, Don
Dobson, Evelyn
Dobson, Fred W.
Dobson, Lena
Dobson, Una
Dobson, Wendy K.
Dockerill, D.
Dohan, Nancie J.
Doherty, Mary
Donald, Rina
Donaldson, D.
Donaldson, J.
Donaldson, Len
Donaldson, Molly
Donohue, Hazel M.
Doole, J.
Doole, L.
Dooley, Brent
Dooley, Robert A.
Doroghy, Miriam
Dorsey, John
Dorsey, Shirley
Dorst, Adrian
Doughton, B.
Doughton, D.
Douglas, Aileen C.
Douglas, Paul
Dow, Anne
Dow, Douglas D.
Downes, Heather M.
Downes, W.
Downey, Phyl
Drabitt, Aaron
Drent, Rudolf H.
Drew, Betty
Drift, Daryl
Drimmelsen, Bruce
Drinnal, Warren

Drury, G.
Dubord, Karen
Dueck, Albert
Duggan, Bill
Dulisse, Jakob
Dunbar, David
Dunlop, Florence
Dunlop, Glen
Dunlop, H.P.
Dunnaway, Al
Dupas, Pauline
Dupilka, Allan
Dupilka, Reba
Dupuis, Linda
Durance, Eva
Durham, Helen
Durkee, Art
Durkee, Marilyn
Durrell, Linda
Duvall, A.J.
Dwight, J.
Dyck, N.
Dyer, Michael J.
Dyer, Orville
Earle, S.
Easthope, C.
Eastman, Donald S.
Easton, M.D.L.
Eaton, Bill
Ebel, G.R.A.
Eccles, Brian
Eckert, Chris
Eckhert, Cameron
Eckhert, K.
Edge, C.
Edge, J.
Edgell, Michael C.R.
Edie, Alan
Edward, Anne M.
Edward, F.H.
Edwards, Barry F.
Edwards, D.
Edwards, R. Yorke
Edwards, W.
Egely, Mary E.
Egger, K.
Eldridge, George
Eldridge, Ros
Ellames, Peggy
Ellingsen, Virginia H.
Elliot, D.E.
Elliot, John
Elliot, K.
Elliot, P.W.
Elliot, Ritchie
Elliott, Diane
Elliott, J.A.
Elliott, Kyle
Elliott, P.W.
Elliott, Peter
Ellis, David W.
Ellis, M.
Ellison, Maurice
Ellison, Veronica
Elphinstone, D.
Elsasser, Irma
Elston, Alice L.
Elston, Marjorie
Elston, Michael
Elston, Suzanne M.
Embleton, Nonie
Emes, Carolyn
Emes, Hanna
Emes, Ken
Emm, Bruce
Emmanuel, Klaus
Emrich, Vivian
Ende, Judy
Enderwick, Peter D.
Engelsleft, E.
Engelsleft, Faye
Engelstonft, Christian
Enns, K.S.
Ensing, Chris
Erasmus, G.
Erasmus, J.
Erhardt, Randy
Erickson, Harriette
Erickson, M.M.
Erickson, Roger
Erickson, Wayne
Erskine, Anthony J.
Erskine, R.C.
Erwin, A.
Erwin, B.
Escher, Sharon
Esouloff, Lorna
Esralson, Lloyd
Esser, Melody
Esser, Ralph
Etzkorn, Jerry
Evanich, J.
Evans, Darcy
Evans, J.
Evans, Norman
Evans, Ralph
Evans, W.J.
Everman, B.W.
Exworthy, R. June
Faasse, Tambrey
Faga, Laura L.
Fahr, Brian
Fair, Brent
Fairley, J.
Fairley, Stephanie
Fallis, Mary
Fallis, Mike
Falls, J. Bruce
Fanjoy, Emilee
Fannin, John
Farland, Dave M.
Farmer, F.
Farquhar, Rochelle
Farr, D.R.
Farrell, Martin
Farrell, Rita
Farrell, Rita
Farrer, A.
Fatt, Norman
Faulkner, Marion
Fawkes, Patrick
Fellowes, Meg
Fenneman, Jamie
Fenwick, A.A.
Ferguson, Elaine
Ferguson, Robert S.
Field, May
Fillion, D.B.
Findlay, R.
Finley, W.L.
Finnestad, Donna
Finnigan, Rory
Fisher, D. Ross
Fisher, Faith
Fisher, H. Dean
Fisher, Jennifer
Fisher, Len
Fisher, R.
Fisk, Roy
Fitch, John
Fitch, P.
Fitz-Gibbon, Joyce
Fitzpatrick, F.
Flahaut, Martha R.
Fleming, J.H.
Fleming, K.J.
Flery, N.
Fletcher, C.S.
Fletcher, F.
Flett, J.A.
Fleury, H.
Fleury, Norman
Flewin, M.
Flood, M.J.
Floyd, Anthony
Floyd, Doug
Flynn, B.T.
Fogg, Irma
Folbegg, Joyce
Folker, Ron
Fomataro, Mark
Foottit, Michael K.
Foottit, Robert G.
Forbes, Butler
Forbes, Susan
Forbes, Ted
Forbush, H.
Force, Michael P.
Ford, Bruce S.
Ford, Shane
Ford, Victor
Forer, A.
Forsek, Gabriella
Forsman, Eric D.
Forster, Craig
Forty, Thelma
Foskett, Ann
Foster, Anthony
Foster, D.F.
Foster, E.
Foster, F.
Foster, G.
Foster, J. Bristol
Foster, Jack W.
Foster, K.B.
Foster, Lee A.
Foubister, M.
Fowle, J.G.
Fox, G.
Fox, L.
Foxall, Roger
Francis, George
Frank, Floyd
Franklin, June
Fraser, David F.
Fraser, Douglas P.
Fraser, H.
Fraser, J.
Fraser, Kitsy
Fraser, M.
Fraser, William.
Frederick, Bruce S.
Freebairn, Tom
Freeman, Cheryl
Freeman, J.
French, F.G.
French, J.E.
Frich, W.
Friedenberger, Karen
Friend, G.B.
Friis, Laura K.
Frith, Wendy
Froese, Dave
Froimovitch, Mark J.
Frost, D. Lorne
Fry, Kathleen
Fryer, Grace
Fryer, Ralph
Fryer, Ron
Fucco, M.
Fuhr, Brian L.
Fuller, A.D.
Fuller, Audrey
Funk, Linda
Funk, Maureen L.
Fusco, L.
Futur, G.
Fyles, Helen
Fyles, J.
Gaelick, N.F.
Gage, Peter
Gagnon, R.
Galbraith, Bob
Galbraith, Florence
Galbraith, J. Douglas
Galbraith, Marjorie
Gales, B.
Galt, Betty
Gardener, Heather
Gardiner, Mark
Gardner, Barbara
Gardner, Elliott
Gardner, Gerry
Gardner, Nora
Gardner, W.E.
Garlick, Ella
Garman, E.
Garnett, J.E.
Garnier, Don
Garnier, Hattie
Garrett, C.B.
Gartshore, Mary
Gaskin, David
Gaskin, Jeff
Gaskin, John
Gasser, Ellen
Gaston, Anthony J.
Gates, Bryan R.
Gates, T.
Gayner, F.
Gebauer, Martin
Geernaert, Tracee O.
Gehlen, Phil
Gehring, Keith L.
Geist, Val
George, D. Val
Gerein, Elsie
Gerein, Ralph
Germyn, D.
Gerow, Helen
Gerry, E.C.
Gibbard, A.J.
Gibbard, Bob
Gibbard, Fern
Gibbard, H.J.
Gibbard, Les A.
Gibbard, Robert T.
Gibbard, Violet
Gibbs, Richard E.
Gibson, C.W.
Gibson, Carlen
Gibson, Daniel D.
Gibson, Dolores L.
Gibson, Doris E.
Gibson, George G.
Gibson, Ian
Gibson, W.H.
Gifford, Kit
Gifford-Brown, Janet
Giggs, Mae
Giliberti, T.
Gilker, Kim
Gillespie, Grahame E.
Gilley, M.
Gillie, Marg
Gillies, Cameron
Gillingham, D.W.
Ginnis, J.
Giovanella, Carlo A.
Gissing, Alwin
Glendenning, R.
Godbe, Loran
Godbout, D.
Godfrey, Dudley
Godfrey, Geoff
Godfrey, Judy
Godfrey, R.B.
Godfrey, W. Earl
Godidek, Mickey
Godider, H.
Godin, Tom
Godkin, Sharon
Godwin, A.
Goerke, Alice
Goerke, Richard
Goff, D.
Goldsmith, Len
Gonzales, B.
Good, Ed
Goodacre, Brian
Goodman, Betty
Goodman, Jim
Goodmanson, H.P.
Goodwill, J.E. Victor
Goodwill, Margaret E.
Goodwin, A.
Goodwin, Lance
Goossen, J. Paul
Gordon, Hilary
Gordon, Orville
Gorham, S.W.
Gornall, Fred A.
Gornall, J.
Gorsuch, Cecilia V.
Goshaluk, Ted
Gosling, A.G.
Gotz, B. Max
Gotz, R.
Goudie, Douglas W.
Gough, Carl
Gould, Dulce
Goulet, Louise
Gow, Bertha
Goward, Trevor
Grabowski, Tony J.
Graham, Douglas J.
Graham, Jim
Graham, Sheila
Grant, G.
Grant, James
Grant, Peter R.
Grant, Tunie
Grass, Al C.
Grass, Jude F.
Grasser, E.
Gravelle, David L.
Gray, Alex
Gray, E.
Gray, Sandra
Gray, T.G.
Green, Charles de Blois
Green, David M.
Green, Henry
Green, M.M.
Greene, R.K.
Greenfield, Tony
Greenwood, Eric
Greenwood, Gwen
Gregory, M.S.
Gregory, Patrick T.
Gresham, A.B.
Greuner, K.
Greyell, Bob
Griffiths, Anabel M.
Griffiths, Deirdre
Griggs, Tamar
Grinberg, Helmet
Grinnell, Dick
Gronau, Christian W.
Gronau, Steffi G.
Groneau, Aileen
Groseth, Bob
Groseth, Janet
Gross, A.
Groth, J.G.
Groves, J.
Groves, Sarah
Gruener, Karl L.
Gruenig, Art
Grunberg, Helmut
Guepel, Geoff
Guernsey, Vera
Guiguet, Charles J.
Guiguet, M.L.
Guillon, Frank E.
Guppy, Cris
Gurr, Ray
Gustafson, Barbara
Gustavus, Monica
Guterson, Shannon
Guthrie, Doreen
Guthrie, P.
Gyug, Les
Haas, Norma
Hack, Alda
Hack, F.W.
Hack, George
Hackman, Sue
Haddow, Douglas J.
Haddow, Suzanne
Haering, Penny
Hagen, Barry
Hagen, Patricia
Hagenstein, W.M.
Haggart, Lea
Haggerstone, B.
Haggert, Leona
Hague, John
Haig-Brown, Anne
Haig-Brown, Roderick
Haines, A.M.
Halaz, Gabor
Hale, W.
Haley, Gordon
Haley, Kathy
Halfnights, B.
Halford, Stephen
Halhead, Wm.
Hall, Alan
Hall, J.
Hall, Ken
Hall, T.B.
Halladay, D. Raymond
Halman, J.
Halsall, Leah
Halverson, Larry R.
Hamel, Peter J.
Hamel, Steve
Hamel, Tracey
Hames, A.M.
Hames, Michael
Hamill, S.
Hamilton, Anthony N.
Hamilton, Daphne
Hamilton, I.
Hamilton, John
Hamilton, K.
Hamilton, Marla
Hamilton, Roberta
Hammell, Terry
Hammer, H.B.

Hammond, Bob
Hammond, Elsie
Hammond, Jack
Hammond, Jo
Hamnett, Ron
Hancock, David
Handley, Catherine
Hanna, Jack
Hannah, K.
Hannam, May
Hannon, Susan
Hanrahan, Christine
Hanrahan, Tom
Hansen, J.A.
Hansen, Lillie
Hansen, Ole
Hansen, Vicki
Hansvall, Erling
Hansvall, Louise
Hanwell, Barbara
Haras, Willie
Harbell, O.G.
Harcombe, Andrew P.
Hardie, M.
Hardie, W.
Harding, Martha
Hards, Jennifer
Hardstaff, Lynn
Hardy, George A.
Harestad, Alton S.
Hargrave, A. Nairn
Hargreaves, D.
Hargrove, Ken
Hargrove, Marian
Harling, M.
Harlock, F.
Harlow, R.C.
Harlow, Susan L.
Harman, Barry C.
Harper, Bill
Harper, Charles
Harper, Fred E.
Harpur, Bill
Harpur, Lorri
Harriet, V.E.
Harrington, R.F.
Harris, A.E.
Harris, Alice
Harris, Christopher G.
Harris, Doreen
Harris, Elizabeth
Harris, L.
Harris, Margaret
Harris, Nancy
Harris, Ross E.
Harrison, Bruce
Harrison, Mary
Harrison, P.
Harrold, C.G.
Hart, A.M.
Hart, Carole
Hart, F. Gordon
Hart, F.E.
Hart, John
Hart, Kit
Hart, Laura
Hart, Mark
Hart, Sue
Hartland, G.
Hartman, F.
Hartwell, Sharon
Harvalias, Irene
Harvath, O.
Harvey, Dwight
Harvey, Merle
Harvey, William
Harwell, Clay M.
Harwell, M.
Hasbell, O.G.
Haskell, K.
Hassell, Sharon
Hastings, D.C.
Hatfield, John
Hatland, G.
Hatler, David F.
Hatter, Ian
Hatter, James
Hausen, Jergen
Hawes, David B.
Hawes, Myrnal A.L.
Hawkes, Bill
Hawkes, Marian
Hawksley, Janet
Hawley, Connie
Hay, Heather
Hay, Robert B.
Hayduk, Mike
Hayduk, Wendy
Hayes, H.
Hayes, Maryann
Hayes, Richard
Hayman, Tom
Haynes, J.
Hayton, Barbara
Hayton, Murray
Hazeldine, W.R.
Hazelwood, Grant W.
Heakes, Todd
Hearn, Dorothy
Hearn, Ed B.
Hearn, T.
Hearne, Margo E.
Hebert, Samantha
Hedley, A.F.
Hedley, Lex A.
Heimstra, H.T.
Heinemann, D.
Heinrich, Ralph
Heizig, J.
Heller, E.
Helm, Charles
Helm, Chris
Helset, Roy
Henderson, C.
Henderson, Joyce
Henderson, M.
Henderson, N.
Henderson, Phil S.
Hendrick, Russell
Hengeveld, Pamela
Hennan, Ed G.
Henry, John
Henry, M.G.
Henson, Gary
Henson, Simon
Hentze, Nathan
Hepburn, J.
Heppner, D.
Heriot, Joan E.
Herring, Les
Herring, Sarah
Herring, Wendy
Hersey, F. Seymour
Herzig, Kim
Herzig, R. Jerry
Herzog, Pat
Hesse, Hilde
Hesse, Werner H.
Hetherington, Anne E.
Hett, M.
Heybroek, William J.
Hickey, E.
Higgins, J.C.
Higgins, John D.
Hill, Bill
Hill, Cecilia L.
Hill, L.G.
Hill, Les
Hill, Lil
Hill, Pat
Hill, Phyllis
Hill, Verna
Hillary, Edward
Himmer, Stefan
Hindshaw, H.H.
Hind-Smith, John
Hippner, D.
Hitchcock, Gordon
Hitchmough, John
Hladly, D.
Hoar, Carol
Hoar, Richard
Hobson, J. Fred
Hobson, James
Hobson, Keith
Hobson, Mark
Hochachka, W.
Hocking, Jennifer H.
Hodgson, H.M.J.
Hodgson, K.
Hoek, Evi
Hoek, Mel
Hogen, Betty
Hogg, Edward
Hogg, Lori
Holdom, C. Martin W.
Holland, Aton Holmes
Holland, George P.
Holland, M.M.
Holland, S.
Hollander, J.
Hollands, Grant R.
Hollick-Kenyon, Tim
Hollington, Jack
Hollington, Madge
Holm, Margaret
Holm, R.O.
Holman, John H.
Holman, Muriel
Holmes, Jean
Holmes, Marc
Holmes, Norman
Holmes, R.
Holms, Maureen
Holroyd, E.
Holroyd, Geoff
Holt, Beryl H.
Holt, John C.
Hooch-Antink, Carl
Hooch-Antink, Tracy
Hooge, Lois
Hook, H.G.
Hook, R.
Hoon, Gay
Hooper, Bob
Hooper, Gordon N.
Hooper, Gwennie
Hooper, Joan
Hooper, Mary
Hooper, Ronald
Hooper, Tracey D.
Hope, Herbert
Hopkins, Jennifer M.
Hopkinson, Bubsie
Horn, W.
Horne, Doreen
Horton, Keith
Horvath, D.
Horvath, Otto
Horwood, Brenda
Horwood, Clive
Horwood, Dennis
Hosford, Harold
Hosie, Suann
Houlden, Marie
Houston, Angela
Houston, Bob
Houston, G.F.
Houston, R.
Hovender, Frank
Howard, D.
Howard, Steve
Howatson, Charles
Howden, Patrick F.
Howe, Ann
Howe, Robert W.
Howell, J.C.
Howell, Steven
Howie, Richard R.
Howie, S.
Howley, Connie
Hrycun, Dean
Hubbard, J.P.
Hudson, Janet
Hudson, Jean
Hudson, Keith
Huff, L.L.
Hughes, William M.
Hulfman, Bruce
Humphries, Dianne
Humphries, Donna
Hunn, Eugene
Hunt, Gary
Hunt, H.K.
Hunter, D.
Hunter, Gary
Hunter, Michael
Hurley, John B.
Hustead, Eileen
Hustead, Jack
Hutchins, Jack
Huxley, Bill
Huxley, Mae
Huxley, W.
Hyde, George
Ibera, Carlos
Ikona, Katherina
Ikona, Richard
Ingles, Joan
Ingram, Dave
Inkster, Mary
Innes, Douglas W.
Innes, Marian
Ireland, John
Ireland, Teresa
Irvine, D.
Irvine, K.
Irving, B.
Irving, E. Bruce
Irving, P.
Irwin, Alan
Irwin, Barbara
Iverson, J.
Jaarsma, Evelyn
Jaarsma, William
Jack, Ian
Jackson, Andrew
Jackson, Anne
Jackson, B.
Jackson, Harry
Jackson, J.
Jackson, M.
Jackson, Mary F.
Jackson, Nell
Jacobs, A.E.
Jacobs, Karen
Jacobson, Tom
Jacobson, W.
Jakimchuk, R.D.
James, Ann
James, M.
James, Paul C.
James, Ross D.
Jamieson, C.
Jamieson, S.
Janes, Russell W.
Janes, Tim
Janz, Douglas W.
Janzen, Pat
Jarosch, Conrad
Jarvis, Janice
Jeal, Margaret
Jeal, Mary
Jean, Cheryl
Jeffers, Pat
Jellicoe, Janice
Jenkins, Jane
Jenkins, Mildred M.
Jenkins, P. Ray
Jenkins, Ron
Jenniks, Peter
Jennings, Hugh A.
Jennings, Peter
Jensen, B.A.
Jensen, Dale
Jensen, Eve
Jensen, Joyce
Jensen, P.
Jerema, Richard S.
Jobin, Leo
Jobin, M.
Johannesen, K.
Johns, Dawn
Johns, Vic
Johnson, A.
Johnson, B.
Johnson, Daryl
Johnson, Donald A.
Johnson, Esther
Johnson, Fran
Johnson, Gordon
Johnson, Joseph W.
Johnson, Mary B.
Johnson, R.E.
Johnson, Scott
Johnson, Stephen R.
Johnson, Stuart
Johnson, Win
Johnston, Ann
Johnston, Jean
Johnston, Joan
Johnstone, Harry
Johnstone, S.
Johnstone, W.E.
Johnstone, W.H.
Johnstone, W.M.
Jolly, Malcolm
Joly, Stephen
Jones, Anne M.N.
Jones, Dave
Jones, Edgar T.
Jones, J. Donald
Jones, Lindsay
Jones, M.
Jones, Peter
Jones, Ruth
Joy, Jeffrey
Jsersel, E.
Jukimchuk, R.D.
Jurgens, P.
Jury, Douglas
Jyrkkanen, Jorma
Kaiser, Dale
Kaiser, Gary W.
Kalman, John
Kalmbach, Richard
Kalum, N.
Karger, Fritz
Karup, Anthony G.
Karup, Lisa
Kaufman, C.L.
Kautesk, Brian M.
Keating, B.
Keating, R.
Keatry, B.
Keber, Lara
Keddie, Grant
Keechyer, N.L.
Keen, Rev. J.H.
Keer, Stephanie
Keith, M.
Keith, Stuart
Keizer, Jasper
Kelleher, Kevin E.
Kelleher, M.N.
Kellogg, L.
Kelly, Colleen
Kelly, Phil
Kelly, W.M.
Kelso, Dr. J.E.H.
Kennedy, D.M.
Kennedy, Ken
Kennedy, Peter
Kenning, John
Kentland, E.
Keranen, Eric
Kerby, O.
Kergin, R.
Kerr, Deborah
Kerr, Elspeth M.
Kerr, Joan
Kerry, Ray
Kertland, E.
Killins, Glyn
Kime, Doris
Kime, Frank
Kimm, Jason
Kimm, Jeremy V.
Kimm, Lindsey
Kimmins, R.A.
Kimpton, Fran
Kinder, B.
King Roderick
King, A.G.
King, Anna
King, Bryan
King, David
King, Don
King, Frances
King, Freeman
King, Janet
King, Jean
King, Joan
King, Roderick
Kinsey, Sandra
Kirbyson, John
Kircner, Mary
Kirkpatrick, F.
Kirkpatrick, T.
Kirkvold, Sherry
Kiryck, Walter
Kising, J.D.
Kitchen, Rip
Klagh, A.B.
Klassen, Barry A.
Klauke, Richard
Klein, H.C.
Kline, Kelly
Kline, L.H.
Klugh, A.B.
Knapton, Richard W.
Knezevich, Gladys

Knezevich, John
Knight, Helen
Knight, J.
Knight, K.
Knight, Pat
Knowles, Anne
Knowles, H.
Knowles, Ken
Knowles, R.H.
Koch, Anne Marie
Koch, Linda
Koechl, Rick
Koenig, Dan F.
Kohn, G.
Kohnert, E.
Kolotello, Dorothy
Kolotello, F.
Koochin, Ed
Koochin, George
Kooistra, Clare
Koran, J.
Korol, J. Burke
Korompai, Marika
Korten, H.
Kovak, Steve
Krabbe, Nils
Kragh, W. Douglas
Krebs, Charles J.
Krebs, John R.
Kremsater, Laurie
Kreuger, Nancy
Kriese, Kevin
Kroning, Angela
Kubica, Greg
Kubica, Lee
Kuch, Dennis
Kuchar, Peter
Kucheron, Nancy
Kurata, T.B.
Kurjata, Nancy
Kurjata, Peter
Kuzmaniuk, Harvey
Kyostia, Derek
La Vigne, M.
Laboida, Ted
Lacey, E.
Lacey, L.
Ladbury, Joan
Ladds, Jim
Ladds, L.
Lafreniere, Elsie
Laing, A.M.
Laing, E.M.
Laing, H.W.
Laing, J.M.
Lake, Barbara E.
Lall, A.
Lamb, R.E.
Lambert, G.
Lambert, Marilyn A.
Lambert, Phil
Lambie, C.
Lambie, David
Lambie, John A.
Lambie, M.
Lambie, Vi
Lambin, Xavier
Lamont, B.
Lamont, Peter
Lance, Arthur N.
Land, J.
Landbury, Bob
Landry, Gaeten
Langevin, Marlene
Langley, A.
Langley, G.
Langston, J.B.
Lanley, George
Lansdowne, Edith
Lansdowne, Fenwick
Lanyon, Susan
Laresser, Cindy
Larson, P.A.
Lau, Alfred
Law, George
Law, Laird
Lawrence, Jim R.
Lawrence, Joanne
Lawrence, Susan
Lawrenson, Lee
Lawson, M.
Leather, E.E.
Leavens, D.
Leckie, B.
Leckie, C.P.
Ledum, Tim
Lee, J. Charlene
Lee, Jeff
Lee, Joyce
Lee, Martin C.
Lee, Sandi
Leech, Hugh B.
Leedham, Tim
Lees, Sybil
Legg, Peter
Leighton, Douglas A.
Leighton, Myriam
Leinor, Betty
Lellis, C.
Lemkie, E.
Lemon, Enid K.
Lemon, Moira
Lenfesty, Jack
Lensey, Ruth
Lesowski, J.M.
Lesowski, N.J.
Leung, Maria
Leupin, Ernest
Leusher, R.G.
Leveson-Gower, Heather
Levesque, Paul
Levitt, Pat
Lewale, J.
Lewis, B.
Lewis, Craig
Lewis, Molly
Lewis, Moray
Lewis, Ryan
Ley, Lynette
Liddel, S.A.
Lima, Steven
Linburg, Anne
Linburg, Richard
Linburg, Wayne
Lincoln, Robert
Lindley, W.
Lindstrom, Dick
Lines, Molly
Linn, Mel
Linnman, Kathi
Linsey, Ruth
Linsey, Ted
Lisicin, Christine
Lisicin, P.
Littau, Paul
Little, B.
Little, R.S.
Livingston, Daryl
Livsey, Ruth
Lloyd, E.
Lloyd, Kevin
Lock, Norm
Lockhart, Roy
Lodge, Terry
Lofroth, Eric C.
Lofts, George
Logan, Phyllis
Logan, Vernon
Logan, Wes
Long, Iris
Long, W.S.
Loosmore, B.
Loosmore, D.
Lorimer, J.
Lortie, P.
Lothian, Betty
Lott, David
Lott, Joan
Lougheed, Lynn
Lovatt, Nancy
Lovatt, R.
Low, David J.
Low, Gladys
Low, S.
Low, W.A.
Low, W.R.
Lowden, Bob
Lowrey, Terry
Lowrey, Ursula
Luce, John
Lunam, Betty
Lunam, James
Luney, H.D.
Luscher, Robert E.
Lusthoff, Donna
Luton, Tony
Luz, Michael
Lydiard, J.M.
Lynion, Margaret
Lyon, Alan M.
Lyons, Ches
Lyons, Tex
Macartney, Malcolm
MacBean, Eric
MacCarter, Alex
MacColl, M.C.
MacDonald, Alan
MacDonald, Brian
MacDonald, Bruce A.
MacDonald, Eloise
MacDonald, Heather
MacDonald, Ian
MacDonald, Jack
MacDonald, Murray
MacDonald, Ross
MacDonald, S.
MacDonald, Stewart D.
MacDonald, W.
MacEachern, Kathleen
MacFarlane, Mair
MacFarlane, Roderick
MacGillivray, Alice E.
MacGillivray, Bruce
MacGillivray, E.R.
MacGregor, Jo
MacGregor, L.
Macguire, W.S.
Machmer, Marlene
Machover, Bob
MacIntosh, Jim
Mack, Eva
Mack, Jim W.
Mack, Mary
MacKay, D.
Mackay, Violet
Mackenzie, Alister
Mackenzie, C.J.G.
MacKenzie, D.I.
MacKenzie, Hue
Mackenzie, J.H.
MacKenzie, Jo Ann
Mackenzie, Orma
MacKenzie, R.
Mackenzie, Tom
MacKenzie-Grieve, Margaret
MacKenzie-Grieve, Robert
Mackie, Al
Mackie, Theda
MacLachlan, D.
MacLean, C.
MacLeod, A.S.
MacLeod, Alan M.
MacLeod, Allan L.
MacLeod, P. Ray
MacNall, Roger
Macpherson, A.
MacPherson, Kevin
MacPherson, Peggy
MacRae, Kenneth
Madden, Sue
Madsen, N.
Magee, K.
Maher, John
Mahony, Nancy
Maisel, G.L.
Malenfant, Bell
Maloff, Diana V.
Maltet, Leo
Mann, Dick
Mann, Miki
Manson, M.
Marchant, Chris
Marcus, Norman
Marcussen, Dave
Mark, David M.
Mark, Mavis
Mark, Murray
Mark, Pat
Mark, Thomas C.
Mark, W.
Marko, Nancy
Marks, G.
Marshall, G.W.
Marshall, Krista
Marshall, Lee-Ann
Marshall, M.
Marshall, Rick
Marshek, Larry
Martin, Bea
Martin, David
Martin, Jack
Martin, Joan
Martin, Kathy
Martin, Nancy
Martin, Patrick W.
Marven, Derrick
Mason, Diane
Mason, Irene
Matheson, Jeff
Matheson, M.
Mathews, N.
Mathewson, Deanna
Mathewson, W.M.
Matkoski, Wayne
Matsushita, Mas
Matsushita, P.
Matthews, H.
Matthews, Nigel
Maurer, April L.
Maxwell, J.
Mayall, June
Mayall, Ken
Mayfield, Hylda
Mayfield, M.
Mayfield, P.
Maynard, F.
Maynard, R.
Mayo, Ron
McAlary, Eric
McAlary, Lois
McAllister, David
McAllister, Nancy
McAllister, Pat
McAllister, Peter B.
McBean, Eric
McCabe, Bernadette
McCabe, Elinor B.
McCabe, Thomas T.
McCabe, Y.
McCammon, James W.
McCann, Gerry
McClarnon, John
McClarnon, S.
McClellan, M.
McClure, Walt
McColl, W.
McCord, Bill
McCord, Dorothy
McCormack, Sue
McCrory, Wayne
McCulloch, H.L.
McDiarmid, Ray
McDonald, B.
McDonald, Ed
McDonald, Jim
McDonald, Marlene
McDonald, Sue
McDonald, W.
McEachem, J.A.
McFarland, David
McFeat, Miggs
McGavin, G.A.
McGavin, Rosemarie
McGhee, Carolyn
McGhee, Jack
McGill, Allan S.
McGregor, Ian
McGregor, Jack
McGrenere, Barbara
McGrenere, Mike
McIlwrath, J.
McIlwrath, K.C.
McIlwrath, T.
McInnis, Nelson
McIntosh, D.
McIntosh, John D.
McIntosh, Lorne D.
McKay, George M.
McKay, H.
McKay, Nellie
McKay, W.A.
McKean, Donna
McKennon, B.
McKie, Mike
McKim, C.
McLachlan, Sandy
McLaren, Karen A.
McLaren, W.C.
McLaren, William D.
McLarnon, J.
McLaughlin, Ronald T.
McLean, C.W.
McLean, Sue
McLeannan, Bette
McLeish, I.
McLellan, D.
McLellan, M.E.
McLennon, Bette
McLeod, Anne C.
McLeod, C.F.
McLeod, E.J.
McLillan, D.
McLintock, B.J.
McLure, S.
McMackin, Ed
McMackin, Patrick
McMann, Michael
McMillan, Susan
McNabb, Carol
McNall, Faye L.
McNall, Michael C.E.
McNally, George
McNicholl, Martin K.
McNobb, C.
McPherson, Kevin
McRae, Douglas
McTaggart-Cowan, D.A.I.
McTaggart-Cowan, Ian
McTaggart-Cowan, Joyce
McTaggart-Cowan, M.
McTaggart-Cowan, R.
Meads, Darlene
Meany, John
Medynski, Jody
Megan, Andy
Mehling, C.
Meiklejohn, Barbara
Meiklejohn, Michael
Melburn, M.
Melke, Avoca
Meller, Brian
Mellett, Mary
Melnychuck, A.
Menzies, Charles R.
Menzies, Helen
Meredith, C.B.
Meredith, M.C.
Merilees, June
Merilees, William J.
Merkens, M.
Merven, D.
Metcalf, E.W.
Meugens, Arthur L.
Meugens, M.M.
Mewhart, S.
Meyer, H.
Meyer, R.
Michaluk, Libor
Michener, Beth
Michener, M.
Mickel, Allison
Mickel, Tom
Mickey, Lil
Mico, Michelle
Middlemass, A.D.
Millar, J.S.
Miller, Alder
Miller, Carl
Miller, Don
Miller, E.P.
Miller, H.R.
Miller, Hazel
Miller, Hettie
Miller, Lynn
Miller, M.D.
Miller, Mike
Miller, R.G.
Miller, Robert
Miller, Roxie
Miller, Shane
Miller, Stella
Milligan, D.
Milligan, Jack A.
Milligan, K.
Milligan, Tony
Milliken, Eileen
Millikin, Rhonda

Mills, Alex
Mills, Eric L.
Mills, J.
Mills, M.C.
Mills, M.L.
Mills, Marjorie
Millward, Peter
Milne, H.
Milnes, Lynne
Milsted, Jack
Minshull, Catherine
Misener, Ken
Misener, Sharon
Mitchell, Aubrey
Mitchell, F.S.
Mitchell, Gail
Mitchell, George J.
Mitchell, Marjorie
Mitchell, Suzann
Mlieczk, Louis
Mobbs, Rosa
Moen, H.L.
Mogensen, Faye
Mollet, L.
Mollet, M.
Momeyer, Ericka
Monberg, Don
Monchak, D.
Moncrieff, R. Scott
Moncton, L.
Monty, Guy L.
Moody, E.
Mooney, Amber
Mooney, Lynn
Moore, B.
Moore, D.
Moore, Elaine
Moore, G.
Moore, Huber
Moore, Iris
Moore, Jackie
Moore, Richard
Moore, Trudy
Moores, Karen
Moores, N.
Morehen, Cy W.
Morford, W.R.
Morgan, Art
Morgan, B.
Morgan, Jeff
Morgan, Jim
Morgan, Joanna
Morgan, Ken
Morgan, Susan
Morgenstern, Bruce
Morgenstern, Pearl
Morley, B.
Morley, J.P.B.
Morley, W.J.
Morrell, Michael
Morris, George
Morris, Irene
Morris, Jackie
Morris, Mary
Morris, Michael S.
Morris, Rick
Morris, W.
Morrison, G.
Morrison, Jim
Morrison, Ken P.
Morrison, Mary
Morrison, Terry
Morse, R.G.
Mortimer, Joan
Morton, Diana T.
Morton, Keith
Morton, Norma J.
Mosedale, W.
Moser, Wendy
Moses, Richard
Moss, W.W.
Mossip, David H.
Mottishaw, J.B.
Mouat, I.
Mouat, S.
Mould, Alice
Mould, Frank
Mounteney, David
Mowbray, C.
Moyle, Gail
Mueller, R.
Muir, Alexander
Muirehead, N.
Mulchinock, J.P.
Mulder ten Kate, Nanny
Müller, Emily
Müller, Heidi
Müller, Lynae
Müller, Naomi
Müller, Nerida
Müller, Nevelle
Müller, Sarah
Müller, Scott
Mundie, E.M.
Mundie, Isabel
Mundy, K.
Munn, David R.
Munn, Ruth
Munro, Alan
Munro, David A.
Munro, Ian H.
Munro, Kathleen
Munro, Ken
Munro, Sylvia
Munro, W.H.
Munro, William T.
Murcklen, Joanne
Murcklen, Ken
Murdoch, Bob
Murehen, C.W.
Murie, O.J.
Murphy, G.R.
Murphy, Glen
Murphy, K.
Murphy, Maggie
Murphy, Michael
Murphy, Tim
Murray, Anne
Murray, Linda
Murray, Susan
Muse, S.
Myer, Ron
Myers, Steven J.
Myres, M. Timothy
Nancarrow, P.
Neaga, Bob M.
Neale, Eve J.
Neff, R.
Nehls, Harry B.
Neill, Betty
Neill, Sandra
Neilsen, C.E.
Neilsen, D.
Neily, Wayne
Nelson, Judy
Nelson, R. Wayne
Nelson, S.
Nero, Robert W.
Nestman, C.
Neville, Heather
Neville, John
Newcombe, W.A.
Newell, David S.
Newell, George
Newell, Janet
Newman, J.
Newson, E.
Newson, Fran
Newson, Marion
Newson, Verna
Newton, Amy D.
Newton, M.I.
Nicholson, Dean
Nicol, Adeline
Nicol, Gwen M.
Nicolai, Brian
Nielsen, Bruce
Nielsen, C.E.
Nielsen, K.J.
Niles, D.M.
Niquidet, Til
Noble, Leigh
Noel, Doug
Noel, Phyllis
Nol, Erica
Nomme, K.
Nordan, Harold C.
Nordan, J.
Norman, Donna
Norman, Hugo
Norman, Joan
Norris, Andrea R.
Norsworthy, Marjorie
North, Elizabeth
North, Lois
Nott, Phil R.
Nowosad, June
Noye, Sandra
Nyberg, Brian
Nye, Doris F.
Nygren, Edward L.
Nyhof, Mark
Nykfork, Elsie D.
O'Brian, Art
O'Brian, Derek
O'Callaghan, Denny
O'Callaghan, Marilyn
O'Connor, F.
O'Dwyer, J.
O'Shaughnessy, Kerri
O'Shaughnessy, M.
O'Shea, Michael
Oakley, Lisa
Oates, Steve
Obana, J.
Oberg, Evelyn
Octing, B.
Odlum, G.C.
Ogilvie, Duncan
Ogilvie, Hugh
Ogilvie, Robert
Ogle, Steve
Ohanjanian, Penny
Oke, Denise
Oldaker, R. Frank
Oliphant, Lynn
Olmstead, John
Olsen, Brent
Olsen, Carl
Olsen, Stan
Omand, Bob
Omand, Cheryl
Omand, J.
Oppenheim, Claire
Oppenheim, David
Oray, A.
Orcutt, Lowell
Orians, Gordon H.
Orser, Elsie
Orser, Ray
Osmond-Jones, E.J.
Oswald, J.P.
Oswaldski, Bob
Ouellet, Henri
Outerbridge, Joan
Overton, Joy
Owen, Rob
Owen, Sue
Pacsay, Joe
Padgett, Fran
Page, D.
Page, Linda
Page, Rick
Page, Ruth
Palleroni, A.
Palmateer, Calvor
Palmer, D.
Palmer, Greg
Palmer, J.
Palmer, Mary
Palmer, Merton
Palmiere, Myrna
Parish, Pat
Parker, Bill
Parker, Joan
Parkin, Tom W.
Parkington, Jackie
Parley, Frank
Parmenter, Louise
Parris, Lucy
Parsons, D.
Parsons, Robert
Parsons, Ron
Partington, Geoffrey W.
Partington, Jackie
Pasieka, Marilyn
Pastrick, Mary
Pastrick, Trudy
Paterson, Aurora M.
Patrick, Robert
Pattenaude, Mary
Pattenden, B.
Patterson, Colleen
Patterson, Gerry
Patterson, James S.
Patterson, Lloyd
Patterson, Rory
Pattison, Lyn
Pattison, Stan
Paul, Adrian B.
Paul, Andy
Paulson, Dennis R.
Pavlick, Leon
Payne, D.E.
Pearce, David
Pearce, Marguerite
Pearce, W.
Pearse, Theed
Pearson, Arthur
Peck, Jess
Peck, W.J.
Peden, Alex E.
Pedley, David
Pedley, Lynne
Peet-Leslie, Mary
Pendergast, Bruce
Pendray, Tom
Perchall, Jennifer
Perchall, L.
Perkins, Ann
Perkins, Glen
Perrin, Janne J.
Perrin, Murielle
Perrin, Peter W.
Perry, Jim
Person, E.V.
Pertson, Diane
Pertson, Paul
Petapiece, V.
Peter, Marsico
Peters, J.
Peters, Vi
Peterson, Bruce R.
Peterson, Dan
Peterson, Gretchen
Peterson, Lawrence
Petrar, Brian J.
Pett, M.
Petty, G. Bernice
Peve, Armand
Peyton, Leonard J.
Phillip, Connie
Phillips, Allan R.
Phillips, Howard
Phillips, Roy W.
Phillipson, G.
Phinney, Mark
Phippen, Stan
Phipps, Barbara
Phipps, R.W.
Pickens, Leonard C.
Picker, Cynthia
Pickering, P.
Pickering, Verna
Pickford, M.E.
Pielous, Chris
Pierce, B.
Pincott, Sylvia
Pirillo, Vera
Pitelka, Frank A.
Pitt, Bill
Pitt, Margaret
Pitt, Rae
Pitts, R.
Piuze, Jean
Pizzey, A.
Plath, Tom
Plaxton, W.
Pohlman, Paul
Poirier, Evelyn
Pojar, Jim
Pojar, Rosamund
Pojar, V.
Poll, David M.
Pollard, A.L.
Pollock, Harold
Polson, John E.
Pond, Lee
Porcher, Alf
Porcier, A.C.
Porteous, A.
Porter, David M.
Porter, Marian M.
Postrick, Mary
Poulin, Andre
Powell, Douglas
Power, Damian
Power, Rosemary
Powers, Gerry
Poynter, Ann
Poynter, G. Allen
Poynter, Helen
Prather, Vaughn
Pratt, H.B.
Pray, T.A.
Preece, W.H.A.
Prehara, Bea
Prehara, George
Prestash, L.
Preston, Allan
Preston, Margaret
Preston, Michael I.
Preston, W.B.
Price, Colin
Price, Karen
Price, Michael
Prier, M.
Prince, Ron
Prince, Shirley
Prior, Paul
Prior, Roy
Provo, Brian
Puckering, P.J.
Punnett, Walter
Purdy, J.
Purdy, Margaret
Purdy, Mary
Pursell, M.
Pursell, Norman
Pustey, Walter
Quadvelieg, R.
Quinton, Tony
Quirk, John
Racey, A.
Racey, S.
Rack, Kurt
Rack, Marilyn
Rae, G.
Rae, William S.
Raedeke, K.
Raginsky, Nina
Railton, Jessop
Raine, Walter
Rainer, E.
Rainville, C.J.
Ramsay, D.L.
Ramsay, Leah R.
Ramsay, M.A.
Ramsay, R.
Ramsay, Terri A.
Rana, J.
Randall, T.E.
Rankin, Leo J.
Rankin, S.
Ranson, J. Philip
Rathbone, Phil
Rathbone, Sandy
Rathbone, Tony
Rathwell, K.
Rattray, Jim
Rawlings, H.
Rawson, Phil
Ray, N.
Raymond, Anne
Raynor, C.
Read, Gloria
Read, S.
Redford, W.
Reed, Annette
Reed, Norm
Reese, William
Reeves, Paul
Reich, Judy
Reid, D.B.
Reid, Emerson
Reid, Trish
Reidemann, S.
Reifel, George C.
Reimchen, Tom E.
Reiter, Jean
Rempel, Dan
Renfroe, Janet
Rennie, Jack
Rennie, Marg
Ressel, Kathy
Rettberg, Lynne
Reusch, R.
Revdell, J.F.
Reynolds, Grant
Reynolds, Sheila
Rhynas, P.
Rich, Harold

Richards, E.
Richards, Frank
Richardson, A.
Richardson, Dave
Richardson, Diane
Richardson, Dorothy
Richardson, Francis
Richardson, Frank
Richardson, Gillian
Richardson, Howie
Richardson, Trevor
Richburg, S.
Ricker, Nancy
Ricker, William E.
Riley, J. Harvey
Rinehart-Pidcock, Dirk T.
Ringer, Dawna-Lea
Rising, James D.
Ritcey, Clara
Ritcey, Frank
Ritcey, Ralph W.
Ritchie, David
Rithalfer, Margaret
Ritz, C.W.
Roach, Anthony J.
Roach, J.A.
Robb, Joanna
Robbie, Shirley
Rober, S.
Roberts, Anna L.
Roberts, Bill
Roberts, Gina
Roberts, Hanna
Roberts, J.A.
Roberts, Leila G.
Roberts, Naomi
Roberts, Stan
Roberts, Syd
Roberts, Ulysses
Robertson, Bob
Robertson, Dorothy
Robertson, Ian
Robertson, Keith
Robertson, Lesley
Robertson, R.
Robertson, Terrance
Robilliard, G.
Robins, Neil
Robinson, Ann
Robinson, David
Robinson, Gary
Robinson, J.
Robinson, J.W.
Robinson, L.
Robinson, Pam
Robinson, Robin D.
Robinson, S.A.
Robinson, Steve H.
Robinson, T.
Robinson, W.
Robinson, Wilma F.
Robson, Louise
Robson, Mark
Rocchini, Helen
Rockwell, Diane
Rockwell, I. Laurie
Rodney, Helen
Rodway, C.M.
Rodway, Heidi
Rodway, Joy A.
Rodway, Michael S.
Roe, Nicholas A.
Roechitz, Manfred
Roemer, Hans
Rogers, Andrea
Rogers, J.E.H.
Rogers, Ruth J.
Rogers, Thomas H.
Rohrmoser, Robin
Rohwer, Sievert A.
Roland, Weiand
Rollins, Percy
Rom, D.
Roper, Alan
Roschitz, Manfred
Roscovich, F.
Rosenberg, E.J.
Ross, D.A.
Ross, Gail
Ross, Glenda
Ross, Greg
Ross, Sandra
Ross, Sue
Rothschild, N.C.
Routledge, David
Routledge, Harold
Routledge, Maureen
Rowan, W.
Rowe, J.F.
Rowlins, H.
Rowlins, R.
Royal, John
Ruckles, Gwen
Rungren, Peggy
Runyan, Bruce
Runyan, Craig S.
Rushton, Charles E.
Rushton, Gerry
Russel, Lorne
Russel, Noel
Russel, Robert
Russell, C.
Russell, L.S.
Rutherglen, Ron
Rutledge, Harold
Rycroft, F.
Ryder, Glenn R.
Ryder, John
Ryder, June M.
Rye, Darrell
Rykes, Petrus
Sadowski, J.
Salkeld, Shirley
Salmons, S.
Salt, W.R.
Samper, Cristian
Sanders, E.
Sandham, Chris
Sandham, Shirley
Sandiford, Peter
Sanford, L.C.
Santonocito, R.
Saphir, Starr
Sarell, Mike
Sarles, Jack G.
Sarles, Rosamond
Sars, Karl
Sather, Michael
Satterfield, Doreen
Satterfield, Joy
Satterfield, Ron
Satterfield, W.
Satterly, Jack
Saturn, J.B.
Saunders, Eve
Saunders, Lowell
Saunders, M.J.
Saunders, P.
Saunders, W.E.
Sauppe, Barry
Savage, Howard
Savale, Bill
Savard, Jean-Pierre L.
Sax, M.
Saxon, Greg
Sayles, Pat
Scarf, D.
Scarf, L.
Scarfe, Ann
Scfareler, F.W.
Schadeli, Anneli
Schadeli, Claire
Schadeli, Heidi
Schadeli, Maura
Scheer, R.
Schellenberg, Al
Schick, W. Jack
Schiek, Jim
Schlegl, Katie
Schmidt, Chris
Schoen, Ellen
Schofield, Al
Schouten, Madelon A.
Schoutz, A.C.
Schueler, Fred
Schuerholz, B.
Schug, Stefan
Schultz, Zella M.
Schurrman, Suzanne
Schutz, Allan C.
Schwab, F.E.
Schweers, Gorden
Schweers, Guthrie
Schweers, Judy
Scolton, G.
Scott, Brian
Scott, Gregg
Scott, Lorraine
Scudder, G.G.E.
Sealy, Spencer G.
Searing, Gary
Seaton, B.
Sedgewick, Barbara M.
Seel, K.E.
Sekhon, K.
Self, Brian
Selk, Jim
Seminoff, Corinne
Semple, John
Sendall, Bill
Sendall, Micky
Sendy, J.
Seney, Nan
Seon, G.E.
Sephton, Janet
Sereno, Elisa
Sergeant, D.E.
Sessions, K.
Severson, G.
Sewell, Richard
Sewell, Rosamond
Shadick, John
Shand, Monica
Shank, I.
Shank, J.
Shank, M.
Sharp, T.J.
Sharples, Robert P.
Shatwell, Alan
Shaw, B.J.
Shaw, Chris
Shaw, Janice
Shaw, Jim
Shaw, Maureen
Shaw, Osborne
Shearman, K.S.
Sheath, S.
Sheehan, Brian
Sheer, C. Mowbray
Shelford, J.C.
Shepard, Chris
Shepard, Michael G.
Shephard, J.
Sheppard, S.
Sheppard, Stephan R.J.
Sherman, F.A.
Sherman, Ken
Sherodd, R.
Sherrett, Jack
Sherrington, Peter
Shields, N.
Shippen, S.
Shipway, R.
Shliek, Ryan
Shopland, Lois
Short, Henry
Short, Terry M.
Shouldice, Bill
Shrieves, C.
Shunter, M.
Shutz, Alan
Sibley, Jim
Siddle, Chris
Siddle, Sonja
Sidhu, Mel
Sigsworth, Jack
Sigsworth, Joan
Silkens, Ed
Silkens, Thelma
Silkie, G.P.
Silver, Rod
Silvey, Joe
Simcoe, Alan
Simcoe, Sandy
Simon, David
Simp, H.M.
Simpson, B.
Simpson, Fred A.
Simpson, G.B.
Simpson, Herbert M.
Simpson, L.
Simpson, R.
Simpson, Winnifred
Sinclair, Brent
Sinclair, Dean
Sinclair, Pam
Sing, Edward C.
Sing, Kay
Sirk, George P.
Sirk, Lauren
Skaar, Stephen M.
Skei, Arnold
Skinner, Andrew
Skipp, Richard L.
Skoba, W. Fred
Skwarok, Denise
Slater, Bill
Slater, Brian
Slater, Glen
Sleik, Ryan
Slessor, N.
Slingerland, M.
Slingerland, Richard
Sloan, B.
Sloan, John
Slocombe, June
Slocombe, Mabel
Slocombe, Maeford
Slocombe, Scott
Slupianek, Barbara
Smailes, Angela
Small, Doe
Small, S.O.
Smart, Gail
Smedley, L.
Smith, Alex
Smith, Barney
Smith, Bernice
Smith, Brent
Smith, Bruce
Smith, Bud
Smith, C.A.
Smith, C.C.
Smith, Craig
Smith, Cyndi M.
Smith, Cyril
Smith, D.A.
Smith, Daphne J.
Smith, David W.
Smith, Dean
Smith, Douglas
Smith, Elizabeth
Smith, Franklin J.
Smith, G.W.
Smith, George
Smith, Gertrude E.
Smith, Glen
Smith, H.
Smith, Jack
Smith, Jamie N.M.
Smith, Jeanne
Smith, Joan
Smith, K. Marian
Smith, K.S.
Smith, Karen M.
Smith, Kathleen
Smith, Kay
Smith, Linda
Smith, Lorraine C.
Smith, Margaret
Smith, Marjorie
Smith, N.
Smith, Pat
Smith, Patricia M.
Smith, Risa
Smith, Susan M.
Smith, Vic
Smith, W.J.
Snell, Les
Snuppe, B.
Sojka, G.
Solecki, C.
Solecki, Daphne
Soluhub, Claire
Somerville, Bonnie
Somerville, Brett
Somerville, Sandi
Sopkinson, S.H.
Sopuck, Leonard
Sorapure, Peter
Soulsby, Gladys
Souppe, B.
South, S.
Sowden, Margaret J.
Sowerby, Tom
Spalding, D.J.
Spalding, David A.E.
Sparling, D.B.
Sparling, Douglas
Speechly, Win
Spence, Doris
Spence, Gordon
Spencer, G.J.
Spencer, Michael
Spicer, Chris
Spicer, Janet
Spitler, Gail
Spitman, Bernie
Spitman, Prue
Stace-Smith, Don
Stace-Smith, Tine
Stacey, Pam J.
Stachera, S.
Staender, Gilbert F.
Stafford, Bess
Stainer, John E.R.
Stainer, W.
Stanley, Denis
Stanley, Elsie
Stanwell-Fletcher, John F.
Stanwell-Fletcher, Theodora C.
Statton, M.B.
Steel, K.E.
Stein, Robert C.
Stenger, J.
Stephan, Jamie
Stephens, F.
Stephens, P.
Stepniewski, M.
Sterling, D.
Sterling, Tom
Stevens, D.
Stevens, Tom
Stevens, William R.
Stevenson, Bruce
Stevenson, D.
Stevenson, George
Stewart, Andrew C.
Stewart, Ann
Stewart, Bill
Stewart, Bob
Stewart, David
Stewart, Geoff E.
Stewart, Ian J.
Stewart, Irene I.
Stewart, Julie
Stewart, R.W.
Stewart, Ronald M.
Stewart-Burton, J.P.
Stirling, David
Stirling, Ian D.
Stirling, J.
Stirling, Ruth
Stirling, Tom
Stiven, A.E.
Stoba, Fred
Stoba, Rosalind
Stoba, William F.
Stocks, J.P.
Stone, Jack
Stoneberg, R.P.
Stoner, E.A.
Straley, G.
Stratman, Andy
Stratton, B.
Stratton, M.B.
Streaker, C.P.
Street, Hazel
Street, Jim A.
Street, Laurie
Strickland, D.
Stroctman, Anita
Strois, Jacques
Strom, Gary
Strong, W.D.
Stuart, A.
Stuart-Smith, Kari
Stubbs, Bob
Stubbs, J.
Sturney, Ray
Stushnoff, Brian G.
Sugden, B.A.
Sugden, L.G.
Sullivan, Atherton
Sullivan, M.G.
Summers, Ken R.
Sunderland, Graham
Surroughs, Betty
Surroughs, Bill

Sutherland, Doreen
Sutherland, S.
Suttill, Dennis L.
Suttill, Kaye
Sutton, Derek
Sutton, Kathy
Swan, Bill
Sweeney, Renee
Swift, Pat
Swindle, G.F.
Switzer, Jim
Symon, Margaret
Symons, P.
Szymkowiak, Linda
Tacker, C.L.
Tagles, Robin
Tait, Doreen
Tait, Jacques
Tait, Mary
Talevi, Teo
Tamasi, J.
Tamke, Vern
Tansdell, Jane
Tarnocai, K.
Tattrie, Vida
Taveli, Teo
Taylor, Adam
Taylor, Dudley
Taylor, Frank
Taylor, G.W.
Taylor, Gwen
Taylor, Keith
Taylor, M. Elizabeth
Taylor, S.
Taylor, Sabrina
Taylor, W.L.
Teal, Aileen
Teal, M.
Telosky, Howard A.
Tener, J.S.
Terrane, Brian
Terrane, Janice
Thacker, Beatrice
Thevarge, Chris
Thomas, Barry
Thomas, Lee
Thomas, Vince
Thommasen, Harvey
Thompson, Anne
Thompson, Bob
Thompson, D.L.
Thompson, Glen
Thompson, K.
Thompson, M.C.
Thompson, Sandra
Thompson, T.
Thompson, Tony
Thomson, B.
Thomson, David
Thomson, H.
Thomson, Ken
Thorn, C.
Thorn, E.
Thornton, Donna
Thorpe, Steve
Thunstrom, Elizabeth
Tidball, J.
Tiernan, John
Tillay, E.W.
Tillenius, N.
Timbs, Linda L.
Tinney, Art
Tinney, Jean
Tobish, Thede
Todd, Jack
Todd, T.
Tolmie, J.W.
Tomlinson, V.
Tomlinson, W.V.
Tommson, Harvey
Tompa, Frank S.
Tonge, Peter
Tonge, Win
Toochin, Barb
Toochin, John
Toochin, Michael
Toochin, Rick
Torrer, A.
Tose, Frank
Townsend, Bill
Townsend, E.W.
Trask, Doug
Trask, Linda
Travers, Charles
Travers, Ruth E.
Trefry, Colin
Tremblay, Roger
Trent, Robert
Trim, Marjorie
Trotter, Brenda
Trotter, Charles A.
Truman, Clayton
Tuck, Cynthia
Tuck, Jim
Tuck, W.J.
Tucker, J.
Tull, C. Eric
Tull, C.W.
Turcotle, M.Y.
Turcotte, Jean-Pierre
Turcotte, Yves
Turnbull, Douglas
Turnbull, H.L.
Turnbull, J.C.
Turnbull, James D.
Turner, I.G.
Turner, Margaret
Turney, Barb
Turney, Laurence
Tutt, J.
Tweit, B.
Tyson, Danny
Tyson, Rick
Udvardy, Miklos D.F.
Uldall-Ekman, Ronald
Uliet, Bob
Ulke, Titus
Underhill, J.E.
Upshall, Muriel
Upton, Dennis H.
Vallee, Anne
Van Damme, Linda M.
Van de Vint, G.
Van den Berg, Duanne
Van den Driesche, Ruth
Van der Pol, Hank
Van Drimmelen, Ben
Van Hove, John
Van Kerkoerle, Peter
Van Maurik, Maria
van Streppen, S.
Van Tets, Gerard F.
Van Tighem, Kevin J.
Van Tine, Leo
Vance, Horace R.
Vandry, A.
Vanstall, Katherine
Vasington, Lynn
Veale, David
Venables, E.P.
Verbeek, Nicolaas A.M.
Verbrugge, William
Vermeer, Kees
Vernon, Nyky
Victor, W.J.
Viken, Audrey J.
Vincent, Jim
Vincent, Pat
Virnig, Mike
von der Linde, Jost
von Haartman, Lars
von Sacken, Barbara
von Schuckmann, Bruno
von Trebra, Charlotte
Vooys, John
Vyse, Frances
Wacheshen, Tony
Wade, Carson
Wade, Kenneth
Wade, Lynn
Wahl, Terry R.
Wainman, A.W.
Wainwright, Eileen
Wainwright, Margaret
Wainwright, Mary
Wainwright, Pat
Wainwright, Roy
Wait, Sally
Waite, Donald E.
Waite, Jean
Waite, Ron
Wakelam, Richard
Waldon, Mary E.
Waldon, R.N.
Walkem, D.
Walker, Foster
Walker, Frank
Walker, H.D.
Walker, P.S.
Walker, Rob
Walker, Ronald P.
Walker, Stephen
Walker, W.
Wallace, Nicole
Waller, Sam
Walsh, L.
Walsh, R.H.
Walters, W. James
Walton, Dr. J.H.
Walton, E.
Walton, Ken
Warburton, J.
Warburton, Tannis
Ward, Ken J.
Ward, P.
Ware, Ruben
Warne, Judy
Warren, David
Warren, Morgan
Water, Ross G.
Waterfield, Ruth
Waterhouse, Louise
Waterhouse, Michaela J.
Waterman, Angeline
Waters, G. Ross
Watkins, J.
Watmough, D.
Watson, Betty
Watson, E.
Watt, Bill
Watts, Brad F.
Watts, Emily
Watts, Marilyn
Watts, Sid
Waye, Les
Wayne, R.
Weaver, Betty-Jo
Weaver, Rob
Webb, Deirdre
Webb, Dick
Webb, J.
Webb, R.
Webb, Scott
Webb, Sue
Weber, Robin A.
Weber, Wayne C.
Weber, Wendy J.
Webster, H.R.
Webster, J. Dan
Webster, Marg
Weeden, J.
Weeden, Robert B.
Weeks, Florence
Wege, Rita
Weismiller, Art
Weismiller, Dianne
Welch, Shirley
Wellborn, Jean
Wells, Hillie
Wells, Joyce
Wells, Lucille
Wells, Pat
Wells, Terry
Wendle, E.
Wendle, Joseph
Wershler, Roy M.
Wessel, J.C.O.
West, G.C.
West, Rick
Westendorf, R.
Westerborg, Betty
Westheuser, H.
Westinghouse, A.
Weston, Elizabeth
Weston, Inez
Weston, Tom
Wetmore, Stephen P.
Wetselaar, Henri
Whalen, Paul
Whalley, Sylvia
Whately, Barbara
Whebell, Yvonne M.
Whelton, B.
White, Barbara
White, Florence
White, Jim
White, Ken
White, Mildred V.
White, Morrison
White, Ted
White, W.
Whitehead, George
Whitelaw, D.M.
Whitelaw, Virginia R.
Whiteside, Evelyn
Whitman, Doug
Whitman, S.D.
Whitney, Carl L.
Whitney, Roy
Whittaker, Jean
Whittaker, Michael
Whittaker, S.
Whittington, Bruce
Whittington, M.
Whittington, P.
Wicky, Dale
Wiebe, Karen L.
Wiedman, Roy
Wiggs, A.J.
Wight, Gordon
Wigle, Michael
Wigmore, Ruth
Wilby, D.
Wilcox, Vi
Wilk, A.L.
Wilkes, Brian
Wilkinson, B.
Wilkinson, George
Wilkinson, Henry
Will, John
Williams, D.R.
Williams, D.S.
Williams, Dorothy
Williams, Eldred
Williams, Ian H.
Williams, Imelda
Williams, Jack E.
Williams, L.
Williams, M.
Williams, M.Y.
Williams, Margaret
Williams, Marlene
Williams, Murray
Williams, Nancy
Williams, P. Ray
Williams, Paul H.
Williams, R.W.
Williams, Raechelle
Williams, Sydney
Willies, Gordon
Willies, Karen
Willock, T.A.
Willox, Mavis E.
Willson, Charlie
Willson, Janet
Wilso, Janice
Wilson, Bill
Wilson, C.
Wilson, Don
Wilson, Dora
Wilson, Douglas J.
Wilson, J.W.
Wilson, Jim C.
Wilson, L.
Wilson, Mary Frances
Wilson, O.E.
Wilson, Q.
Wilson, Ron
Wilson, W.
Winchester, Neville
Winchester, Nick
Windsor, Evelyn
Winson, J.W.
Winstone, Mary
Winterbottom, J.M.
Wiren, George
Wise, Betty
Wiseley, Allen N.
Wisnia, Jim
Withers, Lois
Witt, Lois
Wittrin, Martin P.
Wojnowski, Jul
Wolf, Carl
Wolfe, J.
Wolfe, Michael
Womersley, Marcus
Wong, Kent
Wood, Chauncey
Wood, Christopher S.
Wood, D.
Wood, Frank W.
Wood, L.
Wood, Mari
Wood, Peter
Wood, Phil
Wood, R.
Wood, Sarah
Wood, Trudy
Woods, Carolyn
Woods, John G.
Woods, Marcia E.
Woods, R.
Woodsan, John
Woodsworth, Joy
Woolgar, David
Woolgar, Marie
Woollett, J.
Workman, Dan
Workman, H.
Workman, W.
Worthen, Chas.
Worthington, Eileen
Wraith, A.
Wrenshall, Anne
Wright, Gwen W.
Wright, Harley
Wright, Kenneth G.
Wright, Lynn
Wright, P.F.
Wright, Peter
Wright, Rachelle
Wright, Richard T.
Wy, M.
Wyborn, Margriet
Wyman, E.J.
Wynja, Mark
Wynker, Connie
Wynn, James
Wynne, S.
Wysocki, Roger
Yacyslyn, Brooks
Yaremko, Leslie
Yaremko, Mitch
Yarwood, J.D.
Yellowlees, Lou A.
Yellowlees, Mary
Yellowlees, Robin
Yorke, Paul
Young, C.A.
Young, Darren
Young, Kathy
Young, Myrna
Youngs, K.
Yule, Ian
Yule, J.
Yunker, Mark
Zach, Peter
Zapf, Rick
Zapotoczny, Emil
Zapotoczny, Heather
Zeeman, Thomas
Zettergreen, Barry
Zimmerman, Ellen
Zogaris, Stamatis
Zoyesz, C.A.
Zroback, Ki
Zurowski, B.
Zurowski, Tim
Zwickel, Fred C.

CONTRIBUTING AUTHORS

Dennis A. Demarchi was born and raised in Kamloops, British Columbia. He graduated from the University of British Columbia in 1966, where he majored in plant science and zoology. In 1970 he obtained his Master of Science degree in Range Sciences from the University of Idaho; his thesis was on the effects of cattlegrazing on California bighorn sheep forage along the Chilcotin River, near Williams Lake, B.C.

Dennis joined the British Columbia Fish and Wildlife Branch in 1969 and was posted to Fort St. John as a wildlife biologist. In 1970 he moved to Victoria to work on provincial wildlife/livestock issues. For the past 25 years he has worked in Victoria as part of an inventory team, collecting landform, soil, vegetation, and wildlife information on mapping projects across the length and breadth of the province.

One of the products that he developed for British Columbia, the ecoregion classification, provides a way of understanding the province's diverse wildlife resources when used in combination with the Ministry of Forests' biogeoclimatic zone classification. It has become the standard for determining the ecosystem representativeness of B.C.'s protected areas.

In 2000, Dennis retired from his position as the Provincial Habitat Classification Specialist with the Resources Inventory Branch (RIB) of the Ministry of Environment, Lands and Parks. He was responsible for maintaining wildlife habitat, classification, and mapping standards for the RIB's inventory projects.

Dennis received the Professional Biologists Association's Award of Excellence in 1991 for his development of the British Columbia ecoregion classification. He also received the 1993 Shikar Safari International Annual Award as British Columbia's Wildlife Officer of the year for his work on the ecoregion classification and wildlife habitat mapping. He is a member of both the British Columbia Institute of Agrologists and the British Columbia Association of Professional Biologists.

Dennis was responsible for the section "The Environment" in Volume 1.

Richard R. Howie was born in Vancouver, British Columbia, and graduated from the University of British Columbia in 1970 with a bachelor's degree in agriculture, specializing in wildlife management. He then spent 24 years in the field of conservation education in national and provincial parks across Canada. In 1994, he became a Habitat Protection Biologist with the B.C. Ministry of Environment, Lands and Parks in Kamloops, where he currently works to preserve habitats for all species of fish and wildlife.

Rick has served as a subregional coordinator for *American Birds,* has coordinated Christmas Bird Counts in the Kamloops-Shuswap area for the past 22 years, and has participated in several provincial Breeding Bird Surveys each year since 1975. He prepared the first status reports for the Spotted, Burrowing, and Flammulated owls in the province and is involved in regional studies of Ospreys, Long-billed Curlews, and Lewis's Woodpeckers. He is preparing manuscripts for books on *The Birds of Kamloops, Bird-finding in the Kamloops Area,* and *Raptors of British Columbia.*

Rick Howie is a Registered Professional Biologist of British Columbia and is a member of many organizations, including the American Ornithologists' Union, Cooper Ornithological Society, Raptor Research Foundation, Association of Field Ornithologists, American Birding Association, and Society of Canadian Ornithologists.

Rick prepared the draft account for the Western Meadowlark.

Chris Siddle was born in Mission, British Columbia, and graduated from the University of British Columbia in 1973 with a Bachelor of Arts degree in English literature.

He served as regional editor for *American Birds,* providing quarterly summary reports on birds in the province (and Yukon Territory) between 1990 and 1993. He regularly participates in Breeding Bird Surveys, Christmas Bird Counts, and the British Columbia Nest Records Scheme. He has prepared reports on the status of the Lewis's Woodpecker, Purple Martin, Le Conte's and Nelson's Sharp-tailed sparrows, and warblers for the B.C. Ministry of Environment, Lands and Parks.

He has served as a regional editor for the *Birds of British Columbia* project since the mid-1980s and is currently preparing manuscripts for books on *Birds of the North Peace River Area, British Columbia Birds: A Complete Guide, Raptors of British Columbia,* and *Bird-finding in the Okanagan Valley.*

Chris prepared draft accounts for the Nelson's Sharp-tailed Sparrow, Baltimore Oriole, Bullock's Oriole, Common Redpoll, and Hoary Redpoll.

Linda M. Van Damme was born in Ontario and graduated in 1974 with a nursing diploma from St. Clair College in Chatham, Ontario. She currently holds a licence with the Registered Nurses Association of British Columbia and is employed as a home care nurse in the interior community of Nelson.

Since 1979, Linda has contributed her bird observations to *The Birds of British Columbia* project and the British Columbia Nest Records Scheme and has served as a regional reviewer of draft species accounts for Volumes 3 and 4.

Linda's publications include the species account for the Sky Lark in *The Birds of North America* and a status report on the Bobolink in B.C. She is working on *British Columbia Birds: A Complete Guide, Big Eyes* (Owls) *in the Creston Valley,* and *The Birds of Creston Valley.* Her research interests include the winter ecology of raptors in the Creston valley and food habits of owls in the province.

She is a member of the American Ornithologists' Union, Cooper Ornithological Society, Raptor Research Foundation, Association of Field Ornithologists, Society of Canadian Ornithologists, and the Federation of B.C. Naturalists.

Linda prepared draft accounts for the Black-throated Sparrow, Harris's Sparrow, Bobolink, and Brambling.

ABOUT THE AUTHORS

From left, top: Ian McTaggart-Cowan, Gary W. Kaiser, Andrew C. Stewart, Michael C.E. McNall, and John M. Cooper. Bottom: R. Wayne Campbell and Neil K. Dawe (Victoria, 21 August 2000; Mark Nyhof)

R. Wayne Campbell was born in Edmonton, Alberta. His fields of interest include zoogeography, feeding and breeding ecology of raptors, marine bird populations, and conservation of birds. He graduated from the University of Victoria in 1976 and received his M.S. degree from the University of Washington in 1983.

After high school, he worked as a seasonal naturalist with the British Columbia Parks Branch, and in 1969 joined the staff at the University of British Columbia as Curator of the Cowan Vertebrate Museum in the Department of Zoology. Over the next 4 years, he established the Photo-Records File, a system to document the occurrence of rare vertebrates in the province, and took over administrative and financial responsibilities for the British Columbia Nest Records Scheme, which he continues to oversee today. He also became very active on the executives of several conservation organizations, including the British Columbia Waterfowl Society and the Vancouver Natural History Society.

In 1973 he moved to the then British Columbia Provincial Museum as Curator of Ornithology. Over the next 20 years, he conducted wildlife inventories of remote areas of the province, including the first complete census of breeding seabird colonies. In addition, he amassed an enormous provincial vertebrate data base that includes details for nearly 2 million specimen and sight records, 300,000 breeding records, and hard copies of nearly 10,000 published and unpublished articles on amphibians, reptiles, birds, and mammals.

Wayne has written over 375 scientific papers, popular articles, and government reports, and has co-authored almost 40 books on higher vertebrates. He has served as British Columbia coordinator for the North American Breeding Bird Survey since 1976, and as a select member of the national ornithology group for the Committee on the Status of Endangered Wildlife in Canada since 1981. He belongs to 23 professional and natural history organizations and is a life member and elected member of the American Ornithologists' Union, life member of the Cooper Ornithological Society, and honorary life member of the Vancouver Natural History Society.

He has received numerous honours and awards for lecturing, writing, and conservation activities. In 1989 he received the Award of Excellence in Biology from the Association of Professional Biologists of British Columbia, and two years later the Lifetime Service Award from the Federation of British Columbia Naturalists. He was appointed to the Order of British Columbia in 1992 and recently received the Commemorative Medal of Canada in recognition of the 125th anniversary of the Confederation of Canada.

Wayne retired from 28 years of government service in May 2000. He is presently a director of WBT Wild Bird Trust of British Columbia and has actively participated in establishing the nation's first provincial wildlife data centre in Victoria.

Neil K. Dawe was born in New Westminster, British Columbia. After graduation from high school his interest turned to the world of finance, and banking became his vocation for the next 7 years. In 1970 he returned to the University of British Columbia, where his interest in natural history was inspired by the enthusiasm and encouragement of Wayne

Campbell. This interest grew into a career in wildlife and habitat research and management.

In 1971 he worked as a seasonal naturalist for the British Columbia Parks Branch. Later that year, he became Chief Naturalist at the George C. Reifel Bird Sanctuary, where he developed interpretation and education programs for the British Columbia Waterfowl Society until 1975. Since 1975, he has worked for the Canadian Wildlife Service on Vancouver Island, managing their National Wildlife Areas and Migratory Bird Sanctuaries and working to protect migratory birds and their habitat.

Neil is a member of a number of professional, conservation, and scientific organizations, including the Association of Professional Biologists of British Columbia, the Society for Conservation Biology, the American Association for the Advancement of Science, and the Federation of British Columbia Naturalists, and is an honorary life member of the Arrowsmith Naturalists. He co-founded the Brant Wildlife Festival and co-chaired the festival for its first five years. He was appointed a member of the Healthy Community Advisory Commission for Parksville, British Columbia. He also served as the first Regional Vice-President of the Federation of British Columbia Naturalists for Vancouver Island and has sat on the executives of the Vancouver Natural History Society, the Mid Island Wildlife Watch Society, the Rosewall-Bonnel Land Trust Society, and the Mount Arrowsmith Biosphere Reserve Foundation.

Neil has been the recipient of Environment Canada's Regional Citation of Excellence Award for his work in promoting the value of wildlife to Canadians and for the Brant Wildlife Festival, plus their National Citation of Excellence Award for outstanding achievement in advancing the goals of conservation and protection. He has also received the Federation of British Columbia Naturalist's Outstanding Service Award, and the Ina Mitchell Award from the Tourism Association of Vancouver Island, for bringing wildlife and people together on Vancouver Island. He is especially proud of being the recipient of Ruth Masters' *Hero Spoon* in 1997.

Much of his work over the past 20 years has involved studies on bird use of estuaries and the restoration and creation of estuarine marsh habitat. His primary interests now, focus on the two limiting factors affecting conservation of migratory birds and other wildlife on the Earth today: human population growth and economic growth.

Neil has written over 75 scientific, technical, and popular papers and articles on birds, ecology, and the environment. He recently authored *The Hummingbird Book and Feeders* and is a co-author of the popular children's book *The Bird Book and Bird Feeder.*

Ian McTaggart-Cowan, born in Edinburgh, Scotland, is a career biologist and educator with special concentration on the systematics, biology, and conservation of birds and mammals. He graduated from the University of British Columbia in 1932 and earned a Ph.D. degree from the University of California (Berkeley) in 1935. He has been awarded D.Sc. degrees by the University of British Columbia and the University of Victoria, LL.D. degrees by the University of Alberta and Simon Fraser University, and a Doctor of Environmental Studies degree by the University of Waterloo.

He was the biologist at the Provincial Museum in Victoria from 1935 to 1940, when he joined the faculty of the University of British Columbia. During 35 years there, he established and taught courses in vertebrate zoology, undertook research in ornithology and mammalogy, and guided the studies and research of some 100 graduate students while serving successively as professor, head of the Department of Zoology, and dean of the Faculty of Graduate Studies. His studies took him to 6 continents and resulted in more than 300 publications, 110 television programs, and 12 teaching films.

His public service related to vertebrate zoology, conservation, and education includes 7 years on the National Research Council of Canada, where he was the first chairman of the Advisory Committee on Wildlife Research, and 14 years on the Fisheries Research Board of Canada. It also includes serving as chairman of such bodies as the Environment Council of Canada, the Advisory Committee on Whales and Whaling, The Habitat Conservation Trust Fund of the Province of British Columbia (19 years), the Board of Governors of the Arctic Institute of North America, The Wildlife Society, and the Pacific Science Association; as vice chairman of the International Union for the Conservation of Nature and Natural Resources; as a member of the Select Committee on National Parks for the United States Secretary of the Interior; and as a director of the Nature Trust of British Columbia and of the National Audubon Society.

Ian McTaggart-Cowan has received numerous honours and awards for his contributions to biology and conservation, including: Officer of the Order of Canada, the Order of British Columbia, Fellow of the Royal Society of Canada, the Leopold Medal of the Wildlife Society, the Fry Medal of the Canadian Society of Zoologists, the Einarsen Award in Conservation from the Northwest Section of the Wildlife Society, the Dewey Soper Award from the Alberta Society of Professional Biologists, the Canadian Environmental Achievement Award (1992), the Doris Huestis Spiers Award of the Canadian Society of Ornithologists (1997), and the Barsby Trophy of the British Columbia Wildlife Federation (1999).

John M. Cooper, born in New Westminster, British Columbia, is a career wildlife biologist. His early interest in birds and the natural world was stimulated by his parents, Jack and Louise Cooper. Each spring, for two decades, the Cooper family travelled throughout British Columbia and Alberta, often with close friends Lorne Frost and Glen Ryder, in search of birds and their nests. His passion for birds, wilderness, and environmental issues was born from those experiences.

John obtained his B.Sc. degree from the University of British Columbia in 1978 and his M.Sc. degree from the University of Victoria in 1993. He worked as a consulting biologist from 1978 to 1981, then joined the Royal British Columbia Museum as the ornithology technician. At the museum he realized his dream of working with Wayne Campbell, and travelled to many remote regions of the province to inventory birds and other wildlife. After 10 years at the museum, John returned to private business with his own consulting company. Since 1991, he has worked on a wide variety of projects for clients in industry, government, and the conservation movement, mainly environmental impact assessments, mitigation of wildlife/development conflicts, wildlife management, biomonitoring, and wildlife inventory.

John is a member of many scientific and conservation organizations but is most active as a fundraiser for Ducks Unlimited Canada, an organization that gave him his first biology-related employment in 1975. For 19 years he has directed or assisted in fundraising activities in Victoria to aid Ducks Unlimited's wetland enhancement programs. His most poignant career moment came in 1993: he joined a Sierra Club research trip to the Tatshenshini River, his most beloved wilderness area, and while he was on the river, the provincial government announced the creation of the new Tatshenshini park.

John has written over 100 technical reports, popular articles, scientific papers, and books on birds and other wildlife. His most significant research projects include a long-term study of breeding shorebirds on the Queen Charlotte Islands, breeding ecology and the effects of logging on woodpeckers, conservation of the endangered Northern Leopard Frog, and an inventory of rare raptors.

Gary W. Kaiser grew up in Ottawa where the Macoun Field Club promoted his early interest in natural history. He is a graduate of Carleton University (B.Sc. 1966, M.Sc. 1972) and joined the Canadian Wildlife Service in 1968. He moved to British Columbia as the regional population biologist in 1974 and has led projects on waterfowl, shorebirds, and seabirds for the past 25 years. In 1984 he undertook a banding project in Sabah, Malaysia, and for 10 years offered annual workshops and training programs in the Philippines, Australia, Colombia, and Peru and has seen many of the participants become important conservationists in their own countries.

Recently, Gary has led two major projects in British Columbia. The first involved obtaining funding from the *Nestucca* oil spill to exterminate rats on Langara Island. The Langara Island Seabird Habitat Recovery Project was successfully completed by the summer of 1999 when the rats were extirpated and the island once again held large numbers of ground nesting birds. The second was to address the biology of the Marbled Murrelet, which had just been added to the list of Nationally Threatened birds.

He marked his retirement from government service in 2000 with the publication of *Seabirds of the Russian Far East,* for which he was the English-language editor. He published a paper on theories for the origin of flight in that same year and is currently working on the physics of the exceptionally fast flight of Marbled Murrelets, the biomechanics of their underwater flight, and the biomolecular evidence for their place in the evolutionary tree. In September 2000, he retired from Environment Canada and undertook the associate curatorship of the Cowan Museum of Vertebrate Zoology at the University of British Columbia.

Andrew C. Stewart was born in Vermilion, Alberta, and grew up in the Peace River region of northeastern British Columbia. There, in the areas around Fort St. John, he spent his formative years pursuing outdoor explorations and developing an early, life-long interest in birds and other wildlife. After graduating from the British Columbia Institute of Technology in 1973, Andy moved to Victoria where he began his career in wildlife habitat inventory with the B.C. Ministry of Environment, Lands and Parks. With the exception of a 1-year stint in Williams Lake, he has been based in Victoria while working extensively throughout British Columbia on a variety of field projects. During the course of his 27 years in government, he has undertaken habitat projects in most regions of the British Columbia, including many of the province's last remaining wilderness areas. Although a comparative newcomer to *The Birds of British Columbia* project, Andy has been a long-term contributor to the data bases upon which this project is based. He brings a wealth of knowledge and field experience to the project as well as biological expertise with respect to British Columbia's northern regions.

Andy is a member of a number of professional and scientific organizations, including the Association of Professional Biologists of British Columbia, Association of Field Ornithologists, Western Bird Banding Association, Cooper Ornithological Society, Raptor Research Foundation, Wilson Ornithological Society, Waterbird Society, Wildlife Society, and the American Ornithologists' Union. He is also a member of the Victoria Natural History Society, Federation of British Columbia Naturalists, and British Columbia Field Ornithologists.

Andy is presently carrying out two bird projects on southern Vancouver Island involving research on the ecology of urban-nesting Cooper's Hawks and on the invasion movements of the Steller's Jay. He has published several articles and scientific papers resulting from these studies and has several other scientific works in progress.

Michael C.E. McNall, born in Wingham, Ontario, spent much of his early life hunting, fishing, and studying nature. He obtained a diploma in Wildlife Management from Sir Sandford Fleming College, Ontario, in 1971, joined the Ornithology Department of the Royal Ontario Museum, and spent the next 3 years on field expeditions in the West Indies, British Isles, Netherlands, Iceland, and throughout North America.

While at the Royal Ontario Museum, he was inspired by artist Terry Shortt to carry out his own research. In 1975 and 1976, with guidance and support from Henri Ouellet and Stewart MacDonald of the National Museum of Canada, he carried out a behavioural study of Parasitic and Long-tailed jaegers in the Canadian Arctic.

After his arctic experience, Michael moved to Victoria, and in 1980 joined the staff of the Royal British Columbia Museum. He has travelled throughout the province collecting data for this book and is currently the Ornithology Collections Manager. In this role, he prepares exhibits and gives talks on bird behaviour and conservation.

Outside of ornithology, Michael is vice president of the Vancouver Island Arms Collectors Association.

INDEX

IAN McTAGGART-COWAN, O.C., O.B.C., Ph.D., LL.D., F.R.S.C.

Clockwise from upper left: Clinton, B.C., 1932; home office, Saanich, 2000; University of British Columbia, Vancouver, 1951; Mayne Island, 1976; Shaw Creek, 1950; Thompson Pass, 1945; insets: Chain Islets, 1989; "the rower," circa 1950.

THE LAST WORD ...

There is another factor which enters into the plans. You are now about the only freelance collector in the West who is competent to do museum collecting, and is familiar with the technique, and as an old apostle we want you to help pass on some of the tradition to a disciple. We have a young man in view who has been recommended to me from several different sources. His name is Ian McTaggart Cowan of North Vancouver, now a third-year student at University of B.C. I met him at Winson's place in Huntingdon last fall, and Kenneth Racey and Allan Brooks spoke highly of him, also Professors Spencer and MacLean Fraser of the department of zoology at the University.

... They say his forebears were naturalists, and he has camped and hunted all his life. Spencer says he is one of the best shots in B.C. and is a go-getter in the field. I had only a short conversation with him last fall and was much taken by him ... I think that Cowan is the real thing ...

– *Excerpt from a letter to Hamilton M. Laing from Dr. Rudolph M. Anderson (at Canadian National Museum), 26 December 1929*
(Courtesy of British Columbia Archives)

Over the last 20 odd years, Ian McTaggart-Cowan has provided dedicated leadership, guidance, inspiration, and stability to this project – what we, in its later stages, have affectionately called *the albatross*. Since the publication of Volumes 1 and 2, Ian has lived through his ninth decade. Over that time he has learned to master a computer, written species accounts and other chapters, proofed the galleys, acted as project manager, and liaised with our steering committee. Leading by example, he has been a mentor to us all. There is no doubt that "Cowan is the real thing."

We have made much of the significant efforts of the volunteers who have contributed to the success of this project, but, without question, the volunteer who has made the greatest contribution of all has been Ian, the only non-salaried author involved in the project. So, now that *the albatross* has flown, we simply want to say, *thanks Ian, thanks from us all.*

R. Wayne Campbell
Gary W. Kaiser
Neil K. Dawe
Andrew C. Stewart
G.E. John Smith
John M. Cooper
Michael C.E. McNall

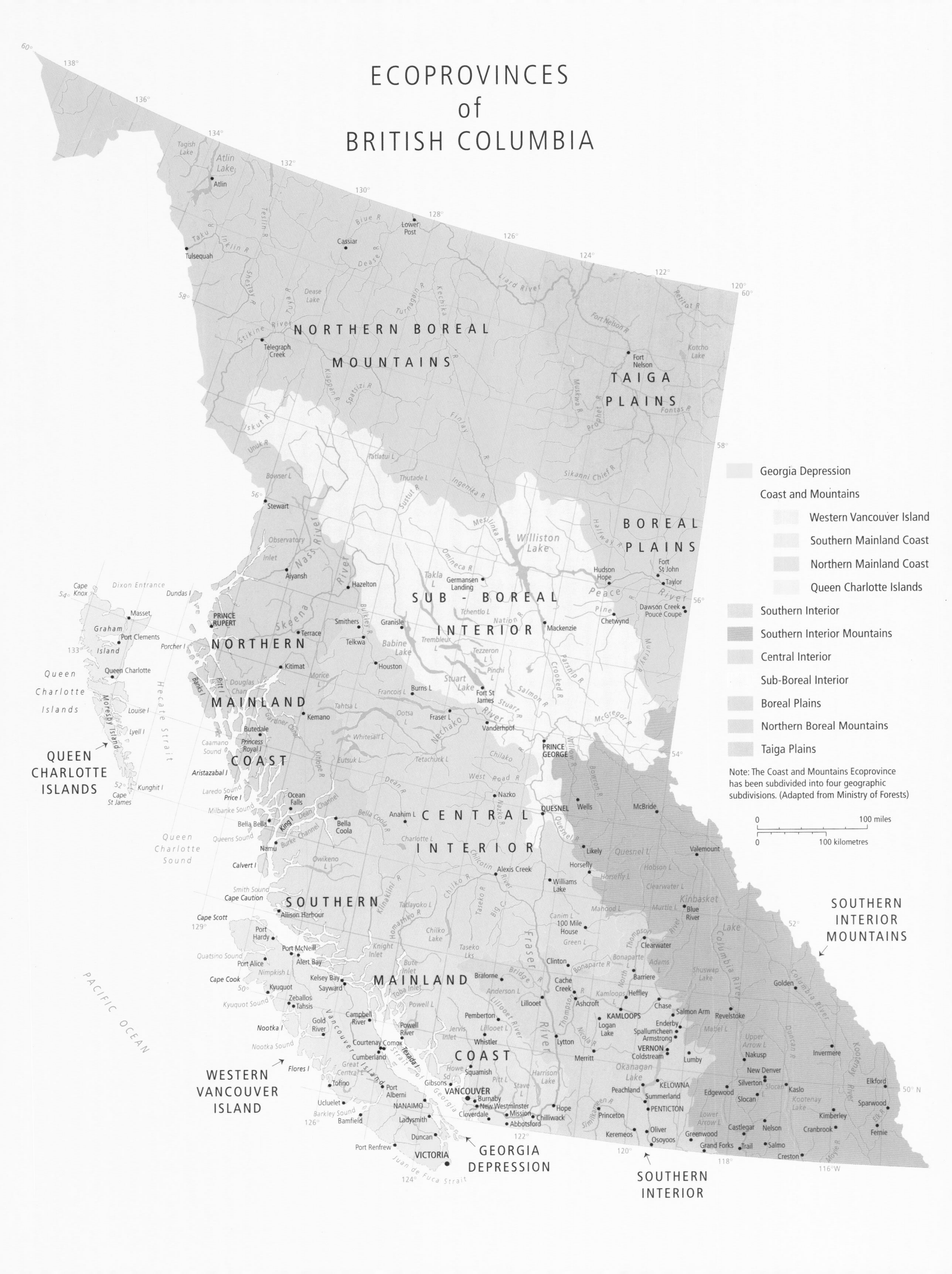

ECOPROVINCES
of
BRITISH COLUMBIA
Georgia Depression
Coast and Mountains
Western Vancouver Island
Southern Mainland Coast
Northern Mainland Coast
Queen Charlotte Islands
Southern Interior
Southern Interior Mountains
Central Interior
Sub-Boreal Interior
Boreal Plains
Northern Boreal Mountains
Taiga Plains
Note: The Coast and Mountains Ecoprovince has been subdivided into four geographic subdivisions. (Adapted from Ministry of Forests)
100 miles
100 kilometres
NORTHERN BOREAL MOUNTAINS
TAIGA PLAINS
BOREAL PLAINS
SUB - BOREAL INTERIOR
NORTHERN MAINLAND COAST
QUEEN CHARLOTTE ISLANDS
CENTRAL INTERIOR
SOUTHERN MAINLAND COAST
SOUTHERN INTERIOR MOUNTAINS
WESTERN VANCOUVER ISLAND
GEORGIA DEPRESSION
SOUTHERN INTERIOR
PACIFIC OCEAN